ORTHOMOLECULARISM

Principles and Practice

R.A.S. HEMAT
MB; BCh, FRCSI, Dip. Urol. London

authorHOUSE™
1663 LIBERTY DRIVE, SUITE 200
BLOOMINGTON, INDIANA 47403
(800) 839-8640
WWW.AUTHORHOUSE.COM

© 2004 R.A.S. HEMAT, MB; BCh, FRCSI, Dip. Urol. London All Rights Reserved.

No part of this book may be reproduced, stored in a retrieval system, or transmitted by any means without the written permission of the author.

First published by AuthorHouse 11/16/04

ORTHOMOLECULARISM: PRINCIPLES AND PRACTICE   ISBN: 1-4184-6202-0 (e)
BY R.A.S. HEMAT   ISBN: 1-4184-6203-9 (sc)

Printed in the United States of America
Bloomington, Indiana

This book is printed on acid-free paper.

Although all reasonable care has been taken during the preparation of this edition, the publisher, editor and author cannot accept responsibility for any consequences arising from its use or from the information it contains. The publisher-author is not responsible (as a matter of product liability, negligence or otherwise) for any injury resulting from any material contained herein. This publication contains information relating to general principles of orthomolecularism, which should not be construed as specific instructions for individual patients. Readers are encouraged to confirm the information contained herein with other sources. Readers are advised to check the products information for each drug they plan to use for their patients to be certain that information contained in this book is accurate. Orthomolecularism is an ever-changing science. There is always possibility of human error and medical sciences will continue to change. Neither the publisher nor author warrants that the information contained herein is in every respect accurate or complete, and they are not responsible for any errors or omissions or for the results obtained from use of such information.

All right reserved. No part of this publication may be reproduced or transmitted in any form or any means, electronic or mechanical including photocopy, recording, or any information storage and retrieval system, without permission in writing from the author. For information, address urotext@urotext.com.

There is a poverty of medical evidence to support the majority of present day medical practices. Orthomolecularism offers the real hope of reversing and preventing diseases. Orthomolecularism is the intervention for disease by varying the concentrations of substances normally present in the human body. Many diseases are caused by molecular imbalances. The term orthomolecularism derives from the words [ortho- + molecular] relating to or aimed at restoring the optimal concentrations and functions at the molecular level of the substances (e.g., vitamins) normally present in the body. Orthomolecularism involves every organs of the body. The term is used today lacks the range to convey the total extent of the field that it covers. Orthomolecularism should not be established as an independent field of medicine as it exists in every medical and surgical subspecialty. I write the books I wish I had, covered the basics, books that I noticed my students need. My students taught me so much about how people need to read and hear things so they make sense. I saw them read and followed my directions, so I learnt what did not work and fixed it. This book is for both the traditional medical education and the integrated medical education systems. We must carefully assess the information presented to us, to move along a continuum of thought regarding cost, practice and outcome. I am convinced this book will make orthomolecularism a lot simpler, a lot clearer and a whole easier. I hope that the solid information contained in this book will allow anybody to approach any medical professional with a logical and convincing description of the excellence of this system. This book collates and summarises, and displays information in the rapidly changing field of orthomolecularism, which encompasses medicine, surgery, psychiatry, oncology, basic science, pathobiology, etc. The reader is advised to consult the instruction and information material included in the package insert of each drug or therapeutic agent before administration. Orthomolecularism rests strongly on the following areas of scientific knowledge: 1- nutrition, 2- biochemistry, 3- cell biology, 4- physiology, 5- general medicine, 6- immunology, 7- allergy, 8- endocrinology, 9- pharmacology, 10- toxicology, 11- gastroenterology, 12- parasitology, 13- nephrology, 14- physical medicine and manipulation therapies, 15- dentistry, 16- veterinary science, 17- food science, 18- agriculture, 19- climatology, and 20- medical politics. Orthomolecularism is integrated medicine. The free radical chemistry was established field in 1950s. In 1983, it was suggested that mitochondria might serve as the biologic clock. Only about 15% of medical interventions are supported by solid scientific evidence. I had too many questions about the application of medical conventional biochemistry that I found it as Orthomolecularism during my years of teaching integrated medicine as a lecturer at university. Integrated medicine let you see the person as a cell, and to treat the cell. Orthomolecularism fits within the concept. I believe it is unethical and irresponsible to fail to advise the public. I fear that many public members will continue to tell patients that pharmaceuticals are waste of time and will dangerously misinforming public. Diet and exercise alone may fail to normalise a disease in the majority of patients, and pharmacotherapy is necessary. At one time, doctors commanded and decided virtually all treatment options for a patient, with no obligation to consider the patient's values or decisions. The published evidence is heavily biased. In this context, meta-analysis becomes a tool of authoritarianism, replacing the use of judgement with the improper use of statistical analysis. Good judgement requires good information, but there are forces that would over-rule individual judgement as to whether published information is applicable to certain patients. There is no perfect solution, and no decision is risk-free. The history of science is the history of struggle against entrenched error.

2004

ORTHOMOLECULARISM: PRINCIPLES AND PRACTICE is an authoritative guide.
Visit http://www.urotext.com
For more information and other books.

Urotext
Urology and health
www.urotext.com

BASICS   7

FUNCTIONAL MEDICINE	8
PHARMACOTHERAPEUTICS AND PHARMACOKINETICS	13
BIOETHICS	16
THE CELL	18
STRUCTURE AND FUNCTION OF RECEPTORS	34
STRESSORS	48
THE FREE RADICALS	64
THE CYTOCHROME P450 SYSTEM	81
HEAT SHOCK PROTEINS	84
CYTOKINES	87
APOPTOSIS	96
ALVEOLAR EPITHELIAL CELLS	115
RENAL PHYSIOLOGY	120
FAT AND MUSCLE PHYSIOLOGY	144
SMOKING	146
ALCOHOL	150
RESPONSES TO SURGERY	164
HOST RESPONSE TO INJURY	166
CONSEQUENCES OF TRAUMA	171
POSTINJURY PAIN	173
THE SPINAL CORD INJURY	175
INFLAMMATION AND INFECTION	181
BIOGERNOTOLOGY	210
HOMEOSTASIS	229
DIABETES MELLITUS	235
NEUROPATHIC PAIN	252
ONCOLOGY	262
THE BIOLOGY OF METASTASIS	277
PARANEOPLASTIC SYNDROMES	283
WOUND HEALING	291
ANDROLOGY	297

NUTRIENTS   307

FOOD COMBINATIONS	308
PROTEINS	344
POTASSIUM	359
FLAVNOIDS	368
ANTIOXIDANTS	374
COENZYME Q_{10} (CoQ_{10})	384
VITAMIN E	389
SELENIUM	391
GLUTAMINE	395
GLUTATHIONE	401
NADH	407
VITAMIN C (ASCORBIC ACID)	408
ALPHA-LIPOIC ACID	416
MAGNESIUM	422
WATER	431
LIPIDS	441
THE BRAIN CHEMICAL-MOOD AND MENTAL FUNCTIONING	453

APPLIED 461

ORTHOMOLECULARISM	462
BIOCHEMICAL INTERVENTION	464
SLEEP DISORDERS	467
NEUROTOXINS	474
ENVIORNMENTAL DISORDERS	482
AUTISM	501
PSYCHIATRY CONDITIONS	503
DIABETES AND PSYCHIATRIC ILLNESS	508
EXCITOTOXINS	515
TOXIC LIVER INJURY	519
SEPSIS AND ALI-ARDS	523
ADRENAL INSUFFICIENCY	526
BONE HEALTH	527
NICOTINE	528
MICROWAVE	529
ELECTROMAGNETIC FIELDS AND HUMAN HEALTH	531
WATER	533
DEMENTIA	534
SULPHITES	542
FLUORIDATION	547
AIR	549
CHRONIC FATIGUE SYNDROME	558
DIETARY INTERVENTION	566
INTERVENTION FOR CANCER	572
ANTIOXIDANTS FOR CANCER CHEMOTHERAPY, AND CANCER CHEMOPREVENTION	578
NUTRITION, WASTING, INVOLUNTARY WEIGHT LOSS (IWL) AND PROTEIN-ENERGY MALNUTRITION (PEM)	584
DIETARY INTERVENTION FOR CHRONIC DISEASE PREVENTION	605
CACHEXIA IN IMMUNOCOMPROMISED PATIENTS	608
NUTRITIONAL SUPPORT IN CRITICALLY ILL PATIENTS	611
SURGICAL METABOLISM AND NUTRITION	612
OBESITY	613
LIPID DISORDERS	623
HYPOGLYCAEMIA	632
DYSAUTONOMIAS	636
UBIQUINONE	637
RHABDOMYOLYSIS	638
CAERULOPLASMIN, COPPER AND NUTRITION	640
POTASSIUM DEFICIENCY	644
THYROID DYSFUNCTION	645
NUTRITIONAL AND METABOLIC EFFECTS OF ALCOHOLISM	647
THE ELEMENTS AND INFLAMMATORY RESPONSE	650
IRON ASSOCIATED PROBLEMS	651
SOME GASTROINTESTINAL TRACT PROBLEMS	652
CENESTHOPATHY	655
ANDROPATHY	665
GYNOPATHY	689

APPENDICES 703

INDEX 741

BASICS

FUNCTIONAL MEDICINE

Functional Medicine is the field of health care that employs laboratory assessment and early intervention to improve physiological, emotional/cognitive, and physical function. This health care approach focuses attention on biochemical individuality, metabolic balance, ecological context and unique personal experience in the dynamics of health. Evaluating organ "function" versus organ "pathology" is one of the principles of functional medicine. Tests are non- or minimally invasive, using samples of stool, urine, saliva, blood, and hair and focus on how well the body is doing its job in six important areas: digestion (gastrointestinal), nutrition, detoxification/oxidative stress, immunology/allergy, production and regulation of hormones (endocrinology), and heart and blood vessels (cardiovascular). The goal of Functional Medicine is to optimise health by improving biomechanical (structural), biochemical and emotional/psychological function of the individual through changes in lifestyle, environment, and nutrition. Functional medicine aims to determine what kinds of internal or external stressors cause the abnormal function which in turn causes the often very deceptive symptoms. Intravenous administration of nutrition and medications is nothing new in standard medicine. The benefit of nutritional substances depends solely upon the degree to which they reach the cells and tissues in need. The use of intravenous administration is known in medicine to be the most reliable delivery system and for this reason, should be combined with intelligently applied and safe nutritional approaches. Intravenous nutrition provides nutrients directly into the blood stream. Chronic disease always compromises the digestive capacities of those affected. IV and IM nutrition bypass absorption and utilisation blocks at the stomach or intestinal level allowing for large doses to gain access to the circulation. The use of nutrients and herbs should be the first defense in the prevention and treatment of disease, not the last resort. Supplements taken orally may not raise blood levels high enough to affect more serious illness and symptoms for a number of important and well-proven reasons: 1- There is a physiologic limit of all oral nutrients. For example, only 35% of oral calcium is absorbed by mouth-regardless of the form of calcium consumed. Beta-carotene has 25% oral limit; the more one takes orally, the less is absorbed. Only 25% of oral Vitamin C is absorbed-regardless of how much you eat or take by vitamin supplement. 2- Deficiencies of stomach acid (hydrochloric acid and pepsin) and pancreatic enzymes (amylases, lipases, proteases) may inhibit the absorption of a variety of nutrients. For example, most minerals (i.e., iron, zinc and calcium) depend upon adequate levels of stomach acid. 3- Deficiency of pancreatic enzymes can inhibit the proper assimilation and absorption of proteins (by not breaking them down into their constituent amino acids and peptides), fats and carbohydrates. As we age, stomach acid levels decline progressively. The average 50-year-old person has little or no HCl left due to the destruction of the stomach's parietal cells that produce hydrochloric acid. 4- Auto-immune conditions such as multiple sclerosis, lupus and rheumatoid arthritis and digestive conditions such as ulcerative colitis, irritable bowel syndrome, Crohn's disease, feel allergies, yeast overgrowth and leaky gut syndrome may either reduce one's absorption of nutrients or increase one's needs beyond what diet and oral supplements alone provide. Since Hepatitis C is often accompanied by an autoimmune component and/or digestive disturbances, it is fair to assume that many people with HCV are not maximally absorbing nutrients orally. Nutritional substances exert their beneficial function in a number of ways, ranging from chemically simple to complex. Nutrients exert some of their physiological effects at the cellular or tissue level. Nutrients are needed for a countless number of important functions relating to the health and maintenance of body tissues. Nutrients must reach the cells and tissues in need to exert their essential benefits. When very high levels of nutrients are achieved in the blood, as is capable through intravenous administration, nutrients may gain direct access into tissues and cells via mass action. The typical nutrient drip takes between 45 minutes to 2 hours to complete. The immediate and ultimate effects of IV-nutrition vary from individual to individual. It is not uncommon for one receiving IV-nutrition to experience greater energy, reduce paid and alleviation or elimination of symptoms. Others experience no improvements in overall symptoms until many IV treatments are received. The effects of IV-nutrition can be compared to those of exercise; the overall benefits of exercise are not always realised during the performance of the exercise. Benefits of anaerobic or aerobic exercise occur during rest and sleep. Likewise, the benefits of IV-nutrition occur because the cells of the body are exposed to 100% of the nutrition in the IV. Some of the nutrients stay in the body only hours while others stay in the body for years. Like cumulative exercise, the effects of IV-nutrition are cumulative over time. Like each individual workout, each IV-infusion produces benefits on top of the previous infusion-enhancing health benefits exponentially. Not all of the symptoms and disease states are caused by nutritional imbalances. Disease may increase one's nutritional requirements. Nutritional problems, whether a result of or a cause of disease have impact on health. Intravenous nutrition is a reasonable step towards correcting nutritional problems that have taken years to develop. Oral absorption of nutrients is limited to 50-60%.

DIAGNOSTIC BIOTECHNOLOGIES

Health is related to the additive oxidative stress imposed on the individual's antioxidant defense system by exogenous and endogenous oxidant stressors. Diagnosis is a process of logical deduction, but in practice most doctors work by pattern recognition. Scientific knowledge together with good biochemical assays can lead to proper treatment strategy. Treatment application would require biochemical assays to confirm a positive results and adjustment of treatment strategy. Imbalanced functioning defence system leads to to general health problems, disorder and disease. The defence system consists of: 1- manager (nervous system), 2- regulator (endocrine system), 3- executor (immune system), and 4- executor (cellular enzymatic systems). The quantity of a substance can be: none (0%), trace (<10% of normal), low (10-74% of normal), normal (100%), Above normal (101-125% of normal), high (126-200% of normal), very high (>200% of normal). The analysis of any result can be normal, abnormal, or pathological. A complete evaluation includes establishment of the reliability and validity of the measurement, ascertainment of diagnostic accuracy, assessment of safety, efficacy, and effectiveness, and most important, improvement in patients treated. Uncontrolled release of untested diagnostic technology may lead to widespread use without clear evidence of benefit to patients. Computed axial tomography (CT) and MRI enjoy widespread acceptance despite the absence of rigorous evaluations. The assessment of diagnostic technology are as follow: 1- Technologic capability, before clinical evaluation can begin, the diagnostic technology must be accepted by regulatory bodies and that parameters of safety, set, should be tested by inventors, physicists, biochemists, and manufacturers. 2- Range of possible uses, to provide important diagnostic information in number of clinical situations. 3- Diagnostic accuracy, should give more precise and exacting information that allows the safety of therapeutics. 4- Impact on medical practitioners, the technology should allow health care workers to be more confident of their diagnoses and therapy, and increase their skill. 5- Therapeutic impact, the therapeutic decision made should be altered as a result of application of the technology. 6- Patient outcome, application of the technology results in benefits for patient. Diagnostic accuracy: Measurements that are not reliable cannot be valid. Reliability is obtaining the same measurement when the same phenomenon is measured on different occasions by the same (intraobserver) or different (interobserver) observer. Test or technology should satisfactorily answer certain questions in the clinician's mind such as: 1- Are the results of the study valid? What were the results? Will the results help in caring the patients? Accuracy is the degree to which a measurement represents the true value of the attribute that is being measured. A diagnostic measurement must be able to distinguish between diseased and nondiseased or differentiate different levels of severity of a condition. Gold standard describes a method, procedure, or measurement that is widely accepted as being the best available. Sensitivity is the probability that those in the population who have the disease are correctly detected by the test. Specificity reflects the probability that those in the population without the disease are correctly identified as free of disease. Positive predictive value is defined as the proportion of the population with positive results who truly diseased. Negative predictive value is the converse. An increase in sensitivity leads to lower specificity. When prevalence of a disease drops, the positive predictive value drops and the negative predictive value increases. The receiver operating characteristic (ROC) is described by plotting the sensitivity (true positive rate) on the y-axis against specificity (false-positive rate) on the x-axis. The area under the ROC curve represents the probability that test correctly identifies 2 subjects as normal or abnormal when one is randomly chosen from the abnormal group. The likelihood ratios (LRs) estimate the probability that disease is present at any level of a test result. LR for a given diagnostic test gives the odds that the test result comes from a person who has the disease for which the test was ordered. Working out LRs is confusing because clinicians must alternate between odds and probabilities. Bias in the assessment of diagnostic biotechnology: Work bias occurs if the results of the diagnostic test alter the subsequent investigations of patients. Test review bias occurs when the observer is aware of the diagnosis at the time of reporting the test results. Diagnostic review bias occurs if the results of the test in question are used in part to determine the find diagnosis (gold standard). The danger is that the results of accuracy will be correlated with the gold standard and perhaps not with the disease it was intended to measure. Interpretation-skill bias occurs when interobserver variations are present. The characteristics of a diagnosis test may change over time because of technologic improvements or innovations. Methods of technologic assessment: Efficacy is the extent to which a specific intervention, procedure, or regimen produces a beneficial result under ideal conditions. Efficacy is determined by the results of a randomised controlled trial. Effectiveness is the extent to which a specific intervention, procedure, or regimen, when used in real life, does what it is intended to do for a definite population. Safety is a primary issue when examining a new diagnostic technology for assessment. Safety is defined as the condition that a drug, device, or procedure presents an acceptable risk of harm, injury, or loss when the technology is used for a specific indication. Diagnostic tests can be evaluated by using randomised controlled trials, metanalysis before-after design studies, and natural history of pretechnology and posttechnology intervention. Clinical trials may be too cumbersome, impractical, or expensive for regular evaluation of diagnostic technologies. Problems include the need for a large number of patients, the need for preliminary use of the technology in practice for clinicians to develop expertise in interpretation of the results, and rapid developments in technology that may make the results of a trial obsolete by time they appear. Metanalysis is a process of using statistical methods to combine the results of different studies. Metanalysis is a developing field with considerable promise not only for technology assessment but also for many other applications in scientific research and clinical medicine. The full integration of a new biotechnology into clinical practice is jeopardised and a potential disservice is done to both clinician and patients. Potential biases in assessing new biotechnology include work-up bias, uninterpretable test results, interpretation-skill bias, temporal changes, influence of clinical factors on test interpretation, absence of a definitive reference test, test-review bias, and diagnostic review bias. The number of red blood cells (RBCs) is determined by age, sex, altitude, exercise, diet, pollution, drug use, tobacco/nicotine use, kidney function, etc. The test measures of the oxygen carrying capacity of the blood. Optimal values for an adult male are 4.70-5.25 million/mm^3 and for an adult female are 4-4.50 million/mm^3. The number of RBCs are increased in: 1- chronic respiratory insufficiency (emphysema, respiratory distress, living at a high altitude, and cystic fibrosis, 2- Non-respiratory (adrenal cortical hyperfunction, polycythaemia vera, anabolic metabolism). The number of RBCs are decreased in: 1- iron deficiency, 2- vitamin B_6, B_{12}, and/or folic acid deficiency, 3- chronic disease, e.g., liver dysfunction, kidney dysfunction, 4- hereditary anaemia, 5- free radical pathology, 5- toxic metals, and 6- catabolic metabolism. Haemoglobin (Hb) contains the iron, which carries the oxygen to the cells. The Hb level indicates the amount of intracellular iron. Hb is the most abundant protein found within the RBC. A Hb value above or below the average may not necessarily be a problem. Hb must be evaluated with the haematocrit (HCT), RBC, and the RBC indices (MCV) to determine if there is fact anaemia and the type of anaemia. The causes of low Hb may need serum iron studies, globulin levels, uric acid, caeruloplasmin (copper), and ferritin (iron stores) to be determined. Optimum values for an adult male is 14-15 g/dl and for an adult female is 13.5-14.5 g/dl. Hb is increased in: 1- dehydration (prolonged or severe diarrhea), 2- emphysema, severe asthma, and other forms of long-standing respiratory distress, 3- macrocytosis (deficiency of B_6, B_{12}, folic acid, or hypothyroidism), 4- adrenal cortex overactivity, 5- polycythaemia vera, 6- high altitude adaptation, 7- splenic hypofunction, 8- exogenous testosterone. Hb is decreased in: 1- digestive inflammatory diseases, 2- free radical pathology, 3- adrenal cortical hypofunction, 4- hereditary anaemia, 5- haemodilution (pregnancy, oedema), 6- haemorrhage, 7- deficiency (protein malnutrition, iron, copper, vitamin C, B_1 (thiamine), B_{12}, folic acid), 8- chronic disease, and 9- bone marrow insufficiency (infiltration with tumour or tuberculosis, toxic or drug induced hypoplasia). Anaemia is not a disease, but a term indicating insufficient haemoglobin to deliver oxygen to the cells. It is always a secondary phenomenon. The haematocrit (HCT) represents the volume of RBCs in 100 ml of blood. The HCT is one of the most precise methods of determining the degree of anaemia or polycythaemia. A low HCT and Hb usually indicates decreased production, excessive loss, or destruction of RBCs. Optimum values in an adult males is 42-48% and in an adult female is 39-45%. An elevated HCT may be due to spleen hyperfunction, and a reduced HCT may indicate low thymus function. The mean corpuscular volume (MCV) relates to the average size of the RBC. Optimum value is 87-92 cuMicrons. The MCV is increased in: 1- hereditary anaemia, 2- megaloblastic anaemias (pernicious, folic acid deficiency, B_{12} deficiency), 3- reticulocytosis (acute blood loss response), 4- artifact (aplasia, myelofibrosis, hyperglycaemia, cold agglutinins), 5- liver disease, 6- hypothyroidism, 7- drugs (anti-convulsants), and 8- zidovidune treatment (AIDS). The MCV is decreased in: 1- copper deficiency, 2- hypochlorhydria, 3- vitamin C insufficiency, 4- vitamin B_6 anaemia, 5- rheumatoid arthritis, 6- toxic effects of lead and other toxic elements, 7- hereditary (thalassemias, sideroblastic), 8- iron deficiency, 9- post-splenectomy, and 10- haemolytic anaemia. Because anaemia due to folic acid and B_{12} deficiency is difficult to differentiate without more sophisticated tests, any supplementation of B_{12} should always be accompanied by folic acid as well, and vice versa. Folic acid and B_{12} should be considered in all cases of nerve inflammation, nerve degeneration, blood sugar problems, nerve irritation and vegetarian diets. B_6 and Mg need should be considered whenever P.M.S. is present. The amount of haemoglobin in a single RBC is indicated by the mean corpuscular haemoglobin (MCH). It is a variation of the MCV measurement. The MCH is increased in and decreased in the same conditions as the MCV. The average haemoglobin concentration per unit volume (100 ml) of packed red cells is indicated by mean corpuscular haemoglobin concentration (MCHC). The MCHC is elevated in spherocytosis but not in pernicious anaemia. MCHC is increased in and decreased in the same conditions as the MCV. Platelets are actually fragments of cells that participate in clotting. Platelets are increased in: 1- reactive (infection, acute blood loss, disseminated carcinoma, splenectomy, free radical pathology), 2- thrombocythaemia (polycythaemia vera, myeloproliferative disorders, chronic granulocytic leukaemia, haemolytic anaemia, myelosclerosis). Platelets are decreased in: 1- decreased production (marrow depression, marrow infiltration, megaloblastic anaemia, congenital), 2- increased destruction (immunologic, drugs such as chemotherapy or heparin, dilution due to overhydration, coagulation disorders, hypersplenism, platelet aggregation, rubella, liver dysfunction, idiopathic thrombocytopenic purpura [ITP]). The random distribution of weight (RDW) stands for random distribution of RBC weight. This is an electronic index, which may help clarify if an anaemia has multiple components. Optimal range is 13. The RDW is increased in: 1- B_{12} and pernicious anaemia, 2- folic acid anaemia, 3- iron deficiency anaemia combined with other anaemia, 4- haemolytic anaemia, 5- transfusions, 6- sideroblastic anaemia, and 7- alcohol abuse. The RDW is decreased in: 1- iron deficiency anaemia, 2- vitamin B_6 anaemia, and 3- rheumatoid arthritis. γ-Glutamyltransferase (GGT) is a microsomal enzyme that is widely distributed in human tissues involved in secretory and absorptive processes, especially the bile canaliculi and brush border of renal tubules. It is also present in heart, pancreas, lungs and seminal vesicles. It is located in cell membranes, where its role is transmembrane amino acid transport as γ-glutamyl amino acids. Most of the activity in blood appears to originate primarily from the hepato-biliary system, and GGT is elevated in any type of liver disease. The mechanism of GGT increase in diseased human liver is not well understood but is thought to be microsomal enzyme induction due to drugs or alcohol, whereas high molecular weight forms of GGT may represent the release of cell membrane fragments rich in GGT into the circulation. GGT is more sensitive than alkaline phosphatase (ALP), 5'nucleotidase and the aminotransferases in detecting obstructive jaundice, cholangitis and cholecystitis. Its rise occurs earlier than with these other enzymes and persists longer. GGT activity in the neonatal period (0-4 weeks) is approximately 5-7 times higher than the upper limit of the adult reference range. Thereafter the activity declines, and after 7 months reaches that of the adult. Plasma GGT measurement is of value in the differential diagnosis of certain liver disorders, e.g., extrahepatic biliary atresia, primary sclerosing cholangitis and progressive familial intrahepatic cholestasis. The liver can function adequately on 20% of liver tissue; thus early diagnosis by laboratory methods is difficult. With severe liver cell damage, the prolonged prothrombin time will not change with ingestion of vitamin K. Proteins are the most abundant compounds in the serum. Amino acids are the building blocks of all proteins. Proteins are the building blocks of all cells and body tissues. Amino acids are the basic components of enzymes, many hormones, antibodies and clotting agents. Proteins act as transport substances for hormones, vitamins, minerals, lipids and other materials. Proteins help balance the osmotic pressure of the blood and tissue. Proteins play a major role in maintaining the delicate acid-alkaline balance of the blood. Finally, serum proteins can serve as a reserve source of energy for tissues and muscle. There are 4 major types of globulins, each with specific properties and actions. Total protein represents the sum of albumin and globulins. Total protein may be elevated due to: 1- chronic infection, 2- adrenal cortical hypofunction, 3- liver dysfunction, 4- collagen vascular disease (rheumatoid arthritis, systemic lupus, scleroderma), 5- hypersensitivity states, 6- sarcoidosis, 7- dehydration (diabetic acidosis, chronic diarrhoea, etc.), 8- respiratory distress, 9- haemolysis, 10- cryoglobulinaemia, 11- alcoholism, 12- leukaemia. Total protein may be decreased due to: 1- malnutrition and malabsorption, 2- liver disease, 3- diarrhoea, 4- severe burns, 5- loss through the urine in severe kidney disease, 6- low albumin, 7- low globulins, 8- pregnancy. Albumin is synthesised by the liver using dietary protein. Its presence in the plasma creates an osmotic force that maintains fluid volume within the vascular space. Albumin is a predictor of health; low albumin is a sign of poor health and a predictor of a bad outcome. Albumin levels may be elevated in: 1- dehydration, 2- congestive heart failure, 3- possibly poor protein utilisation, 4- glucocorticoid excess (medications with cortisone effect, the adrenal gland overproducing cortisol, or tumour produces extra cortisol-like compounds), 5- congenital. Albumin levels may be decreased in: 1- dehydration, 2- hypothyroidism, 3- chronic debilitating diseases (e.g., RA), 4- malnutrition-protein deficiency, 5- dilution by excess H_2O (drinking too much water [polydipsia], or excess administration of IV fluids), 6-renal losses (nephrotic syndrome), 7- protein losing-enteropathy (protein is lost from the GIT during diarrhoea), 8- skin losses (burns, exfoliative dermatitis), 9- liver dysfunction (the body is not synthesising enough albumin and indicates very poor liver function). Globulins are proteins that include γ globulins (antibodies) and a variety of enzymes and carrier/transport proteins. The specific profile of the globulins is determined by protein electrophoresis (SPEP), which separates the proteins according to size and charge. There are 4 major groups that can be identified: γ globulins, β globulins, α-2 globulins, and α-1 globulins. Once the abnormal group has been identified, further studies can determine the specific protein excess or deficit. The γ fraction makes up the largest portion of the globulins, and antibody deficiency should be related to it when the globulin level is low. Antibodies are produced by mature B lymphocytes called plasma cells, while most of the other proteins in the α and β fractions are made in the liver. The globulin level may be elevated in: 1- chronic infections (parasites, some cases of viral and bacterial infection), 2- liver disease (biliary cirrhosis, obstructive jaundice), 3- carcinoid syndrome, 4-rheumatoid arthritis, 5- ulcerative colitis, 6- multiple myelomas, leukaemias, 7-Waldenstrom's macroglobulinaemia, 8- autoimmunity (systemic lupus, collagen diseases renal dysfunction [e.g., nephrosis]). The serum globulin level may be decreased in: 1- nephrosis, 2- α-1 antitrypsin deficiency (e.g., emphysema), 3- acute haemolytic anaemia, 4- liver dysfunction, and 5- hypogammaglobulinaemia/agammaglobulinaemia. The proper albumin to globulin (AG) ratio is 2:1. When <1.7, there is may be a need for increasing stomach acidity. When >3.5 there may be a need for stomach acidity and pepsin. The AG ratio may be elevated in: 1- hypothyroidism, 2- high protein/high carbohydrate diet with poor nitrogen retention, 3- hypogammaglobulinaemia (low globulin), 4- glucocorticoid excess (medications with cortisone effect, the adrenal gland overproducing cortisol, or tumour produces extra cortisol like compounds, low globulin). The AG ratio may be decreased in liver dysfunction.

BUCCAL MUCOSA AS A ROUTE FOR SYSTEMIC DRUG DELIVERY

Peroral administration of drugs has disadvantages such as hepatic first pass metabolism and enzymatic degradation within the GIT, that prohibit oral administration of certain classes of drugs especially peptides and proteins. Transmucosal routes of drug delivery (i.e., the mucosal linings of the nasal, rectal, vaginal, ocular, and oral cavity) offer distinct advantages over peroral administration for systemic drug delivery. These advantages include possible bypass of first pass effect, avoidance of presystemic elimination within the GIT, and, depending on the particular drug, a better enzymatic flora for drug absorption. The mucosa of oral cavity is relatively permeable with a rich blood supply, it is robust and shows short recovery times after stress or damage, and the virtual lack of Langerhans cells makes the oral mucosa tolerant to potential allergens. Delivery of drugs within the oral mucosal cavity is classified into 3 categories: 1- sublingual delivery, which is systemic delivery of drugs through the mucosal membranes lining the floor of the mouth; 2- buccal delivery, which is drug administration through the mucosal membranes lining the cheeks (buccal mucosa); and 3- local delivery, which is drug delivery into the oral cavity. The epithelium of the buccal mucosa is about 40-50 cell layers thick, while that of the sublingual epithelium contains somewhat fewer. The turnover time for the buccal epithelium is 5-6 days. The thick buccal mucosa is 500-800 μm, while the mucosal thickness of the hard and soft palates, the floor of the mouth, the ventral tongue, and the gingivae measure at about 100-200 μm. The keratinized epithelia contain neutral lipids like ceramides and acylceramides, which have been associated with the barrier function. These epithelia are relatively impermeable to water. The non-keratinized epithelia, such as the floor of the mouth and the buccal epithelia, do not contain acylceramides and only have small amounts of ceramide. They also contain small amounts of neutral but polar lipids, mainly cholesterol sulfate and glucosyl ceramides. The permeabilities of the oral mucosae decrease in the order of sublingual greater than buccal, and buccal greater than palatal. The sublingual mucosa is relatively thin and non-keratinized, the buccal thicker and non-keratinized, and the palatal intermediate in thickness but keratinized. The cells of the oral epithelia are surrounded by an intercellular ground substance, mucus, the principle components of which are complexes made up of proteins and carbohydrates. In stratified squamous epithelia found elsewhere in the body, mucus is synthesised by specialized mucus secreting cells like the goblet cells, however in the oral mucosa, mucus is secreted by the major and minor salivary glands as part of saliva. Up to 70% of the total mucin found in saliva is contributed by the minor salivary glands. At physiological pH the mucus network carries a negative charge (due to the sialic acid and sulfate residues), which may play a role in mucoadhesion. At this pH mucus can form a strongly cohesive gel structure that will bind to the epithelial cell surface as a gelatinous layer. Saliva is the protective fluid for all tissues of the oral cavity. It protects the soft tissues from abrasion by rough materials and from chemicals. It allows for the continuous mineralisation of the tooth enamel after eruption and helps in remineralisation of the enamel in the early stages of dental caries. The flow rate of saliva depends on the time of day, the type of stimulus, and the degree of stimulation. The daily salivary volume is between 0.5-2 liters. The passive drug transport across the oral mucosa occur via paracellular and transcellular routes. The intercellular spaces and cytoplasm are hydrophilic in character, and the cell membrane is lipophilic in nature. The lipophilic compounds have low solubilities, the intercellular spaces pose as the major barrier to permeation of lipophilic compounds and the cell membrane acts as the major transport barrier for hydrophilic compounds. Sublingual dosage forms are of 2 different designs, those composed of rapidly disintegrating tablets, and those consisting of soft gelatine capsules filled with liquid drug. The sublingual route is capable of producing a rapid onset of action making it appropriate for drugs with short delivery period requirements with infrequent dosing regimen. The buccal mucosa is less permeable and is thus not able to give a rapid onset of absorption, more suitable for a sustained release formulation). The buccal mucosa is more fitted for sustained delivery applications, delivery of less permeable molecules, and perhaps peptide drugs. Enhancers used to improve drug permeation in other absorptive mucosae have been shown to work in improving buccal drug penetration. The superficial layers and protein domain of the epithelium may be responsible for maintaining the barrier function of the buccal mucosa (low flux). Bioadhesive polymers are defined as polymers that can adhere onto a biological substrate. The term mucoadhesion is applied when the substrate is mucosal tissue. Normally, hydrogels are crosslinked so that they would not dissolve in the medium and would only absorb water. When drugs are loaded into these hydrogels, as water is absorbed into the matrix, chain relaxation occurs and drug molecules are released through the spaces or channels within the hydrogel network. In a more broad meaning of the term, hydrogels would also include water-soluble matrices that are capable of swelling in aqueous media, these include natural gums and cellulose derivatives. These pseudo-hydrogels swell infinitely and the component molecules dissolve from the surface of the matrix. Drug release would then occur through the spaces or channels within the network as well as through the dissolution and/or the disintegration of the matrix. The buccal mucosa offers several advantages for controlled drug delivery for extended periods of time. The mucosa is well supplied with both vascular and lymphatic drainage and first-pass metabolism in the liver and pre-systemic elimination in the GIT are avoided. The area is well suited for a retentive device and appears to be acceptable to the patient.

PHARMACOTHERAPEUTICS AND PHARMACOKINETICS

Pharmacotherapeutics directly relates to core drug knowledge and core patient variables such as life span and gender. In pharmacotherapy, drugs are used to prevent, diagnose, relieve, treat, or cure disease because a drug is a substance that interacts with a living organism to produce a biologic response. This is termed the therapeutic use of a drug. A response to drug results from a biochemical or physiologic interaction between the drug and a cellular component (receptor) to which a drug binds to produce its effects. Exceptions to this include antacids, free radicals, osmotic diuretics, and chelators of heavy metals. Potency-the product of a drug's affinity and efficacy-reflects its overall ability as an agonist to stimulate a receptors. Each drug-induced response has a threshold. Drugs produce multiple predictable biologic effects. The safety of a drug is partly indicated by its dose-response curve. A gradual slope indicates that a relatively large change in dose will produce a relatively small change in effect. A steep slope means that a small change in dose will produce very large changes in drug response. The drug with a steep dose-response curve will be difficult to administer safely. If indicated, monitoring the plasma level of a compound yields information about the efficacy and safety of a drug and allows it to be more closely controlled in terms of dosage, scheduling, and route of administration. Pharmacokinetics is the ways in which the body processes drugs, is a mathematical science. The 4 basic components of pharmacokinetics are absorption, distribution, metabolism, and excretion (abbreviated as ADME). Absorption is the process by which a drug moves from its site of administration into the venous or lymphatic circulation. Any drug (except those injected intravenously) must first be absorbed before it can produce an action within the body. Bioavailability describes the fraction of the administered dose that reaches the systemic circulation and produces effects. Distribution is the delivery of the drug into any and all body compartments it can penetrate. Drug elimination usually begins with hepatic metabolism and is followed with renal excretion. Metabolism (also known as biotransformation) is the alteration or changing of a drug to more ionised or water-soluble and less lipid-soluble forms called metabolites. Most drugs are renally eliminated either as drug metabolites or unchanged drug molecules. ADME processes are interrelated and involve drug passage across cell membranes.

Drug routes:

Route	Pattern	Advantage	Disadvantage
Intravenous	Immediate effect. No absorption.	Accommodate large doses.	Adverse reactions. Unsuitable for oily solution. Risk of infection. Not for self-administration.
Intramuscular	Prompt absorption of aqueous solution. Slow or sustained absorption of depot.		Depends on local blood flow. Painful. Increases CPK. Contraindicated with anticoagulants.
Subcutaneous	Prompt absorption of aqueous solution. Slow or sustained absorption of depot.	Self-administration. Suitable for implantation.	Depends on local blood flow. Painful. Risk of necrosis.
Per ora (PO)	Absorption.	Convenient Economical Safe	Requires adherence. Erratic bioavailability. Dysphagia.
Topical	Absorption is slow.	Acceptability Convenient	Few drugs available. Systemic absorption.
Inhalation	Absorption is rapid.	Convenient	Requires nebuliser. Cough or wheezing.
Rectal	Absorption.	Local or systemic effect.	Uncomfortable. Systemic effect.

Drugs administered in a semisolid or solid form must first disintegrate and dissolve into a solution before absorption can occur. Enteric coating of capsules or tablets delays disintegration; these forms are formulated specifically to disintegrate in the alkaline environment of the small intestine. The route of administration for a drug is primarily determined by its chemical properties (e.g., solubility in water or lipids) and the therapeutic or treatment objective (e.g., rapid onset of action, localised effect) for its use. Drugs administered enterally can be absorbed in the oral or gastric mucosa, or in the small or large intestines, such as fluids and nutrients (e.g., tube feedings). Per ora (PO) is convenient, inexpensive, and relatively safe and simple. This route involves one of the most complicated pathways to the tissues. Although some drugs are absorbed from the stomach, most drugs are absorbed from the duodenum by a passive process. Drugs go through the portal circulation to the liver before reaching the general circulation (first-pass phenomenon). First-pass metabolism may significantly limit a drug's effectiveness by extensively breaking down the drug before it ever reaches the general circulation, e.g., 90% of nitroglycerin is eliminated by the hepatic first-pass effect. Drugs will lose much of their active dose before ever entering the general circulation. Sublingual and buccal drugs involve placement of drug under the tongue or in the cheek, which allows the drug to diffuse directly into the capillary network and enter the general circulation, bypassing the first-pass phenomenon. Drugs administered parenterally bypass the GIT. Parenteral routes include intravenous (IV), intramuscular (IM), subcutaneous (SC), intradermal (ID), and intra-articular (IA) and intrathecal. Drugs absorbed from suppositories in the lower rectum enter vessels that drain into the inferior vena cava, thus bypassing the liver. The rectal route prevents decreased stomach pH, intestinal enzymes, or the hepatic first-pass phenomenon. Suppositories tend to move upward in the rectum into a region where veins that lead to the liver (e.g., superior haemorrhoidal vein) predominate. Consequently, only about 50% of a rectal dose can be assumed to bypass the liver. Like the sublingual route, the rectal route facilitates drug absorption with drugs that produce nausea or vomiting in a patient who is already vomiting. Solubility is the ability of a drug to dissolve and form a solution. Two primary factors in drug absorption are drug solubility and ability of the drug to move through cell membranes. A drug already in solution will be absorbed faster than the same drug in a solid form, which must first disintegrate and dissolve. Lipid solubility is essential for any drug that must diffuse across the cell membrane because the membrane is partially composed of lipids. Lipid solubility is influenced by the cellular environment at the absorptive site. Many drugs are lipid soluble as well as water-soluble and can dissolve in body fluids and pass through cell membranes. A molecule dissociates into ions when it dissolves in water or other liquids. In dissociating, the molecule gives up or accepts a proton, thus converting some molecules into charged particles (i.e., ions). Ionised drugs are poorly absorbed, because they do not diffuse easily across lipid membranes. Nonionized drug molecules are lipid soluble and easily pass through lipid membranes. The pH absorptive environment determines the extent to which a molecule will ionise. Lipid-soluble molecules (e.g., alcohol, carbon dioxide, fatty acids, and oxygen) readily move across most biologic membranes after dissolving in the lipid matrix of the membrane. Water-soluble drugs penetrate the cell membrane through aqueous channels. Most drugs are distributed through diffusion. A carrier (e.g., insulin) is needed to cross the cell membrane, but energy is not required. The rate of movement depends on 3 factors: 1- the differences in drug concentrations, 2- the amount of readily available carrier, and 3- the speed at which the carrier binds and release the transported substance. Filtration is the passage of a drug through the pores of a semipermeable membrane. Osmotic or hydrostatic pressure drives the molecules down a concentration gradient. Generally, active transport is more rapid than passive diffusion. An energy source is required to move molecules across the cell membrane in active diffusion, because this movement takes place against a concentration gradient. During pinocytosis the cell membrane surrounds and engulfs a substance on its outer surface, forms a membrane-covered vesicle, and carries it inside the cell, e.g., proteins, fat-soluble vitamins. Bioavailability is the fraction of administered drug that reaches the systemic circulation. The extent of drug absorption may be calculated by plotting plasma concentrations of the drug versus time and measured by the area under the curve (AUC). The curve reflects the extent of drug absorption. Factors that influence bioavailability are first-pass hepatic metabolism, solubility of the drug, chemical instability in gastric pH, and drug formulation. Distribution is the process by which drug molecules leave the bloodstream and are transported by body fluids to sites of action where the drug produces its effects. The drug is also distributed to body tissues where no effects are produced. Factors influence the distribution of drugs include the volume of distribution, cardiac output, regional blood flow and capillary permeability, degree of plasma protein binding and drug reservoirs or storage sites, drug concentration, tissue affinity, and physiologic barriers. Drugs that are highly lipid soluble and highly tissue bound have a high volume distribution (Vd) and therefore a lower serum level. Women have a higher percentage of body fat than men do, causing women to have a greater Vd for highly lipophilic drugs (e.g., diazepam, trazodone). This results in a lower peak plasma level but a longer half-life. Women have a lower Vd for highly hydrophilic agents (e.g., alcohol, aminoglycoside antibiotics). In the case of ethanol, this means that an equal amount per kg of body weight will produce higher levels in body fluids, delivering higher ethanol levels to brain cells and thereby producing greater intoxication. The quantity of drug that enters the tissue is directly proportional to the perfusion of that tissue. For example, blood flow to the brain, liver, and kidney is greater than that of the skeletal muscles followed by adipose tissue. Drugs circulate in the plasma either bound or unbound to plasma protein, usually albumin. Bound drug molecules are pharmacologically inactive and they remain that way until released from the protein. Only the unbound (free) portion of the drug is available to cross membranes and produce an effect. The binding is reversible. Albumin is the most abundant plasma protein; acidic drugs bind strongly to albumin. The large size of plasma proteins (molecular weight about 68,000) keeps it within the bloodstream, and therefore bound drug molecules cannot leave the bloodstream to reach sites of action or to undergo metabolism and excretion. Plasma proteins serve as reservoirs for drugs. Basic drugs as well as the protease inhibitors have a high affinity for α-1 acid glycoproteins (AAG). Many alkaline drugs are bound to AAG. Plasma protein binding is expressed as a percentage. Hypoalbuminaemia may alter the level of free drug. Examples of tightly bound drugs are warfarin and propranolol (90% and 93%, respectively). The percentages of loosely protein-bound drugs usually range between 10% and 35%. Competitive binding is a property that accounts for clinically important drug interactions. Two drugs competing for the same protein sites will displace each other in some cases, with serious consequences. For example, aspirin will displace warfarin, thereby increasing free warfarin and possibly causing bleeding tendencies. Drugs may be distributed to bone, fat, plasma proteins, or other tissues for storage. Once bound, these storage areas can serve as drug reservoirs for prolonged periods of time, e.g., fat-soluble drugs (lipophilic drugs) have a high affinity for adipose tissue. They are usually released slowly into the bloodstream only after the drug administration stops. The drug concentration level may complicate therapeutic objectives, e.g., antibiotics are not typically distributed to abscesses and exudates, glands (e.g., prostate) tend to be impermeable to most antibiotics, and drug distribution to tumours tends to be unpredictable. Affinity is the chemical force that drives certain atoms to unite with certain others to form compounds and attract molecules to specific sites. The blood-brain barrier is a selective mechanism that opposes the passage of most ions and large molecular compounds from the blood to the brain tissue. This structure of capillaries is characterised by very tight junctions through which most drugs cannot pass. Consequently, drugs must be lipid soluble or have a transport system to leave the bloodstream and reach a site of action within the brain. There are special transport mechanisms present in the brain capillaries for glucose, amino acids, amines, purines, nucleosides, and organic acids. The thick membranes of the placenta do not create an absolute barrier to the passage of drugs from mother to foetus. Steroids, narcotics, anaesthetics, and some antibiotics are easily transported across the placental membrane, producing their effects in the foetus. Metabolism is the process by which the body changes the chemical structure of a drug. The liver is the major site for drug metabolism, but not the only site; other sites are GIT, lungs, kidneys, and skin. Xenobiotic (i.e., foreign non-food) substances or chemicals (e.g., most drugs, environmental pollutants, food additives, fillers, dyes, and contaminants) initiate the process of metabolism because they are foreign or alien to the organism. From metabolism and biotransformation, a drug (or chemical) is converted into another chemical, called a metabolite. Sometimes the metabolites are also pharmacologically active. Enzymes break down drugs and their resulting metabolites. Hepatocytes contain the enzymes that transform or convert drugs. Drugs may require a one- or two-phase set of hepatic transformations before they can be excreted. Phase I reactions most frequently involve the cytochrome P-450 system (also called microsomal mixed function oxidase). Phase I reactions are oxidation reactions and usually convert the parent drug to a metabolite sufficiently polar so it can then be renally excreted. Phase II reactions are conjugation reactions and considered synthetic. They involve the combination of a drug or its metabolite with an equal portion of an endogenous water-soluble substance (e.g., glucuronic acid). These conjugates are highly polar and rapidly excreted in the urine and faeces. Other metabolic sites exist (e.g., GI mucosa, kidney, lung, brain, skin) where different metabolic enzymes carry out biotransformation phases I and II. The hepatic microsomal enzyme system is responsible for metabolising and biotransforming drugs. This enzyme system is known as the cytochrome P-450 (CYP-450) system, which carries out drug-metabolising activity. This system is important for the metabolism of many endogenous compounds (e.g., steroids, lipids) and for the detoxification of exogenous substances (e.g., drugs, chemicals). The CYP-450 system contains 12 identified families of heme-containing isoenzymes; they occur in packets known as microsomes and are localised in the endoplasmic reticulum (or microsomal fraction) of numerous cells in the liver, intestine, and throughout the body. CYP3A4 is the most abundant of all isoenzymes (20% to 60% of total) and it has the broadest substrate (substance acted upon and changed by an enzyme) of any known iso-enzyme. A drug that is described as a CYP inducer increases the amount of the isoenzyme. An inducer of CYP3A4 isoenzyme (e.g., carbamazepine, phenobarbital, phenytoin) may cause a significant decrease in the blood level of CYP3A4 substrate drugs (e.g., alprazolam, fentanyl, midazolam). A drug that is described as a CYP inhibitor decreases the isoenzyme. Concomitant administration of a CYP2D6 iso-enzyme inhibitor (e.g., cimetidine, fluoxetine, paroxetine) may cause a distinct elevation of a CYP2D6 drug (e.g., amitriptyline, desipramine, haloperidol). A drug that may be a CYP inhibitor without being a substrate is a moderate inhibitor. Circadian reabsorption of 70-80% of the bile may cause the drug or its metabolites to be returned to the systemic circulation. Enterohepatic cycling may prolong the biologic activity and delay the eventual disposition of the drug. Any drug that can cause hepatotoxicity has the potential to prolong a person's metabolic response to any other drug. The metabolic inhibition that results from the inhibition or poisoning of the microsomal enzymes. Numerous drugs and drug classes can cause hepatic dysfunction, necrosis, injury, and hepatotoxicity (e.g., anticonvulsants, NSAIDs, and antibiotics). Some drugs stimulate or induce the hepatic microsomal enzyme system, which increases the liver's capacity to metabolise other drugs (stimulation). Some drugs are self-inducers, such as alcohol, carbamazepine, phenobarbital, phenytoin, and rifampin. When a drug is known to be a metabolic inducer, the metabolism of any concurrent drugs may be affected. Excretion is the removal of a drug (or its metabolites) from the body. Drug excretion occurs mainly through the kidneys into the urine, hepatic, respiratory, and biliary systems, the GIT, and the extrahepatic structures (sweat, salivary glands, skin). Clearance is the rate of disappearance of the drug molecules from the circulation. Drugs with a slow clearance rate are removed more slowly, increasing the circulating blood levels as half-life is increased. Protein binding capacity and body weight may affect drug clearance rates. Renal clearance is governed by the processes of filtration, secretion, and reabsorption. Most drugs or their metabolites are efficiently excreted in the urine by glomerular filtration (GFR). GFR (125 ml/min) is normally about 20% of the renal plasma flow (600 ml/min). Low-molecular-weight molecules (i.e., <15 angstroms) readily pass through the glomeruli and, along with water. Any free drugs will appear in the filtrate, whereas protein-bound drugs will remain with the unfiltered serum proteins. Tubular secretion is an active process that occurs in the proximal tubule. Renal clearance is the sum of both routes (GFR and RTS). Acids and bases are secreted through the proximal tubules. Renal ubular secretion (RTR) is dependent on active carriers. As the filtrate moves through the renal tubule, 2 types of reabsorption may occur, namely active or passive. Most drugs are eliminated from the body through the liver or kidney and when the patient has liver damage, some impairment of hepatic excretion may be expected. Some drugs are eliminated in other ways and this is termed extrahepatic excretion. For example, macromolecules (e.g., heparin, amphotericin B) are engulfed and digested by phagocytic cells (e.g., hepatic Kupffer cells). This is termed reticuloendothelial system clearance. Volatile and gaseous drugs (e.g., alcohol, anaesthetics, solvents) may be excreted unchanged by the lungs through respiration or excreted through the skin. The steady state is the evenly distributed concentration of a drug, occurs when the administration rate equals the rate of drug elimination. More frequent dosing of a drug does not create steady state. Steady state is dependent on the half-life of the drug. Half-life is the time needed for the plasma concentration of a drug to be reduced by 50%. Steady state is achieved in approximately 5 half-lives. Drugs with short half-lives are rapidly cleared from the body; drugs with longer half-lives stay in the body for a longer duration of time.

PATHOBIOLOGY

Pathophysiology is the biomedical science of the mechanisms of development and elimination of pathological processes. This terminology was first used in Russia in 1878 as "pathological physiology". In the 1930s pathophysiology acquired the significance of a medical discipline and was included in the medical education curricula. Modern pathophysiology is an integrative, fundamental biochemical science. The term pathophysiology is better to be replaced by the term pathobiology. The aim of pathobiology: 1-Identification of the mechanisms underlying pathological processes at various levels of the organism, under different conditions, and study their relationships. 2- Determination of the specific and common characteristics of different pathological processes. 3- Constitute significant trends in fundamental research. The final biological outcome of pathological processes is the disturbance of certain functions. Pathobiology represents a functional pathology of various structures, from intracellular events to complex systemic relationships at the highest levels. The pathogenesis of a pathological process is not confined to the injuries to the morphological structures and physiological systems. Along with the destructive phenomena, quite a different process takes place, the formation of a new relationships between the involved structures, emergence of the secondary endogenous mechanisms of pathological processes inherent in the altered structures. Result entirely in a new pathodynamic organisations and pathological systems which underlie the corresponding syndromes of functional disorders. Modern pathobiology elaborates the general concept of endogenization pathological processes. Biochemistry and biophysics obtain an ever increasing body of data about the endogenous mechanisms of pathological process development at the molecular, membranous and cellular levels. Modern pathobiology also studies the opposing phenomena, such as the mechanisms of prevention and elimination of the pathological process, and mechanisms responsible for compensation and recovery of disordered functions, i.e., sanogenesis. Knowledge of both classes of mechanisms-pathogenetic and sanogetic is needed for the development of principles and means of pathogenetic therapy and pathogenetic prophylaxis. Modern pathobiology is a bridge between the fundamental theoretical disciplines (biochemistry, physics, physiology, etc.) and clinical medicine. It lays down the foundation of the physician's biochemical mentality. The characteristic objects of pathobiology investigation are the typical pathological processes in tissues: oedema, inflammation, microcirculation disorders, hypoxia, energy deficit, degeneration, dystrophic changes, ischaemia, mitochondrial disorders etc. The typical processes realised at the membrane level; intracellular signalling, secondary messengers and genome systems are known, due to successes attained in biochemistry, biophysics and molecular biology. The boundary between experimental and clinical pathobiology has become less distinct. Clinical physiology is physiology at the patient's bedside. Physiological mechanisms ensure normal functioning, homeostasis, adaptation, recovery, rehabilitation and heath maintenance, while pathophysiological mechanisms proper underlie the development of pathological processes, diseases and functional disorders. Modern pathobiology has a new development in the form of space pathophysiology, ecology, environmental pathophysiology, pathophysiology of work and sport, adaptation pathophysiology, etc. Pathobiology is characterised by one main topic "dysregulation and its correction".

BIOETHICS

The development of a procedural structure for decision-making not only facilitates collective decision-making by health professionals, the family and other involved parties, but also opens the decision-making process to public scrutiny, a necessary prerequisite for the development of uniform standards. Practical skills should include the ability to present a conceptual framework that students can grasp and build on, and to create a teaching milieu, which fosters self-criticism and personal development by probing, questioning, stimulating and bringing forth self-criticism about one's own values often evokes anxiety, self-doubt and sometimes hostility. The teacher of medical ethics must know how to respond to such reactions sensitively and humanely in order to be able to conduct a meaningful class discussion and to psychologically counsel and support participants without imposing personally held values. The necessity for both classroom and bedside components to the teaching programme inevitably means that the teaching team should include both those professionals skilled in teaching the basic of medical ethics, and clinicians with the cognitive and clinical skills required to analyse, make and implement moral decisions during the course of everyday medical practice. The teaching team should be multi- and inter-disciplinary, co-ordination and control should be undertaken by a person or persons with a broad and liberal approach, and knowledge of the subject. The analytical and procedural techniques can be applied to a wide variety of medical problems at beginning, middle or end of life, highlighting points of conflict, questions log-held assumptions and convictions, and attempts to separate subjective from objective aspects of the case under consideration. Basic medical curricular goals include training that facilitates the ability to: 1- identify moral issues in bioethics; 2- obtain valid consent; 3- determine patients competence and enhance competence where possible; 4- proceed appropriately when dealing with partially competent or incompetent patients; 5- identify the appropriate course of action when a patient refuses treatment; 6- understand the boundaries of confidentiality; 7- deal with the terminally ill; 8- deliberate with patients and colleagues on a wide variety of ethical issues; 9- participate in rational decisions concerning the allocation of scare medical resources. Education must lead to competent medical practitioners, the medical education environment must provide and teach high standards of care; the manner of care provision must be acceptable to both the doctor and the patient, cost-effective medicine is necessary in order to be able to meet all needs, and the means of achieving this without compromising patient advocacy must be debated; and the educational system must be concerned with interpersonal relationships and structured in a way that will prevent dehumanisation. These are not easy goals to achieve, and the selection, teaching and training of health care professional is in a constant state of flux. Bioethics has the potential to build on, refine and extend common sense and traditional clinical judgement, but it is important to retain a critical outlook on the structure and content of bioethics education to ensure its evolution and transformation as an intellectual discipline. Every medical decision involves human beings, both as decision makers and as those who have to live with the decisions. Difficult decisions often have to be made under far-from-ideal circumstances, and involve choices, which can never be value-free. Because ethical decisions are frequently not ideal, the result is that often a tragic choice has to be made. Medical ethics has the potential both to benefit medical practice and to do harm to it. Teachers of medical ethics should enable health professionals to frame moral issues in appropriate terms, to produce the main arguments on either side, and to situate the issue in the context of the history and community in which it has arisen. The effective practice of medicine requires a combination of scientific knowledge and artistry. The art with which medicine is practised refers to the wisdom prudence, compassion, patience, sensitivity to ethical issues and devotion with which doctors apply scientific knowledge to the care of their patients. Bioethics re-orientate medical practice towards the humanitarian component of human life. The commercialisation and corporation of medicine within the current economic paradigm of thinking is turning medicine into a medical-industrial complex in which economic and conflict-of-interest considerations dominate. To ensure that those who enter the health profession remain motivated by valued virtue will require the convergence of forces to overcome the greed and guile of others who would drive medicine towards a deceitful and extreme form of exploitation of man by man. This will require recognition that individuals' preferences, values and perceptions are pervasively influenced by the peer group and by the social and institutional structures within which they operate, and that implementing sound ethical values will require collective action to change the incentive structures in existing institutions. What we teach must reflect what we want and what we do. Viewing medicine and ethics as hermeneutics may contribute to progress. The original definition of hermeneutics as the art of interpreting and the science of interpreting has evolved into a way of understanding and explaining human existence. Considering medicine as a hermeneutic enterprise in which metaphorically the patient is conceived as a multifaceted text to be interpreted biologically, personally, as a social being, and also in a broader context, requires not only scientific and social knowledge, but also the understanding and empathy, which comes from moral experience.

QUALITY OF LIFE (QOL)

QoL is a state of well-being that includes: 1- ability to perform everyday activities that reflect physical, psychological, and social well-being, and 2- patient satisfaction with levels of functioning and the control of disease and/or treatment-related symptoms. QoL should provide information about the actual levels of functioning and patient evaluation of quality of life. QoL is more than physical health status, clinical symptoms, or functional ability. Health-related QoL focuses on the effects illness or treatment on QoL and to distinguish these from aspects beyond the realm of health care, such as education, income, and quality of the environment. No one instrument of QoL is ideal for all situations in which QoL outcomes can be assessed. QoL may be a mixture of ordinary common sense plus a reasonable knowledge of pathobiology and clinical reality. Karnofsky performance status is an unidimensional indicator of the ability to carry on normal activity and so is not a measure of QoL. Elements of QoL may include health and functional domain (health, pain, sex, etc.), family domain (children, family health, etc.), psychological/spiritual domain (satisfaction, faith, etc.), and social and economic domain (emotional support from friends, education, job, etc.). QoL is commonly viewed as a subjective evaluation of the experience of life and, as such, is dependent on the individual's own perspective, expectations, and values. Objective indicators of QoL are externally observed by a third party and may include blood pressure, haemoglobin levels, tumour size, income, or years of education.

PATIENT EDUCATION

Patient education is a planned learning experience using a combination of methods such as teaching, counselling, and behaviour modification techniques that influence patient's knowledge and health behaviour. There is no single way to conduct patient education. The purpose of patient education is: 1- to understand the diagnosis, 2- to understand treatment options, 3- to understand responsibilities as part of treatment, 4- to understand symptoms that need to be reported, 5- to comply with treatment procedures, 6- to adapt to ongoing limitations, and 7- to cope with the reality of a terminal condition. The rationales of patient education include 1- patient's rights, 2- professional standards, 3- legal and mandates, 4- benefits to patients, and 5- benefits to society. The basic steps for conducting patient education include: 1- assessing educational needs, 2- planning the teaching process, 3- implementation, and 4- evaluation. In spite of the large volume of patient education materials available, materials designed to reach special populations are often difficult to find. Developing such materials requires special attention to cultural differences as well as sophisticated understanding of the language of the target audience (connotation, common expressions, colloquial terms), and reading level. Locally produced materials make less use of more complex technologies such as audio- and video-tapes, computer programs, computer technologies that provide interactive video capabilities, and illustrations. The pace of presentation is very important; the pace of an audio or video program is not directly controlled by the listener or viewer. Effective patient teaching is dependent upon process and content. Accurate, effective instructional materials are essential.

THE CELL

Energy is the capacity to make things happen, to do work. DNA molecules contain instructions from assembling each new organism from lifeless molecules that contain carbon, hydrogen, and few other kinds of substances. DNA is the molecule of inheritance in organisms. Its instructions for reproducing traits are passed on from parents to offspring. Atomic building blocks are charged particles called protons and electrons, and no charged particles called neutrons. An atom's core region, or nucleus, consists of some number of protons and (except hydrogen) neutrons. Protons carry a positive charge. The atomic number refers to the number of protons in the nucleus, it differs for each element, e.g., the hydrogen atom has one proton, and its atomic number is 1. Atomic weight (mass) is the total number of protons and neutrons in the nucleus; e.g., the mass number of a carbon atom (6 protons and 6 neutrons) is 12. The isolation of intact mitochondria followed improvements in centrifuge technology in the 1940s, and the introduction of isotonic sucrose preparation media to prevent osmotic lysis. The basic structure of mitochondria was established by electron microscopy. It was shown by controlled disruption that most of the respiratory enzymes responsible for oxygen uptake were attached to the inner mitochondrial membrane (which is extensively folded to form the mitochondrial cristae); while the soluble enzymes catalysing the Krebs cycle and fatty acid oxidation pathways were confined to the internal matrix space. The inner membrane is a major permeability barrier within the cell, but the outer membrane contains a protein called porin, which renders it largely permeable to molecules <1500 daltons. An important group of enzymes that metabolise ATP, including myokinase, creatine kinase, and the nucleoside diphosphate kinases, are trapped in the inter-membrane space, between the inner and the outer membranes. The citric acid cycle (Krebs cycle) is a central part of the chemistry, which converts food components like carbohydrates, fats, and proteins into ATP and thus provides energy to carry out biological functions. The function of the citric acid cycle is to convert the 2 carbons of acetyl CoA to CO_2, ATP, NADH and $FADH_2$. Since this cycle is a major part of the energy production system, its control is important so as to balance the cycle's activity and the body's needs. For example, NADH is an inhibitor of the enzyme for step 3, which oxidises the now secondary alcohol to a ketone. If there is lots of NADH, the cycle is slowed down to reduce formation of more. If ATP is being used up by the body's energy needs, the concentration of ADP increases and it activates that same enzyme. If there's lots of ATP, it inhibits the enzyme for step 1, which combines oxaloacetate with acetyl CoA to form citrate. These control mechanisms adjust the cycle's activity so that it keeps in step with the energy needs of the body, that is the function of the respiratory chain. The H atoms are removed from the NADH and $FADH_2$ molecules, making NAD^+ and FAD again available for the citric acid cycle. The H atoms are effectively split into protons and electrons. This process takes place in a membrane, so that the chemistry on one side is kept separate from the chemistry on the other side. The protons are deposited on one side, making that side more acidic. The electrons are used to reduce oxygen to water on the other side of the membrane. This latter reaction also consumes protons, so that the reduction of oxygen to water makes its side of the membrane less acidic. Oxidative phosphorylation general features are the passage of protons through a channel, which contains the enzyme complex called ATPase, provides the energy needed to make ATP from ADP and phosphate. When the results of the respiratory chain and oxidative phosphorylation are added up, the oxidation of a molecule of NADH to NAD^+ yields 3 ATP molecules, and the oxidation of a molecule of $FADH_2$ to FAD yields 2 ATP molecules. In eukaryotic cells the Krebs cycle occurs in the liquid part of the mitochondria (matrix) while the electron transport chain occurs in the inner membrane (cristae). The Krebs cycle converts pyruvate to CO_2 and reducing energy (NADH and $FADH_2$) and phosphorylated energy (GTP).

Pyruvate + 2 GDP + 2 H_3PO_4 + 4 H_2O + 2 FAD + 8 NAD^+ → 6 CO_2 + 2 GTP + 2 $FADH_2$ + 8 NADH

The reduced energy can be used to generate ATP using the electron transport chain in the presence of oxygen. The Krebs cycle is a cyclic process in that oxaloacetate reacts with acetyl CoA to form citrate, which starts a series of several other reactions. The final reaction in the series involves the regeneration of oxaloacetate. NAD and FAD, both are rather small nucleotide molecules that are electron carriers. FAD can transfer its reducing power to flavoprotein. The reduced and oxidised forms of NAD are not clear; the oxidised NAD is called NADox (NAD^+); or NAD; reduced NAD is called NADred (NADH; and $NADH_2$). The actual net gain of ATP is unknown but must be regarded as <36. Frequently 36 ATP are quoted because it is known that in eukaryotic cells that the reduced NAD formed by glycolysis in the cytoplasm must be actively transported across the mitochondrial membrane to be made available to the electron transport chain. The cost of such active transport is one ATP for each NADH transported. Thus, the net gain for each cytoplasmic NADH is only 2 ATP rather than 3. However, we should also consider the active transport of other molecules as well (pyruvate, phosphate, Mg^{++} etc.).

Energy balance per glucose molecule:

Source	Aerobic	Anaerobic
2 ATP used in glycolysis	-2 ATP	-2 ATP
4 ATP formed in glycolysis	+4 ATP	+4 ATP
2 $NADH_2$ formed in glycolysis (via electron transport).	+6 ATP	
8 $NADH_2$ formed in Krebs cycle (via electron transport).	+24 ATP	
2 GTP in Kerb's cycle	+2 ATP	
2 $FADH_2$ in Krebs cycle (via electron transport).	+4 ATP	
Total:	38 ATP	2 ATP

Myokinase: ATP + AMP → ADP + ADP.
Creatine kinase: ATP + creatine → ADP + creatine phosphate.
Cytidine diphosphate kinase: CDP + ATP → CTP + ADP (many similar enzymes exist).
Oxygen uptake that is dependent on the presence of ADP and phosphate is termed coupled respiration. The system of mitochondrial enzymes and redox carrier molecules that ferry reducing equivalents from substrates to oxygen are collectively known as the electron transport system, or the respiratory chain. The respiratory chain can be dissected into 4 large multi-subunit complexes containing the principal respiratory carriers, named Complex I to Complex IV. These substantial protein "icebergs" float in the sheet of inner membrane lipids, often presenting one face to the mitochondrial matrix and another to the inter-membrane space. Other membrane bound enzymes such as the energy-linked transhydrogenase (ELTH) are also present, which fulfil ancillary roles. Except for succinate dehydrogenase (complex II) all these complexes pump protons from the matrix space into the cytosol as they transfer reducing equivalents (either hydrogen atoms or electrons) from one carrier to the next. Complexes I, II, and IV all are "pump" protons (i.e., H^+) into the mitochondrial space between the inner and outer mitochondrial membrane, establishing a proton gradient across the inner mitochondrial membrane. As the protons pass through Complex V, the osmotic energy of the gradient is converted into chemical energy, in the form of ATP. In addition to succinate dehydrogenase, several other enzymes contribute electrons to directly to ubiquinone. Acyl CoA dehydrogenase is a soluble flavoprotein involved in fatty acid oxidation in the mitochondrial matrix space. It feeds its reducing equivalents to ubiquinone via an intermediate low molecular weight electron transferring flavoprotein (ETF). Glycerol phosphate oxidase is another flavoprotein that also feeds reducing equivalents to ubiquinone. It is an inner membrane protein, but the relevant active centre faces outwards and reacts with its substrate in the cytosolic compartment, not in the matrix space. There are 6 distinct types of poison, which may affect mitochondrial function: 1- Respiratory chain inhibitors (e.g., cyanide, antimycin, rotenone and TTFA) block respiration in the presence of either ADP or uncouplers. 2- Phosphorylation inhibitors (e.g., oligomycin) abolish the burst of oxygen consumption after adding ADP, but have no effect on uncoupler-stimulated respiration. 3- Uncoupling agents (e.g., dinitrophenol, CCCP, FCCP) abolish the obligatory linkage between the respiratory chain and the phosphorylation system, which is observed with intact mitochondria. 4- Transport inhibitors (e.g., atractyloside, bongkrekic acid,

NEM) either prevent the export of ATP, or the import of raw materials across the mitochondrial inner membrane. 5- Ionophores (e.g., valinomycin, and nigericin) make the inner membrane permeable to compounds, which are ordinarily unable to cross. 6- Krebs cycle inhibitors (e.g., arsenite, and aminooxyacetate), which block one or more of the TCA cycle enzymes, or an ancillary reaction. Cyanide Blocks cytochrome oxidase (complex IV) and prevents both coupled and uncoupled respiration with all substrates, including NADH, succinate and ascorbate + TMPD. Treatment of mitochondria with ultrasonic vibrations under appropriate conditions tears open the membranes and leads to the formation of tiny inside-out vesicles derived from the inner membrane, containing trapped cytochrome-c originating from the inter-membrane space. Sub-mitochondrial particles may retain partial coupling between electron transport and ADP phosphorylation.

The mitochondria respiratory chain:

Components	Location	Function
NADPH/NADP	Matrix space (separate cytosolic pool).	Mobile carrier.
NADPH + NAD$^+$ → NADH + NADP$^+$	Membrane spanning protein.	Proton pump $2H^+/2e^{-1}$.
NADH/NAD	Matrix space (separate cytosolic pool).	Mobile carrier.
NADH dehydrogenase (complex I).	Membrane spanning multi-subunit protein.	Proton pump $4H^+/2e^{-1}$.
Succinate dehydrogenase (complex II)	Membrane spanning multi-subunit protein.	No Proton pumping.
Ubiquinol/ubiquinone	Dissolved in the inner membrane lipids.	Mobile carrier.
Ubiquinol:cytochrome c reductase (complex III).	Membrane spanning multi-subunit protein.	Proton pump $4H^+/2e^{-1}$.
Cytochrome c (ferrous/ferric).	Inter-membrane space.	Mobile carrier.
Cytochrome c oxidase (complex IV).	Membrane spanning multi-subunit protein.	Proton pump $2H^+/2e^{-1}$.
F0/F1 ATPase (ATP synthetase).	Membrane spanning multi-subunit protein.	Proton pump $3H^+/ATP$.

Enzyme of oxidative phosphorylation:

Complex	Enzyme	KDa	Polypeptides
I	NADH dehydrogenase (or) NADH-CoQ reductase.	800	25
II	Succinate dehydrogenase (or) Succinate-CoQ reductase.	140	4
III	Cytochrome C-CoQ oxidoreductase.	250	9-10
IV	Cytochrome oxidase.	170	13
V	ATP synthase.	380	12-14

The physiological function of ATPase is the synthesis of ATP, using the energy stored in the transmembrane pH and potential gradients. Normal operation of the respiratory chain creates both a pH differential and a voltage gradient of 30,000,000 volts/m across the inner mitochondrial membrane. This demands highly specific transport proteins to control the movement of small ions across the membrane, and to prevent the dissipation of the various gradients. The majority of inner membrane carriers are antiporters, which exchange one molecule for another. Symports transport 2 molecules in the same direction. In a very few cases (e.g., for calcium uptake) the carrier simply allows the charged ion to traverse the membrane. This constitutes an electrical (or electrogenic) uniporter. The membrane potential exerts its full effect, leading to substantial concentration gradients if the reaction were able to reach equilibrium. Muscle contraction apparently requires a very high ATP/ADP ratio, and it is difficult to maintain a low myofibrillar ADP because it will not diffuse quickly enough back to the mitochondria at low ADP levels. Diffusion rate is directly proportional to metabolite concentration. Active muscles contain creatine phosphokinase (CPK) and large amounts of creatine and creatine phosphate. These compounds equilibrate with the adenine nucleotide pools. It is thought that the high concentrations of these highly diffusible energy carriers increase the maximum energy transport rate. One of the most important asymmetric transporters is the aspartate carrier, which plays a key role in the re-oxidation of glycolytic NADH by the malate-aspartate cycle. This cycle is necessary because the inner membrane is not permeable to either NAD or NADH. The component reactions must revolve twice for each molecule of glucose oxidised by the cell. The 2 enzymes involved, malate dehydrogenase (MDH) and glutamate oxaloacetate transaminase (GOT) are among the most active in the body. Both enzymes exist in mitochondrial and cytosolic variants, but all 4 proteins are coded by nuclear genes. There are about 2000 copies of the mitochondrial genome in a typical cell, so that despite its small size it comprises about 0.5% of the DNA mass. Mitochondrial DNA replication appears to be less accurate than nuclear copying, and it is far from clear how consistency is maintained between these multiple copies, or how their replication is synchronised with the remainder of the cell. There are substantial differences between the proteins present in mitochondria from different tissues, reflecting the tissue specific patterns of nuclear gene expression. Protein turnover, however, seems to be fairly slow and mitochondrial protein composition does not respond very quickly to dietary or hormonal stimuli. The total number of mitochondria per cell can be changed (for example, through muscle activity) over the course of several weeks. Mitochondrial proteins are usually more basic than their cytosolic counterparts. Protein folding is temporarily prevented in the cytosol by binding to the chaperone protein hsp70, while 2 further mitochondrial chaperones hsp60 and hsp70 supervise re-folding of the imported proteins after they have entered the matrix space. ATP hydrolysis is required in the cytosol and in the mitochondria for successful import of functional proteins, in addition to the membrane potential and pH gradient. This ATP requirement is probably associated with protein folding. The mitochondrial genetic information is very densely packed. All the hundreds of other mitochondrial proteins, including DNA polymerase, RNA polymerase, amino acid activating enzymes and all the ribosomal proteins are coded by nuclear genes and imported from the cytosol. The human mitochondrial genome (circular chromosome) is only 16.6 kb and among the smallest in the animal kingdom. About 150 different types of hereditary mitochondrial defect have been reported, and mitochondria play an important part in several common conditions. Mitochondrial DNA is maternally inherited through the egg cell cytoplasm, but many of the inherited defects map to the nuclear genome since the majority of mitochondrial proteins are imported from the cytosol. Mitochondrial defects may be confined to a limited range of tissues, and, unexpectedly, they may change substantially as the patient ages.

Reaction catalyzed by SOD: $2\ O_2^- + 2\ H^+ \rightarrow HO_2 + O_2$

The O_2^- substrate for SOD is generated indirectly in the oxidation of epinephrine at alkaline pH by the action of oxygen on epinephrine. As O_2^- builds in the solution, the formation of adrenochrome accelerates because O_2^- also reacts with epinephrine to form adrenochrome. Toward the end of the reaction, when the epinephrine is consumed, the adrenochrome formation slows down. If observed for long times, the adrenochrome disappears and brown, insoluble products form in the solution. These brown products are closely related to the brown pigments of the skin and to the pigments that form when fruit is cut open and exposed to dioxygen. SOD reacts with the O_2^- formed during the epinephrine oxidation and therefore slows down the rate of formation of the adrenochrome as well as the amount that is formed. Because of this slowing process, SOD is said to inhibit the oxidation of epinephrine. The SOD in a cell extract is the total SOD determined by the activity of both the CuZnSOD (cytosolic) and the MnSOD (mitochondrial) enzymes. Both enzymes react with the same rate constant with O_2^- so they are identical in their ability to inhibit oxidation of epinephrine. The smooth ER (SER) is not extensive in neurons; rather, it is enriched in cells that synthesize or modify large amounts of steroids. Under aerobic conditions the end product of glycolysis is pyruvic acid. The next step is the formation of acetyl coenzyme A (acetyl CoA), which is the initiator of the citric acid cycle. In carbohydrate metabolism, acetyl CoA is the link between glycolysis and the citric acid cycle.

Pyruvic acid + CoA + NAD^+ $\rightarrow$ acetyl CoA + NADH + H^+ + CO_2

This reaction may be called the oxidative decarboxylation of pyruvic acid to acetyl CoA. NAD^+ coenzyme is used to remove 2H's and 2e's from pyruvic acid. The reduced form, NADH, initiates the respiratory chain to regenerate NAD^+. Decarboxylation, which will be observed twice more in the citric acid cycle, is the removal of the carboxylic acid group and subsequent conversion into CO_2. This oxidative decarboxylation of pyruvic acid is catalysed by the enzyme complex-pyruvate dehydrogenase. The term complex is used because 3 enzymes and 5 coenzymes are involved. The specific enzyme, pyruvate dehydrogenase contains thiamine pyrophosphate (TPP) as coenzyme. The pyruvic acid becomes attached to positively charged nitrogen in the 5 membered ring of TPP. This is unstable and the carboxyl group is lost as CO_2 in decarboxylation reaction. The lipoic acid attached to dihydrolipoyl transacetylase through amide linkage with lysine in the protein chain of the enzyme. This is oxidation-reduction (the disulfide bond is reduced) and the acetyl group is transferred from TPP to the sulfur on lipoic acid (a thioester is formed). The final 2 reactions catalysed by dihydrolipoyl dehydrogenase, involve the regeneration of the disulfide bond in the 5 membered ring in lipoamide. FAD causes the oxidation and formation of the disulfide bond to form the ring. Finally $FADH_2$ reacts with NAD^+ in the electron transport chain. The NADH plus H^+ formed in the last reaction initiates the electron transport chain sequence.

THE CELL WALL

Substances move across cell membranes as follows: 1- Diffusion is an unassisted movement of solutes from a region of higher to lower concentration. It depends on the concentration gradients. 2- Osmosis is the movement of water across a selectively permeable membrane in response to concentration gradients, pressure or both. 3- Passive transport is diffusion of a solute through the interior of a membrane protein that does not require an energy input to serve as a channel or carrier. It is a reversible changes in a membrane protein's shape allow a solute to diffuse through the protein's interior, occurs when a solute binds to the protein. 4- Active transport is pumping of solutes with or against their concentration gradient, through the interior of a membrane protein that requires an energy boost to operate. Transport of one kind of solute is coupled with transport of another kind in the opposite direction. Carrier protein receives an energy boost from ATP. 5- Oxygen, carbon dioxide, and other small molecules with no net charge diffuse across the lipid bilayer. Ions and large, water-soluble molecules such as glucose are actively or passively transported across through the interior of membrane proteins. 6- Exocytosis, vesicle (small membrane-bound sac) moves to plasma membrane, fuses with it, then the contents released outside. 7- Endocytosis, vesicle forms at surface of plasma membrane, sinks into cytoplasm, then the contents released inside. The methods of intercellular (IC) communication fall into 2 broad groups: 1- those in which cells make direct physical contact by means of adhesion molecules, 2- those involving indirect communication via a variety of soluble compounds (extracellular signalling proteins, or ESPs). ESPs interact with receptors located on the outer surface of the target cell (transmembrane receptors). Some other types of soluble messengers, such as corticosteroids, pass straight through the membrane and team up with their receptors in the cellular cytoplasm. Once an ESP has bound on to its specific receptor, one or more IC signal transducing pathways are activated or blocked. At the end of these pathways lies the on- or off-switch, for one-target genes. Activation or inhibition will result in a change of cell state, or expression of a particular function of that cell. Changes of state include cell growth, differentiation, division and suicide, or the altered set up and expressions of cell surface molecules and receptors. Expression of functions includes the production and release of soluble factors. So far the best understood receptor types are the tyrosine kinase receptors, the G-protein linked receptors and the cytokine receptors. There are 2 remarkable features of cytokines: their functional redundancy and extensive pleiotropy. Most cytokines are pleiotropic, i.e.; a single cytokine exhibits a wide range of different biological effects in various tissues and cells. Different cytokines can act on the same cell type to mediate similar effects; this is the phenomenon known as redundancy. The IL2R is made up of at least 3 distinct subunits (α, β and γ). The IL2Rα contains a very short cytoplasmic region. Human IL2Rβ is a 75 kD glycoprotein structurally unrelated to IL2Rα. The IL2Rβ is also a functional component of IL15R. Human IL2Rγ is a 64 kD glycoprotein that can be associate with IL2Rβ in the presence of IL2 in lymphoid cells expressing the high-affinity IL2R. The predirected mature form of the human IL2Rγ also consists of the EC region, followed by a single transmembrane region. The cytoplasmic region of IL2Rγ is considerably shorter than that of IL2Rβ. The IL2Rγ chain is also a functional component of the IL4R, IL7R, IL9R, and IL15R, and is now referred to as the common γ (γc) chain. Such shared components (subunits) have also been reported for other cytokine receptors. IL-2 and other cytokines induce rapid phosphorylation of intracellular proteins including the receptors themselves. Neither the IL2Rβ nor the IL2Rγ chains contains an obvious protein tyrosine kinase (PTK) motif within the identified cytoplasmic region essential for signal transduction. Instead, the cytoplasmic regions of both IL2Rβ and IL2Rγ can interact with and activate multiple non-receptor type PTKs to elicit the cellular tyrosine phosphorylation that lead to cellular responses such as proliferation. Nuclear proto-oncogenes, including c-fos, c-jun and c-myc, are potentially critical targets for proliferative signals mediated by growth factor receptors. These nuclear proto-oncogene products function as regulators of gene transcription, thereby regulating cellular responses such as proliferation. IL2 stimulation induces the expression of nuclear proto-oncogenes (c-fos, c-jun and c-myc) as well as of cell cycle regulating genes. The bcl-2 oncogene and the pim-1 proto-oncogene are also target genes for IL2 signals. Induction of the c-myc and bcl-2 genes by IL2 appears to be mediated by 2 distinct pathways. Following the stimulation of Jak-PTKs, there is activation of a family of latent transcription factors Stat proteins (signal transducers and activators of transcription). The pim-1 gene promoter contains a Stat-binding site and is likely to associate the Stat 5 protein after IL2R activation. Cytokine signalling may be controlled at different levels: 1- ligand-receptor interaction has to be correct in terms of affinity and set-up; 2- the intracellular structures of the receptor dictate which of the signalling molecules such as PTKs or Stats are to be recruited or activated, 3- the specificity of signalling by a given cytokine may also be dependent on the expression levels of signalling molecules, including receptors in different cell types; and 4- in the nucleus, the binding of a transcription factor to a target gene depends on affinity and availability. The genes encoding the ligand IL2 and the receptor subunits IL2Rα, IL2Rβ, ILRγ are mapped on chromosomes 4q, 10q, 22q and Xq. Dysregulation of a component or components may cause immunological disorders or malignancies under certain circumstances.

CELL STRUCTURES AND FUNCTION

Cell structure and function begin with proteins and instructions for building the proteins themselves are contained in DNA. The nucleus serves 2 function: 1- control access to hereditary instructions contained in DNA; and 2- keeps DNA separated from all substances and metabolic machinery in the cytoplasm. Nuclear envelope is a 2 membrane thick; each is a lipid bilayer studded with various proteins. Pores composed of proteins span both bilayers. The pores are passageways through which substances can move into and out of the nucleus in controlled ways. Nucleolus, is one or more dense irregular masses appear in the nucleus. Nucleoli are sites where the protein and RNA subunits of ribosomes are assembled, then ship out into cytoplasm. Endoplasmic reticulum (isolation, modification, transport of proteins and other substances) begins at the nucleus and curves through the cytoplasm; it has rough and smooth region. Rough ER has many ribosomes; it is abundant in cells that specialised in secreting digestive enzymes and other proteins, the polypeptide chains enter spaces inside it. Smooth ER is free of ribosomes. It is the main site of lipid synthesis in many cells. Mitochondria (aerobic energy metabolism) liberate energy stored in glucose and other substances (e.g., glutamine) then use it to form many ATP molecules. Metabolic machinery in mitochondria uses oxygen to extract far more energy from glucose than can be done by any other means. Oxygen is inhaled for mitochondria. Each mitochondrion has an outer membrane facing the cytoplasm, and inner membrane with many inward folds called cristae. The double membrane system creates two compartments that are used in ATP formation. Mitochondrion has their own DNA and some ribosomes. They divide on their own. Golgi bodies (modification, distribution of materials), are flattened interconnected sacs in which lipids and proteins are modified. An enzyme in a Golgi vesicle may attach a phosphate group to a particular protein. Among these vesicles are lysosomes, the cell's main organelles of digestion. Different enzymes in lysosomes can break down virtually all proteins, polysaccharides, nucleic acids and some lipids. The cytoskeleton, are components of the cytoplasms function in the internal organisation, shape and motion of cells. Microtubules are the most predominant component, consist of tubulin subunits linked in parallel rows. Some parts of the cytoskeleton appear only at certain times in a cell's life, such as the spindle, which moves chromosomes about. Because the cytoskeleton is viscoelastic, it provides a continuous mechanical coupling throughout the cell that changes as the cytoskeleton remodels. This can influence ion channel activity at the plasma membrane of cells and may conduct mechanical stresses from the cell membrane to internal organelles. Cytoskeleton also provides a large negatively charged surface on which many signalling molecules localise in response to activation of specific transmembrane receptors.

DNA STRUCTURE AND FUNCTION

The bases of DNA (Deoxyribonucleic acid) are adenine, guanine, thymine and cytosine. Two strands of nucleotides twist together like a spiral stairway, they form a double helix. Hydrogen bonds connect the bases of one strand to bases of the other. As a rule, adenine pairs (hydrogen bonds) only with thymine, and guanine only with cytosine. Before cell division, its DNA is replicated with the help of enzymes and other proteins. At other times in the cell's life, the 2 strands unwind in certain regions only, so that the cell gains access to particular proteins. There is only one DNA molecule in a chromosome. Through interactions between histones and DNA, a chromosome can coil back on itself again and again. The coiling greatly increases its diameter. All chromosomes contain DNA. DNA contains only 4 nucleotides. Each nucleotide has a five-carbon sugar, which has a phosphate group attached to the fifth carbon atom of its ring structure. It also has one of 4 kinds of nitrogen-containing bases attached to the first carbon atom. The nucleotides differ only in which base is attach to that atom. The path leading from genes to proteins has 2 steps, called transcription and translation. In transcription the double-stranded DNA is unwound at a gene region in the nucleus. Nucleotide bases in DNA serve as a structural pattern (template) for assembling a strand of ribonucleic acid from the cell's pool of free nucleotides. Then RNA molecule is assembled on the exposed bases of one of the strands. The RNA is then shipped to the cytoplasm, where translation occurs. In translation, RNA directs the linkage of one amino acid after another, in the sequence required to produce a specific kind of polypeptide chain, then the chains become folded into the three dimensional shapes of proteins. Protein synthesis requires 3 classes of RNA, messenger RNA (mRNA), ribosomal RNA (rRNA), and transfer RNA (tRNA). Transcription differs from DNA replication in three key aspect: A- Only one region of one DNA strand, not the whole strand serves as the template; B- Different kinds of enzymes (RNA polmerases) are used; and C- Transcription results in a single strand of nucleotides. A gene's base sequence is vulnerable to permanent changes, which is called mutations. These changes are the source of genetic variation in populations. Translation proceeds through 3 stages: initiation, chain elongation, and chain termination.

PURINE AND URATE

Nucleosides-purines or pyrimidines combined with ribose, are components of coenzymes (NAD$^+$, NADP$^+$, ATP, UDPG, etc.), ribonucleic acid (RNA) and deoxyribonucleic acid (DNA). The purines, adenine and guanine are constituents of both types of nucleic acid (DNA and RNA). RNA is in dynamic equilibrium with the amino acid pool, but DNA, once formed, is metabolically stable throughout life. The pyrimidines are catabolised to CO_2 and NH_3, and the purines are converted to uric acid. Minor amounts of purines and pyrimidines are excreted unchanged in the urine. The purines used by the body for nucleic acid synthesis may be derived from the breakdown of ingested nucleic acid, e.g., meat, or may be synthesised in the body from small molecules de novo: 1- condensation of pyrophosphate with phosphoribose to form phosphoribosyl pyrophosphate (PRPP); 2- the amino group of glutamine is then incorporated into the ribose phosphate molecule and pyrophosphate is released; 3- glycine is next added to phosphoribosylamine; 4- after many complex steps purine ribonucleotides (purine ribose phosphates) are formed; 5- ribose phosphate is split off releasing purine; and 6- purines may follow one of two pathways: i- reused for nucleic acid synthesis, or ii- oxidised to urate. The liver enzyme xanthine oxidase oxidises both hypoxanthine and xanthine. Uric acid is formed by the breakdown of purines and by direct synthesis from 5-phosphoribosyl pyrophosphate (5-PRPP) and glutamine. In human uric acid is excreted in the urine. In the kidney, uric acid is filtered, reabsorbed, and secreted. Normally, 98% of the filtered uric acid is reabsorbed, and the remaining 2% make up about 20% of the amount excreted. The remaining 80% come from tubular secretion. The uric acid excretion on a purine-free diet is about 0.5 g/24 h and on regular diet about 1 g/24h. Excretion is decreased in renal disease, and production is increased in leukaemia and pneumonia because of increased breakdown of uric acid-rich white blood cells. Colchicine does not affect uric acid metabolism; it relieves gouty attacks by inhibiting the phagocytosis of uric crystals by leukocytes. Adrenal glucocorticoid hormones and their synthetic derivatives increase uric acid excretion. Insoluble urate may, like calcium precipitate in tissues. If this takes place in the kidney, renal damage can result. Hyperuricaemia may be due to a primary lesion of purine metabolism or be secondary to a variety of other conditions. Urate is the end product of purine metabolism. The poor solubility of urates that man is prone to clinical gout and renal damage by urate. Excretion of urate: 75% of the urate leaving the body is excreted in the urine and 25% passes into the intestine, where it is broken down by intestinal bacteria (uricolysis). The urate filtered at the renal glomerulus is probably almost completely reabsorbed in the tubules and most urinary urate is derived from active tubular secretion. Urinary excretion of urate is slightly lower in males than females; this may contribute to the higher incidence of hyperuricaemia in men. The solubility of urate in plasma is limited. Precipitation in tissues may be favoured by a variety of local factors of which the most important are probably tissue pH and trauma. Local inflammation due to urate precipitation produces an increase in leucocytes in the area, and that lactic acid production by these cells lowers the pH locally. Women are more prone to hyperuricaemia, may be that renal excretion of urate is affected by sex-hormone levels. Two factors may account for the high incidence of clinical gout: 1- alcohol, increases lactic acid production, which inhibits urate excretion; and 2- high meat diet, contains a high proportion of purines. Principle of treatment of hyperuricaemia: 1- reduce purine intake; 2- increase renal excretion of urate, probenecid and salicylates in large doses, if renal function is normal; 3- reduce urate production, allopurinol (hydroxypyrazolopyrimidine), acts as a competitive inhibitor of the xanthine oxidase; and 4- colchicine, does not affect urate metabolism. Causes of secondary hyperuricaemia: 1- malignancy, can cause acute renal failure due to tubular blockage by crystalline uric acid; 2- starvation and tissue damage, acidosis is probably inhibit renal excretion of urate, aggravating the hyperuricaemia; and 3- glomerular failure, causes retention of urate as well as retention of other waste products of metabolism. The major active metabolite of allopurinol is oxypurinol, which is cleared quite slowly by renal elimination (giving oxypurinol a half-life of 24 hrs in normal individuals). Some of the serious toxicities appear to be dose-dependent and increased in incidence by renal insufficiency, include full-blown allopurinol hypersensitivity syndrome, which has a mortality rate of 20%, but also may include significant hepatic disease. In patients with normal renal function, the goal of the uric acid lowering therapy is to drop the serum uric acid to below 6.0 mg/dl, a point below which monosodium urate becomes supersaturated at physiologic pH in normal tissues. Renal excretion is the predominant mechanism for clearing body stores of uric acid, which is normally done in a manner that directly adapts to increases in serum urate. But a decreased GFR constrains the amount by which urate can be eliminated from the serum to lessen (or maintain) total body urate stores. Thus, increasing the allopurinol dose beyond the recommended maximum for marked renal insufficiency (100 mg for creatinine clearance of 30 ml/min, 200 mg for creatinine clearance of 60 ml/min) could decrease new formation of uric acid in cells but would fail to efficiently decrease total body urate stores. The approach generally fails to reduce serum urate to below 6.0 mg/dl. Increased intestinal secretion and bacterial uricolysis have been claimed to occur in renal failure and may account for the rise in urate is less than that of urea. Diuretics inhibit excretion of urate. Xanthinuria is a rare inborn error, probably inherited as autosomal recessive disease. Plasma and urinary urate levels are very low. Increased urinary excretion of xanthine may lead to the formation of xanthine stones. Xanthine oxidase/dehydrogenase degrades the purines, xanthine and hypoxanthine, to uric acid. Xanthine dehydrogenase/oxidase has two forms: 1- xanthine dehydrogenase, requires nicotinamide adenine dinucleotide (NAD), produces uric acid and reduced nicotinamide adenine dinucleotide; and 2- xanthine oxidase, requires oxygen, produces uric acid and superoxide. Under certain conditions, including ischaemia-reperfusion and hypoxia, there is increased conversion of the dehydrogenase form to the oxidase form. Increased conversion to the oxidase form promotes production of ROS. Increased cell destruction leads to an excess of purines with increased availability of substrate for xanthine dehydrogenase/oxidase. An extreme example occurs during the initiation of chemotherapy in myeloproliferative malignancies. Adenine triphosphate (ATP) is another source of purines. During hypoxia there is increased degradation of ATP to adenine monophosphate (AMP), which in the absence of oxygen cannot be recycled to ATP. Excess AMP is degraded to adenosine, hypoxanthine and finally uric acid. Uric acid and purines are increased in cerebrospinal fluid from patients with brain ischaemia from ATP degradation and cell damage. Hypoxia and ischaemia induce the expression of xanthine dehydrogenase/oxidase. Cytokines such as interferon induce the expression of xanthine dehydrogenase/oxidase. Hypoxia-reperfusion, ischaemic conditions, and cytokines thus will promote both the formation of xanthine dehydrogenase/oxidase and the conversion of xanthine dehydrogenase to the oxidase form. With the oxidase form the formation of uric acid from purine is coupled with production of reactive oxygen radicals. The resulting burst of free radicals may damage not only the organ in which they are from but, if free radical metabolites are released into the circulation, also remote organs. Production of ROS such as superoxide and hydrogen peroxide occurs as by-products during the enzymatic degradation of purines to uric acid by xanthine oxidase. AMP and GMP (the purines) are metabolised to the same product, xanthine. Xanthine is thought to be metabolised in the peroxisome. Primates, birds, reptiles, and insects only degrade xanthine to uric acid, while other vertebrates metabolise xanthine to allantoin. It can even be metabolised further in some animals, with marine invertebrates metabolising it all the way to ammonia. Xanthine oxidase is a molybdenum-utilising enzyme that exists predominantly as a dehydrogenase, which utilises NAD^+ as the electron acceptor, but can be converted to an oxidase that utilises molecular oxygen as the electron acceptor. The human xanthine oxidase enzyme has never been demonstrated to be peroxisomal. Urate is a potent antioxidant. Allantoin is further broken down; it is first metabolised to glyoxylate and urea by allantoicase. The urea can then be broken down to ammonia by urease. The subcellular localisation of these enzymes in the pathway seems to vary between species. A deficiency in xanthine oxidase (also known as xanthine dehydrogenase) results in a condition known as xanthinuria type I. It is biochemically characterised by high levels of xanthine, and decreased levels of uric acid. This has been observed to result in xanthine renal stones, and crystalline deposits on the kidney or skeletal muscle.

TRANSCRIPTION FACTORS

Up to 10% of human genes may encode transcription factors. Before a mature mRNA molecule reaches the cytoplasm of the cell to be translated into protein, a number of highly regulated processes must have taken place: transcription initiation, elongation, termination, splicing, polyadenylation and nuclear export. The initiation of the DNA sequences around the start of the gene are recognised by a number of nuclear proteins termed transcription factors, which is divided into 2 groups: 1- Group consists of a series of 20-30 proteins, found in all cells, which complex with RNA polymerase II (Pol II) at the transcription start site. They are collectively termed basal machinery or general transcription factors (GTFs). These proteins are required for the initiation of transcription of all protein-encoding genes and consequently they cannot be considered as tumour-specific targets. 2- Gene- and cell-specific proteins that have 2 main functional domains. One domain specifically recognises and binds to DNA sequences within the gene regulatory elements, while the second interacts with the basal machinery to regulate the efficiency of transcription initiation. The upstream factors (activating factors) either stimulate transcription initiation or interfere with repressors, depending on the nature of their interactions with the general factors. Activators increase the chances of transcription initiation by encouraging the binding of Pol II holoenzyme through interactions with their activation domain, while repressors act via their repression domain to destabilise complex formation at the start site, making transcription initiation less likely. Interactions between some upstream factors and the general machinery may also be ensured via non-DNA binding co-activators or co-repressors, which can also be specific to certain transcription factors or promoters. The binding sites for upstream factors often lie close to the gene transcription start site, in a region termed the gene promoter. All nuclear DNA are packaged into higher order structures through the binding of histones to form chromatin, but the precise architecture of these structures is thought to be quite distinct in transcriptionally active regions of the genome. Thus transcriptionally active chromatin is thought to be more open allowing easier passage of Pol II, and both general and gene-specific transcription factors are thought to contribute to this local reorganisation of chromatin which can lead to the establishment of patterns of gene expression that are stable through several rounds of cellular division. The relative stiffness of short DNA segments (up 500 bp) means that even in the promoter region, bending or kinking of the DNA is required to bring factors into close proximity. This may occur through the binding of general chromatin proteins such as the high mobility group (HMG) proteins, but may also be achieved by the interaction of gene-specific architectural factors, or DNA binding of certain upstream factors and general transcription factors (GTFs). The bent DNA structures caused by the anticancer drug, cisplatin for example, are specifically recognised by the protein HMG1 and this has been shown to contribute to drug efficacy by preventing recognition of the modified DNA by the cellular repair machinery. After being synthesised, most upstream factors translocate to the nucleus, but some are sequestered in the cytoplasm bound to an inhibitor protein until released on receipt of an appropriate signal. Once bound to DNA, the upstream factor needs to interact with the basal machinery and possibly with other proteins such as cofactors or other upstream factors (cross talk). NFκB is a factor involved in the control of expression of inflammatory response genes, which is held in the cytoplasm in complex with its inhibitor IκB. A number of stimulants, including phobrol esters and dsRNA, can inactivate this complex, allowing NFκB to translocate to the nucleus. The anti-inflammatory action of aspirin is to maintain the IκB/NFκB complex, thus NFκB is unable to translocate to the nucleus and activation of the inflammatory response genes is thereby prevented. Cyclosporin prevents nuclear entry of the NF-AT, upstream factor required for expression of the IL2 gene during T-cell activation. The oncoprotein Fos can only bind to DNA when dimerised with Jun to form a protein complex often referred to as AP1. Ligand binding controls dimerisation and DNA binding, but the precise role is subtly different for each family member. There are already drugs used in clinical practice, which specifically target upstream transcription factors, namely aspirin in inflammatory disease and cyclosporin for the immunosuppression of transplant patients. Although initially thought to act as completely independent upstream factors, it has become increasingly clear that the nuclear receptors also act to modulate the activity of other upstream factors, in particular the Fos/Jun AP1 complexes. This interaction is called cross talk, takes the form of mutual transrepression such that non-DNA bound nuclear receptors, particularly retinoic acid receptors (RARs), can repress the activity of DNA-bound AP1, and vice versa. For the cross talk interactions to occur, the RARs must bind ligand. AP1 activity is linked to cell proliferation and the induction of genes involved in tumour metastases, processes that can be ablated by administration of retinoids. Tomoxifen is antitranscription agent. The p53 protein is another upstream transcription factor (TF) and it plays a pivotal role in the cell by monitoring the genome for damage and hence activating expression of a number of genes leading to either DNA repair or cell death, depending on the extent of the damage. When TP53 is mutated or deleted, a damaged cell has a greater chance of continuing to proliferate and ultimately forming a tumour. Biochemically, p53 has 3 major domains: an N-terminal transcription activation domain, a central sequence-specific DNA binding domain and a C-terminal domain, which interacts, with the central domain to autoregulate DNA binding. The majority of TP53 mutations found in human tumours map to the central DNA binding domain, and up to a third may result in too tight an interaction with C-terminal autoregulatory domain leading to a failure to activate p53. There are a number of cellular genes whose expression is upregulated during tumourigenesis, such as the extracellular proteases (e.g., cathepsin D, stromelysin-3, collagenases) which are believed to contribute to the metastatic potential of a tumour. The multidrug resistance gene, MDR1, which encodes a membrane pump capable of removing from the cell a wide range of toxic compounds used as chemotherapy agents. The c-erbB-2 proto-oncogene, is a receptor tyrosine kinase expressed in 25-30% of solid tumours, and is associated with poor prognosis and a reduced response to conventional therapy. Tumour cells that overexpress c-erbB-c have acquired an additional upstream factor. The normal counterparts of many oncogenes are transcription factors whose proper role is in the control of normal cell growth. NF-κB was first identified as a regulator of the expression of the κ light-chain in murine B-lymphocytes but has subsequently been found in many different cells. Several different NF-κB proteins have been characterised. The activated form of NFκB is a heterodimer, which usually consists of 2 proteins, a p65 subunit and p50 subunit. Other subunits, such as rel, relB, v-rel, and p52, may also be part of activated NFκB. In unstimulated cells, NFκB is found in cytoplasm and is bound to IκBα and IκBβ, which prevent it from entering the nuclei. When these cells are stimulated, specific kinases phosphorylate IκB, causing its rapid degradation by proteasomes. The release of NFκB from IκB results in the passage of NFκB into the nucleus, where it binds to specific sequences in the promotor regions of target genes. NFκB induces the synthesis of IκBα, which enters the nucleus to bind to activated NFκB and carry the NFκB to the cytoplasm, thereby terminating the activation of gene expression. Many stimuli activate NFκB including cytokines, activators of protein kinase C, viruses and oxidants. NFκB regulates the expression of many genes involved in immune and inflammatory responses. It is not the only transcription factor involved in regulating these genes, and it is frequently functions together with other transcription factors, such as activator protein 1 (AP-1) and the nuclear factor of IL-6 (NF-IL6), that are also involved in the regulation of inflammatory and immune genes. Products of the genes that are regulated by NFκB also cause the activation of NFκB. The proinflammatory cytokines IL-1β and TNF-α both activate and are activated by NFκB. NFκB increases the expression of the genes for many cytokines, and adhesion molecules in chronic inflammatory diseases. One such gene is that for iNOS. Cyclooxygenase-2, regulated by NFκB is responsible for the increased production of prostaglandins and thromboxane in inflammatory diseases. In all chronic inflammatory diseases, adhesion molecules recruit inflammatory cells, such as neutrophils, eosinophils, and T lymphocytes, from the circulation to the site of inflammation. NFκB regulates the expression of several genes that encode adhesion molecules such as ICAM-1, VCAM-1, and E-selectin. IL-1β and TNF-α influence the severity of inflammatory disease, possibly by the persistent activation of NFκB. Aspirin and sodium salicylate inhibit the activation of NFκB. The antiinflammatory cytokine IL-10 inhibits the action of NFκB, through an effect on IκBα. The free NFκB passes into the nucleus, where it binds to κB sites in the promotor regions of genes for inflammatory proteins such as cytokines, enzymes, and adhesion molecules. Glucocorticoid-receptor complexes bind the p65 subunit of NFκB, and thus prevent activation of inflammatory genes. Synthesis of IκBα is stimulated by the binding of glucocorticoid-glucocorticoid-receptor complexes to a glucocorticoid response element in the promoter region of the IκBα gene. The targeted disruption (or knockout) of the p65 component of NFκB is lethal because of the associated development abnormalities, whereas the lack of the p50 component results in immune deficiencies and increased susceptibility to infection.

ENERGY PATHWAYS

Oxidation is the combination of a substance with O_2, or loss of hydrogen, or loss of electrons. Oxidative phosphorylation is the transfer of hydrogen from NADH to flavoprotein, which is associated with the formation of ATP from ADP, and the further transfer along the flavoprotein-cytochrome system generates 2 or more molecules of

ATP per pair of protons transferred. Under anaerobic conditions (anaerobic glycolysis), a block of glycolysis at the phosphoglyceraldehyde conversion step might be expected to develop as soon as the available NAD^+ is converted to NADH. However pyruvic acid can accept hydrogen from NADH, forming NAD^+ and lactic acid. In this way, glucose metabolism and energy production can continue for a while without O_2. The lactic acid that accumulates is converted back to pyruvic acid when O_2 supply restored, and NADH transfers its hydrogen to flavoprotein-cytochrome chain. During aerobic glycolysis, the production of ATP is 19 times as great as it is under anaerobic conditions. The net production per mole of blood glucose metabolised aerobically via Embden-Meyerhof pathway and citric acid cycle is 36-38 mol (2 + [2 X 3] + [2 X 3] + [2 X 12]). Oxidative deamination of amino acids occurs in the liver. An imino acid is formed by dehydrogenation, and this compound is hydrolysed to the corresponding keto acid, with the liberation of ammonia. Amino acids can take up NH_3, forming the corresponding amide. The reverse reaction occurs in the kidney, with the liberation of NH_3 into the urine. The NH_3 reacts with H^+ to form NH_4^+, thus permitting more H^+ to be secreted into the urine. Products of metabolisms of sugars, carbohydrate, amino acids, and lipids are used as carbon sources for tricarboxylic acid (TCA) cycle in mitochondrial matrix. NADH and $FADH_2$ (products of TCA cycle and of β oxidation of fatty acids), which occurs in mitochondrial matrix, need reducing equivalents into electron transport chain (ETC) located in mitochondrial inner membrane. This culminates in passage of these reducing equivalents from cytochrome oxidase to oxygen. At 3 locations within the ETC, energy acquired from oxidation is used to pump H^+ outward from the matrix to space between inner and outer mitochondrial membranes, creating an electrochemical H^+ gradient. H^+ falling down their electrochemical gradients pass through a pore in ATPase. This complex mediates phosphorylation of ADP and P_i bound to its subunit, forming bound ATP. Energy of the H^+ electrochemical gradient is used to release tightly bound ATP to form free ATP in mitochondrial matrix. This ATP is subsequently exchanged for extramitochondrial ADP over the adenine nucleotide translocase.

Energy pathways:

1- ATP delivered energy nearly to all metabolic reactions.
2- ATP produced by aerobic respiration begins in the cytoplasm as a result of glycolysis.
3- In aerobic respiration oxygen is the final acceptor of electrons stripped from glucose, and joins with H^+ to form water.
4- The pathway for aerobic respiration includes: a- glycolysis (break down of glucose), b- the Krebs cycle (pyruvate is broken down to carbon dioxide and water, coenzymes pick up electrons liberated by the breakdown reactions), c- electron transport phosphorylation. The unbounded hydrogen H^+ ions and electrons are used to form many ATP molecules. Net energy yields 36 or more ATP molecules.
5- PGAl (phosphoglyceraldehyde) marks the start of the energy-releasing steps of glycosis. Glycolysis starts with an energy investment of 2 ATP. A glucose molecule is partially broken down and yield 2 pyruvate, 2 NADH, and 4 ATP, 2 of the later will be invested to get the reaction going, so only 2 ATP is the net energy yield. This takes place in the cytoplasm. During enzyme-mediated reactions, 2 NADH form after each PGAl gives up 2 electrons and a hydrogen atom to 2 NAD^+. Each PGAl combines with inorganic phosphate (P_i) present in the cytoplasm, then donates a phosphate group to ADP, results in formation of 2 ATP. The H_2O then formed, with the resulting intermediates (2 molecules of 3-phosphoenolpyruvate, or PEP), are rather unstable. Each gives up a phosphate group to ADP. So we started with 6 carbons (glucose) and ended by 2 molecules of pyruvate, each with a 3-carbon backbone.
6- The second stage takes place in the mitochondria. Pyruvate is converted to a molecule that can enter a cyclic pathway, the Krebs cycle, and ends in that glucose is broken down into carbon dioxide and water. This end in production of 10 coenzymes (8 NADH and 2 $FADH_2$) and 2 ATP.
The Krebs cycle has 3 functions: A- Hydrogen and electrons from substrates are transferred to NAD^+ and FAD, which thereby become NADH and $FADH_2$. B- Substrate-level phosphorylations produce 2 ATP. C- Intermediates are juggled back to the form of oxaloacetate. Cells have only so much oxaloacetate, and it must be regenerated to keep the cyclic reaction going.
7- The third stage also takes place in the mitochondria. An inner membrane divides a mitochondrion into 2 compartments: A- The inner compartment is the site of the Krebs cycle, in which many coenzymes are loaded up with hydrogen and electrons (Electron transport systems). B- The channel proteins are embedded in the membrane for the coenzymes delivery, which drives ATP formation.
8- Coenzymes deliver electrons to a transport system. Operation of the system sets up H^+ concentration and electric gradients across the membrane. H^+ flows down the gradients, through channel proteins. Energy associated with the flow drives the formation of ATP from ADP and unbound phosphate. Oxygen withdraws electrons and combines with H^+ to form water.
9- Anaerobic electron transport uses an organic or inorganic compound (not oxygen) as the final electron acceptor.
10- Electron transport phosphorylation occurs at transport system and channel proteins (ATP synthases) embedded in the inner mitochondrial membrane. Each transport system consists of enzymes and other proteins (including cytochrome molecules) that operate in sequence.

The electron transport phosphorylation occurs in the inner mitochondrial membrane via ATP syntheses. The reactions begin in the inner compartment, when NADH and $FADH_2$ give up hydrogen and electron to the transport system. Hydrogen follows the gradient and flows back into the inner compartment, through channel proteins. The flow drives ATP formation from ADP and unbound phosphate. The chemosmotic theory is a process in which the concentration and electric gradients across a membrane drive ATP formation. Free oxygen withdraws electrons from the transport system, and then combines with H^+ to form water molecules. Its action pulls electrons through the system and keeps ATP production going. The third stage commonly produces 32 ATP. This brings total net yield of the aerobic pathway to 36 ATP for each glucose molecule. So, in aerobic respiration, glucose is completely broken down to carbon dioxide and water. NAD^+ and FAD accept hydrogen and electrons stripped from substrate of the reactions and deliver them to an electron transport system. Oxygen is the final acceptor of these electrons. 95% of ATP is produced by oxidative phosphorylation through a chemosmotic mechanism within mitochondria. Under normal resting condition within cells, mitochondria do not contain very much stored Ca^{2+}. The energy available from ATP hydrolysis decreases as ATP falls and as ADP and P_i increase. The ATP concentration increases and ADP and P_i decrease near mitochondria, which decrease the rate of phosphorylation even though more ATP is needed to maintain a higher workload. Mitochondria function in calcium transport includes: 1- Regulates cytosolic free Ca^{2+} concentration. 2- Protects the cytosol against damage due to high Ca^{2+}. 3- Regulates mitochondrial matrix free Ca^{2+} to control the level of activation of Ca^{2+}-sensitive metabolic processes.

Co-enzymes and energy pathways:

Glucose		2NAD
Pyruvate conversion	Preceding Krebs	2NADH
Krebs cycle	$2FADH_2$	6NADH
Total coenzymes	$2FADH_2$	10NADH

FATTY ACID OXIDATION

Peroxisome is any of the microbodies present in vertebrate animal cells, especially the liver and kidney cells, which are rich in the enzymes peroxidase, catalase, D-amino acid oxidase, and, to a lesser extent, urate oxidase. Fatty acids are a major source of energy for most tissues. The first step in their utilization is by β-oxidation of fatty acyl-CoAs. In this process the size of the fatty acylCoA is reduced in a sequence of steps. Repeat of the cycle shortens the fatty acyl-CoA 2 carbons known as the β-oxidation spiral. Catalase converts the potentially toxic hydrogen peroxide produced by acyl-CoA oxidase to water and oxygen. Acyl-CoA oxidase is flavoenzyme that introduces a double bond between the a and b carbons of the acyl-CoA and passes the electrons to oxygen, producing hydrogen peroxide (H_2O_2). There are at least 3 human acyl-CoA oxidases, named for substrate specificity. These include a straight chain acyl-CoA oxidase, a branched chain oxidase, and a pristanoyl-CoA oxidase. The fatty acyl-CoA synthetases ligate CoA to a free fatty acid. This step requires ATP and Mg, as well as the CoASH. There are a 2 human peroxisomal synthetases, a very long chain fatty acyl-CoA synthetase (VLCFA), and a long chain fatty acyl-CoA synthetase (LCFA). There are links between peroxisomal β-oxidation, peroxisome assembly, cell death, and cell proliferation in the liver. 10% of patients initially diagnosed with a disorder of peroxisome have a deficiency of an β-oxidation enzyme. Peroxisomal acyl-CoA oxidase (ACOX) is the first enzyme of the fatty acid β-oxidation pathway, which catalyses the desaturation of acyl-CoAs to 2-trans-enoyl-CoAs. ACOX donates electrons directly to molecular oxygen, thereby producing hydrogen peroxide. The marked induction of this hydrogen peroxide-producing enzyme in the liver is implicated in the oxidative DNA damage and hepatocarcinogenesis. There are two major differences between mitochondrial and peroxisomal β-oxidation: specificity and mechanism. The mitochondria oxidize short, medium, and most long chain fatty acids, while peroxisomes oxidize some long chain, but mostly very long chain fatty acids. Peroxisomal and mitochondrial enzymes are encoded by distinct genes. Mitochondria utilize an acyl-CoA dehydrogenase to convert the acyl-CoA to enoyl-CoA. This enzyme transfers the electrons to FAD, and then into the electron transport pathway. However, peroxisomes utilize an acyl-CoA oxidase, which finally transfers the electrons to oxygen, producing hydrogen peroxide, a product converted to water and oxygen by catalase. The mechanism of the rest of the β-oxidation pathway is identical between the mitochondria and peroxisomes. The peroxisome handles the β-oxidation of many substrates. These include very long chain fatty acids (VLCFAs; both saturated and unsaturated), some long chain fatty acids, and long chain dicarboxylic acids. Peroxisomes can also oxidize the side chains of eicosanoids, which are molecules important in short-range signaling and are derived from arachidonic acid (AA). The eicosanoid family includes prostaglandins (involved in the regulation of cAMP, and therefore implicated in numerous pathways), thromboxanes (involved in blood clotting), and leukotrienes (involved in muscle contraction). Peroxisomal β-oxidation also plays a role in bile acid synthesis. Fatty-acid β-oxidation in peroxisomes proceeds via a similar mechanism as in mitochondria and involves the sequential action of acyl-CoA oxidase, bifunctional enzyme (with enoyl-CoA-hydratase and 3-hydroxyacyl-CoA dehydrogenase activity), and finally 3-ketoacyl-CoA thiolase. A fourth enzyme involved is fatty acyl-CoA synthetase. Separate measurement of each enzyme activity of peroxisomal β-oxidation is difficult, for the following reasons: 1- the activities of the corresponding mitochondrial β-oxidation enzymes are much higher than the activity of the peroxisomal enzymes; 2- the elimination of mitochondrial enzymes is a complicated task; and 3- the activity of each peroxisomal enzyme in fibroblasts is below the limit of detection when conventional methods are used. Most patients with isolated enzyme deficiencies may be diagnosed on the basis of lack of the enzyme protein. The accumulation of very-long-chain fatty acids and decreased VLCFA oxidation is common in these patients. In the body, fatty acids are broken down to acetyl-CoA, which enters the citric acid cycle. Fatty acid oxidation begins with activation of the fatty acid, and this occurs both inside and outside the mitochondria. Active fatty acids formed outside the mitochondria cross the mitochondrial membrane by a process that requires carnitine, a lysine derivative that stimulates fat oxidation. β-Oxidation occurs in mitochondria results in 2-carbon fragments are serially split off the fatty acid. Unsaturated fatty acids are β-oxidised in the normal pathway as far as possible. However, polyunsaturation or unsaturation at odd positions of the acyl-CoA will produce a molecule that the major pathway cannot utilise as a substrate. A number of enzymes exist in the peroxisome to convert these molecules to appropriate substrates, which can be shuttled into the normal pathway. Certain types of polyunsaturated fatty acids (PUFA) may enter the β-oxidation spiral and produce a D3,5-dienoyl-CoA. This product is unable to enter the normal β-oxidation spiral, because the acyl-CoA oxidase adds a double bond at the 2 position, but cannot do this because of the double bond at the 3 position. Therefore, D3,5-D2,4-dienoyl-CoA isomerase converts the D3,5-dienoyl-CoA to a 2,4-dienoyl-CoA. This enzyme is both peroxisomal and mitochondrial. 2,4-dienoyl-CoA reductase converts the 2,4-dienoyl-CoA into 3-enoyl-CoA. These enzymes are NADPH dependent, and both mitochondrial and peroxisomal forms exist. 3-Enoyl-CoAs is produced by the normal β-oxidation of odd-position unsaturated fatty acids, as well as through the breakdown of PUFA. The D3-enoyl-CoA is converted to a D2-enoyl-CoA, a molecule that is an intermediate of the normal β-oxidation spiral. Isocitrate dehydrogenase is not specifically required to convert an acyl-CoA to a substrate for the β-oxidation pathway, but is necessary for the regeneration of NADPH for use by 2,4-dienoyl-CoA reductase in the auxiliary β-oxidation pathway, and HMG-CoA reductase (an enzyme involved in cholesterol biosynthesis). Isocitrate dehydrogenase contributes the 2-oxoglutarate required by PAHX, an enzyme in the α-oxidation pathway. Peroxisomal β-oxidation of fatty acids originally described in 1976, is catalysed by 4 enzymes: acyl-CoA oxidase, bifunctional enzyme (with enoyl-CoA-hydratase and 3-hydroxyacyl-CoA dehydrogenase activity), 3-ketoacyl-CoA thiolase, and fatty acyl-CoA synthetase. Peroxisomal β-oxidation proceeds from enoyl-CoA through D-3-hydroxyacyl-CoA to 3-ketoacyl-CoA by action of the D-3-hydroxyacyl-CoA dehydratase/D-3-hydroxyacyl-CoA dehydrogenase bifunctional protein, also called D-bifunctional protein (DBP). The oxidation of bile acid precursors may be catalysed by the D-bifunctional protein. Peroxisomes contain 2 sets of β-oxidation enzymes, which differ in substrate specificity. Straight-chain acyl-CoA oxidase (SCOX) is responsible for the oxidation of VLCFAs. Branched-chain acyl-CoA oxidase (BCOX) is involved in the β-oxidation of pristanic acid and of the bile acid intermediates trihydroxycholestanoic acid (THCA) and dihydroxycholestanoic acid (DHCA). D-bifunctional protein (DBP) forms and dehydrogenates D-3-hydroxyacyl-CoAs. L-bifunctional protein (LBP) only produces L-hydroxy intermediates. DBP is involved in the degradation of VLCFAs, as well as of the branched-chain fatty acids, pristanic acid, and DHCA/THCA. Sterol carrier protein-2 (SCP2) contains both a thiolase domain and a sterol carrier-protein domain and is the key enzyme in the β-oxidation of pristanic acid and DHCA/THCA. The peroxisome is required for proper synthesis of bile acids. Bile acids are derived from cholesterol via a number of steps, which are compartmentalised in the cytosol, the mitochondria, the ER, and the peroxisome. The steps of bile acid synthesis which take place in the peroxisome involve the metabolism of the side chain of trihydroxycholestanoic acid (THCA) or dihydroxycholestanoic acid (DHCA), precursors of bile acids The steps mirror those of the normal β-oxidation pathway, and utilise specific isoforms of this pathway. These reactions produce choloyl-CoA or chenodeoxycholoyl-CoA from THCA and DHCA, respectively. These products are then converted to their glycine or taurine conjugates by another peroxisomal enzyme, then transported out of the peroxisome. The final peroxisomal step of this pathway involves the replacement of the CoA of the β-oxidation product (either choloyl-CoA or chenodeoxycholoyl-CoA) with glycine or taurine, resulting in glyco/taurocholate. This reaction is performed by bile acid CoA:amino acid N-acyltransferase, an enzyme that has been found to terminate in Ser-Gln-Leu in humans. The β-oxidation pathway is cyclic. The product, 2 carbons shorter, is the input to another round of the pathway. If, as is usually the case, the fatty acid contains an even number of C atoms, in the final reaction cycle butyryl-CoA is converted to 2 copies of acetyl-CoA: 1- Acyl-CoA dehydrogenase catalyses oxidation of the fatty acid moiety of acyl-CoA, to produce a double bond between carbon atoms 2 and 3. Very long chain Acyl-CoA dehydrogenase is membrane bound; the others (short, medium, long) are in the mitochondrial matrix. FAD is the prosthetic group that functions as electron acceptor for Acyl-CoA dehydrogenase. A glutamate residue extracts a proton, facilitating transfer of 2 electrons-with H$^+$ (a hydride) to FAD. The reduced FAD accepts a second H$^+$, yielding $FADH_2$. $FADH_2$ of acyl CoA dehydrogenase is reoxidised by transfer of 2 electrons to an electron transfer flavoprotein (ETF), which in turn passes the electrons to CoQ of the respiratory chain. 2- Enoyl-CoA hydratase catalyses stereospecific hydration of the trans double bond produced in the 1st step of the pathway, yielding L-hydroxyacyl-CoA. 3- Hydroxyacyl-CoA dehydrogenase catalyses oxidation of the hydroxyl in the β position (C3) to a ketone. NAD$^+$ is the electron acceptor. 4- β-Ketothiolase (β-Ketoacyl-CoA thiolase) catalyses thiolytic cleavage. Proposed mechanism: A cysteine S attacks the β-keto C. Acetyl-CoA is released, leaving the fatty acyl moiety in thioester linkage to the cysteine thiol. The thiol of HSCoA displaces the cysteine thiol, yielding fatty acyl-CoA (2 C shorter). A membrane-bound trifunctional protein complex with 2 subunit types expresses the enzyme activities for steps 2-4 of the β-oxidation pathway for long chain fatty acids. Equivalent enzymes for shorter chain fatty acids are soluble proteins of the mitochondrial matrix.

β-Oxidation pathway:
Fatty acyl-CoA + FAD + NAD^+ + HS-CoA → Fatty acyl-CoA (2 C shorter) + $FADH_2$ + NADH + H^+ + acetyl-CoA.

Fatty acid oxidation is a major source of cellular ATP. $FADH_2$ of Acyl CoA dehydrogenase is reoxidised by transfer of 2 electrons-via ETF to CoQ of the respiratory chain. H^+ ejection from the mitochondrial matrix that accompanies transfer of 2 electrons-from CoQ to O_2 leads via chemiosmotic coupling to production of 1.5 ATP. ADH is reoxidised by transfer of 2 electrons-to the respiratory chain complex. Transfer of 2 electrons-from complex I to O_2 yields 2.5 ATP. Acetyl-CoA can enter Krebs cycle, where the acetate is oxidised to CO_2, yielding additional NADH, $FADH_2$, and ATP. Genetic abnormalities and mutations plasma membrane fatty acid transport had been identified. Symptoms vary depending on the specific genetic defect but may include hypoglycaemia during fasting, fatty degeneration of the liver; heart and/or skeletal muscle defects, and maternal complications of pregnancy. Hereditary deficiency of medium chain acyl-CoA dehydrogenase (MCAD), the most common genetic disease relating to fatty acid catabolism, can lead to sudden death in infants (SIDS). The final round of β-oxidation of a fatty acid with an odd number of carbon atoms yields acetyl-CoA and propionyl-CoA. Propionyl-CoA is converted to the Krebs cycle intermediate succinyl-CoA, by a pathway involving vitamin B_{12}. Most double bonds of naturally occurring fatty acids have the cis configuration. As carbon atoms are removed 2 at a time, a double bond may end up in the wrong position or wrong configuration for the enoyl-CoA to be a substrate for enoyl-CoA hydratase. β-Oxidation of very long chain fatty acids also occurs within peroxisomes. FAD is electron acceptor for peroxisomal acyl-CoA oxidase, which catalyses the first oxidative step of the pathway. The resulting $FADH_2$ is reoxidised in the peroxisome producing hydrogen peroxide:
$FADH_2 + O_2 \rightarrow FAD + H_2O_2$
The peroxisomal enzyme catalase degrades H_2O_2 by the reaction:
$2 H_2O_2 \rightarrow 2 H_2O + O_2$
These reactions produce no ATP. Once fatty acids are reduced in length within the peroxisomes they may shift to the mitochondria to be catabolised all the way to CO_2. Carnitine is also involved in transfer of fatty acids into and out of peroxisomes. During carbohydrate starvation, oxaloacetate in liver is depleted because it is used for gluconeogenesis. This impedes entry of acetyl-CoA into Krebs cycle. Acetyl-CoA then is converted in liver mitochondria to ketone bodies, acetoacetate and β-hydroxybutyrate. Three enzymes are involved in synthesis of ketone bodies: 1- β-Ketothiolase, the final step of the β-oxidation pathway runs backwards, condensing 2 acetyl-CoA to produce acetoacetyl-CoA, with release of one CoA. 2- HMG-CoA synthase catalyses condensation of a third acetate moiety (from acetyl-CoA) with acetoacetyl-CoA to form hydroxymethylglutaryl-CoA (HMG-CoA). 3- HMG-CoA Lyase cleaves HMG-CoA to yield acetoacetate plus acetyl-CoA. β-Hydroxybutyrate dehydrogenase catalyses inter-conversion of the ketone bodies acetoacetate and β hydroxybutyrate. Ketone bodies are transported in the blood to other tissue cells, where they are converted back to acetyl-CoA for catabolism in Krebs cycle. 3-methyl-branched fatty acids (such as phytanic acid) are taken up the diet. These compounds cannot be degraded by the normal peroxisomal β-oxidation pathway, because the 3-methyl blocks the dehydrogenation of the hydroxyl group by hydroxyacyl-CoA dehydrogenase. The solution for this dilemma is the α-oxidation pathway. The first 2 steps of α-oxidation consist of the ligation of phytanate to phytanoyl-CoA, and the hydroxylation of phytanoyl-CoA. The long chain fatty acyl-CoA synthetase is capable of ligating CoA to phytanic acid. Phytanoyl-CoA hydroxylase (PAHX) converts phytanoyl-CoA to α-hydroxylphytanoyl-CoA. The enzymes, thioesterase and α-hydroxy acid oxidase exist in peroxisomes. There are actually at least two thioesterases (PTE1 and PTE2) and 3 α-hydroxyacid oxidases (HAOX1, HAOX2, and HAOX3), all of which are at least partly peroxisomal enzymes. The hydroxyphytanoyl-CoA lyase enzyme would convert α-hydroxyphytanoyl-CoA directly to pristanal. Pristanal is converted to pristanic acid by the action of an aldehyde dehydrogenase. Pristanic acid then enter the peroxisomal β-oxidation pathway. The family of glycerolipids includes plasmalogens and platelet-activating factor (PAF). Plasmalogens are components of cellular membranes, make up approximately half of the heart's phospholipids, but are the most abundant in nervous tissue. They make up 80-90% of the ethanolamine phospholipid class in myelin. Plasmalogens might be effective antioxidants in the membrane. The peroxisome is involved in the synthesis of glycerolipids.

EPITHELIAL MEMBRANE FLUIDITY AND TRANSPORT

Fluidity is inverse of viscosity, the property describing the resistance of a fluid to movement. The membrane is constituted of an oriented, 2 dimensional, viscous solution of amphipathic proteins and lipids. Membrane fluidity represents a set of parameters including structural information (order and packing of acyl chains) as well as dynamic information (rotational and translational diffusion). Fluidity (and/or composition) affects the activity of numerous proteins. Freedom of protein conformational changes can be regulated by lipid fluidity. Major determinants of membrane fluidity are: 1) extrinsic determinants regroup environmental factors; 2) intrinsic determinants, which quantify the membrane composition. Any manipulation that induces an increase in lipids molal volume, such as an increase in temperature, or replacement of a saturated by an unsaturated fatty acid, will result in an increase in membrane fluidity, and conversely. The length and the unsaturation of fatty acyl chains of phospholipids, the nature of their polar head group and, as a function of this head group, the pH and the concentration of ions (mainly divalent cations), determine the temperature at which pure hydrated phospholipids undergo an endothermic transition from the gel to the liquid crystalline state. These transitions are associated with a large increase in bilayer fluidity. In complex mixtures of lipids such as those found in biological membranes, some lipid species may enter the gel phase while others are still in a liquid crystalline state. Such phase separations phenomena in which gel and fluid phase coexist modify the lateral organisation and thus the environment of membrane proteins. For membrane proteins sensitive to the physical state of surrounding lipids, the liquid to gel phase transition is generally associated with large changes in activity. There are differences exist in the lipid composition of apical and basolateral originating from the proximal tubule of the kidney, for example. The cholesterol content is higher in brush border than in basolateral membranes. Cholesterol decreases the fluidity. The low fluidity of apical membranes relative to that of basolateral membranes constitutes a characteristic feature of epithelial cells. Tight junctions that encircle the apex of epithelial cells, play a major role in the maintenance of lipid polarity between the apical and basolateral membrane regions. Whereas lipids in the cytoplasmic membrane leaflet diffuse freely all around the cell, intact tight junctions prevent the mixing of lipids localised on the outer leaflet. Phospholipid distribution between inner and outer leaflets is highly asymmetrical with sphingomyelin accounting for 3/4 of the outer leaflet phospholipids. Membrane fluidity can modulate transport processes at different levels. The fluidity state of the plasma membrane of epithelial cells depends on physiological/pathological conditions. These changes affect transport functions. Maturation and aging decrease the fluidity of epithelial cells, while ischaemia and gentamicin increases it. Lipid peroxidation decreases membrane fluidity.

Major determinants of membrane fluidity

Extrinsic	Effect	Intrinsic	Effect
Temperature	Increase	Cholesterol	Decrease
Pressure	Decrease	Sphingomyelin	Decrease
Cell volume	Increase	Protein/lipid	Decrease
Cations (divalent)	Decrease	Sat./unsat.	Decrease
H^+	Decrease	Chain length	Decrease

EPITHELIAL CELL POLARITY AND DISEASES

Polarised epithelial cells line many organs in the body and facilitate specific vectorial functions by means of asymmetric insertion of specific proteins into apical, basal, or lateral plasma membrane domains. An epithelial cell depends upon cues from the extracellular environment to generate an orientation and to establish its polarised domains. The apical and basolateral surfaces of polarised epithelia are equipped with markedly distinct populations of channels, carriers, and pumps. The physiological properties of an individual epithelial cell type are determined not only by its census of transport proteins but also by manner in which these proteins are segregated between the apical and basolateral portions of the plasmalemma. An epithelial cell must be able to establish distinct surface domains, to target newly synthesised transport proteins to their appropriate sites of functional residence, and to retain them there following their delivery. Attachments to the substratum occur through interactions between integrins and the extracellular matrix. This divides the plasmalemma into 2 nonequivalent regions: 1- the adherent surface, which is in direct apposition to the basement membrane; and 2- the free surface. Attachment initiates the process of polarisation. Polypeptides that are bound to the apical domains of confluent epithelia are restricted to the free surfaces of newly attached individual epithelial cells. Proteins normally associated with the basolateral plasmalemma domains, are distributed over both the attached and free surface. The complete segregation requires contact between neighbouring epithelial cells, which is mediated by calcium-dependent adhesion molecules, or cadherins. Cadherins are transmembrane proteins whose ectodomains participate in calcium-dependent homotypic interactions. The cadherins of neighbouring epithelial cells bind to one another and communicate this binding event to the cell interior directly with the catenins, a family of proteins that interact directly with the cadherin cytoplasmic tails. The establishment of cell-cell contact nucleates the assembly of the subcortical cytoskeleton, to provide structural stability to the lateral plasmalemma, the subcortical cytoskeleton interacts directly with numerous basolateral transmembrane proteins. Cadherin-mediated cell-cell contact is a prerequisite for the formation of tight junctions, whose establishment delimit the boundaries of the newly formed apical membrane. Prior to establishment of cell-cell contact, precursors of the apical membrane appear to form in intracellular vocuolar compartments. The membranes that delimit these vocuoles display lumen-facing microvilli and are enriched in polypeptides destined for apical insertion. It has been suggested that the formation of cell-cell contact catalyses the fusion of this preformed apical domain. In this manner the interaction between neighbouring cells is able to initiate the biochemical differentiation of both the basolateral and the apical surfaces. Cell-cell contact is required for the maintenance of the polarised state. Disruption of cell-cell contact usually results in the rapid loss of polarity. The biochemical identities of the membrane domains become lost as proteins diffuse and mingle in the plane of bilayer. Proximal tubule cells of the kidney provide an interesting exception to this pattern. Epithelial cells are constantly synthesising, internalising, recycling, and degrading their membrane proteins. The outer leaflets of the apical membranes in many epithelial cell types are heavily enriched in glycosphingolipids (GSLs). Newly synthesised GSLs appear to cocluster as they pass through the trans-Golgi network (TGN) on their way to the apical surface. The most homologous members of the P-type ATPase family are the Na-K-ATPase and the gastric H-K-ATPase, which are both heterodimeric complexes composed of $\approx$100 kDa α-subunits and 55 kDa β-subunits. The α-subunits are oriented polytopic, spanning the bilayer 10 times. The β-subunits are oriented as type II membrane proteins, with a single membrane span and a cytoplasmically disposed NH_2-terminus. Whereas Na-K-ATPase is basolateral in most epithelial cell types, nongastric H-K-ATPase isoforms present in the apical membrane of renal collecting tubule epithelial cells. While residues in the COOH-terminal half of the α-subunit determine the identity of the β-subunit with which it will assemble, the NH_2-terminal half of the α-polypeptide dictates targeting in polarised cells. Interactions with GSLs may play an important role in Na-K-ATPase sorting. Cytoskeletal interactions play an important role in stabilising pumps within their appropriate membrane domains and in preventing their rapid internalisation. Newly synthesised proteins are delivered to specific apical or basal and/or lateral domains where they may be stabilised by complex formation with adhesion or cytoskeletal proteins. The polarised segregation of proteins is further maintained by the presence of tight junctions in the apical lateral regions of the cell, which effectively prevents the mixing of apical membrane proteins with those of the basolateral domains. The establishment and continuous maintenance of polarity in epithelia is essential to the integrity and function of most epithelial organs, and this is especially critical in the kidney where the primary functions of reabsorption and secretion are made possible by the differential polarised insertion of specific transporters and related proteins into apical membranes lining the renal tubule lumen or basal domains adjacent to the interstitium and blood space. The high degree of morphological segmentation of the nephron is accompanied by the differential expression of polarised membrane proteins, transporters, and channels in all the different nephron segments. For instance, the Na^+-K^+-$2Cl^-$ transporter is characteristic of the thick ascending limb of the loop of Henle, whereas the vasopressin-sensitive water channels (aquaporin-2, AQP2) is characteristic of the apical membrane of the collecting tubules. In eukaryotic cells, newly synthesised proteins destined for the plasma membrane are synthesised in the rough endoplasmic reticulum, translocated to the cis-Golgi, undergo pottranslational modifications in the medial Golgi stacks, and finally reach the trans-Golgi network (TNG), which is the major sorting center of the cell. At this point, proteins bearing mannose 6-phosphate (M6P) are sorted into M6P-receptor-bearing vesicles, which are destined for lysosomes, whereas proteins destined for secretory vesicles or the plasma membrane are sorted by less well-characterised signal-recognition pathways. Clathrin-coated vesicles containing specific TNG adaptor protein complexes bud from the TNG, and further specificity and regulation of fission and fusion is thought to be conferred by rab proteins. Polarised plasma membrane proteins are sorted at the level of the TGN into separate discrete vesicle population destined for specific apical or basolateral plasma membrane domains. Abnormal trafficking and membrane insertion of functionally important proteins has been defined in several disease states, and the list is rapidly growing. Understanding the basic cell biology underlying intracellular protein and vesicle transport processes is of crucial targeting. Faulty intracellular delivery and polarisation of proteins, when not fatal to the cell or organism, can produce serious disease, such as: 1- Cystic fibrosis. 2- Congenital goiter with hypothyroidism. 3- Familial hypercholesterolaemia. 4- I cell disease. 5- Hyperoxaluria, in which a mutation in alanine-glycoxalate aminotransferase results in the unmasking of a cryptic site and results in mistargeting of this enzyme to mitochondria instead of to the peroxisomes as in normal cell. 6- Renal diseases, the normal function of the kidney depends on differential polarisation of membrane transporters and enzymes and on the appropriate vectorial transport of fluid and ions. Thus defects in targeting and polarisation of membrane proteins have deleterious consequences on renal function leading to disease. Inappropriate secretion into the nephron tubule lumen instead of reabsorption results in the formation of fluid-filled cysts. Permanent cell loss usually results from protracted ischaemia in brain and heart, whereas the kidney can often restore structure and function to a large extent. One hallmark of ischaemic injury to renal epithelial cells is a disruption of their structural organisation, including disorganisation of the cytoskeleton. In the proximal tubule, the polarised distribution of the normally basolateral Na^+-K^+-ATPase is modified, and appears on the apical plasma membrane at the level of brush border. The at least partial loss of cell polarity resulting from ischaemia is probably related to the breakdown of the tight junctional barrier that normally segregates apical and basolateral plasma membrane domains of these cells. Ischaemia followed by reperfusion results in the depolymerization of microtubules in epithelial cells of the kidney, notably proximal tubule cells. The effect on microtubules is dependent on reperfusion and may be mediated by a sudden rise in intracellular calcium levels.

SPHINGOLIPIDS

When a fatty acid is linked to the amide group of the long-chain base, the resultant lipid is referred to as ceramide. The primary alcoholic function at carbon one of ceramide is the site of attachment for phosphocholine-producing sphingomyelins and saccharides producing a multitude of glycosphingolipids. When a monosaccharide is present (glucose) the lipid is referred to as a cerebroside. There may be as many as 15-20 sugars in the oligosaccharide chain include neutral glycosphingolipids, acidic glycosphingolipids (gangliosides, where one or more sialic acids present and sulfatides where sulfate group present). The ceramide portion of a sphingolipid is highly hydrophobic. The composition diversity of membrane lipids may facilitate a lateral phase separation of different lipid components, resulting in clusters of lipids rich in ceramide in the case of sphingolipids that create distinct microenvironments within the cell membrane (hot spots). Factors that promote clustering include the content of sialic acid, the presence of Ca^{2+} ions, the ceramide content, and proteins such as caveolin (VIP21). Ceramide is a key intermediate in the formation of most of the biologically important sphingolipids. Ceramide is formed via the de novo synthetic NAD dependent route by the unsaturation at the 4,5 positions of the sphingoid amine by another oxidoreductase. The sphingomyelin synthase enzyme localised at the cis and medial Golgi membranes is responsible for the transfer of phosphocholine from phosphoatidylcholine to ceramide. Sphingomyelin synthase pathway regulates level of the precursor ceramide and may have important implications for cell-signalling phenomena. Glucosylceramide is formed from ceramide and UDP-glucose by UDP-glucoseceramide glycosyltransferase. Glucosylceramide is the precursor of most of the mammalian glycosphingolipids. Glucosylceramide can be further glycosylated within the Golgi or can be directly transported to the plasma membrane. Ceramide can be phosphorylated to ceramide-1-phosphate by calcium-dependent ceramide kinase or bacterial diglyceride kinase. The main catabolic pathway of sphingolipids is through endosomal/lysosomal pathways. Glycosphingolipids are degraded via one or more glycosidases, and ceramide is degraded by an acidic ceramidase. The latter reaction results in the formation of sphingosine and a fatty acid. Free sphingosine is released from the lysosome by an uncharacterised pathway. Free sphingosine is further catabolised outside of the lysosome. Free sphingosine may be reacylated forming ceramide or phosphorylated by sphingosine kinase to form sphingosine phosphate. The terminal catabolism of sphingosine forms ethanolamine phosphate, a fatty aldehyde, and monomethyl and dimethyl sphingosines. The subcellular sites of metabolism and the subcellular distribution of sphingolipids are important determinants in understanding the cellular biology of these lipids. Enzymes associated with long-chain base synthesis and ceramide formation are present in the smooth endoplasmic reticulum. Acyl-CoA:sphingoid base acetyltransferase is localised in mitochondria. The degradation of sphingomyelin produces phosphocholine and ceramide. The sphingomyelinase activities include: 1- acidic, lysosomal sphingomyelinase, 2- zinc dependent serum sphingomyelinase, 3- neutral, magnesium-dependent, membrane sphingomyelinase, 4- neutral Mg^{2+}-independent sphingomyelinase (myelin sheath), 5- Mg^{2+} and dithiothreitol-stimulated neutral sphingomyelinase (nuclei of ascites cells), 6- neutral sphingomyelinase localised to chromatin and nuclear envelop of liver cells, and 7- alkaline sphingomyelinase in the bile and intestine. Both sphingomyelin and glycosphingolipids are highly enriched in the outer leaflet of the plasma membrane. Sphingolipids flip from the outer to the inner leaflet. The multidrug resistance protein may serve as a membrane "flipase" for placement of glycosphingolipids such as glucosylceramide in the outer leaflet. Tumour cells that express high levels of multiple-drug resistance protein are characterised by high contents of glucosylceramide. Ceramide is the product of sphingomyelin hydrolysis. Several sphingolipid metabolites are implicated in signalling events including ceramide, sphingosine, sphingosine-1-phosphate, and ceramide-1-phosphate. Ceramide regulates NFκB signalling and apoptosis. Effectors that increase cell ceramide levels cause cell growth arrest, differentiation, or apoptosis by cytokines, environmental stressors, and pharmacologic agents. These include NFκB activation, COX transcription, heat-shock protein transcription, selectin expression, and cytokine secretion. The growth factors increase sphingosine and sphingosine-phosphate. Multiple activators associated with the generation of cellular sphingolipids include TNFα, IL-1β, IL-2, γ-IFN, NGF, hormones (glucocorticoids, progesterone, 1,25(OH)$_2$ vitamin D$_3$, bradykinin, retinoic acid), ET-1, antigens (FAS ligand), environmental stressors (ionising radiation, heat shock, ischaemia, hypoxia, peroxide), pharmacologicals (vincristine), Ara-C, phorbol esters, calcium channel blockers, protein kinase C inhibitors, and infective agents (virus, bacteria, bacterial toxins). Ceramide is recycled to sphingomyelin by sphingomyelin synthase. The biological consequences of sphingomyelin degradation and ceramide generation can often be replicated by manoeuvres that augment or inhibit ceramide levels. Phospholipase C and protein kinase C activation may abrogate biological effects secondary to ceramide generation such apoptosis. Arachidonic acid, the product of phosphlipase A_2 is a lipid mediator of sphingomyelinase. Arachidonate formation precedes sphingomyelin hydrolysis. Other polyunsaturated fatty acids may activate neutral sphingomyelinase. Glutathione, a tripeptide, serves a major role in protecting cells against oxidative stress. Reduced glutathione is an antioxidant that is depleted in cell death. Glutathione regulates ceramide formation. Both reduced and oxidised glutathione inhibit the neutral Mg^{2+}-dependent sphingomyelinase. Sphingomyelin hydrolysis and apoptosis can be prevented by replention of intracellular glutathione. Sphingosine-1-phosphate is rho dependent and mediates stress-fiber formation. Signalling events following sphingosine-1-phosphate binding include: 1- Activation of mitogen-activated protein kinases, phospholipase C, phospholipase D, and potassium channels. 2- Inhibition of adenyl cyclase and mobilisation of intracellular calcium. 3- Inhibition of ceramide-induced apoptosis. 4- Inhibition of cell motility. 5- Neurite retraction. 6- Platelet activation. 7- Induction of P-cadherin transcription. 8- Mitogenesis. Sphingosine-1-phosphate can mobilise Ca^{2+} from intracellular stores. Ceramide and sphingosine stimulate cellular phosphatase activity against Rb protein and induce cell cycle arrest at the G_1/S transition. Ceramide-activated protein kinase (CAP kinase) may directly activate Raf and mitogen-activated protein kinase activation in response to TNFα. Death receptors such as the TNF and CD95 receptors activate death effector domain-containing enzymes, the caspases (cystein protease that cleaves after aspartic acid). Activation is mediated by binding to the cytoplasmic portions of these receptors via adaptor proteins such as TRADD, TRF-2 and FADD. Death receptor-independent stimuli such as etoposide and irradiation induce caspase activation at other sites. Inducer caspases (caspase 8) appear to act upstream of the mitochondrial release of cytochrome C into the cytosol, which in association with Apaf-1 and caspase-9 subsequently process and activate executioner caspases (caspase 3, 6, 7). Both neutral and acidic sphingomyelinases involve in cell death pathways. Ceramide formed secondary to exogenous sphingomyelinase (outer membrane leaflet), within acidic cellular compartments, and via activation of a neutral sphingomyelinase (inner membrane leaflet) is implicated in the cell death response. Apoptotic stimuli induce both very rapid and slowly rising ceramide levels. The verocytotoxin elaborated by pathogenic strains of E Coli (O157:H7) associated with haemolytic uraemic syndrome and haemorrhagic colitis binds to globotriaosylceramide (ceramide trihexoside). Gb3 is expressed on platelets, but verotoxin does not affect platelet aggregation. Adult glomeruli poorly express Gb3, whereas Gb3 is highly expressed in the glomeruli of paediatric patients. Gb3 levels can be induced in response to multiple cytokines (TNF, IL-1β, and lipopolysaccharide). Gb3 can also be induced by viral infection, e.g., EBV and HIV. Gb3 is a Burkitt lymphoma-associated antigen. Viral prodromes may render adult patients susceptible to haemolytic uraemic syndromes (reported in HIV). Thrombotic disorders such as haemolytic uraemic syndrome may be mediated in part by sphingosine-1-phosphate released from activated platelets. Sphingosine-1-phosphate stimulates mesengial cell proliferation. Ceramide is a potential mediator of vascular relaxation, act by inducing endothelial nitric oxide synthase (eNOS). Ceramide increases NOS activity. Bradykinin induces eNOS and raises cell ceramides levels. Ceramide levels may rise in ischaemic injury.

STRUCTURE AND FUNCTION OF RECEPTORS

Most drugs function by binding to receptors, with notable exceptions including antibiotics and anaesthetics. Endocytosis of fluid is called pinocytosis (cell drinking); phagocytosis of a large multimolecular particle is phagocytosis (cell eating). The engulfed vesicle either fuse with lysosome for degradation and release of its contents into the intracellular fluid, or, travels to the opposite side of the cell to release its contents by exocytosis. Intracellular signalling is achieved through myriad molecules that interact specially with their respective docking sites, or receptor proteins. In endocrine functions, circulatory hormones bind to high-affinity receptors in their target tissues, and induce a cascade of events that culminate in their defining effects on cell structure, function and growth. In paracrine functions, effector molecules exert more local effects by binding to nearby receptors to produce coordination and synchronisation of function within a cluster of cells. Subsequent to binding of ligand, the receptor serves a transducing function, and signal to intracellular portions of itself, or other associated proteins, that ligand has been bound. This in turn activates specific effector pathways resulting in metabolic and structural changes in the cell that constitute the signature effect of the ligand. Hormone receptors may be both on membranes, particularly for peptides, and within cells, particularly for steroid hormones, and that second messengers take many forms.

CARGO-TYPE RECEPTORS
Cargo-type receptor, such as the LDL receptor, resides on the plasma membranes of cells and serves as a mechanism whereby the cell can take up essential nutrients (cholesterol). After circulating LDL particles bind to the LDL

receptor, the receptor-ligand complexes migrate laterally in the plasma membrane to cluster within clathrin-coated pits. Subsequently, these coated pits invaginate, and give rise to discreet clathrin-coated vesicles that contain within their interior the LDL particles. Within 90 seconds after internalisation, clathrin is shed from these vesicles, and the now uncoated vesicles migrate and fuse with the endosomal compartment. Endosomes are acidic relative to the cytosol and have an internal pH of 5.2. This acidity is critical to the processing of the endocytosed receptor-ligand complex, and in the case of the LDL receptor, and its cargo substrate LDL, acidification of the vacuole causes release of the LDL particle from the LDL receptor. Once this occurs, cholesterol is transferred from the endosomal compartment to lysosomes. Lysosomes are more acidic than endosomes, with a steady-state pH of 4.5. This acidity, in concert with a newly discovered carrier protein within the lysosomal membrane, facilitates the transfer of cholesterol from this compartment, making it available for cellular metabolism. The LDL receptor is ritualised, vesicles containing the emptied receptor bud from the endosome, and recycle to the plasma membrane.

Messenger to membrane receptors:

First messenger	1- Opens or closes specific channels in the membrane to regulate the movement of ions into or out of the cell. 2- Transfer the signal to an intracellular chemical messenger (second messenger).
Second messenger	Triggers pre-programmed biochemical events within the cell.

The extracellular portion of the LDL receptor consists of a series of cysteine-rich repeats, which form the actual LDL binding site. Other features of the protein include an EGF precursor homology domain that is essential for dissociation of LDL in the acidic endosome, a single transmembranous spanning region, and a small domain in the cytoplasmic tail that function as an internalisation motif. Reduction of cellular cholesterol, sensed within the endoplasmic reticulum, leads to an increase in mRNA encoding the receptor, and ultimately in the synthesis of LDL receptors that serve to enhance the LDL-binding capacity of the cell. These types of receptors serve to generate specific ligand-triggered responses. General features of these types of receptors include an extracellular ligand-binding site that, when occupied, causes a conformational change in the cytosolic portion of the receptor molecule. This conformational change subsequently activates a variety of intracellular signalling pathways that is ultimately manifested in the ligand-specific effect on the cell.

Channel regulation of EC messengers:

1- By binding of an EC chemical messenger to a specific membrane receptor that is in close association with the channel.
2- By changes in electrical current in the plasma membrane.
3- By stretching or deformation of the channel.

Lipoprotein receptors: Lipoprotein BE receptor is a single protein with 839 amino acids. The ligands for these receptors are apo B-100 and apo E. The receptor is distributed in 5 domains with independent contribution to the activity: 1- The first domain, rich in cysteine with 292 amino acids, is the binding site for apo B-100. 2- The second domain, with 400 amino acids has a homology to EGF and certain coagulation factors such as Stuart factor and C protein. 3- The third domain, with 58 amino acids is glucose rich. It has important role in receptor orientation on the cell membrane. 4- The fourth domain, with 22 amino acids is the only transmembrane part of the receptor. 5- The fifth domain with 50 amino acids is intracytoplasmic and is needed for LDL endocytosis with submembranous cytoskeleton proteins namely clathrins. This domain contains the signal that directs the receptor to clathrin-coated pits. For this to occur requires 807 residue to an aromatic amino acid. Tyrosine is needed for endocytosis. Mechanisms of uptake: 1- synthesis in ribosomes of a receptor precursor, which migrates later to the Golgi apparatus for glycosylation. 2- Appearance of the receptor in the coated pits, as membrane aggregates due to random distribution. 3- Coated pits interact with LDL forming endocytic vesicles coated with clathrin-coated vesicles. 4- Vesicles lose clathrin and fusion form larger vesicles (endosomes or receptosomes. 5- At pH lower than 6.5, LDL leaves the receptors and return to the surface (receptor recycling). 6- LDL is taken up by lysosomes, its proteins hydrolysed and cholesterol esters hydrolysed into cholesterol and fatty acids.

ENZYME LINKED RECEPTORS

A- Receptors with intrinsic kinase activity: The intrinsic activity is within the receptor molecule itself. Receptors of this class respond to circulating hormonal stimuli to provoke the cell to undergo a wide array of shifts in metabolism, which ultimately leads to growth and/or differentiation of the individual cells (insulin receptor, IGFR, EGFR, FGFR, PDGFR). They hold in common an extracellular domain that is responsible for binding the circulating ligand (e.g., insulin), a single transmembranous sector, and most characteristically, an internal domain that has intrinsic tyrosine kinase activity. The insulin receptor α subunits are external to the cell, and contain the actual insulin binding sites, and tyrosine kinase domains are found in the β chains within the cell interior. Binding of insulin to the α subunit results in a conformational change in the protein that is transmitted through the membrane-spanning segment to the intracellular β subunits. These changes in the quarternary structure of the receptor activate tyrosine kinase domains in the β subunits, which then phosphorylate multiple tyrosine residues in the corresponding trans β subunits. Termination of these insulin-provoked signals is in part achieved through the receptor-mediated endocytotic pathway. Autophosphorylation of the insulin receptor (IR) triggers internalisation of the receptor-ligand complex through clathrin-coated vesicles. After fusion with endosomes, insulin is released from its receptor by the acidity of the compartment, and is degraded. Tyrosine phosphatases present in the endosomal membrane dephosphorylate the cytoplasmic tail of the receptor. The EGFR, exist intrinsically within the membrane as monomers. They differ from IR in that ligand binding to their extracellular domains initially results in a dimerization of 2 of these monomers. Subsequent to this, activation of the tyrosine kinase domains of the dimerized receptors leads to a cascade of signalling events generally similar to that described for the insulin receptor. B- Receptors with associated tyrosine kinase activities: There exists a second subclass of enzyme-associated receptors that do not possess intrinsic tyrosine kinase activity but rather have an associated partner protein that itself has tyrosine kinase activity (e.g., receptors binding prolactin, growth hormone, and numerous cytokines). Ligand binding to this type of receptor induces dimerization, followed by the noncovalent attachment of a nonreceptor tyrosine kinase, such as those of the Src family. C- Receptors with intrinsic protein tyrosine phosphatase activity: When this receptor protein undergoes dimerization through cell-cell interactions (which function as nonsoluble ligands), its phosphatase domain is activated, leading to dephosphorylation of proteins within signalling cascades and by this mechanism, activation of T and B lymphocytes. D- Receptors with intrinsic serine/threonine kinase activities: This receptor exists as a single polypeptide chain with one membrane-spanning domain. The cytosolic portion of the molecule contains a domain with intrinsic serine/threonine kinase activity (TGFβR). TGFβ responses include growth inhibition and apoptosis. Mutations that disrupt function of this signalling pathway have the important property of stimulating tumour growth in various tissues, thus implicating this receptor and its signalling pathway as potential tumour suppressant elements.

RECEPTORS LINKED TO G PROTEINS
Over 1,000 such receptors have been identified (e.g., receptors bind adrenaline, ACTH, LH, parathyroid hormone, glucagon, vasopressin, and calcium-sensing receptor). The calcium sensing receptor (CaR) is responsible for maintenance calcium homeostasis through its effects in parathyroid gland and in the kidney. Mutations in the CaR results in 3 genetic diseases of calcium metabolism: familial hypocalciuric hypercalcaemia, neonatal severe hyperparathyroidism, and autosomal dominant hypocalcaemia. The G protein receptors have a characteristic secondary structure that includes an extracellular amino terminus, 7 transmembranous sectors, and a cytosolic carboxyl terminus (seven-pass serpentine transmembranous structure). The carboxyl terminus of G protein when phosphorylated, inactivates the receptor as a means of desensitising the system to further ligand stimulation. These receptors utilise G proteins to activate their effector pathways. Once ligand binds to the extracellular face of the receptor, there is a resulting change in the conformation of the cytosolic portion of the molecule that permits binding of a G protein complex. The latter consists of 3 subunits, termed α, β and γ. A single cell typically has multiple different G protein coupled receptors on its plasma membrane. Under basal conditions, one molecule of GDP is bound to the α subunit. Association of the G protein complex with its receptor results in release of GDP from the α subunit. Subsequently, a molecule of GTP binds to the α subunit, provoking release of the heterotrimeric G protein complex from its receptor. The released heterotrimeric G protein dissociates into an isolated α subunit and a γ-β heterodimer. The α subunit then diffuses laterally on the cytosolic face of the plasma membrane and activates (or inhibits) one, or several, effector molecules, which include adenylyl cyclases, phospholipases, phosphodiesterases, PI_3 kinases and ion channels. These effector molecules, in turn, activate one, or several, signalling pathways that amplify the ligand signal, and generate the cellular responses characteristic of the ligand. Termination of the signalling activity of the isolated α subunit occurs when its molecule of GTP is hydrolysed to GDP and Pi. Once this occurs, the α subunit reassembles with the β-γ pair to result in an inert, heterotrimeric complex that awaits a repeat of the activation of cycle.

ION CHANNEL RECEPTORS
Change in membrane potential serves as the initial signal effector. More details below (Page 40).

INTRACELLULAR RECEPTORS

These receptors are members of the steroid hormone receptor superfamily (cortisol, steroid sex hormones, thyroid hormone, 1,25 dihydroxtvitamin D_3, and retinoids), termed orphan receptors. Members of this superfamily of receptors function as transcription regulatory elements, they share highly conserved DNA binding motif that contains 2 zinc finger domains. The cortisol receptors are found in the cytosol, and enter the nucleus after the steroid is bound. Oestrogen receptor is constitutively situated in the nucleus and is bound to DNA. The receptors are inactive in the absence of ligand, and exist in a complex with an inhibitory protein. Ligand binding results in dimerization of the receptor and release of the associated inhibitory protein. The receptor-ligand complex then induces transcription of its target genes. Within 30 minutes of binding ligand, these receptors initiate a primary response leading to the synthesis of a group of proteins that are transcription regulatory elements themselves. Subsequently, these factors initiate a second wave of transcription, leading to the translation of proteins, which produce the cellular responses that typify the action of the specific ligand. In some circumstances these receptors can activate transcription in the absence of ligand, where the receptor protein is activated by growth factors, which stimulate kinase pathways, leading to phosphorylation of the receptor.

MECHANISMS OF SIGNAL TRANSDUCTION

Signal transduction at the cellular level is the movement of signals from outside the cell to inside. The movement of signals can be simple passing in the form of small ion movement, either into or out of the cell, e.g., the acetylcholine class. These ion movements result in changes in the electrical potential of the cells that, in turn, propagates the signal along the cell. More complex signal transduction involves the coupling of ligand-receptor interactions to many intracellular events. These events include phosphorylations by tyrosine kinases and/or serine/threonine kinases. Signal transducing receptors are of three general classes: 1. Receptors that penetrate the plasma membrane and have intrinsic enzymatic activity include tyrosine kinases (e.g., PDGF, insulin, EGF and FGF receptors), tyrosine phosphatases (e.g., CD45 [cluster determinant-45] protein of T cells and macrophages), guanylate cyclases (e.g., natriuretic peptide receptors) and serine/threonine kinases (e.g., activin and TGF-b receptors). Receptors with intrinsic tyrosine kinase activity are capable of autophosphorylation as well as phosphorylation of other substrates. 2. Receptors that are coupled, inside the cell, to GTP-binding and hydrolysing proteins (termed G-proteins). These receptors have a structure that is characterised by 7 transmembrane spanning domains and termed serpentine receptors, e.g., the adrenergic receptors, odorant receptors, and certain hormone receptors (e.g., glucagon, angiotensin, vasopressin and bradykinin). 3. Receptors that are found intracellularly and upon ligand binding migrate to the nucleus where the ligand-receptor complex directly affects gene transcription. The proteins encoding receptor tyrosine kinases (RTKs) contain 4 major domains: 1- An extracellular ligand binding domain. 2- An intracellular tyrosine kinase domain. 3- An intracellular regulatory domain. 4- A transmembrane domain. The amino acid sequences of the tyrosine kinase domains of RTKs are highly conserved with those of cAMP-dependent protein kinase (PKA) within the ATP binding and substrate binding regions. Based upon the presence of these various extracellular domains the RTKs have been sub-divided into at least 14 different families.

Common classes of RTKs:

Class	Examples	Structural features
I	EGFR, NEU/HER2, HER3.	Cysteine-rich sequences.
II	IGF-1R	Cysteine-rich sequences, characterised by disulfide-linked heterotetramers.
III	PDGR, c-Kit	Contain 5 immunoglobulin-like domains, contain kinase insert.
IV	FGFR	Contain 3 immunoglobulin-like domains, contain kinase insert acidic domain..
V	VEGFR	Contain 7 immunoglobulin-like domains, contain kinase insert.
VI	HGFR, SCR	Cysteine-rich sequences, characterised by disulfide-linked heterotetramers. The HGFR is a proto-oncogene.
VII	Neutrophin receptor family and NGFR	Contain no or few cysteine-rich domains, NGFR has leucine rich domain.

Several SH2 containing proteins that have intrinsic enzymatic activity include phospholipase C-g (PLC-g), the proto-oncogene c-Ras associated GTPase activating protein (rasGAP), phosphatidylinositol-3-kinase (PI-3K), protein phosphatase-1C (PTP1C), as well as members of the Src family of protein tyrosine kinases (PTKs). There are numerous intracellular non-receptor protein tyrosine kinases (PTKs) that are responsible for phosphorylating a variety of intracellular proteins on tyrosine residues following activation of cellular growth and proliferation signals. The Src protein is a tyrosine kinase first identified as the transforming protein in Rous sarcoma virus. Subsequently, a cellular homologue was identified as c-Src. Numerous proto-oncogenes were identified as the transforming proteins carried by retroviruses. The second family is related to the Janus kinase (Jak). Most of the proteins of both families of non-receptor PTKs couple to cellular receptors that lack enzymatic activity themselves. This class of receptors includes all of the cytokine receptors (e.g., the interleukin-2 (IL-2) receptor) as well as the CD4 and CD8 cell surface glycoproteins of T cells and the T cell antigen receptor (TCR). The insulin receptor (IR) has intrinsic tyrosine kinase activity but does not directly interact, following autophosphorylation, with enzymatically active proteins containing SH2 domains (e.g., PI-3K or PLC-g). IRS-1 is IR substrate protein, which contains several motifs that resemble SH2 binding consensus sites for the catalytically active subunit of PI-3K. IRS-1 may act as a docking or adapter protein to couple the IR to SH2 containing signalling proteins. The receptors for the activins and TGFs-β have intrinsic serine/threonine kinase activity. One nuclear protein involved in the responses of cells to TGF-β is the proto-oncogene c-Myc, which directly affects the expression of genes harbouring Myc-binding elements. The most commonly known non-receptor serine/threonine kinases are cAMP-dependent protein kinase (PKA) and protein kinase C (PKC), and the mitogen activated protein kinases (MAP kinases). Protein kinase C may be involved in modulating steroidogenesis in the adrenals in addition to the classical cAMP-dependent protein kinase pathway. Protein kinase C (PKC) was originally identified as a serine/threonine kinase that was maximally active in the presence of diacylglycerols (DAG) and calcium ion. It is now known that there are at least 10 proteins of the PKC family. The phosphorylation of various proteins, by PKC, can lead to either increased or decreased activity, e.g., the phosphorylation of the EGFR by PKC that down-regulates the tyrosine kinase activity of the receptor, limiting the length of the cellular responses initiated through the EGFR. MAP kinases are also called ERKs for extracellular-signal regulated kinases. On the basis of in vitro substrates the MAP kinases have been variously called microtubule associated protein-2 kinase (MAP-2 kinase), myelin basic protein kinase (MBP kinase), ribosomal S6 protein kinase (RSK-kinase, i.e., a kinase that phosphorylates a kinase) and EGF receptor threonine kinase (ERT kinase). All of these proteins have similar biochemical properties, immuno-cross reactivities, amino acid sequence and ability to in vitro phosphorylate similar substrates.

Phospholipids as second messengers (<1 min):

1- The phosphoinositide system is stimulated by a ligand (hormone or growth factor).
2- The ligand bins to a specific receptor on the surface of the cell.
3- The receptor is coupled to a G-protein or tyrosine kinase-linked receptor, which translates the signal by activating a phospholipase C inside the cell.
3- The activated phospholipase C then hydrolyses a phosphorylated phosphatidylinositol, phosphatidylinositol-4,5-biphosphate (PIP_2). This generates 2 second messengers, inositol-1,4,5-triphosphate (IP_3) and diacylglycerol (DAG).
4- IP_3 and DAG activate separate, interacting pathways within the cell.
5- The 2 pathways work in variety of mechanisms, to produce signalling system capable of responding rapidly to a variety of external signals and controlling a wide range of cellular processes.
6- The pathways perform 2 main function: A- bringing the second messengers to directly activate the cell, or B- providing the second messengers to directly activate the cell.
7- Once the second messengers are generated 2 things may happen: A- IP_3 is released into the cytosol and acts to mobilise intracellular calcium stores from ER; and B- DAG stays in the plane of the membrane and activates a PKC.
8- The 2 messengers are metabolised in a number of ways, which terminates the signal and can regenerate PIP_2.

Maximal MAP kinase activity requires that both tyrosine and threonine residues are phosphorylated. This indicates that MAP kinases act as switch kinases that transmits information of increased intracellular tyrosine phosphorylation to that of serine/threonine phosphorylation. MAP kinases are activated by an additional class of kinases termed MAP kinase kinases (MAPK kinases) and MAPK kinase kinases (MAPKK kinases). Ultimate targets of the MAP kinases are several transcriptional regulators e.g., serum response factor (SRF), and the proto-oncogenes Fos, Myc and Jun as well as members of the steroid/thyroid hormone receptor super family of proteins. Phospholipases and phospholipids in signal transduction: Phospholipases and phospholipids are involved in the processes of transmitting ligand-receptor induced signals from the plasma membrane to intracellular proteins. The generation of DAG occurs in response to agonist activation of various phospholipases. The principal mediators of PKC activity are receptors coupled to activation of phospholipase C-g (PLC-g). PLC-g contains SH2 domains that allow it to interact with tyrosine phosphorylated RTKs. Activation of PLC-g leads primarily to the hydrolysis of membrane phosphatidylinositol bisphosphate (PIP2) leading to an increase in intracellular DAG and inositol trisphosphate (IP3). The released IP3 interacts with intracellular membrane receptors leading to an increased release of stored calcium ions. Together, the increased DAG and intracellular free calcium ion concentrations lead to increased activity of PKC. PLD action on PC leads to the release of phosphatidic acid, which in turn is converted to DAG by a specific phosphatidic acid phosphomonoesterase. PLA_2 hydrolyses PC to yield free fatty acids and lysoPC, both of which have been shown to potentiate the DAG, mediated activation of PKC. Of medical significance is the ability of phorbol ester tumour promoters to activate PKC directly. This leads to elevated and unregulated activation of PKC and the consequent disruption in normal cellular growth and proliferation control leading ultimately to neoplasia. Phosphatidylinositol-3-Kinase (PI-3K) is tyrosine phosphorylated, and subsequently activated, by various RTKs and receptor-associated PTKs. PI-3K is a heterodimeric protein containing an 85 kDa and

110 kDa subunits. The p85 subunit contains SH2 domains that interact with activated receptors or other receptor-associated PTKs. The 85 kDa subunit is non-catalytic, the 110 kDa subunit is enzymatically active. PI-3K, associates with and is activated by the PDGF, EGF, insulin, IGF-1, HGF and NGF receptors. PI-3K phosphorylates various phosphatidylinositols at the 3 position of the inositol ring. This activity generates additional substrates for PLC-g allowing a cascade of DAG and IP3 to be generated by a single activated RTK or other protein tyrosine kinases. There are 3 different classes of G-Protein coupled receptors: 1- G-protein coupled receptors that modulate adenylate cyclase activity. One class of adenylate cyclase modulating receptors activate the enzyme leading to the production of cAMP as the second messenger, e.g., β-adrenergic, glucagon, and odorant molecule receptors. Increases in the production of cAMP leads to an increase in the activity of PKA in the case of β-adrenergic and glucagon receptors. The α-type adrenergic receptors are coupled to inhibitory G-proteins that repress adenylate cyclase activity upon receptor activation. 2- G-protein coupled receptors that activate PLC-g leading to hydrolysis of polyphosphoinositides (e.g., PIP2) generating the second messengers, diacylglycerol (DAG) and inositoltrisphosphate (IP3). This class of receptors includes the angiotensin, bradykinin and vasopressin receptors. 3- A novel class of G-protein coupled receptors are the photoreceptors. This class is coupled to a G-protein termed transducin that activates a phosphodiesterase that leads to a decrease in the level of cGMP. The drop in cGMP then results in the closing of a Na^+/Ca^{2+} channel leading to hyperpolarisation of the cell. The activity of G-proteins with respect to GTP hydrolysis is regulated by a family of proteins termed GTPase activating proteins, GAPs. The proto-oncogenic protein, Ras, is a G-protein involved in the genesis of numerous forms of cancer (when the protein sustains specific mutations). Hormone receptors are proteins that effectively bypass all of the signal transduction pathways. All of the hormone receptors are bi-functional, capable of binding hormone as well as directly activating gene transcription. These hormones are capable of freely penetrating the hydrophobic plasma membrane. Upon binding ligand the hormone-receptor complex translocates to the nucleus and bind to specific DNA sequences termed hormone response elements (HREs), resulting in altered transcription rates of the associated gene. Phosphatases may function as anti-oncogenes or growth suppressor genes. The loss of a functional phosphatase involved in regulating growth promoting signals could lead to neoplasia. There are 2 broad classes of protein tyrosine phosphatases (PTPs): 1- transmembrane enzymes, which contain the phosphatase activity domain in the intracellular portion of the protein; and 2- intracellularly localised enzymes. Other phosphatases that recognise serine and/or threonine phosphorylated proteins also exist in cells. These are referred to as protein serine phosphatases (PSPs), and are at least 15 in number.

ROLE OF MEMBRANE TRAFFICKING IN PLASMA MEMBRANE SOLUTE TRANSPORT

Cells can rapidly and reversibly alter solute transport rates by changing the kinetics of transport proteins resident within the plasma membrane (acute regulation of plasma membrane). This can be brought about by reversible phosphorylation of the transporter, by the regulated exocytic insertion of transport proteins from the intracellular vesicles into the plasma membrane and their subsequent regulated endocytic retrieval. Agonists such as hormones or neurotransmitters can cause rapid increases in the rate of transmembrane solute movement. Subsequent removal of these agonists usually causes a rapid return of the transport rate to basal unstimulated levels. Hormone stimulation can lead to a decrease in solute transport. Parathyroid hormone (PTH) produces rapid inhibition of Na^+-coupled inorganic phosphate (P_i) reabsorption across the brush border membrane of proximal tubule cells in the kidney, for example. The rate of solute flux across the plasma membrane is proportional to the turnover rate (mol/s) of the individual transport and the number of that transporter present in the plasma membrane. Alterations in the kinetic properties of individual transporters or activation of quiescent transport proteins can be effected by reversible covalent modification of the transporter. An increase in the number of plasma membrane transporters synthesised may be expected to lead to an increase in the number of those proteins in the plasma membrane. Hormones or other agents can increase the number of membrane transporters by causing the recruitment of preexisting transporters (carrier, pumps, or channels) into the plasma membrane. Recruitment occurs by regulating the traffic of cytoplasmic vesicles, whose limiting membranes contain transporter proteins, to the plasma membrane, where they fuse and insert the transport protein into the membrane. Plasma membrane transport proteins are integral membrane proteins and such must reside within a membrane throughout their cellular existence. Transport proteins are found in membrane vesicles during their biosynthetic passage from the endoplasmic reticulum, through the Golgi to the plasma membrane, as well as during their degradative removal from the plasma membrane via endosomes to lysosomal compartments (biosynthesis, trafficking, and degradation). Glucose transporters are localised within the cytoplasm of insulin-sensitive cells under basal conditions. After insulin stimulation, there is a marked redistribution of glucose transporters from intracellular sites to the plasma membrane. The activation energy, for water transporter is <6 kcal/mol for channel-mediated transport, compared with >10 kcal/mol for lipid-mediated water transport. The α-carbonic anhydrase (CA)-rich cells are involved in acid secretion. The presence H^+-ATPase in endocytic membrane vesicles allows for acidification of the intravesicular milieu. Transcellular movement of Cl^-

across epithelial cells is brought about by the concerted actions of several basolateral transporters accumulating Cl⁻ in the cytoplasm, which then exits the cell through a cAMP-regulatable Cl⁻ channel (CFTR) located in the apical membrane. Ion channels are capable of allowing the passage of 10^6 ions/s, and so the number of copies per cell necessary to fulfill the transport requirements of the cell is relatively low. Acidification of intracellular organelles (endosomes, lysosome, etc.) is brought about by the concerted actions of an electrogenic vacuolar-type H⁺-ATPase and a counter ion conductance to maintain electroneurtrallity secretion. The CFTR increases in cellular cAMP levels inhibit endocytic activity in coloncyte and pancreatic cells. Increases in cAMP stimulate exocytosis in cells expressing wild-type CFTR but not in cells expressing mutant CFTR. After stimulation, one would expect a loss of protein and transport function intracellular membranes and a concomitant increase in protein and transport function in the plasma membrane. Subsequently, after removal of the stimulus, the reverse should by observed. The exocytic insertion of transporter-containing vesicles into the plasma membrane may lead to an increase in the surface area of the membrane. Hormone stimulation leads to a decrease in membrane transport by stimulating endocytic retrieval of the transporter. Regulated trafficking of transporters in either direction can occur to either increase or decrease plasma membrane transport and depends on the stimulus-transporter cell system. Increases in plasma membrane transport brought about by insertion of membrane vesicles containing transporters may be accompanied by increases in plasma membrane surface area. Increases in intracellular cAMP lead to an increase short-circuit current. Increases in membrane conductance consistent with increased surface area rather than an increase in membrane conductance without changes in membrane area. Adjacent to the cytoplasmic surface of the plasma membrane is a dense network of cytoskeletal components that help maintain the structure and shape of the cell. The F-actin barrier is under agonist regulation, disappearing during exocytic stimulation. Reorganisation of the actin net during stimulation is associated with several transport systems. Subplasmalemmal F-actin constitutes 10% of total cellular protein. The rearrangement of the actin cytoskeleton is brought about by cycles of monomeric G-actin first polymerising to form filamentous F-actin, which in turn depolymerizes to form monomeric G-actin. Phosphorylation of G-actin and actin-binding proteins promotes dissociation of F-actin networks by altering the affinity of actin-binding proteins for F-actin. Direct serine phosphorylation of actin PKA diminishes the ability of actin to polymerise. In contrast, serine and threonine phosphorylation of actin by protein kinase C (PKC) results in an increase in the ability for actin to polymerise. This PKA-mediated exocytosis of membrane vesicles would be enhanced by breakdown of the actin network at the site of insertion. Second messengers such as diacylglycerol and Ca^{2+} can regulate actin structures. Diacylglycerol can stimulate the nucleation of actin filaments at the membrane. Increased in intracellular Ca^{2+}, brought about by the actions of the inositol polyphosphate signalling pathway, activate proteins that sever and nucleate actin filaments. This process may be expected to promote actin network disassembly, removing the actin barrier that prevents fusion of exocytic vesicles with the plasma membrane. Actin filaments and microtubules interact with a variety of other proteins such as kinins, dynein, and myosins to provide transport between cellular domains. Actin microfilaments are involved in endocytic events. In nonploarized cells, the Na⁺-K⁺-ATPase is distributed randomly, but, in polarised epithelial cells, the Na⁺-K⁺-ATPase is found in the basolateral membrane in the kidney, intestine and exocrine glands. Polarisation is brought about by the subsequent endocytic removal of the Na⁺-K⁺-ATPase from the apical domain, with stabilisation of the Na⁺-K⁺-ATPase in the basolateral membrane preventing its removal. Cytoskeletal interactions play a role in reducing the lateral mobility of proteins. Colchicine is a microtubule-disrupting agent, significantly inhibits the ADH-induced osmotic water flow. Microtubules are involved in trafficking of water channels to the apical membrane but that, once inserted, microtubules are not involved in maintaining the presence of water channels in the membrane. Colchicine inhibits cAMP-mediated increases in Cl⁻ secretion from the colon. Although the Na⁺-K⁺-ATPase is normally restricted to the basolateral domains of epithelial cells, the H⁺-K⁺-ATPase is a component of the apical surface of the parietal cells of the gastric epithelium. The low-density lipoprotein (LDL) receptor possesses a coated pit localised sequence. In the epithelial tissues of intestine and liver, the LDL receptor is targeted basolaterally, whereas, in the proximal tubule of the kidney, the LDL receptor is targeted to the apical brush-border domain. The LDL receptor bears 2 tyrosine-containing determinants that target the receptor to the basolateral domain in polarised cells.

ION CHANNELS

Ion channels constitute a class of proteins that is ultimately responsible for generating and orchestrating the electrical signals passing through the thinking brain, the beating heart, and contracting muscle. Defective ion-channel proteins are responsible for many diseases, e.g., cystic fibrosis. Ion channels are macromolecular protein tunnels that span the lipid bilayer of the cell membrane. Approximately 30% of the energy expanded by cells is used to maintain the gradient of sodium and potassium ions across the cell membrane. They are more efficient than enzymes, small conformational changes change (gate) a single channel from closed to open, allowing up to 10 million ions to flow into or out of the cell each second. A few picoamps of current are generated by the flow of highly selected ions each time the channels opens. Since ion channels are efficient, their numbers per cell are relatively low; a few thousand of given type are usually sufficient. Ion channels are usually classified according to the type of ion they allow to pass-Na, K, Ca, or Cl, although some are less selective. They may be gated by extracellular ligands, changes in transmembrane voltages, or intracellular second messengers. The direction in which ions move through a channel is governed by electrical and chemical concentration gradients. Ions flow passively through ion channels down a chemical gradient. Electrically charged ions also move in an electrical field, just as ions in solution flow to one of the poles of a battery connected to the solution. The point at which the chemical driving force and the electrical driving force are exactly balanced is called the Nernst potential (or reversal potential). Above or below this point of equilibrium, a particular species of ion flows in the direction of the dominant force. Electrophysiologic concepts are simplified by recalling the Nernst potential of the 4 major ions across the plasma membrane of cells. These are approximately as follows: Na, +70 mV; K, -98 mV; calcium, +150 mV; and chloride, -30 to -65 mV. If a single sodium-selective channel opens in a cell in which all other types of channels are closed, the transmembrane potential of the cell will become ENa (+70mV). If single potassium channels opens, the cell's membrane potential will become EK (-98mV). Because cells have an abundance of open K channels, most cells' transmembrane potentials (at rest) are approximately -70 mV, near EK. A cell with one open Na channel and one open K channel, each with the same conductance will have a transmembrane potential halfway between ENa (+70 mV) and EK (-98 mV), or -14 mV. Ion channels are both potent and fast, and they are tightly controlled by the gating mechanisms of the cell. Most ion channel proteins are composed of individual subunits or groups of subunits, with each subunit containing six hydrophobic transmembrane regions S1 through S6. The Na and calcium channels comprise a single (α) subunit containing 4 repeats of the 6 transmembrane-spanning motifs. Potassium selective channel, Kir (**K** channel, **i**nward **r**ectifier) is open in the steady state. Kir channels are known as inward rectifier because they conduct current much more effectively into the cell than out of it. The subunits in Kir channels lack S1 to S4 segments present in Kv channels. Kir channels have a deceptively simple surrounding the conserved H5 pore. Pore formation by different combination of subunits, direct gating of G proteins, and interactions with other proteins adds considerable complexity to the behaviour of the Kir channels. Under physiological conditions, calcium and sodium ions flow into the cells and depolarise the membrane potential, whereas potassium ions flow outward to repolarise the cell towards EK. Factors controlling mechanisms of gating are voltage, time, direct agonist, G protein, and calcium. Factors controlling mechanisms of modulation are increases in phosphorylation, oxidation-reduction, cytoskeleton, calcium, and ATP.

WATER TRANSPORT ACROSS CELL MEMBRANES

A family of water-transporting proteins (water channels, aquaporins) has been identified, consisting of small hydrophobic proteins expressed widely in epithelial and nonepithelial tissues. Although there remain major unresolved questions concerning the physiological significance of water channels and the cell biology of vasopressin-stimulated water permeability, there have been remarkable advances in the identification, molecular cloning, and structure-function analysis of water-transporting proteins, also referred to as water channels or aquaporins. The osmotic water permeability coefficient (P_f; in cm/s) is defined as the net flow of volume across a membrane, in response to hydrostatic or osmotic driving force ($P_f > 0.01$ cm/s, for a single membrane). The osmotic gradient dominates over the hydrostatic pressure in most biological situations, generally independent of osmotic gradient size and direction except for some complex membrane barriers. The parameter P_f provides a quantitative measure of water flow through single channels. P_f is related to pore radius and length. The activating energy is generally >10 kcal/mol for water movement by a channel-independent solubility-diffusion mechanism and, <6 kcal/mol for water movement through aqueous pores. The high activation energy for water movement through lipid may be related to the formation and breaking of hydrogen bonds as water moves between aqueous and membrane phases. The activation energy depends on the nature of the rate-limiting barrier for water movement and the energetics of water-pore interactions. The ratio of osmotic-to-diffusional water permeability may provide useful information about the presence of a facilitated water-transporting pathway. The solute selectivity of the water pathway, which is the ability of small solutes (protons, urea, NH_3, CO_2) to move through the water-transporting pathway may provide a lower limit to effective pore size. A membrane with high osmotic water permeability coefficient ($P_f > 0.01$ cm), low activation energy (<6 kcal/mol), and osmotic-to-diffusional water permeability (>1) is likely to contain water channels. Osmotically induced water flow causes cell swelling or shrinking. Channel-forming integral protein of 28 kDa (CHIP28) is expressed widely in various epithelial and capillary endothelia, including proximal tubule and thin descending limb in kidney. A second water channel expressed selectivity in kidney collecting duct apical membrane (WCH-CD), is vasopressin-inducible water channel by its mutation in congenital nephrogenic diabetes insipidus (NDI) and its vasopressin-dependent membrane targeting. WCH-3 is a protein with 55% amino acid identity to WCH-CD. WCH-3 is expressed only in kidney, and like WCH-CD, displays transcript upregulation in response to dehydration. The water channels identified to date belong to a family of small hydrophobic channel-forming membrane proteins of which major intrinsic protein (MIP) is the prototype.

Hydropathy plots of these proteins are similar, suggesting up to 6 transmembrane helical segments. Homology in amino acid sequence between the first and second halves of each protein suggests that they arose from tandem intragenic duplication of a three-transmembrane segment. The gene loci of the human water channels dispersed to different chromosomes in single copy. The CHIP28 has been localised to chromosome 7p14, WCH-CD to chromosome 12q13, and MIWC to chromosome 18q22. Interestingly, the WCH-CD gene is located near the MIP gene at chromosome 12q13. The WCH-CD transcript is expressed only in the kidney, and its level of expression is strongly upregulated in dehydration. Because vasopressin receptors are present in collecting duct principal cells, elevations in cAMP levels during dehydration might act on the cAMP response and AP-1 elements to enhance WCH-CD gene expression. Mutations in water channels CHIP28 and WCH-CD have been detected in humans. It is not known whether these individuals manifest subtle clinical abnormalities under physiological stress or whether upregulation of other water channels can functionally compensate for the lack CHIP28 expression in some tissues. CHIP28 water channel contains 6 translayer α-helices that pack together to form the aqueous channel. CHIP28 is a 28-kDa glycoprotein forms tetramers in membranes; each monomer contain six putative helical domains surrounding a central aqueous pathway and function independently as a water-selective channel. The amino and carboxy termini are in a cytosolic orientation. CHIP is synthesised at the endoplasmic reticulum. CHIP28 hydrophobic regions (Hrs) are potential membrane-spanning domains. There is fully functional protein at the plasma membrane and in the Golgi vesicle fraction. Only mutations in the vasopressin-sensitive water channel have been shown to cause human disease. Most cell membranes probably have adequate water permeability through membrane lipids to support volume regulation and other housekeeping functions. Certain cell plasma membranes, such as those in secretory and absorptive epithelial and endothelial cells, may require a high or regulated water permeability to facilitate the vectorial transport of fluids across cell layers. High water permeability in proximal tubule is required for the near-isosomotic reabsorption of glomerular filtrate. The renal concentrating mechanism relies on high water permeability in the thin descending limb of Henle and vasa recta, low water permeability in the ascending limb of Henle, and vasopressin-regulated water permeability in the collecting duct. A physiological important water channel should be increased cellular water permeability and, when absent or mutated, should result in abnormal physiology. Osmotically induced water movement in the lungs is rapid. High osmotic water permeability in lung may be important for air space fluid replacement to offset insensible respiratory losses and for reabsortion of fluid in alveolar oedema and in the neonatal lung. Airway water permeability is relatively high, consistent with the expression of MIWC in airway basolateral cell membrane.

Some tissue distribution of water channels:

System	Site	Water channel
Respiratory:		
	1- Tracheal epithelium:	CHIP, AQP-5, GLIP, MIWC.
	2- Bronchial epithelium:	MIWC.
	3- Alveolar endothelium and epithelium:	CHIP.
Cardiovascular:		
	1- Myocardium:	CHIP.
	2- Capillary endothelium:	CHIP.
Digestive:		
	1- Colon, crypt epithelium:	CHIP.
	2- Colon, villus epithelium:	MIWC, GLIP.
Genitourinary:		
	1- Kidney, proximal tubule:	WCH-3.
	2- Kidney, thin descending limb of Henle:	CHIP.
	3- Kidney, collecting duct apical membrane:	WCH-CD.
	4- Kidney, collecting duct basolateral membrane:	MIWC, GLIP.
	5- Urinary bladder (transitional epithelium):	GLIP.
Reproductive:		
	1- Testis, efferent ductules, seminal vesicles, prostate:	CHIP.
	2- Uterus:	CHIP.
	3- Placenta:	CHIP.

THE EXTRACELLULAR CALCIUM-SENSING RECEPTOR

The system governing extracellular calcium (Ca^{2+}_o) homeostasis maintains near constancy of Ca^{2+}_o so as to ensure continual availability of calcium ions for their numerous intracellular and extracellular roles. The intracellular ionised calcium concentration (Ca^{2+}_i), varies substantially during intracellular signalling via key second messenger, Ca^{2+}_o remains nearly invariant. The Ca^{2+}_o-sensing receptor (CaR) permits Ca^{2+}_o to function in a hormone-like role as an extracellular first messenger through which parathyroid, kidney, and other cells communicate with one another via the CaR. Virtually all physiological processes utilise intracellular and/or extracellular calcium (Ca^{2+}) ions in some fashion. Calcium ions within the cytosol (e.g., the cytosolic free calcium ion concentration or Ca^{2+}_i) of essentially all cells act as a key intracellular second messenger and as an enzymic cofactor, coordinating and controlling cellular functions as diverse as muscular contraction, hormonal secretion, glycogen metabolism, cellular differentiation, proliferation, and motility. The basal level of Ca^{2+}_i, usually on the order of 100 nanomolar (nM), is about 10,000-fold lower than the extracellular ionised calcium concentration (1 mM). Ca^{2+}_i can undergo large and rapid elevations to well over 1 μM upon cellular activation, due to release of Ca^{2+} from intracellular stores and/or uptake of extracellular calcium ions. The level of Ca^{2+}_o measured in the blood, in contrast, remains nearly invariant under normal circumstances, fluctuating from its mean value by only a few percent. Ca^{2+}_o is ultimately the source of all intracellular calcium ions, and also participates in numerous essential functions, such as clotting of the blood, maintenance of skeletal integrity, intercellular adhesion and regulation of neuromuscular excitability. The three Ca^{2+}_o-regulated, calciotropic hormones (1,25-dihydroxyvitamin D3 [1,25(OH)$_2$D3], calcitonin [CT], and PTH) act on their various effector tissues, principally intestine, kidney, and bone, which are responsible for normalising Ca^{2+}_o by altering the transport of calcium ions into or out of the extracellular fluid in response to these hormones. Raising Ca^{2+}_o inhibits bone resorption and stimulates bone formation. Increasing Ca^{2+}_o exerts actions on osteoblasts and osteoclasts. Elevating Ca^{2+}_o promotes the chemotaxis and proliferation of osteoblast precursors, thereby potentially expanding the pool of mature osteoblasts at sites of new bone formation. High Ca^{2+}_o inhibits osteoclast formation in the presence of osteoblasts, likely through paracrine interactions that could potentially limit the initiation of bone resorption in sites where new bone is being formed. Normal skeletal turnover entails a sequence of osteoclastic resorption of small packets of bone over several weeks, which must then be replaced by osteoblastic formation of new bone over about 3 months. Disruption of this normally precise, spatiotemporal coupling of bone resorption to subsequent bone formation, with an excess of the former relative to the latter, contributes importantly to the development of osteoporosis in the aging population. Changes in Ca^{2+}_o modulate the functions of selected cells involved in mineral ion homeostasis. At a local level, calcium ions resorbed from bone by osteoclasts may feed back to inhibit osteoclastic action as well as to stimulate the proliferation and chemotaxis of osteoblast precursors, thereby potentially coupling bone resorption to its subsequent formation. Elevated levels of Ca^{2+}_o affect multiple parameters of renal function, several of which are relevant to mineral ion metabolism. Ca^{2+}_o exerts additional actions on renal function that are less clearly related to mineral ion metabolism but modulate other homeostatic mechanisms. High levels of Ca^{2+}_o exert a diuretic action that contributes to the volume depletion that can be encountered in hypercalcaemic individuals. Hypercalcaemia likewise produces abnormalities in water balance, reducing maximal concentrating ability, probably by both inhibiting vasopressin action in the collecting duct and by permitting "wash out" of the hypertonic medullary interstitium that normally underlies vasopressin-elicited reabsorption of water. In duodenum, an important site of vitamin D-stimulates intestinal calcium absorption, elevates levels of Ca^{2+}_o in combination with 1,25(OH)$_2$D3 increase calbindin production. The CaR has a large amino-terminal extracellular domain (600 amino acid) that plays an important role in sensing Ca^{2+}_o. It is followed by the receptor's second major structural domain, the 7 membrane-spanning helices that are the signature of the GPCRs. The CaR's intracellular domains couple the activated receptor to stimulation of various guanine nucleotide regulatory (G) proteins. CaR activates phospholipase C and inhibits adenylate cyclase, CaR also activates phospholipases A_2 and D. CaR activates the mitogen-activated protein (MAP) kinase pathway that transduces extracellular signals to the nucleus. CaR-mediated regulation of calciotropic hormone secretion enables endocrine control of systemic Ca^{2+}_o homeostasis. Direct CaR-mediated regulation of renal, intestinal, and skeletal function may enable local control of systemic Ca^{2+}_o homeostasis. The CaR acts like a loop diuretic, it inhibits the cotransporter through an indirect, second messenger-mediated pathway, most likely involving a metabolite of AA liberated by PLA_2. The CaR may regulate second messenger pathways and, in turn, transport of NaCl, K^+, Mg^{2+} and Ca^{2+} in CTAL. The hormones that stimulate cAMP accumulation, such as parathyroid hormone, enhance the reabsorption of Ca^{2+} and Mg^{2+} through the paracellular pathway by augmenting the lumen-positive transepithelial potential, Vte, as a result of stimulating the activity of the Na-K-2Cl cotransporter and an apical K^+ channel. The CaR, also situated on the basolateral membrane, promotes arachidonic acid (AA) formation by activating PLA_2. AA is metabolised through the P-450 pathway to form an active metabolite, probably 20-HETE, which inhibits the apical K^+ channel and, perhaps, the Na-K-2Cl cotransporter. Both actions diminish overall activity of the cotransporter, and, in turn, paracellular transport of Ca^{2+} and Mg^{2+}. The CaR inhibits adenylate cyclase, thereby reducing hormone- and cAMP-stimulated divalent cation transport. The CaR's presence in the epithelial cells of the small and large intestines is involved in calcium absorption.

ANDROGEN RECEPTOR (AR)

The action of androgens within target cells is transduced by low-abundance intracellular ARs. The AR is a member of a superfamily of nuclear transcription factors that mediate the action of steroid hormones. This gene superfamily includes specific receptors for oestrogen, progesterone, glucocorticoid, mineralocorticoids, vitamin D, retinoic acid, and thyroid hormone, as well as many orphan receptors. When activated by ligand binding, these transcription factors bind to specific DNA sequences on target genes, hormone response elements (HREs), and regulate the transcriptional activity of those genes. The receptor proteins contain 3 common domains: 1- highly conserved central domain that is responsible for DNA binding, 2- N-terminal region that is involved in trascriptional activation, and 3- C-terminal domain that binds ligand. The AR was first described in 1969 and cloned DNA (cDNA) in 1988. The single copy AR gene is localised on the human X chromosome between q11 and q13. The entire gene contains 8 exons, with a total length of 90 kb. The transcription initiation site is located 1116 base pairs upstream from the AR translocation initiation site. The promoter contains a GC box that may play a role in transcription initiation. The translated human AR is a 918-amino acid protein with a molecular weight 100-110 kd. Homology within the N region is highest between androgen, glucocorticoid, and progesterone receptors (75-80%). At the distal end of the C region and within the hinge region (D region), there is a bipartite nuclear targeting sequence (amino acids 617-633) that is responsible for androgen-dependent nuclear import of the AR. The C-terminal steroid-binding domain (E region), is predominantly hydrophobic and consists of 253 amino acids. There is a high methionine content within this region that may be important for ligand binding. ARs are hormonally regulated, and fluctuations in their levels are a contributing factor in determining androgenic responsiveness of tissues and genes. ARs are primarily autoregulated in the prostate, with androgens upregulating the receptor protein level. ARs are nuclear proteins that are activated by ligand binding. Androgens mediate a number of diverse responses through the androgen receptor, a 110 kD ligand-activated nuclear receptor. The androgen receptor can be activated by 2 ligands, testosterone and dihydrotestosterone, which bind to the AR with different affinities. The AR acts as a transcriptional modifier of a variety of genes by binding to an androgen response element. The AR is a member of the steroid hormone receptor family. These receptors regulate gene transcription by interacting with a specific DNA sequence in a ligand-dependent manner. The AR is a 110 kD nuclear protein, which consists of approximately 918 amino acid residues. AR consists of transactivation (6-10), DNA binding (11-16), nuclear localisation (17-19), dimerization (20-22), and ligand binding domains. AR is present in most tissues. AR expression is modified during foetal development, sexual development, aging, and malignant transformation. Regulation of AR levels may occur anywhere along the path from AR gene transcription to post-translational modification. A variety of factors, including androgens, are implicated in modulating the AR protein and mRNA expression. In carcinoma of the prostate (CaP), development of an androgen independent state is associated with a heterogeneous AR expression. Dedifferentiation may result in expression of factors that modulate the AR expression. In addition to androgen, several other hormones and growth factors can regulate the AR expression. FSH increases the level of mRNA of AR in the Sertoli cells. Growth hormone, prolactin and EGF increase the ARmRNA levels in prostatic cells. Androgens can modulate expression of a variety of growth factors in the prostate stroma. Many androgen responsive genes have promoters that consist of several regulatory elements, which suggests that an androgen specific response may be due to a combination of interacting transcription factors. Steroid specific expression can be mediated at both the gene and cellular levels. At the gene level, the context of the HRE may be critical for orchestrating a specific steroid response. Response elements in the proximity of the HRE may provide binding sites for accessory transcription factors. At the cellular level, tissue specific expression of steroid receptors will limit the steroid response only to those cognate receptors. Progestagens, estradiol and anti-androgens may compete with androgens for binding to the AR and may stimulate both cell growth.

MEMBRANE PHOSPHOLIPID

The membrane phospholipids asymmetry is ubiquitous. In general terms, the outer leaflet of eukaryotic plasma membranes is formed predominantly with the cholinephospholipids (sphingomyelin and phosphatidylcholine [PC]), whereas the majority of the aminophospholipids (phosphatidylserine [PS]) and phosphatidylethanolamine [PE]). Asymmetry is generated by the activity of ATP-dependent aminophospholipid translocase that specifically transports PS and PE between bilayer leaflets. Membrane lipid asymmetry is of a major physiologic importance, because it showed that cells invest energy to catalyse lipid movement in order to maintain a specific transmembrane phospholipid distribution. Loss of symmetry, especially the appearance of PS at the cell surface, is associated with many physiologic and pathologic phenomena. Asymmetric orientation of phospholipids in blood platelets is rapidly lost upon influx of Ca^{2+} during their activation, suggests a critical role for PS in thrombosis. Apoptosis and tumourigenic cells express relatively large amounts of outer-leaflet PS that may serve as a trigger for macrophage recognition and promote the cells' phagocytosis. At least 3 distinct activities are involved in the regulation of membrane lipid sidedness. Two energy-requiring activities seem to work in concert to maintain a nonrandom transbilayer phospholipid orientation. Inhibition of these activities stops lipid movement. Influx of Ca^{2+} into the cytoplasm activates a scrumblase activity that results in rapid transbilayer phospholipid mixing that leads to a nearly symmetric distribution of phospholipids across the membrane bilayer. Aminophospholipid translocase is an ATP-dependent aminophospholipid generates and maintains asymmetry through the transport of specific lipids across the cell's membrane. This activity is distinguished by its ability to transport PS and PE from the outer to inner leaflet of plasma membranes against the concentration gradient. This process consumes one molecule of ATP/molecule of lipid transported. This activity is abrogated when cytoplasmic Ca^{2+} levels reach micromolar concentrations. Aminophospholipid translocase activity has been observed in intracellular chromaffin granule membranes and endoplasmic reticulum. ATP-dependent floppase: A less specific ATP-requiring floppase transports both aminophopholipids and cholinephospholipids from the inner to the outer leaflet. Outward movement is abrogated by ATP depletion, sulphydryl oxidation, and histidine modification, indicating that this process is also energy- and protein-dependent. Because rapid inward translocation of aminophospholipids does not accelerate outward migration of all phospholipids, both processes may be mediated by independent mechanisms. The maintenance of membrane lipid asymmetry is stable and resistant to the mechanical stresses. Lipid scramblase: Platelet plasma membranes harbour a Ca^{2+}-dependent mechanism that can rapidly move phospholipids back and forth between the two membrane leaflets (flip-flop), leading within minutes to a loss of membrane lipid asymmetry. Because the influx of Ca^{2+} abrogates aminophospholipid translocase activity, Ca^{2+}-dependent loss of membrane phospholipid asymmetry is not corrected. Ca^{2+}-induced scramblase activity has been found in RBCs and other cells, but its activity is usually lower than blood platelets. Scramblase activity requires the continuous presence of cytoplasmic Ca^{2+}. Lipid scramblase is bidirectional, and all major lipid classes move back and forth at comparable rates. Lipid scramblase does not require hydrolyzable ATP. Lipid scrambling is the result of a complex between phosphatidylinositol 4,5-biphosphate and Ca^{2+}. Loss of membrane lipid asymmetry is often accompanied by blebbing and subsequent shedding of lipid-symmetric microvesicles from the cell surface. Lipid scrambling and calpain activation are required for shedding of microvesicles. The synchronous and cooperative action of the aminophospholipid translocase and the nonspecific floppase contribute to the generation and maintenance of membrane phospholipid asymmetry, whereas lipid scramblase activity results in its collapse. At physiologic (low) cytoplasmic Ca^{2+} concentration, PS activate lipid scramblase and block the cooperative action of translocase and floppase, leading to randomisation of phospholipids across the membrane lipid bilayer. Cytoskleton proteins assist in the maintenance of membrane phospholipid asymmetry by selective interacting with aminophospholipids. The interaction between PS and cytoskeletal proteins is thermodynamically weak. Spherocytic erythrocytes fully conserve lipid asymmetry despite markedly diminished levels of spectin. The cytoskeleton's function in the maintenance of lipid asymmetry presumably not of major importance. Unidirectional transport of PC is catalysed by a member of P-glycoprotein family. P-glycoprotein encoded by mdr1 gene is abundantly expressed in drug-resistant tumour cells, where it nonselectively expels lipid-soluble compounds from the inner to the outer membrane leaflet. Ca^{2+}-induced lipid scrambling involves nonspecific flip-flop of all lipid classes. Platelets are unlikely to a specific mechanism different from that of red blood cells.

THE REGULATION OF MEMBRANE PHOSPHOLIPID ASYMMETRY

Membrane lipid asymmetry is regulated by the cooperative activities of 3 transporters: 1- The ATP-dependent aminophospholipid-specific translocase, which rapidly transport PS and PE from the cell's outer-to-inner leaflet; 2- The ATP-dependent nonspecific lipid floppase, which slowly transports lipids from the cell's inner-to-outer leaflet; and 3- The Ca^{2+}-dependent nonspecific lipid scramblase, which allows lipids to move randomly between both leaflets. The translocases are targets for Ca^{2+} that directly regulates the transporter's activities. Elevated intracellular Ca^{2+} induces PS randomisation across the cell's plasma membrane by providing a stimulus that positively and negatively regulates scramblase and translocase activities. At physiologic Ca^{2+} concentrations, PS asymmetry is promoted because of an active translocase and floppase but inactive scramblase. Increased cytosolic Ca^{2+} can also result in calpain activation, which facilitates membrane blebbing and the release of PS-expressing procoagulant microvesicles. The appearance of PS at the cell's outer leaflet promotes coagulation and thrombosis by providing a catalytic surface for the assembly of the prothrombinase and tenase complexes and marks the cell as a pathologic target for elimination by phagocytes. Recognition of the PS-expressing targets can occur by both antibody-dependent and direct receptor-mediated pathways. Haemostasis and thrombosis: Membrane phospholipids propagate the proteolytic reactions that result in thrombin formation by promoting the assembly of coagulation factors on their surface. The most important pathway of coagulation is initiated by tissue factor, an integral membrane protein expressed on the surface of activated or disrupted cells. Tissue factor interacts with factor VII or VIIa, and this complex rapidly converts the zymogen factor IX, and factor VII itself into their active forms. Activity of tissue factor/factor VIIa complex is effective in the absence of anionic phospholipids, activity is increases by PS. In the prothrombinase complex, binding of factor Va to an anionic lipid surface promotes Ca^{2+}-dependent binding of factor Xa, which converts prothrombin to thrombin. In both complexes, PS is the most effective anionic phospholipid. PS is equally important in promoting anticogulatant protein C pathway that provides feedback inhibition of thrombin formation. Protein C effectively inactivates factor Va when both are bound to the same lipid surface, which leads to disassembly of the prothrombinase complex. Surface exposure of PS in platelet membranes provides for efficient propagation and control of the haemostatic process. Specific protein receptors for factors Va and VIIIa may, in addition to PS binding sites, be present in cellular membranes. PS exposure is critical to coagulation. Platelets do not contain tissue factor, they cannot initiate the coagulation cascade. Ca^{2+}-ionophore is the most effective followed by complement membrane attack complex C5b-9, collagen plus thrombin, collagen, and thrombin; are observed in that order for the extent of lipid-symmetric microvesicle shedding from the platelet surface. Platelet activation after vascular damage involves adhesion to subendothelial structures and aggregation of platelets to form a primary haemostatic plug at the wound site. The exposure of PS on aggregated platelets restricts and controls thrombin formation at the site of injury by providing a catalytic membrane surface for both procoagulant (tenase and prothrombinase) and anticoagulant (protein C) reactions. Increased amounts of circulating microvesicles have been observed in patients suffering from various disorders associated with secondary activated coagulation. Platelet microvesicles bind to and activate neutrophils. The leukocyte adhesion molecule L-selectin binds PS. Activation and aggregation of platelets at the site of injury could recruit leukocytes to the site of inflammation via the binding of platelet-exposed by PS to leukocyte L-selectin. Microparticles released from other cells may also contain tissue factor activity and initiate undesired coagulation in the circulation. Procoagulant activity of RBCs: Increased procoagulant activity associated with loss of lipid asymmetry has been observed in erythrocytes incubated in hyperglycaemic buffers, in platelets from diabetic patients, and in vesicles shed from reversibly sickled cells by repeated hypoxia-induced sickling. Procoagulant activity of WBCs: The prothrombinase activity of monocytes is enhanced by endotoxin activation, a result that supports the notion that their ability to contribute to thrombin formation is important in the thrombotic events associated with inflammation and healing. The prothrombinase activity of activated monocytes can be enhanced by a reduction in their ATP levels and reduced by inhibition of protein synthesis. Inhibition of protein synthesis would limit expression of the lipid scramblase and prevent PS exposure. Because stimulated monocytes also release microvesicles that express tissue factor activity, they have the capacity both to initiate and propagate coagulation in response to an inflammatory stimulus. Complement activation of endothelial cells: There is a possibility that complement-induced release of microvesicles into the circulation could contribute to inflammation-associated disseminated intravascular coagulation. Complement-induced loss of lipid asymmetry in cell-derived microvesicles may be an intrinsic property of complement pore formation because incubation of lipid vesicles with C5b-9 causes transbilayer lipid exchange between both leaflets. These pore-forming peptides promote fast flip-flop of lipids in liposomes. Neoplasia and thrombosis: The principle is increased expression of tissue factor activity and the presence of a tumour-specific cysteine protease that direct factor X independent of the tissue factor pathway. Tumour cells also release microvesicles that catalyses prothrombinase activity a process similar to phospholipid scrambling and tissue factor and PS in the tumour cells and their shed microvesicles could facilitate platelet-independent initiation and propagation of coagulation, and it may be responsible for fibrin deposits often seen in solid tumours. Role of PS in cell-cell recognition: Cell clearance depended on the amount of exogenously-inserted PS, and it occurs when the cells contained only about ~1 mol% of the PS analogue. Normally cells do not expose significant amounts of PS, but pathologic cells seem to have undergone lipid rearrangements that result in PS exposure (reorientation). Macrophages bind PS-expressing deoxygenated sickle cells and leukaemic cells by a mechanism that is PS-dependent. Exposure of this lipid in aging RBCs presumably contributes to their removal from the circulation. Stored RBCs also suffer from the gradual appearance of PS at the outer surface in an amount proportional to the duration of storage. Because aging cells progressively lose ATP-dependent enzymatic activities, both Mg^{2+}-ATP-dependent aminophospholipid translocase and the Ca^{2+} pump will be affected. This condition leads to increased cytoplasmic Ca^{2+} levels that stimulate lipid scramblase and suppress aminophospholipid translocase. Aminophospholipid transport activity decreases upon storage of RBCs and platelets. Because oxidation affects the activity of membrane lipid transporters, age-related alterations in the cells' redox state may also contribute to PS exposure and cell recognition. Apoptosis is accompanied by exposure of PS at the cell's outer surface. PS exposure is one of the earliest manifestations of apoptosis, and that it proceeds DNA fragmentation, plasma membrane blebbing, and loss of membrane integrity. The process is Ca^{2+}-dependent and involve bidirectional, nonspecific flip-flop of phospholipids. Apoptotic cell membrane lipid asymmetry is compromised by the combined actions of an activated scramblase and inhibited translocase. PS receptors in macrophages: Inflammatory macrophages can recognise PS-expressing apoptotic lymphocytes via a specific PS receptor that is inhibited by liposomes containing phosphatidyl-L-serine but not by other anionic phospholipids, including phophatidyl-D-serine.

CELL VOLUME

To survive, cells have to avoid excessive swelling or shrinkage. Alterations of cell volume play a crucial role in the regulation of cell function: 1- Epithelial cell transport may lead to cell swelling, which then triggers volume regulatory mechanisms modifying transcellular transport. 2- Insulin swells hepatocytes by activation of Na^+-K^+-$2Cl^-$ cotransport and Na^+/H^+ exchange, glucagon shrinks cells by activation of ion channels. The respective volume changes participate in the regulation of cellular protein and glycogen metabolism by these hormones. 3- Growth factors and expression of ras oncogene activate $Na^+-K^+-2Cl^-$ cotransport and Na^+/H^+ exchange, leading to the respective cell swelling. 4- Hepatocyte swelling triggers a hepatorenal reflex decreasing renal blood flow (RBF). The osmotic equilibrium permanently is compromised by alterations of EC osmolarity and by cellular transport and metabolism of osmotically active substances. Upon cell swelling most cells release cellular electrolytes. The resulting decrease of intracellular osmolarity achieves a regulatory cell volume decrease (RVD). Cell shrinkage triggers uptake of electrolytes into the cells, leading to regulatory cell volume increase (RVI). Cellular osmolarity is modified by transport of organic substances and metabolic formation or disposal of osmotically active substances, contributing to RVD and RVI, respectively. Cell swelling leads to the activation of K^+ channels and anion channels. The K^+ channel is highly K^+ selective and slightly inwardly rectifying, i.e.; depolarisation decreases the single channel conductance. The K^+ channel is Ca^{2+}-sensitive and upon cell swelling it is probably activated by an increase of intracellular Ca^{2+} activity. The ion channels inactivated by cell shrinkage appear to be distinct from ion channels activated during cell swelling. The activation of the Na^+/H^+ antiporter surpasses the activation of the Cl^-/HCO_3^- antiporter, leading to intracellular alkalinization during cell shrinkage. The net effect of the Na^+/H^+ exchange and $Na^+-K^+-2Cl^-$ cotransport and Cl^-/HCO_3^- exchange is cellular gain of NaCl. Sodium entering the cell via the $Na^+-K^+-2Cl^-$ cotransport or via the Na^+/H^+ antiport is rapidly exchanged for K^+ by the Na^+/K^+-ATPase. Thus, the ion transporters eventually serve to accumulate cellular KCl. Both, the $Na^+-K^+-2Cl^-$ cotransport and the Na^+/H^+ antiporter are modified by phosphorylation and both are most likely attached to the cytoskeleton. The phosphorylation site serves to modify the sensitivity or the respective carrier to alterations of cell volume. Cellular formation and accumulation of these osmolytes are stimulated by cell shrinkage; their release is stimulated by cell swelling. Cell shrinkage activates the degradation of proteins to amino acids and of glycogen to glucosephosphate and their metabolites. The sodium-coupled transport of organic substances, such as amino acids (e.g., glutamine) and glucose, leads to the cellular accumulation of osmotic activity, if the entry across the apical cell membrane surpasses the extrusion at the basolateral cell membrane. Cell swelling leads to delayed activation of K^+ channels in the basolateral cell membrane of renal proximal tubules during stimulated sodium coupled transport. This activation of K^+ channels leads to a hyperpolarisation of the cell membranes and thus further consequences for transepithelial transport. Insulin increases the cell volume by stimulation of the $Na^+-K^+-2Cl^-$ cotransport and Na^+/H^+ antiporter. Glucagon reduces the cell volume by activation of ion channels. The action of glucagon on cell volume may be mediated by cAMP. Furosemide blunts the swelling effect of insulin and leads to a proportionate blunting of the antiproteolytic action of the hormone. Growth factors activate $Na^+-K^+-2Cl^-$ cotransport and Na^+/H^+ antiport. The products of ras oncogenes are GTP binding proteins, which are resistant to inactivation by GTPase. Ras oncogenes are expressed in variety of tumours and expression of ras oncogenes leads to transformation. Activation of $Na^+-K^+-2Cl^-$ cotransport and Na^+/H^+ antiport are pre-requisites for the proliferation of the ras oncogene-expressing cells. Cell injury leads to cell swelling. The liver cell swelling triggers hepatorenal syndrome. Cells suddenly exposed to hyposmotic media, initially swell like more or less perfect osmometers but within minutes retain almost their original cell volume; this is called regulatory cell volume decrease (RVD). Upon sudden exposure to hyperosmotic media, the cells shrink like osmometers and display within minutes a volume regulatory increase (RVI), which brings back cell volume largely to the starting level. RVD and RVI may differ among different cell types but in general involve the activation of ion transport systems in the plasma membrane. RVD is largely achieved by a release of cellular K^+, Cl^-, and HCO_3^-. Cell swelling apparently leads to activation of Ba^{2+}- and quinidine-sensitive K^+ channels in parallel to anion channels. Swelling may open stretch-activated nonselective cation channels, which allow passage of Ca^{2+} into cell. The increase of intracellular Ca^{2+} then may activate Ca^{2+}-sensitive K^+ channels. Ca^{2+} activation of K^+ channels may not only be the mechanism allowing RVD; K^+ channels do not require an increase of intracellular Ca^{2+} activity. K^+ channels are directly activated by cell membrane stretch. RVI is achieved by activation of Na^+-H^+ exchange and $Cl^--HCO_3^-$ exchange. One of the most important challenges for cell volume homeostasis is the cumulative uptake of osmotically active substances, such as amino acids. Na^+-dependent amino acid transports in the plasma membrane can build up intracellular/extracellular amino acid concentration gradients of up to 20 fold. Na^+ entering the cell together with the amino acid is extruded in exchange for K^+ by the electrogenic Na^+-K^+-ATPase, which in turn triggers volume regulatory K^+ efflux. Insulin stimulates Na^+-H^+ exchange, $Na^+-K^+-2Cl^-$-cotransporter, and Na^+-K^+-ATPase, leads to cellular accumulation of K^+, Na^+, and Cl^- and consequently cell swelling. Glucagon activates Na^+-K^+-ATPase. Glucagon may lead to depletion of cellular Na^+, K^+, and probably Cl^- resulting in cell shrinkage. Insulin activates both loop diuretic-sensitive $Na^+-K^+-2Cl^-$-cotransporter and amiloride-sensitive Na^+-H^+ exchange, whereas glucagon stimulates cellular K^+ release through K^+ channels. Cell shrinkage and K^+ channel opening occurs when H_2O_2 is generated intracellularly during the oxidation of monoamines. Inhibition of $Na^+-K^+-2Cl^-$ cotransporter occurs in response to oxidative stress in vascular endothelial cells, which also would tend to shrink the cells. Cell proteolysis is under the control of amino acids and hormones, such as insulin and glucagon, but the underlying mechanisms remained obscure. Hyposmotic cell swelling inhibits proteolysis, whereas hyperosmotic cell shrinkage stimulates protein breakdown. Disruption of microtubules abolishes the antiproteolytic action of hyposmotic, amino acid- or insulin-induced cell swelling. Hyposomotic, amino acid- and insulin-induced cell swelling leads to a rapid alkalisation of intracellular acidic vesicular compartments, including lysosomes and prelysosomal endocytotic compartments. Cell shrinkage increases acidification of these compartments. Cell swelling not only inhibits proteolysis but simultaneously stimulates protein synthesis. Cell shrinkage triggers the inhibition of protein synthesis and stimulation of proteolysis. Alteration of amino acid metabolism after anisosmotic cell volume changes can at least in part be explained by parallel alterations of mitochondrial matrix volume. Cell swelling increases the polymerisation state of β-actin and increases the stability of microtubules. This accompanied by an increase of the mRNA levels for β-actin and tubulin. Microtubules play an important role in transducing some metabolic alterations in response to changes of cellular hydration. Cellular hydration increases mRNA levels for β-actin, tubulin, ornithine decarboxylase, and c-jun. Alterations of cellular hydration will influence membrane stretch, membrane-bound signalling systems, the cytoskeleton, protein phosphorylation, the ionic interior of the cell as well as the extent of macromolecular crowding in the cytosol. The cellular responses to alterations of cell water require an intact cytoskeleton. Cell swelling increases intracellular Ca^{2+} concentration, decreases pH, hyperpolarises the cell membrane, and stimulates IP_3 formation. Cell swelling inhibits glycolysis. Cellular hydration may interfere with activity of protein kinases and phosphatases, and changes in protein phosphorylation may trigger not only volume regulatory responses, but also the volume-dependent alterations in cellular metabolism and gene expression. Most cell membrane is highly permeable to water, which follows any osmotic gradient. Cell volume is of course compromised by any alteration of extracellular osmolarity. The cell accumulates a number of osmotically active substances such as amino acids and is forced to decrease intracellular electrolyte concentration to create an osmotic counterbalance. Cellular osmolarity is increased by conservative uptake of substrates such as amino acids or glucose, and by the breakdown of proteins, glycogen or triglycerides, which are osmotically less active than the sum of their constituent parts. The release of substance and formation of proteins, glycogen or triglyceride decreases cellular osmolarity. Influence of cell volume on metabolism: Some cells when shrunken stimulate the formation of osmolytes (substances designed to create intracellular osmolarity), include: 1- polyols, such as sorbitol, glycerol and inositol, 2- methylamines, such as glycerophosphorylcholine. Cell shrinkage stimulates the breakdown of macromolecules such as proteins and thus stimulates the production of monomers, which are osmotically, more active. Cell swelling stimulates protein and glycogen synthesis and inhibits proteolysis and glycogenolysis. Cell swelling inhibits glycolysis and stimulates flux through the pentose phosphate pathway, enhancing the availability of NADPH, glutathione (GSH) formation and efflux into blood. Cell swelling stimulates glycine oxidation, ketoisocaproate oxidation and lipogenesis from glucose. The mRNA for phosphoenolpyruvate carboxykinase, a key enzyme for gluconeogenesis, is decreased by cell swelling. Influence of cell volume on cytoskeleton: Cell swelling stabilises the microtubule network, stimulates actin polymerisation, and increase mRNA for β-actin and tubulin. These effects are reversed by cell shrinkage. Influence of cell volume on lysosomal pH: Cell swelling leads to alkalinization of acidic cellular compartments, whereas cell shrinkage enhances the acidity in these compartments. Influence of cell volume on specific cellular functions: Volume regulatory activation of ion channels modifies the electrical driving force and thus transepithelial transport, cell volume may modify the incorporation and/or activation of transporters. Growth promoters stimulate the Na^+/H^+ exchange and some growth factors have been described to stimulate Na^+-K^+-$2Cl^-$ cotransport. Expression of ras oncogene leads to growth factor-independent cell proliferation, by shifting the set point for cell volume regulation towards greater cell volume. Cellular alkalinization by Na^+/H^+ exchange favours cell proliferation. Increase of pH in lysosomes, stimulation of protein synthesis and inhibition of proteolysis stimulate cell proliferation.

STRESSORS

Adaptation to high salinity, and the response to hyper- and hypo-osmotic shocks are short- and long-term. Each has a different set of molecules as distinctive players, whose role is to maintain a physiological concentration of intracellular proteins with a functional configuration. A common mechanism for counteracting an increase in the external salinity is the intracellular accumulation of compatible solutes, so called because they do not interfere with cellular functions despite their high concentrations. Organic solutes play a role in the response to osmotic shock. There are 2 osmoadaptation mechanisms present in bacteria: mechanosensitive ion channels (MSC) and volume-activated channels (VAC). Adaptation to high temperature or salinity, for example, requires a number of intracellular mechanisms, and it also involves the cell membrane. The central event unchained by a stressor impacting on a cell is protein denaturation, which in turn elicits the stress response. The stress response involves the cell membrane and intracellular mechanisms, e.g., the cell may exclude the extracellular stressor (excess NaCl) and accumulates compatible solutes, such as glycerol. Among the intracellular mechanisms is the increase in the stress or heat-shock proteins (Hsp), including molecular chaperones whose central role is to assist in the folding and re-folding of polypeptides, as they are produced in the ribosome and as they are unfolded because of the stress, respectively. Molecular chaperones are a means to abate irreversible protein denaturation. Hsp belong to several families according to their molecular mass. Peptidyl-prolyl cis-trans isomerase (PPIase) plays a role in protein folding in eukaryotes and bacteria. PPIases catalyse rotation of the peptidyl-prolyl bonds that are required for protein folding, accelerating it and thus making it compatible with the rapid pace of other intracellular activities. Abnormal proteins must either be converted to normality or eliminated, lest they interfere with cellular functions. Abnormal proteins may aggregate, form precipitates, and be toxic, all factors that conspire against cell physiology. Conversion of proteins to normality is mediated by molecular chaperones, whereas elimination of molecules beyond repair is carried out by proteases. A protein, or fragments destined for degradation is tagged by ubiquitin and digested by the proteasome. Proteins in a cell may be normal or abnormal, and both must be degraded at one time or another. Normal proteins are more or less stable and long-lived, depending on their type and role. Proteins that regulate gene transcription, cell cycle and division, DNA repair, and metabolic pathways are needed temporarily and are short-lived. There are several causes for the presence of abnormal proteins inside a cell. Stress tends to denature most proteins, even if they are structurally normal. Even in the absence of stress a cell may contain abnormal proteins due to gene mutations, or to deficiencies in the post-transcriptional or post-translational mechanisms. A combination of stress and genetic or synthetic abnormalities may be deadly. In eukaryotes, the proteasome is a major cellular tool for degrading proteins that relies on ubiquitin for selecting its targets. Membrane lipids have been implicated in the ubiquitin-dependent proteasome-mediated proteolysis induced by heat shock, which indicates once more the importance of the cell membrane in the stress response. Aerobic organisms such as most known eukaryotes have mitochondria where O_2 is reduced to H_2O_2 with generation of the energy-rich compounds necessary for the cellular activities. Also, small amounts of toxic forms of oxygen, ROS, are generated in the mitochondria: O_2^- (superoxide), and OH^- (hydroxyl radical). In the normal cell, accumulation of toxic oxygen species does not occur because there are mechanisms for their elimination. Imbalances between ROS production and ROS elimination may happen leading to ROS accumulation, which can cause oxidative stress with damage to proteins, lipids, and nucleic acids. ROS not only cause oxidative stress, which is characterised by activation of some stress genes, but they also repress many genes, as other stressors do. This ROS-induced gene down-regulation has profound consequences upon the cell, above and beyond those typical of stress-gene induction. Oxidative stress is, in fact, one of the leading mechanisms of aging and cell death. Thus mitochondria are central players in the cell's life not only because they produce energy from O_2, but also because they have the potential for generating dangerous levels of toxic oxygen derivatives. Free radicals have an unpaired electron in an outer orbit. The energy generated by this unstable atomic state is released via reactions with surrounding molecules, which results in molecular damage. The mechanisms available to the cell for counteracting the effects of ROS are varied. The majority of the biosphere is cold by comparison with the temperature that is pleasant to humans (25-27°C), or with that which is optimal for human cells in general to grow, divide, and function, namely 37°C. While the classical stress response, namely that induced by stressors such as heat, increase or decrease in pH or salinity levels, chemicals, etc., is characterised by protein denaturation, the cold-stress response is not. Protein denaturation is not a major effect caused by the stressor cold. Prominent features of the cold-stress response are: 1- Stabilisation of the secondary structure of nucleic acids with ensuing inhibition of DNA replication, gene transcription, and mRNA translation; 2- Decrease in the activity of many enzymes with the consequent slow-down of metabolism; 3- Decrease in membrane fluidity, which tends to impede transport across it; and 4- Formation of crystalline ice, which if unchecked damages intracellular structures and, ultimately, causes cell death.

OXYGEN SENSING AND MOLECULAR ADAPTATION TO HYPOXIA

Molecular oxygen is the central fulcrum upon which life processes depend. Bacteria can flourish in environments that are essentially devoid of oxygen, and are much more metabolically malleable than man. Although many prokaryotes possess aerobic respiration, they also utilise anaerobic respiratory chains that do not require oxygen. In bacteria, coproporphrinogen oxidase and protoporphyrinogen oxidase are synthesised in the presence of oxygen. Under aerobic conditions, bacteria use molecular oxygen as the electron acceptor for both reactions. Anaerobically, coproporphrinogen oxidase uses NADP as the electron acceptor in a reaction that also requires methionine and ATP. Under anaerobic conditions, protoporphyrinogen oxidase is obligatory coupled to the cell's anaerobic respiratory chain. Any compound that can serve as a terminal electron acceptor (e.g., nitrate) in this chain permits the oxidation of porphrinogen. Aerobic organisms produce the bulk of their ATP by the oxidation of reduced substrates coupled to the phosphorylation of ADP to ATP, with oxygen serving as the terminal electron acceptor. Oxygen tension is an important environmental and developmental signal for the regulation of cell growth and differentiation in most organisms. During the course of aerobic respiration, molecular oxygen is generally reduced by 4 electron to give water. Some of the oxygen consumed during respiration is not completely reduced to water but is only partially reduced to the highly reactive intermediates, superoxide (O_2^-.), hydrogen peroxide (H_2O_2), and hydroxyl (OH.). Reactive oxygen species (ROS) are partially reduced forms of oxygen, produced independently of respiration during oxygen-consuming reactions in the cytosol. The production of ROS increases proportionally with partial oxygen pressure. These highly unstable reactive compounds have been shown to mutate DNA, oxidise proteins, and damage membranes, i.e., oxygen toxicity or oxidative stress.

BACTERIA

Enterobacteria, such as E coli, are facultative anaerobes that can grow in the presence or absence of oxygen. They can produce energy by aerobic respiration, anaerobic respiration, or fermentation, depends on the availability of oxygen or other electron acceptors and on the presence of fermentable carbon sources. Both aerobic and anaerobic pathways are arranged asymmetrically across the cytoplasmic membrane and generate a proton-motive force that in turn drives ATP synthesis by membrane-bound ATP synthase. Escherichia coli possess several different types of respiratory chains, which use different electron acceptors and terminal oxidoreductases. Glucose is completely oxidise to CO_2 by the citric acid cycle with the concomitant transfer of electrons to the cellular quinone pool (ubiquinone or menaquinone) via NADH, succinate, or lactate. Electrons are transferred from these quinones to cytochrome b and then to one of two terminal oxidases. At atmospheric oxygen concentrations, cytochrome bo is used. At limiting oxygen concentrations, cytochrome bd is used. When oxygen is available in abundance, cytochrome bd may function both as an oxidase and oxygen scavenger. Cytochrome bd may mitigate oxidative stress by protecting cells from reactive species. When grown aerobically, cells contain about 300 molecules of cytochrome bo and 200 molecules of cytochrome bd. During anaerobic or microaerophilic growth, the level of cytochrome bo drops 150-fold, while cytochrome bd increases 3-fold. So, in the absence of oxygen, cytochrome bd is predominant, but in the presence of oxygen, both terminal oxidases are present with the electron flow restricted to cytochrome bo. In the absence of oxygen and the presence of an appropriate alternative electron acceptor (e.g., dimethyl sulfoxide [DMSO]), E coli cells can produce one of at least 5 alternative oxidoreductases for anaerobic respiration (fumarate reductase, 2 nitrate reductases, DMSO/TMAO reductase, or TMAO reductase). The 2 nitrate reductase catalyse the conversion of nitrate to nitrite. Anaerobically, the citric acid cycle is converted to a noncyclic form, and most of the reduced substrate (glucose) used for energy is converted to ethanol, acetate, H_2O, and CO_2. These cells choose the most energetically favourable pathways for energy production. Enteric bacteria are found in soil and sewage, in the intestinal tracts of mammals, and in other parts of the mammalian body. Under the anaerobic conditions in the gut, their exposure to ROS is minimal. However, when outside of the gut, these organisms are exposed to air and can encounter reactive oxygen species (ROS) produced from at least three sources: 1- aerobic respiration, 2- environmental agents capable of generating free radicals when taken up by bacterial cells, and 3- phagocytic cells that release a respiratory burst of superoxide and other reactive oxygen radicals when they encounter bacterial cells. Like other organisms, enteric bacteria defend themselves against oxidative stress by producing the antioxidant enzymes superoxide dismutase, hydroperoxidase, glutathione reductase, and enzymes involved in DNA repair and metabolism. Oxygen affects bacterial gene expression. Several proteins involved in sensing oxygen or oxidative stress have been identified. Three of these are transcription factors that undergo reversible oxidation-reduction: 1- Fnr, functions in anaerobic cells, is activated when Fe^{3+} is reduced to Fe^{2+}. 2- SoxR, responds to oxidative stress brought about by chemicals that aggregate superoxide or by nitric oxide (NO), is activated when its iron-sulfur centers are oxidised. 3- OxyR, activated by exposing bacterial to H_2O_2. E coli flavohaemoglobin HMP, function as an electron gate that can transfer electrons from NAD(P)H to either its own heme moiety or to iron sulfur clusters in other proteins. In aerobic cells, it would transfer electrons intramolecularly to its own heme moiety, in anaerobic cells; it would transfer electrons to the iron-sulfur clusters of other proteins, like Fnr and SoxR. FixL is the only bacterial protein that is known to sense oxygen per se. This protein defines as a new class of hemoproteins, the heme-based sensors, ideally suited to sense oxygen and transduce its presence via kinase activity. Bacterial haemoglobins fall into 2 general categories, located in the cytosol: 1- dimeric hemoproteins composed of two single heme domain polypeptide; and 2- monomeric flavohemoproteins composed of a single heme binding domain and a single flavin binding domain. The first type is VGB, has been proposed to function in oxygen storage and diffusion, when overexpressed in E coli it enhances growth under oxygen-limiting conditions. The second type is HMP, has been found in E coli, candida norvensis. The NH_2-terminal regions of these proteins bind heme; the COOH-terminal region has a FAD-binding domain and is related to proteins in the ferredoxin-$NADP^+$ reductase (FNR) family. The E coli flavohaemoglobin can function as an oxidase and a reductase. As a reductase, it can use a variety of electron acceptors, including Fe^{3+} citrate, ferricyanide, and dihydropterine. The function of the flavin in HMP is to transfer electrons from NAD(P)H to the heme moiety, maintaining it in the Fe^{2+} state required for oxygen binding. When oxygen is bound to the heme, FAD is maintained in its oxidised state. At low oxygen tension, the heme moiety becomes fully reduced. If excess reductant [i.e., NAD(P)H] is present, FAD also becomes more reduced.

MULTICELLULAR ORGANISMS

Oxygen transport is optimised by tight regulation of ventilation and the red cell mass. In the carotid body, the neuronal stimulus to respiration in response to hypoxia depends importantly on rapid inhibition of conductance through potassium channels as well as on the relatively fast induction of the gene encoding tyrosine hydroxylase, the rate-limiting step in the synthesis of the neurotransmitter dopamine. In mammals, hypoxia stimulates erythropoiesis by upregulating the production of erythropoietin. Hypoxia induces the expression of genes encoding a number of cytokines, most notably vascular endothelial growth factor (VEGF), which appears to play an important role in wound healing, response to ischaemic injury, and tumour pathogenesis. Tissue oxygenation affects the regulation of a number of genes expressed in endothelial cells, including certain growth factors such as platelet-derived growth factor-β, interleukin-1 α, interleukin-8, and endothelin, as well as adhesion molecules such as vascular cell adhesion molecule-1, intercellular adhesion molecule-1, and endothelial leukocyte adhesion molecule-1. At cellular level, alterations in oxygen tension impact critically on the regulation of intermediary metabolism, affecting genes encoding enzymes responsible for glucose transport, glycolysis, and gluconeogenesis. Enzymatic adaptation to hyperoxia depends on the upregulation of appropriate genes that detoxify ROS. Most ligands, including polypeptide hormones and growth factors (erythropoietin), and variety of smaller molecules (catecholamines, steroids, thyroid hormones), act solely as messengers, having no function independent of their interaction with receptor. Oxygen is known to bind to and react with heme proteins. Oxygen transport depends on the circulation of haemoglobin-containing cells that enable oxygen unloading into tissues at relatively high O_2 tension. Erythropoietin (Epo) is glycoprotein hormone required for the proliferation and differentiation of erythroid cells. Epo production is markedly enhanced by hypoxia. The oxygen sensing mechanism is independent of transcription and translation. Thus physiologically relevant oxygen sensor appears to be localised in the plasma membrane as opposed to somewhere in the cell interior. However, this single Ca^{2+}-independent K^+ channel may not be sufficient for membrane depolarisation. The oxygen sensor for Epo production depends on an interaction between cytochrome P-450 and cytochrome P-450 reductase with the reduction of oxygen to superoxide. In cells containing hydrogen peroxide, more reactive oxygen compounds such as hydroxyl radical and singlet oxygen can be formed. The generation of these reactive oxygen intermediates (ROI) is catalysed by free iron via the Fenton reaction. Genes that are inducible by hypoxia are also upregulated by desferrioxamine and other strong chelators of iron. It is likely drastic reduction in intracellular free iron lowers the level of ROI, thereby mimicking a hypoxic environment. The neutrophil-macrophage cytochrome b558 functions as a NAD(P)H oxidase, converting oxygen to superoxide. The oxygen tension in mitochondria is far too low and too dependent on metabolic fluctuations to make it useful site for oxygen sensing. Mitochondria are a major source of superoxide (O_2^-) owing to inefficient transfer of electrons in the respiratory chain. Mitochondrial production of superoxide is dependent on vicissitudes in the cell's metabolic activity. The presence of abundant mitochondrial specific superoxide dismutase is likely limit egress of superoxide to the cytosol. The carotid body mitochondria are involved in oxygen sensing. It seems likely that most cells share a common oxygen-sensing apparatus. The sensor is likely to be a cytosolic, membrane bound, multisubunit b-like cytochrome that binds and reduces it to superoxide, which serve as chemical signals that impact on transcription factors such as HIF-1 that regulate oxygen-responsive genes. Genes that are induced by hypoxia appear to share a common sensing mechanism, a common mode of signal transduction, and perhaps a common transcription apparatus. CO may engender activated oxygen compounds via its ability to reduce heme groups.

$$CO + 2F^{3+} + H_2O \rightarrow CO_2 + 2Fe^{2+} + 2H^+$$

This reaction provides an alternate way of regenerating ferroheme in the sensor. CO may have an indirect effect on oxygen sensing. By inhibiting, via high affinity binding, cytochrome oxidase in the respiratory chain, a build up of NADH would again result in more rapid regeneration of the sensor's ferroheme groups, thereby enabling a continued flux of superoxide production. ROI are continuously generated as side products of electron transfer reactions. The principle ROI species include hydrogen peroxide, superoxide, hydroxyl radical, and singlet oxygen. When present in excess, these compounds are toxic to cells, causing lipid peroxidation, DNA damage, as well as protein cross-links and sulfydryl oxidation. A number of genes are induced by toxic concentrations of ROI many of these genes help protect the cell from oxidant damage. Less marked increases in levels of ROI are seen in cells exposed to ultraviolet light γ rays and also following stimulation of cells with certain cytokines and other ligands, such as tumour necrosis factor and interleukin-1 (the generation of ROI may explain why TNF and IL-1 blunt the hypoxic induction of the Epo gene). There is a chemical link between alterations in intracellular oxygen concentration and responsive changes in the structure and function of appropriate transcription factors. The most logical and best worked out mechanism for such signalling is by means of oxidation-reduction (redox) modification of protein sulfydryl groups. The protooncogenes fos and jun are members of a multigene family that can be rapidly and transiently induced by a wide range of stimuli in an equally wide range of cells and tissues. The protooncogenes fos and jun are members of a multigene family that can be rapidly and transiently induced by a wide range of stimuli in an equally wide range of cells and tissues. The AP-1 transcription complex is a heterodimer composed of c-Fos and c-Jun proteins or homodimers of c-Jun, which interact with each other at the leucine zipper domains. Redox chemistry is a critical determinant of the formation of AP-1 complexes. Site directed mutagenesis localises the critical redox-sensitive sites to a single conserved cysteine residue in the DNA binding domains of Fos and Jun. This cysteine residue is directly involved in DNA binding. When it is replaced, DNA binding is retained, but, activation is abolished. The c-fos and c-jun family of genes are inducible by hypoxia, which are accompanied by an increase in AP-1 activity as well as by increased expression of Ref-1. NFκB, among the well characterised transcription factors, non has a broader biologic repertoire than NFκB, a trimer composed of 2 DNA binding proteins and an inhibitory subunit (IκB), which keeps the complex sequestered in the cytosol in an inactive form. NFκB activated by dissociation of IκB from the complex, allowing the DNA binding dimer to travel to the nucleus, where it rapidly induces expression of a number of genes imported in inflammation and immune responses. A variety of signals are known to activate NFκB, including cytokines (TNF-α and IL-1), lipopolysaccharide, phorbol ester, mitomycin C, calcium ionophore, lectins, and ultraviolet light. All of these agonists increase intracellular levels of ROI and deplete the cell of reduced glutathione, thereby triggering a common redox-signalling pathway. NFκB can be activated by hypoxia. Heat shock proteins and heat shock factor: HSPs may serve as molecular chaperones that not only protect stressed proteins from denaturation but, under nonstress circumstances, may facilitate protein folding, processing, and transit between cell compartments. The rapid expression of HSP genes depends on the stress-induced activation of heat shock transcription factors (HSF), which binds to cogene cis-elements in their promoters. Under nonstress conditions, HSF resides as a monomer in the cytosol. Upon exposure to stresses that induce HSPs, HSF aggregates into a trimer, is phosphorylated, and is translocated to the nucleus. The activation of HSF is highly dependent on redox status of the cell. Both HSF and HSPs are induced transiently by hypoxia but consistent with a second wave of HSP expression with reoxygenation. E2A Basic Helix-loop-Helix Proteins (bHLH), are transcriptional factors form either homo- or heterodimers. The basic regions bind to specific DNA response elements while the helix-loop-helix regions are required for dimerization. Redox chemistry appears to play a critical role in the regulation of these complexes. Hypoxic induction of biologically important genes depends in a major way on the activation of HIF-1; a heterodimer composed of 2 bHLH transcription factors. Hypoxia inducible factor-1 (HIF-1), these genes are activated by a common set of transcription factors. The kinetics of activation of HIF-1 by hypoxia closely mimic those for the induction of Epo mRNA. The activation of HIF-1 appears to be complex. Epo has a physiological importance and a higher degree of biological specificity than other cytokines. Epo is a 30.4 kDa glycoprotein hormone. The basal production of Epo is increased as much as 1000-fold in response to hypoxic stress such as anaemia or hypobaric atmosphere. The hormone travels to haematopoietic tissues where it binds to receptors on erythroid progenitor cells, protecting them from apoptosis and enabling them to proliferate and differentiate into functioning RBCs. An increase in red cell mass may relieve the hypoxic stress. Epo mRNA is localised to a subset of peritubular interstitial cells (fibroblasts) in the interface between the cortex and outer medulla of the kidney. The peritubular cells would become hypoxic if the O_2 supply to the metabolically active tubular cells were compromised. In the foetus, the liver is the major site of Epo production. In adults, the liver produce little Epo, but during hypoxia, it contributes up to 33% of the total Epo production. Hypoxic induction of Epo gene expression, both in the kidney and liver cells, depends on increased transcription. The Epo 3'-enhancer is composed of three interacting parts. Specific nuclear binds to these half sites in Epo 3'-enhancer and has a marked impact on hypoxic induction and tissue specificity. Hypoxia induces carotid body type I cells to polarise. The hypoxic signal is transmitted by release of the neurotransmitter dopamine to primary terminals of the afferent fibers of the sinus nerve. Tyrosine hydroxylase (TH), the rate-limiting enzyme in dopamine synthesis, is activated by hypoxia, resulting in amplified transmission of the hypoxic signal, via the sensory sinus nerve to the brain stem. Exposure to modest degrees of hypoxia inhibits K^+ current accompanied by membrane depolarisation and increased intracellular Ca^{2+} concentration. This increase in Ca^{2+} concentration may initiate a signal transduction pathway that effects both release of neurotransmitter and induction of TH. Nuclear receptor may play a role on TH gene transcription. Regulation of angiogenesis is critical in development and also in adaptation to episodes of local hypoxia. Effective wound healing depends on the proliferation of new blood vessels to maintain an adequate supply of oxygen and nutrients to the metabolically active repair tissue. Angiogenesis plays a critical role in the pathogenesis of a variety of pathological states ranging from cancer to diabetic and sickle cell retinopathies. Recovery from myocardial ischaemia and infarction is favoured by neovascularization. Hypoxia effects variable induction of different proteins in different cell types, such as PDGF (types A and B), placental growth factor (PLGF), fibroblast growth factors, transforming growth factor-β1, and vascular endothelial growth factor (VEGF). VEGF is distinguished from the other cytokines by being expressed in nearly all cells at much higher levels than in endothelial cells, its receptor is restricted to endothelial cells, and its expression is markedly enhanced by hypoxia. Hyperbaric hypoxia or CO (which induces hypoxia by impairing oxygen release to tissues), upregulate VEGF and may contribute to the pathogenesis of tumour growth. It is likely that angiogenesis is regulated in a paracrine fashion. The hypoxic induction of both VEGF and Epo, along TH and certain glycolytic enzymes depends on the activation of the HIF-1 transcription factor. Thus, these factors are likely to share a common mechanism for oxygen sensing signal transduction, and transcriptional regulation. The VEGF gene appears to be regulated at both the transcriptional and posttranscriptional levels. After hypoxia, VEGF mRNA rise 12-fold while, transcription increases only 3-fold. VEGF mRNA is also stabilises by hypoglycaemia, which like hypoxia is a consequence of tissue ischaemia. The cell's viability depend on the orderly utilisation of its nutrients, primarily glucose and fatty acids, in concert with oxygen. When and where the availability of glucose and oxygen are adequate, anaerobic glycolysis provides pyruvate, which is then processed by the oxygen-dependent tricarboxylic (Krebs) cycle, generating generous amounts of the cell's metabolical currency, ATP. When glucose levels are limiting, gluconeogenesis is turned on, primarily by activation of 3 other rate-limiting enzymes, glucose-6-phosphatase; fructose-1,6-biphosphatase; and biphosphatase, and phosphoenolpyruvate carboxykinase. During hypoxia, ATP yield from the glycolytic pathway is 18-fold lower than that from the Kerbs oxidative pathway, so the glucose consumption must increase substantially. The absorptive surfaces from the gut epithelial cells and kidney tubular cells express a cotransporter that is energetically driven by a sodium electrochemical gradient, enabling transport of glucose across a concentration gradient. In contrast, glucose transport into and across other types of cells depends on channels that operate via facilitated diffusion. Glucose transporters (GT) are responsible for transporting six-carbon sugars and other carbon compounds into the cell. There are several different types of glucose transporters. Bacteria contain several unique types of glucose transporters, such as proton symporters, substrate-binding transporters and group translocation systems. Mammalian cells employ at least 2 types of transporters, Na^+-dependent cotransporters and facilitative transporters. The majority of glucose transporters belong to a family of proteins, called GLUT. There are 7 glucose transporters identified to date, GLUT1-5, 7 and SGLT1, encoded by different genes and are expressed in various tissue: 1- GLUT1 (erythrocyte), contributes most to overall homeostasis, also responsible for basal or constitutive glucose transport in wide variety of cells, and composes 5% of the entire erythrocyte membrane; 2- GLUT2 (hepatocytes, pancreatic cells, intestine and kidney), mediates high-capacity low-affinity transport, and transepithelial transport (basolateral membrane); 3- GLUT3 (brain), appears to be responsible for basal or constitutive glucose transport in wide variety of cells; 4- GLUT4, regulates insulin-dependent uptake by muscle and fat cells; 5- GLUT5 (intestine, adipose, muscle, brain and kidney), involved in intestinal absorption of fructose and other hexoses; 6- SGLT1 (kidney, intestine), mediates Na^+-dependent active transport; concentrates across apical epithelial membranes; and 7- GLUT7, (hepatocytes and other gluconeogenic tissues), mediates flux across endoplasmic reticulum membrane. The precise structure is not known for any of the GLUT proteins. The GLUT proteins all contain approximately 500 amino acids, with GLUT1-5 exhibiting 39-65% sequence identity. They all share a similar structure, but GLUT-1 is the one for which the most information is known. The GLUT proteins contain 12 transmembrane helices, with the amino and carboxyl termini situated on the cytoplasmic face. Prolonged oxidative stress has a significant effect on GLUT1 and GLUT4. GLUT1 expression is increased when the cell is exposed to low-grade oxidative stress for an extended period of time, while GLUT4 is reduced. Oxidative stress impairs GLUT4 translocation, but has no effect on GLUT1 translocation. Glucose transporters are an integral part of every cell and when their function is inhibited or augmented, it can prove disastrous for the cell. Considering the fact that GLUT-1 responds to a number of environmental stimuli in addition to hypoxia, it is likely that its regulation depends on a complex array of cis-acting elements and transcription factors. GLUT-1 promoter responses to both hypoxia and mitochondrial inhibitors. This enhancer contains an atypical HIF-1 binding site. Hypoxia appears to increase steady-state GLUT-1 mRNA levels by increasing mRNA stability. The effect of hypoxia on expression of glycolytic enzymes varies among cell types and may contribute to stress adaptation. Endothelial cells maintain viability during prolonged hypoxic exposure much better than a number of nonendothelial cells types such as fibroblasts, smooth muscle cells, and renal tubular cells. The ability of endothelial cells to tolerate hypoxia better than other cells may be due in part to maintaining a greater level of high-energy phosphates, ATP and GTP. Oxygen sensing transduction in multicellular organisms includes: 1- protein phosphorylation; 2- redox; and 3- HIF-1. Hypoxic induction of physiologically relevant genes, in multicellular organisms, including: 1- Epo (enhancer, promotor, nuclear receptor, RNA stability); 2- tyrosine hydroxylase; 3- VEGF; and 4- gene involved in glucose metabolism. Changes in oxygen tension appear to be sensed by heme proteins, with subsequent transfer of electrons along a signal translocation pathway, which may depend on reactive oxygen species. These heme-based sensors are generally two-domain. Some are hemokinases, while others are flavohemoproteins [flavohaemoglobins and NAD(P)H oxidases]. Hypoxia-dependent kinase activation of transcription factors in nitrogen-fixed bacteria bears a striking analogy to the phosphorylation of hypoxia inducible factor-1 (HIF-1). Redox chemistry appears to play a critical role both in the trans-activation of oxygen-responsive genes in unicellular organisms as well as in the activation of HIF-1. Inappropriate delivery of oxygen to cells underlies the pathobiology of many diseases. Aberrant activation of the homeostatic mechanisms that normally balance oxygen supply and demand may also have pathological consequences, examples being angiogenesis and diabetic retinopathy. Erythropoietin is normally produced by specialised cells in kidney and liver. HIF-1 is capable of binding to erythropoietin. Oxygen-regulated genes include genes involved in glucose transport

(Glut-1 and Glut-3), gluconeogenesis (PEPCK), vascular growth factor (VGF), vasomotor regulation (iNOS, endothelin), iron metabolism (transferrin), coagulation (tissue factor), retroviral transposons (VL30) and catecholamine synthesis (tyrosine hydroxylase). Aryl hydrocarbon receptor nuclear translocator (ARNT) deficiency reduces hypoxic gene regulation, results in failure of cells to produce HIF-1.

RESPONSES TO CELL STIMULATION

Changes in cell metabolism including energy production occur in response either to conditions external to the cells (oxygen deprivation), or to cellular signalling mechanisms (primary messengers) such as hormones, neurotransmitters, or growth factors. All the primary messengers except steroid hormones function by binding to specific receptors on the cell's plasma membrane, setting up a sequence of events leading to release of a variety of second messengers in the cytoplasm of the cell. Among these intracellular mediators of the primary message are cAMP, inositol, 1,4,5-triphosphate (IP_3), diacyl glycerol, and intracellular free Ca^{2+}. Many other second messengers have been proposed including glutamate, cyclic ADP ribose, and cGMP. An increase in one or more second messengers leads to the activation of various kinases that subsequently phosphorylate a variety of proteins on serine, threonine, or tyrosine residues, one of these proteins could function as a messenger. Depolarisation of the plasma membrane potential opens Ca^{2+} channels, causing a rapid influx of Ca^{2+} that is supplemented by Ca^{2+}-induced Ca^{2+} release from sarcoplasmic reticulum (SR). Second messengers could signal the mitochondrion to modify the metabolic rate at the same time that it or another messenger mediates the cytosolic and/or nuclear changes mandated by the primary messenger. Of all these second messengers only Ca^{2+} is known to interact directly with mitochondria. Because Ca^{2+} sequestered rapidly by mitochondria, it has a direct access to the likely sites of action of such a mediator: enzymes mediating the tricarboxylic acid (TCA) cycle and thus the production of NADH and $FADH_2$. Along the pathway of oxidative phosphorylation, other mechanisms controlled by Ca^{2+}, include: 1- amino acid catabolism, 2- fatty acid oxidation, 3- stimulation of electron transport through a Ca^{2+}-induced increase in matrix volume, 4- Ca^{2+}-induced decrease in inhibition of the ATPase through Ca^{2+}-sensitive inhibitor protein, 5- activation of the adenine nucleotide translocase, and 6- Ca^{2+} plays a role in gluconeogenesis. Under pathological conditions in which the plasma membrane becomes permeable to Ca^{2+} or cytosolic ATP levels fall significantly (e.g., ischaemia), Ca^{2+} can rise significantly in some tissues. When inward leakage of Ca^{2+} increases, a new steady state may be reached between influx and efflux through the Ca^{2+}-ATPase pump and $Ca^{2+}/2Na^+$ exchanger. Under physiological conditions, pulses of Ca^{2+} are often observed following depolarisation of the plasma membrane or exposure of the cell to a primary messenger. The driving force for Ca^{2+} transport is diffusion down its electrochemical gradient. Influx via uniporter is energetically downhill. Because Ca^{2+} sequestration is extremely rapid, the uniporter may be one of the fastest gated pores known. The Na^+-dependent Ca^{2+} efflux mechanism is $Ca^{2+}/2Na^+$ exchanger, which can be inhibited by a wide range of inhibitors, including verapamil. Na-independent transport dominates in liver and kidney. The Na^+-independent efflux mechanism of Ca^{2+} transport is an active mechanism. The Na^+-dependent efflux mechanism of Ca^{2+}-transport may also be an active mechanism. The mitochondrial permeability transition (PT) is characterised by a sudden increase in the permeability of the mitochondrial inner membrane to small ions and molecules, leading to complete collapse of the membrane potential and colloid-osmotic swelling of mitochondrial matrix. Under pathological conditions PT can be a source of irreversible injury. Swollen mitochondria are hallmark of cells undergoing necrosis. Inducers of PT include: 1- excessive amounts of endogenous oxaloacetate, acetoactate, and P_i; 2- agents that create oxidative stress by oxidising pyridine nucleotides and/or depleting matrix glutathione; and 3- phospholipase A_2 (PLA_2) reaction products such as lysophospholipids. Pore opening is inhibited by: 1- antioxidant, 2- reducing agents, and 3- inhibitors of PLA_2. Pore opening is reversible by: 1- chelating Ca^{2+}, and 2- reductant dithiothreitol. The most dramatic inhibitive and pressive peptide is cyclosporin A (CSA). Important endogenous inhibitors of PT include: 1- intramitochondrial ADP, 2- reduced pyridine nucleotides, 3- spermine, 4- acidic conditions, and 5- high membrane potential. A few seconds are required by individual mitochondrion to complete the transition. Opening of the transition pore destroys the mitochondrial membrane potential and therefore ADP phophorylation; the resulting energy deprivation may result in cell injury and death. Ischaemia and hypoxia manifestations include: 1- decreased mitochondrial membrane potential, 2- decreased ATP and Pcr, 3- acidification, 4- increased intracellular P_i from adenine nucleotide breakdown, and 5- disruption of the plasma membrane. Ca^{2+} remains unchanged up to the point of cell death, which occurs about 30 min after ATP depletion. Anoxic cells may be rescued by fructose, or by stimulation of glycolysis. Depletion of mitochondrial but not cytosolic reduced glutathione (GSH) kills the cell via loss of ATP, as it induces PT. Reperfusion is accompanied by generation of ROS, release of AA (suggesting PLA_2 activation). Reperfusion of irreversibly injured cell results in swollen mitochondria and a decreased rate of oxidative phosphorylation. Reperfusion or reoxygenation after prolonged hypoxia is more clearly associated with the PT, manifesting a burst of ROS, a sudden surge of Ca^{2+} influx and greatly increased tissue damage, often marked by swollen mitochondria. Ca^{2+} activates: 1- dehydrogenases associated with TCA cycle, 2- electron transport, 3- the F1-ATPase, 4- adenine nucleotide translocase, 5- activates

PDH and KGDH, leading to an increase rate of NADH and $FADH_2$. Increases in NADH stimulate oxidative phosphorylation. The efflux mechanisms have the primary control of Ca^{2+} transport. The primary role of mitochondrial Ca^{2+} transport is likely to be sequestering cytosolic Ca^{2+} to activate Ca^{2+}-sensitive metabolic steps and to control matrix Ca^{2+} for mediation of metabolic control. A secondary purpose may be to protect the cytosol against hypercalcaemia under a limited range of pathological circumstances. The mitochondrial Ca^{2+} efflux mechanisms may also function to minimise the risk of opening the PT pore under hypoxic conditions. Because Ca^{2+} in mitochondrial matrix is at lower energy than Ca^{2+} in external space, its extrusion requires energy.

ATP DEPLETION

As ATP production by mitochondrial oxidative phosphorylation accounts for more than 90% of total oxygen consumption, so mitochondrial dysfunction could result in organ failure. This may be related to NO, which inhibits mitochondrial respiration either directly or through effects upon the antioxidant status (GSH) of tissue. Tissue oxygen tensions are raised in humans during sepsis, which suggesting reduced ability of the organs to utilise oxygen. Mitochondrial dysfunction, reducing ATP generation, is linked to glutathione deficiency. Consequences of reduction of ATP production are: 1- active cells cease to function; 2- anaerobic glycolysis leads to the production of pyruvate and accumulation of lactic acid; 3- impairment of sodium pump, so Na is retained in the cell and potassium escapes from it; 4- ribosome become swollen as fluid accumulates in the sacs of the rough endoplasmic reticulum; 5- ribosomes become detached from the rough endoplasmic reticulum and protein synthesis is reduced; and 6- mitochondria become swollen. Two main problems to chemotherapy are toxicity to normal cells and failure to kill cancer cells. Both problems stem from the indirect mechanism by which both chemotherapeutic drugs and irradiation kill cells. They damage both normal and cancer cells, and this damage is then translated through multiple steps into cell death, likely through activation of caspases and apoptosis. When these steps are compromised the therapy fails. Another potential approach involves bypassing the defective part of the pathway to restore the signal triggered by chemotherapeutic drugs or by the oncogenic transformation itself, which can be thought of as proapoptotic signal that is present only in transformed cells. Bcl-2 directly or indirectly prevents the release from mitochondria of cytochrome c, which along with ATP may facilitate a change in Apaf-1 structure to allow procaspase-9 recruitment and processing. Bcl-2 can protect cells after much cytochrome c has been released. Bcl-2 can protect by preventing mitochondrial disruption. The Bcl-2 family is regulated by cytokines and other death-survival signals at different levels. Early events: After ATP depletion that can be caused by hypoxia, ischaemia or hypoglycaemia there may be complete recovery, death, or secondary damage (about 7 days after the initial ATP depletion). A major factor controlling the cell damage is the magnitude and duration of ATP depletion, which depends on irreducible ATP consumption, metabolic reserves and physiological readjustments. ATP consumption can be reduced during the initial episode by cooling and/or cessation of function, which can be automatically triggered by adenosine. The xanthine oxidase inhibitor (allopurinol) and other administered antioxidants will conserve cellular ATP, the former encourages hypothanthine recycling, and the later saves the energy used to keep glutathione reduced, in addition to its direct antioxidant benefits. The early phases of damage are at about 0-0.5 and 0.5-12 hours after initial ATP depletion. During ATP depletion there is: 1- an automatic reduction of function such as loss of consciousness, reduction of bile secretion, reduction of cellular membrane pumping; and 2- changes in inorganic biochemistry such as increases in extracellular potassium and intracellular calcium. ATP itself is broken down to adenosine, inosine and hypoxanthine which are not strongly charged, and which cross plasma and mitochondrial membranes. Hypoxanthine may be oxidised by xanthine oxidase. At about 0.5 h protein denaturation in cytoplasm is visible histologically. Protein synthesis is markedly inhibited, to at least 80% of normal levels. At about 12 h there may be death with evidence of cell necrosis, possibly with inflammation, complement activation and leukocyte infiltration. In necrotic cells there is lysosomal acid hydrolase action and with inflammation, and oxidation from leukocytes. Later events: At about 12-48 h, there is physiological and biochemical quiescence during that new treatments might be successful: 1- cellular ATP, energy charge and extracellular fluid hypoxanthine would have returned to near normal; 2- there is cellular and mitochondrial swelling and increases in extracellular fluid levels of normally cytoplasmic enzymes; and 3- cytoplasmic changes, visible as swelling, include changes in ATP content and apparently decreased efficiency of oxidative phosphorylation that may persist for up to 24 h. About 48-168 h in the sequence, there is: 1- a fall in cellular ATP concentrations and phosphocreatine; 2- markedly increased hypoxanthine output. Irreversibly damaged cells retain the ability to phosphorylate purine nucleotides but cannot restore concentrations to control levels. Since there is mechanisms for increased ATP consumption, this period of secondary energy failure is probably due to increased demand for ATP as in the heat stress or increased work by the heart. Changes in protein synthesis and breakdown, includes: 1- The synthesis of the heat shock proteins is increased, thus reduction of synthesis of non-heat shock proteins may be even more marked. 2- After ATP depletion early reductions in proteolysis, the short-lived cell components, plasma membranes and cytoplasm, do show rapid hitological and functional changes. 3- The control systems, especially receptors and rate-limiting initial enzymes are known to be short-lived whereas the housekeeping enzymes, mitochondria and lysosomes have lives of 1 or more weeks. 4- The membrane trafficking proteins only last about 2 days (about 300 round journeys). 5- The adenosine-dependent inhibitory protective systems have receptor half-lives of 45 min-16 hours. 6- The α and β adrenergic receptors in the presence of antagonist have lives of 1-3 h. Heat shock proteins (HSP), are synthesised after hypoxia. ATP driven proteolysis occurs at neutral pH. Protease activity is also stimulated by increased cytosolic calcium concentrations. Peptide fragments stimulate HSP ATPase activity, whereas intact cell proteins can inhibit this ATPase activity. 4 ATP + GTP are needed to split every peptide bond and 4 ATP + GTP are needed to synthesise that bond, the total loss is 8 ATP + GTP per peptide bond. Proteolysis occurs 48 h or more after primary ATP depletion, the prolonged release of fragments of structural proteins such as troponin and myosin is delayed, slow and prolonged. Such proteolysis is consistent with the marked and generalised aminoacidaemia and aminoaciduria often seen in sick newborn in the first week of life. There is neurotoxicity from excitatory amino acids. Similar energy (NAD and ATP) consuming but slower mechanisms repair DNA after early inhibition, the poly (ADP-ribose) transferase-dependent mechanism is inhibited by a reduction of cellular ATP. Severely damaged cells assess their damage and if repairs are too costly, they rapidly remove themselves by apoptotic cascade mechanisms. Poly (ADP-ribose) transferase, when attached to DNA breaks and adds NAD to itself and can lower NAD to about 10% of control preventing ATP regeneration and causing death. In apoptosis, there is late ATP-driven lysis of Poly (ADP-ribose) transferase. In ATP depletion, function is reduced, repair and renewal are inhibited, there is progressive loss of control, but there is no adequate evidence of persistent and marked mitochondrial failure.

MITOCHONDRIAL PERMEABILITY TRANSITION

Ischaemia/reperfusion injury is a complex phenomenon involving cell-to-cell interactions, tissue architecture, and zonal structure, as well as changes in individual cells. Mitochondrial permeability transition (MPT) leads to necrotic and apoptotic cell death as a result of ischaemia/reperfusion. Anaerobic glycolysis and ATP hydrolysis during ischaemia rapidly decreases tissue pH, which protects against hypoxic cell killing in heart, liver, kidney, vascular, and endothelial cells. The recovery from acidosis to physiologic pH after reperfusion precipitates necrotic cell killing. The onset of the MPT and subsequent ATP depletion are causative events in pH-dependent necrosis to reperfused hepatocytes. Cyclosporine A (CsA) prevents MPT-induced mitochondrial depolarisation, inner membrane permeabilisation, ATP exhaustion after reperfusion and blocks the necrotic cell killing. Fructose prevents the necrotic cell death by supplying ATP via glycolysis. Preservation of 15-20% of normal is sufficient to prevent necrotic cell killing in hypoxia/ischaemia. Glycine prevents MPT-dependent reperfusion injury, but protects after ATP depletion by preventing plasma membrane failure. Cytoprotection by fructose and glycine is downstream of the MPT because neither fructose nor glycine block mitochondrial depolarisation and inner membrane permeabilisation induced in situ by reperfusion or an MPT-inducing calcium ionophore. Ischaemia/reperfusion injury can lead to apoptosis in liver and other organs. Necrotic cell death occurs after reoxygenation at pH 6.2 (stimulates ischaemia) in anoxia. Fructose plus glycine treatment during reperfusion prevents the necrotic cell killing, and apoptosis occurs instead. CsA is a specific blocker of MPT and prevents apoptosis, whereas tacrolimus does not. Onset of the MPT leads to mitochondrial swelling, outer membrane rupture and release of cytochrome C into cytosol. The interaction of cytochrome C with apoptosis protease activating factor-1 (APAF-1) leads to caspase 9 activation, which then activates caspase 3. APAF-1 and cytochrome C-dependent activation of caspase 9 requires ATP or the less abundant dATP. ATP acts a switch between apoptosis and necrosis. Fructose is a glycolytic substrate that promotes ATP formation after reperfusion. Reperfusion with fructose or fructose plus glycine increases ATP and prevents necrotic cell death, but instead promotes apoptotic cell death. Glycine is a cytoprotective amino acid that prevents plasma membrane permeabilisation, which occurs down-stream of ATP depletion. ATP is required for development of MPT-dependent apoptosis after ischaemia/reperfusion injury to hepatocytes. Fructose promotes apoptosis after ischaemia/reperfusion. Fructose causes a profound decrease of inorganic phosphate (Pi) in aerobic cells, as sugar phosphates are formed by the action of fructokinase and subsequent steps in glycolysis. Pi promotes the MPT, and a decline of Pi and the consequent suppression of the MPT by fructose treatment may be the reason that fructose retards TNF-α-induced apoptosis. Fructose-induced Pi depletion may responsible for the delayed onset of MPT-inducing calcium ionophore. ATP at 27% to 36% of normal levels is sufficient to support apoptosis.

MITOCHONDRIAL DYSFUNCTION

Mitochondrial permeability transition in the inner membrane causes swelling of mitochondria and release of cytochrome C, which progresses to autophagy and apoptosis. The cellular concentration of ATP, OFR, and cell necrosis may follow depending on the stimulus. Disturbance of cell function of CFS is best described as a mitochondrial dysfunction or energy production dysfunction, with an associated intracellular acidosis. Mitochondrial dysfunction and oxidative stress have been implicated in cellular senescence, apoptosis, aging and aging-associated pathologies. Mitochondria produce ATP through oxidative metabolism to provide cells with energy under physiological conditions. Mitochondria are also the major cellular source for generation of reactive oxygen species (ROS), while they are one of the main targets of ROS-induced oxidative damage. Alterations in mitochondrial structure and function are early events of apoptosis and mitochondria appear to be a central regulator of apoptosis in most somatic cell systems. As ATP production by mitochondrial oxidative phosphorylation accounts for >90% of total oxygen consumption, so mitochondrial dysfunction could result in organ failure. This may be related to NO, which inhibits mitochondrial respiration either directly or through effects upon the antioxidant status (GSH) of tissue. Tissue oxygen tensions are raised in humans during sepsis, which suggesting reduced ability of the organs to utilise oxygen. Telomeres consist of tandem repeats of the TTAGGG sequence that cap the ends of chromosomes, protecting them from degradation and fusion. The length of telomere repeats is primarily maintained by active telomerase, which is composed of Telomerase RNA (TR) and a catalytic subunit telomerase reverse transcriptase (TERT). Telomere shortening and erosion lead to chromosome end-to-end fusions and genomic instability, causing replicative senescence in human cells and possibly aging. Maintenance of telomere length is essential for bypassing senescence and crisis checkpoints in cancer cells. Severe telomere shortening may also become a signal that triggers apoptosis. Mitochondrial dysfunction generates ROS and leads to chromosomal instability via telomere attrition. Telomere attrition occurs prior to apoptosis and may serve as an intermediate step between mitochondrial dysfunction and apoptosis. Telomere shortening may signal apoptosis. Oxidants damage DNA, breaks polyguanosine sequences in telomere repeats, and causes telomere shortening, cell cycle arrest and replicative senescence. ROS can also oxidize proteins, especially during chronological process of aging. The ROS generated by compromised mitochondria may oxidize proteins necessary for telomere maintenance, such as telomerase, TRF1 or Ku, resulting in genomic instability. The inhibitory effects of oxidant scavenger NAC on telomere attrition and cell death suggest that ROS are important mediators that link mitochondrial dysfunction and telomere shortening and loss, genomic instability, and apoptosis as well. The anti-apoptotic protein bcl-2 binds to mitochondrial membranes and prevents apoptosis by acting as an antioxidant. Bcl-2 might serve as an endogenous ROS scavenger as well to prevent ROS produced by mitochondria from damaging telomeres and generating chromosome instability. Status epilepticus (SE) results in significant cerebral damage and an increased risk of subsequent seizures, associated with a characteristic pattern of neuronal loss, especially affecting the hippocampus. SE is associated with mitochondrial dysfunction and loss of brain glutathione (GSH), which could be due to reversible excitotoxic cell damage as a result of free radical, especially peroxynitrite damage. Status epilepticus (SE) is a medical emergency, which can result in permanent neurological and mental disability. The hippocampus is especially vulnerable with cell loss in the hilus, CA1 and CA3 regions, but relative sparing of the dentate granule cells and the NADPH-diaphorase positive, nitric oxide synthase (NOS) expressing CA2 cells. Excitotoxicity has been widely studied in a variety of other human neurodegenerative diseases including Huntington's disease, in which NOS expressing cells are similarly spared, Parkinson's disease and motor neuron disease. Calcium influx into the cell, activation of apoptotic pathways, oxidative damage via the production of free radicals, nitric oxide (NO) production and mitochondrial dysfunction are all believed to play a role. Reduced glutathione (GSH) appears to be an important factor in determining molecular susceptibility to oxidising species. Respiratory chain inhibition itself can then contribute to further local free radical production, thereby possibly exacerbating damage. With NO, defects of complex II-III and IV with sparing of complex I are most commonly documented in brain. The biochemical abnormalities and histological evidence of neuronal damage may last up to 44 h after status, with neuronal loss in most studies maximal from 72 h onwards. Neuroprotective strategies for status induced neuronal damage are possible. There is a direct evidence of mitochondrial dysfunction associated with neuronal death following SE, suggesting that oxidative stress play a critical role. It may be that, in seizure associated cell death, a combination of local differences in free radical production, glutathione levels, and NO production contribute to the selective neuronal vulnerability. A dysregulation in the storage/release of dopamine (DA), and its auto-oxidation at physiological pH, could play a decisive role in the neurotoxicant-mediated mitochondrial failure and neurodegenerative mechanisms associated with MPTP-toxicity. The oxidation of DA leads to the formation of insoluble melanin-like polymers in a nonenzymatic autocatalyzed mechanism. MPP^+, the active metabolite of the Parkinsonism-inducing compound MPTP, is produced by the action of monoamine oxidase B and accumulated by dopaminergic neurons through the dopamine-reuptake system. Inside the neurons MPP^+ can be accumulated in catecholaminergic vesicles or in mitochondria by a mechanism depending on the membrane potential. Inside the mitochondria MPP^+ reduces the mitochondrial respiration rate and the NADH-dehydrogenase activity (NADH-DH) of the respiratory chain complex I. MPP^+ acts at hydrophobic sites. MPP^+ seems to inhibit complex I, both in state 4 (without ADP) and in state 3 (with ADP) of respiration, acting at the same site of rotenone, between NADH-DH and coenzyme Q, without affecting complex II activity. The inhibition of the mitochondrial chain occurs when mitochondria are activated in complex I (i.e., with pyruvate/malate) and it is reflected in a decrease in oxygen consumption and an increase in cytoplasmic lactate, both of which can be reverted by succinate. MPP^+ depletes dopamine (DA) at a low range than that required for complex I inhibition. Dopamine itself, or via its autooxidation products, is implicated in radical production, neurotoxic processes, and apoptotic neurodegeneration, and thus in the pathogenesis of Parkinson's disease. The oxidation of DA leads to the formation of insoluble melanin-like polymers in a nonenzymatic autocatalyzed mechanism. Many injuries to the brain result in the uncontrolled release of neurotransmitters, including DA. DA is quickly taken up, metabolized or auto-oxidized in the presence of O_2, so that, DA, its metabolites or oxidative products can accumulate in side surviving cells. High concentrations of DA may induce intracellular calcium increase and DNA-laddering in brain neurons, without significantly altering mitochondrial membrane potential. MPP^+ also inhibits α-ketoglutarate dehydrogenase complex, decreases mitochondrial DNA content, opens the mitochondrial permeability transition pore and releases cytochrome C. Inhibition of mitochondrial complex I and the depletion of ATP supplies results in a loss of mitochondrial transmembrane potential. DA and MPP^+ may have a synergistic effect on mitochondrial function. DA plus MPP^+ cause a significant increase in mitochondrial membrane potential and mitochondrial swelling. A failure in the adenine nucleotide translocator/voltage-dependent anion channel may precede mitochondrial hyperpolarization and swelling. DA and Daox cause a similar potentiation of the effects of MPP^+. Dopamine released by MPP^+ may play a key role in MPP^+-enhanced generation of OH^- free radical. The oxidation of DA can be catalyzed by Daox itself and could enhance the MPP^+ redox cycling, so that a local increase in free radicals could be enough to cause a whole mitochondrial dysfunction. Hepatitis C virus (HCV) is the most common cause of viral hepatitis. Chronic HCV-infection can lead to severe sequelae, most notably liver cirrhosis and hepatocellular carcinoma. The histological findings in livers of HCV-infected patients vary widely and likely represent a complex interaction between viral and host factors. Characteristic histopathological lesions include portal tract inflammation, especially lymphoid aggregates, steatosis, liver fibrosis, and even cirrhosis in advanced disease. Ultrastructural findings are correlated with several markers of lipid peroxidation, a consequence of oxidative stress. Mitochondria of HCV-infected cells demonstrate irregular shapes, often with thinned and fragmented cristae. The finding of altered mitochondrial structure in HCV-related liver disease would help to explain the steatosis that is so prevalent in this disease. All mitochondrial alterations are associated with a depletion of tissue glutathione stores. The loss of the antioxidant glutathione could be due to excessive oxidant production. The oxidative stress would also be responsible for the lipid peroxidation. In conditions of cellular stress, mitochondria are a source, as well as the target, of reactive oxygen species. Mitochondrial dysfunction or failure is an important factor in cytotoxicity, in part, by causing the cell to undergo apoptosis. Mitochondrial dysfunction leads to a release of proapoptotic factors such as cytochrome c and apoptosis-inducing factor (AIF) from the mitochondria. These factors promote activation of caspase proteases, a family of proteases that have been strongly implicated in apoptosis. The caspases then cause proteolytic cleavage of death substrates culminating in cytotoxicity. Reactive oxygen species are associated with disease activity in chronic HCV. This process might be enhanced by nonspecific production of oxygen radicals by monocytes or other inflammatory cells. The expression of the death receptor Fas is significantly higher in HCV core antigen positive hepatocytes than in uninfected cells. Fas expression in hepatocytes is upregulated in areas of more severe liver inflammation. When HCV-specific T-cells recognise HCV antigens via the T cell receptor and MHC antigens, they become activated and express Fas ligand, which in turn will induce the formation of a death complex in Fas-expressing hepatocytes. This death complex then activates caspase-8 followed by downstream effector caspases, which eventually lead to apoptotic cell death. Activation of death receptors like Fas causes release of cytochrome c from mitochondria mediated by a Bcl-2 interacting protein named Bid. Fas activation causes cellular glutathione depletion. The mitochondrial dysfunction would be expected to lead to oxygen radical formation and lipid peroxidation. Antioxidants should benefit and provide protection from mitochondrial injury mediated by oxidative stress. High dose of α-tocopherol (vitamin E) improves aminotransferase levels. Statins inhibit synthesis of mevalonate, a precursor of ubiquinone, which is a central compound of the mitochondrial respiratory chain. The main adverse effect of statins is a toxic myopathy possibly related to mitochondrial dysfunction. Lactate/pyruvate ratios are significantly high in patients treated by statins. Ubiquinone serum levels are low in statin-treated patients. Statin therapy is associated with high blood lactate/pyruvate ratio, which suggests mitochondrial dysfunction.

HCV could induce mitochondrial dysfunction by:

1- Direct effects of viral protein on mitochondria.
2- Inducing an intracellular oxidative stress causing secondary mitochondrial dysfunction.
3- Immune-mediated activation of cell death pathways such as the Fas system.
4- The associated oxidative stress from inflammatory cells, which could target mitochondria.

NEURONAL DEGENERATION AND MITOCHONDRIAL DYSFUNCTION

Mitochondria are the seat of a number of important cellular functions, including essential pathways of intermediate metabolism, amino acid biosynthesis, fatty acid oxidation, steroid metabolism, and apoptosis. Oxidative phosphorylation (OXPHOS) generates most of the cell's ATP, and any impairment of the organelle's ability to produce energy can have catastrophic consequences, due to the primary loss of ATP, and the indirect impairment of downstream functions, such as the maintenance of organellar and cellular calcium homeostasis. Deficient mitochondrial metabolism may generate reactive oxygen species (ROS) that can destroy the cell. Mitochondrial DNA (mtDNA) is a tiny 16.6-kb circle of double-stranded that encodes 13 polypeptides, all of which are components of the respiratory chain/OXPHOS system, plus 24 genes, specifying 2 ribosomal RNAs (rRNAs) and 22 transfer RNAs (tRNAs) that are required to synthesise the 13 polypeptides. About 850 polypeptides, all encoded by nuclear DNA (nDNA), are required to build and maintain a functioning mitochondrial. These proteins are synthesised in the cytoplasm and are imported into the organelle, where they are partitioned into the mitochondrion's 4 main compartments, the outer mitochondrial membrane (OMM), the inner mitochondrial membrane (IMM), the intermembrane space (IMS), and the matrix, located in the interior (the organelle's cytoplasm). Of the 850 proteins, approximately 75 are structural components of the respiratory complexes and at least another 20 are required to assemble and maintain them in working order. The 5 complexes of the respiratory chain/OXPHOS system are all located in the IMM including complexes I (NADH ubiquinone oxidoreductase), II (succinate ubiquinone oxidoreductase), III (ubiquinone cytochrome c reductase), IV (cytochrome c oxidase), and V (ATP synthase). There are also 2 electron carriers, ubiquinone (coenzyme Q), located in the IMM, and cytochrome c, located in the IMS. There are thousands of mtDNAs in each cell, with approximately 5 mtDNAs per organelle. Organellar division and mtDNA replication operate independently of the cell cycle, both in dividing cells (such as glia) and in postmitotic nondividing cells (such as neurons). Upon cell division, the mitochondria (and their mtDNAs) are partitioned randomly between the daughter cells (mitotic segregation). The number of organelles varies among cells, depending in large part on the metabolic requirements of that cell. Skin fibroblasts contain a few hundred mitochondria, whereas neurons may contain thousands and cardiomyocytes tens of thousands of organelles. Most mitochondrial diseases due to maternally inherited mutations in mtDNA are recessive, i.e., a very high amount of mutated mtDNA must be present (>70% of the total population of mtDNAs) in order to cause overt dysfunction. Diseases associated with mtDNA mutations are typically heterogeneous and are often multisystemic. Many mitochondrial disorders are encephalo-cardiomyopathies. Mitochondrial diseases share a number of features including lactic acidosis, massive mitochondrial proliferation in muscle (resulting in ragged-red fibers [RRFs]), and cytochrome c oxidase (CCO) deficiency. Some mtDNA mutations that tend to cause rather selective neuronal degeneration. In neurodegenerative diseases, specific populations of neurons die. Apoptosis operates via 2 pathways mitochondrion-mediated or receptor-mediated but mitochondrion-independent. In the mitochondrion-dependent pathway, an external insult (e.g., elevated cytosolic calcium) acts to cause release of cytochrome c, which is located in the IMS. Cytosolic cytochrome c can then bind APAF-1 (apoptotic protease-activating factor 1), which then binds to the inactive form of caspase-9. This initiator caspase complex, or apoptosome, can now activate a cascade of events, beginning with activation of downstream effector caspases, such as caspase-3, followed by activation of caspases further downstream, ultimately resulting in the hallmarks of apoptosis (condensation of nuclear and cytoplasmic contents, nDNA fragmentation, cell blebbing, and autophagy of membrane-bound bodies). Mitochondrion-mediated activation of caspase-9 can also occur via external (extracellular) receptor-mediated signals (e.g., TNF, growth factor deprivation, etc.) to target various ligands, e.g., Bad, Bax, Bik, Noxa-to the mitochondrion, thereby causing cytochrome c release and binding of APAF-1. Under other circumstances, a separate mitochondrion-independent pathway also operates. It is well known that muscle biopsies from patients with classic mitochondrial disorders, show little or no evidence of necrosis, fiber loss, elevated circulating creatine kinase, or inflammation. Markers of increased ROS have been found in muscle biopsies from patients with mitochondrial disease. Elevated ROS are injurious to mitochondria. There are at least 2 reasons why apoptosis is not occurring in muscle: 1- APAF-1 is either present at very low levels or missing entirely in skeletal muscle. In the absence of APAF-1, cytosolic cytochrome c has no partner with which to bind in order to activate caspase-9 and induce apoptosis. 2- An inhibitor of apoptosis called ARC (apoptotic repressor with caspase recruitment domain), which is expressed almost exclusively in skeletal and heart muscle, not only inhibits apoptosis in these tissues but can also protects mitochondria from free radical damage. Impairment of mitochondrial energy metabolism is the key pathogenic factor in a number of neurodegenerative disorders.

IRON

Iron is indispensable for life, serving co-factor for many enzymes, either nonheme or hemoproteins. Hemoproteins are involved in a broad spectrum of crucial biologic functions including oxygen binding (haemoglobins), oxygen metabolism (oxidases, peroxidases, catalases, and hydroxylases), and electron transfer (cytochromes). Therefore heme is formed in almost all-living systems, except for a few obligatory anaerobes and certain unicellular organisms auxotrophic for porphyrins and/or heme. All animal cells can synthesise heme, with the exception of mature erythrocytes and perhaps some other cells at the very end of their differentiation pathways. In the late 1940s and early 1950s, it was shown that glycine and succinyl CoA were the source of all the heme. At the same time, in vitro heme synthesis was demonstrated in RBCs of birds and human reticulocytes. The heme biosynthesis pathway and its subcellular compartmentation are probably identical in all mammalian cells. Heme biosynthesis involves 8 enzymes, 4 of which are cytoplasmic and 4 are localised in the mitochondria. The first step occurs in the mitochondria and involves the concentration of succinyl CoA and glycine to form 5-aminolevulinic acid (ALA), catalysed by ALA synthase (ALA-S). The next 4 steps of the biosynthetic pathway take place in the cytosol, ALA dehydrogenase (ALA-D) converts two molecules of ALA to a monopyrol porphobilinogen (PBG). Two subsequent enzymatic steps convert 4 molecules of PBG into the cyclic tetrapyrole uroporphyrinogen III, which is then decarboxylated to form coproporphyrinogen III. The final 3 steps of the biosynthetic pathway, including the insertion of ferrous iron into protoporphyrin IX by ferrochelatase, occur in the mitochondria. Rapid rates of heme biosynthesis occur in liver and erythroid cells where large amounts of heme are needed not only for mitochondrial cytochromes but also as prosthetic groups for cytochrome P450 and haemoglobin respectively. 85% of organism's heme is synthesised in erythroid cells whose total number is considerably lower than that of hepatocytes, this because haemoglobin is the most abundant hemoprotein containing as much as 70% of the total iron content of a normal adult. Both iron metabolism and the regulation of heme synthesis are different in haemoglobin-synthesising as compared with nonerythroid cells. Iron is transported between sites of absorption, storage, and utilisation by transferrin, which has a specific relationship to haemoglobin-synthesising cells. The iron transport machinery in erythroid cells is an integral part of the heme biosynthesis pathway. There are 2 different genes for ALA-S, one of which is expressed ubiquitously while the expression of the other is specific to erythroid cells. No tissue-specific isozyme is known for ALA-D but there are subtle differences in 5' untranslated regions (UTRs) in housekeeping and erythroid ALA-DmRNAs. PBG deaminase (PBG-D) exists in two isoforms, one being present in all cells whereas the other is expressed only in erythroid cells. ALA-S condenses glycine with succinate upon decarboxylation. ALA-S is a homodimer residing on the matrix side of the inner mitochondrial membrane with absolute specificity for glycine. Succinyl-CoA is an intermediate in the tricarboxylic acid cycle and provides the major source of energy for the entire porphyrin biosynthesis pathway. Human ALA-S2 synthesised in the cytoplasm contain N-terminal leader peptide (49 amino acids) required for the transport to the mitochondria. ALA-D is composed of 8 identical 36 kD subunits and contains 4 catalytic sites. The enzyme requires Zn^{2+} (1 atom/subunit) and intact sulphydryl groups for activity. No tissue specific isozyme is known for ALA-D. PBG-D catalyses the stepwise deamination and condensation of 4 molecules of PBG to yield an extremely unstable tetrapyrole intermediate, preuroporphyrinogen (hydroxymethylbilane). In human the PBG-D gene is assigned to chromosome 11 (11q23-11qter). PBG-D exists in 2 isoforms: ubiquitous (44 kD) and erythroid-specific (42 kD). A dominantly inherited partial deficiency of the PBG-D causes acute intermittent porphyria (AIP) characterised by acute, life-threatening attacks of abdominal pain, motor and sensory neurological deficits, and psychiatric symptoms. AIP includes decreased hepatic PGB-D activity, increased hepatic ALA-S activity, and dramatically increased urinary excretion of ALA after treatment with phenobarbital. The terminal 3 enzymes of the heme biosynthetic pathway are located in mitochondrion in association with the inner membrane, and this arrangement may suggest that 3 enzymes may form a multienzyme complex with accompanying substrate channelling. Coproporphyrinogen (CPO) catalyses oxidative decarboxylation of the 2 propionate groups at positions 2 and 4 of CPO III. CPO is localised in the intermembrane space of mitochondria. CPO gene contains 2 polyadenylation sites, and an alternative use of different polyadenylation signals seems to be responsible for the production of CPOmRNA with different 3' ends identified in placenta and fibroblasts. Protoporphyrinogen oxidase (PPO) is an integral protein of the cytoplasmic side of the inner mitochondrial membrane with the active site facing the intermembrane space. Hemin inhibits PPO activity by 50%, but the concentration of heme is too high to consider the effect physiologically relevant. The induction of erythroid differentiation is associated with a significant increase in PPO enzyme activity. PPO gene is localised on chromosome 1q23. Ferrochelatase, old name was heme synthase. The enzyme has 2 substrates, protoporphyrin IX and ferrous iron, and 2 products, heme and 2 protons. Ferrochelatase is associated with complex I of the mitochondrial electron transport chain and suggesting that Fe^{2+} is produced via a NADH-dependent ferric-iron-reducing system. Solubilized ferrochelatase is stimulated by fatty acids and phospholipids, suggesting that local mitochondrial environment play a role in the overall enzyme activity. Histidine residues of ferrochelatase are ligands of the ferrous substrate but do not rule out the involvement of other ionic ligands (gultamate). Heme inhibits ferrochelatase in noncompetitive manner; it is unclear whether overall heme synthesis or cellular iron metabolism is modulated by this effect of heme. Human ferrochelatase gene is located on chromosome 18q21.3. Erythroid cells can take up iron only via the transferrin receptor-pathway. Although nonerythroid cells can also take up iron from nontransferrin sources, under normal conditions they acquire iron from transferrin. Released ferric iron is reduced to the more malleable and soluble ferrous form. Membrane carriers for ferrous iron may be involved in the transmembrane transport of Fe^{2+}. Other possibility for the reticulocyte-endosomal iron transport may be the proton ATPase, which is responsible for endosomal acidification. Iron after its release from endosomes, becomes available for mitochondrial heme synthesis, for the insertion into iron-dependent proteins and enzymes, and for storage in ferritin. Iron may complexes to citrate, sugars, some amino acids, pyridoxal and nucleotides. When iron reaches the outer mitochondrial membrane, it is entrapped by ligands and transferred across the inner membrane to ferrochelatase. Ferrochelatase is located on the inner mitochondrial membrane with its active site facing the mitochondrial matrix. Because only the reduced form of iron can be processed by ferrochelatase, reduction of iron must occur at some point after its release from transferrin. In nonerythroid cells enlargement of the intracellular transit iron pool stimulates the synthesis of ferritin and the opposite scenario develops when this pool is depleted of iron. Iron-dependent regulation of both ferritin and transferrin receptor occurs posttranscriptionally, and is mediated by virtually identical iron-responsive elements (IREs). IREs are cis-acting nucleotide sequences, forming stem-loop structures are recognised by trans-acting cytosolic RNA-binding proteins known as regulatory proteins IRPs. IRPs are of two types, IRP-1 and -2, shares homology with mitochondrial aconitase. IRP-2 shares 60% overall amino acid identity with IRP-1. IRP-2 functions solely as an RNA-binding protein because it lacks aconitase activity, and regulation of IRP-2 by iron is mediated by specific proteolysis. When cellular iron becomes limiting, the IRP-1 is recruited into the high-affinity binding state. The binding IRP-1 with IREs of the ferritin mRNA represses the translation of ferritin, whereas association of IRP-1 with IREs of transferrin receptor mRNA stabilises this transcript against ribonucleases. On the other hand, the expansion of the labile iron pool inactivates IRE-P1 and leads to a degradation of IRP-2, resulting in an efficient translation of ferritin mRNA and rapid degradation of transferrin receptor mRNA. IRE/IRP-dependent mechanism is involved in the coordinated regulation of iron uptake and storage by all cells. Erythroid cells appear to possess distinct control mechanisms to satisfy their needs for iron and to specifically handle the metal. Iron modulates transferrin receptor expression in nonerythroid cells, in a negative feedback manner at the level of mRNA stabilisation that affects steady-state transferrin receptor mRNA levels. Transferrin receptor expression is regulated differently in erythroid cells. It is possible that erythroid cells may express a unique transferrin receptor isoform, which may be subject to different control mechanisms. In haemoglobin-synthesising cells iron appears to be specifically targeted towards mitochondria.

Fe-transferrin cycle:

1- Transferrin attaches to specific receptors on the cell surface by a physicochemical interaction.
2- The transferrin-receptor complexes are internalised by the cells enclosed within endocytic vesicles.
3- Iron is released from the transferrin within the endocytic vesicles by a temperature- and energy-dependent process, which is endosomal acidification.
4- Iron is transported to intracellular sites of utilisation and/or storage in ferritin, but this aspect of iron metabolism includes the nature of the elusive intermediary pool of iron and its cellular trafficking.
5- The iron-free apotransferrin, which remains attached to the receptor returns to the cell surface, where the apotransferrin is released from the cells.

Iron and heme metabolism:

	Erythroid	Nonerythroid
Iron		
1- Iron source	Transferrin (Tf).	Transferrin + non-Tf Fe
2- Tf receptors	Differentiation increases.	Proliferation decreases.
3- Tf receptors + Fe	Little change	Decrease.
4- Tf receptors Regulation	Transcriptional	Primarily mRNA stability
5- Effect of heme on Fe uptake from Tf	Inhibits	No effect
6- Fe overload	Mitochondria	Cytosol ferritin (never mitochondria).
Heme		
1- content	Non-covalent associated with globin.	Covalent binding to cytochromes.
2- Major function	Very high.	Trace.
3- Control of synthesis	O_2 transport.	e^- transport.
4- Effect on ALA-S	Fe from Tf	ALA synthase
	Translational induction by Fe (IRE in 5' UTR).	Feedback repression by heme (no IRE).
5- Heme Oxygenase	mRNA decreased during erythroid differentiation.	Induced by heme.

In erythroid cells, the vast majority of iron released from endosomes must cross both the outer and inner mitochondrial membranes to reach ferrochelatase. The chemical nature of iron that accumulates in mitochondria is unknown but iron does not appear to be in the form of ferritin. The only other living system capable of accumulating excessive iron in a nonferritin form is yeast. A certain proportion of this nonheme iron accumulated in the mitochondria is in the form readily for heme synthesis. Heme is quickly transported out of the mitochondria to combine with globin chain in the cytosol; this suggesting that the majority of iron taken up by erythroid mitochondria can leave the organelle only after iron is inserted into protoporphyrin IX. When heme synthesis in inhibited in erythroid cells, very little or no iron accumulates in cytosolic ferritin. In nonerythroid cells, iron in excess of metabolic needs ends up in ferritin. Mitochondria in nonerythroid cells do not accumulate nonheme iron, even in severely iron overload individuals. Four factors plays a role in the pathogenesis of mitochondrial iron in sideroblastic anaemias: 1- iron is specifically targeted toward erythroid mitochondria; 2- iron cannot be used because of the lake of protoporphyrin IX; 3- there is a lack of heme-the negative regulator of iron uptake; and 4- iron can leave mitochondria only after being inserted into protporphyrin IX. Mature erythrocytes contain virtually all their iron in haemoglobin. Iron is required to maintain the stability of ALA-S2mRNA. Iron deficiency decreases ALA-S activity. ALA-S is induced first and ferrochelatase last. There is significant ferrochelatase activity already in early erythroblasts. At all stages during erythropoiesis the contribution of ferrochelatase to the overall heme synthesis exceeds that of ALA-S. Heme appears to be essential for the induction of the heme biosynthetic pathway in erythroid cells. Erythroid differentiation is also associated with a significant increase in transferrin receptors. During the maturation of reticulocytes to erythrocytes, the cells lose all components of the haemoglobin-synthesising system, including the transferrin receptors. Erythroid differentiation physiologically triggered by erythrpoietin leads to an early transcriptional induction of all heme pathway enzymes occurring coordinately with the induction of transferrin receptors. The availability of transferrin-bound iron limits erythroid heme synthesis. Overall haemoglobin synthesis rate appears to be controlled by the capacity of erythroid cells to acquire iron from transferrin. An earlier cessation in porphyrin synthesis would likely lead to mitochondrial iron accumulation, leading to an increased risk of free radical damage. Breakdown of internal membranes, including those of mitochondria, is accomplished in part by erythroid 15-lipoxygenase (LOX). Erythroid 15-LOX is unique in its ability to attack phospholipids and is the main factor in mitochondrial degradation during reticulocyte maturation. The mRNA for erythroid 15-LOX is the second most abundant reticulocyte mRNA after globin mRNA. Proteolysis, receptor shedding, and LOX-induced destruction of mitochondria in the termination of haemoglobin biosynthesis processes are triggered by some kind of feedback mechanism that may be linked to the concentration of haemoglobin in reticulocytes. An increased demand for haemoglobin, usually sensed as a deficiency of tissue oxygen, stimulates renal production of erythrpoietin (Epo), which in turn enhances erythroid differentiation. Epo-mediated increase in the production of immature erythroid cells does not appear to increase the formation of haemoglobin (and heme) per cell, suggesting that even under normal conditions the heme formation per cell is maximal. Many substances foreign to the body (xenobiotics) cause a marked increase in the amounts of the cytochrome P450 in the liver that is associated with the increase in heme synthesis per hepatocyte, do not increase the production of heme in erythroid cells. Although xenobiotics may have some primary inducing effect on hepatic ALA-S1, many chemical inducers are believed to increase ALA-S1 by depleting free or uncommitted heme pool in hepatocytes. 20% of newly formed heme is directly converted to bile pigments, while 80% is used for the formation of hemoproteins. The depletion of cellular heme levels in hepatocytes can occur due to: 1- increased rate of heme degradation; 2- enhanced utilisation of heme for the production of hemoproteins, e.g., cytochrome P450; and 3- inhibition of heme synthesis, or various combinations of the above. An increase in hepatocyte free or loosely bound heme leads to transcriptional induction of heme oxygenase, an enzyme responsible for heme degradation. Hepatocytes maintain adequate heme levels by a combination of synthetic and degradative mechanisms. Haemoglobin-synthesising cells have evolved mechanisms for upregulating, rather than downregulating, heme levels. Neither the intracellular iron pathway nor an immediate substrate for ferrochelatase has been identified in hepatocytes. Iron synergizes the drug-dependent induction of ALA-S1. Iron has a role in accumulation of uro- and hepato-carboxyl porphyrins in the liver known to occur in porphyria cutanea tarda (PCT). NOS, which catalyses five-electron oxidation of L-arginine to citrulline and NO, is a cytochrome P-450 type hemoprotein, activating cytosolic guanylate cyclase via its binding to the heme present in this enzyme. Many of NO's intracellular targets are represented by proteins containing Fe-S clusters. NO inhibits purified human ferrochelatase as well as its activity in hepatocytes, and macrophages. NO interacts with the [2Fe-2S] cluster. The RNA-binding activity of IRP can also be modulated by NO. NO mediates inhibition of haemoglobin expression. NO inhibits gene expression by preventing the binding of NF-E$_2$ to DNA. NO-induced inhibition of haemoglobinization may play a role in the pathogenesis of anaemia of chronic disease, a condition associated with a high level of inflammatory cytokines, which stimulate NO production. There is a chemical similarity between carbon monoxide (CO) and NO, CO may be a physiological regulator akin to NO. CO can act as a neuronal or epithelial messenger, presumably by binding to iron in the heme moiety of guanyl cyclase to produce cyclic GMP. In mammalian cells the only source of CO is heme following its cleavage by heme oxygenase. Iron is absorbed in ferrous state. Most of dietary iron is in the ferric form. Gastric secretin dissolves iron and reduce it to the Fe^{2+} form. Ascorbic acid and other reducing substances in diet facilitate conversion of ferric to ferrous iron. Heme is also absorbed. The phytic acid found in cereals reacts with iron to form insoluble compounds in the intestine, as well as phosphates and oxalates. Pancreatic juice inhibits iron absorption. Iron absorption is an active process. Most of the absorption occurs mainly in duodenum and adjacent jejunum. The mucosal cells pass part of the iron directly into the bloodstream, but much of it is bound to apoferritin. Apoferrtin combines with iron to form ferritin. Apoferritin is a globular protein made of 24 subunits. Iron forms micelle of ferric hydroxyphosphate, and in ferritin, the subunits surround this micelle. The ferritin molecule can contain as many as 4000 atoms of iron. 70% of the iron in the body is in haemoglobin, 3% in myoglobin, and the remainder in ferritin. Ferritin is the principal storage form of iron in tissues, although hemosiderin also stores iron. Iron absorption into the bloodstream is increased when body iron stores are depleted or when erythopoiesis is increased, and decreased in the opposite conditions.

ACUTE METHEMOGLOBINEMIA

75% of oxygen carried in the body is bound to haemoglobin. Iron in the ferrous (Fe^{2+}) state is the central element in the heme portion of haemoglobin, binds with oxygen. The conjugated structure of the heme molecule in the Fe^{2+} state gives blood its characteristic red colour. Methaemoglobin is haemoglobin in which the iron has been oxized from the ferrous to the ferric (Fe^{3+}) state is incapable of binding oxygen. The inability of methaemoglobin to bind oxygen increases the affinity of the remaining normal haemoglobin for oxygen, further diminishing oxygen release to the tissues. The body maintains a balance between haemoglobin oxidation and reduction within the RBCs. Oxidation occurs when an oxygen atom is gained or a hydrogen atom is lost. Reduction occurs when an oxygen atom is lost or a hydrogen atom is gained.

Methaemoglobin symptoms:

Methaemoglobin	Symptoms
0-2%	None,
3-15%	Slate-gray skin colour.
15-20%	Asymptomatic cyanosis. Chocolate brown coloured blood.
20-45%	Symptomatic cyanosis. Headache, dyspnoea, gastric upset, lethargy, dizziness, and syncope.
45-60%	Decreased level of consciousness. Metabolic acidosis, tachypnoea.
60-70%	Seizures, cardiac arrhythmias, coma.
>70%	High incidence of death.

The glycolytic and glucose-6-phosphate dehydrogenase (G6PD) pathways are essential regulatory processes in glucose metabolism. The glycolytic pathway produces NADH, which is the primary initiator of methaemoglobin reduction. NADH requires the enzyme methaemoglobin reductase-I to catalyse the conversion of 90% of methaemoglobin to haemoglobin. The G6PD pathway produces NADPH, which is the secondary mechanism of methaemoglobin reduction. NADPH requires the enzyme methaemoglobin reductase-II to facilitate 10% of methaemoglobin reduction. A methaemoglobin level >2% results from disruption of the normal enzymatic reducing pathways. Methaemoglobinaemia may be genetically acquired or acutely induced. Acute methaemoglobinaemia occurs when the capacities of NADH or NADPH reduction pathways are overpowered by exogenous oxidising agents. The rate of haemoglobin oxidation then exceeds the rate of its reduction, and methaemoglobin is produced in greater than normal quantities. Symptoms are intensified in persons with preexisting conditions of anaemia or cardiopulmonary pathology. A significant diagnostic hint of methaemoglobinaemia is the chocolate-brown appearance of both venous and arterial blood. Exposure to air does not redden the colour of the blood. The diagnosis is confirmed when the application of 10% potassium cyanide solution forms cyanomethaemoglobin, which causes the blood to turn bright red. With any cyanotic patient, an airway should initially be secured and oxygen administered. Cyanosis from methaemoglobinaemia is often asymptomatic. Cyanosis due to methaemoglobinaemia does not reverse quickly with oxygen administration. Arterial blood gas results are not always accurate in methaemoglobinaemia. Pulse oximetry is based on the total absorption spectra of haemoglobin types. Methaemoglobin and oxyhaemoglobin absorb light at the same wavelength. So the SaO_2 may be decreased although not indicated. The pulse oximeter reading may decrease during methylene blue administration, because the dye is interpreted by the oximeter as reduced haemoglobin and causes a false-low reading. Nitrates and nitrites oxidise haemoglobin to methaemoglobin. Amyl nitrate is often abused to enhance sexual pleasure. Methaemoglobinaemia is one of the most dangerous complications of nitrate abuse. Benzocaine suppresses the erythrocytes' reduction capabilities in normal and reductase-deficient patients. Benzocaine is absorbed from mucous or pulmonary membranes; methaemoglobin is produced in proportion to the absorbed dose. Methaemoglobin should be suspected in patients who suddenly become cyanotic after receiving topical anaesthesia. Intervention: Causative agent should be discontinued. Methaemoglobin levels up to 30% may resolve spontaneously in 2-3 days. Methylene blue acts as an electron donor to facilitate methaemoglobin reduction. A slow intravenous injection of 1-2 mg of methylene blue/kg in a 1% solution should be given over 5 minutes. If the patient has not responded to methylene blue administration within 1 hr, subsequent doses should be considered carefully. Lack of response to methylene blue may indicate a deficiency of G6PD (2-3%), and may even induce acute haemolytic episode. The side effects of methylene blue include blue-greenish urine, GIT distress, and hypotension. With a total dose of 7 mg/kg, toxic effects such as precordial pain, dyspnoea, restlessness, haemolytic anaemia, and dysuria may occur. Ascorbic acid works directly but slowly to reduce methaemoglobin, used mostly in congenital methaemoglobinaemia particularly

G6PD deficiency. Ascorbic acid can be given orally as 500 mg or 1-2 g intravenously every 6 hr. Activated charcoal and gastric lavage may be indicated for oral ingestion. In rare circumstances exchange transfusions or hyperbaric oxygen therapy may be considered.

Exogenous causes of methaemoglobinaemia:

1- Local anaesthetic agents	Benzocaine, lidocaine, prilocaine, procaine.
2- Nitrates and nitrites	Nitroglycerine, nitroprusside, silver nitrate, volatile nitrites (Amyl nitrite), bismuth subnitrate, nitrate-rich foods (spinach, broccoli, carrots), nitrate preservatives, contaminated well water, nitrous gas.
3- Aromatic amino- and nitroso-compounds	Aniline dyes (ink, shoe polish), phenazopyridine.
4- Sulfonamides	Dapsone.
5- Miscellaneous	Naphthalene, antimalarials.

THE FREE RADICALS

When two atoms combine to form a molecule, some of their electrons are shared between them. The molecule is flexible. Catalytic transformations can be either oxidative or reductive. Catalytic asymmetric oxidations include epoxidation, dihydroxylation, and aminohydroxylation. Most of the oxidants give rise to by-products that must be either disposed of recycled. The ideal oxidant is molecular oxygen atoms of the O_2-molecule to the substrate. Examples of dioxygenase enzymes are cyclooxygenase and catechol oxidase. Products typically obtained from such dioxygenations are peroxides (from olefins) or carboxylic acids (from catechols). The monoxygenation of an organic substrate with molecular oxygen is usually performed by monooxygenases, where only one of the oxygen atoms is transferred to the substrate and other one is reduced to water by a co-reductant. Hydrogen peroxide is a safe, readily available and cheap reagent. Catalysts are needed, which transfer oxygen from H_2O_2 to the substrate without effecting a disproportionation or a homolytic cleavage of the H_2O_2 molecule. In nature, peroxidases perform this desired cleavage of the H_2O_2 molecule. In many peroxidases, the catalitically active iron center is coordinated by the 4 pyrrole nitrogen atoms of its heme ligand plus an axial imidazole donor. Mn activates hydrogen peroxide in a catalytic manner. L-ascorbic acid is co-ligand that renders extremely high catalytic activity of Mn. A combination of Mn with ascorbic acid is also active in the oxidation of alcohols with hydrogen peroxide. Secondary alcohols are transformed to ketones, and primary alcohols are directly converted to the corresponding carboxylic acid. Hydrogen peroxide is an extremely advantageous source of oxygen. All living matter and synthetic organic materials within the biosphere share a common condition and fate; they are thermodynamically unstable with respect to CO_2 and H_2O_2 products of reaction with O_2. Dioxygen (O_2) is a diradical, possessing a triplet ground-state electronic configuration (two unpaired electrons), whereas almost everything else possesses singlet ground state (paired electrons only). Once dioxygen is activated, its reaction with organic substrates is difficult to control. This is because the diradical nature of dioxygen facilitates the formation of highly reactive and nonselective radical intermediates and radical-chain processes. Because of the thermodynamic instability, such oxidations, once begun, are often highly exothermic and the resulting temperature increases further decrease selectivity. Dioxygen has 4 oxidising electron equivalents. Reductive activation of O_2 can also be accomplished by in situ introduction of an external supply of electrons such as with the effective reduction of O_2 by the coenzyme NADH in oxidative phosphorylation. The catalyst (iron and copper) facilitates direct O_2 activation and substrate oxidation without the generation of nonselective radicals. Oxygen is incorporated directly into the substrate, the sole source of the electrons needed to the reductive activation of O_2. Although free radicals are essential part of normal body processes, too many are bad for us. An atom contains a nucleus, around which electrons move usually in pairs. A free radical is any atom or molecules that contains one or more unpaired electrons, i.e., electrons alter the chemical reactivity of an atom or molecule usually making it more reactive than the corresponding nonradical. The hydrogen radical (H$\cdot$, the same as a hydrogen atom), contains one proton and one electron, this is the simplest free radical. Free radical chain reactions often started by removing H$\cdot$ from other molecules. The growth of anaerobic microorganisms is inhibited and often can be killed by exposure to concentrations of 21% oxygen, the current atmospheric level. Mitochondria make over 80% of the ATP the cells need. All aerobic species suffer injury if exposed to oxygen at concentrations higher than 21%. Human breathing pure oxygen for as little as six hours can develop chest soreness, cough, and sore throat. Longer periods of exposure damage the alveoli of the lung. Radiation causes the splitting of one of covalent bonds in water to generate highly reactive radicals called hydroxyl radical, OH$\cdot$.

H-O-H → H$\cdot$ + OH$\cdot$

Hydroxyl radicals are always bad, as they attack proteins, lipids, and DNA within the cells to start off radical chain reactions, which can propagate for many years and cause damage. Some radicals can be useful. The free radical NO$\cdot$ helps to control blood pressure and certain cells in the brain, NO$\cdot$ is involved in memory. NO$\cdot$ also is made by WBCs, to help them kill foreign organisms. The body also makes superoxide radical, by adding a single electron to oxygen.

$O_2 + e^- \rightarrow O_2^{-}$

Autooxidizable molecules are adrenaline, noradrenaline, dihdroxyphenylalanine (DOPA), and reduced folates. These autooxidizable molecules are slowly chemically oxidised by oxygen and the oxygen is reduced to O_2^{-}. Superoxide radical can also be made deliberately, it is made by WBCs. Excess O_2^{-} and NO^{-} can lead to formation of OH^{-} and other toxic species. The superoxide dismutase (SOD) enzyme helps to get rid of superoxide by converting it to hydrogen peroxide. Hydrogen peroxide is then disposed of by other enzymes, catalase and glutathione peroxidase. The latter enzyme disposes of H_2O_2 by using it to oxidize a substrate found in all cells, glutathione (GSH). Iron and copper ions are powerful catalysts of free radical reactions because their variable oxidation numbers (Fe^{3+}, Fe^{2+}, Cu^{2+}, Cu^{+}) allow them to transfer single electrons. The human body needs these metals for the activity of many proteins, but they must be handled carefully to minimise free radical damage. Iron and copper ions are transported around the body attached to proteins. Keeping iron and copper ions protein-bound helps prevent them from catalysing free radical damage. Reactive free radicals destroy the membrane via lipid oxidation, by targeting the polyunsaturated fats (PUFAs).

$PUFA + OH^{-} \rightarrow PUFA^{-}$ (PUFA radical)

$PUFA^{-} + O_2 \rightarrow PUFAO_2^{-}$ (PUFA peroxyl radical)

$PUFAO_2^{-} + PUFA \rightarrow PUFA^{-} + PUFAO_2H$ (Lipid peroxide)

$PUFAO_2^{-} + \alpha\text{-TOH} \rightarrow \alpha\text{-TO}^{-} + PUFAO_2H$

The fat soluble vitamin E (α-TOH) from the food dissolves in cell membranes and can stop the chain reaction. The poorly reactive α-TO^{-} is recycled to α-TOH by a water soluble vitamin from food, vitamin C. Several plant phenolics such as the flavonoids can also scavenge $PUFAO_2^{-}$ radicals in the same way as α-TOH. Minute amounts of the selenium are needed for the enzyme glutathione peroxidase to destroy hydrogen peroxide. Glutathione peroxidase catalyses the formation of oxidised glutathione (GSSG), in which two GSH molecules are linked to a disulphide bridge. Reduced glutathione (GSH) is a tripeptide (glutamic acid-cysteine-glycine) that has multiple metabolic functions, which also involves NADPH. Polyunsaturated-rich cooking and margarines can easily oxidize to peroxides, causing unpleasant smells and tastes-rancidity, staleness, warmed-over flavours, and potential toxicity. Antioxidants both natural as vitamin E and synthetics such as butylated hydroxyanisole (BHA) and butylated hydroxytoluene (BHT) are often added to prevent this. BHA and BHT act in the same way as vitamin E scavenging the intermediate $PUFAO_2^{-}$ radicals and block chain reaction of lipid peroxidation.

OXIDATIVE STRESS

There are only 2 basic types of chemical reactions: oxidation and reduction. An oxidation reaction removes negative charge by removing electrons. A reduction reaction adds negative charge by adding electrons. For example, Iron + Oxygen (with electrons to contribute) = Iron reacted with Oxygen (electrons contributed to iron) or: Iron + oxygen + water = rust. Rust is mixed iron oxide and iron hydroxide. Iron reacted with oxygen is known as "rust". Iron is changed into something entirely different through this redox reaction with oxygen. Oxygen has the ability to accept electrons from atoms of other elements, thus combining with these other atoms, forming a new compound. Oxygen has this action with many different elements and molecules. Molecules are made from more than one atom. Anywhere oxygen comes into contact with an atom or molecule that has electrons to contribute (they are not tightly bound in the atom or molecule), oxygen will take those available electrons and transform that atom or molecule into a compound called an oxide. Oxygen was discovered in 1775. Organic free radicals were demonstrated in 1900. In 1934, Haber and Weiss proposed that hydrogen peroxide could yield hydroxyl radicals (OH.) in the presence of iron salt acting as a catalyst agent as previously suggested by Fenton. In 1963, attention was attracted to the possible toxicity of pressured oxygen in divers, and suggested the use of some antioxidants for protection in 1966. In 1968, vitamin E that was discovered in 1922 was recognised to be an essential vitamin because of its potential antioxidant properties. One year later, superoxide dismutase (SOD), enzyme present in erythrocytes and capable of destroying superoxide anion radical, was discovered. When not correctly metabolised, oxygen gives rise to the formation of reactive oxygen species (ROS) including free radical species (superoxide anion, hydroxyl radical, peroxyl and alkoxyl radicals) or not (hydrogen peroxide, singlet oxygen, hydrohalous acids). Physiological metals such as iron or copper acting as catalyst agents are strongly associated with the free radical chemistry. ROS are cytotoxic agents as a result of their ability: to inactivate proteins, to oxidize lipoproteins, to promote DNA strand scission, to degrade carbohydrates, to induce haemolysis and overall to initiate lipid peroxidation processes. Reactive nitrogen metabolites (free radical nitric oxide) and also sulphur-centered radical have been discovered. If a free radical activity suddenly increases as the consequence of either a primary (e.g., excess radiation exposure) or a secondary (e.g., tissue damage by trauma) event, the antioxidant defence will be weakened, and in case of too prolonged ROS production, rapidly overwhelmed. Oxidative stress occurs because there is profound disturbance in the proxidant-antioxidant balance in favour of the former leading to potential for damage. ROS have been implicated in over 100 pathobiological conditions, from cancer, arthritis, organ transplantation to AIDS, aging and neurodegenerative diseases. Free radicals are implicated in Parkinson's disease, taking vitamin E prevents heart attacks, and the anti-aging cream contains free radicals scavengers. Identifying oxidative stress in human disease, can be separated into nine categories: 1- Detection of oxidised biological material (lipid damage, protein and lipoprotein damage, DNA damage): Lipid present in high quantity in tissue, unsaturated fatty acids, phospholipids and cholesterol esters are the primary targets for free radicals resulting in the formation of lipoperoxidases (LOOH) that can be detected by gas chromatography-mass spectrometry or by high performance liquid chromatography (HPLC) separation coupled with luminescence detection. Proteins are also sensitive to oxidative damage that leads to alterations in their primary, secondary and tertiary conformation, which is reflected by changes in electrophoretic migration or in the fluorometric spectrum of the protein, by oxidation of thiol group. Reaction of ROS with amino groups such as lysyl residues leads to the formation of a carbonyl group measured by the reaction with dinitrophenylhydrazone as assessed by fluorescence spectroscopy. Interaction of free radicals, and more especially hydroxyl radical, with bases of DNA gives rise to some derivatives (8-hydroxy-guanine, thymine glycol) that can be detected in urine. Determination of DNA damage is time consuming and not easy to use in clinical situations. 2- Salicylate hydroxylation: Based on the ability of OH. to attack the benzene rings in aromatic molecules and to produce hydroxylated products that can be quantified by the HPLC procedure. 3- Measurement of endogenous antioxidants: Human plasma does not contain enzymatic antioxidants other than superoxide dismutase (SOD), but is rich in small molecular antioxidants being divided into: i- water soluble (uric acid, ascorbic acid, bilirubin, glutathione), and ii- lipid soluble (α-tocopherol, ubiquinol-10, α- and β-carotene). Red blood cells are very reach in enzymatic antioxidants (superoxide dismutase, glutathione peroxidase, catalase). Vitamin E loss occurs in trauma patients after admission to hospital. Antioxidant determination appears to be indicative of the individual for oxidative stress. 4- Determination of total antioxidant capacity of blood. 5- Evidence of free iron: Transition metals most particularly iron, play a key role in initiating free radical chemistry through the Fenton reaction. Almost all iron is physiologically stored in ferritin, lactoferrin and transferrin. Under the action of superoxide anion radical, it has been shown that iron could be released as free or in low molecular weight forms from stock proteins and then be able to participate in a free radical reaction. Haemoglobin in the presence of hydrogen peroxide can also release its iron, which is capable of initiating lipid peroxidation. 6- Neutrophil activation: PMNs are important source of activated oxygen species in biological systems. Activated PMNs also release into the extracellular medium enzymes such as elastase and myeloperoxidase. Abnormal concentrations of such enzymes have been found in patients with septic shock. 7- Bioluminescence. 8- Electron spin resonance (esr): Two radicals can be detected by esr in blood, plasma or tissue: i- ascorbyl radical, derived from the one-electron reduction of ascorbate has a prolonged life duration because of the delocalisation of the unpaired electron on the whole molecule, and ii- nitric oxide radical (NO.). 9- Antioxidant therapy: Antioxidants are defined as any substance that, when present at low concentration compared with those of the oxidizable substrate, significantly delays or inhibits oxidation of that substrate. They act as molecules: a- removing ROS after their generation (SOD, catalase, glutathione peroxidase), b- scavenging free radicals (glutathione, vitamin C, carotenoids, flavonoids), c- blocking the chain propagation of lipid peroxidation (vitamin E, probucol, lazaroids), d- inhibiting free radicals generating enzyme (allopurinol), e- increasing endogenous antioxidant capability of the cell (N-Acetylcysteine), and f- complexing iron (desferrioxamine, flavonoids). The use of allopurinol, for xanthine oxidase inhibiting, significantly decreases the incidence of poor early renal function.

ISCHAEMIA

Directly or indirectly, all cellular work is done at the expense of ATP, formed during glycolysis in the cytoplasm (a minor part) or during oxidative phosphorylation in the mitochondria (the major part). The central damaging event in ischaemia is ATP depletion. In the absence of ATP, ion gradients are dissipated, intracellular communication is halted, and macromolecules assemblies are broken down in reactions that are thermodynamically spontaneous. Although cell death is the end result of ATP depletion, however, such depletion may not be the only or, the direct cause. After anoxia, active transport or biosynthetic work can be done only at the expense of the available stores of phosphocreatine (PCr) and ATP, and of the ATP that is formed during anaerobic glycolysis. During glycolysis, two ATPs are formed from one glucose with the simultaneous release of two lactate and two H^+. After the sudden interruption of circulation, energy stores are depleted. PCr approaches zero during the first minute, and ATP is close to its minimal value after 2 minutes, e.g., brain. Glycolytically produced ATP is available to do transport work. Mitochondrial ATP production is almost nil in the absence of oxygen and probably poor even if some oxygen supply persists. Ischaemia inflicts damage to the mitochondria. Provided that ischaemia can be terminated, prolonged ischaemia can be followed by extensive and surprisingly rapid restoration of mitochondrial function. Most mitochondria can resume activity after 60 min of ischaemia. After complete ischaemia, changes in extracellular ion concentrations have two major phases: 1- Phase I, lasting 1.5-2 min, extracellular potassium rises slowly, extracellular calcium, sodium, and chloride remain unchanged or increase slightly. 2- Phase II, potassium increases abruptly, and calcium, sodium, and chloride decrease precipitously. Tissue impedance increases, reflecting a reduction in ECF volume to about 50%. The beginning of phase 2 occurs when the PCr and ATP concentrations approach zero. Membrane permeability to all participating ions is increased, i.e., shock opening of ion gates. These fluxes are unlikely to occur via the classical voltage-dependent channels that mediate spike discharge. The ATP that is present or made available during ischaemia is probably used mainly to pump ions against their thermodynamic gradients. Thus, if the amount of glycolytically produced ATP is reduced by perischaemic hypoglycaemia or increased by hyperglycaemia, the duration of phase I is shortened or prolonged, respectively. Ischaemia leads to an increased tissue lactate content, which is already significant after 5 sec and approximately linear during the first 60 sec. In complete ischaemia, anaerobic glycolysis is the dominating cause of acidosis during ischaemia. Hypoglycaemia is associated with reduced, and hyperglycaemia with increased tissue lactate content. During severe incomplete ischaemia, the lactate acidosis depends on 2 factors: the plasma glucose concentration and the amount of glucose delivered by the residual blood flow. Lactate accumulations of comparable severity when arterial hypoxia is combined with a reduction in cerebral perfusion pressures, e.g., due to arterial hypotension. Intracellular pH (pH_i) values of 5.5 or lower may be reached under unfavourable conditions. Changes in extracellular pH (pH_e) and PCO_2 are also greater in hypoglycaemic than normoglycaemic individuals. During ischaemia, pHe is rapidly reduced. When recirculation provides O_2 for resumed oxidation phosphorylation, it also sets the stage for normalisation of pH_i and pH_e. Hydrogen is rapidly extruded when the Na^+_e/Na^+_i gradient is restored. Such overregulation is most easily explained by persisting stimulation of the Na^+/H^+ antiporter by activation of phospholipase C and C-kinase. Postinsult alkalosis may also occur after transient ischaemia. Mechanisms that cause membrane dysfunction and cell death are, calcium, osmolysis, free radicals, and acidosis: 1- Calcium: Ischaemia causes increased calcium influx into cells because depolarisation and transmitter release open voltage-sensitive Ca^{2+} channels and agonist-operated calcium channels, and because energy failure prevents ATP-dependent calcium extrusion. Loss of ATP combined with anoxia prevents the mitochondria from sequestering Ca^{2+} in the absence of ATP, sequestration of Ca^{2+} by endoplasmic reticulum is also compromised. Since calcium influx enhances breakdown of proteins and lipids, it will adversely affect cytoskeletal and membrane functions, and cause catabolism and eventually, cell death. A breakdown of cytoskeletal components and loss of ordered membrane functions could occur in the absence of a rise in Ca^{2+}_i, and be due to a decreased ATP. Since lipolysis and proteolysis occur in the absence of a rise in Ca^{2+}_i, such a rise may merely accelerate breakdown of structure. Lipolysis during shortlasting ischaemia is due to agonist-dependent activation of phospholipase C rather than to Ca^{2+} activation of phospholipase A_2. 2- Free radicals: The important production sites of such radicals as the superoxide and hydroxyl species are the mitochondrial respiratory chain and the sequences catalysed by cyclooxygenase and lipoxygenase. However, radicals also are formed during auto-oxidation of many compounds (catecholamines) and in the xanthine oxidase reaction. Several ischaemic events favour a spurt of free-radical formation, e.g., those causing oxidation of polyenoic free acids release and reuptake of catecholamines and oxidation of hypoxanthine by xanthine oxidase. Stimulation of phospholipase C and ATP shortage cause lipolysis and accumulation of free fatty acids, the substrates for cyclooxygenase and lipoxygenase during recirculation. Calcium-activated proteolysis may cause a conversion of xanthine dehydrogenase to xanthine oxidase, allowing postischaemic oxidation of hypoxanthine with production of superoxide. Iron in the form of Fe^{3+} is normally relatively tightly bound to proteins such as ferritin and transferritin. Reduction of superoxide may cause the iron to be released and chelated to compounds of lower molecular weight to yield prooxidant species. Such iron-chelated species increases during recirculation (postischaemia). 3- Acidosis is due to the anaerobic metabolism of glucose to lactate in the ischaemia that leads to oedema. Acidosis enhances free-radical reaction, and acidosis may enhance such reactions either by converting superoxide to its hydrated, more lipid soluble, and more prooxidant form, or by releasing iron from its binding sites in transferrin-type proteins. Cells that possess coupled antiporters for $2Na^+/H^+$ and Cl^-/HCO_3^- exchange may regulate pH_i at the expense of volume regulation. In other words oedema may couple to acidosis. Within cells actin exists in either a monomeric form designated G-actin or in polymeric form termed filamentous or F-actin. G-actin exists as a diffuse pool of readily available monomers concentrated at areas of high actin utilisation. Filaments mediate actin-dependent cellular processes. They are arranged into higher-order forms including a cortical actin network (actin gel), stress fibers (loose nonparallel bundles), and tight parallel bundles seen in microvilli or filopodia (fingerlike surface membrane extensions). These F-actin structures have different functions, lengths, and cellular location. Their generation and maintenance is a dynamic energy-dependent process requiring actin assembly and disassembly. Specificity and the spatial and structural properties of the actin cytoskleton are mediated by an ever-increasing group of proteins, called actin effector proteins (either actin-associated proteins or actin-binding proteins). Effector proteins can be divided on the basis of their primary functions into polymerisation effectors, structural proteins, and motors. In cells the ratio of G-actin to F-actin is about 1:1, even though the thermodynamic equilibrium in the cytoplasm for actin polymerisation markedly favours polymerisation. High concentrations of G-actin are necessary to provide the substrate for rapid polymerisation. Maintenance of high G-actin concentrations and the state of actin assembly depend on 4 broad groups of polymerisation effector proteins. Sequestration proteins bind to G-actin and inhibit polymerisation by reducing the effective or free G-actin concentrations. Capping proteins, by binding to the filament end, stabilise either the barbed end (rapid-growth end) or the pointed end. Severing proteins mediate filament cleavage and nucleation proteins may mediate and enhance actin filament formation. The coordinated interactions of these effector proteins result in the continual dynamic remodelling of the cytoskeleton and in the ability to respond rapidly to external and internal stimuli. Filament growth at the barbed end requires displacement of capping protein and the presence of free (unsequestered) G-actin above its critical concentration for polymerisation. Many actin polymerisation effector proteins have multiple functions. Actin cytoskeletal-surface membrane interactions mediating such diverse cellular events as cell polarity, endocytosis, exocytosis, cell division, cellular migration, cell adhesion, signal transduction, and ion channel activity are part of an ever-growing list of cellular processes dependent on precise polarisation and regulation of assembly and disassembly. Ischaemia and the actin cytoskeleton: Ischaemia-induced alterations in actin cytoskeletal effector proteins are in part responsible for the rapidly occurring duration-dependent alterations in cellular actin that occur during ATP depletion or ischaemic cell injury. The release of G-actin from the monomer sequestering protein thymosin-β_4 is in part responsible for enhanced actin polymerisation during ischaemia or cellular ATP depletion. During cellular recovery, as ATP increased, the actin cytoskeleton returns to its normal polymerisation state accompanied by resequestration of G-actin by thymosin-β_4. Disruption of ezrin binding to microvilli surface membranes during ischaemic injury may in part be responsible for microvilli collapse, which is related to the dephosphorylation of ezrin and its dissociation from F-actin. Myosin I alterations during ischaemic injury may also play a role in the microvilli collapse. Spectrin plays an important role in organising and regulating the membrane cytoskeleton. During reversible ATP depletion, spectrin dissociates from actin and the membrane linker protein ankyrin, becomes Trinton X-100 soluble and redistributing throughout the cytosol. Thus previously tethered surface membrane proteins, such as a Na^+-K^+-ATPase, are released and free to diffuse within the bilayer through an open tight junction resulting in loss of surface membrane polarity. During ATP repletion, the dissociated spectrin is reutilised in surface membrane repair. During ischaemia and ATP deletion, renal proximal tubule cells detach. Ischaemia-mediated inactivation of R-ras results in reduced integrin adhesion. Growth factors may mediate this effect through signal transduction pathways involving actin-regulated processes.

POSTISCHAEMIA

The postischaemic injury processes that lead to cell dysfunction and death are multifactorial in nature and include oxidant generation, elaboration of inflammatory mediators, infiltration of leukocytes, Ca^{2+} overload, phospholipid peroxidation and depletion, impaired nitric oxide metabolism, and reduced ATP production. Extracellularly generated oxidants enter endothelial cells and interact with low-molecular-weight iron chelates and/or heme-associated iron. Ferritin does not promote the formation of hydroxyl radical in the presence of physiological intracellular buffers. Leukocytes may play an important role in genesis of ischaemia-reperfusion injury. Leukocytes can produce reactive O_2 metabolites, release cytotoxic enzymes, and adhere to and occlude microcirculatory vessels. Leukocytes stimulates the release of cytotoxic arsenal, as well as their participation in the development of capillary no-flow, many contribute to postischaemic cell damage. Leukocyte elastase may degrade basement membrane and facilitate diapedesis and extravasation in response to proinflammatory mediators. Ischaemia stimulates the release of leukocyte elastase. Leukocyte-generated OFR contributes to membrane injury, microvascular dysfunction, and tissue necrosis during reperfusion. Neutrophil-dependent injury can occur in organ sites distant from the tissue subjected to the ischaemic insult. Neutrophil sequestration in the lungs leads to the development of noncardiogenic pulmonary oedema. Leukocyte adhesion to endothelium involves several different adhesion molecules expressed by both endothelium and granulocytes, including the leukocyte β-integrins (CD11/CD18), a member of the immunoglobulin gene superfamily (I-CAM-1), and the selectin family (L-selectin). Spontaneously, leukocyte rolling on unstimulated endothelium is mediated by P-selectin. The initial adhesive interaction between neutrophils and stimulated endothelium involves upregulation of adhesion molecules (E-selectin and P-selectin) on the surface of the endothelium adjacent to sites of inflammation. Neutrophilic L-selectin appears to mediate the initial adhesive interaction (leukocyte rolling and reversible sticking) between nonactivated neutrophils and the stimulated endothelium by interacting with P-selectin and/or E-selectin. The initial adhesion mediated by L-selectin is downregulated by release of this leukocyte-associated selectin. CD11/CD18 binding to ICAM-1 is required for transendothelial migration and potential oxidative metabolism and neutrophil granulation. Early leukocyte rolling and adhesion following ischaemia and reperfusion appear to be mediated by P-selectin. E-selectin plays an important role in the more delayed adherence reactions induced by ischaemia-reperfusion. Cytoskeleton, microfilament-rich endothelial cell lamellopodia extend luminally and envelope leukocytes as they emigrate. The accumulation of leukocytes in damaged tissues is mediated by locally generated soluble factors that stimulate adherence and chemotaxis. The principal chemoattractants are derived from arachidonic acid (AA) metabolism, superoxide-modified plasma lipids, endothelial cells, and complements activation. Activation of the complement cascade of plasma proteins results in the formation of the biologically active peptides C3a and C5a. These peptides cause increased capillary permeability and vasodilation and stimulate leukocyte chemotaxis, CD18-receptor upregulation, adherence to endothelium, OFR production, and degranulation. Complement activation is potentiated by ischaemia. Reperfusion causes a rapid decrease in systemic levels of factor B, a component of the alternative complement cascade. Alternate pathways activation may be modulated by regulatory proteins following cleavage of factor B. Influx of Ca^{2+} into postischaemic cells may cause phospholipase activation and mitochondrial dysfunction. Phospholipids are the principal structure component of cell membranes. The fatty acyl composition of the phospholipids modulates the function of membrane receptors, enzymes, and ion channels and therefore a critical determinant of cell viability. The content and composition of phospholipid fatty acids may be altered by ischaemia and reperfusion, which causes oxidant-mediated lipid peroxidation, activation of phospholipases, and decreased activity of both reacylation enzymes and de novo phospholipid synthesis. The resulting changes in membrane phospholipids may lead to increased Ca^{2+} permeability, integral protein failure, and cell death. Plasmalogens are the predominant type phospholipids in electrically active tissue, such as cardiomyocytes and neurons, may function as a repository of AA for eicosanoid biosynthesis, as a preferential targets of free radical-mediated lipid peroxidation, or by modulating membrane enzyme and ion-channel function. Cytochrome-c oxidase, the terminal enzyme complex of the mitochondrial respiratory chain, is composed mainly of linoleic acid. Because ischaemia and reperfusion causes peroxidation of linoleic acid, this may adversely affect cytochrome-c oxidase function. Glucose-6-phosphatase and Ca^{2+}-ATP also appear to be dependent on an intact phospholipid bilayer. Ongoing free radical mediates lipid peroxidation, phospholipase activation; decreases activity of the reacylation enzymes lysophosphatidylcholine acyltransferase and fatty acid coenzyme A synthase; or decreases availability of fatty acylcoenzyme A due to the lack of phosphorylated adenine nucleotides. Reperfusion is associated with an increase in phospholipid stearic acid content. These compositional changes, in concert with net phospholipid depletion, may adversely affect membrane function and cell viability during reperfusion. Reperfusion of ischaemic tissue leads to eicosanoid synthesis. Stimulated neutrophils may be a source of eicosanoids. Leukotriene synthesis may stimulate neutrophils by inducing thromboxane A_2 (TXA_2) production. Intravascular coagulation may limit delivery of O_2 and removal of metabolic waste products during reperfusion, and contributes to capillary occlusion and localised no-reflow. Prolonged reperfusion injury may be the result of ongoing leukocyte sequestration, phospholipid peroxidation, depletion, and remodelling, or impaired adenine nucleotide resynthesis. Degradation of phosphorylated adenine nucleotides during ischaemia leads to the depletion of energy stores and the formation of the xanthine oxidase substrates hypoxanthine and xanthine. One of the enzymes that mediate nucleotide catabolism is adenosine deaminase, which catalyses the deamination of AMP to IMP (inosine monophosphate). Deamination of the AMP to IMP impairs ATP resynthesis and functional recovery during reperfusion of ischaemia. Necrosis appears to be the degree of ATP catabolism past IMP to dephosphorylated metabolites during ischaemia. Nucleotide dephosphorylation leads to the formation of lipid soluble bases that can diffuse out of cells during reperfusion, resulting in precursor depletion. The toxicity of NO is thought to be mediated by binding to cytosolic or mitochondrial iron-containing enzymes, which may interfere with cytochrome function and DNA synthesis; or by formation and decomposition of peroxynitrous acid to form hydroxyl radicals, nitrite, and nitrate. The decrease in NO activity during reperfusion may be related to inactivation by the superoxide anion. NO decrease leukocyte sequestration and necrosis. IL-1 may potentiate leukocyte-mediated injury by causing lysozyme release; and stimulating superoxide production, and neutrophil chemotaxis and adherence to endothelium; possibly by upregulating endothelial adhesion molecule expression. TNF-α may participate in postischaemic injury by stimulating leukocyte OFR production, upregulating leukocyte and endothelial adhesion molecule expression, and blocking NO production and acetylcholine-dependent vasodilation. IL-6 is elevated during ischaemia/reperfusion. TGF-β is an angiogenic peptide that inhibits neutrophil adhesion to endothelial cells, deactivates macrophages, and tends to oppose the actions of TGF-α. The pathobiological cascade activated by ischaemia and reperfusion include: 1- complement activation, 2- eicosanoid, 3- cytokines, 4- NO biosynthesis, 5- OFR, 6- cytoskeletal alterations, 7- adenine nucleotide depletion, 8- altered Ca^{2+} and phospholipid metabolism, 9- leukocyte activation, 10- endothelial dysfunction, and 11- change in DNA and RNA metabolism. Reperfusion of ischaemic tissues is necessary to restore its normal function, however, it paradoxically initiates poorly understood processes that lead to tissue injury and organ dysfunction. Impaired mitochondrial function and increased production of O_2^- are very common reperfusion-associated events.

MITOCHONDRIAL DNA (mtDNA)

Mitochondrial DNA (mtDNA) is more susceptible than nuclear DNA to mutations. The oxidative phosphorylation (OXPHOS) system consists of 5 protein-lipid enzyme complexes located in the mitochondrial inner membrane complexes I, II, III, IV (cytochrome c oxidase) and V (ATP synthase). The enzymes contain flavins, CoQ_{10} (ubiquinone), iron sulfur clusters, hemes, and protein-bound copper. Complexes I and II collect electrons from the catabolism of fats, proteins and carbohydrates and transfer them to CoQ_{10}, complex III and complex IV. The human mtDNA is contained within mitochondria, which are located in the cytoplasm. The mDNA is a 16569 nucleotide pair, double stranded, circular molecule that codes for 2 ribosomal RNAs (rRNA), 22 transfer RNAs (tRNA), and 13 polypeptides that, together with polypeptides coded by the nuclear DNA, form complexes I, III, IV, and V. Only complex II is encoded entirely by the nuclear DNA. The cytoplasmic location of the mtDNA is associated with a unique inheritance, which refers to the transmission of mtDNA exclusively from a mother to her children. Normal and mutant mtDNA sequences differ only at the nucleotide or nucleotides that have mutated. Somatic mtDNA mutations are characterised by spontaneous mtDNA rearrangements, point mutations, and free-radical-mediated nucleotide modifications that increase with age in a variety of tissues. Free-radical-mediated damage to the mtDNA results from its susceptibility to oxygen radical production, lack of protective histones, and poor mtDNA repair mechanisms. Ageing is associated with a decline in exercise capacity, accumulation of complex IV-deficient fibres in various muscle groups, and declining OXPHOS function in variety of human tissues. The age related increase in somatic mtDNA mutations might be important during aging as well as cell death in variety of degenerative diseases. Increased concentrations of lactate, pyruvate, and alanine, and/or a generalised aminoaciduria can be important diagnostic clues to the presence of an OXPHOS disease. OXPHOS diseases' muscle biopsy is characterised by proliferation of subsacrolemmal mitochondria (ragged-red fibers), detected by Gomori trichrome stain, succinate dehydrogenase stain, and by abnormal cytochrome c oxidase (complex IV) histochemical reaction. Structurally abnormal mitochondria containing paracrystalline inclusions, condensations of mitochondrial creatine kinase between the mitochondrial inner and outer membranes may be seen by electron microscopic ultrastructural study. Pathogenetic mtDNA may be more prevalent among adults than in children. mtDNA mutations are detected in a third of the adults, and 4% of children. Mutations in nuclear OXPHOS genes may account for a larger percentage of paediatric OXPHOS disease cases. There are 3 classes of pathogenetic mtDNA mutations: 1- mtDNA rearrangement, 2- mtDNA point mutations in tRNA or tRNA genes, and 3- missense mutations that change an amino acid.

REOXYGENATION INJURY

When oxygen is reintroduced after hypoxia-ischaemia, arachidonic acid (AA) is metabolised, and OFRs are produced when prostaglandin G_2 (PGG_2) is converted to prostaglandin H_2 (PGH_2). Purine metabolism halts and hypoxanthine accumulates during asphyxia. Xanthine oxidase uses oxygen in the metabolism of hypoxanthine, which produces OFRs. Non-enzymatic electron transfer in the reaction between oxygen and monoamines, ascorbate, semiquinones and haemoglobin are examples of autooxidation. Semiquinone, is a free radical derived from quinones or quinone imines by addition of a single H atom to a molecule. Quinones, are any benze derivatives in which 2 hydrogen atoms are replaced by 2 oxygen atoms. Ubiquinones (Coenzyme Q) are any group of related quinone with isoprenoid units in the side chain (ubiquinones), occurring in the lipid fraction of mitochondria and serving along with cytochromes, as an intermediate in electron transport. They are similar in structure and function to the vitamin K_1. Respiring cells are capable of reducing molecular oxygen (dioxygen, O_2) by 4 electrons to water (H_2O) by way of cytochrome C oxidase, without forming any free intermediary species of reduced O_2. O_2 can be reduced by only one, two, or three electrons giving rise to the superoxide radical (O_2^-.), hydrogen peroxide (H_2O_2), and the hydroxyl radical (OH.), respectively. The superoxide radicals can be formed both enzymatically and nonenzymatically. Enzymatic pathways, occurs in the cytosol, in which xanthine oxidase (XO) catalyses the oxidation of hypoxanthine to xanthine and uric acid, with the formation of superoxide radical (O_2^-.). Both the xanthine dehydrogenase and the oxidase can convert xanthine to uric acid, but only the xanthine oxidase can at the same time transfer an electron to molecular oxygen to form the superoxide radical. Polymorphonuclear leukocytes and other phagocytic cells contain an NADPH oxidase, which catalyses the reduction of O_2 to superoxide radical. The enzyme is inactive in the resting state of the cells, but can undergo receptor-mediated activation when the cells adhere to bacteria or other cells. Nonenzymatic formation of O_2^-. can take place through autooxidation of semiquinones and other redox-labile compounds (e.g., Fe^{2+} chelates). Quinones of both endogenous and exogenous origin can undergo one-electron reduction through a number of NADH or NADPH-linked flavoenzymes, e.g., mitochondrial NADPH dehydrogenase or the mitochondrial NADH-cytochrome b 5 and NADPH-cytochrome P-450 reductase, giving rise to semiquinones that are readily oxidised by O_2 with the formation of superoxide radical. Normally the extent of O_2^-. production due to this type of redox cycling is relatively small, <2% of the total O_2 consumed by respiration, and originates from some components of the mitochondrial electron-transport system in the region of ubiquinone. The extent of O_2^-. production can greatly increase, in the presence of certain exogenous agents, especially quinones (e.g., Adriamycin) can constitute a major source of O_2^-. generation. Removal of O_2^-. takes place by dismutation, either spontaneously or through superoxide dismutase (SOD):

$$2O_2^-. + 2H \rightarrow O_2 + H_2O_2 \qquad [1]$$

There are 2 isoenzymes of SOD, a Cu-Zn enzyme in the cytosol, and Mn enzyme in the mitochondria. Cu-Zn enzyme catalyses the reaction:

$O_2^- + QH.^+ \rightarrow O_2 + QH_2$ (Q is a quinone) [2]

O_2^-. can serve as an electron donor for mobilisation of Fe from ferritin:

ferritin-$Fe^{3+} + O_2^-. \rightarrow$ Ferritin + $Fe^{2+} + O_2$ [3]

which may be a physiological function of O_2^-.

Hydrogen peroxide formed through SOD is removed by catalase and various peroxidases:

$2H_2O_2 \rightarrow 2H_2O + O_2$ [4]

or by glutathione (GSH) peroxidase

$H_2O_2 + 2 GSH \rightarrow 2H_2O + GSSG$ [5]

Glutathione peroxidase is located partly in the mitochondria and partly in the cytosol.

Formation of hydroxyl radical (OH.) may occur nonenzymatically through the Fenton reaction:

$Fe^{2+} + H_2O_2 \rightarrow Fe^{3+} + OH^- + OH.$ [6]

The sum of reaction [3] and reaction [6] gives the Haber-Weiss reaction:

$O_2^-. + H_2O_2 \rightarrow OH^- + OH.$ [7]

The effects of oxygen-radical generation on cell components and function, may be primary or secondary. The primary effects are those to perferryl-radical-initiated lipid peroxidation, result in lipid peroxidation and ends in DNA strand-breaks. Peroxidation of membrane phospholipids may result in severe cell injury owing to disruption and lysis of intracellular organelles and of the cell as a whole. Hydroxyl radical OH. effects are not dependent on Fe per se, but the formation of the OH. itself through the Fenton and Haber-Weiss reactions requires Fe. The hydroxyl radical is a powerful oxidant of certain amino acid residues and purine and pyrimidine bases in proteins and nucleic acids, respectively, thereby causing enzyme inactivation and DNA strand-break. Secondary effects consist of a disturbance in cellular Ca^{2+} homeostasis, leading to an increase in cytosolic Ca^{2+} and a consequent activation of cytosolic proteases and phospholipases. The latter effects in turn cause breakdown of the cytoskeleton and activation of enzymes, such as xanthine and NADPH oxidase, which give rise to further oxygen-radical formation. Intracellular Ca^{2+} is sequestered in the endoplasmic reticulum (ER) and in the mitochondria. The Ca^{2+} gradients in the cytosol in relation to the extracellular space and ER are maintained by Ca^{2+}-pumping ATPase, whereas the sequestration of Ca^{2+} by the mitochondria operates via an electrogenic uniporter located in the mitochondrial inner membrane and control mechanism, involving various hormones and other factors, such as inositol lipids, the function of which is to regulate the activities of Ca^{2+}-dependent cytosol enzymes such as adenylate cyclase, certain proteases, and phospholipases. Oxidative stress may upset the intracellular Ca^{2+} distribution by depleting the level of GSH through glutathione peroxidase. This results in: 1- release of the Ca^{2+} from the ER, due to inactivation of the Ca^{2+}-ATPase; and 2- release of Ca^{2+} from the mitochondria, due to an oxidation, via glutathione reductase and nicotinamide nucleotide transhydrogenase, of the mitochondria NAD(P)H, and consequent activation of a mitochondrial Ca^{2+}-efflux mechanism. As a result, the cytosolic Ca^{2+} level will increase. Increased cytosolic Ca^{2+} also may result from an influx of extracellular Ca^{2+}, as caused by an inactivation of the plasma membrane Ca^{2+}-ATPase, which also may occur when the cellular GSH is depleted. In the case of ischaemia, the increased in cytosolic Ca^{2+} will have the following effects: 1- An activation of cytosolic Ca^{2+}-dependent proteases. One of these proteases causes a breakdown of the cytoskeleton, leading to severe damage, blebbing of the cell surface. Another cytosolic protease activated by Ca^{2+} is calpein, which causes a conversion of xanthine dehydrogenase to xanthine oxidase, responsible for reoxygenation injury. 2- An activation of phospholipase A_2, with consequent release of AA from phospholipids, promoting the synthesis of PGs, leukotrienes, and thromboxanes. Leukotrienes can stimulate adhesion of WBC to the endothelium. In damaged endothelium, this can produce an activation of the O_2^-. producing NADPH oxidase. Thus, the increased cytosolic Ca^{2+} results in enhanced O_2^-. production by both enzymatic pathways, xanthine oxidase and NADPH oxidase. Primary and secondary effects of oxygen radical form a highly interconnected synergistic network of reactions. Thus, in a complex system such as a cell, a tissue, or a whole organism, it is difficult to predict the final outcome. Living cells possess powerful enzymatic and nonenzymatic mechanisms, by which they remove or prevent the formation of ROS. SOD, with its localisation in those cellular compartments, mitochondria and cytosol, where O_2^-. may be formed, is generally recognised as the key enzyme responsible for the removal of this radical. Catalase, peroxidases, and glutathione reductase function to remove H_2O_2. Glutathione reductase is located in the same cellular compartments as SOD, to remove the H_2O_2 produced by SOD. Glutathione reductase also reacts with organic peroxides and thus may play an important role in the regulation of the arachidonic acid (AA) cascade. Also exists special phospholipid hydroperoxide glutathione peroxidase, for detoxification of peroxidised membrane phospholipids. Nonenzymatic protective devices against oxygen-derived and other free radicals include the classical antioxidants α-tocopherol, β-carotene, retinol, ascorbic acid, uric acid and bilirubin. The reduced form of ubiquinone may serve as a built-in antioxidant in the mitochondria and other biologic membranes. The dimeric flavoprotein of DT diaphorase widely distributed in tissues and occurring mainly in the cytosol, with minor fractions of the enzyme located in mitochondria and the endoplasmic reticulum. It is unique among known NAD(P)H-oxidising flavoproteins in being a two-electron transforming quinone reductase, and plays as such a major role in preventing one-electron reduction of exogenous quinones by other enzymes to autoxidable semiquinones and concomitant O_2^-. generation. Two major pathways considered responsible for the reoxygenation injury: 1) involves xanthine oxidase, and 2) NADPH oxidase. Common to the 2 pathways is the instrumental involvement of an increased cytosolic Ca^{2+} level in the oxygen-radical generation. The use of Ca^{2+}-channel blockers may be too late at this stage if the increase in cytosolic Ca^{2+} occurs during ischaemia. An iron chelator such as deferoxamine blocks lipid peroxidation and hydroxyl-radical formation. It may penetrate the cellular and intracellular membranes too slowly to prevent certain types of damage, for example nuclear DNA strand-breaks caused by OH, or Fe binding to DNA, and an ensuing inhibition of protein synthesis. Allopurinol and related inhibitors are obvious relevance in the case of xanthine oxidase. The 2 pathways may closely interact with each other.

NITRIC OXIDE

Nitric oxide (NO) is synthesised in the body by: 1- endothelium, is responsible of vasodilator tone; 2- central nervous system, is a neurotransmitter; 3- peripheral nervous system, mediates some form of neurogenic vasodilatation and regulate various GIT, respiratory, and genitourinary tract functions; and 4- generated by activated macrophages, it is likely to have a role in nonspecific immunity. NO contributes to the control of platelet aggregation and the regulation of cardiac contractility. Nitric oxide released from nonadrenergic, noncholinergic terminals may contribute to the regulation of blood flow and pressure. Nitroglycerine and sodium nitroprusside act after their conversion into NO. The reaction of NO with the ferrous iron in the heme prosthetic group of the soluble guanylate cyclase in vascular smooth-muscle cells increases the concentration of cGMP, leading to vascular relaxation. Haemoglobin, a potent inactivator of NO, binds to it by a similar mechanism. Nitric oxide also inhibits platelet aggregation by a mechanism dependent on cGMP and synergizes with prostacyclin, which inhibits the aggregation of platelets by increasing their concentrations of cAMP. NO also inhibits platelets adhesion. Platelets themselves generate NO, which acts as a negative-feed-back mechanism to inhibit platelet activation. NO may also be involved in the interaction of leukocytes with vessel walls, since it inhibits leukocyte activation. The L-arginine-NO pathway is responsible for the relaxation of the corpus cavernosum and thus the development of penile erections in humans. Inhibiting the synthesis of NO induces hyperactivity in the urinary bladder and decreases bladder activity. Activated macrophages synthesise nitrite and nitrate. This generation is dependent on L-arginine, its inhibition is responsible for the cytotoxicity of these cells against tumour cells and bacteria. The biochemical basis for the cytotoxicity induced by NO depends on the combination of NO with iron-containing moieties in key enzymes of the respiratory cycle and of the synthesis of DNA in the target cells. The L-arginine-NO pathway constitutes a primary mechanism of defense against tumour cells and intracellular microoganisms. NO is a colourless gas that, in the absence of oxygen, dissolves in water. In air, NO reacts rapidly with oxygen to form nitrogen dioxide, a brown gas capable of inducing tissue damage. At low concentrations, NO is fairly stable, even in the presence of oxygen. This stability, together with the very high affinity of haemoglobin for NO (about 3000 times that of oxygen), allows NO gas to be given by inhalation, since it will combine with haemoglobin before combining with oxygen. In water, ultrafiltrate, and plasma, NO is oxidised to nitrite, which is stable for several hours. In whole blood, nitrite concentrations are low, and those of nitrate are 100 times higher. The endogenous synthesis of nitrite is markedly enhanced in diarrhoea and fever and doubles during exercise. Plasma concentrations of nitrite and nitrate are elevated in patients with septic shock. NO is also rapidly oxidised to higher oxides of nitrogen, nitrosates molecules containing sufhydryl groups, such as glutathione, cysteine, and albumin. The kinetics of the association and dissociation of these molecules, retaining or yielding NO, may vary such that some of them may act as carrier or as biological sinks for NO. NO interacts with other heme-containing proteins, including myoglobin, the prosthetic group of the soluble guanylate cyclase, and enzymes containing iron-sulfur centers. The fate of NO in the body is likely to be complex. NO is an important mediator of homeostatic processes and host defense mechanisms. Inhaled NO proved beneficial in the treatment of pulmonary hypertension and, infant and ARDS. The biological chemistry of nitric oxide can largely summarised by 3 principal reactions: 1- binding to ferrous heme in guanylate cyclase to activate cGMP; 2- elimination of oxyhaemoglobin; and 3- formation of toxic oxidants by reacting with superoxide. These 3 reactions are rapid under physiological conditions. NO is a ubiquitous intracellular messenger; modulating blood flow, thrombosis, and neural activity. NO is important for nonspecific host defense, helping to kill tumours and intracellular pathogens. NO also can be highly damaging when produced in 10- to 100-fold higher concentrations after only a few minutes, during pathological events like cerebral ischaemia. NO is so useful as an intracellular messenger because it can diffuse through most cells and tissues with little consumption or direct reaction. The toxicity of NO is more likely to result from the diffusion-limited reaction of NO with superoxide (O_2^-.) to produce the powerful and toxic oxidant peroxynitrite ($ONOO^-$). NO is an uncharged molecule composed of 7 electrons from nitrogen and 8 electrons from oxygen. Molecular oxygen contains 2 unpaired electrons and clearly, tissue survives containing between 20-200 µM dissolved oxygen. The majority of biological molecules contain bonds filled with 2 electrons, which makes them unreactive with NO. NO only reacts rapidly with a select range of molecules that have orbitals with unpaired electrons, free radicals, and with transition metals like heme iron. NO does not react rapidly with most biological molecules at the dilute concentrations, the reaction with oxygen to form nitrogen dioxide is not likely to be significant in vivo, because diffusion of blood vessels is far more rapid and removes NO by reaction with oxyhaemoglobin in RBCs.

.O=O.   oxygen

.N=O   nitric oxide ↔ 1/2 O=N-N=O   dinitrogen dioxide

:N≡N: molecular nitrogen

NO remains as a free radical rather than dimerizing to form dinitrogen dioxide. NO is intermediate between nitrogen, with a bond order of 3, and oxygen, with a bond order of 2. NO produced in an endothelial cell will diffuse out far faster than it will react with most intracellular component. NO binds to and activate guanylate cyclase. If it diffuses into a RBC, NO will be eliminated by oxyhaemoglobin (Hb-O_2) to form nitrate (NO_3) plus methaemoglobin. The reaction with superoxide to form peroxynitrite is normally limited by the micromolar concentrations of superoxide dismutase in cells. Copper, zinc superoxide dismutase has only a slight effect on the loss of nitric oxide. When enough NO is added to react with all of the haemoglobin, then NO is as stable in the presence of methaemoglobin as a buffer alone. Plasma itself does not consume NO. The diffusion of NO is sufficiently rapid that NO produced in tissues is more likely to diffuse into vascular compartment and be destroyed by haemoglobin than to react with normal constituents in tissues. Clearly oxygen is able to diffuse over 100 μm from a blood vessel to reach mitochondria in tissues. Therefore, NO which has a 1.4-fold larger diffusion coefficient at 37°C than oxygen, can diffuse over 100 μm back to a blood vessel within a few seconds. If a tank of pure NO is allowed to leak into air, a cloud of lethal and highly reactive orange-brown nitrogen dioxide forms by the following reaction:

2 .NO + O_2 → 2 .NO_2

Nitrogen dioxide is a strong and toxic oxidant, 2 nitrogen dioxides on 2 additional NO forms dinitrogen trioxide (N_2O_3), a strong one- and two-electron oxidant. The N_2O_3 can also react with water to give nitrite molecules:

NO + NO_2 → N_2O_3 + H_2O → 2NO_2^- + 2H^+

Nitrite is a major decomposition product of NO in simple buffers. Albumin or other proteins, when added to NO, the apparent half-life of NO increases, because nitrogen dioxide reacts directly with the protein rather than with a second NO. Half-life is not a constant value and is inversely proportional to the concentration of NO, so the half-life becomes much longer as NO becomes dilute. Atmospheric chemists have long known that the oxidation of nitric oxide is an insignificant process. Concentrations of NO that cause vasodilation are not highly toxic. Superoxide reacts with NO 7,000 times faster than cytochrome c. Superoxide dismutase is less efficient than NO as a scavenger of superoxide. The constant removal of NO by haemoglobin and other reactions will allow superoxide dismutase to successfully reduce peroxynitrite formation by limiting the accumulation of NO. The steady-state concentration of any molecule is a balance between the rate of formation and the rate of removal. Without oxyhaemoglobin or some other drain to remove NO, the concentration of NO will rise to the micromolar range, at which nitrogen dioxide will be quantitatively formed. When either haemoglobin or myoglobin is present, even when packaged in RBCs located several microns away in blood vessel, the half-life of NO will be much shorter. The half-life is largely determined by the time needed to diffuse to a blood vessel. Sodium nitroprusside attacks thiols and releases cyanide, cyanide can also inhibit cell proliferation. Sodium nitroprusside also can form complexes with many metalloproteins, including superoxide dismutase. Nitrosothiols release only minute amounts of NO in metal free buffers, nitrosothiols is a much better thiol-modifying agents than NO itself. Activation of cAMP-dependent kinase by cGMP accounts for the inhibition of myocyte proliferation. Nitrothiols have been proposed to be a major form of EDRF, they are vasodilatory, with plasma half-lives of 40 min. Nitrothiols in serum are not inactivated by haemoglobin in RBCs. NO is rapidly interchangeable with nitrosonium ion (NO^+). NO does not directly nitrosate organic molecules without a strongly oxidising cofactor to accept an electron and balance the following reaction:

.NO + RSH → RSNO + H^+ + e^-

For example, NO bound to ferric heme iron can nitrosate phenolics, thiols, and secondary amines. Free nitrosonium itself will not exist long in solution because it will react immediately with water to form nitrite. Nitrosonium ion moiety is probably the most common transfer agents. Thiols such as glutathione are widespread antioxidants and are continuously being oxidised to thiyl radicals. A rapid and biologically likely reaction to form nitrosothiols:

RS. + .NO → RSNO

Nitroxyl anion has been suggested to be a potential short lived EDRF. Nitroxyl anion will react rapidly with up to 2 NO to form nitrous oxide and nitrite:

NO^- + NO → $ONNO^-$

$ONNO^-$ + NO → NO_2 + NO_2^-

So NO can be rapidly scavengered by nitroxyl anion. When NO is dilute, the intermediate $ONNO^-$ can decompose to form hydroxyl radical:

$ONNO^-$ + H^+ → NO_2 + OH.

ONNO⁻ has many similarities with peroxynitrite, except that it will be a one-electron oxidant, whereas peroxynitrite can be both a one- and two-electron oxidant. Nitroxyl anion reacts with molecular oxygen, which is usually present in much higher concentrations than NO, to form peroxynitrite anion:
NO⁻ + O$_2$ → ONOO⁻
Peroxynitrite will cause vessel relaxation and increase cGMP in smooth muscle in concentrations that are 50- to 1000-fold higher than NO. NO activates guanylate cyclase by binding to ferrous heme, which it does with a much higher affinity than oxygen. The mechanism by which NO is released from guanylate cyclase is not known. Local concentration of NO must be rapidly removed to prevent saturation of guanylate cyclase. Because membranes provide no greater barrier to NO than to oxygen or carbon dioxide, diffusion away from a single cell producing NO is more significant than reactions of NO within the cell. NO will diffuse in and out of a cell membrane thousands of times in a second. Diffusion also allows for the rapid summation of NO production by several cells within a local region. Haemoglobin in blood vessels will capture NO diffusing through tissues. When a local group of cells is producing NO, the net concentration of NO within a cell can be much greater from the contribution of surrounding cells than from its own internal production. A red blood cell can reduce the internal concentration of NO in a NO-producing cell because the oxyhaemoglobin will greatly reduce the reentry of NO into cell. Endothelium can make 10- to 40-fold more NO than needed to activate guanylate cyclase, but the majority of the nitric oxide will be lost to the vascular compartment. Even after NO has diffused into the smooth muscle it can rapidly diffuse back along the diffusion gradient to the RBC. The intermediate duration of NO and its rapid diffusion through most tissues allows NO to integrate and modulate complex physiological processes. The modulating effects of NO are illustrated by the release of NO from endothelium in response to turbulent flow. A potential complication of NO modulation of vascular tone is the development of atherosclerosis, which tends to develop at sites of low turbulence. Macrophages accumulate in subendothelial spaces, ingesting oxidised lipoproteins to form atherosclerotic lesions. Peroxynitrite modifies low-density lipoproteins into a form recognised by the macrophage scavenger receptor, such vessels show greater release of nitrite as well as upregulation of the inducible form of nitric oxide synthase, indicating that NO is being inactivated more quickly. Nitrotyrosine, a marker of peroxynitrite oxidation of proteins, is wide spread in human atherosclerotic lesions. NO does not directly attack DNA unless it is allowed to oxidise to higher nitrogen oxides. The production of NO by activated macrophages will inactivate iron/sulfur centers in tumour cell mitochondria. NO will also bind reversibly to iron/sulfur proteins. The inactivation of iron/sulfur centers is mediated by peroxynitrite. Activated macrophages also produce superoxide, so the inactivation of tumour cell mitochondria may well be mediated by peroxynitrite. The inactivation of mitochondria by peroxynitrite caused by activated macrophages. NO can reversibly inhibit enzymes containing transition metals or free radical intermediates in their catalytic cycle. NO in micromolar concentrations will reversibly inhibit catalase and cytochrome P-450. It can inhibit ribonucleotide reductase, a critical enzyme for synthesis of DNA precursors that contains a tyrosine radical. Inhibition of DNA synthesis by NO inhibits viral replication. Large continuous fluxes of NO are necessary to inhibit ribonucleotide reductase and are most likely to occur only with severe sepsis or in cells near activated macrophages. Enormous amounts of oxygen are required to maintain synthesis of NO in micromolar concentrations. NOS requires 2 oxygen per NO produced. If the half-life of NO is 1 sec, then 170 nmol O$_2$/min would be needed/gram of tissue to maintain NO at a steady-state concentration of 1 μM NO. NO in the micromolar range can also reversibly inhibit cytochrome-c oxidase, which may transiently increase the leakage of superoxide from the electron transport chain. If the internal NO concentration approaches that of mitochondrial superoxide dismutase, the superoxide would largely react with the NO and the peroxynitrite formed would irreversibly injure mitochondria. Peroxynitrite is the ugly side of NO. A major mechanism of injury associated with the production of NO in vivo is due to its diffusion-limited reaction with superoxide to form peroxynitrite. Superoxide is produced by the reduction of molecular oxygen by one electron, which has two unpaired electrons:
.O-O. + 1e⁻ → :O-O. (O$_2^-$)
Consequently, superoxide still has one unpaired electron that rapidly combines with NO, with the release of 22 kcal/mol.
:O-O. + .N-O → :O-O=N-O
Peroxynitrite is not a free radical because the unpaired electrons on superoxide and on NO have each combined to form a new chemical bond. Peroxynitrite is approximately 6 times faster than the scavenging of superoxide with copper, zinc superoxide dismutase at physiological ionic strength. NO is the only known biological molecule production in high enough concentrations under pathological conditions to outcomplete endogenous superoxide dismutase for superoxide. A large flux of superoxide is produced by aerobic metabolism. 1-5 % of total oxygen consumption is estimated to be reduced to superoxide. The concentration of superoxide is kept remarkably low by superoxide dismutase. The stability of peroxynitrite allows a greater opportunity for it to diffuse through a cell to find a target. At pH 7.4 only 20% of peroxynitrite will be protonated to form peroxynitrous acid. Peroxynitrous acid is a strong oxidant that can react with biological molecules by a number of complex mechanisms. It is particularly efficient at oxidising iron/sulfur centers, zinc fingers, and protein thiols. It can also produces the products expected

for hydroxyl radical attack. Thus peroxynitrite leaves the hallmarks of oxidation traditionally used to implicate free radicals, including formation of protein carbonyls, salicylate, and 8-hydroxyguanidine, and formation of HO.. However, the direct reaction of peroxynitrite with specific moieties is more likely to account for its toxicity. Peroxynitrite is toxic by more direct oxidative mechanisms. The underlying chemistry of peroxynitrite is sufficiently complex. Superoxide dismutase can also inhibit the formation of peroxynitrite by directly scavenging superoxide before it reacts with NO. Extensive nitration takes place around foamy macrophages in human atherosclerotic lesions. Tyrosine nitration is found in human lung biopsy and autopsy samples with sepsis, pneumonia, or ARDS, peroxynitrite is the most likely source. The superoxide dismutase combines rapidly with peroxynitrite and directs a more selective nitration of tyrosines on certain proteins. The vast number of alternative targets in biological materials for peroxynitrite makes the superoxide dismutase-peroxynitrite complex more selective. NO contrasts with most intracellular messengers because it diffuses rapidly and isotropically through most tissues with little reaction but cannot be transported through the vasculature due to rapid destruction by oxyhaemoglobin. The rapid diffusion of NO between cells allows it to locally integrate the responses of blood vessels to turbulence, modulate synaptic plasticity in neurons, and control the oscillatory behaviour of neuronal networks. NO is removed within seconds in vivo by diffusion over 100 μm through tissues to enter RBCs and react with oxyhaemoglobin. The direct toxicity of NO is modest but is greatly enhanced by reacting with superoxide to form peroxynitrite.

PATHOBIOLOGY OF NO

NO was discovered in 1979. NO markedly inhibits human platelet aggregation, and induces vasorelaxation, both are mediated by intracellular cyclic GMP (cGMP). The mechanism of the so called nitrovasodilator (nitroglycerine, isoamyl nitrite, and sodium nitroprusside) was shown to be attributed to the generation of NO in vascular smooth muscle, in which NO activates cytosolic guanylate cyclase, thereby stimulating the intracellular biosynthesis of cGMP. cGMP then triggers a cascade of intracellular events that culminate in a reduction in calcium-dependent vascular smooth muscle tone. Eight years after the finding that exogenous NO relaxes vascular smooth muscle, endogenous NO was identified as being the EDRF responsible for mediating endothelium-dependent vasorelaxation. NOS inhibitors cause a sustained hypertension that is reversible by administration of excess L-arginine. NO is an essential modulator of vascular smooth muscle tone, systemic vascular resistance, and regional blood flow. cGMP acts as an intracellular amplifier and second messenger to rapidly lower intracellular free calcium levels and to inactivate myosin light chain kinase. The first endogenous vasorelaxant shown to stimulate the formation of NO was acetylcholine. Autacoids or locally-acting tissue hormones including bradykinin, histamine, and serotonin are also capable of causing endothelium-dependent vascular smooth muscle relaxation. Bradykinin and histamine are well recognised autacoids that cause local vasodilation, increased vascular membrane permeability, pain, and itching. The vasodilator action of bradykinin may also be attributed to locally released prostacyclin. The chemical properties of NO enable NO to act as a true autacoid or local tissue hormone. NO is a water soluble gas, and biologically active NO concentrations range from 1 to 100 nM. NO is a small lipophilic molecule that rapidly permeates biological membrane barriers, but its very short half-life (3-5 s) limits the area within which newly synthesised NO can be biologically active. NO is inactivated by rapid oxidation to nitrite (NO_2^-) and, under certain conditions, nitrate (NO_3^-), both of which are essentially inactive as vasodilators at low concentrations. Tissue hormones are released into the local circulation and interact with selective receptors on vascular endothelial cells. These receptors are coupled to inward calcium channels, a signal transduction mechanism often referred to as pharmacomechanical coupling. Shear force, via uncharacterised mechanoreceptors, represents a signal for enhanced production of NO in vascular endothelial cells is calcium- and L-arginine-dependent but prostaglandin-independent. NO formation by shear forces could be a major mechanism for local antithrombosis. Stimulation of nonadrenergic-noncholinergic (NANC) neurons results in the biosynthesis and release of NO. NANC neurons innervate both vascular and nonvascular smooth muscle. NANC neurons innervate the highly vascular corpus cavernosum of the penis. NANC neuronal stimulation results in profound relaxation of the sinusoidal arteriolar bed, thereby causing penile erection, and this physiological response is mediated by NO and cGMP. The NOS in the vascular endothelial cells is membrane-bound, whereas the NOS in neuronal tissue is cytosolic, but both isoforms are activated by calcium. Inducible NOS are subjected to appropriate immunologic stimulation, iNOS does not appear to be regulated by calcium although contain tightly bound calmodulin. iNOS plays a major pathophysiological role in expressing the cytotoxic function of activated macrophages. Cell for cell, activated macrophages can generate several orders of magnitude more NO than can vascular endothelial cells. The main function of macrophage-derived NO is target cell cytotoxicity. Target cells may include tumour cells and invading bacteria, viral particles and other microorganisms. The cytotoxic mechanism of action of NO appears to be a reaction between NO and iron contained in numerous enzymes involved in mitochondrial function and cell proliferation to cause cytostatic and cytocidal effects. NO may also react with superoxide anion generated locally to form the reactive peroxynitrite anion, which can cause tissue injury. Macrophage-derived NO local vasodilation and inhibition of thrombosis, would favour the onset of local inflammatory reactions in order to complement macrophage function. NOS can be induced not only in tissue macrophages, but also in vascular smooth muscle cells. Substances such as lipopolysaccharide, interferon-γ, TNF, and IL-1 induce NOS in vascular smooth muscle cells and provide the biosynthesis and release of large quantities of NO. The consequence is a vasodilation that can lead to marked hypotensive shock. Thus chemical inhibitors of iNOS may be useful clinically to treat systemic hypotension caused by iNOS activity. NO must translocate from the endothelial cells to the underlying smooth muscle cells and exogenous nitrovasodilators must reach the vascular smooth muscle cells and undergo metabolic transformation to NO. Once inside the smooth muscle cell, NO binds rapidly to the heme prosthetic group of cytosolic guanylate cyclase to form a nitrosyl-heme complex, which is the species that activates guanylate cyclase. NO generated by nNOS in NANC neurons diffuses into target cells such as vascular and nonvascular smooth muscle, to trigger cGMP formation and initiate target cell responses. Nitric oxide is an important intra-and inter-cellular mediator. NO activates guanylate cyclase, which leads to increased production of cyclic guanosine monophosphate (cGMP) and to the relaxation of smooth muscle and decreased aggregation of platelets. NO is an important neuronal mediator, may kill bacteria, regulates penile erection and may be diabetogenic by destroying islet cells. The half life of NO in biological systems is of the order of 10-30 s, during which time it can diffuse to a distance of 200-600 μm. In the presence of haemoglobin the half life may even shorter. Most of the nitrite/nitrate in plasma or urine derives from endogenous NO, if the intake of nitrite in the diet is strictly restricted. Increased NO production is associated with progressively increased concentrations of nitrate and methaemoglobin. Nitrate is the main metabolite of NO in blood and urine. NO seemed to be metabolised partly to nitrate and partly to nitrosyl haemoglobin (HbNO). The measurement of nitrite/nitrate in plasma or urine can be used to evaluate endogenous NO production if proper attention is paid to the confounding effect of their alimentary intake. Not all of the effects of NO are mediated through cGMP formation. Nitric oxide can react with heme groups, iron centers, thiol groups, and many other functionalities. The NOS isoforms have considerable homology (about 50-60%), but greater homology of each isoform (about 90%) are found between species. The enzyme is quite complex in that it uses heme as a prosthetic group and flavin adenine dinucleotide, tetrahydrobiopterin, and nicotinamide adenine dinucleotide phosphate as cofactors or cosubstrates. Calcium and/or calmodulin can activate NOS-1 and NOS-3, while NOS-2 has no apparent calcium dependence since calmodulin is tightly associated with this isoform. Both NOS-1 and NOS-3 have been called constitutive or housekeeping enzymes, but their expression and synthesis can be regulated under various conditions. However NOS-2 is probably absent in most cells until induced by endotoxin and/or various cytokines. Some cytokines and corticosteroids can inhibit the induction of iNOS or NOS-2 and perhaps some of their anti-inflammatory properties could be attributable to these effects. When agonist such as acetylcholine acts on receptor of the endothelial cell, it causes an increase in intracellular free Ca^{2+} by opening membrane calcium ion channels and/or by increasing release of calcium from stores in the endoplasmic reticulum. An increase in intracellular free Ca^{2+} can also result from increase in shear stress on the endothelial cell. The increase in Ca^{2+} and intracellular calmodulin activates the enzyme nitric oxide synthase (NOS), an oxygenase that catalyses the oxidation of guanidinium nitrogen of L-arginine to NO, with L-citrulline as a by-product. The reduced form of nicotinamide adenine dinucleotide phosphate is a required cosubstrate, being oxidised to the oxidised form of nicotinamide adenine dinucleotide phosphate, and flavin adenine dinucleotide; both flavin mononucleotide and tetrahydrobiopterin are necessary cofactors. The NO diffuses from the endothelial cells to the adjacent smooth muscle cells of the vessel activates soluble guanylyl cyclase (G-cyclase), resulting in an increase concentration of cyclic guanosine monophosphate (cGMP) formed from guanosine triphosphate (GTP). The increase in cGMP leads to relaxation of the smooth muscle by unknown mechanisms, but may be by activation of cGMP-dependent protein kinase (G-kinase) as an early step, leading to a reduction in intracellular Ca^{2+} and/or a reduced sensitivity of the contractile proteins to Ca^{2+}. The constitutive isoforms of NOS are regulated by the level of intracellular calcium. Extracellular hormones or other chemical agents interact with extracellular receptors to open calcium channels and thereby allows the influx of calcium. Calcium binds to calmodulin and this complex then binds to NOS to cause enzyme activation. Enzyme-bound calmodulin facilitates the transfer of electrons from NADPH to the flavoprotein domain of NOS. These electrons are used to reduce the iron to the ferrous state so that it can bind to oxygen, which is incorporated into the substrate, arginine, to generate NO plus citrulline. Another cofactor, tetrahydrobiopterin, is required for NOS catalytic activity although its precise function remains unknown. iNOS does not require calcium for activity since calmodulin is permanently bound. The inflammatory process results in the induction of iNOS in the inflamed tissues, and endotoxin can result in iNOS induction in vascular smooth muscle and other cell types. A negative feedback effect of NO on NOS is to inhibit further NO biosynthesis. NO inhibits monocyte adhesion, platelet aggregation, and vascular SMC proliferation. NO is synthesised by stereospecific oxidation of the terminal guanidino nitrogen of the amino acid L-arginine by NOS. Asymmetric dimethylarginine (ADMA) is an endogenous competitive inhibitor of NOS. A number of cell types elaborate ADMA including endothelial cells. ADMA is an endogenous inhibitor of endothelial NOS. ADMA accumulation is responsible for cardiovascular disorders such as hypertension and atherosclerosis, and ESRD. ESRD is associated with elevated ADMA levels (6-fold) and reduced levels of nitrosothiols in plasma. L-arginine restores endothelial function independently in haemodialysis.

OXIDATIVELY MODIFIED PROTEINS IN CHRONIC RENAL FAILURE (CRF)

Glucose and other reducing sugars can react in the body with proteins and with nucleic acids like DNA and RNA to form complex compounds. These glycosylation reactions are known generally as the Maillard reaction. It is the cause of the yellow brown colour (called Amadori products), that develops when foods containing proteins and sugars are heated in air during cooking. The high concentrations of proteins modified by the Maillard reaction may be responsible for the development of cataracts and atherosclerosis in diabetics. Oxidative stress is the disruption of the equilibrium between the generation of oxidants and the activity of antioxidant systems. The major source of oxidants is afforded by circulating polymorphonuclear (PMN) neutrophils and monocytes when activated. Excessive ROS may damage normal structures. NADPH complex is dormant in the resting phagocyte. Following activation, NADPH assembles its distinct molecular components in an electron transfer chain capable of performing an univalent reduction of molecular oxygen in superoxide anion (O_2^-). Superoxide anion is converted by SOD into H_2O_2, which produces single oxygen and hydroxyl radicals in the presence of iron. Superoxide anion and hydrogen peroxide are capable to kill microorganisms and are used by phagocytes as precursors for the production of more powerful oxidants (peroxynitrite). H_2O_2 is able to cross plasma membranes, reacts with intracellular iron to form hydroxyl radicals by Haber-Weiss cycle. Hydroxyl radicals trigger peroxidation of cell membrane lipids, promote protein aggregation, and cause damage leading to mutation and cleavage of DNA. Phagocytic cells also have unique capacity to produce chlorinated oxidants caused by myeloperoxidase, which in the presence of chloride (Cl^-) converts hydrogen peroxide into hypochlorus acid (HOCl). Hypochlorus acid is the most toxic and the most reactive species formed by phagocytic cells. HOCl targets the membrane proteins and their thiol groups; intracellular enzymes of nucleotides and cytochromes. The latter action leads to inhibition of the respiratory chain. HOCl may react with endogenous amines (R-NHI) to generate chloramines (RNH-Cl), also named long-lived oxidants (OFRs have extremely short life span). Oxidative stress presents in the CRF patient in combination with a chronic antioxidant deficiency. Proteins are elective targets of oxidant-mediated injury, which lead to functional alterations and progressive loss to their metabolic, enzymatic, or immunologic properties. These alterations affect their primary, secondary and tertiary structure. Moderate oxidation of proteins leads to augmentation of hydrophobicity and favours their catabolism by the multicatalytic or proteasome complex. Intense oxidation generates insoluble products that resist proteolysis. Lysine, arginine, proline, and histidine are prone to generate carbonyls. The formation of carbonyls involves cations of the redox cycle such as Fe^{2+} and Cu^{2+}, which have binding sites on proteins and may transform amino acid residues in carbonyls in the presence of hydrogen peroxide and superoxide anion. Advanced oxidation protein products (AOPP) are of 2 types high molecular weight (HMW-AOPP) and low molecular weight (LMW-AOPP). HMW-AOPP peak is mostly due to albumin, which forms aggregate likely resulting from disulfide bridges and/or dityrosine cross-linking. LMW-AOPP peak contains albumin in its monomeric form. Uraemia per se may induce a state of oxidative stress. AOPP plasma levels are increased at a mild stage of CRF and gradually increased with the progression of renal failure. AOPPs are more accurate oxidative stress marker than oxidised lipid products. AOPP are uraemic toxins and as well cytokine-like mediators between neutrophils and monocytes. AOPP stimulates neutrophil respiratory burst. AOPP-mediated biological effects involve a ligand-receptor type of interaction. The accumulation of reactive carbonyl compounds (RCOs) derived from carbohydrates and lipids and the subsequent carbonyl modifications of proteins is carbonyl stress. Uraemia is characterised by the simultaneous accumulation of irreversible nonenzymatic protein modifications with carbohydrates and lipids, i.e., AGEs/ALEs. A variety of RCOs are generated during lipid peroxidation of polyunsaturated fatty acids including glyoxal, malondialdehyde, hydroxynonenal, and acrolein. RCOs such as glyoxal, methylglyoxal, acrolein, and glycoaldehyde are generated during the myeloperoxidase-catalysed metabolism of amino acids. These RCOs are highly reactive with proteins and form AGEs and ALEs on proteins. Under carbonyl stress AGEs and ALEs are derived from lipids accumulate in plasma and tissue proteins. Evidence of a uraemia-associated oxidative stress include: 1- increased serum ratios of oxidised to total ascorbate, 2- increased serum ratios of oxidised serum albumin (nonmercaptoalbumin) to reduced albumin (mercaptoalbumin), 3- increased serum ratios of oxidised to reduced glutathione (GSH), 4- increased serum levels of advanced oxidation protein products (AOPPs) and protein carbonyls, 5- decreased serum activity of GSH-dependent enzymes, 6- increased levels of the lipid peroxidation marker malondialdehyde-lysine in the plasma protein of haemodialysis patients, and 7- accumulation of dialyzable oxidants in plasma. Under oxidative stress, proteins are modified either directly by ROS resulting in oxidised amino acids or indirectly by RCOs generated by autoxidation of a variety of sources. ROS targets carbohydrates and lipids yielding RCOs involved eventually in the formation of AGEs/ALEs. The raised carbonyl stress and AGEs/ALEs in uraemia may be the consequences of an increased production of RCOs by oxidative stress, proved by raised serum levels of pentosidine and oxidative markers (dehydroascorbate, AOPP). The oxidative stress cannot explain all of the mechanisms of raised RCOs, as they are driven from nonoxidative chemistry as well and accumulate without changes in glycinaemia or lipaemia. Enzymes and enzymatic pathways contribute to the detoxification of RCOs includes aldose reductase, aldehyde dehydrogenases, and glyoxalase pathway. Redox enzymes (reduced GSH, NAD(P)H are important elements for the activity of these pathways. RCOs, methylglyoxal, glyoxal, etc., react reversibly with thiol group of GSH and are subsequently detoxified by glyoxalases into lactate and GSH. A decrease in GSH and NAD(P)H can result in augmented levels of a wide range of RCOs. GSH concentration in RBCs and serum activity of GSH-dependent enzymes of uraemic patients is significantly low. RCOs might be derived from the decrease of thiol concentration (GSH, NAD(P)H, mercaptoalbumin) in uraemia. The decrease of thiol concentrations in uraemia might be derived from consumption of thiol to detoxify ROS generated under uraemia-associated oxidative stress. Also, a nonoxidative pathway may reduce thiol concentration in diabetes. The polyol pathway is activated by hyperglycaemia and consumes NAD(P)H for the reduction of glucose to sorbitol, which is catalysed by aldose reductase. The decrease in NAD(P)H for GSH reductase consequently results in the decrease of GSH. RCOs are derived from both oxidative and nonoxidative chemistry. A decrease in renal clearance of RCOs in uraemia may also be involved in raised carbonyl stress. Dialysis-related amyloidosis is a serious bone and joint destruction associated with CRF. It develops in 2 stages: 1- The preclinical stage, characterised by histologic evidence of β_2-microglobulin (β_2-m) deposits in the joints several years before the onset of clinical or radiological signs. Amyloid deposits are not surrounded by macrophages or evidence of bone destruction. 2- The clinical stage, characterised by arthralgias, carpal tunnel syndrome, and radiologically visible bone cysts. The amyloid deposits are surrounded by macrophages and bone resorption is observed. Levels of AGEs in arterial tissues are higher in dialysis patients than in normal subjects. AGEs/ALEs staining in the thickened neointima corresponds with that for protein carbonyls, a biomarker of oxidative protein damage. AGEs/ALEs are identified in vascular lesions regardless the type of stimulation of the vascular injury (metabolic or mechanical). Colocalization of AGEs/ALEs and protein carbonyls in the vascular tissue indicates a wide range of chemical modifications in vascular matrix proteins. Peritoneal membrane alteration with PD is characterised by interstitial fibrosis, disappearance of mesothelial cells, vascular wall thickening, vasodilation and increased angiogenesis. AGEs/ALEs are inert surrogate markers for carbonyl stress, actively play roles in the pathogenesis of protein modification with RCOs. Proteins modified with AGEs/ALEs exhibit several biological activities including: 1- Initiation of a range of inflammatory responses, e.g., stimulation of monocyte chemotaxis. 2- Secretion of inflammatory cytokines from macrophages. 3- Stimulation of collagenase secretion from synovial cells. 4- Stimulation of osteoclast-induced bone resorption. 5- Proliferation of vascular SMC. 6- Stimulation of aggregation of platelets. These biological activities may account for bone and joint destruction. Uraemic arthropathy may result from carbonyl stress modification of amyloid deposits, followed by secondary localised inflammatory responses. Uraemia and aging are important factors enhancing carbonyl stress. RCOs react covalently with tissue matrix proteins and alter their structures. RCOs induce the functional alteration of proteins. RCOs capable of Schiff base formation with lysine residues of vitamin D receptor inhibit the vitamin D receptor interaction with vitamin D response element. RCOs are biologically active and initiate a variety of cellular responses. Carbonyl stress mediates the intracellular signalling by multiple pathways including: 1- Upon interaction with the receptor for AGEs, trigger the signalling involving P21^{ras}, mitogen-activated protein kinases (MAPKs), and NFκB. 2- Exposure of cell to glyoxal activates protein kinases such as -c-Src and increases intracellular tyrosine phosphorylation of several cellular proteins. 3- Hydroxynonenal causes a capping of EGFR that mimics the capping induced by EGF, and downstream signalling pathways involving MAPKs, which may play a role in oxidative stress-linked-induced apoptotic cell death. 4- Hydroxynonenal triggers oxidative stress-linked-induced apoptotic cell death by activating caspase-3 through a Fas-independent but in a GSH-dependent redox pathway. Homocysteine, similar to cysteine, but with an additional methylene group in its side chain, is a sulfur amino acid not normally inserted into the protein backbone, which is positioned at a crucial branch point of the metabolism of the essential amino acid methionine. Methionine can be transformed intracellularly into the nucleoside S-adenosylmethionine (Ado-Met), a sulfonium compound possessing a highly reactive methyl group. The demethylated product, the thioether S-adenosylmethionine (AdoHcy), represents a competitive inhibitor of transmethylation reactions. AdoHcy is hydrolysed to adenosine and homocysteine. Homocysteine is rapidly removed through: 1- Transsulfuration to cystathionine by β-synthase. Cysteine is formed subsequently by a cystathioninase. 2- Remethylation to methionine using methyltetrahydrofolate (MTHF) as a cosubstrate. Vitamin B_6 is a cofactor in transsulfuration and B_{12} in remethylation. Homocystinuria is inherited connective tissue disease characterised by elevated blood levels of homocysteine and homocystine, the oxidised product, in the urine. Homocysteinaemia is associated with cardiovascular risk, and mortality of coronaropathy is strictly related to homocysteine levels. A higher folate intake is associated with a significant reduction of cardiovascular risk. Homocysteine toxic effects on the endothelium, vascular SMC, platelets, and coagulation cascade include: 1- oxidation and nitrosylation, 2- acylation, and 3- hypomethylation. Oxidation of homocysteine generates homodimer or mixed disulfides, hydrogen peroxide and other ROS (lipid peroxidation). Vitamin C and vitamin E prevent alterations in endothelial cell function caused by acute hypercysteinaemia. Homocysteine can react with NO and form S-nitroso-homocysteine. Homocysteine inhibits the scavenging of ROS by inhibiting glutathione peroxidase. Acylation of homocysteine generates homocysteine thiolactone, a highly reactive anhydride. The thiolactone ring opens up, and carbonyl group reacts with the free amino group of lysine or with terminal amino group of proteins. Hypomethylation of homocysteine causes hydrolysis of AdoHcy to slow down. Therefore AdoHcy accumulates and inhibits transmethylation reactions. In CRF the AdoMet/AdoHcy ration is significantly reduced both in patients on conservative treatment and in patients on haemodialysis. This is highly favourable for transmethylation reactions. Homocysteine causes changes at the molecular level through multiple mechanisms including oxidation and hypomethylation. The repair of damaged proteins causes the formation of racemised residues as byproducts of the reaction. AdoHcy inhibition in uraemic patient targets include: 1- cellular growth (carcinogenic), 2- lymphocyte chemokinetic response, 3- insulin release, 4- interferon synthesis, 5- norepinephrine uptake and release, 6- catecholamine degradation, 7- conversion of phosphatidylethanolamine to phosphatidylcholine, 8- brain histamine content, and 9- serotonin and dopamine turnover.

ERYTHROCYTE FREE RADICAL AND ENERGY METABOLISM

The erythrocyte retains only a cytosol and plasma membrane. The human erythrocyte is biconcave, maximal in surface area, 6-8 μm in diameter, 2 μm thick and has life span of 100-130 days during which it travels 400 km. Glycolysis and the oxidative pentose phosphate pathway generate NADH and NADPH to reduce methaemoglobin (MetHb) and the antioxidant glutathione. NADH is needed for reducing continuously produced MetHb to oxygen-carrying haemoglobin, and NADPH for reducing the glutathione present in high concentration in the erythrocyte. Mature erythrocytes consume 1.25 mmol of glucose per liter of cells each hour. 10% enters the oxidative pentose pathway. Lactate is produced at 2.3-2.5 mmol/l of cells per hour. Glucose consumption in the reticulocyte is markedly higher than in the mature erythrocyte. ATP turnover in erythrocytes is 2 mmol/l of cells per hour. ATP half-life is 1 hour. ATP consumption in erythrocytes takes place in and on the plasma membrane, with <10% accounted for by cytosolic processes. The main consumers in erythrocytes are ion transport processes (30%). RBCs possess high activities of the most important antioxidant enzymes. Circulating RBCs are mobile free radical scavengers and provide antioxidant protection to other tissues and organs. Erythrocytes are rich in the important antioxidant glutathione (glutamate-glycine-cysteine). The most reactive group, the sulfhydryl group on the cysteinyl side chain, acts as an electron donor and thus as a nucleophilic center, reducing agent and free radical scavenger. The structure of glutathione was elucidated 70 years ago. More than 99% of intracellular glutathione exists in reduced form (GSH). Glutathione peroxidase catalyses the reduction of hydrogen peroxide and other hydroperoxides. In the process GSH is converted to GSSG, which is then reconverted to GSH by glutathione reductase. The reducing equivalents for the glutathione reductase reaction (NADPH + H$^+$) are derived from the oxidative pentose phosphate pathway, whose key enzyme is glucose-6-phosphate dehydrogenase (G6PDH). Glutathione protects haemoglobin, other proteins and membrane components from oxidation by means of the glutathione peroxidase reaction. Each of the 4 subunits of glutathione peroxidase (GPx) contains a selenium atom. Glutathione is synthesised in 2 enzyme-catalysed steps in mature RBC, the reactions of γ-glutamylcysteine synthetase and GSH synthetase. Each of the 2 reaction steps consumes one molecule of ATP. Cysteine is the rate-limiting substrate in glutathione synthesis. GSH degradation occurs extracellulary and involves γ-glutamyltranspeptidase and dipeptidases. Both enzymatic reactions proceed by binding of these enzymes to the outer surface of the cell membrane and depend on glutathione export from the intracellular to extracellular space. A pro-oxidant shift results from changes in cellular metabolism (especially energy metabolism), higher flux in catecholamine metabolism and permanent leukocyte activation. The following systems generate a whole range of reactive oxygen species, which then exert their effects in the cell: 1- mitochondria, 2- enzyme cascades such as AA or catecholamine catabolism, 3- individual enzymes such as xanthine oxidase or cytochrome P450, and 4- altered iron metabolism. An intermediate position between lipophilic and hydrophilic antioxidants is occupied by the glutathione system. Despite the high efficiency of these defense systems, adverse reactions do occur. Oxidant exposure damages macromolecules, cellular structures and system. Tissue hypoxia in anaemic state is accompanied by an absolute increase in free radical production triggered by: 1- changes in cellular metabolism, especially energy metabolism; 2- higher flux rates in catecholamine metabolism and permanent activation of leukocytes. At the end-stage of purine catabolism (xanthine oxidase reaction) is the main source of free radicals. The bulk of free radicals are generated in the early reoxygenation phase. Hypoxia conditions unlike anoxia, lead to increased free radical production. An inadequate tissue oxygen supply per se contributes to free radical stress in anaemia. Tissue hypoxia forces a change in catecholamine metabolism, i.e., it increases catecholamine production. After exerting their effect, catecholamines are degraded by monoamine oxidase in a biochemical reaction that generates oxygen radicals. Hypoxic modification of cellular metabolism is accompanied by the release of a range of inflammatory mediators. This activates leukocytes, which release various radicals and oxidants via enzymatic cascades. Invading leukocytes are one the main sources of oxidants in many forms of organ-specific hypoxia and ischaemia. The permanent tissue hypoxia in chronic anaemia results in continuous activation of leukocytes, which both migrate into tissue and circulate in the activated state. Under conditions of haemodialysis in uraemic there is also increased leukocyte activation by the dialysis membrane. Antioxidant system is markedly weakened by anaemia. The erythrocytes represent a major component of the antioxidant capacity of the blood. Erythrocyte contribution is in regenerating consumed redox equivalent, via the oxidative pentose phosphate pathway and glutathione reductase. Erythrocytes are largely responsible for regenerating consumed antioxidants in whole blood. The erythrocyte cytosol is connected to the plasma through membrane transport systems enabling it to influence the thiol status of plasma and hence its antioxidant capacity. Lipid peroxidation products are elevated in uraemic serum. Lipid peroxidation leads to changes in membrane viscosity and, through protein modification, to changes in the cytoskeleton. Lipid peroxidation processes take place not only in cell membranes but also in circulating lipoproteins. This results in an increase in oxidised lipoprotein species, and these, especially oxidised LDL, are key factors in the development of arteriosclerosis. Persistent leukocyte activation and free radical damage to the cells themselves reduce the rate of phagocytosis, which is associated with increased susceptibility to infection. ROS play a role in the pathogenesis of CRF. ROS have a role in pathogenesis of glomerular diseases such as minimal change disease and membranous nephropathy, postischaemic or toxic ARF and pyelonephritis. Renal cells, neutrophils or other cells in circulation may be the sources of ROS. Glomerular mesangial cells may produce ROS when these cells interact with complement 5b-9-membrane complexes. Oxygen free radicals are the molecules, which contain unpaired electrons. They are called as oxidants molecules or reactive oxygen species. Reactive oxygen metabolites are constant products of normal aerobic cell metabolism. A defect in antioxidant system in glomerular diseases causes nephrotic syndrome. Glomerular toxicity of hydrogen peroxide, activation of proteolytic capacity of metalloproteinases, inactivation of circulatory inhibitors of some proteolytic enzymes, sustained leukocytic recruitment into the glomerulus result in continuous generation of ROS and reduction in single nephron GFR due to ROS have been incriminated in glomerular damage related to oxidative stress. Antioxidant deficiencies affect the course of nephrotic syndrome negatively and, both low plasma selenium levels and reduced GSH-Px activities. Supplementation of antioxidants such as selenium may have beneficial effects on antioxidative system.

THE OZONE

The atmosphere is a mixture of gases, water vapour and particles, held to the earth by gravity. The lowest part of the atmosphere is called the troposphere. The ozone is a form of the oxygen we breathe and need to stay alive. The oxygen we breathe is found throughout the atmosphere and contains two oxygen atoms (O_2) but ozone contains three (O_3). The ozone layer is more concentrated in the stratosphere between 10 and 50 km above the surface of earth, and only 3 mm thick. The ozone layer makes up <1 in 100,000 parts of the atmosphere. It is produced from oxygen in the stratosphere in a series of photochemical reactions stimulated by sunlight, especially ultraviolet radiation. It is then broken up again by the activity of ultraviolet radiation. Thus ozone is constantly being produced and broken down in the stratosphere by the action of the sun. The process of forming and destroying ozone absorbs almost all the ultraviolet radiation from the sun, with different wavelength. UV-C is the shortwave radiation that is lethal to living things and is easily absorbed by the ozone layer. Longer wavelength UV-A is relatively harmless. In the middle is UV-B that is less dangerous than UV-C. The ozone layer is important in determining the climatic and weather patterns of the world. The concentration of ozone is measured in Dobson units. One hundred Dobson units are the equivalent of a layer of ozone 1 mm thick at the sea level. A number of human-made chemical compounds released into the atmosphere are responsible for the destruction of the ozone layer. Over 200 different destruction processes have been identified but the worst compounds are chlorofluorocarbons (CFC), carbon tetrachloride, methyl chloroform and halons (used mainly for fire fighting). In the lower atmosphere CFCs remain inert and do not break down quickly. Once in the stratosphere, ultraviolet radiation decomposes them. Each chlorine atom released is capable of destroying tens of thousands of ozone molecules. The chlorine in the chemicals is converted to a form that is much more sensitive to sunlight so that when sunlight returns in spring, the destruction of the ozone is very rapid. CFCs became very popular with industry because of their special qualities: they are stable in the lower atmosphere, do not smell or affect the colour of liquids like deodorants and are not toxic, they are used for blowing plastic foam for food containers and propelling liquids from aerosols, they also make excellent cleaning fluid for electrical components and coolants in refrigerators and air-conditioning system. Chlorofluorocarbons (CFCs) have been one of the major chemicals responsible for the depletion of the ozone layer. This depletion of the ozone layer leads to an increase in skin cancers and cataracts. A 1% decline in ozone levels translates into a 3.5% increase in the incidence of squamous cell carcinomas and a 2.1% increase in the incidence of basal cell carcinomas. Under Montreal protocol (1987), the production and consumption of CFCs has been banned. In vivo exposure to ozone produces an increase in a 70-kDa heat shock protein. Ozone is a ubiquitous environmental pollutant, and acute exposure to this oxidant gas causes a number of concentration-dependent alterations in the mammalian respiratory tract. Oxidative stress produced by ozone, leads to, lipid peroxidation, generation of free radicals, and alterations in antioxidants. Aldehydes, hydrogen peroxide, organic radicals, and hydroxyl radicals are among the toxic mediators generated by ozone. Hydrogen peroxide and hydroxyl radical have been reported to induce members of the HSP 70 family in macrophages and other cells. Other oxidative stressors such as hydroxyquinoline, ischaemia-reperfusion, anoxia-reoxygenation, and γ-irradiation induce the expression of HSPs. States of imbalance in oxidants/antioxidants may underlie the observed inductions in HSP 70 family of gene products. Ozone is thought to exert its toxic effects either by interacting directly with cell membranes or generating secondary products that are often more reactive than ozone itself. A function for 72-kDa HSP has been proposed under both physiological and pathological conditions. Transient expression during development suggests a role in cell differentiation and proliferation. 70-kDa HSP binds to proteins to prevent denaturation and to facilitate their recovery after stress. HSP 70 has a protective mechanism in cells responding to environmental stresses.

CARBON MONOXIDE POISONING

Carbon monoxide (CO) is a colourless, odourless, toxic gas that is a product of incomplete combustion. Motor vehicles, heaters, appliances that use carbon based fuels, and household fires are the main sources of this poison. The true incidence of CO poisoning is not known, since many non-lethal exposures go undetected. The most common sources for CO include car exhaust fumes, smoke from fires, furnaces, gas-powered engines, wood stoves, and paint removers containing methylene chloride. CO from motor-vehicle exhaust is the single most common cause of CO poisoning deaths. Epidemics of CO poisoning commonly occur during winter months and sources include misuse of non-electric heating or cooking devices as well as snow-obstructed motor vehicle exhaust systems. It is well known that urban environments contain higher ambient CO concentrations, primarily due to automotive emissions, and that non-smoking city dwellers commonly have carboxyhaemoglobin (COHb) levels in the 1-2% range. Tobacco smoke is also a significant source of CO, containing approximately 4% CO; smokers have been observed to have COHb levels typically in the 4-5% range and as high as 9%. Methylene chloride (MC) deserves a special mention because it is contained in many paint removers and its vapours are readily absorbed through the lungs. Once it reaches the circulation, MC is converted into CO in the liver. CO combines preferentially with haemoglobin to produce COHb, displacing oxygen and reducing systemic arterial oxygen (O_2) content. CO binds reversibly to haemoglobin with an affinity 200-230 times that of oxygen. Consequently, relatively minute concentrations of the gas in the environment can result in toxic concentrations in human blood. Possible mechanisms of toxicity include: 1) a decrease in the oxygen carrying capacity of blood; 2) an alteration of the dissociation characteristics of oxyhaemoglobin, further decreasing oxygen delivery to the tissues, 3) a decrease in cellular respiration by binding with cytochrome a_3; and 4) binding to myoglobin, potentially causing myocardial and skeletal muscle dysfunction.

CHLORINE

The standard household bleach is 5.25% sodium hypochlorite solution (NaOCl). When NaOCl and an acid are mixed, chlorine gas and water are released. Chlorine gas reacts with the water to form hydrochloric and hypochlorous acids. Hydrochloric acid causes inflammation that may, along with nascent oxygen release, be one of the mechanisms of tissue damage by chlorine. Symptoms following exposure to chlorine include irritation of the eyes, nose, and throat; dizziness; cough; and chest pain or constriction. Severe exposure may cause pulmonary oedema, bronchiolar and alveolar damage, and pneumomediastinum. When bleach is mixed with ammonia-containing compounds, monochloramine (NH_2Cl) and dichloramine ($NHCl_2$) are formed, which may produce tearing, respiratory tract irritation, and nausea. These compounds decompose in water to hypochlorous acid and free ammonia gas; the former combines with moisture forming hydrochloric acid and toxic nascent oxygen; the latter is a respiratory and mucous membrane irritant and can cause pulmonary oedema and pneumonia.

THE CYTOCHROME P450 SYSTEM

The P450 system is the mother of all drug reactions. Not all drugs in the same class have the same effect on the P450 enzymes. The basics of the enzyme system help to understand the complex drug interactions that occur. The P450 system is a group of enzymes responsible for the metabolism of many drugs. Although there are over 30 different enzymes in the P450 system, only 6 account for the metabolism of nearly all clinically important drugs. Two of the 6 are critically important for drug metabolism. The P450 enzymes are located on the endoplasmic reticulum of the liver, small intestine, lungs, kidneys, blood vessels, and almost every cell of the body. These enzymes become active after drug absorption and distribution when the drug moves through the cell membrane into the cytoplasm. The enzymes interact with the drug to produce metabolites. Sometimes the enzymes turn the drug action on; sometimes it is the active metabolite produced that causes the desired effect; and sometimes the enzymes turn the drug action off. For example, after ingestion, Cozaar (losartan) is inactive until it meets a specific enzyme in the system (the 2C9 enzyme), which turns it into an active metabolite, subsequently reducing blood pressure. The 2 enzymes most important for drug metabolism are CYP 3A4, and CYP 2D6. The CYP 3A4 enzyme is needed to metabolise antihistamines, antibiotics, lipid-lowering drugs, antihypertensives, protease inhibitors, and antifungals. It metabolises about 50% of all clinically useful medications. It is the most abundant and clinically significant enzyme, and is actually composed of 4 subsystems. The CYP 2D6 enzyme metabolises selective serotonin reuptake inhibitors (SSRIs), pain relievers, β-blockers, and some other medications. It is involved with the metabolism of about 30% of all clinically useful medications, and is the second most abundant enzyme system. Alcohol and cigarettes are examples of chemicals that alter typical interactions with drugs. Alterations in these processes are called induction and inhibition. Cruciferous vegetables, such as broccoli or cabbage, may also function as either inducers or inhibitors of drug reactions. Induction occurs when the action of an enzyme increases after exposure to particular drugs or toxins. Inducers decrease the efficacy of drugs. For example, a patient with gastroesophageal reflux disease takes omeprazole 20 mg daily and then smokes soon after taking the medication. The cigarette acts as an inducer, producing lots of CYP 1A2 enzymes. Because of this, the omeprazole is quickly destroyed and the patient continues to be symptomatic because he essentially did not receive the medication. Inhibition occurs because 2 or more drugs compete for the same enzyme. Inhibition slows down metabolism, so individuals may experience an increased incidence of toxicity. For example, if verapamil could not get to the 3A4 enzyme because another drug was using it, it would hang around and keep lowering the blood pressure because it wasn't turned off. A classic inhibition reaction occurs when grapefruit juice is taken in conjunction with a calcium channel blocker. If verapamil is taken, but the 3A4 system needed for metabolism is saturated with the grapefruit, the drug can not be turned off. Increased drug effect may be associated with increased risk of toxicity as a result.

Other important enzymes in the P450 system include:

Enzyme	Action
CYP 2C19	Metabolises proton-pump inhibitors, nonsteroidal anti-inflammatory drugs (NSAIDs), β-blockers.
CYP 2C9	Metabolises sulfonylureas, NSAIDs, S-warfarin, sildenafil citrate (Viagra).
CYP 1A2	Metabolises acetaminophen, R-warfarin, theophylline, caffeine, diazepam (Valium), verapamil.
CYP 2E1	Metabolises acetaminophen, ethanol, and dextromethorphan. While it metabolises only a small fraction of drugs, it plays a significant role in activation and inactivation of toxins.

CYTOCHROME P4502E1

Cytochrome P-4502E1 (2E1) is physiologically involved in the metabolism of lipids and ketones (starvation, DM, obesity). The most significant role of 2E1 is its adaptive response to high blood ethanol levels with a corresponding acceleration of ethanol metabolism. Cytochrome P-4502E1 is inducible not only by ethanol, but also by some endogenous substrates, e.g., acetone. The associated free radicals production contributes to liver injury in the alcoholic. The induction of 2E1 results in production of acetaldehyde, which is a highly reactive and toxic metabolite. 2E1 induction is associated with proliferation of the endoplasmic reticulum, which is accompanied by enhanced activity of other cytochrome P-450. Hydroxyl radicals (OH) originate from iron-catalysed degradation of H_2O_2. Cytochrome P-4502E1 exhibits a unique ability to potentiate an iron-catalysed Fenton-type reaction, and increases rates of hydroxyl radical-mediated metabolism of substrates such as ethanol and dimethyl sulfoxide. There is increased rate of microsomal lipid peroxidation by microsomes or liposomes enriched in 2E1. Basal levels of 2E1 are 10-20 times lower in lung and kidney than liver. Cytochrome P-4502E1 is induced by fasting. Cytochrome P-4502E1, reconstituted with cytochrome b_5 and NADPH-cytochrome P-450 reductase metabolises lauric, stearic, oleic, linoleic, linolenic, and arachidonic acids. Alcohol consumption enhances fatty acid ω-oxidation via induction of cytochrome P4504A1. Ethanol promotes oxidative stress via increased generation of oxygen- and ethanol-derived free radicals at the microsomal level, especially through the intervention of the ethanol-inducible 2E1. 2E1 reduces O_2 to superoxide anion and H_2O_2 in the absence of substrates for hydroxylation and even in their presence. Membranous lipids are not the sole targets of free radicals attack. The increased microsomal generation of active oxygen derivatives could also contribute to ethanol toxicity through radical-mediated interactivation of metabolic enzymes, including 2E1 itself. Cytochrome P4502E1 induction results in increased oxidation of ethanol to acetaldehyde, a highly reactive and toxic compound. The capacity of mitochondria of alcohol-fed subjects is decreased to oxidise acetaldehyde, and is associated with unaltered or even enhanced rates of ethanol oxidation. Acetaldehyde's toxicity is due, in part, to its capacity to form protein adducts. This protein binding results in an antibody response, enzyme inactivation, decreased DNA repair, and impairment of the capacity of the liver to utilise oxygen. Acetaldehyde promotes glutathione depletion, free radical-mediated toxicity, and lipid peroxidation. Cytochrome P-4502E1 initiates lipid peroxidation, and elimination of hydroperoxides. Lipid peroxidation reflects tissue damage and plays a pathologic role, for instance by promoting collagen production. Iron overload in chronic alcohol consumption is a result of increased iron uptake by hepatocytes. Acute ethanol intoxication leads to iron exposure that accentuates the changes of lipid peroxidation and the glutathione status of the liver cell. The crucial role of 2E1 and its induction in the pathogenesis of alcoholic liver injury is through free radical and acetaldehyde generation and promotion of adduct formation and peroxidation. 2E1 inhibitors (e.g., disulfiram, diallyl sulfone) may be beneficial in preventing or improving ethanol-induced liver disorders, for instance, diallyl sulfide is a component of garlic oil, exerts a potent inhibitory effect on the induction of colon and liver cancer induced by chemical carcinogens, and also inhibits 2E1 activity. The association between alcohol misuse and an increased incidence of upper GUT and respiratory tract cancers, may be, in part, due to the effect of ethanol on the 2E1-dependent activation of procarcinogens to carcinogens. The intestines are the chief portals of entry for tobacco smoke and dietary carcinogens. Nitrosamines are metabolised by 2E1 even at low concentrations, with a resulting increased mutagen production after 2E1 induction. Uncoupling of oxidation with phosphorylation in mitochondria damaged by chronic ethanol consumption generates acetaldehyde. The combination of ethanol with tranquillisers and barbiturates also results in increased drug concentrations in the blood, sometimes to dangerously high levels as commonly observed in successful suicides. The metabolic tolerance to ethanol that develops in the heavy drinker is due to ethanol-inducible 2E1 in the MEOS. 2E1 induction is associated with increased release of toxic oxygen free radicals and the highly reactive metabolite acetaldehyde, both playing a significant role in the pathogenesis of alcoholic liver injury. The heavy drinker is vulnerable to industrial solvents, anaesthetic, commonly used medications, analgesics (acetaminophen), and carcinogens. Induction of 2E1 spills over to other microsomal P-450s, such as those involved in the degradation of retinol.

HMG-CoA REDUCTASE INHIBITORS

The cytochrome P450 (CYP450) system is often implicated in clinically significant drug interactions and subsequent adverse drug reactions. CYP3A4 is one of the major isoenzymes of the CYP450 group. Drug-drug interactions can occur when drugs that are metabolised and/or inhibited by CYP3A4 are taken concomitantly. When a medication metabolised by this enzyme system (i.e., a substrate) is taken simultaneously with an agent that decreases the activity of the same enzyme system (i.e., inhibitor), the result is often increased concentrations of the substrate, which increases the potential for adverse drug reactions. Inducers, by increasing the number of enzymes available for metabolism, may increase the metabolism of substrates, generally leading to a decreased drug effect. Drug interactions between certain HMG-CoA reductase inhibitors (statins) and drugs that inhibit CYP3A4 may result in an increased concentration of the statin and possibly myopathy and varying degrees of rhabdomyolysis (injury to the plasma membrane of skeletal muscle, resulting in leakage of its components into the blood or urine). The cytochrome P450 Enzyme System (CYP450 enzyme system) is a key pathway for drug metabolism. Many lipophilic drugs must undergo biotransformation to more hydrophilic compounds to be excreted from the body. The majority of drugs undergo phase I metabolism (e.g., oxidation, reduction) by CYP450 enzymes. The most common phase I reaction is oxidation, which involves the insertion of an oxygen atom into the compound to form a polar hydroxyl group. Other drugs undergo phase 2 metabolism and are directly conjugated without CYP450 biotransformation. The cytochrome P450 superfamily of haemoproteins can be divided into families, subfamilies, and/or single enzymes. The letters "CYP" represents "cytochrome P450", followed by an Arabic numeral denoting the family, a letter representing the subfamily (when 2 or more exist) and another Arabic numeral designating the individual gene within the subfamily (e.g., CYP2D6). Each enzyme is termed an isoform (or isoenzyme), since it is derived from a different gene. The CYP3A4 isoenzyme accounts for nearly 60% of the total CYP450 in the liver and 70% in the intestine. Extrahepatic expression of CYP3A4 is important in the biotransformation of many drugs. Extensive metabolism by CYP3A4 in the gastrointestinal tract contributes to the poor oral bioavailability of many drugs. Many clinically significant drug interactions have resulted from the inhibition or induction of CYP3A4. A large number of drugs are either substrates, inducers, or inhibitors of CYP3A4; coupled with the increased awareness of the role of CYP3A4 in drug metabolism. P-glycoprotein, an ATP-dependent drug transport protein, also plays a role in the secretion/elimination of certain drugs. P-glycoprotein reduces the oral bioavailability of certain drugs by detecting and expelling the drugs as they enter the intestinal plasma membrane, thus decreasing their intestinal absorption. There is coordination between CYP3A4 and P-glycoprotein in the plasma membrane of the intestine. A drug entering the intestinal epithelium may be effluxed (pumped) back into the intestine via P-glycoprotein and/or metabolised by CYP3A4 with the residual native drug absorbed across the epithelium. The activity of P-glycoprotein is subject to induction and inhibition. Rifampicin decreases the bioavailability of cyclosporine A not only by inducing CYP3A4, leading to increased cyclosporine metabolism, but also by inducing P-glycoprotein, leading to decreased oral cyclosporine absorption. The most serious adverse effects associated with the statins involve the skeletal muscle and may range from muscle aches and weakness to myopathy. Myopathy is defined as muscle aches or weakness in conjunction with increases in creatine phosphokinase (CK) values to values that are >10 times the upper limit of normal (ULN) to, rarely, rhabdomyolysis (actual skeletal muscle destruction). Toxic levels of a statin can cause direct injury to the plasma membrane of the skeletal muscle. The leakage of myoglobin into the blood or urine can subsequently cause ARF and death. Statins that are substrates of CYP3A4 have the greatest potential for interacting with drugs known to inhibit the CYP450 system, increasing the concentrations of substrate and the potential for adverse drug interactions. Lovastatin, simvastatin, and atorvastatin are substrates of CYP3A4 (also they may inhibit CYP2C9 to a minor extent). Fluvastatin is significantly metabolised by CYP2C9 and cerivastatin is metabolised by 2 isoenzymes, CYP3A4 and CYP2C8. In contrast, pravastatin is not extensively metabolised by either CYP3A4 or CYP2C9. These differences in drug metabolism among the statins may account for their different drug-interaction profiles. All inhibitors of CYP3A4 such as itraconazole, erythromycin, and grapefruit juice can produce interactions with lovastatin. Grapefruit juice has been shown to increase the bioavailability of many other drugs metabolised by CYP3A4. Grapefruit juice increases the serum concentrations of simvastatin and simvastatin acid when taken concomitantly. The coadministration of itraconazole with simvastatin has been associated with rhabdomyolysis. Itraconazole significantly increases the serum concentration of simvastatin and its active metabolite, simvastatin acid. The concomitant administration of cyclosporine with simvastatin in heart transplant patients causes a reduced metabolic clearance of simvastatin and buildup of active metabolite. Grapefruit juice causes a significant increase in AUC, time of peak concentration, and elimination half-life of atorvastatin acid and a metabolite, atorvastatin lactone. Grapefruit juice prolongs the time to peak pravastatin concentration. Cerivastatin is metabolised by CYP3A4 and CYP2C8. The concomitant administration of fluvastatin with drugs that are CYP2C9 substrates may lead to an increase in serum concentrations of these drugs. The concomitant administration of fluvastatin with S-warfarin, may inhibit S-warfarin metabolism and produce marked increases in prothrombin time measurements and bleeding. Warfarin is metabolised by CYP2C9. The concomitant administration of fluvastatin with certain oral hypoglycaemics may result in poor glycaemic control and low blood glucose concentrations. The coadministration of fluvastatin with certain nonsteroidal anti-inflammatory agents may increase the frequency and severity of adverse drug reactions associated with this class of drugs, such as gastritis and nephrotoxicity. Pravastatin is not extensively metabolised by CYP3A4, CYP2C8, or CYP2C9. If administered with protease inhibitors, HMG-CoA reductase inhibitors that are CYP3A4 inhibitors may increase the possibility for drug interactions in this population.

HEAT SHOCK PROTEINS

Heat shock proteins appeared to involve the phenomenon of thermotolerance. The stress proteins are divided into: the heat shock proteins (HSPs) and the glucose-regulated proteins (GRPs). The GRPs are expressed in response to glucose starvation, calcium ionophores, and anoxia. The HSPs are group of stress proteins ranging in molecular mass from 8-110 kDa.

HSPs:

Name	size	Localisation	Functions
Ubiquitin	8 kDa	Cytosol/nucleus.	Nonlysosomal degradation pathways.
HSP27	27 kDa	Cytosol/nucleus.	Regular of actin cytoskeleton; molecular chaperone; cytoprotection.
Heme oxygenase	32 kDa	Bound to ER extends to cytoplasm.	Degradation of heme to bilirubin; resistance to oxidant stress.
HSP 47	47 kDa	ER.	Collagen chaperone.
HSP 60	60 kDa	Mitochondria.	Molecular chaperone.
HSP 70 family	72 kDa	Cytosol/nucleus.	Highly stress inducible; involved in cytoprotection against diverse agents.
	73 kDa	Cytosol/nucleus.	Regulation of steroid receptor activity.
HSP 110	110 kDa	Nucleolus/cytosol.	Protects nucleoli from heat stress.

Heme oxygenase (HO) catalyses the first and rate-limiting step in the oxidative degradation of heme to bilirubin. It is a major stress protein (HSP 32), which is induced by multiple stresses, particularly oxidant stress. HO exists in two isomers: HO-1 is highly stress inducible, and HO-2 expressed constitutively. The HSP 70 family posses ATP-binding and protein-binding domains thought to be important for their roles in protein folding, maturation, and transport. HSP 73 is expressed consitutively in the cytosol and nucleus and is thought to function as a chaperone under basal conditions. HSP 72 is highly stress inducible, plays a critical role in stress response-mediated cytoprotection. Stress proteins are induced by thermal and nonthermal forms of cellular stress. Nonthermal inducers include, oxidant injury, inflammation, and ischaemia-reperfusion. The stress proteins are regulated primarily by changes in gene transcription. Transcriptional regulation of HSP 70 is mediated by a transcription factor heat shock factor (HSF). HSF is present as a large cytosolic pool of inactive monomers. In response to cellular stress, HSF trimerizes and translocates to the nucleus to bind the DNA consensus sequence in the HSP 70 promoter termed the heat shock element (HSE). Cytoplasmic accumulation of denatured or abnormally folded proteins is a key proximal signal for initiation of the stress response. HSF is complexed with HSP 70 under basal conditions, thus rendering HSF inactive. During stress, denatured proteins are bound by existing intracellular pools of HSP 70, causing a relative depletion of unbound HSP 70. The decreased level of intracellular HSP 70 shifts the equilibrium between HSF and HSP 70, liberating HSF to trimerize, translocate to the nucleus, and activate HSP 70 transcription via high-affinity binding with the HSE. When the level of the newly synthesised HSP 70 reaches some critical level, the equilibrium between HSF and HSP 70 is restored, and HSF activation is attenuated. HSF activity is regulated by HSP 70. Transcriptional regulation of HO-1 is more complex than that of HSP 70. The HO-1 promoter contains several potential inducer-enhancer elements. The dominant stress-responsive element in the HO-1 promoter appears to be a 10-bp sequence. This stress response element contains the binding site for the activator protein-1 (AP-1) family of transcription factors. Stress proteins have an important cytoprotective role during inflammation and injury.

Effects of cellular stressors:

Agent	Stress protein induced	Note
Heat	HSP 72, 110, 90, 27	Most universal signal.
Sodium arsenite	as above, HO	Inhibitor of energy metabolism.
Oxidants	Heme oxygenase (HO)	Less consistent induction of other stress proteins.
Ischaemia/reperfusion	HSP 72	Possibly via generation of oxidants.
Inflammation	HSP 72, HO	Endotoxin and TNF.
Metabolic stress	Variable	Anoxia, ischaemia, and energy depletion.
Herbimycin	HSP	Tyrosine kinase inhibitor.
Haemoglobin	HO	
Restraint	HSP 72 and 27	Mediated by neural-humoral axis.

Induction of stress proteins protected against the lethal effects of endotoxin and attenuated endotoxin-mediated cardiovascular dysfunction. Induction of HO-1 by haemoglobin protects against endotoxin-mediated shock and organ injury. Stress proteins transiently stabilise and refold damaged intracellular proteins and prevent intracellular protein aggregation during times of stress. HO-mediated protection against oxidant stress involves several products derived from the enzymatic conversion of heme to bilirubin by HO. These products include ferritin (expressed in response to increased levels of free ion), bilirubin, and CO. Ferritin acts as an indirect antioxidant by sequestering free iron that could otherwise participate in hydroxyl radical formation via the Haber-Weiss reaction, whereas bilirubin acts as a direct antioxidant. Carbon monoxide functions as a cellular messenger that modulates vascular tone and cell proliferation. During stress gene expression characterised by the rapid expression of stress protein, whereas the expression of various nonstress proteins is transiently inhibited. Induction of the stress response inhibits endotoxin-mediated iNOS mRNA expression. Increased production of NO during proinflammatory states significantly contributes to acute injury. Induction of stress response inhibits TNF-α-mediated NFκB nuclear translocation, via stabilisation of the NFκB inhibitory protein I-κB. Stress response inhibits TNF-α-mediated degradation and dissociation of I-κB. Increased expression of HSF inhibits expression of prointerleukin-1β gene, allowing ongoing expression of beneficial patterns of gene expression. The stress response is a highly conserved cellular defense mechanism defined by the rapid and specific expression of stress proteins, with concomitant transient inhibition of nonstress protein gene expression.

CRP

C-reactive protein (CRP) is released by the body in response to acute injury, infection, or other inflammatory stimuli. C-reactive protein is a leading blood marker of systemic inflammation. The activation of the acute phase response is signaled by IL-6, which produces proteins such as fibrinogen, CRP, and serum amyloid A that lead to inflammatory reactions. Conditions associated with an elevated CRP include heart disease/atherosclerosis, strokes, obesity, dental disease, blood sugar disorders, AD, arthritis, cancer, viral diseases, smoking tobacco, use of estradiol, hidden bacterial infections, and ageing. High CRP levels are a better indicator than either total cholesterol, low density lipoprotein cholesterol, or homocysteine in predicting the risk of a heart attack, as well as of death in the first month after coronary-artery bypass surgery. CRP is present in lesions that form on blood vessel walls, and is strongly associated with the rupture of these lesions, which can lead to dangerous blood vessel clots. The prevalence of coronary artery disease increases 1.5 fold for each doubling of CRP level. During 1821-1902, pathologists hypothesized that inflammation was the causative factor in the atherogenic process. Decades later, it was confirmed that increased monocytes (WBCs critical in early plaque development) and macrophages (mononuclear phagocytic cells capable of scavenging and ingesting dead tissue and degenerated cells) are present, particularly at points of plaque rupture. CRP and several other inflammatory markers may be elevated many years prior to a coronary event. CRP causes cells in the arteries to produce higher levels of plasminogen activator inhibitor-1 (PAI-1), which inhibits the breakdown of clots, and is also a strong risk marker for heart disease, especially in diabetics. CRP activation of PAI-1 in aortic cells causes lesions in the arteries that ultimately lead to formation of plaque and blood clots. Cholesterol levels alone could significantly underestimate a patient's risk level. CRP and PAI-1 are linked to diabetes and insulin-resistant syndrome. CRP damages the blood vessel wall by blocking a critical protector protein and inhibiting nitric oxide (NO). CRP has multiple, independent effects that cause heart disease, and activation of PAI-1 may not be related to the NO inhibition. With very subtle changes in CRP levels, even within the normal range, there's a huge increase in cardiovascular disease risk. CRP binds with LDL cholesterol (a union that increases stickiness and increases vascular adherence), which prepare LDL cholesterol for uptake by macrophages and increasing the formation of foam cells. Macrophages, gorged with fats contained in blood, become bloated and develop into foam cells. When they have reached their maximum load, they explode, discharging their fatty contents into the blood vessel wall at the site of injury. The presence of added fat signals the need for more macrophages to clean up the mess. They stuff themselves, explode, and the cycle starts anew. By causing LDL cholesterol to oxidize into a more reactive, abrasive form, CRP becomes an initiator in this vicious cycle. Elderly who have the herpes simplex I virus have twice the risk of having a heart attack or dying from heart disease as those never infected by the virus. The cytomegalovirus, herpes, and hepatitis B and C infection end up as arterial disease. Bacteria may gain entry into the heart via immune cells, most likely activated in the process of clearing infections from the respiratory passages. The bacteria most suspected of initiating coronary problems are C pneumoniae, P aerogenes, E endocarditis, S aureus, E faecalis, C albicans, V streptococcus, and H pylori. Leukocytosis is associated with an increased coronary risk by diminishing blood flow to the heart muscle and encouraging blood clot formation. The higher the white blood cell count, the higher a patient's risk of death from a heart attack or of developing congestive heart failure. C pneumoniae is a specific microbial antigen that causes inflammation and atherosclerotic cells to proliferate. Bacterial infections may not induce significant increases in the CRP, and inflammation that induces CRP may not be due to bacteria. Human fat cells, especially those that form around the abdomen, release the pro-inflammatory cytokine TNFα, IL1, and IL6 induces low-grade systemic inflammation, and persons with excess body fat are likely to have higher levels of CRP. Waist-to-hip ratio is positively associated with raised CRP levels, independent of body size. Obesity is partly an inflammatory disorder, and body fat promotes inflammation. Being obese increases the risk of diabetes, heart disease, and other disorders. CRP levels are generally elevated in overweight children as well as adults. Insulin resistance, syndrome X, and diabetes are all associated with increased levels of CRP, and increased risk of coronary artery disease. High CRP levels have been found in patients with the inflammatory brain disorder Alzheimer's disease. Patients with arthritis and cancer tend to have high CRP levels. Individuals with periodontal disease have elevated CRP levels, which may be the result of chronic infection or inflammation of the gums. It may also reflect inadequate levels of antioxidants. Inflammatory effects from periodontal disease, a chronic bacterial infection of the gums, cause oral bacterial byproducts to enter the bloodstream and trigger the liver to make acute phase proteins such as CRP that inflame arteries and promote blood clot formation. Patients with influenza A, a flu virus, tend to have much higher levels of CRP. Tobacco smoke raises CRP levels, and remains elevated in ex-smokers. The increased risk of cardiovascular events seen with female hormone therapy is related to an initial increase in CRP levels after starting hormone replacement therapy. The median CRP approximately doubled with age, from approximately 1 mg/l in the youngest decade to approximately 2 mg/l in the oldest, and tend to be higher in females. Intervention: High intake of vitamin E reduces CRP levels. There are at least 8 forms of vitamin E; the most common form available is α-tocopherol, which is not necessarily the most effective. HMG-CoA reductase inhibitor, pravastatin (12-24 weeks) significantly reduces the level of CRP. Pravastatin may be acting as an anti-inflammatory. Red yeast rice contains a naturally-occurring statin (lovastatin) as well as other cholesterol-lowering compounds, some with antioxidant effects. Polyphenolic compounds present in virgin olive oil also have anti-inflammatory and antioxidative effects in cardiovascular disease. The pro-inflammatory immune cytokines may be suppressed by DHA, the hormone DHEA, vitamin K, and nettle leaf extract.Ê CRP causes depletion of vitamins A, C, and E, as well as carotenoids, zinc, and selenium, which are nutrients for cardiac health. Reducing refined carbohydrates and high-glycaemic foods, engaging in moderate physical activity, and losing weight are essential.

FIBRINOGEN

Fibrinogen is a protein that plays a key role in blood clotting. Fibrinogen is a powerful predictor of stroke including fatal and nonfatal strokes, first time strokes, and haemorrhagic and ischaemic strokes. Fibrinogen's association with increased mortality is probably directly related to its ability to promote thromboses, or clots, by causing platelets to clump inside blood vessels. This is one of the main mechanisms underlying ischaemia and heart attack. Exercise, quitting smoking, and certain medications have been shown to lower fibrinogen in the short term. Elevated fibrinogen is a major risk factor for coronary heart disease (heart attacks) and cerebrovascular disease (strokes), which together account for about 60% of deaths in the elderly. Fibrinogen may possibly be the major risk factor, exceeding the contributions of homocysteine, cholesterol and other lipids in the pathogenesis of these diseases. Elevated fibrinogen levels have also been associated with a number of other diseases, including cancer, diabetes and hypertension. Fibrinogen levels rise about 25 mg/dl per decade of age. Antioxidants can reduce fibrinogen levels and the presumably the inherent risk of hyper-fibrinogenaemia-related diseases.

CYTOKINES

Cytokines (CKs) are glycopeptides (5-50 kDa) that act in picomolar concentrations in paracrine, autocrine and endocrine ways in the regulation of inflammation, cell-growth and -differentiation. CKs may act in synergism or antagonism and display countless, unrelated and often contradictory bioactivities depending on the nature of the target cells and the context in which the particular CK is present. Many CKs have overlapping effects.

Major cytokines:

Cytokine	Group	MW (KDa)	Producing cells	Target cells and functions
IL1α	2	17.5	Many cells, macrophages, lymphocytes.	Many cells, lymphocytes, hepatocytes. Increase fever and acute phase protein synthesis. Increase thymocyte and T cell activation and B cell growth differentiation, and immunoglobulin secretion.
IL1β	2	17.5	Epithelial cells, endothelial cells.	Cynovial cells. Increase fever and acute phase protein synthesis. Increase thymocyte and T cell activation and B cell growth differentiation, and immunoglobulin secretion.
IL2	1	15-20	T-lymphocytes.	Increase growth and differentiation of T- and B-lymphocytes, NK-cells.
IL3	1	14-30	T-lymphocytes.	Increase growth of Haematopoietic stem cells, basophils, and mast cells.
IL4	1	15-19	T-lymphocytes.	T- and B-lymphocytes, haematopoietic precursors. Increase differentiation of B cells and Th2 cells, increase IgG, and IgE synthesis, decrease proliferatory Th1 cell and MØ function.
IL5	1	45	T-lymphocytes.	Increase growth and differentiation haematopoietic precursors, eosinophils, B-lymphocytes.
IL6	2	26	Many cells, macrophages, lymphocytes, epithelial cells, endothelial cells.	Many cells, lymphocytes, bone marrow precursors, and hepatocytes. Increase acute phase protein synthesis, increase thymocyte and T cell activation, increase B cell growth, differentiation and Ig production.
IL7	1	20-28	Bone marrow and thymic stromal cells.	Growth of pre-T- and B-cells. Increase growth and differentiation of pre-T cells and mature T cells.
IL9	1	32-39	T-lymphocytes.	T-lymphocytes, erythrocyte and mast precursors. Increase T cell and mast cell activation. Increase IL-4-induced IgE and IgG expression.
IL10	1	35-40	T-lymphocytes, B-lymphocytes, macrophages.	T-lymphocytes, macrophages, B-cell precursors. Decrease Th1, NK cell and MØ function including cytokine synthesis/release. Increase B cell and mast cell proliferation.
IL11	1	23	Stromal cells.	Haematopoietic stem cells, magacryocytes, and plasmacytomas. Increase megalocyte growth.
IL12	1	35	Monocytes, B-lymphocytes.	T- and NK-cells. Increase NK cells, CTL and Th1 generation; increase IFNγ production by NK cells and T cells. Increase NK and ADCC activity, costimulates T cell proliferation.
IL13	1	9-17	T-lymphocytes.	B-lymphocytes, NK-cells. Growth factor for B cells, increase IgM, IgE, and IgG4 synthesis, increase T cell membrane CD32 and MHC class II Aq expression, decrease monocyte/MØ functions including proinflammatory cytokine synthesis.
IL14	1	60	T-lymphocytes, follicular dendritic cells.	Haematopoietic stem cells. Increase B cell proliferation and memory B cell generation, decrease immunoglobulin synthesis.
IL15	1	14-15	All cells.	T- and B-lymphocytes, NK-cells.
IL16		17	T cells, brain, thymus, spleen and pancreas.	Chemotactic for CD4⁺ cells, induces proliferation of T cell.
IL17		30-38	CD4⁺ T cells.	Increase epithelial, endothelial, and fibroblastic cells to secrete IL6, IL8, and GM-CSF.
IL18		18	Monocytes/MØ.	Increase INFγ production by T cells, increases NK activity.
TNFα	2	17	Macrophages, neutrophils, NK-cells.	Many different cells. Expressed as cell surface homotrimer, shed in soluble form by enzymatic cleavage, increase fever and septic shock, cytotoxic for many tumour cell types.
TNFβ	2	20-25	T-lymphocytes.	Many different cells. Secreted as homotrimer or complexed with LT-β and expressed as cell surface heterotrimer, involved in lymphoid tissue organogenesis.
IFNα	-		Leukocytes.	Many different cells.
IFNβ	2		Many cells, fibroblasts.	Many different cells.
IFNγ	2	40-70	T-lymphocytes, NK-cells.	Many different cells. Antiviral, increase MØ and NK cell function, increase MHC class I and II surface antigen expression.
NTGFβ	-		Many cells.	Many different cells.
G-CSF		18-22	BMC stromal cells and monocytes/MØ.	Precursor and mature granulocytes. Increase growth, differentiation and activation.
M-CSF		45-90	T cells and monocytes/MØ.	Precursor and mature MØs. Increase growth, differentiation and activation.
GM-CSF		22	T cells and monocytes/MØ.	Mature granulocytes and MØs. Increase growth, differentiation and activation.
LIF		46	T cells, myelomonocytic lineages.	Increase acute phase protein synthesis; increase MØ differentiation and haemopoietic stem cell proliferation.

At the molecular level, overlapping CKs either share identical receptor associated signal transducer or their distinct receptors, intrinsically, activate overlapping sets of intracellular messengers. CK gene disruption reveals that lack of a single CK is either: 1- incompatible with life; 2- associated with fertility dysfunction; 3- results in quite unexpected and complex disorders; or 4- causes minor disease. X-linked severe combined immunodeficiency is due to a multiple CK-deficiency syndrome caused by mutations in the gene encoding the signal transducing γ-chain of the IL-2, IL-4, IL-7, and IL-15 receptor system. Overactivity of one or more cytokines is pathogenic or even fatal, and overactivity of some CKs is believed to be involved in the pathogenesis of a variety of disorders. The CK-auto-antibodies (CK-aAb), are the single most important serum proteins that bind the CKs and specifically interfere with measurements and biological activities of the corresponding CKs. CK-aAb are exchangeable in the extracellular phase, which may exceed local inflammatory or tumour tissue volumes considerably. Some CK-aAb may potentiate killing of tumour cells via their binding to cell membrane associated CKs and subsequent triggering of Fc dependent effector mechanisms. The aAb are frequently present in healthy individuals, and the aAb and their corresponding CKs are concentrations of CK-aAb and CK/CK-aAb complexes, bind to and taken up the cells that under physiological conditions are competent to handle and inactivate them. The identity of cytokines is determined by their physiochemical as well as biological properties. Bioassays are an essential component in cytokine characterisation. The detection sensitivity of bioassays is often greater than immunoassays. Bioassays assess intact, biologically active cytokine. Immunoassays measure degradation products, inactive pro-cytokines or inactive isomers.

STATs AND CYTOKINES

Signal transducers and activators of transcription (STATs) are latent cytosolic transcription factors that are activated by Janus family (Jak) tyrosine kinases after cellular stimulation by cytokines. There are 4 Jak kinases and 7 STAT proteins. Cytokines bind to specific receptor chains, including their dimerization or oligomerization. Jak kinases (associated with receptor chains) respond by phosphorylating each other and also tyrosine residues on the receptor cytoplasmic domains, creating phosphotyrosine docking sites for the Src homology 2 domains of STAT proteins. After their phosphorylation, bound STAT proteins dissociate from the receptor and dimerize forming homo- or hetero-dimers. These have different binding specificities for DNA, and they presumably differ in their potency as transcriptional activators. Additional levels of regulation are probably also provided by selective interactions of different STATs with other accessory transcription factors. The variety of regulatory mechanisms may explain why many cytokines activate the same STATs nevertheless induce expression of different genes. This classification depends on the functional ability of various cytokines to promote growth of primitive haematopoietic progenitor cells. Cytokines are soluble peptidic mediators that mediate interactions between immunocompetent and haematopoietic cells and between the immune and the neuroendocrine systems. They are produced by activated cells and exert their biological activities upon binding to specific receptors expressed on target cells. The major cytokines have characterised into: 1- Group one, produced by one or a small number of specialised cell types and have target restriction, often with autocrine activity. 2- Group two, produced by many different cells, within and outside of the immune system, and act on many target cells in the body. IL1s, IL6 and TNFs, are pro-inflammatory cytokines, the production of which is induced not only by immunological stimuli, but also by bacterial endotoxins, as well as by physical and chemical irritants. 3- Group three, produced at a site of an inflammatory reaction and attract cells, such as a lymphocytes, neutrophils or monocytes. They allowing sustaining immune responses, and known as chemokines. Chemokines activity is exerted by members of the cytokine network, such as IL8, MIPα. Many cytokines do the same job. Most cytokines have different, and sometimes opposite activities. Most of cytokines are synthesise for a short period after the mitogenic or antigenic stimulus, with exception of two: A- IL1β, produced under a precursor form. The precursor is then cleaved intracellularly by a specific enzyme, called IL1β-converting enzyme (ICE) present in many cells independently of the presence of IL1. IL1β is then secreted in the supernatant under its active form. B- TNFα, anchored in the membrane of the macrophages and neutrophils under an active form. The cells that express TNFα can destroy other cells and induce inflammatory reactions. Upon cell activation, metalloproteinase cleaves the membrane-associated TNFα allowing the release of the cytokine in the milieu. TNFα then exerts its biological activities at a distance.

CYTOKINES IN IMMUNE RESPONSES

The first step of an immune response is the interaction of a lymphocyte with an antigen, followed by subsequent lymphocyte activation. All antigens, to initiate an immune reaction, have to be presented to T-lymphocytes. When a foreign antigen penetrates the body, e.g., bacterium, it does not bind in its native form to the antigen receptors expressed on T-cells (the T-cell receptors TCR). The first step leading to antigen recognition is antigen endocytosis or phagocytosis by specialised cells that digest antigenic proteins into peptides and associate the resulting peptides to MHC Class II molecules. Antigen presenting cells (APC) are diverse. T-helper lymphocytes, equipped with a receptor recognising a peptide, interact with the molecular complex formed by the peptide presented in the groove of the acid MHC molecules. T-cell-produced IFNγ increases the number of MHC molecules at the surface of APC. APC-produced IL1 activates the proliferation of T-cells as well as induces the expression of receptors for IL2. IL12 orientates the T-cell response by inducing the production of IL2 and IFNγ. T-lymphocytes (Tc) and natural killer (NK)-cells are capable of killing target cells. Tc are equipped with a receptor that binds antigen CD4-positive cells are usually TH-cells and CD8-positive lymphocytes are often Tc-cells. Antigen taken up from the extracellular milieu, or resulting from active immunisation, will activate TH-cells whereas intracellular antigen activate Tc. NK-cells, do not recognise antigen in a conventional sense, but are equipped with receptors that do not have the high diversity of TCR. The NK receptors allow the interaction of NK-cells with monomorphic determinants expressed on different cells, especially when they are cancerous or infected by a virus. To exert their cytotoxic activity, Tc lymphocytes require the presence of MHC Class I molecules, presenting a peptide recognised by their TCR, while

NK-cells are active to eliminate MHC Class I-negative-or weakly positive-target cells, as are many human tumours. IL2 is the major growth and differentiation factor of NK-cells, IL12 and IFNγ acting as co-factors to increase their cytolytic potential. A third cellular population can be induced to become cytotoxic by IL2 and IFNγ. Monocytes activated by these cytokines become cytostatic, or even cytotoxic, towards target cells. Once the first signal is given by a cell-to-cell interaction, via specific receptors, IL2 stimulates the precursors of all types of cytotoxic cells to proliferate and differentiate into effector cells. Depending on the cell type, other cytokines act as cofactors. The generation of cytotoxic cells, follows the same rules whatever cell is concerned, only the nature of the initiating signal differs. Primary antibody production: Antibody production is the property of B-lymphocytes. Generated in the bone marrow, using membrane immunoglobulin as an antigen receptor, B-lymphocytes circulate in the body and accumulate in the secondary lymphoid organs, e.g., spleen, lymph nodes. The B-cell antigen receptor binds native antigen. B-cells that express membrane IgM and IgD bind antigen that initiates their activation. The cells are then driven from a resting G_o phase to an active G_1 phase, followed by DNA synthesis during the S phase, mitosis and terminal differentiation to IgM-producing plasma cells. After antigen triggering, IL4 pushes B-cells to the G_1 phase, then IL5 allows the G_1/S transition followed by B-cell proliferation. Finally IL6 acts as a late differentiation factor to promote efficient plasma-cell maturation. Secondary antibody responses and isotypic regulation: If during a primary antibody response, the immunoglobulin isotype predominantly produced is IgM, secondary encounters with the same antigen induces isotype switching resulting in the production of antibodies of the IgG, IgA or IgE isotypes. Isotypes switching requires genetic rearrangements that take place in activated B-cells eliminating the gene coding for the IgM heavy chain. Isotype switching takes place under certain conditions and in specific places of the body, e.g., allergens induce IgE production, which, upon binding to high affinity receptors and basophils, initiate allergic reactions. Cytokines are the major factors that govern isotype switching and orientate the antibody responses towards IgE, IgA or certain IgG subclasses.

CYTOKINES IN INFLAMMATION

Inflammation is the physiological response to aggressive agents, from mechanical, chemical or biological (infections). Always starting as a local reaction, followed by systemic phenomena. Aimed at controlling and eliminating the aggressive agents, the inflammatory process can become pathogenic if hypertensive or directed towards self-components. Inflammation is characterised by a local reaction that includes the activation of the clotting process, plasmatic extravasation, complement activation, the activation of chemotaxis of leucocytes and monocytes and the production of various mediators, especially cytokines. The production of TNFα, followed by IL1β, IL6, IL8 and other cytokines. In inflammation, IL1 has activities similar to TNFα while IL6, IL8 and IL11, act as co-factors, with IL6 and IL11 increasing the production of acute phase proteins while IL8 is chemotactic for neutrophils. A typical inflammatory reaction starts with the induction of IL1 and TNF of the production of prostaglandins. This can occur in almost all cells of the body and reflects the neosynthesis of phospholipase, cyclo-oxygenase and lipo-oxygenase. In conjuction with platelet activating factor (PAF) induced in certain cells, prostaglandins increase vascular permeability and the production of toxic free radicals. Vascular endothelial cells play a crucial role in inflammation. Being at the interface between blood and tissues, they control the adhesion and migration of blood cells to the inflamed tissues. IL1 and TNF increase the expression of adhesion molecules at surface of endothelial cells, facilitating blood cell adhesion. In conjunction with IL8, these cytokines increase vascular permeability allowing cell and protein extravasation. Activated monocytes and neutrophils, attracted by IL8, MIP-1 and other chemostatic factors, aggravate this process by producing thromboxane and PAF. Cytokines are major factors in these systemic effects because of their interactions with neuroendocrine systems, helping to resolve infection or, if hyperreactive, responsible for deleterious pathological effects, that may lead to death. Glucocorticoids are potent immunosuppressive molecules and induce the apoptosis of non-mature thymocytes. Neuropeptides, such as serotonin, catecholamines, bradykinin, somatostatin or endorphins play a major role in inflammation by increasing vascular permeability and under certain conditions, may modulate antibody production. Systemic effects of cytokines: IL1, TNF, IL6 and IL8 have pyrogenic activities. IL1 effect results from a central effect of the cytokine, which induces eicosanoid synthesis and the production of corticotropin-releasing factor. Cytokines produced by cerebral cells, modify behavioural functions such as somnolence, anorexia or depression. IL1 behaves as a true brain modulator, produced and acts on the hypothalamus increasing the duration for slow-wave sleep, inducing anorexia and reduced interest in social activities. Cytokines are not only immunoregulatory molecules, but are involved in the general homeostasis of the organism. The immune system produces substances that mediate neuroendocrine activities. Proinflammatory cytokines have a dual role in to induce expression of a variety of genes and the synthesis of proteins that in turn induce acute and chronic inflammation. Whereas on the other hand, they serve as alarm systems, which lead to increase in a variety of defense mechanisms, especially immunologic and haematologic responses. IL-1β levels correlates with survival, whereas higher TNFα levels correlates with disease severity. Although IL1β and TNFα plasma content correlates with body temperature, plasma levels also fluctuates with transient increases that does not correspond to changes in the individual patient's clinical condition or severity of illness. IL1 and TNF cause dramatic alterations in carbohydrate and energy metabolism. IL1 reduces the redox state in the hepatocyte by inducing the local production of NO. IL1 as a mediator of sepsis induces skeletal muscle catabolism. There may be an important role for proinflammatory cytokines within minutes of the initiation of inflammatory diseases, as an alarm system to signal the cascade of inflammatory diseases, as an alarm system to signal the cascade of inflammatory events to follow. Within minutes to hours, profound antagonist processes begin to dominate the host response and control the proinflammatory systemic response. These systemic mechanisms may have developed to produce large excesses of antagonists to prevent local inflammatory products from spilling over into the systemic system and resulting in a generalised activation of the inflammatory system in an uncontrolled manner. Therefore, 30- to 100,000-fold greater capacity in the plasma is evidently necessary to ensure that the intense acute inflammatory cytokines remains localised.

REGULATION OF CYTOKINE PRODUCTION

TH1 and TH2, play a major role in orienting immune response by the set of cytokines they produce. TH1-cells produce IFNγ, IL2 and lymphotoxin. TH2-cells release IL4, IL5 and IL6. TH1 cytokines are involved in cell mediated rather than in humoral responses, while TH2 cytokines are known as supporting antibody production and inflammatory reactions. Although most cytokines are synthesised either by TH1- or TH2-cells, a few mediators are produced by both, which are haematopoietic growth factors (IL3, GM-CSF) that do not directly interfere with immune response. IL2 is a growth factor for T-cells, NK-cells and activates monocytes. IFNγ is a potent co-factor for the activation of NK-cells and monocytes/macrophages, potentiating the effect of IL2. Natural Killer (NK)-cells, lack a classical antigen receptor, but bind to target cells via a non-polymorphic receptor, which recognises monomorphic determinants on various molecules. Upon binding to their targets, NK-cells are activated and their lytic machinery will allow the release of cytolytic granules and eventually the elimination of the target cells. NK-cells can also be activated by other receptors. The interaction of NK-cells with IgG sensitised target cells results in a very potent cytotoxic activity towards the targets, a phenomenon known as antibody dependent cellular cytotoxicity (ADCC). In both cases, triggering of NK-cells via their Fcγ receptors results in the production of cytokines. As cytotoxic T-cells, NK-cells produce lymphotoxin, which is involved in their lytic potential, as well as IL2 and IFNγ which have an autocrine growth and differentiation activity. NK-cells also respond to IL2 signalling by proliferating and by producing other cytokines, such as IFNγ. Thus NK-cells act as TH1-cells. Mast cells are equipped with receptors for immunoglobulins. Mast cells and basophils express various receptors for different Ig isotypes. They have a high affinity IgE receptor (FcεRI) responsible, when cell-bound IgE is aggregated by specific antigen, for mast cell degranulation provoking the release of the performed inflammatory mediators, such as histamine, serotonin, or PAF, which mediate the allergic reaction. Mast cells have been shown to behave as regulatory cells. Their activation through IgE receptor or type III IgG receptor (FcγRIII) induces the production of TH2 cytokines such as IL4, IL5, IL6 and IL13 making the mast cells a player in the regulation of the production of IgE, their triggering molecule. The release of TNFα and IL1β by activated mast cells reinforces their importance in inflammation. Upon stimulation by bacterial endotoxins, or by antigen-antibody complexes binding to activating IgG (FcγRI and FcγRIII) or IgA (FcαR) receptors, circulating or resident macrophages are triggered to produce IL1β, TNFα and IL6, the cytokines responsible for inflammatory reactions. Macrophages are also immunoregulatory cells, in particular, they produce IL12, an heterodimeric protein, which major function is to induce the production (by T- and NK-cells) of IFNγ and IL2. This leads to the increase of their cytotoxic potential and to a deviation of the immune response towards TH1-cells, since one of the activities of IFNγ is to inhibit the activation of TH2-cells. Macrophages have a dual role: producing inflammatory cytokines, they are major effectors of inflammatory reactions producing IL12, they influence the profile of the immune response to the antigens they present to T-lymphocytes. The factors involved in the selection of the profile of the cytokines produced by T-helper cells are cytokines themselves, produced by cells in the microenvironment. T-precursor (Tp) cells produce IL2, which sustained their own growth. IFNγ is critical factor for TH1 differentiation. It is produced by NK-cells in response to stimulus delivered by antigen-antibody complexes and by IL12. IFNγ and TGFβ can act in synergy with IL12 and IFNγ for the development of TH1-cells. The nature of the cytokine induced by various stimuli is crucial factor, which controls the profile of the immune responses by these stimuli. There is integration between TH1- and TH2- cells and cytokines in physiological and pathological responses. The cytokine network is as follow: 1- Cytokines are produced by activated cells from numerous origins. Besides immunocompetent cells, which produce cytokines involved in immune and non-immune reactions, many cells produce cytokines that play a role in inflammation and general homeostasis. 2- Cytokines are produced for a short period after cell stimulation. The stimuli are diverse:

antigens for T- and B-lymphocytes, immune complexes for NK-cells, mast cells, or macrophages, bacterial products or other cytokines. 3- Not all cells produce cytokines. 4- Under non-pathological situations, cytokines act at a short range, do not circulate in the body fluids, may act in a paracrine as well as endocrine manner. 5- Cytokines have pleiotropic and redundant activities, may synergize or inhibit each other, induce the production of other cytokines.

THE CYTOKINE RECEPTORS

Many cytokines receptors belong to the haemopoietin receptor family characterised by a 200 amino acid extracellular conserved domain with 4 positionally conserved cysteine residues and a WSXWS motif. The intracellular C terminal half of the extracellular domain of some of these receptors shows an evolutionary linkage to the fibronectin type III modules found in a series of cell surface molecules with adhesive properties. Members of the haemopoietin receptor family are expressed as a dimer with a ligand binding subunit α chain and β chain common different cytokine receptors. The interleukin 6 receptor family: IL6, leukaemia inhibiting factor (LIF), IL11, ciliary neurotrophic factor (CNTF), oncostatin M (OSM), cardiotrophin, share the gp130 molecule as a common β chain transducing signal. gp130 consists of 918 amino acids and its extracellular region comprises units of a fibronectin type III module. gp130 is found in almost all tissues. Binding of cytokine to its receptor α chain induces homodimerization of gp130 or heterodimerization of the gp130 with α chain of the receptor. When IL6 binds to the α chain of the IL6R, the complex then associates with gp130 allowing it to dimerize. The IL3-IL5-GM-CSF receptor family: High affinity receptor for IL3, IL5 and GM-CSF consists of a cytokine-specific α subunit and a common β subunit. The β subunit, a glycoprotein of about 120-240 kDa has two segments of the conserved motif of the haemopoietic receptors. The IL2 receptor family: The receptor for IL2, IL4, IL7, IL9, IL13 share a common γ chain belonging to the haematopoietic family. The high affinity of IL2 binding site consists of 3 polypeptide chains, the α, β, and γ chains. The cytoplasmic region of IL2Rγ contains sequences homologous to the SRS homology region, which activate tyrosine kinases from the JAK and STAT families. IL2Rγ is required for receptor mediated internalisation and increases cytokines binding affinity. The TNF receptor family: Activation of these receptors involves receptor trimerization. The TNF-RI is responsible for signalling cytotoxicity and the induction of several genes whereas TNF-RII is capable of signalling proliferation of primary thymocytes. TNF-RII significantly reduces the TNF concentration required for cell killing. TNF-RII regulates the rate of TNF association with TNF-RI possibly by increasing the local concentration of TNF at the cell surface through rapid ligand association and dissociation and ligand passing mechanism. TNF associates more rapidly with TNF-RII than TNF-RI. The ability to induce death is unique to this family of receptor and is well established for TNFα, LTα and FasL. The p55 TNF-R and Fas share a 65 residue homology region in the cytoplasmic domains, which are crucial for apoptotic death activity. IL1 receptors: IL1-R type I and type II are members of the immunoglobulin superfamily and are structurally related to each other. These receptors are recognised by IL1α and IL1β. IL1β has a greater affinity for the type II receptor, whereas IL1α has a higher affinity for type I receptor. IFNα/β receptor: This belongs to the interferon receptor family, which also includes the two IFNγ receptor subunits and the IL10 receptor. Generation of soluble cytokines receptors: Two mechanisms have been proposed for the formation of these soluble cytokines receptors: 1) Proteolytic cleavage of the membrane anchored receptor as shown with IL2R, TNFR, and IFNXR. TNFα receptors occur via enzymatic cleavage of cell membrane receptors with elastase. 2) Alternative splicing of cytokine receptor mRNA, which often deletes the exon coding of the transmembrane region of the protein as in the case for IL4, IL5, and IL7. For soluble receptors such as IL6Rs the two mechanisms coexists. Soluble cytokines are present in serum and urine from healthy individuals. Soluble cytokine receptors inhibit cytokine functions. Soluble cytokine receptors may act as specific carriers of cytokines and prolong their life by preventing urinary excretion or by protecting the cytokine against proteolytic degradation. Signal transduction via cytokine receptors: The binding of cytokine ligands to members of the cytokine receptor superfamily typically results in receptor homo- or hetero-dimerization and activation of protein tyrosine kinase such as members of the Src or Janus (Jak) families, which are physically associated with the receptor. Then Stat factors are rapidly tyrosine phosphorylated and subsequently dimerize and translocate to the nucleus, where they can activate transcription.

CYTOKINES AND CANCER

Cytokines can be tumour growth inhibitors, e.g., TNFα and interferons with cytotoxicity towards tumour cell. Cytokines can be tumour growth factors; e.g., IL6 is a growth factor for RCC, and prostate carcinoma. Tumour cells release cytokines, which may act as autocrine growth factors. Tumour progression is associated with difference in the sensitivity of tumour cell to cytokines. Cytokines modulate the phenotype of tumour cell. Cytokines may play a role in the promotion of tumour metastasis. IL12 inhibits cancer cell motility and invasion by upregulation of E-cadherin cell surface adhesion molecule. IL6 promotes tumour metastasis and invasion by increasing motility and decreasing adherence of cancer cell. TGFβ increases the metastatic potential of tumour cell. Cytokines are clinical prognostic markers; elevated cytokine serum levels are associated with poor outcome in patients with cancer, e.g., RCC. Cytokines are predictor factors for chemoimmunotherapy; endogenous expression of TNFα has been correlated with resistance to the cytotoxic effects of TNFα in some tumour cells. Secretion of IL6 by advanced cancer is associated with an increase in resistance to inhibitory factors such as IL1β, TNF or TGF. This may explain the correlation observed between high pretherapeutic serum IL6 levels and poor response to IL2 therapy in patients with RCC. The role of cytokines in predicting response to immunotherapy may be extended to chemotherapeutic agents. Cytokines have a role in the paraneoplastic syndrome; increased production of cytokines may contribute to disturbances in calcium homeostasis in some malignancies. Administration of anti-IL6 antibodies decreases the thrombocytosis often observed in patients with RCC. Cytokines have a role in cachexia, TNF, IL1, IL6, and IFNγ role has been emphasised in cachexia. IL6 reduces adipose lipoprotein lipase activity and this may contribute to the loss of body fat stores associated with some cases of cancer cachexia. Cytokine profile pattern switches during tumour progression, cytokines released by CD4 T-cells could be divided into 2 groups: TH_1 cytokines which include IL2, IFNγ, TNFβ, or TH_2 cytokines such as IL4, IL5, IL6, IL10. During tumour progression there is a switch towards a TH_2 phenotype. Cytokines are antitumour agents; IL2 was the first immunological agent that demonstrated an antitumour effect by activating immune effectors. The measure of IL6 and CRP levels before treatment has shown a high predictive value for response to immunotherapy in RCC. Objective clinical response rates of 10-15% have been reported after administration of IFNα in RCC. Retinoic acid and IFNγ act synergistically, induced objective clinical response in 30% of patients with RCC.

CYTOKINE REGULATION OF THE MACROPHAGE SYSTEM
Cells of the macrophage (MØ) lineage or mononuclear phagocyte system (MPS) are more complex than other myeloid cells. MPS includes strikingly different end cells, even specialised derivatives that are neither phagocytes nor mononuclear cells, e.g., osteoclasts. MØ also differ from other myeloid lineages in that, cells leaving the bone marrow may continue to proliferate, mature, and become functionally activated in various tissues. Almost every organ of the body contains mononuclear phagocytes that possess unique characteristics for that location, while still sharing common features with analogous cells in other organs. Constituent cells should be produced from common progenitors under the influence of the same primary growth factors. The end cells belonging to the system share at least some cardinal features. These include the capacity of non-immune and immune (opsonized particles) phagocytosis, the ability to collaborate with lymphocytes in generating specific immune responses, and production of secretory products including cytokines, termed monokines. The following major cell types are classically assigned to the mononuclear phagocyte system: 1) Monoblasts, promonocytes, monocytes, and MØ of the bone marrow. 2) Monocytes of the peripheral blood. 3) Tissue MØ including kupffer cells, alveolar MØ histiocytes (MØ of connective tissue), serosal MØ, synovial type A cells, and MØ of other organs, e.g., gut, genitourinary tract, and microglia. 4) Osteoclasts, bone-resorbing cells, which are not phagocytic, characterised by the enzyme tartrate-resistant acid phosphatase (AcP). 5) Dendritic cells, which are poorly phagocytic but very efficient antigen-presenting cells to resting T lymphocytes, and are critically dependent on granulocyte-macrophage (GM)-colony stimulating factor (CSF), one of the key factors responsible for MØ production. A widely held scheme of the mononuclear phagocyte production is as follows: a) haematopoietic stem cells form common neutrophil-MØ progenitor cells; b) these cells further differentiate, forming MØ progenitors that finally produce the first morphologically recognisable precursor cells, the monoblasts; c) monoblasts further divide to form monocytes that leave bone marrow and after passage for 1-2 days through peripheral blood, reach various tissue compartments; and d) in those tissues, while still having the capacity to undergo one or two divisions, they finally mature to become MØ or other specialized cells of the system. Very primitive progenitor cells, the high proliferative potential colony-forming cell (HPP)-CFC are pluripotential; they are able to proliferate under the exclusive influence of CSF-1. Progenitors committed to MØ may arise at several different levels of differentiation and the biopotential progenitors finally producing MØ may give rise not only to neutrophils but also eosinophils, mast cells, erythroid cells, and magakaryocytes. MØ is one of the earliest appearing lineages of haematopoiesis. Three different factors induce MØ formation from progenitor cells independently of each other: a) CSF-1 [also known as MØ (M)-CSF]; b) GM-CSF; and c) IL3. They could be termed primary MØ growth factors. The early synergistic factors include steel factor (SLF) and IL1; late synergistic factors include factor increasing monocytopoiesis, TNFα, and possibly other molecules. Each growth factor acts only after binding to a specific receptor. Colony stimulating factor-1, a large dimeric cytokines with homology to SLF and PDGF binds to c-fms, a dimeric receptor member of the immunoglobulin superfamily with intrinsic tyrosine kinase activity. The CSF-1 gene is located on chromosome

3, and is composed of 10 exons. There are at least 3 different (NH_2-terminal identical) protein forms of CSF-1, two of which (glycoprotein and proteoglycan) are soluble and one glycoprotein) membrane bound. It is not known whether these different forms of CSF-1 posses different physiological roles, and why such a complex system exists for just one factor. Both GM-CSF and IL3 belong to the haematopoietin family, and their receptors belong to the family of haematopoietin receptors. These receptors have unique α-subunits specific for each growth factor and serving as low-affinity receptors. The α-subunits have no transmembrane domains and to transduce a signal associate with β-subunits posing both transmembrane and intracellular domains. β-Subunits that after binding to α-subunits form high-affinity receptors, are the same for GM-CSF and IL3. These receptors do not have intrinsic tyrosine kinase activity and do not have immunoglobulin domains. It is therefore possible that 2 different biochemical pathways can induce MØ differentiation from progenitor cells: the first initiated by CSF-1, and the second initiated by GM-CSF/IL3. MØ differentiation is described as being effected by primary growth factors acting sequentially: IL3 followed by GM-CSF followed by CSF-1. Colony stimulating factor-1, which is constitutively produced by ubiquitous cells such as fibroblasts and endothelial cells, therefore may play a major role. Granulocyte-macrophage-CSF has low levels of constitutive production and is probably less significant, and IL3, which is produced only by stimulated T cells, is likely to play a marginal role. Colony stimulating factor-1 is the only primary MØ growth factor detectable in peripheral blood in the steady state and that acts in an endocrine manner. One of the CSF-1 forms may act as an endocrine factor, the others as paracrine factor, and that the membrane-bound form primarily receptor-bearing cells during cell-to-cell contact. Membrane-bound CSF-1 may be shed from the membrane and reach intercellular fluids and peripheral blood. Both GM-CSF and IL3 have only one known protein form, but GM-CSFs of different origin may differ in glycosylation. Granulocyte-macrophage-CSF, similar to CSF-1, is produced by many different cell types, but not detected in peripheral blood. IL3 is most likely produced exclusively by helper T cells, and this constitutes a potential restriction to the biological role of this molecule. These growth factors regulate proliferation and maturation of cells belonging to the system and also support their survival. MØ have been found to produce CSF-1 and GM-CSF themselves, i.e., these factors can act in an autocrine manner. Autocrine growth factor production may allow for autonomous survival and even growth of MØ in situations such as entry to foci of infection. Absence of exogenous growth factors in such foci would limit local host cells, while self-stimulation of relatively mature cells would allow them to perform their functions without endangering systemic homeostasis. At least two MØ products, TNF-α and IL1α, stimulate release of primary MØ growth factors, creating a positive feedback loop. Such loops are most likely triggered by infection, since bacterial products such as lipopolysaccharide (LPS) are potent inducers of the release of both TNFα and ILα. Vitamin D_3 also serves as a stimulator of CSF-1 production. Major candidates for negative humoral regulation are soluble growth factor receptors that bind MØ growth factors and prevent them from stimulating MØ formation. Negative feedback may operate because receptor-positive populations expanded under the influence of MØ growth factor would be expected to release/shed more soluble receptors, bind more growth factor, and in turn, diminish factor availability to cells. The organism functions in a very complex environment and has developed similarly complex mechanisms to protect itself. Each growth factor is capable of inducing terminal differentiation of MØ may independently stimulate its own factor-dependent subpopulation of MØ, and this population would then in turn regulate the production and tissue concentration of this particular growth factor. The MØ lineage is a sum of CSF-1-dependent, GM-CSF-dependent, and IL3-dependent subpopulations. The major role of peritoneal MØ is to recruit appropriate cells to the peritoneal cavity and not to exert antibacterial activity by themselves. Two opposing roles have been assigned to MØ in tumour growth. First, which exert tumouricidal action through several mechanisms such as TNFα release, direct cytotoxicity, and antibody-dependent cellular cytotoxicity. Second, as a major source of so-called fibrogenic cytokines, they may support the formation of tumour stroma and stimulate tumour vascularization. The role of CSF-1-dependent MØ is in providing the tumour with factors necessary for stroma formation and tumour vascularization, and then absence of MØ could delay tumour growth. CSF-1-dependent MØ plays a role in formation of tumour stroma and vascularization but not in natural defense against tumour. MØ are prominent in atherosclerotic lesions from early stages of fatty streak formation, consisting of endothelium-adherent monocytes, which become foam cells as a result of lipid uptake once they enter the intima. Subsequently, they interact with ECM, fibroblasts, and medial smooth muscle cells, which are stimulated to migrate to the intima. Lesions expand as a result of platelet adhesion and deposition of plasma proteins and matrix and local proliferation of MØ and other cells, counterbalanced by apoptosis. In addition to occlusion and neovascularization, MØ metalloproteinases and other lytic products are thought to contribute to plaque rupture and local haemorrhage, which may be fatal. Recruitment of monocytes and persistence of MØ depends on monocytes-specific chemokines and adhesion molecules interacting with modified/oxidised lipoproteins and other potential ligands. CSF-1 plays a key role in this type of monocyte-restricted inflammatory process and that the scavenger receptor, is responsible for adhesion/retention of these cells, as well as uptake of lipoprotein ligands and foam cell formation. The type A I and II scavenger receptors (SR-A I, II) have been implicated in MØ foam-cell formation. CSF-1 accounts for some, but not all, SR-A expression in tissues.

The macrophage (MØ) lineage is more complex than other myeloid lineages of haematopoietic cells and includes strikingly different end cells such as Kupffer cells, alveolar MØ, histocytes, serosal MØ, synovial type A cells, microglia, osteoclasts, and possibly dendritic cells. The MØ lineage is split into CSF-1-dependent and CSF-1-independent cells that are largely independently regulated. Both CSF-1 and GM-CSF are responsible for transition of cells of the MØ lineage from bone marrow to blood, and from blood to tissues, and have a critical extramedullary role. Regulation of the MØ system by CSF-1 is complex. They play a regulatory role in various tissue reactions including responses to bacterial infection, neoplasia, and atherosclerosis.

CHEMOKINES

Chemokines (chemotactic cytokines) are structurally related molecules participate in activation of leukocytes, and play a critical role in controlling the movement of these cells during inflammation. Chemokines are 8-12 KDa proteins sharing 20-70% homology in amino acid sequence, and divided into 4 families including CXC (α-chemokine), CC (β-chemokine), C (γ-chemokine) and CX$_3$C (δ-chemokine). The CXC chemokines, with the first 2 cysteine separated by amino acid residue can be further subdivided into 2 groups based on the presence of the amino acid motif, ELR (glutamic acid-leucine-arginine, at the N-terminus, e.g., IL8, GRO, ENA-78, GCP). Most ELR chemokines are angiogenic, whereas non-ELR chemokines are angiostatic. CC chemokines lack amino acid residue between the first 2 cysteines (e.g., MIP-1α, MIP-1β, RANTES, eotaxin). γ-Chemokines retain only the second and the fourth cysteine residues of the four-cysteine motif. δ-Chemokines have 3 amino acids between the first 2 cysteines. Fractalkine (neurotactin) is a membrane-bound chemokine and is expressed mainly in the brain, lung and heart. Fibroblasts are ubiquitous cells that provide much more than a source of scaffolding on which other cells function and migrate. Fibroblasts when activated by substances released during tissue injury or derived from infectious microorganisms or by other environmental factors produce chemokines that initiate the recruitment of bone-marrow-derived cells.

Major chemokines:

Chemokines	MW (KDa)	Producing cells	Target cells and functions
IL8	6-8	Monocytes, T-lymphocytes.	Neutrophils, T-lymphocytes. Chemotactic and activating for neutrophils, promote angiogenesis.
GROα	7-11	Monocytes, epithelial and endothelial cells, and tumour cells.	Chemotactic and activating for neutrophils, promote angiogenesis and growth of certain tumours.
IP-10	10-11	Endothelial cells, monocytes and fibroblasts, thymic and splenic stromal cells.	Chemotactic for activated T cells. Inhibit endothelial cell proliferation.
SDF-1	10	Stromal cells, liver, muscle.	Stimulates growth of pre-B cells. Chemotactic for monocytes and T cells.
MIG	14-15	IFNγ treated monocytes and macrophages.	Chemotactic for tumour-infiltrating lymphocytes.
MCP-1	11-17	Many cell types.	Chemotactic for T cells, induces chemotaxis and activation of monocytes.
MCP-2	7.5-11	Many cell types.	Chemotactic for T cells, induces chemotaxis and activation of monocytes.
MCP-3	11	Many cell types.	Chemotactic for T cells, induces chemotaxis and activation of monocytes.
MIP-1α	10	Many cell types.	Chemotactic for T cells and monocytes, inhibits proliferation of haemopoietic stem cells.
MIP-1β	10	Many cell types.	Chemotactic for T cells and monocytes, inhibits proliferation of haemopoietic stem cells.
RANTES	10	Many cell types.	Chemotactic for T cells, eosinophils and monocytes.
Eotaxin	8-9	Endothelial cells, alveolar macrophages, lung, intestine, heart, thymus, spleen, kidney, and liver.	Eosinophil chemotaxis.
Lymphotactin	10	Thymocytes, activated T cells.	T cell chemotaxis.
I-309	10-16	T cells, mast cells.	Neutrophils chemotaxis.

The ability to recruit specific populations of leukocytes during inflammation is the role of a family of cytokines called chemokines. Chemokines are small polypeptides that range in size from 7-10 kDa and have significant sequence identity at the amino acid level. Most chemokines have been divided into 3 different groups called the CXC, CC, and C families, so designated because of a conserved amino acid sequence of cysteines at the NH$_2$-terminal end of the protein. The CXC chemokines (α chemokines) posses 2 cysteines separated by a single unconserved amino acid residue. This amino acid sequence is important for signal transduction involved in neutrophil chemotaxis and may also play a role in chemokine stimulation of angiogenesis. The CXC chemokines have 20-50% amino acid homology, and their genes are located on human chromosome 4. The CC chemokines (β chemokines) have 2 adjacent NH$_2$-terminal cysteines. They exhibit 28-45% amino acid homology, and the encoding genes are located on the human chromosome 17. The C chemokines family contains only one member and has a single NH$_2$-terminal cysteine. The C chemokine gene localises to human chromosome 1. Most of the CXC chemokines are specifically chemotactic for neutrophils, such as IL8. IL8 is mainly chemotactic for neutrophils, but also stimulates the migration of T cells and basophils. Fibroblasts lack the tight junctions and keratin synthetic ability of epithelial cells and do not produce factor VIII or display the morphological characteristics of endothelial cells. Fibroblasts consist of subsets of cells, much like lymphocytes. Fibroblasts from different anatomic regions display characteristic phenotypes. Fibroblasts can be activated to display new functions. Chemokines exert their function through a family of 7 transmembrane receptors, G protein-coupled. They transduce their signals through heterotrimeric G proteins, making cellular responses to chemokines. The sentinel cells are likely to be the mesenchymal cells that, when stimulated by bacterial products such as LPS, generate chemokines capable of recruiting the inflammatory cells. Tissue fibroblasts, in concert with other mesenchymal cells, appear to function as key regulators of the inflammatory process in many tissues. The autoinducers are N-acyl homoserine lactone-based structures that bind and activate transcriptional activator proteins leading to the induction of gene expression. Induction of proteolytic elastases in Pseudomonas permits rapid tissue destruction providing nutrients and space for the colonising bacteria. Some autoinducer molecules stimulate in human epithelial cells the synthesis of the chemokine IL8. Pseudomonas autoinducer are powerful stimulators of IL8 and other cytokines in several types of primary human fibroblasts. Secretion of autoinducer by bacteria activate the regional host mesenchymal cells to synthesise IL8 and perhaps other chemokines, which in turn recruit immune cells to protect the tissue undergoing bacterial colonisation. Autoinducer and LPS from the outer cell wall of the bacteria may participate in the promotion of chemokine synthesis. Continued stimulation of the fibroblasts without eradication of the infectious agent may lead to persistent colonisation and chronic inflammation. Fibroblasts in a tissue are initially signalled by bacterial products such as LPS and autoinducer molecules to synthesise chemokines. This initiate recruitment of the professional haematopoietic cells (T lymphocytes, macrophages, and granulocytes) in an attempt to resolve the infection. Infiltrating immune cells further activate the fibroblasts through CD40/CD40L interactions with further chemokine production. Depending on the magnitude of these interactions and clearance of the invading microbe, either acute or chronic inflammation may occur with attendant tissue repair or fibrosis, respectively. Products of microorganisms are the initial alarm signal for resident mesenchymal cells to produce chemokines. The second defense layer for the eukaryotic host are activation receptors such as CD40 on regional mesenchymal cells. CD40 is a 50 kD integral membrane protein, which is a member of the TNFα receptor superfamily originally described on B lymphocytes and subsequently on antigen-presenting haematopoietic cells such as macrophages and dentritic cells. This receptor and its natural ligand CD40L are critical for activation of and production of antibody by B lymphocytes. CD40 on macrophages and dendritic cells functions as major pathway for activating these cells to produce proinflammatory cytokines (e.g., IL1) and chemokines (IL8). CD40L is present on activated T lymphocytes and is also displayed by mast cells, eosinophils, and basophils. Interaction between CD40L-expressing immune cells and CD40-bearing cells results in the activation of the CD40-displaying cells causing augmented expression of adhesion molecules, co-stimulatory molecules (e.g., B7) and cytokines. Human epithelial and endothelial cells and fibroblasts are capable of displaying CD40, especially after exposure to IFNγ. Triggering of CD40 with ligands for CD40 increases expression of adhesion molecules, permitting attachment of haematopoietic cells and their entry into the tissue. IL8 is one of the abundant chemokines secreted by fibroblasts and endothelial cells after CD40-mediated activation. Recruitment of neutrophils and T lymphocytes to sites of fibroblast activation would permit further interaction between these cells. This is potentially important amplification scheme that would help eradicate the invading microorganism, leading to acute inflammation and tissue repair. Alternatively, it could lead to a state of chronic inflammation, with possible fibrotic consequences, as occurs in diseases such as rheumatoid arthritis, and idiopathic pulmonary fibrosis. PGE_2 is the principle prostanoid synthesised by fibroblasts. PGE_2 appears to participate in the inflammatory response in a complex manner by its effects on immunocomponent cells. Prostaglandin-endoperoxide H synthases (PGHSs) are bifunctional enzymes involved in the conversion of AA first to PGG_2 through cyclooxygenation, which is ordinarily expressed at low levels but is inducible by mitogens, cytokines, and serum. Fibroblasts, irrespective of anatomical site of origin, express PGHS-1 at relatively high levels, and this activity accounts for basal PGE_2 production. The human PGHS-1 promoter contain 2 identifiable NF-κB sites, which are thought to be used in the induction of PGHS-2 gene transcription. PGE_2 can bias the commitment of naive T lymphocytes away from the Th1 phenotype and towards Th2 and can also influence B lymphocyte behaviour. The prostanoid can increase IL5 synthesis in Th2 lymphocytes, and plays important role in mast cell activation. Thus, anatomic regions where fibroblasts produce high levels of PGE_2 may exhibit a characteristic immunological environment predicted on the high levels of this prostanoid. The participation of fibroblasts in the inflammatory response relates to their ability to receive and respond to complex molecular cues from the pericellular cytokine and growth factor milieu. Several target genes encoding proteins proximate to cell trafficking, prostanoid biosynthesis, cell adhesion, and ECM synthesis have been identified in fibroblasts. Many of these fibroblast genes are potential targets for activation through NF-κB/Rel. NF-κB family function in very different ways depending on the cell type. Quiescent fibroblasts express low levels of RelB and that the protein is bound to IκBα and is therefore inactive. PDGF, TNFα, and serum up-regulate DNA binding of NF-κB and RelB in the fibroblast. Fibroblast's ability to function as a structural element and as a vital immunoregulatory cell suggests that the fibroblast should be considered as sentinel cell.

APOPTOSIS

Cell death exhibiting a conserved pattern of morphological alterations is a common and critical component of embryogenesis. Apoptosis is a form of cell death with morphological features quite distinct from necrosis. Apoptosis is often subtle and rapid event and is easily missed in routine light microscopic preparations. Apoptosis is clearly pathological, and difficult to define it functionally. Programmed cell death is primarily developmental and is part of the normal life of metazoan. Such programmes exist within the genome of the organism, at least implicitly as epigenetic phenomena, but are differently expressed according to some differentiation within individual groups of cells. The trigger that initiates the programme must also be within the genome of the whole organism and not an external. Apoptosis occurs in a wide variety of pathological conditions and in situations that appear to be concerned with the control of tissue size, where it is seen to counteract the mitotic activity to some extent. Apoptosis is a response to physiologic and pathologic stresses that disrupt the balanced rates of cells generation and elimination. Carcinogens in the diet can promote apoptosis of cells instigating epithelium turnover. In response to ultraviolet irradiation, epidermal cells initiate apoptotic death rather than waiting to slough. Ionising radiation, hyperthermia, introduction of toxic agents, high chemotherapeutic agents, elicit apoptotic response. Apoptosis is induced in at least population of sensitive cells by most of the cytotoxic anticancer drugs in current use. Apoptosis is an energy-dependent process and can be delayed by the inhibition of RNA or protein synthesis. An early biochemical event is the production, activation or release of endonuclease(s) that cut the DNA at linker regions between nucleosomes to form fragments of double-stranded DNA. Apoptosis is a destructive and irreversible event, and it may be that it serves to prevent repair mechanisms that might otherwise save a dangerously damaged cell. In the average human about 500,000 neutrophils die every second and are programmed to do so, but, their survival can be extended by CSF and it is thus apparent that although cells may contain programmes for their own death they do so. Apoptosis is characterised by both cytosolic and nuclear shrinkage. Apoptosis is a well-controlled mechanism by which cells undergo a carefully coordinated series of enzymatic reactions to activate their own destruction. The extracellular matrix (ECM) provides a mechanical scaffold for cells and tissues, and also helps regulate cell proliferation, differentiation, and survival. ECM proteins such as integrins, matrix metalloproteinases, and collagen are involved in regulating tumour growth, invasiveness, and angiogenesis. Collagen type-IV facilitates tumour cell invasion and migration, and types I, II, III, IV, and V regulate cell proliferation as well as migration. Collagen type-I, a structural protein that is organized into a complex network of fibrous triple helical molecules, is one of the major elements of the interstitial ECM, which lies beneath the basement membrane. Denatured or proteolyzed forms of collagen are also known to have biochemical functions, and the collagen environment has been observed to change as tumours become invasive. The endothelial cells of tumour blood vessels divide very rapidly and are much more dependent on growth factors. The endothelial cells, even the ones providing blood supply to the tumour, are nonmalignant, genetically stable cells and, therefore, much less likely to develop drug resistance. Many different types of angiogenesis inhibitors have been identified such as IFN-α; IL-12; antibodies against the vascular endothelial growth factor (VEGF), and their receptors; and traditional chemotherapeutic drugs including taxanes, camptothecin analogues, antimetabolites, anthracyclines, and platinum drugs. These drugs likely inhibit endothelial cell growth in a similar manner to the way in which they inhibit proliferation and renewal of other rapidly dividing cells, such as those in the bone marrow, hair follicle, and gut. Because of their low turnover rate, mature quiescent cells of the normal vasculature are not affected by the chemotherapeutic agents. Chemotherapy is given at maximal tolerated doses, requiring long periods without chemotherapy to allow for bone marrow recovery. This recovery period, however, provides the tumour endothelial cell compartment with an opportunity to repair the damage inflicted by chemotherapy. Antiangiogenic scheduling or metronomic dosing is a way to attack the tumour vasculature by administering lower doses of chemotherapeutic drugs on a more frequent basis, without periods of rest and for longer periods of time. Cellular insults such as hypoxia, ionising radiation, or DNA damage activate p53, which regulates a complex biochemical pathway leading to cell growth arrest or apoptosis. p53 is the most frequently mutated gene in human cancers, and its inactivation allows cells to actively proliferate and evade cell death induction. p53 becomes stabilised and accumulates after ionising radiation and DNA breakage. In normal nonstressed cells, p53 is rapidly degraded by the ubiquitin-proteasome pathway. Ionising radiation leads to phosphorylation of p53. Oncogenes, such as myc, ras or β-catenin, are cell regulatory proteins whose function becomes disrupted through mutation or other mechanisms, increasing their activity and allowing the cell to undergo uncontrolled proliferation. p53 must be inactivated for cells to become cancerous. On the other hand, overexpression of p53 confers chemoresistance to human cancer cells, protecting them from etoposide-mediated cell death. Wild-type p53 activates Fas transcription, while mutant forms of p53 inhibit transcription. Mutant p53 inhibits the induction of Fas-expression in response to etoposide. Causes of cell death, may be: 1- Internal, e.g., i- genetic error, ii- deprivation of essentials, e.g., O_2, vitamins; and iii- loss of blood supply (ischaemia). 2- External, e.g., i- physical, trauma, radiation, cold, heat; ii- chemical, ischaemia; and iii- microbial. A free radical is a molecule containing an odd number of electrons. If a free radical reacts with a non-radical, another free radical must be produced. Hence a chain reaction may commence, and even more reactive can be formed. Consequences of reduction of ATP production are: 1- active cells cease to function; 2- anaerobic glycolysis leads to the production of pyruvate and accumulation of lactic acid; 3- impairment of sodium pump, so sodium is retained in the cell and potassium escapes from it; 4- ribosome become swollen as fluid accumulates in the sacs of the rough endoplasmic reticulum; 5- ribosomes become detached from the rough endoplasmic reticulum and protein synthesis is reduced; and 6- mitochondria become swollen. The nuclear chromatin in apoptotic cells undergoes condensation, whereas the plasma membrane and mitochondria remain morphologically intact. Apoptosis is a term invented to describe the scattered, apparently random cell deaths in healthy tissue. Morphology of apoptosis includes: cell shrinkage, plasma and nuclear membrane blebbing, organelle relocalisation and compaction, chromatin condensation, and production of membrane-enclosed particles containing intracellular material known as apoptotic bodies. Cytotoxic agents or other conditions lead to rapid cellular disintegration, processes, which typically involve cellular swelling, organelle dysfunction, mitochondrial collapse, and the spillage of cellular contents into the extracellular milieu. The lack of cellular debris generated by apoptotic cells appears to be important. They display no obvious change in plasma membrane integrity, even while drastic biochemical alterations are underway, allowing apoptotic cells to be cleared without triggering an inflammatory response. Apoptosis is characterised by the loss of contact with neighbouring cells in tissues and usually occurs in isolated single cells. Criteria for determining apoptosis, include: 1- Morphological, detachment of the cell from its neighbours, cytoplasmic condensation resulting in cell shrinkage, and condensation of chromatin around the nuclear perimeter. 2- Biochemical, the activation of endogenous endonuclease that attacks the cell's genomic DNA at the linker regions that connect the nucleosomal units, produces DNA fragments in size multiple of 185-200 base pairs. In pathological cell death, DNA is randomly degraded. Apoptosis plays a pivotal role in a wide variety of physiologic phenomena.

Morphological changes of apoptosis and necrosis:

Features	Viable cell	Necrosis	Apoptosis	Apoptosis with secondary necrosis
Cell size	Normal	Increased	Decreased	Increased
Detachment of cells	None	Sheets of cells	Individually	Individually
Plasma membrane	Intact	Permeable	Intact	Permeable
Trypan blue and EH staining	Negative	Positive	Negative	Positive
Chromatin pattern	Normal	Normal	Condensed and fragmented	Condensed and fragmented

CELLULAR MECHANISMS OF APOPTOSIS

Cellular mechanisms of apoptosis remains unknown, but possibly involving Ca^{2+}/Mg^{2+}-dependent endonucleases. Zn^{2+} is known to inhibit both apoptosis and DNA fragmentation also inhibits Ca^{2+}/Mg^{2+} nuclease activity. A cascade of apoptosis-associated genes, overexpressions of several genes are important for cell death. c-myc stimulates specific intracellular signals for growth and death, at the same time. Expression of c-jun and c-fos, and TP53, are associated with apoptosis. High expression of bcl_2 inhibits the process, bcl_2 shields against oxidative damage. TNF-α is an important inducer of apoptosis in the immune system. Cytokines are also important regulators of apoptosis in lymphoid cells. IL-4 triggers apoptosis in immune B cells. IL-6 inhibits the apoptotic response of myeloid cells. Apoptosis involves a coordinated pathway beginning with a loss of contact between apoptotic cells and their neighbouring cells. Detachment from adjacent cells isolates the dying cells as they simultaneously begin shedding specialized surface elements such as cell-cell junctions and microvilli. The cell surface initially adopts a smooth contour, but soon immature glycan structures, once embedded in mature surface glycoproteins start to appear. As cytoskeletal integrity deteriorates, apoptotic cells are released from the basement membrane. The cell undergoes a rapid, yet selective export of water and ions without a corresponding loss of macromolecules or organelles. The direct consequent of this disturbed homeostasis is cytoplasmic condensation and compaction of organelles. The result is an overall shrinkage of cell volume and an abrupt rise in intracellular density. The nucleus reduces in size as chromatin condenses into dense granular caps around the nuclear envelope. Apoptotic bodies are engulfed by neighbouring cells and degraded. Macrophages recognise specific vitronectin receptors and quickly engulf apoptotic cells. Recognition is mediated by integrin peptide and lectin-carbohydrate interactions between the receptors on the phagocytic acceptor cell and the apoptotic body. After ingestion, the apoptotic body membrane ruptures and organelles become nondescript as lysosomes digest the cell debris into residual material within a few hours. Phagocytes target apoptotic bodies so rapidly that even large numbers of dying cells do not induce an inflammatory response, that is classified into 3 mechanisms: 1- Type I cell death, heterophagy, the lysosomes of neighbouring cells break down the apoptotic body; 2- Type II cell death, autophagy, has been found only in some developing systems, the indigenous lysosomes of the dying cell degrade the apoptotic bodies; and 3- Type III death characterised by macrophage digestion of apoptotic bodies.

BIOCHEMISTRY OF APOPTOSIS

The biochemical events, are less clearly defined, because apoptosis occurs in cells that are scattered rather than in one concentrated location. The c. elegans cell death gene system, such as ced-3, is a well-characterised cysteine protease known as IL-1β converting enzyme (ICE) homologues, is the only endonuclease implicated in apoptosis, acts on cell corpse and is expressed by a phagocytic cell within the organism. Transglutaminases are a group of Ca^{2+}-dependent enzymes that catalyse the post-translational coupling of amines (including polyamines) into proteins and the crosslinking of proteins via γ- glutamyl-lysine bridges when the amine is a peptide-bound lysine residue. The enzyme appears to be activated by elevations of the cytosolic Ca^{2+} concentration. Protein crosslinking stabilises apoptotic cells and bodies, preventing leakage of the intracellular contents into the extracellular milieu. Transglutaminase is extensively crosslinks membrane and cytosolic proteins and a Ca^{2+}- and Mg^{2+}-dependent endonuclease, accumulating immediately before the onset of apoptosis. Through mediation by a Na^+-K^+-cotransporter the transglutaminase may limit leakage of intracellular material into the extracellular space. The Ca^{2+}- and Mg^{2+}-dependent endonuclease is responsible for the irrevocable nuclear DNA fragmentation. The p53 protein increases in response to DNA damage leading to G_1 growth arrest until DNA repairs is accomplished. c-myc leads to cell proliferation. Apoptosis-specific protein (ASP) is not found in viable or necrotic cells. ASP is an apoptotic marker, ASP expression is localised to cytoplasm and remains easily detectable after DNA degradation and death. ASP soon after induction of apoptosis and may be up-regulated immediately prior to or simultaneously with DNA fragmentation. Bcl_2 blocks ASP expression, revealing that bcl_2 obstructs apoptosis before ASP expression and close to the onset of the apoptosis. Cancer cells are relatively resistant to apoptosis.

MITOCHONDRIA AND APOPTOSIS

Bax, a mammalian cell death protein that targets mitochondrial membranes, can induce mitochondrial damage and cell death even when a family of intracellular cysteine proteases known as caspases are inactivated. At least 3 general mechanisms are known, by which mitochondria triggers apoptosis, and their effects may be interrelated, including: 1- disruption of electron transport, oxidative phosphorylation, and ATP production; 2- release of proteins that trigger activation of caspase family proteases; and 3- alteration of cellular reduction-oxidation (redox) potential. Ceramide disrupts electron transport at the same step in cells as well as in isolated mitochondria. Ligation of Fas also leads to a disruption in cytochrome c function in electron transport. A drop in ATP production occurs relatively late in apoptosis. Bcl-2-inhibitable nuclear condensation and DNA-fragmentation are dependent on the presence of mitochondria. During apoptosis cytochrome c (cyto c) is released from mitochondria and this is inhibited by the presence of Bcl-2 on these organelles. Cytosolic cyto c forms an essential part apoptosome, which is composed cyto c, Apaf-1, and procaspase-9. Once cyto c is released, this commits the cell to die by either a rapid apoptotic mechanism involving Apaf-1-mediated caspase activation or a slower necrotic process due to collapse of electron transport, which occurs when cyto c is depleted from mitochondria, resulting in a variety of deleterious sequelae including generation of OFR and decreased production of ATP. Superoxides and lipid peroxidation are increased during apoptosis induced by myriad stimuli. However, generation of ROS may be a relatively late event, occurring after cells have embarked on a process of caspase activation. Multiple stimuli such as Bax, oxidants, Ca^{2+} overload, active caspases, and perhaps ceramide can trigger mitochondria to release caspase-activating proteins, among which are cyto c and possibly other proteins such as AIF and intramitochondrial caspases. Mechanisms of release of caspase-activating protein from mitochondria, include: 1) Osmotic disequilibrium leading to an expansion of the matrix space, organellar swelling, and subsequent rupture of the outer membrane. 2) Opening of channels in the outer membrane, thus releasing cyto c from the intermembrane space of mitochondria into the cytosol. Cyto c activates caspases by binding to Apaf-1, inducing it to associate with procaspase-9, thereby triggering caspase-9 activation and initiating the proteolytic cascade that culminates in apoptosis. Cells in which mitochondria have ruptured are at risk for death through a slower nonapoptotic mechanism resembling necrosis because of loss of the electrochemical gradient across the inner membrane ($\Delta\Psi m$), production of ROS, and declining ATP production. In many apoptosis scenarios, the mitochondrial inner membrane potential ($\Delta\Psi m$) collapses, indicating the opening of a large conductance channel known as the mitochondrial permeability transition (PT) pore. Opening of this nonselective channel in the inner membrane allows for an equilibration of ions within the matrix and intermembrane space of mitochondria, thus dissipating the H^+ gradient across the inner membrane and uncoupling the respiratory chain. PT pore opening results in a volume dysregulation of mitochondria due to the hyperosmolality of the matrix, which causes the matrix space to expand. Because the inner membrane with its fold cristae possesses a large surface area than the outer membrane, this matrix volume expansion can eventually cause outer membrane rupture, releasing caspase-activating proteins located within the intermembrane space into the cytosol. In the open configuration, water and solutes enter the matrix, causing matrix swelling and outer membrane disruption, leading to release of cyto c and other proteins. Bcl-2 can prevent PT, while Bax induces both apoptosis and PT. Oxidants and pathological elevation of cytosolic Ca^{2+}, can induce rupture of the outer membrane of mitochondria and release of caspase-activating proteins. Hyperpolarisation of the mitochondrial inner membrane may disrupt the outer membrane. Bcl-x_L suppresses this transient hyperpolarisation. Hyperpolarisation could be due increased export of protons into intermembrane space that may result in protonation of weak acids. These can then freely diffuse across the inner membrane and are trapped when the protons are lost. As these metabolites accumulate, osmolality increases and water enters, resulting in matrix space expansion and eventually rupture of the outer membrane, thereby releasing all contents of the intermembrane space. Although rupture of the outer membrane causes release of cyto c, caspases, AIF, and sometimes generation of ROSs, only a subset of mitochondria appear to be affected. Many (but not all) Bcl-2 family proteins reside in the mitochondrial outer membrane, anchored by a hydrophobic stretch of amino acids located within their COOH-termini, with the proteins oriented toward the cytosol. Bcl-2 and Bcl-x_L suppress release of sequestered matrix Ca^{2+} induced by uncoupled of respiration. Bcl-2 and Bcl-x_L enhances proton extrusion from mitochondria and increases mitochondrial Ca^{2+} buffering capacity. Bcl-2 and Bcl-x_L may communicate functionally or physically with inner membrane proteins that govern ion transport, such as components of the PT pore, or ion-transporting proteins that control volume regulation of the matrix space independently of the PT. Bcl-2 and Bcl-x_L may regulate the pH of the intermembrane space, causing increased rates of proton extrusion from mitochondria. Although the outer membrane of mitochondria apparently is not polarised, its close apposition to the inner membrane at the junctional complexes may affect proteins located there. Apoptosis is a mode of cell death used by multicellular organisms to eradicate cells in diverse physiological and pathological settings. Diverse cellular stresses trigger caspase activation by promoting release of mitochondrial components, including cytochrome c, into the cytoplasm. In turn cytochrome c promotes the assembly of a caspase-activating complex (apoptosome). Caspases (cysteine aspartic acid-specific proteases) are typically found in cells as inactive precursors, being converted into cytopaths at the onset of apoptosis by adaptor proteins that aggregate dormant caspase molecules. At least 14 caspases have been found, a subset of which are involved in the regulation of apoptosis; the remainder are thought to be involved in the processing of pro-inflammatory cytokines. Upon activation, initiator caspases activate the effector caspases by restricted proteolysis. Efflux of cytochrome c from the mitochondria is a pivotal event in apoptosis as it drives the assembly of a high-molecular weight caspase-activating complex in the cytoplasm, termed the mitochondrial apoptosome. Uncoupling of mitochondrial oxidative phophorylation is commonly observed during apoptosis, resulting in a loss of mitochondrial transmembrane potential. Several heat shock proteins (Hsps) might interfere with assembly of the apoptosome, either through interaction with cytochrome c, or through interaction with Apaf-1, inhibitor of apoptosis proteins (IAPs) might interfere with caspase activation events downstream of apoptosome assembly by directly binding to certain caspases. Smac/Diabio, which is also released from mitochondria during apoptosis, might facilitate caspase activation in this pathway by neutralising IAP function.

CASPASES

Caspases are all expressed as proenzymes (30-50 kD) that contain 3 domains: an NH_2-terminal domain, a large subunit (~20 kD), and a small subunit (~10 kD). Caspases are among the most specific of proteases, with an unusual and absolute requirement for cleavage after aspartic acid. A subset of caspases (effectors) is responsible for the cellular changes that occur during apoptosis and provide insights into the mechanisms that they employ. One role of caspases is to inactivate proteins that protect living cells from apoptosis. For example is the cleavage of I^{CAD}/DFF45, an inhibitor of the nuclease responsible for DNA fragmentation, CAD (caspase-activated deoxyribonuclease). In nonapoptotic cells, CAD is present as an inactive complex with I^{CAD}. During apoptosis, I^{CAD} is inactivated by caspases, leaving CAD free to function as a nuclease. But I^{CAD} is required for both the activity and inhibition of this nuclease. Bcl-2 proteins are cleaved by caspases. It appears that cleavage not only inactivates these proteins, but also produces a fragment that promotes apoptosis. Caspases contribute to apoptosis through direct disassembly of cell structures, as illustrated by the destruction of nuclear lamina, a rigid structure that underlies the nuclear membrane and is involved in chromatin organisation. Lamina is formed by head-to-tail polymers of intermediate filament proteins called lamins. During apoptosis, lamins are cleaved at a single site by caspases, causing lamina to collapse and contributing to chromatin condensation. Caspases also reorganise cell structures indirectly by cleaving several proteins involved in cytoskeleton regulation, including gelsolin, focal adhesion kinase (FAK), p21-activated kinase 2 (PAK2). Cleavage of these proteins results in deregulation of their activity. Caspases cut off contracts with surrounding cells, reorganise the cytoskeleton, shut down DNA replication and repair, interrupt splicing, destroy DNA, disrupt the nuclear structure, induce the cell to display signals that mark it for phagocytosis, and disintegrate the cell into apoptotic bodies. Caspases are regulated by opposing effects of activators and inhibitors. A signal apparently initiates 3 pathways involving: 1- cofactors (cyto c relocation from mitochondria to cytoplasm), 2- initiators caspases (relocation of caspase-8 to a receptor complex); and 3- inhibitors. Caspase-8 is associated with apoptosis involving death receptors. Caspase-9 is involved in death induced by cytotoxic agents. Activation of initiator caspases requires binding to specific cofactors, a mechanism commonly observed with proteases. This binding is triggered by a proapoptotic signal and mediated through one of at least 2 distinct structural motifs that reside in both the caspase prodomain and its corresponding cofactor. Activation of procaspase-8 requires association with its cofactor FADD (Fas-associated protein with death domain) through the DED (death effector domain) while procaspase-9 activation involves a complex with the cofactor APAF-1 through CARD (caspase recruitment domain). Activation of caspase-9 also requires cyto c and deoxyadenosine triphosphate, indicating that caspase activation may require multiple cofactors. The cofactors serve to bring 2 or more caspase precursors in close proximity, allowing for intermolecular autoproteolytic activation. The cofactors facilitate activation by changing the conformation of the precursors either directly or by removing an inhibitor. FADD-like ICE inhibitory proteins (FLIPs) are similar in sequence to procaspase-8, except that they lack essential catalytic residues. These proteins probably compete with procaspase-8 for binding to its effector, FADD, thus preventing caspase activation. Compartmentalisation of caspases and their cofactors is likely to be another way of regulating caspase activation. The precise caspase targets of the IAPs remain elusive. IAP proteins (inhibitors of apoptosis) are a large family. IAPs inhibit apoptosis through inhibition of effector caspases. Two opposite types of diseases that involved deregulation of apoptosis are those involving excessive apoptosis, causing damage to normal tissues, and those in which apoptosis is prevented, allowing malignant tissues to grow. Inhibition of caspases can functionally rescue cells from death. Peptidyl caspase inhibitors are effective in animal stroke, myocardial ischaemia-reperfusion injury, liver disease, and traumatic brain injury. Two classes of major drugs work by inhibiting proteases, angiotensin converting enzyme (ACE) inhibitors and HIV protease inhibitors. Two main problems to chemotherapy are toxicity to normal cells and failure to kill cancer cells. Both problems stem from the indirect mechanism by which both chemotherapeutic drugs and irradiation kill cells. They damage both normal and cancer cells, and this damage is then translated through multiple steps into cell death, likely through activation of caspases and apoptosis. When these steps are compromised the therapy fails. Oncoproteins that deregulate the cell cycle can activate caspases and induce apoptosis. Chemotherapeutic drugs induce apoptosis indirectly. They inflict cell damage that is then translated through several poorly understood steps into activation of caspase-9. In drug-resistant cells, the apoptosis fails because of defects in signalling pathways (antiapoptotic events, such as p53 mutation or overexpression of bcl-2) that lead to caspase activation. Therapeutic opportunities may lie in bypassing these defects. One possibility is to activate the death receptor complexes, resulting in activation of their corresponding initiator caspase. Another potential approach involves bypassing the defective part of the pathway to restore the signal triggered by chemotherapeutic drugs or by the oncogenic transformation itself, which can be thought of as proapoptotic signal that is present only in transformed cells.

DEATH RECEPTORS

Development and homeostasis of multicellular organisms are controlled by cell proliferation and differentiation, and the elimination of cells that are unnecessary or deleterious. Apoptosis is the result of an active cellular response that can be elicited by a variety of stimuli such as growth factor deprivation, a molecular damage that does not cause severe loss of integrity, or by triggering of specific cellular receptors such as the TNFR1 or Fas/Apo-1. There is growing evidence that not all Fas positive cells are susceptible to apoptosis induction. Fas-mediated apoptosis undoubtedly involves a delicate balance of receptor/ligand interactions and that these may be modulated by soluble proteins. Death receptors-cell surface receptors that transmit apoptosis signals initiated by specific death ligand-play a central role in instructive apoptosis. These receptors can activate death caspases within seconds of ligand binding, causing an apoptotic demise of the cell within hours. Death receptors belong to the TNFR gene superfamily, which is defined by similar, cysteine-rich extracellular domains. The death receptors contain in addition a homologous cytoplasmic sequence termed the death domain. Death domains typically enable death receptors to engage the cell's apoptotic machinery, but in some instances they mediate functions that are distinct from or even counteract apoptosis. The best characterised death receptors are CD95 (also called Fas or Apo1) and TNFR1 (also called p55 or CD 120 a). The p75 nerve growth factor (NGF) receptor also contains a death domain. The ligands that activate these receptors, with the exception of NGF, are structurally related molecules that belong to the TNF gene superfamily. CD95 ligand (CD95L) binds to CD95; TNF and lymphotoxin α bind to TNFR1; Apo3 ligand (Apo3L, also called TWEAK) binds to DR3; Apo2 ligand (Apo2L, also called TRAIL) binds to DR4 and DR5. Apaf-1 is a mammalian CED-4 homologue. The products of the mammalian Bcl-2 gene family are related to CED-9 but include two groups of proteins that either inhibit or promote apoptosis. The nematode Caenorhabditis elegans has been good model organism for studying cell death. Three C. elegans gene products are essential for apoptosis: CED-3 and CED-4 promote apoptosis, whereas CED-9 inhibits apoptosis. CED-3 is a caspase, that is, a cysteine protease that cleaves certain proteins after specific aspartic acid residues, it exists as a zymogen, which is activated through self-cleavage. CED-4 binds to CED-3 and promotes CED-3 activation, whereas CED-9 binds to CED-4 and prevents it from activating CED-3. Normally, CED-9 is complexed with CED-4 and CED-3, keeping CED-3 inactive.

Apoptosis stimuli cause CED-9 dissociation, allowing CED-3 activation and thereby committing the cell to die by apoptosis. CD95 and CD95L play an important role mainly in 3 types of physiologic apoptosis: (i) peripheral deletion of activated mature T cells at the end of an immune response; (ii) killing of targets such as virus-infected cells or cancer cells by cytotoxic T cells and by natural killer cells; and (iii) killing of inflammatory cells at immune-privileged sites such as the eye. Evidence for the biological role of CD95 comes from certain mouse strains and from human patients who have defective genes for CD95 or CD95L. Such mutations can lead to accumulation of peripheral lymphoid cells and to a fatal autoimmune syndrome characterised by massive enlargement of lymph nodes. CD95 and CD95L are implicated also in pathological suppression of immune surveillance, namely, elimination of tumour-reactive immune cells by certain tumours that constitutively express CD95L. Like other TNF family, CD95L is a homotrimeric molecule. Because death domains have a propensity to associate with one another, CD95 ligation leads to clustering of the receptors death domains. Upon recruitment by FADD, caspase-8 oligomerization drives its activation through self-cleavage. Caspase-8 then activates downstream effector caspases such as caspase-9, committing the cell to apoptosis. TNF is produced mainly by activated macrophages and T cells in response to infection. By engaging TNFR1, TNF activates the transcription factors NFκB and AP-1, leading to induction of proinflammatory and immunomodulatory genes. In some cell types, TNF also induces apoptosis through TNFR1. TNF rarely triggers apoptosis unless protein synthesis is blocked. Inhibition of NFκB and/or JNK/AP-1 pathway sensitises cells to apoptosis induction by TNF. TNF trimerizes TNFR1 upon binding, inducing association of the receptors' death domains. Subsequently, an adapter termed TRADD (TNFR-associated death domain) binds through its own death domain to the clustered receptor death domains. TNFR-associated factor-2 (TRAF2) and receptor-interacting protein (RIP) stimulate pathways leading to activation of NFκB and JNK/AP-1, whereas FADD mediates activation of apoptosis. TRAF2 and RIP activate the NFκB-inducing kinase (NIK), which in turn activates the inhibitor of κB (I-κB) kinase complex, IKK. IKK phosphorylates I-κB, leading to I-κB degradation and allowing NF-κB to move to the nucleus to activate transcription. TRAF2 may not be essential for NF-κB activation by TNF, alternatively, there may be another TRAF family member that binds to TRADD and NIK and substitutes for TRAF2. TRAF2-deficient cells are totally lacking in JNK activation in response to TNF, demonstrating a critical role for TRAF2 in this response. RIP is required for coupling TNFR1 to NF-κB. TRAF2 also binds to cIAP1 and cIAP2 (cellular inhibitor of apoptosis-1 and -2), with anti-apoptotic activity. FADD couples the TNFR1-TRADD complex to activation of caspase-8, thereby initiating apoptosis. DR3 triggers responses that resemble those of TNFR1, namely NF-κB activation and apoptosis. Like TNFR1, DR3 activates NF-κB through TRADD, TRAF2, and RIP and apoptosis through TRADD, FADD, and caspase-8. DR3 binds to Apo3L. Apo3L activates NF-κB through TRADD, TRAF2, RIP, and NIK, and triggers apoptosis through TRADD and FADD. Apo3L closely resembles TNF. Apo3LmRNA is expressed constitutively in many tissues. DR3 transcripts are present mainly in spleen, thymus, and peripheral blood and are induced by activation in T cells. Apo3L-Dr3 and TNF-TNFr1 interactions probably have distinct biological roles. Like CD9L, Apo2L transcription is elevated upon stimulation in peripheral blood T cells. Apoptosis induction by Apo2L requires caspase activity. Overexpression of DR4 or DR5, which bind to Apo2L, triggers apoptosis. The unique set of decoy receptors (DcRs) competes with DR4 and DR5 for binding to Apo2L. DcR1 is a glycosyl phosphatidylinositol (GPI)-anchored to cell surface protein that resembles DR4 and DR5, but lacks a cytoplasmic tail. DcR1 appears to function as a decoy that prevents Apo2L from binding to its death receptors. DcR2 (also called TRAIL-R4 or TRUNND) resembles DR4 and DR5, but it has a substantially truncated cytoplasmic death domain. DcR2 acts as decoy that competes with DR4 and DR5 for binding to Apo2L. The genes encoding DR4, DR5, DcR1, and DcR2 map together to common ancestral gene. The idea of targeting specific death receptors to induce apoptosis in tumours is attractive, because death receptors have direct access to the caspase machinery. Death receptors initiate apoptosis independently of the p53 tumour suppressor gene, which is inactivated by mutation in more than half of human cancers. TNF causes a severe inflammatory response syndrome that resembles septic shock, mediated mainly by induction of proinflammatory genes in macrophages and endothelial cells through NF-κB activation. Apo2L may be a safer agent, Apo2l induces activation of NF-κBκ weakly, and activation requires doses that are considerably higher than doses of TNF that activate a strong NF-κB response. Many tissues constitutively express the Apo2LmRNA. DR4 and DR5 are expressed in normal tissues and in many types of tumour cells, whereas DcR1 and DcR2 are expressed frequently in normal tissues but infrequently in tumour cells.

THE Bcl-2 PROTEIN FAMILY

With the discovery of the anti-apoptotic function of the bcl-2 oncogene, raised threshold that apoptosis represents a central step in tumourigenesis. Bcl-2 family members are essential for maintenance of major organ systems, and mutations affecting them are implicated in cancer. CED-9 and Bcl-2 proved to be functional and structural homologues, and their survival function is opposed either by close relatives such as Bax. At least 15 Bcl-2 family members have been identified in mammalian cells and several others in viruses. All members posses at least one of four conserved motifs known as Bcl-2 homology domains (BH1 to BH4). Pro- and anti-apoptotic family members can heterodimerize and seemingly titrate one another's function, suggesting that their relative concentration may act as rheostat for the suicide program. Heterodimerization is not required for pro-survival function. Bcl-2 resides on the cytoplasmic face of the mitochondrial outer membrane, endoplasmic reticulum (ER), and nuclear envelope and may register damage to these compartments and affect their behaviour, perhaps by modifying the flux of small molecules or proteins. Although the COOH-terminal hydrophobic domain of Bcl-2 is important in membrane docking, its deletion does not abrogate Bcl-2 survival function. Only a fraction of Bcl-x_L resides on membranes, and Bax is cytosolic. Bcl-2 and its relatives dock on specific proteins on each organelle. The pro-survival proteins may function by directly inhibiting the ability of CED-4-like molecules to activate caspases. CED-9 and Bcl-x_L can bind to CED-4, which also bind CED-3 and stimulates its activation. Bcl-x_L binds also to CED-4-like portion of Apaf-1, whereas procaspase-9 binds to its NH_2-terminal caspase recruitment domain (CARD). Bcl-x_L may inhibit the association of Apaf-1 with procaspase-9 and thereby prevent caspase-9 activation. Bcl-2 directly or indirectly prevents the release from mitochondria of cyto c, which along with ATP may facilitate a change in Apaf-1 structure to allow procaspase-9 recruitment and processing. Bcl-2 can protect cells after much cyto c has been released. Bcl-x_L, Bcl-2 and Bax form channels in lipid bilayers, and those created by Bax and Bcl-2 have distinct characteristics, including some ion selectivity. Bcl-2 can protect, by preventing mitochondrial disruption. Bax and Bax-like proteins might mediate caspase-independent death via channel-forming activity, which could promote the mitochondrial permeability transition (PT) or puncture the mitochondrial outer membrane. The Bcl-2 family is regulated by cytokines and other death-survival signals at different levels. Several pro-survival genes are induced transcriptionally by certain cytokines, and bax is induced in some cells as part of the p53-mediated damage response. The signal from the receptor seems to be transduced by phosphoinositide 3-kinase. The physiological roles of Bcl-2, include: 1) Bcl-2 protects against diverse cytotoxic insults, such as γ- and ultraviolet-irradiation, cytokine withdrawal, dexamethasone, and cytotoxic drugs; and 2) Bcl-2 protects poorly against apoptosis of lymphocytes induced by ligation of the receptor CD95. Thus, in at least the lymphoid system, the major CD95-induced pathway, which activates caspase-8, bypass the Bcl-2-inhibitable step common to most stress pathways. CD95 may also trigger alternative pathways. Bax is not essential for p53-dependent apoptosis. The Bcl-2 family can modulate cell cycle progression. Under suboptimal growth conditions, Bcl-2 promotes exit into quiescence and retards re-entry into cycle. This effect is genetically separable from its survival function. T-cell expressing Bcl-2 make less IL2, the cytokine required for their progression into S phase, because of reduced nuclear translocation of the transcription factor NFAT. NFAT transit requires calcineurin, which Bcl-2 may sequester on cytoplasmic membranes. The cell cycle-inhibitory effect may be to reduce the oncogenic impact of Bcl-2. Apoptosis normally eliminates cells with damaged DNA or an aberrant cell cycle. Synergy between myc and bcl-2 in tumourigenesis has been noted. All pro-survival bcl-2-like genes are potentially oncogenic, and some mutations probably increase their expression indirectly. Bax is mutated in human GIT cancer and some leukaemias. Loss of bax reduces apoptosis and increase tumourigenicity but only about half as much as loss of p53. Thus, bax is not the only gene responsible for p53-driven apoptosis. A major arm of the mammalian DNA damage response involves a suite of protein kinases, distantly related to the intracellular signalling molecule phosphatidylinositol 3-kinase (PI 3-kinase), ATM, and DNA-dependent protein kinase (DNA-PK). Both ATM and DNA-PK trigger a plethora of cellular responses that include activation of cell cycle checkpoints and growth arrest, repair, and apoptosis. One pivotal target of ATM, possibly DNA-PK, is the tumour suppressor p53, a transcription factor normally maintained in abeyance at low levels through interaction with the Mdm-2 protein that signals its degradation. Mdm-2 is itself a target for DNA-PK. DNA damage-induced phosphorylation of either p53 or Mdm-2 prevents the two proteins from interacting, thus stabilising and activating p53. Lesions leading to elevated Mdm-2 also lead to p53 activation. Two cellular responses to p53 activation are well described-growth arrest (in cell cycle stages G_1 and G_2) and apoptosis. p53-induced growth arrest after certain types of DNA damage is irreversible; although alive, such cells are genetically dead and thus constitute no further neoplastic risk. p53-mediated growth arrest proceeds through induction of the cyclin-dependent kinase (cdk) inhibitor p21. p53 is also implicated in cell responses to a variety of insults that do not involve DNA damage, such as metabolite deprivation, physical damage, heat shock, hypoxia, and expression of oncogenes (e.g., myc and E1A). The processes of cell renewal and cell death are linked. Cells harbouring activated oncogenes exhibited a high amount of cell death. Deregulated expression of myc genes is frequent in cancer. Ectopic expression of Myc is sufficient to drive many cells into the cycle in the absence of external mitogens. However, in certain circumstances, Myc also promotes apoptosis. The mitogenic and proapoptotic properties of c-Myc are genetically inseparable. Both require an intact NH_2-terminal transcriptional activation domain. DNA binding and dimerization domains, and interaction with the Myc partner protein Max. Cancers arise through inactivation of growth-suppressive pathways, in particular lesions affecting the activity of Rb, which restrains activation of genes through necessary for progression through G_1 and into S phase. Phosphorylation of Rb, by a pathway involving cyclin D, cdk4, and the cdk inhibitor p16INK4, eliminates its growth-restraining action, thus allowing cells to enter the cell cycle. Cells lacking Rb exhibit deregulated entry into the cell cycle. No myc gene has been shown to act directly as a tumour suppressor. Apoptosis limits Myc's oncogenic activity. Myc-induced apoptosis is suppressed by survival signalling pathways such as that triggered by IGF-1 in fibroblasts or IL3 in myeloid cells. In the case of IGF-1, a discrete antiapoptotic pathway routes through Ras, PI 3-kinase, and the serine-threonine kinase Akt/protein kinase B, ultimately impacting on Bad, a key modulator of the Bcl-2 family. The sensitisation to apoptosis afforded by oncogenes explains the remarkable sensitivity of many primary tumours to anticancer agents when compared with their normal counterparts. p53 is widely accredited as guardian of replicative normalcy and general factotum of cellular cataclysm. Oncogene-generated activity candidates are the components of the intracellular apoptosome-the complex whose assembly is central to the activation of apoptosis, Apaf-1, holo-cyto c (released from mitochondria), deoxyadenosine triphosphate, and pro-caspase 9. Activation or accumulation of any of these apoptosome components could generate the requisite oncogene-dependent sensitisation. The mechanism by which Bcl-2 family members impact on cell proliferation is unclear, however, growth inhibition by Bcl-2 requires critical residues in the NH_2-terminal BH4 domain of the Bcl-2 that are dispensable for suppression of apoptosis. Interference with CD95 signalling pathway also inhibits cell proliferation. Interference with the action of FADD, a critical intermediary for at least some of the apoptotic signalling initiated by CD95, results in a profound impairment in cell proliferation that is p53. Ras proteins are key transducers of mitogenic signals through their activation of Raf-MAP kinase pathways, a fact attested by the high frequency of Ras-activating mutations in human cancer. Ras proteins are also involved in transducing survival signals from a receptor such as IGF-1R to downstream effectors such as PI 3-kinase, Akt, and Bad. Oncogenic mutation of Ras appears able to simultaneously activate cell proliferation and suppress the concomitant apoptosis-a potentially catastrophic combination. Ras triggers precisely the opposite responses-a profound p53-dependent growth arrest, frequently accompanied by apoptosis, both of which actions are mediated by the Ras kinase. Like Myc, Ras bears the seeds of its own destruction.

PHAGOCYTOSIS

The rapid removal of cellular material by professional phagocytic cells (macrophages) and neighbouring epithelial and mesenchymal cells is one of the most diagnostic features of apoptotic cell death. The orderly removal of potentially hazardous intracellular contents prevents secondary tissue injury due to inflammation and the direct effects of intracellular catabolic enzymes. The release of solubilized DNA and cytoskeletal/nuclear matrix proteins as a consequence of apoptotic cell death due to improper phagocytosis might contribute to the circulating antibodies to these molecules observed in patients with systemic lupus erythematosus and other autoimmune diseases. Apoptotic cell recognition seems to be mediated by at least two different mechanisms: The first is the recognition and binding of the apoptotic cell by and integrin heterodimer ($\alpha v \beta 3$) expressed by the phagocytic cell termed the vitronectin receptor. The mechanism does not appear to involve direct binding of the integrin to the apoptotic cell. The expression of thrombospondin can be induced by p53. Surface antigen CD36 appears to play a role in facilitating the vitronectin receptor-mediated clearance of apoptotic cells. The other mechanism takes the advantage of the changes in phospholipid asymmetry seen in old and dying cells. Neutral phospholipids such as sphingomyelin and phosphatidylcholine, are present in both leaflets of the plasma membrane, while anionic phospholipids (most notably phosphatidylserine) are found on the inner surface. During apoptosis this asymmetry is abolished, resulting in exposure of phosphatidylserine (PS) on the outer leaflet of the plasma membrane. Macrophages (and potentially adjacent cells) are capable of directly binding surface exposed PS, and liposomes containing PS can block the uptake of apoptotic cell in a number of important situations. The oxidised LDL receptor is responsible for the recognition of PS exposed on the outer plasma membrane leaflet of apoptotic cells.

TRIGGERING OF APOPTOSIS

Apoptosis triggers, include: 1- Growth factor deprivation; 2- cytokines (TNF, TNFRs, and Fas); 3- Glucocorticoids and retinoids; 4- cytotoxic chemicals and toxins; and 5- physical damage. Suppression of apoptosis by cytokines and other survival factors (growth factor) is a critical component of cytokine action. Growth factor withdrawal is one of the most common signals for apoptosis, which occurs upon withdrawal of androgens from prostatic epithelial cells. Apoptosis can also be induced by eliminating cell-cell contact, indicating that integrins and other adhesion molecules are capable of delivering survival signals as well, this explains why nonmalignant cells are not found outside their tissues of origin. Migration from the site of growth factor production may lead to apoptotic cell death due to growth factor withdrawal. The key aspects of tumour metastasis are that it requires the tumour cell to survive outside of its normal microenvironment. Cytokine induced apoptosis is particularly common in transformed cells and may represent a physiological anti-tumour defense mechanism. Among the cytokines that are now known to be capable of triggering apoptosis, most attention has been TNF binding to two distinct surface receptors, termed TNFR1 and 2, which are members of a gene family that includes the nerve growth factors receptors. These polypeptides possess homologous cysteine-rich regions in their extracellular domains. Interestingly, the TNFR1 and TNFR2 exhibit no significant homology in their intracellular domains. Most of the activities of TNF appear to be mediated by TNFR1 whereas TNFR2 seems to play a more selective role in immune regulation. TNF can trigger apoptosis in a variety of susceptible cells, most of which are either activated or transformed, mediated via TNFR1. Mutagenesis studies have identified an 80-amino acid sequence in the receptor, termed the "death domain", which is required for the response. This domain binds a 34 kDa protein, called TNFR1-associated death domain protein (TRADD), which is capable of activating the downstream events observed following TNF stimulation when it is overexpressed in cells. Binding to TNFR1 results in its oligomerization and the activation of protein kinase C, the generation of ROS, the activation of the transcription factor NFκB, and the stimulation of endosomal sphingomyelinase, which results in the production of ceramide. Ceramide can induce apoptosis in TNF-sensitive cells, by sphingomyelin hydrolysis. Fas/APO-1 (CD-95) is a member of the TNF receptor superfamily; engagement with specific antibodies, or upon exposure to its natural ligand transduces an extremely rapid synchronous apoptotic signal to a variety of cell types. This signal appears to be delivered via a cytoplasmic 'death domain' which is homologous to a corresponding stretch of amino acids within the carboxyterminus of the p55 TNF receptor. The Fas pathway of apoptosis may mediate apoptosis initiated by other primary signals. TGF-β is cytotoxic, can induce apoptosis in many different cell types, and augment Fas-mediated apoptosis. TGF-β may alter bcl_2 expression independent of p53 expression or alterations in bax expression. TGF-β regulation of apoptosis may be related to the effects of the cytokine on cell cycle progression. Glucocorticoids induce cytolysis, and apoptosis. Glucocorticoids action requires glucocorticoid receptor (GR) binding. Glucocorticoid receptor levels are rate limiting for cytolysis. GR-mediated alterations in transcription are required, glucocorticoid-induced "death genes" is at the very least linked to apoptosis. Glucocorticoids alters cellular metabolism and intracellular ion homeostasis. Glucocorticoids can trigger the uncoupling of electron transport from ATP production, leading to decrease mitochondrial membrane potential, ATP depletion, and the production of OFR. Mitochondrial membrane potential collapse in conjunction with oxidative stress appears to promote the release of Ca^{2+} from intracellular pools and the subsequent stimulation of a Ca^{2+} influx resulting in a sustained increase in the cytosolic Ca^{2+} concentration that is involved in triggering protease endonuclease activation and cell death. Retinoids are common triggers for apoptosis, via a mechanism involving tissue transglutaminase activation. Most of the biological effects of retinol have been shown to be exerted by two retinoids, 9-cis and all-trans retinoic, via activation of two different receptors, the retinoic acid receptor (RAR) and the retinoid X receptor (RXR). Like glucocorticoids, retinoids induce apoptosis via repression of the expression of some of the same genes shown to be targeted by glucocorticoids in dividing cells, including myc and AP-1. Cytotoxic chemical induces apoptosis. γ irradiation, heat shock, cold shock and u.v radiation, damage-induced apoptosis represents a physiological mechanism for removal of extensively damaged cells that may prevent the expansion of clones possessing genetic lesion.

Key genes associated with apoptosis:

Gene	Function
c-myc	Mediates cell proliferation, especially through the G1 checkpoint of the cell cycle
p53	Mediate cell cycle arrest in G1 phase in response to DNA damage
Rb	Functions as a cell cycle checkpoint control in G0/G1, interacts with p53 and is modulated by protein phosphorylation
bcl_2	Functions to protect cells from undergoing apoptosis
bcl_x	A member of the bcl_2 family that produces two alternatively spliced transcripts, the larger protein, $bcl-x_l$, acts similarly to bcl_2, the smaller protein, $bclx_s$, suppresses bcl_2
bax	A member of the bcl_2 family that dimerizes with bcl_2 and antagonises its function.

SIGNALLING FOR APOPTOSIS

Signalling for apoptosis include: 1- Ca^{2+}; 2- PKC; 3- cAMP; 4- ceramide; and 5- redox. Increase in intracellular Ca^{2+} might be involved in triggering apoptosis. Glucocorticoid-stimulated apoptosis is associated with Ca^{2+} influx. The Ca^{2+} increase is responsible for the initiation of apoptosis. Ca^{2+} causes endonuclease activation as well as many of the morphological changes that are typical of apoptosis in the endoplasmic reticulum. Ca^{2+}-ATPase inhibitor can also trigger the morphological and biochemical events of apoptosis in thymocytes. Rapid, sustained Ca^{2+} increases precede the cytolysis of the targets of cytotoxic T-lymphocytes. Elevation of the intracellular Ca^{2+} level might promote apoptosis by directly stimulating the enzymatics activities of proteases, phospholipases and/or endonucleases responsible for mediating cellular demise in apoptosis. DNA fragmentation and death in apoptotic cells, require sustained Ca^{2+} elevations. Calmodulin expression is linked to apoptosis in the prostate following withdrawal of androgen. Ca^{2+} influx in the triggering of apoptosis has come from studies with specific Ca^{2+} channel blockers, which abrogate apoptosis in the regressing prostate following testosterone withdrawal. Possible targets in the nucleus for PKC include AP-1, a transcription factor composed of Fos/Jun dimers, which serves as a common target for signal transduction through Ras activation and may involve activation of the Ras GTP exchange protein, Vav. The effect of cAMP involves protein phosphorylation changes that lead to apoptosis. cAMP stimulates DNA fragmentation. The effects of cAMP are mediated by protein kinases and phosphatases. Sphingomyelin (SM) hydrolysis leading to the production of the second messenger ceramide. Two different sphingomylinases appear to be involved, one of which is a plasma-membrane associated, diacylglycerol (DAG)-independent Mg^{2+}-dependent, possessing a neutral pH optimum, the other is an acidic form of the enzyme, which may be localised to lysosomes. Ceramide triggers apoptosis by DAG or cAMP. TNF-induced apoptosis may be mediated via the activation of sphingomyelin pathway in variety of cells. Ceramide participate apoptosis induced by the Fas ligand, growth factor and serum factor withdrawal, shock, and u.v. ionising radiation. Ceramide modulates protein kinase(s) and or phosphatase(s). Membrane bound ceramide activated protein kinase (CAPK), autophosphorylates on serine residues and phosphorylates the myelin basic proteins (MBP). Ceramic can activate mitogen-activated protein kinase (MAPK), a family of serine/threonine kinases that are downstream of Raf-I in the Ras pathway. Oxidants cause multiple intracellular alterations, including elevation of cytosolic Ca^{2+}, energy depletion and oxidation of glutathione, NADPH, protein thiols and lipids. Necrosis is the end point of such a dramatic disturbance in cell homeostasis. Low-level exposure of cells to oxidants triggers apoptosis rather than necrosis. Oxidised low-density lipoproteins and lipid hydroperoxides are other oxidants reported to induce apoptosis. Following their production, intracellular oxidants have the potential to trigger apoptosis in several ways, including direct oxidative damage to DNA, and the oxidation of intracellular proteins. Oxidation of the critical sulfhydryls present in Ca^{2+} transport systems located in the mitochondria, ER, plasma membrane may promote Ca^{2+} increases in the cytosol. Protein oxidation changing nuclear gene transcription that the apoptotic pathway is activated.

APOPTOSIS IN IMMUNITY

Clonal deletion of autoreactive precursors and mature lymphocytes via antigen-receptor-mediated apoptosis appears to be a predominant means of ensuring self-tolerance. Cell death is massive within the thymus, and it has been estimated that up to 90% of precursor cells never emerge in the periphery. Immature precursors derived from the bone marrow emigrate to the thymus, where they begin a series of differentiation events that serve to generate T-cell receptor diversity. T-lymphocytes recognise peptides derived from intracellularly cleaved proteins that are bound to, and 'presented' by the polymorphic products of the major histocompatability complex (MHC). The majority of total intrathymic apoptotic deaths occur by default. T-cell receptor-induced apoptosis is blocked by inhibitors of mRNA or protein synthesis suggesting that changes in gene expression may be involved. The TCR-mediated apoptosis appears to be a form of "autocrine suicide", where FasL produced by the cell binds to its own surface Fas receptor and triggers cell death. TCR-induced expression of FasL is blocked by Ca^{2+} chelators, cyclosporin A. B-cells go through two major waves of antibody affinity maturation. The first of these occurs during the early differentiation of pre-B-cells in the spleen and bone marrow when rearrangement of the immunoglobulin genetic loci is initiated. Exogenous antigen or antibodies to surface immunoglobulin can induce tolerance in early B-cells. Peripheral deletion of T-cells not only facilitates the removal of unnecessary cells, but may also represent a back-up means of establishing self-tolerance in T-cells, which have escaped surveillance mechanisms in the thymus. Peripheral deletion may be somewhat distinct from TCR-induced apoptosis in the thymus. B-cell deletion is involved in both the further functional developments of mature B-cells (affinity maturation), as well as the elimination of B-cell at the termination of the immune response. The former is dependent upon interactions between antigen specific "helper" (CD4+) T-cells and B-cells within lymph node structures known as germinal centers. Germinal centers are structures that develop within 4-8 days following antigen administration that have subdomains, termed the dark zone, basal light zone, apical light zone, and follicular mantle, which contain proliferating centroblast precursors undergoing somatic hypermutation that enhances immunoglobulin binding affinities. These cells proceed to the light zone, where they are subjected to selection processes akin to positive and negative selection in the thymus. Cell death within the basal light zone is extensive and fragmented, condensed chromatin, characteristic of apoptotic cell death, can be observed throughout this section of the germinal center. Cells that survive this initial selection migrate to the apical light zone, where cell survival and functional differentiation are promoted by interaction with specific antigens presented by follicular dendritic cells. Defective cell death may be involved in the pathogenesis of systemic lupus erythematosus (SLE), in which Fas is overexpressed on T and B-cells in the peripheral blood. It is possible that the release of solubilized nuclear material produced by apoptotic cells, which subsequently undergo secondary necrosis may stimulate autoantibody production that contributes to disease severity. Peripheral T-cells from SLE patients may express higher levels of bcl_2. Apoptosis is also associated with various forms of diabetes. A factor triggers apoptosis in pancreatic islet cells via a Ca^{2+}-dependent mechanism. Therapeutic induction of apoptosis may represent a novel treatment strategy for the removal of autoimmune T and B-cells.

APOPTOSIS IN THE CNS

The development of the nervous system is characterised by the generation of an excessive amount of neurons and glial cells, which migrate and form numerous connections. NGF protects developing neurons from dying. Although most neurons probably die by necrosis in cerebral ischaemia, some of them die by apoptosis. There is apoptosis-associated DNA fragmentation in epileptic brain damage. Chronic inhibition of superoxide dismutase produces apoptosis in spinal neurons, which is related to the mutations in the Cu/Zn superoxide dismutase gene detected in some families with an autosomal dominant form of amylotrophic lateral sclerosis (ALS). Apoptosis is also likely to be important mechanism in HIV infection that involves the loss of both white matter and grey matter structures. Ca^{2+} mediates the neurotoxicity of heavy metals as well as other toxicants. Intracellular Ca^{2+} overload due to the excessive stimulation of excitatory amino acid receptors and enhanced Ca^{2+} influx through membrane channels appears to play an important role in ischaemic brain damage. The calcium ion plays a critical role in the ability of gultamate or related compounds to induce neuronal death by receptor overstimulation. Increases in intracellular Ca^{2+} can lead to the activation of degenerative enzymes, such as phospholipases, proteases and endonucleases and to mitochondrial dysfunction and perturbation of cytoskeletal organisation. Ca^{2+} overload may play a critical role in ischaemia as well as in various neurodegenerative disorders. NO triggered by a Ca^{2+} rise caused by glutamate receptor overstimulation in these neurons may be lethal to neighbouring cells. The mechanism of NO-induced cell death has been reported to be related to DNA damage and the subsequent activation of poly (ADP-ribose) synthetase, leading to energy depletion. The proto-oncogene bcl_2 prevents neural cell death induced by glutathione depletion by decreasing the net cellular generation of reactive oxygen species, although it is unclear whether bcl_2 protection in neuronal cells is exclusively related to apoptotic cell death.

APOPTOSIS IN CANCER

Cell loss can also have profound effects on the growth rates of tumours. There are established roles for apoptosis in all facets of cancer. Variety of chemicals capable of acting as tumour promoters stimulates marked liver hyperplasia. Removal of these compounds lead can to rapid and extensive apoptosis, which precedes a decrease in liver mass DNA content. Apoptosis is particularly pronounced in preneoplastic lesions. Tumour promoting phorbol ester is capable of suppressing apoptotic cell death in a variety of cell types. Thus, suppression of apoptosis by chemical carcinogens may represent an important component of their ability to promote oncogenesis. Endogenous survival signals also appear to contribute to oncogenesis. The sensitivity of primary tumours to growth factor withdrawal-induced apoptosis mirrors the cytokine dependencies of the normal tissue from which they were derived. Survival of the glandular cells of the prostate is actively controlled by the male sex steroid, testosterone. Castration-induced testosterone withdrawal triggers rapid involution of the gland, a response which is accompanied by the accumulation of cells with apoptotic morphology, and is accompanied by extensive oligonucleosomal DNA fragmentation due to the activation of the Ca^{2+}/Mg^{2+}-dependent endonuclease. Androgen ablation therapy can trigger apoptosis in prostatic tumours. However, the loss of steroid responsiveness is a key indicator of tumour progression in prostate cancer, which may be related in part to the fact that these cells no longer depend upon steroids because they have developed alternative mechanisms to suppress cell death. Possibilities include the mutational activation of oncogenes or tumour suppressor genes and/or acquisition of the capacity to respond to alternative survival signals provided by metastatic microenvironments. Cytokines such as IGF-1 and EGF, which can possess apoptosis-suppressing activities, may be involved in tumour progression adhesion to the extracellular matrix and/or neighbouring cells may provide important survival signals, as disruption of these interactions can trigger apoptosis in most normal and some malignant cells. It appears likely that the aggressiveness of tumour cells, which display anchorage-independent growth, is due in part to their ability to survive without these survival signals, which are critical for the viability of normal cells. The inhibition of apoptosis can promote neoplastic transformation. Tumourigenesis occur in a multistep process involving the acquisition of multiple genetic lesions. Tumour progression and metastasis cannot occur without the induction of new blood vessels within the primary tumour (angiogenesis), which involve a balance between stimulators and inhibitors of endothelial cell survival. The two major growth and survival factors are: VEGF and bFGF; and the plasminogen fragment known as angiostatin and IFNβ. Endothelial cell survival is also dependent upon integrin-dependent interactions with the extracellular matrix. Metastasis involves the migration of tumour cells outside of their normal microenvironment. Cytokine-induced apoptosis, a response that involves the production of NO by the tumour cells. Metastatic variants are nearly completely resistant to this effect, due to a defect in the expression of the enzyme iNOS. The defective expression of iNOS is responsible for their apoptosis resistance. Apoptosis resistance may also be engendered by extrinsic survival signals, including cytokines such as IGF-1. Chemotherapeutic agents are capable of triggering apoptosis in both normal and transformed cells, involving both Proteolysis and oligonucleosomal DNA fragmentation. Oncogones and tumour suppressor genes control cellular chemosensitivity via their capacities to regulate apoptosis. The first oncogene to be implicated in chemosensitivity was the bcl_2 oncogene. p53-dependent and independent pathways for apoptosis can co-exist, although p53-independent DNA damage responses can be observed under some conditions. Ionising radiation induces basically 2 types of tumour cell death: interphase and mitotic (reproductive) death according to the time required for cell death after irradiation. Reproductive cell death is characterised by its selectivity for cells within association with the mitotic cycle, whereas interphase cell death does not depend on cell division and occurs before the cell enters the cell cycle. Tumours from the same histological group and the same stage of development are extremely heterogeneous in their sensitivity to radiotherapy. The main challenge for radiotherapeutic cancer treatments is to understand how to use relatively low doses of radiation to selectively activate tumour cell death by apoptosis while protecting normal cells. Interphase death showed increased sensitivity to radiation-induced DNA double-strand breakage (DSB) when compared with cells that undergo mitotic death. More than 90% of DNA single-strand breaks (SSB) and some of DSB were repaired within 1 hr post-irradiation, yet the irradiated cells were nonetheless committed to apoptosis, which was initiated in a stochastic fashion over the ensuing several hours. This could be explained by 2 possibilities: a) radiation-induced apoptosis, which is triggered by an initial event and once triggered proceeds to its inevitable conclusion independently of the cell's attempt to repair these initial radiation-induced lesions; and b) because of a defect in the slow phase of DNA repair, the unrepaired or misrepaired lesions can initiate the signal cascade leading to apoptosis, alternatives, which at this time both remain viable. The DNA damage response is a genetic program initiated by p53 and other transcriptional regulators, which can lead to cell cycle arrest or cell death. p53 status is a major determinant in radiation-induced cell death. p53 is involved in spontaneous and radiation-induced apoptosis in the GIT. p53-independent mechanisms may also produce radioresistant tumours. Polypeptides Bcl_2 family can also prevent radiation-induced cell death regardless of p53 status. Both p53 and Bcl_2 appear to act within the same molecular pathway, p53 can downregulate Bcl_2 and upregulate bax expression indicates that p53 may act, at least in part, via these mechanisms. Another alternative pathway for radioresistance involves the ras oncogene. In radiation-induced cell death, Ca^{2+} increases, which is a critical step in the signalling of apoptosis. Release of intracellular Ca^{2+} stores may be sufficient to facilitate apoptosis in these cells. Radiation can also trigger secondary responses mediated by growth factors and cytokines, including TNF, PDGF-B, and bFGF-B. TNF can protect normal bone marrow progenitor cells against radiation, it sensitises tumour cells to the insult. Both TNF and radiation can activate the sphingomyelin-ceramide pathway, which results in induction of early response genes related to apoptosis. A variety of pharmacological agents can modulate the efficacy of radiation therapy. Radiation-induced apoptosis can also be modified by membrane-active agents such as vitamin E, and TEMPOL, a non-specific free radical scavenger, which can completely block apoptosis in some irradiated cells. Another classical radiomodifier, 3-aminobenzamide, can either induce or inhibit radiation-induced apoptosis. Fractionation of dose and hypoxia play an important role in tumour radiosensitivity. Radiobiological factor, tumour hypoxia, inhibits induction of apoptosis by a factor of about 2.5. Increasing the time between 2 radiation treatments can lead to the recovery of some cells or the appearance of cells with a resistant phenotype. The overall contribution of apoptosis to radiation-induced tumour regression is still unclear. The chemotherapeutic identified as apoptosis-inducing includes etoposide, VM26, m-AMSA, dexamethasone, vincristine, cis-platinum, cyclophosphamide, paclitaxel, 5'-fluorouracil, and adriamycin. Tumour hypoxia, ionising radiation, and hormone withdrawal in hormone-dependent tumours have been shown to cause apoptosis. Proteolytic cleavage of poly (ADP-ribose) polymerase (PRAP), is the result of activation of a specific protease and that it preceded endonuclease activation and DNA fragmentation. This PRAP protease has evolved as a centerpiece in the study of apoptotic mechanisms. Bcl-2 overexpression inhibits apoptosis in response to several chemotherapeutic agents. Reduced Bax, a pro-apoptotic homologue of Bcl-2, is associated with poor response to combination chemotherapy and worse survival in patients with metastatic cancer. Other modulators of apoptosis interact with chemotherapy-induced cell death. For example the cytotoxicity of several chemotherapeutic agents is inhibited by haematopoietic growth factors, such as granulocyte-macrophage colony-stimulating factor, granulocyte colony-stimulating factor, or IL6. It is not yet ascertain what fraction of tumour cell death is effected through apoptotic mechanisms. Chemotherapy-induced cell death proceeds through 3 distinct general phases: i- phase I, an insult-generating mechanism. Chemotherapeutic agents interact with a specific target such as DNA, RNA, and microtubules causing target injury or dysfunction; ii- phase II, signal transduction. The cell is able to decipher and assess the specific injury to the chemotherapy target. For example, DNA-damaging agents may use the c-Abl tyrosine kinase to induce cell cycle arrest in a p53-dependent mechanism; iii- phase III, induction of apoptosis. The cells undergo the orderly breakdown of macromolecules through the operation of proteases, endonucleases, transglutaminases, and possibly lipases. If one component of the cell receive an insult, the factory has to make a decision as to whether to slow down other production facilities to allow repair. If the injury is severe and leads to irreversible damage, then the decision may be to permanently close down the production facility and move viable functions to other existing and better-suited production units so as to minimise the losses of the mother company. Most chemotherapeutic agents induce damage to a major component of the cell (act in phase I). Cancers can be segregated into 2 groups: those that are prone to apoptosis and those that are resistant. Most solid tumours are intrinsically resistant to apoptosis, which could be due to mutations in p53. Other mutations may exist that also allow these cancer cells to escape suicide. Many cancer cells are at least intrinsically resistant to apoptosis as normal cells. There are no windows of opportunity to selectively kill cancer cells without killing the host tissues. Possible mechanisms that make cancer cells resistant to chemotherapy: 1- a cancer cell has a defect in one arm of induction of apoptosis, such as a mutation in p53, and the cell becomes resistant to activators of that arm (DNA-damaging agents); 2- a cancer cell has one or more mutations or defects in the common signalling components of phase II, but otherwise the downstream apoptotic machinery is intact; 3- a cancer cell has mutation or a change in a gene that provides a selective survival advantage (over expression of Bcl-2); and 4- the cancer cells have either a defect in the apoptotic machinery or are unable to respond to damage by undergoing apoptosis.

APOPTOSIS IN ISCHAEMIA/REPERFUSION

Reperfusion leads to oligonucleosomal DNA fragmentation, this late cell death may contribute to tissue pathology. Brief periods of ischaemia followed by reperfusion lead to apoptosis in the renal tubular cells of the kidney. Apoptosis is detectable by 12 hours after reflow and is not detected after 24 hr of chronic ischaemia. Even when periods of ischaemia are brief, secondary tissue damage may still occur via induction of apoptosis.

APOPTOSIS IN THE KIDNEY

Dramatically enhanced rates of programmed cell death were observed at day 12 of embryonic development in normal embryos. bcl_2 levels are dramatically and rapidly upregulated in the developing human kidney within the induced metanephrogenic mesenchymal cells which differentiate into renal vesicles and nephrons. Bcl_2 are high in all renal neoplasms, supporting a role for apoptosis suppression in the development and progression of renal tumours. Partial ureteric obstruction, induces hydronephrosis and apoptosis. Cell suicide is triggered in response to pathogenic invasion to halt the spread to neighbouring cells, when they have exhausted their overall usefulness within the organism, including human. Types of cell death, are: 1- physiological; and 2- pathological. Physiological cell death is characterised by nuclear changes such as: a- chromatin condensation, b- fragmentation, c- margination, d- internucleosomal DNA cleavage, e- cytoskeletal disruption, f- cell shrinkage, g- membrane blebbing, which leads to fragmentation of the dying cell and subsequently engulfed by neighbouring cells or professional macrophages. Inappropriate apoptosis, can be either: 1- excessive apoptosis, ischaemic damage; or 2- insufficient apoptosis, as autoimmune syndromes, cancer, infections. When p53 is dysfunctional, apoptosis fails to occur in part because of failed cell-cycle checkpoint control and the loss of transcriptional activation of death-promoting genes. Overexpression of death-suppressing proteins such as Bcl_2, retards the normal suicide process, leading to cell accumulation.

APOPTOSIS AND OSTEOPOROSIS

Bone turnover begins by conversion of an inactive or quiescent skeletal surface to a remodelling site, a process referred to as activation. Activation involves proliferation of new blood vessels needed to bring recruited resorbing cells to the remodelling site and retraction of the flat, pavement-like cells that cover quiescent surfaces to expose the mineralised bone surface. The recruited cells become multinucleated osteoclasts, which attach to the newly exposed bone surface with a ring of contractile proteins sealing off a subosteoclastic resorption compartment. Lysosomal enzymes, hydrogen ions, and collagenase are secreted through the microvilli of the ruffled underside of the osteoclasts and begin to excavate a resorption cavity or bay. Osteoclasts are motile cells, capable of resorbing more than just the cavity within which they are identified. After an osteoclast digs a cavity, it may detach from bone and move on to a new resorption site. When the osteoclasts have moved away, osteoblasts are drafted to reconstitute the previously resorbed cavity with new bone. Bone formation begins to occur while bone resorption advances. Intermediate between the end of bone resorption and the beginning of bone formation is the reversal phase, when mononuclear phagocytes smooth out the jagged erosion bays. During this phase, the old bone is coated by a thin layer of cement substance, a collagen- and mineral-poor matrix rich in glycosaminoglycans, glycoproteins, and acid phosphatase, to which the new osteoblasts attach. New osteoblasts assemble only at sites where osteoclasts have recently been active, a phenomenon referred to as coupling. BMU digs through the bone. By this means, the adult skeleton, containing approximately 35 million bone structural units or BSUs (one from each previous BMU), is almost completely regenerated every 10 years. In healthy human adults, 3 to 4 million BMUs are initiated per year and approximately 1 million are operating at any moment. That amounts to a skeletal replacement of 0.027% per day or one new BMU every 7 seconds (with considerable regional variation). Basic multicellular units (BMUs) are temporary anatomic structures of remodelling or turnover that is carried out by a battalion of juxtaposed osteoclasts and osteoblasts. Regions containing fatty (yellow) marrow may turn over every 20 years, whereas the cancellous bone of the ilium, containing haematopoietic (red) marrow, is exchanged at roughly 3-year intervals. Growth factors and cytokines produced in the bone marrow control development of both osteoblasts and osteoclasts. The production and responsiveness to these agents are modulated by systemic hormones. The precursors of osteoblasts are pluripotent mesenchymal stem cells, presumably present in both red and yellow marrow, which may also give rise to chondrocytes, myocytes, fibroblasts, adipocytes, and cells that are necessary to support haematopoiesis, whereas the precursors of osteoclasts are haematopoietic cells of the monocyte-macrophage lineage. In bones containing yellow marrow (the majority of the adult skeleton), osteoclast precursors must travel in the bloodstream from their origin in red marrow to newly activated remodelling sites as directed by currently unknown signals. Bone morphogenetic proteins (BMPs), members of the TGF-β superfamily, the same family of polypeptides that is involved in endochondral bone formation during embryogenesis, participates in fracture healing, and can induce ectopic bone formation. BMP2, BMP4, and their receptors are not only associated with organogenesis and repair but are also expressed in postnatal bone marrow and are required for the formation of both osteoblasts and osteoclasts. BMPs stimulate transcription of the gene encoding Cbfa1 (core-binding factor a1), a transcription factor that can induce osteoblast differentiation by activating osteoblast-specific genes to produce a characteristic molecular repertoire including alkaline phosphatase, osteopontin, bone sialoprotein, type I procollagen, and osteocalcin. Cbfa1 is also known as osteoblast specific transcription factor 2 (Osf2). Lack of Cbfa1 prevents osteoblast development and also leads to a paucity of osteoclasts. Continued bone resorption with interrupted bone formation is demonstrated by the absence of a sclerotic rim around lytic lesions in multiple myeloma and the low serum bone alkaline phosphatase level typical of the humoral hypercalcaemia of malignancy. Cell renewal and cell death may seem to be opposing and contradictory, but they are inexorably linked in bone, coupled and balanced just as tightly as are bone resorption and formation. After osteoclasts finally stop resorbing bone, they die by apoptosis and are quickly removed by phagocytes during the early portion of the reversal phase. Over 90% of bone cells are osteocytes, the fate of as many as 30% of the originally assembled osteoblasts, osteocytes are long-lived cells, surviving for decades. Osteoporosis is characterised by low bone mass and microarchitectural deterioration of the skeleton leading to an increased risk of fracture after minimal trauma. These fractures result from the accelerated loss of bone that occurs in women after natural or surgically induced menopause. Oestrogen loss (postmenopausal osteoporosis) disrupts local regulatory control of cytokines in the bone marrow, provoking an inappropriate increase in bone remodelling far in excess of that necessary to repair microdamage and replace old BMUs. Increased production of cytokines such as IL-1, IL-6, and TNF after oestrogen deficiency mediates the increase in osteoblast and osteoclast numbers and, consequentially, the increased frequency of activation. Loss of oestrogen's effect to promote apoptosis of osteoclasts would prolong the lifespan of osteoclasts and increase their numbers two- to three-fold, thus accounting for the perforation of trabeculae and grinding away of endocortical margins. With oestrogen deficiency, osteoclasts erode deeper than normal cavities, creating a biomechanically weakened structure. If they break through the trabeculae, the surface on which new bone would be formed is lost. The accelerated loss of bone that occurs after oestrogen deficiency is not only from an increase in osteoclast number and lifespan, but also from a premature reduction in the work hours of the osteoblasts. The increased bone remodelling that follows oestrogen deficiency induced by gonadotropin-releasing hormone analogues (GnRH) in humans is also associated with an increase in osteocyte apoptosis. The accumulation of apoptotic osteocytes caused by loss of oestrogen could increase bone fragility even before significant loss of bone mass because of the impaired detection of microdamage and repair of substandard bone. Similar effects on cytokine production and bone turnover follow loss of gonadal function in either sex. Involutional osteoporosis, which develops in both sexes after the age of 60, occurs more frequently in women than in men, indicating that even in the older age group oestrogen deficiency continues to play a significant role. Older women have >3 times as many hip fractures than do older men. An increase in the distance between adjacent trabeculae accounts for 68% of the reduction in cancellous bone with advancing age, whereas only 23% are attributed to diminished trabecular width. The increase in trabecular spacing is predominantly the result of oestrogen deficiency, whereas the decrease in trabecular width is the result of defective osteoblast production relative to the demand and the subsequent decrease in bone formation. With increasing age, osteoblast production becomes progressively less capable of refilling resorption cavities, a bone deficit associated with a reciprocal increase in bone marrow adipose tissue. This may result from an overexpression of the transcription factor peroxisome proliferator-activated receptor 2 (PPAR2) and its ligand, prostaglandin G_2, that redirect pluripotent mesenchymal stem cells toward adipocytes instead of osteoblasts and perhaps to changes in insulin-like growth factors (IGFs) and their binding proteins. The resultant loss of bone from incomplete refilling is presently irreversible, because each bone remodelling cycle is a transaction that, once consummated, is irrevocable. Osteocyte death in cancellous bone, indicated by absence of lactic dehydrogenase activity, increases in prevalence with age in the femur but not in lumbar vertebrae, probably because of the higher rate of bone remodelling in the spine. Glucocorticoid-induced osteoporosis is the third leading cause of osteoporosis after the postmenopausal and involutional variety. Chronic glucocorticoid therapy causes the accumulation of apoptotic osteocytes and lining cells. Glucocorticoid-induced osteonecrosis may actually be osteocyte apoptosis, a cumulative and unrepairable defect that would uniquely disrupt the mechanosensory role of the osteocyte-canalicular network and thus promote collapse of the femoral head. Oestrogen promotes osteoclast apoptosis more than enough to reduce osteoclast numbers greatly and is effective in preventing bone loss in both early and late postmenopausal women. Oestrogen therapy must, however, be taken chronically to reduce fractures and, unfortunately, the side effects of breast tenderness, fluid retention, and withdrawal bleeding along with concern about the risk of breast cancer limit the long-term acceptance of this approach for many women. Drugs that could decrease the prevalence of osteoblast apoptosis would expand the pool of mature osteoblasts at sites of new bone formation and allow these cells more time to make bone. The molecular basis of the dependency of osteoclastogenesis on osteoblast precursors has recently been discovered by the identification of a membrane-bound TNF-related cytokine, RANK (receptor activator of NF-β) ligand, that is expressed on committed preosteoblastic cells. RANK ligand is also known as osteoclast differentiation factor (ODF) or osteoprotegerin ligand (OPGL). The absolute requirement for RANK ligand in osteoclastogenesis has been revealed by several experiments. Lack of RANK ligand results in osteopetrosis, and administration of recombinant RANK ligand produces hypercalcaemia, increases bone turnover, and loss of bone.

APOPTOSIS IN CARDIAC DISEASE

Apoptotic cells either shrink or do not change size, whereas necrotic cells typically swell. The plasmalemma of apoptotic cells remains intact, but this external cell membrane of necrotic cells promptly ruptures. Most organelles (mitochondria, sarcoplasmic membrane, and so forth) of apoptotic cells remain morphologically preserved, whereas these same structures soon rupture or disintegrate in necrosis. The nucleus of apoptotic cells typically partitions its chromatin marginally and then the nucleus is cleaved into separate membrane-bound particles (apoptotic bodies), while the necrotic nucleus randomly clumps its chromatin as the nucleus itself disintegrates. Histologically, necrosis is characterised by all the consequences of cellular disintegration including tissue acidosis and the rapid chemotactic attraction of lymphocytes and neutrophils. By contrast, apoptotic cells quickly exteriorise molecules of phosphatidylserine from their normal internal location, thereby signalling macrophages (and even like cells such as neighbouring myocytes) to engulf them, leaving the apoptotic focus of cell death conspicuously devoid of inflammation. If phagocytosis fails, for whatever reason, apoptotic cells eventually disintegrate, releasing their internal contents, and then become indistinguishable from necrosis. The TUNEL stain identifies any nucleus containing broken strands of DNA. Any focus of apoptotic cell death in human myocardium has at least the following multiple diagnostic morphological features: 1) an apoptotic focus is distinctively separated by a sharp visible boundary from viable myocardium; 2) the apoptotic focus has a homogenous appearance, totally devoid of any evidence of inflammation; 3) the focus is filled with relatively intact myocytes mixed with fibroblasts, macrophages, and neural and vascular elements, any number of which may stain positively for apoptosis; 4) throughout the apoptotic focus there are numerous macrophages containing either a few or many ingested apoptotic bodies; 5) in some apoptotic areas, large groups of apoptotic bodies aggregate in sheets or pools in the intercellular space, owing to local overwhelming of phagocytic capacity; and 6) nearly every apoptotic focus also contains intermixed nonapoptotic cells, and the negative staining of their nuclei makes a useful contrast with the TUNEL-positive nuclei of the apoptotic cells. Apoptosis may be either beneficial or harmful depending upon coexisting circumstances. Apoptotic cell death is devoid of nearly all secondary harmful features of necrotic cell death. Thrombotic thrombocytopenic purpura (TTP) is characterised clinically as episodic widespread occlusion by platelet masses within small arteries and capillaries. An associated paradox included in the name of this disease comes about because so many platelets are consumed by this extensive multifocal thrombosis that a deficit of circulating platelets coexists, resulting in purpura and other bleeding problems. Virtually every organ is affected by the episodic bouts typical of TTP, but notably affected are the kidneys, brain, adrenals, and the heart. There are variable morphologic coexistence of apoptosis and necrosis in myocardial infarction. There is no single clinical picture or course applicable to all human myocardial infarctions. Cell death in the first couple of days of myocardial infarction may be entirely by apoptosis, in the next few days a mixture of apoptosis and necrosis. Many factors contribute to this initial clinical instability, including the nature of the coronary obstruction (spasm may wane, platelet clumps can disintegrate, partial or intermittent unhinging of a fragment of plaque can occur), or the level of myocardial metabolic demand (compounded by many autonomic and neurohumoral influences upon the heart); arrhythmias can come and go, the level and effectiveness of available collateral circulation can fluctuate, and countless other events such as fever, medications, and high or low blood pressure can each and all continually change the clinical picture. Brief renal ischaemia causes apoptosis but prolonged ischaemia produces necrosis.

THE ROLE OF APOPTOSIS IN INTESTINAL DISEASE

Apoptosis is a stereotypic form of cell death, defined on morphological criteria that is under tight genetic control. Introduced in 1965, the word apoptosis, which is derived from ancient Greek and means "leaves falling in autumn," invokes the idea that death is required for life. Unlike necrosis, which occurs only in pathological situations, apoptosis is a normal and essential part of growth and development that acts to counterbalance cell division and remove unwanted or damaged cells. It is therefore not surprising that inappropriate regulation of apoptosis can lead to disease. Failure of apoptosis can allow the development of cancer. In apoptosis the cell first detach from its neighbours and shrinks. The nucleus shrinks as well and, under electron microscopy, the nuclear chromatin can be seen to condense around the edge of the nucleus. Cell surface features, such as microvilli, are lost, and blebs of plasma membrane develop, gives the cell the appearance of boiling. Necrotic cells swell then burst because of loss of integrity in the plasma membrane early in the necrotic process, with consequent failure of intracellular ionic homeostasis. The escape of intracellular contents in necrosis invariably provokes an inflammatory response, which results in further damage to neighbouring healthy cells. Apoptosis prevents the escape of intracellular contents. Plasma membrane integrity is maintained until after phagocytosis. Prevention of the premature rupture of the apoptotic cell is assisted by the activation of γ-glutamyl transferase, which forms cross-links between membrane proteins, forming protein shells that are insoluble in detergent. The mechanism that attracts and facilitates the action of phagocytes could be redistribution of plasma membrane phosphatidylserine to the outer leaflet of the plasma membrane bilayer might be an indicator of apoptosis to surrounding cells. This redistribution of lipid within the leaflets of plasma membrane appears to be widespread in apoptosis and can be regulated by 2 genes that are known to inhibit apoptosis, bcl-2 and abl. A numbering system has been developed in which the epithelial cell at the base of the crypt is designated "1" and the cells are then numbered up the sides of the crypt. In the small intestine, spontaneous apoptosis has a maximum incidence at cell position 4-6. This implies that approximately 10% of the stem cells are undergoing apoptosis at any one time. Percentage of all crypt cells, apoptosis rare (<1%), in only one of every 5 to 10 crypt sections. In the colon, the stem cells are located lower down in the crypt, at cell positions 1 and 2, while apoptosis is at the higher cell positions 4-6. Crypt cells migrate up to the tip of the villus where they are shed at a rate of 1400 cells/villus/day. In the human small intestine, at the villus tip, epithelial cells first shrink to form dome-like profiles, leaving large vacuoles underneath. A lymphocyte often migrates into this vacuole. Then chromatin condensation develops in the epithelial cell just before it is shed into the lumen. The barrier function of the epithelium is maintained by neighbouring cells extending processes under the shed cell and forming tight junctions with its neighbours. Cells shrink and loosen from their neighbours do not develop chromatin condensation until the point of cell shedding. DNA fragmentation occurs while the cell is still histologically normal. Bcl-2, a gene that functions to suppress apoptosis, is expressed only at the base of the colonic crypt and is barely expressed in the small intestine. Bax, another member of the bcl-2 gene family, accelerates apoptosis and is expressed most strongly at the surface of colonic epithelium and base of small intestinal crypt epithelium. There are important differences between apoptosis in the crypt and the villus tip. In the crypt the apoptotic calls loosen from their neighbours and undergo phagocytosis in situ. At the villus tip, although the apoptotic cells loosen from their neighbours the barrier function of the epithelium is maintained, rather than undergoing phagocytosis, the cells are shed. This suggests that apoptosis at the villus tip is a specialized process and not typical of the generality of apoptosis. Activated neutrophils kill microorganisms by the generation of OFR. These reactive species induce apoptosis in intestinal epithelial cells at low concentration, where cellular damage is mild. At high concentration, cellular membranes break down rapidly through lipid peroxidation reactions, and necrosis results. Cytotoxic T cells induce apoptosis in target cells by 2 distinct pathways. First, by the secretion of perforin, a pore-forming protein, which is inserted into the plasma membrane of the target cell and allow granzyme B to pass through it into the target cell. The mechanism by which granzyme B induces apoptosis is not fully understood. However, it has been demonstrated that it activates CPP32, one of the interleukin-1β-converting enzyme (ICE)-like cystein proteases, and a novel 55-kDa protein, designated FLICE, which has homology to both FADD and the ICE/CED-3 family of cystein proteases, which are part of the effector mechanism of apoptosis. Secondly, ligation of CD95 ligand on T cells to CD95 on target cells will induce apoptosis by activating ICE-like proteases, initiated by the generation of ceramide through the activation of an acidic sphingomyelinase. Apoptosis of both target epithelial cells and lymphocytes will be found in inflammatory reactions in the GIT, e.g., Helicobacter pylori-induced gastritis. Abnormalities in apoptosis could give rise to inflammatory bowel disease. Shigella flexneri can evade immune attack by inducing apoptosis in host macrophages. Clostridium difficle causes detachment of intestinal epithelial cells, resulting in their apoptosis. Non-steroidal anti-inflammatory drugs (NSAIDs) induce apoptosis in the intestinal epithelium. Testicular cells constitutively express CD95 ligand, which induces apoptosis in mature T cells (which express CD95) entering the testis. This probably explains the mechanism of immune-privilege in the testis. Cells are constantly exposed to mutagens and divide billions of times in the life span of human being. Therefore the chances of a cell developing a malignant mutation and becoming a fatal carcinoma should be high. Only one in three humans develop cancer. Apoptosis is one of these anti-cancer mechanisms. For a cell to become malignant, not only must it acquire mutations to give the cell unregulated growth but also mutations to prevent destruction of the mutant cell by apoptosis before it can undergo clonal expansion. Small intestinal cancer is at least 100% less common than colorectal cancer. In the small intestine, chemical carcinogen-induced apoptosis occurs within maximum frequency at the level of stem cells within the crypt. A stem cell with a potentially malignant mutation will be deleted immediately. By contrast, in the colon, the maximum incidence of carcinogen-induced apoptosis is at cell position 5-10, above the positions of the stem cells, which are at position 1-2. These data suggest that mutant stem cells in the colon have a greater chance of avoiding deletion by apoptosis than mutant stem cells of the small intestine. Bcl-2 is expressed in 75% of colonic adenomas and in a lower proportion of colorectal carcinomas. Bcl-2 expression disappears during progression from adenoma to carcinoma. Deletions of chromosome 18q, the locus of the bcl-2 gene, occur in 69% of colorectal cancers. There is inverse relationship between bcl-2 and p53 expression. p-53 will be mutant and non-functional. Wild-type p53 and some mutants down-regulate bcl-2 by binding to a transcriptional silencer element within the bcl-2 promoter. Mutation of p53 occurs in 86% of colorectal cancers and occurs most frequently at the transition between adenoma and carcinoma. The function of p53 is to prevent the accumulation of genetic damage within the cell. In response to DNA damage, oncogene activation, infection with certain viruses or tissue hypoxia p53 levels rise. If the damage had been mild, the cyclin-dependent-kinase inhibitor p21 is activated; this halts the cell cycle and allows the repair to take place. If, on the other hand, genomic damage is severe, the cell is induced to under apoptosis. In the intestine, p53 plays an important role in the regulation of apoptosis when transfected with wild-type p53. Functional p53 is essential for the induction of apoptosis at the base of small and large intestinal crypt 4.5h after irradiation. Tumours contain areas of hypoxia because of inadequate blood supply. Defective apoptosis plays a particularly interesting role in hypoxic tumours, which can lead to the selection of p53-mutant clones. In normal tissue, hypoxia induces p53, causing the death of hypoxic cells through apoptosis. In areas of hypoxia within a tumour, p53 mutant cells will tend to replace p53 wild-type cells and lead to the progression of the tumour's malignant phenotype. NSAIDs can prevent or reduce the risk of developing colorectal cancer. Individuals taking one aspirin per day have a reduced risk of developing colorectal cancer up to 50%. The mechanism of induction of apoptosis by NSAIDs is controversial. NSAIDs inhibit an enzyme cyclooxygenase (COX), which is sometimes also referred to as prostaglandin endoperoxide synthase. There are 2 isoforms of COX. COX-1 is constitutively expressed in the GIT and is responsible for the maintenance of tissue integrity. COX-2, on the other hand, is usually undetectable in resting conditions but can be induced by growth factors or cytokines. 43% of adenomas and 90% of carcinomas express COX-2. COX-2 may be important in the development of colorectal cancer from adenomas. NSAIDs may regulate bcl-2 expression. COX-2-overexpressing cells have increased adhesion to ECM compared to controls. COX-2 expression activates matrix metalloproteinase-2, which enhances the invasiveness of cells and their potential to form metastases. Both irradiation and the anti-cancer chemotherapeutic agents induce apoptosis in intestinal epithelium. These agents could be separated into 3 classes depending on the site of apoptosis: 1- isopropylmethane-sulfonate, bleomycin and andriamycin affected cells at positions 4-6 (the probable stem cells); 2- bischloroethinylnitrosourea, actinomycin D, cyclophosphamide, and cycloheximide affected cells at positions 6-8 (the early transit cells); and 3- while mechloroethane, triethylenethiophosphoramide, vincristine, 5-flourouracil, hydroxyurea, and methotrexate affected cell positions 8-11 (the late transit cells). Apoptosis may account for the enteropathy, which sometimes occurs after chemotherapy. Radiation induces less apoptosis in p53-null cells or cells expressing bcl-2.

PROSTATE CELL APOPTOSIS

Cell death may occur passively by necrosis or actively by apoptosis. Apoptosis has a central role in the development of the prostate, in the normal process of prostate glandular self-renewal and in the neoplastic processes that occur after androgen ablation. Prostate cell apoptosis under normal physiological conditions may be interpreted as a default mechanism or a dormant process requiring direct activation. Epithelial cells are prevented from initiating apoptosis by critical survival-growth factors, which evident after androgen withdrawal when glandular epithelial cells but not basal epithelial or stromal prostate cells die, causing the prostate gland to involute and decrease in size. Within the initial 24 hrs after androgen ablation, prostatic tissue death commences via apoptosis with DNA degradation. Apoptosis in these cells is proliferation independent and proceeds without the cells leaving the G_0 phase of the cell cycle or proceeding through a defective cell proliferation stage. Apoptosis may be brought about by a range of diverse stimuli in the prostate cell, including cytokines, glucocorticoids, ligation of cell surface death signalling transmitting receptors and subnecrotic damage by toxins, hyperthermia or radiation in the presence of androgen. The majority of CaP cells maintain the sensitivity to androgen that explains why certain metastatic forms of prostate cancer may be managed for moths and even years after androgen ablation. Despite these beneficial effects patients eventually have relapse to an androgen insensitive state in which the outcome is uniformly fatal. CaP cells that escape androgen ablation maintain the capacity to undergo apoptosis. Chemotherapeutic agents are capable of arresting androgen insensitive prostate cells proliferating at various phases of the cell cycle and subsequently stimulating their apoptosis. Events leading to apoptosis (intracellular and cell surface functions) include: 1- receipt of apostate, 2- recognition and intracellular signal transduction of the stimulus, 3- downregulation of anti-apoptotic and cell cycle regulatory proteins, 4- recruitment of the effector apoptotic machinery, and 5- removal of the apoptotic body. Cell surface death receptors-ligand systems including Fas and TNFR-1 contain the death domain (80 amino acids) that is required for signal transduction during apoptosis of the prostate. Fas associated death domain FADD (recruited by Fas-Fas ligand) activates caspases 8 and 10 to initiate apoptosis. The Fas-FADD complex is also associated with RIP, a death domain containing serine-threonine kinase. Prostatic epithelia may express soluble Fas and Fas receptor. The Fas pathway of apoptosis have a significant biological role in normal tissue homeostasis, the hormone-refractory response and cancer progression. CaP often proves not to be immunoreactive. Fas mediated apoptosis is down-regulated in the development of metastatic cancer. CaP cells have decreased potential to undergo Fas induced apoptosis, while locally invasive CaP cells maintain sensitivity. TNF mediates cell death by binding its receptor TNFR-1, which recruits the death domain TRADD. TRADD recruits FADD in a TNF-α dependent process to form the death signalling complex merging the apoptotic pathways of TNFR-1 and Fas. TNF-α sensitises cells to γ-radiation induced apoptosis via ceramide mediated pathway. Inhibition of ceramide production is associated with radioresistance in CaP.

Effect of Apoptotic stimuli:

Apoptotic stimulus		Effect
1- TNF-α		Increases ceramide, sensitiser to γ-radiation. Affects mitochondria/cytochrome C. Caspases involved are 8 and 6.
2- Fas		Affects mitochondria/cytochrome C. Generates ROI. Decreases bcl-2. Caspases involved are 7 and 8.
3- TGF-β		Decreases bcl-2. Caspase involved is 1.
4- Androgen ablation		
	Sensitive cells	Increases Fas receptor expression. Induces TGF-1. Increases calcium influx. Decreases bcl-2 and increases Bax.
	Resistant cells	Decreases bcl-2 and increases Bax. Inhibit Crm A.
5- Radiation		Increases ceramide. Caspases involved are 3 and 7.
6- Chemotherapy		
	Camptothecin	Sensitises to Fas. Cytochrome C release/change in mitochondrion membrane potential. Induces peroxide and superoxide. Inhibit Z-VAD.
	Staurosporine	Cytochrome C release/change in mitochondrion membrane potential. Inhibit Bcl-2. Caspases involved are 3, 7.
	Sodium phenylacetate	Cytochrome C release/change in mitochondrion membrane potential. Inhibit Bcl-2. Caspases involved are 3 to 7.
	Bicalutamide	Inhibit Bcl-2. Caspases involved are 3 and 7.
	Phosphmevalonate decarboxylase	Cytochrome C release/change in mitochondrion membrane potential. Induces ROI/amino acid inhibition. Caspase involved is 3.
	Paclitaxel	Caspases involved are 3, 7 and 10.
	Lovastatin	Caspase involved is 7.
	Etoposide	Inhibit Bcl-2. Caspase involved is 3.
7- Gene therapy		
	Caspase 7	Induce calcium influx. ΔMφ. Z-AVD inhibition.
	Caspases 3 + 7	Caspases involved are 3, 7 and 10.
8- NO		Induces ROI/amino acid inhibition. Decreases Bcl-1. Caspase involved is 3.
9- BMD188		Cytochrome C release/change in mitochondrion membrane potential. Induces ROI/amino acid inhibition. Caspase involved is 3.
10- Diethylmaleate		ΔMφ. Induces ROI/amino acid inhibition. Caspase involved is 3.
11- Thapsigargin		Increase calcium influx. Caspases involved are 3, 7 and 10.
12- FTY 720		Caspase involved is 3.

There are mitochondrial changes in membrane integrity that take place before the manifestation of the biochemical features of apoptosis. Opening mitochondrial mega-channels between the inner and outer membrane by permeability transition marks the point of no return and commitment of the cell to death. Certain apoptotic molecules that normally stored in the mitochondrial intermembrane space are released into the cytosol, including Apaf-1-cytochrome c that forms a ternary complex with and activates procaspase 9, a death signalling protease apoptosis inducing factor, a flavoprotein that travels to the nucleus and is responsible for nuclear disassembly and, in certain cell types, the proforms of caspases 2, 9 (death signalling proteases) and 3 (death effector protease). Intracellular biochemical signalling pathways include calcium influx, ROI generation, glutathione depletion and ceramide production may occur in response to extracellular signalling for apoptosis, or induce apoptosis directly. Prostate calcium signalling has an important role in the signalling events during the response to hormone therapy and drug induced apoptosis. Androgen ablation causes a rapid elevation in intracellular calcium in androgen dependent cells only after hormone withdrawal. The depletion of glutathione (endogenous thiol antioxidant) and the generation of ROI have important roles in the early signalling of apoptosis. Replenished glutathione and use of scavengers prolong the onset of apoptosis. The intracellular trafficking of glutathione within the cell is an important event for determining the susceptibility of the cell to apoptosis. Bcl-2 expression is associated with the intracellular relocation of glutathione to the nucleus and prevents the functioning of certain death proteases (caspases). Manipulating intracellular trafficking of glutathione may induce and increase CaP susceptibility to chemotherapy and radiation induced apoptosis. Glutathione depletion increases cytotoxicity and sensitises Bcl-2 over expressing human prostate tumour cell to radiation induced apoptosis. Sphingomylinase activation is an immediate response to plasma membrane damage associated with ionising radiation. Digestion of membrane lipid by sphingomylinase generates ceramide. The cell death proteases (caspases) are family of cyteine dependent proteases (ranging between 10 to 20 kDa). Caspases implicated in cell death may loosely classified as initiators and executioners. Limited proteolysis to the active form releases the prodomain and interdomain linker region. Prodomain size has functional importance. Cytochrome C release from the mitochondria serves as a critical step in the activation of the downstream caspases, e.g., caspase 9. Caspases cleave structural protein involved in cell architecture, and functional proteins involved in cell cycle regulation and DNA repair. The activation of DNA repair enzymes, such as poly(ADP-ribose) polymerase (PARP) and activation of a caspase activated deoxyribonuclease (CAD), serves as a crucial event in the commitment of the cell to undergo apoptosis. Caspase 3 mediated cleavage of $p21^{Waf-1}$ converts cancer cells from growth arrest to apoptosis, leading to acceleration of chemotherapy induced apoptotic process in these cancer cells. The prostate expresses caspases 3 and 1. Basal cell expresses low caspase 3 compared with glandular secretory cells. High caspase 3 in the luminal secretory cells correlates with high cell turnover. Caspases are involved during prostate progression after castration. The continued growth of androgen sensitive tumours after androgen withdrawal may result from altered apoptotic signalling mediated by caspase inhibition. There is increased Bcl-2, Bcl-X1 and Mcl-1 expression during CaP progression. The mechanisms by which Bcl-2 family member inhibit apoptosis are multifunctional. Bcl-2 family exerts anti-apoptotic effects downstream in the apoptotic cascade by physically interacting with caspase proforms. Bcl-2 may serve as a molecular chaperone preventing caspase activation. Bax may disrupt such complexes, leading to caspase activation and apoptosis. Elevated Bax-to-Bcl_2 ratio corresponds with the onset of prostate epithelial cell apoptosis. p53 Mutation is the most common genetic mutation associated with human malignancy. CaP progression is partially due to the failure of cells to die by apoptosis. Chemical-induced dimerisation of transfected caspase proforms may activate caspases in cancer cells, which may starts a proteolytic caspase cascade to commits the cell to die by apoptosis. Identifying caspase expression with androgen resistance enables therapies that up-regulate caspase expression and down-regulate anti-apoptotic protein expression pharmacologically or by gene transfer, which may serve as adjuvant strategy before androgen withdrawal. CaP cells use multiple pathways to evade apoptosis, bypassing these events and inducing the process via downstream activation of the caspases serve as an attractive strategy for treating CaP. Prostate carcinogenesis and apoptosis are related. The deregulation of apoptosis contributes to initiation, metastasis and progression to androgen insensitive state. The effectiveness of treatment of CaP depends on its ability to induce apoptosis in CaP cells.

Caspase recruitment:

1- Fas receptor or TNFR-1 engaged by soluble ligand, or directly.
2- Recruitment of caspases 2, 8 and 10 through FADD and TRADD associated death domains.
3- Mitochondrial membrane disruption with mega-channel opening and release of cytochrome C, proforms of caspases 3 and 9, and AIF1.
4- Increased Ca^{2+} and ROS.
5- Cytochrome C activates caspase 9.
6- Downstream effector caspases, e.g., 3 and 7, commit cell to death.
7- Bcl-2 may inhibit apoptosis at level of mitochondria and directly at caspases.

APOPTOSIS OF LEUKOCYTES

Apoptosis in both excessive and reduced amounts has pathological implications. Apoptosis is initiated by a number of different stimuli, including DNA damage, toxins, or extracellular signals. Apoptosis is an essential component of development and cellular regulation. Abnormal regulation of apoptosis can lead to disorders such as cancer, lymphocytes depletion in AIDS, and degenerative diseases. Uraemia is associated with alterations in host defense mechanisms, which increase the risk of infection and malignancy. Regulation of the apoptotic process is complex and involves several cellular pathways. PMNLs have the shortest half-life among leukocytes. Mature PMNLs spend 12 hrs in the bloodstream, after which time they migrate into normal tissues or are drawn by chemotactic stimuli to inflamed tissues. PMNLs undergo apoptosis and are recognised and engulfed by tissue-derived macrophages. PMNLs undergoing apoptosis are dysfunctional. Polyamines are uraemic retention solutes that are generated by intestinal bacteria and include spermine, spermidine, putrescine, and cadevrine. These toxins attenuate PMNL apoptosis. Urea is generated during amino acid breakdown, but is a weak uraemic toxin. It participates in the generation of cyanide and protein carbamylation. These toxic compounds lead to PMNL dysfunction. Arginine, creatinine and urea participate in the generation of aminoguanidine compounds, such as hydroxyurea, which are toxic to different cell types and may be apoptogenic. Impaired homocysteine metabolism in CRF results in hyperhomocysteinaemia. Homocysteine is a pro-oxidant, induces leukocyte apoptosis. There is retention of apoptogenic molecules in the serum of patients with CRF. G-CSF, IL-2, IL-1β, TNFα, glucocorticoids, and complement component C5a inhibit PMNL apoptosis (survival factors). IL-10 is anti-inflammatory molecule, promotes PMNL apoptosis (death factor). Apoptosis-inducing molecules are not significantly cleared by dialysis. Solutions enriched with effective osmolytes such as mannitol or amino acids result in cell apoptosis. Superoxide is a free radical that is derived from molecular oxygen by the addition of a single electron. PMNLs possess a reduced NADPH oxidase that produces superoxide radical when the cell is activated. An electron transfer from NADPH to molecular oxygen generates superoxide anion, which is rapidly dismutated through superoxide dismutase (SOD) to form hydrogen peroxide. Hydrogen peroxide reacts with chloride to generate hypochloric acid promoted by myeloperoxidase (MPO) enzyme of the PMNL azurophilic granules, and in extracellular space, following cellular degranulation. Hydrogen peroxide is detoxified by catalase or glutathione peroxidase. The direct interaction between superoxide and hydrogen peroxide can lead to the generation of highly reactive hydroxyl radicals. A metal-ion-dependent pathway (Haber-Weiss reaction) is responsible for the generation of hydroxyl radicals. The targets of free radicals and ROS include membrane polyunsaturated fatty acids, lipoproteins, proteins, and DNA. Oxidative stress in patients with CRF is multifactorial in origin and is due to an increased production of pro-oxidants, mainly ROS generated by activated PMNLs, and improper antioxidant defense mechanisms. The mitochondria are the main cellular source of superoxide and other ROS. Cytochrome c, a mitochondrial heme protein that is primarily involved in electron chain transport, can leak out into the cytosol, where it combines with a putative ICE-like protease. This complex activates caspase-9, which in turn activates caspase-3, the effector death molecule. Hydroxyl radicals may result in lipid peroxidation of the outer membrane of mitochondria, with consequent increased membrane permeability and leakage of cytochrome c into the cytosol. Bcl-2 may exert an antioxidant effect primarily as a free radical scavenger. Bcl-2 is poorly expressed in mature PMNLs. Bcl-2 blood concentration is reduced in patients undergoing HD. Endotoxic shock is associated with upregulated expression of inflammatory cytokines, sometimes secondarily to the release of bacterial lipopolysaccharides. TNF induces hepatocyte apoptosis during septic shock.

ALVEOLAR EPITHELIAL CELLS

In normal conditions, the alveolar space is free from fluid except for a very thin layer that forms a subphase between the surfactant layer and the alveolar epithelium. The lung air spaces are lined by an epithelial monolayer that in normal conditions is exposed to relatively high oxygen tension (100 mmHg) when compared to other tissues. The large changes in alveolar O_2 tension result in a progressive injury of the components of the alveolar wall, leading to increased protein permeability, pulmonary oedema and compromised gas exchange. Na may enter alveolar type II (ATII) cells through a variety of different pathways including Na cotransporter: Na-H antiporter, Na-phosphate, Na-amino acid, Na-glucose symport, Na-K-Cl cotransporters, and diffusion through Na channels. Along with the apical Na channels, basolateral Na-K-ATPase represents the major protein involved in transepithelial Na transport in alveolar cells. Na-K-ATPase, localised at the basolateral side of ATII cells, extrudes actively the Na from the cell and therefore tightly regulates the intracellular Na concentration. The α_1 subunit of Na-K-ATPase is the catalytic component that translocates Na ions from the cell and K ions into the cell, and is usually produced in excess. The β_1 subunit allows insertion of functional Na pumps in the plasma membrane and in several tissues is the rate-limiting step in the formation of functional Na pumps. Oxygen is an indispensable therapeutic agent used extensively for the correction of arterial and tissue hypoxaemia in patients with cardiac and pulmonary disease. However, its usefulness is limited by its well-known toxicity on the pulmonary system. Prolonged exposure to 100% O_2 causes atelectasis and the formation of pulmonary oedema leading to death from respiratory failure. Exposure to hyperoxia results in increased intracellular production of partially reduced oxygen species (PROS) by pulmonary tissue and lung cells. These species cause lipid peroxidation, protein inactivation and destruction of nucleic acids. In normal conditions, lung cells have the capacity to scavenge physiological concentrations of PROS inasmuch as they possess large amount of antioxidant enzymes such as superoxide dismutase, catalase, and glutathione peroxidase. During hyperoxia, the production of PROS can exceed the antioxidant defenses of lung cells and lead to progressive damage of the alveolar epithelium, increased endothelial and alveolar permeability and formation of protein-rich alveolar oedema. Among lung cells, alveolar type II cells are resistant to oxidant injury because of high intracellular antioxidant levels, which play a key role in the reparative phase of lung oxidant injury. Acute hyperoxic injury (the lung oxidant injury) occurs in 2 sequential phases: An early acute-exudative phase, marked by a disruption of alveolar epithelium; and a late reparative phase, characterised by ATII cell proliferation and restoration of the alveolar epithelium. A pulmonary oedema is present in the acute phase, i.e., a reduction of lung liquid clearance and of unidirectional Na flux out the alveolar space, and Na-K-ATPase activity is reduced by half in ATII cells. Acute severe hyperoxia has an opposite impact on protein and gene expression. Hyperoxia alters the membrane proteins that lead to a decrease of Na pump activity. Hyperoxia increases gene expression of the 2 subunits of Na-K-ATPase, which likely maintains or preserves tissue function. This increase in gene expression likely participates in the normalisation of Na-K-ATPase activity observed during the reparative phase of the acute injury. The decrease in alveolar O_2 availability may be observed either in pathologic situations such as respiratory distress syndrome, or in environmental conditions associated with decreased barometric pressure, such as that which occurs during a rapid ascent to high altitude. Exposure to altitude hypoxia potentially induces acute lung injury with pulmonary oedema in healthy subjects. Pulmonary oedema appears to be the consequence of an increase in microvascular permeability due either to an increase in capillary pressure in some areas of the lung or to the local release of mediators. Surfactant is composed of phospholipids (90%) and proteins (10%), and is important for the maintenance of normal alveolar function and gas exchange. IFNγ, dexamethasone, and the combination of the two all increase the content of surfactant protein-A (SP-A) in alveolar cells. TNFα decreases mRNA levels and expression of both SP-A and SP-B. TNFα decreases surfactant phospholipid synthesis by human ATII cells. NO generation, secondary to PGE_2 production is responsible for the TNFα-induced inhibition of phosphatidylcholine synthesis. IL1α up-regulates the expression of SP-A. PMN are capable of impairing surfactant function and of degrading the major apoprotein. The effects on surfactant function are mediated largely if not exclusively by oxidant radicals. Transepithelial fluid transport by alveolar epithelial cells is the target of acute inflammation. Cytokines increase the rate of alveolar fluid clearance. EGF increases the capacity of ATII cells to transport Na, and TGFα increases alveolar fluid clearance. TNFα increases alveolar fluid clearance and mediates in part the increase in alveolar fluid clearance. ATII cells constitutively express class I HLA molecules. It might be possible that when an acute inflammatory reaction is initiated, IFNγ secretion increase the expression of HLA class II molecules by ATII cells. ATI cells constitutively express ICAM1. ICAM1 is used by lymphoblasts and alveolar macrophages to adhere to epithelial cells. ATII cells may play an important role in creating distinct immunologic environment within the lung through the secretion of GM-CSF. ATII metabolise AA through the cyclooxygenase and lipoxygenase pathways. ATII cells secrete PGE_2, PGI_2 and $PGF_{2\alpha}$. PGE_2 and PGI_2 inhibit many functions of macrophages, lymphocytes and neutrophils. ATII cells are important source of cytokines in the alveolar space. TGFβ inhibits lymphocyte proliferation and promotes the transcription and the secretion of pro-inflammatory cytokines such as TNFα and IL1. ATII expresses IL6 in the fibrotic lung. IL1β and TNFα act synergistically to increase IL6 secretion by ATII cells. Chemokines are members of a continuously growing family of molecules with potent chemoattractant activities for inflammatory cells. C-C chemokines are chemoattractant for monocytes (monocytes-inflammatory protein [MIP]-1), chemoattractant activity for neutrophils (macrophage-inflammatory protein [MIP]-2) or eosinophils and basophils (RANTES). Monocyte chemoattractant protein (MCP-1) is expressed in fibrotic lung. The host response to tissue injury, the acute phase response, is a highly coordinated series of physiologic reactions involving almost every major organ system. Marked changes in concentrations of plasma glycoproteins termed acute phase plasma proteins occur in the course of the acute phase of inflammation. The latter result from changes in the expression of specific genes in the liver and are the consequences of the action of inflammatory mediators (such as cytokines) and hormones (such as glucocorticosteroids) on the hepatocytes. The acute phase response may take place in extra-hepatic cell types, notably epithelial cells, and may be regulated by cytokines as observed in hepatocytes. ATII cells might be the sites of an acute phase reaction in the alveolus. ATII cells secrete NO. Proliferation of type II pneumocytes and their differentiation into type I pneumocytes is the basis of an adequate repair of the alveolar epithelium after injury. Dysfunction of ATII cells may promote the development of lung fibrosis. Altered fibroblasts emerging during fibrotic lung injury release soluble factors capable of inducing cell death and net loss of ATII cells. ATII cells release heat-stable peptides that protects endothelial cells against apoptosis induced by TNF. Two types of pulmonary oedema have been recognised depending on whether capillary permeability was altered or not. Whatever its type, pulmonary oedema proceeds by 2 stages of increasing severity: 1- interstitial oedema, during which the excess fluid is confined into the parenchyma, accumulating in peribronchovascular cuffs, and finally 2- alveolar oedema, when the capacity of the interstitium is overwhelmed, an internal drowning that excludes parts of the lung from gas exchange and may be immediately life-threatening. Alveolar type II pneumocytes transport Na and fluid in the apical to the basolateral direction. Protein concentration increased into airspeces to high values, developing an osmotic counter-pressure of >50 cmH_2O, which is less insufficient to step water absorption. AQP1 is sufficiently abundant in the alveolar region, more abundant in endothelial than in epithelial cells. The short term stimulation of transepithelial Na transport is probably the consequence of a direct effect of cAMP on Na channel open probability. Growth factors and cytokines may directly or indirectly affect alveolar Na transport and fluid absorption. TNFα produces an immediate increase in alveolar fluid clearance. When the permeability of the alveolar epithelial barrier to proteins is severely increased, the epithelium is unable to reabsorb alveolar oedema. Exposure to high levels of oxygen has well-known pulmonary and neurologic toxic effects. Hyperoxia increases the production of ROS in the lungs that result in cellular damage, pulmonary oedema and alveolar epithelial destruction. Exposure to 100% O_2 is lethal if prolonged, but during this there is increase in antioxidant enzymes and in surfactant synthesis and secretion. During recovery from hyperoxic lung injury there is an increase of alveolar epithelial Na and fluid transports. The increase in Na transport could be due to increased expression and activity of Na channels and of Na-

K-ATPase. During lung injury, exchange areas may not be ventilated either because they are flooded or because oedema in the bronchial walls affects small airway opening, resulting in gas trapping and eventually in atelectasis. The barrier and transport properties of the epithelium are preserved during acute hypoxia. Alveolar epithelium, even in aerobic conditions, displays a high glycolytic activity. A difference of concentration of only 1 mmol/l Na across the epithelium will result in a pressure difference of about 25 cmH$_2$O, which is the maximum osmotic pressure that can develop lung proteins. Increasing the passive permeability of the alveolar epithelial barrier may affect fluid movement. As large as five-fold increases in albumin permeability do not alter alveolar fluid removal. Preservation of the transport properties of the epithelium correlate with clinical improvement in patients suffering from various pulmonary oedema.

ALLERGY

Clinical manifestations of food allergy may be primarily gastrointestinal, and include vomiting, oedema of the lips and throat, diarrhoea, blood in stools, malabsorption, and failure to thrive. However, food allergies may also affect other body systems. Dermatologic manifestations such as atopic dermatitis and urticaria commonly occur. Oral allergy syndrome, triggered by contact with fresh fruits and vegetables, consists of oropharyngeal pruritus and local angioedema. Respiratory manifestations of the syndrome may include wheezing, rhinitis, and otitis. Anaphylactic shock may occur in severely food-allergic children. Pulmonary haemosiderosis, haemolytic anaemia, and thrombocytopenia occur rarely. Erythromycin and cephalosporins are associated with the development of allergy when administered during the first 2 years of life only. A potential mechanism by which antibiotics administered in infancy could promote the development of allergy is through the effect of oral antibiotics on the gastrointestinal microflora: 1- antibiotics causes impaired lymphoid development and disordered immune responses that can be prevented by exposure to microflora products; and 2- antibiotic use during infancy promotes a shift in the Th1/Th2 balance toward Th2-dominant immunity. A 7-day course of kanamycin in infant results in elimination of intestinal Gram-negative bacteria, increased IgE and IL-4 levels, and decreased IFN-γ and IL-12 synthesis. Exposure to Mycobacterium tuberculosis may, by modification of immune profiles, inhibit atopic disorder (asthma) as it lowers serum IgE levels, and cytokine profiles biased toward a Th1 response. Gram-negative bacterial outer membrane (lipopolysaccharide [LPS], also known as endotoxin) may play an important role in asthma. LPS is present in high concentrations in organic dusts, air pollution, and household dusts. Inhaled LPS can exacerbate airway inflammation and airflow obstruction in allergic asthmatics. Allergic subjects are more sensitive than nonallergic subjects to the bronchoconstrictive properties of inhaled LPS. Prior allergen exposure significantly augments the inflammatory response to inhaled LPS. Not everyone develops airflow obstruction after inhaling LPS. TLR4 receptor polymorphism (mutation in the extracellular domain of the receptor that binds LPS) is significantly associated with those individuals who are resistant to the bronchoconstrictor effects of high doses of inhaled LPS. LPS exposure in early life could protect against the establishment of atopy, as it induces the innate immune system to generate IL-12, which promotes Th1 as opposed to proallergic Th2 responses to inhaled or ingested antigens. High levels of LPS exposure promote the development of a nonallergic Th1 phenotype. Antibiotic usage in the first 2 years of life might contribute to the inception of allergy. At least 80-90% of patients with atopic dermatitis is colonised with S aureus, whereas only 5% of healthy subjects are colonised. The susceptibility of atopic dermatitis skin to colonization with S aureus may be due to a lack of an intact stratum corneum in atopic dermatitis either uncovering receptors for staphylococcal adhesins, or reducing levels of free fatty acids and polar lipids with antibacterial activity in the stratum corneum. Most atopic dermatitis patients express specific IgE antibodies directed to the staphylococcal toxins found in their skin. These anti-S aureus toxin IgE antibodies are functional as they bind to high-affinity IgE receptors on basophils to induce histamine release, which can contribute to the prominent pruritus associated with atopic dermatitis. Initial definitions of asthma focused on a primary defect in airway smooth muscle as the cause of asthma. Airway inflammation has more recently replaced the concept of a primary defect in airway smooth muscle contraction as a key mechanism in the development of airway hyperreactivity and asthma. There is no primary genetic abnormality of airway smooth muscle in asthmatics. The most effective therapy for asthmatics namely, corticosteroids inhibits airway inflammation and airway hyperreactivity but has no direct effect on airway smooth muscle contraction. A cardinal feature of asthma is the development of airway hyperresponsiveness. Asthmatic subjects develop airway narrowing in response to low concentrations of diverse pharmacologic (e.g., methacholine), physiologic (e.g., histamine), or physical (e.g., exercise) stimuli. Any process that results in thickening of the airway wall can lead to the development of persistent airway hyperresponsiveness. Contraction and shortening of smooth muscle in the thickened airway leads to a much greater narrowing of the airway and to resultant airflow obstruction. The components of the thickened airway in asthma include airway oedema, extracellular matrix components, and increased airway smooth muscle thickness. CD4+ lymphocytes play a key role in the development of airway responsiveness as depletion of CD4+ lymphocytes abrogates induction of airway responsiveness. Depletion of mast cells has a variable effect on reducing eosinophilic inflammation and airway responsiveness. A variety of inflammatory mediators present in the airways of asthmatics (e.g., histamine, leukotrienes, platelet-activating factor, endothelin, and eosinophil major basic protein), released by mast cells and eosinophils, could either directly constrict airways via stimulation of specific receptors on airway smooth muscle, or indirectly constrict airways through induction of the release of other contractile agonists. Airway inflammation induces airway hyperreactivity independent of mediators of inflammation acting on smooth muscle cells to alter their excitability. Airway inflammation induces alterations in the airway wall geometry that result in greater narrowing of the airway for a given stimulus. The thickening of the airway smooth muscle layer may be due to smooth muscle hyperplasia (i.e., an increase in cell number) and/or smooth muscle hypertrophy (i.e., an increase in cell size). A common intracellular signalling pathway (possibly the mitogen-activated protein kinase/extracellular signal-regulated kinase pathway, and the phosphatidylinositol 3-kinase pathway) may mediate the response of airway smooth muscle to the divergent stimuli. Airway smooth muscle can be induced to express adhesion molecules (ICAM, VCAM), which allow for the direct binding of recruited inflammatory cells (T lymphocytes, eosinophils) to airway smooth muscle. This allows for a direct effect of these adherent leukocytes on airway smooth muscle function. Airway smooth muscle expresses receptors for interleukins, interferons, tumour necrosis factor (TNF), and CD40, allowing it to respond to a variety of proinflammatory cytokines and mediators, including TNF, IL-1, IL-4, IL-5, and IL-13. Airway smooth muscle is an important source of mediators, cytokines, chemokines, extracellular matrix, matrix metalloproteinases, and growth factors; all of which can contribute to the pathogenesis of airway inflammation and, potentially, airway remodelling. Respiratory viruses in particular, rhinovirus, respiratory syncytial virus, influenza virus, and parainfluenza virus are important precipitants of asthma. Rhinovirus has been implicated in the majority of viral-induced wheezing episodes in children and in 50% of viral exacerbations of asthma in adults. Rhinovirus may either directly affect airway smooth muscle to induce airway hyperresponsiveness, or indirectly affect airway smooth muscle by effects on neurotransmission (increased parasympathetic activity), effects on epithelium (release of cytokines, chemokines), effects on endothelial cells (permeability changes), or effects mediated by recruited inflammatory cells (releasing cytokines and mediators).

Mediators of allergies

Name	Source	Mechanism
Histamine	Mast cells (early phase), basophils (late phase).	Vasodilation, plasma leakage, granular secretion.
PGD_2	Mast cells, platelets, fibroblasts.	Vasodilation.
Tryptase	Mast cells.	
TAME-esterase	Plasma/granular kallikrein, mast cell tryptase.	Vasodilation.
LTB_4	Mast cells, neutrophils.	Eosinophil chemotaxis, neutrophil activation.
LTC_4 and LTD_4	Mast cells, eosinophils, neutrophils.	Vasodilation, increased blood flow.
Kinins	Plasma.	Vasodilation, increased capillary flow.
PAF	Macrophages, neutrophils, eosinophils.	Vasoconstriction/ vasodilation, eosinophil and neutrophil chemotaxis.
ECP, EPO, MBP	Eosinophils.	Epithelial damage.

Rhinovirus is capable of directly infecting airway smooth muscle cells, which is associated with the development of airway smooth muscle hyperresponsiveness to acetylcholine and impaired relaxation responses to β-adrenergic receptor stimulation. The ability of rhinovirus to induce airway smooth muscle hyperresponsiveness is dependent on rhinovirus binding to the adhesion receptor ICAM-1 on the surface of airway smooth muscle and does not require the virus to enter and infect the airway smooth muscle cell, which activates a signalling cascade that downstreams results in activation of an inhibitory G protein (Gi). Activation of Gi by rhinovirus results in inhibition of cAMP accumulation in smooth muscle. Rhinovirus binding to ICAM on airway smooth muscle induces the initial release of IL-5 by airway smooth muscle, which subsequently binds in an autocrine manner to IL-5 receptors expressed on the same airway smooth muscle cell, inducing the release of IL-1β. Airway smooth muscle is both a contractile and secretory cell contributing to airway inflammation. Airway remodelling is characterised by the development of airway wall thickening, subepithelial fibrosis, mucus metaplasia, myofibroblast hyperplasia, myocyte hyperplasia and hypertrophy, and epithelial damage. The normal bronchial epithelium is a stratified structure consisting of a columnar layer composed of ciliated and secretory cells supported by basal cells. In asthma, the epithelium shows evidence of activation linked to structural damage and goblet cell metaplasia. Evidence of epithelial injury in asthma includes epithelial expression of transcription factors, autacoid mediators, chemokines, and growth factors that sustain ongoing inflammation. Asthmatics airway epithelium is more susceptible to oxidant-induced apoptosis. The epidermal growth factor receptor (EGFR) serves a central role as a primary regulator of epithelial cell function. In asthma, there is increased expression of EGFR both in damaged and morphologically intact epithelium, the extent of which correlates with sub-basement membrane collagen thickness. The disease-related increase in epithelial EGFR expression in asthma is not matched by increased proliferation of epithelial cells to replace columnar epithelial cells that have been shed. The lack of epithelial proliferation and repair may be due to the strong epithelial expression in asthma of a negative regulator of epithelial proliferation $p21^{waf}$. $p21^{waf}$ is an inhibitor of the cyclin-dependent kinase (CDK), which regulates the progression of the cell cycle from G1 into S phase and from G2 into mitosis. The CDK inhibitor $p21^{waf}$ is a negative regulator of G1 cyclins and can be induced in epithelium by stress, injury, and TGF-β, which may explain its potential role in airway remodelling. Communication between the epithelium and the subepithelial fibroblast sheath functions as a trophic unit to regulate airway growth and branching during embryogenesis. Reactivation of this epithelial-mesenchymal trophic unit in asthma has been proposed to mediate remodelling of the airway in asthma. Signalling between the epithelium and myofibroblasts (which lie directly beneath the epithelium in the lamina reticularis) involves the release of growth factors with effects on myofibroblast proliferation and function. Epithelial cells are a source of several important growth factors, including FGF-2, PDGF, TGF-β2, and IGF-1. Chronic rhinosinusitis and asthma are common problems encountered by physicians in both primary and speciality care. Infectious sinusitis may be more characterised by influx of neutrophils, whereas noninfectious sinusitis may be more characterised by eosinophils.

Eosinophil secondary granule proteins:

Protein	Name	Biological activities
MBP	Major basic protein.	1- Triggers the granulation of mast cells and basophils. 2- Neutralises heparin. 3- Bactericidal. 4- Increases smooth muscle reactivity.
ECP	Eosinophil cationic protein.	1- Potent neurotoxin. 2- Inhibits lymphocytes. 3- Releases histamine from mast cell. 4- Bactericidal. 5- Causes voltage-insensitive, ion-nonselective toxic pores in the membranes of target cells, and these pores may facilitate the entry of other toxic molecules.
EDN	Eosinophil derived neurotoxin.	1- Potent neurotoxin. 2- Inhibits lymphocytes. 3- Potent RNase activity.
EPO	Eosinophil peroxidase.	1- Kills microorganisms and tumour cells, in the presence of H_2O_2 and Halide. 2- Releases histamine from the mast cells. 3- Inactivates leukotrienes. 4- Damages respiratory epithelium promotes bronchospasm, in the presence of H_2O_2 and Halide.

In infectious inflammation, in addition to more neutrophils, there are increased amounts of IL-8 in both sinus tissue and lavage fluid. IL-8 levels correlate with the radiographic extent of disease. Infectious sinus inflammation may increase symptoms of asthma by a number of possible mechanisms, including nasobronchial reflexes, postnasal drip (PND), and pharyngobronchial reflexes. Noninfectious inflammation may be marked by significant bilateral sinus mucosal thickening, nasal congestion, facial pressure or fullness sensation, PND, and a decrease or loss in the patient's sense of smell. These patients have a high incidence of nasal polyposis, and at least 50% of these patients have associated asthma. Approximately 1 in 5 of these patients will have associated aspirin and nonsteroidal anti-inflammatory drug (NSAID) sensitivity, resulting in increasing asthma symptoms as well as possible life-threatening angioedema and anaphylaxis. Sinusitis produced bronchial hyperresponsiveness is due to the direct passage of inflammatory mediators from the upper to the lower airways via aspiration. Corticosteroids exert an anti-inflammatory effect in asthma because they switch off the expression of multiple genes important to the pathogenesis of asthma, including cytokines, chemokines, adhesion molecules, receptors, and proinflammatory enzymes. Corticosteroids enter cells and bind to glucocorticoid receptors in the cytoplasm. Occupied glucocorticoid receptors rapidly translocate to the nucleus where they bind to glucocorticoid response elements in promoter regions of genes. Corticosteroid effects are on transcription factors (NF-κB and AP-1), and on histone deacetylases, which regulate gene transcription. Transcription factors that are important in asthma include NF-κB and AP-1, both of which are activated in the airways of asthmatics. One mechanism by which glucocorticoids regulate NF-κB expression is by regulating the expression of IκB, which normally keeps NF-κB in an inactive state in the cytoplasm, and not available in the nucleus to induce the expression of NF-κB-regulated proinflammatory genes. In response to cell stimulation with proinflammatory cytokines, IκB kinase phosphorylates IκB, which targets the IκB to the proteasome for degradation. NF-κB dimers in the cytoplasm are, thus, freed from the degraded IκB and can translocate to the nucleus to activate gene transcription. Corticosteroids work by inhibiting histone acetyltransferase (and, thus, corticosteroids prevent unwinding of DNA). Under resting conditions, nuclei of inflammatory cells contain tightly wound DNA with a closed chromatin structure that excludes binding of transcription factors so that the gene is silent or repressed. When the cell is stimulated, transcription factors bind to large protein coactivator molecules in the nucleus, which act as molecular gene switches. The binding of transcription factors to coactivator molecules activates histone acetyltransferase, which changes the charge of the core histone proteins around which DNA is wound. Viral respiratory infections are a very common trigger for acute exacerbations of asthma; in fact, most acute asthma episodes begin as an upper respiratory infection (URI). Drugs that inhibit leukotrienes help in the treatment of asthma. The goals of acute asthma therapy include correcting significant hypoxaemia, reversing airflow obstruction, and decreasing the likelihood of recurrence of asthma symptoms. Gastroesophageal reflux disease (GERD) is common in patients presenting with a number of respiratory complaints, including chronic cough. The incidence of symptomatic GERD in patients with asthma averages 57%. GERD also appears to be more common in a number of other pulmonary conditions, including chronic bronchitis, idiopathic pulmonary fibrosis, recurrent pneumonia, and cystic fibrosis. 50% of patients with GERD have eosinophils in biopsy specimens of the oesophagus "eosinophilic oesophagitis". Eosinophilic bronchitis without asthma may produce cough. Most frequently allergic response occurs at a mucosal surface, an important interface between the external environment and internal milieu. Irrespective of the organs in which they manifest, almost all of the above diseases share a single triggering mechanism, which underpins adverse responses to specific environmental allergens. This factor is present in the serum and is called immunoglobulin E (IgE). IgE, like the other antibody classes, is present in the blood, it is usually present in greater amounts and, more importantly, it is directed against specific environmental allergens. Any inhaled, injected or ingested allergen is presented to the immune system by the antigen-presenting cells (APCs) located at sites of interface between the external and internal environment, where it is recognised as foreign. IgE antibodies are secreted by plasma cells derived from activated B cells. Prostaglandins and leucotrienes derive from AA through different pathways and this is where the use of aspirin or leukotriene antagonists play a role in allergic response. The circulating inflammatory cells bind to the endothelium of mucosal vessels by interacting with adhesion molecules and counter ligands expressed freely on inflammatory cells and endothelial cell surfaces. This is followed by trans-endothelial migration, chemotaxis and chemokine involvement between the cell and matrix proteins as well as additional complex interaction with chemokines. The rolling of circulating eosinophils on the endothelium is mediated by P-selectin, whereas neutrophil rolling is mediated by E-selectin that interact with carbohydrate ligands-Sialyl Lewis^x and Sialyl Lewis^a. It is impossible to accurately group cytokines according to unique tissue sources or biological activities. Cytokines may be grouped according to those that are predominantly mononuclear phagocyte-derived (monokines) or T-lymphocyte-derived (lymphokines) and then according to those that predominantly mediate humoral, cell-mediated, or allergic immunity or, alternatively, are immunosuppressive. T cells that differentiate in response to MHC class II of antigen presentation are called T helper cells (T_H: CD4$^+$) and release an array of cytokines that orchestrate different types of inflammatory response. These are in contrast to CD8$^+$ cytotoxic T cell (Tc) that utilises MHC class I in response to antigen, are more promiscuous and serve to kill infected or cancerous cells or suppress immune responses by a targeted attack on CD4$^+$ T cells. MHC class I glycoproteins are expressed on the surface of almost nucleated cells. The expression of MHC class II molecules is mainly confined to cells directly involved in immune responses, i.e., dendritic cells, macrophages, B cells, activated T cells, and Langerhans cells. Glucocorticoids (GC) are very effective as anti-allergic therapy and the molecular mechanisms involved in suppression of inflammation are blocking many of pathways activated in allergy. Steroids may have direct inhibitory actions on several inflammatory cells including macrophages, eosinophils, T-lymphocytes, mast cells, dendritic cells, neutrophils, epithelial cells, submucosal gland and goblet cells (mucous secretion).

RENAL PHYSIOLOGY

OXYGEN AND RENAL METABOLISM

Oxygen extraction across the kidney is so low and mixed renal venous blood so oxygenated that it is often assumed that under normal circumstances kidney cells are plentifully supplied with more than enough oxygen to perform their work. The bulk of this supply is directed to the renal cortex, where it maximises flow-dependent clearance of wastes. The supply of O_2 to the renal medulla barely exceeds its O_2 utilisation. The medullary pO_2 is about 10 mmHg. The sharp corticomedullary gradient of O_2 is most easily explained by organisation of vessels in the medulla, which disposed in hairpin loop pattern to allow countercurrent exchange between the descending and ascending limbs for urine concentration. As a by-product of this arrangement, O_2 diffuses directly from the arterial to the venous branches of the vasa recta, thus reducing oxygen supply to the deeper portions of the medulla. The high O_2 requirement for active transport by the mTAL and the constraint imposed on medullary blood flow. Medullary hypoxia is the unavoidable price we pay for the ability to concentrate our urine. Ambient pO_2 in the medulla is normally close to the critical pO_2 (5-15 mmHg) at which the cytochrome oxidase of mitochondria in intact cells becomes reduced and respiration is slowed. The enzymes responsible for gluconeogenesis (glucose-6-phosphatase;

fructose 1,6 diphosphatase; and phosphoenolpyruvate carboxykinase) are limited to the proximal tubule. Oxidative metabolism, mainly fat, is the major source of energy for proximal tubular cells of the cortex. The enzymes of glycolysis, such as hexokinase, phosphofructokinase, and pyruvate kinase, predominate in distal structures of the nephron: medullary ascending limb, cortical thick limb, distal convoluted tubule, and entire collecting duct. Because of the high requirement for energy necessary to reabsorb approximately 20% of filtered sodium (Na) in the thick ascending limb, which can be supplied by oxidative metabolism of substrates, mitochondrial density is particularly high in the medullary thick limb but low in the medullary collecting duct. Mitochondrial respiration is exceptionally low in the inner medulla, where the ambient pO_2 is lowest in the body, and glycolysis predominates. Transepithelial sodium transport that depends on Na-K-ATPase, in the proximal tubule of the kidney, permits 3 mEq of Na to be transported for ever mM of ATP hydrolysed. The permeability of water along the proximal tubule is 10,000 higher than along the distal tubule. It is not surprising that proximal tubular cells have a striking propensity to develop cytoplasmic oedema in response to a decrease in energy stores. The cortex most often responds to any renal insult by a decrease in renal blood flow. Proximal tubular cells are not remarkably protected from hypoxia by a reduction in transport activity. By contrast, renal blood flow to the medulla usually increases in response to the threat of hypoxia, while a decrease in transport activity regularly protects against hypoxic injury to the medulla. Variety of heat-shock proteins, are found in higher concentration in cells lining medullary tubules than in those of the cortex. Incomplete renal ischaemia causes selective injury to mTALs. Combinations of multiple renal insults to a radiocontrast agent, produce selective necrosis of mTALs associated with acute renal failure. Chronic renal insults associated with prolonged renal vasoconstriction, induce selective degeneration and atrophy of mTALs in zones remote from oxygen. Reduction of transport reabsorption activity, along medullary tubules, and mTAL in particular lowers O_2 requirements. Inhibition of active transport by furosemide protects mTAL from hypoxic damage. Reduction of GFR diminishes the delivery of urine for solute reabsorption in the mTAL and blunts or prevents medullary hypoxic injury. Hypoxic mTAL injury observed after systemic hypotension and renal hypoperfusion is confined to open tubules, while collapsed nephrons are protected. Medullary hypoxia damage occurs only when reabsorptive work is active and that reduction of GFR is likely to prevent such injury. Rapid and profound reduction in metabolic activity is the response to lack of O_2. Hypoxic cells exhibit patterns of self-regulation of mitochondrial respiration at the level of key enzymes like aconitase, oxidases, and oxygenases. The hydrolysis of ATP in hypoxic cells increases the concentration of internal Mg^{2+}, close chloride exit channels in thick ascending limb cells, thus inhibiting active transport. In cells capable of glycolysis, a decrease in cellular pH produced by accumulation of lactic acid in the presence of hypoxia therefore acts as a break on cellular metabolism. Acidosis also decreases Na-K-ATPase activity. Cellular acidosis affords protection from hypoxic necrosis.

Liberation of adenosine during ischaemia, might protect medullary cells from ischemic injury by:

1- Increase the supply of O_2 by improving vasa recta flow.
2- Decrease the work of transport in thick ascending limbs.
3- Further decreases the necessity for active transport by depressing glomerular filtration rate via an exaggerated tubuloglomerular feedback and a decrease in renal cortical blood flow.

Functions of the kidneys:

1- Regulation of water and inorganic-ion balance.
2- Removal of metabolic waste products from the blood and their excretion in the urine.
3- Removal of foreign chemicals from the blood and their excretion in the urine.
4- Gluconeogenesis.
5- Secretion of hormones (erythropoietin, renin, 1,25-dihydroxyvitamin D_3).

The critical pO_2 of cells is the level at which O_2 tension in the surrounding medium becomes rate limiting. The rate of O_2 diffusion through highly proteinaceous solutions like cytoplasm is quite slow, cellular hypertrophy greatly increases the critical pO_2 and cellular susceptibility to hypoxia. The critical pO_2 is also raised by proliferation and hypertrophy of mitochondria within cells and by increase in the rate of O_2 utilisation. Hypertrophic mTAL cells have a critical pO_2 more than 3 times as high as smaller mTAL cells. While renal hypertrophy has long been known to predispose to glomerular injury, renal tubular cell hypertrophy, especially in the renal medulla might confer additional susceptibility to hypoxic injury and contribute to a vicious circle of progressive interstitial disease. The renal regulatory mechanisms protect medullary oxygenation by either increasing O_2 supply (blood flow) or decreasing O_2 demand. Leading factors for this mechanisms are derivatives of arachidonic acid (AA), especially PGE_2, adenosine and purine metabolites, and nitric oxide (NO). Prostaglandin E_2 (PGE_2) is produced by collecting duct cells and interstitial cells of the papilla. PGE_2 inhibits active transport by the medullary thick limb and reduces transport-related O_2 consumption by mTAL cells. PGE_2 enhances medullary blood flow. Chronic interstitial nephritis of analgesic nephropathy, characterised by fibrosis that is especially marked in the renal medulla and papilla, progressing to papillary necrosis, which contributes to ARF, conferred by NSAIDs. Prostaglandins (PGs) manufactured in the papilla or inner medulla may be carried by vasa rectal blood to their neighbours in outer medulla to spare thick ascending limb cells from ischaemic damage. Adenosine is endogenous modulator of ischaemic injury. Hydrolysis of ATP in excess of its synthesis is thought to lead to an increase in the intracellular concentration of AMP, which exists the cell via specific membrane transporters. AMP is hydrolysed to adenosine by 5'-nucleotidase located at the external face of the plasma membrane of most cells. In low concentrations, adenosine activates inhibitory receptors, which reduce the activity of adenylate cyclase and diminish the burden of cell work in excitable and secretory tissues. In higher concentrations, adenosine diminishes the rate of transport-associated respiration of mTAL cells. Adenosine enhances tubuloglomerular feedback. Nitric oxide (NO), increases vasa recta flow and decreases active transport by mTAL. The susceptibility to acute renal failure (ARF) conferred by the liberation of iron pigments into circulation, as in haemolysis or rhabdomyolysis, is probably due in part to the binding of renal NO by iron-containing pigments. Radiocontrast agents decreases medullary pO_2 by inhibiting NOS, leading to ischaemia. The loop diuretic furosemide improves medullary pO_2 without altering O_2 tension in the cortex. The proximal tubular diuretic, acetazolamide, produces little change in estimated pO_2 in the medulla or the cortex. Water diuresis increases pO_2 in the renal medulla, promotes excretion of PGE_2 in young but not in elderly subjects, and may reflect the effect of aging on compensatory mechanisms designed to protect the kidney against hypoxic injury.

THE GLOMERULUS

The renal corpuscle consists of glomerulus and a Bowman's capsule. The filtration barrier in the renal corpuscle consists of 3 layers: 1- The capillary endothelium of the glomerular capillaries is perforated by large fenestrae. 2- A basement membrane, is a gel-like acellular meshwork of glycoproteins and proteoglycans. 3- The single-celled layer of epithelial cells, called podocytes. Podocytes posses foot processes, which embedded in the basement membrane. Slits between adjacent foot processes constitute the path through which the filtrate travels to enter the Bowman's space. These slit do not offer completely passageways, because: 1- the foot processes are coated by a thick layer of extracellular material (glycosialoproteins), which partially occludes the slits, and 2- extremely thin diaphragm bridge the slits at the surface of the basement membrane. Blood in the glomerulus is separated from Bowman's space only by a thin set of membranes, which permit the filtration of fluid from the capillaries into the space. Bowman's capsule connects at the side opposite the glomerulus with the first portion of the tubule, into which the filtered fluid flows. Cells within the renal corpuscle are: 1- capillary endothelium, 2- podocytes, and 3- mesangial cells. The kidneys process blood by removing substances from it and by adding substances to it.

Substances handled by filtration and reabsorption:

Substance	Filtered	Excreted	% reabsorbed
Water	180 L	1.8 L	99 L
Sodium	630 g	3.2 g	99.5 g
Glucose	180 g	0	100 g
Urea	56 g	28 g	50 g

THE TUBULE

The tubule is made up of a single layer of epithelial cells resting on a basement membrane, with one common feature is the presence of tight junction between adjacent cells. A number of initial collecting tubule join end to end or side to side to form larger cortical ducts. All the cortical collecting ducts then run downward to enter the medulla and become outer medullary collecting ducts and then inner medullary collecting ducts. The inner medullary ducts merge to form several hundred large ducts, the last portions of which are called papillary collecting ducts, each of which empties into a calyx of the renal pelvis. The last portions of the tubule, the collecting duct system, are not part of the nephron. Nephrons are categorised according to the locations of their renal corpuscles in the cortex: 1- In the superficial cortical nephrons renal corpuscles are located within 1 mm of the capsular surface of the kidneys. 2- In midcortical nephrons renal corpuscles are located in the midcortex, deep to the superficial cortical nephrons but above the next category. 3- In juxtamedullary nephrons renal corpuscles are located just above the junction between cortex and medulla. One major distinction among these 3 categories is the length of Henle's loop. All superficial cortical nephrons have short loop, which make their hairpin turn above the junction of outer and inner medulla. All juxtamedullary nephrons have long loops, which extend into the inner medulla, often to the tip of a papilla. Midcortical nephrons may be either short-looped or long looped. The beginning of the thick ascending limb in the longest loops marks the border between outer and inner medulla, the thick ascending limbs are found only in the cortex and outer medulla. There are 2 million nephrons in the 2 human kidneys. The kidney receives a supply of sympathetic noradrenergic neurons, to the afferent and efferent arterioles, the juxtaglomerular apparatus, and many portions of the tubule. There is no significant parasympathetic innervation. There are some dopamine-containing neurons. There are 3 important generalisation to be drawn from the table: 1- Because of the huge GFR the filtered quantities are enormous, generally larger than the amounts of the substances in the body. The body contains 40 L of water, but the volume of water filtered each day is 180 L. If reabsorption of water ceased but filtration continued, the total plasma would be urinated within 30 min. 2- Reabsorption of waste products, such as urea, is relatively incomplete, so that large fractions of their filtered amounts are excreted in the urine. 3- Reabsorption of most useful plasma components, e.g., water, is relatively complete. The tubular cells may extract organic nutrients from the glomerular filtrate or peritubular capillaries and metabolise them as dictated by the cells' own nutrient requirements, e.g. the synthesis of the ammonia from glutamine and the production of bicarbonate. The biochemistry of the renin-angiotensin system is a complex: 1- About 50% of the renin in plasma is the inactive prohormone form of renin (prorenin) initially synthesised by the kidneys, and it is possible that this protein can be activated to renin by enzymes in peripheral tissues. 2- Clinically important situations exist in which changes in the concentrations of angiotensinogen or angiotensin-converting enzyme occur, which may significantly increase or reduce the generation of angiotensin II at any giving plasma concentration of renin. Contraceptives may cause a large increase in plasma angiotensinogen, whereas lung disease may reduce angiotensin-converting enzyme activity in the pulmonary capillaries. 3- In several tissues an enzyme exists, which splits angiotensin II to yield the heptapeptide angiotensin III, which is also active biologically.

Terminology of the tubular segments

Segments	Terms
Proximal convoluted tubule. Proximal straight tubule.	Proximal tubule
Descending thin limb of Henle's. Ascending thin limb of Henle's. Thick ascending limb of Henle's (contains macula densa near end).	Henle's loop
Distal convoluted tubule	
Connecting tubule. Cortical collecting tubule. Inner medullary collecting duct (last portion is papillary duct).	Collecting-duct system

Some intrarenal paracrine agents:

1- Angiotensin II.
2- Eicosanoids, oxygenated derivatives of arachidonic metabolism.
3- Kinins, specific kallikrein produced and secreted by the distal tubules cells splits plasma kinnogen to yield lysl bradykinin, which cleaved again within the kidneys to yield bradykinin. Bradykinin can exert potent effects on the renal vasculature, renin secretion, and ion transport.
4- Dopamine.
5- Endothelin, secreted by vascular endothelium and cells of several tubular segments, and is a potent vasoconstrictor.
6- Growth factors, cause the kidney to increase in size and changes in development such in embryogenesis and compensatory renal hypertrophy following loss of part or all of a kidney.
7- Adenosine.
8- Nitric oxide.
9- Cytokines, which mediate immune function intrarenally but may also, influence the renal vasculature and tubules under normal physiological conditions.

RENAL BLOOD FLOW AND GLOMERULAR FILTRATION

Given a normal haematocrite of 0.45, the renal plasma flow (RPF) = 0.55x1.1 L/m = 605 ml/min. GFR is 125 ml/min, so 605 ml of plasma that enters the glomeruli via the afferent arterioles, 125/605, or 20%, filters into Bowman's capsule, the remaining 480 ml passing via the effrent arterioles into the peritubular capillaries. This ratio GFR/RPF is known as filtration fraction. The high glomerular pressure (60 mmHg) is crucial for glomerular filtration. The low peritubular capillary pressure (20 mmHg) is equally crucial for the tubular reabsorption of fluid. RBF is determined mainly by the mean arterial pressure and the contractile state of the smooth muscle of the renal arterioles. The renal cortex receives 90% of the renal blood flow. The paucity of medullary blood flow is due to the high resistance offered by the vasa recta. The route that filtered substances take through: 1- fenestrate in the glomerular capillary endothelial layer, 2- basement membrane, 3- slit diaphragms, and slits between podocyte foot processes. The membranes of the renal corpuscle provide passage to the movement of molecules with molecular weights (MW) less than 7000 and almost hindrance to plasma albumin (MW 70,000), which is not 100%. So the glomerular filtrate contains extremely small quantities of albumin, on the order of 10 mg/L or less, which is 0.02% of the concentration of albumin in plasma. Many normally occurring plasma peptides and small proteins are filtered to a significant degree. Electrical charge is the second variable determining filterability of macromolecules. For any given size, negatively macromolecules are filtered to a lesser extent, and positively charged macromolecules to a greater extent, than neutral molecules. The surface of all the components of the filtration barrier contain fixed polyanions, which repel negatively charged macromolecules during filtration. The diseases that cause renal corpuscles to become "leaky" to protein do so by eliminating negative charges in the membranes. Low-molecular-weight solutes that completely filterable are partially bound to large plasma proteins. 40% of the plasma calcium is protein-bound, and so the calcium concentration of the glomerular filtrate is 60% of that in plasma. The rate of fluid movement, by filtration, in any of the body's capillaries is determined by the hydraulic permeability of the capillaries, their surface area, and the net filtration pressure (NFP) acting across them. Filtration coefficient (K_f) is used to denote the product of the hydraulic permeability and the area. Since the filtrate is essentially protein free, the filtration process removes water but not protein from the plasma, thereby increasing the protein concentration and hence, the oncotic pressure of the unfiltered plasma remaining in the glomerular capillaries. Mainly because of this large increase in oncotic pressure the net filtration pressure decreases by a large amount from the beginning of the glomerular capillaries to the end, 17 mmHg. The GFR is not fixed but many show marked fluctuations in differing physiological states and in disease. The major cause of decreased GFR in renal disease is a decrease in the number of functioning nephrons. The glomerular-capillary hydraulic pressure (P_{GC}) reflects the interplay of the renal arterial pressure, afferent-arteriolar resistance (R_A), and efferent-arteriolar resistance (R_E): 1- A change in renal arterial pressure will tend to cause a change in P_{GC} of the same direction. 2- At any given renal arterial pressure, an increase in R_A will tend to lower P_{GC}, simply by causing a greater loss of pressure between the renal arteries and glomerular capillaries. 3- Changes in R_E tens to cause changes in P_{GC} opposite to those caused by changes in R_A. An increase in R_E tends to elevate P_{GC}. Hydraulic pressure in Bowman's capsule (P_{BC}), obstruction anywhere along the tubule or in the external portions of the urinary system, increase the tubular pressure everywhere proximal to the occlusion, all the way back to Bowman's capsule. The result is to decrease GFR. Oncotic pressure in glomerular-capillary plasma (II_{GC}), is identical to arterial oncotic pressure only at the very beginning of the glomerular capillaries, then progressively increases along the glomerular capillaries as protein-free fluid filters out of the capillary, concentrating the protein left behind. Anything that causes a steeper rise in II_{GC} will tend to lower average net filtration pressure and hence GFR. The RBF remains relatively constant in response to changes in mean renal arterial pressure in the range between 85-200 mmHg. Two intrarenal mechanisms responsible for autoregulation: 1) a myogenic mechanism, and 2) tubuloglomerular feedback. The luminal concentrations of the Na and Cl ions in the Henle's loop are always lower than those of plasma, and there is a direct relationship between the luminal flow rate and concentrations of the ions, i.e., the higher the flow rate the higher the sodium chloride concentrations. These increased concentrations in turn cause the macula densa to reabsorb more Na and Cl, and it is this increased reabsorption that somehow causes increased production of the vasoconstrictor that acts on the afferent arterioles. Adenosine constricts afferent arterioles, in contrast to its vasodilator effects on other vascular beds in the body. Decreased stretch of afferent-arteriolar smooth muscle causes relaxation, i.e., less contraction and decreased flow through the macula densa causes less adenosine to be produced by the JGA. Tubuloglomerular feedback actually causes GFR to go down, whereas during autoregulatory responses, it keeps GFR from changing. The afferent and efferent arterioles are richly supplied with sympathetic neurons, which release norepinephrine, via α-adrenergic receptors to cause constriction of both sets of arterioles. The sympathetic nerves stimulate renin secretion and the resulting increase in angiotensin II also causes renal arteriolar constriction. Input from baroreceptors in the veins and cardiac chambers probably has greater reflex sympathetic effects on the renal circulation than does input from the arterial baroreceptors. Input from the peripheral chemoreceptors or from higher brain centers can also trigger increased sympathetic outflow to the kidneys. Angiotensin II functions in the kidney as hormone and paracrine agent, is a very powerful vasoconstrictor, it acts on both the afferent and efferent arterioles to raise renal vascular resistance, and thereby reduce GFR. The combined afferent and efferent constriction causes a rise P_{GC}, whereas the decrease in GFR causes an increase in the average II_{GC}. Angiotensin II reduces K_f, by the action on the glomerular mesangial cells. Higher concentrations of angiotensin II cause larger falls in GFR. Angiotensin-induced renal haemodynamic changes can be expected whenever renin secretion is significantly elevated. The renin-secreting granular cells of the juxtaglomerular apparatus, acts as intrarenal baroreceptors, monitoring the pressure or vascular volume within the afferent arterioles and varying their secretion of renin inversely with these parameters. Adenosine increases NaCl reabsorption by the macula densa, elicits inhibition of renin release. When the flow and hence, Na and Cl concentrations in the macula densa are low, renin secretion is stimulated. Thus the macula densa mechanism causes increased renin secretion when there is a decrease in GFR and/or an increase in fluid reabsorption by the proximal tubule. When a decrease in arterial blood pressure is the cause of a decreased GFR, the macula densa and intrarenal baroreceptors cooperate, i.e. they both stimulate renin secretion and vice versa. Sympathetic neurons and in the immediate vicinity of granular cells, and these neurones exert a direct stimulatory effect on renin secretion via β_1-adrenergic receptors on the granular cells. Increased activity of the renal nerves can lower the GFR, increase proximal tubular fluid reabsorption. The net effect will be to lower fluid delivery to the macula densa, which stimulate renin release. Angiotensin II exerts a direct inhibitory effect on renin secretion by the granular cells. Angiotensin II constricts renal arterioles and arterioles in most organs and tissues. An increased plasma angiotensin II concentration will increase total peripheral resistance and raise arterial blood pressure. Angiotensin II causes the tubular reabsorption of sodium to increase, which causes retention of sodium in the body. Prostaglandins, PGE_2 and PGI_2, are vasodilators. In normal unstressed state they are produced in concentrations too low to influence the renal arterioles. Increased activity of the renal nerves or increased angiotensin II stimulates the kidney to synthesise and release large amounts of these vasodilator prostaglandins, which then act locally on both afferent and efferent arterioles. The end result is that much of the vasoconstrictor actions of norepinephrine and angiotensin II are concentrated by the vasodilator actions of the prostaglandins, and the renal resistance changes much less than would otherwise have occurred. 3 major factors reducing renal blood flow: 1- the decrease blood pressure, 2- the renal sympathetic nerves, 3- angiotensin II. The renal arterioles have receptors for more than 25 vasoactive neurotransmitters, hormones, and paracrine agents, and there are more than 20 vasoactive substances produced by cells of the renal corpuscles alone.

Chemical messengers that influence the renal vasculature:

Vasoconstrictors	Vasodilators
Norepinephrine and epinephrine.	PGE_2, PGI_2 (prostacyclin).
Angiotensin II.	Atrial natriuretic factor (ANF).
Antidiuretic hormone (ADH, vasopressin).	Nitric oxide.
Thromboxane A_2.	Dopamine.
Leukotrienes.	Bradykinin.
Endothelin.	

2 reasons for the presence of protein in plasma but not in glomerular filtrate: 1- Solutes, dissolved in water, not protein, and the presence of the plasma proteins causes the plasma water to occupy only 95% of the total plasma volume. 2- The plasma proteins cause a Donnan equilibrium to exist between these fluids and this affects the concentrations of ions. The filtration process that occurs in renal corpuscle is often called ultrafiltration to denote the fact that proteins are excluded. Accordingly K_f is often called the ultrafiltration coefficient. Loop acting drugs block Na and Cl reabsorption by the macula densa and so actually eliminate the feedback signal. Increased plasma concentrations of ADH, K, and calcium, inhibit renin release. Increased plasma hydrogen ion concentration stimulates renin release.

RENAL CLEARANCE

Clearance is extremely useful for evaluating renal function both laboratory and clinically. The clearance of a substance is the volume of plasma from which that substance is completely cleared by the kidneys per unit time. The clearance of glucose is zero. Clearance is defined not as mass excreted but rather as the volume of plasma supplying that mass per unit time. The volume of plasma completely cleared of inulin is a measure of GFR, i.e.; the volume-excreted inulin is equal to the volume of plasma filtered. The clearance of protein is zero, because it is not filtered. Creatinine is freely filterable and not reabsorbed; therefore a volume of plasma equal to that of the GFR is completely cleared of creatinine. But small amount of creatinine is secreted, so the clearance is greater than the Inulin clearance. Para-aminohippurate (PAH) clearance measures the effective renal plasma flow (ERPF) and is approximately 85-90% of the total renal plasma flow. Urea, like inulin, is freely filterable, but 40-60% is reabsorbed. The clearance of a filterable substance is less than that of inulin definitely proves reabsorption but does not disprove secretion, secretion may be masked by a greater rate of reabsorption. 50% GFR reduction can be counterbalanced by doubling of plasma creatinine, so filtered creatinine is again normal. Normal plasma urea concentration varies widely, depending on protein intake and changes in tissue catabolism, and that urea is reabsorbed to a variable degree. It is common to measure plasma creatinine and to use it as indicator of GFR, clinically. It is not completely accurate for 3 reasons: 1) Some creatinine is secreted. 2) There is no way of knowing exactly what the person's original creatinine was when GFR was normal. 3) Creatinine production may remain completely unchanged. When GFR fall, creatinine retention would occur until a new steady state established.

TUBULAR REABSORPTION AND SECRETION

Diffusion arises from random molecular motion and requires an electrochemical gradient for net movement to occur. Lipid-soluble substances can diffuse across the lipid portions of a cell's plasma membranes, whereas ionic diffusion is restricted pretty much to water-filled channels created by the proteins of the plasma membrane. Facilitated diffusion can produce net movement of a substance only down its electrochemical gradient. The transport depends on the interaction of the substance with specific membrane proteins, called transporters, which facilitate its movements. The initial event in facilitated diffusion is the binding of the substance to the transporter, followed by conformational change in the transporter that causes the translocation of the substance across the membrane. The substance then separates from the transporter. Because of the interaction with membrane proteins, facilitated diffusion manifests specificity, saturability, and competition. Primary active transport, the transported molecule also interacts with membrane transports and exhibits specificity, saturability, and competition. Primary active transport produces net uphill transport, i.e. net transport against an electrochemical gradient. Membrane bound ATPase not only splits the ATP to provide energy but also is a component of the actual carrier mechanism. The primary active transporters are Na-K-ATPase, H-ATPase, H-K-ATPase, and Ca-ATPase. Two or more substances interact simultaneously with the same specific membrane transporter, and both are translocated across the membrane. One of the substances undergoes only net downhill transport, whereas the other manifests net uphill movement against its electrochemical gradient. Yet the latter occurs without input of metabolic energy directly into the transport process. Rather the direct source of energy is the energy liberated by the simultaneous downhill movement of the other transported substance. The substance moving uphill is said to undergo secondary active transport because the active transport is not directly linked to ATP hydrolysis the way primary active transport is. Paracellular reabsorption can be by diffusion or solvent drag. Diffusion requires that an electrochemical gradient exists for the substance in the reabsorptive direction and that the tight junctions be permeable to the substance. Solvent drag requires the flow of water through relatively leaky tight junctions. Transtubular (transepithelial) potentials are simply the algebraic sum of the individual luminal-membrane and basolateral-membrane potentials. Both potentials are oriented cell-interior negative, i.e.; the cell interior is negative relative to both the lumen of tubule and the interstitial fluid. But the magnitudes of the 2 plasma-membrane cell-negative potentials may be different. In the proximal tubule, the transtubular potential is very small, a few mV lumen-negative relative to the interstitial fluid in the early portions and a few mV lumen-positive in the middle-to-late portions. In the thick ascending limb of Henle's loop transtubular potential is always significantly lumen-positive and therefore is a force for paracellular reabsorption of cations (Na^+, K^+, Ca^{2+}). The transtubular potential in most of the more distal segments is significantly lumen-negative and is therefore a force for paracellular reabsorption of anions, chloride being the only one of importance. Transcellular (across the cell), in which the reabsorbed substance must cross 2 plasma membranes in its journey from tubular lumen to interstitial fluid, the luminal (apical) membrane-separating the luminal fluid from the cell cytoplasm and the basolateral (or contralateral) membrane-separating the cytoplasm from the interstitial fluid. Lipid-soluble substances can traverse both membranes (and cytosol) by diffusion, and net passive reabsorption occurs by this route, simultaneously with the paracellular route, when a favourable transtubular electrochemical gradient exists for substance. Transcellular movement of poorly lipid-soluble substances is active. In the principal cells of the cortical collecting ducts, sodium reabsorption is active and transcellular, down its electrochemical gradient from the lumen membrane into the cytoplasm through sodium channels. The sodium is then actively transported against its electrochemical gradient across the basolateral membrane into the interstitial fluid using Na-K-ATPase, found only in the lateral membrane. The net unidirectional reabsorption of sodium in this tubular segment occurs because: 1) there is asymmetry of the luminal and basolateral membranes-sodium channels in the former and Na-K-ATPase pumps in the latter, and 2) energy is supplied for the basolateral membrane Na-K-ATPase. The net diffusion of sodium across the luminal membrane depends on the existence of a favourable electrochemical gradient cytoplasmic Na<luminal Na and/or electrical potential difference oriented so that the cell interior is negative relative to the lumen. The basolateral Na-K-ATPase pump creates this electrochemical gradient across the luminal membrane by keeping the cytoplasmic Na low and the cell interior negatively charged, just as plasma membrane Na-K-ATPase pumps do in all cells of the body. The reabsorption of glucose occurs in the proximal tubule. Glucose moves uphill from the lumen across the luminal membrane into the cytoplasm by cotransport with sodium, which is secondary active transport since the energy utilised to drive the uphill movement of glucose across the luminal membrane is derived from the simultaneous downhill movement of sodium along its electrochemical gradient via the cotransporter. After entry into the cell, the glucose then exits across the basolateral membrane by sodium independent facilitated diffusion, this downhill movement being driven by the high-glucose concentration achieved in the cell by the action of the luminal cotransport process. The entire process of glucose reabsorption depends on the primary active Na-K-ATPase pump in the basolateral membrane. Amino acids, inorganic phosphate or sulfate, or a variety of organic nutrients can use the same principle, they too undergo secondary active reabsorption by being cotransported with Na across the luminal membrane in precisely the same manner. Movement from the interstitium into capillaries follow the same type of dynamics that occur across all

capillaries in the body. Net diffusion is one mechanism for entry into the peritubular capillaries. The net filtration pressure across the peritubular capillaries always favours net movement into the capillaries, for 2 reasons: 1) the peritubular capillary hydraulic pressure is generally quite low (20 mmHg) because the blood entering the peritubular capillaries has already to flow through the afferent arterioles, glomeruli, and efferent arterioles; and 2) the oncotic pressure of the plasma entering the peritubular capillaries is higher than that of arterial plasma because the plasma proteins are concentrated by loss of protein free filtrate during passage of the glomerular capillaries. How much this oncotic pressure exceeds that of arterial plasma is determined by the filtration, the greater the filtration fraction the more concentrated the protein and higher the oncotic pressure. The oncotic pressure decreases as protein free fluid enters the peritubular capillary, i.e., as absorption occurs, but would not go below 25 mmHg (the value of arterial plasma) even if all fluid originally filtered at the glomerulus were absorbed. Transport maximum per unit time of the renal tubule is due to saturation of the membrane proteins responsible for the transport. Normal persons do not excrete glucose in their urine because all filtrated glucose in a completely normal person is absorbed. Because of insulin deficiency in diabetes mellitus, the person's glucose may rise to extremely high values, the filtered load of glucose becomes great enough to exceed the T_m, and glucose appears in the urine. Glucose began to be excreted in the urine before the true T_m of 375 mg/min is reached. The plasma concentration at which glucose first appears in the urine is known as the renal plasma threshold for glucose. The appearance of glucose in the urine before the T_m is reached is called splay. Not all nephrons have the same T_m for glucose. The plasma glucose in normal persons never becomes high enough to cause urinary excretion of glucose, because T_m for glucose is much higher than necessary for normal filtered loads normally completely reabsorbed, none appearing in the urine. Tubular secretory processes transport substances across the tubular epithelium into lumen, i.e., in the direction opposite to tubular reabsorption, and constitute a second pathway into tubule, the first pathway being glomerular filtration. Substance diffusion out of the peritubular capillaries into the interstitial fluid, from which it makes its way into the lumen by crossing either tight junctions (the paracellular route) or, in turn, the basolateral and luminal membranes of the cell (the transcellular route). Passive secretion, either paracellular or transcellular-can occur by diffusion. Active secretion, always transcellular, its requirement is the same as active reabsorption. Among the most important secretory processes are those for potassium and hydrogen ions. The potassium entering across the basolateral membrane via the Na-K-ATPase pumps can diffuse across the luminal membrane. Only rarely, if ever, does any transported substance manifest purely unidirectional flux across the tubule totally unopposed by a flux in the other direction. Pump-leak systems, in which the active transport system (the pump) creates a diffusion gradient that opposes its own action by favouring back-diffusion. The leak component of epithelial pump-leak is a very important determinant of the maximal concentration gradients that can be established across the epithelial layer. The more permeable the epithelium is to sodium, the more difficult it will be for the active reabsorption mechanism to decrease luminal sodium concentration below interstitial-fluid sodium concentration. An active secretory mechanism is less able to raise the luminal concentration of the transported substance above its interstitial fluid concentration when the permeability of the epithelial layer to that substance is very high. For most mineral ions and many organic molecules, the major route for the leak in pump-leak systems is paracellular. Leaky epithelia include the proximal tubules, small intestine and gall bladder. Tight epithelia include the distal convoluted tubules and collecting ducts. In order to excrete waste products adequately the GFR must be very large. The proximal tubule has the primary role of recovering these filtered loads by marked reabsorption, i.e., mass reabsorber. 50-100% of the filtered loads is reabsorbed, depending on the substance. Similarly, with one major exception (potassium), the proximal tubule is the major site of solute secretion. Thick ascending loop of Henle reabsorbs relatively large quantities of the major ions and, to a lesser extents water, together they entering the more distal segments are relatively small. Beyond Henle's loop, the tubular segments do fine tuning for most substances, determining the final amounts excreted in the urine by adjusting their rates of reabsorption and, in a few cases, secretion.

RENAL HANDLING OF ORGANIC SUBSTANCES

The proximal tubule is the major site for reabsorption of large quantities of organic nutrients filtered, include glucose, amino acids, acetate, Krebs cycle intermediates, certain water soluble vitamins, lactate, acetoacetate, β-hydroxybutyrate, etc. Characteristics of organic substances reabsorption: 1- They are active, they can reabsorb their respective solutes against electrochemical gradients. Reabsorption can be 100% complete. 2- The uphill step is across the luminal membrane, usually via cotransport with sodium. 3- They manifest T_{ms} that are usually well above the amounts normally filtered. 4- They manifest specificity, the amino acid transporters is one for arginine, lysine, and ornithine, another for glutamate and aspartate, and so on. Shared pathways allow for competition among those substances utilising any given pathway. 5- They are initiated by a variety of drugs and diseases. The proximal tubule is the major site for protein reabsorption. There is a very small amount of protein in the glomerular filtrate. The normal concentration approximates 10 mg/L, about 0.02% of plasma albumin concentration (5 g/L). Yet because of the huge volume of fluid filtered per day, this concentration is not negligible.

Total filtered protein = GFR X filtered concentration of protein
= 180 L/day X 10 mg/L = 1.8 g/L

If none of this protein were reabsorbed, the entire 1.8 g would be lost in the urine. Almost all the filtered protein is reabsorbed so that the excretion of protein in the urine is normal only 100 mg/day. So any large increase in filtered protein resulting from increased glomerular permeability can cause the excretion of large quantities of protein. Endocytosis at the luminal membrane, is the initial step in protein reabsorption, it is energy requiring process triggered by the binding of filtered protein molecules to specific receptors on the luminal membrane. The rate of endocytosis is increased in proportion to the concentration of protein in the glomerular filtrate until a maximal rate of vesicle formation, and thus the T_m for protein reabsorption is reached. The pinched off intracellular vesicles resulting from endocytosis merge with lysosome, whose enzyme degrade the protein to low molecular weight fragments, mainly individual amino acids, that exit the cells across the basolateral membrane into the interstitial fluid, from which they gain entry to the peritubular capillaries. Growth hormone (mw = 20,000) is approximately 60% filterable, i.e., relatively large fractions of these smaller plasma proteins are filtered and then degraded in tubular cells. The kidneys are major sites of catabolism of many plasma proteins, including polypeptide hormones, and decreased rates of degradation occurring in renal disease may result in elevated plasma hormone concentrations. Small polypeptides, such as angiotensin II, are handled differently than proteins although the end result is the same, catabolism of the peptide and preservation of its amino acids. They are completely filterable at the renal corpuscles and are catabolised mainly into amino acids within the proximal tubular lumen by peptidases located on the luminal plasma membrane, then reabsorbed. Urea is the primary end product of protein catabolism, passively reabsorbed, driven by concentration gradients across the tubule. Urea is freely filtered at the renal corpuscle; its concentration in Bowman's capsule is identical to its concentration in peritubular capillary plasma. As the fluid flows along the proximal convoluted tubule, water reabsorption occurs, increasing the concentration of any intratubular solute, such as urea. The concentration of urea in the tubular lumen becomes greater than in the peritubular capillary plasma. Diffusion of urea from tubular lumen to interstitial fluid and then into peritubular capillaries. Urea reabsorption is completely dependent on water reabsorption. Approximately 50% of the filtered urea is reabsorbed by the proximal tubule. Henle's loop, the distal convoluted tubule, the cortical collecting duct, and medullary collecting duct, all are relatively impermeable to urea. In the inner medulla, high luminal urea concentration drives urea reabsorption via facilitated diffusion urea transporters in both the apical and basolateral membranes out the collecting duct into the medullary interstitial fluid. About another 10% of the filtered load of urea is reabsorbed at this site. The active secretory pathway for organic anions in the proximal tubule has a relatively low specificity, i.e., a single kind of transporter is responsible for the secretion of all the organic anions accounts for kidney's ability to eliminate from the body so many drugs and other foreign environmental chemicals. Some of the other organic anions secreted by the proximal tubule can undergo other forms of transport both in the proximal tubule and in more distal segments. The major form of uric acid in plasma is ionised urate. Urate is not protein bound and so is freely filterable at the renal corpuscles. Urate undergoes active tubular secretion, mainly in the proximal tubule, and actively reabsorbed mainly in the proximal tubule. The rate of tubular reabsorption is normally much greater than the rate of tubular secretion. The secretory process is the one that is homeostatically controlled to maintain relatively constancy of plasma urate. In gout there is: 1) decreased filtration of urate secondary to decreased GFR, 2) excessive reabsorption of urate, and, 3) diminished secretion of urate. Active transport system of proximal tubule for organic material is relatively nonspecific in that it transports a large number of foreign and endogenous substances, which compete with each other for transport, and it manifests a T_m limitation. Some organic cations are not only actively secreted by proximal tubules but may also undergo other forms of tubular handling, mainly passive reabsorption or secretion. Many weak acids undergo passive tubular secretion when the urine is highly alkaline but passive tubular reabsorption when it is acidic. The tubular epithelium is mainly a lipid barrier; highly lipid-soluble substances can penetrate it fairly readily by diffusion. One of the major determinants of lipid-solubility is the polarity of a molecule, the more polar, the less lipid-soluble. A weak acid exists as a polar ion in alkaline solution and as nonpolar molecule in acid solution. The diffusible form of weak acids is generated in acidic fluid, whereas the diffusible form of weak bases is generated in alkaline fluid. Non protein-bound substance concentration in Bowman's capsule is identical to that in peritubular-capillary plasma. Glomerular filtrate pH is identical to that of peritubular-capillary plasma. As the filtered fluid flows along the tubule, water is reabsorbed, and this removal of solvent concentrates creating lumen to plasma concentration gradients favouring net reabsorption by diffusion, as described for urea. Secretion of hydrogen ions into lumen lowers the luminal pH and favouring the generation of weak acid, which can then diffuse along its concentration gradient from lumen to peritubular plasma. So water reabsorption is one factor that helps create the concentration gradient required for passive reabsorption of weak acid, but luminal acidification, by generating the diffusible form weak acid of a nondiffusible substance is even more important in creating the gradient. The quantitative relationship is determined by the pK of the acid and the pH of the tubular fluid. An acid with a pK well below the lowest tubular fluid pH of 4.4 would always exist within the tubule almost entirely in the anionic (nondiffusible) form and thus always undergo relatively little passive reabsorption.

An organic acid with a pK 6 would exhibit a marked increase in the degree of passive reabsorption if intratubular fluid pH were lowered from 7 to 5, since the nonionized (diffusible) form would go from 10% to 90%. pH changes are greatest in the more distal tubular segments; these segments are the major site for such a pH-dependent passive transport. Renal tubular handling a weak bases, would have the concept, when tubular fluid is highly acidic, the generation of weak base from a substance is favoured. The weak base cannot diffuse out of the lumen because of its charge, but the lowering of intraluminal substance favours net passive secretion of such a substance from peritubular capillary plasma into the lumen. Thus, weak bases are reabsorbed when the urine is alkaline but may be passively secreted when it is acid. Active secretory mechanisms also exist in the proximal tubule for the anionic and cationic forms of many weak acids and bases. Depending on urine flow rate and changes in luminal pH along the tubule, these weak acids and bases may be actively secreted into proximal tubular lumen followed by either passive secretion of the nonionized forms in the subsequent tubular segments.

Na, Cl, AND WATER HANDLING

Control of the renal excretion of Na, Cl, and water constitutes the most important mechanism for regulation of the body content of these substances. Urinary water excretion can be varied physiologically from approximately 0.4 L/day to 25 L/day, depending on whether one is lost in the desert or participating in a beer drinking contest. Na, Cl, and water are all freely filterable at the renal corpuscle because they are not bound to protein, more than 99% reabsorbed. Most renal energy utilisation goes to accomplish this enormous reabsorptive task. The proximal tubule reabsorbs 65% of the filtered Na, the thin and thick ascending limbs of Henle's loop together 25%, the distal convoluted tubule and collecting duct system together most of the remaining 10%, the final urine contains less than 1% of the total filtered Na. Under physiological control by neural, hormonal, and paracrine inputs, the exact amount of Na is homeostatically regulated. The primary active transport of Na from cell to interstitial fluid by the Na-K-ATPase pumps in the basolateral membrane. These pumps keep the intracellular Na concentration very low and the inside of the cell negatively charged with respect to the lumen. Therefore, luminal Na enter the cell passively, along their electrochemical gradient. The critical step for chloride in each tubular segment is from lumen to cell. Luminal membrane chloride transports serve essentially the same active function for chloride as the basolateral membrane Na-K-ATPase pumps do for sodium. Water reabsorption occurs in proximal tubule (65%), the descending thin limb of Henle's loop (10%), and the collecting duct (>24%). Na and water reabsorption occur in the proximal tubule to the same degree. Both are also reabsorbed in Henle's loop but the limbs involved are entirely different for the 2 substances, and the percent of sodium reabsorbed for the loop as a whole is always greater than that of water. Na reabsorption but not water reabsorption occurs in the distal convoluted tubule. Both occur in the collecting duct system, and the percentage of water reabsorbed may vary enormously, depending on the person's water balance. Water reabsorption occurs by simple net diffusion through the lipid bilayer and/or water channels in plasma membranes of the tubular cells and the tight junctions between the cells. The net diffusion is caused by differences in osmolarity between lumen and interstitial fluid created by the reabsorption of solutes. Osmolarity is an inverse measure of water concentration. The epithelium of the proximal tubule and descending thin limb of Henle's loop always has a very high water permeability, that of ascending limbs of Henle's loop (both thin and thick) and distal convoluted tubule is always relatively water-impermeable, and that of the collecting duct system can be regulated so that its water permeability is either very high or very low. The differences in water reabsorption beyond the proximal tubule permit the kidneys to dissociate their reabsorption of water from that of solutes. The urine osmolarity (the measure of the total concentration of solute molecules in a solution) can vary over a wide range, from extremely hypoosmotic (dilute) to extremely hyperosmotic (concentrated), compared to plasma. The human kidney can produce a maximal urinary concentration of 1400 mOsm/L, almost 5 times the osmolarity of plasma. The sum of the urea, sulfate, phosphate, other waste products, and small number of non-waste ions excreted each day normally averages approximately 600 mOsm/L. Therefore, the minimal volume of water in which this mass of solute can be dissolved equals 0.43 L/day. This volume of urine is known as the obligatory water loss and changes with different physiological states, such as increased tissue catabolism as during fasting and trauma, releases much solute and so increases obligatory water loss. If we could produce urine with an osmolarity of 6000 mOsm/L, the obligatory water loss would only be 100 ml of water, and survival time would be greatly expanded. Proximal tubule, a large fraction of the filtered Na enters the cell across the luminal membrane via cotransport with organic nutrients and phosphate, the luminal concentrations of which decrease rapidly. The rest of the sodium movement from the lumen into the cell in the early proximal tubule is mainly via countertransporter with hydrogen ions. In the early proximal tubule bicarbonate is the major anions reabsorbed with Na, and luminal bicarbonate also fall markedly. A major fraction of chloride reabsorption in the proximal tubule is by paracellular diffusion. Along the early proximal tubule, the reabsorption of water, driven by reabsorption of sodium plus its cotransported solutes and bicarbonate, causes chloride concentration in the tubular lumen to increase substantially above that in the peritubular capillaries. Then, as the fluid flows through the middle and late proximal tubule, this concentration gradient, maintained by continued water reabsorption, provides the driving force paracellular reabsorption by diffusion. There is an active chloride transport in the later proximal tubule. It employs parallel Na-H and Cl-base countertransporters. Chloride transport into the cell is powered by the downhill countertransport of organic bases (including formate and oxalate) that arise in the cell by dissociation of their respective acids. Simultaneously, the hydrogen ions generated by the dissociation are actively transported into the lumen by Na-H countertransporters. In the lumen the hydrogen ions and organic bases recombine, and the nonpolar acid molecules then diffuse across the luminal membrane into the cell, where the entire process is repeated. The recycling allows a small number of hydrogen ions to be involved in the reabsorption of a large number of sodium and chloride ions. The recycling dependent on the basolateral-membrane Na-K-ATPases to establish the gradient for sodium that powers the luminal Na-H countertransporter. The proximal tubule always has a very high permeability to water. The osmolarity of the freshly filtered tubular fluid at the very beginning of the proximal tubule is the same as that of plasma and interstitial fluid. Then, as solute is reabsorbed from the proximal tubule, the movement of this solute out of the lumen lowers osmolarity (i.e., raises water concentration) compared to interstitial fluid. This osmotic gradient from lumen to interstitial fluid causes net diffusion of water from the lumen across the plasma membranes and/or tight junction into the interstitial fluid. Diuresis simply means increased urine flow, and osmotic diuresis denotes the situation in which the increased urine flow is due to an abnormally high concentration in the glomerular filtrate of any substance that is reabsorbed incompletely or not at all by proximal tubule. As water reabsorption begins in this segment, secondary to sodium reabsorption, the concentration of the unreabsorbed osmotic diuretic builds up and its osmotic presence retards the further reabsorption of water here (and downstream as well). The failure of water to follow Na causes the sodium concentration in the proximal tubular lumen to fall below that in the interstitial fluid, this concentration difference drives a net passive diffusion of sodium across the epithelium back into the lumen. Thus osmotic diuretics inhibit the reabsorption of both water and sodium. Osmotic diuresis can occur in persons with uncontrolled diabetes mellitus; the unabsorbed glucose acts as an osmotic diuretic. Henle's loop, as a whole reabsorbs more sodium and chloride (25% of the filtered loads) than water (10% of the filtered water). The ascending limb does not reabsorb sodium or chloride, but it quite permeable to water and reabsorbs it. The ascending limbs (both thin and thick) reabsorb sodium and chloride, but they are quite impermeable to water, and reabsorb no water. The movement of sodium chloride out of the ascending limbs and into the interstitial fluid raises this fluid's osmolarity, which causes water reabsorption, by diffusion, from the water permeable descending limb. The major luminal entry step for sodium and chloride is via Na-K-2Cl cotransporters, and Na-H countertransporters. 50% of total Na reabsorption in this segment is by paracellular diffusion in the thick ascending limb. The ascending limb of Henle's loop is called a diluting segment. Because Henle's loop as a whole has reabsorbed more solute than water, the fluid leaving the loop to enter the distal convoluted tubule is hypoosmotic (more dilute) compared to plasma. Distal convoluted tubule and collecting duct system, the major luminal entry step in the reabsorption of sodium and chloride by the distal convoluted tubule is via Na-Cl cotransporters. In the collecting ducts, the principal cells reabsorb sodium, the luminal entry step being via sodium channels. The reabsorbing cells for chloride are the type-B intercalated cells, by a luminal-membrane Cl-HCO_3 countertransporter, and paracellular diffusion driven by the lumen-negative potential. As fluid flows through the distal convoluted tubule and sodium chloride reabsorption proceeds, no water is reabsorbed results in more hypoosmotic, so the distal convoluted tubule functions as a diluting segment. The water permeability of the collecting duct system both the cortical and medullary portions is subject to physiological control by a hormone. Beyond the loop of Henle, tubular segments are tight epithelia and so there is very little back-leakage of sodium from interstitium to tubular lumen despite the large electrochemical gradient favouring diffusion. As the hypoosmotic fluid entering the collecting duct system from the distal convoluted tubule flows through the cortical collecting ducts, water is rapidly reabsorbed by diffusion. This is because of the large difference in osmolarity between the hypoosmotic luminal fluid and the isoosmotic (300 mOsm/L) interstitial fluid of the cortex. The cortical collecting duct is reabsorbing the large volume of water that did not accompany solute reabsorption in the ascending limbs of Henle's loop and distal convoluted tubule, i.e., the cortical collecting duct undoes the dilution carried out by the diluting segments. The tubular fluid, which leaves the cortical collecting duct to enter the medullary collecting duct, is isoosmotic compared to cortical plasma. In the medullary collecting duct, the tubular fluid becomes more and more hyperosmotic in its passage through the medullary collecting ducts. The interstitial fluid of the medulla is very hyperosmotic and is what causes water to diffuse out of the lumen into the interstitial fluid. In the absence of vasopressin, or ADH the water permeability of the collecting duct is very low and little if any water is reabsorbed from these segments, the result being a water diuresis. In the presence of high plasma concentrations of ADH, the water permeability of the collecting ducts is very great and only a very small volume of maximally hyperosmotic urine is excreted. ADH acts in the collecting ducts on the principal cells. The receptors for ADH are in the basolateral membrane of the principal cells, and the binding of ADH by its receptors results in the activation of adenylate cyclase, which catalyses the intracellular production of cAMP, which induces the migration of the intracellular particle aggregates to the luminal membrane and the insertion into the membrane of protein channels through which water can diffuse. In the absence of ADH, these channels are withdrawn from the luminal membrane by endocytosis. The complex sets up the medullary interstitial hyperosmolarity is called the countercurrent multiplier system, takes place in the long loops of Henle that like the collecting ducts extend into the medulla. Reabsorption of sodium and chloride together from thick ascending limb is active, whereas reabsorption from the ascending thin limb is passive. Fluid leaving the proximal tubule is isoosmotic to plasma (300 mOsm/L). As water enters from the descending limb the interstitial osmolarity is maintained at 400 mOsm/L because of continued sodium chloride transport out of the ascending limb. Thus the osmolarities of the descending limb and interstitium are equal and both are higher than that of the ascending limb. Water diffuses out of the descending limb until descending limb and interstitium have the same osmolarity. Fluid leaves the loop via the distal convoluted tubule, and new fluid enters the loop from the proximal tubule. Although a gradient of only 200 mOsm/L is maintained across the ascending limb at any given horizontal level in the medulla, there is a much larger osmotic gradient from the top of the medulla to the bottom (312 mOsm/L versus 700 mOsm/L). In other words, the gradient of 200 mOsm/L established by active ion transport has been multiplied because of the countercurrent flow (i.e., flow in opposing directions through the 2 limb) within the loop. In humans, the maximal value reached in the interstitial fluid at the tip of the papilla is 1400 mOsm/L, and this is why the maximal osmolarity of the final urine is 1400 mOsm/L. Countercurrent exchange, the vasa recta is a unique hair-loop anatomy of the medullary blood vessels, characteristic of the medullary circulation without which the countercurrent multiplier system could not operate, which run parallel to the loops of Henle and medullary collecting ducts. The loop of Henle, which creates the gradient and is, a multiplier. The hairpin-loop blood vessels minimise losses of solute or water from the interstitium by diffusion. The vasa recta are not perfect countercurrent exchangers, a change in the blood flow to the medulla, either too much or to little, will reduce the interstitial gradient and, so, the maximum hyperosmolarity of the urine. Plasma Na concentration (P_{Na}) may change considerably in several pathological conditions, and these changes can influence Na excretion by altering the filtered load of Na. Sodium reabsorption is more important than that of GFR. The reflexes that control GFR and Na reabsorption are initiated by 2 major types of sensors: 1- Extrarenal baroreceptors, like the carotid sinuses, in arteries, cardiac chambers, and great veins. 2- The renal juxtaglomerular apparatuses, specifically the intrarenal baroreceptors and macula densa, which control the secretion of renin. The efferent portions of these reflexes are the renal sympathetic nerves and multiple hormones, including the renin-angiotensin system and the adrenocortical hormone aldosterone. Both GFR and tubular reabsorption are also influenced directly by the hydraulic pressure and composition of the blood perfusing the kidneys (oncotic pressure). Glomerulotubular balance, the mechanisms responsible for adjusting tubular reabsorption to GFR are completely intrarenal, requires no external neural or hormone input. Glomerulotubular balance is a second line of defence preventing changes in renal haemodynamics per se from causing large changes in sodium excretion. GFR autoregulation prevents GFR from changing too much in direct response to changes in blood pressure, and glomerulotubular balance blunts the sodium-excretion response to whatever GFR change does not occur in response to altered blood pressure and other GFR-altering inputs. Aldosterone is produced by the zona glomerulosa of the adrenal cortex. Aldosterone stimulates 2% of the total filtered sodium reabsorption by the principal cells of cortical collecting duct. Thus aldosterone controls the reabsorption of 0.02 X 26,100 mmol/d = 522 mmol/d, this approximately 30 g NaCl/day, an amount considerably more than the average person eats.

Total filtered Na^+/d = GFR X P_{Na} = 180 L/d x 145 mmol/L = 26100 mmol/d

Aldosterone exerts its effect by combining with intracellular receptors and stimulating, in the nucleus, synthesis of mRNA, which then mediate translation of specific proteins. The effects of these proteins are to increase the activity and/or number of luminal-membrane sodium channels and basolateral-membrane Na-K-ATPase pumps. Aldosterone secretion is controlled by: 1- adrenocortical hormone (ACTH), 2- increased plasma potassium concentration, 3- angiotensin II, 4- natriuretic factor (inhibits secretion). Peritubular-capillary Starling factors: 1- Primary changes in peritubular-capillary Starling factors (hydraulic pressure and oncotic pressure) influence renal interstitial hydraulic pressure (RIHP). 2- An increased RIHP produced in this manner to reduce sodium and water reabsorption, mainly in the proximal tubule, and a decreased RIHP increases fluid reabsorption. An increased RIHP causes back-leak of reabsorbed fluid from interstitial fluid across the tight junctions into the tubule. Thus this effect does not alter the cellular transport mechanisms for Na and water but rather reduces the net reabsorption achieved by these mechanisms, especially the leaky proximal tubule. Increased RIHP may inhibit the Na transport mechanisms, either by direct effect on the tubular cells or by stimulating the release of paracrine agents that then act on the cells. An increase in peritubular capillary hydraulic pressure reduces favouring movement of interstitial fluid into capillaries, which causes fluid to accumulate in the interstitium, thereby raising RIHP. A decrease in peritubular capillary oncotic pressure does the same. Thus RIHP is directly proportional to peritubular capillary hydraulic pressure and inversely proportional to peritubular capillary oncotic pressure. Fluid loss ends with 3 changes that then lower GFR: 1- Increased constriction of the afferent and efferent arterioles, induced by angiotensin II. 2- Decreased arterial hydraulic pressure. 3- Increased arterial oncotic pressure. When expansion of the extracellular volume, GFR goes up a small amount and so RIHP, which reduces fluid reabsorption, there occurs: 1- Decreased plasma oncotic pressure, due to dilution of plasma proteins. 2- Increased arterial pressure. 3- Renal vasodilation, secondary to decreased activity of the renal sympathetic nerves and a decreased angiotensin II. Direct tubular effects of renal sympathetic nerves, the renal sympathetic nerves indirect effect is by stimulating renin secretion, and by decreasing renal interstitial hydraulic pressure. The direct effect is by actions on the tubular cells themselves. Angiotensin II enhances Na reabsorption indirectly by stimulating the secretion of aldosterone and by decreasing renal interstitial hydraulic pressure. Angiotensin II acts directly on tubular cells themselves to stimulate Na reabsorption. Effects of renal nerve stimulation 1- Stimulates renin secretion via a direct action on β_1-receptors of granular cells. 2- Stimulates Na reabsorption via a direct action on tubular cells, e.g., proximal tubule. 3- Stimulates afferent arteriolar constriction (α- adrenergic receptors). The results: a- Decrease of GFR and RBF. b- The increased renal resistance decreases peritubular capillary hydraulic pressure and the increased filtration (GFR/RBF) increases peritubular capillary oncotic pressure. Pressure natriuresis, an increased in arterial pressure inhibits Na reabsorption and a decrease in pressure stimulates it. The major mechanisms of pressure natriuresis are completely intrarenal: 1- inhibition of renin release, with loss of angiotensin II's tonic paracrine stimulation of sodium reabsorption, 2- increased production of renal paracrine agents that inhibit Na reabsorption, 3- increase in renal interstitial hydraulic pressure. Atrial natriuretic factor (ANF), secreted by distension of the cardiac atria acts directly on the inner medullary collecting ducts to inhibit Na reabsorption, and indirectly inhibit the renin-angiotensin-aldosterone pathway. It acts directly on the adrenal cortex to inhibit angiotensin-induced aldosterone secretion. It increases GFR via effects on the renal arterioles, which contributes to the increased Na excretion. ADH and aldosterone secretions are stimulated when plasma volume is reduced. Other hormones such as cortisol, oestrogen, growth hormone, thyroid hormone and insulin all enhance Na reabsorption, whereas glucagon, progesterone, and parathyroid hormone all decrease it. The secretion of these hormones is not reflexly controlled, but increase in hormone level that exert influence on Na reabsorption and excretion. Abnormal Na retention, in congestive heart failure, the heart contractility is too low to maintain the cardiac output required for the body's metabolic requirements manifests decreased GFR and increased activities of the renin-angiotensin-aldosterone system and renal sympathetic nerves. Renal filtration is almost always increased, causes decreased renal interstitial hydraulic pressure. All these contribute to the almost complete reabsorption of sodium. The atrial distension of congestive heart failure increases ANF plasma concentration opposes this Na reabsorption. The person who has a lower than normal cardiac output and hence a lower arterial pressure at any given plasma and extracellular volume. The liver disease cirrhosis, and the kidney syndrome nephrosis, tends to produce Na retention. All these oedematous conditions are sometimes termed diseases of secondary hyperaldosteronism because they are usually associated with increased secretion of aldosterone secondary to increased angiotensin II. Primary hyperaldosteronism is characterised by persistent oversecretion of aldosterone because of a primary adrenal defect, usually an aldosterone-producing tumour. Water excretion is the difference between the volume of water filtered and the volume of reabsorbed. The major regulated determinant of water excretion is the rate of water reabsorption. Total body water is regulated by reflexes that altered the secretion of ADH. ADH is a peptide produced by a discrete group of hypothalamic neurons whose cell bodies are located in the supraoptic and paraventricular nuclei, and whose axons terminate in the posterior pituitary, from which ADH is released into the blood. The most important of the inputs to these neurons are from cardiovascular baroreceptors and osmoreceptors. A decreased extracellular volume induces increased ADH secretion, a reflex mediated by neural input to the ADH secreting neurons from venous, cardiac, and arterial baroreceptors. The opposite occurs when decreased cardiovascular pressure. ADH is able to exert direct vasoconstrictor effects on arteriolar smooth muscle, result in increased total peripheral resistance, which helps raise arterial blood pressure independently of the slower restoration of body fluid volumes. Renal arterioles and mesangial cells also participate in this constrictor response, and so a high plasma concentration of ADH, quite apart from its effect on tubular water permeability, may promote retention of both Na and water by lowering GFR. ADH enhances the ability of aldosterone to stimulate Na reabsorption by cortical collecting duct. Angiotensin II stimulates the secretion of ADH. ADH is also elevated in the diseases, like congestive heart failure, characterised by secondary hyperaldosteronism with oedema. The receptors that initiate the reflexes controlling ADH secretion are osmoreceptors in the hypothalamus, receptors responsive to change in osmolarity. The hypothalamic cells that secrete ADH receive neural input from osmoreceptors. Decreased osmolarity inhibits ADH secretion. The hypothalamic cells are the true integrators, whose rate of activity is determined by the total synaptic input to them. A simultaneous increase in plasma volume and decrease in body fluid osmolarity cause strong inhibition of ADH secretion. A simultaneous decrease in plasma volume and increase in osmolarity produce very marked stimulation of ADH secretion. When the plasma volume and osmolarity are both decreased, the osmoreceptor influence predominates over that of the baroreceptor when changes in osmolarity and plasma volume are small to moderate. A very large change in plasma volume will take precedence over decreased body fluid osmolarity in influencing ADH secretion, under such conditions water is retained in excess of solute and the body fluids become hypoosmotic. The ADH secreting cells receive synaptic input from many other brain areas. Thus ADH secretion and hence urine flow can be altered by pain, fear, and a variety of other factors, including drugs such as alcohol, which inhibits ADH release. Diabetes insipidus is characterised by a constant water diuresis, as much as 25 L/d, due to lost ability to produce ADH as a result of damage to the hypothalamus. Thus, collecting duct permeability to water is low and unchanging regardless of extracellular osmolarity or volume. In inappropriately large secretion of ADH, there is decreased plasma osmolarity because of the excessive reabsorption of pure water. Sweat is a hypoosmotic solution containing mainly water, Na, and chloride. Therefore, sweating causes both a decrease in extracellular volume and an increase in body fluid osmolarity. The renal retention of water and Na helps to compensate for water and salt lost in sweat. The centre of thirst is located in the hypothalamus (very close to ADH producing areas). Thirst is a subjective feeling, which drives one to obtain and ingest water, is stimulated by reduced plasma volume and by increased body fluid osmolarity, and stimulates ADH production. Dryness of the mouth and throat causes profound thirst, which is relieved by merely moistening them. Angiotensin II is another factor that stimulates thirst by a direct effect on the brain, by which thirst is stimulated when extracellular volume is decreased. Salt appetite consists of 2 components: 1- hedonistic appetite, and 2- regulatory appetite. Humans seem to have hedonistic appetite for salt. Human can survive quite normally on less than 0.5 g/d salt.

POTASSIUM BALANCE

The resting membrane potentials of nerve and muscle are directly related to the ratio of intracellular to extracellular K^+ concentration. Raising the extracellular K^+ concentration lowers the resting membrane potential, thus increasing cell excitability. Extracellular K^+ concentration is a function of 2 variables: 1- the total amount of K^+ in the body, which is determined by the relative rates of K^+ intake and excretion, and 2- the distribution of this K^+ between the extracellular and intracellular fluid compartments. About 98% of total-body K^+ is located within cells because of the Na-K-ATPase plasma pumps, which actively transport K^+ into cells. Epinephrine and insulin cause increased K^+ uptake by certain cells, e.g., muscle cell through stimulation of plasma membrane Na-K-ATPase. During exercise or trauma, K^+ moves out of the exciting muscle cells or the damaged cells, which raises extracellular K^+ concentration. At the same time epinephrine secretion is increased by adrenomedullary, stimulates K^+ uptake by other cells partially offsets the outflow from the exercising or damaged cells. After a meal, insulin causes the movement of the ingested and absorbed K^+ into cells. A negative feedback system for opposing elevation in plasma K^+ concentration exists, as well. An increase in extracellular fluid hydrogen ion concentration (acidosis) is associated with net K^+ movement out of cells, whereas alkalosis causes net K^+ movement into them, i.e., hydrogen ions moving into cell during acidosis and out during alkalosis, with K^+ doing the opposite. K^+ is freely filtered at the renal corpuscle. 5-15% of the filtered K^+ (35-100 mmol/d) is excreted. The renal tubules are capable of reabsorbing and secreting K^+. The proximal tubule absorbs 55% of filtered K^+, and the thick ascending limb of Henle's loop about 30%. The reabsorption in the proximal tubule is mainly by paracellular diffusion, the concentration gradient for which is created. Reabsorption in thick ascending limb is partially active (Na-K-2Cl cotransporter), and partially by paracellular diffusion. K^+ reabsorption by both is ultimately dependent upon Na^+ reabsorption. The rest of the tubule, low K^+ intake or depletion, the distal segments reabsorb K^+. Normal or high K intake, the distal convoluted tubule and cortical collecting duct both manifest net K^+ secretion. The great majority of the excreted K^+ is K^+ that was secreted by the distal convoluted tubule and cortical collecting duct. The cortical collecting duct contribution is much greater than that of the distal convoluted tubule. In the principal cell the aldosterone regulates reabsorbers of Na^+ and ADH regulates reabsorbers of water-that secrete K^+. The intercalated cells type A reabsorb K^+. Diuretics that inhibit Na^+ reabsorption by the proximal tubule or thick ascending limb of Henle's loop also inhibit K^+ reabsorption at these sites, the reasons should be clear from the Na^+ dependence of K^+ reabsorption by these segments. Osmotic diuretics interfere with K^+ reabsorption by these segments, and this is one reason for the marked urinary loss of K^+ suffered in patients with uncontrolled diabetes mellitus. K^+ is secreted by the principal cells of the cortical collecting duct, involves active transport of K^+ into the cell across the basolateral membrane and passive exit across the luminal membrane, which is mediated by Na-K-ATPase pumps creating a high intracellular K^+ concentration. The resulting increase in intracellular K^+ concentration enhances the gradient for K^+ movement into the lumen and raises K secretion. Aldosterone enhances tubular K^+ secretion by the principal cells. The aldosterone-secreting cells of the adrenal cortex are sensitive to the K^+ concentration of the extracellular fluid bathing them. Increased extracellular K^+ concentration increases plasma aldosterone concentration, which stimulate K^+ secretion by the cortical collecting duct and thereby eliminates the excess K^+ from the body. Lowered extracellular K^+ does the opposite. The basolateral-membrane Na-K-ATPase pumps, increases basolateral K^+ transport into the cell, hence increasing intracellular K^+ concentration and the gradient for movement into the lumen. Aldosterone increases luminal-membrane permeability to K^+ by increasing the activity and/or number of K^+ channels in the luminal membrane, so that the concentration gradient for diffusion is more effective in driving K^+ from cell to lumen. A person on high sodium diet has a low plasma aldosterone concentration because of decreased renin secretion, and this will tend to decrease K^+ secretion. A person on high Na^+ diet has both decreased GFR and reduced proximal Na^+ reabsorption, resulting in increased fluid delivery to the more distal tubular segments. The increased fluid flow through the cortical collecting duct tends to increase K^+ secretion. The decreased aldosterone caused by increased Na^+ intake can increase Na^+ excretion without producing significant K^+ retention, as the net result is that the effects on K^+ secretion and high fluid delivery counterbalance each other. This is the same explanation of Na^+ depleted persons, as the high aldosterone will tend to increase K^+ secretion, but opposed by low fluid delivery to the cortical duct, and hence reduce K^+ secretion. K^+ excretion is almost always increased in persons undergoing osmotic diuresis or treatment with diuretics that act on the proximal tubule, loop of Henle, or distal convoluted tubule. The K^+ loss may cause severe K^+ depletion. This is due to decreased K^+ reabsorption and increased K^+ secretion by the cortical collecting duct. The increased fluid flowing into collecting duct per unit time as a result of inhibition of Na^+ and water reabsorption, derives increased K^+ secretion and hence excretion. The elevated aldosterone in heart failure or secondary hyperaldosteronism does not cause K^+ hypersecretion. The use of diuretics in these patients cause marked increases in K^+ secretion and excretion. K-sparing diuretics can simultaneously block aldosterone's stimulation of K^+ secretion. Another class of K-sparing diuretics blocks Na^+ channels in the cortical collecting duct, preventing Na^+ entry from lumen to cell and effectively prevents the basolateral-membrane Na-K-ATPase pumps from transporting either Na^+ or K^+. Primary acid-base disturbances are a major cause of secondary K^+ imbalance because they cause marked changes in K^+ secretion and excretion. Alkalosis induces increased K^+ secretion and excretion, i.e., increased urinary excretion of K^+ and K-deficiency. Alkalosis stimulates the basolateral-membrane Na-K-ATPase pumps in the collecting duct. Acidosis, does just the opposite.

HYDROGEN ION AND BICARBONATE REGULATION

A huge quantity of CO_2 (15000-20000 mmol) is generated daily as the result of oxidative metabolism and yields hydrogen ions via the following reactions:

$CO_2 + H_2O \leftrightarrow H_2CO_3 \leftrightarrow HCO_3^- + H^+$

Hydrogen ions generated via these reactions during passage of blood through the tissues are reincorporated into water when the reactions are reversed during passage of blood through the lungs. The body also produces acids both organic and inorganic termed nonvolatile acids or fixed acids, to distinguish them from those produced from CO_2. These acids include: 1- sulfuric acid and phosphoric acid, generated during catabolism of proteins and other organic molecules containing sulfur and phosphorus, and 2- organic acids, such as lactic acid, ketone bodies, and others. Dissociation of all these acids yields anions and hydrogen ions. Where diet is high in protein, the generation of nonvolatile acids usually predominates, daily production of 40-80 mmol of hydrogen ions. Vegetarian diet, metabolic utilisation of hydrogen ions, with production of an equivalent amount of bicarbonate, predominates. The GIT secretion leaving the body, such vomiting contains a high concentration of hydrogen ions and so constitutes a source of net loss. The other GIT secretions are alkaline and contain a higher concentration of bicarbonate than exists in plasma, such as diarrhoea, i.e., gain of hydrogen ions as a result of loss of a bicarbonate ion from the body. The urine constitutes the fourth source of net hydrogen ion gain or loss. The kidneys normally excrete the 40-80 mmol of hydrogen ion generated by high protein diet; they excrete the required amount of bicarbonate. The kidneys adjust their excretion of H^+ and bicarbonate to compensate for any net retention or elimination of CO_2, for any increase in the metabolic production of H^+, and for any increased loss of H^+ or bicarbonate via the GIT. The kidneys create an abnormal H^+ concentration by excreting too little or too much H^+. Between their generation and their elimination, most hydrogen ions are buffered by extracellular and intracellular buffers, i.e., disappear. The extracellular fluid pH of 7.4 corresponds to an actual H^+ concentration of only 40 nmol/L. Without buffering interposed between generation and excretion, the daily net production rate (40-80 mmol) of only the nonvolatile acids amounting to millions of nanomoles (1 millimole = 1 million nanomoles) would cause huge changes in pH. Buffering doesn't actually eliminate the H^+ from the body or retain them; this is the function of the kidneys. The only extracellular buffers are phosphates and proteins, including haemoglobin, all are in equilibrium with one another, and a change in one buffer pair will be associated with changes in the other. The intracellular buffers account for 50-90% of the buffering of excess H^+. The kidneys contribute to the homeostasis of extracellular fluid H^+ concentration by regulating plasma bicarbonate concentrations. In alkalosis, the kidneys excrete large quantities of bicarbonate in the urine, thereby raising plasma H^+ concentration. In acidosis, the kidney does not excrete bicarbonate in the urine but instead add new bicarbonate to the blood, thereby lowering the plasma H^+ concentration. The renal addition of new bicarbonate to the blood is associated with the excretion of an equal amount of H^+ in the urine. Thus, to compensate for acidosis, the kidneys excrete acid urine and alkalinise the blood, in response to alkalosis; they excrete bicarbonate containing alkaline urine and acidify the blood.

Filtered HCO_3^-/day = GFR X $pHCO_3^-$ = 180 L/d X 24 mmol/L = 4320 mmol/d

Bicarbonate reabsorption is an active process, involves the tubular secretion of H^+. Hydrogen ion secretion occurs mainly in the proximal tubule, thick ascending limb of Henle's loop, and the collecting duct system (type A intercalated cells, not the principal cells). Within the cells a H^+ and a hydroxyl ion are generated from water. The hydrogen ion is actively secreted into the tubular lumen; the hydroxyl ion left behind inside the cell combines with

CO_2 to form a bicarbonate, in a reaction catalysed by carbonic anhydrase. The bicarbonate moves downhill across the basolateral membrane into the interstitial fluid and then into the peritubular capillary blood. The net result is that, for every H^+ secreted into the lumen; a bicarbonate enters the blood in the peritubular capillaries. Active transport of H^+ across the luminal membrane from cell to lumen is achieved by 3 distinct luminal membrane transporters: 1- A primary active H-ATPase, exists in all the hydrogen ion secreting tubular segments. 2- The proximal tubule and thick ascending limb of Henle's loop possess large numbers of Na-H countertransporters. 3- The type A intercalated cells of the collecting duct, H-ATPase and H-K-ATPase, which simultaneously moves H^+ into the lumen and K^+ into the cell, both actively. The basolateral membrane downhill exit step for bicarbonate is via Cl-HCO_3 countertransporters or Na-HCO_3 cotransporters, depending on the tubular segment. Once in the tubular lumen, a secreted H^+ combines with a filtered bicarbonate to form carbonic acid. The overall result is that the bicarbonate filtered from the blood at the renal corpuscle has disappeared, but its place in the blood has been taken by the bicarbonate that was produced inside the cell. The bicarbonate appears in the peritubular capillary is not the same bicarbonate that was filtered. The H^+ secreted into the lumen is not excreted in the urine. It has been incorporated into water. Any secreted H^+ that combines with bicarbonate in the lumen to cause bicarbonate reabsorption does not contribute to the urinary excretion of hydrogen ions. Throughout the tubule, intracellular carbonic anhydrase is involved in the reaction generating hydrogen ion and bicarbonate. In the proximal tubule, carbonic anhydrase is also located in the luminal cell membranes, and this carbonic anhydrase catalyses the intraluminal decomposition of the very large quantities of carbonic acid formed in this tubular segment. When there is an increased filtered load of bicarbonate, whether caused by an increased plasma bicarbonate concentration, the proximal tubule automatically reabsorbed approximately 80%. The type A-intercalated cells of the collecting duct system reabsorb bicarbonate, type-B intercalated cells in the cortical collecting duct secrete bicarbonate. The mass of bicarbonate leaving the kidneys via the renal veins exceeds that entering the kidneys via the renal arteries. There are 2 mechanisms by which the kidneys add new bicarbonate to the body: 1- Secretion of H^+, that instead of causing bicarbonate reabsorption, are excreted in the urine, combined with nonbicarbonate buffers supplied by filtration. 2- Catabolism of glutamine to yield ammonium (NH_4^+) followed by excretion of the NH_4^+ in the urine. In the case of bicarbonate reabsorption, the secreted H^+ combines with filtered bicarbonate and is incorporated into water. In contrast, in the case of the addition of new bicarbonate to the blood, the secreted H^+ combines with noncarbonate buffers (phosphate) in the lumen and is excreted. The bicarbonate generated within the tubular cell and entering the plasma constitutes a net gain of bicarbonate by the blood, not merely a replacement for a filtered bicarbonate. Thus, when a secreted H^+ combines in the lumen with a filtered buffer other than bicarbonate, the overall effect is not merely bicarbonate conversation but rather addition to the blood of a new bicarbonate, which raises the bicarbonate concentration of the blood and alkalinises it.

Normal contribution of tubular segments to renal H^+ balance:

Proximal tubule.	A- Reabsorbs most filtered bicarbonate (80%). B- Produces and secrete ammonium.
Thick ascending limb of Henle's loop.	Reabsorbs second largest fraction of filtered bicarbonate (10-15%).
Distal convoluted tubule and collecting duct system.	A- Reabsorbs all remaining filtered bicarbonate as well as any secreted bicarbonate (type-A intercalated cells). B- Produces titratable acid (type-A intercalated cells). C- Secretes bicarbonate (type-A intercalated cells).

The renal contribution of new bicarbonate to the blood is accompanied by the excretion of an equivalent amount of buffered H^+ in the urine. Glomerular filtration of H^+ makes no significant contribution to H^+ excretion because the concentration of free hydrogen ion at a pH 7.4, the pH of the glomerular filtrate is less than 10^{-7} M. Even multiplying this figure by 180 L/d, one comes up with less than 0.1 mmol filtered/d. Filtered phosphate is normally the most important nonbicarbonate urinary buffer. There is 4 times more dibasic (HPO_4^{2-}) than monobasic ($H_2PO_4^-$) phosphate in plasma. Not all filtered HPO_4^{2-} is available for buffering, 75% of filtered phosphate is reabsorbed. The reabsorption of phosphate considerably limits the supply of HPO_4^{2-} remaining in the tubule and available for buffering. Glutamine metabolism bears most of the compensatory burden. As a result of insulin deficiency in diabetes mellitus, such patient may become extremely acidotic because of the production of large quantities of acetoacetic acid and β-hydroxybutyric acid, which at plasma pH almost completely dissociate to yield anions (acetoacetate and β-hydroxybutyrate) and H^+. These anions are filtered at the renal corpuscle but are only partly reabsorbed because they are present in large enough quantities to exceed the renal reabsorptive for them. So they are available in the tubular fluid to buffer a portion of the hydrogen ions being secreted by the tubules. Only half of these anions can be used as buffers. A H^+ secreted by the tubule can buffer on 2 general fates, depends on the pKs of each buffer-pair reaction and on the concentrations of each buffer present: 1) it can combine with filtered bicarbonate, in which case the overall process accomplishes bicarbonate reabsorption (conservative process), and 2) it can combine with filtered nonbicarbonate buffers such as phosphate (contributes new bicarbonate to the body and simultaneously excretes hydrogen ions, thereby alkalinising the body). Once most of the filtered bicarbonate has been reabsorbed, the secreted H^+ combine with the other buffers. The proximal tubule secretes very large numbers of H^+ and almost all of these go to achieve bicarbonate reabsorption, 80% of filtered bicarbonate is reabsorbed proximally, where the pH of the luminal fluid falls less than 1 pH unit, and only a small amount of H^+ is picked up by phosphate and organic buffers. Most of the hydrogen ions secreted by the thick ascending limb of Henle's loop also go to reabsorb bicarbonate. 5 to 10% of filtered bicarbonate normally remains by the beginning of the distal convoluted tubule, the hydrogen ions secreted by this segment and the collecting-duct system can reabsorb this bicarbonate and then create the low pH required for titration of nonbicarbonate urinary buffers. Should a large amount of bicarbonate escape proximal and loop reabsorption, most of the hydrogen ions secreted by these more distal segments, too, would be expended in reabsorbing bicarbonate rather than in titrating urinary buffers. The distal portions of the tubule have tight epithelia; there is little paracellular leakage of H^+ from the lumen to the interstitium or of bicarbonate from interstitium to lumen despite very large concentration gradients favouring such leakages. At a luminal pH of 4.4 there is a 1000-fold concentration difference for H^+ between the renal interstitial fluid and tubular lumen.

GLUTAMINE CATABOLISM AND AMMONIUM (NH_4^+) EXCRETION

The cells of the proximal tubule extract glutamine from both the glomerular filtrate and peritubular capillary blood and hydrolyse it to glutamate ion and NH_4^+. Glutamate is then metabolised to α-ketoglutarate, with the liberation of another NH_4^+. The α-ketoglutarate metabolised either to glucose or to CO_2 and water yields 2 bicarbonates (HCO_3^-).
1 glutamine → $2NH_4^+ + 2HCO_3^-$
The NH_4^+ is actively secreted into the lumen and excreted, whereas the HCO_3^- moves into the peritubular capillaries and constitutes new bicarbonate. It is absolutely essential for the NH_4^+ produced from glutamine to be excreted rather than to enter the blood for the bicarbonate released into the blood truly to constitute a net gain of HCO_3^- to the body. Where the NH_4^+ to enter the blood along with the HCO_3^-, the 2 substances would rapidly be incorporated into urea or glutamine in the liver, with the disappearance of the bicarbonate from the blood. Renal contribution of a new bicarbonate to the blood is the same regardless of whether it is achieved by H^+ secretion and excretion on buffers or by glutamine metabolism with NH_4^+ excretion. The tubular handling of NH_4^+ beyond the proximal tubule is actually a very complex sequence of transport events. Most of NH_4^+ formed by proximal tubule does not end up being excreted, and that the actual percentage excreted is under physiological control. Control of renal glutamine metabolism and NH_4^+ excretion, there are several homeostatic controls over the production and tubular handling of NH_4^+. The renal metabolism of glutamine is subject to physiological control by extracellular pH. A decreased in extracellular pH stimulates renal glutamine oxidation by the proximal tubule whereas an increase does just the opposite. Thus, an acidosis, by stimulating renal glutamine oxidation, causes the kidneys to contribute more new bicarbonate to the blood, thereby counteracting the acidosis. An alkalosis inhibits glutamine metabolism, resulting in little or no renal contribution of new bicarbonate via this route. Acidosis influences NH_4^+ transport in ways that enhance excretion, whereas alkalosis does the opposite. Acidosis increases renal NH_4^+ synthesis and excretion, whereas alkalosis does the opposite.

RENAL COMPENSATION ACIDOSIS AND ALKALOSIS

In pulmonary insufficiency or hyperventilation, carbon dioxide is retained, and the resulting increase in arterial PCO_2 drives, by mass action, the formation of more H^+, with a resulting respiratory acidosis. It is the kidneys' job to cause bicarbonate increase via: 1) production and excretion of NH_4^+ increased by acidosis, and 2) the increase in both PCO_2 and extracellular pH stimulates renal-tubular H^+ secretion so that all filtered HCO_3^- is reabsorbed and increased amounts of secreted H^+ are left over for the formation of titratable acid. Respiratory alkalosis, raise plasma pH due to CO_2 loss, is just the opposite. Respiratory alkalosis is the result of hyperventilation, in which the person eliminates CO_2 faster results in lowering arterial PCO_2 and raising pH, increases in extracellular pH reduces tubular H^+ secretion so that HCO_3^- reabsorption is not complete. In addition HCO_3^- secretion is stimulated. Therefore bicarbonate is lost from the body. There is no titratable acid in the urine and little or no NH_4^+ in the urine since the alkalosis inhibits NH_4^+ production and excretion. The primary cause of metabolic acidosis is either the addition to the body of increased amounts of any acid other than carbonic acid or, alternatively, the loss from the body of bicarbonate. The kidneys' compensation is to raise the plasma HCO_3^- concentration back toward normal, thereby returning pH toward normal. The kidneys must reabsorb all the filtered HCO_3^- and contribute new HCO_3^- through increased formation and excretion of NH_4^+ and titratable acid. The mass of HCO_3^- filtered is reduced proportionally to the decreased plasma HCO_3^-, and less H^+ needs to be secreted to accomplish total reabsorption of the filtered bicarbonate. The situation is just the opposite in metabolic alkalosis. The load of filtered bicarbonate is so great that more bicarbonate escapes reabsorption, also increased bicarbonate secretion occurs in metabolic alkalosis. Then, bicarbonate is lost in the urine, plasma bicarbonate is decreased, and pH decreased toward normal. In extracellular volume contraction, chloride depletion, and the combination of aldosterone excess and potassium depletion, there is oversecretion of hydrogen ion (and sometimes NH_4^+ as well), producing one of 2 general results: 1- the kidneys may generate a metabolic alkalosis, or, 2- the kidneys may not generate the metabolic alkalosis but may fail to compensate as usual for an existing metabolic alkalosis. In the presence of extracellular volume contraction, aldosterone is increased, stimulate H^+ secretion. This action is mainly on type-A intercalated cells, is distinct from aldosterone's stimulation of sodium reabsorption and potassium secretion. The urine becomes somewhat acid. Aldosterone stimulates relatively small H^+ secretion. Potassium depletion, weakly stimulates tubular H^+ secretion and NH_4^+ production. The combination of these 2 mechanisms stimulates tubular H^+ secretion markedly (NH_4^+ also increased), thereby causing the development of metabolic alkalosis. This phenomenon is important because the combination of a markedly elevated aldosterone and potassium depletion occurs in a variety of clinical situations, i.e., extensive use of diuretic drugs. Extracellular volume contraction per se, via multiple mechanisms stimulates reabsorption of HCO_3^-. This helps to maintain the alkalosis once the high-aldosterone/K-depletion combination has generated it. If the diuretics have also produced chloride depletion as well, the person will be quadruply in trouble because chloride depletion causes excessive secretion of H^+. Another large potential source of bicarbonate is protein catabolism to NH_4^+ and HCO_3^-. The NH_4^+ will not dissociate, to any significant degree, to yield hydrogen ions because the pK of the $NH_4^+ \leftrightarrow NH_3$ (ammonia) + H^+ reaction is very high, so protein catabolism yields net HCO_3^-. NH_4^+ and HCO_3^- formed by protein catabolism are rapidly combined to form urea in the liver, and so the HCO_3^- disappears. Urea formation undoes the alkalinization effects of protein catabolism. NH_4^+ donates no proton at physiological pH. The NH_4^+ incorporated into urea/or glutamine, and these reactions utilise bicarbonate from the plasma. It is this disappearance of bicarbonate that causes the acidosis. Acidosis also causes a marked increase in hepatic glutamine synthesis, thereby supplying to the kidneys the additional glutamine required for increased renal glutamine metabolism.

CALCIUM AND PHOSPHATE BALANCE

The amount of ingested calcium is equal to the calcium lost in urine, faeces, and sweating combined. A large fraction of ingested calcium is not normally absorbed from the intestine and simply leaves the body along with faeces. Only 60% of the plasma calcium is filterable, the remainder are protein-bound. 60% of calcium reabsorption occurs in the proximal tubule and the remainder in the thick ascending limb of Henle's loop, distal convoluted tubule, and collecting ducts. Calcium reabsorption in the proximal tubule and thick ascending limb of Henle's loop is largely passive and paracellular, and electrochemical forces driving it are dependent directly or indirectly on Na reabsorption. Calcium reabsorption in the more distal segments is active and transcellular. Calcium is not secreted. An increase or decrease in urinary calcium excretion can be induced by administrating or withholding salt, respectively. Acidosis markedly inhibits calcium reabsorption by unknown mechanism, and hence increased calcium excretion. Alkalosis does the opposite. Approximately 99% of the total-body calcium are contained in bone, which is a collagen-protein framework on which calcium phosphate is deposited in a crystal structure known as hydroxyapatite. The plasma calcium (5 mEq/l or 2.5 mmol/l) exists in 3 forms: A- 45% is in ionised form (Ca^{2+}), the only biologically active form in nerve, muscle, etc. B- 15% is complexed to anions with relatively low molecular weights, such as citrate and phosphate. C- 40% is reversibly bound to plasma proteins

PARATHYROID HORMONE (PTH)

PTH exerts at least 4 effects on calcium homeostasis: 1- Increases the movement of calcium from bone into extracellular fluid by stimulating bone resorption. 2- Stimulates the activation of vitamin D, and this hormone then increases intestinal absorption of calcium. 3- Increases renal tubular calcium reabsorption, by the action on the distal convoluted tubule, thus reduces urinary calcium excretion. 4- Reduces the proximal tubular reabsorption of phosphate. When PTH induces bone resorption, both calcium and phosphate are released. Activated vitamin D enhances the intestinal absorption of both calcium and phosphate. Plasma phosphate dose not actually increases because of PTH's inhibition of tubular phosphate reabsorption. Plasma phosphate may decrease when PTH levels are elevated, while facilitates further bone resorption. Primary hyperparathyroidism causes enhanced bone resorption leading to bone thinning and the formation of completely calcium-free areas (cysts). Plasma calcium increases and plasma phosphate decreases. The increased plasma calcium is deposited in various body tissues, including the kidneys, where bone may be formed. The urinary calcium excretion is increased despite the fact that tubular calcium reabsorption is enhanced by PTH. The reason is that elevated plasma calcium concentration induced by the nonrenal effects of parathyroid hormone causes the filtered load of calcium to increase even more than the reabsorptive rate. Vitamin D_3 or cholecalciferol is formed by the action of ultraviolet radiation on 7-dehydrocholesterol in the skin. A second source of vitamin D is that ingested in food, specially in plants. The form of food vitamin D differs only slightly in structure from vitamin D_3. Vitamin D_3 enters the blood and is hydroxylated in the 25 position by the liver and then in the 1 position by the kidneys (proximal tubular cells). The end result is 1,25-$(OH)_2D_3$. 1,25-$(OH)_2D_3$ is a hormone, since it is made in the body. 1,25-$(OH)_2D_3$ stimulates active absorption of calcium and phosphate from intestine, enhances bone resorption. The major blood concentration of 1,25-$(OH)_2D_3$ control point is the second hydroxylation step that occurs in the kidney. This step is stimulated by parathyroid hormone, and both hormones contribute to the restoration of the plasma calcium to normal. Calcitonin is a peptide hormone secreted by parafollicular cells within the thyroid gland. Calcitonin lowers plasma calcium by inhibiting bone resorption. Its secretion is controlled by calcium concentration of the plasma supplying the thyroid gland, increased calcium causes increased calcitonin secretion. Thyroidectomized persons with no detectable plasma calcitonin have no effect on their plasma calcium concentration, which indicate that calcitonin has no role in regulating plasma calcium. High levels of cortisol can induce negative calcium balance by decreasing GIT absorption of calcium while increasing its renal excretion. Growth hormone increases urinary calcium excretion, increases GIT absorption, with net result a positive calcium balance. Approximately 5-10% of plasma phosphate is protein bound, so 90-95% is filterable at the renal corpuscle. 75% of the filtered phosphate is actively reabsorbed in the proximal tubule as cotransport with Na. A diet low in phosphate induces, overtime an increase in the rate of phosphate reabsorption, a diet high in phosphate does the opposite. Whenever PTH is increased or decreased, reabsorption of phosphate is inhibited or stimulated. Other hormones such as inulin increase it and glucagon decreases it. 1,25-$(OH)_2D_3$ may play minor roles. Parathyroid hormone inhibits calcium reabsorption by proximal tubule, because it inhibits Na^+ reabsorption. PTH's inhibition of proximal tubular hydrogen secretion via the effect on the luminal-membrane Na-H countertransporter, and thereby bicarbonate reabsorption. The result is increased extracellular fluid H^+ concentration (acidosis), which is known to displace calcium from plasma protein and from bone. Thus free calcium concentration raises. PTH secretion is decreased when plasma phosphate decreases, because the latter causes a rise in plasma ionised calcium concentration, the lowered PTH would result in greater phosphate reabsorption. Despite the decrease in parathyroid hormone, the formation of 1,25-$(OH)_2D_3$ is increased when plasma phosphate decreases. Phosphate is an important stimulator of the 1-hydroxylation step in the formation of 1,25-$(OH)_2D_3$.

REGIONAL RENAL BLOOD FLOW IN NORMAL AND DISEASE STATES

Endothelial cells release numerous other vasoactive substances in response to various humoral agents or physical stimuli such as shear stress. Endothelial cells possess active enzymatic mechanisms for the formation of several AA metabolites including PGI_2, PGE_2, TXA_2, and cytochrome P-450 metabolites that exert both vasoconstrictor and vasodilator effects. Endothelial bound angiotensin converting enzyme (ACE, Kininase II, peptidyl-dipeptidase A) acts on systemically delivered or locally formed ANGI. Endothelin (ET) is a 21-amino acid peptide that exerts multiple effects on a variety of tissues, from a large prepropeptide precursor, proETs of 38-39 amino acids are cleaved by endopeptidases. Upon exposure to a phosphoramidon-inhibitable metalloprotease, these give rise to the active 21-amino acid peptides. There are three isopeptides encoded on separate gene, are very similar. Endothelin-1 is released by endothelial cells, and ET-3 is found in a variety of other cell types and does not appear to circulate. Endothelin-1 is not stored but rather is released constitutively in response to various stimuli. In particular, endothelial release of ET is stimulated by thrombin, bradykinin, ATP, PAF, TGF-β1, and other cytokines and by physical stimuli such as shear stress. Endothelin induces powerful and long-lasting VSM cell contractile responses in various systems with particular potent actions on pulmonary, mesenteric, and renal vascular beds. The sustained effects is due, in part to the irreversibility of ET binding to ET receptors leading, in part to sustained increases in cytosolic Ca^{2+}. At least 2 ET receptor subtypes ETA and ETB have been identified, and both seem to utilise G protein-coupled activation mechanisms to enhance Ca^{2+} mobilisation and activation of PKC. The ETA and ETB receptors are expressed in VSM and mesangial cells, while ETB receptors are also found on endothelial cells as well as on vasa recta and loops of Henle. In the main branch of the renal artery, ET-1 exerts a much more potent vasoconstrictor action than ET-3. The kidneys receive between 20-25% of the total cardiac output (5 L/d), although their combined weight is less than 1% of the total body weight. Nephron reduction is associated with less nitric oxide (NO) formation in the kidney and excessive ET-1 generation in response to protein traffic and reabsorption. The high renal blood flow has a dual function: 1- Serves to provide blood for filtration since an adequate glomerular blood supply is necessary if the normal excretory function of the kidney is to be maintained. 2- The renal circulation provides oxygen and nutrients to the renal parenchyma. The kidney reacts to alterations in systemic haemodynamics and in turn interacts with the systemic regulatory mechanisms in order to maintain total and regional renal blood flow at a level consistent with optimal renal function. In man total blood flow has been estimated at approximately 1.1 L/min, which is equal to a total daily renal blood flow of 1640 L/d. Mean glomerular capillary pressure is around 50 mmHg or about half of the mean systemic arterial blood pressure. This positive filtration pressure is opposed by a combination of the capsular hydrostatic pressure and the osmotic gradient across Bowman's capsule, resulting in a final positive filtration pressure of about 10 mmHg. The entire body plasma volume filtered over 60 times/day, 99% is subsequently reabsorbed into the peritubular capillaries. Renal autoregulation is independent of hormonal or neurological influences and is thought to be primarily due to a combination of afferent and efferent arteriolar vasoconstrictor activity. Tubuloglomerular feedback loop is mediated by the detection of alterations in chloride ion concentration by the macula densa. Changes in the chloride ion composition of the tubular fluid in the region of the macula densa result in alterations in the glomerular haemodynamics of that nephron. More than 90% of blood entering the kidney goes to supply the renal cortex, resulting in a cortical perfusion rate of between 500 and 850 ml/min per 100 g tissue, while the remainder serves to supply the capsule and the renal adipose tissue. Renin, an acid protease, is released from the juxtaglomerular apparatus in response to Na depletion and alterations in circulating volume. Renin releases the decapeptide angiotensin I by cleavage of the α2-globulin angiotensinogen. Angiotensin I is rapidly converted to angiotensin II by the plasma-converting enzyme kinase II. Angiotensin II is a week vasoconstrictor. The endocrine hormone atrial natriuretic peptide (ANP) is produced by the cardiac atrial in response to increased effective circulating volume. The actions of ANP are: a- vasodilatation, b- reduction in vasopressin release-natriuresis, and c- diuresis. The latter effect of ANP is due to dilatation of the intrarenal vasculature. ANP tends to redistribute the renal blood flow towards the outer cortex and this may contribute to its natriuretic action. Decreased collecting tubule Na^+ reabsorption could in part be a consequence of increased medullary blood flow and a reduced corticomedullary solute gradient. Receptors for ANP are located in glomeruli and other areas of the renal vasculature. A further natriuretic peptide, urodilantin, with a similar structure and range of activities has been isolated from the urine. It is synthesised within the kidney. Urodilantin is found in the urine but not in the plasma and has a natriuretic potency similar to ANP. ANP also antagonises the renal vasoconstricting actions of norepinephrine by dilating the afferent arterioles. Endothelin is a potent vasoconstrictor peptide, synthesised by vascular endothelial cells, whose actions are thought to be mediated by leukotrienes. The renal sympathetic innervation is connected primarily to the afferent renal arterioles and it is thought that their effect is mediated by α-adrenergic receptors and to lesser extent postsynaptic α2-adrenergic receptors. Mild stimulation of the renal nerves leads to a reduction in the blood flow to the superficial nephrons and increases the blood flow to the juxtamedullary nephrons, which in turn leads to Na^+ retention. Direct renal nerve stimulation acting via β1-adrenergic receptors at the juxtaglomerular apparatus, releases renin with secondary production of angiotensin. Renal vasoconstriction can occur in response to a decreased discharge in the baroreceptor nerves. Hypoxia is also a stimulus for renal vasoconstriction, but only when arterial pO_2 falls less than 50% of its normal value. The response is mediated via chemoreceptors, which stimulate the vasomotor centre to produce renal vasoconstriction when the renal nerves are intact. Catecholamines constrict the renal vessels. Decreases in renal blood flow have also been observed during exercise and to a lesser extent on rising from the supine to the standing position. PGE_1 and PGE_2 are vasodilators and are produced by the kidney in increased amounts when the renal perfusion is threatened. Inhibition of renal prostaglandin synthesis, can lead to large falls in renal blood flow in individuals who have vasoconstrictor stimuli to the kidney, such as occur when effective circulating volume is decreased. Prostaglandins (PGs) modulate the alterations in renal medullary haemodynamics, which are observed in response to changes in renal perfusion pressure. The antivasoconstrictor role of the renal prostaglandins is a cortical effect. Thus effective circulating volume leads to increased cortical PGs synthesis. Vasoconstrictor stimuli to the kidney lead to increased renal cortical synthesis of vasodilator PGs (PGE_1 and PGE_2) so that generally the renal blood flow (RBF) remains adequate for glomerular filtration since efferent arteriolar vasoconstriction occurs to maintain filtration pressure, unless the mean blood pressure falls below 80 mmHg. Below this level the renal blood flow falls dramatically, renal function is impaired and unless there is a prompt restoration of the effective circulating volume there is a danger of acute renal failure. 15% of nephrons have long loops of Henle, which pass into the medulla (juxtaglomerular nephrons), while 85% have short loops barely reach the medulla (cortical nephrons). The O_2 consumption of the human kidney is about 18 ml/min. There is significant juxtaglomerular shunting of O_2 within the juxtaglomerular vasculature. This most likely occurs between the interlobular vessels, which are arranged in countercurrent fashion. If the renal blood flow is reduced, the arteriovenous O_2 difference does not increase until the cortical flow is down to about 150 ml/min per 100 g tissue. This is because the blood flow determines the rate of filtration and >50-80% of the O_2 consumption is used for sodium reabsorption. Excessive generation of OFRs may lead to cell swelling, increased capillary permeability and medullary hyperaemia. Toxic levels of ammonia accumulate in the kidney and renal vein during ischaemia, these are produced by the action of the glutaminase enzyme system in the tubular cells at the corticomedullary junction. This enzyme converts glutamine to glutamic acid, leading to the production of free ammonia. During ischaemia, glutaminase is not initially inhibited and production of ammonia continues, leading to toxic levels. Ammonia is nephrotoxic, causing inhibition of renal uptake of oxygen and P-aminohippuric acid and reducing creatinine clearance. Urinary alkalinization inhibits the main part of the glutaminase enzyme system and therefore prevents potentially toxic effects of ammonia accumulation during ischaemia. During reperfusion the hypoxanthine is converted to xanthine by xanthine oxidase and this results in OFRs generation. Nephron reduction is associated with less NO formation in the kidney and excessive ET-1 generation in response to protein traffic and reabsorption. The increased renal vascular permeability for albumin associated with diabetes is due to increase metabolism of glucose to sorbitol and is prevented by inhibition of aldose reductase. Basement membrane thickening of glomerular capillaries is associated with increased metabolism of glucose and galactose to their respective polyols, sorbitol and galactitol. The pathophysiological significance of the decrease in vascular resistance, associated in blood flow in diabetes, is that arterial pressure will be transmitted further downstream into terminal arterioles and capillaries. An increase in microvascular pressure would tend to increase the rate of vascular permeation of plasma constituents that normally permeate the microvasculature any change in the permeability characteristics of the vessel wall per se. Protein compensation, prostaglandin metabolism and ANP have been reported to modulate both the renal haemodynamic and structural changes with diabetes.

EICOSANOIDS

In 1930 it was reported that a substance in seminal fluid contracts smooth muscle in the human uterus. In the mid 1930s, the term "prostaglandin" (PG) was given for the putative humoral agonist. In the 1950s, this unique agent PG was isolated and crystallised from sheep prostate gland. All cells but erythrocytes produce these substance and their congeners, but the misnomer "prostaglandin" has persisted. PGs belong to a larger family of related, oxidised, 20-carbon, unsaturated fatty acids, the eicosanoids. In 1967 PGE_2 was discovered to be synthesised in the renal medulla. Eicosanoids are generated from polyunsaturated 20-carbon (eicosa-) oxidation products of the essential fatty acids linoleate, arachidonate (AA). Essential fatty acids (i.e., eicosanoids) are required for normal growth cell membrane integrity, regulation of cholesterol metabolism, and maintenance of the integument. Essential fatty acid deficiency results in dermatitis, hepatic steatosis, delayed wound healing, increased thrombosis, water loss by evaporation, mitochondrial oxidation, decreased metabolic rate and susceptibility to infection. This syndrome may be prevented by 10 ml of sunflower soybean oil given orally three times a day under basal conditions. Each eicosanoid precursor generates its own prostanoid series. Linoleate-derived prostanoids contain one double bond, AA-derived prostanoids contain 2 double bonds, and linolenate-derived prostanoids contain 3 double bonds. The eicosanoid family includes the prostanoids (PGs and thromboxanes [TX]), leukotrienes (LT), and monohydroperoxy and monohydroxy fatty acids (lipoxins). Eicosanoids precursors (e.g., AA) are stored in the plasma membrane as acyl moieties of phospholipids. The membrane-bound acylhydrolases, phospholipases A_1, A_2, C, and D, are Ca^{2+}-dependent enzymes activated by the ubiquitous intracellular calcium-regulating protein calmodulin. The phospholipases are inhibited by cAMP, lipomodulin, a similar compound stimulated by adrenal corticosteroids, lipocortin. Activation of adenylatecyclase by PGE_2 or PGI_2 and subsequent production of cAMP regulates eicosanoid production in many tissues through negative feedback by reducing intracellular Ca^{++}. Corticosteroid-induced lipocortin competes for the Ca^{++}-binding sites on phospholipase A_2, deactivates the enzyme, and reduces the production of all eicosanoids. By this mechanism, some of the antiinflammatory activities of the corticosteroids may be mediated by suppression of lipoxin, oxygen-free radical and LT synthesis in macrophages, monocytes and leukocytes. The eicosanoid synthetic (AA oxidation) cascade contains 2 major metabolic pathways: the cyclo-oxygenase/peroxidase (prostaglandin endoperoxide synthase) pathway yields the prostanoids (PGs and TXs) and the lipoxygenase pathway yields monohydroperoxy- and monohydroxy-unsaturated fatty acids (LTs and lipoxins). The multienzyme cyclo-oxygenase/peroxidase complex is located within the plasma membrane. The endoperoxide PGG_2 is synthesised in a concerted reaction that removes a hydrogen at carbon-13 of AA to form a cyclopentane ring. Sequential hydroperoxidation of PGG_2 at carbon-11 yields the pivotal endoperoxide PGH_2. PGH_2 is rapidly rearranged by specific isomerases into the various prostanoids. The action of prostacyclin synthetase on PGH_2 produces prostacyclin (PGI_2). TX synthetase produces TXA_2, which is degraded to the inactive metabolite TXB_2. Reductases and isomerases produce PGD_2, PGE_2, and $PGF_{2\alpha}$. PGH_2 also decomposes nonenzymatically to a mixture of PGD_2, PGE_2, and $PGF_{2\alpha}$. Through the action of 9-keto reductase, PGE_2 is metabolised to $PGF_{2\alpha}$. 15-hydroxyprostaglandin dehydrogenase transforms PGE_2, $PGF_{2\alpha}$, PGA_2, and PGI_2 to their keto analogues. These and other eicosanoids are further altered by a variety of degrative enzymes that reduce the double bond at carbon-13, induce β-oxidation, reduce the 9-keto group, reduce the 9α-hydroxy group, and induce ω-hydroxylation at carbons -19 and -20. Lipoxygenases catalyse the oxidation of AA to hydroxy and hydroxyperoxy-unsaturated fatty acids (the lipoxins) such as 12-hydroxyperoxyeicosatetraenoic acid (12-HPETE). A peroxidase reaction reduces HPETE to hydroxyeicosatetraenoic acid (HETE) while generating oxygen free radicals (OFRs). Highly reactive OFRs contribute to tissue injury in inflammatory reactions by lipid peroxidation that disrupt cell membrane integrity, a process inhibited by superoxide dismutase, catalase and peroxidase. 5-lipoxygenase and oxygen convert AA to 5-hydroxyperoxyeicosatetraenoic acid (5-HPETE), which is subsequently converted to the progenitor of the LTs, LTA_4. LTA_4 possesses a reactive, three-membered oxygen-containing epoxide ring that may react with water at the

C-12 position to open the epoxide bridge and yield LTB_4 or it may react with glutathione at the C-6 position to yield 3 thionly peptide derivatives, LTC_4 (containing the tripeptide glutathione [serine, glycine, and cysteine]), LTD_4 (containing glycine and cysteine) and LTE_4 (containing only cysteine), because the latter 3 compounds account for the activity of slow-reacting substance A, they are important components of several immune-mediated inflammatory response. The LTs, first discovered in leukocyte preparations (leuko-), contain 3 conjugated or alternating double bonds (-triene). Each LT derived from AA has a fourth double bond in addition to the 3 conjugated ones, hence the subscript. Monokines and lymphokines orchestrate the body's immune and inflammatory responses to foreign antigen by interacting with specific cell surface receptors in their target tissues. Binding of a monokine or lymphokine to its receptors activates phospholipase, increases the free AA pool, and induces the synthesis of lipoxins and LTs, that may act as intracellular second messengers to mediate the inflammatory and immune responses. Polypeptide hormones interact with membrane-bound receptors linked to guanine nucleotide regulatory (G) proteins. G proteins activate phospholipases A_2, and C, also known as phosphodiesterase or acyl hydrolase. Phospholipase C removes 1,2-diacylglycerol and inositol-1,4,5-trisphosphate from phosphatidylinositol-4,5-bisphosphate. Inositol-1,4,5-triphosphate mobilises Ca^{++} from the endoplasmic reticulum. Binding of 1,2-diacylglycerol to protein kinase C increases the affinity of this enzyme for 2 other cofactors, phosphatidylserine and Ca^{++}. Ca^{++} liberated from the endoplasmic reticulum by the action of inositol-1,4,5-trisphosphate thus acts synergistically with protein kinase C to promote Na^+/H^+ exchange and to activate phosphorylases, which in turn produce active phosphoproteins. Diacylgyceride lipase release AA from 1,2-diacylglycerol, initiating PG synthesis. PGs down regulate the actions of polypeptide hormones by increasing cytosolic Ca^{++}, activating inhibitory G_i proteins, reducing the activity of adenylate cyclase and directly reducing cAMP by activating cAMP phosphodiesterase. The same cell membrane contains receptors that through stimulatory G_s proteins activate adenylate cyclase, lower cytosolic Ca^{++} and inhibit PG synthesis. This system of multiple negative feedback loops assures that intracellular eicosanoids, protein kinase, cytoplasmic Ca^{++}, cGMP, and cAMP levels are rapidly buffered to oscillate around a set point. PGs exert their biologic activities by reacting with specific cell membrane receptors. Second messengers cAMP or changes in cytosolic Ca^{2+} carry out the cellular reactions initiated by PG-ligand binding. The biologic activities of the same eicosanoids may vary from tissue to tissue. Because they are synthesised at or near their sites of action, eicosanoids are autacoids. Rather than being stored intracellularly eicosanoids are rapidly metabolised within the kidney. Locally synthesised eicosanoids, presented to adjacent receptors in high concentrations are largely degraded or metabolised in situ within seconds to minutes, and raise blood and urine levels minimally. Linoleate and linolenate-derived prostanoids, play minor roles in renal physiology whereas the AA-derived prostanoids are by far the most abundant and physiologically active renal eicosanoids. PGI_1 is the most abundant renal prostanoid and PGE_2 and TXA_2 are present in lesser amounts. Prostanoid synthesis varies locally within the kidney and axially along the nephron. The LTs and lipoxins, particularly 12- and 15-HETE are synthesised in both cortex and medulla, participation in inflammatory responses and mesangial cell contraction. Cyclo-oxygenase inhibitors, decrease renal vascular resistance. Nonsteroidal anti-inflammatory agents may induce sodium retention (oedema or hypertension) or hyporeninaemic hypoaldosteronism (hyperkalaemic metabolic acidosis). Rarely, after 3 to 6 months of use, the nonsteroidals may induce nephrotic-range proteinuria or acute renal failure associated with tubulointerstitial inflammation. Papillary necrosis very rarely follows the prolonged high-dose intake of nonsteroidals. PGE_2 induces a natriuresis, both through afferent arteriolar dilation and through direct inhibition of sodium and chloride transport in the proximal and distal nephron. Inhibition of PG-mediated natriuresis by cyclo-oxygenase inhibition may lead to sodium retention, oedema and hypertension. Concurrent administration of a cyclo-oxygenase inhibitors with a loop diuretic decreases or completely eliminates the diuretic's haemodynamic effects without decreasing its natriuretic potency under basal conditions. Cyclo-oxygenase inhibition may reduce the natriuresis of furosemide, thiazide, or spironolactone by >75% and GFR by 20-40%. Renin is stored in secretory granules in smooth muscle cells of the media of afferent arterioles and to a lesser extent within efferent arterioles, glomerular mesangium, and the goormaghtigh cells. Cyclo-oxygenase inhibitors quickly reduce plasma renin activity regardless of salt intake. Diminution of renal perfusion pressure, effective circulating volume, afferent arteriolar wall tension, or JGA cell stretch; raises cytosolic K^+ and decreases cytosolic Na^+, decreases Na^+-for-Ca^{++} transcellular exchange, lowers cytosolic Ca^{++}, and releases stored renin. Prostagladins do not mediate the baroreceptor-stimulated release of renin. Vasoconstriction is a prominent feature of early acute renal failure of various aetiologies. It is certainly prudent to avoid nonsteroidal anti-inflammatory agents during the initiation phase of acute renal failure or in clinical situations in which acute renal failure is a significant risk. Chronic renal failure, if protein consumption by patients with renal impairment is not curtailed, the solute load per surviving nephron increase in proportion to nephron loss. The surviving nephrons respond to a solute load by increasing glomerular filtration pressure and single nephron GFR (hyperfiltration) through constriction of the efferent arterioles in order to maintain the volume and composition of the extracellular fluid. This normal adaptive mechanism maintains nitrogenous solute homeostasis until 60% of the functioning nephrons have been lost and the functional renal reserve (i.e., the different between basal (fasting) GFR and the maximal GFR induced by administration of a protein load) has been expended. This adaptation to nephron loss may itself cause progressive renal failure. Glomerular hypertension and hyperfiltration in the functionally intact nephron remnant leads to capillary endothelial injury, loss of permselectivity of the basement membrane, and filtration of macromolecules that are phagocytized by the mesangium. Subsequent overproduction of mesangial matrix (basement membrane-like material) may lead initially to focal/segmental glomerulosclerosis and later to global glomerular obsolescence, focal ischaemic tubular atrophy and interstitial fibrosis. Glomerulosclerosis is manifested clinically by proteinuria, hypertension, and progressive loss of renal function ending in end-stage renal failure. Following a protein meal, arginine-stimulated glucagon induces renal PG synthesis. Nitrogenous solute clearance may thereby be critically dependent on intact PG synthesis in hyperfiltering surviving nephrons. Dilating the efferent arterioles, constricting the afferent arterioles and contracting the mesangium may be the mechanism by which nonsteroidal anti-inflammatory agents further reduce GFR in patients with renal insufficiency. The effect of cyclo-oxygenase inhibition traditionally thought to be deleterious in patients with renal insufficiency might actually prove to be renal-sparing. A nonsteroidal anti-inflammatory agent like dietary protein restriction, calcium channel blockers and ACE inhibitors could ultimately protect damaged kidneys from progressive glomerulosclerosis. The renal fate of a patient with chronic renal insufficiency treated with a nonsteroidal anti-inflammatory agent is difficult to predict. The effects of nonsteroidals on GFR in patients with interstitial nephritis, hypertensive nephrosclerosis and diabetic glomerulosclerosis are usually mild. Eicosanoids mediate the formation and target activities of monokines and lymphokines as well as antigen expression and inflammatory responses. Corticosteroids induce the production of lipocortin a substance, which like lipomodulin competitively occupies the active Ca^{++}-binding sites on phospholipase. Subsequent reduction of the free AA pool non-specifically reduces the production of all eicosanoids.

ORGANIC OSMOLYTES

Organic osmolytes are directly involved in the osmoadaptation of renal medullary cells. The 4 main organic osmolytes in mammalian kidney, sorbitol, myo-inositol, betaine, and glycerophosphorylcholine, change their intracellular content parallel with the extracellular tonicity. Each osmolyte exhibits an individual distribution pattern along the corticopapillary axis, suggesting site-specific functions. The synthesis of sorbitol is enhanced by elevated glucose concentrations. Besides the regulation by the extracellular Na, cellular glycerophosphorylcholine concentration seems to depend on intracellular urea concentration. Cells accumulating organic osmolytes to balance extracellular hypertonicity respond to a decrease of extracellular hypertonicity. The furosemide diuretic effect is apparent after 30 min, with breakdown of the corticopapillary concentration gradient; results in decrease tubular sorbitol content. The decrease of intracellular osmolytes after hypotonic stress is predominantly caused by an efflux of all organic osmolytes finished within 30 min, except for glycerophosphorylcholine. Cell swelling triggers anion channel of papillary collecting duct cells after furosemide diuresis. During hypotonic stress the renal medullary osmoadaptation is mainly regulated by osmolyte efflux. Because of their rapid reduction by 46% within 15 min, the role of organic osmolytes as a fast system in osmoregulation is evident. In contrast to furosemide diuresis, a change of the osmolyte content in outer medulla is not seen when acute or chronic water diuresis is induced. Acute furosemide diuresis induces efflux of all osmolytes. Water diuresis reduces inner medullary and does not change outer medullary osmolyte contents.

RENAL FAILURE FROM MITOCHONDRIAL CYTOPATHIES

Renal disease may be the first sign of mitochondrial cytopathy, or it may appear together with neurological and neuromuscular signs. Findings of hyperlactaemia and reduced enzymatic activity on the respiratory chain in tissue biopsies are of diagnostic significance in mitochondrial cytopathy. Mitochondrial cytopathies are polymorphic diseases occurring secondary to altered oxidative mitochondrial function. They are clinical syndromes that affect all mitochondrial nucleated cells, leading to the multisymptomatic clinical expression of disease in different organs. Mitochondria possess DNA, which is susceptible to pathological mutations. The respiratory chain consists of 5 multienzymatic complexes that function as electron transporters: complex I (NADH-CoQ reductase), complex II (succinyl-CoQ reductase), complex III (ubiquinone-cytochrome c reductase), complex IV (cytochrome c oxidase), and complex V or ATPase, and synthesise mitochondrial ATP. Two genomes intervene in the mitochondrial respiratory chain: the nuclear genome for 90% of the proteins of that it consists, and the mitochondrial genome for the remaining 10%. In man, the mitochondrial genome is formed by a double-chain circular molecule with 16,589 pairs of bases. Human mitochondrial DNA has 2 alleles, each mitochondrion containing 2-10 DNA copies, so each cell, with its mitochondria, contains hundreds or thousands of mitochondrial genomes. The zygote receives genetic nuclear information from the ovocyte and the spermatozoon while the mother transfers information to mitochondrial DNA, although a small quota of paternal mitochondria can be inherited. Genetic diseases due to errors in the mitochondrial genome are inherited from the mother, who can transmit the defect to all her offspring, whether male or female. Only daughters can then transmit this defect to their offspring. During mitosis the mitochondria redivide in the offspring's cells, which is called "mitotic segregation". Therefore, if the maternal cell contains 2 types of mitochondria that are differentiated from each other by their respective genomes, after numerous cell divisions the daughter cells may contain only one type of mitochondrial DNA, or they may retain 2 types, thus becoming heteroplasmic. The presence of mitochondrial DNA affected by mutations can be expressed in an oxidative phosphorylation deficiency, disturbed cytoplasmic and mitochondrial oxidative-reductive balances, a reduction in the mitochondrial ketone bodies, and an increase in the β-hydroxybutyric/acetoacetic ratio, while cytoplasmic pyruvate/lactate balance is shifted towards the lactate with hyperlactaemia. The tissue expresses the pathological morphotype when the percentage of mutated mitochondrial DNA reaches a threshold level that varies depending on the organ affected by the mutation. In elderly, the oxidative metabolism of tissue is diminished and it is particularly vulnerable to the effects of mitochondrial DNA mutation. Mitochondrial diseases are heterogeneous and can affect any age. The first sign of mitochondrial disease is often nephropathy, tubular and interstitial lesions being encountered during mitochondrial cytopathy, which in newborns and infants may present as the Fanconi's syndrome, evolving into terminal renal failure. Reasons for CoQ deficiency in mitochondria: 1) decreased CoQ levels, decreased or insufficient biosynthesis or increased catabolism; and 2) increased demand for CoQ, bioenergetics (decreased activity of the respiratory chain, or increased affinity for CoQ of respiratory enzymes); or as an oxidant (oxidative stress, aging). Fanconi syndrome is characterised by an impairment of proximal tubular reabsorption, leading to urinary losses of amino acids, glucose, proteins, phosphate, uric acid, bicarbonate, potassium and water. Glomerular disease has been observed in a few patients with mitochondrial cytopathies, e.g., segmental glomerular sclerosis. Renal disease in patients with mitochondrial cytopathies consists of tubulointerstitial nephritis, diffuse interstitial fibrosis with tubular atrophy and sclerosed glomeruli within the area of interstitial fibrosis. Impaired proximal tubular functions may lower blood lactate and increase urinary lactate. An enzymatic deficiency of the respiratory chain results in an increase in reducing equivalents (NADH-FADH) in both mitochondria and cytoplasm and the functional impairment of the Krebs cycle. In the mitochondrion, the excess of NADH induces the transformation of acetoacetate to 3-hydroxybutyrate with a concomitant increase of the ratio 3-hydroxybutyrate/acetoacetate. Similarly, in the cytoplasm, pyruvate is transformed to lactate resulting in an increased molar ratio of lactate/pyruvate with a secondary increase of blood lactate. These abnormalities are more pronounced during the post absorptive period or after an exercise, when more NAD is required to oxidise glycolytic substrates. Similarly, because of the functional impairment of the Krebs cycle, ketone body synthesis increases due to the channelling of Acetyl-CoA towards ketogenesis. The increase in ketone bodies after meal is paradoxical as it normally decreases following insulin release. Therefore, a persistent hyperlactacidaemia, an impaired redox status and a paradoxical ketonaemia are highly suggestive of a respiratory chain defect. All humans contain several mitochondria. mtDNA copies in each mitochondrion are identical, a condition named homoplasmy. When a mtDNA mutation arises, it creates a mixture of normal and mutant molecules called heteroplasmy. Heteroplasmy is almost constant in case of mitochondrial cytopathy. During mitotic cell division, mitochondria are randomly partitioned to daughter cells. If normal and mutant mtDNA are present in mother cells, some daughter cells may drift towards purely mutant mtDNA or purely normal DNA while others remain heteroplasmic. The phenotype is defined not only by the nature of the mutation but also by the amount of mutated mtDNA in a given tissue. In cells with both wild-type and mutant mtDNA, the phenotype depends on the proportion of mutant mtDNA molecules and the extent at which the cell type relies on mitochondrial function. As the phenotype of a given tissue may vary during the course of the disease according to the proportion of mutant/wild-type mtDNA, clinical symptoms related to an affected organ may worsen or improve and even disappear with time, while other organs become involved. In mitochondrial disorders, children in the same family may not be all clinically affected. Any mode of inheritance may be observed in mitochondrial diseases: autosomal recessive, dominant, X-linked, maternal or sporadic. Point mutations of mtDNA result in amino-acid substitutions and mutations of mRNA and tRNA. Most of them are maternally inherited and heteroplasmic. Mitochondrial DNA depletions due to a decreased number of mtDNA copies have been also described in cases of lethal infantile respiratory, liver and kidney disease with an autosomal recessive mode of inheritance. No satisfactory treatment is presently available to alter the course of mitochondrial disorders. The treatment is mainly symptomatic. It includes avoidance of drugs that are known to interfere with the respiratory chain and may precipitate hepatic failure. Lactate acidosis may be exacerbated by exercise or intercurrent infection and should be treated with slow infusion with sodium bicarbonate. Patients with complex III deficiency may be improved with vitamin K_3 (40-160 mg/day) or coenzyme Q_{10} (80-300 mg/day). Patients with complex I deficiency myopathy may be treated with riboflavin. The prevention of oxygen radical damage is the rationale for ascorbate administration. High lipid and low carbohydrate diet in patients with complex I deficiency. Supplements of sodium bicarbonate, potassium, vitamin D, phosphorus and water may be necessary in patients with proximal tubular losses.

FAT AND MUSCLE PHYSIOLOGY

Most people find it difficult to maintain exercise regimens for long periods of time. To couple exercise with its metabolic benefits would require a detailed understanding of the signalling pathways that regulate energy production, storage, and expenditure in skeletal muscle. The activity of adenosine monophosphate-activated kinase (AMPK) is increased by muscle contraction as well as other factors that increase the ratio of AMP to ATP in the cell, i.e., energy sensor. The energy-regulating hormones, such as leptin, as well as commonly used therapeutic agents, such as metformin and rosiglitazone, can modulate AMPK activity. Activating AMPK shifts the cell's metabolic activities away from substrate storage toward oxidative substrate disposal to create ATP; e.g., the suppression of cholesterol, fatty acid, and triglyceride synthesis, and the activation of fatty acyl-CoA β oxidation, glucose uptake, and glycogenolysis. AMPK's several isoforms derived from combinations of forms of its 3 subunits: α, β, and γ. The 2 chief isoforms in muscle are AMPK α1 and AMPK α2; the latter isoform is generally more likely to be regulated by exercise. AMPK's activity is increased by elevation in cellular AMP concentration, acute exercise, exercise training, and the activity of an upstream kinase termed AMPK kinase (AMPKK). AMP appears to activate some but not all isoforms of AMPKK. Metformin appears to increase the activity of AMPK directly, while rosiglitazone does so by increasing the cellular concentration of its allosteric activator, AMP. Inhibition of acetyl-CoA carboxylase (ACC) as a result of phosphorylation by AMPK drops the level of malonyl-CoA, and this disinhibit carnitine palmitoyl transferase I resulting in the movement of acylated fatty acids into mitochondria to undergo β oxidation. Short-duration exercise of varying intensities increases AMPK activity. Prolonged (3.5 hrs) exercise of low intensity leads to a progressive increase in AMPK activity, but not that of ACC, which peaked after 1 hr and then declined. This dissociation between AMPK activity and lipid oxidation may be important to provide energy to muscle during prolonged exercise. There is no correlation in the exercising muscle between AMPK activity and ACC phosphorylation and glucose uptake. With short, intense exercise, carbohydrate appears to be the chief fuel for energy, and malonyl-CoA content in the muscle declines. With prolonged, low-intensity exercise, there is a progressive shift towards lipid oxidation for energy. Exercise-associated decrease in glycogen synthase activity occurs in association with increased AMPK activity. AMPK promotes glycogen breakdown by inhibiting glycogen synthase activity and activating glycogen phosphorylase. Although net glycogen breakdown occurs with exercise, there is actually an increase in both glycogen synthesis and glycogenolysis (i.e., increased turnover). AMPK may actually increase glycogen synthase activity in the context of exercising muscle. Exercise-induced increase in glycogen synthase activity is not necessarily associated with the increase in AMPK activity. Exercise increases AMPK activity, which increases glucose transport, thereby enhancing glycogen synthesis. Increased muscle glycogen content might then feed back to inhibit AMPK activity. Adipocyte depots comprise a key endocrine organ. Adiponectin is an adipocyte-secreted insulin-sensitising and anti-atherosclerotic hormone. Plasma adiponectin levels correlate inversely with adiposity and fasting blood glucose levels and low adiponectin levels may precede declines in insulin sensitivity in humans. Adiponectin improves hepatic insulin sensitivity; decreases rate of plasma glucose entry and reduce the activity of the gluconeogenic enzyme phosphoenolpyruvate carboxykinase (PCK). Hepatocyte metabolic signalling molecules occur, with activation of adenosine monophosphate-activated kinase (AMPK), a sensor of cellular energy stores, and protein kinase B, a key intermediate of insulin action. This effect of adiponectin resembles the effects of the thiazolidinediones and metformin. Thiazolidinediones modulates circulating levels of adiponectin. Adiponectin expression is highly regulated in abdominal fat, the amount of which is inversely related to insulin sensitivity. Increasing abdominal obesity is associated inversely with plasma adiponectin levels. Adiponectin's diurnal variations are roughly inverse to those of leptin and insulin, peaking at early hours of the morning and declining during the midafternoon. Adiponectin secretion is increased by glucocorticoids and proliferator-activated receptor gamma (PPARγ) agonists, potentiated by testosterone, and decreased by oestrogen and prolactin. Adiponectin is initially synthesised by adipocytes in an inactive, high-molecular-weight form. Subsequent insulin action may be needed to modify the protein to a secreted, hexameric form. In the circulation, a putative serum reductase may alter the protein to a trimer, which, at its target cells, may be proteolytically cleaved to its most bioactive form. The final cleavage step may determine varying levels of activity in different target tissues. For instance, the globular head domain alone is highly bioactive in muscle cells but not in hepatocytes. Adiponectin increases hepatic insulin sensitivity with respect to glucose metabolism, lowers plasma insulin levels, and improves plasma triglyceride and free fatty acid (FFAs) levels. Adiponectin deficiency induces insulin resistance. Plasma adiponectin levels increase with age, decrease with adiposity, decrease during puberty, and are higher in women. Adiponectin gene polymorphisms are linked to type 2 diabetes. Lower circulating adiponectin levels diminish the permissive effect of the hormone on insulin sensitivity in muscle and liver. There is also some evidence of linkage between the adiponectin polymorphism and coronary artery disease (CAD). Resistin, adipsin, TNF-α, plasminogen activator inhibitor-1 (PAI-1), and adiponectin are adipocyte-secreted factors that clinically relevant actors in the drama of insulin resistance and the metabolic syndrome. The adiponectin gene is a strong candidate for an important contributor to the metabolic syndrome. A major metabolic defect associated with type 2 diabetes is the failure of peripheral tissues in the body to properly utilise glucose, thereby resulting in chronic hyperglycaemia. Adipose tissue and skeletal muscle are the primary targets of insulin-stimulated glucose uptake. Insulin is an anabolic hormone that stimulates glucose and fatty acid uptake into adipose cells, promoting triglyceride (TG) synthesis. Although adipose cells are storage depots for TG and predominantly contain lipid droplets, they possess other organelles typical of most cells. Insulin has powerful effects on glucose transport in adipose tissue. In the basal state without insulin, glucose transport is very low. In the presence of insulin, glucose transport is rapidly stimulated. Vesicles containing glucose transporters are mobilised to the plasma membrane by insulin stimulation, thereby effecting glucose transport into the cell. In the absence of insulin, glucose transporters are located intracellularly. Following insulin receptor binding and signal transduction, glucose transporters are recruited from the intracellular pool to the plasma membrane, exposing functional glucose transporters to the extracellular medium containing glucose. On termination of the insulin stimulus, the glucose transporters are recycled from the plasma membrane into the intracellular pool in readiness for the next insulin stimulus. Glucose transport molecules transport glucose by facilitative diffusion down concentration gradients, in contrast to energy-dependent uptake of glucose in the gut or kidney. GLUT4 is the predominant contributor to insulin-regulated glucose transport activity in adipose cells. The structure of the glucose transporters is similar, and they have a region where carbohydrate can bind the protein. In metabolically active cells where the glucose concentrations approximate zero, external glucose binds the transporter that is internalised and thereby glucose enters the cell. The concentration gradient is the major determinant of glucose flux. In the basal (insulin-free) state, the cell-surface plasma membrane fraction contains no glucose transporters, whereas an intracellular fraction of low-density microsomes contains all the glucose transporters. During the insulin-stimulated state this distribution is reversed, with insulin rapidly (within 10 minutes) inducing an increase in glucose transporters in the plasma membrane and a decrease in the intracellular fraction. Insulin acts by increasing the glucose transport rate of each transporter, increasing the number of functional glucose transporters, or by a combination of both mechanisms. Other hormones can interact with insulin to regulate glucose transport. Adenosine and hormones that activate Gs (a GTP-binding protein that activates adenylate cyclase), in conjunction with insulin, boost insulin-stimulated glucose transport, whereas lipolytic hormones, isoproterenol, and hormones that activate Gi (a GTP-binding protein that inhibits adenylate cyclase) inhibit insulin-stimulated glucose transport. In contrast to insulin, these additives do not change the number of glucose transporters present on the plasma membrane. Morphologically, the basal state glucose transporters have a punctuate cytosolic distribution in adipose cells, whereas in the presence of insulin, glucose transporters are redistributed to the cell surface. Glucose transporters are associated with vesicles near the trans-Golgi network and endosomal compartment in the basal state, with no surface expression.

Glucose transporters:

Transporter	Specificity
GLUT1	Ubiquitously distributed; exhibits constitutive transport activity.
GLUT2	Present in gut, liver, and pancreatic islets.
GLUT3	Present in the central nervous system and brain.
GLUT4	Present in insulin-responsive tissues, skeletal muscle, adipose tissue, and heart.

GLUT4 is co-localised with vesicles containing: 1- VAMP2 (vesicle-associated membrane protein involved in the docking and fusion of glucose-transporter vesicles with the plasma membrane), and 2- M6PR (mannose-6-phosphate receptor, which is a marker for endosomes and the trans-Golgi network). In the presence of insulin, the GLUT4 transporter moves out of its specialised compartment to the cell surface, where it may be associated with cell component proteins from other trafficking pathways. In addition to adipose tissue, skeletal muscle is also a peripheral target for insulin action. Insulin stimulates glucose uptake into muscle through increased glucose transporter activity, and this process, like in adipose cells, involves insulin-stimulated translocation of intracellular glucose transporters to the T-tubules or plasma membrane. Muscle contraction also stimulates glucose uptake, and the effects of insulin and contraction are additives, which occurs through activation of parallel but distinct pathways. In general, the magnitude of insulin-stimulated glucose transport is proportional to the number of expressed glucose transporters. In diabetes mellitus insulin-stimulated glucose transport is markedly attenuated, glucose transporter number may be reduced by 50% or more, and there is a decreased translocation of the transporters to the cellular surface in the presence of insulin. Adipose tissue serves a storage function, with adipose cells predominantly containing large lipid droplets with a thin rim of cytoplasm and other organelles, and secretory function. There is intercellular communication between adipose and muscle cells, raising the intriguing possibility that adipose cells may be secretory organs. GLUT4 and leptin (a secreted protein) are detected in intracellular compartments of the adipose cells in the basal state, in the presence of insulin GLUT4, transporters are translocated to the plasma membrane and are visible, whereas decreased amounts of leptin are located intracellularly because leptin is secreted. There are several genes associated with insulin resistance in adipose cells, including IRS-1, a signalling molecule; PI3CK, another signalling molecule; GLUT-4; PPAR-γ, a transcription factor known to be involved in the development of adipose cells; and FABP4 (fatty acid binding protein 4), a major protein expressed in adipose cells. There was also a marked increase in the gene for interleukin 6 (IL-6), which is an adipose cell secreted protein, raising the intriguing possibility that IL-6 may play an important role in the development of insulin resistance.

Clinical implications:

1- Insulin mediates glucose uptake into adipose tissue and skeletal muscle through GLUT4 glucose transporters.
2- Vesicles containing GLUT4 glucose transporters are mobilised to the plasma membrane by insulin stimulation, thereby effecting glucose transport into the cell. In the absence of insulin, glucose transporters are recycled intracellularly.
3- Insulin acts by increasing the glucose transport rate of each transporter, by increasing the number of functional glucose transporters, or by a combination of both mechanisms.
4- Insulin resistance in adipose cells is associated with a decrease in GLUT4 transporter number and activity.
5- Exercise can be beneficial in the treatment of diabetes because contraction-induced glucose uptake occurs in distinct, insulin-independent pathways.
6- Adipose cell-secreted proteins may play a role in the development of in vivo insulin resistance.

SMOKING

Tobacco smoking is the leading preventable cause of death (40%). In 1950, it was estimated that an individual adult consumed 3500 cigarettes per year. The effects of tobacco smoke are multifactorial and complex. Nicotine is a tertiary amine composed of a pyridine and pyrolidine ring. Tobacco smoke contains mainly the isomer levorotatory S-nicotine. Nicotine is a weak base with a pH of 8.0. One third of the nicotine molecule is in its unionised form that readily crosses cell membrane. Inhaled cigarette smoke is acidic at a pH of 3.5 and not readily absorbed until it reaches the small airways and alveoli. The pH of smoke from pipes and cigars is alkaline at 8.5 and is well absorbed by buccal mucosa. The small airways and alveoli absorb nicotine rapidly, regardless of the pH of the smoke. In the bloodstream, nicotine crosses the blood-brain barrier by way of passive diffusion and active transport. Nicotine dose-dependent effects include: 1- low-dose effects are mediated by direct stimulation of CNS, and 2- high-dose effects are mediated by direct actions on the peripheral nervous system. Nicotine induces desquamation and cell death to the endothelial cells. The endothelial dysfunction is reversible. Von Willebrand's factor antigen and NO are abnormal in smokers. PGI_2 synthesis is inhibited by nicotine and, damaged PGI_2 promotes vasoconstriction and platelet aggregation. Thromboxane A_2 is increased by nicotine. Catecholamine increases once nicotine is absorbed followed by increase in free fatty acids and possible increase intracellular lipid deposition. Catecholamines increase blood pressure and heart rate, which result in an increase in myocardial oxygen demand. Cigarette smoke contains 2-6% carbon monoxide (CO), which can cause hypoxia. The high oxygen extraction ratio of the myocardium makes it more sensitive to hypoxia. CO increases vascular permeability as a result of hypoxia. Nicotine produces vasospasm and increases platelet aggregation that can decrease myocardial oxygen supply. Passive smoking promotes arteriosclerotic plaque formation. Smoking elevates serum cholesterol, phospholipids, triglycerides, LDL, and VLDL cholesterol. Smoking decreases serum HDL cholesterol. Smoking adversely affects blood fluidity, with haematocrit; fibrinogen, and WBC count are elevated in chronic smokers and normalises with abstinence. Smoking increases platelet activation >100-fold. Smoking increases the production of PDGF, which is an atherogenic promoter of smooth muscle cell growth. Smoking increases fibrinogen and factor VII, which leads to progression of atherosclerosis and acute thrombosis due to thrombogensis. Smoking one cigarette increases coronary vascular resistance and decreases coronary flow velocity. Cigarette smokers have 2-3 times more risk to develop stroke than non-smokers; this could be due to intimal hyperplasia. Mortality is higher among smokers after myocardial infarction (MI). Smoking cessation results in a 50% reduction of risk from coronary heart disease within the first year, and the risk equalises to that of non-smokers after 10 years. Patients who stop smoking more than 6 weeks before surgery have reduced postoperative respiratory complications and improved pulmonary function. Cigarette smoking decreases exhaled NO level. Elevated blood viscosity, haematocrit, blood cell filterability, plasma fibrinogen, and WBC normalises after 8 weeks of smoking cessation. The risk of stroke diminishes after smoking cessation, and the risk of stroke reverts to the level of nonsmokers within 5 years. Nicotine creates tolerance, physical dependence, and withdrawal symptoms. Nicotine withdrawal symptoms include craving, irritability, anxiety, difficulty concentrating, increased appetite, and sleep disturbances. Smoking cessation may take 3-4 attempts before successful abstinence is maintained. Almost 90% of those attempts to stop smoking have one relapse. Glucocorticoids are released in response to stress. Glucocorticoids may play part in the promotion of excess gastric acid production or the destruction of gastric mucosal defenses. Peptic ulcer disease (PUD) is twice as likely to develop in smokers as in nonsmokers. Smoking is related to poor ulcer healing and high rates of ulcer recurrence. Alcohol can injure the mucosal barrier. Coffee, including decaffeinated forms stimulates acid secretion. In 1984, cigarette-initiated fires were reported in 67,000 cases in USA and resulted in 1570 deaths and 7000 serious injuries. 80% of the world's smokers live in developing countries. 500 million people alive today will die by tobacco. In the year 2000 Four million people died from illnesses related to tobacco. By 2030, 10 million people will die each year, 70% of them in developing countries. Cigarette smoke contains >4000 toxic gaseous or particular compounds. Nicotine causes a release of adrenal catecholamines resulting in vasospasm and subcutaneous hypoperfusion with a simultaneous increased oxygen demand. Vasoconstriction may also follow a decrease in production of prostacyclin induced by nicotine. Increased platelet adhesiveness and formation of microthrombi are induced by nicotine, and platelets may be rendered hyperaggregable for as long as 2 hrs after a short smoking period. The net effect of these processes is hypoxia at the cellular level. Nicotine inhibits the function of erythrocytes, fibroblasts, and macrophages. Carbon monoxide has a high affinity for haemoglobin, reducing the amount of oxygen to be carried by the molecule, and reduces oxygen supply to the tissues. Cigarette smoking inhibits human lung fibroblast proliferation and chemotaxis. Hydrogen cyanide inhibits oxidative enzymes and oxygen delivery at the cellular level. Cadmium depresses the procollagen production in fibroblasts. Smoking 20 cigarettes a day results in tissue hypoxia for most of each day. 30 minutes after smoking one cigarette, there is a mean decline in subcutaneous oxygen tension of 32%. The cross-linking of the procollagen molecules to form a stable triple helix is dependent on the conversion of proline to hydroxyproline by prolyl hydroxylase and lysine to hydroxylysine by lysyl hydroxylase. These pathways in particular require molecular oxygen. Age is not negatively correlated with the amount of collagen deposited. The synthesis of subcutaneous collagen in smokers is specifically impeded, indicating impaired wound healing process.

Some additives and smoke constituents:

Substance	Action
Acetaldehyde	Enhances addiction by working synergetically with nicotine.
Acetone	Toxic solvent.
Ammonia	Enhances addiction by boosting absorption of nicotine.
Arsenic	
Cadmium	Carcinogenic.
Carbon monoxide	Toxic.
Cocoa	Masks the tobacco taste, bronchodilator allowing smoke inhalation deep into the lung.
Formaldehyde	Carcinogenic.
Mercury	
Nitrosamines	Carcinogenic.
Polonium-210	Carcinogenic.

BIOMOLECULAR DAMAGE OF CIGARETTE SMOKE

Cigarette smoking contributes to much of mortalities worldwide. In 1985, smoking accounted for 87% of lung cancer deaths, 82% of chronic obstructive pulmonary disease (COPD) deaths, 21% of coronary heart disease (CHD) deaths, and 18% of stroke deaths. Free radicals in cigarette smoke, or the production of ROS by recruitment and activation of phagocytes in the lung, may be the major contributory factors to cigarette smoking-related diseases. Gas-phase cigarette smoke contains 1×10^{15} radicals per puff, primarily the alkyl, alkoxyl, and peroxyl type. Nitric oxide (nitrogen monoxide, NO) is present in cigarette smoke in amounts up to 500-1000 ppm. NO reacts quickly with superoxide radical (O_2^-) to form peroxynitrite ($ONOO^-$) and with organic peroxyl radicals to give alkyl peroxynitrites (ROONO). The first biological fluids that come into contact with inhaled cigarette smoke are the respiratory tract lining fluids (RTLFs). RTLFs contain a variety of antioxidants and serve to protect the underlying respiratory tract epithelial cells against damage. When the lung is injured and exudate processes across the epithelial lining cell layer are initiated, RTLFs become more plasma-like in composition. Cigarette smoke-induced injury to underlying respiratory tract epithelial cells occurs by: 1- direct toxic interaction, 2- damage to the cells by toxic reaction products of cigarette smoke with RTLFs, and 3- reactions subsequent to activation of inflammatory-immune processes initiated by process 1 and 2. Human alveolar epithelial cells exposed to gas-phase cigarette smoke exhibit decreased cell attachment and cell proliferation, and reduced glutathione (GSH) concentration. Increased damage to respiratory tract epithelial cells and subsequent increased permeability expose the circulatory system to toxicants derived from cigarette smoke, resulting in oxidative damage or other modifications to lipoprotein, especially low-density lipoprotein (LDL). Cigarette smoke causes changes in the electrophoretic mobility of LDL, and enhances uptake of LDL by macrophages. There is increased in circulating products of lipid peroxidation (F_2-isoprostanes) in the plasma of cigarette smokers. Alterations in plasma lipids and lipoprotein metabolism include decreased high-density-lipoprotein (HDL) concentrations, alterations of cholesterol transport, and decreased plasma lecithin-cholesterol acyltransferase (LCAT). Cigarette smoke induces adhesion-promoting processes between leukocytes and endothelium, oxidised lipoprotein can mediate interactions of circulating neutrophils with platelets and endothelium. Cigarette smoke influences endothelial injury by activating neutrophil and monocyte selectin expression. Vitamin E concentration in smokers' epithelial lining fluid is decreased and concentration of ascorbic acid in blood plasma is low. GSH is markedly elevated in epithelial lining fluids of cigarette smokers and in lung tissue that suggest a response to chronic oxidative stress. Cigarette smoke causes complete depletions of ascorbic acid and ubiquinol-10 after 6 and 9 puffs respectively, whereas urate and α-tocopherol are only depleted by 25% to the initial concentrations. Consumption of ubiquinol-10 and α-tocopherol occurs after ascorbate is nearly completely utilised. Ascorbate may be the first strong reductant in plasma to react with cigarette smoke oxidants and affords considerable protection to unsaturated lipids. These lipid-soluble antioxidants (CoQ_{10} and α-tocopherol) react with peroxyl radicals and terminate propagation events in the lipid peroxidation pathway. Ubiquinol-10 protects LDL more efficiently against lipid peroxidation than does α-tocopherol. Ubiquinol-10 concentrations in plasma are 1 μmol/l and represent only small fraction of the overall lipid-soluble antioxidant defenses. Ubiquinol and α-tocopherol protect plasma lipids from cigarette smoke-induced oxidative damage primarily by terminating peroxyl radical propagation reactions. Cigarette smoke is capable of inducing the release of iron from ferritin, which may accelerate the decomposition of 18:2OOH (<1 μm/l in nine puffs) through Fenton-like reaction. NO of gas-phase cigarette smoke may counteract the effects of the radical-initiating species in cigarette smoke by terminating propagation reactions during lipid peroxidation. NO can inhibit lipid peroxidation by rapidly scavenging peroxyl radicals. Nitrogen-containing lipid derivatives resulting from interactions of lipid peroxyl and alkoxyl radicals with NO. β-carotene decreased by 30%, 50%, and 60% at 9, 18, 27 puffs of cigarette smoke, respectively. Lycopene is more susceptible to cigarette smoke-induced depletion. Cigarette smoke-mediated singlet oxygen ($^1O_2$) production is via reaction of NO with H_2O_2. $ONOO^-$ is formed by reaction of NO with O_2^-, and reacts with H_2O_2 to form $^1O_2$. The activation of phagocytes in the lung of cigarette smoker can yield high amounts of hypochlorite (CLO^-) and H_2O_2 that react to form $^1O_2$. Standard measures of protein oxidation and modification are the loss of free protein sulfhydryl groups and the formation of protein-bound carbonyl groups. The basal concentration of carbonyl groups in human plasma is 50 μmol/l (0.5 μmol/g protein) and concentration of protein sulfhydryl groups of 500 μmol/l. Cigarette smoke-induced protein modification (450 μmol/l) occurs to a large extent than does lipid peroxidation (<1 μmol/l). Cigarette smoke induces iron mobilisation from ferritin. Formation of protein carbonyls begins immediately with cigarette smoke exposure to plasma, CoQ_{10} and ascorbate do not inhibit it. GSH decreases protein carbonyl formation. GSH does not oxidise to GSSG or S-nitrosoglutathione (GSNO) during exposure to cigarette smoke, these species convert to GSH by reductants. GSH reacts with aldehydes in cigarette smoke. Protein sulfhydryl groups in plasma react with aldehydes in cigarette smoke. The majority of sulfydryl groups in plasma proteins are associated with albumin. Several enzymes contain sulfydryl groups that are critical for functional activity. Cigarette smoke decreases LCAT activity, which is an important component of reverse cholesterol transport, and like creatine kinase (CK), contains sulfhydryl groups. Plasma CK activity is decreased by 30% after nine puffs of cigarette smoke. Nitrogen oxides such as NO and NO_2 are capable of oxidising sulfhydryl groups to disulfides via thiyl radical intermediates. NO_2 inactivates α_1-proteinase inhibitor (α_1-PI) enzyme, which may be important in cigarette smoke-induced diseases such as emphysema. Pulmonary surfactant can be damaged by cigarette smoke constituents including NO_2 and $ONOO^-$. The damage may involve tyrosine nitration. GSH, ascorbate, and urate inhibit the formation of 3-nitrotyrosine and dityrosine. NO_2 is a likely candidate for a nitrating agent. NO is oxidised in aqueous solution to nitrite (NO_2^-), which proceed via a steady state concentration of NO_2 and dinitrogen trioxide (N_2O_3). Cigarette smoke particulates (tar phase) that are deposited in the respiratory tract, contribute directly to the damaging of biomolecules (most cigarette smoke carcinogens are in the tar phase) and may act synergistically with the gas phase of cigarette smoke to produce an increased oxidant burden on the respiratory tract. Tar-phase, lipid-soluble phenolic constituents may diffuse through the respiratory tract epithelium into the systemic circulation and perhaps undergo redox cycling. Carcinogens in cigarette smoke include polycyclic aromatic hydrocarbons (PAH), aza-arenes, N-nitrosamines, aromatic amines, heterocyclic aromatic amines, aldehydes, organic compounds, and inorganic compounds. Cigarette smoke can induce the release of iron from metal-binding proteins and concentrations of free iron in the respiratory tract substantially increased in smokers. Oxidative damage to proteins or lipids can still occur in the presence of antioxidants. Reductants such as GSH and ascorbate may in fact exacerbate oxidative damage to biomolecules that is catalysed by metal ions. Scavenging of acrolein by GSH protects plasma proteins from cigarette smoke-induced damage, but GSH-acrolein adduct itself is toxic. Neutrophil and macrophage accumulation and activation, and their subsequent production of inflammatory oxidants such as hydrogen peroxide, hypochlorous acid, and a variety of reactive nitrogen species, contribute to lung (and plasma) indexes of oxidative injury during smoke inhalation. Cigarette smoke-derived NO has a bronchodilatory and vasodilatory effect, and can modulate the inflammatory response first by inhibiting leukocyte adhesion, and second by direct inhibition of NADPH oxidase function in neutrophils. NO may exacerbate inflammatory conditions via the action of proinflammatory prostaglandins. NO can rapidly react with tyrosine. NO can inactivate ribonucleotide reductase through a mechanism involving tyrosine radical scavenging by NO. After a puff of cigarette smoke is inhaled, the components of the smoke are absorbed into RTLF and continue to react with biomolecules in these fluids and the underlying epithelial cells for an unknown period of time. The α- and β-unsaturated aldehydes in cigarette smoke react by a nonradical mechanism with proteins resulting in the formation of Michael addition products with protein -SH and $-NH_2$ groups. The formation of nitrated and oxidised forms of tyrosine by cigarette smoke strongly suggests that free radical reactions of nitrogen oxides and other ROS may play a role in cigarette smoke-induced protein modification. These types of modifications may result in altered activity of critical enzymes, membrane receptors, and transport proteins (e.g., membrane ion channels), and would interfere with cell signalling pathways (e.g., tyrosine phosphorylation). CoQ_{10} and α-tocopherol afford protection against smoke-induced damage to plasma lipids, but do not inhibit protein modifications associated with cigarette smoke aldehydes with the exception of GSH. These antioxidants may attenuate some of the adverse effects associated with cigarette smoking including a variety of respiratory and cardiovascular diseases.

TOBACCO SMOKE CARCINOGENS AND LUNG CANCER

Pulmonary carcinogens include: A- Polycyclic aromatic hydrocarbons (PAH) such as benzo[a]pyrene, benzo[b]fluoranthane, benzo[j]fluoranthane, dibenzo[a,i]pyrene, indenol[1,2,3-cd]pyrene, dibenz[a,h]anthracene, and 5-methylchrysene. B- Aza-arenes such as dibenz[a,h]acridine, and 7H-dibenzo[c,g]carbazole]. C- N.Nitrosamines such as N-nitrosodiethylamine, and 4-(Methylnitrosamino)-1-(3-pyridyl)-1-butanone (NNK). D- Organic compounds such as 1,3-butadiene, and ethyl carbamate. E- Inorganic compounds such as nickel, chromium, cadmium, polonium-210, arsenic, and hydrazine. Environmental tobacco smoke (ETS) is a cause of lung cancer. The mainstream smoke emerging from the mouthpiece of a cigarette is an aerosol containing 10^{10} particles/ml. 95% of the smoke (vapour phase) is made up of gases, mainly nitrogen, oxygen and carbon dioxide. The particulate phase of smoking contains at least 3500 compounds and most of the carcinogens. There are 55 carcinogens in cigarette smoke that have been evaluated. Other carcinogens may also be present such as PAHs, multiple alkylated and high molecular-weight compounds are not completely characterised with respect to their carcinogenicity. Among PAHs, benzo[a]pyrene (BaP) is a potent lung carcinogen. Of these, polycyclic aromatic hydrocarbons and the tobacco-specific nitrosamine 4-(methylnitrosamine)-1-(3-pyridyl)-1-butamine are likely to play a major role. Nickel, chromium, cadmium, and arsenic are all present in tobacco, and a percentage of each is transferred to mainstream smoke. Cigarette smoke is also a tumour promoter. Carcinogens are enzymatically transformed to a series of metabolites as the exposed organism attempts to convert them to forms that are more readily excreted. The initial steps are usually carried out by cytochrome P450 enzymes, which oxygenate the substrate. Some metabolites produced by the P450s react with DNA or other macromolecules to form covalent binding products known as adducts. This is called metabolic activation, other reactions are detoxification pathways. Cigarette smoke contains free radicals and induces oxidative damage in humans. The gas phase of freshly generated cigarette smoke contains up to 600 μg of NO. The particulate phase contains stable free radicals, quinone-hydroquinone complex held in a tar complex. Peroxynitrite generated from NO and superoxide anion, might be involved in DNA single-strand breakage. Peroxynitrite is induced by aqueous cigarette smoke fractions. The gas phase of cigarette smoke causes lipid peroxidation of human blood plasma. Both whole-cigarette smoke and gas-phase cigarette smoke cause formation of carbonyls in human plasma.

THE RENAL RISKS OF SMOKING

Renal cell cancer developing dose relationship is 1.1 for moderate and 3.8 for heavy smokers. For urothelial cancer, smoking is the single most important risk factor (1.5-3) apart from industrial exposure to carcinogens. Heaviest smokers risk for urothelial cancer is ranged from 4.7 to 7.9. Smoking causes premature mortality. The basic problem is the addictive property of nicotine in tobacco smoke. The prevalence of smoking in adolescent is increasing. Haematuria tends to be more prevalent in smokers. In diabetic patients smoking increases mortality two- to three-fold, mainly as a result of cardiovascular complications. Smoking is an independent risk factor for diabetic nephropathy. Smoking aggravates insulin resistance in healthy smokers. Smoking predictor of significant renal artery stenosis as well as atheroembolic renal disease. Cessation of smoking is beneficial for renal function in diabetic patients. In patients with IDDM albuminuria improves significantly when patients stopped smoking. Progression of diabetic nephropathy is 53% of current smokers, and only 33% of patients who has stopped smoking and 11% of non-smokers. Smoking at the time of lupus nephritis, is an independent risk factor for the more rapid development of ESRD. Smoking is a predictor of significant renal artery stenosis. Smoking and/or nicotine have 3 major known effects on the kidney: 1) on renal haemodynamics, 2) on water diuresis and electrolyte excretion, and 3) on the proximal tubule. Nicotine increases GFR, Na^+ and Cl^- excretion and urine flow, mediated by release of catecholamines from the adrenal medulla, which exerts their effect via β-adrenergic receptors. In insulin-treated diabetics, there is a significantly higher prevalence of glomerular hyperfiltration (41% vs. 18%) in the smokers than in the non-smokers. Vasopressin increases GFR inappropriately via modulating the activity of the tubuloglomerular feedback, probably as a result from change in urea recycling. Smoking is associated with changes in proximal tubular function, including increased excretion of N-acetyl-β-glucosaminidase (NAG) and impairment of organic cation transport. During smoking there is transient increase of blood pressure and heart rate, which actions are mediated via sympathetic activation and vasopressin release. The haemodynamic effects of one cigarette only last approximately 30 minutes and are related to nicotine. The intermittent and persistent increase in blood pressure induced by smoking may well contribute to progression of diabetic and non-diabetic renal disease. Smoking may indirectly damage the microvasculature (including the glomerulus) in the kidney through its known effects on platelet function, thromboxane metabolism and endothelial cell function. Chronic smoking increases the production of TXA_2, a potent vasoconstrictor. The deleterious synergism between activation of platelets and diversion of AA metabolites to proaggregatory vasoconstrictors may be of particular importance in lupus nephritis and other inflammatory renal diseases. Increased frequency of endothelial cell death by nicotine results in enhanced transendothelial leakage of macromolecules. Transepithelial leakage of macromolecules, particularly of low-density lipoproteins, is important in the genesis of atherosclerosis. Human plasma exposed to cigarette smoke is toxic to endothelial cells. Oxidative stress is the result of free radicals present in tobacco smoke. Cigarette smoke induces the adhesion of human monocytes to endothelial cells, due to activation of PKC, leading to increased surface expression of the adhesive ligand CD11b on monocytes and its receptors ICAM-1 and ELAM-1 on endothelial cells. There are increased levels of von Willebrand factor antigen in smokers. The number of desquamated endothelial cells in the circulation increases significantly after smoking only one cigarette. Smoking-induced endothelial cell injury and interaction of white blood cells with endothelial cell injury and interaction of WBCs with endothelial cells therefore could be involved in immune and non-immune damage to the kidney. Cigarette smoking and passive smoking impair endothelial cell-dependent vascular dilation in healthy children and adults. Endothelial cell dependent and independent relaxation induced by endogenous vasodilators are impaired in smokers as compared to non-smokers. Disturbed active vasodilation is also deleterious for coronary arteries and might be important in the genesis of altered renal haemodynamics of smokers. The costs resulting from smoking are enormous. This includes medical costs and social costs due to mortality and morbidity. Smoking lowers the plasma level of vitamin C, but not vitamin E. Smokers have significantly higher plasma concentrations of lipid peroxides. The integrated function between vitamins C and E in the arterial wall appears to be crucial for controlling oxidation and protecting LDL from oxidation. In smokers, a pro-oxidant/antioxidant imbalance exists in the arterial wall, and significant decrease in vitamin E and C levels, along with a significant increase in tissue lipid peroxides.

ALCOHOL

Alcohol is a drug that produces a dual effect on the body: A primary depressant effect that lasts a relatively short time, and a weaker agitation of the central nervous system that persists about six times as long as the depressant effect. In a quiet, nonsocial environment, the excitatory influence may be impaired, and the sedation and drowsiness produced by the drugs are then readily perceived as depression of the CNS. In a social setting, where there is a great deal of sensory input, the effects of low doses of alcohol may be perceived as stimulation. With prolonged or chronic drinking, the presence of the dual effects depends on the time that elapses between drinking episodes. Blood Alcohol Level (BAL) is the amount of alcohol in the bloodstream, recorded in milligrams of alcohol per 100 milliliters of blood, or milligrams percent. General Effect of BAL include: 1- Reduces activity of CNS, reduces muscle tone, loss of fine motor coordination, and induces a staggering "drunken" gait. 2- The eyes may appear glossy and pupils may be slow to respond to stimulus. At high doses pupils may become constricted. 3- At intoxicating doses, alcohol can decrease heart rate, lower blood pressure and respiration rate, and result in decreased reflex responses and slower reaction times. 4- Skin may be cool to the touch (but the user may feel warm), profuse sweating may accompany alcohol use. 5- Loose muscle tone, loss of fine motor coordination, odour of alcohol on the breath, and a staggering "drunken" gait. Since the depressant effects of chronic drinking are greater, they will be significant for the first 2 hours after the last drink. As the time since the last drink increases, the longer-lasting agitation effect becomes dominant. This effect eventually leads to morning drinking to calm the drinker. The morning after hangover and shakiness is due to the residual CNS agitation. This agitation can be temporarily counter-balanced by more drinking because of its dominant depressant effect. Thus, a vicious circle is in motion. Withdrawal symptoms eventually include restlessness, shakiness, confusion, hyperventilation, hallucination, and convulsions. The chronic loss of calcium and magnesium, general malnutrition, and dehydration contribute to these symptoms. Symptoms are usually far more dangerous than those after withdrawal from the opiates or other drugs to which physical dependence may be developed. Withdrawal symptoms begin within the first 24 hours after the last drink, reach their peak intensity within 2-3 days, and disappear within one or two weeks. During the first day of withdrawal, there may be headaches, anxiety, involuntary twitching of muscles, tremor of hands, weakness, insomnia, and nausea. During the next 48 hours, the symptoms become progressively more intense. There may be a fall in blood pressure; fever; delirium characterised by disorientation, delusions, visual hallucinations; and convulsions similar to those exhibited in grand mal epileptic seizures. The fever, delirium, and convulsions are the most serious symptoms and have proved fatal in a number of instances. Aside from withdrawal itself, alcohol has a pervasive effect on the body's GIT, liver, bloodstream, brain and nervous system, heart, muscles, and endocrine system. Alcohol, unlike other drugs, can be utilised by the body as a source of energy. This supply of calories often suppresses appetite, leading to dietary deficiencies that may be responsible in part for the pathologic conditions seen in chronic alcoholism. Alcoholic drinking related specific medical conditions: A- Early alcoholic drinking: 1- loss of control of eye muscles, 2- hypoglycaemia, 3- gastritis, 4- increased susceptibility to infections, 5- cardiac arrhythmia, 6- anaemia, 7- constant flushing of facial oedema, 8- peripheral neuritis, 9- pancreatitis, 10- increase in blood alcohol level, 11- withdrawal signs, 12- fatty liver, and 13- hypertension. B- Chronic alcoholic drinking: 1- liver damage, 2- Korsokoff syndrome (vitamin B deficiency), 3- brain damage, 4- cardiomyopathy, and 5- cancer of the tongue, mouth, pharynx, hypopharynx, oesophagus, and liver. Alcoholism is a multi-factorial psychiatric disorder, with both psychosocial and biochemical/genetic factors. Stress influences ethanol consumption and relapse in abstinent alcoholics. The drinking pattern, particularly frequent drunkard, has an influence on the occurrence of alcohol-related disorders, such as liver disease, pancreatitis, upper gastrointestinal and neurological disorders (polyneuropathy). Alcoholism, one of the most common chronic disorders in the western world, causes or promotes a plethora of diseases and injuries. The social health care costs of harmful alcohol consumption are enormous. Ethanol is a hydrophilic and lipophilic substance, it may harm nearly every organ. The high amounts of alcohol drunk by frequently intoxicated alcoholics per day may injure the gastrointestinal tract directly and may also harm the liver as a detoxifying organ. Complications during alcohol withdrawal, particularly delirium and seizures, occur more frequently in alcoholics with high alcohol consumption. The frequent heavy drinkers more frequently attempt suicide. The economic implications of alcohol abuse in surgical patients are tremendous. Alcohol abusers have a threefold risk of post-operative morbidity after surgery.

Special effects of increasing BAL:

Level	Effect
0.02.	Mellow feeling. Slight body warmth. Less inhibited. Slight euphoria and loss of shyness.
0.05.	Noticeable relaxation. Less alert. Less self-focused. Coordination impairment begins.
0.08.	Drunk driving limit. Definite impairment in coordination and judgement. Slight impairment of balance, speech, vision, reaction time, and hearing. Euphoria. Judgement and self-control are reduced, and caution, reason and memory are impaired.
0.10.	Noisy. Possible embarrassing behaviour. Mood swings. Reduction in reaction time. Significant impairment of motor coordination and loss of good judgement. Speech may be slurred; balance, vision, reaction time and hearing will be impaired. Euphoria.
0.15.	Impaired balance and movement. Clearly drunk. Gross motor impairment and lack of physical control. Blurred vision and major loss of balance. Euphoria is reduced and dysphoria is beginning to appear. Dysphoria (anxiety, restlessness) predominates, nausea may appear.
0.30.	Many lose consciousness.
0.40.	Most lose consciousness; some die.
0.50.	Breathing stops. Many die.

Common postoperative morbidity related to alcohol: 1- disseminated and focal infections (pneumonia, cystitis, abscesses), 2- cardiopulmonary insufficiency, 3- bleeding episodes, and 4- delayed healing with risk of anastomotic leak and wound failure. The most frequent complications are infections, cardiopulmonary insufficiency, and bleeding episodes. The pathogenesis is suppressed immune capacity, subclinical cardiac dysfunction, and haemostatic imbalance, respectively. Heavy alcohol consumption is associated with abnormalities of the menstrual cycle, sterility, miscarriages, and loss of female sexual characteristics, all of which are affected by the hormonal balance in women. Acute alcohol intake leads to decreased levels of testosterone in normal healthy men. Alcohol ingestion causes an acute elevation of the total testosterone levels in premenopausal women using oral contraceptives, but occurs during the mid-cycle phase among non-users as well. Increased androstenedione to testosterone conversion in the liver is caused by the alcohol-mediated elevation in the $[NADH]:[NAD^+]$ ratio. This may be relevant in the development of hyperandrogenism and loss of female sexual characteristics associated with heavy alcohol consumption. Ethanol may inhibit retinoic acid synthesis by providing competition for alcohol dehydrogenase-catalysed retinol oxidation, i.e., blocking the oxidation of retinol by inhibiting ADH. Ethanol may increase destruction of the retinoic acid by increased P450 enzyme activity, i.e., stimulating the hydrolysis of retinoic acid by cytochrome P450. Ethanol might have an indirect effect in reducing retinoic acid concentrations in liver cells by increasing the mobilisation of vitamin A to peripheral tissues. The decrease in tissue all-trans-retinoic acid concentrations could interfere with normal retinoid signal transduction by causing a functional down-regulation of

RAR activity. RARs act as regulators of AP-1/responsive genes because RARs bound to retinoic acid can combine with the c-Fos/c-Jun complex and sequester it, thereby preventing it from binding to the AP-1 binding site. In the absence of retinoic acid, RARs can no longer bind to the c-Fos/c-Jun complex; thus, AP-1 can bind to DNA sequence motifs, resulting in the transactivation of target genes and cell proliferation. Such a mechanism could, in part, be responsible for alcohol-induced cell injury as well as malignant transformation. A genetic contribution to alcoholism has been firmly established. It has been evaluating that a roughly 40% rate of lifetime alcohol dependence among individuals who started drinking at age 14 or younger, compared to a roughly 10% rate in individuals who started drinking at age 20 and older. Alcohol misusers report an earlier age of onset of drinking. The early exposure to alcohol causes an increase in alcohol use and misuse in subsequent years through an, as yet, unknown mechanism. Alcohol is only one of many substances likely to be misused by alcoholics and that underlying disorders may promote the use of alcohol as well as other addictive substances. Alcohol consumption may be secondary to (or symptomatic of) underlying disorders that predate the onset of drinking and perhaps contribute to the development of alcohol problems. The age of onset of smoking is earlier in alcoholics than in moderate drinkers. Alcohol abuse is defined as consumption of at least 60 g of ethanol/day, corresponding to 5 drinks/day, for at least 6 month. One standard drink contains 12 g of ethanol, corresponding to 360 ml (12 oz) of beer, 180 ml (60 oz) of wine, and 45 ml (1.5 oz) of spirits. Operations on asymptomatic abusers are associated with postoperative morbidity. The alcohol dependent patient may develop withdrawal symptoms postoperatively. Alcohol abuser has threefold-increased risk of postoperative morbidity. Acute ethanol intoxication can cause urinary retention in patients with BPE. Ethanol relaxes smooth muscle, decreases contractile strength and is a potent anaesthetic for CNS. High blood levels of ethanol decreases detrusor contractility, increases bladder capacity with decreased voided volume and increased residual urine volume, and subsequently impairs voiding efficacy. Alcohol inhibits triggering of the sacral micturation reflex, resulting in overdistension of the bladder and subsequently in an increase bladder tonus. Ethanol suppresses Ca^{2+} channels on the smooth muscle membrane, specifically by suppressing the influx of calcium through the voltage-dependent and receptor-operated calcium channels. Ethanol inhibits the responses induced by direct receptor stimulation and by membrane depolarisation of the detrusor muscle. Ethanol affects the activity of membrane-bound enzymes such as adenylate cyclase and Na^+/K^+-ATPase. Adenylate cyclase may be involved in the detrusor relaxation induced by isoproterenol (β stimulator). Ethanol may increase adenylate cyclase activity by acting at a locus on or near the G-protein and can increase adenylate cyclase activation by itself or with isoproterenol and G-unit activator. Ethanol excreted from the kidney may influence the detrusor function. In human, the ratio of urine alcohol to blood alcohol is 1.35.

BIOLOGICAL WATER

Alcohol and water compete with each other on target membrane molecules, specifically, lipids and proteins near the membrane surface. The basis for the ability is the hydrogen bonding capability of both compounds. Alcohol's amphiphilic properties give it the capability to be attracted simultaneously to both hydrophobic and hydrophilic membrane targets. Alcohol displaces water leading to conformational consequences to involve alcohol-water interactions on glycoproteins and glycolipids in the membrane. Alcohol has antifreeze properties when mixed with nonbiological water. Carbon chain length of alcohols affects water's ability to freeze. Nonbiological water has no opportunity to hydrogen bond to organic molecules. Water molecules have large permanent dipole moments and they are polarizable. Hydrogen bonding can occur between any 2 molecular groups as long as one can donate a proton and another can accept it. The common proton donors in living systems are carboxyl, hydroxyl, amine, or amide groups. The common proton acceptors in living systems are atoms of oxygen and nitrogen. Water is both a donor and an acceptor. Water organisation is influenced by repulsive forces, such as these found near hydrophobic domains of membrane lipid and protein. Proteins are able to gel (imbibe) large volumes of water and thus restrict its flow. The intricate network of protein scaffolding inside and on the surface of cells imposes deterministic, rather than random, constraints on water flow and movement of diffusible molecules. This may be important in organelles where water is trapped in small compartments, such as the cell nucleus, endoplasmic reticulum, lysosomes, endosomes, the Golgi apparatus, secretory vesicles, peroxisomes, and mitochondria. The hydration layer at the cell surface is 3 nm thick, which is the same thickness as the membrane itself. On a membrane surface, especially where there are extracellular projections of sugar domains by glycolipids and glycoproteins, water clusters organise in gel-like fashion around the projecting moieties. For lipids, the predominant structural components of membranes, alcohol's action could involve either direct or indirect changes in the solvation of lipid polar head groups, electrostatic repulsions among lipid head groups, and the hydrophobic interactions among alkyl chains. Water becomes an active participant in intermolecular interactions. Most of the components of membrane, such as phospholipids, sialoglycolipids and proteins, have microdomains that attract hydrogen bonding with water. Under normal conditions in membrane phospholipids, there are 7-20 water molecules per molecule of lipid. Total hydration is affected by polar group identity, polar group methylation, the physical status of the hydrocarbon chain, chain heterogeneity, and the mixing of lipid species. Phosphatidylcholine micelles have three states of water: 1- tightly bound water (<14 molecules of water present in each molecule of lipid), 2- loosely bound water (>14 molecules of water), and 3- bulk water (>30 molecules of water permolecule of lipid). Water holds the polar head groups of lipids in place and also serves to partially offset the large electrostatic repulsion of electronegative phospholipid head groups. The water molecules, in turn, become transformed from bulk water to being bound or constrained. Proteins imbibe water like sponge. Water binding sites in proteins fall into two categories: 1- the side-chain polar groups (OH groups on serine, β-carboxyl groups on aspartic acid, and ε-amino groups on lysine); and 2- polypeptide chain groups that are polar (NH imido groups and CO carbonyl groups). Proteins have ample capability to absorb, and organise water. Water is an allosteric ligand, as provided by: 1- hexokinase releases 100 molecules of water in the process of binding glucose; 2- ion channels hydrate upon opening, and this is impaired as water is depleted from their vicinity; 3- as cytochrome A binds an electron, cytochrome oxidase is associated with an addition 10 molecules of water; and 4- four oxygen bind to haemoglobin together with 60 additional water molecules. The oxygen affinity of haemoglobin varies linearly with the chemical potential of water in the bathing medium. Each human haemoglobin molecule is solvated with a cluster of 60-75 molecules of water. Glycoproteins attract large volumes of water (95%). Degree of hydration is very important in the adhesive properties of mucus and in regulating mucocillary transport. The use of surfactant therapy at the sol phase in the respiratory tract is to make mucus less viscous and less adhesive. The carbohydrate moieties of biologically active glycoproteins play a major role in the stabilisation and oriented immobilisation of these proteins. Gangliosides provide ample opportunity for hydrogen bonding of water because of their various hydroxyl, carbonyl, carboxyl and amide groups. Phospholipids are the major structural lipids of membranes. Phospholipids have a polar head group (glycerophosphoryl ester moiety) and a nonpolar hydrocarbon tail (2 chains of esterified fatty acids). In the presence of water, the polar head groups forms hydrogen bonds with water, whereas the hydrocarbon tails interact among themselves and avoid water. Gangliosides can comprise 10 mol% of the plasma membrane lipid. Gangliosides concentrate in the brain and synapses. They project sugar residues into the aqueous extracellular space, and they form hydrogen bonds with each other and with water. Small mol fractions of ganglioside can have profound effects on the molecular organisation of membrane phospholipid. Gangliosides have binding affinity for bacterial toxins, viruses, IFN, IL-2, and several glycoproteins. Gangliosides modulate function of various proteins secondary to interaction with Ca^{2+}, e.g., enzymes, receptors, ion channels, and cell adhesion molecules residing on the cell surface. Gangliosides are intimately associated with serotonin receptors. Gangliosides present a massive target for attracting both water and alcohol. The smallest gangliosides, GM1, when mixed with phosphtidylcholine in fully hydrated bilayers, has a head group cross sectional area of 100Å. The sugars project straight out from the bilayer surface at least 12Å beyond the boundary of the phospholipid head groups (22Å) above the glycerol backbone. This space is filled with water. All other gangliosides species are bigger and have multiple branched sialic residues. Gangliosides may interact with each other, especially if ethanol influences dispersion. The effectiveness of gangliosides as targets depends on their accessibility at the surface. Alcohol may alter the surface exposure of gangliosides and sugar residues of glycoproteins. The displacement of hydrogen-bonded water is part of the mechanism of alcohol-induced intoxication. Intoxication may be associated with physical changes in the membrane surface environment that partially hide ganglioside-reactive groups. Gangliosides potentiate the actions of NGF and binding of GM1 to the cell surface glycoprotein (tyrosine kinase receptor, Trk), which is thought to be the initial step in the intracellular transduction pathway of NGF. Gangliosides are associated with the receptor protein for TSH. Gangliosides inhibit PKC. Gangliosides may potentiate the activity of cAMP-dependent protein kinase. Acute ethanol can decrease the level of brain sialic acid, the acidic sugar that is a key component of gangliosides and many glycoproteins. The carboxyl group of sialic acid could be the prime target for alcohol-induced displacement of hydrogen-bonded water. Gangliosides are the source for the reduced sialic acid in lipid. Acute Alcohol ingestion decreases sialylated glycoconjugates. Alcohol decreases contents of sialic acid in synaptosomal gangliosides and glycoproteins. Alcohol impairs synthesis of sialoglycoprotein in stomach mucosa. Alcohol-induced desialylation should decrease bound water. The sugar of complex lipids and alcohols may share some similar properties. The molecular mechanisms of sweet taste perception are due to hydrogen-bonding domains and a hydrophobic domain-separated by appropriate distances. Water is repelled from hydrophobic domains and attracted to the hydrophilic domains of sweeteners. Mobility of water around a sweetener molecule correlates with the strength of opposing hydrophobic and hydrophilic domains. Mobility also correlates with perceived sweetness and with effectiveness of the ligand action on sweet taste receptors. Acute alcohol increases the mobility (fluidity) of the acyl chain region of the membrane, which need not be entirely direct. Tolerance to alcohol-induced intoxication is attributed to changes in membrane composition (e.g., increased cholesterol content) that make it more difficult for ethanol to fluidise the membrane interior. In transmembranous receptors, the proteins fold and coil, with their hydrophobic surfaces in contact with the hydrophobic interior of lipid bilayers. The hydrophobic forces that anchor the protein are reinforced by water at the exofacial surfaces whenever there is an exposed and charged amino acid. At the lipid-protein interface, hydrogen-bonded water can serve as a bridge to link the oxygen of lipid carboxyl groups to the NH group of positively charged amino acids. The well-known marker of alcoholism, transferrin is an N-glycosylated glycoprotein in which the sialic acid residues are changed by chronic alcohol. The nature of alcohol interaction with protein is not stereospecific. When ethanol binds at the surface of membrane, it perturbs the acyl chain region. Alcohol concentration affects where alcohol binds within plasma membranes. During acute intoxicating phase, ethanol levels are several-fold higher in the brain than in the blood, and differential distribution occurs within the plasma membrane. Alcohol-induced displacement of water should weaken the strength of interlipid interactions that are bridged by hydrogen-bonded water. Changes in water structure, especially displacement from hydrogen-bonded sites, could cause allosteric effects in membrane bound enzymes and receptor proteins. Intracellular hydration would likely be altered. Water and many small molecules that are dissolved in it have high membrane permeability. The disordering ethanol effects in the interior of the membrane depends on the oligosaccharide composition and not on the sialic acid. Alcohol displaces hydrogen-bonded water from membrane surfaces. Increasing the concentrations of GM1 causes increased alcohol-induced disordering of the membrane at the surface, intermediate and interior regions. Lipid-protein interfaces, receptor surfaces, and internal pockets of water in proteins are also potential sites for ethanol interactions that could alter membrane function. Given the multiplier effect of water clustering, perturbations of hydrogen-bonded water could easily be transferred from the membrane-water interface to both the membrane interior and the protein domains. There are highly mobile charges that are highly dependent on the physical state of protein-bounded water. Ion channel function seems to be influenced by water distribution. The mechanism of alcohol could be on the water structure that is associated with ion channels. The hydroxyl group of water would be attracted to the carboxyl group of gangliosidic acid residues, and simultaneously by oxygen of water would be attracted to NH groups of adjacent protein. Alcohol cannot bind both sites at the same time, as it has only a hydroxyl group to be attracted to NH groups of protein or sialic acid. Alcohol's carbon tail is insert into hydrocarbon domains of membrane. The monopolar hydrogen bonding of alcohol would break the seal of water that normally helps to maintain the conformational relationships between ganglioside and protein. Alcohol's displacement of water could also occur at NH groups within the protein that are adjacent to a hydrophobic domain. Water displacement (at the surface or in the interior) could alter the conformation and receptor properties of the protein. Gangliosides are electrostatically attracted to complementary electrical charge groups of receptor protein. The exofacial boundary of the lipids and the ganglioside-protein interface are likely to hold large volumes of structured water clusters (e.g., major synaptic membrane components). Extracellular water is considered to be organised by and to help organise the conformation of surface lipids and proteins. The normal structure in membrane is lamellar. Stability of lamellar structure depends on fatty acid chain length, degree of double bonding, relative sizes of polar and hydrocarbon tail groups, temperatures, and presence of perturbing drugs, such as anaesthetics and alcohol.

ALCOHOL-INDUCED OXIDATIVE STRESS

Cytochrome P-450 2E1 (CYP2E1) induction is related to oxidative stress. Ethanol consumption is able to induce an oxidative stress in the liver and in extrahepatic tissues, linked to an imbalance between the pro-oxidant and the antioxidant systems in favour of the former. CYP2E1, which is induced by administration of a large ethanol dose or after chronic ethanol ingestion, has a high oxidase activity and plays a crucial role in the microsomal generation of reactive oxygen species (ROS), which have the capability of initiating membranous lipid peroxidation. CYP2E1 is also involved in the production of ethanol-derived free radicals during ethanol oxidation. These free radicals have the property to bind covalently to proteins forming adducts able to induce autoantibodies. These antibodies may represent markers of the production of ethanol-derived free radical adducts and contribute to the hepatotoxicity of ethanol in promoting immune mechanisms of liver injury. Ethanol induces an oxidative stress either by enhancing the production of ROS and/or by decreasing the level of endogenous antioxidants. Blood markers of oxidative stress, such as oxidised proteins and lipid peroxides, are increased by ~20% in alcoholic patients, whereas plasma concentrations of α-tocopherol, a marker of antioxidant defence, are depressed. Tobacco smoking, which is very often associated with alcohol consumption, also produces free radicals and thus participates in the generation of oxidative stress. During ethanol intoxication, pro-oxidant species are generated in hepatic tissue at various subcellular sites: in the endoplasmic reticulum by CYP2E1, in the cytosol by xanthine oxidase and in the mitochondria by the respiratory chain. In addition, phagocytes recruited and activated by tissue injury will also contribute to this production. CYP2E1 does not appear to be the main factor responsible for the oxidative stress occurring during human chronic alcoholism. Free radicals from other sources may therefore contribute significantly to the generation of this oxidative stress. Glutathione (GSH) is a tripeptide that exists in high concentrations (mM range) in all cells. GSH is not homogeneously distributed within cellular organelles. 80-85% of total cellular GSH is found in the cytosol, manufactured exclusively in the cytosol from amino acids, such as glutamate, cysteine, and glycine, through the action of 2 sequential ATP-dependent enzymatic reactions catalysed by γ-glutamyl-cysteine synthetase and GSH synthetase. 10-15 % of the total GSH pool is found in mitochondria. Oxygen intermediates may act as single transducers and represent a versatile cellular control mechanism for gene regulation. Mitochondrial electron transport is required for the activation of NF-κB and subsequent gene production by TNF. In conditions of increased oxidant stress in mitochondria, such as ischaemia-reperfusion, chronic ethanol ingestion, TNF-induced cytotoxicity, and bile acid retention in cholestasis; mitochondrial GSH status is a critical factor in determining loss of mitochondrial function and cell viability, as well as transcription factor activation and gene regulation. TNF is a 17-kDa polypeptide exerts a pleiotropic array of strikingly different cellular reactions, depending on its concentration and the target cell type. Despite its central role in infection and immunity, TNF exerts cytotoxicity effects against a variety of cell types. Mitochondria are not only the source of oxidative stress but are also a target of ROS. ROS generated from mitochondria participate and cause the initiation and propagation of peroxidative phenomena associated with necrotic cell death and gene regulation. TNF evokes apoptosis and necrotic cell death, depends on the length of action. Ceramide is a lipid-signalling moiety able to perform divergent actions depending on its concentration and the environment in which it is produced. Ceramide can be generated by sphingomyelin (SM) hydrolysis at various subcellular sites by the action of at least 2 types of sphingomyelinases (SMases). Neutral SMase located in the vicinity of the plasma membrane hydrolyses SM located in the outer leaflet of the plasma membrane, whereas acidic SMase is located in the endosome and/or lysosome compartment, which generates a distinct functional pool of ceramide. The action of the neutral SMase increases the amount of ceramide directly involved in the activation of membrane-associated protein kinases, including ceramide-activated protein kinases and cytosolic heterotrimeric protein phosphatase 2A. The ceramide pool generated through the activation of the acidic SMase can regulate gene expression by activation of NF-κB and is involved in the chain of events that culminate in apoptosis in response to several stimuli. Interaction of TNF receptor subunit p55 (TR55) evokes an array of intricate signals that trigger downstream responses. Activation of neutral sphingomyelinases (SMase) by interaction of the cytoplasmic neutral Smase (NSD) with the protein factor associated with neutral Smase (FAN) increases ceramide pool involved in activation of the ceramide-dependent enzymes cytosolic heterotrimeric protein phosphatase 2A (CAP) and ceramide-activated protein kinase (CAPK), which participate in the chain reactions leading to activation of mitogenic-activated protein kinases (MAPK) and activation of AA release that is may be responsible for inflammatory effects of TNF. Activation of acid SMase requires a different domain of the receptor, which corresponds with the death domain by interacting with TNF-receptor-associated through death domain (TRADD). Whereas interaction of TRADD with the signal transducer Fas-associated death domain (FADD) results in cell death, the association of other signal transducers, receptor-interacting protein (RIP) and the ring finger protein (TRAF2), results in activation of NF-κB, responses that diverge downstream to TRAF2. The enzyme phosphocholine-phospholipase C (PC-PLC) appears to be involved in activation of acid SMase through diacylglycerol (DAG). This enzyme generates a functional distinct pool of ceramide, which evokes activation of protein kinase C, thus phosphorylating IκB proteases and resulting in activation of NF-κB. Ceramide pool acts on mitochondria, enhancing production of ROS. Generation of ROS can also activate NF-κB, either by serine phosphorylation (Ser-P) of NFκB followed by its hydrolysis by the proteasome or tyrosine phosphorylation (Tyr-P). Ceramide may interact with mitochondria leading to generation of ROS. Ceramide exerts a direct effect in mitochondria, including sphingolipids as inducers of oxidative stress. Activation of the acid SMase generates a ceramide pool that is involved in the production of ROS. TNF activation of NF-κB requires mitochondrial ROS participation. TNF-induced activation of NF-κB. TNF activation of ceramide of protein kinase C, activates NF-κB. NF-κB becomes activated on its release from IκB either by serine phosphorylation followed by degradation via proteasome or by phosphorylation in tyrosine residues without apparent degradation of IκB. Ceramide produced through the acid SMase in mediating the cytotoxicity of TNF and other cytokines by increasing the generation of ROS and identify mitochondria as a key target of ceramide, leading to generation of ROS by interrupting the electron transport chain at complex III. Oxidative stress is generally considered the result of an imbalance between two antagonistic forces, ROS and antioxidants, in which the effects of the former predominates. In mitochondria, hydrogen peroxide can increase if the activity of hydrogen peroxide-metabolising enzyme, GSH peroxide, is low relative to that of MnSOD. Chronic alcohol intake depletes mitochondrial GSH, without changes in the cytosolic GSH pool. This effect is specific for liver; renal mitochondrial GSH is unchanged. Circulating levels of TNF and other cytokines are increased in patients with acute alcoholic hepatitis and chronic alcohol liver disease. Hepatic mitochondria from ethanol impairs capacity to release GSH from cytosol pool would exhibit increased susceptibility to the effects of prooxidants, generated directly (peroxides) or indirectly (cytokines) by ethanol metabolism. As the level of ROS derived from the mitochondria increased, damage to mitochondrial ATP synthesis through effects on lipids, proteins, or mitochondrial DNA and effects on NF-κB to activate chemokines production would supervene. Ethanol impairs mitochondrial GSH transport. Mitochondria do not contain γ-glutamylcysteine synthetase of GSH synthetase, indicating that mitochondria cannot synthesise their own GSH. Mitochondria generate ROS as byproducts of molecular oxygen consumption in the electron transport chain. Most cellular oxygen is consumed in the cytochrome-c oxidase complex of the respiratory chain, which does not generate reactive species. Glutathione (GSH) in mitochondria is the only defence available to metabolise hydrogen peroxide. A small fraction of the total cellular GSH pool is sequestered in mitochondria by the action of a carrier that transports GSH from the cytosol to the mitochondrial matrix. Mitochondria are subcellular targets of cytokines, especially TNF; depletion of GSH in this organelle renders the cell more susceptible to oxidative stress originating in mitochondria. Ceramide generated during TNF signalling leads to increase production of ROS in mitochondria, Chronic alcohol intake deplete GSH in mitochondria due to effective operation of the carrier responsible for transport of GSH from the cytosol into the mitochondrial matrix. This critical carrier displays functional characteristics distinct from other plasma membrane GSH carrier, such as its ATP dependency, inhibitory specificity, and the size class of mRNA that encode the corresponding carrier, suggesting that the mitochondrial carrier of GSH is a gene product distinct from the plasma membrane transporters. NF-κB is a pivotal transcription factor in chronic inflammatory disease. Genes may determine a patient's susceptibility to the disease and disease's severity, but environmental factors, often unknown, may determine its course. Once established, a chronic inflammatory process appears to take on a momentum of its own. The vicious circle may be suppressed by glucocorticoid or immunosuppressive therapy, but there is no curative treatment for any chronic inflammatory disease.

ALCOHOL IN SUPPLEMENTS

Alcohol is used in the medical field as an indispensable solvent because water alone cannot absorb all the beneficial properties from plant materials. Many homeopathic companies manufacture remedies that contain up to 60-70% alcohol by volume. Fermented products should not be higher than 40 proof alcohol (20% ethanol), because once the alcohol level reaches ~20%, yeast is naturally destroyed and alcohol production stops. Alcohol in small doses: 1- Has a stimulating effect on the vasomotor center of the heart, increases blood flow to the arteries and reduces coronary risk. 2- Benefits peripheral circulation. 3- Prevents water retention. 4- Increases secretion of gastric fundus. 5- Increases the defensive capabilities of the immune system. 6- Lowers the carcinogenicity of nitrosamines in foods. Individuals with a history of alcohol abuse, psychiatric patients on medication and patients subject to seizures should not use products that contain alcohol.

ALCOHOL AND IMMUNE SYSTEM

Chronic alcoholics are more prone to infections with a variety of pathogens, have decreased ability to fight against infections, and have an increased risk of developing cancers, particularly those of the head, neck, and upper gastrointestinal system. While malnutrition, vitamin deficiency, and advanced liver cirrhosis can contribute to some of the immune abnormalities in chronic alcoholics, alcohol itself is a potent modulator of the immune system. Both acute and chronic alcohol uses can affect the immune system at the level of innate or acquired immune responses. Altered inflammatory neutrophil, leukocyte, and macrophage functions after acute or chronic alcohol use contribute to impaired host defence against microbial infections. In addition, the humoral and cellular components of the specific immune system can be equally damaged by alcohol use. Impaired B lymphocyte functions, and increased levels of certain types of immunoglobulins at the expense of others, contribute to the inappropriate immune defence. Furthermore, the impairment of cellular immune responses is pivotal in increased susceptibility to various infections after either acute or chronic alcohol use. During the process of inflammatory response, phagocytic cells, such as neutrophils and macrophages, have a major role in locating, ingesting, and killing microorganisms that invade the body. This complex process involves recruitment of phagocytic cells from the bloodstream to the site of the inflammation by chemotactic agents such as activated complement components (C5a), leukotrienes (LTB4), or various proteins belonging to the chemokine family. Alcohol use can affect this process at several levels. Migration of neutrophils and monocytes from the bloodstream involves adherence and migration through the vascular endothelium at the site of infection, phagocytosis of the microorganism, and intracellular destruction of the pathogen by proteolytic enzymes in the phagolysosomes or via toxic oxygen-derived radicals. In addition to neutrophils, phagocytic monocytes and macrophages are also affected by alcohol use. Acute alcohol inhibits monocyte phagocytic functions, antimicrobial activity, and expression of FcR-type II, which is involved in phagocytosis of antibody-coated particles. Impaired macrophage phagocytic functions as well as abnormal neutrophil leukocyte adherence and chemotaxis are likely to contribute to impaired local antimicrobial defence after alcohol use. Altered production (decreased) of the oxygen radicals, superoxide anion and hydrogen peroxide, after alcohol exposure could be a mechanism undermining antibacterial immune defence. Conversely, overproduction of reactive oxygen radicals has been implicated as a potential pathomechanism for alcohol-induced liver damage. Ethanol can also inhibit gene expression for inducible nitric oxide synthase (iNOS). The adverse effect of ethanol on reactive oxygen radical production may cause dual damage to the host. First, ethanol may inhibit induction of ROS and NO in alveolar macrophages where these mediators play a crucial role in microbial killing. Second, ethanol appears to increase reactive oxygen radical production in the liver where these mediators can mediate or contribute to direct tissue damage. Infection with intracellular pathogens is prevalent in chronic alcoholics. Ethanol-exposed macrophages have been shown as more susceptible to Legionella pneumophilia infection, which is a typical intracellular Gram-negative bacillus. Ethanol augments intracellular survival of Mycobacterium. Alcohol use (potentially both acute and chronic) is likely to increase host susceptibility to HIV-1 infection and to contribute to an accelerated progression of HIV disease. It has been proposed that the modulatory effects of alcohol on the immune system may play a role, not only in increased risk of initial infection, but also in the rapid progression of HIV-1 disease. The percentage of $CD8^+$ T cells significantly increased among the heaviest drinkers between 2 to 5 years post-seroconversion. There is a potential for greater immune changes in those HIV-1 positive patients who also consume alcohol. Ethanol may accelerate the development of AIDS by disrupting cytokine production. Alcohol consumption promotes clinical progression and liver damage in patients with chronic hepatitis C infection, which could be due to IL-12 and inflammatory cytokines. Ethanol decreases lymphoid cell number. Acute ethanol increases apoptosis of thymocytes and human blood mononuclear cells. Ethanol-exposed lymphocytes have a reduced capacity to undergo proliferation and differentiation in response to an antigenic challenge, which may be due to direct effect of ethanol on protein kinase C (PKC), impaired accessory cell/monocyte function, or reduced expression of MHC II antigens on lymphocytes. The overall immune alterations after both acute and chronic alcohol exposures are consistent with decreased cellular immune responses. The immunological abnormalities after both chronic and acute alcohol consumption appear to be consistent with a decreased Th1-type immune response based on reduced antigen-specific T cell proliferation, and increased antibody and autoantibody levels. Ethanol results in decreased IL-4-induced B cell proliferation and IL-4-induced Ig class switching, while IL-2-induced B cell proliferation is not affected by ethanol. Alcohol consumption is associated with increased morbidity and mortality related to malignancy. The activity of NK cells, which are involved in destruction of virus-infected cells and prevention of tumour development and metastasis, is important in anti-cancer defence. Chronic alcoholics decrease the frequency of activated NK cells. Cytokines, immunoregulatory proteins produced by lymphoid cells, have a capacity to affect the functions of both lymphoid and non-immune cell types. IL-2 is one of the most important cytokines promoting T cell growth, survival and proliferation. Ethanol probably affects T cell utilisation of IL-2. IFN-γ, in concert with the macrophage-derived IL-12, is thought to be crucial for induction of Th1-type, cellular immune response. In addition to cytokines, ethanol has been shown to affect the production of non-protein inflammatory mediators, particularly the cyclooxygenase products. The biological relevance of increased (PGE_2) after alcohol exposure is complex, it can inhibit monocyte antigen presentation capacity, inflammatory cytokine production, as well as T cell proliferation. Acute alcohol consumption may significantly modulate responses to subsequent challenges to the immune system, whether it is a bacterial, viral pathogen or trauma injury. Alcoholics, especially those with clinical evidence of liver injury, have been observed to have an increased incidence of infections, which were associated with a higher mortality, e.g., BCG. Chronic large consumption of alcohol significantly impaired organ (liver) clearance of the infection, caused fragmentation of granuloma formation, and decreased skin test responsiveness. The BCG infection in both alcoholics and non-alcoholic is associated with a significant increase in both CD4 and CD8 splenic T lymphocytes. T lymphocyte modulation of the host response to infection may be altered in the alcoholic, a selective reduction of 47% may be observed in $CD4^+$ T lymphocytes in the alcoholics. TNF and IFN are associated with a Type 1 (Th-1) cytokine response that contributes to a cell-mediated defense against intracellular infections (i.e., TB), whereas IL-4, IL-5, IL-9 and IL-13 are typically associated with the activation of Type 2, (Th-2) cytokine responses against extra cellular infections. IL-10 increases as IL-4 decreases during a Th-1 activation. A failure of IL-10 to increase or IL-4 to decrease would suggest an inadequate Th-1 response. The host immune response to a BCG infection is significantly altered in excess alcohol consumer (10 g/kg/d). Following infection with BCG, mRNA levels for both the proinflamatory cytokines (TNF, IFN) are increased significantly. IL-10 modulates the Type 1 cytokine production, preventing an excessive response to infection. BCG infection increases TNF, IFN, and IL-10 and decreases IL-4.

ALCOHOL'S HARMFUL EFFECTS ON BONE

The process of skeletal growth and maturation involves 3 general phases: 1- growth and modelling, 2- consolidation, and 3- remodelling. Achieving an optimal peak bone mass during adolescence may reduce a person's risk for developing osteoporosis (i.e., bone loss with fracture) later in life. Long-term alcohol administration to young, rapidly growing rats significantly reduced bone growth, volume, density, and strength. The longitudinal growth rate and the rate of proliferation of cells in the growing region near the ends of long bones stop during long-term alcohol administration. Alcohol may disrupt the action of hormones, vitamins, and local growth factors balance that regulate the distribution of calcium between blood and bone by affecting the hormones that regulate calcium metabolism, as well as the hormones that influence calcium metabolism indirectly (e.g., steroid reproductive hormones and growth hormone). Short-term alcohol consumption increases PTH secretion, possibly by causing calcium to leave body fluids and flow into cells. Levels of PTH declines sharply until the end of the drinking period and rise over the next 9 hrs. Urinary calcium excretion increases during the first 3 hrs and subsequently decrease. Long-term heavy drinking is associated with hypocalcaemia but normal PTH levels. Calcitonin levels increase only briefly during acute and short-term alcohol consumption. The alcohol-induced decrease in activated vitamin D results in decreased absorption of calcium, but calcium levels quickly return to normal following abstinence. Osteoporosis can develop in post-menopausal women as well as in men with inadequate gonadal function. Alcoholic men frequently have decreased levels of testosterone, and female alcoholics experience increased metabolic conversion of testosterone to estradiol. Oestrogen replacement reduces a woman's risk of developing postmenopausal osteoporosis. Moderate alcohol (i.e., no more than one drink per day for women) consumption increases oestrogen levels in the blood. There is a relationship between the consumption of large quantities of alcohol and bone loss. Growth hormone, secreted by the pituitary gland, is important in bone growth and remodelling. Growth hormone exerts its effects largely through IGF-1, which is produced in the liver and other organs. Alcohol reduces the IGF-1 levels, return to normal after 6 months, but bone deficiencies resulting from alcohol consumption continue, possibly through a mechanism independent of growth factors. Remodelling occurs in small, circumscribed areas scattered on the surface of the bone. Osteoclasts erode a cavity on the bone surface in a process known as resorption. When the resorption cavity is complete, the osteoclasts disappear, the floor of the cavity is smoothed off, and a thin layer of matrix or cement is deposited. The osteoblasts fill the newly formed cavity with new bone. Local imbalance of bone remodelling can occur when osteoclasts erode cavities that are too deep or when osteoblasts lay down layers of new bone, which are too shallow. Alcohol reduces bone formation and increases bone resorption. Alcohol consumption leads to delayed and impaired osteoblast activity (decrease in osteocalcin levels) associated with normal osteoclast function. Microscopically decreased trabecular bone volume, decreased numbers of osteoblasts, and decreased rates of bone formation, indicate impaired bone formation and mineralization, along with other characteristics indicative of osteoporosis. Chronic alcohol consumption has major harmful effects on bone development and maintenance at all ages. Alcohol reduces peak bone mass and can result in relatively weak adult bones that are more susceptible to fracture. Alcohol and bone effects modulate and are modulated by hormones, including PTH, calcitonin, and growth hormone, as well as by other substances, such as vitamin D. Bone is a living tissue that continues to undergo change and replacement (i.e., remodelling) even after a person has attained full stature. Long-term alcohol consumption can interfere with bone growth and remodelling, resulting in decreased bone density; and may be exerted directly or indirectly through the many cell types, hormones and growth factors that regulate bone metabolism. Bone consists of living cells encased in a hard matrix of protein fibers and calcium crystals. There are 2 types of bone: 1- Cortical bone, which is dense and thick, forms the outer layer of bones and the shafts of the long bones of the arms and legs. 2- Cancellous bone, which is a porous meshwork of thin plates (i.e., trabeculae), occurs mostly within the ends of long bones and in the vertebrae. Alcohol affects both types of bone, although the most dramatic changes occur in cancellous bone. At the end of puberty, after bones stop growing lengthwise, they continue to increase in mass. Women experience an accelerated reduction in bone mineral density following menopause. As aging bones weaken, they reach a point (i.e., fracture threshold) at which even minor stress can cause fractures. A high peak bone mass should withstand a longer duration and greater level of bone loss before reaching the fracture threshold. Peak bone mass can be influenced by hormonal, nutritional, environmental, and lifestyle factors, including tobacco and alcohol consumption. A significant proportion of the adolescent population may be at risk for alcohol's harmful effects on bone including reduced bone growth, volume, density, and strength. Those effects may significantly decrease bone mass. The decreased bone mass that occurs from early, long-term alcohol consumption could result in increased fracture and early onset of osteoporosis. Bone is a major storage depot for calcium and other minerals. The small intestine absorbs calcium from ingested food, and the kidneys excrete excess calcium. An adequate concentration of calcium in the bloodstream is required for the proper functioning of nerves and muscle. Alcohol may disrupt the calcium balance by affecting the hormones that regulate calcium metabolism as well as the hormones that influence calcium metabolism indirectly (e.g., steroid reproductive hormones and growth hormone). Parathyroid hormone (PTH) stimulates the activity of osteoclasts. Osteoclasts dissolve small areas of bone, releasing calcium into the blood. PTH inhibits the excretion of calcium by the kidney and activates vitamin D, which promotes the absorption of calcium from the intestine. The resulting increase in calcium levels eventually inhibits further PTH production. Short-term alcohol consumption increases PTH secretion, possibly by causing calcium to leave body fluids (e.g., blood) and flow into cells. Levels of PTH declines sharply until the end of the drinking period and rose over the next 9 hrs. Urinary calcium excretion increases during the first 3 hours and subsequently decreased. Long-term heavy drinking is associated with hypocalcaemia but normal PTH. Moderate drinking is generally defined as no more than 2 standard drinks per day for men and one per day for women. A standard drink is generally considered to be 12 ounces of beer, 5 ounces of wine, or 1.5 ounces of 80-proof distilled spirits. Alcohol administration impairs the ability of the parathyroid glands to increase PTH production in response to the presence of hypocalcaemia. Specialised cells in the thyroid gland produce calcitonin, a hormone that protects the skeleton from calcium loss by inhibiting osteoclast activity. Calcitonin increases the deposition of calcium in bone and lowers the level of calcium in the blood. Calcitonin levels increase only briefly during acute and short-term alcohol consumption. Vitamin D increases intestinal absorption of dietary calcium and has a function in normal bone metabolism. Vitamin D is formed in the skin through the action of sunlight and occurs in foods such as liver, eggs, and milk. The vitamin becomes physiologically active only after chemical modification in the liver and kidneys. Alcoholics normally have low levels of activated vitamin D, along with low levels of the proteins that bind with vitamin D to protect it during transport within the blood. The alcohol-induced decrease in activated vitamin D results in decreased absorption of calcium, although calcium levels quickly return to normal following abstinence. Osteoporosis can develop in post-menopausal women as well as in men with inadequate gonadal function. Alcoholic men frequently have decreased levels of the male steroid hormone testosterone, and female alcoholics experience increased metabolic conversion of testosterone (produced in the ovaries and adrenal glands) to the female steroid hormone estradiol. Because oestrogen deficiency is a major contributing factor for the development of osteoporosis, alcohol might indirectly affect bone through oestrogen. Moderate alcohol consumption (i.e., no more than one drink per day for women) may increase oestrogen levels in the blood. Growth hormone, secreted by the pituitary gland, is important in bone growth and remodelling. Growth hormone exerts its effects largely through IGF, which is produced in the liver and other organs. Levels of IGF-1 are significantly reduced by alcohol. Bone remodelling occurs in small, circumscribed areas scattered on the surface of the bone. Osteoclasts erode a cavity on the bone surface in a process known as resorption. When the resorption cavity is complete, the osteoclasts disappear, the floor of the cavity is smoothed off, and a thin layer of matrix or cement is deposited. Bone-forming cells (i.e., osteoblasts) fill the newly formed cavity with new bone. Local imbalance of bone remodelling can occur when osteoclasts erode cavities that are too deep or when osteoblasts lay down layers of new bone, which are too shallow. Alcohol administration reduces bone formation and increases bone resorption. Alcoholic osteoporosis is characterised by decreased bone formation but normal levels of resorption. Alcohol consumption leads to delayed and impaired osteoblast activity associated with normal osteoclast function. A decrease in osteocalcin levels in response to alcohol administration suggests that alcohol decrease osteoblastic activity. There is diminished osteoblast numbers and osteoblast function in humans as a result of alcohol consumption. Overall, alcohol appears to suppress osteoblast function in adults, resulting in decreased bone formation. A decreased trabecular bone volume, decreased numbers of osteoblasts, and decreased rates of bone formation, indicating impaired bone formation and mineralization, along with other characteristics indicative of osteoporosis. Reduced of trabecular surface covered by active osteoblasts as a result of alcohol, suggests an inhibition of osteoblast proliferation. Cortical thickness is reduced by 52% in alcoholic. Chronic alcohol consumption has major harmful effects on bone development and maintenance at all ages.

$$CH_3CH_2OH + NAD^+ \xrightarrow{ADH} CH_3CHO + NADH + H^+$$

$$CH_3CH_2OH + NADPH + H^+ \xrightarrow{MEOS} CH_3CHO + NADP^+ + 2H_2O$$

$$\left[NADPH + H^+ + O_2 \xrightarrow{NADPH\ Oxidase} NADP^+ + H_2O_2 \right.$$
$$+$$
$$\left. H_2O_2 + CH_3CH_2OH \xrightarrow{Catalase} 2H_2O + CH_3CHO \right.$$

$$\left[Hypoxanthine + H_2O + O_2 \xrightarrow{Xanthine\ Oxidase} Xanthine + H_2O_2 \right.$$
$$+$$
$$\left. H_2O_2 + CH_3CH_2OH \xrightarrow{Catalase} 2H_2O + CH_3CHO \right.$$

ADH = alcohol dehdrogenase, MEOS = microsomal ethanol-oxidizing system

ALCOHOL-HISTAMINE INTERACTIONS

Histamine is biologically a very active compound that participates in intracellular signalling and is characterised as a neurotransmitter. Histamine cannot cross the BBB easily. The histamine is divided into peripheral and central pools. The main stores of histamine at the periphery are the mast cells, basophils, and enterochromaffin-like cells. In the brain, histamine is localised in histaminergic neurones; small amounts of histamine reside in mast cells and possibly in the endothelium of blood vessels. In the periphery, histamine produces contractile effects on smooth muscles, capillary dilative action, and stimulant effect on gastric secretion. Central histamine is involved in various physiological functions including the regulation of neuroendocrine and CVS, sleep-wakefulness, thermoregulation, feeding, drinking, learning and memory, cerebral vascular regulation, analgesia, etc. Central histamine is involved in various pathological functions including motion sickness, epilepsy, Alzheimer's disease, morphine dependence, etc. Histamine acts through 3 types of receptors, H_1, H_2 (postsynaptic), and H_3 (presynaptic, mediating the autoinhibition of histamine synthesis and release). Histamine and ethanol-metabolic pathways have common enzymes, aldehyde dehydrogenase (ALDH) and aldehyde oxidase. Histamine is synthesised from L-histidine by histidine decarboxylase. Histamine is removed by histamine-N-methyltransferase (HMT) by conversion to tele-methylhistamine. The highly active ethanol metabolite, acetaldehyde interfere with histamine degradation by competition with N-tele-methylimidazole and imidazole acetaldehyde for these enzymes. H_2 receptor antagonist increases (12-18%) the peak blood-ethanol concentration and increases (8-15%) the area under the blood-alcohol concentration curve, may be as a result of inhibition of the δ-ADH in the stomach depending on the timing of alcohol ingestion, dose, relationship to food intake and the varying ethnic backgrounds of persons. Histamine is involved in ethanol-induced injury of the stomach and intestine, bronchial asthma, and flushing. Exposure of the intestinal mucosa to ethanol induces histamine release from the intestinal mast cells and a dramatic jejunal protein loss, reflecting microvascular injury, H_1 and H_2 receptor antagonists partly prevents this action. The metabolite of ethanol, acetaldehyde is a major factor in alcohol-induced asthma, causing bronchoconstriction indirectly through endogenously released histamine in asthmatic persons.

ALCOHOLIC BEVERAGES AS A SOURCE OF OESTROGENS

The excessive consumption of alcoholic beverages is associated with numerous serious medical, social, and legal problems. Alcoholic beverages contain numerous substances in addition to alcohol, which determine a beverage's taste, colour, and aroma. Alcoholic beverages are made from many plants and plant by-products that contain phytoestrogens. There are 2 main factors contribute to the feminisation in men with alcoholic cirrhosis: 1- prolonged exposure to the phytoestrogens contained in alcoholic beverage, and 2- the impaired ability of the alcohol-damaged liver to adequately metabolise and excrete many compounds, including phytoestrogens. Phytoestrogens found in milled by-products and oils made from various grains, hops, corn, and rice exhibit biological activity. In men, small amounts of oestrogens are produced in the testes. Nonsteroidal oestrogens, are also known as phytoestrogens, are produced by certain plants. Steroidal oestrogens exhibit greater affinity and specificity for oestrogen receptors (ER) than do nonsteroidal phytoestrogens. Phytoestrogens bound less strongly to the ER and are less able to compete for binding to the receptors. Phytoestrogens interact with ER in the cytosol, as well. Alcoholic beverages may produce oestrogenic effects in moderate drinkers. Oestrogen increases levels of HDL cholesterol.

ALCOHOL AND THE CARDIOVASCULAR SYSTEM

Alcohol's effects on the CVS can take opposite forms, depending on how much is consumed, by whom, when, and how. Alcoholism is associated with cardiomyopathy, arrhythmias, hypertension, and strokes. Moderate drinking can not be achieved by simply averaging the number of drinks consumed. Consuming seven drinks on one night will not have the same effects as consuming one drink each day of the week. Major plasma lipids circulate as macromolecular complexes, lipoproteins. High levels of HDL and low levels of LDL is desirable, HDL <35 mg/dl and LDL, 160 mg/dl. LDL:HDL ratio should be 2.5:1 and 4.5:1. LDL contains large concentrations of cholesterol and phospholipids. Some cholesterol is released, where it plays an important role in maintaining cell membranes and synthesising steroids. NFκB regulates the transcription of VCAM-1, ICAM-1, and ELAM-1. These cellular adhesion molecules recruit WBC. Recruited monocytes adhere to the endothelium, then move into the subendothelial space and differentiate into macrophages that phagocytose oxidised LDL, forming foam cells, subsequently produce cytokines. Alcohol may disrupt the NFκB function. The quantity of antioxidants consumed in wine may not reach sufficiently high plasma levels to prevent the oxidation of LDL. Dealcoholised red wine changes the composition of platelets and decreases the likelihood of thrombosis. HDL inhibits LDL oxidation. Alcohol may block hydroxymethylglutaryl coenzyme A (HMG-CoA) reductase, so suppressing cell proliferation and cholesterol. Cell proliferation results from cellular signalling. Rapture of the plaque into the vessel, allows platelets to become in contact with collagen and other exposed subendothelial compounds become activated and operate in conjunction with other clotting factors to form a thrombosis, and culminating in the formation of thrombin. Thrombin interacts with platelet membrane receptors, resulting in stimulation of phospholipase C. This enzyme mediate platelet aggregation through the formation of IP_3 and diacylglycerol. The former mobilises ionised Ca^{2+} from intracellular stores, and the latter activates protein kinase C. Calcium and PKC stimulate platelets to form TXA_2, which is a powerful stimulator of platelet aggregation and activation. Alcohol's antithrombotic effects may be related to platelet granule secretion and inhibition of TXA_2 production. When alcohol consumption is chronic, platelet function is significantly reduced, and clotting time increases. Even after alcohol intake ceases, these effects persists for several weeks. There is a positive association between alcohol consumption and fibrinolytic activity. There is increased plasma t-PA levels in heavy drinkers. The plasma t-PA levels increase following moderate consumption of alcohol with dinner. An increase in triglyceride level is positively correlated with PA-1 plasma levels, indicating a predisposition to thrombosis and atherogenesis. Elevated triglyceride levels resulting from heavy alcohol consumption may further stimulate PAI-1 genetic expression, resulting in the inhibition of fibrinolysis and thus increasing the risk for acute cardiac events. The biochemical basis of alcohol-induced cardiomyopathy includes: 1- High blood concentrations of alcohol reduce the oxygen supply to the cardiac muscle and interfere with aerobic metabolism in the heart. The result is decreased ATP and phosphocreatine levels. 2- Alcohol alters the permeability of the sarcoplasmic reticulum to Ca^{2+}. 3- Alcohol has a negative effect on the integrity and function of myosin. 4- Alcohol reduces the synthesis of cardiac proteins in both the actin-myosin complex and in the mitochondria, especially in alcoholic with hypertension. 5- Acetaldehyde and free radicals may contribute to decreased protein synthesis as well. 6- Alcohol increases the expression of c-myc, which can promote apoptosis, resulting in muscle cell loss. Holiday heart syndrome is characterised by specific ECG changes that are hallmarks of cardiac conduction abnormalities. Ventricular arrhythmia is associated with sudden death. Alcohol abuse can cause atrial fibrillation; other factors such as hypertension, CAD and valvular diseases may precipitate it as well. Increased thickening and scarring of connective tissue in heart muscle is a source of the disturbance in ventricular rhythm by impeding electrical conduction. Alcohol-induced arrhythmias may be caused by a reduction in the threshold for ventricular fibrillation. Electrolyte disturbances, cardiac muscle hypoxia, and increased levels of catecholamines from cardiac nerves are other mechanisms contributing to the ventricular arrhythmias. Possible mechanisms of alcohol-induced hypertension include: 1- increased activity of sympathetic nervous system, 2- increased plasma levels of catecholamines, 3- decreased sensitivity of the baroreceptors located in arterial walls, and 4- decreased levels of ionised Mg^{2+} in plasma. Alcohol-induced intracerebral haemorrhage is more pronounced in hypertensive than normotensive subjects. Both chronic heavy drinkers and binge drinkers are at an increased risk for subarachnoid haemorrhage. Women appear to be especially sensitive to an increased risk of haemorrhagic stroke.

Effect of alcohol on CVS:

Beneficial	Harmful
1- Reduction of atherosclerosis. 2- Protection against blood clot formation, which protects against heart attack and atherosclerotic ischaemic stroke. 3- Promotion of blood clot dissolution, which protects against heart attack and atherosclerotic ischaemic stroke.	1- Increased risk of cardiomyopathy. 2- Increased risk of arrhythmia. 3- Increased risk of hypertension. 4- Increased risk of haemorrhagic stroke.

ALCOHOL'S IMPACT ON RENAL FUNCTION

A cell's function depends on: 1- receiving a continuous supply of nutrients, 2- eliminating metabolic waste products, and 3- the existence of stable physical, and chemical conditions in the ECF bathing it. Alcohol is one of the numerous factors that can compromise kidney function. It can interfere with kidney function directly, through acute or chronic consumption, or indirectly, as a consequence of liver disease. Alcohol compromising renal function include: 1- the GBM becomes abnormally thickened and characterised by cell proliferation; 2- enlarged and altered cells in the renal tubules; 3- renal hypertrophy; 4- reduced renal function; and 5- nephromegaly. Alcohol disrupts hormonal control that governs renal function. Alcohol produces urine flow within 20 minutes of consumption, resulting in urinary loss of fluid with consequent increase in concentration of serum electrolytes. A raising BAL disrupts and suppresses ADH secretion into the blood. Neither acute nor chronic consumption of alcohol directly causes significant changes in serum Na. Alcohol reduces excretion of K by the kidneys. Low phosphate occurs acutely in hospitalised alcoholic patients. Causes of low phosphate levels include: 1- phosphate deficiency; 2- alkalosis due to prolonged rapid breathing; 3- insulin administration; 4- excessive excretion in urine; and 5- Mg deficiency. Alcohol also may lead to excessive phosphate levels by altering muscle cell integrity and causing the muscle cells to release phosphate. Chronic alcoholism leads to hypomagnesaemia by: 1- increases Mg excretion in the urine; 2- disrupts Mg absorption from the GIT; and 3- alcohol or its metabolites may directly affect renal tubular Mg exchange. Alcohol induces hypocalcaemia. Chronic alcohol consumption may cause fluid and electrolyte retention, thus increases the body fluid volume contributing to hypertension. Alcohol's influence on blood pressure may be attributable to its effects on the production of hormones that act on the kidneys to regulate fluid balance, and also vasoconstricts blood vessels. Most of the metabolic reactions are highly sensitive to the acidity of the surrounding fluid. Alcohol can hamper the regulation of acidity. Alcohol-related acid-base disturbances may be due to: 1- low levels of phosphate; 2- slow respiratory rate and reduction of respiratory centre sensitivity to CO_2; 3- severe elevation of glucose metabolism (lactate); 4- alcoholic ketoacidosis; and 5- liver diseases. Alkalosis may be present in 71% of patients with established liver disease. Excessive alcohol consumption can have profound effects on the kidneys and their function in maintaining the body fluids, electrolyte, and acid-base balance. Malnourished cirrhotic patients tend to have low levels of urea and creatinine. Hepatorenal syndrome can occur without any apparent precipitating factor. A hospital-related event can trigger the syndrome. Hepatorenal syndrome is not related to structural damage and is instead functional in nature. Major clinical features of hepatorenal syndrome include a marked decrease in urine flow, almost no Na excretion and, usually hyponatraemia and ascites. Liver transplantation is one of two options available as a treatment for hepatorenal syndrome. The other option is transjugular intrahepatic portosystemic shunt (TIPS), under local anaesthesia, in which a bypass between 2 veins inside the liver is created by way of the jugular vein.

THE HEPATOPULMONARY SYNDROME

The hepatopulmonary syndrome (HPS) is a disease entity that is seen in association with liver disease and is one of the many extrahepatic manifestations of liver failure, with a reported incidence of 13-47%. The triad of liver disease, hypoxaemia, and intrapulmonary vascular dilations characterises HPS. The degree of hypoxaemia is defined as an arterial oxygen tension of <70 mmHg or an alveolar-arterial oxygen gradient of >20 mmHg. Pulmonary vascular dilations are mandatory for the evolution of this disease process and ultimately result in the development of hypoxaemia. The normal pulmonary capillary diameter is 8-15 μm. Dilations of the capillary and precapillary vessels up to a diameter of 500 μm occurs in HPS. The intrapulmonary vascular changes are located primarily at the base of the lungs and these finding help to explain the clinical features of orthodoexia and platypnea. These vascular alterations result in an intrapulmonary right-to-left shunting of blood and contribute to the resultant hypoxaemia of the syndrome. Portopulmonary venous anastomosis (between oesophageal varices or coronary veins and the pulmonary veins) occurs in HPS. The pathophysiology of HPS has been attributed to intrapulmonary, precapillary, and capillary vascular dilations, and found in close proximity to the alveoli. Because of this abnormal dilation of the pulmonary vessels, blood flowing only adjacent to the alveolus becomes appropriately oxygenated. The result is the inability of oxygen to diffuse to the center of these abnormally dilated vessels and couple to haemoglobin, which results in an apparent right-to-left intrapulmonary shunt. The hyperdynamic circulation and the resultant low pulmonary vascular resistance in this disease state result in the rapid transit of blood through the lungs and potentiate the transit of deoxygenated blood to the systemic circulation. As a result of the vascular dilations being at the base of the lungs, a ventilation-perfusion mismatch is augmented. Large vessels at the base of the lungs, spider angiomata, or direct arteriovenous communications act as true anatomic shunts. The mechanism of hypoxaemia in patients with liver failure is multifactorial. Restrictive pulmonary defects secondary to pleural effusion and ascites may also contribute to hypoxaemia in patients with liver failure. Elevated erythrocytes levels of 2,3-DPG can cause the right shift of oxyhaemoglobin dissociation curve, respiratory and metabolic alkalosis may counteract this effect in the patients with liver failure. Altered production of endothelin may account for the vasoconstriction noted in portopulmonary hypertension, and increased circulating NO or NOS may account for the vasodilation in the HPS. Endotoxin, TNFα, somatostatin analogue (octreotide), glucagon, PGI_2, and angiotensin-2 are potential progenitors for the genesis of the vascular alterations noted in HPS.

EFFECTS OF ALCOHOL USE IN DIABETICS

Alcohol consumption by diabetics can worsen blood sugar. Long-term alcohol use in well-nourished diabetics can result in excessive blood sugar levels. Long-term alcohol ingestion in diabetics who are not adequately nourished can lead to dangerously low blood sugar levels. Heavy drinking, particularly in diabetics, also can cause the accumulation of certain acids in the blood that may result in severe health consequences. Alcohol consumption can worsen diabetes-related medical complications, such as disturbances in fat metabolism, nerve damage, and eye disease. The two most common forms of diabetes are type 1 and type 2 diabetes, with type 2 diabetes accounting for at least 90% of all cases. Type 1 diabetes is an autoimmune disease. In most patients, the disease develops before age 40, primarily during childhood or adolescence. In those patients, the immune system attacks β-cells, which produce insulin. Most importantly, insulin leads to the uptake of the glucose into muscle and fat tissue and prevents glucose release from the liver, thereby lowering blood sugar levels (e.g., after a meal). Insulin deficiency leads to major impairment of the body's regulation of carbohydrate, lipid, and protein metabolism. People with type 2 continue to produce insulin in early disease stages; however, their bodies do not respond adequately to the hormone. Obesity, inactivity, and cigarette smoking may worsen genetically determined insulin resistance. Insulin resistance does not immediately lead to overt diabetes, because the patient's pancreatic β-cells initially can increase their insulin production enough to compensate for the insulin resistance. Ultimately, insulin secretion declines to levels below those seen in nondiabetics. Thus, whereas type 1 diabetes is characterised by a complete lack of insulin production, type 2 is characterised by reduced insulin production plus insulin resistance. In people with either type 1 or type 2 diabetes, single episodes of acute alcohol consumption generally do not lead to clinically significant changes in blood sugar levels. Long-term (i.e., chronic) alcohol consumption in well-nourished diabetics results in increased blood sugar levels (i.e., hyperglycaemia). Haemoglobin A1c (HbA1c), a blood component that reflects blood sugar control over the past 2 to 3 months (the higher the HbA1c levels, the higher were the average blood sugar levels). C-peptide, a molecule that is produced together with insulin, because C-peptide is more stable than insulin, it can be detected longer in the blood and is frequently measured as a reflection of the average insulin secretion. The fasting blood sugar levels of the drinking type 2 diabetics are significantly higher than those of the nondrinking type 2 diabetics. HbA1c levels (and, by extension, average blood sugar levels) are significantly higher in drinking type 2 diabetics than in nondrinking type 2 diabetics. C-peptide levels, and thus insulin production, are significantly lower in both groups of diabetics than in non-diabetics. Acute alcohol administration leads to increased insulin resistance in diabetic, elevated insulin secretion cannot compensate for increased insulin resistance. The same mechanism might occur in chronically drinking diabetics. Chronically drinking diabetics may show worse compliance with their dietary and pharmacological treatment regimens, which also may result in uncontrolled blood sugar levels. Alcohol consumption in the fasting state can induce a profound reduction in blood glucose levels (i.e., hypoglycaemia). Hypoglycaemia can have serious, even life-threatening, consequences, because adequate blood sugar levels are needed to ensure brain functioning necessary to provide energy to the brain: 1- breakdown of glycogen, or glycogenolysis, and 2- production of glucose, or gluconeogenesis. Generally, the glycogen supply is depleted after 1-2 days of fasting. Thus, a person who has been drinking alcohol and not eating for 1 or more days has exhausted his or her glycogen supply. Alcohol metabolism in the liver actually shuts down the process of gluconeogenesis and thus the second line of defense against hypoglycaemia. Alcohol consumption can lead to hypoglycaemic unawareness in both diabetics and nondiabetics. A hypoglycaemic person normally experiences several warning symptoms, such as palpitation or tachycardia. A person in a state of hypoglycaemic unawareness, may not notice or recognise hypoglycaemic warning signs (sweating, weakness, shakiness, nervousness, and palpitation) and is therefore at increased risk of severe hypoglycaemia. Diabetics who have consumed alcohol, particularly those with type 1 diabetes, experience a delayed glucose recovery from hypoglycaemia, suggesting an alcohol-related impairment in the counter-regulatory response to hypoglycaemia. The combination of alcohol-induced hypoglycaemia, hypoglycaemic unawareness, and delayed recovery from hypoglycaemia can lead to deleterious health consequences (incontinence, inability to follow simple commands, perseveration, disorientation, and impairment of recent memory). Ketoacidosis is characterised by excessive levels of certain acids called ketone bodies (e.g., acetone, acetoacetate, and β-hydroxybutyrate) in the blood. Ketoacidosis is caused by complete or near-complete lack of insulin and by excessive glucagon levels. Insulin and glucagon regulate the conversion of fatty acids into larger molecules (i.e., triglycerides), which are stored in the fat tissue. In the absence of insulin, the triglycerides are broken down into free fatty acids, which are secreted into the bloodstream and delivered to the liver. The liver normally re-incorporates free fatty acids into triglycerides, which are then packaged and secreted as part of very low-density lipoproteins (VLDL). Under the influence of excess glucagon, some of the free fatty acids are converted to ketone bodies and secreted into the blood. Ketoacidosis typically occurs in patients with type 1 diabetes who completely lack insulin. In rare cases, however, the condition also may affect people with type 2 diabetes. Heavy alcohol consumption (i.e., 200 g of pure alcohol, or approximately 16 standard drinks, per day) can cause ketoacidosis in both diabetics and nondiabetics. Poor food intake can lead to depleted glycogen levels. Furthermore, continued alcohol metabolism results in diminished gluconeogenesis. Both the depletion of glycogen and diminished gluconeogenesis lead to lower blood sugar levels. As blood sugar falls, insulin secretion is reduced as well. Because insulin restrains glucagon secretion, lower insulin secretion allows increased glucagon secretion, setting the stage for the development of ketoacidosis. Vomiting can lead to dehydration and a reduced blood volume, which, in turn, increases the levels of catecholamines. Catecholamines further decrease insulin production and increase glucagon production. Abnormalities in the levels and metabolism of lipids are extremely common in people with either type 1 or type 2 diabetes and may contribute to those patients' risk of developing cardiovascular disease. Alcohol consumption can exacerbate the diabetes-related lipid abnormalities; heavy drinking can alter lipid levels even in nondiabetics. Alcohol can induce several types of lipid alterations, including elevated triglyceride levels in the blood (i.e., hypertriglyceridaemia), reduced levels of low-density lipoprotein (LDL) cholesterol, and elevated levels of high-density lipoprotein (HDL) cholesterol. Hypertriglyceridaemia is an important risk factor for cardiovascular diseases. Moreover, elevated triglyceride levels can cause severe inflammation of the pancreas (i.e., pancreatitis). Pancreatitis may interfere with the production of insulin, thereby potentially worsening control of blood sugar levels and making hypertriglyceridaemia a particularly serious complication in diabetics. Heavy drinking (i.e., >140 g of pure alcohol, or approximately 12 standard drinks, per day) can cause alcohol-induced hypertriglyceridaemia in both diabetics and nondiabetics. Abstinence from alcohol leads to normalisation of the triglyceride levels, unless the person has an underlying genetic predisposition for hypertriglyceridaemia. Alcohol stimulates the generation of VLDL particles in the liver, which are rich in triglycerides. Alcohol may inhibit VLDL particle breakdown. Alcohol may enhance the increase in triglyceride levels in the blood that usually occurs after a meal. LDL cholesterol is strongly related to cardiovascular disease and stroke and has been called bad cholesterol. LDL cholesterol levels tend to be lower in alcoholics than in nondrinkers, but the LDL cholesterol in alcoholics exhibits altered biological functions and may more readily cause cardiovascular disease. Reduced levels of vitamin E in alcoholics actually may have harmful long-term effects. HDL cholesterol has a protective effect against cardiovascular disease and is called good cholesterol. This protective effect results at least partly from a process called reverse cholesterol transport, in which HDL particles carry cholesterol from blood vessel walls and other sites back to the liver, where it is broken down and subsequently eliminated from the body. Two subtypes of HDL (HDL_2 and HDL_3) are particularly effective in this reverse cholesterol transport. Alcohol increases in HDL_2 and HDL_3 levels. Consumption of more than 3 alcoholic drinks/day results in elevated blood pressure in both men and women compared with nondrinkers. Peripheral neuropathy is a condition in which nerves are damaged that extend from the spinal cord to control muscle function (i.e., motor nerves) or that transmit various sensations-such as touch, pain, temperature, and vibration-back to the spinal cord and brain (i.e., sensory nerves). The symptoms (tingling, burning, pain, and numbness) occur most commonly in the legs and feet and are frequently worse at night. Numbness in particular can be a serious problem. Patients may not sense lesions, such as cuts or ulcers, which may then become seriously infected and result in amputation of the affected limb. Diabetes and alcohol consumption are the 2 most common underlying causes of peripheral neuropathy. The prevalence of symptomatic peripheral neuropathy is greater in men who consume at least 2-4 alcohol-containing beverages every night. Alcohol use also is associated with an inability to sense vibrations from a tuning fork (i.e., absent vibratory perception). Alcohol and diabetes can enhance each other's effects in terms of causing nerve damage. Neuropathy, in addition to other factors (e.g., vascular disease in the penis or altered hormone levels), also may contribute to erectile dysfunction in diabetic men. The nerves that control erection are part of the autonomic nervous system. Many impotent diabetic men also have lower than normal levels of the sex hormone testosterone in their blood. Alcohol reduces blood levels of testosterone and may thereby further exacerbate the existing hormonal deficit. Both neuropathy and vascular disease likely play significant roles in erectile dysfunction in diabetic men. Diabetic eye disease (i.e., retinopathy) is another troublesome tissue complication of diabetes and one of the leading causes of blindness. Heavy alcohol consumption (10 pints of beer/week) may increase a person's risk for developing this disease. Alcohol consumption in men (but not in women) is associated with more severe retinopathy. Alcohol consumption and its associated health consequences may interact with or alter the effects of several medications used to treat diabetes, including chlorpropamide (unpleasant, disulfiram-like reaction), metformin (liver damage), troglitazone (impair liver function).

RESPONSES TO SURGERY

The stress of surgery starts before, continues during and subsides after an operative procedure and is the end result of a variety of stimuli. A rise in the levels of circulating cortisol is one of the earliest consequences of skin incision. ACTH from the hypothalamus is secreted via release of corticotrophin releasing factor hormone (CRH). CRH is secreted in the median eminence and transported in the portal-hypophyseal vessels to the anterior pituitary where it stimulates ACTH release. Vasopressin and vasoactive intestinal peptide (VIP) also increase ACTH release. The nonactivated glucocorticoid complex tightly bound to several heat shock proteins (HSP), which seem to inhibit DNA binding activity. After steroid binding, HSP hetero-complex facilitates the transport of the receptor to the nucleus and dissociates from the receptors, thereby exposing the DNA binding site. Insulin levels in the immediate postoperative period are low, due to suppression of insulin secretion by elevated levels of catecholamines. Later, insulin concentration begins to rise but the effect on peripheral tissue is blunted by the surge of cortisol, glucagon and catecholamines serving to maintain the post-traumatic hyperglycaemia (insulin resistance). The adrenal medulla is encapsulated by the cortex; it offers a unique site for catecholamine-glucocorticoid interaction. Glucocorticoids synthesised in the cortical zona fasciculata reach the medulla by the intra-adrenal portal system and stimulate the synthesis of enzymes responsible for conversion of tyrosine to adrenaline. The catecholamine level peaks at 24-48 hours after injury, following which it tends to decrease towards baseline. Catecholamines exert their diverse effects through transmembrane adrenergic receptors (at least 9 distinct adrenoreceptors). These receptors act via intracellular second messenger, adenylate cyclase or phospholipase C. Catecholamines remain persistently elevated, they suppress the expression of cytokine genes and inhibit cytokine production. The period immediately following surgery is often termed the ebb phase, during which there is a decrease in resting energy expenditure and core temperature where blood glucose may fall below normal level, this might not occur if the operative procedure involves little blood loss. During the flow phase of adrenergic-corticoid phase, there is increase in the levels of circulating catecholamines, glucocorticoids, and glucose, and the metabolic rate and body temperature rise. This slowly merges into anabolic phase if host compensatory responses prevail over the acute insult. The transition phase between catabolism and anabolism is characterised by a reduction in nitrogen excretion and spontaneous diuresis and often is termed the corticoid withdrawal phase. During the late anabolic phase there is progressive reaccumulation of protein followed by deposition of fat in the adipose tissue. Surgically traumatised patients are hypermetabolic with a slight increase in resting energy expenditure (REE) early after surgery. The REE is the balance between increases caused by the catabolic process and decreases due to tissue depletion caused by malnutrition and cachexia. The total energy requirement may not increase and may in fact below normal levels during the first postoperative week. Systemic infection and fever may also increase the REE by 10-25%; fever per se increases the metabolic rate by 7% for each half-degree centigrade above normal. Patients with peritonitis and empyema exhibit 30-50% increase in REE. TNF and prostaglandin may contribute to increased metabolism. Cytokines liberated by inflammatory cells are partly responsible for the hyperglycaemia of injury. TNFα stimulates hepatic amino acid transport and IL1β augments hepatic gluconeogenesis via greater alanine utilisation. Fibroblast, macrophages and leucocytes increase demand for uptake of glucose. In these metabolically-active tissues, the anaerobic conversion of glucose to lactate provides the main energy; lactate is then recycled in the liver by the Cori cycle. The resting skeletal muscle takes up only a fraction of glucose in this state despite persistent hyperinsulinaemia (insulin resistance). Such a resistance to glucose uptake further increases blood glucose, which stimulate the release of insulin. Insulin resistance is due to elevated levels of catecholamine, glucocorticoids and glucagon. Catecholamines stimulate glucagon secretion and glycogenolysis. By stimulating lipolysis, catecholamines increase circulating fatty acids, which subsequently decrease glucose uptake. Glucocorticoids increase gluconeogenesis by releasing substrates like amino acids and lactate and by stimulating hepatic gluconeogenic enzymes, and decreasing the number and affinity of insulin receptors on cell surface. Glucocorticoids appear to inhibit the ability of insulin to mediate the recruitment of glucose transport protein from the cell interior to the cell surface. Following injury the daily urinary loss of nitrogen increases 2-3 times signifying net proteolysis. TNFα and IL1β augment protein loss. Changes in cell hydration and volume act as potent signals that modify proteolysis include liver cell swelling, inhibit proteolysis and simultaneously stimulate protein synthesis. Glutamine and alanine constitute over 50% of circulating amino acid released after injury, whereas each compromises only 6% of muscle protein. As regards the intracellular free amino acids in muscle, glutamine has the highest concentration. Under the influence of the stress hormones and IL1β this intramuscular glutamine store soon becomes depleted. During sepsis, however, cytokines diminish gut uptake of circulating glutamine and also inhibit mucosal glutaminase activity. Lack of glutamine reduces crypt cell numbers thereby increasing transcellular permeability, predispose to bacterial translocation from the intestinal lumen, which may initiate gram-negative bacteraemia, with the potential risk of developing multiple system organ dysfunction. Glutamine is not added to parenteral preparations because it is unstable in solution, especially during sterilisation and rapidly degrades to ammonia. Synthesis of binding molecules like albumin and transferrin decrease during stress, whereas proteins, which combat infection and injury, are produced in plenty. Acute phase proteins are glycoprotein, mediate a variety of functions. C-reactive proteins activate: 1- the complement cascade; ceruloplasmin, which is a carrier of protein and is essential for enzymes involved with free radicals and hormone synthesis; 2- the protease inhibitor α1-antitrypsin, which limits the tissue damage of proteases in local inflammation; 3- fibrinogen and other clotting factors, which enhance thrombus formation; and 4- fibronectin and α-2 macroglobulin, which modify fibroblast function and thereby influence wound healing. IL6 is the primary mediator of altered hepatic protein synthesis and the final common mediator of the cytokine cascade initiated by endotoxin, TNFα and IL1β. Glucocorticoids augment cytokine effects on acute phase protein synthesis. Fat metabolism, is the preferred fuel, since continued proteolysis would deplete muscle protein, the only reserve of protein. The hormone sensitive adipocyte lipase increases the rate of free fatty acids (FFA) and glycerol turnover. The FFA are oxidised by cardiac and skeletal muscles to produce energy while glycerol contributes to modest gluconeogenesis lipoprotein lipase (LPL) is a lipolytic enzyme, which is bound to the capillary endothelium and is capable of hydrolysing triglycerides bound to very low density lipoproteins and chylomicrons in serum. TNFα, IL1 and IL6 inhibit LPL activity, thereby decreasing total body fatty stores, this increases triglyceride-rich circulating lipoproteins. Triglyceride-rich lipoproteins are part of the host defence mechanism against endotoxaemia. Despite the introduction of new generations of antibiotics the overall incidence of sepsis after elective surgery remains in the region of 5-10%. Immune depression may be due in part to blood loss and cellular hypoxia. Blood transfusion can also contribute to this immune depression. The amount of blood transfused determines the increase in the risk independently of the severity of underlying disease. Defects in neutrophil chemotaxis, phagocytosis, lysosomal enzyme contents and respiratory burst; reduction of the ratio of T helper to T suppressor cells; the lowering of the number of circulating natural killer (NK) cells are most pronounced within 6-48 hr of operation. The initial event in the generation of the immune response is uptake and degradation of foreign antigens by monocyte/macrophage. The antigens are then presented by the macrophages to the T helper cells. The T helper cells will only recognise the antigen when they are presented on macrophage cell surface in close proximity to major histocompatiblity (MHC) class II antigens (HLA-DR in humans). In appropriate circumstances, T helper cells are activated and a number of cytokines are released. Th1 cells, which release IL2 and IFNγ modulating the cell mediated effector response and the Th2 subclass, which liberate IL4, 5, 6 and 10 thereby influencing B cells development and augmenting immunoglobulin production. Laparotomy reduces the percentage of circulating HLA-DR positive monocytes, which is due to an actual down-regulation of HLA-DR at the cellular level rather than the percentage of HLA-DR positive monocytes being reduced by an influx of immature monocytes. Under the influence of injury, haemorrhage and endotoxin, the monocytes liberate a number of cytokines, amongst which PGE_2 and TGFβ. Exposure to endotoxin enhances the expression of TGFβ gene in the macrophages. PGE_2 has a differential effect on the T helper cells; it inhibits the Th1 cells and stimulates Th2 subset. The Th2 cells are also stimulated by TGFβ. As a result, an accelerated release of IL10 takes place, which is known to down-regulate the expression of vital HLA-RD antigen. IL4 is produced by activated Th2 cells, also upregulated the release of IL10. IL4 and IL10 are also capable of suppressing the Th1 cells. TNFα augments PGE_2 release, while high levels of IL6 prolongs the release of TGFβ. The down-regulation of the macrophages after a traumatic insult may be beneficial in the early phase. Exaggerated neutrophil activation is an early event in those patients destined to develop postsurgical sepsis. Postoperative neutrophil dysfunction may in reality be a physiologic immune modulation in as much as it does not contribute any further to potentially destructive effects of endotoxin and microbes. The host response to injury or the sepsis phenomena is mediated by OFR. The genetic constitution of a patient may influence the immunological perturbations following surgery. The secretion of IL1 and TNFα is genetically linked with the MHC locus. Patients express high CD16 have an increased tendency to develop postoperative sepsis. The number of circulatory lymphocytes immediately is decreased after surgery because of Fas-mediated apoptosis. The change in the number of peripheral lymphocytes correlates with the degree of surgical damage. IL-6 delayed neutrophil apoptosis in a large population of surviving neutrophils has a greater collective capacity for superoxide production. The number of neutrophils immediately increases up to approximately 160-220% of the original on day 1 after all the surgical procedures performed, without a notable difference between major and minor surgery. A delay in neutrophil apoptosis might cause acute respiratory dysfunction syndrome (ARDS) after these surgical procedures. Traumatic wounds; bee stings; and bacterial, viral and parasitic infections all produce early phase reactants that lead to the efficient recruitment of targeted inflammatory cells to the efficient site, with the production of cytokines and growth factors that precisely control the immune system response. These secreted protein mediators include cytokines (IL-1-16), growth factors such as TGF-β, and chemokines such as macrophage inflammatory protein-1 α and β, and RANTES. Post-surgery period is characterised by large fluid shifts; temperature changes, coagulation disturbances, and increased concentrations of catecholamines and stress hormones. Inflammatory response to reperfusion may be largely responsible for systemic vasodilatation during rewarming, fever, postoperative bleeding, and reperfusion injury to the heart, brain, kidney, and gut, leading to either permanent organ injury or transient dysfunction. Neutrophil-endothelial cell adhesion is central to postischaemic reperfusion. Initially, oxygen free radicals (OFR) are released from endothelial cells in reperfused organs, such as the heart and lung, leading to alterations in endothelial cells that promote early neutrophil targeting, activation of neutrophils in transit through these organs, local activation of serum completely, and induction of cytokine synthesis. Initial neutrophil attachment to the endothelium results in translocation of P-selectin from intracellular vesicles to the cell membrane and initiation of platelet activating factors (PAF) synthesis. These changes lead to more neutrophil adhesion, neutrophil activation, and neutrophil adhesion protein (CD11/CD18) integrin expression via endothelial membrane-bound PAF. Protein inflammatory cytokines such as IL-1β and TNFα, and cytokines such as IL-8, are newly synthesised, largely within endothelial cells, in reperfused organs within hours, and induce increased expression of adhesion molecules on endothelial cells and other tissue-specific cells (myocytes, glomerular or renal tubular epithelium, pulmonary alveolar lining cells). In turn this result in tighter neutrophil attachment via intercellular adhesion molecule-1 (ICAM-1), transmigration of neutrophils into the interstitial space, and release of large amounts of free radicals. Nitric oxide is produced by endothelial cell NOS may control many of these processes by inducing vasodilation, regulating leukocyte recruitment, scavenging endothelial-generated oxygen radicals, and preventing upregulation of neutrophil CD11/CD18. NO production is impaired after ischaemia/reperfusion, and inhibitory controls over leukocyte migration, oxygen radicals, and neutrophil adherence are mitigated. Postoperative blood loss may result from multiple factors: 1- direct and indirect effect on platelet function; 2- a reduction in capillary leak typically induced by proinflammatory cytokines and complements; and 3- inhibition in cytokine-induced coagulopathy and fibrinolysis.

HOST RESPONSE TO INJURY

The host response to injury involves a complex set of interactions of different parameters: 1- cytokines, 2- chemokines, 3- hormones, 4- neurotransmitters, 5- cell types, and 6- organs. Cytokines are indispensable for inflammation, the immune response, activation of endothelial cells, and coordination of the steps leading to wound healing and tissue repair. Severe injury complicated by infection or secondary injuries can lead to "systemic inflammatory response syndrome" (SIRS), which is diffuse "whole body" inflammation often associated with multiple organ system failure (MOSF). TNF and IL-1 appear to be two key cytokines that mediate SIRS, together with IL-6, and IL-8. TNF has both beneficial and deleterious effects, depending upon the amount produced and the milieu, which it is acting. This proinflammatory event is counterbalanced by other cytokines (IL-4, IL-10, TGF-β), the production of which is stimulated by TNF. The systemic levels of TNF and IL-1 are only transiently elevated, there are circulating inhibitors such as soluble TNF receptors, IL-receptor antagonists. Once SIRS and secondary tissue damage due to activated neutrophils are established, it may be too late to see a beneficial effect. Polymorphonuclear leukocytes (PMNs) accumulate early at sites of injury, release mediators such as platelet-activating factor (PAF), oxygen free radicals (OFR), and arachidonic acid (AA) metabolites. Activation of PMN function is associated with adult respiratory distress syndrome (ARDS). Colony-stimulating factors (CSF), are group of glycoproteins, with the following biological properties: 1- stimulating the proliferation and differentiation of cells; 2- acts on mature neutrophils to enhance their response to physiologic stimuli; 3- enhances binding of chemotactic peptide on peripheral blood neutrophils; 4- increases antibody-dependent cellular cytotoxicity; 5- increases phagocytosis by neutrophils; 6- increases superoxide anion generation; and 7- may affect the production of other cytokines. Platelet activating factor (PAF), is a phospholipid produced by a variety of cell types, such as blood cells, endothelial cells. PAF is a potent inflammatory mediator, associated with hypertension, increased pulmonary vascular resistance, bronchoconstriction, cardiac dysfunction, and increased microvascular permeability. PAF levels increase following trauma and ischaemia. PAF results in elevated levels of TNF, IL-1, and IL-2. The complement system is made up of proteins that mediate inflammation and promote microorganism ingestion by phagocytes, that can be initiated by antibody-antigen binding (classic pathway) or directly by a variety of substances (alternative pathway), in a variety of pathophysiologic states, including multisystem trauma. Complement activation results in neutrophil upregulation of CD11/CD18 associated with increase in vascular permeability. Complement activated neutrophils may then release oxygen-derived free radicals. Nitric oxide (NO), is vasodilator that inhibit platelet aggregation, suppress PMN adhesion and activation. Suppression of NO release during trauma may be an adaptive mechanism to maintain the blood pressure. The synthesis of NO may be interrupted by OFRs and cytokines. Oxygen free radicals (OFRs) are formed from a variety of sources, including catecholamine oxidation, AA metabolism, neutrophil NADPH oxidase, and the xanthine-xanthine oxidase system. There is an acute rise of OFRs formation with trauma. Transforming growth factor-beta (TGF-β), is found in a variety of cell types, including mononuclear and polymorphonuclear leukocytes and endothelial cells. Elevated level of TGF-β in trauma is related to increased septic events, and mediate post-trauma immunosuppression. Ischaemia is the result of three well-defined mechanisms: 1- changes in calcium homeostasis, 2- free radical reactions, and 3- inflammatory reactions. Mechanisms of cell death after temporary ischaemia include: 1- ATP decreases to near zero, formation of AMP; 2- shift of sodium ions (Na^+), water (H_2O) and Ca^{2+} from the extracellular into the intracellular space (cytosolic oedema); 3- potassium (K^+) leakage from the intracellular into extracellular space; 4- breaks down membrane phospholipids (PL) into free fatty acids (FAA), particularly AA; 5- conversion of xanthine dehydrogenase to xanthine oxidase (XO), priming the cell for the production of the OFR; and 6- hypoxia results in anaerobic metabolism and lactic acidosis. Normally, intracellular Ca^{2+} to extracellular Ca^{2+} gradient is 1:10,000, i.e., 0.1 μmol:1 mmol. Calcium regulators include calcium/magnesium-ATPase, the endoplasmic reticulum (ER), mitochondria (M), and AA. With stimulation, different cells respond with an increase in Ca^{2+}, either by release of bound Ca^{2+} in the ER or influx of Ca^{2+} or both. During complete ischaemic anoxia, the level of energy (phosphocreatine, Pcr and adenosine triphosphate, ATP) decreases to near zero in all tissues at different rates, depending on stores of oxygen and substrate, it is fastest in the brain ($\approx$5 min), and slower in the heart and other vital organs. This energy loss causes membrane pump failure, which causes a shift of Na^+, H_2O and Ca^{2+} from the extracellular into the intracellular space (cytosolic oedema), and K^+ leakage from the intracellular into extracellular space. Increase in Ca^{2+} activates phospholipase A_2, which breaks down membrane phospholipids (PL) into FAA, particularly AA. Increase in Ca^{2+} also activates proteolytic enzymes, such as calpain, which may disrupt the cytoskeleton (CS), and possibly the nucleus. In mitochondria, hydrolysis of ATP to adenosine monophosphate (AMP) leads to an accumulation of hypoxanthine (HX). Increased Ca^{2+} may enhance conversion of xanthine dehydrogenase (XD) to xanthine oxidase (XO), priming the cell (e.g., neurone) for the production of OFR (O_2^-). Excitatory amino acid neurotransmitters (EAA), particularly glutamate and aspartate, increase in the extracellular fluid (ECF), which activates n-methyl-d-aspartate (NMDA) and non-NMDA receptors, thereby increasing calcium and Na^+ influx and mobilising stores of Ca^{2+}. Increase extracellular K^+ activates EAA receptors by membrane depolarisation. Glycolysis during hypoxia results in anaerobic metabolism and lactic acidosis. Until all glucose is used (20 min in the brain), this lactic acidosis plus inability to wash out CO_2, results in a mixed tissue acidosis which adversely influences neuronal viability. Without reoxygenation, cells progress via first reversible, later irreversible structural damage, to necrosis of all cells, homogeneously, at specific rates for different cell types. During reperfusion and reoxygenation, lactate and molecular breakdown products can create osmotic oedema and rupture of organelles and mitochondria. Recovery of ATP and Pcr and of the ionic membrane pump may be hampered by hypoperfusion as a result of vasospasm, cell slugging, adhesion of neutrophils (granulocytes), and capillary compression by swollen cells, which also help to protect neurons by absorbing extracellular potassium. Capillary leakage results in interstitial (vasogenic) oedema. Increased concentrations of at least 4 free radicals species that breakdown membranes and collagen, worsen the microcirculation, and possibly also damage the nucleus may be formed: superoxide (O_2^-) leading to hydroxyl radical (.OH) via the iron-catalysed Fe^{3+} to Fe^{2+}, Haber-Weiss/Fenton reaction, free lipid radicals (FLR) and peroxynitrate (ONOO$^-$). Superoxide may be formed from several sources: 1- directly from AA metabolism by cyclooxygenase, 2- by the XO system; 3- via quinone-mediated reactions within and outside the electron transport chain (from Mt); and 4- by activation of NADPH-oxidase in accumulated neutrophils in the microvasculature or after diapedesis into tissue. Increased superoxide leads to increased hydrogen peroxide (H_2O_2) production as a result of intracellular action of superoxide dismutase (SOD). H_2O_2 is controlled by intracellular catalase. Increased superoxide further leads to increased .OH, because of conversion of H_2O_2 to .OH via the Haber-Weiss/Fenton reaction, with iron liberated from mitochondria. This reaction is prompted by acidosis. The hydroxyl radical and peroxynitrite damage cellular lipids, proteins, and nucleic acids. AA increases activity of: 1- cyclooxygenase pathways to produce prostaglandins (PGs), including thromboxane A_2, 2- the lipoxygenase pathway to produce leukotrienes, and 3- the cytochrome P450 pathway. These products act as neurotransmitters and signal transducers in neurone and glia and can activate thrombotic and inflammatory pathways in the microcirculation. Neuronal injury can signal IL-1 and other cytokines to be produced and trigger endogenous activation of microglia, with additional injury. In addition tissue and/or endothelial injury, especially associated with necrosis, can signal the endothelium to produce adhesion molecules (ICAM, E-selectin, P-selectin), cytokines, chemokines, and other mediators, triggering local involvement of systemic inflammatory cells in an interaction between blood and damaged tissue. Reoxygenation restores ATP through oxidative phosphorylation, which may result in massive uptake of Ca^{2+} into mitochondria, which are swollen from increased osmolality. Mitochondria loaded with bounded Ca^{2+} may self-destruct by rupturing and releasing free radicals. Increased Ca^{2+} by itself and by triggering free radical reactions may result in lipid peroxidation, leaky membranes, and cell death. During ischaemia and subsequent reperfusion, Ca^{2+} loading signals a wide variety of pathologic processes. Proteases, lipases, and nucleases are activated, which may contribute to activation of genes or gene products (ICE or p53) critical to the development of apoptosis, or inactivation of genes or gene products normally inhibiting this process. Activation of iNOS by Ca^{2+} lead to production of NO, which can combine with superoxide to generate peroxynitrite (ONOO$^-$). Peroxynitrite and radical hydroxyl both can lead to DNA injury and apoptosis, or protein and membrane peroxidation and necrosis. The following works to lessen the damage: 1- nerve growth factor (NGF), 2- nuclear immediate early response genes (IERG) such as heat shock protein, free radical scavengers (FRSs), and 3- adenosine, and other endogenous defenses (ED). Longer ischaemia may trigger apoptosis, and very long arrests and reperfusion are followed by membrane damage, in part due to radical reactions. During reperfusion and reoxygenation there is: 1- osmotic oedema and rupture of organelles and mitochondria, 2- capillary leakage results in interstitial oedema, and 3- superoxide releases. These products can activate thrombotic and inflammatory pathways in the microcirculation. In addition tissue and/or endothelial injury, especially associated with necrosis, can signal the endothelium to produce cytokines, chemokines, and other mediators, triggering local involvement of systemic inflammatory cells. Reoxygenation restores ATP through oxidative phosphorylation that may result in massive uptake of Ca^{2+} into mitochondria, which are swollen from increased osmolality, may self-destruct by rupturing and releasing free radicals. Activation of nitric oxide synthase by Ca^{2+} lead to production of NO, which can combine with superoxide to generate peroxynitrite (ONOO$^-$). OFR induces the following: 1- DNA injury and apoptosis; and 2- lipid peroxidation, leaking membrane and cell death; and 3- damage cellular lipids, proteins and nucleic acid. The reaction of nitric oxide (NO) with superoxide, producing peroxynitrite. Peroxynitrite has a pKa of 6.8. Under physiologic conditions, protonation will form peroxynitrous acid, which will rearrange and form nitrate and a proton. Nitrate is unreactive. Thus one possible interaction of NO with superoxide is that scavenger, to form something nontoxic (nitrate). Peroxynitrite and peroxynitrous acid, are highly reactive. Peroxynitrous acid reacts like hydroxyl radical and in the process liberates nitrogen dioxide, both are extremely reactive. NO can either be damaging or protective in oxidative injury. NO in the vascular system is normally produced by endothelial cell. It is a small molecule and is very diffusible. It is not a gas but a dissolved nonelectrolyte. NO induces vascular smooth muscle relaxation. In ischaemia-reperfusion NO is a potent inhibitor of platelet aggregation and neutrophil adherence, and it is highly antithrombotic. Superoxide production, decrease NO and therefore potentially increase vascular contraction, remove a mechanism of the inhibition of platelet aggregation and neutrophil adherence, and therefore promote vascular coagulation. Ischaemia-reperfusion and other types of injury will actually upregulate NO formation. NO can be formed in haemorrhagic shock. Nitric oxide is protective in the intestine and vascular integrity in oxidative injury in the kidney and liver. NO can also be damaging, contributing to oxidative injury in ischaemic perfusion or to endogenous NO formation in the heart. Inhibition of NO has a protective effect or opposite effect. Nitric oxide renders the cell more sensitive to oxidative injury. The half-life of NO is 4 s, it diffuses a distance of 160 μ during that time, and it reacts with oxygen. Oxygen free radicals are not one specific compound; they are a group of substances that are formed during ischaemia, when oxygen becomes unavailable as the terminal electron acceptor in the electron transport chain. A free radical is any molecule that has an unpaired electron in its outer shell. In biological systems, there are a number of free radicals produced, such as superoxide and hydroxyl radicals. Hydrogen peroxide, although itself not an oxygen radicals, participates in a series of reactions with free radical. Nitric oxide is a free radical. Free radicals may also serve as sources of injury to the heart and brain. Free radical reactions plays a role in cardiac dysfunction after reperfusion of the temporarily ischaemic liver. The liver itself has a great deal of xanthine dehydrogenase, which during ischaemia can be converted to xanthine oxidase. The latter could go into the systemic circulation, generate free radicals, and cause dysfunction and injury at distant sites. The intestine, which is extremely rich in xanthine dehydrogenase, may also be important in cytokine activation. The mucosal breakdown that may take place in the postischaemic intestines may lead to further loss of the intestines' integrity. There may be protein leakage into the gastrointestinal tract. The barrier functions of the intestines against the intestinal flora may be decreased leading to sepsis and multiple organ failure. Increased in intracellular calcium lead to increase in the activity of serine proteases, which is at least one of the mechanisms for conversion of xanthine dehydrogenase to xanthine oxidase. The initial production of superoxide during ischaemia is due to xanthine oxidase. There may be contribution by neutrophils, but xanthine oxidase is more important. Decreasing superoxide upon reperfusion can be achieved by: 1- vitamin C, which can be reductant or oxidant, i.e., anti- or pro-oxidant; 2- allopurinol, can inhibit xanthine oxidase; 3- superoxide dismutase, is equally effective as allopurinol; and 4- vitamin E and ubiquinol (reduced CoQ_{10}) seem to play an important role in increasing the antioxidant capacity, in facing not only acute ischaemia/reperfusion-induced challenge, but also some chronic effects, such as low-density lipoprotein oxidation and ensuing cardiovascular diseases. Antioxidants, used before or during reperfusion, have been shown to efficiently prevent reoxygenation injury.

MITOCHONDRIA

The whole point of life is to keep ATP relatively constant. ATP function must be balanced with pathways, which re-synthesise ATP during large changes in metabolic rate. The primary signals for activating metabolic pathways are end products of ATP utilisation. Cytochrome oxidase is almost always saturated with oxygen. Mitochondria have drawn particular attention because of their involvement in producing OFRs via the oxidoreductive cycling in the electron transport chain, in sequestering calcium, and in generating ATP. Under physiological conditions, the production of the reactive oxygen species (ROS) by mitochondria is relatively small (2 nmol H_2O_2/min/mg protein in the heart), mitochondrial enzymatic and nonenzymatic antioxidant defence can provide adequate protection against potential oxidative damage. Ischaemia and I-R cause an imbalance of energy supply, indicated by a degradation of adenine nucleotides, accumulation of purine byproducts, a shift from aerobic to anaerobic metabolism. The irreversible damage of the mitochondria impair the oxidative phosphorylation, exhibit significant depression of ADP-stimulated state 3 respiration and RCI following hypoxia-reoxygenation.

Oxidant production:

Normal conditions	Pathological conditions
1- Leak from mitochondrial respiratory chain.	1- Increased leak from respiratory chain.
2- Neutrophils.	2- Pathologically active neutrophils.
3- Catecholamine auto-oxidation.	3- Increased catecholamine auto-oxidation during stress response or inotrope therapy.
4- Background radiation.	4- Exposure to high levels of radiation.
	5- Transition metal ion-catalysed reactions (iron, copper).
	6- Xanthine oxidase.
	7- Prostaglandin metabolism.
	8- Drugs (bleomycin, xenobiotics).
	9- NO interaction with superoxide.
	10- Inhaled oxidants, such as ozone, exhaust fumes, high-inspired oxygen concentrations.
	11- Calcium mediated.

Electron flow from NADH to iron-sulphur protein (complex I) and/or to cytochrome c (complex III) might be blocked as a result of ischaemia. Upon reperfusion, electrons instead of being transferred to oxygen through cytochrome c oxidase, can leak out of the respiratory chain reducing oxygen to O_2^- and subsequently H_2O_2 and .OH. The produced ROS might cause further damage to the mitochondria and enhance electron leakage, thus forming a vicious cycle. Increased levels of ADP and inorganic phosphate in the post-ischaemia are well known stimulants of mitochondrial respiration. Ischaemia-reperfusion may be attributed to the following factors: 1- Several well-identified pathways of free radical generation, e.g., xanthine oxidase, polymorphonuclear neutrophil and catecholamine, are located in the cytosol or endothelial cells of the capillary, thus having only limited access to the mitochondrial respiratory chain. Their potential to cause damage may be reduced by the cytosolic antioxidants, such as Cu-Zn SOD, GPX, glutathione and ascorbate. 2- Mitochondria contain high levels of enzymatic and nonenzymatic antioxidants, such as Mn SOD, GPX, glutathione and α-tocopherol that can readily eliminate ROS produced within the respiratory chain. Their effectiveness in protecting against oxygen species may compromise. 3- Ischaemia-reperfusion, in which blood-free perfusate is used, may be protected by the blood-borne free radical scavengers. Leakage of proton occurs at the mitochondrial inner membrane. It is known that in state 4 (ADP-independent) respiration moderately damaged mitochondria will consume more oxygen than functionally intact mitochondria. Mitochondrial susceptibility to free radical damage is dependent upon respiratory substrate. Succinate dehydrogenase located on the outer surface of the mitochondrial inner membrane may make the mitochondria accessible to exogenous free radical inhibition. Mitochondrial NADH oxidase (complex I) and succinoxidase (complex II) are inhibited by oxygen radical generating system. The electron transport chain and the ATP synthase complex are damaged by OFR. State 4 and state 3 respirations with succinate are inhibited by the free radicals. Mitochondrial oxygen consumption in the presence of succinate and ADP (state 3) is a measure of both electron transport and the ATP synthase enzyme complex. Mitochondria respiring on succinate alone indicating the function of electron transport per se. The most likely sites of damage by O_2^- are the cross membrane translocates for various substrates and high-energy phosphates, thereby limiting the availability of reducing equivalents and phosphate acceptors for oxidative phosphorylation.

PROSTAGLANDIN E_2 IN TRAUMA

The lung participates in multiple metabolic and endocrine activities. The lung is a target for damage and failure after injury. TNFα, IL8 and NO produced locally in the lung or transiting systemically from the liver to the lung cause ARDS and lung function impairment. The associated presence of direct pulmonary injury may exacerbate this process. Injury and infection compromise respiratory function, disrupt the crucial metabolic activities of the lung such as PGE_2 degradation. So, PGE_2 escapes and become available peripherally to cause pain, hypertension, and immune dysfunction. Type II phospholipase A_2 (PLA_2) is active in lung epithelium, alveolar macrophages, and Kupffer cells. PGE_2 produces multiple effects in different cell types and in the same cell through interactions with one of the 4 classes of PGE_2 receptor subtypes. Macrophages stimulated by lipopolysaccharide (LPS) up-regulate the expression of the PGE_2 receptors, EP2 and EP4, which induce an increase in cAMP. cAMP mediates the decrease in TNFα production by the macrophage. The increased levels of PGE_2 correlates with augmented COX-2 expression. PGE_2 is significantly elevated in the plasma of injured patients. Bacterial LPS triggers the production of TNFα, IL1, and IL6 during sepsis. LPS induces COX-2 expression and increases PGE_2 production by monocytes and macrophages. cAMP regulates COX-2 gene activation. NFκB levels are important for PGE_2 production by macrophages. IL-1, IL-6, and TNF-α increase PGE_2 formation by mesangial cells, macrophages, and fibroblasts. These cytokines induce COX-2 activity and COX-2 gene expression. IL-1 and TNF-α increase the expression of PLA_2. PAF is a potent AA derivative secreted in response to stimulation by LPS. PAF is linked to acute lung injury, abnormal liver glycogenolysis, and increased portal pressure and the stimulation of neutrophils and macrophages. PAF stimulates the production of PGE_2 and the increased expression of COX-2 in macrophages and neutrophils. The anti-inflammatory cytokines IL-10 inhibits COX-2 expression in macrophages and neutrophils. PGE_2 contributes to the demise of an injured patient, such as massive vasodilation, pain, and immune system dysfunction. PGE_2 is a potent immunosuppressive agent that modulates the activity of T cells, macrophages, and granulocytes. T-lymphocyte function is depressed as a result of burn injury. PGE_2 alters T-cell function by depressing IL-2 release. Stimulated macrophages that produce PGE_2, are autocrine targets of PGE_2 action. PGE_2 contributes to the neutropenia in burn patient by depressing granulocyte production in the bone marrow. PGE_2 reduces G-CSFmRNA expression, which is mediated by increases in cAMP. Indomethacin or ibuprofen improves haemodynamic parameters, restores T-cell function and activation, restore GM-CFU growth. Ibuprofen (10 mg/kg) improves cardiac output and mean arterial pressure and reduced plasma TXB_2 levels. Ibuprofen reduces illness without decreasing plasma TNFα or IL-6 levels. Ibuprofen reduces tachycardia, oxygen consumption, and lactic acidosis. Indomethacin and ibuprofen are nonspecific inhibitors of COX and reduce the activities of both COX-1 and COX-2. Selective COX blockade may be better option. COX-2 inhibitor improves neutrophil counts and reduces mortality related to burn injury with infection. The lung has the capacity to remove 90-98% of circulating PGs on first pass through the pulmonary circuit. PGE_2 is degraded to inactive metabolites. The initial step is the oxidation of the 15-(S)-hydroxyl group by the enzyme prostaglandin 15-hydroxyprostaglandin dehydrogenase (PGDH), which renders PGE_2 inactive. The 2 enzymes of PGDH are oxidised by NAD or NADP. PGDH metabolises PGs of the E, F, and A series; and prostacyclin. PGDH is highly concentrated in the lung, kidney, and uterus. PGDH is localised to the epithelial cells in the lung including the epithelial cells of respiratory bronchioles and type II alveolar cells. Pulmonary endothelial cells produce PGE_2, but do not contain PGDH and do not metabolise PGs. The metabolism of PGs in the stomach, skin, kidney, and colon is regulated by PGDH. PGE_2 is protective to the stomach; the integrity of the gastric mucosa depends on the balance between PG synthesis and degradation. An imbalance in the system results in gastric erosion and ulcer formation. The PGDH is localised in the cortex, and in the proximal and distal tubules of the kidney. PGDH is small labile protein (M, 28000) with a half-life of an hour. PGDH is elevated in the uterus and ovaries during pregnancy. PGDH activity in the lung is reduced by hyperoxia. Transpulmonary metabolism of PGs decreases during the early stages of oxygen exposure. Oxygen inactivates PGDH. Antioxidants, vitamin E (600 IU/kg) or 100 ppb selenium reverse the hypoxia-induced inhibition of PGDH activity reversed by hypoxia-induced inhibition of PGDH activity in lung tissue. PGDH is vulnerable to damage by free radicals. High oxygen concentration ventilation may be detrimental to PG metabolism in the lung by depressing PGDH activity. Linoleic acid and AA inhibit PGDH. PAF present in human milk and placenta inhibits PGDH activity. PAF is an AA metabolite that is linked to hypotension and multiple organ failure during sepsis, reduces PGDH in lung epithelial cells. Burn injury with wound infection depresses PGDH expression in the presence of increased COX-2. Specific COX-2 inhibitors may be contraindicated in pregnant women or children. AA metabolites synthesised via COX-2, primarily PGD_2 and PGJ_2 may have anti-inflammatory properties. Selective COX-2 inhibition may exacerbate lung inflammation. Pharmacologic enhancement of PGDH-induced PGE_2 degradation would reduce PGE_2 without reducing beneficial prostanoids such as PGD_2.

CONSEQUENCES OF TRAUMA

Organ failure following trauma was first recognised in the late 1960s. Critically injured patients are at high risk for developing complications including SIRS and MODS (multiple organ dysfunction syndrome) that often result in death or profound disability. Acute inflammation follows all types of tissue trauma. The inflammatory process lessen haemorrhage, removes debris from injured areas, limits invasion of infecting organisms, and promotes healing. If the inflammatory response is not contained, widespread systemic inflammation occurs in tissue and organs remote from the initial injury. The acute inflammatory response is generally a self-limiting (8-10 days), nonspecific physiologic response that begins almost immediately after the initial insult and prepares tissue for healing including, a vascular response, phagocytosis, release of biochemical mediators, activation of plasma protein systems, and the immune response. Initially, arterioles near the site of injury constrict briefly, then dilate, secondary to serotonin release, to increase blood flow and nutrient delivery to the injured area. Concurrently, capillary and venule endothelial cells retract, creating spaces at junctions between cells. Retraction allows leukocytes to squeeze out of the cell (diapedesis) and phagocytize invading organisms and cellular debris. The net effect of this process is increased capillary permeability and movement of blood cells and plasma proteins into injured tissue, surrounding tissue, causing interstitial oedema and stasis in the microcirculation. This localised inflammatory response results in the classical redness, swelling, heat, pain and loss of function (manifestations of inflammation). Normally antiproteases, circulating albumin, vitamin C and E, RBCs, and phagocytic cells limit cellular damage from the inflammatory process by scavenging toxic oxygen metabolites, and by phagocytosis. Causes of SIRS and MODS in trauma patient include: 1- intense uncontrolled activation of inflammatory cells and mediator release; 2- direct damage of vascular endothelium; 3- altered immune cell function; 4- hypermetabolism; 5- maldistribution of circulatory volume to organ systems; and 6- release of inflammatory mediators from damaged or necrotic tissue. Normally, neutrophils limit tissue injury and defend against invading microorganisms by margination, diapedesis, chemotaxis, and phagocytosis. During the SIRS, intense uncontrolled neutrophil activity damages host cells and causes organ damage, vascular injury, oedema, thrombosis, and haemorrhage in multiple organ systems via the release of cytotoxic biochemical mediators. Neutrophilic function is also affected by other circulating mediators that intensify systemic inflammation. Pulmonary alveolar macrophages produce OFRs and proteolytic enzymes that destroy alveolar epithelial cells leading to ARDS. Mast cells are tissue-based leukocytes located primarily adjacent to blood vessels. Mast cells release multiple mediators that have local and systemic inflammatory effects. Lymphocytes adhere to microvascular endothelium and produce ILs that activates other inflammatory cells. Endothelial cells are metabolically active, important in anticoagulation, and manufacture chemotactic agents that attract neutrophils to areas of inflammation and endothelial injury. TNF may precipitate organ injury after trauma and infection by causing generalised endothelial injury, fibrin deposition, a procoagulant state, decreased organ perfusion, and increased capillary permeability. TNF induces fever, hypotension, hyperglycaemia progressing to hypoglycaemia, and hypertriglyceridaemia. DIC, interstitial pneumonitis, ATN, and necrosis of the GIT, liver and adrenal glands have been linked to excessive TNF activity. IL1 induces vascular congestion, capillary leakage, altered coagulation, leukocyte adherence to the endothelium, and the synthesis of two vasodilators PGI_1 and PGI_2. IL1 increases the synthesis of acute phase reactant proteins and causes muscle catabolism, hypotension, fever, tachycardia, diarrhoea, acute lung injury, and leukopenia. PAF affects the heart, vascular system, coagulation, platelets, and the lungs by promoting platelet aggregation, microvascular stasis and ischaemia, and vascular permeability. Arachidonic acid is found in all cells except RBCs. Eicosanoids increase vascular reactivity and permeability, and favour the activation of inflammatory cells. Eicosanoids cause vascular instability and maldistribution of blood flow. Hypoxia, ischaemia, endotoxins, catecholamines, and tissue injury activate AA metabolism, producing metabolites from both the cyclooxygenase and lipoxygenase pathways. All leukotriens (LT) and TXA_2 enhance capillary membrane permeability and increase vascular leakage. LTB_4 stimulates leukocyte adherence to vascular endothelium and tissue injury. TXA_2 enhances neutrophils aggregation, vasoconstriction, and pulmonary bronchoconstriction. Proteases are released from activated inflammatory cells to wall off injured areas and aid in the digestion of foreign debris. Proteases are destructive to healthy tissue, and their uncontrolled release can cause organ damage. SIRS is characterised by generalised inflammation in organs remote from the initial insult. Organ dysfunction, such as acute lung injury and acute renal failure, is a complication of SIRS. The transition from SIRS to MODS is associated with: 1- failure to control the source of inflammation or infection; 2- a persistent perfusion deficit; 3- flow-dependent oxygen consumption; and 4- the presence of necrotic tissue. MODS is a clinical syndrome of progressive physiologic dysfunction of several interdependent organ systems. Dysfunction of at least 2-organ systems occurs in 7-15% of critically ill patients. In trauma patients, dysfunction of 2-organ systems is associated with a 41-67% mortality rate, with mortality rates ranging from 60-100% with failure of 3- or more-organs. The inflammatory response peaks in 3-4 days postinjury and resolves within 7-10 days. Hypermetabolism may last for 14-21 days. About 25-40% of patients dies during hypermetabolism. Death may occur in 21-28 days after the initial insult. MODS survivors may require prolonged, extensive rehabilitation and may develop chronic generalised polyneuropathy and chronic lung disease. ARDS occurs 24-72 hrs after the initial injury. Not all ARDS patients develop MODS. Mediators associated with ARDS include AA metabolites, OFRs, proteases, TNF, and ILs. These mediators damage the pulmonary vascular endothelium and the alveolar epithelium, depress surfactant production, cause pulmonary hypertension, and increase lung water. Pulmonary failure results in refractory hypoxaemia, altered airway mechanics, and bronchoconstriction. The patients require intubation and mechanical ventilation. Translocated gut bacteria are not adequately detoxified or cleared by hepatic macrophages. In response to enteric organisms, hepatic macrophages greatly increase their production of TNF, which further intensifies systemic inflammation. Bacterial translocation has also been associated with the use of antibiotics, antacids, and histamine blockers. Hypoxia as well as reperfusion injury, damages hepatocytes and vascular endothelium. Ischaemic hepatitis (shock liver, posttraumatic hepatic insufficiency) is associated with centrilobular hepatocellular necrosis and is directly related to severity and duration of physiologic shock. Ischaemic hepatitis either resolves spontaneously or progresses to hepatic failure or encaphalopathy. Acalculus cholecystitis manifests 3-4 weeks following the initial insult and is attributed to increased TXA_2 and ILs in splanchnic microcirculation and absence of enteral intake. 50% of patients with acalculus cholecystitis have gallbladder gangrene. Acute renal dysfunction may result from renal hypoperfusion or from direct damage to renal tubular cells (ATN). The frequent use of nephrotoxic drugs critical illness also intensifies the risk of ATN. The clinical management of SIRS/MODS should achieve the following goals: 1- control or elimination of the source of inflammation; 2- maintenance of tissue oxygenation; 3- nutritional and metabolic support; 4- support for individual organs; and 5- effective pain control. Antibiotics initially may contribute to SIRS by triggering the release of endotoxin, an inflammatory mediator, upon cell death. Altered oxygen extraction may lead to anaerobic metabolism and increased lactate production. Escalating serum lactate levels increasing from 2 to 8 mmol/L are associated with a 90% mortality rate. Decreasing oxygen need by pain control and sedation, mechanical ventilation, temperature control, and rest is essential. Oxygen delivery may be further increased by positive end-expiratory pressure, maintenance of adequate haemoglobin levels, and by optimising cardiac output by increasing preload or myocardial contractility or by reducing afterload. The trauma patient experiences extreme alterations in carbohydrate, protein, and fat metabolism. MODS is a clinical syndrome of progressive physiologic dysfunction of organ systems. Trauma patients are at high risk for SIRS and MODS due to circulatory shock with tissue hypoxaemia, tissue injury, and infection. Trauma patients may experience both primary and secondary MODS. Activation of neutrophils (PMNs) and subsequent adherence of PMNs to inflamed vascular endothelium is an important step in the development of MODS. The first step in this process is initiated by a group of glycoprotein adhesion molecules, selectins. L-selectin is present on PMNs under basal conditions, and E- and P-selectin are expressed on activated endothelium. P-selectin expression appears rapidly after endothelial activation by histamine, complement, and thrombin. E-selectin expression follows endothelial cell activation by endotoxin (LPS), TNF-α, or IL-1β, and is transcription-dependent. The appearance of E-selectin is consequently slower, taking some 30 minutes before it becomes evident, and peaking at 4-6 hours. The ligands for these endothelial adhesion molecules are carbohydrates, specifically the Lewis^x (CD15) and the sialyl Lewis^x on PMNs, monocytes, and eosinophils. The result of selectin-mediated PMN-endothelial cell adhesion is a slowing or rolling of PMNs as they pass over the endothelium under normal flow conditions. This permits the subsequent PMN integrin-endothelial cell ligand interaction, resulting in firm adhesion and diapedesis. After activation and particularly up-regulation of integrin receptors, selectin receptors are shed from the cell surface and are measurable in the circulation. The underlying pathophysiology of sepsis and accompanying MODS after trauma involves an early humoral mediators response followed by a cellular phase involving intravascular margination and activation of PMNs. The initial immediate consequence of injury is humoral response includes the activation of the various plasma protein cascades, resulting in the elaboration of coagulation and fibrinolytic factors, activated complement fragments, histamine, and bradykinin. Followed by the elaboration of cell-derived mediators, including NO, lipid products (platelet activating factor and prostaglandin), and cytokines (TNF, IL1, IL6, and IL8). The net result of this systemic proinflammatory humoral milieu is the activation of cellular elements, including circulating and tissue-fixed phagocytes and vascular endothelium. For subsequent organ injury to occur, activated PMNs need to migrate in the postcapillary venules, adhere to the inflamed endothelium, and migrate into interstitium. Injury occurs when activated PMNs release toxic proteases and elaborate OFRs during the process of degranulation and respiratory burst while still attached to the endothelium or on subsequent diapedesis into the interstitium. This requires three steps: 1) slowing of the PMNs, causing them to roll along the endothelium, 2) firm adherence, and 3) diapedesis. The latter two steps are mediated by PMN adhesion molecules of the β2 integrin family (CD18/CD11a or LFA-1, and CD18/CD11b or MAC-1). Their ligands on endothelium are ICAM-1 and ICAM-2 that belong to the IgG superfamily. For the adhesion and diapedesis to occur, the following are required: 1- both endothelium and PMNs require activation by proinflammatory mediators before adhesion molecules become expressed and adhesion can occur; 2- the normal lamina flow of PMNs over vascular endothelium needs to be slowed to allow subsequent integrin-mediated adhesion to occur. This slowing or rolling effect is mediated also by selectins. The selectins are a family of glycoproteins having carbohydrate-binding domains homologous to those on lectins and other domains resembling EGF and complement binding proteins. L-selectin is expressed on PMNs under basal conditions and will loosely attach to its ligand on activated endothelium. E- and P-selectin are expressed on activated endothelium. P-selectin is expressed rapidly after stimulation of endothelium by thrombin, complement, or histamine. P-selectin expression may be cytokine-dependent. E-selectin expression follows endothelial cell activation by LPS, TNFα, or IL1, and is slower because it is transcription-dependent. E-selectin expression has been shown to precede organ dysfunction. Elevated P-selectin levels have been detected in patients with acute lung injury and levels seemed to correlate with outcome. Injured patients express elevated levels of E-selectin and possibly P-selectin after injury and resuscitation. These patients may be at increased risk for subsequent death, and potentially for infectious complications and organ dysfunction. The lack of correlation between selectin activation and ISS emphasises the limitation of current injury scoring tools in identifying the high-risk patient. Elevated E-selectin levels after trauma may identify a group of patients at a higher risk of death and complications. Patients without elevated levels of selectin would seem to be at low risk for infectious complications, organ dysfunction, and death. Postinjury elevation of selectin levels are a marker for endothelial activation and may be a marker for sustained SIRS. PMN-endothelial cell interaction seems to be at the core of many of the systemic deleterious effects of injury. Sepsis and MOSD continue to cause delayed morbidity and mortality after major injury.

POSTINJURY PAIN

Trauma will evoke intensity-dependent increases in firing rates of small, lightly myelinated (Aδ) or unmyelinated (C) afferents. Electrical activation of Aδ nociceptors produces a short-lasting, pickering sensation (first pain), whereas activation of C fibers results in a poorly-localised, burring sensation (second pain). If stimulus produces a local injury as in a tissue crush or incision, two events are observed to occur: 1- The normally silent sensory afferent displays a persistent bursting discharge that continues for extended interval (minutes to hours) after the injuring stimulus is removed. 2- The stimulus intensity required for activating the otherwise high-threshold afferent may fall significantly, such that otherwise moderately intensive stimuli will be highly effective. The origin of ongoing activity after injury is the terminal region of the sensory afferent and have two sources: 1) afferent terminals that are in the vicinity of the injury may develop spontaneous activity, as a local damage to the terminal, which may result in an increase in local sodium channel activation; 2) a tissue-injuring stimulus will lead to the release of local factors that directly activate the local terminals of afferents and facilitate the discharge of the afferent in response to otherwise submaximal stimuli. Stimuli will evoke intensity-dependent increases in firing rates of small afferents, and this corresponds to the psychological report of pain sensation in humans. Such stimuli may result in local injury and the subsequent elaboration of active products that both directly activate the local terminals of afferents, enervating the injury region, and facilitate their discharge in response to otherwise submaximal stimuli. Many small fibers contain one or more neuropeptides, including, but not limit to, substance P (SP) calcitonin gene-related peptide (CGRP), cholecystokinin, galanin, somatostatin, and the excitatory amino acids (EAAs) aspartate and glutamate. Activity proceeds orthodromically to the spinal cord and invades local collaterals (antidromic activity). The orthodromic traffic reaches the spinal cord and may serve to produce sufficient local depolarisation in the dorsal horn that an antidromic action potential is generated in the terminals of an adjacent sensory axon. The antidromic activity generated by a local axon reflex and by the spinal component invades the distal terminal region to locally release neuropeptides (SP, CGRP). Local injury and the released hormones serve to activate local inflammatory cells. Hormones such as bradykinin, prostaglandins, and cytokines, or K^+/H^+ released from inflammatory cells and plasma extravasation products result in stimulation and sensitisation of free nerve endings. Acute activation of small sensory afferent axons evokes autonomic responses and pain behaviour. This effect is mediated by the activation of spinifugal projection neurons. These projection systems travel both ipsilaterally and contralaterally in ventrolateral aspect of the spinal cord, projecting supraspinally into the medulla, mesencephalon, and diancephalon. Medullary projections serve to activate spino-bulbo-spinal reflexes that influence autonomic tone. Other projections into the mesencephalon and thalamus are assumed to contribute to the perceptual and complex emotive and discriminative components of the pain state. Supraspinal projections to the ventrobasal thalamus send projections to the sensory cortex. Areas in the anterior cingulate cortex known to receive input from several thalamic nuclei, notably the nucleus submedius, may be activated in the presence of noxious input. The submedius is the recipient of nociceptive-specific input. The pain pathway appears to preserve several properties of the stimulus: the anatomical sign (localisation) and the intensity. The intensity of the stimulus is mirrored either by the specific neuronal population activated nociceptive specific (NS) cells and/or the frequency of the discharge as with wide dynamic-range (WDR) neurons. The pain pathway reflects the connectivity by which afferent traffic generated by tissue injury reaches higher centers and the conscious state. The excitation of dorsal horn neurons evoked by small afferent input is subject to modulation by a number of receptor systems within the spinal cord.

Factors released after tissue injury that influence primary afferent fibers:

1- Amines:
:   Histamine (mast cells, basophils, and plates) and serotonin (mast cells and platelets) are released by a variety of stimuli including mechanical trauma, tissue damage, thrombin, collagen, epinephrine, AA derivative (leukotrienes and prostanoids)

2- Kinin:
:   Bradykinin released by physical trauma is synthesised by a cascade, which triggered by activation of factor XII by agents such as kallikrein and trypsin. Bradykinin evokes activity through specific bradykinin receptors (B1/B2) located on small afferent terminals

3- Lipidic acids:
:   AA released into the cytosol by the activation of PLA_2 is converted by lipoxygenase or cyclooxygenase (prostanoids) into a large family of diffusible lipidic acids. A number of prostanoids, including PGE_2, can directly activate C fibers. PGI_2 and TXA_2 and several leukotrienes can markedly facilitate the excitability of C fibers. These effects are also mediated by specific membrane receptors.

4- Cytokines
:   Interleukins and $TNF\alpha$ are formed as part of the inflammatory reaction exert powerful sensitising effects on C fibers.

    IL-1 may sensitise C fibers through a prostaglandin intermediary. $TNF\alpha$ acts directly upon axon membranes to form voltage Na channels.

5- Primary afferent peptides:
:   CGRP and SP are found in and released from the peripheral terminals of C fibers, produce degranulation of mast cells, local cutaneous vasodilatation, and plasma extravasation, that heighten the inflammatory response to injury.

6- H^+/K^+:
:   Low pH ($\uparrow H^+$) directly stimulate small afferent terminals through a proton channel, activate the local axon reflex, and result in the local release of afferent peptides, which enhances response to mechanical stimuli. Wounds and inflammatory fluids are acidic as compared with normal tissue.

The graded response properties of the WDR system and its strong projection into the sensory thalamus suggests its role in defining aspects of the sensory discriminative components of the perceptual condition evoked by a high-intensity stimulus. Dorsal horn WDR neurons can be inhibited by transient activation of large primary afferents. Such inhibition is mediated by a local spinal circuit that produces primary afferent depolarisation (PAD), which exerts an inhibitory effect upon the terminals of adjacent afferents. This serves to reduce the amount of neurotransmitter released from the afferent fibers in response to a fixed input. PAD is mediated by release from the local interneurons of the inhibitory amino acid, γ-aminobutyric acid (GABA). Repetitive stimulation of C, but not A fibers, produces a gradual increase in the frequency discharge until the neurone is in a state of virtually continuous discharge (wind-up). Organ convergence depends on the spinal level, a neurone in the nucleus proprius may be activated by: 1) stimulation of the sympathetic afferent; 2) stimulation of the splanchnic nerve; 3) distension of hollow viscera (bladder, intestine, etc.); and 4) bradykinin. The same WDR neurone can be excited by cutaneous or deep (muscle) input applied within the dermatome that coincides with the segmental location of these cells. Visceral pain has its substrate in the viscerosomatic and musculosomatic convergence onto dorsal horn neurons. The components of spinal systems involved in the postinjury pain state, may be classified into: 1- Primary systems: i) primary afferent C fibers, these release peptide (SP, CGRP and ESSa); and these evoke excitation in second order neurons, ii) spinal facilitation and behaviourally defined hyperalgesia. 2- Secondary systems: i) prostaglandins, cyclooxygenase is found in the spinal dorsal horn, ii) nitric oxide (NO), small dorsal root ganglion cells as well as some postsynaptic elements contain NOS. NO facilitates terminal release of glutamate. The time course of events in human surgery is more extended and the postoperative phase is complicated by the presence of a frank injury that will lead presumably to high levels of afferent traffic. In human surgery, there is an extended second phase. Whether the injury is induced iatrogenically (postoperative), as a result of trauma, or in a chronic state (cancer), the conditioning of the sensory system with uncontrolled afferent input gives rise to prominent changes in system function. These effects include: 1- Acute activation of the autonomic nervous system leads to hypertension and myocardial oxygen consumption. 2- Release of stress hormones that increase metabolic activity and may divert resources from processes requiring healing. 3- Influence gastrointestinal motility leading to increased likelihood of stasis. 4- C-fiber input leads to suppression of the immune competency.

THE SPINAL CORD INJURY

Spinal cord injury (SCI) is one of the most catastrophic and devastating injuries known to man. Males sustain spinal cord injury more than females; the ratio is 4 males to one female. Deaths within the first few months following injury are due to respiratory complications and after the first year are due to renal failure. UP to 67% of all deaths among chronic paraplegics have been attributed to renal failure. 90 % of the patients dying of renal failure are managed chronically with indwelling catheters and renal deterioration occurs secondary to pyelonephritis, reflux, hydronephrosis and stones. Catheter drainage of the shocked bladder reduced mortality from 100% to 50%, improvement of urodynamics in the early 1970s allowed early identification of patients at risk for developing problems with their upper urinary tracts. It was noticed that sustained detrusor pressures in excess of 40 cm H_2O will produce damage to the kidney and early intervention in this group of patients is mandatory. SCI in children is a specific clinical phenomenon as children are in the growing process. The main feature of SCI in children are: 1- radiographs do not show the boney injury; 2- many of the injuries are complete SCI; 3- many involve the upper thoracic spine; and 4- the duration of spinal shock is short. Pressure sores, spinal deformity and hip dislocation are typical complications.

PATHOBIOLOGY

Mechanisms of injury include vehicular and high-velocity missile (gunshot). The majority of injury occurring between the ages of 11-30 years. The cervical spine being the most common (C4-to-C7, 48%), followed by high thoracic (T3-to-T6, 13%) and low thoracic (T10-to-T12, 18%). Most SCIs are preventable. Over 70% of motor vehicle accidents are due to human factors and >90% of the spinal cord injuries that result from motor vehicle accidents occur in unrestrained passengers. Alcohol is a major contributing factor. The actual process of damage can be considered a two-part injury sequence. The primary injury is the tissue destruction due to the mechanical impact or penetrating injury. The damage may be to neural tissue or to vascular structures. The injury results in a series of physiologic and biochemical events. The release of endogenous opiate peptides after acute SCI contributes to the pathobiology of secondary injury. There is increased κ-receptor binding after SCI. Gangliosides are complex acidic glycolipids present in high concentrations in CNS cell membranes. Gangliosides augment neurite outgrowth, induce regeneration and sprouting of neurons, promote healing and regeneration of damaged nervous tissue. The monosialotetrahexosylganglioside (GM-1) enhances the ultimate function in the white matter tracts rather than gray matter. Hypoxaemia may complicate SCI. It may result from airway obstruction, pneumonia, respiratory arrest, aspiration pneumonia, shock, fat or air embolism, neurogenic pulmonary oedema, or high SCI. Spinal cord blood flow (SCBF) is decreased pottrauma. SCBF is autoregulated in manner similar to the brain. In the patients with severe SCI, the presence of spinal shock may lead to hypotension and bradycardia, which reduce perfusion pressure. Mild temperature reductions may provide a protective benefit without adverse complications. Patients with acute trauma are often hypothermic, and it is not clear to what temperature it is optimal to rewarm these patients when they have associated CNS injury. High glucose levels may aggravate ischaemic injury by increasing lactic acidosis. Lactic acidosis can contribute to cellular injury via several mechanisms: 1- At low pH levels, acidosis can cause endothelial injury, resulting in swelling and rupture of the astrocytes, leading to a reduction in vascular integrity and collateral blood flow. 2- Denaturation and inactivation of proteins and enzymes may also occur, which hamper energy production and cellular function. 3- The acidosis can also promote the reduction of iron from the ferric to the ferrous state, leading to the generation of free radicals. The degree of cellular lactic acidosis is reduced by decreasing the cellular stores of glucose. By withholding glucose (but avoiding hypoglycaemia), the amount of anaerobic glycolysis is limited during ischaemia, thereby limiting the amount of lactic acidosis that occurs. Hyperglycaemia may worsen neurologic outcome following global ischaemia. Hyperglycaemia also aggravates focal ischaemic injury. Hyperglycaemia may also be deleterious in SCI. In the patient with acute SCI, surgery may be necessary for decompression in a neurologically deteriorating patient, for reduction and stabilisation when conservative management fails, and also for treatment of other life-threatening conditions. Pathobiologic response to spinal cord injury include: 1- Physical forces: Acute spinal cord injury most frequently involves axial or translational forces applied to the spinal column, with subluxation or fracture of the bony processes, disruption of the ligamentous elements, and secondary damage to the spinal cord itself. Although the spinal cord is relatively strong anatomically, it is very vulnerable physiologically. Immediate autodestructive process results in acute, widespread necrosis of its substance, with eventual replacement by fibrosis and scar tissue. 2- Anatomical changes: The spinal cord may be deformed by direct injury, with subsequent axonal swelling, central haemorrhage, oedema, and white matter necrosis. 3- Physiologically, immediate loss of autoregulation, with ischaemia, hypoxia, and infarction result. 4- Biochemical changes: A release of biogenic amines and leukotactic factors from the injured cells is noted. Serum endorphin levels are dramatically elevated. Lipid peroxidation occurs after disruption of myelin lipid bilayers and may enhance the generation of free radicals, PGI_2 and thromboxanes. Endoperoxidases may be the final common pathway for the injury process. The effect of high-dose steroids likely relies on inhibition of endoperoxidase activity.

Pathobiologic response to spinal cord injury:

Time	Anatomic	Physiologic	Biochemical
Immediate	Cold deformation.	Loss of evoked potentials.	Lipid peroxidation, free radicals formation.
5 min	Axonal swelling.	Vasoconstriction.	
15 min		Decreased gray and white matter blood flow.	Increased thromboxane levels, increased tissue norepinephrine levels.
30 min	Central haemorrhages.	Ischaemia.	Profound tissue hypoxia.
4 h	Blood vessel necrosis, white matter oedema.		
8 h	Central haematoma formation.		
24 h	White matter necrosis.		

Primary injury occurs at the time of mechanical trauma. Only small number of nerve axons in the brain and spinal cord are immediately destroyed by impact injuries. In the spinal cord, most of the nearly 20 million nerve fibers remain after injury. These axons often have reduced myelin thickness and cannot impulses reliably. The cellular repair apparatus for these axons is often intact because it is usually located at a distance from the injury. Penetrating injury (missiles and fragments) can produce extensive damage. Nonpenetrating injury (compression, stretch or shear forces) however is more common. The gray matter is more susceptible to injury, because of its higher metabolic rate that makes it more dependent on blood supply. In the gray matter of the spinal cord, the mechanical trauma results in a central zone of contusion with petechial haemorrhage. As early as 1 hr after trauma, central chromatolysis, microvacuolation, and ischaemic injury are seen in the anterior horn cells. There is a loss of ATP production with subsequent disruption of membrane homeostasis and ATP-dependent functions. This results in immediate loss of axonal transport, disruption of energy metabolism, and altered neural function, leading to depression of consciousness or coma. Secondary injury is mediated by microvascular and subsequent biochemical forces that result by the injury. These effects are responsible for the majority of damage occurring after nonpenetrating injury. Secondary injury is a self-propagating cascade of autodestruction. Secondary injury appears to begin following a latent period of several hours and then continues for days until injury is complete (6 days). Pathologic changes seen include haemorrhage, oedema, axonal and neuronal necrosis with demylination followed by cyst formation and infarction. These biochemical processes are intertwined in a complex way. The excitatory amino acids (EAA) play a key role in the initiating injury. They are normally stored in presynaptic vesicles, are transiently present during synaptic activity, and then are quickly taken back up into the presynaptic area. With injury, they are released in large quantities, and their reuptake is impaired. EAAs produce cell swelling, vacuolisation, seizure activity, and neuronal death. The most frequently studied EAA is glutamate, binds 3 classes of receptors. The kainate and quisqualte (AMPA) receptors are associated with monovalent cation channels (Na^+, K^+). The N-methyl-D-aspartate (NMDA) receptor is linked to the monovalent cation channels as well as a divalent cation channel that allows Ca^{2+} influx when activated. This NMDA receptor is also regulated by binding of Zn^{2+} and Mg^{2+} (which block the ion channel) and has binding sites for phencyclidine-like compounds (e.g., ketamine), MK-801, dextrophan and dextromethorphan. The other major EAA is aspartate, and it appears to bind only the NMDA receptor. It is postulated that EAAs released with injury activate the ion channels, producing shifts in Na^+ and K^+ (disrupting membrane function), as well as Ca^{2+} (calcium influx raises intracellular calcium to toxic levels). The NMDA receptor is coupled to a membrane channel permeable to Na^+ and Ca^{2+}. Glutamate binding is facilitated by glycine. Competitive and noncompetitive antagonists may bind to the NMDA receptor, inhibiting the opening of the ion channel. The noncompetitive antagonists bind within the ion channel, thereby impeding ion flux through the channel. These compounds include Mg^{2+}, Zn^{2+}, and other compounds that bind to the phencyclidine site within the channel (phencyclidine, ketamine, MK-801, dextrophan, and dextromethorphan). Ketamine, a noncompetitive NMDA-receptor antagonist improves neurologic outcome and decreases infarct size, but increases cerebral metabolic rate for oxygen ($CMRO_2$), cerebral blood flow, and intracranial pressure, which is a major concern in associated head injury. The abnormal accumulation of calcium in neural cells may occur by several mechanisms: 1- disruption in membrane function and permeability may allow passive entry of calcium; 2- calcium may enter through receptor-controlled channels linked to NMDA receptors; or 3- calcium may enter through voltage-sensitive channels. The effects of raised intracellular calcium are: 1- Exacerbate vasospasm or promote hyperperfusion by causing myosin phosphorylation and subsequent contraction. 2- Calcium activation of phospholipases (A_1, A_2, and C) can destroy cell and mitochondrial membranes, liberating free fatty acids such as arachidonic acid (AA). The release of AA promotes the synthesis of prostaglandins, thromboxanes, and leukotrienes, which may further damage the cell. 3- Cell membrane failure causes ionic shifts with loss of neural activity; mitochondrial membrane injury can result in paralysis of energy generation. 4- Activation of calcium-dependent proteases can lead to degradation of neurofilament proteins and membrane receptors. Calcium channel blockers for SCI failed to demonstrate a beneficial effect. Steroids, inhibit lipid peroxidation, stabilise lysosomal membranes, and modify oedema production. Activation of membrane phospholipids and lipases results in the release of fatty acids, notably AA. AA is stored as component membrane phospholipids. Its release is normally limited by endogenous glucocorticoids, which inhibit phospholipase action. AA can promote the synthesis of prostaglandins (PGEs, PGFs, PGIs, prostacyclin) and thromboxanes (A_2) from endoperoxides via the cyclooxygenase pathway. Arachidonic acid can form leukotriens (A_4, B_4, slow-reacting substance of anaphylaxis C4, D4) from hydroperoxy and hydroxy fatty acids via the lipoxygenase pathway. The products of AA metabolism effects, including: 1- diminish vascular flow by causing vasoconstriction and platelet aggregation; 2- produce oedema from an increase in vascular permeability; and 3- stimulate an inflammatory response by attracting neutrophils. Methylprednisolone (30 mg/kg i.v bolus, followed by 5.4 mg/kg/h X 23 hr) results in a significant improvement in motor neurologic scores that may persist for 1 year after injury. Methylprednisolone has to be administered within 8 hrs of injury to achieve efficacy. Free radicals are released into the cytoplasm during ischaemia from damaged neurons, astrocytes, endothelial cells, and vascular smooth muscle. Free radicals may be produced by catecholamine oxidation, mitochondrial leak, dislocation or decompartmentalization of intracellular iron (transferritin, ferritin), and extravasation of haemoglobin, hematin, and other iron compounds. Arachidonic acid (AA) promotes the formation of the superoxide radical during its conversion to endoperoxides by lipoxygenase. The peroxynitrite anion radical is formed in endothelial cells from NO as a result of EAAs acting on the NMDA receptor. It is also formed as a result of the action of cyclo-oxygenase and lipoxygenase. Leukocytes are potent generators of free radicals. Free radicals also released during reperfusion. Free radicals produce damage to neurons, astrocytes, and vascular structures (endothelial cells and vascular smooth muscle) by peroxidation of cell membranes. Cells of the CNS are especially prone to free radical damage because they contain high levels of polyunsaturated fatty acids. Methylprednisolone is antioxidant. Vitamin E alone is not effective in situations where rapid and extensive free radical removal is required because the active form must be regenerated each time it donates an electron to combine with an oxygen radical. Vitamin C is water-soluble and therefore its effects on membrane events are limited. Superoxide dismutase has beneficial effects in CNS injury. The levels of H_2O_2 increases within 30-min post-spinal cord injury, remains elevated for over 11 hrs, and then decline. Because H_2O_2 can cross the cellular membrane, the H_2O_2 levels measured in dialysates may reflect the intracellular levels of H_2O_2. Therefore, the postinjury increase in H_2O_2 concentrations in the dialysates may indicate that intracellular H_2O_2 concentrations increase. .OH has very short lifetime (nanoseconds). Generation of both O_2^- and .OH in the spinal cord block electrical conduction and induce cell death. The possible lengthy production of .OH from the long-lasting increased H_2O_2 and O_2^- may cause damage. The relatively long-lasting increases of O_2^- and H_2O_2 may indicate that ROS production is not the immediate consequence of impact injury as excitatory neurotransmitter release is, but rather results from subsequent secondary processes. A variety of agents may mediate secondary damage upon CNS injury, including excitatory neurotransmitters, neuropeptides, eicosanoids, and free radicals. The relatively long-lasting production of ROS extends into a time such that removal O_2^- and H_2O_2 may be a realistic treatment strategy (antioxidant) for reducing secondary injury.

PAIN

The psychological issues have tremendous importance in the experience and expression of pain. Psychological factors interact with physiological factors to affect any of the pains. Acute pain applied to one region of the body is known to attenuate the perception of acute pain administered to a different body area. While several studies found that acute pain threshold is raised in chronic pain subjects; others showed that these subjects exhibit a decreased pain threshold compared to pain free subjects. A critical level of chronic pain must be perceived in order to induce an elevation in acute pain threshold. An elevation in pain threshold can be seen in spinal cord injured patients, above the level of injury, only if these patients are suffering from chronic pain and only if spinal transection is functionally complete. In contrast, in the same patients, thresholds for warmth and cold perception are unchanged. The area of the body to which chronic pain is projected is considerably larger in these patients than in those with partial spinal cord damage. Pain continues to be a significant management problem in people with spinal cord injuries. Pain following spinal cord injury includes musculoskeletal, visceral, neuropathic and other types of pain. Neuropathic pain is divided into 2 subdivision depending on the basis of origin, neuropathic at level and neuropathic below level pain. The neuropathic pain at the level is further divided into radicular and central, to indicate the presumed site of the lesion responsible for pain generation. 34-94% of the patients experience pain following SCI, and describe it as severe. Pain has a direct bearing on the ability or inability of the spinally injured person to regain his or her optimal level of activity. Pain arising from both musculoskeletal and nervous structures can occur at the level of the lesion. Pain arising from both spinal cord pathology and visceral structures can occur below the level of the lesion. Different types of pain have the same features. Burning, tingling and aching could equally describe root pain, or diffuse pain below the level. Burning, tingling or aching could also refer to segmental pain or to the pain associated with syringomyelia. Burning pain has been termed sympathetic, but yet this pain may not be related to sympathetic nervous system function. Headache or reflex sympathetic dystrophy are more indirectly related to SCI. Musculoskeletal pain arises from damage or overuse in structures such as bones, ligaments, muscles, intervertebral discs and facet joints, or mechanical pain due to damage to spinal structures, e.g., the acute pain that occurs prior to spinal stabilising operations. Musculoskeletal pain can be identified by location (at or above SC lesions) and by pain features (dull, aching, worse with activities, eased by rest). Visceral pain can be identified by location (abdomen) and by pain features (dull, poorly localised, cramping, related to visceral function or pathology). Neuropathic pain describes pain that occurs following damage to the central or peripheral nervous system. Neuropathic pain can be identified by site (region of sensory distribution) and by features (sharp, shooting, electric, burning, stabbing). Site is easy to distinguish on clinical grounds and provides a simple method of further distinction, which is readily and reliably determined: 1- Neuropathic at level pain, can be further divided into radicular or central. Pain arising from nerve root damage may be suggested by neuropathic features with characteristics such as increased pain in relation to spinal movement. The pain may be due to direct damage to the nerve root during the initial injury or it may be secondary to spinal column instability and impingement by facet or disc material; or due to pathology within the spinal cord or other parts of the CNS. The pain is bilateral in distribution. At level pain should include 2 segments, above and below the level of SCI because input from several segments may be disrupted or disturbed following injury at any particular level. 2- Neuropathic below level pain is a term that has been consistent with the terminology used to identify neuropathic at level pain. Neuropathic below level pain should include pain that is described by the words burning, tingling, aching, shooting, stabbing, present at least 3 segments below the level of injury and is more likely to be diffuse. 3- Other pain types: These types of pain are specific pains that are a consequence of SCI, such as syringomyelia, headache associated with dysreflexia, compressive mononeuropathies and reflex sympathetic dystrophy (complex regional pain syndrome). Enkephalin-synthesising neurons in the superficial laminae of the spinal and trigeminal dorsal horn are critical components of the endogenous pain-modulatory system. These neurons display intracellular oestrogen receptors, suggesting that oestrogen can potentially influence their enkephalin expression. Oestrogen has an acute effect on spinal opioid levels in areas involved in the transmission of nociceptive. The steroid hormone oestrogen has a remarkably wide range of actions in the human body. Apart from its effects on sexual differentiation and behaviour, it modulates bone formation, cardiovascular function, metabolic rate, water and salt balance, and haemostasis. Studies on human subjects have demonstrated that the subjective experience of pain varies during the menstrual cycle and a meta-analysis of the majority of studies on experimentally induced pain indicates that there are gender differences in pain sensitivity, with females displaying greater sensitivity. Oestrogen and progesterone modulate spinal cord δ and κ opioid receptor systems as well as the spinal content of the endogenous opioid dynorphin. Like most other effects of oestrogen, this effect is thought to be mediated by the binding of oestrogen to intracellular oestrogen receptors that, in turn, bind to specific gene sequences known as oestrogen response elements, thereby inducing transcriptional activity of that particular gene. The promoter region of the enkephalin gene contains a putative binding site for oestrogen receptors, which implicates that oestrogen induces transcription of the enkephalin gene by way of nuclear oestrogen receptors. Oestrogen has the ability to rapidly induce an increase in the amount of enkephalin mRNA in the lumbar spinal cord. Oestrogen may have direct genomic action on enkephalin gene transcription. Oestrogen increases the expression of progesterone receptors as well as muscarinic receptors. Both these receptors have been shown to be involved in the regulation of the expression of enkephalin and other neuropeptides in several regions of the brain. The spinal cord expression of the enkephalin gene is sensitive to fluctuating oestrogen levels. The oestrogen-induced increase in enkephalin gene expression takes place in the superficial part of the dorsal horn. The enkephalinergic neurons in the superficial dorsal horn constitute a crucial component of the endogenous pain control system. They are local circuit neurons that mediate pre- and post-synaptic inhibition onto nociceptive relay cells, and they are activated by primary nociceptive and non-nociceptive afferent fibers as well as descending fibers from pain modulating regions of the brain stem. The levels of enkephalins in the spinal cord are directly related to the pain threshold, with increasing levels of enkephalins being associated with less pain sensitivity. Hence, the rate of opioid peptide synthesis and release by the dorsal horn enkephalinergic neurons is likely to be a determining factor for the pain threshold and perceived pain intensity. Intra- and inter-individual differences in pain sensitivity may be influenced by oestrogen. Furthermore, because of the short time span of the oestrogen-induced increase of spinal enkephalin transcription, which differs from the more sustained effect seen in the hypothalamus, changes in pain sensitivity could follow physiologically occurring fluctuations in oestrogen levels, as seen during the menstrual cycle. However, there is clearly no simple relationship between oestrogen levels and pain threshold, since pain threshold is likely to be influenced by a variety of biological and other factors, such as, in the normal cycling individual the changes in progesterone levels are probably important. Oestrogen and progesterone may affect the sensitivity to pain from different organs differently. Changes in hormonal status could affect pain threshold, which in turn could be important in the development of chronic pain.

SPINAL CORD INFARCTION

Infarction of the spinal cord is uncommon in comparison with cerebral stroke yet is as disabling. The principal vascular supply to the human spinal cord is the anterior spinal artery, which originates from the vertebral arteries. Along its discontinuous course the anterior spinal artery receives input from 6-9 radical arteries in variable locations. The major radicular artery, the arteria radicularis magna anterior of Adamkiewicz, supplies the lumbar cord and conus medullaris. The anterior spinal artery syndrome, in which the infarction overcomes the anterior two-thirds of the cord. These patients present with the abrupt onset of weakness and, below the level of the lesion, flaccid paraplegia, areflexia, loss of spinothalmic perception of pain and temperature, and autonomic deficits compromising sphincter flaccidity, atonic urinary bladder and paralytic ileus. Transient radicular or back pain may herald these findings. Spasticity eventually ensues, with exaggerated deep tendon reflexes, Babinski responses, and clonus, except when ischaemia involves anterior horn cells, nerve roots, or the cauda equina, in which case mixed or lower motoneuron deficits may occur. Autonomic dysfunction may accompany anterior spinal artery infarctions. The urologic consequences of complete suprasacral lesions usually include detrusor hyperreflexia and striated sphincter dyssynergia, and if above T6, smooth sphincter dyssynergia may also occur. Orthostatic hypotension may occur from lesions above the greater splanchnic nerve (T4-9). Sexual dysfunction, impairment of vasomotor and pseudomotor tone and piloerection may occur below the level of the lesion, leading to impaired thermoregulation. Disinhibited sympathetic neurons of the intermediolateral cell column may mount an exaggerated response to mildly noxious stimuli such as a distended bladder, resulting in the paroxysmal, generalised hypersympathetic state known as spinal dysautonomia. Spinal cord hemisection or Brown-Sequard syndrome originates from a lesion obstructing the sulcocommissural artery. Infarction of the posterior third of the spinal cord is quite rare and can result from interruption of the posterior spinal arteries or from deficient collateral perfusion in the setting of diffuse arteriosclerosis. Spinal venous infarction is rare; it can be haemorrhagic or ischaemic. Spinal TIAs, manifest as painless paraparesis or quadriparesis (drop attacks) that may be sporadic or associated with postural changes but without loss of consciousness or cranial localising features. This is most likely to occur in patients with foramenal stenosis during cervical or lumbar extension, which maximally compromises the intervertebral foramina through which pass spinal radicular arteries (intermittent spinal claudication). Spinal cord infarction shares with cerebral infarction a common systemic vascular differential diagnosis that includes hypoxia and ischaemia, cardiogenic thromboembolus, vasculitis, atherosclerosis, arteriovenous malformation, collagen and elastin disorders, sickle cell disease, polycythaemia, hypercoagulability, paradoxical embolism via a patent foramen ovale, and cocaine use. The spectrum of aetiologies for spinal cord stroke is as diverse as for cerebral stroke and the degree of recovery is generally poor in comparison with cerebral stroke. Common causes including cervical spondylosis, cervical spinal trauma or sprain, and sustained neck rotation as might occur during yoga exercises. The relative hypovascularity of the midthoracic spinal cord might be an important determinant in some patients.

AUTONOMIC DYSREFLEXIA

Autonomic dysreflexia (AD) is defined as an acute syndrome of massive disordered autonomic response to a specific stimulus seen in patients with spinal cord injuries above the level of splanchnic outflow. It occurs in 30-85% of patients with quadriplegia and high paraplegia, 60% for women and 46% for men. Prevalence associated with cervical injuries is 60% compared to 20% with thoracic injuries. This syndrome is characterised by excessive sweating, flushing of the face, congestion of the nasal passages, pounding headache, intermittent hypertension (diastolic and systolic), piloerection and bradycardia or tachycardia. Spinal cord-injured patients can experience a life threatening exaggeration of the sympathetic response to stimulation. This acute syndrome was reported in 1890, the syndrome occurs in patients with unopposed sympathetic discharge, due to excessive sacral afferent feedback to the spinal cord, seen in patients with lesion above the sympathetic splanchnic outflow from the cord in spastic lesions above T.V1 but also seen with lesion at T.V5 and as low as T.V8. The level of the injury would not predict who would or will not have elevated blood pressure. Viability of the distal cord segment is necessary to elicit this condition. It may occur at any time after neurogenic shock has subsided up to as late as 15 years after injury. Although the frequency of attacks diminishes with time the severity remains unchanged. AD is also known as autonomic hyperreflexia, spinal poikilopoiesis, paroxysmal neurogenic hypertension, autonomic spasticity, paroxysmal hypertension, autonomic reflex, sympathetic hyperreflexia, mass reflex and the neurovegetative syndrome. It is usually due to a distended bladder caused by blocked catheter or to poor bladder emptying as a result of detrusor-sphincter dyssynergia or distended bowel. The distension of the bladder or the bowel or a painful stimulus can result in reflex sympathetic overactivity below the level of the spinal cord lesion causing vasoconstriction and systemic hypertension. The carotid and aortic baroreceptors are stimulated, the signals travel via the glossopharyngeal nerve to the tractus solitarius in the medulla where the vasoconstrictor center is inhibited and the vagal center is stimulated, this respond via the vasomotor center with increased vagal tone results in bradycardia but the peripheral vasodilation that would normally have relieved hypertension dose not occur because stimuli can not pass distally through the injured cord. Hypertension may cause seizures, intracerebral haemorrhage or death. Without treatment intracranial haemorrhage, risking stroke, myocardial infarction, and ocular haemorrhage that may lead to blindness may occur. An important challenge for the home care nurse is to prevent autonomic dysreflexia. During voiding 78% of spinal cord patient have significant hypertension. There is no correlation of AD with length of injury, maximum voiding pressure or bladder capacity. Urodynamics are helpful to detect symptomatic and asymptomatic AD. Significant elevation in blood pressure can occur without the symptoms of AD.

PATHOPHYSIOLOGY
The classical description of AD was presented in 1956. Impulses travel along afferent pathways, and ascend along the spinothalamic tracts and posterior columns. Reflex motor outflow via neurons in the lateral horns causes spasm of the pelvic viscera, arteriolar spasm, pilomotor spasm and sweating. The major splanchnic outflow of the sympathetic emanates from T5 to L2. The sudden increase in blood pressure caused by the reflex arteriolar spasm is appreciated by receptors in the aortic, carotid sinus and cerebral vessels. Afferents from the aortic arch and the carotid sinus course in the tenth and ninth cranial nerves, respectively, and the efferents to the sinoatrial node in the tenth cranial nerve. The result is marked bradycardia with minimal correction of hypertension, can be explained by Poiseuille's formula, which states that the pressure in a pipe is linearly proportional to the rate of flow and inversely proportional to the fourth power of the diameter of the tube. The head and neck are hot and red, and perspire freely. Haemodynamic changes in addition to hypertension and bradycardia associated with autonomic crisis include increased cardiac output, stroke volume, systemic vascular resistance, and pulmonary artery pressure. Anything that can cause discomfort to the neurologically intact person can trigger AD in a patient with spinal cord injury. All stimuli acting below the level of injury can evoke the stereotypic response such as bladder distension, retention of urine, clamping of the Foley catheter, bladder calculi, UTI, acute cystitis, epididymyitis, loaded colon, barium enemas, anal fissure, flatulence, acute abdominal conditions (acute appendicitis), ejaculation, labour, uterine contraction and foetal movement, procedures (cystoscopy, cystometry, percutaneous nephrolithotomy), detrusor-sphincter dyssynergia, cleansing enemas, pressure sores, external temperature changes, scratching the soles of the feet, skin lesions (ingrowing toe-nails or sunburn), tight clothing; shoes or leg bag straps, placement of tourniquet during extremity surgery, distension of the renal pelvis, pressure on the testicles and glans, intraoperative manipulation of the renal pelvis, bladder irrigation, ileal conduit loopography, sexual intercourse in men and women, intracavernosal injections, bladder percussion, renal or biliary colic, vesicoureteral reflux, manipulation of an indwelling catheter, introduction of vaginal speculum, and testicular torsion. AD during labour is expected in two-thirds of women with spinal cord lesions above the T6 level and concurrent intraventricular haemorrhage. During distension of the renal pelvis the stimulus transmits through the T12 afferent pathway to the spinothalamic tract and the posterior columns of the spinal cord. When the transmission is interrupted reflex activity may stimulate the sympathetic ganglia, which in turn stimulates the arterioles to constrict and causes severe hypertension. The interruption of central excitatory and inhibitory impulses to the distal spinal cord results in defective regulation of body temperature, blood pressure and sweating. Anxiety, unusual fear or apprehension, chills, flushing above the level of injury and coolness below it and stuffy nose are common symptoms. Paradoxical hypertension, the absence of symptoms of dysreflexia does not exclude significant elevations in blood pressure, and morbidity increases with hypertension. Blood pressure should be taken with the patient supine and not sitting. Headache is severe in 56% of the patients, of occipital, bitemporal and bifrontal locations. Headache may result from passive dilatation of cerebral vessels or increased circulating PGE_2. Digital occlusion of the carotid arteries will stop the headache. Sweating, blotching of the skin and chest, bradycardia, tachycardia, arrhythmias, vasoconstriction, cutis anserina, changes in level of consciousness, visual field defects, changes in skin and rectal temperature also are common with autonomic dysreflexia. Ocular findings may include, conjunctival congestion, lid lag, Horner's syndrome, mydriasis, and oculosympathetic spasm. Myocardial infarction (MI) is a serious problem that may result from autonomic dysreflexia. The resulting sympathetic response is profound vasoconstriction, producing a rapid rise in blood pressure is exaggerated because the parasympathetic response is partially disabled. The parasympathetic response is limited to vagal slowing of the heart rate and vasodilation, flushing, and

diaphoresis above the level of spinal cord injury. The absence of symptoms of dysreflexia does not exclude significant elevations in blood pressure. 43% of patients would have no symptoms or awareness of hypertension. Elevated blood pressures may not be detected if blood pressure had been obtained while the patient in sitting in a wheel chair. The morbidity is primarily associated with hypertension and includes confusion, visual disturbances, unconsciousness, seizures, hypertensive encephalopathy, cerebrovascular accident with residual weakness, haemorrhages (subdural and retinal) and apnoea. The differential diagnosis includes pheochromocytoma, migraine, cluster headaches, posterior fossa neoplasms, toxaemia of pregnancy, and essential hypertension. In autonomic dysreflexia, blood pressure returns to normal after removal of the triggering stimulus. Chest x-ray may reveal cardiac enlargement. ECG findings may include premature atrial and ventricular contractions, increase amplitude of T waves, prominent U waves, first and second degree heart block and arteriovenous junctional escape. Performance of a cystometrogram before the onset of labour may be helpful to identify patients at risk. AD may be prevented by proper medical care for the spinal cord injury patient. The patient and family should be well educated regarding the causes, presentation, prevention and first aid for autonomic dysreflexia. How to intervene: 1- Rapidly identifying and removing the cause of the dysreflexia and lowering the blood pressure are the first priorities. The systolic blood pressure can reach 300 mmHg. 2- Remove anything that may be stimulating the patient, such as constricting bed linens or sneakers worn to decrease foot drop. 3- Sit the patient up or elevate the head of the bed to 90°, while keeping the feet down, to promote orthostatic reduction of blood pressure. 4- Assess the abdomen for evidence of distended bladder and conclude the need for catheterisation. Insert the catheter after liberally lubricating it with Xylocaine to reduce stimulation. Drain the bladder slow while monitoring the patient. 5- Check blood pressure every few minutes, a blood pressure above 200/100 requires monitoring at least every 15 minutes until blood pressure and heart rate are under control. Heart rate and blood pressure should be monitored every four hours for 24 hours. 6- If after the bladder fully drained, no change in the blood pressure, constipation is then the primary factor. Repeat abdominal examination for bowel distension. 7- Gently insert 1 oz of Xylocaine into the rectum. Manually check for faecal impaction and remove all accessible stool. 8- If blood pressure remains high, order hydralazine hydrochloride 20-40 mg iv (lower BP and raise the heart rate). Sublingual nifedipine 10-20 mg is alternative treatment. 9- Continue check the blood pressure every 2-3 minutes until returns to normal. 10- Patient must learn how to prevent another episode by maintaining a bowel and bladder emptying program at home, instructed about other possible triggers and reminded to call for help immediately if experience any of the warning signs. 11- Blood pressure should be carefully monitored in patients with injuries at or above T6 who undergo various diagnostic tests and procedures, even if they are asymptomatic. 12- Instruct the patient not to drink several glasses of fluid at one time, which can distend the bladder.

INFLAMMATION AND INFECTION

Invading microbes are constantly subject to host immune surveillance and avoidance of detection by neutralising antibodies is an additional necessity for survival. The human body has physical, chemical and cellular defences against invasion by viruses, bacteria and other disease's agent. During early stages of invasion, white cells and plasma proteins take part in an inflammatory response, non-specific counter-attack, phagocytic white cells ingest invaders. Plasma proteins promote phagocytosis and some destroy invaders directly. If invasion persists, some white blood cells make immune responses. If the foreign or abnormal molecule triggers an immune response, it is called an antigen. Antibodies are molecules that bind to a specific antigen and tag it for destruction. There are 3 lines of defense against pathogens: 1- Barriers at body surface (non-specific targets), a) intact skin, mucosal cover of other body surfaces; b) infection-fighting substances in body fluid as tears, saliva; c) normally harmless bacterial inhabitants of body surfaces that outcomplete pathogenic visitors; d) flushing effect of body fluids as urine, diarrhoea and tears. 2- Non-specific responses (non-specific targets): a) inflammation, i- fast-acting white cells, as basophil, neutrophils, eosinophils, ii- macrophages, also take part in immune responses, iii- complement proteins, blood-clotting proteins; b) organs with phagocytic functions, as lymph nodes, spleen; c) immune response (specific targets): i- white blood cells, macrophages, T-cells, B-cells, ii- communication signals as interleukins, and chemical weapons as antibodies, complement proteins. The inflammatory response begins with changes in blood flow to the besieged tissue: 1- Substances released by pathogens and dead or damaged body cells makes the capillary leaky. Histamine and some other substances released from the mast cells. The vessels become engorged with blood, so the tissue reddens and become warmer. Circulating white cells enter the tissue and destroy invaders. 2- Plasma proteins also enter the tissue. Swelling and pain follow. Voluntary movements aggravate the pain and tend to be avoided, this promote tissue repair. Within hours, neutrophils squeeze out through gaps in the blood vessel walls and swiftly go to work. Monocytes arrive later, differentiate into macrophages, engulfing pathogens and secreting several substances such as interleukins. IL1 signals the brain region that controls body temperature to raise the setpoint on the body's thermostat. A fever of about 39°C may develop, it promotes an increase in host defence activities and also makes the body too hot for the functioning of most pathogens. IL-1 induces drowsiness, which reduces the body's demands for energy. The complement proteins induce lysis of pathogens and attract phagocytes. Blood clotting proteins help repair damaged blood vessels. About 20 kinds of complement proteins are circulating in the blood in inactive form. When activated they induce a cascade of reactions. Some complement proteins joins together to form pore complexes, which are structures with an interior channel, become inserted into the plasma membrane of many pathogens and induce lysis. Lysis is a gross structural disruption of the plasma membrane and leads to cell death. Lysozyme molecules diffuse through the pores, and when they reach the bacterial cell wall, they destroy it. Some activated complement proteins promote inflammation. The invader ends up with a complement coat, which adheres to the phagocytes and this is like putting a delicious meal on for dinner. Macrophages and other antigen-presenting cells process and display fragments of antigen with their own MHC markers. Lymphocytes have receptors that can bind to antigen-MHC complexes. Binding is the start signal of an immune response. Immune response begins by recognition of antigen. Interleukins and other signals among white blood cells drive the responses. Effector helper T and cytotoxic T cells, as well as effector B cells and antibodies, carry out the response. Secondary cells are set aside for secondary responses. They directly destroy cells infected with intercellular pathogens, tumour cells, and cells of tissue or organ transplants. When antibody binds to antigen, toxins, may become neutralised, pathogens may be tagged for destruction, or the attachment of pathogens to body cells may be prevented. In active immunisation, vaccines provoke an immune response, with production of effector and memory cells. In passive immunisation, injections of purified antibodies help patients through an infection. An allergy is an immune response to generally harmless substances. An autoimmune response is an attack by lymphocytes on the body's own cells. An immune deficiency is a weakened or nonexistent capacity to mount an immune response. AIDS is syndrome of disorders that follow infection by HIV virus. The virus cripples the immune system, and the body becomes highly susceptible to usually harmless infection and some rare forms of cancer. Anaemia in severe infection is associated with reduced erythropoiesis due to toxaemia and non-availability of iron for haemoglobin (Hb) synthesis. While Hb measures larger deficits in the functional iron compartment, serum ferritin (SF) reflects residual iron stores. Nevertheless, neither can reflect the iron status of individuals during infection. Changes occurring in Hb, SF and serum transferrin receptor (sTfR) concentration during mild and severe respiratory infections include: 1- infection tends to aggravate anaemia by blocking iron utilisation (increase SF level), 2- increase in synthesis of apoferritin by the membrane-bound polysomes with subsequent release into circulation as an acute phase response to infection, 3- accumulation of iron in storage tissues during acute/chronic inflammatory conditions, 4- tissue ferritin synthesis is induced probably to accumulate the iron that returns to storage sites from plasma with the onset of infection/inflammation, 5- shift of plasma iron to tissues is responsible for the block in tissue iron release in these conditions, 6- metabolic changes occurring during infection could lead to iron-deficient haemopoiesis, although the total body iron content may not have changed significantly, 7- metabolic redistribution of iron from circulation to tissues appears to play a major role in altering the iron status during infection, 8- decrease in intestinal absorption of exogenous iron, and 9- SF significantly rise with the onset of infection, confirming the acute phase response to infection. These changes tend to explain how anaemia in infection is a result of increased iron deprivation to the erythropoietic tissue, which enhances the expression of transferrin receptor. Infection-induced release of cytokines like TNF, IL-1, β and γ interferon are known to suppress colony forming units-erythroid (CFU-E) development, leading to suppression of erythropoiesis resulting in anaemia. Reduction in erythropoiesis may reflect lowered receptor concentration. Persistence of anaemia with high sTfR levels even 30 days after the acute phase of the severe infection is of concern.

PATHOBIOLOGY OF INFLAMMATION AND INFECTION

Many pathogenic microorganisms specifically target host-cell receptors that normally mediate intercellular interactions, attachment to substrata, or receive hormonal, cytokine and other environmental signals. In eukaryotic cells, signal transduction involves a complex array of interactive molecules and second messengers that control multiple cellular functions. Two distinct mechanisms of eukaryotic transmembrane signal transduction are recognised: 1- unimolecular receptor/effector systems that span the plasma membrane with the ligand-binding domain exposed at the cell surface and enzymatic effector domain at cytoplasmic face; and 2- multi-component systems with independent but coupled receptors, transducers and effector molecules. Signals originating at the membrane are passed on to cytosolic molecules in a cascade of activating events and eventually to the nucleus. Multiple receptor systems can interact and synergize with each other to regulate cellular events. Binding of appropriate stimuli to cell surface receptors of phagocyte leads to increased consumption of oxygen concomitant with the formation and secretion of superoxide anion by a plasma membrane associated NADPH-oxidase. The oxygen metabolites not only can initiate cytotoxic injury, but can also directly alter the biochemical and biophysical properties of structural proteins of tissue, including elastin and collagen, as well mucopolysaccharides. Cells can eliminate OFRs by specific enzyme scavengers, thereby limiting formation of hydroxyl radicals. The stimulus, its secretory products, the target cell, and the extracellular environment can modulate the type and magnitude of oxidants generated by phagocyte. Nitric oxide, a short-lived, highly reactive gas generated by lightning and pollution, is generated by mammalian and other eukaryotic cells. Dissolved in the aqueous cellular fluid, it diffuses through membranes, and is so reactive that it disappears within moments of its production. Superoxide reacts with NO generating peroxynitrate ($ONOO^-$), which can further give rise to hydroxyl radical through ($ONOO^- + H^+ \rightarrow HO^· + NO_2$) that is not dependent on redox active transition metal ion. NO promotes the ADP ribosylation of actin in human neutrophil, which is associated with reversible inhibition of neutrophil adhesion to a fibronectin-coated surface. NO can limit the degree of endothelial activation and inhibit monocyte adhesion, which may contribute to some properties within the vessel wall.

EUKARYOTIC SIGNALLING PATHWAYS

Receptor tyrosine kinase pathway (RTK)/Ras pathways have tyrosine kinase activity directly associated with the receptor as a part of the receptor molecules. The membrane-located proto-oncogene product Ras is central to this pathway. Ras is a low-molecular-mass GTP-binding protein (small G-protein) whose activation involves conversion from GDP to GTP-bound (activated) form. Binding of the ligand results in receptor clustering, activation of cytoplasmically located tyrosine residues at the C terminus. This leads to binding of other proteins such as Grb-2/Sos complex (cytosolic complexes with affinity for phosphorylated receptor as well as Ras) to the receptor and to plasma-membrane-located Ras. Sos activates Ras via exchange of GDP for GTP, leading to activation of other proteins downstream in the cascade, which include cytoplasmic serine/threonine kinases. Direct activation of the nuclear factor NF-κB may occur via Raf-1, which can phosphorylate IκB, the inhibitor of NF-κB. The superfamily of small GTPases include 5 classes of proteins: Ras (primarily affects cell growth and development), Ran (involved in nuclear protein import), Rab and Arf (monitor and direct vesicular movement) and the Rho subfamily (involved in regulation of the actin cytoskleton). Non-receptor tyrosine kinase (NRTK) pathways, activation requires receptor aggregation and non-covalent association with other membrane-located proteins such as Src. Full activation of Src family kinases requires 2 opposing enzyme activities: the removal of C-terminal phosphate and phosphorylation of tyrosine proximal to the active site. G-protein pathways are utilised by many classes of hormones and involve β-helical receptors. These membrane-associated trimers consist of α, β, and γ subunits. The Gβ and Gγ subunits control the function of Gα subunit. The Gα binds GTP, results in activation of other effector enzymes to generate second messengers such as cAMP, 1,2 diacylglycerol (DAG) and inositol triphosphate (IP_3) or to regulate ion channels. These second messengers are involved in activating protein kinases directly or via Ca^{2+}/calmodulin complexes, leading to activation of cellular events. Pathways used by adhesion molecules, integrins have short cytoplasmic tails without catalytic domains that associate with cytoskeletal proteins such as α-actinin, talin, vinculin, paxillin and tensin in complexes and affect cytoskeletal arrangement via actin microfilaments. The extracellular domains have Ca^{2+}-binding sites. Several integrins interact with matrix proteins and thus have the capacity of forming a physical link between the extracellular matrix and cytoskeleton. Two principal mechanisms by which integrins are activated to transduce signals involve conformation change and receptor clustering. Conformation-dependent activation is exemplified by thrombin-mediated activation of gpIIb/IIIa integrin expressed on platelets. Thrombin induces intracellular signalling leading to modulation of the conformation of gpIIb/IIIa, which is then, able to bind fibrinogen leading to platelet aggregation. This has called inside out signallings. Signalling transduction via integrins may reside, at least in part, in cytoskeletal rearrangement, which may physically bring together molecular complexes by forming focal adhesions. Inhibition of clustering, which is achieved via cytoskeletal rearrangement, also inhibits tyrosine kinases. Stimulation of integrins may be translated into a variety of intracellular signals. Monocytes adherence to fibronectin or collagen activates transcription of inflammation mediators such as IL1β and IL8. Integrins may also act in concert with other pathways to enhance or dampen signals. Activation via growth factor receptors requires adherence of cells to ECM via integrins. Protein tyrosine phosphatases (PTPases) appear to be involved in regulation of cell adhesion; PTPases may also activate Src kinases by dephosphorylating the negative-regulatory C-terminus phosphotyrosine. FGFR and tyrosine phophatases have Ig motifs in their extracellular domains. Cadherins bind to intracellular proteins called catenins. Tyrosine phosphorylation of catenins or other molecules leads to the detachment of cell-cell contacts mediated by cadherins and results in disruption of epithelial integrity. TNF-α, interferon, IL1 and vitamin D3, engaging with receptor results in activation of sphingomyelinase that cleaves sphingomylin to release ceramide and phosphocholine. Ceramide stimulates protein kinases and protein phosphatases leading to activation of NF-κB and c-jun. Janus kinases (JAKs) can activate Ras signal pathway or directly phosphorylate tyrosine in cytoplasmic proteins STATs (signal transducers and activators of transcription), which migrate to the nucleus and affect transcription of specific genes. Neisseriae gonorrhoeae induce endocytosis in host epithelial or endothelial cells (parasite-directed phagocytosis), involve tyrosine phosphorylation of host protein and changes intracellular second messengers. Actin polymerisation is a frequent requirement. E coli are associated with characteristic attaching and effacing (AE) lesions on adhesion to host epithelial cells, resulting in rearrangement of host cytoskeleton beneath adherent bacteria. Signal transduction occurs include tyrosine phosphorylation and elevation of intracellular levels of second messengers, such as Ca^{2+} and IP_3 in host cells and resulting in rearrangement of host-cell microfilament. LPS-mediated signal transduction mechanisms result in the synthesis and release of inflammation mediators. Direct interactions of micro-organisms with eukaryotic receptors often occur at the normal host ligand recognition site and may utilise the ligands of recognition motif, e.g., RGD (Arg-Gly-Asp), a recognition sequence on many ligands of integrins. Within RGD motif, aspartic acid is important for recognition by integrins. Microbial adhesion to integrins may also occur by a bridging or sandwich mechanism involving natural ligands of integrin receptors such as ECM proteins. Type 1 fimbriae of E coli binds mannoside chains of matrix proteins. It interacts with mannosyl residues on CD11/CD18 ($\alpha x\beta 2$) integrins of phagocytes. Candida albicans adhere to host epithelial cells via RGD-dependent mechanisms and the ligands on the host epithelial cells are C3bi integrins. Opa proteins of N. gonorrhoeae are a family of structurally related proteins, which mediate interactions with neutrophils and some also interact with epithelial cells. N. Gonorrhoeae interact specifically with CD66-expressing cells. The adhesion is to the N-terminal domain of the molecule, which is largely conserved between different members CD66. Leucocyte-adhesion-deficiency syndrome (LAD) is an inherited disorder involving the $\beta 2$ chain (CD18) of the leucocyte cell adhesion molecules LFA-1, CR3 and CR4, and results in a low number or absence of normal receptors on the cell surface. Patients with LAD develop severe bacterial infections involving oral, respiratory and urogenital mucosa, as well as the skin and the intestine. The infections seem to result from impaired adhesion-dependent functions such as chemotaxis, aggregation and CR3-dependent phagocytosis. Epithelial cells that line mucosal surfaces are an important mechanical barrier that separates the host's internal milieu from the external environment, as most microbes found in the environment, including commensals, do not enter epithelial cells. Stimulation of epithelial cells by microorganisms or TNFα, IL1, causes the increased expression and secretion of cytokines with chemoattractant and proinflammatory functions. In addition epithelial cells can produce additional physiologic mediators that may act in an autocrine/paracrine function on neighbouring epithelial cells or other mucosal cells. NO is produced by NOS, affects blood flow and mucosal inflammation. Increased NO production by epithelial cells, contributes to increase in apoptosis. Chlamydia infection is characterised by acute inflammation that is exacerbated upon re-infection, ultimately leading to tissue damage and scarring. Although Chlamydia resides within epithelial cells and does not invade deeper layers of the mucosa, the infection initiates an inflammatory response that is key for the development of disease. Chlamydia upregulates epithelial cell expression and secretion of IL8, GM-CSF, and IL6. IL1α released from epithelial cells after Chlamydia-induces cell lysis can amplify the inflammatory response by stimulating additional cytokine production by non-infected neighbouring cells. E coli induces actin reorganisation and membrane pedestal formation, and stimulate low level epithelial cell IL8 responses. The epithelial cell inflammatory gene program includes the expression and production of proinflammatory cytokines, prostaglandin H synthase (PGHS)-2 and prostaglandins, NOS2 and NO, and increased cell surface expression of ICAM-1. Each of the rapidly upregulated genes is a target gene of the transcription factor NF-κB and each of the products encoded or produced as a result of upregulated expression of these genes likely plays a role in the host's resistance to the invading microbe and host mucosal defence. Mucosal infections with microbial pathogens can result in a spectrum of disease manifestations that range from mild and self-limited to fulminant and lethal, or can be chronic and debilitating. The latter may occur in response to repetitive infections and eventually lead to organ dysfunction.

OXIDANTS AND MICROBIAL PATHOBIOLOGY

Many biochemical reactions vital to normal aerobic metabolism of human and microbial cells require the transfer of 4 electrons to molecular oxygen to form H_2O. Molecular oxygen has the capacity to undergo sequential univalent reduction to form other oxygen intermediates with different toxicities prior to the generation of H_2O. The addition of one electron to O_2 yields the superoxide radical (O_2^-), which at physiologic pH rapidly reduces itself to form the divalent oxygen reduction product, hydrogen peroxide (H_2O_2). Trivalent oxygen reduction occurs via the reaction of H_2O_2 with O_2^- to produce the hydroxyl radical (OH^-). At physiological pH, this reaction is of little biologic importance unless a transition metal catalyst (Fe^{3+}) is present to enhance the reaction rate, yielding OH^- via the Haber-Weiss-Fenton reaction. NO. is not a classic product of O_2 reduction, instead, its formation in mammalian cells is dependent on a group of enzymes termed nitric oxide synthases (NOS). Factors known to modulate iNOS levels include a number of cytokines, microorganisms, and microbial products, consistent with the importance of iNOS in host defence and inflammation. Many bacterial species are also capable of generating NO under conditions of low oxygen tension via nitrite reductases. Once formed, NO. has the ability to act as an oxidising agent alone or interact with O_2^- to generate peroxynitrate ($ONOO^-$). Superoxide is a moderately reactive compound capable of acting as an oxidant or reductant in biologic systems, diffuse for considerable distances before it exerts its toxic effects. Extracellularly generated O_2^- can gain access to intracellular targets via cellular anion channels. These targets include bacterial enzymes; particularly those involved in biosynthesis of branched amino acids (e.g., NADH-bound lactic dehydrogenase). At low pH, such as at site of inflammation or inside the phagosome, O_2^- becomes protonated to form HO_2^-. Because of its neutral charge, HO_2^- is more membrane permeable and more likely to react with itself to form H_2O_2. Hydrogen peroxide is more reactive oxidant than O_2^-, and readily diffuses across cell membranes. Potential sources of H_2O_2-mediated damage of cellular constituents include the oxidation of cellular membranes and enzymes, DNA damage and mutagenesis, and the inhibition of membrane transport processes. Killing E coli by H_2O_2 is bimodal in that low and high concentrations of H_2O_2 are more lethal than intermediate concentrations. Low concentration killing is attributed to DNA damage mediated by the interaction of H_2O_2 with F^{3+} to form the toxic ferryl radical, an intermediate product in the formation of OH^-. Exposure of E coli to these low concentrations of H_2O_2 induces a postive response, which confers increased resistance to subsequent H_2O_2 exposures by an enhanced ability to carry out recombinational DNA repair. At high concentration killing, which does not require iron or an electron source, is not due to DNA damage but may involve the oxidation of a separate cellular target. The mechanism by which OH^- and other oxidants may cause cell injury at sites distant from their formation is via the initiation of a free radical cascade. Oxidation of unsaturated fatty acids within a lipid membrane can produce peroxyl radical, which in turn can react with other nearby lipid molecules to generate additional lipid radicals, that then react with other unsaturated lipids, setting up a free chain reaction. In human, intracellular iron is predominantly complexed to ferritin in a relatively noncatalytic form. Almost all extracellular iron is tightly bound to host binding proteins (transferrin and lactoferrin) in forms unable to catalyse OH^- formation. Neutrophil lactoferrin may function to trap iron from ingested microorganisms. Thus, via its interaction with phagocytes, lactoferrin may prevent iron-catalysed oxidant formation, thereby limiting inflammatory tissue injury. Lactoferrin also binds lipopolysaccharide, an important compound mediating toxicity in sepsis. Lactoferrin disrupt lipopolysaccharide priming of phagocytes for O_2^- production. The Pseudomonas aerogenosa secretory product, Pseudomonas elastase and other host-derived proteases present at sites of inflammation are known to cleave transferrin and lactoferrin into lower-molecular-weight iron chelates. Pseudomonas elastase-cleaved transferrin and, to a lesser extent, lactoferrin are capable of catalysing OH^- formation when a source of O_2^- and H_2O_2 is concurrently present. Pseudomonas elastase and other protease-cleaved transferrin to enhance oxidant-mediated pulmonary artery endothelial cell injury via OH^- generation. As microorganisms require iron for growth and replication, their mechanisms of iron acquisition and storage have evolved to fulfill these needs. Intracellular bacterial iron is primarily complexed to ferritin-like iron storage proteins. To acquire iron from the extracellular environment, aerobic and facultative anaerobic bacteria, and fungi, synthesise diverse low-molecular-weight Fe^{3+}-scavenging ligands collectively termed siderophores. These compounds posses a high affinity for iron, which is probably important at sites of infection, where the availability of iron for bacteria is extremely limited due to competition from host iron-binding protein. P. aeruginosa synthesises and secretes 2 types of siderophores: pyochelin and pyoverdin. Additional potential sources of catalytically active iron related to host-microbe interactions include iron released from haemoglobin through the action of the bacterial toxin haemolysin, host cell exposure to bacterially derived iron reduction compounds such as pyocyanin, and/or the release of intercellular iron from damaged mammalian or bacterial cells into the microenvironment, it is necessary that extracellularly generated iron catalysts remain in close proximity to the cell in order to facilitate OH^--mediated injury, given the limit diffusibility of OH^-. Coincident with their production of O_2^- and H_2O_2, stimulated human phagocytes release 1 or 2 distinct peroxidases from their cytoplasmic granules. In the case of neutrophils and monocytes, this enzyme is myeloperoxidase (MPO), whereas for eosinophil it is eosinophil peroxidase (EPO). The interaction of MPO and EPO with H_2O_2 forms hypohalous acids (HOX, where X = halide). Myeloperoxidase is a glycoprotein, consists of a pair of glycosylated heavy(α)-light(β) protomers, each of which contains an iron atom. EPO is an $\alpha\beta$ glycoprotein, similar in structure to hemi-MPO. These enzymes are cationic, thus allowing them to stick to cell surfaces and perhaps enhancing their potential for cell injury by increasing the local concentration of hypohalous acid at the target cell membrane. Hypohalous acids are potent oxidants known to have several cytotoxic effects on mammalian and bacterial cells. Cell membrane integrity may be violated by membrane peroxidation and the oxidation and/or decarboxylation of membrane proteins. Nitric oxide is cytostatic or cytotoxic for both prokaryotic and eukaryotic cells. The primary mechanism of injury involves the interaction of NO with iron-containing moieties in key enzymes of the respiratory cycle and with DNA synthesis leading to mutagenesis in target cells. The formation of nitrosothiol groups on proteins can lead to interactivation of enzymes or changes in protein function. These groups can react further to cross-link sulfydryl groups and thus initiate a chain reaction. NO and its derivatives can form toxic alkylating agents by reacting with secondary amines. $ONOO^-$ is generated by the reaction of O_2^- with NO^-. Its ability to directly oxidize sulfydryl groups and DNA bases, catalyse iron-independent membrane lipid peroxidation, and react with metals or metalloproteins (SOD) to form the toxic nitronium ion (NO^{2-}), suggesting that $ONOO^-$ plays a more important role than its precursor NO in mediating cytotoxicity. $ONOO^-$ can undergo homolytic cleavage to form OH^- by an iron-independent mechanism. Aerobic microorganisms are continually exposed to endogenous sources of toxic oxygen species as a consequence of aerobic metabolism. Univalent electron reduction of molecular O_2 generates such species, as superoxide, hydrogen peroxide and hydroxyl. These toxic oxygen species also can be generated as by-products of reactions involving glucose oxidase, xanthine oxidase, thiol groups and flavins. Microbial exposure to ultraviolet or γ-irradiation induces O_2^- production. Anaerobic organisms are particularly susceptible to oxidants derived via the above mechanisms, as they often lack the antioxidant defence mechanisms observed in aerobic organisms. E coli, and other microorganisms (streptococcus pneumoniae), generate extracellular O_2^- and H_2O_2. H_2O_2-producing lactobacillus spp. inhibit Neisseria gonorrhoeae and human immunodeficiency virus (HIV), suggesting a nonspecific antimicrobial defence mechanism resulting from the presence of lactobacilli in the normal vaginal flora. Women with bacterial vaginosis, H_2O_2-producing lactobacilli are notably absent from the vaginal flora. Microorganisms are also continually exposed to endogenously produced NO· through dentrification, that involves the transformation of oxyanions of nitrogen to N_2, mainly under conditions of reduced oxygen tension or strict anaerobiosis. Dentrification is controlled by a number of metalloenzymes, of which nitrite reductase has been identified as the enzyme responsible for the conversion of nitrite to NO·. Two nitrite reductases among dentrifying bacteria: a tetraheme cytochrome cd located in the periplasma of gram-negative organisms, and Cu-containing protein bound to the cytoplasmic membrane of gram-positive organisms. The locations of these enzymes may limit the potential toxicity of the endogenously produced NO·, as NO is subsequently rapidly reduced by a cytoplasmic membrane-associated NO· reductase. When phagocyte encounters a microorganism, the latter is surrounded by a portion of the phagocyte membrane, which then invaginates forming a district phagosome. This increased phagocyte oxygen consumption and initiates a complex biochemical signalling system, which activates a unique membrane-associated NADPH-dependent oxidase complex. This enzyme univalently reduces O_2 to O_2^-, which is then secreted into the phagosome, where O_2^- dismutases to H_2O_2. These toxic compounds may also leak extracellularly as the phagosome is closing. Following phagocytosis, microorganisms are subjected to further insult as phagocyte primary cytoplasmic granules fuse with the phagosome. In addition to MPO, these granules contain mainly hydrolases (such as, acid hydrolases, lysozyme, neutral proteases, deoxyribonucleases), which are probably responsible for the decomposition of killed organisms. Secondary cytoplasmic granules fuse with the external plasma membrane before the primary granules do, thereby secreting their contents (lactoferrin, lysozyme, and vitamin B_{12}-binding protein) extracellularly. The membranes of these secondary granules also contain a number of functionally significant proteins, including CD11b/CD18, the formyl-methionyl-leucyl-phenylalanine receptor, and cytochrome b_{558}. The fusion of these granules with the plasma membrane serves to reinforce or sustain various cellular responses. Defects in the NADPH-oxidase complex results in a lack of phagocyte O_2^- production. The NADPH-oxidase requires the assembly of its membrane and cytosolic components for generation of the respiration burst. The genetic defects in this enzyme observed among chronic granulomatous disease (CGD) patients, are characterised by their localisation to the membrane or cytosol. These persons suffer from recurrent pyogenic infections with organisms that are normally rapidly killed by oxidants: Staphylococcus aureus, enteric gram-negative rods, Candida spp, typically begin in infancy and recur throughout childhood and adolescence. Oxidants are not critical for neutrophil-mediated killing of P. aeruginosa. Neutrophils from CGD patients have the same capacity as normal neutrophils to kill P. aeruginosa. Without the presence of ambient O_2, neutrophils are unable to generate O_2^- and H_2O_2. However, their abilities to kill P. aeruginosa under aerobic and anaerobic conditions appear to be similar. This is in contrast to finding with S. aureus, in which neutrophils are unable to kill the organism under anaerobic conditions. Phagocyte-derived H_2O_2 may also be converted intra- or extra-cellularly to HOCl and the longer-lived chloramines in the presence of chloride and myeloperoxidase. MPO can catalyse the reaction of O_2^- and HOCl to form OH·. All of these compounds are known to have a number of cytotoxic effects. The interaction of exogenous H_2O_2 at low concentrations with intracellular Fe^{2+} in E coli results in DNA damage mediated by the ferryl radical. Bacterial exposure to higher H_2O_2 concentrations can result in killing by separate oxidative mechanism. NO· has been recognised as another phagocyte-derived oxidant involved in microbicidal activity. The inducible NOS can be induced by a number of cytokines and lipopolysaccharides. The primary microbial effect of phagocyte-derived NO· appears to involve intracellular pathogens. NO· is having microbiostatic and/or microbial activity against pathogens. NO· play an important role in containing the extent of infection and decreasing the overall organism load. TNF-α and granulocyte-macrophage colony-stimulating factor stimulate human macrophages to restrict the growth of virulent Mycobacterium avium by a mechanism involving NO· production. Nitric oxide may also contribute to the killing of staphylococci by neutrophil cytoplasts (anucleate, granule-poor, motile cells) which rapidly took up and kill the bacteria. Several antimicrobial agents used in the treatment of clinical infections, in addition to blocking key enzymes and other metabolic functions of microorganisms, produce reactive oxygen intermediates that are capable of damaging other biomolecules. β-lactam antibiotics (penicillins and cephalosporins) have been shown to oxidatively damage DNA and deoxyribose in the presence of iron and copper salts, consistent with an OH·-mediated mechanism. The polyunsaturated structure of the polyene antifungal antibiotics (amphotericin, nystatin, and natamycin) gives them the propensity to oxidize to form peroxy radicals and thiobarbituric acid-reactive aldehyde fragments. These interactions can then lead to the generation of other oxygen-centered radicals capable of inciting further microbial injury. A number of compounds undergo rapid redox cycling under aerobic conditions, potentially resulting in an additional source of extracellular oxidants for microbial encounter. These compounds are univalently reduced to free radicals by cellular systems. In the presence of O_2, these reduced molecules are then reoxidised, with the resulting transfer of that electron to O_2, forming O_2^- and H_2O_2, the latter via O_2^- dismutation, e.g., adriamycin, bleomycin, and nitrofurantoin. The P. aeruginosa pyocyanin is a phenazine-derived pigment that undergoes redox cycling to induce both intra- and extra-cellular O_2^- and H_2O_2 production from O_2 in both eukaryotic and prokaryotic cells. This process contributes to cell death through the diversion of electron flow from normal biologic pathways into those leading to toxic oxidant generation.

OFRs chemical reactions:

Reaction	Formula
Haber-Weiss reaction	$O_2^- + Fe^{3+} \rightarrow O_2 + Fe^{2+}$
	$H_2O_2 + Fe^{2+} \rightarrow .OH + OH^- + Fe^{3+}$
	$O_2^- + H_2O_2 \rightarrow OH^- + OH^- + O_2$
Myeloperoxidase (eosinophil)	$H_2O_2 + HX \rightarrow HOX$ (hypohalous acids) $+ H_2O$ (X = halide)
Nitric oxide synthase	L-Arginine $\rightarrow$ L-citruline $+ NO^-$
Peroxynitrate formation/decomposition	$NO^- + O_2^- \rightarrow ONOO^-$
	$ONOO^- + H^+ \rightarrow ONOOH$
	$ONOOH \rightarrow OH^- + NO^-$
GSH	
Peroxidase	$2GSH + H_2O_2 \rightarrow G\text{-}S\text{-}S\text{-}G + 2H_2O$
Reductase	$G\text{-}S\text{-}S\text{-}G + 2NADPH \rightarrow 2GSH + 2NADP$
Catalase	$2H_2O_2 \rightarrow 2H_2O + O_2$
SOD	$2O_2^- + 2H^+ \rightarrow H_2O_2 + O_2$

MECHANISMS OF MICROBIAL DEFENCE AGAINST OXIDANTS

Evidence of phagocyte-derived oxidants: 1- Some microorganisms secrete toxins to kill the phagocyte before they can be killed by it, e.g., leukocidin by staphylococcus aureus, and clostridium septicum toxin. 2- Pathogenic mucoid strains of P. aeruginosa synthesise alginate, an exopolysaccharide. Alginate has the ability to scavenge reactive oxygen intermediates, suppress leukocyte function, and promote bacterial adhesion. 3- P. aeruginosa also require a unique glucose-dependent pathway for phaygocytosis by macrophages, this may enhance its pathogenicity in the bronchoalveolar space where concentrations of glucose and other carbohydrate are low. 4- Acidification within the phagocytic vacuole is an important process to maximise the spontaneous dismutation of O_2^-, hydrolase activity, and phagosome-lysosome fusions. Legionella pneumophilia and Toxoplasma gondii can inhibit this acidification process. 5- Activation of protein tyrosine kinases is an important signalling mechanism in many cells, including macrophages. By interfering with host signalling pathways, will modify the host immune response, e.g., Yersinia spp. Specific defence against oxidants: A- Non-enzymatics: 1- Inhibition of protein kinase C activity in macrophages, resulting in suppression of respiratory burst ultimately of O_2^- production, also impedes macrophage chemotactic locomotion and IL1 production, e.g., Legionella pneumophila. 2- Antioxidant scavengers unique to specific pathogens have also evolved to protect microorganisms from phagocyte-derived oxidants. Mannitol in high concentrations has the ability to scavenge reactive oxygen species, its production by Cryptococcus neoformans may be a protective mechanism by which the organism protects itself from oxidative killing by host phagocytes. 3- Accumulation of mannitol may influence C. neoformans tolerance to heat and osmotic stresses and its phagogenicity. 4- The formation of heat shock proteins (HSP), can be induced by increased temperature and/or oxidant exposure (which may include phagocytosis) as a means of protection against both heat and oxidant damage. HSP may play a role in the regulation of antioxidant enzyme production in E coli. 5- Several stresses also increase the production of antioxidant enzymes such as SOD. B- Enzymatic: 1- Glutathione serves as a substrate for the H_2O_2-removing enzyme glutathione peroxidase. It can then be redox cycled via glutathione reductase for further H_2O_2 removal. GSH is also an OH^- scavenger. Eukaryotic cells depleted of GSH exhibit increased susceptibility to oxidant-mediated killing. GSH depletion is involved in HIV replication. GSH reductase-negative E coli mutants do not demonstrate an increased susceptibility to H_2O_2-mediated stress compared with the isogenic parenteral strain. GSH may facilitate the deactivation of E coli aconitase and other-containing dehydrogenases that been oxidatively inactivated by O_2^-. There may be several types of GSH-metabolising proteins in bacteria, which serve a similar purpose, and their distribution may even vary within a single sepsis. 2- The antioxidant action of the peroxidase enzymes is to catalytically convert H_2O_2 to H_2O and O_2. Nearly all aerobic and facultatively anaerobic microorganisms, with the exception of the Streptococcus spp. synthesise at least one form of catalase and/or peroxidase. The majority of obligate anaerobes lack this capability. The catalases of E coli, are hydroperoxidase I (HPI) and hydroperoxidase II (HPII). HPI is a bifunctional catalase-peroxidase contains 2 protoheme IX groups associated with a tetramer of identical 80-kDa subunits and is localised in the periplasmic space. HPII is a monofunctional catalase consists of 84.2 kDa subunits and is found solely in the cytoplasm. HPI is synthesised in response to oxidative stress (H_2O_2), whereas HPII is produced in response to nutrient depletion as occurs in the stationary growth phase. Neutrophils easily kill low- but not high-catalase-producing Staphylococcus aureus strains. Catalase-deficient E coli mutants exhibit an increased susceptibility to phagocyte-mediated killing. 3- SOD, production is a key defence strategy aimed at the elimination of O_2^-. Not only does this decrease the possibility of direct O_2^--mediated toxicity, but also it prevents O_2^--mediated reduction of iron and subsequent OH generation via the Haber-Weiss reaction. The eukaryotes, several Haemophilus spp, E coli, and some higher fungi predominantly produce CuZnSODs, homodimers with 2 noncovalently linked identical subunits containing one atom each of copper and zinc. All bacteria, produce either FeSOD, MnSOD, or both. The metal content of both isoenzymes varies between 1 and 2 atoms per dimer. Most SODs are cytoplasmically located, although a few are located on or secreted through the cytoplasmic membrane. FeSOD predominates in anaerobic organisms whereas MnSOD is more commonly found among aerobic organisms. Microbial SOD regulation and genetics, is dependent on a number of environmental stimuli. FeSOD, encoded by the sodB gene, is produced constitutively in E coli grown aerobically or anaerobically but is upregulated when grown aerobically in the presence of nitrate. MnSOD becomes the predominant form when the cell is exposed to oxidative stress. FeSOD provides E coli with the first line of defense against O_2^- and MnSOD is subsequently recruited in circumstances of increased oxidative stress.

ROLE OF OXIDANTS IN VIRAL INFECTIONS
HIV infection is associated with proinflammatory state in the host, resulting in high levels of circulating cytokines, including TNF-α, IL-1α, IL-1β, IL-2, IL-6, IFNα, and IFN-γ. Cytokines can activate HIV replication in the infected host cell directly, cytokine activation of phagocytes and other cells can also stimulate oxidant production. Oxidants also have direct effects on HIV replication. H_2O_2 increases replication of the HIV. Cytokines direct activation of HIV is mediated by the induction of NF-κB, an ubiquitous transcription factor that is recognised by HIV promoter. HIV gene expression enhances T-cell susceptibility to H_2O_2-induced apoptosis. HIV-infected cells may be sensitive to oxidant stress, due to low levels of GSH, the main intracellular defense against oxidants. HIV-infected patients demonstrate decreased GSH levels in blood and peripheral blood mononuclear cells relative to those in normal patients, and this decrease becomes more pronounced with advanced disease. In patients with symptomatic AIDS, GSH concentrations in CD8 and CD4 T cells are 62% and 63%, respectively, of those found seronegative controls. The greatest decreases in GSH levels are seen in those patients with advanced infection. Decrease in intracellular GSH levels leave the infected cell susceptible to the direct effects of oxidants, and also it leads to increased NF-κB expression, resulting in further activation of HIV replication. HIV-mediated modification of host antioxidant enzymes may be an important component in mediating ongoing HIV infection and the ultimate progression to severe immunodeficiency. This process may be further altered in the presence of opportunistic pathogens. HIV expression can be decreased by treatment of the cells with GSH, glutathione ester, or N-acetylcysteine. Each of these compounds increases intracellular thiol concentrations and, as a result, inhibit NF-κB and ultimately HIV expression. There are no significant differences in CD4 cell counts, viral load, or proviral DNA frequency. The oxidant scavenger ascorbate also suppresses HIV replication in chronically and acutely infected T cells. Oxidants may also be involved in the pathogenesis of other viral infections. As in HIV-infected cells, H_2O_2 effectively induces synthesis of viral antigens in several lymphoid cells that harbour the Epstein-Barr virus genome. Levels of vitamin E in plasma are notably low in patients with chronic liver disease. Vitamin E is an important cell membrane antioxidant, which acts as a free radical scavenger. Approximately 50% of healthy women posses polyamine oxidase and/or diamine oxidase in their cervical mucus. These enzymes act on spermine and spermidine (polyamines present in seminal fluid) to generate H_2O_2 and reactive aldehydes, which are likely to exert local mutagenic effects. These transformed cervical cells may exhibit prolonged survival in the presence of HPV (Human papiloma virus) infection through HPV suppression of apoptosis in the keratinocytes. Polyamine oxidation occurring in the cervical environment of sexually active women. Viral-host cell interactions in relation to oxidant production also appear to be important in the pathogenesis of influenza. Although exposure to the virus leads to neutrophil activation and generation of respiratory burst, the neutrophil response is atypical with regard to calcium fluxes, phospholipase C activation, and release of H_2O_2.

UNFAVOURABLE CONSEQUENCES OF OXIDANT PRODUCTION
At sites of infection, host-derived oxidants not only place the offending organisms under oxidative stress but also cause stress to neighbouring host tissues. These oxidants are derived primarily from phagocytes, can be produced by

other cell types via induction by redox-active agents. Tissue injury at sites of infection may be the result of the host inflammatory response to the pathogen rather than cytotoxic components of the microorganism. Many aspects of acute and chronic inflammatory tissue injury appear to be mediated by oxidants released by neutrophils and other phagocytes. This process is enhanced by adherence of the phagocyte to the target cell surface with subsequent movement of phagocytes from the blood to sites of inflammation require a complex signalling system involving a family of polyproteins termed selectins. Selectins are synthesised by endothelial cells and stored in their secretory granules. When endothelial cells are activated by compounds such as thrombin or histamine (released in response to inflammation), the granules fuse with the outer membrane to expose the selectins on the cell surface. Phagocytes recognise these proteins, and this promotes their adherence to the endothelium and primes them for degranulation. Following leukocyte activation, phagocyte-derived proteins termed integrins bind to their respective receptors on the endothelial cell. This interaction strengthens adhesion and directs the migration of the phagocyte beneath the endothelium. Thus this process targets phagocytes to areas of inflammation, where oxidant-mediated tissue injury may results. Phagocyte-derived H_2O_2 can also be indirectly lead to inflammatory tissue injury by upregulating selectin expression on endothelial cells and promoting further neutrophil localisation. Inflammatory tissue injury may also result via oxidant induced cellular production of proinflammatory cytokines. The production of cytokines may potentiate further cellular oxidant release. Many of these interactions are mediated through the transcription-regulating factor, NF-κB. Endotoxin-induces NF-κB activation via the production of cytokines-induced chemoattractant by macrophages. Oxidants can also activate NF-κB, promoting the production and release of cytokines such as IL1 and TNF-α. Joint inflammation can also be induced by bacterial products, immune complexes, and crystals, which recruit and activate phagocytes to form ROS. This process can result in tissue destruction via oxidant interactions with host proteoglycans, collagen, and elastin. Peroxynitrite inhibits epithelial cell ion channels, that could contribute to diffusion barrier disruption under conditions in which both O_2^- and NO are present concurrently. MPO-derived oxidants released in response to a microbial stimulus may also contribute to inflammatory tissue injury directly via their toxic effects and indirectly by their ability to inactivate serine protease inhibitors such as α_1-antitrypsin. These antiproteases play a critical role in limiting the activity at local sites of inflammation. NO may also have antioxidants properties. Bacterially derived lipopolysaccharide induces NO production in endothelial cells, may contribute to the vasodilation and hypotension observed in septic shock. Like microorganisms, host cells have evolved a complex system to defend themselves against oxidant injury, such as, synthesis of a manganese-containing enzyme (MnSOD) induced in the mitochondrial matrix under conditions of increased oxidative stress specific cytokine stimulation, or heat shock. Since the H_2O_2 formed by the dismutation of O_2^- is also cytotoxic, eukaryotic cells have developed various mechanisms for its removal analogues to those found in prokaryotic microorganisms. Preventing the formation of O_2^- and H_2O_2 is the primary mechanism by which cells can limit the formation of other potent oxidants such as OH and the MPO-derived oxidants. Hydroxyl radical generation via the Haber-Weiss reaction can also be controlled by limiting the availability of redox-active iron catalysts through the formation of less active iron complexes such as extracellular lactoferrin and transferrin and intracellular ferritin. Heme oxygenase mRNA expression in mammalian cells is increased following cell exposure to oxidant stress. Heme oxygenase decreases the availability of intracellular iron capable of participating in the Haber-Weiss reaction by catalysing the conversion of free heme to bile pigments. These bile pigments in turn exert antioxidant effects. The extent of oxidant-mediated cytotoxicity observed at sites of inflammation is dependent on the balance between host- and microorganism-derived prooxidant forces. When this balance is in favour of the prooxidants, microbial and host cell cytotoxicity results, leading to clinical manifestation such as the sepsis syndrome, acute respiratory distress syndrome.

REGULATION OF BACTERIAL VIRULENCE GENE EXPRESSION BY THE HOST ENVIRONMENT
Bacterial pathogens must adapt to a wide variety of changing environment conditions during the infection cycle. Upon encountering the host, pathogens must express specific gene products to persist and proliferate in the appropriate location and to circumvent host defenses. During the course of the infection, the host environment may change markedly. All pathogens must also ensure successful transmission, which may entail prolonged periods in external environments or adaptation to an intermediate host of a different species. All bacteria regulate gene expression in response to different environment signals, a property that is crucial to their ability to compete with other organisms. These virulence genes are subject to complex regulatory mechanisms that ensure expression in the appropriate host environment. The general metabolic economy of the bacterial cell selects against unnecessary expenditure of energy. Surface molecules that localise the bacteria to a specific site in the host have to be altered if the organism is to change its location or be transmitted to a new host. Invasive pathogens need one set of genes to gain access to the host and different set to proliferate and circumvent host defenses in the tissues. Some pathogens can downregulate or switch antigens during the course of infection. Signals control of virulence gene expression is, simple physical and chemical factors: temperature, osmolarity, O_2, CO_2, pH, ROI and RNI, nutrient availability, and

inorganic ion concentrations. Many of these factors change in the transitions from the external environment to the host, between different locations in the host, and between different hosts. Bacteria are able to respond to a complex environment by integrating the signals from a variety of these cues to sense their location and express the appropriate genes. Temperature is a particularly useful signal for pathogens since the host temperature is relatively constant and elevated with respect to ambient environment. Temperature-sensitive gene expression allows the organism to discriminate between the host and the outside world, and certain genes required to produce disease. Elevation in temperature also induces the heat shock stress response, a prominent regulatory circuit important for many pathogens. Invasive pathogens are exposed to the toxic effects of ROI and RNI produced by phagocytic cells, inducing crucial resistance and repair mechanisms. Inorganic ion concentrations are used by pathogenic bacteria to sense their location. Iron plays a key regulatory role, since virtually all available iron on mucosal surfaces and in the tissues is bound to host proteins. Low iron induces a variety of virulence traits including toxin production in uropathogenic E coli. An octapeptide pheromone has been reported in Staphylococcus aureus, accumulates in the post-exponential phase of growth, signalling the cells to express secreted virulence proteins and repressing the production of surface proteins. Conditions that control virulence gene expression in pathogens also stimulate global regulatory changes in many genes common to both pathogens and commensal bacteria. Efficient entry of organism into epithelial cells requires production of secreted invasion proteins, which act both on the epithelial cell surface and internally to remodel the cell membrane into ruffles, leading eventually to macropinocytosis of the bacteria. The protein secretion is stimulated by contact with host cells.

ROLE OF COMPLEMENT

Inappropriate activation and excess production of complement may contribute to tissue injury in many diseases. C5a is capable of amplifying the inflammatory responses via its chemoattractant properties, its induction of granule release in phagocytic cells, and its ability to induce phagocyte generation of oxygen metabolites. Neutrophils store C3, C7, and C6, which can be released under specific stimulatory conditions. Monocytes and macrophages produce all components of the terminal complement pathway. Human cells are protected from complement by membrane proteins, which inhibit activation and its effects at several stages of the pathway. These include: 1- decay-accelerating factor (DAF, CD55); 2- membrane cofactor protein (MCP, CD46); 3- complement receptor 1 (CR1, CD35); and 4- CD59.

ACUTE PHASE PROTEINS

Acute phase proteins are synthesised almost exclusively in the liver and most are glycosylated. They serve important functions in restoring homeostasis after infection and inflammation. These include haemostatic functions (fibrinogen), microbicidal and phagocytic functions (complement components, C-reactive protein), antithrombotic properties (α1-acid glycoprotein), and antiproteolytic actions that are important to contain protease activity at sites of inflammation (α2-macroglobulin, α2-antitrypsin, and α1-antichymotrypsin). Acute phase proteins can be divided into two groups: 1- Type I, are induced by IL1α, IL1β, TNF-α and TNF-β. Type I include SAA (serum amyloid A), CRP (C-reactive protein), complement C3, haptoglobin, and a1-acid glycoprotein. 2- Type II, are induced by IL6, LIF (leukaemia inhibitory factor), IL11, OSM (oncostatin M), CNTF (ciliary neurotrophic factor), and CT1 (cardiotrophin-1). Type II includes fibrinogen, haptoglobin, α1-antichymotrypsin, α1-antitrypsin, and α2-macroglobulin. In general, IL6-like cytokines synergize with IL1-like cytokines in the induction of type I acute phase proteins, whereas IL1-like cytokines have no effect on, or even inhibit, the induction of type II acute phase proteins. Signal transduction: IL1-like cytokines: activation of TNF/IL receptors initiates the conversion of membrane sphingomyelin to ceramide via sphingomyelinase. Ceramide-activated protein kinases connect to several signalling pathways, which ultimately lead to activation and translocation factors AP-1 (activating protein-1, c-jun/c-fos heterodimer) and nuclear factor NF-κB. NF-κB is activated and translocated to the nucleus after phosphorylation and degradation of the inhibitory subunit I-κB. In addition, IL-1 signal connects to the mitogen activated protein (MAP)-kinase pathway, ultimately activating transcription factor NF-IL-6. Many type I acute phase protein genes contain NF-κB, NF-IL-6 and AP-1 response elements in their promotor regions. Activation of the IL6 receptor complex activates JAK tyrosine kinases. Tyrosine phosphorylation of STAT proteins (signal transducers and activators of transcription), including STAT3, also known as APRF (acute phase response factor), induces STAT protein homo- and hetero-dimerization and translocation to the nucleus, where they bind to their response elements. IL6 signal transduction pathway activates the MAP-kinase pathway, connecting the IL6 and IL1 signalling pathways and converging on NF-IL6, another IL6 response element. The initial trigger leads to rapid activation of transcription factor NF-κB, resulting in increased expression of cytokines. Binding to cytokine receptors initiates signalling events, leading to activation of several transcription factors. These factors bind to their response elements in the promotor regions of acute phase genes. Transcription of acute phase genes is induced and acute phase proteins are secreted. In the extracellular milieu, acute phase proteins function to restore homeostasis. Fibrosis is a local inflammatory process in the tissue, characterised by increased deposition of extracellular matrix components most notably collagen. Fibrogenesis is an extremely complex process in which locally produced cytokines and growth factors act on several cell types. A key event in the development of fibrosis expression of TGFβ, which is expressed by many cell types. TGFβ induces the expression of PDGF receptors on activated cells, rendering them responsive to the mitogenic effect of PDGF. Proteolytic events play an important role in the inhibition and perpetuation of fibrosis. TGFβ is secreted as an inactive latent precursor. Activation of latent TGFβ into active TGFβ is protease-mediated, probably by plasmin provided by endothelial cells. Plasmin actively is regulated by plasminogen activators, in particular urokinase type plasminogen activator (uPA), which convert inactive plasminogen into active plasmin. The plasminogen activators are under the control of plasminogen activator inhibitor type 1 (PAI-1). TGFβ is a powerful profibrogenic factor, directly stimulates matrix gene transcription and synthesis, while it inhibits interstitial matrix degradation via inhibition of matrix metalloproteinase (MMP) synthesis and increased synthesis of TIMPs (tissue inhibitor of metalloproteinases). MMPs are synthesised as active precursors (proMMPs), which are activated by plasmin.

Acute-phase proteins:

Positive acute-phase proteins:

	A- Complement system.	1- C3, C4, C9.
		2- Factor B.
		3- C1 inhibitor.
		4- C4b-binding protein.
		5- Mannose-binding lectin.
	B- Coagulation and fibrinolytic system.	1- Fibrinogen.
		2- Plasminogen.
		3- Tissue plasminogen activator.
		4- Urokinase.
		5- Protein S.
		6- Vitronectin.
		7- Plasminogen-activator inhibitor 1.
	C- Antiproteases.	1- α_1-Protease inhibitor.
		2- α_1-Antichymotrypsin
		3- Pancreatic secretory trypsin inhibitor.
		4- Inter-α-trypsin inhibitors.
	D- Transport proteins.	1- Ceruloplasmin.
		2- Haptoglobin.
		3- Haemopexin.
	E- Participants in inflammatory responses.	1- Secreted phospholipase A_2.
		2- Lipopolysaccharide-binding protein.
		3- IL1R antagonist.
		4- GCSF.
	F- Others.	1- C-reactive protein.
		2- Serum amyloid A.
		3- α_1-Acid glycoprotein.
		4- Fibronectin.
		5- Ferritin.
		6- Angiotensinogen.

Negative acute-phase proteins:
Albumin, Transferrin, Transthyretin, α_2-HS glycoprotein, AFP, TBG, IGF-I, Factor XII.

Cytokines modulate both the proliferative responses and the matrix-producing processes. Some of those known to induce acute phase proteins are locally released during the development of fibrosis. Acute-phase changes reflect the presence and intensity of inflammation, and they have long been used as clinical guide to diagnosis and management. Acute phase changes may be divided into changes in the concentrations of many plasma proteins, known as the acute phase proteins, and a large number of behavioural, physiologic, biochemical, and nutritional changes. Acute phase protein is one whose plasma concentration increases (positive acute phase) or decreases (negative acute phase) by at least 25% during inflammatory disorders. The changes in the concentrations of acute-phase proteins are due largely to changes in their production by hepatocytes. Cytokines are intracellular signalling polypeptides produced by activated cells. Most cytokines have multiple sources, multiple targets, and multiple functions. IL6, IL1β, TNFα, IFNγ, TGFβ, and possibly IL8 are the chief stimulators of the production of acute-phase proteins. IL6 is the chief stimulator of the production of most acute-phase proteins. Lipopolysaccharide causes the production of other cytokines capable of stimulating the production of acute-phase proteins. IL11, LIF, OSM, CNF,

CT1 may have actions similar to those of IL6. Cytokines operate both as a cascade and as a network in stimulating the production of acute-phase proteins. The effects of cytokines on target cells may be inhibited or enhanced by other cytokines, by hormones, and by cytokines-receptor antagonists and circulating receptors. IL6 enhances the effect of IL1β. Glucocorticoids generally enhance the stimulatory effects of cytokines on the production of acute-phase proteins, whereas insulin decreases their effects on the production of some acute-phase proteins. Fever is representative of the neuroendocrine changes that characterise the acute-phase response. Although several cytokines may induce fever, IL6 produced in the brain stem is required for final steps leading to fever. Stimulation of the production of arginine-vasopressin by IL6 can explain the hyponatraemia that occurs during some inflammatory disorders. Neural mechanisms have been implicated in anorexia as well as cytokines, vagal afferent is required for the induction of anorexia by IL1β and lipopolysaccharide. Increased plasma leptin concentrations occur in inflammation, probably in response to stimulation of adipocytes by cytokines, and may also contribute to anorexia. Pathogenesis of anaemia in chronic diseases include decreased responsiveness of erythrocyte precursors to erythropoietin, decreased production of erythropoietin, and impaired mobilisation of iron from macrophages. Hypoferraemia results largely from the sequestration of iron in macrophages by apoferritin produced in response to the inflammation-associated cytokines IL4 and IL13. The thrombocytosis of inflammation appears to be caused by IL6. Cachexia, the loss of body mass occurs during severe chronic inflammatory disease, results from decrease in skeletal muscle, fat tissue, and bone mass. IL1β, IL6, TNFα, and IFNγ all contribute to these processes.

Other acute-phase phenomena:

A- Neuroendocrine changes:
 1- Fever, somnolence, and anorexia.
 2- Increased secretion of CTRH, corticotropin, and cortisol.
 3- Increased secretion of arginine vasopressin.
 4- Decreased production of IGF-I.
 5- Increased adrenal secretion of catecholamines.
B- Haematopoietic changes:
 1- Anaemia of chronic disease.
 2- Leukocytosis.
 3- Thrombocytosis.
C- Metabolic changes:
 1- Loss of muscle and negative nitrogen balance.
 2- Decreased gluconeogenesis.
 3- Osteoporosis.
 4- Increased hepatic lipogenesis.
 5- Increased lipolysis in adipose tissue.
 6- Decreased lipoprotein lipase activity in muscle and adipose tissue.
 7- Cachexia.
D- Hepatic changes:
 1- Increased metallothionin, iNOS, heme oxygenase, manganese superoxide dismutase, and tissue inhibitor of metalloproteinase-1.
 2- Decreased phosphoenolpyruvate carboxykinase activity.
E- Nonprotein plasma constituents changes:
 1- Hypozincaemia, hypoferraemia, and hypercupraemia.
 2- Increased plasma retinol and glutathione concentrations.

IL6 increases the production of the metal-binding protein metallothionein, with consequent increased zinc binding and hypozincaemia. IL1β and TNFα decrease the expression of growth-hormone receptors on hepatocytes, with subsequent decreased responsiveness to growth hormone and low plasma concentrations of IGF-I. C-reactive protein has many pathophysiologic roles in the inflammatory process. C-reactive protein binds phosphocholine and thus recognises some foreign pathogens as well as phospholipid constituents of damaged cells. C-reactive protein activates the complement system when bound to one of its ligands and can also bind to phagocytic cells. C-reactive-protein initiates the elimination of targeted cells by its interaction with both humoral and cellular effector systems of inflammation. C-reactive protein induces inflammatory cytokines and tissue factor in monocytes. C-reactive protein prevents the adhesion of neutrophils to endothelial cells by decreasing the surface expression of L-selectin, to inhibit the generation of superoxide by neutrophils. C-reactive-protein stimulates the synthesis of IL1R antagonist by mononuclear cells. Several acute-phase proteins initiate or sustain inflammation. Complement activation leads to chemotaxis, plasma protein exudation at inflammatory sites, and opsonization of infectious agents and damaged cells. GCSF increases the inflammatory response by increasing the numbers of granulocyte precursor in bone marrow and by activating mature granulocytes. The antioxidants haptoglobin and haemopexin protect against OFR. Wound healing is influenced by 2 acute-phase proteins: 1- fibrinogen causes endothelial cell adhesion, spreading, and proliferation, all critical to tissue repair; and 2- haptoglobin aids in wound repair by stimulating angiogenesis. A decrease in plasma concentration of transthyretin may be pro-inflammatory, as this negative acute-phase protein inhibits IL1 production by monocytes and endothelial cells. Somnolence associated with inflammatory states may reduce demands for energy. The adaptive value of fever has been attributed to both enhancement of immunity and the stabilisation of cell membranes. Glucocorticoids maintain the haemodynamic stability during severe illness and can modulate the immune and inflammatory responses to different noxious stimuli. The increase in plasma lipid concentrations may provide nutrients to cell involved in host defense and substrates for the regeneration of damaged membranes. Circulating Lipoproteins can bind to and decrease the toxic effects of lipopolysaccharide, thus participating in the defense against microbes. Increased production of heme oxygenase and manganese superoxide dismutase may limit oxidant-mediated tissue injury, whereas inhibitor of metalloproteinase-1 antagonises the destructive effects of metalloproteinases. When extreme, cytokine-induced changes associated with the acute-phase response can be fatal, as in septic shock. Elevated serum amyloid-A concentrations in some chronic inflammatory conditions, leads to secondary amyloidosis. Both anaemia and hypoalbuminaemia due to inflammation are common among hospitalised patients. Measurements of plasma or serum C-reactive protein can help differentiate inflammatory from non-inflammatory conditions and are useful in managing the patient's disease, since the concentration often reflects the response to and need for therapeutic intervention. Serum amyloid A concentrations parallel those of C-reactive protein. Serum amyloid A is a more sensitive marker of inflammatory disease, but assays are not widely available. The erythrocyte sedimentation rate (the rate at which erythrocyte fall through plasma) depends on the plasma concentration of fibrinogen. The erythrocyte sedimentation rate (ESR) has the advantage of familiarity, and simplicity. The ESR is an indirect measurement of plasma acute-phase protein concentrations and can be greatly influenced by the size, shape, number of erythrocytes, and by other plasma constituents such as immunoglobulins. As a patient's condition worsens or improves, the ESR changes relatively slowly, whereas C-reactive protein concentrations change rapidly. Patients with plasma C-reactive protein concentrations >100 mg/l, 80-85% have bacterial infections. The ESR increases steadily with age, but plasma C-reactive protein concentrations do not. Normal values of C-reactive protein concentrations range from 2 mg/l or less to 10 mg/l, due to low-grade processes such as gingivitis or trivial injury. Measurement of cytokines in plasma is difficult, because of their short plasma half-lives and the presence of blocking factors. Plasma IL6 concentrations are elevated in patients with many inflammatory diseases, but except for the rapidity with which change occurs, measurement of plasma IL6 concentrations has no apparent advantage over measurement of plasma C-reactive protein.

ROLE OF PROTEASES AND OTHER INFLAMMATORY PROTEINS

The inflammatory processes are modulated by a number of different cellular and plasma including serine proteases, metalloproteinase, cationic proteins, acid hydrolyases, myloperoxidase, and lactoferrin. Proteinases, in addition to their role in cellular cytotoxicity, can influence monocyte motility and chemotaxis, modulate cytokines responses, contribute to mononuclear cell proliferation, or induce target cell apoptosis. Thrombin, in addition to its role in leukocyte growth control, exerts a potent proinflammatory effect through its modulation of leukocyte chemotaxis and adherence properties, also it has a direct effect (chemoattractant) on leukocyte migration, through modulation of gene expression and cytokine release. Serine proteases are stored in secretory granules of neutrophils, cytolytic T lymphocytes and natural killer cells. The neutrophil granular serine proteases (cathepsin G, elastase, proteinase-3) can directly degrade vascular basement membrane and are thought to play an important role in mediating tissue in a number of diseases. Urokinase plays a crucial role in tissue remodelling through its ability to degrade matrix proteins, laminin and fibronectin and to activate latent matrix metalloproteinases and collagenases. Factor X, binds vascular epithelial cells, induces release of PDGF-like molecules. Metalloproteinases (collagenase, gelatinase), are family of zinc- and Ca^{2+}-dependent endopeptidases that are active in the degradation of extracellular matrix. Neutrophil metalloproteinases are localised in the secretory granules and are released secondary to lysosomal degranulation at inflammatory sites. Matrix metalloproteinase-9 derived from macrophages can solublize the elastin of the mature elastic fibres in arterial wall. The activity of metalloproteinase is regulated by plasma proteins including a-macroglobulin and cell-derived tissue inhibitors of metalloproteinases (TIMP-1 and TIMP-2). A cationic antimicrobial protein (CAP37) has a bacterial role, implicated in inflammation because of its effect on the recruitment and activation of monocytes and its effect on endothelial cells and fibroblasts. Lactoferrin is an iron binding protein, which is found in external secretions, can affect various immunological functions, including antibody synthesis; production of IL1, IL2, and TNF-α; natural killer-cell cytotoxicity, complement activation, and lymphocyte proliferation. Lactoferrin binds to LPS and thereby decreases the production of LPS-mediated TNF-α, production.

INFLAMMATORY RESPONSE AND CASCADE

In adults most inflammation responses resolve with development of scar tissue. Inflammation is a complex tissue response to injury in which a network of chemical signals initiates and maintains the host response by stimulating endothelial cells, inducing leukocytes recruitment and activating phagocytic leukocyte functions. Persistent injury or inflammation stimulation may result in pathological inflammation and fibrotic repair. Initially, platelets aggregate and release inflammatory mediators, including TGF-β. The role of TGF-β in leukocyte recruitment and activation is that it increases integrin expression and adhesion, is chemicotactic, induce type IV collagenase, auto-induces TGF-β and activates growth factor/cytokines production. The platelet products interact with vascular leukocytes and endothelial cells to bring about slowing and adhesion of leukocytes at the inflammation site. Signalling molecules, including cytokines and growth factors, mediate localised adherence to the vascular wall through adhesion receptor interactions. Adherence of the WBC to the vessel wall precedes the extravasation of these cells into the tissues where they participate in the inflammatory response. The recruitment of leukocytes at sites of inflammation involves: 1- Rolling phenomenon in which leukocyte selectins, carbohydrate lectin receptors, interact with low affinity ligands on the endothelium to slow the movement of the white blood cells; 2- Exposure to chemoattractants or cytokines induces up regulation and/or activation of adhesion molecules on the leukocytes and endothelial cells to promote a firmer attachment; 3- The leukocytes commence migration between the endothelial cells towards the site of tissue inflammation. Much of this interaction is mediated by the cellular adhesion molecules, selectins and integrins. Integrins composed of 1 α and β subunit, includes the β2 heterodimers involved in cell-to-cell interactions and the β1 integrins that promote cell-to-matrix and/or cell-to-cell interactions. The α5β1 receptor is specific for the ECM molecule, fibronectin, and interacts with its arginyl-glycyl-aspartic acid domain. Interaction of leukocytes with matrix proteins or endothelium is regulated by inflammatory molecules. Following adhesion, the blood cells emigrate into the tissues aided by secretion of tissue degrading enzymes such as collagenase. Migration from the vessels at the site of adherence then proceeds by the process of chemotaxis in response to chemotactically active secreted factors. Complement fragments (C5a), products of the coagulation and kinin systems, platelet products and cytokines are responsible for inducing directed migration. Cytokines and growth factors bind to their corresponding receptors on neutrophils, lymphocytes and monocytes, to initiate signal transduction with enhanced biochemical, endocytic and synthetic functions. In the absence of activating factors monocytes undergo apoptosis and are rapidly cleared. Activated macrophages release a network of biologically active molecules, including prostaglandins, ROM, nitrogen intermediates, enzymes and cytokines, which orchestrate an inflammatory response. ROM and hydrogen peroxide may also contribute to tissue damage, as do nitrogen intermediates. Whereas IL-1, TNF-α, PDGF and TGF-β promote inflammatory response, these same cytokines contribute to tissue repair as the inciting agent or injury is resolved. Tissue repair is dependent on the accumulation and activation of fibroblasts, which are responsible for the deposition of ECM molecules and the generation of scar tissue. Cytokines may directly and indirectly influence fibroblasts recruitment and proliferation by stimulating the fibroblasts to produce PDGF, which then acts in an autocrine or paracrine fashion to regulate cell growth. TGF-β indirectly regulates fibroblast proliferation, but directly stimulates matrix production. Under normal circumstances, once the inflammatory response begins to subside, cytokines regulation of fibroblast activation and proliferation progresses to tissue repair, which occurs with minimal scarring. If antigens persist or repetitive injuries occur, prolonged inflammation becomes a chronic inflammatory lesion with the potential for substantial tissue damage. Scarring may involve into fibro-obstructive pathology. Inhibitors of leukocyte adhesion as well as cytokines antagonists could potentially minimise damaging inflammatory consequences. Synthetic fibronectin peptides are capable of blocking inflammatory cell adhesion. Superoxide is capable of rapid reaction with NO, reducing its effectiveness as a vascular relaxant. Potential sources of free radicals are numerous, including xanthine oxidase, mitochondrial enzymes, cytochrome p-450, cyclooxygenases, lipoxygenases, and the leukocyte superoxide-generating NADPH oxidase. Phagocyte NADPH oxidase plays a major part in phagocyte defense against disease processes, catalysing the one-electron reduction of oxygen to superoxide on the outer surface of the cell with the simultaneous oxidation of cytosolic NADPH. Redox centers involved in electron transfer by this enzyme include FAD and a low-potential cytochrome b_{558}. At least 5 protein components are required for NADPH oxidase function. Although cytosolic components of NADPH oxidase are present in endothelium, its expression is at low level and is unlikely to be a major source of endothelium-derived superoxide. The phagocytic NADPH oxidase is a well-characterised, multicomponent enzyme capable of generating large amounts of superoxide in response to a variety of stimuli. NADPH oxidase has a very low affinity for NADH. A contribution of the phagocyte NADPH oxidase to endothelial oxidant generation may be unlikely. Endothelial oxidant generation could be of major clinical importance, reaction of superoxide with endothelium-derived NO may lead to inhibition of vasorelaxation, whereas oxidative processes themselves may play a major role in the pathogenesis of athrosclerosis. Traumatic wounds, bee stings, and bacterial, viral and parasitic infections all produce early phase reactants that lead to the efficient recruitment of targeted inflammatory cells to the efficient site, with the production of cytokines and growth factors that precisely control the immune system response. These secreted protein mediators include cytokines (IL-1-16), growth factors such as TGF-β, and chemokines such as macrophage inflammatory protein-1 α and β, and RANTES.

Post-surgery period characterised by large fluid shifts; temperature changes, coagulation disturbances, and increased concentrations of catecholamines and stress hormones. Inflammatory response to reperfusion may be largely responsible for systemic vasodilatation during rewarming, fever, postoperative bleeding, and reperfusion injury to the heart, brain, kidney, and gut, leading to either permanent organ injury or transient dysfunction. Neutrophil-endothelial cell adhesion is central to postischaemic reperfusion. Initially, OFR are released from endothelial cells in reperfused organs, such as the heart and lung, leading to alterations in endothelial cells that promote early neutrophil targeting, activation of neutrophils in transit through these organs, local activation of serum completely, and induction of cytokine synthesis. Initial neutrophil attachment to the endothelium results in translocation of P-selectin from intracellular vesicles to the cell membrane and initiation of platelet activating factors synthesis. These changes lead to more neutrophil adhesion, neutrophil activation, and neutrophil adhesion protein (CD11/CD18 integrin) expression via endothelial membrane-bound platelet activating factor. Protein inflammatory cytokines such as IL-1β and TNFα, and cytokines such as IL-8, are newly synthesised, largely within endothelial cells, in reperfused organs within hours, and induce increased expression of adhesion molecules on endothelial cells and other tissue-specific cells (myocytes, glomerular or renal tubular epithelium, pulmonary alveolar lining cells). In turn, this result in tighter neutrophil attachment via ICAM-1, transmigration of neutrophils into the interstitial space, and release of large amounts of free radicals. NO produced by endothelial cell NOS may control many of these processes by inducing vasodilation, regulating leukocyte recruitment, scavenging endothelial-generated oxygen radicals, and preventing upregulation of neutrophil CD11/CD18. NO production is impaired after ischaemia/reperfusion, and inhibitory controls over leukocyte migration, oxygen radicals, and neutrophil adherence are mitigated. Sympathetic efferents are involved in the pain of inflammation. Thus the control of these fibers is a matter of considerable importance. Both primary afferent and sympathetic fibers are involved in the production of, and sequelae that follow, inflammation. The release of peptides from primary afferents and norepinephrine (NE) and prostaglandins from sympathetic efferents are significant contributors to the pain, swelling and plasma extravasation that occurs in inflammation. Glutamate has an important role in the pain that accompanies inflammation. Glutamate may also regulate sympathetic efferents locally since ionotropic receptors are expressed by peripheral postganglionic sympathetic fibers. Glutamate released into the extracellular fluid in the inflamed state could activate more postganglionic axons than in the normal state. This in turn might release more NE and prostaglandins as well as other substances from postganglionic sympathetic terminals, which would enhance the changes in the peripheral vascular system that are under the control of sympathetic efferents. It is also possible that the increased labelling could simply reflect an increase in the cytosolic pool of receptors, such as occurs in receptor endocytosis. Postganglionic sympathetic axons express glutamate receptors. Prominent sources of glutamate in the peripheral extracellular fluid are plasma or reticuloendothelial elements such as macrophages. Other major sources are the primary afferent fibers that travel in close proximity to the sympathetic axons. These fibers use glutamate as their primary transmitter, and activity in peripheral sensory fibers releases glutamate into the peripheral extracellular fluid. Thus, in the inflammatory state, which is characterised by plasma extravasation, infiltration of macrophages, increased glutamate levels in the epidermal cells and increased activity in primary afferent fibers, each or all of these phenomena could lead to enhanced levels of glutamate in the inflamed region and thus heighten local sympathetic activity. The signal(s) that might lead to the increase in the numbers of sympathetic axons that express glutamate receptors, include NGF, substance P and calcitonin gene-related peptide (CGRP). Adrenergic receptors have been demonstrated on primary afferent neurons and capsaicin-induced inflammatory pain is ameliorated by adrenergic blockers. The pain-related behaviour that accompanies some other types of inflammation is partially dependent on an intact peripheral sympathetic system. The sympathetic activity has an important role in the nociception that characterises certain types of inflammation. Certain neurochemicals, including glutamate, influence cardiac activity by direct interactions with sympathetic terminals in the heart. Inflammation leads to a considerable increase in the proportions of sympathetic axons that express ionotropic glutamate receptors. This suggests that glutamate-induced modulation of postganglionic activity may be enhanced during inflammation. Increased activity in sensory fibers in inflamed skin may release excess glutamate into the extracellular space. Due to the increased number of sympathetic efferents containing glutamate receptors, the probability that more of these fibers would be excited by glutamate is greatly enhanced. This in turn could lead to an increased release of NE and other substances in postganglionic efferents such as prostaglandins and purines, which in turn could excite an increasing number of sensory axons. This positive feedback loop could result in the exacerbation of both the sensory and the vascular components of inflammation.

NITRIC OXIDE

The inflammation is characterised by hyperaemia, increased microvascular permeability to plasma proteins and fluid, and enhanced recruitment of leukocytes to the site of injury. Mast cells lie in close proximity to the microvascular and, on stimulation, are a rich source of numerous inflammatory mediators, such as vasoactive amines and proteases and agents that are synthesised including leukotriens, PGs, PAF, and variety of cytokines. Mast cells have been implicated in diseases such as IgE-mediated anaphylaxis, ischaemia/reperfusion of various organs, rheumatoid arthritis, and atherosclerosis. Prevention of mast cell activity could be a potential therapeutic strategy in controlling inflammation. NO have profound anti-inflammatory properties in various acute inflammatory conditions, including ischaemia/reperfusion and cardiac anaphylaxis. Mast cell has also been implicated as playing an important role in the associated inflammation. NO reduces histamine release, suggested a potential link between NO and mast cell reactivity. Mast cell stabilisation reduces ischaemia/reperfusion-induced leukocyte rolling, leukocyte adhesion and emigration, and vascular dysfunction. NO can reduce neutrophil recruitment, and attenuate the profound increase in microvascular permeability associated with direct activation of perivenular mast cells. After their displacement from the mainstream of blood, leukocytes make initial contact with endothelium via leukocyte rolling mediated by adhesion molecules (e.g., selectins) found on both leukocytes and endothelial cells. Early recruitment of rolling leukocytes is by P-selectin, which is mobilised by various mast cell-derived mediators, including histamine, leukotriene C4, and oxidants. Mast cell-derived mediators, including histamine mediates P-selectin-dependent leukocyte rolling. NO completely inhibit the enhanced flux of rolling leukocytes associated with mast cell activation. NO has no direct inhibition of P-selectin expression or effect on leukocyte/endothelial cell, P-selectin/ligand interactions. NO functions directly to prevent histamines release from the mast cells and thereby inhibit the recruitment of leukocyte rolling. NO directly reduces histamine release from mast cells. Once leukocyte rolling is established, firm adhesion of the leukocytes to the venular endothelium is next step in leukocyte recruitment during an inflammatory response. PAF and CD18 cause the mast cell-induced leukocyte adhesion during inflammation. Mast cells activated with CMP 48/80 promote leukocyte adhesion to endothelium primarily via PAF. Microvascular dysfunction may be at least in part related to increased mast cell reactivity. NO has a profound inhibitory effect on mast cell-induced sequelae of inflammation. Mast cells are leukocytes situated in connective tissues, characterised by the presence of numerous basophilic granules. Its action is due to: 1- the release of the contents of the preformed granules, which may contain various substances connected with the inflammatory process and the fibrosis; 2- the release of newly formed substances such as a metabolites of AA, thromboxane, prostaglandin. The gas nitric oxide (NO) is highly reactive free radical, that play a vital role in many biological events including regulation of blood flow, platelet function, immunity and neurotransmission. NO tends to exist as a gas and is poorly soluble in water. The small size (molecular weight 30D) and high lipid solubility of NO makes it readily membrane-permeable. Once within the cell it may bind to transition metals such as Fe and Cu in enzymes. In blood, NO is rapidly inactivated by its reaction with haemoglobin. α-Tocopherol, the principle component of vitamin E, may also reduce nitrite to NO. NO is formed from acid in the presence of ascorbate. NO is formed in cells when amino acid L-arginine, in the presence of molecular oxygen, is catalytically converted to L-citruline by NOS. NADPH is cosubstrate for NOS, and flavin mononucleotide, flavin adenine dinucleotide, heme, and tetrahydrobiopterin are cofactors. Type-1 NOS (neuronal, nNOS) and type-3 NOS (endothelial, eNOS) are generally constitutively expressed in certain cells and produce small amounts of NO in response to agonists or physiological stimuli that raise intracellular Ca^{2+} concentrations (Ca^{2+}-dependent). The third isoform (type-2 or inducible NOS, iNOS) is generally not present in resting cells but is expressed after induction by variety of cytokines, or by bacterial products such as lipopolysaccharide. Once expressed, iNOS can produce large amounts of NO over a long period of time in a Ca^{2+}-independent manner. The low amounts of NO that are released after transient activation of constitutive NOSs in nerve or in endothelial cells are thought to be involved in neurotransmission and in regulation of vascular tone. The high rate NO producing iNOS, which is present for example in activated WBC, seems to be geared towards host defence. Anti-inflammatory glucocorticoids inhibit the expression of iNOS. Urinary excretion of NO_3^- increases in subjects with fever and diarrhoea. Macrophages produce increased amounts of NO_2^- and NO_3^-, and the cytotoxicity of macrophages against tumour cells is dependent on the presence of L-arginine. NO may inhibit the growth of the tumour cells and various pathogens including bacteria, virus, helminths, protozoa and fungi. Many biological functions of NO seems to be related to its binding to iron-containing enzymes, resulting in activation or inactivation of the enzyme, the cytotoxicity induced by NO is thought to be an effect of NO on iron-containing moieties in key enzymes involved in the respiratory cycle and in the synthesis of DNA in the target cells. Human macrophages may produce NO if activated with appropriate immunological stimuli. NO may cause vasodilation such as that observed in sepsis can results in hypotension. An increased local blood flow in tissues may promote leukocyte infiltration, and NO may also directly modulate leukocyte adhesion to endothelial cells. NO may bind to and activate proinflammatory enzymes such as cyclo-oxygenase and certain metalloproteases, which in turn may contribute to tissue damage. NO may also react with other agents and produce toxic compounds that cause tissue injury. NO has the capacity to react rapidly with superoxide anion to form peroxinitrite, which is a much powerful oxidant than NO itself. Highly toxic oxidants are also formed when peroxynitrite is protonated, such as hydroxyl radical and nitrogen dioxide. NO may serve to protect against superoxide and peroxynitrite-dependent lipid peroxidation by acting as a radical scavenger. NO may reduce mast cell-dependent inflammation. NO is present in low concentrations in exhaled air of humans, and increases in adult asthmatics. The NO excretion is much larger in the upper airways than in the lower airways. NO may be trapped in the lower airways and the lungs following reactions with, e.g., haemoglobin or other compounds that react rapidly with NO. The high local NO concentrations in the nasal airways and in the sinuses may help to protect against airborne infectious agents. Cigarette smoke contains high levels of NO, which may be contributing to vasodilation in the lower airways upon smoke inhalation. NO producing cells in the gastrointestinal tract include potganglionic parasympathetic neurons, and endothelial and epithelial cells. In neuronal cells, NO is thought to act as the principal noradrenergic, noncholinergic (NANC) relaxant transmitter. iNOS is upregulated in the presence of infection or cytokines. NOS immunoreactive neurons have been found in the ureter and in the renal pelvis. NANC relaxation of urethra is mediated by NO. The NO contained in the sample gas reacts with an excess of ozone (O_3) to produce NO_2 with electron in an excited state (NO_2^*). NO_2^* changes back to the ground state (NO_2) while emitting electromagnetic radiation in the 600-3000-nm wavelength range. A continuous production of NO takes place in the acidic stomach through chemical reduction of nitrite present in swallowed saliva. This was the first evidence of non-enzymatic NO production in humans. Stomach NO may be involved in local defence against swallowed pathogens and in regulation of superficial mucosal blood flow and mucus production. Luminal concentrations of NO are increased in the lower airways of asthmatic children, in the colon of patients with inflammatory bowel disease, and in the urinary bladder of patients with cystitis. Local steroid treatment reduces orally exhaled NO levels in asthmatic children. Only small amount of NO are released from infected urine. NO levels are higher in chemically induced cystitis despite negative urine cultures. NOS expression in the mucosa of the urinary bladder may be induced either directly by bacterial toxins or indirectly by cytokines present in the bladder after BCG treatment. The antitumour effect of BCG treatment is partly due to the increased mucosal production of NO, which is known to have cytostatic effects. Urgency symptoms may be due to inflammatory diseases or non-inflammatory functional disorders, and bladder biopsies are often required to distinguish between the 2 groups of patients. With a molecular weight of 30 D, NO is certainly one of the smallest biological molecular mediators. NO production is a part of an effective host response to infection. NO production by NOS2 is stimulated by proinflammatory cytokines such as IFNγ, TNFα, IL1, and IL2, as well as by microbial products such as LPS and lipoteichoic acid. Infections in humans is associated with significant increases in systemic NO production, when determined by measurement of NO end-oxidation products (nitrite and nitrate) in plasma and urine. High output NO synthesis is indeed part of the antimicrobial armamentarium of human macrophages. Inducible NOS and the phagocyte NADPH may be co-stimulated by inflammatory stimuli (IFNγ). Simultaneous production of reactive nitrogen and oxygen intermediates may lead to the formation of a variety of antimicrobial molecular species, each with distinct stability, compartmentalisation, and reactivity. NO radical ($NO^·$) itself, potentially important NO congeners include peroxynitrite ($OONO^-$), S-nitrosothiols (RSNO), nitrogen dioxide (NO_2^-), dinitrogen trioxide (N_2O_3), dinitrogen tetroxide (N_2O_4), and dinitrosyl-iron complexes (DNIC). Peroxynitrite may be formed from the rapid interaction of $NO^·$ and superoxide (O_2^-), from the combination of hydrogen peroxide (H_2O_2) and nitrous acid (HNO_2), which would exist in equilibrium with nitrite (NO_2^-) within an acidified phagolysosomal vacuole, or from the interaction of nitroxyl (NO^-) and oxygen. S-nitrosothiols such as S-nitrosoglutathione can be formed from $NO^·$ and reduced thiols in the presence of an electron acceptor. The potent oxidant NO_2^- can be formed by the autooxidation of $NO^·$, or possibly by the oxidation of NO_2^- by myeloperoxidase and H_2O_2. Additional potent nitrosating agents can arise from the autooxidation of NO (N_2O_3, N_2O_4), or from the interaction of NO with thiols and nonheme iron (DNICs). Interactions between reactive oxygen and nitrogen intermediates provide a molecular basis for synergy between the respiratory burst and synthesis of NO. Peroxynitrite can have greater cytotoxic potential than $NO^·$ or O_2^- alone. The combination of H_2O_2 and $NO^·$ appears to posses particularly potent antibacterial activity, possibly resulting from increased formation of the potent oxidant hydroxyl radical (·OH) in the presence of iron species or from generation of singlet oxygen ($^1O_2$). NO has been shown to be capable of reducing Fe^{3+} complexes, providing a mechanism for enhancement of the Fe^{2+}-catalysed Haber-Weiss reaction. NO can also inhibit antioxidant metalloenzyme such as catalase, thereby limiting H_2O_2 disproportionation. The combination of reactive oxygen and nitrogen intermediate may be antagonistic in some circumstances. NO protects mammalian cells against oxidant injury, perhaps by forming iron-nitrosyl complexes (making iron less available for catalysis of prooxidant reactions), inhibiting the respiratory burst oxidases, directly scavenging radical species such as ·OH, or inducing the expression of protective stress regulons. NO may also antagonise oxidant membrane injury by terminating lipid peroxidation reactions. NO production may simultaneously enhance the antimicrobial function of the respiratory burst while protecting tissues from oxidant injury. Interactions with O_2^- reduce concentrations of $NO^·$, which may actually diminish antimicrobial activity toward microbes more sensitive to $NO^·$ or another NO congener, than to peroxynitrite. The diffusion of NO across membranes resembles that of oxygen, with the exception that oxygen is more lipophilic. Although superoxide (O_2^-) does not appear to enter bacterial cells to a significant extent, its congener peroxynitrite can pass through membranes, probably as peroxynitrous acid (HOONO); greater activity for lipid and proteins may nonetheless limit its effective diffusion into microbial target cells relative to $NO^·$. Peroxynitrite may also be formed in situ within target cells if both $NO^·$ and O_2^- are present. S-nitrosoglutathione (GSNO) appears to be recognised as a substrate by the periplasmic enzyme γ-glutamyltranspeptidase, with subsequent conversion to S-nitrosocysteinyl-glycine. This nitrosated dipeptide in turn is imported into the bacterial cytoplasm across the inner membrane by a specialized dipeptide permease (Dpp). DNA is an important target of RNI (reactive nitrogen intermediates). NO can deaminate DNA. NO_2^- and peroxynitrite can also oxidatively damage DNA, resulting in abasic sites, strand breaks, and a variety of other DNA alterations. NO-related DNA damage has been shown in intact bacteria. NO interactions with proteins can involve reactive thiols, heme groups, iron-sulfur clusters, phenolic or aromatic amino acid residues, tyrosyl radicals, or amines. NO might directly releases iron from metalloenzymes and promote iron depletion. NO-related inhibition of metabolic enzymes may constitute an important mechanism of NO-related cytostasis. Actions of RNI on metabolic enzymes or membrane transporters may lead to dissipation of transmembrane electrochemical proton-motive force. NO-heme interactions can result in the inactivation of other heme proteins, such as catalase and cytochrome systems. Ribonucleotide reductase, a non-heme metalloenzyme essential for DNA synthesis, has been implicated as a major target of NO in tumour cells. Thiols are among the most important protein targets of NO, and S-nitrosylation is favoured over N-nitrosylation under physiologic conditions. Nitrosylation of thiols may modify protein function per se, facilitate subsequent modification such as ADP-ribosylation, or accelerate disulfide formation between vicinal thiols. Nitration of tyrosine residues has received particular attention because this protein modification can be produced by peroxynitrite, although myeloperoxidase may also catalyse tyrosine nitration in the presence of NO_2^- and H_2O_2. Tyrosine nitration may disrupt pathways involving tyrosine phosphorylation and modify protein function or turnover NO can be associated with membrane damage. Peroxynitrite mediates lipid peroxidation of liposomes, which does not require iron. NO_2^- can induce lipid peroxidation. NO has been shown to exert immunomodulatory effects including effects on immune cell adherence and function, cellular proliferation, and cytokine production. Mechanisms of microbial resistance to RNI appear to overlap considerably with antioxidant defences. Staphylococci contain low concentrations of GSH and appear to be susceptible to NO. The molecular interactions of S-nitrosoglutathione and glutathione have similar transformations within microbial cells. A complex set of reactions, dependent upon both relative glutathione concentration and the presence of oxygen, produces a mixture of products including oxidised thiol, ammonia, and nitrite. Homocysteine is another low molecular weight thiol compound, which has been implicated in resistance to S-nitrosothiols. Inhibition of glutathione peroxidase by homocysteine with subsequent oxidative inactivation of NO equivalents has been proposed as a mechanistic explanation in mammalian cells, however this may not be the case in microbes. Microbial systems that repair oxidative injury appear to be similarly involved in repairing nitrosative injury. Glucose-6-phosphate dehydrogenase, which provides a major source of reducing equivalents (NADPH) used to regenerate thiols and other antioxidants, is similarly involved a defenses against both ROI and RNI. Cu, Zn-superoxide dismutase (SODc) may protect against both oxidative and nitrosative stress by removing periplasmic superoxide and limiting peroxinitrite formation. Peroxynitrite production from NO and O_2^-, antagonised by the presence of bacterial superoxide dismutase, might constitute a significant component of the antimicrobial host defense. E coli have implicated specific antioxidant regulons, which is resistant to NO-related antimicrobial activity. The oxyR regulon, originally described as a hydrogen peroxide-inducible genetic system, has been conferred resistance to S-nitrosothiols in E coli. Induction of oxyR-regulated genes by S-nitrosothiols is dramatically augmented in glutathione-deficient cells. Pathogenic E coli are more adapted to survival in phagocytic cells. Pyocyanin, a phenazine pigment produced by Pseudomonas aeruginosa, has been reported to inhibit NO in vitro, but the pathophysiological significance of this action is uncertain. Pathogenic microorganisms might resist NO-related antimicrobial activity by avoiding phagocytosis and avoiding or suppressing stimulation of NO synthase. NO may act in concert with reactive oxygen species to damage microbial DNA, proteins, and lipids. Microbial defenses against oxidative and nitrosative stress regulons, scavengers, detoxifying enzymes, repair systems, and strategies to subvert or avoid host phagocytes.

GLUCOCORTICOIDS

The primary purpose of the glucocorticoids, including cortisol (hydrocortisone), is to mobilise the body to resist infection. Cortisol is for intestinal disease and corticosterone serum disease. Most processes that resist infection impair fight or flight. Potassium loss is the most serious aspect of intestinal diseases, so the electrolyte capabilities of cortisol, but not corticosterone, are oriented toward conserving K. Low cell K reduces adrenal synthesis of cortisol, but not corticosterone. Sodium, water, glucose, amino acids, chloride, hydrogen ion, copper, and numerous others are controlled by cortisol such as to survive during intestinal disease. Endotoxin of gram-negative bacteria induces secretion of huge amounts of ACTH, which is the chief mediator of cortisol. A glucocorticoid response modifying factor (GRMF) and IL-1, raises the effective threshold of cortisol. Cortisol (hydrocortisone) and corticosterone steroids control a large number of enzymes, hormones, and processes, most of which could enhance growth of pathogens or make the adverse symptoms worse. Glucocorticoids mobilise immunity by declining their serum concentration. Cortisol may prevent proliferation of T-cells by rendering the IL-2 producer T-cells unresponsive to IL-1, and unable to produce the T-cell growth factor. But, immune cells take over their own regulation, but at a higher set point. Activated macrophages start to secrete IL-1, which synergistically with CRF increase ACTH, T-cells also secrete glucosteroid response modifying factor [GRMF] as well as IL-1, both of which increase the amount of cortisol required to inhibit almost all the immune cells. Cortisol even has a negative feedback effect on IL-1. The suppressor cells are not affected by GRMFs. Cortisol operates by changing the nucleus commands to send RNA for production of enzymes, etc., in almost every case and the various diffusion steps take an hour or more to complete. Cortisol and aldosterone can induce euphoria. When a process must move in the same direction for both immunity and fight or flight, a different hormone system controls it for stress. The most dangerous digestive diseases produce a protein poison, which stimulates c-AMP hormone in such a way that the intestines cannot remove water from their contents and thus cause diarrhoea. Since K in food and the 2.5 g or so secreted with digestive fluids can only move into the blood stream passively, this causes a large loss of K. A wide range of processes are stimulated by cortisol, each of which increase the ability to resist K and water wasting intestinal disease. The greatest urgency during diarrhoea is to prevent loss of K, since there is no storage of K in any cell. In cells, 88% of the K is in free solution. One of cortisol's functions conserves potassium, it tends to move K inversely into the cells. In order for K to move into the cell, cortisol inversely moves out an equal number of sodium ions. Cortisol consistently causes alkalosis of the serum. Potassium is primarily blocked from loss in the kidneys by a drastic decline of aldosterone during dehydration. As osmotic pressure rises during dehydration; aldosterone undergoes a drastic decline. Sodium depletion does not affect cortisol. The Na retaining hormone, 18-hydroxy 11-deoxycorticosterone (18OH DOC) acting on the kidneys is strongly dependent on ACTH. Cortisol also acts as a diuretic hormone, which is probably due to inverse stimulation of antidiuretic hormone (ADH or arginine vasopressin [AVP]). Cortisone significantly inhibits insulin secretion. The inversed stimulation of insulin by corticosterone would lower serum glucose and thus deny glucose to pathogens. A sudden withdrawal of glucose by insulin in a K deficiency can lower serum K enough to be lethal. Cortisol induces hypoglycaemia, which could be indirect effect caused by inverse inhibition of amino acid degradation. The intestinal brush border disaccharide enzymes are inversely inhibited by cortisone. Sucrose and fructose induces copper deficiency. Glucocorticoids inversely lower amino acids in the serum. They do this by inversely stimulating collagen formation, increasing amino acid uptake by muscle, and stimulating protein synthesis. Cortisol also inversely inhibits protein degradation, which denies amino acids to bacteria. Collagen is useful in repair of infected tissue. Loss of collagen from skin by cortisol is ten times greater than from all other tissue. Thus the skin can be a reasonably safe source of energy during stress and be rapidly repaired during damage preliminary to or caused by infection. Lowering serum amino acid or even tissue damage repair during intestinal disease should be not nearly so advantageous. DOC acts in the opposite direction for collagen, and thus tends to cancel cortisol's effect. Denying amino acids to bacteria could be very advantageous in a serum infection. 40% of the protein synthesis is in the intestines, much of it for synthesis of IgA. IgA acts as an inert, nonlethal coating on bacteria to prevent adhesion to intestinal walls, and is the predominant immunoglobulin in the human intestine. Cortisol may inversely stimulate IgA precursor cells in the intestines. Cortisol also inversely stimulates IgA in serum, and also IgM, but not IgE. Cortisol has an opposite effect on liver than it has on muscle, which could be to provide a small amount of maintenance amino acids when the muscles are withdrawing them from the blood and possibly also to provide liver amino acids for IgA. Cortisol inversely activates luteinizing hormone. Sodium, K, and Cl make strong bases and acid so that any unilateral movement by any of them has considerable implications in H^+ control. Cortisol inversely inhibits gastric acid secretion. Hydrogen ion interferes with K excretion at the kidneys. Gastric secretion carries 0.6 g of K/day into the stomach. Corticosterone has a much greater effect on gastric acid secretion than cortisol. Some leucocyte enzymes have a pH optimum lower than serum. Cortisol's only direct effect on the H^+ excretion of the kidneys is to inversely inhibit excretion of ammonium ion by inactivation of renal glutaminase enzyme. Glutaminase splits ammonia off of the amino acid glutamic acid, and this provides ammonium ion to take the place of K for excretion. Cortisol's presence is necessary for the other hydrogen ion excretion regulator to operate. When K is deficient, the kidneys fail to absorb Cl and the serum tends toward alkalosis. The net effect of glucocorticoids is to inversely acidify the serum. Chloride (Cl) is intimately involved with K loss because when the cell loses K to take the place of serum losses and Na migrates in, Cl must also be excreted as the only ion that has a chance of maintaining serum pH. In a K deficiency chloride is lost. The immune system is very sensitive to copper (Cu) availability. Resistance to infection is reduced by Cu deficiency. A reduction in neutrophils is the first symptom of a Cu deficiency in children. Many copper enzymes are inversely inhibited (up to 50% of their total potential) by cortisol. This includes lysyl oxidase, an enzyme that is used to cross link collagen and elastin. DOC acts in the same direction as cortisol for lysyl oxidase. Immunity is the inverse shutdown of superoxide dismutase by cortisol, since this copper enzyme is almost certainly used to inversely permit superoxide to poison bacteria. The safest way to transport Cu to the immune system is by the transport protein, caeruloplasmin. This avoids Cu toxicity when Cu availability to the cells from the liver is increased, since caeruloplasmin Cu is not in equilibrium with the serum. Caeruloplasmin is used by the immune cells as a source of Cu during infection. Several antigens raise plasma caeruloplasmin. Stress requires extra Cu, and at high ACTH levels epinephrine is used for this purpose. Transporting Cu as the ion is not so important for denying Cu to pathogens during digestive disease, which is probably why DOC inversely loses Cu from the liver and inhibits liver uptake somewhat, thus providing the immune cells with free Cu to supplement the caeruloplasmin source. Patients with Wilson's disease, in whom caeruloplasmin cannot be synthesised, are not prone to infection. These patients cannot transport Cu to the bile excretory proteins either, so their cells are already loaded and even overloaded with Cu. Cortisol causes an inverse 4- or 5-fold decrease of metallothionein, a Cu storage protein, which may be to furnish more Cu for caeruloplasmin synthesis. Cortisol has an opposite effect on α-aminoisobutyric acid than on the other amino acids. The gram-negative bacteria have lipopolysacharride (endotoxin) on their cell wall. Some endotoxin erodes off the wall and more is released into the blood stream when polymorpholeucocytes eject debris from bacteria, which they have engulfed. The lipid A part of the molecule stimulates the hypothalamus to secrete large amounts of CRF. An amount of endotoxin that causes no other symptoms than a mild fever causes a 6-fold rise in ACTH, from CRF stimulation. Some responses to endotoxin are fever, creation of IFN by spleen cells as well as division of spleen cells, synthesis of IL-6, activation of complement by three mechanisms, creation of hypotension, stimulation of adherence and oxidative processes of neutrophils, activation of a burst of activity in macrophages in extremely small amounts, proliferation and maturation of B-cells, low serum glucose, metabolic acidosis, and numerous other functions. Most of these responses are mediated by TNF secreted by macrophages and they last only the first couple of hours. TNF evolves after the endotoxin assault on ACTH evolved. In the absence of TNF and GIFN, bacteria proliferate very rapidly. Lipid A fraction of endotoxin enhances local IgA response to mucosally applied antigen (toxin), at least when lipid A and antigen are associated on a liposome carrier. GRMFs' secretions are stimulated by endotoxin. The body forces endotoxin to mount a preliminary quick response even before the antigens can activate a response, and then quickly turns it off again assisted by a cachectin (TNF) half-life of only 6 minutes. Antidiuretic hormone (ADH) quickly rises 20-fold in only 15 minutes. The release of endotoxin by phagocytosis may be the reason why glucocorticoids inhibit digestion but not uptake of bacteria by macrophages. Those glucocorticoid hormones cause the immune cells to rise to a peak of activity at low concentrations and then decline again at increasing concentrations. Tumour necrosis factor also stimulates ACTH production by a direct effect on the pituitary, possibly an advantage the first few hours, especially if the shutdown of ACTH is rapid. A glycoprotein produced by T-cells called glucocorticoid response modifying factor (GRMFs, also GAF) along with IL-1 has the power to inhibit the response of immune cells to cortisol. GRMFs affect most of the physiological processes affected by cortisol other than immune cell activity. Vitamin C (ascorbic acid) supplements may be useful against stress effect, blocking a rise in corticosterone resulting from stress. Continuing rising of corticosterone would affect this advantageous effect of vitamin C. Refraining from coffee, tea, soft drinks or cocoa may be advantageous because of an effect on cortisol by caffeine. Cortisol is not a medicine, it is a hormone, a hormone whose effects ramify through multiple functions in most of the cell groups in the body. Arthritics have normal cortisol level. Very virulent diseases are endotoxin involved include cholera, typhoid, pneumonia, salmonella, campylobacter, and meningitis.

TROPISM
Tropism is the phenomenon by which commensal and pathologic bacteria are restricted to certain hosts' tissues and cell types. Tropism helps to explain the diversity of bacteria known to cause certain diseases. The architecture of the outer surface of bacteria modulates the interaction with environment, including the ability of the bacteria to interact with only certain cells and tissues. Irreversible adherence to the surface is the most significant mechanism by which bacteria colonize and/or invade the host. The negative surface charges of the tissue cells and bacteria, as well as diffuse ion clouds in the area, repulse adhesion. Bacterial fimbriae or polymers, which are of a much smaller diameter, allow bacterial adherence that might not otherwise occur, with the fimbriae reaching cell surface receptors for firm adherence to the cell surface. Once adherent, the normal flow of body fluids, such as mucus or urine, does not wash away the bacteria. Bacterial growth then reaches a critical mass for parenchymal invasion. The term adhesin is a class designation of any structure leading to bacterial adhesion to a cell or tissue. Adhesion may also occur because of a nonspecific characteristic, such as hydrophobicity, but more frequently it is due to a specific receptor-lectin interaction by means of surface hair-like appendages termed fimbriae or pili. Fimbriae are not the only means of attachment, since carbohydrate polymers can also overcome the repulsive forces. The adhesins include fimbriae, adherence pedestals and the afimbrial adhesins, such as polymers, polysaccharides, lipoteichoic acid and high molecular weight proteins. In the urinary tract infections it appears that adhesion to urothelial cells by fimbriae or pili of E coli (class II adhesion in the kidney and class III adhesin in the bladder) is most often the initiating event. Streptococcus Pneumonia, bind to immobilised fibronectin but not to the soluble fibronectin. The adherence of pneumococci is within the C-terminal epithelial binding site compared to most bacterial binding to fibronectin by other organisms, which require recognition of an N-terminal domain of fibronectin. Pneumococci bind poorly to

ciliated cells of the airways, and by strong binding to basement membrane components, which become exposed following injury to respiratory epithelium induced by the organisms pneumolysin or by previous viral infection. Haemophilus influenzae is a normal inhabitant of the pharynx but can cause invasive infections. Organisms with type B capsules are associated with more severe diseases. Encapsulated and unencapsulated H. influenzae colonize oropharyngeal epithelial cells, most frequently because of bacterial fimbriae but also by means of nonfimbrial high molecular weight adhesion. H. influenzae adhesion by uncapsulated nonfimbriated strains to the ECM enhances degradation of ECM with bacterial penetration through the basement membrane, and adherence to immobilised laminin, fibronectin and various collagens. The plasminogen appears to be activated to plasmin by tissue plasminogen activator. Plasminogen activation leads to formation of proteolytic plasmin activity when bound to the bacteria, which leads to the degradation of fibrin and noncollagenous glycoproteins of the ECM and basement membranes by which invasive disease occurs. The bacterial capsule, which is composed of a polymer of ribose and ribotol-5-phosphate, improves intravascular survival and host resistance. Encapsulated Haemophilus tends to be the most pathogenic. In general following the adherence of bacteria to vascular cells, activation of neutrophils occurs and via pro-inflammatory cytokines, such as IL1 and TNF, endothelial injury follows with the release of proteases and glycanases as well as OFRs. All of which further damages the endothelium causing capillary leak as well as intravascular thrombosis. Inflammation has also been correlated with resistance to phagocytosis and to complement mediated killing by human serum. Capsulation is a major virulence factor for many organisms. Neisseria gonorrhoeae freshly isolated have bacterial fimbriae, N. gonorrhoeae has the ability to switch between phenotypic expression of the fimbriae and no expression (phase variation). An effective vaccine to prevent gonorrhoea has not been devised. Receptors for the fimbriae of gonococci are present in human tissues, such as the urethra, cervix, endometrium, cornea, intestine, stomach and midbrain meninges. The adherence is mediated by a receptor having protein characteristic. Normal flora may prevent colonisation of certain tissues although the receptor is present. Receptor present in the intestine can explain how gonococci can colonize and infect the human rectum, and receptors in the cornea explain the conjunctivitis with ulceration that may ensue in newborn infections. The disease is due to tissue specific, cell specific, tropism and host tropism, occurs only in man. Parasitism is a relation in which one species benefits at expense of the other. Commensalism may be defined as a relationship in which one species derives benefit and the other is unharmed. Symbiosis is the biological association of 2 or more species to their mutual benefit. Amphibiotic organisms may be either beneficial or disease causing, depending on context. The gram-negative organism, Helicobacter pylori colonise the stomach of human and other primates. H pylori are obligate parasites, with no free-living form in nature yet identified, and that natural infection is specific for primates, including humans. Failure to eliminate the parasite implies that the cost to the host is greater than benefit. This may be due to high costs (e.g., protection against lethal diseases), or that both cost and benefit are relatively low. Environmental phenomena affecting H pylori include gastric pH, nutrient limitation, and physical removal via peristalsis. Bacterial secretion of proinflammatory molecules may result in greater tissue injury and permitting nutrient release. Similarly, secretion of Na-methyl-histamine would stimulate the host to increase acid production when pH in a particular locale is too high for optimum microbial function. A consequence of these interactions can be epithelial cell proliferation and apoptosis, gastric secretion, and antigenic stimulation of lymphoid cell populations. Metabolic products and physiological alterations from the host can signal the microbe in multiple ways, including changes in pH, nutrient supply, and physical removal. The ultra-low biological activity of H pylori LPS and the expression of host epithelial cell Lewis antigens on the bacterial cell surface are example of the organism limiting its constitutive signalling to the host. The ability of the organism to modulate Lewis expression may permit optimal adaptation to particular host microenvironments. H pylori are highly diverse at the genetic level. Individuals may be simultaneously colonised by multiple H pylori strains.

DRUG INTERACTIONS AND ANTI-INFECTIVE THERAPIES

Drug interactions are classified as pharmacokinetic, pharmacodynamic, or occasionally. Pharmacokinetic interactions occur when the precipitant drug modifies the object drug's absorption, distribution, metabolism, and excretion. Pharmacodynamic interactions occur when the precipitant drug changes the effects of the object drug, such as when an aminoglycoside antibiotic potentiates the neuromuscular blocking effects of succinylcholine or a nondepolarizing muscle relaxant. Anticholinergics, opiates, and food slow gastric emptying, which may delay the time to peak level and reduce the peak concentration of many drugs, while not usually reducing the area under the plasma drug concentration curve. Conversely, prokinetic drugs such as erythromycin, cisapride, or metoclopramide accelerate gastric emptying and delivery of drugs to the small intestine, often resulting in a shorter time to peak level and a greater peak concentration, again without necessarily changing the net absorption. In about 10% of people who take digoxin, as much as 40% of the drug is converted by upper intestinal bacteria (Eubacterium lentum and perhaps others) to inactive digoxin reduction products. The growth of those bacteria can be inhibited by orally administered antibiotics, resulting in greater bioavailability of pharmacologically active digoxin.

Coadministration of oral neomycin, erythromycin, or tetracyclines has resulted in digitalis toxicity, and diminished digoxin metabolism in the gut may persist for months following cessation of the antibiotic. Gastric pH is important in the solubility or chemical stability of some oral antimicrobials (azole antifungals and β-lactam antibiotics). The bioavailability of these drugs may be altered by antacid therapy. Cationic (especially magnesium or aluminium but also, to a lesser degree, calcium) antacids, sucralfate (sucrose aluminium sulfonate), or perhaps kaolin-pectin form insoluble chelates with certain antibiotics including tetracyclines, fluoroquinolones, and perhaps lincosamides, reducing the absorption of the antibiotic. Most drugs are reversibly bound to some degree to plasma proteins, (albumin). Biotransformation of drugs refers to the oxidation or other enzymatic modification of a compound to one that is more polar and thus more easily excreted either by the liver or kidney. A second phase that may occur is hepatic conjugation of the oxidised drug by glucuronidation, acetylation, or sulfation. The oxidative phase is a function of hepatic or intestinal cytochrome P450 (mixed-function oxidase) enzymes, which are versatile in their substrate specificity. Many P450 enzymes can be inhibited by precipitant drugs that may or may not also be substrates for the enzymes. This inhibition may result in decreased metabolism of other drugs, causing their accumulation and perhaps toxicity. The inhibition or induction of certain P450 isoforms by drugs is generally regarded as the most important mechanism of drug interaction, and many of these involve anti-infective agents. The P450 families or subfamilies that are most involved in adverse drug interactions are CYP1A2, the CYP2C subfamily, CYP2D6, and the CYP3A subfamily (principally CYP3A4). There are as much as 10-fold interindividual variations in P450 subfamily content. Decreased glomerular filtration results in accumulation of drugs that do not have an alternative route of excretion; to the extent that drug interactions are often dose or concentration related, renal failure can predispose a patient to adverse drug interactions. Interference with renal excretion of drugs can cause drug interaction by competition for renal tubular secretion or by altered tubular reabsorption. Aluminium- or magnesium-containing antacids can decrease the bioavailability of fluoroquinolones as object drugs by as much as 90%. Sucralfate (sucrose aluminium sulfonate), calcium (antacids, calcium supplements, milk, yoghurt), iron, zinc, and bismuth may result in a more modest but still significant decrease in absorption of fluoroquinolones as object drugs. The most important interaction of fluoroquinolones as precipitant drugs is the ability of enoxacin, ciprofloxacin, and to a lesser extent, norfloxacin to inhibit the metabolism of theophylline by CYP1A2, resulting in theophylline accumulation and toxicity. Enoxacin as well as ciprofloxacin, but not other fluoroquinolones, inhibit CYP1A2 metabolism of caffeine, prolonging its half-life four-fold to five-fold and increasing its maximum concentration. Trimethoprim and sulfamethoxazole are precipitant drugs in interactions mediated by alterations of hepatic biotransformation, competitive renal excretion, or pharmacodynamic effects. Trimethoprim inhibits sodium channels in the distal tubules of the kidneys. Trimethoprim has caused hyperkalaemia in the elderly, in patients with AIDS, and when added to therapy with potassium-sparing diuretics or potassium supplements, including salt substitutes. Trimethoprim inhibits renal tubular excretion and increases toxicity of amantadine, dapsone, digoxin, methotrexate, procainamide, and zidovudine. Sulfa drugs, including sulfisoxazole and trimethoprim-sulfamethoxazole, enhance warfarin-induced anticoagulation. Sulfa drugs cause modest displacement of warfarin from plasma protein binding sites. Sulfamethoxazole may inhibit CYP2C9, the major cytochrome P450 isoenzyme responsible for metabolism of the more potent S-warfarin. Trimethoprim-sulfamethoxazole in combination with methotrexate may cause bone marrow suppression. Sulfa drugs may displace methotrexate from plasma protein binding sites resulting in transiently higher levels of unbound methotrexate. Trimethoprim competes with methotrexate for renal tubular elimination. When trimethoprim is added to chronic methotrexate therapy resulting in inhibition of dihydrofolate reductase can result in acute megaloblastic anaemia. Trimethoprim may have a pharmacodynamic effect of causing a rise in serum creatinine level in patients on cyclosporine. Macrolide antibiotics differ in their abilities to bind to and inhibit cytochrome P450 isoforms, especially CYP3A4. Erythromycin and troleandomycin bind strongly to and inhibit CYP3A4. Clarithromycin have intermediate binding affinity to CYP3A4. Azithromycin and dirithromycin do not bind CYP3A4 and are associated with the fewest adverse interactions. Ethanol decreases the absorption of erythromycin ethylsuccinate, and food modestly decreases the absorption of all macrolides except clarithromycin and enteric-coated erythromycins. Cimetidine inhibits microsomal P450 metabolism and has been associated with transient reversible deafness due to high-dose erythromycin. A few pharmacokinetic interactions in which macrolides are the precipitant drug are common or dangerous. The object drugs in these interactions include astemizole, terfenadine, carbamazepine, cisapride, clozapine, cyclosporine, digoxin, ergot alkaloids, pimozide, tacrolimus, theophylline, and warfarin. Cisapride also metabolised by CYP3A4. Erythromycin may lead to elevated tacrolimus levels and toxicity in organ transplant recipients. Azole antifungals, ketoconazole, itraconazole, and fluconazole alternate the solubility and absorption of the azole, and modify P450 biotransformation of either the azole or the other drug in the interacting pair. Ketoconazole and itraconazole are soluble at acid pH but only 10% soluble at pH 6. Ketoconazole or itraconazole should be given at least 2 hrs before antacids or sucralfate. Fluconazole does not depend on gastric acidity or food for its dissolution or absorption. Precipitant interactions involving azole antifungals result from their inhibition of cytochrome P450 enzymes. Azole antifungals inhibit the CYP3A4 metabolism of cyclosporine and tacrolimus, which can result in toxicity. The metabolism of orally administered midazolam, triazolam, and other anxiolytics that are oxidised by CYP3A4 are inhibited by ketoconazole or itraconazole therapy more than by fluconazole. Rifampin is the most potent and broadly specific inducer of cytochrome P450 isoenzymes, including CYP1A2, the CYP2C subfamily, and CYP3A4. Aluminium hydroxide antacids, ketoconazole, or pyrazinamide can reduce the oral bioavailability of rifampin. When therapy with a rifamycin is discontinued there is usually a washout period of 1-3 weeks during which cytochrome P450 metabolism returns to baseline. The digoxin dose should be decreased by 50% when rifampin is discontinued. Metronidazole has the ability to precipitate a disulfiram-like reaction when taken concurrently with ethanol; metronidazole should be used cautiously in patients concurrently using oral tinctures or intravenous medications containing ethanol (e.g., diazepam, nitroglycerine, phenobarbital, phenytoin, trimethoprim-sulfamethoxazole, and homeopathy tinctures). Metronidazole enhances the toxicity of fluorouracil and of the effect of warfarin.

THE ENTERIC PATHOGENS

The intestinal epithelium, in addition to its absorptive and digestive properties, represents an efficient barrier against the commensal flora and pathogens. The mucus layer, intestinal peristalsis, and array of innate antibacterial factors such as lactoferrin, lysozyme, and cryptdins, a family of short hydrophobic antibacterial peptides produced by Paneth cells in intestinal crypts are involved in the defence process. Entry of enteroinvasive bacteria into the intestinal epithelial cell is key to a successful invasive process. There are essentially 2 major mechanisms of bacterial internalisation: 1- The zippering process corresponds to tight envelopment of the bacterial body by the mammalian cell membrane, involving a surface bound bacterial protein binding an adherence molecule of the mammalian cell surface with high affinity, e.g., the invasin (Inv) of Yersinia, or internalin A of Listeria monocytogenes. 2- The trigger process corresponds to bacteria inducing massive cytoskeletal changes in the mammalian cell underneath its site of contact, thereby causing a ruffling process that internalises the bacterial body in a macropinocytic vacuole. Adherent microorganisms bind to the apical pole of the intestinal epithelium, whereas invasive microorganisms disrupt and invade the epithelium. Cross talks leading to alteration of the epithelial cell actin cytoskeleton appear as a recurrent theme during entry and dissemination into epithelial cells. Other cross talks alter the trafficking of cellular vesicles and induce changes in the intracellular compartment in which they reside, thus creating niches favourable to bacterial survival and growth. There are 2 main kinds of bacteria in the intestinal tract: aerobic and anaerobic. Another major group of organisms in the intestine are the yeast and fungi. The single-celled organisms, protozoa in a normal intestinal tract are found in a natural balance that is healthy. It is estimated that there are 500 or more different species of bacteria in the average human intestinal tract. It's estimated that there are about 10-100 trillion cells of bacteria in the intestinal tract at any one time. There are about a 100 trillion human cells in the entire human body. Thus, 10-50% of the total cells is due to bacteria in a normal individual who is not on antibiotics. Bacteria constitute about 50% of the content of faeces. Oral antibiotics reduce the total population of anaerobic bacteria by a factor of 1,000 including beneficial bacteria (Lactobacilli). Lactobacilli are present in yoghurt. As the good bacteria are killed off, the potentially harmful bacteria increase rapidly. The harmful bacteria translocate out of the intestinal tract into the lymph nodes surrounding the intestinal tract. From these lymph nodes, these bacteria are then strategically placed to cause new infections throughout the body. Antibiotics kill off all the normal bacteria and result in the proliferation of yeast. Also the yeast may be stimulated by many of the antibiotics. If antibiotics are given at an early age, the Candida in the intestinal tract increases by 130-fold. The hormone cortisone increases Candida in the intestine by 8-fold. Largely because of the overuse of antibiotics, the incidence of disseminated candidiasis has changed from a rare occurrence prior to 1960 to the fifth most common organism encountered in hospital acquired infections. The bad bacteria and yeast produce chemical byproducts, which get absorbed from the intestinal tract into the blood. From there, they circulate throughout the body to all the tissues and are eventually filtered out of the body into the urine. Also the yeast cells may convert to their more invasive colony form. The yeast imbed themselves into the lining of the intestinal tract, which is facilitated by the secretion of yeast digestive enzymes at the point of attachment. The intestinal lining is thus digested by a variety of yeast enzymes including phospholipase A_2, catalase, acid and alkaline phosphatases, coagulase, keratinase, and secretory aspartate protease. The secretory aspartate protease may destroy the lining of the intestinal tract and may also digest the IgA and IgM antibodies produced by the body to attack the yeast. The destruction of this gastrointestinal lining may be the reason for the abnormal secretin response. Some of the intestinal cells probably die as a result of this attack. As a result of multiple yeast attaching to the intestinal lining, the lining may appear like Swiss-cheese on a microscopic level (ulcerations). Normally, undigested food molecules would not be able to pass through this intestinal lining. However, because of the ulcers (holes) in the intestinal wall, undigested food molecules pass through. This phenomenon is called the leaky gut syndrome. A major consequence of the leaky gut syndrome is much greater susceptibility to food allergies. The undigested food is recognised as an invader by the immune system and as a consequence, antibodies of both the IgE and IgG types may start to be produced. After a while both behavioural and allergic reactions may occur after eating certain foods. Newborns infants have extremely low values of dihydroxyphenylpropionic acid-like compound (DHPPA-like compound) in urine since newborns are not colonised with intestinal germs. In older children, the values are much higher. In children with autism, values may be extremely high. The antifungal treatment does not decrease in DHPPA-like compound. Some of the common species of Clostridium are Clostridium tetani that causes tetanus, Clostridium botulinum that causes the food poisoning botulism, and Clostridium perfringens and Clostridium difficile that cause diarrhoea. Clostridium perfringens, Clostridium novyi, Clostridium bifermentans, Clostridium histolyticum, Clostridium septicum, and Clostridium fallax may all cause gangrene. Since other species of Clostridium must be processed in an oxygen free environment, many hospital laboratories do not have the capability to identify these organisms. The exception is Clostridium difficile, which is identified by the toxin it produces in the stool rather than by the isolation of the organism itself. Clostridium difficile overgrowth of the intestinal tract causes pseudomembranous colitis, which is frequently associated with the use of oral antibiotics. This indicating that this organism is resistant to many of the common antibiotics such as penicillin, ampicillin, tetracyclines, cephalosporins, chloramphenicol, and others. This organism is usually treated with either metronidazole or vancomycin followed by a replenishment of the intestine with Lactobacillus acidophilus. Since many bacteria can genetically transfer drug resistance to other similar species and even unrelated species, most likely that multiple species of Clostridia may now be resistant to the most common drugs. Individuals with tetanus have extreme sensory sensitivity and may need to be placed in dimly lighted-quite rooms. The toxins produced by several different species of Clostridia (tetani, botulinum, barati, and butyricum) are very similar biochemically and therefore antibodies produced against one Clostridium toxin may react against the tetanus toxin. Also the gene for the tetanus neurotoxin is located on a plasmid, a piece of naked DNA that can be easily passed on to different species of Clostridia and may be even to other species of bacteria and which would confer on the new species the ability to make tetanus toxin. Eukaryotic and prokaryotic interactions is two way involving a complex and dynamic cross talk between species. Enteric pathogens disrupt the tight junction (TJ) by either altering the cellular cytoskeleton or by affecting specific tight junction proteins. Tight junction regulation via the cytoskeleton may occur indirectly through changes in the perijunctional actomyosin ring or directly through changes in specific TJ proteins. Disruption of specific TJ proteins can result from degradation by bacterial derived proteases or by biochemical alterations such as phosphorylation or dephosphorylation. Clostridium difficile is an anaerobic bacterium that causes antibiotic associated pseudomembranous colitis. C difficile's enterotoxins A and B causes TJ disruption by inducing: 1- structural changes (degradation of filamentous actin and increases levels of soluble actin, resulting in cell rounding), and 2- functional changes (decreases TER [transepithelial electrical resistance] and increases flux of paracellular markers) changes. These structural and functional changes appear to be due to toxin induced modifications of the Rho family of proteins, resulting in inactivation of the Rho GTPases and filamentous actin degradation. Associated with C difficile toxin induced depolymerisation of actin is the movement of TJ proteins such as ZO-1 and occludin away from the TJ and into the cytoplasm of the cell. C difficile may affect TJ proteins directly. Enteropathogenic Escherichia coli (EPEC) adheres to the surface epithelial cells and induces accumulation of cytoskeletal proteins underneath the site of attachment to produce the characteristic attaching and effacing lesion. Effector molecules are proteins that once inserted into host cell trigger a broad range of cellular events including alterations in electrolyte secretion, disruption of the TJ barrier, and inflammation. The permeability defect that occurs at the level of the TJ occurs via calcium and myosin light chain kinase (MLCK) dependent process. EPEC infection of intestinal epithelial cells stimulates phosphorylation of the 20 kD myosin light chain (MLC20) by MLCK, accounting in part for the EPEC induced TJ disruption. MLC20 phosphorylation by MLCK stimulates cytoskeletal contraction, including that of the perijunctional actomyosin ring. This event causes a drop in TER and increase in paracellular permeability. EPEC alters occuldin, a transmembrane TJ protein that is important for barrier formation. Phosphorylation of occludin is required for its localisation to the TJ complex. Following infection by EPEC, occludin shifts from TJ into the cytosol of the cell, an event that is accompanied by protein dephosphorylation. Reversal of these changes in occludin and the return of TER to baseline occur following elimination of the infection with gentamicin. Inhibition of serine/threonine phosphatases prevents EPEC induced changes in both occludin and TER. Enterohaemorrhagic E Coli (EHEC) also lowers TER, increases the paracellular flux of mannitol, and alters ZO-1 disruption via MLCK and conventional PKCs pathways. The conventional PKCs do not participate in the EPEC associated disruption of TJs. Bacteroides fragilis disrupts TJs by proteolytic degradation of TJ proteins. B fragilis enterotoxin (BFT) or fragilysin is a metalloprotease with a zinc binding motif and has the ability to hydrolyse several proteins including gelatin, fibrinogen, and actin. This toxin causes a decrease in TER and increase in paracellular permeability resulting in rounding cells and loss of microvilli. Cleavage of E-cadherin occurs near the plasma membrane by BFT. The initial cleavage is ATP independent and directly caused by BFT, the second part of the cleavage is ATP dependent caused by host proteases. Proteolysis of the intracellular domain of E-cadherin may disrupt β-catenin which links E-cadherin to α-catenin, therefore causing actin disruption. Like BF, Vibrio cholera secretes a zinc binding metalloprotease, termed haemagglutinin protease (HA/P), which has the ability to act on a wide variety of substrates, including cholera toxin, and E1 Tor cytolysin/haemolysin toxin. Clostridium perfringens is an anaerobic bacterium that causes GIT illness and may also cause antibiotic associated diarrhoea. C perfringens enterotoxin (CPE) has a C terminal binding domain, while the biologically active portion of the toxin is localised to the N terminus. Exposure of the intestinal cells to CPE results in tissue damage followed by fluid and electrolyte secretion. After binding to the cell surface, CPE remains associated with the plasma membrane and increases membrane permeability. CPE induces disruption of TJ fibrils resulting in decreased TER and increased paracellular flux. The intestinal epithelium has a remarkable capacity for fluid and electrolyte absorption. Under normal condition 8-9 L of fluid enter the gut on a daily basis and all except for 100-200 ml/d are reabsorbed. Under the tight control of various neurotransmitters, hormones, inflammatory mediators, and intraluminal contents, exist to carry out this function. With Cl secretion, paracellular movement of Na follows, resulting in luminal accumulation of NaCl, which provides gradient for the diffusion of water. Chloride enters the basolateral membrane via the actions of the Na-K-2Cl cotransporter. K^+ channels, also basally situated, provide the exit route of transported K^+. The apically located cAMP dependent cystic fibrosis transmembrane conductance regulator (CFTR) is responsible for the majority of apical chloride secretion. There are also calcium activated chloride channels (CaCC) that secrete Cl in response to increased intracellular calcium levels. Various pathogens, including V cholera, E coli, Salmonella, Campylobacter jejuni, Shigella dysenteriae, and Pseudomonas aeruginosa, mediate Cl secretion in a cAMP dependent fashion, which accomplished by elaboration of enterotoxins. Increased in intracellular cAMP levels leads to the activation of cAMP dependent protein kinase and subsequently CFTR. The final result is stimulation of electrogenic Cl⁻ secretion and massive diarrhoea. Cholera toxin may also indirectly affect intestinal secretion by inducing the enteric nervous system to release increased levels of 5-hydroxytryptamine (5HT). Under physiological conditions, increased intracellular cGMP can also lead to phosphorylation and activation of CFTR by cGMP dependent protein kinase II, resulting in Cl and bicarbonate secretion. Guanylate cyclase is responsible for the generation of GMP. Guanylin and uroguanylin are endogenous small peptides that stimulate chloride and bicarbonate secretion in the intestine through activation of GC-C. Several bacteria elaborate heat stable toxins (ST) that share considerable homology with guanylin and uroguanylin. ST binding to GC-C increases intracellular cGMP and initiates a signalling cascade, leading to phosphorylation of CFTR followed by Cl and bicarbonate secretion. Intracellular calcium levels are generally very low (100 nM) and are under tight physiological control. Changes in intracellular calcium concentrations are transient, even in the continued presence of agonist. Calcium is sequestered into 3 compartments, the extracellular space, mitochondria, and non-mitochondrial intracellular stores. Neurohormonal substances such as acetylcholine, serotonin, carbachol, and bradykinin can increase intracellular calcium by altering the permeability of these stores. Binding of ligand to receptor results in activation of the membrane associated phospholipase C. The phospholipase C enzyme hydrolyses phosphatidyl inositol 4,5-bisphosphate (PIP_2), releasing inositol 1,4,5-trisphosphate (IP_3) and diacylglycerol (DAG). IP_3 then acts to increase intracellular calcium levels while DAG activates PKC. Apically located channels (CaCC) are activated by increases in cytosolic calcium. Regulation of these channels is complex involving phosphorylation by the calmodulin dependent protein kinase CaMKII and possibly PKC. Non-bacterial microbial pathogens, most notably rotavirus, a common cause of severe diarrhoea in children, also exploit a calcium-based mechanism to stimulate intestinal secretion. Increased intracellular calcium through both release from intracellular stores and through the plasma is mediated through activation of phospholipase C and IP_3 mobilisation. Bacterial infection of host intestinal epithelial cells can also upregulate the expression of host derived products that, through autocrine or paracrine effects, also stimulate intestinal, primarily Cl, secretion. Galanin is a neuropeptide released from the nerve endings of the enteric nervous system. Galanin modulates intestinal motility, and induces Cl secretion in human intestinal epithelium through activation of the galanin-1 receptor (Gal-1R). Gal-1R expression is upregulated in an NFκB dependent fashion and that such upregulation plays a role in fluid secretion on bacterial infection. Nitric oxide (NO) functions in the intestinal epithelium include: 1- regulation of barrier function and antimicrobial activity, and 2- stimulating intestinal epithelial chloride secretion by increasing intracellular cGMP levels. Infection by S dublin, S flexneri, enteroinvasive E coli (EIEC), and Shigella flexneri increases intracellular cGMP and electrogenic chloride secretion. These responses can be attenuated by inhibitors of NFκB and tyrosine kinases. Prostaglandins are the products of arachidonic acid metabolism through the cyclooxygenase pathway. PGE_2 production is catalysed by the enzyme prostaglandin H synthase (PGHS) and stimulates chloride secretion in epithelial cells. Most the invasive pathogenic bacteria, but not non-pathogenic bacteria, induce the expression of PHGS-2 and its products PGE_2 and $PGE_{2\alpha}$ resulting in Cl secretion. Cyclooxygenase 2 (COX-2) induction by S dublin and EIEC also results in increased PG production, which in turn, contributes to Cl secretion. The host cell response to infection by enteric pathogens is capable of contributing to the intestinal secretory process, and ultimately diarrhoea, through a variety of mechanisms. The advantage to the host is the potential flushing of the intestinal lumen and resultant clearing of pathogenic microbes. The advantage of this process to the microbe is the increased potential for transmission to additional hosts. Intestinal pathogens are capable of inciting inflammation in the GIT mucosa. The common final response is the release of cytokines, chemokines, and the recruitment of inflammatory cells. EPEC infection activates NFκB, mitogen activated protein (MAP) kinases, tyrosine kinases, and PKCs, all of which are involved in inducing the inflammatory response. Enteric pathogens elicit expression of a characteristic profile of cytokines from epithelial cells that recruit effector cells to the site to clear the infecting organisms. After bacterial adherence or invasion, epithelial cells respond by secreting or expressing a characteristic pattern of cytokines, adhesion molecules, and major histocompatibility complex (MHC) class II molecules. These molecules, whose expression is regulated by various nuclear transcription factors and protein kinases, recruit a wide variety of effector cells, including neutrophils, monocytes, lymphocytes, and eosinophils to the site of infection. IL8 is secreted basolaterally by the intestinal epithelium in response to pathogenic bacteria or specific proinflammatory cytokines. IL8 gradients are responsible for neutrophil migration through the ECM and to the apical surface epithelial cells. Once at the subepithelial surface, further signals are needed to complete transepithelial migration such that neutrophils will be strategically positioned adjacent to the pathogens, e.g., pathogen elicited epithelial derived chemoattractant (PEEC) of S typhimurium infection. This appears not to induce superoxide production or degranulation. Transepithelial migration of neutrophils contributes to intestinal epithelial Cl secretion. Neutrophils release 5'AMP, which is converted to adenosine by the apical membrane enzyme 5' ectonucleotidase. Interaction of adenosine with the intestinal adenosine receptor $A2_b$ stimulates the secretion of Cl through a Gα and cAMP dependent mechanism. NFκB controls the expression of essentially all proinflammatory cytokines, chemokines, immune receptors, and cell surface adhesion molecules including IL-1β, TNF-α, IL-6, IL-8, IL-12, iNOS, ICAM-1, VCAM-1, T cell receptor α, and MHC class II molecules. Both invasive and non-invasive enteric pathogens, but not commensal flora, trigger the inflammatory cascade through activation of NFκB. NFκB regulates apoptosis. The magnitude of NFκB activation in the crypt cells of the intestine of patients with Crohn's disease and ulcerative colitis correlates with disease activity as does the expression of cytokines including IL-8 and IL-6. Phosphorylation of NFκB RelA/p65 subunit is necessary for the induction of NFκB dependent transcription. TNF-α induced phosphorylation is mediated by casein kinase II. Activation of NFκB by cytokines, such as TNF-α and IL-1, is initiated by their interaction with cognate receptors located in the cell membrane. In these situations, NFκB inducing kinase (NIK) communicates with the intracellular domain of these receptors triggering a complex array of downstream signalling events that ultimately activate IKK and, in turn, NFκB. Bacteria employ cytokine independent pathways to activate NFκB. The MAP kinase (MAPK) family may be involved in membrane cytosolic, cytoskeletal, and nuclear processes, they commonly target transcription factors that impact a diverse array of cellular functions, including cytokine production. There are 3 parallel MAPK cascades, ERK, JNK, and p38 that can simultaneously or independently activated. The cascade is initiated by activation of MAPK kinase kinase (MAPKKK). One of the best defined means of MAPKKK activation is through receptor tyrosine kinases. Receptor autophosphorylation caused by ligand binding induces the recruitment of SH2 containing adaptor proteins, which subsequently recruit guanine nucleotide exchange factors near the membrane, promoting GTP binding to Ras. GTP bound Ras binds the protein kinases Raf-1 and B-Raf thus increasing their protein kinase activity. Raf-1 and B-Raf then phosphorylate MAPKK, also called MAP/ERK (MEK), which subsequently phosphorylates and activates the MAP kinases. Downstream ERK may be involved in the regulation of NFκB through MEK, which has been shown to activate both IKK-α and IKK-β. The protein kinase C (PKC) family consists of 11 identified members divided into 3 groups based: 1- The conventional PKCs (cPKC-a, β1, β2, and γ) are activated in calcium and DAG dependent manner. 2- Novel PKCs (nPKC-δ, ε, η, φ, and μ) are calcium independent but DAG dependent. 3- Atypical PKCs (aPKC-ζ and λ/ι) are DAG and calcium independent and, in contrast with conventional and novel PKCs, do not respond to phorbol esters. The different isotypes have signal and tissue specificity. Bacteria may use the host cellular apparatus for invasion, attachment, or synthesis of bacterial proteins. Pathogens utilise many inducers of the inflammatory cascade. Infection of epithelial monolayers with strains of invasive bacteria such as Salmonella dublin, enteroinvasive E coli and others results in the expression of proinflammatory cytokines such as IL8, monocyte chemotactic protein 1, granulocyte macrophage-colony stimulating factor, and TNFα. Invasion is not necessary for induction of inflammation.

INTERACTION OF PATHOGENS WITH THE ENDOTHELIUM
The capability of bacteria to produce and release toxin may contribute to its ability to cause disease. After entering the bloodstream or being in the area of the capillary beds, virtually all pathogens and their products come in close contact with the endothelium. Pathogens and pathogen-derived products may use preformed endothelial signalling pathways by binding to cell-surface receptors or alter endothelial function by attacking membrane integrity.

Endothelial-matrix and intercellular junctions may be altered by pathogens. Specific pathogen surface structures mimic endogenous ligand binding to endothelial surface structures thereby allowing adhesion to or uptake into the endothelium. Fibronectin-binding protein A from S. aureus, for example, facilitates adhesion and temporal uptake in endothelial cells by binding to cell surface fibronectin. Expression of fibronectin-binding protein A by non-invasive bacterium causes cellular uptake of this bacterium by the endothelium. Beside fibronectin on endothelia, S. aureus attaches to host ECM components including fibrinogen, vitronectin, as well as elastin thereby demonstrating the complex nature of the adhesion process. These bacterial adhesion proteins may also function as potent inhibitors of leukocyte recruitment to the site of inflammation. S. aureus secrets extracellular adherence protein efficiently blocks β_2-integrin-dependent neutrophil recruitment. Adhesion of pathogens to endothelial cells activate their signalling pathways. Simple attachment of Chlamydia pneumoniae or Bartonella henselae to endothelial cells induces lipopolysaccharide (LPS)-independent activation of protein kinases within minutes. The bacterial outer membrane proteins (OMP) contribute to this rapid activation. Highly aggressive bacteria perturb membrane integrity of endothelial cells by exotoxins, which are secreted in the host environment. The endothelium is a major target of pore forming bacterial exotoxins, e.g., S aureus α-toxin and E coli hemolysin (HlyA). Pore forming exotoxins are amphipathic proteins, which insert and oligomerize into the membrane by interacting with lipid chains or nonpolar segments of integral membrane proteins, thereby creating an inner hydrophilic activity. These toxins punch small holes in the plasma membrane of eukaryotic cells, thus allowing fluxes of ion but not of macromolecules. These toxins activate expression of endothelial adhesion molecules on the cell's surface with subsequently enhanced endothelial-leukocyte interaction promoting the inflammatory response. Some bacterial exotoxins bind to endothelial cell receptors resulting in toxin translocation into the cytosol. Enterohemorrhagic E coli (EHEC) causes postdiarrheal haemolytic-uremic syndrome (HUS). Glomerular capillary as well as microvascular cortical renal endothelial cells are highly susceptible for Shiga toxin induced cell damage. The cellular biology of pneumolysin-endothelial interaction has been elucidated. Many host cells including endothelial cells recognize pathogen-associated molecular pattern (PAMP) by pattern recognition receptors, which clearly can distinguish between self and conserved microbial structures shared by different microbes. Viral protein also capable of activating endothelial cells. A few bacteria persist and replicate in the endothelium, e.g., Chlamydia pneumoniae, Listeria, Reckettsiae. Many pathogens enter endothelial cells as a transient process, e.g., S pneumoinae, Neisseria menigitides. Different viruses infect and replicate within the endothelial cells, e.g., Kaposi's sarcoma associated herpesvirus. Proteins derived from the pathogens can directly affect the cytosketelton of endothelial cell. Pathogens induce ruffles at the cell membrane, the formation of actin tails and pedestals in non-phagocytic cells to induce their uptake. Inside the host cell, bacterial pathogens may either remain within a vacuole, e.g., Chlamydia, or escape into the cytoplasm, e.g., Listeria. Almost all of these interactions require alterations of the actin cytoskeleton. Pathogens release effector molecules into the extracellular space. Many bacteria affect the microfilament system by modulation of small GTP-binding Rho proteins, which act as molecular switches in the regulation of the cytoskeleton. N meningitides enters endothelial cells by interaction of bacterial pili with a cellular receptor leading to Rho protein dependent microvilli formation. Recruitment of a tyrosine receptor and subsequent phosphorylation of the cytoskeletal protein cortactin promote bacterial invasion into endothelial cells. Intracellular replection within endothelial cells results in profound alterations of endothelial metabolism and endothelial factors may interfere with bacteria. Chlamydia induce delivery of spingomyelin, cholesterol and phospholipids from the Golgi apparatus to the chlamydial inclusion. HIV-related damage to the endothelium contributes to AIDS-related vasculitis in the central and peripheral nervous system, and manifestation of bacilliary angiomatosis. Kaposi's sarcoma-associated herpesvirus (KSHV) alters cellular pattern of gene expression. Mediator release, leukocyte recruitment, pro-coagulant activity, and the breakdown of endothelial barrier function are responses of the cell upon pathogen-related stimulation. Activation of NFκB-dependent gene transcription contributes significantly to the endothelial pro-inflammatory response. The response of the endothelium to inflammatory stimuli may differ grossly among different microvascularture sites of the body. Upon stimulation by a pathogen or pathogen components (LPS, liptoeichonic acid) as well as by pro-inflammatory mediators produced by the host defense, an array of inflammatory mediators is produced by endothelial cells. E-selectin, ICAM-1, or VCAM-1 are upregulated within hours by gene-transcription after pro-inflammatory stimulation. These molecules act in a complex concert with their counterparts expressed on leucocytes and on platelets mediating cell-cell adhesion. Pro-inflammatory stimulation leads to expression of the inducible NOS2 resulting in Ca-independent libration of large quantities of NO into the vascular and perivascular space. Besides regulating vascular tone, these large amounts of NO may contribute to the killing of pathogens. The release of cytokines and vasoactive compounds by the endothelium in the vascular space has systemic as well as paracrine effects. Endothelium triggered by infection with for example Chlamydia or viruses loses its anticoagulant properties and becomes a procoagulant surface. Increased tissue factor expression by the pathogen-activated endothelium, tethering of monocytes bearing tissue factor and leukocyte-derived microparticles together may activate the coagulation cascade. The serine protease thrombin activates the G protein leading to hyperpermeability, adhesion molecule expression and cytokine production. Activation of platelets by pathogens contributes to the formation of platelet-bacteria thrombi on the endothelial surface. Activation of the coagulation cascade may bring up a molecular net fishing and immobilizing invading pathogens. Activation of the plasminogen system may be activated by some microorganisms to counteract the host response, e.g., E coli. Bacterial plasminogen receptors immobilize plasminogen on the bacterial surface, thereby enhancing plasminogen activation and turning the bacteria into a proteolytic organism. This proteolytic activity may counteract the host's fishing strategy and promotes invasion of pathogens by degradation of extracellular matrices. Hyperpermeability with subsequent oedema formation is a hallmark of acute inflammatory reactions. Viral proteins, endogenous mediators of the inflammatory response (e.g., TNFα) or leucocyte-derived agents (OFR, proteases) attack the endothelium finally resulting in barrier dysfunction. Most of the activated signalling pathways converge at the phosphorylation of myosin light chain leading to increased actin-myosin interaction. Endothelial cells contract as muscle cells do, and paracellular gaps open, thereby breaking down the permselective endothelial barrier. Bacterial or host proteases promote endothelial hyperpermeability by directly degrading cell-cell contacts or cell-matrix adhesions. Thus, the exposed sub-endothelial matrix becomes accessible to bacterial adhesins and a vulnerable target of bacterial proteases promoting bacterial colonialisation of host tissue. Endothelial dysfunction contributes to the development of mulyiple organ failure. Loss of barrier function is a prerequisite for lung oedema formation in adult respiratory distress syndrome.

BOWEL DYSBIOSIS
The maintenance of stable host flora equilibrium involves host defenses, environmental factors, and bacterial interactions. These host mechanisms include acid secretion, intestinal peristalsis, and a properly functioning ileocaecal valve. In addition, intestinal immunoglobulin secretion and a mucosal barrier preventing bacterial adherence are of vital importance in minimising pathogenic bacterial populations. Intestinal motility is another major host defense against bacterial overgrowth. There is an average of 109 microbes per ml in saliva. Anaerobic bacteria colonize at the gumline and between the teeth. Both Streptococci and Lactobacilli play a major role in dental caries by converting oligosaccharides from the diet into dextran and levan, which cause plaque formation on the teeth. Fluoride and vitamin B_6 reduce the numbers of these bacteria in the mouth, reducing the potential for cavities. The stomach harbours few bacterial species, and those that survive the highly acidic pH of gastric juices are primarily acid tolerant Lactobacilli and yeasts, which are constituents of the oral flora. The duodenum has a very low bacterial count of an average of 105 organisms per ml of intestinal juice, which are the same oral type found in the mouth and stomach. HCl kills bacteria by dissolving plasma membranes. If normal acidity is maintained in conditions of gastritis or gastric ulcers, microbial presence would remain at a low level. Helicobacter pylori is a microaerophillic gram negative bacilli, which is uniquely adapted to survive in the stomach mucosa, and causes chronic gastritis and peptic ulcers by impairing the mucosal integrity of the stomach, and is treated with antibiotics, berberine sulfate, deglycyrrhizinated licorice, bismuth, bentonite, and L-glutamine. The small intestine is relatively free of bacterial populations until the ileum approaches the ascending colon, where some faecal type bacteria are believed to migrate. Dysbiosis of the small intestine results a number of factors, including surgical trauma, low acidity, environmental and emotional stress, and altered intestinal transit, which can facilitate the overgrowth of bacteria and contribute to malnutrition. Reduced mucosal immunity (Secretory IgA) can further contribute to increasing potentially pathogenic bacterial populations of the bowel. The small intestine is the primary site of food absorption, and the metabolism of microbes can interfere with this process. Bacteria can prevent the uptake of vitamin B_{12} in the distal ileum, causing B_{12} deficiency and megaloblastic anaemia. The lost dietary B_{12} can be recovered bound to bacteria in the faeces of patients with bacterial overgrowth. Urinary concentrations of indole and indican, which are degradation products of tryptophan, are increased in patients with small bowel overgrowth. Amino acid absorption can be impaired, while faecal nitrogen is increased and serum proteins lowered. The bacterial overgrowth can contribute to the dysfunction of mucosal cells, impairing protein and carbohydrate absorption. In blind loop syndrome, a portion of the small intestine is dilated and does not empty during peristalsis, as does the rest of the intestine. Bacteria remain and propagate in this area causing enteritis and penetration of the bacteria into the intestinal epithelium, where obligate anaerobes rather than facultative anaerobes tend to grow. Malabsorption is prominent in this disease, with bacterial deconjugation of bile acids preventing fatty acid and monoglyceride absorption, causing steatorrhea. Decreased HCl, digestive enzyme deficiency, and imbalances in short and long chain fatty acids can all contribute to bacterial overgrowth of the small and/or large bowel. Malabsorption, maldigestion, and malnutrition can all result from disturbance or disruption of the bowel microbial ecology. Identifying potential pathogenic bacteria is not an easy task. The faeces contain about 400 billion bacteria per gram of dry weight, which accounts for about 40% of faecal volume. These organisms are representative of the population of the ascending, transverse, and descending colon as well as the rectum. There are 400-500 different species of bacteria in the human intestinal tract at any given time. Within a week, a breast-fed

infant develops a faecal flora primarily consisting of Bifidobacterium, Lactobacillus, Escherichia, Streptococcus, and Staphylococcus. As the diet changes, the intestinal flora also changes. Gram negative anaerobes (Bacteroides) are observed when meat is introduced into the diet. When bacterial overgrowth occurs in the small intestine, changes in the colonic flora also occur. Bifidobacterium is greatly reduced and there may be an increase in aerobes such as Enterobacteriaceae, Enterococcus, Staphylococcus, Proteus, and Clostridium. The major increase is within the Enterobacteriaceae family and includes Klebsiella and Enterobacter. The overgrowth of this faecal flora seems to be associated with an increased pH of gastric juice. Opportunistic infection and complications in the small intestine are interrelated. Clostridium perfringens produces a number of exotoxins and an enterotoxin, which cause food poisoning, cellulitis, antibiotic associated diarrhoea, and may be associated with rheumatoid arthritis. Clostridium difficile causes antibiotic-associated diarrhoea (e.g., clindamycin and ampicillin), and pseudomembranous colitis by producing a toxin that promotes necrosis and ulceration of the colonic mucosa. Antibiotics eliminate facultative anaerobes, which compete with Clostridium difficile for growth. Saccharomyces boulardii, Lactobacillus and Bifidobacterium are biotherapeutic agents that have been used successfully to prevent antibiotic-associated diarrhoea caused by Clostridium difficile infection. Facultative anaerobes create a reduced environment by utilising oxygen, therefore creating a hospitable environment for anaerobes. Metabolic products of anaerobic bacteria (and facultatives to some extent) are short chain fatty acids such as acetic, propionic and butyric acid, which can inhibit bacterial proliferation by entering the bacterial cell at low pH and inhibiting metabolism. Some bacterial strains produce antibiotic substances that can directly reduce populations of potential pathogenic bacterial species. Bacteriocins are examples of antibiotic substances that are produced by some bacterial species. E coli produces a substance known as colicine, which prevents growth of other species, and can autoregulate its own population. Diet has significant influence on microbial populations in the bowel. Extremely oxygen sensitive anaerobes as well as Bifidobacterium disappeared in individuals consuming a chemically defined, non-bulk diet. The major changes observed were an increased number of aerobes, primarily E coli, and a decrease of Lactobacillus in these individuals. Subjects with significant amounts of meat in the diet have many more gram negative, non-spore-forming anaerobes such as Bacteroides in their faeces, whereas subjects who consumed a vegetarian diet have a higher proportion of Streptococci and Enterobacteriaceae. Bacteria may have mutagenic activity, which may further increase procarcinogenic activity. Procarcinogenic substances such as food preservatives, dyes, additives, or pollutants can be transformed into carcinogenic substances by bacterial enzymes β-glucuronidase, β-galactosidase, β-glucosidase, nitro-reductase, azoreductase, 7-α dehydroxylase, and cholesterol dehydrogenase. Bacterial β-glucuronidase hydrolyses glucuronidase, contributing to the toxicity of compounds, which have been previously detoxified by liver glucuronidation. Bacterial nitroreductase reduces nitrogenous compounds to aromatic amines via nitrosamines and n-hydroxy compounds, which are suspected mutagens. Lactobacilli may influence the metabolic activity of colonic flora and inhibit cancer potential. Lactobacillus feeding reduces faecal amounts of β-glucuronidase, azoreductase, and nitroreductase. Refined sugar significantly increases in gut transit time, fermentative colonic bacterial activity, and intestinal bile acids in healthy subjects. The lengthened transit time caused by an increase in secondary bile acid concentration may be associated with the development of colorectal cancer. High sugar intake and increased gut fermentation are also associated with and implicated in the formation of gallstones and Crohn's disease. There is increased psychomorbidity in hospitalised patients with irritable bowel syndrome. Anxiety is associated with increased bowel frequency, and depressed patients tend to be constipated. Emotional reactions trigger changes in intestinal motility and enzyme, bile, and/or mucin secretion. Mucin is an important substrate for many intestinal microorganisms. Intestinal transit time can alter symbiotic relationships of the bowel microflora. Bacterial presence stimulates defensive factors in the intestinal wall. Continued use of antibiotics reduces the number of normal flora, which can cause susceptibility by reducing the defenses of the intestinal wall and enabling adherence and growth of pathogenic bacteria. Penicillin, clindamycin, and metronidazole eliminate certain obligate anaerobes, allowing gram negative facultative anaerobes to thrive and overpopulate, then translocate from the gastrointestinal tract. Facultative anaerobes such as Escherichia coli, Klebsiella pneumoniae, Proteus mirabilis, and Pseudomonas aeruginosa are likely to translocate to the mesenteric lymph nodes; and translocation is especially promoted by abnormally high numbers of these organisms. Patients who receive chemotherapy in addition to antibiotics are especially at higher risk. Translocation has also been known to occur apart from trauma or any other known aetiology. Candida ingested can translocate within 3 hrs of inoculation into the blood stream. Translocation occurs in healthy individuals. Delayed transit time and achlorhydria are 2 common manifestations in individuals with bacterial overgrowth. Antibiotics clearly alter the bowel flora, creating an environment favouring proliferation and adhesion of pathogens to the bowel mucosa. Broad spectrum antibiotics generally do not disturb the anaerobic bacteria. Bacteria may transfer antibiotic resistance to each other as well as secrete substances with antibiotic activity. Many individuals harbour bacterial overgrowth without symptoms. Clinical features in symptomatic individuals can include megaloblastic anaemia, weight loss, malnutrition, diarrhoea, constipation, bowel cramping, and chronic joint pain. Rare presentations may include night blindness, abnormal bleeding, hypocalcaemic tetany, osteomalacia, peripheral neuropathy, and oedema. Yoghurt is a coagulated milk product that results from fermentation of lactic acid in milk by Lactobacillus bulgaricus and Streptococcus thermophilus. Other lactic acid bacteria (LAB) species (L acidophilus and L casei) can be combined with L bulgaricus and S thermophilus. During fermentation, vitamins B_{12} and C are consumed and folic acid is produced. The differences in other vitamins between milk and yoghurt are small and depend on the strain of bacteria used for fermentation. Although milk and yoghurt have similar mineral compositions, some minerals, e.g., calcium, are more bioavailable from yoghurt than from milk. In general, yoghurt also has less lactose and more lactic acid, galactose, peptides, free amino acids, and free fatty acid than does milk. The potential therapeutic effects of LAB and yoghurt, including their immunostimulatory effect, are due primarily to yoghurt-induced changes in the gastrointestinal (GI) microecology. Increased amounts of LAB in the intestines can suppress the growth of pathogenic bacteria, which contributes in turn to reduced infection and heightened anticarcinogenic effects. The immunostimulatory effect of LAB also depends on the degree of contact with lymphoid tissues while the bacteria are transiently colonising the intestinal lumen. Cytokine production, phagocytic activity, antibody production, T cell function, and natural killer (NK) cell activity increase with yoghurt consumption. The ability to perform phagocytosis and kill microbes, including bacterial pathogens, is a major effector function of macrophages. Bacteria bind to complement components and the bacterium-complement complexes bind complement receptors on the surface of macrophages. The bacterium-antibody complex then binds the macrophages via the Fc receptor and phagocytosis begins. Yoghurt consumption and oral administration of LAB stimulate the host immune system. Muramyl dipeptide (MDP) is a main constituent of peptidoglycan in the cell wall of pathogenic and nonpathogenic bacteria such as LAB. MDP stimulates macrophages to release IL-1, which is needed for activation of T lymphocytes and induces IFN-γ production by lymphocytes. MDP stimulates the production of IL-1, IL-6, and TNF-α by monocytes as well as that of IL-4 and IFN-γ by lymphocytes. Peptides produced from the fermentation of milk may also contribute to the immunoenhancing effect of yoghurt. Glutathione is important for detoxification of endogenous and exogenous carcinogen and free radicals and in regulation of immune function. Depletion of cellular glutathione was reported to suppress mitogenic response of lymphocytes, to prevent lymphocytes from entering the S phase in the cell cycle, and to decrease antibody-dependent cellular cytotoxicity and spontaneous cell-mediated cytotoxicity and IL-2 induced LAK activity. Whey protein (glutathione) decreases the incidence of infections and neoplastic diseases and increases longevity. Milk fermentation results in a complete solubilization of calcium, magnesium, and phosphorus and a partial solubilization of trace minerals.

BIOGERNOTOLOGY

Lifestyle activities are necessary for successful ageing. Lifestyle consists of 2 components: 1- life conduct, expressed in personal choices; and 2- the opportunities available to realise these choices. Cognitive functioning in old age may be facilitated by an active lifestyle. While many age differences in memory performance are attributed to intellectual activity and educational level, declining cognitive performance in old age may specifically relate to both cognitive style (attitudinal and motor-cognitive flexibility), specific lifestyle variables (the absence of an active lifestyle or a breakdown of family ties), socio-economic status and social interaction. Perceived memory change is influenced by ageing, whereas memory capacity and memory anxiety are more influenced by social factors. Activity and frequent contact with friends and family increase the chance for good memory capacity. The hallmark of cellular aging is the failure of senescent cells to initiate the DNA synthesis during the progression of cell cycle. IL-2 plays a critical role in immune function, the expression of IL-2 decreases with age, and the decrease in expression parallels the age-related decrease in immune function. The failure of maintenance and repair mechanisms underlies the ageing process. The term gerontogenes does not refer to a tangible physical reality of real genes for ageing, but refers to an emergent functional property of a number of genes, which influence ageing. Free radical scavenging and antioxidant gene is part of the gerontogene family. Antiageing replacement therapy should include calorie restriction, antioxidants, and gene therapy. Aging is a process characterised by several changes that include a reduced capacity to use oxygen along with impaired cardiocirculatory capacity and respiratory adaptation; deterioration of the nervous system (a decrease in the form, width, and rate of conduction of evoked potential); and degeneration in muscle mass characterised by a reduction in muscle fiber diameters and by a qualitative and quantitative alteration in muscle fibers. In human skeletal muscles there is a significant decline in total SOD activity during aging and a significant increase in MnSOD activity. Skeletal muscle GSHtot and GSH levels remains unchanged with age indicating that membrane transport of GSH into the cell, which plays a critical role in determining tissue GSH levels, does not change during aging. Protein carbonyl content tends to increase during aging. Protein oxidative damage is linked to the rate of aging and contributes to this process. Consequences of slowing-down of protein synthesis include: 1- Decrease in the availability of enzymes. 2- Inefficiency in the removal of intracellular damage. 3- Accumulation of abnormal and defective molecules. 4- Inefficient intra- and inter-cellular communication. 5- Decrease in the production of hormones and growth factors. 6- Decrease in production of antibodies. 7- Decrease in the production of ECM.

PATHOBIOLOGY

Ageing has many facets. Ageing is not only an important biological issue, but is also an important social and emotional issue that affects almost all aspects of our lives. Not all parts of the body become functionally exhausted with age. Degenerative changes fall into 3 main categories: rapid, negligible and gradual. Biogernotology describes age-related changes in cells, tissues and organisms. It is highly complex implicates both genetic and epigenetic stochastic causative factors. There are no any apparent deterministic, predictive or universal biomarkers of ageing. There are age-related changes in dopaminergic, nonadrenergic, serotoninergic, cholinergic, neuropeptidic, GABAergic, and amino-excitatory systems. Glandular atrophy, vascular changes and fibrosis are some of the direct effects of ageing in the endocrine glands. Altered hormonal responsiveness is also the basis of some of the major age-related impairments and diseases, include diabetes, osteoporosis, hyperthyroidism, masked hypothyroidism, breast cancer, and various disorders associated with cognition and motor skills. Ageing of the hypothalamus leads to a reduced capacity for thermoregulation and then to increased sensitivity to sudden changes in environmental temperatures. A failing immune system makes the body vulnerable to all kind of opportunistic infections and allows cancerous cells to develop into cancers. With ageing, it is difficult to make the generalisation that the changes in the immune system are totally deteriorative. Some of the mechanisms expected to be crucial in ageing, are those involve in maintaining: 1- The structural (anatomical) and functional (physiological) integrity of the nuclear and mitochondrial genome. 2- The accuracy and speed of transfer of genetic information from genes to gene products. 3- The turnover of defective and abnormal macromolecules. 4- The efficiency of intracellular and extracellular communication and responsiveness. Increased adverse reaction in old age is a sign of failing mechanisms of drug uptake, absorption, clearance and detoxification. Altered cellular responsiveness is one of the critical aspects of cellular ageing. The loss of a cellular function as important as division, along with a large number of other physiological, biochemical and molecular changes, which precede and/or accompany this loss, is one of the fundamental determinants of ageing. Instability of the genome is an important aspect of the failure of homeostasis, as the survival and continued existence of any lifeform depends upon the stability of its genome. Various physical, chemical and biological factors are continuously challenging the DNA in all cells of the body. At the same time several DNA repair system operate in the cell, which counteract the effects of various DNA-damaging agents. Major DNA-damaging agents include solar UV radiation, background ionising radiation, a wide range of chemicals in food and in the environment, and several endogenous agents such as aldehydes, reactive oxygen species and other free radicals, which are the result of metabolic pathways. In addition to that, damage to DNA in a cell can also result from several spontaneous chemical changes in the DNA, such as those due to hydrolysis, deamination, methylation, demethylation and glycation, along with the errors that can occur during DNA duplication and repair, as a result of innate limits on the accuracy of any biochemical processes. The biological consequences of damage to DNA are similar, irrespective of the nature of the damaging agent. Single-strand breaks of DNA increases during ageing. There is age-related increase in DNA crosslinks, dicentric chromosomes, aneuploidy, polyploidy, loss of centromeric tandem repeats, and shorter linker regions between nucleosomes. An increase in free-radical-induced oxidative damage, particularly the formation of 8-hydroxy-2'-deoxyguanosine (8OHdG) in the nuclear and mitochondrial DNAs occurs in various organs of ageing. One of the most common results of DNA damage is the causation of mutations and epimutations. Although the genomic instructions of life are written in the language of nucleic acids, the life is actually lived in the language of proteins. A decline in the rate of total protein synthesis is one of the most common age-associated biochemical changes that has been observed in a wide variety of cells, tissues, organs. Even though the bulk of protein synthesis slows down with age, the total protein content of a cell generally increases because of an accumulation of abnormal proteins during ageing. Age-related changes in protein synthesis are regulated both at the transcriptional and pre-transcriptional levels in terms of the availability of individual mRNA, specificity and stability of a protein. One of the reasons for the inactivation of enzymes can be their oxidative modification by OFRs and by mixed-function oxidation (MFO) systems or metal catalysed oxidation (MCO). Structural alterations introduced into proteins by oxidation can lead to the aggregation, fragmentation, denaturation, and distortion of secondary and tertiary structure, increasing thereby the proteolytic susceptibility of oxidised proteins. Glycation is one of the most prevalent covalent modifications in which the free amino groups of proteins react with glucose forming a ketoamine called Amadori product. This is followed by a sequence of further reactions and rearrangements producing the so-called advanced glycosylation end products (AGEs). The long-lived structural proteins such as crystallins, collagen and basement membrane proteins are more susceptible to glycation. An increase in the levels of glycated proteins during ageing has been observed in a wide variety of systems, such as nerves, aorta and skin collagen. There is increased membrane rigidity of erythrocytes during ageing due to increased methylation of proteins. Detyrosination of micromolecules may be important during ageing. Alterations at the level of protein synthesis and their post-synthetic modifications can have global detrimental effects on the maintenance and survival of cells, tissues, organs and organisms leading to ageing and death. Age related decline in protein turnover is generally due to a decrease in the proteolytic activity of various lysosomal and cytoplasmic proteases. There is slower transcription, reduced rates of synthesis and altered patterns of post-synthetic modification. There are increased levels of expression and activities of TIMP and trypsin inhibitor during ageing of human fibroblasts. This will also lead to a decrease in the activities of proteases leading to decreased protein degradation during ageing. One of the common features of cells from senescent tissues is the accumulation of abnormal proteins. Aging is associated with a progressive reduction in almost all-physiological functions. Senescent organisms show an impaired responsiveness to environmental stimuli and to physiological stress. Immune system impairment in aging contributes to increased mortality from infections, autoimmune diseases, and cancer in the elderly. At the cellular level, aging results in the inability of cells to proliferate. Senescent cells resemble cells that are arrested at the G_1-S boundary of the cell cycle. A blockage in the expression of the c-fos gene, which leads to changes in the AP-1 (active gene regulatory protein), along with changes in CREBP (cyclic AMP response element binding protein) and CTF (CAAAT transcription factor) transcription factor complexes, all may contribute to the loss of cellular proliferation in senescent cells. Age-related changes in the activity of several cell cycle regulatory components such as Rb (retinoblastoma protein), Cdk2 (cyclin-dependent kinase 2), p21 (inhibitor of cyclin-dependent kinases), and p53 (tumour suppressor protein) seem also to participate in the mechanism of proliferation arrest in aging. Senescent cell remains metabolically active for long periods of time, but there are progressive changes in cell structure and function. Protein degradation together with protein synthesis takes place continuously in all cells. Small modifications in the balance between these processes allow cells to rapidly adapt to changes in the extracellular environment. Continuous turnover of proteins inside cells, abnormally synthesised proteins or proteins incorrectly modified are also eliminated from the cells by the proteolytic systems.

Mechanisms of maintenance and repair:

A- Whole-body level:

    1- Neuronal and hormonal responsiveness.
    2- Immune defences.
    3- Thermoregulation.
    4- Repair, wound healing and regeneration.

B- Tissue level:

    1- Neutralising and removing toxic chemicals.
    2- Tissue regeneration.
    3- Cell replacement and turnover.

C- Cellular level:

    1- Stability of the differentiated state.
    2- Regulation of cell proliferation.
    3- Stability of the cellular milieu (viscosity, ion balance, pH).
    4- Cell communication.

D- Molecular level:

    1- Stability of the genome.
    2- Fidelity of genetic information transfer.
    3- Turnover and degradation of macromolecules.
    4- Stress-protein synthesis.
    5- Scavenging of free radicals.

Intracellular proteases also participate in many other fundamental cell processes including cell differentiation, cell cycle progression, antigen presentation, and intracellular traffic of proteins. Cells with impaired protein degradation are less able to eliminate abnormal or damaged proteins. Accumulation of those altered proteins may cause a variety of further problems within cells. Defects in the processes of intercellular communication (i.e., antigen presentation, hormone and peptide secretion) will make those cells more vulnerable to the attack of exogenous agents and unable to develop a coordinated systemic response. These changes may contribute to common phenotypes of senescent cells, including the inability to progress through the cell cycle. The mechanisms responsible for aging may be catastrophic (aging is the result of damage accumulation throughout the cells' lifespan) or genetic (aging is due to a genetic clock that causes age-related phenotypes in cells). The effect of extracellular agents during the cellular life span modulates the severity of the age-related alterations. Several proteolytic systems participate in intracellular protein degradation. They mainly differ in their intracellular localisation and in the regulation of their activities. Although all of these degradation pathways are present in all cells, their activity varies from tissue to tissue and also depends on environmental conditions. Aging characteristics of senescent cells include: 1- The proliferative potential of cells depends on the number of cell doublings undergone and is independent of the chronological age of the cells. 2- There is a correlation between the proliferative potential of cells from different species and the average life span of that species. 3- Cells from patients with premature aging disorders show reduced proliferative potential. 4- Senescent cells have enlarged, flattened morphology and fail to replicate their DNA in response to normal growth stimuli. 5- There is an inverse relationship between the age of a donor and the proliferative potential of their cells. There is a decrease in a non-acidic protease activity with age. Calpains, or calcium-dependent proteases, constitute the other major cytosolic proteolytic system; their activity is tightly regulated by intracellular calcium levels. Micromolar calcium concentrations activate micro-calpain, but millimolar calcium concentrations are required for milli-calpain activation. Calpains partially degrade membrane and cytoskeletal proteins and several membrane-associated enzymes. Two of the best characterised substrates for the calpain system are the band 3 protein (an anion exchanger) in the erythrocyte membrane and protein kinase C. After translocation to the cell membrane and limited autolysis, calpains are activated. Calpains are responsible for the degradation of certain membrane and cytoskeletal proteins. An increased degradation of band 3 protein correlates with a higher translocation of calpains from the cytosol to the cell membrane. A diminution of thiols in senescent cells, as a consequence of free radical reactions and oxidative damage, may cause enhanced calpain translocation to the cell membrane. The increase in degradation of membrane proteins modifies the stability of the cell membrane and reduces cell life span in aged humans, e.g., the abnormal proteolysis and neuronal degeneration in Alzheimer's disease may be, in part, a consequence of calpain activation.

Age-related changes in protein degradation in the nervous system:

A- Changes in protease activity:
    1- Increase of calpain activity in spinal cord.
    2- Increase in calpastatin levels in brain.
    3- Increase in calpain and cathepsin D activity in brain.
    4- Decrease in calpain levels but increase in calpain activity in Alzheimer's disease patients.
    5- Increase in levels of cathepsin D in brain.
    6- Increase of brain calpain activity depends on the cerebral area analysed.
    7- Increase in cathepsin D, E, and B activities and decrease in cathepsin L activity in brain.
B- Changes in substrate susceptibility to proteases:
    1- Microtubule-associated proteins in brain are more susceptible to cathepsin D-like proteases.
    2- Phosphorylation and carbonyl modification of neurofilament proteins reduce their proteolytic susceptibility.
    3- Age-related modifications of synaptosomal membrane proteins inhibit their proteolysis.
    4- Binding of aluminium to neurofilament proteins inhibits their proteolysis.
    5- Peroxidation of myelin membrane proteins increases their susceptibility to proteolysis.

Lysosomes are organelles that contain a powerful mixture of proteases, peptidases, and other hydrolases capable of degrading most intracellular and extracellular macromolecules. The proteases responsible for the degradation of proteins inside lysosomes are called cathepsins. Lysosomes are one of the major proteolytic systems affected by age. In most cells there is an age-related increase in number and size of lysosomes. Primary lysosomes are able to completely degrade the protein content of several kinds of intracellular vesicles after membrane fusion. Fusion of lysosomes with endosomes, vesicles containing several extracellular and plasma membrane proteins, is fairly well characterised. A large number of intracellular proteins can be degraded in lysosomes by a process called macroautophagy. In this process, complete regions of the cytoplasm, including cytosolic proteins as well as entire organelles, are surrounded by a membrane to become double-membraned structures known as autophagosomes or autophagic vacuoles. The proteins inside autophagosomes are then completely degraded after fusion with lysosomes. Factors inhibiting this process include amino acids, insulin, cell swelling, and disruption of the cytoskeleton. During senescence, there are severe changes in the autophagosome/lysosomal system. A decrease in autophagosome formation as well as a decrease in the degradative activity of lysosomes would result in the slower degradation of substrate proteins. Since autophagic vacuole formation rates decrease in senescent cells, and rates of elimination of autophagic vacuoles also decrease with age, reductions in autophagosome formation and fusion with lysosomes both contribute to the age-related decline in macroautophagy. The specific degradative mechanisms of lysosomes are impaired in senescent cells. The important role of lysosomes in bulk protein turnover in liver, kidney, brain and fibroblasts make them mainly responsible for the age-related decrease in protein degradation in those organs. The CNS is one of the tissues that show the most dramatic changes in protein degradation with age. In Alzheimer's disease, proteins of the cell membrane and cytoskeleton are abnormally processed and accumulate in the brain. Acid hydrolases accumulate in atrophic and degenerating neurons, and their release to the extracellular space may contribute to senile plaque formation. The typical amyloid deposits in Alzheimer's disease are normally produced in an acidic compartment by noncysteine proteases, and then they are eliminated by lysosomal cysteine proteases. The liver is one of the organs in which a decrease in total protein degradation occurs with age. During aging there is a progressive loss of muscle mass that originates because of an imbalance between synthesis and degradation of proteins. The protein synthesis/degradation imbalance in senescence is only evident after induction of a catabolic state such as in starvation, denervation atrophy, cancer, acidosis or trauma, and reflect a failure to rapidly restore muscle protein after those circumstances. In the eye lens the proteolytic removal of damaged proteins may play an important role in maintaining the lens transparency. Normal lenses have the ability to increase ubiquitin conjugation activity in response to oxidative stress, but that response is impaired in senescent lenses. No changes in levels of ubiquitin conjugating enzymes in lenses were detected with age, but changes in their ability to respond to oxidative stress were found. The age-related decrease in the ability to mount an ubiquitin-dependent response upon oxidation may contribute to the accumulation of damaged proteins in the old lenses. Other proteolytic systems also decline with age in lens.

MITOCHONDRIA AND AGEING

Mitochondrion means, Greek mitos, a thread, and chondros, a grain. The primary functions of mitochondrion are the conservation of oxidatively derived energy and its utilisation for ATP synthesis, namely, oxidative phosphorylation. Progressive age-dependent damage in mitochondrial genomes and functions is an important contributor to human aging. Mitochondrion is the site of oxidation-reduction in the cell. There are 3 distinct pigments cytochrome a, b and c that undergo oxidation-reduction changes in a determined sequence, which bridges dehydrogenase and oxygenase leading to the concept of respiratory chain.

$$NADH \rightarrow Flavoprotein \rightarrow Cytochromes\ b \rightarrow c \rightarrow a \rightarrow a_3 \rightarrow O_2$$

Mitochondria are centres of energy metabolism. Mitochondria carry out the tricarboxylic acid cycle and the β-oxidation pathway for fatty acids. These degradative sequences essentially remove hydrogen from metabolic fuels with the release of CO_2 and transfer it through coenzymic carrier to the respiratory chain in the mitochondrial inner membrane. The chain passes the electron sequentially through complex I (NADH dehydrogenase) or complex II (succinate dehydrogenase), coenzyme Q (CoQ), complex III (ubiquinol: cytochrome c oxidoreductase), cytochrome c and complex IV (cytochrome oxidase) to give oxygen and water. The released energy is used to pump protons out of the mitochondrial inner membrane, creating an electrochemical gradient. The energy stored in this gradient is the driving force for complex V (ATP synthetase), which is associated with the inner membrane to condense ADP and P_i to make ATP. Mitochondria posses an independent genetic system necessary for the morphogenesis of the energy transduction system consisted of the respiratory chain and ATP synthetase. The human mitochondrial DNA (mtDNA) is remarkable for both its small size of 16,596 bp, compared to the nuclear DNA (nDNA) of ≈109 bp and for the most economically packed genomes so far characterised in the whole biosphere. Mitochondrial DNA is located inside of mitochondrial inner membrane where reactive oxygen species (ROS) always escape from the respiratory chain arising mtDNA mutation. mtDNA has inefficient repair systems compared with those in nDNA. Therefore, the mutated mitochondrial genomes tend to accumulate in a cell especially in the stable tissues such as nerve and muscle, which consist of postmitotic cells. A number of human neuromuscular diseases have structural abnormalities of mitochondria as their common feature, and these appear to be caused by mutations in mtDNA. mtDNA random mutations that occur throughout human life in the population of mtDNA molecules of each cell are major contributor to the gradual loss of cellular bioenergy capacity within tissues and organs associated with general senescence and diseases of aging. Mitochondrial electron-transport chain is composed primarily of 4 complexes: 1- complex I, NADH-ubiquinone oxidoreductase; 2- complex II, succinate-ubiquinone oxidoreductase; 3- complex III, ubiquinone-cytochrome c oxidoreductase; and 4- complex IV, cytochrome oxidase. These complexes together with complex V, proton-motive ATP synthetase, are all embedded in the mitochondrial inner membrane. A decreased numerical density of mitochondria in the fibroblasts from old individuals seems to be balanced by the increased size of individual mitochondria. Mitochondrial myopathy, encephalomyopathy, and cardiomyopathy could be regarded as the disease associated with premature aging. Defective mitochondrial respiratory chain encoded by human mtDNA with deletion would enhance oxygen free radical formation, resulting in more accumulation of hydroxyl radical damage; such vicious cycle of the oxidative damage and mutation in the mtDNA seems to result in those changes to be synergistic and exponential. More efficient and complete repair was reported among tumour cells, could be due to efficient repair system or to suppressed amount of oxidative damage and may have relevance to their high rate of cellular replication that requires substantial amount of bioenergy supplied by intact mitochondria. In human mtDNA, there are 2 kinds of deletions and addition, namely, the small ones in the noncoding region and the large ones spanning several genes. There is no satisfactory explanation for generation of large deletion except the double-strand break and rejoining model that is suggested by size distribution pattern of type B deletions. Production of reactive oxygen species is enhanced by both endogenous factors such as inherited point mutations, and exogenous factors such as toxins, drugs, or viral infections. Damages of macromolecules due to reactive oxygen species are regulated by protooncogene products. Bcl-2/Bax levels are differentially affected by p53 protein. Hydroxyl radical adduct of guanosine, 8-hydroxydeoxyguanosine (8-OH-dG) results in random point mutations and double-strand separation leading to a long stretch of single-stranded DNA, then deletion. Defective electron-transport chain encoded by deleted mtDNA enhances hydroxyl radical formation resulting in more accumulation of 8-

OH-dG and ions in mitochondrial matrix. High concentration of ions facilitates double-strand separation. Such a vicious cycle of OFR damage and deletion in mtDNA seems to result in those changes being synergistic and exponential. The intermediates of oxygen reduction remain within the active site of cytochrome oxidase until the final reaction stage of water is achieved, probably for protection against cellular intoxication. The active sites of complex IV and III, consisting of cytochromes a and b, respectively, play a crucial role in cellular energy production, and protection against cellular oxidative damage. A large quantity of ROS is expected to generate from the genetically defected sites of cytochrome oxidase and/or cytochrome b, or with too much oxygen supply over enzymatic capability to dispose ROS. All the patients harbouring severe point mutations in the cytochrome oxidase subunit genes or in the cytochrome b gene express most severe clinical phenotype, such as fatal infantile cardiomyopathy died at age 1. Mitochondrial DNA locates inside mitochondrial inner membrane where ROS continuously leak from the respiratory chain, hence being directly susceptible to attack by ROS, despite the cellular defences against damage from ROS, such as superoxide dismutase and glutathione peroxidase. Mitochondria consume ≈ 90% of the cell's oxygen. During the life of an individual, a vast sum of the redox energy produced in mitochondria is converted to activate the molecular oxygen to ROS that seems to be utilised in the process of cellular apoptosis. mtDNA could be one of major targets of hydroxyl radicals, in addition to mitochondrial enzymes and lipids. Hydroxyl radical produces single-strand breaks in a plasmid forming endonuclease sensitive sites, hydroxyl radical scavengers are able to prevent the damage. The breakdown of mtDNA by ROS is expected during the life of an individual. Consistent with oxygen damage in muscle, oxidative damage in human brain to mtDNA shows marked age-dependent increases. Decline in the mitochondrial respiratory chain activity associated with age or premature aging due to the mitochondrial diseases, lead to the expression of the geriatric process, such as mental retardation, progressive decline of heart performance, muscle weakness, and clinical signs of degenerative diseases. Insufficient repair system and oxidative repair damage are the major cause of the progressive mtDNA instability associated with age. The mtDNA's oxidative damage results in cumulative increase in somatic mutations in mtDNA leading to bioenergetic deficit, cell death, and aging. Oxidative damage in mtDNA with age induces local DNA structure alteration, leading to double-strand separation, thus resulting in double-strand breaks by hydroxyl radical. The mutation in human genome seems to cause imbalance of the inherent mitochondrial antioxidant apparatus, especially that built-in active center of cytochrome oxidase and that of the protooncogene products. The cumulative accumulation of oxidative damages and somatic mitochondrial gene mutations will lead to a mosaic of cells ranging in their bioenergetic capacity from normal to partially or grossly defective. A progressive fall in the efficiency of the oxidative phosphorylation occurs in the tissues. Biologically active quinones harbour unique capability to form multiple bypaths of electron flow in respiratory chain: 1- bypass complex I by succinate or glycerol 3-phosphate → CoQ → complex III; 2- bypass complex III by CoQ → complex IV; 3- bypass complex IV by CoQ → molecular oxygen forming H_2O_2; or 4- bypass the mtDNA-encoded proton channel of complex V with CoQ proton cycle. Ubiquinone has a membrane-stabilising effect in the inner mitochondrial membrane. Coenzyme Q reduces the amounts of lactate and pyruvate in cerebrospinal fluid, and their increases during aerobic exercise. Succinate and CoQ had been successful in treatment of respiratory failure. Several oxidation-reduction substances have been used for the treatment of the mitochondrial diseases. Riboflavin had been successful in treatment of NADH-CoQ reductase-deficient myopathy. A fundamental therapy of aging and degenerative diseases in postmitotic cells would be a gene therapy to replace mutated mtDNA or to attenuate the control of protooncogenes on ROS damage.

REGULATION OF NFκB IN IMMUNE SENESCENCE

Aging results in a significant decline in immune function that has been directly or indirectly linked to increased susceptibility to infections, autoimmunity, and cancer. Age-associated decreases in both humoral and cell mediated immune responses. The 2 main cellular components of the immune system mediating specific immune responses, i.e., T lymphocytes and B lymphocytes, exhibit functional deficits with advancing age. Phagocytic activity decreases with age. Total immunoglobulin production is unchanged with age, while serum immunoglobulin components and specificities, show age-related alterations. Serum concentrations of IgG and IgA increase, whereas the concentration of IgM is either unchanged or decreases with age. The production of antibodies directed towards self antigens such as thyroglobulin, nuclear proteins, and DNA increases with aging. Aging results in a few changes in the phenotypic profiles of T lymphocytes. The organ responsible for T cell maturation and function, the thymus, undergoes considerable involution with age. Memory cells may differ from naive cells in terms of activation requirements and cytokine production. T cell function in elderly humans is characterised by decreases in classic tests of T cell mediated immunity such as mixed lymphocyte reactions (MLR) and delayed type hypersensitivity responses (DTH). Cytotoxic T cell (CTL) generation and activity decline with age. Activation of transcription factors is an important mechanism for the transmission of extracellular signals from the cytoplasm to the nucleus. There is a decrease in the induction of transcription factors for the resulting decline in signalling and proliferation observed in the aged. Defects occur immediately following receptor ligation, defects also occur in downstream signalling events of which the induction of transcription factors may be vital to the events resulting in aberrant nuclear signalling. T cell activation results in the activation of several important transcription factors including Activator protein-1 (AP-1), Nuclear factor of activated T cells (NFAT), and Nuclear factor kappa B (NFκB), which play key roles in mediating transcription of genes involved in the generation of T cell immune responses. NFAT is a transcriptional activator that is a specific target for signals from the antigen receptor and binds to sites in the IL-2 promoter. This complex is largely restricted to T lymphocytes and is thought to be responsible for the T cell-specific inducibility of IL-2. Binding sites for NFAT have also been observed in the regulatory regions of genes for other cytokines such as IL-4, GM-CSF, and TNF-α indicating that NFAT may regulate the expression of cytokine genes other than IL-2. NFκB was originally identified as a B lymphocyte nuclear factor binding to a site in the immunoglobulin kappa light chain enhancer. NFκB is activated by a myriad of agents including cytokines like IL-1 and TNF-α; bacterial LPS; viral infection and certain viral proteins (e.g., HTLV-1 Tax, LMP1 of EBV); antigen receptor cross linking on T and B lymphocytes; calcium ionophores; phorbol esters; UV radiation and others. The genes regulated by NFκB family of transcription factors are just as diverse as the activators and include those involved in immune function, inflammatory response, cell adhesion, cell growth, and cell death. First characterised in mature B and plasma cells as a nuclear protein that binds specifically to a 10-bp sequence in the kappa intronic enhancer, NFκB has been demonstrated in virtually all cells. In most cells with the exception of mature B cells, macrophages and some neurons, NFκB remains dormant in the cytoplasm bound to its inhibitor, IκB. Cellular activation by various agents leads to the dissociation of the inhibitor and the translocation of the free NFκB to the nucleus. The speed of induction and its ubiquitous expression makes NFκB an ideal regulator of rapid-response genes. NFκB family of transcription factors are homo- and hetero-dimeric complexes formed from combinations of members of the Rel family of proteins. There are 5 mammalian members of the Rel family of proteins include c-Rel, NFκB1 (p105/p50), NFκB2 (p100/p52), RelA (p65), and RelB. These 5 members of the Rel family proteins may form almost any possible combination of homo or heterodimers. The classic and most well studied NFκB molecule is a heterodimer of p50/p65 subunits. This heterodimer is the most abundant complex and is found in virtually all cell types. Heterodimers of p50/p65 are rapidly translocated to the nucleus following cellular activation and bind the consensus sequence 5'GGGRNNYYCC3'. Each of the heterodimers exhibit unique properties including cell type specificity, DNA binding site preference, differential interactions with IκB isoforms, differential activation requirements, and kinetics of activation, thus being capable of regulating gene expression in a uniquely specific manner. IκB proteins regulate the cellular location, DNA binding, and transcriptional properties of NFκB/Rel family of proteins. The IκB family of proteins include IκB-α, IκB-β, IκB-γ, IκB-δ, (generated from alternative splicing of the NFκB1 gene, NFκB1, NFIκB2), IκB-ε and the predominantly nuclear protein Bcl-3. All IκB proteins have in common a conserved domain containing 5-8 repeats of the erythrocyte protein ankyrin. IκB binds to the NFκB dimer and masks its nuclear localisation signal thereby sequestering NFκB in the cytoplasm of the cell. The NFκB/IκB complex itself cannot bind to DNA, however disassociation of IκB from NFκB that can be achieved with various activating agents will produce an NFκB dimer that is capable of translocating to the nucleus and binding DNA. IκB α is the most extensively studied and most abundant IκB family member and unlike IκB-β, is primarily involved in regulating the rapid and transient activation of NFκB. IκB-α is a 37 kd protein, which can be structurally divided into a 70 amino acid N-terminus, a central section of 205 amino acids composed of 6 ankyrin repeats, and an acidic 42 amino acid C-terminus that contains a PEST (pro-glu-ser-thr) sequence, a motif correlated with rapid protein turnover IκB-α performs several critical functions including cytoplasmic retention of NFκB in resting cells, release of NFκB in response to activating signals, and inhibition of DNA binding by NFκB. The classic cytoplasmic NFκB complex contains a p50/p65 dimer bound to one IκB-α. IκB-α can have differential affinity for the various NFκB dimers. Newly IκB-α may enter the nucleus and regulate NFκB activity, resulting in a transient response. Activation of NFκB results from the degradation of IκB-α and subsequent translocation of the active NFκB dimer to the nucleus. NFκB can be activated by a wide variety of physiological and non-physiological stimuli including cytokines, mitogens, viruses and viral products, oxidative stress, and chemical agents such as phosphatase inhibitors and ceramide. In T cells, almost any stimuli capable of activation result in NFκB induction. Signals may directly activate the IκB kinase complex (IKK) or may first activate the NFκB inducible kinase (NIK). This kinase complex may also phosphorylate the p105/p65 complex, as well. IKK1/IKK2 then phosphorylates the NFκB-IκB-α complex on IκB-α, which is followed by ubiquitination, proteasomal degradation and the nuclear translocation of NFκB. In the nucleus, NFκB induces the transcription of several immune response genes. N-acetyl cysteine may inhibit the activation of NFκB. Phosphorylation of IκB-α is required for an additional modification of

IκB-α, i.e., ubiquitination. Mutations in IκB-α that block phosphorylation block ubiquitination. Transactivating NFκB dimers can induce their own inhibitor, IκB-α, which then binds to cytoplasmic dimers to restore the inhibited state and reestablish cytoplasmic pools of NFκB/IκB-α complexes. NFκB dimers are regulated at the level of cytoplasmic retention and at the level of synthesis. Absence of the c-rel gene has a large impact on immune function. Both mature B and T cells exhibit proliferative defects in response to various activating agents. The defect in T cell proliferation correlates with a lack of IL-2 production as IL-2 levels in c-rel deficient T cells are 50-fold lower than that in wild type T cells. Production of other cytokines in response to stimulation is also affected, 2- to 3-fold lower amounts of IL-5, GM-CSF, TNF-α, IFN-γ when compared to wild type T cells. Dysregulation of transcription factors appears to accompany aging in several cell types and may contribute to the age-related changes in cellular function. Aging may affect post translational processes required for activation of transcription factors rather than affecting constitutive levels.

T CELLS AND AGING

T cell activation is initiated when an antigenic peptide is recognised by the antigen receptor of T cells. This recognition event promotes sequential activation of a network of signalling molecules such as kinases, phosphatases, and adaptor proteins that couple the stimulatory signal received from T cell receptor (TCR) to intracellular signalling pathways. Activation of T cells results in a transient increase in intracellular free calcium ion concentrations $[Ca^{2+}]_i$. The rise in $[Ca^{2+}]_i$ results from the release of intracellular stores and also from the influx of extracellular Ca^{2+}. The age-related changes in calcium signal mobilisation may result from alterations in production of IP_3. A common feature of the antigen receptor-mediated signalling is the activation of PLC-g, resulting in the hydrolysis of phosphoinositide lipids and the production of IP_3 and DAG. Production of these second messengers in turn leads to the increase in the intracellular free calcium ion concentrations $[Ca^{2+}]_i$ and the activation of PKC, respectively. There is selective alteration in PKC isoenzymes in T cells during aging, which might contribute to alterations in intracellular signalling events. Aging alters the activation of Ras/MAPK cascade that leads to cytokine gene expression and T cell function. The age-associated decline in the immune system is reflected by a sum of dysfunction and dysregulation of the immunologic responses. Many of these defects implicate a deficit in the functional properties of T cells. Contributors to immune senescence may include diminished proliferative response of T cells to antigenic/mitogenic challenges, altered cytokine expression, accumulation of the hyporesponsive memory T cells, and decreased calcium mobilisation. T cells undergo a series of genetically programmed processes, first maturation into functionally competent cells and later development into hyporesponsive senescent cells. Changes in signal transduction machinery are one of the underlying causes of the age-related decline in T cell function. Deterioration of the immune system with aging (immunosenescence) is believed to contribute to morbidity and mortality in man due to the greater incidence of infection, as well as possibly autoimmune phenomena and cancer in the aged. Factors contributing to T cell immunosenescence may include: a- stem cell defects; b- thymus involution; c- defects in antigen presenting cells (APC); d- aging of resting immune cells, disrupted activation pathways in immune cells; and f- replicative senescence of clonally expanding cells. The term immunosenescence designates that deterioration of immune function seen in the elderly that is believed to manifest in the increased susceptibility to cancer, autoimmune phenomena and infectious disease of the aged. Immunosenescence is clinically relevant for protection against infectious disease in the elderly. Healthy life may be extended by preventing these infections using interventions designed to prevent, ameliorate or reverse immunosenescence. Immunosenescence may contribute to decreased pathology in elderly individuals, as in the lesser degree of acute rejection seen in clinical corneal and kidney transplantation. Despite declining immune function, aging is also associated with increased autoimmune phenomena. The requirement for peripheral control of potentially damaging autoreactivity could be dysregulated in aging. Thus, immunodeficiency on the one hand could be reconciled with increased autoimmunity on the other by postulating a compromised cellular regulatory activity with age. Dysregulated haematopoiesis is seen in elderly individuals, raising the possibility that multiple lesions are responsible for altered immune function in the aged. Haematopoiesis may be compromised because of a severely reduced capacity to produce GM-CSF and because lower numbers of progenitor cells are present in the BM. Aging is associated with reduced numbers of committed haematopoietic progenitor cells. The phenomenon of thymic involution begins very early in life, even before puberty, and progressively continues. The replacement of thymic parenchyma with adipose tissue is a discontinuous process, reaching a maximum at around 50 years of age in humans and thereafter not progressing further. The amount of non-fatty material in the thymus may not decrease further after the age 50. Secretion of the important immunoactive hormone, thymulin, continues throughout life, blood thymulin levels decrease with age. Lower levels of thyroid hormones and insulin may be responsible for lower thymulin levels. The ability to generate new T lymphocytes after chemotherapy is inversely related to the patients' age, probably an indirect indication of thymic involution. During the first year of recovery after chemotherapy, the CD4 cells in adults also mostly carry memory markers, but in children they carry markers of naive T cell. The prime source of reconstituting cells in adults is from peripheral expansion of pre-existing CD4 T cell subsets, which survived conditioning, and not by thymus-dependent generation of new T cells. CD8 cell generation is thought to be extrathymic. A similar phenomenon may be observed in HIV infection, where antiviral therapy results in an increase of naive CD4 cells only if some were still present before initiation of therapy. In bone marrow transplant (BMT) patients, even as long as 5 years after transplantation, CD4 cell counts are still depressed and cells with a naive phenotype are also rare. In human peripheral blood stem cell transplantation, only recovery of the $CD4^+$ $CD45RA^+$ population, but not the $CD45RO^+$ population, is thymus-dependent. Mature T cells are also subject to aging processes, either of the type affecting post-mitotic cells (when quiescent) or of the replicative senescence type (during clonal expansion for effective immune responses). Decreased T cell responses in the elderly may be due to decreased T cell function, decreased accessory cell function, or both. Lipopolysaccharide (LPS)-stimulated monocytes from the elderly produce less G-CSF, GM-CSF, IL 8, TNF-α, and MIP-1-α as well as less IL 1β. At least a subset of APC in the elderly retains optimal function, dendritic cells (DC). DC are able to present antigen, to inhibit apoptosis, and stimulate proliferation in pre-senescent T cells. Damage to the cytoskeleton paralleling aging may have profound effects on cell function. In the case of T cells this may be relevant even to the earliest stages of T cell activation. In T cells, signalling through the TCR, CD4, CD8 or the IL 2R resulted in lowered protein tyrosine kinase activity in cells from old subjects. Age-associated inactivation of proteosome function may be attributed to the effects of oxidative damage, which can be partly prevented by hsp90. Hsp90 levels are themselves decreased with age, in T cells. NF-AT may exert negative regulatory, not stimulatory, effects on the immune response. T cells require stimulation via the antigen-specific TCR for activation. Once stimulated, T cells must transcribe T cell growth factor (TCGF) genes, secrete growth factors, upregulate TCGF receptors and respond to the cytokines. One result of poorer T cell function is decreased cytokine production. Cytokine secretion may be affected by other (non-pathological) parameters such as exercise. IL 6 levels may be a good overall biomarker of health in aging because plasma levels are correlated with functional status. Chronic stress (caring for a demented spouse) results in significantly lower antibody titers, as well as IL-1 and IL-2 production in the elderly caregivers. Increasing levels of expression of cellular amyloid precursor protein (APP) by human lymphocytes are significantly positively associated with increasing age. Age-associated alterations in cytokine production are not determined solely by the subset changes, but by alterations within each of those subsets. Decreased soluble IL-2R secretion has been noted in elderly and may be of greater significance than possible IL-2 secretion defects. IL-8 is produced in elderly males. Increases in TNF-α in elderly may be directly relevant for decreased T cell responses. The major lymphoid organs from which T cells can be obtained are skin and gut. Autocrine proliferative capacity relies upon the stimulation of growth factor secretion, upregulation of the growth factor receptor and correct signal transduction. Telomere length in sperm DNA does not decrease with increasing age, suggesting that a mechanism for maintaining telomere length may be active in germ cells but not somatic cells, i.e., telomerase. Telomerase activity is detectable at very low levels in normal human T and B cells, and it increases greatly after mitogenic stimulation. Only a small proportion of T cell clones manifest full telomerase function and are effectively immortal. Telomerase may not be the only factor determining telomere length and cell survival. Expansion of CD8 cells is common, but it may be the rarer CD4 expansions that are observed at increasing frequency in the aged. In some disease states, e.g., RA, CD4 cells may show striking oligoclonal expansions. CD8-positive subpopulation contains suppressor cells. The alterations in proportions of different T cell subsets may be more marked in $CD8^+$ than in $CD4^+$ cells of aged humans. There are clearly more $CD4^+$ $CD45RO^+$ "memory" cells and less $CD45RA^+$ naive cells in elderly individuals. In human, most T cells are $CD7^+$, but the frequency of CD7-negative cells increases with age. Increased proportions of $CD7^-$ cells are found in situations of chronic antigenic stimulation in vivo, e.g., in RA and in kidney transplant recipients. CD28 is perhaps the closest to a biomarker of aging found for human lymphocytes. Proliferative senescence could play a role in the eventual loss of the memory cells. Some of the manifestations of aging on the immune system are related to downregulated apoptosis. Vaccination may have a less long-lasting effect in elderly. Responses to secondary antigens may normalise after boosting in elderly, but the improved response is not sustained for the same duration as in the young. The precise clinical relevance of T cell immunosenescence is hard to define. The general state of health is not the only factor contributing to depressed immunity. The prime cause of death is infectious disease in the elderly over 80's (pneumonia, influenza, gastroenteritis, bronchitis). The major predictor of mortality in the elderly is lung function. Immunogerontological parameters may be affected by many outside influences rather than aging per se. Nutritional status may play a significant role in exacerbating immunological differences. Vitamin C supplementation enhances the mitogenic responses of lymphocytes, and vitamin E supplementation enhances lymphocyte proliferative responses and IL 2 production in elderly people. The immunosuppressive effects of PGE_2 are predominantly mediated by increasing cAMP levels; therefore agents (insulin and chromium) which decrease cAMP levels may enhance lymphocyte responses. Vitamin E supplementation for 4 months improves a number of clinically-relevant indices of cell-mediated immunity in the healthy elderly, such as antibody responses to hepatitis B and tetanus vaccines, but without increasing autoantibody titers. High-dose vitamin E significantly enhances lung virus clearance in elderly. The effects of vitamin E may be attributable to its antioxidant or some other function. Not only vitamin but also trace element-deficiencies in the elderly may contribute to immunodeficiency. Levels of selenium decrease in lymphoid tissues with increasing age, selenium supplementation reverse it. In elderly people, selenium supplementation enhances lymphocyte proliferative responses. Zinc is necessary for the function of many hormones and enzymes, including those known to affect immune responses (e.g., testosterone). Zinc supplementation improves immune function. Zinc enhances age-associated lowered levels of IFN-α production in the aged. Secretion of the Th1-type cytokines IFN-γ and IL2 decreases during zinc deficiency, whereas Th2-type cytokines (IL4, IL6, IL10) are not affected. Oral zinc supplementation results in a recovery of thymic function. Low levels of activity of the zinc-dependent hormone thymulin are not dependent on the state of the thymus itself, but on decreased zinc saturation of the synthesised hormone. Melatonin supplementation is associated with increased plasma zinc levels in elderly in the absence of exogenous zinc supplementation. Melatonin may have direct effects on lymphocytes, which express mRNA for melatonin receptors. Age-associated changes in secretion of growth hormone (GH) and related hormones, releasing factors and binding factors may contribute to immunosenescence. Administration of low-dose GH to elderly adults for 6 months results in an increase in IGF-1 levels (which are reduced in aging) and an improvement in some physiological parameters, such as muscle strength. The native steroid dehydroepiandrosterone (DHEA), which, like most steroids, has immunomodulating activity. Whereas levels of cortisol increase with age in both men and women, in general the levels of DHEA decline with age. DHEA augments the decreased capacity of T cells to produce IL2 and IFN-γ. DHEA decreases the spontaneous secretion of IL6 and IL10 and reverses their hypersensitivity to endotoxin-stimulated release of both IL6 and IL10 in elderly. DHEA enhances IL2 production, significantly augment serum IGF-1 and decrease IGFBP-1, which may contribute to immune enhancement. DHEA increases monocytes, as well as increases mitogenic responses of both T cells and B cells, and enhances the numbers and activity of NK cells. Dihydrotestosterone (DHT) downregulates IL4, IL5 and IFN-γ production but does not affect IL2. Together, DHEA and DHT supplementation may alter the cytokine profile. There are an increasing number of cells showing evidence of mitochondrial dysfunction with aging, in terms of depolarised membrane potential and decreased mitochondrial mass. Reduced glutathione levels in the plasma are decreased in elderly. The age-associated reduction of GSH levels do not correlate with increased susceptibility of lymphocytes to oxidative damage, may due to a predominance of memory cells, which are more resistant to oxidative damage. Interventionist approaches with antioxidants remain attractive because of their cheapness and easiness. Caloric restriction slows down the age-related increase in mutations, perhaps via effects on DNA polymerase-α. Caloric restriction reduces levels of a marker of oxidative DNA damage. Caloric restriction lowers the nutritionally-driven insulin exposure, which lowers overall growth factor exposure. Caloric restriction reduces body temperature, as a result of decreasing energy expenditure. Mitogen-stimulated proliferation, NK activity, and antibody production, all are reduced with Caloric restriction. Caloric restriction decreases the age-associated increased constitutive serum levels of IL6 and TNF-α.

AGING OF THE NMDA RECEPTOR COMPLEX

Aging causes functional declines in many organs of the body, including the brain. The earliest cognitive dysfunctions that humans experience with ageing (fifth decade of life) are a decline in learning and memory performance. These declines in memory can range in severity from benign senescent forgetfulness, in which individuals have trouble accessing new and old information, to the degenerative disorder, Alzheimer's disease, which induces dementia and severe declines in cognitive functions. One subtype of the glutamate receptors, the N-methyl-D-aspartate (NMDA) receptor, is expressed in high density in cortical and hippocampal regions and is very important in the initiation steps of learning and memory. NMDA receptors are involved in the performance of spatial, working, and passive avoidance memory tasks and in long-term potentiation (LTP), a cellular phenomenon that is believed to be involved in at least some types of memory. The NMDA receptors appear to be more vulnerable to the aging process than other glutamate receptors and show declines in their binding densities, electrophysiological functions, and influence on other transmitter systems. The NMDA receptors are present at high density in the cerebral cortex and hippocampus. The NMDA receptor has an absolute requirement of co-agonism for channel activity; both glutamate and glycine must occupy their binding sites for channel activation. The age-associated changes in the NMDA binding site are not homogeneous across brain regions. Aging may alter the affinities of certain binding sites on the receptor complex or changes the ability of the binding sites to influence the channel. Changes with aging in the electrophysiological characteristics of the NMDA receptor appear to be highly region-specific. The NMDA receptor complex during aging probably contributes to age-related dysfunctions that occur in other transmitter systems.

BONE DISEASES

Alterations in the development of osteoclasts and/or osteoblasts underlie the pathobiology of bone diseases that is associated with bone destruction including osteoporosis, which can result from the bone loss that occurs after menopause, during aging, or after administration of such immunosuppressive agents as glucocorticoids, and arthritis. Osteoclast differentiation from haematopoietic progenitors depends on activation of a receptor, called RANK (receptor activator of NF-κB signalling), transcriptional regulation by NF-κB is important for osteoclast development. Stromal/osteoblastic cells express on their surface RANKL, the ligand for RANK. Bone-resorbing hormones, PTH, IL-1 and TNF, promote osteoclast differentiation by stimulating the synthesis of RANKL in stromal/osteoblastic cells. RANKL then engages RANK on osteoclast progenitors to stimulate differentiation. This process also requires expression of macrophage colony-stimulating factor (M-CSF) by stromal/osteoblastic cells, which binds to specific receptors on osteoclast progenitors. Stromal/osteoblastic cells also secrete a soluble RANK decoy called osteoprotegerin (OPG) that acts as a potent inhibitor of osteoclast formation by binding to RANKL and preventing its interaction with RANK on osteoclast progenitors. Bacterial lipopolysaccharide (LPS), an intensely inflammatory substance, can substitute for RANKL and M-CSF in stimulating osteoclastogenesis and bone resorption directly by binding to a protein called Toll-like receptor 4 to activate NF-κB in osteoclastic cells. IL-17, a T-cell-derived cytokine may act in combination with RANKL, independent of NF-κB activation. IL-12 induces the secretion of an undefined soluble T-cell product that strongly synergizes with IL-18 to inhibit osteoclastogenesis. T cells may play a role in the increased osteoclast formation and bone loss caused by oestrogen deficiency. Available therapies for osteoporosis, such as oestrogens, selective oestrogen receptor modulators (SERMs), bisphosphonates, and calcitonin, are anticatabolic, i.e., they prevent bone loss by inhibiting resorption and suppressing remodelling. RANKL synthesis is restricted to cells in an early phase of osteoblast differentiation, a time at which collagen synthesis has begun, but before the onset of osteocalcin synthesis. The bone morphogenetic proteins (BMPs), especially BMP-2 and BMP-4, exert potent prodifferentiation effects on early osteoblast progenitors. The most common forms of osteoporosis are diseases of bone remodelling, in which the production and life span of osteoclasts and osteoblasts are altered in such a way as to result in excessive bone resorption and/or inadequate bone formation. The pathobiology of osteoporosis resulting from sex-steroid deficiency or glucocorticoid excess involves increased osteoblast apoptosis. Sex steroids, bisphosphonates, and PTH activate intracellular signalling pathways that prevent apoptosis in osteoblasts. Both androgen and oestrogen rapidly induce Src activation via either the androgen or oestrogen receptor (ER); the Src activation in turn culminates in the generation of ERK-dependent antiapoptosis signalling. The bisphosphonate alendronate also stimulates Src/ERK-dependent antiapoptotic pathways in osteoblasts, and this effect appears to be exerted via connexin 43 hemichannels. PTH inhibits osteoblast apoptosis via cAMP-dependent protein kinase A (PKA) activation. PKA in turn phosphorylates the proapoptotic protein Bad to inactivate its death function and also phosphorylates a transcription factor called CREB to activate its activity. New protein synthesis, along with bad phosphorylation, is required for the apoptosis-inhibiting action of PTH. PTH-induced suppression of osteoblast apoptosis is a transient phenomenon. Short-term, but not long-term, exposure of marrow-derived osteoblast progenitors to PTH causes an increase in IGF and Cbfa1. Cbfa1 is one of the principal transcription factors responsible for the genetic determination of the osteoblast phenotype during differentiation from primitive mesenchymal progenitors. Bone mass in women and men at older ages, when fractures are most common, reflects a combination of the peak bone mass achieved as a young adult and subsequent bone loss. Both of these factors are under genetic control. Skeletal fragility and susceptibility to fracture are complex, polygenic traits that will be affected by variation in many different genes. Fractures are a direct manifestation of bone fragility and are therefore a key component of an osteoporosis phenotype. Fracture risk is a function not only of skeletal parameters but also of numerous nonskeletal factors, such as neuromuscular function, vision, balance, body size, and muscle strength. Type I collagen makes up the major proportion of the protein content of bone. Risk factors for osteoporotic fractures have several important roles to play in the investigation, prevention, and treatment of osteoporosis. The bone mineral density (BMD) measurement can be used to make a diagnosis of osteoporosis, perhaps its most important use is as an indicator of subsequent risk of fractures. A history of a low-trauma fracture, or a fracture sustained after the age of 50, is one of the strongest and most consistent risk factors for subsequent fracture (2- to 6-fold), and, for certain types of fracture (e.g., hip and vertebral), is associated with an increased risk that is independent of BMD. Identifying a high-risk target group can be done with any one or more risk factors, including a BMD measurement or a combination of BMD and other risk factors. Elderly women with very low levels of serum estradiol and vitamin D (<30 nmol/l), a high urinary deoxypyridinoline/creatinine ratio, and low levels of serum IGF-1 have an increased risk of hip and vertebral fractures (30-70% each). In elderly men, a low baseline serum vitamin D (<30 nmol/l) level, a high urinary deoxypyridinoline/creatinine ratio, low levels of serum IGF-1 and higher levels of serum osteocalcin, are each associated with a 30-70% increased risk of any fracture. It is important to treat vitamin D deficiency to prevent wintertime bone loss and a possible increase in fracture risk. Diets rich in animal protein and poor in vegetable intake deliver a net acid load to the body that is sufficient to induce a low-level sustained metabolic acidosis. This acid-base imbalance is exacerbated by age-related decline in ability of the kidney to excrete acid. The alkaline salts of calcium in bone are an important reservoir of base that is released into the circulation in response to systemic acid load. Significant bone loss is induced by metabolic acidosis. Elderly persons with impaired kidney function are thus more susceptible to a chronic metabolic acidosis resulting from a high-protein diet, and this may be a factor in age-related bone loss. Bone loss from chronic acidosis may be prevented by increasing dietary base intake. Dietary base intake reduces urinary calcium loss, decreases markers of bone resorption, and increases markers of bone formation. Alfacalcidol (ALF) supplementation inhibits bone resorption markers, causes modulation of cytokines, and reduces pain in subjects with osteopenia. Reduced extracellular pH directly stimulates osteoclasts to resorb bone. Most of the bone mineral release that occurs in response to acidosis results from osteoclast activation, which increases resorption pit formation in bone (with the organic matrix being destroyed at the same time). A small drop in pH causes a tremendous burst in bone resorption. The role of bone in acid-base balance is complex. The skeleton is a giant ion exchange column loaded with an alkali buffer, as 80% of body carbonate, 80% of body citrate, and 35% of body sodium are contained in solution within the hydration shell of bone and are released in response to metabolic acid. The increased incidence of osteoporosis with age may represent, in part, the results of a life-long utilisation of the buffering capacity of the basic salts of bone for the constant assault against pH homeostasis. Animal proteins and cereals are rich sources of phosphoric and sulphuric acid, which are recognised as "acid ash" foods. On a Western diet, adult humans produce roughly 1 mEq of acid/day and 2 mEq of Ca/day is required to buffer it; i.e., 15% loss of inorganic bone in an average individual over a decade. Humans become more acidic as they age; the more acid precursors consumed, the greater the degree of systemic acidity. With increasing age, overall renal function declines and acidity increases. Insufficient intake of dietary protein has been implicated in the pathogenesis of osteoporosis, and protein supplementation has been shown to improve the clinical outcomes of hip fractures. Isocaloric protein undernutrition is associated with alterations of somatotropic and gonadotropic axes and results in decreased bone mineral density (BMD) and bone strength of cancellous and cortical bone. Protein undernutrition, with normal calorie intake, reduces IGF-1 levels independently of growth hormone secretion. The resistance of both IGF-1 and growth hormone in target organs is associated with early depression of bone formation. The inhibition of the sex steroid secretion may explain, in part, the increase in bone resorption. Lower protein intake is significantly associated with greater bone loss at the spine and femur skeletal sites. Vitamin D insufficiency/deficiency is a major problem in the elderly population and is prevalent in certain ethnic groups. There is a positive link between the consumption of alkali-forming foods (in particular fruits and vegetables) and indices of bone health.

Causes of protein-energy malnutrition in nursing homes:

1- Medications (e.g., digoxin, theophylline, antipsychotics).
2- Emotional problems (e.g., depression).
3- Anorexia.
4- Late-life paranoia.
5- Swallowing disorders.
6- Oral problems.
7- Nosocomial infections (e.g., tuberculosis, Helicobacter pylori, Clostridium difficile).
8- Wandering and other dementia-related behaviours.
9- Hyperthyroidism/hypercalcaemia/hypoadrenalism.
10- Enteric problems (e.g., malabsorption).
11- Eating problems.
12- Low-salt, low-cholesterol diets.
13- Stones (cholelithiasis).

MALNUTRITION

Advancing age is associated with a remarkable number of changes in body composition. Sarcopenia accounts for age-associated decreases in basal metabolic rate, muscle strength, and activity levels, which, in turn, cause the decreased energy requirements in the elderly. Sarcopenia is a direct cause of age-related decrease in muscle strength. Reduced muscle strength in the elderly is a major cause of disability and may also account for the high prevalence of falls among the institutionalised elderly. Causes of Sarcopenia: 1- Reduced physical activity. 2- Aging of the CNS and resulting loss of motor units. 3- Changing endocrine function, including decreased circulating growth hormone, decreased testosterone production in both men and women, and decreased estradiol production in women. Changing endocrine function with age may also result in a different pattern of fat distribution, with increased visceral fat content. Inadequate intake of energy and/or protein may also contribute to sarcopenia. The recommended daily allowance (RDA) for protein of 0.8 g/kg body weight per day may be inadequate to meet the needs of elderly people. Advancing age is associated with a dysregulation in appetite and thirst. Aging is associated with decreased thirst, even when dehydrated. The elderly remain hypophagic and maintained their lower body weight after resumption of an unrestricted diet. Involuntary weight loss that may be experienced by elderly people because of illness, trauma, depression, or other reasons could result in permanent weight loss. More food or even a nutritional supplement may not result in body weight gain in elderly people. Involuntary weight loss among elderly people increases mortality risk. Involuntary weight loss may result in a BMI that places an elderly person at a greatly increased risk of premature death, and weight-gain strategies using protein-calorie supplements or providing increased dietary energy may not be effective for underweight elderly people. Increased mortality risk is seen in men with a body mass index (BMI) <23.5 kg/m^2 and in women with a BMI <22 kg/m^2. Depression and adverse drug effects are the most common causes of weight loss. Adults of all ages are candidates for resistance exercise. Elderly, hypertensive patients should be carefully evaluated before beginning a strength training program. Resistance training should be directed at the large muscle groups important in everyday activities, including those centered at the shoulders, arms, spine, hips, and legs. The amount of weight lifted should increase as strength builds (2-3 weeks). Patients should be instructed to inhale prior to a lift, exhale during the lift, and inhale as the weight is lowered to the beginning position. Patients should avoid performing the Valsalva manoeuvre. Any device that provides sufficient resistance to stress muscles (e.g., Weight-stack or compressed-air resistance machines) beyond levels usually encountered may be used. Effective host response to trauma, surgery, or sepsis involves a complex interaction between systemic and local immune function and tissue repair processes. Dietary supplementation with the semiessential amino acids arginine enhances T-cell-mediated immune function. Arginine abrogates the immune suppression associated with trauma. Arginine enhances reparative collagen synthesis in elderly population. Wound breaking strength during the early stages of healing is directly related to collagen synthesis. A progressive gain in tensile strength is an important factor in determining the effectiveness of tissue repair. Wound strength is retarded with increasing age. There is increased wound total protein synthesis in young subjects. Arginine has powerful secretagogue effects, which may explain its immunosuppressive actions. IGF-1 induced by arginine is one of the principal mediators of the growth promoting actions of growth hormone. The increase in T-lymphocyte activity induced by arginine may be indirectly responsible for the induction of collagen synthesis, possibly through their influence on wound fibroblast activity.

ANOREXIA

The physiologic anorexia of aging may involve alterations in both the peripheral satiation system and the central feeding drive. This dysregulation of the regulation of food intake places older persons at a major risk of developing severe protein energy malnutrition when diseases intervene. The regulation of food intake in humans is modulated by a variety of sociologic, psychologic, physiologic, and nutritional factors. Specific genetic differences determine the total food intake and amount of food ingested at an individual meal. Alterations in body mass index occur over the life span, with the greatest mass (and obesity rates) occurring during the middle decades (40-60 y) with a decline thereafter. From the ages of 20-90 y, there is a linear decline in food intake. Old females eat 30% fewer calories than young females, and old males eat 43% fewer than young males. This decline in energy intake is predominantly due to a decrease in fat calories with a small increase in the percent of calories ingested as carbohydrate. Weight loss in older persons seems to be predominantly secondary to the decline in food intake. There is a decline in total energy expenditure (TEE) with aging, which is accounted for by a decline in physical activity and by a decreased resting metabolic rate (RMR). The factors leading to the decline in RMR with aging are multifactorial and include a decline in Na$^+$-K$^+$-ATPase activity, a decline in triiodothyronine, a decrease in the postreceptor effect of norepinephrine, a decrease in food intake, and a decrease in the number of type II muscle fibers. Older persons eat 55% less fat and 40% less carbohydrate. Older persons are less likely to snack between meals. When given a yoghurt preload, older persons may tend to overeat compared to younger persons, suggesting a failure of the normal energy-sensing mechanisms that operate in younger persons. After an overnight fast, older persons are less hungry and had less desire to eat. Older persons eat smaller meals and tend to eat their meals at a slower pace. Once food is ingested, a variety of messages from the stomach and intestine, levels of circulating nutrients, and availability of stored nutrients interact to signal the state of satiety or hunger to the brain. There are clear declines in olfaction and less obvious alterations in taste sensation with aging. Flavour enhancers may reverse the decreased enjoyment of food in some older persons. With aging, there is a decline in gastric emptying of large meals that has been associated with satiation. Also, there is a decreased adaptive relaxation of the fundus of the stomach to food resulting in more rapid antral filling. Adaptive relaxation occurs secondarily to nitric oxide release. There is a decrease in nitric oxide synthase with aging. Cholecystokinin is the best studied of the gastrointestinal satiating hormones. Amylin is a peptide hormone released from the islets of Langerhans in response to a meal. In humans amylin levels increase from middle age to old age suggesting a possible role for amylin in the anorexia of aging. Insulin is unlikely to play a role in the anorexia of aging. Leptin decreases food intake and increases metabolic rate. In humans, increases in adiposity are, in general, accompanied by an increase in leptin levels. Leptin levels are higher in women than men. There is an increase in leptin levels at middle age in women, followed by a decline in older women. Leptin levels decline in females with aging, but increase in males. Testosterone levels decline with age in males. Testosterone is strongly related to the development of sarcopenia (muscle loss) in this population. Leptin has a role in the physiologic anorexia of aging in males, but not in females. Dynorphin, an endogenous opioid that produces its effects through the kappa opioid receptor, appears to play an important role in producing the feeding drive for fat. Decline in the opioid feeding system with aging may be responsible for the decline in fat intake that occurs in old humans. Declining nitric oxide may play both a peripheral and central role in the pathogenesis of the anorexia of aging. Corticotrophin-releasing factor (CRF) is one of the most potent centrally acting anorectic agents. Depression is one of the most common causes of weight loss in older persons and depression increases CRF levels, suggesting a role for CRF in this pathologic cause of anorexia and weight loss in older persons. Megesterol acetate reverses the low β-endorphin levels. Patients with idiopathic age-related anorexia have low β-endorphin levels in the cerebrospinal fluid.

SKIN AGING AND PHOTOAGING

Cell senescence limits cell divisions in normal somatic cells and may play a central role in age-related diseases. Damage to human skin due to ultraviolet light from the sun (photoaging) and damage occurring as a consequence of the passage of time (chronologic or natural aging) are considered to be distinct entities. Photoaging is caused in part by damage to skin connective tissue by increased elaboration of collagen-degrading matrix metalloproteinases, and by reduced collagen synthesis. Matrix metalloproteinase levels rise fibroblasts as a function of age, and oxidant stress underlies changes associated with both photoaging and natural aging. Histologic and cellular markers of connective tissue abnormalities are significantly elevated in the 6th and 8th decades. Increased matrix metalloproteinase levels and decreased collagen synthesis/expression are associated with this connective tissue damage. Topical application of 1% vitamin A for 7 days increases fibroblast growth and collagen synthesis, and concomitantly reduces the levels of matrix-degrading matrix metalloproteinases. Naturally aged, sun-protected skin and photoaged skin share important molecular features including connective tissue damage, elevated matrix metalloproteinase levels, and reduced collagen production. Vitamin A treatment reduces matrix metalloproteinase expression and stimulates collagen synthesis in naturally aged, sun-protected skin, as it does in photoaged skin. The ability of human skin to rejuvenate itself diminishes with the passage of time, resulting in increased fragility. This increased fragility reflects both reduced growth of skin cells and loss of collagenous connective tissue. Oxidative damage plays a central role in cellular aging. Cellular responses to growth signals and oxidative stress are mediated, in part, by growth-factor-activated and stress-activated MAP kinases. Extracellular-signal-regulated kinase protein is reduced by 60% in old skin. Cyclin D2, which is regulated by extracellular-signal-regulated kinase and functions to promote cell cycle progression is reduced by 50% in old skin. Transcription factor c-Jun, which is activated by stress-activated MAP kinases and promotes expression of connective-tissue-degrading matrix metalloproteinases is elevated 2-fold in old skin. Treatment of old skin with vitamin A (retinol) for 7 days stimulates extracellular-signal-regulated kinase activity, consistent with its demonstrated ability to stimulate cell growth in old human skin. Telomeres play a role in cellular aging and might contribute to the genetic background of human aging and longevity. Telomerase activity has replicative immortality to skin fibroblasts, and prevents or reverses the loss of biological function of senescent cell populations. Shortening of telomeres occurs with cell proliferation and correlates well with ageing in humans. Telomerase is a ribonucleoprotein, and is the body's most widely studied mechanism for extension of telomeres to circumvent cellular ageing. Premature aging of the skin is a prominent side-effect of psoralen photoactivation, a therapy used for a variety of skin disorders. The aged appearance of skin following repeated exposure to solar ultraviolet (UV) irradiation stems largely from damage to cutaneous connective tissue, which is composed primarily of type I and type III collagens. Single exposure to UV irradiation causes significant loss of procollagen synthesis in human skin. Expression of type I and type III procollagens is substantially reduced within 24 hrs after a single UV exposure, even at UV doses that cause only minimal skin reddening. Daily UV exposures over 4 days result in sustained reductions of both type I and type III procollagen protein levels for at least 24 hrs after the final UV exposure. UV inhibition of type I procollagen synthesis is mediated in part by c-Jun, which is induced by UV irradiation and interferes with procollagen transcription. Pretreatment of human skin in vivo with all-trans retinoic acid inhibits UV induction of c-Jun and protects skin against loss of procollagen synthesis. Following exposure to oxidative stress of various kinds, the release of low molecular weight antioxidant (LMWA) from the skin is significantly enhanced. This may suggest a physiological mechanism of the skin to cope with oxidative stress. During the aging process or following exposure to oxidative stress a significant decrease in the levels and activity of the water-soluble LMWA, while no change and even a slight increase occurs in the lipophilic LMWA. Along with the reduction in total water soluble antioxidant activity there is an accumulation of oxidised adducts, both on the surface of the skin and in deeper layers. The skin is increasingly exposed to ambient UV-irradiation, thus increasing its risk for photooxidative damage with long-term detrimental effects like photoaging, which is characterised by wrinkles, loss of skin tone, and resilience. Photoaged skin displays prominent alterations in the cellular component and the extracellular matrix of the connective tissue with an accumulation of disorganised elastin and its microfibrillar component fibrillin in the deep dermis and a severe loss of interstitial collagens, the major structural proteins of the dermal connective tissue. The unifying pathogenic agents for these changes are UV-generated ROS that deplete and damage non-enzymatic and enzymatic antioxidant defense systems of the skin. As well as causing permanent genetic changes, ROS activate cytoplasmic signal transduction pathways in resident fibroblasts that are related to growth, differentiation, senescence, and connective tissue degradation. Aging skin is a factor of many things-genetics, sun exposure, environmental insults, stress, and more. Proteasome is downregulated during replicative senescence as well as in aged cells, possibly resulting in the accumulation of modified proteins. UV-induced DNA photoproduct levels increase with age. Extracellular lipids of the stratum corneum, which are composed of cholesterol, fatty acid, and ceramides, are essential for the epidermal permeability barrier function. With damage to the barrier, a decreased capacity for epidermal lipid biosynthesis in aged epidermis results in an impaired repair response. Mevalonic acid is an intermediate after the rate-limiting step in cholesterol biosynthesis, which is catalysed by 3-hydroxy-3-methylglutaryl coenzyme A reductase. Topical application of mevalonic acid may enhance barrier recovery in aged skin, which is accompanied by acceleration of cholesterol synthesis from mevalonic acid and stimulation of the whole cholesterol biosynthesis.

ANDROPAUSE

Earth is hosting a rapidly aging population. Androgen decline in ageing male (ADAM) is a clinical entity characterised biochemically by a decrease in serum androgen and other hormones such as growth hormone, melatonin and dehydroepiandrosterone (DHEA). Clinically it is manifested by fatigue, depression, decreased libido, erectile dysfunction, and alterations in mood and cognition. Health problems affecting the elderly are cerebrovascular and ischaemic heart disease, cancer, respiratory conditions, Alzheimer's and other dementias, and diabetes. Age associated alterations in hormone levels in general and androgen in particular in men is significant. In men, corticosteroid and estradiol production remains fairly constant throughout life, leptin is altered in men with hypotestosteronaemia (changes in fat distribution). Leptin levels can be decreased by androgen supplementation, which improve obesity. The ADAM is characterised by: 1- Diminished sexual desire and erectile quality, especially nocturnal erection. 2- Changes in mood, decrease in intellectual activity, spatial orientation ability, fatigue, depression and anger. 3- Decrease in lean body mass with associated diminution in muscle volume and strength. 4- Decrease in body hair and skin alterations. 6- Increase in visceral fat. Serum testosterone decreases with age, at 1% per year after age 50 years. Sex hormone-binding globulin (SHBG) increases with age, resulting in decrease bioavailability of testosterone (free and albumin bound fractions). Dehydroepiandrosterone (DHEA and its sulfate [DHEAS]) are week androgens secreted by the adrenal glands. Decrease in DHEA and DHEAS is a constant feature of advancing age. By the 5th decade of life DHEA levels decrease (<30%). The declining levels of DHEA parallel a decrease in well-being. Supplemental DHEA results in improvement of QoL parameters. Penile erection had been documented in human male foetuses and castrated subjects. Serum testosterone assessment should be done between 8-11 a.m. Goals of treatment are restoration of sexual function, libido, and sense of well-being. Androgen replacement can prevent osteoporosis and optimise bone density, maintain virilisation, improve mental acuity and restore normal growth hormone levels, especially in elderly males. Low testosterone levels are associated with potentially unfavourable changes in triglycerides and high density lipoprotein cholesterol. Exogenous androgen modestly elevates low density lipoprotein cholesterol. The relationships between androgens and CVS risk factors are complex. Hepatoxicity induced by exogenous testosterone includes hepatocellular adenoma, cholestatic jaundice and haemorrhagic liver cysts may occur with androgen suplementation. Injectable, dermal and oral agents without methyltestosterone are safe in regard of hepatic carcinoma. Yearly LFT are advisable. Testosterone ester may result in supraphysiological levels of dihydrotestosterone. Parental androgen does not provide normal circadian patterns of serum testosterone and the injection is uncomfortable. Supraphysiological levels of serum testosterone occur following injections, which may result in breast tenderness or gyaencomastia. Aqueous testosterone requires frequent administration and is unsatisfactory for chronic testosterone replacement. Testosterone enanthate and cypionate induces supraphysiological levels as high as 1400 ng/ml within 72 hrs after administration that decreases over 14-21 days reaching baseline at 21 day. Transdermal testosterone therapy is expensive but may provide physiological approach to testosterone replacement. Transdermal testosterone replacement improves testosterone levels, sexual function, libido and nocturnal penile tumescence response, with maintenance of normal haematocrite, lipid profile, PSA and prostatic volume. Serum sex hormones have no relation to the development of CaP. There is no effect of exogenous androgens on PSA or prostate volume. Testosterone promotes growth of an established adenocarcinoma of the prostate. There is no credible evidence that prostatic biopsies are indicated before initiation of androgen supplementation. A rapid increase in PSA or detection of abnormalities by DRE after androgen supplementation is indication for evaluation. Sleep apnoea may be exacerbated during the administration of exogenous testosterone. Androgen supplementation results in enhancement of aggressiveness.

RENAL FUNCTION CHANGES

There is decreased renal blood flow, altered tubular function, altered endocrine function, and decreased GFR with ageing. There is increased load per remaining nephron. Hypertension causes changes in the elasticity of the arteriolar wall. There are changes in distribution of renal blood flow. The decrease in blood flow in the cortical nephrons may result in an increase in blood flow to the remaining nephrons so that there is a wash-out effect, which reduces the kidney's capacity to concentrate urine. The ability of the kidney to maintain sodium is affected. Physiologic response to sodium deprivation is slower than that of a younger person. When placed on a sodium-restricted diet, the time for sodium excretion is nearly doubles that of the younger individual. There is decreased response to aldosterone in tubules, which causes an increase in sodium excretion and may predispose the geriatric patient to the effect of hyponatraemia. Acid-base balance is altered as the cortical mass of the kidney shrinks, which causes decreased capacity of the cortical cells to generate ammonia from glutamine. Once the kidney loses its ability to generate ammonia, metabolic acidosis becomes a problem. The response to antidiuretic hormone in the older adult is reduced. Erythropoietin production in the kidney is also altered with ageing. This decline may lead to chronic anaemia and increased demands on the CVS secondary to oxygen carrying capacity of the blood. GFR is a process driven by hydrostatic forces within the glomerular capillaries that strain solutes and solvents through a semipermeable glomerular basement membrane. With age, the number of functioning nephrons decreases, sclerosis of the glomerular apparatus occurs, and there is a decline in the renal vasodilator response. A high carbohydrate and high fat diet have little effect on glomerular structural changes, whereas protein intake markedly alters the glomerulus. Glomerular filtration and glomerular blood flow rise with a high protein diet. A protein rich diet causes hyperfiltration and glomerular sclerosis caused by an increase in glomerular pressure and blood flow over time. Age-related glomerular sclerosis alone posses little threat to a person's overall health. Because older individual has less renal reserve, any additional stress will cause a marked decrease in overall renal function. Diseases such as hypertension and diabetes mellitus, and high protein diets increase glomerular pressure and glomerular filtration, thus resulting in further glomerular damage with a decrease in GFR. Serum creatinine, an index of GFR, depends on diet and muscle mass, with production being equal to renal function. Differences in production or excretion of creatinine may affect the serum creatinine level even if renal function is normal. Serum creatinine levels that are normally low may lead one to conclude that GFR is increased. Creatinine is formed as a result of the nonenzymatic dehydration of muscle creatine. Synthesis of creatine is performed primarily in the liver and is selectively uptaken by muscle. Because creatine is taken up by muscle, muscle mass also an important consideration in determining the creatine pool and ultimately the serum creatinine. Age and gender also determine muscle mass, men have a larger muscle mass than women, which accounts for a higher creatinine clearance in women. The muscle mass decreases with aging, with corresponding decrease in creatinine clearance. High quantities of meat in the diet will yield significant amounts of creatinine, which in turn will increase GFR. In older persons, especially those with renal disease, a high protein diet can be detrimental. Eliminating protein content from the diet can decrease creatinine excretion by 10-30%. The decline in creatinine clearance associated with aging is intrinsically renal in origin. A decrease in GFR increases the half-life and the potential for drug toxicity. The decreased ability of the kidney to concentrate urine may lead to dehydration, hypernatraemia, hyponatraemia and hyperkalaemia. Dehydration is a common problem in older persons and may result from inadequate fluid intake during distress and illness and increased insensible loss. Skin turgor is not a reliable indicator in the older adult. Impaired tubular function increases the renal threshold for glucose, making the presence of glycosuria an unreliable measure in the older adult with diabetes. Hyponatraemia and hypernatraemia may result from inadequate food or fluid intake. Hypernatraemia may occur in the patient undergoing a contrast medium study. It is imperative that the patient's renal status be assessed before the study so that the patient is adequately hydrated. Hyperkalaemia in the older patient may result from depressed aldosterone levels. When aldosterone is depressed, renal excretion of potassium is diminished. There is a tendency in the older adult to develop acidosis, which contributes to the rise in serum potassium. Potassium-sparing diuretics and angiotensin-converting enzyme inhibitors should therefore be administered with caution. Serum potassium levels should be monitored frequently. Changes in mental status may result from changes in fluid and electrolyte balance. Hypertension and diabetes mellitus alter glomerular pressure and filtration, the older individual is at greater risk for developing renal insufficiency.

ORAL ANTICOAGULANT

Oral anticoagulant drugs are recommended for primary prevention of thromboembolic events in patients with chronic atrial fibrillation, recent myocardial infarction, and prosthetic heart valves. The highest annual incidence of fatal and major bleeding is 0.8% and 2.0%, respectively, in patients on warfarin and for atrial fibrillation. In patients treated with warfarin after a recent myocardial infarction, the incidence of fatal and major bleeding is 0.2% and 0.5% per year, respectively. In patients with prosthetic heart valves on warfarin treatment is 1.4% and 5.2% for fatal and major bleeding, respectively. The mean incidence of fatal and major bleeding in patients on warfarin is 0.5% and 1.7% per year, respectively. The mean incidence of fatal and major bleeds in patients on placebo was 0.1% and 0.7% per year, respectively. The mean incidences of fatal and major bleeding during aspirin treatment are 0.2% to 0.8% per year, respectively. Reasons for these low incidence may be less intensive anticoagulant regimes, better control of anticoagulant therapy due to the introduction of the international normalised ratio, and careful pretreatment evaluation of risk factors for bleeding. Bleedings can be: 1- fatal, 2- major, and 3- minor. A bleeding episode requiring admission to hospital or blood transfusion is defined as major. A bleeding episode that is neither fatal nor major is defined as minor. Intracerebral haemorrhage is defined as endpoint, fatal. Comparison of incidences of bleeding complications to oral anticoagulant therapy, decreases over time in the incidence of major and fatal bleedings. Warfarin has a well-defined and stable pharmacokinetics, and a better bioavailability than dicoumarol and phenoprocoumon. The short half-life of warfarin is especially convenient during the period of adjustment of treatment and in cases of overdosing. INR is defined as the PT ratio that would be obtained if the WHO reference thromboplastin is used to perform the PT test. The responsiveness of a thromboplastin to reduction of the vitamin K dependent clotting factors is expressed as an international sensitive index, ISI. Thromboplastin has an ISI of 1.0. The ISI is a determinant for the conversion from PT ratio to INR. Risk factors for major bleeding: 1- age over 65 years, 2- history of stroke; 3- history of gastrointestinal bleeding, 4- serious illness, 5- atrial fibrillation, 6- age >75 and arterial hypertension (160/95 mmHg), 7- intensive anticoagulation (INR 7.5), and 8- long-term anticoagulation treatment. Untreated arterial hypertension and intensive anticoagulation are well defined risk factors for intracranial bleeding, 2-9% per year during anticoagulant therapy in patients with ischaemic cerebrovascular disease. 71% of the fatal bleedings are intracerebral. Ischaemic cerebrovascular disease may increase the risk of intracerebral bleeding. Aspirin significantly increases bleeding in the early post-operative period. Aspirin acts in 2 ways: 1- inhibits platelets aggregation by acetylating cyclo-oxygenase, thus inhibiting the production of TXA_2; and 2- inhibits the production of prostacyclin (PGI_2) by vascular endothelium. Following a single 300 mg aspirin tablet the inhibition of TXA_2 production lasts for several days, while recovery of prostacyclin production occurs within hours.

PAIN

Rheumatoid arthritis (RA) is an autoimmune disorder affecting 2% of the adult population. Although the peak incidence of RA occurs between the ages of 35-45, the prevalence of the disorder increases with age, due to the aggregate of early- and late-onset cases. RA is characterised by intermittent flare-ups of aching/burning joint pain. If untreated, the disease can ultimately result in irreversible joint damage, disability, and premature death. Arthritis pain varies in intensity, onset of flares can be unpredictable, and the impact of arthritis pain can be disabling. These aspects of the disease are the most salient for patients. Stress is viewed as a dynamic and ongoing interaction between the person and environment, in which specific demands on the individual tax or exceeds his/her resources to adapt. Coping refers to the process of adapting to these demands by managing or altering the situation (problem-focused coping) or by regulating the emotional response to the situation (emotion-focused coping). When RA pain is mild, middle- and older-aged adults report a greater tendency to catastrophize and engage in prayer/hoping efforts than younger adults. Older adults exhibit greater expertise in using coping strategies, which minimise health threats and stress associated with chronic pain. Older adults may better understand, through their own experiences and observations of others, that severe pain is harder to control and warrants more effective coping responses than dwelling on hopeless aspects of pain and suffering. Religious involvement is the coping strategy most frequently reported by older adults. Headache is independently associated with depression in the elderly. Headache prevalence is ranked as one of the most frequent complaints in the elderly. Depression in late life is widely acknowledged as a serious public health problem with considerable morbidity and mortality. Depression in the elderly is related to several factors. The association of migraine and depression is well established in young populations. Headache frequency is an important predictor of the presence of depression in the elderly.

LABORATORY CHANGES IN ELDERLY

Aging decreases immune competence, reducing the sensitivity of homeostatic mechanisms. The older patient is at increased risk for iatrogenic and nosocomial complications. Older patients are known to have platelet-rich plasma with enhanced platelet aggregation. There is an increase in plasma fibrinogen, activated factor VII, VIII, β-thromboglobulin, thromboxane A_2 in platelets, plasminogen activator inhibitor 1, fibrinopeptide A, prothrombin activation fragments F1 and F2, and D-dimer levels. There is a decreased plasma plasminogen, decrease in the number of platelet prostacyclin and thromboxane A_2 receptors. There is increase of the fibrinolytic system to balance the procoagulant activity. Ecchymosis resulting from capillary fragility, usually present on the arms and legs, can be a normal finding. Longevity is described in 2 ways life span and life expectancy. Life span is obtained by averaging the ages of the oldest 5% and 10% of the population, is the maximum age to which an individual can live (100-120 years). Life expectancy, the average number of years a person born in certain year can anticipate living. Educational background has impact on health risks and disease. Within a single individual, the aging of specific systems varies according to a complex interaction of inheritance; hormonal and immunologic regulation; and environmental stressors such as drug use, infection, diet, and physical activity. The existence of stable, chronic diseases in older adults may be difficult to recognise. 10% of elderly are diabetics. Cellular aging may be considered to be intrinsic (genetically programmed). DNA have a limited capacity to replicate itself, therefore cell proliferation is limited. Aged cells are less sensitive to growth factors, they have a diminished capacity to respond to mitogens. As cells age, they synthesise less of their own growth factors, such as IL-2; and the pre DNA synthesis G_1 phase of its life cycle is lengthened. The free radicals are atoms or molecules with one or more unpaired electrons in their outer orbitals. They are generated as byproducts of a number of cellular metabolic activities including mitochondrial electron transport, prostaglandin synthesis, cytochrome P450 activity, and the oxidative bursts of macrophages and neutrophils. Free radicals can oxidise and thereby damage DNA, lipids, and proteins. Age-associated decline in levels of scavenging enzymes, due to declining lean body mass, and changes in nutritional status that lower levels of the antioxidants may be related to numerous changes with age, including diminished immune responsiveness and higher risks for malignant transformation and atherogenesis. Age-related decline in reserve capacity, places the elderly individual closer to the threshold for dysfunction. As long as the body is essentially unstressed, homeostatic mechanisms can adequately cope with minor fluctuations. Many organs shrink with increasing age, e.g., thymus, spleen, kidney, liver and brain. The GFR declines by 8 ml/min/1.73 m^2 each decade of life. Increased glomerular pressures/flow induces cell injury and platelet aggregation, leads to permeability changes, and glomerular hyperfiltration. These changes lead to glomerular sclerosis. Aging, glomerular sclerosis, primary renal disease, and renal ablation, each reduce nephron number. All these factors lead to systemic hypertension. Systemic hypertension, diabetes, and ad libitum feeding, all increase glomerular pressures/flow. Reduction in the length and volume of proximal tubules occur and may affect tubular transport. There is a 5% decline in maximum renal concentrating capacity for each decade after age 50. Despite decreasing renal disease, blood pH, $PaCO_2$, and bicarbonate values do not significantly vary from those observed in young adults. Partial obstruction of the urethra is found in 20-30% of men >65 y and in 11-14% of incontinent women. The consequent of stagnation of urine in the bladder predisposes these individuals to an abrupt increase in UTIs after ages 65-70 y. Aging causes many differences in metabolism that would increase the RDA above that set on the basis of observing healthy younger adults. Ageing leads to diminished secretion of gastric acid, with or without accompanying decline in availability of intrinsic factor (IF). The absorptive capacity of the small intestine is also reduced. There is a decline in macronutrient and micronutrient absorption. All calcium salts are more soluble in an acidic medium.

Elderly vs. Young renal functions:

Function/test	Elderly	Young
Serum creatinine	0.84 mg/dl	0.81 mg/dl
Urinary creatinine excretion	1259 mg/24 h.	1862 mg/24 h.
Creatine clearance	97 ml/min/1.73 m^2.	140 ml/min/1.73 m^2.
Inulin clearance	75 ml/min/1.73 m^2.	125 ml/min/1.73 m^2.
PAH clearance (effective renal plasma flow).	289 ml/min.	649 ml/min.
Concentrating ability (maximum urine osmolality after 12 hrs water deprivation).	882 mOsm/kg H_2O.	1109 mOsm/kg H_2O.
Diluting ability (minimum urine osmolality after water loading).	92 mOsm/kg H_2O.	52 mOsm/kg H_2O.
Urine acidification (minimum urine pH after acid loading).	4.85	4.96

Relative achlorhydria reduces calcium absorption. Prevalence of Helicobacter pylori increases (>80%) with age and correlates strongly with the presence of acute and chronic gastritis. Vitamin B_{12} levels decline with advancing age. Suboptimal levels of cobalamin may produce neuropsychiatric manifestations without the development of the classic peripheral blood picture of a magloblastic anaemia. Lactose intolerance development is age-related. Declines in renal cortical tissue restrict the body's ability to maintain calcium homeostasis by limiting the availability of the renal enzyme 1α-hydroxylase, which is responsible for catalysing the synthesis of the active form of the vitamin, calciferol. Achlorohydria inhibits ferric iron absorption, or to a marginal status of vitamin C, which enhances intestinal uptake of nonhaeme iron. Serum iron level decreases with age, and serum ferritin as well as bone marrow iron stores increases, indicating impaired uptake of iron by erythrocyte precursors. Serum transferrin level does not change in elderly despite their low iron levels, which may be due to limited hepatic synthesis of transferrin or to development of a new set point for regulation that perhaps is mediated by altered membrane receptors. Bone marrow cellularity decreases 50% during the first 30 years of life, then stabilises until age 70 years, and subsequently decreases an additional 40% during the following decade, which could be due to increased bone marrow fat. Arterial PaO$_2$ decreases 5% every 15 years starting in the 30s, and PaCO$_2$ increases 2% per decade after age 50. Age-related organ changes affecting the CVS are mostly anatomic and include fibrosis; calcification; and the deposition of lipofucin, amyloid, and cholesterol in arteries and arterioles and in the heart muscle, its lining and covering, and valves. Levels of total homocysteine (tHcy) are increased, and are higher in men than women. tHcy may be damaging to the vascular endothelium and directly involved in the pathogenesis of vascular disease. Infections are less likely to produce fever in the aged, which may be due to depressed neutrophil function; or depression of the specific hypothalamic response to endogenous pyrogen (IL-1) and a generalised suppression of the hypothalamus or the autonomic system. Hepatic mass declines by 28%, and hepatic blood flow decreases 25-35%, between the age of 30 and 75 years. Hepatic conversion vitamin D to 25-OHD declines. A gross indicator of plasma protein concentration is the erythrocyte sedimentation rate (ESR), which rises in adults at a rate of 0.22 mm/hr per year because of the corresponding small increases in globulins and fibrinogen. The change in ESR is not related to mortality. The total protein range decrease slightly with age. Serum albumin levels decline by 10-15% between the ages of 30 to 80 years. Total serum protein concentration is constant after the age of 55, because increases in acute phase proteins (e.g., α_1-acid glycoprotein) and globulins are offset by a decrease in albumin. Decreased albumin causes an increase in the competition among different drugs for the remaining protein-binding sites, and tends to enlarge the active (non-protein-bound) fraction of a drug. Hepatic enzymes may increase, decrease, or show no change during aging. The volume of muscle is smaller in the elderly. The thyroid becomes more nodular with age. An abnormal TFT should never be attributed to ageing per se. The incidence of both hypothyroidism and hyperthyroidism rises in the elderly. Postmenopausal women taking thyroid hormone are at risk of developing osteoporosis. Oestrogen therapy reverses the adverse effects of thyroid hormone on bone mineral density. Major problems exist in defining reference ranges for the elderly in general.

Hepatic enzymes changes with aging:

Enzyme	Male >60 years	Female >60 years
Acid phosphatase	15-20% increase (prostatic component).	10% increase.
Alanine aminotransferase	Decrease but remains 10-15% higher than female.	Decreased.
Alkaline phosphatase	At 70 years male values is greater than female values.	
Amylase	Falls to 40%, and at 80 years at 20% of total.	
Aspartate aminotransferase	Slight increase, but 10% higher than female.	Increases, but 10% less than in males.
Catalase	Slight decrease.	Decreases by 5% below male value.
Creatine kinase	7-10% declines (physical activity).	Declines, but 10-15% less than in males.
γ-Glutamyltransferase	5-10% increase.	Increase, but 10% less than male value.
Glutathione reductase and peroxidase	Unchanged.	Unchanged.
Lactate dehydrogenase	10% increase.	Similar to male trend.
Lipase	Slight increase.	Similar to male trend.
Lecithin cholesterol acyltransferase (LCAT).	Remains constant.	Similar to male trend.

POLYPHARMACY IN THE ELDERLY

One of the most significant factors in caring for the elderly is widespread polypharmacy-the administration of multiple medications to an individual. Polypharmacy is a critical safety issue for patients. The elderly are at risk for adverse effects of polypharmacy. They frequently take multiple drugs for multiple diseases, are prescribed medications by a number of providers, and may receive care in many different settings. They may also have sensory deficits leading to nonadherence to a medication regimen due to confusion or other problems. 70-80% of geriatric patients experiences side effects 2-3 times more frequently than do younger persons. 10% of the elderly will have reactions if they take 1 drug; the rate can reach 100% with use of additional medications. Factors that affect the rate and degree of absorption of medications in the elderly include slowed gastric motility, decreased hydrochloric acid production, decreased effective surface area available for absorption in the intestines, and frequent coadministration of other drugs that affect absorption. Changes in percentage of body water, body fat, and muscle tissue also affect drug distribution. If drugs are lipophilic, they become stored in fat and may not be available to the body. There may be decreased blood supply to tissues and alterations in protein binding in the elderly patient. These factors affect the half-life of the drugs. The elderly may have low albumin levels that can enhance a drug's potential for adverse drug effects. Metabolism in the liver renders drugs inactive. There may be decreased hepatic mass with aging, and decreased regional blood flow can result in a decreased clearance of drugs through first-pass effects. Phase 1 metabolism in the liver involves oxidation by the P450 system, as well as reduction and hydrolysis, which convert a drug to more polar chemicals, thus preparing the drug for elimination. There are multiple opportunities for drug interactions in this phase because of the P450 reactions. Excretion is affected in the elderly if there are changes in kidney function. The net effect of age-related renal changes is the accumulation of drugs that are inactivated primarily by the kidneys (e.g., digoxin, antibiotics, antihypertensives, and antiarrhythmics). Creatinine clearance should always be calculated to ensure that medications will not pose a problem for patients. This is especially important for all potentially nephrotoxic drugs being prescribed. Clinicians writing the order would be required to review the patient record, provide rationale for every drug choice, and make recommendations about how long the drugs should be taken. Heparin is a drug with chronobiological implications. The anticoagulant effects of periodic doses of heparin are decreased in the morning. Therefore, the dosage prescribed for the morning should be increase, and decrease the dosage in the evening. Women have higher drug-concentration levels with antibiotics, with variances based on their menstrual cycles. Blood pressure is lowest between 2 am and 4 am. At 6 am, blood pressure begins to rise dramatically. Thus, increased stress on the heart in the early morning when blood pressure begins to rise is associated with increased likelihood of having a myocardial infarction. Transmural myocardial infarctions occur between 6 am and 12 noon. Histamine is released by mast cells in the late afternoon. Patients are more likely to have allergic reactions in early afternoon or evening. Drugs with allergic properties should be given earlier in the day. Antihistamines are more effective when taken at 7 pm than when taken in the morning. Everyone has a drop in lung forced expiratory volume between 10 pm and am. People with asthma have up to 50% greater decrease in forced expiratory volume than individuals without asthma. Individuals sleep better when their body temperature is lower (chronopersonality). Melatonin is released at night, and has a sedating effect that relaxes the body much like alcohol. Epileptic seizures occur more frequently when body temperature is higher. Aspirin is a very effective antiplatelet drug, but should be given in the morning. Bioavailability in the morning is twice as high as in the afternoon. Digoxin should be taken in the morning. Side effects increase by 40% when taken at night. Diuretics work best from 6 am to 12 noon, and don't work as well in the afternoon or evening. The absorption of ibuprofen is greater in the morning, but is also associated with more intolerance if taken at this time. It is better to take NSAIDs in the late afternoon or evening.

HOMEOSTASIS

FLUID AND ELECTROLYTE

The patient's fluid and electrolyte status depends on weight changes, serum electrolyte concentrations, and blood pH and Pco_2. Therapy must recognise: 1- the magnitude of the volume deficit present, 2- the pathogenesis and treatment of abnormal sodium concentration, 3- assessment of potassium requirement, 4- management of acid-base disturbance, and 5- presence of ongoing obligatory fluid losses. The normal response to the stress of surgery is to conserve water and electrolyte. Increased catabolism will deliver more potassium to the circulation, so maintenance of potassium is not essential for several days postoperatively. Fever will increase insensible losses. Volume deficits is best estimated on the basis of acute changes in weight or from clinical estimates. Deficit of <5% of body water will not be detected and 15% of body water will be associated with severe circulatory compromise. If the serum Na concentration is normal, fluid losses have been isotonic, if hyponatraemia is present, means more Na than water has been lost. Initial replacement should be with isotonic saline solutions. If Hypokalaemia exists at normal pH, the management for example, a serum K concentration of 2.5 mEq/L at pH 7.4 suggests a 20% depletion of total body K. A normal human has a K capacity of 45 mEq/kgBW. For a normal 75 Kg man, total potassium capacity is 45 X 75 = 3375 mEq. The deficit is 20% of this, or 675 mEq, and this amount must be considered in therapy calculations. Half of the calculated deficits should be replaced in a 24 hr period, with subsequent reassessment of the clinical situation. ECF is divided into two compartments: 1- plasma water (20% of ECF or 5% of body weight), and 2- interstitial fluid (75% of ECF, or 15% of body weight). The total body water compromises 45-60% of body weight. The total body water is divided into ICF and ECF compartments. IC water represents two thirds of total body water, or 40% of body weight. The plasma proteins, mainly albumin, account for the high colloid osmotic pressure of plasma, that is an important determinant of the distribution of fluid between vascular and interstitial compartments, as defined by the Starling relationships. Plasma water contains more protein than interstitial fluid. The kidneys keep the volume and osmolality of body fluid constant within a few percentage points despite wide variations in intake to salt and water. The osmolality (total solute concentration) of all the body compartments is identical, 290 mOsm/kgH$_2$O. In ECF, salts of sodium account for most of the osmolality, whereas in ICF salts of potassium are chiefly responsible. The kidneys can reduce urine volume and raise urine solute concentration 4-fold above plasma (1200-1400 mOsm/kgH$_2$O). If water intake is high, the kidneys can excrete a large volume of dilute (50 mOsm/kgH$_2$O) urine. A 1-molar (M) solution contains 1 gram molecular weight of a compound dissolved in 1 liter of fluid; 1 equivalent (Eq) of an ion is equal to 1 mole (mol) multiplied by the valance of the ion. 1 g NaCl = 17 mEq Na$^+$. The clinical manifestation of volume depletion is chiefly caused by hypernatraemia, which can depress the CNS, resulting in lethargy or coma. Muscle rigidity, tremors, spasticity, and seizures may occur. Treatment of volume depletion involves replacement of enough water to restore the plasma sodium (P_{Na}) concentration to normal.

$\Delta Na = (140 - P_{Na}) \times TBW$

The Δ Na is the total mEq of Na in excess of water. Divide Δ Na by 140 to obtain the amount of water required to return the serum Na concentration to 140 mEq/l. Combined water-electrolyte deficits are corrected by restoring volume and the deficient electrolytes. Clinical manifestations of volume overload include oedema of the sacrum and extremities, jugular venous distension, tachyponea, increased body weight, and elevated pulmonary artery and central venous pressure. A gallop rhythm would indicate cardiac failure. Inappropriate secretion of ADH syndrome may occur with acute head injury, some cancers, and burns; characterised by hyponatraemia, concentrated urine, elevated urine sodium concentration, and a normal or mildly expanded ECF volume. The serum Na^+ values may drop below 110 mEq/l and produce confusion and lethargy. Restriction of water intake may be sufficient to correct the abnormality. For rapid correction of hyponatraemia furosemide and isotonic saline infused at a rate equal to the urine output may be indicated. The potassium in ECF constitutes only 2% of total body K; the remaining 98% is within body cells. The serum K^+ is determined primarily by the pH of ECF and the size of intracellular K^+ pool. With EC acidosis, the majority of the excess hydrogen is buffered intracellularly by an exchange of IC K^+ for EC H^+; this movement of K^+ may produce dangerous hyperkalaemia. Alkalosis has an opposite effect. In the absence of acid-base disturbance, serum K^+ reflects the total body pool of K^+. A loss of 10% of total body K^+ drops the serum K^+ from 4 to 3 mEq/l at a normal pH. Renal excretion of K is regulated by aldosterone levels. Adrenal insufficiency may produce hyperkalaemia through impaired renal excretion. Hyperkalaemia is a treatable problem. Hyperkalaemia results from severe trauma, burns, crush injury, renal insufficiency, marked catabolism from other causes, and Addison's disease. Nausea, vomiting, colicky abdominal pain, and diarrhoea may occur. Early ECG changes include peaking of T waves, widening QRS complex, and depression of ST segment. Platelets counts >1 million/μl may elevate the serum K, since the ion is liberated from platelets as they are consumed during clotting. Hyperkalaemia is treated initially by intravenous infusion of 100 ml of 50% dextrose solution containing 20 units of regular insulin to lower EC K^+ by promoting its IC transport in association with glucose. Intravenous $NaHCO_3$ solution to lower serum K^+ as acidosis is corrected. Calcium gluconate to transiently reverse cardiac depression from hyperkalaemia without changing the serum potassium concentration, as calcium will antagonises the tissue effect of potassium. The cation exchange resin sodium polystyrene sulfonate orally or by enema at a rate of 40-80 g/d, is a slower method of control. The cation exchange resin will bind K in the intestine in exchange for Na. It can be given with sorbitol to induce osmotic diarrhoea and enhance the rate of potassium removal. Dialysis is indicated if hyperkalaemia is a result of renal failure. Hypokalaemia associated with alkalosis may result from IC shift of K^+ in exchange for H^+, or renal wasting of K^+. Hypokalaemia is manifested by decreased muscle contractility and muscle cell potential develop, and in extreme cases death may result from paralysis of the muscles of respiration. Urine K^+ excretion of >30 mEq/24 hr associated with a serum K^+ <3.5 mEq/l indicates renal potassium wasting. Potassium should be given orally, otherwise intravenously. Intravenous solutions should not exceed 40 mEq/l. In moderate to severe hypokalaemia (<3-3.5 mEq/l), K^+ can be administered at a rate of 20-30 mEq/hr. Many of the causes of K deficiency will also result in magnesium depletion. The normal serum calcium rage is 4.2-5.2 mEq/l. Hypocalcaemia occurs in hypoparathyroidism, hypomangesaemia, severe pancreatitis, chronic or ARF, severe trauma, crush injuries, and necrotizing fasciitis. Clinical manifestations include: hyperactive deep tendon reflexes, a postive Chvosteck sign, muscle and abdominal cramps, carpopedal spasm, rarely convulsions, and a prolonged QT interval on ECG. It should be treated by intravenous calcium gluconate or calcium chloride. Hypercalcaemia presented by fatigability, muscle weakness, depression, anorexia, nausea, and constipation. Long-term hypercalcaemia may impair renal concentrating mechanisms, resulting in polyuria and polydipsia and in metastatic deposition of calcium. Severe hypercalcaemia can cause coma and death (>12 mg/dl) and should be regarded as a medical emergency. With severe hypercalcaemia, intravenous isotonic saline will expand ECF, increase urine flow, enhance calcium excretion, and reduce the serum level. Intravenous furosemide and sodium sulfate, increase renal calcium excretion. Mithramycin is useful in cases associated with metastatic cancer. Calcitonin is indicated in patients with impaired renal and cardiovascular function. Dialysis is indicated in renal failure. Adrenal corticosteroids are useful in sarcoidosis, vitamin D intoxication, and Addison's disease. The normal plasma Mg is 1.5-2.5 mEq/l. Hypomagnesaemia occurs with poor dietary intake, intestinal malabsorption of ingested Mg, or excessive losses from the gut (diarrhoea, nasogastric suction). Clinically it is manifested as those of hypocalcaemia. In moderate Mg^{2+} deficiency, oral replacement is adequate. In more severe deficits, intravenous administration of 40-80 mEq of $MgSO_4$/l is indicated. Hypokalaemia should be corrected as well. Hypermagnesaemia occurs in renal disease, it is rare in surgical patients. Hypermagnesaemia is presented as lethargy and weakness. ECG changes as hyperkalaemia. When the serum level reaches 6 mEq/l, deep tendon reflexes are lost; with levels above 10 mEq/l coma and death may occur. Intravenous isotonic saline will increase the rate of renal excretion. Slow intravenous calcium to antagonise some of the neuromuscular actions. Dialysis is indicated in severe renal failure. Clinical manifestations of hypophosphataemia appear when the serum phosphorus levels falls to ≤1 mg/dl. Lassitude, fatigue, weakness, convulsion, and death may occur. RBCs haemolyse, oxygen delivery is impaired, and WBC phagocytosis is depressed. Chronic phosphate depletion leads to osteomalacia. Hyperphosphataemia develops in renal disease, after trauma, or with marked tissue catabolism. It is usually asymptomatic, the serum calcium is decreased.

ACID-BASE ABNORMALITIES

Hydrogen ions generated from metabolism are buffered intracellulary by protein, e.g., by haemoglobin; or by the carbonate/carbonic acid system. The renal tubules regulate plasma bicarbonate by: 1- reabsorption of the filtered bicarbonate in the proximal tubule; 2- hydrogen ions are secreted as titratable acid to regenerate the bicarbonate that was buffered when these hydrogen ions were initially produced and to provide a vehicle for excretion of about one-third of the daily acid production; and 3- the kidney excretes hydrogen also as ammonium ion by regenerating bicarbonate initially consumed in the production of these hydrogen ions. Volume depletion, increased Pco_2, and hypokalaemia all favour enhanced tubular reabsorption of HCO_3^-. Chronic disturbances allow the full range of compensatory mechanisms to come into play, so that blood pH may remain near normal despite wide variations in the plasma bicarbonate or blood Pco_2. Over 80% of the carbonic acid during respiratory acidosis resulting from the increased Pco_2 is buffered by intracellular mechanisms, about 50% by intracellular protein and another 30% by haemoglobin. An acute increase in the Pco_2 from 40 mmHg to 80 mmHg will increase the plasma bicarbonate by only 3 mEq/l. Renal compensation raises plasma bicarbonate by increased renal excretion of ammonia ion, which enhances acid excretion and regenerates bicarbonate that is returned to the blood. Chronic respiratory acidosis is generally well tolerated until severe pulmonary insufficiency leads to hypoxia. Rapid correction of the condition will rapidly drop Pco_2 and the compensated respiratory acidosis may be converted to a severe metabolic alkalosis. Acute hyperventilation with respiratory alkalosis may be an early sign of bacterial sepsis. Chronic respiratory alkalosis occurs in pulmonary and liver disease. The renal response to chronic hypocapnia is to decrease the tubular reabsorption of filtered bicarbonate. As the bicarbonate concentration falls, the chloride concentration rises. This is the same pattern seen in hyperchloraemic acidosis, and the two can only be distinguished by blood gas and pH measurements. Generally, chronic respiratory alkalosis does not require treatment. During metabolic acidosis that is due to excessive bicarbonate (severe diarrhoea, diuretics, CA inhibitors, ureterosigmoidostomy), the decreased in plasma bicarbonate concentration is matched by an increase in the serum chloride, so that the anion gap (the sum of chloride and bicarbonate concentrations subtracted from the serum sodium concentration) remains at the normal level, below 15 mEq/l. Metabolic acidosis due to increased acid production (renal failure, lactic acidosis, ketoacidosis) is associated with increased anion gap exceeding 15 mEq/l. The lungs compensate by hyperventilation, which returns the hydrogen ion concentration towards normal by lowering the blood Pco_2 (10-15 mmHg). The amount of sodium bicarbonate required to restore the plasma bicarbonate concentration to normal can be estimated by subtracting the existing plasma bicarbonate concentration from the normal value of 24 mEq/l and multiplying the resulting number by half the estimated total body water. It is better to raise the plasma bicarbonate concentration by 5 mEq/l initially and then reassess the clinical situation. The long-term management can be either as supplemental sodium bicarbonate tablets or by dietary manipulation. Metabolic alkalosis is the most common acid-base disturbance in surgical patient. It can be due to loss of gastric secretions rich in hydrochloric acid; volume depletion; or potassium depletion. HCl secretion by the gastric mucosa returns bicarbonate ion to the blood. Gastric acid, after mixing with ingested food is reabsorbed in the small intestine, so that there is no net gain or loss of hydrogen ion. If secreted hydrogen ion is lost through vomiting or drainage, the result is a net delivery of bicarbonate into the circulation. The kidneys are able to excrete the excess bicarbonate load.

24 hr volume and electrolyte content of GIT fluid losses:

	Na^+	K^+	Cl^-	HCO_3^-	Volume
Gastric juice, high in acid	20-30 mEq/l	5-40 mEq/l	80-150 mEq/l	0	1000-9000 ml
Gastric juice, low in acid	70-140 mEq/l	5-40 mEq/l	40-120 mEq/l	5-25 mEq/l	1000-2500 ml
Pancreatic juice	115-180 mEq/l	3-8 mEq/l	55-95 mEq/l	60-110 mEq/l	500-1000 ml
Bile	130-160 mEq/l	3-12 mEq/l	90-120 mEq/l	30-40 mEq/l	300-1000 ml
Small bowel drainage	80-150 mEq/l	2-8 mEq/l	60-125 mEq/l	20-40 mEq/l	1000-3000 ml
Distal ileum and caecum drainage	40-135 mEq/l	5-30 mEq/l	20-90 mEq/l	20-40 mEq/l	1000-3000 ml
Diarrhoea	20-160 mEq/l	10-40 mEq/l	30-120 mEq/l	30-50 mEq/l	500-17000 ml

The excess bicarbonate load cannot be completely excreted. Some of the filtered bicarbonate escapes reabsorption in the proximal tubule and reaches the distal tubule, it promotes potassium secretion and enhanced potassium loss in the urine. The urine pH will be either neutral or alkaline, because of the presence of bicarbonate. If K depletion is severe, Na is reabsorbed in exchange for hydrogen ion. This may result in the paradoxically acid urine. Potassium depleted patients should be given KCl, since chloride depletion is another hallmark of this condition and potassium given as citrate or lactate may not correct the K deficit. Adequate volume of saline solution will diminish tubular Na reabsorption, and the kidneys can excrete the excess bicarbonate. Combined metabolic acidosis and respiratory alkalosis, can arise in patients with septic shock or hepatorenal syndrome. Combined metabolic and respiratory acidosis occurs in cardiopulmonary arrest.

THE ANION GAP

The anion gap is an approximate measurement of ions, which are molecules with a charge, either negative or positive. There are always more unmeasured anions than cations, and thus the anion gap equation, $(Na^+ + K^+) - (Cl^- + HCO_3^-)$, is always greater than zero. Some of the unmeasured cations (~7mmol/l) include calcium, magnesium, and most other minerals. Unmeasured anions (~24 mmol/l) include proteins like albumin, and phosphates, sulfates, etc. The anion gap is increased when there are excessive anions/acids in the blood, which could be either due too much acid production or insufficient removal of acids (either through the lungs, stomach, or kidneys). Excess acids lead to a rapid respiratory rate (extra CO_2), an inability to hold ones breath (the acid build up forces ones to exhale), low blood pressure (due to vasodilatation), fatigue, poor appetite, etc. The high anion gap indicates that the electrical charge of the fluids are too negative compared to the inside of the cell. The charge across cell membranes is required for many enzymes and energy production, so a reduced charge may result in less energy production (oxidative phosphorylation and ATP). A high anion gap may also indicate a functional need for alkaline minerals. The electrical potential between the inside of the cell and the outside of the cell is basis for nearly all transactions that occur with in the cell. Common causes of an elevated Anion Gap include: 1- Ketoacid overproduction due to fat metabolism (diabetes, alcohol, and starvation). 2- Lactic Acid overproduction due to respiratory failure, genetic defects of enzymes of carbohydrate metabolism, nutritional deficiencies that impair the bodies ability to metabolize lactic acid (B vitamins, especially vitamin B_1). 3- Inability to excrete acids (sulfate and phosphate) due to renal disease (usually with an elevated BUN and creatinine). 4- Dehydration. 5- Medications such as salicylates causes a metabolic block. 6- Toxins such as ethylene glycol, methanol, paraldehyde, propyl alcohol. The anion gap is decreased by free radical pathology due to overproduction of alkaloids. Other causes associated with a reduced anion gap are: 1- Alkalosis for any reason [a- hyperchloraemic acidosis (excess chloride), b- multiple myeloma, c- hyponatraemia, d- hypoalbuminaemiaÊ (can increase the amount of free blood calcium), e- bromide ingestion (displaces chloride), f- uncalculated blood cations (calcium, magnesium)]. 2- Lithium toxicity (can be due to effects on sodium). 3- Primary hypothyroidism. 4- Renal disease (due to the loss of the cations sodium and or potassium). 5- Polymyxin B.

D-LACTIC ACIDOSIS

Lactate is the end product of the cytosolic metabolism of glucose, and its accumulation in the blood signals an increase in production or a decrease in utilisation, or both. The liver is a major site for removal of lactate, so abnormalities in the metabolism of lactate by mitochondria in hepatocytes and other cells may contribute to clinical conditions in which overproduction and underuse of lactate occur. Lactic acidosis in all settings is associated with a high mortality. Treatment and correcting its underlying cause, ensuring adequate oxygen delivery to tissues, reducing oxygen demand through sedation and mechanical ventilation, and alkalise the blood and clear the lactate would manage the condition. The higher the lactate, the more acidotic the patient and the worse the prognosis. Lactic acidaemia or lactataemia is elevated lactate without acidosis. Low, asymptomatic lactataemia (2-5 mmol/l) does not appear to predict the development of more severe lactataemia. Patients with the most severe form of lactic acidaemia (>10 mmol/l) are generally symptomatic, often are acidotic, and have a high mortality rate. A rare but potentially fatal clinical condition, D-lactic acidosis, is seen in patients with a shortened (or by passes) small intestine and an intact colon. D-Lactic acidosis provides insight into the role of intestinal flora in normal human metabolism and serves as a model for demonstrating how altered intestinal flora can produce disease in humans.

Anion gap acidosis (normal range is 9-14 mEq/l):

Anion gap $= \Delta = Na^+ - (Cl^- + HCO_3^-)$
$= 140 - (105 + 24) = 11$

Most organic acid production occurs in the caecum, the major site where bacteria and appropriate substrate for fermentation coexist. The fermentation of one molecule of a hexose sugar produces 2 molecules of organic acids. Organic acids may be used for oxidative metabolism by mucosal cells of the colon. Locally produced butyric acid is a major source of energy for colonic mucosal cells. Organic acids seem to be absorbed at the site of production. High rates of organic acid production depend on the number, site, and metabolism of the bacteria in the GIT, the amount of incompletely digested polysaccharides and disaccharides as substrates for bacterial fermentation, and the length of time the bacteria and substrate remain in contact in an environment favourable for endogenous acid production. Non-spore forming gram-positive organisms normally found in the colon ferment the carbohydrates producing organic acids. This leads to a progressive decrease in intraluminal pH, which alters the intestinal microenvironment favouring the overgrowth of lactobacilli and other organisms at the expense of normal gut flora such as Bacteroides. Lactobacilli possess the enzyme DL-lactate racemase, resulting in the production of D-lactate from L-lactate through racemization by the altered intestinal flora. Abnormal colonic flora is thought to be central to the pathogenesis of the syndrome of D-lactic acidosis. Trimethoprim sulfamethoxazole, the consumption of Lactobacillus tablets or fermented dairy products, and the use of medium-chain triglyceride may precipitate D-lactic acidosis. Other sources of D-lactate include some pickled or fermented foods such as yoghurt and sauerkraut. A racemic mixture of L- and D-lactate is also used as an agent in various processed foods. Lactobacillus acidophillus, L. fermenti, L. buchneri, and Streptococcus bovis are the main bacteria that produce D-lactate in both animal and humans. L-lactate is a normal product of glycolysis and an intermediary of human metabolism. It is rapidly metabolised by the L-isomer-specific lactate dehydrogenase. The D-isomer accumulates in the blood because it cannot be metabolised. D-lactate is metabolised to pyruvate by D-2-hydroxy acid dehydrogenase (D-2-HDH). D-2-HDH is an intramitochondrial flavoprotein of wide-spread distribution in mammals, with high activities in the kidney cortex and liver. D-2-HDH can metabolise D-lactate rapidly and will prevent accumulation of D-lactate in patients receiving hypertonic sodium lactate solution when they undergo peritoneal dialysis. D-Lactate half-life is 20-40 minutes. The total body clearance rates suggest that L-isomer is metabolised at 5 times the rate of the D-isomer. D-lactate acidosis may be the result of a combination of overproduction of D-lactate and inhibition of its metabolism by high L-lactate concentrations. The half-life of D-lactate in the blood of healthy humans is 21 minutes. The increase in half-life probably reflects the saturation of D-lactate metabolism. Only minimal D-lactate is excreted in the urine. Organic acids absorbed from GIT are delivered to the liver through the portal vein. During episodes of acidosis when colonic fermentation produces organic acids at a rate exceeding their clearance, organic acids compete for oxidation. Organic acids that cannot be converted to pyruvate enter human metabolism after conversion to acetyl coenzyme A (acetyl CoA). They can be oxidised and converted to ketone bodies or fatty acids. Accumulation of acetyl CoA will lead to inhibition of pyruvate dehydrogenase and accumulation of D-lactate. Only D-lactate will accumulate during prolonged fasting or starvation when insulin levels are low, a greater availability of fatty acids may lead to a lower rate of oxidation and accumulation of D-lactate. Administration of insulin to diminish fatty acid levels and enhance D-lactate clearance may be a treatment option during an episode of acidosis. Cerebellar symptoms in D-lactic acidosis may be especially prominent because the normal cerebellum contains minimal pyruvate dehydrogenase reserve over that needed for normal metabolism. Congenital and acquired disorders of pyruvate metabolism often manifest with CNS symptoms. D-lactate produces its CNS manifestations by altering pyruvate metabolism. The pyruvate dehydrogenase complex in the cerebellum is especially sensitive to thiamine deficiency. The patient with increased erythrocyte translocase level, suggests a thiamine deficiency. Nutritional deficiencies in patients with a short bowel may have an adjuvant role in the pathogenesis of D-lactic acidosis. D-lactic acidosis is often a transient, self-limited disorder that responds to carbohydrate restriction and intravenous rehydration. Trimethoprim-sulfamethoxazole promotes the overgrowth of resistant D-lactate-producing organisms at the expense of other competing bowel flora. D-lactic acidosis is due to a combination of an increased production by abnormal colonic flora, inhibition of D-lactate metabolism by L-lactate and other organic acids such as free fatty acids. D-Lactic acidosis should always be suspected in a patient with unexplained acidosis and acute neurologic dysfunction who has previously undergone surgical resection of the bowel, resulting in a shortened small intestine with an intact colon.

Intervention for D-Lactic acidosis

Treatment	Outcome
1- Antibiotics	Decreases lactate-producing bacilli in the intestine.
2- Low-carbohydrate diet	Decreases substrate available for bacterial fermentation.
3- Reversal of jejunoileal bypass	Prevents a large load of partially digested carbohydrate from reaching the colon.
4- Treatment of thiamine deficiency	Increases availability of pyruvate dehydrogenase.
5- Insulin	May enhance D-lactate metabolism by decreasing levels of competitors for metabolism, predominantly free fatty acids.

CHRONIC METABOLIC ACIDOSIS

Chronic metabolic acidosis is a process whereby an excess nonvolatile acid load is chronically placed on the body due to excess acid generation or diminished acid removal by normal homeostatic mechanisms. Significant decreases in plasma HCO_3^- concentration are frequently associated with only small decreases in blood pH because of respiratory compensation. While respiratory compensations typically do not return blood pH to 7.4, they frequently return blood pH to values close to or within the normal range. Conditions with an excess nonvolatile acid load also may be associated with a normal serum HCO_3^- concentration. Diet induced changes in net acid production from 0 to 150 mEq/d produces minimal changes in the plasma HCO_3^- concentration. The limits for normal serum HCO_3^- concentration are 22-31 mEq/l. Causes of variability in the serum HCO_3^- concentration include: 1- a subject's serum HCO_3^- concentration varies during the day; 2- ingestion of food leads to secretion of acid into the stomach, generating HCO_3^-, which is added to the bloodstream, increasing the serum HCO_3^-; later this HCO_3^- secreted into the intestine to neutralise the acid, returning serum HCO_3^- to normal; and 3- variable loss of CO_2/HCO_3^- from the blood samples. Common causes of chronic metabolic acidosis, include: 1- GIT HCO_3^- loss; and 2- renal tubular acidosis (RTA); renal insufficiency (RI). An average 70 kg adult ingests a diet that generates about 70 mEq/d acid. This acid is excreted in the form of 40 mEq/d NH_4^+ and 30 mEq/d of titratable acid. This balance is associated with a normal serum HCO_3^- concentration of 24 mEq/l. Increase in endogenous acid production may be related to an increase in dietary protein intake. Initially, renal NH_4^+ and titratable acid excretion rates continue, leading to a net positive acid balance and a decrease in the serum HCO_3^- concentration. This decrease in serum HCO_3^- concentration and blood pH then activates homeostatic mechanisms that serve to defend acid-base balance. Urinary NH_4^+ excretion and titratable acid excretion increase to return acid-base balance toward normal. The increase in urinary NH_4^+ and titratable acidity is not of a magnitude equal to the increase in endogenous acid production, leading to a positive acid balance in the body. The serum HCO_3^- concentration stabilises because of a continuous release of alkali from bone. Aging can cause a significant metabolic acidosis. Aging is associated with a progressive loss of nephrons, which results in a progressive decrease in GFR. Metabolic acidosis associated with aging is likely attributable to the age-related decrease in renal function. The 2 common complications of chronic metabolic acidosis are bone demineralisation and muscle wasting. Chronic metabolic acidosis is a common clinical problem in elderly patients.

Net acid excretion = titratable acidity + urinary NH_4^+ - urinary HCO_3^-

Homeostatic response to metabolic acidosis: 1- Increased renal tubular acid secretion, this adaptation occurs along the entire nephron. In the proximal tubule, rates of H^+ secretion into the luminal fluid are greater in chronic metabolic acidosis than in more severe acute metabolic acidosis. Chronic metabolic acidosis is associated with increased activities of the apical membrane Na/H antiporter and the basolateral membrane Na/HCO_3 cotransporter. In the proximal tubule and the ascending thick limb of Henle's loop, apical membrane Na/H antiporter is mediated by an isoform of the Na/H antiporter gene family, NHE-3. These results likely apply to the adaptations in Na/H antiporter activity in the proximal tubule and thick ascending limb. The adaptation in the cortical collecting duct is mediated by changes in the relative activities of type A and type B intercalated cells, with acid ingestion increasing the relative activity of the type A intercalated cells (secretes H^+ in the luminal fluid), and alkali ingestion increasing the activity of type B intercalated cells (secretes HCO_3^- into the luminal fluid). This occurs in the absence of cell death and/or cell proliferation. 2- Increased renal tubular citrate absorption. Citrate is a molecule of intermediary metabolism that at pH 7-7.4 possesses 3 negatively charged carboxyl groups. Its metabolism generates three HCO_3^- ions. In response to chronic metabolic acidosis, urinary citrate levels markedly decrease. Decreases urinary citrate levels increases the risk of nephrolithiasis and nephrocalcinosis. Citrate is freely filtered at the glomerulus and its urinary excretion is determined by the amount of citrate absorption in the proximal tubule. Citrate is reabsorbed from the luminal fluid on a Na/citrate transporter that couples the transport of 3 Na^+ ions with one citrate anion. This leads to accumulation of citrate within the proximal tubule cell. Cytoplasmic citrate is then metabolised to HCO_3^- and either CO_2 and water or glucose, in the mitochondria. Citrate is transported into mitochondria on a citrate/malate exchanger (tricarboxylic acid transporter) and is then metabolised within the tricarboxylic acid cycle, leaving the mitochondria as malate. Malate is then converted within the cytoplasm to oxaloacetate, then to phosphoenolpyruvate, and finally to glucose. Citrate produced by the mitochondria in the liver, exits mitochondria in exchange for malate on tricarboxylic acid exchanger. Cytoplasmic citrate is then metabolised by ATP citrate lyase to acetyl CoA and oxaloacetate. The oxaloacetate is then metabolised to phosphoenolpyruvate and then glucose. 3- Increased NH_3/NH_4^+ synthesis: NH_3/NH_4^+ is synthesised in the renal proximal tubule by a sequence of events that includes glutamine transport into the cell on a Na/glutamine coupled transporter, glutamine is transported into mitochondria, and then the metabolism of glutamine takes place. Chronic metabolic acidosis leads to increase in the activities of the Na/glutamine cotransporter, glutaminase, glutamine dehydrogenase, and phosphoenolpyruvate carboxykinase, all of which contribute to glutamine metabolism and ammonia synthesis. In the proximal tubule cells there is increase in the mRNA of glutaminase, glutamate dehydrogenase, and phosphoenolpyruvate carboxykinase. Acidosis-induced enhanced muscle breakdown is mediated by activation of an ATP-dependent pathway involving ubiquitin and proteasomes. Metabolic acidosis inhibits albumin synthesis. Muscle wasting seen with aging may be attributable to acidosis. 4- Increased renal tubular growth, chronic metabolic acidosis is associated with increased renal growth. Renal hypertrophy has been associated with progression of renal disease. Polycystic kidney disease may represent one clinical disorder in which complete correction of metabolic acidosis is essential. 5- Increased bone alkali release. Metabolic acidosis directly regulates bone function by mobilising Ca^{2+} and alkali. Metabolic acidosis directly inhibits renal Ca^{2+} absorption, leading to Ca^{2+} removal from the body. Acidosis leads to the physicochemical dissolution of bone, causing release of HCO_3^- together with Na^+ and K^+ and small amount of Ca^{2+}. Chronic metabolic acidosis leads to increases in the activity of osteoclasts and decreases in the activity of osteoblasts. Ca^{2+} is reabsorbed at 3 major sites along the nephron. In the proximal tubule, Ca^{2+} absorption parallels NaCl and volume absorption and is largely passive. In the thick ascending limb, Ca^{2+} absorption is also largely passive, driven by the lumen-positive voltage. The distal nephron appears to provide a site where, although only a small fraction of filtered Ca^{2+} is reabsorbed, it is highly regulated. Chronic metabolic acidosis leads to mobilisation of alkali from bone, which helps to defend blood pH. These processes occur secondary to the ability of acidosis to physico-chemically dissolve bone, to increase osteoclast activity, to inhibit osteoblast activity, and to directly impair renal distal nephron Ca^{2+}. Increases in protein intake increase urinary Ca^{2+} excretion. This effect of protein is likely mediated by the increased endogenous acid production associated with ingestion of protein. The incidence of hip fractures (related to osteoporosis) is high in industrialised countries where the population eats a high-protein diet (the acidosis-osteoporosis hypothesis). Changes in intracellular pH may modulate cell function. There is an increase in the mRNA of AP-1 (c-fos, c-jun), junB and erg1. Acidosis increases expression of NHE-1 and NHE-3. Acidosis increases urinary corticosterone excretion. Acidosis increases secretion of glucocorticoids, and glucocorticoids are required for acidosis to increase Na/H antiporter activity and protein catabolism. The presence of glucocorticoids is required for the acid-induced adaptations. Intervention: The treatment of metabolic acidosis is relatively simple. HCO_3^- or citrate can be used as an oral alkali. HCO_3^- can cause a feeling of bloating due to the reaction of gastric acid with HCO_3^- to form CO_2. Citrate, can enhance aluminium absorption. Aluminium absorption is a major problem in patients with renal insufficiency. Na^+ is more likely to expand ECF and intravascular volume. This will result in worsening hypertension in patients with RI. Volume expansion increases Ca^{2+} excretion, which can further contribute to bone demineralisation, nephrocalcinosis, and nephrolithiasis. This volume expansion is not indicated in patients with serum HCO_3^- <22 mEq/l. Alkali treatment in combination of Ca^{2+} could lead to milk alkali syndrome, especially in elderly. Small doses of $KHCO_3$ sufficient to neutralise endogenous acid production, is indicated.

DIABETES MELLITUS

Diabetes Mellitus (DM) is a chronic disorder of glucose homeostasis that, over time, damages multiple end-organs. DM is a principal risk factor in thromboembolic stroke. The dyslipidaemia of DM involves alteration in composition and concentration of various lipoproteins, including hypertriglyceridaemia. Glucose auto-oxidation leads to the formation of free radicals that have been incriminated in the activation of coagulation. Glucose transporters (GT) are an integral part of every cell and when their function is inhibited or augmented, it can prove disastrous for the cell. Diabetic patients (Type II) are characterised by insensitivity to insulin, a decrease in the amount of GLUT4 translocation, decreased glucose transport, increased glucose in the plasma and either an increased ascorbate requirement or a decreased amount available inside the cell. These complications are all connected to each other to some extent; the majority of these conditions may be the total or partial result of the reduction of GLUT4.

PATHOBIOLOGY OF DM

Supra-physiological concentration of glucose has a toxic effect on the endothelium. The two most common forms of diabetes are type 1 and type 2 diabetes, with type 2 diabetes accounting for at least 90% of all cases. Type 1 diabetes is an autoimmune disease. In most patients, the disease develops before age 40, primarily during childhood or adolescence. In those patients, the immune system attacks β-cells, which produce insulin. Most importantly, insulin leads to the uptake of the glucose into muscle and fat tissue and prevents glucose release from the liver, thereby lowering blood sugar levels (e.g., after a meal). Insulin deficiency leads to major impairment of the body's regulation of carbohydrate, lipid, and protein metabolism. People with type 2 continue to produce insulin in early disease stages; however, their bodies do not respond adequately to the hormone. Obesity, inactivity, and cigarette smoking may worsen genetically determined insulin resistance. Insulin resistance does not immediately lead to overt diabetes, because the patient's pancreatic β-cells initially can increase their insulin production enough to compensate for the insulin resistance. Ultimately, insulin secretion declines to levels below those seen in nondiabetics. Thus, whereas type 1 diabetes is characterised by a complete lack of insulin production, type 2 is characterised by reduced insulin production plus insulin resistance. Diabetes mellitus is the most frequent metabolic disorder with incidence of 1-5% in general population. All types of diabetes mellitus are characterised by hyperglycaemia, a relative or absolute lack of insulin, and the development of diabetes-induced vascular changes (microvascular disorders). The microvascular disorders involves especially the retinal vessels, the renal small arterioles, venules and capillaries, and the microvessels supplying peripheral nerves, generalised atherosclerotic damage to large blood vessels. Atherosclerosis accounts for 60% of all diabetes-related deaths. The simultaneous occurrence of diabetes mellitus and hypertension greatly accelerates the atherosclerotic process, also other factors play important role, such as compromise defective lipid metabolism, obesity, hyperinsulinaemia, and insulin resistance as well as changes in platelet aggregation and haemostasis. Each of the factors is an independent risk factor for the development of coronary heart disease in non-diabetics. Hyperglycaemia and hyperinsulinaemia play a unique role in the pathogenesis of diabetes-associated atherosclerosis. The early indicator for the development of diabetic nephropathy, namely microalbuminuria, is also a good predictor for cardiovascular morbidity in diabetes and macroangiopathy. Mechanisms by which Hyperglycaemia contributes to the pathogenesis of athrosclerosis include: 1- elevated plasma levels of glucose, which can induce functional cellular changes by direct stimulation of cells; 2- hyperglycaemia leads to long-lasting cellular damage and chronic alterations of the vessel wall. Hyperglycaemia and hyperinsulinaemia, which present in patients with NIDDM, appear to contribute to the pathogenesis of diabetic atherosclerosis. 80% of patients with NIDDM are overweight. Obesity leads to a relative tissue resistance to the action of insulin and subsequent hyperinsulinaemia, which may be directly involved in the pathogenesis of hypertension. There is a close association between hyperinsulinaemia and the occurrence of atherosclerosis vascular changes. Patients with hyperinsulinaemia has a higher incidence of other atherosclerosis risk factors, including high triglyceride plasma concentrations, low HDL cholesterol levels, elevated total cholesterol, and hypertension. Structural cohesion of the cells in the arterial wall is achieved by matrix proteins, endothelial cells and smooth muscle cells. Atherosclerosis changes in the vessel wall result from a global sclerosing process, which includes increases in the extracellular matrix, and proliferative changes, such as excessive cell growth mainly of smooth muscle cells. The early stages of atherogenesis are characterised by damage to endothelial cells resulting in an increased permeability and functional changes of the affected epithelium. Concomitantly, there is an activation of smooth muscle cells by vasoactive substances and growth factors with migration of these cells into the intima and increased proliferation. Further more, there is an increased production of matrix proteins in the vessel wall. There are different mechanisms by which hyperglycaemia and hyperinsulinaemia can contribute to this process of cellular damage with subsequent proliferation and matrix formation.

REGULATION OF INTESTINAL SUGAR TRANSPORT
The epithelial cells lining the small intestine have a finite life span of few days to several weeks. Intestinal cells arise from actively dividing stem cells in the crypt. These new cells gradually mature as they migrate upward along the upper crypt regions, the crypt/villus junction, and the villus walls until they are finally exfoliated from the villus tip. Cells acquire new functions during their journey. When uptake rate is normalised per unit length (cm) of intestine, then uptake per mg is multiplied by intestinal weight/cm. Nonspecific adaptations often arise at the tissue level by changes in intestinal length or intestinal weight. Increases in food intake under energy-demanding conditions, leads to adaptive upregulation of transport capacity by either specific or nonspecific mechanisms, thereby tending to restore some reserve capacity. Mechanisms of regulation: A- Nonspecific mechanisms are mainly due more area per cell or more cells: 1- Changes in surface area, increases in intestinal surface area mediated by increases in height of microvilli produced by a high calorie diet. 2- Change in electrochemical gradient for Na^+ may be responsible for changes in Na^+-dependent nutrient uptake across the brush-border membrane. 3- Changes in plasma membrane lipid composition of the entrocyte, such as that induced by diet may change the passive permeability of the membrane to various nutrients. Moreover, the lipid environment of transporters may change, and if there is a corresponding change in membrane fluidity as well, this may alter the activity of transporters themselves and eventually affect rates of carrier-mediated transport. The effect of dietary lipid may be more pronounced on the lipid composition of the brush-border membrane than on the basolateral membrane. This mechanism may also be involved in changes in patterns of sugar uptake during development of the small intestine. 4- Changes in ration of transporting to nontransporting cells, the entrocytes lining the intestinal mucosa are in different stages of maturity. The relationships among the rates of cell division, cell maturation, cell migration, and cell exfoliation are usually in a steady state, and entrocytes would by expected to assume a new function or lose an existing function at a certain site during their migration from crypt to villus tip. The ration of transporting cells to nontransporting cells also changes during gut development. Different transporters are expressed at different ages of the enterocyte and/or at different sites along the crypt-villus axis. 5- Other nonspecific adaptive mechanisms include, changes in transmembrane potential induced by alterations in electrochemical gradient of ions other than Na^+ and changes in paracellular permeability. B- Specific mechanisms are mainly due to either more transporters or faster turnover rate: 1- Changes in turnover number of transpoters, the maximum velocity (V_{max}) of transport rate can be changed by changing the turnover number of each transporter molecule. Hyperglycaemia induces a 3.4 fold increase in the V_{max} of glucose, but only a 1.5-fold increase in site density of transporters in the basolateral membrane of entrocytes. Short-term hyperglycaemia increases the turnover number of preexisting transporters already in the membrane, perhaps by switching transporters from inactive to active states. 2- Changes in affinity constants. 3- Changes in site density of transporters. Glucose, galactose, or fructose transport increases with dietary carbohydrate due to an increase in number of transport sites. In the case of fructose, the increases in transporter number are clearly paralleled by increases in GLUT5mRNA levels. In the case of glucose and galactose, changes in transporter number seem independent of changes in SGLT1mRNA levels. Basolateral transport can also be influenced by dietary carbohydrate. Sugar transport across both poles of the entrocyte is regulated by dietary carbohydrate. Dietary fiber can be divided into 2 categories: a- insoluble fiber, comes from lignin, cellulose, and some hemicelluloses, while soluble fiber is made up of pectins (although some pectins may be insoluble) and other

soluble polysaccharides (guar gum, oat bran); b- soluble pectins and gums are branched hydrophilic compounds that form viscous solutions and delay gastric emptying. The insoluble fibers cellulose and lignin have little effect on gastric emptying and motility and, instead, are used mainly as laxatives to accelerate colonic transit. Fiber improves glucose tolerance and may even decrease cholesterol and phospholipid synthesis in the small intestine. Increases in soluble and insoluble dietary fiber at the expense of digestible carbohydrate would decrease intestinal sugar transport, because non-nutritive fiber would dilute the level of carbohydrate in the diet. V_{max} is unchanged with cellulose supplementation but is lower with guar-gum supplementation. Both soluble and insoluble fibers induce increases in absorption of glucose. Fiber causes adaptive reductions in rates of sugar absorption. Chronic consumption of diets supplemented with fiber also induces nonspecific changes in transport rates. Dietary fiber affects rates of intestinal cell turnover, entrocyte migration along the crypt/villus axis, entrocyte life span, villus appearance, thickness of smooth muscle, microbial density, and rates of microbial metabolism. Dietary fat increases caloric content. A hypercaloric diet consisting of food typically found in fast food restaurants and containing a much greater energy density than control diets enhances active glucose transport/mg of the intestine, and increases intestinal weights. A high calorie diet increases sugar uptake by a nonspecific mechanism involving changes in intestinal mass. A high calorie galactose-rich diet increases intestinal surface area by increasing villus height and crypt depth and by increasing the length of microvilli of entrocytes. Foods with high caloric densities may also increase by unknown mechanisms V_{max} of sugar transport without changes in intestinal mass. Starvation often leads to decrease in intestinal weight, mucosal thickness, and mucosal protein, which decrease intestinal nutrient transport/cm of intestine. Total parenteral nutrition deprives the gut of luminal nutrition while maintaining total body energy and nitrogen balance. Luminal nutrition may be required to maintain glucose transporters. Glucose absorption increases after intestinal resection, as a result of mucosal hyperplasia of the remnant intestine. Absorption normalises to intestinal weight decreases or does not change, but absorption normalises to intestinal length increases. The main mechanism for adaptive increase in sugar transport after intestinal is a nonspecific increase in intestinal mass. Resection enhanced rates of crypt cell proliferation typically result in hyperplasia of the remnant small intestine. Intestinal resection induces mucosal hyperplasia and in some cases a decreased uptake/gm. After a massive resection or after a short period following a more modest resection, there may be fewer transport sites per villus because the villi are occupied by a greater number of immature cells. The main mechanism underlying adaptation to the physiological demands of pregnancy and lactation is a nonspecific increase in intestinal mucosal mass, accompanied by slight decreases in sugar uptake/mg of intestine. Obesity increases intestinal growth, which augments absorption of sugars and all other nutrients. Diabetes mellitus enhances intestinal glucose transport. The small intestine recruits more cells to transport glucose, and those cells can either be cells along the crypt/villus axis that normally would not be transporting glucose or cells in the distal ileum that normally also would not be transporting glucose. An increase in intestinal mass is the main mechanism for non-specifically enhancing transport of glucose and all nutrients. The human foetal intestine can actively transport glucose and display a typically adult decrease in the rate of glucose absorption from the duodenum to the ileum. Fructose transport becomes significant only after the initial ingestion of solid foods. Shortly after birth there are more Na^+/glucose cotransporter/cell. Aging impairs absorption of carbohydrate. Sugar transport per unit intestinal tissue declines with age because of fewer functional transporters per cell, or because of a change in ratio of transporting to nontransporting cells. Pancreatic glucagon, insulin, and EGF affect intestinal sugar transport by different mechanisms. Glucagon affects mainly the brush border membrane potential, while EGF affects membrane fluidity. Insulin mechanism is not known.

BIOCHEMICAL DERANGEMENTS IN DIABETES MELLITUS

A- Polyol pathway, glucose is converted to sorbitol by aldose reductase, sorbitol is further metabolised by sorbitol dehydrogenase to fructose, with production of reduced nicotinamide adenine dinucleotide hydrogenase (NADH). Sorbitol consumption is very slow process, as a result hyperglycaemia causes accumulation of sorbitol in the cell, with several consequences: 1- sorbitol acts as osmolyte; 2- sorbitol can act as a competitive inhibitor of myoinositol uptake by the cells resulting in disturbed inositol signalling; 3- the increased utilisation of NADH in the polyol pathways leaves the tissue more susceptible to oxidative stress; and 4- activation of polyol pathways depletes the NADPH reserves in the cells. B- Diacylglycerol-protein kinase C pathway, protein kinase C (PKC) β II isoform is a ubiquitous intracellular enzyme, and can be activated by hyperglycaemia. PKC produces a variety of cellular responses, including proliferation, concentration, calcium influx, and response to other growth stimuli. DAG, an intermediate in glycerolipid metabolism, primarily regulates the cellular activity of PKC. Activation of PKC results in translocation of kinase from a cytosolic pool to a membranous pool. α-Tocopherol (vitamin E) can prevent the increase in DAG and PKC activities. This may have important implication for the management of diabetes and its complications. Diabetes mellitus stimulates the PKC activity by increasing de novo synthesis of DAG from intermediate products of glycolysis without altering the inositol phosphate levels. C- Nonenzymatic glycosylation, glucose interact with reactive amino groups in proteins to form early glycosylation end products. These products undergo slow chemical rearrangement leading to advanced glycosylation end products (AGEs). Advanced glycosylation end products interfere with multiple cellular functions and matrix protein dynamics. Binding sites for AGEs exist on endothelium, mononuclear phagocytes, mesengial cells, and smooth muscle cells. In the endothelium AGEs enhance procoagulant activity and permeability across the monolayer, increases oxidative stress with activation of transcription factor NFκB, which is linked with transcriptional activation of cytokines and adhesion molecules. AGEs are chemotactic and induce the synthesis of growth factors such as PDGF in human monocytes, TNF-α and IL1α. These cytokines in turn disrupt the function of mesenchymal and endothelial cells. Formation of AGEs on the extracellular matrix can initiate the cross-linking of collagen and decrease enzymatic degradation. There is increased lipid peroxide levels, lower antioxidant levels, and increased free radical activity in diabetes mellitus. Glucose auto-oxidation and transitional metal-catalysed oxidation are also suspected.

EFFECTS OF FREE RADICALS IN DIABETES

Diabetes mellitus is a metabolic disease characterised by hyperglycaemia and is associated with a series of late complications, such as vascular and metabolic abnormalities. The oxidative stress is greatly increased due to prolonged exposure to hyperglycaemia in diabetes. Increased free radical generation, which results from increased non-enzymatic glycosylation, glucose autoxidation, and alterations in polyol pathway activity, may exert a modulator effect on the level of oxidative stress in diabetes. Reduced capacities of antioxidant defense systems in diabetes also increase the oxidative stress. Diabetes mellitus is a very complex chronic disease with syndrome of hyperglycaemia. The cause of diabetes mellitus is not fully understood. The oxidative stress is significantly increased in diabetes because prolonged exposure to hyperglycaemia increases the generation of free radicals and reduces capacities of antioxidation defense systems. Some biochemical pathways are strictly associated with hyperglycaemia (non-enzymatic glycosylation, glucose autoxidation, polyol pathways) can increase the production of free radicals. Non-enzymatic glycosylation of protein ensues exposure to hyperglycaemia. Initially, glucose undergoes a nucleophilic addition reaction with proteins to form the Schiff base. Ketoamine (early glycosylation product) is chemically reversible and dissociates when blood glucose level returns to normal. However, it subsequently undergoes an Amadori compound further reactions, rearrangements, dehydration and cleavage irreversibly results in the formation of brown, insoluble, crosslinking complexes called advanced glycosylation end-products (AGEs). Amadori products could form H_2O_2 via two pathways: 1- The 1,2-enolization pathway, which lead to 3-deoxyglucosone formation under anaerobic conditions. In the presence of a suitable electron acceptor, enolization forms H_2O_2 and glucosone. 2- The 2,3-enolization pathway, which leads to 1-deoxyglucosone and the putative 1,4-deoxyglucosone. Under oxidative conditions, the 2,3-enediol generates H_2O_2 and carboxymethyllysine. The 3-deoxyglucosones are highly reactive intermediate in the non-enzymatic glycosylation and a potent cross-linker responsible for the polymerisation of proteins to AGEs. AGEs tend to accumulate on long-lived macromolecules in tissues, which results in abnormalities of cell and tissue functions. AGEs contribute to increased vascular permeability in both micro- and macro-vascular structure by binding to a specific macrophage receptor. This process induces the synthesis and secretion of cytokines such as TNF and IL-1, which causes endothelial dysfunction and induces free radicals. Monosaccharides and fructose-lysine can spontaneously reduce molecular oxygen under physiological conditions. The reduced oxygen products formed in the autoxidative reaction are superoxide, hydroxyl radical, and hydrogen peroxide. These radicals can damage lipids, as well as proteins, through cross-linking and fragmentation. Free radicals also accelerate the formation of advanced glycosylation end-products, which in turn generate more free radicals. This process is called as glucose autoxidation. Exposure to elevated glucose levels increase intracellular sorbitol and fructose content due to aldose reductase and sorbitol dehydrogenase activity. Oxidation of sorbitol to fructose is coupled to reduction of $NAPD^+$ to NADPH. An increase in $NAD^+/NADH$ ratio is linked to O_2^- formation via the reduction of prostaglandin G2 (PGG_2) to prostaglandin H2 (PGH_2) by prostaglandin hydroperoxidase that use NADH or NADPH as a reducing cosubstrate. Antioxidant defense system is compromised in diabetic patients. The mechanism by which the antioxidant reserve is reduced is not clear. Protein damage due to the protein glycosylation may be a mechanism that lowers the activities of primary antioxidant enzymes. In addition, GSH deficiency may result from depletion of NADH in polyol pathway. IDDM diabetes may be caused by the autoimmune β-cell destruction in pancreatic islets. The destruction of β-cells results in deficiency and finally total loss of insulin secretion. Hydrogen peroxide, nitric oxide, and superoxide are toxic to the human pancreatic islets. Vitamin E suppresses the nitric oxide toxicity in the pancreatic islet cells. β-cells are prone to be destroyed by free radicals because of the low antioxidant enzyme nature. Immune-effect cells such as macrophages, T cells, nature killer cells and B cells are believed to produce free radicals that causes damage to β-cells. There are two mechanisms of free radicals in β-cells destruction: 1- Infiltrating macrophages produce superoxide as primary source of free radicals. The superoxide can be further converted to more active radical, hydroxyl radical that attacks cellular membrane and cause DNA breaks. The consequence is cells death if cells fail to repair the damage. DNA repair enzymes, especially the activation of poly (ADP-ribose) synthetase, deplete the NAD levels in cells, inhibiting proinsulin synthesis and, in addition, causing cells more sensitive to free radicals. NAD and Na supplement can increase cellular NAD level so that elevates the efficiency of DNA repair and prevent decrease of proinsulin level. 2- Cytokines are released by T cells, macrophages, NK cells in the insulitis and induce the formation of intracellular free radical causing selective damage to β-cells. IL-1 is the major factor in the damage of β-cells. Interferon γ (IFNγ) and TNF are released by macrophages during the insulitis. These cytokines induce intracellular free radicals in endothelial cells, fibroblasts, and β-cells. Two types of free radicals are induced from these cells superoxide and NO. IL-1 can induce the production of nitric oxide synthase (NOS), which is the enzyme in charge of the synthesis of NO. NO inhibitory effect is on β-cells mitochondria function. NIDDM is a heterogeneous disorder, i.e.; very different pathologic events result in the same clinical symptom. Genetic abnormalities, or environmental factors, or obesity that may induce β-cells malfunction and/or insulin resistance can cause mild hyperglycaemia, which further develops to NIDDM. In patients with essential hypertension and NIDDM, the oxidant stress is high and the plasma levels of NO are low. The mean lipid peroxide value and mean plasma glutathione peroxidase activity are significantly higher in diabetic women. The activities of SOD, glutathione peroxidase, glutathione reductase and vitamin E levels are significant low in the plasma in NIDDM, with no significant change in catalase activity. Endothelial cells play an important role in vascular relaxation because they continuously produce NO by NOS through incorporation of molecular oxygen into L-arginine, resulting in the formation of NO and L-citrulline. NO, as a potent endogenous nitrovasodilator, modulates vascular tone by increasing the production of cGMP in smooth muscle cells. Endothelium-dependent vasodilation is abnormal in both IDDM and NIDDM, which is caused by decreased release or activity of NO. Many biochemical pathways associated with hyperglycaemia, such as non-enzymatic glycosylation, glucose autoxidation and polyol pathway increases the production of free radicals. Superoxide anion exhibits endothelium-dependent vasoconstrictive properties by inactivation of basal release of NO and by stimulation of endothelium-derived vasoconstrictor prostanoids. Prolonged hyperglycaemia leads to an alternative metabolism of glucose through the polyol pathway. The effect of an increased polyol pathway is an increased cytosolic NADH/NAD$^+$ ratio. Such an altered redox state may influence the availability of tetrahydrobiopterin (BH$_4$), an essential cofactor for NOS. During BH$_4$ depletion NOS is "uncoupled", leading to increased superoxide, rather than NO production. BH$_4$ supplementation may improve impaired endothelium-dependent vasodilatation. Aldose reductase inhibitors restore endothelium-dependent relaxations. Aldose reductase requires NADPH as a cofactor and, thus, an increased flux through the aldose reductase pathway leads to depletion of NADPH. Because NADPH is required for generation of NO from arginine, the depletion of NADPH is a possible explanation for impairment of endothelial cell function in diabetes. The endogenous antioxidant enzymes such as glutathione reductase also require NADPH, and NADPH depletion increases generation of reactive oxidants during hyperglycaemia. Because glutathione is one of the most important protective factors against oxidative damage, its depletion is responsible for increased susceptibility to tissue injury. The abnormalities of proteins and lipids are the major causes for diabetic complications. In diabetic patients, extracellular long-lived proteins, such as collagen, elastin, laminin, are the targets of free radicals. The glycoproteins are further broken into fragments by free radicals. These are associated with the development of complications in diabetes such as cataracts, microangiopathy, atherosclerosis and nephropathy. Diabetes is almost always related to the changes in plasma lipoproteins. There are multiple abnormalities of lipoprotein metabolism in VLDL, LDL and HDL in diabetes. The oxidised lipoproteins are then rapidly internalised by macrophages, which in turn convert to cholesterol-loaded foam cells. The formation of the foam cells is a key mechanism in atherosclerotic lesion. Treatment with antioxidants, vitamin E and probucol decrease the oxidation of lipoproteins without decrease of blood glucose. Free radicals react with polyunsaturated fatty acid (PUFA) to form peroxides. Peroxidation of PUFA results in the degradation of lipids and releases malnodialdehyde as final products. The breakdown of lipid components causes characteristic changes in lipid rich cell component (lipid peroxidation). Reduced SOD and GPx in β-cells, render them more susceptible to free radical damage. Normally, SOD performs the dismutation of superoxide to hydrogen peroxide. And GPx converts hydrogen peroxide, the product of superoxide dismutation, into water. The changes in pancreatic antioxidant enzymes may reflect susceptibility of β-cells to free radical destruction. The oxidative stress is greatly increased due to prolonged exposure to hyperglycaemia in diabetes. Vitamin C and vitamin E are the known free radical scavengers because they inhibit glucose autoxidation and reduce the covalent linking of glucose to serum proteins. Diabetic patients have at least 30% lower circulating vitamin C and vitamin E concentrations. The cellular intake of vitamin C and vitamin E is inhibited by hyperglycaemia and promoted by insulin. Cross-linking AGE-protein leads to cells dysfunction and increase vascular permeability. During hyperglycaemia, aldose reductase is activated and part of the glucose is metabolised to sorbitol. As sorbitol does not diffuse readily across cell membranes, it accumulates intracellularly and may cause damage. Aldose reductase may be a target for prevention of endothelial cell dysfunction. Diabetes mellitus is a chronic disease characterised by hyperglycaemia with a lot of serious complications. Endothelial dysfunction, lipid and lipoprotein oxidation, and long-live proteins damage are the sequences of increased oxidative stress.

VASCULAR DISEASE IN DM
Adherence of monocytes to the endothelium followed by directed migration into the subendothelium culminates in the formation of foam cells. Lipid peroxidation of the unsaturated fatty acids present in LDL results in formation of oxidised LDL. Oxidised LDL is a chemoattractant for monocytes, promote the expression of certain adhesion molecules such as VCAM and ELAM. Lipoproteins trigger the immune response and promote the formation of autoantibodies. Low-density lipoprotein-containing immune complex, avidly taken up by macrophages, induces intracellular accumulation of cholesteryl ester in macrophages and their subsequent activation. This results in the release of TNF-α and IL1 from macrophages, which in turn influence the behaviour of smooth muscle cells (SMCs) and endothelium. Hyperglycaemia increase the growth of VSMCs. Insulin stimulates the proliferation of VSMCs. Endothlin-I (ET-I), vasoactive peptide, and mitogens in VSMCs are elevated in patients with diabetes mellitus. Hyperglycaemia and insulin stimulate secretion of ET-I by the endothelium. NO is a mediator of the vasodilatory effect of insulin. IGF-I inhibits the cytokine-induced production of NO by preventing the induction of NOS. Endothelial dysfunction in diabetes mellitus is the reflection of an astounding interplay of multiple metabolic pathways. Supra-physiological concentration of glucose has toxic effects on the endothelium. High glucose levels can induce single-strand breaks in DNA and increased DNA repair, which can interfere with gene expression. High levels of glucose result in increase levels of mRNA for fibronectin, laminin, and collagen type IV in endothelial cells. Elevated glucose levels hamper the migration of endothelial cells. Insulin and IGF-I receptors on the endothelium transport insulin unidirectionally by receptor-mediated transcytosis. Chemical changes in the blood vessel basement membrane: 1- increased hydroxylysine/hydroxylysine-disaccharide; 2- decreased heparan sulfate; 3- increased collagen type IV; 4- decreased lysine; and 5- decreased laminin. A characteristic feature of diabetic microangiopathy is a thickened basement membrane, which involves vascular and nonvascular tissues. Basement membrane components, 7S collagen and laminin have been identified in the serum of diabetic patients. The increased permeability of the vessel wall in diabetes mellitus has been attributed to activation of the polyol pathway and to the formation of AGEs in tissues. The initial haemodynamic changes in early DM are increased flow and pressure of the capillary blood. As the disease progresses, autocoagulation in various microvascular beds is lost. These changes expose microvessels to haemodynamic stresses that may play a role in subsequent thickening of the basement membrane and increased permeability of the vessel wall. Capillary hypertension is more pronounced in individuals who are at risk of developing nephropathy. Rheological factors may contribute to ischaemia by promoting atherosclerosis, thrombosis, and obstruction of microcirculatory flow distal to atherosclersosis stenosis.

DM, new therapeutics:

Agents/methods	Mechanism
1- Growth factors.	↑ Wound healing.
2- Aminoguanidine.	↓ AGEs formation.
3- Sorbinil.	↓ Aldose reductase activity.
4- Antioxidants.	↓ Free radical formation.
5- Biosynthetic growth hormone.	↑ Wound healing.
6- Glucose sensors and insulin delivery systems.	Better glucose control.
7- Pancreatic transplantation.	Endogenous insulin supply.

A variety of haemorheological abnormalities have been associated with diabetes mellitus. Plasma and blood viscosity increases from 10% to 50% in DM, and in persons with DM who have complications affecting the major organs. Erythrocytes from patients with DM display reduced deformability and are prone to increased aggregation. Hyperglycaemia alters the physiochemical properties of proteins in the erythrocyte membrane, which may involve nonenzymatic glycosylation of these proteins. There is decreased sialyation of glycophorin A of the erythrocyte membrane, which may further interfere with the surface charge of the erythrocyte. The resultant decrease in repulsion force provides a possible explanation for the increased erythrocyte aggregation that is characteristic of DM. The metabolism of glutathione is impaired, which in turn impairs erythrocyte defenses against oxidative stress. The endothelium synthesises less prostacyclin, but more tissue factor and von Willebrand's factor in DM. There is increase in plasma levels of multiple coagulation factors such as fibrinogen, factor VII, and factor VIII. Hyperglycaemia reduces the biological activity of antithrombin-III (AT-III) and result in a drop in levels of thrombin-antithrombin complex. Levels of protein C and protein S are also depressed. High levels of insulin in non-IDDM are associated with elevated levels of plasminogen activator inhibitor (PAI), where they are normal or slightly elevated in IDDM. There is hypersensitivity of platelets in DM to aggregating agents such as ADP, collagen, arachidonic acid, and thrombin. Platelets release the content of α-granules, platelet factor 4, and β-thromboglobulin at increased rate in DM. There is increased number of large activated platelets circulate in DM. Glycosylation of the platelet membrane reduces its fluidity. Glucose auto-oxidation leads to the formation of free radicals that have been incriminated in the activation of coagulation. Intermittent administration of insulin and self-monitoring of blood glucose levels, is nonphysiological and demands compliance of patients with the treatment regimen and continuous active participation of health care provider. There is erratic absorption of insulin after subcutaneous route. The new approaches to insulin delivery include continuous intraperitoneal insulin infusion, nasal, rectal, and transdermal routes. Pancreatic transplantation results in favourable glucose metabolism, decreases hypoglycaemic episodes, and normalises glycosylated haemoglobin levels. Diabetic angiopathy has a central role in the development of almost all the major complications of DM. Patient education and active participation in the management is essential for a successful outcome.

Haemostastic abnormalities in DM:

Cellular abnormalities:
- A- Endothelium
  - 1- ↓ prostacyclin.
  - 2- ↓ tissue factor production.
- B- Platelets
  - 1- ↑ hypersensitivity to agonists.
  - 2- ↑ aggregation.
  - 3- ↓ membrane fluidity.
  - 4- ↓ platelet volume.

Coagulation abnormalities:
- A- Coagulation factors
  - 1- ↑ fibrinogen.
  - 2- ↑ factor VII.
  - 3- ↑ factor VIII.
  - 4- ↑ von Willebrand's factor.
- B- Coagulation inhibitors
  - 1- ↓ antithrombin III activity.
  - 2- ↓ heparin co-factor II activity.
  - 3- ↓ thrombin-antithrombin complex levels.
  - 4- ↓ protein C levels.
- C- Fibrinolysis abnormalities
  - 1- ↑ plasminogen activator inhibitor-I levels.

MITOCHONDRIAL CYTOPATHY

Mitochondrial multisystem disorders are characterised by various neuromuscular dysfunctions, in which several pathogenetic mutations of mitochondrial DNA (mtDNA) are involved. DM-Mt3243 is a NIDDM of maternal transmission, reported 1992. The disease has an earlier onset of DM, affect younger patients than NIDDM. NIDDM with a family history have a similar pathophysiology to that of DM-Mt3243 patients. Acetaldehyde accumulation may damage mtDNA, inducing DM. Thus, NIDDM with family history of DM, some patients may have pathogenic mtDNA mutations or deletions, and therefore have similar pathophysiology to DM-Mt3243 patients. Oxidative phosphorylation in the mitochondria may play an important role in insulin secretion; mutant mtDNA in the pancreatic β cell may interfere with the normal process of insulin secretion. Consequently, the patients secrete less insulin and therefore need extrinsic supplementation. In addition, they may suffer severe hyperglycaemia for some time over the course of the disease, in which oral hypoglycaemic treatment becomes ineffective, necessitating insulin treatment. Insulin administration induces endoneural hypoxic effect in peripheral nerves. In DM-Mt3243, bioenergry production is insufficient due to mitochondrial dysfunction, which possibly makes the peripheral nerves vulnerable to hypoxia. Abnormal mitochondria in the small arteries may worsen the hypoxia in peripheral nerves following insulin treatment. Subsequently, the damage to peripheral nerves responsible for vasomotor regulation may disrupt microcirculation in the legs or cause an arterio-venous shunt, eventually leading to oedema of the leg. Ambiguous psychiatric disorders can be frequently observed. Recurrent headache is known complication of MEALS (90%), of unknown aetiology. DM-Mt3243 induces irreversible damage in the affected tissues. Two months of treatment using CoQ reduces the symptoms in terms of fatigability, parathesia in the legs, palpitation and chest discomfort, constipation, and sleep disturbances. The disappearance of leg oedema may be due to improved circulation in the legs, or to anti-aldosterone effect of CoQ.

CELLULAR MECHANISM IN ATHEROSCLEROSIS

The late atherogenesis is reached only after years of or decades of complex cellular interactions. The process of plaque development involves endothelial cells, smooth muscle cells, and macrophages. Atherosclerotic lesions begin to develop at sites of excessive proliferation of smooth muscle cells in the intima. These cells migrate from the media, their original site, into the intima, where they proliferate and thus cause thickening of the vessel wall. Growth of the atherosclerotic plaque is caused by an increase in the number of cells and by an elevated formation of extracellular matrix. Then lipids accumulate in the smooth muscle cells, macrophages, and extracellular matrix. Secretion of paracrine factors by endothelial cells may further enhance such growth processes. Growth factors that stimulate the proliferation of smooth muscle cells such as PDGF are expressed by smooth muscle cells and can also be released by them. The greater endothelial permeability allows higher concentrations of angiotensin II to reach the intima. Endothelial cells express angiotensin-converting enzyme (ACE). Endothelin formed by endothelial cells also induces proliferation of smooth muscle cells, whereas the vasorelaxing substances released from endothelial cells have rather an antiproliferative action, such as prostaglandin, NO. Damaged endothelium loses its ability to produce the above vasorelaxing substances and instead secretes more vasoactive substances. This imbalance between proliferative and antiproliferative substances in the vessel wall leads to increased cell proliferation. Macrophages migrates from the bloodstream are also involved in the release of cytokines and growth factors in the vessel wall. These macrophages first become attached to stimulated endothelial cells by adhesion molecules and then penetrate the wall and accumulate in the area of the intima. Since they are a rich source of cytokines and other growth factors, they are important mediators of the proliferative processes that occur in the course of atherogenesis. Effects of hyperglycaemia on endothelial cell function include: 1- increased release of vasoactive hormones, e.g., endothelin, prostanoids; 2- decreased release of vasodilatory substances, e.g., NO; 3- increased expression of basement membrane components, e.g., fibronectin, collagen IV; 4- increased permeability for macromolecules; 5- increased adhesive proteins for leukocytes and platelets; 6- decreased secretion of tPA with reduced fibrinolytic capacity; and 7- delayed cell replication and cell death. Endothelial cell damage in patients with diabetes mellitus is also reflected by markedly elevated plasma concentrations of von Willebrand's factor and factor VIII. Damage to endothelial cells initially affects the release of vasorelaxing substances. Vasorelaxation is disturbed in patients with diabetes mellitus, probably because of a decrease in the release of vasorelaxing substances from damaged endothelial cells. Growth factors such as PDGF, which are secreted by stimulated endothelial cells, also have a vasoconstrictor action. In diabetics, the ration of vasorelaxing to vasoconstricting substances in the vessel wall is shifted towards vasoconstriction. Glucose also exert direct toxic effects on the endothelial cells, reduces the replication rate of endothelial cells, increases the permeability of the endothelial cell layer in diabetes mellitus and thus leads to a greater influx of substances from the circulating blood into the intima and media. Endothelial cells can actively contribute to local enhancement of coagulation. NO and prostacyclin act as inhibitor of thrombocyte activation. Endothelial damage can affect the role of the endothelial cells in fibrinolysis. The expression of plasminogen-activating factors by damaged endothelial cells is markedly reduced. Intact endothelial cells are important for the integrity of the vessel wall, and the interactions with cells in the bloodstream. Endothelial cell dysfunction can lead to accelerate intravascular blood coagulation. The adhesion of thrombocytes and WBCs can also be augmented by stimulation of endothelial cells. Activated endothelial cells express adhesion molecules for leukocytes and thrombocytes on their surface. Both the increased expression of adhesion molecules on the surface of endothelial cells and alterations of the monocytes in the bloodstream contribute to the enhanced adhesion of these cells to the vessel surface. Endothelial cells are also able to express and secrete matrix proteins and thus contribute to the sclerotic changes of the vessel wall. Elevated glucose concentrations increase the expression of collagen IV and fibronectin in endothelial cells and also enhance the activity of enzymes involved in collagen synthesis.

ENDOTHELIAL CELLS AND GROWTH FACTORS

Vascular endothelial cells regulate the contraction of the smooth muscle cells in the vessel wall, and also involved in the increased migration and proliferation of these cells in atherogenesis. Endothelial cells are capable of expressing: 1) PDGF, which is assumed to have a paracrine action on the smooth muscle cells, 2) IL-1 and TNF, which are involved in the stimulation of smooth muscle proliferation. Monocytes and macrophages migrating into the vessel wall are another source of growth factors. Low density lipoprotein (LDL) can also increase the surface expression of adhesion molecules on endothelial cells. Adherent monocytes migrate into the vessel wall, are rich source of cytokines and other growth factors play an important role in mediation of the proliferative processes that occur in the course of atherogenesis. These monocytes have surface for AGEs and are activated by binding to glycosylated proteins in the vessel wall. Activation leads to an increased release of cytokines and growth factors including PDGF, ILs, TNF, and TGF-β. Insulin is a growth factor for muscle cells, and stimulates DNA synthesis. Hyperinsulinaemia and hyperglycaemia directly affect the cellular processes involved in the development of diabetic microangiopathy.

INTRACELLULAR EFFECTS OF HYPERGLYCAEMIA

1- Effect of glucose on intracellular signal transduction: Persistent hyperglycaemia increases insulin-mediated glucose transport, non-insulin-dependent glucose uptake and intracellular glucose concentrations in different cell types. Excessive intracellular glucose is intracellularly metabolised via sorbitol. Sorbitol pathway reduces intracellular myoinositol uptake and lowers the activity of Na-K-ATPase. The de novo synthesis of diacylglycerol formed by the splitting of membrane-bound phosphatidyl inositol 4,5-biphosphate (PIP_2), which together with elevated free intracellular Ca activates protein kinase C (PKC). PKC is an important intracellular mediator of extracellular physiologic stimuli such as vasoconstrictive hormones and mitogenic substances for smooth cells, endothelial cells and other cell types. In smooth muscle cells, activation of PKC leads to an increased proliferation, DNA synthesis, and induction of growth factors and matrix proteins. The greater endothelial permeability caused by high glucose concentrations is associated with the activation of this enzyme system. High glucose concentrations in diabetes mellitus can directly activate a central step in intracellular signal transduction. The gene expression for matrix molecules in endothelial cells remains elevated even after correction of hyperglycaemia, this is called cellular memory for elevated glucose concentrations, which means that the activation of this enzyme system persists for some time after removal of the stimulus. 2- Non-enzymatic glycosylation of proteins and cell stimulation, long lasting changes of the vessel wall associated with diabetes mellitus appear to be mainly due to the non-enzymatic glycosylation of proteins and membrane constituents and the resulting activation of surrounding cells. Non-enzymatic glycosylation of protein starts with the attachment of an aldehyde or ketone moiety to a free amino acid, resulting in the formation of a so called Schiff's base. The final steps of non-enzymatic glycosylation are irreversible and induce long-lasting structural changes in the involved molecules. These changes affect both extra- and intra-cellular proteins, e.g., modification of haemoglobin. The changes in extracellular proteins primarily involve matrix constituents and membrane proteins and have an indirect effect on cellular processes, which results from the binding of glycosylated proteins to specific receptors on the surface of adjacent cells in the vessel wall. Specific receptors for non-enzymatically glycosylated proteins are present on monocytes and macrophages. Interaction of these cells with AGEs stimulates the release of TNF-α, IL-1, and IGF-I. The secretion of these cytokines is sufficient to stimulate the proliferation of endothelial cells, mesangial cells, and smooth muscle cells. The interaction of endothelial cells with AGEs leads to an increased procoagulant activity on the surface of these cells. Glycosylation of the mesangial cell matrix leads to substantial decrease in cell proliferation, which is probably due to reduced attachment of these cells to glycosylation matrix proteins. Glycosylation of type IV collagen reduces binding of NC1 domains, impairing the lateral linkage of these molecules that form a matrix under physiological conditions. Glycosylation of laminin decreases polymerisation and reduces binding to type IV collagen. Glycosylation markedly reduces binding of heparan sulfate to the matrix of the vessel wall. This lack of heparan sulfate proteoglycan reduces the permeability of the basal membrane and may also have a role in the excessive matrix production of adjacent cells. The structural changes resulting from increased matrix production contribute to the disturbed vasorelaxation in diabetes mellitus, and a direct effect on vasorelaxing substances. NO released from the endothelial cells is inhibited by AGEs in a dose-dependent fashion. The non-enzymatic glycosylation of vascular wall proteins leads to a vacious circle in patients with diabetes mellitus. Glycosylated proteins are resistant to proteases. The metabolic changes seen in diabetes mellitus have short-, medium- and long-term effects on the cellular elements of the vessel wall. The short-term stimulation of cells by hyperglycaemia and hyperinuslinaemia is rapidly reversible. The long-term influences appear to induce structural alterations, which in turn lead to the activation of cells such as monocytes and macrophages. Two long-term changes can be distinguished: 1- the intracellular processes susceptible to high glucose concentrations, i.e., the formation of diacylglycerol and the glycosylation of enzymes; and 2- the formation of AGEs proteins that bind to and thus activate cells of the vessel wall. The intracellular effects of hyperglycaemia, includes: 1- direct effects on intracellular signal transduction, and 2- long lasting changes in cellular function.

ADVANCED GLYCATION END PRODUCTS (AGEs)

Aldoses, and in particular glucose, react non-enzymatically with the amino group of macromolecules to produce a chemically heterogeneous group of glycated molecular species. The consequence of the interaction of AGEs with the cells dramatically modifies their functions and is associated with the production of an oxidant stress. The generation of reactive oxygen intermediates (ROI) can be considered either a deleterious factor or step that produces cell-signalling molecules that induce the activation of cell-responsive elements and result in gene transcription. The formation of AGEs is observed in diabetes mellitus and ageing both associated with vascular disorders. AGEs form by the interaction of an aldose with NH_2 of proteins, and the subsequent Amadori rearrangement leads to complex molecules. Hyperglycaemia is one of the major causes of vascular dysfunction. Mechanisms related to hyperglycaemia include sorbitol hypothesis, diacylglycerol-protein kinase pathway, non-enzymatic glycation and alteration of redox potential. Glucose is capable of reducing molecular oxygen to form, by nucleophilic addition to protein amino groups, an aldimine or Schiff base. AGE formation from Amadori products requires the presence of molecular oxygen and does not proceed under anaerobic conditions. The process is catalysed by transition metals and inhibited by reducing compounds, such as ascorbate. This chemical pathway is irreversible and is called glycation. The Amadori products lead, after intra- and inter-molecular rearrangements, to a new class of molecules called Maillard products or AGEs. Concomitantly, glucose, like other α-hydroxyaldehydes and AGEs per se generates, by auto-oxidation, H_2O_2 and hydroxyl radicals. This oxidation, catalysed by transition metals, causes structural and molecular damage to proteins. Amadori products reacting with ROIs form N-carboxymethyllysine (CML) and erythonic acid by oxidative fragmentation. Glucose, like other circulating sugars, such as galactose or fructose, is capable of reducing molecular oxygen to form, by nucleophilic addition to protein amino groups, an aldimine or Schiff base. The Schiff base intermediates may arrange in more stable products called Amadori products. Amadori products by isomerization lead to α-ketoaldehyde (1-amino-deoxy-2-cetose) compounds, such as deoxyglucosone, which are highly reactive dicarbonyl compounds. AGE formation from Amadori products requires the presence of molecular oxygen and does not proceed under anaerobic conditions. The process is catalysed by transition metals and inhibited by reducing compounds, such as ascorbate. This chemical pathway is irreversible and is called glycation. The Amadori products lead, after intra- and inter-molecular rearrangements, to a new class of molecules called Maillard products or AGEs. Glucose, and other α-hydroxyaldehydes and AGEs generate, by autooxidation, H_2O_2 and hydroxyl radicals. This oxidation, catalysed by transition metals, causes structural and molecular damage to proteins. Aminoguanidine prevent the development of irreversible AGE formation. The amino group in aminoguanidine reacts with the glucose-derived reactive intermediates to form, a triazine compound. Aminoguanidine does not inhibit AGE formation but prevents the protein-protein or protein-lipid cross-linking. Aminoguanidine modulates the activity of diamine oxidase and the inducible form of NOS. Glycation of Ig occurs on Fab and Fc fragments. Glycation of IgM is two-fold greater than that of IgG. Albumin can be glycated at multiple sites. Endothelial cells ingest and degrade glycated albumin according to the degree of glycation. AGE modified fibrinogen and fibrin are less sensitive to plasmin, and could lead to the fibrin accumulation. The glycation of anti-thrombin III produces a significant decrease in its heparin-catalysed thrombin-inhibiting activity. AGE-apo-B levels are found to be about two-fold higher in diabetic patients, and 24-fold higher in patients with end-stage renal disease. The existence of primary amino groups on phospholipids supports the possible direct reactivity between glucose and lipids to form AGEs. The severity of diabetes is associated with lipid oxidation. Oxidised LDL concentration is related to the level of glycation ion, which leads to the uptake of LDLs by macrophage scavenger receptors and to the formation of foam cells. The AGE- and oxidised-AGE-LDLs are immunogenic. Antibodies could increase the plasma clearance of glycated-LDLs by macrophages. Glycated HDLs have decreased affinity for their specific membrane sites on fibroblasts, which have a decrease in their capacity to induce reverse transport of cholesterol. Receptors for AGEs (RAGE) have different molecular weights: 36-83 KD for a macrophage cell, 60-90 KD for liver cell, and 30-50 KD for renal tissue. The gene for RAGE is localised on chromosome 6 in the human MHC class III region. The RBC is a target for the glycation process, because of its long half-life in the blood. The glycated haemoglobins are more negatively charged and their functional activity differs in their capacity to bind molecular oxygen. Other intracellular proteins of RBCs are also glycated, such as D-aminolaevulinate dehydratase. The glycation of the enzyme decreases its activity. A loss in RBC membrane fluidity is observed in diabetic patients and is correlated with glycation of membrane proteins. The glycation of spectrin, a major RBC membrane protein, which occurs in the cytoskeleton network, may account for reduced RBC deformability. AGE interactions with monocyte macrophage and T lymphocytes are mediated by RAGE. Macrophages internalise and degrade AGE-modified soluble proteins. AGEs, abundant in most tissues and fluids in diabetic patients, through the induction of cytokines (IL-1β, TNF-α) and growth factors (PDGF, IGF-1) may participate in tissue remodelling. AGE-activated T cells may contribute, in co-operation with primed macrophages, to tissue damage, particularly to the development of atherosclerotic lesions. There is an increased glycation of platelet membrane protein in diabetes. Most of the glycated platelets could be incorporated into thrombi. Modification of platelet functions, including increased fibrinogen binding and increased aggregation in response to ADP, could participate in modification of platelet functions in diabetes. Thickening of basement membrane is a widespread phenomenon in diabetes and ageing, affecting mainly, but not exclusively, the microvasculature. AGE accumulation on collagen results in the stiffing of collagen fibres and modification in the mechanical properties of the arterial wall. AGE accumulation in the subendothelial matrix could play a major role in the modification of the properties of the vascular wall and in early development of atherosclerosis in diabetes. The RAGE forms an integral membrane protein, which represents a cell-surface binding site for AGE. The AGE-endothelial interaction induces an oxidant stress, which is mediated by activation of the transcription factor, NF-κB, and enhancement of endothelial VCAM-1 expression. Binding RBCs to ECs results in a range of cellular changes, particularly increase IL-6 secretion. Increased IL6 could mediate the increase in fibrinogen concentration observed in diabetic patients. AGEs are efficient inactivators of NO by a quenching phenomenon. Vascular SMC proliferation is a characteristic abnormality of atherosclerosis that could be accelerated by the NO quenching effect of AGE-modified matrix and vascular proteins in diabetic subjects. Together, both effects lead to an increase in systemic arterial pressure. Vascular SMCs are exposed to subendothelial and circulating AGEs by an increase in permeability. Vascular SMCs and nerves in the vessel wall showed constitutively high levels of RAGE expression unchanged with ageing or by a vascular disease. When endothelium is damaged, the balance between growth promoters and inhibitors of vascular SMC proliferation shifts towards cell hyperplasia. AGE albumin exerts potent angiogenic effect on human ECs by stimulating their migration and tube formation. The binding of AGE albumin to EC monolayers is followed by internalisation and subsequent segregation either to a lysosomal compartment or to endothelial-derived matrix after transcytosis.

$$O_2 \rightarrow O_2^- \rightarrow H_2O_2 \rightarrow OH \rightarrow H_2O$$

AGEs produce ROI. Aminoguanidine prevents glycoxidation of the Amadori products and the oxidative modification of LDLs. Vitamin E reduces lipid peroxide levels in diabetes. Vitamin C in diabetes inhibits early protein glycation, but its effect on athrosclerosis is unknown. The oxides of nitrogen in cigarette smoke cause oxidation of macromolecules and deplete antioxidation. Haemochromatosis is a high risk factor for heart, may be related to iron accumulation in cells and mitochondria. The relaxing effects of NO are antagonised by superoxide and other free radical. ROIs can alter cell function of the blood vessels resulting in leucocyte accumulation, increased permeability, coagulation initiation and eventually thrombosis. AGE-modified protein induces a cellular-oxidant stress activating the transcription factor NFκB and cytokine production. Oxidants stress mediated by AGE-RAGE interactions is central to the development of vascular hyperpermeability. The endothelium-erythrocyte interaction is abnormal in malaria, diabetes mellitus, and sickle cell anaemia. In diabetes mellitus the increased adhesion RBCs is correlated with the extent and severity of the vascular lesions and with the blood level of glycated haemoglobin HbA1, mediated by the AGEs present on erythrocyte membranes and by the receptors for AGEs on endothelium.

DIABETIC NEPHROPATHY

Diabetes mellitus is an endocrine disease that is diagnosed by chronic hyperglycaemia. Diabetes mellitus (DM) is a chronic syndrome of impaired carbohydrate, protein, and fat metabolism. Type I diabetes induces higher incidence of microangiopathy than type II. Type II induces macroangiopathy leading to atherosclerosis. Both types are associated with micro- and macro-angiopathy. Diabetic nephropathy encompasses a complex of structural changes including renal hypertrophy, thickening of basement membranes, and progressive glomerular and tubulointerstitial accumulation of ECM components. Local proliferation of both macrophages and myofibroblasts likely contributes to the tubulointerstitial infiltrate. Glomerular and tubular hypertrophy is one of the earliest structural alterations of diabetic nephropathy. Amino sugars are physiologically synthesised through the hexosamine biosynthetic pathway. In this pathway, fructose-6-phosphate is converted to glucosamine-6-phosphate (GlucN-6-P) by glutamine:fructose-6-phosphate-amidotransferase (GFAT), which uses glutamine as an amino donor. Most glucose taken up by the cells is metabolised through glycolysis; only 2-5% enters hexosamine pathway. Hyperglycaemia induces production of TGF-β1, a prosclerotic cytokine involved in diabetic nephropathy. TGF-β1 induction is mediated by the hexosamine pathway.

PATHOBIOLOGY OF DIABETIC NEPHROPATHY

In euglycaemia, ROS increases the concentration of toxic aldehydes. NO levels are reduced, converting aldose reductase to a higher activity form. Glutathione levels are unaffected, and the reactive aldehydes are detoxified. In hyperglycaemia, ROS increases, and the reactive aldehydes are then detoxified. Hyperglycaemia increases intracellular ROS and lipid peroxidation. The enzyme aldose reductase converts various aldehydes to inactive alcohols. ROS may activate aldose reductase by reducing NO levels. Increased intracellular glucose concentration results in increased enzymatic conversion of glucose to the polyalcohol sorbitol, with concomitant decreases in NADPH and glutathione. In cells in which aldose reductase activity is sufficient to deplete glutathione, hyperglycaemia-induced oxidative stress is augmented. Sorbitol is oxidised to fructose by enzyme sorbitol dehydrogenase (SDH). In cells in which SDH activity is high, this may result in an increased NADH:NAD$^+$ ratio (reduced form to oxidised form ratio). Intracellular hyperglycaemia may damage tissue by PKC activation. Hyperglycaemia increases DAG content. Increased DAG activates PKC, primarily the β and δ isoforms. Activated PKC increases the production of cytokines and ECM, the fibrinolytic inhibitor plasminogen activator inhibitor (PAI-1), and the vasoconstrictor endothelin-1; PKC is also a mediator of VEGF activity. The result is thickening, vascular occlusion, increased permeability, and activation of angiogenesis. ROS activate PKC in vascular endothelial cells. H_2O_2 activates PKC. ROS activate phospholipase D, which hydrolyses phosphatidylcholine to produce DAG. ROS increase DAG through de novo synthesis resulting from ROS inhibition of the enzyme glyceraldehyde phosphate dehydrogenase. Diabetes increases mean retinal circulation time. Diabetes increases GFR. Intracellular hyperglycaemia damages susceptible tissues by increasing the formation of advanced glycation end products (AGEs). AGEs can arise from autooxidation of glucose to glyoxal, decomposition of the Amadori product to

3-deoxyglucosone, and fragmentation of glyceraldehyde-3-phosphate to methylglyoxal. These reactive dicarbonyls react with amino groups of proteins to form AGEs. Intracellular production of AGE precursors damages target cells by 3 mechanisms: 1- Intracellular protein glycation (e.g., increase in macromolecular endocytosis) alters protein function. 2- ECM modified by AGE precursors has abnormal functional properties. 3- Plasma proteins modified by AGE precursors bind to AGE receptors on adjacent cells such as macrophages, thus inducing receptor-mediated ROS production that activates NF-κB and expression of pathogenic gene products, including cytokines and hormones. There are 5 AGE-binding proteins: i- The receptor for advanced glycation end products (RAGE), ii- p60, iii- p90, iv- galcetin-3, and v- the scavenger receptor type II. RAGE is a member of immunoglobulin family, whose ligation generates ROS and activates the pleiotropic transcription factor NF-κB. The type II macrophage scavenger receptor binds AGEs and mediates their uptake by endocytosis. On macrophages and mesangial cells, binding of AGE-modified protein to specific receptors stimulates production of TNF-α IL-1, IGF-1, and GM-CSF at levels that increase proliferation of smooth muscle cells and increase matrix production. On endothelial cells, binding of AGE-modified protein to specific receptors induces changes in gene expression that are procoagulatory and increases expression of leukocyte-binding VCAM-1. Increased glucose flux through polyol pathway decreases the NADH:NADP$^+$ ratio, thus reducing glutathione reductase activity and increasing oxidative stress.

AGE-binding proteins:

Protein	MW	Function	Features
p60 (AGE-R1)	50 kD	AGE uptake and degradation.	Homologous to the oligosaccharyl transferase.
p90 (AGE-R2)	80 kD	Cell activation.	Homologous to PKC and FGFR3.
Galectin (AGE-R3)	32 kD	Cell activation.	
RAGE	35 kD	Cell activation.	Member of the Ig superfamily.
Scavenger receptor type I	220 kD	AGE uptake and degradation.	Homotrimeric protein with 5 domains.
Scavenger receptor type II	170 kD	AGE uptake and degradation.	Homotrimeric protein with 5 domains.

Increased sorbitol flux through this pathway increases the NADH:NAD$^+$ ratio, thereby blocking glycolysis at the level of triose phosphates and increasing formation of α-glycerol phosphate, a precursor of DAG. Aldose reductase activity is reduced in several major target cells damaged in diabetes, such as endothelial cells. ROS may activate aldose reductase, induce DAG, activate PKC, induce AGE formation, and activate NF-κB. Insulin resistance may damage arteries indirectly by exacerbating known risk factors for vascular damage. Insulin resistance alone or combined with hyperinsulinaemia is associated with atherogenic changes in plasma lipoproteins, increased PAI-1, and hypertension (syndrome X). Insulin resistance in vascular cells may promote damage by inhibiting antiatherogenic gene expression. Selective resistance to insulin action in the phosphatidyl inositol 3-kinase signalling pathway may reduce antiproliferative NO production and interfere with insulin's inhibitory effect on TNF and angiotensin II, a stimulation of PAI-1 and ICAM expression. Hyperglycaemic memory refers to the development of retinopathy during posthyperglycaemic normoglycaemia. The mechanism of hyperglycaemic memory include: 1- Induction, hyperglycaemia-induced increases in ROS that may be a consequence of increased reducing equivalents generated from increased glucose metabolism flowing through the mitochondrial electron transport chain. Increased ROS cause cellular dysfunction and induce mutations in mitochondrial DNA. 2 Perpetuation, occurs as a result of mDNA mutation by hyperglycaemia-induced ROS, which would encode defective electron transport chain subunits. These defective subunits would cause increased ROS production by the electron transport chain at physiologic concentrations of glucose and glucose-derived reducing equivalents. Hyperglycaemia may cause damage at a cellular level in both glomerular and tubular locations, often preceding overt dysfunction. The polyol pathway enzymatic reactions include: 1- The reduction of glucose to sorbitol by the aldose reductase; and 2- Oxidation of sorbitol to fructose by the action of sorbitol dehydrogenase. Aldose reductase activity exists in both the cortex and outer medulla. Aldose reductase is greatest in the medulla at the inner stripe of the outer medulla, the inner medulla, and at the papillary tip. Fructose, the second product of polyol pathway is increased several fold in tissues with an activated polyol pathway and can contribute to non-enzymatic fructosylation of proteins and provide 3-deoxyglucosone, the precursor to AGEs. Decreased NADPH may compromise reduction of glutathione in oxidatively stressed cells. Nonenzymatic glycosylation of protein begins with the covalent attachment of glucose to reactive amino groups at a rate determined by blood glucose concentrations. These early glycation products (Schiff bases and Amadori products) can serve as precursors to AGEs with complex glucose-derived cross-linking altering the structure and function of cells and supporting matrix. Plasma levels of Amadori albumin are increased in type 1 diabetes and are associated with nephropathy classified by persistent albuminuria. Nonenzymatic glycation of reactive amino groups increases the rate of free radical production nearly 50-fold. Trace metal-catalysed oxidation of glucose can form ROS. The oxidative modification of carbohydrates, lipids and proteins that follows can produce glycoxidation and AGEs such as pentosidine, carboxymethyllysine, carboxyethyllysine, malondialdehyde-lysine and 4-hydroxynonenal and acrolein-protein adducts. Free radicals may arise from multiple pathways including autoxidation of glucose; H_2O_2 generated from the oxidation of enedisols from Amadori products or O_2^- formed by the mitochondrial oxidation of NADH to NAD^+ or in the formation of prostaglandins, as PGH synthase reaction utilises NADH and NADPH with the generation of O_2^-. Glomerular and tubule cells can generate ROS with potential to impose a free radical stress. Levels of H_2O_2 may be decreased by the action of catalase or glutathione peroxidase. Glutathione peroxidase requires an effective cellular level of reduced glutathione. Dicarbonyl species formed without oxidation include methylglyoxal, which forms nonoxidatively from triose phosphates and 3-deoxyglucosone, which is formed by the nonoxidative 1,3-enolation of Amadori adducts and from fructose. Dicarbonyls, methylglyoxal and 3-deoxyglucosone contribute to glycoxidation products, and the lipid dialdehyde products of the oxidation of unsaturated fatty acids in membrane and lipoprotein phospholipids, which contribute to lipoxidation adducts that are most effective substrates for aldose reductase with a substrate specificity for a recombinant form of the enzyme higher than that glucose. High glucose exhibit activation of intracellular signalling pathways, up-regulation of the TGF-β system, and a consequent overproduction of the ECM molecules that leads to glomerulosclerosis, tubulointerstitial fibrosis, and renal failure. Glycated proteins arise from a condensation reaction, driven by the ambient glucose concentration, in which a free sugar covalently attaches to a protein at reactive $-NH_2$ groups. Glycation proceeds through the formation of a labile Schiff base adduct, which then undergoes an intramolecular Amadori rearrangement to become a stable glucose-modified protein. The reaction occurs slowly, and the degree and duration of hyperglycaemia influence the amount of glycated protein. Amadori-modified proteins may further evolve through a series of spontaneous rearrangement, dehydration, and polymerisation reaction to become AGEs. The glycated proteins alter the permeability properties of the glomerular capillary wall, and transport across the glomerular filtration barrier and into the mesangial space. Endothelial cells take up circulating glycated proteins, which may lead to endothelial dysfunction and contribute to diabetic microangiopathy. All vasoactive hormones (ANGII, endothelin) exert profound growth stimulatory actions on many renal cells, especially in the presence of high glucose. Hypertrophy of the proximal tubule mainly accounts for the increase in kidney size in diabetes because glomeruli alone account for <10% of the kidney's volume. ANGII exerts proliferative effects on some renal cells (mesangium cells, distal tubular cells), and mediate hypertrophy of proximal tubules. Continuous production and release of growth factors during the diabetic state cause renal growth as an integral early part of diabetic nephropathy and may explain why patients with diabetes have a higher incidence of renal cancer. Growth factors and peptide in diabetes mellitus include: 1- TGF-β, 2- PDGF-β, 3- IGF-I, 4- insulin, 5- HGF, 6- TNF-α, 7- FGF, 8- VEGF, 9- leptin, 10- ANGII; and 11- endothelin 1. Leptin is produced by adipocytes and reduces food intake through interacting with specific hypothalamic receptors. Leptin stimulates proliferation and synthesis of TGF-β in glomerular endothelial cells. Leptin leads to proteinuria and segmental glomerular sclerosis. Leptin increases TGF-β receptor type 2 expression on mesangial cells (paracrine action). Hyperleptinaemia is part of the syndrome in type 2 diabetes; increased leptin concentrations may modulate growth of glomerular cells in diabetic nephropathy. Renal cells, under normal conditions, have a very low turnover rate of <1%. After stimulation with a growth factor or cytokine, dormant cells actively enter the G_1-phase in which cells increase their size and stimulate protein and mRNA synthesis to prepare for DNA replication.

AGNII and diabetic nephropathy:

1- AGNII induces renal hypertrophy and cell proliferation.
2- AGNII stimulates ECM synthesis.
3- AGNII inhibits ECM degradation.
4- AGNII stimulates cytokine production (TGF-β1, VEGF).
5- AGNII stimulates superoxide production
6- AGNII increases glomerular capillary pressure and permeability.
7- AGNII induces mesangial cell contraction leading to reduction of filtration surface area.

Mesangial cells exposed to high glucose actively enter the cell cycle, expressing c-fos, c-jun, and Egr-1. High glucose stimulates expression of $p27^{kip1}$ (bound to CDK2) involved the activation of PKC and partly depends on induction of TGF-β. ANGII induces $p27^{Kip1}$ expression in renal cells. MAP-kinase phosphorylates $p27^{Kip1}$ and increases its stability. High glucose increases $p27^{Kip1}$ protein expression. ACE inhibitor prevents glomerular hypertrophy in diabetes mellitus, by interfering with the expression of selected CKI. TGF-β transforms a mitogenic stimulus into hypertrophy. High glucose and/or TGF-β cause sustained proliferation of some renal cell types such as tubulointerstitial fibroblasts. Most renal cells undergo hypertrophy in diabetes mellitus, but limited cell population within the kidney may proliferate. TGF-β leads to down-regulation of cyclin D and induces $p16^{INK4}$. Lovastatin suppresses glomerular TGF-β1 expression in 4 days and urinary albumin excretion in 4 weeks; it reduces glomerular volume in 2 weeks, and suppresses serum cholesterol level in diabetic within 6 months. HMG CoA reductase is a rate-limiting enzyme in a cholesterol biosynthetic mevalonate pathway. HMG CoA reductase inhibitor suppresses the expression of chemokines and cytokines, induces apoptosis of mesengial cells, inhibits cell proliferation, interferes with intracellular signalling, reduces ECM production and modulates ECM protein degradation system. From the mevalonate pathway, several isoprenoids are produced as an intermediate stage. These isoprenoids are enrolled in post-translational modification and membrane localisation of small GTP binding proteins such as Ras and Rho family proteins. Localisation to plasma small GTP binding proteins are followed by intracellular signalling and activation of various transcription factors. HMG CoA reductase inhibitor may reduce the concentration of isoprenoids intracellularly, thus reducing functional small GTP binding proteins and interfere with intracellular signalling.

Classification of diabetic neuropathy:

Diffuse Neuropathy:
    1. Distal symmetric sensorimotor polyneuropathy.
    2. Autonomic neuropathy:
        A- Sudomotor neuropathy.
        B- Cardiovascular autonomic neuropathy.
        C- Gastrointestinal neuropathy.
        D- Genitourinary neuropathy.
Focal Neuropathy:
    1- Cranial neuropathy.
    2- Radiculopathy/plexopathy.
    3- Entrapment neuropathy.

DIABETIC POLYNEUROPATHY

Diabetic peripheral neuropathy is a complex group of disorders. Diabetic neuropathy is one of the most common manifestations of diabetes and potentially its most debilitating. Diabetic polyneuropathy is the most common type of diabetic neuropathy. It presents primarily with sensory symptoms but may have a prominent autonomic component. It is this type of neuropathy that is associated with secondary complications of neuropathy. The prevalence of diabetic neuropathy ranges from 5% to 100%. The prevalence increases with the duration of diabetes, so that 25 years after the initial diagnosis of diabetes, the prevalence is 50%. The prevalence is similar for insulin-dependent (Type 1) and non-insulin-dependent diabetes (Type 2). Distal symmetric polyneuropathy, the most common type of diabetic neuropathy, is an anatomically diffuse process primarily affecting sensory and autonomic fibers. The longest nerves are affected first, with symptoms typically beginning insidiously in the toes and then advancing proximally up the legs. Both small- and large-diameter fiber sensory neurons can be involved. A compression neuropathy often results in demyelination with the axon left relatively intact. Sensory loss follows a radicular pattern. When the neuronal cell body dies the condition is called neuronopathy. Involvement of small-fiber sensory neurons may result in loss of normal pain and temperature sensation, which can predispose the patient to injury, ulceration, and chronic infection. Large sensory fiber degeneration leads to loss of vibratory and proprioceptive sensation, which is not clinically significant until the neuropathy is advanced. Motor symptoms are generally mild. Autonomic nervous system involvement is common and present in up to 40% of Type 2 diabetic patients at the time of diagnosis. Acute painful neuropathy is a variant of the distal sensory polyneuropathy, and is characterised by the presence of severe and often unrelenting pain in the legs that is usually worse at night. Abnormal sensitivity to innocuous stimuli such as bed sheets or clothing is sometimes reported, and the neurological examination is often unimpressive. Proximal motor neuropathy is a syndrome known as amyotrophy, as well as femoral neuropathy or sacral plexopathy. Mononeuropathy occurs in variety of isolated nerve lesions. Oculomotor nerve palsies are the commonest among cranial nerves. Other mononeuropathies include trunk or thoracoabdominal neuropathy and peripheral compressive neuropathies. Thoracoabdominal neuropathy occurs infrequently and almost exclusively in older patients. They may present with the abrupt onset of chest or abdominal pain with sensory loss in a dermatomal pattern, suggesting spinal root involvement. Motor involvement may occur and can lead to herniation through the weakened muscle of the abdominal wall. Focal lesions of peripheral nerves occur at common sites of nerve entrapment or compression, such as the carpal tunnel (median nerve), tarsal tunnel (posterior tibial), and elbow (ulnar nerve). Diabetic neuropathy (DN) is a group of clinical syndromes that alone or in combination affect distinct regions of the nervous system. It is a heterogeneous disorder that encompasses a wide range of abnormalities affecting proximal and distal peripheral sensory and motor nerves as well as the autonomic nervous systems. The major effect of diabetes is on the small unmyelinated or thinly myelinated C and A δ nerve fibers that subserve autonomic function and thermal and pain perception. Once autonomic neuropathy sets in, life can become quite dismal and the mortality rate approximates 25% to 50% within years. Patients with small-fiber dysfunction most commonly present with burning feet. Diabetes remains the most common cause for neuropathy, but heavy metal poisoning, amyloid, vasculitis, autoimmune disease, collagen vascular disease, chemotherapeutic drugs, Chagas' disease, and idiopathic and familial causes need to be considered as well. Autonomic neuropathy can affect virtually every system in the body.

Clinical aspect of Neuropathy:

Autonomic	Small fiber	Large fiber
1- Abnormal heart rate. 2- Postural hypotension. 3- Neuropathic diarrhoea. 4- Gastroparesis. 5- Abnormal sweating. 6- Erectile dysfunction. 7- Retrograde ejaculation.	1- Cutaneous hyperesthesia. 2- Burning pain. 3- Paraesthesias. 4- Lancinating pain. 5- Loss of pain sensation. 6- Loss of temperature sensation. 7- Foot ulceration. 8- Loss of visceral pain.	1- Loss of vibration sensation. 2- Loss of proprioception. 3- Loss of reflexes. 4- Slowed nerve conduction velocities.

PATHOBIOLOGY OF DIABETIC NEUROPATHY

There are several different mechanisms that contribute to the pathobiology mechanism of diabetic neuropathy. These include direct metabolic compromise, ischaemia, oxidative stress, and loss of neurotrophic support. Glucose is converted to sorbitol in cells by the activity of the aldose reductase enzyme. The accumulation of sorbitol results in a decrease in intracellular levels of myoinositol and taurine to the point that they are rate-limiting for intracellular metabolism. The reduced levels of myoinositol are associated with decreased Na^+-K^+-ATPase activity and slow nerve conduction velocities. Vascular perfusion of peripheral nerves is impaired in diabetic patients. The reduction in nerve blood flow, reduced endoneural blood flow and nerve ischaemia are the primary contributors to the pathogenesis of diabetic neuropathy. The endoneural ischaemia is characterised by vascular basement membrane thickening, platelet aggregations, endothelial cell hyperplasia, and occluded vessels. The conversion of glucose to sorbitol and subsequently to fructose results in the depletion of NADPH and NAD^+ stores in the cell, which makes the cell more vulnerable to ROS. Ischaemia also induces ROS, further increasing the oxidative stress and causing nerve injury. Neurotrophic factors are proteins that promote the development, survival, and maintenance of specific neuronal populations. Levels of NGF are significantly reduced, and its retrograde axonal transport from target tissues to the neuronal cell bodies, where it exerts its effects, is impaired. Reduced expression of NGF in the skin of diabetic patients correlates with the presence of early signs of small fiber sensory neuropathy. Neurotrophin-3 or the insulin-like growth factors play a role in the pathogenesis of diabetic neuropathy. The pathobiology of diabetic autonomic neuropathy (DAN) is multifactorial. The neuropathic progression is dynamic, with nerve degeneration and regeneration occurring spontaneously and simultaneously. Cells adapt to osmotic stress by inducing the formation of organic osmolytes, which allows them to maintain their normal volume while avoiding accumulation of toxic concentrations of ions. These organic osmolytes include sorbitol, myo-inositol, taurine, and glycerophosphoryl choline. Hyperosmotic stress normally increases the concentration of all these osmolytes through different mechanisms. Hyperosmotic stress normally increases the expression of the enzyme aldose reductase, which in turn converts glucose to sorbitol. It also increases the Na^+-dependent transport of myo-inositol and taurine into cells. The resulting accumulation of intracellular sorbitol in the absence of osmotic stress produces a reciprocal decrease in levels of myo-inositol and taurine to the point that they become insufficient for normal intracellular metabolism. Myo-Inositol and taurine depletion has been associated with reduced Na^+/K^+-ATPase activity and slowed nerve conduction velocity. Aldose reductase inhibitors restore myo-inositol and taurine levels and improve motor conduction velocity. Aldose reductase inhibitors may also improve nerve conduction velocities and protect small sensory fibers from degeneration in humans. Nerve injury may result from direct metabolic effects of hyperglycaemia on nerve fibers or mediated by local nerve ischaemia. Evidence of diabetic local vascular disease includes basement membrane thickening, endothelial cell proliferation, and vessel occlusions. When glucose is converted to sorbitol and sorbitol is subsequently converted to fructose, the oxidation/reduction status of the cell is altered with loss of reduced nicotinamide-adenine dinucleotide phosphate (NADPH), and glutathione stores in the cell. This may impair the cell's ability to detoxify ROS. Hyperglycaemia promotes formation of ROS by auto-oxidation of glucose and formation of AGEs. The sorbitol pathway may accelerate this process because fructose is a more potent reactant than glucose. Ischaemia secondary to vascular disease induces oxidative stress in the nerve; increasing the production of ROS and inducing nerve injury. Antioxidant therapy ameliorates signs of diabetic neuropathy, such as slowed nerve conduction velocities and reduced nerve blood flow, and may improve pain in painful diabetic neuropathy. Hyperglycaemia-induced depletion of NADPH stores, with subsequent loss of detoxification mechanisms, further exacerbates the damage resulting from ischaemic oxidative stress. Neurotrophic factors are involved in the development, maintenance, and regeneration of responsive elements of the nervous system. Nerve growth factor (NGF) is a protein that promotes the survival of sympathetic and small-fiber neural crest-derived sensory neurons in the peripheral nervous system. In diabetes, both NGF production and retrograde transport appear to be impaired, and abnormal expression of NGF in skin keratinocytes correlates with early manifestations of small-fiber sensory neuropathy. Neurotrophic factors appear to play a significant role in mediating the cell's defense against oxidative stress. Deficiency in the availability of these factors could further exacerbate the injury caused by oxidative stress. Antioxidant treatment also enhances NGF action. Aldose reductase inhibitors may offset the intracellular metabolic consequences of hyperglycaemia, while antioxidants directly limit damage from ROS. NGF may increase the resistance of the cell to injury from oxidative stress and at the same time promote regeneration of damaged nerves. Oxidative stress, "hot" mitochondria, and nitric oxide (NO) play important role in mediating nerve damage. There is an association among oxidative stress, mitochondrial (Mt) membrane depolarisation (MMD), and induction of apoptosis. Glucose increases ROS, which induces an initial Mt membrane hyperpolarisation followed by MMD. In turn, MMD is coupled with cytochrome-c release and cleavage of caspases. Hyperglycaemia is associated with oxidative stress-induced neuronal and Schwann cell death, and targeted therapies aimed at regulating ROS bursts may prove effective in therapy of diabetic neuropathy. Early during the course of evolving diabetes, there is an increase in NO and its derivatives NO_2 and NO_3, which are collectively known as Nox (referred to as noxious). Under normal conditions, the synthesis of small amounts of NO is catalysed cNOS, which facilitates vasodilation and inhibits the aggregatory tendencies of platelets. Under inflammatory conditions, iNOS is activated and large quantities of NO are formed. NO may be responsible for nerve damage, neuronal apoptosis, and other types of cell death. Nox may also act as an oxidative stressor, increasing the production of ROS, which may also derive from overheated mitochondria, xanthine oxidase, and NADPH among others. Interaction of NO with superoxide radical (SO_2^-) results in the formation of peroxynitrate (ONOO), which nitrotyrosylates proteins, thereby impairing their function. Glucose-induced oxidative stress and secondary depletion of NO at sites both within the peripheral nerve trunks and the endoneural microvasculature and more proximally in the sympathetic ganglia. The depletion of NO at these sites has been suggested to lead both directly and indirectly to nerve blood flow deficits, nerve hypoxia, and disruption of nerve mitochondrial function and energy production. Neuropathy is found in 11% of patients who are euglycaemic. Functional disturbances in the diabetic dermal microvasculature include: 1- decreased microvascular blood flow, 2- increased vascular resistance, 3- decreased tissue PO_2, and 4- altered vascular permeability characteristics. Diabetes disrupts vasomotion, the rhythmic contraction exhibited by arterioles and small arteries. Unmyelinated C-fibers are assumed to be damaged in DN, contributing to abnormalities in cutaneous blood flow. Warm thermal sensation is a functional measure of C-fiber function in the periphery, and the impairment of this is paralleled by a reduction of vasomotion indicating an interaction between small unmyelinated C-fiber function and vasomotion. The susceptibility to foot ulceration is the consequence of impaired microvascular perfusion of skin and subcutaneous tissue and a concomitant decrease in vascular supply to small C-fibers subserving pain and warm thermal perception, allowing increased heat and tissue injury in an ischaemic limb. Postischaemic hyperaemia is thought to be endothelium-dependent, largely the result of the release and action of endothelial-derived NO and PG. AGE in vessels and connective tissues, accumulation of sorbitol in nerves, alterations in PKC metabolism, or other metabolic factors that remain poorly understood, are likely to precede structural changes in the vessels. Blood flow is a sensitive marker of C-fiber dysfunction, precedes development of diabetes mellitus. Epidermal nerve fibers (ENFs) are the distal projections of small dorsal root ganglion (DRG) neurons that pierce the dermal-epidermal basement membrane and penetrate the epidermis, often to the stratum corneum. Here these thinly myelinated or unmyelinated fibers, which are not assessed by routine nerve conduction studies, can be visualised by immunostaining with antibodies to the protein gene product (PGP). They subserve heat, pain, and autonomic functions and may be preferentially affected in diabetes. Reduced numbers of ENFs are found in diabetic neuropathy and other small-fiber neuropathies.

INTERVENTION

The clinical course of the syndrome varies depending on the particular type of neuropathy. Positive sensory symptoms of polyneuropathy may include numbness, tingling, sharp lancinating pains, dull aches, burning pain, tightness in the foot or calf, hyperesthesia, and allodynia. The long-term clinical course is one of slow but relentless progression. Sensory loss due to neuropathy results in a 7-fold increase in the risk of acquiring a foot ulcer. Chronic foot ulcers are associated with small-fiber sensory dysfunction. Because the wound heals slowly and there is an increased risk of infection, the extremity is prone to gangrene and possible amputation. Small-fiber sensory neuropathy predisposes the patient to traumatic injury, painless burns, neuropathic osteoarthropathy, and painless blisters. Asymptomatic patients with neuropathy are still at significant risk for complications, such as ulceration or cardiac arrhythmias and sudden death. The evaluation of the diabetic patient should always be centered about the feet. They should be inspected for signs of ulceration, callous formation (especially in areas subject to high pressure), cracks in the skin, and hair loss. The proper evaluation of a patient at risk for or with suspected diabetic neuropathy should involve a complete symptom history, a thorough neurologic examination and foot examination, and electrophysiological testing. Pharmacologic intervention is used to treat the pain of neuropathy. Neuropathic pain may also improve rapidly with tightening of glycaemic control; hyperglycaemia is thought to interfere with opioid receptors. Analgesic use is best restricted to NSAIDs. Narcotics may exacerbate features of autonomic neuropathy. Tricyclic antidepressants effectively control neuropathic pain but pain relief is not dependent on mood elevation. They act in the CNS, blocking the reuptake of norepinephrine and serotonin at central synapses that are involved in pain inhibition. Tricyclics provide >50% pain relief in 30% of patients with neuropathic pain. Nortriptyline and imipramine tend to have less anticholinergic side effects than amitriptyline, and imipramine has the lowest potential for causing urinary retention. Doxepin is the least cardiotoxic. Older patients should have an electrocardiogram and all patients should be questioned about urinary retention and glaucoma prior to starting these drugs. Capsaicin is a topical medication applied over the painful areas. Capsaicin depletes the neuropeptide substance P at the terminals. Substance P plays an important role in pain transmission. Capsaicin most common side effect is a burning pain or a transient worsening of the neuropathic pain that lasts for the first few days of treatment. Pharmacologic intervention may be useful in the treatment of autonomic neuropathy and its complications. Gastroparesis may be treated by having the patient take frequent small meals, a liquid diet, and metoclopramide (a dopamine antagonist) 10 mg, 30 minutes before meals. Diabetic diarrhoea may be due to bacterial overgrowth or abnormal motility. Aldose reductase converts glucose to sorbitol in cells. Aldose reductase inhibitors (ARI) are thought to be promising agents not only for the treatment of diabetic polyneuropathy but also for other complications of diabetes. α-Lipoic acid (ALA) is a potent lipophilic free-radical scavenger that has demonstrated effectiveness in preventing neuropathic abnormalities. Strategies to inhibit oxidative burst include: 1- IGF-I, can block induction of ROS and/or stabilise the mitochondrial membrane potential. This in turn is associated with inhibition of apoptosis. 2- Reduction of ROS generation in neuronal Mt by upregulation of manganese superoxide dismutase, or downregulation of nNOS, prevents neuronal degeneration. 3- Upregulation of uncoupling proteins that stabilises the mitochondrial membrane potential blocks induction of caspase cleavage. Diabetic neuropathy is a common complication of diabetes that may be disabling and even contribute to mortality. Diabetic neuropathy is a common and sometimes disabling problem that may result in premature death or loss of a limb. Diabetic neuropathy affects >50% of patients with a history of diabetes >25 years' duration. The presence of neuropathy significantly increases the risk of foot ulcerations and infections, which may lead to amputations. Autonomic neuropathy (25%) is a very serious and often overlooked component of diabetic neuropathy. Cranial nerve palsies most often associated with diabetes affect the third nerve, followed in frequency by the fourth and sixth cranial nerves. They may present with eye pain, diplopia, and, in the case of third nerve involvement, ptosis. The pupillary light response is typically spared. Cranial neuropathies generally affect older patients and usually reverse themselves naturally within 3-6 months. Thoracolumbar neuropathy/radiculopathy is also more common in older diabetic patients. This syndrome presents with intermittent chest or abdominal pain in a nerve root distribution and may be associated with a dermatomal pattern of sensory loss. Compression neuropathies, such as carpal tunnel syndrome and ulnar nerve entrapment, are more common in diabetic than in nondiabetic patients. Diabetic autonomic neuropathy may be presented by loss of sweating in the feet, sexual dysfunction, bladder abnormalities, gastroparesis, orthostatic hypotension, and diabetic diarrhoea. Loss of the normal sympathetic innervation to the vascular supply of the feet may contribute to the development of foot ulcers. Decreased sudomotor function in the feet leads to drying and cracking of the skin, fissures formation, and infection and ulcer formation. Lubrication of the feet is important to prevent this from happening. Patients with symptomatic autonomic neuropathy and reduced heart rate variability have a significantly reduced 10-year survival rate. Regional sympathetic myocardial denervation in diabetes is associated with abnormal patterns of myocardial blood flow. In patients with postural hypotension or known cardiac disease, it might be advisable to measure heart rate variability. Cardiovascular autonomic neuropathy (CAN) is a common complication of diabetes. The onset of loss of heart rate variability (HRV) is a predictor of premature death and is associated with a mortality of 25-50% within 5-10 years. Loss in HRV is reversible with glycaemic control, graded exercise, antioxidants, and a variety of other agents including α-blockers, ACE inhibitors, aldosterone antagonists, and calcium-channel blockers. The polyol pathway; that is, high tissue glucose levels may lead to an accumulation of sorbitol and fructose within the nerves, associated with myo-inositol depletion, decreased Na^+-K^+-ATPase activity, and axoglial disjunction in peripheral nerves. Aldose reductase is the first end rate-limiting enzyme of the polyol pathway that catalyses the reduction of glucose to sorbitol. Overall, the incidence of neuropathy 25 years after the diagnosis of diabetes is 20% and the prevalence is 50%. In patients with excellent glycaemic control, the prevalence is <20%, whereas in those with poor control it is >60%. Better control of blood glucose for longer periods would be necessary to achieve meaningful improvement in nerve function. The insulin pump permits even more meticulous glycaemic control; meticulous control of blood glucose is feasible, that it can be sustained for several years. Hypoglycaemia markedly increases the severity of axonal degeneration in experimental diabetic neuropathy. A lifetime of recurrent hypoglycaemia may result in progressive cognitive and intellectual deterioration due to cumulative effects of small, subclinical insults to the brain. Too rapid glycaemic control may precipitate an acute, painful neuropathy in humans. Rigorous glycaemic control exerts beneficial effects on neuropathy types other than polyneuropathy. The most physiologic means of achieving glycaemic control is pancreas transplantation; it may restore motor-nerve conduction velocity and amplitude to normal and prevents development of structural nerve abnormalities that may be due to improvement in the uraemic component of neuropathy. Even with normal glucose metabolism and insulin secretion, it may take >3 years before a significant difference is noticed. Meticulous control of hyperglycaemia can slow the progression of mild somatic and autonomic neuropathy. The best means of controlling or improving neuropathy is pancreas transplantation, but this is an expensive and potentially dangerous procedure, applicable to only small numbers of patients, and the beneficial effects are small and are achieved only after several years of normal glucose metabolism. Hyperinsulinaemia in type 2 diabetes is the result of an abnormal incretin signal from the gut, while the insulin resistance is a secondary protective phenomenon. Chronic exaggerated stimulation of the proximal gut with fat and carbohydrates may induce overproduction of an unknown factor that causes impairment of incretin production and/or action, leading to insufficient or untimely production of insulin so that glucose intolerance develops. A hormone overproduced in the proximal foregut in diabetic patients might not directly increase the production of insulin, but rather counteract the action of insulin, thus inducing insulin resistance (IR) and only secondarily hyperinsulinaemia. Type II diabetes mellitus is a heterogeneous disorder. Roux-en-Y gastric bypass (GBP) and biliopancreatic diversion (BPD) restore insulin sensitivity, but the possibility of an additional incretin-mediated effect on insulin secretion cannot be ruled out. IGF-1 has a potent hypoglycaemic effect, effectively lower blood glucose concentrations in patients with type I or type II diabetes. Decreased levels of IGF-1 have been documented in patients with type II diabetes mellitus. GBP increases IGF-1 levels only in morbidly obese patients with diabetes and not in non-diabetic subjects. Leptin may directly affect glucose and fat metabolism. Leptin-induced increase in fatty acid oxidation could also improve glucose uptake and influence insulin sensitivity indirectly through the brain and sympathetic nervous system or by changes in the concentration of serum fatty acids and glucose flux in the liver. Leptin levels decrease rapidly after GBP and BPD without correlation with postoperative BMI. Body fat composition is not the only factor that regulates leptin levels.

NEUROPATHIC PAIN

Neuropathic pain is not a single entity or diagnosis; rather, it represents a variety of syndromes representing certain neurogenic signs and symptoms. Neuropathic pain syndromes, such as painful diabetic neuropathy (PDN), trigeminal neuralgia, and postherpetic neuralgia (PHN), may have different neural mechanisms and demonstrate different responses to pharmacologic and nonpharmacologic treatments. Glial cells are found in conjunction with neurons. Microglia and astrocytes are generally considered the supporting cells in the CNS. Glial cells release neuroactive substances, such as excitatory amino acids, nerve growth factor, and enkephalins. They encapsulate synapses and can regulate the uptake of neuroactive substances. Glial cells are also found within the dorsal horn of the spinal cord and their activation may induce pain. Glial transmission is thought to occur by propagation of calcium waves, gap junctions, and proexcitatory cytokines. Astrocytes can communicate with neurons, microglia, and oligodendroglia via calcium waves. The calcium waves are mediated by extracellular ATP. The microglia change shape in response to calcium. Glial activation and pro-inflammatory cytokine release occur during exaggerated pain states. Uninjured afferents may develop spontaneous activity when injury is caused to nearby afferents. Nerve injury induces Wallerian degeneration and the inflammation drives the sensitivity of nearby afferents. The uninjured afferents adjacent to injured neurons are believed to be causing the neuropathic pain. The bone cancer pain is initiated by cancer cells. The pain impulses result in reorganisation of the spinal cord and massive proliferation of astrocytes (probably resulting in central desensitisation). The tumour cell causes differentiation of the osteoclast into a hypertrophied tumour osteoclast. The osteolysis is acidic and causes pain. Neutralising the osteoclasts may improve bone cancer pain. Osteoprotagan (OPG) interferes with the differentiation of osteoclasts into tumour osteoclasts and causes senescence. OPG reduces bone destruction and reduces ongoing pain; OPG reduces the neurochemical changes induced in the spinal cord. The acute diabetic neuropathies include acute distal sensory neuropathy, acute thoracic radiculopathy, acute lumbar radiculoplexopathy, and insulin neuritis. The chronic diabetic neuropathies include distal sensory neuropathy, distal small fiber neuropathy, small fiber type, and demyelinating neuropathy. Diabetes damages both small unmyelinated and large myelinated fibers. There is axonal atrophy, reduced DRG neural volume, and myelin splitting in the dorsal root as a result of diabetic neuropathy, but no associated Wallerian degeneration, fiber loss, or regeneration. The functional changes include large-fiber nerve conduction velocity slowing, ischaemic hypoxia, and reduction in axonal transport but no ectopic activity. Neuropeptide levels of substance P (SP), N-methyl-D-aspartate (NMDA), and α-amino-3-hydroxy-5-methyl-4-isoxazolepropionic acid (AMPA) increase. The painfulness may be correlated with the amount of active axonal degeneration. The regeneration of sprouts is the source of ectopic pain generation. Spontaneous pain has been correlated with progressive loss of A-δ and C fibers. Hyperglycaemia itself may be a cause of painful diabetic neuropathy because better glucose control can relieve pain. There may be an exaggerated polyol pathway flux. The nerves are damaged by the accumulation of sorbitol and other by-products of the polyol pathway as a result of myo-inositol deficiency. Aldose reductase, which reduces sorbitol accumulation, can prevent myo-inositol deficiency; and diabetic neuropathy can be blocked by aldose reductase. Another aetiology may be a neurotrophin deficiency such as NGF. Postherpetic Neuralgia is caused by the reactivation of the varicella zoster virus (VZV), resulting in chronic burning pain and allodynia. Antidepressants and antiepileptics are considered the first-line treatment for postherpetic neuralgia. Selective serotonin reuptake inhibitors (SSRIs) seem to be less effective for neuropathic pain. Gabapentin is recommended for postherpetic neuralgia because of its superior efficacy, tolerability, and safety. Topical agents such as capsaicin (deplete SP), lidocaine cream 2.5%, and lidocaine patch 5% may have a role. Sympathetic blocks may be used to alleviate acute pain, accelerate healing of lesions, and prevent the development of postherpetic neuralgia. Intrathecal steroids for intractable postherpetic neuralgia may have a role. Polypharmacy provides greater reduction of pain than a monotherapy. The combination is equally effective and has no more significant side effects than monotherapy. Neuropathic pain is defined as pain caused by a primary lesion or dysfunction in the nervous system. Most of the anticonvulsant drugs have at least some ability to block sodium channel conduction (very weak evidence), and this is often cited as a likely antiepileptic mechanism of action because of the presumed mechanism of action of carbamazepine and phenytoin.

Pharmacologic properties of anticonvulsants:

Anticonvulsant	Possible mechanisms	Comment
First-generation anticonvulsants:		
1- Carbamazepine	1- Slows recovery rate of voltage-gated Na$^+$ channels. 2- Minor Ca^{2+} channel antagonist effect. 3- Chemically related to tricyclic antidepressants.	Induces hepatic enzyme activity.
2- Valproate	1- Slows recovery rate of voltage-gated Na+ channels, limits repetitive firing. 2- Increases GABA levels through effects on various enzymes.	
Second-generation anticonvulsants:		1- Second-generation drugs have wide therapeutic to toxicity windows, except for felbamate, which may cause fatal aplastic anaemia and hepatic failure. 2- Unlike first-generation drugs most do not induce hepatic enzyme activity.
1- Gabapentin	1- Increases brain GABA levels. 2- Binds to α-2-δ subunit of voltage-gated Ca^{2+} channel. 3- Inhibits branched chain AA transferase.	1- Gabapentin is not protein bound and is not metabolised. 2- Gabapentin is the only drug excreted intact through the kidney.
2- Lamotrigine	1- Limits repetitive firing of Na$^+$ channels by slowing of recovery rate. 2- Inhibits neurotransmitter release (glutamate, aspartate, GABA, acetycholine).	
3- Felbamate	1- Blocks voltage-gated Na$^+$ channels. 2- Inhibits responses to NMDA-evoked activity. 3- Enhances GABA-evoked activity.	
4- Topiramate	1- Blocks voltage-gated Na$^+$ channels. 2- Affects activity at kainate and AMPA subtype of the glutamate receptor. 3- Has positive modulatory effect on GABA receptors.	

It is possible that these drugs have efficacy because of an undefined selectivity for certain kinds of sodium channels. The tricyclic antidepressants, which have proven efficacy in neuropathic pain, are fairly good sodium channel blockers. The relief of neuropathic pain that occurs with tricyclic antidepressants does not result from their antidepressant properties. There is also no evidence that tricyclic antidepressants are more effective for one type of pain than another. Gabapentin treatment is associated with significant pain relief for sensory pain (e.g., throbbing pain, shooting pain, cramping pain, burning pain), but not gnawing pain. Gabapentin treatment is associated with significant improvement in all of the affective pain items (i.e., tiring, sickening, fearful, punishing). Gabapentin has analgesic activity against a broad range of sensory and affective pain qualities in patients with postherpetic neuralgia and painful diabetic neuropathy. Gabapentin reduces continuous pain (e.g., burning), intermittent pain (e.g., sharp, shooting), and dynamic allodynia (stimulus-evoked pain), and increases cold pain thresholds (e.g., thermal thresholds) in patients with peripheral and central neuropathic pain. Gabapentin reduces continuous pain, intermittent pain, and dynamic allodynia in patients with neuropathic cancer pain. Pharmacotherapy for the treatment of neuropathic pain includes 3 general categories of medications, specifically, opioids, nonopioids, and adjuvant analgesics. Opioids produce analgesia through the pre- and post-synaptic interaction with μ-receptors within the spinal cord dorsal horn and CNS. In the case of neuropathic pain, the μ-receptor sensitivity is down-regulated; as a consequence, μ-receptor stimulation may require large doses of opioid compounds, leading to adverse effects such as sedation, constipation, nausea, and tolerance. Nonopioid analgesics, which include first- and second-generation NSAIDs and acetaminophen, can be used for treating mild to moderate pain in patients with cancer. When used in conjunction with other agents, including opioids, antidepressants, corticosteroids, and anticonvulsants, additional pain relief may be obtained. Adjuvant analgesics, such as antidepressants and anticonvulsants, are defined as medications that have a primary indication other than for pain, but that are also effective analgesics. Traditional models of neuropathic pain treatment are based on the use of tricyclic antidepressants (e.g., amitriptyline, nortriptyline, and desipramine) for continuous pain-burning pain.

Adverse effects of anticonvulsants:

	Systemic effects	Neurotoxic effects	Rare Idiosyncratic Reactions
First generation:			
1- Carbamazepine	Nausea, vomiting, diarrhoea, hyponatraemia, rash, pruritus, vision, fluid retention.	Drowsiness, dizziness, blurred or double vision, lethargy, headache.	Agranulocytosis, Stevens-Johnson syndrome, aplastic anaemia, hepatic failure, dermatitis or rash, serum sickness, pancreatitis.
2- Valproic acid, Divalproex, sodium	Weight gain, nausea, vomiting, hair loss, and easy bruising.	Tremor	Agranulocytosis, Stevens-Johnson syndrome, aplastic anaemia, hepatic failure, dermatitis or rash, serum sickness, pancreatitis.
Second generation:			
1- Gabapentin	Gastrointestinal upset, oedema (non-pitting).	Somnolence, fatigue, ataxia, dizziness.	
2- Lamotrigine	Rash, gastrointestinal upset.	Dizziness, tremor, ataxia, diplopia, headache.	Stevens-Johnson syndrome.
3- Felbamate	Anorexia, nausea, aplastic anaemia, hepatic failure.	Irritability, insomnia, headache.	
4- Topiramate	Fatigue, gastrointestinal upset, renal calculi.	Cognitive difficulties, tremor, dizziness, ataxia, headache.	

PAIN CONTROL

Pain is a fact of modern life and can be a direct result of day-to-day interaction with people, jobs, physical activity, ageing, sedentary lifestyle and relationships. Patients who suffer from chronic pain are at a disadvantage, since they do not usually get a well balanced, multidisciplinary regimen. Medical models of chronic pain management have failed to produce a reliable, effective treatment. The immune system produces mediators, which can either transmit pain signals to the brain or cause inflammation. Those with chronic pain have elevated levels of destructive inflammatory mediators that must be suppressed, as they set up a chronic pain feedback loop with the brain. An interaction between the brain and the immune system causes pain signals to remain permanently turned on resulting in further chronic pain. Some of the most recognisable substances that cause pain include prostaglandins, cytokines (immunotransmitters), and peptides such as substance P, glutamate and nitric oxide. The principal substances that produce inflammation include histamine, kinins, leukotrienes, complement and PGs. Prostaglandins (PGs) and leukotrienes are compounds that are produce via the metabolism of fats in the diets. They can be either good or bad. The bad ones are very destructive, having roles in increasing cancer, heart disease and inflammation. The high amounts of saturated fat give rise to the production of the bad prostaglandins and leukotrienes. The enzymes that are involved in the production of the destructive PGs and leukotrienes (LTs), called cyclooxygenase-2 (COX-2) and lipoxygenase (LOX), are the targets of both medical and nutritional intervention in order to suppress these inflammatory substances. NSAIDs and steroids have significant side effects that can occur with long-term use. NSAIDs induce nausea, liver disorders, kidney damage, ulcers, skin rashes, and bone marrow suppression. Their long-term use can worsen the arthritis since they interfere with GAG (glycosaminoglycan) synthesis. The steroid associated side effects include infection, delayed wound healing, peptic ulcer, Cushing syndrome, osteoporosis and fractures, muscle weakness, diabetes and psychosis. The nutritional approach for the treatment of chronic pain is a comprehensive approach, including: 1- nutrients that inhibit the immune system's inflammatory cytokines, 2- nutrients that inhibit the bad PGs and LTs, 3- nutrients that increase endorphins (the natural pain-relieving substances in the brain), 4- nutrients that are involved in the repair of tissue once it is damaged from inflammation, 5- nutrients that provide antioxidant support, since inflammation that damages tissues gives rise to increased free radicals, and 6- nutrients that can assist in relieving the stress and anxiety associated with chronic pain conditions. The lipoxygenase inhibitors include taxifolin, turmeric, boswellia, perilla seed extract, ω-3 fatty acids, and pycnogenol. The COX-2 inhibitors include CAPE (caffeic acid phenethyl ester), resveratrol, stinging nettle, willow bark, turmeric, and ω-3 fatty acids. The nutritional agents suppress the inflammatory cytokines include stinging nettle, vitamin D (inhibits IL-1 and IL-12), ω-3 fatty acids (reduce IL-1 and IL-6), vitamin E (increases levels of IL-11, IL-6), curcumin (decreases IL-1 and IL-8), pycnogenol (decreases NO), and quercetin (inhibits IL-8). Repair Nutrients include NAC (helps boost glutathione), glucosamine (helps rebuild hyaluronic acid and cartilage), chondroitin sulfate (rebuilds cartilage and is anti-inflammatory, SAMe (s-adenosyl-L-methionine, is important for cartilage synthesis, is analgesic and has antidepressant activity), niacinamide (assists in chondrocyte activation, an important element in cartilage synthesis), antioxidants (Taxifolin, vitamin C, tocotrienols/tocopherols, folic acid, carotenoids, CoQ_{10}, α-Lipoic acid, polyphenols, selenium, zinc, manganese, copper). The nutrients for treating the depression, anxiety and stress include 5-HTP (boosts serotonin levels), St. Johns wort (increases serotonin levels), Rhodiola rosea (helps maintain serotonin levels and is adaptogenic), SAMe (boosts serotonin levels), phosphatidylserine (decreases destructive cortisol levels that are elevated during chronic stress), kava kava, valerian and passionflower (reliable sedative and anti-anxiety). The comprehensive approach should use anti-inflammatories, antioxidants, antidepressants and tissue repair nutrients together for future direction towards complete rehabilitation. There are many paths and directions to an illness, and there is no single drug or nutrient has the capability of addressing the many pathways to the condition.

ANALGESICS-INDUCED PAPILLARY NECROSIS

Nonsteroidal antiinflammatory drugs (NSAIDs) block the synthesis of cyclo-oxygenase (COX) products of arachidonic acid (AA), which have a critical modulatory role on renal haemodynamics, renal epithelial cell fluid and ion transport, and the synthesis and action of renal hormones. Withdrawal of NSAIDs usually leads to prompt reversal of the acute renal failure (ARF). During plasma volume depletion, counterregulatory renal prostaglandins are released that counteract vasoconstrictors and normalise RBF. NSAIDs taken under these circumstances blunt this counterregulatory response and intensify the renal vasoconstriction leading to ARF. In congestive heart failure and hepatic failure with ascites, there is circulating neurohumoral vasoconstrictors, and the use of NSAID may lead to ARF by augmenting arteriolar constriction. Eicosanoids or oxygenated metabolites of AA exert modulatory influences on many ion transport sites along the nephron, that is interrupted by NSAID use, which leads to a wide variety of disorders of ion transport with the result of the sodium retention. Natriuresis rapidly ensues once the drug is discontinued. NSAIDs antagonise both thiazide and loop diuretics. Hyperkalaemia is the second major electrolyte disorder that accompanies NSAID use. NSAID-induced hyperkalaemia seldom occurs in the absence of other defects on potassium homeostasis. NSAID action is the suppression of prostaglandin-mediated renin release leading to a state of hyporeninaemic hypoaldosteronism. Acute interstitial nephritis syndrome with or without minimal-change glomerulopathy occurs after 2-18 months of NSAID therapy and may sufficiently serve as to require dialysis. Most cases are reversible and are characterised pathologically by a mononuclear cell infiltrate of lymphocytes and plasma cells. The mechanism of action may be related to a reactive non-COX product of AA metabolism. The long-term use of NSAIDs can cause renal papillary necrosis and renal insufficiency. Renal papillary necrosis induced-by NSAIDs is enhanced by caffeine. Neonatal renal failure deaths were reported with 150-400 mg of indomethacin/d for 2-11 weeks during pregnancy. NSAIDs induces ARF and serious fluid and electrolyte disorders. Renal papillary necrosis and chronic renal insufficiency can occur secondary to prolonged use of NSAIDs. Neonatal renal failure and renal death may occur from use during pregnancy. NSAIDs on the market today inhibit PGH synthase, the enzyme involved in the conversion of AA to PGG_2 and PGH_2. PGG_2 and PGH_2 are the 2 intermediates that subsequently result in the formation of PGE_2, $PGF_{2α}$, TXA_2, and PGI_2. PGH synthase is composed of 2 isoenzymes having somewhat different reaction properties (PGHS-1 and PGHS-2). PGHS-1 is expressed constitutively in almost all tissues, whereas PGHS-2 is expressed in response to cytokines, growth factor, or tumour promoters. Any manoeuvre that decreases medullary blood flow would lead to ischaemia of the deeper parts of the kidney and, thus, could result in papillary necrosis. Ischaemia could still occur easily in the face of normal blood flow if the affinity of haemoglobin for oxygen is decreased. An agent well known to affect oxygen affinity is 2,3-diphosphoglyceric acid (2,3-DPG). The salicylates decrease erythrocytes 2,3-DPG in a dose-dependent manner could easily lead to renal ischaemia and infarction. Phenacetin causes methaemoglobinaemia and enhances sulfhaemoglobin formation, either of which could cause renal ischaemia and papillary necrosis. The reason for multiple vasodilators synthesised by the kidney that act locally in a paracrine or autocrine fashion is to prevent medullary tissue oxygenation from falling below a PO_2 of 10-20 mmHg. This low medullary PO_2 is always present if urinary concentrating mechanisms are operating in a normal fashion, but it is obvious that further additional stress could easily result in papillary necrosis. The medullary thick limb synthesises the vasodilator, NO, and inhibition of its synthesis results in medullary infarction. Another vasodilator, made by the distal tubule is urodilantin. Adenosine is released with ischaemia. Adenosine causes medullary vasodilation and cortical constriction. High corticomedullary concentrations of the drug could be directly toxic to renal tubular cells in general and papillary cells in particular. The concentration of the drug at the papillary tip is several-fold higher than that seen in the cortex. The salicylates inhibit gluconeogenesis and decrease protein synthesis probably due to an inhibition of amino acyl-t-RNA synthetase. Cell ATP levels are decreased. Inhibition of glucose-6-phosphatase activity and altered membrane permeability. Free fatty acids (FFAs) cause Ca efflux from kidney mitochondria, to increase cytosolic Ca, which could irreversibly damage cells. The metabolites of acetaminophen bind to proteins of the renal parenchyma and deplete the cell of glutathione. Free radical formation also results from acetaminophen and its more stable metabolites, 3,5-dimethyl acetaminophen. Indomethacin causes an increase in the papillary concentration of sphingomyelin, phosphatidylcholine, phosphatidylserine, phosphatidylethanolamine, and phosphatidylinositol. The daily use of single agent at common doses is not likely to induce analgesic nephropathy in patients with normal renal function. Caffeine is ingested regularly either in the form of coffee, its congeners (tea, cola, chocolate), or in over-the-counter preparations. Caffeine inhibits phosphodiesterase enzyme; increases cAMP levels; directly stimulates calcium-activated Ca^{++} release channels, resulting in an increase in the cytosolic Ca^{++} concentration.

MULTIPLE DISCIPLINARY INTERVENTION FOR DIABETIC NEUROPATHY

The patient is the ultimate loser. There is a vast amount of sound medical research that is ignored due to lack of or limited profit potentials. The first step in the conversion of the essential fatty acid (EFA) linoleic acid is γ-linolenic acid (GLA), is broken in diabetics, due to deficiency of the enzyme δ-6-desaturase and/or δ-5-desaturase; resulting in shortage of GLA and its metabolites, prostacyclin (PGI_2) and prostaglandins (PGs). The aim of analgesia is to reduce pain and inflammation by stopping production of the PGs. Incorrect use of analgesic creates break in the conversion of GLA to PGI_2, resulting in removing some pain, but worsening the underlying neuropathy by exacerbating a prostacyclin deficiency. Aspirin inhibits prostacyclin production for 3-4 hrs, piroxicam for 3-4 days. One aspirin every 3 days maximising PGI_2 production and minimising COX-2 prostanoid TXA_2, which causes hypertension. The very low levels of PGI_2/PGs in diabetics make RBCs fragile and unable to be deformed and squeezed into capillaries, resulting in endoneural hypoxia especially in the extremities (irritation and pain). Agents promote PGI_2 (prostacyclin): 1- GLA, 2- pentoxifylline, 3- ACE-inhibitors, 4- EPA (eicosapentaenoic acid) from fish oil, 5- Ginkogo biloba extract, 6- vitamin C, and 7- vanadium. Aspartame metabolises into formaldehyde and methanol. Within 6-8 weeks GLA makes RBCs more deformable, regenerate veins/capillaries (increases endoneural capillary by 22%) and encourage nerve growth. Evening primrose oil (EPO) provides AA for the production of PGI_2, and stimulates COX-1 expression in some tissues. Ascorbyl-GLA is 40 times as efficacious as GLA as a treatment for neuropathy. GLA plus ascorbate is >70% as efficacious as ascorbyl-GLA. Ascorbate is excreted from the body every 5 hrs, thus it is advised to take it together with GLA. The synergy between GLA and racemic α-lipoic acid (1.3:1 ratio) results in correcting motor nerve conduction velocity and endoneural blood flow, i.e., reverses the effects of the broken neurotrophic mechanisms that correlate with diabetic neuropathy. Pentoxifylline (Trental, 400 mg) is used for peripheral vascular disorders (PVD) including diabetes-induced PVD. In diabetes it acts much like prostacyclin because it causes PGI_2 to be liberated by the body. Trental have a more immediate effect, EPO is slower, but has a more complete spectrum of effects in managing the fatty acid deficiency that diabetics suffer from. Pentoxifylline suppresses the production of TNF-α, which is a major cause of vicious circles of insulin resistance (obesity → TNF-α → insulin resistance → more insulin → more obesity → more TNF-α → more insulin resistance). Excess levels of TNF-α accounts for >40% of insulin resistance. High TNF-α levels cause COX 2 overexpression that may cause increase in levels of TXA_2 of NIDDM resulting in hypertension and attendant vascular complications. Ginkgo Biloba (Egb) is preferred to Trental, it is as efficacious in its production of PGI_2 and is a powerful antioxidant preventing retinopathy and treating macular degeneration. ACE-inhibitors are also used as protective against renal and cardiac complications of diabetes. Vanadium also suppresses TNF-α-induced insulin resistance. DM is a free-radical-associated disease; diabetics have significantly accelerated levels of oxidative stress. 200 mg of α-lipoic acid qid for 4 months results in improvement of cardiac autonomic neuropathy in NIDDM, protects against oxidative stress-induced insulin resistance, and improves and may prevent diabetic neuropathy. Thioctic acid (600 mg) is superior to NGF in the induction of neurotrophic support. ALA is a coenzyme essential for energy production, a superb antioxidant that recycles vitamins C, E and important enzymatic antioxidant glutathione, and antihyperglycaemic agent too. Hyperglycaemia significantly diminishes glutathione levels lowering defences against oxidative stress. NAC is a precursor of GSH; dietary NAC inhibits the development of functional and structural abnormalities of peripheral nerves in diabetes. NAC reduces lipoprotein (α) by 24% in NIDDM. α-Lipoic acid increases intracellular GSH and recycles it. Ginkgo biloba extract has SOD liberating properties. SOD and GSH prevents hyperglycaemia-induced free radical cell damage and may help reduce the diabetic vascular complications. Vitamin E has a synergistic effect with GLA, protects prostacyclin, anticoagulant, may restore depressed PGI_2, dilates blood vessels, and has the ability to bind oxygen. Vitamin E acts as an anticoagulant in the veins by preventing clotting of platelets. Vitamin E prevents haemolysis of RBC. α-Tocopherol provides protection against electrophiles-extremely reactive radicals implicated in LDL-cholesterol oxidation and coronary heart disease. Vitamin E reduces protein glycosylation. Carnitine is an amino acid. Acetyl-L-carnitine (ALC) may be the most useful in treating neuropathy and repairing neural damage. Chromium affects cellular absorption and makes a more effective use of the insulin. Biotin levels is significantly reduced in DM. Biotin may have a role in the prevention and treatment of peripheral neuropathy. Niacin (B_3) restores electrical polarity of blood cells, and vasodilate capillaries. Diabetics over-excrete inositol. Inositol plays a role in the fat metabolism, and may prevent nerve fibers from excess glucose. Taurine deficiency may lead to diabetic cardiomyopathy. Taurine supplementation reduces total proteinuria and albuminuria by 50%, attenuates diabetes-induced hyperglycaemia, prevents arrhythmias, stabilises platelets, lower high blood pressure, retards cholesterol-induced atherogenesis. Taurine reduces renal oxidant injury with decreased lipid peroxidation and less accumulation of AGEs within the kidney. Under the prominent oxidative stress in the body, taurine is converted to taurine chloramide, which reduces TNF-α overexpression hence, increase insulin sensitivity. The aim of dietary intervention for diabetics: 1- To make RBCs more deformable (change of lipid characteristics). 2- To induce angiogenesis. 3- To encourage nerve growth. 4- To increase oxygen delivery capacity of circulation. The vascular endothelium is a highly active metabolic and endocrine organ, secretes prostacyclin, TX and other essential substances. The vascular endothelium is easily damaged. Zinc, taurine and magnesium are protective and critical nutrients for maintenance of endothelial integrity. GLA from EPO is best. ALC, GLA + aLA and NAC must be taken between meals on an empty stomach (1 hr before or 2 hrs after eating) and some nutrients must be taken together (e.g., GLA + vitamin C, GFT chromium + Niacin).

Direction to control PN pain:

1- GLA (130 mg) + α-lipoic acid (100 mg) on an empty stomach.
2- GLA (130 mg) + vitamin C (500 mg) + NAC (600 mg) on an empty stomach later in the day.
3- Taurine (1500 mg) + magnesium orotate/aspartate for at least 2 weeks.
4- Antioxidants.
5- Trental (400 mg bid), taper gradually to zero once EPO has the same effect (2-3 months).
6- Vitamin E (400 iu bid) with fatty meal.

Mixture for maintenance-level antioxidants:

Antioxidant	Dose
1- β-carotenoids	15,000 iu (Mixed carotenoids).
2- Vitamin A	10,000 iu.
3- Vitamin C	2000 mg bid + cofactors (e.g., quercitin).
4- Vitamin E	400 iu bid.
5- Selenium	200 µg daily.
6- NAC	600 mg daily.
7- α-lipoic acid	100 mg bid.
8- Ginkogo Biloba extract	120 mg daily.
9- B-complex	

Ascorbate is transported into cells via 2 mechanisms. Ascorbate can be brought into the cell directly through an undetermined Na^+-dependent transporter. Ascorbate can also be oxidised outside the cell, allowing it to be transported by several of the glucose transporters into the cell, where it is reduced to ascorbate again. DHA has a structure very similar to that of glucose, this allows several of the GLUT family to act as DHA transporters. DHA uptake and its subsequent reduction can account for an increase in ascorbate accumulation of 5-20 fold within of minutes of insulin stimulation (ascorbate recycling), allows the cell to respond quickly to oxidative stress. Diabetic patients tend to have a condition termed micro-scurvy, meaning that the cells have a vitamin C deficiency. The amount of vitamin C entering the cells of diabetic patients through the GLUT proteins can be reduced in 2 different ways. If there is decreased GLUT4 translocation to the membrane, less DHA will be able to enter the cell via the transporter. If the flux of glucose into the cells is increased, as it is in diabetes, then less vitamin C will be able to enter the cells. Acupuncture has been shown to offer relief from chronic pain (neuropathy). Biofeedback is a technique that helps a person become more aware of and learn to deal with the body's response to pain. Biofeedback emphasizes relaxation and stress-reduction techniques. With guided imagery, a person thinks of peaceful mental images, such as ocean waves. Patient's condition can be eased with these positive images. Chromium is needed to make glucose tolerance factor, which helps insulin improve its action. A deficiency in magnesium may worsen the blood sugar control in type 2 diabetes, interrupts insulin secretion in the pancreas, increases insulin resistance in the body's tissues, and contributes to certain diabetes complications. Vanadium is a compound found in tiny amounts in plants and animals. Vanadium normalizes blood glucose levels in type 1 and type 2 diabetes, modestly increase in insulin sensitivity, and decreases their insulin requirements.

INSULIN RESISTANCE SYNDROME

Insulin Resistance (IR) is central to the pathogenesis of cardiovascular metabolic syndrome, which is known by a number of other names, including Syndrome X, Reaven's syndrome, the Deadly Quarter, metabolic cardiovascular syndrome, atherothrombogenic syndrome, and cardiovascular dysmetabolic syndrome. Metabolic syndrome arises from a number of causes, including predetermined genetic factors such as IR as well as acquired or lifestyle characteristics such as obesity, physical inactivity, and high carbohydrate diets (>60% total calories). Increased IR is inversely related to decreased urinary uric acid clearance, which leads to a dramatic increase in the rate of gout. More than 50% of women with polycystic ovary syndrome are insulin resistant; treatment with insulin-sensitising medications can lead to a resumption of ovulation, return of fertility, and reduction in hirsutism. Acanthosis nigricans, hyperpigmentation of the skin often in the neck and axilla, is also correlated with IR. This finding is most common in children and young adults with IR and DM risk. Apple-shaped body types, or central abdominal obesity, are made up of metabolically active fat and associated with high insulin levels, IR, high mobilisation rate of free fatty acids, and increased appetite. IR helps to increase insulin levels and promote fat storage. IR and its resulting metabolic syndrome contribute to a prothrombotic and proatherogenic state. Plasminogen activator inhibitor (PAI-1) produced by the liver and endothelial cells, inhibits fibrin degradation by plasmin and enhances clot formation. PAI-1 increased levels are found in atherosclerotic lesions. High levels of triglycerides, very-low-density lipoprotein, and oxidised low-density lipoprotein stimulate the production of PAI-1. PAI-1 levels correlate with increased body mass and high plasma insulin levels, whereas plasma insulin levels are reduced when endogenous insulin levels are reduced by exercise, weight loss, and/or insulin-sensitising medications such as metformin and the thiazolidinediones (TZDs). Hyperinsulinaemia leads to increase renal sodium resorption, potentially expanding circulating volume and increasing vascular resistance leading to hypertension. Other cardiovascular effects include increased vascular smooth muscle proliferation, greater responsivity to angiotensin II, and greater sympathetic activation. Endothelial dysfunction is correlated with decreased NO production and peripheral vasodilatation in the muscle tissue. Refining foods result in ill-health. The causes of syndrome X are, may be 2-fold and the decline in testosterone with age, and sugar and refined carbohydrates. The principal features of syndrome X are hypertension, abnormal glucose tolerance, increased VLDL triglyceride levels, decreased HDL cholesterol levels, obesity and hyperinsulinaemia/insulin insensitivity. These are all cardiovascular risk factors (CAD). Hyperinsulinaemia is the common factor, linked with each of the six risk factors, and reflects decreased insulin sensitivity. There is a positive association between hyperinsulinaemia and hypertension. Hyperinsulinaemia increases sympathetic neural activity and, as a consequence, blood pressure. Increased insulin levels is associated with increased circulating catecholamine concentration. Increased very low-density lipoprotein triglyceride and decreased HDL cholesterol concentrations are features of syndrome X and non-insulin dependent diabetes mellitus (NIDDM). An elevated LDL/HDL ratio is predictive of CAD in diabetic subjects. Increased VLDL triglyceride in NIDDM is primarily due to increase in hepatic synthesis. VLDL triglyceride production rates are correlated with insulin levels. Insulin has a direct stimulatory effect on hepatic VLDL triglyceride synthesis. A combination of hyperinsulinaemia and decreased insulin sensitivity, particularly at the liver, is required for the sustained increase in circulating VLDL triglyceride levels to develop in syndrome X. Impaired adipose tissue insulin sensitivity in NIDDM and the increase adipose tissue mass in obesity are responsible for the increase on circulating, non-esterified fatty acid levels in these conditions, and this increases the supply of substrate for hepatic VLDL triglyceride synthesis. Lipoprotein lipase is the principal enzyme involved with catabolism of VLDL triglyceride. There is a deficiency of lipoprotein lipase in NIDDM. Decreased HDL cholesterol levels in NIDDM are due to an increased rate of clearance, primarily mediated by hepatic lipase. Fasting insulin levels correlate with the rate of HDL cholesterol clearance. There is an inverse correlation between circulating HDL cholesterol levels and whole body insulin resistance. Acute elevation of triglyceride levels can induce insulin insensitivity. Circulating non-esterified fatty acid (NEFA) and triglyceride levels change simultaneously. An increase in circulating NEFA levels will impair insulin sensitivity, and vice versa. Upper body obesity and lower body fat distribution are features of male and female sexes, respectively. An increased waist to hip ratio is a useful index of upper body obesity. Increased visceral fat stores can be distinguished from those with increased abdominal subcutaneous fat. Only the former have an association with increased plasma triglycerides and blood glucose levels, following an oral glucose load, and are associated with syndrome X, NIDDM, hypertension, decreased plasma HDL cholesterol and increased plasma triglyceride levels. Insulin insensitivity is a feature of obesity but peripheral insulin insensitivity is more marked in obese subjects with upper versus lower body fat distribution. There is an increase in the circulating levels of plasminogen activator inhibitor 1 (PAI-1) activity, which is a CAD risk factor. Insulin has a direct atherogenic role. Insulin level is a sensitive but not a specific test. Triglyceride levels are well correlated with insulin levels, so if TG is elevated, it's likely insulin level is as well. Cigarette smoking increases IR, as does inactivity and obesity. 80% of the body's insulin-mediated glucose uptake, takes place in muscle and are enhanced by physical activity. Exercise reduces IR by 40%, and the effects persist for up to 48 hrs after the activity. Exercise aids in weight loss, reduces blood pressure, and improves lipids. Regular aerobic physical activity is one of most effective therapies to help decrease IR and prevent the development of DM. Weight loss improves insulin sensitivity and lowers blood pressure. Eating a mixture of protein, carbohydrate, and fat at each meal reduces glycaemic load. Eating frequent, high-fiber, small meals, foods with a low glycaemic index, and smaller serving sizes reduces glycaemic load. Vitamin E 400-1000 IU/day may improve insulin sensitivity and lipids as well as offer antioxidant activity. Chromium supplementation, a trace mineral, may improve insulin sensitivity at a dose of 200-1000 µg/day. Folate and vitamin B_6 supplements can help reduce homocysteine levels, a proatherogenic substance. Dyslipidaemia and hypertension must be aggressively treated to minimise risk of cardiovascular disease. Aspirin counteracts the proinflammatory and prothrombotic effects of IR. Intervention: 1- weight loss (wholesome unrefined food), 2- medium protein and restricted carbohydrate diet, 3- increase in physical activity, exercise, 4- nutriceuticals-vanadium, chromium, vitamin E, and 5- endocrine evaluation (hormone therapy-which may include testosterone). Pancreas transplant patients exhibit marked and sustained hyperinsulinaemia with putative tissue resistance to insulin action.

Directions to control DM and PN:

1- Trental 400 mg bid.
2- Before breakfast (empty stomach):
      GLA (130 mg) + vitamin C
      ALC (1000 mg)
      NAC (600 mg).
3- At breakfast meal:
      CoQ_{10} (120 mg) daily.
      GTF chromium (200 µg) daily.
      Taurine (1500 mg) for 2 weeks then tapers it.
      α-Lipoic acid (100 mg).
      Antioxidants.
4- Before the lunch meal (empty stomach:
      GLA (130 mg) + α-lipoic acid (100 mg).
5- At lunch:
      Biotin (5 mg).
      Vitamin E (400 iu).
      α-Lipoic acid (100 mg).
      Antioxidants.
6- At dinner meal:
      Mg-orotate/aspartate (100 mg elemental Mg).
      α-Lipoic acid (100 mg).
      Antioxidants.

Effects of insulin resistance on dyslipidaemia development:

Normal Effects of Insulin on Lipids	Effects of insulin resistance on Lipids	Resulting dyslipidaemia
1- Suppresses nonesterified fatty acid (NEFA) release from adipose tissue.	Less NEFA suppression, converted by liver to triglycerides.	Elevated triglycerides.
2- Suppresses hepatic very-low-density lipoprotein (VLDL)-triglyceride secretion.	Increased VLDL-triglyceride secretion postprandial.	Elevated triglycerides, formation of small, dense highly atherogenic LDL.
3- Activates lipoprotein lipase (LPL), reducing postprandial triglycerides, and causes transfer of cholesterol to high-density lipoprotein (HDL).	Reduced activation of LPL, causing elevated triglycerides, decreased transfer of cholesterol to HDL.	Low HDL, elevated triglycerides.

THE GLYCAEMIC INDEX

The glycaemic index is a useful tool that measures how fast a particular food is likely to raise the blood sugar. In case of hypoglycaemia or that it may occur during exercise, it would be preferred to eat carbohydrates that raise the blood sugar quickly. Eat extra carbohydrate with a lower glycaemic index and longer action time is preferable when blood sugar is expected to drop during a few hours of mild activity. The glycaemic index numbers are based on glucose, which is the fastest carbohydrate available except for maltose. Glucose is given a value of 100; other carbs are given a number relative to glucose. Faster carbs (higher numbers) are great for raising low blood sugars and for covering brief periods of intense exercise. Slower carbs (lower numbers) are helpful for preventing overnight drops in the blood sugar and for long periods of exercise. For example, for white bread, simply multiply the numbers below by 1.42, i.e., glucose would have a glycaemic index of 142. The impact a food will have on blood sugars depends on many other factors like ripeness, cooking time, fiber and fat content, time of day, blood insulin levels, and recent activity. Using the glycaemic Index as just one of the many tools available to improve glycaemic control. The glycaemic index of food is a ranking of foods based on their immediate effect on blood glucose (blood sugar) levels. The GI ranks carbohydrates according to their effect on blood glucose (sugar) levels. For the same amount of carbohydrate, foods with a lower GI raise blood glucose less than those with higher GI values.

Calculating the GI:

Food type	Carbohydrate (g)	% Total Carbohydrate	GI	Contribution to Meal GI
150ml orange juice	13	24	46	24% x 46 = 11
2 wheatflake biscuits	21	38	69	38% x 69 = 26
150ml milk	7	13	27	13% x 27 = 4
1 slice toast	13	24	70	24% x 70 = 17
Total	54			Meal GI = 58

GI:

BEANS
Baby lima 32
Baked 43
Black 30
Brown 38
Butter 31
Chickpeas 33
Kidney 27
Lentil 30
Navy 38
Pinto 42
Red lentils 27
Split peas 32
Soy 18

CEREALS
All Bran 44
Cheerios 74
Corn Bran 75
Cornflakes 83
Flakes 80
Muesli 60
NutriGrain 66
Oatmeal 66
Puffed Wheat 76
Puffed Rice 90
Rice Bran 19
Rice Chex 89
Rice Krispies 82
Shredded wheat 69

MILK PRODUCTS
Chocolate milk 34
Ice cream 50
Milk 34
Pudding 43
Soy milk 31
Yogurt 38

COOKIES
Crackers 74
Oatmeal 55
Shortbread 64
Vanilla Wafers 77
Rice cakes 52
Rye 63

DESSERTS
Cake 67
Banana bread 47
Bran muffin 60
Danish 59
Sponge cake 46

FRUIT
Apple 38
Apricot, canned 64
Apricot, dried 30
Apricot jam 55
Banana 62
Banana unripe 30
Canteloupe 65
Cherries 22
Dates, dried 103
Fruit cocktail 55
Grapefruit 25
Grapes 43
Kiwi 52
Mango 55
Orange 43
Papaya 58
Peach 42
Pear 36
Pineapple 66
Plum 24
Raisins 64
Strawberries 32
Strawberry jam 51
Watermelon 72

GRAINS
Barley 22
Brown rice 59
Buckwheat 54
Bulger 47
Chickpeas 36
Cornmeal 68
Couscous 65
Hominy 40
Millet 75
Rice, instant 91
Rice parboiled 47
Rye 34
Sweet corn 55
Wheat, whole 41
White rice 88

JUICES
Agave nector 11
Apple 41
Grapefruit 48
Orange 55
Pineapple 46

SWEETS
Honey 58
Jelly beans 80
M & M's Choc peanut 33
Skittles 70
Snickers 41

PASTA
Brown rice 92
Gnocchi 68
Linguine, durum 50
Macaroni 46
Macaroni and cheese 64
Spaghetti 40
Vermicelli 35
Vermicelli, rice 58

Carbohydrate foods that breakdown quickly during digestion have the highest glycaemic indexes. Their blood sugar response is fast and high. Carbohydrates that breakdown slowly, releasing glucose gradually into the blood stream, have low glycaemic indexes. Low GI means a smaller rise in blood sugar and can help control established diabetes. Low GI diets can help people lose weight and lower blood lipids. Low GI diets can improve the body's sensitivity to insulin. High GI foods can help re-fuel carbohydrate stores after exercise. Low GI diets are easy to apply in practice. Substituting half of a day's total carbohydrate with low GI instead of high GI foods on average results in a 15-unit reduction in the overall diet GI. High carbohydrate foods at each meal should be considered, low GI food should be chosen instead of a high GI food for each meal (e.g. pasta or sweet potato instead of potato, oatmeal instead of a high GI breakfast cereal, apples, oranges and pears instead of tropical fruits, Basmati rice instead of Calrose rice). Low GI diets can improve both glycaemic control and quality of life in diabetics. Low saturated fat is important. The GI of a mixed meal or diet can be predicted when the GI values and carbohydrate content of foods in the meal are known. The prediction is best when meals are not high in fat, as high fat content tends to reduce the impact of carbohydrate foods on blood glucose levels. Determining the GI of a food with precision is not easy. Because the GI is a reflection of the combined effect of several physiological processes, it must be determined using human subjects (in vivo testing). To determine a food's GI rating, measured portions of the food containing 50 g (or 25 g in some cases) of carbohydrate are fed to 10 healthy people after an overnight fast. Finger-prick blood samples are taken at 15-30 minute intervals over the next 2 hours. These blood samples are used to construct a blood glucose response curve for the 2-hr period. The area under the curve (AUC) is calculated and reflects the total rise in blood glucose levels after eating the test food. The GI rating (%) of the test food is calculated by dividing the AUC for the test food by the AUC for the reference food (white bread or glucose) and multiplying by 100. The average of the GI ratings from all ten subjects is published as the GI of that food. In vitro (test tube) methods are also being used to estimate the GI of a food, but are in fact crude estimates for true GI values. One of these in vitro tests measures rapidly available glucose (RAG) (i.e. the glucose released within 20 minutes of the start of an incubation of food plus enzymes). The RAG test cannot distinguish the effect on GI of adding viscous fibre or acid to a food, and does not distinguish effects that alter gastric emptying like acidity, osmolality or concentration of sugars. It also fails to pick up subtle effects on the degree of gelatinisation with small changes to moisture and heating. Optimal GI function plays a fundamental role in the full spectrum of health and illness. With a total surface area of 300-400 m^2 (about the size of a tennis court)-the gut mucosal barrier plays an important dual role: 1- protecting the systemic circulation from the entry of undesirable bacteria and toxins, and 2- allow essential nutrients free passage into the bloodstream. The barrier can become too leaky once the mucosal layer is damaged by chronic or acute illness, allowing pathogens to seep through and contaminate the rest of the system (translocation). The resulting bacterial translocation can make patients more susceptible to organ failure. Microbes such as lactobacillus prevent the colonization of their pathogenic cousins, but they are also more sensitive, being easily decimated by antibiotic treatment. As much as 70- 80% of the body's immunoglobulin-producing cells are produced within the gut. IgA is a primary immunoglobulin that acts like a sentry, guarding against the intrusion of potentially damaging foreign agents. IgA prevents the adherence of bacteria and viruses to the mucosal epithelium to defend against systemic invasion. The balance of the natural gut microflora (yeast and bacteria), digestion and absorption, faecal IgA and other crucial parameters of gut health can be evaluated by noninvasive GIT assessment. Intestinal permeability test is a urine analysis, which evaluates the integrity of the gut mucosal barrier, utilising a specialized lactulose/mannitol challenge drink. The recovery percentage of each marker, combined with the ratio of lactulose to mannitol, provides an indication of permeability and malabsorption. Both of these assessments are important clinical tools for uncovering the gut source of many chronic symptoms and illnesses. The metabolism of carbohydrates, proteins and fats into energy is referred to as oxidation. Energy is formed and released at different stages during 2 cycles: glycolysis and the citric acid cycle. If carbohydrates and amino acids are oxidised too slowly (slow oxidation) in one cycle or too quickly in another cycle (fast oxidation), energy production is reduced. The most common symptoms of a fast or slow oxidative rate are fatigue, emotional duress of some type, lowered resistance to infections, a low body temperature, gall bladder or liver problems, and being over or under weight. The rate of oxidation and energy production in turn affects the individual mental, emotional, behavioural, and in some cases, physical characteristics. The slow (ketogenic) oxidation tends to be of the alkaline, hypoactive quality. Slow oxidisers tend to have very little appetite, an aversion to heavy proteins and fats, low but steady energy levels, depression, digestive problems due to lack of hydrochloric acid production, calcium deposits, poor fat metabolism, apathy, lethargy, repressed emotions, introversion, belching, premature aging, and often feel cold. A slow oxidiser often does not eat a heavy breakfast. High intensity, short duration exercise is poorly tolerated, and for the slow oxidiser prefers low intensity and long duration (aerobic). Slow oxidisers have problems metabolising carbohydrates (hyperglycaemia). Slower oxidisers have lower levels of blood lipid (cholesterol, triglyceride) and citric acid cycle intermediates and higher levels of pyruvate and lactate. They tend to be able to hold their breath for a relatively long period and have a relatively lower pulse rate. The fast (glucogenic) oxidation characteristics tend to be of the acid, hyperactive quality. Fast oxidisers tend to have strong appetites, crave and do well on heavy proteins and fatty foods, tend to get hyper yet feel exhausted underneath, feel anxious, nervous, jittery, have severe emotional ups and downs, feel too warm, irritable, impatient, are competitive and usually extroverted. Fast oxidisers tend to have low blood sugar (reactive hypoglycaemia) and higher levels of blood cholesterol and triglyceride and citric acid cycle intermediates. They tend to be unable to hold their breath a long period (one can consider the fast oxidiser functionally anaemic due to low oxygen capacity in the blood) and have a relatively faster pulse rate. Exercise should be of high intensity and short duration (anaerobic) if normal or underweight but aerobic (walking, biking, etc.) if overweight. Bilirubin is commonly found in the urine. In general meal that is predominantly carbohydrates should be avoided. All trans-fats (hydrogenated vegetable oils) should be avoided. Common pesticides, paints, and chemicals can disturb the energy producing abilities of tissues. The metabolites of protein-derived sugar are stored in the liver as glycogen and are converted to glucose when sugar derived directly from carbohydrates in the diet runs out. The gradual digestion of protein keeps an adequate and continuous glycogen (and thus blood sugar) reserve. Alcohol depletes glycogen storage in the liver causing an increase in blood sugar. Alcohol also increases the demand for carbohydrates (by being directly broken down into acetyl CoA of the tricarboxylic acid cycle) and the resultant nutrients needed to metabolise it. The diet contains enough of the correct type of protein at every meal (about 1 g/Kg/2.2 pounds of ideal body weight a day). An easy way to calculate the amount is to divide the ideal body weight by 15 to get the number of ounces of cooked meat to be consumed per day. Example: 150 pound Ideal body weight = 10 ounces).

Foods:

Foods	Fast Oxidiser	Slow Oxidiser
Recommended	1- Eat a full breakfast. Eat frequently. 2- Fats/proteins: all meats (especially beef, lamb, and venison), fish (especially tuna and salmon) and fowl, especially high fat, high purine (adenine) types: such as anchovies, brains, meat gravies, soups, heart, herring, caviar, kidney, liver, sweetbreads, mussels, sardines, tuna, and meat extracts. 3- Foods with moderate purine content include meat, shellfish (clams, crabs, lobster, oysters, shrimp), asparagus, cauliflower, spinach, lentils, yeast, whole grain breads and cereals, beans, peas, mushrooms, and peanuts. 4- Nuts and seeds: almonds, walnuts, peanuts, peanut butter, and sunflower seeds. 5- Carbohydrates: cauliflower, beans, peas, lentils, broccoli, barley, corn, sprouted grains (sprouting destroys the phytates that bind calcium). 6- Supplements: Your supplement should contain vitamin A, vitamin C, vitamin E, vitamin B_{12}, niacinamide, calcium pantothenate, bioflavonoids, choline, inositol, calcium, phosphorus, iodine, zinc, and carnitine.	1- Eat a light breakfast (that contains protein) and restrict calcium. 2- Proteins: low fat, low purine variety such as selected fish, chicken, turkey, eggs, and low fat dairy. 3- Carbohydrates: vegetables. 4- Supplements: emphasise activated vitamin B_1, B_2, and B_6, niacin, and potassium citrate, magnesium citrate and chloride, copper, manganese aspartate, and iron. PABA, vitamin C and D, and chromium.
Allowed in moderation	1- Proteins: milk, buttermilk, cottage cheese, eggs. 2- Vegetables: root vegetables (carrots, beets, yams, potatoes, radishes, onions), lettuce, green peppers, cabbages, pickles, cucumbers, and tomatoes.	1- Whole fruits, lean beef, lamb, natural and whole grains, breads and cereals, cold-processed non-hydrogenated vegetable oil.
Avoid	1- Sweets and starches:- simple carbohydrates like glucose, maltose, fruit juices, honey, corn syrup, highly glycaemic foods like white bread, white rice, soft drinks, catsup, and meals consisting mainly of starches and sugars. 2- Miscellaneous: spices, sauces, alcohol, and caffeinated drinks such as coffee, colas or tea. 3- Supplements: limit vitamin B_1 (thiamine) and vitamin B_3 (niacin) because they increase Coenzyme A and accelerate carbohydrate oxidation; vitamin B_2 (riboflavin) and vitamin B_6 (pyridoxine) because they increase the breakdown of amino acids leading to a faster citric acid cycle activity and more CO_2 generation; glucogenic amino acids (Alanine, glycine, and serine), and citrates.	1- High fat or high purine proteins: fatty red meat, salmon, tuna, herring, anchovies, high purine proteins such as liver, caviar, meat concentrates, artichoke hearts, and modest purine containing foods such as beans, peas, lentils, cauliflower, spinach, and asparagus. 2- Fatty foods: lard, butter, oils, fatty meats, nuts, avocado, high fat pastries low in flour such as cheese cake, Danish, torts, peanuts, and peanut butter. High fat content dairy products like cheese and cream. 3- Carbohydrates: sugars, fruit juices, alcoholic beverages, and meals consisting mainly of starches and sugars.

ONCOLOGY

There has been a significant increase in the incidence of cancer during the past 100 years. The immune system is an integrated body system that is compromised of various organs, tissues, cells, and cell products, which mediate host defense against pathogenic cells (e.g., neoplasms), microorganisms (e.g., bacteria, virus, fungi, etc.) or foreign substances (e.g., bacterial toxins). Protective immune responses occur as a result of interaction of vast array of effectors cell types including T and B lymphocytes, NK cells, macrophages, dentritic cells and granulocytes. These effector cells arise from precursor cells that originate from either haematopoietic or primary lymphoid organs such as the bone marrow and thymus. The development and function of the immune system depends on the cellular production of and response to intracellular cytokines. Cytokines act as autocrine, paracrine, and juxtacrine regulators of cell function. Cytokines such as IL-1, IL-6, TNF-α exhibits an endocrine mode of action. When cytokines interact with their receptors, these receptors initiate intracellular signal cascades that ultimately determine the overall cellular response. Chemokines play major roles in guiding leukocyte migration into within tissue sites, including inflamed tissues. Neoplasms may be benign or malignant. The early the malignant tumours appear, the more immature they are, the more malignant are their growth and the worse the prognosis and cure for malignant tumours. Malignant tumours should be diagnosed early, before infiltration and metastasis, and treated by radical surgery. The general practitioner (GP) is responsible for early diagnosis. TNM classification is anatomical classification system developed by UICC (international union against cancer) based on anatomical extent of the disease as determined by all diagnostic methods findings observed at surgical exploration and results of pathological examination of the operative specimens, the tumour is then staged. Essentials for staging are: complete Hx and Ex, Biopsy, X-rays examination may be useful for staging or patient management, CT scan, special X-ray, multichemistry screen may be useful for future staging system or research, panendoscopy, studies of immune competence. Characteristics of tumour include exophytic, superficial, moderately infiltrating, deeply infiltrating, ulcerated, extends to overlies bone, and destruction of the bone (pathological). Cancer can have several methods of extensions: spread within the wall, spread into the regional lymphatics, direct invasion of adjacent organs, haematological spread. Grading: Broder's histopatholgical classification is an index of malignancy based on the fact that the more undifferentiated or embryonic the cells of a tumour, the more malignant is the tumour.

TNM classifications:

TNM	Meaning	Description
cTNM	Clinical staging.	Based on information from the history and physical examination, laboratory and imaging tests, endoscopy and biopsy, and surgical exploration. cTNM uses all information available before the initiation of definitive intervention.
pTNM	Pathologic staging. Includes: resected tumour (pT), lymph nodes (pN), and distant metastasis (pM).	Based on information obtained before intervention, supplemented by information from surgery and pathologic specimen, i.e., after pathologic review. pTNM is determined after surgery when the true extent of the disease is known and treatment decisions can be made.
rTNM	Retreatment staging.	Based on information available after a disease-free interval or at the time of a second look surgery, i.e., at the time of retreatment.
aTNM	Autopsy staging.	Based on information available at the time of postmortem examination, i.e., on autopsy. aTNM may help to answer about tumour's response to treatment, recurrence patterns, and the extent of disease at the time of death.

The higher the grade, the faster the tumour growth and worse is the prognosis:

Grade I	Well differentiated, contain one-fourth undifferentiated cells.
Grade II	Moderately differentiated, contains one half undifferentiated
Grade III	Poorly differentiated, contains three fourth undifferentiated cells.
Grade IV	Undifferentiated or dedifferentiated or anaplastic, all cells are undifferentiated.

NEOPLASIA

Cancer is defined as a disease where normal cells change and undergo rapid and uncontrolled division leading to the development of a malignant growth. Tumour cells characterised by the up-regulation of oncogenes and the down regulation of tumour-suppressor genes. Cancer development requires the cumulative actions of multiple events: 1- the induction of DNA mutation in somatic cell (initiation); 2- the stimulation of tumourigenic expansion of cell clone (promotion); and 3- the malignant conversion of the tumour into cancer (progression). In up to 50% of all clinical cancers, metastases are present by the time of diagnosis. Adhesion molecules determine cell shape and polarity. Growth factors may act on the cell of origin (autocrine function), nearby cells (paracrine function), or distant cells (endocrine function). Binding of the specific growth factor to its receptor (GFR) causes the receptor to send a signal into the cell, activating other molecules in the cytoplasm and the nucleus. Messenger RNA (mRNA) made in the nucleus from template DNA (transcribed), contains the genetic information that directly codes for specific proteins. In the nucleus, DNA is organized into regulatory segments (promoters, enhancers) and protein coding sequences (open reading frames; ORFs). ORFs are further organized into exons and introns; the former code for structural units, protein, and RNA, whereas the latter are spliced out during RNA processing, allowing for diversity in genetic response. Communication between the nucleus and other parts of the cell is bidirectional. Mutations in DNA produce altered or lost genes. Signal bases may be changed (point mutations), or segments of DNA may be lost (deletions) or rearranged (translocation). Dominant mutations result in increased production or lack of regulation of molecules and a gain in function. Recessive mutations result in loss of a molecule (due to premature termination of transcription) or structural problems. Dominant/negative mutations produce an abnormal molecule that is able to complex with a normal counterpart, but the unit in inactive. Growth factor receptors have an extracellular (EC) domain, an α helical transmembrane segment, and an intracellular domain. The EC domain binds ligand; the transmembrane helix anchors the protein; and the intracellular (IC) domain (a tyrosine kinase) phosphorylates proteins on tyrosine residues using ATP. The signal transduction pathway composed of the phosphatidylinositol/ITP$_3$/DAG/Ca^{2+} system, protein kinase C, and the GTP-binding proteins (the ras proto-oncogenes), carries signals from GFRs to the nucleus. When ligand binds to GFR, phosphatidylinositol is split by phospholipase C into inositol triphosphate (ITP$_3$) and diacylglycerol (DAG). ITP$_3$ releases Ca^{2+} from IC vesicles, which activates a number of enzymes and structure-motility proteins. It also directly stimulates a number of membrane proteins, including GFRs and protein kinase C. Activated protein kinase C phosphorylates proteins on serine and threonine, which either stimulates or inhibits their activity. Stimulation of protein kinase C is responsible for the increase in c-fos and c-myc expression seen after many GFs bind to their GFRs. GFRs stimulate protein kinase C either by directly activating PLC, which raises membrane DAG levels, or by activating G proteins, which subsequently stimulate PLC and increase membrane DAG. Proteins that bind to GTP are called G proteins; binding activates the protein. Activation stops when GTP is converted to GDP, a process stimulated by GTP-activating protein. The principal G proteins involved in signal transduction following GFR stimulation belong to the ras family, which are proto-oncogenes encoded by c-Ha (Harvey)-ras, c-Ki (Kirsten)-ras, or N-ras.

Levels of cancer prevention:

Primary prevention	Relate to initiation of cancer.	Decreases the vulnerability of a healthy individual or population to illness or dysfunction through health promotion strategies and specific protection recommendations. The avoidance of exposure to carcinogens (pollutants, chemicals, radiation, viruses), tobacco use, changes in diet, and the administration of specific agents (antioxidants).
Secondary prevention	Relate to promotion of cancer.	Defines and identifies high-risk individuals and populations, including those with precursor lesions or syndromes, and consists of early diagnosis, early detection, screening, and treatment of early stages of disease. The prevention of promotion by cessation of smoking, changing diet, and the administration of specific agents (antioxidants).
Tertiary prevention	Relate to progression of cancer.	Minimises morbidity resulting from permanent or irreversible disease by preventing complications. Tertiary prevention consists of arresting, removing, or reversing a premalignant lesion to prevent recurrence or progression to cancer.

Incidence of cancer in human:

Cancer	Male	Female
Breast	-	26%
Prostate	19%	-
Lung	22%	10%
Colon and rectum	14%	16%
Urinary tract	9%	4%
Leukaemia and lymphomas	8%	7%
Uterus	-	11%
Ovary	-	4%
Oral	4%	2%
Skin	3%	2%
All other	18%	14%

When ligand binds to GFR, cytoplasmic ras protein binds GTP, translocates into the membrane, and then shuttles to PLC or other as yet unidentified effectors and activates them. They in turn activate PKC. The signal decays when ras GTP-activating protein activates hydrolysis. Point mutations at 12, 13, or 61 alter critical amino acids at the GTP binding site, so GTP-activating protein is unable to stimulate hydrolysis of bound GTP, and the ras protein produces a continuous proliferative signal. Oncogenes cause malignant transformation by overproduction of a normal or mutant oncoprotein. Grading and staging of neoplasms are attempts to describe the degree of malignancy and the dissemination of the cancer. Virtually every cell type in the body is capable of transforming into a malignant cell. 5-10% of patients diagnosed with cancer each year are found to have a malignancy from an unknown primary site. Patients and their families who are facing cancer from an unknown source present unique challenges. Although most staging classifications are based on the anatomic extent of disease, other criteria are included for specific malignancies. The objectives of staging including: 1- to provide the necessary information for individual treatment planning, 2- to give prognostic information, 3- to assist in treatment evaluation, 4- to facilitate the exchange of information and comparative statistics among treatment centers, and 5- to stratify individuals who may be eligible for clinical trials. The non-solid tumours do not confirm to solid tumour staging principles because of their disseminated nature.

EPITHELIAL-MESENCHYMAL TRANSITIONS (EMT) IN CANCER PROGRESSION

Epithelia form continuous sheets of tightly adhering cells. Epithelia grow in aggregates with little cell motility, characterised by specialized organelles, tight junctions, adherns junctions and desmosomes that are responsible for the tight intercellular contracts between the individual cells. Distinct proteins are expressed on the basolateral or apical surface and free diffusion of membrane protein to all surfaces of the cells is inhibited. Polar epithelial cells have evolved special mechanisms that allow the transport of membrane proteins to either the apical or basolateral surface. Adherens junction are specialized structures containing the transmembrane cell adhesion molecule E-cadherin that recognises and binds E-cadherin present on the neighbouring cells in a Ca^{2+}-dependent manner. E-cadherin is a 120-kD transmembrane glycoprotein, of which an extracellular 80 kD tryptic fragment can be released in the presence of Ca^{2+}. E-cadherin is the prototype of a family of Ca^{2+}-dependent cell adhesion molecules, close relatives are N-cadherin (neurones) and P-cadherin (placenta and epithelia), M-cadherin (muscle), OB-cadherin (osteoblasts), and LI-cadherin (liver). The cytoplasmic portion of E-cadherin interacts with the α-, β-, and γ-catenins (plakoglobin). Other cytoskeletal proteins (vinculin, α-actinin, radixin, ezrin, and moesin) are located on the cytoplasmatic side of the junctional complex that anchor the actin-containing filaments. The integrity of this junctional complex is critical for the maintenance of the functional characteristics of epithelia. Its disruption leads to the dissociation of epithelial sheets, a change in cell morphology to a more flattened shape, and to an increased motility of the cells. A controlled loss of epithelial character concomitant with dissociation and increased motility of the cells is a prerequisite for normal morphogenic processes. Loss of epithelial differentiation and the acquisition of mesenchymal characteristics, like the ability to move, correlate with the malignancy of carcinoma cells. Hemidesmosomes are located on the basal surface of epithelia, and are junctions contacting the basement membrane. The basement membrane separates the epithelial cell compartment from underlying mesenchymal cells and is formed by both cell types. Characteristics constituents of basement membranes are laminin, collagen IV, nidogen/entactin and basement proteoglycan. The mesenchymal reticular lamina, which lies below the basement membrane contain collagen type I and III as well as fibronectin. The epithelial receptors for basement membrane components are located on the basal surface of the plasma membrane. Mutations giving deregulated expression of such receptors as well as the formation of autocrine loops are observed in carcinoma cells and contribute to the formation of tumours. Mesenchymal cells in contrast to epithelial cells are generally loosely associated, and are surrounded by extracellular matrix (ECM). The major adhesive interactions of mesenchymal cells occur with cell substrate. Mesenchymal cells produce particular integrin receptors, such as $\alpha_5\beta_1$, which enables them to adhere to the extracellular matrix within the connective tissue. Mesenchymal cells also express specific ligands for various epithelial receptor tyrosine kinases, such as HGF or KGF, and thereby play an important role in the control of epithelial growth, morphogenesis and differentiation. E-cadherin is important for maintenance of morphology and adhesion in differentiated epithelia and also for the acquisition of epithelial characteristics early in embryogenesis. The integrity of the adherens junction also requires functional catinen molecules. The N-terminus of β-catenin interacts with α-catenin, and via α-catenin the cytoskeletal proteins bind to the complex. γ-Catenin (plakoglobin) is similar in structure to β-catenin, and it interacts in an analogous manner with E-cadherin and α-catenin. The ligand for the c-met tyrosine kinase is called scatter factor (SF), also called hepatocyte growth factor (HGF) as it can induce growth of hepatocytes. SF/HGF has a unique structure closely resembles proteases like plasminogen. SF/HGF has no catalytic activity. All known biological activities of SF/HGF are mediated by the c-met receptor. The c-met receptor is widely expressed on epithelial cells, both during development and in the adult, whereas SF/HGF is produced usually by mesenchymal cells. SF/HGF can induce EMT. SF/HGF activates the intrinsic morphogenic potential of various epithelial cell types, but does not instruct the cells to form one particular structure. SF/HGF induces morphogenesis or motility of epithelial cell, and also invasiveness into collagen matrices. The different morphogenic programs can only be induced by SF/HGF in cells with intact epithelial characteristics, including expression of E-cadherin and functional adherens junctions. SF/HGF induces metastasis to the lymph nodes. Morphogenesis requires an intact epithelial program, whereas metastatic behaviour is observed in cells that have already undergone EMT. Various molecular mechanisms can interfere with the integrity of the adherens junction and can induce epithelial cells to assume mesenchymal characteristics, which correlates with dedifferentiation of the cells and increased metastatic potential. In human carcinoma, the loss of epithelial morphology is an important prognostic marker and correlates with a poor outcome of the disease. The loss of epithelial differentiation is particularly prominent at invasive fronts of the tumours, where the carcinoma cells break into the surrounding mesenchymal tissues. The invading cells can lose their epithelial appearance, become spindle-shaped and fibroblast-like with reduced number of desmosomes. The prominent change in morphology of malignant carcinoma cells is associated with a loosening of intracellular adhesion. This is a consequence of a functional disturbance of cell-cell contacts. The malignant cells down-modulate intercellular adhesiveness, the loss of intercellular adhesion is a crucial step in carcinoma progression. Carcinoma cells with a well-differentiated epithelial morphology do not invade a collagen matrix, and produce E-cadherin, whereas carcinoma cells with a fibroblast-like morphology are mobile, invade collagen matrices and have lost E-cadherin. More than 90% of all malignant tumours are carcinoma, and thus of epithelial origin. Aberrant growth and the ability to invade the underlying tissues are intrinsic properties of the fatally altered cells. Loss of epithelial morphology and the acquisition of mesenchymal characteristics are typical for carcinoma cells late in tumour progression and correlate with metastatic potential. In carcinoma, downregulation of E-cadherin correlates with poor survival rates.

PRINCIPLES OF GENETICS AND CANCER

Genes are the fundamental units of heredity. They are the DNA molecules that contain the instructions for cell growth, division, and structure. The length of genes is commonly reported in terms of 1000 bp or kilobase pairs (kb). Genetic information from DNA is copied to a molecule of ribonucleic (RNA), known as mRNA. The information encoded in the base sequence of mRNA is translated to the amino acid sequence of a protein. Each amino acid is coded by a sequence of three bases, called a codon. The sections of the gene that contain the protein code are termed exons and the intervening sections are called interons. Mutation takes place at the time of cell division, if a mistake is made in the synthesis of DNA. One or both of the daughter cells will have a different DNA code at the mistake point. Mutations that occur in germ cells (sperm and ovum) are called germ line mutations and transmit the mutant gene to the individual's offspring. Mutations that occur in other tissues are called somatic mutations and may alter the growth of that tissue but are not transmitted to the person's offspring. Mutations in germ line cells can predispose to cancer in the individual's offspring, where mutations in somatic cells lead to cancer in the organ involved. Nonsense mutations occur when a base pair is completely lost. Since the amino acid code is in units of three bases, the loss of one base leads to a shift in the reading frame so that all subsequent amino acid codes will be wrong and incorrect amino acids will be inserted into the polypeptide chain (frameshift mutation), e.g., BRCA1 and BRCA2 genes. Each chromosome is divided into two arms (long and short) by a constricted region called a centromere. Under the microscope individual genes cannot be visualised, but using special stains bands can be seen at different distances from the centromere. Bands may contain many genes. The short arm is p (petite), and long arm is q. Sometimes in meiosis homologous chromosomes may exchange sections of DNA and this event means that there is an exchange of genetic material. This is called recombination, or crossing over, and the resulting chromosome has a new combination of genes. The recombined sequence of genes will transfer to the new offspring a mixture of the genes from the grandfather and grandmother. Crossing over is an important concept in oncology to understand because some genetic testing for inheritance of cancer-related genes is done with linkages analysis. It is sometimes possible to identify sections of DNA called markers that are not associated with disease, but are located close to the disease gene (i.e., linkage). Linkage analysis can be used to determine whether an individual has inherited the disease gene. When the disease gene is dominant the parent of either gender with this gene may donate an autosomal disease gene. When inheritance is recessive, both parents must donate a disease gene because the disease will not be expressed unless both genes are abnormal. The frequency of disease expression is called penetrance. The basic cause of cancer is damage to specific genes that are responsible for regulating the cell's growth. Tumour suppressor genes are recessive at the cellular level, i.e.; both genes must be abnormal. The p53 tumour suppressor gene is the most common gene mutation in cancer, being found in half of all cancers. The protein from a normal p53 gene halts the cell cycle in the G1 phase, before DNA replication occurs. This allows the cell to repair any DNA damage. If the DNA damage is too great to be repaired, the p53 protein induces apoptosis. When p53 is mutated, cells do not repair the DNA damage and continue to replicate. Oncogenes are mutant forms of proto-oncogenes; genes that are involved in basic mechanisms regulating cell growth. Their mutant forms lead to protein that stimulates growth and division. Oncogenes are often dominant at the cellular level, meaning only one copy is required for the tumour development to occur. When DNA replication occurs within the nucleus of a cell, this process is monitored by a system that is designed to detect mistakes in replication before the new DNA is completed. If error occurs when small loops emerging from the double helix cause two strands not to be precisely aligned. The misalignment leads to short, repeated segments of DNA termed micro-satellites.

p53

Under normal conditions, very little p53 protein is found in most cells, which is not essential for normal cell function but is needed in special circumstances, prevents the cell from progressing into the S phase and replicating cellular DNA, so, it acts as a checkpoint and consequently blocking cell proliferation. The p53 gene has been mapped to chromosome 17q13 in the human genome and has been shown to be composed of 11 exons. This autosomal gene encodes a 53 kDa nuclear phosphoprotein of 393 amino-acids, which acts as a transcription factor and participates in 2 separate but independent processes during the cell cycle, DNA repair and apoptosis. The wild type p53 (i.e., normal type) acts not by interfering with progression of cells through the cell cycle but by acting as a checkpoint and consequently blocking cell proliferation. Cells with mutated p53 may die, or survive with a corrupted genome, with the development of a less stable cell, thus increasing the chance of developing cell clones with more malignant phenotypes. Mutation results in the loss of p53 function. Cellular location of p53 during cell division: 1- about 95% of the cells shows strong nuclear staining whilst their cytoplasm is completely negative (immunocytochemical staining), 2- during mitosis p53 is dissociated from DNA and appears in the cytoplasm, and 3- once cell division is completed, p53 re-enters the nucleus and bind to the DNA. During G1 phase of the cell cycle, p53 protein accumulates in the cytoplasm and migrates to the nucleus at the beginning of the S phase. Once bound to DNA, p53 exerts its effect by inducing the transcription of another regulatory gene, producing a 21 kDa protein and termed wild-type p53-activated fragment 1 (WAF1). This protein forms a complex with G1 cyclin and cyclin-dependent kinase-2 (CDK2) protein that normally serve to drive the cell past the G1 checkpoint in the cell cycle. By blocking kinase activity, the complex prevents the cell from progressing into the S phase and replicating cellular DNA, therefore, these cells are either delayed in the G1 phase for DNA repair or die by apoptosis. This can be seen clearly when cells with wild type p53 are exposed to stimuli, which damage DNA, where there is a concomitant increase in cellular p53. Apoptosis is a critical mechanism by which unwanted or defective cells are eliminated. p53 induction prevents cell proliferation and allows repair of damaged DNA or allows cells to die by series of orchestrated steps, i.e., apoptosis. Mutation of p53 occurs in a major proportion of various human cancers, such mutation results in the loss of p53 function. Cells with mutated p53 may die, or survive with a corrupted genome, with the development of a less stable cell, thus increasing the chance of developing cell clones with more malignant phenotypes. The mutations of p53 may appear as point mutations, allelic loss, and rearrangements or as a complete deletion, which is different from that in other tumour-suppressor genes. p53 can be used as a marker to select patients for more aggressive regimens of chemotherapy. Although inherited p53 mutations are present in all somatic cells, malignant transformation is limited to certain organs and target cells. p53 mutations are capable of initiating the process of malignant transformation. The location of mutations within the p53 gene is similar to that of somatic mutations in sporting tumours. There is no evidence of an organ or target cell specificity of p53 germline mutations. For some neoplasms at target sites p53 mutations are a necessary but not sufficient event and that additional factors are required to initiate carcinogenesis, e.g., cigarette smoking, alcohol abuse. Chronic inflammatory states may also be responsible in determining organ-specific effects. The types of p53 germline mutations includes deletions (10%), splice mutations (4%), and insertions (2%). G:C → A:T transitions are most frequent, followed by G:C → T:A and A:T → G:C transversions. About 50% of the G:C → A:T transitions are at CpG sites. Most neoplasms have in common a high incidence (47-73%) of G:C → A:T transitions, of which 45-75% is located at CoG sites, suggesting a common mechanism of tumourigenesis. This may occur spontaneously or be factor mediated, e.g., through OFRs or NO, produced by NOS, in the setting of chronic inflammatory states. G:C → A:T transition mutations may also result from an enzymatic deamination by DNA methyltransferase in the absence of S-adenosylmethionine. Similar to somatic mutations, p53 germline mutations are located in highly conversed regions of exons 5 to 8, with clusters at codons 248 and 273. Somatic cells from affected family members carry a p53 mutation in one allele whereas the other allele has retained the wild-type sequence. p53 germline mutations often show a loss of the wild-type allele. A certain degree of tissue specificity may exist with respect to the loss of the wild-type p53 allele. Mechanisms of p53 induction: 1- Irradiation, exposing of normal cells to ionising irradiation results in p53 induction. Radiation therapy is more effective in killing those cells lacking wild-type p53 compared with their counterparts carrying the normal gene. 2- Agents, which damage DNA result in an increased expression of p53, such as distillates of tar from cigarette smoke, doxyrubicin. 3- Cytokine stimulation, leads to significant induction of p53 and consequent increase in cellular apoptosis, e.g., TNF-α. Inactivation of p53, include: 1- Viral inactivation, protein in virally infected cells form a complex with p53, leading to inactivation. 2- Oncogene inactivation, products of oncogenes can lead to the inactivation of p53. 3- Negative feed-back mechanism, p53 once overexpressed inhibits further synthesis of itself by negative feed-back. Location of cellular p53 can be either nuclear or cytoplasmic. mAbs in an immunocytochemical staining technique has shown that: 1- about 95% of the cells shows strong nuclear staining whilst their cytoplasm is completely negative; 2- during mitosis, p53 is dissociated from DNA and appears in the cytoplasm, i.e., cells at various stages of mitosis, shows no staining attached to the chromosomes but the cytoplasm is strongly positive; and 3- once cell division is completed, p53 re-enters the nucleus and bind to the DNA. Effects of antimitotic drugs on p53, such as mitomycin, cisplatin, include: 1- shows no staining of p53 in either the nucleus or cytoplasm, which may indicate that cellular p53 is being continuously generated and degraded; and 2- at high drug concentrations protein synthesis is inhibited, resulting in the prevention of regeneration of new p53 molecules. The mutation and overexpression of p53 has been reported in a significant proportion of human testis, prostate and bladder cancers. The overexpression of p53 has been correlated with a poor rate of survival, could be used as indices of tumour aggression. p53 mutation may correlate with a late event of cancer. Wild-type p53 is only detected in cells under stress, whereas in tumour cells, mutated p53 is constitutively expressed in large quantities. The cellular heat shock protein HSP70 is important for the stabilisation of p53. The levels of serum anti-p53 antibodies correlates with the overexpression of mutated p53 in their tumour. In all tumour eliciting anti-p53 antibody there is a circulating complex between p53 and HSP70 molecule. These observations implies that the use of HSP70 with mutated p53 peptides for vaccination may be immunologically more effective than p53 alone. Insertion of wild-type p53 with mutated p53 may lead to increased sensitivity of the tumour cells to cisplatin in cell line. Overexpression of p53 in tumour cells can lead to their increased radiosensitivity.

APOPTOSIS

Apoptosis is type of cell death that is highly regulated and critically important for maintaining homeostatic cells number. Disruption of normal cell death mechanisms contributes to the pathogenesis of some diseases, including cancer. Expression of oncogenes, such as bcl-2, and inactivation of tumour suppressor genes, such as p53, reduces the susceptibility of cells to undergo apoptosis. Apoptosis can be rapidly and reliably quantified in acquired tissue samples. A cell can die by 2 mechanisms: necrosis or apoptosis. Necrosis is a passive process resulting from direct physical or chemical damage to the plasma membrane or disturbances in the osmotic. Extracellular fluid enters the cell and results in swelling, cell lysis, and subsequently an inflammatory response. Necrosis usually affects groups of cells and consequently tends to disrupt normal tissue architecture. Apoptotic cell death is a normal physiologic process that functions in concert with cell proliferation in the maintenance of homeostatic cell populations. It is a highly regulated process, characterised by a loss of cell volume and loss of junctional contacts with neighbouring cells. The chromatin in apoptotic cells typically becomes highly condensed and eccentrically positioned within the nucleus, and ultimately the cell degenerates into smaller membrane-bound "apoptotic bodies". The apoptotic bodies are rapidly engulfed within 1-3 h by neighbouring cells or macrophages such as that an inflammatory response is not elicited. The alterations in nuclear structure observed in apoptotic cells are usually preceded by endonucleolytic DNA cleavage that initially results in the generation of large 50-300 kb fragments and subsequently to characteristic DNA "ladders" of apoptosis, which is the result of internucleosomal DNA cleavage mediated by Ca^{2+}/Mg^{2+}-dependent endonuclease. With individual steps of the ladder occurring as integer multiples of 200 nucleotides. The histological features of apoptosis include cell shrinkage, loss of junctional contacts, "halo" surrounding the cell, chromatin condensation and margination within the nucleus, and significantly the lack of inflammatory infiltrate. Apoptosis is an essential process for normal homeostasis and development of multicellular organism. Deregulated expression of the bcl-2 oncogene is frequently implicated in this process in that bcl-2 protein is detected in a wide variety of human neoplasms and the only known function of bcl-2 is the initiation of apoptotic cell death. Bcl-2 overexpression may represent an alternative molecular basis for achieving multidrug resistance. Commonly utilised chemotherapeutic agents have been shown to be potent inducers of apoptotic cell death. Bcl-2 protein blocks DNA fragmentation in response to dexamethasone. The p53 tumour suppressor gene are among the most common genetic alterations occurring during multistep carcinogenesis, result in the inhibition of apoptotic cell death induction mediated by wild-type p53 protein in response to radiation and certain chemotherapeutic agents. Whether the assessment of p53 in individual tumours is predictive of patient response to therapy or other clinical endpoints remain controversial and appears to be context dependent. The proliferative rate of most malignancies does not exceed that of their normal tissues of origin so that the growth of these neoplasms may largely be determined by the rate of cell accumulation. This rate of accumulation is influenced by the rate of a complexly regulated process, apoptosis. Apoptosis is an essential process for the normal homeostasis and development of multicellular organisms. The determination of the susceptibility of individual malignancies to undergo apoptosis in response to therapy may provide useful information for predicting patient response and for design of more effective treatment strategies. Informative biomarkers such as bcl-2 protein may provide important information regarding individual tumour cell dynamics for treatment and prognostic purposes.

OXYGEN FREE RADICALS (OFR)

OFR are continuously generated in cells exposed to aerobic environment, and removed by antioxidant defence systems. Cancer development is a microevolutionary process that requires the cumulative actions of multiple events: 1- the induction of DNA mutation in somatic cell (initiation); 2- the stimulation of tumourigenic expansion of cell clone (promotion); and 3- the malignant conversion of the tumour into cancer (progression). OFR can stimulate cancer development at all 3 stages. Free radicals are molecules with one or more unpaired electrons. An important feature of free radical reactions with non-radicals is that they result in new radicals, which leads to chain reactions. Molecular oxygen reacts easily with free radicals, to become radicals themselves, the OFR. Many other radical and non-radical molecules play an important role in mediating OFR-related effects. The reaction of OFR with biomolecules gives rise to organic radicals that can propagate the oxidative damage. Thus, the peroxidation of membrane lipids to organic peroxyl radicals initiates a chain reaction that may explain many membrane-mediated effects of OFR. Nitric oxide (NO) may react with O_2^- to form the reactive peroxynitrite anion, $ONOO^-$ and OH. NO can act as a chain-breaking antioxidant against lipid peroxidation. Non-radical reactive oxygen metabolites include H_2O_2, $HOCl$, O_3 and the single O_2.

[A] $$O_2 \xrightarrow{e^-} .O_2^- \xrightarrow{e^-} H_2O_2 \xrightarrow{e^-} .OH \xrightarrow{e^-} H_2O$$

[B] $$L\text{-Arg} \rightarrow NO. \xrightarrow{.O_2^-} ONOO^- \rightarrow .OH$$

Oxidative stress arises either from the overproduction of OFR or from the deficiency of antioxidant defence or repair mechanisms, and results in reversible or irreversible tissue injury. Activated leucocytes generate $.O_2^-$ and HOCl, which represent an important source of OFR in situ. These OFR not only mediate the killing of the target cells but also induce oxidative stress in adjacent tissue cells. Activated neutrophils stimulate mutagenesis, and oxidative stress from chronic inflammation favours cancer development in many organs. Chronic inflammation contributes to one third of the world's cancer. Cancer induction by chronic inflammation is frequently observed in ulcerative colitis, urinary bladder cancer (induced by Schistosoma haematobium infections) and hepatocellular carcinoma (induced by viral hepatitis). Increased formation of oxidative DNA damage has been found in chronic hepatitis. Smoking, the major cause of bronchogenic carcinoma by exposes the bronchial epithelium to OFR.

Major exogenous causes of oxidative stress:

Causes of oxidative stress	Oxygen free radicals
Tobacco smoke	NO., .OH
Ultraviolet light	.OH, organic radicals
Fatty acid in food	Lipid peroxides
Iron and copper ions	.OH
Ethanol	Lipid peroxides

There are other several carcinogens in the tobacco smoke, including nitrosamines, and polycyclic aromatic hydrocarbons, such as bezo(a)pyrene. The antioxidant defence systems, include: 1- endogenous enzymatic antioxidants, 2- endogenous non-enzymatic antioxidants, and 3- exogenous antioxidant molecules, such as: α-tocopherol, ascorbic acid, β carotene. The action of uric acid, which is a purine base, is 2-fold: a- it is like DNA itself, a substrate for scavenger of OFR, and b- by complexing iron ions, it prevents the generation of OFR in the Fenton reaction. Glutathione (GSH), is a tripeptide with a reactive sulphydryl group, acts on multiple levels of the antioxidant defence: a) as a scavenger of free radicals such as $.O_2^-$, .OH and lipid hydroperoxides, b) as substrate for the antioxidant enzyme GPX, and c) in the direct repair of oxidative DNA lesions. α-Tocopherol, is lipophilic, it is one of several isomers included in the term vitamin E. It protects biological membranes from lipid peroxidation. Thus it may prevent membrane-mediated effects of OFR, that play a role in free radical chain reaction induced by lipid peroxidation, in Ca^{2+}-mediated apoptosis, and in Ca^{2+}-mediated tumour promotion. Ascorbic acid can replace the superoxide anion in the iron-catalysed Fenton reaction, which might actually increase OFR production from oxygen metabolites. Vitamin C deficiency is associated with increased oxidative DNA damage in the human. The high plasma levels of uric acid in human may serve for endogenous ascorbic acid. The additional supply of vitamin C appears to have no effect on cancer incidence, either beneficial or harmful. Mechanisms of OFR-related mutagenesis: 1- Tumour initiation, the first step in carcinogenesis requires a permanent modification of the genetic material in one cell. The number of oxidative hits to DNA is about 10,000/cell/day in the human. The DNA damage from OFR is continuously removed by specific and non-specific repair mechanisms. Higher doses of OFR increase the chance that the DNA lesions may not be effectively countered by DNA repair. 2- DNA base modifications, structural damage is always caused or accompanied by chemical DNA damage and vice versa. The hydroxyl radical reacts with all components of DNA molecule, the deoxyribose backbone, the purine bases and the pyrimidine bases. The chemical alteration of the deoxyribose elements can cause the release of purine or pyrimidine bases, producing abasic sites, which have been shown to be mutagenic. .OH attack on the duplex DNA results in radicals adducts with purine or pyrimidine bases that yield a variety of end products. Many of the OFR-related DNA base modifications result in a replicative block, some can induce point mutations by base misreading at replication. 8-Hydroxy-guanine (8-OH Gua) can produce GC to TA transversions as a result of 8-OH-Gua-Ade mispairing, GC to TA transversions are frequently detected in the Ras oncogene. GC to TA transversions in the p53 tumour suppressor gene have been observed. OFR-related point mutations may lead to initiation as a first step of carcinogenesis as well as participate in tumour progression at alter stage. 3- DNA helix alterations, including: a- helical distortion, b- single-strand breaks, c- double-strand breaks, d- interstrand crosslinks, and e- chromosomal aberrations. OFR can promote proliferation in mammalian cells after the initiation of cancer development by radiation or mutagenic chemicals. While high levels of oxidative stress inhibit proliferation by cytotoxic effects, distinct low levels can stimulate cell division and promote tumour growth. OFR can selectively promote the growth of initiated cells, while having a toxic effect on the normal cell population. OFR can induce large increases in cytosolic Ca^{2+}, through the mobilisation of intracellular Ca^{2+} stores and through the influx of extracellular Ca^{2+}, may regulate the transcription of genes involved in cell growth and proliferation through a direct effect of Ca^{2+} on the gene level, or though an indirect action. The activation of Ca^{2+}-dependent protein kinases (PKC) is mediated by the OFR-related increase in intracellular Ca^{2+}. The activation of PKC and other protein kinases leads to phosphorylation and activation of S6-kinase, which is involved in the acquisition of growth competence. Both PKC and S6-kinase can regulate the activity of transcription factors via multiple phosphorylation cascades and may thus mediate many of the effects of OFR on cell proliferation. Progression include accelerate growth, escape from immune surveillance, tissue invasion and the formation of metastases. Most of these changes involve additional DNA lesions. Normal cell cycle controlled by the p53 protein, and by a cell cycle arrest that allow DNA repair before replication (OFR lead to accumulation of DNA lesions). Cell lacking functional p53 proceeds with cell divisions and thus permit DNA damage to be carried in the following generations. Thus the absence of p53 can cause continued chromosome rearrangement from the initial DNA damage. p53 protects against carcinogenesis from spontaneously generated OFR. Most cancers stimulate an immune response of variable intensity in their host organisms. OFR from activated leucocytes can cause: 1- chronic inflammation that does not eliminate the tumour cells, but enhances tumour progression; 2- apoptotic cell death through intracellular Ca^{2+} redistribution; 3- high levels of OFR induce cell death from direct cytotoxicity; and 4- too low production of OFR, and/or OFR-related cell death mechanisms are blocked in the tumour cell, the oxidative stress contributes to cancer progression by further DNA damage and growth stimulation. The bcl-2 proto-oncogene protects cancer cells from death through apoptosis induced by OFR, so it could be possible that cells that overexpress bcl-2 may escape elimination by immune system through their resistance against OFR-induced apoptosis.

DNA REPAIR MECHANISMS

1- Repair of specific DNA lesions: DNA mismatch glycosylase, recognises and removes the adenosine inserted opposite to the 8-OH-Gua. DNA lesion is kept at an acceptably low levels. The DNA glycosylase endonuclease III recognises thymine glycol and broad selection of other oxidative and non-oxidative base modifications. Several endonucleases without glycosylase activity recognise abasic sites resulting from direct free radical damage, are essential for the removal of replication block induced by OFR-related single-strand breaks. 2- General repair mechanisms, are associated with histone and non-histone proteins that afford a relative protection against nuclease and free radical-mediated attack. DNA damage from OFR stimulates poly(ADP-ribose) polymerase to produce ADP-ribose polymers, that temporarily attract and detach histones from DNA. This important mechanism may have additional implications: A- the resulting changes in DNA accessibility for transcription factors may represent a mechanism that regulates gene expression in response to oxidative stress; B- in high levels of DNA damage, bulky DNA lesions are removed by the excision nuclease (exinuclease) system. The DNA mismatch repair system recognise the base-pairing anomalies within the DNA helix induced by oxidative stress. The NAD depletion by poly(ADP-ribose) polymerase becomes important enough to interfere with ATP synthesis. As ATP depletion can induce apoptosis, poly(ADP-ribose) polymerase activation may be an alternative pathway of OFR-related apoptosis; and C- rapid depletion of NAD pools by poly(ADP-ribose) polymerase is a defensive mechanism to prevent NADPH-driven Fenton reaction in oxidative stress. OFR should be recognised as an important class of carcinogens that stimulate cancer development at multiple stages. Both ultraviolet light and ionising radiation of higher energy (X-rays, γ-radiation) stimulate mutagenesis by induces DNA damage via the generation of OFR such as hydroxyl radicals, as well as from direct induction of free radicals in the biomolecules. OFR from lipid peroxidation reactions are responsible for the association between fat intake and cancer. For fatty acids ingested with meat, iron (Fe^{2+}) is an important cofactor that enhances the production of OFR. Copper (Cu^+) is as effective as iron as a catalyst in Fenton reaction, and by direct interactions of copper with the bases of DNA. Ethanol is a major cancer risk factor that may in part act through free radicals generated during its metabolism. The intervention with antioxidants is a double-edged sword that might enhance the effect of the oxidant stress, even perfect protection against OFR by a well-balanced antioxidant action might actually stimulate cancer development through the improved survival of tumour cells. Supplementing antioxidant in a well-nourished population has not had a consistent effect on cancer incidence, either beneficial or harmful.

CYTOKINES IN CANCER

Cytokines are intercellular messengers through which tumour cells and host tissues can communicate. Among the actions of the cytokines within a tumour are stimulation of parenchymal growth, induction of a supporting stroma, ingress of new blood vessels, chemoattraction of host immune cells and dispersal of tumour cells. Tumour growth and metastasis relies on the autocrine and paracrine actions of cytokines at every stage: 1- the development of a blood supply depends on the expression of vascular permeability factor/vascular epithelial growth factor (VPF/VEGF) as well as platelet-derived endothelial-cell growth factor (PD-ECGF), the FGFs and TNF-α; 2-

proliferation of the tumour parenchyma is orchestrated by autocrine loops including EGF, TNF-α and TNF-β1 in concert with EGFR, and IGF-1; 3- stromal components such as fibroblasts proliferate in response to paracrine effects of tumour-cell-derived chemokines promote a mononuclear-cell infiltrate; and 4- the primary tumour may even have endocrine effects on distant metastases via the action of inhibitors of angiogenesis such as angiostatin. Macrophages are capable of producing a wide range of cytokines that may promote tumour cell proliferation directly or through effects on neovascularization and stroma formation. Cytokines are divided into autocrine ones and paracrine group: 1- Autocrine cytokines include: a- mitogens, e.g., EGF, TGF-α, TGF-β1, IGF-1, IGF-2; b- motogens, e.g., autotoxin (ATX), SF, other motility factors. 2- Paracrine cytokines include: a- angiogenic factors (VPF/VEGF, FGFs, PD-ECGF, TNF-α); b- antiangiogenic factors (Angiotensin, PF4); c- mitogens (PDGF); d- cytokines regulating Matrix metalloproteases [MMPs] (TNF-α, IL-1β, TGF-β); and e- cytokines regulating monocytes (MCP-1, M-CSF). The role of MMPs in malignancy is emigration of tumour cells and the immigration of vascular endothelial cells and monocytes. Monocyte chemoattractants and proliferation factors, most epithelial tumours contain infiltrate of T-cells and macrophages, compromising a significant proportion of the tumour mass. Macrophages and T-cells are cytotoxic to tumour cells. Macrophages are rich source of cytokines, they may even help in tumour growth. Too few macrophages and the tumour fail to thrive because of a lack of growth factors. On the other hand, too many macrophages or too much activation results in tumour destruction. Monocytes within tumours accumulate in response to specific attractive signals.

ENVIRONMENTAL FACTORS

Only a small fraction of cancers can be attributed to germline mutations in cancer-related genes. Examples of environmental factors that have been associated with increased cancer risk in the human population include chemical and physical mutagens (e.g., cigarette smoke, heterocyclic amines, asbestos and UV irradiation), infection by certain viral or bacterial pathogens, and dietary non-genotoxic constituents (e.g., macro- and micro-nutrients). Among molecular targets of environmental influences on carcinogenesis are somatic mutation (genetic change) and aberrant DNA methylation (epigenetic change) at the genomic level and post-translational modifications at the protein level. Changes in environment may be associated with major shifts in cancer prevalence. Comparison of p53 mutations in prostate cancers of Japanese and Western populations have also led to identification of different mutation patterns, suggesting that different aetiologic factors are involved. Mutations in oncogenes have also been associated with environmental exposure. Concurrent with the well-documented activities of environmental genotoxic agents at the genomic level is their effect on cell surface receptors and various signalling cascades. The formation of reactive oxygen species (ROS) result in altered cellular redox potential and lead to respective activation of protein kinases and subsequent changes in transcription factors. The putative mechanisms by which changes in redox potential alter protein kinases include the formation of disulfide bonds between selective cysteines on signal-transducing molecules, which results in protein dimerization; such a mechanism leads either to protein activation or to inactivation. ROS and redox potential can be considered the primary cytoplasmic changes that regulate protein kinases. The same kinases also can be activated by alterations within cell surface receptors. Cross-linking of receptors (e.g., EGFR, IGFR) and subsequent trans-phosphorylation have been shown to occur in response to physical, chemical and cytokine stimuli. UV irradiation efficiently causes dimerization of receptors, as shown for IGFR and EGFR. In the nuclei, UV irradiation causes DNA lesions that lead to the formation of pyrimidine dimers and subsequent signature mutations, that coincide with activation of DNA-damage, related signalling cascades, as documented for c-jun N-terminal kinases (JNK). Within the cytosol, UV irradiation has been implicated in the activation of various signalling cascades including protein kinase C (PKC), mitogen activated protein kinase (MAPK), and JNK with changes in downstream effectors, as shown for transcription factors cAMP-response element binding protein (CREB), jun, fos, p53 and NFκB. Activation of JNK stress kinase occurs within minutes of UV irradiation, and last from 30 min to 24 h depending on the type (i.e., UVB, UVC) and dose (i.e., amount and dose rate) of irradiation. JNK activation by UV irradiation occurs as a result of JNK phosphorylation by its upstream kinases concurrent with inactivation of redox-sensitive JNK inhibitor. Activated JNK phosphorylate key regulatory transcription factors including c-jun, ATF2 and p53. Such phosphorylation contributes to the activity and stability of JNK-associated proteins, which under non-stressed growth conditions are targeted by JNK for ubiquitination and subsequent degradation. It is the duration and magnitude of activity of stress-activated kinases and respective transcription factors that dictates whether the damaged cell will undergo growth arrest or apoptosis. In general, dietary components relevant to cancer can be divided into 3 major categories: (i) dietary constituents that are carcinogenic including aflatoxins, heterocyclic amines, N-nitroso compounds, polycyclic aromatic hydrocarbons and trihalomethane; (ii) dietary factors that promote tumour development (tumour promoters) including diverse chemical classes, such as phorbol ester derivatives, non-TPA type tumour promoters, chlorinated hydrocarbons (from industrial or agricultural sources), alcohol and salt (sodium chloride). Increased ingestion of fats and/or calories markedly enhances tumour promotion; and (iii) dietary components can also improve cellular defense mechanisms. For example, many bioactive compounds found in plants increase expression/induction of crucial detoxification enzymes, particularly glutathione synthetase, glutathione transferase and glucuronyl transferase, resulting in decreased bioavailability of potentially DNA-damaging carcinogens. The carcinogenic potential of Helicobacter pylori infection is implicated in the endogenous synthesis of nitric oxide (NO) in macrophages by the induction of NO synthase and by its genotypes, particularly the cytotoxin-associated gene A (cagA), which encodes a high molecular weight immunodominant antigen, and vacA (vacuolating cytotoxin A). Possible mechanisms by which HPV participates in the development of cancer have been shown to depend on the activity of the two viral oncoproteins encoded by the E6 and E7 genes. E6 and E7 form complexes with several cellular proteins involved in cell cycle and growth control. E6 binding to wild-type p53 promotes its degradation through the ubiquitination pathway and abrogates its transcriptional activities. E7 binds preferentially to unphosphorylated Rb, participates in cell growth regulation, and disrupts or alters cellular complexes that form between Rb and transcription factors E2F or myc. Importantly, continuous expression of HPV oncoproteins is required for inactivation of these tumour suppressor proteins. HPV genes interact with transcription factors, including AP1, NF1, Oct-1, Sp1, GRE and YY1. Molecular mechanisms that underlie E6 and E7 activities also include other cellular regulatory pathways. HPV infection by itself is not sufficient for development of cervical cancer. It is the combination of HPV infection and accumulated genomic alterations that drives malignant progression. Germline mutations play a key role in the relative risk of cancer predisposition, but for the most part, they are limited to those that meet the criteria of familial cases. Polymorphism in metabolic activation and detoxification enzymes are believed to play important roles in the acquisition of susceptibility to environmental factors. It is further believed that genetic makeup, which differs among individuals, as reflected in the ability to cope with certain carcinogens, plays an important role in determining the susceptibility and development rate of the multistep carcinogenesis process. Electromagnetic fields (EMF) either as direct current (DC) or alternating current (AC), with a field strength in the range of 0.1-0.15 V/m applied for 20 min up to several hours have been reported support bone and wound healing. In cancer therapy, high electrical fields above irreversible electrical breakdown of cell membranes have been used to induce retardation of tumour growth. Avascular tumours can remain in the body in a dormant status for up to 10 years. A single electrical field pulse resulted in an increase of the mitogenic activity of strength of 500 V/m represents the optimum beneath which no effect and above which irreversible membrane breakdown. The stimulation of mitogenic activity is accompanied by a rise of intracellular RNA, indicating gene transcription. Ca^{2+} signals have been frequently implicated in the control of the G_1, the G_2/M transition, and several events of the mitosis phase of the cell cycle. Ca^{2+} may directly initiates cell proliferation. In nonexcitable tissues Ca^{2+} release from intracellular stores is preferentially induced by inositol trisphosphate ($InsP_3$) acting on $InsP_3$-sensitive release channels levels. ROS affect various molecular components of the cell, such as fatty acids, proteins, and DNA, and an excess of ROS inevitably leads to cell degeneration and death. ROS are able to serve as secondary messengers affecting gene expression, as long as their production does not overrun a toxic threshold. The underlying mechanism apparently involves activation of tyrosine kinases and inhibition of phosphatases. Growth-related genes such as c-fos, c-myc, c-jun, and egr-1 are activated by a rise in Ca^{2+} and others by the production of ROS. The rise in intracellular ROS and the Ca^{2+} transient are involved in promotion of mitogenic activity. Oxidants promote cancer cell growth.

HORMONES AND CARCINOGENESIS

In general pathology, there are two fundamental concepts concerning the cause of a disease: aetiology and pathogenesis. Aetiology refers to the agent that evokes the disease (bacteria, radiation), pathogenesis refers to the mechanisms through which the causative agent evokes the histopathological lesions, the clinical symptoms and the laboratory signs of the disease. Carcinogenesis is a multi-step process involving many mutations in genes regulating growth control, resulting in increased stimulation (oncogenes) or removal of inhibition (tumour suppressor genes). Carcinogenesis is a continuous process, affecting cellular growth control. Mutations occur at cell divisions with a certain frequency. Mutagens increase the risk of mutation via a direct effect on the genes. Both steroid and peptide hormones could be important in carcinogenesis. Hormones are not only co-carcinogens, but also the most important carcinogens by increasing the risk of mutations in their normal target cells and at the same time stimulating the growth of the mutated cells. The racial differences in the incidence of CaP are paralleled by differences in the testosterone levels. Hormones are important regulators of growth. By stimulating proliferation, hormones may increase the risk of mutation and at same time stimulate the replication of mutated cell. Thus hormones are complete carcinogens. Androgens contribute to the age-associated increase in prostatatic carcinoma (CaP) by increasing oxidative stress. Androgens induce production of ROS and cause prolonged AP-1 and NF-κB DNA-binding activities. Antioxidant vitamin C plus E blocks both androgen-induced DNA-binding activity and production of ROS. Long-term α-tocopherol supplementation reduces CaP incidence by 32% and CaP mortality by 41% in cigarette smokers. Lycopene and selenium supplementation reduces CaP risk. AP-1 and NF-κB are ubiquitous protein complexes that mediate cellular response to various external signals by binding to distinct DNA sites. AP-1 and NF-κB are redox regulated and their DNA-binding activities are sensitive to ROS. AP-1 is ubiquitous transcriptional activator, mediates responses to external signals by regulating the expression of genes involved in growth, differentiation, and stress responses. It is composed of members of the Jun and Fos families that form homodimers (c-Jun:c-Fos) that bind to a specific DNA response element. Redox regulation of AP-1-binding activity occurs through a conserved cysteine residue in the DNA-binding domain of c-Fos and c-Jun proteins. ROS-induced injuries to DNA, proteins, and lipids may be important carcinogenic mechanisms. Damage associated with ROS has been demonstrated in CaP. Transcriptional activation by nuclear receptors, like androgen receptor, is influenced by interactions with other transcription factors. AR-induced transcriptional activity can be inhibited by members of the Jun and Fos families or can be enhanced by c-Jun. AR inhibits c-Jun-induced AP-1 transcriptional activity. c-Jun can interact with the DNA- and ligand-binding domains of the AR. NF-κB is a ubiquitous transcriptional factor involved in growth, differentiation, and stress response. It is a dimer of members of the Rel/NF-κB family. NF-κB exists in an inactive form bound to IκB proteins in the cytoplasm and thus does not require de novo protein synthesis for activation. Upon activation, IκB is degraded and the active NF-κB complex is translocated to the nucleus. Vitamin C plus E diminish androgen-induced AP-1 DNA-binding activity. H_2O_2 and superoxide radical elevate AP-1. Androgen-induced NF-κB activity is elevated much later than AP-1 DNA-binding activity. NF-κB is not sensitive to superoxide radical but induced by H_2O_2. NF-κB DNA-binding activity is decreased with high dose of antioxidants.

Antioxidants:

Endogenous factors	Endogenous enzymes	Nutritional factors
Thiols (glutathione).	GSH reductase.	Ascorbic acid (vitamin C).
Haem proteins	GSH reductase	Tocopherols (vitamin E).
Coenzymes Q	GSH peroxidases (GPX/PHGPX)	β-carotene/retinoids.
Bilirubin	Superoxide dismutase	Selenium-essential dietary component of peroxidase.
Urates	Catalase	Methionine or lipotropes for choline biosynthesis.

Antioxidant nutrient repair systems:

1- Reduction of disulphides, peroxides, quinones by GSH (with GSH reductase and GSH peroxidase).
2- Radical scavenging by the GSH/ascorbate/tocopherol system.
3- Reduction of soluble peroxides by GSH peroxidase (GPX).
4- Reduction of phospholipid membrane peroxides by GSH phospholipid peroxidase (PHGPX).
5- Phospholipid membrane integrity, methylation of ethanolamine to choline, to prevent electron loss.
6- Defence against NO nitosation in stomach by ascorbate.

An extract mixture from eight different herbs, PC-SPES is used by patients with prostate cancer at different stages as a supplement or alternative to their traditional therapy. PC-SPES seems to have substantial clinical effects in CaP patients, and its mechanism of action appears to be hormonal in nature. PC-SPES exhibits some oestrogenic effects, including loss of libido and extreme breast enlargement. Rare but potentially serious side effects include venous thrombosis, pulmonary embolism, and allergic reactions. PSA may drop by ≥80% in patients receiving PC-SPES, and ≥50% decrease in tumour volume. PC-SPES promotes apoptosis, decreases the expression of the androgen receptor, and reduces PSA levels in an androgen-dependent and androgen-independent CaP. Vitamin E has potent antioxidant properties and may also protect against cancer by enhancing immune functions, lowering the activity of protein kinase C (involved in regulating cellular proliferation), and inducing apoptosis. This vitamin inhibits the growth of CaP. Side effects of large doses of vitamin E (>800 IU) include, among others, delayed blood clotting and an increased requirement of vitamin K in vitamin K-deficient patients. Higher selenium levels are associated with a reduced risk of prostate cancer. Selenium may affect several types of anticarcinogenic activities including antioxidant protection, carcinogen metabolism, immune enhancement, and apoptosis. Selenium works synergistically with vitamin E to inhibit carcinogenesis, and vitamin E reduces the oxidative damage seen in selenium deficiency. Selenium may be toxic above 400 μg/day, resulting in characteristic garlicky breath and brittle fingernails and hair. Tomatoes are the major source of lycopene. It reduces cellular proliferation, which may be associated with a higher risk of prostate cancer. Lycopene was found in relatively high concentrations in the prostate. Dietary carotenoids might be related to prostate function and disease processes, particularly reducing prostate cancer risk. Spaghetti sauce, tomato soup, salsa, ketchup, and tomato paste may be better sources of bioavailable lycopene than the fresh tomatoes. Lycopene reduces the risk of total prostate cancer by 35% and aggressive prostate cancer by 53%. Genistein is a phytochemical found in soybeans. Foods that contain large amounts of soy are tofu, soy milk, and miso. Genistein may be a chemopreventive agent in prostate cancer. Genistein is a potent inhibitor of protein-tyrosine kinase and topoisomerase II, enzymes, which are crucial to cellular proliferation. Genistein is also an inhibitor of angiogenesis and several steroid metabolising enzymes, such as aromatase and 5 α-reductase.

POLYCYCLIC AROMATIC HYDROCARBONS AND HETEROCYCLIC AMINES IN THE DIET

Heterocyclic amines (HCAs) are promutagens or procarcinogens. They are activated by N-oxidation step involving the cytochrome P4501A2 (CYP1A2) and O-acetylation by the enzyme N-acetyltransferase NAT2 to produce an arylamine, which can then form adducts with DNA. HCAs are powerful mutagens and are carcinogenic. HCAs consists of amino-imidazo moiety that is postulated to arise from the cyclisation of creatine (of which muscle contains about 0.4% of wet weight), and a pyridine or pyrazine derivative that is probably generated by the reaction of a hexose with an amino acid by the Maillard reaction. A reaction mixture of creatine, a single amino acid or a dipeptide, and a single hexose sugar generates highly mutagenic reactants when heated. HCAs are generated by all types of muscle meats, e.g., red meat (beef, lamb) or white meat (fish, poultry or chicken). Muscle meats contain the synthetic precursors from which heterocyclic amines are generated. Creatinine that is enriched in the meat crust and in pan residues during cooking, produces more mutagenic products than creatine. Proteins do not appear to be reactive, although the dipeptide carnosine (rich in meat) reacts to produce a mutagenic product during high-temperature cooking. In meat, glucose-6-phosphatase are the main sugars present at about half the molar concentrations of creatine and free amino acids. Low glucose beef patties contain very little mutagenic activity after frying, but this activity can be restored by the addition of small amounts of glucose before frying. Excess sugars inhibit the production of mutagenically active products; and potato, starch, glucose and bread crumbs have all inhibit the mutagenic activity in fried beef patties when present in sufficient amounts, probably due to competitive reactions of other Maillard reaction products with creatinine. The presence of oils may enhance HCAs.

The most potent HCA in muscle meats:

Name	Abbreviation
2-amino-1-methyl-6-phenyl-imidazo[4,5-b]pyridine	PhIP
2-amino-3,8-dimethylimidazo[4,5-f]quinoxaline	MeIQx
2-amino-3,4,8-trimethylimidazo[4,5-f]quinoxaline	DiMeIQx
2-amino-3-methylimidazo[4,5-f]quinoline	IQ
2-amino-9H-pyrido[2,3-b]indole (or 2-amino-a-carboline)	AαC

Highest temperatures (grilling, pan-frying and barbecuing) cooking may increase HCAs. Heterocyclic amines are procarcinogens that may be converted to carcinogens through oxidation and acetylation. The oxidative enzyme P4501A2 might facilitate the unfavourable transformation of HCAs to the corresponding adduct-forming arylamine. Similar induction levels of P4501A2 have also been observed with tobacco smoke. Cruciferous vegetables are rich source of indole-3-carbinol, which is a very powerful inducer of P4501A2 activity. Thus induction of cytochrome per se is not necessarily good or bad for health. Green or black tea, polyphenols from tea-epigallocatechin gallate and theaflavine gallate reduces mutagenic effects of HCAs. Vegetable factors, pyrrole pigments, vitamin A, anthraflavic acid, ellagic acid, conjugated linoleic acid (CLA), chlorophyllin, dietary fibre and parsley reduce mutagenic effects of HCAs. Polycyclic aromatic hydrocarbons (PAHs) are derivatives of 2 or more fused benzene rings. Since PAHs are ubiquitous, it is fortunate that only a few appear to be carcinogenic. The polyaromatic hydrocarbons source is from the melted fat that drips onto hot coals and is pyrolized at the prevailing high temperature. The PAH is then deposited on the meat as the smoke rises. Cereals are the major contributors of PAH (35%), followed by oils and fats (34%), with meats contributing only 4% of the total dietary exposure. Fruit, sugar, root vegetables and other vegetables together contribute about 24-25%. A vegetarian might receive 3-9 mg/day of PAHs, while extremely heavy meat diet might consume 6-12 mg/day. Cereal foods and vegetables may be major contributors to an individual's total dietary exposure to PAH. High consumption of fruit and vegetables, and high cereal intakes may be associated with lower risks of cancer.

TELOMERASE

Vertebrates have special structure at the end of their chromosomes, known as telomeres, which are composed of 5-15 kb pairs of a guanine-rich hexameric repeat, (TTAGGG)n. Telomeres provide chromosomes with stability and protect them from exonucleolytic degradation. In normal somatic cells there is a progressive degradation of telomeres with aging. Cells lose a small amount of DNA at their terminal ends with each replication. The 3' to 5' (leading) strand of the parent DNA is copied in a continuous manner, but 5' to 3' (lagging) strand of the parent DNA is copied discontinuously. Each fragment is primed with an RNA primer, which is subsequently degraded. The fragments are then ligated by DNA-repair enzymes that operate behind the replication fork. The 3' end of the lagging strand is incompletely copied and is lost. The cell can afford to lose only a finite number of these telomeres before significant sequences of the parent DNA are lost, resulting in chromosomal instability and cell death. Telomerase is necessary for indefinite cell division in most immortal cells, but apparently unnecessary for the normal function of most somatic tissues. Telomere functions can be broadly distributed into 3 areas: chromosome positioning, protection, and replication. Germ-cell telomeres are maintained despite multiple rounds of replication. The telomere sequence is synthesised by a ribonucleoprotein called telomerase. Cellular immortality is believed to be a critical step in tumourigenesis. The activation of the enzyme telomerase is tightly associated with cellular immortality and cancer. Telomerase expression is detected in a majority of tumours, but is absent in most somatic tissues and correlates to clinical outcome in a number of cancer types. Telomerase expression is associated with the stage of differentiation but not necessarily with the rate of cell proliferation. Inhibition or absence of telomerase may result in cell crisis in cancer cells and tumour progression in cancer patients. The activation of the enzyme telomerase by cancer cell is to achieve cellular immorality. Activation of telomerase to compensate for telomere loss and maintain chromosomal integrity is an important factor in tumour's progression. Cellular immortality in germ-line cells is one of the most important criteria for the progression of species. Although telomeres may not mediate chromosome segregation, the separation of telomeres during cell division creates a special problem for the segregation. DNA polymerase can only move in one direction (5' to 3'). Thus once the RNA primer on the leading strand at the end is removed; there is no way to back-fill the missing base pairs. Left unaltered, the next replication event will result in a daughter chromatid that has been shortened by a few base pairs. Telomeres exist in at least two general forms, each of which appears to solve the problem of replicating chromosome ends in a different manner. The ends of the chromosome are prevented by the ability of an enzyme, called telomerase, to synthesise more such repeats at each end, which is mediated by a corresponding RNA template within the telomerase holoenzyme. The RNA transposition intermediate is converted into end DNA by reverse transcriptase. Telomeres participate in meiotic pairing, in meiotic and mitotic chromosome segregation, and in the organisation of the nuclear architecture. Telomere-deficient chromosomes can be transmitted both in meiosis and mitosis. Telomere defect can impair mitotic chromosome segregation. Wild-type telomeres are normally associated on sister chromatids, until metaphase and that the defective telomeres prevent a critical process in telomere separation. Thus, a specific element of the telomere repeat is required in cis to either mediate chromatid separation or to prevent persistent association. The construction of telomeres created a solution to the problem of DNA replication, while creating a problem for chromatid and chromosome separation. The solution to that problem appears to require both the telomeric DNA sequences themselves and the proteins acting at the telomere. Telomerase may be the most specific and universal cancer-marker. Most advanced human neoplasms contain telomerase activity. These telomerase-expressing cancer cells may arise either from telomerase-active stem cells involved in tissue remodelling or by the reactivation of telomerase activity in normally telomerase-silent somatic cells. Telomerase activity has been demonstrated in various human tumours such as, prostate cancer. Not all tumours express telomerase activity. The telomerase-positive tumours are larger and more aggressive and led to a decrease in patient's survival. A tumour biopsy may not manifest telomerase activity because: 1- the tumour cells remain mortal or are quiescent, 2- there may be inhibitors or RNases in the sample that interfere with the assay, and 3- cells may maintain their telomerase in the absence of telomerase activity. Telomerase RNA expression is one of the initial events in tumourigenesis but RNA levels does not correlate with telomerase activity. Telomerase activity is detected only in advanced tumours. Telomerase activity may not be obligatory for primary tumour formation. Reduction in telomere length has been reported in different tumours. Tumours of all grades express telomerase activity. All of the grade I tumours, 92% of the grade II tumours, and 83% of the grade III tumours are positive for telomerase. TRAP (Telomeric repeat amplification protocol) assay for telomerase activity may be useful in the detection of low-grade (I and II) bladder cancers in assays involving voided urine. Vitamin D receptors are present in numerous normal and malignant cells that are not involved in calcium or phosphorus metabolism. Vitamin D receptors have also been demonstrated in human prostate. Vitamin D may be capable of causing differentiation in a subset of tumours, perhaps by its ability to suppress telomerase activity. Telomerase inhibitors may be effective agents against certain types of cancer, either through pharmacological inhibition, transcriptional repressors, or genetic intervention, may interrupt telomerase activity in immortalised tumour cells and may cause these cells to reassume a finite life span. The cell can afford to lose only finite number of these telomeres before significant sequences of the parent DNA are lost, resulting in chromosomal instability and cell death. Telomerase activity may be a useful marker for tumour immortality and may be a sensitive tool for detection of the presence of malignancy.

ABNORMAL CELL CYCLE REGULATION

The cell cycle is the process by which cells replicate, and is driven by a set of enzymes and regulatory proteins that compose the cell cycle engine. The cell cycle consists of an initial growth phase (G_1), DNA replication (S), a gap phase (G_2), and mitosis (M), after which the cell may differentiate or enter the resting stage (G_0). The cycle is driven by a number of positive and negative regulatory phosphorylation and dephosphorylation events, involving protein kinases, protein phosphatases, cyclins, cyclin-dependent kinases, and cyclin-dependent kinase inhibitors that ultimately impinge on the activity of transcription factors. Unreplicated or damaged DNA blocks the progression of the cell cycle at checkpoints, including a late G_1 checkpoint regulated by the dephosphorylated retinoblastoma protein (pRb) and a late G_2 checkpoint regulated by the phosphorylation of cyclin-dependent kinase 1 complexed with cyclin B. Many cell cycle regulator genes may be considered proto-oncogenes or tumour suppressor genes, and point mutations, amplifications, deletions, or rearrangements involving their loci, especially those in the Rb pathways, are associated with various tumours. The initiation of pRb phosphorylation will arrest the cell cycle at the G_1 checkpoint. Inhibition of pRb phosphorylation may also occur after DNA damage, somehow sensed by the tumour suppressor gene p53. The p53 protein then transactivates p21, which blocks the function of the cyclin-CDK complexes. The products of the p21 family share a common N-terminal domain for the binding to and inhibiting the kinase activity of CDK-cyclin complexes. Both p21 and p57 bind to proliferating cell nuclear antigen (PCNA) through a separate C-terminal domain, inhibiting DNA replication. Ki-67 antigen is a high molecular weight nonhistone protein absent in quiescent G_0 cells, and thus is useful as a human nuclear proliferation marker. Ki-67 antigen is expressed even when DNA synthesis is blocked and cells are arrested in G_1-S or G_2-M. Carcinogenesis entails loss of cell cycle checkpoint control, with loss of control of genomic integrity, disturbances in the cell death pathway, and altered cell-cell and cell-matrix interactions. The cell cycle may go into overdrive and lead to cancer if pRb function is lost or the G_1-S check-point constraint is breached by loss of p16 activity, thereby effectively increasing cyclin D1-CDK4 activity. The various kinases are activated and deactivated in cascades of phosphorylation and dephosphorylation involving other kinases and phosphatases.

HYPERCOAGULABILITY

Patients presented with idiopathic venous thromboses in the absence of underlying risk factors may have cancer. Many of these malignancies may be at their early stages, and they escape detection by routine low sensitivities screening techniques. Hypercoagulability may be associated with a variety of malignancies of almost every organ system in the body, making diagnostic evaluation of such patients difficult and costly. Identifying such tumours at earlier stages may prove more fruitful and have favourable therapeutic and financial implications than discovery at a more advanced stage. In 1872, Armand Trousseau described a relationship between malignancy and hypercoagulability (an increased incidence of deep venous thromboses of the extremities in patients with visceral malignancies). Venous thromboses associated with cancer are frequently migratory, may involve superficial as well as deep veins, and affect unusual sites including the face, arms, and chest. Thrombosis associated with cancer tends to be refractory to standard anticoagulation therapy. Thromboembolic disease is often the earliest manifestation of malignancy. Haemostasis can be viewed as a balance between a series of cross-linked proteolytic reactions finally resulting in a fibrin clot and the degradation of this clot by a series of fibrinolytic reactions. A major inhibitor of the coagulation process, antithrombin III and its association with thrombin is thought to be a sensitive and specific indicator of the activation of the coagulation system. α-2-Antiplasmin is one of the major inhibitors of fibrinolysis, and the presence of plasmin-α-2-plasmin complexes is thought to be a sensitive indicator of the fibrinolytic state. There is an abundance of fibrin and platelet aggregation products on the surface of many tumour cells, which suggest local activation of the coagulation system. Tumour cells and their products interact with the blood's thrombin and plasmin-generating systems. Tumour cells and their products interact with platelets, leukocytes, and endothelial cells. Mechanisms for activation of the coagulation system in malignancy include direct activation of factor X by tumour cell proteases, release of thromboplastin-like substances, and destruction and necrosis of endothelial cells with subsequent exposure of the highly thrombogenic subendothelial tissue and tissue factors. Many tumour cells express the transmembrane protein tissue factor, which, when exposed to circulating factor VII, facilitates activation of factor X, leading to thrombin generation and eventual deposition of fibrin. Other tumour-derived activators of factor X include mucin and cysteine proteases, both of which are abundant on visceral tumours. Tumour cells secrete plasminogen activator inhibitor, which reduces steady-state fibrinolysis and contributing to the procoagulant state. Several types of tumour secrete potent fibrinolytic tissue type plasminogen activator. Tumours can possess fibrinogenic and fibrinolytic activity concomitantly, and it is the delicate balance between the 2 processes that determines the net effect on activation of the procoagulatory pathways. Interactions between tumour cells and monocytes mediated by adhesion molecules on the surfaces of both cell types could be responsible for inducing the expression of tissue factor. Direct interaction of tumour cells with platelets induces platelet aggregation. Factor V gene factor (factor V Leiden) increases the risk of venous thromboembolism in healthy individuals. This gene is present in 5% to 10% of Caucasian individuals. The ability to generate adhesive and proteolytic factors is essential for tumour invasion, implantation, angiogenesis, and growth. There is a reduction in morbidity associated with the hypercoagulable state once this anticoagulation therapy started. A careful history, physical examination, routine blood chemistries, chest x-ray, and measurement of the erythrocyte sedimentation rate (ESR) followed by other modalities if indicated may diagnose cancer in 10.5% of patients with idiopathic deep vein thromboses. Venous thrombosis refractory to anticoagulant therapy is a particularly sensitive indicator of an underlying malignancy (17-23%). The more aggressive approach (measurement of CEA levels, CT scanning, ultrasonography, and endoscopy) yields a higher incidence of cancers detected at initial presentation. There is a significantly higher risk of cancer in patients with documented thromboembolic disease. Despite oral warfarin therapy with International Normalised Ratio (INR) levels between 2.0 and 4.0, some patients with known malignancies will continue to present with recurrent superficial and deep venous thromboses. These patients should receive more aggressive anticoagulation with intravenous or subcutaneous heparin. Treatment of the underlying cancer with appropriate regimens, including chemotherapy and radiation is accompanied by a reduction in tumour mass and resolution of the underlying hypercoagulable state.

THE BIOLOGY OF METASTASIS

The onset of metastasis represents the beginning of an irreversible terminal stage in the life and suffering of most patients that even modern systemic therapeutic approaches cannot halt at best palliation. In advanced disease, systemic therapy offers at best palliation but rarely prolongs survival and even more rarely cures, for most solid tumour, with the exception of testicular cancer. The metastasis is a coordinated, multiple process encompassing the detachment of tumour cells from the primary site to the development of tumourigenic lesion in a secondary organ. The metastatic cascade is initiated by tumour-cell motility, adherence, and proteolysis leading to invasion. This complex process is determined by both random and selective events. Tumourgenesis and metastasis are 2 consecutive processes, being governed by equal as well as by different mechanisms. Tumour cells characterised by the up-regulation of oncogenes and the down regulation of tumour-suppressor genes might be subject to further cellular changes, resulting in a metastatic phenotype. The vast majority of tumour cells are eliminated by shear forces during passage through the capillary bed or by immune defense mechanism or because they lack certain functional requirements for successful implantation at the secondary site. Up to 50% of all clinical cancers, metastases are present by the time of diagnosis. The probability of metastatic dissemination increases with increasing tumour size. The observation of a hypercoagulable state in a cancer patient led to an inquiry into the role of the fibrinolytic system. The surgical observation of tumour vascularity highlighted the importance of angiogenesis. The significance of local regional spread of disease is exemplified in renal cancer, in which local spread of disease to the regional lymph nodes dramatically changes the overall survival from 87% for localised disease to 57% for regional disease. In contrast, in prostate cancer, surgery or radiotherapy can be effective for the treatment of regional disease, but effective therapy is needed for metastatic disease. 5 year survival of patients with distant disease is 29% as compared with 85% for regional disease. Metastasis is regulated by, cell-cell and cell-extracellular matrix (ECM) receptors, proteolytic enzymes necessary for invasion of the basal membrane, the vasculature, different organs, motility factors essential for migration through the tissues, receptors required for organ-specific invasion, growth factors, which will support the growth and development of tumour microcolonies in the secondary organ, and angiogenesis-specific factors. The sequential steps of the metastatic cascade are as follow: A) local invasion: 1- formation of primary tumour; 2- angiogenesis; 3- attachment, proteolysis, and cell motility; and 4- intravasation. B) Circulating tumour cells: 1- homotypic interaction; 2- heterotypic interaction; and 3- coagulation abnormalities. C) Extravasation. D) Secondary tumour formation. Tumour angiogenesis is a process during which tumour cells interact directly with endothelial cells. Cell-ECM interactions have a major impact on gene regulation, cytoskeletal structure, differentiation, and growth control. Alterations in apoptosis and in cellular interactions are features of neoplastic transformation. Detachment, motility, and invasion require alterations in the normal cell-ECM contacts. Anoikis (homelessness) is caused by disruption of the interaction of normal epithelial cells with ECM leading to apoptosis. Protection against anoikis is conferred by transformation with oncogenes and can be alleviated by scatter factor. The transition from an epithelial to a mesenchymal phenotype shares many properties of embryonic cells. During development the first epithelial cells appear at the blastula stage. They are polarised cells, interconnected with each other by adherens junctions, tight junctions and by desmosomes. During gastrulation the three germinal layers, namely the embryonic endoderm, the embryonic mesoderm and the embryonic ectoderm, develop from the primitive ectoderm. Whereas the former two layers keep their epithelial phenotype, epithelial-mesenchymal transition (EMT) leads to the development of the first mesoderm. In EMT, epithelial cells become elongated and mobile and lose their polarity and firm cellular junctions. The transformation event from a nonmalignant to a malignant mimics the losses of polarisation, motility, and cellular contacts that occur during EMT. The main cause of mortality in tumour patients is the development of secondary tumours, i.e., metastases. Often at time of primary excision, specialized tumour cells have left the primary lesion to settle and grow in a more distant tissue, establishing micrometastases. Most tumours develop from a single cell that has been affected by a somatic mutation, leading to immortalization or to a change in the capacity to interact with other cells. Highly proliferative metastatic cells will develop from the primary neoplasm. The distribution of metastatic lesions in the body does not follow statistical parameters. The liver and lung tend to be invaded more frequently, but there is a preference for less vascularized organs to be targets of metastatic cells, such as bone marrow and brain. During normal nonmalignant processes, cells respect their boundaries, whereas metastatic cells are characterised by their ability to escape from their tissue organisation and interact actively with different cell types. The basal membrane is composed of collagen, laminin, glycoproteins, and proteoglycans and cannot be crossed by cells without being damaged. Several molecules involved in tumour progression and metastasis have been found: 1- ECM receptors, 2- proteolytic enzymes, 3- motility factors, 4- cell-cell adhesion molecules, 5- growth factors and cytoplasmic proteins, and 6- angiogenesis regulating factors. ECM receptors, intgrins are heterodimeric molecules composed of an α-chain and a β-chain. Integrins are capable of integrating the cytoplasmic cell network, connecting the cytoskeleton with components of the ECM. The integrin adhesion molecules are such a big family that is expressed during development, lymphocyte circulation, and tumour progression. The 67-kDa non-integrin high affinity laminin receptor has been localised on the surface of numerous cell types, including fibroblasts, chondroblasts, and monocytes, and is up-regulated in malignant, invasive tumour cells. Elastin, laminin and collagen type IV are known ligands. Unlike other cell-surface receptors, this laminin receptor is not a transmembrane molecule but can be immobilised in association with other integral membrane proteins. Its ligand interaction occurs only in the absence of galactosugar. When galactosugar bind to the lectin site, the laminin receptor changes its molecular folding, releases the ECM ligand, and is concomitantly shed from the cell surface. The transmembrane glycoprotein CD44 is identical to ECM receptor III, fibronectin, collagen and hyaluronidase receptor, and others. Lymphocyte homing receptors permit memory and activated lymphocytes to enter the vascular system and invade tissues. Lymphocytes are known to up regulate CD44 expression during activation and differentiation. CD44 integrates ECM components with the cytoskeleton. The cytoplasmic tail of CD44 is also involved in signal transduction, regulating cell migration and/or proliferation. CD44 plasma membrane domain seems to be required for outside-in signalling. At the N-terminus, CD44 interacts with disaccharide units of hyaluronic acid. The up-regulating of CD44 isoforms has been reported to correlate with an adverse progression in kidney carcinoma. Down-regulation of CD44v surface expression accompanied the dedifferentiation of urothelial tumour cells. Early tumour detection and the establishment of criteria that predict the dissemination and invasive potential of tumour cells are goals. Soluble molecules from the serum, urine, or synovial fluids of patients have successfully been used for tumour assessment and/or treatment monitoring. Proteolytic enzymes, capable of digesting ECM and stromal components surrounding the primary tumour tissue and the vessels level the pathways of metastatic cells. Matrix metalloproteases (MMPs), serine proteases (cathepsins), heparanases, hyaluronidases and proteoglycanases belong to this group of enzymes. The activity of these proteases is regulated by a fine adjustment of activators and inhibitors, which induce the transition from an inactive to an active enzymatic stage and block proteolytic activity. Degradation of the ECM is one of the initial steps in the invasion of metastatic tumour cells. Substrate-specific proteases also play an essential role in organogenesis, trophoblast implantation, and wound healing. Proteases suffer complex posttranslational modifications before they become active. Consequently, immunohistochemical analysis cannot distinguish between the active form and the inactive form. Tumour cell motility can be influenced by ECM fragments, e.g., from laminin, fibronectin, tenascin, osteopontin, thrombospondin, and hyaluronic acid, as well as by growth and motility factors surrounding the primary tumour area. Motility factors, also known as motogenic cytokines, influencing the nondirectional movement of cells: AMF (autocrine motility factor), SF/SGF (scatter factor/hepatocyte growth factor) and ATX (autotaxin). SF/SGF can act as a mitogen, as a motility factor (motogen). The transmembrane tyrosine kinase and proto-oncogene c-met is the receptor for SF/SGF. The AMF receptor is a guanosine triphosphate (GTP)-binding protein with homology to the tumour-suppressor gene p53 and is highly expressed at the leading edge of migrating cells. Integrins mediate cell-cell and cell-ECM contacts in a heterotypic as well as homotypic manner. Major cellular ligands for integrins are members of the immunoglobulin superfamily, e.g., ICAM-1 and VCAM-1. Integrins might be altered as well by posttranslational modifications, thereby influencing the receptor-ligand specificity and, subsequently, the organ-specific invasion of metastatic cells. The immunoglobulin superfamily of adhesion molecules, interact with integrins but also among themselves, like CD31, a molecule involved in leukocyte homing. These molecules are used by circulating leukocytes on their way through the vascular and lymphatic system as well as by invasive tumour cells during invasion and migration. The selectin family, L-selectin, P-selectin, and E-selectin are involved in the homing of leukocytes, namely, during their initial docking and rolling on the endothelial cells. P- and E-selectin are expressed by endothelial cells, whereas L-selectin is present on leukocytes. Complex sugar structures have been identified as selectin ligands, sialyl-Lewis α and x, often presented on proteoglycans as carrier molecules, P- and E-selectin also act as tumour cell receptors, probably through sialyl acid ligands. Herewith, tumour cells make use of the same extravasation mechanisms as leukocytes. E-cadherin is a transmembrane molecule providing Ca^{2+}-dependent homotypic cell-cell interactions between epithelial cells. The loss of E-cadherin expression or function correlates with an aggressiveness and dedifferentiation of many carcinomas. E-cadherin appears very early during vertebrate development.

Metastatic cells properties:

1- Intracellular junctions are no longer strictly regulated during tumour progression.
2- Adherens junctions as well as desmosomal components are redistributed in the cell, leading to a distortion in polarisation and intracellular communication, which might result in mobile and invasive cells.
3- Invasive cells break through the naturally established barriers between different tissues and become invasive.
4- Cells that capable of breaking through the basal membrane and entering the bloodstream or the lymphatics will have the potential to generate metastases.
5- Tumour cells secrete proteases, which in a more or less specific manner degrade the ECM in its single components, opening holes through which cells can migrate.
6- Migration through the ECM and endothelium of local blood and lymphatic vessels is controlled by cell-surface receptors.
7- Tumour-cell-surface receptors interact with ECM ligands and with endothelial cell receptors.
8- After crossing the basal membrane of the primary tumour, the surrounding ECM, and the endothelial basal membrane, invasive tumour cells may enter the bloodstream or the lymphatics. They well aggregate either with each other or with blood cells to survive as emboli.
9- As soon as tumour-cell emboli get entrapped in capillaries, such as those of the lung, they may break through the lining endothelium, invade the surrounding ECM, and settle to form micrometastatic foci, which might stay dormant in the body. Only those capable of proliferation will develop into secondary tumours.
10- From these secondary lesions, tertiary tumours may arise, generated by metastatic cells even more aggressive and invasive than before.

E-cadherin has been shown to be relevant marker for tumour cell invasion in several clinical studies. A correlation between dedifferentiation processes, an unfavourable clinical prognosis, and a reduction in E-cadherin expression has been described in several tumours. Growth factors are functional only when bound to their corresponding receptors. Therefore, it is of crucial importance that the expression of both, receptor and factor be regulated. The TIAM-1 (T-lymphoma invasion and metastasis) is a cytoplasmic gene linked to tumour progression. It is homologous to GDP-GTP exchangers such as rho-like proteins. The oncogene products ras, rho, and rab are active when bound to GTP and modulate cell motility through signal-transduction processes, which modulate cytoskeletal organisation. The metastatic cascade depends on neovascularization at two stages: 1- metastatic cells can leave the primary tumour when it is properly vascularized; and 2- a metastatic lesion can survive and grow to a clinically relevant size only if it is connected to nutritive resources and if degradation products are removed. Lung micrometastases are not vascularized and have, although highly proliferative, a high rate of apoptosis. They will stay dormant unless they are neovascularized. Tumour angiogenesis is modulated by angiogenic factors produced either by the tumour or by the surrounding stroma cells. An interplay of both stimulatory and inhibitory factors is probably the most likely situation. The following are considered as being stimulatory factors: aFGF and bFGF (acidic and basic FGF), angiogenin, VEGF (vascular endothelial growth factor), IL-8, TGF-α and TGF-β, TNF-α, perlecan, and SF/HGF. The following are considered as being inhibitory factors: angiostatin, IFN-α and IFN-β, thrombospondin, TIMP-2, and heparinase. The regulation of these factors is yet largely unknown. IFN-α, used for the treatment of highly vascularized tumours, inhibits bFGF expression and, therefore, angiogenesis. Angiostatin is a cleavage product of plasminogen, a serine protease present in nonvascularized tumours. In an animal model it has been shown that surgical elimination of the primary tumour reduces the level of circulating angiostatin, inducing neovascularization in the proximity of metastatic lesions, which consequently start to grow and this suggests a broad distribution of tumours that cannot be detected clinically. These micrometastases are held in an equilibrium between cell proliferation and death, staying in a kind of "steady state" until angiogenesis is induced by the reduction of angiogenesis inhibitors. Development of the vascular system during embryogenesis is governed by mechanisms similar to those occurring during progression. The main concern of patients suffering from a tumour disease is the development of metastases. The need to resect the primary tumour, followed usually by chemotherapy and/or radiotherapy, has been a matter of debate. Due to the incision into the nonmalignant tissue, wounds are set, which could induce a number of changes resulting in secondary tumour growth. Enhanced growth of micrometastases after surgical resection of the primary tumour has been observed. Resection of the primary tumour leads to relapse-free survival of almost 50% of patients if no metastases have been detected at the time of surgery. Following resection, diagnostic markers may then determine the necessity for and the dose of chemo- and radio-therapy to be delivered. Resistant tumour cells with an increased metastatic capacity could be selected by radiation and/or cytostatics. Tumour markers allow a clear-cut diagnosis only when a massive tumour load has developed but cannot be applied for the detection of minimal residual disease. High levels of MCA (mucin-like-associated antigen) are associated with mammary and kidney tumours. AFP (α-fetoprotein) can be elevated in liver and testicular tumours, as can β-HCG (human β-chorionic gonadotropin) in testicular tumours, and PSA (prostate-specific antigen) in prostate carcinomas. All these markers are currently in use for clinical follow up of patients after surgery, but they are usually not indicative enough to be used as prognostic factors when tumour growth is not obvious. After surgical resection of the primary tumour, the early detection of micrometastatic growth is of crucial importance. The results might overburden the patients, causing unnecessary psychological constrains as well as social and legal problems. Any single metastatic lesion follows its own criteria under specific selective pressures. All the properties of a metastatic cell, e.g., its particular adhesion, invasion, and growth capacities, may develop in parallel independently. Adhesive interactions between like cells (homophilic), different cell types (heterophilic), or cell and the extracellular matrix exert multiple influences on the metastatic process. Normal cells secrete ECM components around themselves; tumour cells produce less ECM. There are 4 broad classes of cell-adhesion molecules (CAMs): A- Cadherins, which are calcium-dependent and mediate cell binding, mediate homophilic interactions. E-cadherin expression may contribute to cell detachment, dedifferentiation, and/or invasion in cancer progression. B- Immunoglobulin super family, mediate both homophilic and heterophilic interactions. They are expressed on many types of cells. C- Selectins, are expressed on blood cells and endothelial cells and are involved in heterophilic interactions, thus play a role in leukocyte adhesion. D- Integrins, mediate interactions with the ECM by binding multiple ECM proteins. Integrins interact extracellulary with ECM components and intracellulary with the cytoskeletal proteins actin, talin. Protease remodelling of the tumour microenvironment is a key step in invasion process. The plasminogen-activator system is implicated in clot formation, interaction of cells with the endothelial, and tumour invasion. The proenzyme plasminogen is involved in the final steps of the extrinsic fibrinolytic pathways. Plasminogen is converted to plasmin by one of two different serine proteases, tissue plasminogen activator (tPA) or urokinase plasminogen activator (uPA), and the activated plasmin then degrades fibrin and fibrinogen. Secreted pro-uPA binds to a uPAR (uPA receptor) on the same cell or a neighbouring cell. Enzyme bound at the cell surface is activated to cleave plasminogen, unless inhibited by PA inhibitors. The receptor is internalised and its ligand is degraded in intracellular lysosomes and recycled to the cell surface, such trafficking may permit repositioning of uPAR to change the area of localised proteolysis. The zinc metalloproteinases family share several properties, including secretion as a proenzyme and requirement of cleavage for activation and inhibition by metal chelators and by tissue inhibitors of metalloproteinases (TIMPs). There are 11 members of the group. The enzymatic activity is pericellular and tightly regulated, although the nature of the cellular receptors and endogenous activators are incompletely defined.

Metastasis is regulated by:

1- Cell-cell and cell-extracellular matrix (ECM) receptors.
2- Proteolytic enzymes necessary for invasion of the basal membrane.
3- The vasculature, and different organs.
4- Motility factors are essential for migration through the tissues.
5- Receptors required for organ-specific invasion.
6- Growth factors, which will support the growth and development of tumour microcolonies in the secondary organ.
7- Angiogenesis-specific factors.

ANGIOGENESIS

The development of new capillaries from preexisting blood vessels is required in normal situations such as wound healing and development but is critical to the continued growth of both primary and metastatic tumour deposits. Angiogenesis contributes to the outgrowth of occult micrometastases that have potentially established by the time of cancer diagnosis and therapy. Tumour angiogenesis is a complex process comprising numerous molecular and cellular mechanisms that are involved in the generation of the expression of angiogenic factors, their mode of action at the vascular endothelial cells up to the outgrowth of new capillaries, and their penetration into the tumour. Vascular endothelial growth factor (VEGF) is expressed by many solid tumours and induces angiogenesis at the vessels adjacent to the tumours. The receptors for VEGF are present in large amounts on the surface of endothelial cells in these vessels, whereas they are only poorly expressed in the remaining vasculature. VEGF receptor expression is cell-specific and occurs almost exclusively in endothelial cells. The angiogenic process is complex, resembling tumour-cell invasion in certain characteristics: 1- the basement membrane is dissolved at a postcapillary venule; 2- endothelial cells proliferate near the venule and migrate toward the tumour; 3- canalisation, branching, and formation of vascular loops complete a functional connection; 4- progressive arteriolization, in which smooth-muscle cells proliferate and encase a developing artery; 5- the vasculature of the tumour bed has a unique, poorly organized composition; 6- the blood therefore coming in direct contact with the tumour; 7- tumour may also be found within the endothelial lining; 8- the postcapillary venules are devoid of basement membrane (BM) and are a recognised site of cancer-cell intravasation; and 9- the venous system outnumbers the arterial system, and the pressure in the venules is disproportionately lower than that in the arterial system. This combined with the increased interstitial pressure in the tumour bed, result in a more sluggish circulation and, hence, in a tendency toward vessel-wall collapse. As endothelial cell migrates across the ECM it undergoes a change in morphology from a cuboidal to an elongated form. The composition of the ECM is contributory to morphology, proliferation, and differentiation. Tumour vascularization may reflect the degree of activity of angiogenic inhibitors as well as stimulators. Many angiogenic factors are heparan-binding growth factors. Angiogenesis-stimulating factors are produced by the primary tumour in excess of inhibitors. At the primary tumour site the stimulatory effect is greater and neovascularization occurs, but distally, systematically released inhibitor suppresses neovascularization at metastatic foci.

COLONIZATION

The outgrowth of tumour cells at distant sites remains one of the steps in the metastatic cascade that are "open" for therapeutic development in a large number of cancer patients. The colonization response of metastatically component tumour cells is qualitatively different from that of nonmetastatic tumour cells, which is due to the production and responsiveness to growth factors present either in the microenvironment or in the circulation, leading to more aggressive metastatic phenotype. Colonization is involved in the outgrowth of micrometastases. TGF-β is inhibitory to the colonization of many nonmetastatic cells but either lacks inhibitory activity or is actually stimulatory for metastatic tumour cells. In prostatic carcinoma, angiogenesis is related to tumour growth and to metastatic potential. The density of microvessels within prostate carcinoma is almost doubled in tumours from patients with metastases as compared with those from patients without metastases. Invasive bladder carcinoma, tumour vascularity is significantly correlates with patients' survival. Hypoxia has a promoting effect on VEGF expression. Hypoxia seems to increase VEGF mRNA stability rather than substantially inducing transcription. A switch from normal p53 to mutated p53 causes an increase in he levels of VEGF mRNA expression. Renal cell carcinomas express high levels of VEGF mRNA as assessed by in situ hybridisation. Expression is especially accentuated adjacent to areas of necrosis. These levels are 3-13 fold higher than those measured in the adjacent normal kidney tissue, even tumours of small size overexpress VEGF mRNA. Most of the hypervascular RCCs also overexpress mRNA for placenta growth factor (PlGF), which is shown to act synergistically with VEGF.

BONE METASTASIS

Metastatic bone pain can be attributed to damage done by direct tumour involvement or to an indirect effect of osteoclast-mediated bone resorption. Causes of pain include: 1- release of prostaglandins, bradykinin, substance P, and histamine, which stimulate nearby nerve endings; and 2- stretching of the periosteum by the tumour, weakening of the bone causing fractures, and tumour growth into surrounding tissues. Metastatic spread to bone is primarily due to the haematogenous spread of cancer cells from a primary tumour. Prostate cancer (CaP) exhibits a preference for certain organs. This preference may be due to local growth factors or hormones at the sites of metastases, physical or chemical characteristics that promote adherence of cancer cells to the endothelial surface only at specific, and chemotactic factors that diffuse from the sites of metastases. Substances released during normal bone resorption act as chemotactic factors in the case of cancer metastatic to bone. This tumour involvement is typically seen in the axial skeleton, with most metastases noted in the spine, ribs, pelvis, and proximal parts of the femur or humerus. Pathological fractures and pain most often occur at these sites. Osteoclast-mediated bone resorption, leads to weakening of the bone, fractures, and pain. The process of bone remoulding itself is complex. In patients with cancer there can be disruption in the balance of bone resorption and formation. Metastatic cancer cells secrete substances, including parathyroid hormone-related protein, TGF-α, IL-1, IL-6, TGF-β and prostaglandins, which induce the proliferation and activity of osteoclasts directly or through osteoblasts. When osteoclasts are activated by these substances, a net loss of bone at the site of increased activity is because of inadequate bone formation by osteoblasts or inability of the cavity to attract osteoblasts. In some cases there can be a net of increase in bone, i.e., osteosclerosis metastases, associated with malignancies as CaP. Lytic bone lesions resulting from a net of loss of bone are more common and cause the greatest morbidity, including hypercalcaemia, fractures, and pain. The bone is the most common metastatic site for solid tumours. The most frequently involved skeletal sites for solid tumours are the vertebrae, pelvis, ribs, femur, and skull. IL-1 and TNF as local factors together with tumour-produced mediators that stimulate osteoclasts play a crucial role in the complex process of direct destruction of bone by metastatic spread. Mechanism of tumour metastasis at a distant bone involves complex steps including: 1- progressive growth at the primary site; 2- vascularisation; 3- invasion of the circulation; 4- detachment; 5- embolisation; 6- survival in the circulation; 7- arrest at the metastatic site; 8- extravasation; 9- evasion of host defenses; and 10- progressive growth. Malignant cells can act directly that leads to bone destruction. Radiologically, bone metastases can be lytic, blastic, or mixed. Prostate cancer (CaP) cells produce osteoblastic stimulatory factors with a corresponding sclerotic radiographic appearance. These osteoblastic lesions are prone to fracture because of the increased production of unstable bone matrix. Elevated alkaline phosphatase levels in these patients result from increased osteoblastic activity. Bone metastases morbidity includes 1- pain, 2- impaired mobility, 3- hypercalcaemia, 4- pathological fracture, 5- spinal cord or never root compression, and 6- bone marrow infiltration with resulting disruption of haemopoiesis. Pain from bone metastasis develops gradually during a period of weeks or months, progressively become more severe. The severity of pain does not correlate with the extent of metastatic bone disease. Pain is worse at night or exacerbates by weight-bearing activity. Microfracture can occur at the site of metastasis resulting in bone distortion. Many nerves are found in the periosteum, and others enter bones via blood vessels. The stretching of periosteum by tumour expansion, mechanical stress of the weakened bone, nerve entrapment by the tumour, or direct destruction of the bone with consequent collapse are possible cause of pain. The complex, mixed pathobiology of the pain, and its difficulty to control may be exacerbated by movement or weight bearing. Spinal cord compression occurs when an unstable vertebra (dorsal, lumbar, and cervical spine) is displaced posteriorly on the spinal cord. A patient's vague complaints of back pain and leg weakness associated with bone metastasis must be investigated to exclude metastasis. Tumour-induced hypercalceamia occurs when calcium is released during bone breakdown caused by metastasis, overwhelming the kidney's capacity to filter it properly or to maintain calcium homeostasis. The associated hypercalciuria, volume depletion and resulting dehydration leads to poor renal perfusion, reduced GFR, and a compromised calcium excretion, which causes a further increase in plasma calcium. Rib fractures and vertebral collapse results in loss of height, kyphoscoliosis, and restrictive lung disease. Bone metastasis may be diagnosed by radiography, bone scan, CT scan, and MRI. 30-40% change in bone density occurs before bone metastasis can be identified in cancellous bone by radiography. Bone scan is the most sensitive method of detecting bone metastasis.

Turnover/replacement time of some tissues:

System	Turnover/replacement time
1- Integumentary:	
Epidermis	30 days
Basal cells	Nadir: 21 days. Re-epithelialization: 28-31 days.
Endothelial cells	unknown.
2- Blood:	
RBCs	120 days
Granulocytes	6-10 hr in blood, 2-3 days in tissue.
Lymphocytes	100-300+ days
Platelets	5-10 days
3- Respiratory tract:	
Tracheal epithelium	50 days
4- GIT:	
Oral mucosa	10-14 days
Stomach	3-9 days
Small intestine	1.5 days
Colon	10 days
Skin	20 days
5- GUT:	
Urinary bladder	50 days
Testis	20 days
Lung alveolar cells	10-30 days
6- Eye:	
Cornea	7 days.

Bone scan may not be sensitive to detect purely osteolytic metastasis. CT scan or MRI should be used to detect lytic metastasis. MRI is useful for imaging the vertebral and spinal canal, including evaluation of the bone marrow. In general, the presence of bone metastasis indicates progressive and an incurable disease, with median survival of 2.1 years in breast cancer, when compared to 1.6 years in patients with additional organ involvement. CaP, the median duration of controlled disease by androgen blockade is 4 years in the presence of bone Metastasis only, with good performance status. Metastasis involving 66% of the long bones places the patient at a higher risk for fracture and may require prophylactic orthopaedic intervention followed by radiotherapy. Bone pain that is worsening by activity can occur before fracture. The femur is the most common site of pathological fractures. Pain management with analgesics or adjuvant therapy decreases pain, increases mobility, and enhances the patient's adaptation to pain from bone metastasis. NSAIDs should be used in combination with opiods because NSAIDs have additive analgesic effects and lessen GI and CNS side effects of opiods without compromising analgesia. For localised bone pain, external radiotherapy to the lesion may reduce pain in up to 80% of cases, over 1-2 weeks. 10-20% of patients may develop pain flare shortly after administration of radiopharmaceuticals, which disappears within 48 hrs. So, radiopharmaceuticals are contraindicated in patients at risk for spinal cord compression. Strontium-89 is chemically similar to calcium. Pain control using radiopharmaceuticals may last up to 16 weeks. Weekly platelet and WBC count should be monitored for 8 weeks because of accompanying bone marrow suppression. A variety of bisphosphonates are available, and i.v. Amino-bisphosphonate (pamidronate disodium) is indicated for osteolytic bone metastasis together with adjuvant therapy. The bisphosphonates are pyrophosphate analogues characterised by a phosphate-calcium-phosphate bond that renders them resistant to hydrolysis by phosphates for long periods. Bisphosphonates bind strongly to exposed bone mineral around osteoclasts, then internalised by the osteoclasts and cause disruption of biochemical processes involved in bone resorption and destruction of the osteoclast. Bisphosphonates inhibit the generation of new osteoclasts and disturb the production of osteoblast-stimulated coupling factors and bone-resorbing cytokine release from macrophages adjacent to the bone surface. The aminobisphosphonates induce apoptosis. Aredia, 90 mg i.v. infusion/4 h every 3-4 weeks in conjunction with adjuvant therapy may treat osteolytic bone metastasis, prevent and mange hypercalcaemia.

PARANEOPLASTIC SYNDROMES

Paraneoplastic syndromes are disorders of host organ function occurring at a site remote from the primary tumour and its metastases. The result is at least one syndrome for every tumour type, and every host target organ described as being perturbed by some tumour. The majority of syndromes occur in relation to specific tumours and specific mediators. It has been estimated that paraneoplastic syndromes will be present at diagnosis in 7-10% of patients with malignancy, but many as 50% of patients may experience such a syndrome. These syndromes may not be specific for malignancy. Prostate cancer, secreting corticotropin can cause adrenal hyperplasia followed by high levels of glucocorticoids and resulting diabetes, potassium loss, hyperkalaemic alkalosis, mania, and coma. Paraneoplastic syndromes reflect communication between tumour cells and host cells. Tumours that invade and die in nonapoptotic fashion may release autoantigens and set the stage for a host of paraneoplastic host responses. From the urological ridge, the kidney and excretory organs arise as well as internal sexual apparatus. Tumours of adult differentiated tissues that have arisen from these embryological areas may share the same synthetic capacities with other cells from the same anlage. When a tumour cell is highly undifferentiated, it may be dividing at a stage at which its ability to produce the specific products of the same tissues terminally differentiated cells may be impaired. In prostatic cancer, when Gleason's stage increases in number and differentiation decreases, the amount of PSA may also drop, even as the less differentiated tumour progresses. A certain degree of differentiation seems necessary for endocrine paraneoplastic products to be produced, but also increasing cell mass is needed for the level of the product to exceed that present under normal circumstances and create paraneoplastic syndromes in tissue sites at a distance from the tumour. Tumours arising from tissues of the urogenital ridge anlage would be expected to produce products resembling those of normal kidneys, ovary, prostate, adrenal, endometrium, and bladder. The kidney produces paraneoplastic syndromes secondary to excess secretion of prostaglandins, 1,25-dihydroxy vitamin D, EPO, renin, and cholecalcification. Renal carcinomas produce trophic polypeptides resembling parathormone, ACTH, glucagon, and IL1. Small cell neuroendocrine tumours of bladder and prostate may produce hypercalcaemia and hyperadrenocorticism. Prostate cancer accompanied by diffuse haemorrhage due to the secretion of plasmin or tissue plasminogen activator. Benign tumours within the urogenital ridge can produce EPO in the fashion that it is produced by a normal kidney or a renal cell cancer (RCC). The mesenchymal anlage, when it becomes neoplastic, might be expected to create paraneoplastic syndromes involving overproduction or underproduction of RBCs and WBCs procoagulants and antibodies and/or cytotoxic T cells. Apudomas can develop in multiple sites in many organs from residual neuroactive cells, tend to secrete products that are characteristic of the amine precursor uptake and decarboxylating cells, including bombesin, kallikerins, and serotonin derivatives. Tumours that express glycoprotein synthetic capacities for example trophoblastic or testicular neoplasms usually can manufacture large amounts of the nonspecific β unit, but the ability to manufacture the specific α subunit appears limited. Several behaviours forbidden in normal organisms may occur in tumourigenesis. Tumour cells proliferate and extend at their peripheries rather than along a central cord as in normal embryogenesis. There are open intratumoural vascular spaces lacking epithelium, in which RBCs and WBCs and plasma can be found directly in contact with tumour cell walls and clusters of dying cells. This slow sinusoidal circulation produces the tumour blush found during tumour arteriography. Lacking the normal protective endothelium, RBCs circulating through the tumour vascular spaces come into direct contact with tumour cell surface and tumour cell secretions, which appear to be damaging leading to microangiopathic haemolytic anaemia. Altered RBC may be found as antigenic by normal host immunocytes and reacted to by the secretion of RBCs-directed antibody globulins, then removed at a rapid rate by conventional splenic mechanisms giving rise to haemolytic anaemia. Some of the angiogenic substances produced by tumour also escape through the venous circulation and may act at great distances producing paradoxical vascular proliferations. The central areas of a tumour lack well-developed vascular systems and soon become anoxic and necrotic. Such ischaemic necrosis is nonapoptotic in character. Random nonapoptotic tumour cell death consequently leaves many intact macromolecules in their original physiologically and antigenically active forms. Circulating antibodies, or cytotoxic lymphocytes, may be produced by the normal immune system in response to a tumour-associated material sharing antigenic epitopic determinants with normal tissue. The resulting antibodies and cells may be cytotoxic to the tumour and, more importantly for paraneoplasia, to bystander normal cells. Tumours that produce parathormone-like endocrine substances must rely on normal bystander osteoclasts to dissolve osseous bone and release calcium resulting in hypercalcaemic syndromes. Octreotide blocks the receptors for neuroendocrine mediators in carcinoids have shown not only decrease in symptoms but also decrease in the actual size of the tumour, suggesting that the tumour growth could be based on autocrine paraneoplastic stimuli. Exposure to exogenous or endogenous carcinogens is common to the stem cells of some tissues, such as inhaled tobacco or radioactive carcinogens in the upper respiratory tract. Even low doses of most carcinogens are highly cytotoxic, giving rise to triple simultaneous impacts: 1- causes cell death, 2- promoting rapid proliferation in remaining normal cells, and 3- providing ligands for DNA mutations that may lead to neoplastic transformation.

THE CANCER CACHEXIA SYNDROME

Cachexia is based on the Greek words "kakos", meaning bad, and "hexis", meaning state of being. The cancer cachexia syndrome encompasses a wide range of metabolic, hormonal, and cytokine-related abnormalities that result in a wasting syndrome. The clinical manifestations, including anorexia, early satiety, weight loss, weakness, easy fatigue, impaired immune function, tissue wasting, and poor performance status. It is not resolved by forced caloric intake. There are at least 3 possibilities for cachexia: 1- increase in metabolic needs generated by the nutritional demands of the tumour; 2- increased energy expenditure in the cancer patient; and 3- maladaptive metabolism. Decreasing blood sugar that results from both hypermetabolism of the tumour and the host, together with decreased nutrient intake related to anorexia or the anatomic changes produced by tumour of various sites also play a role. Insulin resistance state is characterised by a decreased uptake and use of glucose, especially in muscle, and a tendency towards gluconeogenesis and lipolysis. The lipolytic component of insulin resistance is characterised by excess fatty acid oxidation, which does not decrease even with increasing fat and caloric intake. Increasing rates of glycerol and free fatty acid turnover occur. The result is that cancer patients deplete fat stores, develop hypertriglyceridaemia and decreased lipoprotein lipase (LPL) levels. The lower LPL levels result in failure of production of free fatty acids (FFA) and monoacylglycerol and, thus, decrease adipocyte triglyceride synthesis, which depends on these fatty acids. The frequency of anorexia in cancer varies from 15-40% at presentation and up to 80% in those with advanced disease. Anorexia may result from pain, mechanical obstruction of the GIT, from nausea induced by chemotherapeutic drugs, or from psychological factors. Weight loss is associated with complaints of weakness, fatigue, loss of energy, and inability to perform the tasks of everyday life. Patients loss of weight may result from factors that decrease food intake (nausea, vomiting), direct tumour encroachment on the GIT, iatrogenic causes such as radiotherapy and cytotoxic drugs, and pain and/or emotional distress. Cachexia can impact on quality of life (QoL) on several levels. Profound muscle weakness leads to loss of physical function and deterioration of performance status. Fatigue and weakness may impair a patient's ability to perform even simple activities of daily life such as dressing, preparing meals, and eating. The physical fatigue and changes in body image that may occur with cachexia may contribute to depression and decreased social interactions. Suspected mediators of cachexia include: 1- cytokines, TNF-α, IL-1, IL6, IFN-α and -γ; 2- hormonal factors, insulin, growth hormones, epinephrine, ACTH, IGF; 3- tumour factors, proteoglycan; and 4- satiety factors, leptin, glucagon-like factor II, and urocortin. Early diagnosis and intervention may well be the foundation of treatment. Approaches to treatment include: 1- treat associate symptoms, 2- increase nutritional support, 3- pharmacological approaches, stimulate appetite, reverse abnormal metabolism, e.g., anabolic steroids, metoclopramide, megesterol acetate, and 4- nutritional counselling. Tumour-induced weight loss (TIWL) is a common cause of morbidity and mortality in patients with advanced cancer. This cachexia is similar to the metabolic changes seen in a variety of inflammatory and traumatic insults characterised by inefficient substrate use, alterations in energy balance, and the acute-phase protein response (APPR). These changes seem to be driven by proinflammatory cytokines, alterations of the neuroendocrine axis, and tumour-derived catabolic factors. Steroids and megestrol may improve appetite but do not affect the cachectic process. Agents that influence the inflammatory metabolic state, such as nonsteroidal anti-inflammatory drugs (NSAIDs) and ω-3 fatty acids from fish oil, show more promise, particularly in combination with nutritional supplementation. Weight loss is seen in a large proportion of cancer patients. The fundamental difference between TIWL and other forms of weight loss is the lack of reversibility with feeding. This seems to be due to metabolic changes in cachexia, produced by the tumour or by the host in response to the tumour. Weight loss implies a loss of energy balance with an increase in energy expenditure, a reduction in energy intake, or both. Energy intake is reduced in cancer patients. The resting energy expenditure of cancer patients vary between <60% and >150% of the predicted. Cachectic cancer patients exhibit relative glucose intolerance and insulin resistance with an increased rate of glucose production and recycling via lactate (the Cori cycle). Whole body protein turnover is increased in the majority of advanced cancer patients, with energy cost of about 100 Kcal/d. Loss of skeletal muscle protein is a prominent feature of TIWL, i.e., a reduction in the rate of muscle protein synthesis producing net protein breakdown. The balance of liver export proteins is altered in many cancer patients such that while albumin synthesis remains unchanged, fibrinogen synthesis rates are significantly increased. These changes occur on a background of a decrease in the circulating concentration of albumin (a negative acute-phase protein) and an increase in the concentration of fibrinogen (a positive acute-phase protein). During an inflammatory response, there are altered demands for amino acids. In cancer patients, feeding stimulates the synthesis not only of the negative acute-phase protein albumin (as seen in normal individuals) but also the positive acute-phase protein fibrinogen. A tumour constitutes a new organ requiring its own sustenance and, thus, increases demand for nutrients and causes weight loss if nutrients are not forthcoming. However, the presence and severity of cachexia does not correlate with the size of the tumour, since cachexia is often seen early in the course of cancer and manifestations such as alterations in appetite and nutrient metabolism cannot easily be explained by this hypothesis. Several proinflammatory cytokines, including TNF-α, IL-1-β, IL-6, and IFNG, have been implicated in cachexia. These cytokines lead to anorexia, weight loss, an APPR, protein and fat breakdown, rises in levels of cortisol and glucagon and falls in insulin levels, insulin resistance, anaemia, fever, and elevated energy expenditure. Release of prostaglandins is a major step in the signalling pathway leading to muscle protein breakdown in normal tissues. Prostaglandins (PGs) may mediate the actions of most proinflammatory cytokines. Hormones such as cortisol, glucagon, and adrenaline in humans produce features of cachexia such as protein loss, an APPR, increased energy expenditure, and glucose intolerance. With cancer there is elevated levels of cortisol and glucagon. A 24-kd glycoprotein proteolysis-inducing factor (PIF) activates the ATP-ubiquitin-dependent proteolytic pathway and induces the nuclear transcription factors NFκB and STAT3, resulting in cytokine and acute-phase protein synthesis. PIF is expressed in tumour cells from patients with significant weight loss but not in those who are reasonably weight stable. The best way to treat cancer cachexia is to cure the cancer, but this remains a rare achievement among adults with advanced solid tumours. Prednisolone at a dose of 5 mg 3 times daily and dexamethasone 3-6 mg daily may improve appetite. Methylprednisolone given intravenously at a dose of 125 mg daily will improve QoL. Steroids can be associated with a number of adverse effects such as worsened glycaemic control and catabolism. The progestational agents' megestrol and medroxyprogesterone (240-1600 mg/day) may improve appetite. Megestrol improves appetite with no changes in nutritional parameters or QoL. These agents have a number of side effects, including an increased incidence of venous thrombosis and peripheral oedema, reduced response to chemotherapy, and a trend to poorer survival. Any weight gained by TIWL patients taking megestrol tends to consist of fat and water. Pentoxifylline inhibits the production of TNF-α, resulting in improved well being. Melatonin of 20 mg daily given in the evening may influence TNF-α production. Hydrazine (60 mg/tds) inhibits phosphoenolpyruvate carboxykinase, an enzyme responsible for gluconeogenesis from lactate (the Cori cycle), but with no effect on cancer patients. Indomethacin 50 mg bd stabilises Karnofsky performance status. Ibuprofen 400 mg tds reduces levels of acute-phase proteins, IL-6, and cortisol and normalises whole-body protein kinetics to some extent in cachectic cancer patients. Ibuprofen also reduces levels of acute-phase proteins and resting energy expenditure. Megestrol together with ibuprofen may stabilise quality of life (QoL) and weight in cachectic patients with advanced cancer. Eicosapentaenoic acid (EPA) affects a number of potential mediators of cachexia, including cytokine production, PIF, and the APPR. Fish oil prolongs survival in cancer patients. 2 g of EPA and 600 kcal/day significantly increase weight and lean body mass in cancer patients as a result of improved appetite and stabilisation of APPR and downregulation of a number of inflammatory mediators. TIWL is a challenging clinical problem. TIWL remains a significant cause of morbidity and mortality in malignant disease. An anti-inflammatory agent such as fish oil or an NSAID in combination with oral nutritional supplementation represent the more promising treatments at present. The treatment of cachexia is, on its own, not going to cure anyone of cancer.

FEVER

Fever is an increase of temperature above the daily variation of low, early morning temperatures and higher, late afternoon readings. In children, the daily swings can be high (1.2°C/day), whereas in older adults, the daily variation can be only 0.3°C/day. Hyperthermia is the increased in core temperature due to dysregulation of normal thermoregulation. Characteristic of hyperthermia is the failure of common antipyretics to lower the increased temperature, whereas in fever antipyretics are highly effective. Core temperature is defined as the temperature of the blood at the hypothalamic level. The brain and the liver have a higher temperature, about 38°C, whereas the skin and peripheral lymph nodes are maintained at a lower temperature. Axillary temperature tends to be about 1°C less (36°C) than the core temperature. Oral and rectal temperatures reflect core temperature. Oral readings are lower probably because of mouth breathing, which is particularly important in patients who have respiratory infections and rapid breathing. Freshly voided urine temperatures also can reflect core temperature. The thermoregulatory center located in the anterior hypothalamus regulates internal temperature at about 36.5°C, primarily by its ability to balance heat production and peripheral heat loss. Pyrogenic cytokines have the ability to increase the hypothalamic set-point to febrile levels, cause fever whether produced peripherally or in the CNS. Pyrogenic cytokines cause fever only by their effects on the hypothalamic thermoregulatory center and not by peripheral effects. Pyrogenic cytokines first enter the area of the blood-brain barrier near the anterior hypothalamus called the organ vasculosum, results in a release of PGE_2 which, in turn, leads to an increased set-point. The increased set-point then triggers the processes of heat conservation (vasoconstriction) and heat production (shivering and metabolism). In fever, the hypothalamic set-point is increased and triggers the vasomotor center to commence vasoconstriction. Blood is shunted away from the periphery to the internal organs essentially decreasing or even ceasing the usual heat loss. As a result, there is a steady increase in blood temperature, 2-3°C. Fever is hardly an isolated event. It can be part of a broader host-response called the acute phase response. Fever is primarily associated with infectious diseases, but the febrile response can be a prominent component of inflammatory and immunologically mediated diseases and frequently accompanies certain malignancies, e.g., renal cell cancer (RCC). During acute phase response, increases in the total and relative numbers of circulating young neutrophils often occur. Hepatic acute phase proteins include antiproteases, haptoglobin, several complements, fibrinogen, ceruloplasmin, and ferritin, are increased. There is also a 100- to 1000-fold increase in special acute phase proteins such as C-reactive protein (CRP) and serum amyloid A protein. The acute phase response is a systemic, generalised reactions, although most disease processes that induce it are localised. The role of acute phase protein is to help contain pathogens and their toxins and to inactivate microbial proteases and highly reactive O_2 metabolites. IL1, TNF, IL6, IL11, Oncostatin M, CNTF, cardiotropin-1, and LIF posses the ability to stimulate the hepatocyte to increase gene expression for various acute phase proteins as well as suppress the synthesis of several "house-hold" proteins. These latter proteins include lipoprotein lipase, albumin, and cytochrome P450. Several RCC cells will produce IL1 and IL6. Solid tumours may induce the production of pyrogenic cytokines from "by stander" cells such as local endothelium or "infiltrating cells" such as monocytes. These likely to account for the fever associated with tumour growth in which either an intermediate cytokines-inducing substances is released from the tumour cells or a local inflammatory process takes place due to tumour cell death or ischaemia. In cells derived from bladder transitional cell carcinomas, the autocrine growth factor effect of intracellular adhesion molecule-1 is, in part, due to constitutive production of IL-1α. Inhibitors of cyclooxygenase are potent antipyretics. In the brain, Acetaminophen is oxidised by the p450 cytochrome system and the oxidised form inhibits cyclooxygenase activity. There is no difference between oral Aspirin and Acetaminophen in reducing fever in humans. Aspirin use in children is contraindicated because of the association with Reye's syndrome. Chronic high dose antipyretic therapy such as Aspirin or NSAIDs used in arthritis does not reduce normal core body temperature. Corticosteroids are also effective antipyretics, act at two levels: 1- similar to the cyclooxygenase inhibitors, corticosteroids reduce PGE_2 synthesis by inhibiting the activity of phospholipase A_2; and 2- corticosteroids also block the transcription of the mRNA for the pyrogenic cytokines. The most common cause of fever in cancer patients is infection, followed by paraneoplastic syndrome, drug reaction, and rarely adrenal insufficiency. 30% of cancer patients developed fever and 5% have no explanation other than paraneoplastic syndrome. Paraneoplastic fever is most common in certain types of malignancies including Hodgkin's disease, and RCC. The most effective therapy for paraneoplastic fever is the use of NSAIDs. PGE_2 induces cAMP, which increases IL1-induced IL6. Histamine triggers adenyl cyclase via H2 receptor can increase cAMP. Thus, macrophagic cells induced by IL1 to make IL6 produce more IL6 in the presence of histamine. IL1 and IL6 induction is enhanced by the presence of PGE_2, blocking endogenous PGE_2 synthesis reduces IL1 induced IL6 and other cytokines.

ENDOCRINE/METABOLIC SYNDROMES OF CANCER

The most common cause of endocrine syndromes of cancers (ectopic hormone syndromes) is cancer production of protein hormones or hormone precursors. The endocrine syndromes of cancer are caused by one of three mechanisms: 1- cancer production of protein hormones and hormone precursors, 2- cancer production of cytokines, or 3- cancer metabolism of precursor steroids to bioactive steroids. The cells of the body produce peptide hormones are derived from the neural crest embryologically share two properties: 1- peptide hormone synthesis, storage and secretion; and 2- amine metabolism and uptake (APUD). Many of the neoplasms associated with ectopic hormone production may be derived from the neural crest. Peptide hormones are produced in small quantities by most or all normal tissues and act in paracrine fashion. Cellular transformation caused by oncogene expression is associated with increased expression of one or more genes coding for protein hormones. Ectopic ACTH syndrome: Most or all normal tissues produce a precursor ACTH molecule, which has been shown to act in paracrine fashion for several tissues. Carcinoma produces these same precursors, often in much greater concentrations. Some neoplasms convert the pro-ACTH to biologically active ACTH, producing clinically apparent Cushing's syndrome. Clinically, the ectopic ACTH syndrome may usually be distinguished from the pituitary-dependent Cushing's disease by the following: 1- serum and urine cortisol concentrations are usually increased in cancer-induced Cushing, but are not in a high-normal or moderately increased range in Cushing's disease; 2- plasma ACTH and ACTH precursor levels are usually increased in cancer-induced Cushing's, and high-normal or slightly increased in Cushing's disease; 3- hypokalaemia is common in cancer-induced Cushing's, but is uncommon in Cushing's disease; 4- ACTH and cortisol are usually not suppressed by large doses of dexamethasone in cancer induced Cushing's, but are greatly suppressed by such treatment in Cushing's disease; and 5- corticotropin releasing hormone (CRH) stimulation increases ACTH and cortisol secretion in Cushing's disease, but generally produces no change in cancer-induced Cushing's. Hypercalcaemia and cancer: Hypercalcaemia as a manifestation of cancer is a relatively common finding. In 15% of RCC, hypercalcaemia exists, and also hypophosphataemia presents. 1% of prostate cancer, carcinoma of the penis, and carcinoma of the bladder may produce hypercalcaemia. Treatment of patient with modest doses of indomethacin may return blood calcium levels to normal. Parathormone-related protein (PTH-RP), is a 141-amino acid protein homologous with PTH in the amino terminus, both PTH and PTH-RP bind to the PTH receptor. The PTH gene is located on chromosome 11. The PTH-RP gene is expressed at low levels in many normal tissues. Mechanisms of hypercalcaemia associated with cancer: 1- cancer production, PTH-RP, 1,25 OH-D, cytokines, Parathormone; and 2- primary hyperparathyroidism. There is no single mechanism by which calcium homeostasis is disturbed in cancer. Many patients who develop hypercalcaemia do not have skeletal metastases. PTH-related protein (PTHrP) is structurally similar to parathyroid hormone and mimics much of the endogenous substance's activity, including increasing renal reabsorption of calcium. TGF-α stimulates osteoclasts activity. PGEs are potent stimulators of bone resorption by osteoclasts, are released by cancer cells. TNF (released by macrophages), lymphotoxin (produced by T-lymphocytes), and IL-1 (secreted by monocytes), all have osteoclastic effects. Elevated calcium concentration causes a tubular defect in the kidney that results in the loss of urinary concentrating ability and polyuria, both of which promote dehydration. Profound cellular dehydration and hypotension may develop from electrolyte and volume depletion by reduced proximal reabsorption of sodium, magnesium, and potassium. In some patients, renal failure may develop. Bone loss due immobilisation and lack of physical activity, are common in oncology patients. Hypercalcaemia develop after only 4 weeks of bed rest. Inappropriate use of diuretics, poor diet, and general physiological wasting are also contributing to hypercalcaemia. Hypercalcaemia causes nonspecific neurological, musculoskeletal, cardiovascular, and GIT symptoms primarily by decreasing the excitability of nerve cells and the contractility of smooth, striated, and cardiac muscle cells.

Effects of tumours on the skeletal and calcium homeostasis:

Osteolytic bone disease	It is due to an increase in osteoclastic bone resorption. It is responsible for catastrophic consequences in the patient with malignant disease, such as intractable bone pain, susceptibility to fracture, hypercalcaemia, nerve compression syndromes, and spinal cord compression, e.g., myeloma.
Osteoblastic bone disease.	The increased bone formation is due to the tumour cells stimulating osteoblast activity. Carcinoma of urinary tract, Hodgkin's disease, carcinoma of the breast.
Mixed osteolytic and osteoblastic lesions.	Solid tumours of the skeleton.
Hypercalcaemia	Associated with a marked increase in osteoclastic bone resorption, increased renal tubular calcium reabsorption, and impairment in GFR.
Hypocalcaemia	Occurs in multiple osteoblastic and in tumour lysis syndrome.
Oncogenic osteomalacia	Osteomas, haemangiopericytomas, CaP, and osteomalacia superimposed on metastatic bone disease.

Careful monitoring of serum calcium levels in cancer patients is essential. Treatment begins with hydration to promote urinary Ca excretion, inhibit bone resorption, and reduce the entry of Ca into the ECF. Mortality from hypercalcaemia without treatment is 50%. Hydration stimulates renal excretion of Ca and Na by increasing the GFR, and it interferes with Ca reabsorption in the proximal tubule. Renal function must be evaluated by before initiation of any large fluid infusion. Dialysis is indicated in CRF. Teaching of the patient must focus on prevention, and possible complication of their illness. Hypercalcaemia often heralds the final stages of cancer. The prevention and treatment of hypercalcaemia can help control the patient's symptoms and thereby preserve the QoL as well as prolongation of survival. Tumours cause multiple effects on the skeleton and on calcium homeostasis. There are 2 distinct syndromes for hypercalcaemia, the humoral hypercalcaemia of malignancy (HHM), and hypercalcaemia based on widespread osteolysis. In multiple osteoblastic lesions hypocalcaemia, the extracellular calcaemia is presumably utilised for the mineralisation of new bone, which is being rapidly formed. In tumour lysis syndromes, the tumour releases intracellular ions such as phosphate, K and uric acid. Increased intracellular phosphate release lead to a rapid fall in serum calcium. Chorionic gonadotropin, human chorionic gonadotropin (HCG) is a glycoprotein biochemically related to human thyrotropin, LH, and FSH. All 4 hormones are composed of α subunit, which is identical in amino acid sequences, and β subunit, which is unique to each and which counters biological and immunological specificity, e.g., RCC. There are 2 sources of β core: 1- direct secretion by trophoblast cells or carcinoma; and 2- degradation of intact HCG by the kidney with excretion into urine. Erythropoietin production by tumours, both carcinomas and sarcomas causes polycythaemia. The hypophosphataemia syndrome appears to be caused by tumour production of a protein, which inhibits renal cell reabsorption of phosphorus, occurs in 7% of prostate cancer. This syndrome may be cured by surgical excision of the neoplasm. Renin production by cancers:

Wilms' tumour produces pro-renin. After tumour removal, prorenin and renin values decreases to normal. Renin production can be caused by tumours arising in tissues outside the kidney, e.g., paraganglioma of adrenal origin. Hypoglycaemia is associated with 10% of adrenal carcinoma.

PARANEOPLASTIC SYNDROMES OF THE KIDNEY

Malignant disease is associated with a wide variety of derangements in renal function and electrolyte homeostasis, which leads to a clinically significant worsening of health status and rarely may lead to the patient's death. The paraneoplastic syndromes of the kidney are the following categories: 1- glomerular diseases, e.g., glomerulonephritis; 2- microvascular involvement; 3- fluid and electrolyte abnormalities; 4- structural abnormalities resulting from tumours products such as myeloma protein, e.g., ARF in 50% secondary to infections, ARF and CRF related to the toxic effects of light chains of immunoglobulins and Tamm-Horsfall proteins on renal function, and hypercalcaemia induces reversible ARF; and 5- renal tubular obstruction by tumour products, e.g., Bence-Jones proteinuria (myeloma-kidney), light chains and Tamm-Horsfall proteins make up the proteinaceous casts that obstruct the distal tubule. Prostate cancer, bladder cancer, RCC, and Wilms' tumour can induce membranous nephropathy. It is uncertain whether malignancy is the cause of membranous nephropathy when the 2 diseases coexist. Benign tumours also have been associated with membranous nephropathy. Renal microvascular involvement in association with neoplasia takes two forms: 1- thrombotic microangiopathy (TMA). Haemolytic uraemic syndrome, usually occur as a complication of chemotherapeutic agents. Disseminated intravascular coagulation (DIC) from sepsis, and specific malignancies including prostate carcinoma; and 2- vasculitis. Hyponatraemia is defined by serum sodium ≤130 mEq/l, in the presence of hypervolaemic, euvolaemic, or hypovolaemic ECV state. Those with malignant ascites or oedema of any cause are typically of the hypervolaemic group, which is characterised by a urinary Na. The euvolaemic group is due to the syndrome of inappropriate secretion of antidiuretic hormone (SIADH), e.g., malignancies involving the thorax and CNS. Hyponatraemia and hypouricaemia coexists in SIADH, serum urate ≤4 mg/dl. Natriuretic factor in the plasma of patients with intracranial diseases, tumours of the CNS and extracranial malignancies might account for the cerebral salt wasting syndrome of patients with intracranial diseases.

HAEMATOLOGICAL PARANEOPLASTIC SYNDROMES

Cancer may indirectly affect both the cellular elements of the blood as well as the coagulation system. Iron deficiency anaemia may be related to bleeding mucosal tumours, nutritional deficiencies, marrow invasion by tumours, and coexistent inflammatory disorders or endocrine disease. Chemotherapy and radiation can also suppress bone marrow function and result in anaemia. A normochromic or slightly hypochromic RBC morphology, low serum iron, normal to increased ferritin levels and iron stores, are the characteristics of iron deficiency anaemia. Anaemia of chronic inflammation is related to IL-1, TNF, and TGF-β produced or stimulated by the neoplasm mediate low serum erythropoietin (EPO). Most often erythrocytosis is not a manifestation of solid malignancy, but rather it is a phenomenon associated with myeloproliferative disorders or as a result of a physiological compensation for hypoxaemia. Leukocytosis, most likely due to cytokine release or secondary to the effects of IL-1 or granulocyte colony-stimulating factor released or induced by the presence of tumour. Thrombocytopenia is frequently the result of antitumour therapy. Rarely an autoimmune thrombocytopenia occurs in association with a solid malignancy. Thrombocytosis may be due to GI bleeding, coexisting inflammatory disorders, chemotherapy-induced myelosuppression, and tumour cytokine release such as IL-6 and thrombopoietin. Factors contribute to thrombosis are: fibrinopeptide A, postsurgical state, immobility, procoagulant effects of chemotherapy, coexistent atherosclerotic disease, procoagulant mediators released by tumour cells such as sialic acid, tissue factor that activate haemostatic mechanisms after disruption of endothelium may gain direct access to circulation blood by abnormal tumour vascularity or direct release of tumours cells into the circulation, cytokines with procoagulant activity. Abnormal activation of the haemostatic system can also present as haemorrhage when coagulation factors and platelets are secondarily depleted by ongoing consumption, e.g., DIC. Anaemia is common but easily treated with transfusion if symptomatic and may now be palliated with recombinant EPO in selected circumstances. The thrombotic and haemorrhagic manifestations of DIC are perhaps the most frequently troublesome paraneoplastic syndrome associated with cancer. A general misunderstanding exists between patients and their oncologists about the impact of anaemia. Many cancer patients with mild-to-moderate anaemia suffer markedly reduced functional capacity during chemotherapy, even long before they receive transfusion. Anaemia is a common complication of myelosuppressive chemotherapy that results in decreased functional capacity and QoL for cancer patients. The typical incidence of chemotherapy-induced anaemia is 30-80% with a single-agent use. The frequency and severity of anaemia in cancer patients are variable. The main mechanisms involved in the development of chemotherapy-induced anaemia are direct bone marrow damage and renal impairment leading to deficient production of erythropoietin (EPO). In addition to the myelosuppressive effects of chemo- and radio-therapy, cancer-related anaemia can result also from inflammatory responses and from activation of the immune system (anaemia of chronic disease, AIDS). It is a consequence of release of IL-1 and TNF and some IFNs. RBC survival is shortened, possibly under the influence of anaemia-inducing substances released from tumour cells or from cells involved in immune or inflammatory responses. There is failure of erythropoiesis as a result of suppression of erythroid progenitor cells, to impaired iron utilisation and to inadequate erythropoietin production. rhEPO compensates for the relative endogenous erythropoietin deficiency, and overcome the suppression of erythroid progenitor cells. Cisplatin inhibits renal erythropoietin production, and cumulative dose leads to progressive decrease in renal function with increasing defective erythropoietin production. The free radicals induced by therapeutic radiation by therapeutic radiation and some chemotherapy drugs need to be activated by oxygen before they can fatally damage tumour cells.

Coagulation factors:

Factor	Name	Normal range
I	Fibrinogen	142-366 mg/dl
II	Prothrombin	80-120%
III	Tissue factor, tissue thromboplastin (extrinsic prothrombin activator)	80-120%
IV	Calcium	8.5-10.5 mg/dl
V	Proaccelerin, accelerator globulin	50-150%
VII	Proconvertin, serum prothrombin conversion accelerator (SPCA).	60-140%
VIII	Antihaemophilic globulin (AHG), antihaemophilic factor (AHF)	60-150%
IX	Plasma thromboplastin component (PTC), Christmas factor	60-150%
X	Stuart-Power factor	60-150%
XI	Plasma thromboplastin antecedent (PTA)	60-135%
XII	Hageman factor	50-150%
XIII	Fibrin stabilising factor (FSF)	Present

Common haematological paraneoplastic syndromes:

RBCs disorders:
    A- Anaemia:
        1- Anaemia of cancer.
        2- Autoimmune haemolytic anaemia.
        3- Microangiopathic haemolytic anaemia.
    B- Erythrocytosis
WBCs disorders:
    A- Leukocytosis:
        1- Leukemoid reactions.
        2- Eosinophilia and basophilia.
    B- Neutropenia   Autoantibody mediated.
Platelet disorders:
    A- Thrombocytopenia:
        1- Autoimmune thrombocytopenia.
        2- Thrombotic thrombocytopenic purpura and Haemolytic uraemic syndromes.
    B- Platelet functional disorders   Hyperfunctional and hypofunctional.
    C- Thrombocytosis
Haemostatic disorders:
    A- Thrombotic:
        1- Trousseau's syndrome.
        2- Non-bacterial thrombotic endocarditis.
        3- Budd-Chiari syndrome and portal vein thrombosis.
    B- Haemorrhagic:
        1- Disseminated intravascular coagulation.
        2- Amyloidosis.
        3- Acquired von Willebrand disease.

A hypoxic state may encourage tumours to release angiogenic factors, which promote the formation of new blood vessels around the tumour, thereby facilitating tumour growth and metastasis. Hypoxia stimulates cellular expression of certain proteins that inhibit apoptosis, and a low oxygen level can lead to selective pressures on tumour cells that result in the evolution of therapy-resistant forms. Glutathione (GSH) and its related enzymatic system constitute an important cellular detoxification system, especially for RBCs. Cisplatin (CP) binds to GSH and may alter the enzyme-substrate interaction between GSH and GSH peroxidase, making RBCs much more sensitive to oxidative stress. Haemoglobin (Hb) level after radio-chemotherapy is a significant prognostic factor. The subnormal EPO most likely the result of the general decline in protein biosynthesis and/or increased protein catabolism characteristic of cachexia. TNF-α may be the primary mediator of cachexia. Suppression of RBC precursors in the bone marrow by cytokines such as TNF-α is the primary mechanism. Anaemia will continue to affect large numbers of cancer patients, leading to a decrease in functional capacity and QoL. The causes of anaemia in patients with malignancy are multifactorial including: 1- Anaemia of chronic disease, bone marrow infiltration, and myelosuppressive effects of chemotherapy. 2- Causes seen in patients without cancer, such as bleeding, haemolysis, infection, and nutritional deficiencies. Inflammatory cytokines might contribute to its pathogenesis, including an increase in IL-1, TNF-α, and IFNG. The impact of anaemia on the lives of patients with cancer is from a quality-of-life perspective and outcomes perspective. Each patient with cancer and anaemia should undergo an evaluation for other reversible causes of anaemia, e.g., drug, nutritional deficiency (i.e., ferritin, folate, and B_{12} levels), haemolysis (i.e., LDH, bilirubin, haptoglobin, and reticulocyte count), or comorbidities such as renal disease (i.e., serum creatinine). The most common symptoms reported at the time of transfusion are lethargy, tiredness, and breathlessness. A decrease of 1 g/dl in the Hb concentration results in a rise of 20 μg/l in the serum ferritin concentration. Serum ferritin contains virtually no iron and is secreted from the reticuloendothelial cells. Its function is unknown. Intracellular ferritin is part of the iron storage system and is not released into plasma unless there is cellular damage. A raised serum ferritin may be the result of increased iron stores but is more likely to come from intracellular release. Transfusion-associated cardiomyopathies and infections appear to be explained by blood-borne infection and other comorbidities and not the iron load. The marrow requires 30-40 mg of iron each day to sustain normal red cell production and this is derived from the iron recycled from effete red cells. When erythropoiesis is stimulated by rhEPO therapy, this supply will be insufficient. An extra 40-50 mg iron a day will be required. The stores cannot deliver this and i.v. iron therapy is required (200-300 mg of i.v. iron per week). In the maintenance phase of rhEPO therapy as little as 100 mg of iron a week may be appropriate depending upon the percentage of hypochromic cells in the circulation or transferrin saturation if that is not available. The intravenous iron therapy should aim at keeping the percentage of hypochromic cells near normal level (<5%). Oral ferrous iron therapy has side effects and poor compliance. Milder degrees of chemotherapy-induced anaemia may be associated with decreased QoL and that rhEPO-induced increase in haemoglobin levels is associated with improvements in QoL independent of disease response. rhEPO treatment (150 IU/kg/day, 3 times a week) is effective and safe in children with chemotherapy-induced anaemia. A good response would be achieved within 4 weeks. Epoetin is contraindicated in patients with uncontrolled hypertension or during pregnancy or lactation. Adverse events should be monitored included hypertension, seizures, thrombotic events, and allergic reactions.

CANCER PAIN

Cancer pain characteristics, syndromes and pathobiologies are very heterogeneous. Patients with chronic cancer-related pain usually experience fluctuations in pain intensity. Often, these fluctuations occur as discrete transitory flares of pain. When these flares interrupt a tolerable background pain, they are commonly described as "breakthrough pains" or "incident pains". In the cancer setting, breakthrough pain (BP) typically interrupts a background pain that is generally well controlled by opioid therapy. Incident pain is usually understood to be a subtype of BP induced by movement or some other voluntary action of the patient. BP is a prevalent phenomenon associated with adverse outcomes. 64% of inpatients with cancer pain report breakthrough pain, and 90% of ambulatory cancer patients experience pain with movement. The presence of BP reduces the likelihood of a satisfactory response to opioid therapy. Breakthrough pains are heterogeneous and must be distinguished from recurrent acute pains and other clinically insignificant fluctuations in the intensity of chronic pain. Patient may experiences multiple pains during the day of rapid onset. Precipitating factors are often identified, but pains occur without warning about half the time. In general, the occurrence of BP is a marker of a more severe pain syndrome. BP is also associated with relatively more severe pain-related functional impairment and psychological distress. The onset time of breakthrough pain for most patients is measured in minutes. The time required reaching peak concentration after oral administration of a supplemental analgesic dose is far longer. For many patients, particularly those whose pains are both unpredictable and rapid in onset, the utility of conventional therapy may be limited. The phenomenon of BP is associated with adverse functional and psychological outcomes. The assessment of BP is difficult, given its heterogeneity and lack of predictability. BP should be recognised as a common problem, which is often associated with other aversive experiences. Interventions should attempt to provide comfort, support the patient's sense of control over the pain, and address the concurrent physical and psychosocial burdens that contribute to impaired quality of life. Pain is a common problem in cancer patients, and morphine is the most common analgesic in advanced disease. Cancer patients commonly tolerate high doses of morphine without respiratory depression. Pain is a physiological antagonist of the respiratory depressant effects of opioids; reduction in the painful stimulus may precipitate abrupt respiratory depression if the opioid dose is not decreased. The dose of morphine becomes excessive only after the neurologic damage. Patients with decompensated chronic liver disease have reduced metabolism and clearance of morphine. Hepatic encephalopathy may be precipitated if the morphine dose is not reduced. Another potential factor is impaired respiratory muscle function as a result of spinal cord compression and radiation therapy to the spine. The patient should improve immediately after one dose of 0.4 mg of naloxone given intravenously. Complex regional pain syndrome I (CRPS I), also called reflex sympathetic dystrophy (RSD) can occur notably in an extremity and is characterised by regional severe pain, swelling, vasomotor changes (changes in skin temperature and colour) and a reduced range of motion. Signs and symptoms increase during or after effort, for example when using an affected hand for activities of daily living. They occur in an area much larger than the area of primary injury or operation and include the area distal to the primary injury. Pain is a leading symptom in CRPS I. The pain level that occurs with CRPS I can vary widely. The pain changes often during the day and night, but its intensity can also be constant. The intensity of the pain may increase during or after loading of the affected extremity, or owing to changes in the environmental temperature. Evoked pain is generally described as hyperalgesia, hyperpathia, or allodynia. One single pain recording does not seem to be sufficient, especially if it only concerns the momentary pain. Momentary pain is greatly influenced by precipitating activities and the time of day. Also, it is often impossible to rate the pain; pain not only changes in magnitude, but it can also manifest itself in different forms. One week seems sufficient to rate the changes in pain manifestation and it is not too long to forget worsening or lessening of the total pain presentation.

Paraneoplastic syndrome:

Syndrome	Causes
Endocrine	
ACTH/Cushing's.	Lung, adrenal cancer.
Inappropriate ADH secretion	Lung cancer
Hypercalcaemia	Prostate, kidney, lung, breast, myeloma
Adult onset rickets	Soft tissue/ bone tumours
Hypoglycaemia	Insulinomas, sarcomas, hepatomas.
Neurogenic	
Encephalomyelitis	Lung cancer
Opsoclonus	Lung, breast, uterine cancers. Paediatric neuroblastoma.
Neuropathies and neuromuscular syndrome	Myeloma, breast, lung, ovarian cancers, thymomas
Cerebellar atrophy	Breast, ovarian cancers
Haematologic	
Erythrocytosis	RCC, Wilms' tumour, hepatoma.
Granulocytosis	GI, lung, brain tumours, melanoma
Eosinophilia	Lymphomas, melanoma, brain tumours.
Thrombocytosis	Leukaemias, lymphomas
Disseminated intravascular coagulation (DIC)	Prostate, pancreas, lung, stomach cancers.
Migratory thrombophlebitis	Prostate, GI, breast, ovarian cancers.
Dermatologic	
Acanthosis nigricans	Gastric, abdominal cancers.
Leser-T related keratoses.	Lymphomas, GI cancers.
Bazex's acrokeratosis	Head and neck, GI, lung cancers.
Flushing	Carcinoid tumour.
Dermatomyositis	All types of cancer.
Pruritus	Lymphomas
Miscellaneous	
Fever	RCC, lymphoma, hepatoma.
Nonmetastatic hepatic dysfunction	RCC.
Hypertrophic osteoarthropathy	Lung cancer, tumours metastatic to chest.
Amyloidosis	RCC, myeloma, lymphoma.
Renal and electrolytes	
Uric acid nephropathy	Large cell lymphoma, Burkitt's lymphoma, acute leukaemias, response to chemotherapy or radiotherapy, small lung cancer, testicular cancer.
Rapid tumour lysis syndrome	Burkitt's and undifferentiated lymphomas, acute lymphoblastic leukaemia, non-Hodgkin's lymphomas, myeloproliferative syndromes, and small cell lung cancer.

WOUND HEALING

Anaesthetists already use drug, which enhance oxidant production (high inspired oxygen concentrations and halothane) or posses antioxidant properties (mannitol, propofol). Oxidants are produced under normal conditions both accidentally by leakage from the respiratory chain and deliberately to serve useful biological roles. Under pathological conditions oxidants production contributes to many disease states and in some cases may be the underlying cause. When oxidant activity exceeds antioxidant capacity the resulting oxidative stress can damage lipids, proteins, DNA, and connective tissue. Oxidative stress occurs in the following conditions: 1- reperfusion injury (brain, myocardium, gut, kidney, and transplanted organs); 2- sepsis/inflammation; 3- shock/trauma; 4- pulmonary toxicity; and 5- halothane hepatitis. Oxidants are extremely reactive, short lived and difficult to detect in biological systems (hydroxyl radicals believed to have a half-life of 10^{-6}s. MRI can delineate macroscopic oxidant activity in acute lung injury by showing high-intensity spots where the lung damage has occurred. Useful biological roles of oxidants include: a- controlled release of energy within the respiratory chain; b- microbial killing; c- tumour surveillance; d- intracellular signal transduction; and e- DNA biosynthesis. Heat shock proteins assists in folding nascent proteins to their tertiary structures as they emerge from the ribosomal apparatus under normal conditions. As cells are subject to thermal or oxidant stress, rapid upregulation of heat shock proteins occurs with suppression of normal protein synthesis and subsequent production against further stress. Ischaemic preconditioning is a short period of ischaemia that offers protection against subsequent ischaemic insults and this appears to be biphasic, with short-term protection probably mediated through adenosine, and longer protection (>24 hr) mediated through heat shock proteins. Antioxidants and HSPs repair DNA damage and restore protein homeostasis. Exogenous antioxidants strategies for antagonising oxidant stress include: a- preventing initial oxidant production; and b- stopping oxidants begetting further oxidants through chain reaction. With acute disease, timing of antioxidant intervention is important, particularly with reperfusion injury. Neutrophils can produce oxidants through 3 pathways, which in turn they use to kill ingested microbes. This killing role may be extended to abnormal or mutant cells during immune surveillance, which may be particularly important in the prevention of cancer. Oxidants are intracellular signals upregulating defences during oxidant stress (increases expression of HSP) and under normal physiological conditions. Oxidants are involved in protein synthesis through their interactions with redox-sensitive recognition sites. Exogenous antioxidants may interfere with this process by: 1- Altering the intracellular level of available oxidant, they may indirectly reduce protein synthesis coupled to oxidant stress. Since these proteins would have served a protective role then exogenous antioxidant may paradoxically increase the effects of oxidant stress. 2- Antioxidant responsive element (ARE), which on exposure to redox antioxidants increases the genetic expression of antioxidant enzymes (NADPH reductase, glutathione-S-transferase). Exogenous antioxidants may not act on the ARE to upregulate defensive enzymes in this way. 5-10% of oxygen enters mitochondria is used for oxidants formation. Endogenous antioxidants may also work synergistically. Vitamin E and C are believed to act synergistically, with free radical bearing vitamin E in cell membranes being regenerated by aqueous-phase vitamin C. Increasing membranous vitamin E without regulating vitamin C levels may not produce the treatment effect anticipated. Excess antioxidant activity could, paradoxically, protect mutagenic DNA from natural destruction, leading to tumour promotion. Endogenous antioxidant defence is internally regulated and in many cases orchestrated by a number of complementary biologically active molecules. Exogenous antioxidants during oxidant stress may improve outcome but equally may produce indifferent or even damaging effects.

Effects of oxidant:

Biological target	Damage
1- Proteins	A- Dysfunctional enzymes.
	B- Ultrastructural disruption.
2- Lipids	A- Disruption of cellular membranes and organelles.
	B- Chain reaction producing further oxidants (lipid peroxidation products).
	C- Damage to circulation lipids
3- DNA	A- Dysfunctional genes.
	B- Mutations.
4- Connective tissue	A- Direct: oxidation of collagen and lipopolysaccharides.
	B- Indirect: through elastase and collagenase released from activated neutrophils.

Endogenous antioxidant:

Blood	Cytoplasm	Membranes
Vitamin C	Superoxide dismutase (SOD)	Vit E
Urate	Glutathione peroxidase (GPX)	Phospholipid hydroperoxide glutathione peroxidase (PHGPX)
Haemoglobin	Glutathione-S-transferase (GST)	
Vit A	Glutathione	
Glutathione		

The mechanisms for deleterious effects include suppressing upregulation of more appropriate endogenous defences, conversion of low-toxicity oxidant to higher toxicity oxidant (superoxide to hydrogen peroxide when using SOD alone) and pro-oxidant activity of vitamin C in the presence of free iron. During ischaemia, oxidants may be derived from the organ in question, from blood elements or from the vascular endothelium of the blood vessels supplying it. The principal sources of free radicals in reperfusion injury are xanthine oxidase (XO), activated neutrophils, AA metabolites, free transition metal ions (iron, copper) and NO interaction with superoxide. During ischaemia, vascular endothelial xanthine dehydrogenase is converted to XO, which in the presence of molecular oxygen after reperfusion acts on hypoxanthine to produce superoxide radicals. Activated neutrophil provides superoxide and singlet oxygen through their respiratory burst. XO and iron-mediated free radicals may appear over a longer time course particularly associated with prolonged ischaemia and necrosis. Vascular endothelium is a highly active metabolic site produces many mediators of acute inflammation, elements promoting free radical production and leucocyte adhesion. Free iron is a potent source of cerebral radicals, and traumatic release of free iron from injured brain cannot be contained by the limited transferrin concentrations in CSF. Histidine antagonises singlet oxygen, protects against reperfusion injury. During ischaemia early XO and iron-mediated radicals are followed by late, neutrophil-derived free radicals. Free radicals cause lipid membrane disruption, which in turn releases excessive free cytosolic calcium and AA metabolites, leading to chemotaxis and neutrophil activation. Complement is activated by free radicals and itself generates further oxidants and chemotactic agents as well as producing direct cell damage through membrane attack complexes, which again release AA substrate. Auto-oxidative injury is a frequent physiologic abnormality that persists following trauma. Auto-oxidative injury has not been reported to occur secondary to blood transfusion, blood loss, or glucocorticoids. Trauma induces auto-oxidative receptor injury to intracellular and cell surface receptors. Trauma induces auto-oxidative reduction of Fc and complement receptors. This defect correlates with the development of nosocomial infections in trauma patients and provides a biological basis for the addition of exogenous antioxidants or strategies to increase endogenous antioxidant levels as therapy shortly after severe trauma. Critically ill trauma patients have auto-oxidative receptor injury, which is closely linked with the development of nosocomial infections. Nosocomial infections occur in 87% of patients with auto-oxidative injury. Oxidants mediate intestinal reperfusion injury and antioxidant therapy in man has been shown to be effective. Oxidants are the molecular trigger and phospholipase A_2 the enzymatic trigger for post-reperfusion gut injury. XO probably triggers the mucosal reperfusion injury. Pretreatment with the xanthine oxidase inhibitor allopurinol prevents intestinal reperfusion injury. Agents interfering with leucocyte adherence also protect against reperfusion injury. Allopurinol and desferioxamine improve cure and relapse rates in acute colitis and peptic ulceration. In critically ill patients with stress ulceration, free radical-mediated damage secondary to mucosal ischaemia and reperfusion is a major contributory factor. The incidence of endoscopically proven gastric ulceration is 3% when allopurinol is administered within 3 days of admission to ICU, 4% with dimethyl sulphoxide, and 22% if no treatment is giving. It is probable that mucosal villous degradation allows translocation of proteolytic enzymes, bacteria and endotoxin, with adverse cardiopulmonary effects. This villous damage can occur even in the presence of apparently adequate intestinal blood flow, through extravascular short-circuiting of oxygen at the base of villi.

Factors associated with wound healing:

Factors	Function
A- Haemostatic factors:	
1- Fibrin, plasma fibronectin	Coagulation, chemoattraction, adhesion, scaffolding for cell migration.
2- Factor XIII (fibrin-stabilising factor)	Induced chemoattraction and adhesion.
3- Circulatory growth factors	Regulation of chemoattraction, mitogenesis, fibroplasia.
4- Complement	Antimicrobial activity, chemoattraction.
B- Platelet-derived factors:	
1- Cytokines, growth factors	Regulation of chemoattraction, mitogenesis, fibroplasia.
2- Fibronectin	Early matrix, ligand for platelet aggregation.
3- Platelet activating factor (PAF)	Platelet aggregation.
4- Thromboxane A (TXA)	Vasoconstriction, platelet aggregation, chemotaxis.
5- Platelet factor 4	Chemotactic for fibroblasts and monocytes, neutralises activity of heparin, inhibits collagenase.
6- Serotonin	Induces vascular permeability, chemoattractant for neutrophils.
7- Adenosine dinucleotide	Stimulates cell proliferation and migration, induces platelet aggregation.

The lung is exposed to higher concentration of oxygen than other body organs, particularly when higher inspired oxygen concentrations are administered during anaesthesia or intensive care. Lung injury may be produced if hyperoxia increases free radical generation (through semiquinones) and antioxidant defences are reduced (steroid-mediated reduction in SOD and catalase). Many of the predisposing factors to lung injury will produce free radicals through activated neutrophils, free transition metals and prostaglandin synthesis and endothelial NO. Neutrophils are particularly implicated since large numbers of neutrophils are observed in the lung vasculature in ARDS, while neutrophil depletion is protective. Free radicals may produce direct damage but they have such a short half-life as alternative mechanism. Factors should be determined in patients with soft tissue injuries include: 1- type of the wounds (e.g., abrasion, or contusion); 2- the cause of injury; 3- patient's age, 4- the site of the injured tissues; 5- degree of contamination of the injured area; 6- associated injuries; and 7- the general condition of the patient. Contusion and swelling require ice packs for 24 hrs, rest and elevation. General principles of wound healing can be divided into 3 distinct but overlapping phases: 1- haemostasis and inflammation; 2- proliferation; and 3- maturation or remodelling. Failure or prolongation in one phase may result in delay of healing or nonclosure of the wound. Wound healing failures remain a significant clinical problem with large impact on health care costs. All injuries result into motion an orderly sequence of events that are involved in the healing response characterised by the movement of specialised cells into the wound site. Platelets and inflammatory cells are the first cells to arrive, and they proved signals known as cytokines needed for the influx of connective tissue cells and a new blood supply. The fibroblast is the connective tissue cell responsible for collagen deposition that is needed to repair the tissue injury. Collagen is the most abundant protein, accounting for 30% of the total protein in the human body. If too much collagen is deposited in the wound site, normal anatomical structure is lost, function is compromised and fibrosis occurs. Conversely, if an insufficient amount of collagen is deposited, the wound is weak and may dehisce. Wound is a disruption of normal anatomical structure and function. Healing is the complex and dynamic process that results in the restoration of anatomical continuity and function. The 3 basic responses following injury are resolution, regeneration, and repair. Resolution defines the situation when there is temporary loss of architecture, e.g., oedema, when tissue is temporarily disrupted by the influx of fluid and plasma but after a short time the oedema resolves and the structure and function return to exactly the condition that existed before the injury. Regeneration occurs when there is loss of structure and function, replacement of that structure by regenerating exactly what was before injury. Epidermis can partially regenerate after injury. The pathological response to tissue injury is opposite to that of the normal repair response. In fibrosis there is excessive deposition of connective tissue, e.g., Peyronie's disease. In dehiscence, there is insufficient deposition of matrix and the tissue is weakened to the point where it can fall apart. Contraction is part of the normal process of healing but, if there is excessive contraction, it is known as the pathological process called contracture. All dermal wounds heal by 3 mechanisms: epithelialization, contraction, and matrix deposition. Matrix deposition is the process in which collagen, proteoglycans and attachment proteins are deposited to form a new extracellular matrix. Adult skin consists of 2 tissue layers: a keratinized stratified epidermis and an underlying thick layer of collagen-rich dermal connective tissue providing support and nourishment. Appendages such as hairs and glands are derived from, and linked to, the epidermis but project deep into the dermal layer. Because the skin serves as a protective barrier against the outside world, any break in it must be rapidly and efficiently mended. As soon as the skin and dermis are disrupted and blood leaks into the site of injury, the cascade of healing is set into motion. Platelets give off the clotting factors needed to control the bleeding and loss of fluid and electrolytes and cascade of cytokines that initiate the healing response. Platelet is the initial most important cell to set the healing response in motion. The fibrin clot traps RBCs, controls bleeding, once it is coated with fibronectin binds various cytokines that are released at the time of injury. A temporary repair is achieved in the form of a clot that plugs the defect, and over subsequent days steps to regenerate the missing parts are initiated. Inflammatory cells and then fibroblasts and capillaries invade the clot to form a contractile granulation tissue that draws the wound margins together, meanwhile the cut epidermal edges migrate forward to cover the denuded wound surface. The fibrin network also provides an initial matrix to give the wound site some stability and initiates the healing response. The fibrin in the clot is usually organised parallel with the lines of tension in the wound, providing a meshwork by which inflammatory cells, and subsequently fibroblasts are attracted to the wound site and attach to this matrix. Neutrophils are the predominant cell in the wound within 24 hrs after injury, to remove foreign material and bacteria that may present in the wound site. Bacteria give off chemical signals, attracting these inflammatory cells. The neutrophils will engorge themselves until they are filled with bacteria and, subsequently, these cells will be removed by macrophages. By 48 hrs after trauma macrophages, are the predominant cells in the wound. They are derived from fixed tissue monocytes that originate from peripheral monocytes normally present in connective tissue matrix. Macrophages give off chemical signals include, platelet derived growth factor (PDGF), transforming growth factor (TGF)-β. The mast cell releases granules filled with histamine, prostaglandin metabolites and tumour necrosis factor (TNF)-α. The mast cell is the major contributor in keloids. A host cell of polypeptide factors is released by these inflammatory cells. At least 13 individual types of collagen have been identified but, type I is predominant in scar tissue of the skin. After transcription and processing of the collagen mRNA, it is attached to polyribosomes on the endoplasmic reticulum where the new collagen chains are produced. The procollagen molecule is then secreted into the extracellular spaces where it is further processed. Hydroxyproline in collagen is important because it gives the molecule its stable helical conformation. Fully hydroxylated collagen has a higher melting temperature, whereas when hydroxyproline is not present in collagen. The collagen gets to the extracellular space where it undergoes further processing by cleavage of the procollagen N- and C-terminal peptides. As the collagen matures and becomes older, more and more of these intramolecular and intermolecular cross-links are placed in the molecules for strength and stability. Dermal collagen on a per weight basis approach the tensile strength of steel, in normal tissue it is a stronger molecule and highly organised. In contrast, collagen fibers formed in scar tissue are much smaller and have a random appearance; scar tissue is always weaker and will break apart before the surrounding normal tissue. Finally, in the process of collagen remodelling collagen degradation occurs. Specific collagenase enzymes in fibroblasts, neutrophils and macrophages clip the molecule at a specific site through all chains, and break it down to characteristic three-quarter and one-quarter pieces. These collagen fragments undergo further denaturation and digestion by other proteases. Growth factors are essential, specifically to the starting and stopping of each of the many cell activities by which the wound is healed. Most skin lesions are healed rapidly and efficiently within a week or two, although the end product is neither aesthetically or functionally perfect. Epidermal appendages that have been lost at the site of damage do not regenerate, and when the wound has healed there remains a connective tissue scar where the collagen matrix has been poorly reconstituted. The formation of a clot serves as temporary shield protecting the denuded wound tissues and provides a provisional matrix over and through which cells can migrate during the repair process. The clot consists of platelets embedded in a mesh of cross-linked fibrin fibers derived by thrombin cleavage of fibrinogen, together with smaller amounts of plasma fibronectin, vitronectin, and thrombopondin. The clot serves as a reservoir of cytokines and growth factors that are released as activated platelets degranulated. This early cocktail of growth factors provide chemotactic cues to the wound site, initiates the tissue movements of reepithelialization and connective tissue contraction, and stimulates the characteristic wound angiogenic response. Neutrophils and monocytes are attracted to wound sites by a huge variety of chemotactic signals. Growth factors released by degranulating platelets, but also cues diverse as formely methionyl peptides cleaved from bacterial proteins and the by-products of proteolysis of fibrin and other matrix components. Members of the selectin family of adhesion molecules expressed by the surface of endothelial cells lining capillaries at wound site allow rapid but light adhesion to neutrophils and monocytes. So that leukocytes are slowed and pulled from rapid circulation in the blood, followed by tighter adhesions and arrest mediated by the β_2 class of integrins, lead to diapaedesis, whereby the activated leukocytes crawl out between endothelial cells into the extravascular space.

Growth factors:

Growth factor	Source	Primary target cells and effect
EGF	Platelets.	Keratinocyte motogen and mitogen. Stimulates epithelialization, proliferation.
TGF-α	Macrophages, Keratinocytes.	Keratinocyte motogen and mitogen. Induces angiogenesis, epithelialization.
HB-EGF	Macrophages	Keratinocyte and fibroblast mitogen
FGFs1, 2, 4	Macrophages and damaged endothelial cells.	Angiogenic and fibroblasts mitogen. Stimulates proliferation, angiogenesis.
FGF7(KGF)	Dermal fibroblasts.	Keratinocyte motogen and mitogen.
PDGF	Platelets, macrophages, keratinocytes.	Chemotactic for macrophages, fibroblasts, macrophages activation, fibroblast mitogen, and matrix production. Induces proliferation.
IGF1	Plasma, platelets.	Endothelial cell and fibroblast mitogen.
VEGF	Keratinocytes, macrophages.	Angiogenesis.
TGFβ1 and -β2	Platelets, macrophages.	Keratinocyte migration, chemotactic for macrophages and fibroblasts, fibroblast matrix synthesis and remodelling. Increases matrix synthesis, proliferation.
TGFβ3	Macrophages.	Antiscarring.
CTGF	Fibroblasts, endothelia.	Fibroblasts, downstream of TGFβ1.
Activin	Fibroblasts, keratinocytes.	Unknown.
IL1β	Neutrophils.	Early activators of growth factor expression in macrophages, keratinocytes, and fibroblasts.
TGFα	Neutrophils.	Similar to the ILs.

Neutrophils normally begin arriving at the wound site within minutes of injury to clearing the initial rush of contaminating bacteria, also as a source of pro-inflammatory cytokines. Unless a wound is grossly infected, the neutrophil infiltration ceases after a few days, phagocytosed by tissue macrophages. Macrophages continue to accumulate at the wound site by recruitment of blood-borne monocytes and are essential for effective wound healing, if macrophages infiltration is prevented, then healing is severely impaired. Macrophages tasks include phagocytosis of any remaining pathogenic organisms and other cell and matrix debris. In unwounded skin, the basal keratinocyte layer attaches to a carpet of specialised matrix, the basal lamina. The keratinocyte's primary anchoring contacts are hemidesmosomes, which bind to laminin in the basal lamina by way of $\alpha_6\beta_4$ integrins and have intracellular links with the keratin cytoskeletal network. The hemidesmosome attachments have to be dissolved and leading edge keratinocytes have to express new integrins, primarily the $\alpha_5\beta_1$ and $\alpha_v\beta_6$ fibronectin/tenascin receptors and the $\alpha_2\beta_1$ collagen receptor, and relocalize $\alpha_2\beta_1$ collagen receptors, in order to grasp hold of, and crawl over, the provisional wound matrix and underlying wound dermis. Forward locomotion involves contraction of intracellular actinomysin filaments that insert into the new adhesion complexes. It is unclear which cells lead the keratinocytes' forward march. The chief fibrinolytic enzyme is plasmin, which is derived from plasminogen within the clot itself and can be activated either by tissue-type plasminogen activator (tPA) or urokinase-type plasminogen activator (uPA). Both of these activators are upregulated in the migrating keratinocytes. Various members of the matrix metalloproteinase (MMP) family, each of which cleaves a specific subset of matrix proteins, are also up-regulated by wound-edge keratinocytes. MMP-9 (gelatinase B) can cut basal lamina collagen (type IV), and anchoring fibril collagen (type VII), and is thought to be responsible for releasing keratinocytes from their tethers to the basal lamina. MMP-1 (interstitial collagenase) is up-regulated only in those basal keratinocytes that migrated beyond the free edge of the basal lamina, suggesting that cell-matrix interactions may control aids keratinocyte crawling by cutting collagens I and III at sites of focal adhesion attachment to the dermal substratum. MMP-10 (stromelysin-2) has wider substrate specificity and is also up-regulated by keratinocytes at the wound margin, but its expression is increased in situations of impaired healing. Once the denuded wound surface has been covered by a monolayer of keratinocytes, epidermal migration ceases and a new stratified epidermis with underlying basal lamina is re-established from the margins of the wound inward. Suprabasal cells cease to express integrins and basal keratins and instead undergo the standard differentiation program of cells in the outer layers of unwounded epidermis, probably due to contact inhibition arising from mechanical cues. EGF and TGF-α, heparin binding epidermal growth factor (HB-EGF), all acting as ligands for the EGF receptor, and are released in abundance at a site of injury. EGF activates the small GTPase Rac, which mediates lamellipodial extension. EGFs are the chief epidermal wound regulators with keratinocyte growth factor (KGF), or FGF7, which acts specifically on keratinocytes. KGF is up-regulated more than 100-fold within 24 hours by dermal fibroblasts at the wound margin, possibly in response to pro-inflammatory cytokines. TGF-β1 and some pro-inflammatory cytokines appear to stimulate expression of some of the integrin subunits that facilitate keratinocyte migration. Although the actin cytoskeleton is critical for crawling motility of adult keratinocytes, the keratin cytoskeletal network would supply essential cell and tissue strength during such strenuous epithelial movements. As with integrins, keratins that are normally basally restricted appear suprabasally in keratinocytes at the wound margin. The embryonic epidermis does not need intermediate filament support during the repair process. If an adult skin wound is deeper than the level of hair bulbs in the dermis so that no remnants of hair follicles remain, and the repairing epithelium does not regenerate hairs, or sweat glands. During embryogenesis, the dermal connective tissue fibroblasts supply permissive and instructive signals that govern the positions and types of hairs and other cutaneous appendages that will differentiate from the overlying epidermis. Adult wound epidermis fails to regenerate hairs, not because it is unable to respond to hair-inducing signals but because it does not receive such signals from the underlying wound dermis. As an early response to injury, resident dermal fibroblasts in the neighbourhood of the wound begin to proliferate, and then 3 or 4 days after the wound insult they begin migration into the provisional matrix of the wound clot where they lay down their own collagen-rich matrix. Many of the growth factors present at a wound site can act either as mitogens or as chemotactic factors for wound fibroblasts, such as platelet-derived growth factor (PDGF) and TGF-β. TGF-β-related growth factors Activin are induced in the proliferative fibroblasts of a wound margin and in the adjacent wound-edge keratinocytes. Connective-tissue growth factor (CTGF) is expressed at high levels by wound fibroblasts. Just as wound-edge keratinocytes have to adjust their integrin profile before migration, dermal fibroblasts, which normally lie in a collagen-I-rich matrix, must down-regulate their collagen receptors and up-regulate integrins that bind fibrin, fibronectin, and vitronectin in order to crawl into the clot. Fibroblasts read and act according to dual signals from their matrix surroundings and from the growth factor. Fibroblasts may use a fibronectin conduit to lead them into the fibrin clot. Fibronectin is a good substratum for cell migration. By about a week after wounding, the wound clot will have been fully invaded and all but replaced by activated fibroblasts that are stimulated by TGF-β1 and other growth factors to synthesise and remodel a new collagen-rich matrix, at his stage. A proportion of the wound fibroblasts transform into myofibroblasts, which express α-smooth muscle actin and resemble smooth muscle cells in their capacity for generating strong contractile forces. This conversion is triggered by growth factors such as TGF-β1. A number of growth factors at the wound site are potent stimulators of fibroblast-driven gel contraction and presumably signal granulation tissue contraction. Potential "stop" signals for wound contraction stimulate the loss of resistance after wound has closed. Within minutes of release from resisting forces, fibroblasts activate cAMP signal transduction pathway, which involves influx of extracellular Ca^{2+} ions and production of phosphatidic acid by phospholipase. Subsequently, PDGF and EGF receptors on the cell surface become desensitised and the relaxed cells return to a quiescent state similar to that existing before the injury. Apoptosis occurs in some of the wound fibroblasts, probably the myofibroblasts, after wound contraction has ceased. The wound connective tissue is known as granulation tissue because of the pink granular appearance of numerous capillaries that invade the wound neodermis. VEGF, also called vascular permeability factor, is induced in wound-edge keratinocytes and macrophages, possibly in response to KGF and TGFα, and synchronously at least one of its receptors, flt-1, is up-regulated by endothelial cells at site of injury. VEGF expression fails at the wound site and healing is impaired in diabetes mellitus. Endothelial cells must up-regulate $α_vβ_3$ integrins to respond to any wound angiogenic signal. $α_vβ_3$ is expressed transiently at the tips of sprouting capillaries in the granulation tissue. Capillary morphogenesis is dependent on tightly regulated proteolysis of the matrix surrounds during the invasion phase. As embryo develops, its skin becomes densely innervated by a plexus of sensory and sympathetic nerves serving the blood vessels and cutaneous appendages as well as supplying sensation. The sensory nerve termini are exquisitely sensitive to signals released after injury, resulting in transient nerve sprouting at the site of an adult skin lesion and more dramatic, permanent hyperinnervation after wounding of neonatal skin. The wounding-induced signal controlling this nerve overgrowth may be nerve growth factor (NGF), and because NGF is up-regulated after exposure to any of the TGFβ isoforms, it is tempting to consider nerves as another indirect target for TGFβ at wound site. Sparsely innervated regions of the body tend to heal poorly and lacking low-affinity NGF receptor. Connective-tissue contraction closes embryonic as well as adult wounds, but in embryos there is no apparent conversion from fibroblast to myofibroblast and neither is there a significant angiogenic response, no sign of a connective-tissue scar, and the repair is perfect. There is a strong correlation between the age of onset of scarring and the first stage in development when a noticeable inflammatory response is raised after wounding. In the embryo, TGFβ1 is expressed transiently and at low levels after injury, but at the adult wound site it is present at high levels for the duration of healing and beyond. TGFβ3 down-regulates the other two TGFβ isoforms. The mannose-6-phosphate (M-6-P)-IGFII receptors directly applied to wounds prevent scarring, and TGFβ antibodies neutralise TGFβ1 and β2 at the time of wounding reduce scarring. Growth factors and cytokines are the mediators for the various processes of wound healing. High levels of bacteria lead to up-regulated inflammatory mediators, up-regulated matrix metalloproteinases (MMPs), and decreased levels of endogenous growth factors in chronic wounds. Disruption of the wound healing process leads to a chronic non-healing wound. High bacterial levels can interfere with the orderly progression of wound healing and interrupt the process because of the effects on endogenous growth factors. In chronic wounds, repetitive trauma, ischaemia, and infection increase proinflammatory cytokines, increase MMPs and decrease the levels of growth factors. Tissue repair is interrupted, and failure of the wound to heal occurs, resulting in a chronic wound. Degradation of growth factors occurs in the presence of significant quantities of bacteria. The growth factors levels may drop up to 69% for gram-negative species and to 76% for gram-positive species. Further enhancement of bacterial degradation of the growth factors occurs in the presence of fibroblasts. Bacteria secrete proteases, which by cleaving complex molecules, can degrade and inactivate growth factors. When bacteria and fibroblasts are present together, enhanced degradation occurs, due to a synergistic effect of bacteria on fibroblasts or MMPs. Bacteria are the causative factor and fibroblasts are an affected enhancer of growth factor degradation. The presence of tissue components (fibroblasts) enhances and increases growth factors degradation. Bacterial proteases and MMPs synergistic effect degrades growth factors in wound tissue, and this mechanism could account for the degradation of bFGF, GM-CSF, KGF-2, and TGF-β2. Control of bacterial burden should be accomplished before use of exogenously applied growth factors to avoid degradation in the chronic wound environment.

ANDROLOGY

Sexual determination is a 3 process that must occur in fixed, sequential manner (chromosomal, gonadal, and phenotypic sex): 1) chromosomal sex is determined at the time of fertilisation by the union of maternal and paternal haploid gametes; 2) the chromosomal sex then dictates the gonadal sex by directing the formation of testes or ovaries; 3) the phenotypic sex is the result of testicular hormones acting upon the undifferentiated genitalia and other target tissues, whereas in the absence of these hormones the female phenotype is produced. Errors in the sex-determination pathway are not lethal to the individual but are detrimental to the species by making procreation impossible. The testicular-determining factor (TDF) acts upon small number of cells to initiate the development of a testis from the genital ridge, and the embryonal testicular cells then amplify the message of sexual differentiation by secreting hormones to create the male phenotype. Without the Y chromosome TDF is not present. Analysis of XX males revealed that TDF genes must by transferred from the Y chromosome to the X chromosome during paternal meiotic crossover. Androgens induce the virilization of the genitalia and multiple extragenital sites in utero and again at puberty to create the recognisable features of the male phenotype. Locally in the testis, testosterone promotes maturation of the developing spermatogenic tubules. It is also secreted into the foetal circulation as blood vessels grow into the developing testis, and through this distribution it exerts its masculinizing effects upon receptive extragonadal tissues. Testosterone directly causes virilization of the Wolffian duct into the epididymis, vas deferens, and seminal vesical. Development of the prostate and conversion of the genital analg into male genitalia are the result of the action of DHT, not testosterone. These structures can also be abnormally virilized to varying degrees in cases of female pseudohermaphroditism due to an excess of adrenal androgen production secondary to errors in adrenal biosynthesis of glucocorticoids and mineralocorticoids. Leydig cells develop from the undifferentiated mesenchymal cells of the testis under the direction of the Sertoli cells and produce testosterone almost immediately. Testosterone synthesis by the human foetal testis is independent of gonadotropin control during this time and up to 18 weeks. Following birth, foetal Leydig cells are replaced by adult Leydig cells, which remain relatively quiescent until the onset of puberty.

THE PROSTATE

The prostate gland is not an endocrine gland; but comprised of endocrine-sensitive tissue whose embryonic development, postnatal growth, maintenance and even unregulated growth are hormone responsive. The hypothalamic-pituitary-gonadal axis is a self-regulating system comprised of: 1- the hypothalamus, 2- the pituitary gland, and 3- the testes. The function of this system is to produce sufficient circulating gonadal steroids (androgen) to ensure appropriate target organ responses, such as, sexual maturation, maintenance of the sexually mature phenotype, and fertility. The integrating function of the hypothalamus is mirrored by its location at the base of the brain, bounded anatomically by the optic chiasm anteriorly, by the mamillary bodies posteriorly, and the thalamus superiorly. The most inferior portion of the hypothalamus is the medial eminence, from which the pituitary stalk arises. Gonadotropin-releasing hormone (GnRH), also known as luteinizing hormone-releasing hormone (LHRH), is a peptide hormone that is synthesised in the hypothalamus. LHRH is transported to the anterior lobe of the pituitary gland by means of a portal venous system, where it controls the synthesis and secretion of the gonadotropin hormones, luteinizing hormone (LH) and follicle-stimulating hormone (FSH). The suprachiasmatic nucleus, median eminence, and arcuate nucleus contain the most concentrated region of GnRH-releasing neurons, and most involved in the homeostatic control of the reproductive hormone axis. GnRH is a decapeptide synthesised in the neurosecretory neurons of the medial basal region of the hypothalamus, transported by axoplasmic flow to the axon terminals in the hypothalamic median eminence. The median eminence is highly vascularized by a portal circulation system derived from the superior and the inferior hypophyseal arteries. This capillary system is unique in the fenestration of its endothelial cells, which allow the free passage of macromolecules, including GnRH, without interference from a functional blood-brain barrier. Once secreted in the area of the median eminence, GnRH is carried by this vascular system along the pituitary stalk to the adenohypophysis (anterior pituitary) and neurohypophysis (posterior pituitary). GnRH is secreted into the regional vascular system once every 70-90 minutes. The half-life of GnRH in the circulation is 2-5 minutes. The pituitary is exposed to high concentrations of GnRH for very brief periods of time. The pulsatile nature of the signal conveyed by GnRH, not just the absolute concentration of GnRH, appears to be essential for LH and FSH production by the pituitary. A variety of neurotransmitters have been found in cells synapsing on GnRH neurons, including norepinephrine, dopamine, serotinin, γ-aminobutyric acid (GABA), and opioid peptides (endorphins). Both endogenous and exogenous opiates suppress GnRH secretion. When GnRH is released in a pulsatile fashion by the hypothalamus, the amplitude and frequency of this pulse result in LH and FSH into the peripheral circulation by the cells in the adenohypophysis. The glycopeptides LH and FSH are each composed of 2 chains, a shared common α chain, which is also shared with thyroid-stimulating hormone (TSH) and human chorionic gonadotropin (hCG), and a unique β chain. Although both LH and FSH are secreted in a cyclical fashion, the longer half-life of FSH results in a more constant peripheral blood concentration. By contrast, LH, which is more rapidly metabolised, has a peak and trough pattern with a cycle time of about 2 hours. After puberty, the periodic fluctuation in LH levels becomes independent of the sleep-wake cycle. Factors control gonadotropin secretion: 1- Trophic influence of GnRH. 2- Gonadal steroids, including androgens, oestrogen and prolactin, have the capacity of inhibiting the secretion of FSH and LH. The effect of androgen occurs at the post-GnRH receptor site. FSH and LH levels continue to rise up to 100 days after orchiectomy. All androgens are capable of inhibiting LH and FSH. 3- Oestrogens, estradiol is an oestrogen that can be synthesised by the testis but is derived primarily from the peripheral conversion of androgen (aromatisation). As men age, an increase in the total plasma estradiol levels is observed, although increased binding of estradiol by elevated serum testosterone-oestrogen-binding globulin (TeBG) maintains a relatively constant level of free estradiol so that there may not be a significant change in the steady-state inhibition of LH and FSH secretion by oestrogens. The free testosterone levels fall with age. Estradiol exhibits 1000-fold greater potency than testosterone in the inhibition of LH and FSH secretion by the pituitary gland. Whether or not there is a synergistic effect of estradiol and androgen in the regulation of LH or FSH is unknown, either compound is effective in suppressing LH production. Oestrogen has direct effects on prostatic tissue. 4- Gonadal peptides, the testicles are capable of producing gonadal peptide hormone, which is capable of inhibiting FSH secretion "inhibin". Sertoli cells produce the inhibin. Two subtypes exist, sharing a common α subunit and a unique β subunit, inhibin A and inhibin B. Combinations of the 2 β subunits appear to increase FSH secretion and have been termed activins. Inhibin secretion is stimulated by LH and FSH or both. 5- Prolactin has direct effects on prostatic tissue. Prolactin is involved in regulation of LH levels. Hyperprolactinaemia is associated with: a- suppressed LH levels, by inhibiting GnRH, b- lowered testosterone levels and hypogonadism, and c- impotence, through a direct effect on the CNS. Treatment with agents that lower prolactin levels may result in reversal of impotence in men with hyperprolactinaemia, but not androgen. Testosterone synthesis by the testis, steroidogenesis is undertaken by Leydig cells, characterised histologically by the presence of long, slender structures, the crystalloids of Reinke, which presumably are the site of androgen synthesis. Testosterone is synthesised from cholesterol by similar pathways in both testes and the adrenal cortex. Cholesterol is either synthesised from acetate or derived from LDL-cholesterol carried in the circulation and then is converted to pregnenolone by an oxidation reaction, catalysed by P450 scc. P450 scc are part of the cytochrome P450 system, which require oxygen as a substrate, and are located in the inner membrane of mitochondria. This mitochondrial membrane enzyme system is fairly tightly regulated and appears to be the site at which LH exerts its stimulatory effect on testosterone synthesis. Pregnenolone has 2 major metabolic fate: a- In the adrenal cortex, pregnenolone is hydrolysed to 17-hydroxypregnenolone, which in turn is converted to dehydroepiandrosterone (DHEA). DHEA is then sulfated to its conjugate, dehydroepiandrosterone sulfate (DHAES). B- In the testis, DHEA is converted to either androstenediol or androstenedione; both are the final substrate for the synthesis of testosterone. Testicular steroidogenesis by Leydig cells is controlled by the interaction of circulating LH with its receptor. The LH receptor is found in the cytoplasm of Leydig cells. The conversion of cholesterol to pregnenolone on the inner mitochondrial membrane is regulated by LH. The binding of LH to its receptor results in the production of cAMP. cAMP stimulates the synthesis of androgens by at least 2 mechanisms: 1- increased cAMP accelerates transport of cholesterol from lipid stores in the cytoplasm to mitochondria, and 2- cAMP activates Leydig cell protein kinase, which activates a secondary messenger that catalyses the cholesterol to pregnenolone reaction. Adrenal steroidogenesis, including adrenal androgen, is controlled by ACTH. ACTH interacts with specific receptor, in similarity by which LH interacts with its receptor in Leydig cells. The binding of ACTH to its receptor activates adenylate cyclase to convert ATP to cAMP. Elevated cAMP serve as a secondary messenger that activates a cAMP-dependent protein kinase, which in turn appears to activate an unknown factor that catalyses the mitochondrial conversion of cholesterol to pregnenolone. After orchiectomy or treatment with a GnRH analogue, human intraprostatic dihydrotestosterone (DHT) levels are lowered only 30-40% of normal, despite a 90-95% reduction in serum testosterone levels. The average plasma testosterone level in healthy human males is 700 ng/ml. Under physiologic conditions, Leydig cells account for 95% of testosterone production in adult men. About 5% of circulating testosterone is derived from the adrenal production of androgens, such as androstenedione, and their subsequent conversion in peripheral tissues, including prostate, to testosterone. Minor sleep-related increase in testosterone levels can be observed in young adults, which disappears with advancing age. Gonadal testosterone then released into the spermatic veins and from there into systemic circulation. Other testicular androgens compromise <20% of the testosterone concentration. Two features of the testosterone pool make it the most physiologically important androgen, that its relatively high serum concentration and its potency as an androgen. Although DHT appears to be 1.5-2 fold more potent than testosterone, its plasma concentration is less than one-tenth that of testosterone. Testosterone is relatively water insoluble, transported in the bloodstream bounded to plasma proteins (TeBG, albumin and corticosteroid-binding globulin (CBG), accounting for 57%, 40%, and <1%, respectively). Over 98% of testosterone are bound to plasma proteins, only 2-3% of the total testosterone pool is free, accounting for 12 ng/ml of testosterone. The rate of testosterone metabolism correlates with the amount of albumin-bound and free testosterone. TeBG-bound testosterone is biologically inactive. Exogenous administration of testosterone results in a 2-fold decrement in plasma concentrations of TeBG and potentially increasing the free testosterone level. Exogenous oestrogen results in a 5-10 fold increase in TeBG, causing increased binding of testosterone, effectively lowering the free testosterone level. Although testosterone bound to TeBG is unavailable for metabolism or binding by end organs, testosterone bound to albumin dissociates fairly readily and, along with the small amount of circulating free testosterone, comprises the pool of testosterone that can either exert its effects on end organ cells or be further metabolised. Four of the major metabolic pathways for testosterone include: a) peripheral aromatisation to oestrogen; b) hepatic conversion to nonandrogenic steroids, which then undergo "c) and d)"; c) hepatic conjugation to glucouronides or sulfates; and d) reduction to a more androgenic metabolite, DHT. The majority of oestrogens found in men are produced in peripheral tissues by the aromatisation of androstenedione and testosterone. The tissues in which the aromatisation reactions are believed to occur include adipose tissue, brain, and the breast. 75-90% of blood oestrogen is derived from testosterone and androstenedione. 10-25% of the total oestrogen pool is derived from the testes by Sertoli cells, after FSH stimulation. Free testosterone enters the cell by passive diffusion. In certain tissues such as the brain, pituitary and kidney, the majority of testosterone is bound unchanged by the AR. In the prostate, seminal vesicle, epididymis, liver, adrenals, and skin, testosterone serves as a prohormone; whose metabolic fate is the conversion by 5α-reductase to DHT. The androgen receptor (AR) has a 2-3-fold higher affinity for DHT than testosterone. DHT is primarily an intracellular androgen. DHT has a 5-fold higher prostatic tissue concentration than testosterone. Thus, DHT appears to be the primary intracellular ligand for the AR. The enzyme 5α-reductase, located on the endoplasmic reticulum and nuclear membrane, irreversibly reduces the unsaturated C4-5 bond in testosterone, thereby producing DHT. Two different isozymes of 5α-reductase exist: a) type 1, expressed in newborn skin and scalp and after birth in the liver; b) type 2, located in the accessory sex glands, including the prostate, seminal vesicles, epididymis, the foreskin, and in the liver. Regulation of 5α-reductase is not completely understood but appears to be tissue specific and dependent on androgen. Short-term androgen deprivation does not affect the expression of 5α-reductase in the prostate, and long-term androgen ablation results only in small decrease in 5α-reductase expression. When fenesteride is administered to normal men, DHT tissue level can be decreased to 70-90% of normal. 5α-reductase-2 activity appears to be concentrated in stromal cells. Whereas 5α-reductase activity has been shown to be higher in benign prostatic enlargement (BPE) than in normal prostatic tissue. DHT levels are not higher than normal in BPE. DHT can be reversibly converted to

3α-androstenediol. DHT can also be converted to 3β-androstenediol, which is rapidly and irreversibly converted to water-soluble inactive triol steroids as a final metabolic sink for DHT. Androgens regulation of normal functions of the prostate, the activated AR acts as a transcription activator, resulting in increased transcription of specific androgen target genes. The prostate is stimulated to grow and is maintained in size and secretory function by androgens. At puberty, androgens cause body and sexual maturation. Through adult life, androgens are required for the maintenance of the accessory sex glands. Although androgen stimulation prolongs DNA synthesis, once prostatic cell numbers return to baseline, DNA synthesis also falls to baseline levels (negative gene regulation), the mechanisms are not fully understood. There is a strong endocrine component to the genesis of BPE and carcinoma of the prostate (CaP). Human ejaculate comprised of seminal plasma and spermatozoa. Seminal plasma accounts for 99% of ejaculate volume and is compromised primarily of seminal vesicle secretions (85%) and prostatic secretions (15%). The prostate is comprised of extensive glandular epithelium composed of secretory epithelial cells, basal cells, and neuroendocrine cells embedded in a stromal compartment consisting of ECM and a variety of cell, including fibroblasts, angiolymphatic cells, and smooth muscle cells. Secretory epithelial cells are most affected by androgen deprivation, with an up to 90% decrease in cellular numbers and 80% decrease in tissue volume. Basal cells account for <10% of epithelial cells but may represent a prostatic stem cell. Neuroendocrine epithelial cells, although relatively small in number, almost certainly play an important role in prostatic growth. During early foetal development, the AR localises to mesenchyme only, and as the foetus matures, the AR can be found expressed on both stromal and epithelial cells. In the mature normal adult prostate, the AR is found primarily in the epithelial compartment. The stroma retains 5α-reductase activity. In both BPE and cancer, 5α-reductase-2 activity appears to be more concentrated in stromal cells. Oestrogen receptors are present in both human prostatic epithelial and stromal cells, but they appear to exist in a higher concentration in stromal cells. Oestrogens have other direct effects on prostatic tissue that are synergistic to those of androgens. Oestrogens may have a direct cytotoxic effect on prostatic cells. The prolactin receptor exists in high abundance in prostatic tissue. Patients with BPE have higher prolactin serum levels than control patients. The role of prolactin in prostatic growth is not fully understood. The actions of androgens within target cells are transduced by low abundance intracellular androgen receptor (AR). The AR is a member of a superfamily of nuclear transcription factors that mediate the action of steroid hormones. This gene superfamily includes specific receptors for oestrogen, progesterone, glucocorticoid, mineralocorticoids, vitamin D, retinoic acid, thyroid hormone, and orphan receptors. On activation by ligand binding, these transcription factors bind to specific DNA sequences on target genes, referred to as hormone response elements (HREs) and regulate transcriptional activity of those specific genes. The steroid receptor proteins contain 3 common domains: 1- a highly conserved central domain, responsible for DNA binding; 2- a variable N-terminal region, involved in transcriptional activation; and 3- C-terminal domain that binds ligand. The androgen receptor was first described in 1969. The single-copy AR gene is localised on the human X chromosome between q11-q13. The entire gene contains eight exons with a total length of 90 kb. The translated human AR is a 918 amino acid protein with a molecular weight between 100 and 110 kDa. Androgen receptors are nuclear proteins that are activated by ligand binding. This initiates a complex cascade of events, which includes a conformational change, phosphorylation, dissociation from heat shock proteins, dimerization, DNA binding of zinc fingers to androgen response element (ARE), interaction with accessory factors, other transcription factors, and RNA polymerase to alter transcriptional activity of a specific gene. In the prostate, free testosterone diffuses into cells and is rapidly converted via 5α-reductase to DHT. Although the AR can be activated by binding to either the androgen, the affinity for DHT is greater than the affinity for testosterone. Unliganded AR is primarily localised to the cell nucleus and that androgens diffuse into the nucleus where they bind to available receptors. The binding of androgen by its receptor induces an allosteric conformational change of the receptor, which in turn results in hormone-dependent, DNA-independent phosphorylation of the receptor at several sites and dissociation from heat shock proteins. Dissociation from these associated factors allows the AR to dimerize, which is essential for binding to AREs. Various genes have been identified that are regulated by androgen, including the gene encoding PSA, a prostate-specific protease, that is now widely used as a serum marker for human prostate cancer. Although androgens directly influence the transcription of this gene, posttranscripitional regulation via stability of the RNA product is believed to be a major mechanism of androgen's effect. The final responses of interaction of tropic peptide hormones such as LH and ACTH, with their receptors, include: 1- Early responses: ion transport, carbohydrate metabolism, steroidogenesis. 2- Late responses: RNA and protein synthesis, cell growth, DNA synthesis, and cell division.

ERECTION

Reoxgenation of hypoxic or ishaemic tissue presents a new and generally deleterious stress to tissues, most notably by the generation of reactive oxygen metabolites (ROM). In penile tissue, prostaglandins may have several important physiological roles. Prostacyclin (PGI_2) is the most abundant cyclo-oxygenase metabolites produced and may play a key role in corporal blood homeostasis via its inhibition of platelet aggregation and adhesion of inflammatory cells to the vascular wall. The production of PGI_2 is modulated by oxygen tension, suppressed by hypoxia due to inhibition of prostaglandin H synthase (PGHS); O_2 is a substrate for this enzyme, a dioxygenase. Nocturnal penile tumescence and rigidity testing (NPTR) reflects the integrity of the corticospinal efferents to the penis, when nocturnal erections are appropriate neuroeffectors and corporal regulators of penile haemodynamics are intact. Three to five erections per night are normal. At puberty, 20% of total sleep time may be spent with an erection. In the second decade of life the average duration of nocturnal erection is 38 minutes. For adult males the average duration is 27 minutes. Nocturnal erectile rigidity diminishes with age. 50-80% of men seeking urological advice has organic dysfunction. 70% of men under 35 years of age having psychogenic cause and 85% of men over 50 years of age having organic impotence. Coital frequency similarly varies with age: 75% of men in their seventh decade report coitus once a month, weekly coitus is reported among 37% of patients 61-65 years old, and 28% of patients 66-71 years old. The age-related changes in penile sensitivity, response to penile ischaemia, and somatosensory evoked potentials. Diabetes under treatment is associated with a three-fold increase in the probability of complete impotence. Complete impotence is more prevalent among men taking hypoglycaemic (26%), antihypertensives (14%), vasodilators (36%), and cardiac drugs (28%), and dehydroepiandrosterone. There is no correlation between impotence and obesity. Concurrent smoking amplifies the risk of complete impotence among patients with heart disease, hypertension; and intensifies drug effects, such as cardiac drugs, vasodilators, and hypertensives. Most of the pathobiology of ED is related to cell injury and formation of OFRs. NO is a radical that can lead to cell injury in the presence of other radicals. Oxidative and nitrous stresses lead to lipid peroxidation. Psychological stress increases OFRs. OFRs are involved in vascular pathology. Trauma is associated with increase in OFRs. Substances released by the sinusoidal endothelium directly alter penile smooth muscle tone to maintain flaccidity or promote erection. The endothelium releases NO in response to cholinergic stimulation. NO diffuses to smooth muscle via paracrine mechanism, stimulates guanylate cyclase, which results in an increase in intracellular accumulation of cGMP and cavernous smooth muscle (CSM) relaxation. Endothelin-1 is a potent vasoconstrictor that is synthesised by corporal endothelial cells. At low oxygen tensions found in the flaccid penis nitric oxide synthase (NOS) activity is inhibited, following cavernous artery relaxation oxygenated blood is delivered to the corpus cavernosa. In the oxygen-enriched corpora the autonomic dilator nerves and endothelium of the sinusoids increase synthesis and release of NO. PGE synthesis is regulated by oxygen tension. TGF-β synthesis is increased under hypoxic conditions. In the absence of periodic oxygenation (ischaemia) from nocturnal erections, growth factors and cytokines released by the endothelium are altered either quantitatively or qualitatively. The result is a gradual conversion of CSM and extracellular matrix (ECM), remodelling these tissues yields corporal fibrosis. Cavernous arteriolar flows decrease with age, but an excellent erection can be maintained over a wide range of penile inflows. As intracorporal pressure increases diastolic flow approaches zero or may actually reverse direction in the corporal artery during rigidity. The dynamics of veno-occlusion are the critical factor in the aging erectile response. Vasculogenic components of erection: the main areas engorged with blood during an erection are corpora cavernosa and to a lesser extent, the corpora spongiosum. The function of corpus cavernosa is pure erectile. An erect penis holds about 8 times as much blood as does a flaccid one. When excitement or direct stimulation occurs, the arterioles dilate, the cavernous bodies fill with blood, the veins are constricted, trapping the blood in, and an erection occurs. In a full erection, the average pressure in the corpus cavernosa is about 90-100 mmHg and in the glans is 40-50 mmHg. During erection, the sinusoid and arterioles relax, allowing more blood to flow into the corpora, distending the sinusoids spaces and compressing the subtunical venous plexus against tunica albuginea. Psychological components of erection: A- The brain, the limbic system, the location of the libido is responsible for a variety of automatic and somatic responses. Sensory sexual stimuli such as sight, hearing, touch, smell and taste all are integrated in the brain and can stimulate erection. Many men are turned on by the smell, touch, or sound of a lover. B- Hypothalamus releases LHRF that stimulates the pituitary to secrete LH, which act on testes to secrete testosterone. Testosterone influences: a- men's desire for sexual activity (libido), and b- men's aggression, which is related to sexual intercourse. Parasympathetic sends impulses to: a) penile arteries to expand, and b) corpora cavernosa to fill with blood. Hormonal balance: 1- testosterone (350-800 ng/dl), remains normal until the age of 70, then decline. It is not known how it affects an erection. 2- LH (5-18 µIU/ml), stimulate testicular production of testosterone. 3- Prolactin (<15 ng/ml) in large amount blocks the effectiveness of testosterone and decreases libido. Erection involves the coordinated activity of at least 4 neuropharmacologic events: 1- Decrease in α-adrenergic neurotransmission in association with possible local inhibition of α-adrenergic activities. 2- Release of NO in a rapid and profound relaxation of cavernous smooth muscle (CSM). 3- The activation of endothelial cells by parasympathetic neurotransmission (acetylcholine) stimulates the release of NO, which diffuses into the sinusoidal smooth muscle. 4- The activation of the postganglionic dilator neurotransmitter, which are nonadrenergic, noncholinergic (NANC) neuroeffectors, such as NO, VIP, and calcitonin gene-releasing peptide (CGRP). In the flaccid penis, the smooth muscle of the cavernous arteries, arterioles, and sinusoids is contracted. Although vascular resistance to inflow is elevated in the flaccid state, venular outflow is unrestricted. Regulation of venous outflow from the penis is a passive phenomenon. Venous outflow during erection is limited dynamically by distension of the sinusoids compressing the subtunical venular plexus against the inner layer of the tunica albuginea, and structurally by the differential stretching of the 2 primary layers of the tunica across which the emissary vein exit. The tone of smooth muscle fibers is controlled by the autonomic nervous system (the thoracolumbar sympathetic and sacral parasympathetic systems). Stimulation of the sacral parasympathetic outflow conveyed by the pelvic and cavernous nerves has elicited penile erection. The sacral parasympathetic nucleus (SPN) contains different groups of cells: sacral preganglionic neurons sending axons into the pelvic nerve, interneurons, and neurons projecting to the hypothalamus. Stimulation of hypogastric nerves elicits penile erection in spinal-cord-injured patients. The SPN contains serotonergic terminals. Dense 5-hydroxytryptamine$_{1A}$ (5-HT1A) and 5-HT1B serotonergic receptor-subtype binding sites are present in the SPN. Synaptic contacts exist between serotonergic fibers and preganglionic SPN neurons. Dopaminergic fibers also run in the SPN. SPN also receives a dense neuropeptide Y (NPY) innervation. Cholecystokinin neurons present in the intermediate gray and the dorsal horn of the lumbosacral spinal cord contribute fibers to the SPN. Specific oxytocin receptors exist in the SPN. Sympathetic fibers running in the paravertebral sympathetic chain to sacral levels reach the penis through the pudendal nerves. Visceral primary afferents conveyed by the hypogastric nerve may in turn reach interomediolateral (IML) neurons. Neurons in the lumbosacral dorsal gray commissure are activated by both somatic and visceral afferent inputs and receive dense peptidergic (VIP, substance P, somatostatin, encephalon) and serotonergic innervation. The perineal muscles are innervated by the pudendal nerves, whose motoneurons are located in Onuf's nucleus, a group of neurons in the ventral horn of the lower lumber and sacral spinal cord. Pudendal motoneurons concentrate androgens. During the perinatal period in the male, androgens are responsible for the correct specificity of motoneuron death. In adult males, androgens are involved in the soma size, dendritic extension, and number of gap junctions. The dorsal nerve of the penis (DNP) conveys sensory information from the glans penis and perigenital skin to the spinal cord. Reticulospinal and coeruleospinal descending pathways from brain stem and pons distribute many fibers to the pudendal motoneurons and SPN neurons. Serotonergic descending pathways to the sympathetic and parasympathetic nuclei and motoneurons originate in raphe nuclei. Dopaminergic neurons from the A11 cell group project in the thoracolumbar sympathetic and sacral parasympathetic nuclei. The paraventricular nucleus of the hypothalamus (PVN) provides descending innervation to the IML in the thoracolumbar spinal cord and contributes fibers to the L5-L6 spinal cord. This innervation is vasopressinergic and oxytocinergic. The command of erections occurring during visual sexual stimulation, during masturbation, or during episodes of rapid-eye-movement (REM) sleep could differ. Reflexive erections depend on the disinhibition of the spinal lumbosacral circuits, when penile reflexes occurring during copulation involve the integration of both excitatory and inhibitory processes by the nervous system, activated by supraspinal centers. Stimulation of peripheral neural afferents elicit penile reflexes. Dorsal horn and dorsal gray-commissure neurons at the lumbosacral spinal cord are activated by DNP stimulation. Repetitive stimulation of the nerve elicits a reflexive erection that is abolished by pelvic nerve section. Glans erections and cups are due to engorgement of the corpus spongiosum and bulbospongiosus (BS) muscles activity. Flips are elicited by corpus cavernosum erection and ischiocavernosus (IC) muscles activity. Reflexive erections represent a complex activation of automatic projectile pathways to the corpus cavernosum and corpus spongiosum combined with somatic pathways to the IC and BS muscles. Reflexive erections occur as a plateau phase of intracavernous pressure increase that remains infrasystolic, over which brief and sharp suprasystolic intracavernous pressure increases are superimposed. Supraspinal structures participate in the modulation of the frequency of spinal rhythms. Penile reflex responses decrease after sexual exhaustion. The intraspinal organisation of this inhibitory effect is both inhibitory and excitatory influences from the brain to the spinal cord and the presence of spinal pacemakers responsible for reflexive erections. At least 2 spinal mechanisms regulating the reflex exist: 1) the starter, would be responsible for response latency, 2) the generator, would decide the number and intensity of responses. A large population of neurons in the intermediate layers of the spinal cord, from the region around the central canal to the interomediolateral (IML) column, and extending from the lower thoracic to the sacral levels. This neural network may represent the morphological basis of coordination between sympathetic and parasympathetic nuclei and motoneurons. The transmission from the flaccid to the erectile state involves factors controlling the tone of corpus cavernosum and penile vascular smooth-muscle cells (SMC). The relaxation necessary for erection requires the removal of contractant factors, the release of relaxant factors, or, most probably, both of the above. Each corporal body is composed of cavernous smooth (CSM), fibrous tissue matrix, and endothelium-lined vascular spaces, surrounded by a fibrous cover, the tunica albuginea. The tunica of the corpora cavernosa is bilayered with multiple sublayers. Elastic fibers provide an irregular lattice network on which collagen fibers rest. The inner layer of collagen is oriented circularly adjacent to the cavernous tissue. Morphologic changes in erectile tissue in patients with diabetes, are changes in ultrastructure of cavernous nerves with diffuse thickening of Schwann and perineural basement membranes, endothelial cell dysfunction. Morphological changes in patients with moderate arterial insufficiency are changes in the number and morphology of mitochondria, cytoplasmic, cytoplasmic vacuolisation and endothelial cell alterations. Morphological changes in patients with severe arterial disease are irregular contour, fragmentation of the smooth muscle; and loss of basement membrane. The tunica plays an important role in the mechanical of erection because it is essential for penile extensibility, rigidity, compliance and veno-occlusion; it is composed mainly of thick collagen bundles and elastic fibres. The reduction of the elastic fibres in Peyronie's disease can decrease the structural resilience of the tunica. The transmission from the flaccid to the erectile state involves factors controlling the tone of corpus cavernosum and penile vascular smooth-muscle cells. The relaxation necessary for erection requires the removal of contractant factors, the release of relaxant factors, or, most probably, both of the above. Contraction of penile erectile tissues: 1- Noradrenaline and α-adrenoceptors, noradrenaline acting on post junctional α-adrenoceptors. The number of α-adrenoceptors is 10-fold that of β-adrenoceptors. There is a functional predominance of α_2-adrenoceptors, although a contribution of α_2-adrenoceptors to the contraction induced by noradrenaline and electrical stimulation of nerves cannot be excluded. At least 3 subtypes of α_1-adrenoceptors exist, α_{1A}, α_{1B}, α_{1D}. There is increasing evidence that an additional α_1-adrenoceptors subtype with low affinity for prazosin (α_{1L}), which is not fully characterised. Noradrenaline-mediated contraction may require synergistic receptor interactions at the second-messenger or receptor-protein level. 2- Endothelins, ET-1 released from endothelial cells might act as a paracrine/autocrine factor. Binding sites for ET-1 have been demonstrated in the deep penile artery and the circumflex veins and throughout the cavernous tissue. At least 2 distinct ET receptors in corporal membranes: A) high affinity for ET-1 and ET-2 and low affinity for ET-3, and B) less abundant type with high affinity for ET-1, ET-2 and ET-3. The contractions induced by ET-1 may be dependent both on transmembrane calcium flux, and on the mobilisation of inositol triphosphate (IP_3)-sensitive intracellular calcium stores. ET-1 might change the sensitivity to Ca of the corporal smooth muscle. The ability to increase both cytosolic and nuclear Ca levels suggests that ET-1 may function not only as a long-term regulator of corporal smooth-muscle tone but also as a modulator of the contractile effect of other agents, e.g., noradrenaline, or as a modulator of cellular proliferation and phenotypic expression. ET-1 has a vasodilator action (increase in corporal pressure) at low doses but vasoconstriction action at high doses. ET-3 has mainly vasodilator effects, which may be due to activation of ET_B receptors on the endothelium and local release of NO. ET_B receptors partly located on the inhibitory nerves that mediate relaxation via activation of the L-arginine/NOS pathway. ET-1 and ET_A receptor binding has been found to be increased in diabetic cavernosal tissue. Relaxation of penile erectile tissues: 1- Nitric oxide (NO), both acetylcholine and neuronally mediated relaxation in corpus cavernosum involves the release of NO or a NO-like substance. The endothelium and/or the nerves innervating the corpus cavernosum may be the source of the NO involved in erection. NO synthesised from L-arginine by NOS in the endothelium and NANC nerves then diffuse to smooth muscle to exert its activity. Binding of NO to soluble guanylate cyclase induces the formation of cGMP from GTP and the accumulation of cGMP leads to relaxation. 2- Vasoactive intestinal polypeptide (VIP) and related peptides, NO is crucial for parasympathetic vasodilatation by regulating peptide release and second-messenger systems for VIP and acetylcholine. VIP may contribute to non-adrenergic; noncholinergic (NANC) mediated corpus cavernosum relaxation and that its mechanism of relaxation is dependent on prostanoids and involved the generation of NO. The production of prostanoids can be modulated by oxygen tension and suppressed by hypoxia. Prostanoids may be involved in contraction (PGE_2, TXA_2) and relaxation (PGE_1, PGE_2) of the corpus cavernosum and penile vasculature as well as in the inhibition of platelet aggregation and white cell adhesion (PGI_2). PGs and $TGF-\beta_1$ may have a role in the modulation of collagen synthesis and the regulation of fibrosis of the corpus cavernosum. Androgens (Testosterone) are necessary for the normal development of the penis and their deficiency results in significant structural abnormalities. Although androgen receptors (AR) in the penis decrease after puberty, they usually do not disappear completely. Castration results in loss of libido and in ED. Testosterone enhances libido, frequency of sexual acts and sleep-related erections. Signal transduction in cavernous smooth muscle, in the resting state the level of sarcoplasmic free Ca^{2+} amounts to about 120-270 nM, whereas in the extracellular fluid (ECF) the level of Ca^{2+} is in the range of 1.5-2 mM. This 10,000-fold gradient is maintained by the cell-membrane Ca^{2+} pump and the Na^+/Ca^{2+} exchanger. Neuronal or hormonal stimulation can open Ca^{2+} channels, resulting in Ca^{2+} entry to the sarcoplasm down its concentration gradient. Increase in the level of free sarcoplasmic Ca^{2+} by a factor 2-3 to 550-700 nM triggers myosin phosphorylation and subsequent smooth muscle contraction. The intracellular free Ca^{2+} ($[Ca^{2+}]_i$) binds to the thin filament-associated protein troponin, in the smooth muscle it binds to calmodulin. This Ca-calmodulin complex activates myosin light-chain kinase calmodulin complex activates myosin light-chain kinase (MLCK) by association with catalytic subunit of the enzyme. The active MLCK catalyses the phosphorylation of the regulatory light chain subunits of myosin (MLC_{20}). Phosphorylated MLC_{20} activates myosin-ATPase, thus triggering cycling of the myosin heads (cross-bridges) along the actin filaments, resulting in concentration of the smooth muscle. A decrease in the level of ($[Ca^{2+}]_i$) induces a dissociation of the Ca-calmodulin MLCK complex, resulting in dephosphorylation of the MLC_{20} by myosin light-chain phosphatase and in relaxation of the smooth muscle. Ca-sensitising agonists are mediated by GTP-binding proteins that generate proteins that generate protein kinase C or arachidonic acid (AA) as second messengers. These inhibit the myosin light-chain phosphatase (MLCP), thus increasing MLC_{20} phosphatase by basal-level activity of MLCK. The resulting myosin phosphorylation and subsequent smooth-muscle contraction without a change in ($[Ca^{2+}]_i$). Calcium-depolarisation occurs in the presence of ($[Ca^{2+}]_i$) higher than that required for the activation of the MLCK, activates the Ca-calmodulin-dependent protein kinase II, which then reduces the affinity of MLCK for Ca-calmodulin by phosphorylation of a specific site. The resulting decrease in the activity of MLCK leads to an increase in myosin dephosphorylation by basal-level activity of MLCP and to subsequent smooth-muscle relaxation. Ca^{2+}-independent contractions are probably mediated via receptor-operated membrane-bound GTP proteins that activate Ca^{2+}-independent protein kinase C (I-PKC). Electromechanical coupling, Ca^{2+} depolarise the membrane potential, opening voltage-graded L-type Ca^{2+} channel. Thus Ca^{2+} enters the sarcoplasm driven by the concentration gradient and triggers contraction. Changes in the membrane channels other than Ca^{2+} channels include β-adrenergic agents or atrial natriuretic factor (ANF) activates K^+ channels by involving the intracellular second messenger cGMP or cAMP. These activate protein kinases G and A, respectively, which then phosphorylate K^+ channels, thus hyperpolarising the cell membrane. This hyperpolarisation then inactivate the L-type calcium channels, resulting in a decrease Ca^{2+} influx and subsequent smooth muscle relaxation. The major mechanisms of pharmacomechanical coupling-induced smooth muscle contractions are the release of IP_3 and the regulation of Ca^{2+} sensitivity.

Sexual responses related conditions:

Condition	Related condition
Hypoactive sexual desire disorder	Anxiety
Sexual aversion disorder	Body image disturbance
Sexual arousal disorder	Decisional conflict
Orgasmic disorder	Fear
Premature ejaculation	Grieving, dysfunctional
Dyspareunia	Health maintenance, altered
Vaginismus	Health-seeking behaviour
Sexual dysfunction due to a general medical condition	Pain
Substance-induced sexual dysfunction	Disturbance
Exhibitionism	Powerlessness
Fetishism	Role performance, altered
Frotteurism	Self-care deficit
Paedophilia	Self-esteem disturbance
Sexual masochism	Sensory/perceptual alterations
Sexual sadism	Sexual dysfunction
Voyeurism	Sexual patterns, altered
Transvestic fetishism	Social interactions, impaired
Gender identity disorder	Spiritual distress

Pharmacomechanical coupling mechanisms of smooth-muscle relaxation are mediated via the intracellular cyclic nucleotide/protein kinase/second-messenger system. β-adrenergic agonists activates protein kinase A (PKA) and, to a lesser extent, protein kinase G (PKG). ANF acts via membrane-bound guanylate cyclase, whereas NO acts via the soluble form of guanylate cyclase, both generate cGMP, which activates PKG and to a lesser extent, PKA. The protein kinases activate the cell-membrane Ca^{2+} pump, leading to a decrease in ($[Ca^{2+}]_i$) in the sarcoplasm and to subsequent relaxation. cAMP and cGMP are inactivated by phosphodiesterases (PDEs) by hydrolytic cleave of the 3'-ribose-phosphate bond. There are 5 PDE isoenzymes, in smooth muscle cells. Ca/calmodulin-stimulated PDE (PDE1), cGMP-stimulated PDE (PDE II), cGMP-inhibited PDE (PDE III), cAMP-specific PDE (PDE IV), and cGMP-specific PDE (PDE V). The existence of syncytial tissue triad confers a plasticity, adaptability, and flexibility to erectile function. The gap junctions represent a diverse family of related proteins, known as connexins. The union of these connexins across the extracellular space provides an aqueous intercellular channel, and thus, partial cytoplasmic continuity between adjacent cells. Connexin 43 appears to be the predominant gap junction protein expressed in human corporal smooth muscle. The presence of gap junctions, in concert with the autonomic nervous system, myogenic intracellular signal transduction mechanisms, and/or electronic current spread (both hyperpolarising and depolarising waves through gap junctions), confers a plasticity, adaptability, and flexibility to erectile physiology that may well account for the diversity in regulation and function of penile erection. AGE role in erectile dysfunction, diabetes mellitus is one of the most frequent causal factors of erectile dysfunction. The prevalence of ED in diabetic men is 35-75%. The development of autonomic neuropathy and alteration of endothelial cell-mediated mechanisms that control penile smooth tone could result in ED in diabetes. Diabetes-related abnormality in the synthesis of NO may result in the diminished relaxation. The longer duration of the illness the less pronounced the neurogenic relaxation. Ageing and castration, along with diabetes contribute to erectile dysfunction (ED) by decreasing the NOS content or activity or both in penile tissue. There is a significantly lower value for glycosylated haemoglobin and plasma glucose in potent diabetic compared with impotent patients. Nonenzymatic glycosylation (glycation) of proteins, often referred to as the Maillard reaction, has been proposed to play a role in age and diabetes-related processes by forming protein and DNA adducts and cross-links. These cross-links may contribute to ED by scavenging NO, which is needed for erection. Maillard reaction is a process by which a reducing sugar attaches to an amino group of an amino acid residue and then undergoes rearrangement to form a ketoamine-linked sugar (Amadori product). Amadori products have been demonstrated in collagen, myelin, albumin, haemoglobin, and low-density lipoproteins (LDL). Amadori products undergo further reactions leading to the formation of intermediates that again react with proteins to form stable protein cross-links, chromophores, and fluorophores called advanced glycation end products (AGEs). The effect of insulin on endothelium-dependent relaxation is quenching of NO by an accumulation of AGEs in subendothelial collagen. AGEs could change cellular function or they could generate free radicals. The major factors that govern the rate of formation of the AGEs are glucose concentration and duration of exposure to glucose. The specific AGE pentosidine in penile corpus cavernosum tissue and penile tunica albuginea tissue was found to increase with age in cadaver as well as living penis. Pentosidine levels increases 4-6 fold from puberty to the age of 100 years. Mean pentosidine levels are higher in the tunica than in the corpora. It is not known whether pentosidine itself is directly associated with erectile dysfunction, but its formation is usually accompanied by extensive tissue modification. Between 40-70 years the probability of complete ED tripled from 5 to 15%; the probability of moderate ED doubled from 17 to 34% while the probability of minimal ED remained constant at 17%. By the age of 70 years, only 32% portrayed themselves as being free of ED. Collagen is the major component of the corpus cavernosum. Of the total collagen, 40% is type I collagen and 40% is type IV collagen. Type IV collagen is distributed in the basement membrane of the cavernosal tissue as well as in the trabeculae of the cavernosal spaces. Type III collagen, 20%, is present in the corporal fibrous network. The tunical collagen content is divided among type I and III in equal amounts.

Urotext
Urology and health
www.urotext.com

NUTRIENTS

FOOD COMBINATIONS

The goal of eating is to supply the system with the nutrients, vitamins, minerals, amino acids, essential fatty acids, enzymes, fiber bulk and other required essentials. Current foods is a mess of chemicals including pesticides, radioactive isotopes, bleaching agents, heavy metals, aromatics, extenders, emulsifiers, softeners, thickeners, hydrogenators, curing agents, buffers, deodorisers, sprout inhibitors, fungicides, sweeteners, conditioners, stabilisers and preservatives. There are almost 15,000 substances that are added to the foods we eat. Diets rich in whole and unrefined foods, like whole grains, dark green and yellow/orange-fleshed vegetables and fruits, legumes, nuts and seeds, contain high concentrations of antioxidant phenolics, fibers and numerous other phytochemicals that may be protective against chronic diseases. A diet abundant in phytochemically-rich foods beneficially affects lipoproteins, decreases need for oxidative defense mechanisms and improves colon function. There is a protective relationship between consumption of fruit and vegetables-particularly cruciferous and green leafy vegetables and citrus fruit and juice-and ischaemic stroke risk. Nutrition is a primary issue, there is an international consensus regarding the optimum diet for the prevention of both coronary heart disease and cancer. There is a mismatch between the attitude of the public, who appear willing to accept dietary advice from professionals, and the reluctance on behalf of these professionals to fulfil this role. The contribution that diet could make is significant and, the role of nutrition needs to be put firmly on the health care agenda. Clinical nutrition is concerned with the diagnosis and treatment of diseases that affect the intake, absorption, and metabolism of dietary constituents and with the promotion of health through the prevention of diet related diseases. Diseases of clinical nutrition encompass the most common causes of mortality including obesity with its co-morbidities of hypertension, diabetes, dyslipidaemias, increased risks of cardiovascular disease, cancers, pulmonary failure; GIT disorders related to inadequate nutrient absorption; eating disorders; and malnutrition associated with chronic illness and surgical trauma. Despite the prevalence of nutritional disorders in clinical medicine and increasing scientific evidence on the significance of dietary modification to disease prevention, the majority of doctors are untrained in the relationship of diet to health and disease. Standardisation of curricula for nutrition education of medical students and trainees is relevant to the cost-effective integration of nutritional concepts into medical practice. Diet can play a major role in cancer prevention. About 50% of cancer incidence and 35% of cancer mortality, represented by cancers of the breast, prostate, pancreas, ovary, endometrium, and colon, are associated with Western dietary habits. Cancer of the stomach, currently a major disease in the Far East, relates to distinct, specific nutritional elements such as excessive salt intake. Fiber should be increased to 25-35 g/day for adults. The geographic distribution of colon cancer is in areas that had high prevalence rates of rickets-regions with winter ultraviolet radiation deficiency. Generally due to a combination of high or moderately high latitude, high-sulfur content air pollution (acid haze), higher than average stratospheric ozone thickness, and persistently thick winter cloud cover (combination of latitude, climate, and air pollution that prevents any synthesis of vitamin D during winter). Breast cancer death rates rise with distance from the equator and are highest in areas with long vitamin D winters. Colon cancer may be prevented with regular intake of calcium in the range of 1,800 mg/day, in a dietary context that includes 800 IU per day (20 micrograms) of vitamin D3. Intake of 800 IU/day of vitamin D may be associated with enhanced survival rates among breast cancer cases. The consumption of even relatively small amounts of fish is a favourable indicator of the risk of several cancers, especially of the digestive tract. It is difficult to trust that the vitamins and minerals we need are still in the food we eat. Vitamin E is a fat-soluble antioxidant, protects the fatty parts of the cells from damage. Vitamin E is almost completely destroyed by food processing and by cooking, making a nonsense of claims that we get more than enough in our diets. The natural form of E is d-α-tocopherol is more biologically active than the synthetic version, dl-α-tocopherol. Vitamin E labelled as acetate or succinate means that it has been stabilised. Vitamin E can help RBCs survive in premature infants. 100 iu of vitamin E or 2 g of vitamin C give a 20% greater resistance to sunburn. 100 mg of vitamin E/day for one month reduces the traces of fatty oxidation in the urine by 27%. Vitamin C is water-soluble, not stored in the body. Our need for vitamin C varies tremendously with stress, infection and injury. Vitamin C is found in high concentrations in the adrenal glands. Vitamin C helps the immune system kill viruses. Aspirin interferes with the uptake of vitamin C by cells of the immune system. Vitamin C is destroyed by heat. Bioflavonoids may help its absorption. Selenium is generally unevenly distributed in the earth's crust. Trace elements are vital to health because they are necessary integral components of many body enzymes and hormones. Enzymes and hormones are required for body functions and processes and so play a vital health-giving role. Selenium protects the millions of cells in the body, and could possibly slow down the whole body's ageing process. It further enhances its action synergitically by vitamin E, C, and A. Selenium can help: 1- maintain health by protecting against cell damage caused by free radicals produced by normal body processes, radiation, and pollution; 2- enhance the body's immune system, thereby improving the resistance to diseases; 3- act as an anti-inflammatory agent for arthritis and rheumatism; 4- protects against ageing by acting as an antioxidant together with vitamin A, C, and E. Selenium is a trace element, bread is a major source of it. Canadian wheat has high selenium. In Europe, selenium gone missing from soil depleted by industrial-scale farming and imblanced use of fertilisers. Low intake of selenium is associated with cataracts, increased infections, inability to detoxify properly, fertility dysfunction (in men), heart disease and changes that may lead to cancer. It forms one of the 4 most powerful antioxidant enzymes. β-Carotene is the most well-known carotene, or carotenoid. But there are more than 20 of them. Carotenes help the immune system, protect against cancers and gives us enhanced protection against harmful radiation from the sun. Lycopene in tomatoes is linked with prevention of prostate cancer. Men eating a lot of tomatoes and tomato paste have low rates of prostate cancer. Lycopene is made available also by cooking. Lecithin is a special type of fat, rich in the liver, where it is made, and the brain. It is mainly made up of two B vitamins, choline and inositol. Lecithin splits up the fats we eat and enable them to be mobilised and used for energy instead of being laid down and being allowed to accumulate in the body. Zinc and essential fatty acids protect against prostate cancer. Fatty acids are extremely vulnerable to light, heat, and spoilage by oxygen. Oleic acid comes from olive oil, prevents heart disease as helps to stop cholesterol being damaged by oxygen in the body. Oxidised cholesterol is thought to be a prime cause of damage to blood vessels. Olive oil contains phenol, chemicals that add more protection. Mediterranean's diet reduces the risk of death by 50%, reduces the risk of cancers by 61% and reduces the risk for combination of death. Magnesium is needed by hundreds of vital enzymes and is important in blood sugar regulation and the correct firing of muscles. Calcium alone, at the rate of 1000-1500 mg/day can reduce bone loss in post-menopausal women by 40%. There are 16 B vitamins making up a family whose main role is to help convert carbohydrates into energy. B_6 is depleted by use of contraceptive pill. B_6 is known as pyridoxine, is needed for protein metabolism and nervous system function. It is essential for the production of serotonin, which affects mood, behaviour and sleep patterns. B_6 can be toxic if taken a massive doses 2000-6000 mg/day. Niacin or vitamin B_3 lowers blood cholesterol levels and helps the body dealing with toxins such as caffeine. B_{12} is anti-fatigue and anti-anaemia vitamin. All the B complex are water-soluble, some are vulnerable to heat and light. B complex is destroyed during stress. Folate is an important cofactor in the transfer of one-carbon moieties and plays a key role in DNA synthesis, repair, and methylation. The role of folate includes prevention of macrocytic anaemia, prevention of cardiovascular disease and neural tube defects and cancer prevention. CoQ_{10} starts to decline in the age of 20s, often leaving us seriously deficient by the end of middle age. CoQ_{10} is present in food items and absorbed to a significant degree. The mean daily consumption of CoQ_{10} is 3-5 mg. The major contributions arise from meat and poultry representing 64% of the CoQ_{10} intake. Dietary CoQ_{10} may contribute to the plasma CoQ_{10} concentration. The soy bean oil based, gelatine-coated capsules containing CoQ_{10} has a good bioavailability, producing a 3-fold increase in blood concentrations after 6 week's supplementation with 90 mg/day. Resynthesis of Pcr in skeletal muscles after exercise appears to require ATP produced through mitochondrial oxidative phosphorylation, there is no recovery of Pcr when ischaemia is continued after exercise. CoQ_{10} acts in its reduced form as an antioxidant, preventing initiation and/or propagation of lipid peroxidation in biological membranes and in serum low-density lipoprotein. CoQ_{10} does not improve glycaemic control nor diminish the insulin requirement in patients with IDDM and therefore can be taken freely without risk of hypoglycaemia. The bacteria in our large intestine weigh around 3 lb; large enough to be considered an organ in their own. These bacteria are friendly. They help the digestive process and synthesise vitamins including B_1, B_2, B_6, B_{12}, folic acid, biotin, niacin, pantothenic acid and vitamin K. They improve the availability of minerals and stimulate the immune system, by producing natural antibiotics. They also keep the environment of the intestine unfavourable to candida as well as other pathogens. Gut bacteria break down some of the fibre in our food to produce a fatty acid called butyric acid. This type of fat also found in butter, is an important fuel for the cells that line the intestines. Butyric acid is thought to be one of the reasons why high-fibre diet protects against colon cancer. Pineapple contains an enzyme, bromelain that digests protein. Red grape juice is made from purple, not red, grapes. Grape juice is effective as aspirin as anticoagulant and is good for the prevention of heart attacks and strokes. The mortality profile of prostate cancer mirrors that of breast cancer. Cranberry juice reduces bacteria and WBCs in the urine of these with UTI. Soy protein, as well as many other vegan proteins, are higher in non-essential amino acids than most animal-derived food proteins, resulting in increased glucagon production. Acting on hepatocytes, glucagon promotes (and insulin inhibits) cAMP-dependent mechanisms that down-regulate lipogenic enzymes and cholesterol synthesis, while up-regulating hepatic LDL receptors and production of the IGF-I antagonist IGFBP-1. The insulin-sensitising properties of many vegan diets-high in fiber, low in saturated fat should amplify these effects by down-regulating insulin secretion. Decrease circulating IGF-I activity should impede cancer induction, lessen neutrophil-mediated inflammatory damage, and slow growth and maturation in children. Low-fat vegan diets may be especially protective in regard to cancers linked to insulin resistance--namely, breast and colon cancer as well as prostate cancer; conversely, the high IGF-I activity associated with heavy ingestion of animal products may be largely responsible for the epidemic of Western cancers in wealthy societies. Increased phytochemical intake is also likely to contribute to the reduction of cancer risk in vegans. Regression of coronary stenoses has been documented during low-fat vegan diets coupled with exercise training; such regimens also tend to markedly improve diabetic control and lower elevated blood pressure. Risk of many other degenerative disorders may be decreased in vegans; although reduced growth factor activity may be responsible for an increased risk of haemorrhagic stroke. Phytotherapy is the treatment of health conditions with plant-derived medicine. Thirty percent of modern drugs remain plant-derived substances. Alcohol medicinally known, as ethanol is a natural substance, which is, produced everyday by the body. Glycerol or glycine is less stable, has dehydrating action on the body, and may give headaches, nausea and vomiting. Alcohol in phytotherapy is 0.001-0.006%. Epigallocatechin gallate (EGCG) or crude green tea extract inhibits the growth of cancer cells, lung metastasis, and urokinase activity. Increased consumption of green tea is associated with decreased numbers of axillary lymph node metastases among premenopausal patients with stage I and II breast cancer and with increased expression of progesterone receptor (PGR) and oestrogen receptor (ER) among postmenopausal ones. Increased consumption of green tea correlates with decreased recurrence of stage I and II breast cancer. The recurrence rate is 16.7% or 24.3% among those consuming ≥ 5 cups or ≤ 4 cups per day, respectively, for stage I and II breast cancer, and the relative risk of recurrence is 0.564 after adjustment for other lifestyle factors. No improvement in prognosis is expected for stage III breast cancer. Soy isoflavones are not only antioestrogenic/proestrogenic, but they also have antioxidant, antiproliferative, anti-inflammatory, and pro-apoptotic effects in cancer cells, which may provide additional mechanisms for their potential anticarcinogenic effects. Soy foods and/or soy isoflavones should be a high priority in patients with breast cancer or in individuals at high risk for breast cancer. Soybeans are unique among the legumes because they are a concentrated source of isoflavones. Isoflavones have weak oestrogenic properties and the isoflavone genistein influences signal transduction. Soyfoods and isoflavones have potential role in preventing and treating cancer and osteoporosis. Soy or isoflavones may reduce the risk of prostate cancer. The weak oestrogenic effects of isoflavones and the similarity in chemical structure between soybean isoflavones and the synthetic isoflavone ipriflavone, increases bone mineral density in postmenopausal women. Selenium as an essential trace element is required for synthesis and activity of glutathione peroxidase, which protects against free radical damage to cell and tissues. Dry beans and soybeans are nutrient-dense, fiber-rich, and are high-quality sources of protein. Protective and therapeutic effects of both dry bean and soybean intake have been documented. Dry bean intake has the potential to decrease serum cholesterol concentrations, improve many aspects of the diabetic state, and provide metabolic benefits that aid in weight control. Soybeans are a unique source of the isoflavones genistein and diadzein, which have numerous biological functions. Soybeans and soyfoods potentially have multifaceted health-promoting effects, including cholesterol reduction, improved vascular health, preserved bone mineral density, reduction of menopausal symptoms, salutary effects on renal function and lower prevalences of certain cancers. Increased intake of whole grains and higher fiber intake may protect against CHD. Nut consumption may offer protection against IHD. Frequent nut consumption is associated with a reduced risk of both fatal coronary heart disease and non-fatal myocardial infarction. Diet rich in olive oil may attenuate the acute procoagulant effects of fatty meals, which might contribute to the low incidence of IHD. Conjugated metabolites of typical flavanol- and flavonol-flavonoids participate in the antioxidant defense in blood plasma. Therefore, the intake of vegetables, fruits and tea rich in flavonoids may help to prevent oxidative damages in the blood. Their metabolic conversion begins in the intestinal mucosa where the activity of uridine-5'-diphosphoglucuronosyltransferase (UGT) is at its highest. Both flavonoids accumulated mostly as glucuronide and sulfate conjugates in blood plasma after oral administration. Cancer chemopreventive agents include NSAIDs such as indomethacin, aspirin, piroxicam, and sulindac, all of which inhibits cyclooxygenase (COX). COX catalyses the conversion of AA to pro-inflammatory substances such as prostaglandins, which can stimulate tumour cell growth and suppress immune surveillance. COX can activate carcinogens to form that damage genetic material. The process of chemical carcinogenesis can be divided into 3 general stages, and chemopreventive agents have been categorised according to the stage that they inhibit. Resveratrol inhibits cellular events associated with tumour initiation, promotion, and progression. By inhibiting the COX-1, resveratrol inhibit promotion. Its inhibitory activity is less than NSAIDs, but greater than aspirin. Resveratrol inhibits the hydroperoxidase activity of COX-1. Resveratrol reduces pedal oedema both in the acute phase (3-7 hrs) and in the chronic phase (24-1444 hrs). Resveratrol inhibits free radical formation. Resveratrol induces quinone reductase activity, which is capable of metabolically detoxifying carcinogens. Resveratrol has been found in at least 72 plants species, a number of which are components of the human diet, such as mulberries, peanuts, and grapes. Relatively high quantities are found in the latter. Fresh grape skin contains about 50-100 μg of resveratrol/g, and the concentration in red wine is in the range of 1.6-3 mg/L. Appreciable amounts are also found in white and rose wines. Resveratrol plays a role in the prevention of heart disease because it has been reported to inhibit platelet aggregation and coagulation, alter eicosanoid synthesis, and modulate lipoprotein metabolism. Resveratrol, a phytolaxin found in grapes and other food products have cancer chemopreventive activity. Resveratrol acts as an antioxidant and antimutagen, anti-initiation activity; it mediates anti-inflammatory effects and inhibits cyclooxygenase and hydroperoxidase functions (anti-promotion activity). Resveratrol inhibits the development of preneoplastic lesions, and inhibit tumourigenesis. The term "alternative cancer therapy" refers to clinically unproven treatments that are used in place of conventional cancer treatments, while the term "complementary therapy" is attributed to such treatments when they are used as adjuncts to conventional therapy. These treatments may enhance, interfere, or have no interactions with standard cancer therapy. Types of complementary and alternative medicine (CAM) may include: 1- Orthomolecular medicine, such as the use of products that may be used as nutritional and food supplements, such as ultra-high doses of magnesium, CoQ_{10} (an antioxidant), carnitine (an amino acid), melatonin, or vitamins, when investigated for therapeutic or preventive purposes. 2- Pharmacologic therapies, such as metabolic therapies and immunoaugmentative therapies as used by CAM practitioners or the public, including antineoplastons (medium- or small-size peptides and amino acid derivatives that are taken orally or injected, and are said to form a defense against cancer), enzyme therapies, the Revici system (which focuses on improving the balance between anabolic and catabolic activities, using injectable selenium, calcium, and copper), or 714X (a camphor rich in nitrogen) therapy. 3- Herbal medicine, such as echinecea, gingko biloba, and other herbals. 4- Mind-body medicine, such as transcendental meditation, imagery, hypnosis, biofeedback, music therapy, yoga, spirituality, and biological effects of consciousness. 5- Biofield therapy, such as energy healing and intentional effects on living systems. 6- Bioelectromagnetics, such as diagnostic and therapeutic application of electromagnetic (EM) fields, including pulsed EM fields, magnetic fields, direct current fields, and artificial light therapy. 7- Alternative medical systems, which include traditional oriental medicine (acupuncture, herbal medicine, oriental massage, and qi gong, or vital energy); ayurveda (India's traditional system of medicine that places equal emphasis on body, mind, and spirit, and strives to restore the innate harmony of the individual); other traditional medical systems such as those developed by Native American, Aboriginal, or African cultures; homeopathy; and naturopathy. 8- Manipulative and body-based systems, such as chiropractic, osteopathic, or unconventional applications of integrated conventional and physical therapies, including massage therapy. As cancer incidence rates and survival time increase, use of CAM will likely increase. Women, younger patients, and those who had undergone cancer surgery or who are poor are more likely to use alternative treatments. Fruits, vegetables, and other plants contain many potentially cancer-preventive compounds. Cinnamon proved to be a potent neutraliser of microbes, particularly of common food-borne microbes such as Escherichia coli (E. coli). Postmenopausal women who consumed 40 g of ground flaxseed every day for three months reduced their total cholesterol levels by 6%. Raspberries are a good source of ellagic acid, a compound believed to inhibit cancer growth. The ellagic acid content of raspberries is mostly maintained when the berries are processed into jams or spreads. Sunflower seeds are rich in vitamin E and fibre. Men who eat nuts at least twice per week have a 46% lower chance of dying suddenly of a heart attack compared to men who rarely or never eat nuts. Peanuts, almonds, walnuts, and Brazil nuts are all good nuts to try. Consuming a serving of peanut butter five times per week lower the risk of type 2 diabetes (20%) compared to eating little or no peanut products. Dried plums, or prunes, are an excellent source of boron and fiber. Fermenting the cabbage appears to produce isothiocyanates, which demonstrate anti-carcinogenic properties. Once the enzymes on the surface of the small intestine are damaged or impaired, carbohydrates would be available to intestinal bacteria and yeast to multiply in a vicious circle. Damaged surface of the small intestine leads to impaired digestion and malabsorption of disaccharides. Since the sugar is not broken down and absorbed, it becomes available for fermentation by bacteria and yeast, which overgrow in the presence of abundant sugar. Toxic by-products of bacteria and yeast injure the lining of intestine and enzymes necessary for carbohydrate digestion and absorption. Excessive mucus may also be produced as the body tries to protect itself.

Carbohydrates types:

Type	Criteria	Subtype
1- Monosaccharides	Single sugar	Glucose, fructose, and galactose.
2- Disaccharides	Require splitting into the single sugars by intestinal wall enzymes.	Lactose (found in milk), sucrose (table sugar), maltose and isomaltose (found in corn syrup, malted foods, and candies).
3- Polysaccharide	Amylose contains straight and unbranched chains of sugars while amylopectin is branched. Amylopectin is harder to digest and therefore especially a problem when there is damage to the intestinal enzymes.	Amylose and amylopectin. Fiber is a starch for which man does not have the intestinal enzyme to digest.

Impaired digestion of sugars have been found in celiac disease (gluten enteropathy), soy-protein intolerance, cow's milk protein intolerance, diarrhoea in infancy and children, intestinal parasite infections (Giardia), cystic fibrosis, and Crohn's disease. Lactose intolerance represents a common form of this condition in which lactase, the enzyme that breaks down milk sugar (lactose) is damaged, impaired, or absent. Often there is a combination of affected enzymes. Some starchy foods that were assumed to be digested completely are, in fact, incompletely digested by most healthy people. Autism, functional diarrhoea, irritable bowel syndrome, ulcerative colitis are as well types of impaired digestion of sugars disease. The following proteins should be avoided: processed meats such as hot dogs, bologna, turkey loaf, spiced ham, breaded fish, canned meat if they contain starches such as whey powder, lactose, sucrose, etc. The following vegetables should be avoided: grains such as arrowroot, barley, buckwheat, bulgur, corn, millet, oats, rice, rye, triticale, or wheat. No flour, germ, pasta, starch, or cereal products from these. Potatoes (white or sweet), yams or parsnips. Beans (sprouts, soybeans, mung, faba and garbanzo). Amaranth flour, Jerusalem artichoke flour or powder, quinoa flour, or other grain substitutes such as cottonseed, tapioca, sago. Seaweed, margarine, chocolate, carob. The following fruits should be avoided: canned fruits. Dried fruit that has been glazed with corn syrup or sugar such as many brands of banana chips. Molasses, ketchup, agar-agar, carrageenan, jams, jellies. The following nuts should be avoided: roasted peanuts or peanuts in salted mixtures. Beernuts, glazed nuts, etc. The following beverages should be avoided: cow, goat, soy, rice, coconut milk products. Instant coffee or tea, postum, coffee substitutes. Beer, wine, tonic water, soda, etc. Eating small, frequent meals, avoiding fats, and eating less fiber would relieve mild symptoms of gastroparesis. Erythromycin speeds digestion, metoclopramide speeds digestion and help relieve nausea, other drugs may help regulate digestion or reduce stomach acid secretion, which may indicate in severe gastroparesis. Sitting or standing slowly may help prevent the light-headedness, dizziness, or fainting associated with blood pressure and circulatory problems. Raising the head of the bed or wearing elastic stockings may also help. Some people may benefit from increased salt in the diet and treatment with salt-retaining hormones. Others may benefit from high blood pressure medications. Physical therapy can help when muscle weakness or loss of coordination is a problem. Smoking significantly increases the risk of foot problems and amputation. Some antigen, e.g., viruses, fungi, foods, chemicals, parasites, environmental, and bacteria cause antibodies to be formed that cross-react with some organs of the body such as skin, thyroid, kidneys, joints, etc. This causes damage to these organs, which release chemicals from the interior of the cells (secondary antigens) to which more antibodies are, formed that cause a secondary autoimmune reaction. Free radicals increase the affinity of antibodies for their antigen. Free radicals causing an oxidative redox potential in the affected tissues also facilitate the binding of other components of the immune system. Avoiding all foods containing gluten, which are wheat, rye, barley, oats, and short grain rice is a good idea. The use of large amounts of certain nutrients (zinc, manganese, chromium, selenium, calcium, magnesium, multiple B vitamins, vitamin E, EPA and vitamin A in cod liver oil, alpha lipoic acid, etc.) compensate for inefficiency in absorption of nutrients because of the leaky bowel syndrome, for increased needs due to the necessity of repairing damaged tissues, and for the inefficiencies in metabolism of diseased tissues. Dietary fibre binds metal ions. Increase of fibre intake could lead to reduced intestinal absorption of certain minerals. Inulin intake does not inhibit mineral absorption. Inulin intake exerts its actions on mineral absorption following fermentation in the colon. In the colon, dietary inulin is totally fermented by the natural micro flora to yield short-chain fatty acids (SCFAs) and lactic acid (lowering the pH of the colon lumen by 1-2 units). Any inulin-mineral complexes are degraded during fermentation, liberating the minerals for bio-absorption. The solubility/bio-availability of many mineral salts, e.g. Ca-phosphate, is markedly enhanced in the dilute acidic environment. SCFAs, especially butyrate, seem to stimulate proliferation of colon epithelial cells, thereby increasing the absorptive capacity of the epithelium. Increased mineral absorption, especially calcium phosphate, may be particularly relevant for post-menopausal women and elderly, as in these groups it may prove effective in preventing or delaying the onset of osteoporosis. SCFAs from colonic inulin fermentation stimulate water and electrolyte absorption, which is important to patients suffering or recovering from acute diarrhoea. St John's wort, ginkgo, and saw palmetto now have sufficient clinical studies to consider orthodox use. There is still insufficient evidence to draw definitive conclusions about the efficacy of individual herbs and supplements for diabetes and other diseases; however, they appear to be generally safe. Known or potential drug-herb interactions exist and should be screened for. Echinacea, when used beyond 8 weeks, could cause hepatotoxicity and therefore should not be used with other known hepatoxic drugs, such as anabolic steroids, amiodarone, methotrexate, and ketoconazole. Echinacea lacks the 1,2 saturated necrine ring associated with hepatotoxicity of pyrrolizidine alkaloids. Nonsteroidal anti-inflammatory drugs (NSADs) may negate the usefulness of feverfew in the treatment of migraine headaches. Feverfew, garlic, Ginkgo, ginger, and ginseng may alter bleeding time and should not be used concomitantly with warfarin sodium. Additionally, ginseng may cause headache, tremulousness, and manic episodes in patients treated with phenelzine sulfate. Ginseng should also not be used with oestrogens or corticosteroids because of possible additive effects. Since the mechanism of action of St John wort is uncertain, concomitant use with monoamine oxidase inhibitors and selective serotonin reuptake inhibitors is ill advised. Valerian should not be used concomitantly with barbiturates because excessive sedation may occur. Kyushin, licorice, plantain, uzara root, hawthorn, and ginseng may interfere with either digoxin pharmacodynamically or with digoxin monitoring. Evening primrose oil and borage should not be used with anticonvulsants because they may lower the seizure threshold. Shankapulshpi, an Ayurvedic preparation, may decrease phenytoin levels as well as diminish drug efficacy. Kava when used with alprazolam has resulted in coma. Immunostimulants (e.g., Echinacea and zinc) should not be given with immunosuppressants (e.g., corticosteroids and cyclosporine). Tannic acids present in some herbs (e.g., St John wort and saw palmetto) may inhibit the absorption of iron. Kelp as a source of iodine may interfere with thyroid replacement therapies. Licorice can offset the pharmacological effect of spironolactone. Numerous herbs (e.g., karela and ginseng) may affect blood glucose levels and should not be used in patients with diabetes mellitus. Every time a healthy person combines acids and starches he is making trouble for his digestion, he is getting less value from his foods, and he is hurting himself. Inadequate absorption of food causes degeneration of tissue. For perfect metabolism combining foods high in starches with food high in proteins or fats in the same meal should not take place. The combination of high protein and high starches inhibits the absorption of all the nutritive factors of foods and results in an unnecessary burden upon the entire digestive apparatus. Many illnesses are due to deficiencies of certain essential food factors, e.g., vitamins and minerals. These deficiencies produce degeneration of certain tissues, and this degeneration result in loss of resistance and susceptibility to infections. It is possible to eat large quantities of nutritious foods and get no benefit at all from them if other foods are eaten at the same time, which interfere with the proper digestion of vitamin and mineral bearing foods. Eating cheese, which is rich in calcium, during alkaline digestive process in the small intestine, only very little (if any) of that calcium will be available to for absorption. The calcium will make a chemical combination with the alkali and become non-absorbable, it will pass through and out of the body unused, which would lead to calcium deficiency. But if this food reaches the small intestine when an acid condition is present, much of the calcium will be utilised. Eating carbohydrates (starches and sugars), the small intestine environment becomes alkaline, and a condition is created by that essential factors in other foods cannot be used. Carbohydrates may interfere with the digestion of certain proteins in the stomach, and partially digested protein becomes toxic material. Imperfect splitted large protein molecules may be absorbed into circulation as macromolecules, which then initiate a cascade of immunologic reactions that can cause symptoms and disease. Protein is a compound composed of hydrogen, oxygen, and nitrogen present in the body and in foods that form complex combinations of amino acids. Protein is essential for life and is used for growth and repair. Foods that supply the body with protein include animal products, grains, legumes, and vegetables. Proteins from animal sources contain the essential amino acids. Proteins are changed to amino acids in the body. Proteins eaten with carbohydrates may become toxic due to incomplete digestion, systemic absorption of the allergy producing and poisonous amines. There are 2 distinctly different types of digestion: 1- acid digestion for proteins (meat, fish, eggs, and cheese) and, 2- an alkaline digestion for carbohydrates (sugars and starches). Proteins are digested largely in the stomach, by the gastric juice, which is acidic in reaction. One of the most important constituents of the gastric juice is hydrochloric acid (HCl). Another important ingredient of gastric juice is pepsin, which splits protein only in an acid medium. In other words the stomach must be acid in order to digest protein. Carbohydrates, on the other hand, are not digested in the stomach, but are digested largely in the small intestine, principally by the pancreas secretions, which are alkaline. Amylase splits the starch only in an alkaline medium. The carbohydrates inhibit the secretion of hydrochloric acid and combine with some of the free HCl in the stomach. Fats leave the stomach largely unchanged and upon entering the small intestine cause the gall bladder to empty bile into the small intestine. Bile emulsifies the fat and releases fatty acids, which can neutralise alkaline secretions in the small intestine. If these fatty acids are produced in the intestine while carbohydrates are being digested there, the alkaline secretions that are part of the carbohydrate digestion will be neutralised, and the action of the amylase will be inhibited. The undigested carbohydrates will be left free to ferment and produce gas. So, pure fats (butter, cream, and bacon fat) must not combine with high starches (potatoes, rice, pasta, bread, cereal, sweets) at any one meal. Neither fats nor starches have to be eliminated or restricted, they should be eaten at different times. High starches and pure fats are incompatible. Exclusive high protein and fat diet allows greater absorption of foodstuffs when eaten in the proper combinations, no gas and a distinct simplification of putrefactive organisms in the intestine, no constipation, and no increase in blood pressure. Fats and proteins are an excellent combination. Since protein is digested largely in the stomach by acids, and since the pepsin only works in an acid medium, when pepsin and protein get into the small intestine, if fats are being digested there at the same time and they have liberated enough fatty acid to acidify the intestine and prolong the action of pepsin so that the digestion of the protein would be carried further. Carbohydrates (starches and sugars) are digested by alkalies. Naturally, if any acid is combined with carbohydrates it will tend to neutralise the alkaline digestive juices they need. The more acid present, the more alkaline secretion will be required to neutralise the acid before it can begin to digest the carbohydrates. So, combining acids and carbohydrates must not take place. Buttermilk, orange juice, lemon juice, grapefruit juice or vinegar should not be taken at any meal; also including high starches and sugars. Orange juice with cereals, toast or other carbohydrates are incompatible. Orange juice should be taken alone or with protein foods only. Combining high proteins (meat, fish, eggs or cheese) with high starches (potatoes, cereals, breads, and sweets) at the same meal is not good. Carbohydrates interfere with the digestion of the proteins, and the proteins make more difficult the digestion of the carbohydrates. Carbohydrates inhibit the secretion of HCl in the stomach. Carbohydrates combine with the free hydrochloric acid in the stomach. Both of these actions, by lessening the amount of hydrochloric acid in the stomach, interfere with the digestion of proteins, which must have that acid. Conversely, if proteins are being digested in the stomach and there is more acid there for the carbohydrates to combine with (pick up and take along to the small intestine), then it will require just so much more alkaline secretion from the pancreas to neutralise the extra acid before it acts on the carbohydrates. When the starch meal entered the small intestine comparatively little alkali would be required to neutralise the acid it had picked up in the stomach, but when the mixed meal reached the small intestine just twice as much alkaline pancreatic secretion would be needed to neutralise its acid before starch digestion could begin. When the mixed meal is eaten, the proteins in it are digested under difficult conditions, as the acidity is reduced. The starches cut the acidity to one-third less, producing imperfectly split up proteins-the large toxic protein molecule. When high proteins and high carbohydrates are mixed, there is not enough acid to digest the protein part readily, and too much acid to digest the starch readily. The bad effects of this abuse are not always immediately apparent. The digestion of youth has abundant juices but improper food combination places an extra burden. By the time we reach mid adulthood impaired digestion is evident. The chemically impaired digestion contributes to an increasing deficiency of food elements, which in turn, leads to more tissue degeneration. Minor disturbances are directly created, serious diseases are made more probable, and one more obstacle is raised to be able to live a full life of glowing health. Nearly all foods contain some starch elements and some protein elements. Meat does contain carbohydrate-glycogen. This is a carbohydrate was eaten by the animal and then metabolised and stored in its muscles. Little digestion, if any, is required to make this sugar ready to be absorbed, it is ready to be absorbed as soon as it is liberated from the protein of the meat. The amount of protein in starchy vegetables is small in proportion (negligible quantity) and presents none of the difficulties in digestion, which result from combining large quantities of high protein with high starches. Herbivorous animals, such as the cow or sheep, eating only vegetable food, have specialized on alkaline digestion. Carnivorous animals, such as lions or wild dogs, have specialized acid digestion. Humans have both types of digestion. Humans can not get enough protein from vegetable sources. When we eat meat, we should chew it as little as possible; but when we eat vegetable, we should chew well and thoroughly. Improperly digested proteins split up into intermediate or large protein molecules that are actually toxic, e.g., histamine. Histamine is irritant and vasodilator associated with allergies such hay fever, asthma, eczema, coryza, migraine headaches and general malaise. Mixed diets produce more histamine in the system. Histaminase splits up histamine destroying its toxic effect. Mixed meal requires more histaminase to control the symptoms than when eating proteins only or carbohydrates only. The adrenal gland regulates the normal amount of histamine produced in the body; but the years of improper food habits lead to certain deficiencies and degenerations. The combination of excess histamine and food deficiencies depletes the adrenal glands, the control is lost, allergic reactions appear more readily with possible serious bodily degeneration.

Food combinations:

1- Eat all kinds of meats, fish, eggs, leafy vegetables, citrus fruits (and carbohydrates only if must) as the safest way to avoid deficiencies.
2- Do not combine pure fats (butter, cream or bacon) with high starches (potatoes, cereals, breads, cakes or sweets) in any one meal.
3- Do not combine acids (citrus juice, vinegar, buttermilk) with high starches at any one meal.
4- Do not combine high proteins (meats, fish eggs, cheese) with high starches at any one meal.
5- Eat fats freely with proteins and acid solutions.
6- Be sure you get enough of each essential nutritional element as follows:
A- Meat, fish, fowl and eggs: One serving of each, or two servings of one per day with butter or other fat.
B- Milk, buttermilk, or cheese: Two glasses of raw organic milk or buttermilk, or two and one-half ounces of cheese a day (or one glass of milk or buttermilk plus an ounce or more of cheese).
C- Raw, low-starch fruits and raw green and yellow vegetables: Two servings a day or one large salad bowl a day.
D- 1 or 2 tablespoonfuls of a plain cod liver oil, or its equivalent in other fish liver oils, or their concentrates in capsules. But if you use capsules, then be sure to take plenty butter fats and cream; your liver must have fats, if it is going to make bile for you.
E- If you are a carbohydrate eater, supplements with yeast or other equivalent vitamin B complex. Other natural fats and oils may also be necessary, as the fact remains that natural fats and oils are absolutely necessary in ample quantities for natural, healthy metabolism.

Digesting food is the most energy consuming function. Maintaining the chemical balance between the stomach acid and the body alkalinity takes a lot of chemical energy. Energy is required to produce huge amounts of digestive enzymes. The body has to process the food into energy for immediate use or for storage. Fat is the bank the body uses when it can't use all the calories consumed (hips, thighs, and waistline). Eating frequent smaller meals prevents the ups and downs of the blood sugar level, i.e., craving less sugar. A large meal, and thus a large assault of the immune system, could cause many symptoms of an activated immune system including fatigue, joint aches, flu-like symptoms, headaches, etc. Hydrochloric acid consists of the protons (H^+) and chloride ions (Cl^-). Large quantities of acidic H^+ ions are formed and secreted into the stomach, the alkaline HCO_3^- ions are also formed in large quantities and delivered to the blood in slow but equal fashion.Ê This results in what is called the alkaline tide after eating, i.e., eating helps the metabolism shift toward an alkaline state. Many enzymatic and hormonal body processes require a specific alkaline environment. Digestive enzymes reduce undigested food levels and the chance of stimulating an immune reaction, and help to get more energy out of the food.

SPHINGOLIPIDS IN FOOD

Sphingolipids are constituents of most foods. Both complex sphingolipids and their digestion products (ceramides, sphingosines) are highly bioactive compounds that have effects on cell regulation. The amino group of the sphingoid base is usually substituted with a long-chain fatty acid to produce ceramides. The fatty acids vary in chain length, 14-30 carbon atoms; sphingolipids accounts for a substantial portion of the very long-chain fatty acids. Sphingolipids are located in cellular membranes, lipoproteins, skin, etc. Sphingolipids function as second messengers for growth factors, cytokines, differentiation factors, $1\alpha,25$-didroxy-cholecalciferol and toxins (γ-radiation). PDGF induces sphingomyelin hydrolysis to ceramide (by sphingomyelinase), which is further metabolised (by ceramidase and sphingosine kinase) to sphingosine and sphingosine-1-phosphate. TNF-α activates only sphingomyelinase that results in ceramide accumulation. Sphingosine 1-phosphate is a potent mitogen, inhibits apoptosis. Sphingosine and ceramide inhibits growth and/or induce apoptosis. All organs are capable of de novo sphingolipid biosynthesis. Exogenous sphingolipids are required for the growth of mammalian cells with defects in serine palmitoyl-transferase, the initial enzyme of sphingolipid biosynthesis. Sphingosine and ceramide affect cell growth, differentiation and apoptosis in most types of cells. Normal intestinal cells undergo rapid turnover, except in cancer in which there is loss of normal growth arrest and apoptosis. Digestion of sphingolipids to ceramide and sphingosine may reduce the risk of colon cancer. Sphingomyelin feeding suppresses the conversion of adenomas to adenocarcinomas. Sphingomyelin affects many aspects of cholesterol transport and metabolism and vice versa, cholesterol efflux from cells, the conversion of cholesterol to bile acids, cholesterol esters and other metabolites, the regulation of β-hydroxyl-β-methyl glutarate (HMG)-CoA reductase, and proteolysis of serol regulatory element binding proteins. Induction of sphingomyelin turnover as part of cell signalling (TNF-α) increases cholesterol esterification, which provides a relatively unexplored link between cell signalling events and cholesterol homeostasis. Cholesterol and other lipids can also alter sphingomyelin metabolism. Essential fatty acids deficiency reduces the formation of the skin ceramides. Sphingomyelin influences atherosclerosis as follows: 1- sphingomyelin affects LDL binding and utilisation by cells, 2- hydrolysis of LDL sphingomyelin by an extracellular Sphingomyelinase that is enriched in atherosclerotic lesions alters the aggregation state of the particle and promotes foam cell formation by macrophages, 3- oxidised lipoproteins stimulate the growth of vascular smooth muscle cells and human blood monocytes via triggering of the sphingomyelin signalling pathway, 4- elevation of sphingomyelin in aortic lesions in which this lipid can account for 70% of the total phopholipid, 5- the ration of sphingomyelin to phosphatidylcholine increases five-fold in VLDL hypercholesterolaemia. Changes in sphingomyelin content with aging have been seen in many tissues. Ceramide can inhibit cell growth and induces apoptosis. Sphingomyelinase of sphingolipid signalling pathway activity can be affect by $1\alpha,25$-dihydroxycholecalciferol, unsaturated fatty acids and glutathione. Dietary (ω-3) polyunsaturated fatty acids (PUFA) suppress the formation of ceramide and diacylglycerol. Ceramide induces pancreatic β cells apoptosis. Free sphingoid bases inhibit insulin-induced glucose uptake and oxidation by adipose cells, ceramide down-regulates GLUT4 gene transcription in adipocytes, and sphingolipids may alter insulin action at the level of cell membrane. A number of microorganisms produce secondary metabolites that disrupt sphingolipids metabolism. Many microorganisms, microbial toxins and viruses bind to cells via sphingolipids. Many bacteria utilise sphingolipids to adhere to cells, e.g., E coli (galactosylceramide), Haemophilus influenza (gangliotetraosylceramide and gangliotriosylceramide), Helicobacter pylori (gangliotetraosylceramide, gangliotriosylceramide, sulfatides and G_{M3}), Candida albicans (asialo-G_{M1}), and HIV-1 gp120 (galactosylceramide). Synthetic sphingolipids are effective in inhibiting the binding of bacteria and viruses, sphingolipids in food may compete for cellular binding sites and facilitate the elimination of pathologic organisms from the intestine. Some glycosphingolipids may participate in disease induced by microorganisms (Campylobacter jejuni may lead to Guillian-Barre or Miller Fisher syndrome). Sphingolipids are found in all eukaryotic and some prokaryotic cells and therefore most foods contain varying amounts. They are digested in the intestinal tract, and the bioactive metabolites ceramide, sphingosine are released. Sphingolipids are not ordinary fats, but are functional ingredients with structural and regulatory functions. Sphingolipids are effective at low concentrations, the intake of 0.3-0.4 g/day does not significantly add to the energy content of the diet. They are located mostly in membranes, showing a stabilising effect on membrane structures, and on the function membrane-bound proteins that contain a glycosyl the effective phosphatidyl-inositol anchor (such as the folate in soy are receptor). As receptors and ligands, they are involved in interactions between cells, and between cells and the underlying matrix; they also serve as binding sites for bacteria, toxins, and, or viruses. The sphingolipid metabolites are lipid second messengers in the signal transduction pathway of growth factors (e.g., PDGF), cytokines (e.g., TNF-α), chemotherapeutics such as daunorubicin, and vincristine, and cellular stresses such as γ-irradiation. These agonists induce the hydrolysis of complex sphingolipids, resulting in the release of the highly bioactive metabolites ceramide, sphingosine, sphingosine-1-phosphate and possibly others. This is crucial for the response of the cells in which bioactive compounds accumulate because ceramide and sphingosine are growth inhibitors, and/or induce cell death, while sphingosine-1-phosphate inhibits apoptosis and is a potent mitogen (opposite effects). Most of the sphingolipids consumed per year are derived from dairy products (containing 160 mmol/kg in milk and more than 1300 mmol/kg in cheese), followed by meat products (400-500 mmol/kg) and eggs (2410 mmol/kg). In milk, sphingolipids are located in the membranes of the milk fat globule, and a substantial amount remains in the milk after the skimming process. Fruits and vegetables contain less sphingolipids (<100 mmol/kg), but soybeans are an excellent source (>2400 mmol/kg). The structures of sphingolipids vary considerably with the type of food, which make sphingolipids structurally the most diverse class of membrane lipids. Enzymes of sphingolipid digestion (sphingomyelinases, glucosylceramidases and ceramidases) are found in the small intestine, and remove the headgroups and the fatty acids. The metabolites are absorbed rapidly and utilised for the synthesis of complex sphingolipids, or they are degraded further to fatty acids via fatty aldehydes. Most of the sphingolipid metabolites are retained in the intestinal mucosa; only a small amount is transported into the body via lymph or blood. The digestion of complex sphingolipids is incomplete and about 25% of the administered dose is excreted via the faeces. Sphingomyelin affects cholesterol synthesis, esterification, and bile acid synthesis. Cholesterol increases VLDL sphingomyelin via up-regulation of the synthesis, and down-regulation of degradation. Sphingomyelin inhibits lipid peroxidation in LDL. Dietary sphingolipids may suppress early and late stages of colon carcinogenesis.

Phosphatidylserine:

1- Improves concentration and short-term memory.
2- Effectively alleviates depression-especially in elderly.
3- Increases brain α-waves by 15-20%.
4- Helps stabilise brain wave patterns in epileptics.
5- Increases intelligence.
6- Prevents the decline in Learning capacity that occurs with age.
7- Prevents the decline in the number of brain dendrites that occurs with age.
8- Improves mood, especially in elderly persons.
9- Involved in myelin sheath repair.
10- Increases the number of neurotransmitter receptor sites.
11- Stimulates the release of the brain neurotransmitter dopamine.
12- Improves reflexes.
13- Counteracts cortisol that rises during intensive exercise and during stress.
14- Enhances the function of nerve growth factor (NGF).
15- Increases production of the brain neurotransmitter acetylcholine.
16- Enhances brain glucose metabolism.

PHOSPHATIDYLSERINE

Working in the cell membrane milieu, phosphatidylserine (PS) is a nutrient that supports membrane proteins crucial for homeostasis, maintenance, and specialized cell functions. PS is found most concentrated in the brain, where its relative abundance reflects its involvement in specialized nerve cell functions such as chemical transmitter production and release, receptor action, and synaptic activity. PS supports EEG integration, the HPAA (hypothalamic-pituitary-adrenal axis), and circadian rhythms of hormone release. PS benefits measurable cognitive functions, which tend to decline with age; these include memory, learning, vocabulary skills and concentration, as well as mood, alertness, and sociability. PS is a phospholipid, ubiquitous in membranes and obligatory for all the cells of the body. The in vivo synthesising of PS from precursors is multistep biosynthesis and is energetically costly. Beginning around midlife, the brain's higher functions of memory, learning, semantic manipulation, and concentration (collectively referred to as cognition) measurably begin to fade. Over the adult life span, individuals who are otherwise healthy can lose as much as half of their cognitive capacities. Such progressive and insidious loss of the brain's higher functions can have a telling effect on personal productivity, damage self-esteem, and bring considerable distress to many aging adults. Phosphatidylserine (PS) is one of the 5 phospholipids essential to the functionality of all the body's cells. Phosphatidylserine is most notably found in the cell membrane of neurons, comprising about 7-10% of its lipid content. The phosphatidylserine currently available over the counter is derived from soy. PS enhances neuronal membrane function and hence cognitive function, especially in the elderly. Ingestion of apoptotic cells by macrophages induces TGF-β secretion, resulting in an anti-inflammatory effect and suppression of proinflammatory mediators. This means that apoptotic cell enhances the resolution of acute inflammation. This enhancement appeared to require phosphatidylserine (PS) on the apoptotic cells and local induction of TGF-β. Engulfment of these apoptotic cells is thought not only to remove them from the tissues but also to provide protection from local damage resulting from release or discharge of injurious or proinflammatory contents. Ingestion of apoptotic cells actively suppresses production of proinflammatory growth factors, cytokines, and chemokines (e.g., GM-CSF, MIP2, IL-1β, KC, IL-8, and TNF-α, and eicosanoids). The potential anti-inflammatory effect of recognition and uptake of apoptotic cells may explain the quiet, noninflammatory nature of apoptotic cell removal during development and tissue remodelling. It also seems likely that it is not only intact apoptotic cells that can induce the anti-inflammatory response, since membrane fragments that express PS on their surface may have a similar effect. The longer-term effects of TGF-β production, in response to apoptotic cells during resolution of inflammation, are contributions to wound healing and fibrosis. PS is essential to the healthy functioning of the human brain where it affects an assortment of nerve cell functions, including: conduction of nerve impulses; accumulation, storage and release of neurotransmitters; the activity and number of receptors involved in synaptic discharge; and the biological maintenance of cellular housekeeping functions. Oral PS slows halts, or in many cases, even reverses cognitive degeneration due to age-related cognitive decline (ARCD), and dementing illnesses like Alzheimer's disease. PS is extremely bioavailable and crosses the blood-brain barrier with ease. Once in the brain, the PS molecule as a unit merges smoothly into the nerve cell membrane where it is available to facilitate cell-level energy and homeostasis, as well as enhance neurotransmitter production, release, and action. PS also serves as a precursor reservoir for the related phospholipids, phosphatidylethanolamine and phosphatidylcholine. Phosphatidylserine restores acetylcholine synthesis; mechanisms through which the action on cholinergic systems might take place, e.g., stimulation of the high affinity choline uptake. PS counteracts physiological and pharmacological suppression of humoral immune response. PS is a necessary cofactor for protein kinase C (PKC) activation, and changes in the synthesis of PS. PS participates in the mechanism(s) involved in the transmembrane signalling of IL-1. PS may reverse the physiological decline of the humoral immune response induced by the ageing process (50 mg/kg, in drinking water), and raises specific antibody titers to levels back to normal. PS 800 mg/d for 10 days would significantly blunt the ACTH and cortisol responses to physical exercise, without affecting the rise in plasma GH and PRL. Physical exercise significantly increases the plasma lactate concentration. Chronic oral administration of phosphatidylserine may counteract stress-induced activation of the hypothalamo-pituitary- adrenal axis in man. PS enhances neural events involved in the encoding or consolidation of new information into memory. Cortisol decreases significantly the level of phosphatidylcholine, phosphatidylethanolamine, and sphingomyelin levels in liver microsomes. Some of the microsomal enzymes that are affected by corticosteroids or reduced adrenal function require phospholipids for full activity. By affecting the lipid composition of the membranes, corticosteroids may regulate or modulate the activity of the lipid-requiring enzyme systems. β-Lipid is an intracellular liposome delivery system composed of the same 4 essential phospholipids that surround every cell in the body. Sphingomyelin (SPM) serves as a surface receptor for immunoglobulins and some bacteria and is an activator of epithelial growth receptors. Phosphatidyl ethanolamine (PE) plays an important role in myelin structure of nerve endings in the brain. Phosphatidyl choline (PC) is the most abundant component of all cell membranes. It helps the liver to effectively eliminate viruses, pollutants, and pharmaceuticals. PC helps slow the aging process by protecting the cell membranes from damage. Phosphatidyl serine (PS) stimulates the production of a brain messenger chemical that helps regulate memory by increasing the availability of glucose in the brain. PS stimulates production of dopamine and protein kinase C (PKC) and protects against stress-induced behavioural changes. Phosphatidylserine (PS) is a nutrient, which supports membrane proteins crucial for homeostasis, maintenance, and specialized cell functions. PS is found most concentrated in the brain. PS supports EEG integration, the HPAA (hypothalamic-pituitary-adrenal axis), and circadian rhythms of hormone release. PS is a phospholipid, ubiquitous in membranes and obligatory for all the cells of the body. Present in common foods in small amounts, PS may be a semi-essential nutrient. Beginning around midlife, the brain's higher functions of memory, learning, semantic manipulation, and concentration (collectively referred to as cognition) measurably begin to fade. Over the adult life span, individuals who are otherwise healthy can lose as much as half of their cognitive capacities, as measured from tests related to everyday tasks that rely on cognitive skills. Phosphatidylserine (PS) is one of the 5 phospholipids essential to the functionality of all the body's cells. PS facilitates membrane-to-membrane fusion, a central process in nerve transmitter release; and by activating cell surface receptors supports signal transduction, the process through which the cell responds to chemical signalling substances. Cognitive decline with aging is to some degree inevitable (Age-related cognitive decline, or ARCD). Age-associated mental impairment correlates roughly with progressive decline of nerve net density in the hippocampus and the cortex. Total, or near-total, breakdown of mental functions is neither strictly age-related nor an inevitable accompaniment of aging. Heavy metals and free-radical stressors can destroy brain tissue. Smoking and alcohol intake are major negative factors. One alcoholic drinking binge can initiate memory impairment that continues for days afterward, and drinkers can develop a dementia as debilitating as other dementias. Sustained emotional stress impairs memory. Nutrient deficiencies can impair the higher mental faculties, and food additives can be sources of excitotoxins (e.g., monosodium glutamate, aspartame) that destroy neurons over the long-term. Substance abuse (cocaine, amphetamines, psychedelics, and to some extent marijuana) impairs the brain's higher functions. PS can benefit cognitive functioning in subjects over the age of 50 who are in relatively early stages of cognitive decline. Phosphatidylserine appears to exert an action in 2 distinct contexts: one relating to the cognitive effects of vigilance, attention, and short-term memory, and the other relating to behavioural aspects such as apathy, withdrawal and daily living. PS improves the quality of life and contribute in keeping the patients within their own families and social background. PS improves adaptability to the environment, which can have an important impact on the quality of life (QoL) of the patients. Epileptics could possibly benefit from PS, but only through chronic administration. PS stabilises cognitive decline, stabilises resting brain metabolism and boosts brain activation. PS can benefit brain dysfunctions other than the strictly cognitive. PS can have beneficial effects on mood. Phosphatidylserine modulates the cortisol release, conserves hypothalamic function and benefit the aging hypothalamus-pituitary-adrenal axis (HPAA), restores the daily rhythm of TSH secretion to a level comparable with the young male adult, and spikes the growth hormone release (activation of dopamine metabolism in the pituitary gland by PS). PS works in nerve cell membranes, and helps optimise a variety of functions indispensable at the level of the single nerve cell. These encompass homeostatic (basic, survival-type) processes; maintenance (renewal, repair, scavenging); and specialized processes unique to the nerve cell. The PS phospholipids are one of five phospholipid classes, the others being phosphatidyl-cholines (PC), phosphatidyl-ethanolamines (PE), phosphatidyl-inositols (PI); and the sphingomyelins. Each of phosphatidyl PL molecules has a head group that contains phosphorus and one other chemical subgroup, which in the case of PS is serine. A 3-carbon backbone, structurally identical to glycerol, is attached to the head group. Extending out from the glycerol backbone are two tails, each of which is a fatty acid. The sphingomyelin phospholipids have a different molecular structure: they do not have the glycerol backbone, and carry only one fatty acid tail. PS controls: 1- Entry of nutrient substances into the cell; and clearance of wastes. 2- Movements of charged atoms for impulse conduction. 3- Reception of molecular messages from outside the cell. 4- Transformation of messages into metabolic response. 5- Cell movement, shape changes, flattening or expansion. 6- Cell-to-cell recognition and communication. The ion pumps, transport molecules, enzymes, and receptors, which manage these master-switch activities are all primarily protein in nature. All are built into or onto the phospholipid (PL) membrane matrix, and all require PL for full functional capacity and optimal activity. PS is ubiquitous in the outer cell membrane, in membranes of secretory vesicles, and in the mitochondrial membrane system (where it also serves as a metabolic reservoir for PE and PC via enzymatic decarboxylation). Function of PS includes: 1- Maze learning, other adaptive behaviours are partially rejuvenated. 2- Glucose utilisation efficiency, synaptic efficiency improves. 3- Abnormal EEG patterns are reversed. Other phospholipids, oleic acid, serine amino acid do not substitute for PS. 4- Structurally, decline of nerve network density is reversed. 5- Nerve transmitters are boosted-acetylcholine, catecholamine turnover, tyrosine hydroxylase activity, and dopamine release. 6- Lagging circadian and hormonal rhythms are reset. PS can confer protection from toxic free-radical attack. PS may have antioxidant effects. PS enhances cellular detoxification capacity, due to its overall enhancement of membrane-based cell functions. PS is present in every cell type in the body. PS facilitates the recycling of old cells. Membrane enzymes (amino-PL translocases) transpose PS from its usual position in the inner half of the membrane to the outer half. This apparently acts as a signal to circulating immune cells to remove the aged red cell from the circulation. PS is also involved in membrane phenomena linked to bone matrix formation, testicular function, signal transduction in the heart, and secretion by the adrenal glands. PS is an orthomolecule, a nutrient present in most common foods in small amounts. PS is a semi-essential nutrient, the body can make this phospholipid only through a complex series of reactions and with substantial investment of energy.

Membrane proteins known to require PS.

1-Sodium-potassium ATPases, regulating Na/K flux.
2- Calcium and Magnesium ATPases, others.
3- Protein kinases, mediating one type of signal transduction.
4- Adenyl cyclases, regulating another type of transduction.
5- NADPH-cytochrome reductase and other mitochondrial complexes.
6- Proteins of secretory vesicles.
7- Receptors for N-Methyl-D-Aspartic acid and other nerve transmitters.

Lecithins are normally a good dietary source of phospholipids. Phosphatidylserine has good bioavailability by the oral route. Following oral dosing PS reaches the blood at about 30 minutes, and immediately also begins to build up in the liver and the brain. As PS is being absorbed into the intestinal lining cells, the PS headgroup remains intact but some of the tails (particularly those at position 2) are removed. The fatty acid tails in the 2 position are hydrolysed away by digestive enzymes. Directly following absorption the tails are re-added by re-acylation enzymes. The most active form of PS in membranes is the lyso-form, which lacks a tail in the 2-position. Lyso-PS can act synergistically with NGF, a small protein that enhances nerve cell renewal in the brain. In the brain's gray matter, much of position 2 is occupied by the long-chain, ω-3 fatty acid docosahexaenoic acid (DHA). Hypertensive, diabetic, or alcoholic may be impaired in making DHA from ω-3 precursors, and such individuals should probably supplement their diet directly with DHA. While PS appears to be a highly effective substance for conserving the intellect, it is not a magic bullet, and may not accomplish miracles by itself. Rather, its membrane-based action mechanisms lend PS to compatibility with other nutrients that are safe and have different mechanisms of action, such as Ginkgo biloba extract and acetyl-l-carnitine. PS is also fully compatible with vitamins, minerals, antioxidants, and other nutrients. PS is not contraindicated with respect to drugs. PS improves the QoL for the elderly, especially conserving memory, learning, concentration, and other higher mental capacities in the face of advancing age. PS enhances neural events involved in the encoding or consolidation of new information into memory. Physical stress induces increase in plasma epinephrine, NE, ACTH, cortisol, GH and prolactin (PRL), with no significant change in plasma dopamine and glucose. Pretreatment with both 50-75 mg PS significantly blunts the ACTH and cortisol responses to physical stress. Diminished activity of glucose-6-phosphatase may be partly due to a low level of phosphatidylcholine.

CHOLINE

Choline is one of the "lipotropic" B vitamins. Choline is widely available in food but is sensitive to water and may be destroyed by cooking, food processing, improper food storage, and the intake of various drugs, including alcohol, oestrogen, and sulfa antibiotics. Choline is easily absorbed from the intestines and crosses the blood-brain barrier into the spinal fluid to be involved directly in brain chemical metabolism. Humans can synthesise choline from the amino acid glycine. Choline is probably also manufactured by intestinal bacteria. Choline as phosphatidylcholine, is a basic component of soy lecithin and thereby helps in the emulsification of fats and cholesterol in the body, by helping form smaller fat globules in the blood and aiding the transport of fats through the smaller vasculature and in and out of the cells. Choline is combined with fatty acids glycerol and phosphate to make lecithin, an important part of cell membranes. Choline is also an integral part of the neurotransmitter acetylcholine. Its availability preserves the integrity of the electrical transmission across the gaps between nerves, and this helps the flow of electrical energy within the nervous system. It is also important to the health of the myelin sheaths covering the nerve fibers. Choline helps the liver and gallbladder function and is vital to brain chemistry, as it seems to aid thinking capacity and memory. When choline is depleted, fat metabolism and utilisation may be decreased, leading to fat accumulations, loss of cell membrane integrity and the effects on the myelin covering of the nerves.

COPPER

Copper (Cu)[3] ions serve as important catalytic cofactors in redox chemistry for proteins that carry out fundamental biological functions that are required for growth and development. Cu participates in reactions that result in the production of highly ROS. Hydroxyl radicals may be responsible for devastating cellular damage that include lipid peroxidation in membranes, direct oxidation of proteins, and cleavage of DNA and RNA molecules. Cu may manifest its toxicity by displacing other metal cofactors from their natural ligands in key cellular signalling proteins. Dietary Cu is absorbed across the mucosal intestine (primarily stomach and small intestine). Once transported into intestinal mucosal cells, Cu is transported across basolateral membrane. Cu homeostasis is coordinated by several proteins to ensure that it is delivered to specific subcellular compartments and Cu-requiring proteins without releasing free Cu ions that will cause damage to cellular components. Cu is delivered to specific molecules or subcellular compartments through highly controlled pathways by forming complexes with small cytosolic proteins known as Cu chaperones. Cu-deficient microcytic hypochromic anaemia can be corrected by dietary Cu and not Fe. Cu supplementation lowers the abnormal Fe accumulation and increases Fe absorption into cells. A cup of cooked mushrooms contains almost a milligram of copper, a trace mineral that may play a role in guarding against osteoporosis. There is no recommended daily intake for copper, but 1.5-3 mg/day is a good goal. Copper is a brain stimulant and destroys histamine, the elevated serum (and presumably brain) copper level probably accounts for many symptoms of histapenia (histamine low), including the low blood histamine level. Medical conditions that may be associated with excess copper (Cu) include: biliary obstruction (reduced ability to excrete Cu), liver disease (hepatitis or cirrhosis), and renal dysfunction. Symptoms associated with excess Cu accumulation are muscle and joint pain, depression, irritability, tremor, haemolytic anaemia, learning disabilities, and behavioral disorders. It is important to rule out contamination from permanent solutions, dyes, bleaches, swimming pool/hot tub water, and washing hair in acidic water carried through copper pipes. In the case of contamination from hair preparations, other elements (aluminum, silver, nickel, titanium) are usually also elevated. Sources of excessive Cu include contaminated food or drinking water, excessive Cu supplementation, and occupational or environmental exposures. Insufficient intake of competitively absorbed elements such as zinc or molybdenum can lead to, or worsen Cu excess. Confirmatory tests for copper excess are a comparison of Cu in pre vs. post provocation (D-Penicillamine, DMPS) urine elements tests and a whole blood elements analysis. Caeruloplasmin can also be useful in copper retention syndromes. Copper excess causes brain dopamine levels to rise in low histamine schizophrenia. Paranoia is also associated with elevated copper. Copper oxidises catecholamines such as dopamine and therefore propagates neurotoxin formation. Copper is also a powerful free radical generator and has been shown to be elevated within the substantia nigra of Parkinsonian brains. Copper is an essential mineral that is a component of several important enzymes in the body and is essential to good health. Copper is found in all body tissues. Copper deficiency leads to a variety of abnormalities, including anaemia, skeletal defects, degeneration of the nervous system, reproductive failure, pronounced cardiovascular lesions, elevated blood cholesterol, impaired immunity and defects in the pigmentation and structure of hair. Copper is involved in iron incorporation into haemoglobin. It is also involved with vitamin C in the formation of collagen and the proper functioning in CNS. The best studied enzymes that have been found to contain copper are superoxide dismutase (SOD), cytochrome C oxidase, catalase, dopamine hydroxylase, uricase, tryptophan dioxygenase, lecithinase and other monoamine and diamine oxidases. Copper is a brain stimulant and destroys histamine, the elevated serum (and presumably brain) copper level probably accounts for many symptoms of histapenia (histamine low), including the low blood histamine level. Behavioural symptoms in high-copper histapenia include paranoia and hallucinations in younger patients, but depression may predominate in older patients. Schizophrenics may have high blood copper, as seen in histadelia, with low urinary copper (due to retained copper) as well as low blood zinc. Intervention includes zinc, manganese, vitamin C, niacin, vitamin B_{12}, and folic acid, would slowly reduces high blood copper and symptoms are slowly relieved in several months' time. Risk factor for histapenia is severe childhood hyperactivity. Histidine, High/increased protein diet, Mn, EFAs, folic acid in conjunction with vitamin B_{12} (cobalamin) injections raises the blood histamine, vitamin B_3 (Niacin), vitamin C (Ascorbic acid), Zn and Mn with vitamin C remove copper from the tissues. Copper destroys histamine and therefore as copper levels decrease, histamine levels should return towards normal.

Cu-binding proteins:

Protein	Function	Deficiency
Cu/Zn SOD	Free radical detoxification.	Oxidative damage of cell.
Cytochrome c oxidase	Electron transport in the mitochondria.	ATP deficiency (myopathy, ataxia, seizures).
Lysyl oxidase	Crosslinking of collagen and elastin.	Connective tissue diseases (vascular rupture and torsion). Loose skin and joints, emphysema.
Dopamine β-hydroxylase	Catecholamine production.	Hypothalamic imbalance (hypothermia, hypotension, dehydration, somnolence).
Tyrosinase	Melanin production	Depigmentation.
Peptidylglycine monooxygenase	Bioactivation of peptide hormones.	Malfunction of several peptide hormones.
Ceruloplasmin	Ferroxidase, Cu transport.	Anaemia.
Factor V, VIII	Coagulopathy.	Bleeding tendency.
Angiogenin	Angiogenesis	Defective blood vessel development.
Metallothionin	Cu-sequestration.	Cu toxicity.
Prion protein	Cu binding properties.	Insomnia.
β-amyloid precursor protein.	Unknown.	Familial Alzheimer's disease.
Hephaestin	Iron egress from intestines.	Sex-linked anaemia.

ZINC

Zinc (Zn) is a small hydrophilic, highly charged species that cannot cross biological membranes by passive diffusion. Specialised transpoters are required for both its uptake and release. Zinc transport-1 (ZnT-1) is localised to the basolateral membrane, allow Zn absorption/retention. ZnT-2 may involve in Zn efflux or uptake into vesicles in intestine, kidney, and testis. ZnT-3 is involved Zn uptake into vesicles in neurons and testis. ZnT-4 is in mammary gland and brain. The divalent cation transporter 1 (DCT1) is regulated by iron and exhibits transport activity for a number of trace elements including Zn. Zinc (Zn) is an essential element that is required in numerous biochemical processes including protein, nucleic acid and energy metabolism. Zn is an obligatory co-factor for numerous enzymes including alcohol dehydrogenase, carbonic anhydrase, and superoxide dismutase. Low zinc may be the result of poor dietary intake, digestive dysfunction, malabsorption syndromes, chronic diarrhea, or excessive tissue levels of copper or iron. Many possible dysfunctional conditions may be associated with zinc inadequacy. These include impaired taste or smell, poor night vision, fatigue, skin disease (dermatoses), sexual dysfunction, growth retardation in children and (partial) alopecia. Conditions associated with low Zn include maldigestion, celiac disease, chronic hepatitis, sickle cell anaemia, renal dialysis, cancer, anorexia, obesity and Wilson's disease. Low Zn has also been noted in premature birth babies and their mothers, as well as mothers of infants with spina bifida. Zn is commonly low in diabetics, and in association with ADD/ADHD and autism. Reported symptoms of Zn deficiency include: fatigue, apathy, hypochlorhydria, decreased vision and dysgeusia, anorexia, anaemia, dermatitis, weak/brittle nails and hair, white spots on nails, alopecia, impaired would healing, sexual dysfunction (males), and hypogonadism. Zn competes for absorption with Cu and iron, Cd, Pb and mercury. They are also potent Zn antagonists. Zn deficiency can be caused by malabsorption, chelating agents, poor diet, excessive use of alcohol or diuretics, metabolic disorder of metallothionein metabolism, surgery, and burns. Zinc imbalance is associated with CNS disorders such as schizophrenia and autism and several other pathologies. Zinc is an essential trace mineral. The functions of zinc are enzymatic. There are over 70 metalloenzymes known to require zinc for their functions. The main biochemicals in which zinc has been found to be necessary include enzymes and enzymatic function, protein synthesis and carbohydrate metabolism. Zinc is a constituent of insulin and male reproductive fluid. Zinc is necessary for the proper metabolism of alcohol, to get rid of the lactic acid that builds up in working muscles and to transfer it to the lungs. Zinc is involved in the health of the immune system, assists vitamin A utilisation and is involved in the formation of bone and teeth. High concentrations of Zn are found in brain hippocampus, which may indicate that it acts as a neurotransmitter. Low zinc levels at these sites could reduce the inhibition of neuron activity, thus leading to abnormal behaviour. Zinc deficiency can result in irritability, anger episodes, impaired immune function, acne, stunting of growth, poor taste sensitivity, and impaired wound healing. There is a high incidence of zinc deficiency in ADD, autism, depression, schizophrenia, and bipolar disorders. Zinc deficiency is hard to confirm since no single laboratory test is always low. Absorption of dietary Zn into the bloodstream is usually about 35-45% efficient, but malabsorption syndromes can reduce Zn uptake to about 10-15%. Once in the bloodstream, Zn concentrations are controlled by the metal-binding protein, metallothionein. Many individuals with zinc deficiency appear to have a metallothionein disorder. Patients with an overproduction of pyrroles (pyrroluria) also develop zinc deficiencies. Zinc deficiency usually responds well to supplementation. Zinc deficiency can be corrected, but not cured. If treatment is discontinued, zinc deficiency usually will re-emerge with all symptoms gradually returning. Zinc deficiency, like diabetes, requires life long treatment. Zinc is essential in nerve development, intellectual function, serotonin formation, the regulation of mood, and the prevention of oxidative damage. Zinc deficiency is common. There is no specific disease associated with zinc deficiency, but many general signs and symptoms can point to it. It tends to occur in the elderly, when zinc intake is inadequate, when there are increased losses of zinc from the body, when copper exposure is high, or when the body's requirement for zinc increases. Symptoms and signs include weak appetite, sparse head hair, abnormal tastes in mouth, sensations of unpleasant tastes, reduced sense of taste, unpleasant smell sensations, grooves across fingernails, white spots on fingernails, hang nails, inflamed cuticles, anorexia, tinnitus, weak sexual desire, nasal polyps, darker/redder skin colour, depression, aphthous ulcers, psoriasis, boils, abscesses, carbuncles, hair loss, adult acne, and being a light sleeper, The healing time of surgical wounds is reduced by 43% with zinc sulfate at 50 mg tid. Topically applied zinc oxide enhances the regeneration of epithelial tissue on leg ulcers, and reduces inflammation and bacterial growth. Zinc status should be evaluated in men with decreased serum testosterone. Zinc supplementation is essential in the treatment of low sperm count especially in the presence of low testosterone levels. Both sperm count and testosterone levels are expected to rise in men with initially low testosterone levels. Zinc deficiency weakens the immune system. High concentrations of zinc are found in the inner ear. A deficiency of protein or zinc can reduce the amount of vitamin A released from the liver. Zinc is required in order to transport vitamin A from the liver to the retina. 30-50mg daily zinc dramatically reduces certain body odours. Zinc may also reduce perspiration and sweaty feet. Copper to zinc ratio is significantly high in patients with lymphoma or acute and chronic leukaemias. Zinc increases sperm count and motility as well as raising testosterone levels when low. Zinc deficiency is frequently associated with alcoholism, due to a lower intake of food. Zinc, manganese and vitamin B_6 have been helpful in treating osteochondrosis (Perthes disease). Diarrhoea causes a loss of zinc and therefore digestive diseases or gastrointestinal surgery that result in diarrhoea are often associated with a deficiency. Kryptopyrrole combines with vitamin B_6 and zinc, resulting in their excretion in the urine. Gastric acid secretion plays an important role in the regulation of zinc absorption. By reducing stomach acid levels, H2 blockers might interfere with the absorption of iron, zinc and perhaps other minerals. Drinking coffee causes a significant loss of several vitamins and minerals, including vitamins B and C, calcium, iron, and zinc.

COPPER/ZINC IMBALANCE

Nutrient metals from the diet are: 1- incorporated into blood if blood levels are depleted, 2- transported into cells if cellular levels are inadequate, or 3- excreted if blood and cell levels are sufficient or overloaded. When this system fails to function properly, abnormal levels of trace metals can develop in the brain and other parts of the body. One of the most common trace-metal imbalances is elevated copper and depressed zinc (the optimal plasma or serum ratio is 0.7-1). Copper and zinc are regarded as neurotransmitters and are in high concentrations in brain hippocampus. Elevated copper and depressed zinc is associated with hyperactivity, attention deficit disorders, behaviour disorders, and depression. Autism and paranoid schizophrenia have elevated blood copper levels in addition to other biochemical imbalances. 80% of hyperactive patients and 68% of behaviour-disordered patients have elevated blood copper levels. In many cases, symptoms may be provoked by consuming chocolate (rich in copper) or food dyes rich in hydrazines, which lower blood zinc levels. Many high-copper patients (often labelled depressives) experience severe PMS, are intolerant to oestrogen, and may have a family history of postpartum depression. Also they have a high incidence of acne, eczema, sensitive skin, sunburn, headaches, poor immune function, and white spots under their fingernails. Elevated copper/zinc ratios can be especially serious for persons with low blood histamine (over-methylation). This combination of imbalances may be associated with anxiety, panic disorders, paranoia, and in severe cases hallucinations. Copper and zinc levels are regulated by metallothionein, which is a short linear protein composed of 61 amino acids. Metallothionein deficiency results in abnormal levels of nutrient metals (such as copper, zinc, and manganese) and toxic metals (such as cadmium, mercury, and lead). A high ratio of copper/zinc may cause zinc deficiency and could lead to violent behaviour. High copper levels indicate a zinc deficiency. Less physical affection (or more physical neglect) can contribute to greater aggression. There may be a high copper/zinc ratio among assaultive people. Reduction in sugar intake may improve aggressive/delinquent behaviour. Over 80 enzymes are known to require zinc as part of their prosthetic groups. These include alcohol dehydrogenase, carbonic anhydrase, DNA and RNA polymerases, and carboxypeptidase. Zinc is found in high concentrations in the prostate gland, sperm cells and the eyes, where it presumably plays an important but unknown function. Phytate contained in certain breads combines with zinc; it prevents its absorption. Zinc deficiency during the gestational period may be related to development of schizophrenia. Zinc is required in alcohol dehydrogenase in alcohol metabolism. If zinc/copper ratio is greater than 2.5/1 copper absorption is decreased. Daily ingestion of 150 mg zinc produced overt copper depletion with anaemia in some patients. Pharmacologic doses of zinc (100-300 mg daily) for several weeks can impair immune response, and could lower HDL cholesterol.

PYRIDOXAL-5-PHOSPHATE (P5P)

Many B_6 supplements are the inactive pyridoxine HCl form. The active form of vitamin B_6 is pyridoxal-5-phosphate or P5P, this active form allows for the best absorption. P5P is synthesised primarily in the liver from pyridoxine with the help of enzymes; this requires vitamin B_2, zinc and magnesium for their activity. P5P is associated with numerous enzymes, many of which are involved in amino acid metabolism. This necessary process produces the neurotransmitters dopamine, noradrenaline, GABA and as well as the haemoglobin in red blood cells. Breast-fed infants, autistic, elderly persons on a poor diet, and women on oestrogen-containing oral contraceptive pills, PMS, pregnant and nursing mothers, the elderly, and babies at risk of SIDS (sudden infant death syndrome), are all suffer from P5P deficiency. Vitamin B_6 can result in irreversible CNS damage in premature babies, who are more susceptible than full-term babies to SIDS. A lack of P5P can predispose a surviving premature infant to atherosclerosis in later life. P5P benefits some persons with carpal tunnel syndrome (CTS). Symptoms of P5P deficiency: 1- depression, 2- nervousness, 3- irritability, 4- slow learning, 5- poor dream recall, 6- dizziness, 7- fatigue, 8- cracks around mouth and eyes, 9- dermatitis and acne, 10- inflamed eyes, 11- facial oiliness, 12- stillbirths from deficiency during pregnancy, 13- decreased lymphocytes, 14- decreased vitamin C levels, 15- PMS, 16- increased sensitivity to sound, 17- water retention, 18- decreased resistance to infection, 19- impaired wound healing, 20- poor appetite, 21- AM nausea/vomiting, 22- dental cavities, 23- hair loss, 24- impaired calcium utilisation, 25- decreased absorption of copper, 26- decreased iron status, 27- decreased vitamin B_{12} absorption, 28- arthritis, 29- muscular weakness, 30- neuritis, 31- carpal tunnel syndrome (CTS), 32- temporary limb paralysis, 33- numbness and tingling in the limbs, 34- anaemia, 35- elevated homocysteine levels, 36- seizures, 37- hypoglycaemia, 38- low glucose tolerance, and 39- abdominal pain. P5P Helps to relieve premenstrual syndrome, useful with carpal-tunnel syndrome (CTS), helps boost immunity, protective against atherosclerosis, and may be protective against forms of cancer. Carpal tunnel syndrome (CTS) may result from long-term repetitive motions of the hands and wrists, such as from typing. But, this does not explain the frequent occurrence of CTS with non-motion-related conditions, such as pregnancy. Many people with CTS have vitamin B_6 deficiencies. 100 mg of vitamin B_6 three times per day is the recommend dose, with improvement in 2-3 months. Pregnant and lactating women should not take >100 mg of vitamin B_6/day, or 75 mg of P5P. Zinc and vitamin B_6 are critical nutrients for brain function, fertility, and regulation of the menstrual cycle, maintenance of pregnancy and for the production of digestive enzymes. In addition, zinc is needed for producing stomach acid that helps to break down protein. Zinc deficiency is associated with white marks on the fingernails and stretch marks. Pyrrolurics do better on a low protein diet. Nausea, flatulence and bloating can be the result of inadequate digestion of protein. Red meat is a rich source of protein and Zinc. Zinc is essential for the mechanism of taste and smell. Pyrroluria often starts in teenage years and is more common in females. The development of the male reproductive organs requires more zinc. Amino acids are also needed for the production of neurotransmitters. For neurotransmitters to function properly, the brain needs an adequate supply of important fatty acids. The vitamin B_6 is a cofactor for normal cortisone activity in tendons and synovium. Both B_2 and B_6 are very effective in treating carpal tunnel syndrome within 12 weeks. Vitamin B_6 exists naturally in 3 forms: pyridoxine, pyridoxal phosphate and pyridoxamine phosphate. B_6 conversion process can be interfered with by metabolic disorders, diseases or drugs. Pyridoxine hydrochloride as a supplement, under normal conditions, is converted to the phosphorylated pyridoxal within the cells. Under abnormal conditions the conversion process may be interrupted and a deficiency results. The phosphorylation enzyme, pyridoxal kinase, is preferentially activated by zinc and/or magnesium. Subnormal activity of this enzyme, low levels of zinc or magnesium, or impaired membrane transport can cause a deficiency of P5P. P5P is 10 times more effective than pyridoxine HCl. Common conditions associated with decreased plasma levels of P5P include: 1- immunological disorders, 2- renal disorders, 3- atherosclerosis, 4- hepatic biliary disorders, 5-

pregnancy/lactation, 6- PMS, and 7- cancer. Patients with cirrhosis have an increase in alkaline phosphatase and degradation of P5P. Intestinal absorption of P5P is decreased in patients with celiac disease. Common drugs associated with decreased plasma levels of P5P include: 1- oral contraceptives, 2- hydrazines, 3- cycloserine, and 4- L-dopa, and 5- penicillamine. Pyridoxal phosphate is necessary for amino acid metabolism and many human metabolic sequences. P5P is a coenzyme for transamination steps (transfer of an amino group). Various hyper-aminoacidurias are often found in cases of functional deficiency of pyridoxal phosphate. Pyridoxal phosphate is required in many human metabolic sequences including: 1- amino acid metabolism, 2- carbohydrate metabolism, 3- heme biosynthesis, 4- sphingolipid biosynthesis and degradation, 5- neurotransmitter biosynthesis, 6- collagen formation, and 7- glucocorticoid action.

SESAME LIGNANS (SESAMIN AND EPISESSAMIN)

The sesame lignans (sesamin and episessamin) are compounds commonly found in refined sesame seed oil. Sesame lignans (sesamin and episessamin) have multiple physiological functions including antioxidant activity, anticarcinogenicity, antihypertensive effects, and may alleviate hepatic injury caused by alcohol. Sesame lignans (sesamin and episessamin) have hypocholesterolaemic effects, they inhibit cholesterol absorption from the small intestine, reducing 3-hydroxy-3-methylglutaryl CoA reductase activity in liver microsomes. Sesame lignans (sesamin and episessamin) inhibit marked changes of the ω-6/ω-3 fatty acid ratio by reducing polyunsaturated fatty acid content. Sesame lignans (sesamin and episessamin) do not accumulate in liver, but get absorbed by the lymphatic.

OMEGA-3 FATTY ACIDS

Over 60% of the dry weight of the brain is composed of lipids whose role in the CNS is structural (e.g., neuronal membranes) or functional (e.g., membrane-bound receptors and associated neurotransmitter functioning). Essential fatty acid metabolism can influence many aspects of brain development, including neuronal migration, axonal and dendritic growth, and the creation, remodelling, and pruning of synaptic connections. Humans cannot synthesise certain essential fatty acids. There is a 31% increase in the ratio of having mild to severe depression symptoms among infrequent (<once a week) fish consumers compared with frequent (at least once a week) users. The rates of depression are high and increasing in parts of the world (e.g., United States and Western Europe) where changes in agriculture and food technology have shifted diets away from ω-3 fatty acids toward the physiologically competitive ω-6 fatty acids (from commercial and processed vegetable oils). The depression is often comorbid with various medical disorders, such as cancer, diabetes, cardiovascular disease, and inflammatory disorders, all may be related to impaired fatty acid and phospholipid metabolism. The flaxseed oil supplement has a distinct mood-elevating effect. A higher consumption of ω-3 fatty acids correlated with less severe symptomatology of schizophrenia. Attention-deficit hyperactivity disorder (ADHD) has also been associated with a deficiency in essential fatty acids. Clinical signs consistent with a deficiency of essential fatty acids include: excessive thirst, frequent urination, dry skin, and dry hair. ω-3 fatty acids may also be helpful in the treatment of dementia. Fish consumption is inversely related to incident dementia. A low-fat diet supplemented with flaxseed (30 g/day of ground flaxseed) appears to reduce the growth of prostate cancer cells. Flaxseed supplement for >30 days leads to a significant decrease in mean total testosterone, free androgen index, serum cholesterol, and PSA. The prostate cancer cells are dividing much less rapidly and undergo apoptosis much more quickly as a result of flaxseed supplement (lignins).

CONJUGATED LINOLEIC ACID

Conjugated linoleic acid (CLA) refers to a group of positional and geometric isomers of the ω-6 essential fatty acid linoleic acid (cis-9, cis-12, octadecadienoic acid). CLA is formed when reactions shift the location of one or both of the double bonds of linoleic acid in such a manner that the two double bonds are no longer separated by two single bonds. Several dozen different CLA isomers are possible depending on which double bonds are relocated and the resultant isomeric reconfigurations. CLA, unless otherwise specified, should be construed to indicate a mixture of isomers. Synthesizing and isolating each unique CLA isomer from vegetable oils is a more difficult and expensive process than generating a mixture of CLA isomers. CLA is formed, for example, as a result of rumen gut microbial isomerization of dietary linoleic acid and desaturation of oleic acid derivatives, or as subsequent to bacterial δ-9 desaturase enzyme activity in cows. CLA concentrations in dairy products is 2.9-8.92 mg CLA/g fat, of cheeses is 3.59-7.96 mg CLA/g fat. CLA content of cultured dairy products is 3.82-4.66 mg CLA/g fat. However, significant variation of CLA content of cow's milk products occurs. CLA content of milk fat can be influenced by directly manipulating the type of dietary supplements fed to dairy cows. Human production of CLA from linoleic acid is insignificant. The amount of CLA in human adipose tissue is directly related to milk fat intake. Adipose and lung tissues contain the highest concentrations of CLA. The majority of commercially available CLA is in the form of free fatty acids and not the triglyceride-bound CLA occurring in food. Supplementing the diet with CLA might generate favorable changes in body composition in some human subjects, e.g., reduced body fat. The greatest changes in body fat mass reduction and lean body mass gain is expected with no less than 3.4 g daily dose. CLA may increase lipolysis and β-oxidation of fatty acids, and reduce deposition of fatty acids in adipose tissue. CLA may increase energy expenditure, which is sufficient to account for lower body fat stores. CLA does not cause destruction of fat cells. Fat-mass decrease by CLA may be due to apoptosis of adipose tissue cells, or reduction of adipocyte size. CLA supplementation results in a decrease in adipose tissue weight. CLA significantly reduces LDL, HDL, and total cholesterol. CLA's anticancer activity might be partially a result of CLA-inducing lipid peroxidation. CLA's anticancer activity might be a result of modifying eicosanoid production, e.g., decrease arachidonic acid production, reduction in the release of leukotriene B4 (LTB4) and a reduction of serum PGE_2 levels. CLA might have temporary effect on the oestrogen-mediated mitogenic pathway. CLA's ability to decrease tumour mass might be a result of inducing apoptosis. A dose of greater than 3 g/d of CLA appeared to produce the most favorable results in altering body composition. Adverse effects reported after CLA administration in human subjects include gastrointestinal complaints and fatigue. Along with a reduction in total and LDL cholesterol, reductions in HDL and increases in Lp(α) have been reported. This suggests some lipid fractions improved while others worsened. CLA may accelerate some autoimmune processes.

Ethanol results in decreased liver retinoic acid concentrations by:

1) Increasing the mobilisation of vitamin A from hepatic stores.
2) Blocking the oxidation of retinol by inhibiting alcohol dehydrogenase (ADH).
3) Stimulating the hydrolysis of retinoic acid by cytochrome P450.

VITAMIN A

Dark adaptation has been used as a tool for identifying patients with subclinical vitamin A deficiency. Serum vitamin A concentrations >1.4 µmol/l predict normal dark adaptation 95% of the time. Other causes of abnormal dark adaptation include zinc and protein deficiencies. In the absence of zinc deficiency or severe protein deficiency, dark adaptation is a reliable and highly reproducible functional indicator of vitamin A nutritional status. With regard to the toxicity of retinoids, the biological activity of carotenoid metabolites must be better understood in terms of their possible beneficial as well as harmful effects. Ethanol can compete with retinol for alcohol dehydrogenase, which catalyses retinol oxidation to retinaldehyde, which then can be further oxidised to retinoic acid. Abnormal oxidation products of carotenoids can cause toxicity and may increase the incidence of lung cancer. Alcohol ingestion can result in abnormal gene expression. Retinoic acid exerts profound effects on cellular growth and differentiation. The antiproliferative effect of the retinoic acid receptors is through an interaction with the activator protein 1 (AP-1) complex made up of c-Fos and c-Jun. AP-1 mediates signals from several growth factors, inflammatory peptides, oncogenes, and tumour promoters, usually resulting in cell proliferation. AP-1 induced gene transcription can be inhibited by nuclear retinoic acid receptors (RAR and RXR) when bound to retinoic acid. Ethanol may inhibit retinoic acid synthesis by providing competition for alcohol dehydrogenase-catalysed retinol oxidation. Ethanol may increase destruction of the retinoic acid by increased P450 enzyme activity. Ethanol might have an indirect effect in reducing retinoic acid concentrations in liver cells by increasing the mobilisation of vitamin A to peripheral tissues. The decrease in tissue all-trans-retinoic acid concentrations could interfere with normal retinoid signal transduction by causing a functional down-regulation of RAR activity. RARs act as regulators of AP-1/responsive genes because RARs bound to retinoic acid can combine with the c-Fos/c-Jun complex and sequester it, thereby preventing it from binding to the AP-1 binding site. In the absence of retinoic acid, RARs can no longer bind to the c-Fos/c-Jun complex; thus, AP-1 can bind to DNA sequence motifs, resulting in the transactivation of target genes and cell proliferation. Such a mechanism could, in part, be responsible for alcohol-induced cell injury as well as malignant transformation. Serum concentrations of vitamin A after a physiologic dose of vitamin A reach higher peaks in old people than in young people. Elderly people taking vitamin A supplements in amounts greater than the recommended dietary allowance tends to accumulate more retinyl esters. The longer the individuals took the vitamins containing vitamin A, the greater the tendency for concentrations of potentially toxic retinyl esters to be high. Pumpkin is an excellent source of β-carotene, the precursor to vitamin A.

SEROTONIN

A neurotransmitter occupies the synapse between 2 or more nerve cells (neurons) and thereby allows the triggering of tiny electrical currents in adjacent cells. Serotonin is a neurotransmitter that conveys the positive sensations of satiety, satisfaction and relaxation. It regulates appetite and when converted to melatonin helps us to sleep. A deficiency of serotonin in the brain causes endogenous depression upsets the appetite mechanism and may lead to obesity or other eating disorders such as anorexia and bulimia nervosa and may be responsible for insomnia. Serotonin is a calming, analgesic-like substance, which is secreted in response to carbohydrate and sugar consumption. Sugar addiction may be a misguided attempt to replenish serotonin in the system. A small carbohydrate-rich meal increases the level of serotonin in the brain, and this in turn increases the amount of protein in relation to carbohydrate eaten at the subsequent meal. If tryptophan is given before a meal a similar result may be anticipated. The phenomenon of carbohydrate craving (sugar addition), found in many people on a reducing diet based on a high protein diet, may therefore be the result of reduced serotonin, due to high protein intake. Individuals with serotonin deficiency become depressed. Selective serotonin reuptake inhibitor (SSRI) agents are effective, include fluoxetine (Prozac), sertraline (Zoloft), paroxetine (Paxil), and fluvoxamine (Luvox). In downregulation, there is a decrease in postsynaptic receptors because the brain is flooded with serotonin and the brain adjusts to that level of chemical stimulus; a reduction in serotonin release is produced because of autoreceptors on the serotonin neurone that indicate too much serotonin and shut production down. The increase in serotonin acutely sets something in the body into action that causes an adaptation within the brain that reduces depression. SSRIs treat a lot more than depression, such as those conditions that are part of anxiety disorders: obsessive-compulsive disorder, panic disorder, bulimia, or generalised anxiety disorder, and a whole host of other problems. There are anatomic differences in the brain and serotonin receptors also vary. Most of the cell body neurons that produce serotonin occur in the discrete portion of the brain stem called the medium raphe and extend through specific pathways throughout the brain. The action produced by the drug depends upon the pathway used. The pathway used also explains the side effects that are seen. The current thinking is that the antidepressant effect happens via a pathway that terminates in the prefrontal cortex and related compensatory effects then occur. A second pathway terminates in the basal ganglia structures. The hippocampus and limbic cortex is another system involved to integrate emotional responses to cortical responses. The amygdala is the source of emotional activity and drugs that act here would help in regulating panic and panic disorders, and may be helpful in treating posttraumatic stress disorders. Other pathways may explain why SSRIs are helpful in treating eating disorders, or produce insomnia, nausea, or sexual dysfunction. Multiple serotonin systems exist then, and it is hard to affect some pathways, but not others. A person treated for depression with SSRIs may respond well for a few years, and then, for no discernible reason, the depression is not helped. A long-term increase in serotonin may result in downregulation in the total amount of dopamine produced. Dopamine is involved in producing feelings of reward, so the person seems to feel a lack of satisfaction, or that they are not appreciated, because of the lack of dopamine. SAMe boosts serotonin levels. L-tryptophan is depressed in the plasma of 80% of CFS patients, suggesting that brain serotonin levels may be depressed. Tryptophan is the dietary precursor of serotonin, a neurotransmitter intimately connected with mood. A low tryptophan diet may cause relapse in recovering depressives, while low tryptophan concentrations may rise when depression remits. Tryptophan supplementation provides a mild degree of analgesia and may be especially effective for the subset of chronic pain patients with a disorder of serotonergic transmission. Tetrahydrobiopterin (BH_4) is an essential cofactor for dopamine, serotonin and nitric oxide (NO) formation. Deficiency of BH_4 occurs in a number of inherited metabolic disorders and also in neurodegenerative disorders such as Parkinson's disease. When higher doses of BH_4 are used dopamine and serotonin status can also be restored. Tryptophan is converted to serotonin in the presence of vitamin B_6 (Pyridoxine). When there is a deficiency of vitamin B_6, tryptophan may be transformed into excessive xanthurenic acid, which may cause cancer (bladder), attack the pancreas and cause diabetes. Tryptophan supplementation may have adverse reactions and should be administered under the supervision. Natural sources of tryptophan include soya protein, brown rice (uncooked), cottage cheese, fish, beef, liver, lamb, peanuts, pumpkin, sesame seeds and lentils. Milk and cheese contain tryptophan and this is why a glass of warm milk before bedtime initiates sleep. Warm milk combined with a tablespoon of glycerine is an ideal sleeping agent. Bananas and dates are also providing tryptophan. Other good sources of tryptophan are chlorella or other green or blue algae tablets taken at bedtime to induce sleep (via serotonin production). Serotonin is the precursor of melatonin, a hormone produced by the pineal gland. Decreased brain serotonin has been associated with insomnia, depression, anxiety, panic attacks, hallucinations, suicidal attempts, hostility and psychopathic conditions. Aspartame inhibits the carbohydrate-induced synthesis of serotonin. Serotonin is an important component of the feedback system, which helps limit the consumption of carbohydrate to appropriate levels by blunting the carbohydrate craving. The amino acid tyrosine, derived from phenylalanine, reduces the amount of tryptophan that can cross the blood-brain barrier for utilisation in serotonin production. The decrease of serotonin by aspartame administration has considerable relevance to depression and other psychological problems. The rapid lowering of brain serotonin can precipitate clinical depressive symptoms in untreated individuals vulnerable to major depression. Psychological stress may lead to increased cortisol levels, interfering with serotonin synthesis causing endogenous depression. There is a highly significant correlation between sugar consumption and the annual rate of depression; a correlation does not necessarily imply aetiology. Depression may be caused by frequent consumption of caffeine or sucrose (sugar). An extended period of physical or psychological stress will produce stress hormones such as cortisol and adrenaline, which can interfere with the synthesis of the brain neurotransmitter, serotonin. It takes 60 mg of tryptophan to produce 1 mg of niacin. Hence, niacin deficiency may also be responsible for depression. The absorption of tryptophan can be accelerated by consuming refined carbohydrates, such as sugar. Insulin speeds up the absorption of amino acids other than tryptophan. A person low in serotonin will be inclined to consume greater amounts of sugar in an attempt to increase serotonin production and this may lead to sugar addiction. Sugar addiction can lead to insulin resistance. High levels of insulin cause receptors for insulin to shut down by means of down-regulation. Insulin resistance starts first as mild insulin resistance leading to hypoglycaemia (also called hyperinsulinism), then reactive hypoglycaemia, more severe insulin resistance, which causes unstable concentrations of blood glucose, and finally more complete insulin resistance, causing diabetes over time. Hyperinsulinism blocks the utilisation of adipocytes as a source of energy, thus causing obesity. It also causes to dump magnesium into the urine, upsetting the delicate balance of intracellular magnesium and calcium ions that regulate blood pressure, thereby contributing to hypertension. Hypoglycaemic fluctuations in blood sugar levels causes the body to produce excess adrenaline, which functions to convert glycogen (stored sugar) into glucose in an attempt to stabilise the supply of glucose to the brain. The overproduction of adrenaline, known as the fight/flight hormone, can cause nervousness, panic attacks, anxiety, phobias, extreme mood swings and bouts of aggression and many other symptoms of hypoglycaemia. Depressant drugs, such as alcohol, tranquillisers, benzodiazepines, sleeping pills may temporarily counteract the effects of adrenaline, these are however very addictive and this helps to explain how hypoglycaemia may lead to alcohol or drug addiction. Most drug addicts have been found to be hypoglycaemic. Insulin resistance may also interfere with the absorption of other essential amino acids such as phenylalanine and tyrosine, which are precursors of important brain neurotransmitters, such as dopamine and norepinephrine. An error in norepinephrine synthesis has been associated with attention deficit and hyperactivity disorder (ADHD), because the person is bombarded with irrelevant information and cannot concentrate. Norepinephrine is a neurotransmitter that may block out any irrelevant information from the brain and helps a person (especially young children) to concentrate on the task at hand. Thus, ADHD may be a consequence of insulin resistance and hypoglycaemia. Hypoglycaemia and/or IR may result in a dysfunction of dopamine metabolism. Dopamine conveys the sensation of pleasure and many addictive drugs, e.g., as heroin and cocaine, increase the amount of dopamine, by blocking (inhibiting) the reabsorption (reuptake) of dopamine by brain cells. This causes increased levels of dopamine, which is felt by the addict as a high and as a feeling of great pleasure. The presence of excess dopamine in the brain causes the down-regulation of dopamine receptors as a defence against superfluous dopamine. Receptors for dopamine are reduced and the person becomes dependent on the heroin, cocaine or any other addictive drug to artificially obtain normal levels of dopamine. ω-3 essential fatty acids (fish oil) may help to restore brain cell membranes.

MELATONIN

Melatonin is a non-toxic molecule over a very wide range of doses even when given for years. Melatonin reduces the severity of septic shock in premature newborn infants. Melatonin treatment in these children reduces the clinical measures (white blood cell count, C reactive protein, etc.) of sepsis and improves their survival. Pharmacological melatonin improves the outcome of infants who experience asphyxia during birth. Melatonin reduces the signs of tardive dyskinesia in adults. Co-treating cancer chemotherapy patients with melatonin prolongs their survival, improves their general well being, and reduces the severity of myelosuppression and thrombocytopenia induced by chemotherapeutic agents. Melatonin is effective in slowing the progression of Alzheimer's disease. Melatonin is available in edible foods, e.g., tart cherries, and therefore can easily be included in a healthy diet. Melatonin (N-acetyl-5-methoxytryptamine) is a superior and ubiquitously-acting free radical scavenger and antioxidant for several reasons including the following: 1) Melatonin directly detoxifies a variety of oxygen and nitrogen-based free radicals and non-radical reactants; 2) the metabolites that are formed when melatonin scavenges radicals are themselves radical scavengers (thus, there is an antioxidant cascade that greatly increases the efficacy of melatonin as a scavenger; 3) melatonin stimulates a number of antioxidative enzymes including superoxide dismutase, glutathione peroxidase and catalase, which rid the cell of toxic agents; 4) melatonin inhibits one prooxidative enzyme, i.e., nitric oxide synthase (NOS); 5) melatonin stimulates the production of another important intracellular antioxidant, glutathione; 6) melatonin increases efficiency of mitochondrial electron transport thereby likely reducing the free radical generation; 7) melatonin stimulates ATP product that enhances the ability of cells to repair free radical damage; 8) melatonin is readily absorbed via any route and enters all cells and subcellular organelles; 9) melatonin has no known prooxidative activity unlike some other classical antioxidants. Drugs that deplete melatonin include β-blockers, NSAIDs, steroids, nicotine, alcohol, caffeine, sleep aids and anti-anxiety medications. Fluoxetine (Prozac) may lower melatonin levels. Low melatonin may contribute to insomnia, sleep/wake disorders, or PMS. Some forms of depression are associated with low melatonin levels. Low levels have also been implicated in increased risk for coronary heart disease. A disturbance in the circadian rhythm of melatonin may influence other hormones such as thyroid, testosterone, and oestrogen. Melatonin influences other vital functions including cardiovascular and antioxidant protection, endocrine function, immune regulation and body temperature.

INOSITOL

IP_6 (Inositol Hexaphosphate) suppresses lipid peroxide production and absorbs excess iron ions, having an effect on heart disease, liver dysfunction, and dermatitis; prevents kidney stone formation and cholesterol deposition, and has immunity-enhancing and anticancer actions. IP_6 totally inhibits iron's ability to catalyse the formation of hydroxyl radicals, and by a process of chelation gets rid of it and its ability to form free radicals that cause oxidation and aging. The bran or germ that comprises 10% of whole rice is removed during the polishing process. However, rice bran is an important source of rice oil and other phytochemicals, which possess antioxidative and disease-fighting properties. IP_6 is the major form of phosphorylated inositol present in foods, constituting 1-5% by weight of most cereals, nuts, oilseeds, legumes and grains. It occurs at 9.5-14.5% by weight in rice bran. Antioxidative polyphenols in rice bran include ferulic acid, its esterified derivatives (oryzanols), tocopherols, and other phenolic compounds. Increased consumption of rice and its products would result in improved health, with reduction in heart disease, renal stones and some forms of cancer. IP_6 and its parent molecule, inositol has protective role as antioxidants in oxidative stress. Consumption of whole grains, vegetables and fruits is linked to reduced cancer risk for both colonic and mammary cancer. Fiber is not the sole anticarcinogen since other substances in fiber-containing foods also exert protective influences on cancer. Foods rich in phytate (IP_6) but poor in fiber, such as cereals and grains, correlates better with reduced risk of colon cancer than phytate-poor fiber foods such as fruits and vegetables. IP_6 is a strong chelating agent and certain metals are known to promote cancer through generation of reactive free radicals from oxidation of fats. IP_6 also plays an important role in regulating cell proliferation and differentiation. Inositol hexaphosphate (IP_6) prevents carcinogen-induced tumour development, and interferes with growth of pre-formed, transplanted tumours. Myo-inositol as a chemopreventive agent is another phytochemical with low toxicity and ability to inhibit carcinogenesis in various organs. IP_6 is rapidly absorbed by cells and metabolised to lower phosphates and inositol. Lower inositol and phosphates may mediate cancer inhibition, have metal chelating activity and may interfere with tumour formation by suppressing metal catalysed oxidation of fats, and plays a central role in signal transduction and cell transformation triggered by growth factor or tumour promoter. IP_6 reduces hyperlipidaemia (high levels of cholesterol and triglycerides in blood) and protect against cardiovascular disease (CVD). Bengal gram, a bean species rich in IP_6, is associated with reduced hypercholesterolaemia. Dietary IP_6 lowers the zinc/copper ratio, a marker of hypercholesterolaemia, without significantly affecting levels of other minerals in serum. IP_6 inhibits rises in hepatic total lipids and triglycerides. Higher supplementary levels of dietary IP_6 depress accumulation of lipids. IP_6 inhibits platelet aggregation and to enhance inflammatory responses of neutrophils in response to microbial stimuli. IP_6 inhibits calcification in the aorta and lipid peroxidation in ischaemic kidneys. IP_6 is naturally present in human urine where normal levels fluctuate between 0.5 to 5.0 mg/l. 1-3% of oral doses are excreted in the urine with an associated reduced risk of developing renal stones, phytate can interfere with formation of calculi (crystals) of calcium oxalate and phosphate.

Summary of phospholipids as second messengers (<1 min):

1- The phosphoinositide system is stimulated by a ligand (hormone or growth factor).
2- The ligand bins to a specific receptor on the surface of the cell.
3- The receptor is coupled to a G-protein or tyrosine kinase-linked receptor, which translates the signal by activating a phospholipase C inside the cell.
3- The activated phospholipase C then hydrolyses a phosphorylated phosphatidylinositol, phosphatidylinositol-4,5-biphosphate (PIP_2). This generates 2 second messengers, inositol-1,4,5-triphosphate (IP_3) and diacylglycerol (DAG).
4- IP_3 and DAG activate separate, interacting pathways within the cell.
5- The 2 pathways work in variety of mechanisms, to produce signalling system capable of responding rapidly to a variety of external signals and controlling a wide range of cellular processes.
6- The pathways perform 2 main function: A- bringing the second messengers to directly activate the cell, or B- providing the second messengers to directly activate the cell.
7- Once the second messengers are generated 2 things may happen: A- IP_3 is released into the cytosol and acts to mobilise intracellular calcium stores from ER; and B- DAG stays in the plane of the membrane and activates a PKC.
8- The 2 messengers are metabolised in a number of ways, which terminates the signal and can regenerate PIP_2.

Ingestion of 120 mg/day of IP_6 reduces the urinary risk of kidney stone development. Inositol phopholipids present in plasma membranes (inositol occurs ubiquitously in cell membranes in conjugation with lipids as phosphatidylinositol) and plays a critical role in transmission of signals elicited by growth factors and mitogens acting at the cell surface. IP_6 is the only known dietary source of inositol phospholipids. During cell stimulation, these molecules are converted by special enzymes (PI kinases and phospholipase C) to inositol triphosphate (IP_3) and diacylglycerol, which act as second messengers inside cells. IP_3 also plays a role in cell-to-cell communication and can be generated from IP_6 via a salvage pathway. Anti-tumour compounds such as genistein and quercetin act by inhibiting PI kinases and lowering IP_3 concentration in tumour cells leading to cellular differentiation and death. A nuclear inositol-lipid pathway with its signal-transduction components located and acting in the nucleus may be important in switching cell programming from a proliferative to a differentiative state. IP_6 is the dominant inositol phosphate in insulin-secreting cells of the pancreas where it influences secretion of the hormone by modulating activity of a calcium channel. Other rice components include: 1- Polyphenols from edible plants are a rich source of antioxidative compounds with chemopreventive activity. 2- Ferulic acid is a ubiquitous polyphenol, which is formed from metabolism of 2 amino acids (phenylalanine and tyrosine), occurring primarily in the bran fraction of plant seeds. Natural ferulic acid is commercially extracted and purified from rice bran oil. The compound has strong antioxidant potential, protecting skin cells from light or radiation-induced damage and preserving foods from spoilage due to lipid peroxidation. Ferulic acid inhibits liver carcinogenesis and reduces development of oral lesions induced by chemical carcinogen. Ferulic acid prevents and dissolved platelet aggregation related to thrombosis. An ester of DL-α tocopherol (vitamin E) and ferulic acid prevented facial hyperpigmentation by suppressing melanogenesis induced by UV light. 3- Rice germ showed a chemopreventive effect in large bowel carcinogenesis. 4- Rice bran oil is rich in polyunsaturated fatty acids, lowers blood lipid levels and has a role in preventing atherosclerosis, and prevents hypercholesterolaemia. In combination with safflower oil (7 parts to 3), it exhibits a blending effect, yielding greater reduction of serum cholesterol than either oil alone. Rice bran oil is rich in non-saponifiable matters such as steryl ferulates, which have growth-promoting vitamin like activity. They consist of a mixture of ferulic acid esters called oryzanols. γ-Oryzanol protects rice bran oil from oxidation and inhibits peroxidation of lipids mediated by iron or UV irradiation. Its triterpene alcohol components, cycloartenyl and 24-methylene-cycloartenyl ferulates are effective in the treatment of arteriosclerosis.

TAURINE

Taurine (2-aminoethanesulfonic acid) is a conditionally-essential amino acid that is not utilised in protein synthesis, but found free or in simple peptides. Taurine was first discovered in 1827, and in 1975 taurine's significance in human nutrition was identified. Taurine plays an important role in numerous physiological functions. It is derived from methionine and cysteine metabolism. While conjugation of bile acids is perhaps its best-known function, this accounts for only a small proportion of the total body pool of taurine in humans. The taurine molecule contains a sulfonic acid group, rather than the carboxylic acid moiety found in other amino acids. Taurine is one of the most abundant free amino acids in many tissues, including skeletal and cardiac muscle, and the brain. Vitamin B_6 deficiency impairs taurine synthesis, the 3 pathways for the synthesis of taurine from cysteine require pyridoxal-5'-phosphate (P5P), and the active coenzyme form of vitamin B_6, as a cofactor. Cysteine sulfinic acid decarboxylase (CSAD) converts both cysteine sulfinic acid into hypotaurine, and cysteic acid into taurine. Humans have relatively low CSAD activity, and therefore possibly lower capacity for taurine synthesis. Taurine comprises over 50% of the total free amino acid pool of the heart. It has a positive inotropic action on cardiac tissue, and lowers blood pressure. Accumulation of intracellular calcium leads to cellular death. Taurine may both directly and indirectly help regulate intracellular Ca^{2+} ion levels by modulating the activity of the voltage-dependent Ca^{2+} channels, and by regulation of Na^+ channels. Taurine also acts on many other ion channels and transporters. When an adequate amount of taurine is present, calcium-induced myocardial damage is significantly reduced. Taurine functions as a membrane stabiliser and may prevent suppression of membrane-bound Na-K-ATPase. Taurine can protect the heart from neutrophil-induced reperfusion injury and oxidative stress. This antioxidative property of taurine is due to reduction of respiratory burst activity of neutrophils. Taurine alleviates physical signs and symptoms of congestive heart failure (CHF). Taurine may reverse ECG abnormalities such as S-T segment changes, T-wave inversions, and extra systoles in arrhythmias. Taurine is an effective agent for the treatment of heart failure without any adverse effects. Taurine may prevent rapid progress of congestive heart failure. The liver forms 2-4 g of bile acid pool that has 10 enterohepatic cycles/day, with the terminal ileum serving as the main absorption site for the enterohepatic recycling of 80% of these acids. Bile acids function as a detergent for emulsification and absorption of lipids and fat-soluble vitamins. The intestinal bacteria form the secondary bile acids deoxycholic acid and lithocholic acid, from the primary bile acids, cholic acid and chenodeoxycholic acid, respectively. These secondary bile acids conjugate through peptide linkages with either glycine or taurine to form bile salts. Taurine conjugation of bile acids has a significant effect on the solubility of cholesterol, increasing its excretion, and administration of taurine reduces serum cholesterol levels in human subjects. Much of the insult of cystic fibrosis (CF) is in the ileum, with the enterohepatic recycling of bile acids malabsorbed as well. Taurine (13 g/day) decreases the severity of steatorrhea associated with many CF. Taurine is an antioxidant, which mediates the chloride ion and hypochlorous acid concentration, and protects the body from potentially toxic effects of aldehyde release. Taurine neutralises hypochlorous acid, a potent oxidising substance, and attenuates DNA damage caused by aromatic amine compounds. Taurine contains a sulfonic acid moiety rather than carboxylic acid, it does not form an aldehyde from hypochlorous acid, rather a relatively stable chloroamine compound. Taurine inhibits bacterial intestinal translocation and may protect against endotoxaemic injury. Acamprosate is a synthetic taurine analogue, has a chemical structure similar to that of γ-aminobutyric acid, and acts via several mechanisms affecting multiple neurotransmitter systems, and by modulation of calcium ion fluxes. 50% of alcoholic patients relapse within 3 months of treatment, acamprosate increases abstinence rates and durations of abstinence for 6-12 month post-treatment periods. Ethanol reduces ionic transfer through alterations in the cationic paracellular pathway, the monovalent cation pump, and the antiport system. Caution should be used in extrapolating the effects of acamprosate to taurine or other taurine analogues. The retina contains one of the highest concentrations of taurine in the body. Taurine deficiency induces changes in the photoreceptor cells of the retina, and further depletion can result in permanent retinal degeneration. Taurine (1-2 g/day) for one year may show clinical evidence of improvement in patients with retinitis pigmentosa (RP). Taurine's limited diffusibility across the blood-brain barrier may be the main factor restricting the antiepileptic effect of this compound. There is a decreased concentration of taurine in the cerebral spinal fluid of patients with advanced symptoms of Alzheimer's disease when compared to age-matched controls. Both plasma and platelet taurine levels are depressed in insulin-dependent diabetic patients; oral taurine supplementation raises these levels to normal. Taurine reduces platelet aggregation in diabetic patients in a dose-dependent manner. Taurine (2-aminoethanesulfonic acid) was discovered about 175 years ago. Abundant in human mother's milk but nearly absent in cow's milk. Taurine is a critical metabolite for phagocytosis. In the CNS, taurine helps to regulate the activity of two neurotransmitters, GABA and glutamic acid. In liver tissue, it combines with activated cholesterol (cholyl-Coenzyme-A) to produce taurocholic acid, a primary bile acid that assists uptake of dietary lipids. Taurine regulates the flux of potassium, calcium and magnesium into cells while limiting cellular levels of sodium. The anti-arrhythmic action of taurine appears to be connected with its abundance in heart tissue in humans. Taurine is magnesium-sparing for the body. Normalised or healthy blood Mg levels may decrease blood platelet aggregation and associated vascular disorders. Assessment of taurine levels can be accomplished by amino acid analysis of fasting blood plasma or 24-hour urine. Fasting blood plasma represents the metabolic level rather than a transient dietary influx. When low, inadequacy is certain. When the level is normal in plasma, primary metabolic precursor cysteine must then be checked. Low cysteine or cystine suggests limited taurine during periods of increased need. When high in plasma, one must check for blood cell haemolysis, leukocytolysis, infection, or toxicity that causes rupture of cell membranes. Deficient urine taurine (with normal renal clearance) means deficiency. Elevated urine taurine can also mean deficiency due to wasting. Urinary taurine wasting typically occurs with elevated β-alanine and possibly occurs with elevated β-aminoisobutyric acid. In renal tubules, β-alanine blocks reabsorption of taurine. Elevated β-alanine occurs in the following conditions: A- maldigestion and increased uptake of dietary peptides (anserine, carnosine), B- infection or intestinal dysbiosis in which bacterial production is abnormally increase, and C- catabolism of DNA and RNA as occurs with tissue necrosis or malignancy. Catabolism of tissue produces both β-alanine and β-aminoisobutyric acid. Structurally, hair is about 14-15% cysteine and about 5% by weight sulfur. Low hair sulfur results often are found to correlate with methionine deficiency, cystinuria, or urinary taurine wasting, all of which are diagnosed by 24 hr urine amino acid analysis. Both methionine deficiency and urinary cysteine wasting imply limited reserves or capacity for taurine formation. Conditions consistent with taurine insufficiency include: 1- Signs of Mg deficiency (muscle cramps, lower backache, constipation, fatigue, depression). 2- Hypersensitivity to chlorine, bleach or chlorinated water. 3- Cardiac arrhythmia. 4- Steatorrhea. 5- Elevated cholesterol. 6- Abnormally enhanced inflammation during infections. 7- Frequent infections, leukocytolysis or leukopenia. 8- Maldigestion. 9- Seizures.

METHYL-SULFONYL-METHANE (MSM)

MSM (methylsulfonylmethane, also known as dimethyl sulfone) has generated numerous anecdotal reports of its benefits in cases of allergies, arthritis and joint pain, nutrition for the skin, hair and nails; and support of the health of the GIT. MSM is a bioavailable source of sulfur. Many of the benefits derived from onions, garlic and the cruciferous vegetables, such as cabbage and broccoli, may come from the sulfur, which these supply to the body. The sulfur-bearing amino acids methionine, cysteine and taurine are very important in maintaining normal metabolism and in supplying the building blocks for the production and repair of the skin, cartilage, ligaments and tendons. MSM is a stable source of sulfur, which can be derived from plants grown either on land or in the sea. Marine sources include algae and phytoplakton. There is a sulfur cycle in the biosphere in which sulfur is taken up from the soil by plants, is released into the atmosphere as the highly volatile dimethyl sulfide, which in turn is oxidised in the upper atmosphere to dimethyl sulfoxide (DMSO), which then becomes the atmospheric source of MSM. DMSO and MSM return to the soil via the rain, and then the sulfur cycle repeats itself. Plants in their fresh state thus contain a quantity MSM when grown on sulfur-rich soils, although most of the compound found in plant foods may be lost by improper handling and storage. Sulfur compounds have antioxidant properties role in the body and help to transport methyl groups for various purposes. The story of MSM actually dates back to the early 1960's, when the compound was first prominently mentioned as being of potential importance in health and healing. In the case of osteoarthritis and similar joint and ligament injuries, MSM may work through several different nutritional mechanisms. For instance, it was discovered in the 1930's that sufferers from arthritis often have below normal levels of cystine (a metabolite of cysteine) in their fingernails. This can lead to brittle or soft nails and can be an indication of either inadequate sulfur in the diet or a poor ability to manipulate dietary sulfur to match the body's needs. Sulfur is well represented in the human organism because it is required for the repair of joint tissues and for the construction of connective tissues more generally. MSM itself is abundant in our bodies. 85% of the sulfur found in living organisms is provided by MSM and related compounds. For instance, the circulatory system of an adult human contains about 0.2 parts per million MSM. MSM may improve GIT health. Rheumatoid arthritis, which is an autoimmune disease, is strongly associated with the passage of toxins and certain proteins through the wall of the gut and into the blood stream, i.e., leaky gut syndrome. MSM may improve allergies, constipation, and problems with parasites. Common to all of these are problems in the health of the intestinal wall. Organic sulfur is a precursor to the biosynthesis of the amino acid taurine; an important element is the production of bile. The lining mucosal cells of the digestive tract have high turnover rate such that the whole layer of surface cells may be renewed in 3-4 days. An inability to manufacture adequate building blocks (in this case, glucosamine "amino-sugars') will cause the intestinal wall to "thin" and allow toxins and not fully digested proteins into the blood stream. A lower than normal sulfur content in the gut wall is associated with many autoimmune issues and may play a contributory role. Microorganisms in the gut lining may be responsible for incorporating sulfur from MSM into sulfur-bearing amino acids, with a positive benefit to this essential aspect of the metabolism. MSM may thus play a role in improving this aspect of gut health. MSM may reduce the impact upon the health of the intestinal tissues of various toxins. Similar protective benefits have been found with other tissues, as well.

Phytochemicals:

A- Chemicals with cGMP phosphodiesterase inhibitor activity:	1- Caffeine. 2- Theobromine. 3- Theophylline.
B- Chemical with cAMP phosphodiesterase inhibitor activity:	1- Catechin. 2- Matairesinol. 3- Amentoflavone (5-10 x papaverine). 4- Caffeine. 5- Cyanidin-chloride. 6- Hyperoside. 7- Kaempferol. 8- Kurarinone. 9- Naringenin. 10- Orientin. 11- Quercetin. 12- Rutin. 13- Theobromine. 14- Theophylline. 15- Vitexin.
C- Chemical with phosphodiesterase inhibitor activity:	1- Arctigenin. 2- L-Acetoxypinoresinol. 3- Agathisflavone. 4- Amentoflavone. 5- Biflorin. 6- Cupressiflavone. 7- Hinokiflavone.
D- Biological activities of amentoflavone:	1- Aldose-reductase inhibitor. 2- Antibradykinic. 3- Anti-inflammatory. 4- Antileukaemic. 5- Antiperoxidant. 6- Antiulcer. 7- Antiviral. 8- Fungicide. 9- Pesticide. 10- Phosphodiesterase inhibitor. 11- cAMP-phosphodiesterase inhibitor (5-10 x papaverine).
E- Some plants containing amentoflavone (flavonoid):	1- Hypericum perforatum (Clusiaceae), St. John's wort, 100-500 ppm. 2- Salix alba (Salicaceae), whilte willow, 120 ppm 3- Ginkgo biloba (Ginkgoaceae), ginkgo 3.8-5 ppm.

CHOCOLATE

Chocolate comes from beans harvested from the cocoa tree (Theobroma cacao), native to Central America. The beans are fermented, dried, roasted and pressed to give chocolate liquor that consists of an even mix of cocoa particles and cocoa butter. The chocolate liquor is then pressed to extract more of the cocoa butter, leaving between 10 and 25%; the remaining paste is cooled, ground and sifted to give cocoa powder. A combination of cocoa and cocoa butter with other ingredients produces different forms of chocolate such as dark, milk, and white chocolate. Some of the nutrients in chocolate include magnesium, phosphorus, potassium, calcium, iron, zinc, copper and manganese. Also chocolate contains substantial amounts of antioxidants in the form of complex mixtures of phenolic compounds, and that these may have beneficial effects on health. Cocoa powder has equivalent or greater antioxidant activities than many fruits and vegetables. Tea also contains high levels of antioxidant compounds. Phenolic antioxidants inhibit the oxidation of low density lipoprotein (LDL) cholesterol. Phenols in milk chocolate are higher than that in black and green teas. It is 20 times higher than in tomatoes, twice the level of that in garlic and over three times higher than in grapes. Dark chocolate provides more than twice the level of phenols as milk chocolate per serving, while white chocolate has no antioxidant content. Cocoa powder extract is a potent antioxidant for LDL oxidation. The 3 forms of chocolate, cocoa have the highest levels of phenols, followed by baking chocolate and milk chocolate. Chocolate is high in copper. Chocolate is high in cadmium and nickel. Cocoa beans may be high in both cadmium and lead. Cadmium as well can be introduced to the cocoa during processing, possibly by contact with galvanised containers. The saturated fats in chocolate blocks the toxicity of oils rich in linoleic acid and its odd proteins seem to have an anabolic action.

CRANBERRIES

Cranberries have long been recognised as a good source of vitamin C. As well as preventing scurvy, vitamin C has important antioxidant properties, which may help protect against cardiovascular disease and cancer. Cranberries also contain high levels of flavonoids and phenolic acids. Classes of cranberry flavonoids include the procyanidins (the pigments that give cranberries their rich red colour), found also in chocolate and apples. On average, cranberry juice has been shown to contain about 32 mg of procyanidins per serving. Cranberries contain quinic, malic and citric acids; some of these acids may reach the urine unchanged and provide a hostile environment in which the bacteria cannot grow. Eschericia coli attach themselves to the cells lining the urinary tract and then multiply, causing symptoms such as increased frequency and urgency of urination, irritation and lower back pain. Daily consumption of a low-calorie cranberry juice cocktail reduces bacteriuria, and dysuria (50%) often experienced by elderly women. Cranberry juice acts to promote urinary tract health by inhibiting bacterial adherence to mucosal surfaces. It appears that cranberry juice contains at least 2 different inhibitors: 1- fructose, common to many fruit juices, and 2- the second factor appears to inhibit certain adhesins (P-fimbriae) of some pathogenic strains of E coli. While orange juice, pineapple juice and cranberry juice exhibited antiadhesin activity against type 1 fimbriated E coli, containing a mannose-sensitive adhesin, only cranberry juice contained the mannose-resistant adhesin inhibitor. Cranberry juice inhibits the sialic acid-specific adhesion of H. pylori to human gastric mucus and to human erythrocytes. Cranberry juice has a moderately high concentration of oxalate. Cranberries are rich in several different types of flavonoids, including anthocyanins, flavonols, and proanthocyanins. The antioxidant-like properties of these flavonoids may keep strokes and heart attacks away by helping to prevent fatty plaques from clogging your arteries.

COFFEE

Acute intake of coffee increases blood pressure; the pressor response is strongest in hypertensive subjects. Serum uric acid has been found to be related to risk of gout, coronary heart disease, diabetes mellitus and hypertension. Caffeine's diuretic properties have inverse association to serum uric acid concentrations. Despite the pharmacological similarities between tea and coffee, epidemiological differences occurred. The number of cups of coffee drunk daily is directly proportional to the prevalence of rheumatoid factor positively. High levels of potassium (K) may reduce the risk of hypertension and stroke. Potassium is water soluble, so leaches into water during cooking. A boiled potato, for example, loses at least half its K to the water it's boiled in. Caffeine increases genetic damage in cells from individuals with Alzheimer's disease (AD). Caffeine increases amyloid β protein levels in cells. High levels of amyloid β protein can aggravate the symptoms of AD. Caffeine is found abundantly in coffee, tea, and soda. Coffee and individual coffee constituents stimulate cholecystokinin release, enhance gallbladder contractility, inhibit gallbladder fluid absorption, decrease cholesterol crystallisation in bile, and increase intestinal motility. Coffee diterpines may down-regulate the hepatic low-density lipoprotein receptor and decrease 3-hydroxy-3-methylglutaryl CoA reductase activity. Patients with symptomatic gallstone disease are sometimes advised to avoid coffee. Coffee appears to cause heartburn in certain individuals by diminishing lower oesophageal sphincter pressure, a disorder that may share a common pathogenesis with gallstone disease. Coffee sensitivity is common in individuals with functional dyspepsia, a condition that may be due in part to heightened visceral nociception. Consumers of large amounts of coffee show less health-seeking behaviour than coffee abstainers. There is a positive association with caffeinated soft drinks and increased diagnosis of gallstones with subsequent cholecystectomy. However, a consumption of ≥4 cups of coffee/day is associated with >25% risk reduction of cholecystectomy. Consumption of decaffeinated coffee or tea is not associated with risk, possibly due to the lower amount of caffeine in these beverages. Caffeine and other methylxanthines may prevent bile cholesterol supersaturation by stimulating ileal bile acid absorption, increasing hepatic bile acid uptake, decreasing serum oestrogen levels, increasing sex-hormone-binding globulin concentrations, and increasing thermogenesis and reducing body fat stores. Other ingredients in coffee may contribute to the inverse relation. Coffee may act through the effects of Mg, K, or niacin, which are coffee constituents and are inversely associated with cholecystectomy. Coffee contains an insoluble hemicellulose fiber that may decrease the colonic absorption of deoxycholic acid. Coffee contains antioxidants such as tocopherols and caffeic acid, capable of inhibiting ROM, which appear to precede cholesterol crystallisation. These and other coffee components may be lost during industrial processing of decaffeinated coffee. Removal of diterpines by filtering may modulate hepatic cholesterol metabolism, possibly by down-regulating the activity of sterol regulatory element-binding proteins. Moderate intake of caffeinated coffee may play a role in the prevention of symptomatic gallstone disease. Drinking coffee causes a significant loss of several vitamins and minerals, including vitamins B and C, calcium, iron, and zinc. Coffee percolated in an aluminium pot contained a large amount of dissolved aluminium, because of coffee's acidity. Drinking coffee with iron rich foods can reduce iron's toxic effects. Consuming stimulants such as caffeine appears to increase the risk of bruxism; a sleep disorder characterised by nighttime teeth grinding or jaw clenching that can damage teeth.

SCOPOLETIN

Scopolia have properties like those of hyoscyamus and belladonna and is used as anticholinergic. Scopoletin is 7-hydroxy-6-methoxycoumarin; a plant growth factor derived from the root of scopolia carniolica or scopolia japonica. Scopoletin is a reactive oxygen intermediate (ROI) scavenger. TOGT-mediated glucosylation is required for scopoletin accumulation in cells surrounding TMV lesions, where this compound could both exert a direct antiviral effect and participate in ROI buffering. Tobacco salicylic acid and pathogen-inducible UGTs (TOGTs) act on the hydroxycoumarin scopoletin and on hydroxycinnamic acids. Plant UDP-Glc:phenylpropanoid glucosyltransferases (UGTs) catalyse the transfer of Glc from UDP-Glc to numerous substrates and regulate the activity of compounds that play important roles in plant defense against pathogens. The hypersensitive response (HR) is characterised by localised cell and tissue death at the site of infection and is associated with the induction of intense metabolic alterations, resulting in confinement of the pathogen. One of the earliest responses underlying HR cell death in plants is the increase in the production of ROIs, giving rise to the so-called oxidative burst. The cells surrounding the HR lesion actually are stimulated strongly without being destined to die, and they produce a large set of defense responses that contribute to the efficient restriction of pathogen spread. TOGT is involved in scopoletin glucosylation by regulating the solubility, biological activity, and transport of compounds within the cell; glucosyltransferases are crucial in the maintenance of cellular homeostasis. Scopoletin is a reactant of peroxidases. Scopolin is coumarin glucoside, which displays strong blue fluorescence, accumulates to very high levels compared with other phenylpropanoids (apart from chlorogenic acid) in the ring of stimulated cells surrounding TMV lesions. Scopolin and scopoletin play an important role in plant disease resistance. Scopoletin accumulation could be part of the mechanism of virus restriction in plants. Phenolic compounds display antiviral activity. Scopoletin may represent an antioxidant in living cells surrounding HR lesions besides the possible antiviral role of scopoletin, this molecule likely represents a potent antioxidant in plant defense responses, together with scopolin, which can act as a source of scopoletin after its hydrolysis by glucosidases. Scopoletin, together with other antioxidant systems, plays a role in the regulation of ROI accumulation in living cells surrounding necrotic lesions, either as a substrate of peroxidases or as a direct ROI scavenger. Phenolic compounds display antiviral activity, especially against animal viruses. For example, phenolic moieties, and particularly compounds containing 4-hydroxycoumarin residues, are inhibitors of the Human immunodeficiency virus integrase required for virus replication. Glucosylation is necessary for the accumulation of phenylpropanoids. The coumarin scopoletin have an inducible nitric oxide synthase (iNOS) inhibitory effect. Non-specific spasmolytic action of scopoletin can be attributed, at least in part, to its ability to inhibit the intracellular calcium mobilisation. Scopoletin is one of the phytoalexins in tobacco. Scopoletin, a naturally occurring fluorescent component of some plants and a proven plant growth inhibitor, is a known reactant with peroxidase. Although NADH is also a peroxidase substrate, it cannot compete effectively for the oxidised forms of the peroxidase enzyme. On the other hand, scopoletin stimulates the oxidation of NADH by the H_2O_2 system, apparently by forming a phenoxyl radical, which then oxidises NADH to NAD· radicals. Scopoletin is widely used in a peroxidase assay for H_2O_2.

XERONINE

Xeronine is a relatively small alkaloid that is physiologically active in the picogram range. Noni fruits have a negligible amount of free xeronine, but contain appreciable amounts of the precursor of xeronine, i.e., proxeronine. Proxeronine's molecular weight is relatively large, about 16,000. Noni fruits also contain the inactive form of the enzyme that releases xeronine from proxeronine. Taken Noni juice on an empty stomach, the critical pro-enzyme escapes digestion in the stomach and enters the intestines, where the chances are high that it may become activated. The pepsin and acid in the stomach will destroy the enzyme that liberates xeronine. It is recommend to drink 100 ml of Noni juice half-hour before breakfast. At this time the juice will pass rapidly through the stomach and into the intestines, where it may be converted into the active enzyme. The primary function of xeronine may be regulating the rigidity and shape of specific proteins. The action of ginseng, bromelain and Noni in making a person feel well is probably caused by xeronine converting certain brain receptor proteins into active sites for the absorption of the endorphin, the well-being hormones. To obtain the maximum effect of the active ingredient in Noni, it is recommended that not to drink Noni juice with coffee, tobacco or alcohol. The combination lowers the potentially beneficial effect of xeronine, and gives some unexpected side effects. The green fruit has more of the potentially valuable components and less of the undesirable flavor. Xeronine attaches to the surface of certain proteins (e.g., enzymes, cell receptors or cell membrane channels) and stabilises their structure, thus enhancing their efficiency. Xeronine can be found in every healthy cell in plants, animals or micro-organisms. This substance is short-lived. The extract of Tahitian Noni is effective to inhibit the growth of tumours. Tahitian Noni juice, made from the extract of the noni fruit, Morinda citrifolia is not a medicine, but a fruit juice.

MOLYBDENUM

Molybdenum (Mo) is widely distributed in nature, the crustal abundance being 1.5 mg Mo/kg. Molybdenite (MoS_2) is the major source for industrial production of molybdenum compounds. Mo is ubiquitous in food and water as soluble molybdates. Mo-containing enzymes are found in many plants and animal organisms. In human and animal tissues the enzymes xanthine dehydrogenase (XD)/oxidase (XO), aldehyde oxidase (AO) and sulphite oxidase (SO) require molybdopterin as cofactor and part of the enzyme molecule. In molybdopterin Mo is bound by two S atoms to the pterin. Mo is an essential component of flavin- and Fe-containing enzymes. This evaluation covers those forms of Mo, which are found naturally in food and water, as well as soluble molybdates added to foods. Good food sources of Mo are sorghum, leafy vegetables (levels depending on soil content, those grown on neutral or alkaline soil are rich in Mo, those grown on leached acid soil are Mo deficient, legumes (beans), grains (cereals, wheat germ), organ meats (liver, kidney), milk and eggs. Some 40% of Mo in cereals are lost on milling. Fruits, root vegetables, and muscle meat are poor sources. High concentration is in shellfish. Soft tissue of fish contains about 1 mg Mo/kg, vascular plants 0.03-5 mg Mo/kg. Mo levels in drinking water range from 0-68 µg/l, but usually does not exceed 10 µg/l. There are no reliable estimates of human requirements for Mo and no recommended intake has been established. Intakes estimate for breast fed infants (aged 0-3 m) varies, typically, from 0.1-0.5 µg/kgBW/day, children (from weaning to 3 years) from 5-7 µg/kgBW/day, and adolescents/adults from 1.5-2.5 µg/kgBW/day. In mining areas with contaminated drinking water (levels up to 400 µg/l) intakes from food plus 2 L water can reach about 1000 µg Mo/day. Mo deficiency in humans is unknown under normal dietary conditions. Mo-deficiency is suggestive of functional deficiency of XO activity, e.g. a doubling of xanthine excretion, a 20% decrease in uric acid excretion after purine load. Decrease in AO activity was noted because of nicotinamide metabolism abnormalities. A human syndrome suggestive of Mo deficiency occurs in prolonged total parenteral feeding in association with intolerance of cysteine and methionine, manifested by irritability, tachycardia, tachypnoea, nightblindness, encephalopathies and coma. Biochemical indicators are low tissues SO and XO, raised plasma methionine, reduced plasma uric acid, high excretion of thiosulphate, xanthine and hypoxanthine, low excretion of inorganic sulphate. Intervention requires reduction of protein intake especially S-containing amino acids. Clinical symptoms may totally be eliminated by administration of 300 µg ammonium molybdate/day (equivalent to 147 µg Mo/day). A similar symptom complex is seen in the short-bowel syndrome and after ileal resection for Crohn's disease with faecal loss of 350-530 µg Mo/day, requiring 500 µg Mo parenterally/day for correction. Human Mo deficiency can be seen in a rare autosomal recessive syndrome in infants, where there is a defective hepatic synthesis of Mo-pterin cofactor. This disease is associated with abnormal faeces, feeding difficulties, neurological and developmental abnormalities, mental retardation, encephalopathy and ectopy of the lens. The urinary levels of sulphite, thiosulphate and S-sulpho-L-cysteine are increased and urinary sulphate levels decreased. Death occurs by age 3. This condition is not ameliorated by dietary Mo supplementation because it is the result of a defective gene. Low intakes of Mo have been claimed to be associated with oesophageal cancer, where low serum, hair and urine Mo levels is found. Keshan disease (myocardial defects associated with Se deficiency) may be linked to low cereal and drinking water levels of Mo, as the incidence was reduced by using Mo fertilisers. However, high Mo levels in rice, wheat and soya combined with high tissue and hair levels were found in some Keshan endemic areas. Mo deficiency may be associated with reduced conception rate, increased abortion rate, and increased perinatal-mortality. Xanthine dehydrogenase (XD) irreversibly converts tissue purines, pyrimidines, pteridines and pyridines by oxidative hydroxylation to uric acid. Its normal action is that of a dehydrogenase, but when reacting with O_2, during proteolysis, freezing/thawing or in the presence of reactive -SH reagents it changes into Xanthine oxidase (XO), which produces oxygen free radicals known to be involved in tissue damage following physical injury, reperfusion, injury by toxins or Mo excess. Avian XD is stable; hence birds excrete uric acid. Allopurinol oxidises metabolically to alloxanthine, which inhibits XD. Reduced XD activity is associated with xanthinuria, low urinary uric acid, and high blood xanthine levels, high urinary and blood hypoxanthine levels, renal calculi and depositions in muscles with myopathy. Low Mo intake reduces tissue XD activity, however the intake variations from normal diet are insufficient to exert an effect on XD activity, which can cause overt clinical changes. Low XD activity can also be due to low protein intake or hepatoma, while high XD activity can be due to high protein intake, low vitamin E status, administration of interferon or administration of agents stimulating interferon release. It is not known whether high Mo intake stimulates tissue XD activity. Aldehyde oxidase is structurally and chemically similar to XO, has a similar tissue distribution and shares some substrates, e.g., aldehydes, substituted pyridines, pyrimidines, quinolines and purine derivatives. Its principal metabolic role is unknown. Sulphite oxidase (SO) is a haem-containing molybdoprotein located in the intermembraneous space of mitochondria. SO converts sulphite to sulphate. Sulphite derives metabolically from S-amino acids, e.g., cysteine, methionine. SO occurs in the liver of man and other species. Silicates inhibit the absorption of dietary molybdates. Absorbed Mo rapidly appears in the blood loosely attached to the erythrocytes, specifically bound to 2-macroglobulins. Mo crosses the placenta. Sulphate reduces the utilisation of Mo by some tissues and increases the urinary Mo excretion. Mo is reabsorbed by the renal tubules but this reabsorption is reduced by S-containing and by acid proteins. The reabsorbed Mo deposits in liver, lung, bone, and skin. Water-soluble Mo compounds and Mo in herbage and green vegetables are absorbed by man from 40-50%. The absorption rate from drinking water may be the same as from food. 25% of absorbed Mo appears rapidly in the blood loosely attached to the erythrocytes, specifically bound to 2-macroglobulins, normal blood levels being 2-6 µg/l whole blood or 0.55 µg/l serum. In man, the highest levels appear in kidney, liver and bone, raised levels appear also in adrenals, fat and omentum. There is no bioaccumulation, tissue levels rapidly returning to normal once exposure stops. Increased exposure at the work place or through drinking water is balanced by increased urinary excretion. Serum levels of Mo rise in liver functional defects, hepatitis, hepatic tumours, and after certain drugs. Raised blood levels are seen in uraemia, rheumatic disorders and CVS disease. Human liver contains 1.3-2.9 mg Mo/Kg dry matter, kidney 1.6 mg/Kg dry matter, lung 0.15 mg/Kg dry matter, brain and muscle 0.14 mg/Kg dry matter, hair 0.07-0.16 mg/Kg. Signs of human Mo toxicity are diarrhoea, anaemia, immaturity of erythrocytes, uricaemia. Thiomolybdate at levels of 5 mg Mo/kgBW causes in experimental animals diarrhoea, anaemia and skeletal lesions. Mo released from surgical metal implants can induce a delayed type of hypersensitivity with PUO and ANA-ve systemic lupus erythematosus (positive lymphocyte transformation test). Chronic small doses of molybdate inactivate the glutaminases of brain and liver causing a decrease in ammonia release. Small amounts of molybdate impair the intestinal utilisation of carotenes and reduce vitamin A status. The evidence for anticarcinogenicity in man is contradictory and inconclusive. Mo accumulates in teeth and dental enamel. In humans high Mo intakes occur with industrial exposure or through food. It is associated with raised XD activity, uricaemia, uricosuria and a higher incidence of gout. In areas with high geological Mo levels the human XO level is increased. Biochemical changes include hypoalbuminaemia, a rise in globulins, and raised serum bilirubin as sign of hepatotoxicity. It may be associated with oesophageal cancer. In an area in Armenia, where the population is exposed to a high dietary intake of Mo for geophysical reasons from soil levels of 77 mg Mo/kg and 39 mg Cu/kg, aching joints and gout-like symptoms have been reported. Haemodialysis reduces Mo serum level from 2.7 µg/dl to 1.4 µg/dl (normal 0.02-0.13 µg Mo/dl).

ENZYMES

Enzymes are substances that speed up chemical reactions. Enzyme is a protein molecule that catalyses chemical reactions of other substances without itself being destroyed or altered upon completion of the reactions. Enzymes are classified into 6 main groups: oxidoreductases, transferases, hydrolyases, lyases, isomerases, and ligases. There are at least 4 nutritional causes of mental problems including: 1- Protease deficiency (difficulty digesting protein). 2- Sugar intolerance, i.e., the inability to digest disaccharides (common sugars like sucrose) to simple sugars. 3- Hypothyroidism. 4- Junk-food diet laced with sugar. Protease deficiency leads to a build-up of excess alkaline reserves because there is inadequate digested protein to supply enough acidity. The person becomes anxious and sighs in an attempt to restore the acid-base balance. Everyone is sugar intolerant to some degree. Sucrose intolerance can lead to mental and emotional problems, such as panic attacks, depression, insomnia, mood swings that can progress to a bipolar disorder, and a tendency towards irritable, aggressive or violent behaviour. There are herbs, which have a stimulating effect on thyroid function. Sugar intolerance can lead to severe mental and emotional problems. Patients may have panic attacks, horrible nightmares, mood swings, and violent behaviour and be diagnosed as schizophrenic, bipolar or manic and so on. Enzymes catalyse nearly all the reactions that occur in living things. Enzymes accelerate reactions. Enzymes exhibit a high degree of specificity. With one exception, all enzymes are proteins. General principles of enzyme catalysis include: 1- enzymes concentrate reactants from solution, 2- enzymes reduce the order of complexity of reactions by making them multi-step processes, 3- enzymes provide an isolated environment for reaction, 4- enzyme active sites bind substrates so precisely that they align them for maximum reactivity, 5- enzymes provide reactive functional groups that participate in and facilitate the reaction, 6- enzymes act by lowering the energy of activation facilitating the formation of the transition, and 7- enzymes facilitate the formation of the transition state by binding specifically to it. Inhibition of enzymes includes: 1- competitive inhibition, and 2- non-competitive inhibition. Chymotrypsin is a serine endopeptiase that preferentially cleaves peptide bonds on the carboxyl side of amino acids with bulky hydrophobic residues, particularly tyrosine, tryptophan, phenylalanine and leucine. It is secreted by the pancreas as the inactive proenzyme chymotrypsinogen. Chymotrypsin is one of a large family of related proteases. Chymotrypsin cleaves amides and esters. Many reactions require enzymes to use co-factors or coenzymes to carry out catalysis. Most cofactors cannot be synthesised de novo in humans. They must be ingested as vitamins. Co-factors can be directly involved in catalysis or act as carriers of substrates. Co-factors can be either diffusible molecules/co-substrates such as NAD^+ or permanently bound or attached co-factors such as biotin. Enzyme activity and metabolism can be regulated by many different mechanisms. Many enzymes and pathways are regulated by reversible phosphorylation on threonine, serine and tyrosine residues. The phosphate can be removed by phosphoprotein phosphatases. The two isoforms of the enzyme that phosphorylates glucose regulate the utilisation and storage of glucose. In many cases, metabolic pathways are regulated directly through allosteric enzymes. Allosteric enzymes are oligomers. Allosteric enzymes are generally composed of multiple subunits and contain multiple catalytic and regulatory sites that are cooperative. Allosteric enzymes have distinct regulatory sites, which influence their activity. Binding of the allosteric ligands, ATP, cause conformational changes that are transmitted through the subunit-subunit contacts to all the rest of the molecule. In general, enzymes show pH optima, which reflect the ionisation of functional groups in the active site required for activity. The pH dependence of chymotrypsin reflects the active site residues. Catalysis involves a set of 3 amino acids arranged precisely aligned with each other and with the specificity pocket. Determination of enzyme activities is an important diagnostic tool in the clinical laboratory. Damaged cells release enzyme molecules into the circulation, e.g., creatine kinase (CK). Different organs contain different isoforms of CK. Many proteases (e.g., chymotrypsin) and esterases (e.g., acetylcholine esterase) contain serine in their active center. This serine forms a catalytic triad with 2 other residues, histidine and aspartic acid. Proteases belong to one of 4 broad families: 1- the serine proteases, such as chymotrypsin, in which catalysis involves formation of an acyl-enzyme with the acyl group on the serine; 2- the cysteine proteases, such as papain, with a some-what similar mechanism involving an acyl-cysteine; 3- the acid proteases, such as pepsin, with an acid pH optimum, and two aspartate residues involved in the mechanism; and 4- the metalloproteases, such as carboxypeptidase, which use a metal ion, usually zinc. There are other hydrolytic enzymes, such as lipases and acetylcholine esterase, with mechanisms closely related to that of the serine proteases. Examples of the mammalian serine proteases, such as trypsin, chymotrypsin, elastase, the enzymes of the blood clotting system, and many other proteases with specific roles in control of systems. These enzymes generally belong to 2 classes defined by their specificity, 1- the chymotryptic enzymes, which cleave at the carboxyl side of amino acids with large hydrophobic side chains-phe, tyr, trp, leu, ile; and 2- the tryptic enzymes, which cleave at the carboxyl side of basic amino acids, lys and arg. There are also elastases, which attack the structural protein elastin at alanine and valine residues. These enzymes typically are synthesised in inactive forms, which require activation by cleavage of a peptide bond near the NH_2-terminus. These enzymes act by forming and hydrolysing an ester on a serine residue. In most of the blood-clotting proteases such as Factors VII, IX, X, XI and XII, there is a large amino-terminal portion, which remains attached by disulfide bonds and is important in keeping the active protease bound on membrane surfaces to act on the next protease that is similarly bound. The final protease, thrombin, does not have its amino terminal domain attached by a disulfide bond and goes free in the plasma to attack fibrinogen and generate clots. The serine proteases act on both esters and amides. The serine displaces the alcohol or amine part of the substrate to form an acyl-enzyme, and water then displaces the serine to yield the acid product and free enzyme. An ester and an amide of the same acid yield the same acyl-enzyme and have the same rate constant for deacylation; but the acyl-enzyme accumulates only when the substrate is an ester or acyl-imidazole. Trypsin and chymotrypsin work well on esters and amides of single amino acids, as long as the α-amino group is blocked. The imidazole in effect is a much stronger base, facilitating proton removal from the serine. Deacylation is considered to be essentially the reverse of acylation. Protein inhibitors of serine proteases such as soybean trypsin inhibitor undergo the reaction, including formation of an acyl-enzyme, but have many interactions with the protease, so that the first product does not diffuse away and water has no room to attack the acyl-enzyme. Metabolic disease are not transmittable to another person, i.e., non-infectious. Cancer is a chronic metabolic disease. Most tumours are composed of a mixture of cancer and non-cancerous tissue. Cancer is a deficiency disease aggravated by the lack of essential food compounds. These compounds are part of nitriloside family, which occurs abundantly in over 200 edible plants and found virtually in every part of the world. It is particularly prevalent in the seeds of those fruits in Prunus Rosacea family (bitter almond, apricot, blackthorn, cherry, nectarine, peach, and plum), but also contained in grasses, maize, sorghum, millet, cassava, linseed, apple seeds, etc. Nitriloside is not food or drug, but may be considered as necessary food factors. Many people are alive but not really living and the QoL is devastated by impaired mental capacity, or life-support machine dependency, or requirement of round-the-clock care. The QoL is more important than the quantity. Cancer and trophoblast are the same. The trophoblast in pregnancy exhibits all the classical characteristic of cancer. It spreads and multiplies rapidly as it invades into the uterine wall preparing a place where the embryo can attach itself for maternal protection and nourishment. The trophoblast is formed as a result of chain reaction starting with another cell the diploid totipotent. 80% of totipotent cells are located in the ovaries or testes serving as a genetic reservoir for future offspring. Wherever the body is damaged, either physical trauma, chemical action, or illness, oestrogen and other steroid hormones appear in high concentration, serving as stimulators or catalysts for cellular growth and body repair. These steroids stimulate the totipotent cells to produce trophoblast. Anything that causes damage to the body can lead to cancer. The appearance of trophoblast proliferation, invasion, extension, and metastasis takes place. Promotion of trophoblast activity is the beginning of cancer. Oestrogen is the fodder on which carcinoma grows. Women using contraceptive pills-especially those containing oestrogen undergo irreversible breast changes and become 3 times more cancer-prone than women who do not. The body floods the areas of the trophoblast with β-glucouridase (BG), which inactivate all oestrogen. The trophoblast dies and the benign polyp or other benign tumour remains. Trophoblast cells produce chorionic gonadotrophic hormone (CGH). No other cell produces CGH. If CGH is detected in the urine, it indicates that there is pregnancy or cancer in female patient, and only cancer in male. Measuring CGH in the urine is 95% accurate for detecting cancer. Although cancer cells are foreign, they scape the lethal attentions of the immunological system. Cancer cells are vital part of the life cycle (pregnancy and healing), and provided with an effective means of avoiding WBCs. Trophoblasts are surrounded by a thin protein coating, which carries a negative electrostatic charge, i.e., pericellular sialomucin coat. WBC also carries a negative charge, and this polarity repeal cancer and WBCs, i.e., protect trophoblast. The cancer or trophoblast cell is non-antigenic because of the pericellular sialomucin coat. Trypsin and chymotrypsin are important in trophoblast destruction. These enzymes exist in their inactive form (zymogens) in the pancreatic glands. In the small intestine these zymogens are connected to their active form and then absorbed. When they reach the trophoblast, they digest the negatively-charged protein coat. The cancer then exposed to the attack of WBCs and it dies. Monocytes are more destructive of cancer cells than lymphocytes. Vitamin B_{17} releases almost all the pancreatic enzymes for absorption into the blood stream where they work on the cancer cells. By week 8 the infant pancreas begins to function, that is why the trophoblast cells are destroyed and stop to grow. There is almost complete absence of carcinoma in the duodenum. Diabetics are 3-times more likely to develop cancer than non-diabetic. Tumour may not be a disease, but is the symptom of the disease. Tumours are mixtures of malignant and benign cells, and most tumours have only a small percentage of cancer cells. Laetrile stops metastasis, improves general health, inhibits the growth of small tumours, provides relief from pain, and act as cancer prevention. Amygdalin inhibits the appearance of lung cancer metastasis. β-Glucouridase (BG) is activated by secretions of the mouth and stomach causes a minute amount of cyanide and benzaldehyde to be released in these locations. Scurvy is extreme form of vitamin C deficiency. A lesser form may not reveal the classic symptoms of scurvy but could manifest itself as fatigue, susceptibility to infection, and other non-fatal maladies. Scurvy is the final collapse of the organism, a pre-mortal syndrome. Cyanide is the only chemical substance that has been misunderstood.

Serine proteases of physiological importance:

Protease	Cleavage Site	Function
Trypsin	Arg-X, Lys-X	Digestion
Chymotrypsin	Tyr-X, Phe-X, Trp-X	Digestion
Elastase	Val-X	Tissue degradation
Thrombin	Arg-Gly	Blood coagulation
Factor Xa	Arg-Ile, Arg-Gly	Blood coagulation

Cyanide is synonymous with poison. Amygdalin was discovered in 1830, and was listed in the pharmacopoeia by 1834. Amygdalin means like almond. Amygdalin may exist in several different crystalline forms, depending on the number of water molecules. Amygdalin cryst is known as laetrile, is more soluble than any of the other forms. Aspirin is 20 time more toxic than equivalent amount of laetrile. The toxicity of aspirin is cumulative. Laetrile is less toxic than sugar. Kernels provide excellent food material, rich in protein and minerals. Laetrile is safe and effective in the treatment of cancer. Vitamin B_{15} (panganic acid) detoxifies the liver as a trans-methylating agent, and increases the O_2 uptake potential to the tissues. The surgical treatment can be life-saving particularly where intestinal blockages must be relieved to prevent death from secondary complications. Surgery has the psychological advantage of visibly removing the tumour and offering temporary comfort of hope. Cutting into the tumour (surgery)-even a biopsy-can aggravate the condition. It causes trauma to the area, which triggers the healing process that in turn brings more trophoblast cells into as a by-product of the process. If malignant tissue is left behind, it becomes encased in scar tissue as a result of surgery. Consequently, the cancer tends to become insulated from the action of pancreatic enzymes, which are essential for exposing trophoblast cells to the surveillance action of the WBCs. No relation between intensity of surgical treatment and duration of survival has been found in verified malignancies. Simple excision of cancers has produced essentially the same survival as radical excision. Once cancer has metastasised to a second location, surgery has almost no survival value. X-rays induce cancer by: 1- physical damage to the body, which triggers the production of trophoblast cells as part of the healing process; 2- weaken or destroy the production of WBCs, which constitute the immunological defence mechanism. There is little or no evidence that medication improves the patient's chances for survival. WBCs count is reduced by X-ray therapy, which leaves the patient susceptible to infections (pneumonia) and other diseases as well. Longer life brings human to the age in which cancer most strikes (5th decade on). The trophoblast cells are produced in the body as a result of a chain reaction involving oestrogen. Oestrogen always is present in large quantities at the site of damaged tissue, possibly serving as an organiser or catalyst for body repair. Many factors are involved to control the growth of these trophoblast cells, but most direct-acting of them appear to be the pancreatic enzymes and nitriloside (vitamin B_{17}), a unique compound that destroy the cancer cells while nourishing and sustaining all others.

DIARY PRODUCTS

pH is measure of an environment's acidity or alkalinity. The more acidic the solution, the lower the pH. For example, a pH of 1 is very acidic; a pH of 7 is neutral; a pH of 14 is very alkaline. Calcium (Ca) is the body's most abundant mineral. Its primary function is to help build and maintain bones and teeth. Calcium is also important to heart health, nerves, muscles and skin. Calcium helps control blood acid-alkaline balance, plays a role in cell division, muscle growth and iron utilisation, activates certain enzymes, and helps transport nutrients through cell membranes. Calcium also forms cellular cement called the matrix (ground substance) that helps hold cells and tissues together. It is well established that an acute oral dose of Ca rapidly suppresses bone resorption. Milk is a good source of magnesium as well as calcium. Magnesium (Mg) deficiency leads to an increase in serum PTH, which promotes Ca release from bone. There is increased bone mineral density, as well as relief of back pain and movement restrictions by Mg supplementation in osteoporotic patients. Bone mineral density is greater in people with high intakes of Mg and K. Ca supplementation is clinically, as well as biologically, important. Milk clots in the gastrointestinal tract, making its digestion slower than that of aqueous solutions. Longer transit time may enhance the total absorption of nutrients. Dietary Ca may inhibit Mg reabsorption in the nephron because of the decrease in PTH. This hormone enhances Mg uptake in the distal convoluted tubule of the nephron. Increase in serum Ca or Mg each inhibit both Ca and Mg reabsorption in the kidney, which is attributed to an extracellular Ca^{2+}/Mg^{2+}-sensing receptor located in the cortical thick ascending limb of the loop of Henle. But, the enrichment of high Ca skim milk with Mg has no additional impact on serum PTH or bone resorption than high Ca skim milk alone. Milk and other dairy products contribute about 4-10% of total energy worldwide. Milk has been suggested as a risk factor of atherosclerosis and coronary heart disease because it is a source of cholesterol and saturated fatty acids. Milk contains conjugated linoleic acid (CLA), which may have hypolipidaemic and antioxidative and thus antiatherosclerotic properties, calcium, which may protect from hypertension, and folic acid, vitamin B_6 and B_{12}, which contribute to lower homocysteine levels. Dairy consumption may be related to colorectal cancer risk. Milk fats and particularly saturated fats might increase cancer risk. Ionised Ca or calcium phosphate might reduce colon cancer by binding secondary bile acids and free fatty acids, primarily deoxycholic and lithocholic acids, thereby reducing their effective toxic dose to the colonic epithelial cells and preventing their stimulatory effects on proliferation of the intestinal mucosa. Calcium reduces the colonic content of diacylglycerol formed by bacteria, which may activate cellular transduction pathways and has been postulated to increase proliferation in the colonic epithelium. The major part (97-98%) of milk lipids is triglycerides or esters of fatty acids. The remainder comprises phospholipids (0.22-1%), sterols, free fatty acids and variable quantities of liposoluble vitamins (A, D, E and K). Around two-thirds of the fatty acids in milk are saturated. Polyunsaturated fatty acids (PUFA) make up less than 4% of milk fat. Dietary fat may promote colon cancer by increasing bile acid and fatty acid excretion in the colonic lumen. The microbial flora hydrolyses bile acids to form secondary bile acids. Free fatty acids (FFAs) and ionised secondary bile acids may damage colonic epithelial cells and thus induce crypt cells. Insulin is an important growth factor of colonic epithelial cells and a mitogen of tumour cell growth. Chronic hyperinsulinaemia consequent to insulin resistance increases colon cancer risk. The butyric acid may inhibit proliferation and induce differentiation in a wide range of tumour cells. Dietary butyric acid occurs exclusively in the lipid fraction of milk and its derivatives. Butyrate can likewise result from fermentation of dietary fibre by the microflora of the colon that has not been digested by intestinal enzymes. Dietary butyric acids are rapidly absorbed by the intestine and are largely metabolised in the liver. It is unlikely that butyric acid from milk is involved in colon carcinogenesis in any way similar to that of butyrate produced by the colonic microflora from fibre fermentation. Conjugated linoleic acid (CLA) inhibits the proliferation of colorectal, breast and skin tumour cells, and decreases the number of aberrant crypt foci. CLA could have a cancer-protective effect by modifying the fluidity of cell membranes, reducing the synthesis of prostaglandins and/or stimulating the immune response. Lactobacillus bulgaricus, one of the various microbial species (Lactobacillus bulgaricus, Streptococcus thermophilus, Lactobacillus acidophilus, Bifidobacterium bifidus, etc.) that can be used in the preparation of fermented products, suppresses toxins produced by putrefactive bacteria in human intestines. Some strains of Bifidobacteria and Lactobacilli bind the apical surface of colonic epithelial cells without injuring them, which could explain their protective action in gastrointestinal diseases. By protecting the surface epithelium, Bifidobacteria may thus facilitate repair and reduce irritation and epithelial permeability to electrolytes. Ingestion of L. acidophilus significantly reduces the excretion of mutagens following consumption of meat heavily browned or burnt by cooking at high-temperature. Milk containing L. bulgaricus or L. casei activate the lymphocytes and macrophages. Yoghurt increases production of INFG, a cytokine with anti-proliferative properties that is able to activate natural killer (NK) cells. Lactoferrin, a glycoprotein that participates in the transport and storage of iron, has a bacteriostatic effect by binding the iron that is necessary for microbial development. Lactoferrin activates NK cells and stimulates lymphokine-activated killer (LAK) cells, inhibit the development of solid tumours of the colon and of metastases from a melanoma, and may reduce significantly the incidence and number of adenocarcinomas of the large intestine. Lactoferrin is the milk component with cancer-chemopreventive properties (0.02% in cows' milk) and it is destroyed by heat treatments used for milk conservation. Folate is associated with a reduced risk of colorectal neoplasia. Folates represent an important B vitamin, participating in one-carbon transfer reactions required in many metabolic pathways, especially purine and pyrimidine biosynthesis (DNA and RNA) and amino acid interconversions. Folate is central to methyl-group metabolism, may influence both methylation of DNA and the available nucleotide pool for DNA replication and repair. Folate concentrations in cow's milk is 5-10 µg/100 g. Fermented milk contains slightly higher amounts of folate, sometimes double, depending on the starter culture used. Most cheese varieties contain 10-40 µg of folate/100 g. Ripened soft cheese may contain up to 100 µg/100 g. There is no evidence of the effect of milk folates on colorectal cancer risk. Insulin-like growth factors (IGFs), IGF-I and IGF-II, present in mammalian milk, play an important role during GIT development. IGF-1 may have direct effects on colorectal carcinogenesis, by stimulating cell proliferation, and by inhibiting apoptosis. But, IGF-I is less stable in the intestine of adult. 2% fat-milk was found to be significantly protective for rectal cancer. There is no significant associations between low-fat and high-fat dairy products and colorectal cancer risk with the exception of a significant risk increase in rectal cancer in women associated with an intake higher than 16 g/day of cream, milk desserts, and ice cream, ice milk and sherbert cheese. There may be a risk increase associated with intake of high-fat dairy products. There may be an inverse association between Ca and colon cancer. It is unlikely that calcium supplementation can substantially lower colorectal epithelial cell proliferation rates, but it may normalise the distribution of proliferating cells within colon crypts. A higher consumption of Ca may reduce the risk of colorectal cancer. In short-term, calcium and vitamin supplementation does not reduce cell kinetics of the colon, but in long-term (>1 y), Ca supplementation suppresses rectal epithelial cell proliferation. Calcium supplementation may be associated with a modest but not significant reduction in the risk of adenoma recurrence. There may not be association between vitamin D intake (nor Ca) and colorectal cancer risk. Calcium could play an indirect role in the development of prostate cancer, through a reduction in plasma levels of the active form of vitamin D (1, 25-(OH)2 vitamin D). The consumption of dairy products and in particular milk may be associated to a modest reduction in colorectal cancer risk. There is no evidence of either reduction or increase of colorectal cancer risk specifically associated with consumption of cheese or yoghurt.

SOME BONE HEALTH ISSUES

As we age, several conditions tend to become more troubling because of changes in the body's repair mechanisms. Among these is osteoporosis. The age-related slowing in repair mechanisms that fosters this condition, fortunately, often can be compensated for with appropriate supplementation. Senile osteoporosis is a paediatric disease. Osteoporosis is Latin for porous bones. It is a progressive condition in which the bones gradually lose their strength and density. As a living tissue, the bone is continuously remodelled as it renews itself, responds to damage, and so forth. It constantly is both releasing and absorbing new calcium. As is true of all other tissues, the bone renews itself with a turnover of its cells over time. Bone loss results when the balance of the constructive and destructive processes is tipped from equilibrium towards a loss of calcium and other bone components. Women are far more susceptible than are men and suffer about 80% of the injuries caused by this condition. By the age of 65, men on average have lost approximately 9% of their bone mass, whereas women have lost 26% of their bone mass on reaching this age. Most discussions of osteoporosis focus on the loss of calcium, which is a major error in analysis. Calcium may not be the most important in preventing demineralisation. The bone consists of both inorganic mineral components and organic components. Osteomalacia is the technical name for the softening of the bones, which results from a lack of calcium in the diet. In osteoporosis, not only calcium and other minerals, but also the non-mineral bone matrix, which consists of collagen and proteins, is disrupted. Many factors have been suggested as leading to elevated rates of bone loss. Caffeine, especially that derived from coffee and soft drinks, elevates the rate at which the body loses calcium, magnesium and other minerals; in green and black teas, flavonoids appear to offer countervailing beneficial effects. Soft drinks are sources of both carbonic and phosphoric acids. Smoking lowers a woman's oestrogen levels. Corticosteroids cause osteoporosis. Thyroid hormones induce osteoporosis. Alcohol, heavy drinking damages the liver (hence may impair vitamin D metabolism) and is a source of empty calories, which can displace mineral-rich foods; although, red wine contains resveratrol and flavonoids that enhance bone health. A high intake of animal proteins supplies calcium-leaching phosphorous and sulfur-containing amino acids, and increasing protein intake is usually linked to increasing urinary losses of calcium; nevertheless, the role of protein consumption in osteoporosis is disputed. Osteoporosis in Western societies is probably the cumulative result of a number of factors. Here are a few of the areas in which intervention appears to be helpful. Vegetables, such as lettuce, tomato, cucumber, arrugula, onion, garlic, wild garlic, common parsley, Italian parsley, and dill, have a significant positive effect upon bone mineral density. Along with alcohol, caffeine and carbonated soft drinks, sugar may be a prime offender when it comes to leeching calcium from the body. Exercise against resistance, which may include the force of gravity, should be undertaken at least three times per week. Swimming, in this regard, is less beneficial than walking, jogging, vigorous bike riding, etc. Mild resistance training is another good choice. The benefits of simple exercise are often found to be equal to that of hormone replacement therapy in preventing bone loss. Calcium is usually suggested to be supplemented at the rate of 800 to 1200 mg per day. Calcium citrate is often recommended. However, a strong case might be made for calcium hydroxyapatite, which is the form of calcium actually found in bone tissue. It produces a more prolonged calcium balance than do soluble calcium salts. It can cause the bone osteoblasts to become receptive to its components and to build bone tissue. Hydroxyapatite provides both the organic and inorganic constituents found in bone. Hydroxyapatite microcrystals are made of calcium, phosphorous, oxygen and hydrogen; the trace minerals zinc, strontium, silicon and iron; and proteins, amino acids and aminoglycans. It has been proposed that the bone-building process is enhanced by the presence of the proteins (the organic matrix) or that the microcrystalline structure provides a large surface area from which the minerals may be released from the organic matrix in the intestines. Vitamin D is required for the absorption of calcium. A good level of intake is approximately 400 IU daily from all sources, but can be toxic in larger quantities. Magnesium is required for the proper metabolism of vitamin D. Magnesium is as important as calcium in preventing osteoporosis. Magnesium (as citrate or aspartate), 350-600 mg/day, should be supplemented in conjunction with vitamin B_6 because this combination helps to keep calcium soluble and prevent its deposition in the soft tissues. Vitamin B_6, 25-50 mg/day, in conjunction with folic acid (400-800 µg/day) and vitamin B_{12} (100-300 µg) serves to control the levels of the compound homocysteine, which is usually elevated in individuals with osteoporosis. Homocysteine prevents collagen from properly cross-linking to form a stable bone matrix. Vitamin K (150-500 µg/day) is required for the proper structuring of osteocalcin, the major non-collagen protein found in bone. All connective and hard tissue conditions depend upon the production of collagen, proteoglycans (PG) and glycosaminoglycans (GAGs). Perhaps 50% of all the collagen protein of the body is found in the bones. Another way of viewing bone tissue is to see it as mineralised connective tissue. Glucosamine and MSM (methylsulfonylmethane, contains sulfur) provide the two most crucial nutrients needed for the synthesis of the bone matrix consisting of collagen, PG and GAGs. Ipriflavone, which is a semisynthetic isoflavone, has been shown to be metabolised to some degree (about 10%) in the body to daidzein, one of the isoflavones found in soybeans. Daidzein's mild oestrogenic effect in bone tissues, which may decrease the decalcification of the bone, is thought to be one of the sources of soy's protective effects against osteoporosis. However, ipriflavone itself exerts no oestrogen-like effects. The rapid metabolization of ipriflavone leading to its quick clearance from the body is often cited as a factor in its safety. Fosamax (alendronate) poses risks of liver damage. 200-600 mg of ipriflavone three times per day along with 1000 mg calcium leads to a 5.8% average increase in bone density over a twelve month period.

Vitamins and minerals:

1- Vitamin C (for collagen production) 500-2000 mg/day
2- Boron (important if magnesium is lacking) 1-3 mg/day
3- Copper (to balance zinc) 1 mg/day
4- Manganese 15-25 mg/day
5- Silica 100-1000 mg/day
6- Zinc (monomethionine) 15 mg/day
7- Glucosamine (helps form bone matrix) 300-1500 mg/day
8- MSM 100-1000 mg/day

PANTETHINE

Pantethine is a derivative of vitamin B-5; it is the stable disulfate form of pantethine, which provides the active part of coenzyme A molecules (CoA). The actions of vitamin B-5 (pantothenic acid) require a sulfhydryl (SH) group to be donated from cysteine and attached to the molecule. Thus, pantethine (has SH group) is more metabolically active than is pantothenic acid. Coenzyme A is a cofactor for >70 distinct biochemical pathways in the body, including fatty acid oxidation, carbohydrate metabolism, the synthesis of acetylcholine, and Phase II detoxification in the liver. Pantethine promotes proper lipid metabolism. Pantethine may direct the primary precursor of cholesterol into the β-oxidation (where fats are used to produce energy) or into another energy-producing cycle of the cell. Pantethine improves the metabolism of undesirable cholesterol, triglycerides, low density lipoprotein (LDL) cholesterol and apolipoprotein B (Apo-B) while leading to an increase in the desirable high density lipoprotein (HDL) cholesterol and apolipoprotein A (Apo-A). Pantethine favourably influences platelet lipid composition and cell membrane fluidity, resulting in reduction of excessive aggregation. Pantethine is important to acetylation reactions inside the cells. Within cells, the attachment or removal of acetyl groups allows the production of acetyl-CoA from coenzyme A and many other molecules. The acetyl groups are what make vinegar sour. Acetylation reactions utilising acetyl-CoA are an important component of the liver's Phase II detoxification system. As a stable precursor to CoA, pantethine may offer support to this detoxification pathway. Pantethine protects the clarity of the lens of the eye against the assault of several compounds, which are known to produce cataract. Pantothenic acid is known as an anti-stress vitamin, which is necessary for proper adrenal functioning.

TRIMETHYLGLYCINE (TMG)

Methylation is directly related to many diseases, including cancer, heart disease, liver disease, and neurological disorders. Furthermore, methylation also appears to play a significant role in the aging process in general. Methylation can be enhanced or inhibited through diet, lifestyle factors (such as smoking, drinking and taking birth control pills that decrease methylation), and direct supplementation, primarily with folic acid, B_{12} and trimethylglycine (TMG) is recommended. Methylation is the process by which methyl groups attach to different substances in the body, working to either protect or transform them. Methyl groups convert homocysteine, a toxic amino acid, which can cause heart disease and vascular disease, to methionine, a beneficial amino acid. Methionine produces SAM-e (S-adenosylmethionine) a natural anti-depressant and methyl donor. Elevation of SAM-e is beneficial in both the prevention and treatment of a variety of liver disorders, including those caused by alcohol stress. Methylation is a naturally occurring process; however, the presence of methyl groups is inversely correlated with the aging process. Trimethylglycine (TMG) is most effective methylation enhancing compound, and is commonly known as betaine, glycine betaine, or oxyneurine. TMG is not the same as betaine HCl, which acts as a stomach acidifier and is not practical due to stomach irritation at the doses required to enhance methylation metabolism. There are many benefits of methylation include: 1- Helping prevent heart disease, cancer, liver disease, depression and perhaps slowing down the perennial human condition of aging are some of the many possibilities. 2- Lowering homocysteine, protecting DNA, and producing SAM-e are the three ways that methylation works to improve health. When homocysteine levels are tested they may indicate certain problems with folate metabolism. There are many risk factors for heart disease including modifiable factors such as hyperlipidaemia (LDL 130 mg/dl or greater), hypertension (140/90 mmHg or greater), smoking, diabetes, obesity (especially abdominal), and physical inactivity. Unmodifiable factors include advanced age, male gender, and genetic predisposition (family history of heart disease). There are many nutrients that, in general, promote healthy cardiovascular function. Also, there are some specific nutrients that affect specific forms of cardiovascular disease (considering that the pathophysiology of one form of heart disease can be completely different from another). For example, the therapeutic antihypertensive effect of magnesium is very effective for arteriosclerosis, but may not do much for someone suffering from atherosclerosis caused by elevated levels of homocysteine. It is probably easier to utilise a range of nutrients including: 1- Silicon is an essential component of connective tissue and found in significant quantities in the aorta. Silicon resists atherosclerosis by reducing the permeability of the arterial wall. This directly results in the reduction of atheromatous plaques. Dietary sources of silicon are polymerised and must first be converted into OrthoSilicic Acid prior to absorption. 2- Trimethylglycine, folic Acid, B_6 and B_{12} combinations to reduce elevated homocysteine. An increase in the daily intake of folic acid and B_6 results in a 50% reduction in coronary hearth disease. Long-term use of folic acid (for at least 15 years) has a 75% reduction risk of colon cancer. 3- Magnesium, potassium, taurine combinations for reduction on hypertension. 4- CoQ_{10}, vitamin E are both excellent for antioxidant effect. CoQ_{10} provides superior LDL oxidation resistance. Those are taking cholesterol-inhibiting HMG-CoA reductase inhibitors should particularly use CoQ10 since it causes reduced levels of Co-Q10 in plasma. 5- Flavonoids are potent inhibitors of LDL. Flavonoids have an excellent effect on various disorders including hypertension, hypercholesterolaemia, and vascular fragility. Some flavonoids are poorly absorbed and subject to intestinal degradation, but they are nonetheless conditionally essential in heart disease. Both grape tannins and green tea catechins increase HDL and reduce LDL. Whole soy products, including soy proteins enriched with isoflavones are also important for inhibiting the formation of atherosclerotic plaque. Healthy individuals absorb only 9-21% of soy isoflavones. The amount of soy isoflavones in the Asian diet is estimated to be 20-80 mg daily. Once the hyperplasia and hypertrophying takes place in the arterioles, then it may be irreversible. Consumption of juice rich in folic acid (150 μg) per day results in a 11% drop in 30 days.

CHLORELLA

Chlorella is one of the earliest instances of a functional food, i.e., a food with special nutritive properties. The story of chlorella growth factor (CGF) goes back to the 1950's, when by using electrophoresis a substance was separated from a hot water extract of chlorella. This special growth factor was found to be especially rich in DNA. In fact, chlorella contains 17 times the amount of DNA found in sardines, probably the next richest natural source of these nucleic acids. Chlorella is approximately 10% RNA and 3% DNA. RNA and DNA may be the long life factors. The internal machinery of the cells revolves around RNA and DNA. As the body ages, the production of these compounds slows down. Many habits and environmental conditions also reduce the rate of production of RNA and DNA. These include smoking, alcohol, pollution, poor eating habits, and so forth and so on. Cell growth factor provides the basic building material for the immune system, tissue repair, the production of enzymes, proper bowel function, etc. The chlorella algae, unlike most other green foods, have a tough cell wall that is not easily digested. Without special processing, digestibility is only on the order of 50%. The cell wall of chlorella may need to be ruptured to afford the greatest nutritional benefit, yet components of the chlorella cell wall provide important health benefits. These cell walls bind to toxins and supply high grade fiber. Compounds found in the cell wall also activate sensing components of the immune system. Chlorella is one of the richest sources the green plant pigment chlorophyll. In its various forms, both water-soluble and fat-soluble, chlorophyll has many uses such as control purification and unwanted odours in the body, and an aid to healing. In the fat-soluble form found in fresh plants, chlorophyll positively modulates the production of haemoglobin, red blood cell production, and influences menstrual blood flow. The similarity of the chlorophyll molecule to the haeme portion of the haemoglobin molecule makes it useful to the body in the area of blood maintenance. Chlorophyll is antioxidant. As a deep green food, chlorella has an alkalinising effect upon the body. Some of this effect is indirect in that chlorella promotes the growth of lactobacilli in the intestines and thus helps to improve overall bowel health. With its content of easily assimilated vitamins, minerals and various plant factors, chlorella is a concentrated green food.

CHRYSIN

Chrysin (5,7-di-OH-flavone) belongs to bioflavonoids. These are ubiquitous compounds in the vegetable kingdom, present in the cells involved in photosynthesis. Over 4000 of these compounds have been identified from both higher and lower plants and the list constantly expands. Clinical interest in the bioflavonoids was triggered in the early 1930's. They were collectively termed vitamin P, in view of the fact that they possessed vitamin C sparing activity. Flavonoids are essential for the treatment of thyroid and other hormonal disorders. They also possess antioxidant and antimicrobial properties and are of particular interest because several of them have been shown to be antimutagenic and anticarcinogenic. They exhibit a wide spectrum of pharmacological activities. Including stimulation of oxidative mitochondrial phosphorylation and inhibition of several enzymatic activities. They are powerful inhibitors of both 5-lipoxygenase and cycloxygenase, interfering with the metabolism of arachidonic acid (AA). Flavonoids competitively inhibit the activity of hyaluronidases enzymes that depolymerize hyaluronic acid, which can be involved in allergic effects, migration of cancer cells, and permeability of the vascular system[1]. Flavonoids reduce the relative risk of mortality from coronary heart disease (CHD) in elderly men. Chrysin was originally isolated from the heartwood of the plant genus Pinaceae and from the bark of Dolichandrone falcata. It has also been isolated from Passiflora plants such as P. coerulea (has sedative effect) and P. incarnata. P. incarnata (maracuja "passion flower") has diverse biological effects. The methyl derivative, of chrysin, methylchrysin present as the glucoside in the buds of Populus spp. Salicaceae is used as a diuretic. Chrysin has also been obtained synthetically. It's known biological effects include antioxidant action, anti-inflammatory action, antiviral (including anti-HIV) action, vasodilatory effects and anxiolytic action. The compound has also been proven to have a potential role in drug metabolism and the chemoprevention of carcinogenesis. Chrysin has been shown to catalyse penicillin biogenesis in Penicillium chrysogenum, with the addition of chrysin to the germinating culture. In terms of chemical nomenclature, Chrysin is 5,7-Dihydroxy-2-phenyl-4H-1-1benzopyran-4-one (5,7-dihydroxyflavone) with molecular formula of $C_{15}H_{10}O_4$. Chrysin is insoluble in water, soluble in alkali hydroxide solutions and slightly soluble in alcohol, chloroform and ether. Biological activity of chrysin includes: 1- Anti-anxiety effects, with no associated sedation or muscle relaxation. 2- Anticonvulsant action, chrysin is a ligand for both central and peripheral benzodiazepine (BDZ) receptors. 3- Antioxidant action and anti-inflammatory effects, the antioxidant effects of chrysin linked to its antiinflammatory action by the inhibition of nitric oxide synthase (iNOS) expression in macrophages. In cytokine stimulated cells, activation of the NFκB by oxidative stress is crucial for the increase of iNOS gene expression. This activation therefore involves a redox-sensitive step. Chrysin (50 μmol) prevents the activation of NFκB. Chrysin may inhibit the secretion of lysosomal enzymes and AA release from membranes in neutrophils. Chrysin inhibits degranulation of mast cells, a prelude to inflammation. Chrysin inhibits HIV-1 activation. Flavonoids with hydroxyl groups inhibit P-450 enzyme activity, while those without hydroxyl groups stimulate activity. P-450 isoenzymes are involved in carcinogen activation. Chrysin, like some other flavonoids with the 7-hydroxyl group, is a potential chemopreventive agent. Chrysin is an effective antioxidant phytonutrient with unique neuropharmacological effects and potential utility in the management of conditions related to oxidative stress.

GINKGO BILOBA

Also called the "maidenhair tree", the Ginkgo tree possesses fan-shaped leaves and bears a foul-smelling inedible fruit; even the seeds are edible only after roasting. Extracts are made from the leaves, although Oriental medicine also employs the seeds. The Ginkgo tree today is a common ornamental planting world wide due to its ability to withstand pollutants and harsh conditions. The extract often is recommended for improving memory and reaction time, for improving circulation, and for protecting against free radical damage. Ginkgo biloba also is suggested in traditional practices for improving the physiologic effects of other herbs and nutrients. Chemical analysis reveals several important biologically active compounds are present in the Gingko biloba leaves such as: 1- Gingko flavone glycosides or gingko heterosides (which are flavanoid molecules unique to ginkgo to which sugars are attached). 2- Several terpene lactone molecules unique to gingko (ginkgolides and bilobalide), the most important of which may be Ginkgolide A and Ginkgolide B. 3- Organic acids. Gingko's benefits for the brain are perhaps the best known of its properties. The extract improves many aspects of brain function. Several different mechanisms likely are involved in these benefits. Much of the decline in cognitive abilities of ageing can be traced to a decrease in blood flow to this most vital of organs, e.g., cerebral vascular insufficiency. Ginkgo extracts markedly improve the circulation to the brain as a whole and especially within certain areas of the brain. These extracts enhance both the uptake of glucose and the utilisation of oxygen by the brain and other nerve tissues, and thus increase the amount of energy available for mental functioning. The rate of signal transmission by brain nerve cells is improved with Ginkgo extracts and the synthesis of some of the neurochemicals used to transmit nerve signals, such as acetylcholine, is increased. Ginkgo powerfully protects the brain against the ravages of free radicals. It is quite active against the damaging substances peroxides. Ginkgo also may improve the production of glutathione. These extracts are powerful antioxidants and free radical scavengers, and therefore they protect against damage to the fats that are found in the blood, the brain and all cell membranes. Ginkgo extracts improve the blood flow through the artery, which serves the heart, improve circulation more generally throughout the body and increase the clearance of toxins from the system. Ginkgo protects the body against platelet-activating factor (PAF). PAF was only discovered in 1972, and since that time scientists have come to realise that it is involved in disturbances ranging from internal blood clots (often leading to heart attacks and strokes) to allergic reactions to asthma to declining brain function. PAF has wide-ranging effects within the body. By inhibiting PAF, Ginkgo influences platelet function, inhibits excessive aggregation and adhesion, and helps to control degranulation (release of allergic and inflammatory components). Ginkgo increases the tone of the venous system; which helps to clear of toxic metabolites that accumulate in times of insufficient oxygen supply. Ginkgo acts on the vascular endothelium to enhance the release of relaxing factors. The extract helps to stabilise cell membranes and to promote the scavenging of free radicals in the brain and nerve cells. Brain tissues have a high percentage of unsaturated fatty acids, something that make them very susceptible to free radical damage. Ginkgo biloba extracts should contain both flavonoid glycosides (usually 24%) and terpenes consisting of ginkgolides and bilobalides (6%). Any product, which does not contain both the flavonoids and, especially, the terpenes, is unlikely to be effective.

VITAMIN B_3

Vitamin B_3 is made in the body from the amino acid tryptophan. On the average 1 mg of vitamin B_3 is made from 60 mg of tryptophan, about 1.5%. Since it is made in the body it does not meet the definition of a vitamin; these are defined as substances that can not be made. It should have been classified with the amino acids, but long usage of the term vitamin has given it permanent status as a vitamin. The 1.5% conversion rate is a compromise based upon the conversion of tryptophan to N-methyl nicotinamide and its metabolites in human subjects. Women pregnant in their last trimester convert tryptophan to niacin metabolites three times as efficiently as in non-pregnant females. Oestrogens stimulate tryptophan oxygenase, the enzyme that converts the tryptophan into niacin. The increased longevity in women in comparison to men is the result of greater conversion of tryptophan into niacin under the stimulus of their increase in oestrogen production. Vitamin B_3 exists as the amide, in nicotinamide adenine dinucleotide (NAD). Pure nicotinamide and niacin are synthetics. Niacin was known as a chemical for 100 years before it was recognised to be vitamin B_3. It is made from nicotine, a poison produced in the tobacco plant to protect itself against its predators. When the nicotine is simplified by cracking open one of the rings, it becomes the immensely valuable vitamin B_3. Plants use one type of antioxidants and animals use another type. There is a wide overlap and the same antioxidants such as vitamin C are used by both plants and animals. The catecholamines, of which adrenaline is the best known example, and the aminochromes, of which adrenochrome is the best known example, are intimately involved in stress reactions. Vitamin B_3 is a specific antidote to adrenaline, and the antioxidants such as vitamin C, vitamin E, β-carotene, selenium and others protect the body against the effect of the free radicals by removing them more rapidly from the body. Vitamin B_3 is very helpful, especially in cardiovascular-induced forms of dementia as it reverses sludging of the RBCs and permits proper oxygenation of the cells of the body. About 10% of arthritics have allergic reactions to the solanine family of plants (sugar, potatoes, tomatoes and peppers). Vitamin B_3 has anti-cancer properties. Niacin, niacinamide and nicotinamide adenine dinucleotide (NAD) are interconvertable via a pyridine nucleotide cycle. NAD, the coenzyme, is hydrolysed or split into niacinamide and adenosine dinucleotide phosphate (ADP-ribose). Niacinamide is converted into niacin, which in turn is once more built into NAD. The enzyme that splits ADP is known as poly(ADP-ribose) polymerase, or poly(ADP) synthetase, or poly(ADP-ribose) transferase. Poly(ADP-ribose) polymerase is activated when strands of deoxyribonucleic acid (DNA) are broken. The enzyme transfers NAD to the ADP-ribose polymer, binding it onto a number of proteins. The poly(ADP-ribose) activated by DNA breaks helps repair the breaks by unwinding the nucleosomal structure of damaged chromatids. It also may increase the activity of DNA ligase. This enzyme cuts damaged ends off strands of DNA and increases the cell's capacity to repair itself. Damage caused by any carcinogenic factor, radiation, and chemicals, is thus to a degree neutralised or counteracted. Diet is a major risk factor in cancer, diet has both beneficial and detrimental components. About 20 mg/day of niacin will prevent pellagra in people who are not chronic pellagrins. The latter may require 25 times as much niacin to remain free of pellagra. Vitamin B_3 may increase the therapeutic efficacy of anti-cancer treatment. Niacinamide increases the toxicity of irradiation against tumours, it enhances blood flow to the tumour. Nicotinamide also enhances the effect of chemotherapy. Niacin may offer some cardioprotection during long-term adriamycin chemotherapy. In cancer there is increased destruction of nicotinamide, thus making N-methyl nicotinamide less available for the pyridine nucleotide cycle.

MANGANESE

Manganese (Mn) activates several enzyme systems and supports the utilisation of vitamin C, E, choline, and other B-vitamins. Inadequate choline utilisation reduces the acetylcholine synthesis, causing conditions such as myasthenia gravis (loss of muscle strength). Manganese and zinc therapy can reduce copper levels and therefore manganese and/or zinc may be of therapeutic value in the treatment of symptoms linked to excess copper.

Manganese:

1- Important for normal skeletal growth and development.
2- Essential for glucose utilisation.
3- Lipid synthesis and lipid metabolism.
4- Cholesterol metabolism.
5- Pancreatic function and development.
6- Prevention of sterility.
7- Important for protein and nucleic acid metabolism.
8- Activates enzyme functions.
9- Involved in thyroid hormone synthesis.

Arginine sources:

Sunflower	8.2%
Carob	5.5%
Butternut	5%
Watermelon seed	4.4%
Peanut	3.7%
Chaya and sesame	3.5%
Soy	4.7%
White lupine	3.1%
Watercress	5%
Fenugreek, mustard, Indian fig	2.7%
Almond and velvet bean	2.6%
Bean sprouts, brazil nut, and chives	2.5%
Garlic	06-2.2%

Manganese metabolism is similar to that of iron. It is absorbed in the small intestines and while the absorption process is slow, the total absorption rate is exceptionally high about 40%. Excess Mn is excreted in bile and pancreatic secretion. Only a small amount is excreted in the urine. Liver and kidneys are the primary meat source of Mn. Wheat germ; legumes, nuts, and black tea are good plant sources. Manganese deficiency symptoms include ataxia, fainting, hearing loss, weak tendons and ligaments. Manganese deficiency can be a cause of diabetes. Manganese deficiency impairs glucose metabolism and reduces insulin production. Deficiency has been linked to myasthenia gravis. Manganese deficiency has been associated with cancer, rheumatic conditions, rickets, morning sickness, jaundice, and diabetes. Symptoms and side-effects of Mn deficiency include infertility, impaired glucose metabolism, diseases of the skeletal structure, and impaired growth, pancreatic dysfunction, hypertension, atherosclerosis, reduced protein metabolism, reduced immune function, ataxia, selenium deficiency, depressed activity of mammary glands in nursing mothers, and mitochondrial abnormalities. Excessive ingestion of iron, combined with hypochlorhydria, can cause an imbalance in the Mn/Fe ratio. Manganese overload is generally due to industrial pollution. Workers in the Mn processing industry are most at risk. Well water rich in Mn can be the cause of excessive Mn intake and can increase bacterial growth in water. Manganese poisoning has been found among workers in the battery manufacturing industry. Symptoms of toxicity mimic those of Parkinson's disease (tremors, stiff muscles) and excessive Mn intake can cause hypertension in patients older than 40. Excess Mn interferes with the absorption of dietary iron. Long-term exposure to excess levels of Mn may result in iron-deficiency anaemia. Increased Mn intake impairs the activity of copper metalloenzymes. Significant rises in Mn concentrations have been found in patients with severe hepatitis and post-hepatic cirrhosis, in dialysis patients and in patients suffering heart attacks. In the presence of high levels, dopamine levels are reduced. Manganese influences the copper and iron metabolism. Oestrogen therapy may raise serum Mn concentration. Glucosteroids alter the Mn distribution in the body. Manganese toxicity can cause renal failure, hallucinations, as well as diseases of the CNS. Symptoms of increased Mn levels include psychiatric illnesses, mental confusion, impaired memory, loss of appetite, mask-like facial expression and monotonous voice, spastic gait, neurological problems, impaired thiamine (B_1) metabolism, iron deficiency, and increased demand for vitamin C and copper. Vitamin C improves cellular exchange of Mn. Manganese poisoning can be treated successfully with chelation therapy. Elevated calcium and/or phosphorus intake suppresses absorption of Mn.

HERBAL VIAGRA

4 g of L-arginine/day is sometimes recommended to address erectile dysfunction (ED). This 4 g of arginine could be obtained from 50 g of sunflower seed. 60 mg of zinc/day is recommended for infertility dysfunction (ID), this could be found in a kilo of sunflower seed. Garlic is one of the best sources of free arginine, which might be much more potent than sunflower seed. Without NO, erections are impossible. 2.2% free arginine of garlic is more effective than sunflower 8.2% bound arginine. Arginine is essential for normal sperm production. Too much arginine may aggravate herpes, unless it is balanced with lysine. The three Gs of China are ginseng, ginkgo, and goat weed. Ginkgo stimulates both cerebral and peripheral circulation, contains some amentoflavone (phosphodiesterase inhibitor), and arginine. Ginkgo biloba extract, 80 mg tds, may be the ideal treatment of ED with good result in 9 months (78%). Fababeans is the best dietary sources of L-dopa, may induce priapism.

PROTEINS

The primary structure of peptides and proteins refers to the linear number and order of the amino acids present. The convention for the designation of the order of amino acids is that the N-terminal end (i.e., the end bearing the residue with the free a-amino group) is to the left (and the number 1 amino acid) and the C-terminal end (i.e., the end with the residue containing a free a-carboxyl group) is to the right. In general proteins fold into 2 broad classes of structure termed, globular proteins or fibrous proteins. Globular proteins are compactly folded and coiled, whereas, fibrous proteins are more filamentous or elongated. The α-helix is a common secondary structure encountered in proteins of the globular class. Tertiary structure refers to the complete three-dimensional structure of the polypeptide units of a given protein. Proteins with multiple polypeptide chains are termed oligomeric proteins. The structure formed by monomer-monomer interaction in an oligomeric protein is known as quaternary structure. Haemoglobin, contains two α and two β subunits arranged with a quaternary structure in the form, $α_2β_2$. Proteins also are found to be covalently conjugated with carbohydrates. These modifications occur following the synthesis (translation) of proteins and are, therefore, termed post-translational modifications. These forms of modification impart specialised functions upon the resultant proteins. Proteins covalently associated with carbohydrates are termed glycoproteins. Glycoproteins are of two classes, N-linked and O-linked, referring to the site of covalent attachment of the sugar moieties. N-linked sugars are attached to the amide nitrogen of the R-group of asparagine; O-linked sugars are attached to the hydroxyl groups of either serine or threonine and occasionally to the hydroxyl group of the modified amino acid, hydroxylysine. There are extremely important glycoproteins found on the surface of erythrocytes. There are at least 100 blood group determinants, most of which are due to carbohydrate differences. The most common blood groups, A, B, and O, are specified by the activity of specific gene products whose activities are to incorporate distinct sugar groups onto RBC membrane glycosphingolipids as well as secreted glycoproteins. Structural complexes involving protein associated with lipid via noncovalent interactions are termed lipoproteins. Their major function in the body is to aid in the storage transport of lipid and cholesterol. Several forms of familial hypercholesterolaemia are the result of genetic defects in the gene encoding the receptor for low-density lipoprotein

(LDL). These defects result in the synthesis of abnormal LDL receptors that are incapable of binding to LDLs, or that bind LDLs but the receptor/LDL complexes are not properly internalised and degraded. The outcome is an elevation in serum cholesterol levels and increased propensity toward the development of atherosclerosis. A number of proteins can contribute to cellular transformation and carcinogenesis when their basic structure is disrupted by mutations in their genes. These genes are termed proto-oncogenes. For some of these proteins, all that is required to convert them to the oncogenic form is a single amino acid substitution. A lot of claims are being made for many of the ergogenic aids on the market. Supplement can be divided into those correct deficiencies, those that raise nutrients to optimal levels, and those that alter cell and tissue functions. It is important to figure out which prevent or reduce damage and improve the rate of recovery from workouts, and which improve exercise anabolic effects. The lines between these functions are not always distinct. Many athletes still fail to take advantage of antioxidant supplements to protect against the downside of workouts. α-Lipoic acid may be the most useful of all the antioxidants as an ergogenic and recovery aid. Proteins are made up of amino acids. The essential amino acids are isoleucine, leucine, lysine, methionine, phenylalanine, threonine, tryptophan and valine. The semi-essential ones are arginine and histidine. All of the others can be produced from these ten amino acids. The branched-chain amino acids (BCAAs) are so named because of their structural peculiarities of having branched side chains off the main part of the molecule. The protein balance in the body-also called the nitrogen balance since proteins, but neither carbohydrates nor fats, contain nitrogen-is largely controlled by the liver and other regulatory mechanisms. Excess protein is simply broken down into carbohydrates and urea wastes and can place a burden upon the liver and the kidneys. The BCAAs valine, isoleucine and leucine make up roughly one third of muscle tissue. These amino acids, especially valine, are used up in very large amounts during intensive exercise. This may be a result of their use for fuel in the muscles inasmuch as their nitrogen component can be snipped off and the rest of the molecule used to produce pyruvate, an energy source. The BCAAs also can be used to make glutamine and alanine, two more important amino acids. The BCAAs, especially leucine, seem to be able to spare muscle from being broken down during exercise to supply energy. Leucine, further, may increase growth hormone (GH) release. HMB (β-hydroxy β-methylbutyrate) is a metabolite of leucine. The BCAAs must be taken in very large amounts (2-10 g) to have any significant effect. Whey protein is rich in all three branched-chain amino acids, so consuming a quality whey protein may be the preferred way to consume the BCAAs. Both GH and IGF-1 increase the synthesis of protein in the muscles, and IGF-1 can decrease the rate at which protein is broken down in the tissues, as well. IGF-1 may be partially responsible for increasing blood flow into the muscle, and this would also be important for tissue repair. L-arginine, L-ornithine, L-tryptophan, and L-glycine are GH releasers. Injecting these amino acids produce dramatic results, but oral supplements often have less consistent results. Excessive dosages of amino acids taken by mouth can cause stomach upset and diarrhoea. Typical dosages for L-arginine and L-ornithine are of 2-10 g, and 250 mg-6.75 g for glycine. α-Ketoglutarate forms of arginine and ornithine maintain muscle mass and protein synthesis during severe trauma, i.e., the effects of GH release. L-ornithine α-ketoglutarate (OKG) has superior GH releasing effects. It must be taken with plenty of water to avoid diarrhoea. Dosages begin at 1 to 2 g and range up to 10 g at one time. Individuals with herpes simplex virus should balance arginine intake with 1.25 g or more lysine taken at a different time of the day. GH release may cause a rise in the blood sugar. L-ornithine α-ketoglutarate may raise glutamine levels better than ingesting glutamine itself. Glutamine is a significant factor in the body's ability to withstand stress. Glutamine has potent anti-catabolic effects that can prevent muscle wasting. Glutamine is perhaps the most potent anti-stress and anti-catabolic member of the amino acids. The branched-chain amino acids valine, leucine and isoleucine appear to improve athletic performance by acting as building blocks for the production of glutamine by the body. Glutamine is important for immune function, digestive health and the transport of nitrogen as part of the physiologic system for removing ammonia from the brain and other tissues for disposal. The amino acid L-carnitine is absolutely essential to the body's ability to metabolise fats for energy. L-carnitine supplementation should be considered for elevated triglyceride and other blood lipids, 500 mg-4 g/day in divided doses. L-carnitine may interfere with sleep. The amino acid taurine may mimic some of the effect of insulin in the body, include improving the uptake of glucose and other amino acids by the cells. This is the anabolic aspect of taurine. Taurine is also an important anti-catabolic agent. 500 mg three times a day may help as much as 20% in reducing muscle breakdown. L-tyrosine can be used as an alternative to caffeine one hour before workouts. This amino acid is an important precursor to both the hormones produced by the thyroid and neurotransmitters produced in the brain and nerves. L-tyrosine improves mood and energy. Oxidative stress rises dramatically during intense exercise. This stress has been linked to immune suppression, reduced performance, and injury. Potentially, there is oxidative damage in both water-based tissues, such as the blood serum, and fat-based tissues, such as all cell membranes. Moreover, there is the potential for damage to the mitochondria inasmuch as it is the mitochondria that must process glucose and fats to produce most of the energy generated in the body. Resistance exercise such as weight training causes countless small tears in the muscle (known as microtrauma) and other forms of damage to tissues. Workouts exhaust cellular supplies of glucose (stored as glycogen) and various amino acids. α-Lipoic acid increases the cellular uptake and oxidation of glucose by 50% in adult diabetic patients. Efficient burning of blood sugar (glucose) for fuel is essential for the normal production of energy in muscles for both diabetics and normal individual. Replenishment of glycogen in the muscles is reduced by insulin resistance (IR), which makes it very difficult for an athlete to recover from workouts and for the diabetic to be healthy. Increased glycogen stores in muscles improve endurance and increase the fullness of the muscles, which is why athletes spend so much time reaching for carbohydrates before and after workouts. The ability to properly handle carbohydrates is important for energy production in the brain. Too much carbohydrate will release so much insulin that GH release will be blocked. Excessive aerobics performed in the form of running, dancing or any other form where impact forces are imposed on the body are potentially dangerous. When forces imposed on the body exceed the structural integrity of the muscle, connective tissue or bone, injury must occur. Anaerobic exercise such as weight training/strength building must also be performed safely in slow controlled movements, which are a requirement in order to minimise impact/deceleration forces. Excessive aerobics may burn lean muscle tissue. Excessive aerobics in combination with very low body fat levels may lead the overstrained female to a loss of the monthly menstrual cycle and possibly the onset of premature bone loss. Excessive aerobic or anaerobic exercise may be so overdone that the individual's daily "exercise fix" may lead to psychological hang-ups manifesting in anorexia, bulimia or other symptoms of such compulsive behaviour. Aerobics, for all of its benefits has limitations. In order to strengthen the muscular structures of the body, to enhance the structural integrity of the connective tissues, the joints and the bones, aerobics definitely falls short. While working in an aerobic pathway, the muscles, working against minimal or zero resistance, are contracting with little of their potential force output. Under such conditions a muscle can continue work for lengthy periods of time without stimulating any meaningful strength gain. Anaerobic exercise by contrast requires much higher muscular force production. In anaerobic exercise fatigue is induced in the muscle faster than the muscle can compensate. Soon, while working against a sufficient resistance, your reduced strength level will no longer be enough to allow you to continue against that level of resistance. Such exercise, if progressive, does have the potential to stimulate strength gain. Aerobics can not be the be-all or end-all of a sensible exercise program. The failure to understand this point has led to the physiological and psychological trauma of a very large number of sincere but misdirected participants. Neither vitamin C nor vitamin E is abundant in those parts of the body that need the most protection, e.g., collagen-based tissues. Inside the cell, where the first steps in the burning of sugar take place (glycolysis), there is very little vitamin C or vitamin E. The most important antioxidant inside the cell with a role in protecting against oxidative damage resulting from glycolysis is glutathione. Vitamin C and vitamin E also do little to protect the mitochondria. α-Lipoic acid is often called the universal antioxidant because it scavenges a variety of free radicals and regenerates both vitamin C and vitamin E, regenerates glutathione in the cytoplasm of the cell, and it even regenerates CoQ_{10} within the mitochondria. Endocrine hormones may be derived from amino acids, peptides, or sterols and act at sites distant from their tissue of origin. But, the latter definition has begun to blur as it is found that some secreted substances act at a distance (classical endocrines), close to the cells that secrete them (paracrines), or directly on the cell that secreted them (autocrines). For example insulin-like growth factor-I (IGF-I) behaves as an endocrine, paracrine, and autocrine. Hormones are normally present in the plasma and interstitial tissue at concentrations in the range of 10-7M to 10-10M. Sensitive protein receptors in target tissues sense the presence of very weak signals. Carrier proteins for peptide hormones prevent hormone destruction by plasma proteases. Carriers for steroid and thyroid hormones allow these very hydrophobic substances to be present in the plasma at concentrations several hundred-fold greater than their solubility in water would permit. Carriers for small, hydrophilic amino acid-derived hormones prevent their filtration through the renal glomerulus, greatly prolonging their circulating half-life. Activation of these receptors by hormones (the first messenger) leads to the intracellular production of a second messenger, such as cAMP, which is responsible for initiating the intracellular biological response. Steroid and thyroid hormones are hydrophobic and diffuse from their binding proteins in the plasma, across the plasma membrane to intracellularly localised receptors. The resultant complex of steroid and receptor bind to response elements of nuclear DNA, regulating the production of mRNA for specific proteins. With the exception of the thyroid hormone receptor, the receptors for amino acid-derived and peptide hormones are located in the plasma membrane. Receptor structure is varied: 1- Some receptors consist of a single polypeptide chain with a domain on either side of the membrane, connected by a membrane-spanning domain. 2- Some receptors consist of a single polypeptide chain that is passed back and forth in serpentine fashion across the membrane, giving multiple intracellular, transmembrane, and extracellular domains. 3- Some receptors are composed of multiple polypeptides. Subsequent to hormone binding, second messengers and phosphorylated proteins generate appropriate metabolic responses. The main second messengers are cAMP, Ca^{2+}, inositol triphosphate (IP_3), and diacylglycerol (DAG). Proteins are phosphorylated on serine and threonine by cAMP-dependent protein kinase (PKA) and DAG-activated protein kinase C (PKC). Additionally a series of membrane-associated and intracellular tyrosine kinases phosphorylate specific tyrosine residues on target enzymes and other regulatory proteins. For most hormones, plasma membrane receptors is transduced to the interior of cells by the binding of receptor-ligand complexes to a series of membrane-localised GDP/GTP binding proteins known as G-proteins. When G-proteins bind to receptors, GTP exchanges with GDP bound to the a subunit of the G-protein. The Ga-GTP complex binds adenylate cyclase, activating the enzyme. The activation of adenylate cyclase leads to cAMP production in the cytosol and to the activation of PKA, followed by regulatory phosphorylation of numerous enzymes. Stimulatory G-proteins are designated Gs, inhibitory G-proteins are designated Gi. Other class of peptide hormones induces the transduction of 2 second messengers, DAG and IP_3. Hormone binding is followed by interaction with a stimulatory G-protein, which is followed in turn by G-protein activation of membrane-localised phospholipase C-g, (PLC-g). PLC-g hydrolyses phosphatidylinositol bisphosphate to produce 2 messengers: IP_3, which is soluble in the cytosol, and DAG, which remains in the membrane phase. Cytosolic IP_3 binds to sites on the endoplasmic reticulum, opening Ca^{2+} channels and allowing stored Ca^{2+} to flood the cytosol. Ca^{2+} activates numerous enzymes, many by activating their calmodulin or calmodulin-like subunits. DAG has 2 roles: it binds and activates PKC, and it opens Ca^{2+} channels in the plasma membrane, reinforcing the effect of IP_3. Like PKA, PKC phosphorylates serine and threonine residues of many proteins, thus modulating their catalytic activity. The atrial natriuretic factor (ANF), a peptide secreted by cardiac atrial tissue, is much like other peptide hormones in that it is secreted into the circulatory system and has effects on distant tissue. The principal site of ANF action is the renal glomerulus, where it modulates the rate of filtration, increasing Na^+ excretion in the urine. The receptors for the natriuretic factors are integral plasma membrane proteins, whose intracellular domains catalyse the formation of cGMP following natriuretic factor binding. Intracellular cGMP activates a protein kinase G (PKG), which phosphorylates and modulates enzyme activity, leading to the biological effects of the natriuretic factors. Many amino acid and peptide hormones are elaborated by neural tissue, with ultimate impact on the entire system. Releasing hormones (factors) are synthesised in neural cell bodies of the hypothalamus and secreted at the axon terminals into the portal hypophyseal circulation, which directly bathes the anterior pituitary. These peptides initiate a cascade of biochemical reactions that culminate in hormone-regulated, whole-body biological end points. Cells of the anterior pituitary, with specific receptors for individual releasing hormones, generally respond through a Ca^{2+}, IP_3, PKC-linked pathway that stimulates exocytosis of pre-existing vesicles containing the various anterior pituitary hormones. The pituitary hormones are carried via the systemic circulation to target tissues throughout the body and generate unique biological activities. This is controlled by feed-back-forward mechanism.

Growth hormone abnormalities:

GH-deficient dwarfs	Lack the ability to synthesise or secrete GH, respond well to GH therapy.
Pygmies	Lack the IGF-1 response to GH but not its metabolic effects
Laron dwarfs	Have low levels of circulating IGF-1. The defect is due to inability to respond to GH by the production of IGF-1.
Gigantism	The production of excessive amounts of GH before epiphyseal closure of the long bones.
Acromegaly	Excessive GH after epiphyseal closure, acral bone growth leads to its characteristic features.

Thyroid abnormalities:

Cretinism	Hypothyroidism in the embryo, characterised by multiple congenital defects and mental retardation.
Graves' disease	Thyroid stimulating autoantibodies (TSAb) activate the human thyroid TSH receptor, leading to hyperthyroidism. TSAbs bind to the TSH receptor and mimic the TSH stimulation of the gland by increasing intracellular cAMP.

The stimulatory substance growth hormone releasing hormone (GRH), and the inhibitory substance somatostatin (SS), both products of the hypothalamus, control pituitary growth hormone (GH) secretion. Under the influence of GRH, GH is released into the systemic circulation, causing the target tissue to secrete IGF-1. The principal source of systemic IGF-1 is the liver, although most other tissues secrete and contribute to systemic IGF-1. Liver IGF-1 is considered to be the principal regulator of tissue growth. IGF-1 secreted by peripheral tissues is generally considered to be autocrine or paracrine in its biological action. The longer positive feedback loop, involving IGF-1 regulation at the hypothalamus, stimulates the secretion of somatostatin (SS, also called growth hormone-inhibiting hormone, GIH), which in turn inhibits the secretion of growth hormone by the pituitary. Human placental lactogen (hPL), GH, and prolactin (Prl) comprise the growth hormone family. All have about 200 amino acids, 2 disulfide bonds, and no glycosylation. Mature GH (22,000 daltons) is synthesised in acidophilic pituitary somatotropes as a single polypeptide chain. In humans, growth hormone promotes gluconeogenesis (hyperglycaemic), and promotes amino acid uptake by cells. The growth hormone is lipolytic, inducing the breakdown of tissue lipids and thus providing energy supplies that are used to support the stimulated protein synthesis induced by increased amino acid uptake. The glycoprotein hormones are the most chemically complex family of the peptide hormones. All members of the family are highly glycosylated. Each of the glycoprotein hormones is an ($\alpha:\beta$) heterodimer, with the a subunit being identical in all members of the family. The biological activity of the hormone is determined by the β-

subunit, which is not active in the absence of the α-subunit. The molecular weight of the gonadotropins FSH, LH, and CG is about 25,000, whereas that of the thyroid tropic hormone TSH is about 30,000. All members of the glycoprotein family transduce their intracellular effects via the receptor, G-protein, adenylate cyclase, second-messenger system. The gonadotropins (LH, FSH and CG) bind to cells in the ovaries and testes, stimulating the production of the steroid sex hormones oestrogen, testosterone (T), and dihydrotestosterone (DHT). Human chorionic gonadotropin (hCG) is a placental hormone. The production of hCG increases markedly after implantation; its appearance in the plasma and urine is one of the earliest signals of pregnancy and the basis of many pregnancy tests. cAMP causes increased secretion of thyroid stimulating hormone (TSH) by thyrotropes. Chronic stimulation of the TSH receptor at the basal membrane of thyroid follicles causes an increase in the synthesis of a major thyroid hormone precursor, thyroglobulin. Thyroglobulin produced on rough endoplasmic reticulum has a molecular weight of 660,000. It is glycosylated and contains more than 100 tyrosine residues, which become iodinated and are used to synthesise T3 and T4. Thyroglobulin is exocytosed through the apical membrane into the closed lumen of thyroid follicles, where it accumulates as the major protein of the thyroid and where maturation takes place. A Na^+/K^+-ATPase-driven pump concentrates iodide (I^-) in thyroid cells, and the iodide is transported to the follicle lumen. There it is oxidised to I^+ by a thyroperoxidase found only in thyroid tissue. The addition of I^+ to tyrosine residues of thyroglobulin is catalysed by the same enzyme, leading to the production of thyroglobulin containing monoiodotyrosyl (MIT) and diiodotyrosyl (DIT) residues. Mature, iodinated thyroglobulin is taken up in vesicles by thyrocytes and fuses with lysosomes. Lysosomal proteases degrade thyroglobulin releasing amino acids and T3 and T4, which are secreted into the circulation and carried by glycoprotein known as thyroxin-binding globulin. Thyroid hormones act by binding to cytosolic receptors very similar to steroid hormone receptors. In adults, the ligand receptor combination binds to thyroid hormone response elements in nuclear DNA and is responsible for up-regulating general protein synthesis and inducing a state of positive nitrogen balance. In the embryo, thyroid hormone is necessary for normal development. Corticotropin releasing hormone (CRH) induces rapid secretion of adrenocorticotropic hormone (ACTH, also called corticotropin) and a variety of other peptides from corticotropes of the anterior pituitary. ACTH, a 39 amino acid peptide, is the main physiologically active product of CRH activity. ACTH is derived by post-translational modification from a 241 amino acid precursor known as pro-opiomelanocortin (POMC). The processing of POMC involves glycosylations, acetylations, and extensive proteolytic cleavage at sites shown to contain regions of basic protein sequences. In human embryos and in pregnant women, the intermediate lobe is active and leads to the production of endorphins and enkephalins. These same endorphin-producing pathways are active in other neural tissue, and since they bind to the opiate receptors in other parts of the brain they are assumed to represent natural opiate-like analgesic compounds. The biological role of ACTH is to stimulate the production of adrenal cortex steroids, principally cortisol and corticosterone. The mechanism of action of ACTH involves activation of adenylate cyclase, elevation of cAMP, and increased PKA activity of adrenal cortex tissue. The main effect of these events is to increase the activity of the side chain-cleaving enzyme, which converts cholesterol to pregnenolone. The principal physiological effect of ACTH is production of the glucocorticosteroids. The nonapeptides oxytocin and vasopressin are the principal hormones of the posterior pituitary. Both are synthesised as prohormones in neural cell bodies of the hypothalamus and mature as they pass down axons in association with carrier proteins termed neurophysins. The axons terminate in the posterior pituitary, and the hormones are secreted directly into the systemic circulation. Vasopressin is also known as antidiuretic hormone (ADH), because it is the main regulator of body fluid osmolarity. The secretion of vasopressin is regulated in the hypothalamus by osmoreceptors, which sense water concentration and stimulate increased vasopressin secretion when plasma osmolarity increases. The secreted vasopressin increases the reabsorption rate of water in kidney tubule cells, causing the excretion of urine that is concentrated in Na^+ and thus yielding a net drop in osmolarity of body fluids. Vasopressin binds plasma membrane receptors and acts through G-proteins to activate the cAMP/PKA regulatory system. Vasopressin deficiency leads to diabetes insipidus, which characterised by watery urine and polydipsia. Oxytocin secretion in nursing women is stimulated by direct neural feedback obtained by stimulation of the nipple during suckling. Its physiological effects include the contraction of mammary gland myoepithelial cells, which induces the ejection of milk from mammary glands, and the stimulation of uterine smooth muscle contraction leading to childbirth. Parathyroid hormone (PTH, molecular weight 9,500) is synthesised and secreted by chief cells of the parathyroid in response to systemic Ca^{2+} levels. The Ca^{2+} receptor of the parathyroid gland responds to Ca^{2+} by increasing intracellular levels of PKC, Ca^{2+} and IP_3; this stage is followed, after a period of protein synthesis, by PTH secretion. PTH acts by binding to cAMP-coupled plasma membrane receptors, initiating a cascade of reactions that culminates in the biological response. The body response to PTH is complex but is aimed in all tissues at increasing Ca^{2+} levels in extracellular fluids. PTH induces the dissolution of bone by stimulating osteoclast activity, which leads to elevated plasma Ca^{2+} and phosphate. In the kidney, PTH reduces renal Ca^{2+} clearance by stimulating its reabsorption; at the same time, PTH reduces the reabsorption of phosphate and thereby increases its clearance. PTH acts on the liver, kidney, and intestine to stimulate the production of the steroid hormone 1,25-dihydroxycholecalciferol (calcitriol), which is responsible for

Ca^{2+} absorption in the intestine. Calcitonin (CT) is a 32-amino acid peptide secreted by C cells of the thyroid gland. Calcitonin relieves the symptoms of osteoporosis, but the mechanism of action remains unclear. CT induces the synthesis of PTH, which leads to increased plasma Ca^{2+} levels. CT reduces the synthesis of osteoporin (Opn), a protein made by osteoclasts and responsible for attaching osteoclasts to bone. Thus, it appears that CT elevates plasma Ca^{2+} via PTH induction and reduces bone reabsorption by decreasing osteoclast binding to bone. The principal role of the pancreatic hormones is the regulation of whole-body energy metabolism, principally by regulating the concentration and activity of numerous enzymes involved in catabolism and anabolism of the major cell energy supplies. Insulin is a member of a family of structurally and functionally similar molecules that include IGF-1, IGF-2, and relaxin. The tertiary structure of all 4 molecules is similar, and all have growth-promoting activities, but the dominant role of insulin is metabolic while the dominant roles of the IGFs and relaxin are in the regulation of cell growth and differentiation. Insulin secreted by the pancreas is directly infused via the portal vein to the liver, where it exerts profound metabolic effects. In most other tissues insulin increases the number of plasma membrane glucose transporters, but in liver glucose uptake is dramatically increased because of increased activity of the enzymes glucokinase, phosphofructokinase-1 (PFK-1), and pyruvate kinase (PK), the key regulatory enzymes of glycolysis. Insulin generates its intracellular effects by binding to a plasma membrane receptor, which is the same in all cells. The receptor is a disulfide-bonded glycoprotein. One function of insulin (aside from its role in signal transduction) is to increase glucose transport in extrahepatic tissue by increasing the number of glucose transport molecules in the plasma membrane. Insulin stimulates lipogenesis, diminishes lipolysis, and increases amino acid transport into cells. Insulin modulates transcription, altering the cell content of numerous mRNAs. It stimulates growth, DNA synthesis, and cell replication, effects that it holds in common with the IGFs and relaxin. Like insulin, glucagon lacks a plasma carrier protein, and like insulin its circulating half-life is also about 5 minutes. The principal effect of glucagon is on the liver, which is the first tissue perfused by blood containing pancreatic secretions. Glucagon binds to plasma membrane receptors and is coupled through G-proteins to adenylate cyclase. The resultant increases in cAMP and PKA reverse all of the effects insulin has on liver. Prolactin (PRL) is produced by acidophilic pituitary lactotropes. Prolactin is the lone tropic hormone of the pituitary that is routinely under negative control by prolactin inhibiting hormone (PIH), which is now known to be dopamine. Decreased hypophyseal dopamine production, or damage to the hypophyseal stalk, leads to rapid up-regulation of PRL secretion. PRL initiates and maintains lactation in mammals, but normally only in mammary tissue that has been primed with oestrogenic sex hormones.

Gastrointestinal hormones:

Hormone	Location	Major Action
Gastrin	Gastric antrum, and duodenum	Gastric acid and pepsin secretion
Cholecystokinin (CCK)	Duodenum, jejunum	Pancreatic amylase secretion
Secretin	Duodenum, jejunum	Pancreatic bicarbonate secretion
Gastric inhibitory peptide (GIP)	Small bowel	Enhances glucose-mediated insulin release; inhibits gastric acid secretion
Vasoactive intestinal peptide (VIP)	Pancreas	Smooth muscle relaxation; stimulates pancreatic bicarbonate secretion
Motilin	Small bowel	Initiates interdigestive intestinal motility
Pancreatic polypeptide (PP)	Pancreas	Inhibits pancreatic bicarbonate and protein secretion
Enkephalins	Stomach, duodenum, gallbladder	Opiate-like actions
Substance P	Entire gastrointestinal tract	Physiological actions uncertain
Bombesin-like immunoreactivity (BLI)	Stomach, duodenum	Stimulates release of gastrin and CCK
Neurotensin	Ileum	Physiological actions unknown
Enteroglucagon	Pancreas, small intestine	Physiological actions unknown

SPECIALISED PRODUCTS OF AMINO ACIDS

The majority of tyrosine that does not get incorporated into proteins is catabolised for energy production. One other significant fate of tyrosine is conversion to the catecholamines. The catecholamine neurotransmitters are dopamine, norepinephrine, and epinephrine. Tyrosine is transported into catecholamine-secreting neurons and adrenal medullary cells where catecholamine synthesis takes place. The first step in the process requires tyrosine hydroxylase, which like phenylalanine hydroxylase requires tetrahydrobiopterin as cofactor. The hydroxylation reaction generates DOPA (3,4-dihydrophenylalanine). DOPA decarboxylase converts DOPA to dopamine, dopamine β-hydroxylase converts dopamine to norepinephrine and phenylethanolamine N-methyltransferase converts norepinephrine to epinephrine. Within the substantia nigra and some other regions of the brain, synthesis proceeds only to dopamine. Within the adrenal medulla dopamine is converted to norepinephrine and epinephrine. Tryptophan serves as the precursor for the synthesis of serotonin (5-hydroxytryptamine, 5-HT) and melatonin (N-acetyl-5-methoxytryptamine). An increased uptake of tryptophan in the diet will lead to increased brain serotonin content. Serotonin is present at highest concentrations in platelets and in the GIT. Lesser amounts are found in the brain and the retina. Serotonin containing neurons have their cell bodies in the midline raphe nuclei of the brain stem and project to portions of the hypothalamus, the limbic system, the neocortex and the spinal cord. After release from serotonergic neurons, most of the released serotonin is recaptured by an active reuptake mechanism. The function of the antidepressant, Prozac is to inhibit this reuptake process, thereby, resulting in prolonged serotonin presence in the synaptic cleft. The function of serotonin is exerted upon its interaction with specific receptors and their subtypes (5HT1, 5HT2, 5HT3, 5HT4, 5HT5, 5HT6, and 5HT7). Most of these receptors are coupled to G-proteins that affect the activities of either adenylate cyclase or phospholipase Cg (PLCg). The 5HT3 class of receptors are ion channels. Some serotonin receptors are presynaptic and others postsynaptic. The 5HT2A receptors mediate platelet aggregation and smooth muscle contraction. The 5HT2C receptors are suspected in control of food intake. The 5HT3 receptors are present in the GIT and are related to vomiting. Also present in the GIT are 5HT4 receptors where they function in secretion and peristalsis. The 5HT6 and 5HT7 receptors are distributed throughout the limbic system of the brain and the 5HT6 receptors have high affinity for antidepressant drugs. Melatonin is derived from serotonin within the pineal gland and the retina, where the necessary N-acetyltransferase enzyme is found. The pineal parenchymal cells secrete melatonin into the blood and cerebrospinal fluid. Synthesis and secretion of melatonin increases during the dark period of the day and is maintained at a low level during daylight hours. This diurnal variation in melatonin synthesis is brought about by norepinephrine secreted by the postganglionic sympathetic nerves that innervate the pineal gland. Norepinephrine interaction with β-adrenergic receptors leads to increased levels of cAMP, which in turn activate the N-acetyltransferase required for melatonin synthesis. Glutathione (GSH) serves as a reductant, is conjugated to drugs to make them more water soluble, is involved in amino acid transport across cell membranes (the γ-glutamyl cycle), is a part of the peptidoleukotrienes, serves as a cofactor for some enzymatic reactions and as an aid in the rearrangement of protein disulfide bonds. The sulfhydryl of GSH can be used to reduce peroxides formed during oxygen transport. The resulting oxidised form of GSH consists of two molecules disulfide bonded together (abbreviated GSSG). The enzyme glutathione reductase utilises NADPH as a cofactor to reduce GSSG back to two moles of GSH. Hence, the pentose phosphate pathway is an extremely important pathway of erythrocytes for the continuing production of the NADPH needed by glutathione reductase. 10% of glucose consumption, by erythrocytes, may be mediated by the pentose phosphate pathway. The γ-glutamyl cycle is a group transfer mechanism of amino acid transport that requires more energy input, it is rapid and has a high capacity. The cycle functions primarily in the kidney, particularly renal epithelial cells. The enzyme γ-glutamyl transpeptidase is located in the cell membrane and shuttles GSH to the cell surface to interact with an amino acid. Reaction with an amino acid liberates cysteinylglycine and generates a γ-glutamyl-amino acid, which is transported into the cell and hydrolysed to release the amino acid. Glutamate is released as 5-oxoproline and the cysteinylglycine is cleaved to its component amino acids. Regeneration of GSH requires an ATP-dependent conversion of 5-oxoproline to glutamate and then the 2 additional moles of ATP that are required during the normal generation of GSH. One of the earliest signals that cells have entered their replication cycle is the appearance of elevated levels of mRNA for ornithine decarboxylase (ODC), and then increased levels of the enzyme, which is the first enzyme in the pathway to synthesis of the polyamines. The polyamines are highly cationic and tend to bind nucleic acids with high affinity. The polyamines may be important participants in DNA synthesis, or in the regulation of that process. The function of ODC is to produce the 4-carbon saturated diamine, putrescine. At the same time, SAM-e decarboxylase cleaves the S-adenosylmethionine (SAM-e) carboxyl residue, producing decarboxylated SAM-e (S-adenosymethylthiopropylamine), which retains the methyl group usually involved in SAM-e methyltransferase activity. Spermidine synthase catalyses the condensation reaction, producing spermidine and 5'-methylthioadenosine. A second propylamine residue is added to spermidine producing spermine. Vasodilators, such as acetylcholine, do not exert their effects upon the vascular smooth muscle cell in the absence of the overlying endothelium. When acetylcholine binds its receptor on the surface of endothelial cells, a signal cascade, coupled to the activation phospholipase C-γ (PLCγ), is initiated. The PLCγ-mediated release of inositol trisphosphate, IP_3 (from membrane associated phosphatidylinositol-4,5-bisphosphate, PIP_2), leads to the release of intracellular stores of Ca^{2+}. In turn, the elevation in Ca^{2+} leads to the liberation of endothelium-derive relaxing factor (EDRF) the free radical diatomic gas, nitric oxide (NO), which then diffuses into the adjacent smooth muscle. Within smooth muscle cells, NO reacts with the haeme moiety of a soluble guanylyl cyclase, resulting in activation of the latter and a consequent elevation of intracellular levels of cGMP. The net effect is the activation of cGMP-responsive enzymes, which lead to smooth muscle cell relaxation. The coronary artery vasodilator, nitroglycerin, acts to increase intracellular release of NO and thus of cGMP. NO is formed by the action of NO synthase, (NOS) on the amino acid L-arginine. NOS is a very complex enzyme, employing 5 redox cofactors: NADPH, FAD, FMN, haeme and tetrahydrobiopterin. NO is a highly reactive free radical (lasting only 2-4 seconds) and interacts with oxygen and superoxide. NO is inhibited by haemoglobin and other haeme proteins which bind it tightly. NO produces cGMP, which acts to inhibit platelet aggregation. Creatine is synthesised in the liver by methylation of guanidoacetate using SAM-e as the methyl donor. Guanidoacetate is formed in the kidney from the amino acids arginine and glycine. The phosphate of ATP is transferred to creatine, generating creatine phosphate, through the action of creatine kinase. The reaction is reversible such that when energy demand is high (e.g., during muscle exertion) creatine phosphate donates its phosphate to ADP to yield ATP. Both creatine and creatine phosphate are found in muscle, brain and blood. The amount of creatinine produced is related to muscle mass and remains remarkably constant from day to day. Creatinine is excreted by the kidneys and the level of excretion (creatinine clearance rate) is a measure of renal function.

POLYAMINES

Polyamines are part of a class of proteins called biogenic amines. The polyamines are organic compounds, such as putrescine, spermidine, and spermine that are growth factors in both eukaryotic and prokaryotic cells. They are synthesised in cells in pathways that are very highly regulated. The actual function of these compounds is not entirely clear. As cations, they do bind to DNA, and structurally, they represent compounds with cations that are found at regularly spaced intervals (unlike, say, Mg^{2+} or Ca^{2+}, which are point charges). If synthesis of polyamines is blocked, then cell growth is stopped or profoundly slowed. Exogenous polyamines restore the growth of these cells. Cell growth is inhibited at the early stage of polyamine accumulation, which may be due to the inactivation of ribosomes through the replacement of Mg^{2+} on magnesium-binding sites by polyamines. At later stage of polyamine accumulation, a decrease in ATP content takes place. This is followed by swelling of the mitochondria, which may be a symptom of the subsequent cell death. Polyamines are essential for the maintenance of the high metabolic activity of a normal functioning and healthy body. Cell growth requires certain amounts of polyamines. Polyamines are found in very high amounts in the tissues of severely injured trauma patients and in food that have been morphologically shocked by excessive processing, such as rapid freezing. Polyamines exist in vegetables, grains, fruits, sprouts, meats, and seafood. The concentration of polyamines inside the cell is tightly regulated. The range of cellular polyamine concentration is determined at the lower limit by their absolute requirement for cell growth, and at the upper limit by their potential toxicity. Polyamines are also synthesised in the body, or by the action of the bacteria in the gut, synthesising polyamines from dietary amino acids. The process of producing polyamines by bacteria begins long before the food is eaten. Frozen, canned and otherwise tainted foods are loaded with polyamines long before they hit the gut. When manufactured by the body, polyamines are derived from the amino acid ornithine (Orn), through the actions of the enzyme ornithine decarboxylase, or ODC. Almost all tissues can manufacture polyamines, but the liver makes the vast majority of them. Ornithine is a non-essential amino acid and the body can make ornithine from other amino acids through the ornithine cycle. This pathway converts either of 2 other amino acids, arginine (Arg) or citrulline into ornithine through conversion into intermediates such as arginosuccinic acid. These highly controlled conversions are the result of a variety of specialized enzymes. Arginine is one of the most versatile amino acids in animal cells, serving as a precursor for the synthesis not only of polyamines but also of proteins, nitric oxide, urea, proline, glutamate. Arginine is critical for the synthesis of creatine, a major source of high-energy phosphate for regeneration of energy production in muscle. The production of arginosuccinic acid, a key intermediate in the production of ornithine (and thus a major factor in the synthesis of polyamines) is genetically linked to ABO blood type. The gene for the enzyme that manufactures arginosuccinic acid, called arginosuccinate synthase (ASS), lies adjacent to the ABO gene on 9q34 and studies have shown their linkage to strongly correlate.

The ornithine cycle:

> 1- Arginine or citrulline is converted into ornithine through conversion into intermediates such as arginosuccinic acid.
> 2- Polyamine synthesis begins with the conversion of ornithine into the polyamine putrescine by the action of the enzyme ornithine decarboxylase (ODC).
> 3- Putrescine can then be converted to 2 other polyamines, spermine and spermidine, each of which have slightly different effects in the body.
> 4- Because both spermidine and spermine are made from putrescine, and putrescine is made from the amino acid ornithine by the enzyme ornithine decarboxylase (ODC), blocking ODC is usually sufficient to block the synthesis of all 3 polyamines.

ODC is a rate limiting enzyme, i.e., lots of ODC, lots of polyamines; and vice versa. Cell growth and differentiation are dependent on precise control of the levels of polyamines inside the cell. ODC is one of those ephemeral enzymes that don't last very long in the cell (very short half-life); protein-splitting enzymes degrade it very rapidly. This short half-life of ODC gives the cell a way to rapidly change polyamine synthesis. Polyamines are essential to cellular proliferation and differentiation. Kids have high polyamines, and bodybuilders think they need high levels as well. Polyamines accumulated in the small bowel are largely obtained from the food consumed. The size and electrical charge of the polyamines permit them to interact with huge molecules such as DNA and RNA, and pass through phospholipid membranes and compartments with ease. There is an intimate relationship between polyamines RNA and insulin. Insulin, whose primary effect once inside the cell is to activate growth, does this by providing stimulation to the protein synthesising factory of the cell, the ribosomes. Ribosomes act on instructions from mRNA, which carries the blueprint for that particular type of protein coded in the DNA. Polyamines seem to stabilise and amplify the message contained in mRNA, which serves to increase the protein produced from it. So, inside the cell, polyamines work to increase growth by 2 separate mechanisms: 1- a direct influence on specific growth promoting genes, and 2- enhancement of the production of the various cell proteins needed for growth. By this mechanism polyamines amplify the effects of DNA and insulin by acting to stabilise mRNA. This results in more the synthesis of larger amounts of protein. Human milk is very high in polyamines, particularly spermine. Human milk provides substantial amounts of spermine and spermidine to newborns and infants, which could potentially modulate the maturation of the infant's intestines. During the first week after birth, putrescine levels in human milk remain very low and vary little, while spermidine and spermine concentrations rise markedly during the first 3 days, reaching levels that are 12 times higher, respectively, than the values measured on the first day. The polyamine concentration of powdered milk formulas is 10 times lower than in human milk. Low maternal protein intake decreases the activity of ornithine decarboxylase and consequently the levels of polyamines in the placenta, resulting in reduced foetal growth. Polyamines, particularly putrescine, in sufficient amounts are important in maintaining the healthy structure and function of the intestinal mucosa, a function that seems to also require vitamin D. In infants, polyamines are very important growth facts. Cancer cells are voracious consumers of polyamines. The strategy of polyamine deprivation shows great promise as a new horizon in cancer treatment, especially cancers that are in themselves hormonally sensitive, such as prostate and breast cancer. α-Difluoromethylornithine (DFMO), shows great promise in prostate cancer, which inhibits ODC, so blocks the ability of cancer cells to benefit from polyamines. High polyamine levels inhibit the anti-cancer response of the body through specialized anti-tumour NK (Natural Killer cells). Deprivation and lowering of polyamines, through blocking their manufacture or uptake from the intestines, on the other hand, increases NK cell activity. Polyamines, secreted by the tumour itself as well as absorbed through the GIT, could be considered as growth factors for the cancer, and also as natural immunosuppressive factors as well. Vitamin B_6 has a profoundly stimulating effect on ODC and consequently on polyamine synthesis. Changes in ODC activity and polyamine synthesis and changes in the expression of the blood group antigens may be the most important biological markers of colonic precancer. Polyamines made by gut bacteria or present in food itself are an important stimulus to tumour growth. When polyamines are systematically blockaded by the use of drugs that inhibit ODC and the elimination of all external sources by use of a polyamine-free diet and decontamination of GIT, the number of metastases is significantly reduced. High polyamine levels act as promoters of other tumours. The concentration of polyamines in spoiled food can be toxic. Fish tissue, which is more perishable than animal tissue, is very susceptible to microorganism invasion. This is why freshly caught fish stored at moderate temperature (60°F) will remain unspoiled for 1 day or less. A condition called scombroid poisoning is associated with polyamine toxicity. It is resulted from widespread consumption of scombroidea species, mackerel, tuna, bluefish, and skipjack, and the association with a seafood poisoning. The rapidly moving fish, like mahi-mahi (yellow-fin dolphin), sardines, anchovies, and herring are subject to a rather rapid microbial decomposition. All of these fish have a relatively high content of the amino acid histidine in their tissues. Bacterial decomposition of the fish converts the histidine to histamine. Histamine can reach concentrations of up to fairly high concentrations without the development of off-flavours that would cause it to be rejected. In most cases the histamine will not produce the observed toxicity-histamine has a relatively low oral toxicity. However, putrescine is also encountered in fish, which enhances the effects of histamine and causes a violent allergic reaction. These chemically are stable and will not be reduced by cooking, freezing, or other processing. The symptoms of scombroid poisoning can begin within 10 minutes and up to 2 hrs after consumption of the tainted fish. Most of the acute symptoms are gone within 16-24 hr. The symptoms resemble a severe allergic reaction and may include a facial flush, tightness of chest, sweating, nausea, vomiting, tingling, body rash (hives or urticaria), severe headache, shortness of breath, dizziness, throbbing, thirst, and diarrhoea. Paraquat, the herbicide used to spray on marijuana plants is a substantial lung carcinogen. Its cancer causing abilities have in part to be the result of huge increases of polyamines in the lungs. Most dietary lectins are potent inducers of polyamine production in the gut. This is probably the result of the intestinal cells synthesising large amounts of polyamines in an effort to repair the damage caused by the lectins. Lectins typically damage the mucosal microvilli. Many lectins cause growth increases in several organs including the liver, pancreas and spleen. These organ enlargements are the result of a huge influx of polyamines into the organs. Wheat germ lectin induces significant polyamine production. Incorporating wheat germ lectin into the diet reduces the digestibility and utilisation of dietary proteins. As a result of its binding and uptake by the cells of the small intestine, wheat germ lectin induces extensive polyamine-dependent growth of the small bowel tissue by increasing its content of proteins, RNA and DNA. An appreciable portion of the absorbed wheat germ lectin is transported across the gut wall into the systemic circulation, where it deposits in the walls of the blood and lymphatic vessels. Wheat germ lectins also induce growth of the pancreas. These same effects have been shown to occur with several bean and legume lectins as well. In adults, low protein diets tend to increase ornithine decarboxylase activity in the liver, probably to make up for the loss of polyamines normally available from food or made by the gut bacteria. Many type O's who have been on high lectin diets experience higher rates of autoimmune disease, obesity and allergies. Internal production of polyamines (i.e., made by the liver) may increase the risk of autoimmunity, and many of the grain lectins can have significant effects on internal polyamine synthesis by the pancreas, liver and small intestine. Polyamines are typically found in fermented foods such as cheese, beer, sauerkraut and yeast extracts. The polyamines are thought to be produced from amino acids by fermentation by enzymes formed by the microorganisms. Polyamines are also found in foods, which through processing have had the structural integrity of their tissues shocked or damaged through food preparation such as quick freezing or canning. The polyamines putrescine, spermidine and spermine are essential for cell renewal and, therefore, are needed to keep the body healthy. We need enough polyamines to help growth and healing, but not so much as to slow down the immune systems, and change the metabolism of the tissues. Polyamines temporarily rise in the first stages of fasting, which is probably a response on the part of the liver to the lack of dietary sources. Polyamine synthesis has a positive effect on albumin levels. Albumin is used to assess the long-term nutritional status of patients since it reflects body protein stores for the last month. The reference range is 3.5-5.2 g/dl. Levels above 4.8 probably indicate higher polyamine levels, levels below 4 low levels. The indican test gauges the level of indoles in the intestine by looking for the product of their metabolism, indican, in the urine. Large amounts of indican are usually a sign of high bacteria counts in the upper intestine. Large amounts of bacteria in the upper intestines produce large amounts of polyamines. Putrescine and one of the secondary polyamines, cadaverine, are responsible for much of the odour characteristic of halitosis. High levels of polyamines in the mouth inhibit the migration of white blood cells to areas of infection and inflammation. A common symptom of high levels of polyamines is headaches resulting from eating fermented foods such as wine, beer or sauerkraut. Polyamines enhance the effect of histamine (usually present in the form of histidine containing foods, such as red wine) in the diet. The malabsorption that results from the incomplete breakdown of animal protein in type A will serve as a very tempting source of amino acids for the intestinal bacteria. In gratitude for this free meal they will synthesise huge amounts of polyamines. Increase the production of polyamines by the cells of the intestines, pancreas and liver would result from the grain lectins in type O. Both guar gum and pectin lead to the appearance of cadaverine and to elevate putrescine concentrations in the caecum. The intestinal microflora are a major source of polyamines in the contents of the large intestine. Bacteroides, fusobacteria and anaerobic cocci can synthesise high amounts of putrescine and spermidine. The shock to the tissues of many foods processed by canned or flash-frozen manner allows the release of many polyamines either prior to pasteurisation of upon de-thawing. High levels of polyamines have been found in quick frozen cucumbers, especially when sealed in non-perforated packaging.

Polyamines:

Polyamine	Source
Putrescine	Aged or sharp cheeses, potatoes, canned/frozen vegetables, or certain fruit products, such as oranges and tangerines, fermented soy sauce (containing wheat), and shrimp, especially the packaged and frozen types.
Spermidine	Mature cheeses, fermented soybeans, fermented tea, Japanese sake, domestic mushrooms, potatoes and fresh bread.
Spermine	Cereals (other than bread), canned or frozen vegetables, meat products, red meat and poultry.

Putrescine, cadaverine, spermidine, agmatine and spermine are detected in different concentrations, depending on the type of sprouts. In pre-packed retail products the total polyamine content is higher than in home-grown samples. Pollution-induced stress in many grains raises their polyamine content. Arsenic has been shown to increase polyamine synthesis. Foods that inhibit ODC include walnuts, curcumin (found in turmeric), green tea, pomegranates, guava, broccoli leaves, plantain, black currant fruit, bilberry, elderberry, grapes, onion, garlic, dill, tarragon, and chives. A handful of walnuts thrown into a salad twice daily inhibits excess ODC activity. Two common carotenoids found in many yellow and red vegetables, xanthophylls and canthaxanthin, reduce polyamine levels in the oral cavity and GIT. Vitamin E seems to upregulate polyamine metabolism, probably by its direct effects on ODC. The ability of vitamin E to increase levels of polyamines may account for its more positive effects in Alzheimer's disease. Much of the nerve damage in Alzheimer's results from oxidative stress, or the generation of free radicals in the brain. Vitamin E may help reverse this effect by raising brain polyamine levels to allow the nerve cells to survive the neurodegenerative process. Milk thistle or schizandra are bioactive because they raise polyamines in the liver, which might be rationale to use in liver disease but not in cancer. Ornithine and methionine are precursors of polyamines. Ornithine is the direct precursor to putrescine; methionine and cysteine are intermediates in the synthesis of polyamines. Amino acids promote growth hormone and insulin, perhaps through increased polyamine secretion. The relatively low methionine levels in some phytochemicals, such as soy, may limit the synthesis of polyamines necessary for tumour growth. Methionine promotes intestinal carcinogenesis. Zinc deficiency causes significantly higher plasma levels of the polyamines spermidine and spermine. Copper in high concentrations increases the levels of polyamines. Alterations in polyamine metabolism are associated with ischaemic and traumatic brain injury. Polyamines may play a multifaceted detrimental role following ischaemia reperfusion. Polyamines are major intracellular modulators of inward rectifier potassium channels and certain types of NMDA and AMPA receptors. Alterations in polyamines could have major effects on ion homeostasis in the CNS, especially K, and thus account for the observed injury after cerebral ischaemia. There is also evidence suggesting a role for polyamines in apoptosis. Polyamines play important roles in a number of cellular processes such as replication, transcription, and translation. Presumably these roles are exerted by specific interactions, which can only be mediated by the cationic polyamines with their characteristic, unique, and flexible charge distributions. The importance of the polyamines in cell function is reflected in a strict regulatory control of their intracellular levels. Adequate cellular polyamine levels are achieved by a careful balance between biosynthesis, degradation, and uptake of the amines. The polyamine biosynthetic pathway consists of 2 highly regulated enzymes, ornithine decarboxylase and S-adenosylmethionine decarboxylase, and 2 constitutively expressed enzymes, spermidine synthase and spermine synthase. The biological half-lives of the 2 regulatory enzymes ornithine decarboxylase and S-adenosylmethionine decarboxylase (5-60 min) are among the shortest known for mammalian enzymes, allowing the cell to rapidly change the cellular polyamine levels. The polyamine degradation pathway consists of the highly regulated enzyme spermidine/spermine N1-acetyl transferase and the constitutively expressed polyamine oxidase. Also, cells are equipped with an efficient transport system for utilisation of exogenously derived polyamines. The biosynthesis of polyamines is increased by a great variety of physiological growth stimuli, and polyamine deficiency results in an arrest of cell proliferation, which can be reversed by supplementation with external polyamines. Polyamine deficiency can also, under certain circumstances, result in apoptosis. The constitutive overproduction of ornithine decarboxylase has been observed in many types of cancer cells, the ornithine decarboxylase gene appears to be of central importance in the regulation of cell growth. When the ornithine decarboxylase gene is transfected into cells and overexpressed, the cells go through malignant transformation. Polyamine biosynthesis is activated at the G_1 to S and G_2 to M transitions of cell cycle. Ornithine decarboxylase and S-adenosylmethionine decarboxylase are regulated at both transcriptional and translational levels during the cell cycle. The role of the polyamines in cell cycle regulation is dependent on oncogenes, tumour suppressor genes, and cyclins with their associated kinases. The polyamine biosynthesis activated at the G_1 to S transition infers a role for the polyamines in DNA replication. Polyamine depletion affects DNA replication negatively, presumably by reducing the rate of DNA elongation. The G_1 to S transition and the G_2 phase progression are affected only after several cell cycles. DNA replication requires the presence of a number of proteins, which have a direct or indirect role for the incorporation of the nucleotides into DNA as well as for the structural integrity of the chromatin. Topoisomerases are required to release various stresses in DNA that might affect the progress of the replication machinery. In most cell types, the biochemical characteristics of apoptosis include the activation of endogenous calcium and magnesium dependent endonucleases, leading to fragmentation of the chromosomal DNA. DNA is destabilised in polyamine-depleted cells. The diamine putrescine and the polyamines spermidine and spermine are ubiquitous substances found virtually in all cells from higher prokaryotes and eukaryotes. These highly charged, low molecular weight and polycationic substances are essential for cell growth and differentiation and their intracellular concentrations increase during periods of rapid cell proliferation. To meet the needs of division processes and protein synthesis, the intracellular concentration of polyamines are critically regulated by several enzymes including meanly ornithine decarboxylase (ODC) and S-adenosyl-methionine decarboxylase (SAM-DC). In the de novo synthesis pathway, ODC catalyses the decarboxylation of the amino acid ornithine to form putrescine. This is the first rate-limiting step in polyamine synthesis. The synthesis of ODC activity and of ODC mRNA can be enhanced by several hormones and growth factors including insulin, gastrin, and nerve growth factor and epidermal growth factor. The turnover of ODC is the fastest among eukaryotic enzymes, and any change in the rate of ODC synthesis will be rapidly transmitted to the amount of enzyme protein and thus to polyamine synthesis. Polyamines by means of various feedback mechanisms control their own cellular levels. Increasing polyamine levels exert a feedback regulation on ODC activity by a direct inhibition of ODC mRNA translation. A second rate-limiting enzyme in polyamine synthesis is S-adenosyl methionine decarboxylase (SAM DC), which decarboxylates S-adenosyl methionine, providing aminopropyl groups for spermidine and spermine synthesis. Putrescine is converted into spermidine by the action of an aminopropyltransferase called spermidine synthase. A second aminopropyltransferase, termed spermine synthase, adds an additional propylamine moiety to spermidine forming spermine. Regarding the catabolic pathway, putrescine can be catabolised by several enzymes, the much frequently analysed being diamine oxidase. Putrescine is also a precursor in the biosynthesis of γ-aminobutyric acid (GABA) via a pathway that does not involve L-glutamic acid. GABA formation from putrescine represents a major process in some developing neural tissues. Rapidly dividing tissues such as the intestinal epithelium are also dependent upon exogenous sources of polyamines supplied by food, secretions and microbial flora. The apical surfaces of intestinal cells are exposed to variable luminal contents, and has a unique advantage for regulating its polyamine levels. Diet can supply sufficient amounts of polyamines to support at least intestinal cell renewal and growth. The major sources of putrescine are fruit, cheese and non-green vegetables. Meat is the richest source of spermine. More than 80% of putrescine can be converted into other polyamines including cadaverine and into non-polyamine metabolites, mostly to amino acids. Diamine oxidase (DAO) is the enzyme responsible for controlling the bioavailability of putrescine. At higher endoluminal concentrations, absorption of polyamines seems to occur following a passive diffusion process especially when mucosal permeability is increased. Genistein, a tyrosine kinase inhibitor inhibits EGF-stimulated polyamine uptake, which indicate that tyrosine phosphorylation plays a critical role in the hormonal stimulation of the uptake. The effect of EGF on polyamine uptake may be due to a translocation of intracellular proteins into the membrane and mobilisation of a specific transporter. Human milk contains substantial quantities of polyamines mainly spermine and spermidine, with much less putrescine. Milk contains polyamine oxidases, which can induce a decrease in polyamine concentration over time. However, spermine and spermidine concentrations markedly increase during the first 3 days of lactation. The peak concentrations of polyamines occur in the colostrum and milk between the first week of lactation in cow. The concentrations of polyamines in artificial powdered formulas are much lower than in human milk. Over 250 nonprotein amino acids have been identified in plants. A number of these compounds are intermediates in the synthesis and catabolism of the protein amino acids. Over 10,000 alkaloids have been isolated from plants and their structures elucidated. In higher plants, polyamines also influence developmental processes and play an important role in the response to abiotic stress. Spermidine and spermine are generated from putrescine by the addition of aminopropyl groups derived from decarboxylated S-adenosyl Met. The rate-limiting step in the formation of putrescine in animals and most fungi is the decarboxylation of Orn by Orn decarboxylase (ODC). ODC activity is regulated both at the translational level and via controlled, ATP-dependent proteolysis by the 26S proteasome, at least in mammalian and yeast (Saccharomyces cerevisiae) cells. ODC is an enzyme with rather short half-life, varying between 30-120 min in eukaryotic organisms. In some plants, putrescine is also synthesised via the decarboxylation of Arg by Arg decarboxylase and subsequent degradation of the generated agmatine. In some plant systems, polyamines may act to store organic nitrogen being the sole source of nitrogen. Polyamine and ethylene biosynthesis is related by the common precursor S-adenosylmethionine, and the distribution of this compound can have important physiological implications. Polyamines are also bound to hydroxycinnamic acids and thus possibly affect the processes of differentiation, flowering and maturation. Polyamines have an effect on the resistance to virus and fungus in some plants. The plant cell wall is one of the most important compartments in relation to the polyamine catabolism, emphasising the increase in the conjugation of polyamines in the cell wall during the cell senescence. Polyamines (putrescine, spermidine and spermine) are aliphatic nitrogenous compounds that at present are considered growth regulators because they have some demonstrated effect on cellular growth, division and differentiation at low concentrations. Polyamines, due to polycationic nature, can be bound to negatively charged molecules such as nucleic acids, proteins or phospholipids altering genic expression, the activity of some enzymes and also modifying the fluidity and permeability of the biological membranes.

Elements of the feedback loop:

> Polyamines rise; polyamines induce production of more antizyme; antizyme inhibits and demolishes ODC; polyamine synthesis declines.

Antizyme (AZ) is the central element in a feedback loop that controls cellular polyamines. Labile proteins are important and important proteins are labile. AZ3 is expressed only in the testes and is there restricted to the post-meiotic stages of spermatogenesis. Cells make, transport, and destroy polyamines. AZ controls and limits polyamine accumulation by impeding the first and second of these processes. AZ inhibits and sometimes destroys a key polyamine biosynthetic enzyme, ornithine decarboxylase (ODC), and inhibits cellular uptake of polyamines. The polyamines are biologically ubiquitous small alkylamines bearing multiple amine groups that carry positive charge at physiologic pH. Biosynthesis begins with the enzymatic decarboxylation of ornithine by ODC to produce putrescine (diaminobutane). Successive addition by distinct enzymes of aminopropyl groups to putrescine results in spermidine and spermine. Polyamines are present at multimillimolar concentrations within all cells. Polyamine biosynthesis can in effect be reversed in a catabolic series of oxidations, leading to the net conversion of spermine to spermidine and then to putrescine. The rate-limiting step for this back-conversion pathway is spermidine/spermine acetyltransferase. AZ has a very high affinity for the ODC monomer, so formation of the enzymatically inactive AZ:ODC heterodimer is strongly favoured. In addition to biosynthesis, uptake from the cellular environment can provide polyamines, transport of which is also elaborately regulated. AZ inhibits uptake, synthesis, transport, and catabolic destruction are all in play. AZ3 mutation may underlie a specific form of heritable male infertility. Antizyme, a moniker for 'anti-enzyme for ornithine decarboxylase (ODC). There are at least 3 independently conserved antizyme isoforms among vertebrates, AZ3 is testis specific; and its expression is restricted to a late stage in sperm production. Mutations that reduce polyamine biosynthesis or drugs that inhibit their biosynthesis halt cell growth and eventually cause cell death and translation. Suboptimal concentrations of Mg^{2+} diminish translation. Spermidine is known to have one vital and unique biochemical function-it provides an aminobutyl group for a post-translational modification, termed hypusination, which is limited to a single lysyl residue of one protein. Polyamines have been shown to gate the inward rectifier recurrent of an ion channel and to have effects on chromatin condensation and transcriptional regulation. Although cells need polyamines, failure to limit their amount can be disastrous. Antizyme production depends on polyamine levels. ODC is enzymatically active only as a homodimer. Monomers of ODC interact with each other only weakly, but antizyme has high affinity for the ODC monomer. The site of antizyme binding is placed so that it obstructs the ODC homodimer interface. Consequently, two antizyme molecules convert one homodimer to a pair of enzymatically inactive antizyme–ODC heterodimers. ODC also acts catalytically to direct the proteasome to degrade the enzyme. The proteasome is the main neutral protease of the cell, eliminating proteins that have outlived their usefulness, or that might become a danger to the cellular community because they are mutant or misfolded. Antizyme is released to participate in subsequent cycles of ODC destruction. ODC is a substrate for the proteasome even if antizyme is absent, with a half-life of an hour or two; antizyme reduces this half-life to minutes. Polyamines hold prostate cells in the G_1 phase of the cell cycle by inducing antizyme. This depletes cyclin-dependent kinases whose action is required for progression from G1 into the DNA synthetic phase of the cell cycle. As polyamines are not general cell growth inhibitors, prostate cells might prove to have (or lack) some special component that determines the unusual response of these cells. Polyamine pools is coordinated with pronounced changes in cell state. Disrupting polyamine homeostasis is difficult because biosynthesis, catabolism and import/export all affect the pool. Antizyme is widely dispersed in eukaryotes, and is found in vertebrates, insects, nematodes and fungi. Antizyme modulates polyamine transport into cells. Polyamines can be determined in urine and in RBC.

Polyamine functions:

1- Act as counter-ions for negative charges on RNA and DNA.
2- Enhance biosynthesis of DNA, RNA and proteins.
3- Mediate gating of K^+ rectifier channel.
4- Donate modifying groups in hypusination.
5- Act as odorants.
6- Function during transcriptional regulation.
7- Function as constituents of bis-glutothionylspermidine (trypanothione), which might have some of the functional roles of glutathione in trypanosomes.

Patients with prostate cancer have a significant reduction of polyamines during pelvis radiotherapy, followed by the maintenance of low levels in patients with favourable outcome. In high proliferating tissues (small intestine and spleen) a statistically significant decrease of these molecules within few hours after irradiation well demonstrates cell loss and block of proliferation. The levels and the time when modifications appear to depend on the proliferation level and turnover time of the tissues. Treatment with ionising radiation and chemotherapy produces an early and more severe reduction of polyamines than the chemotherapy alone. The usefulness of the determination of polyamines in the monitoring the effects of cytotoxic agents in humans is well documented. Polyamines significantly decrease during the early phase of the injury in tissues with high proliferative activity (small intestine, spleen) whereas do not show any modification in kidney. Polyamines and ethanol interact with the glutamate/NMDA receptor (NMDAR). Polyamines potentiate the function of NMDARs. Polyamine levels in the CNS are highest during development, and the NMDARs expressed in developing brain are highly sensitive to the potentiating effects of these polyamines. Polyamines potentiating NMDAR function in the adult CNS may contribute to hyperexcitability states such as seizures and to excitation-induced neurotoxicity. Polyamines are implicated in unrestricted cell growth and division, as in some cancers later in life. The NMDAR is a multisubunit receptor that responds to glutamate by fluxing Ca^{2+}. It has several physiological functions, which include contributing to increased survival of functional neurons during development (neurotrophic effects) and the synaptic plasticity associated with learning and memory (e.g., the long-term potentiation of synaptic transmission after high-frequency stimulation in hippocampus). Pathologically, hyperactivation of NMDARs is well known to lead to seizures and Ca^{2+}-mediated excitotoxicity in neurons. NMDARs contain several modulatory sites include: 1- the glutamate/NMDA agonist binding site, 2- a coagonist site for glycine, 3- proton sites that are inhibitory to channel function, and 4- Mg^{2+} binding site that inhibits Ca^{2+} flux until the membrane environment is depolarised. Polyamines also modulate the NMDAR, both positively and negatively, at several sites. Ethanol is acutely inhibitory to NMDAR function at concentrations within the pharmacological range (10–100 mM). Ethanol may modify the glycine site negatively. The presence of ethanol may alter NMDAR function by indirect effects on the polyamine sites, e.g., by inhibiting ornithine decarboxylase. Polyamines are implicated in both behavioural and neuropathological consequences of alcohol withdrawal. Polyamines may be released during ethanol withdrawal as part of a cascade of excitotoxic reactions. The polyamines are important player in adaptive responses to ethanol and also are involved in generating the hyperexcitation and neurotoxicity that accompany alcohol withdrawal. The diamine putrescine, the triamine spermidine, and the tetramine spermine are ubiquitous in plant tissues. Heat stress is a major factor limiting the productivity and adaptation of crops. The polyamines physiological role in nitrogen fixation might be associated to a large extent with the changes of metabolite and O_2 transport across mitochondrial and symbiotic membranes. The naturally occurring polyamines, spermidine, spermine and their precursor putrescine are aliphatic polycations present ubiquitously in all living cells. The intracellular concentration of the polyamines is highly regulated by the modulation of enzymes that are involved in both their synthesis and catabolism. The two highly regulated enzymes, ornithine decarboxylase (ODC) and S-adenosylmethionine decarboxylase (SAMDc), constitute the control points in the biosynthesis pathway, whereas the catabolic pathway is controlled predominantly by the activity of spermidine/spermine N^1-acetyltransferase (SSAT). The polyamine transport system depends on energy supply and is able of transporting polyamines against significant concentration gradient. The range of cellular polyamine concentrations is determined at the lower limit by their absolute requirement for cellular proliferation, and at the upper limit by their potential toxicity. Drugs interfering with polyamine biosynthesis have considerable potential for use as therapeutic agents, e.g., α-DFMO. The first research publication describing polyaminated compounds in cancer tissues occurred in 1958. Between 1960 and 1985, there were more than 5,000 research publications about polyaminated small molecules and compounds, including 11 multi-chapter research review books and six extensive review publications. The major natural, biologically active polyamines identified in mammalian cells/tissues include putrescine (PUT), spermidine (SPD), spermine (SPM), and possibly some of the polyamine metabolites such as N-acetyl spermidine (N-acetyl SPD), N-acetyl spermine (N-acetyl SPM), and hydrogen peroxide (H_2O_2). Catabolism may occur via polyamine oxidase (PAO), which converts N-acetyl-SPD and N-acetyl-SPM to putrescine and SPD, respectively, and yields H_2O_2 and acetoamidopropanol (ap). ODC is located in both the cytoplasm and the nuclei of mammalian cells. Therefore, the polyamine system is compartmentalised and may play different roles in the cytoplasm versus the nucleus. In the 1970s, difluoromethylornithine (DFMO) was established to be a very potent mechanism based inhibitor of ODC. Because the natural polyamines are small, completely water soluble, aliphatic carbon chains with multiple positive charges, they bind both specifically and non-specifically to numerous macromolecules such as DNAs, RNAs, membrane proteins, soluble proteins, enzymes, and many small, negatively charged polyphosphorylated molecules in the cytoplasm and nucleus. Polyamines bind to many proteins (non-specific). Elevated intracellular polyamine levels correlate with alterations in histone acetylation and deacetylation in normal and cancer cells. During cellular embryogenesis, differentiation, and oncogenesis, genes appear to be turned off and turned on by the acetylation-deacetylation of their associated histones. Many histone acetylases and deacetylases can be regulated by gene promotors and gene inhibitors, which, in turn, may be controlled by various polyamines.

Genomes controlled by myc oncoprotein:

Genes	Type	Function
1- Induced:		
	Cyclins A, D, E	Activate cyclin dependent kinases in cell cycle.
	Cyclin dependent kinase	Phosphorylates Rp protein in cell cycle.
	cdc 25A and B phosphatase.	Dephosphorylate and activate cyclin dep kinase2
	FADD and caspase 8	Stimulate the apoptosis pathway.
	Bax	Stimulates the apoptosis pathway.
	Dead box helicase	DNA synthesis.
	RNA polymerase I	RNA synthesis.
	Carbamoyl P synthetase	Synthesis of nucleic acids.
	eIF-4E and 2α	Translation factors
	Ornithine decarboxylase	The rate limiting enzyme in polyamine synthesis.
2- Repressed:		
	$p27^{Kip-1}$	Cyclin dependent kinase inhibitor in cell cycle.
	Growth Arrest and DNA	Upon cell damage, inhibits cell proliferation.
3- Damages (GADDs):		
	Growth Arrest Specific (GASs)	Inhibitor of cell proliferation.

These systems are modified in cells depleted of polyamines, in quiescent cells upon initiation of cell proliferation, and in cancer cells when compared to their non-malignant counterparts. Several protein kinases have been identified as enzymes, which may be regulated by polyamines, including polyamine dependent protein kinase, nuclear protein kinase NII, mammary gland polyamine responsive protein kinase, self phosphorylating polyamine stimulated protein kinase, and casein kinase II (CKII). Polyamines have linkages to the actions of EGF, TGFβ, TNFα and HGF. In addition, polyamine linkages have also been established for several oncogenes including: NFκB, c-myc, c-jun, and c-fos; and cancer suppressor genes including p53, Rb, $p21^{WAF1/CIP1/SDI1}$, and $p27^{Kip1}$. Because the myc oncoprotein and the protein products of the above cancer suppressor genes are directly involved with the cell cycle and apoptosis, polyamines may be linked to key mechanisms involving mammalian cell proliferation, differentiation, and apoptosis. There are molecular linkages between polyamines and apoptosis. Increases in the polyamine pathway in mammalian cells correlates with decreased mitochondrial membrane potential (MMP). This results in the release of mitochondrial cytochrome c, which stimulates caspase 8 cascade activation down to caspase 3. Activation of caspase 3, in turn, is considered to activate the key endonucleases and proteases that directly cause apoptosis. The polyamines are postulated to affect mitochondrial transmembrane potential via increased activity of PAO and/or increased intracellular levels of putrescine (PUT) and/or H_2O_2 (catabolic pathway products). CKII activity is increased several fold in most types cancers and induced by with various growth factors (e.g., EGF). CKII (and other protein kinases) activates myc oncoprotein (which is a nuclear transcription factor) by phosphorylation of several serine residues in the nuclear entry, dimerization, DNA binding, and transactivation domains. The myc oncoprotein may be increased by both transcriptional and translational means in cancer cells. Fast growing normal and cancer cells have elevated levels of ODC, polyamines, and CKII. Many cancer cells have amplified myc oncogene/oncoprotein. Phosphorylated myc oncoprotein could then be increased via transcription and translation control mechanisms, which, in turn, would induce the genes for cyclins A, D, and E, cyclin dependent kinase 2, cdc 25A and cdc 25B phosphatases, and ODC. Phosphorylated myc oncoprotein also repressed the cyclin dependent kinase inhibitor gene $p27^{Kip1}$. Increased cyclins and/or inhibition of cyclin dependent kinase inhibitors result in increased phosphorylation of Rb. Once a cell enters S phase, it will continue through M to produce 2 new daughter cells. The polyamine catabolic system enzyme, polyamine oxidase (PAO), may be increased in some cancer cells, which could allow for increased intracellular levels of H_2O_2. The oxidation system of H_2O_2 could initiate a decreased mitochondrial membrane potential (MMP) and lead to cytochrome c induced apoptosis. In addition, myc oncoprotein induction of the bax gene would increase the bax protein in cells with amplified myc. Bax protein can enter the mitochondrial membrane and cause cytochrome c leakages from the mitochondria, which in turn, can stimulate the cytochrome c to APAF to caspase 9 to caspase 3 pathway of apoptosis. The effect of polyamine inhibitor (DFMO) on cancer cell: 1- amplifies myc induced rapid cell proliferation without causing apoptosis, 2- amplifies myc induced rapid cell proliferation and apoptosis together, 3- apoptosis can occur without amplified myc, and 4- polyamines are involved in both cell proliferation and apoptosis in cancer cells. High levels of intracellular polyamines correlate with high grades/stages of many human cancers. There is a direct correlation of increased metastatic potential of the cancers and increased negative prognosis for the patients. Reduction of intracellular levels of polyamines causes a decrease in oncogenesis by most carcinogenesis. Prevention of human skin, stomach, colorectal, lung, prostate, and breast cancers is possible using DFMO. However, it is extremely difficult to deplete 100% of the polyamines from cancer tissue when that cancer tissue can obtain polyamines from the surrounding normal tissues. But, decreasing intracellular levels of polyamines in normal tissues interferes with the oncogenesis mechanisms of toxic chemical carcinogens.

WHEY PROTEIN

Athletes have been long enjoying the benefits of speciality nutrients like creatine, glutamine and branched chain amino acids (BCAA) before they became widely used dietary supplements. Whey proteins consist of a diverse group of proteins with a wide range of biological applications. Industrially, whey is one of the main by-products of the dairy industry and is produced during cheese and casein manufacturing. Cow's milk contains about 6.25% protein, of which 20% is whey and 80% is caseinate. Initially considered a waste product, whey is now considered a functional food with remarkable applications ranging from protein supplementation to immune support. There is need of protein supplementation for nitrogen balance during exercise, particularly resistance training, approximately 2.0 g of protein per kilograms per day (2.0g/kg/d). Increased protein helps counteract the increased oxidation of amino acids caused the metabolic needs of exercise. If amino acid oxidation is unchecked by protein supplementation, this can lead to catabolism of muscle tissue. Amounts above approximately 2.0 grams of protein per kilograms per day (2.0g/kg/d) do not have a positive impact on nitrogen balance. Whey is possibly the best biologically active protein. Biological value (BV) refers to the amount of body protein that can be replaced by food protein in adult humans. Protein efficiency ratio is a standard used to assess growth-potential of protein, calculated by measuring the weight gain of an adult animal caused by 1 g of a food protein. The highest BV of Whey is, in part, because it contains an extremely high concentration of essential amino acids. The daily requirements of essential amino acids would be met by the consumption of 28.4 g of caseinate or 17.4 g of egg protein. But, only 14.5 g of whey protein would fulfil the daily requirement of essential amino acids. Whey protein supports muscle growth (protein synthesis) because it is the richest natural source of BCAAs (isoleucine, leucine and valine), important components of muscle tissue. BCAAs are depleted from muscle tissue following strenuous exercise, which can cause catabolic loss of protein. BCAAs suppress proteolysis, which is the breakdown of muscle tissue. BCAAs are important as a metabolic fuel for muscle and other tissues by their involvement in the alanine cycle (a process of energy production from BCAAs). BCAAs have been known to enhance performance in moderate as well as intense exercise by improving endurance. No other source of protein provides as much BCAAs as Whey. Whey is also a good source of several vitamins and minerals, such as calcium and B-vitamins. Overall, whey protein is very effective in protecting precious lean body mass (LBM) but also supports the growth of new muscle tissue. Whey protein offers antioxidant activity, especially scavenging lipid oxidation catalysts. Most of the antioxidants found in Whey are present in low molecular weight fractions of the protein (approximately 500-3,000) and are water soluble. Although hydrolysing does increase the quantity of low molecular weight compounds, the hydrolysis process will deactivate the antioxidants. Unfortunately, even normal processing may deactivate the enzymes itself. The immunomodulatory properties of whey: As a potent source of lactalbumin, Whey offers the highest natural source of immunoglobulins, which are known to be important for proper immune function. Mother's milk is rich in immunoglobulins. Bovine serum albumin is very rich in glutamylcysteine groups, compounds that are very rare in both animal and plant proteins. Each 100 g of whey protein contains approximately 2.8 g of L-cysteine. Both glutamylcysteine and other cysteine compounds have been known to stimulate the production of glutathione, which is necessary for lymphocyte proliferation. Dietary cysteine from whey is more efficient in raising glutathione than free cysteine. Hydrolysed whey protein (often referred to as Whey peptides) is thought to be the second, more sophisticated generation of Whey proteins. Infants can develop allergies to proteins found in whey, the strongest antigen being the β-lactoglobulin protein. This allergenic effect can be reduced by treating the whey with the enzyme trypsin. The hydrolysis process itself will cause a loss of glutamine. Because the hydrolysis of proteins requires the application of certain enzymes and a reduction of the pH, some of the glutamine will be converted to glutamic acid. Approximately, one third of the glutamine will be lost during the hydrolysis process. The reduction of pH, required for the hydrolysis of whey reduces the activity of certain antioxidant. These low molecular weight antioxidants are deactivated, yet they account for 90% of the antioxidant benefit attributed to whey. Hydrolysis of Whey protein results in the formation of peptide fractions, which are easier to absorb and, for some individuals, reduce allergenic reaction to the proteins. However, if an individual is allergic to some of the proteins found in whey, a partial hydrolysis will still cause problems. The immune system of these individuals will be compromised by the antigenic load from whey. The growth-hormone (GH) secreting benefits caused by oral glutamine occur when glutamine is administered singularly. If other amino acids are present, they may competitively inhibit the glutamine from rapid absorption and significantly reduce GH-secretion. Glutamine utilises some of the same absorption pathway as other amino acids. It is likely that Glutamine absorption will be inhibited by the presence of other amino acids. It is impossible to get the same growth hormone secretion by drinking 2 cups of skim milk, which contains about 2 g of glutamine. Although, the addition of glutamine to whey may beautify a label, it does nothing to enhance the functional attributes of the whey protein. Whey protein naturally contains a low concentration of phenylalanine. An essential amino acid, phenylalanine is not manufactured in the body and is required in the diet. As early as 1972, the potential application of Whey protein for the treatment of phenylketonuria (PKU) was recognised. Individuals with PKU lack or have reduced amounts of the enzyme phenylalanine hydroxylase. Phenylketonurics accumulate toxic levels of phenylalanine and often require dietary control of phenylalanine. Whey is rich in phenylalanine. Muscle tissue, however, contains very low quantities of phenylalanine (only tryptophan and histidine are found in lower amounts) and therefore excessive enrichment of phenylalanine to Whey is not warranted. Whey protein significantly lowers plasma and liver cholesterol and also plasma triacylglycerols. The hypocholesterolaemic effect of whey protein is associated with a decrease in very-low-density-lipoprotein cholesterol. At the high dietary protein concentration, whey protein reduces the faecal excretion of bile acids. The cholesterol-lowering effect of whey protein may be caused by inhibition of hepatic cholesterol synthesis. Milk growth factors enriched from cheese whey ameliorate intestinal damage by methotrexate, which suggests clinical applications for the treatment of intestinal mucositis. Lactose is a simple carbohydrate found in milk that can cause gastrointestinal irritation to individuals that lack adequate lactase. A good source of whey protein contains a very high concentration of protein, with a minimum of lactose.

Biological value of common proteins:

Type of protein	Biological value
Whey protein	104
Eggs (whole)	100
Egg albumin	88
Beef	80
Caseinate	77
Soy	74
Rice	59
Beans	49

POTASSIUM

Potassium reduces urinary calcium excretion (potentially resulting in a more positive calcium balance), and that potassium deprivation stimulates bone resorption (potentially causing a more negative calcium balance). The protein to potassium ratio in the diet predicts net acid excretion via the urine and that, in turn, net acid excretion via this route predicts calcium excretion. Diet rich in fruit and vegetables induces alkaline environment. Diets rich in fruit and vegetables are associated with a significant fall in blood pressure compared with baseline measurements; they also result in a fall in urinary calcium. Potassium depletion results in hyperplasia of renal tubular and interstitial cells, and potassium repletion induces rapid regression of hyperplasia. Apoptosis participates importantly in this reduction of cell number, tubular and interstitial apoptosis may ocuur during K depletion as well. Plasma and not serum K is more accurate, since serum can give inaccurately high readings. But, analysis of serum within half an hour would prevent the disparity. This inaccuracy results when blood stands at low temperatures and K leaks from the cells and also there can be cell losses if the blood is handled roughly. Serum K will be higher than plasma by 0.2 mEq/l in normal subjects. Plasma K can be anonymously high because of K losses from platelets during rheumatoid arthritis. A difference of 0.4 mEq/l can occur when platelets release K. As levels increase, the first ECG change is tall peaked T waves. The QT interval is normal or diminished. As K levels continue to rise, the PR interval becomes prolonged, then the P wave amplitude decreases. The QRS complex widens into a sine wave pattern, with subsequent cardiac arrest. Some symptoms are irregular/tachycardia, paralysis of limbs, hypotension, convulsions, coma, cardiac arrest, black or bloody stool, diarrhea, confusion, breathing difficulty, and vomiting. The kindeys can excrete 26,000 mg of K in a day if necessary. The kidneys heal up slowly beyond the age of 50. Diabetes causes over 30% of renal failure for unknown reasons, and K elevates after eating 100 g of glucose, unlike normal people who show a slight decline. About 30% are related to hypertension, and 15% of cases result from atherosclerotic renovascular disease.The latter together with glomerular derangement may often be from a copper deficiency combined with a high salt intake. Analgesia are implicated as a risk factor in acquiring chronic fatigue syndrome (CFS), but not methylsulfonylmethane (MSM). Pain also can cause increased K losses. Systemic lupus erythematosis (SLE) causes extensive damage to renal tubules (66%), resulting in chronic high plasma K, which does not respond to aldosterone. Other causes of hyperkalaemia are crush injury, severe burns, haemolysis, hyperkalaemic periodic paralysis (during paralysis episodes), acute tumour destruction after chemotherapy, transfusion of haemolyzed (old) blood, Addison's disease (rare), hypoaldosteronism (very rare), severe dehydration, and respiratory acidosis. The latter is a failure of the lungs to remove CO_2 leaving behind carbonic acid. Since acid (H^+) interferes with K excretion, i.e., K can build up. Respiratory acidosis can result from failure of the lung to remove CO_2 because of bronchitis, asthma or airway obstruction. Excessive dreaming could be an indication of high CO_2 in the air. Angiotensin-converting enzyme (ACE) inhibitors, NSAIDs, and potassium sparing diuretics prevent K excretion. The combined use of NSAIDs and β-adrenergic blockers may increase the risk of life-threatening hyperkalaemia through their suppressive effect on the renin-aldosterone system. Angiotensin receptor blockers and antiinfective agents as well may be involved. Digitalis glycosides, digoxin, and oleander inhibits the Na/K cell wall pumps. Arginine (nuts and chocolate), hypertonic solutions, and fluoride salts cause cell leakage. Fluoride damages the kidneys and inhibit the thyroid, and act synergistically with aluminum to produce Alzheimer's disease. Fluoride strengthens the bones but reduces resorption, which may imply an occasional increased problem with muscle spasms from hyperkalemia if calcium should be low. Heparin, α-adrenoreceptor stimulants, succinylcholine, and β-adrenergic nerve blocking agents also produce hyperkalaemia. The kidneys are a critical target for cadmium (cd), the half-life of cadmium in the body is 17 years. Cd poisoning rarely causes severe hyperkalaemia directly unless the kidneys have been damaged or muscles destroyed as well. The kidneys can be damaged by a K deficiency, especially if hypertension is also involved. A modest K deficiency can cause irreversible renal damage in the collecting tubules, which is where the K is excreted. When the cell K is low the body attempts to make the plasma K low also. Excess Ca can also damage the renal distal tubules. In chronic renal failure, aldosterone secretion is often high and more K is excreted per unit of decreased glomerular filtration. Persistent hypertension can damage the glomerulus, which is where the fluids are initially filtered into the kidney tubules. Progression of Immunoglobulin A (IgA) nephropathy (IgAN) can be muted by increasing ingestion of ω-3 oil in food. There are medications that can damage renal tubules. Aristolochic acid found in certain plants and botanicals is toxic to the kidneys and is a potential carcinogen. This chemical can cause serious renal damage. The use of aristolochic acid-containing product may be linked to increased risk of renal cancer. Patients with renal failure who are using diuretics have a higher death rate but also a higher rate of subsequent chronic renal failure requiring dialysis. The death rate is especially high in patients who would fail to quickly respond to the diuretics with an increased flow of urine. Heavy metals like lead, arsenic, and mercury, carbon tetrachloride (used in the dry cleaning industry), pesticides, fungicides, and hydrocarbons can cause damage to the kidneys. Lead may prevent K retention and is the main cause of gout. Timalol combined with hydrochlorothiazide and amiloride can cause gout, as well. The lead poisoning makes the aldosterone system insensitive to K concentration and increases the K content of the plasma. The blood lead content is no indicator of toxicity and the status must be obtained with an EDTA mobilization test. Ethylenediaminetetraacetic acid chelator of lead has sucessfully increased uric acid excretion. Gout can be triggered by the same agents that cause K losses such as fasting, surgery, and K-losing diuretics. A potassium deficiency can increase urate levels in the blood. Urate renal stones form during gout in a fifth of the cases. Making the urine less acid with K-citrate or Na-bicarbonate is the treatment for urate stones. Toluene in automobile enamel reducers is also a poison, which can trigger gout. Swelling of the prostate tissue (enlargement) is most often caused by a deficiency of linoleic or linolenic acids (ω-6 and ω-3 oils) induced by hydrogenating vegetable oils. Zinc deficiency induced by excessive intake of copper, also a cause of prostatic enlargement. If calcium intake is normal but phosphorus is too low, renal calcium citrate stones can form. If phosphorus is much too high (drinking soft drinks), renal calcium phosphate stones should form. Incidence of renal stone: calcium oxalate (65-75%), calcium phosphate (5%), uric acid (7%), cystine (2%), and ammonium phosphate (2%) stones. Uric acid stones are less probable when K is adequate in the diet. Medications that can cause hyperkalaemia include cyclosporine, lithium, heparin, enalapril, triampterene, amiloride, spironolactone, and trimethoprim. Indomethecin by inhibiting PG stimulation of renin and captopril by prevention of angiotensin II can also produce hyperkalaemia. Intravenous aldosterone will help the kidneys get rid of excess potassium. Intravenous calcium, intravenous glucose and insulin works for about an hour. If the hyperkalaemia is caused by acidosis, sodium bicarbonate will help to correct it. Drinking salt (sodium chloride) water as a vehicle for the bicarbonate is helpful. Increased urine flow increases K loss, so part of the effect of the salt solution may be from this phenomenon. While Na is a good antidote for high serum K, kidneys that have been conditioned by a prior low Na intake can excrete an additionally larger amount of K from the collecting ducts than kidneys that have had a prior large intake. Drinking extra water should be helpful, and essential if dehydrated. However, drinking huge amounts of water (gallon) if severely dehydrated without sodium chloride is very dangerous and can kill from hyponatraemia. Potassium tends to be correlated with other essential nutrients. Celery is high in potassium per calorie. Patients on dialysis have low copper and zinc in their serum as well as too much retinol. No vitamin capsules have adequate amounts of the macronutrients such as calcium, magnesium, sodium, chloride, phosphate, and amino acids, so imbalances are possible. If vitamin B_1 (thiamin) is deficient when K is adequate, the wet heart disease of beriberi is possible. Sulfites degrade vitamin B_1 in the intestines. Sulfites intefere with K excretion. Wine, beer, dried fruits should be avoided during hyperkalaemia. Ethanol increases K excretion. When the protein in meat, eggs, and milk is burned for energy the nitrogen in it degrades to ammonium, uric acid, and urea. Damaged kidneys can have trouble excreting these wastes also. 0.6 g of protein/KgBW slows down progression of renal disease. The kidneys activate an enzyme that degrades glutamic acid during K deficiency. This could be an adaptation to conserve potassium by interfering with excretion at the excretion site. Meat, milk and eggs foods provide amino acids lysine and methionine, which plants are low in. They also provide vitamin B_{12}. Glucosamine furnishes glutamine for ammonium synthesis in the kidneys, as ammonia production goes up and potassium excretion goes down when ingesting glutamine. Acids that is absorbed but not metabolized interfere with K excretion since hydrogen ion (H^+) competes with K at the excretion site. Possible examples of such mechanism include acetic acid, cherries and cranberries. But, citric acid is part of the Kreb's cycle. Perspiration removes excess K. Enemas also cause large K losses. Leaving water in contact with the mucous membranes for a long time increases the losses. Glycyrrhetinic acid interferes with degradation of aldosterone, and increases cortisol. Sodium bicarbonate is often used as baking powder causes large potassium loss. Hyperventilating gets rid of CO_2 (carbonic acid) to an abnormal extent. Fats and oils contain little water and therefore little K. Molasses is high in K. Oils are extracted by solvents (chlorinated molecules), and always contain traces of the solvents. Animal fats have very little of the oil soluble linoleic and linolenic acid in them in addition to being low in K. Hydrogenated oils are almost in the same class as animal fat since the linoleic and linolenic acid (the omega oils) are destroyed. Preserving food has two major drawbacks in regard to potassium. First, the sodium displaces potassium in the food cells, and the potassium is discarded with the fluid in those cases where the food is immersed or loses liquid. A glance at the food tables where salt treated frozen or canned goods are shown will confirm this. This can cause fairly considerable loss of K. The second problem is that an increase in sodium above its optimum intake causes a greater excretion of K. The adverse affect on health from sodium (or chloride) may not be essentially related to reduced cell K, but it is from the hormone imbalance that one or both create in hormones. The bones store up to 45% of the body's Na. Freezing of vegetables is done by boiling the vegetables to kill the enzymes (blanching), and then moving the vegetables around the plant in water filled troughs or flumes to the subsequent processing and packaging steps. By the time they arrive at a store, they have lost anywhere from 20-50% of their K and some much more if the liquid is drained off, and other water-soluble vitamins are similarly affected. The whole body K is significantly lower in older arthritics. Potassium presents in the saliva of patients. Potassium in the body cells is not often determined for patients because of the enormous cost of the equipment (scintillation counter, using K isotopes, K-40). Other methods for determining cell potassium involve biopsies, balance studies, and isotope dilution studies. When cell K is low the serum K is usually low also. Serum K measurement can give incorrectly high results. Plasma K measurement can also be anonymously high because of K losses from platelets during rheumatoid arthritis (RA). 80% of people with rheumatic heart disease have low blood plasma content of K. The hormone most involved in K excretion, aldosterone, is significantly increased by fear and anxiety. 1.5-3 g of K supplements per day have a complete healing effect of arteries. Potassium citrate can prevent arterial lesions, but KCl has a somewhat lesser protective affect. Serum K does not improve with 1 g of K/day unless Mg is also supplemented. Gouty arthritis is characterised by deposition of sodium urate crystals in cartilage, especially in the feet. Gout is linked to lead poisoning. The lead poisoning makes the aldosterone system insensitive to K concentration and increases the K content of the blood plasma. Ethylenediaminetetraacetic acid (EDTA) chelator of lead increases uric acid excretion. Timalol combined with hydrochlorothiazide and amiloride may induce gout. Potassium losing diuretics may trigger an attack of gout. Adding a K supplement removes the gout. Gout can be triggered by the same agents that cause K losses such as fasting, surgery, and K losing diuretics. A K deficiency can increase urate levels in the blood. Low cell K can inhibit the insulin response independently of serum K. Glucose intolerance develops exclusively associated with lower insulin secretion rather than cellular response to insulin. It is possible that K deficiency can cause hypothyroid secretion and that a severe deficiency can cause hyperthyroid secretion. Chloride wasting starts when 20 g of K out of 150 are gone. Most of the Cl reabsorption occurs in the ascending limb of the Henle tubule via the Na-K-2Cl cotransporter and most of the Cl reabsorption in the distal tubule is by thiazide sensitive Na-Cl cotransporter. These transporters are inhibited during a K deficiency. Several enzyme systems in the kidneys are affected by a K deficiency, e.g., increased activity of enzyme that converts glutamine to ammonia. The ammonium ion has a positive charge and is about the same size as K. The ammonium is may be synthesized in the mitochondria of the proximal tubule cells, excreted in part by the sodium/hydrogen ion exchanger (NHE-3), then reabsorbed by the Na-K-Cl cotransporter, and then brought to the collecting duct and excreted. Active excretion of K ceases in the renal tubules after 2 days on a low K diet. Urinary excretion of Ca, Mg and phosphate is higher during a K deficiency. Loss of intracellular K is replaced by Na, which may explain the increased oedema when K is repleted. Adequate K is necessary for protein synthesis. Potassium may be essential to defense against pathologicl bacteria. Muscular strength is directly related to K intake. A K deficiency seems to be most destructive to the tissues that derive from the middle layer of the embryo, include all the connecting tissues, the heart, the blood vessels, the kidneys and WBCs. Potassium deficiency alone causes a higher mortality during stroke. RA may be essentially a chronic K deficiency condition. It may be that some genetic difference like sexual hormones, or differences in secretion of other hormones such as the glucocorticosteroid response modifying factors (GRMFs), or some other imbalance with other nutrients such as copper affect who and when arthritis strikes. The symptoms of a K deficiency take a long time to heal, and a deficiency should be avoided with almost the same urgency as a water deficiency (dehydration). Also, a deficiency of, e.g., 40-50 g of K would take a fairly long time to be completely corrected by food. Potassium as KCl act very fast, but raises blood pressure, so such a supplement probably not be used for hypertensive and Mg and vitamin B_1 must be added. But, vitamin B_1 supplements may be dangerous for arthritics since heart disease is more common among rheumatoid arthritics and osteoarthritis. Potassium is odorless, colorless, and, in the usual concentrations, tasteless. There is no way to detect a deficiency and cell content can not even easily be assessed in the body by modern analytical procedures. Whole body cell content is virtually invisible. Potassium ramifies through every cell and process in the body, has no storage, and has a dangerous dependence on its precise control for nerve impulse transmission. The terrible pains associated with RA, along with the actual physical disability, weak joints, loss of energy, and other systemic symptoms that accompany them, cause an enormous loss of productivity. RA has no obvious clear association with any culture. Most of the people who have pains in the joints have them because of arthritis. The pains usually strike first in the outer joints like wrists, carpels, fingers or joints with a history of injury. Load bearing joints are also vulnerable. The pain is most likely in the early morning. It is often accompanied by stiffness. Arthritis is a disease largely associated with humans, present throughout the body and can affect kidneys, pericardium of the heart, and connecting tissue. RA has few externally observable symptoms, especially in early stages. There are no known consistent biochemical changes in RA except a lower cellular K content than normal, and a somewhat higher plasma copper content along with a protein, which binds the copper in the serum (caeruloplasmin). The Na/K-ATPase activity is lower in erythrocyte membrane and lower than in normal, osteoarthritis, or gout. The steroid hormone dehydroepiandrosterone sulfate (DHEA) is statistically lower in arthritics as is cortisol and pregnanediol, even though ACTH is higher, as is aldosterone. Aldosterone stimulates excretion of K and has a positive feedback. There is lower glycosylation of immune peptides during arthritis. Epithelial sodium channels, α and β types are higher than normal in rheumatoid arthritis but not in osteoarthritis. Arthritis sometimes has fatigue associated with it. The Dead Sea water has a reputation for healing arthritis lasting up to 3 months. The ocean contains potassium in about the same concentration as blood fluid. An anion is a negatively charged substance that neutralizes the positive charge of an ion like K. Honey is extremely low in K. A vinegar like ferment may be helpful for arthritis. The acid hydrogen ion (acetate) interferes with K at the excretion site. Sulfite destroys vitamin B_1. Vitamin D is necessary for Mg reabsorption in the kidneys. Magnesium is necessary for powering some of the electrolyte pumps, and may have an indirect affect on K. Allergens may damage the kidneys' ability to retain K. A decline in cortisol during a K deficiency may stimulate the allergic response. A diet free of chemicals, milk, meat, sugar, and low in fat, with increased unprocessed vegetables, would reduce the allergens, which may be helpful in treating RA. Glucose-insulin-potassium (GIK) therapy may suppress TNF, which is thought to produce some of the symptoms of arthritis. Too much K can have some undesirable side effects, such as muscle spasms or cramps. The spasms are from too low a calcium coupled with too high a K. Symptoms of an acute toxicity are listlessness, mental confusion, numbness, tingling of limbs, a sense of weakness, a cold gray pallor, hypotension, and bradycardia. During surgery the release of cortisone and other steroids into the blood stream causes a release of K into the plasma too rapid for the kidneys to clear. Simple measurement to deal with hypokalaemia include: 1- rest, 2- intake of sugar, preferably glucose, e.g., honey for rapid absorption (glucose increases insulin, which in turn moves K into the cells to associate with glycogen), 3- intake of dilute salt water (hypotonic) containing bicarbonate of soda (to interfere with hydrogen ion excretion at the potassium site), and 4- hyperventilating. Increased urine flow increases K loss. Aldosterone is reduced during dehydration. Aldosterone is excellent for increasing survival in shock. Licorice inhibit the enzyme that degrades aldosterone and cortisol. Licorice has a chemical, glycyrrhetinic acid, in it that interferes with degradation of aldosterone. There may be renal tubule damage in 30% of rheumatoid arthritis. Systemic lupus erythmatosis (60%) has caused such extensive damage to renal tubules. Potassium supplements are most dangerous in impaired renal function. Both hypokalaemia and hyperkalaemia can induce periodic paralysis. Heavy carbohydrates meal can trigger it in case of hypokalaemia. Individuals drink much wine, lemon juice, or vinegar fermented with sulfur dioxide with their meals are at risk for beri-beri and the diet can vary widely as to vitamin B_1. Dying cells may not be able to reabsorb K during the acute phases and thus cause death from this and the adjacent hyperkalaemia, e.g., disionic cardiopathy. Administering the K in conjunction with glucose sugar and insulin (polarizing solution or GIK), may help K to enters the cell to be tied up with glycogen. The insulin may speed movement across the cell wall because of its effect on a glucose-potassium pump. Hypokalaemia reduces insulin secretion. The unpredictability for heart disease in RA, may be partly because some of the disease is caused or accentuated by copper or magnesium or vitamin C deficiencies. Adding magnesium to GIK make the therapy MAGIK. The Na/K pump depends on inositol. There is 100% mortality in heart attack during K deficiency in the presence of excess phosphate (soft drinks). Anxiety results in K deficiency, probably because of low aldosterone. Copper deficiency is a much more plausible explanation for high cholesterol and maybe a vitamin C deficiency than increased dietary intake of cholesterol. Aldosterone is normally constantly destroyed by liver enzymes as fast as it is produced. Aldosterone accumulates when the clearance of blood through the liver is reduced in cardiac disease, and partly because of lack of exercise. It is difficult to restore the body's potassium with food alone. Excess chloride can induce hypertension. Death of heart cells from a K deficiency is prevented by a thiamin deficiency, so supplementing vitamin B_1 prematurely could be dangerous. Potassium is low in muscular dystrophy. Sulfur dioxide rapidly destroys vitamin B_1 at the pH of the intestines. Celery has twenty times the K content of wheat per calorie, and eighty times of refined wheat. Bamboo shoots and green coconut water are also very high. The serum only holds about 2% of the total body's potassium, and it is possible that there is an excess K in the cell and a low level in the blood. During anabolic/anaerobic condition, there is excess intra-cellular K with a relatively decreased serum K. Choosing agents like calcium and magnesium that drive the potassium out of the cell may be preferred rather than giving more K. Serum K values range from 3.5-5 mmol/l while the concentration inside the red blood cell is at least 15-20 times this amount. Causes of hyperkalaemia: 1- Renal insufficiency. 2- Adrenal cortex underactivity, with aldosterone insufficient production, the body loses Na in the urine in exchange for K. The adrenal corticosteroids also favour the anabolic over catabolic balance. There is less K in the cell and more in the serum. Symptoms include fatigue (especially feeling tired in the morning), crave salt or salty foods, low systolic blood pressure, food and/or environmental sensitivities, weakness or tiredness after colds, stress or exercise, sweat easily with exertion, dyspnea after very minimal exertion even though the person is in shape, recurrent colds and upper respiratory tract infections easily, etc. 3- Metabolic acidosis, the potassium is pumped out of the cell in exchange for the excess hydrogen ion (acid). 4- Bradycardia. 5- Massive tissue destruction, K is lost and the blood and then excreted. Although K is increased in the serum, K may be needed because the largest store of K is inside the cell. 6- Diabetes without adequate insulin. 7- Catabolic/dysaerobic state. Causes of hypokalaemia: 1- Diarrhoea and/or vomiting. 2- Adrenal Cortex overactivity, with excess aldosterone production the body reclaims Na from the urine in exchange for excreting K. A relatively high Na with a relative decreased serum K with adrenal cortex overactivity. 3- Several types of anaemia. 4- Metabolic alkalosis, the K is pumped into the cell in exchange for the hydrogen ion (acid). 5- Diuretic use. 6- Familial periodic paralysis. 7- Diets high in refined foods due to lack of K in the diet. 8- Hypertension. 9- Insulin use, K moves into cells when there is use of glucose or build-up of protein. 10- Anabolic/anaerobic states. Potassium is not found as a pure metal in nature but is always combined with other substances, the most common form being potassium chloride. The alkaline, metallic element potassium (K) belongs to group IA of the periodic table and has an atomic weight of 39.1. It is a soft, silvery metal, which is rapidly oxidised. K is widely distributed in silicate rocks and also occurs in salt beds and seawater. Potassium is not found as a pure metal in nature, being easily ionised to the cation (K^+) through the loss of the outer electron, and always combined with other substances, e.g., chloride salt (KCl). Good food sources of K include milk, fruits (especially oranges, prunes, apples, pears, peaches, and bananas), vegetables (especially broccoli, carrots, tomatoes and potatoes), fish, shellfish, beef, liver, chicken and turkey. Levels in food vary from up to 1740 mg/100 g of edible portion in soya beans, 1000 mg in parsley, 400 mg in banana and 350 mg/100 g of edible portion in beef. The potassium content of water varies widely. The potassium values of mineral water vary between 1.4 mg/l and 611 mg/l. Natural spa waters have K levels varying from 1.88 mg/l to 12830 mg/l. The total daily dose of KCl depends on the amount of K lost, but the maximum is generally 4.8 g/day. Potassium chloride is used as fertiliser and plant nutrients. Salt substitutes (Lo Salt) contains 9 mmol K/g. Most sports drinks contain K to replace losses in sweat induced by exercise. There is little evidence for its inclusion as the losses can normally be replaced by potassium in foods following exercise. A Reference nutrient intake (RNI) of 3500 mg/day (90 mmol/day) of potassium has been recommended for males and females over 15 years old. Higher intakes of dietary K are associated with lower blood pressures and fewer strokes, in part mitigating the effect of Na. In the elderly, K depletion is common, usually because of dietary deficiencies and frequent drug therapy. Measurements of K tissue levels can be done by: 1- electron microprobe X ray analysis, 2- dilution of radioactive isotopes of potassium (^{40}K), and 3- measuring total body K. Potassium is the third most abundant element in the body (after calcium and phosphorus) and constitutes about 5% of the mineral content of the body. Total body K is about 45-55 mmol/kg of bodyweight. A 70 kg adult man contains about 135 g of K. Of the 135 g body burden, approximately 84 g is in the muscle and 15 g is in the skeleton. Potassium is also stored in fat, blood, CNS, intestines, liver, lung and skin. The extracellular K concentration is a function of two variables: 1- total body potassium content; and 2- the relative distribution of potassium between the extracellular and intracellular fluid compartments. 85-90% of ingested K is absorbed. The vast majority of intestinal K absorption occurs in the small intestine, with the contribution of the colon being trivial. Absorption in the human small intestine is thought to take place mainly by passive diffusion in response to electrochemical gradients, reaching equilibrium between plasma and lumen in the jejunum and ileum. Only the colon (and the distal nephron) can modify K^+ transport in response to variations in K status. The proximal colon secretes K^+ via a transcellular pathway, with active K^+ uptake mediated by Na^+-K^+-ATPase and passive exit via a conductance pathway. The distal colon can also both secrete and absorb K^+. KCl is much more readily absorbed by the GIT than K contained in food. The bioavailability of K is high, with 90-95% of ingested K being utilised in normal metabolic pathways. This is for 2 reasons: 1- potassium salts are completely soluble because they are wholly ionic; and 2- few dietary components alter the digestive utilisation of K. The duodenum and jejunum can absorb K even more rapidly than water. As water is absorbed, the concentration of K decreases. Potassium is transported mainly in ionic form in the extracellular liquid. Potassium is secreted by the renal tubules. More K is found in the urine than is filtered at the glomerulus. Secretion is accomplished by an ion exchange mechanism, in which K of the tubular cells is exchanged for the Na of the glomerular filtrate. Secretion of up to 130 µEq K/minute occurs after large oral doses (5 g) of KCl. The major excretory route of K is via the kidneys and urine. The amount of K entering the glomerular filtrate each day is about 800 mEq, whereas daily intake is only about 100 mEq. Therefore, to maintain normal body K balance only 1/8 of the total daily tubular load of K can be excreted. The digestive juices contain relatively large amounts of K but most are reabsorbed and therefore loss in the faeces is small (about 9 mEq/day). Approximately 98% of the total body K is located intracellularly, where the concentration can be 30 times that of the extracellular concentration. The concentration in the ECF is a critical determinant of neuromuscular excitability. The extracellular and intracellular balance of K is maintained by the enzyme K^+-exchanging ATPase, found in the plasma membrane of virtually all cells. Potassium is a cofactor for a number of enzymes including glycerol dehydrogenase, mitochondrial pyruvate carboxylase, pyruvate kinase, vitamin B_{12}-dependent diol dehydratase, L-threonine dehydratase, adenosine triphosphatase and aminoacyl transferase. Potassium is required for the secretion of insulin by the pancreas. Potassium is also involved in phosphorylation of creatine, in carbohydrate metabolism and protein synthesis. Potassium deficiency or hypokalaemia is defined as plasma K concentration below 137 mg/l^{-1} (3.5 mmol/l^{-1}). Levels of 98 mg/l^{-1} (2.5 mmol/l^{-1}) or less may be associated with severe K depletion (total body K loss). Hypokalaemia can induce growth retardation, with pronounced decrease in circulating somatomedin C and concomitant inhibition of protein synthesis. Hypokalaemia can also predispose to hypertension. Hypokalaemia results mainly from crash diets, diarrhoea, diabetic acidosis, vomiting, intense and prolonged sweating, body burns and heavy urine losses induced by diuretic drugs. Potassium has protective effect against vascular damage and stroke. Oral K supplements lower blood pressure in humans, 120 mEq/day of KCl is well tolerated and decreases blood pressure in mild hypertensives. During ageing, Na enters the cells and K leaves them and the rates of cancer increase; patients with hyperkalaemic diseases (e.g., Parkinson's, Addison's) have reduced rates of cancer. Potassium is an anticarcinogenic agent. Acute human toxicity from K can also result from accidental ingestion/overdose. Individuals witty compromised renal or cardiac function may die of accidental oral overdose of KCl. A single dose of a potassium salt containing 80-100 mg K/kgBW can be toxic. Acute human toxicity resulting from ingestion of K is not common mainly severe gastritis, and the necrotic gastric mucosal lining may slough off resulting in tight antrial stenosis, which may necessitate partial gastric resection. Vomiting within 30 minutes, and sweating and breathless, which few hours later may lead left ventricular failure with cyanosis and widespread lung crepitations. Nausea may last for 3 days. Discomfort in the epigastrium experienced by the subjects after ingestion of large dose of potassium salts is the result of the effect of K^+ on sensory nerve endings causing severe burning pain. Hyperkalaemia (sub-chronic toxicity) can develop as a result of combination of K supplements and the K-sparing diuretics, this account for the tenting of T-waves seen on ECG and may contribute to the tachycardia. Serum K rises during exercise and may reach 6.8 mmol/l compared to 5 mmol/l, the upper limit of the normal range. Elevated K can cause an imbalance of the transmembrane K gradient and a lowering of resting potential across muscle cell membranes with a consequent impairment of membrane excitability. Increased extracellular K also causes shortening on the cardiac action potential and an increased velocity of repolarization, with a reduction of the Q-T interval and tenting of the T wave of ECG. Further lowering of the resting membrane potential causes a decreased upstroke velocity of the action potential and results from an even more elevated K^+ level and a slowing on intraventricular conduction, and widening of the QRS complex on the ECG. The atrial muscle is particularly sensitive to an alteration in extracellular K and to its depolarisation effects but the SA and AV nodes are relatively resistant at higher concentrations of K. The P-R interval widens and an irregular cardiac rhythm supervenes with eventual asystole. Chronic toxicity has also been seen with slow-release potassium tablets, such as nausea and abdominal cramp after meals. Focal ulceration with subsequent oedema, which further contributes to injury by trapping subsequent tablets at the site of injury could be the possible mechanism. Adverse effects can occur even after 12 weeks including stomach pains, nausea and vomiting, diarrhoea and bright red blood in the stools. The severity of potassium salts associated ulceration of the gut depends on factors such as the formulation of the treatment and gut transit time. There is an association between potassium chloride (KCl) supplements and GIT disturbance, small-bowel ulceration, erosions and ulcerations. The range of GIT disorders including upper gastrointestinal ulceration and perforation, and related deaths. Breakdown of K homeostasis is associated with changes in acid-base balance, respiratory and heart rates, hyperkalaemia and hypernatraemia. Death may occur from respiratory failure (with or without convulsion) accompanied by gastroenteritis, dehydration of most organs and renal tubular necrosis. 2.4 g KCl (1.25g K) three times a day for 7 days (total daily dose 3.7g K) can induce considerable mucosal pathology, with erosions, gastric ulcers, inflammatory lesions and bleeding at endoscopy. Cyto-adaptation to KCl treatment does not seem to occur. In the elderly, depletion or excess of K is frequent. Although no adverse effects have been reported with the use of high K diets in healthy people, increasing dietary K in the elderly must be considered with caution (renal disease, adrenal insufficiency, acidosis, insulin deficiency, or digitalis intoxication). In addition, certain drugs predispose to hyperkalaemia, including K-sparing diuretics, β-adrenergic blockers, ACE inhibitors. Potassium chlorate ($KClO_3$) is less toxic than KCl. There are no data on the carcinogenicity of K itself. Potassium has negative carcinogenicity. Excess KI has a thyroid tumour-promoting effect but KI per se does not induce thyroid tumours. As with bromate ($KBrO_3$), iodide rather than K is thought to be the active component, via its effects on thyroid metabolism. There may be a link between increased urinary Na^+ and/or K^+ level and urinary alkalinity to the development of urothelial hyperplasia and urinary bladder tumour promotion. Alkalinuria and kaliuresis may result in simple epithelial hyperplasia, and in the long-term to papillary/nodular hyperplasia, papillomas and transitional cell carcinomas of the urinary bladder. Potassium hydrogen carbonate is able to induce urinary bladder cancer without prior initiation by another chemical but KCl is only a weak tumour promotor and could only induce a few (pre)neoplastic lesions. $KBrO_3$ cause lipid peroxidation in the kidney and subsequently produces 8-hydroxydeoxyguanosine (8OHdG) in DNA. Dose-dependent $KBrO_3$ may induce renal cell tumours, thyroid follicular tumours and mesotheliomas on the tunica vaginalis testis. Potassium bromate is mutagenic. Potassium chloride is a weak tumour promotor, only inducing a few neoplastic changes. There is no evidence for neurotoxicity, carcinogenicity or genotoxicity of K in humans. The dietary administration of large doses of K results in death due to ECG abnormalities and cardiac arrest.

THE HORMONAL BASIS FOR ELECTROLYTE CONTROL

The medical establishment must understand the physiological basis for nutritional intervention. Excess potassium (K) is the main problem in metabolic shock. Aldosterone, deoxycorticosterone (also called cortexone, 11 desoxycorticosterone, DOCA, or DOC), 18 hydroxy 11 deoxycorticosterone (also designated 18 OH-DOC), and 16 α-18 dihydroxy 11 deoxycorticosterone (DOH-DOC), all must conserve sodium (Na) in order to be called a mineralocorticoid. Sodium makes up most of the cations of blood plasma. 580 g of Na and 36 g of K are filtered each day. All but the 1-10 g of Na and the 1-3 g of K likely to be in the diet must be reabsorbed. Sodium must be reabsorbed in such a way as to keep the blood volume exactly right and the osmotic pressure correct; K must be absorbed in such a way as to keep serum concentration as close to 4.8 mEq/l (about 190 mg) as possible. The pool of K in the cells is fifty times as large as outside it. Potassium moves passively in counter flow to Na in response. Potassium is secreted twice and reabsorbed 3 times before the urine reaches the collecting tubules. When K in the serum is higher than 4.8 mEq/l, the zona glomerulosa of the adrenal gland secretes more aldosterone and K is excreted into the end of the tubules and the collecting ducts. Aldosterone also reverses K out flow in the last part of the colon only and increases Na absorption throughout the colon. In the ascending colon chloride enters with the Na while in the descending colon there is only a K-Na exchange. The amount of aldosterone secreted is a function of the serum K as probably determined by sensors in the carotid artery, pressure in the carotid artery, the inverse of the Na intake as sensed via osmotic pressure, anxiety, and of the angiotensin II (ANGII) formation. ANGII is regulated by the rennin from the kidneys. Depletion of either K or Na activates secretion of rennin, but in K depletion aldosterone is suppressed. ANGII acts synergistically with K, and the K feedback is virtually inoperative when no ANGII is present. A portion of the regulation resulting from ANGII must take place indirectly from decreased blood flow through the liver due to constriction of capillaries. When the blood flow decreases so does the destruction of aldosterone by liver enzymes. The K feedback to aldosterone is virtually inoperative when no angiotensin is present. When ANGII drops out in order to correct the situation, it leaves behind a somewhat enhanced K serum concentration, which also tends to reduce pressure at serum contents of K above 4.8 mEq/l, and causes Na to start to decline by the same failure to stimulate aldosterone. ACTH has some stimulating effect on aldosterone probably by stimulating DOC formation, which is a precursor of aldosterone. The primary function of ACTH is to inversely mobilise the body's defenses against intestinal disease. Aldosterone is increased by blood loss, pregnancy, and possibly by other circumstances such as physical exertion, endotoxin shock, and burns. Carotid artery pressure, pain, posture, emotion (anxiety, fear, and hostility) including surgical stress produce an unknown messenger hormone that stimulates aldosterone secretion. Aldosterone operates by diffusing to the nucleus to produce a mRNA and the various steps take about an hour to come completely on stream. Too high a serum content of K has very adverse effects on nervous transmission. Individuals with an anxiety neurosis can have as high as 4 times the secretion as normal and these with schizophrenia have a low secretion. K feedback is the main regulation of aldosterone in normal diet and health. The slope of the response of aldosterone to serum K is almost independent of Na intake. Aldosterone is much increased at low Na intakes. The K is strongly regulated at all Na intakes by aldosterone when the supply of K is adequate, which it usually is in primitive diets. Na pump, which secretes K into the distal loop of the tubules, along with the nature of the K feedback, make aldosterone certainly a hormone for unloading K. As much as 26 g of K can be unloaded per day by healthy people accustomed to a large intake. Aldosterone makes the Na in the bones available. The bone contains nearly half the body's Na. The body depends considerably on aldosterone to keep the serum Na retained and normal. DOC stimulates the collecting tubules to continue to excrete K in much the same way that aldosterone does but not like aldosterone in the end of the distal convoluted tubules. Kallikrein is augmented by DOC and suppressed by aldosterone. If Na becomes very high, DOC also increases urine flow. DOC has about 1/20 of the Na retaining power of aldosterone and may be as little as 1% of aldosterone at high water intakes. DOC has about 1/5 the K excreting power of aldosterone. A rise in serum K causes a rise in DOC secretion. Angiotensin has little effect on DOC, but DOC causes a rapid falls in rennin, and therefore ANGI, the precursor of ANGII. Therefore, DOC must be indirectly inhibiting aldosterone since aldosterone depends on ANGII. ACTH has more effect on DOC than aldosterone. During dehydration, aldosterone virtually disappears and rennin and angiotensin rise high. Potassium supplements are very dangerous during dehydration. DOC's primary purpose is to regulate electrolytes. It has other effects on copper enzymes, proteins and connective tissue, which may be used by the body to survive during K wasting intestinal diseases. When K becomes low, the first thing happens is that excretion of K from the far end of the renal tubules and collecting tubules declines (within 24 hrs and stops in 2 days). Under low sodium intake 18-OH DOC is increased in serum. There is a marked increase in serum 18-OH DOC after injection of insulin and this may be due to the hypokalaemic tendency after a rise in insulin, which in turn would make the serum more acidic. 18 OH-DOC may lower urine pH. 18 OH-DOC's primary purpose may stimulate hydrogen ion or ammonium excretion. Hydrogen ion's interference with K excretion depends on the K cell or plasma content. The large affect that ACTH has on 18 OH-DOC revolves primarily around keeping immune enzymes at a low pH during infection. Insulin counters high serum K only at low K intakes. At high intakes the affect of insulin stays normal. 18 OH-DOC acts primarily by blocking aldosterone's effect on potassium, and must have aldosterone to assist it. ANGII has very little effect on 18 OH-DOC and is ambiguous nor does serum K above 4.8 mEq/l (187 mg). Under low Na intake, 18 OH-DOC rises in the serum. ACTH causes a marked increase in 18 OH-DOC up 20-fold, probably by a generalised affect on the zona fasciculata of the adrenal cortex where 18 OH-DOC is synthesised. At hypokalaemic situations, the intracellular K is usually decreased and this depresses heart contraction. The intracellular K is much more important than serum K on the strength of heart contractions. When cardiac contraction strength decreases from low K status, it should be imperative to contract the capillaries in order to make sure that blood pressure does not drop. The relaxation of capillaries takes place by K between 4-8 mEq/l serum content is some kind of an adaptive circumstance rather than an inherent characteristic of pre capillary blood vessels. ANGII causes a drop in 18 OH-DOC, which may be a negative feedback phenomenon. Modern man eats only starchy, salty refined food. DOH-DOC increases the Na to K ratio in urine. The ascending colon increases water absorption under c-AMP stimulation, opposite the effect in the descending colon. DOH-DOC's greatest effect on Na in the colon because it is here where it would be most advantageous to unload Na in order to keep water loss in the kidneys at a minimum. DOH-DOC decreases ANGII in the vicinity of 4.8 mEq/l and then considerably increases it if the intracellular K becomes low. The cell status is maintained largely by controlling the serum. DOC is associated with increased synthesis of collagen and may tend to increase the thickness of the arterial walls with time and decrease their elasticity. Aldosterone is low in arthritics. The amount of K to heal rheumatoid arthritis must be 3.5 g/day or more because this is the amount that permitted slow improvement of a man across a 3-month time span. DOC secretion is increased 10 times by the end of the pregnancy. Potassium is ominously low in the diet. Potassium supplements are very important for recovery from heart disease. Heart disease can be caused by K deficiency or vitamin B_1 deficiency. Psychic stress stimulation of aldosterone, profuse perspiration, excessive vomiting, eating Na carbonate or bicarbonate (because hydrogen ion is excreted at the same site as K), laxatives, diuretics, licorice, hyperventilating, enemas, shock from burns or injury, hostile or fearful emotions, and very high or low Na intakes all increase K losses, some massively. Reduced appetite is associated with a sedentary life.

POTASSIUM-TO-SODIUM RATIO

The sodium pump in a resting cell used almost a quarter of all the energy available. Insulin regulates the Na/K pump; the energy required is lower than those to affect glucose uptake in cells. Insulin raises the intracellular pH, thus making the cell interiors more alkaline. The pH is not constant, but is a physiological variable. The pH level is involved with regulation of glycolysis, and to some extent, cell division. Every enzyme is affected by pH. The regulation of intracellular pH is via the Na/H exchange pump, whereby Na leaks back into the cell due to a difference in its energy gradient. The electrochemical potential (free-energy gradient) provides the energy to move a proton, which is acid or a hydrogen ion (H^+) out of the cell. If the extracellular Na is lowered by replacing it with magnesium or sucrose, the Na/H exchange pump no longer works. Further, if the Na is lowered below that point where the insulin stimulated it, it should make the pH more acidic. By changing the extracellular Na, the action of insulin on glycolysis can be converted from stimulation to inhibition. Lowering extracellular Na inhibits glycolysis without adding any foreign chemicals. The sodium/calcium exchange pump moves 3 Na ions into the cell in exchange for one calcium ion (Ca^{2+}) going out. The Na/Ca exchange is electrically not neutral; the membrane potential affects it. A higher membrane potential tends to make that pump move Ca out of the cell. At each cycle, that pump moves a positive charge in. Lowering the intracellular Na moves Ca out. Inhibition of the Na/K pump leads to accumulation of Na inside the cell, and decreases the plasma membrane potential. Both of these effects make the sodium/calcium exchange pump less active. The Na/Ca pump is very sensitive to a slowdown of the sodium/potassium pump. Dietary Na increases intracellular Ca in the muscle cells and the resultant is vasoconstriction. Potassium would stimulate the Na/K pump and, thus, indirectly, through the resulting increase in Na-Ca exchange decrease the intracellular Ca and allow the muscles in the blood vessel walls to relax. The Na-Ca exchange is involved with the action of digitalis-increased strength of contraction of cardiac muscle. The potassium/sodium imbalance is present in every cell in the body. The potassium/sodium imbalance also is a cause of insulin resistance. The potassium/sodium imbalance, which contributes to insulin resistance also is associated with an abnormal metabolism of carbohydrates and fatty acids, and may contribute to causing Type 2 diabetes. Vegetarians rarely develop hypertension, because potassium/sodium ratio is extremely high in their diet. Slaughtered meat has a high potassium/sodium ratio (>3:1), because most of the Na is in the blood, which is eliminated. As a result of osmotic equilibrium, the sum of the Na and K inside the cell is very close to constant (within about 2%). The higher the K Factor (potassium), the better the food or diet. Soybeans have a K Factor of 340. Corned beef hash has a K Factor of about 0.37. The K Factor of the total diet is the important number. A K Factor above 4 is a better goal. Our ancestors had a K Factor of about 16 to 1. The caveman forebears got around 11,000 mg of K daily and about 700 mg of Na (Palaeolithic diets have about 16 times more K than Na, whereas modern civilised diets have about 1.6 times more Na than K). Nowadays this 11,000 mg has shrunk to 2,500 mg of potassium, and the sodium intake has increased from 700 mg to 4,000 mg (This is a K Factor of 0.6). Unprocessed fruit or vegetable has 20-100 times as much potassium as sodium (high K Factor). This is because the Na-K pumps in their cells work to keep K in. Potatoes are one of the richest sources of K. A big baked potato have about twice as much K as a banana but a lot of people will add salt to their baked potato. Most of the processed foods have huge amounts of salt added, and in great many cases, large amounts of potassium have been depleted. Typical of this is polished rice; three-fourths of the K disappears, and bout three-quarters of the K is removed from the wheat berry by reducing it to white flour. Cow's milk has a K Factor of 2.8. Salmon, tuna, sardines or any marine fish are quite high in K and quite low in Na. The Na-K pumps in the kidneys are different than those in other tissues and they function to return Na back into the bloodstream. Food processors refine foods in such a way as to reduce K and then they add salt to satisfy our cravings and taste preferences. The Na-K pump is essential to every cell in the body; it is a form of ATP. The pumps are proteins in the cell membranes that pull K into the cell and push Na out, which generates a voltage between the inside and the outside of the cell. The pumping action against the gradient is brought about by a protein commonly called Na-K-ATPase, which functions by conformational changes that can attract or repel one of the ions or the other. The conformation change by the protein requires energy from ATP. When the potassium-to-sodium balance is not right, the body is aging biochemically as a result of stiffening the membrane and decreasing the flow of nutrients into the cell (e.g., amino acids, glucose). If the voltage further decreases, channels and pores in the membrane may malfunction. Eventually, various illnesses develop. The potassium-sodium imbalance is linked to a number of other diseases, and among these is osteoporosis. Excess Na simply sucks Ca out of the bones; a high-salt diet probably may be a greater risk for osteoporosis than a low-calcium diet. The Ca pump removes Ca from inside the cell by allowing some of the Na back in, drawn by the charge differential. The intracellular environment has a negative charge that attracts ions such as Na. The membrane protein, calcium-ATPase, could be called the calcium pump is like a wheel that loads up 3 sodium ions on one side and one calcium ion on the other. This wheel also is powered by ATP. Since Ca has a charge of plus two (valence), the net effect is that the negative charge on the cell interior is reduced and Ca is removed from the cell. There is a relationship to asthma, gastric cancer, ulcers, and to age-related cognitive memory decline. Salt is a hidden component of much processed food, processed food accounts for up to 75% of the salt intake. About 15% of salt intake comes from the saltshaker (one teaspoon of salt is about 2,500 mg). The potassium/sodium imbalance affects every cell in the body and can lead to a range of ailments, ranging from hypertension to cataracts and possibly even erectile dysfunction. The more salt one eats, the more calcium the body loses in urine. The serum Ca level is tightly regulated because it has a major impact on the heart's electrical rhythm and other vital physiological processes. When a lot of salt is eaten, the immediate impact of the resulting urinary loss of Ca is a transient drop in serum calcium that has to be immediately corrected. The parathyroid hormone aids the efficiency of Ca absorption from the diet, but it also has the unfortunate effect of leaching Ca from the bone mineral. The K rich diets are protective in two ways: 1- the increased K intake helps the kidneys to excrete salt more efficiently, so that salt can't promote as much Ca loss in the urine, and 2- the negatively charged organic molecule associated with that K can be metabolised to release bicarbonate, which has an alkalinising flirt on the body that protects bone from the adverse impact of acid-generating proteins. Unless the diet is extremely high in calcium, reducing dietary Ca doesn't have much impact on the amount of Ca in the urine-and a low-calcium diet actually can make matters worse by increasing the intestinal absorption of the dietary compound oxalate. This may lead to the formation of calcium oxalate renal stones. Individuals whose diets have low dietary potassium-to-sodium ratios are 3 times more likely to develop renal stones than those with high potassium-to-sodium ratios. Salty diet appears to exacerbate asthma in males. Salty diets can make exercise-induced asthma worse. Inefficient function of the Na-K pumps can make the bronchial tubes more sensitive to bronchoconstrictors such as those released by exposure to allergens. Heavy salting of foods is practised to prevent microbial contamination of food in societies that lack refrigeration. Moulds or bacteria frequently generate mutagens in salt-preserved food, which is due to lack of refrigeration. High-salt diets can be directly damaging to the stomach lining, increase the ability of the Helicobacter pylori to colonize the stomach lining. Gastric cancer is second only to lung cancer as the chief cause of cancer mortality worldwide. A lot of vegetarians make up for the lack of animal products and fat in their diet with an increased intake of salty condiments. Individuals with the lowest blood levels of K have the highest risk of stroke. One cup of cubed acorn squash has almost 900 mg of K.

FLAVNOIDS

For about 25 years, the flavonoids were referred to as vitamin P, because they were considered the "permeability" factor. About 4,000 flavonoids have been identified, and they are part of an even larger class (polyphenols). The mix of flavonoids, like other nutrients, varies from plant species to species. Along with their common antioxidant properties, flavonoids display some striking differences. Some function as mild oestrogens, some inhibit tumour growth by preventing angiogenesis, and a few even appear to promote free radicals and cancer. Both pine bark and grape seed extracts are rich in a group of flavonoids called proanthocyanidins, or condensed tannins. There are about 250 different proanthocyanidins in plants and, as a group, they constitute one of 12 subcategories of flavonoids. Grape seeds, along with their pulp and skin, are also a rich source of proanthocyanidins, though the mix of proanthocyanidins is a little different from that of pine bark. Tea is rich in a variety of gallate compounds, but the most active antioxidant among them appears to be epigallocatechin gallate. It accounts for about 30% of the flavonoids in green tea. The gallates are powerful antioxidants that prevent the oxidation of low-density lipoprotein, a risk factor for coronary heart disease. Epigallocatechin gallate can inhibit antibiotic-resistant Staphylococcus aureus. Gallic acid, a related compound, has potent anticancer properties. Soy flavonoids (genistein and diadzein) appear to reduce the risk of prostate and breast cancers. Ellagic acid, related more closely to gallates than to proanthocyanidins, is a potent antioxidant and cancer inhibitor. The citrus flavonoids, Hesperidin, quercetin, rutin (a sugar of quercetin), and tangeritin are abundant and effective. Hesperidin raises blood levels of the good high-density lipoprotein and lowers the bad low-density lipoprotein and triglycerides. It also possesses significant anti-inflammatory and analgesic effects. Quercetin reduces inflammation associated with allergies, can inhibit the growth of head and neck cancers, and may stop reverse transcriptase (HIV replication). Tangeretin induces apoptosis in leukaemia cells, but does not harm normal cells. The flavonol quercetin exhibits a wide range of biological activities, such as antioxidative, anticarcinogenic and enzyme-inhibiting activities. Quercetin has protective effect on cardiovascular disease. The main dietary sources of quercetin are onions, tea, berries, and apples. Quercetin concentrations of 74-146 mg/kg have been found in lingonberries, in black currants the concentration is 52-122 mg/kg, and a 30 mg/kg in bilberries. Quercetin is mainly present in plants as glycosides and different plants contain different quercetin glycosides.

SILYMARIN

Flavonoids belong to the family of the benzo-γ-pyrones. More than 4000 different flavonoids are currently known; they are ubiquitous. Multiple biological effects of flavonoids include antiinflammatory, antiallergic, antihaemorrhagic, antimutagenic, antineoplastic and hepatoprotective activities. The biological and pharmacological effects of flavonoids are due to modulation of certain enzymes (hexokinase, aldose reductase, phospholipase C, protein kinase C, cyclo-oxygenase, lipoxygenase, myeloperoxidase, NADPH oxidase, and xanthine oxidase) and their antioxidant activity. A single flavonoid can inhibit one enzyme at a certain concentration while inhibiting another enzyme at a 100-fold higher concentration. Some flavonoids, including quercetin and silibinin can protect cells and tissues against the effects exerted by reactive oxygen species. Their antioxidant activity results from the scavenging of free radicals and other oxidising intermediates, from the chelation of iron or copper ions, and from inhibition of oxidases. Flavonoids from Silybum marianum have positive effect on intact liver cells or cells not yet irreversibly damaged, and also stimulate their regenerative capacity after partial hepatectomy. Antihepatotoxic activity was also demonstrated for kolaviron, a defatted alcoholic extract of the seeds of Garcinia kola, and for Garcinia biflavonones. Other flavonoids extracted from Baccharis trimera were reported to protect against hepatic damage; hispidulin appeared to be the most active compound. Quercetin was demonstrated to exert some ameliorative effects on tissue damage induced by cigarette smoke. The flavonoid silymarin and one of its structural components, silibinin, are substances with documented hepatoprotective properties. Silymarin and silibinin mechanisms of action include: 1- antioxidants, scavengers and regulators of the intracellular content of glutathione; 2- cell membrane stabilisers and permeability regulators that prevent hepatotoxic agents from entering hepatocytes; 3- as promotors of ribosomal RNAsynthesis, stimulating liver regeneration; and 4- as inhibitors of the transformation of stellate hepatocytes into myofibroblasts, the process responsible for the deposition of collagen fibres leading to cirrhosis. The key mechanism that ensures hepatoprotection appears to be free radical scavenging, anti-inflammatory and anticarcinogenic properties. Silymarin exerts hepatoprotective effects in acute viral hepatitis, poisoning by A phalloides, toxic hepatitis produced by psychotropic agents and alcohol-related liver disease, including cirrhosis, at daily doses ranging from 280-800 mg, equivalent to 400-1140 mg of standardised extract. Silymarin is absorbed by the oral route and that it distributes into the alimentary tract (liver, stomach, intestine, and pancreas). It is mainly excreted as metabolites in the bile, and is subject to enterohepatic circulation. Toxicity is very low; silymarin is devoid of embryotoxic potential. Silymarin is a flavonolignan, the most well known compound of the flavonoids, extracted from the seeds and fruit of Silybum marianum (Compositae) and in reality is a mixture of 3 structural components: silibinin, silydianine and silychristine. The structure of the constituents of silymarin was clarified in the 1960s. Silymarin and silibinin have been found to provide cytoprotection and, above all, hepatoprotection. Silymarin is used for the treatment of numerous liver disorders including: 1- degenerative necrosis and functional impairment, 2- to antagonise the toxin of Amanita phalloides and provides hepatoprotection against poisoning by phalloidin, galactosamine, thioacetamide, halothane and carbon tetrachloride, 3- to protect hepatocytes from injury caused by ischaemia, radiation, iron overload and viral hepatitis. Silymarin is as active as quercetin and dihydroquercitin, and more active than quercitin in terms of antiperoxidant activity. Silymarin reduces the loss of lactate dehydrogenase (LDH), increases oxygen consumption, reduces the formation of lipid peroxides, reduces Ca^{2+}, and increases the synthesis of urea in hepatocyte. Oxidative stress is defined as structural and/or functional injury produced in tissues by the uncontrolled formation of pro-oxidant free radicals. Silibinin protects hepatocytes from cell damage induced by erythromycin, amitriptyline, and nortriptyline. Silymarin may increase the stability of the erythrocyte membrane. Silymarin increases erythrocytes resistance against haemolysis. Glutathione (GSH) in the liver exerts important protective activity against chemically induced oxidative stress. Ethanol or paracetamol are peroxidation inducers, which produce marked GSH depletion in the liver. Silymarin antioxidant property acts as a scavenger of the free radicals that induce lipid peroxidation and also influences enzyme systems associated with glutathione and superoxide dismutase. Silymarin inhibit linoleic acid peroxidation catalysed by lipoxygenase, and protects liver mitochondria and microsomes against the formation of lipid peroxides. Silymarin reduces plasma levels of cholesterol and low-density lipoprotein (LDL) cholesterol. One of the mechanisms that can explain the capacity of silymarin to stimulate liver tissue regeneration is the increase in protein synthesis in the injured liver. The application of silymarin significantly reduces apoptosis, skin oedema, depletion of catalase activity, and induction of cyclo-oxygenase and ornithine decarboxylase activity. This effect provides protection against photocarcinogenesis. The molecular bases of the anti-inflammatory and anticarcinogenic effects of silymarin might be related to the inhibition of NFκB, which regulates the expression of various genes involved in the inflammatory process, in cytoprotection and carcinogenesis. Silymarin may act by modulating the activation of regulating substances of the cellular cycle and of mitogen-activated protein kinase. Silymarin improves hepatic fibrosis. Colchicine and silymarin prevent completely peroxidation of lipids, Na^+,K^+- and Ca^{2+}-ATPase. Silymarin inhibits completely the loss of glycogen. Silymarin can inhibit the hepatic cytochrome P450 (CYP) detoxification system (phase I metabolism), which could explain some of the hepatoprotective properties. Silymarin, together with other antioxidant substances, could contribute towards protection against free radicals generated by enzymes of the CYP system. Silymarin and silibinin inhibit the absorption of toxins, preventing them from binding to the cell surface and inhibiting membrane transport systems. Silymarin renders cell membranes more resistant to lesions. Silibinin probably also acts on the nucleus, where it appeared to increase ribosomal protein synthesis by stimulating RNA polymerase I and the transcription of rRNA. The stimulation of protein synthesis is an important step in the repair of hepatic injury and is essential for restoring structural proteins and enzymes damaged by hepatotoxins. Silymarin is not soluble in water and is usually administered in capsules as a standard extract (75% silymarin). Its elimination half-life ranges from 6 to 8 hours. Silibinin and other components of silymarin are rapidly conjugated with sulfate and glucuronic acid in the liver. They undergo enterohepatic circulation: intestinal absorption, conjugation in the liver, excretion in the bile, hydrolysis by the intestinal flora, and reuptake in the intestine. Silibinin is excreted in minimal quantities in the urine during the 48 hrs following oral (2-5%) or intravenous (8%) administration. On the contrary, biliary excretion is fairly high during the same period (about 40-45% after oral administration of up to 20 mg/kg, and about 80% after intravenous administration). Acute, subacute and chronic toxicity of silymarin is very low; the compound is also devoid of embryotoxic potential. Silymarin (140 mg tds) reduces complications, reduces the duration of hospital stay and promotes recovery in patients with acute viral hepatitis. Resolution of subjective symptoms of patients with chronic liver disease is expected achieve in 63% of cases treated with high doses of silymarin (560 mg/day); AST diminished on average by 36%, ALT by 34% and GGT by 46%.

Action of silymarin and silibinin in xenobiotics:

A- Inhibition of haemolysis.
B- Protection against liver glutathione depletion and lipid superoxidation.
C- Inhibition of O_2 consumption.
D- Reduction of enzyme loss and morphological alterations.
E- Inhibition of lipid peroxidation.
F- Prevention of hepatotoxicity.
G- Prevention of chronic liver damage.
H- Neutralisation of lipid peroxidation.
I- Antihepatotoxic effects (reduction in liver alterations).
J- Hepatoprotection.
K- Inhibition of toxic effects on protein synthesis.
L- Reduction of glutathione depletion.
M- Reduction of enzyme loss.
N- Neutralisation of lethal effects.
O- Prevention of increase in liver enzymes and of reduction in coagulation factors.

Histological examination would reveal an improvement in portal inflammation, parenchymal alterations and necrosis. The daily oral dose of silymarin range from 280 to 800 mg. This is equivalent to 400-1140 mg of standardised extract containing 70% silymarin. The recommended dosage for active disease is 140 mg of silymarin (200 mg of extract) three times daily. At higher dosages (>1500 mg/day) silymarin may have a laxative effect due to an increase in secretion and bile flow. Moderate allergic reactions have also been reported. Silymarin exerts hepatocyte membrane stabilisation and permeability regulation, stimulation of ribosomal RNA synthesis promoting liver regeneration, and prevent the transformation of stellate hepatocytes into myofibroblasts, which are responsible for the deposition of collagen fibres. These properties induce protection against the hepatotoxic effects of a number of xenobiotics, such as Amanita phalloides toxins, ethanol and psychotropic compounds. Silymarin, an antioxidant flavonoid complex derived from the herb milk thistle (Silybum marianum). Silymarin can reduce insulin resistance (adult-onset diabetes, hypertension, and hypercholesterolaemia) and diabetic complications. Insulin resistance is constantly fairly high in patients with NIDDM and hepatic cirrhosis. This metabolic disorder is partly related to increased blood glucose levels due to reduced glucose uptake by the liver. Hyperglycaemia promotes hyperinsulinaemia and this, together with the decreased hepatic degradation of the insulin molecule, may lead to insulin resistance in the target tissues. Silymarin stabilises as well as lowers glucose levels, decreases glucosuria, decreases glycosylated haemoglobin levels significantly, decreases serum malondialdehyde (marker of lipid peroxidation). Silymarin flavonoid, silybinin, prevents the accumulation of fibronectin protein in renal cells. Fibronectin is one of the principal causes of renal injury in diabetics. Silybinin's protective effect is the result of its antioxidant properties. Silymarin inhibits EGFR, a type of tyrosine kinase receptor that promotes tumour growth. Silymarin significantly inhibits cell growth and proliferation. Other antioxidant-like green tea polyphenols, EGCG (epigallocatechin gallate), quercetin, curcumin and genistein, also result in similar inhibitory effects.

OLIGOMERIC PROANTHOCYANIDINS (OPCs)

Bioflavonoids are complex organic plant compounds found mostly in fruits, vegetables and certain tree barks. They are powerful antioxidants, free-radical scavengers, and influence the body immune response to inflammation, allergy and infection. There are more than 20,000 different types of bioflavonoids, of which OPCs are considered the most potent antioxidants. Oligomeric proanthocyanidins (OPCs) are very powerful non-toxic water-soluble bioflavonoids, often referred to as Pycnogenols. OPCs are derived from one or more of a combination of grape seed extract and/or pine bark extract. OPCs have a powerful free-radical scavenging activity. OPCs are 20 times more powerful than vitamin C and 50 times more powerful than vitamin E. They are non-toxic and cross the blood-brain barrier. Grape seed extract is a superior source of OPC, containing 92% active ingredients and pine bark contains 84% of the active OPC ingredient. OPC potentiates vitamin C and regenerates vitamin E. Some of the causes of free radicals include environmental pollution, food additives, ketogenic diet (high fat), excessive alcohol, smoking and passive smoke inhalation, burns, infection, stress, radiation and nutrient deficiencies. Oligomeric proanthocyanidins (OPC) is the name of a particular molecule, which is ubiquitously found throughout the plant kingdom. Oligomeric proanthocyanidins were first extracted in 1947 from the red skins of peanuts, and later from pine bark (1950). OPC are pairs and triples of one particular molecule, called flavan-3-ol. The single flavan-3-ol molecule is called catechin. Catechin could be a precursor of OPC. OPC preserve and protect collagen and elastin. Collagen and elastin contribute to the elasticity of all these tissues. OPC protects plants against oxidation, against rancidity and against aging. Both grape seed and pine bark extracts share OPC as the major constituent and because also the catechin part is very similar. The pine possesses a high level of monomers (flavanol singles) of the catechin type. The Grape contains more oligomers, while the epicatechin dominates amongst the monomers. As histidine decarboxylase inhibitors, pycnogenols lower the histamine level in the aortic endothelium, which protect it against the vascular permeability alterations occurring early in the atherogenic process. Procyanidins have capillary protective action, which is probably a pluricentric mechanism, based on radical quenching and antioxidant effects and on the inhibition of some key enzymes of the microvascular endothelium and extravascular matrix. Procyanidins are effective in peripheral venous insufficiency with rapid duration of action. OPC are strong agents in counteracting spontaneous mutation of DNA both at mitochondrial and nuclear level. This effect could be due, at least in part, to the antioxidant properties of procyanidins and it could be a rational basis for their potential use in chemoprevention of several pathological situations. OPC and catechin bound to insoluble elastin markedly affect its rate of degradation by elastases. In normal conditions vascular permeability is precisely regulated by mechanisms, which involve among others macromolecules of the extracellular matrix of the vascular wall. Permeability for a given substance varies according to the anatomical localisation of the vessel determining also its structure and composition. In some pathological conditions, such as inflammation or diabetes, permeability can be abnormally increased. OPC prevents the permeability increase induced by collagenase. OPC may increase the resistance of the tight junctions of arteriolar capillaries. The effect of OPC on the vasculatory constituents (venolymphatic problems) offers an effective and well-tolerated alternative to hormonal treatment of premenstrual syndrome. It prolongs and reinforces with the passage of time the intensity of the symptoms and their duration. OPC have effects on capillary resistance in hypertension and in certain nephropathies. Light vision is a major element of the visual function. However, it varies from one subject to another. Many factors may modify it in the absence of any retinal pathology (age, fatigue, and stress). OPC improves the visual performances and visual adaptation.

GREEN TEA

Fatty tissue, like a tumour, requires an increased blood supply to grow (angiogenesis). Angiogenesis inhibitors prevent new fat accumulation; trigger shed significant amounts of fat, without affecting vessels elsewhere. Theanine and green tea catechins enhance the effectiveness of at least some forms of chemotherapy. The amino acid L-theanine and epigallocatechine gallate (EGCG) are bioavailable orally, induce apoptosis of several tumour cell lines, inhibit chemically induced carcinogenesis, are chemosensitizing agents, induce relaxation, in humans, and are associated with a reduced occurrence of some cancer types. Green components may inhibit angiogenesis, reduce cell signal pathways involved in HER-2/neu, reduce toxicity of certain chemotherapeutic agents, and inhibit TNF-α. The 4 major catechins in green tea are epigallocatechin gallate (EGCG), epigallocatechin (EGC), epicatechin gallate (ECG), and epicatechin (EC). Catechins are members of the flavan-3-ol class of flavonoids. Catechins are polyphenolic compounds that provide the antioxidant activity of green tea, and accounts for 8-30% of the dry weight of green tea. L-theanine constitutes between 1% and 2% of the dry weight of tealeaves, and exists only in the free (non-protein) form. L-theanine gives tea its characteristic, slightly brothy, unami taste. EGCG induces apoptosis in certain types of cancer cells, including carcinoma of the prostate, stomach cancer, epidermoid carcinoma, and keratinocyte carcinoma. Green tea polyphenols induce apoptosis and inhibit the growth of lung cancer. EGFG inhibits colorectal cancer cells, breast cancer cells, and fibroblasts. Catechins may have anti-angiogenesis activity. EGCG inhibits tumour cell invasion by inhibiting matrix metalloproteinases in lung carcinoma cells. Green tea polyphenols inhibits ornithine decarboxylase; an enzyme overexpressed in prostate cancer and apparently involved in regulation of growth and development in normal and dysplastic prostate cells. EGCG induces apoptosis and induces cell cycle arrest in prostate cancer. Theanine could increase the effectiveness of several anthracycline chemotherapy agents including doxorubicin (adriamycin), idarubicin (idamycin), and pirarubicin. Theanine inhibits the efflux of anthracycline agents from several tumour cell lines, so the agent is active over a longer period of time thereby enhancing tumouricidal activity. Doxorubicin with theanine leads to 68% tumour weight reduction. P-gp (P-glycoprotein) pump inhibition, a well-known mechanism for chemosensitisation, does not appear to be the mechanism of action of theanine. Theanine is most effective at dose of 10 mg/kg/d. Green tea powder, EGC, and EGCG enhance the effectiveness of anthracycline chemotherapy agents. Green tea polyphenols (EGCG) inhibit P-gp pump activity. EGCG and green tea powder both appears to significantly enhance chemotherapeutic efficacy. EGCG inhibits the efflux of doxorubicin by 19.5%, i.e., 1.5-fold increase in concentration of doxorubicin in tumour cells. Green tea catechins especially EGCG inhibits P-gp pump activity and enhances the effectiveness of vinblastine in multidrug-resistant tumour cells. Theanine decreases the concentration of doxorubicin chemotherapy in the heart (by 44%). Theanine reduces the concentration of pirarubicin in the liver (by 21%) and in the heart (25%). Theanine does not increase chemotherapy concentrations in healthy tissue nor inhibit the effectiveness of chemotherapy on tumours. Green tea catechins inhibit TNF-α gene expression and release. Over-production of TNF-α is implicated in the cachexia symptoms of cancer patients. Theanine stimulates the alpha waves in the brain. Alpha waves are typically elevated in individuals who are in a state of calm relaxation such as meditation. In anxiety, alpha wave activity is almost nonexistent. Cancer patients experiencing stress and anxiety from the diagnosis been given and treatment undergoing. Theanine consistently creates a sense of relaxation 30-40 min after oral ingestion. Theanine has an apparent role in the formation of γ-aminobutyric acid (GABA). The amino acid also appears to cross the blood brain barrier and increase serotonin and dopamine concentrations in the brain. 200 mg theanine increases the production of alpha waves in the brain as an index of increased relaxation. Green tea consumption is a component of Asia diets (5 cups a day in Japan). Drinking 5 cups a day of green tea decreases risk of axillary node metastases in premenopausal breast cancer patients and the postmenopausal women increases the expression of oestrogen receptors (ER) and progesterone receptors (PR). Green tea consumption is associated with a lower risk of cancer including colon, pancreatic, and rectal. EGCG scavenges free radicals and protects against DNA damage. Green tea catechins inhibit tumour growth, induce carcinogenesis, inhibit matrix metalloproteinase activity, and reduce intracellular communication in Her-2/neu. L-theanine and green tea catechins are chemosensitizing agents. Green tea polyphenols may also have a protective effect on Parkinson's disease. Selective cell death of dopaminergic neurons in the midbrain (substantia nigra) and marked decrease in dopamine neurotransmitter produced by these neurons are characteristic of Parkinson's disease. Polyphenols inhibit the uptake of dopamine or the neurotoxin MPP^+ by blocking dopamine transporter (DAT), which could be the protective effect in Parkinson's disease by blocking the DAT-dependent uptake of environmental neurotoxin. Increased consumption of green tea correlates with decreased recurrence of stage I and II breast cancer. The recurrence rate is 16.7% or 24.3% among those consuming $\geq$ 5 cups or $\leq$ 4 cups per day, respectively, for stage I and II breast cancer, and the relative risk of recurrence is 0.564 after adjustment for other lifestyle factors. Increased consumption of green tea is associated with decreased numbers of axillary lymph node metastases among premenopausal patients with stage I and II breast cancer and with increased expression of progesterone receptor (PGR) and oestrogen receptor (ER) among postmenopausal ones. Green tea has favourable alterations on carbohydrate (glucose) metabolism. Green tea and its active catechins influence both carbohydrate absorption and metabolism. Green tea and its most popular catechin, EGCG (epigallocatechin gallate), have insulin-enhancing action. EGCG is the prime mover of glucose into cells, and potentiate insulin's effects on promoting the entry of glucose into cells. Green tea lowers blood glucose and increases blood antioxidant markers. Green tea improves renal function. The biological activities of catechins are related with the affinity for cell membranes and the hydrogen peroxide formation during oxidations. Compound found in green tea, reactivates the dying skin cells. Energising dying skin cells might help to improve the skin's condition. Exfoliating with the green tea bags when washing face may improve the skin condition. The urinary bladder benefits from drinking green tea. A compound (type of catechin) found in green tea inhibits the growth of bladder cancer. Green tea is significantly more effective than the black tea. Black and green tea improve the risk factors for heart disease by both hypolipaemic and antioxidant mechanisms and possibly a fibrinolytic effect.

LUTEIN/ZEAXANTHIN CAROTENOIDS

Lutein/Zeaxanthin carotenoids (a family of brightly coloured pigments in plant foods) are less common than their better-known relative, β-carotene, but they appear to offer more and greater benefits to the body. In particular, lutein and zeaxanthin have been shown clinically to support visual function and to protect the health of eye tissues during aging. Lutein and zeaxanthin may protect the cardiovascular system and maintain normal cell differentiation in the tissues of the breast, cervix, colon and skin. The eyes are especially prone to certain types of oxidative damage. Age-related macular degeneration (AMD) is a typical result of the aging process, as the formation of cataracts. Diet may play a significant role in these age-related degenerations. Lutein and zeaxanthin are important antioxidants used by the body for a number of physiological functions. Lutein is a carotenoid, which does not supply vitamin A activity to the body. It is chemically distinctive in that it lacks part of the terminal ring structure of the other carotenoids. Like its close relative zeaxanthin, lutein is what is termed a xanthophyll carotenoid. Both of these related carotenoids are better antioxidants than is β-carotene under normal oxygen conditions. Lutein is the more important of the two. Lutein can be metabolised into zeaxanthin and is therefore the more essential carotenoid. Zeaxanthin has been shown to be present in the center of the macula. Lutein and zeaxanthin are usually found together in leafy green vegetables, such as kale, broccoli, spinach and mustard greens. One of the primary functions of lutein and zeaxanthin is to provide protection against oxidative and free radical damage. These yellow-coloured carotenoids are found in high concentrations within the macula lutea (the yellow spot in the center of the retina) and in smaller amounts throughout the retina and the eye lens. They are also concentrated in the skin, breast and cervical tissues. These stores, however, appear to diminish with age if not regularly replenished through dietary means. Age-related macular degeneration (AMD) is the leading cause of irreversible blindness in people over age 65. The exact cause of AMD is not yet known. A daily intake of 6 mg/day of lutein leads to a 43% lower risk of developing AMD. Lutein and zeaxanthin may contribute to the density of macular pigment-the component of the eye, which typically absorbs and filters out 40-60% of damaging near-ultraviolet blue light (near-UV blue light) which strikes the retina. The denser the pigment, the more the inner retina is protected from light-induced damage. Lutein/zeaxanthin also helps limit blue light damage to the inner retina by inhibiting lipid peroxidation and by neutralising free radicals. Lutein and zeaxanthin are important in reducing changes in the opacity of the eye lens during ageing. The consumption of spinach, which is an excellent source of lutein and zeaxanthin, leads to a much lower level of such eye lens changes than the consumption of other vegetables, such as carrots, sweet potatoes and winter squash, which contain primarily β-carotene and very little lutein. Eating foods rich in lutein-particularly kale and spinach reduces the risk of developing macular degeneration. The intake of carotenoids other than β-carotene, i.e., α-carotene, lutein and lycopene has been inversely correlated with the risk of developing cataracts. Protection most likely comes from the scavenging of free radicals. Oxidative/free radical damage to the eye lens may play an important part in the development of cataracts. Lutein/zeaxanthin prevent peroxidation in the lens, thus limiting damage to the opacity of this tissue. However, there is no evidence that lutein/zeaxanthin can help to reverse an existing cataract. As is true of vitamin E, but not of other carotenoids, lutein is found in blood plasma at levels correlated with the amount of cholesterol. Although lutein is relatively minor as a component of LDL cholesterol in comparison with vitamin E, its antioxidant protective effects may be ten times greater. This may provide another clue to the French paradox in which the consumption of large amounts of saturated fats has not led to elevated rates of heart disease among the French. Lutein is associated with higher high-density lipoprotein cholesterol (HDL) serum levels. Lutein and zeaxanthin may be helpful in preventing irregular cell growth in various tissues of the body, such as the skin, breast, cervix and large intestine. Lutein, like lycopene and a number of other carotenoids, would appear to influence the inner regulation of cell growth and repair.

Food sources of lutein:

Vegetable (1/2-cup serving)	Lutein Content (mg)
Kale	2,190
Collard greens	1,630
Spinach, raw	1,020
Broccoli	190
Leaf lettuce	180
Green peas	170
Brussels sprouts	130
Corn	78
Green Beans	74
Carrot	26
Tomato	10

ANTIOXIDANTS

The use of vitamin supplementation has surged for treatment and prevention of a wide range of diseases in the past few years. Free radicals and antioxidants are among the important discoveries of the past 100 years. All of the major diseases confronting people today are caused by or aggravated by free radicals. Antioxidants may help us live better. Oxidative stress is a situation when there is a serious imbalance in the ratio of free radicals to antioxidants. Too many free radicals and too few antioxidants lead to oxidative stress. At high levels, reactive oxygen species can be damaging to cells and may contribute to cellular dysfunction and disease. Antioxidant is a nutrient that protects body components against undesirable chemical reactions. There are significant health benefits from dietary antioxidants. Diseases linked to excess free radicals include: 1- Ageing, 1954. 2- Cancers. 3- Coronary heart disease. 4- Autoimmune diseases. 5- Rheumatoid arthritis. 6- Alzheimer's disease. 7- Cataracts. 8- Parkinson's disease. Antioxidant nutrients play a role in the prevention and treatment of a variety of chronic diseases, ranging from asthma to cardiovascular disease (CVD) and cancer. Two forms of chemical reactions, oxidation and reduction, occur widely. Oxidation is the loss of electrons, and reduction is the gain of electrons. Oxidation and reduction reactions always occur in pairs, i.e., when one atom or molecule is oxidised, another is reduced. Highly reactive molecules can oxidise molecules (i.e., remove electrons from molecules) that were previously stable, and may cause them to become unstable species, such as free radicals. A free radical is a chemical species with an unpaired electron that can be neutral, positively charged, or negatively charged, and most are very reactive. In free radical chain reactions, the radical product of one reaction becomes the starting material for another, propagating free radical damage. There are 3 steps to free-radical chain reaction: initiation, propagation, and termination. Without termination by an agent such as an antioxidant, a single free radical can damage numerous molecules. The 4 common oxygen metabolites in biologic systems that are free radicals (superoxide anion, hydrogen peroxide, hydroxyl radical, and singlet oxygen) can be generated via a number of mechanisms, including normal physiologic processes and processes resulting from external factors. The excited molecule generated by photosensitisation reactions transfers the increased energy to molecular oxygen, creating singlet oxygen, which then can attack other cell components. A dietary antioxidant is defined as a substance in foods that significantly decreases the adverse effects of reactive species, such as reactive oxygen and nitrogen species, on normal physiological function in humans. The major antioxidant constituents of fruit, vegetables, and beverages are the phenolic phytochemicals, synthesised by the shikimate pathway from tyrosine and phenylalanine, usually as O-glycosides and O-methyl conjugates. Factors affect the bioavailability of these compound include the biologic barriers such as degradation by gut microflora and enzymes, decomposition in the gut lumen, first pass metabolism by the liver, and the physico-chemical properties of the compound (molecular weight, partition coefficient, and pKa). The reaction of a compound with oxygen, or wherever a molecule loses an electron during a chemical reaction is called oxidation. An electron is a negatively charged elementary particle of atoms and molecules. Coenzyme is a cofactor that combines with an enzyme and helps the enzyme function. Vitamin E was once believed to be a mystery vitamin without deficiency symptoms. Antioxidants work together in concert (antioxidant synergism). The reason for the synergism is that some antioxidants are more effective against some free radicals, whereas other antioxidants are more effective against other free radicals. Synergism makes every link in the chain strong. Sulfur compounds are protective against the effects of radiation on the body. Antioxidants can regenerate other antioxidants. After antioxidant neutralises a free radical, it becomes a weak free radical. Both α lipoic acid and Pycnogenol can regenerate used vitamin C, which in turn can regenerate used vitamin E This means that α-lipoic acid and Pycnogenol extend the usefulness of vitamins C and E. Many antioxidants are nutrients removed during food processing. Every single cell suffers about 10,000 free radical blows per day. Over a typical 70 year life span, the body generates an estimated 17 tons of free radicals. Inflammation involves superoxide anion free radical. Cataracts are caused by free radicals reacting with the proteins in the eye lens. As the ozone layer in the atmosphere diminishes, we are exposed to more ultraviolet energy from the sun. Sunlight including ultraviolet rays, generates free radicals when they react with proteins in the lens. High glucose levels generate free radicals and can damage the lens. The body uses free radicals to destroy germs. Free radicals are needed for energy production. Tobacco smoke and smog increase free radicals in the body. The increase in oxygen consumption required during heavy exercise increases free radical formation. Free radicals damage the particles that carry cholesterol in the blood and turn good cholesterol into bad cholesterol. Cholesterol is a fat and not soluble in blood, it is carried by lipoproteins. The cholesterol carried by Low-density lipoprotein (LDL) is bad cholesterol. The cholesterol carried by high-density lipoproteins (HDL) is good cholesterol. HDL carries cholesterol away from cells. Cholesterol deposits seem to form only when LDL becomes damaged by oxidation (oxidised LDL). Oxidised LDL can infiltrate the artery lining and initiate a series of events that trap the cholesterol in the oxidised LDL, attract white blood cells, and form a deposit. LDL becomes oxidised only when the amount of antioxidants is insufficient to protect the LDL against oxidation. Oxidised LDL is a sign of very low antioxidant level. The amount of LDL and the balance between antioxidants and free radicals all important factors in forming oxidised LDL. Cholesterol deposits don't cause a heart attack. Vitamin E, and especially Pycnogenol, has protective anti-aggregation effect on blood platelets, which are critical factors in the blood clotting process. They are effective against the damage to platelets from stress and smoking. The antioxidant nutrient Pycnogenol is a mild hypotensive. Pycnogenol acts to maintain adequate NO levels so blood vessels can relax. Antioxidant nutrients, especially Pycnogenol reduce inflammation. Pycnogenol regenerates vitamin C, which in turn regenerates vitamin E. Daily 400-800 IU of natural vitamin E reduces heart attacks by 77%. Daily 100 IU of vitamin E or more slows the formation of cholesterol deposits. Stress increases production of adrenaline. Adrenaline activates the blood platelets so that they clump together and form a blood clot. Pycnogenol blocks the effect of adrenaline on blood platelets. Pycnogenol is effective against increased platelets aggregation caused by smoking. 100 mg of Pycnogenol can achieve the same desire effect on blood platelets in smokers as 500 mg of aspirin, this due Pycnogenol's effects on the enzyme 5-lipoxygenase, rather than cyclooxygenase (the enzyme that aspirin inhibits). Pycnogenol does not increase bleeding tendency as does aspirin. Free radicals mutate DNA. Cancerous tumours generate their own free radicals and promote still more mutations and abnormal cells. Free radicals can damage cell membranes and inactivate the sensory mechanisms in the membranes that limit cell growth and reproduction. Free radicals can suppress the immune system, inactivating the body's defense against cancer. Antioxidants can stop or slow each of the steps in cancer development. Antioxidants can reduce the chances of metastasis and boost the immune system. Antioxidants may have a role in apoptosis, which helps eliminate mutated cells from the body, antioxidants help cancer cells commit suicide. Antioxidants inhibit several tumour promoters and the activation of some pre-carcinogens into true carcinogens. Pycnogenol inhibits monooxygenase enzyme from converting benzo[a]pyrene into epoxide, which is a true carcinogen. Pycnogenol boosts the levels of cytokines IL6, IL10 secreted by Th2 cells. These cytokines decrease during HIV infection and leads to progressive defects in T- and B-cell functions. These cytokines are important in the body's resistance to cancers. Pycnogenol increases the activity of natural killer cell. The primary function of carotenoids, an important class of antioxidants, is to scavenge free radicals, particularly singlet oxygen produced by photosensitisation. β-carotene increases the activity of monocytes. β-carotene increases the production of TNF, which is a cancer cell killer. Antioxidants increase the killing power of WBCs and protect them from excess free radicals. Pycnogenol, quercetin, and other antioxidants reduce the activity of ICAM-1, VCAM-1 adhesion molecules, preventing the attachment of cancer cells. Adhesion molecules are also involved in inflammation, allergies, and atherosclerosis. Vitamin E supplements cut prostate cancer incidence by 32%. A certain amount of oxidative function is necessary for proper health. When the level of toxic reactive oxygen intermediates (ROI) overcomes the antioxidant defenses of the host, resulting in an excess of free radicals and a state called oxidative stress. These free radicals can induce local injury by reacting with lipids, proteins, and nucleic acids. The interaction of free radicals with cellular lipids leads to membrane damage and the generation of lipid peroxide byproducts. Cells contain a number of antioxidants that have various roles in protecting against free radical reactions. Despite the actions of antioxidant nutrients, some oxidative damage will occur, and accumulation of this damage throughout life is believed to be a major contributing factor to aging and disease. Aging is the process that reduces the number of healthy cells in the body. Different organs seem to age at different rates in different people. Free radical damage to the cell membranes can impair the cell's ability to transport nutrients into the cell and waste products out. Centenarians (people aged 100 years or older) have substantially higher levels of antioxidants and lower levels of free radicals in their blood, compared with people between the ages of 70-99 years old. Vitamin C and E have been associated with a lower risk of cataracts. Lutein is the only carotenoid found near the lens. The lens is bathed in a fluid rich in glutathione. Vitamin C, α-lipoic acid, and N-aceylcysteine increase glutathione and are very important antioxidants. Antioxidants creams are absorbed and retained by the skin. Sunburn is inflammation caused by free radicals, triggered by ultraviolet rays in sunlight. Antioxidants increase the skin's resistance to free radicals and inflammation. Vitamin E, vitamin C, selenium, α-lipoic acid, and ferulic acid (an antioxidant found in Pycnogenol) have each been shown to improve fertility. Antioxidants improve sperm motility. 200 IU of vitamin E can improve fertility by 30%. Pycnogenol prevents β-amyloid from accumulating in brain cells. Vitamin E activity is shared by 8 different compounds: A- 4 of which are members of tocopherol family (α-, β-, γ-, δ-tocopherol), B- 4 of which are members of the tocotrienol family (α-, β-, γ-, δ-tocotrienol). The human body selects for the natural d-α tocopherol form of vitamin E over all others, though others play important roles in health. Synthetic vitamin E, which is identified by the term dl-α is not assimilated or retained as well as the natural form, natural vitamin E is active as twice as potent. The current RDA of vitamin E is 15 IU daily far below the amount needed to reduce the risk of heart disease. Pycnogenol consists of about 40 or so compounds. Vitamin C 2-4 g daily reduces cold symptoms by a third. Cancer patients may benefit from large amounts of vitamin C (10 g or more daily). Vitamin C decreases pain and increases life expectancy in cancer patients. The first symptoms of vitamin C deprivation are fatigue and irritability, large amounts of vitamin C supplements relieve fatigue in patients. There are 2 classes of carotenoids, the carotenes and the xanthophylls. Carotenes are hydrocarbons, while xanthophylls also contain oxygen. β-carotene is a very effective antioxidant against singlet oxygen. There is no vitamin A in carrots or any other vegetables, only carotenoids that is converted into vitamin A in the body. Vitamin A is found only in animals. Strict vegetarian may have trouble getting optimal amounts of vitamin A. Lycopene is one of the more important dietary carotenoids. Men who eat 10 or more lycopene-rich tomato meals weekly have a 45% reduced risk of developing prostate cancer. Diets high in lycopene is associated with a 48% reduction in heart attacks compared with diets low in lycopene. Lutein is a very important carotenoid found in many leafy green vegetables, alfalfa, marigold petals, and egg yolks. Zeaxanthin is found in corn. Lutein and zeaxanthin are essential for vision. Lutein protects vitamin E from oxidation in LDL. The term bioflavonoids, or flavonoids, cover thousands of nutritional substances that have a common basic structure. The structure of flavonoid compounds makes them easy to donate electrons to other molecules, and thus they are usually excellent antioxidants. Flavonoids prevent capillary permeability. Some common bioflavonoids are quercetin, rutin, hesperidin, genistein, and diadzein. β-carotene and tobacco smoke interact to increase activator protein-1 (AP-1) production, which may explain the increased risk of lung cancer as AP-1 overexpression is associated with squamous metaplasia in lung tissue. CoQ_{10} helps to convert food to energy, is a powerful antioxidant, and may help prevent the recurrence of breast cancer. α-lipoic acid is a very powerful antioxidant help to regenerate numerous antioxidants, including vitamin C, vitamin E, and glutathione. High blood sugar levels generate large numbers of free radicals. These radicals account in part for complications of diabetes. α-lipoic acid helps in 2 ways, by lowering blood sugar levels a little and by scavenging free radicals. NADH is a complex compound built around vitamin B_3 (niacinamide, nicotinamide). Like CoQ_{10} and α-lipoic acid, NADH plays a key role in converting food to energy. NADH is a powerful antioxidant. Glutathione as a powerful antioxidant is sulfur-containing tripeptide formed in the body from 3 amino acids: cysteine (a sulphur-containing amino acid), glutamic acid, and glycine.

Recommended comprehensive antioxidants program:

Antioxidants	Daily dosage
Vitamin E	400-800 IU.
Vitamin C	500-4,000 mg. 25 g and more may cause loose stools.
Selenium	100-200 mg. Do not exceed 600 mg daily.
Mixed carotenoids	15-25 mg. Do not exceed 25 mg daily
Vitamin A	8,000-12,000 IU. Do not exceed 25,000 IU daily.
CoQ_{10}	90-600 mg.
Pycnogenol	25-100 mg.
α-lipoic acid	25-100 mg.
NADH	5-10 mg.
NAC	300-600 mg. Do not exceed 600 mg daily.
Lycopene	5 mg.
Lutein	5 mg.
Grape seed extract	50-100 mg.

Glutathione serves as a substrate for many enzymes, such as selenium-containing glutathione peroxidases that reduce free radical reactions. Glutathione can regenerate most other antioxidants, but not NADH. NAC (N acetylcysteine) reduces symptoms of the flu, and is a cancer-preventing compound. The estimated average requirement (EAR) is the intake value that is estimated to meet the requirements of a defined indicator of adequacy in 50% of the population. The recommended dietary allowances (RDA) are the dietary intake level that is sufficient to meet the nutrient requirements of nearly all individuals in the group. These requirements must base on other markers of deficiency (e.g., prevention of scurvy), i.e., the tolerable upper intake level (UL). The UL is not intended to be a recommended level of intake, but represents the highest level of intake that is unlikely to have any adverse health effects in most individuals. The upper limits may be indicated only for a short periods. Origin of ROS: 1- Homolytic scission of water by ionising radiation. 2- Leaking of electrons from membranes and reduction of O_2. 3- Futile cycling of CYPs. 4- Activation of CYP2E1. 5- Reduction of tissue O_2 by Fe^{2+}/Fe^{3+} and other metal redox systems. 6- Activation of leukocytes in inflammation. 7- Redox cycling of quinones. 8- Prostaglandin biosynthesis. Mechanisms of ROS toxicity: 1- Oxidation of vital thiol compounds to disulphides. 2- Loss of tissue GSH. 3- Impairment of energy generation (ATP, NADH, NADPH). 4- Inhibition of Ca^{2+} transport and electrolyte homeostasis. 5- Oxidation of cytochromes. 6- DNA strand cleavage. 7- Initiation and promotion of mutations and carcinogenesis. Reactive oxygen species (ROS), comprising of superoxide anion radical (O_2^-.), the peroxide anion (O_2^{2-}), and singlet oxygen ($^1O_2$), are highly reactive entities, produced from molecular oxygen (O_2) by gain electrons, or the realignment of the electron spins. The hydroxyl radical (OH), the most highly reactive species of ROS, is formed by dismutation of peroxide catalysed by Fe^{2+}. The hypochlorite ion (OCl^-) is yet another highly reactive oxygen species, which together with the former ROS, is produced by leukocytes to kill invading microorganisms. Electron leakage occurs continuously from the mitochondrial membranes and endoplasmic reticulum, and from various cytochrome P450 in the catalysis of microsomal oxygenations, especially with CYP2E1 that acts as a ROS generator to oxidise resistant chemicals such as benzene and ethanol. ROS are the mediators of inflammation. ROS interaction with platelets, neutrophils, macrophages and other cells leads to the synthesis of eicosanoids and the activation and release of various cytokines, and propagating the inflammatory process from one organ system to another. This results in tissue oxidative stress and multiple system organ failure. Fasting and exposure to ether anaesthetic agent induces CYP2E1 and resulting in tissue oxidative stress by depletion of glutathione (GSH). ROS-induced oxidative damage is characteristic of acute inflammation and chronic inflammatory diseases. The most common targets for autoxidative tissue injury and disease are the biological membranes. These comprise phospholipids and cerebrosides, short-chain and long-chain saturated, unsaturated and polysaturated fatty acids. The short-chain acid, butyrate, has a protective effect against ROS damage, inflammation and cancer, possibly by ROS scavenging or rapid production of energy. The most fundamental membrane function is electron conductance, which appears to depend to a large extent on the presence of choline, a charged molecule, and a zwitter ion $(OCH_2CH_2N(CH_3)_3$, which may facilitate membrane electron transport by Grottus conduction.

Antioxidants functions:

1- Immune Function, Vitamin C $\geq$200 mg/day increases antimicrobial and natural killer cell activities, lymphocyte proliferation, chemotaxis, and delays dermal sensitivity. Treatment of the common cold with vitamin C ($\geq$1 g/day) is associated with reduced duration of cold symptoms. Vitamin C 600 mg/day to 1 g/day is beneficial for preventing colds. vitamin E enhances cellular immune function. Carotenes are associated with enhancing immune function (results may be attributed to vitamin A).
2- Cancer, diets rich in fruits and vegetables are associated with a lower risk of incurring a number of common cancers. Low serum concentrations of β-carotene increase the risk of certain types of cancer, including those of the breast, lung, stomach, intestine, prostate, colon, uterus, ovary, and cervix. Higher consumptions of β-carotene and vitamin C improve survival of prostate cancer. 5 years supplementation with β-carotene 14 mg, α-tocopherol 20 mg, and selenium 50 mg lead to a 9% reduction in cancer mortality, and a 21% reduction in mortality from stomach cancer. Plasma concentrations of all the antioxidants studied except α-tocopherol are significantly reduced in women with cervical neoplasia. Women with the greatest serum concentrations of lycopene have a significantly reduced risk for developing cervical dysplasia. β-carotene may have a beneficial effect for treating oral leukoplakia, but does not have any benefit for treating patients with cervical dysplasia. A diet high in β-carotene, vitamin E, and calcium has been found to be protective against breast cancer. The dietary antioxidants cryptoxanthin, lycopene, lutein, and zeaxanthin protect against breast cancer. Supplemental β-carotene is associated with a significant reduction in the risk of CaP in men with low baseline plasma concentrations of β-carotene. Lycopene, along with other compounds in tomato products, has been found to reduce the risk of cancer of the prostate, pancreas, and the GI tract. Vitamin E, selenium, and lycopene are chemopreventive for prostate cancer.
3- Smoking, antioxidants (β-carotene) may actually increase the risk of all cancer in male-smokers (lung, prostate). β-carotene is susceptible to oxidative damage from alcohol and the gases in cigarette smoke, which may lead to the formation of harmful byproducts. Vitamin C protects against the harmful effects of β-carotene in smokers.
4- Secondary prevention, selenium significantly decreases total cancer mortality and the incidence of lung, colorectal, and prostate cancer. β-carotene may play a protective role for women who already have breast cancer.
5- Treatment, alkylating agents involves the intentional generation of free radicals to cause cellular damage and necrosis of malignant cells. Antioxidants can be beneficial in the treatment of cancer, either as sole agents or as adjuncts to standard radiation and chemotherapy protocols. Antioxidants prolong survival and reduce some of the adverse events associated with chemotherapy. CoQ_{10} for the treatment of breast cancer is promising. Antioxidants have synergistic effects with a number of anticancer medications.

Deficiency of dietary choline or dietary lipotropes (B_{12}, folate, pyridoxal, glycine, PO_4^{3-}) leads to ROS production, lipid peroxidation, tissue injury, malignancy and death. Dietary cholesterol has little affect on blood cholesterol levels. Dietary cholesterol decreases blood cholesterol by enhancing cholesterol 7α-hydroxylase activity and promoting bile acid production. High dietary fiber keeps the bile acid in the stool and hence removes the negative feedback of cholic acid on cholesterol 7α-hydroxylase. Most cholesterol is synthesised in the body where it is needed, from fatty acids that may be derived from dietary carbohydrate. Cholesterol is relatively inert, but some of its polyunsaturated fatty acid esters are highly peroxidisable, and form peroxyl radicals capable of converting cholesterol to toxic oxidation products. Reduction of low-density lipoprotein (LDL) cholesterol by 35% leads to a 40% decrease in adverse CHD events. Nutritional prophylaxis in CHD should focus on prevention of lipid peroxidation, by using antioxidants rather than cholesterol removal. Oestrogens act as antioxidant cardioprotectants. LDL oxidative damage that leads to atherogenesis is inhibited by 17β-oestradiol. Oxidised low-density lipoprotein (OxLDL) exerts proliferation and apoptosis in vascular cells, depending on its concentration and the time of exposure. NADPH oxidase is a major source for O_2^- formation. OxLDL stimulates vascular O_2^- formation. OxLDL induces proliferation at low and apoptosis at higher concentrations. Both effects are mediated by O_2^- formation, with NADPH oxidase being a major source for O_2^-. The endogenous enzymes and other factors act by:
1- scavenges ROS, 2- reduces toxic disulphides, and 3- reduces soluble and membrane-bound peroxides. The vital

repair requires NADH, NADPH and antioxidant vitamins to function efficiently. Ascorbic acid, α-tocopherol and GSH, interacts as a complex system to reduce ROS and other oxidants. Disease is a sequential series of interactive phenomena. Prevention of disease can be achieved by arresting one or more of the critical stages of those phenomena. The phenomena linking organ failures are inflammation and cytokines, and beside aggressive rescue and resuscitation by perfusion, nutrition and appropriate medication, the use of antioxidants to arrest ROS production is indicated. Gastric cancer is attributed to Helicobacter pylorii infections, surgical vagotomy, stress and trauma mediated by cytokines. This pathobiological phenomena progressing from stress to inflammation of the gastric mucosa, to achlorhydra, bacterial invasion of the gastric mucosal barrier, depletion of antioxidant protectants, formation of nitrosamines in gastric lumen, mutations and malignancy. H. Pylorii infection is highly resistant to antimicrobial treatment. Ascorbic acid by inhibiting nitrosation, aids the antibiotics to eliminate the microbial overgrowth and restore the normal mucosa. Vitamin C enhances the effect of endothelium-derived nitric oxide (NO). Supplementation with vitamin C normalises vascular function in patients with coronary artery disease and associated risk factors, including hypercholesterolaemia, hyperhomocysteinaemia, hypertension, diabetes and smoking. Treatment of hypertensive patients with ascorbic acid lowers blood pressure.

BIOMEDICAL ASPECTS OF ANTIOXIDANTS

Oxygen-derived free radicals are the by-products of aerobic metabolism. A free radical is an atom or molecule with one or more unpaired electron(s). This electron imbalance causes high reactivity, creating other free radicals by chain reactions. Oxygen-derived free radicals (OFRs) and other pro-oxidants are important mediators in signal transduction and play a vital role in the production of biologically active and essential compounds. OFRs are toxic, inflicting damage upon cells by promoting the oxidation of lipids, proteins, and DNA to induce peroxidation, modification, and strand break. Humans are exposed to environmental sources of prooxidants such as cigarette smoke, ultraviolet (UV) radiation and oxidising agents. Antioxidants are chemical compounds or substances that inhibit oxidation. OFRs interact with the lipid component of LDL particles and cause this lipid to become peroxidatively modified. Polyunsaturated fatty acids contained within the LDL particle especially susceptible to attack by free radicals species. These reactive compounds are capable of abstracting one of the double allylic hydrogen atoms on the carbon atom between the double bonds of the fatty acid.

$$LH + R^\cdot \rightarrow L^\cdot$$
$$L^\cdot + O_2 \rightarrow LOO^\cdot$$
$$LOO^\cdot + LH \rightarrow LOOH + L^\cdot$$
$$LOOH + Fe^{3+} \rightarrow LOO^\cdot + H^+ + Fe^{2+}$$
$$LOOH + 2GSH \rightarrow LOH + GSSH + H_2O$$

Where $R^\cdot$ is free radical, LH is polyunsaturated fatty acid; $L^\cdot$ is the pentadienyl radical, $LOO^\cdot$ is peroxyl radical, LOH is alcohol, GSH is selenium dependent glutathione peroxidase.

LDL is susceptible to modification by oxidation, and the oxidatively modified LDL triggers foam cell formation. Oxidation of LDL results from interaction with metal ion such as Cu^{2+} and Fe^{2+}, with cigarette smoke, with arterial cells including endothelial, macrophage and smooth muscle cells. Iron is a major generator of free radicals. At physiological levels, the free radicals generated by iron are essential for cell regeneration, while an excess amount of iron could be potentially detrimental to the cell. Oxidative stress stimulates chemotactic responses in macrophages and promotes smooth muscle proliferation and differentiation. The pathological consequence of lipid free radical and peroxide productions is athrogenesis. These reactive substances react with apolipoprotein B (apo-B), the sole protein component of LDL, and promote its oxidation (oxidatively modified forms of LDL). The oxidation of lipids in LDL is characterised by the conversion of polyunsaturated fatty acids into lipid hydroxyperoxides with conjugated double bonds (conjugated dienes). Further oxidation of LDL causes the conjugation of lysine resides of apo-B with lipid peroxide products such as malnodialdehyde, and ultimately the fragmentation of the protein. Ascorbic acid (vitamin C) reacts with hydrogen peroxide (H_2O_2), the hydroxyl radical ($^\cdot OH$), peroxyl radical ($ROO^\cdot$) and singlet oxygen ($O_2^{\cdot-}$), to form the semidehydroascorbic radical ($A^\cdot$) and dehydroascorbate (A):

$$AH^- + {^\cdot OH} \rightarrow H_2O + A^\cdot$$
$$AH^- + O_2^{\cdot-} + H^+ \rightarrow H_2O_2 + A^\cdot$$
$$AH^- + ROO^\cdot \rightarrow RH + A^\cdot$$
$$AH^- + H_2O_2 + H^+ \rightarrow 2H_2O + A$$

Ascorbic acid acts in the plasma to scavenge free radicals, dissipating these reactive species before they can react with biological membranes and lipoproteins. Elderly subjects receiving 400 mg of vitamin C for 1 year, experience a 13% decrease in serum peroxide levels. Ascorbic acid has the ability to regenerate the activity of lipid-soluble antioxidants, such as γ-tocopherol and β-carotene, by interacting with biological membranes at the aqueous-lipid interphase. α-Tocopherol is ineffective in cardiac tissue that contains low level of ascorbic acid. Ascorbic acid lowers plasma total cholesterol by promoting its conversion to bile acids. Chronic ascorbic acid deficiency may result from increased vascular fragility secondary to a deficiency in collagen synthesis. Ascorbic acid lowers blood pressure, possibly by influencing prostaglandin (PGs) synthesis. α-Tocopherol (vitamin E) is the most abundant and the most active isomer in vitamin E family in humans. Once absorbed from the gut, vitamin E is transported to target tissues via the lipoproteins. LDL carries 50% of circulating vitamin E. An average of six molecules of α-tocopherol are present in each LDL particle. Vitamin E functions by denoting hydrogen to fatty peroxyl radicals, thereby halting lipid peroxidation.

α-Tocopherol → α-tocopherol· + LOOH

α-Tocopherol· + LOO· → LOO-α-tocopherol

Vitamin E is the most effective antioxidant for reducing lipid peroxidation. Vitamin E is the strongest contribution to the inverse relationship between serum antioxidant concentrations and ischaemic heart disease. α-Tocopherol supplementation decreases the severity of atherosclerosis and promotes the regression of diet-induced atherosclerotic lesions. Vitamin E inhibits oxidative LDL modification, but fails to prevent dietary cholesterol-induced atherosclerosis. Vitamin E supplementation of 600 mg/day protects against smoking-induced LDL oxidation. α-Tocopherol radical is reduced by vitamin C, resulting in the regeneration of the metabolically active form of vitamin E. Selenium is another antioxidant that is involved in vitamin E function. Selenium is an essential element that is necessary component of the enzyme glutathione peroxidase (GSH). Selenium dependent glutathione peroxidase functions to inactivate lipid peroxides. Carotenoids including β-carotene, γ-carotene and lycopene, are lipid soluble, carries within lipoprotein particles. 1 mol of LDL contains 0.29 mol of β-carotene, 0.12 mol of α-carotene and 0.16 mol of lycopene. Carotenoids are thought to be highly efficient quenchers of singlet oxygen ($^1O_2$·). β-Carotene is capable to trap more lipid free radicals than α-tocopherol, the latter is capable to trap a maximum of 2.

CAR + LOO· → LOO-CAR·

LOO-CAR· + LOO· → LOO-CAR-OOL

LOO-CAR-OOL + LOO· → (LOO)$_2$-CAR-(OOL)$_2$

(LOO)$_2$-CAR-OOL· + LOO· → (LOO)$_2$-CAR-(OOL)$_2$

Where CAR is β-carotene.

A single β-carotene molecule eliminates up to 1000 singlet oxygens before it is oxidised and loses its antioxidant properties. Foams cells are the major constituents of the fatty streak associated with developing atherosclerosis. Oxidised LDL is responsible for the lipid loading of macrophages in arterial walls resulting in the formation of foam cells. A cocktail of antioxidants is more effective than one individual antioxidant in isolation. All antioxidants can potentially become pro-oxidants subject to their concentrations. Antioxidants interfere with the production of free radicals and/or inactivate them once they are formed. Vitamin E protects against lipid peroxidation and prevents the loss of cell membrane fluidity. β-Carotene enhances natural killer (NK)-immune cell functions such as killing of tumour cells, and increases the secretion of TNF from human monocytes.

ANTIOXIDANTS AND RESPIRATORY DISEASES

Ozone (O_3) and nitrogen dioxide (NO_2) have high risk of oxidant injury. Epithelial lining fluid (ELF) and lung tissue plays an important role in protecting the respiratory system against oxidant damage. Uric acid is the major low-molecular-weight antioxidant in upper respiratory tract fluids. It is co-secreted with lactoferrin into the upper airways, and is closely associated with mucin. In plasma, urate is the most potent scavenger of ozone, and it has the same action in the respiratory tract. Urate can chelate transition metals and this may contribute to its antioxidant activity. Ascorbate contributes to antioxidant protection in the lung, especially lower respiratory tract. Ascorbate protects against the bronchoconstriction induced by exposure to both ozone and nitrogen dioxide, and increases bronchial responsiveness observed during upper respiratory tract infection. Glutathione in the lower respiratory tract normal levels are >50 times those in plasma. Glutathione is synthesised within cells lining the respiratory tract and subsequently released into ELF. Oxidised glutathione can then be recycled by glutathione reductase. α-Tocopherol is low in ELF, where the aqueous environment has a little lipid content. Most of the α-tocopherol is likely to be associated components of surfactant, which constitutes the main lipid components present in ELF. Normally ELF contains very little transition metals, but in the presence of lung inflammation, metals may be released from damaged cells, and binding proteins will then serve an important antioxidant function (lactoferrin, transferrin, albumin, caeruloplasmin [ferroxidase activity]), catalase and superoxide dismutase [SOD]). Albumin has copper-binding activity and rich content of sulphydryl groups. Mucin consists of a core glycoprotein rich in threonine and serine with a number of cysteine-rich domains that are involved in disulphide formation. These disulphide bridges are important in giving mucin its characteristic consistency. The abundant thiol groups provide mucin with the capacity to directly scavenge many oxidants. Mucins have transition metal-binding properties. Mucins are excreted in increased amounts following exposure of airways to pollutants and other irritants and are likely to provide a substantial contribution to the antioxidant protection of upper airways. Asthma is prevalent in areas with higher levels of ambient ozone, with outdoor ozone pollution exacerbating symptoms. Antioxidant defences of ELF and lung tissue plays an important role in protecting the respiratory system against oxidant damage. Once these antioxidants are depleted, the bronchial epithelium undergoes oxidative injury, resulting in enhancement of eicosanoid metabolism and subsequent generation of products of the AA pathway, which contribute to the development of early symptoms and decrement in lung function. Damage to the bronchial epithelium results in generation of tachykinins (e.g., substance P), which contribute to the early symptomatic responses and changes in lung function. Exposure to ozone leads to rapid depletion of urate in ELF. Uric acid is the most potent lung antioxidant, followed by ascorbic acid and finally glutathione. Cigarette smoking is a major risk factor for the development of pulmonary disease, including emphysema, chronic bronchitis and lung cancer. Cigarette smoking contains substantial quantities of free radicals in both gas and tar phase including superoxide and NO, which may combine to produce peroxynitrites, the highly damaging hydroxyl radical, tar semiquinone free radicals and various xenobiotic electrophiles. Cigarette smoking have increased numbers of pulmonary inflammatory cells that will provide a secondary source of increased free radical production, and circulating leukocytes have an increased oxidative burst, which will make significant contribution to oxidative damage in the airways. Cigarette smoking is associated with increased levels of lipid peroxidation products in plasma, exhaled breath, and lung tissue. Elevated levels of DNA and protein damage products can be detected that may contribute to smoking-induced cancer and emphysema. In emphysema, tobacco smoke inactivates the anti-protease α_1-antitrypsin, which opposes the action of neutrophil elastase. Smoking is associated with reduced antioxidant levels in various body fluids, which is due to a combination of reduced dietary intake of fruit and vegetables and repetitive sessions of oxidative stress. Smokers have low plasma vitamin E status to both a reduction in the ability to absorb α-tocopherol and increased clearance of the freshly absorbed vitamin E. Iron concentrations of smokers increase in alveolar macrophages, ELF and upper lobes. This is due to iron present in the cigarette, and increased release of iron from its binding proteins as a result of biochemical disturbances induced by cigarette smoke. The superoxide and hydrogen peroxide react in the presence of iron to produce the hydroxyl radical (oxidative stress). Cigarette smoke is a rich source of oxidants, contains 500 mg/kg NO and 10^{15} radicals per puff that contribute to oxidative stress. Only 20% of smokers develop chronic obstructive pulmonary disease (COPD). Airways inflammation is a key factor in asthma, characterised by the presence of eosinophils, T lymphocytes, and mast cells in the airways. These cells produce a wide range of inflammatory mediators including free radicals. Vitamin C that protects against bronchoconstriction is reduced in asthmatic; also selenium, glutathione and retinol are reduced in ELF. Magnesium influences muscle contraction and is related to wheezing and airway hyperactivity in asthma. Selenium deficiency in premature infants is associated with glutathione peroxidase deficiency. Antioxidant enzymes are also reduced that is associated with increased susceptibility to lung damage. Corticosteroids induce antioxidant enzymes. Premature babies have reduced serum transferrin resulting in circulating non-protein-bound redox reactive iron. Leakage of this fluid into alveolar fluid results in a pro-oxidant reaction with the formation of hydroxyl radicals. The caeruloplasmin is reduced in premature infants. High ascorbate concentration in neonates inhibits ferroxidase activity of ceruloplasmin. In cystic fibrosis (CF) there is 1000-fold increase in lung neutrophil numbers as a result of heightened immune response from repetitive pulmonary infections. Once activated, these neutrophils are a major source of free radicals. CF patients are placed under increased oxidative stress, have insufficient antioxidant defences resulting in oxidative lung damage. This would exacerbate the progressive decrease in lung function. ARDS is associated with massive infiltration of the lungs by neutrophils with consequent oxidant injury. There is ELF glutathione deficiency in ARDS, with systemic increase in antioxidant turnover. Depletion of antioxidants increases tissue sensitivity to oxidant injury in a variety of respiratory diseases. Intravenous administration of N-acetylcysteine (NAC) increases glutathione in plasma, RBCs and ELF. Detrimental oxidative process in the body can be initiated by: 1- pro-oxidants present in the food and the environment, 2- activation of immune system due to infection, 3- neutrophils migrating to the site of infection, 4- smoking and drug toxicities. The cyclooxygenase-independent, free radical-catalysed oxidation of the polyunsaturated fatty acid (PUFA) results in arachidonic acid (AA). The superoxide ion (O_2^-) and hydrogen peroxide (H_2O_2) are 2 prominent free radicals that can catalyse the conversion of AA to isoprostanes. Isoprostanes are 8 racemic diastereomers, with the potential to form 64 isomeric structures, and are referred as F_2 isoprostanes. Of these, 8-epi-prostaglandin $F_{2\alpha}$ (8-iso-PGF$_{2\alpha}$) is the most abundant in human tissues and fluids. Monocytes and platelets have the ability to synthesise 8-iso-PGF$_{2\alpha}$ via both cyclooxygenase-catalysed and free radical-catalysed mechanisms. Prostanoids are formed via enzyme-catalysed reactions, while isoprostanes are produced via an enzyme-independent free radical-catalysed process. Prostanoid formation requires the fatty acid substrate to be cleaved from the parent phospholipid by phopholipase A_2 before it can be acted upon by cyclooxygenase, while isoprostanes are formed while AA is still esterified to its parent phospholipid, after which it is cleaved by phospholipases. Isoprostanes are found in human fluids at higher concentrations than prostanoids. Isoprostanes are chemically stable, whereas TXA_2 half-life is 30 s after which it rearranges to the biologically inactive TXB_2. Isoprostanes does not induce platelet shape-change. Isoprostanes are the most accurate indicator of oxidative stress and lipid peroxidation. Increased free radical production is one of the main causes of tissue damage during coronary reperfusion. Coronary reperfusion is associated with increased urinary level of 8-iso-$PGF_{2\alpha}$. Urinary 8-iso-$PGF_{2\alpha}$ is elevated in smokers. Urinary analysis of isoprostanes can provide an indication of oxidative stress. Cystic fibrosis and asthma have elevated isoprostanes levels. Administration of oxygen can cause pulmonary oedema. After acute and chronic hyperoxia, increased oxygen radical formation can occur, leading to toxic injury. Liver cirrhosis is associated with increased free radical production and decreased levels of antioxidants, e.g., vitamin E, glutathione and selenium. Cigarette smoke contains large quantities of oxidants. ROS can act as signalling molecules in the regulation of gene expression. Extracellular radicals can initiate cell signalling, and intracellular radicals can act as second messengers. Antioxidants can affect the binding ability of transcription factors to DNA, can modify redox-sensitive sites on signalling molecules, and can alter phosphorylation reactions. The antiatherosclerotic effect of ω-3 PUFA eicosapentaenoic acid (EPA) and docosahexaenoic acid (DHA) results from increasing glutathione peroxidase activity. Apoptosis is reduced by antioxidants. Excess iron induces cellular oxidant injury via catalyse the generation of OFRs. Excess iron increases hepatic F_2 isoprostane levels and reduces levels of the antioxidants α-tocopherol, β-carotene and ascorbic acid. Selenium acts as antioxidant by being an essential component of active glutathione peroxidase. CF patients have elevated RBC 8-iso-$PGF_{2\alpha}$, and depressed β-carotene levels. Hypertensive patients have low endogenous antioxidants. Daily administration of zinc sulphate, ascorbic acid, α-tocopherol, and β-carotene can reduce systolic blood pressure. Low dose aspirin inhibits TX in placental cells of pre-eclamptic women. Oxidative stress is an inevitable feature of life, induced by reactive forms of oxygen released during normal respiration, by the oxidative burst of the macrophages in response to infection, and by variety of exogenous agents including cigarette smoke and ionising radiation.

Prostanoid vs isoprostane:

Prostanoid	Isoprostane
Formed via enzyme-catalysed reactions	Produced via an enzyme-independent free radical-catalysed process.
Requires the fatty acid substrate to be cleaved from the parent phospholipid by phospholipase A_2 before it can be acted upon by cyclooxygenase.	Formed while AA is still esterified to its parent phospholipid, after that it is cleaved by phospholipases.
Are found in human fluids.	Are found in human fluids at higher concentrations.
Chemically unstable.	Chemically stable.
Stimulate constriction of vascular smooth muscle.	Stimulate constriction of vascular smooth muscle.
Induce platelet shape-change.	Does not.

BIOMEDICAL ASPECTS OF COENZYME Q AND α-TOCOPHEROL

Hydroperoxides are high in erythrocyte membranes at birth and significantly decreases at 3 and 72 h after birth. In the erythrocyte membranes, CoQ_{10} content shows an opposite behaviour with respect to the plasma compartment. A normal delivery means a strong stress, which is characterised by an increase of free radicals. In the case of premature newborns, this oxidant damage could be exacerbated due to the probable immature mechanisms of antioxidant defence. The erythrocyte membranes of at term newborns show the highest amount of hydroperoxides at birth. Erythrocyte membranes levels of α-tocopherol in at term newborns amount tripling after 72 h of life. The fact could be attributed to its utilisation in a moment of strong oxidative stress. Oxidation of LDL, the major cholesterol-carrying protein in human blood plasma, is implicated as an important early event in atherogenesis. Oxidation of the lipid moiety of LDL is generally held to proceed, and to some extent cause, the modification of apolipoprotein B_{100}, which ultimately results in uncontrolled uptake of the lipoprotein by cells via the scavenger receptor to form lipid-laden or foam cells. Antioxidants modulate expression of certain inflammatory genes via alteration of the activation potential of the redox transcription factors, NFκB, and AP-1. α-TOH, biologically and chemically the most active form of Vitamin E, and quantitatively the major lipid-soluble antioxidant in extracts from human LDL. α-TOH does not act as a conventional chain-breaking antioxidant; rather α-TOH facilitates the transfer of radical reactions from the aqueous phase into LDL, and its one-electron oxidation product, α-TO can react as a peroxidation chain transfer agent causing formation of lipid hydroperoxides. (N.B., Bisallylic hydrogens refer to the PUFA, i.e., the most readily oxidised lipid moieties of the lipoprotein. Linoleic and arachidonic acid contain one and three pairs of bisallylic hydrogens, respectively). α-TOH's activity is the ability to react readily with the chain-carrying lipid peroxyl radical (LOO˙), thereby breaking the chain of lipid peroxidation. Alternatively, α-TOH can react directly with the peroxidation initiating radical (ROO˙). In either case the relatively unreactive α-TO˙ is assumed to rapidly react with another available radical to yield non-radical products (NRP). Each molecule of α-TOH destroys 2 radicals.

LOO˙ + α-TOH → LOOH + α-TO˙
ROO˙ + α-TOH → ROOH + α-TO˙
LOO˙ + α-TO˙ → NRP
ROO˙ + α-TO˙ → NRP

This conventional mode of lipid peroxidation and action of α-TOH has been verified extensively in homogeneous solution, micelles and liposomes.

LOO˙ + LH → LOOH + L˙
L˙ + O_2 → LOO˙

$CoQ_{10}H_2$-free lipoproteins and ascorbate- and $CoQ_{10}H_2$-free plasma are exposed to a variety of oxidants. α-TOH makes lipoproteins oxidizable, it aids the transfer of radical reactions from the aqueous phase into the lipoprotein particle. Once a LDL contains a radical species it is present predominantly as α-TO˙. α-TO˙ is relatively free to move within LDL, it cannot readily escape from the lipoprotein, due to the lipophilic phytyl tail of the vitamin. α-TOH is regenerated with concomitant formation of a carbon-centred lipid radical (L˙). L˙ will react rapidly with molecular oxygen forming the lipid peroxyl radical (LOO˙), which will be scavenged rapidly by α-TOH to generate α-TO˙ and result in the formation of a lipid hydroperoxide (LOOH). So, α-TOH makes the lipoprotein oxidizable and α-TO˙ can act as chain-carrying radical resulting in substantial amounts of LOOH being formed without consumption of α-TOH. In order for α-TO˙ to be destroyed, a suitable reducing agent must be present, or alternatively, a second radical must enter the confined space of an oxidising LDL particle, in which case radical-radical termination occurs with formation of NRP and consumption of α-TOH. Once a radical enters LDL, it is present predominantly as α-TO˙, because this is the most stable radical that can be formed and α-TO˙ cannot readily leave LDL.

Composition of human LDL:

Component		Weights	mol/LDL
1- Protein:			
	Apolipoprotein B_{100}	22 ± 1.9%	1
2- Lipids:			
	Phospholipids	22.3 ± 3.9%	700
	Phosphatidylcholine		450
	Bisallylic hydrogens		375
	Cholesterol	9.6 ± 0.7%	600
	Cholesteryl esters	42.2 ± 3.8%	1600
	Cholesteryl linoleate		880
	Cholesteryl arachidonate		95
	Bisallylic hydrogens		1165
	Triglycerides	5.9 ± 2.7%	180
	Bisallylic hydrogens		50
	Total bisallylic hydrogens		1590
3- Antioxidants:			
	α-TOH		6-12
	γ-TOH		0.5
	Ubiquinol-10		0.5-0.8
	Lycopene		0.2-0.7
	β-Carotene		0.1-0.4

As a consequence, α-TO˙ becomes the chain-transfer agent. Although the chain transfer activity of α-TO˙ is retarded compared to that of the conventional, more reactive, chain-carrying LOO˙, the overall effect of α-TOH can still be pro-oxidant due to its dual activity as a chain- and phase-transfer agent. The conversion by the co-antioxidant of the lipophilic α-TO˙ into harmless aqueous radicals or NRP, which prevents LOOH-formation. A variety of natural reductants can act as co-antioxidants, including $CoQ_{10}H_2$, ascorbate, albumin-bound bilirubin, and the tryptophan metabolite 3-hydroxyanthranilic acid. $CoQ_{10}H_2$ is the first antioxidant consumed when LDL is exposed to oxidants such as aqueous and lipophilic peroxyl radicals, transition metals (Cu^{2+}, Fe^{3+}), activated human neutrophil, unstimulated monocytes/macrophages, hypochlorite, singlet oxygen, and peroxynitrite. While $CoQ_{10}H_2$ is present, formation of oxidised lipids is markedly suppressed except in the case of hypochlrorite where lipid peroxidation is a minor reaction. In unsupplemented human subjects only every second LDL particle contains, on average one molecule of $CoQ_{10}H_2$. However dietary supplementation of human subjects with 100-300 mg/d of coenzyme Q results in increased concentrations of $CoQ_{10}H_2$ in plasma and all its lipoproteins. Interestingly, supplementation does not alter the redox ratio of $CoQ_{10}H_2$ to ubiquinone-10 (CoQ_{10}) in LDL (and plasma), which remains constant with 80% of the total coenzyme Q present as ubiquinol-10 ($CoQ_{10}H_2$, the reduced form of CoQ_{10}). This suggests that there is sufficient reducing potential available to keep circulating coenzyme Q in the reduced (co)-antioxidant active form. It is assumed that lipid peroxyl radicals ($LOO^\cdot$) and $\alpha\text{-}TO^\cdot$ move freely within though do not readily escape from oxidising lipoprotein particles. LDL from most healthy people contains, on average, <1 molecule $CoQ_{10}H_2$/particle. $CoQ_{10}H_2$ can scavenge $\alpha\text{-}TO^\cdot$, resulting in the formation of $\alpha\text{-}TOH$ and the ubisemiquinone radical ($CoQ_{10}^{\cdot-}$). $CoQ_{10}^{\cdot-}$ can under go one of two reactions: 1- it may scavenge an additional $\alpha\text{-}TO^\cdot$, a more favourable than of $CoQ_{10}H_2$ with $\alpha\text{-}TO^\cdot$; 2- $CoQ_{10}^{\cdot-}$ may auto-oxidise to yield the charged superoxide anion radical ($O_2^{\cdot-}$), which is forced to leave the LDL particle. Whether the putative $O_2^{\cdot-}$ formed dismutates (into H_2O_2 and O_2) or perhaps reduces a second molecule of $\alpha\text{-}TO^\cdot$ is not known. Plasma $CoQ_{10}H_2$, total coenzyme Q and the coenzyme Q redox status are slightly lower and the levels of $\alpha\text{-}TOH$ slightly higher. $\alpha\text{-}TOH$ does not appear to be limiting in advanced atherosclerotic plaques while $CoQ_{10}H_2$ may be. Megamitochondria (MG) are often intimately related to the generation of free radicals. MG is irregular shaped mitochondria with poorly developed cristae, may be the result of suppression of the dividing process of mitochondria. Ethanol, hydrazine and chloramphenicol (CP) induce the formation of MG. This formation of MG decreases the body weight and the weight of the liver, remarkably increases the level of lipid peroxidation, and increases the activity of xanthine oxidase. Ethanol-induced chronic liver hepatic injury has been correlated to free radical production besides acute one. CP induces MG since it suppresses protein synthesis in mitochondria resulting in the disturbance in their living dividing process. Aromatic nitro compounds, such as nitrofurantoin, metronidazole and CP, undergo enzymatic one electron reduction to the corresponding radicals, and these radicals react with molecular oxygen resulting in the formation of superoxide. Allopurinol partly suppressed the hydrazine- and CP-induced formation MG. α-tocopherol and CoQ_{10} partly suppress the hydrazine-induced formation of MG. The reduced form of coenzyme Q, ubiquinol, has been shown to act as an antioxidant against free radical-mediated oxidations in membranes and lipoproteins. Ubiquinol is consumed much faster than α-tocopherol and the induction period produced by ubiquinol-10 is much shorter than that by α-tocopherol.

$$UQ_{10}H^\cdot + O_2 \rightarrow UQ_{10} + HO_2^\cdot$$
$$HO_2^\cdot + UQ_{10}H_2 \rightarrow UQ_{10}H^\cdot + H_2O_2$$

The lower antioxidant activity of ubiquinol-10 than α-tocopherol against lipid peroxidation, in spite of its higher reactivity toward peroxyl radical than α-tocopherol, may be explained by the different behaviour of the ubiquinol semiquinone radical and α-tocopheroxyl radical. The former gives hydroxyperoxyl radicals by reaction with oxygen, which may attack the substrate to induce chain initiation, whereas the latter scavenges another radical to give a stable adduct product. Vitamin C present in the aqueous phase efficiently reduces vitamin E radicals located within the membrane and LDL particle to regenerate vitamin E and also to inhibit vitamin E radical-induced chain initiation. Therefore, vitamin C and E inhibit the lipid peroxidation synergistically. Ubiquinol reduces vitamin E radicals. Ubiquinol and vitamin C compete in the reduction of vitamin E radicals. Vitamin C does not reduce ubiquinol semiquinone radical efficiently, they must function independently. The side chain of lipophilic antioxidant is important for its incorporation and retainment in biomembranes and lipoproteins. The mobility of ubiquinol-10 between the membranes is restricted. Ubiquinol-10 spared vitamin E in the same membranes but not in different membranes. Ubiquinol and vitamin E are localised in the same lipophilic domain of lipoproteins and membranes and act as a potent radical-scavenging antioxidant. Ubiquinol may contribute to the reduction of α-tocopherol radicals formed when α-tocopherol scavenges radicals, which results in the sparing of α-tocopherol and the consumption of ubiquinol. Ubiquinol is rapidly auto-oxidised by a chain mechanism to give ubiquinone and hydrogen peroxide. The kinetic chain length of ubiquinol auto-oxidation depends on the conditions. The fate of radicals derived from ubiquinol is also important in determining the antioxidant capacity.

$$UQ^\cdot + O_2 \rightarrow UQ + O_2^{\cdot-}$$

The interaction between ubiquinol, vitamin E and vitamin C play an important protective role against oxidative stress. Coenzyme Q is uniquely designed as an electron and proton carrier within the lipid phase of membranes. This unique chemistry has diverse application to important functions in all cellular membranes. Coenzyme Q functions in the plasma membrane electron transport involved in activation of signalling protein kinases related to gene activation for cellular proliferation. An extremely basic function of coenzyme Q in medication and preservation of antioxidant function in cells, acts to remove oxygen radicals, and reduce tocopheryl radicals and semihydroascorbate back to tocopherol and ascorbate, respectively. Coenzyme Q acts as antioxidant because it can be reduced by metabolic supply of NADH or NADPH in the cell to form the hydroquinone. Electron transport across the plasma membrane induces massive proton release by cells. Most of the proton release activated by electron transport at the plasma membrane is based on activation of Na^+/H^+ exchange. This proton movement might be based on anisotropic oxidation of coenzyme Q. Added coenzyme Q may reach interior sites and not act exclusively at the plasma membrane. The reduced coenzyme Q can reduce tocopheryl radical. Coenzyme Q acts to reduce extracellular ascorbate free radical back to ascorbate. These antioxidant functions for CoQ are evidence that coenzyme is present in all cellular membranes, not just mitochondria. NADH cytochrome b_5 (Nb_5R) is located on the cytosolic side of all endomembranes including the plasma membrane and is known primarily for its function in reduction of the non-heme iron fatty acid desaturase. It is also related to the p450-based detoxification as an alternative source of electrons from NADPH. Nb_5R can reduce coenzyme Q_{10}. The reduction of coenzyme Q in the plasma membrane provides a source of electrons for the ascorbate free radical outside the cell. NADH and NADPH can be a source of electrons for regeneration of antioxidants. In addition to its protonophoric role in energy coupling are the redox control of gene expression and cell growth and a primary role in antioxidant protection. Reasons for CoQ deficiency in mitochondria: 1- Decreased CoQ levels, decreased or insufficient biosynthesis or increased catabolism. 2- Increased demand for CoQ may be: a- bioenergetics (decreased activity of the respiratory chain, or increased affinity for CoQ of respiratory enzymes); b- oxidant (oxidative stress, aging). Ischaemia decreases the content of semiquinones of CoQ_{10} in the mitochondria. Changes in the characteristics of CoQ_{10}, flavoproteins and iron-sulfur centers in mitochondrial membranes, produced by prolonged ischaemia, are the underlying pathology in the reperfusion or re-oxygenation damage of cells. The enzyme responsible for the reduction of UQ to UQH_2 in animal tissue has not been identified. About 60% of plasma CoQ_{10} is found associated with LDL. LDL is drastically involved in the onset of atherosclerosis. The redox status of CoQ_{10} in plasma may change in acute or chronic pathological conditions, where a certain degree of peroxidative insult is present. Vitamin E, β-carotene or selenium reduces platelet aggregability and decreases the serum concentration of β-thromboglobulin and thromboxane in humans. CoQ_{10} could modulate haemostasis, and affect the expression of vitronectin receptor on the platelets. Inhibition of platelet vitronectin receptors may contribute substantially to the success of antioxidant therapy in an expanding array of cardiovascular diseases.

COENZYME Q_{10} (CoQ_{10})

In 1957 an orange molecule from the mitochondria of beef heart was isolated and named ubiquinone. Several years later, the molecule's specificity for 2 mitochondrial enzymes NADH and succinate dehydrogenase, which elucidated its function as a coenzyme were demonstrated. CoQ_{10} is an endogenously synthesised provitamin that serves as a lipid-soluble electron carrier in the mitochondrial electron transport. The alternative names ubidecarenone and ubiquinone, meaning ubiquitous quinone, allude to the presence of CoQ_{10} in all cells. Coenzyme Q exists in several forms. CoQ_{10} is prevalent in humans, with high endogenous concentrations found in the heart, liver, kidneys, and pancreas. Current CoQ_{10} supplements is manufactured by the fermentation of beets and sugarcane. Supplementation with CoQ_{10} is common in Europe, Russia, and Japan. CoQ_{10} is an essential electron and proton carrier that functions in the production of biochemical energy in aerobic organisms. CoQ_{10} also has antioxidant and membrane stabilising properties that serve to prevent the cellular damage that results from normal metabolic processes. The structure of CoQ_{10} consists of a quinone ring attached to an isoprene side chain. It contains 82.08% carbon, 10.51% hydrogen and oxygen and its molecular weight is 863.37. CoQ_{10}'s formula is $C_{59}H_{90}O_4$. CoQ_{10} is produced in all mammalian tissues. The quinone ring is synthesised from the amino acids tyrosine and phenylalanine, and the polyprenyl side chain is synthesised from acetyl-CoA. CoQ_{10} or ubiquinone is a redox component of the respiratory chain, which may be involved in the pathogenesis of cancer. Since prooxidants may promote tumourigenesis, ubiquinone supplementation in cancer could be relevant. CoQ_{10} can provide rapid protective effects in patients with acute myocardial infarction (AMI) when administered within 3 days of the onset of symptoms. CoQ_{10} is an essential electron and protein carrier in ATP synthesis in the mitochondrial inner membrane. CoQ_{10} is required for transmembrane electron transport that activates signals in the cell, which stimulate cell growth. The reduced form of CoQ_{10} acts as a lipophilic antioxidant, preventing initiation and/or propagation of free radicals and lipid peroxidation in biological membranes, and is the only known lipid soluble antioxidant that cells can synthesise de novo. CoQ_{10} and possibly other components of the mevalonate pathway, such as dolichol and dolichyl phosphate, could be important in various disease and senescence mechanisms. Ageing cells contain altered mitochondria that are less able to fulfill their energy requirements so that a general lowering of homeostasis and increased susceptibility occurs. When oxidative phosphorylation decreases, disease symptoms appear and cell degenerates resulting in energy production decreases further. Between the ages of 20-30 and 60-90, there are large and significant decreases in the activities of complexes I (NADH) and IV (CoQ_{10}) of oxidative phosphorylation by 59% and 47% respectively. Alteration and decline of respiratory chain enzyme activity decreases the maximal rate of ATP formation in aging cells, forcing the cells to adapt to a declining availability of energy for biosynthesis and repair.

The concentration of CoQ_{10} falls with increasing age in all tissues. As CoQ_{10} levels decrease dolichol levels increase, indicating a shift in the regulation of the related pathways of dolichol, CoQ_{10}, and cholesterol synthesis. Dolichol destabilises membranes and increases fluidity and permeability. This shift in the pathway could alter the role of CoQ_{10} in signalling for cell growth, and a reduction of CoQ_{10}'s mitogenic properties could indirectly lead to accumulation of DNA damage and reduction of cell viability. There is a drastic decreases in activity in human from a defect in complex I activity. CoQ_{10} is the rate-limiting compound of the activity of complexes I and III but not of complexes II and IV. Lowered ATP synthesis results both directly and indirectly from the shift in mevalonate regulation, and not from an actual lack of CoQ_{10} in the mitochondria. Increasing the concentration of CoQ_{10} within the mitochondrial inner membrane will cause an increase in the production of ATP due to CoQ_{10} being the rate limiting compound for complex I. CoQ_{10} concentration within the mitochondrial inner membrane can control the efficiency of oxidative phosphorylation. CoQ_{10} improves contractility and ejection fraction in heart failure, and significantly increase myocardial function and work capacity in normal sedentary people and in patients with mitochondrial disease. Caloric restriction leads to low blood glucose levels, which in turn stimulate the release of glucagon. Glucagon has a range of effects on different pathways, including the mevalonate pathway. Increased glucagon levels inhibit glycolysis by lowering the level of the intermediate fructose-2,6-bisphosphate, which is an inhibitor of fructose-1,6-phosphatase and an activator of phosphofructokinase-1. Glucagon also inhibits pyruvate kinase, so that pyruvate is prevented from entering the citric acid cycle, and the resulting accumulation of phosphoenol pyruvate favours gluconeogenesis. Caloric restriction maintained over long period of time reduces acetyl-CoA due to fatty acid metabolism, and there would not be an increase in the level of precursors of the mevalonate pathway. An increased level of glucagon itself is sufficient to inhibit HMG-CoA reductase, and thereby would actually decrease the level of mevalonate. Mevalonate is at a major control point in this pathway, and is converted into farnesyl pyrophosphate, common precursor to the cholesterol, ubiquinone, and dolichol synthetic pathways. A decreased level of the substrate farnesyl pyrophosphate would lead to a shift in the production of the 3 end products, and to a shift in the dolichol/ubiquinone ratio. CoQ_{10} decreases blood pressure possibly by decreasing oxidative stress and insulin response in patients with known hypertension receiving conventional antihypertensive drugs. CoQ_{10} decreases lipoprotein concentration in patients with acute coronary disease. Lipid peroxidation in LDL is corroborated by increased proportion of oxidised form of CoQ_{10}. Elevated apoB-bound cholesterol is an indication of lipid peroxidation. Antioxidants significantly increase the concentration of vitamin E in serum and increase the resistance of LDL to undergo Cu^{2+}-catalysed oxidation, reduce formation of lipid peroxides, and reduce relative electrophoretic mobility. The mitochondrial membrane-phospholipid (MMP) injury changes of peripheral lymphocytes in patients with heart failure can be used as an injury indicator of myocardia, and are related to the long-term prognosis. Protection and repairment of MMP injury can improve the life-quality and prolong the life-span of the patients. Statins inhibit synthesis of mevalonate, a precursor of ubiquinone that is a central compound of the mitochondrial respiratory chain. The main adverse effect of statins is a toxic myopathy possibly related to mitochondrial dysfunction. Ubiquinone serum levels are lower in statin-treated patients than in untreated hypercholesterolaemic patients. Lactate/pyruvate ratios are significantly higher in patients treated by statins than in untreated hypercholesterolaemic patients. Statin therapy can be associated with high blood lactate/pyruvate ratio suggestive of mitochondrial dysfunction. CoQ_{10} may have a beneficial role in ischaemia-reperfusion injury. CoQ_{10} administered either as an additive to cardioplegia or as long-term preoperative oral supplementation has been reported to ameliorate myocardial injury after cardiac operations. CoQ_{10} that is involved in mitochondrial ATP, is also a powerful antioxidant. CoQ_{10} pretreatment improves myocardial function after ischaemia and reperfusion. This results from a tripartite effect: 1- higher concentration of ATP and phosphocreatine, initially and during reperfusion, 2- improved myocardial aerobic efficiency during reperfusion, and 3- protection of creatine kinase from oxidative inactivation during reperfusion. The heart is the most susceptible of all the organs to premature aging and free radical oxidative stress. Free radical damage has a role in the progression of numerous degenerative diseases, particularly cardiovascular disease. This may be the result of acute ischaemia-reperfusion injury, endothelial damage of hyperhomocysteinaemia, as well as chronic oxidative damage secondary to lipid peroxidation. The heart is also receptive to the benefits of targeted phytonutrients and antioxidants. Phytonutrients such as the natural flavonoids and carotenoids found in fresh fruits and vegetables or vitamins C, E, and β-carotene have powerful antioxidant effects. In addition, minerals like selenium and nutrients such as CoQ_{10} will minimise free radical risk and optimise a favourable outcome from the ubiquitous presence of oxidative stress on the cardiovascular system. The B complex, particularly folic acid, B_{12}, and B_6 are also essential in the prevention of hyperhomocysteinaemia, another major risk factor for the circulatory system. Measures to minimise accumulation of heavy metals in the body, especially iron and copper, which are capable of initiating adverse free radical reactions, will also help to alleviate oxidative stress. Plasma CoQ_{10} levels is inversely related to metabolic demand. Definite levels of CoQ_{10} are also found in white and red blood cell components, as well as in platelets. Plasma and erythrocyte CoQ_{10} has a well assessed antioxidant role. Erythrocytes previously enriched with exogenous CoQ_{10} were found more resistant to a haemolysis induced by a free radical initiator. Heart failure is always characterised by an energy depletion status, as indicated by low intramyocardial ATP and CoQ_{10} levels. The biosynthetic pathway of the CoQ_{10} polyisoprenoid side chain, starting from acetyl-CoA and proceeding through mevalonate and isopentenylpyrophosphate, is the same as that of cholesterol. Statin lowers both LDL-C and apo B plasma levels together with the plasma and platelet levels of CoQ_{10}, and that CoQ_{10} therapy prevents both plasma and platelet CoQ_{10} decrease, without affecting the cholesterol lowering effect of statin. CoQ_{10} improves cardiac haemodynamic response to exercise in patients with CHF. CoQ_{10} have antioxidative role in the plasma. The pretreatment with CoQ_{10} may have a protective role during routine coronary bypass grafting by attenuating the degree of peroxidative damage. Radical-mediated lipid peroxidation proceeds via similar mechanisms in isolated LDL and VLDL. Efficient LDL antioxidants are also likely to be effective protective agents for VLDL. The ratio of α-tocopherol to CoQ_{10} in VLDL is close to that of LDL. These lipoproteins may transport some CoQ_{10} to extrahepatic tissues, as they do tocopherol. Most of the CoQ_{10} associated with VLDL is present in its reduced, antioxidant active form, ubiquinol-10. The small amounts of ubiquinol-10 in VLDL provide the lipoprotein lipids with a highly efficient antioxidant protection. The reduced CoQ_{10} levels may contribute to the defective serum antioxidant activity and the increased peroxidative damage in uraemic patients on chronic haemodialysis. The LDL/ubiquinone ratio is likely to be a risk factor for atherogenesis, and administration of ubiquinone to patients at risk might be needed. The 3-hydroxy-3-methylglutaryl coenzyme A (HMG CoA) reductase inhibitor is thought to prevent atherosclerosis, however, it also inhibits ubiquinone production. Ubiquinol-10 ($CoQH_2$, the reduced form of CoQ_{10}) is a potent antioxidant present in human low-density lipoprotein (LDL). Supplementation of humans with ubiquinone-10 (CoQ, the oxidised coenzyme) increases the concentrations of $CoQH_2$ in plasma and in all of its lipoproteins. Intake of a single oral dose of 100-200 mg CoQ_{10} increases the total plasma coenzyme content by 80 or 150%, respectively, within 6 h. Long-term supplementation (three times 100 mg CoQ_{10}/day) results in 4-fold enrichment of $CoQH_2$ in plasma and LDL with the latter containing 2.8 $CoQH_2$ molecules per LDL particle (on day 11). Oral supplementation with CoQ_{10} increases $CoQH_2$ in the plasma and all lipoproteins thereby increasing the resistance of LDL to radical oxidation. CoQ_{10} and vitamin B_6 increase IgG and T4-lymphocytes. Defective sperm function in infertile men has been associated with increased lipid peroxidation and impaired function of antioxidant defenses in spermatozoa. Higher intracellular concentrations of CoQ_{10} may represent a mechanism of protection of the spermatozoa. In varicocele patients, this mechanism could be deficient, leading to higher sensitivity to oxidative damage. CoQ_{10} may result in improvement in sperm functions in selective patients. There is a correspondence between a low CoQ_{10} level and spontaneous abortion. There is an increase in the plasma CoQ_{10} level in relation to the contractile activity of the uterine muscle, mainly in the third trimester. There is a significant correlation with sperm count and with sperm motility, and seminal plasma CoQ_{10} levels. CoQ_{10} is present in small amounts in a number of foods and its highest concentration occurs in organ meats like the heart, kidney and liver. It is present also in sardines, mackerel and peanuts. One would have to eat over 7 pounds of peanuts, or 3 pounds of mackerel, or 6 pounds of beef daily to obtain 90 mg of CoQ_{10}. CoQ_{10} is manufactured by the body but this is a rather complex 17 step process requiring vitamin B_2, vitamin B_3, vitamin B_6, folic acid, vitamin B_{12}, vitamin C, pantothenic acid and several trace minerals. As we age, however, our ability to manufacture CoQ_{10} diminishes considerably. Optimal nutrition include optimal levels of CoQ_{10} may be beneficial in any disease state, including cancer. CoQ_{10} is a prescription drug in Japan, the most prescribed drug in that country and it is estimated to be taken by over 50 million people daily. As early as 1974, the Japanese government approved CoQ_{10} for the treatment of chronic heart failure (CHF), and approximately 250 CoQ_{10}-containing preparations are available in Japan for the treatment of cardiovascular disease indications. CoQ_{10} is involved in the mitochondrial electron transport chain leading to synthesis of ATP and on it possesses antioxidant and membrane-stabilising properties. Low endogenous levels of CoQ_{10} can be the result of inadequate nutritional intake, a genetic defect in biosynthesis, or a depletion of vitamins, trace elements, or other precursors necessary for CoQ_{10} biosynthesis. CoQ_{10}, a lipid-soluble benzoquinone with a 10 isoprenyl unit side chain, is structurally similar to vitamin K. CoQ_{10} biosynthesis is multifold, with the isoprenyl side chain deriving from mevalonate, the benzoquinone ring structure from tyrosine, and condensation of these structures through polyprenyl transferase enzyme activity. The primary regulation of CoQ_{10} biosynthesis is the 3-hydroxy-methylglutaryl coenzyme A (HMG-CoA) reductase reaction, which is similar in cholesterol synthesis. 30 mg of CoQ_{10} in healthy subjects can significantly increase the mean peak blood level to about 1 µg/ml within an average of 6 hours, with a second peak occurring again at 24 hours after dosing. 100 mg of CoQ_{10} 3 times/day achieve a mean steady-state level of 5.4 µg/ml, which may be up to 7 times higher than endogenous levels of CoQ_{10}. The plasma half-life of CoQ_{10} is approximately 34 hours and is indicative of the low clearance rate from plasma. CoQ_{10} is a lipophilic molecule and therefore can display variability in absorption depending on the formulation. The slow absorption of CoQ_{10} from the GIT can be attributed to the coenzyme's high molecular weight and low water solubility. The highest bioavailability of CoQ_{10} is with the soybean oil-only formulation. Vegetable oil and vitamin E increase solubility of CoQ_{10} and yield more efficient absorption rates. The extent of hepatic metabolism is unknown, and the excretion of CoQ_{10} is mainly through the biliary tract, with over 60% of an oral dose recovered unchanged in the faeces. A significant deficiency of CoQ_{10} is expected in patients with hypertension. This may be due to a deficiency in the activity of succinate dehydrogenase CoQ_{10} reductase in leukocytes. Deficient activity of this enzyme can result in decreased levels of CoQ_{10}. Side effects of CoQ_{10} occur in <1% of patients, include nausea, epigastric pain, diarrhoea, heartburn, and appetite suppression. Asymptomatic elevations in serum lactate dehydrogenase and hepatic enzymes may be observed and may occur with oral dosages of CoQ_{10} in excess of 300 mg/day. A drug interaction may exist between HMG-CoA reductase inhibitors (e.g., simvastatin, pravastatin, lovastatin) and CoQ_{10} because the coenzyme is a byproduct of the cholesterol biosynthetic pathway. Use of HMG-CoA reductase inhibitors (statins) may result in the diminution of CoQ_{10} blood levels due to interruption of synthesis. Simultaneous administration with CoQ_{10} can result in a significant increase in CoQ_{10} blood levels without opposing the lipid-lowering effect of an HMG-CoA reductase inhibitor. Oral antidiabetic agents (i.e., acetohexamide, glyburide, phenformin, and tolazamide) inhibit CoQ_{10} enzymes (e.g., NADH and succinate dehydrogenase), thus reducing serum CoQ_{10} levels. Reduced insulin requirements are observed in patients with diabetes who are taking CoQ_{10} because it exerts favourable effects on ATP, which in turns acts as a chemical energy carrier in the biosynthesis of insulin. CoQ_{10} inhibits catecholamines and downregulates insulin receptors.

The mechanisms/functions of CoQ_{10} include:

1- It is an essential component in the synthesis of ATP and exhibits both antioxidant and membrane-stabilising properties.
2- Present in the inner mitochondrial membrane, CoQ_{10} serves as an electron transport carrier during the processes of respiration and oxidative phosphorylation, and it has involvement in the manufacture of ATP.
3- CoQ_{10} directly regulates NADH and succinate dehydrogenase, enabling reversible reactions between these enzymes in the mitochondrial electron transport chain.
4- CoQ_{10} may prevent the depletion of metabolites necessary for the resynthesis of ATP.
5- CoQ_{10} may inhibit lipid oxidizability (peroxidation), both initiation and propagation, whereas vitamin E inhibits only propagation.
6- Membrane-stabilising properties, due to the reduction of free radicals that may cause damage to structural proteins and lipids found in membranes.
7- Stabilisation of calcium-dependent slow channels.
8- Inhibition of intracellular phospholipases.
9- Alteration of prostaglandin metabolism.
10- Positive inotropic effect, similar to the effects of digoxin.
11- CoQ_{10} corrects mitochondrial leak of electrons during oxidative respiration.

The concomitant use of warfarin and CoQ_{10} should be avoided due to the risk of thrombotic complications. CoQ_{10} is structurally related to vitamin K and subsequently possesses procoagulant effects. CoQ_{10} decreases INR. CoQ_{10} supplementation in patients with hepatic insufficiency or biliary obstruction may increase serum CoQ_{10} levels because this molecule is metabolised in liver and excreted primarily through the biliary tract. Peak blood levels of CoQ_{10} occur in 5-10 hours. The elimination half-life of ubidecarenone is 34 hrs. The recommended i.v of CoQ_{10} dose is 1.5 mg/kg once daily. 100 mg daily of CoQ_{10} for 14 days prior to surgery, followed by postsurgical administration of 100 mg daily for 30 days may be recommended for cardiac surgery patient for myocardial preservation. CoQ_{10} increases exercise duration in chronic stable angina patients (0.7-1 µg/ml). With administration of 100 mg tds, the levels would be steady at 5.4 µg/ml, after 4 days. Blood levels of CoQ_{10} would increase 3-fold after 2 months of treatment with doses of 300 mg daily in patients with mitochondrial disease. Ubidecarenone is absorbed slowly from the GIT, due to the high molecular weight of ubidecarenone and its water solubility. After oral absorption or i.v administration, ubidecarenone is taken up by chylomicrons. Most of an exogenous dose is distributed to the liver and incorporated into very-low-density lipoprotein. Endogenous CoQ_{10} is found in relatively high concentrations in the heart, liver, kidney, and pancreas. Intracellularly, most of the CoQ_{10} (40-50%) is found in the mitochondrial inner membrane. Other intracellular distribution sites include cytosolic 5-10%, microsomal 15-20%, and nucleus 25-30%. After incorporation of CoQ_{10} into VLDL in the liver, it subsequently concentrates in various tissues such as adrenals, spleen, kidney, lung and myocardium. CoQ_{10} appears to have a low plasma clearance. CoQ_{10} has shown promise in the treatment of several cardiovascular and noncardiovascular disorders, including CHF, hypertension, angina, and periodontal disease. Add-on use can be recommended in patients with CHF who are responding poorly to conventional regimens (e.g., digoxin, diuretics, ACE inhibitors). Endogenous deficiency of CoQ_{10} occurs in a variety of disorders/conditions, including cancer, CHF, hypertension, chronic haemodialysis, mitochondrial disease, and periodontal disease. CoQ_{10} protects the myocardium against functional and structural changes induced by ischaemia and reperfusion. CoQ_{10} appears to be directly responsible for the partial or complete regression of breast cancer, and reduces morphine doses. CoQ_{10} is a fat-soluble quinone that is synthesised intracellularly and participates in a variety of essential cellular processes. CoQ_{10} is found primarily in the inner mitochondrial membrane, and highest concentrations in the human body are in the heart, liver, kidney and pancreas; the total body content ranges from 0.5-1.5 g. CoQ_{10} has a significant role in mitochondrial electron transfer and the synthesis of ATP. It serves as a mobile electron carrier in the mitochondrial electron-transfer process of respiratory and oxidative phosphorylation. The coenzyme regulates NADH and succinyl dehydrogenases, and enables reversible interactions between the NADH dehydrogenase, succinate dehydrogenase, and cytochrome b-c1 portion of the mitochondrial electron transport chain. CoQ_{10} may have direct membrane-stabilising properties and is an antioxidant and free radical scavenger (scavenge lipid peroxidation-produced free radicals). CoQ_{10} improves pulmonary function and exercise performance in patients with chronic lung disease. Patients with hypoxaemia at rest tend to have lower CoQ_{10} levels, and those with exercise-induced hypoxaemia also have reduced level. Under these low level conditions, some organs such as heart, liver and skeletal muscles may become hypoxic. CoQ_{10} improves oxygen transport to muscles during exercise, as a result of the improved $PaCO_2$ and increased cardiac output. CoQ_{10} improves left ventricular ejection fraction, stroke volume, clinical symptoms, and functional status. CoQ_{10} improves QoL and reduces number of patients' readmission to hospital for worsening of heart failure. CoQ_{10} inhibits lipid peroxidation initiated by doxorubicin, scavenging of doxorubicin induced free radicals, and/or corrects deficiency of CoQ_{10}. CoQ_{10} decreases systolic and diastolic blood pressure. CoQ_{10} is effective in the treatment of warfarin-induced alopecia. CoQ_{10} may improve mitochondrial respiration in hair roots.

Potential therapeutic uses of CoQ_{10} in cardiovascular, include:

1- Angina pectoris.
2- Unstable anginal syndrome.
3- Myocardial preserving agent during chemical thrombolysis.
4- Myocardial preserving agent for cardiac surgery.
5- CHF.
6- Toxin-induced cardiotoxicity (adriamycin).
7- Essential and renovascular hypertension.
8- Ventricular arrhythmia.
9- Mitral valve prolapse.
10- Prevents oxidation of LDL.

CoQ_{10} and myocardial failure:

1- Key substance in biological energy production (ATP), needed for both muscle contraction and relaxation.
2- Depletion of CoQ_{10} may be crucial in the development of heart failure due to an increased demand on the respiratory chain.
3- Possible restoration of the function of the exchanged myocytes in the energy-starving heart via CoQ_{10} supplementation.
4- Protective substance against toxic myocardial damage from metabolic inhibitors (anthracyclines).
5- Protective effects on myocardial ischaemia and reperfusion during open heart surgery (antioxidant and antiischaemic effects).

CoQ_{10} is a lipid soluble benzoquinone, which has properties potentially useful in preventing or attenuating the damage associated with ischaemia-reperfusion. CoQ_{10} is directly involved in energy transduction and aerobic ATP production as it transports electrons in the respiratory chain, as well as couples the respiratory chain to oxidative phosphorylation. The antioxidant actions of CoQ_{10} are not limited to the mitochondria but are applicable to any other cell membrane containing CoQ_{10}. CoQ_{10} pretreatment effectively increases myocardial CoQ_{10} levels leading to an improved tolerance to myocardial reperfusion injury. CoQ_{10} increases myocardial high energy phosphate production, preserves myocardial aerobic efficiency, and decreases the oxidative stress after an ischaemic insult. Via an antioxidant mechanism, CoQ_{10} also protects endothelial cells against I/R injury, thereby preserving endothelium-dependent and endothellum-independent vasorelaxation along with recovery of myocardial function. CoQ_{10} may be protecting myocardial and arterial smooth muscle cell function via OH scavenger action. Left ventricular function depends on the operational capacity of myocardial cells to generate the energy to expand and contract. Insufficient myocardial contractive forces often contribute significantly to CHF. Heart failure is an energy-starved heart. It is important to treat both the molecular and cellular components of the heart when managing CHF, e.g., the biochemistry of pulsation, fluid retention etc. CoQ_{10} has a significant effect upon electron transfer within the respiratory chain and supports intramyocardial energy at the cellular level. Because O_2 based production of energy takes place in cellular mitochondria, CoQ_{10} concentrations in myocardial cells is ten-fold that in the brain or the colon. The myocardium requires elevated ATP supports. Decrease in CoQ_{10} could precipitate a decrease in oxidative phosphorylation of the mitochondrial respiratory chain, thus making the tissue more susceptible to free radical attack. If CoQ_{10} levels evaluation is not possible, increasing the dosage of CoQ_{10} according to clinical symptoms must be considered. Treatment of congestive heart failure with CoQ_{10} reduces hospitalisation by at least 20%. Triggers for headaches and migraines are: chocolate and cola drinks, oranges, citrus fruits, peanuts and peanut paste, green beans and peas, Cow's milk dairy, and MSG. CoQ_{10} appears to be a good migraine preventive, it reduces incidence of migraine headaches by 50%. Moderating the consumption of cow's milk significantly reduces the incidence of antisocial behaviour.

VITAMIN E

Since vitamin E was discovered over 70 years ago, a number of species-dependent deficiency symptoms of vitamin E have been reported. The essentiality of vitamin E for humans was established in the late 1960s. The recognition was largely derived from clinical investigations involving premature infants. Studies of children and adults with specific causes of fat malabsorption and patients with familial isolated vitamin E deficiency syndrome have conclusively shown that neurologic dysfunction is associated with vitamin E deficiency. Vitamin E is an essential nutrient necessary for the optimal development and maintenance of human nervous system integrity and function. Mitochondrion, the intracellular organelle that produce ATP, constitute the greatest source of steady state oxidants. The mitochondrial electron transport system consumes >85% of all the oxygen used by the cells, and it is estimated that between 1% and 5% of the oxygen consumed by mitochondria is converted to superoxide, hydrogen peroxide, and other ROS under normal physiological conditions. Proximal to a large flux of ROS, mitochondrial DNA are particularly susceptible to oxidative damage and mutation because it lacks protective histones and an effective repair system. Mitochondrial DNA, for example, has a 16-fold higher oxidised base than nuclear DNA, and the accumulation of mitochondrial DNA damage products increases with age. In addition to oxidative damage to lipid and protein in mitochondria, many studies have found that increased oxidative lesions, deletions, point mutations, and aberrant forms in mitochondrial DNA of postmitotic tissues on aging. Mitochondrial respiratory chain defects and DNA mutations are 2 key contributors to human aging and neurodegenerative diseases. Dietary vitamin E markedly reduces hydrogen peroxide production in the mitochondria. Disruption of mitochondrial ultrastructure is one of the earliest pathologic events resulting from vitamin E depletion and suggests a role for ROS in initiating the degenerative process. Oxidative damage is a consequence of excessive oxidative stress, insufficient antioxidant potential, or both. Oxidative damage induced by ROS is increasingly implicated as an important contributor in the pathogenesis of many degenerative diseases, including cancer, cardiovascular diseases, and aging. Increasing evidence indicates that superoxide plays a central role in the generation and action of other ROS. In the presence of superoxide dismutase, superoxide is readily converted to hydrogen peroxide. Hydrogen peroxide in turn can be reduced to water by the activity of glutathione peroxidase or catalase. However, under the condition where the removal capacity of hydrogen peroxide is low and in the presence of heavy metal ions, the highly reactive hydroxyl radicals can be formed via the iron-catalysed Haber-Weiss reaction. Hydroxyl radical is the most oxidising radical known in biologic systems. Furthermore, superoxide may react with another important free radical modifier, nitric oxide (NO), to form highly reactive free radical peroxynitrite. Formation of peroxynitrite has been implicated in a variety of free radical-induced tissue injuries. These include inhibition of mitochondrial electron transport chain, thereby leading to more ROS generation. The protonated form of peroxynitrite can also decompose to reactive nitrogen dioxide and hydroxyl radicals. Vitamin E has long been recognised as the major lipid-soluble chain-breaking antioxidant that prevents free radical-initiated peroxidative tissue damage. Vitamin E may act as a biologic modifier independent of its antioxidant property. Vitamin E is capable of regulating mitochondrial generation of superoxide and other ROS. A certain level of vitamin E is required to prevent mitochondrial electron leakage, or to downregulate superoxide generation systems. Vitamin E regulates superoxide generation in human neutrophils and monocytes. Vitamin E or oxidative stress status mediates the activation and gene expression of protein kinase (PKC), tumour growth factors, transcription factor activator protein-1 (AP-1), TGFβ1, NFκB, and other related transcription factors, as well as the diacylglycerol-PKC pathway. These factors are known to play important roles in mediating a number of pathophysiologic events including platelet adhesion and aggression and mural thrombosis, vascular smooth muscle cell proliferation, apoptosis, glomerulosclerosis, tumour angiogenesis, vascular hyperglycaemia, and abnormal retinal blood flow. Vitamin E may attenuate the development of cancer, cardiovascular disease, aging, and other degenerative disease by regulating mitochondrial generation of ROS. The pulmonary system is the major target of oxygen toxicity. Prolonged exposure to 100% oxygen causes atelectasis and the formation of pulmonary oedema, leading to death from respiratory failure. Alveolar surfactant, the prime target of oxidants in lung tissue, is supplemented with vitamin E during its assembly in type II pneumocytes. The classic LDL receptor pathway involves LDL binding to specific surface receptors followed by the formation of clathrin-coated pits, coated vesicle, endosomes, and lysosomes. Type II pneumocytes are able to adapt vitamin E uptake independently from cholesterol uptake. Because vitamin E is a constituent of all lipoprotein classes, it is taken up together with cholesterol during internalisation of lipoproteins. Cholesterol is a major component of the alveolar surfactant and has to be taken up by type II pneumocytes from plasma lipoproteins while only a small portion (1-7%) is synthesised de novo. HDL is the primary source of vitamin E for type II pneumocytes. The rate of uptake of vitamin E by this cell type might be regulated by the expression of SR-B1. Vitamin E is the name given to a group of 8 fat-soluble compounds. The esterified vitamin E is hydrolysed and absorbed as efficiently as α-tocopherol. Vitamin E is incorporated into the lipid portion of cell membranes and other molecules, protecting these structures from oxidative damage and preventing the propagation of lipid peroxidation. Vitamin E may be protective effects against cancer, heart disease, and complications of diabetes. Vitamin E is necessary for maintaining a healthy immune system, and it protects the thymus and circulating white blood cells from oxidative damage. Vitamin E works synergistically with vitamin C in enhancing immune function. In the eyes, vitamin E is needed for the development of the retina and protects against cataracts and macular degeneration. Vitamin E deficiency occurs in people with chronic liver disease and fat malabsorption syndromes, such as celiac disease and cystic fibrosis. Vitamin E deficiency leads to nerve damage, lethargy, apathy, inability to concentrate, staggering gait, low thyroid hormone levels, decreased immune response, and anaemia. Marginal vitamin E deficiency may be much more common and has been linked to an increased risk of CVD and cancer. The UL for vitamin E may be 1,000 mg/day. Vitamin E supplements is often recommended, as it is impossible to obtain a high intake of vitamin E without consuming a high-fat diet. Reported adverse effects of vitamin E include increased risk of bleeding, diarrhoea, abdominal pain, fatigue, reduced immunity, and transiently raised blood pressure. Very high doses of vitamin E may be pro-oxidant (i.e., acting as free radicals), especially in smokers. Vitamin E exerts antioxidant effects in combination with other antioxidants, including β-carotene, vitamin C, and selenium. Vitamin C can restore vitamin E to its natural reduced form. Vitamin E is necessary for the action of vitamin A and may protect against some of the adverse effects of excessive vitamin A, because inorganic iron destroys vitamin E, the two should not be taken simultaneously. Cholestyramine, mineral oil, and alcohol may reduce the absorption of vitamin E. Vitamin E may safely be given to patients receiving warfarin. Modified dialyzer with vitamin E provides more effective antioxidant defense than peroral administration of vitamin E. In uraemic condition there is defective antioxidant production and increased susceptibility to plasma lipid oxidation, oxidative damage to proteins and nucleic acids, and accumulation of molecules with pro-oxidant function such as homocysteine and some reactive carbonyls deriving from the non-enzymatic oxidation of glucose and lipids.

Mechanism of vitamin E in artherosclerosis:

Site	Mechanism
1- LDL	Inhibits its oxidation.
2- Lipoproteins	Inhibits thrombin generation and assembly.
3- Endothelial cell	A- Potentiates prostacyclin synthesis.
	B- Up-regulate the expression of cytosolic phospholipase A_2 and cyclooxygenase.
	C- Inhibits agonist-induced monocyte adhesion.
	D- Attenuates cell-mediated LDL oxidation.
	E- Decreases endothelial expression of adhesion molecules induced by oxidised LDL.
4- Smooth muscle	Inhibits proliferation.
5- Platelets	Inhibits platelet aggregation.
6- Neutrophils	Reduces leukotriene synthesis.
7- Monocytes	Reduces monocyte adhesion.

Haemodialysis weakens the antioxidant defenses by: 1- leakage and consumption of hydrosoluble antioxidants (vitamin C, free thiol) during dialysis; 2- consumption of liposoluble/lipoprotein-associated antioxidants (vitamin E, CoQ_{10}); 3- changes in the lipid consumption of biological fluids and cell membranes; and 4- a deficit in cofactors and damage to antioxidant enzymes such as extracellular glutathione peroxidase and HDL-associated paraoxonase. The oral supplementation with vitamin E increases the level of vitamin E, PUFA and GSH; but has no effect on triglyceride (TG). There is increased susceptibility to lipid peroxidation during haemodialysis. Vitamin E supplementation inhibits lipid peroxidation, protect plasma thiols and GSH from oxidation, and participate in the scavenging of ROS. Vitamin E is a scavenger of lipid hydroperoxides and regulates both enzymatic and nonenzymatic pathways of lipid oxidation. Vitamin E may control the endothelial component of the atherogenic process (endothelial cells, resident macrophages, SMC). Vitamin E modulates ROS productions of neutrophils exposed to dialysis membranes. Vitamin E-modified dialyzers can lead to overall beneficial effects against oxidative stress. Chronic fat malabsorption due to biliary atresia, intestinal resection, Crohn's disease, or pancreatic insufficiency is the most common cause of low levels of the highly fat-soluble vitamin E. Resection of the terminal ileum interrupts the enterohepatic circulation and prevents reabsorption of bile salts. Intestinal resection can also lead to gastric hypersecretion and small intestine bacterial contamination, which in turn lead to the deconjugation of bile salts. Both these processes cause bile salt precipitation, which decreases micelle formation for lipid absorption. Two autosomal recessive conditions can cause vitamin E deficiency. A gene defect in a microsomal triacylglycerol transfer protein essential for hepatic very low-density lipoprotein (VLDL) synthesis occurs in a β-lipoproteinaemia. Mutations in the α-tocopherol transfer protein, which normally incorporates α-tocopherol into plasma VLDL, produce autosomal recessive vitamin E deficiency (AVED). Posterior column deficits, axonal peripheral neuropathy, and ataxic gait disturbance are constant features of chronic vitamin E deficiency. Vitamin E inhibits peroxidation of polyunsaturated fatty acids in photoreceptor membranes. Vitamin E deficiency is associated with photoreceptor degeneration and retinal lipofuscin accumulation. Without vitamin E, end-products of lipid peroxidation may result in lipofuscin accumulation. This, in turn, may lead to sequestering of vitamin A esters and thereby result in visual impairment. Vitamin E prevents free radical damage to the nervous system. Chronic vitamin E deficiency results in ataxia, neuropathy and bilateral centro-cecal scotomata. Progesterone dissolved in vitamin E is absorbed very efficiently, and distributed quickly to all of the tissues. Vitamin E and oestrogen act in opposite directions on the clot-removing enzymes. Vitamin E facilitates clot removal, by activating proteolytic enzymes. $CoQ_{10}H_2$ is closely linked to vitamin E and serves to generate the reduced (active) α-tocopherol form of the vitamin. High intake of vitamin E reduces CRP levels. Vitamin E may block glutamate toxicity. Vitamin E is found abundantly in the membranes of cells, where it protects membrane lipids from oxidation by free radicals. Consuming vitamin E regularly lowers incidence of dementia. More than 60% of all demented patients have subnormal vitamin E levels. Vitamin E is reduced the most in Alzheimer patients with lower bodyweight and body mass. Vitamin E combined with other antioxidant vitamins may slow the course of Alzheimer's disease and Parkinson's disease. Vitamin E seems to upregulate polyamine metabolism, probably by its direct effects on ornithine decarboxylase (ODC). The ability of vitamin E to increase levels of polyamines may account for its more positive effects in Alzheimer's disease. ω-3 fatty acids and vitamin E may increase RBC membrane fluidity and potentially reduce diabetic symptoms due to impaired cell deformability such as intermittent claudication. Unsaturated fats destruction of vitamin E will lead to the destruction of vitamin A. Vitamin E is likely to reduce the intensity and frequency of epileptic seizures. Vitamin E suppressed the nitric oxide toxicity in the pancreatic islet cells. β-Cells are prone to be destroyed by free radicals because of the low antioxidant enzyme nature. Vitamin C and Vitamin E are the known free radical scavengers because they inhibit glucose autoxidation and reduce the covalent linking of glucose to serum proteins. The combination of vitamin E and selenium significantly increases sperm motility and the overall percentage of normal spermatozoa. High-dose vitamin E significantly enhances lung virus clearance in elderly.

SELENIUM

Selenium (Se) metabolism is dynamic. Se-methylselenocysteine is a major constituent of selenized garlic and has several advantages as a chemopreventive form of Se compared with selenomethionine. Hydrogen selenide is a key metabolite, formed from inorganic sodium selenite (oxidation state +4) via selenodiglutathione (GSSeSG) through reduction by thiols and NADPH-dependent reductases and released from selenocysteine by lyase action. Hydrogen selenide provides Se for synthesis of selenoproteins after activation to selenophosphate. The known functions of Se as an essential element in animals are attributed to ~12 known mammalian selenoproteins, all containing selenocysteine, specifically incorporated through a unique co-translational mechanism. Monomethylated forms of Se is a critical class of Se metabolites having powerful effects on carcinogenesis, while lacking some of the toxic effects produced by other forms such as inorganic selenite. The monomethylated selenium compounds are effective in vitro at very low concentrations to give chemopreventive effects (apoptosis and cell cycle arrest) in transformed cells. Se-methylselenocysteine serves as a reservoir that provides a steady stream of monomethylated Se so that a critical level is maintained and cell growth is inhibited. Apoptosis, by causing deletion of carcinogen-initiated cells and suppression of clonal expansion of a transformed cell population, is an attractive mechanism for chemoprevention. Apoptosis is an important mechanism for the anticancer effects of Se. A chemopreventive mechanism based on induction of apoptosis, separate from toxic effects and independent of a functional p53, strengthens the case for Se chemoprevention in the human population. Four different types of reactions by which Se might modify proteins including: A- type 1, formation of selenotrisulfide bonds (S-Se-S); B- type 2, formation of selenenylsulfide bonds (S-Se); C- type 3, catalysis of disulfide bond (S-S) formation or its reversal; and D- type 4, formation of diselenide bonds (Se-Se). Inactivation of key proteins such as protein kinase C is clearly relevant to chemoprevention. Protein activating reactions involving scission of disulfide or cysteinyl mercaptide linkages in proteins are also potential chemopreventive mechanisms to consider for Se metabolites such as methylselenol. An interesting possibility, particularly with regard to Se-induced apoptosis, might be direct activation of a cysteine protease having the active site cysteine blocked by zinc. The higher affinity of methylselenolate for Zn could result in release of the cysteinylthiolate-bound Zn and activate the enzyme. Se inactivation of transcription factors such as AP-1 and NF-κB that are known to be modulated by redox control mechanisms was described as a mechanism by which Se compounds inhibit cell growth at micromolar levels. Transcriptional factor (NF-κB) modulation by Se may be relevant to gene expression and chemopreventive mechanisms. Se induces normalisation of regulatory pathways that are perturbed in early stages of carcinogenesis, as its effects are more pronounced in early stages of transformed cancer cells. Redox-regulated transcription factors can have two states, ON and OFF, that differ by the oxidation state of a cysteine residue. In the case of OxyR and some other factors, this involves the presence or absence of an intramolecular disulfide bond between two cysteine residues. The formation of a disulfide by oxidation is reversed by cellular disulfide-reducing enzyme systems such as glutaredoxin and thioredoxin. Having a transcription factor poised between the opposing pathways is the basis for a dynamic mechanism that makes activation a transitory process. Oxidised or reduced forms of Se react with protein thiols or disulfides and form more reactive intermediates. Se catalysis of the reversible redox changes in redox-regulated proteins (type 3 mechanism) facilitates resetting of the basal state. By shortening the length of time that a transcription factor would be in the ON configuration, less transcriptional activity would be observed. In this mechanism, Se is serving as a redox catalyst that links the cellular redox poise to the critical targets, rather than causing bulk changes in peroxide levels and the cellular redox potential. The selenoprotein is a homodimeric flavoprotein using NADPH as electron donor and reduces 5,5'-dithiobis (2-nitrobenzoic acid) (DTNB) as well as catalysing thioredoxin-dependent reduction of insulin, similar to other mammalian thioredoxin reductases. A conformational change affecting interaction of thioredoxin reductase with other molecules could be important with regard to triggering cell signalling in response to oxidative stress. Selenocysteine in thioredoxin reductase is an oxidant sensor controlling cell signalling pathways. Since hydrogen selenide is rapidly oxidised by oxygen, the Se compounds catalysed oxygen-dependent oxidation of the thioredoxin system and NADPH. Selenite increases the activity of thioredoxin reductase (measured with insulin as substrate) in human cancer cell lines and in rats fed supranutritional levels of selenite. The inhibitory effects of selenite on cell growth involve NADPH depletion, competitive inhibition of the thioredoxin system, and oxidation of structural cysteines in thioredoxin to disulfides. The chemopreventive effects of Se are due to these inhibitory effects of Se on the purified thioredoxin system. The in vivo effects of Se in the nutritional to supranutritional range on thioredoxin reductase activity differ from other selenoenzymes in two ways: (i) increased activity with excess Se; and (ii) a decline in activity with continued high level Se administration. Thioredoxin is overexpressed in many forms of cancer, is secreted by tumour cells and stimulates tumour growth while decreasing apoptosis. Se is a component of enzymes in two major redox systems of the cell, namely the glutathione and the thioredoxin systems. Glutathione peroxidase (GPX1) is decreased while thioredoxin reductase is increased in any cancer. Functions to consider as chemopreventive mechanisms of Se include cell signalling and redox regulation of transcription factors or reactivation of oxidatively inactivated proteins. Diselenides are frequently observed in organoselenium chemistry as stable products that arise from more unstable forms of Se. A selenocysteine residue in a protein could be inactivated by forming a diselenide bond after reaction with a small Se moiety such as a methylated Se metabolite. The stability or resistance of the diselenide bond to cleavage would be expected to vary, depending on such factors as accessibility of the diselenide in the protein and the nature of the second Se moiety. Cessation of Se supplementation would diminish metabolite production (diselenide), allowing an opportunity for removal of the diselenide block and recovery of activity. Se modulates many enzyme activities besides classic selenoproteins like GSH peroxidase. These alterations involve increased activity of some enzymes and decreased activity of others. New selenoproteins of unidentified function such as the 15 kDa human selenoprotein found in prostate and other tissues, and expressed at lower levels in some cancers, are of great interest. A selenoprotein could have a role in the regulation of signalling pathways through catalysis of thiol/disulfide exchange. Epigenetic tumour suppressor gene silencing through methylation of cytosine occurs in many transformed cells. Mammalian cells contain an enzyme that catalyses demethylation, apparently releasing the methyl group in the form of methanol. A possible role for Se in the chemistry of demethylation could involve initiation of the process through nucleophilic attack on a carbon of the cytosine ring or facilitation of the process by serving as a methyl group acceptor. The ability of Se compounds to release zinc from tightly bound zinc-sulfur clusters in metallothionein has been demonstrated and may involve catalysis of redox chemistry as well as thiol/disulfide exchange. The potential for Se to release zinc from clusters in zinc fingers of transcription factors or signalling proteins may have importance as a potential chemopreventive mechanism. Cell cycle cdk2 or cell signalling protein kinases may be targets of Se metabolites. A novel chemopreventive mechanism is proposed involving Se as a catalyst of the reversible cysteine/disulfide transformations that occur in a number of redox-regulated proteins, including transcription factors, effectively limiting the period of time such proteins are in the activated state. Selenoproteins hold promise for a number of chemopreventive mechanisms. Selenoproteins such as mammalian thioredoxin reductase may prove to have a role to play throughout the range of Se-mediated cancer prevention. Selenium (Se) prevents cancer. Apoptosis can be triggered by micromolar levels of monomethylated forms of Se independent of DNA damage and in cells having a null p53 phenotype. Cell cycle protein kinase cdk2 and protein kinase C (PKC) are strongly inhibited by various forms of Se. Inhibitory mechanisms involving modification of cysteine residues in proteins by Se have been proposed that involve formation of Se adducts of the selenotrisulfide (S-Se-S) or selenenylsulfide (S-Se) type or catalysis of disulfide formation. Selenium may facilitate reactions of protein cysteine residues by the transient formation of more reactive S-Se intermediates. Se reversibly catalyses cysteine/disulfide transformations that occur in a number of redox-regulated proteins, including transcription factors. A time-limited activation mechanism for such proteins, with deactivation facilitated by Se, would allow normalisation of critical cellular processes in the early stages of transformation. The metabolism of selenium is dynamic. Hydrogen selenide is a key metabolite, formed from inorganic sodium selenite (oxidation state +4) via selenodiglutathione (GSSeSG) through reduction by thiols and NADPH-dependent reductases and released from selenocysteine by lyase action. Methylation is a major pathway for Se metabolism in microbes, plants and animals, but demethylation back to inorganic selenium can occur. Methylation of Se produces less toxic forms. Monomethylated forms of Se have emerged as a critical class of Se metabolites having powerful effects on carcinogenesis. Hydrogen selenide provides Se for synthesis of selenoproteins after activation to selenophosphate. There are ~12 known mammalian selenoproteins, all containing selenocysteine, specifically incorporated through a unique co-translational mechanism. Se-methylselenocysteine is a major constituent of selenized garlic and has several advantages as a chemopreventive form of Se compared with selenomethionine. Apoptosis, by causing deletion of carcinogen-initiated cells and suppression of clonal expansion of a transformed cell population, is an attractive mechanism for chemoprevention. Apoptosis can be triggered by Se independent of DNA damage and in cells having a null p53 phenotype. Apoptosis is an important mechanism for the anticancer effects of Se. Se-induced alterations in cell cycle proteins associated with G1/S phase and decreased DNA synthesis. Growth inhibition caused by Se-methylselenocysteine was coincident with a marked decrease in cdk2 kinase activity and impeded progress through S phase. There are 4 different types of reactions by which Se might modify proteins including Type 1, formation of selenotrisulfide bonds (S-Se-S); Type 2, formation of selenenylsulfide bonds (S-Se); Type 3, catalysis of disulfide bond (S-S) formation or its reversal; Type 4, formation of diselenide bonds (Se-Se). Non-specific incorporation of Se into proteins occurs through substitution of selenomethionine for methionine. In type 1 and type 2 mechanisms, where Se inactivates thiol proteins by bonding to the sulfur to form an adduct. Proteins having regulatory cysteines can form Se adducts. In the type 3 mechanism Se is a catalyst and is not taken up in the protein. The type 3 mechanism (Se-catalysed disulfide bond formation/reversal) is related to thiol/disulfide interchange, long regarded as a potentially useful function for Se. Se can facilitate intramolecular disulfide bond formation in protein kinase C, leading to inactivation. Inactivation of key proteins such as PKC is clearly relevant to chemoprevention. Protein activating reactions involving scission of disulfide or cysteinyl mercaptide linkages in proteins are also potential chemopreventive mechanisms to consider for Se metabolites such as methylselenol. Se inactivates transcription factors such as AP-1 and NF-κB. Se effects are more pronounced in early stages of transformed cancer cells. Se might be considered as an agent inducing normalisation of regulatory pathways. The formation of a disulfide by oxidation is reversed by cellular disulfide-reducing enzyme systems such as glutaredoxin and thioredoxin. Oxidised or reduced forms of Se react with protein thiols or disulfides and form more reactive intermediates. Se catalysis of the reversible redox changes in redox-regulated proteins (type 3 mechanism) facilitates resetting of the basal state. Se serves as a redox catalyst that links the cellular redox poises to the critical targets, rather than causing bulk changes in peroxide levels and the cellular redox potential. Reaction of an oxidised Se metabolite with a protein thiol forms an activated selenenylsulfide intermediate, which is attacked by a second thiol to form an intramolecular disulfide. Release of the methylselenolate and spontaneous oxidation of the methylselenolate helps drive the coupled reactions. The strongly nucleophilic selenolate ion opens the disulfide bond and forms an activated selenenylsulfide intermediate that undergoes facile reaction with GSH to form the glutathione protein mixed disulfide. The glutathione moiety facilitates recognition of the protein mixed disulfide by a thiol-disulfide oxidoreductase (TDOR) such as thioltransferase (glutaredoxin) or protein disulfide isomerase, allowing further reduction to protein dithiol with formation of GSSG. NADPH-linked reduction of GSSG by glutathione reductase couples protein disulfide reduction to the cellular reducing systems. The small size and nucleophilicity of methylselenolate allows it to open the protein disulfide and form a selenenylsulfide derivative, which undergoes facile reaction with glutathione, forming the glutathione selenenylsulfide derivative and facilitating reduction of the glutathione-protein mixed disulfide by thioltransferase or protein disulfide isomerase, giving the active (reduced) form of the factor. Protein disulfide isomerase is a relatively abundant cellular protein present in millimolar concentrations in the lumen of the endoplasmic reticulum. It has a relatively low catalytic efficiency, but the combined effects of micromolar levels of a Se catalyst and a thiol-disulfide oxidoreductase may have physiological significance for in vivo disulfide metabolism. Thioredoxin is overexpressed in many forms of cancer, is secreted by tumour cells and stimulates tumour growth while decreasing apoptosis. Se is a component of enzymes in 2 major redox systems of the cell, namely the glutathione and the thioredoxin systems. Glutathione peroxidase (GPX1) is decreased while thioredoxin reductase is increased in all cancer. Both thioredoxin and thioredoxin reductase offer interesting possibilities for inhibition by antitumour agents. A selenocysteine residue in a protein could be inactivated by forming a diselenide bond after reaction with a small Se moiety such as a methylated Se metabolite (CH_3-Se). CH_3-Se interacts with glutathioneyl mixed disulfides. Thioredoxin reductase activity is declined with continued exposure to relatively high dietary Se levels. Sustained exposure of cells to high levels of Se, generating reactive Se intermediates would bring about diselenide formation leading to inhibition of thioredoxin activity over time. Se status modulates many enzyme activities besides classic selenoproteins like GSH peroxidase. The normalisation of enzyme activities observed with Se supplementation might involve effects of Se on redox-regulated transcription factors. Se induces much larger increase in the leukocyte phospholipid GSH peroxidase compared with GSH peroxidase activity. A possible role for Se in the chemistry of demethylation could involve initiation of the process through nucleophilic attack on a carbon of the cytosine ring or facilitation of the process by serving as a methyl group acceptor. Se compounds releases zinc from tightly bound zinc-sulfur clusters in metallothionein. The potential for Se to release zinc from clusters in zinc fingers of transcription factors or signalling proteins may have importance as a potential chemopreventive mechanism. Cell cycle cdk2 or cell signalling protein kinases may be targets of Se metabolites. Selenoproteins hold promise for a number of chemopreventive mechanisms. Most of the human clinical trials with selenium have been with selenium yeast. The nutritional yeasts are totally dead cells, there are a few strains of pernicious yeast and these have live cells. Yeast selenium-containing compounds include: 1- Selenomethionine accounts for no more than about 20% of all selenium-containing compounds. 2- Selenocysteine, seleno-methylselenocysteine and selenomethionine. 3- Selenophosphates, triphenylphosphine selenide, diselenides, triselenides and other organic selenium compounds (make up 40-50% of all the selenium compounds in yeast). Selenium and the GSH/GSSH redox couple enhance delivery of Zn to cells, and sequestering of mercury and other heavy metals. Prostaglandins (PGs) require selenium for their synthesis; PGs deficiency may be a source of schizophrenia. Selenium and glutathione are essential to the formation of phospholipid hydroperoxide glutathione peroxidase, an enzyme present in spermatids, which becomes a structural protein comprising over 50% of the mitochondrial capsule in the mid-piece of mature spermatozoa. Higher tissue selenium content, which is indicative of dietary intake, is associated with lower prostate cancer risk. Selenium is necessary for formation of the deiodinase enzyme, which converts T4 into T3. Selenium (Se) and vitamin E prevent lipid peroxidation and EPA and DHA upregulate CuZnSOD. Selenium deficiency exaggerates copper toxicity. Ions are atoms, which have an electrical charge because one or more electrons have been removed from or added to them. One selenium compound is not the same as another. There are many different selenium-containing compounds that have various biochemical actions. The element selenium is not useful to the body, but the selenium ions are to a degree. The form of selenium, which is most useful in the body, is selenium that is incorporated as an integral part of an organic molecule, such as a selenoprotein. Not all sulfur compounds are sulfur. Some of the known sulfur-containing compounds include cysteine, methionine, MSM, SAMe, biotin, lipoic acid, and N-acetyl cysteine (NAC). Not all nitrogen compounds are nitrogen. Some of the known nitrogen-containing compounds include lysine, nitric oxide (NO) and the polyamine growth factor spermine. Selenium compounds vary in their suitability for nourishing the body. Different selenium compounds are metabolised in various ways. They can affect the metabolic pool of selenium compounds and metabolic pathways differently. Different selenium compounds have diverse functions in the body. Different selenium compounds vary in their cancer-protection and even cancer-destroying abilities. An oxygen free radical (OFR) can react with a lipid molecule to form a lipid peroxide, which like hydrogen peroxide, is not a free radical but a reactive oxygen species (ROS) that can cause harm to body components, especially cell membranes. The selenium-containing compounds glutathione peroxidases repair phospholipid peroxide damage in membranes, thus breaking the peroxidation chain reactions that damage cells. The selenium-containing compounds glutathione S-transferases repair epoxide damage in DNA, thus repairing abnormal DNA and preventing mutations, which can lead to cancer. Various selenium compounds have several other biochemical functions as well. The selenium peroxide reducing factor is a family of 4 glutathione peroxidases. The selenium epoxide-reducing factor is a large family of cellular enzymes called glutathione S-transferases. On the periodic chart, selenium is found in group 16 (formerly known as group VIA) just beneath sulfur. Both sulfur and selenium atoms have outer electronic shells that contain 6 electrons. Thus, their chemical reactions are very similar. Selenium, however, is in the next lower period and is a metalloid, whereas sulfur is a non-metal. As a metalloid, selenium can behave as a metal by donating electrons during a chemical reaction and behave as a non-metal by accepting electrons. Selenium atoms have a ground state electronic shell structure that gives selenium the versatility to readily accept or donate electrons, and makes the selenium atom an ideal catalytic center and a great semiconductor. The selenium atoms in organic selenium-containing compounds can function as redox centers. Selenium and sulfur have similar atomic sizes, bond energies, ionisation potentials, electron affinities, and electronegativities. This may explain why selenium is sometimes incorporated into proteins by plants and animals. Selenium is the only trace element that genes specify to form an amino acid, which is then incorporated into proteins. Selenocysteine is the keystone compound needed to make several selenium-containing proteins (selenoproteins). The selenium that most often gets into proteins by accident is in the form of selenomethionine, which can readily be incorporated by accident in place of methionine. Selenocysteine and other selenoamino acids are formed this way, too, but to a lesser degree. The difference between methionine and selenomethionine is that the methionine molecule contains an atom of sulfur, whereas selenomethionine molecules have an atom of selenium instead. Soil selenium varies widely from region to region, and thus, a given plant will have differing amounts of selenomethionine. Selenomethionine-specific proteins including thiolase, β-galactosidase and certain muscle proteins in humans are inadvertent selenium-containing proteins formed randomly by accident. Humans cannot make selenomethionine directly from inorganic selenium. As long as there are selenium ions available in the metabolic pool, the DNA specifies for selenocysteine production. About 80% of the body's selenium may be present as selenocysteine. The selenium compounds that circulate in the blood are of three types: 1- selenoproteins and other selenium-containing organic compounds; 2- organic compounds that inadvertently contain selenium; and 3- inorganic selenium compounds. There are at least 35 selenoproteins. There are 8 selenoproteins in arterial walls, 8 selenoproteins in brain tissue and 9 selenoproteins in testis. Selenium yeast has a measured level of organically bound selenium compounds. Nearly all selenium compounds contribute to both the body's nutritional needs for selenium and its defense against cancer, but through several different mechanisms. Increasing amounts of selenium compounds results in increasing cancer protection until toxic limit is reached. Selenium-methylselenocysteine proved to be all effective compound. The most powerful anti-cancer compounds are the allyl phosphine selenides. Individuals with higher levels of selenium compounds in their blood have less cancer than those having lower selenium levels. Some of the standard nutritional functions of selenium include the formation of selenoproteins such as the glutathione peroxidases, which protect cell membranes against free radical damage. At low selenium intake levels, the body cuts back on glutathione peroxidase production to help make selenium available for conversion into other selenoproteins. Other selenoproteins are involved in immune system enhancements and this too is an anti-cancer action. Some of the selenium compounds several anti-cancer mechanisms involve prevention (membrane protection, DNA protection, etc.) and others (epoxide repair, immune stimulation, apoptosis, etc.) involve cancer cell destruction. The anti-cancer effect of selenium compounds is not related to formation of selenoproteins. The anti-cancer effects of selenium compounds are not related to the tissue accumulation of the compounds. Selenium's anti-cancer action is independent of its nutritional role and toxicity. Selenium supplements include high-selenium (selenized) yeast, selenomethionine, inorganic selenium, and more. The production of selenium anti-cancer compounds from inorganic selenium compounds (chiefly selenates and selenites) is inefficient. The inorganic selenium first has to be converted into compounds that can enter the selenium metabolic pool and then be converted into specific organic anti-cancer selenium compounds. The toxic effects from intake of inorganic selenium occur at quantities far less than those from organic selenium compounds. The benefits of selenomethionine occur over a longer time frame and do not produce rapid results. Most of the selenomethionine intake can be temporarily incorporated non-specifically into proteins in place of methionine, and not be available to the selenium pool again until that protein is degraded by the body. Selenomethionine can be converted into selenocysteine readily, and the latter is the building block for the selenoproteins. Selenomethionine is significantly less toxic than inorganic selenium. Selenium yeast provides compounds that are direct anti-cancer compounds not having to go through the selenium metabolic pool. The effects of selenium yeast are both immediate and long-term. Selenomethionine accounts for about 20% of all selenium-containing compounds. Another group of selenium-containing compounds includes selenocysteine, seleno-methyl selenocysteine and selenomethionine. There are other groups of compounds make up about 40-50% of all the selenium compounds in yeast. These may include selenophosphates, triphenylphosphine selenide, diselenides, triselenides and other very interesting organic selenium compounds.

GLUTAMINE

Glutamine is produced in the skeletal muscle from glutamate and ammonia catalysed by glutamine synthetase. Glutamine comprises 60% of the free intracellular amino acids in the skeletal muscle. Glutamine 10 g three times a day may improve survival rate and decreases complications related to bone marrow transplant in cancer patients. Glutamine 10 g three times a day is effective in preventing myalgia and arthralgia associated with paclitaxel. Glutamine decreases methotrexate toxicity and increases methotrexate concentration in cancer cells during treatment (0.5 g/kgBW/day). Glutamine decreases intestinal permeability caused by NSAIDs. Oral stomatitis and mouth pain is reduced in cancer patients by oral suspension of glutamine twice daily swish and swallow (2 g). Glutamine together with human recombinant growth factor (HGH) increases body weight, increases lean muscle mass, and decreases body fat in patients with short bowel syndrome (SBS). Glutamine restores impaired permeability of the intestine. Glutamine (5 g/500 ml enteral feeding) shorten ICU and hospital stays. Glutamine preserves hepatic function in cancer patients undergoing marrow transplantation. Glutamine enhances T-cell function. Glutamine may be contraindicated in hypersensitivity to glutamine, liver diseases, Reyes syndrome, cirrhosis and hepatic insufficiency (ammonia induced encephalopathy and coma). Glutamine supplementation in trauma patients decreases incidence of pneumonia, sepsis, and bacteraemia. Glutamine is the fuel for intestinal cells, can be used to treat ulcers, food allergies, leaky gut syndrome, reflux disease, and diarrhoea. Glutamine promotes the release of growth hormone. Glutamine is a marker for aging. Glutamine decreases cachexia. Glutamine is muscle enhancer and recovery aid for athletes. The body is depleted of glutamine stores during trauma, hypercatabolism, immunodeficiency, malnutrition or extreme stress. Glutamine decreases tumour activity. Glutamine causes an increased in T-cell DNA synthesis. Glutamine is released from the skeletal muscle into circulation during metabolic stress, trauma, and surgery. Glutamine muscle concentration is affected by injury, sepsis, prolonged stress, and starvation. Plasma glutamine levels decrease after severe organ injuries partly because the small intestine uses it faster than it can be produced by skeletal muscle. Hormones, concentrations of electrolytes, and branched-chain amino acids influence a transport mechanism allowing glutamine to be released by crossing the cell membrane. Arginine stimulates lymphocyte immune responses and wound healing. Glutamine stimulates renal production of arginine by raising plasma concentrations of the precursor citrulline. Glutamine is manufactured in the liver from glutamate, cysteine, and glycine. Glutamine acts as a powerful antioxidant protecting hepatocytes. Glutamine preserves glutathione liver stores and protects the liver from free radical damage. Glutamine diminishes electrolyte and water loss during acute bouts of diarrhoea. Glutamine functions as a dual transporter crossing the cell membrane attached to sodium. Glutamine stimulates electroneutral absorption of NaCl. Increasing the amount of glutamine in the diet increases the ability of T-lymphocytes to respond to mitogenic stimulation. Increasing oral glutamine may promote the T-cell immune response. Glutamine increases the production of IL-2 and IFNγ that is important for optimal lymphocyte proliferation. Glutamine supplementation causes an increase in T-cell synthesis postoperatively. L-glutamine supplementation increases glutathione (GSH) levels, decreases pro-inflammatory PGE_2 production, increases NK activity by 2.5 times and may result in 40% reduction in tumour growth of breast cancer. Serum levels of glutamine are reduced in gastric carcinoma. Glutamine increases cytotoxicity (300%) of methotrexate and 5-FU while protecting the patient from GIT adverse effects. Glutamine protects lymphocytes and intestinal integrity in patients with advanced oesophageal cancer. Glutamine protects against gut integrity and T-cell functioning during radiochemotherapy. Glutamic acid is an aliphatic amino acid that is degraded in the body to form levoglutamine (glutamine). Glutamic acid and levoglutamine are used as dietary supplements. Glutamine is the most abundant amino acids in blood and tissue fluids. Glutamine is synthesised in skeletal muscle, released into the bloodstream, and transported to the main utilising organs: intestine, kidney, immune cells, and brain. Glutamine provides substrate for protein synthesis and precursors for nucleic acid biosynthesis. Glutamine is an important carrier of nitrogen and carbon and a precursor for gluconeogenesis in the liver and takes part in the acid-base homeostasis in the kidney and the liver. Glutamine is the principle fuel used by intestinal mucosa, where it accounts for 35% of total metabolic requirements of the entrocytes. Gut glutamine requirements are increased in several states of disease such as sepsis and surgery. A lack of glutamine promotes mucosal atrophy, an increase in intestinal permeability and bacterial translocation, and a reduced synthesis of glutathione, which is an important factor for antioxidant defense. Glutamine-enriched enteral nutrition up-regulates intestinal glutaminase activity and stimulates gut glutamine utilisation, at the same time, it improves growth and repair of the small bowel and colonic mucosa and reduces bacterial translocation. Oral therapy and prophylaxis with glutamine reduces macroscopic and microcirculatory inflammatory activity. There is a negative correlation between faecal pH of the lower ileum and leukocyte adherence, the lower the pH the higher the adherence. Glutamine normalises the reduced faecal pH induced by ileitis. Glutamine normalises leukocyte rolling velocity. Glutamine stimulates mucosal protein synthesis, preserves gut barrier function by augmenting immune responses, has effect on radical scavenger production especially glutathione, preserves intestinal blood flow by dilation of submucosal arteries, or promotion of secretion of peptide hormones such as glucagon, which might affect intestinal mucosa. The beneficial effect of glutamine prophylaxis could be due to an increase of glutathione reserve. Oxidised glutamine provides substrate for the synthesis of purines and pyrimadines needed for DNA, RNA, mRNA, and in the kidney, glutamine is involved in acid-base balance through ammonia production. Glutamine supplementation limits skeletal muscle wasting, reduce diarrhoea and malabsorption, enhance immune host defense, and reduce the incidence of opportunistic infections. Glutamine contains 2 nitrogen moieties. Much of the nitrogen transport from the skeletal to the visceral tissues is done by glutamine. Glutamine is a primary fuel for rapidly dividing cells including enterocytes, colonocytes, lymphocytes and fibroblasts. During periods of increased metabolic stress, glutamine is freely released from skeletal muscle and intracellular glutamine concentrations fall >50%. Glutamine synthesis can not keep up with the higher requirements during stress. Cancer cachexia is marked by massive host skeletal muscle glutamine depletion. Glutamine is an essential component of lymphocyte cell division. Natural killer (NK) cells are cytotoxic lymphocytes capable of killing tumour cells as well as producing other cytokines. Tumours do not grow well in hosts with high NK cell activity. Optimal functioning of lymphocytes, including NK cells, is dependent on adequate supplies of glutamine and glutathione. Oral glutamine supplementation leads to decrease tumour growth through NK cell activity, which is mediated by glutathione suppression of PGE_2. Radiation enteritis mucosal injuries include destruction of crypt cells, decreased villous height, and ulceration and necrosis of the GI epithelium. Glutamine is a primary cellular fuel for entrocytes, a precursor for nucleotides needed for cell regeneration and a source of glutathione. Glutamine reduces bacterial translocation. Exogenous glutamine maintains the gut barrier by decreasing the incidence of mucosal ulcerations and increasing the number and the height of intestinal villi. Glutamine increases tumour cell sensitivity by decreasing tumour intracellular glutathione. Glutamine improves nutritional status, decreases intestinal injury, decreases bacterial translocation, reduces endotoxaemia and improves survival during chemotherapy. Glutamine maintains normal cardiac glutathione levels decreasing cardiotoxicity. Increased glutamine concentration causes an increase in the proportion of polyglutamated methotrexate that results in improved retention within the tumour (3-fold increase of methotrexate). During periods of severe metabolic stress or catabolic insult 20-40 g of glutamine may be required to maintain homeostasis. Glutamine is more effective when administered via enteral route. Cancer cells do not absorb antioxidants as efficiently as healthy cells. The liver injury of veno-occlusive disease (VOD) occurs because of free radical damage to the hepatic endothelial cells after radiation or chemotherapy. Glutamine and vitamin E may have a role in the treatment of VOD. Vitamin E along with vitamin C and glutamine act on a cellular level by protecting cell membranes and preventing lipid peroxidation. Vitamin E improves mucositis; lessening chemotherapy induced alopecia and reversing certain precancerous conditions. The provision of antioxidants nutrients and precursors such as vitamin A, C and E, selenium, copper, zinc, and N-acetyl cysteine (NAC) may offer protection to healthy cells against oxidative injury by antineoplastic therapy. Glutamine combined with antioxidants may sensitise malignant cells therapy making conventional therapy more effective with the added benefit of increasing patients tolerance to dose escalating therapies. Glutamine is a neutral gluconeogenic amino acid that can be synthesised by virtually all tissues. Rapidly replicating cells (entrocytes, lymphocytes, macrophages, and fibroblasts) remove glutamine from the circulation and utilise its five-carbon chain for energy. Metabolism of glutamine to α-ketoglutarate and subsequent complete oxidation via the Kreb's cycle yields 30 mol of ATP per mole of glutamine. Glutamine double nitrogen (α-amino group and amide group) allows glutamine to function as a nitrogen shuttle between various organs. Glutamine provides nitrogen for the synthesis of purines and pyrimidines and other amino acids. Intestinal glutamine provides precursors and nitrogenous end products for hepatic gluconeogenesis and for urea synthesis of intraluminal ammonia that may assist in the transport of sodium and water. Glutamine is a precursor to glutathione, a potent antioxidant important to limiting tissue damage from free radical attack. Glutamine is the primary oxidative fuel for both the enterocyte and the colonocyte and is necessary for the maintenance of intestinal structure in both normal and stressed states. Subjects receiving glutamine supplementation both before and during abdominal radiation have increased crypt cell proliferation and improve gut integrity. A daily dose of 5-10 g before onset of chemotherapy or radiation treatment, with an increase to 20-30 g during actual therapy is recommended. Glutathione levels in the liver fall within 24 hrs of injury. Replacement is accomplished primarily by replacing glutamine, NAC and selenium. Enteral glutamine enhances glucose absorption. Daily dosages are divided throughout the day to increase direct contact to the enterocytes. Glutamine stops alcohol sugar craving. Glutamine is a major fuel and important nitrogen source for enterocytes and plays a key role in maintaining mucosal cell integrity and gut barrier function. Although ammonia derived from glutamine is excreted as ammonium via the kidneys or converted to urea in the liver, with the liver also mopping up ammonia from both the systemic and portal circulations. Glucosamine furnishes glutamine for ammonium synthesis in the kidneys, as ammonia production goes up and potassium excretion goes down when ingesting glutamine. A combination of vitamin C and L-glutamine three times a day helps alcoholics to say "no" to alcohol.

Indications for L-Glutamine:

1- Growth hormone secretagogue (more muscle/less fat). Growth hormone releasers do not cause or create masculine features such as excessive hair growth or huge muscles in females.
2- Stronger immune system.
3- More energy, less fatigue and better mood.
4- Crohn's disease, leaky gut syndrome, colitis, and diarrhoea.
5- Blood sugar control.
6- More agile brain.
7- Healthier intestines.
8- Heavy stress (including strenuous exercise) or recovering from injury or other trauma.
9- Helpful as adjunct therapy in the treatment of addictions such as alcoholism.
10- Especially popular with bodybuilders, and with those who wish to perk up their physical and mental energy.

Glutamine has been one of the most intensively studied nutrients in recent years. Glutamine is effective against catabolic stress. Glutamine supplementation improves organ function, and survival, or both. Glutamine is a critical nutrient for the gut mucosa and immune cells. Glutamine-enriched diets are safe and effective in catabolic patients. Intravenous glutamine increases plasma glutamine levels; exerts protein anabolic effects; improves gut structure and function; and reduces important indices of disease, including infection rates and length of hospital stay. Glutamine is the most abundant free amino acid in the human body. In catabolic stress situations, such as after surgical operations or trauma and during sepsis, glutamine is rapidly transported to organs and to blood cells. This results in an intracellular depletion of glutamine in the muscles and the ensuing catabolic wasting effect. Glutamine depletion decreases the proliferation of lymphocytes, possibly by arresting a critical phase of the cells' growth cycle. Glutamine is a precursor for the synthesis of glutathione and stimulates the formation of heat-shock proteins. Glutamine plays a crucial role in the stimulation of intracellular protein synthesis. Glutamine deficiency causes a necrotizing enterocolitis, which is an inflammation of the small intestine and colon leading to cell death-increases the mortality due to bacterial stress. Parenteral glutamine reduces nitrogen loss and causes a reduction of the mortality rate. In surgical patients, glutamine evokes an improvement of several immunological parameters. Glutamine exerts a nutritional (trophic) effect on the intestinal mucosa, decreases the intestinal permeability, and thus may prevent the bacterial translocation. Glutamine is an important metabolic substrate of rapidly proliferating cells (e.g., intestinal). It influences the cellular hydration (molecular water content) state and has multiple effects on the immune system, intestinal function, and protein metabolism. Glutamine is an indispensable nutrient supplement. In the brain, glutamine is a substrate for the production of both excitatory and inhibitory neurotransmitters (glutamate and γ-aminobutyric acid [GABA]). Glutamine is an important source of energy for the nervous system. Glutamine users often report more energy, less fatigue, and better mood. Glutamine plays a part in maintaining proper blood glucose levels and the right pH range. Glutamine is osmoregulator in various tissues. Glutamine constitutes 50% of all amino acids in the serum and >60% of free amino acids within the body. When glutamine was first discovered, it used to be called intestinal permeability factor. It is by far the most important nutrient for intestinal health. Glutamine is the chief source of energy for enterocytes. Most glutamine in the diet (and most dietary glutamate and aspartate) is metabolised by the intestines, both to serve as intestinal fuel and also to produce glutathione, nitric oxide, polyamines, nucleotides and the amino acids alanine, citrulline and proline, making these available to the rest of the body. Glutamine can also be used to treat colitis, Crohn's disease, leaky gut syndrome, ulcers and diarrhoea in doses of up to 20 g/day. Monosodium glutamate-a flavour enhancer originally manufactured from seaweed is the most abundant natural source of glutamate. Monosodium glutamate (MSG) is the sodium salt of glutamic acid just as sodium ascorbate and calcium ascorbate are salts of ascorbic acid. MSG raises glutamate levels. Glutamate is the term used interchangeably with glutamic acid. Glutamate is an anionic amino acid or the anionic form of glutamic acid. Glutamine differs from glutamate in that it has been formed from glutamate and ammonia, and thus has an extra nitrogen it can easily donate whenever nitrogen might be needed. The enzyme that catalyses the addition of ammonia to glutamate is called glutamine synthase. The biosynthesis of glutamine is the process through which the body eliminates excess ammonia. As we will see later, glutamine synthase is of incredible importance in brain function. Survival may be dependent on this enzyme, and on the glial cells that secrete it. MSG is the sodium salt of glutamic acid, while glutamate is the ionic form of glutamic acid. Abundant sodium of MSG causes Na/K imbalance, dehydration, and disturbances in the constriction and dilation of blood vessels. Dehydration alone is enough to cause the kind of dull headache that some Western patrons of Chinese restaurants have complained about. Those prone to migraines should avoid MSG and aspartame and all of us should avoid these compounds in large doses. Young children with immature nervous systems are most susceptible to MSG damage, not the elderly. Stress increases the permeability of the blood-brain barrier to exogenous glutamate. Serum levels of glutamine are in the range of 390-650 mg/dl for adults compared to 18-98 range for glutamate. Children have higher upper values for both glutamate and glutamine, 140 mg/dl and 730 mg/dl respectively. Glutamine readily crosses the blood-brain barrier. Neurons take up glutamine and convert it to glutamate or GABA (through the additional step of decarboxylating glutamate). Some glutamate is used for energy, some for synthesis of glutathione and niacin and, some as neurotransmitter. After either glutamate or GABA is released into the synaptic junction, the supportive cells (the glia) take up the glutamate or GABA and resynthesize glutamine-detoxifying ammonia in the process. The glutamate that is not converted to glutamine is used by the glia as a source of energy and also to produce energy nutrients alanine and α-ketoglutarate, which are then released to the neurons. Very little glutamine is released by the brain in contrast to muscle and adipose tissue, which donate a lot. If the glia are dysfunctional due to reduced aerobic metabolism or the release and/or activity of the glial glutamine synthase is inhibited in any way (free-radical damage, toxins, certain drugs), lethargy and cognitive dysfunction can be the result. This too is one of the phenomena takes place in the aging brain. On the one hand, glutamate excitotoxicity damages or destroys some neurons leading to deficiencies in memory and learning. On the other hand, excess GABA can lead to lethargy. Excess ammonia not detoxified through sufficient glutamine synthesis by the glia leads to further neural damage. During stroke, the dying neurons release glutamate, which then unfortunately can cause more neuron death. Furthermore, ischaemic episodes damage the glutamate receptors so that the glutamate can't work as a neurotransmitter. Without glutamate, there is no memory and no learning. Ampakines amplify the glutamate signal through a yet unknown mechanism possibly by rebuilding glutamate receptors. In healthy people, ampakines enhance cognitive performance. Glutamate is also deficient in those diagnosed with schizophrenia. However, most likely, there are many neurotransmitters are out of balance in neurological disorders. When the CNS is aroused, some of the glucose is converted to glutamate. The other source of glutamate is, of course, glutamine. An abundant supply of glutamine makes it easier for the brain to maintain neurotransmitter balance by increasing the production of glutamate when required for alertness, learning, memory, and the production of GABA when its inhibitory properties are needed. Glutamate is the chief excitatory neurotransmitter. It is essential for learning and both short- and long-term memory. Problems arise only if the normal process of glutamate removal and conversion to glutamine malfunctions and an excess of this excitatory neurotransmitter builds up in the synaptic junctions. Excess glutamate causes excessive influx of calcium ions into the neurons causing excitotoxicity and ultimately even death of the neurons. It also destroys glutathione-a crucial brain-protective antioxidant. Low levels of brain glutathione are associated with neurodegenerative disorders. Glutathione depletion further leads to neuronal death. CNS damage is associated with excess glutamate in Alzheimer's, AIDS patients, and cancer patients. The AIDS virus inhibits glutamate uptake by the glia. Cancer may start with brain dysfunction and in those who have suffered a severe brain injury. Very high fever or artificially induced hyperthermia can also result in excess glutamate release, leading to seizures. The use of glutamine as a free amino acid has never been associated with any form of brain damage. Glutamine is in fact abundantly produced in the brain as a vital defense against ammonia and also against excess glutamate. The main defense against glutamate excitotoxicity is the synthesis of glutamine by the glia, or more specifically, astroglia or astrocytes. They are most abundant types of cells in the CNS exhibiting high amounts of glutamine synthase. Proinflammatory cytokines IL-1β and TNF-α inhibit the induction of glutamine synthase. These pro-inflammatory cytokines are released after a brain injury and in neurodegenerative disorders. Thus, neuronal death may occur because the inflammatory process interferes with the conversion of glutamate into glutamine. Controlling inflammation can prevent glutamate excitotoxicity by protecting the glia. The anti-inflammatory hormones (glucocorticoids) induce glutamine synthase. Excess cortisol, however, can inhibit the uptake of glutamate by the glia. Bioflavonoids, such as the catechins in green tea or proanthocyanidins in grape seed extract, can help protect against the excitotoxic injury due to glutamate build-up. So can uric acid one of the endogenous antioxidants and the amino acid taurine. It seems that the brain can produce its own taurine. It might be an extra precaution to take supplemental taurine, if MSG is regularly consumed. Certain B vitamins, including methylcobalamin (one of the active forms of vitamin B_{12}), are likewise protective. A ginkgo extract as well as one of its constituents, ginkolide B, protects against glutamate excitotoxicity by reducing the rise in calcium ions. Thus, it would be advisable to include ginkgo biloba in the supplement regimen, and drinking green tea or taking green tea extract, as well as eating berries or taking bilberry extract, in order to obtain a good dose of flavonoids for general neural protection and prevention of neurodegenerative diseases. Retinal damage in diabetes is also partly due to excitotoxic glutamate build-up. This is due to insufficient conversion of glutamate to glutamine probably due to the malfunction of glial cells (both insufficient or excessive glucose levels can lead to cell dysfunction; diabetics also show higher levels of free radicals). Alcohol also inhibits glutamine synthase, which explains at least in part the neurotoxicity of alcohol. Certain drugs including many anti-epileptic drugs likewise inhibit glutamine synthase and this may be partly responsible for their toxic side effects. Vigabatrin (anti-epileptic drug) raises both GABA and glutamine while decreasing glutamate. Toxic residues from the compounds formerly used for bleaching white flour may contribute to the increase in neurodegenerative diseases, by inhibiting glutamine synthase. Neurotoxins might impair the quick conversion of glutamate to glutamine; thus avoiding them may enhance the production of glutamine synthase. Malnutrition can likewise lead to glial malfunction and thus to the inability of glia to remove excitatory neurotransmitters (glutamate and aspartate) from the synaptic junctions. Glutamine is the primary source of energy for the various cells of the immune system, including T cells and macrophages. Strenuous exercise, viral and bacterial infections, and stress and trauma in general cause glutamine depletion that starves the immune cells. Up to 40 g of glutamine a day can be used to sustain the immune system of AIDS patients or cancer patients undergoing bone marrow transplantation. Glutamine is a substrate for glutathione, a tripeptide amino acid that acts as the master of antioxidants and also helps enhance the immune function. It is important to reduce stress as much as possible. Stress hormones may interfere with glutamine metabolism in the immune cells. Relaxation and DHEA supplementation might prove to be very helpful in addition to glutamine. In its role as a carbon donor, glutamine is muscle food, helping to replenish glycogen. Strenuous exercise such as weight lifting causes micro-injuries to the muscle. By donating nitrogen, glutamine helps build proteins and repair the muscle as well as help build up more muscle. Part of its muscle-building action may be due to its ability to induce the release of growth hormone (secretagogue), 2-3 g after workout is particularly recommended. Long-term users of anti-inflammatory steroids tend to suffer from muscle atrophy. The concomitant use of glutamine has been shown to prevent most of this muscle loss. Glutamine serves the anabolic needs of the whole body. Since it can very easily donate nitrogen, it functions as a nitrogen shuttle, delivering nitrogen wherever it is needed. Very ill patients suffer both a decrease in glutamine levels and muscle loss. The use of glutamine has been documented to aid the survival of severely ill surgical and burn patients. It also speeds up wound and burn healing, and improves recovery in general. Glutamine is an important source of fuel for the heart muscle. Glutamine is converted to glutamate, which then enters the Krebs cycle to produce ATP. Glutamine serves as a substrate for the synthesis of a special type of β-endorphin, glycyl-l-glutamine. This dipeptide appears to be important for the regulation of blood pressure and prevention of cardiorespiratory depression. Glycyl-l-glutamine is also important for the immune response, since it enhances the activity of the natural killer (NK) cells. So, If the brain has a faulty glutamine/glutamate/GABA metabolism, we can expect the development of cardiovascular dysfunction as well. The neural control of cardiovascular function relies on glutamate and GABA. Glutamine can enter the Krebs cycle and serve as a non-carbohydrate source of energy. Together with alanine, glycine, serine and threonine, glutamine is an important gluconeogenic amino acid-in fact the primary one. Glutamine is catabolised in the liver to provide more glucose. During hypoglycaemia the kidneys can contribute as much as 25% to whole-body glucose production. The kidneys are especially equipped to process glutamine due to its importance in the detoxification of ammonia. Calorie-restricted dieters are at risk of losing muscle mass more than fatty tissue. The metabolically active muscle mass helps keep us slender, strong and fit, extra glutamine can help dieters lose girth around the waist while preserving muscle mass. Many alcoholics appear to suffer from hypoglycaemia. Diabetics have an abnormal glutamine metabolism. Glutamine gluconeogenesis increases the production of glucose from glutamine (and also from alanine, an amino acid in the same family), which is probably related to the diabetes-related excess levels of the serum glucose-raising pancreatic hormone, glucagon. This excessive breakdown of glutamine into glucose in diabetes occurs without any supplementation since muscle and fatty tissues release so much glutamine in response to the endocrine pathology. Diabetics also show other enzymatic abnormalities in relation to glutamine including poor function of the retinal glia (glia are cells that have various supportive functions in the nervous system including detoxifying ammonia through the production of glutamine). Thus, the diabetic retina is prone to damage through glutamate excitotoxicity since the glia are not converting enough glutamate to glutamine. The use of high doses of antioxidants, including vitamin E and various polyphenols is beneficial as well as supplementation with taurine. Taurine is amino acid that seems to be very helpful to diabetics. Glutamine is frequently used as an adjuvant treatment of advanced cancer. It prolongs survival by slowing down catabolic wasting, beneficial for the depleted immune system, and helps preserve intestinal function as well. Glutamine can be given to cancer patients without stimulating tumour growth or metastasis. Glutamine increases the natural killer cells activity (2.5 times greater), rises in glutathione levels (25%), and decreases inflammatory prostaglandins (PGE_2). PGE_2 fuels tumour growth. Glutamine lowers the toxicity of methotrexate augmenting its effectiveness against inflammatory breast cancer. The glutamine dose used in conjunction with methotrexate should be 5g/kg/day. High doses of glutamine (30 g/day) are recommended as adjuvant therapy for sickle cell anaemia. Glutamine is used as part of the treatment for AIDS. Glutamate excitotoxicity arises only under certain pathological conditions such as stroke, extremely high fever, certain viral infections, the presence of neurotoxins, or severe inflammation. It can be due to excess release of glutamate by the neurons (stroke) and/or to glial malfunction where the glia are incapable of secreting enough glutamine synthase in order to convert glutamate to glutamine. Glutamine (GLN) is the most abundant amino acid in the body. Glutamine is essential amino acid that serves not only as a primary respiratory fuel but as a necessary substrate for nucleotide synthesis in most dividing cells. GLN is required by enterocytes, lymphocytes, fibroblasts, and rapidly growing tumours. In kidney tissue with oxidative stress, GLN becomes rate limiting in glutathione (GSH) synthesis. The disruption of interorgan glutamine flux by progressive tumour growth may contribute to host cachexia. Dietary glutamine to the cancer-bearing host nearly doubles the tumouricidal action of methotrexate (MTX) while reducing mortality and morbidity. Glutamine is the most abundant amino acid in the blood, comprising 50% of the whole body pool of free amino acids; and 5% resides in skeletal muscle, with most of the remaining stores in the liver. Glutamine transports almost one third of circulating amino acids and nitrogen; it is also the principle carrier of nitrogen from skeletal muscle to visceral organs. Concentrations of GLN in blood and skeletal muscle decrease markedly after injury and catabolic states, may be depleted by >50%, while plasma levels fall 20-30%. Gut is a major producer of glutathione (GSH) synthesis and this can be stimulated three-fold with oral GLN. Gut GSH production is decreased with tumour growth. Glutamine-mediated recovery of depressed levels of GSH in lung, liver, kidney, heart, gut, and muscle after radiation and chemotherapy has been demonstrated. Glutamine is the principal fuel utilised by most rapidly growing tumours. Such tumours have high glutaminase activity, the principal enzyme of GLN degradation. Advanced malignant disease results in muscle GLN depletion and weight loss. As the tumour grow, it becomes a major GLN consumer, behaving as a "GLN trap" that may contribute to host GLN and GSH depletion. Glutamine not only increases intracellular MTX tumour concentration but also increase the tumouricidal effectiveness as demonstrated by decreased tumour volume. All host tissues have significantly elevated GSH content whereas the tumour tissue levels are significantly decreased. The dose-limiting toxic effect of radiotherapy on the gut is not acute mucosal injury but the late reactions of fibrosis, stricture, and necrosis, mediated in part by vascular injury and are responsible for significant morbidity and mortality. Histological parameters of chronic radiation injury are ulcerations, epithelial atypia, serosal thickening, vascular sclerosis, fibrosis, thickening of the intestinal wall, lymph congestion, and ileitis cystica profunda. GLN during abdominal or pelvic XRT may accelerate healing of the irradiated bowel, prevent injury, and decrease the long-term complications of radiation enteropathy. Glutamine increases the content of GSH in liver and has been shown to exert a protective effect against oxidant injury. Glutamine-enriched diet decreases tumour GSH content while increasing gut GSH levels. Glutamine is a primary fuel utilised by rapidly dividing host cells, including lymphocytes. GLN depletion has a significant effect on host cell immunity, particularly immunological function of the gastrointestinal tract stimulating secretory IgA and decreasing bacterial translocation. GLN is necessary for the growth and maximal function of T cells and natural killer (NK) cells. NK cells, a subpopulation of cytotoxic lymphocytes present in normal individual, are capable of spontaneous cytolytic activity against a variety of tumour cells. Unlike T cells, tumour cell killing is mediated by non-MHC restricted mechanism, allowing NK cells to act spontaneously at first contact. IL-2 stimulates the activation and proliferation of NK cells, which enhances NK cytotoxicity. Activated NK cells also participate in other host functions through their production of cytokines such as interferon. NK cell may be the first line of defence against the blood-bourne phase of tumour metastases. The role of GLN in lymphocyte activity is to provide both nitrogen and carbon for precursor synthesis of purines and pyrimidines and also for energy production. The rate of GLN use by lymphocytes is markedly in excess of these precursor requirements. In other host tissues GLN is used in part for GSH synthesis. A GLN deficit affects the generation of lymphokine-activated killer (LAK) cells. Upregulation of GSH via supplemental GLN will improve antitumour NK activity and suppress tumour growth in both sarcoma and carcinoma. Interferons and IL-2 have not been useful clinically as anticipated. The modest response of cancer patients to cytokine therapy may reflect variations in depletion of host GLN and GSH stores associated with poor nutritional status or tumour progression in these patients. GLN can enhance both radiation and chemotherapy toxicity is by alteration in GSH metabolism. Glutathione is a tripeptide that is ubiquitous and acts in a protective role against radiation- and chemotherapy-related injury in tumour tissue. GSH plays a central role in calcium metabolism, leukotriene biosynthesis, thyroid metabolism, and membrane and channel function and nutrition. Glutathione is also important in the protection of critical cellular molecules. Depletion of GSH in host tissues occurs during shock, sepsis, endotoxaemia, multiple trauma, haemorrhagic shock, and malnourished. Toxicity of target tissue is a result of depletion of tissue GSH concentration and protein alkylation. Depletion of >70% of tissue GSH is associated with irreversible cellular damage. GLN can bypass the acidotic block in the GSH-recycling enzyme, oxoprolinase, in host but not in tumour cells. Uptake and metabolism of extracellular GSH requires the oxidation of an intracellular GSH molecule. In cells receiving radiation or chemotherapy, intracellular GSH regeneration is slowed because the pH-sensitive oxoprolinase enzyme is inhibited in the resulting acidotic environment. This results in depletion of intracellular GSH levels in the tumour. In normal cells, the presence of abundant GLN can regulate the γ-glutamyl transferase and glutaminase enzymes, providing additional glutamate to bypass the blocked oxoprolinase enzyme. These enzymes in the tumour cell are not upregulated by supplemented GLN. If no radiation or chemotherapy is given, tumour GSH remains unchanged while host GSH stores increase and tumour growth decreases, possibly through GSH-mediated upregulation of the immune system. Reduced NK cell activity in cancer is associated with increased prostaglandin (PGE_2) synthesis. PGE_2 has profound effects on cellular immunity and may influence host-tumour interplay. Tumour cell PGE_2 production might constitute another mechanism whereby tumours subvert immune surveillance. Inhibition of PGE_2 synthesis by indomethacin or flurbiprofen is associated with decreased tumour growth. GSH is also an inhibitor of prostaglandin synthesis. GLN will upregulate GSH, improve NK activity, and suppress tumour growth possibly through the action of GSH on PGE_2 synthesis. Glutamine may be superior to other antioxidants, which unlike GLN may also enhance tumour GSH, increasing tumour resistance to therapy. The low cost of GLN, its ease administration, and lacks of toxicity make it an ideal adjunct to radiation and chemotherapy. GLN when given with radiation or chemotherapy protects the host and actually increases the selectivity of therapy for the tumour. Glutamine is an important fuel for the intestinal mucosa. Glutamine is extracted at all locations of the human intestine with the highest fractional extraction in the jejunum. This extraction of glutamine is associated with release of the metabolic products of glutamine degradation in the enterocyte: ammonia, alanine, and citrulline. Glutamine extraction in ileum and jejunum is associated with the release of citrulline, glutamate, alanine, and a small but significant amount of arginine. The jejunum contributes to citrulline release. Impaired glutamine metabolism by the gut may result in decreased citrulline release and consequently impaired arginine production by the kidneys. In addition to arginine production from citrulline in the kidney, the gut itself releases arginine. The release of glycine (aminoacetic acid) and taurine in the jejunum and especially the ileum most likely reflects the deconjugation of bile acids in the gut and the reuptake of glycine and taurine derived from deconjugation. Ammonia production by the intestine is long ascribed to urease activity of intestinal bacteria. Ammonia production is predominantly a metabolic process in the proximal bowel. In patients who are nutritionally depleted, colonic ammonia production is reduced with 75%. The reduction in ammonia release is associated with a decrease in glutamine extraction. Malignant cells, like other rapidly dividing cells, oxidize glutamine for energy supply at a rapid rate. A high tumour load may, result in a decreased availability of glutamine for other tissue. Human carcinomas rely predominantly on glucose for their energy. Tumour cells exhibit a high rate of aerobic glycolysis, but a substantial part of the glucose is converted to lactate and not completely oxidised. Glutamine extraction in the ileum and most likely in the jejunum is dependent on arterial glutamine concentrations. Glutamine is more important substrate for the proximal intestine than for the distal gut. Nutritional depletion results in decreased arterial glutamine concentration, which in turn results in diminished extraction.

GLUTATHIONE

Glutathione (γ-glutamylcysteinylglycine, GSH) is a sulfhydryl (-SH) antioxidant, antitoxin, and enzyme cofactor. Glutathione is water soluble ubiquitous found mainly in the cell cytosol and other aqueous phases of the living system. Glutathione attains millimolar levels inside cells, which makes it one of the most highly concentrated intracellular antioxidants. Glutathione is homeostatically controlled, both inside the cell and outside. Enzyme systems synthesise it, utilise it, and regenerate it as per the γ-glutamyl cycle. Glutathione is most concentrated in the liver (10 mM), where the P450 Phase II enzymes require it to convert fat-soluble substances into water-soluble GSH conjugates, to facilitate their excretion. GSH depletion leads to cell death. Glutathione exists in two forms: 1- the antioxidant "reduced glutathione" tripeptide GSH; and 2- the oxidised form is a sulfur-sulfur linked compound, glutathione disulfide or GSSG. The GSSG/GSH ratio may be a sensitive indicator of oxidative stress. Any substance with great readiness to donate electrons, when present at high concentrations, has greatly enhanced effectiveness as a reductant. This reducing power is most expressed by GSH where its concentrations are highest (as in the liver). Reducing power is the key to the multiple actions of GSH at the molecular, cellular, and tissue levels, and to its effectiveness as a systemic antitoxin (free-radical scavenging, electron-donating, and sulfhydryl-donating capacity). Glutathione is present inside cells mainly in its reduced (electron-rich, antioxidant) GSH form. Glutathione status is homeostatically controlled, being continually self-adjusting with respect to the balance between GSH synthesis (by GSH synthetase enzymes), its recycling from GSSG (by GSH reductase), and its utilisation (by peroxidases, transferases, transhydrogenases, and transpeptidases). Cysteine is generated from the essential amino acid methionine, from the degradation of dietary protein, or from turnover of endogenous proteins. Cysteine and glutamate are combined by the enzyme γ-glutamyl cysteinyl synthetase. γ-Glutamylcysteine combines with glycine to generate GSH (catalysed by GSH synthetase). Excessive accumulation of γ-glutamylcysteine in the absence of its conversion to GSH can lead to its conversion to 5-oxoproline by the enzyme γ-glutamyl cyclotransferase. Increased 5-oxoproline induces metabolic acidosis. Applications of GSH: 1- as cofactor for the GSG-S-transferases in the detoxicative pathways; 2- as substrate for the γ-glutamyl transpeptidases that are located on the outer cell surface and which transfer the moiety from GSH to other amino acids for subsequent uptake into the cell; and 3- for direct free-radical scavenging and as an antioxidant enzyme cofactor. The GSH transferases are a large group of isozymes that conjugate GSH with fat-soluble substances as the major feature of liver detoxification. Glutathione is an essential cofactor for antioxidant enzymes, namely the GSH peroxidases (both Se-dependent and non-Se-dependent forms) and phospholipid hydroperoxide GSH peroxidases. The GSH peroxidases serve to detoxify peroxides (hydrogen peroxide, other peroxides) in the water-phase, by reacting them with GSH. The phospholipid hydroperoxide GSH peroxidases use GSH to detoxify peroxides generated in the cell membranes and other lipophilic cell phases. GSH transhydrogenases use GSH as a cofactor to reconvert dehydroascorbate to ascorbate, ribonucleotides to deoxyribonucleotides, and for a variety of -S-S- ↔ -SH inter-conversions. After GSH has been oxidised to GSSG, the recycling of GSSG to GSH is accomplished mainly by the enzyme glutathione reductase. Glutathione reductase uses as its source of electrons the coenzyme NADPH (nicotinamide adenine dinucleotide phosphate, reduced). Therefore NADPH, coming mainly from the pentose phosphate shunt, is the predominant source of GSH reducing power. Subjects unable to make adequate NADPH may be at increased risk of oxidative damage from GSH insufficiency. GSH makes major contributions to the recycling of other antioxidants that have become oxidised. GSH helps to conserve lipid-phase antioxidants such as α-tocopherol and the carotenoids. The liver seems to have 2 pools of GSH: 1- a fast turnover pool (half-life of 2-4 hrs), and 2- avidly retained pool with a half-life of about 30 hrs. The first corresponds to cytosolic GSH, the second mainly to mitochondrial GSH. Inherited deficiency of the γ-glutamyl cysteine synthetase, the first of the two enzymes necessary for GSH synthesis, has been described in 2 human siblings. They exhibited generalised GSH deficiency, haemolytic anaemia, spinocerebellar degeneration, peripheral neuropathy, myopathy, and aminoaciduria, and severe neurological complications as they moved into their fourth decade of life. Their red cell GSH was <3% of normal, their muscle GSH <25%, and their white cell GSH <50% normal. One of them may have been hypersensitive to antibiotics, having developed psychosis after a single dose of sulfonamide for UTI. Deficiency in GSH synthetase, the second enzyme of GSH synthesis, also is associated with haemolytic tendency and defective central nervous system function. This condition is complicated by the metabolic consequences of an excess of 5-oxoproline, formed as a spillover from the accumulation of γ-glutamylcysteine after its normal synthesis by the first enzyme and its lack of conversion to GSH by the second enzyme. Human hereditary GSH deficiency states are not necessarily lethal, characterised by GSH levels decreased in the plasma, liver, kidney, and other tissues. At the cell level, the damage mostly involved the mitochondria, but nuclear changes were also observed. Lung Type 2 cells showed damage to their lamellar bodies, the vesicles that package lung surfactant and release it to the cell exterior. The mitochondria appeared to be the most susceptible foci in the GSH-depleted tissues. If the mitochondria are unable to make their own GSH, they must import it from the cell cytosol. Dietary ascorbate can protect against the tissue damage that typically results from depletion of GSH, GSH can conserve ascorbate, and ascorbate can conserve GSH. GSH plays a role in diverse biological processes as protein synthesis, enzyme catalysis, transmembrane transport, receptor action, intermediary metabolism, and cell maturation. Redox phenomena are intrinsic to life processes, and GSH is a major pro-homeostatic modulator of intracellular sulfhydryl (-SH) groups on proteins. Many important enzymes (e.g., adenylate cyclase, glucose-6-phosphatase, pyruvate kinase, the transport Ca-ATPases), and at least eight participating in glucose metabolism, are regulatable by redox balance as largely defined by the balance of (2-SH ↔ -S-S-). Other proteins (tubulin of microtubules, thioredoxins, metallothioneins) have -SH groups at or near their active sites, or are otherwise regulated by the ambient redox state. Glutathione's reducing power is used in conjunction with ascorbate and other antioxidants to protect the entire spectrum of biomolecules, to help regulate their function, and to facilitate the survival and optimal performance of the cell. Glutathione's -SH character and its reducing power also set the redox stage for the proteins known as metallothioneins, which are able to bind with heavy metals and other potential sulfhydryl poisons to facilitate their subsequent removal from the body. Metallothioneins are inducible, and their levels are augmented in response to heavy metal overload or related oxidative challenge. GSH/GSSG ratio is normally very high in mitochondria, and their reducing potential highly negative help control the high flux of oxygen radicals from the mitochondria's OXPHOS activities (reducing environment). In the endoplasmic reticulum (ER), the protein biosynthesis does not consistently require a highly reducing environment, the GSH/GSSG ratio is low, and the ER microenvironment is set at a comparatively oxidising point. The antioxidant defense system is sophisticated and adaptive, and GSH is a central constituent of this system. Originating within the mitochondria of aerobic cells is a steady flux of OFRs, unavoidably generated from the processes that utilise oxygen to make ATP. This complex system of enzyme pathways by which the mitochondria use oxygen to break carbon-carbon bonds and produce ATP is called oxidative phosphorylation (OXPHOS). As OXPHOS substrates are processed in the mitochondria, invariably single electrons escape, leaking out of the OXPHOS complex to react with ambient oxygen and generate OFRs. 2-5% of the electrons that pass through the OXPHOS system is converted into superoxide and other oxygen radicals. The continual flux of single electrons to oxygen generates an endogenous oxidative stress in human tissues. Healthy cells homeostatically oppose free radicals through the use of antioxidants. The cumulative damaging effects of oxygen radicals and other oxidants are principal contributors to degenerative diseases, and aging (the progressive loss of organ functions). Oxidative stress originating from outside the body is a feature of life in the modern world (toxic substances). A single puff of cigarette smoke contains trillions of free radicals that burns away the antioxidant vitamins C and E, as well as other nutrients. The cigarette tars are long-lived free radical generators and potent carcinogens. Acetaminophen is a potent oxidant that depletes GSH from the cells of the liver, and by so doing renders the liver more vulnerable to toxic damage. Adriamycin has been used in animal experiments as a model for free radical-induced tissue damage; its foremost threat is to the heart. The halogenated hydrocarbons (halocarbons) are potent oxidants. Halocarbons are ubiquitous, being used in the plastics industry, as industrial and dry cleaning solvents, as pesticides and herbicides, and as refrigerants. The chlorofluorocarbons that currently threaten the ozone layer are one type of halocarbon. Halocarbons contaminate much of the ground water, and can now be detected in adipose tissue of humans from around the globe. They are potent free radical generators in the liver, by way of P450 activation, and they effectively deplete liver GSH. Strenuous aerobic exercise can deplete antioxidants from the skeletal muscles, and sometimes also from the other organs. Exercise increases the body's oxidative burden by calling on the tissues to generate more ATP, which requires using more oxygen and this in turn results in greater production of OFRs. GSH is depleted by exercise, and that for the habitual exerciser supplementation with GSH precursors may be a prudent policy. Some of the other exogenous factors depleting GSH include: 1- Dietary deficiency of methionine, an essential amino acid and GSH precursor. 2- Ionising radiation (X-rays or ultraviolet from sunlight). 3- Tissue injury (burns, ischaemia and reperfusion, surgery, septic shock, or trauma). 4- Iron overload, as in haemochromatosis and transfusional iron excess. Surgery can cause iron release from damaged tissue, and unbound iron catalyses free radical generation. 5- Bacterial or viral infections (HIV-1). 6- Alcohol intake is toxic through a number of differing pathways, some of which are free radical/oxidative in character. Negative lifestyle factors (smoking, alcohol consumption, legal or illegal drug use, emotional stress and life in the fast lane) can converge with environmental stressors to attack the body through related oxidative pathways. Individual cells die in those areas most affected by GSH depletion. Then zones of tissue damage begin to appear; those tissues with the highest content of polyunsaturated lipids and/or the most insubstantial antioxidant defenses are generally the most vulnerable. Localised free-radical damage spreads across the tissue in an ever-widening, self-propagating wave. If this spreading wave of tissue degeneration is to be halted, the antioxidant defenses must be augmented. Repletion of glutathione appears to be central to intrinsic adaptive strategies for meeting the challenge of sustained (or acute) oxidative stress. Normally, GSH is abundant inside cells (at millimolar levels) and relatively lacking outside of cells. One exception is the high concentration of GSH in lower regions of the lungs, where it helps neutralise inhaled toxins (e.g., cigarette smoke) and free radicals produced by activated lung phagocytes. The liver is the organ most involved with the detoxification of xenobiotics, and also is the main storage locale for GSH. Glutathione reaches its highest intracellular concentrations (10 millimolar) in the parenchymal cells (hepatocytes) of the healthy liver. The hepatocytes are highly specialized to synthesise GSH from its precursors or to recycle it from GSSG, as well as to utilise GSH against potential toxicants. Liver GSH stores are sensitive to depletion by malnutrition or starvation, but in the normally functioning liver the major drain on GSH is the activity of the GSH transferase enzymes (GSTs). GSTs are a large family of cytosolic isozymes with a collective broad specificity for endogenous orthomolecules as well as for xenobiotics. They are inducible, exhibiting multiple forms, and differing in their developmental patterns and induciblities. The GSTs constitute 10% of the extractable protein of liver. In classic toxicology, these are the P450-Phase II conjugating enzymes. The role of GSH in liver P450 conjugation activity normally is quite considerable, accounting for up to 60% of all the liver metabolites found in bile, but its outcome is not positive in every instance. Several classes of xenobiotics induce or otherwise activate P450-type enzymes, which generate GSH conjugates that are then more potentially toxic than the parent xenobiotic. Factors that deplete the liver pool of GSH can decrease conjugation and increase the toxicity of xenobiotics. To optimise nutritional support for the liver's detoxification functions, it is more rational to supply other nutrients in addition to GSH. A variety of oral antioxidants would be required for support of the entire antioxidant defense system (including its GSH branches), and non-antioxidant nutrients (phosphatidylcholine, B vitamins and minerals) would lend additional dimensions of support. Glutathione (GSH) in the liver exerts important protective activity against chemically induced oxidative stress. The mean lipid peroxide value and mean plasma glutathione peroxidase activity are significantly higher in diabetic women. The activities of SOD, glutathione peroxidase, glutathione reductase and vitamin E levels are significantly low in the plasma in NIDDM, with no significant change in catalase activity. Because glutathione is one of the most important protective factors against oxidative damage, its depletion is responsible for increased susceptibility to tissue injury. Reduced glutathione is the single most important antioxidant found within cells. Glutathione peroxidase and superoxide dismutase are two of the body's antioxidant enzymes and are, incidentally, inducible by aerobic exercise. Elevated levels of reduced glutathione, glutathione peroxidase, and/or superoxide dismutase indicate the body is diligently battling a heavy free radical load. Glutathione have a positive effect on sperm motility. Selenium and glutathione are essential to the formation of phospholipid hydroperoxide glutathione peroxidase, an enzyme present in spermatids, which becomes a structural protein comprising over 50% of the mitochondrial capsule in the mid-piece of mature spermatozoa. Glutathione (GSH) is an important natural antioxidant found in brain and nerve cells. The brains of Alzheimer patients are considered to be especially vulnerable to attack by free radicals. GSH levels are lower in immune cells and brains of Alzheimer patients. Antioxidants such as glutathione have effect on intercellular trafficking of NO metabolites. The oxygen of the water has been used to oxidize carbon monoxide to carbon dioxide with the liberation of hydrogen. Glutathione may facilitate this cellular oxidation by acting as a hydrogen acceptor. GSH (catalysed by glutathione peroxidase) reduces the peroxide to water but in the process is oxidised to GSSG. The resulting GSSG is reduced by NAD(P)H (catalysed by glutathione reductase). Mitochondrial dysfunction, reducing ATP generation, is linked to glutathione deficiency. Glutathione reduction induces an increase in citrate levels, which can inhibit 2,3 diphosphoglycerate (2,3-DPG). Citrate is a very potent inhibitor of 2,3-DPG in the RBCs, which would further reduce oxygen transfer at the haemoglobin level. Glutathione is also a powerful antiviral and antimicrobial. Glutathione deficiency has a potent pro-viral effect. Glutathione deficiency can augment viral replication. Sulphur containing glutathione is attached to the inside of cell membrane, which helps grasp free radicals that make it across the membrane and into the cell. Glutathione is low in HIV-positive patients.

GLUTATHIONE DEFICIENCY IN DISEASES

GSH depletion is an important contributory factor to liver injury, and enhances morbidity related to liver hypofunction. A significant decrease in cysteine occurs in severe cirrhosis. GSH is decreased in the alcoholic and non-alcoholic liver disease; and GSSG is significantly higher. This decreased GSH and/or increased GSSG could have contributed to liver injury susceptibility and toxic risk in these patients. Acetaminophen intake superimposes on the alcohol-damaged liver, when γ-glutamyl transferase (SGGT) is high GSH will decrease. The lung tissue is particularly at risk from oxidative stressors such as cigarette smoke, atmospheric pollutants, and other inhaled environmental toxins. GSH and GSH-associated enzymes present in the epithelial lining fluid (ELF) of the lower respiratory tract may be the first line of defense against such challenges. GSH deficiencies related pulmonary diseases, including acute respiratory distress syndrome (ARDS), asthma, chronic obstructive pulmonary disease, idiopathic pulmonary fibrosis, and neonatal lung damage. Intravenous N-acetylcysteine for patients with ARDS and sepsis, who have a deficiency of GSH in the ELF in intensive care, regains independent lung function and would be expected to leave the intensive care unit significantly faster. In patients with idiopathic pulmonary fibrosis, GSH concentrations in the ELF are a mere 25% of normal, and may contribute to the pathophysiology of this disease. Intracellular GSH is required for the T-cell proliferative response to mitogenic stimulation, for the activation of cytotoxic T "killer" cells, and for many specific T-cell functions, including DNA synthesis for cell replication, as well as for the metabolism of IL-2 (mitogenic response). Depletion of GSH inhibits immune cell functions; the intracellular GSH of lymphocytes determines the magnitude of immunological capacity. Patients with rheumatoid arthritis (RA) have low blood sulfhydryl (-SH) status, as do patients with Type II diabetes or with ulcerative colitis. Chronic viral infections may trigger GSH depletion in circulating immune cells or GSH/GSSG imbalance. Hepatitis C virus patients have low GSH in their circulating monocytes. Monocyte GSH levels are abnormal in early HIV-1 disease, in advanced disease the GSH levels normalise in monocytes but the GSH/GSSG ratio become abnormal. There is significant reduction in the plasma levels of both cysteine and cystine in subjects with HIV-1 infection. Decrease of GSH in the lung ELF is highly suggestive of a systemic GSH insufficiency in these subjects. The most marked GSH reduction occurs in subjects who are asymptomatic but have CD-4 counts below 400. Both the abnormal cytokine expression and the progression to weight loss seen in HIV-1 disease may be linked (at least in part) to abnormalities in the uptake of GSH precursors by immune cells of HIV-1 subjects, and/or to abnormalities in their synthesis of GSH. The brain is susceptible to free radical attack, it is highly oxygenated that makes it vulnerable to endogenous oxygen radical production, and it has a high proportion of unsaturated lipid, which makes it vulnerable to peroxidation. Brain regions that are rich in catecholamines are exceptionally vulnerable to free radical generation. The catecholamines adrenaline, noradrenaline, and dopamine auto-oxidise to free radicals, or become metabolised to radicals by MAO (monoamine oxidases). One such region is the substantia nigra (SN), where a connection has been established between antioxidant depletion (including GSH) and tissue degeneration. Parkinson's disease (PD) is based primarily in the SN. GSH depletion might have particular significance in PD. The melanized catecholaminergic cells found in large quantities in the SN contain less GSH peroxidase and tend to bind to redox-active metals, which makes them more vulnerable to free radical generation from their easily oxidizable melanin complement. There is an increased level of such metals, especially iron, in PD brains. Indicators of decreased antioxidants in PD tissue include the disappearance of melanin from the SN, the increase of total iron and ferric iron, the marked decrease of GSH in the SN, the decreases in antioxidant enzyme activities, and the substantial increases of lipid peroxidation indicators. A combination of vitamin E (3,200 IU/day) and vitamin C (3,000 mg/day) could slow PD progression. The basal ganglia are exceptionally vulnerable to free-radical overload because they are so rich in dopamine as well as other catecholamines. By blocking dopamine receptors, neuroleptics (antipyschotic drugs) may cause dopamine build-up in the basal ganglia that then increases free-radical production. Glutamate excess may contribute to the free-radical overload in tardive dyskinesia (TD). There is elevated lipid peroxide levels in the CSF of patients maintained on neuroleptics and exhibiting symptoms of TD. Decreasing the severity of TD may be achieved by using high doses of vitamin E, and combinations of antioxidants. Schizophrenia may have a component of free-radical overload. Lipid peroxides are elevated in the blood, and pentane gas (a marker for lipid peroxidation) is increased in the breath of schizophrenics. The enzyme SOD is also increased possibly as an adaptive response to free radical overload, and GSH peroxidase is reduced. The use of antioxidants in schizophrenia should include selenium and GSH precursor nutrients. Down's Syndrome (DS) is associated with increased systemic oxidative stress. There is a 50% overexpression of SOD on chromosome 21 that contributes to heightened fluxes of superoxide in all the tissues. Yet DS does not manifest until after birth; the mother's antioxidant defenses may protect the foetus until delivery. Nutritional antioxidants may conserve DS children's mental resources after birth. DS children are also at greatly increased risk for an Alzheimer's-type dementia as they age. It may be possible that potent nutritional supplementation from birth can delay the onset of dementia in DS subjects. Alzheimer's disease (AD) has both direct and indirect indications of free radical involvement. There is increased lipid peroxides in the temporal and cerebral cortex of patients with AD. Iron is raised and GSH is decreased in the cortical areas; and there is significantly higher lipid peroxide generation. Fibroblast cells have increased susceptibility to free-radical damage; the sites of their increased vulnerability may be the mitochondria. Glutathione metabolism may also be abnormal in AD; GSH is lower in the hippocampus, the primary site of short-term memory initiation, also GSH is decreased in the cortical areas. The vascular endothelial cells are arranged in a single, attenuated layer, and are vulnerable to oxidative challenge. They are continually exposed both to exogenous oxidants that reach the circulation, and to endogenous sources of oxidative challenge such as hydrogen peroxide (produced from OXPHOS fluxes) or activated phagocytic cells. Atherosclerosis is linked to oxidative damage to the vessel wall. There is increased lipid peroxides, decreased GSH peroxidase levels, and lowered levels of the protective eicosanoid prostacyclin (PGI_2) in atherosclerotic arteries. Oxidative stress within atherosclerotic arteries depletes GSH and other antioxidants, and results in a shift in the eicosanoid balance from anti-inflammatory towards pro-inflammatory. GSH can produce coronary vasodilation by its normalising effect on prostaglandin synthesis. Platelets contain millimolar levels of GSH. Oxidative stressors shift the platelet's eicosanoid balance away from PGI_2 and toward TXA_2, resulting in a proaggregatory state. Platelets from diabetics have lower GSH levels and make excess TXA_2, thus having a lowered threshold for aggregation; this may contribute to the increased atherosclerosis seen in the diabetic population. Reduced levels of GSH peroxidase have been found in the platelets of patients with acute myocardial infarction, Glanzmann's thrombasthenia, and the Hermansky-Pudlak syndrome characterised by dysfunctional platelets. Exogenous GSH or combinations of antioxidants can be employed to raise the threshold for platelet aggregation, and so ultimately to protect the endothelium against further damage. Human pancreatic inflammatory states are linked to damage inflicted on the pancreatic tissue by oxygen free radicals. These patients suffer from a depletion of antioxidants. Many show increased lipid peroxidation products in their pancreatic tissue, duodenal juice, and bile. N-acetylcysteine (NAC), a precursor of GSH significantly improves acute pancreatitis patient within 72 hours. The clinical status significantly improves on the second and third day in those patients with combined pancreatic and other organ failure when treated with NAC. Chronic pancreatitis patients have increased serum lipid peroxides, and deficiency in several antioxidants. Antioxidants significantly reduce pain and prevent relapse, independent of the aetiology and acuteness of pancreatitis. Both hepatic iron overload and copper overload feature increased lipid peroxidation and detectable free radical damage at the cell level in Metal storage diseases. Humans with thalassemia and secondary iron overload have a significant reduction in GSH reductase activity. Sickle cell anaemia, the lifespan of the red cell is markedly decreased, from an average 120 days to 17 days. Sickling is associated with increased oxidative stress in the red cell, and depletion of antioxidants, including GSH. Higher GSH concentrations are associated with good health (free of disease), regardless of age. An oral bolus of 15 mg/kg of glutathione raises plasma GSH 2-to-5-fold. The enterocyte cells that line the intestinal lumen absorb GSH via non-energy-requiring, carrier-mediated diffusion, and later export it into the blood. GSH also can be absorbed intact by epithelial cells other than the enterocytes, such as lung alveolar cells, vessel endothelial cells, retinal pigmented epithelial cells, and cells of the kidney's proximal tubule; it seems also to cross the blood-brain barrier. Intact GSH also can be delivered directly into the lungs as an aerosol. Brain endothelial and nerve cells, red blood cells, lymphocytes seem to be incapable of absorbing GSH as the intact tripeptide; rather they must synthesise GSH anew from cysteine (or cystine) that they transport inward from the outside. Here transpeptidase enzymes on the outside surface of the cell assist by removing single amino acids from circulating GSH, some of which are then subsequently absorbed. The essential amino acid L-methionine must be obtained from the diet. But methionine must first be converted to cysteine which itself is then available for synthesis into GSH. This pathway requires many cofactors and may be inactive in neonates and in certain adults, such as patients with liver disease. The activated methionine metabolite known as SAM (S-adenosyl methionine) is effective in raising red cell GSH and hepatic GSH when given orally at 1600 mg/day. SAM has proven clinical benefit against cirrhosis and cholestasis. The sulfur-containing amino acid L-cysteine is the precursor that most limits the synthesis of GSH. When circulating in the blood cysteine readily auto-oxidises to potentially toxic degradation products (hydroxyl radical). Cysteine has excitotoxin activity in the brain, similar to that of the amino acids glutamate and aspartate, and can be toxic to the retina. GSH has none of these liabilities. GSH acts as a reservoir for cysteine. Cysteine is unstable in the blood because the ambient oxygen is high enough to oxidise it, yet its availability limits GSH synthesis. The cystine produced from cysteine oxidation is not significantly taken up into cells other than those of the kidney, and requires energy and enzymatic intervention to be converted to cysteine. The mechanistic solution to this problem may be that once replete with GSH, the liver's cells export it. After GSH exits the liver cell, it can quickly be back-converted to cysteine, which then is used elsewhere for protein synthesis and for the biosynthesis of taurine and other sulfur metabolites. Circulating GSH is safe; it reacts only slowly with oxygen, is less susceptible to auto-oxidative degradation than is cysteine, and is more soluble in the plasma. Certainly as a water-soluble, transportable form of sulfhydryl (-SH) reducing power, GSH is more reliable than circulating cysteine. Liver GSH synthesis is closely linked to overall protein synthesis. Hormones and other vasoactive substances increase GSH efflux into the bile, and this may contribute to the hepatic GSH loss noted under conditions of stress. 80% of the GSH synthesised in the liver is exported from the hepatocytes, and most of this is utilised by the kidneys, which also carry a major toxic burden. Some cells of the body are unable to directly utilise GSH. N-acetyl cysteine (NAC) is a cysteine precursor; it is well absorbed by the intestine, and becomes converted to circulating cysteine by de-acetylation. It seems not to raise GSH levels if they are already within the normal range, but it can raise abnormally low GSH levels back to normal. NAC has antimutagenic and anticarcinogenic properties while also being a potent antioxidant. It may not be the perfect GSH source. GSH precursors with known safety profiles include NAC, glycine, L-glutamine, L-taurine, L-methionine, and S-adenosyl methionine. Glutathione is used as a non-enzymatic reducing agent to help keep cysteine thiol side chains in a reduced state on the surface of proteins, and to prevent oxidative stress in most cells and helps to scavenge free radicals that can damage DNA and RNA. There is a direct relationship between the speed of aging and the reduction of glutathione concentrations in intracellular fluids. Glutathione levels drop and the ability to detoxify free radicals decreases with aging. 90% of GSH that is synthesised within the cell is stored in the cytosol, 10% is stored in mitochondria. GSH act as a cofactor in various enzymatic reactions, and maintenance of sulfydryl redox status. GSH protects against electrophilic xenobiotics and intracellular oxidants. Once synthesised in the liver, GSH is either translocated to plasma or excreted in the bile. The bulk of plasma GSH is in the reduced form (85%), while the remainder is oxidised (15%). GSH is cleared in the kidney through direct glomerular infiltration and a noninfiltration mechanism using the γ-glutamyl transpeptidation reaction. GSH aerosol restores respiratory epithelial surface (RES) oxidant-antioxidant balance in cystic fibrosis and in chronic disease of the upper respiratory tract. GSH levels are necessary for both T and B lymphocyte function and immune function in general. Hepatocyte GSH levels is low in patients with cirrhosis. Ethanol exposed hepatocytes have increased susceptibility to oxidant injury. High dose of reduced GSH protects against cisplatin nephrotoxicity and from oxazaphosphorine (cyclophosphamide, ifosfamide) urotoxicity, without interfering with antitumour efficacy. Glutathione may reduce some adverse effects associated with reperfusion-induced oxidative stress. High doses of reduced glutathione may reduce adverse effects of alcohol on liver function. Reduced glutathione (2400 mg daily) decreases total bilirubin in chronic alcoholic liver disease. GSH is effective in reducing radiotherapy toxicity in various tumour types. GSH improves anaemic status in haemodialysis patients. GSH supplementation in CRF increases RBCs, reduce glutathione, haematocrite, and haemoglobin, with a concomitant decrease in plasma oxidised glutathione and reticulocytes after 120 days of therapy. Aerosolised glutathione (600 mg, daily) reverses oxidant-antioxidant imbalance of idiopathic pulmonary fibrosis, and decreases superoxide release by alveolar macrophages. GSH (600 mg daily) improves semen parameters in patients with dyspermia. Perioperative GSH (200 mg/kg in 100 ml saline) may improve perioperative renal function in coronary by pass patients.

NADH

Nicotinamide adenine dinucleotide (NADH) is the reduced coenzyme form of the vitamin niacin. NADH is an electron donor. NADH is the reduced (electron-energy rich) coenzyme form of vitamin B_3, while NAD is the oxidised (burned) coenzyme form of B_3. NAD and NADH are converted into each other in numerous different metabolic activities. NAD activates alcohol dehydrogenase and acetaldehyde dehydrogenase that are the 2 enzymes needed to detoxify the alcohol into carbon dioxide and water. NADH is the first of five enzyme complexes of the electron transport chain, where much of the ATP bioenergy that runs every biological process. NAD(H) is necessary to oxidise all fats, sugars, and amino acids into ATP bioenergy. There are 3 interlinked energy production cycles: 1- the glycolytic cycle (sugar), 2- Krebs' citric acid cycle (amino acids and fats), and 3- the electron transport side chain. A glycolytic cycle waste end product-pyruvic acid, helps power the Krebs' cycle, while electron sparks released from the step by step slow oxidation that occurs in the Krebs' cycle provide the fuel used by the electron transport side chain to generate much of the ATP bioenergy. NAD(H) is involved in all of these different cycles. NADH captures the electron sparks thrown off during Krebs' cycle oxidation and shuttles them to the electron transport side chain energy production cycle. Each unit of NADH is capable of generating 3 units of ATP energy. NAD(H) is vitamin B_3 (niacinamide) combined with a ribose (5-carbon sugar), a phosphate group and an adenine nucleotide (a DNA component). NAD(H) can also be made from the amino acid L-Tryptophan at the expensive ratio of 60 mg tryptophan for 1 mg B_3. Taking in exogenous (from outside the body) B_3 or NAD(H) may spare the scarce amino acid tryptophan. Tryptophan is the precursor of one of the most important antidepressant neurotransmitters, serotonin. The dopamine formed through L-dopa therapy is prone to auto-oxidation to free radical forms that eventually destroy what few dopaminergic neurons are left. Parkinson's disease begins when the substantia nigra neurone population has dropped to 20-30% of normal. Dopamine is usually made inside the neurons that use it through a 2 step process: 1- The amino acid tyrosine is first converted to L-dopa through an enzyme called tyrosine hydroxylase (TH). 2- L-dopa is then converted to dopamine. The activity of tyrosine hydroxylase, is the rate-limiting controller of dopamine synthesis, and tyrosine hydroxylase activity is considerably lower in Parkinson's patients. L-dopa diminishes the already weakened tyrosine hydroxylase activity, thus further limiting their own L-dopa production and increasing the need for L-dopa supplements. The coenzyme that activates tyrosine hydroxylase-tetrahydrobiopterin (H_4BP) is reduced 50% in the brains of Parkinson's patients. NADH activates the enzyme that helps produce H_4BP. NADH could elevate H_4BP production, tyrosine hydroxylase activity and dopamine production. NADH increases brain dopamine and noradrenaline using brain cells use dopamine to make noradrenaline. Dopamine and/or noradrenaline are frequently diminished in the brains of depressed patients. Through NADH's sparing of tryptophan, more tryptophan would be left to end up as brain serotonin, another neurotransmitter frequently reduced in depressives. The human brain must produce and use 20% of the body's total ATP bioenergy. NADH may halt the progression of Alzheimer's disease, it significantly reverses the cognitive and behavioural problems. The standard dosage of NADH is 5-10 mg, taken with water 30 minutes before breakfast. Animal studies suggest 1000 mg/kg of body weight to be a tolerable dosage. In chronic alcoholism, the cellular NAD/NADH ratio is already in favour of NADH, so NADH would not be appropriate. Coenzyme Nicotinamide adenine dinucleotide (NADH) was discovered in 1943. The more NADH a cell has available, the more energy it can produce. NADH is abundant in meat, less so in vegetables and fruits. NADH activity decreases in tangent with the human aging process. Alzheimer's disease is caused by a steady loss of neurons, or nerve cells, that causes progressive dementia, resulting in a steady decline in mental and physical function. NADH is 25-50% lower in Alzheimer's patients than similar age matched individuals. NADH may prevent the premature death of brain cells, resulting in higher levels of acetylcholine, noradrenaline, and dopamine. NADH is sensitive to light, temperature, oxygen, water and other reactive molecules. To get 10 mg of NADH from normal food, it would be necessary to consume 4 pounds of steak. NADH is safe in levels up to 500 mg/kg of body weight, over 7,000 times greater than the recommended level of 5 mg/day. The NADH:ubiquinone oxidoreductase (Complex I), provides the input to the respiratory chain from the NAD-linked dehydrogenases of the citric acid cycle. The complex couples the oxidation of NADH and the reduction of ubiquinone, to the generation of a proton gradient, which is then used, for ATP synthesis. Mutations in this complex are associated with Leber hereditary optic neuropathy, Melas syndrome, and possibly with Alzheimer's disease. NADH:ubiquinone oxidoreductase catalyses the oxidation of NADH, the reduction of ubiquinone, and the transfer of $4H^+/NADH$ across the coupling membrane.

$NADH + H^+ + Q + 4H^+_N \leftrightarrow NAD^+ + QH_2 + 4H^+_P$

Nicotinamide adenine dinucleotide (NAD) and its relative nicotinamide adenine dinucleotide phosphate (NADP) are two of the most important coenzymes in the cell. NADP is simply NAD with a third phosphate group attached. Because of the positive charge on the nitrogen atom in the nicotinamide ring, the oxidised forms of these important redox reagents are often depicted as NAD^+ and $NADP^+$ respectively. In cells, most oxidations are accomplished by the removal of hydrogen atoms. Both of these coenzymes play crucial roles in this. Each molecule of NAD^+ (or $NADP^+$) can acquire 2 electrons; i.e., be reduced by 2 electrons. However, only one proton accompanies the reduction. The other proton produced as 2 hydrogen atoms are removed from the molecule being oxidised is liberated into the surrounding medium. For NAD, the reaction is thus:

$NAD^+ + 2H \rightarrow NADH + H^+$

NAD participates in many redox reactions in cells, including those in glycolysis and most of those in the citric acid cycle of cellular respiration. NADP is used in many anabolic reactions. NADH functions as an electron donor in many oxidation-reduction reactions. It is a loosely bound cofactor for several hundred enzymes that catalyse reactions ranging from lactate production to cholesterol biosynthesis, and is present at micromolar concentrations in many cell types. NADH protects against free radical damage (H_2O_2). The precursor of the NADH is the vitamin niacin. α-Ketoglutarate increases the absorption of NADH. Each unit of NADH is capable of generating 3 units of ATP energy. NAD(H) can be made in the liver and other cells from vitamin B_3 (niacinamide). Taking in exogenous B_3 or NAD(H) may spare the scarce amino acid tryptophan, which is the least plentiful amino in any normal diet. The human brain must produce and use 20% of the body's total ATP bioenergy, and PET scans of the brains of depressed and demented people frequently show reduced brain energy production. NADH treats depressed patients (93%). Parkinson's patients exhibit dementia as well as neurotransmitter problems, while many Alzheimer's patients exhibit neuromotor dysfunction as well as dementia. NADH halts the progression of Alzheimer's disease, and significantly reverse the cognitive and behavioural problems, even in the worst cases. NADH combats fatigue. NADH in Parkinson's may be accompanied with tyrosine and deprenyl. NADH for depression may be added to DLPA, tyrosine and/ or tryptophan or 5-hydroxy-tryptophan. NADH for Alzheimer's dementia might include acetyl L-carnitine and DMAE or centrophenoxine. In serious fatigue situations, B-complex vitamins, α-lipoic acid, CoQ_{10} and magnesium would be synergistic with NADH.

VITAMIN C (ASCORBIC ACID)

Vitamin C ($C_6H_8O_6$) is a vitamin and an electron carrier. Vitamin C is a reducing substance, an electron donor. When vitamin C donates its 2 high-energy electrons to scavenge free radicals, much of the resulting dehydroascorbate is re-reduced to vitamin C and therefore used repeatedly. After vitamin C donates an electron to a free radical, it becomes what is known as the ascorbyl radical. Scurvy is a common occurrence in disseminated metastatic disease. Vitamin C deficiency (scurvy) results in weakened arteries and bleeding into tissue spaces. Levels of 400 mg/dl vitamin C in the blood can kill cancer cells by a pro-oxidative mechanism. Vitamin C is a water-soluble antioxidant vitamin essential to the body's health. When bound to other nutrients, for example calcium, it would be referred to as calcium ascorbate. Vitamin C is important for tissue integrity and the prevention of invasion by foreign organisms and cancer cells. Vitamin C recharges fat-soluble vitamin E and water-soluble glutathione, allowing them to be reused many times. Diets are deficient in vitamin C as a result of storage and processing. Vitamin C deficiency leads to weakness of the connective tissues. Vitamin C stimulates the absorption of iron; so, drinking orange juice should be avoided at the same meal with iron-rich foods. Vitamin C supplement daily has 28% reduction in risk of developing heart disease. Vitamin C (ascorbic acid) supplements may be useful against stress effect, blocking a rise in corticosterone resulting from stress. Continuing rising of corticosterone would affect this advantageous effect of vitamin C. Vitamin C reduces the oxidation of adrenaline to adrenochrome. Vitamin C is a vital component of every cancer treatment program; it may add 20-30% to survival rate. Vitamin C hastens the healing of wounds by producing healthy granulation tissue and reducing local oedema. Ascorbic acid may be used locally as a 2% dressing, which may possess astringent properties similar to hydrogen peroxide. Ascorbic acid has many important functions. It is a powerful oxidiser and when given in massive amounts; i.e., 50-150 g, intravenously, for certain pathological conditions, and run in as fast as 20 Gauge needle will allow, it acts as a flash oxidiser, often correcting the pathology within minutes. Ascorbic acid is also a powerful reducing agent. Its neutralising action on certain toxins, exotoxins, virus infections, endotoxins and histamine is in direct proportion to the amount of the lethal factor involved and the amount of ascorbic acid given. There are many factors that increase the demand by the body for ascorbic acid: 1- age; 2- habits (smoking, alcohol, playing); 3- sleep; 4- trauma (physical, emotional, biochemical); 5- renal threshold; 6- environment; 7- stress (physiological, emotional, chemical, physical); 8- season of the year; 9- loss in the stool; 10- variations in individual absorption; 11- variations in binders in commercial tablets; 12- body chemistry; 13- drugs; 14- pesticides; 15- body weight; and 16- inadequate storage. The active enzyme l-gulonolactone oxidase is absent from the human liver. A defect or loss of the gene controlling the synthesis of this enzyme in man, blocks the final phase in the series for converting glucose to ascorbic acid. Other recognised genetic diseases in which a missing enzyme causes a pathological syndrome, in man, are phenylketonuria, galactosaemia and alkaptonuria. Ascorbic acid is necessary co-enzyme in the metabolic oxidation of tyrosine. The velocity of the oxidation in this reaction is dependent upon the concentration of vitamin C. Tyrosine is essential in breaking down protein to usable amino acid. Hypoascorbaemia is due to inability of man to manufacture his own ascorbic acid. Ascorbic acid has the capability of entering all cells. Under normal circumstances its presence is beneficial to the cell, however, when the cell has been invaded by a foreign substance, like virus nucleic acid, enzymic action by ascorbic acid contributes to the breakdown of virus nucleic acid to adenosine deaminase, which converts adenosine to inosine. The net result is to lead to purines, which are extensively catabolised. Ascorbic acid also joins with the available virus protein, making a new macromolecule, which acts as the repressor factor. The repressor factor inhibits the multiplication of new virus bodies. The tensile strength of the cell membrane is exceeded by these macromolecules with rupture and destruction. The permeability of the blood-brain barrier can be changed by introducing various toxic agents into the blood circulation. A virus, e.g., adenovirus, gains entrance into the brain through the blood cerebrospinal fluid barrier and/or the blood brain barrier by one of the following mechanisms: 1- electrical charge; 2- chemical lysis of tissue; or 3- osmosis. This reflects the importance of the intercellular cement of the capillary wall in regulating permeability of the blood vessels of the central nervous system. Ascorbic acid repairs and maintains the integrity of the capillary wall. Many times the first degree burn progresses rapidly to the second degree stage and remains as blisters. Still others go on to third degree, which usually is more pronounced on the third plus post burn day. Ascorbic acid may eliminate the fourth stage and the third stage. The pathobiology of a burn wound from the moment of the accident is in a state of dynamic change until the wound heals or the patient dies. Initially there is intravascular agglutination of RBCs into distinctly visible, smooth, hard, rigid, basic masses. Oxygen uptake by the tissues is greatly reduced because of the sludging and therefore reduced rate of flow. Sludging or agglutination results in capillary thrombosis in the area of the burn, extending proximally to involve the large arterioles and venules and thereby creating tissue destruction greater than that originally produced by the burn. Anoxia produces added tissue destruction. After severe burns there is considerable alteration in the metabolism of ascorbic acid. The extent of the abnormality parallels the severity of the burn. There is an increased demand for ascorbic acid in burns especially when epithelialization and formation of granulation tissue are taking place. Vitamin C hastens the healing of wounds by producing healthy granulation tissue and reducing local oedema. Ascorbic acid may be used locally as a 2% dressing, which may possess astringent properties similar to hydrogen peroxide. Absorption of Pseudomonas' exotoxin from the infected burn wound inhibits the bacterial defense mechanism of the reticuloendothelial system. Death can result either from the toxaemia alone or from an associated septicaemia. Vitamin A and D ointment should be used over the area of the burn alternating with the 3% ascorbic acid solution. Ascorbic acid can be given by vein and by mouth. 500 mg/kgBW diluted to at least 18 ml/g vitamin C using sterile distilled water, 5% dextrose in water, saline in water or Ringers solution and for the initial injection, run in as fast as a 20 gauge needle or catheter will carry the flow. Vitamin C solution is repeated every 8 hrs for the first several days, then at 12 hr intervals. Ascorbic acid, by mouth, is given to tolerance, i.e., loose stools. At least one gram calcium gluconate, daily is required, to replace free calcium ions removed in the breakdown chemical action as ascorbic acid goes to dehydroascorbic acid, then to ketogulonic acid and later to oxalic acid as the calcium salt. Ascorbic acid destroys the exotoxin systemically and locally. Ascorbic acid also eliminates pain. There are varying degrees of chronic carbon monoxide poisoning, because of railroads on the highways, smoking, and being too lazy to walk. Small amounts of carbon monoxide, if constantly maintained in the alveoli, can produce serious effects. Carbon monoxide in the inspired air leads to oxygen deficiency in the tissues causing extreme exhaustion. The affinity of carbon monoxide for haemoglobin is roughly 300 times as great as that for oxygen. In addition to active replacement of oxy-haemoglobin the presence of some proportion of carboxy-haemoglobin decreases the dissociability of such oxy-haemoglobin as remains. Carbon monoxide can be released from haemoglobin if the patient is exposed to high pressure of oxygen, 93% along with 7% carbon dioxide. Ascorbic acid in the blood is constantly losing molecules of water. Perfectly dry carbon monoxide and oxygen cannot unite to form carbon dioxide, but carbon monoxide and water may give rise to carbon dioxide in the complete absence of oxygen.

$$CO + H_2O = HCOOH \rightarrow CO_2 + H_2$$

The oxygen of the water has been used to oxidize carbon monoxide to carbon dioxide with the liberation of hydrogen. Glutathione may facilitate this cellular oxidation by acting as a hydrogen acceptor. With the flash oxidation of high dose ascorbic acid, a high concentration of oxygen pulls carbon monoxide from haemoglobin to form carbon dioxide. This rapidly formed carbon dioxide acts with the high oxygen tension to serve the same purpose as when given by mask, further enhancing the chemical action taking place. Ascorbic acid prevents residuals such as paralysis, blindness, interference with sensations, muscle spasms or twitchings, which in some cases can be permanent. 4 g first trimester, 6 g second trimester and 10 g third trimester may be advised to pregnant women. With such requirements haemoglobin levels are much easier to maintain, leg cramps are less, striae gravidarum is seldom encountered (the capacity of the skin to resist the pressure of an expanding uterus varies in different individuals), labour is shorter and less painful, the postpartum haemorrhage is rare, the perineum is remarkably elastic and episiotomy is performed electively (healing is always by first intention), the firmness of the perineum is expected to be similar to that of a primigravida, no cardiac stress, and infants born are expected to be robust with no feeding problems. The only way that oxalic acid can be produced from ascorbic acid is through splitting of the lactone ring, which occurs above pH 5. The reaction of urine when 10 g of vitamin C is taken daily is usually pH 6. Oxalic acid precipitates out of solution only from a neutral or alkaline solution-pH 7 to pH 10. Dehydroascorbic acid is protected in vivo from rapid transformation to the antiscorbutically impotent diketogulonic acid from which oxalic acid is derived. The normal 24 hr urinary oxalate excretions for humans is 14-56 mg. The amount of oxalic acid found in the diabetic patient approximates that found in the urine of a normal person taking 10 g vitamin C/d. Vitamin C (>10 g) is an excellent diuretic, i.e., no urinary stasis; no urine concentration. Methylene blue will dissolve calcium oxalate stones giving 65 mg orally 2 to 3 times a day. The production of histamine and other end products from demonised cell proteins released by injury to cells are a cause of shock. Demonising enzymes from the damaged cells are inhibited by vitamin C. Mechanical damage to a cell results in pH changes, which reverse the cell enzymes from constructive to destructive activity. The pH changes spread to other cells. This destructive activity releases histamine a major shock producing substance. The presence of vitamin C inhibits this enzyme transition into the destructive phase. Shock and stress cause depletion of the ascorbic acid content of the plasma. Optimal primary wound healing is dependent to a large extent upon the vitamin C content of the tissues. Large doses of intravenous vitamin C have a striking influence on the course of mononucleosis. As low as 1.5 g ascorbic acid daily may prevent recurrences of cancer of the bladder. In the presence of ascorbic acid, carcinogenic metabolites do not develop in the urine. The spontaneous tumour formation is the result of faulty tryptophan metabolism while urine is retained in the bladder. Vitamin C may be necessary either directly or indirectly for formation of mast cells, or for their maintenance once formed or both. Ascorbic acid would control myelocytic leukaemia provided 25-30 g is given orally each day. Ascorbic acid assists with hepatic metabolism of barbiturate and other poisons and as a major diuretic flush these compounds out by way of the kidneys. Vitamin C is a regulator of the rate at which cholesterol is formed in the body; deficiency of the vitamin speeding (600%) the formation of this substance in the adrenal glands. Vitamin C may inactivate tetanus toxin. Heavy metal intoxication may resolve with adequate vitamin C therapy. Ascorbic acid destroys virus bodies by taking up the protein coat so that new units cannot be made, by contributing to the break-down of virus nucleic acid with the result of controlled purine metabolism, e.g., virus pneumonia and virus encephalitis. Virus encephalitis is a deadly syndrome and must be treated heroically with intravenous and/or intramuscular injections of ascorbic acid. As much as 12 g can be given in this manner with a 50 ml syringe. Larger amounts must be diluted with dextrose or saline solutions and run in by I.V drip. Amounts like 20-25 g, which can be given with a 100 ml syringe, can suddenly dehydrate the cerebral cortex so as to produce convulsive movements of the legs, i.e., symptomatic epilepsy. This epileptiform type seizure continues for 20 minutes or more and then abruptly stops. Mild pressure on the knees will stop the seizure so long as pressure is maintained. Massive doses of ascorbic acid pulls free calcium ions from the vicinity of the platelets or from the calcium-prothrombin complex as the lactone ring of dehydroascorbic acid is opened. The first sign of calcium ion loss is epistaxis. Ascorbic acid plays an important role in maintaining fluid balance in the body. Vitamin C activates the enzyme arginase, which breaks down the amino acid arginine, resulting in production of urea, which is one key to tissue fluid balance. The simple stress of pregnancy demands supplemental vitamin C. Vitamin C seems especially concerned with mesenchymal tissue. Ascorbic acid is vital to the body tissues in the formation and maintenance of normal intercellular material, especially in the connective tissue, bones, teeth, and blood vessels. Other pathological conditions in which ascorbic acid plays an important part in recovery include cardiovascular diseases, hypermenorrhea, peptic and duodenal ulcers, post-operative and radiation sickness, rheumatic fever, scarlet fever, poliomyelitis, acute and chronic pancreatitis, tularaemia, whooping cough, and tuberculosis. Vitamin C plays a very important role in general nutrition. Deficiency of this substance in sufficient amounts can be a factor in loss of appetite, loss of weight or failure to grow, muscular weakness, anaemia, various skin lesions, and other conditions. When ascorbate acts as a scavenger, dehydroascorbate is formed; but if the ascorbate/dehydroascorbate (AA/DHA) ratio is kept high (the redox potential kept reducing) until the unstable dehydroascorbate undergoes hydrolysis or can be reduced back to ascorbate, the dehydroascorbate will do no harm. A very high dose of ascorbate is virtually nontoxic, which quench almost all unwanted free radicals and oxidants. Mixtures of mineral ascorbates (calcium, magnesium, potassium, zinc, and sometimes sodium) can be used to increase bowel tolerance and for even more clinical effectiveness, but do not clearly demonstrate the increasing bowel tolerance phenomenon. Part of the unexpected benefit at the high dose levels of vitamin C is frequently a feeling of well-being, despite the gas and diarrhoea sometimes produced. Chemical reactions involving free radicals and highly reactive oxidants are necessary in the normal metabolism of cells. Metabolic processes utilising oxygen (aerobic metabolism), which release energy are important examples. Ordinarily, these reactions occur in conjunction with appropriate enzymes or in the proper places within the cells. In general free radical scavenging occurs through complex metabolic pathways involving many steps, which are rate-limited. Deficiencies of nutrients, vitamins and minerals, which make up the enzymes and coenzymes of these systems can slow down or halt certain pathways. Superoxide dismutase can catalyse the conversion of superoxide to O_2 and H_2O_2. Ascorbate, nonenzymatically, also converts superoxide to H_2O_2 but is oxidised in the process to the ascorbate free radical and dehydroascorbate. The ascorbate free radical and the dehydroascorbate are reduced back to ascorbate either by NADH (catalysed by semidehydroascorbate reductase and forming NAD) or reduced glutathione (GSH) (catalysed by dehydroascorbate reductase and forming oxidised glutathione [GSSG]). Some of the peroxide can be converted to oxygen and water by catalase but most will be destroyed by a glutathione-requiring enzyme system. GSH (catalysed by glutathione peroxidase) reduces the peroxide to water but in the process is oxidised to GSSG. The resulting GSSG is reduced by NAD(P)H (catalysed by glutathione reductase). The resulting NAD is reduced back to NADH by way of the Krebs cycle or resulting NADP is reduced back to NADPH by the hexose monophosphate (HMP) pathway. The rate-limiting step in the last series of reactions may be catalysed by glutathione peroxidase and its cofactor selenium, but other substances that could limit all this are the vitamin E, vitamin C, vitamin B_2, vitamin B_3, cysteine, etc. At small amount, vitamin C available is oxidised to dehydroascorbate and then must be reduced back to ascorbate by the pathway described, to be reused as ascorbate. This can be overwhelmed by a toxic pathogen liberating free radicals or by an inflammatory cascade regardless of its cause. If a pathogen produces free radicals at a rate sufficient to exceed the rate at which the host can produce free radical scavengers to protect the immune system, the pathogen will be free to invade and multiply. The more toxic pathogens produce more free radical toxins than just necessary to suppress the immune system. The spill over of free radicals reaches a threshold where an inflammatory cascade in the tissues affected, is initiated. Neutrophils liberate free radicals and highly reactive oxidants both intracellularly and extracellularly in their attempt to destroy pathogens, in the process termed the respiratory burst. The respiratory burst consumes NADPH, which must be continually restored if the respiratory burst is to be maintained. Restoration of NADPH supplies is accomplished by way of the HMP pathway, by various rate-limited enzymatic mechanisms. If the rate-limited enzymatic processes or the limited availability of the antioxidant free radical scavenging mechanisms of the leukocytes, superoxide dismutase, catalase, glutathione peroxidase, and glutathione, fall short of being able to contain and direct free radicals and reactive oxidants toward the pathogen, that failure causes the free radicals to backfire, damage the host itself, and initiate an inflammatory cascade. By neutralising virtually all unwanted free radicals and toxic oxidants, massive doses of ascorbate can be made to protect the immune system to such a degree that early in acute viral diseases, the immune system can usually destroy the pathogen within hours. Massive doses of ascorbate work synergistically with appropriate antibiotics when used against acute bacterial diseases, and broaden the spectrum of the antibiotics considerably. Certain bacteria may do very poorly in the face of massive doses of ascorbate even where antibiotics are not used, e.g., in chronic bronchitis, sinusitis, otitis media, tonsillitis, osteomyelitis, nonspecific urethritis, etc. When the induced scurvy is eliminated by driving tissue levels of ascorbate up above a certain threshold, the immune system usually rapidly eliminates the infection and the affected areas heal. Where allergies in combination with infections play a major role, massive doses of ascorbate are helpful but continuing maintenance doses will be required. In this case, continuing blockade of the allergically-induced inflammatory cascade must be maintained. Topical ascorbate is particularly effective on herpes simplex. In chronic hepatitis, ascorbate in massive doses may control the disease. In conditions where a virus has become well established intracellularly, there are some limitations on the ability of ascorbate to assist the immune system. The AIDS patient who has already suffered a marked suppression of helper T-cells, presents a clinical problem. High tissue concentrations of ascorbate to dehydroascorbate can directly reduce various substances (e.g., disulfides). Vitamin functions of lower doses of vitamin C are frequently potentiated by and work in conjunction with vitamin A, zinc, selenium, bioflavonoids, and other nutrients, which play roles in various defense mechanisms. Whether a reaction will proceed left to right, or in reverse, depends upon the ratio of the oxidised to the reduced members of a redox couple. In conditions resulting from combinations of mechanical derangements, nutritional deficiencies, immune dysregulations, haemorrhage with release of free radical generating iron and copper atoms, and then secondary inflammatory cascades (e.g., degenerative disc disease, degenerative arthritis, rheumatoid arthritis, ankylosing spondylitis, blunt trauma of the spine, etc.), therapeutic effects could be expected proportional to what might result from blocking of the free radicals and the inflammatory cascade. Toxic substances, whose mechanisms of action involve free radical generation, e.g., toxic poisons such as snake bites and spider bites, certain drugs, such as barbiturates, chemotherapeutic agents, narcotics, and powerful oxidising pollutant chemicals, might be neutralised. Psychological symptoms resulting from oxidative products such as adrenochrome and noradrenochrome, would be expected to be ameliorated to a degree by vitamin C and other antioxidants. Tumours invading the body or holding off the immune system by way of free radical toxicity might be expected to respond to varying degrees. The sequence of reactions whereby certain drugs cause haemolysis with glucose-6-phosphate dehydrogenase (G-6-PD) deficiencies is poorly understood. It appears that G-6-PD deficient cells lack a mechanism to regenerate reduced glutathione (GSH) from oxidised glutathione (GSSG), which may result in several biochemical alterations, the final result being haemolysis of the red cells. The maintenance of glutathione in the reduced state (GSH) is probably the most important function of the HMP pathway. It may be that the haemolysis caused by certain drugs is initiated by the drug forming either free radicals or hydrogen peroxide. When peroxides are reduced back to water, GSH is oxidised to GSSG, a reaction catalysed by glutathione peroxidase. Ordinarily the GSSG is reduced back to GSH by NADPH, a reduction catalysed by glutathione reductase. The resulting oxidised NADP is reduced back to NADPH in the first step of the HMP pathway, as glucose-6- phosphate is oxidised to 6-phosphogluconolactone. This critical reaction is catalysed by G-6-PD. G-6-PD deficient cells may be expected to accumulate peroxides, which could then oxidize other red cell components. If the AA/DHA redox potential is kept reducing enough by high enough concentrations of ascorbate, it should directly reduce the GSSG to GSH. Such mechanism may compensate for the lack of G-6-PD. G-6-PD deficiencies have a wide range of clinical severities. There is substantial decrease in the activity of G-6-PD with aging. The erythrocytes of individuals with sickle cell trait and sickle cell anaemia, possess more copper than normal persons. Chemically allergic persons accumulate dehydroascorbate more readily than others because of a deficiency of glutathione peroxidase. Frequently, after the administration of selenium, ascorbate is better tolerated by chemically allergic patients, because selenium augments the glutathione peroxidase activity. The slight increase in the acidity of the urine from ascorbate, and the slight diuresis solubilizes calcium salts. High concentrations of ascorbate, by being bacteriostatic in the urine, should prevent many of the niduses of infection around which oxalate stones frequently form. The increased ascorbate concentration complexes Ca^{2+} and thereby decreases the amount of Ca^{2+} available to complex with oxalate. Clinically, ascorbate in the very large doses described is very effective and safe as part of the treatment of a wide variety of conditions, especially infectious diseases. Acute scurvy can be induced by any stress and is responsible for a high percentage of the secondary complications of many diseases. The magnitude of this scavenging drain on ascorbate is enormous. Ascorbate does not cure AIDS but it will prolong the life of AIDS patients and make their life much more comfortable. Vitamin C is essential in the hydroxylation of proline in the synthesis of collagen. The AIDS patient cannot possibly take enough NAC to provide the amount of reducing equivalents necessary to quench most of the free radicals generated in the AIDS process. NAC provides cysteine in a readily available form is used by the body to make glutathione. Glutathione is low in HIV-positive patients. To reduce some oxidised substance, there is a threshold in the concentration of reducing substance that has to be exceeded before the reaction will proceed. The free radicals activate the immune system. Massive doses of ascorbate prevent allergic reactions, and can terminate it abruptly. Free radicals are an almost universal sign of damage to the body. The immune system is activated by the resulting free radicals. Ascorbate in massive quantities suppresses the humoral immunity of antibodies, and augments the cellular immunity of phagocytic cells. Ascorbate provides reducing power that initiates the respiratory burst (by the reduction of molecular oxygen); and protects the WBCs from those very radicals the white cell makes in its vacuoles to kill the pathogens. WBCs may not be able to continue to produce the good radicals when the radicals leak significantly into the cytoplasm (then becoming bad free radicals) and exceed the ability of their free radical scavengers to neutralise. Stimulating the immune system in HIV-positive patient results in the autoimmune destruction of the T4 cells. Treatment of HIV-positive patient (AZT, DNCB) increases the T4 cells in the early stages of the disease; but hastens the final autoimmune destruction of the T4 cells. This stimulation would sometimes have some beneficial effect on secondary infections such as KS for a period of time. When free radicals dominate in a tissue for long, the small amount of vitamin C ordinarily present is converted to dehydroascorbate (DHA), which has a very short half-life resulting in acute induced scurvy. Patient dying of AIDS has many of the classic signs of scurvy, e.g., nonfunctional WBCs. The vitamin C levels can be conserved only if a reducing redox potential is maintained in those tissues. Vitamin C acts as a pro-oxidant, and destroys cancer cell membranes, and produces hydrogen-peroxide. Cancer cells have 10-100 times less catalase enzymes. Thus vitamin C tends to concentrate in cancer cells when given at high dose, and therefore destroys these cancer cells. A serum level between 350-400 mg/dl of vitamin C may achieve maximal cancer cell destruction. 75 g of vitamin C in an intravenous drip with a mixture of minerals and other vitamins included, and also some amino acids every day for 3 weeks (excluding weekends) may reach that level. Patients on this intravenous treatment programme often become tired due to lot of cancer cell destruction. B_{17} intravenously, also concentrates in cancer cells and can destroy these cells. At the end of such programme the patient provides 2 litres of urine to extract tumour antigen, that is proteins on the cell surfaces of cancer cells, is used to make specific cancer vaccines, which is specific to the tumour, known as dendritic cell therapy vaccines. Dendritic cell primes the native lymphocyte population towards an antitumour response. Augmentative molecular immunology may stimulate the dendritic cell-T lymphocyte network using cancer vaccine. The cancer vaccine is comprised of the following 3 parts: 1- Tumour-specific antigen injections, 2- Cytokine injections [antigen presentation mixture (APM), lymphocyte proliferation mixture (LPM), granulocyte macrophage-colony stimulating factor (GM-CSF), interleukin-2 (IL-2)], and 3- Heat shock protein/tumour-specific antigen injections. Dendritic cells (DCs) are naturally found in almost all tissues of the body and are especially numerous in the skin and in the liver. Dendritic cells are termed "professional antigen presenting cells" as they are the most competent at capturing and presenting antigen. Immature dendritic cells can engulf tumour cells and mature dendritic cells can present antigen. Heat shock proteins (HSPs) are small proteins that associate with tumour antigen and initiate the phase I response by the dendritic cell, i.e., heat shock proteins act as "antigen chaperones" to the dendritic cells. Tumour specific antigens (TSA) are located on the cancer cells and they can be harvested either from the urine or directly from tumour cells. Vitamin C supplementation may also share with vitamin B_{12} the ability to reverse erythrocyte membrane abnormalities seen in CFS and thus improve capillary blood flow. Fifteen minutes postinfusion of vitamin C (15 g), over 80% of the membrane abnormalities would disappear. Vitamin C supplementation also bolsters immune responses including increased neutrophil motility and chemotaxis, increased immunoglobulin levels, and increased lymphocyte blastogenesis in response to mitogens. Vitamin C enhances interferon activity (antiviral activity). Excess vitamin C may drop magnesium levels. Vitamin C completely protects the brain cells from amyloid β toxicity. Vitamin C cures viral infections. There is remarkable absence of the common cold and other viral diseases, psychosomatic disorders, such as ulcerative colitis, peptic ulcer, as well as asthma and hay fever in patients during an active psychotic episode. A 30 years ago it was reported that the thymus is an important endocrine gland. Administering large doses of ascorbic acid orally or sodium ascorbate intramuscularly reduces the incidence of Reye's syndrome associated with aspirin use. Ascorbic acid administered orally to bowel tolerance (just short of producing diarrhoea) has a definite antipyretic effect. Increased bowel tolerance is a sign that the metabolic processes are utilising the additional ascorbate so that it does not reach the rectum and cause diarrhoea. Aspirin increases the body's utilisation of ascorbate. Ascorbate deficiency may be associated with the increased incidence of Reye's syndrome in profoundly ill children administered aspirin. When the ascorbate destroys the free radicals produced by a disease process, the free radicals destroy the ascorbate. Diarrhoea is an osmotic phenomenon. The amounts of ascorbate tolerated reveal a magnitude of free radicals involvement in the disease process, which has not been suspected. All processes involving free radicals affect the bowel tolerance of vitamin C. Bowel tolerance to oral ascorbic acid increases almost directly with the feeling of toxicity in disease processes. With recovery, the bowel tolerance returns to the normal baseline. Vitamin C must be a vital component of every cancer treatment program; it may add 20-30% to survival rate.

THE UNIQUE ASCORBATE FUNCTION

Free radical scavenging is a dynamic process. The rate at which free radicals are formed becomes excessive and causes symptoms when it exceeds the rate of reduction of the free radicals. The free radical scavengers cycle from the reduced form carrying the hydride anion (H^+) with the high-energy electron back to the oxidised form lacking the hydride anion. Most of the free radical scavengers are rereduced and used over and over again. The body can make NAD(P)H available for this purpose only at a limited rate, in a well nourished person. When the need to scavenge free radicals exceeds this rate, then symptoms, damage, and ageing occur. All these free radical scavengers are cycled several times an hour when a person is sick. The NAD(P)H keeps rereducing these free radical scavengers so they are used repeatedly. Taking of the usual amounts of nutrient free radical scavengers only assures that there are no critical deficiencies that would limit this free radical scavenging electron-transfer chain. There is a normal limit to the free radical scavenging ability. Vitamin C is a reducing substance, an electron donor. When vitamin C donates its 2 high-energy electrons to scavenge free radicals, much of the resulting dehydroascorbate is rereduced to vitamin C and therefore used repeatedly. The limiting part in nonenzymatic free radical scavenging is the rate at which extra high-energy electrons are provided through NADH to rereduce the vitamin C and other free radical scavengers. During illness free radicals are formed at a rate faster than the high-energy electrons are made available. High concentrations ascorbate (doses 10-200 g/24 h) provides the electrons necessary to quench the free radicals of almost any inflammation, and reduces NAD(P)H and therefore can provide the high-energy electrons necessary to reduce the molecular oxygen used in the respiratory burst of phagocytes. Vitamin C ($C_6H_8O_6$) is a vitamin and an electron carrier. Dehydroascorbate (DHA) is either further metabolised, releasing more electrons, or is rereduced back to vitamin C to be used over and over again. Other nonenzymatic free radical scavengers such as glutathione and vitamin E function in a similar manner of regeneration. Hypoascorbaemia is common in many pathological conditions. There is a proportionate increased bowel tolerance to oral ascorbic acid with toxicity of disease. Individual who can tolerate orally 10-15 g of ascorbic acid per 24 hrs when well, might be able to tolerate 30-60 g per 24 hrs if he/she has a mild cold, 100 g with a severe cold, 150 g with influenza, and 200 g per 24 hrs with mononucleosis or viral pneumonia. The beneficial effect is achieved only when the redox couple, ascorbate/dehydroascorbate, becomes reducing in the tissues affected by the disease. Radicals are molecules that have lost an electron. When a radical escapes its normal location, it becomes a free radical. These free radicals are very reactive and seize electrons from adjacent molecules. All type of inflammations involves free radicals. Cells injured by free radicals spill free radicals onto adjacent cells injuring them resulting in generating more free radicals. Some free radicals spontaneously decay and others are destroyed by enzymatic free radical scavengers such as superoxide dismutase (SOD) and catalase, which act on free radicals in such a way that they neutralise themselves without the addition of extra electrons. The remainder must be destroyed by the high-energy electrons carried by the nonenzymatic free radical scavengers. Free radicals escaping scavenging, cause symptoms and damage. The free radical scavenger carries the high-energy electron that does the neutralising. The high-energy electron is neutralising the free radical, not the free radical scavenger. Plants store this energy by photosynthesis in carbohydrates, fats, and proteins that are then eaten by man and animals. During metabolism of these substances, this energy is past from one molecule to another in the form of high-energy electrons, which often, but not always, are in association with hydrogens. Together with a high-energy electron, one such hydrogen can be called a hydride anion. As glucose is metabolised, NAD^+ (nicotinamide adenine dinucleotide) is reduced to NADH. The high-energy electron in the hydride anion (H^+) is added to the NAD^+. NAD^+ can be reduced to NADH only at a limited rate by the addition of the hydride anion with its high-energy electron derived from the metabolism of carbohydrates, fats, or proteins. The energy NADH carries must be shared among several other critical functions. Most must be used in the process of oxidative phosphorylation to make ATP, which is used as a source of energy by the various tissues of the body. When phagocytes engulf pathogens into its vacuoles, NADPH (nicotinamide dinucleotide diphosphate, reduced form) provides the high-energy electrons the phagocytes require to oxidise substances (radicals) with which they kill various pathogens, i.e., the respiratory burst. The first oxidising substance, superoxide, (O_2^-), in the respiratory burst is made by the reduction of molecular oxygen (O_2) by NADPH. $NADP^+$ is rereduced back to NADPH in the hexosemonophosphate shunt. NADH and NADPH have a common source of energy and can be made available only at some limited rate. Glucose is metabolised for the source of the high-energy electron, which is a rate-limited process. The glucose comes from the metabolism of carbohydrates, fats, and proteins. As NAD(P)H is used in regenerating free radical scavengers, it gives up the hydride anion with its extra high-energy electron and becomes $NAD(P)^+$ again. As these high-energy electrons are used up within the phagocytes, the phagocytes are unable to produce more oxidising substances within their vacuoles to kill pathogens. Some of the previously made oxidising substances leak from within the vacuoles into the cytoplasm thereby becoming free radicals. Once the high-energy electrons are exhausted and the nonenzymatic free radical scavengers cannot be rereduced, the free radicals damage the phagocytes and interfere with phagocytosis. NAD(P)H reduces oxidised flavin adenine dinucleotide (FAD^+), to reduced flavin adenine dinucleotide ($FADH_2$), and becomes $NAD(P)^+$ again. $FADH_2$ reduces oxidised glutathione (GSSG) to reduced glutathione (GSH). Part of NAD(P)H is from vitamin B_3, and part of $FADH_2$ is from vitamin B_2. The high-energy electrons of reduced glutathione (GSH) can directly reduce some free radicals and dehydroascorbate back to ascorbate. In the process the GSH is oxidised back to GSSG. Two hydride anions are added to the dehydroascorbate reducing it back to vitamin C. The enzyme glutathione peroxidase and its coenzyme selenium catalyse these reactions. Ascorbate ($C_6H_8O_6$ or $C_6H_6O_6H_2$, the bolded and separated H_2 is to emphasise the hydrogens containing the high-energy electrons) differs from dehydroascorbate ($C_6H_6O_6$) in that it has 2 extra hydrogen atoms with 2 high-energy electrons in its molecular structure, which it can donate to reduce free radicals. The high-energy electrons of ascorbate, $C_6H_6O_6H_2$, can directly quench free radicals. But some may reduce tocopheryl quinone (an oxidised form of vitamin E) back to α-tocopherol (vitamin E). Some high-energy electrons are passed to the α-tocopherol and then quench free radicals. Ascorbate ($C_6H_6O_6H_2$) is used as the source of electrons and as the electron carrier. The $C_6H_6O_6H_2$ used in massive doses substitutes for the limited availability of the NAD(P)H. The $C_6H_6O_6$ part of the $C_6H_6O_6H_2$ used this way is thrown away; the $C_6H_6O_6H_2$ is only used for the electrons it carries. Amounts of 30-200 g of $C_6H_6O_6H_2$ provide ample high-energy electrons to directly scavenge the large amounts of free radicals generated in disease processes and provide enough high-energy electrons to rereduce $NAD(P)^+$, FAD^+, GSSG, tocopheryl quinone, etc. back to their reduced forms. The $C_6H_6O_6H_2$/ $C_6H_6O_6$ redox couple is reduced by GSH at the concentrations in which these substances are ordinarily present, when $C_6H_6O_6H_2$ is present in large concentrations, it reduces GSSG to GSH. The usual direction of the redox reaction is reversed and the $C_6H_6O_6H_2$ supplies the high-energy electrons reducing the GSSG. The dehydroascorbate, $C_6H_6O_6$, part of the ascorbate, $C_6H_6O_6H_2$, used this way is excreted rapidly in the urine or metabolised further by the body. Certain breakdown products of dehydroascorbate may supply even more high-energy electrons. The high-energy electron is doing the free radical scavenging. The ascorbate and the high-energy electrons are required by various metabolic steps using glucose. It is the high-energy electrons, not the ascorbate that is most important. Any disease process that involves free radicals can be ameliorated by the high-energy electrons carried by ascorbate when used properly in massive doses. The purpose of dietary free radical scavengers is to replace the scavengers incidentally lost. The free radical scavengers are intermediaries. It is up to other metabolic processes to provide the high-energy electrons with which the free radical scavengers reduce free radicals. When doses of 30-200 g or more per 24 hrs of ascorbate are used, the high- energy electrons carried in on the administered ascorbate adds significantly and decisively to the actual electrons doing the reducing. The ascorbate is not used as the vitamin C where it is rereduced by NAD(P)H and used repeatedly; it is used for the high-energy electrons it carries. In high concentrations ascorbate reduces NAD(P)H and provides the high-energy electrons necessary to reduce molecular oxygen used in the respiratory burst of phagocytes. Damaged mitochondria produce most of the free radicals. Additionally, the free radicals produced by these damaged mitochondria further damage adjacent mitochondria. These damaged mitochondria are unable to make available the extra electrons to rereduce the spent non-enzymatic free radical scavengers. So, the immune system is stimulated by the oxidative redox potential and antibodies, T-cell, NK-cells, etc., which hopefully will rid the body of the offending organisms. The massive amounts of extra electrons carried (ascorbate) will neutralise the free radicals even with this advanced damage to the mitochondria situation. The sicker a person the more ascorbic acid he can take orally without it producing diarrhoea. As ascorbate destroys free radicals, the free radicals destroy the ascorbate, and of the ascorbate that reach the rectum does not cause diarrhoea, i.e., titrating to bowel tolerance. Intravenous ascorbate does not cause diarrhoea. Ascorbate cures acute viral diseases but only block symptoms of allergies. Free radicals are needed for the perpetuation of acute viral diseases but only needed for the symptoms due to allergies. Ascorbate broadens the spectrum of activity of antibiotics against bacterial infections and prevents allergic reactions to the antibiotics. Anaphylaxis is a manifestation of acute induced scurvy. Certain drugs such as the barbiturates are neutralised by massive doses of ascorbate. Pain is minimised and recovery rate is amazingly augmented with ascorbate. WBCs require vitamin C to fight. The number of electrons in the free radical scavengers that exist in the body at any time are not enough to last but for a few minutes unless the scavengers are refuelled with electrons by the mitochondria. Antibodies are turned on by free radicals, which can be controlled by ascorbic acid. A virus can damage enough mitochondria to produce enough free radicals to cause the acute induced scurvy, e.g., cold. Therefore, small doses of vitamin C may prevent many colds. Even after a severe cold has been established, massive doses of ascorbate sufficient to force electrons into the nose and throat will again allow WBCs to kill the viruses. Massive doses of ascorbate stimulate WBCs or allow them to continue fighting while turning off antibodies. Ascorbate inhibits antibodies by neutralising the free radicals that induce them. Ascorbate does not cure AIDS but it will prolong the life of AIDS patients and make their life much more comfortable. Vitamin C is essential in the hydroxylation of proline in the synthesis of collagen. The AIDS patient cannot possibly take enough NAC to provide the amount of reducing equivalents necessary to quench most of the free radicals generated in the AIDS process. NAC provides cysteine in a readily available form is used by the body to make glutathione. Glutathione is low in HIV-positive patients. To reduce some oxidised substance, there is a threshold in the concentration of reducing substance that has to be exceeded before the reaction will proceed. The free radicals activate the immune system. Massive doses of ascorbate prevent allergic reactions, and can terminate it abruptly. Free radicals are an almost universal sign of damage to the body. The immune system is activated by the resulting free radicals. Ascorbate in massive quantities suppresses the humoral immunity of antibodies, and augments the cellular immunity of phagocytic cells. Ascorbate provides reducing power that initiates the respiratory burst (by the reduction of molecular oxygen); and protects the WBCs from those very radicals the white cell makes in its vacuoles to kill the pathogens. WBCs may not be able to continue to produce the good radicals when the radicals leak significantly into the cytoplasm (then becoming bad free radicals) and exceed the ability of their free radical scavengers to neutralise. Stimulating the immune system in HIV-positive patient results in the autoimmune destruction of the T4 cells. Treatment of HIV-positive patient (AZT, DNCB) increases the T4 cells in the early stages of the disease; but hastens the final autoimmune destruction of the T4 cells. This stimulation would sometimes have some beneficial effect on secondary infections such as Kaposi sarcoma for a period of time. When free radicals dominate in a tissue for long, the small amount of vitamin C ordinarily present is converted to dehydroascorbate (DHA), which has a very short half-life resulting in acute induced scurvy. Patient dying of AIDS has many of the classic signs of scurvy, e.g., nonfunctional WBCs. The vitamin C levels can be conserved only if a reducing redox potential is maintained in those tissues.

ASCORBATE FUNCTION AND METABOLISM IN THE HUMAN ERYTHROCYTE

Ascorbate is intimately associated with early phases of reproduction. Pregnancy is a severe biochemical stress. Ascorbate is the natural antihistaminic. The ascorbate passage through the kidney appears necessary for the physiological homeostasis and efficiency of the kidney. High levels of ascorbate in the urine tend to lower its pH and endow it with antibacterial, antiviral, anticancer and healing qualities. A gram of ascorbic acid daily would result in a complete remission of the cystitis. Ascorbate administration is also useful in urolithiasis. Ascorbic acid aided spinal nerve regeneration. Ascorbic acid, or vitamin C, is an important antioxidant in plasma, where it consumes OFRs and helps to preserve α-tocopherol (vitamin E) in lipoproteins. Erythrocytes, as the most plentiful cell in blood, help to preserve ascorbate in the blood plasma. Erythrocytes regenerate vitamin C from its 2 electron-oxidised form, dehydroascorbic acid (DHA). DHA is rapidly taken up by these cells on the abundant glucose transport protein, GLUT1. Intracellular DHA is rapidly reduced to ascorbate by GSH in a direct chemical reaction. Ascorbate, which carries a negative charge at physiologic pH, enters and leaves the cells slowly. Vitamin C prevents oxidation of α-tocopherol in low-density lipoprotein. Intracellular ascorbate can spare and recycle α-tocopherol in the erythrocyte membrane. The ability of erythrocytes to recycle ascorbate, coupled with the ability of ascorbate to protect α-tocopherol in the cell membrane and in lipoproteins, provides a potentially important mechanism for preventing lipid peroxidative damage in areas of inflammation in the vascular bed, such as those involved with atherosclerosis. Newly absorbed ascorbate is distributed to tissues in the blood plasma, in which it is one of the most important antioxidants. The partially oxidised form of ascorbate, termed the (mono)ascorbyl free radical or AFR, may serve as an electron acceptor or donor. Loss of the second electron results in dehydroascorbic acid, which is not an acid. DHA is quite unstable at physiologic pH and temperature, with a half-life of about 6 min. With hydrolysis of the lactone ring, DHA is converted to 2,3-diketo-1-gulonic acid, which is probably irreversible in cells. The erythrocyte has redundant mechanisms to recycle DHA back to ascorbate. Uptake of ascorbic acid by erythrocytes is very slow, with a half-time of hours, and occurs by simple diffusion. Most nucleated cells possess a sodium- and energy-dependent transporter with a high affinity for ascorbate that maintains low millimolar intracellular concentrations of the vitamin. DHA is taken up by erythrocytes and other cells by facilitated diffusion on the glucose transporter. Once it has entered erythrocytes or other cells, DHA is rapidly converted to ascorbate and trapped within the cells. The mechanism by which erythrocytes reduce intracellular DHA to ascorbate (recycling) is primarily GSH- and NADPH-dependent. Excess GSH can chemically reduce DHA to ascorbate. Erythrocytes have enzymes (e.g., thioltransferase glutaredoxin, thioredoxin reductase) that can facilitate GSH-dependent reduction of DHA to ascorbate. The most likely mechanism for either direct or enzyme-mediated GSH-dependent DHA reduction involves nucleophilic addition of the thiyl anion of GSH to carbon-3 of DHA, followed by reduction by another molecule of GSH to form the ascorbate double bond and GSSG. In the presence of glucose, DHA does not affect the erythrocyte content of GSH, which reflects recycling of GSSG to GSH by glutathione reductase when adequate NADPH is available from the hexose monophosphate shunt. Depletion of cellular GSH (50-70%) decreases the ability of glucose-depleted cells to recycle DHA to ascorbate. Ascorbate is one of the primary antioxidants in plasma; its presence is required to prevent lipid hydroperoxide formation in plasma lipoproteins. Ascorbate directly protects against peroxide-mediated oxidation of plasma low-density lipoprotein (LDL). LDL is the primary atherogenic lipoprotein in human plasma, and is cleared by LDL receptors and by a scavenger pathway. If the surface lipids and protein sulfhydryls of LDL become oxidised, its receptor-mediated clearance is markedly impaired, and its uptake occurs largely by the scavenger pathway in monocyte-derived macrophages (lipid-laden foam cells found in atherosclerotic plaques). Ascorbate probably acts synergistically with the tocopherols in protecting against LDL oxidation. Ascorbate can both consume OFRs before they can oxidise α-tocopherol, and can reduce α-tocopherol in LDL in the face of an oxidative stress. Erythrocytes are likely to be an important source of ascorbate in plasma, if nothing else because of their abundance. In addition to release of ascorbate from erythrocytes, the presence of an AFR reductase on the outer surface of the cells could also contribute to ascorbate recycling. One-electron oxidation of ascorbate outside erythrocytes would provide AFR to this enzyme for recycling back to ascorbate using intracellular reducing equivalents. Erythrocyte ascorbate recycling may preserve α-tocopherol in LDL, prevent LDL oxidation, and scavenge oxidants released by monocytes and leukocytes. The latter effect will in turn prevent damage to the vascular endothelium. Erythrocyte cell is equipped with so many different types of defenses against oxygen free radicals. In the cytoplasm, enzymatic defenses against both endogenous and exogenous oxidants include catalase, glutathione peroxidase/reductase, and superoxide dismutase. The primary non-enzymatic antioxidant defenses in the erythrocyte cytoplasm are GSH, ascorbate and others. Defenses against oxidant damage in the erythrocyte plasma membrane involve prevention and reversal of peroxidation of unsaturated fatty acids in the lipid bilayer. In the face of a profound external oxidant stress to the erythrocyte, the plasma membrane is often the initial site of damage; the resulting peroxidation of membrane lipids then causes haemolysis. Ascorbate is more sensitive to oxidation by exogenous H_2O_2 generated by the glucose oxidase system than are GSH and α-tocopherol. α-Tocopherol is the most important antioxidant in the cell membrane; it scavenges peroxide free radicals and converts them to less toxic lipid hydroperoxides. In so doing, it protects the cell membrane and decreases haemolysis. The erythrocyte content of α-tocopherol correlates directly with the resistance of the cell to oxidant-induced haemolysis. CoQ_{10} and membrane protein sulfhydryls may also contribute to protection of the membrane. The GSH-dependent phospholipid hydroperoxidase can reduce membrane lipid hydroperoxides, and can also spare α-tocopherol in cellular membranes. α-Tocopherol may be recycled in the cell membrane from the α-tocopheroxyl free radical. Ascorbate does not directly affect membrane lipid peroxidation, but it may perform this function indirectly by reducing the tocopheroxyl free radical at the aqueous-lipid interface of the membrane bilayer.

ALPHA-LIPOIC ACID

It was observed in 1937 that certain bacteria required a component of potato extract for growth. The so-called "potato growth factor" was, in fact, Alpha lipoic acid. In 1947, it was reported that yeast extracts contained an unidentified compound that allowed Streptococcus faecalis to oxidise the carbohydrate pyruvate to acetate. In 1957, the compound was formally isolated and characterised as α-lipoic acid (ALA). The in-vitro antioxidant function of ALA was investigated in 1939. Inside the cell, α-lipoic acid (ALA) is readily reduced to dihydrolipoic acid. Dihydrolipoic acid is more potent than α-lipoic acid, neutralising free radicals. Free radicals damage membranes, creating capillary fragility, damage proteins resulting in cataracts, and breakdown elastin and collagen that are associated with aging and wrinkles, and cancer. α-Lipoic acid regenerates other antioxidants such as C, E and glutathione, prolonging their existence in the body. Free radicals formation through the oxidation of LDL creates arterial cholesterol deposits associated with atherosclerosis. Vitamin E is transported by LDL and plays a critical role in protecting it against oxidation. High vitamin E concentrations in LDL can be achieved by reducing its free radicals, i.e., by vitamin E recycling. β-Carotene is not active in vitamin E recycling. α-Lipoic acid treats and prevents complications associated with diabetes including neuropathy and cataracts. α-Lipoic acid improves nerve blood flow, reduces oxidative stress, and improves distal nerve conduction in diabetic PN. α-Lipoic acid reduces oxidative stress in diabetic peripheral nerves and improves neuropathy. Glutathione is an important free radical deactivator offering protection against cataract formation, as well as immune enhancement, liver protection, protection against cancer, and heavy metal detoxification. α-Lipoic acid restores the activities of glutathione peroxidase, catalase, and ascorbate free radical reductase in lenses. α-Lipoic acid prevents radiation injury and capillary fragility. α-Lipoic acid functions as cofactor for ATP formation. α-Lipoic acid is found in plants containing mitochondria and in non-photosynthetic plant tissues such as potatoes, carrots, beets, yam, kohlrabi and others. Beef is loaded with α-lipoic acid. α-Lipoic acid is a natural co-factor that is integral to the enzymatic rendering of glucose and oxygen into energy. It is also a unique and powerful antioxidant both in itself and in its action of recharging other antioxidants such as vitamins C and E. Antioxidant scavenges or neutralises free radicals that are caused by oxygen metabolism and toxic substances. Free radical reactions in biological systems relate directly to genetics and immunology, as well as all other biological processes, because free radicals are an unavoidable element in the basic chemistry of living systems and, therefore, are involved in a broad spectrum of disease processes. α-Lipoic acid is distributed generally throughout the cytoplasm, functioning as an antioxidant. α-Lipoic functions as a co-enzyme in various metabolic sequences in the mitochondria and protects from free radicals. α-Lipoic, as an antioxidant, prevents certain free radical, genetic damage and thereby inhibits carcinogenesis. The oxidation of rubber is via the action of free radicals on protein. The rancidity of fats in foods is a damage that is caused by free radical reactions. There are over a 60 trillion cells in the body, and about 10,000 oxidative hits to DNA per cell per day. Damage to DNA is relatively small in comparison what happens to other cellular components such as the mitochondria and the cell walls, the chemical reactions in normal metabolism that need to be buffered by antioxidants are enormous. The antioxidant surveillance of biological structures needs to be constant and very tight otherwise, severe damage occurs. Combination medications containing α-lipoic acid is useful for producing analgesic, anti-inflammatory, antidiabetic, cytoprotective, anti-ulcer, antinecrotic, neuroprotective, detoxifying, anti-ischaemic, liver function regulating, anti-allergic, immune-stimulating and anti-oncogenic effects. Pharmaceutical compositions containing α-lipoic acid are used in combating retroviruses. Combination medications containing α-lipoic acid is used for treating circulatory changes reducing the formation of deposits of thrombocytes in the vascular system, preventing the constriction in blood vessels, particularly in smokers. Combination medications containing α-lipoic acid is used for the treatment of nerve cell and nerve fiber diseases and also for the prophylaxis and treatment of circulatory disturbances associated with diabetes mellitus, hyperpathic polyneuropathy and carpal tunnel syndrome. α-Lipoic acid compounds are used for the protection of lipid-containing substances against oxidation and in pharmaceuticals for the prophylaxis and treatment of diseases in which bioradicals are involved, in particular of coronary, circulatory and vascular diseases. The principal, chemical name of α-lipoic acid is 1,2-dithiolane-3-pentanoic acid. The compound has been reported under various names-most commonly thioctic acid. The chemical formula of α-lipoic acid is $C_8H_{14}O_2S_2$. Vitamins as a term was developed in 1912 to denote substances, which were found to be essential or vital for life but, which an organism does not produce, by oneself, in sufficient quantities and which, therefore, must be acquired from the diet, by either food or by supplementation. Vitamin B_1 was discovered in 1912. Vitamin C was first isolated in 1928, and identified as the curative agent for scurvy in 1932. Vitamin A was discovered in 1940. α-Lipoic acid and compounds like it function within the idea of what has been called a bio-energy supplement. The medical paradigm had shifted from the acute to the chronic. Orthomolecularism is a framework that is probably one of the next, major steps forward in the evolution of medicine. Orthomolecularism postulates that normal metabolic mechanisms function on the basis of the right molecule being at the right place at the right time; and both preventive and curative medical treatments should support the natural metabolic mechanisms so that inherent processes can better work toward restoring proper function. Chronic conditions might be associated with sub-optimal levels of essential nutrients. α-Lipoic acid has the unique ability to alleviate diabetes-induced reduction in intracellular vitamin C levels. For sub-clinical (syndrome X) elevated glucose, a dosage of 200-400 mg/day of α-lipoic acid should be considered; and for general, health maintenance or high carbohydrate diets, 100-200 mg/day should be adequate. In the human body, there are about 200 different types of cells, all of which must have some glucose. If a cell receives too much glucose, it starts converting the glucose into fats for storage; if a cell receives too little glucose, it starts using its proteins and fats for energy. Glucose is the only source of energy for certain specialized cells and is the major fuel used by the brain. It is metabolised differently by different cells, depending mostly on whether the cell-type has mitochondria or not. Because glucose is the major fuel for the brain, the nervous system is highly sensitive to both hyperglycaemia and hypoglycaemia. Because the skeletal muscle tissue constitutes 40-50% of the body weight, that tissue is the major pool of glucose; and resistance to glucose uptake in skeletal muscle cells is believed to be the major cause of Type II diabetes. α-Lipoic acid counters insulin resistance to glucose metabolism in muscle tissue; which probably the main reason why lipoic effectively lowers serum glucose in general. Cells of the nervous system and the liver do not require insulin for glucose transport. Type I diabetes is caused by destruction of the insulin producing, β-cells, of the pancreas; and α-lipoic acid slows the onset of this type of diabetes if detected early enough. In Type II diabetes, there is an adequate amount of insulin, but there is a resistance to its ability to transport glucose across the cell membrane. With its actions as both an antioxidant and coenzyme, α-lipoic acid is used as a liver protectant against a variety of toxins, both natural and artificial. Hyperglycaemia means some malfunction has occurred in one or more aspects of the metabolic apparatus, which is supposed to control the usage of glucose. Dosages of 300 mg or more of α-Lipoic acid, prior to a meal, should be considered for both Type I (insulin dependent) diabetes and Type II (non-insulin dependent) diabetes, and probably there should be no administration after 7 p.m. to avoid hypoglycaemia during sleep. Sub-clinical biological damage from intermittent spikes or a chronic, moderately elevated glucose can be caused by: 1- frequent overeating or a junk food diet that is high in refined carbohydrates, 2- chronic stress, 3- certain therapeutic drugs and common stimulants such a caffeine, 4- some medical disorders, and 5- the aging process. Diabetics suffer from accelerated degenerative diseases and require very rigorous medication, diet, and life-style regimens. Cardiovascular disease accounts for 75% of diabetes-related deaths. The risk of stroke is 2-4 times higher among this population and an estimated 60-65% of persons with diabetes have high blood pressure. Diabetes is the leading cause of new cases of blindness among adults 20-74 years of age; and diabetic retinopathy causes from 12,000 to 24,000 new cases of blindness per year.

Elemental constituents of the human body and their percentages:

1- Oxygen	65%.
2- Carbon	18%.
3- Hydrogen	10%.
4- Nitrogen	3%.
5- Macro elements:	0.72%.
A- Calcium	2%.
B- Phosphorus	1%.
C- Potassium, Sulfur, Sodium, Chlorine, Manganese, Iron, Manganese, Copper, Iodine.	Altogether, >1%.
6- Micro elements:	0.28%
Zinc, Cobalt, Aluminium, Arsenic, Barium, Boron, Bromine, Cadmium, and Trace elements.	

Major functions of cell:

Component	Major functions
Cell membrane	1- A lipid and protein barrier that contains and protects the cell contents. 2- Functions include: transport of chemicals in and out, and movement of the cell. Imbedded in the cell membrane are protein structures, which act as chemical receptors and transporters. 3- The membrane and receptor/transporters are principal sites of oxidative damage, and lipoic acid is active in protecting both the lipid and protein structures.
Cytosol or cytoplasm	1- The interior matrix for the cellular components. 2- Site of metabolism of lipids, carbohydrates, amino acids/proteins, and nucleotides.
Mitochondria	Site of oxidative reactions in the production of energy.
Nucleus	Site of the genetic template, DNA, and the nucleolus, which is the site of RNA processing and ribosome synthesis, which is instrumental in the synthesis of proteins.
Endoplasmic reticulum	Membrane synthesis, synthesis of proteins and lipids for cell organelles and for export, detoxification reactions.
Glycogen granules	Stored glucose.
Golgi apparatus	Modification and storage of proteins for organelles and for export.
Lysosome	Digestion of cellular proteins, carbohydrates, lipids and nucleic acids.
Lipid	Stored fats.

Diabetes is the leading cause of end-stage renal failure (ESRF), accounting for 36% of new cases. About 60-70% of people with diabetes have mild to severe forms of diabetic nerve damage, e.g., impaired sensation in the feet or hands, delayed stomach emptying, carpal tunnel syndrome, and peripheral neuropathy. Severe forms of diabetic nerve disease are a major contributing cause of lower extremity amputations. α-Lipoic increases the storage of glucose in both the liver and muscle; α-Lipoic lowers blood sugar and increases glycogen in liver and muscle. Eumetabolic therapies of diabetes are defined as those, which promote and potentiate a normal physiological pattern of insulin activity. Possible components of a eumetabolic therapy include: 1- aspirin, as a potentiator of glucose-stimulated insulin secretion; GTF (glucose tolerance factor), to directly enhance the efficacy of insulin; 2- weight loss; exercise; and fasting, to help reduce tissue resistance to insulin; 3- mitochondrial metavitamins, to optimise the oxidative disposal of excess substrate; 4- a high-fiber, low-fat diet, which appears superior to traditional diabetic diets as a promoter of glucose tolerance. The symptoms of diabetic neuropathies may be intermittent, with different degrees of severity, and vary according to the individual. Oxidative stress plays a promoting role in developing of long term diabetic late complications; a therapy with adjuvant antioxidants may lead to a regression of diabetic late complications. Long term diabetic late syndrome is a prominent feature of the neuropathies. Excessive, free glucose (i.e., circulating, extracellular glucose in the serum or intra-cellular glucose, which is not in the form of glycogen) binds to and denatures protein structures, i.e., glycation. Glycation causes microvascular deterioration and which, in turn, causes damage in diabetic nerve, vascular, renal, retinal, and other tissue damage. α-Lipoic acid prevents the glycation. α-Lipoic acid normalises myocardial oxygen uptake, myocardial ATP levels, and cardiac output. Lactate and pyruvate production is also normalised. Additionally, α-lipoic acid improves endogenous glycogen in the diabetic heart. α-Lipoic acid acts especially by increasing glucose uptake, glycogen breakdown and glucose oxidation. Hyperglycaemia causes a decrease in the permeability and fluidity of cell membranes. α-Lipoic acid results in a significant increase in membrane fluidity. α-Lipoic acid increases insulin stimulated glucose disposal in Type II diabetes. ALA (α-lipoic acid) significantly enhances the capacity of the insulin-stimulatable glucose transport system and of both oxidative and non-oxidative pathways of glucose metabolism in insulin-resistant skeletal muscle. ALA improves the insulin-mediated glucose uptake in muscle tissue by 62% as well as increases both insulin-stimulated glucose oxidation (33%) and glycogen synthesis (38%); and significantly greater (21%) muscle glycogen concentration. ALA significantly lowers (15-17%) plasma levels of insulin and frees fatty acids. The free radical chemistry plays a significant role in many disease processes, including diabetes, atherosclerosis, neuronal degeneration, infection and immunologic reactions, exposure to toxic substances, and the cumulative deterioration known as the aging process. These chronic disorders cause an array of diseases such as: cataracts and blindness, hypertension, heart disease, cancer, kidney failure, Parkinson's and Alzheimer's, and arthritis, all of which are amenable, in varying degrees, to prevention/treatment by antioxidant compounds. The antioxidants are natural physiological substances. There are many environmental sources of free radicals exist, such as ozone, sunlight, and other forms of radiation, smog, dust, and other atmospheric pollutants.

Free radicals:

Radical	Name	Half-life (Reactivity)
$HO^{\cdot}$	Hydroxyl radical	.0000000001 sec.
$LO^{\cdot}$	Lipid alkoxyl radical	.0000001 sec.
$LOO^{\cdot}$	Lipid peroxyl radical	7 sec.
$L^{\cdot}$	Lipid carbon-centered radical	.000000001 sec.
H_2O_2	Hydrogen peroxide	Minutes
$O_2^{\cdot-}$	Superoxide anion	.000001 sec.
$^{1}O_2$	Singlet oxygen	.0000001 sec.
HQf	Semiquinone radical	Days.
$NO^{\cdot}$	Nitric oxide	~1 sec.

The antioxidant nutrients may have major significance in the prevention of a number of diseases, including cardiovascular and cerebrovascular disease, some forms of cancer and several other disorders, many of which may be age-related. There is a great need for improvement in public awareness of the potential preventive benefits of antioxidant nutrient intake. The free radicals are a major and fundamental biological phenomenon. Of all the types of stress to which biological systems are subjected (e.g., thermal and mechanical stresses, exposure to toxins, immunological challenges, nutritional fluctuations, etc.), the generation of free radicals is probably the single highest stress factor and is responsible for the greatest amount of "wear and tear" on biological damage. Free radicals are generated in all cell types and are highest in the ones which work the hardest and are functionally most critical ones - the nervous system, cardiac and skeletal muscle, liver, etc. The generation of free radicals is an intrinsic aspect of basic metabolism; and these chemical reactions are unavoidable. Most free radicals are generated in the process of combining oxygen and glucose and rendering them into ATP. The function of ALA in the pyruvate dehydrogenase reaction is to act in conjunction with primary enzymes, as an electron carrier in the oxidation of pyruvate. The oxidation of glucose and the concomitant generation of free radicals together with the operation of ALA as an antioxidant, all, take place in and are contained by the mitochondria. The criticalness of controlling the reactions (about 30 steps in all) which are involved in the oxidation of glucose is evidenced by the number of mitochondria per cell that are required. For example, cardiac tissue requires large amounts of energy, which is derived from the oxidation of glucose; and mitochondria comprise about 50% of the volume of cardiac cells. Liver cells also oxidise glucose, and they contain 800-2000 mitochondria per cell. As a general rule, in aging cells, as much as 60% of the mitochondria can be lost. Any agent that would facilitate the conversion of glucose into ATP would lower the over-all plasma glucose, thus reducing glycation of proteins and the free radicals, which result from the oxidation of glucose outside the mitochondrial containment. Other chemical reactions are also major sources of free radicals, including toxic chemicals, radiation, and the diffusion of molecular oxygen; and ALA plays a role in buffering those reactions from damage to cell membranes, all the organelles, the intracellular matrix itself, and the cytosol. Free radical chemistry is complex and depends on the particular metabolic reactions that are evoked by particular stressors or agents, which further depends on the particular tissue. For example, when the 2 free radicals, superoxide and NO, combine, they form another radical called peroxynitrite. That, in turn, causes tissue-damage that apparently is involved in the pathobiology of several human diseases, e.g., emphysema. Peroxynitrite modifies the amino acid, tyrosine by inactivating α-1-antiproteinase, which inhibits an enzyme that degrades certain proteins. Consequently, certain proteins are excessively degraded and disease occurs. α-Lipoic acid effectively protects against this reaction. ALA functions as an antioxidant, and has ability to reactivate other antioxidants, which have, been oxidised. α-Lipoic acid is the antioxidant's antioxidant; ALA reactivates other antioxidants that have been oxidised. Free radical modification of low-density lipoprotein is one of the primary pathologies in cholesterol deposits of atherosclerotic plaques. The recycling of vitamin E and other phenolic antioxidants by plasma reductants (such as ALA) may be an important mechanism for the enhanced antioxidant protection of LDL. ALA influences the maintenance of intra-cellular glutathione, which is one of the main endogenous antioxidant that is responsible for free radical scavenging in all cell types. α-Lipoic acid and other antioxidants are effective in inhibiting the activation of NFκB and other nuclear factors. Aging is an extremely complex biologic with a poor and incomplete understanding of its fundamental molecular mechanisms. Aging is a multifactorial process composed of both genetic and environmental components. All aging changes have a cellular basis. α-Lipoic acid is both water- and fat-soluble. It is active in the circulating plasma, the cell membranes, and throughout the cytosol. It is in the mitochondria and the other organelles of the cells, including the genetic material in the nucleus. ALA passes the blood brain barrier and, therefore, is active in the tissue in which there is the highest free radical activity because neurons have the highest utilisation of oxygen and glucose. α-Lipoic acid recharges oxidised vitamins C and E, which are exogenous antioxidants and must be supplied by a good diet. ALA recharges glutathione, which is one of the principal endogenous antioxidants that are made within cells. ε-Carboxymethyl lysine (CML) represents a general marker of oxidative stress and of long-term damage to proteins in aging, atherosclerosis, and diabetes. CML accumulates in skin, lung, heart, kidney, intestine, intervertebral discs, and particularly in arteries with aging, and an acceleration of this process in diabetes. The oxidative formation of CML can be reduced by ALA, as well as by other antioxidants. In order to be transparent, the cells of the eye lens do not contain any mitochondria; and therefore, they must metabolise glucose anaerobically. Because anaerobic glycolysis is a relatively slow process, this makes the lens more susceptible to the detrimental effects of elevated glucose. Premature cataract is a common feature of diabetics. To protect itself from glycosylation and free radical reactions, the eye lens cells must generate a constant and substantial amount of glutathione (an endogenous antioxidant) to buffer the potential damage. ALA supports that specific type of antioxidant surveillance by both recharging glutathione and compensating for glutathione deficiencies by recharging other antioxidants. ALA increases the levels of glutathione, vitamin C and vitamin E, and also restores the activities of glutathione peroxidase, catalase, and ascorbate free radical reductase in the lenses. Neurons are non-dividing types of cells. Old neurons have a substantial accumulation of lipofuscin or age pigment, which is probably a residue from oxidised and disintegrated mitochondria. About one-half of the mitochondria of neurons are lost in aging. Brain damage commonly occurs in stroke, cardiac arrest, haemorrhage, and head trauma by the combination of initial lack of oxygen to the tissue (i.e., ischaemia) followed by a rapid reoxygenation (i.e., reperfusion). Maximum tissue damage is observed during the reperfusion stage, which is attributed mostly to injury from OFR. ALA treatment leads to dramatic improvement in survival and a reduction of brain damage, which is resulting from ischaemia-reperfusion injury. Chronic elevated glucose in diabetics and insulin resistance syndromes increase the risk of premature heart disease, stroke, and hypertension. High cholesterol in the blood is associated with an increased risk of atherosclerosis. Oxidative modifications of low-density lipoprotein (LDL), in particular, increase its atherogenicity by altering the cell receptors and their uptake of LDL, particularly cells in the intima of blood vessels, which is where the atherosclerotic lesion occurs. Oxidised LDL is taken up by scavenger receptors on monocytes, smooth muscle cells, and macrophages in an uncontrolled process leading to accumulation of lipid and the formation of foam cells. Antioxidants have a protective effect on the oxidation of LDL and a mitigation of atherosclerotic plaques. The atherosclerotic diseases (i.e., heart disease and stroke) cause about 50% of the deaths and cancer causes an additional 20%. Free radicals play a significant role in cancer. Once cancer becomes established, the presently available methods of treatment (i.e., surgery, radiation, and chemotherapy) can be so traumatic that frequently it is more devastating than the disease itself. After about 20 years of the war on cancer and billions of dollars spent on cancer research with essentially no improvement in survival statistics or quality of life (QoL). The possible mechanisms of cancer chemoprevention include carcinogen-blocking activities, antioxidant/anti-inflammatory activities, and antiproliferation/antiprogression activities. Carcinogen-blocking activities encompass inhibition of carcinogen uptake, inhibition of carcinogen formation or activation, deactivation or detoxification of carcinogens, prevention of carcinogen binding to DNA, and enhancement of the level or fidelity of DNA repair. Antioxidant/anti-inflammatory activities include scavenging of reactive electrophiles and OFR, and inhibition of AA metabolism. Antiproliferation/antiprogression activities comprise modulation of signal transduction, modulation of hormonal and growth factor activity, inhibition of aberrant oncogene activity, inhibition of polyamine metabolism, induction of terminal differentiation, restoration of immune responses, enhancement of intercellular communication, restoration of tumour suppressor function, induction of apoptosis, telomerase inhibition, correction of DNA methylation imbalances, inhibition of angiogenesis, inhibition of basement membrane degradation, and activation of antimetastasis genes. Oxidative or free radical damage to DNA is the main factor in causing cells to become cancerous. In living cells ROS or free radicals are formed continuously as a consequence of metabolic and other biochemical reactions as well as external factors. These include oxidative damage to DNA, is important factor in carcinogenesis. Despite extensive repair capacity, oxidatively modified DNA is abundant in human tissues, in particular in tumours. The damaged DNA accumulates with age in both nucleus and mitochondria; and the incidence of cancer accelerates exponentially with aging (>45 years). α-Lipoic acid is active at the DNA site, and operates synergistically with and potentiates other antioxidants. Alpha lipoic acid is unique in that both its oxidised and reduced forms possess antioxidant properties. In its oxidised form, the surface atoms at the end of the molecule form a ring structure known as the dithiolane ring. It is because of a minute particle of disulfide in this ring that ALA is able to perform its attributed functions as an enzyme-catalyst and as an antioxidant. The dithiolane ring is broken when the molecule is reduced, either by enzymes or free radicals. The result is dihydrolipoic acid, which itself is even more aggressive in its antioxidant potency. Since ALA is a medium chain fatty acid, it is readily absorbed and transported across cell membranes. In this way, ALA differs from glutathione, which is the major thiol (sulfur containing) antioxidant in the body. Glutathione cannot be transported across the intestinal tract so glutathione levels cannot be increased by dietary means to increase antioxidant defense from this substance. Dihydrolipoic acid acts directly to destroy certain oxygen species such as superoxide radicals, hydroperoxy radicals, and hydroxyl radicals. ALA is a molecule that connects the activity of antioxidants in the cell membrane with antioxidants in the cytoplasm, strengthening the antioxidant network. The intake of vitamin E and vitamin C should be coupled with ALA supplementation in order to ensure complete cell protection since vitamin E, vitamin C, and ALA work synergistically. ALA plays an important role in antioxidant and vitamin recycling. This process can be viewed as a sort of chain reaction.

Cell types:

Cells without mitochondria:	
	Red blood cells, and some cells of the kidney, eye lens, cornea, and the retina.
	Glucose is converted to lactic acid. α-Lipoic acid helps recharge glutathione for antioxidant scavenging of free radical reactions.
Cells with mitochondria:	
	Brain, muscle, adipose, and liver cells.
	Glucose is converted to pyruvate. α-Lipoic acid is a coenzyme in the further oxidation of pyruvate into ATP or into stored fats. α-Lipoic acid also has antioxidant functions.

Antioxidants are most powerful in their reduced form. When antioxidants come into contact with free radicals, they lose their free radical scavenger fighting abilities and return to their oxidised form. The reduced form of molecules always has an extra electron. The reduced form of ALA, dihydrolipoic acid, is able to donate this electron to the oxidised or antioxidant-inactive form of glutathione and/or vitamin C. The oxidised form of glutathione is called glutathione disulfide. The oxidised form of vitamin C is called dehydroascorbate. When ALA donates the electron to either of these molecules, it serves to regenerate them back to their reduced, potent antioxidant forms known as glutathione and ascorbate respectively. The reduced form of vitamin C (ascorbate), regenerates vitamin E from its oxidised form (chromanoxyl radical to its reduced form tocopherol) by means of a similar process of electron donation. Each time a molecule in its reduced form donates an electron, it returns to its oxidised form. Each time a molecule in its oxidised form receives or accepts an electron it returns to its reduced form. This is known as the redox cycle. The reduced form of the antioxidant works in a similar way in combating free radicals. For example, vitamin E donates an electron to peroxide, a free radical, thus balancing out the unpaired electron in the peroxide molecule to create hydrogen peroxide, a relatively harmless molecule. Now that vitamin E has given up the extra electron it loses its free radical scavenging properties and is in its oxidised state. ALA is found in food stuffs, which have been derived from sources where active energy production is occurring. Mitochondria produce energy for the cell and have abundant fats, protein and enzymes. Cells or tissues that are mitochondria-rich would be expected to have higher sources of ALA. ALA is present in the leaves of plants containing mitochondria and nonphotosynthetic plant tissues, such as potatoes. Another source, which is very high in mitochondria, is red meat.

MAGNESIUM

More than 350 enzymes, aside from metabolic cycles, appear to require and be regulated by concentrations of Mg^{2+} that are well within the physiological range observed in tissues and cells. Magnesium is indispensable for enzyme activity and structural modification of phosphometabolites or channels. Ca^{2+} is a signalling molecule because of the following conditions: a- the free cytosolic concentration is extremely low; b- the concentrations in plasma and cytosolic reservoirs are very high, establishing a large concentration gradient across biological membranes; c- because of the low resting Ca^{2+} concentration, the movement of a few Ca^{2+} molecules can increase or decrease the concentration of Ca^{2+} by several orders of magnitude; and d- the increase of Ca^{2+} concentrations results in specific Ca^{2+} binding to cytosolic proteins that are modified in their 3-D structure and function upon formation of a complex with Ca^{2+}. Mg^{2+} is kept both in the cytosol and in extracellular fluids in the millimolar or submillimolar concentration. Because of this initial large concentration, a total increase or decrease of Mg^{2+} in the cytosol equivalent to that occurring for Ca^{2+} will result in negligible free Mg^{2+} changes. Furthermore, due to the large difference between the radius of hydrated and not hydrated Mg^{2+}, specific coordination to proteins is less likely than that of Ca^{2+}. Circulating Mg^{2+} level is 1.5-1.7 mEq/l in humans. Within the cell, total Mg^{2+} content is distributed almost homogeneously among nucleus, mitochondria and endo-(sarco)-plasmic reticulum. A considerable amount of Mg^{2+}, approximately 4-5 mM, is present in the cytosol as a complex with ATP and other phosphometabolites. The cell senses the cytosolic free Mg^{2+} concentration and adjusts it according to its physiological requirement as a result of changes in energy content or other cations distribution. Hormones (vasopressin, catecholamine) stimulating Mg^{2+} extrusion from different organs or tissues, thereby increasing plasma Mg^{2+}, also increase Mg^{2+} reabsorption at the renal level to prevent a net loss of the cation. Insulin prevents the Mg^{2+} extrusion. The rapid increase in cytosolic Ca^{2+} induced by hormones like vasopressin may be required to activate an entry of Mg^{2+} across the cell plasma membrane and its redistribution within intracellular compartments. Within the cell mitochondria contain large amount of Mg^{2+}, some of which can be rapidly mobilised following the increase in cytosolic cAMP level. Many conditions leading to cAMP increase are accompanied by an efflux of Mg^{2+} from the mitochondria. Several mitochondrial dehydrogenases can increase activity within minutes in the absence of an increase in mitochondrial Ca^{2+}. The catecholamine increasing respiration could be due to the decrease in mitochondrial Mg^{2+} resulting from an increase in cellular cAMP. This in turn will stimulate some dehydrogenases directly and render others more susceptible to the concentrations of Ca^{2+} present in the mitochondrial matrix. The abundance of Mg^{2+} within mammalian cells is consistent with its relevant role in regulating tissue and cell functions. There are more than three hundred and fifty enzymes, aside from metabolic cycles, appear to require and be regulated by concentrations of Mg^{2+}. Large fluxes of Mg^{2+} can cross the cell plasma membrane in either direction following a variety of hormonal and non-hormonal stimuli, resulting in major changes in total and, to a lesser extent, free Mg^{2+} content within tissues, and in a marked variation in the opposite direction of circulating Mg^{2+} level. Intracellular magnesium (Mg_i) is the second most abundant intracellular cation. Mg is essential key cofactor for enzymes involved with transfer of phosphate groups (e.g., ATPases, phosphatases, kinases, etc.). Mg_i modulates membrane receptors, ionic channels and transporters. Activation of FAS on B-cell lymphomas causes an increase in $[Mg^{2+}]_i$ that appears to be required for apoptosis. Homeostasis of plasma concentration of Mg^{2+} in humans is achieved via renal conservation mechanisms and hormonal control of Mg absorption. Changes in the Mg^{2+} plasma concentration occur during alcoholism, CNS injury and diabetes mellitus. A primary defect in $[Mg^{2+}]_i$ handling may be a critical effector of non-insulin dependent diabetes mellitus. Hypomagnesaemia may produce nervous hyperexcitability, tetanic syndrome, and meningo-encephalic syndrome. In excitable cells, the intracellular free Mg^{2+} concentration ($[Mg^{2+}]_i$) is several hundred times lower than expected if distributed passively. Since plasma membranes are permeable to Mg^{2+}, constant extrusion of this ion occurs. Hormones induce massive efflux of Mg_i from cells. Operation of the Na^+/Mg^{2+} exchanger requires that Na^+ bind to the "cis"-side of the transporter protein while Mg^{2+} binds to the "trans"-side. Subsequently, the ions are translocated to the opposite side of the membrane. Na/Mg exchanger has an absolute requirement for ATP, and is voltage-insensitive. Extracellular Na^+ activates Mg^{2+} efflux and extracellular Mg^{2+} activates Na^+ efflux. Two or more extracellular Na^+ ions are required to account for the observed steady-state distribution of $[Mg^{2+}]_i$. The Mg^{2+} transporter has an absolute requirement for ATP. ATP may activate transporters because it may work: 1- as a substrate for ATPases; 2- as a substrate for protein kinases; 3- as a substrate for lipid kinases generating second messengers, (e.g., phosphatidylinositol phosphates); 4- by directly binding to the transporter inducing allosteric effects; 5- by inducing changes of actin cytoskeleton; 6- by chelating polyvalent cations; and 7- by activating ATP-dependent phospholipases. The possible mechanisms by which ATP could activate the Mg^{2+} transporter include: 1- via a direct effect; 2- by being a precursor for phosphoarginine (P-Arg); or 3- by being a precursor for PIP_2. PIP_2 could in turn activate the transporter by: 4- a direct effect; 5- being a precursor of PIP_3; 6- being a precursor of diacylglycerol (DAG); or 7- being a precursor of IP_3, 8- ATP could work by being a substrate of phosphatidylinositol (PI) kinase yielding PIP_2; 9- PIP_2 could directly activate the transporter; 10- PIP_2 could work by being a substrate of phosphatidylinositol 3'-kinase (PI_3 kinase) yielding phosphatidylinositol-3,4,5-trisphosphate (PIP_3) that would in turn activate the transporter; 11- PIP_2 could work by being a substrate of phospholipase C (PLC) yielding diacylglycerol (DAG) that would in turn activate the transporter; 12- PIP_2 could work by being a precursor of inositol 1,4,5-trisphosphate (IP_3) that would in turn activate the transporter. There is a direct correlation between the concentrations of intracellular free K^+ and Mg^{2+} in skeletal muscle. There is a strong correlation between the total intracellular concentrations of Mg and K in red cells, lymphocytes, and skeletal muscle. Mg^{2+} affects K^+ transport in red blood cells and muscle cells. There is a direct correlation between the changes in plasma Mg and K in pathological conditions such as Bartter's syndrome and Alzheimer disease. There is a relationship between the fluxes of Cl^- and Mg^{2+} across the plasma membrane of cells. Extracellular Mg^{2+} and Cl^- interact to modulate the tone and contractility of vascular muscle. Net Mg^{2+} efflux is dependent on net Cl^- efflux in Mg^{2+}-loaded human erythrocytes. Ingestion of supplementary dietary Cl^- may reduce plasma Mg^{2+}. There is a direct correlation between the levels of $[Mg^{2+}]_i$ and intracellular free K^+. The electrochemical gradients of Na^+ appear to be coupled to regulate intracellular Mg^{2+}. Removal of extracellular Mg^{2+} produces a simultaneous equimolar reduction in Na^+ and Cl^- efflux. Magnesium is an essential (macro) mineral in vertebrates with many biochemical and physiological functions including activation of enzymes, involvement into metabolic pathways, regulation of membrane channels and muscle contraction. Despite these important functions, Mg^{++} homeostasis depends on absorption from the gastrointestinal tract (GIT), requirement of the body, and excretion via the kidneys. Paracellular movement of Mg^{++} is only important in leaky epithelia as in the small intestine. The transcellular transport of Mg^{++}, luminal uptake and basolateral extrusion, require membrane proteins that increase the low permeability of the membranes and facilitate the movement of Mg^{++} through these lipid bilayers. Magnesium is an essential mineral in vertebrates and is the fourth abundant cation in the body, within the cell second only to potassium. Patients with a severe short bowel syndrome exhibit very low or even negative net absorption of Mg^{++} resulting in hypomagnesaemia. The abnormality in these patients may be a defect in carrier-mediated transport of Mg (in the small intestine) from low intraluminal concentrations of magnesium. Paracellular pathway (shunt) between the cells consisting of tight junctions and the intercellular space. Magnitude and direction of flow of a solute through this pathway depends on the passive driving forces and the permeability of the shunt for the solute. The transcellular route includes uptake across the luminal and the extrusion across the basolateral membrane. Because hydrophilic compounds such as Mg^{++} permeate very poorly across (lipid) cell membranes of the enterocytes, transport proteins in the luminal and basolateral membrane of epithelial cells are necessary for transcellular transport (absorption) of Mg^{++}. Transcellular intestinal Mg^{++} transport (or absorption) may be regarded as a three-step process, consisting of: 1- entry into the epithelial cell from the lumen, 2- transit through the cytosol, and 3- extrusion from the cell, across the basolateral membrane. The entry of Mg^{++} into the intestinal cell across the brush border or apical membrane requires no metabolic energy, since Mg^{++} moves down a steep electrochemical gradient. Cellular uptake and transcellular transport are not necessarily linked. The transcellular pathway plays only a minor role in net Mg^{2+} absorption under normal conditions, because solvent drag via the paracellular pathway dominates as long as net solute and water movement across the colonic epithelium proceeds from lumen to blood. In contrast, during diarrhoea the colon secretes water and solutes. Under these conditions osmotic water and solvent drag of Mg^{++} through the paracellular pathway probably reverses, leading to Mg^{++} losses and clinically observed hypomagnesaemia. High dietary K^+ intake and, consequently, high luminal K^+ concentrations may decrease the apparent digestibility of Mg^{++}. The reduced Mg^{++} transport at high luminal K^+ concentrations is closely correlated with electrophysiological changes within the lumen epithelium. At higher luminal K^+ concentrations there is a small passive backflow of Mg^{2+} most likely through the paracellular pathway. The transcellular component of Mg^{++} transport, is significantly reduced by high luminal K^+ concentrations. Increasing mucosal K^+ concentrations causes a reversible, concentration-dependent depolarisation of the potential across the apical membrane of the lumen epithelium, which is accompanied by a decrease of the unidirectional mucosa to serosa transcellular Mg^{++} flux and decrease of intracellular Mg^{++}. Extrusion of Mg^{++} is an uphill transport (basolateral extrusion) and is probably coupled to the electrochemical gradient of Na^+. The transcellular transport of Mg^{++}, luminal uptake and basolateral extrusion, require membrane proteins that increase the low permeability of the membranes and facilitate the movement of Mg^{++} through these lipid bilayers (e.g., Na^+/Mg^{++} exchange). Magnesium has an important modulatory role in the control of secretory epithelial cells function. The kidney provides the most sensitive control for magnesium balance. About 80% of the total serum Mg are ultrafilterable through the glomerular membrane. The proximal tubule reabsorbs only a small fraction, 10-60%, of the filtered Mg. A large part (60%) of the filtered Mg is reabsorbed in the loop of Henle. Magnesium reabsorption in the loop occurs within the cortical thick ascending limb (cTAL) by passive means driven by the transepithelial voltage through the paracellular pathway. The superficial distal tubule reabsorbs significant amounts of Mg. Magnesium reabsorption in the distal tubule is transcellular and active in nature. Many hormones and nonhormonal factors influence renal Mg reabsorption to variable extent in the cTAL and distal tubule. Nonhormonal factors may have important implications on hormonal controls of renal magnesium conservation. Dietary magnesium restriction leads to renal magnesium conservation with diminished urinary magnesium excretion. Elevation of plasma Mg or calcium concentration inhibits Mg and calcium reabsorption leading to hypermagnesiuria and hypercalciuria. Loop diuretics, such as furosemide and bumetanide, diminish salt absorption in the cTAL whereas the distally acting diuretics, amiloride and chlorothiazide stimulate Mg reabsorption within the distal convoluted tubule. Metabolic acidosis, potassium depletion or phosphate restriction can diminish magnesium reabsorption within the loop and distal tubule. Mg participates in activation of many enzymes such as ATPases, regulation of channel activities such as K and Ca channels, and because it can establish stable ternary complexes with nucleotides, it is also involved in transcriptional, nucleic acid polymerisation, and translational processes. The recommended daily estimated average requirement for Mg is 265 and 350 mg for adult females and males, respectively. In the proximal tubule, little Mg is transported across the epithelium. In the loop of Henle, specifically the thick ascending limb, transepithelial Mg movement is abundant and principally passive in nature moving between the cells through the paracellular pathway. More distally, in the distal convoluted tubule and connecting segments Mg transport is transcellular and active in nature quite different from that observed in the thick descending limb. Finally, there appears to be no transepithelial Mg transport in the collecting ducts. Approximately 70% of plasma Mg are in the ionic form, Mg^{2+}. The remaining Mg is bound to circulating proteins (essentially albumin) or is complexed with citrate, oxalate and phosphate anions. About 80% of the total serum magnesium are ultrafilterable through the glomerular membrane. With a GFR, of 125 ml/min the filtered Mg amounts to about 140 mmol/day of which the kidney reabsorbs some 80-99%. Accordingly, of the filtered Mg is excreted in the final urine depending on the dietary Mg intake. Many hormones and nonhormonal factors influence renal Mg reabsorption to variable extent in the various nephron segments (not proximal tubule). Hormones, such as PTH, calcitonin, glucagon, and AVP increase magnesium reabsorption within the kidney partly additive with antidiuretic hormone. The major effects of these hormones are within the thick ascending limb of the loop of Henle and the distal convoluted tubule. These hormones stimulate intracellular cAMP formation in the distal tubule. Insulin stimulates Mg^{2+} transport through activation of tyrosine kinase-mediated signalling pathways. Peptide hormones stimulate intracellular cAMP formation and Mg^{2+} uptake. The steroid hormones have important effects on Mg^{2+} uptake in distal tubule cells. Aldosterone potentiates hormone-mediated cAMP formation and Mg^{2+} transport. $1,25(OH)_2D_3$ stimulates Mg^{2+} uptake by cellular mechanisms which are separate from cAMP-mediated pathways. $1,25(OH)_2D_3$ has an additive action with hormones, like PTH, which act through cAMP. The interactions of the various peptide and steroid hormones, prostaglandins, and renal innervations are complex. Adaptation of magnesium transport with dietary magnesium restriction occurs at least in 2 segments, the ascending limb and distal tubule. Chronic Mg depletion in humans is associated with diminished urinary Mg and normal to high excretion of calcium. This may be in part due to acquired PTH resistance in magnesium deficiency. Elevation of plasma Mg or Ca concentration inhibits Mg and Ca reabsorption leading to hypermagnesiuria and hypercalciuria. Inhibition of reabsorption occurs within the thick ascending limb of the loop of Henle and the distal convoluted tubule. In the loop where Mg is passively reabsorbed an elevation of extracellular Mg or Ca may decrease the permeability for these cations in the paracellular pathway so that for any given transepithelial voltage there is less Mg and Ca absorption. The extracellular Ca^{2+}/Mg^{2+} sensing receptor is located on the peritubular side of TAL cells and comprised of 3 major domains: 1- a large extracellular amino-terminal domain of 613 amino acids that is thought to possess the cationic binding sites, 2- a 250 amino acid domain with 7 predicted membrane-spanning segments characteristic of the superfamily of G protein-coupled receptors, and 3- a carboxyl terminal domain of 222 amino acids that likely resides within the cytoplasm and is involved with intracellular signalling processes. Ca^{2+} or Mg^{2+} binds to the extracellular domain initiating a number of intracellular signals. Stimulation of Gi-proteins modulate adenylate cyclase activity and cAMP levels and Gq proteins activate phospholipase C releasing inositol 1,4,5-trisphosphate, cytosolic Ca^{2+}, and cytochrome P-450 metabolites. A Ca^{2+}/Mg^{2+}-sensing receptor has been found in glomeruli, proximal tubules, cortical and medullary thick ascending limbs, distal convoluted tubules, cortical collecting ducts, and outer medullary collecting ducts. The site and location of the Ca^{2+}/Mg^{2+}-sensing receptor(s) have important effects on renal Mg handling. Ca^{2+}/Mg^{2+}-sensing receptor plays an important role in the physiological control of renal Mg handling. Renal Mg reabsorption is influenced by extracellular volume in that volume contraction increases Mg conservation and volume expansion decreases Mg absorption. Changes of Mg reabsorption within the proximal tubule are not markedly altered even though NaCl and water are significantly affected. The basis for these changes is principally associated with alterations of salt and Mg transport within the loop of Henle, both the descending thin limb and thick ascending limb. Hypermineralocorticoidism is accompanied by salt conservation due to enhanced distal tubular NaCl reabsorption. The resulting extracellular volume expansion diminishes proximal and loop of Henle Mg reabsorption leading to renal Mg wasting and in extreme cases to hypomagnesaemia. Any solute that is not absorbed or reaches its maximal tubular transport rate becomes an osmotic diuretic. Osmotic diuretics diminish salt and water reabsorption in the proximal tubule, loop, and distal tubule by increasing the flow rate along these segments. The principal action on Mg handling is in the thin descending and thick ascending limbs of the loop of Henle. Mg wasting in uncontrolled diabetes may be explained, in part, on increased filtered glucose and unreabsorbable ketoacids. Loop diuretics, such as furosemide and bumetanide, diminish salt absorption in the thick ascending limb by virtue of their action on electroneutral Na-2Cl-K cotransport across the luminal membrane. Magnesium, a divalent cation, may be influenced to a greater degree than the monovalent cations, such as Na and K. Chronic usage of furosemide may lead to renal Mg wasting and Mg deficiency. The distally acting diuretics, amiloride and chlorothiazide stimulate Mg reabsorption within the distal convoluted tubule. Amiloride is a magnesium-conserving diuretic. Thiazides are extensively used in the management of diseases due to fluid retention, diabetes insipidus, and nephrolithiasis. Chlorothiazide may stimulate Mg^{2+} transport through changes in the membrane voltage similar to the basis of amiloride actions: 1- blocks Na^+ entry into DCT cells and hyperpolarises the cell; and 2- does not stimulate Mg^{2+} uptake in the absence of a change in voltage. Chlorothiazide increases Mg^{2+} uptake in a dose-dependent fashion. Maximal concentrations of chlorothiazide increase Mg^{2+} transport by 58%. This is associated with hyperpolarisation of the plasma membrane voltage. An increase in the membrane voltage enhances Mg^{2+} uptake into MDCT cells. The clinical use of thiazide diuretics in patients over a long period of time sometimes lead to hypomagnesaemia probably from renal magnesium-wasting. Chlorothiazide inhibits Na-Cl cotransport leading to an increase in urinary NaCl excretion and contraction of the ECF volume. The renin-angiotensin system (RAS) is activated resulting in elevated aldosterone levels and increased potassium and hydrogen ion secretion that may result in hypokalaemia and exacerbation of metabolic alkalosis. Volume depletion, elevated aldosterone, and metabolic alkalosis increase renal magnesium conservation so that it is unlikely that these influences would lead to magnesium-wasting. However, hypokalaemia has been associated with altered renal magnesium handling that may explain some of the cases of potassium deficiency with chronic chlorothiazide usage. Systemic acidosis is associated with renal magnesium-wasting. Acute metabolic acidosis leads to significant increases in urinary Mg excretion. Chronic acidosis also leads to urinary Mg-wasting, which, as with acute acidosis, may be partially corrected by the administration of bicarbonate. Acute and chronic metabolic alkalosis consistently leads to a fall in urinary Mg excretion. Metabolic acidosis and alkalosis act within both the loop of Henle and the distal tubule. Alkalosis changes the permeability of the paracellular pathway so that Mg moves passively through the pathway to a greater degree resulting in greater Mg absorption. Metabolic acidosis of any aetiology would be expected to lead to diminished Mg reabsorption in the loop and distal tubule. Mg wasting with diabetes is due to insulin deficiency but more importantly as a result of keto acidosis often associated with the disease. Hypokalaemia and potassium depletion is associated with diminished Mg absorption within the loop and distal tubule that may lead to renal Mg wasting. One of the hallmarks of hypophosphataemia and cellular phosphate-depletion is the striking increase in urinary excretion of Ca and Mg. The possible mechanisms for the increased renal excretion include: 1- mobilisation of Ca and Mg from bone, 2- suppression of PTH secretion, and 3- aberrant tubular transport. Mg-wasting commonly observed with hypophosphataemia and phosphate depletion could be due, in part, to diminished Mg^{2+} uptake in the distal convoluted tubule. Acid-base changes (H^+ ions) have different effects on Mg transport relative to potassium or phosphate depletion so that the 3 disturbances may act in an additive manner to compromise renal magnesium conservation. The regulation of intracellular Mg availability parallels the molecular control of cell proliferation, and may be also cell differentiation and death. Magnesium deprivation may influence cell cycle control by upregulating the cyclin inhibitor $p27^{Kip1}$ thus influencing cyclin E-dependent kinases. In many neoplastic cells, Mg is higher than in normal counterparts and this high Mg is maintained also against concentration gradient. Severe Mg deprivation causes growth arrest also in tumour cells, while chronic Mg deprivation leads to an adaptation of tumour cells both to growth rate and Mg content. In tumour cells deranged Mg content and distribution is likely due to an inhibition of Mg efflux via the Na-Mg antiport. When differentiation process is induced by receptor mediated stimuli such as IFN-α and ATP, decrease of cell Mg content accompanies with activation of Mg efflux. Transformed cells may thus display high growth rate also because they retain a large amount of Mg. Magnesium stabilises DNA structure, promotes DNA replication and transcription, and influences RNA translation, presumably by acting as allosteric or catalytic modulator of critical enzymes such as topoisomerases, endonucleases or polymerases. Magnesium induces ribosome assembly, thus influencing protein synthesis at a post-translational level. Magnesium regulates the opening-closure of ion channels, which may account for pH changes preceding cell division. Magnesium couples energy metabolism to sustained protein synthesis and DNA duplication. Intracellular Mg content and distribution may vary with physiologic or neoplastic proliferation. Elevation of extracellular Mg above the physiologic concentration could stimulate the proliferation and migration of capillary endothelial cells. Elevation of extracellular Mg is accompanied by consistent increase of $[Mg^{2+}]_i$ within 2-10 minutes. Tumour cells are more resistant to Mg deprivation than are normal cells. The transition from G1 to S phases in proliferating normal cells is strictly dependent on Ca availability. Magnesium deficiency delays transit through the G1 and S phases of normal cells. Extracellular Mg deprivation may affect cell cycle. Magnesium can regulate cell cycle by modulating cyclin expression. An increase of $[Mg^{2+}]_i$ can trigger endonuclease activation in cells, but Mg-dependent endonucleases are not essential for the apoptotic process. Topoisomerase II inhibitors, like etoposide, or intercalating agents, like cytosine-arabinoside (ARA-C), can be used to increase the population of cells committed to apoptosis. Magnesium restores apoptosis by etoposide but not by ARA-C, and neither drug is targeted to endonucleases. Mg can influence apoptosis upstream DNA fragmentation elicited by endonuclease. Mitochondria actively participate to the apoptotic process by various mechanisms such as release of caspase activators, disruption of electron transport and energy metabolism and production of reactive oxygen species and cellular redox potential. Opening of mitochondrial transition pore that control the inner transmembrane potential (channel) induces swelling and consequent rupture of outer membrane leading to the release of various substances that induce apoptosis, i.e., caspase activators. Inhibition of cell proliferation is often considered a prerequisite to differentiative events. In eukaryotic cells concentration of intracellular Mg is remarkably high, yet it is still way below the electrochemical equilibrium. PGE can activate adenyl-cyclase, thus increasing cAMP, which eventually stimulates the Na-Mg antiport.

REGULATION OF INTRACELLULAR FREE MAGNESIUM IN CNS INJURY

Traumatic injury to the central nervous system (CNS) initiates an autodestructive cascade of biochemical and pathophysiological changes that ultimately results in irreversible tissue damage (secondary injury). Magnesium may play a pivotal role in the secondary injury process following CNS trauma, affecting a number of secondary injury factors including neurotransmitter release and activity, ion changes, oxidative stress, protein synthesis, changes in blood flow, oedema and energy metabolism. Secondary injury events, occur between minutes and days after the primary insult, and need not be activated exclusively by direct primary injury to the brain. Secondary injury events may also be initiated in the CNS following peripheral injury, by the action of, amongst others, autacoids that are released into the systemic circulation, by afferent hyperexcitability, or by enhanced neurotransmitter release. Targeting several factors using interventional cocktail or using drugs that affect a number of secondary factors is considered the most likely method of improving outcome. Magnesium cation multifactorial intervention has affects on a number of secondary injury factors. Magnesium decline after trauma is associated with neuronal cell death and functional impairment. Intracellular brain free magnesium declines by up to 60% following moderate traumatic brain injury. Decline in brain intracellular free magnesium concentration is associated with the development of functional deficits in neurologic motor outcome. The decline in free Mg after trauma is reflected in a much smaller decline in tissue total Mg concentration, which is limited to the injury zone and does not extend to non-injured tissue. Magnesium declines after traumatic brain injury results in a decreased ability of the cell to provide sufficient energy to restore disrupted ion gradients and to repair itself. Magnesium declines following indirect neurotrauma correlates with decline in activity of the Na^+/K^+ ATPase and the associated oedema development. Declines in magnesium concentration occur also in spinal cord injury, drug intoxication, and even migraine. The declines in brain free Mg persist for at least 4 days following traumatic brain injury and with increasing severity of injury, out to 7 days. Acute alcohol consumption causes a significant but transient decline in brain free Mg concentration, which does not result in a permanent neurologic deficit. In contrast, chronic alcohol consumption, which is known to cause neurologic deficit, causes a persistent decline in brain free Mg concentration. Measure of free Mg homeostasis may be a convenient prognostic indicator of outcome following severe trauma. Therapies targeting the restoration of Mg homeostasis have demonstrated beneficial effects on neurologic outcome following traumatic brain injury. The traumatic event permits the entry of blood Mg into the brain. Significant improvements in cognitive performance and the motor outcome result with magnesium sulfate administered intravenously at 30 minutes posttrauma. Both magnesium sulfate and magnesium chloride salts improve brain intracellular free magnesium concentration following traumatic brain injury with a resultant improvement in neurologic motor outcome. A single bolus of Mg (within the first 24 hours after trauma) is as effective as repeated administration in improving motor outcome following diffuse brain trauma. Thyrotropin releasing hormone analogues, n-methyl-D-aspartate antagonists, opiate antagonists, adenosine agonists, apovincaminic acid derivatives and phospholipase C inhibitors, all increase brain free Mg concentration after trauma with an associated improvement in neurologic outcome. Magnesium is an important cation necessary for the functioning of over 300 key enzymes involved in energy transformation, protein synthesis, and lipid and nucleic acid metabolism. Any reaction that either produces or consumes ATP has a mandatory requirement for Mg, e.g., enzymes of glycolysis and oxidative phosphorylation. Magnesium is also essential for the stability and normal functioning of the cell membranes. Magnesium depletion increases membrane turnover and fluidity and increases lipid peroxidation. Increased lipid peroxidation is indicative of oxidative stress. Hydrolysis of membrane phospholipids and activation of the free fatty acid cascade is strongly related to the generation of ROS. These reactive oxygen species initiate further cell damage, in part, through peroxidation of membrane components. Peroxidative reactions themselves initiate complex chain-reactions that generate even more free radicals. These free radicals, along with other damaging neurochemical events, play an important role in the pathophysiology of CNS injury. The increase in ROS in the brain soon after CNS injury correlates with formation of cerebral oedema, secondary ischaemia, and impairment of microvascular regulation. The vulnerability of neurons to iron-dependent oxidative injury is an inverse function of the extracellular Mg concentration. Magnesium at high concentration directly inhibits lipid peroxidation, most probably by competing with iron for phospholipid binding sites. At low Mg concentrations, increased cell death may be result of to the combined effect of increased n-methyl-D-aspartate receptor activity, impairment of antioxidant defense, and direct potentiation of the oxidative stress. Combination of Mg and specific free radical scavenging therapies may be of some utility as a treatment intervention following traumatic brain injury. Magnesium administration attenuates oxidative stress following indirect neurotrauma by reducing superoxide anion generation. The membrane-stabilising properties of Mg affect lipid peroxidation and generation of reactive oxygen species, and impact upon the release of neurotransmitters and other mediators/modulators. Both glutamate release and acetylcholine release (after trauma) is reduced by the administration of Mg. While Ca^{2+} increases acetylcholine release during neuronal depolarisation, Mg^{2+} decreases acetylcholine release by stabilising the membranes of presynaptic vesicles. Significant elevations in neurotransmitters, and particularly the excitatory amino acids including glutamate, are among the most important autodestructive responses occurring during the early posttraumatic period after trauma. Glutamate activation of n-methyl-D-aspartate (NMDA) and α-amino-3-hydroxy-5-methylisoxazole-4-propionate (AMPA)/kainate receptors leads to increased intracellular calcium and sodium concentrations, which activates a complex cascade of interactive biochemical alterations and subsequent cell death. NMDA channel is implicated as a critical factor in the development of cellular injury following neurotrauma. Magnesium cation is the endogenous regulator of NMDA channel activity. Magnesium block of the NMDA channel is reduced after neural injury, and that this reduction may be linked to either a decline in Mg levels, or a change in the structure of the NMDA channel. Magnesium is neuroprotective, significantly reduces the NMDA receptor binding capacity in the brain. Magnesium regulates calcium concentration within a cell and modulates posttraumatic neurochemical changes mediated by these elevated intracellular calcium levels. Elevations in intracellular calcium concentration occur through ionotropic receptors, and through second-messenger linked receptors. Neurotransmitters, hormones and even mechanical damage also increase intracellular calcium concentration through activation of phospholipase C and subsequent hydrolysis of phosphatidyl inositol 4,5-bisphosphate into inositol 1,4,5-triphosphate (IP_3). IP_3 is an intracellular messenger with regulatory effects on nerve cell excitability, neurotransmitter secretion, posttetanic potentiation and differentiation, both in physiological and pathological conditions. Magnesium (as an intracellular calcium antagonist) modifies the activity of the IP_3-messenger system mainly by reducing the affinity of the receptor for IP_3. Calcium ion influx and its intracellular redistribution are key events following brain injury. One of the consequences of calcium influx is initiation of pathways involved in breakdown of lipid membrane constituents and subsequent accumulation of free fatty acids, particularly arachidonic acid (AA). At high concentrations, AA can inhibit Na^+/K^+-ATPase activity and induce cerebral oedema, as well as increasing release of neurotransmitters like glutamate and its reuptake. Depletion of intracellular Mg^{2+} reduces incorporation of exogenous AA into tissue phospholipids, perhaps by reducing synthesis of arachidonyl CoA and thus modifying the first phase in the incorporation of exogenous fatty acids into membrane phospholipids. Mg^{2+} binding to its specific binding site on protein kinase C causes reduced activity of this enzyme, intracellular Mg^{2+} depletion may be mediated through protein kinase C activation. Increased PKC-mediated phosphorylation of enzymes that are involved in AA incorporation (CoA synthetase, lysophosphatidylcholine acyl transferase) reduces their activities and subsequently increases the concentration of the free AA and its metabolites. Phospholipase C-activated second messenger pathways may affect Mg homeostasis. PAF synthesised through the actions of phospholipase A_2 significantly contributes to posttraumatic ischaemia, inflammatory changes and the impairment of the cerebral blood flow in much the same way as the products of the AA metabolic cascade. Phospholipase-induced products may chelate Mg. Magnesium sulfate administration increases cerebral blood flow velocity in the normal intact cerebral vasculature, accompanied by increased arterial carbon dioxide tension. Magnesium deficiency sensitises the vascular epithelium to prostanoids and induces rapid calcium-mediated vasospastic responses in cerebral blood vessels. Magnesium decreases vascular resistance and causes vascular dilation. Magnesium reverses pressor effects and cerebral vasoconstriction previously induced by noradrenaline. Magnesium administration dramatically attenuates myocardial dysfunction associated with brain damage and caused by excessive release of catecholamines and calcium ion overload. Mg^{2+} has direct effect on the actions of noradrenaline in the cerebral and peripheral vasculature leading to subsequent vasodilation and decrease in vascular resistance. Magnesium's effect on the cerebral vascular bed may contribute to the neuroprotection observed following stroke. Significant declines in free Mg concentration occur after penetrating and non-penetrating head injury (both direct and indirect) and that such a decline may be an important factor determining posttraumatic course and outcome. Following indirect neurotrauma, these changes are correlated to alterations in oxidative status/antioxidant defense. Dose-dependency of plasma ionised Mg decline and head injury severity implicates that measurement of plasma free Mg should be included in evaluation of patients with traumatic brain injury, planning of treatment and determining prognosis. Loss of cells during necrosis or apoptosis might explain the total tissue losses of Mg. Phospholipase C activity results in decline in free Mg levels in ATP loaded membrane vesicles. Increased adenylate cyclase activity results in loss of total Mg content from mitochondria. Activation of adenylate cyclase results in extrusion of mitochondrial Mg with subsequent loss from the cell down its concentration gradient. Magnesium is an important element for health and disease. Magnesium, the second most abundant intracellular cation, has been identified as a cofactor in over 300 enzymatic reactions involving energy metabolism and protein and nucleic acid synthesis. Approximately half of the total Mg in the body is present in soft tissue, and the other half in bone. Less than 1% of the total body Mg is present in blood. Magnesium is absorbed uniformly from the small intestine and the serum concentration controlled by excretion from the kidney. Magnesium deficiency may cause weakness, tremors, seizures, cardiac arrhythmias, hypokalaemia, and hypocalcaemia. The causes of hypomagnesaemia are reduced intake (poor nutrition or IV fluids without magnesium), reduced absorption (chronic diarrhoea, malabsorption, or bypass/resection of bowel), redistribution (exchange transfusion or acute pancreatitis), neuromuscular hyperexcitability, and increased excretion (medication, alcoholism, diabetes mellitus, renal tubular disorders, hypercalcaemia, hyperthyroidism, aldosteronism, stress, or excessive lactation). Magnesium is a prominent intracellular cation required for the function of hundreds of enzyme systems. Magnesium depletion is observed frequently in hospitalised patients and is usually secondary to renal or intestinal Mg loss. Magnesium therapy appears to improve survival in patients with myocardial infarction. The diagnosis of Mg deficiency is usually made by a low-serum Mg concentration, although the Mg tolerance test may be more indicative of low Mg states. In acutely ill patients, Mg is usually give parenterally; oral Mg may be given for long-term repletion. Approximately 15% of bone Mg (equivalent to 1.5 mmol/KgBW) can be lost, whereas less than one tenth of that amount is available from skeletal muscle. In the human, up to 35% of bone Mg can be lost. In the human, the bone and skeletal-muscle Mg pools can provide an average of 1.7 mmol/KgBW equivalent to 15% of total body Mg. Release of Mg from these stores appears to depend on the presence of hypomagnesaemia, which may also result in small, significant and potentially adverse Mg losses from certain vital organs such as the heart, kidney and brain. The liver and other organs appear not to lose Mg; despite Mg deprivation, although intracellular Mg shifts of importance cannot be ruled out. The body reserve to combat Mg depletion is not designed to protect the extracellular Mg pool and certain critical organs from Mg deficiency. A continuous optimal intake of Mg is needed for good nutrition and health. Supplementation with magnesium aspartate hydrochloride may effectively prevent the diuretic induced disturbances of electrolyte balance. Magnesium aspartate hydrochloride results in a significant rise of the cellular potassium and magnesium content and in a significant decrease of both systolic and diastolic blood pressure. Oral Mg supplementation restores the concentrations of Mg, K and sodium-potassium pumps in skeletal muscle of patients receiving diuretic treatment. A supplemental period of at least 6 months seems to be required before complete normalisation can be expected. Oral Mg supplementation may restore diuretic-induced disturbances in the concentrations of Mg, K and Na-K pumps in skeletal muscle. Patients who have undergone ileal resection are at risk for developing Mg depletion/deficiency because of poor absorption and decreased intake as well as increased endogenous losses. Magnesium repletion is difficult to accomplish because of the cathartic action of most oral Mg supplements at therapeutic doses. Magnesium diglycinate (chelate) may be a good alternative to commonly used Mg supplements in patients with intestinal resection, which is absorbed in part as an intact dipeptide in the proximal small intestine. Binge drinking of ethanol could result in cerebrovasospasm, ischaemia, and rupture of cerebral blood vessels as a consequence of depletion of cerebral VSMC $[Mg^{2+}]_i$. Deficits in $[Mg^{2+}]_i$, O_2, and nutrient delivery could account in part for some of the behavioural actions of alcohol. Ethanol promotes rapid depletion of intracellular free Mg in cerebral vascular smooth muscle cells. Magnesium deficiency is common among chronic alcoholics. Oral Mg supplementation improves metabolic variables and muscle strength in alcoholics. Magnesium induces significant reduction of aspartate-aminotransferase (ASAT), alanine-aminotransferase (ALAT), γ-glutamyl-transpeptidase (GGT), and bilirubin in chronic alcoholic patients. Serum Na, Ca, and P increased significantly, and serum K and Mg increased slightly during magnesium therapy. Muscle strength increased significantly during Mg treatment in chronic alcoholic patients. Even short-term oral Mg therapy may improve liver cell function, electrolyte status, and muscle strength in chronic alcoholics. The cellular ionic consequences of hyperglycaemia may contribute to the increased risk of hypertension and vascular diseases present among subjects with NIDDM, impaired glucose tolerance, or both. Glucose alters intracellular ions, increases cytosolic free Ca, while suppressing intracellular free Mg and pH levels. Insulin secretion requires Mg, Mg deficiency results in impaired insulin secretion, and while Mg replacement restores insulin secretion. Magnesium deficiency reduces the tissues sensitivity to insulin. Subclinical Mg deficiency is common in diabetes. It results from both insufficient Mg intakes and increase Mg losses, particularly in the urine. In type 2, or non-insulin-dependent, diabetes mellitus, Mg deficiency seems to be associated with insulin resistance. Magnesium deficiency may participate in the pathogenesis of diabetes complications and may contribute to the increased risk of sudden death associated with diabetes. Magnesium deficiency may play a role in spontaneous abortion of diabetic women, in foetal malformations and in the pathogenesis of neonatal hypocalcaemia of the infants of diabetic mothers. Magnesium supplementation to patients with type 2 diabetes tends to reduce insulin resistance. Magnesium, the second most abundant intracellular cation, has several critically important roles in the body. In addition to energy production and maintaining electrolyte balance, Mg is essential for normal neuromuscular function as well as Ca and K transport. Magnesium deficiency is closely interrelated to K deficiency and refractory K repletion. In spite of K supplements long term treatment with diuretics may lead to K and Mg deficiencies, which are not detectable using the standard methods of serum analysis. The changes in concentrations of electrolytes and Na-K pumps associated with treatment with diuretics may impair muscle function and K homeostasis and interfere with the distribution of digitalis glycosides. Potassium depletion induced by diuretics leads to a selective decrease in the concentrations of Mg and in the concentration of Na-K pumps in skeletal muscle. Low extracellular Mg^{2+} can induce myocardial injury and subsequent cardiac failure. Ferrylmyoglobin formation is formed considerably before intracellular release of either creatine phosphokinase or lactic dehydrogenase. Cardiac tissue damage, induced by Mg deficiency, is probably involved in the generation of a ferrylmyoglobin radical, which could be prevented by ascorbate. Low extracellular $[Mg^{2+}]_o$ leads to >80% of the oxymyoglobin converted to its deoxygenated form. The level of reduced cytochrome oxidase aa3 also increased about 80% in low $[Mg^{2+}]_o$. The deoxymyoglobin is converted further to a species identified as ferrylmyoglobin. Magnesium may be useful in the prevention of migraine attacks. Dietary Mg may contribute to the reduction of total serum cholesterol, LDL-cholesterol, and triglyceride as well as to the marginal rise in HDL-cholesterol. Citrate directly and isocitrate by prior conversion into citrate exert a protective action by chelating and retaining Mg^{2+} within the mitochondria. If magnesium citrate is to be used in the management of recurrent calcium oxalate nephrolithiasis, they should be administered with meals. Magnesium salts (10 mEq QID or 486 mg Mg/day for 2 weeks) provided with meals lead to more prominent increases in urinary Mg (by 92-96 mg/day) and in citrate (by 218-226 mg/day). Hypomagnesaemia is common and causes weakness in humans. Hypomagnesaemia should be excluded as a cause of muscle weakness because its replacement may improve muscle power. Free radical-mediated injury could contribute to skeletal muscle lesions resulting from Mg deficiency. Hypomagnesaemia is accompanied by significantly lower Mg and greater Ca concentrations in skeletal muscle tissue. Electron microscopy of skeletal muscle tissue revealed ultrastructural changes, including swelling mitochondria and disorganisation of the sarcoplasmic reticulum network. Magnesium deficiency reduces the threshold antioxidant capacity. Erythrocytes are more susceptible to free radical injury during Mg deficiency. Antioxidant drugs and nutrients protect against Mg deficiency-induced myocardial injury.

MAGNESIUM IN ONCOGENESIS AND IN ANTI-CANCER TREATMENT

Over 300 enzymes that influence the metabolism of carbohydrate, amino acids, nucleic acids and protein, and ion transport, require magnesium (Mg). Mg plays roles in fatty acid and phospholipid acid metabolism, that affect permeability and stability of membranes. Mg is central in the cell cycle, and its deficiency is an important conditioner in precancerous cell transformation. Mg is central in the cell cycle; its deficiency may lead to precancerous cell transformation. Environmental factors contribute to most human cancers. Greater morbidity and mortality from cardiovascular disease is directly correlated with water softness and diet. Negative Mg balance contributes (<70% RDA) to the high mortality rate from ischaemic heart disease (IHD) in the USA. Mg, rather than Ca, protects against IHD, myocardial infarcts and sudden unexpected cardiac death caused by arrhythmias. Cancer is second to heart disease as a cause of death in the aged, and thus is more common in regions where more people reach old age. Depressed B-cell and T-cell immunologic function occur with aging. The longer the exposure to environmental agents with oncogenic potential, the greater the risk of developing cancer. The increased longevity of those living in hard water areas might obscure protection by geochemical factors against cell transformation. Oncogenic trace metals as cadmium (Cd), lead (Pb), and nickel (Ni) are found more in soft than in hard water. Stomach cancer is four times more common in the Ukraine where the Mg content of soil and drinking water is low, than it is in Armenia where the Mg content is more than twice as high. In Poland the neoplastic-related deaths are nearly three-fold higher in the one in a low soil Mg area (27%) than in the one with high soil Mg. Also considered aflatoxins from environmental fungi, radiation, use of pesticides, and application to the soil of fertilisers that are rich in phosphates, potassium (K) and nitrogen (that lower Mg and selenium [Se] in the earth), in evaluating regional differences in leukaemia. Inadequacy of Mg and antioxidants are important risk factors in predisposing to leukaemias. Chronic Mg deficiency can cause lymphosarcomas and leukaemia. Also cattle leukaemia is rare in the Orient. Subacute Mg deficiency may cause lymphopoietic neoplasms. Persistent neutrophilic leukocytosis leads to chronic myeloid leukaemia as a result of Mg deficient. Normal Mg intakes may reverse the leukocytosis, except for those progressing to irreversible myeloid leukaemia. Magnesium is essential for cartilage cell maturation. Mg suppresses oncogene-induced large bowel carcinogenesis. Mg inhibits the increased DNA synthesis of the colon epithelium, and excess proliferation. Lead (Pb) is more leukaemogenic in the presence of Mg deficient. Mg inhibits nickel (Ni) carcinogenesis, probably in part from Mg-antagonism of suppression by Ni of the T cell killer activity. Mg may also protect against Ni oncogenicity by inhibiting Ni-induced breaks in DNA strands, which gives rise to abnormal chromosomes and cell transformation. Mg is a non-competitive antagonist of Ni, exerting its effect in the nucleus, but a competitive antagonist of Pb and Cd (Cadmium). Mg, which has a central regulatory role in the cell cycle, which affects transphosphorylation and DNA synthesis, has been proposed as the controller of cell growth, rather than Ca. This has been related to mitosis and division. Mg^{2+} controls the timing of spindle and chromosome cycles by changes in intracellular concentration during the cell cycle. Intracellular Mg falls as cells enlarge, until it reaches a level that allows for spindle formation. Magnesium influx then causes spindle breakdown and cell division. Low intracellular Ca levels can be regulators only when there is adequate free Mg. The metabolic effects of Ca are produced indirectly through its competition with Mg for membrane sites. The processes that are regulated by Ca/calmodulin are Mg-dependent. Activators of lymphoblastic mitosis, mediated by stimulation of Mg-dependent adenylate cyclase, increase cAMP levels-causing Mg influx, which initiates mitosis. Changes in intracellular Mg-binding are associated with different degrees of transformation. Membrane inositol lipids, influenced by Mg at several steps, and also have been implicated in the action of oncogenes. The intracellular Mg, which is present in a small amount as the free cation, or bound to ligands (i.e., the many enzymes it activates), is compartmented to a different degree in different types of cells; it is not influenced by extracellular Mg concentration. In gradually Mg depleted cells, Mg^{2+}-dependent metabolic functions are inhibited in the following order: glycolysis<RNA and DNA synthesis<respiration<protein synthesis, protein synthesis appears to be the most sensitive function affected. Altered membrane phospholipids influence the viscosity of membranes, which is decreased both in Mg-deficient and in cancer cells. Membrane abnormalities of erythrocytes are associated with Mg deficiency. Magnesium low cell membranes are characterised by increased fluidity and permeability. Changed lipid membrane components can reflect changes in lipid metabolism caused by Mg deficiency. Lowered cholesterol and sphingomyelin, and decreased ratios of sphingomyelin/phosphatidyl-choline and cholesterol/phospholipid result from Mg deficiency. Membrane changes of Mg deficiency are caused by altered binding to negatively charged phosphate groups of phospholipids of the membranes. Mg deficiency-induced lymphoma cells have such membrane lipid changes: higher phosphatidyl-inositol and phosphatidyl-choline and decreased cholesterol, which cause decreased membrane viscosity. There is also abnormal enzyme activity affecting phospholipids. Modifications of cell membranes are principal triggering factors in cell transformation leading to cancer. There is much less Mg^{2+} binding to membrane phospholipids of cancer cells, than to normal cell membranes, which occurs in precancerous changes; in the preneoplastic phase, binding of Mg to intracellular membranes is decreased at the same time that cytosolic Mg increases. There is drastic change in ionic flux from the outer and inner cell membranes (higher Ca and Na; lower Mg and K levels), both in the impaired membranes of cancer, and of Mg deficiency. Mg deficiency may trigger carcinogenesis by altering fidelity of DNA replication, and increasing membrane permeability. A mechanism proposed is competition of Mg^{2+} with oncogenes for DNA binding sites, and its prevention of incorporation of incorrect nucleotides during DNA synthesis. Abnormal membrane properties in Mg deficient lymphocytes may give rise to defective differentiation that can lead to lymphoma. The membranes have a smoother surface than normal and decreased membrane viscosity. Malignant tumours have higher Mg levels than do normal tissues, possibly caused by the "capture" of Mg by the tumour, as a result of the high Mg requirement of growing cells. This result in lowering of Mg levels in healthy tissues. Undernutrition causes both humoral and cell mediated immuno-incompetence, as does aging. Mg deficiency depresses cell-mediated immunity. It impairs phagocytic activity, as well as lymphocytic function. Immunologic defects of lymphoma cells include reduction of concavalin A stimulation of phospholipid metabolism, and of thymidine uptake and IgG synthesis, and failure to synthesise interleukins, which are related to proliferation and function of lymphocytes. Mg-deficient cytoxic T-cells, which lyse aberrant cells have defective response to antigens. The metabolism of Mg is interlinked with that of pyridoxine, and with that of zinc-which is also interlinked with vitamins B_6 and E. Deficiencies of each can contribute to cancer. Both Mg and Zn are needed for many enzyme systems; each is needed for nucleic acid and protein synthesis. Vitamin E is antioxidant; free-radical scavenger, with Mg, stabilises cell membranes, and deficiency of E lowers tissue Mg levels. Vitamin B_6 deficiency is associated with decreased cellular Mg and Zn levels, decreased oxidative phosphorylation (Mg dependent), and abnormal tryptophan metabolism (Mg dependent enzymes). Pyridoxine deficiency causes loss of tissue Mg, and its supplementation increases Mg tissue uptake. Pyridoxine deficiency also adversely affects Zn metabolism by decreasing its absorption, and lowering its tissue levels. All participate in cell-mediated immunity. Pyridoxine deficiency, like that of Mg, causes thymic depletion, as well as decreased cell-mediated and humoral immune response. The functions of T- and B-cells are more affected by pyridoxine deficiency than killer cell and macrophage. Low plasma pyridoxal phosphate levels and abnormal tryptophan metabolites: 3-OH-anthranilic acid (OHA) and 3-OH-kynurenine (indicative of disturbed B_6 metabolism), have been detected in patients with breast, bladder, intestines and lung cancers, which are evidence of vitamin B_6 deficiency in patients with malignancies. Bladder tumours recur in patients with abnormal tryptophan metabolism. Families prone to bladder cancer have high levels of an oncogenic tryptophan metabolite, which is associated with abnormal kynureninase (a B_6-dependent enzyme), although pyridoxine deficiency may not be manifested. Kynureninase is also Mg-dependent; a genetic Mg abnormality may be a factor. Zinc deficiency impairs T-cell mediated immunity, predominantly, although B-cells are

also affected. Patients with Zn deficiency caused by alcohol consumption, intestinal inflammatory disease or other causes of malabsorption, also are deficient in Mg, and are prone to develop cancer. Zinc deficiency increases oncogenicity of some chemical agents and decreases that of others. Membrane damage caused by agents that increase free radicals and peroxidise lipids (lipid peroxidation) has been implicated in the decreased immunocompetence of the aged. Low serum levels of vitamin E have been associated with high incidence of lung cancer. Membrane lipid abnormalities in Mg deficient and in neoplastic cells involve peroxidation of unsaturated fatty acids. Vitamin E deficiency intensifies Mg deficiency. Mg depletion from intestinal Mg loss and malabsorption contributes to total body irradiation-associated mortality. Irradiation of the intestines, while treating abdominal or pelvic neoplasms, causes hypomagnesaemia. Radiation-induced proctosigmoiditis may be corrected by $MgSO_4$ infusions. Administration of Mg salts daily has some protective effect against carcinogenesis. Toxicity of cyclophosphamide and other chemotherapeutics may significantly decrease Mg. Hypomagnesaemia occurs in cancer patients, as a result of cisplatin-induced renal tubular defect in Mg reabsorption (distal convoluted segment) persists long after the drug treatment has been stopped. Similar renal tubular wasting occurs with use of vinblastine, bleomycin and cyclosporin. Besides damage to tubular function, cisplatin may also interfere with Mg cellular metabolism. Drug-induced Mg loss can cause arrhythmias in cancer patients, even during the first cycle of treatment (e.g., cisplatin and fluorouracil). The anticancer activity of cisplatin has been linked to its effects on mitochondrial and nucleic acid Mg. Amphotericin B binds Mg to cell membranes, inactivating it, and thereby intensifying Mg deficiency. Mg supplementation, accompanying cisplatin treatment, has not affected tumour growth rates in patients. Intravenous Mg supplementation should be given during the antineoplastic courses, with oral Mg supplementation provided between courses. Rapidly metabolising cancer cells have high Mg requirements. Iatrogenic Mg depletion is effective in treating inoperable cancers, but associated with serious consequences includes acute tumour necrosis resulting in haemorrhage, arrhythmias and stroke. Recurrent tumour with Mg repletion necessitates surgical removal of the remaining tumour and antineoplastic chemotherapy or radiation. Competitive inhibition of Mg within the cancer is preferable, if feasible, deplete cancer cell Mg might be fruitful. Adequate Mg might protect against initiation of precancerous cellular changes. Mg correlation in cancer prevalence is inconsistent. Mg deficiency might be implicated in aspects of pathogenesis and treatment of neoplasms.

WATER

Water is H_2O, hydrogen two parts, oxygen one, but there is also a third thing, that makes it water and nobody knows what it is. Biological molecules are based upon carbon skeleton and polar and non-polar side chain groups. Water is the solvent in which biological molecules are dissolved. The polar character of water permits other polar molecules to dissolve in it and excludes nonpolar (oily) molecules. The polar and nonpolar interactions are major determinants of the structure of biologically important molecules and of cells. Water is a polar molecule. The hydrogen atoms take on a slight positive charge and the oxygen atom takes on a slight negative charge due to unequal sharing of electrons. Hydrogen bonds are the result of the weak bonding between the +ve charge on one water molecule and the -ve charge on a part of an adjacent water molecule. Water forms Hydrogen bonds between adjacent molecules, giving some stability to liquid water. Solid water (ice) is very stable and has the unusual property of being less dense than liquid. Water also forms transient Hydrogen bonds to polar, hydrophilic, groups on proteins, lipids, and carbohydrates and to ions in solution. Hydrogen bonds break and reform very rapidly, so molecules and cellular components can move. Water clustering around ions (e.g., Na and Cl) shields the ion to ion bonds, weakens the ion to ion interaction to about the strength of a hydrogen bond. The ionic bonds in water are weak, and like hydrogen bonds, can easily be broken. As a result, ions can move quite easily through water. Hydrogen bonds can form between water molecules and from water molecules to polar (hydrophilic) groups on biologically important molecules, and between polar groups on biomolecules. The interactions between polar groups on biomolecules such as proteins and nucleic acids and between these polar groups and water play a very important role in determining and stabilising the three-dimensional structure of biomolecules and subcellular structures. Polar Groups found on carbohydrates, lipid, proteins, and nucleic acids include groups like -OH, $-NH_2$, $-PO_4$ and -COOH. Non polar groups are hydrocarbon groups, with no electronegative atoms, e.g., $-CH_3$. Non-polar groups form hydrophobic bonds between each other. Hydrophobic bonds aren't actually bonds-they are formed because water molecules hydrogen bond to each other and not to the nonpolar groups. Thus the nonpolar groups are excluded from the water and forced together. Hydrophobic substances are those that don't dissolve in water because the hydrocarbon groups have no electronegative atoms to create polarity of electrical charge. Ammonia (NH_3) is similar to water but boils at -33.5°C, freezes at -77.7°C, and the solid is denser than the liquid. The carbon skeleton is very stable because of the strong covalent bonds formed by the equal sharing of electrons between carbon atoms. 100 kcal/mole is required to form a C-C bond and about the same amount of energy is released when the bond is broken, e.g., metabolism. 10 or fewer kcal/mole net energy is released when a covalent bond is broken in an organic molecule. Donation of an electron from Na to Cl leaves Na with a +ve charge and Cl with a -ve charge. Water is a diamagnetic substance. Structured water helps to improve absorption of minerals and vitamins, which improves the removal of unhealthy toxins. Structured water also provides antioxidant protection. The magnetised water system is able to greatly reduce acids while increasing oxygenation and therefore, improve digestion and give much better tasting water. Water that comes from wells, lakes, or running streams is naturally charged as it flows through the earth's magnetic field. During the water treatment and transportation phase, this water passes underneath the ground through metallic pipes where the charge dissipates. When water is treated with magnetic fields, the natural energy and balance are restored. The ionising magnetic field increases oxygenation between molecules, slowing the bacteria while improving the quality and taste of the water. As water passes through the magnetic field, the hydrogen ion and any minerals dissolved in the water become charged, which creates a temporary separation of the minerals from the molecular water clusters. The result is improved taste. Chemicals and water are able to change weight by being influenced by magnetic fields. There are more hydroxyl ions created in order to form bicarbonate, calcium, and other alkaline-type molecules. These molecules reduce the acid from the traditional pH level of 7 or greater. Magnetism also has an affect on the bonding angle between the oxygen atom in a water molecule and hydrogen. This causes a hydrogen-oxygen bond angle within the water molecule that is reduced from 104° to 103°. In turn, the water molecule begins to cluster in groups of six or seven, rather than the normal level of 10 to 12. The smaller the cluster, the better the water absorption across the cell walls. Free radicals such as hydroxyl radicals can be introduced into water by techniques such as ultrasonic or ultraviolet radiation, but their lifetime is likely to be very short and health benefits unproven. There are proven health benefits from simply drinking more water and from changing fluid intakes from coffee, tea, alcohol, and hypertonic soft drinks to mineral or tap water. First thing in the morning cup of coffee may better be replaced by a glass of water in order to reduce the higher risk of heart attacks at this time of day. High magnetic field has an insignificant effect on the equilibrium content of dissolved oxygen, but does significantly enhance its dissolution rate. Drinking of oxygenated water does give a transient moderate increases in serum ascorbyl radicals (with unknown consequences), an affect that disappears with regular consumption. Electrolysis changes the isotope ratio, which may have an effect. Magnetically treated water retains a significantly changed effect on fungal spore germination for at least 24 hrs; however other parameters (e.g. reduced dissolved oxygen levels) may be responsible for such effects. Mechanically-induced hydrogen bond breakage, caused by shaking (succussion) when producing homeopathic solutions, has been reported to last for weeks. There is incompatibility of certain polymers in aqueous solution. Anions and cations distribute themselves differently between the phases depending on their affinity for low or higher density water but with the requirements that the phases be electrically neutral and iso-osmotic, so producing an interfacial potential difference, which may aid the partitioning of charged biomolecules. All drinking water contains some natural contaminants. About a third of bottled water contains contaminants too. 90% of all municipalities add chlorine to kill bacteria. Chlorine is linked with the probability of cancer. In ice (1h), all water molecules participate in 4 hydrogen bonds (2 as donor and 2 as acceptor) and are held relatively static. In liquid water, some hydrogen bonds must be broken to allow the molecules move around. The large energy required for breaking these bonds must be supplied during the melting process and only a relatively minor amount of energy is reclaimed from the change in volume. The free energy change (DG=DH-TDS) must be zero at the melting point. As temperature is increased, the amount of hydrogen bonding in liquid water decreases and its entropy increases. Melting will only occur when there is sufficient entropy change to provide the energy required for the bond breaking. The low entropy (high organisation) of liquid water causes this melting point to be high. There is considerable hydrogen bonding in liquid water, which prevents water molecules from being easily released from the water's surface. This reduces the vapour pressure. Boiling cannot occur until this vapour pressure equals the external pressure. This occurs at a consequentially higher temperature. The boiling point of water is over 150 K higher than expected by extrapolation of the boiling points of other Group 6A hydrides compared with Group 4A hydrides. The critical point of water is over 250 K higher than expected by extrapolation of the critical points of other Group 6A hydrides compared with Group 4A hydrides. The critical point can only be reached when the interactions between the water molecules fall below a certain threshold level. Due to the strength and extent of the hydrogen bonding, much energy is needed to cause this reduction in molecular interaction and this requires higher temperatures. Even close to the critical point, a considerable number of hydrogen bonds remain, albeit bent, elongated and no longer tetrahedrally arranged. Water molecules at the liquid-gas surface have lost potential hydrogen bonds directed at the gas phase and are pulled towards the underlying bulk liquid water by the remaining stronger hydrogen bonds. Energy is required to increase the surface area (removing a molecule from well hydrogen bonded interior bulk water to the lesser hydrogen-bonded surface), so it is minimised and held under tension. As the forces between the water molecules are several and relatively large on a per-mass basis, compared to those between most other molecules, the surface tension is large. Lowering the temperature greatly increases the hydrogen bonding causing increased surface tension. If a small drop of water (typically 1 mm diameter) is coated in a fine (typically 20 mm diameter) hydrophobic dust then the drop can roll and bounce without leakage, and the aqueous spheres can even float on water. Capillarity holds the dust at the air-liquid interface with the elasticity being due to the high surface tension. Surfactants lower the surface tension because they prefer to sit in the surface, attracting the surface water molecules in competition to the bulk water hydrogen bonding and so reducing the net forces away from the surface (i.e., the surface tension). In contrast, most cations and anions prefer to be fully hydrated in the bulk liquid water so adding to the attractive forces on the surface water molecules, consequently increasing the surface tension. The viscosity of a liquid is determined by the ease with which molecules can move relative to each other. It depends on the forces holding the molecules together. This cohesivity is large in water due to its extensive three-dimensional hydrogen bonding. Although the viscosity is high, it causes no much difficulty for water to move around within organisms. There is still considerable hydrogen bonding (~75%) in water at 100°C. As effectively all these bonds need to be broken (very few indeed remaining in the gas phase), there is a great deal of energy required to convert the water to gas, where the water molecules are effectively separated. The increased hydrogen bonding at lower temperatures causes higher heats of vaporisation. When water freezes at 0°C, at atmospheric pressure, its volume increases by about 9%. If the melting point is lowered by increased pressure, the increase in volume on freezing is even greater (e.g., 13% at -20°C). The structure of ice (1h) is open with a low packing efficiency where all the water molecules are involved in 4 straight tetrahedrally-oriented hydrogen bonds; for comparison, solid hydrogen sulfide has a face centered cubic closed packed structure with each molecule having twelve nearest neighbours. On melting, some of these ice (1h) bonds break, others bend and the structure undergoes a partial collapse. The high-pressure ices (ice III, ice V, ice VI and ice VII) all expand on melting to form liquid water (under high pressure). It is the expansion in volume when going from liquid to solid, under ambient pressure, that causes much of the tissue damage in biological organisms on freezing. In contrast, freezing under high pressure directly to the more dense ice VI may cause little structural damage. Viscous flow occurs by molecules moving through the voids that exist between them. As the pressure increases, the volume decreases and the volume of these voids reduces, so normally increasing the viscosity. Water's pressure-viscosity behaviour can be explained by the increased pressure causing deformation, so reducing the strength of the hydrogen-bonded network, which is also partially responsible for the viscosity. In water, the cluster equilibrium shifts towards the more open structure as the temperature is reduced due to it favouring the more ordered structure. As the water structure is more open at these lower temperatures, the capacity for it to be compressed increases. Water has about twice the specific heat capacity of ice or steam. As water is heated, much of the energy is used to bend the hydrogen bonds; a factor not available in the solid or gaseous phase. This extra energy causes the specific heat to be greater in liquid water. The presence of this large specific heat offers strong support for the extensive nature of the hydrogen-bonded network of liquid water. The water cluster equilibrium shifts towards less structure and higher enthalpy as the temperature is raised. Solutes have varying effects on properties such as density and viscosity. Solutes interfere with the cluster equilibrium by favouring either open or collapsed structures. Any effect will cause the physical properties of the solution, such as density or viscosity, to change. Supercooled water has 2 phases and a second critical point at about -50°C. As water is supercooled it converts mainly into its expanded form at ambient pressures. Liquid water is easily supercooled. Water, supercooled down to -37.5°C, is sustained in storm clouds and the condensed clouds formed by aircraft. As water is cooled, the cluster equilibrium shifts towards the more open structure with higher viscosity. Lowering the temperature further, which should encourage crystallisation. Methods that break the hydrogen bonding in these clusters, such as ultrasonics, cause the supercooled water to freeze. A possible explanation of the existence of low-temperature-range supercooled water (124-150 K) may be the formation of strands of icosahedral structures. This also may explain the high viscosity and strong (i.e., low specific heat) liquid behaviour of this extremely supercooled water. Solid water exists in a wide variety of stable (and metastable) crystal and amorphous structures. The ability for water to form extensive networks of hydrogen bonds increases the number of solid phases possible. For comparison, hydrogen sulfide has only 4 distinct solid phases. Hot water may freeze faster than cold water; (the Mpemba effect). The degree of supercooling is greater in initially-cold water than initially-hot water. The initially-hot water appears to freeze at a higher temperature (less supercooling) but less of the apparently frozen ice is solid and a considerable amount is trapped liquid water. Initially-cold water freezes at a lower temperature to a more completely solid ice with less included liquid water; the lower temperature causing intensive nucleation and a faster crystal growth rate. Icosahedral clusters do not readily allow the necessary arrangement of water molecules to enable hexagonal ice crystal initiation; such clustering is the cause of the facile supercooling of water. Water that is initially-cold will have the maximum (equilibrium) concentration of such icosahedral clustering. Initially-hot water has lost much of its ordered clustering and, if the cooling time is sufficiently short, this will not be fully re-attained before freezing. Such clustering processes may take some time. It is also possible that dissolved gases may encourage supercooling by increasing the degree of structuring, by hydrophobic hydration, in the previously-cold water relative to the gas-reduced previously-hot water and increasing the pressure as gas comes out of solution when the water starts to crystallise, so lowering the melting point and reducing the tendency to freeze. Also, the presence of tiny gas bubbles (produced on heating) may increase the rate of nucleation, so reducing supercooling. Proton and hydroxide ion mobilities are anomalously fast in an electric field. The thermal conductivity of water rises to a maximum at about 130°C and then falls. Under high pressure water molecules move further away from each other with increasing pressure. The electrical conductivity of water rises to a maximum at about 230°C and then falls. The electrical conductivity of water increases with temperature up to about 230°C due mainly to its increased ionisation producing higher concentrations of the highly conducting H^+ and OH^- ions, which reach maximum concentrations at about 250°C. Above this temperature, for liquid water in equilibrium with the vapour, the density is much reduced (e.g., 0.7 g cm^{-3} at 300°C) and this reduces the ability for ionisation. Pipes burst due to the rapid formation of a network of feathery dendritic ice enclosing water, which then expands on freezing within a now restricted volume to generate the required pressure. Hexagonal ice (ice 1h) is the normal form of ice and snow, as evidenced in the six-fold symmetry in ice crystals grown from water vapour (i.e. snow flakes). Hexagonal ice possesses a fairly open low-density structure, where the packing efficiency is low compared with simple cubic or face centered cubic structures (and in contrast to face centered cubic close packed solid hydrogen sulfide). The crystals may be consisting of sheets lying on top of each other. Ions may destroy the natural hydrogen bonded network of water. Ions cause negligible change to water's bulk structure; these differences due to ionic concentration and the sensitivity of the methods to bulk structural changes. Ions that have the greatest such effect (exhibiting weaker interactions with water than water itself) are known as structure-breakers or chaotropes, whereas ions having the opposite effect are known as structure-makers or kosmotropes (exhibiting strong interactions with water molecules). Strongly hydrated ions considerably increase the difference between the hydrogen bond donating and accepting capacity of the linked water molecules resulting in the breakdown of the tetrahedral network. Anions hydrate more strongly than cations for the same ionic radius as water hydrogen atoms can approach more closely than the water oxygen atoms (but note that most anions are larger than most cations), giving rise to greater electrostatic potential. Also, breaking such hydrogen bonds is relatively slow due to the difficulty in finding a new hydrogen-bonding partner. Small ions are strongly hydrated, with small or negative entropies of hydration, creating local order and higher local density. Large singly charged ions such as I^-, with more positive entropies of hydration, act like hydrophobic molecules, binding to surfaces dependant not only on charge but also on van der Waals forces. They may additionally be pushed on by strong water-water interactions. Such large ions possess low surface charge density. Less large ions (e.g., K^+, Br^-) cause the partial collapse of such clathrate structures through puckering. These ions allow rotations of the water molecule dipole towards the oppositely charged ions, through weak interactions, that would be prevented at truly hydrophobic surfaces and hence produce greater localised water molecule mobility (negative hydration). Larger ions, such as the tetramethylammonium cation, form clathrate structures but do not allow these rotations in the surface water surface (hydrophobic hydration). The collapse, through puckering, of the water clathrate structures surrounding the smallest ions (e.g. Na^+, F^-) is tightly formed as these ions hold strongly to the first shell of their hydrating water molecules and hence there is less localised water molecule mobility (strong or positive hydration) and higher apparent density for the solution water. There is also a less complete cluster structure, due to the hydrogen-bonding defects caused by the inward-pointing primary hydrogen-bonding to anions or disoriented lone-pair electrons and electrostatic repulsion together with weakened hydrogen bonding reducing inward-pointing secondary hydrogen bond donation near cations or acceptance near anions. Such strong hydration round both anions and cations costs the equivalent of 2 hydrogen bonds. Ions possessing high charge density bind larger water clusters more strongly. The entropies of hydration correlate with the tendency for the ion to accumulate in low-density water (LDW) such that a gain in entropy of the ion on solution is countered by a loss in entropy of the water. Ions that are weakly hydrated exhibit a smaller change in viscosity with concentration. Low viscosity is at least partially due to the lower density reducing non-bonded inter-molecular attractions. Arrangement of water around ions is not restricted to the first hydration layer (e.g. magnesium ions) except at high concentrations. The number of hydrating water molecules depends on the ion, the method of determination, and the ionic concentration. The type of ions present in solution control their overall properties; inner sphere ion pairs may be formed between two small ions of high charge density (e.g. CaF_2), where the strong ionic attraction overcomes the hydration shells, or between 2 large ions of small charge density, where there are no strong hydration shells. A small ion of high charge density plus a large counter-ion of low charge density forms a highly soluble, solvent-separated hydrated but clustered ion pair as the large ion cannot break through its counter-ion's hydration shell (e.g. CaI_2 and AgF) but prefers to sit within the disturbed hydrogen bonding at intermediate distance between the ordered but poorly hydrogen bonded strongly-held first hydration shell and the more disordered but strongly hydrogen bonded bulk phase water. Hydrophilic polymers in solution are surrounded by water with varying LDW content; e.g. proteins. Generally this LDW acts to separate such molecules but this process is also dependent on the ions present. LDW is labile and may be abolished by solutes it accumulates due to the micro-osmotic gradients that may be fleetingly established. If both ions accumulate then micro-osmosis destroys the LDW and oscillations may occur. Ions such as sulfate form a hydration shell from small rings of hydrogen bonded water. Such a shell may consist of a symmetrical dodecahedral arrangement of 16 water molecules where each sulfate oxygen is hydrogen bonded to 3 water molecules; these water molecules around a sulfate ion molecules forming small looped chains of 2 or 3 molecules from one sulfate oxygen to another. Such a cluster can form the central part of an icosahedral water cluster ($SO_4^{2-}(H_2O)_{276}$) possessing just four defects. One of the puzzles associated with this ability to interact with a water dodecahedron is the very different aqueous property of sulfate (SO_4^{2-}) and perchlorate (ClO_4^-) ions. Cluster defects are water molecules, within the icosahedral cluster structure, with only 3 rather than 4 hydrogen bonds as the fourth site cannot accept/donate a hydrogen bond from/to either (a) a water molecule already possessing 4 hydrogen bonds; or (b) a bound atom that already possesses 3 hydrogen bonds such as the oxygen atoms in SO_4^{2-}. Dilution and shaking have effect on the health benefits of water, which depends on the presence of a working hypothesis for the mode of action (also magnetic effects). Meta-analysis of 89 placebo-controlled trials failed to prove either that homeopathy was efficacious for any single clinical condition or that its positive clinical effects could entirely be due to a placebo effect, thus leaving the scientific door open both ways. Some of these studies may have failed to avoid bias and that studies using better methodology yielded the less positive effects. There are more negative view concerning the clinical effectiveness of homeopathic remedies. One of the main reasons concerning this disbelief in the efficacy of homeopathy lies in the difficulty in understanding how it might work. If there were an acceptable theory then more people would consider it more seriously. There is a similar strange occurrence to homeopathy in enzyme chemistry where an effectively non-existent material still has a major effect; enzymes prepared in buffers of known pH retain (remember) those specific pH-dependent kinetic properties even when effectively dry; these molecules seemingly having an effect in their absence somewhat against common sense at the simplistic level. Water does store and transmit information, concerning solutes, by means of its hydrogen-bonded network. Changes to this clustering network brought about by solutes may take some time to re-equilibrate. Succussion may also have an effect on the hydrogen bonded network (shear encouraging destructuring) and the gaseous solutes (with critical effect on structuring). Such mechanically-induced hydrogen bond breakage has been reported to last for weeks. Some molecules form larger clusters on dilution rather than the smaller clusters thermodynamically expected. Such biologically-active molecules may cooperatively form icosahedral expanded water networks to surround and screen them by the formation of face-linked icosahedra. Overall the balance is expected to be rather fine between water cluster stabilisation and particle cluster stabilisation, which is due to the hydrophobic effect and the tendency of biologically-active molecules to form a small surface with the water. Hydrocolloids are hydrophilic polymers, of vegetable, animal, microbial or synthetic origin, that generally contains many hydroxyl groups and may be polyelectrolytes. They are naturally present or added to control the functional properties of aqueous foodstuffs. Most important amongst these properties are viscosity (including thickening and gelling) and water binding but also significant are many others including emulsion stabilisation, prevention of ice recrystallization and organoleptic properties. Each of the hydrocolloids consists of mixtures of similar, but not identical, molecules and different sources, methods of preparation, thermal processing and foodstuff environment (e.g., salt content, pH and temperature) all affect the physical properties they exhibit. Hydrocolloids are natural products (or derivatives) with structures determined by stochastic enzymic action, not laid down exactly by the genetic code. They are made up of mixtures of molecules with different molecular weights and no one molecule is likely to be conformationally identical or even structurally identical (cellulose excepted) to any other. All hydrocolloids interact with water, reducing its diffusion and stabilising its presence. Generally neutral hydrocolloids are less soluble whereas polyelectrolytes are more soluble. Interactions between hydrocolloids and water depend on hydrogen-bonding and therefore on temperature and pressure in the same way as water cluster formation. Hydrocolloids may exhibit a wide range of conformations in solution as the links along the polymeric chains can rotate relatively freely. Large, conformationally stiff hydrocolloids present essentially static surfaces encouraging extensive structuring in the surrounding water. Water binding affects texture and processing characteristics, prevents syneresis and may have substantial economical benefit. Hydrocolloids can provide water for increasing the flexibility (plasticizing) of other food components. They can also effect ice crystal formation and growth so exerting a particular influence on the texture of frozen foods. Some hydrocolloids, such as locust bean gum and xanthan gum, may form stronger gels on freeze-thaw due to kinetically irreversible changes consequent upon forced association as water is removed (as ice) on freezing. Hydrocolloids are used to increase viscosity, which is used to stabilise foodstuffs by preventing settling, phase separation, foam collapse and crystallisation. Viscosity generally changes with concentration, temperature and shear strain rate in a complex manner dependent on the hydrocolloid(s) and other materials present. Mixtures of hydrocolloids may act synergically to increase viscosity or antagonistically to reduce it. Many hydrocolloids also gel, so controlling many textural properties. Gels are liquid-water-containing networks showing solid-like behaviour with characteristic strength, dependent on their concentration, and hardness and brittleness dependent on the structure of the hydrocolloid(s) present. Hydrocolloids display both elastic and viscous behaviour where the elasticity occurs when the entangled polymers are unable to disentangle in time to allow flow. Mixtures of hydrocolloids may act synergistically, associating to precipitate, gel or form incompatible biphasic systems; such phase confinement affecting both viscosity and elasticity. Hydrocolloids are extremely versatile and they are used for many other purposes including: 1- production of pseudoplasticity (i.e., fluidity under shear) at high temperatures to ease mixing and processing followed by thickening on cooling, 2- liquefaction on heating followed by gelling on cooling, 3- gelling on heating to hold the structure together (thermogelling), and 4- production and stabilisation of multiphase systems including films. Hydrocolloids gel when intra- or inter-molecular hydrogen-bonding (and sometimes salt formation) is favoured over hydrogen bonding (and sometimes ionic interactions) to water. Often the hydrocolloids exhibit a delicate balance between hydrophobicity and hydrophillicity. Mixtures of hydrocolloids may avoid self-aggregation at high concentration due to structural heterogeneity, which discourages crystallisation but encourages solubility. Hydrocolloids may interact with other food components such as aiding the emulsification of fats, stabilising milk protein micelles or affecting the stickiness of gluten. Different hydrocolloids prefer low-density or higher density water and other hydrocolloids show compatibility with both. As more intra-molecular hydrogen-bonds form so the hydrocolloids become more hydrophobic and this may change the local structuring of the water. Water held by hydrocolloids might act as plasticizer (allowing molecular motion) greatly reduces the glass transition temperature by breaking inter-molecular hydrogen-bonding. Hydrocolloids, together with other dietary fiber, are increasingly being seen as contributing to a healthy diet, having a number of positive health benefits. Hydrocolloids have many other major economic uses such as in the chemicals, oil and cosmetic industries. Hydrogen bonding occurs when an atom of hydrogen is attracted by rather strong forces to two atoms instead of only one, so that it may be considered to be acting as a bond between them. Typically this occurs where the partially positively charged hydrogen atom lies between partially negatively charged oxygen and nitrogen atoms, but is also found elsewhere, such as between fluorine atoms in HF_2^- and between water and the smaller halide ions F^-, Cl^- and Br^-. In water the hydrogen atom is covalently attached to the oxygen of a water molecule, but has an additional attraction to a neighbouring oxygen atom of another water molecule. Increased extent of hydrogen bonding within clusters results in higher NMR chemical shifts with greater cooperativity. The bond strength depends on its length and angle. The dependency on bond length is very important and has been shown to exponentially decay with distance. As water molecules are relatively well separated in solution there is plenty of room for the bending and stretching of these hydrogen bonds away from their preferred structures. The hydrogen bonding patterns are random in water (and ice 1h); for any water molecule chosen at random, there is equal probability (50%) that the four hydrogen bonds (i.e., the two hydrogen donors and the two hydrogen acceptors) are located at any of the four sites around the oxygen. Water molecules surrounded by four hydrogen bonds tend to clump together, forming clusters. Hydrogen bonded chains are cooperative; the breakage of the first bond is the hardest, then the next one is weakened, and so on. Thus unzipping may occur with complex macromolecules held together by hydrogen bonding, e.g. nucleic acids. Such cooperativity is a fundamental property of liquid water where hydrogen bonds are up to 250% stronger than the single hydrogen bond in the dimer. A strong base at the end of a chain may strengthen the bonding further. The cooperative nature of the hydrogen bond means that acting as an acceptor strengthens the water molecule acting as a donor. Water molecule with 2 hydrogen bonds where it acts as both donor and acceptor is somewhat stabilised relative to one where it is either the donor or acceptor of two. Total hydrogen bonding around ions may be disrupted as if the electron pair acceptance increases (e.g., in water around cations) so the electron pair donating power of these water molecules is reduced; with opposite effects in the hydration water around anions. These changes in the relative hydration ability of salt solutions are responsible for the swelling and deswelling behaviour of hydrophilic polymer gels. The hydrogen-bonded cluster size in water at 0°C has been estimated to be 400. Weakly hydrogen-bonding surface restricts the hydrogen-bonding potential of adjacent water so that these make fewer and weaker hydrogen bonds. As hydrogen bonds strengthen each other in a cooperative manner, such weak bonding also persists over several layers and may cause locally changed solvation. Conversely, strong hydrogen bonding will be evident at distance. The weakening of hydrogen bonds is observed when many bonds are broken at superheating temperatures (>100°C) so reducing the cooperativity. The breakage of these bonds is due to the more energetic conditions at high temperature and a related reduction in the hydrogen bond donating ability by about 10% for each 100°C increase. The loss of these hydrogen bonds results in a small increase in the hydrogen bond accepting ability of water, due possibly to increased accessibility. Every hydrogen bond formed increases the hydrogen bond status of two water molecules and every hydrogen bond broken reduces the hydrogen bond status of two water molecules. Bifurcated hydrogen bonds (where both hydrogen atoms from one water molecule are hydrogen bonding to the same other water molecule) have just under half the strength of a normal hydrogen bond and present a low-energy route for hydrogen-bonding rearrangements. Clusters can persist for much longer times. Broken bonds will probably reform to give same hydrogen bond (slow ortho-water/para-water equilibrium process), particularly if the other three hydrogen bonds are in place. If not, breakage usually leads to rotation around one of the remaining hydrogen bond(s) and not to translation away, as the resultant free hydroxyl group and lone pair are both quite reactive. Dissociation is a rare event, occurring only twice a day, i.e., only once for every 1016 times the hydrogen bond breaks. Hydrogen bonding carries information about solutes and surfaces over significant distances in liquid water. The effect is synergistic, directive and extensive. Reorientation of one molecule induces corresponding motions in the neighbours. Solute molecules can sense (e.g., effect each other's solubility) each other at distances of several nanometers and surfaces may have effects extending to tens of nanometers. Where water molecules are next to flat hydrophobic surfaces, and unable to form extensive clathrate structuring, some hydrogen bonds must be broken. This causes the water molecules to collapse into their shallow energy minima due to non-bonded interactions. Although there may be a consequentially increased density in the first water layer, the second and subsequent water shells compensate by forming stronger hydrogen bonds and a less dense structure. Consequences of this include differential solvation properties affecting surface absorption. Just breaking the hydrogen bond in liquid water, leaving the molecules essentially in the same position requires only about 25% of the energy required for breaking and completely separating the bond. The hydrophobic effect of hydrophobic solutes in water (such as non-polar gasses) is primarily a consequence of changes in the clustering in the surrounding water rather than water-solute interactions. Hydrophobic hydration produces an increase in the heat capacity, as the ordered bonds must be bent on increasing the temperature. Thus hydrophobic hydration behaves in an opposite manner to polar hydration, which decreases heat capacity due to their associated disorganised hydrogen bonds being already bent or broken. Water at a hydrophobic surface loses a hydrogen bond, therefore has increased enthalpy. Water molecules compensate for this by doing pressure-volume work, i.e., the network expands to form low-density water with lower entropy. Water is a more reactive environment when the extent of hydrogen bonding is reduced. An open, more hydrogen-bonded network structure slows reactions due to its increased viscosity, reduced diffusivities and the less active participation of water molecules. Water clusters (even with random arrangements) have equal hydrogen bonding in all directions. As such, magnetic, electric or electromagnetic fields that attempt to reorient the water molecules should necessitate the breakage of some hydrogen bonds; e.g., electric fields have been reported to halve the mean water cluster size. Electromagnetic radiation (e.g., microwave) exerts its effect primarily through the electrical rather than magnetic effect. At metallic electrodes, even quite low voltages can have impressive effects on the orientation of the water molecules and the positioning of ions. Ions are attracted or repelled dependent on their charge. Similar orientations may take place at the surface of minerals containing alternating positive and negative charges such that many layers of structured water may be found at the surfaces of complex silicates and a solid (static and non-exchangeable) water layer has been reported at the surface of highly polar metal oxides. High-voltage electric field raises the water activity in bread dough, so ensuring a more efficient hydration of the gluten and treatment of water with magnetic fields of about one Tesla increases the strength of mortar due to its greater hydration. Electric fields lower the dielectric constant of the water, due to the resultant partial or complete destruction of the hydrogen bonded network. Consequentially, the solubility properties of the water will change in the presence of such fields. Magnetic fields may reorient liquid water molecules, weakening and stretching their hydrogen bonds, but very high field strengths are required to reorient water in ice such that freezing is inhibited. The effects of magnets and electromagnetic radiation on the properties of water may long lifetime. There is evidence that water structuring in still deaerated pure water increases over a period of a day or two and clathrates may also persist metastably in water. Electromagnetic fields may perturb in the gas/liquid interface and produce reactive oxygen species. Changes in hydrogen bonding may effect carbon dioxide hydration resulting in pH changes. Thus the role of dissolved gas in water chemistry is likely to be more important than commonly realised. Gas accumulating at hydrophobic surfaces promotes the hydrophobic effect and low-density water formation. The accumulated gas molecules at such hydrophobic surfaces become supersaturating when this surface low-density water is disrupted by electromagnetic effects. Static magnetic effects cause an increase in the ordered structure of water formed around hydrophobes and colloids. The movement through a magnetic field, and its associated electromagnetic effect, that is important for disrupting the hydrogen bonding. Microwave frequencies can give rise to signals audible to radar operators. Unstructured water with fewer hydrogen bonds is a more reactive environment, as exemplified by the enhanced reactivity of supercritical water. If electromagnetic effects do indeed influence the degree of structuring in water, then it is clear that they may have an effect on health. Water is the main absorber of the sunlight. The 13000 billion tons of water in the atmosphere removes about 70% of the radiation, mainly in the infrared region. It contributes significantly to the greenhouse effect ensuring a habitable planet, but operates a negative feedback effect, due to cloud formation, to attenuate global warming. The water molecule may vibrate in a number of ways. The main stretching band in liquid water is shifted to a lower frequency and the bending frequency increases by hydrogen bonding. In liquid water the molecular stretch vibrations shift to higher frequency, on raising the temperature or pressure (as hydrogen bonding weakens the covalent O-H bonds strengthen, causing them to vibrate at higher frequencies) whereas the intramolecular and molecular bend vibrations shift to lower frequencies. Raising the temperature also lowers the intensity of the stretching bands. Combinations of stretching vibrations shift to higher frequency with temperature with this trend reduced when bending vibrations are also combined. Hydration is a general term concerning the amount of bound water but it is poorly defined. Even what is meant by bound is very difficult to explain (or investigate) exactly. Using a simplistic approach to polysaccharide hydration, water can be divided into bound water, subcategorised as being capable of freezing or not, and unbound water, subcategorised as being trapped or not. Unbound water freezes at the same temperature as normal water ($<0°C$ dependent on cooling rate). However some water may take up to 24 hr to freeze. Bound freezable water freezes at a lower temperature than normal water, being easily supercooled. It also exhibits a reduced enthalpy of fusion (melting). Bound water may be divided into non-freezing and freezing. Unbound water may be divided into trapped and bulk. Bound water may be divided into tightly bound (removed by freeze-drying) and loosely bound (removed by centrifugation). The firmly held water not removed by centrifugation, gives the water binding capacity (WBC) whereas the loosely associated water which is not removed by filtration gives the water holding capacity (WHC). Unbound water is removed by filtration. Non-freezing water may be trapped in a glassy state, lowering diffusion by several orders of magnitude and hindering crystal formation. The effects of water on polysaccharide and polysaccharide on water are complex and become even more complex in the presence of other materials, such as salts. Water competes for hydrogen bonding sites with intramolecular and intermolecular hydrogen bonding and may determine the carbohydrate's conformation. The polysaccharide hydration may be polar, weak hydrogen bonding, strong hydrogen bonding, hydrophobic, and the presence of other non-ionic and ionic solutes have effect as well. D-galacturonic acid residues occur in pectins and tightly bind some water molecules. As hydrogen bonding (through donation) is weakened if one of the donor hydrogen bonds of water is hydrogen bonded to a stronger base than water, carboxylates are expected to give rise to a particularly weak hydrogen bond in next shell, so encouraging a local collapse in the hydrogen bonded network. The polysaccharide chain rigidity depends on whether the counter ions are close (leading to possible intermolecular attraction but reducing the intramolecular charge repulsion) or far away. The low potential of the water associated with the ion pairs affects nearby water to encourage a levelling of the water potentials. This causes surrounding water to become more expanded (lower in density). α-L-arabinofuranose occurs in arabinoxylans and cannot form intra-residue hydrogen-bonding due to the position of its alcohol groups and the fluid nature of the furanose ring, which is constantly changing its conformation due to the low potential energy barriers between the large number of conformers with similar potential energies. However, it does form many single hydrogen bonds to water molecules. This hydrogen bonding is weak and destructuring with every carbohydrate hydroxyl groups acting as a donor, and preferably also as a double acceptor. Strong hydrogen bonding (e.g., β-1-4-linked D-xylose) requires two hydrogen bonds from the polysaccharide to the same water molecule, one as donor and one as acceptor for maximum cooperativity. Strong (double) hydrogen-bonded water links often appear to displace single intra-residue hydrogen bonds as they have the similar requirements for suitably oriented vicinal hydroxyls and their presence can reduce the stereochemical demands. Polysaccharides are more hydrophobic if they have intra-molecular hydrogen bonds. Several carbohydrates (e.g., β-1-3-linked D-xylose, β-glucans) may be considered as having two hydrophobic surfaces with a hydrogen-bonding edge. The hydrophobic effect is primarily a consequence of changes in the clustering in the surrounding water rather than water-solute interactions. Local strong hydrogen bonding may be able to create low-density water without assistance. Kosmotropes are very soluble, well hydrated molecules, having no net charge and enforcing extensive hydrogen bonding. They may compensate for the disrupting effects of high ionic concentrations in some natural microorganisms. Kosmotropes are molecules that stabilise the structure of macromolecules in solution. They stabilise polysaccharide junction zone formation in the same way as they are preferentially excluded from their surfaces. This exclusion entropically drives junction formation. Low molecular weight sugars can cause hydrogen-bonding links between polysaccharides by dehydrating the surface. Ions generally have an effect on the structure of water like increased temperature or pressure. Small ions are strongly hydrated, creating local order but destroying the natural hydrogen bonded network. Ions in polysaccharide solutions may behave differently from when in solution by themselves as the polysaccharides are capable of producing relatively stable low-density or high-density aqueous microenvironments. Polysaccharides are classified on the basis of their main monosaccharide components and the sequences and linkages between them, as well as the anomeric configuration of linkages, the ring size (furanose or pyranose), the absolute configuration (D- or L-) and any other substituents present. The most stable arrangement of atoms in a polysaccharide will be that which satisfies both the intra- and inter-molecular forces. The structural non-starch polysaccharides such as cellulose, and xylan have preferred orientations that automatically support extended conformations. Carbohydrates, especially those containing large numbers of hydroxyl groups, are often thought of as being hydrophilic but they are also capable of generating apolar surfaces depending on the monomer ring conformation, the epimeric structure, and the stereochemistry of the glycosidic linkages. Hydrophobicity is affected by the degree of polysaccharide hydration, particularly the amount of intra-molecular hydrogen bonding. Hydrophobicity affects their availability for fermentation in the gut, and their binding to bile acids. Polysaccharides are more hydrophobic if they have a greater number of internal hydrogen bonds, and as their hydrophobicity increases there is less direct interaction with water. Carbohydrates contain alcohol groups that preferentially interact with two water molecules each if they are not interacting with other hydroxyl groups on the molecule. Polysaccharide linkage through the methyl hydroxyl group is more flexible due to the extra degree of freedom in the link. The flexibility of polysaccharide chains depends on the ease of rotation around the anomeric links. Interactions with the aqueous solvent may determine the preferred conformation by disrupting intramolecular hydrogen bonding. Hydration is very important for the three-dimensional structure and activity of proteins. Proteins lack activity in the absence of water. In solution they possess a conformational flexibility, which encompasses a wide range of hydration states, not seen in the crystal or in non-aqueous environments. Equilibrium between these states will depend on the activity of the water within its microenvironment; i.e., the freedom that the water has to hydrate the protein. Water acts as a lubricant, so easing the necessary peptide amide-carbonyl hydrogen bonding changes. The internal molecular motions in proteins, necessary for biological activity, are very dependent on the degree of plasticizing, which is determined by the level of hydration. Thus internal water enables the folding of proteins and is only expelled from the hydrophobic central core when finally squeezed out by cooperative protein chain interactions. Water is required for the protein to show its biological function as, without it, the necessary fast conformational fluctuations cannot occur. Proteins are formed from a mixture of polar and non-polar groups. Water is most well ordered round the polar groups where residence times are longer than around non-polar groups. Both types of group create order in the water molecules surrounding them but their ability to do this and the type of ordering produced are very different. Polar groups are most capable of creating ordered hydration through hydrogen bonding and ionic interactions. This is most energetically favourable where there is no pre-existing order in the water that requires destruction. Non-polar groups promote clathrate structures surrounded by denser water. The energetic optimisation of mutual hydrogen bonded networks between protein, water and ligand is an intrinsic part of the molecular recognition process in enzymes, binding proteins and biological macromolecules generally. Water molecules have also proved integral to the structure and biological function of a dimeric haemoglobin. Protein folding is driven by hydrophobic interactions, due to the unfavourable entropy decrease forming a large surface area of non-polar groups with water. Compatible solutes (osmolytes, e.g., betaine), that stabilise the surface low-density water, will stabilise the protein's structure. Such hydrophobic collapse is necessarily accompanied and guided by (secondary) structural hydrogen-bond formation between favourable peptide linkages in parallel with their desolvation. Water is critical, not only for the correct folding of proteins but also for the maintenance of this structure. The free energy change on folding or unfolding is due to the combined effects of both protein folding/unfolding and hydration changes. There are both enthalpic and entropic contributions to this free energy that change with temperature and so give rise to heat denaturation and, in some cases, cold denaturation. The midpoint temperatures of both heat and cold denaturation may be determined from peaks in the temperature dependence of the heat capacity, where additional heat is being absorbed by the intermediate structures. At ambient temperature, the entropies of hydration of both non-polar and polar groups are negative indicating that both create order in the aqueous environment. However these entropies differ with respect to how they change with increasing temperature. The entropy of hydration of non-polar groups increases through zero with increasing temperature, indicating that they are less able to order the water at higher temperatures and may, indeed, contribute to its disorder by interfering with the extent of the hydrogen-bonded network. Also, there is an entropy gain from the greater freedom of the non-polar groups when the protein is unfolded. Protein stability has been directly tied to the equilibrium structuring of water between low-density and higher density forms. This provides an equivalent but alternative way of looking at the above analysis. Effectively the denaturation is treated as increased solubility of the unfolded form in a manner similar to that given in the treatment of the anomalous solubility behaviour of non-polar gases. There are two rheological properties of particular importance to hydrocolloid science. These are their gel and flow properties. When a force is applied to a volume of material then a displacement (deformation) occurs. If two plates (area, A), separated by fluid distance (separation height, H) apart, are moved (at velocity V by a force, F) relative to each other, Newton's law states that the shear stress (the force divided by area parallel to the force, F/A) is proportional to the shear strain rate (V/H). The proportionality constant is known as the (dynamic) viscosity (h). The viscosity (h) is the tendency of the fluid to resist flow and is defined by: viscosity = shear stress/shear rate; the SI units are Pascal seconds. Increasing the concentration of a dissolved or dispersed substance generally gives rise to increasing viscosity (i.e., thickening), as does increasing the molecular weight of a solute. For water and solutions containing only low molecular weight material, the viscosity is independent of shear strain rate and a plot of shear strain rate (e.g. the rate of stirring) against shear stress (e.g. force, per unit area stirred, required for stirring) is linear and passes through the origin. Many hydrocolloids are capable of forming gels of various strength dependent on their structure and concentration plus environmental factors such as ionic strength, pH and temperature. The combined viscosity and gel behaviour (viscoelasticity) can be examined by determining the effect that an oscillating force has on the movement of the material. The viscosity increases with concentration until the shape of the volume occupied by these molecules becomes elongated under stress causing some overlap between molecules and a consequent reduction in the overall molecular volume with the resultant effect of reducing the amount that viscosity increases with concentration (under stress). At high shear strain rate (and sufficient concentration) molecules may become more ordered and elastic. Shear flow (and its related stress) causes molecules to become stretched and compressed (at right angle to stretch) resulting in isotropic solutions becoming anisotropic. After release from such conditions, the molecules relax back with time (the relaxation time). At low concentrations below the critical value, the shear modulus of hydrocolloid solutions is mainly determined by the loss modulus at low frequencies. Reduced viscosity is the specific viscosity divided by the concentration. Relative viscosity is the ratio of the dynamic viscosity of the solution to that of the pure solvent. As it is a ratio, it is dimensionless having no units. Specific viscosity is one less than the relative viscosity. Life depends on the anomalous properties of water. In particular, the large heat capacity and high water content in organisms contribute to thermal regulation and prevent local temperature fluctuations. The high latent heat of evaporation gives resistance to dehydration and considerable evaporative cooling. Water is an excellent solvent due to its polarity, high dielectric constant and small size, particularly for polar and ionic compounds and salts. It has unique hydration properties towards biological macromolecules (particularly proteins and nucleic acids) that determine their three-dimensional structures, and hence their functions, in solution. This hydration forms gels that can reversibly undergo the gel-sol phase transitions, which underlie many cellular mechanisms. Water ionises and allows easy proton exchange between molecules, so contributing to the richness of the ionic interactions in biology. Water has unusually high melting point. Water has unusually high boiling point. Water has unusually high critical point. Water has unusually high surface tension and can bounce. Water has unusually high viscosity. Water has unusually high heat of vaporisation. Water shrinks on melting. Water has a high density that increases on heating. Water shows an unusually large viscosity increase as the temperature is lowered. Water's viscosity decreases with pressure. Water has unusually low compressibility. Water has a low coefficient of expansion (thermal expansivity). Water has over twice the specific heat capacity of ice or steam. Solutes have varying effects on properties such as density and viscosity. Solid water exists in a wider variety of stable (and metastable) crystal and amorphous structures than other materials. Hot water may freeze faster than cold water. The solubilities of non-polar gases in water decrease with temperature to a minimum and then rise. Proton and hydroxide ion mobilities are anomalously fast in an electric field. Under high pressure water molecules move further away from each other with increasing pressure. Pressure reduces the temperature of maximum density. As water passes through a magnetic field, the suspended particles receive a charge that reduces their tendency to cluster. Chlorine destroys bacteria; it makes the water safe for bathing or drinking. In fact, chlorinated water is one of the most important advances in sanitation. Chlorine is absorbed through the skin and enters the body in vapour form by inhalation and passes into bloodstream. Shower is responsible for 50% of daily exposure to chlorine-the other 50% is from drinking water. When water interacts with solutes and surfaces, it is unavailable for other hydration interactions. The term water activity describes the (equilibrium) amount of water available for hydration of materials; a value of unity indicates pure water whereas zero indicates the total absence of water molecules. It has particular relevance in food chemistry and preservation. The water activity usually increases with temperature and pressure increases. The multi-ingredient nature of food and its processing (e.g., cooking) commonly result in a range of water activities being present. Foods containing macroscopic or microstructural aqueous pools of differing water activity will be prone to time and temperature dependent water migration from areas with high water activity to those with low water activity; a useful property used in the salting of fish and cheese but in other cases may have disastrous organoleptic consequences. Such changes in water activity may cause water migration between food components. Control of water activity (rather than water content) is very important in the food industry as low water activity prevents microbial growth (increasing shelf life), causes large changes in textural characteristics such as crispness and changes the rate of chemical reactions (increasing hydrophobe lipophilic reactions but reducing hydrophile aqueous-diffusion-limited reactions). Water activity is defined as equal to the ratio of the fugacity (the real gas equivalent of an ideal gas's partial pressure) of the water to its fugacity under reference conditions, but it approximates well to the more easily determined ratio of partial pressures under normal working conditions. In some materials (e.g., salts and some sugars) water activity may reduce with temperature increase. At high pressures, water behaves similar to solutions with increasing salt content in that the water activity apparently reduces with increased pressure. Hydration is very important for the conformation and utility of nucleic acids. Hydration is greater and more strongly held around the phosphate groups, due to their rather diffuse electron distribution, but more ordered and more persistent around the bases with their more directional hydrogen-bonding ability. Nucleic acids have a number of groups, which can hydrogen bond to water, with RNA having a greater extent of hydration than DNA due to its extra oxygen atoms and unpaired base sites. In DNA, the bases are involved in hydrogen-bonded pairing. Guanine will hydrogen bond to a water molecule, cytosine will hydrogen bond to a water molecule, adenine will hydrogen bond to a water molecule, and thymine (and uracil, if base-paired in RNA) will hydrogen bond to a water molecule. Phosphate hydration in the major groove is thermodynamically stronger but exchanges faster. Water (H_2O) is the third most common molecule in the Universe (after H_2 and CO), the most abundant substance on earth and the only naturally occurring inorganic liquid. Successful models for water to study must encompass the radial distribution function, the pressure-viscosity and temperature-density behaviour and the effects of solutes. Water dodecahedra have been found in aqueous solutions, and in the gas phase. Dodecahedral water clusters have also been reported at hydrophobic and protein surfaces, where low-density water with stronger hydrogen bonds and lower entropy has been found. Dodecahedral cavities are formed relatively easily in water during molecular simulations. There are more than twenty commercial products on the market that purport to alter the structure of water in order to help maintain or restore health, youth, and vigour. Chemists have long recognised water as a substance having unusual and unique properties that one would not at first sight expect from a small molecule having the formula H_2O. The special properties of water stem from the tendency of its molecules to associate, forming short-lived and ever-changing polymeric units, which are sometimes described as clusters. Water may be one of a loosely-connected network that might best be described as one huge cluster whose internal connections are continually undergoing rearrangement. Water molecules ionise endothermically due to electric field fluctuations caused by nearby dipole librations resulting from thermal effects; a process that is facilitated by exciting the O-H stretch overtone vibration. Water has the molecular formula H_2O a but the hydrogen atoms are constantly exchanging due to protonation/deprotonation processes. These are catalysed by both acids and bases and even when at their slowest (at pH 7), the average residence time is only about a millisecond. Liquid water (H_2O) is the most remarkable substance. We wash in water, fish in water, swim in water, drink water and cook with water, although probably not all at the same time. Pi-water was discovered in 1964 through the study of botanical physiology, which is based on a very minute amount of ferric ferrous salt. Inducing this ferric ferrous salt into a high-energy state and infusing it through a ceramic filter process creates pi-water. A pi particle is a quantum particle smaller than an electron, neutron and proton. The pi meson causes the protons and neutrons to exchange energy states with one another, which create energy and electrical charges of it's own. The amount of ferric ferrous (bivalent and trivalent ferrite) is minute, a quantum particle essentially such a trace amount that it is only the energetic signature of the substance that remains. It possesses high anti-oxidation properties because of its ability to eliminate free radicals. It neutralises the pH of water, bringing it to just above 7. It improves cellular adaptation to stress and stressful environments. Agriculturally, pi-water has been used for cultivating crops without chemicals and for producing larger vegetables. Industrial use has shown that oxidation or rusting was greatly reduced. Livestock and animals, have responded with improved meat quality, a decreased level of anaerobic fermentation, an increase in egg production, an improved taste, and lethargic animals became more active playful and interested in their surroundings. Cooking-soaking meats increases tenderness. Cleaning fruits and vegetables with pi-water removes toxins and prolongs their freshness.

LIPIDS

There is no nutritional substance as controversial as cholesterol, and no substance about which there is more confusion, and widely publicised and no bigger health scandal. Cholesterol has been given so much attention over the last 35 years. Cholesterol is a hard waxy lipid substance, which melts at 149°C (300°F). Cholesterol is essential for health, manufactured in the body from simpler substances (2 carbon acetates), which the cell drives from the breakdown of sugars, fats and even proteins. Cholesterol is precursor of stress hormones. Cholesterol compensates for changes in membrane fluidity, keeping it within the narrow limits required for optimal membrane function. Cholesterol is secreted by glands in the skin, covers and protects the skin against dehydration, cracking, and wear and tear of sun, wind, and water. Chromium with niacin molecules has been patented as cholesterol-lowering agent. By hooking 15-two-carbon acetates (vinegar) end to end to make a 30-carbon chain, then this chain is cyclised and finally 3 carbons are clipped off to produce the 27-carbon cholesterol molecule. Break down fatty acids, sugars, starches, or amino acids in the mitochondria, produces the 2-carbon acetates at each step. Alcohol also provides acetates for cholesterol production. Saturated and monounsaturated fatty acids are the main sources of acetate fragments from fats. Foods rich in refined carbohydrates also produce an excess of acetates. Stress also increases cholesterol production. Only foods from animal sources contain cholesterol, plant foods are cholesterol free. One egg, 1/4 pound of liver, and 1/4 pound of butter each contain about 250 mg cholesterol. About half of the dietary cholesterol is absorbed. The average person's body contains 150 g of cholesterol. Most of this is found in membranes, and about 7 g is carried out in the circulation. Muscle meats and dairy products contain fewer minerals and vitamins, less EFAs, more SaFAs, and no fiber. Eggs are nutrient-dense but contain no fiber. Meat, eggs, and dairy products are poor sources of vitamin C and B_3, ω3 fatty acids, and fiber. Vitamin C is required for synthesis of mucopolysaccharides. Vitamin C is important for tissue integrity and the prevention of invasion by foreign organisms and cancer cells. Vitamin C recharges fat-soluble vitamin E and water-soluble glutathione, allowing them to be reused many times. It also protects many B vitamins and other substances from oxidative destruction. Vitamin C is necessary for the production of the proteins collagen and elstin, especially the hydroxylation of the amino acids lysine and proline. Vitamin C deficiency (scurvy) results in weakened arteries and bleeding into tissue spaces. Diets are deficient in vitamin C as a result of storage and processing. Vitamin C deficiency leads to weakness of the connective tissues. Thickening of arteries by using adhesive repair protein apo(α) made in the liver protects man from early death and scurvy during survival pressure. Apo(α) and its carrier vehicle in the bloodstream, lipoprotein(α) [Lp(α)], are stronger risk factors for CVD than LDL, which if not associated with Lp(α), is only a weak risk factor. Lp(α) resembles LDL except for the apo(α) protein. Increasing ascorbate levels is associated with reduction of apo(α) levels, because less repair protein is necessary for strong connective tissue in the arteries. Indiscriminately lowering cholesterol increases death rate from suicide and cancer. Increased suicide may result from increased aggression. Lowering cholesterol levels reduce the number of receptors for serotonin on brain cell membranes. Lack of receptors might result in decreased suppression of aggression and violence. Lack of fat-soluble antioxidants (anti-cancer), which are transported to the cell by LDL transport may become deficient when LDL levels in the blood decreased. More stress requires more oils and proteins. Oil rich in EFAs and sulphur-rich proteins may be essential to avoid or control the effect of stress. Oils containing many slightly negatively charged cis-double bonds and proteins containing many slightly positively charged sulphydryl groups. Alkylglycerols remove mercury from the body. Alkylglycerols are oil-based chelating agents. Alkylglycerols are not sources of EFAs. Their fatty acid are saturated or monounsaturated, 16 and 18 long. They are linked to glycerol by ether bonds. They contain no phosphate groups, lecithin, choline, or inositol. Essential oil is volatile oil, typically fragrant, which are extracted from botanicals using steam distillation. Essential oils are normally liquid, but in some cases, such as Anise, may be solid, depending on temperature. For commercial purposes expressed oils such as Orange are identified as essential oils, while they technically are not. Essence oil is oil collected in the water distillate during the production and concentration of fruit juices. Upon separation from the water, the remaining oil contains the highly volatile top notes of natural juice. Fixed oils are non-volatile oils derived from plant materials, commonly referred to as vegetable oils. Folded oil is an essential oil that is concentrated by distillation, e.g., removal of terpenes from citrus oils. Terpene is fraction of an essential oil consisting mainly of hydrocarbons, obtained as a byproduct from either concentration of distillation of the oil. Lipids include fatty acids, phospholipids, and alkylglycerols. Fats and oils are triglycerides, which consist of 3 (tri) fatty acid molecules joined to a glycerol (glyceride) molecule. Cell membrane lipids are phospholipids, which chemical arrangement of phosphorus and oxygen called phosphate (phospho), and 2 fatty acids (lipids) are attached glycerol. Fatty acids are members of several different families. A molecule of any solid fat or liquid oil is made up of 1 molecule of glycerol to which 3 fatty acid molecules are attached. Each fatty acid is made of 2 parts: a fatty chain at one end, and an acid group at the other end. The fatty chain is water-insoluble (hydrophobic), oil-soluble, non-polar chain of variable length. This fatty chain is made entirely of carbon and hydrogen atoms, and ends in a methyl ($-CH_3$) group. The acid end is water-soluble (hydrophilic), polar, weak organic acid known as carboxyl (-COOH) group, which dissolves in water but not in oil. Fatty acids can have any number of carbon atoms in their fatty chain, range from 4 carbons (butyric acid) to 24 carbons (fish oils and brain tissue). Formic acid with 1 carbon (bee sting and ant bite) and acetic acid with 2 carbons (vinegar) are also part of this family of compounds. The omega (ω) system is the numbering system that numbers carbon atoms in sequence, starting from methyl end. Fatty acids whose carbons are joined by single bonds are called saturated. Removing 2 hydrogen atoms from it produces unsaturated fatty acids, which differ from saturated fatty acids in shape, physical properties, and chemical properties, such as the ability to react with water, oxygen, hydroxyl (OH) and sulphydryl (SH) groups, and light. Unsaturated fatty acids are less stable and more active chemically than saturated fatty acids, which are relatively stable and inert. Double bonds can occur between any 2 carbons in a carbon chain. Human enzymes can not insert double bonds into position closer than 7 carbons from the methyl (ω) end. Unsaturated fatty acids with 2 or more double bonds are known as polyunsaturated fatty acids. In cis-configuration, both hydrogen atoms on the carbons involved in the double bond are on the same side of the molecule. The hydrogen atoms repel one another, and the fatty carbon chain kinks to take up some of the empty space on the side of the molecule opposite the hydrogen atoms. Trans-fatty acids are more stable than cis-fatty acids. Saturated fatty acids containing <16 carbon atoms provide energy, calories, and heat. All fatty acids produce 9 calories/g, the body prefers to save $\omega 3$ and $\omega 6$ EFAs for vital hormone-like functions. Unsaturated and essential fatty acids are used in the body to construct membranes, create electrical potentials, move electric currents, and burn them to produce energy. Saturated fatty acids (SaFAs) are the simplest fatty acids. They carry the carboxyl (acid) group at one end and the rest of the molecule is fatty material. SaFA carbon chain can be 4-28 carbon atoms long. The longer the fatty part of the molecule, the greater its tendency is to aggregate. SaFAs are sluggish molecules. Short chain SaFAs make up <10% of the total fatty acids found in butter and milk fat. Up to a length of 8 carbons, SaFAs are liquid at room temperature. Up to 10 carbons, they are liquid at body temperature. Above 10 carbons, they are solid at body temperature. Butyric acid helps feed the friendly bacteria that keep the colon healthy. Caprylic acid inhibits the growth of yeasts and candida in the intestine, possibly by incorporating into membranes of yeast cells and then these membrane ruptures killing the yeast cell. Medium chain SaFAs contain 6-12 carbon atoms. The cell use long-chain SaFAs to build cell membranes. Long-chain SaFAs are solid at body temperature. By making platelets more sticky, the aggregating tendency of long-chain SaFAs plays a role in CVD. SaFAs can be deposited within the cells, organs, and arteries along with proteins, minerals, and cholesterol. Refined sugars can be converted into SaFAs in the body. Refined sugars need no digestion and are absorbed rapidly. Starches can also turn into saturated fats. Digestion and absorption of sugars occurs rapidly, and turn into SaFAs in the body. Soft drinks and ketchup contain a huge amount of sugar. Many meat and sausage products are extended with refined starch. Protein-starch mixtures are more difficult to digest, which leads to bloating, intestinal pain, and gas. Sweet fruits contain starches, but also large amounts of sugar, and can lead to fat production in big amount. Bears eat fruit in the fall in order to build up their fat. Hyperglycaemia triggers the pancreas to secrete insulin, which aid glucose to move into the cells. Once in the cytoplasm, the glucose is converted into pyruvate, which is fed into the energy producing (Krebs) cycle within the mitochondria. Excess intracellular glucose stimulates the production of fatty acids. Three fatty acids, hooked to a glycerol molecule, make a fat molecule (triglyceride). These fat molecules are deposited in the cells and organs, or transported to the adipose tissues for storage. Sugar consumption leads to high blood triglyceride levels, which are associated with CVD. During breaking down a 6-carbon glucose molecules to produce energy, one of the steps involves the creation of 2-carbon acetates (vinegar) inside the mitochondria. These acetates are building blocks for both cholesterol and SaFAs. Excess intracellular vinegar is more toxic than excess fats and cholesterol. The brain demands glucose, glutamic acid, or ketones to function. The adrenal glands mobilise the body's stores of glycogen, and also stimulate the synthesis of glucose from proteins and other substances present the body. Wheat and corn allergy is common. Hard fats interfere with insulin function, leading to insulin insensitivity. Stress-caused diseases occur when adrenal gland exhausted, there is inability to respond biochemically to stress. Adrenal gland exhaustion results in hypoglycaemia, which give rise to craving for sweets. Hyperglycaemia-hypoglycaemia-sugar-craving vicious cycle is the result. Fatty degeneration is the deposition of visible fat in places where it is not normally found in healthy individuals. Athrosclerosis, fatty liver and renals, some tumours, obesity, and some forms of diabetes belong to this group. SaFAs decrease oxygen supply to the tissues (hypoxia), choking them by sticking RBCs together and making them less mobile, less able to deliver oxygen to cells. Refined sugars cause hypoxia. Hypoxia is common to many degenerative conditions and to fatty degeneration. Sugars inhibit the functions of immune system, and increase diseases caused by poor immune function, such as cold and flu. Colitis, asthma, behaviour disorders, joint pain and deterioration, and muscle pain develop as a result of sugar. Food allergies can lead to autoimmune diseases, which attack and destroy tissues. Sugars increase the production of adrenaline by 4 times (fight or flight stress). This stress reaction increases the production of both cholesterol and cortisone. Cortisone inhibits immune function. Sugars lack vitamins and minerals required for the metabolism, to mobilise fats and cholesterol, to convert cholesterol into bile acids for removal via the stool, or burn excess fats as heat or increased activity. Decreased metabolic rate is involved in ageing, arthritic diseases, cancer, and CVD, and is another general symptoms of degenerative diseases. Sugars feed candida (yeasts), fungi, other pathological organisms and cancer cells. Sugars interfere with the transport of vitamin C, as both use the same transport system. Vitamin C's immune, virocidal, bacteriocidal, collagen- and elastin-building, and mucopolysaccharide (tissue glue) forming functions are inhibited by sugar. Sugar cross-link proteins, leading to ageing and wrinkles even in young skin. Hard fats, those made from sugars and those that come from food, induce pimples (teenagers). High blood sugar inhibits the release of linoleic acid (LA) from storage in fat tissues, and thereby contributes to EFA deficiency (functionally deficient in LA). Lack of fiber in refined carbohydrates slows down the speed at which foods pass through the GIT. They remain in the colon too long, serve as food for harmful bacteria that produce gas and toxins, and can induce inflammation of the colon (e.g., diverticulitis) and ballooned (diverticulosis). White bread was used in the 1800's to stop diarrhoea because of its reliability in plugging up the colon. Fatty acids were once thought to be nothing more than a stored form of caloric energy. Fatty acids are some of the most essential nutrients in the human diet, critical for cell membrane structure and function and for local "hormonal" signalling. There are a number of diseases that can be improved by fatty acid therapy, including inflammatory disorders, cardiovascular disease, hormonal disorders, auto-immune disorders, arthritis, mental and behavioural disorders, and many cases of senile neurological degeneration. EFAs are transformed into critical local hormones "eicosanoids". One of the most important functions of eicosanoids is to regulate all stages of the inflammatory process. They control initiation, propagation, and termination of the inflammatory process, which is vital to the body's ability to repair and to protect itself immunologically. Many of the chronic inflammatory conditions, which accompany an EFA imbalance, are treated with symptom-specific pharmaceutical drugs such as steroids, prednisone, sulfasalazine, colchicine, aspirin and other aspirin-like drugs. The problem with such drug therapies is that they prevent the formation of "good" anti-inflammatory eicosanoids as well as the "bad" pro-inflammatory eicosanoids, or they shift production of one type of eicosanoid to another. Eicosanoid production can be modified through dietary changes (balancing dietary intake of specific fats, as indicated by testing) and by controlling insulin levels in the circulation. There are a number of clinical conditions where EFA involvement in the pathophysiology is critical, include: 1- Disorders of cell membrane and cell receptor function, which includes many neurological and mental disorders. 2- Any condition involving a chronic inflammatory process, which includes such disorders as atherosclerosis, heart disease, eczema, autoimmune diseases, and irritable bowel syndrome. Eating merely one fatty fish serving per week reduces the risk of a first heart attack by 70%, and modest dietary fatty fish intake (2-3, 100 g servings per week) significantly reduces the risk of death from myocardial infarction by 29% in patients who had already experienced one heart attack. Maintaining a proper balance between the various families of dietary fats (ω-3, ω-6, ω-9, saturated, and cholesterol) may be one of the most important preventative measures a person can take to reduce the likelihood of developing one of the chronic diseases of modern civilisation, such as diabetes, heart disease, obesity, irritable bowel syndrome, or auto-immune disease. EFA testing and therapy reduce both morbidity and mortality associated with these diseases in patients who already have the problems.

TOXICITY OF UNSATURATED VEGETABLE OILS

50 years ago, paints and varnishes were made of soy oil, safflower oil, and linseed (flaxseed) oil. Then paints were made from petroleum, which was much cheaper. The cholesterol focus was just one of the marketing tools used by the oil industry. Radiation and vegetable oils can cause acquired immunodeficiency. Unsaturated oils, especially polyunsaturates, weaken the immune system's function in ways that are similar to the damage caused by radiation, hormone imbalance, cancer, aging, or viral infections. When an oil is saturated, that means that the molecule has all the hydrogen atoms it can hold. Unsaturation means that some hydrogen atoms have been removed, and this opens the structure of the molecule in a way that makes it susceptible to attack by free radicals. When unsaturated oils are exposed to free radicals they can create chain reactions of free radicals that spread the damage in the cell, and contribute to the cell's aging. Rancidity of oils occurs when they are exposed to oxygen, in the body just as in the bottle. Harmful free radicals are formed, and oxygen is used up. Soybean oil, corn oil, safflower oil, canola, sesame oil, sunflower seed oil, palm oil, and any others that are labelled as unsaturated or polyunsaturated are hazardous to the health. Almond oil, which is used in many cosmetics, is very unsaturated. Corn and soybeans have antithyroid effect, causing obesity. Plants produce the oils for protection, not only to store energy for the germination of the seed. Seeds germinate in early spring, so their energy stores must be accessible when the temperatures are cool, and they normally don't have to remain viable through the hot summer months. Unsaturated oils are liquid when they are cold, and this is necessary for any organism that lives at low temperatures. Seeds contain a small amount of vitamin E to delay rancidity. When the oils are stored in the human tissues, they are much warmer, and more directly exposed to oxygen. These oxidative processes can damage enzymes and other parts of cells, and especially their ability to produce energy. The enzymes, which break down proteins are inhibited by unsaturated fats, and these enzymes are needed not only for digestion, but also for production of thyroid hormones, clot removal, immunity, and the general adaptability of cells. On top of the natural toxicity, the plants are sprayed with industrial pesticides, which can concentrate in the seed oils. Obesity, free radical production, the formation of age pigment, blood clotting, inflammation, immunity, and energy production are all responsive to the ratio of unsaturated fats to saturated fats, and the higher this ratio is, the greater the probability of harm there is. Puberty occurs at an earlier age if oestrogen is high, or if these oils are more abundant in the diet. This is probably a factor in the development of cancer. Hormonal imbalances, damage to the immune system, and oxidative damage are the main damage of excess of these oils. Unsaturated oils block thyroid hormone secretion, its movement in the circulatory system, and the response of tissues to the hormone. When the thyroid hormone is deficient, the body is generally exposed to increased levels of oestrogen. The thyroid hormone is essential for making the "protective hormones" progesterone and pregnenolone, so these hormones are lowered when anything interferes with the function of the thyroid. The thyroid hormone is required for using and eliminating cholesterol, so cholesterol is likely to be raised by anything that blocks the thyroid function. The unsaturated oils suppress the immune systems of cancer patients. Unsaturated oils get rancid when exposed to air; i.e., oxidation, and it is the same process that occurs when oil paint dries. Free radicals are produced in the process. This process is accelerated at higher temperatures. The free radicals produced in this process react with parts of cells, such as DNA and protein and may become attached to those molecules, causing abnormalities of structure and function. An excess of unsaturated vegetable oils damages the human body, even if it is organic. Alcoholic cirrhosis of the liver cannot occur unless there are unsaturated oils in the diet. Heart disease can be produced by unsaturated oils, and prevented by adding saturated oils to the diet. Coconut and olive oil are the only vegetable oils that are really safe, but butter and lamb fat, which are highly saturated, are generally very safe (except when the animals have been poisoned). Coconut oil is unique in its ability to prevent weight-gain or cure obesity, by stimulating metabolism. It is quickly metabolised, and functions in some ways as an antioxidant. Olive oil is less fattening than corn or soy oil, and contains antioxidant, which makes it protective against heart disease and cancer. The insecticide lindane when used in dairies induces cancer. Certain cancers are several times more common among corn farmers than among other farmers, presumably because corn requires the use of more pesticides. This probably makes corn oil's toxicity greater than it would be otherwise, but even the pure, organically grown material is toxic, because of its intrinsic unsaturation. Lard is toxic because the pigs are fed large quantities of corn and soybeans. Tropical plants live at a temperature that is close to the normal body temperature. Coconut oil and other tropical oils also contain some hormones that are related to pregnenolone or progesterone. Coconut oil is the least fattening of all the oils. Hard margarine, because it resists oxidation, isn't suppressive to the thyroid gland, and doesn't cause cancer. Butter contains natural vitamin A and D and some beneficial natural hormones. It is less fattening than the unsaturated oils. There is much less cholesterol in an ounce of butter than in a lean chicken breast. A diet rich in fish oil causes intense production of toxic lipid peroxides, which induces azoospermia. Women with breast cancer have very high levels of agricultural pesticides in their breasts. Vegetables, grains, nuts, fish and meats all naturally contain large amounts of polyunsaturated oils, and the extra oil used in cooking becomes a more serious problem. Mayonnaise, pastries, even candies may contain these oils; check the labels for ingredients. Avoid foods which contain the polyunsaturated oils, such as corn, soy, safflower, flax,

cottonseed, canola, peanut, and sesame oil. Unsaturated fats intensify oestrogen's harmful effects. Use coconut oil, butter, and olive oil. Unsaturated fats cause aging, clotting, inflammation, cancer, and weight gain.

COCONUT OIL

There are many types of saturated fat, just as there are different types of polyunsaturated fat. Each has a different effect on the body. The saturated fat in coconut is identical to a special group of fats found in human breast milk, the MCFA. Most of fats in the diet whether they be saturated or unsaturated are in the form of large molecules the long-chain fatty acids (LCFA). The medium-chain fatty acids (MCFA) in coconut are much smaller in size. The large LCFA are digested slowly. MCFAs do not require pancreatic enzymes for digestion and absorbed quickly and channelled to the liver rather than the bloodstream. Medium chain fatty acids (MCFAs) in coconut oil break the virus apart. The MCFAs in coconut oil mimic the fatty acids in the lipid membrane that coats the virus, so the antimicrobial MCFAs are readily absorbed by the unsuspecting virus, weakening its protective membrane until it breaks apart and dies. Coconut oil boosts immune system in cases of viral outbreaks. Coconut oil has the highest level of MCFAs of all palm oils. Lauric acid makes up 53% of its content. Coconut oil kills STDs' microbes and some strains of HIV. It inactivates viruses such as measles, herpes simplex 1 and 2, and disease-causing bacteria such as H. pylori, E coli, salmonella, influenza, and yeast infections. It also kills funguses such as candida, chlamydia, and eczema. Coconut oil must be naturally processed at a temperature below 100°F (40°C) in order to retain its natural antioxidant properties, vitamin E, fatty acids, and enzymes. One litter of virgin coconut oil is produced from 15-17 fresh nuts, even with mechanised aids. The best virgin oil is cold-pressed, without heat being applied or generated. Athlete's performance and endurance improves with MCFA. Because coconut oil produces energy, it stimulates the metabolism supporting thyroid function. This thermogenic or metabolic stimulating effect causes the body to burn more calories. The thermogenic effect of MCFA is almost twice as high as the LCFA. MCFA given over a 6-day period can increase diet-induced thermogenesis by 50%. The thermogenic effect of MCFA over 6 hours is 3 times greater than that of LCFA. Coconut oil contains the most concentrated natural source of MCFA available. Polyunsaturated vegetable oils depress thyroid activity, thus lowering metabolic rate. Eating polyunsaturated oils, like soybean oil, will contribute more to weight gain than other fat known. Unsaturated oils block thyroid hormone secretion, its movement in the circulation, and the response of tissues to the hormone. Polyunsaturated oils are high calorie fats, which encourage weight gain more than any other fats. Lard does not interfere with thyroid function. MCFA is essential for infants, as they can not digest LCFA. MCFA in the breast milk is quickly digested for use as energy. MCFAs improve the absorption of other nutrients, such as minerals (Ca, Mg), vitamins and amino acids. Coconut oil may be useful in protecting against heart disease, breast and colon cancer, liver disease, kidney disease, Crohn's disease, epilepsy, candida, herps, influenza, and numerous other infectious diseases. Coconut oil contains a small amount of the unsaturated oils. The unsaturated oils in some cooked foods become rancid in just a few hours, even at refrigerator temperatures, and are responsible for the stale taste of leftover foods. Coconut oil that has been kept at room temperature for a year has been tested for rancidity, and showed no evidence of it, the other (saturated) oils have an antioxidative effect. To interrupt chain-reactions of oxidation is one of the functions of antioxidants, and it is possible that a sufficient quantity of coconut oil in the body has this function. Coconut oil is unusually rich in short and medium chain fatty acids. Shorter chain length allows fatty acids to be metabolised without use of the carnitine transport system. Mildronate protects cells against stress partly by opposing the action of carnitine. Carnitine promotes the oxidation of unsaturated fats during stress, and increases oxidative damage to cells. Very soluble and mobile short-chain saturated fats have priority for oxidation, because they don't require carnitine transport into the mitochondrion, and thus tend to inhibit oxidation of the unstable, peroxidisable unsaturated fatty acids. There is association between consumption of unsaturated oils and the incidence of cancer. Emulsions of unsaturated oils are used specifically in organ-transplant patients for their immunosuppressive effects. Age pigment is produced in proportion to the ratio of oxidants to antioxidants, multiplied by the ratio of unsaturated oils to saturated oils. Ultraviolet light induces peroxidation in unsaturated fats, but not saturated fats, and that this occurs in the skin. The amount of unsaturated oil in the diet strongly affects the rate at which aged, wrinkled skin develops. The unsaturated fat in the skin is a major target for the aging and carcinogenic effects of ultraviolet light, though not necessarily the only one. Unsaturated fats suppress the metabolic rate, apparently creating hypothyroidism. Unsaturated fats damage the mitochondria, partly by suppressing the respiratory enzyme, and partly by causing generalised oxidative damage. The more unsaturated the oils are, the more specifically they suppress tissue response to thyroid hormone, and transport of the hormone on the thyroid transport protein. The thyroid hormone is formed in the gland by the action of a proteolytic enzyme, and the unsaturated oils also inhibit that enzyme. Similar proteolytic enzymes involved in clot removal and phagocytosis appear to be similarly inhibited by these oils. Brain tissue is very rich in complex forms of fats. Because coconut oil supports thyroid function, and thyroid governs brain development, including myelination, the result might simply reflect the difference between normal and hypothyroid individuals. Lipid peroxidation occurs during seizures, and antioxidants such as vitamin E have some anti-seizure activity. Lipid peroxidation is involved in the nerve cell degeneration of Alzheimer's disease. Aspartic and glutamic acids promote seizures and cause brain damage, and are intimately involved in the process of stress-induced brain aging, and tryptophan is carcinogenic. Some of physiological effects of coconut include antihistamines, antiinfectives/antiseptics, promoters of immunity, glucocorticoid antagonist, nontoxic anticancer agents, etc. The liver defensively retains its cholesterol, rather than releasing it into the blood, when unsaturated oils are used. This could be the explanation that the unsaturated oils are able to slightly lower serum cholesterol. Whenever drugs, including the unsaturated oils are used to lower serum cholesterol, mortality increase, from a variety of causes including accidents, but mainly from cancer. Suppression of the thyroid raises serum cholesterol (while increasing mortality from infections, cancer, and heart disease), while restoring the thyroid hormone brings cholesterol down to normal. The thyroid does not suppress the synthesis of cholesterol, but rather promotes its use to form hormones and bile salts. When the thyroid is functioning properly, the amount of cholesterol in the blood entering the ovary governs the amount of progesterone being produced by the ovary, and the same situation exists in all steroid-forming tissues, such as the adrenal glands and the brain. The generalised protective function of progesterone and its precursor, pregnenolone include antioxidant, anti-seizure, antitoxin, anti-spasm, anti-clot, anti-cancer, pro-memory, pro-myelination, pro-attention, etc. Any interference with the formation of cholesterol will interfere with all of these exceedingly important protective functions. Coconut oil added regularly to a balanced diet, lowers cholesterol to normal by promoting its conversion into pregnenolone. The cholesterol-rich plaques in blood vessels are caused mostly by lipid peroxidation of unsaturated fats, and relate to stress, because adrenaline liberates fats from storage, and the lining of blood vessels is exposed to high concentrations of the blood-borne material. The ability of some of the medium chain saturated fatty acids to inhibit the liver's formation of fat very likely synergizes with the pro-thyroid effect, in allowing energy to be used, rather than stored. Shifting from unsaturated fats in foods to coconut oil involves several anti-stress processes, reducing the need for the adrenal hormones. Hypoglycaemia is a basic signal for the release of adrenal hormones. Unsaturated oil tends to induce hypoglycaemia by: 1- It damages mitochondria, causing respiration to be uncoupled from energy production. 2- It suppresses the activity of the respiratory enzyme (directly, and through its anti-thyroid actions), decreasing the respiratory production of energy. 3- It tends to direct carbohydrate into fat production, making both stress and obesity more probable. The use of coconut oil consistently provides the ability to go for several hours without eating, and to feel hungry without having symptoms of hypoglycaemia. Unsaturated fats, and their breakdown products, interfere with enzymes and transport proteins, which accounts for many of their toxic effects. They probably bind to all proteins, and disrupt some of them, but for some reason their affinity for proteolytic and respiration-related enzymes is particularly obvious. Unsaturated fats are slightly more water-soluble than fully saturated fats, and so they do have a greater tendency to concentrate at interfaces between water and fats or proteins. The fluidity or viscosity of cell surfaces is an extremely complex, and the degree of viscosity has to be appropriate for the function of the cell. In some cells, such as the cells that line the alveoli of the lungs, cholesterol and one of the saturated fatty acids found in coconut oil can increase the fluidity of the cell surface. Stressful conditions create structural disorder in cells. Lipid peroxidation of unsaturated fats weakens the cellular structure of RBCs, causing the cells to be destroyed prematurely. Lipid peroxidation products lower the rigidity of regions of cells considered to be membranes. Red blood cell is more like a sponge in structure, consisting of a skeleton of proteins, which (if not damaged by oxidation) can hold its shape, even when the haemoglobin has been removed. Oxidants damage the protein structure, and it is this structural damage that in turn increases the fluidity of the associated fats. The anti-obesity effect of coconut oil is known. The physiologically functional components of coconuts are found in the fat part of whole coconut, in the fat part of desiccated coconut, and in the extracted coconut oil. Lauric acid, the major fatty acid from the fat of the coconut, lends to nonfood uses in the soaps and cosmetics industry. Lauric acid has unique properties in food use, which are related to its antiviral, antibacterial, and antiprotozoal functions. Capric acid has been added to the list of coconut's antimicrobial components. Natural coconut fat in the diet leads to a normalization of body lipids, protects against alcohol damage to the liver, and improves the immune system's anti-inflammatory response. Functional foods provides a health benefit over and beyond the basic nutrients. As a functional food, coconut has fatty acids that provide both energy (nutrients) and raw material for antimicrobial fatty acids and monoglycerides (functional components) when it is eaten. Desiccated coconut is about 69% coconut fat, as is creamed coconut. Full coconut milk is approximately 24% fat. 50% of the fatty acids in coconut fat are lauric acid. Lauric acid is a medium chain fatty acid, which has the additional beneficial function of being formed into monolaurin in the human or animal body. Monolaurin is the antiviral, antibacterial, and antiprotozoal monoglyceride used by the human or animal to destroy lipid-coated viruses such as HIV, herpes, cytomegalovirus, influenza, various pathogenic bacteria, including listeria monocytogenes and helicobacter pylori, and protozoa such as giardia lamblia. 6-7% of the fatty acids in coconut fat are capric acid. Capric acid is another medium chain fatty acid, which has a similar beneficial function when it is formed into monocaprin in the human or animal body. Monocaprin has also been shown to have antiviral effects against HIV and is being tested for antiviral effects against herpes simplex and antibacterial effects against chlamydia and other sexually transmitted bacteria. The antimicrobial effects of the fatty acids (Fas) and monoglycerides (MGs) are additive, and total concentration is critical for inactivating viruses. The properties that determine the anti-infective action of lipids are related to their structure, e.g., monoglycerides, free fatty acids. The monoglycerides are active; diglycerides and triglycerides are inactive. Of the saturated fatty acids, lauric acid has greater antiviral activity than either caprylic acid (C-8), capric acid (C-10), or myristic acid (C-14). The fatty acids and monoglycerides produce their killing/inactivating effect by lysing the plasma membrane lipid bilayer. The antiviral action attributed to monolaurin is that of solubilizing the lipids and phospholipids in the envelope of the virus, causing the disintegration of the virus envelope. Also, monolaurin's interferes with signal transduction, and another antimicrobial effect in viruses is due to lauric acid's interference with virus assembly and viral maturation. The antiviral aspects of the antimicrobial activity of the monoglyceride of lauric acid (monolaurin) has been reported since 1966. The envelope of the RNA and DNA viruses is a lipid membrane, and the presence of a lipid membrane on viruses makes them especially vulnerable to lauric acid and its derivative monolaurin. The medium-chain saturated fatty acids and their derivatives act by disrupting the lipid membranes of the viruses. The enveloped viruses are inactivated in both human and bovine milk by adding fatty acids and monoglycerides, and also by endogenous fatty acids and monoglycerides of the appropriate length. Some of the viruses inactivated by these lipids, in addition to HIV, are the measles virus, herpes simplex virus-1 (HSV-1), vesicular stomatitis virus (VSV), visna virus, and cytomegalovirus (CMV). Many of these pathogenic organisms are known to be responsible for opportunistic infections in HIV-positive individuals. Monolaurin and its precursor lauric acid are antiviral, antimicrobial, and antiprotozoal. These antimicrobial fatty acids and their derivatives are essentially nontoxic to man; they are produced in vivo by humans when they ingest those commonly available foods that contain adequate levels of medium-chain fatty acids such as lauric acid. The lipid-coated (envelope) viruses are dependent on host lipids for their lipid constituents. Monolaurin does not appear to have an adverse effect on desirable gut bacteria, but rather on only potentially pathogenic microorganisms, with major inactivation of Hemophilus influenzae, Staphylococcus epidermidis and Group B gram positive streptococcus. Potentially pathogenic bacteria inactivated by monolaurin include Listeria monocytogenes, Staphylococcus aureus, Streptococcus agalactiae, Groups A, F & G streptococci, gram-positive organisms, and some gram-negative organisms. Monolaurin is 5000 times more inhibitory against Listeria monocytogenes than ethanol. Helicobacter pylori is rapidly inactivated by medium-chain monoglycerides and lauric acid, with very little development of resistance of the organism to the bactericidal effects of these natural antimicrobials. A number of fungi, yeast, and protozoa are inactivated or killed by lauric acid or monolaurin, e.g., ringworm, Candida albicans, and Giardia lamblia, respectively. Chlamydia trachomatis is inactivated by lauric acid, capric acid, and monocaprin, and hydrogels containing monocaprin are potent in vitro inactivators of sexually transmitted viruses such as HSV-2 and HIV-1 and bacteria such as Neisseria gonorrhoeae. Coconut oil has been shown to be beneficial to the heart. Dietary coconut oil does not lead to high serum cholesterol nor to high coronary heart disease mortality or morbidity. The chemical analysis of the atheroma shows that the fatty acids from the cholesterol esters are 74% unsaturated (41% of the total fatty acids is polyunsaturated) and only 24% are saturated. None of the saturated fatty acids were reported to be lauric acid or myristic acid. There is a causative role for the herpes virus and cytomegalovirus in the initial formation of atherosclerotic plaques and the reclogging of arteries after angioplasty. Herpes virus and cytomegalovirus are both inhibited by the antimicrobial lipid monolaurin, but monolaurin is not formed in the body unless there is a source of lauric acid in the diet. Chlamydia pneumoniae, a gram-negative bacteria, is another of the microorganisms suspected of playing a role in atherosclerosis by provoking an inflammatory process that would result in the oxidation of lipoproteins with induction of cytokines and production of proteolystic enzymes, a typical phenomena in atherosclerosis. All members of the herpes virus family are reported to be killed by the fatty acids and monoglycerides from saturated fatty acids ranging from C-6 to C-14, which include approximately 80% of the fatty acids in coconut oil. Trans fatty acids raise the low density lipoprotein (LDL) cholesterol and lower the high density lipoprotein (HDL) cholesterol in serum. There is a significant positive association between the intake of trans fatty acids and the risk of death from coronary disease. There is no association between intakes of saturated fatty acids, or dietary cholesterol and the risk of coronary deaths. There may be association between breast-cancer incidence and linoleic acid status. The mean fatty acid composition of adipose does not show an association with ω-6 linoleic acid and breast, colon or prostate cancer. However, cancers of the breast and colon are positively associated with the trans fatty acids. The adipose tissue concentration of trans fatty acids have a positive association with postmenopausal breast cancer. Saturated fatty acids raise HDL cholesterol, whereas the trans fatty acids lower HDL cholesterol. saturated fatty acids lower the blood levels of the atherogenic lipoprotein (a), whereas trans fatty acids raise it. Saturated fatty acids conserve the elongated ω-3 fatty acids, whereas trans fatty acids cause the tissues to lose these ω-3 fatty acids. Saturated fatty acids do not inhibit insulin binding, whereas trans fatty acids do inhibit insulin binding. Saturated fatty acids are the normal fatty acids made by the body, and they do not interfere with enzyme functions such as the δ-6-desaturase, whereas trans fatty acids are not made by the body, and they interfere with many enzyme functions such as δ-6-desaturase. Some saturated fatty acids are used by the body to fight viruses, bacteria, and protozoa, and they support the immune system, whereas trans fatty acids interfere with the function of the immune system. Cholesterol is a major support molecule for the immune system, an important antioxidant, and a necessary component of neurotransmitter receptors. The brain does not work very well without adequate cholesterol. The pathway to cholesterol synthesis starts with a molecule of acetyl CoA that comes from the metabolism of excess protein forming ketogenic amino acids and from the metabolism of excess carbohydrate, as well as from the oxidation of excess fatty acids. The degree of saturation of the fat in the diet does not affect the rate of synthesis of cholesterol. Polyunsaturated fatty acids in the diet increase the rate of cholesterol synthesis relative to other fatty acids. Dietary intake of the ω-6 polyunsaturated fatty acid linoleic acid is positively related to coronary artery disease. The elongated ω-3 fatty acids of concern are eicosapentaenoic acid (EPA) and docosahexaenoic acid (DHA). Linolenic acid (the basic ω-3 fatty acid) is not readily converted to the elongated forms in humans or animals, especially when there is ingestion of the trans fatty acids and the consequent inhibition of the δ-6-desaturase enzyme. If the background fat in the diet is high in saturated fat, the conversion is 6% for EPA and 3.8% for DHA, whereas if the background fat in the diet is high in ω-6 polyunsaturated fatty acids (PUFA), the conversion is reduced 40-50%. A diet enriched in saturated but not unsaturated fatty acids reverses alcoholic liver injury in animals that could be due to down-regulation of lipid peroxidation, which was caused by dietary linoleic acid. Dietary saturated fatty acids protect the liver from alcohol injury by retarding ethanol metabolism, and carnitine may be involved. Diets with high linoleic acid may promote acetaminophen-induced liver injury compared to diets with more saturated and monounsaturated fatty acids. Coconut-oil enriched diet decreases white fat stores. Coconut-oil conserves the elongated ω-3 and normalizes the ω-6-to-ω-3 balance. Soybean oils significantly decrease HDL cholesterol. Both fat and calcium are needed by the infant for proper growth. Coconut oil appears to help the immune system response in a beneficial manner. Coconut oil has an inhibitory effect on IL-1 production, which is largely due to a reduced prostaglandin and leukotriene production. Coconut oil does not produce an increase in inflammatory prostaglandin E2 (PGE_2), which is due to a reduction of phospholipd arachidonic acid content. ω-6 oil enhances inflammatory stimuli, but the coconut oil, along with fish oil and olive oil, suppresses the production of IL-1. Monolaurin and the ether analogue of monolaurin reduces adverse reactions to toxic forms of glutamic acid. Lauric acid and capric acid have very potent effects on insulin secretion. Monolaurin induces proliferation of T cells and inhibits the toxic shock syndrome toxin-1 mitogenic effects on T cells. Consumption of coconut is beneficial for individuals with the chronic fatigue and immune dysfunction syndrome known as CFIDS. Coconut oil may protect against the renal necrosis and renal failure produced by a diet deficient in choline (a methyl donor group).

SHEA BUTTER

Crude shea butter is a grey, tallow-like substance, which is extracted from the kernels of the fruit of the shea nut tree (Butyrospermum parkii) by hydraulic pressing or by using specially adjusted screw expellers. The greenish crude fat is usually more or less contaminated with the latex from the fruits, rendering the shea butter rather difficult to refine. Most qualities offered are rather grey in colour, but if proper care is taken in the selection of the nuts, and if adequately refined an attractive white fat can be obtained, which is very suitable for cosmetic applications. Shea Butter is refined, bleached and deodorised to obtain a nice whitish product. Shea butter is a non-toxic and non-irritating material derived from totally renewable natural resources. It is easy to use within cosmetic formulations. Shea butter contains tocopherol (100-150 ppm) and tocotrienol (110-175 ppm). Because of its unique fatty acid composition Shea butter is a soft fat, which readily melts at body temperature, making it a very attractive emollient for many skin care applications, such as baby care products, skin care products, massage creams, make-up. Shea butter is absorbed quickly by the skin, acts as a refatting agent and has good waterbinding properties. As its consistency at room temperature is somewhere between a liquid and solid, it can be used in emulsions as a binding component between solid and liquid components. Shea butter exhibits two polymorphic forms at about 38°C, which makes it more predictable and stable in formulations than other fats with more polymorphic forms, such as cocoa butter. The particular strength of Shea butter lies in its content of unsaponifiables (up to 8%), which gives it very good soothing properties and provide extra sun protection. The unsaponifiables consist of up to 65% terpenic alcohols. Sterols and terpenic alcohols are found almost exclusively as cinnamic acid esters. This is unique to Shea Butter when compared to common oils whose unsaponifiable components are found as free alcohols. It is the high content of cinnamic acid that gives Shea Butter its healing properties. Traditionally the Africans use Shea butter as a medical balm for rheumatism, muscle aches, burns and light wounds, with astonishing results. Shea butter is good in the treatment of cutaneous dryness, dermatitis, dermatoses, solar erythema, burns and other skin irritation. Shea butter seems to induce the capillary circulation in the skin, which in turn increases tissue re-oxygenation and enhances the elimination of metabolic waste products. The presence of esterified terpenic alcohols in Shea butter favour its cutaneous compatibility. The unsaponifiables contain 5-10% phytosterols, which are known to be active in cellular growth stimulation.

OTHER LIPIDS

Unsaturated fatty acids (UFAs) are liquid oils. UFAs differ from SaFAs in that they contain one or more double bonds between carbon atoms in their fatty carbon chain, and for each double bond; they have given up 2 hydrogen atoms. Monounsaturated fatty acids (MUFAs) are unsaturated fatty acids with one double bond. The most important MUFA in nutrition has an 18-carbon chain, and its double bond is always between carbons 9 and 10, called oleic acid (OA). OA is found in olive, almond, peanut, pistachio, pecan, canola, avocado, hazelnut, cashew, and macadamia oils, in the membranes of plant and animal cell structures, and in the fat deposits of most land animals. OA resists damage by oxygen and is fairly stable. In excess OA interfere with essential fatty acids and prostaglandins (PGs). Two 18-carbon UFAs are known as essential fatty acids (EFAs). EFAs may be important in nutrition, and may be vital to health. α-Linolenic acid (LNA) has 3 double bonds. The key components of healing fats are the EFAs linoleic (LA) and α-linolenic acid (LNA). LA is cis-ω6,9-octadecadienoic acid (18:2ω6), which means that there are 18 carbon atoms in the chain, there are 2 double bonds, the double bonds are methylene interrupted, the first double bond starts at carbon atom number 6 counting from the methyl end, and the double bonds are in the cis-configuration. LA is abundant in safflower, sunflower, corn, sesame, and other oils. LNA is cis-ω3,6,9-octadecadienoic acid (18:3ω3). Many of the effects of ω6s and ω3s in the body are opposite. Both LA and LNA are EFAs for humans. Arachidonic acid (AA, 20:4ω6) is not an EFAs, but EFA derivative. EFAs attract O_2, which made EFA-rich oils useful to the paint industry. EFAs are chemically reactive, increases oxidation and metabolic rate. Flaxseed oil (the undenatured version of linseed oil) and hem oil are nutritionally superior to safflower oil because they contain both EFAs (safflower contains only LA) and are chemically more active. EFAs absorb sunlight, which is thousand-fold their ability to react with O_2. EFA molecules repel one another, because they carry slightly negative charges. Thus they do not easily aggregate, they keep membranes fluid, a property that is important in membrane functions.

Symptoms of deficiency:

LA deficiency	LNA deficiency
1- Eczema-like skin eruptions 2- Loss of hair 3- Liver degeneration. 4- Behavioural disturbances. 5- Renal degeneration. 6- Excessive water loss through the skin accompanied by thirst. 7- Drying up of glands. 8- Susceptibility to infections. 9- Wound healing failure. 10- Male fertility dysfunction. 11- Abortions. 12- Arthritis-like conditions. 13- Cardiovascular problems. 14- Growth retardation.	1- Growth retardation. 2- Weakness. 3- Impairment of vision and learning ability. 4- Motor incoordination. 5- Tingling sensation in arms and legs. 6- Behaviour changes. 7- High TGs 8- Hypertension. 9- Platelets adhesiveness. 10- Inflammatory conditions. 11- Oedema. 12- Dry skin. 13- Mental deterioration. 14- Low metabolic rate. 15- Immune dysfunction.

EFAs are weekly basic, and able to form weak hydrogen bonds with weak groups such as the sulphydryl groups found in proteins. EFAs can form phase boundary potentials, like charges of static electricity in a capacitor, which are caught between the water within and outside cells, and the oils within the membranes. These charges can produce measurable bioelectric currents, like the zap when static electricity discharges, which are important in nerve, muscle, heart, and membrane functions. LA and LNA transfer O_2 from air in the lungs via alveolar membranes to blood plasma, across membranes of RBCs to haemoglobin, which then carries O_2 to all cells in the body. At the cellular end, they help transport O_2 from RBCs across the walls of our capillaries, across cell membranes and, inside the cell to precise locations in the mitochondria, which use it in oxidation reactions to produce energy. LA and LNA appear to hold oxygen in the cell membranes, where the O_2 acts as a barrier to viruses, fungi, bacteria, and other foreign organisms that cannot thrive in its presence. LA helps produce haemoglobin from simpler substances. Both EFAs are involved in a process that makes O_2 available to tissues by activating O_2 molecules by way of free radicals or electrostatic forces, regulated by sulphur-containing proteins. EFAs are part of all cell membranes. They help hold proteins in the membrane by the electrostatic attractive forces of their double bonds, and thus they are involved in the traffic of substances via protein channels, pumps, and other special mechanisms. EFAs help maintain the fluidity of membranes. EFAs help creating electrical potentials across membranes, which when stimulated, generate bioelectric currents that travel along cell membranes to other cells, transmitting messages. LA and LNA substantially shorten the time required for fatigued muscles to recover after exercise. They facilitate the conversion of lactic acid to H_2O and CO_2. LA and LNA are involved in exocrine- and endocrine-hormones production. EFAs are precursors of prostaglandins (PGs). EFAs are precursors for even longer and more (5-6 times) unsaturated fatty acids needed by the most active, oxygen-requiring, energy- and electron-exchanging tissues, such as brain, retina, adrenal, and testicular tissues. EFAs are growth-enhancer. At levels above 12-15% of total calories, they increase the rate of metabolic reactions, and the increased rate burns more fat into H_2O and CO_2 and energy, resulting in fat-burn-off and loss of excess weight. EFAs seem to be involved in electron and energy transport. EFAs are involved in the transport (esterification) of cholesterol. LNA and its derivative cal lower elevated blood fats (S. TGs) by up to 65%. Heart tissue requires LA for proper functioning. EFAs exist around chromatin in the chromosomes, where they may play part in maintaining chromosome stability, and may have functions in starting and stopping gene expression. EFAs help energise and govern the movement of chromosomes during cell division by their functions in spindle fiber development, and they form part of the new cell membranes that separate the daughter cells after a cell has divided. EFAs can buffer excess acid as well as excess base in the body, and are the richest source of energy in nutrition. LNA produces smooth, velvety skin, increases stamina, speeds healing, increases vitality, and brings a feeling of calmness. LNA reduces inflammation, water retention, platelet adhesiveness, and blood pressure. LNA inhibits tumour growth. LNA enhances some immune function, reduces pain and swelling of arthritis, and completely reverses PMS in some cases. LNA kills malaria, and treats bacterial infections. LNA is required for brain development. Deficiency of LNA during foetal development and early infancy results in permanent learning disabilities. Linoleic acid (LA) is the one with the highest daily requirement. LA requirement varies from time to time for the same person depending on the level of physical activity and stress, nutritional state, and individual differences. Males require more than females. 1-2% of calories (3-6 g/d) is enough to prevent symptoms of deficiency in most healthy adults, with optimum amount of 9-18 g/d. Obese individuals and those dieting in high saturated fatty acids and olive oil may require even more. Vitamin should be supplied as 1 part for 1500 parts of LA, i.e., 30 IU vitamin E for 18 g LA/day. 2% of daily calories should be LNA. An average adult carries a total of 10 kg of body fat. About 10% of that (1 kg) is LA. Vegetarians contain up to 25% of their total body fat as LA. Patients with degenerative diseases average 8% of their fatty tissue as LA. LA and LNA are essential because of the absence of desaturase enzyme in man. Safflower oil is the richest source of LA. Flax oil is the richest source of LNA. Flax oil is the richest source of ω3s and provides a quick way to make up for long-standing, widespread ω3 deficiency. Long-term exclusive use of flax oil can result in ω6 deficiency within 16-24 months. Contemporary diets cause degenerative conditions. The ω6:ω3 ratio in the brain is about 1:1, in fat tissue is 5:1, and other tissues are 4:1. Hemp is marijuana, but the oil is legal. Eating hemp seeds or drinking hemp seed oil does not produce any intoxication effects. Hemp seeds and oils contain little tetrahydrocannabinol (THC), which cause the intoxication, but traces of THC may be present, enough to show up in urine drug screens. Light, air, and heat destroy LA and LNA. Fresh EFAs-rich oils should be pressed and packed in the dark, in the absence of O_2, and with minimal heat, then stored in opaque containers to prevent contact with light and O_2. Frozen oil remains unspoiled for a long time. Light, the greatest enemy of EFAs, produces free radicals in oil, and speeds up the reaction of oils with O_2 from the air by 1000 times, resulting in rancid oil. Light can induce cross-linking of fatty acid molecules into dimers and polymers that are rubber-like substances. Light-induced free radical chain reactions can break EFAs down into many different kinds of products, including aldehydes, ketones, and other toxic and non-toxic components. Light destroys the vital biological properties of the EFAs. Oxygen breaks down EFAs resulting in rancid oil. Heat, used in deodorising, hydrogenation (margarines), and commercial frying and deep-frying used to make consumer items, destroys EFAs by twisting their molecules from a natural cis-shape to an unnatural trans-shape. All fats and oils are mixtures of Triglycerides (TGs). Most stored body fat is composed of TGs. High serum TGs are risk factor in CVD. When sugars turn into saturated fatty acids, these fatty acids are also carried as TGs. TGs are body's reserve of the vital EFAs, LA and LNA. Beef fat is made of TGs that carry mostly saturated and monosaturated fatty acids in all positions; it contains hardly any EFAs. Completely hydrogenated fats carry 100% SaFAs in all position. Flax oil and safflower oil must carry EFAs in outside positions, because both of these oils contain >70% EFAs. Butter contains many short-chain SaFAs between 4 and 14 carbon atoms in length. Tropical fats contain mostly SaFAs 16 carbons or less in length. Most plant seed oil and animal depot fats have mainly 18-carbon fatty acids in their TGs. Fatty acids in the TGs of fish oils are up to 22 carbon long. In nutrition, the arrangement of fatty acids on the glycerol molecule appears less important than the total amount of each kind of fatty acid present. TGs have several functions: 1- Excellent insulation material, it is more efficient and less wasteful to conserve heat than to keep producing it. 2- Toxic excess sugars is converted into less harmful TGs. 3- Fatty acids and TGs are fuel for all organs except brain. 4- Stores body's reserves of EFAs. Phospholipids (PLs), also known as phosphatides, are the second major class of lipids besides TGs. PLs accounts for 5% of total lipids in foods and body. PLs are similar to TGs in that 2 fatty acids are attached to glycerol backbone. But, the third position holds a phosphate group. TGs are water-insoluble and fat-soluble, and aggregate into oil droplets. The phosphate group of PLs is polar and water-soluble, while its fatty acids are fat-soluble. As a result, PLs spread out in a thin layer over surfaces of water, and forms thin membranes. PLs surround the cell and all cell organelles. PLs have many functions within the membrane: 1- form a barrier, 2- helps hold proteins in place in membranes to fulfill structural, enzymatic, and transport functions, 3-

together with proteins control cellular trafficking, and 4- fluidise cell membranes, enabling proteins within them to move freely around the surface of the cell to perform vital functions. Fat-soluble toxic substances such as alcohol, barbiturates, drugs, and carcinogens can exert toxic effects because they dissolve in and pass through the cell's PL membranes into the cell. Toxins can disrupt cell functions. The middle carbon of glycerol molecules in PLs usually holds EFAs, which, being saturated, is bent. EFAs molecules attract O_2, which discourages infectious organisms from thriving. Several other chemical groups can be attached to phosphate groups, such as choline, inositol, serine, or ethanolamine. Besides PLs and proteins, the membranes contain cholesterol, which fine tunes membrane fluidity under constantly fluctuating conditions of food fat intake. Providing a diet rich in EFAs (fluid) means that more cholesterol (rigid) is built into membranes, i.e., EFAs lowers serum cholesterol levels to balance fluidity. Providing a diet rich in SaFAs (hard) means that more cholesterol is removed from membranes, i.e., SaFAs raises serum cholesterol levels. Membranes also contain vitamin E and carotene. Sulphur containing glutathione is attached to the inside of cell membrane, which helps grasp free radicals that make it across the membrane and into the cell. PLs are important in the single-layer membranes, which form the envelopes that surround the fat and cholesterol (as well as vitamins E and A, and carotene) being transported from GIT to liver, from liver to cells, and from cells back to liver. Examples of single-layer membranes include chylomicron, VLDL, LDL, HDL, and albumin. Chylomicrons, made in the intestinal cells, carry food fats. VLDL are made in the liver. LDL carries fats, cholesterol, and fat-soluble vitamins from the liver to cells. HDL carries cholesterol and fats from cells back to the liver. Lipoprotein(α) [LP(α)] has a single-layered membrane as its envelope. Single-layer LP membranes keep water-soluble fats, cholesterol, and vitamins soluble in the watery bloodstream. Lecithin is PL, which was first isolated from egg yolk. Lecithin supplies choline, which is necessary to make the neurotransmitter acetycholine. Acetylcholine is required for brain and nerve function. In the liver, choline helps utilise fats properly. Lecithin emulsifies fats. All unrefined seed oils contain some lecithin. The richest source of lecithin is soybean oil, of which 2% or more is lecithin containing both EFAs (57% LA, and 5-7% LNA). Lecithin is removed when oils are refined. Eggs and meat from animals raised on commercial feeds contain low EFA fats and lecithins, which provide insufficient EFAs for optimum human health. Lecithin helps keep cholesterol soluble. Lecithin keeps cholesterol isolated from arterial linings, protects it from oxidation, and helps prevent and dissolve gall and renal stones by its emulsifying action on fatty substances. Lecithin is necessary for liver detoxification functions. Poor liver function is a common precursor of cancer. Deficiency of either choline or EFAs may induce cancer. Lecithin increases resistance to disease by its role in the thymus gland. In the thymus gland, EFAs are precursors of several prostaglandins (PGs), as well as being vital as part of the ammunition made by immune cells to kill bacteria. Fatty acid peroxides are used to produce bacteriocidal H_2O_2. Lecithin is a phospholipid that makes up 22% of both the HDL and LDL cholesterol-carrying capacity vehicles in the blood. Lecithin is an important part of membrane PLs, which is involved in electric phenomena, membrane fluidity, and other functions for which EFAs are responsible. Lecithin is an important component of bile.

FAT-LOSS SUPPLEMENTS

5-HTP (5-hydroxy tryptophan) is naturally derived from the seedpods of a West African plant known as Griffonia simplicifolia. In the body, the amino acid tryptophan is converted to 5-HTP, which is converted to serotonin. 5-HTP is a precursor to serotonin, the neurotransmitter that signals the brain to feel happy and content. 5-HTP is used then to induce quality sleep and fight depression, and also help curb cravings for sweets and other carbohydrates. 5-HTP is also believed to cause a feeling of early satiation. Low serotonin levels can illicit cravings for high-sugar foods. CLA (conjugated linoleic acid) may significantly help reduce body fat and increases muscle tissue. CLA significantly shift body composition in favour of fat loss and muscle gain. CLA appears to block fat uptake and then increase the speed of fat burning. CLA has the ability to regulate the metabolism of fat through a fairly complex process via lipoprotein lipase and hormone-sensitive lipase. CLA reduces overall fat mass. CLA dramatically reduces the adipocyte (fat cell) size; the CLA plus the addition of guarana reduce the fat cells by 50% in just 6 weeks. Guarana (Paullinia cupana) is an herb that grows within the Brazilian Amazon rainforest. It contains significant amounts of guaranine (the active constituent that's virtually identical to caffeine) and has thus been used for centuries by indigenous tribes to help reduce hunger, relieve fatigue, and treat obesity. Guarana has the ability to "free" fatty acids into the bloodstream and break down and mobilise them to use for energy. Like caffeine, guarana works by stimulating the adrenal glands to release the hormones epinephrine, norepinephrine, and dopamine, which in turn enhance thermogenesis-the body's ability to free fatty acids and use them for energy production (i.e., fat loss) as well as energy, endurance, and mental clarity. Guarana have mild diuretic effects, so increasing water intake is very important with use of this herb. Chromium (usually found in the form of chromium picolinate) is an essential trace mineral that plays an important role in the body's normal carbohydrate metabolism. Chromium aids insulin in properly shuttling these sugars into cells to be stored as energy. Glucose tolerance could be restored in chromium-deficient patients once they were given a diet rich in chromium. Green tea (Camellia sinensis), black, and oolong teas are all derived from the same plant-amellia sinensis. Green tea or, more accurately, the polyphenols in green tea, appear to activate the body's thermogenic activity, promoting the use of calories as energy. Polyphenols in green tea also appear to help produce natural killer immune cells, which scavenge and fight off bacteria and flush out toxins, which basically means it may protect the body from the free radicals that damage cells and weaken the immune systems. Intense exercise increases free radicals. Green tea is packed with caffeine and catechin polyphenols, both of which increase resting metabolic rate. 7-KETO (3-acetyl-7-oxo-dehydroepiandrosterone) could be the most potent thermogenic enhancer available. 7-Keto increases the body's metabolic rate by way of T3 (thyroid hormone). After the age of 25, the metabolisms begin to slow. The inevitable response is to store the unburned calories as fat, and thus, a harder time keeping unwanted weight off. 100 mg of 7-Ketot twice daily for 8 weeks increases levels of the thyroid hormone T3 (or triiodothyronine) by (within the normal range) 17.88%. T3 is only the most active thyroid hormone and has the greatest effect on metabolic rate. 100 mg of 7-Keto taken twice daily is most effective for weight loss and has positive effects on fat loss.

Nutrient deficiency:

Nutrient	Problem
Biotin	Dermatitis, eye inflammation, hair loss, loss of muscle control, insomnia, muscle weakness.
Calcium	Brittle nails, cramps, delusions, depression, insomnia, irritability, osteoporosis, palpitations, periodontal disease, rickets, and tooth decay.
Chromium	Anxiety, fatigue, glucose intolerance, adult-onset diabetes.
Copper	Anaemia, arterial damage, depression, diarrhoea, fatigue, fragile bones, hair loss, hypothyroidism, weakness.
Essential fatty acids	Diarrhoea, dry skin and hair, hair loss, immune impairment, infertility, poor wound healing, premenstrual syndrome, acne, eczema, gall stones, liver degeneration.
Folic acid (B_9)	Anaemia, apathy, diarrhoea, fatigue, headaches, insomnia, loss of appetite, neural tube defects in foetus, paranoia, shortness of breath, weakness.
Iodine	Cretinism, fatigue, hypothyroidism, weight gain.
Iron	Anaemia, brittle nails, confusion, constipation, depression, dizziness, fatigue, headaches, inflamed tongue, mouth lesions.
Magnesium	Anxiety, confusion, heart attack, hyperactivity, insomnia, nervousness, muscular irritability, restlessness, weakness.
Manganese	Arteriosclerosis, dizziness, elevated cholesterol, glucose intolerance, hearing loss, loss of muscle control, ringing in ears.
Niacin	Bad breath, canker sores, confusion, depression, dermatitis, diarrhoea, emotional instability, fatigue, irritability, loss of appetite, memory impairment, muscle weakness, nausea, skineruptions and inflammation.
Pantothenic acid (B_5)	Abdominal pains, burning feet, depression, eczema, fatigue, hair loss, immune impairment, insomnia, irritability, low blood pressure, muscle spasms, nausea, poor coordination.
Potassium	Acne, constipation, depression, oedema, excessive water consumption, fatigue, glucose intolerance, high cholesterol levels, insomnia, mental impairment, muscle weakness, nervousness, poor reflexes.
Pyridoxine (B_6)	Acne, anaemia, arthritis, eye inflammation, depression, dizziness, facial oiliness, fatigue, impaired wound healing, irritability, loss of appetite, loss of hair, mouth lesions, and nausea.
Riboflavin	Blurred vision, cataracts, depression, dermatitis, dizziness, hair loss, inflamed eyes, mouth lesions, nervousness, neurological symptoms (numbness, loss of sensation, electric shock sensations), seizures, sensitivity to light, sleepiness, weakness.
Selenium	Growth impairment, high cholesterol levels, increased incidence of cancer, pancreatic insufficiency (inability to secrete adequate amounts of digestive enzymes), immune impairment, liver impairment, male sterility.
Thiamine	Confusion, constipation, digestive problems, irritability, loss of appetite, memory loss, nervousness, numbness of hands and feet, pain sensitivity, poor coordination, weakness.
Vitamin A	Acne, dry hair, fatigue, growth impairment, insomnia, hyperkeratosis (thickening and roughness of skin), immune impairment, night blindness, weight loss.
Vitamin B_{12}	Anaemia, constipation, depression, dizziness, fatigue, intestinal disturbances, headaches, irritability, loss of vibration sensation, low stomach acid, mental disturbances, moodiness, mouth lesions, numbness, spinal cord degeneration.
Vitamin C	Bleeding gums, depression, easy bruising, impaired wound healing, irritability, joint pains, loose teeth, malaise, and tiredness.
Vitamin D	Burning sensation in mouth, diarrhoea, insomnia, myopia, nervousness, osteomalacia, osteoporosis, rickets, scalp sweating.
Vitamin E	Gait disturbances, poor reflexes, loss of position sense, loss of vibration sense, shortened RBC life.
Vitamin K	Bleeding disorders.
Zinc	Acne, amnesia, apathy, brittle nails, delayed sexual maturity, depression, diarrhoea, eczema, fatigue, growth impairment, hair loss, high cholesterol levels, immune impairment, impotence, irritability, lethargy, loss of appetite, loss of sense of taste, low stomach acid, male fertility dysfunction, night blindness, paranoia, white spots on nails, wound healing impairment.

L-carnitine is nonessential amino acid, required for the heart to function efficiently, especially during exercise. L-carnitine is fat transporter, needed to transfer fatty acids across cell membranes into the mitochondria, which in turn uses the fat as a primary source of energy. Carnitine may not increase the rate of weight loss but rather could increase the ratio of fat to muscle loss, thus preserving muscle mass while increasing the rate fat is burned. Carnitine deficiency is associated with increased level of fats in bloodstream, which may interfere with our ability to lose bodyfat. Carnitine deficiency may result in lower ATP (muscle energy) levels. Carnitine turnover is accelerated during exercise, and shortages could limit the amount of energy available to muscles. The result is a rapid onset of fatigue and subsequent compromised recovery. Individuals who supplement with carnitine while engaging in intense exercise programs are less likely to experience muscle soreness and fatigue. Carnitine should be supplemented about 30 minutes to 1 hour before exercise. Carnitine (2000-4000 mg, in divided 2 dosages throughout the day) shouldn't be used with protein foods or supplements. Pyruvate (calcium pyruvate) increases resting metabolic rate. Pyruvate (6 g tid) significantly decreases bodyweight and bodyfat (12%), while gaining lean mass with 6 weeks. Citrus aurantium (bitter orange, synephrine) is a powerful thermogenic Chinese herb. Citrus aurantium appears to work by way of its active compound called synephrine, which is a bit like ephedra (an amphetamine-like chemical found in a lot of weight-loss and performance products). Synephrine reduces appetite and boost metabolism, thus stimulating fat loss. Citrus aurantium contains chemicals called amines (tyramine and octopamine), which are not as lipophilic-meaning they do not cross the blood/brain barrier as easily as ephedra, which reduces CNS stimulation and cardiovascular effects. Citrus aurantium stimulates certain receptors (β_3 adrenergic receptors) that help break down fat (lipolysis). Simultaneously, this stimulation causes an increase in the metabolic rate. 4-20 mg of synephrine/day, which usually is provided by supplementing with 200-600 mg of a standardised Citrus aurantium extract (at 3% to 6% synephrine), may be most effective. Coleus forskohlii (coleus) is an ancient Ayurvedic plant and member of the mint and lavender family, which grows in the mountains of Asia. The active ingredient in coleus is forskolin, which plays a major role in a variety of important cellular functions, including inhibiting histamine release, relaxing muscles, increasing thyroid function, and increasing fat-burning activity. Coleus bypasses the adrenergic receptors and goes straight into the cAMP cycle. Forskolin increases adenylate cyclase, which increases levels of cAMP, which is found in fat. cAMP then stimulates another enzyme, hormone sensitive lipase (HSL), to burn fat. Coleus increases thyroid hormone production and release via cAMP to burn more calories. Coleus has abundance of benefits, including bodyfat reduction and lean body mass enhancement. Coleus forskholii may promote weight and fat loss and/or mitigate (or moderate) weight gain in overweight subjects. HCA (Garcinia cambogia, hydroxycitric acid) is an inhibitor or blocker of the enzyme ATP citrate lyase, which is required for the synthesis of fatty acids blocking fat storage. HCA suppresses fatty acid mobilisation and blocks the conversion of carbohydrates to fat, resulting in less fat storage. HCA is active when carbohydrates are over-consumed. Guggul lipid (guggalsterone) reduces bad cholesterols (LDL). Guggul may enhance thyroid function, increasing levels of circulating triiodothyroxine (T3), a thyroxine metabolite known to raise overall metabolism. Yerba mate (maté, Ilex paraguariensis) grows wild in the rainforests of Argentina, Chile, Peru, Brazil, and Paraguay. Yerba mate improves mood and concentration, reduces anxiety, prevents mental fatigue, and prolongs the effects of thermogenesis, along with a multitude of other compelling benefits. There are 196 chemicals in Yerba mate that become active in the body once consumed-including B vitamins; Vitamins A, C, and E; and the minerals calcium, magnesium, iron, potassium, and selenium. Yerba maté has 11 polyphenols, which are powerful antioxidants. But the most important chemical in maté is mateine. Mateine is a xanthine alkaloid. Like other xanthines, it mildly arouses the central nervous system, but unlike other stimulants, it doesn't appear to be addicting, nor does it produce unwanted side effects such as insomnia or nervousness. Interestingly, it actually works as a tonic for the CNS, calming the body and mind. Ephedra, or ma huang (its interchangeable herbal name), is an ancient Chinese herbal form of the powerful stimulant ephedrine. Ephedra (or ma huang) produces fight-or-flight response (sweat, and increases blood flow to the heart, brain, and muscles) due to increased adrenaline and CNS stimulation of β_1, β_2 receptors. This internal act, in turn, raises the body's core temperature-a process called thermogenesis-and help burn fat. Ephedra is a thermogenic and appetite suppressant. The dietary supplements that contain ephedra alkaloids pose a serious health risk.

THE BRAIN CHEMICAL-MOOD AND MENTAL FUNCTIONING

The B vitamins and vitamin C can improve mood and mental functioning, a good balanced multi-vitamin and mineral supplement should be the backbone to any nutritional program. The messengers, i.e, neurotransmitters, include acetylcholine, serotonin, noradrenaline (norepinephrine) and dopamine. Acetylcholine is significantly involved in mental acuity and memory. Vitamin B_5 (pantothenic acid) is necessary for the production of acetylcholine from the nutrient choline, which is provided by the diet, for instance, in the form of lecithin provided by foods such as eggs and soybeans. Phosphatidylserine (PS) is a member of the class of compounds known as phospholipids. PS is not abundant in the normal diet and is found in trace amounts in the usual sources of lecithin. PS induces the production of a number of neurotransmitters and/or to prevent their age-related decline. PS stimulates acetylcholine output and the synthesis and release of dopamine. PS reverses the loss of signal, which marks memory decline. PS may play a role in preparing the membranes of the cell to receive the signals sent by other cells. Cytidine 5'-diphosphocholine, also known as CDP-choline and citicholine, is involved in the synthesis of the phospholipids, which make up the membranes of cells, especially the pathway involving phosphatidylcholine. When taken orally, CDP choline is metabolised to yield the free nucleotide cytidine and choline, and peak plasma levels occur several hours after consumption for delivery to various parts of the body. Both elements readily cross the blood/brain barrier, become incorporated into brain membrane lipids, and increase the production of neurotransmitters in the CNS, including the synthesis of acetylcholine, noradrenaline and dopamine. CDP choline consumption promotes brain metabolism by restoring phospholipid content in the brain and regulation of neuronal membrane excitability and osmolarity (by its effect on ATP-dependent Na-K pumps). CDP-choline also influences the mitochondria or energy factories of the brain cells. It improves memory in the elderly and has been shown to be useful as co-therapy in Alzheimer's disease and Parkinson's disease. S-adenosylmethionine (SAMe) benefits have been recognised for years. These include amelioration of depression and migraine headaches, but also range to improving osteoarthritis, fibromyalgia and liver disorders. SAMe supplementation increases brain levels of serotonin, dopamine and phosphatidylserine, plus it improves the binding of these neurotransmitters to their binding sites. SAMe is effective in elevating mood and brightening subjects' outlook on life. Vitamin B_6, as is true of SAMe, improves serotonin production in the brain. A factor, which is common to the actions of supplements such as PS, CDP-choline and SAMe, is their effect upon the cell membranes of the cells of the brain and nervous tissues. These tissues are largely composed of fats. Increasing the degree of ω-3 fatty acids available to these membranes, especially the amount of DHA (docosahexaenoic acid), improves the fluidity of these membranes and is linked to improvements in mood and mental functioning in general. Improvement in fluidity promotes proper neuronal transmission and brain communication. Docosahexaenoic acid is the primary structural fat of the brain and also in the retina. Neuronal membranes should also be protected from free radical damage through supplementation with vitamin E. Free radicals damage both the membranes and the nuclei of the cells. Acetyl-L-carnitine is perhaps the only nutrient, which has been shown to rejuvenate the membrane of the mitochondria. Acetyl-L-carnitine performs this function best when used in conjunction with ALA. α-Lipoic acid has been shown to improve the energy metabolism of the brain, and CoQ_{10} is also useful in this regard. Nutrients for improving mood and energy are the amino acids L-phenylalanine and L-tyrosine. These amino acids are precursors for the production of the neurotransmitter norepinephrine. The conversion of L-phenylalanine and L-tyrosine to norepinephrine is enhanced by the actions of vitamin B_6. Siberian ginseng appears to increase the monoamine content, i.e., the neurotransmitter content, of the brain. It serves as a natural MOA inhibitor without the side effects and dangers of the drugs normally used in this regard. Depression and insomnia respond well to Siberian ginseng, and the result is an increased feeling of well-being. The flavonoid chrysin, the so-called "Flavone X" sometimes used by athletes. One source of this flavonoid is a member of the passion flower family. Chrysin turns out to be a successful anxiolytic compound, which is very useful for combating stress. This flavonoid also exerts its anti-stress effect without sedation. Ginkgo biloba extracts are perhaps the most widely used of all cognitive enhancers. They have been shown to improve Alzheimer's disease, memory, the ability to concentrate, reaction time, mood, and many other aspects of brain function. Ginkgo extracts may speed recovery even from stroke. Ginkgo extracts markedly improve the circulation to the brain as a whole and especially within certain areas of the brain. However, they do more than just increase circulation. These extracts enhance both the uptake of glucose and the utilisation of oxygen by the brain and other nerve tissues, and thus increase the amount of energy available for mental functioning. The rate of signal transmission by brain nerve cells is improved with ginkgo extracts and that the synthesis of some of the neurochemicals used to transmit nerve signals, such as acetylcholine, is increased. St. John's Wort has a long history of improving not just mood, but wound healing and recuperation. In terms of neurotransmitters, St. John's Wort maintains norepinephrine levels and may act as a natural MAO (monoamine oxidase) inhibitor, but one without side effects. Hypericin, a constituent of St. John's Wort, may be one active ingredient, but it now is know that other ingredients are perhaps even active. St. John's Wort may improve brain levels of serotonin and dopamine along with those of norepinephrine by serving as a selective reuptake inhibitor for all of these. 5-hydroxytryptophan (5-HTP) is very closely related to the essential amino acid tryptophan. Tryptophan helps to alleviate the mood swings, which are associated with premenstrual syndrome (PMS) and as a sleep aid. In fact, tryptophan, which is found in milk, poultry and many other foods as well as being added routinely to infant formulas, is a precursor to 5-HTP. The traditional cup of warm milk before bedtime is helpful for inducing sleep in part because of the tryptophan found in the milk. Tryptophan from the diet initially is taken up by brain cells. These cells transform the tryptophan into 5-HTP. 5-HTP, in turn, is converted into serotonin (5-hydroxytryptamine [5-HT]). In the case of the neurochemistry of sleep, serotonin undergoes yet other conversions in the pineal gland to yield melatonin (N-acetyl-5-methoxy-tryptamine). In 1980s, tryptophan produced in Japan was linked with eosinophilia-myalgia syndrome. No other batches of tryptophan have been associated with eosinophilia-myalgia syndrome. The employment of 5-HTP aims at providing the substrate or metabolic precursor to serotonin while allowing the body itself to regulate the further steps in serotonin metabolism. This is the orthomolecular approach to nutrition. The conversion of tryptophan to serotonin and/or the realisation of serotonin's proper metabolic and physiologic effects depend upon the body having access to adequate amounts of vitamins (such as vitamin B_6) and minerals (such as Mg and Zn). Likewise, diets that lead to insulin resistance and the blood sugar roller coaster can have an unfortunate impact upon serotonin production and regulation. Serotonin is a classic neurotransmitter, fulfils a number of criteria: 1- A neurotransmitter must be produced in neurons, and serotonin is produced both in the brain and throughout the nervous system. 2- Serotonin is stored at the ends of the neurons at the synapses in vacuoles. 3- Upon the proper stimulation, serotonin is released from storage into the synaptic cleft, i.e., the synapse. 4- Once it has crossed the synapse, serotonin binds to a specific receptor on the postsynaptic membrane of the adjacent neuron. 5- Once it has bound to the receptor, serotonin triggers a response. After the receptor has been activated, the serotonin is either broken down or taken up again by the originating nerve cell via reuptake transporters and place back in storage. If the body is producing too little serotonin, there are 3 ways in which this problem can be addressed: A- Monoamine oxidase inhibitor (MAOI) is used to prevent the rapid breakdown of released serotonin and thus make a smaller amount goes further. This mode of action has been suggested as explaining the effects of St. John's Wort. B- A second approach is to prevent the rapid reuptake of serotonin by the originating neurons, such as is the action of Prozac. C- Increase the supply of the basic building block for serotonin. There is a direct relationship between the amount of tryptophan available from the diet and the concentration of serotonin found in the brain. Serotonin deficiency syndrome is a complex of symptoms, which includes depression, anxiety, insomnia, and obesity and migraine headaches. Correcting the serotonin deficiency can be achieved through supplementation with 5-HTP (100 mg), which is both safer and more effective than using drugs, such as Prozac. Supplementation with 5-HTP also appears to be as good or better than supplementation with tryptophan for most purposes. Psychological conditions other than depression are linked to deficiencies in the production of serotonin. Among these are anxiety, insomnia, aggressiveness and some obsessive-compulsive disorders. A substantial fraction of children with attention deficit hyperactivity disorder (ADHD) are deficient in serotonin. Migraine headaches may be related to serotonin levels in the brain. 5-HTP has proven to be useful in many of these conditions. The sleep-enhancing effect of 5-HTP reflects the fact that serotonin is required for the production of the hormone melatonin. 5-HTP might be the preferred preventive measure against migraine attacks. Unfortunately, the amount of 5-HTP necessary to influence weight loss via appetite suppression appears to be quite high, 600-900 mg/day. 5-HTP may be effective as an adjuvant to other diet aids at much smaller dosages (100-200 mg/day). The primary complaint, at least during the initial weeks of use, appears to be transient nausea, which can be minimised by taking the compound with food and/or starting with low dosages and increasing these slowly. Side effects are uncommon at the lower dosages generally suggested to improve sleep and mood. Acetyl L-carnitine is the more stable and more bioavailable form of carnitine. Acetyl L-carnitine characteristics include: 1- provides antioxidant protection for neurons against the superoxide radical, 2- enhances the metabolism of fatty acids in the mitochondria, 3- supports energy production in the heart muscle, 4- easily crosses the blood-brain barrier, 5- is found naturally in the central nervous system, 6- aids in the body in coping with elevated levels of ammonia in the brain and the blood, and 7- may enhance tolerance of exercise and improve physical performance in some individuals. Acetyl L-carnitine is sometimes referred to as vitamin B-T because of its vitamin-like roles. The body needs lysine, methionine, vitamin C, iron, niacin, and vitamin B_6 to produce limited quantities of L-carnitine, mainly in the liver and the kidneys. L-carnitine is found in small amounts in the diet, primarily in red meat. Mutton and lamb are the richest sources, whereas chicken and turkey contain much less; dairy products contain only small amounts of L-carnitine. The typical daily diet of a non-vegetarian supplies approximately 50 mg of L-carnitine. Human beings may have to consume 500 mg or more of this nutrient. In the human body, L-carnitine is concentrated in the heart and the skeletal muscles, and also in the brain and in the sperm. Only the natural L- form should be used; the synthetic D- and DL- forms do not function properly for human metabolism and have severe side effects. Although the L-carnitine is usually referred to as an amino acid, this is technically incorrect inasmuch as there is no amino (NH_2) group present in the molecule. The primary role of L-carnitine in the body is as a biocatalyst or coenzyme. One of the most important functions of L-carnitine is in the oxidation of long chain fatty acids, a process that takes place inside of mitochondria, the energy factories of the cells. This process is known as β-oxidation. L-carnitine acts as a shuttle for bringing fatty acids into the mitochondria and then removing waste afterwards. Fats are the preferred source of fuel for the skeletal muscles, and even more so for the heart muscle. As much as 70% of the energy generated in muscle tissues comes from the oxidation of fats. L-carnitine increases the rate of oxidation of fats in the liver. L-carnitine supplementation during dieting can help to control the negative effects of ketosis (the accumulation of waste products of fat metabolism) in those who are susceptible to this problem. L-carnitine is produced internally by the liver and kidney. L-carnitine reduces muscle soreness related to exercise. 2 g of L-carnitine taken twice per day for 2-4 weeks lead to positive changes in breathing response to exercise. L-carnitine supplementation reduces at least some of the after-effects of strenuous exercise. L-carnitine acts locally in the muscle tissues as an antioxidant. L-carnitine penetrates into the mitochondria, where most of the free radicals are generated through the oxidation of food for energy. L-carnitine may serve to spare antioxidants, such as vitamin C. L-carnitine has positive benefits upon the myocardium of the heart and upon peripheral circulation. Supplemental L-carnitine is associated with significantly higher concentrations of pyruvate, ATP and creatine phosphate in portions of the heart muscle during conditions of extreme stress. L-carnitine is quite useful for improving blood flow. Acetyl L-carnitine is composed of acetic acid and L-carnitine bound together. In the human brain, the acetylation of L-carnitine is a normal event. Acetyl L-carnitine influences neural levels of acetylcholine, a major neurotransmitter responsible for memory and proper brain function. As the brain ages, the ability to synthesise and utilise acetylcholine declines. Acetyl L-carnitine acts as an antioxidant within the brain, stabilises cell membranes, enhances energy production within brain cells, and supports the synthesis and functions of acetylcholine. The brain, which composed mostly of fats, enzymatically converts cholesterol to pregnenolone and DHEA (dehydroepiandrosterone). These are then precursors for various other steroid compounds. At least 150 steroid hormones have been identified in the brain. These have important functions. Pregnenolone is the most potent memory enhancer. Pregnenolone, DHEA and testosterone decrease with age. This decrease may contribute to the age-related deficit in learning and memory. Pregnenolone acts as an antagonist to the receptors, which are stimulated by GABA (γ-amino-butyric acid). GABA leads to relaxation, the slowing of reflexes, even to sleep. This is desirable at the end of the day, but not during it. By antagonising GABA receptors, pregnenolone may lead to greater focus and awareness during the day. Unfortunately, pregnenolone and DHEA can unfavourably influence other hormone levels, e.g., 7-keto DHEA. Huperzine A is an alkaloid that was first isolated from Huperzia serrata. It is found in an extract from a club moss that has been used for centuries in Chinese folk medicine. Huperzine is an extremely potent acetylcholine-esterase inhibitor, i.e., it inhibits the enzyme that breaks down the neurotransmitter acetylcholine. This is of importance in degenerative diseases, such as Alzheimer's. Approximately 58% of patients treated with huperzine (0.2 mg bid) are expected to show improvements in their memory, cognitive and behavioural functions. Vinpocetine is a derivative of the alkaloid vincamine, which is found in small amounts in the seeds of periwinkle (Vinca major) and also in a few other plants, such as voaconga and Crioceras longiflorus. Vinpocetine has 5 main pharmacological and biochemical actions: 1) selective enhancement of the brain circulation and oxygen utilisation without significant alteration in parameters of systemic circulation, 2) increased tolerance of the brain toward hypoxia and ischaemia, 3) anticonvulsant activity, 4) inhibitory effect on phosphodiesterase (PDE) enzyme, and 5) improvement of rheological properties of the blood and inhibition of aggregation of thrombocytes. Vinpocetine (20 mg bid) increases mental alertness, reduces confusion, enhances recall of recent events, diminished anxiety, less depression and fatigue, and greater emotional stability. Many fats, such as conjugated linoleic acid (CLA) and ω-3 fats, have extraordinary nutritive value. Docosahexapentaenoic acid (DHA) is an essential fatty acid (EFAs) derivative from the ω-3 family. Both EPA and DHA play roles in body functions. EPA seems to exert its activity on the cardiovascular system by suppression of arachidonic acid (AA) and increase of beneficial eicosanoids. In times of need, it is rather difficult for EPA to convert into DHA. On the other hand, DHA is a more versatile fatty acid: 1- DHA is easily retroconverted to EPA when needed, and 2- DHA is vital during old age, infancy, pregnancy and is critical for the proper functioning of the nervous system. DHA is selectively incorporated into the synapses of the brain and nervous system tissue. The prevalence of DHA in brain membranes results in a tremendous impact on the fluidity and regulation of cell receptors. The benefits of compounds like DHA and phophatidylserine may be as a result of their capacity to regulate proteins found in cell membranes. It is possible, that mental abnormalities such as dementia and Alzheimer's disease may be related to a lifelong deficiency of EFAs. The most important period in supplying adequate DHA and EPA is during the first 3 months before conception. This is an important time for cell commitment and growth. Because of its importance in nourishing brain and nervous system development, DHA in the maternal diet is a factor in birth weight of newborn. The fatty acids found in mother's milk is 0.2% DHA. It is for this reason that infants fed breast milk are healthier and have better visual acuity. ω-3 fish oils can be present as re-esterified-tryglycerides (rTG), ethyl esters (EE) or free fatty acids (FFAs). The ethyl ester forms of ω-3s display a delayed and reduced incorporation. The FFA form of ω-3s also displays delayed and reduced absorption.

NUTRITIONAL ASPECTS OF DEPRESSION

60 % of the dry weight of the brain is fat. Essential fatty acid (EFA), including ω-3 and ω-6, are important components of nerve cell walls and are involved in neurotransmitter electrical activity and post-receptor phospholipid mediated signal transduction. Neuronal degeneration is found commonly among people with chronic schizophrenia as they have apparent increased phospholipid neuron membrane breakdown, which concentrates in the frontal cortex and other areas of the brain. EFAs offer a means of maintaining brain membrane structure and avoiding brain mass loss. 10%-15% of individuals with schizophrenia has cerebral/brain allergies. Elimination dieting is important for diagnosis. Testing for overpopulated gut microorganisms such as candida albicans is a vital part of the assessment when cerebral allergies are suspected. Hypoglycaemia tends to be an aggravating factor in mental illnesses rather than a causative factor. B_6 and zinc supplementation is 95% successful in the management of pyrroluria-type schizophrenia. Depression is often a symptom of other disorders, as in schizophrenia or manic-depressive reactions. Psychotic depression is characterised by more severe symptoms. Typically, sleep is disturbed, with problems of waking up early in the morning. It may affect appetite and lead to anorexia (pathological loss of appetite) and decreased sex drive. Although exhaustion is the physical aspect, failure to reach one's goals may be related to personality problems. Some unhealthy work environments contribute to depression. If the depression were seen as being caused by psychological aspects or personality problem, a course in psychotherapy would be the most appropriate step. Depression is often caused by ill-health or some metabolic disorder. Any conventional illness can contribute to depression and these should be eliminated in the first place. The thyroid gland controls the rate of metabolism and all chemical processes of the body slow down in hypothyroidism. Low thyroid function may also be an important factor in chronic fatigue and depression. One way of testing hypothyroidism is to take the temperature in the morning before coming out of bed. Temperature below 36.5°C or 97.8°F over a number of days may indicate hypothyroidism. Hypothyroidism may also be the cause of high cholesterol. Thyroxine production depends on a complex range of nutrients. Iodine is one of the precursors of thyroxine. This is contained in kelp and iodised salt. Vitamin A-retinol-and not in the carotene form may be essential in converting iodine into thyroxine. The liver can't convert carotene to vitamin A in the absence of thyroxine or in hypothyroidism. Vitamins B_2, B_3, B_6 and C are required for absorption of iodine. A vitamin B_1 (thiamine) deficiency alone can cause hypothyroidism. Vitamin B_{12} can't be absorbed with a deficient thyroid gland. Copper is required for the production of TSH from the

pituitary. Foods that interfere with the uptake of iodine are: cabbage, kale, Brussels sprout, cauliflower, broccoli, Kohlrabi, turnips, rutabaga, rapeseed, brown (Indian), black, or white mustard, garden cress and radish, soybeans, skins of peanuts, almonds, and cashews. Thus when eating these food frequently one should take extra iodine supplementation. The following chemical substances inhibit iodine uptake; sulfa, anti-diabetic drugs, prednisone, oestrogen, smoking (thyocyanide inhibitor) and fluoride (thyro suppression). Tyrosine-a non-essential amino acid- is a precursor to thyroid, adrenocortical hormones, dopamine, and melanin-pigment found in hair, skin and the choroid of the eye. Vitiligo is the disorder of melanin distribution on the skin and could therefore be related to hypothyroidism. Deficiency of tyrosine may show up as having low body temperature, low blood pressure and restless legs. The body can produce tyrosine from an essential amino acid called phenylalanine. Deficiency of the latter lead to a variety of symptoms includes bloodshot eyes, cataracts and behavioural changes. Phenylalanine is also the precursor (via tyrosine) of dopamine, and then on to norepinephrine and epinephrine (adrenaline)-a deficiency of these may lead to depression indicating that it affects behaviour in a fundamental way. Hypochlorhydria may block the digestive process of amino acids including phenylalanine. Thyroid deficiency may be treated naturally with supplementation of phenylalanine or tyrosine, as well as the treatment for depression. Phenylalanine in large doses-in excess of 3% of diet-an amino acid imbalance may cause tyrosine toxicity. Phenylalanine can aggravate a preexisting pigmented melanoma. Schizophrenia may be due to an error in dopamine metabolism. As phenylalanine is a precursor of tyrosine and then of dopamine, administration of L-dopa (which passes the brain barrier, not dopamine) together with antioxidants may help some schizophrenics. Phenylalanine or tyrosine is contraindicated in individuals taking monoamine oxidase inhibitors (MAO inhibitors). For depressive states 100-500 mg/day of L-phenylalanine. Individuals suffering from phenylketonuria, which is a disease caused by a defective enzyme-phenylalanine hydroxylase-converting phenylalanine to tyrosine are accumulating phenylalanine at toxic levels and should be avoided. Monoamine oxidase (MAO) is an enzyme in the brain, which degrades the monoamine neurotransmitters dopamine, norepinephrine (NE) and serotonin. This enzyme increases in activity with ageing, lowering the levels of the neurotransmitters available to the brain, and make elderly more susceptible to depression. MAO inhibitors can cause hypertensive crises, interact with other depressant, or hypotensive drugs and they react with many foods and beverages such as cheese, protein extracts, soy sauce, pickled herrings, and red wine. Individuals with epilepsy, cardiovascular disease and those with hepatic and renal insufficiency are especially at risk with MAO inhibitors. Some side effects are insomnia, agitation, dizziness, hypotension when in a lying position (sleep), constipation, dry mouth, blurred vision, difficulty in urination, etc. Natural sources of phenylalanine: soybeans, cottage cheese, fish (especially trout), meat, liver, lamb poultry, almonds, Brazil nuts, pecans, pumpkins, sesame seeds, lima beans, chickpeas and lentils. Soybeans and almonds may interfere with iodine uptake. Over 62% of hypoglycaemics suffer from depression and insomnia. If the fall in blood glucose is over 2.8 mm/l in any one-hour or 1.9 mm/l in any half-hour, the brain is starved of glucose with all the pseudo-psychological consequences, including depression. When the brain is suddenly starved of glucose, the pituitary gland sends an urgent message to the adrenal glands to pour adrenaline into the blood stream. Adrenaline is a hormone that rapidly converts glycogen (stored liver sugar) into glucose, thus raising the blood sugar level. However, adrenaline is also the fight/flight hormone, readying the body for quick action in case of danger. Thus the sudden presence of adrenaline in the blood stream wakes up the poor sleeper-usually in the early morning. Thus depression and insomnia are often found together. A strict hypoglycaemic diet is the main remedy against depression. The hypoglycaemic diet consists of 3 hourly, high protein snacks, the avoidance of sugar, coffee, sugary drinks, white rice, white bread and cakes, plus high potency B-complex vitamins, vitamin C, chromium and zinc. Hypoglycaemic symptoms may be overcome by the taking of one tablespoon of glycerine mixed in fruit juice or even water with lemon juice. Glycerine is not recognised by the pancreas as a sugar, so it does not stimulate the over-production of insulin. Fructose has a similar biochemical pathway as glycerine, but excess fructose may raise triglyceride levels. A peaceful night can be obtained with glycerine, vitamin B_1 (thiamine is involved in glucose metabolism), or vitamin B_5 (pantothenic acid). Nicotine, caffeine and alcohol are stimulants (release adrenaline), which destabilise the blood glucose levels. Depression can also be caused by the body's inability to produce the neurotransmitter serotonin, which is normally synthesised in the body from other substances. Tryptophan-an amino acid and building block of protein-is the precursor of serotonin. A low protein diet, typical of hypoglycaemics, causes a tryptophan deficiency. Protein should be avoided for 90 min before and after administration of (L-tryptophan, 4-6 g/day) and the uptake can be improved with sugar. Insulin improves absorption by lowering levels of competing amino acids. Tryptophan is converted to serotonin in the presence of vitamin B_6 (Pyridoxine). When there is a deficiency of vitamin B_6, tryptophan may be transformed into excessive xanthurenic acid, which may cause cancer (bladder), attack the pancreas and cause diabetes. A B_6 deficiency can cause insomnia. Patients on anti-psychotic drugs need higher doses of vitamin B_6. Any drug taking or the presence of toxins will use up all vitamin B_6, so that none is left to convert tryptophan into serotonin. Detoxification is also aided by vitamin C. Tryptophan is also a precursor of vitamin B_3 (niacin), which is so important that the body considers its production to be more important than that of serotonin. It requires 60 mg of tryptophan to produce 1 mg of niacin in case of dietary niacin deficiency. Niacinamide supplementation to schizophrenics may sometime be helpful to liberate the production of serotonin from tryptophan. Vitamin B_3 deficiency can cause insomnia, mood swings, bedwetting in children, crying spells, anxiety, depression and affect the eyesight. Tryptophan supplementation may have adverse reactions and should be administered under the supervision. Natural sources of tryptophan: Soya protein, brown rice (uncooked), cottage cheese, fish, beef, liver, lamb, peanuts, pumpkin, sesame seeds and lentils. Milk and cheese contain tryptophan and this is why a glass of warm milk before bedtime initiates sleep. Warm milk combined with a tablespoon of glycerine is an ideal sleeping agent. Bananas and dates are also providing tryptophan. Other good sources of tryptophan are chlorella or other green or blue algae tablets taken at bedtime to induce sleep (via serotonin production). Serotonin is the precursor of melatonin, a hormone produced by the pineal gland. When the eyes perceive dusk or darkness, it signals the pineal gland to produce this hormone which is closely related to the diurnal cycles of sleep and wakefulness. Melatonin has sedative qualities and helps reduce anxiety, panic disorders and migraines as well as induce sleep. Melatonin is a powerful antioxidant and eliminates free radicals toxic to DNA. Thus sleeping restores the immune system. Melatonin inhibits release of oestrogen thereby reduces risk of breast cancer. A disturbance in the diurnal melatonin production causes depression, rather than the amount of melatonin in the body at a certain time. Exposure to bright, early morning sunlight (between 7.00 AM and 9.00 AM) for at least 15 min is perhaps the most powerful signal that sets the biological clock, thereby washing away depression. When individuals are exposed to artificial light, i.e., light lacking the full spectrum sun light, the body cannot absorb certain nutrients and this contributes to fatigue, tooth decay, depression, hostility, suppressed immune function, hair loss, alcoholism and drug addiction and even cancer. Taurine levels may rise in the pineal and pituitary gland through exposure to full spectrum daylight. Lack of taurine may lead to mental impairment and depression. GABA or γ-amino-butyric acid is essentially an inhibiting neurotransmitter. Neurotransmitters are hormone-like chemicals controlling messages between neurons in the brain. GABA is produced by specialized cells. It fits neatly into receptor molecules of other cells and thereby can act to inhibit release of dopamine from dopamine cells. Dopamine causes intense feelings of pleasure. Thus GABA regulate the release of dopamine, which influences other cells to experience pleasure (or satiety).

The following individual nutrient deficiencies are associated with depression:

A- Vitamins	
	1- Biotin
	2- Folic acid
	3- Pyridoxine
	4- Riboflavin
	5- Thiamine
	6- Vitamin B_{12}
	7- Vitamin C
B- Minerals	
	1- Calcium
	2- Iron
	3- Magnesium
	4- Potassium

Severely depressed individual cannot experience pleasure. Excess dopamine production-Intense pleasurable rewards produce addiction to substances that causes excess dopamine secretion. In cocaine addiction, the reabsorption of dopamine is blocked by dopamine cells, resulting in excess dopamine. This leads to intense pleasure and results in cravings for the same substance. Nicotine, as an addictive substance, acts by occupying the GABA receptor sites on dopamine cells, drowning out GABA, thus causing increased dopamine production and addiction. Ongoing dopamine synthesis causes dopaminergic exhaustion. The amino acids phenylalanine and/or tyrosine are precursors of dopamine, which has been used in the treatment depression. The conversion from dopa to dopamine is dependent on vitamin B_6; a B_6 deficiency can cause depression. Inositol and vitamin B_3 (niacinamide) occupy the same receptors, which may explain why some people feel relaxed and sleepy when taking these nutrients. The body produces GABA from glutamic acid in the presence of vitamin B_6 (pyridoxine). Glutamic acid cannot pass the lipid layer of the brain cell unless in the form of glutamine. When glutamine enters the brain cell it is converted to glutamic acid. In this form it may: 1- Combine with ammonia (a highly toxic end-product of protein) to form glutamine, which is carried to the liver and then excreted as urea in the urine. 2- Combine with vitamin B_6 to form GABA. Glutamic acid itself is an excitatory substance. Glutamine stops alcohol sugar craving. If there is a deficiency of vitamin B_6 there may be an excess of glutamic acid causing anxiety and restlessness. If there is an excess of vitamin B_6, too much GABA is produced causing one to feel tired and depressed. Related to hypoglycaemia is heavy metal intoxication. High levels of lead, mercury and cadmium interfere with the enzymes breaking down glucose into energy within the mitochondrion, which carry out aerobic respiration and where the Krebs cycle is located. Fatigue, insomnia and depression are the classical symptoms, which are indistinguishable from those of hypoglycaemia. Increasing zinc intake prevents heavy metals from occupying substrate molecules in enzymes. Sunflower seeds, oysters and crustaceans have a high zinc content. Foodstuffs containing mercaptan groups or sulphur containing compounds, such as in onions, garlic and eggs have the ability to claw out heavy metals from the body over a period of time. The name mercaptan comes from their ability to react with (seize) mercury. The amino acid methionine plus vitamin B_6 may be the most effective way of detoxifying the body of heavy metals. Anti-oxidant supplementation with vitamins A, E, C and selenium is also helpful. Toxic metals increase free radicals in the body, which have been associated with cancer. Foods may cause mental and behavioural symptoms by a variety of mechanisms including cerebral allergies, food addiction, caffeinism, hypersensitivity to chemical food additives, and reactions to amines in food. The body's response to foreign body causes stress, which over time will lead to exhaustion and overt illness, including depression. Avoidance of the source of allergy is the most important treatment part of treating depression. There are several treatment approaches: avoidance, reduction of total load, rotary diet, desensitisation, neutralisation, nutritional supplements etc. Prostaglandins are very active organic compounds derived from essential fatty acids, cause a range of physiological effects: 1- At very low concentrations they induce contraction of smooth muscles. 2- Thromboxane A_2 causes blood clotting while prostacyclin causes blood vessels to dilate, both (PGE_2) are associated with many degenerative diseases, e.g., arthritis and allergies. 3- The series 1 prostaglandins (PGE_1) are known to prevent platelet adhesiveness, inhibit inflammatory reactions, dilate blood vessels thereby improve blood circulation and control blood pressure, help in weight reduction, improve the effects of insulin, activate T lymphocytes and inhibit abnormal cell proliferation. Allergic individuals have low PGE_1 and the reason is that they may be deficient in cis-linoleic acid in the diet from which it is manufactured. Safflower oil contains 70% of linoleic acid and is therefore a rich source along with poppy seed, sunflower, soybean corn etc. An enzyme, δ-6-desaturase converts cis linoleic acid (cLA) to γ linolenic acid (GLA) requiring the following vitamins and minerals; pyridoxine (B_6), zinc, magnesium, B-complex vitamins and vitamin C and E. It is thought that some people have a deficient D6D enzyme and if this is so they are advised to take Evening Primrose oil as this contain about 10% of GLA. Other plant sources of GLA are borage and blackcurrant. These are all precursors of the PGE_1. Supplementation with the ω-6 essential fatty acids may bring some order into the erratic behaviour of the immune system. ω-6 are warm climate oils such as safflower, sunflower, corn, almond oils etc., and the cold climate oils (ω-3) include linseed, salmon, walnut, wheat germ and soybean. Cold climate oils are more unsaturated and the body need them to produce beneficial PGs. Fish oils contain 2 additional types of ω-3 fatty acids, made from linolenic acid: DHA or docosahexaenoic acid, and EPA or eicosapentaenoic acid. They keep the blood thin, prevent platelet stickiness and are especially recommended to prevent cardiovascular diseases. Fish produce these from plankton in the sea. Flaxseed (Linseed) oil contains 60% ω-3 and 20% ω-6 essential fatty acid. Internal parasites and fungi, especially in hypochlorhydria (low levels of HCl acid, a natural defence barrier to internal parasites) interfere with the absorption of food in the gut. This may produce irritable bowel symptoms, diarrhoea, fatigue, depression, urticaria, arthralgia, uveitis and generally malabsorption of carbohydrates, fats, proteins, vitamins and minerals. This often follows a long period of medication with antibiotics, which tend to kill off friendly flora inside the intestines. Friendly intestinal bacteria produce most of the required vitamins and will make up for any deficiency in the diet. Also pectin in apples and bananas tend to absorb unfavourable bacteria while promoting the growth of beneficial organisms. Excesses of magnesium and vanadium have also been associated with depression. St John's Wort (Hypericum perforatum) has antidepressant effects similar action as the SSRI drugs. It inhibits the reuptake of serotonin in the brain in the treatment of mild to moderate depression. Hypericum is also useful in conditions associated with anxiety, stress, premenstrual syndrome, fibromyalgia or chronic pain. St John's Wort interacts with a number of drugs: 1- it decreases bioavailability of digoxin, theophylline (asthma), cyclosporin (immunosuppressant), and phenoprocoumon (anticoagulant), and 2- it potentiates MAO inhibitors and SSRI. It takes about 4 weeks before the herb becomes effective. Where cerebrovascular insufficiency is a contributing factor of depression, Ginkgo biloba is effective in reducing anxiety and depression. Milk Thistle (Sylibum marianum) will help the liver to accelerate detoxification, in case of toxaemia (toxic overload). Tricyclic anti-depressants are potent anti-histamines, which may explain their effectiveness against psychiatric symptoms associated with allergic reactions. Tardive Dyskinesia (the trembling disease of anti-psychotic drugs) can be improved by vitamin B_3, B_6, C, E and Mg. Lithium for manic-depression should be accompanied with safflower oil, or GLA. Evening Primrose oil is an excellent source of GLA. A long-lasting increase in the availability of serotonin neurotransmitter as a result of SSRI (specific serotonin reuptake inhibitors) at a synaptic receptor site results in a decrease in the number of receptors on the cell surface (so-called downregulation). Each person is a biochemical individual.

Urotext
Urology and health
www.urotext.com

APPLIED

ORTHOMOLECULARISM

Orthomolecularism offers the only real hope of reversing and preventing degenerative conditions. At least 68% of humans die from food-related degenerative conditions. Healthy people from healthy backgrounds become prone to degenerative diseases soon after they adapt to western dietary ways. In the last 90 years, deaths from degenerative diseases have raisen from 15% to 68%. Recommended daily allowances (RDA) are estimates that do not account for increased requirements during pregnancy, lactation, growth, puberty, activity, athletic performance, stress, infection, injury, convalescence, or aging. RDAs are statistical averages, not indications of individual requirements. RDAs are official standards of accepted levels of substandard nutrition, inadequate to maintain optimum health for almost everyone. Over 50% of patients become nutritionally deficient and more ill from eating hospital foods. Some individuals have much higher requirements of these nutrients than others do. Subclinical deficiency of nutrients lowers the body's ability to withstand stress. Subclinical deficiencies are less severe than those that result in the classical symptoms of scurvy, beriberi, etc. Each essential nutrient has its own set of subclinical symptoms. Subclinical deficiency of vitamin C results in tissue weakness, because vitamin C is necessary to build the major potent connective tissue, i.e., collagen. Subclinical deficiency of zinc slows down wound healing. Subclinical deficiency of vitamin A results in susceptibility to infection and poor skin condition, because vitamin A has important function. Pharmaceutical companies have gained approach, have exercised a strong influence over the course content of medical studies and the politics of medical associations. Orthomolecularism attempts to bring the levels of all essential nutrients in the body up to the minimum required level for health, and to optimum level for well-being, vitality, and accomplishment. Orthomolecularism concerns with lowering the concentration of excessive quantities of substances, e.g., glucose, drugs. Drugs may remove, mask, or bury symptoms of diseases caused by other factors, and because symptoms are thus being controlled, the use of drugs destroy the incentive to find root causes and real cures. Drugs cause diseases and side effects. Diuretics deplete potassium. Contraceptive pills result in water retention and behavioural changes, and deplete vitamins B_6, B_{12}, and folic acid. Tranquillisers interfere with cell membrane functions, resulting in lowered vitality, sleep without rest, and increased risk of cancer.

Internal pollution results from environmental poisons:

1- Food	
	A- Pesticides.
	B- Heavy metals.
	C- Toxic synthetics.
	D- Disease-producing organisms.
2- Water	
	A- Chlorine.
	B- Trihalomethanes.
	C- Soil-water-air pollutants.
3- Air	
	A- Dust.
	B- Smog.
	C- Ozone.
	D- Nitrous oxide.
	E- Moulds.
	F- Bacteria.
4- Street	
5- Pharmaceuticals	
6- Food additives	
7- Synthetic substances used in:	
	A- Paints.
	B- Carpets.
	C- Tiles.
	D- Countertops.
	E- Adhesives.
8- Tobacco	
9- Alcohol	
10- Toxic allergic reactions	
11- Normal metabolic functions	

Antibiotics and cortisone both interfere with the immune system, increasing risk of cancer. Antibiotics can also cause rashes, nausea, vomiting, anaemia, joint pain, and yeast infections, as well as failure of intrauterine devices. Pharmaceutical companies often start with a natural remedy or herb that affects human functions, and isolate its most potent active ingredient. Then they use chemical methods to change the ingredient's configuration slightly to make it synthetic and therefore patentable. Then they sell the patented, non-natural substance as a drug under patent-protected profits of as much as 600% or even more for 12-20 years. The patented drugs suppress symptoms of disease. 68% of humans died from 3 conditions, which involve fatty degeneration including CVD (44%), cancer (22%), and diabetes (2%). There are about 50 essential factors essential for living that must come from the environment: 1- essential nutrients (20 minerals, 13 vitamins, 8-11 amino acids, 2 essential fatty acids), 2- energy, 3- water, 4- oxygen, and 5- light. A complete absence of an essential factor increases the risk of degeneration. Toxicity can be reversed by decreased intake of the factor. Degeneration from a deficiency is greater that toxicity from excess. Progressive deficiency of essential nutrients results in a progressive deficiency syndrome. More difficult to identify is a mixture of syndromes resulting from multiple deficiencies of essential nutrients. Malnutrition results from deficiencies, imbalances of nutrients, poor digestion, and poor absorption. In 1900, 1:7 died of CVS, and 1:30 died of cancer. Essential nutrients are about 50 factors. Disease results from an absence of health. Deficiency of essential nutrients may be due to poor food choices, poor digestion, poor absorption, food allergies, intestinal injuries, imbalances in bowl flora, drug interferences with metabolic processes, increased nutrient requirements for drug detoxification, and increased requirements, which could be genetic, congenital, life-style, athletic performance, or environmental conditions.

Body contents and doses of some essential nutrients:

Name	Body content	Therapeutic dose	Toxic dose
LA	1000 g	Up to 60 g/d	Unknown
LNA	200 g	Up to 70 g/d	Unknown
Vitamin A	172 mg (500,000 IU)	50,000-500,000 IU/d	100,000 IU/d
Vitamin B_1 (Thiamine pyrophosphate [TPP])	25 mg	10-100 mg/d	Non-toxic
Vitamin B_2 (Riboflavin)		10-100 mg/d	Non-toxic
Vitamin B_3 (Niacin)		50-3000 mg/d	Transient flush above 75 mg/d
Vitamin B_5 (Pantothenic acid)		50-3000 mg/d	Non-toxic
Vitamin B_6 (Pyridoxine-5-phosphate [P-5-P])		50-500 mg/day	>500 mg/d
Vitamin B_7 (Biotin)		Up to 0.3 mg/d	Non-toxic
Vitamin B_9 (Folic acid)		0.8-5 mg/day	Non-toxic
Vitamin B_{12} (Cobalamin)		0.05-1 mg/d	Non-toxic
Vitamin C	1400 mg	100-100,000 mg/d	Non-toxic
Vitamin D		500-1000 mg/d	25,000 mg/d
Vitamin E		50-1600 mg/d	Non-toxic
Vitamin K		Up to 0.3 mg/d	>0.5 mg/d
Calcium (Ca)	1050 g	800-1200 mg/d	
Phosphorus	1000 g		
Potassium (K)	300 g	Up to 21000 mg/d (not KCl)	25000 mg/d
Sulphur (S)	175 g		
Sodium (Na)	150 g		15000 mg/d
Chlorine (Cl)	140 g		15000 mg/d
Magnesium (Mg)	355 g	Up to 2000 mg/day	>1000 mg/d
Silicon (Si)	30 g	200-1000 mg/d	
Iron (Fe)	4.2 g	Up to 50 mg/d (short-term	5000 mg/d
Fluorine (F)	2.6 g	1 mg/d	>2 mg/d
Zinc (Zn)	2.3 g	Up to 100 mg/d	>300 mg/d
Copper (Cu)	0.125 g	Up to 5 mg/d	20 mg/d
Selenium	0.013 g	0.2-0.8 mg/d	0.9 mg/d (salt)
Manganese (Mn)	0.012 g	10-100 mg/d	
Iodine (I)	0.011 g	Up to 0.5 mg/d	>0.3 mg/d
Molybdenum (Mo)	0.009 g	0.5 mg/d	
Chromium (Cr)	0.0015 g	0.6 mg/d	
Protein	12% of total BW		High protein diet
Tryptophan	0.5 g	1-3 mg/d	

Polished rice is 26-83% mineral-deficient. Corn starch is 31-100% mineral-deficient. Sugar is 83-100% mineral-deficient. Fat-free milk is deficient in Mg, Se, and vitamins A, D, and E. The regular consumption of processed foods lacking essential nutrients. People of poorer nations are better nourished than the majority of affluent. The former consumes less food, but their foods are less processed and more nutrient-dense. The technology of food alteration is carried on with out regard to its effects on human health, and has remained undefined and standarless. For successful therapy, junk foods and toxic substances must be removed from the food supply. Short-term deficiency of vitamin E may make w3 fatty acids more effective killers of the malaria-causing protozoa. The total body content for most vitamins is difficult to assess accurately, because vitamins and minerals degrade rapidly. Mineral supplements are available as inorganic salts, organic acid salts, and amino acid chelates. Inorganic phosphates, sulfates, and chlorides are important for electrolyte balance, organic ascorbates, acetates, and citrates are natural and are more effectively absorbed via GIT than inorganic salts. Mineral amino acid chelates may be absorbed even better. The best formulations combine all 3 forms in appropriate ratios. Specific mineral chelates best with specific amino acids, e.g., Mg with glycine or aspartic acid, copper with histidine, glutamine, or threonine, Mg with arginine, selenium with methionine, and iron with cysteine. Vegetable proteins may contain fewer toxic contaminants such as pesticides that are concentrated in animal tissues, fewer contaminants such as antibiotics, excess fat, hormones, and fewer parasites that affect humans. Phenylalanine and tyrosine can be used as anti-depressants, and for increasing assertiveness. Arginine may decrease atherosclerosis and gallstones.

Major deficiencies:

Calcium	60%
Chromium	40%
Cobalt	89%
Copper	68%
Iron	76%
Mg	85%
Mn	86%
Molybdenum	48%
Phosphorus	71%
K	77%
Se	16%
Stronium	95%
Zinc	78%
Vitamin B1, 2, 3, 6	72-81%
Pantothenic acid	50%
Folacin	67%
Vitamin E	86%
LA	95%
ANL	95%
Protein	33%
Fiber	95%

BIOCHEMICAL INTERVENTION

Innate chemical imbalances, rather than distorted environments or disturbing life experiences, may be responsible for most mental disorders. Biochemical treatment uses natural body chemicals rather than foreign drug molecules. Biochemical individuality focuses on the enormous complexity and variability of molecular biology in human beings. Many diseases result from disordered body chemistry, whether genetic or acquired, intervention is best achieved through correction of these chemical imbalances. Once a patient's chemical anomalies are established through laboratory analysis, The body chemistry may be balanced using biochemical treatment. Most schizophrenics, over the course of their lives, spend several months in mental hospitals, with numerous episodes of recurring mental illness. Current medications offers little hope for a real recovery. 90% of schizophrenics have either histapenia, meaning low histamine (histamine is an essential protein metabolite); histadelia, meaning high histamine; or pyrroluria (disordered vitamin B_6 metabolism) as their principal disorder, with hypoglycaemia often a complicating factor. Many histapenics and histadelics are also pyrroluric. The remaining 10% of the schizophrenic population are afflicted by a variety of splinter disorders, including cerebral allergies, wheat gluten intolerance, homocystineuria (an inborn error of sulfur amino acid metabolism), caeliac disease, polydypsia, prolactin hormone overabundance, and thyroid deficiency. 50% of schizophrenics have histapenia as their major chemical imbalance. Symptoms commonly include paranoia, suicidal depression, auditory or visual hallucinations, religiosity, and sleep disorder. Depressed blood histamine and basophils and elevated serum copper are diagnostic. Intervention revolves vitamin B_3 as either niacin or niacinamide; folic acid; cobalamine (part of the vitamin B12 group); vitamins B_6 and C; zinc; and manganese. Most histapenics experience major improvement within six weeks, but a year of treatment is commonly required before the last symptom (usually paranoia) can be overcome. Biological treatment of schizophrenia is inexpensive, and often results in a complete recovery. Histadelia accounts for 20% of schizophrenics, involves elevated blood histamine. Histadelia is characterized by delusions, severe depression, obsessive/compulsive behavior, and blank-mindedness, and often results in a diagnosis of schizo-affective disorder. Intervention revolves around anti-folates such as calcium, methionine, and the prescription drug Dilantin along with augmenting nutrients. Histadelia treatment requires great patience, because 6-10 weeks are often needed before the beginning of significant improvement. The treatment usually takes twelve months to complete. Pyrrolurics typically have light skin, poor wound healing, absence of dream recall, high internal tension, photosensitivity, severe depression, assaultive behaviour, delusions, and hallucinations. This condition is caused by an overproduction during haemoglobin synthesis of kryptopyrrole, which chemically combines with vitamin B_6 and zinc, resulting in their excretion and a severe deficiency of both of these essential nutrients. Most pyrroluric individuals never develop schizophrenia symptoms. 20% of all schizophrenics have pyrroluria as their primary imbalance. The biological imbalances of depressed patients include elevated histamine, zinc deficiency, copper overload, thyroid deficiency, low histamine, pyrroluria, and many others. Biochemical intervention appears to be about 85% effective in combating depression. Depressed individuals with elevated histamine commonly have headaches, allergies, and obsessive/compulsive/addictive tendencies. Those with zinc deficiency may report poor wound healing, impaired taste acuity, amenorrhea, stress dyscontrol, delayed growth, and/or premenstrual syndrome. Individuals with elevated copper are prone to tinnitus and post-partum depression. Low-histamine depressives often report anxiety/panic attacks, upper body pain, and paranoia. Pyrroluric depressives suffer from rage attacks and severe inner tension. Intervention varies widely, depending entirely on which imbalances are present. Biochemical treatment appears to be about 85% effective in combating depression. Behaviour disorders may be classified into: 1- Type A chemistry individuals are characterized by a high copper/zinc ratio, depressed hair sodium and potassium, and a sensitivity to lead, cadmium, and other toxics. 40% of all behaviour-disordered children exhibit mild or moderate Type A chemistry.

Classifications of bipolar disorder:

Condition	Criteria	Intervention
A- Undermethylation:	1- Congenital. 2- Characterized by low levels of serotonin, dopamine, and norepinephrine, high whole blood histamine and elevated absolute basophils. 3- High incidence of seasonal allergies, OCD tendencies, perfectionism, high libido, sparse body hair, and several other characteristics. 4- In severe cases involving psychosis, the dominant symptom is usually delusional thinking rather than hallucinations. They tend to speak very little and may sit motionless for extended periods. They may appear outwardly calm, but suffer from extreme internal anxiety.	1- Methionine, s-adenosyl-methionine (SAMe), calcium, magnesium, ω-3 essential oils (DHA and EPA), B_6, inositol, and vitamins A, C, and E. 2- Avoid supplements containing folic acid, choline, dimethyl-amino-ethanol (DMAE), copper and histidine.
B- Overmethylation:	1- Biochemically opposite of undermethylation. Elevated levels of serotonin, dopamine, and norepinephrine, low whole blood histamine, and low absolute basophils. 2- Absence of seasonal, inhalent allergies, but a multitude of chemical or food sensitivities, high anxiety that is evident to all, low libido, obsessions but not compulsions, tendency for paranoia and auditory hallucinations, underachievement as a child, heavy body hair, hyperactivity, nervous legs, and grandiosity.	1- Folic acid, B_{12}, niacinamide, DMAE, choline, manganese, zinc, ω-3 essential oils (DHA and EPA) and vitamins C and E. 2- Avoid supplements of methionine, SAMe, inositol, trimethyl-glycine (TMG) and dimethyl-glycine (DMG), tryptophan, phenylalanine, St. John's, wort, tyrosine, copper.
C- Pyrrole disorder (pyrroluria):	1- Genetic stress disorder. 2- Severe mood swings, high anxiety, and depression. 3- Elevated urine kryptopyrroles, a double deficiency of zinc and B_6, and low levels of arachidonic acid. 4- Devastated by stresses including physical injury, emotional trauma, illness, sleep deprivation. 5- Sensitivity to light and loud noises, tendency to skip breakfast, dry skin, abnormal fat distribution, rage episodes, little or no dream recall, reading disorders, underachievement, histrionic behaviors, and severe anxiety.	1- Zinc, B_6, primrose oil, and augmenting nutrients. 2- Avoid histidine, copper and ω-3 fatty acids.

Boys may have episodes of fighting or severe tantrums interspersed with periods of excellent behavior. Girls may be prone to oppositional behavior, mood swings, promiscuity, and non-violent delinquency. 2- Type B chemistry (relatively rare, 0.3%) involves depressed hair copper, pyrroluria, elevated histamine, depressed blood spermine (a protein found in almost all tissues, but first found in sperm), and elevated toxic metals. Type B individuals are very prone to criminality, but respond to biochemical treatment within 7 days. 3- Type C persons exhibit nonviolent delinquent behavior. 80% of Type C persons are slender malabsorbers. Usually impulsive and oppositional, they are seldom able to maintain a valid driver's license. They underachieve in school and have great difficulty keeping jobs. 4- Type D persons exhibit nonviolent delinquent behavior. They have depressed manganese and chromium levels, and clinical studies revealed hypoglycaemia as the principal imbalance. Learning disabilities, hyperactivity, attention deficit disorder (ADD), and dyslexia can be cured by correction of the specific chemical imbalances afflicting each child. Indiscriminate supplementation with multiple vitamins, minerals, and amino acids usually results in worsening of hyperactivity, ADD, and learning disabilities. Nutrients that can adversely impact learning disorders include calcium, folic acid, niacin, pantothenic acid, methionine, histidine, iron, and copper. These same nutrients can be beneficial to children with different body chemistry. Continuation of these medications is required until body chemistry is balanced. In most cases, the amphetamine drugs can be withdrawn after 2-3 months of treatment without adverse effects. Biochemical treatment appears to be about 85%, 70%, and 60% effective for children with learning disabilities, hyperactivity, and ADD, respectively. Many nutritional treatments offered are highly generalized and can produce unpleasant effects. Bipolar disorder is a mental health condition characterized by periods of depression alternate with periods of mania. Mild degrees of bipolar disorder can often go undiagnosed. There is a relationship between ω-3 fatty acids and bipolar disorder. High-dose, concentrated ω-3 fatty acids (EPA and DHA) have effective mood stabilizing and antidepressant effects for many people with bipolar disorder. ω-3 fatty acids are essential components of brain cell membranes, including those of neurotransmitter receptors. ω-3 fatty acids also alter signal transduction and electrical activity in brain cells, and control the synthesis of chemicals such as eicosanoids and cytokines, which may have a direct effect on mood.

NEUROBIOLOGY OF DEPRESSION

The morbidity and mortality associated with depression are considerable and continue to increase. Depression is the fourth major cause of disability worldwide, after lower respiratory infections, perinatal conditions, and HIV/AIDS. 17% of people will suffer from depression during their lifetime. The incidence of depression may be 30-50% in general. Despite the clear decrease in quality of life and decreased productivity associated with depression, it is often underdiagnosed and inadequately treated. Only 25% of depressed patients receive an adequate antidepressant dose and duration of treatment. Depression results from a complex interaction between early life experiences, genetic vulnerability, and more recent environmental factors. Having a family member with depression clearly increases the risk for depression. Stressful life events often precede depression, at least in vulnerable individuals, including divorce, marital problems, assault, or death of a close relative. Disturbances in the monoamine transmitters norepinephrine and serotonin have been demonstrated in depression. Disturbances in glucocorticoid (i.e., cortisol) secretion specifically and hypothalamic pituitary adrenal axis (HPAA) activity generally have also been documented in depressed patients, along with increased synthesis and release of corticotropin-releasing factor (CRF). There is also evidence of disturbed second messenger function in the pathobiology of depression. There is dysfunction in many other neurotransmitter and endocrine systems including glutamate, γ-aminobutyric acid, growth hormone, and thyroid hormones. Depression may represent a common final pathway of multiple underlying pathobiologies. The treatment strategies of depression should be aimed at achieving full remission. The clinical presentation of depression may be pleomorphic, and the diagnosis is often missed unless specifically sought. Many depressive symptoms are nonspecific and occur in both depression and a variety of somatic disorders. Depressive illness is characterised by a broad spectrum of other symptoms such as dizziness, headache, joint pains, fatigue, weakness, muscle pains, abdominal pain, chest pain, back pain, sadness and tearfulness, loss of interest, anxiety/irritability, hopelessness, concentration difficulties, guilt, suicidal ideations, tiredness/fatigue, sleep disturbance, headache, psychomotor changes, GIT disturbances, appetite changes, and/or body aches and pains. Somatic complaints may represent an underlying depression. The perception of pain is under CNS regulation, which processes pain information and modulates pain responses via the descending pathways. Both 5-hydroxytryptamine (5-HT) and norepinephrine (NE) neural circuits directly modulate the descending pathways and compose an integral part of the complex system that controls pain perception. Dual 5-HT/NE reuptake blocking tricyclic antidepressants such as amitriptyline or imipramine is used in the treatment of various pain states. Nontricyclic antidepressants that act as dual serotonin/norepinephrine reuptake inhibitors (SNRIs) possess antinociceptive properties. Dual blockade of 5-HT and NE is much more effective at treating pain than drugs that primarily target either transporter separately. Serotonin-containing nerve terminals primarily originate in the dorsal raphe nucleus located in the midbrain, and project throughout the forebrain, including the prefrontal cortex. Low levels of the major 5-HT metabolites have been detected in the cerebrospinal fluid (CSF) of depressed and suicidal patients. All selective serotonin reuptake inhibitors (SSRIs) are effective antidepressants, but should only be used for short-term (12 weeks). Serotonin biosynthesis in male brains is 52% higher than that in female brains, which may account for the increased prevalence of major depression in females. 5-HT neuronal dysfunction plays a role in the pathobiology of depression. Norepinephrine projections originate from a pontine brain region, the locus ceruleus (LC), and like the raphe neurons, these neurons project to the frontal cortex and limbic system. The urine and CSF of depressed patients is low of NE metabolites. There is increased density of β-adrenergic receptors in cortex of depressed suicide victims. Selective NE reuptake inhibitors are effective antidepressants (i.e., desipramine, reboxetine, maprotiline), but should only be used for short-term (12 weeks). The venoarterial 5-hydroxyindoleacetic acid concentration, the major 5-HT metabolite, is not reduced in the patients with depressive illness. The efficacy of TCAs does not demonstrate that these monoaminergic systems represent the primary deficit in depression. CRF is a 41 amino acid-containing neuropeptide that is the primary physiologic secretagogue of adrenocorticotropic hormone (ACTH) and subsequently cortisol release. CRF also acts as a neurotransmitter in various regions of the brain that modulate affect and emotionality. Dexamethasone (DEX)/CRF test is used to assess HPAA, which invlove administration of 1.5 mg dexamethasone orally at night (23:00 h), and and subjects receive an IV bolus of 100 μg of human CRF at 15:00 h the following day. Patients with HPA axis dysfunction, including many depressed patients, display a paradoxically increased release of ACTH and cortisol. These HPA axis alterations normalize following remission of depression, and clinical remission. The sensitivity of DEX/CRF test for detecting subtle alterations in HPA axis function is 80%. The DEX/CRF test may predicts treatment resistance.

SLEEP DISORDERS

Lack of sleep impairs glucose tolerance, which may increase the risk of carbohydrate metabolism disorders, such as hypoglycaemia. Sleep appears to boost the body's immune system, enhancing its response to viral challenges. Sleep also may increase the activity of cytokines, growth factors that play a role in cellular immune responses. Aiming for 6-8 hrs of sleep each night is essential. Sleep disorders are increasingly prevalent and are associated with significant morbidity. The hypothalamus is a key regulator of sleep and wakefulness, shifting focus away from brainstem and thalamo-cortical systems. This role for the hypothalamus was first recognized with the encephalitis lethargica epidemic (1918–1926), in which affected individuals showed major sleep abnormalities and Parkinsonism. Sleep control systems in the hypothalamus and their interaction with the circadian pacemaker in the suprachiasmatic nuclei (SCN) have been identified. Sleep control systems within the hypothalamus are closely integrated with homeostatic systems. Sleep deprivation alters hormone release, increases body temperature, stimulates appetite and activates the sympathetic nervous system. Sleep oscillates between rapid eye movement (REM) sleep, light sleep and deep slow-wave sleep (SWS); the latter two are collectively called non-REM (NREM) sleep. Sleep-state control (e.g., transitions from NREM to REM) is attributed to reciprocal monoaminergic-cholinergic interactions in the brainstem, whereas the electrophysiological expression of sleep versus wakefulness is attributed to the synchronization and desynchronization of thalamocortical circuits. Serotoninergic, adrenergic and histaminergic activity is high during wakefulness, decreases during NREM stages, and becomes almost silent during REM sleep. In contrast, brainstem cholinergic activity is high during wakefulness and REM sleep. Amphetamine-like stimulants, the most potent wake-promoting agents, increase wakefulness by blocking dopamine reuptake and/or by stimulating dopamine release. Dopaminergic receptor agents have differential effects on sleep depending on their affinity for dopaminergic receptor subtypes and for presynaptic versus postsynaptic receptors. Dopaminergic neurons do not change their activity greatly throughout the sleep cycle. Sleep regulation by dopamine may be as complex as that mediated by cholinergic systems. Descending dopaminergic neurons from the telencephalon to the spinal cord may contribute to abnormal movements during sleep. Dopaminergic cell groups may coordinate sleep with respect to motivated behavior and locomotor activity. Most hypnotics (such as benzodiazepine) increase GABAergic transmission by acting on the $GABA_A$/BZ-Cl-ligand-gated ion channel. This receptor is a pentameric protein that surrounds the trans-membrane chloride channel. Physiologically, cataplexy may represent the atonic component of REM sleep that occurs in isolation from other elements of the narcolepsy tetrad. The effects of sleep loss are frequently described in terms of capacity to perform lengthy, yet cognitively undemanding routine tasks. Performance loss can be largely overcome by compensatory effort after only a single night without sleep. The frontal lobes contribute in performing functions of day-to-day organisation of behaviour, e.g. recent memory, the assessment of novel situations, problem solving, forward planning, articulation, communication, and creativity. Narcoleptics function at the level of 72 h sleep deprivation optimally. Narcoleptic patients are frequently insomniac and γ-hydroxybutyric acid (GHB) consolidates sleep and improves daytime symptoms such as cataplexy. GHB is a GABA metabolite. Primary GHB receptor and $GHBA_B$ mechanisms have been suggested, with secondary effects on dopamine neurotransmission. GHB has a short half-life and, unlike BZ-like drugs, dramatically increases SWS. Histamine is a major wake-promoting neurotransmitter. Human narcolepsy is caused by deficient hypocretin neurotransmission in the lateral hypothalamus. The hypocretin system encompasses two peptides (hypocretin-1 [orexin A] and 2 [orexin B]). Hypocretin neurons have widespread projections, with dense excitatory projections to all monoaminergic and cholinergic cell groups. Hypocretin release is higher in the active period. Hypocretin neurotransmission is also activated by sleep deprivation. Wake-promoting actions of hypocretins are attenuated when histaminergic transmission is impaired. Monoaminergic and hypocretinergic systems work in concert. Hypocretinergic neurons may be more sensitive to metabolic cues such as glucose, leptin and ghrelin. Hypocretinergic, but not histaminergic, neurons are activated by locomotion during wakefulness. Hypocretins may be more important in resisting sleep after sleep-deprivation or during periods of intense activity. Increased hypocretin activity may also mediate some of the effects of sleep deprivation, including deleterious metabolic and autonomic effects (via sympathetic and hypothalamo-pituitary–adrenal axis activation) and antidepressant action (via monoaminergic stimulation). TGF-α and prokineticin-2 are released during the rest period and decrease locomotor activity. TGF-α action on locomotor rhythms is mediated via the EGF receptor. TGF-α is also expressed in a subset of retinal ganglion cells. EGF receptor–positive ganglionic cells may provide direct input from the retina to the hypothalamus and mediate other effects of light, such as masking. Prokineticin-2 receptors are located in selected hypothalamic nuclei. Adenosine has been proposed as a mediator of sleep homeostasis and a link between metabolism and sleep control. Caffeine, an adenosine A1 receptor antagonist, is widely used to induce wakefulness, although its efficacy is low compared to dopaminergic stimulants. Most sedatives are suspected to exacerbate disordered breathing during sleep. DC magnetic fields are produced by the earth and by man-made permanent magnets. Deviations in the earth's natural magnetic field either in strength or direction are cause for concern. More than 60% of all inner spring mattresses and box springs tested show non-uniform magnetic hot spots along, the length of the bed. These hot spots cause a shift in the location of true north so far as the body is concerned, but only for the parts of the body lying on the hot spots. Other objects around the bed may be non-uniformly and strongly magnetized, such as the bed frame, steel furnishings, steel objects stored under or around the bed. Disturbed DC magnetic fields in the sleeping place confuses the body and can cause insomnia, chronic fatigue, and other sleep related problems. DC fields influence enzyme activity, oxygen usage, and the growth rate of cancer cells. Wearing glasses with magnetized frames may induce strong headaches. Stereo headphones use magnets for the speaker and also cause headaches.

Common sleep disorders and their treatments:

Disorder	Incidence	Criteria	Current treatments	Future treatments
Insomnia:	9–15%	Difficulty initiating and maintaining sleep; difficulty initiating sleep in adolescence is often a circadian phase maladjustment called 'delayed sleep phase syndrome'; can be associated with sleep-disordered breathing and other sleep disorders.	Benzodiazepine (BZ) and BZ receptor–acting hypnotics (e.g. zolpidem, zopiclone, zaleplon), sedative anti-depressants and antihistaminergic drugs (e.g. trazodone), melatonin, behavioral treatments.	GABAergic and BZ compounds with improved neural specificity (e.g. VLPO, TMN); hypocretin/orexin antagonists; non-GABAergic hypnotics such as histaminergic H1 receptor antagonists and sedative antidepressants with shorter half-lives; melatonin receptor agonists.
Sleep-disordered breathing (obstructive sleep apnea):	10–20%	Snoring and breathing pauses causing sleep disruption and most often excessive daytime sleepiness; exacerbated by alcohol and sedatives. Associated with airway anatomical abnormalities and/or central control of ventilation; multi-system disorder associated with obesity, hypertension, diabetes mellitus, hyperlipidaemia.	Continuous positive airway therapy (CPAP) or bilevel positive airway pressure (BiPAP) therapy; dental appliances; surgery of the upper airway.	Improved ventilatory devices; 5HT3 and 5HT2A/C antagonists; adenosine A1 receptor agonists.
Restless legs syndrome (RLS):	2–5%	Parasthesias followed by akathisia (urge to move limbs, usually legs); often associated with periodic leg movements (PLM), brief, repetitive muscular jerks of the legs during sleep and wakefulness; pathophysiology possibly due to abnormalities in dopaminergic system and/or brain iron metabolism.	Dopamine agonists; opiates; gabapentin.	Dopamine (D2/3) agonists with improved neural specificity (e.g. A11); hypocretin/orexin compounds; factors influencing central iron metabolism.
REM sleep disorder (RBD):	0.5%	Increased motor activity during REM sleep; behavior multiple causes, often drug induced; frequently predates Parkinsonism.	Clonazepam	GABAergic and BZ compounds with increased neural specificity.
Narcolepsy:	0.02–0.06%	Excessive daytime sleepiness, cataplexy (loss of muscle tone in response to emotions), short REM sleep latency; disturbed nocturnal sleep; often associated with PLM and RBD; HLA association and probable autoimmune aetiology causing hypocretin deficiency.	Amphetamine-like stimulants, Modafinil (sleepiness); REM suppressant anti-depressants (cataplexy); γ-hydroxybutyric acid (GHB, all symptoms).	Hypocretin/orexin agonists or transplantation of hypocretin-producing cells (all symptoms); dopamine uptake inhibitors or histamine H3 receptor antagonists (sleepiness); GHB receptor or $GABA_B$ receptor agonists (all symptoms).

INSOMNIA

Insomnia is a heterogeneous complaints that may involve difficulties falling asleep (initial or sleep onset insomnia), trouble staying asleep with prolonged nocturnal awakenings (middle or maintenance insomnia), or early morning awakening with inability to resume sleep, causing dissatisfaction with day time function. Insomnia may be short term (less than a month), or chronic. 30-88% of elderly suffers from insomnia, 18.6% suffer from early morning awakenings, and 18.6% suffer from difficulty to falling asleep. Daytime sleepiness in elderly is 30.9%. Ageing is associated with increased amount of light sleep (stages 1 and 2 of the sleep cycle), a decreased amount of deep sleep (stages 3 and 4 of the sleep cycle), and less time spent in the rapid eye movement (REM) stage of sleep. Sleep time requirement is 18 hrs for neonates, 11 hrs for young children, 9 hrs for adolescents, and 8 hrs for adults. Sleep time is decreased on average of 27 min/decade from mid-life until the eighth decade. The slow wave sleep of stage 3 and 4 that determines sleep intensity decreased from young adulthood to middle age from 18.9% to 3.4% of total sleep time. Pathological sleep changes tend to interfere with healthy functioning. This may cause attention deficits, delayed reaction, short-time memory difficulties, functional problems and increased risk of fall. Being tired during the day is related to a higher 5-year mortality for both genders aged 65 or more. Normal sleep patterns include tiredness in the mid-afternoon and before bed at night. Primary insomnia is difficulty in initiating or maintaining sleep for a minimum of a month. Sleep problems amongst older patients are generally caused by secondary insomnia. Secondary insomnia may occur as a result of medical, psychiatric, environmental, behavioural, or drug side effects. Medical causes of secondary insomnia include any physically disturbing condition such as pain, thyroid disease, acid reflux, coronary artery disease or pulmonary problems. Sleep related movement disturbance (nocturnal myoclonus, restless leg) is another medical condition that may cause insomnia. Sleep related movement disturbance may be genetic, or secondary to iron or folic acid deficiency anaemia. Anxiety and depression, decreased exposure to bright light, lack of exercise, and drugs (steroids, theophylline, anticancer drugs, β-blockers, caffeine, alcohol and nicotine) are associated with insomnia. Shift-work sleep disorder causes sleep disturbances, which are disorders of sleep timing rather than sleep production. Insomnia may be caused by a patient's sleeping environment. Sleep diary (self-reporting measure) is the primary subjective method. Sleep diaries can provide reliable and valid estimates of sleep parameters. They capture data on the time between sleep onset and the moment at which the EEG displays stage 2 sleep pattern. Sleep diaries are the most valid tools for measuring insomnia in general practice. Benzodiazepine hypnotics cause: 1- Impairment of sleep quality (decrease slow wave sleep), and some potential significant adverse effects. 2- Residual sedation and memory or functional impairment occurring the day following drug administration tends to be associated with hypnotics that have a long half-life or active metabolites. 3- Rebound insomnia is associated with short half-life hypnotics. 4- Increased risk of falls, drowsiness, dizziness, cognitive impairment and motor vehicle crashes. 12% of women and 9% of men over the age of 65 use benzodiazepine. 34% of institutionalised patients use benzodiazepine. Non-benzodiazepine hypnotics are type I selective GABA (γ-aminobutyric acid) receptor agents. Zolpidem and zopiclone have inactive metabolites, short half-lives, and absence of respiratory depressive side effects, resulting in less risk of falls, daytime sedation, driving impairment, tolerance, rebound, respiratory depression, or exacerbation of sleep-disordered breathing. Zolpidem side effects include dizziness, somnolence, nausea, vomiting, and fall. Zopiclone long-term use may cause dependency and withdrawal symptoms such as anxiety and insomnia. Pharmacological treatment should be limited to 4 weeks. Non-pharmacological treatments cause fewer side effects, and can sustain long-term improvements. Non-pharmacological treatments include stimulus control, sleep restriction, sleep hygiene education, cognitive therapy, multi-component therapy, paradoxical intention, and relaxation therapy. Non-pharmacological treatments for primary and secondary insomnia are feasible and effective alternatives to the use of benzodiazepines. Stimulus control is achieved by reducing activities that interfere with sleep, which include 1- explanation to the patient the rationale of each step of the protocol, 2- providing a copy of the instruction to the patient, and 3- encouragement of the patient to comply with the protocol.

Sleep hygiene education:

1- Avoid the use of caffeine containing products, nicotine and alcohol especially later in the day.
2- Avoid heavy meals within 2 h of bedtime.
3- Avoid drinking fluids after supper to prevent frequent nighttime urination.
4- Avoid environments that will make you active after 5 pm.
5- Use the bed only for sleep.
6- Avoid watching TV in bed.
7- Establish a routine for getting ready to go to bed.
8- Set time aside to relax before bed, and utilise relaxation techniques.
9- Create an atmosphere conducive to sleep (comfortable temperature, dark room, and comfortable mattress).
10- Relax when in bed.
11- Get up at the same time every day.
12- Daytime naps should be before 3 pm.
13- Pursue regular physical activity (walking).
14- Avoid vigorous exercise too close to bedtime.

Sleep restriction therapy:

1- Determine the average estimated total sleep time.
2- Restrict the time in bed to the average estimated total sleep time.
3- Each week, determine the patient's weekly sleep efficiency (total sleep/time in bed X 100) from the data obtained from the sleep diary.
4- Increase total time in bed by 15-20 min when sleep efficiency exceeds 90%. Decrease it by 15-20 min when sleep efficiency below 80%. Keep total time in bed the same when sleep efficiency is between 80-90%.
5- Each week, adjust the total time in bed until the ideal sleep duration is obtained.
6- Do not reduce time in bed to below 5 h.
7- Brief midday naps may be permissible, especially in the early phase of treatment.
8- When applying this protocol to the elderly, some recommend reducing the time in bed only when sleep efficiency is below 75%.

Sleep onset and sleep maintenance insomnia among the elderly can be improved with stimulus control. Secondary insomnia responds well to stimulus control after 4 weeks when combined with sleep hygiene and relaxation therapy. Excessive time in bed causes fragmented sleep and seems to be a critical factor in perpetuating insomnia. Sleep restriction requires less time to implement. Sleep hygiene attempts to change patient's lifestyles and environment to optimise sleep quality. Good sleep hygiene improves mood on morning awakenings, improves sleep continuity and depth. Maladaptive sleep beliefs and attitudes are more frequent in older chronic insomniacs than in self-defined good sleepers. Cognitive therapy is an important component of insomnia management and perhaps more important for older insomniacs. The intervention involves preparing a patient for treatment by providing the conceptual framework of relationships between cognition, effect and behaviour. Cognitive therapy lacks sufficient evidence to be recommended as a single treatment. Multi-component therapy is an insomnia treatment that involves combining several different interventions. Multi-component therapy produces better result than no treatment, but it is not always more effective than stimulus control or sleep restriction alone. Paradoxical intention aims to remove performance anxiety by having the patient partake in his or her most feared behaviour (e.g., remaining awake). Relaxation therapy attempts to decrease somatic arousal. Most relaxation therapies (progressive muscle relaxation, autogenic training) can be self-administered by patients.

Stimulus control therapy:

1- Go to sleep only when you feel tired.
2- Use the bed and bed room for sleep and sex.
3- Do not read books or magazines, watch TV, eat or worry while in bed.
4- Leave bedroom if you do not fall asleep within 15-20 min.
5- Return to bed only when you feel sleepy again.
6- Get up at the same time every morning regardless of how much sleep you obtained the night before.
7- Avoid napping

Paradoxical intention:

1-When you are in bed, remain awake as along as possible.
2-Do not partake in activities that are incompatible with sleep (watching TV, lights on).

NARCOLEPSY

Narcolepsy-cataplexy is a disabling neurological disorder that affects 1:2000 individuals. Narcolepsy-cataplexy with hypocretin deficiency is agenuine disease entity. Narcolepsy is a chronic neurological disorder characterised by excessive daytime sleepiness and sleep attacks. Narcolepsy effects equal numbers of men and women and occurs in all racial groups. This sleep disorder usually develops during the late teens or early twenties and continues throughout the individual's life. Early diagnosis and appropriate intervention are critical for reducing the adverse effects of this disorder. Narcolepsy is a chronic (long lasting) disease of the CNS. Narcolepsy is a life-long disease. The symptoms may vary in severity during the patient's lifespan, but they never disappear completely. The role of heredity in humans with narcolepsy is not fully understood. Most narcolepsy does not run in families. Excessive daytime sleepiness (EDS) is the main symptom (100%). The other primary symptoms of narcolepsy include loss of muscle tone (cataplexy), distorted perceptions (hypnagogic hallucinations), and inability to move or talk (sleep paralysis). Cataplexy is present in 65-70% of patients with narcolepsy. Hypnagogic hallucinations may be present in up to 50% of patients with narcolepsy. Sleep paralysis may be present in up to 60% of patients with narcolepsy. Automatic behaviour may occur in 60% to 80% of patients with narcolepsy. Additional symptoms include disturbed nocturnal sleep and automatic behaviour. All of the symptoms of narcolepsy may be present in various combinations and degrees of severity. The onset of narcolepsy begins usually in teenagers or young adults and affects both sexes equally. The first symptom to appear is EDS, which may remain unrecognised for a long time in that it develops gradually over time. The other symptoms can follow EDS by months or years. The prevalence of narcolepsy is similar to that of Parkinson's disease and multiple sclerosis (0.03-0.09% of the population). Narcolepsy may not be considered in the evaluation of patients who complaining of fatigue, tiredness, or problems with concentration, attention, memory, and performance, and other illnesses (seizures, mental illness, etc.). There are abnormalities in the structure and function of hypocretin neurons, in the brains of patients with narcolepsy. These cells are located in the hypothalamus and they normally secrete neurotransmitter substances called hypocretins. The hypocretin neurons and their connections with many different areas of the brain make up the hypocretin system, whose functions are not yet fully known. The hypocretins may be responsible, both directly and by affecting other neurotransmitter systems, for the daytime sleepiness and abnormal REM sleep found in narcolepsy. Patients with narcolepsy have a markedly decreased number of hypocretin nerve cells in the brain. They also have a decreased level of hypocretins in the CSF. Narcolepsy could be an autoimmune disease, similar to other HLA-associated diseases such as multiple sclerosis and ankylosing spondylitis (degenerative disease). HLAs are genetically determined proteins on the surface of white blood cells. They are a part of the body's immune system. Thus, an autoimmune reaction is possible to cause the loss of nerve cells (hypocretin and some other neurons in the brain) in patients with narcolepsy. Unknown factor in the environment (e.g., infection or trauma) may trigger the autoimmune reaction. The normal brain cells are attacked by the body's own immune system, as though these neurons were foreign proteins. As a result, the neurons are damaged and deteriorate (degenerate). Relatives of patients with narcolepsy may have a higher tendency to develop narcolepsy or some sleep-related abnormalities, such as increased daytime sleepiness, increased REM sleep, or others. Excessive daytime sleepiness (EDS) causes the patient to fall asleep or doze off easily. This happens not only in relaxed situations, but also at inappropriate times and places. The daytime sleepiness is present even after normal nighttime sleep. Patients may refer to this symptom by being tired, fatigued, or sleepy, feeling lazy, or having low energy. EDS usually impairs a patient's functioning because it reduces motivation and vigilance, interferes with concentration and memory, and increases irritability. Cataplexy is a sudden, temporary (reversible) loss of muscle tone (firmness) in a person with narcolepsy. Cataplexy may be triggered by emotional stimuli such as laughter, excitement, surprise, or anger, physical fatigue, stress, and sleepiness. Severe attacks of cataplexy may involve all of the skeletal muscles, which can result in a complete body collapse with a fall to the ground and risk of injury. Milder cataplexy involve regional (focal) muscle groups and result in symptoms such as a dropping head, sagging jaw, slurred speech, buckling of the knees, or weakness in the arms. This muscle weakness (loss of tone) can be quite subtle. Consciousness is maintained throughout the episodes of cataplexy but the patient is usually unable to speak. Cataplectic attacks last from a few seconds to several minutes, varies from a few per year to numerous attacks per day that could disable the patient. Hypnagogic hallucinations are dream-like experiences, which occur during the transition from wakefulness to sleep. Hypnopompic hallucinations are those that occur during the transition from sleep to wakefulness. These hallucinations (distorted sensations or perceptions) may involve hearing, vision, touch, balance, or movement. They often incorporate images of the patient's environment into the dream-like images. The hallucinations are frequently vivid, bizarre, frightening, and disturbing for the patients. As a result, the patients may become fearful that they have or will develop a mental illness. Sleep paralysis is a temporary inability to move or talk, which occurs during sleep-to-wake or wake-to-sleep transitions. The duration of these episodes of sleep paralysis may be from seconds to minutes. They can occur at the same time as hypnagogic (or hypnopompic) hallucinations. During sleep paralysis, breathing is mostly maintained. Cataplexy, hypnagogic hallucinations, and sleep paralysis in patients with narcolepsy are referred to as REM related abnormalities because they are caused by REM sleep intrusions into wakefulness. Disturbed nocturnal sleep with frequent awakenings and increased body movements may develop after the onset of the primary symptoms of narcolepsy. Narcolepsy pentad (a set of 5 symptoms) is additional symptom, along with EDS and the REM related abnormalities Automatic behaviour is the performance of routine activities with reduced awareness or to semi-purposeful behaviour, often with the unusual use of words (irrelevant words, lapses in speech). The patient is unaware of this behaviour, which occurs while she/he is fluctuating between sleep and wakefulness. Other complaints associated with narcolepsy may include eye disturbances due to sleepiness, such as blurred vision, double vision, and droopy eyelids. The diagnosis of narcolepsy is based on a clinical evaluation, specific questionnaires, sleep logs or diaries, and the results of sleep laboratory tests. The clinical evaluation includes a detailed medical history and physical examination by a physician. The Stanford narcolepsy questionnaire is a very extensive questionnaire that can provide the physician with valuable information on all symptoms of narcolepsy, but especially on cataplexy. The Epworth sleepiness scale is a brief self-administered questionnaire, which provides an estimate of the degree of daytime sleepiness. The use of sleep logs or sleep diaries for 2-3 weeks is recommended in the evaluation of any patient with EDS. Sleep diaries provide information about the patient's usual sleep patterns (sleep deprivation, irregular sleep/wake pattern, interrupted sleep), alcohol and/or drug use, and common waking behaviours (e.g., internet syndrome). The Internet syndrome is the habit of surfing the Internet until late at night, thereby causing sleep deprivation and daytime sleepiness. Standard sleep laboratory tests (sleep studies) for narcolepsy include polysomnography (PSG) and the multiple sleep latency test (MSLT). PSG is a full night recording of several different functional (physiological) factors (parameters) of a patient's sleep. The PSG is followed the next day by the MSLT, which is a recording of the patient's tendency to fall asleep during the day. Daytime sleepiness is measured in the MSLT by the sleep latency (SL) time. This is the time from the beginning of the recording to the onset of sleep. In healthy individuals, the SL time is more than 10 minutes, whereas in narcolepsy, it could be as short as 0.5 minutes (an almost immediate onset of sleep). In narcolepsy, the REM sleep begins soon after the onset of sleep. Thus, in healthy individuals, the first REM sleep period occurs about 80-120 minutes after the onset of sleep. By contrast, in narcolepsy, the initial REM sleep period usually occurs within 15 minutes of the onset of sleep. Narcolepsy patients will have two or more sleep onset REM periods during the multiple sleep latency test (MSLT) in the daytime. PSG is helpful in excluding other causes of daytime sleepiness, such as sleep apnoea syndrome (SAS); periodic limb movements in sleep (PLMS), and sleep disruptions. Wakefulness test (MWT) is a recording that measures the ability of a subject to stay awake during the day. There are familial forms of narcolepsy without HLA marker and there are sporadic or individual forms that have the HLA marker. Intervention: The treatment of narcolepsy includes drug (pharmacological) and non-drug, or behavioural, therapies. Treatment options should be individualised depending on the severity of the symptoms, life conditions (e.g., type of work or responsibilities) of the patients, and the specific goals (e.g., relief of certain symptoms) of therapy. Optimal management usually takes weeks to months to be achieved. Good treatment management typically produces significant improvement of the symptoms rather than a resolution of all symptoms.

The types, number, and severity of the symptoms determine which drugs are used to treat the narcolepsy. Severe daytime sleepiness may require treatment with high doses of stimulant medication, and sometimes a combination of stimulants may be needed. Trouble sleeping (insomnia) and depression may also require treatment. Improved alertness may be critical throughout the day for most students and working adults, but may be critical only at certain times of the day (e.g., driving times) for other people. Amphetamines (e.g., Dexedrine, Desoxyn, DextroStat, Adderall) and methylphenidate (Ritalin) are generalised CNS stimulants. These medications also increase the activity of the so-called sympathetic part of the nervous system, which can produce undesirable side effects, e.g., elevation of blood pressure, nervousness, irritability, and rarely, fearful distrust (paranoid) reactions, feeling of great happiness (euphoria). These alerting medications can also lead to drug dependency. Pemoline (Cylert) is less effective than the traditional stimulants. This drug has the potential risk of toxic side effects on the liver. Modafinil (Provigil) has alerting effects similar to those of the traditional stimulants. This medication does not affect cataplexy and the other REM sleep symptoms. Modafinil is not a general CNS stimulant like the amphetamines, but the way it works is unknown. It does not have significant effects on the sympathetic nervous system and does not cause mood changes, euphoria, or dependence. Modafinil does not develop tolerance. Headache and nausea are the most commonly reported side effects, and they are usually mild and temporary (transient). These side effects can be reduced, however, by a slow increase from a low initial dose up to the desired dose. Modafinil is usually used in a single daily dose. Switching patients from amphetamines to modafinil may cause the reappearance of cataplexy in patients previously well controlled. Monoamine oxidase inhibitors (MAOIs), such as phenelzine (Nardil) and selegiline (Eldepryl), can also be used for treatment of EDS. Anticataplectic medication is the general name for drugs that are used to treat cataplexy. These drugs may also be used for the other REM related symptoms, such as hypnagogic hallucinations and sleep paralysis. Tricyclic antidepressants (TCAs), used in lower than antidepressant doses, are often effective in controlling cataplexy. These medications act on some neurotransmitter systems to produce suppression of REM sleep and consequently improve the symptoms of cataplexy. The anticholinergic side effects including dry mouth, dry eyes, blurred vision, urine retention, constipation, impotence, increased appetite, drowsiness, nervousness, confusion, restlessness, headache, etc., in some cases, may limit the use of TCAs. TCAs may increase periodic limb movements in sleep, which could further disrupt already disturbed nighttime sleep in narcoleptic subjects. If TCAs are abruptly discontinued a significant worsening of the cataplexy and other REM related symptoms could occur. This rebound phenomenon may appear in 72 hours after discontinuation of the medication and peak in approximately 10 days from the withdrawal. The most frequently used TCAs for the treatment of cataplexy and other REM related symptoms are protriptyline (Vivactil), imipramine (Tofranil), clomipramine (Anafranil), desipramine (Norpramine), and amitriptyline (Elavil). Sedating TCAs such as clomipramine, amitriptyline, and imipramine, should be prescribed for evening use, whereas the alerting ones (protriptyline and desipramine) should be used during the day. Selective serotonin reuptake inhibitors (SSRIs) are also useful in treating cataplexy at doses that are comparable to those used to treat depression, e.g., fluoxetine (Prozac), paroxetine (Paxil), sertraline (Zoloft), citalopram (Celexa), and venlafaxine (Effexor). The most frequently reported side effects of SSRIs are dizziness, lightheadedness, nausea, and mild tremor. Fluoxetine (Prozac) given late in the day may cause insomnia. Sodium oxybate (Xyrem), also known as γ-hydroxybutyrate or GHB, is anticataplectic, which is usually administered in two doses; the first is given at bedtime and the second 4 hours later. It consolidates (unifies) sleep and improves the disturbed nocturnal (nighttime) sleep that is characteristic of narcolepsy. Sodium oxybate is not hypnotic and is not used for insomnia. However, it can cause drowsiness and should only be taken at night. Non-drug treatments include the disease-specific education of the patient and family members and modification of behaviour patterns. Understanding the symptoms of narcolepsy may help relieve some of the frustrations, fears, anger, depression, and resentment of the patients and family members. Behavioural approaches include establishing a regular, structured sleep-wake schedule. Planned naps of 15-30 minutes or longer are usually refreshing and may be beneficial in reducing the patient's daytime sleepiness. Certain dietary restrictions should be observed (e.g., avoidance of large meals and alcohol). Regular exercise and exposure to bright light can improve alertness. Occupational, marriage, and family counselling may help improve the patient's quality of life (QoL). Occupations that require working in different shifts, changing the work schedule, or driving should be avoided. The dangers of driving while sleepy and/or experiencing cataplexy need to be addressed and the patients should be advised to avoid driving with these symptoms. Patients with narcolepsy may be able to drive for short distances at certain times of the day and after taking their stimulant medications. The EDS sometimes may become more pronounced and require additional medication. At other times, cataplexy and/or the other symptoms may decrease or even disappear for a time. Regular follow-ups and adherence to the plan of medications and behavioural treatment should diminish these fluctuations in symptoms and improve the patients' symptoms and quality of life. Narcolepsy is a disabling sleep disorder. Sleep is a vital behaviour of unknown function that consumes one-third of any given human life. Hypocretin neurons are discretely localised in the lateral hypothalamus, but have diffuse projections. Hypocretin deficiency may decrease monoaminergic tone, an abnormality previously suggested to underlie the narcolepsy symptomatology. Possible aetiologies include baroreceptor hypersensitivity, ultradian or circadian rhythm disturbances, receptor abnormalities, immune response modulation defects, genetic defects, neurotransmitter abnormalities, neuronal membrane defects, and neuroanatomic abnormalities; narcolepsy may also be secondary to events such as infection or accident. It is possible that different aetiologies exist for individual components or narcolepsy, or combinations, such as narcolepsy with cataplexy and narcolepsy without cataplexy. Because narcolepsy is an incurable illness with variable treatment results and symptoms that interfere with daily functioning, its impact on quality of life is significant. No one drug successfully treats all the manifestations of narcolepsy. Patients often do not perceive a major relief of symptoms. The sleep attacks can occur several times a day, 2-6 being typical. The sleep attacks vary in length; they can last anywhere from minutes to hours. The sleep attacks are not under the control of the individual; they are irresistible. Cataplexy is virtually pathognomonic of narcolepsy. Cataplectic episodes may last from a few seconds to as long as 30 minutes or longer. 82.6% of patients report cataplexy when not on medication. Attempts to practice rigid emotional control can have untoward results. Marked or moderate neurasthenia would be reported by 75% of cases, and 78% may express anxiety, feelings of loneliness and depressiveness. Negative self-image, underestimation and pessimistic disposition influence also the negative assessment of the quality of life. Excessive daytime sleepiness (EDS) is the most discomforting symptom responsible for career curtailment in the productive age, and for the need to live on a partial or full disability pension. The most primary complaints of narcolepsy patients are fatigue, excessive daytime sleepiness (EDS), and memory and attention deficits. Sustained alertness in Narcoleptics is reduced and changes diurnally; the largest effect is after 10 minutes. Narcoleptics have a 10 minute attention span, and narcolepsy does directly affect memory and function. The diagnosis of narcolepsy can be problematic.

HYPOCRETIN/OREXIN

Neuropeptides were named hypocretins based on the neurotransmitters' hypothalamic origin and their similarity to the gut hormone secretin. Hypocretins may complement monoaminogeric and cholinergic systems as major contributors for the generations of the sleep cycle. The hypothalamus is one of the most complex brain regions at both the functional and neuroanatomical level. The hypothalamus is the vegetative centre regulating body homeostasis, e.g., CVS, autonomic regulation, circadian rhythms, sleep, and reproduction. Hypocretin-containing neurons are exclusively located in the posterior hypothalamus, in a discrete region including and surrounding the perifornical nucleus. The discovery that hypocretin-containing cell bodies were exclusively located in the lateral hypothalamus, an area where lesions are known to dramatically reduce food intake. Intrahypothalamic projections were dense and included arcuate nucleus (ARC), paraventricular nucleus (PVN) and ventromedial nucleus (VMN), structures known to integrate feeding. Limbic system and associated structures (nucleus accumbens, amygdala, septum, basal forebrain area, bed nucleus of the stria terminalis), specific area of the thalamus (paraventricular and reticular thalamus) and the brainstem also received dense fibers. Hypocretin receptor Hcrtr1 is expressed in the VMN, while Hcrtr2 is mostly expressed in the PVN and arcuate nucleus. Human narcolepsy is a disabling disorder that necessitates life-long therapy. Human narcolepsy is primarily a sporadically occurring disorder but familial clustering has been observed since its initial description. There is impaired hypocretin trafficking and processing for the mutant polypeptide, with accumulation in the smooth endoplasmic reticulum and most probably cell death. The absence of hypocretin mutations in most human cases does not indicate a lack of involvement of this system in human narcolepsy. Hypocretin cells are driving monoaminergic and cholinergic tone during the sleep cycle, with maximal and minimal activity during wakefulness and REM sleep, respectively. High hypocretin tone during wakefulness activates both cholinergic and monoaminergic tone. In non-REM sleep, decreased hypocretin activity reduces monoaminergic and cholinergic tone. In REM sleep, the decreased monoaminergic tone reaches such a low level that a desinhibition of cholinergic systems occur. Hypocretin activity may be modulated by the biological clock. Inhibitory GABAergic projections from the preoptic hypothalamus and periaqueductal gray (PAG) to monoamine- and hypocretin-containing cells plays a complementary role in the driving of the sleep cycle. The REM sleep desinhibition observed in hypocretin-deficient narcolepsy is more compatible with the hypothesis that hypocretin tone is depressed during REM sleep. In the sleep disorder, symptoms are best explained by baseline monoaminergic hyperactivity, hyperactivity of cholinergic systems and cholinergic receptor hypersensitivity. Hypocretin systems are activated by starvation while their activity may be reduced by satiety. Hypocretins were initially described as neuromodulator of food intake. Inhibitory leptin receptors have been reported on hypocretin-containing cells. Hypocretin neurons are also activated during hypoglycaemia. Not only hypocretin but also hypocretin receptors expression is modulated by food deprivation. Hypocretin stimulates corticosterone and adrenergic secretion while decreasing plasma growth hormone and prolactin levels. Hypocretin has modulation effect on luteinizing hormone-releasing hormone secretion, which suggests a more minor contribution to the regulation of reproductive functions. These effects generally result in increasing global energy consumption. Hypocretin-1 induces an increase in the respiratory quotient without increasing activity or feeding, indicating an increased metabolic rate. The hypocretin system also modulates vagal tone, stimulating gastric acid secretion, increase blood insulin and blood glucose. Narcolepsy can occur occasionally with normal or elevated hypocretin levels, independently of HLA status and family history, suggesting disease heterogeneity. Most human brains have no preprohypocretin transcripts. Residual gliosis may present.

DOPAMINERGIC ROLE IN STIMULANT-INDUCED WAKEFULNESS

Dopamine transporters play an important role in sleep regulation and are necessary for the specific wake-promoting action of amphetamines and modafinil. The monoamine serotonin (5-HT) and norepinephrine (NE) and histamine (HA), neuropeptides including hypocretin (orexin), and other transmitters including acetylcholine, GABA, and adenosine have been prominently implicated in sleep–wake regulation. Alteration in acetylcholine, 5-HT, NE, or HA are more critically involved in regulating the cortical electroencephalogram (EEG) desynchronization characteristics of wakefulness, whereas dopaminergic activity is thought to mediate motor-related aspects of behaviors. The terminal release of dopamine varies in concert with arousal states. Uncertainties have persisted about the molecular bases of efficacious wake-promoting compounds, such as amphetamines and modafinil. Amphetamines block plasma membrane transporters for dopamine (DA), NE, and 5-HT and inhibit the vesicular monoamine transporter (VMAT2), releasing monoamines from the synaptic vesicles into which VMAT2 pumps them. Modafinil increases wakefulness could be due to activation of α-1 noradrenergic transmission or hypothalamic cells that contain the peptide hypocretin, or that it may work by modulating GABAergic tone. The wake-promoting effect of amphetamine is maintained after severe reduction of brain norepinephrine. Dopamine-specific reuptake blockers can promote wakefulness in normal and sleep-disordered narcoleptic animals better than NE transporter-selective blockers. Intact hypocretin transmission is not required for the wake-promoting effect of modafinil. Both modafinil and amphetamine promote wakefulness primarily by increasing dopaminergic tone and not by stimulating hypocretin transmission. The mechanism by which caffeine induces wake is not certain. Caffeine has a number of pharmacological effects. Caffeine adenosine–dopamine interactions in sleep is a possible mechanism, similar to those implicated in brain locomotor control systems. Adenosine–dopamine interactions are important in determining the stimulant effect of caffeine. Dopamine metabolism and receptor abnormalities also occur in disorders of excessive day-time sleepiness, such as narcolepsy and in normal aging. Human variants at the dopamine transporter (DAT) gene locus could predispose to vulnerability to sleep–wake disorders. Selective DAT inhibitors can have useful clinical applications and low side effect profiles when compared with classical amphetamine-like stimulants. Deficiency in GHRH is associated with significantly less NREM sleep. Impairment of the somatotropic axis could reduce sleep. Dopamine receptors in the hypothalamus inhibit hypothalamic GHRH release. Because growth hormone-releasing hormone (GHRH) promotes sleep, an inverse relationship would be expected between dopaminergic tone and sleep time.

NEUROTOXINS

When toxic matter and undigested food, collected in the intestines as a result of bowel toxaemia, are absorbed from the bowels into the blood stream, the result is the leaky gut syndrome. The undigested molecules act as antigens, foreign substances that provoke an immune reaction. Many of these antigens are similar in structure to normal body components, and the antibodies produced to fight them can destroy healthy tissues. If the defence system of the body does not work effectively, it is likely that pathological conditions will develop. Primary harmful factors are represented by:- 1- Various invading foreign microorganisms such as bacteria, viruses, fungi, yeast and parasites. 2- Their components such as lipopolysaccharides, nucleic acids and peptidoglycans. 3- Toxic products formed in the metabolic process such as mycotoxins and sugar alcohol. Secondary harmful factors are represented by non-living factors: 1- Radiation. 2- Extremes of temperature. 3- Excessive exercise. 4- Trauma. 5- The general environment, particularly in industrial zones. 6- Psychological chronic stress. Thes factors are capable of detrimentally damaging normal body cells either directly or indirectly, e.g., apoptosis or necrosis etc. Some of these factors are able to utilise or damage various substances such as nutrients and enzymes, which the body needs for itself. The loss of these substances will ultimately result in the development of one or another pathological condition, sooner or later. The dying cells will produce damaging free radicals, which oxidise different substances, such as proteins, lipids and nucleic acids. Oxidised substances lose their normal activities and properties and instead, create some form of abnormal process, which contributes to the development of diseases and disorders. Lipopolysaccharides, which are components of gram negative bacteria, are capable of directly damaging cells, which can result in the induction of oxidative stress. α-Toxin (which is a component of the bacteria Staphylococcus aureus) is equally able to directly damage cells by creating a defined size pore in the cell membrane. T2 Toxin (which is a fungal metabolite) is also capable of killing cells by inducing apoptosis. Radiation and some toxic substances prevalent in an industrial environment are equally capable of inducing oxidative stress. Other harmful factors, such as peptidoglycans, which are components of gram positive bacteria are able to excessively activate monocytes and macrophages. Over activation of the natural killer (NK) cells by various foreign organisms, or their fragments, can also lead to the over production of another mediator of oxidative stress-IFNγ, which is an inductor of apoptosis or perforin, which is an inductor of necrosis. A balanced defence system is capable of negating the adverse affects of harmful substances, thus enabling the individual to combat the demands of everyday life whilst maintaining good health. It is essential that the activity of the nervous system to remain in a state of balance otherwise the signals that it transmits will be inaccurate. The nervous system may be forming the fundamental part of the defence system and therefore balanced activity of the nervous system forms the basis for balanced activity of the defence system. The brain, which forms the main part of the nervous system, controls and regulates the functions of all other parts of the defence system and, in particular, the immunological part. It is the latter, which receives signals from the nervous system to modulate its activity according to the circumstances prevailing at any instant. The vast majority of patients are suffering from some form of infection (viral, bacterial, parasitic, fungal or yeast) and associated immunological imbalance, although the type of infection differs. By passing less than a quarter of a millilitre of the urine specimen through high performance liquid chromatography (HPLC) equipment much can be revealed about the state of the patient's condition and will include the following: 1-The presence and probable type of any infection. 2- If damaging free radicals is being generated. 3- The character of the adrenal activity. 4- The character of the activities of the defence system (NK cells, T cells, B cells and macrophages). 5- How the nervous system can affect mood and behaviour. It is necessary to undertake additional assays using the intravenous blood sample. Neurological profile is designed to detect and to identify various chemical abnormalities within the nervous system, which can significantly affect the functioning of the immune and endocrine systems. It also reflects the behaviour and mood patterns of the patient, which in turn, affects all of the physiological processes. Level is the amount or concentration of a particular substance, which is revealed from the assay. Gene expression is the quantity of DNA, which is responsible for the production of the specific protein that one wishes to investigate. The higher the gene expression then the greater the quantity of the substance produced. Measurement of the gene expression, especially for substances of the defence system, may provide a far more reliable picture because the quantity of DNA within the nucleus of a cell remains stable. Detection is the presence of a particular substance, macro-organism or process, without their quantification. Protein Bcl–2 inhibits apoptosis by preventing the release of apoptosis inducing factor (AIF) and cytochrome C from mitochondria. Protein Bax promotes apoptosis by triggering the release of AIF and cytochrome C from mitochondria. If the Bcl–2 level is higher than the Bax level then apoptosis will be prevented. The ratio of Bcl–2 to Bax in the cell can determine whether or not the cell initiates apoptosis or survives. Many cancer cells over express Bcl–2, thus preventing apoptosis and permitting malignant growth to continue. Anti-apoptotic Bcl–2/Pro-apoptotic Bax genes expression is an indicator of the ability of cancer and other abnormal cells to survive. This assay is able to provide information related to the possibility of tumour growth without destruction. p53 is a tumour suppressor gene, which functions as an inducer of apoptosis and is a regulator of the Bcl-2 gene, which expresses protein inhibiting apoptosis. In cancerous cells, the p53 gene is one of the most mutated genes that express mutant forms of the p53 protein in the nucleus. This mutant p53 protein is not able to induce apoptosis of cancerous cells and they will continue to divide, thus increasing cancer growth. Also, cancerous cells with a deleted (absent), for unknown reasons, p53 gene will continue to proliferate, thus increasing cancer growth. It is not possible to convert cells with mutated or deleted p53 gene into cells with a wild (normal) p53 gene. Thus, such cells can only be destroyed or eliminated. Anti-p53 antibodies are an indicator of the ability of cancer cells to survive. This assay provides information relating to the presence of a wild (normal), mutated or deleted p53 gene, which is associated with the suppression of tumour development. Bactericidal permeability increasing protein is a bactericidal compound, which is present in leukocytes. BPI binds endotoxins; it is rapidly released by leukocytes in response to exposure to bacteria, which produce endotoxic lipopolysaccharides (LPS). LPS are highly toxic components present in the cell wall of gram negative bacteria, which include many bacteria responsible for diseases. The presence of LPS in blood will lead to the development of fever, inflammation and septic shock. Bactericidal permeability increasing (BPI) protein level is a marker of the response to pathogenic gram negative bacteria. This assay is able to provide information related to the presence of pathogenic gram negative bacteria. Cathepsin L is an enzyme, lysosomal endopeptidase, which is produced by certain specialised cells, such as macrophages and osteoclasts. This enzyme is also produced by many malignantly transformed cells. Since Cathepsin L has the ability to degrade the proteins of the extracellular matrix, it is assumed to play a crucial role in tumour progression, metastasis and numerous other disorders, where the destruction of the extracellular matrix is the major cause of disease, e.g., rheumatoid arthritis and neurodegeneration. Cathepsin L Level is a marker of the ability of cancer and other abnormal cells to destroy their environment, resulting in disease progression. This assay is able to assist in the prognosis of the progression of tumour growth and also neurodegenerative and some inflammatory processes. Cell viability detection is an indicator of the presence of harmful substances and abnormal cells. This assay is designed to reveal, indirectly, the presence of cytotoxic agents, which damage cells (drugs, preparations, natural toxins etc.) and abnormal cells (fungi and bacterial cells) in the test media. It is quite possible to monitor the effectiveness of any treatment programme embarked upon. Cu/ZnSOD is a soluble cytosolic enzyme, which present in the cytoplasm of virtually all eukaryotic cells. Cells, which are dying or dead cells, release Cu/ZnSOD into the media where it can be tested, thus indicating the level of apoptosis. Liver dysfunction, kidney damage and neurodegenerative pathologies could be associated with increased Cu/ZnSOD level. Cu/Zn Superoxide Dismutase (Cu/ZnSOD) Level is a marker of inflammation, cell destruction and organ dysfunction. This assay is able to provide information relating to the inflammatory processes, associated dysfunctions of the liver and/or kidney, plus destruction of cells of the nervous system. E-cadherin, also known as uvomorulin or cell-CAM120/80, is one of the sub-classes of cadherins, Ca^{2+}-dependent cell adhesion molecules, which are responsible for the selective cell-cell adhesion. There is evidence to suggest that the changed expression of E-cadherin is a common event in cancer progression. The level of this substance can significantly decrease in the number of solid tumours such as lung, breast and hepato-cellular carcinomas, gastric, prostatic and other tumours. Simultaneously, its level in blood could be increased. Such distribution of E-cadherin is associated with the ability of cancer cells to be invasive, which can lead to the development of metastasis. In addition, the elevated E-cadherin level is associated with systemic inflammatory response and multi-organ dysfunction syndrome (MODS). E-cadherin Level is an indicator of the ability of cancer cells to be invasive, plus a marker of inflammation and organs dysfunction. This assay is able to assist in the prognosis of cancer invasion/metastasis and also in the estimation of systemic inflammation and associated multi-organ dysfunction. It is designed for the quantitative determination of soluble E-cadherin. Fas ligand is the special protein molecule, which is able to trigger apoptosis in the Fas sensitive cells. The immune system T cells and B cells, as well as Natural Killer (NK) cells are Fas sensitive cells. On their surface there is a special receptor molecule called Fas or CD95. When Fas ligand interacts with the Fas receptor, the destructive mechanism of the immune system cells is triggered. Activated T cells, B cells and NK cells have an increased number of Fas receptors, which make them more Fas sensitive. There is evidence that some cancer cells produce significant amounts of Fas Ligand. This Fas L is able to trigger destruction of the cells of the immune system, which would normally attack cancer cells. This means that cancer cells avoid destruction whilst some of the cells of the immune system could be destroyed. This phenomenon is known as the immune escape of cancer cells. Fas Ligand (Fas L) gene expression is an indicator of the ability of cancer cells to destroy cells of the immune system. This assay is able to obtain information related to the cancer cell defence mechanism protecting against its own death. Production of interferon gamma (IFN-γ) is a function of T cells (both cytotoxic and the Th1 type) and NK cells. IFN-γ is a lymphokine, which has a number of properties such as antiviral and antiproliferative. IFN-γ is able to activate the macrophages in order to destroy tumour cells by releasing reactive oxygen intermediates and TNF-α. IFN-γ is capable of down regulating (inhibiting) anti-apoptotic Bcl-2 protein followed by the induction of cell destruction. IFN-γ gene expression is a marker of active NK and Th1 cells. This assay can assist in the evaluation of the processes relating to the destruction of abnormal cells including cancer. It has been designed to estimate the IFN-γ gene expression as part of the characterisation of the state of the defence system. IL-1β, originally known as lymphocyte activating factor (LAF), activates T lymphocytes, which then proliferate and secrete IL-2. IL-1β (also known as endogenous pyrogen) is able to mediate oxidative stress, and is produced mainly by macrophages and monocytes. IL-1β gene expression is a marker of active macrophages, Th1 cells, inflammation and oxidative stress. This assay is able to provide information related to the activation of cellular immunity, promotion of inflammation and oxidative stress. It estimates the risk of inflammation and oxidative stress. Interleukin 2 (IL-2) is produced by the naive CD4+ T cells and Th1 cells. IL-2 is a cytokine, which plays a central role in immune responsiveness by promoting the activation and proliferation of lymphocytes that have been primed by antigens (antigen specific T-lymphocyte proliferation). IL-2 stimulates the proliferation of B cells; augments NK cell activity and inhibit granulocyte macrophage colony formation. IL-2 displays anti-tumoural effects. IL-2 gene expression is marker of the active naive T and Th1 cells. This assists in estimating the specific immune response to cancer (anti-tumour immunity), and various types of pathological conditions. It estimates the IL–2 gene expression as part of the characterisation of the Th1/Th2 immunological balance. IL-4 is produced mainly by the Th2 cells, as well as mast cells. This cytokine exerts numerous effects on various haematopoietic cell types. It can promote survival, growth and differentiation of both T and B lymphocytes, mast cells and endothelial cells. IL-4 can inhibit the production of TNF-α and IL-1 and IL-6, by macrophages. IL-4 gene expression is marker of active naive T and Th2 cells. This assay assists in the estimation of the formation of the anti-tumour immunity and also to provide information related to the presence of some bacterial infections. It estimates the IL-4 gene expression as part of the characterisation of the Th1/Th2 immunological balance. IL-5 is a cytokine, which has the property of promoting the development and the activation of eosinophils. It is produced mainly by Th2 cells, in response to the stimulation provided by parasite derived antigens and allergens. In addition, various transformed B cells, Reed-Sternberg cells in Hodgkin's disease and activated eosinophils can express IL-5. Some kind of inflammatory process (such as asthma or various allergic responses) is associated with Th2 cell activation followed by an elevated production of IL-5, together with IL6. IL-5 gene expression is marker of active Th2 cells. This assay assists in the estimation of the formation of the anti-tumour immunity and provides information related to the presence of parasite infections or allergens. It estimates the IL-5 gene expression as part of the characterisation of the Th1/Th2 immunological balance. IL-6 is a cytokine, which is produced by a variety of cells, including macrophages and Th2 cells. Normal cells do not produce this cytokine constitutively, but its expression is induced by a variety of cytokines, bacterial or viral infections. IL-6 acts on a wide range of tissues, exerting growth-induction, growth-inhibition or differentiation, depending on the nature of the target cells. IL6 is involved in other important processes, such as the proliferation of T cells, the induction of IL-2 expression and B cell differentiation. The changed expression of IL-6 is associated with infections, autoimmune, proliferative and neoplastic diseases and inflammatory responses to infections or trauma. IL-6 gene expression is a marker of infection or trauma. This assay provides information relating to both viral and bacterial infections, trauma and cancer development. IL-10 is a cytokine, which is produced by the regulatory CD4 and T cells and has the ability to suppress the functions of macrophages, Th1 cells and NK cells. It is associated with the promotion of Th2 cell development and consequently with the regulation, proliferation and differentiation of B cells. IL-10 has immunosuppressive properties by inhibiting the production of Th1 specific cytokines and the maturation of dendritic cells and is able to exert strong anti-inflammatory activities. IL-10 gene expression is a marker of cellular immunosuppression and infection. This assay assists in the estimation of the level of IL-10 gene expression as part of the characterisation of the Th1/Th2 immunological balance. The expression elevates in parasitic infections (such as schistosoma mansoni, leishmania, toxoplasma gondii and trypanosoma), mycobacterial infections (such as mycobacterium leprae, mycobacterium tuberculosis and mycobacterium avium) and retroviral infections. IL-12, which is also known as the natural killer cell stimulatory factor or the cytotoxic lymphocyte maturation factor, is a cytokine that can have multiple effects upon T cells and NK cells. IL-12 produced by macrophages is associated with the promotion of Th1 cell development, followed by cell-mediated immune response. In addition, IL-12 is a proinflammatory cytokine produced by phagocytic cells, B cells and some antigen-presenting cells, which modulate adaptive immune responses. IL-12 gene expression is a marker of active B cells, macrophages and the stimulation of Th1 cells. This assists in the estimation of the formation of the anti-tumour immunity and provides information related to the presence of bacterial and viral infections. It estimates the IL-12 gene expression as part of the characterisation of the Th1/Th2 immunological balance. Significant elevation of IL-12 occurs in autistic patients, sufferers of multiple sclerosis, autoimmune diseases, cancer and chronic inflammatory reactions. A large number of bacterial and viral infections are capable of changing the expressed levels of IL-12. Lipid peroxidation is the major indicator of oxidative stress, which is associated with the production of damaging free radicals and contributes to the formation of various pathobiological changes in the cells and organs functioning. The lipid peroxidation of the cell membrane is very often followed by necrosis (accidental death) of such cells. Lipid peroxidation products (LPP) Level is an indicator of the destructive processes by free radicals. Matrix metalloproteinases (MMP's) are a family of zinc dependent endopeptidase, which can degrade the major components of the extracellular matrix (ECM). Cancer cells subvert MMP's activity to promote invasion of the surrounding tissues, as well as metastasis to distant tissues. MMP's by releasing growth factors sequestered in the ECM are also thought to promote the growth of these tumour cells once they are metastasised. The inhibition of one of the MMP enzymes can lead to the dissemination of cancer, in other words the suppression of its progression and metastasis. Matrix metalloproteinase 2 (MMP-2) gene expression is an indicator of the ability of cancer to be invasive and form metastasis. This assists in the prognosis of the progression of cancer. Mycoplasma detection/gene expression is an indicator of the presence of an infection. This assay is able to provide information related to the presence of mycoplasma-the DNA particle of the bacterium. Mycoplasma is able to dramatically change the properties of normal cells and to significantly impair the functions of the defence system. Natural killer (NK) cells are large granular lymphocytes, which are the first line vital defence in the body as they can destroy toxic and pathogenic agents without sensitisation by them and before the involvement of the specific cells of the immune system. NK cells use 2 mechanisms to destroy the targets: 1- the antibody-dependent cell cytotoxicity, NK cells are able to kill antibody-coated cells through this mechanism; and 2- natural cytotoxicity, NK cells are able to kill many different targets through this mechanism such as tumour cells and virus infected cells. The cells, which are attacked by NK cells, can die by either apoptosis or necrosis. For many cancers, necrosis can be the most effective method of their destruction. To trigger necrosis and/or apoptosis, NK cells release special substances, e.g., perforin and INF-γ, in the extracellular media or express receptor molecules (specific glycoproteins) on their surface. These receptors can be killer inhibitory or activatory receptors and under special conditions, the inhibitory receptors prevent NK killing and the inhibitory signal dominates the activatory signal. NK cells activity profile is an indicator of the activity of the defence system. This assessment provides information relating to the involvement of NK cells in the effective functioning of the defence system. Cytomegalovirus inhibits any attack by NK cells. Psychological stress factors significantly depress the activities of NK cells. Nitrotyrosine (modified tyrosine) is formed in the presence of peroxinitrite, which is the active metabolite of nitric oxide (NO). NO is produced by various cells in response to the inflammatory or mitogenic stimuli. The biological role of NO is defensive against non-self pathogens. However, enhanced NO production will have toxic effects. The increased formation of NO metabolite peroxinitrite will lead to elevated nitrotyrosine production. Accumulation of nitrotyrosine dramatically affects the functions of proteins, which negatively influence numerous physiological processes. Nitrotyrosine is associated with various inflammatory processes, including septic shock. Nitrotyrosine level is an indicator of the presence of an infection and inflammation. Non-genomic DNA level is an indicator of the presence of a viral infection or mycoplasma. This assay measures the concentration of extracellular (non-genomic) DNA. Non-genomic RNA level is an indicator of the presence of a viral infection. This assay measures the concentration of extracellular (non-genomic) RNA. Telomeres are specific structures found at the ends of chromosomes in eukaryotes. These telomeres protect chromosome ends because chromosomes lacking telomeres undergo fusion, rearrangement and translocation. In somatic cells the telomere length is progressively shortened with each division due to the inability of the DNA polymerase complex to replicate the 5' end of the lagging strand. Telomerase is a unique enzyme, which is a ribonucleoprotein that synthesises and directs the telomeric repeats on to the 3' end of existing telomeres using its RNA component. Telomerase activity becomes suppressed through the ageing process, but activation of telomerase is regarded as essential to most cancer. Telomerase activity is specifically expressed in immortal cells, cancer and germ cells where it compensates for telomere shortening during its DNA replication and this stabilises telomere length. The expression of telomerase activity in cancerous cells is a necessary step for tumour development and growth. This means that there is a specific association of human telomerase activity with cancer and it is usually high in cancer patients. There is clinical evidence to show that patients whose tumours do not display telomerase activity are very likely to eliminate the cancer, quite often naturally. It is considered that the regression of telomerase activity could be one of the mechanisms for cancer regression. Number of cells with active telomerase is an indicator of the presence of cancer cells but not necessarily cancer. This provides information related to the very early oncogenic changes prior to the definitive formation of any tumour and clinical diagnosis. When the tumour is formed, this assay is able to assist in the prognosis of cancer regression or progression. Parasites and their by-products are able to significantly impair the functions of the nervous and immune systems, which can lead to the development of various pathological conditions. Parasites and smears detection is an indicator of the presence of an infection. This assay provides information related to the presence of blood parasites or smears. Protein fractions level is an indicator of organs dysfunction. Obtaining information from this assay about the concentration of total small peptides in blood can have a negative affect upon the nervous and immune systems. Total peptides can serve as an indicator of gut problems. Soluble p185 Her-2 protein is a transmembrane growth factor receptor, which belongs to a class of oncogenes related to tyrosine protein kinase that possess tumourigenic or transformation activity. It is involved in all cell growth and cell transformation. It is possible to detect increased levels of p185 in the serum of individuals who will subsequently develop cancer up to 60 months before clinical diagnosis. Soluble p185 Her-2 protein level is an indicator of the ability of cancer cells to metastasise. This assay provides information related to very early oncogenic changes and is able to assist in the monitoring of tumour spread, postoperative relapse and metastatic risk. Reactive oxygen species may cause severe damage within the cell, however several enzymes have evolved that scavenge and neutralise their effects. Superoxide dismutase, a metallo enzyme, is one key enzyme involved in the protection of cells from damaging free radicals. Any one of three metals will work together with SOD by catalysing the dismutation of superoxide anion into oxygen and hydrogen peroxide; these are manganese, iron and copper. There is evidence to suggest that many disease processes could be associated with extremely low superoxide dismutase activity. Superoxide dismutase (SOD) activity is an indicator of the ability of the defence system to scavenge and neutralise the damaging effects of the reactive oxygen species. This assay provides information related to the destructive oxidative processes in the body. Survivin is a protein, which is produced in many foetal tissues, such as the kidneys, liver, lungs and brain in addition to transformed cell lines and most cancers. Survivin is not found in terminally differentiated adult tissues. Overproduction of survivin in rapidly growing cells may explain its abundance in tumours. Survivin level is a marker of the ability of cancer cells to inhibit their own death. This assay assists in the estimation of cancer cells to survive the normal destructive process and to continue the growth process. Pyruvate kinase (PK) or pyruvic acid kinase is an enzyme, which catalyses the transfer of a phosphate group from phosphoenolpyruvic acid to ADP with the formation of ATP. This reaction is very important in the glycolytic pathway. This enzyme has an absolute requirement for magnesium and is inhibited by calcium. Whilst the conventional form of pyruvate kinase is a tetrameric molecule a, dimeric isoform of PK is over produced by a wide range of different tumours. This appears to be linked to the different metabolic requirements shown by tumour cells, which leads to these cells using a metabolic shortcut to save energy for cell multiplication. In this process the proportion of pyruvate kinase present as the dimer form is increased. Therefore the level of the dimer, namely tumour M2-PK, is increased in tumour cells. Tumour M2-pyruvate kinase (TM2-PK) is marker of tumour cells. Transforming growth factor (TGF-β) is a multi-functional protein, which is produced mainly by mammalian cells. TGF-β is capable of influencing differentiation, proliferation and a variety of cellular functions. This protein generally slows immunosuppressive and anti-inflammatory activities but is able to stimulate the macrophages and monocytes. TGF-β overproduction is linked to tumour growth, perhaps as an indirect result of its other actions. TGF-β gene expression is marker of immunosuppression. This assay assists in the estimation of the level of TGF-β gene expression as part of the characterisation of the state of the defence system. TNF-α, also known as cachetin, is a cytokine, which is produced mainly by monocytes and macrophages. Endotoxins and INF-γ promote the production of TNF-α and it functions as a modulator of the immune response by activating dendritic and Th1 cells. It may play a role in the pathogenesis of many disease states and, in particular, inflammatory diseases. In high quantities, this cytokine acts as a pyrogen and neurotoxin by increasing the permeability of the blood/brain barrier and destroying the serotonin containing cells of the nervous system. TNF-α gene expression is a marker of active macrophages, Th1 cells and neurotoxicity. This assay assists in the estimation of the formation of the anti-tumour immunity and also to provide information related to the presence of pathogenic microorganisms and inflammatory processes. It measures the TNF-α as part of the characterisation of the state of the defence system. TNF-β, also known as lymphotoxin-α, and the TNF-α are 2 closely related proteins that bind to the same cell surface receptors and show many common biological functions. TNF-β plays a central role in lymphoid development and in normal host resistance to infection and to the growth of malignant tumours, serving as an immunostimulant and as a mediator of the inflammatory response. It is produced by activated T, B and NK cells, astrocytes and human myeloma cells. Excessive production has been found to play a very significant role in a number of autoimmune disorders including multiple sclerosis, insulin dependent diabetes and rheumatoid arthritis. TNF-β gene expression is a marker of Th1 cells activity. This assay assists in the estimation of the formation of cell-mediated immunity. Local hypoxia is a potent inducer of vascular endothelial growth factor (VEGF) production from adjacent cells but it is not synthesised in endothelial cells. Expression of VEGF can be induced in macrophages, T cells, astrocytes, osteoblasts, smooth muscle cells, fibroblasts, endothelial cells, cardiomyocytes, skeletal muscle cells and keratinocytes. It is also expressed in a variety of human tumours. The VEGF is the most active angiogenic factor, appears to inhibit the maturation of the dendritic cells, which have fundamental role in the activation of an effective anti-cancer immunity. Abnormally high pre-treatment levels of VEGF are associated with reduced efficacy of IL-2 cancer immunotherapy. In this case, the concomitant administration of anti-angiogenic drugs could enhance the anti-cancer immunotherapy. Therefore information related to the VEGF expression could be used to control the angiogenic processes, which have been proven to stimulate cancer growth not only through stimulation of cancer related neo-angiogenesis, but also by inducing a suppression of host anti-cancer immune response. VEGF is a family of closely related growth factors, which are polypeptides. VEGF is a potent growth factor for blood vessel endothelial cells, showing pleiotropic responses that facilitate cell migration, proliferation, tube formation and survival. It is also one of the most potent permeability factors, so VEGF is a common link of inflammation permeability and angiogenesis. VEGF-β gene expression is a marker of the angiogenic process, which stimulates the growth of tumours. This assay assists in the estimation of the risk factor, which is able to suppress the host anti-tumour immune response and stimulate tumour growth. Yeast and its by-products are able to significantly impair the functions of the nervous and immune systems, which can lead to the development of various pathological conditions. Yeast gene expression is an indicator of the damaging factors. This assay provides information related to the presence of yeast. Glycoprotein 90K/Mac-2BP is mainly produced by the cells of the defence system: the macrophages, monocytes and to some extent by a variety of other cells. There is evidence of its involvement in immune defence mechanisms. Elevated levels of this protein are associated with viral infections and some types of cancer (probably of viral origination). In some cases, the failure to respond to treatment could be connected with increased production of this molecule. Glycoprotein 90K/Mac-2BP is an indicator of the presence of a viral infection.

URINARY NEUROTRANSMITTER ASSAYS IN CANCER PATIENTS

Scientific knowledge together with good biochemical assays can lead to proper treatment strategy. Treatment application would require biochemical assays to confirm a positive results and adjustment of treatment strategy. Imbalanced functioning defence system leads to general health problems, disorder and disease. The defence system consists of: 1- manager (nervous system), 2- regulator (endocrine system), 3- executor (immune system), and 4- executor (cellular enzymatic systems). The quantity of a substance can be: none (0%), trace (<10% of normal), low (10-74% of normal), normal (100%), Above normal (101-125% of normal), high (126-200% of normal), very high (>200% of normal). The analysis of any result can be normal, abnormal, or pathological. Urinary neurotransmitter assays may have a valuable role to play in the assessment and management of cancer patients and patients with other conditions also. There is interaction between the nervous system and immune system: 1- directly via neurotransmitters and neuropeptides, and 2- indirectly via the stimulation of the endocrine system. There are 2 axes: Hypothalamus-Pituitary-Thymus axis (HPT) and Hypothalamus-Pituitary-Adrenal axis (HPA). The dopaminergic systems stimulate immune function via the HPT axis, e.g., the production of neuropeptides such as tyrosine, which is immunostimulatory. While the serotonin and adrenergic systems suppress immune function via the HPA axis via the release of cortisol. There are 3 major subsystems in the CNS that interact with each other and the other neurotransmitter systems, and to an extent regulate each other by feedback including: 1- dopaminergic, 2- 5HT, and 3- adrenergic (noradrenaline). In normal individuals there is a degree of fluctuation in the day to day levels of neurotransmitters and their metabolites. Factors, both endogenous and exogenous interact with the neurotransmitter subsystems affecting production and/or metabolism, e.g., stress, free radicals, nutrition, medications, environmental toxins, gut derived toxins (polyamines, toxins from yeast species). Metabolites build up when neurotransmission is disrupted, e.g., Bufotenin (5HT derivative) is neurotoxic and an agonist at 5HT receptors, Epenin (dopamine derivative) is a dopamine receptor agonist, and 6-hydroxydopamine and 5,6-dihydrotryptamine destroy catecholamine and 5HT neurons respectively. Many neurotoxins facilitate activity at glutamate receptors, i.e., excitatory, which is associated with an elevation of intracellular calcium levels and tends to result in the formation of free radicals, especially the hydroxyl radical that can predispose to cancer by causing oxidative damage to proteins, DNA and lipids. Free radicals impair the activities of neurotransmitter systems, which is reflected in the dominance of their pre-synaptic (intraneuronal) activities rather than their synaptic (extraneuronal) activities. Disruption of balance within the nervous system results in disruption of immune and endocrine function, which in turn results in the development of disease states.

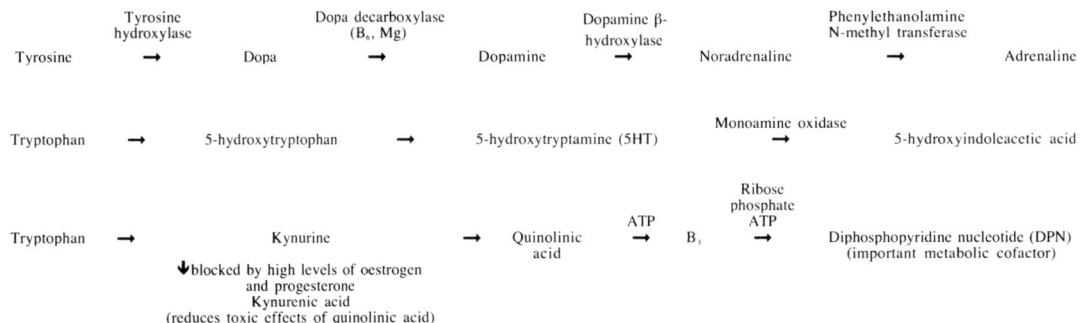

Infections (Chlamydial, Pseudomonas), hypoxic states and others that disrupts respiration and oxidative phosphorylation, or excessive ATP consumption can affect the conversion of quinolinic acid to B_3, resulting in accumulation of quinolinic acid. In cancer there is: 1- a reduction in dopaminergic activity and increase in 5HT activity, 2- the accumulation of abnormal metabolites and neurotoxins, 3- an increase in cortisol levels is associated with raised 5HT activity. This is immunosuppressive with both negative and positive effects in the cancer patient-negative effects include a reduction of NK cell activity, and positive effects include control of IL6 activity. In later stages when 5HT activity drops, control of IL6 activity is reduced and the consequent increase in activity of this cytokine facilitates the development of bone metastases. Cancer patients tends to have more methylated derivatives of dopamine. Many cancer cells have increased methylation of DNA, which tends to cause gene mutations. Treating patients with reuptake inhibitors (SSRI) may be detrimental if there is an overproduction of toxic metabolites in the neurotransmitter subsystem. Overstimulation of immune function may have negative results. An overactive dopaminergic system could favour an autoimmune response and reduce anti-tumour activity if unchecked. Dopaminergic activity (including metabolite activity) stimulates release of growth hormone release hormone and other growth factors, which may stimulate tumour growth. Schizophrenics suffer less cancer than general population, i.e., better immune response. Dopamine metabolites that accumulate in this condition tend not to be methylated derivatives, which are apparent in chronic fatigue syndrome patients who develop more cancers than general population. Muramyl dipeptides (MDPs) activate the central dopaminergic system, with a subsequent stimulation of immune response-possibly via the HPT axis. They also cause an increased turnover of 5HT in the brain. Noradrenaline suppresses tumour necrosis.

URINARY NEUROTRANSMITTER ASSAYS

The function of the nervous system (CNS, PNS) is to control and regulate the immune system, the endocrine system and all other bodily functions; to process the information obtained from the environment and then initiate a reaction accordingly. A healthy nervous system forms the basis for general physical and mental health. The nerve endings, known as dendrites, receive impulses when the sensory cells are stimulated. The impulse is sent along successive nerve cell axons to the brain where the information is processed. A signal is then sent from the brain back down the nerve cell axons to effector cells, such as muscles or glands, for an appropriate response, for example. At the synapse, there are a number of biochemical substances (molecules), which transmit the impulse across the gap (neurotransmitters). For the nervous system to function efficiently the concentration of each neurotransmitter must fall within a specific quantitative range for the time of day or night. Also, other many associated molecules participate in the functioning of a healthy nervous system include: 1- Precursors, substances that are transformed into neurotransmitters under specific conditions. 2- Metabolites, substances that originate from neurotransmitters under specific conditions. They are the products of neurotransmitter utilisation. 3- Non-toxic derivatives, substances that originate from neurotransmitters or precursors under certain circumstances, often through errors in the biochemistry. Derivatives can be agonistic (i.e., increase neurotransmitter action) or antagonistic (i.e., decrease neurotransmitter action). Accumulation of non-toxic derivatives can interfere with the normal functioning of neurotransmitters by preventing them forming a balanced pattern of substances. 4- Neurotoxins, substances that can cause damage and, in some cases, cause irreversible negative effects to the nervous system. Neurotoxins can injure or destroy parts of the nerve cells, or destroy sites of neurotransmitter action. Consequently, the neurotransmitters cannot function properly and so the pattern gradually becomes imbalanced. In a healthy individual, there will be no neurotoxins present in the body. Neurotoxins often originate from yeast, fungal, bacterial, microbial, viral or parasitic infections, as well as from synthetic chemicals in the environment. A balanced pattern of neurochemical substances is the basis of normal functioning channels of communication between the separate bodily systems (healthy individuals). The primary cause of ill health is an imbalanced nervous system, whilst the illness itself is a secondary effect. Neurotoxins include: 1- Bufotenin (N,N–dimethyl–5–hydroxytryptamine) is the methylated toxic derivative of the principal neurotransmitter serotonin. Structurally, bufotenin is very similar to serotonin. This means that bufotenin can act as a serotonergic agonist, which therefore increases serotonin synaptic activity by binding to the same neuronal receptor sites as serotonin. This process effects the metabolism of the principle neurotransmitters leading to thought disturbance, shift of mood, as well as movement, appetite, sleep disturbance and loss of endocrine function. 2- Dihydroxyphenylpropionic acid is an exogenous dopaminergic false neurotransmitter, which imitates dopamine by binding to the same neuronal receptor sites and is produced by bacteria, such as Clostridium in the intestine, not by the body. Dihydroxyphenylpropionic acid suppresses the activity of the dopaminergic subsystem by preventing dopamine from functioning in its usual way. This process can lead to various pathological conditions including autism. 3- Dihydroxytryptamine accumulates when hydroxy radicals attack serotonin molecules and can cause depletion in the levels of serotonin in the nervous system by destroying serotonergic neurones. Problems caused by the process are exacerbated by the slow recovery of serotonin back to normal levels once dihydroxytryptamine has been eliminated from the body. The presence of dihydroxytryptamine can cause a depletion of cortisol and subsequent immune and endocrine abnormalities. 4- 6-Hydroxydopamine (6-OHDA) accumulates when radicals attack dopamine molecules. 6-OHDA is toxic towards the nerve terminals of dopaminergic and noradrenergic nerve terminals, which therefore degenerates the neurones of these two subsystems. This destructive process can lead to locomotion problems similar to that seen in Parkinson's disease. Decreases in self-stimulation, eating, drinking and other behaviours that depend on motor capability can also occur. Urinary 6-OHDA is often associated with the presence of a parasitic infection. 5- Indican and indole-3-propionic acid neurotoxins result from bacterial action on tryptophan in the bowel. Indican and indole-3-propionic acid diminish the bactericidal function of the gastric juices. This can lead to intestinal obstruction, paralytic ileus and obstructive jaundice. The concentration of urinary indican can be used as a qualitative indication of the extent of bacterial infection in the large intestine. 6- Monohydroxyphenylpropionic acid has similar characteristics and implications to that of dihydroxyphenylpropionic acid. 7- Quinolinic acid (QUIN) is known as an excitotoxin, which means it has an excitatory effect on neurones, therefore some neurones may be damaged. The accumulation of QUIN results from impaired tryptophan metabolism in the liver by infectious microorganisms. QUIN can cause the over activation of glutamate receptors in the brain, which leads to the generation of free radicals. The presence of QUIN is therefore associated with the presence of 6-OHDA or/and dihydroxytryptamine. Problems associated with the action of QUIN include thought disturbance, loss of locomotor activity and depression. Parasitic and probably fungal infection as well as physical trauma could be responsible for QUIN accumulation. 8- Tartaric acid is the by-product of yeast metabolism. It is able to adversely affect the general metabolism of the host. The deterioration of neuronal activity and partly immunosuppression with increased production of antibodies is very often associated with the presence of tartaric acid. There are a number of important enzymes that first catalyse the formation of the neurotransmitters from their respective precursors and then catalyse the utilisation of the neurotransmitters into their respective breakdown products (i.e., metabolites, derivatives). The enzymes associated with neurotransmitter formation include phenylalanine hydroxylase, tyrosine hydroxylase, DOPA decarboxylase, phenethanolamine-N-methyltransferase and dopamine-β-hydroxylase. The enzymes associated with the breakdown of the neurotransmitters include catechol-O-methyltransferase and monoamine oxidase. Co-factors must be present in order for these enzyme systems to function efficiently, e.g. vitamins B_6, B_3, magnesium, iron, biopterin, etc. There are some molecules that inhibit the action of these enzymes, e.g., heavy metals. Factors lead to the impairment of the nervous system activities include: 1- accumulation of neurotoxins, 2- bacterial infection, 3- decreased ATP production, 4- decreased nicotinamide adenine dinucleotide (NAD) level, 5- hypoxaemia, 6- fungal infection, 7- impaired blood/brain barrier, 8- impaired gut permeability, 9- increased bacterial activity in the intestinal tract, 10- renal problems, 11- liver problems, 12- magnesium deficiency, 13- malnutrition, 14- over production of free radicals, 15- parasite infection,

16- psychological stress, 17- vitamin C deficiency, 18- vitamins group B deficiency, 19- viral infection, 20- yeast infection.

ENVIORNMENTAL DISORDERS

Environmental pollution and food falsification are an inevitable fact of modern life and are a medical priority. Blood tests do not necessarily reflect tissue levels of nutrients. Hope is the indispensable ally of the physician and the absolute right of the patient. No one alive today is breathing air even remotely as clean as our ancestors of a few dozens generations ago. Cure means that the patient has no symptoms and no disease, this is a successful intervention. Environmental medicine is chemical sensitivity in which an individual reacts to a chemical (simple or complex) to which most other subjects do not react. This can be due to true allergies mediated through the immune system, or chemical intolerance. Humans are biochemically unique and react in different ways to environmental chemicals, e.g., quantity of alcohol. The different reactions are probably inherited differences in the efficiency of various enzymes systems. The more complex cases who react to many environmental chemicals require a much more sophisticated approach including desensitisation, nutritional support, detoxification and the possible use of an environmental unit. Diagnosis is a process of logical deduction, but in practice most doctors work by pattern recognition. Chemical sensitivity exists, and one must consider the possibility of an environmental approach in all cases. The main triggering cause of chemical sensitivity (CS) is air pollution, due to poor indoor air quality. Detection of objective disturbances in the nervous system seem to be the most important to establish a consistent diagnosis of CS. CNS of CS patients is mostly deranged, and neuro-ophthalmological examination may show positive findings, such as: a- disturbance of the visual modulation transfer function, b- abnormal findings in electro-pupillography (an objective estimation of light reaction of the pupil), c- abnormal smooth pursuit eye movement, and d- abnormalities in the accommodation function of eye focusing and in the blood flow of the brain by single photon emission CT (SPECT) and near-infrared oxygen monitoring (NIRO). Intervention for CS should focus on: a) patient education, b) injection therapy, c) nutrition replacement, d) physical therapy, and e) low dose desensitisation and d) selective environmental control. Possible mechanisms of cancer in an ageing population would include xeno-oestrogenic effects, direct damage to DNA and inhibition of apoptosis. The environment is composed of 3 elements: physical, biological, and chemical. The Outbreak of toxic algae (pfiesteria), mould (chytrid), or virus (Norwalk) are biological environmental changes. The widening of the ozone hole over both poles of the globe or global warming would be physical environmental changes. The broad adverse health effects caused by pesticides are chemical environmental changes. There is no single cause for any disease; there is a causal network of factors in the environment and in the body interacting to effect health. Treatment must include physical and non-physical approaches, which are determined on an individual basis. Patients should be offered the opportunity to participate in individual or group sessions depending upon interest, accessibility and availability. The quality of our life, the air we breathe, the water we drink and the food we eat depends on the quality of the soil. The petrochemical solvents are cumulative, aggressive, progressive and carcinogenic in their poisoning effects. Every measure should be taken to prevent exposure and measures to be taken after exposure. Non-drug, non-surgical therapy may be more effective in the treatment of patients with chronic illness than drug or surgical therapy. Multiple modalities are essential to treat patients such as oxygen therapy, intravenous nutrients, avoidance, nutritional supplements, neutralisation, immunotherapy, diet modification, house and workplace environmental changes, and homeopathy. 64% reduction in symptoms at 3-4 months of treatment and 82% reduction at 6 months would be expected irrespective of the diagnosis. The average improvement would be 88-90% at 9 months. Some patients worsen in the second month of treatment presumably from mobilisation of xenobiotics. Many chronic diseases are initially triggered by one group or one substance. The propagation may be by the same, similar substances or even unrelated ones as the chronicity appears. At times it is impossible or unimportant to find the initial triggering agent(s). Eliminating or neutralising the daily propagating agents, maintaining homeostasis by the use of antioxidants, other nutrient therapy and avoidance regimes is necessary to stabilise and/or at times reverse the chronic disease. Environmental pollutants including cigarette smoke, heavy metals, ozone, toxins in crop sprays, etc., are toxic to humans and animals because they generate free radicals in tissues after they enter the body. Free radicals are highly reactive molecules that are often by-products of oxygen (O_2). 95% of the O_2 inhaled is used to produce energy in the form of cellular ATP, some of its spins off as free radicals. These free radicals plunder, mutilate and destroy essential molecules within cells, and contribute to a variety of diseases such as cancer, emphysema, Alzheimer's disease, etc. Antioxidants either directly detoxify or metabolically remove free radicals from cells so their damage is limited. The intracellular accumulation of oxidatively (free radicals) damaged essential molecules damage DNA, which can lead to cancer, destroy cell membranes causing cells death, and destroy proteins leading to weakening of cellular metabolism and energy production. There is strong evidence that aging and many age-related diseases are a consequence of the accumulated oxidative damage. In healthy cells and tissues, free radicals produced by oxidative phosphorylation, biological oxidations, and chemical and drug detoxification reactions are normally insulated from susceptible molecules and enzymes by cell membrane barriers that contain numerous antioxidant molecular species derived from nutrients. Any substantial shift in the local oxidation-reduction balance in response to chemical oxidant exposure, intensified endogenous generation of oxidant molecular species, physical trauma, or infection can directly affect the viability and functioning of the cell, tissue, or organ system. Health is related to the additive oxidative stress imposed on the individual's antioxidant defense system by exogenous and endogenous oxidant stressors. The antioxidant adaptation could be due to myriad of factors that precipitate oxidative stress operates through avenues involving direct tissue damage by free radicals and other activated oxygen species. When the body goes beyond antioxidant adaptation, it can no longer defend itself against stress (e.g., cancer). When pushed beyond its limits (beyond adaptation), the antioxidant system, which is a sophisticated system of aerobic metabolism, reverts to a primitive system of anaerobic metabolism, leaving the body open to disease. Cancer cells may not have the antioxidant adaptation capacity, and have gone beyond the ability to adapt to oxidative stress. And this in part may characterise their transformed state. A trace element (TE) is defined as one, which causes impairment of a vital function, or death, when its intake falls below a certain level, and whose role cannot be undertaken by any other element. The imperative TEs are cobalt, copper, zinc, selenium and iodine. Less important possibly are manganese, boron, molybdenum and chromium. The major nutrients, potassium, phosphorous, sodium, calcium, magnesium, iron, sulphur and chlorine are rarely in short supply. The soil biomass is constantly degraded by modern farming methods. The biomass of the soil is immensely complex and is responsible for producing the nutrients for plants and the grazing animals. One square meter of soil to a depth of 6 cm should contain about 75000 insects and small worms and 200 earth worms plus billions of fungi and single cell organisms. They should weigh about 1 ton per acre and on a permanent field may turn over around 15 ton of soil per year/acre. TE are probably the most important factor in the production of healthy productive ruminants, unless the field of the ruminant is balanced with TE then fertility, resistance to disease and the production of meat, milk, and wool are adversely affected. TE also affect soil life, and plants, and the whole ecosystem depends on these vital elements, which are present in very small amounts. Concentrations of TE in a field depend on factors such as rainfall, altitude, aspect, base rock, temperature and geographical situation. Modern farming methods such as overuse of nitrogen fertilisers, lime, sulphur, and phosphorous plus a variety of herbicide and insecticide chemicals also reduce the TE availability. There is substantial evidence to support the sex hormonal regulation of immune functions, including: 1- the existence of sexual dimorphism in immune response, 2- alteration of immune response by sex steroid replacement, 3- alteration of immune response during pregnancy, and 4- existence of sex steroid receptors in the immune organs that affect T or B lymphocyte differentiation and function. Oestrogenic xenobiotics (i.e., environmental oestrogens), which are one of the endocrine disrupters, have been implicated in a number of human health disorders. These hormonally active agents (HAAs) are derived from a number of relatively common and abundant sources such as pesticides, insecticides, plastics, combustion by products, plants and agricultural products. The peripheral blood lymphocytes (PBL) proliferate in response to IL-2, which is mediated by C kinase activity. HAAs inhibit development in the both T-and B-lymphocyte stem cell compartments (there is evidence for apoptosis in HAAs inhibition of immature lymphocyte development). The cytoplasmic signal-generating system in developing or mitogen-treated lymphocytes are inhibited by HAAs, and the defect occurs at all stages in the sequence of events leading to DNA synthesis, cell proliferation and cell differentiation. Rhinitis, polyps, and recurrent sinusitis have been associated with allergic and environmental aetiologies. Neglected or improperly treated ear, nose, and throat sensitivities result in either irreversible end-organ damage or the spread of sensitivities to other smooth-muscle systems and other body systems such as skin and neurologic systems. The best way in dealing with these patients is to search for environmental triggers and treat them accordingly. Consumption of very hot liquids, deficiency of vitamins C, B_2, PP, A, β-carotene, zinc, molybdenum, selenium and other trace elements can also play an essential role in oesophageal cancer incidence. Increased risk of oesophageal cancer is associated with low social and economic status. Risk of oesophageal adenocarcinoma also depends on tobacco smoking, alcohol and diet. Factors increasing the risk of gastric cancer include low social and economic status, infection of stomach with Helicobacter pylori, influence of nitrites and carcinogenic nitrosamines; heavy metals, asbestos, tobacco smoking, alcohol, high consumption of starch-containing products, salted, pickled and smoked foods, low consumption of fruit and vegetables, and deficiency of some micronutrients (vitamins C and E, β-carotene and iodine). Risk factors of colorectal cancers include a diet high in fat, sucrose and alcohol; and low in vegetables and fruit, fiber, antioxidants, folate and calcium. It is obvious that most of the risk factors of GIT cancers are avoidable. At least 50-95% of human cancers are caused by diet and environmental chemicals. The multiple mechanisms to reverse and heal cancers and metastatic cancers in humans include: 1- Reinstating the processes of cellular dedifferentiation. 2- Genetic reprogramming of mutant p53 cancer-promoting gene back to the wild p53 cancer suppressor gene. 3- Depurating bioaccumulated xenobiotics. 4- Re-establishing gap junctional (connexin) proteins. 5- Utilising enzymes to dissolve the sialoglycoprotein protection about cancer antigen-antibody complexes, to rendering them vulnerable to the host's immune system attack. 6- Inhibition of enterohepatic recirculation of carcinogenic hormones. 7-

Repairing xenobiotic detoxication with special nutrients to include minerals, vitamins, amino acids, essential fatty acids, orphan nutrients, lipotropes, and phytochemicals, as well as. 8- The use of a non-toxic, mineral that provides relief from morphine-resistant cancer pain. Five organ and glandular systems are involved in the regulation of inner ear biochemistry: 1- the adrenal gland, 2- the pituitary gland, 3- the hormonal system, 4- the immune system, and 5- the hypothalamus. Each of these organ systems secretes chemical messengers, which interact with each other and with the chemicals that are being transported to the inner ear. The inner ear is a transducer of mechanical to electrical energy for both hearing and balance functions by means of chemicals present within the perilymph and endolymph. Maintenance of normal hearing and balance therefore is dependent upon the availability of the proper chemicals to perform this transduction task. The inner ear functions as an internal body organ because it is chemically dependent upon many body systems. Common diseases such as recurrent RTIs are symptomatic of a depressed immune system and may be corrected by the use of TE rather than broad-spectrum antibiotics. Trace elements (TE) deficiency is associated with necrotic areas of the liver, damaged intestinal surfaces and deteriorated heart muscle, delayed puberty, reduced libido, females cycle irregularly and poor conception, cystic ovaries, abortion. Sperm efficiency is reduced by a low count, malformation, and reduced motility. All of the above symptoms may be rectified if the TE balance is restored. It is essential to evaluate allergy patients for the presence of adrenal fatigue because many allergy patients present with an increased incidence of low cortisol levels. Many chemicals impair brain function. Toxic heavy metals also impair brain function. When chemicals and toxic heavy metals are combined, they potentiate the toxic effect on the brain and CNS. The toxic effects are cumulative, aggressive and progressive. The loss of brain function resembles accelerated aging. Comprehensive environmental evaluation is one of the most important tools to use in toxic brain patients. Comprehensive treatment program can reverse the symptoms of neurotoxicity. Comprehensive treatment program includes avoidance, glass bottled water, less chemically contaminated food and minerals to augment detoxification, antigen injection treatment, intravenous treatment with antioxidants and vitamins, chelation to eliminate toxic heavy metals, heat deputation, and physical therapy to reduce total toxic load of chemicals and toxic heavy metals. Exposure at home to chlorine and creosol produces persistent CNS impairment, which increases after 3 years. Several years of H_2S exposure is associated with neurobehavioral impairment. There is a considerable increase in reports of illness with inability to work in crewmembers of long distance flights.

Symptoms that may lead to inability to flying civilian air crew members:

A- Mucocutaneous alterations:
    1- Conjunctivitis.
    2- Alopecia.
    3- Disfiguring weeping eczema of uncovered skin (flares up after in-flight spray missions, after entering the aircraft or hotel).
B- Cerebro-organic disorders:
    1- Decrease of intellectual performance.
    2- Increasing loss of memory.
    3- Mix-up of words.
    4- Word finding difficulties.
    5- Narcoleptic attacks alternating enhanced arousal and agitation.
    6- Abrupt nausea followed by black out with collapse from seconds up to minutes (person falls down the stairs or, while car driving, wakes up when the car has come to standstill off the road).
    7- Sudden attacks of clouding of consciousness with complete loss of orientation for 30 to 60 min (persons being at a familiar place suddenly does not know where they are).
C- Movement disturbances:
    1- Walking "like on cotton", a problem with correct estimation of distances (walking into doorframes or missing stair steps).
    2- Clumsy while serving (when pouring the coffee, missing the cup and spattering the passenger).
D- Suspicion of autoimmune diseases:
    1- Multiple sclerosis.
    2- Collagenosis.
    3- Thrombocytopenia.

The occupational stress of the flight crew members may arise from: 1- enhanced psychophysical strain due to irregular way of life, lack of sleep, jet lag, temperature jump, loss of social contacts; 2- cosmic radiation; and 3- insecticide exposure in aircraft and hotels. The illness begins independently from the date of employment. The introduction of pyrethroids for aircraft disinfection is related to outbreak of illness in the air crew members. Organic solvents are one of the major classes of neurotoxic chemicals, which adversely affect neuropsychological functioning. Organic solvent exposure may induce various neuropsychological deficits, including reduced speed of thinking, impaired coordination, decreased concentration, memory, and vocabulary, in addition to psychiatric symptomology including anxiety, panic disorder, and psychosis. Organic solvents are suspected to be causative agents in senile dementia. Classifications of solvent-exposed individuals: 1- Affective syndrome: neuropsychiatric symptoms reported, no objective findings, considered reversible. 2- Mild chronic toxic encephalopathy: neuropsychiatric symptoms, impairments on objective assessment, uncertain reversibility. 3- Severe chronic toxic encephalopathy: severe neuropsychiatric symptoms, impairments more pronounced on objective assessment, usually irreversible. The neuropsychological effects of solvent exposure may be subtle or severe and appear to depend upon a variety of factors including frequency, intensity, duration of exposure, age, educational level, and concomitant illnesses and/or neurotoxic exposures. Chronic chemical exposure at the work place is common. Exposure source varies but mainly from solvents in the ambient air. The main complaints are neurological symptoms (headaches, migraines, short-term memory loss, inability to concentrate, loss of concentration, vertigo, light headedness), fibromyalgia, fatigue, arthritis, arthralgia, musculoskeletal symptoms, respiratory symptoms (shortness of breath, asthma, bronchitis); ENT (hearing loss, tinnitus, hoarseness, laryngeal oedema, dysphasia); GIT (irritable bowel syndrome, malabsorption, diarrhoea); cardiovascular (vasculitis, chest pain, hypertension, cardiomyopathy, angioedema, mitral valve disease); endocrine (ovarian imbalance, PMS, hypothyroid, thyroiditis); eye (cataract, vision loss); and skin rash. Triple camera brain SPECT scan is positive for Neurotoxicity. Intervention has significant 82% improvement rate and consists of a massive avoidance of pollutants in air, food and water. Parental and oral nutrition, sauna therapy and autogenous lymphocytic factor, and air purification. Improvement is noticed after 3-4 months after removal from exposure. Toxic volatile organic hydrocarbons (TVOCs) may pollute indoor air more frequently than outside air. TVOCs in new homes are high, but levels are 1,000 times lower in homes more than three years old. TVOCs share certain characteristics: all are easily vaporised, dissolve in liquids, are transported across cell membranes, and are absorbed by the lungs, skin, and gastrointestinal tract. The effects of TVOCs in homes include symptoms related to the cardiovascular, gastrointestinal, genitourinary, respiratory, neurological, and dermatological systems. Chlorinated compounds are ubiquitous. Chloroform is used in the preparation of pharmaceuticals, artificial silks, insecticides, floor polishes, lacquers, and cleaning solvents. It is a common by-product of water chlorination. Dichloromethane is a major component of paint strippers and degreasers. It is also used as a solvent for oils, fats, and waxes.

Common toxic chemicals and solvent exposure:

1. Trichloroethane (45%).
2. Benzene (24%).
3. Trimethylbenzene (19%).
4. Xylene.
5. Dichloromethane.
6. Chloroform.
7. Trichloroethylene.
8. Tetrachloroethylene.
9. Ethylbenzene.
10. Dichlorobenzene.
11. Aliphatic hydrocarbon (80%) (methylpentane, n-Hexane, cyclopentane, n-pentane).
12. Chlorinated pesticides (100%) (DDE, trans nonachlor, oxychlorodane, heptachlor epoxide, hexachlorobenzene, dieldrin, B-BHC).
13. Food (98%).
14. Mould (93%).
15. Trees (74%).
16. Grasses (74%).
17. Ethanol (55%).
18. Formaldehyde (16-47%).
19. Cologne (96%).
20. Cigarette smoke (64%).
21. Metal (nickel, zinc sulfate).
22. Chlorine (11%).
23. Toluene.
24. Phenol (12.5%).
25. Organophosphate pesticide (16.7%).

1,1,1-Trichloroethane (methyl chloroform) applications include use as a degreaser, dry cleaning agent, and as a solvent in paint products. Trichloroethylene is commonly used in manufacturing computer chips and electrical components and has been used as a degreaser for metals, and as a dry cleaning agent. It can be found in lacquers, printing inks, and paints and has been used as a refrigerant and as an extracting agent in the removal of caffeine from coffee. It has been used in the manufacture of pharmaceuticals, as a metal degreaser, and as a grain fumigant. Tetrachloroethylene exists in food and in dry cleaned clothes present in most patients and houses. Chronic lifetime exposure to benzene, carbon tetrachloride, chloroform, methylene chloride, dinitro-toluene, nitrosamines, and phenols increases the probability of developing cancer. Xylenes are found in a variety of solvents for gums, synthetic resins, rubbers, paints, and inks. They are also used in photographic processes in the manufacture of insecticides and plastics, and are present in degreasing agents, cleaners, and petroleum products. Styrene is used as a solvent for resins and synthetic rubber, and as an intermediate in the chemical synthesis of polymerised synthetic materials. There is an increased exposure in the workplace, home, water source, or diet to TVOCs. Some of these chemicals are found indoors, as emission products from building materials (floor coverings, furnishings, and paint) and consumer products (polishes, cleaners, and solvents), as well as from combustion (cigarette smoking, kerosene heaters, wood stoves). Dichlorobenzene is used as a chemical intermediate in deodorants, disinfectants, insecticides, fumigants, metal polishes, and mothproofing materials. It has applications in lacquers and paint products. The most common application of benzene is in the manufacture of detergents, polymers, pesticides, pharmaceuticals and paint products; it is also found in processed food, petroleum products, and cigarette smoke. Toluene is commonly found in solvents for gums, fats, adhesives, as well as in petroleum products and paint products. Trimethylbenzenes are used as solvents and intermediates for chemical synthesis, in the manufacture of paint thinners, perfumes, dyes, and as motor fuel additives. Toxic chemicals can produce both malignant and nonmalignant tissue damages. There may be a relationship between these cancer-causing hydrocarbons and chemical sensitivity. Heavy exposure to TVOCs may cause illness; e.g., chlorinated solvents can cause severe liver injury and death. There is no safe level of TVOCs; some chlorinated solvents can be mutagenic and carcinogenic, after repeated low level of exposure over a prolonged period. Exposue to nitrogen oxide (NO_2), mainly emitted from gas cooking appliances and motor vehicle exhausts is associated with respiratory symptoms in children and adults. Hig outdoor NO_2 exposure is associated with a raised frequency of lower-responsiveness and a high total IgE concentration. Asthma prevalence is increasing, and exacerbation are the main cause of asthma morbidity and result in very substantial health-care costs. Respiratory viral infections are the main trigger for acute asthma excerbations in children. NO_2 exposure before an infection can impair resistance to respiratory viruses and bacteria through reduction of bacterial clearance and local immunity, and by alteration of macrophage function. There is an association between increased personal exposure to the air pollutant NO_2 and the severity of virus-induced asthma excerbations in children. Factor associated with increased personal exposure include the use of gas appliances, cohabitation with smokers, and travel to school by means other than a car. These viariables explain <30% of the temporal variation in personel exposures. Cigarette smoke emits various oxides of nitrogen. Exposure to domestic cigarette smoke has no significant effects on measured personal NO_2. The potential mechanisms by which NO_2 excerbates asthma in the presence of viral infections include direct effects on the upper and lower airways by ciliary dyskinesis, epithelial damage, increases in pro-inflammatory mediators and cytokines, rises in IgE concentration, and interaction with allergens, or indirectly through impairment of bronchial immunity. NO_2 exposure might increase the susceptibility of respiratory epithelial cells to injury from respiratory syncytial virus. Exacerbate rhinovirus-induced inflammation, which could be mediated by increase in ICAM-1 (the major group rhinovirus receptor) in the presence of air pollution, which could heighten the risk of rhinovirus infection. The major site of NO_2-induced lung injury is the transitional zone (the area beween the terminal bronchioles and alveoli). Personel NO_2 concentrations during cooking hours exceed weekly average personal NO_2 exposure by an order of 3-5 from 21 mg/m^3 to 60-100 mg/m^3 range. Higher personal NO_2 exposure may increase the severity of virus-induced asthma. Some nutrients help remove heavy metals but environmental exposure must be addressed. This includes restrictions on diet and the elimination of environmental factors such as copper tea pots, copper sulphate (Jacuzzi or swimming pool water), bad drinking water, prenatal vitamins, copper IUD's, etc. Drugs such as neuroleptics, antibiotics, antacids, cortisone, tagamet, zantac, diuretics, and birth control pills, etc. may exacerbate copper overload.

PESTICIDES AND BRAIN-FUNCTION CHANGES

Chlorinated hydrocarbon pesticides are strongly correlated with a changed psychological/brain-function condition. These pesticides affect the CNS, maintaining a body burden (neurotic-somato form disorder). Airborne chlorinated pesticides are ubiquitous, resulting in a broad public exposure to potentially hazardous materials. Chlorinated pesticides appear to be associated with a more serious psychological profile in which the patient shows more indications of major variance of behaviour such as depression with obsessive-compulsive patterns of coping, and of symptoms such as emotional sensitivity (an apparent overreaction to emotional stimuli), often with feelings of being rather overwhelmed by adversity sometimes to the point of having persecutory thoughts that lead to unreasonably angry responses. They exhibit central and peripheral nervous system (cognitive, perceptual, motor) symptoms and signs including recent-memory deficits, paraesthesias, headache, dizziness, and motor instability. Evidence of poisoning may appear in subtle behavioural alterations and in emotional or sensorimotor difficulties discernible only when the individual is observed systematically over time in carefully designed assessment situations. At low levels of exposure, the effects may differ in different individuals due to genetic deficits, heavy stress, or other variables. Behavioural changes could promote interpersonal difficulties because of the affected individual's depression, irritability, activity rate, etc. There may be a long latency of reaction (weeks to years) for those persons exposed to chemicals, and reactions may take different forms in different systems. These range from emotional lability to impaired learning ability and can affect the unborn child. Systems Affected by pesticides includes brain/neurological, cardiovascular, respiratory, genitourinary, gastrointestinal, and skin. All water, air, and foods should be controlled to assure a significant reduction of organic chemicals and chlorinated pesticides in individuals' total intake. This would allow for increased removing of the organohalides out of the body into the air, urine, faeces, and through the skin. Patients not only improve symptomatically, but also by objective measurement, under rigid environmentally controlled conditions. Significant clinical improvement requires strict avoidance in intake of chlorinated hydrocarbons in food, air, and water. Additionally, mobilisation of the tissue stores by fasting and neutralisation of free radicals by vitamin intake apparently occur. A regime of vitamin ingestion combined with external hyperthermia could usually remove pesticides from the body fast. Reduction of intake not only assures less stress in the body's immune and detoxification systems, but it also might allow them to function much better in clearing the body burden of pesticides as well as other organic chemicals. Organohalide depresses the activities of the enzymes L-gluconolactone oxidase and dehydroascorbitase along with an increased urinary excretion of ascorbic acid. Environmental control unit (ECU) treatment programs also impact on health of the patient by decrease of other toxins from the blood. Blood pesticide levels may increase after institution of environmental control. Several pesticides may be removed more easily from the body.

IMMUNE-CNS CONNECTIONS IN FOOD AND CHEMICAL SENSITIVITY

Our understanding of the affects of immunity on the brain is still limited. Cytokines (IL1, TNF, IL6, TGFβ, etc) are thought to be the mediators of the immune-CNS connection. The purpose of immune processes acting on the CNS is to elicit sickness behaviour. Such behaviour is believed to mobilise resources to combat infection. Components of sickness behaviour include fever, reduced motor activity, decreased exploratory behaviour and a reduction in food and water intake, a reduction in social and sexual behaviour, and impair learning and memory and a decrease in activities related to body care. When cytokine concentrations are high, act directly on the brain by crossing the blood-brain barrier. At lower concentrations the action of cytokines appears to be mediated by their effects on sensory nerves, responding to immune events producing cytokines. Sensory nerves in the skin may relay to specific brain centres following immune stimulation. Similar considerations apply for the Trigeminal nerve, which provides much of the sensory nerve supply to the upper respiratory tract. In the periphery dendritic cells and other immune cells are intimately associated with vagal fibres. This close association enables cytokines released by immune cells to activate vagal afferent fibres. The sensory branches of the vagus nerve relay to the dorsal vagal complex of the medulla, particularly the nucleus tractus solitarius (NTS). There is subsequent relay from the NTS to the dorsolateral pons and ventrolateral medulla. Projections occur from these sites to the hypothalamus, amygdala, thalamus and cortex. In addition there is an important relay from the NTS to the raphe magnus in the medulla and from there to the dorsal horns of the spinal cord. Activation of this latter pathway facilitates pain sensation. Signalling of the hypothalamus is implicated in fever and loss of appetite. Similarly, the amygdala and hippocampus are implicated in behavioural changes and impaired learning and memory associated with immune activation. A maternal behaviour or TLC, can override some of the effects of immune activation on behaviour. Also the effects of stress considerably overlap those induced by immune events within the nervous system, sometimes acting at the same anatomic sites such as the hippocampus. Immune-brain communication has mainly deployed cytokines such as IL-1 and TNFα or microbial products such as lipopolysaccharide (LPS), but adverse reactions to foods or chemicals, if mediated by immune processes, could also affect the CNS. TGFβ enhances the release of neuropeptides from sensory nerves, similar to that of IL1 on sensory nerve fibres. Antigen affects adjacent sensory nerves, via the action of TGFβ, which could alter local tissue function and signal centres within the CNS. Such a process may for example explain the disturbances in gut motility and symptoms such as malaise and headache reported in some patients with food intolerance. The possibility of adverse reactions involving IgG antibody also acting on sensory nerve fibres, with secondary effects on other tissues is also possible. IgG_1 antibody in particular, on binding to antigen, may induce mediator release from mast cells. This process can occur directly, or by the activation of complement. Mediators derived from mast cells are known to act on sensory C fibres, inducing the release of neuropeptides. Such a process could have wide spread clinical effects.

TOXIC ELEMENTS

Heavy metals are elements of specific weight characteristics. Toxic heavy metals are present in the air, water, soil, and food supply as a byproduct of the industrialised society. Chronic low-level exposure can lead to a wide array of problems, ranging from neuropsychiatric disturbances such as aggressive behaviour, memory loss, depression, irritability, and learning deficits, to physical manifestations such as liver and kidney dysfunction, fatigue, infertility, gout, hypertension, headache, and candida (yeast) infections. Perhaps the two most widespread and significant heavy metal toxins are mercury and lead. Beverages are a source of toxic trace element intake. About a ten times higher lead concentration in wine than in most other beverages. Cocoa was high in cadmium and nickel and some vegetable juices contained high levels of nickel. Wine is the most significant source of lead even if the bottles do not have lead capsules. Consuming half a bottle per day the daily intake of lead would be doubled. Consumption of tea as well as vegetable juices could increase the nickel intake. Children with high lead levels are much more likely to drop out of school, have reading disabilities, and exhibit criminal behaviour. Lead-based paint is major source, which may compose of up to 50% lead. Flakes and microscopic dust from the paint continue to contaminate homes for many years, and can be released in larger amounts during renovations. Lead (Pb) is a cellular toxin, competes with Ca in the body, which can cause various malfunctions in calcium metabolism including a decrease in neurotransmitter release and blockade of Ca channels. The CNS appears to be affected to the greatest degree by Pb toxicity, and this can explain the many neuropsychiatric symptoms associated with exposure to this heavy metal. Chronic exposure to low levels of Pb (<1 mg) over a long period of time can cause serious damage to the brain, kidneys, nervous system and red blood cells. Growing children and pregnant women are at the greatest risk. Hot water dissolves lead more quickly than cold. Chronic exposure to Pb can result in significant accumulation in the brain, soft tissue, and bones. Neuropsychiatric symptoms of chronic lead exposure include: 1- headaches, 2- poor memory, 3- inability to concentrate, 4- attention deficit, 5- aberrant behaviour, 6- irritability, 7- temper tantrums, 8- fearfulness, 9- insomnia, 10- lowered IQ, and 11- difficulty with the reading, writing, language, visual and motor skills. Mercury (Hg) may be more toxic than lead. Mercury is a potent cellular toxin and is known to decrease neurotransmitter production, disrupt important processes within the nerve cells, and decrease important hormones such as thyroid and testosterone. Amalgam fillings contain 50% metallic mercury, and they continuously release mercury vapour during chewing, brushing, or when drinking hot beverages. In addition to amalgam fillings, common sources of Hg include pesticides, laxatives, batteries, paper and pulp products manufacturing, drinking water, and paint products. Mercury can cause neurodegeneration in the brain. The resultant defective microtubule assembly and the aggregation of neurofibrils, found in the brains of Alzheimer's patients. The chronic exposure to Hg may be a potential factor in neurodegeneration in humans that could ultimately be observed as altered behaviour. Neuropsychiatric symptoms associated with Hg toxicity include: 1- insomnia, 2- nervousness, 3- hallucinations, 4- memory loss, 5- headache, 6- dizziness, 7- anxiety, 8- irritability, 9- drowsiness, 10- emotional instability, 11- depression, and 12- poor cognitive function. Blood levels may not accurately reflect the total body burden of toxic metals. High blood levels are usually only found in acute toxic metal exposure, or in individuals exposed to high levels of toxins over a long period of time. In chronic low level exposure, the blood levels may be low due to redistribution of the toxins throughout the body, while bone and other tissue levels remain high. There are arguments over the accuracy of hair analysis due to the possibility of contamination from hair dyes, shampoo, and other factors. Chelating agents bind to heavy metals throughout the body, and then are excreted in the urine, taking the heavy metals with them. In the urine challenge test, a chelating agent is administered and then urine is collected and analysed to determine the amount and type of toxic metals that are excreted. Intervention: DMSA is useful in cases of mercury exposure, while EDTA is useful for Pb toxicity. Both of these agents remove other toxic metals in addition to Pb and Hg. Lead has no taste or smell or colour, so high levels cannot be detected. Prenatal exposure to low Pb levels is associated with striking neurobehavioral deficits, including poorer motor control, reduced attention span and nervous system irritability. Low-dose Pb exposure has a profound impact on brain development. Lead interferes with cellular growth and reduces the brain's ability to transmit neural impulses. Lead can also damage developing cells in the cerebral cortex and frontal lobes of the brain; these regions are linked to motor control and are particularly sensitive to Pb toxicity. Once released, Pb is extremely resistant to break down in the environment. Exposure today still occurs from contaminated air, water, and soil. Lead in drinking water is estimated to constitute about 10-20% of total Pb exposure in children; in infants fed with baby formulas using tap water, this proportion rises to 40-60%. Simple methods for reducing contamination risk in children include washing a child's hands often. Iron and calcium deficiencies increase a child's risk of suffering Pb poisoning. Lead is found in older, oil based paints, glazes on dinnerware, bullets and fishing sinkers, soldered cans, printing inks, plumbing fixtures, stained glass, and cosmetics. Lead in soil will be stirred into the air by foot traffic. Window sashes painted with lead paint generate lead dust when opened and closed. This accumulates on the sills. Lead from exterior siding washed into the soil surrounding older homes and can be found there at high levels. Lead is toxic even at low levels. The body normally contains no Pb. Today Pb level in many people is already significant. Lead can cause serious damage to brain, kidneys, and peripheral nervous system. The effects of Pb include delays in physical and mental development, lower IQ, shortened attention span and increased behavior problems. Lead is more serious for children because of lower body weight. Children have higher exposures because they get lead dust on their hands or eat lead chips that taste sweet. Drinking water pollution as one of the top 5 environmental threats to health. It contains hazards to health such as heavy metals (e.g., Pb), volatile organic compounds (VOCs), pesticides, nitrates (from fertilizer run off), bacteria, viruses and parasites. Fluoride has negative health effects. Chlorine reacts with organic matter to produce dangerous by-products. Asbestos fibers are introduced by 200,000 miles of asbestos pipes. Lead and copper in water are a primary health threat. Lead levels as small as 5 ppb have been shown to cause health problems in children. Lead is leached from older lead distribution pipes. Well water can be contaminated with some of the above and can contain radon too. Heavy metals (arsenic, cadmium, mercury, lead,) are associated with increased disease susceptibility, birth defects, nerve damage, mental retardation, certain cancers. Chlorine by-products are carcinogenic. The skin, the body's largest organ, is permeable to volatile chemicals. At the cellular level, Pb interferes with membrane transport processes and with enzyme functions because it is able to bond to many chemically active sites. The interaction of Pb with sulfhydryl (SH) sites causes most of the toxic effects, which include impaired heme synthesis, inhibition of erythrocyte Na-K-ATPase, diminished RBC glutathione, shortened RBC life span, impaired synthesis of RNA, DNA and protein and impaired metabolism of vitamin D. Lead may also affect the body's ability to utilise the essential elements Ca, Mg, and Zn. Lead is toxic to nerves and at moderate levels of body burden, Pb may have adverse effects on memory, cognitive function, and nerve conduction. High level of Pb in children is associated with a higher incidence of hyperactivity and learning problems. Lead is also toxic to kidneys resulting in disordered renal transport with uricaemia (possibly gout), hyperaminoaciduria, glucosuria and phosphaturia. Excess body burden of Pb is often associated with fatigue, headaches, loss of appetite, insomnia, nervousness, anaemia, weight loss, decreased nerve conduction and possibly motor neuron disorders. Lead exposure includes welding, old leaded paint (chips/dust), drinking water, some fertilisers, industrial pollution, lead-glazed pottery, and newsprint. Confirmatory tests for lead excess are urine elements analysis following provocation, e.g., with intravenous EDTA, DMPS, or oral DMSA. Whole blood analysis for lead only reflects recent or ongoing exposures and may not correlate with total body burden. Increased blood or urine protoporphyrins is a finding consistent with Pb excess, but may occur with other toxic elements as well. Detoxification therapy is by means of chelation. Cadmium (Cd) is a toxic heavy metal with no known metabolic function in the body. Cadmium exerts toxic effects by inhibiting sulfur-bearing enzymes and by displacing enzyme bound zinc or copper. In cells, Cd can inhibit gluconeogenesis and phosphorylation processes. Cadmium's deleterious effects may be slow and not recognised for years before manifestations are apparent. Excessive body burden of Cd is associated with high blood pressure (hypertension) and impaired renal transport with proteinuria and urinary wasting of β_2-microglobulin. Chronic Cd excess can lead to microcytic, hypochromic anaemia. Cadmium can also adversely affect heart, arterial walls, bone and testes. Cadmium excess is also commonly associated with fatigue, weight loss, osteomalacia, and lumbar pain. Inhalation of Cd salts or vapours may produce emphysema. In children, elevated Cd has been correlated with lowered IQ. Smoking and high sugar diets appear to increase Cd levels. Cadmium is found in varying amounts in foods, from 0.04 pg/g for some fruits to 3-5 pg/g in some oysters and anchovies. Cigarette smoking significantly increases Cd intake. Refined carbohydrates have very little zinc in relation to the Cd. Cadmium absorption is reduced by zinc, calcium, and selenium. A confirming test for elevated body burden of Cd is urine analysis following administration of an appropriate chelating agent such as EDTA or sulfhydryl agents (DMSA, D-Penicillamine, and DMPS). Avoiding cadmium (Cd) is essential to preserve zinc and copper and normal thyroid function. Cadmium is a principal toxic metal that disturbs Zn, Cu, and other metals and probably is a major contributor to thyroid disease. Cd sources include: tobacco smoke, burning oil, automobile tire dust, cadmium batteries, canned foods, dried foods, cola drinks, processed coffee, decaffeinated coffee, milk (from galvanised dairy cans), butter, olive oil, lipstick, silver polish residue on eating utensils, metal ice trays, processed meats, pottery, plastic wrappings, wheat gluten, the electric elements that are put directly into containers to heat water for soups, teas, and coffees, and many other sources. Cadmium coating food containers can be dissolved by acid foods and drinks, which can induce acute Cd poisoning. The Cd enters the city's sewage treatment plants, collects, and in time poisons the bacteria digesting sewage and garbage. Cadmium precipitates in salt water to the bottom mud, where it can enter the food chain. Just a little Cd, a half to one part per million in water, is toxic to most bacteria. Cadmium is constantly present in zinc, even the purest. Impure Zn can contain up to 2%. The Zn used for galvanising iron-tin roofs, pails, water storage tanks, iron pipes, gutters, water-softening tanks, maple-sap buckets, cauldrons, barbed wire, chicken fences, wire fences, animal cages, nails, and a host of other articles of iron or steel-is far from pure, generally being the cheapest grade, which contains even more Cd. Rainwater is slightly acid from the dissolved carbon dioxide in the air. Falling on a tin roof, collected by a gutter, stored in a galvanised iron cistern, rainwater will contain Zn and Cd, for it is slightly corrosive. Whenever a slightly acid liquid comes in contact for a time with galvanised metal it dissolves some zinc and cadmium. Soft water is also corrosive, and when it stands overnight in galvanised iron pipes, it dissolves Zn and Cd. Hard water is usually not corrosive. Hot water is much more corrosive than cold. Zinc is not poisonous except in huge doses. The human body contains about 2.2 g of Zn, and there are mechanisms, which keep this amount constant throughout life, unless the diet is low in Zn. Zinc does not accumulate. Cadmium, once in the body, stays, probably for life, in the kidneys, the liver, and the blood vessels. As little as 2 µg daily absorbed and retained results in a body burden of 30 mg in 40 years. Cadmium displaces Zn but does not act beneficially like Zn. The air of some cities contains Cd from fumes spewed out by Zn smelters and refiners and Cu smelters. About 10% of ingested Cd is absorbed into the body from the gut, most of which is excreted in the urine. Almost half of breathed Cd is absorbed from the lungs and retained. A pack of cigarettes contains 16-24 µg of Cd, and a smoker can contaminate a whole roomful of people. Rubber tires, plastics, pigments, plated ware, alloys, insecticides, and solders are some of the things containing Cd. Pigments can be a source, for Cd yellow and Cd red are fast colours; some French lipsticks have it. Cadmium is so ubiquitous in this civilisation that it is very difficult to avoid it. The earliest sign of subtle Cd toxicity is hypertension. Cadmium displaces Zn and slightly poisoned some Zn system that controls blood pressure. When Cd accumulates beyond the subtle poisoning stage, the kidneys and the liver are damaged and blood pressure falls. Nicotine in small doses stimulates nerves; in larger doses it paralyses. Emphysema is characterised by rupture of alveoli of the lungs, making larger scars. Breathing becomes laboured. Pulmonary hypertension appears, straining the heart and leading to heart failure. Emphysema patients also have more Cd in their kidneys and livers than do well people. Cigarette smoke contains Cd, and Cd absorbed directly by the lungs may initiate hypertension there. Nitrilotriacetic acid (NTA)-like the polyphosphate detergents, is a chelating agent, binding metals. Cadmium and methyl mercury are common water pollutants. NTA gets into the water from drains and combines with methyl mercury and cadmium. NTA may allow the metals to get into the body and pass through the placenta into the foetus. The amount of Cd absorbed seems to depend on the amount of Zn also absorb; the more Zn, the less Cd. Cadmium demands Zn. Zinc displaces cd and reverses the spasm induced by it in minor arteries. Zinc also works in angina pectoris. Zinc increases healing of wounds, for it is necessary for the growth of cells. Zinc improves tolerance for alcohol, and has been used with some success in cirrhosis of the liver. Whenever Zn is burned or melted there will be Cd. Incineration is a major source: burning of automobile tires, red and yellow plastic bags, plastic products, paints, discarded automobiles, discarded airplanes, and the parts thereof. Cadmium pollution can be abated by prevention of air pollution with Zn. The largest part of the Cd in the body comes from air-from tobacco smoke, polluted air, and dusts-for most of the Cd inhaled is absorbed directly from the lungs. As little as 1 µg/day retained would build up a body burden of 14.6 mg in 40 years, over a third of the average amount, 38 mg, in all tissues. A pack of cigarettes contains 20 µg, of which 10 µg are absorbed from smoke. So it is easy to get enough Cd to cause illness. Cadmium is a perfect example of an accumulative abnormal and subtly toxic trace metal in the environment causing widespread and serious human diseases, most of which are fatal. The total amount of Cd stored in the livers and kidneys of exposed workers can be nearly 350 mg, the total in those organs of hypertensive persons is only 20 mg, and the total in normal persons is about 12 mg. Smog damages lungs, and may induce sudden death in individuals with existing cardiac problems. Aluminium (Al) can impair cellular energy transfer processes by interfering with phosphate and ATP metabolism. Excess Al can inhibit the formation of α-ketoglutarate and result in toxic levels of ammonia in tissues. Aluminium can bond to phosphorylated bases on DNA and disrupt protein synthesis and catabolism. Neuronal cells are susceptible to long term accumulation of Al, and Al bonding to phosphate can inhibit normal catabolism of neuronal filaments in the CNS. Correlation of elevated Al with degenerative dementia and Alzheimer's disease has been documented. Excessive dietary Al can also form insoluble aluminium phosphates in the GIT and may lead to hypophosphataemia. Possible sources of Al include some antacid medications, Al cookware, baking powder, processed cheese, drinking water, and antiperspirant components that may be absorbed. Many colloidal mineral products have very high levels of Al. Aluminium can effectively complexed and excreted with silicon, a complex of malic acid and Mg, and acetoacetic acid. Aluminium has neurotoxic effects at high levels, but low levels of accumulation may not illicit immediate symptoms. Symptoms of elevated Al may include fatigue, headache and signs of phosphate depletion. Low level of Al exposures may not provoke any immediate symptoms. Aluminium excess should be considered when symptoms of presenile dementia or Alzheimer's disease are observed. Aluminium is elevated in children and adults with low zinc and behavioural/learning disorders such as ADD, ADHD and autism. Individuals with renal problems or on renal dialysis may have elevated Al. Antimony (Sb) is a nonessential toxic element. Antimony has a high affinity for sulfhydryl groups on many enzymes. Excessive exposure to Sb has the potential to deplete intracellular glutathione pools. Antimony is conjugated with glutathione and excreted in urine and faeces. Antimony's deposition in body tissues and its detrimental effects depend upon the oxidation state of the element. Antimony^{+3} affects liver functions, impairs enzymes, and may interfere with sulfur chemistry. If Sb impairs phosphofructokinase (PFK), then purine metabolism may be disrupted, resulting in elevated blood and/or urine levels of hypoxanthine, uric acid and possibly ammonia. Antimony^{+5} deposits in bone, kidney, and in organs of the endocrine system. Antimony spots may result from skin contact with Sb salts and vapours. Symptoms can be variable, including fatigue, myopathy (muscle aches and inflammation), hypotension, angina and immune dysregulation. Food and smoking (e.g., cigarette) are the usual sources of Sb. Cigarette smoke can externally contaminate hair, as well as contribute to uptake via inhalation. Gunpowder (ammunition) often contains Sb. Firearm enthusiasts often have elevated levels of Sb in hair. Other possible sources are textile industry, metal alloys, and some anti-helminthic and anti-protozoal drugs. Antimony is also used in the manufacture of paints, glass, ceramics, solder, batteries, bearing metals and semiconductors. Early signs of Sb excess include fatigue, muscle weakness, myopathy, nausea, low back pain, headache, and metallic taste. Later symptoms include haemolytic anaemia, myoglobinuria, haematuria and renal failure. Transdermal absorption can lead to Sb spots, which resemble chicken pox. Respiratory tissue irritation may result from inhalation of antimony particles or dust. Uranium (U) is a nonessential element that is very abundant in rock, especially granite. It is present at widely varying levels in ground (drinking) water, root vegetables, and present in high phosphate fertilisers. Other sources of include ceramics, some coloured glass, many household products (uranyl acetate) and tailings from uranium mines. Most exposure to uranium (U) comes from natural uranium in ground and drinking water. The U238 isotope of uranium is >99% of naturally occurring uranium. Radioactivity danger from trace quantities of natural U is slight because of its very long half-life (billions of years). The finding of elevated U238 does not imply nor does it rule out exposure to enriched uranium fuel (U235) or to other radioactive isotopes that may be radiation hazards. The major toxicological concern of U238 excess is biochemical rather than radiochemical. Uranyl cations bind tenaciously to protein, nucleotides, and bone, where it substitutes for calcium. Uranium is a reactive element that is able to combine with and affect the metabolisms of lactate, citrate, pyruvate, carbonate and phosphate. Kidney and bone are the primary sites of uranium accumulation, but it also deposits in the liver and spleen. The primary symptom of low level chronic U excess is chronic fatigue. Possible conditions from more severe U contamination include damage to renal glomeruli with disordered renal transport (proteinuria, albuminuria, and hyperaminoaciduria) and haematopoiesis in bone marrow. Published data are sparse, but there appears to be a correlation between U exposure, renal damage and all forms of cancer. Because U is rapidly cleared from blood and deposited in tissues, urine analysis rather than blood analysis may need to be performed to confirm excess exposure to uranium. Hair is sensitive to external contamination with U by shampoos or hair products. The mitochondria accumulate the element As $(CH_2(SH)-CH_2,-CH(SH)-(CH_2)_4-COOH)$. The pyruvate dehydrogenase complex (catalyzes formation of acetyl coenzyme A from mitochondrial pyruvic acid) is inhibited by As^{3+}. Pyruvic acidosis may result; citric acid cycle function and formation of ATP is reduced.Ê The citric acid cycle at the α-ketoglutaric acid dehydrogenase step; and formation of succinyl coenzyme A is impaired.Ê Both of these enzymatic steps require the active thiol, lipoic acid.Ê Arsenic combines with sulfhydryl (-SH) groups. Arsenic may react with any of the enzymes in the body that have sulfhydryl groups. Monoamine oxidase has eight cysteinyl residues with -SH groups.Ê Monoamine oxidase inhibitors are used for severe depression.Ê Other effects of arsenic include irritation of the skin and mucous membranes, chromosomal damage in lymphocytes and erythroblasts in bone marrow with leukopenia, and myocardial capillary damage. Sources of arsenic include contaminated foods (e.g., seafoods), water or medications.Ê Industrial sources are ore smelting/refining/processing plants, galvanizing, etching and plating processes. Tailings from or river bottoms near gold mining areas (past or present) may contain arsenic. Insecticides, rodenticides and fungicides (Na-, K-arsenites, arsenates, also oxides are commercially available). Commercial arsenic products include: sodium arsenite, calcium arsenate, lead arsenate and Paris green (cupric acetoarsenite) a wood preservative. The order of decreasing toxicity for arsenic is arsines, inorganic arsenites, organic arsenoxides, inorganic arsenates, arsenorganics with As valence of +5, and metallic As. Arsines penetrate rubber and well absorbed through the skin, which will become vesicated and blistered during the exposure. Arsines combine with haemoglobin in RBCs, cause haemolysis and cell destruction. Chronic exposures to arsines can result in anaemia. Myocardial failure due to oxygen deprivation can occur in severe cases. Symptoms of arsenic exposure include: 1- Minor exposure may clear skin lesions such as acne. 2- Garlic-like breath occurs with fatigue, malaise, and nausea.Ê 3- Skin and mucus membrane lesions may develop with eczema or allergic-type dermatitis, raindrop areas of lost skin color (hypopigmentation), hair loss, white marks on fingernails, thickening of the skin of the palms and soles (hyperkeratosis), melanosis (eyelids, areolae of nipples, and neck), conjunctivitis, bronchitis, and gingivitis. 4- Excessive salivation, stomatitis, abdominal pain. 5- Jaundice, peripheral neuropathy, polyneuritis, haemolysis, anaemia, leukopenia, cyanosis of the fingers, Raynoud's syndrome.Ê 6- Chronic arsenic exposure has been associated with basal cell carcinomas. 7- Long-term arsenic exposure may lead to the progression or acceleration of carotid artery atherosclerotic disease. urine As levels can and commonly do vary by a factor of 5 from day to day depending on diet.Ê Seafoods and some canned foods may contain variable and high As levels. A reduced L-glutathione (GSH) provocation test can be done using 300 mg of GSH orally, followed by skipping the first am urination and collecting for the next 24 hours including the first urination of the next am provides the urine specimen. Elevated pyruvic acid in serum or plasma, elevated α-ketoglutaric acid in a plasma organic acid analysis, and haematuria are also diagnostic. Excessive arsenic in cells can inhibit mitochondrial processes, especially those related to cofactor activity of lipoic acid. Typical early symptoms of As excess include fatigue, dermatitis, increased salivation and possibly peripheral parasthaesias with tingling or numbness. More advanced symptoms or chronic exposures can lead to muscular weakness, hair loss, and hypopigmentation of skin, anaemia with haemolysis and neuronal degeneration. Even at low levels, As may cause digestive problems, fatigue and skin rashes. Intervention: Arsenic must be methylated using methyl donors such as SAMe, trimethylglycine, dimethylglycine, methionine, etc. Some arsenic is bound to sulfur groups such as glutathione and excreted in the urine or bile. 1- Remove the individual from sources of arsenic. 2- Treat anaemia. 3- Sulfurated methyl groups, supportive therapy (magnesium, B-vitamins, vitamin C, vitamin E, selenomethionine, lipoic acid). 4- Increase fluid intake. 5- Sulfhydryl-group conjugating agents. 6- Cysteine combines with As and will move it around in body tissues but not necessarily clear it from the body. Intervention should be continued until urinary arsenic levels are consistently below those stated above. Bismuth (Bi) is a non-essential element of low toxicity. However, excessive intake of insoluble, inorganic Bi containing compounds can cause nephrotoxicity and encephalopathy. Absorption is dependent upon solubility of the Bi compound, with insoluble Bi excreted in the faeces while soluble forms are excreted in the urine. Sources of Bi include: cosmetics (lipstick), Bi containing medications such as ranitidine Bismuth-citrate, antacids, pigments used in coloured glass and ceramics, dental cement, and dry cell battery electrodes. Symptoms of moderate Bi toxicity include: constipation or bowel irregularity, foul breath, blue/black gum line, and malaise. High levels of Bi accumulation can result in nephrotoxicity (nephrosis, proteinuria) and neurotoxicity (tremor, memory loss, myoclonic jerks, dysarthria, dementia). Dithiol chelating/complexing agents (DMPS, DMSA) markedly reduce Bi levels in liver and kidneys, and increase urinary Bi. Some shampoos and many hair perm dye bleach products place nickel (Ni) into the hair. In blood, Ni binds to albumin, globulin and amino acids, and is deposited in leukocytes. In cells, it binds to mitochondrial and cytosolic proteins. In so doing, it can displace zinc and copper, thereby activating, inhibiting, or dysregulating enzymes. A Ni exposure may hypersensitise the immune system, resulting in inflammatory responses to many environmental substances to which there was formerly little or no response. Possible symptoms of Ni excess include panallergy with rhinitis, sinusitis, conjunctivitis and asthma. Other symptoms may include vertigo, weakness and fatigue, nausea and headache. Nickel contact allergy (nickel itch) or contact dermatitis is not necessarily reflected by elevated Ni. Inorganic tin (Sn) is mildly toxic and may impair liver function by inhibition of the P-450 mixed function oxidase enzyme system. Hence, Sn can have a synergistic effect of rendering organic chemical xenobiotics or drugs more difficult to detoxify. Organic Sn compounds (dimethyl tin, dialkyl tin, and triphenyl tin) are biocidal and can be severely toxic. Exposure to organic Sn compounds may produce headache, muscle ataxia, general fatigue, vertigo and reduced sense of smell. Renal damage may also result. Erythrocyte haemolysis, anaemia and subnormal lymphocytes may occur, causing immune dysfunction. Other conditions include hyperglycaemia, lesions in testes and ovaries, and inflammation or congestion of binary ducts. Boron (B) is sensitive to contamination from hair preparation products. Elevated B may be due to a combination of factors-endogenous excess, external contamination, and maldistribution secondary to toxic excesses. Effects of excess B depend strongly upon chemical form and mode of exposure. Elemental B has low toxicity while borates and boranes can have cumulative neurotoxic effects. Boranes interfere with pyridoxal phosphate-dependent metabolic steps for amino acids. Symptoms may include dizziness, muscular tremors and incoordination. Excess levels of vanadium (V) in the body can result from chronic consumption of fish, shrimp, crabs, and oysters derived from water near offshore oilrigs. Environmental sources of V include processing of mineral ores, phosphate fertilisers, combustion of oil and coal, production of steel, and chemicals used in the fixation of dyes and print. Inhalation of excess V may produce respiratory irritation and bronchitis. Excess ingestion of V can result in decreased appetite, depressed growth, diarrhoea/gastrointestinal disturbances, and nephrotoxic and haematotoxic effects. Pallor, diarrhoea, and green tongue are early signs of V.

Levels of As concern are:

Age	<ppm	<mg/24 hr	<µg/hr
3-5	0.27	0.20	9
6-13	0.27	0.27	12
14-99	0.29	0.16	20
Concern	0.40	0.60	30

NATURAL GAS

Radioactive building materials include gypsum board, granite, slate, and concrete products containing such materials. Radon gas produces radioactive substances. Radioactive products are hazardous waste. Fluorescent lights may contain small amounts of potent radioactive waste. The energy can cause damage to the cell's machinery. Sometimes this damage is manifest as leukaemia or other cancers. Radon is a naturally occurring colourless, odourless gas that comes out of the ground, seeps elements, and moves upstairs. It can be found in soils and rocks containing uranium, graphite, granite, shale, phosphate, and pitchblende. Radon decays into radioactive elements known as radon daughters. Daughters float around in the air attached to dust particles. When breathed in the particles find their way into the bronchial air passages and stick to the surface. There they stay to irradiate the surrounding cells. Radon is second only to smoking as a cause of lung cancer. Breathing in 4 picoCuries/l (pCi/l) of radon is equated to smoking half a pack of cigarettes per day. Outdoor levels are in the range of 0.3-0.7 pCi/l. The average home level is 1.5 pCi/l. However, no level of radon is considered safe. Building a new house should plane to prevent radon from entering the basement by putting a rubber seal between the slab and the block walls. Underlay the slab with a plastic sheet. Six inches of gravel should be placed below the plastic sheet. A sub-slab mechanical ventilation system can be added to give much better assurance that no radon will ever enter the basement. Natural gas, propane gas, and sewer (methane) gas, formaldehyde and other volatile organic compounds (VOC) and gases from combustion are examples of the common toxic gas exposures. The source of fuel gas is leaking pipe fittings, gas valves or defective control equipment. Sewer gas emits from unused drains when the water in the drain elbow has evaporated and no longer blocks the fumes as intended. Formaldehyde gas is originated from a wide variety of building products such as plywood, chip board, particle board, drywall, sealants, calking, carpet padding and paint, as well as treated fabrics. Paints, solvents, dry cleaning, pesticides, sealants, glues, mastics, carpets, household cleaning products are also examples of common VOC. Combustion gases, created when fuel (gas, wood, kerosene, coal) is burned can back up from the stack venturi due to poor air flow up the exhaust stack or back drafts from winds. Toxic gases can cause blurred vision, headaches, nausea, dizziness, coughing, burning eyes, sinus irritation, skin rashes, respiratory illness, light-headedness, concentration problems, depression, and in extreme cases loss of consciousness, and suffocation. Intervention: 1- Unbag dry cleaning and air out in the garage for a day. 2- Keep household cleaning products in well-ventilated area, and replace with safe products. 3- Use no VOC paint, 4- Avoid vinyl products like shower curtains, wall paper and floor coverings. 5- Choose carpet and padding that have a low smell level, and aired out before installing. 6- Run water occasionally into seldom used sinks and floor drains to keep gas trap filled.

BIOLOGICAL CONTAMINANTS

Biological contaminants include fungi (mould, mildew), bacteria, virus, animal dander, cat saliva, cockroach parts and faeces, pollens, dust mites parts and their feces. The natural environment contains all of these materials. There are numerous sources for these contaminants. Pollens originate from plants, and trees. Viruses are transmitted by people and animals. Bacteria cohabitate with people, animals and plants both inside and outside their bodies. Household pets and pests are a source of dander and saliva. Protein in rat and mice urine becomes a potent airborne allergen when it dries. Wet building materials and contaminated air conditioning systems can become breeding grounds for bacterial and fungi. Some contaminants cause allergic reactions including hypersensitivity pneumonitis, allergic rhinitis and asthma. Other contaminants transmit infectious diseases like influenza, measles and chickenpox. Some biologicals like certain fungi produce toxins (mycotoxins) as a by-product of living. Reactions may include sneezing, watery eyes, coughing, shortness of breath, dizziness, lethargy, fever, digestive problems, joint problems, lung damage. Some allergic reactions may occur immediately after exposure. Children, the elderly and people with breathing problems or lung disease are particularly susceptible to biological contaminants. Hydrogen peroxide mixed with water is an efficient way to decontamination. Moulds found in the contaminated area (home or workplace) have a wide range of findings including Cladosporium, Alternaria, aspergillus, stachybotyrus, penicillium and many others. The cause is always water leak (seepage through a sewer pipe to roof leaks, shower and bathtub plumbing leaks, building leaks to faulty construction). The range of time spent in the mouldy buildings is from 3 months to over 2 years. The patient's main symptoms include rhinosinusitis, bronchitis, short-term memory loss, confusion, lack of concentration, chronic fatigue, fibromyalgia, chronic cough and swollen nasal passages. Clinically the patient has a positive stressed Romberg's and Tandem Romberg's. Romberg's test differentiates between peripheral and cerebellar ataxia. An increase in clumsiness in all movements and in the width and uncertainty of the gait when the patient's eyes are closed indicates peripheral ataxia; no change indicates the cerebellar type. Laboratory tests would show sensitivity to moulds on intradermal skin testing, alteration in T and B lymphocytes and cell-mediated immunity, triple camera SPECT brain scan will be positive for neurotoxicity, and autonomic nervous system imbalance (measured by the Iriscorder and/or the heart rate variability machine). Intervention consists of air purification system, massive avoidance of moulds and toxic chemicals, nutrient therapy both oral and intravenous, heat depuration, physical and massage therapy. Broadspread avoidance of toxins, food rotation, nutrient supplementation and sauna are necessary to clear the symptoms. Avoidance of mould exposure alone will not allow them to revert back to normal. Exposure to moulds and mycotoxins indoor induce neurobehavioural and pulmonary impairment. In homes sources include with water intrusion into walls from structural defects, leaky pipes, air-conditioning condensers and icemaker tubes. If moulds are incriminated serum measurements of antibodies to mycotoxins rather than moulds, may provide the best indication, particularly trichothecenes and toxic chemical compounds. These agents have two 6 carbon rings may be a link in understanding the indoor air mystery. Mould exposures indoors are associated with neurobehavioural impairment and evidence of small airways obstruction. There are 3 major mycotoxins related human diseases: 1- Vascular obstruction and necrosis occur, the adrenergic activity of ingested ergot alkaloids from Claviceps species which also stimulate the hypothalamus and midbrain, depress the vasomotor center and act centrally to cause emesis. 2- Alimentary toxic aleukia, the toxins are sesquiterpinoids related to the trichothecenes of Stachybotrys. 3- Liver cancer from consumption of Aflatoxin B1 contaminated food, particularly in malnourished people. Moulds can be grown on agar by uncovering plates or taking air samples obtained with an Anderson sampler. Indoor and outdoor samples should be quantified and compared. There are chemical assays of dust for mycotoxins (aflatoxin, coumadins, zearalenones and trichothocene) using thin layer chromatography on silica gel and identification by charring and by treating with 4 (p-nitrobenzyl) pyridine. Liquid chromatography is used for final separation with identification using mass spectrometer. There is no blood assay for mycotoxins but reactive antibodies can be measured to trichothene, aflatoxin and zearalenones. Mycotoxins, an age-old problem contributes to the adverse effects of indoor air on neurobehavioral functions and airflow. Mould amelioration reduces respiratory problems in school children. Reducing mould toxins in foods decreases hepatocellular cancers. Fungal toxins have irritant, immunosuppressive, neurotoxicological and carcinogenetic effects. Toxic moulds such as penicillium, stachybotrys, cladosporium, aspergillus, and black mould can cause indoor air quality problems leading to allergies and sickness. These problems are a result of airborne mycotoxins and mould spores. Species of mycotoxin-producing moulds include Fusarium, Trichoderma, and Stachybotrys. Mycotoxins are extremely potent toxins produced by some indoor moulds. Mycotoxins are lipid-soluble and are readily absorbed by the intestinal lining, airways, and skin. The most common indoor moulds are Cladosporium, Penicillium, Aspergillus, and Alternaria. The growth of moulds is pervasive throughout the outdoor environment. Given the proper conditions, moulds may also proliferate in the indoor setting. Moulds readily enter indoor environments by circulating through doorways, windows, heating, ventilation systems, and air conditioning systems. Spores in the air also deposit on people and animals, making clothing, shoes, bags, and pets common carriers of mould into indoor environments. Moulds may colonise near standing water. Moulds proliferate in environments that contain excessive moisture, e.g., from leaks in roofs, walls, plant pots, or animal urine. Many building materials are suitable nutrient sources for fungal growth. Cellulose substrates, including paper and paper products, cardboard, ceiling tiles, wood, and wood products, are favourable for the growth of certain moulds. Other substrates such as dust, paints, wallpaper, insulation materials, drywall, carpet, fabric, and upholstery commonly support mould growth. The presence of moulds indicates a long-standing water problem. Moulds can induce acute pulmonary haemorrhage. The quantity of moulds, including the toxigenic fungus Stachybotrys atra is higher in the homes of such patients. Simultaneous exposure to environmental tobacco smoke appeared to increase the risk of acute pulmonary haemorrhage among these infants. Stachybotrys atra requires water-saturated cellulose-based materials for growth in buildings. Infants may be particularly susceptible to the effects of these inhaled mycotoxins because their lungs are growing very rapidly. The toxic spores of Stachybotrys atra induce severe interstitial inflammation with haemorrhagic exudates in the alveoli. Bulk mould must be removed, followed by a thorough cleaning with soap and water. Moulds can trigger asthma episodes in sensitive individuals with asthma. Probably most proteins of non-human origin can cause asthma in a subset of any appropriately exposed population. Notable triggers for these diseases are allergens derived from house dust mites; other arthropods, including cockroaches; pets (cats, dogs, birds, rodents); moulds; and protein-containing furnishings, including feathers, kapok, etc. In occupational settings, more unusual allergens (e.g., bacterial enzymes, and algae) have caused asthma epidemics. Water can enter any home by leaking or by seeping through basement floors. Showers or even cooking can add moisture to the indoor air. As the temperature goes down, the air is able to hold less moisture. This is why, in cold weather, moisture condenses on cold surfaces (e.g., drops of water form on the inside of a window). This moisture can encourage biological pollutants to grow. Water leaks in pipes or around tubs and sinks can provide a place for biological pollutants to grow. The use of exhaust fans in bathrooms and kitchens helps to remove moisture to the outside. Dehumidifiers and air conditioners are useful, especially in hot, humid climates, to reduce moisture in the air, but they may become sources of biological pollutants. Opening doors between rooms increases circulation, which carries heat to the cold surfaces. Carpet can absorb moisture and serve as a place for biological pollutants to grow. Moisture problems and their solutions differ from one climate to another. Moisture problems in school buildings can be caused by a variety of conditions, including roof and plumbing leaks, condensation, and excess humidity. Moisture problems in schools are associated with delayed maintenance or insufficient maintenance, due to budget and other constraints.

AIR IONIZATION, ELECTROSTATIC FIELDS AND MATERIALS

Fresh, rural outdoor air contains 1000-3000 small charged gas particles/cm^2, which are called small ions. Normally, there are 5 positive to 4 negative ions. Ions live for 3-15 minutes and are constantly being made by sun radiation, earth radiations and turbulent water. The human body becomes affected once the number of ions decreases or the ratio of positive to negative is altered. Prior to a thunder storm breaking people feel nervous, irritable, and ill-at-ease. Just before the storm the air is loaded with positive ions. Hot, dry winds are associated with increase in violent behaviour. When the storm breaks everyone feels great. Electrostatic fields or static electricity is stored in non-conducting materials. The human body has to build up thousands of volts. Most synthetic materials build up a positive charge that drains away very slowly. This positive charge on the surface of synthetic carpets, fabrics, and wall coverings and the TV or computer screen will generate a surplus of positive ions in the air. Homes and offices have this condition when the air is dry. This can be in winter and in summer. The indoor air never gets any better. The number of ions in room air is depleted by the AC electric field, by the forced air heating/cooling system, and by tobacco smoke. The only way to replenish ions indoors is to let in outside air (ventilation) or to make them electronically (ionizer). If this is not done positive ions go to surplus and the number of ions falls. Symptoms of low air ionization and high positive ionization were reported in the 1930's, such as migraine, nervousness, tension, anxiety, irritability, breathing difficulties, worsening of asthma, elevated heart rate, insomnia, disturbed digestion, and nausea. 25% of people have severe symptoms, 25% seem unaffected, and 50% have some form of milder symptoms. Children sleeping with a heavily charged stuffed toy would develop sinus problems until the toy is removed or neutralized. Electricity produces both magnetic and electric fields. These together are termed EMF. Magnetic fields are present when current flows to power appliances and lights. Electric fields are present at all times and are produced by the wiring in the walls, floors and ceilings and wires to appliances, etc. AC Magnetic field exposure causes long-term health problems. Children exposed to fields >1 milligauss have twice the risk of leukaemia, >2 milligauss the risk is 3 times; and >3 milligauss 4 times the risk. Adults exposed to fields >2 milligauss have 70% increased risk of developing leukaemia. Electric field exposures cause stress to the body especially at night when resting that can be manifested as insomnia, fitful sleep, night sweats, strains, pulled muscles, heart flutters, waking up tired, constipation, irritability, and bed wetting in children. Intervention: 1- Keep children 10 feet from the TV. 2- Keep clocks, radios, cassette players 3 feet away from sleeping areas. 3- Keep wires and lamp cords away from the sleeping area or most effective. 4- Turn off the circuits identified in the sleeping place evaluation. 5- Keep 3 feet away from computer video display terminal. 6- Stay 3 feet away from appliances, e.g., stove, oven, toaster, dryer, heater. 7- Fix any abnormally high magnetic fields. High frequency electromagnetic fields (HFEMF) come from radio, TV, police, fire, and military communications, microwave, radar, and cellular phones. These energies are pervasive and can be measured every where on the earth. The wireless age is increasing the density of such energies at an unprecedented rate. HFEMF are associated with cataracts, blood composition changes, hormone alterations, chromosomal abnormalities, stress induced reduction in immune system response, and cancers of the pituitary, thyroid, and adrenal glands. Wireless communications signals that are close to the head may increase nonmalignant tumours of the auditory nerve, increase brain cancer, cause chromosome damage and DNA breaks in brain tissue. Living near broadcast towers, cell phone towers and/or are near the transmission paths of the numerous microwave installation that now replace telephone wires in so many places have a negative effect on health and ability to heal. Intervention: 1- Stand away from microwave ovens when they are operating. 2- Avoid living or working near broadcast towers and cell towers. Suburbs and the country are normally better because the sources of these signals are usually in urban areas. Watch out for microwave repeater stations located at high points between large urban centers. The rapid increase in cell phone towers placements has the potential to affect many people. 3- Metal acts as a concentrator of high frequency energy. The metal springs of the bed act as an energy concentrator. Keeping metal away from the sleeping area is prudent. 4- Use shielding materials. DC magnetic fields are produced by the earth and by man-made permanent magnets. Deviations in the earth's natural magnetic field either in strength or direction are cause for concern. More than 60% of all inner spring mattresses and box springs tested show non-uniform magnetic hot spots along the length of the bed. These hot spots cause a shift in the location of true north so far as the body is concerned, but only for the parts of the body lying on the hot spots. Other objects around the bed may be non-uniformly and strongly magnetized, such as the bed frame, steel furnishings, steel objects stored under or around the bed. Disturbed DC magnetic fields in the sleeping place confuses the body and can cause insomnia, chronic fatigue, and other sleep related problems. DC fields influence enzyme activity, oxygen usage, and the growth rate of cancer cells. Wearing glasses with magnetized frames may induce strong headaches. Stereo headphones use magnets for the speaker, and also cause headaches. Intervention: 1- Sleeping with the head to the north optimizes red blood cell production and yields optimal deepness of sleep. 2- Sleep with no steel in or around the bed. This can be accomplished using a futon or a natural latex rubber mattress. A waterbed could be satisfactory, but the water heater subjects the body to high EMF. The vinyl container for the water gives off plasticizer fumes that will add more unwanted stress to the sleeping place. 3- Use a bed frame made of wood. 4- Use magnetic health products cautiously. 5- The needle deviation of liquid filled compass should be no more than 2° to be ideal no more than 10° to be weak.

CARDIOVASCULAR AND RENAL DISEASES

Cardiac allergy is seldom considered in the differential diagnosis of patients with heart disease, which is often misdiagnosed as idiopathic. The heart contains mast cells that release histamine and other allergic mediators when degranulated by antigens to which a patient is allergic. These mediators can have a direct effect on the heart. Histamine produces tachyarrhythmias and decreased cardiac blood flow. Allergic reactions to bee stings produce arrhythmias, electrocardiographic abnormalities, and myocardial infarction, in the absence of pre-existing coronary disease and treatment with drugs with cardiac effects. A number of drugs can produce allergic myocarditis. Elevated homocysteine provokes oxidative stress leading to oxidation of membrane lipids and cholesterol, thus increasing risk for CVD. Homocysteine in blood forms disulfides (homocystine, homocysteine-cysteine, homocysteine-thiols) with concurrent production of superoxide ion. The resulting hydrogen peroxide and hydroxyl radical activity makes homocysteine a cause of endothelial tissue damage, which may cause atherosclerosis. Elevated homocysteine may cause mental depression and Alzheimer's disease. The disordered methylation affects catecholamine metabolism, neurotransmitter chemistry and homocysteine levels, and that homocysteine is symptomatic rather than causal. Assessment of homocyst(e)ine (homocysteine + homocystine) status is done by amino acid analysis of blood plasma or urine. Urine amino acid analysis by HPLC usually measures only the oxidised, disulfide form, homocystine. Plasma HPLC analysis may measure only homocysteine. Non-HPLC usually measures only the oxidised, disulfide form, homocystine. Plasma HPLC analysis may measure only homocysteine. Non-HPLC screening tests may measure blood plasma or serum homocysteine and various disulfides including homocystine, homocysteine-cysteine and homocysteine thioethers. Total serum homocysteine includes these species plus protein-bound homocysteine. Although quantitative amino acid analysis only measures homocysteine and/or homocystine, it is most informative, and also measures methionine, serine, cystathionine and other analytes sensitive to vitamins B_6, B_{12}, and folate. Conditions associated with elevated normal intermediary metabolite of the essential amino acid methionine, homocysteine in cells and body fluids: A- Deficiency or dysfunction of vitamin B_6 or pyridoxal-5-phosphate (P-5-P), the coenzyme for cystathionine β-synthase (the enzyme that catalyzes formation of cystathionine from homocysteine and serine). B- Deficiency of serine. C- Deficiency of vitamin B_{12} or of its cofactor form, methylcobalamin, which methylates homocysteine to re-form methionine. D- Deficiency of folic acid or of its cofactor form, 5-methylcobalamin, which methylates homocysteine to re-form methionine. E- Genetic weakness in the cystathionine β-synthase enzyme (possibly countered with vitamin therapy). F- Genetic weakness in the folate-B_{12} methyl transfer enzymes (5,10-methylene tetrahydrofolate reductase, or 5-methyltetrahydrofolate-homocysteine methyltransferase. Normally during the inspiratory phase of respiration, heart rate increases. This vagus nerve regulated sinus arrhythmia is reduced or abolished by aging, diabetes mellitus and solvent exposure. Exposure to chlorine/creosol reduce the respiratory variation of the heart rate. This effect on vagal function parallel other adverse effects on the brain's regulation of balance, of hearing and choice reaction time, color discrimination, contrast sensitivity and visual field performance. The human body is highly electrically active. Minute currents can be measured from every cell in the body. Currently, the principles of physics is used almost solely for diagnostic rather than curative purposes. X-rays, ultrasound scans, nuclear magnetic resonance and CAT scans are invaluable in the diagnosis of disease. The electromagnetic radiation is a cause of a wide variety of illnesses. The electromagnetic radiation may offer a new way of curing many diseases. The electrical smog, electric pollution may cause a syndrome called electrical sensitivity, and typical symptoms include headache, depression, muscular weakness, incoordination and even blackouts. These symptoms may be caused in electrically sensitive people by many types of electromagnetic radiation including radio- and micro-waves, infrared, normal and ultraviolet light, x-rays, and cosmic radiation. Electrical sensitivity affects at least 1 in 1,000 of the population. Almost all electrically sensitive people are also sufferers from food and/or chemical allergies. Non-ionising electromagnetic radiation is able to cause a wide variety of symptoms especially those related to blood vessels and the brain and nerves. These symptoms include flushing, blushing, palpitations, diarrhea, muscular aches, pins and needles especially in the hands and feet, dizziness, fits and blackouts, disorientation, headaches, noises in the head, depression and suicide, and persistent tiredness unrelieved by rest. Electrical sensitivity may also mimic neurological diseases such as paralysis, epilepsy and multiple sclerosis. Other diseases and symptoms associated with prolonged exposure to magnetic and electric fields include headache, depression, suicide, miscarriage, cancer and leukemia. There is a long latent period, 3-5 years, between exposure to the current and the effect. There may be a further 10 years for a well defined disease to show itself. Electrical activity is absolutely fundamental to life and the state of health of the body can be tested electrically in several ways. It is very difficult to screen magnetic fields but individual devices can be greatly modified to reduce their electric and magnetic radiation. For example, low radiation computer monitors can give out one tenth of the level amount of radiation of standard devices. By understanding the principles of biophysics it has been possible to treat a large number of patients sensitive to non-ionising radiation. Patients who become very ill when they have live close to high voltage powerlines (exposure to 50 Hz), when they move away from this powerful radiation their health improve and electrical sensitivity disappear. Some individuals who use personal radio transmitters such as mobile telephones and deaf aids transmitter may become unable to work due to the radio transmitter making them feel tired all the time. Patients sensitive to marine and aeronautical radar frequencies may become ill when living near airports and coastal marine radar facilities. Intervention for all these cases is consisted of reducing the total load of allergic problem in air, food, water and also the total load of non-ionising electromagnetic radiation from all sources. Cardiovascular involvement is a common feature in environmentally-triggered illness and can present as both large vessel and small vessel disease, Raynaud's-like phenomenon, skin lesions, and vasopastic disease. Foods, inhalants, and chemicals can cause vasculitis. Possible mechanisms include: a- damage to the endothelial barrier, b- activation of leukocytes and platelets, c- initiation of plaque formation, d- stimulation of the inflammatory rresponse, e- direct damage to cardiac and blood vessel tissue, and f- kidney-related hypertension. Disturbances in the peripheral vascular system ultimately lead to extravasation of blood elements outside the vessel, inflammation of the vessel, and reduction and/or occlusion of bloodflow either intermittently or chronically, resulting in ischaemia. Cardiac disturbances include abnormalities in contraction, conduction, excitability, and heart rate. Decreased bloodflow to vital organs can cause intermittent as well as chronic symptoms, some life-threatening. The brain SPECT scan can demonstrate functional pathology. Many of the inflammatory vascular diseases were of unknown aetiology. Capability of incitant definition is significant aid in treatment and allow us to advance further than ever before into the less costly areas of preventive medicine. The development of the ability to measure numerous laboratory parameters such as immunoglobulins, total haemolytic complement and all its components, T and B lymphocytes, blastogenesis and numerous mediators such as kinin, serotonin, and histamine is critical. The spectrum of inflammatory vascular disease (pathobiology of blood vessel change) is a spectrum of reversible to totally irreversible changes. Initially, after insult, localized oedema is well appreciated. However, the much more common but less well appreciated subtle generalized oedema is often ignored (in the extremities and the periorbital areas). Generalized oedema may develop into much more serious disease. The blood vessel wall develops greater leaks or even ruptures as the disease progresses, resulting in red blood cell extravasation with bruising, purpura, and petechiae. Later, as the fibrinolytic system becomes depleted, clots occur and palpable purpura develop. Eventually, if the vessel is injured enough, ischaemic necrosis can occur. If the area survives, granulomatous healing may occur. Atherosclerotic plaque formation can occur in other people after severe inflammation develops. In inflammatory vascular disease, two different cell types occur: 1- The lymphocytic type infiltrate, will occur around the vessel wall, first as a mild form of vasculitis then as irreversible inflammatory disease. 2- The leukocytoclastic form appears to be more acute and perhaps more devastating initially. Here, the leukocytes occur around the vessels and then break up. The debris is incorporated into cells, thus the name leukocytoclastic. Vasculitis may be of necrotizing or non-necrotizing variety. The non-necrotizing varieties is more easily preventable than the necrotizing and should be diagnosed and treated early. Internal disruption of the homeostatic mechanism, which causes vessel damage may occur from any of several pathways. Fatal cardiac arrhythmia occurs in teenager after inhalation of a fluorocarbon spray. Fatal arrhythmia also occur to furniture removal substance (environmental incitants). Cardiac arrhythmias and myocardial infarction may be ascribed to the hypersensitive mechanisms, the clinician will be able to define triggering agents of cardiac disease if he will only look for them. No matter what internal mechanism is disrupted, the search for triggering agents should be carried out before these secondary mechanisms become autonomous and cause irreversible changes in the tissue. There are basic principles to be considered for incitant: 1- four-day avoidance before incitant challenge; 2- any trigger is possible including pollens, dusts, molds, terpenes, foods, or chemicals; 3- responses can occur anywhere from five seconds to fourteen hours after a challenge; 4- the chemical environment must be well-controlled in a less polluted manner; and 5- most of the reactions are not IgE mediated, though this does frequently occur. When considering the response of the cardiovascular system to incitants, one must perceive that any part of the system can be involved. Some patients may just exhibit heart involvement while others will show just venous involvement or small or large peripheral arterial involvement. A patient may have phlebitis at one time and spontaneous bruising at another, or they may occur together. Reynaud's phenomenon or hypertension may be present with cardiac arrhythmia or they may be isolated. Many patients with vascular headaches may be sensitive to multiple foods and chemicals. Occupational renal diseases are preventable. Heavy metals cause acute tubular necrosis (ATN) when injected intravenously, lead and cadmium may induce chronic interstitial nephritis in individuals exposed heavily in the workplace over many years. The chronic interstitial nephritis caused by long-term occupational exposure to lead must be differentiated from the transient acute proximal tubule reabsorptive defect (Fanconi syndrome) caused by the acute lead poisoning. Hypertension and gout accompany occupational lead nephropathy but are absent from cadmium-induced renal disease, which is characterised by calcium wasting, osteomalacia, and renal stones. Minute dose of mercury can cause glomerular disease manifested as the nephrotic syndrome in sensitive individuals. Mercury, arsenic, chromium and uranium may lead to chronic interstitial nephritis in victims who survive acute tubular necrosis. Inhalation of organic compounds such as carbon tetrachloride can also produce acute tubular necrosis, in large dosage. Long-term low-level inhalation of industrial solvents induce a variety of glomerular diseases. Exposure to petroleum products (light hydrocarbon nephropathy) induces eosinophilic lysosomal bodies in the cells of proximal tubules, which is associated with the development of chronic renal disease. Occupational exposure to silica may cause of anti-neutrophile cytoplasmic antibody (ANCA) positive Wegener's granulomatosis. This immunologically-mediated vasculitis of lungs and kidneys represents an unusual manifestation of exposure to silica and may be present in the absence of pulmonary silicosis. Glomerulonephritis may also occur in association with the intense immunological manifestations of silicoproteinosis in the presence of severe silicosis. Interstitial nephritis is difficult to diagnose because of the absence of heavy albuminuria. Occupational renal diseases can lead to end-stage renal disease.

CHELATION

Chelation is the binding of a metal ion by an organic (i.e., carbon based) molecule. All weak organic acids are chelators, including ascorbic acid, acetic acid, (vinegar) lactic acid (produced with exercise), and malic acid (apple acid), etc. Oral chelation can not be compared to the dramatic anti-aging benefits of intravenous chelation therapy. Penicillamine, DMPS, DMSA, EDTA, aspirin and tetracycline are all orally effective chelators. Haemoglobin is chelated iron. Chlorophyll is chelated magnesium and the body utilises chelated Mg far more efficiently than free form Mg. Plants are the source of chelated minerals. Mineral water contains free form minerals and is not a preferred source of minerals. EDTA binds to trace metals and retards the oxidative degradation of vitamin C and other nutrients in the food. Aging relates to an increasing deficiency of essential trace minerals and an excessive accumulation of toxic metals. No one chelator can handle all of the toxic metals absorbed daily from the water, food and air. EDTA has an impact on vascular disease, reduces clotting problems, reduces iron and heavy metal stores already in the body and results in reduced free radicals. EDTA may substitute for Zn in zinc-deficiency, enhance the absorption of heparin, vitamin B_{12}, and increase the blood levels of antibiotics. EDTA prevents excess iron absorption (depending on other factors such as pH, etc.). Oral EDTA decreases the circulating levels of free iron and copper. EDTA protects against many toxic metals including nickel, cadmium (as long as dose is high enough), vanadium (which is linked to diabetes), cobalt, iron, and copper overload. Ethylene-diamine-tetra-acetic acid (EDTA) is a synthetic amino acid first used in the 1940's for treatment of heavy metal poisoning. EDTA was considered possibly effective in the treatment of occlusive vascular disorders caused by arteriosclerosis in 1960s. More than 50% of bypass surgeries are failures. The new cell growth is a fertile ground for the growth of the plaque obstructions that characterise atherosclerosis. Gene therapy shows promise of reducing this failure rate. The gene therapy approach involves soaking the blood vessel prior to grafting with a short segment of DNA that blocks the genetic machinery needed to form the new cells. The average person is receiving about 15-50 mg of EDTA in their diet, and that it is in every lavender tube used for blood tests because of its widely recognised anti-coagulant activity. Other oral chelators include garlic, chlorella, alginates, or vitamin C, or any of the sulfhydryl containing amino acids such as cysteine or methionine containing products. The death rate from heart attack, stroke and other cardiovascular diseases had dropped 20% in the past decade. Despite this progress, heart attacks remain the number-one killer in the west, strokes the number-three killer. Not all plaques are equal. The most dangerous ones are soft and consist largely of a pool of cholesterol covered by a thin fibrous cap. The stress of blood flow can tear the caps and cause a blood clot. These plaques are often inflamed and the inflammation can further weaken the thin cap. However, the harder, more stable, calcium-rich plaques can also trigger a stroke or heart attack by blocking the flow of blood so much that blood pools behind it. Stagnant blood is more likely to clot. Strokes can be triggered by soft plaques in the thoracic aorta. When the turbulence and vibrations become severe enough, a tear in the carotid artery can occur. That tear can launch a blood clot to the brain. Individuals with apolipoprotein E (apoE) tend to have elevated levels of LDL and are at increased risk of developing heart disease. Those with this gene variant respond well when they consume a low-fat diet. Those who have the normal gene variant, called apoE3, show much less reduction in LDL levels when they adhere to a low-fat diet. Patients with LDL subclass pattern B are at higher risk for developing diabetes and heart disease, and have elevated blood levels of triglycerides-fats in the blood that come from food. Patients with pattern B respond much better to low-fat diets than do patients who have the larger, less dense LDL particles (pattern A trait). It is essential to combine physical activity with a low-fat diet in the treatment of elevated LDL cholesterol levels. Leptin, a by-product created by one of the obesity genes, is a hormone that regulates food intake and body weight. A diabetes gene codes for the protein that forms cell receptors for leptin. Defects in the production of either protein may lead to a tendency toward weight gain and obesity. Inflammation in the circulating blood may play an important role in triggering heart attacks and strokes by activating blood-clotting mechanisms, which in turn can slow down or stop blood flow. Inflammation is the body's natural response to injury and blood clotting is often part of that response. During the inflammatory process, C-reactive protein is produced in the blood. Inflammation can limit the effectiveness of clot busting drug therapy, e.g., thrombolysis, which is the first line of treatment for patients suffering a heart attack. Heart attack patients with high levels of C-reactive protein respond more slowly to thrombolytics. Anti-inflammatory drugs may improve the effectiveness of anticlotting treatment in patients with high levels of C-reactive protein. High cholesterol levels can cause an inflammatory response in the circulating blood. People who smoke cigarettes have elevated levels of C-reactive protein. Having increased inflammation in their blood may one of the reasons smokers are at a much higher risk of death from heart disease and stroke than nonsmokers. Men who smoked fewer than 10 cigarettes a day have a 30% higher risk of death from heart disease or lung cancer than nonsmokers do. Those who smoked 10 or more cigarettes a day have an 80% higher incidence of death from these diseases. EDTA prolongs prothrombin time, which could be a synergistic influence with other anti-clotting, anti-platelet substances such as garlic, ginkgo, polysaccharides, and EPA. Lowering of the prothrombin activity may occur in 50% of patients. EDTA lowers cholesterol level. Oral EDTA inhibits the absorption of iron and copper. EDTA in body fluids binds copper and iron so tightly that they are largely unavailable for biological activities including catalysing free radical reactions. EDTA-bound iron is poorly utilised by the erythropoietic system. Iron presented to the cell in the form of an EDTA chelate, is unavailable for incorporation into haemoglobin. This strongly suggests that it is not available to catalyse free radical pathology. EDTA enhances the absorption of heparin, and vitamin B_{12}. EDTA in products ought to be at least 500 mg, preferably 800 mg or more. EDTA enhances lead absorption from the gut in acute lead poisoning. EDTA causes a great increase of urinary lead excretion without modifying its intestinal absorption. Oral EDTA must be evaluated versus penicillamine and DMSA, both of which are far more expensive and clearly more toxic. Removal of lead should be by using of natural therapies including vitamin C, garlic, zinc, and calcium. Oral EDTA is a vitally important additional method of increasing lead removal from the body. These natural adjunctive deleading therapies provide significant protection to patients because, for example, by aggressively administering zinc, any potential for zinc deficiency associated with oral EDTA therapy is entirely avoided. Gradually accumulating levels of toxic metals such as lead and cadmium in tissues, especially in the brain (particularly the pituitary gland) with aging is of great concern. EDTA has been documented to be an effective food preservative that prevents oxidative damage to food by free metals. EDTA prevents the copper catalysed oxidation of vitamin C. EDTA protects against many toxic metals including nickel, cadmium (as long as dose is high enough), vanadium (which links to diabetes), cobalt, iron, and copper overload. EDTA helps lower the body burden of internally deposited fission products. EDTA showed significant protection against the toxicity of cisplatin. Most heart attacks that happen in patients before, during or after their IV chelation directly relate to the inflammatory and thrombotic factors. Chelation contains many useful nutritional compounds and may not rely only on EDTA or any chelating effects for its well-documented benefits. Chelation therapy, like bypass surgery and angioplasty, is based upon a scientific rationale and is of measurable benefit to patients. Coronary bypass surgery appears neither to prolong life nor to prevent myocardial infarction in patients who have mild angina or who are asymptomatic after infarction in the 5-year period after coronary angiography. The operation does not cure patients, it is scandalously overused, and its high cost drains resources from other important areas of need. EDTA is also recommended for emergency treatment of hypercalcaemia and the control of ventricular arrhythmias associated with digitalis toxicity. EDTA improves blood flow and relieves symptoms associated with the disease in greater than 80% of the atherosclerotic patients. EDTA chelation therapy impedes each step of the disease process. EDTA by binding ionic metal catalysts and removing them from the body, reduces subsequent abnormal production of OFR reactive molecules and molecular fragments, which react destructively with other molecules. Thus, reducing pathological lipid peroxidation of cell membranes, DNA, enzyme systems and lipoproteins and allowing the body's natural defensive mechanisms to halt and often reverse the disease process. Oxidative modification depends on low concentrations of copper or iron. Cardiac bypass surgery does not save lives or even prevent heart attacks. Among patients who suffer from coronary-artery disease, those who are treated without surgery enjoy the same survival rates as those who undergo open-heart surgery. Eliminating the performance of inappropriate procedures may lead to reductions in health care expenditures or to improved patient outcomes. The free radical concept provides a scientific basis for treatment and prevention of the major causes of disability and death, including arteriosclerosis, dementia, cancer, arthritis and numerous other diseases. EDTA chelation therapy, nutritional supplementation, physical exercise and moderation of health destroying habits all have common therapeutic mechanisms, which reduce free radical causes of age-related diseases. In USA, EDTA was originally approved by the FDA in July 1953 under a version of the Federal Food, Drug and Cosmetic Act, which required that the drug be shown safe, i.e., that the benefits outweigh the risks. In 1962, the Act was amended so that any new drugs must be proven both safe and effective before they could be introduced into interstate commerce. Three agents have been used in the treatment of lead poisoning: 1- dimercaprol (2,3-dimecaptopropanol), known as British anti-lewisite (BAL); 2- penicillamine; and 3- edetate (EDTA) calcium disodium. BAL is an old agent, first made available during World War II. It is the chelating agent of choice for mercury, arsenic, and gold poisoning, but it also is known to chelate lead. BAL is injected parenterally, after which high levels in the blood are maintained for 2 hours; then is excreted within 6-24 hours. Therefore, in acute poisoning, it is recommended that injections be repeated every 4 hours in doses of 3-5 mg/KgBW. Approximately half the injected dose is excreted through the biliary system. Hepatic injury does require reduction in BAL dosage, although the drug is well tolerated in the presence of renal failure. Combination BAL-EDTA therapy is recommended in all cases of severe acute intoxication. d-Penicillamine was first used in the treatment of Wilson's disease in 1956. It can be used orally over long periods of time. Penicillamine given parenterally has been shown to be nearly as effective as parenteral EDTA in promoting the urinary excretion of lead. Small doses of EDTA are efficacious in treatment of chronic lead intoxication and quite safe even in advanced renal disease. Oral administration of EDTA in the human has been shown to effectively promote lead excretion and has been used for treatment. But, oral EDTA may increase lead absorption. There is increased faecal lead content after oral EDTA therapy, Which suggesting that faecal as well as urinary excretion is promoted. EDTA should not be given orally in the treatment of acute lead poisoning or during periods of excessive exposure, as the lead concentration in the intestinal tract may be high. When used orally, doses must be high (i.e., 2-4 g/d in divided doses) to effectively promote lead excretion. Lead nephropathy is a late consequence of lead exposure and usually does not become clinically apparent for 10-20 years. But, anaemia is always present in the untreated patient, and the erythrocyte δ-amino-levulinic acid dehydratase (ALA-D) activity is depressed. The level of lead in bone is characteristically high, and the abnormal tissue stores can be demonstrated by the administration of EDTA. A dose of 1g given intravenously, the normal individual seldom excretes more than 0.2 mg. The patient with lead poisoning usually excretes >1 mg in the first 24 hours. When renal insufficiency is advanced and lead exposure has been very remote, it may be necessary to collect urine for 3 days to demonstrate a significant increase over basal lead excretion. In the absence of renal failure, most of a 2 gm dose given orally at bedtime will be excreted in the overnight urine, which can be collected and analysed for lead. The response is almost as good as for intravenous administration; an excretion of 0.5mg/l or more in the overnight urine should be considered significant. Illegal alcohol is a source of lead. Lead nephropathy is characterised by focal interstitial fibrosis with tubular degeneration and nephrosclerosis in the absence of urinary infection, obstruction, or other disease. Ethylenediaminetetra-acetic acid (EDTA) of binds Pb^{+4} ions forming a relatively stable and not-toxic complex. EDTA may form complexes with other ions. So, it is giving as calcium chelate ($CaNa_2EDTA$) to prevent of loss of Ca from the body. The EDTA can been given by different routes: intravenous, oral, and inhalation. Lead absorbed through the intestine does not concentrate in the liver but is distributed throughout the whole body. Treatment with $CaNa_2EDTA$ by mouth does not modify the intestinal absorption of lead but increases its elimination in urine. Pb-EDTA complex is excreted by the same route as EDTA itself. $CaNa_2EDTA$ passes into the extravascular space where it binds lead ions, but does not penetrate the cells. $CaNa_2EDTA$ does not mobilise lead from bone, in which probably the metal, bound to other ions, is fixed as a stable organic compound. Lead is fixed by the osteons in the bones. EDTA may increase the intestinal absorption of lead. Ca-EDTA passes through the body unchanged. It is excreted via the kidney (>95%) by both glomerular filtration and tubular excretion. The turnover time from the blood is approximately one hour after intravenous administration and one and a half-hours after intramuscular injection. It quickly mixes with almost all of the body water except that it does not pass into the red cells and passes relatively slowly into the spinal fluid compartment. Ca-EDTA is poorly absorbed from the gastrointestinal tract (a maximum of 5%) and practically not all through the skin. Because EDTA inhibits blood clotting so well, by tying up calcium, it is routinely added to blood samples that are drawn for testing purposes. Repeated administrations of a weak synthetic amino acid (EDTA) gradually reduce atherosclerotic plaque and other mineral deposits throughout the cardiovascular system by literally dissolving them away. Oral EDTA chelation refers to the ingestion of the EDTA per mouth. EDTA removes toxic metals from the blood. Ageing continuously accumulates toxic metals: lead, mercury, aluminium, iron, cadmium, arsenic, and others. These toxins invite an increased risk for various diseases, especially heart disease. EDTA normalises the distribution of most metallic elements in the body. EDTA helps prevent heart attack, stroke, varicose veins, and more. EDTA makes stronger bones and reduces cholesterol by improving calcium and cholesterol metabolism. 6 tablespoons a day of EDTA-based chelation formula would reduce cholesterol. EDTA lowers serum calcium, it also stimulates the parathyroid gland to produce parathormone. This hormone is responsible for removing calcium from places such as the inside of arteries and depositing it in the right places, such as bone. IV chelation makes the individual physiologically younger because it moves calcium from the arteries and makes the bones stronger. EDTA reduces the risk of osteoporosis. Oral chelation provides many, but not all, of the benefits of IV therapy. Overall, the differences in benefits are more those of degree, speed, convenience, and cost per dose than of quality. Intravenous (IV) EDTA chelation has a direct and powerful effect on the body almost instantaneously. An IV EDTA session usually lasts about 3-4 hrs, during which about 1,500-3,000 mg of EDTA (plus vitamin C and other nutrients) are administered. Generally about 20-50 sessions are needed depends on the individual's condition. Oral EDTA is significantly cheap depending on one's intake. Only 5-10% of an oral dose of EDTA is absorbed into the bloodstream, compared with 100% of an IV dose. With continuous daily intake, the amounts add up and may achieve similar benefits as IV chelation. Over the course of 5-6 weeks, regular use of oral EDTA can be as beneficial as a single IV EDTA session. Oral EDTA is appropriate for preventing or delaying the onset of the many complications of the diseases related to atherosclerotic plaque build-up, including heart disease, heart attack, stroke, hypertension, PVD, mental disorders, and erectile dysfunction. It is important to supplement the diet with a good multi-mineral supplement while taking EDTA, to avoid any mineral deficiency, especially in zinc. When a free radical reacts with small amount of toxic metal, instead of a simple chain-reaction of one new free radical being produced, there could be millions of new free radical produced from this one impact, in fraction of a second. Oral chelation fights free radicals. During a heavy metal detoxification program, it is very important to have a high protein diet as the sulfur bearing amino acids in the protein will greatly facilitate detoxification. Increasing psyllium intake to have 2-3 bowel movements per day. 90% of the mercury is eliminated via the stool. Chlorella removes toxic metals from the connective tissue. Chlorella is an algae, which has protein high levels of chlorophyll. Garlic, MSM and cilantro can detoxify mercury. 2-3 cloves of garlic (crushed to release its active ingredients) a day enhances sulfur stores. MSM is a form of sulfur, which aid the removal of mercury. Cilantro helps mobilise mercury out of the tissue so the chlorella can bind to it and allow it to be excreted from the body. The detoxification process works better with toxic metals than no metals at all. Enzymes have certain binding sites that require a metal for them to perform their function as a catalyst. Minerals deficiency in magnesium, sodium, zinc and other, reduces the effect of detoxification. N-Acetyl-Cysteine (NAC) is very important component of a good mercury detoxification program. EDTA is better taken on an empty stomach on a daily basis.

MAGNESIUM DEFICIENCY IN PATIENTS WITH CHEMICAL SENSITIVITY

Magnesium (Mg) deficiency is probably a wide spread problem, which is primarily due to poor dietary intake. Increased phosphate also contributes to magnesium deficiency states, e.g., high phosphate containing soft drinks. Others who have poor protein or high fat metabolism tend to be more likely afflicted with magnesium deficiency. Some disease states such as advancing age, diabetes mellitus, alcoholism, heart failure, or use of diuretics, may cause Mg depletion. Hypoxia and drugs such gentamicin, cylclosporin and angiotensin can also deplete the body of Mg. Excess vitamin C may drop magnesium levels. Malabsorption of Mg is associated with gastrointestinal problems such as functional bowel disorder, ulcerative colitis, and Crohn's disease. Acute Mg deficiency can occur after epinephrine, cold stress, and stress of serious injury or extensive surgery. Many patients with chemical sensitivity have poor dietary intake, poor protein metabolism, or fall within the aforementioned conditions. Ethanol and carbon tetrachloride disturbs liver collagen. Exposures to some pesticides can disturb muscle and nerve physiology resulting in muscular spasm and tetany. Most chlorinated hydrocarbons are lipophilic, and exposures can lead to disturbances in membrane stability. Mg plays an integral role in membrane stability. It is a co-factor in many metabolic reactions. Mg counteracts the diuretic inducer of catecholamine. Serum and RBC measurements of Mg may not reflect a true state of Mg depletion unless it is extremely low. The best way to define deficiency was by Mg challenge. The symptoms of Mg deficiency includes back and neck pain, fine tremors, muscle spasm, anxiety and nervousness, spastic vascular phenomena, ventricular arrhythmia, and fatigue. Intravenous Mg challenge appears to be a more accurate assessment of total body magnesium status (0.2 mEq/kgBW magnesium chloride or magnesium sulfate over a 4 hour period). Other symptoms of Mg deficiency include coarse twitching, hypertonicity, carpal pedal spasm, generalised convulsions, tetany, weakness, confusion, personality changes, nausea, anorexia, lack of coordination, GI disorders, alopecia, swollen gums, skin lesions, lesions of small arteries, high blood pressure, strokes, drowsiness and ventricular arrhythmia. Reaction to Mg includes extreme weakness, inability to move, metallic taste, heavy heart. Immediate severe reactions may be as high as 8%.

AUTISM

There is one autistic infant for every 150. There has never been any research conducted on the toxicity of bolus doses of mercury given to infants, yet. The toxicity of mercury is known for more than a century. Methylmercury (inorganic) has long been associated with serious neurological disorders, demyelinating diseases, gut disease, and visual damage. The mercury in vaccines is in the form of thimerosal, which is 50 times more toxic than methylmercury. Thimerosal is organic mercury. Once it is in nerve tissue, converted irreversibly to its inorganic form. Injected mercury is far more toxic than ingested mercury. Mercury accumulates in brain cells and nerves. There's no blood-brain barrier in infants. Infants don't produce bile, which is necessary to excrete mercury. Thimerosal is converted to ethylmercury, an organic form that has a preference for nerve cells. Once in the nerve cells, mercury is changed back to the inorganic form and becomes tightly bound. Mercury remains for years, like a time-release capsule, causing permanent degeneration and death of brain cells. In the late 1930s, autism was identified as a new type of mental disorder, which is the exact time were thimerosal introduced into vaccines. Mercury in thimerosal is stored in the gut, liver and brain, and becomes very tightly bound to the cells. Once inside the cell, mercury can either immediate cell damage or become latent and cause the onset of autism, brain disorders, or digestive chaos years later. α-Lipoic acid crosses the blood-brain barrier and soak up mercury. Standard chelators such as DMSA can handle mercury in the circulation. There is a major synergistic adverse effects in the gut from the combination of measles and rubella vaccines. Mercury induces oedema in the intestinal wall. Mercury induces nodular hyperplasia, which is due to immune response and an autoimmune response. Mercury affects the Krebs' cycle. Defective functioning of metallothionein protein (MT) is a distinctive feature of autism. This abnormality results in impaired brain development and extreme sensitivity to toxic metals and other environmental substances. Abnormal levels of copper and zinc in blood indicates defective functioning of metallothionein (MT) proteins. In humans, MT proteins regulate blood levels of these metals, detoxify mercury and other heavy metals, and assist in neuronal development. Metallothionein is directly involved in neuronal development and maturation of the brain and GIT, and the timing of environmental insults is critically important. By age three, these systems may have sufficiently matured so that environmental toxins can no longer provoke autism. There are 4 primary types of MT protein: MT-I and MT-II are found in cells throughout the body, MT-III restricted primarily to the brain and MT-IV to squamous epithelial cells in the intestines. Metallothionein functioning involves: 1- induction of thioneine, 2- pre-loading with Zn atoms, and 3- redox reactions in which Zn may be displaced by other metals. Metallothionein proteins are induced by Zn, Cu, Cd, and other toxic and nutrient metals. In addition, MT can be induced by injury, emotional stress, and/or nuclear radiation and is an important antioxidant system in the body. A primary mechanism for Zn loading and metal binding is the glutathione (GSH), glutathione disulfide (GSSH) redox couple. Autism can be caused by either: 1- a genetic MT defect, or 2- a biochemical abnormality, which disables MT protein. Mechanisms with the potential for disrupting MT functioning include severe Zn depletion, impaired synthesis of GSH, toxic metal overload, a pyrrole disorder, and a sulfur amino acid abnormality. A defect of MT-I or MT-II would suggest a therapy concentrating on achieving homeostasis of metals in blood. A MT-III defect might indicate the need for regulation of glutamate chemistry in the brain. A MT-IV defect would suggest a focus on restoring proper GIT function. Metal-metabolism abnormalities observed in ADHD, behaviour disorders, and mental illness may result from biochemical imbalances, which impair MT functioning, rather than a direct genetic MT defect. Metallothionein (MT) does not promote the large overgrowths of unfriendly organisms in the intestines so often associated with the use of lipoic acid. Each molecule of MT requires 7 atoms of zinc (Zn) for proper functioning. Premature synthesis of MT at intestinal mucosa can temporarily prevent Zn transport into the bloodstream, resulting in severe irritability. Most of the body's MT is induced by zinc, with glutathione (GSH) needed for loading apo-MT with zinc and glutathione disulfide (GSSH) required for redox exchange. Selenium and the GSH/GSSH redox couple enhance delivery of Zn to cells, and sequestering of mercury and other heavy metals. Zn-MT is a magnet for these toxic metals. MT proteins are composed of 14 amino acids and zinc. The large amounts of cysteine required for MT synthesis can be supplied in the form of oral GSH, which breaks down in the GIT with minimal side effects. MT-promotion therapy is recommended only for patients with disturbed metal metabolism. Key laboratory tests include serum copper, plasma zinc, and serum caeruloplasmin. In healthy individuals, the Cu/Zn ratio usually ranges between 0.8 and 1.2, and the amount of free copper (unbound by caeruloplasmin) ranges from 5 to 25 µg/dl. There is elevated ammonia levels in patients with autism/PDD and related disorders. Stool comprehensive parasitology (aerobic bacteria, yeast, and parasites) and urine organic acid test is recommended prior to the initiation of any therapy. A good trial of the gluten-free, casein-free diet (at least 6 months) is highly recommended. Restoration of good levels of friendly bacteria is mandatory. Aggressive supplementation with Zn (Zinc Loading) and augmenting nutrients for 4-8 weeks is recommended. Plasma zinc levels should be >100 µg/dl to minimize irritability side effects. A daily mg dosage of Zn may be made equal to weight (lbs.) plus 15-20 mg. Pyridoxal-5-Phosphate (P5P), manganese gluconate, and vitamins C and E should be added. Also, taurine may be used for patients with seizure tendencies. These followed by GSH, Se, and the amino-acid MT. 85% of autistic patients have severely elevated Cu/Zn ratios in blood (1.78), suggesting a disorder of metallothionein (MT), and a short linear protein responsible for homeostasis of Cu and Zn. The average Cu/Zn ratio is 1.15 in a normal population. 30% may of these autistic exhibit a pyrrole disorder (urine kryptopyrrole level of 79 µg/dl) associated with severe zinc deficiency. Severe zinc depletion can result in improper induction and functioning of MT. Overall, >99% of the autistic subjects would exhibit clear evidence of a metal-metabolism disorder. An inborn error of metal-metabolism may be a fundamental cause of autism. The metallothionein (MT) family is a group of 4 cysteine-rich metal binding proteins, which are induced and regulated in response to metal toxicity, cellular stress, neuronal development, and inflammation. The primary consequences of MT disorders in a newborn include: 1- abnormal Cu and Zn levels in blood and hippocampus, 2- impaired neuronal development, especially in the first 30 months of life, which could result in incomplete maturation of the GIT and brain, 3- absence of MT's protective detoxification of Cd, Pb, Hg, and other heavy metals, resulting in greatly-increased vulnerability to these toxins, and 4- impaired immune function. Most of the body's MT is induced by zinc, with glutathione (GSH) needed for loading apo-MT with zinc and glutathione disulfide (GSSH) required for redox exchange. Selenium and the GSH/GSSH redox couple enhance: 1- delivery of Zn to cells, and 2- sequestering of mercury and other heavy metals. The equilibrium constants for binding of MT to heavy metals are remarkably large, with the net result that Zn-MT is a magnet for these toxic metals. MT proteins are composed of 14 amino acids and zinc. Many autism-spectrum patients are unable to efficiently cleave dietary proteins into the individual amino acids needed for MT synthesis. Aggressive Zn loading must precede full-scale MT promotion therapy for best results, which is introduced gradually. Each molecule of MT requires 7 atoms of Zn for proper functioning. If MT is formed too rapidly at the intestinal mucosa, temporary severe zinc depletion in the periphery and brain can occur, resulting in irritability and other side effects. The Metallothionein protein benefits to autism-spectrum patients including: 1- elimination of toxic metals, 2- protection against future toxic exposures, 3- normalisation of the GIT, 4- improved behaviour control, 5- improved immune function, and 6- enhanced development of brain neurons and synaptic connections. Multiple insults can result in multiple disabilities. If serious toxic insults are avoided until after age 3, the brain and GIT may have sufficiently matured so that autism is no longer possible. Metallothionein functioning involves: 1- induction of thioneine, 2- pre-loading with Zn atoms, and 3- redox reactions in which Zn may be displaced by other metals. SOD-GP ratio is elevated in autism similar to Down's syndrome, in which elevated SOD (superoxide dismutase) activity leads to greater production of hydrogen peroxide and as a consequence to greater activity of GP (glutathione peroxidase). Autistic children have progressed than schizophrenics towards loss of their ability to adapt to endogenous oxidative stress. Intermediary metabolism of molecular oxygen plays an important role in mental diseases. Vast majority of inherited mental deficiencies-Down's syndrome, autism-may be under considerable oxidant stress, which is the proximate mediator of their symptoms.

PSYCHIATRY CONDITIONS

Depression and somatization are often comorbid states. The brain is connected to the rest of the body, either hard-wired through the central nervous system (CNS) or soft-wired through neurohormonal pathways. Major depression is highly prevalent and a major cause of disability. Depression will rank second only to ischaemic heart disease in terms of disability by 2020. Depression also influences the morbidity and mortality of a number of somatic illnesses. Depression and chronic pain share synaptic monoamine underpinnings. Antidepressants that affect more than one monoamine system (e.g., dual reuptake inhibitors of serotonin and norepinephrine) seem to possess a greater ability to both affect chronic pain states and demonstrate higher rates of symptom remission in major depression. The diagnosis of depression can be delayed or missed altogether. The pain associated with depression (69%) is commonly represented by headache, back pain, or nonspecific musculoskeletal complaints. Anxiety is often comorbid with depressed mood. Irritable bowel syndrome is markedly aggravated in depressed patients. Depression may amplify the intensity or distress attributable to an existing complaint or serve as the basis for any one of a number of pseudonymous diagnoses aliases for depression that may be specific. Chronic or functional pain, irritable bowel syndrome, disequilibrium, hypoglycaemic episodes, hormone imbalances, fibromyalgia, and the like may reflect the diagnosis of depression. Depressed diabetics tend to have poorer blood glucose control. Stroke victims suffering from depression do more poorly in rehabilitation programs. Depression after an acute myocardial infarction is associated with higher mortalities independent of the severity of cardiovascular function. Depression may reduce the life expectancy of older adults. Hospitalised patients with comorbid depression are less likely to survive life-threatening illnesses. Serotonin and norepinephrine (NE) are associated with depression. Serotonin and NE also modulate pain sensitivity via the descending pain pathway. Descending modulatory pathways mediated by serotonin, norepinephrine, and γ-amino butyric acid (GABA) may be used to limit the intensity of pain signals arriving in the brain from the body via spinal pathways. Dysregulations in the hypothalamic-pituitary-adrenal (HPA) axis and reductions in brain-derived neurotrophic factor (BDNF) are common in depressed patient. BDNF levels in the hippocampus are reduced by increased levels of endogenous glucocorticoids often seen in depressed patients. BDNF appears to retard neuronal atrophy under stress, and a loss of such protection may inhibit accommodation and response to stress that involves both neuron preservation and neurogenesis. Antidepressants may increase levels of BDNF. The goal of depression treatment is robust, sustained response-complete and sustained remission of depression symptoms. Failure to achieve the robust level of symptom remission is associated with disability, failure to achieve normal social functioning, and risk of relapse. Depression is a common condition after MI (20-27%). Even mild depression is associated with increased mortality. The pathobiology of post-MI depression is poorly understood. Atherosclerosis and depression may be linked via common mechanisms, such as altered serotonin metabolism, decreased intake of ω-3 fatty acids, and immune activation. Vitamins and minerals were first used to treat illnesses unrelated to nutrient deficiency in the 1920s. Malnutrition and improper nutrition could place a person at risk, directly causing or contributing to the development of disease and psychiatric disorders.

Symptoms of depression:

Emotional symptoms (mood)	Physical symptoms
1- Sadness and tearfulness.	1- Tiredness/fatigue.
2- Loss of interest.	2- Sleep disturbances.
3- Anxiety/irritability.	3- Headaches.
4- Hopelessness.	4- Psychomotor activity changes.
5- Concentration difficulties.	5- GIT disturbances.
6- Guilt.	6- Appetite changes.
7- Suicidal ideation.	7- Body aches and pains.

Neurotoxicity encompasses injury from chemicals to the brain and to peripheral nerves. As the brain is the body's master controller, primary effects on it affect many bodily functions. Neurotoxicity is the damage to the brain and/or the peripheral nervous system from toxic chemicals. These chemicals include alcohol, lead, insecticides, solvents, lead, mercury, cadmium, car exhaust, formaldehyde, chlorine, phenol, rotten egg gas, ammonia, polyvinyl chloride (PVC) and polychlorinated biphenyls (PCBs), and thousands of others. Symptoms for the brain toxicity are short-term memory loss, loss of circulation, imbalance, and altered feeling or moods (anxiety, depression, confusion, anger, extreme fatigue, short term memory loss, vertigo, and flu-like symptoms with lack of concentration). For the peripheral the symptoms are numbness, tingling, loss of sensation and movement. The best test for the peripheral are the neurometer, and nerve conduction test. For the brain, the triple camera SPECT brain, the computerised balance, the pupillography, the heart rate variability, and a battery of psychiatric tests. Treatment consists of a massive avoidance of pollutants in air, food and water including specific toxic substances that the patient is commonly around, intravenous and/or nutrition, and injection therapy for the substance the patient has become sensitive to. These include food, biological inhalants and some chemicals, moderate temperature sauna, massage and exercise under environmentally controlled conditions. Often immune modulation is necessary using the patient's own autogenous lymphocytes. Mental symptoms of avitaminosis sometimes are observed long before any physical symptoms appear. The functioning of the brain is affected by the molecular concentrations of many substances that are normally present in the brain. Orthomolecular psychiatry is varying the concentrations of substances normally present in the human body, which may control mental disease. It is likely that the brain is more sensitive to changes in concentration of vital substances than any other organs. Orthomolecular psychiatry is the achievement and preservation of good mental health by the provision of the optimum molecular environment for the mind, especially the optimum concentrations of substances normally present in the human body, such as the vitamins. Biochemical and genetic arguments support the idea that orthomolecular therapy may be the preferred treatment for many mentally ill patients. The brain is affected by its molecular environment. The functioning of the brain is dependent on the molecular environment of the mind, i.e., composition and structure. Molecules of many different substances are essential to prevent/intervene with mental disease. Mental diseases are usually associated with physical diseases, results from a low concentration or changes of any one of the following substances in the brain: thiamine (B_1), nicotinic acid or nicotinamide (B_3), pyridoxine (B_6), cyanocobalamin (B_{12}), biotin (H), ascorbic acid (C), folic acid, L-glutamic acid, uric acid, and γ-aminobutyric acid (GABA). The rate of an enzyme-catalysed reaction is approximately proportional to the concentration of the reactant, until concentrations that largely saturate the enzyme are reached. The saturating concentration is larger for a defective enzyme with decreased combining power for the substrate than for the normal enzyme. The functioning of the brain and nervous tissue is more sensitively dependent on the rate of chemical reactions than the functioning of other organs and tissues. B_{12} deficiency may lead to mental disease. Nicotinic acid and nicotinamide are nontoxic to humans (up to 200 mg). Mental symptoms (depression) accompany the physical symptoms of vitamin-C deficiency disease (scurvy). Schizophrenics have a great power of oxidising N,N dimethyl-p-phenylenediamine, which is due to a smaller concentration of ascorbic acid. Poor diet and increased tendency to chronic infectious disease of the patients increases the rate of metabolism of ascorbic acid. Schizophrenics have an increased metabolism of ascorbic acid, presumably genetic in origin, and that the ingestion of massive amounts of ascorbic acid has some value in treating mental disease. Other vitamins (thiamine, pyridoxine, and folic acid) and other substances (zinc ion, Mg ion, uric acid, tryptophan, L-glutamic acid, and others) influence the functioning of the brain. L-glutamic acid is an amino acid that is present at rather high concentration in brain and nerve tissue and plays an essential role in the functioning of these tissues. Psychosis associated with pernicious anaemia may manifest itself in a patient for several years before the other manifestations of this disease become noticeable. The functioning of the brain and nervous tissue is probably more sensitively dependent on molecular composition than is that of other organs and tissues. Mental manifestations of avitaminosis: 1- Schizophrenia results from poor nutrition, e.g., pyridoxine, niacinamide, and ascorbic acid. 2- Shifting the equilibrium for the reaction of apoenzyme and coenzyme to give the active enzyme. Vitamins serve as coenzymes in the brain. Increase in concentration of the coenzyme can counteract the effect of the decrease in the value of the combining constant and lead to the formation of enough of the active enzyme to catalyse effectively the reaction for instant of conversion of methylmalonic acid to succinic acid. Many individuals may require a considerably higher concentration of one or more coenzymes than other people do for optimum health, especially for optimum mental health. Some vitamins are known to serve as coenzymes for several enzyme systems. 90% or more of the protein is converted to active enzyme. Intervention for schizophrenia involves the use of vitamins (megavitamin therapy) and minerals; the control of diet, especially the intake of sucrose; and, during the initial acute phase, the use of conventional methods of controlling the crisis, such as the phenothiazines. I g of ascorbic acid, I g of niacinamide, 50 mg of pyridoxine, and 400 I.U of vitamin E four times a day had been recommended. Other vitamins may also be given. A larger intake, especially of niacinamide or niacin may be prescribed; the usual amount seems to be about 8 g a day after an initial period on 4 g a day. Ascorbic acid is an important substance necessary for optimum functioning of many organs. Chronic psychiatric patients would benefit from the administration of ascorbic acid, e.g., depressive, manic, and paranoid symptoms-complexes, schizophrenia. More than 76% of schizophrenic patients are deficient in ascorbic acid. Pyridoxine, vitamin B_6 is used in the treatment of schizophrenia in amounts of 200-800 mg a day by many orthomolecular psychiatrists, derivatives of this vitamin are known to be the coenzymes for over 50 enzymes. Pyridoxine is involved in tryptophan-niacin metabolism. Pyridoxine may potentiate the action of nicotinic acid. A deficiency in cyanocobalamin (vitamin B_{12}), whatever its cause, leads to mental illness as well as to such physical manifestations as anaemia. 61% of mentally ill patient have a subnormal or pathologically low concentration of vitamin B_{12}, <150 pg/ml (the normal range is 150-1,300 pg/ml.). Also mental illness may accompany some genetic diseases, such as methylmalonic aciduria, which can be controlled only by achieving a serum concentration of cyanocobalamin far greater than normal. Some vitamins, including vitamin C, are not known to be transformed into a coenzyme. Mental disease, usually associated with physical disease, results from a low concentration in the brain of any one of the following vitamins: thiamine (B_1), nicotinic acid or nicotinamide (B_3), pyridoxine (B_6), cyanocobalamin (B_{12}), biotin (H), ascorbic acid (C), and folic acid. There is evidence that mental function and behaviour are also affected by changes in the concentration in the brain of any of a number of other substances that are normally present, such as L-glutamic acid, uric acid, and γ-aminobutyric acid. Humans can not synthesise ascorbic acid and that accordingly requires it in the diet. There is the possibility that for some persons the cerebrospinal concentration of a vital substance may be grossly low at the same time that the concentration in the blood and lymph is essentially normal. The mechanisms of enzyme-catalysed reactions in general involve: A- the formation of a complex between the enzyme and a substrate molecule, and B- the decomposition of this complex to form the enzyme and the products of the reaction. The rate determining step is usually the decomposition of the complex to form the products, or, more precisely, the transition through an intermediate state of the complex, characterised by activation energy less than for the uncatalyzed reaction, to a complex of the enzyme and the products of reaction, with a rapid dissociation. The rate of an enzyme-catalysed reaction is approximately proportional to the concentration of the reactant, until concentrations that largely saturate the enzyme are reached. The saturating concentration is larger for a defective enzyme with decreased combining power for the substrate than for the normal enzyme. For such a defective enzyme the catalysed reaction could be made to take place at or near its normal rate by an increase in the substrate concentration. The functioning of the brain and nervous tissue is more sensitively dependent on the rate of chemical reactions than the functioning of other organs and tissues. Significant improvement in the mental health of many persons might be achieved by the provision of the optimum molecular concentrations of substances normally present in the human body. A deficiency of vitamin B_{12}, whatever its cause (pernicious anaemia; infestation with the fish tapeworm Diphyllobothrium, whose high requirement for the vitamin results in deprivation for the host; excessive bacterial flora, also with a high vitamin requirement, as may develop in intestinal blind loops), leads to mental illness, often even more pronounced than the physical consequences. The mental illness associated with pernicious anaemia (a genetic defect leading to deficiency of the intrinsic factor [a mucoprotein] in the gastric juice and the consequent decreased transport of cyanocobalamin into the blood) often is observed for several years in patients with this disease before any of the physical manifestations of the disease appear. Vitamin B_{12} deficiency, whatever its origin, may lead to mental illness. Nicotinic acid (niacin), when its use was introduced cured hundreds of thousands of pellagra patients of their psychoses, as well as of the physical manifestations of their disease. The use of nicotinic acid and nicotinamide for the treatment of mental disease is essential. Nicotinic acid and nicotinamide are nontoxic (the lethal dose, 50% effective [LD50], is not known for humans, but probably it is over 200 g; the LD50 for rats is 7.0 g/kg for nicotinic acid, and 1.7 g/kg for nicotinamide), and their side effects, even in continued massive doses, seem not to be commonly serious. The advantages of nicotinic acid are the following: it is safe, cheap, and easy to administer, and it is a well-known substance that can be taken for years, if necessary, with only small probability of incidence of unfavourable side effects. Mental symptoms (depression) accompany the physical symptoms of vitamin-C deficiency disease (scurvy). Many schizophrenics have an increased metabolism of ascorbic acid, presumably genetic in origin, and that the ingestion of massive amounts of ascorbic acid has some value in treating mental disease. Other vitamins (thiamine, pyridoxine, folic acid) and other substances (zinc ion, magnesium ion, uric acid, tryptophan, L-glutamic acid, and others] influence the functioning of the brain. L-Glutamic acid is an amino acid that is present at rather high concentration in brain and nerve tissue and plays an essential role in the functioning of these tissues. It is normally ingested (in protein) in amounts of 5-10 g/day. It is not toxic; large doses may cause increased motor activity and nausea. L-Glutamic is more effective than its sodium or potassium salts. The effective dosage is usually between 10-20 g/day (given in three doses with meals), and is adjusted to the patient as the amount somewhat less than that required to cause hyperactivity; improvement in personality and increase in intelligence have been reported for many patients with mild or moderate mental deficiency by several investigators. A deficiency disease may be localised in the human body in such a way that only some of the manifestations usually associated with the disease are present (a localised cerebral avitaminosis or other localised cerebral deficiency disease). There is the possibility that some human beings have a sort of cerebral scurvy, without any of the other manifestations or a sort of cerebral pellagra or cerebral pernicious anaemia. Every vitamin, every essential amino acid, every other essential nutrilite represents a molecular disease. The localised mental deficiency diseases are molecular diseases, compound molecular diseases, involving not only the original lesion, the loss of the ability to synthesise the vital substance, but also another lesion, one that causes a decreased rate of transfer across a membrane, such as the blood brain barrier, to the affected organ, or an increased rate of destruction of the vital substance in the organ, or wine other perturbing reaction. A physiological abnormality such as decreased permeability of the blood brain barrier for the vital substance or increased rate of metabolism of the substance in the brain may lead to a cerebral deficiency and to a mental disease. Diseases of this sort may be called localised cerebral deficiency diseases. The morbidity and mortality associated with depression are considerable and continue to increase. Depression is the fourth major cause of disability worldwide, after lower respiratory infections, perinatal conditions, and HIV/AIDS. 17% of people will suffer from depression during their lifetime. The incidence of depression may be 30% to 50% in patients depending on the specific medical condition. Despite the clear decrease in quality of life (QoL) and decreased productivity associated with depression, it is often underdiagnosed and inadequately treated. Diet plays a major role in controlling schizophrenia and other types of mental conditions. 60% of individuals who suffer from schizophrenia have allergies to wheat and/or dairy. 4% of schizophrenia is the result of wheat allergies alone. These individuals tend to have symptoms that worsen after eating the offending food and tend not to respond favorably to medications. Wheat and dairy contain types of proteins that mimic endorphins, mood elevating hormones, which are naturally produced in the body. For the vast majority of people, this doesn't cause a problem as these proteins to have an effect; they need to cross the blood brain barrier. With susceptible individuals, their blood brain barrier allows wheat and dairy proteins to pass through. When this happens, there is an increased effect of endorphins, which can induce schizophrenic-like symptoms. Eating high carbohydrate meals and snacks often leads to high insulin output by the pancreas leading to hypoglycaemia. Hypoglycaemic symptoms may include sugar-craving, fatigue, sweating, trembling, slurred speech, anxiety, visual disturbances, headaches, irritability, rear, strong temper and palpitation. Blood sugar imbalances can significantly aggravate mental symptoms, such as depression, anxiety, fear, and anger. Caffeine and alcohol can also aggravate sugar imbalance. When vitamins B levels are sub-optimal, there is increased risk for depression, anxiety and other mood disorders. With schizophrenia, it is likely that the requirements for vitamins B are drastically increased. Vitamins B are extremely important for the regulation of neurotransmitters, such as serotonin and dopamine. Vitamins B are also responsible for optimal sugar metabolism. Zinc, manganese and magnesium are also nutrients that are extremely important for the regulation of neurotransmitters. These minerals are important co-factors to enzymes that exist throughout the body. ω-3 fatty acids have been found to be deficient in about 60% of schizophrenic. These fatty acids are essential for integrity of cell membranes, cellular energy, formation of enzymes and maintaining inflammatory mediators under control. Omega-3 fatty acids improve cognitive function. Heavy metal toxicity and environmental toxins play a crucial role in the pathobiology of mental diseases. In general, there are 3 major types of schizophrenia, based on biochemical imbalances: 1- Histapenia (low histamine), 2- Histadelia (high histamine) and, 3- Pyroluria (elevated kryptopyrrole). Regardless of the type of treatment that you utilize, perhaps the biggest obstacle is adherence, and this is especially true with schizophrenia. It is important to be focused and persistent when a patient's symptoms are in remission. Treatment also needs to be long-term, at least 6-12 months, before deciding whether or not any protocol is having an effect. A significant number of diseases including mental disease may have an infectious disease component to the pathological process. For example, the link between psychopathology and infectious disease include with Lyme disease, syphilis, babesiosis, ehrlichiosis, mycoplasma pneumonia, toxoplasmosis, borna virus, AIDS, CMV, herpes, and other unknown infectious agents. Disease begins with vulnerability and an exposure to one or more stressors. The vulnerability may commonly include genetic and/or increased vulnerability as a result of chronic stress. As a result of these and other vulnerabilities, the microbe more easily penetrates the host's defences and an initial infection may then occur. The course of the infection most relevant to psychiatry would be chronic, low-grade, persistent infections or the persistence of the infectious agent in the inactive state. Neural injury may occur by a variety of mechanisms, which include vasculitis, direct cell injury, inflammation, autoimmune mechanisms and excitotoxicity. This injury leads to a vicious cycle of disease resulting in dysfunction of associative and/or modulating centers. Injury to associative centers more commonly causes cognitive symptoms, while injury to modulating centers more commonly causes emotional and allocation of attention disorders. In some cases, the infection and injury may occur before birth. The majority of patients demonstrating psychiatric symptoms in response to infectious disease are infected by Lyme disease and quite often co-infected with other agents. Psychiatric syndromes caused by infectious disease include depression, OCD, panic disorder, social phobias, variants of ADD, impulse control disorders, bi-polar disorders, eating disorders, dementia, various cognitive impairments, psychosis and a few cases of dissociative episodes. In the more obvious cases, symptoms are present with cognitive, neurological and physical signs. Laboratory tests are sometimes, but not always, able to confirm the diagnosis. Enzymes are substances that speed up chemical reactions. There are at least 4 nutritional causes of mental problems including: 1- Protease deficiency (difficulty digesting protein). 2- Sugar intolerance, i.e., the inability to digest disaccharides (common sugars like sucrose) to simple sugars. 3- Hypothyroidism. 4- Junk-food diet laced with sugar. Protease deficiency leads to a build-up of excess alkaline reserves because there is inadequate digested protein to supply enough acidity. The person becomes anxious and sighs in an attempt to restore the acid-base balance. Everyone is sugar intolerant to some degree. Sucrose intolerance can lead to mental and emotional problems, such as panic attacks, depression, insomnia, mood swings that can progress to a bipolar disorder, and a tendency towards irritable, aggressive or violent behaviour. There are herbs, which have a stimulating effect on thyroid function. Sugar intolerance can lead to severe mental and emotional problems. Patients may have panic attacks, horrible nightmares, mood swings, and violent behaviour and be diagnosed as schizophrenic, bipolar or manic and so on. Elevated prostaglandin (PG) levels have been observed in, for example, pre-menstrual syndrome (PMS). PGE_2 is a central nervous system depressant. PGs require selenium for their synthesis; it's believed PGs deficiency may be a source of schizophrenia. Endorphins, discovered in 1975, are substances secreted in the brain. They have a pain-relieving and stress-relieving effect similar to morphine. Endorphin molecules lock onto receptors in the brain to remove the perception of pain. Gluten molecules are molecularly similar in shape to endorphins and thus can create the same stimulatory/suppression activity. Certain dairy proteins have been shown to have similar qualities. Serine is an amino acid, which plays a critical role in maintaining blood sugar levels. It has a vital part in the production of the myelin sheath. Serine metabolism may be abnormal in psychotics. Prolactin is a hormone also known as luteotropic hormone, which is secreted by pituitary gland and induces lactation. Excess prolactin has been connected with pre-menstrual syndrome and a host of extreme mental states that can occur with it. Tranquillisers can increase prolactin levels. Patients who undergo dialysis regularly may be exposed to high levels of aluminium in dialysis fluids and medicines, which induce dialysis encephalopathy. A progressive mental degeneration manifested by tremors, convulsions, psychosis and other changes in speech and behaviour follow. 1-2% of interferon users manifests psychosis or suicidal behaviour. Histamine is a chemical produced by the breakdown of histidine in the body tissues. It is released in allergic reactions and causes dilation of capillaries, decreased blood pressure, increased release of gastric juice, fluid leakage forming itchy skin and hives, and tightening of smooth muscles of the bronchial tube and uterus. Histadelia, more common in males, is characterised by elevated blood levels of histamine. It is estimated that 15-20% of schizophrenics are probably histadelic. The treatment of histadelia requires great patience because 6-10 weeks are often needed before the beginning of significant improvement. The treatment usually takes 12 months to complete. Folic acid increases depression in histadelic patients and a trial of folic acid could be used to distinguish between histapenics and histadelics. Copper is part of the enzyme histaminase, which is involved in the metabolism of histamine. Copper levels may be low to normal in patients with histadelia. Histidine, which is more common in animal proteins, should be avoided as it can be converted into histamine. Methionine supplements lower blood levels of histamine by increasing histamine breakdown. Histadelics are often prone to obsessions, compulsions, and addictions (OCD). Histadelics are often chronically and suicidally depressed, often having allergic rhinitis/hay fever, excess perspiration and warm skin, excess mucous (histamine can cause additional mucus production), and joint pain/swelling/stiffness. Histadelics may have strong sexual desire or weak sexual desire, poor pain tolerance or good pain tolerance, excess/abundant saliva in mouth or very dry mouth, good tolerance of cold and poor tolerance of heat or good tolerance of heat and poor tolerance of cold. Histamine speeds up metabolism producing a tendency towards hyperactivity. Histadelics may suffer from phobias or lack of phobias. Being highly motivated, histadelics tend to work compulsively. A hard-driving personality, histadelics tend to work compulsively. Good creativity/imagination, histadelics are often highly creative. 30-70 % of people with schizophrenia has pyrroluria. In some people, one specific pyrrole, 2,4-dimethyl-3-ethylpyrrole, is produced in excess and excreted in the urine. 2,4-dimethyl-3-ethylpyrrole interacts with vitamin B_6, Zn, and Mg leaving these nutrients deficient. Pyrrole disorder individuals have problems forming serotonin, dopamine, GABA, norepinephrine, and glycine. Under-methylation syndrome often involves seasonal variations in depression, obsessive-compulsive behaviour, inhalant allergies, and frequent headaches. In severe cases involving psychosis, the dominant symptom is usually delusional thinking rather than hallucinations. They tend to speak very little and may sit motionless for extended periods. They may appear outwardly calm, but suffer from extreme internal anxiety. Most OCD patients with both obsessive thoughts and compulsive actions are in this category. Associated with under-methylation, which results in low levels of important neurotransmitters such as serotonin, dopamine and norepinephrine. Under-methylation (high histamine) results in low levels of important neurotransmitters such as serotonin, dopamine and norepinephrine. Treatment focuses on the use of antifolates such as calcium, methionine, SAMe, magnesium, zinc, TMG, ω-3 oils, B_6, inositol, and vitamin A, C and E. The dose of inositol is 500-1000 mg. Choline is anti-dopaminergic and often makes undermethylated patients worse. Also bad are DMAE, copper and folic acid. Three to six months of nutrient therapy are necessary to correct this chemical imbalance. Symptoms will return if treatment is stopped. Vitamin B_6 is needed for proper brain detoxification. Zinc is essential in nerve development, intellectual function, serotonin formation, the regulation of mood, and the prevention of oxidative damage. In order for vitamins to be utilised by the body, they must first be converted into their active coenzyme forms. Kryptopyrrole binds to pyridoxine (vitamin B_6) and zinc and makes them unavailable for their important roles of co-factors in enzymes and metabolism. Individuals with pyrroluria disorder produce too much of a by-product of haemoglobin synthesis called kryptopyrrole (KP) or haemepyrrole. Kryptopyrrole has no known function in the body, and is excreted in urine. The effect of pyrroluria can be mild, moderate, or severe depending on the severity of the imbalance. Most individuals show symptoms of zinc and/or B_6 deficiencies, which include poor stress control, nervousness, anxiety, mood swings, severe inner tension, episodic anger (an explosive temper), poor short-term memory and depression. Most pyrrolurics exhibit at least two of these problems. These individuals cannot efficiently create serotonin since vitamin B_6 is an important factor in the last step of its synthesis. Pyrroluria is detected by chemical analysis of the abnormal pyrroles in urine detectable as a purple (on testing paper) metabolite called the mauve factor. Normal value is <10 µg of KP/dl. Persons with 10-20 µg/dl are considered borderline pyrroluric and may benefit from treatment. Persons with levels >20 µg/dl are considered to have pyrroluria, especially if the above symptoms are present. The chemical analysis for KP is difficult due to the tendency for this chemical to decompose, and no vitamins or minerals should be taken for 2 days before the test. Sometimes it is necessary to repeat the urine test. Pyrrolurics are stress intolerant and especially vulnerable to cumulative stress over many days. Pyrroluric patients are prone to relapses, especially during illness, injury, or emotional stress. Vitamin B_6, zinc, niacinamide, pantothenic acid, manganese, vitamins C and E, ω-6 fatty acids and cysteine are essential in treating pyrroluric. Pyrroluria is a genetic condition resulting in an abnormality in haemoglobin synthesis. The excess amount of kryptopyrrole (2,4-dimethyl-3-ethylpyrrole) is found in the urine of individuals with pyrroluria. The simplest test for the presence of pyrroluria is using Erhlich's reagent. This chemical is added to the urine, which turns a mauve colour if significant amounts of kryptopyrroles are present. A more accurate test directly measures urine kryptopyrrole. Pyrroluria is a form of schizophrenic porphyria, similar to acute intermittent porphyria where both pyrroles and porphyrins are excreted in the urine in excess. This abnormality leads to a higher excretion of vitamin B_6 (pyridoxine) and zinc in the urine, with deficiencies or borderline deficiencies common. Changes in fatty acid metabolism often lead to low levels of arachidonic acid (a ω-6 fatty acid). The presence of pyrroluria can have a profound effect on mental and physical health and was first discovered in relation to schizophrenia. Pyrroluria is not a recognised biochemical abnormality by the orthodox psychiatric community. Psychiatric disorders strongly associated with alcohol disorders include other drug abuse or dependence, major depression, simple phobia, antisocial personality disorder, tobacco dependence, and pathological gambling. The brain consumes about 120 g of glucose daily, which corresponds to an energy input of about 420 kcal. The brain accounts for some 60% of the utilization of glucose by the whole body in the resting state. Glucose ingestion may affect cellular immunity by increasing serum insulin, which complexes with mitogens for binding sites on lymphocytes. Glycerol can be converted into pyruvate or glucose in the liver. One tablespoon of Glycerine mixed in a glass of water and a dash of lemon for taste may help in withdrawing from sugar. Chronic use of alcohol by mental patients undergoing pharmacotherapy with neuroleptics enhances the vulnerability of these patients to TD. Tardive dyskinesia (TD) is a syndrome of potentially irreversible, involuntary, dyskinetic movements, which may develop in patients who have been treated with antipsychotic medications. Patients with TD suffer from repetitive and uncontrollable repetitive movements that can interfere greatly with their quality of life. The supplementation of vitamin E, choline, lecithin, dimethylanimoethanol (DMAE), manganese, Evening primerose oil (EPO) may help controlling the condition. Individuals with TD have higher levels of the amino acid phenylalanine. Phenylalanine (PHE) stimulates the production of cholecystokinin (CCK), which induces satiety (stops hunger) and control appetite, control overweight obesity. Supplementing with branched-chain amino acids (BCAA), including valine, isoleucine, and leucine, could reduce excess phenylalanine in people with this disorder. Vitamin C (4 g/day), vitamin B_3 (4 g/day), vitamin B_6 (800 mg/day), and vitamin E (1,200 IU/day) may prevent TD altogether. Quercetin has a protective role against haloperidol-induced orofacial dyskinesia. The atypical neuroleptic quetiapine improves symptoms within 2 weeks after starting and stops completely TD within 10 weeks. Schizophrenic who eat more ω-3 fatty acids in their normal diet have less severe symptoms. Both schizophrenic symptoms and TD improve over a 6 week period. Vanadium has an insulin-like effect. Vanadium does not restore plasma insulin levels but may spare pancreatic insulin. Elevated vanadium has been reported in the plasma of patients with mania and depression and in the hair of patients with mania. Manic patients may respond significantly better to lithium than to ascorbic acid and EDTA. Vitamin C may possibly reduce symptoms of bipolar disorder due to its ability to the detrimental effects on vanadium on RBC Ca^{2+}-K^+-ATPase activity by reducing vanadate^{+5} to the vandyl ion^{+4}, the latter being much less effective inhibitor of Na^+-K^+-ATPase activity than the former. Both manic and depressed patients would be better following a single 3 g dose of vitamin C.

Glycerol (glycerine) → Dihydroxyacetone phosphate → Glyceraldehyde 3-phosphate → Pyruvate or Glucose.

DIABETES AND PSYCHIATRIC ILLNESS

The prevalence of type 2 diabetes has increased dramatically over the past decade. Impaired glucose tolerance is a condition that often evolves into type 2 diabetes. Patients with type 1 and type 2 diabetes have similar microvascular (retinopathy, nephropathy, and neuropathy) and macrovascular (myocardial infarctions, cerebrovascular disease, and peripheral vascular disease) complications. Individuals with diabetes are at least 2-4 times more likely to suffer from premature ischaemic heart disease, myocardial infarction, heart failure, ischaemic cerebrovascular disease, and peripheral vascular disease, and demonstrate a higher mortality rate from these conditions compared to euglycaemic patients. Cardiovascular events in this population are associated with a poor rate of recovery and a high rate of disability when compared to the nondiabetic population. The severity and duration of hyperglycaemia is clearly accountable for the microvascular complications in both type 1and type 2 diabetes. The clinical combination of hypertension and diabetes carries a particularly poor prognosis. The obesity seen in type 2 diabetes is central in distribution, with a large amount of abdominal or visceral fat that is rich in highly metabolically active adipocytes that are responsible for formation and release of toxic substances (e.g., TNF-α, plasminogen type activator inhibitor-1 [PAI-1], IL-6, serum amyloid A, and asymmetric dimethylarginine, an endogenous nitric oxide synthase inhibitor) and that play a role in atherogenesis. Leptin and resistin are adipocyte hormones associated with insulin resistance and obesity. High leptin levels are found in patients with obesity and type 2 diabetes, most likely a consequence of a metabolic maladaptation. The novel cytokine resistin can be modified by peroxisome proliferator-activated receptors (PPARs) ligands. Insulin resistance and central obesity are also accompanied with an increased flux of free fatty acids and a large pool of very-low-density lipoprotein, with a characteristic dyslipidaemia of elevated triglycerides, and low high-density lipoprotein (HDL) cholesterol. Impaired fibrinolysis includes a high production and activity of PAI-1, together with low tissue plasminogen activity. The management of individuals with type 2 diabetes needs to be early and aggressive, and a global approach. Treatment of dyslipidaemia reduces cardiovascular morbidity and mortality, and aspirin therapy is also beneficial. Healthy lifestyle changes are important in disease prevention and remain the cornerstone for treatment of type 2 diabetes. Diet and exercise alone fail to normalise hyperglycaemia in the majority of patients, and pharmacotherapy is necessary. Some poison in food may be causing diabetes. In spite of years of experience in injecting insulin into the body to thwart IDDM defect, diabetics have poor health. Since insulin, as is the case with most hormones, has a half-life in the blood stream, the concentration in the body varies over wide extremes. When large amounts of insulin enter the body, serum sugar content drops precipitously, which associated with a sudden upsurge in cortisol secretion. The result is the degradation of many proteins in order to produce glucose from them. The connective tissue is especially degraded. Insulin indirectly may increase hydrogen ion or acidity (decreasing pH), which could be an important part of health degradation. When glucose enters the cell, it takes potassium with it, which result in a dangerous, potentially lethal low serum K. The serum K must be kept at 4.8 mEq/l (187 mg/l) for nerve impulses to travel. Individuals with K deficiency secrete less insulin. High serum concentrations of cortisol cause greater excretions of K. Arthritics have low whole body K content. Low insulin may cause greater excretion of K and Mg because of affects on the energy furnished to their cell wall pumps. Total body K is always depleted in diabetic ketoacidosis (DKA) but serum/plasma K may be low, normal or high. It may be disadvantageous to use potassium chloride. Hyperglycaemia results in glycosylation of protein, which is the cause of health problems during diabetes. The acidity of cell fluid of diabetics, which shows up even when the blood serum is not abnormally acidic, may be a large part of those health problems. Thus, it may be disadvantageous to eat foods that have unmetabolizable (not burned) anions, which many fruits and some vegetables have. This should include vinegar and citrus fruit, even though both the acetate and citrate are at least partly metabolised. Chilli pepper comes from a family of plants many of whose members have very vicious poisons. Capsaicin and resiniferatoxin kill human skin cancer cells. Lectins, plant proteins that bind certain carbohydrates, have been proposed as making certain cells susceptible to autoimmune destruction. Dioxin is the compound in agent orange is linked to health effects. Another poisons, which may yet be shown to predispose to diabetes, are olanzapine, clozapine, risperidone, quetiapine and ziprasidone. The metabolic abnormalities in patients using olanzapine ranging from mild blood sugar problems to diabetic ketoacidosis and coma. Diabetic ketoacidosis (DKA) is a serious condition in which a person experiences an extreme rise in blood glucose level coupled with a severe lack of insulin, which results in symptoms such as nausea, vomiting, stomach pain and rapid breathing. Untreated, DKA can lead to coma and even death. There is an increased requirement for nutrients in normal pregnancy, and also increased loss. There is also an increased insulin-resistant state during pregnancy mediated by the placental anti-insulin hormones oestrogen, progesterone, human somatomammotropin; prolactin; and cortisol. Gestational diabetes is associated with excessive nutrient losses due to glycosuria. Specific nutrient deficiencies of chromium, magnesium, potassium and pyridoxine may potentiate the tendency towards hyperglycaemia in gestational diabetic women because each of these 4 deficiencies causes impairment of pancreatic insulin production. During juvenile diabetes, there are lower levels of reduced glutathione, ceruloplasmin oxidase activity, Zn, Cu and Na. Zinc and arachidonic acid (AA) lower glucose via improvement in insulin sensitivity. Copper (Cu) depletion doubled glucose in the blood. Build-up of Cu in the kidneys of diabetics is responsible for the renal damage. Diabetics probably absorb Cu twice more readily than normal individuals. The pancreas can be irreversibly destroyed by Cu deficiency, but the isles of Langerhan are not affected. There is negative correlation between Cu in drinking water and onset of juvenile diabetes. Adequate Cu could help prevent insulin dependant diabetes. Once insulin dependant diabetes sets in handling Cu may be a problem and may be part of the difficulty with poor health afflicting diabetics. Inositol may increase sensitivity to insulin; inositol is required for the electrolyte cell wall pumps. Diabetes is becoming an explosive epidemic. Type 2 diabetes is associated with combination of insulin resistance (IR) and diminished β-cell function. The percentage overweight is increasing in about one third of the population (1%/annum), 31.7% in men and 34.9% in women. Obesity is related to insulin resistance, which leads to the metabolic syndrome, or so-called syndrome X, with a number of components that lead to disease that causes a great deal of morbidity and mortality. Insulin resistance is associated with hyperinsulinaemia and decrease in insulin sensitivity. The insulin resistance tends to get worse as the BMI, or as the overweight/obesity increases. Components of metabolic syndrome include insulin resistance, hyperinsulinaemia, central obesity, hypertension, glucose intolerance, increased serum VLDL triglyceride, decreased serum HDL cholesterol, hyperuricaemia, increased risk of macrovascular disease. The inability of insulin to let glucose into the cell and getting glucose utilised, oxidised, and stored, means that the β cell of the pancreas is putting out 5- to 6-fold more insulin/day in these individuals. Abdominal or android fat distribution (fat distribution in the central or upper body) is significantly more risky for health than gluteal femoral or lower body or gynecoid obesity. Fat distribution also increases the amount of insulin that is circulating. Because with fatness and intra-abdominal fat and fat in the liver, less insulin is degraded by as the insulin passes through the liver, and more of it can flux into the peripheral circulation. Obesity is also related to hypertension risk. As the weight or the total fat burden increases, hypertension increases accordingly (BMI of 31 and above). Dyslipidaemia is high triglyceride/low HDL cholesterol phenotype, and characterised also by dense low-density lipoprotein (LDL) cholesterol particles. Dyslipidaemia even slight is very atherogenic. Part of the reason for this increased cardiovascular risk with insulin resistance (IR) is due to the hypercoagulability that occurs in this condition. Hypercoagulability in IR include: 1- increased PAI-1, 2- increased fibrinogen, 3- increased von Willebrand factor, 4- increased factor X, and 5- increased platelet aggregation. PAI-1 is important in clearing clots from the circulation. Increased fibrinogen is important in enhancing clotting, as is the case with increased von Willebrand factor, increased factor X, and increased platelet aggregation, which occurs in this condition. Depression causes either a diminished or an increased appetite in different individuals. So it can cause subsequent weight loss or subsequent weight gain. Successful antidepressant treatment may reverse such changes over the short term.

Main causes of disability:

Both sexes	Female	Male
1- Major depression.	1- Major depression.	1- Alcohol use.
2- Alcohol use.	2- Schizophrenia.	2- Road traffic accidents.
3- Road traffic accidents.	3- Road traffic accidents.	3- Major depression.
4- Schizophrenia.	4- Bipolar disorder.	4- Self-inflicted injuries.
5- Self-inflicted injuries.	5- Obsessive-compulsive disorder.	5- Schizophrenia.
6- Bipolar disorder.	6- Alcohol use.	6- Drug use.
7- Drug use.	7- Osteoporosis.	7- Violence.

Sometimes antidepressants can cause weight gain changes that do not normalise weight or appetite. The rates of weight-related adverse events may vary across antidepressant classes and within classes. Antipsychotic weight gain is a growing problem. Once the weight gain has occurred, it is very difficult to lose. Possible mechanisms of antidepressant-induced weight gain include: 1- Central CNS effect on specific receptors such as $5-HT_{2c}$, H_1, which leads to changes in appetite regulation. 2- Increased food intake/decreased satiety. 3- Sedative effects. 4- Decreased caloric expenditure. 5- Shift in food preference. 6- Dry mouth/throat. 7- Increased intake of caloric beverages. IR means that the liver increases production of glucose; in the presence of muscle inability to respond appropriately to insulin, the net result is diminished glucose uptake. The visceral accumulations of fat can be seen with computed tomography (CT) scans. Fat is not a passive tissue. This adipose tissue is metabolically active. It is resistant to insulin and, therefore, it is undergoing accelerated lipolysis, generating free fatty acids. So instead of muscle tissue being stimulated to take up glucose, it is inhibited. Instead of glucose production being inhibited by insulin in the liver, free fatty acids stimulate it. There is huge interdigitations of fat with muscle tissue (juxtaposition). Type 2 diabetes is under insulin-stimulated conditions in the muscles, with significant decrease in glucose metabolism. This excessive lipolysis from this rich fat store that is associated with obesity certainly has a significant deleterious effect in muscle. The increased free fatty acid oxidation, the augmentation of gluconeogenesis as a result of exposure of the liver to those increased free fatty acid levels, results in hepatic gluconeogenesis being augmented, and this is precisely what drives up fasting plasma glucose levels. Anything that drives the accumulation of additional adipose tissue, excessive fat, has significant health consequences that derive from all of the consequences of the IR state and the insulin resistance syndrome (IRS). Most antipsychotic medications can cause weight gain, with some medications, especially clozapine and olanzapine, associated with the highest levels of weight gain. Antipsychotic-induced weight gain was first report with chlorpromazine in 1954 (4%-17%), and phenothiazines was added in the 1970s. Ziprasidone (Geodon) is weight-neutral drugs. Olanzapine (Zyprexa) increases weight, increases insulin levels, and raises low-density lipoprotein (LDL) levels. Antipsychotic medications can increase weight and increase levels of adiposity. Increasing adiposity increases the risk of type 2 diabetes and hyperglycaemia, and this occurs via increases in insulin resistance. Diabetes mellitus is actually a group of metabolic disorders. The common theme is hyperglycaemia, but this can occur due to defects in insulin secretion, insulin action, or both. Major depressive disorder and minor depressive disorders have been associated with an up to 2-fold increase in the risk of diabetes mellitus, and first reported in 1914. Not all antipsychotic medications produce this association to the same degree, e.g., high-potency drugs such as haloperidol, seem not to do this as much as low-potency phenothiazines. Clozapine is associated with new onset diabetes mellitus, exacerbations of existing diabetes, impaired glucose tolerance, and complications such as diabetic ketoacidosis (DKA). DKA is a severe metabolic disorder characterised by hyperglycaemia, ketonaemia, and acidosis, with a rapid and unpredictable onset. Diabetes mellitus, or even impaired glucose tolerance or impaired fasting glucose is associated with increased risk for both micro- and macro-vascular disease. Microvascular disease is the risk of retinopathy, nephropathy, and neuropathies. Macrovascular disease is atherosclerosis, with the risk of myocardial infarction and stroke. The microvascular disease risk may not increase significantly until the glucose levels is associated with diabetes mellitus or impaired control. The risk of myocardial infarction and stroke increases progressively and continuously with increasing glucose levels. There is no clear threshold for microvascular disease. As fasting glucose levels rise, the risk of myocardial infarction rises, which can achieve >3 times the chance of a myocardial infarction while still having fasting glucose below the impaired threshold. As fasting glucose rises, the thickness of the atherosclerotic plaque in the carotid artery is measurably increasing. Moving up from a more or less ideal plasma glucose level of 62 mg/dl or 4 mmol/l, progressively and continuously rising risk of cardiovascular events, there is no clear threshold. Retinopathy starts >7 years before clinical diagnosis. Glucose dysregulation can lead to short-term and long-term complications, including the risk of DKA. Antipsychotic treatment is associated with changes in glucose regulation, and this is potentially interacting with disease and lifestyle-related abnormalities. Olanzapine (Zyprexa) causes the most weight gain on average, with at least 6 kg after a year. Risperidone (Risperdal) and quetiapine (Seroquel) are intermediary. Ziprasidone (Geodon) is the only one that does not cause any average weight gain or weight loss, it is weight-neutral. There are some people who will stay thin on olanzapine, others will gain weight. There is behavioural problems related to weight. There is a strong relationship between weight gain and noncompliance to medication. It is difficult to predict who will gain weight. The mechanism of weight gain could be related to increased food intake, not metabolism. Weight gain is not a dose-related adverse effect. Weight gain starts within the first 6 weeks, and plateau between 3 months and 1 year. Pancreas transplant patients exhibit marked and sustained hyperinsulinaemia with putative tissue resistance to insulin action. Diabetic neuropathies are a family of nerve disorders caused by diabetes. Diabetic neuropathies can be classified as peripheral, autonomic, proximal, and focal. Each affects different parts of the body in different ways. 1- Peripheral neuropathy causes either pain or loss of feeling in the toes, feet, legs, hands, and arms. 2- Autonomic neuropathy causes changes in digestion, bowel and bladder function, sexual response, and perspiration. It can also affect the nerves that serve the heart and control blood pressure. Autonomic neuropathy can also cause hypoglycaemia unawareness, a condition in which people no longer experience the warning signs of hypoglycaemia. 3- Proximal neuropathy causes pain in the thighs, hips, or buttocks and leads to weakness in the legs. 4- Focal neuropathy results in the sudden weakness of one nerve, or a group of nerves, causing muscle weakness or pain. Any nerve in the body may be affected. 50% of diabetics have some form of neuropathy, but not all with neuropathy have symptoms. The highest rates of neuropathy are among people who have had the disease for at least 25 years. Nerve damage is likely due to a combination of factors: 1- metabolic factors, such as high blood glucose, long duration of diabetes, possibly low levels of insulin, and abnormal lipids levels, 2- neurovascular factors, leading to damage to the blood vessels that carry oxygen and nutrients to the nerves, 3- autoimmune factors that cause inflammation in nerves mechanical injury to nerves, such as carpal tunnel syndrome, 4- inherited traits that increase susceptibility to nerve disease, 4- lifestyle factors such as smoking or alcohol use. Symptoms depend on the type of neuropathy and which nerves are affected. Symptoms may include: 1- numbness, tingling, or pain in the toes, feet, legs, hands, arms, and fingers, 2- wasting of the muscles of the feet or hands, 3- indigestion, nausea, or vomiting, diarrhoea or constipation, 4- dizziness or faintness due to a drop in postural blood pressure, 5- problems with urination, erectile dysfunction or vaginal dryness, 6- weakness, weight loss, and depression. Neuropathy is diagnosed on the basis of symptoms and a physical examination. Measuring blood pressure and heart rate, muscle strength, reflexes, and sensitivity to position, vibration, temperature, or a light touch is essential. Nerve conduction studies, electromyography (EMG), quantitative sensory testing (QST), nerve or skin biopsy, and ultrasound may be indicated as confirmatory tests. Hyperglycaemia suppresses immune response and delays wound healing. Herbs with hypoglycaemic effects (possibly enhancing insulin sensitivities in hypoglycaemics) include avocado, bitter lemon, aloe vera, fenugreek, gymnema, stevia, garlic, prickle-pear, cactus, curcumin, green tea, holy basil, glucomannan. Holy basil: (ocimum basilicum) holy basil, like sweet (culinary) basil comes from India. Sweet Basil (Ocimum basilicum) has very important medicinal properties-notably its ability to reduce blood sugar levels. It also prevents peptic ulcers and other stress related conditions like hypertension, colitis and asthma. Basil is also used to treat cold and reduce fever, congestion and joint pain. It's known to reduce the stress hormone, cortisol, which in turn reduces the fat, the body stores in the abdomen and around the waist. Burdock root (Arctium lappa), dandelion root (Taraxacum officinalis) and Jerusalem artichoke (Helianthus tuberosa) contain inulin, a polyfructosan or fructose oligosaccharides that exert beneficial effects on blood sugar control. The consumption of these roots, as vegetables or as teas, may be effective in reducing postprandial hyperglycaemia. Hospitalized patients with hyperglycaemia, poor insulin control, have poorer prognoses. In-hospital hyperglycaemia is a common finding and represents an important marker of poor clinical outcome and mortality in patients with and without a history of diabetes. Patients with newly diagnosed hyperglycaemia have a significantly higher mortality rate and lower functional outcome than patients with a known history of diabetes or normoglycaemia. Glucose, in a specific, dose- and time-dependent manner, can directly and coordinately alter intracellular ions, increasing cytosolic free calcium, while suppressing intracellular free magnesium and pH levels. High insulin levels make the intracellular Mg decrease and the Ca increase, causing hypertension. High insulin flushes out Mg but not Ca. Hyperglycaemia suppresses immune response and delays wound healing. Acute hyperglycaemia predicts increased risk of in-hospital mortality after ischaemic stroke in nondiabetic patients and increased risk of poor functional recovery in nondiabetic stroke survivors.

Foot care instruction:

1- Clean the feet daily, using warm-not hot water and a mild soap. Avoid soaking the feet. Dry them with a soft towel; dry carefully between the toes.
2- Inspect the feet and toes every day for cuts, blisters, redness, swelling, calluses, or other problems.
3- Use a mirror (laying a mirror on the floor works well) or get help from someone else if you cannot see the bottoms of your feet. 4- Notify your health care provider of any problems.
5- Moisturise the feet with lotion, but avoid getting it between the toes.
6- After a bath or shower, file corns and calluses gently with pumice stone.
7- Each week or when needed, cut the toenails to the shape of the toes and file the edges with an emery board.
8- Always wear shoes or slippers to protect the feet from injuries. Prevent skin irritation by wearing thick, soft, seamless socks.
9- Wear shoes that fit well and allow the toes to move. Break in new shoes gradually by wearing them for only an hour at a time at first.
10- Before putting the shoes on, look them over carefully and feel the insides with your hand to make sure they have no tears, sharp edges, or objects in them that might injure your feet.

Hyperglycaemia is common and involves up to 50% of the acute stroke patients. Hyperglycaemia is associated with a poor outcome in terms of mortality and neurological recovery. Hyperglycaemia exacerbates the ischaemic lesions and is associated with an increase of the oedema and size of the infarct, as well as a decrease in the cerebral blood flow. Elevated plasma glucose levels (glycaemia >8 mmol/l) predict poor prognosis, irrespective of age, severity, or stroke sub-type. Avoid sugar in immunodepression, wound healing, burn victims, skin grafts, and bacterial infections. Healthy lifestyle changes are important in disease prevention and remain the cornerstone for treatment of type 2 diabetes. Diet and exercise alone fail to normalise hyperglycaemia in the majority of patients, and pharmacotherapy is necessary. Biguanides (e.g., Metformin) enhances hepatic insulin sensitivity, improves impaired vascular reactivity in diabetes, improves β-cell function and/or improves skeletal muscle glucose uptake, and improves PAI-1 levels and endothelial function. Biguanides has favourable lipid effects, and the improved glycaemic control is not associated with weight gain. Its mild and transient GIT side effects, including abdominal discomfort and diarrhoea, are common, occurring in 20% to 30% of patients. Metformin does not increase insulin secretion; thus, hypoglycaemia is rare when it is used as monotherapy. Sulfonylureas enhance of insulin secretion via binding to a specific sulfonylurea receptor on pancreatic β-cells. The average improvement of HbA1c value is 1.5% to 2.0% points. The major adverse effect is hypoglycaemia, which is more commonly observed with the longer-acting sulfonylureas. When glycaemic goals are not met, combination therapy is preferred. The risk of hypoglycaemia is higher in the elderly, in those with a reduced glomerular filtration rate or congestive heart failure, and/or in patients taking angiotensin-converting enzyme (ACE) inhibitors such as captopril. Meglitinides are nonsulfonylurea insulin secretagogues, which work by closing the ATP-dependent K channel, and require the presence of glucose for their action. Their glycaemic improvement is similar to sulfonylureas. These drugs are rapidly absorbed (0.5-1 hr) and rapidly eliminated (half-life <1 hr). Because of their pharmacokinetic behaviour, they cause a more physiological, rapid, and brief release of insulin, with nateglinide having a quicker insulin return to preprandial levels. As is the case with sulfonylureas, weight gain and hypoglycaemia remain the most common side effects. Thiazolidinediones or PPAR-γ analogues enhance insulin sensitivity, which is mediated by decreasing fat mobilisation, thus lowering circulating free fatty acid levels. Thiazolidinediones bind to the novel nuclear peroxisome proliferator activated receptor (PPAR-γ), leading to increased glucose transporter expression. While PPAR-γ receptors are expressed in hepatocytes, haematologic and vascular cells, the main target for the PPAR-γ ligands appears to be fat tissue. These agents change adipocytes into more effective energy storage cells and increase adiponectin levels resulting in favourable metabolic changes. Thiazolidinediones, when used as monotherapy, have a less dramatic improved glycaemic effect. These agents improve the overall lipid profile by reducing triglyceride levels by 10-20% and increasing HDL cholesterol levels by 5-10%. They also increase low-density lipoprotein (LDL) cholesterol levels by 10-15%, predominantly the large buoyant particles, which are less susceptible to oxidation and therefore less atherogenic. They improve endothelial function by normalising endothelium dependent vasoactivity. These agents inhibit vascular smooth-muscle proliferation and migration, decrease expression and activity of matrix metalloproteinase-9 in macrophages, and inhibit endothelial cell mitogenesis. Directly or indirectly, thiazolidinediones appear to have an antiatherogenic effect by inhibiting smooth-muscle cell proliferation, cell-cycle progression, and the cell response to injury. They also decrease production of cytokines and inhibit chemokines that attract inflammatory cells to the vessel wall. Neointimal tissue proliferation after coronary restenosis can also be reduced with troglitazone. The drug appears to have a myocardial protective effect in response to catecholamine stimulation and myocardial ischaemia. Improvement in PAI-1 synthesis, directly by affecting vessel walls and indirectly by improving hyperinsulinaemia, is also described. Adverse effects include liver toxicity, and worsening of CHF. α-Glucosidase inhibitors inhibit the ability of enzymes (maltase, isomaltase, sucrase, and glucoamylase) in the small intestinal brush border to break down oligosaccharides and disaccharides into monosaccharides. By delaying the digestion of carbohydrates, they retard glucose entry into the systemic circulation, blunting the postprandial rise in plasma glucose levels. The postprandial plasma glucose level improves without increasing plasma insulin levels. They are effective in those individuals accustomed to a high carbohydrate intake. Adverse effects include GIT side effects (bloating, abdominal discomfort, diarrhoea, and flatulence) in 30% of patients, and elevated serum aminotransferase levels. Acarbose is contraindicated in patients with inflammatory bowel disease, a plasma creatinine concentration of more than 177 mol/l (>2.0 mg/dl), or cirrhosis.

Effect of diabetic pharmacotherapy on HbA1c levels:

Drug	Decreasing HbA1c
Biguanides	1.5% to 2.0% points
Meglitinides	0.7-1 points.
Sulfonylureas	1.5% to 2.0% points.
Thiazolidinediones or PPAR-γ analogues	0.6% points.
α-Glucosidase inhibitors	0.5% to 0.7% points.

There are two main physiological failures in NIDDM: 1- impaired insulin secretion, and 2- increased cellular uptake. NIDDM humans given a 1000 mg ALA would experience 50% improvement in insulin-stimulated glucose disposal. ALA supplementation may prevent diabetes (70%). α-Lipoic acid (ALA) is effective against diabetic neuropathies (i.e., polyneuropathy, and retinopathy). ALA reduces plasma oxidation, whole body oxidation (as measured by urinary isoprostanes) and LDL-oxidation. The sustain-release format offers the advantage of reduced gastrointestinal irritations caused by the strong dose. Furthermore, the sustained-release format may result in better conversion of ALA into DHLA (dihydrolipoic acid), which takes place intracellularly and not in the plasma. L-glutamine consumption reduces the hyperglycaemia and hyperinsulinaemia caused by high fat diets. Chromium is a popular mineral supplement. Chromium is essential for optimal insulin action. Another source of bioavailable chromium is Colostrum, which is a rich source of low molecular weight chromium compounds (LMWCr's) known for their bioavailability. Chromium supplements (i.e., chromium picolinate) and chromium-rich foods (colostrum) should definitely be included in the diabetic diet. Alpha lipoic acid (Thioctic Acid) is a sulfur containing, vitamin-like substance. It plays an important role as the necessary cofactor in two vital, energy producing reactions in the production of cellular energy (ATP). α-Lipoic acid is not considered a vitamin because presumably either the body can usually manufacture sufficient levels or it is acquired in sufficient quantities from food. However, a relative deficiency can occur in certain situations and its supplementation exerts benefits beyond its role in normal metabolism. α-Lipoic acid is an effective antioxidant. It is unique in that it is effective against both water- and fat-soluble free radicals. ALA helps prevent arteriosclerosis and other health destructive effects of sugars. In protecting the liver from aldehyde-induced changes ALA helps reduce symptoms associated with candidiasis and the chemical sensitivities and poisoning many people are vulnerable to in everyday exposure to the acrylamides found extensively in plastics and adhesives, hexane in auto exhaust and heavy metals such as lead, mercury and aluminium. Detoxication benefits also include protection against damage from radiation and anaesthesia-induced hepatitis. A complex interaction occurs in the body among ALA and other antioxidants, primarily vitamins C and E, and glutathione. ALA produces positive effects in the treatment of AIDS. ALA supplementation increases plasma ascorbate, total glutathione, and T helper lymphocytes and T helper/suppressor cell ratio. The lipid peroxidation product malondialdehyde is also decreased. Discovered in the 1930's and extracted in 1957, ALA is a unique free-radical protector for cells because it is the only such nutrient, which is both fat- and water-soluble. ALA has excellent bioavailability and can easily travel across cell membranes to fight free-radicals both inside and outside the cell. Once inside the cell, ALA is broken down to dihydrophilic acid, an even more potent neutraliser of free-radicals. Many of the beneficial effects of ALA may be due to its ability to regenerate glutathione, a potent amino acid antioxidant, which in turn is a powerful immune enhancer, liver protector, and heavy metal detoxifier. Because it is a sulphur compound, ALA can bind and help eliminate heavy metals such as copper, iron, mercury and cadmium; risk factors for a wide range of degenerative diseases. There is usually little ALA in the body. There are just enough for certain metabolic functions like turning carbohydrates and fats in to blood sugar. ALA inhibits the growth of HIV and inhibits growth more effectively than the NAC (N-acetyl cysteine). ALA inhibits activation of a gene in the AIDS virus that allows it to reproduce. ALA increases the amount of glutathione inside cells. In AIDS, ALA supplement increases glutathione levels (100%), vitamin C levels (90%), T4 cells (66%), and other evidence of oxidative stress is decreased (70%). A combination of ALA and AZT inhibits reverse transcriptase activity more than either drug alone. Nerve damage is one of diabetes' most devastating complications and affects >50% of all diabetics. ALA can partly restore diabetic nerve function after only 4 months of high-dose oral treatment. The elevated blood sugar levels associated with diabetes can distort normal metabolism and may, in turn, increase oxidative stress and free-radical production. Most diabetics are subject to high oxidative stress even when they control their blood sugar. ALA has a statistically significant improvement in the sympathetic system and no change for the parasympathetic. Symptoms due to autonomic nerve disorder are decreased with ALA. ALA is the second nutrient proven effective for treating diabetic neuropathy. Primrose oil can reverse diabetic damage to peripheral nerves. Primrose oil, and other sources of γ-linolenic acid (GLA) such as borage or blackcurrant seed oil, helps balance the body's prostaglandin levels and reduce inflammation. ALA may also be effective for treating peripheral nerve damage. Diabetics with autonomic neuropathy are 5 times more likely to die early than diabetics with healthy nervous systems. A damaged nervous system may render them more susceptible to heart disease, poor wound healing and infection.

STRESS

Elderly who experienced chronic stress tended to have higher blood levels of IL-6. IL-6 is an inflammatory protein that has been linked to aging and disease. Counteracting stress can be achieved by reserving time for long walks, soothing music, or chats with good friends. The psychological stress response is seen after any injury. Stress generates a hypermetabolic catabolic state through the same neuropathway initiated by physical stress. Psychological stress: 1- activates hypothalamus-pituitary-adrenal (HPA) axis, 2- increases catechols, 3- causes hypermetabolism, 4- causes immune dysfunction, 5- decreases wound blood flow, and 6- is correctable with adequate pain/stress management. Emotional structures are located mainly in the limbic system. The hypothalamus, in particular, with its connections to both central and peripheral nervous structures and neuroendocrine integrated systems, is concerned with the organisation of motivated behavioural and endocrine responses. In humans, mental stress is known to induce pronounced and reproducible activation of the sympathoadrenal system, with elevation of plasma epinephrine and norepinephrine concentrations and subsequent metabolic consequences identical to those produced by a wound insult. Stress is a common experience of daily living. Regardless of whether the stresses are physical or psychological, similar responses are activated to maintain homeostasis in the body. Acute stress enhances intestinal epithelial permeability to macromolecules by mechanisms involving corticotropin-releasing hormone (CRH) and mast cells. Repeated stress induces a colonic barrier defect, enlarges epithelial cell mitochondria, and mucosal. Stress can augment the mucosal inflammatory response to hapten, as well as reactivate colonic inflammation. Mast cells are important effector cells in mucosal innate immunity and are involved in mucosal perturbations induced by stress. Mast cells are important mediators in the process leading to stress-induced mucosal abnormalities. Acute stress induces release of CRH. On activation, mast cells release a large variety of mediators, many of them affecting mucosal function, nerves, and immune cells. Mast cells are critically involved in neutrophil recruitment and activation in C. difficile toxin A-induced inflammation. Mast cells increase epithelial permeability and mucus depletion, which facilitates bacterial-epithelial cell interaction. Mast cells may mediate immune activation by release of chemokines and cytokines. Stress-induced mast cell factors may recruit additional inflammatory cells to the lamina propria. Bacterial adherence/penetration into epithelial cells may trigger these cells to synthesise chemokines specific for enteric infections. The intestinal mitochondrial changes are induced by uncoupling of oxidative phosphorylation and associated with an enhanced epithelial permeability, involving rearrangement of F-actin and tight junction proteins. The number of autophagosomes in mitochondria and cytoplasm increases after exposure to 10 days stress. Mitochondrial permeability transition in the inner membrane causes swelling of mitochondria and release of cytochrome C, which progresses to autophagy and apoptosis. The cellular concentration of ATP, OFR, and cell necrosis may follow depending on the stimulus. There is a decreased number of mucus-containing goblet cells. There is also a reduction in the protective capacity of the mucus indicated by an increased number of bacteria within the mucus and adhesion of bacteria to the epithelium. Acute stress stimulates intestinal mucus secretion by a mast cell-dependent mechanism, which may enhance the mucosal barrier acutely, protecting a leaky epithelium against invasion. Corticosterone is increased 10-fold in response to acute stress. In chronic stress there may be a habituation of the HPA-axis. 2 weeks of stress causes inhibition of CRH transcription in the paraventricular nucleus of the hypothalamus and blunted plasma corticosterone response. The mucosal changes may not be entirely dependent on activation of the HPA axis but may be caused in part by parasympathetic modulation, and peripheral CRH receptors. Diminished HPA response may compromise the ability of the host to counteract mucosal inflammation. Decreased basal cortisol levels have been found in patients with rheumatoid arthritis, fibromyalgia, and in diarrhoea-predominant IBD. Chronic ongoing psychological stress in a naïve host can induce prolonged intestinal barrier dysfunction, impair mucosal defense to luminal bacteria, and initiate mucosal inflammation.

EXCITOTOXINS

Excitotoxins play a critical role in the development of several neurological disorders including migraines, seizures, infections, abnormal neural development, certain endocrine disorders, neuropsychiatric disorders, learning disorders in children, AIDS dementia, episodic violence, Lyme borreliosis, hepatic encephalopathy, specific types of obesity, and especially the neurodegenerative diseases such as ALS, Parkinson's disease, Alzheimer's disease, Huntington's disease, and olivopontocerebellar degeneration. Various excitotoxins are added to the food supply, such as monosodium glutamate (MSG), hydrolysed vegetable protein, and aspartame. Since 1948 the amount of MSG added to foods has doubled every decade. Food additives have nothing to do with preserving food or protecting its integrity. They are all used to alter the taste of food. Monosodium glutamate, hydrolysed vegetable protein, and natural flavouring are used to enhance the taste of food so that it tastes better. Aspartame is an artificial sweetener. Excitotoxins are present in almost all processed foods. Excitotoxins are synergistic. Liquid forms of excitotoxins, as occurs in soups, gravies and diet soft drinks are more toxic than that added to solid foods. This is because they are more rapidly absorbed and reach higher blood levels. Excitotoxins are added in disguised forms, such as natural flavouring, spices, yeast extract, textured protein, soy protein extract, etc. Excitotoxins are substances, usually acidic amino acids, which react with specialised receptors in the brain in such a way as to lead to destruction of certain types of neurons. Glutamate is one of the more commonly known excitotoxins, MSG is the sodium salt of glutamate. This amino acid is the most commonly used normal neurotransmitter in the brain. Glutamate, as a neurotransmitter, exists in the extracellular fluid only in very, very small concentrations, 8-12 uM. When the concentration of this transmitter rises above this level, the neurons begin to fire abnormally. At higher concentrations, the cells undergo a specialised process of delayed cell death known as excitotoxicity, i.e., they are excited to death. The effects of excitotoxin food additives generally are not dramatic. Some individuals may be especially sensitive and develop severe symptoms and even sudden death from cardiac irritability, but in most instances the effects are subtle and develop over a long period of time. The individual may never develop a full-blown disorder, but a very mild form of the disease. Food borne excitotoxins may be harmful to those suffering from strokes, head injury and HIV infection and certainly should not be used in a hospital setting. In 1957, MSG excitotoxicity was reported for the first time. In 1969, MSG was reported to be associated with obesity, multiple endocrine deficiencies including TSH, growth hormone, LH, FSH, and ACTH, and discrete lesions of the arcuate nucleus as well as less severe destruction of other hypothalamic nuclei. Later studies indicated that the damage by monosodium glutamate was much more widespread, including the hippocampus, circumventricular organs, locus cereulus, amygdala- limbic system, subthalamus, and striatum. Glutamate is the most important neurotransmitter in the hypothalamus. Glutamate, and other excitatory amino acids, attach to a specialised family of receptors (NMDA, kainate, AMPA and metabotrophic) which in turn, either directly or indirectly, opens the calcium channel on the neuron cell membrane, allowing Ca to flood into the cell. If unchecked, this Ca will trigger a cascade of reactions, including free radical generation, eicosanoid production, and lipid peroxidation, which will destroy the cell. With this Ca-triggered stimulation, the neuron becomes very excited, firing its impulses repetitively until the point of cell death, hence the name excitotoxin. The activation of the calcium channel via the NMDA type receptors also involves other membrane receptors such as the zinc, magnesium, phencyclidine, and glycine receptors. When brain cells are injured they release large amounts of glutamate from surrounding astrocytes, and this glutamate can further damage surrounding normal neuronal cells. This appears to be the case in strokes, seizures and brain trauma. But, food born excitotoxins can add significantly to this accumulation of toxins. Examples of disguised harmful additives names include hydrolysed vegetable protein, vegetable protein, textured protein, hydrolysed plant protein, soy protein extract, caseinate, yeast extract, and natural flavouring. When these excitotoxin taste enhancers are added together they become much more toxic than is seen individually. Excitotoxins in subtoxic concentrations can be fully toxic to specialised brain cells when used in combination. Processed foods on supermarket shelves, especially frozen or diet foods, may contain 2, 3 or even 4 types of excitotoxins. Many of the commercial soups, sauces, and gravies containing MSG are very dangerous to nervous system health, and should especially be avoided by those either having one of the above mentioned disorders, or who are at a high risk of developing one of them. They should also be avoided by cancer patients and those at high risk for cancer, because of the associated generation of free radicals and lipid peroxidation. Red meats, cheeses, and pureed tomatoes, all have higher levels of glutamate. A whole tomato is safer than a pureed tomato; tomato plant contains several powerful antioxidants known to block glutamate toxicity. Hydrolysed vegetable protein may contain at least 2 excitotoxins, glutamate and cysteic acid. Hydrolysed vegetable protein is made by a chemical process, which breaks down the vegetable's protein structure to purposefully free the glutamate, as well as aspartate, another excitotoxin. Growing lists of excitotoxins are being discovered, including several that are found naturally, e.g., L- cysteine. Homocysteine, a metabolic derivative, is also an excitotoxin. Elevated blood levels of homocysteine have been implicated in neurodevelopmental disorders, especially anencephaly and spinal dysraphism (neural tube defects). The protective mechanism of action is associated with the use of the prenatal vitamins B_{12}, B_6, and folate when used in combination. Alzheimer's patients have elevated levels of homocysteine. Alzheimer's patients have widespread destruction of the retinal ganglion cells. Neurological diseases associated with excitotoxic injury are also associated with accumulations of toxic free radicals and destructive lipid oxidation products. For example, the brains of Alzheimer's disease patients have been found to contain high concentration of lipid peroxidation products and evidence of free radical accumulation and damage. One of the early changes in Parkinson's disease is the loss of one of the primary antioxidant defense systems, glutathione, from the neurons of the striate system, and especially in the substantia nigra. Accompanying this, is an accumulation of free iron within the globus pallidus and the substantia nigra, which is one of the most powerful free radical generators known. The neurons within the substantia nigra are especially vulnerable to oxidant stress because the catabolic metabolism of the transmitter-dopamine can proceed to the creation of very powerful free radicals. Dopamine is auto-oxidised to peroxide, which is normally detoxified by glutathione. In the presence of high concentrations of free iron, the peroxide is converted into hydroxide. As the hydroxide radical diffuses throughout the cell, destruction of the lipid components of the cell takes place, i.e., lipid peroxidation. The generation of the powerful peroxynitrite radical produce serious injury to cellular proteins and DNA, both mitochondrial and nuclear. Iron accumulation in Parkinson's disease is primarily localised in the neuromelanin granules. There is also dramatic accumulation of aluminium within these granules. The aluminium may displace the bound iron, releasing highly reactive free iron. Low concentrations of aluminium salts can enhance iron-induced lipid peroxidation. Individuals with Parkinson's disease also have defective iron metabolism. In Parkinson's disease there is reduction in complex I enzymes within the mitochondria. Complex I enzymes are sensitive to free radical injury. These enzymes are critical to the production of cellular energy. When cellular energy is decreased, the toxic effect of excitatory amino acids increases dramatically. ALS, (amyotrophic lateral sclerosis) is also associated with similar free radical damage, most likely triggered by toxic concentrations of excitotoxins, which plays a major role in the disorder. There is lipid peroxidation product accumulation within the spinal cords of ALS patients as well as iron accumulation. Glutamate acts on its receptor via a nitric oxide (NO) mechanism. Overstimulation of the glutamate receptor can produce an accumulation of reactive nitrogen species (RNS), resulting in the generation of several species of dangerous free radicals, including peroxynitrite. Free radical scavengers (antioxidants), have successfully blocked excitotoxic destruction of neurons. Vitamin E may block glutamate toxicity. Vitamin E combined with other antioxidant vitamins may slow the course of Alzheimer's disease and Parkinson's disease. A combination of D-α-tocopherol and ascorbic acid in high doses reduce progression of the Parkinson's disease by 2.5 years. Tocotrienol may have even greater benefits, especially when used in combination with other antioxidants. Vitamin E also may slow the course of ALS as well, especially in the form of D-α-tocopherol. Excessive glutamate stimulation triggers a chain of events that in turn sparks the generation of large numbers of free radical species, both as nitrogen and oxygen species. These free radicals induce damage cellular proteins (protein carbonyl products) and DNA. The most immediate DNA damage is to the mitochondrial DNA, which controls protein expression within that cell and its progeny, producing rather profound changes in cellular energy production. Chronic free radical accumulation would result in an impaired functional reserve of antioxidant vitamins/minerals and enzymes, and thiol compounds necessary for neural protection. Chronic unrelieved stress, chronic infection, free radical generating metals and toxins, and impaired DNA repair enzymes all add to this damage. There are 4 main endogenous sources of oxidants: 1- Those produced naturally from aerobic metabolism of glucose. 2- Those produced during phagocytic cell attack on bacteria, viruses, and parasites, especially with chronic infections. 3- Those produced during the degradation of fatty acids and other molecules that produce H_2O_2 as a by-product. 4- Oxidants produced during the course of P450 degradation of natural toxins. Iron is one mineral heavily promoted by the health industry, and is frequently added to many foods, especially breads and pastas. Copper is also a powerful free radical generator and has been shown to be elevated within the substantia nigra of Parkinsonian brains. There is a direct connection between excitotoxicity and free radical generation in a multitude of diseases and disorders such as seizures, strokes, brain trauma, viral infections, and neurodegenerative diseases. Free radicals prevent glutamate uptake by astrocytes as well, which would significantly increase extracellular glutamate levels. This creates a vicious cycle that will multiply any resulting damage and malfunctioning of neurophysiological systems, such as plasticity. The blood-brain barrier (BBB), a system of specialised capillary structures designed to exclude toxic substance from entering the brain. The brain, even in the adult, has several areas that normally do not have a barrier system, called the circumventricular organs. These include the hypothalamus, the subfornical organ, organium vasculosum, area postrema, pineal gland, and the subcommisural organ. The hypothalamus controls center for all neuroendocrine regulation, sleep wake cycles, emotional control, caloric intake regulation, immune system regulation and regulation of the autonomic nervous system. Careful regulation of blood levels of glutamate is very important, since high blood concentrations of glutamate would be expected to increase hypothalamic levels as well. Arcuate nucleus is a small hypothalamic nucleus controls a multitude of neuroendocrine functions, as well as being intimately connected to several other hypothalamic nuclei. High concentrations of blood glutamate and aspartate (from foods) can enter the so-called protected brain by seeping through the unprotected areas, such as the hypothalamus or other circumventricular organs. Chronic elevations of blood glutamate can even seep through the normal BBB when these high concentrations are maintained over a long period of time. One of the earliest and most consistent findings with exposure to MSG is damage to arcuate nucleus. Conditions associated with brain barrier disruption include: hypertension, diabetes, ministrokes, major strokes, head trauma, multiple sclerosis, brain tumours, chemotherapy, radiation treatments to the nervous system, collagen-vascular diseases (lupus), AIDS, brain infections, certain drugs, Alzheimer's disease, and as a consequence of natural aging. When the barrier is dysfunctional due to one of these conditions, foods containing high concentrations of excitotoxins (glutamate and aspartate) will increase brain concentrations to toxic levels as well. Multiple sclerosis (MS) has an exacerbation of symptoms; the BBB near the lesions breaks down, leaving the surrounding brain vulnerable to excitotoxin entry from the blood, i.e. diet. But, not only is the adjacent brain vulnerable, but the openings act as points of entry, eventually exposing the entire brain to potentially toxic levels of glutamate. Barrier disruption has been demonstrated in the case of Alzheimer's disease. Not only can free radicals open the BBB, but excitotoxins can as well. Free radical scavengers block the opened barrier. Excitotoxin damage is heavily dependent on the energy state of the cell. Cells with normal energy generation systems are very resistant to such toxicity. When cells are energy deficient, as a result of hypoxia, starvation, metabolic poisons, hypoglycaemia, they become infinitely more susceptible to excitotoxic injury or death. Even normal concentrations of glutamate are toxic to energy deficient cells. In many of the neurodegenerative disorders, e.g., Huntington disease and Alzheimer's disease, neuron energy deficiency often precedes the clinical onset of the disease by years. In Parkinson's disease, one of the early deficits of the disorder is an impaired energy production by the complex I group of enzymes within the mitochondria of the substantia nigra. Complex I system is very sensitive to free radical damage. Co-enzyme Q_{10} restores energy production, but does not prevent cellular death. But when combined with niacinamide, both cellular energy production and neuron protection is possible. A combination of CoQ_{10}, acetyl-L carnitine, niacinamide, riboflavin, methylcobalamin, and thiamine are beneficial for neurodegenerative disorders. Acetyl L-carnitine enters the brain in higher concentrations and also increases brain acetylcholine, necessary for normal memory function. Phosphatidyl serine, Ginkgo biloba, vitamin B_{12}, folate, magnesium, vitamin K and several others are also important. Mitochrondial dysfunction may reflect that some are more vulnerable to excitotoxin damage than others, but it does not explain injury in those with normal cellular metabolism. When the blood sugar falls (reactive hypoglycaemia), the body responds by releasing a burst of epinephrine from the adrenal glands, in an effort to raise the blood sugar. This felt as nervousness, palpitations, tremulousness, and profuse sweating. The brain is one of the most glucose dependent organs known, since it has a limited ability to metabolise other substrates such as fats. Several of the neurodegenerative diseases are related to either excessive insulin release, e.g., Alzheimer's disease, or impaired glucose utilisation, e.g., Parkinson's disease and Huntington's disease. Most Alzheimer's patients have low blood sugars, and high CSF insulin levels, i.e., brain hypoglycaemia. High number of Parkinson's and ALS patients suffers from reactive hypoglycaemia. Avoiding gluten containing products, such as bread, crackers, cereal, pasta, etc., means also avoiding products that are high on the glycaemic index, i.e., that produce reactive hypoglycaemia. Also, all of these food items are high in free iron. Clinically, hypoglycaemia will worsen the symptoms of most neurological disorders. Severe hypoglycaemia can, in fact, mimic ALS both clinically and pathologically. It is also known that many of the symptoms of Alzheimer's disease resemble hypoglycaemia, as if the brain is hypoglycaemic in isolation. One of the effects of hypoxia is a massive release of glutamate into the space around the neuron. This results in rapid death of these sensitised cells. As we age, the blood supply to the brain is frequently impaired, either because of atherosclerosis or repeated syncopal episodes, leading to short periods of hypoxia. Hypoglycaemia produces lesions very similar to hypoxia and via the same glutamate excitotoxic mechanism. Diabetics suffering from repeated episodes of hypoglycaemia associated with over medication with insulin would develop brain atrophy and dementia. Another cause of isolated cerebral hypoglycaemia is impaired transport of glucose into the brain across the BBB. Glucose enters the brain by way of a glucose transporter, and in several conditions this transporter is impaired, including aging, arteriosclerosis, and Alzheimer's disease. Prolonged elevation of the blood sugar produces a down-regulation of the glucose transporter and a concomitant brain hypoglycaemia is exacerbated by repeated spells of peripheral hypoglycaemia, which is common to type I diabetics. With aging, several of the energy deficiency syndromes, such as mitochondrial injury, impaired cerebral blood flow, enzyme dysfunction, and impaired glucose transportation, develop simultaneously. This greatly magnifies excitotoxicity, leading to accelerated free radical injury and a progressively rapid loss of cerebral function and profound changes in cellular energy production. Chronic free radical accumulation would also result in an impaired functional reserve of antioxidant vitamins/minerals, antioxidant enzymes (SOD, catalase, and glutathione peroxidase) and thiol compounds necessary for neural protection. Chronic unrelieved stress, chronic infection, free radical generating metals and toxins, and impaired DNA repair enzymes all add to this damage. It is estimated that the number of oxidative free radical injuries to DNA number about 10,000 a day in humans. Under conditions of cellular stress this may reach several hundred thousand. Normally, these injuries are repaired by special DNA repair enzymes. AS we age these repair enzymes decrease or become less efficient. Also, some individuals are born with deficient repair enzymes from birth, e.g., xeroderma pigmentosum. In Alzheimer's patients there is a significant deficiency in DNA repair enzymes and high levels of lipid peroxidation products in the affected parts of the brain. The hippocampus of the brain, most severely damaged in Alzheimer's dementia, is one of the most vulnerable areas of the brain to hypoglycaemia as well as hypoxia. That also makes it very susceptible to glutamate/free radical toxicity. When cells are exposed to glutamate they develop certain inclusions (cellular debris) that not only resembles the characteristic neurofibrillary tangles of Alzheimer's dementia, but are immunologically identical as well. Glutamate toxicity leads to a condition of receptor loss, which is characteristic of neurodegeneration. This receptor loss produces a state of disinhibition, which magnifies excitotoxicity during the later stage of the neurodegenerative process. The brain contains one of the highest concentrations of ascorbic acid in the body. Certain areas of the brain have very high concentrations of ascorbic acid, such as the nucleus accumbens and hippocampus. The lowest levels are in the substantia nigra. These levels seem to fluctuate with the electrical activity of the brain. Amphetamine acts to increase ascorbic acid concentration in the corpus striatum (basal ganglion area) and decrease it in the hippocampus, the memory imprint area of the brain. Ascorbic acid plays a vital role in dopamine production as well. There are links between the secretion of the glutamate neurotransmitter by the brain and the release of ascorbic acid into the extracellular space. The other neurotransmitters do not have a similar effect on ascorbic acid release. The ascorbic acid and glutamate exchange places. High concentration of ascorbic acid in the diet may inhibit glutamate release, lessening the risk of excitotoxic damage. Ascorbic acid has neutralising effect on the free radical. Ascorbic acid modulates the electrophysiological as well as behavioural functioning of the brain. It also attenuates the behavioural response to amphetamine, which is known to act through an excitatory mechanism. This is due to binding of ascorbic acid to the glutamate receptor. With ageing there is a decline in brain levels of ascorbate. When accompanied by a similar decrease in glutathione peroxidase, H_2O_2 accumulates and hence, elevated levels of free radicals and lipid peroxidation. With ageing the extracellular concentration of ascorbic acid is decreased, and the capacity of the brain ascorbic acid system to respond to oxidative stress is impaired. In terms of its antioxidant activity, vitamin C and E interact in such a way as to restore each others active antioxidant state. Vitamin C scavenges oxygen radicals in the aqueous phase and vitamin E in the lipid, chain breaking, phase. The addition of vitamin C suppresses the oxidative consumption of vitamin E almost totally, probably because in the living organism the vitamin C in the aqueous phase is adjacent to the lipid membrane layer containing the vitamin E. When combined, the vitamin C is consumed faster during oxidative stress than vitamin E. Once the vitamin C is totally consumed, vitamin E begins to be depleted at an accelerated rate. N-acetyl-L-cysteine and glutathione can reduce vitamin E consumption as well, but less effectively than vitamin C. The real danger is when vitamin C is combined with iron. This is because the free iron oxidises the ascorbate to produce the free radical dehydroxyascorbate. α-Lipoic acid acts powerfully to keep the ascorbate and tocopherol in the reduced state (antioxidant state). As we age, we produce less of the transferrin transport protein that normally binds free iron. As a result, elderly have higher levels of free iron within their tissues, including brain, and are therefore at greater risk of widespread free radical injury. The development of the brain is a very complex process that occurs in a spatial and temporal sequence, which is carefully controlled by biochemical, structural, as well as neurophysiological events. Even subtle changes in these parameters can produce ultimate changes in brain function that may vary from subtle alteration in behaviour and learning to autism, attention deficit disorder and violence dyscontrol. Glutamate plays a vital role in the development of the nervous system, especially as regards neuronal survival, growth and differentiation, development of circuits and cytoarchitecture. Deficiencies of glutamate in the brain during neurogenesis can result in maldevelopment of the visual cortices and may play a role in the development of schizophrenia. Excess glutamate can cause neural pathways to produce improper connections. Excess glutamate during embryogenesis reduces dendritic length and suppress axonal outgrowth in hippocampal neurons. Glutamate can produce classic toxicity in the immature brain even before the glutamate receptors develop. High glutamate levels can also affect astroglial proliferation as well as neuronal differentiation. It appears to act via the phosphoinositide protein kinase C pathway. During brain development there is an overgrowth of neuronal connections and cellularity, and that at this stage there is a peak in brain glutamate levels, whose function is to remove excess connections and neuronal overexpression, i.e., pruning. Glutamate excess during synaptogenesis and pathway development cause abnormal connections in the hypothalamus that can lead to later endocrinopathies. In general, toxicological injury in the developing foetus carries the greatest risk during the first 2 trimesters. But, this is not so for the brain, which undergoes a spurt of growth that begins during the third trimester and continues at least 2 years after birth. Dendritic growth is maximal in the late foetal period to 1 year of age, but may continue at a slower pace for several more years. Neurotransmitter development also begins during the late foetal period but continues for as long as 4 years after birth. This means that alterations in dietary glutamate and aspartate are especially dangerous to the foetus during pregnancy and for several years after birth. The developing brain's susceptibility to excitotoxicity varies, since each brain region has a distinct developmental profile. The type of excitotoxin also appears to matter. For example, kianate is non-toxic to the immature brain but extremely toxic to the mature brain. The glutamate agonist, NMDA, is especially toxic up to postnatal day 7 while quisqualate and AMPA have peak toxicity from postnatal day 7 through 14. L-cysteine is a powerful excitotoxin on the immature brain. Myelination can also be affected by neurotoxins. Excitotoxic substances affect dendrites and neurons more than axons. During the myelination process, each fiber tract has its own spatiotemporal pattern of development, accompanied by significant biochemical changes, especially in lipid metabolism. Aspartame's methanol component breakdown product is formaldehyde. The aspartate moiety undergoes spontaneous racemization in hot liquids to form D-aspartate, which has been associated with tau (τ) proteins in Alzheimer's disease. The immature brain is four times more sensitive to the toxic effects of the excitatory amino acids as is the mature brain. This means that excitotoxic injury is of special concern from the foetal stage to adolescence. There is evidence that the placenta concentrates several of these toxic amino acids on the foetal side of the placenta. Consumption of aspartame and MSG containing products by pregnant women during this critical period of brain formation is of special concern and should be discouraged. Many of the effects, such as endocrine dysfunction and complex learning, are subtle and may not appear until the child is older. Prenatal exposure to MSG causes normal simple learning but showed significant deficits in complex learning, accompanied by profound reductions in several forebrain neurotransmitters. In human this would mean that during infancy and early adolescence learning would appear normal, but with entry into a more advance education level, learning would be significantly impaired (ADD and ADHD). Neonatal glutamate could severely injure hippocampal neurons and dendrites, which may impair discriminative learning. Neonatal exposure to MSG causes significant alterations in neuroendocrine function that can be prolonged.

TOXIC LIVER INJURY

The liver is usually the target of the immune system because of the production of drug metabolites that bind to proteins that are degraded to peptides presented on the cell surface with major histocompatibility complex class I. This leads to sensitization, which allows T cells to recognize the drug metabolite bound to liver protein (e.g., CYP) targeted to the hepatocyte surface. It is to understand the role of the extrinsic and intrinsic pathways, the intracellular triggers of stress, the participation of the membrane permeability transition (MPT), and the contribution of apoptosis and necrosis as well as the potential for switching from one mode of cell death to the other. The clinical appearance of hepatitis is a consequence of cell death mediated by either the extrinsic immune system (e.g., cytotoxic T cells) or intracellular stress. Intracellular stress can lead to apoptotic or necrotic cell death, depending on the extent of mitochondrial involvement and the balance of factors that activate and inhibit the Bcl_2 family of proteins and the caspases. Drug metabolites can undergo or promote a variety of chemical reactions, including covalent binding, depletion of reduced glutathione, or oxidative stress with consequent effects on proteins, lipids, and DNA. These chemical consequences can directly affect organelles such as mitochondria, cytoskeleton, endoplasmic reticulum, microtubules, or nucleus or indirectly influence these organelles through activation or inhibition of signaling kinases, transcription factors, and gene expression profiles. These changes lead to either triggering of the necrotic or apoptotic process or sensitization to the lethal action of cytokines of the immune system intrinsic to the liver. There are 2 ways in which some drugs produce metabolites that cause cell death independent of the extrinsic immune system: 1- the reactive metabolites can disrupt the balance of factors that favour survival, leading to direct loss of viability, or 2- they can modify this balance so as to render liver cells susceptible to the lethal effects of the intrinsic immune system, e.g., TNF, produced by activation of resident inflammatory cells in the liver. The outcome can range from maintenance of viability to apoptosis or necrosis depending on the drug, the extent of exposure to reactive metabolites, and a variety of environmental and genetic factors that modulate drug metabolism, transport, defense, regeneration, cytokines, and prodeath and prosurvival genes. An important feature of apoptosis is the requirement for ATP to initiate the execution phase. The swelling of the necrotic cell is a consequence of profound loss of mitochondrial function and resultant ATP depletion, leading to loss of ion homeostasis, including volume regulation, and increased Ca_i^{2+}; the latter activates a number of nonspecific hydrolases (i.e., proteases, nucleases, and phospholipases). The initiation of cell death begins either at the plasma membrane with the binding of TNF or FasL to their cognate receptors (i.e., death receptors) or within the cell. The latter is due to the occurrence of intracellular stress in the form of biochemical events such as oxidative stress, redox changes, covalent binding, lipid peroxidation, and consequent functional effects on mitochondria, endoplasmic reticulum (ER), microtubules, cytoskeleton, or DNA. Intracellular stress either directly affects mitochondria or can lead to effects on other organelles, which then send signals to the mitochondria to recruit participation in the death process. The initiation of cell death by both death receptors (extrinsic trigger) and intracellular stress (intrinsic trigger) usually involves the participation of mitochondria with the release of proapoptotic proteins from the intermembrane space. When death receptors are engaged, adaptor proteins (FADD, TRADD) bind to the cytoplasmic tail of the receptors. Procaspase 8 is then recruited and self-cleaves to release caspase 8, which in turn cleaves Bid, a proapoptotic Bcl_2 family member. tBid then causes the translocation of Bax to the mitochondria and the aggregation of Bax and Bak. Bax and Bak promote the permeabilization of the mitochondria to release cytochrome c, Smac, AIF, and some procaspases sequestered in the intermembrane space. Cytoskeletal stress (e.g., anoikis) releases Bmf (sequestered with myosin motors), which translocates to mitochondria and promotes permeabilization. Microtubular stress releases Bim (sequestered with dynein motors) in a similar fashion. Bcl_2 and Bcl_{XL} prevent permeabilization. Various Bcl_2 family members influence the balance between proapoptotic (Bax, Bak) and antiapoptotic (Bcl_2, Bcl_{XL}) family members at the level of the mitochondria. The balance of action of these opposing members of the Bcl_2 family determines the permeability of the mitochondria. Signals transduced by activation of death receptors or intracellular stress disrupt the balance at the mitochondrial gateway in favour of release of intermembrane mitochondrial proteins into the cytosol. Intracellular stress in various organelles leads to permeabilization of the mitochondria with the release of intermembrane proteins such as cytochrome c and Smac. Cytochrome c associates with apaf-1 and ATP. This complex binds procaspase 9 to form the apoptosome, resulting in self-cleavage to release caspase 9, which initiates the executioner phase by cleaving caspase 3. Mitochondria also release Smac, which binds and thus removes IAP protection against caspases. ER stress is unique in releasing a sequestered caspase (caspase 12) that intercalates downstream of mitochondria to activate caspase 9 directly independent of the apoptosome. The availability of ATP is required for continued propagation of the apoptosis cascade. The MPT pore opening in the presence of sustained ATP will lead to apoptosis but in the absence of a critical threshold level of ATP will lead to necrosis. The pore consists of the outer membrane voltage-dependent anion channel (VDAC) and the peripheral benzodiazepine receptor, the inner membrane adenine nucleotide translocase (ANT), matrix cyclophin (binds cyclosporin A, which inhibits channel opening), and cytosol hexokinase and creatine kinase. Some of the Bcl_2 family directly bind to pore constituents, such as ANT, and thus may regulate the pore-open versus pore-closed configuration. The pore opening is promoted by Ca^{2+} and oxidative stress. The pore contains functional dithiols, whose oxidation in response to thiol-disulfide redox changes promotes pore opening. Extensive opening of the pore depolarizes mitochondria and results in profound ATP depletion. Thus, pore opening is a critical and perhaps universal event in necrotic cell death. The proapoptotic Bcl_2 members may selectively permeabilize the outer membrane, allowing the release of intermembrane proteins. Apoptosis depends on the participation of caspases. Caspases can be divided into initiators and executioners; caspases exist as zymogens, which are activated either by self-cleavage or cleavage by other caspases. The initiator caspases include caspase 8 and caspase 9, which is activated (self-cleavage) by interaction with cytochrome c/apaf-1 adaptor complex (so-called apoptosome). Thus, the release of cytochrome c is of major importance in initiating the formation of the activating complex. ATP is required for the proper confirmation of the apoptosome. Caspase 9 then initiates apoptosis by cleaving caspase 3. This executes the final dismantling of the cell with the participation of other caspases, such as 6 and 7, which are cleaved by caspase 3. Intracellular stress promotes apoptosis at the level of the mitochondria to cause the release of cytochrome c and the activation of caspase 9 to initiate the cascade. Other intermembrane proteins that are released from mitochondria and and may have considerable importance include: 1- apoptosis-inducing factor, which activates DNA fragmentation, and 2- Smac, which binds caspase inhibitory proteins (i.e., inhibitor of apoptosis proteins [IAPs]), removing these inhibitors from interacting with caspases. Malfolding or protein overloading induced by toxins, oxidative stress, viral protein, or mutant protein accumulation in the ER up-regulates production of caspase 12. Caspase 12 is sequestered on the cytoplasmic face of the ER, where it binds and activates cytoplasmic caspase 7, resulting in cleavage of procaspase 12. Caspase 12 is then released into the cytosol, where it binds and cleaves procaspase 9 to initiate apoptosis independent of mitochondria or the apoptosome. An important means of protecting against apoptosis is through: 1- Inhibition of caspases. This is achieved by attack on the cysteine thiol in the active site by NO, reactive oxygen species, or the redox effect of profound depletion of reduced glutathione (GSH). 2- Specific protein interactions, which inhibit caspases (inhibitors) including IAP family members and heat shock proteins (HSPs). 3- FLIP inhibits activation of caspase 8 by acting as a decoy for binding to FADD. 4- Bcl-2 and Bcl-XL inhibit mitochondrial permeabilization. 5- Phosphatidylinositol 3-kinase (PI_3-kinase)/Akt promotes survival by phosphorylation of caspase 9 and activation of NF-κB. Phosphorylation of IκBa kinase (IKK) leads to proteosomal degradation of IκBa, releasing NF-κB for translocation to the nucleus. IKK is activated by TNFR signaling, oxidative stress, and PI_3-kinase. Hepatocytes, although expressing abundant TNFR1, are resistant to the lethal effects of TNF. When TNF engages TNFR1, multiple signaling systems are activated-several promote cell death and involve activation of caspase 8 or stress-activated kinases, whereas another signaling pathway involves activation of NF-κB and protection. The NF-κB transcriptional up-regulation of survival genes then protects against apoptosis while enhancing inflammation through the up-regulation of cytokines, chemokines, and adhesion molecules.

Inhibition of NF-κB unmasks the lethal effects of TNF through the apoptotic cascade. This may give explanation as to how can toxins sensitize the hepatocyte to the lethal action of TNF. Another mechanism of sensitization to TNF is GSH depletion. An important factor is compartmentation of GSH in mitochondria and cytosol. TNF increases oxidative stress in mitochondria, which is believed to be a result of the translocation of GD3 sphingolipids to mitochondria and blockage of electron transport. Selective mitochondrial GSH depletion by chronic alcohol intake appears to render hepatic mitochondria incapable of detoxifying the TNF-induced burden of reactive oxygen species (ROS). MPT opening occurs with resultant collapse of the mitochondrial membrane potential, ATP depletion, and necrosis. The redox effects of mitochondrial GSH depletion directly sensitize the MPT pore to opening. Antioxidants protect against this TNF-induced necrosis occurring with selective mitochondrial GSH depletion, underscoring the role of enhanced mitochondrial oxidative stress. TNF-induced apoptosis may occur as a result of acute, moderate GSH depletion, which is due to the redox changes in thiol-disulfide status as a consequence of GSH depletion. GSH depletion activates JNK and p38 kinase through redox inhibition of thioredoxin and GSH S-transferase Pi, which normally binds to and inhibits stress kinases. GSH depletion inhibits NF-κB-dependent transcription. The mitochondria are the most important source of ROS, mainly because of auto-oxidation of ubisemiquinone at the level of complex III. This occurs to some extent under physiological, aerobic conditions and explains why profound depletion of GSH alone is lethal for aerobic cells. It is enhanced by blockage of electron transport, as occurs with TNF action or as consequence of loss of cytochrome c, or because of increased electron flow, as occurs with the uncoupling action of some drugs. Other important cellular sources of ROS include CYP2E1, peroxisomal oxidases, iron and copper overload, and NADPH oxidase of phagocytic cells. The biochemical consequences of oxidative stress include lipid peroxidation, protein thiol oxidation, DNA oxidation, and changes in the GSH/GSSG redox buffer. The consequences of these chemical changes may include MPT pore opening, activation of redox-sensitive kinases and transcription factors, such as AP-1, NF-κB, and p53. This may lead to alterations of gene expression and either promote or protect against cell death. The effects of oxidative stress are complex and contradictory. Massive oxidative stress, particularly in mitochondria, induces necrosis. Lesser exposure may be sufficient to trigger apoptosis. Aside from initiating apoptosis, oxidative stress may occur as a downstream secondary event due to uncoupling of mitochondria, cytochrome c release, or the nearly universal massive export of cell GSH that accompanies apoptosis. In addition, oxidative stress may protect against apoptosis in some circumstances as a consequence of the inhibition of caspases by ROS or the activation of NF-κB and increased expression of survival genes. Fatty liver predisposes to oxidative stress presumably by amplifying the capacity for free radical chain reactions. The most important consequence of oxidative stress in the pathogenesis of liver disease is the promotion of inflammation through the activation of transcription of cytokines, chemokines, and adhesion molecules. Liver injury is encountered in response to exogenous sources, such as infiltrating inflammatory cells, TNF action, or sometimes CYP2E1 induction or storage of unsaturated fatty acids. Cyclosporin A and ursodeoxycholate inhibit the MPT opening. The use of antioxidants such as tocopherol and silymarin or GSH precursors such as S-adenosylmethionine, N-acetylcysteine, and GSH esters may be effective against oxidative stress or GSH depletion. Pentoxifylline may be effective against cytokines and death receptors. Protective cytokines such as IL10, MCP-1, or IL6 may be of value in certain circumstances by down-regulating toxic cytokines, promoting survival gene expression or liver regeneration.

The Apoptotic-Antiapoptotic Molecular Machinery:

Prodeath	Prosurvival
1- Death receptors (e.g., Fas, TNF-R1).	1- NF-κB.
2- Caspases.	2- PI$_3$-kinase/Akt.
3- Bid, Bax, Bak, Bmf, Bim.	3- Bcl-2, Bcl-XL (NF-κB responsive survival genes).
4- Mitochondrial proteins.	4- Caspase inhibitors.
5- Cytochrome c.	5- FLIP.
6- Smac.	6- IAPs (NF-κB responsive survival genes).
7- AIF.	7- HSPs.
8- Stress kinases.	8- NO (iNOS [NF-κB responsive survival genes]).

ALCOHOL-INDUCED LIVER DISEASE (ALD)

Alcohol-induced liver disease (ALD) results from the dose- and time-dependent consumption of alcohol. The disease process is characterized by early steatosis, inflammation, and necrosis (steatohepatitis), and in some individuals, ultimately, it progess to fibrosis and cirrhosis. O_2-derived prooxidants play a central role in alcohol-induced liver injury. Cytosolic Cu/Zn-SOD or mitochondrial Mn-SOD prevents alcohol-induced liver injury. O_2^- cannot directly attack ethanol to form the α-hydroxyethyl radical. The reduction of O_2^- by SOD forms H_2O_2, H_2O_2 and transition metals can lead to formation of hydroxyl radicals (Fenton reaction). H_2O_2 can react with Cl^- via myeloperoxidase in neutrophils to make hypochlorous acid ($HOCl^-$). Although O_2^- is not a potent prooxidant per se, it appears to be a key starting point of oxidative stress during alcohol exposure. Reactive nitrogen species (RNS) are similar to ROS, it is likely that NO^- serves as the parent molecule for other RNS in vivo. Dysregulation of vascular tone after acute alcohol and during alcohol-induced cirrhosis is mediated, in part, by a decrease in NO^- production. NO^- is antiapoptotic in hepatocytes and is required for normal liver regeneration. NO^- is chain-breaking antioxidants by reacting with lipid peroxyl radicals. NOS-inhibitor N(G)-nitro-L-arginine methyl ester exacerbates liver injury. Thus, arginine supplementation may reverse ALD liver damage. High levels of NO^- play damaging role in ALD by production of RNS, such as $ONOO^-$. RNS derived from NO^- cause nitration reactions (formation of 3-nitrotyrosine) and nitrosation reactions (formation of nitrosothiol), as well as oxidation reactions during alcohol exposure. Reactive intermediates formed during $ONOO^-$ degradation can cause one electron oxidation reactions with ethanol, leading to the formation of hydroxyethyl radicals. A key component in the progression of alcohol-induced liver injury is inflammation, both endogenous proinflammatory cells (Kupffer cells) and infiltrating inflammatory cells (neutrophil) are involved. If inappropriate stimulated inflammatory cells cause damage to normal tissue rather than detoxification. Alcohol primes Kupffer cells and monocytes, increasing phagocytic and bactericidal activities, and cytokine production. Oxidant production from Kupffer cells in liver leads to activation of oxidant-sensitive signaling pathways, which prime the cells to LPS activation. In hepatocytes, the stimulus of oxidant production caused by alcohol leads to a sensitization of the cells to cytotoxic signaling through apoptotic/necrotic signaling pathways. NAD(P)H oxidase is invloved in ALD. Damaging oxidant in ALD is dependent on the production of O_2^- and NO^-, and $ONOO^-$ may play a critical role in the initiation and progression of liver damage by increasing RNS production. Other sources of oxidants in inflammatory cells include xanthine oxidase and myeloperoxidases. The xanthine oxidase inhibitor, allopurinol is protective against liver damage caused by alcohol, which may be mediated by preventing O_2^- production after hypoxia/reoxgenation. Myeloperoxidase also can form NO^--derived RNS that can mediate tissue damage. There are also many potential sources of prooxidants, 2 major suspected sites in hepatocytes are the ethanol-inducible CYP2E1 and mitochondria. In microsomes CYP2E1 is coupled loosely with cytochrome reductase, leaking electrons to oxygen to form O_2^-, or to catalyze lipid peroxidation. CYP2E1 overexpression enhances oxidative stress in the cells, either directly, by producing prooxidants, or indirectly, by altering/enhancing the response to prooxidants from other sources. The reduction of O_2 to H_2O by mitochondria is not tightly regulated, it is estimated that 1-2% of O_2 consumption by mitochondria leads to the formation of O_2^-. An 80 Kg human would produce 215-430 mmol O_2^- a day from mitochondria. Alcohol increases the yield of O_2^- from this cellular component. Elevated prooxidant production by mitochondria increases the net yield of prooxidants in the hepatocyte, and leads to direct damage of mitochondrial proteins and DNA, which can exacerbate mitochondrial aging and stimulate mitochondrial-mediated apoptotic pathways. Alcohol depletes mitochondrial glutathione levels, which increases the response of hepatocytes to apoptotic stimuli. Oxidative stress may play a role in the transformation of stellate cells into myofibroblasts, the critical matrix-depositioning cell in fibrotic liver, i.e., nonparenchymal cells prooxidant production. Oxidant-dependent signaling from other cells (cytokines) contributes to stellate cell transformation and collagen deposition. Oxidant production within the cell also may be involved in their transformation. Any changes in the cell that favour prooxidant formation or disfavour antioxidant defenses can lead to oxidative stress. Alcohol can cause modification to the cell that lead to a shift in this delicate balance independent of increased prooxidant production per se. The mobilization of free iron caused by alcohol leads to an increase in transition-metal catalysis to potent oxidants (Fenton reaction). There is a strong link between dysregulated iron homeostasis (e.g., haemochromatosis) and alcohol-induced liver disease. Iron supplementation enhances damage caused by alcohol. Alcohol causes hypoxia in pericentral regions of the the liver lobule. Cells undergoing hypoxia are prone to oxidative stress. Hypoxia increase prooxidant production via hypoxia/reoxygenation and shifts the cellular balance to favour oxidative stress. A decrease in cellular oxygen tension exacerbates the impaired mitochondrial electron flow caused by alcohol by decreasing the delivery of O_2 to the mitochondria. Hypoxia decreases the ability of the liver cells to detoxify free radicals. There exists a host of proteins and systems involved in the antioxidant network. This family does not directly intercept prooxidants, but serves instead as ancillary reductants and maintains the catalytic activity of antioxidants proteins or small molecules. The depletion by ethanol of both cytosolic and mitochondrial energy supplies can thus indirectly impair cellular antioxidant defenses. Oxidative stress causes cellular injury by chemical modification of biological molecules, which alters and/or interfere with normal biologic processes and be toxic directly to the cell. Modified biological molecules may stimulate the host immune response and cause an autoimmune-like disease. Antibodies against such oxidatively modified proteins occurs in ALD. Proxidants can confer highly specific changes in a cell at concentrations well below observable chemical damage. The amount of oxidants produced during liver ischaemia/reperfusion may be too low to cause significant direct chemical changes early in the disease. Prooxidants can mediate and/or amplify their signal by modifying signaling cascades within the cell. Oxidant-sensitive signaling cascades include small molecules (Ca^{2+}), stress-activated protein kinases (c-Jun), transcription factors (NFκB), and modulators of apoptosis signaling (Bcl-2). Ethanol modulates the signal of many of these cascades. Ethanol alters activation of stress-activated protein kinases in cell types of the liver, which may be mediated in part by oxidant production. The activation of NFκB is linked to oxidant production. The effect of NFκB activation has different biologic effects in different cell types. NFκB is a key mediator in the inflammatory response in Kupffer cells by up-regulating proinflammatory genes (TNFα). NFκB inhibtion causes apoptosis. Many prooxidants are regulated by enzymes, channels, or pumps, are enzymatically degraded, are in rapid flux within the cell, and are specific in action. Alcohol increases circulating LPS and these increases appear to stimulate an inflammatory response during ALD. Inflammatory cells appear to be primed to activation by alcohol. Some proteins appear to have both damaging (proinflammatory) and protective (antiapoptotic) roles in ALD, such as iNOS and NFκB. Oxidants also modify intracellular signaling through activation/modulation of cytokine expression. Specifically, the production of proinflammatory cytokines (TNFα) from Kupffer cells is stimulated by activation of oxidants-sensitive NFκB. Prooxidants enhance the production of signaling cascades/molecules, which then mediate their effects. TNFα appears to be a critical link between prooxidant production in inflammatory cells and hepatocytes.

TNFα and other cytokines mechanisms contributing to steatosis:

1- TNFα direct effects include increased FFA release from adipocytes in periphery, increased lipogenesis in hepatocytes, and inhibition of β-oxidation of fatty acids.
2- TNFα can directly increase ROS formation by impairing mitochondrial electron flow, leading to formation of O_2^{-}.
3- Indirectly, the oxidation of lipids by ROS/RNS can further impair β-oxidation of fatty acids, and further damage mitochondria.
4- IL1 and IL6 and other cytokines may impair transport and secretion of triglycerides (TG).

Allopurinol prevents the accummulation of lipid droplets in the hepatocytes. Steatosis caused by alcohol is considered the result of NADH redox inhibtion of mitochondrial β-oxidation caused by alcohol metabolism. TNFα and other cytokines can influence lipid metabolism both in liver and periphery. TNFα increases free fatty acid release from adipocytes in the periphery, increases lipogenesis in hepatocytes, and inhibits β-oxidation of fatty acids. Prooxidant production stimulated by TNFα in hepatocytes could impair mitochondrial electron flow and cause lipid peroxidation, processes that also could slow the metabolism of fat by mitochondria. IL1 and IL6 may impair transport and secretion of triglycerides. The net consequence during alcohol exposure is that cytokines increase the supply of fatty acids to liver while simultaneously impairing the ability of the hepatocytes to metabolize and secrete them. These effects work in tandem with alcohol-induced shifts in the NADH redox state to lead to steatosis. The enhancement of the cell's own intrinsic antioxidant capacity should be a goal of antioxidant therapy for ALD. Antioxidants may not be effective in human owing to the kinetics of antioxidant reactions. Multiplication of rate constant of the reaction by the achievable concentrations in vivo yields the predicted rate of disappearance of the prooxidant caused by the antioxidant without toxicity concerns, which may limit using such doses in human. Nevertheless, there are antioxidants that have sufficiently rapid rate constants to be pharmacologically relevant. S-adenosyl methionine protects mitochondrial glutathione pools against alcohol by normalisation of the microviscosity of the mitochondrial inner membrane. Some classic antioxidants may mediate protection in vivo, not by directly scavenging prooxidants, but by stimulating intrinisic antioxidant production through activation of the antioxidant response element on genes.

SEPSIS AND ALI-ARDS

Sepsis syndrome is a continuum that ranges from infection with inflammation to shock and differs by degree of severity. Sepsis is the systemic inflammatory response to infection and is manifested by at least 2 of the following: hypo- or hyper-thermia; tachycardia; tachypnoea or hypocapnia; and leukopenia, leukocytosis, or bandaemia. Severe sepsis is associated with organ dysfunction from hypoperfusion. The presence of hypotension despite adequate fluid resuscitation constitutes septic shock. Similar to the sepsis syndrome, acute lung injury (ALI) and acute respiratory distress syndrome (ARDS) are a continuum of a single pathologic process of acute respiratory failure, with the spectrum differing by degree of severity. Acute lung injury is characterised by acute development of bilateral infiltrates on chest radiographs, pulmonary capillary wedge pressure of 18 mmHg or less or the absence of clinically evident left atrial hypertension, and a ratio of partial pressure of arterial oxygen to the fraction of inspired oxygen (PaO_2:FiO_2) of 300 or less. In response to noxious stimuli, the host simultaneously activates an interactive network of pathways that act in synergy to increase the probability of survival. The host defense response to insults is similar regardless of the tissue involved and consists of a complex cascade involving integration of inflammation, coagulation (intravascular clotting and extravascular fibrin deposition), immune response modulation, tissue repair, and activation of the hypothalamic-pituitary-adrenal (HPA) axis to produce endogenous corticosteroids. Activation of the sympathetic system to release catecholamines and increase hepatic production of acute-phase cytokines, as they work in synergy to either eliminate the noxious stimulus or repair tissue damage. Both severe sepsis and ALI-ARDS involve an uncontrolled host defense response leading to inflammation, endothelial damage, enhanced coagulation, diminished fibrinolysis and fibroproliferation to produce microthrombi, and relative adrenal insufficiency. The early pathologic mechanisms that cause sepsis and ALI-ARDS are similar. The primary difference is that ALI-ARDS is confined to the alveolocapillary units of the lungs, whereas sepsis is a systemic process that involves many organs. The development of ALI-ARDS occurs over several days. The early pathogenesis of ALI-ARDS may be different in patients with causes other than sepsis. In all cases of ALI-ARDS, the reversible pathologic mechanism involves pulmonary vasoconstriction, microthrombi within alveolocapillaries, interstitial oedema, and reduced production of surfactant to raise pulmonary artery pressure. Balancing the production of proinflammatory and antiinflammatory cytokines establishes tissue homeostasis and coordinates the host defense response. There is a delayed inflammatory response in sepsis and ALI-ARDS, which is mediated by the eicosanoid pathways. Through mediation of other transcription factors, corticosteroids suppress the synthesis of phospholipase A_2 and cyclooxygenase to reduce the production of prostanoids, PAF, and NO, which contribute to the delayed inflammatory response. Corticosteroids protect the host from an overwhelming inflammatory response that if left uncontrolled will have deleterious effects. Restoration and maintenance of proinflammatory-antiinflammatory cytokine homeostasis by corticosteroids are paramount for survival. The production of endogenous corticosteroids is essential for coordinating the host defense response and maintaining tissue homeostasis. During the development of sepsis or ALI-ARDS, proinflammatory cytokines such as IL-1β, IL-6, and TNF-α are released from lymphocytes, activated macrophages, and endothelial cells. These cytokines are generated by various tissues during sepsis but are produced locally within the alveolocapillary units during ALI-ARDS. Together these cytokines promote endothelial leukocyte adhesion, stimulate the release of proteases and arachidonic acid (AA) metabolites from neutrophils, induce coagulation, suppress fibrinolysis, downregulate adrenergic receptor sensitivity, and activate the HPA axis to produce corticotropin. The HPA axis responds in a graded manner according to the degree of stress, with each of the 3 cytokines independently stimulating the HPA axis, but when combined, they act synergistically. The secretion of cortisol from the adrenal glands is important because it acts as an inhibitory feedback to interrupt the inflammatory process of the host defense response. Also, cortisol inhibits the release of catecholamines and decreases hepatic production of acute-phase proinflammatory cytokines. Cytokines and corticosteroids modulate gene expression by altering the rate at which a gene is transcribed into mRNA. For this process to occur, proteins called transcription factors are needed to recognise the DNA sequences that are to be transcribed. Cytokines and corticosteroids enhance the expression and activity of transcription factors. Specifically, proinflammatory cytokines, in addition to bacteria, bacterial products, reperfusion injury, and oxygen free radicals (OFR), activate NF-κB within macrophages, lymphocytes, and endothelial cells to cause the production of more acute proinflammatory cytokines and leukocyte adhesion molecules; activate iNOS; downregulate adrenergic receptor sensitivity; and reduce corticosteroid receptor binding affinity. Activation of iNOS to enhance nitric oxide (NO) formation and reduced adrenergic receptor activity produce vasodilatation. NF-κB activity is higher in nonsurvivors of sepsis and ARDS and correlates with severity of illness. Corticosteroids directly inhibit NF-κB by preventing DNA binding and indirectly inhibit NF-κB by activating inhibitor [kappa]B-[alpha] (IκB-α), which sequesters NF-κB in the cytoplasm of macrophages, lymphocytes, and endothelial cells. The overall result is a reduction in the production of proinflammatory cytokines and leukocyte adhesion molecules, inhibition of iNOS, and reversal of adrenergic receptor desensitisation. Inhibition of iNOS and enhanced adrenergic receptor activity increase myocardial contractility and produce vasoconstriction, both of which elevate blood pressure. Activation of IκB-α by corticosteroids also regulates apoptosis and promotes the production of the antiinflammatory cytokines IL-4 and IL-10 and the cytokine antagonists IL-1 receptor antagonist and soluble TNF receptor. Nonsurvivors of sepsis and ARDS exhibit persistent and exaggerated serum and alveolar proinflammatory cytokine concentrations compared with those of survivors. Nonsurvivors have lower concentrations of antiinflammatory cytokines. If the noxious stimulus is not removed, continuous corticotropin secretion eventually produces cortisol deficiency. Under normal conditions, the amount of cortisol produced by the adrenal glands is a mean of 5.7 mg/m^2 of body surface area/day (equivalent to 8-13 mg/day for the average person). During conditions of high stress such as sepsis or ALI-ARDS, the amount of cortisol produced may be 150-200 mg/m^2/day (equivalent to 225-440 mg/day for the average person) for several days. Serum cortisol concentrations may reflect the severity of illness during sepsis and ALI-ARDS. A single cortisol concentration may inadequately reflects the physiologic status of the patient because adrenal function during sepsis and ALI-ARDS is dynamically dependent on the host defense response. The continuous activation of the HPA axis by proinflammatory cytokines

(IL-1β, IL-6, and TNF-α) during the host defense response may produce relative adrenal insufficiency because the adrenal glands are unable to function at maximal capacity for prolonged periods. Although TNF-α stimulates the HPA axis to produce corticotropin-releasing hormone (CRH) and corticotropin, it blocks, in a concentration-dependent manner, the stimulatory effect of corticotropin. In patients with severe sepsis or ARDS serum cortisol concentrations decline gradually over several days, but serum corticotropin concentrations are not affected. Normally cortisol suppresses the HPA axis, but the inhibitory action of cortisol is impaired during severe sepsis, thus allowing unabated HPA axis stimulation by proinflammatory cytokines. Eventually a vicious cycle is created because endogenous cortisol production is restricted to the degree that NF-κB is not suppressed and proinflammatory cytokines are inadequately regulated. Cortisol production is only minimally enhanced in nonsurvivors of severe sepsis after the corticotropin stimulation test, which reflects that the adrenal glands are functioning at their maximum stimulatory capacity. The adrenal glands are no longer capable of producing cortisol when they are stimulated. Although the frequency of occult adrenal failure is less than 3% in critically ill patients, relative adrenal insufficiency occurs in 16.3-55% of patients with sepsis or ALI-ARDS and in 39.4-66.7% of patients with shock. Ischemic injury, the formation of microthrombi, and localised haemorrhage predispose the adrenal glands to bilateral damage. Cortisol is produced by the adrenal glands and immediately secreted into the serum where it circulates 95% bound to corticosteroid-binding globulin (CBG). During the first 7 days of the inflammatory response, CBG concentrations are markedly decreased owing to increased metabolism by cytokine-induced elastases, decreased hepatic production, and increased capillary leakage. Although unbound cortisol concentrations are increased, the availability of cortisol at the target tissue is reduced because CBG acts as a carrier protein to facilitate cortisol delivery. Hepatic metabolism of endogenous cortisol may be enhanced if the unbound fraction increases. Peripheral cortisol resistance occurs in patients with sepsis and ALI-ARDS because proinflammatory cytokines (IL-1β, IL-2, IL-6), antiinflammatory cytokines (IL-4), and phospholipase A_2 produce a concentration-dependent reduction in cellular receptor affinity to corticosteroids. The stress-dose administration of hydrocortisone may reduce the systemic inflammatory response associated with severe sepsis and septic shock. Higher infection rates are associated with corticosteroids therapy. Therapy with corticosteroids is associated with early shock reversal or blood pressure elevation. Serum concentrations of the proinflammatory cytokines IL-6 and IL-8 are significantly reduced within 24 hrs of starting stress-dose hydrocortisone. Serum concentrations of the late inflammatory mediators phospholipase A_2 and neutrophil elastase are reduced within 48 hrs of starting stress-dose hydrocortisone. But, IL-8 concentrations are not significantly different. The antiinflammatory cytokine IL-10 is not altered by stress-dose hydrocortisone. Elevated heart rate and body temperature, two manifestations of systemic inflammatory response syndrome, and serum cortisol concentration may return to normal during hydrocortisone administration but may trend upward after discontinuation of hydrocortisone. Although mortality rates are 34.7% in the hydrocortisone group and 22.6% in the placebo group, the median daily scores of sepsis-related organ failure assessment decreases significantly after 3 days of hydrocortisone, and remains constant in the placebo group. Respiratory function may improve quicker with hydrocortisone administration, resulting in substantially fewer days of mechanical ventilation. Also hydrocortisone therapy may reduce the frequency of posttraumatic stress disorder in patients surviving sepsis. Similar to sepsis, serum and alveolar concentrations of the antiinflammatory cytokine IL-4 are not altered by stress-dose methylprednisolone. Promising results are reported with stress-dose administration of corticosteroids to patients with vasopressor-dependent septic shock. The mechanism of action of stress-dose administration of corticosteroids is likely multifactorial. Reducing the inflammatory response and enhancing the antiinflammatory response restores cytokine homeostasis in sepsis and ALI-ARDS. For cases of septic shock, upregulation of adrenergic receptors to increase catecholamine sensitivity accelerates the discontinuation of vasopressors. Peripheral cortisol resistance may be reversed with exogenous corticosteroid administration. Exogenous corticosteroid administration may reset the HPA axis. Although adrenocortical insufficiency is common in patients with sepsis, beneficial effects of stress-dose corticosteroid therapy are observed regardless of adrenocortical status. The most important goals in the treatment of sepsis and ALI-ARDS are to reduce mortality and morbidity. Stress-dose corticosteroids cannot be recommended for severe sepsis without shock. The pathobiology of ALI-ARDS is similar to that of severe sepsis as both involve an uncontrolled host defense response. Patients with extensive burns, fungal infections, uncontrolled diabetes mellitus, acute pancreatitis, immunosuppression, or recent GI haemorrhage should not receive stress-dose corticosteroid therapy. Stress-dose corticosteroid therapy may increase the rate of occurrence of GI haemorrhage and secondary infections. The rate of occurrence of GI haemorrhage seemed to be increased with stress-dose corticosteroid therapy in patients with septic shock rather than in those with ALI-ARDS. Corticosteroid therapy in septic shock may have greater detrimental effects because hypotension and vasopressor therapy already compromise GI blood flow.

ADRENAL INSUFFICIENCY

Adrenal insufficiency is often misdiagnose. The functions of the medulla and cortex are distinctly different. DHEA is a well-known sex hormone precursor and immune system enhancer. Pregnenolone is a precursor to DHEA, with a predilection to cascade down to progesterone. Stress is defined in a physiological context as any factor that acts to destroy homeostasis. More pecisely, it is the body's response to any factors that threaten its ability to maintain homeostasis. Many women, however, approach menopause in a state of chronic emotional and nutritional depletion, which affects optimal adrenal function. A secondary side effect of this stress response is decreased mental clarity due to the hippoocampus's chronic exposure to cortisol. The stress response is part of the larger general adaptational syndrome. This syndrome is composed of 3 phases: a- alarm, b- resistance, and c- exhaustion. These reactions ultimately cause the pituitary to release ACTH, which in turn causes the adrenals to release stress-related hormones, such as adrenaline. This short-lived phase is followed by the resistance phase, which allows the system to continue its adaptation to stress long after the alarm phase effects have worn off. Corticosteroids, such as cortisol, mediate this response. However, if prolonged, stress reaches the final phase, which is exhaustion. Secondary disorders can cause adrenal corticol deficiency, which include nonsteroidal anti-inflammatory drugs and pituitary disease. Even oversecretions of the medulla can affect cortical function. A stressor creating a fight-or-flight response releases adrenaline. The most common presenting symptoms of chronic primary adrenal cortical insufficiency are weakness, fatigue, weight loss, and anorexia. Low blood pressure, salt cravings, gastrointestinal symptoms, and hyperpigmentation can also be seen in these patients, Non- specific symptoms are commonly seen in marginally deficient patients include poor mental clarity, decreased sexual function, decreased libido, and just not feeling right. A three-month supply of hydrocortisone (5mg or less q.i.d.) may be recommended for clinical repletion with hydrocortisone (possesses both glutacorticoid and mineralocorticoid activity) and carefully monitoring the patient with regular monthly visits. After 90 days, add nutraceuticals for a week to 30-day period. Then gradually reduce and eliminate the hydrocortisone over a 2-4 weeks. If the gland is still not operating appropriately, further drug repletion may be necessary. Careful examination of the chronically ill patient often reveals significant adrenal insufficiency. Serial saliva testing is the easiest and most convenient way to diagnose adrenal cortical deficiency. Early morning, noon, dinner, and PM samples are obtained by soaking a cotton ball with saliva and sent to the laboratory. The best screening test is to measure a serum cortisol before, and 30-60 minutes after, an IV or IM injection of 0.25mg synthetic ACTH. A normal response would be a rise in the serum cortisol level of two to three times baseline, or a peak response no less than 15 µg/100ml. If DHEA levels are low, replacement is required. Normal ranges for salivary levels are 3-10 ng/ml, with 8-10 ng/ml reflecting the best immune function. Rheumatoid arthritis, fibromyalgia, and even polyneuropathy respond to DHEA more than other drugs. The patient must: a- Stop smoking; b- Discontinue any street drugs; c. Decrease alcohol consumption; d- Decrease intake of fats, salt, and sugar; e- Decrease caffeine consumption, f- Exercise aerobically at least 3 hours a week; g- Use stress-reduction techniques such as prayer, meditation, soothing music, funny movies, h- Guided imagery, or biofeedback, i- Get professional help to deal with anger and rage, and deal with psychospiritual issues, which surround work, family, care-giving, self-esteem, and body image, which may help reduce cortisol levels; and j- Limit exposure to extremely low-frequency magnetic energy fields. If secretory IgA levels are low, repair the probable intestinal permeability with L-glutamine, an amino acid often taken from muscles, lung and bone to replenish the supply for the enterocytes and coloncytes. Whey product and eggs are high in immunoglobulin IgA. Melatonin improves secretary IgA. ω-3 and ω-6 oils (ratio 4: 1) and pantothenic acid (vitamin B_5) is recommended. 2-5 g of potassium a day is recommended. Vitamins C (1-2 g b.d) and B_6 (100-300 mg/day) are vital nutrients for adrenal function. Zinc (50 mg/day) and magnesium (500 mg/day) are recommended. Ginseng 1-2 g/day provides tonic effects (adaptogen), contain 25-50 mg of ginsenosides. Ginseng increases the responsiveness of the adrenal gland and the ability to control the gland secretion. It is adviced to use 90-day courses of ginseng, and alternate this with licorice derivatives. Licorice has aldosterone-like properties. The herb ashwagandha may be helpful. Glycyrrhetinic acid exhibits regulatory action on the adrenal gland. Soluble adrenal fractions, a nutraceutical, can stimulate a sluggish gland to become more productive. Increase this product until there are symptoms of nervousness or difficulty sleeping. Then reduce the dose slightly until there are no more undesirable side effects. Both tyrosine and phenylalanine restore epinephrine levels. Phenytalanine is decarboxylated to phenylethylamine (PEA), which has amphetamine-like stimulant properties (found in high concentrations in chocolate). Phenylalanine is also hydroxylated to tyrosine, which eventually forms epinephrine. Tyrosine, which is necessary for the formation of norepinephrine, is low in depressed patients. Supplementation increases levels of 3-methoxy-4-hydroxyphenethylene glycol (MHPG) in the urine. This is probably the principle breakdown product of norepinephrine in the CNS, and may provide a marker to determine which amino acid to supplement. Heavy metal burdens must be removed with an appropriate chelating agent. Coenzyme Q_{10} and vitamin A may help correct abnormal exhaustion. Repeat the salivary testing every 90 days until the situation is corrected. Plasma renin assess the need for mineral corticoid replacement therapy. Blood pressure should be measured in both arms, in lying, sitting, and standing positions. Causes of failure of treatment include: 1- too low dose, 2- too short treatment intervals, 3- poor patient compliance, 4- external causes have not been eliminated, 5- IgA levels did not improve, 6- inadequate treatment of chronic anxiety and/or depression, 7- food allergies, and 8- bioavailability, e.g., adequate hydrochloric acid, gut dysbiosis.

BONE HEALTH

Heavy consumers of soft drinks (with or without sugar) spill huge amounts of calcium, magnesium and other trace minerals into the urine. The more mineral loss, the greater the risk for osteoporosis, osteoarthritis, hypothyroidism, coronary artery disease, high blood pressure and a long list of degenerative diseases generally associated with premature aging. Soft drinks are harmful because the substances added to the carbonated/distilled/filtered water are really the problem. These include sugar or artificial sweeteners (especially aspartame), caffeine, dyes, artificial flavours, and phosphoric acid. Soft drinks induce acidosis and osteoporosis may result due to loss of calcium from the bone (part of the buffering system of the body, which maintains the acid-base balance of the body critical for life). Soft drinks can also upset the GIT function. Distilled water leaches inorganic minerals rejected by the cells and tissues, so they can be excreted. This is a good property of distilled water, and helpful to the body. Flavour can be added to distilled water by adding 1-2 tablespoons of raw apple cider vinegar per gallon of water. Vinegar is an excellent solvent and aids in digestion. Lemon juice is another good flavouring agent, and has cleansing properties as well. Disability is a general term used to represent the interactions between individuals with a health condition and barriers in their environment. The term disability is operationalized as self-reported activity limitations or use of assistive devices or equipment related to an activity limitation. Osteopenia is a condition similar to osteoporosis except the reduction in bone mass is not as severe. Arthritis and other rheumatic conditions affect quality of life in many ways and are key items of personal interest to individuals with these conditions. As the leading cause of disability, arthritis is a leading cause of difficulty in performing personal care activities and thereby a leading cause of loss of independence. Coping difficulties, depression, anxiety, and low self-efficacy are recognised as major personal and emotional problems among persons with arthritis. Because arthritis is a leading cause of chronic pain, monitoring these mental health outcomes can help assess the success of applied interventions. Health impact of arthritis include: 1- arthritis is the leading cause of disability, 2- arthritis limits the major activities (e.g., working, housekeeping, school), 3- arthritis trails only heart disease as a cause of work disability, 4- arthritis limits the independence of affected persons and disrupts the lives of family members and other caregivers, 5- quality-of-life measures are consistently worse for persons with arthritis (i.e., difficulty in performing personal care activities), 6- arthritis has a sizeable economic impact, 6- arthritis, like other chronic pain conditions, has an important negative effect on a person's mental health, and 7- persons with certain forms of arthritis have higher death rates than the general population (e.g., rheumatoid arthritis are at greater risk of premature death from respiratory and infectious diseases). The activity limitations of arthritis also indirectly affect health and independence by decreasing physical activity, increasing weight, and placing persons at higher risk for all the adverse outcomes of those risk factors. Rheumatoid arthritis is a chronic, inflammatory disease of the body that produces its most prominent manifestations in joints, often leading to joint pain, stiffness, and deformity. Osteoporosis is bone disease characterised by a reduction of bone mass and a deterioration of the microarchitecture of the bone leading to bone fragility. Health impact of osteoporosis is increased risk of fractures (1 in 3 women and 1 in 8 men aged 50 years and older will experience an osteoporotic-related fracture in their lifetime). Chronic low back pain is described in different ways, such as the occurrence of back pain lasting for more than 7-12 weeks, back pain lasting beyond the expected period of healing, or frequently recurring back pain. Health impact of low back pain (LBP) includes: 1- activity limitation, 2- impairment, and 3- disability. Back pain occurs in 15-45% of people each year, and 70-85% of people has back pain some time in their lives. Work-related risk factors account for 28-50% of the low back problems in an adult population. There is a popular belief that arthritis is part of normal aging that a person can do nothing about it, and that it affects only old persons. Bone mineral density (BMD) has been identified as one of the primary predictive factors for osteoporosis-related fractures. Osteoporosis is defined as a BMD value that is more than 2.5 standard deviations below that of an average young adult. The proportion of adults aged 50 years and older with osteoporosis in the total femur region is 10% (16% in women and 3% in men). Osteoporosis is a major risk factor for hip fracture. Virtually all persons with a hip fracture are hospitalised for treatment. Two-thirds of persons who fracture a hip do not return to their prefracture level of functioning. Interventions that reduce the rate of osteoporosis should have a marked impact on the rate of hip fractures. Increasing BMD by 5% may decrease the risk of fractures by 25%. Vertebral fractures are the most common fracture due to osteoporosis. About 30-50% of women and 20-30% of men will experience vertebral fractures in their lifetime. Most of these fractures cause little difficulty and go unrecognised. However, 33% of the fractures will be diagnosed clinically, 8% will require hospitalisation, and about 2% will require long-term nursing care. The most common symptom of vertebral fractures is back pain, which is reported in about half of the cases.

People with these fractures are more likely to have difficulty performing activities of daily living, such as bending, reaching above the head, and walking. Changes in the outward appearance of people experiencing these fractures are a loss of height and the development of a humped back. Interventions that reduce the number of persons with osteoporosis should reduce the rates of vertebral fractures. Normal daily activities can place sufficient stress on the vertebra to cause fractures. Persons who are overweight and persons who frequently bend over or lift heavy objects are at risk for low back injuries. Occupations that require repetitive lifting, particularly in a forward bent and twisted position, place employees at especially high risk. Other risk factors for low back injury include exposure to vibration produced by vehicles or industrial machinery, prolonged vehicle driving, and certain sports activities. Predictors of back problems may include diminished lumbar flexibility, trunk muscle strength, and hamstring elasticity. Osteoporosis increases the risk of vertebral compression, which may account for the increase in reported low back pain in older females. Increased age also is associated with back pain. Persons who have experienced back problems in the past are at increased risk for future injury. Educational and behavioural interventions also can relieve symptoms and reduce disability. Interventions for osteoporosis and fractures can be designed to prevent the development of the disease, reduce further bone loss after the occurrence of the disease, and lessen the risk of fractures. Opportunities for primary prevention occur throughout the lifespan and include programs to promote exercise, avoid smoking, reduce excessive alcohol consumption, and improve nutrition, particularly the amount of calcium and vitamin D in the diet. Bone can be lost at the rate of 2-4% per year in menopause women. Older persons, even those who have had a fracture can benefit from treatment to prevent further bone loss or restore some lost bone to decrease the risk of subsequent fractures. Ergonomic approaches would be effective in preventing chronic LBP as well. The overall benefits of exercise, nutrition, and lifestyle changes on an individual's health and well-being would certainly justify efforts in this area. The goal of intervention is to prevent illness and disability related to arthritis and other rheumatic conditions, osteoporosis, and chronic back conditions.

NICOTINE

Cigarette smoking especially accelerates atherosclerosis in the coronary arteries, the aorta, the carotid and cerebral arteries, and the large arteries in the peripheral circulation. Predictors of atherosclerosis include: 1- hypertension, 2- serum cholesterol level <297 mg/100 ml, 3- glucose intolerance, 4- left ventricular hypertrophy on ECG, and 5- smoking. History of smoking is the most predictive factor of the development of intermittent claudication. Smoking is associated with: 1- acute myocardial infarction, 2- sudden death and stroke, 3- aggravation of stable angina pectoris, vasospastic angina, intermittent claudication, 4- rethrombosis after thrombolysis, and 5- restenosis after angioplasty. The most important constituents of cigarette smoke are nicotine, aromatic hydrocarbons, sterols and oxygenated isoprenoid compounds, aldehydes, nitriles, cyclic ethers, and sulfur compounds. Pharmacology and metabolism of nicotine: Nicotine is a tertiary amine composed of a pyridine and pyrrolidine ring. Tobacco contains both a levorotatory S-nicotine and an R-isomer in quantities up to 10% of the total nicotine present. Nicotine is a weak base with a pK_a of 8. Almost 30% of nicotine is nonionised and can readily cross cell membranes. Nicotine binds to acetylcholine receptors at the autonomic junctions and the brain. Nicotine cholinergic receptors exist in the brain, autonomic ganglia, and the neuromuscular junction. Nicotine is distilled from burning tobacco, carried proximally, and deposited in the small airways and alveoli. On inhalation, nicotine is rapidly distributed to body tissues. It takes 10-20 s for nicotine to pass through the brain. Nicotine levels then fall, due to uptake by peripheral tissues and later to elimination of nicotine from the body. The arterial levels of cigarette smoking exceed venous levels by 6- to 10-fold. Nicotine plasma concentration in smokers range between 10^{-5} mol/l and 10^{-8} mol/l. The elimination half-life of nicotine during the use of tobacco or nicotine products averages 2-3 hrs. Nicotine levels accumulate over 6-8 hrs during regular smoking. But there is a very long terminal half-life, 20 hrs or more, reflecting the slow release of nicotine from body tissues. Nicotine crosses the placenta freely to the amniotic fluid and the umbilical cord blood of neonates. Nicotine is metabolised by the liver and to a lesser extent by the lung. The renal excretion depends on urinary pH and flow. 70-80% of nicotine is metabolised to cotinine and the rest to nicotine-N'-oxide. Cotinine is metabolised to 3'-hydroxycotinine and is abundant metabolite found in urine. Both nicotine and cotinine undergo further N-glucuronidation. Cotinine has longer half-life than nicotine (16 hrs compared to 2 hrs respectively). There is no relationship between cotinine concentrations and metabolic clearance of nicotine. The nicotinic cholinergic receptors are located in the hypothalamus, hippocampus, thalamus, midbrain, brain stem, and areas of the cerebral cortex. Stimulation of nicotinic receptors activates several CNS neurohormonal pathways, leading to the release of acetylcholine, norepinephrine, dopamine, serotonin, vasopressin, growth hormone, and ACTH. Most of the effects of nicotine on the CNS are from: 1- direct action on brain receptors, or 2- activation of afferent nerves of chemoreceptors in the carotid body or the lung. The carotid chemoreceptor is most sensitive to low levels of nicotine. Peripheral mechanisms include: 1- catecholamine release from the adrenals, 2- enhancement of release of catecholamines from vascular nerve endings, or 3- causes release of electrical stimulation-evoked neurotransmitters from sympathetic nerves in blood vessels. The expression of nicotinic cholinergic receptors may be regulated by cAMP-protein kinase system. Enhanced nicotine binding does not involve synthesis of new receptor proteins, but rather decreased turnover rates of nicotinic receptors. Biological activity of nicotine includes: 1- delays wound healing, 2- reproductive disorders (low birth weight, prematurity, spontaneous abortion, foetal neurotoxicity), 3- peptic ulcer disease, 4- oesophageal reflux, and 5- cardiovascular diseases. Nicotine is not carcinogenic but is a precursor of nitrosonornicotine and other nitrosamines, which contribute to tobacco-related cancer. Cardiovascular effects of nicotine may be due to stimulation of cardiac sympathetic nerves and by adrenaline release from the adrenal medulla. Nicotine increases arterial pressure and total body oxygen uptake (7%). Nicotine increases myocardial oxygen consumption. Nicotine increases plasma free fatty acid levels by stimulating lipolysis in adipose tissue, increasing the myocardial uptake and oxidation of FFA, which increases myocardial oxygen consumption. Nicotine constricts certain vascular beds via vasopressin release. Nicotine dilates skeletal muscle by increasing cardiac output or/and release of epinephrine at nerve endings. Initially, blood flow increases in the large coronary vessels, then decreased by α-adrenergic vasoconstriction of coronary vessels. During regular cigarette smoking, the sympathetic nervous system is activated 24 hrs a day. Atherosclerosis is a disease of the intima of large arteries, which causes luminal narrowing, thrombosis, and occlusion associated with ischaemia of the end organ. First monocytes attach to the endothelium in response to endothelial injury. Platelets subsequently adhere, even to minimally injured endothelial cells. Injured endothelial cells stimulate intimal proliferative lesions and, subsequently, smooth muscle proliferation. Intimal thickening proceeds to fibrous plaque, and ultimately to a complicated plaque. Endothelial cell injury is ultrastructurally characterised by cell swelling, cytoplasmic vacuolation, increased irregularity of the luminal surface, mitochondrial swelling, and subendothelial oedema. Endothelial cell injury occurs in the uterine arteries of smoking women. Features of endothelial injury in umbilical cord of children of smoking mothers include: 1- endothelial cell swelling, 2- luminal surface projections, and 3- extensive nuclear folding with basal fibroblasts. Nicotine may be the component of cigarette smoke that causes endothelial injury. The endothelial intercellular clefts, with plasmalemmal vesicle system are the ultrastructural basis of arterial endothelial permeability to blood components. Increased permeability of endothelial cells changes in the intercellular material transport, and the stimulation of secretory and cell migratory activities are all considered being early signs of cell damage. Low levels of nicotine stimulate cytoskeletal protein synthesis and polymerisation. Higher nicotine doses inhibit protein synthesis and ultimately lead to cellular destruction. The formation of an atherosclerotic lesion is initiated by damage to the endothelium and interaction of blood components with the underlying vascular tissue. This leads to the smooth muscle cells in the arterial media to change from contractile to a synthetic phenotype and migrate into the intima, where they multiply by division and deposit extracellular matrix components. Further development of the lesion involves lipoprotein accumulation, cell death, calcification, and intramural thrombosis. Nicotine promotes conversion of arterial smooth muscle cells (SMC) from a contractile to a synthetic phenotype. Nicotine and cotinine are mitogenic for human vascular smooth muscle cells (VSMC). Nicotine and cotinine elevate levels of bFGF, and decrease the levels of $TGF\beta_1$. $TGF\beta_1$ has mitogenic stimulatory effects at low concentrations; at high concentrations $TGF\beta_1$ inhibits SMC proliferation. Aromatic hydrocarbons and polyphenol-containing glycoproteins of tobacco smoke may act together with or independently of nicotine to influence proliferation of arterial SMC. The VSMC migration may occur through activation of the mitogen activated protein (MAP) kinase pathway. Nicotine increases DNA synthesis associated with cellular proliferation in these endothelial cells. Nicotine may contribute to the acceleration of hypertension by contributing to vasoconstriction or by promoting interactions between platelets and endothelial lining. Nicotine reduces renal blood flow, which may promote renal ischaemia and aggravate hypertension in a patient with marginal renal blood flow. The net result is turbulent blood flow, which may accentuate vascular injury. The narrowed arterial lumen may cause a turbulent flow that damages RBC and platelets, leading to the release of ADP, which promotes platelet aggregation. Cigarette smoking increases TXA_2 urinary excretion. Epinephrine stimulates platelet aggregation and thrombus formation by α-adrenergic mechanism. Nicotine does not influence levels of coagulation factors. Nicotine may inhibit prostacyclin synthesis. Prostacyclins play an important role in vascular homeostasis as a local vasodilator and have antiplatelet aggregation effects. Nicotine may produce an imbalance between platelet aggregating TXA_2 and antiplatelet prostacyclins. Smoking increases plasma fibrinogen levels. Nicotine may increase arterial uptake of fibrinogen as a result of defective clearance of fibrinogen from the vessel wall from the vasoconstrictive effects of nicotine on the vasovasorum. Nicotine in combination with other products of combustion from cigarette smoke may produce a hypercoagulable state, which may contribute to accelerated atherosclerosis and to acute cardiac events.

MICROWAVE

Because the body is electrochemical in nature, any force that disrupts or changes human electrochemical events will affect the physiology of the body. Introduction into the human body of molecules and energies, to which it is not accustomed, is much more likely to cause harm than good. The Nazis, for use in their mobile support operations, originally developed microwave radiomissor cooking ovens to be used for the invasion of Russia. The Soviet Union ban the use of microwave ovens in 1976. The Soviets issued an international warning on the health hazards, both biological and environmental, of microwave ovens and similar frequency electronic devices. Radio frequency and microwave radiation sources in USA are increasing at 15% per year. Microwave radiation is odourless and invisible and therefore hard to detect. Microwave oven radiation is present whenever a microwave oven is turned on. The microwave energy causes the water molecules in the food to vibrate rapidly. This rapid vibration produces heat which, in turn, cooks the food. It can also penetrate through living tissue, which is why exposure is harmful to our health. Microwaves are a form of electromagnetic energy, like light waves or radio waves, and occupy a part of the electromagnetic spectrum of power, or energy. Microwaves are very short waves of electromagnetic energy, which travel at the speed of light (186,282 miles/s). Microwaves are used to relay long distance telephone signals, television programs, and computer information across the earth or to a satellite in space. Every microwave oven contains a magnetron, a tube in which electrons are affected by magnetic and electric fields in such a way as to produce micro wavelength radiation at about 2450 MHz or 2.45 GHz. This microwave radiation interacts with the molecules in food. All wave energy changes polarity from positive to negative with each cycle of the wave. In microwaves, these polarity changes happen millions of times every second. Food molecules especially the molecules of water have a positive and negative end in the same way a magnet has a north and a south polarity. The microwave oven has a power input of about 1000 watts of alternating current. As these microwaves generated from the magnetron bombard the food, they cause the polar molecules to rotate at the same frequency millions of times a second. All this agitation creates molecular friction, which heats up the food. The friction also causes substantial damage to the surrounding molecules, often tearing them apart or forcefully deforming them, i.e., structural isomerism. Many terms are used in describing electromagnetic waves, such as: 1- Wavelength determines the type of radiation, i.e. radio, X-ray, ultraviolet, visible, infrared, etc. 2- Amplitude determines the extent of movement measured from the starting point. 3- Cycle determines the unit of frequency, such as Hz, or cycles/s. 4- Frequency determines the number of occurrences within a given time period (usually 1 second). The number of occurrences of a recurring process per unit of time, i.e., the number of repetitions of cycles per second. Radiation is the electromagnetic waves emitted by the atoms and molecules of a radioactive substance as a result of nuclear decay. Radiation causes ionisation, which is what occurs when a neutral atom gains or loses electrons. In simpler terms, a microwave oven decays and changes the molecular structure of the food by the process of radiation. Heating the baby bottle in a microwave can cause slight changes in the milk. In infant formulas, there may be a loss of some vitamins. In expressed breast milk, some protective properties may be destroyed. Microwaving baby formulas converts certain trans-amino acids into their synthetic cis-isomers. Synthetic isomers, whether cis-amino acids or trans-fatty acids, are not biologically active. Further, one of the amino acids, L-proline, is converted to its d-isomer, which is known to be neurotoxic and nephrotoxic. Artificially produced microwaves, including those in ovens, are produced from alternating current and force a billion or more polarity reversals per second in every food molecule they hit. Eaten food processed through the microwave oven causes changes in the blood, including: 1- haemoglobin levels decreased, 2- white cell levels increased, 3- cholesterol levels increased, and 4- lymphocytes decreased (degeneration). Food cooked in microwave ovens has cancerous effects on the blood. Atoms, molecules, and cells hit by this hard electromagnetic radiation are forced to reverse polarity 1-100 billion times a second. There are no atoms, molecules or cells of any organic system able to withstand such a violent, destructive power for any extended period of time, not even in the low energy range of milliwatts. Of all the polar natural substances, the oxygen of water molecules reacts most sensitively. This is how microwave cooking heat (thermic effects) is generated, friction from this violence in water molecules. Structures of molecules are torn apart, molecules are forcefully deformed, called structural isomerism, and thus become impaired in quality. This is contrary to conventional heating of food where heat transfers conventionally from without to within. Cooking by microwaves begins within the cells and molecules where water is present and where the energy is transformed into frictional heat. The athermic effects of microwaves can also deform the structures of molecules and have qualitative consequences, e.g., weakening of cell membranes. As a result of the force involved, the cells are actually broken, thereby neutralising the electrical potentials, the very life of the cells, between the outer and inner side of the cell membranes. Impaired cells become easy prey for viruses, fungi and other microorganisms. The natural repair mechanisms are suppressed and cells are forced to adapt to a state of energy emergency, they switch from aerobic to anaerobic respiration. Instead of water and carbon dioxide, the cell poisons hydrogen peroxide and carbon monoxide are produced. Microwaving creates new compounds, called radiolytic compounds, which are unknown fusions not found normally. Radiolytic compounds are created by molecular decomposition (decay) as a direct result of radiation. Carcinogens are formed in virtually all foods microwaved. Carcinogens in microwaved food: 1- d-Nitrosodienethanolamines are formed in microwaving prepared meats, 2- free radicals (highly reactive incomplete molecules) are formed in microwaved plants, especially root vegetables, 3- thawing frozen fruits converts their glucoside and galactoside containing fractions into carcinogenic substances, 4- extremely short exposure of raw, cooked or frozen vegetables converts their plant alkaloids into carcinogens, 5- microwaving milk and cereal grains converts some of their amino acids into carcinogens, and 6- decrease in nutritional value, marked acceleration of structural degradation leading to a decreased food value of 60-90% (vitamin B complex, vitamin C, vitamin E, essential minerals and lipotropics factors, degradation of nucleo-proteins in meats, damaged to plant alkaloids, glucosides, galactosides and nitrilosides). The microwave sickness was first described in Russia in 1950's. The first signs are low blood pressure and slow pulse. The later and most common manifestations are stress syndrome (chronic excitation of the sympathetic nervous system) and high blood pressure. This phase also often includes headache, dizziness, eye pain, sleeplessness, irritability, anxiety, stomach pain, nervous tension, inability to concentrate, hair loss, plus an increased incidence of appendicitis, cataracts, reproductive problems, and cancer. The chronic symptoms are eventually succeeded by crisis of adrenal insufficiency and ischaemic heart disease. Certain diseases among consumers of microwaved foods: 1- Lymphatic disorders, causing a degeneration of the immune potentials of the body to protect against certain forms of neoplastics. 2- An increased rate of cancer cell formation in the blood. 3- Increased rates of stomach and intestinal cancers. 4- Higher rates of digestive disorders and a gradual breakdown of the systems of elimination. Continually eating food processed from a microwave oven causes long term-permanent-brain damage by depolarising or demagnetising the brain tissue (shorting out electrical impulses in the brain). The human body cannot metabolise the unknown by-products created in microwaved food. Male and female hormone production is shut down and/or altered by continually eating microwaved foods. Microwaved foods cause stomach and intestinal neoplasms, e.g., rapidly increased rate of colon cancer. Eating microwaved food causes loss of memory, concentration, emotional instability, and a decrease of intelligence. The use of artificial microwave transmissions for subliminal psychological control (brainwashing) has also been proven.

ELECTROMAGNETIC FIELDS AND HUMAN HEALTH

X-rays, ultraviolet (UV) light, visible light, infrared light (IR), microwaves (MW), radio-frequency radiation (RF), and magnetic fields from electric power systems are all parts of the electromagnetic (EM) spectrum. As the frequency rises the wavelength gets shorter. The frequency is the rate at which the electromagnetic field goes through one complete oscillation (cycle) and is usually given in Hertz (Hz), where one Hz is one cycle per second. Power-frequency fields in the US vary 60 times per second (60 Hz), and have a wavelength of 5,000 km. Power in most of the rest of the world is at 50 Hz. Broadcast AM radio has a frequency of around 10^6 (1,000,000) Hz and a wavelength of around 300 m. Microwave ovens have a frequency of 2.54×10^9 Hz, and a wavelength of about 12 cm. X-rays have frequencies above 10^{15} Hz, and wavelengths of <100 nm. The interaction of biological material with an electromagnetic source depends on the frequency of the source. At the very high frequencies characteristic of vacuum UV and X-rays (<100 nanometers), electromagnetic particles (photons) have sufficient energy to break chemical bonds. This breaking of bonds is termed ionisation, and this part of the electromagnetic spectrum is termed ionising. Non-ionising electromagnetic sources can produce biological effects. Many of the biological effects of ultraviolet (UV), visible, and infrared (IR) frequencies depend on the photon energy, but they involve electronic excitation rather than ionisation, and do not occur at frequencies below that of infrared (IR) light (below 3×10^{11} Hz). The electromagnetic sources produce both radiant energy (radiation) and non-radiant fields. Radiation travels away from its source, and continues to exist even if the source is turned off. In contrast, some electric and magnetic fields exist near an electromagnetic source that is not projected into space, and that cease to exist when the energy sources is turned off. Ionising electromagnetic radiation carries enough energy per photon to break bonds in the genetic material of the cell, the DNA. Severe damage to DNA can kill cells, resulting in tissue damage or death. Lesser damage to DNA can result in permanent changes, which may lead to cancer. A principal mechanism by which radiofrequency radiation and microwaves cause biological effects is by heating (thermal effects). This heating can kill cells. If enough cells are killed, burns and other forms of long-term and possibly permanent tissue damage can occur. Cells that are not killed by heating gradually return to normal after the heating ceases. It is possible to produce thermal effects even with very low levels of absorbed power. One example is the "microwave hearing" phenomenon; these are auditory sensations that a person experiences when his head is exposed to pulsed microwaves such as those produced by radar. The "microwave hearing" effects is a thermal effect, but it can be observed at very low average power levels. The electric fields associated with the power-frequency sources exist whenever voltage is present, and regardless of whether current is flowing. The electrical fields from power lines may penetrate most buildings, and the electrical currents induced in the body by power line electrical fields may be greater than those induced by power line magnetic fields. Power-frequency electric fields can exert forces on charged and uncharged molecules or cellular structures within a tissue. These forces can cause movement of charged particles, orient or deform cellular structures, orient dipolar molecules, or induce voltages across cell membranes. Power-frequency magnetic fields can also cause biological effects via the electric fields that they induce in the body. The non-thermal biological effects of non-ionising electromagnetic sources are a result of a direct interaction between the field and the organism (e.g., photochemical events like vision and photosynthesis); and thermal effects are a result of heating (e.g., heating with microwave ovens or IR light). In the USA magnetic fields are often still measured in Gauss (G) or milligauss (mG), where: 1,000 mG = 1 G. In the rest of the world and in the scientific community, magnetic fields are measured in tesla (T), were: 10,000 G = 1 T, 1 G = 100 microT (µT), and 1 µT = 10 mG. Electric fields are measured in volts/meter (V/m). Actual magnetic fields depend on distance, voltage, design and current; actual electric fields are affected only by distance, voltage and design (not by current flow). Appliances that have the highest magnetic fields are those with high currents or high-speed electric motors (e.g., vacuum cleaners, microwave ovens, electric washing machines, dishwashers, blenders, can openers, electric shavers). Appliance fields decrease rapidly with distance. Electric trains can be a major source of exposure, as power-frequency fields at seat height in passenger cars can be as high as 60 µT. Children living near certain types of power lines (high-current distribution lines and high-voltage transmission lines) may have higher than average rates of leukaemia, brain cancers and/or overall cancer. Cancer is initiated by damage to the genetic information of a cell (DNA). Agents that cause such injury are called genotoxins. Genotoxins may affect many types of cells, and may cause more than one kind of cancer. Non-genotoxic (epigenetic) agents can contribute to the development of cancer, even though they may not be able to cause cancer by themselves. Epigenetic agents (non-genotoxic carcinogens) affect carcinogenesis indirectly, by increasing the probability that other genotoxic agents will cause genotoxic injury, or that genotoxic injury caused by other agents will lead to cancer. An epigenetic agent might inhibit repair of potentially-genotoxic damage, affect the DNA in such a way as to make it more vulnerable to genotoxic agents, allow a cell with genotoxic injury to survive, or stimulate cell division in a previously non-dividing cell that had genotoxic injury. The actions of epigenetic agents may be tissue-specific. Cancer promoters are a specific class of epigenetic agents. The majority of agents that are known to be carcinogenic in humans are genotoxins.

The biological effects the electromagnetic spectrum:

1- The ionising radiation portion, where direct chemical damage can occur:
   X-rays, "vacuum" ultraviolet light
2- The non-ionising portion of the spectrum, which can be subdivided into:
   A- The optical radiation portion, were electron excitation can occur (ultraviolet light, visible light, infrared light).
   B- The portion where the wavelength is smaller than the body and heating via induced currents can occur (microwaves and higher-frequency radiofrequency radiation).
   C- The portion where the wavelength is much larger than the body and heating via induced currents seldom occurs (lower-frequency radiofrequency radiation, power frequencies fields and static fields).

There are biological effects other than genotoxicity and promotion, which might be related to cancer. Especially, agents that have dramatic effects of cell growth, on the function of the immune system, or on hormone balances might contribute to cancer without meeting the classic definitions of genotoxicity or promotion. Suppression of the immune system in humans is associated with increased rates of only certain types of cancer, particularly lymphomas. Power-frequency fields might suppress the production of hormone melatonin (cancer-preventive activity). The electric fields and static magnetic fields may affect melatonin production. Melatonin levels are decreased in some cancer patients, e.g., breast cancer. Melatonin may inhibit the induction of breast cancer by chemical carcinogens; and inhibition of melatonin production may enhance the induction of breast cancer by chemical carcinogens. Static (DC) magnetic fields can affect the reaction rates of chemical reactions that involve free radical pairs. Traffic exhaust contains known carcinogens, and traffic density is correlated with cancer incidence. Pacemaker function can be affected by power-frequency fields. Fields strong enough to interfere with pacemaker function clearly could exist in some occupational settings, and might even exist in some non-occupational settings. Pacemaker users who work or live in environments where there is equipment capable of causing significant electromagnetic interference should bring this to the attention of the doctor who implanted the pacemaker. Pacemaker users would also be advised to exercise some caution when in the close vicinity of high voltage transmission lines, particularly lines with voltages of 230 Kv and above. The same precaution is probably applicable to implanted defibrillators, and might be applicable to other implanted biomedical devices. Measurements of power-frequency magnetic fields must be done with a calibrated gauss meter in multiple locations over a substantial period of time, because there are large variations in fields over space and time. Fortunately, the magnetic field is far easier to measure than the electric field. A syndrome, now called sensitivity of electricity or electrosensitivity first appeared in Norway in the early 1980's among users of VDTs. The phenomena of electrical hypersensitivity cannot be explained by any known mechanisms, as the threshold for known interactions are at least 50 times higher than actual exposures levels. Initial reports were largely of a transient skin reaction, but in more recent years the syndrome has included central nervous system, respiratory, cardiovascular and digestive symptoms. Certain individuals experience a variety of health symptoms, which they attribute to exposure to electric or magnetic fields from sources such as power lines, household appliances, visual display units (VDUs), light sources, mobile telephones and mobile phone base stations. This perceived sensitivity to electromagnetic fields has the general name electromagnetic hypersensitivity or EHS. Regardless of the science, the public controversy will continue. The public concern is sustained by uneven reporting by the mass media, by the inability of scientists to guarantee that no risk exists, and by statements from scientists and government officials that more research is needed. This public concern is further encouraged by lay-oriented books that allege that there has been a conspiracy to conceal the health risks of power-frequency fields and others.

WATER

Chemicals from the plastic bottle may have leached into water. Plastic can release artificial oestrogens into food and water. Increased oestrogen concentration in the body may increase the risk of breast cancer and cause early puberty in children. When bottled water is stored in a hot environment, for example a hot warehouse, the amount of artificial oestrogens released is even greater. Steam distillation of water involves boiling water to the point of evaporation and then condensing it by cooling the vapours to a liquid state. The end product is water that is free of minerals and toxic inorganic wastes. The concentrated minerals and toxic inorganic wastes remain in the boiler. They are drained out when it is convenient to clean. The water is also filtered to remove organic toxins. Steam distillation with filtration is an excellent method of obtaining pure water that contains no dissolved minerals or inorganic toxins, harmful microorganisms, organic toxins, or heavy metals such as lead, cadmium and mercury.

Water content human body:

Brain	83%
Kidney	82%
Heart	79%
Lungs	80%
Bones	22%
Blood	90%

Carbon filters come in all shapes and sizes. They're one of the oldest and least expensive options for purifying water. Most carbon filters utilise a special form of carbon called activated carbon. Water easily passes through an activated carbon filter, which provides an almost unbelievably large surface area (125 acres per pound). Activated carbon is used in both the solid block and granular forms. It takes water longer to pass through block carbon, which makes this form more effective at absorbing contaminants. Activated carbon filters are best suited for removing organic pollutants such as insecticides, herbicides and PCB's. They can also remove many industrial chemicals and chlorine. Activated carbon will not remove most inorganic chemicals, dissolved heavy metals (e.g., lead) or biological contaminants. Many manufacturers use activated carbon in conjunction with other processes, such as ceramic filters or ultraviolet light. Carbon filters provide a fertile breeding ground for bacteria. Carbon filters lose their effectiveness over time. It essential that the filter be changed every 1-12 months. Rust, dirt, parasites like Cryptosporidium and Giardia lamblia, and others can be easily removed from drinking water by forcing the water through the very fine pores in ceramic material. Ceramic filters, however, are not effective at removing organic pollutants or pesticides, as carbon filters are. They can be repeatedly cleaned by simply scrubbing the outside of the ceramic material, and need to be paired with a carbon filter. Ozone (O_3) differs from normal oxygen in that it contains 3 atoms of oxygen instead of just 2. This extra oxygen atom makes it highly unstable and reactive. When ozone gas bubbles through water, it quickly and very efficiently kills bacteria, viruses, algae and parasites. Ozone is thousands of times more potent than chlorine, and doesn't produce any harmful by-products like chlorine does. It does not remove heavy metals, minerals or pesticides. Ozone has a very short half-life, dissipates almost instantly and offers no residual purifying power. Ozone filters are expensive. When microorganisms like bacteria and viruses absorb UV light, certain chemical reactions are triggered, which kill the organism. But, it is not effective at killing all types of organisms (such as oocysts), it has no effect on heavy metals, pesticides or other contaminants, and the water must be relatively clear to begin with for the light to be able to penetrate. Ion exchange system is nothing more than a water softener. Softening hard water may improve laundry and prolong the life of hot water heater. Water softeners don't purify water. KDF water filtration technology uses granules of copper and zinc alloy. The copper and zinc molecules act like the different poles on a battery. As contaminated water passes through the granules, certain contaminants are drawn toward the zinc while others with a different charge migrate to the copper. Additional chemical reactions take place, which release ozone and other compounds that kill bacteria and other organisms. KDF filtration effectively removes chlorine and its by-products, as well as heavy metals, but cannot remove pesticides and other organic contaminants. Reverse osmosis (RO) is a process whereby soft water (only) is forced through a semipermeable synthetic membrane. It was originally used to transform saltwater into freshwater. RO systems may remove 90-98% of heavy metals, viruses, bacteria and other organisms, organic and inorganic chemicals. For every gallon of pure water produced, 3-8 additional gallons of water (contains concentrated contaminants) get washed down the drain. They are ideal breeding grounds for bacteria. RO systems require a minimum water pressure of 40 psi to work properly, and steps must be taken to insure the integrity of the membrane, which has to be replaced every few years. The membrane will also degrade in the presence of chlorine and turbid water. Nanofiltration is primarily used by beverage manufacturers and some municipal water plants as a sophisticated method of softening water. Nanofiltration is quite expensive. Water distillation provides the purest and safest water available. Distillation is a fairly simple process. Water is heated until it boils and turns to steam. The boiling action kills the various bacteria and other pathogens, and as the steam rises it leaves behind waste material, minerals, heavy metals and other heavier contaminants. The steam is then cooled and returns to water. Some organic chemicals, like by-products of chlorine, boil at lower temperatures than water and will rise to mix with the water vapour. These chemicals can easily be removed by using a carbon filter on the water either before it enters or after it leaves the distiller. Preheating the water in a separate vented chamber drives off these chemicals before the water is boiled and turned into steam. Minerals and other solids accumulate in the bottom of the boiling chamber, which must be cleaned out periodically. They can be easily cleaned in a couple of minutes using plain water or a mild acid, like vinegar. It does take a little longer to distil water. Distillation system will remove every kind of bacteria, virus, parasite, and pathogen, as well as pesticides, herbicides, organic and inorganic chemicals, heavy metals (dissolved or otherwise) and even radioactive contaminants. Although it requires more electricity than other methods, it doesn't waste any water. Any water supply can be used to fill the boiling chamber with water (from a river, lake, pond or swimming pool) and have an endless supply of clean water. In the long run bottled water will be far more expensive and inconvenient, and while the risk of contamination may be less than your tap water, there's far too little regulation to make it a permanent, viable solution. The Doulton water filter combines ceramic/carbon filtration. First, through ceramic filtration, it removes rust, dirt, iron, insoluble lead, and parasites. Then the water moves through the block carbon, which removes chlorine, odour, bad taste, and turbidity. Carbon effectively removes organic pollutants like herbicides and pesticides. Finally, unlike other ceramic/carbon filters, the Doulton has a heavy-metal removing compound called titanium zeolite that removes any lead that's dissolved in the water. The ideal water for the human body should be slightly alkaline and this requires the presence of minerals like calcium and magnesium. Water filtered through a solid charcoal filter is slightly alkaline. Ozonation of this charcoal filtered water is ideal for daily drinking. Distilled water is an active absorber and when it comes into contact with air, it absorbs carbon dioxide, making it acidic. Distilled water tends to dissolve substances with which it is in contact. A poor diet may be partially to blame for the waste accumulation. Meats, sugar, white flour products, fried foods, soft drinks, processed foods; alcohol, dairy products and other junk foods cause the body to become more acidotic. Stress; whether mental or physical can lead to acidosis. The longer one drinks distilled water, the more likely the development of mineral deficiencies and acidosis. Those who supplement their distilled water intake with minerals are not as deficient, but still not as adequately nourished in minerals as their non-distilled water drinking counterparts even after several years of mineral supplementation.

DEMENTIA

The neural stem cells exists in the developing nervous system and in the nervous system of the adult. In the adult brain, new neurons are generated primarily in 2 regions: 1- the subventricular zone and, 2- subgranular zone of the hippocampal dentate gyrus. This neurogeneration of new neurons can increase under pathologic conditions, such as brain ischaemia that typifies a stroke. Stroke leads to a marked increase in the proliferation of neural progenitor cells in the subventricular zone and to the migration of these recently generated neuroblasts into the damaged striatum. Stroke can lead to the proliferation of new neuronal cells that migrate into the zone of ischaemic injury, differentiate, and express markers of mature striatal neurons. Neuronal progenitor cells proliferate in response to growth factors, which are associated with angiogenesis. bFGF, PDGF, TGFβ, VEGF, erythropoietin, and EGF are included in this group. Several of these growth factors are transactivated by hypoxia-inducible factor, and their levels are increased in ischaemic brain. During adult neurogenesis, neuroblasts, glial precursors, and endothelial precursors form tight proliferative clusters, which are grouped around small capillaries, creating a vascular niche in which neurogenesis proceeds in the context of vascular recruitment and remodeling. After angiogenic stimulation, the endothelial cells synthesize and release brain-derived neurotrophic factor, which promotes the survival, migration, and differentiation of neuronal precursors. Astrocyte-derived mediators also appear to regulate the migration of neuronal precursors in the subventricular zone. The self-renewal program for precursors involves the suppression of apoptosis by retinoic acid, erythropoietin, and Notch 1. Brain morphogenic protein can promote apoptosis. Stress, depression, glucocorticoids, excitotoxicity mediated by the N-methyl-D-aspartate receptor, and inflammation all suppresses neurogenesis. Exercise and environmental stimulation increase neurogenesis, whereas depression suppresses it. Neurogenesis has functional consequences; e.g., the hippocampus-dependent learning promotes the survival of new neurons. Neurogenesis supports cognition and memory throughout life. Vasculopathies mat perturb neurogenesis and contribute to vascular dementia. Dementia is a syndrome of progressive cognitive decline, which is severe enough to cause a significant impairment in social or occupational performance. This is often accompanied by behavioural and personality changes, and symptoms may include 1- impairment of memory, language, visual-spatial skills, abstraction, calculation and judgement, and 2- delusion, hallucinations, depression, mania and anxiety. The diagnosis of dementia must be made with care, and impairment should be present for at least 6 months before the diagnosis is made. The prevalence of dementia rises markedly with age, from <10% in people aged 65-70 years to 20-48% in those over 70. The care of these patients will have major medical, social and economic implications. Alzheimer's disease (AD) and vascular dementia are the 2 most common forms of dementia. AD occurs in 55-65% of patients with dementia aged over 65 yrs old but only 34% of cases in patients under 65. Vascular dementia is found in 10-20% of cases in both age groups, including multi-infarct dementia, frontotemporal dementia (FTD), and diffuse Lewy body disease (DLB). AD and vascular dementia often coexist in 10% of dementia cases. 13% of all dementias are potentially reversible, 21% of patients <65 y.o and 5% of patients >65. The commonest causes of reversible dementia are drug intoxication and depression.

Causes of dementia:

1- Degenerative:
   AD, Pick's disease, Huntington's disease, diffuse Lewy body dementia, Parkinson's disease, progressive supranuclear palsy, amytrophic lateral sclerosis, late-onset metachromatic leucodystrophy, adrenoleucodystrophy.
2- Vascular:
   Multi-infarct dementia, subdural haematomas
3- Infective:
   AIDS-related dementia, herpes encephalitis, post-encephalitic dementia, syphilis, toxoplasmosis, subacute sclerosing panencephalitis.
4- Normal pressure hydrocephalus
5- Traumatic:
   Head injury, dementia pugilistica (Boxer's dementia).
6- Neoplastic:
   Primary brain tumours, metastatic tumours.
7- Transmissible:
   Creutzfeldt-Jakob disease.
8- Inflammatory:
   Disseminated lupus erythematosus, giant cell arthritis, polyarteritis nodosa, multiple sclerosis.
9- Toxic:
   Alcoholic dementia, heavy metal poisoning (lead, mercury), solvent abuse, carbon monoxide, drugs (hypnotics, tranquillizers, barbiturates, neuroleptics).
10- Metabolic:
   Hypothyroidism, Cushing's disease, Addison's disease, uraemia, hepatic failure, prolonged hypoglycaemia, vitamin B_{12}, B_1, and B_6 deficiency, Wilson's disease.
11- Pseudodementia:
   Depression.

Reversible metabolic dementia such as thyroid disorders and vitamin B_{12} deficiency are also common but have a smaller chance of full recovery. Distinguishing between depression and true dementia can be difficult. Depression may lead to poor performance in tests of memory and intelligence as well as self-neglect and social withdrawal. Depression may be termed pseudodementia. There is no reliable predictor of who will respond to treatment. 40-50% of patients improves and lasts only as long as the drugs are taken. Careful assessment should be undertaken after 2-4 months of treatment. Tau (τ) protein is a major constituent of the microtubular system of neurons, where it acts to promote microtubular assembly and stability. Six isoforms are present in the human adult brain, which are produced by alternative mRNA splicing from a single gene, chr17q21. These τ protein isoforms are phosphorylated in the healthy adult brain, but in AD they become hyperphosphorylated and form neurofibrillary tangles (NFTs). CSF τ protein concentrations are significantly higher in patients with AD than in healthy age-matched controls or patients with other neurological diseases. The raised CSF τ protein concentrations may reflect either neuronal and axonal degeneration or the formation of NFTs. NFTs are formed from paired helical filaments of hyperphosphorylated τ proteins, and amyloid or senile plaques composed of Aβ peptides. The sensitivity and specificity of CSF τ protein to distinguish AD from normal ageing is >80%, but not for dementia. High CSF τ protein concentrations have been reported in vascular dementia, FTB, DLB and corticobasal degeneration. CSF τ protein is not useful in distinguishing normal ageing and depression from true dementia. Aβ peptides are cleavage products of amyloid precursor protein (APP), and fibrillar aggregates of these peptides are the major constituents of senile plaques. One of the earliest pathological features of AD. The Aβ peptides are generated as soluble peptides during normal cellular metabolism and are secreted into extracellular spaces and into biological fluids. Creutzfeldt-Jakob disease belongs to a family of rare fatal neurodegenerative diseases known as transmissible spongiform encephalopathies (TSEs), which also include animal disease such as scapie and bovine spongiform encephalopathy (BSE). These diseases are characterized by neuropathological evidence of neuronal loss, astrocytosis and spongiform change. Patient's brain would have deposition of a partially protease-resistant isoform (PrP^{SC}) of a host-encoded sialoglycoprotein called prion protein (PrP^C). PrP^C is coded for by the prion protein (PRNP) gene on chromosome 20 and is expressed throughout the body, but in particularly high levels in neuronal cells, where it is found as glycosylphosphatidylinositol-anchored cell-surface protein. 85% of CJD occur sporadically, and 10-15% arise due to mutations in the PRNP gene. The variant form of CJD (vCJD) affects younger patients and has a more prolonged clinical course than sporadic CJD (spCJD). vCJD is causally linked to BSE. Patients tend to present with early and persistent psychiatric symptoms, and definitive neurological features such as ataxia and cognitive impairment develop relatively late. spCJD presents with a rapidly progressive dementia, often with neurological features such as myoclonus and ataxia. The age at onset is 45-75 y, the peak age at onset being 60-65 y. The disease course is rapid and most patients die within 6 months. MRI is useful for CJD investigation. Patients with spCJD have hyperintense signal changes in the putamen and caudate head of the basal ganglia. Patients with vCJD have a distinctive distribution of symmetrical hyperintensity of the pulvinar nucleus of the thalamus. Several neuropsychological tests are used to evaluate patients with cognitive impairment. The most widely used of these is the mini-mental state examination (MMSE), which is a series of tests designed to assess the patient's orientation, registration, attention and calculation ability, recall and language. It is marked on a scale from 0 to 30, with 30 representing a perfect score and less than 24 considered evidence of sufficient cognitive impairment to diagnose dementia. These clinical assessment scales have many limitations, and subjects with poor educational attainment or depression in the absence of dementia may score <24. The MMSE does not assess all aspects of cognitive function equally, and many frontal lobe functions are not adequately assessed. Because it is not ethical to perform repeated lumber punctures in healthy age-matched subjects, it is not possible to study changes of $A\beta$ 42 in normal ageing. CSF $A\beta$ 42 levels decline with time in patients with AD. Alzheimer's disease is not a normal part of aging. Alzheimer's disease a common form of dementia among older people. Alzheimer-type dementia is different from other dementias, but it overlaps with them, and with age-related and stress-related changes in other organs. Criteria of AD: 1- death of neurons (increase of glial cells), 2- amyloid plaques (extracellular), associated with a particular variant of apolipoprotein E, the epsilon 4 allele, 3- fibrillary tangles (intracellular, or remaining after the rest of the cell has disappeared), and 4- amyloid in blood vessels. Functional and biochemical observations: 1- The mitochondrial energy problem, cytochrome oxidase and its regulation; body temperature/pulse-rate cycle disturbance; lipid peroxidation; respiratory defect; altered amino acid uptake; memory impairment; dominance of the excitatory systems vs. the inhibitory adenosine/GABA/progesterone/pregnenolone system. Increased calcium uptake, which is associated with lipid peroxidation and cell death. Increased cortisol and DHEA. 2- Deposit of abnormal proteins, such as transthyretin-amyloid; albumin binding of PUFA, vs. transport of thyroid and retinol. β-Glucuronidase increases, depositing oestrogen in cells. 3- Abnormally phosphorylated (tau) τ proteins, association with the variant form of Apo E; τ microtubule organising proteins, microtubules are involved in transporting cholesterol; phosphorylation, by the kinase systems, regulated by PUFA; the intermediate filaments are generally stress-associated. 4- ApoE, in cytoplasm, involved in cholesterol delivery for pregnenolone synthesis, as in the adrenal; its expression regulated by thyroid. Regulation of the side-chain cleaving enzymes; regulation of the cholesterol intake and conversion to pregnenolone by the endozepine receptor/GABA receptor, modified by progesterone. Ginkgo Biloba is the most ancient of all trees, the leaves are used to increase circulatory oxygenation to brain cells, anti-aging properties, Alzheimer's disease, asthma, depression, attention deficit disorder, blood clots, dementia, kidney disease, memory loss, respiratory disease, senility, stress, tinnitus, and vascular disease. Ginkgo biloba improves function in memory. Ginkgo biloba slows down cognitive decline in patients with Alzheimer's dementia. Phosphatidylserine improves memory decline and also slows Alzheimer's disease progression. Phosphatidylserine is useful in the treatment of mild to moderate deterioration of cognitive functions, with consistent improvement of depressive symptoms, memory and behaviour. Approximately 10% of alzheimer patients engage in binge eating. More than 10% of patients tried to eat inedible things like soap, faeces, tea bags, or paper towels. Alzheimer patients take longer to ingest, chew, and swallow food. Sometimes patients need to be told to swallow their food. They leave behind substantial food. Poor eating habits may aggravate alzheimer's disease pathology. Alzheimer patients avoid green and yellow vegetables and fish. They tend to eat sweets and foods high in fat prior to dementia onset. Antioxidants are especially important for alzheimer patients, because the brains of patients are under increased oxidative attack by free radicals. Vitamin E, a fat-soluble antioxidant, delays the progression of alzheimer's disease. Fish are rich in ω-3 fatty acids. Omega 3 fatty acids, found in the membrane of brain cells, are rapidly oxidised by free radicals in AD, so patients need a rich supply of ω-3 fatty acids in their diet. Residents of Nigeria have extremely high levels of ω-3 fatty acids in their bloodstream, and, an extremely low rate of Alzheimer's disease. N-acetyl-l-cysteine (NAC) is converted in the body to glutathione, a key antioxidant. NAC protects nerve and brain cells from injury induced by hydrogen peroxide, and protects brain cells from the toxic effects of amyloid β protein. 50 mg/KgBW of NAC may increase cognitive abilities of AD patients. Caffeine is found abundantly in coffee, tea, and soda. Caffeine increases genetic damage in cells, and increases amyloid β protein levels in cells. AD patients should avoid excessive caffeine consumption. Calcium levels are depleted in the bloodstream of individuals with Alzheimer's disease. Calcium serum levels seem to fall even before signs of Alzheimer's disease appear. Alzheimer patients do not absorb calcium normally. Alzheimer patients have some sort of calcium deficiency. The incidence of AD is markedly lower in areas where calcium level is elevated in the drinking water. Carnosine prevents the oxidation of glutathione by free radicals. Carnosine is decreased in the brains of people with AD. Carnosine markedly reduces the toxic effects of amyloid β protein in the cell. High levels of cholesterol in the bloodstream is linked to alzheimer's disease. Cholesterol may play a direct role in formation of amyloid β deposits. High density cholesterol (HDL) decreases the toxic effects of amyloid β protein. HDL encourages the microglia to degrade amyloid β protein. Individuals with alzheimer's disease tend to have lower than normal levels of serum HDL cholesterol. Alzheimer patients have normal serum levels of CoQ_{10}. However, CoQ_{10} is dramatically reduced in muscle tissue from alzheimer patients. Copper is found abundantly around senile plaques in alzheimer brains, copper may play a causative role in plaque formation. Copper greatly enhances the aggregation of amyloid β protein into deposits. Alzheimer patients need to limit excessive copper intake. Using copper-bottom cooking pots is a significant dietary source of copper. Dehydroepiandrosterone (DHEA) is an important human hormone. DHEA protects brain cells from free radicals and increases the survival of brain cells during and after injury. DHEA may reverse memory impairment. Serum DHEA levels are generally lower in individuals with alzheimer's disease. Individuals who consume high amounts of saturated fats have almost twice the risk for alzheimer's disease, compared with people who consume low amounts of saturated fats. Regular fish eating have a dramatic-reduced risk for alzheimer's disease. Fish are a rich source of ω-3 fatty acids, which have been found useful in the treatment of individuals with alzheimer's disease. Fluoride, an element added to drinking water to promote tooth health, may also help prevent alzheimer's disease. Fluorine competes with aluminum for absorption in the intestines, high aluminum content in public drinking water may be a risk factor for alzheimer's disease. Individuals with alzheimer's disease may need to take folic acid supplements or eat more foods, rich in folic acid. Folic acid may be especially useful in alzheimer's disease treatment, because it lowers blood homocysteine. Recommended daily intake for folic acid is 200 μg daily for men and 180 μg daily for women. Foods rich in folic acid include: asparagus, bananas, beans, beats, brewer's yeast, fortified grain products, and wheat germ. Folate (an anionic form of folic acid) levels are decreased in alzheimer patients. Folic acid is critical for the development of red blood cells, for normal growth and development, and for the maintainence of the nervous system. RBC folate levels correlate with cognitive abilities. Folate levels are also reduced in the CSF of individuals with alzheimer's disease. Glucose is a sugar extracted from grape juice. The sugar used at the dinner table is sucrose, which is broken down in the intestine to glucose. Glucose is used by brain cells to produce energy. Glucose can enhance memory. Glucose drinks can substantially increase memory in AD's patients. Insulin is responsible for the memory improvement. Thus, memory is improved because the high levels of blood glucose induce higher levels of blood insulin. Glutathione (GSH) is an important natural antioxidant found in brain and nerve cells. The brains of alzheimer patients are considered by experts to be especially vulnerable to attack by free radicals. GSH levels are lower in immune cells and brains of Alzheimer patients. Amyloid β protein found in alzheimer brains, lowers GSH content in brain cells. Homocysteine is a sulfur-containing amino acid that increases the risk for stroke and atherosclerosis, and may be a risk factor for AD. Homocysteine levels are abnormally elevated in the bloodstream of individuals with AD. Serum homocysteine elevates very early in the disease, even while there is only very mild memory impairment. High serum homocysteine is associated with poor memory. Homocysteine itself is toxic to brain cells, and aggravates the toxic effects of amyloid β protein in brain cells. High serum homocysteine is associated with low levels of folic acid in the blood. Folic acid supplementation may be an inexpensive way to reduce elevated serum homocysteine levels in demented patients. Hyperphagia is a condition characterized by excessive eating, often accompanied by weight gain. One-third of Alzheimer patients are hyperphagic and engage in binge eating. They may try to eat inedible things, like soap or flowers. Patients with hyperphagia also tend to develop sweet cravings. Hyperphagia is more common among the severely-demented. Excess iron can stimulate the formation of free radicals. Iron is abnormally elevated in the brains of Alzheimer patients. The basal ganglia contain the highest levels of iron in the brain, and patients with AD or Huntington's disease, have extremely high basal ganglia iron. Iron increases the aggregation of amyloid β protein into insoluble deposits. Iron may aggravate the generation of radicals, which induced initially by amyloid β protein. A mutation in the haemochromatosis gene, which elevates iron levels, is significantly prevalent in individuals with AD. Magnesium is an essential mineral for humans. Magnesium activates enzymes, aids in bone growth, aids in nerve conduction, and assists in muscle contraction. Magnesium levels are low in almost all areas of the brain and other cells in the body, in individuals with AD. Serum Mg levels are also low in individuals with AD; and is elevated paradoxically inside RBCs. Both Mg and calcium are necessary to keep amyloid β protein in a dissolved, non-aggregated state. 300 micromoles of Mg can decrease the aggregation of amyloid β protein into deposits, and 1 millimole of Mg completely blocks the rise of calcium inside brain cells, when the calcium rise is induced by amyloid β protein. Foods rich in Mg include: almonds, bananas, codfish, flounder, leafy, green vegetables, mackerel, molasses, nuts, wheat germ, and whole wheat bread. Melatonin is an important hormone, involved in the regulation of sleep and 24 hr bodily rhythms. Melatonin is also an extremely powerful antioxidant. Serum melatonin level is decreased in Alzheimer patients, and in the CSF. Normal individuals secrete only small quantities of melatonin during the day, and large quantities at night. In contrast, Alzheimer patients secrete more melatonin during daylight hours than normal subjects. Alzheimer patients secrete less nightime melatonin than normal. The more severely-demented patients have the lowest secretion of nighttime melatonin. 3-6 mg of melatonin at bedtime improves sleep quality, decreases sundown behavior in several alzheimer patients, reduces daytime sleepiness, improves mood especially in the one patient with severe dementia. Melatonin slows progression of AD. Melatonin prevents damage to mitochondrial DNA, induced by amyloid β protein. Melatonin decreases lipid peroxidation, induced by amyloid β peptides. Omega 3 fatty acids are found especially in the tissues of cold water fish. Docosahexaenoic acid, (DHA), and eicosapentaenoic acid (EPA), are the predominant ω-3 fatty acids . ω-3 fatty acids reduce the risk of stroke and heart disease and may alter the risk for dementia. Subjects who consumed fish and have high blood values of DHA, have a 40% less risk of developing AD. ω-3 fatty acids are lower than normal in many different tissues in individuals with AD. Serum DHA levels, in RBC membranes, in mitochondrial membranes, and in the brains are low of Alzheimer patients. High blood DHA may even prevent the development of alzheimer's disease. Nigerians consume lots of fish, elderly Nigerians do not show brain changes associated with aging, such as plaques or tangles. Nigerians have extremely high blood levels of DHA and EPA, the highest in the world. Alzheimer patients need more ω-3 fatty acids. DHA and EPA are found in fish oil capsules, tuna, mackerel, sardines,and herring. ω-3 fatty acids differ from ω-6 fatty acids because the placement of a methyl chemical group is different. Omega 3 fatty acids reduces the growth of cancer cells; whereas, ω 6 fatty acids accelerate the growth of cancer cells. ω-6 fatty acid consumption increases dementia risk. Heating vegetable oils destroys the vitamin E content. Normally,vitamin E protects oils and fats from oxidation. However, once vitamin E is destroyed, ingested oils are less protected from oxidation. The brains and tissues of individuals with Alzheimer's disease are particularly prone to excessive oxidation of fats. Pycnogenol powerfully protects endothelial cells from the toxic effects of amyloid beta protein. Alzheimer patients seem to have a markedly lower secretion of saliva. Many Alzheimer patients (80%) experience dysphagia to food and liquids, more than 23% have severe dysphagia. Dysphagia increases as dementia becomes more severe. The AD process itself may destroy cells in the section of the brain that controls swallowing. But, dysphagia could be as a result of decreased saliva flow. The neuroleptics (Loxapine) are known to cause dysphagia. Dysphagia have been linked to lower bodyweight in Alzheimer patients; and, lower bodyweight has been linked to shorter survival time. Some antidepressant drugs, used to treat Alzheimer patients, can cause increased sweet cravings. The brain cells of Alzheimer patients may need more glucose in order to function normally; this increased need, in turn, causes increased sweet cravings. Glucose is the fuel used by brain cells. Thiamine (vitamin B_1) vital functions include promoting normal growth, maintaining the nervous system, and keeping mucous membranes healthy. Thiamine serum levels of alzheimer patients are significantly low. Tofu consumption has been linked to AD. Tooth loss increases the risk for AD. Demented patients have more unrestorable, non-functioning teeth. The increased incidence of dental caries and unrestorable teeth in Alzheimer patients may be related to the decreased saliva flow found in patients. The pH of saliva discourages the proliferation of bacteria in the mouth. In addition, immunoglobulins in saliva protect against bacterial-induced dental caries. Tryptophan is an important amino acid, which is converted in the brain to serotonin, an important brain transmitter. Serum levels of tryptophan are greatly reduced in the alzheimer patients, but not in the brain. Up to 3 g daily tryptophan significantly improves mental function in certain individuals with alzheimer's disease. An overactive immune system in Alzheimer patients activates the enzyme indoleamine 2,dioxygenase, which breaks down tryptophan. Vitamin A functions include assisting in bone growth, preserving night vision, aiding in healing, and building the body's resistance to infection. Vitamin A serum levels are decreased in individuals with AD. Foods rich in vitamin A include asparagus, carrots, endive, liver, milk, spinach, squash, and tomatoes. Vitamin B_{12} (cyanocobalamin) is important for growth and development. Without sufficient vitamin B_{12}, individuals develop anaemia, which is characterized by profound fatigue, weakness in the arms and legs, and irreversible nerve damage. High levels of serum homocysteine or methylmalonic acid occur frequently in patients with vitamin B_{12} deficiency and are used to confirm an initial diagnosis of vitamin B_{12} deficiency. Serum methylmalonic acid and homocysteine are elevated in individuals with AD. Frequently, vitamin B_{12} can be depleted in the CSF of AD patients, while still normal in the bloodstream. Intravenous vitamin B_{12} improves memory, emotion, and social communication in individuals with AD. The improvements correlates with elevated levels of vitamin B_{12} in the CSF. Consuming vitamin C regularly, reduces risk for dementia. Consuming both vitamin C and vitamin E regularly dramatically lower the risk for dementia. Serum levels of vitamin C are consistently reduced in individuals with AD. Alzheimer patients dislike and avoid green-yellow vegetables, which are extremely rich in vitamin C. Vitamin C completely protects the brain cells from amyloid β toxicity. Vitamin C prevents the aggregation of amyloid β protein into deposits, when iron is used to prime aggregation. Foods rich in vitamin C include broccoli, cabbage, grapefruit, green peppers, lemons, oranges, potatoes, spinach, strawberries, tomatoes, and watercress. Vitamin D (cholecaciferol) serum levels are below normal in the individuals with AD. Female Alzheimer patients have decreased bone mass, compared with age-matched, healthy elderly women. Foods rich in vitamin D include cod-liver oil, herring, mackerel, salmon, sunlight, and vitamin-D-fortified milk. Vitamin E is found abundantly in the membranes of cells, where it protects membrane lipids from oxidation by free radicals. Consuming vitamin E regularly lowers incidence of dementia. More than 60% of all demented patients have subnormal vitamin E levels. Vitamin E is reduced in Alzheimer patients with lower bodyweight and body mass. Vitamin E slows the progression of AD, and allows patients to live at home longer, before having to be admitted to a nursing home. Vitamin E protects brain cells from the toxic effects of amyloid β protein. Vitamin E prevents oxidative damage to creatine kinase, which is easily oxidised by amyloid β protein. Vitamin E decreases lipid peroxidation, lipids are also readily oxidised by amyloid β protein. Foods rich in vitamin E include almonds, broccoli, canola oil, corn, fortified cereals, peanuts, spinach, walnuts, and wheat germ. More than 70% of people with AD show weight loss at some stage of their illness. Patients in the more severe stage of the disease seem to show the greatest weight loss. Alzheimer patients with weight loss have 5 times the risk of early death, compared with patients of normal bodyweight. Unexplained weight loss can be an early warning sign of alzheimer's disease. The reason for weight loss is unknown. Megestrol acetate (Megace), a synthetic derivative of progesterone, safely and effectively increases appetite and stabilizes weight in dementia patients. Zinc is an important mineral; which is needed for growth, wound healing, and as a cofactor for many enzymes. Zinc increases the aggregation of amyloid β protein. Zinc induces the aggregation most at acidic pH. Zinc is not recommended for AD patients. Most people are slightly demented now and then, when they are very sleepy or tired, or sick, or drunk, or having a hormone imbalance or extreme anxiety state. When the body temperature is very much below normal, mental functioning is seriously limited. Every moment of malfunction probably leaves its structural mark. Early or late, it is good to prevent the functional errors that lead to further damage, and to give the regenerative systems an opportunity to work. Ordinary healthy sleep is an example of restorative, protective inhibition. The energy charge, including levels of ATP, creatine phosphate, and glycogen storage, regulates many restorative enzyme systems. Restoration of brain energy metabolism is the function of sleep. The conditions that stimulate adenosine release from neural tissue represent either increases in metabolic demand (activation of excitatory receptors) or decreases in metabolic supply (hypoxia, ischaemia, and hypoglycaemia). In the brain, adenosine-stimulated increases in potassium conductance produce hyperpolarisation, thereby reducing neuronal responsiveness. Adenosine release is triggered globally in response to changes in cerebral energy homeostasis. A number of findings provide indirect support for the hypothesis that glycogen stores are depleted during waking and restored during sleep. Reduced availability of glycogen to astrocytes must increase adenosine release. Because ATP concentration is 100-fold greater than AMP concentration, a minute decrease in cellular energy charge is translated into a large proportional increase in extracellular adenosine concentration. Energy to resist stress makes quiescence possible, and prevents the deterioration of cells, of the sort that occurs in aging. Oestrogen production is facilitated when tissue is cooler, and it lowers body temperature. Oestrogen and the endorphins act together in many ways, and naloxone (the antagonist of morphine and the endorphins) raises body temperature and in other ways opposes oestrogen. Naloxone improves the symptoms of demented people, quickly, and dramatically, improve the mental clarity especially of a 60 year old woman who had used oestrogen. It, like clonidine (the anti-adrenaline drug), is a good candidate for controlling the hot flashes and other symptoms of menopause. In various degenerative brain conditions, blood clotting has been implicated either as a cause or a complication. The platelets of Alzheimer's patients are less viscous, and lipids extracted from the brain are more fluid, and contain 30% less cholesterol than normal. In general, lipid peroxidation causes cellular viscosity to increase, apparently by causing cross-linking of proteins. The extracellular matrix (ECM) is a major factor in the function and stability of brain cells. Any factor producing oedema tends to disrupt the ECM. Seizures are promoted by oestrogen, by unsaturated fats, and by lipid peroxidation, and cause an increase in the size of the free fatty acid (FFAs) pool in the brain. Prolonged seizures cause nerve damage in certain areas, especially the hippocampus, thalamus, and neocortex. Dementia is produced by prolonged seizures. Prenatal exposure to oestrogen, to oxygen deficiency, or to unsaturated fats decreases the size of the brain at birth. There is apparently a requirement for saturated fats during development. Under the influence of oestrogen, or unsaturated fats, brain cells swell, and their shape and interactions are altered. Memory is impaired by an excess of oestrogen. Oestrogen and unsaturated fat and excess iron kill cells by lipid peroxidation, and this process is promoted by oxygen deficiency (hypoxia). The foetus and the very old have high levels of iron in the cells. Oestrogen increases iron uptake. Oestrogen treatment produces elevation of FFAs in the blood, and lipid peroxidation in tissues. This tends to accelerate the accumulation of lipofuscin (age-pigment). Lipofuscin is a yellow to brown, granular, iron-negative lipid pigment found particularly in muscle, heart, liver and nerve cells undergoing slow, regressive change and accumulating in lysosomes with age, being the product of oxidation and polymerisation of the membrane lipids of autophagocytosed organelles. Lactic acid, the production of which is promoted by oestrogen, lowers the availability of CO_2, leading to impairment of blood supply to the brain. Oestrogen stimulates cell division, but can also increase the rate of cell death. Unsaturated fatty acids can also stimulate or kill. Both oestrogen and unsaturated fats promote the formation of age-pigment. Besides increasing the FFA concentration, oestrogen possibly depresses the level of cholesterol. Oestrogen causes massive alterations of ECM, and seems to promote dissolution of microtubules, as calcium does. Unsaturated fats increase calcium uptake by at least some brain cells. Unsaturated fats, like oestrogen, increase the permeability of blood vessels. The unsaturated fat causes oedema of the brain, inhibits choline uptake, blocking acetylcholine production. Progesterone is a nerve growth factor, produced by glial cells (oligodendrocytes). It promotes the production of myelin, protects against seizures, and protects cells against free radicals. It protects before conception, during gestation, during growth and puberty, and during aging. It promotes regeneration. Its production is blocked by stress, lipid peroxidation, and an excess of oestrogen. Aspirin protects against iron toxicity, clot formation, and reduces lipid peroxidation while blocking prostaglandin formation. Aspirin and other antiinflammatory drugs, taken for arthritis, have been clearly associated with a reduced incidence of Alzheimer's disease. Aspirin reduces the formation of prostaglandins from arachidonic acid (AA). Unsaturated fatty acids, but not saturated fatty acids, are signals which activate cell systems. A deficiency of polyunsaturated fatty acids (PUFA) leads to altered rates of cellular regeneration and differentiation, a larger brain at birth, improved function of the immune system, decreased inflammation, decreased mortality from endotoxin poisoning, lower susceptibility to lipid peroxidation, increased basal metabolic rate and respiration, increased thyroid function, later puberty and decreases other signs of oestrogen dominance. When dietary PUFA are not available, the body produces a small amount of unsaturated fatty acid (Mead acids), but these do not activate cell systems in the same way that plant-derived PUFAs do, and they are the precursors for an entirely different group of prostaglandins (PGs). In a variety of cell types, vitamin A functions as an oestrogen antagonist, inhibiting cell division and promoting or maintaining the functioning state. It promotes protein synthesis, regulates lysosomes, and protects against lipid peroxidation. Just as stress and oestrogen-toxicity resemble aging, so does a vitamin A deficiency. While its known functions are varied, the largest use of vitamin A could be for the production of pregnenolone, progesterone, and the other youth-associated steroids. One of vitamin E's important functions is protecting vitamin A from destructive oxidation. Unsaturated fats destruction of vitamin E will lead to the destruction of vitamin A. The increased lipid peroxidation of old age represents a vicious circle, in which the loss of the antioxidants and vitamin A leads to their further destruction. To produce pregnenolone, thyroid, vitamin A, and cholesterol have to be delivered to the mitochondria in the right proportion and sufficient quantity. Normally, stress is balanced by increased synthesis of pregnenolone, which improves the ability to cope with stress. Lipid peroxidation, resulting from the accumulation of unsaturated fatty acids, iron, and energy deficiency, damages the mitochondria's ability to produce pregnenolone. When pregnenolone is inadequate, cortisol is over-produced. When progesterone is deficient, oestrogen's effect is largely unopposed. When both thyroid and progesterone are deficient, even fat cells synthesise oestrogen. Vegetable oil suppresses the thyroid, increasing oestrogen. Oestrogen and calcium depolymerize microtubules. Microtubule transport for Apo E, transthyretin, thyroid, and cholesterol for pregnenolone synthesis is disrupted. Transthyretin and Apo E accumulate unused, and deposit in blood vessels, around nerves, and in cytoplasm. Pregnenolone and progesterone deficiency (aggravating thyroid deficiency) causes memory loss, destabilisation of nerve cells, failure of myelin formation, and excess cortisol synthesis. Free radicals and calcium cause multiple cell injuries including nerve-death. Oestrogen is released by elevated β-glucuronidase. Imbalances of other steroids, including cortisol and DHEA, develop as cells compensate for pregnenolone deficiency, causing shifts in balance of glial cells. Hypothyroidism, oestrogen excess, free unsaturated fats cause increased vascular permeability and brain oedema, protein leakage, and alteration of the matrix. The enzymes of glycolysis are bound to the structure of the cell when they are not in use, and that they are desorbed under the conditions that required abundant glycolysis. The cell water undergoes a transition under conditions, which include increased temperature, reduced oxygen, or nervous or hormonal stimulation. Activation of glycolysis is usually explained by the availability of regulatory substances such as ammonia, phosphate, and NAD, some of these basic regulatory molecules bind to structural components of the cell. Hormones and other factors stabilise or destabilise RNA, and that during some of these events relevant enzymes bind to the RNA. The cell's structure is sensitively reshaped constantly by processes that incorporate some of the environment in establishing each new stability. The structure of the cell may be developed very largely on the basis of information received from the environment, i.e., epigenetically. During the terminal stress that produces apoptosis, these enzymes make (which makes specific cuts in the DNA chain) confetti of the genome. Poisons, such as oestrogen, unsaturated fatty acids, or even radiation, produce different effects at different doses. Low doses typically stimulate cell division, larger doses produce changes of cell type and altered states of differentiation, and finally, adequate doses produce apoptotic cell death. Growth factors of various sorts can prevent apoptosis. It is increasingly clear that it represents excessive stress and deficient resources. The involvement of the genetic apparatus in differentiation and radical adaptation suggests that the (epigenetic) resources of cells are unlimited. The changes that are known to be produced by the poisons that we are habitually exposed to are exactly the changes that occur in the aging brain. As people moved north and developed new ways of living, their consumption of unsaturated fats increased, their brain size decreased, and they aged rapidly. Flaxseed was a staple of their diet. Even living in the tropics, there are many possibilities for diets rich in signal-disrupting substances, including iron, and in high latitudes there are opportunities for reducing our exposure to them. As a source of protein, milk is uniquely low in its iron content. Potatoes, because of the high quality of their protein, are probably relatively free of toxic signal-substances. Many tropical fruits, besides having relatively saturated fats, are also low in iron, and often contain important quantities of amino acids and proteins. The saturated fats in chocolate blocks the toxicity of oils rich in linoleic acid and its odd proteins seem to have an anabolic action. The random production of free radicals, rather than acting only by way of genetic damage or protein cross-linking, is also able to act as a signalling process. An excess of unsaturated fatty acids itself constitutes a massive distortion of the regulatory systems, but it also leads to distortions in the eicosanoid system and the increasingly uncontrolled production of free radicals, and to changes in energy, thyroid activity, and steroid balance. Intervention should concentrate on a correction of the signalling process, rather than genetic surgery, transplantation, etc., which is the pessimistic implication of the doctrine that oxidative damage is simply a matter of wear and tear, somatic mutations, and cross-linking. Those problems are reparable, and the emphasis should be on the production of energy and the avoidance of the conditions, which allow the undesirable signals to accumulate. The absence of cancer on a diet lacking unsaturated fats, the increased rate of metabolism, decreased free radical production, resistance to stress and poisoning by iron, alcohol, endotoxin, alloxan and streptozotocin, etc., improvement of brain structure and function, decreased susceptibility to blood clots, and lack of obesity and age pigment on a diet using coconut oil rather than unsaturated fats, indicates that something very simple can be done to reduce the suffering from the major degenerative diseases, and that it is very likely acting by reducing the aging process. Alzheimer's-type disease in old people has a long developmental history. The toxicity of oestrogen and of the unsaturated fats has been known for most of the twentieth century. Iron interacts with oestrogen and unsaturated fats in ways that can change restraint and adaptation into sudden self-destruction, apoptotic cell death. Regenerative age-regressing occurs when many circumstances are just right; e.g., taking a trip to the mountains in the spring with friends can optimise several basic regulatory systems. In the human foetus at 6 months of development, there are about twice as many brain cells as there are at the time of birth, and in old age the number of cells in the brain keeps increasing with age, so that at the age of 90 the amount of DNA in the brain (37 g) is about 50% greater than at the age of 16-20 (23 g). In the aged brain, glial cells multiply while neurons die. In the foetus, the cells that die are apparently nerve cells that haven't yet matured. Examples of the degenerative diseases include Alzheimer's dementia, epileptic dementia, arthritis, osteoporosis, depression, hypertension, hardening of the heart and blood vessels, diabetes, and some types of tumour, immunodeficiencies, reflex problems, and special atrophic problems, including clearing of amyloid and mucoid deposits. The factors that are known to reduce the brain size at birth are also factors that are involved in the degenerating brain in old age or Alzheimer's disease: lack of oxygen, excess unsaturated fats or deficiency of saturated fats, oestrogen excess, progesterone deficiency, and lack of glucose. A lack of carbon dioxide is probably harmful in both. Inflammation and blood clots may be factors in the aging brain, and bleeding with vascular spasm is sometimes a contributing factor to brain damage in both the old and the foetal brain. Endotoxaemia may be a factor in nerve degeneration only during adult life, but it is sometimes present during pregnancy. The brain, like the other organs, stops growing during development when the food supply is used up. The size, complexity, and intelligence of the brain represent a very large part of the information contained in the organism. The structure of the brain is determined at an early point in life. The structure of the brain goes into an entropic deterioration during the process of aging. The brain cells are in a vital developmental process at all times, and that the same things that injure the brain of a foetus also injure the brain of an aging person. Learning is anti-entropic. The directed flow of energy generates the structures. Regeneration involves a production of entropy. Much of the concept of entropy has derived from the study of water, as it changed state in steam engines, etc. Cancer cells, like egg cells, have higher water content than the differentiated; functioning cells of an adult, and the water is less rigidly ordered by the cellular molecules. Oestrogen has a special place in relation to the water in an organism. It is intimately involved with the formation of the egg cell, and wherever it operates, it increases both the quantity of water and, apparently, the disorder of the water. The way oestrogen promotes regeneration is by promoting water uptake, stimulating cell division, and erasing the differentiated state to one degree or another, providing a new supply of stem cells, or cells at the beginning of a certain sequence of differentiation. Degenerative aging seems to be produced by various types of contamination of the energy supply. Unsaturated fats, interacting with an excess of iron and a deficiency of oxygen or usable energy, redirect the developmental path. Endotoxin poisoning tends to intervene during stress and aging, exacerbating the trend begun under the influence of other factors, e.g., glucose, iron and calcium. The environment can be supportive, but it can also divert development from an optimal course. The demented women have a much lower rate of progestogen use, and a much higher incidence of hysterectomy, which interferes with natural progesterone production. Exposure to oestrogen in middle-age may increase the risk of Alzheimer's disease in old age, which medical progestogens offer some protection against it. Magnesium and sodium antagonise calcium in certain situations. The toxicity of linoleic acid, linolenic acid, AA is potentially prevented by the Mead acids, and their eicosanoid derivatives, which behave very differently from the familiar PGs, can be drastically reduced by dietary changes. Formation of PGs, prostacyclin, thromboxane is blocked by aspirin and other antiinflammatory drugs. Adenosine is a free radical scavenger, and protects against calcium and glutamate excitotoxicity. Adenosine is structurally very similar to inosine. Adenosines scavenge hydroxyl radicals and prevent posttraumatic epilepsy. It also appears to protect against the relative hyperventilation that wastes carbon dioxide, and endotoxin can interfere with its protective action. Guanosine, in this same group of substances, might have some similar properties. Thymidine and cytidine, which are pyrimidine-based, are endogenous analogs of the barbiturates, and like them, they might be regulators of the cytochrome P450 enzymes. Uridine, in this group, promotes glycogen synthesis, and is released from bacteria in the presence of penicillin. Iron as regulator of mRNA stability, haeme synthesis; reacts with reductants and unsaturated oils, to produce free radicals and lipid peroxides; its absorption is increased by oestrogen, hypothyroidism, anaemia or hypoxia. Glutamate and aspartate, (excitotoxins), and GABA, (an inhibitory transmitter) have metabolic links with each other, with ammonia, and with stress and energy metabolism. Endorphins are stress-induced, laterally specific, involved in oestrogen action, antagonised by naloxone and similar anti-opiate drugs. The endorphins may cause or sustain some of the symptoms of aging. Naloxone appears to be a useful treatment for senility. Endotoxin has antimitochondrial action, and causes elevation of oestrogen. It synergizes with unsaturated fats, and naloxone opposes some of its toxic effects. Structural stability of proteins and lipid-protein complexes requires urea and cholesterol. Camphor, adamantane, and the antiviral drug amantadine, probably have water-structuring effect (form cages of water molecules around themselves), and amantadine, which is widely used as a therapy in Parkinson's disease, has an anti-excitotoxic action. Post-traumatic syndrome (PTSD) differs from the majority of other diagnostic categories as it includes in its criteria the presumptive cause of the trauma. Stress hormones may influence the development of PTSD through complex and simultaneous interactions on memory formation and retrieval. Patients who received cognitive behaviour therapy (CBT) reported less intense PTSD symptoms, and particularly less frequent and less avoidance symptoms. Early provision of CBT in the initial month after trauma has long-term benefits for people who are at risk of developing PTSD. Brain cells regeneration and improved brain function is associated with methylcobalamin, phosphatidylserine and vinpocetine. Vinpocetine is an excellent vasodilator and cerebral metabolic enhancer. The mechanism of action include: 1- selective enhancement of brain circulation and oxygen utilisation, 2- increases tolerance of the brain towards hypoxia and ischaemia, 3- anticonvulsant activity, 4- inhibitory effects of phosphodiesterase (PDE) enzyme, and 5- improvement of rheological properties of the blood and inhibition of aggregation of thrombocytes. The increased cerebral blood flow associated with vinpocetine therapy is thought to contribute to its neuroprotective properties. Some of the disorders may be preventable or treatable with oral methylcobalamin including Parkinson's disease, peripheral neuropathies, Alzheimer's disease, muscular dystrophy and neurological aging. High doses of methylcobalamin (30-40 mg daily) are needed to regenerate neurons, as well as the myelin sheath that protects axons and peripheral nerves. Methylcobalamin protects against neurotoxicity by enhancing brain cell methylation. Other synergistic methylation enhancing nutrients would include vitamin B_6, trimethylglycine, and folic acid. Dopamine producing brain cells become toxic with the excessive release of glutamate from neurons. Methylcobalamin 5-20 mg daily may enhance the therapeutic benefit of Sinemet in Parkinson's patients over a longer duration of time. Methlycobalamin improves intellectual functions such as memory, emotions and communication with other people. Methylcobalamin improves the activity of helper T-cells. Dopamine deficiency causes a disinhibition (overactivity) of the subthalamic nucleus and that this may result in excitotoxic injury to the substantia nigra. Tardive dyskinesia (TD) is a neurological problem associated with antipsychotic or neuroleptic medications. Some abnormal movements characteristic of TD include grimacing, sticking out the tongue, smacking and sucking of the lips, and sometimes, rapid movements of the arms and legs. TD affects more than 1:4 older patients annually. Symptoms may start as early as one month after commencing medications. Vitamin B_6 (300 mg/day) improves various movement scores (i.e., parkinsonism, dystonia, and dyskenetic movement) in the third week. Higher doses (>400 mg/day) may have greater advantages for TD. Several other nutrients such as vitamin E, lecithin, tryptophan, vitamins B_3, vitamin C, and essential fatty acids have also demonstrated benefit in TD. A significant improvement in TD is expected with melatonin (10 mg/day).

SULPHITES

Preservatives are small molecules that are added to foods to stop or slow the degree of a deterioration reaction. Any additives is prohibited where the same results could be achieved by good manufacturing practice. Chemical preservatives should only be used as part of a holistic strategy (along with freezing, blanching, dehydration, cooking, salting, and packaging) to deliver the highest quality, safest foods possible. Sulfites are various forms of sulfurous acid. Sulfites enjoy widespread use as preservatives in foods, beverages, and pharmaceuticals. Saccharomyces cerevisiae are resistant to sulfites during winemaking. Sulfites, in gaseous or aqueous forms, are widly used as preservatives in foods and beverages. Sulfites prevent enzymatic and non-enzymatic browning reactions. As, an antimicrobial agent, sulfites prevent growth of microorganisms, or in wine fermentations, selectively inhibit undesirable organisms. Sulfites are also used as bleaching agents and conditioners in other foods. Sulphites are antioxidants, antimicrobials, and bleaching agents. In wine, sulphites prevent or minimize oxidation by inhibiting the grape enzyme polyphenol oxidase, and by direct reaction with oxygen and oxygen-derivatives such as hydrogen peroxide. Because sulphites are unstable, wine makers monitor free and total levels and make adjustments as necessary during processing. Only undissociated sulfurous acid (H_2SO_3) possesses significant antimicrobial activity and traverse microbial cell membranes. The lower the pH of a sulfite-containing wine, the more sulfurous acid is present. The bound forms of sulfite appear not to have antimicrobial activity.

SO_2 Sulfur dioxide (gas)
$SO_2 + H_2O = H_2SO_3$ Sulfurous acid or molecular sulfite
$H_2SO_3 = H^+ + HSO_3^-$ Bisulfte ion (pKa 1 = 1.77)
$HSO_3^- = H^+ + SO_3^{-2}$ Sulfite (pKa 2 = 7.2)

At wine pH, the bisulfite ion is the predominant form of free sulfite, regardless of whether gaseous sulfur dioxide or potassium metabisulfate is added. The bound sulfites or sulfonates, do not possess the antimicrobial and antioxidant properties of free sulfites. Acetaldehyde, a yeast fermentation product, forms stable reaction product with sulfite called 1-hydroxyethanesulfonate. Over time, free sulfite levels decrease in wine due to the reversible formation of bound species, and eventually, to irreversible oxidation to sulfate. Sulfites prevent or minimise oxidation-derivatives such as hydrogen peroxide. Antioxidants work by becoming oxidised more readily than the compounds they protect. In wine, this antioxidant function protects phenols. Sulfites are unstable compounds, so the winemakers monitor free and total levels to make adjustments as necessary during processing. Only undissociated sulfurous acid (H_2SO_3) possesses significant antimicrobial activity, because they traverse microbial cell membrane. The lower the pH of a sulfite-containing wine, the more sulfurous acid is present. The bound forms of sulfite appear not to have antimicrobial activity. Sulfite does not inhibit certain non-Saccharomyces yeasts indigenous to grape, especially the red varieties, e.g., Zygosaccharomyces, Torulaspora, Bettanomyces, and Schizosaccharomyces. Other yeasts such as Klooeckera, Candida, Pichia and Hansenula are significantly more sensitive to sulfite than Saccharomyces. Millimolar concentrations of sulfite cause a rapid depletion of ATP content of yeast at low pH. Inhibition of glycolysis at the step of glyceraldehyde-3-phosphate dehydrogenase is largely responsible for the decrease in ATP generation. Glyceraldehyde-3-phosphate dehydrogenase catalyses the conversion of glyceraldehyde-3-phosphate to 1,3-diphosphoglycerate, which in a subsequent step is converted to 3-phosphoglycerate with the concomitant production of ATP. Alcohol dehydrogenase catalyses the reduction of acetaldehyde to ethanol by NADH during alcoholic fermentation. The NAD^+ regenerated by acetaldeyde reduction is an obligate electron acceptor in the earlier reaction catalysed by glyceraldehyde-3-phosphate dehydrogenase. Thus, when alcohol dehydrogenase is inhibited, NAD^+ is no longer produced, and this in turn prevents the oxidation of glyceraldehyde-3-phosphate, indirectly causing depletion of ATP. Sulphite can combine with anthocyanin pigments to form a colorless adduct. This bleaching action is exploited in the manufacture of maraschino cherries where high levels of sulfite are deliberately added to bleach natural color at an early stage in the process. Sulfite is a normal metabolite in humans, other mammals, plants, and in microorganisms. In humans, sulfite is produced as an intermediate in the metabolism of the sulfur-containing amino acids methionine and cysteine, which are liberated during digestion of sulfur-containing proteins. The sulfite formed is oxidised to sulfate by the mitochondrial enzyme sulfite oxidase. Depending on the diet, 1.5 to 2.5 g of sulfite oxidase-generated sulfate is excreted daily in the urine of normal adults. Sulfur dioxide is a major urban pollutant resulting from combustion of sulfur-containing fossil fuels.

$S + O_2 = SO_2$

The gaseous sulfur dioxide released to the atmosphere may become hydrated to form sulfurous acid. In either form, it oxidises readily to form sulfuric acid. Sulfuric acid is stronger than sulfurous acid, and dissolves in rainwater constitutes one form of acid rain. Human exposure to acute air pollution and to less severe episodes is generally correlates with an increase in mortality and morbidity. Other components of air pollution that endanger human health include nitrogen oxides, ozone, and smoke. Sulfites are widely used food preservatives, and their interactions with food components and with the microorganisms they inhibits-and do not inhibit-are important as these interactions determine when and where sulfite use will be effective. Sulfite is a potentially toxic but normal metabolite in animals, plants, and microorganisms. The basis for human sulfite hypersensitivity is not well understood, although to the sulfite added to foods and beverages has been clearly established to be the causative agent. As no cure is available, the most sensible course is avoidance. Proteins or other molecules with which sulfite reacts to elicit hypersensitivity in humans may have counterparts in yeast. Sulfite reductase is the yeast enzyme that consumes sulfite in the methionine and cysteine biosythetic pathway. Glutathione is sulfite-detoxifying compound. Acetaldehyde production by yeast is an important means of detoxifying sulfite, not the only mean. The sulfite pump mediates efflux from yeast cells. The sulfite sensitivity of mutant yeast cells may be secondary to defect caused in part, by its impaired abitility to metabolise glucose. Multi-drug transporters are well-known in organisms ranging from bacteria to humans. In structure, and almost certainly in mode of action the sulfite pump (e.g., Ssu 1) differs from this class of transporters. The multi-drug transporters all contain an ATP-binding domain, which allows direct use of the energy released by ATP hydrolysis to drive efflux of a broad range of toxic compounds from the cell. In contrast, Ssu 1 does not have an ATP-binding domain, and likely uses energy obtained indirectly to pump sulfite out of the cell. The Ssu 1 protein may transport other compounds as well. Ssu 1 protein may recognise different compounds. Cells overexpressing Ssu 1 are more resistant to the structuraly related compound, such as selenite. Sulfur dioxide is a useful preservative but its use is declining in foods because of its antinutritional effects (thiamin destruction) and allergic reactions in small population. Sulfur dioxide is a toxic, chocking gas that is almost never found in foods. When the gas is mixed with water a variety of salts of sulfurous acid form, e.g., sulfurous acid (is a hydrated sulfur dioxide gas), bisulfite ions and sulfite ions. Most foods are in the pH of 3-7 and the most prevalent compounds are bisulfite (sulfite) ions. There is little of free SO_2 in the food. Sulfites are effective inhibitors of Maillard and enzymic browning. In most of these cases the sulfite bound to a carbonyl intermediate group and prevent them from polymerisation to form a brown pigment. These reactions depens on sulfite acting as nucleophile to the electron-poor carbon of the carbonyl group to form a C-sulfonate. Once the sulfite is depleted, the adduct breaks down and the preservative effect is lost. Sulfite ion can also be oxidised, which is irreversible reaction. It consumes oxygen and so limit the oxidation of other compounds. The chemical reactivity of preservatives is rarely specific, and additive may react in an uncontrolled way with other food molecules present, e.g., thiamin molecule. Sulfites are banned in foods rich in thiamin, e.g., meat products. Sulfite ions are absorbed into the circulation and very rapidly oxidised to sulfate ion by the enzyme sulfite oxidase. The sulfate is excreted in the urine. Deficieny of sulfite oxidase tends to be catastrophic. Sulfur dioxide is harmful in large quantities and even small amounts can trigger an attack in a certain class of asthmatics. Low pH foods are especially high risk for producing gaseous SO_2. Sulfites are chemicals used to keep certain farm produce fresh longer. In some patients with asthma, ingesting sulfites may lead to an acute and sometimes life-threatening asthmatic attack. The possibility for this reaction is the inhalation of sulfur dioxide generated in the stomach by the interaction between the sulfites and HCl. Sulfur dioxide (SO_2) may act as a carcinogen with benzo[*a*]pyrene (BaP) in the respiratory tract. Sulfite is the physiological form of SO_2. Sulfite bind to cellular protein and nuclear DNA, and form sulfonate compound. Once formed, this compound may be unable to leave the cell. A grain of corn is almost 70% starch. When heated in solution, the starch granules absorb water and eventually gelatinise into a paste. Food starches include thick soups, sauces, pie fillings and yougarts. They help suspend fats and proteins in gravies, salad dressings and puddings. The first step in corn refining is a 2-day soak in hot water laced with sulfur dioxide, i.e, steeping. Then the kernels are ground into a slurry, centrifuged and sreened to separate the oil, protein and starch. The starch is washed and dried, looks pure and white, but carries traces of the original sulfur dioxide. Corn starch may contain up to 50 ppm (parts per million) of sulfur dioxide. Starch bonds are built of sugar-like molecules. If the bonds are broken into smaller pieces by acids or enzymes, the starch become slightly sweet. Maltodextrin and polydextrose fall into this category of slightly sweet starches. The shorter the bonds, the sweeter the starch until finally sugars in the form of glucose syrup, commonly used as corn syrup. If the corn syrup is crystallized, dextrose is the outcome. If it is dried, corn syrup solids are the outcome. All of these forms of corn syrup have effective sulfur oxide concentrations about half that of starch. The sulfur dioxide is lowered as the processing continues. Corn syrup and dextrose are not as sweet as the sucrose in cane or beet sugar. Fructose is slightly sweeters than sucrose. High fructose corn syrup is economically maufactured at 42%, 55% and 90% fructose are so highly refined that much of the sulfur dioxide is left behind and the effective concentrations are usually low enough to be ignored. Corn starches and syrup are the most common sulfured ingredients in modern foods. The slightly sweet starches like maltodextrin are used as bulking and texturing agents. This exremely important for low sugar and fat free products. Cola bottle have mixture of high fructose corn syrup and/or sugar. Sulfur is no stranger to the vineyard, sulfur is the number one pesticide in California. All this sulfur enhances the tendency of grapes to be naturally sulfited at around 2.5 ppm. Most table grapes are treated with sulfur dioxide to extend their shelf life, doubling the sulfur preservatives before they make it to the market. Grape juices and concentrates are widely used as sweeteners in preserves and other products. The juices are divided into 2 categories, purple concord and clear white. The purple juices have a modest sulfur oxide content not to different from field grapes. The white grape juice is normally preserved with added sodium bisulfite at levels exceeding wine. Wine vinegar (and apple cider vinegar) are also treated with added sulfites and wind up with sulfur levels not too different from wine. Most raisins are sun-dried grapes without any added sulfites. But since it takes over 4 pounds of grapes to make a pound of raisins, the natural sulfur in grapes is concentrated to give raisins a sulfite level about the same as corn syrup. About 5% of raisins are not just sun-dried, but sulfured like dried apricots. These golden raisins are soft, light colored and highly preserved. Wine starts out with grapes; so wine is going to have sulfites no matter what, then most wine makers add sulfites along the way to control fermentation. Organic rules allow sulfur on the grapes and sulfur dioxide in the wine. Most white wines are more heavily preserved than red wines and inexpensive wine blends are the worst. Air oxidises sulfites to sulfate, which takes many weeks for a bottle of wine. But, after a few days, mold appears on the surface of the wine and renders it undrinkable. Most of the processed potatoes contain sulfites. Potatoes have a slight temperature problem. If they are stored at temperatures below 45^0F, the starches can convert to sugars and they can turn brown. If they are frozen, browning becomes a major problem. Any product that contains a frozen potato has a serious color problem and that is why sulfites are used in frozen potatoes. Dehydrated potatoes have similar problems that are solved by sulfite preservatives (sulfur dioxide to bleach the final product, giving it a nice white appearance like other flours). French fries are the kings of processed potatoes and most fries are prepared in a potato factory and frozen before being sent to the local fish and ships/burger stand or supermarket. At the factory, potatoes are cooked under pressure until the skin softens. When the pressure is suddenly released, the skin is jolted loose and it easily peels off in a stream of water jets. The potato skin is used either as cattle feed or to produce methane gas to help run the potato plant. The peeled potatoes are inspected and sent to the cutting machines. The cutting machines are centrifugal pumps that accelerate the potatoes to 50 mph and shoot them into stationary blades. The french fries are inspected by automated machinery for size, spots and color. The french fries that pass are blanched in vats of hot water. If necessary, a sugar dip improves the sugar content of the potatoes. Then the fries are dried to controlled water content by blasts of hot air. Finally, they are "par fried" for about a minute and a half in very hot cooking oil. At this point, they are blast frozen and boxed for shipment. Somewhere in this process, sulfites are added because french fries wind up with a sulfite concentration of about 13 ppm. For a super-sized bag, this means 2730 µg of sulfur oxide. Dehydrated potatoes have a cousin, potato flour. Potato flour is a sulfited starch. Ordinary hamburger buns contain high fructose corn syrup. Potatoes are a good source of starch just like corn. Some french fries are coated with potato starch to keep them crisp for longer periods. Sulfites are a preservative for fish. Sulfites are not allowed on red meat. Gelatin is pure protein processed to promote the gelling of liquids. Most of the gelatin produced is made from pigskin, although cattle hide and bones are also used. The first step in making gelatin is a softening soak in sulfur dioxide and water. Gelatin is used in many foods to build body and improve texture. Lowfat yogurts use lots of gelatin to make up for the missing milkfat. Gelatin typically has an effective sulfur oxide concentration of 45 ppm. An alternative to gelatin is fruit pectin. Cheese contains sulfites.

Food content of sulfites:

Food	Serving	Delay (hr)	SOx (µg)
Brown Potatoes	1 dish	24	560
Toast with Grape Jelly	2 slices	12	250
Pancakes with Syrup	1 stack	12	240
Lowfat Yogurt	8 ounces	24	200
French Fries	1 large bag	24	2275
Chocolate Shake	1 large Shake	12	480
Super-size Cola	22 ounces	18	320
Fat Free Salad dressing	2 tablespoon	12	150
Wine Salad Dressing	2 tablespoon	12	400
Lemon Juice concentrate	1 teaspoon	12	770
Inexpensive White Wine	1 wine glass	12	7500
Quality Red Wine	1 wine glass	12	3000
Instant Potatoes	2 scoops	18	2100
Pizza	4 slices	12	1600
Canned Potato Soup	1 cup	18	420
Dried Fruit	4 pieces	12	3500
Candy	26 pieces	12	160

All cheeses contain low levels of sulfite created naturally during the aging process. Milk is separated into solid curds and liquid whey by adding a little acid and enzyme borrowed from a cow's stomach to make cheese. As the mixture is pressed, salt is added, the whey is drained and the curds become more firm. Cheese may be categorized as fresh, soft, hard and dry. Fresh cheeses are not aged. Soft cheeses are aged but remain creamy either naturally or by processing. Hard cheeses include cheddar, jack, mozzarella, swiss and gouda. Dry cheeses are strongly aged with very low moisture like parmesan and dry jack. Sour cream is similar to fresh cheeses, but yogurt is free of sulfites unless corn starch and gelatin are added. Cottage cheese contain 50 µg of SOx, and cheddar 84 µg of SOx, for example. Sourdough bread using yeast that is aged have very small amount of sulfite. The sulfur oxide concentration of eggs is about 1 ppm. Some of the sulfites are lost or converted during cooking eggs. Scrambling the eggs exposes more of the egg to heat and atmospheric oxygen, and wind up with a slightly lower sulfite concentration. Sulfur preservatives can be dissolved as ions or bound to organic molecules and the human body reacts similarly, but not identically, to all forms. The headache is a measure of the swelling caused by sulfites and sulfur dioxide. Sulfite swells body tissue. Red wine contains histamines. The alcohol in any wine can cause migrains. The levels of sulfites may vary from brand to brand and from year to year. Food diary is good idea to determine the sensitivity. All foods on this planet contains natural sources of sulfur oxide. The worst offenders are onion and garlic. Raw onion contains about 1 ppm and raw garlic 4 ppm. Most other foods are under 1 ppm of effective sulfur oxide. Pork is rated at 1 ppm and a large serving might be 8 ounces or 240 g. Upon digestion thiamin (B_1) is a very weak source of sulfur dioxide. Biotin (B_7) is an incredibly strong sulfite. Most biotin in the human diet is produced by intestinal flora. Taurine is celebrated in trendy energy drinks. Red bull contains 1000 mg of taurine resulting in more sulfur oxide than a glass of wine. MSM (sulfur supplement) is bad news for people that are sensitive to sulfites. Food color is poorly digested and little sulfite is actually released within the body. Both asulfame-k and saccharine contain sulfur dioxide. Drugs can contain sulfite preservative like food. Some are called sulfa drugs, some contain sulfur and oxygen. Sulfonamide contains a sulfur dioxide group. About 15% of the drug is digested to release sulfur dioxide and cause a headache. A common sulfa drug is celebrex, 200 mg tablet contains sulfite equal to that of a glass of red wine.

Food to avoid that sulfite in it:

Everybody	Instant potatoes, dried fruits, lemon juice concentrate, molasses, taurine, biotin, fast food, french fries, fast food pizza, all food with a label that declares sodium sulfite, bisulfite, metabisulfite, sulfur dioxide or E220.
Infants	Formula made with corn syrup, corn syrup solids or maltodextrin, cereal made with dried fruit, juice sweetened with white grape juice, jams sweetened with white grape juice.
Children	Super size cola, gummy worms, coconut candy and fruit roll-ups.
Adults	Wine, wine vingar, guacamole, coconut drinks, sulfa drugs and MSM.

Free SO_2 is measure in mg/l, which is approximately 1 ppm. Both units are used interchangeably in the wine industry. The effectivess of free SO_2 decreases as pH increases. Sulfite is added to the wine industry using one of 3 simple methods: 1- sulfite powder (dissolved in water), 2- Campden tablets (crushed and dissolved in water), and 3- 10% dilute sulfite solution. More that 50% of the sulfite becomes free SO_2 to protect wine. Sulfite and other sensitivies should be considered dangerous and must be sorted out. About 5% of all asthma sufferers are quite sensitive to sulfites. Sulfur oxides: Na_2SO_3 (sodium sulfite), $NaHSO_3$ (sodium bisulfite), $Na_2S_2O_5$ (sodium metabisulfite), and SO_2 (sulfur dioxide gas). The first 3 are true sulfur salts and K may replace Na in some formulas. When dissolved in water, they dissociate to free sulfite (SO_3) ion. This combination of one sulfur and 3 oxygen atoms is a strong antioxidant and explains the food preservative mechanism of sulfites. If O_2 is available, sulfite ions readily oxidise to become sulfate SO_4 ions. Sulfur dioxide lacks the third oxygen atom, but humans response to SO_2 mimics the response to sulfites, probably because it may be oxidised to become a sulfite. SO_2 is air polllutant gas. In foods sulfur can easily combine with organic molecules taking the place of carbon and oxygen groups. When carbon foods or tree wood is burned, CO_2 (carbon dioxide) is produced. When sulfur is burned, SO_2 (sulfur dioxide) is produced. SO_2 gas is used to prevent discoloration when drying fruits such as apricots. SO_2 is an important ingredient in caramel color, which is produced in the same way as true caramel. Diet colas are nearly all water and caramel colour. Regular colas have sweetner, high fructose corn syrup. A gram is the weight of 10 drops of water. The PPM means parts per million, and refers to the weight of the preservative as a part of the total weight of fruit. For example, 220 ppm means that if the fruit is divided into one million parts, 220 of those parts would biologically effective sulfur oxide. Ppm numbers calculte weights.

Food weight X ppm = SOx in μg, i.e, for 220 ppm concentration of effective suldur oxide in dried fruit, a gram bite would be preserved with 4 X 220 = 880 μg of SOx.

Caramel colour has a sulfur oxide concentration of 0.34 based on the total weight of the cola. 355 ml can of cola means that 0.34 times 355 to get the amount of caramel color SOx. The sugar weight of cola is 40 g, the sweetner will most likely be high fructose corn syrup. This syrup has a code name HFS and a sulfur oxide concentration of 1.3 ppm based on the weight of the syrup. So, carmel color (CCL) 121 μg (355 X 0.34) + High fructose syrup (HFS) 52 μg (40 X 1.3) = 173 μg (total micrograms sulfur oxide in a can of cola). Sulfur dioxide has a number of unwanted environmental effects: 1- Acid rain is formed when SO_2 mixes with and dissolves in the water in clouds forming dilute sulfuric acid. Acid rain causes lake and soil acidification, forest die-off, and corrosion of stone and metalwok. 2- SO_2 contributes to the formation of acidic aerosols, which can cause haze to form over large regions. Haziness reduces average temperatures in the affected areas. 3- SO_2 and related pollutants are linked to a number of human diseases. Sulfite reduction is energetically more favourable than sulfate reduction. At a biochemical level, this is manifested by the ATP demanding activation of sulfate to adenosine-5'-phosphosulfate (APS) by ATP-sulfurylase, which is followed by APS reduction to form sulfite and AMP. Sulfite is directly suitable as an electron acceptor for sulfate reducing bacteria (SRB). SRB constitute a diverse group of prokaryotes that contribute to a wide variety of essential functions in many anaerobic environments. SRB are classified into 4 groups: 1- Gram-negative mesophilic SRB, 2- Gram-positive spore-forming SRB, 3- Gram-negative thermophilic SRB, and 4- thermophilic archaeal SRB. All these groups are characterised by their use of sulfur oxyanions such as sulfate, sulfite, or thiosulfate as the terminal electron acceptor for anaerobic respiration. The production of endospores as an effective reproducing and survival strategy is a phenomenon observed in prokaryotic as well as in eukaryotic species.

Dormant spores are not hazardous, but spore germination, outgrowth and proliferation could result in food spoilage and toxin production, in turn causing food poisoning. Spore heat resistance is both complex and multifactorial, and it is unquestionable that the spore cortex peptidoglycan, the dehydration, minerlization, and dipicolinic acid content of the spore core are involved in spore heat resistance. Enzymes involved in primary alcohol degradation under anaerobic conditions are divided into 3 different groups, alcohol dehydrogenases (ADHs): 1- NAD(P)-independent alcohol dehydrogenase, mediate the oxidation of methanol via formaldehyde to formate. 2- NAD(P)-dependent alcohol dehydrogenase. 3- Methyltransferases, in the degradation of methanol, corrinoid proteins (corrinoids are cobalt-containing co-factor) play a role in methyl transfer processes. Part of methanol is oxidised to CO_2 to generate the reducing rquivalents needed for the reduction of the remainder of the methanol to methane. In homoacetogenic bacteria (AB), several enzymes are involved in acetate synthesis, such as formate dehydrogenase, a corrinoid protein, and carbon monoxide dehydrogenase. The formation of acetate from methanol is only possible if either carbon containing compounds more oxidize than methanol-such asformate, CO, or CO_2-are present. Sulfate reduction results in the production of hydrogen sulfide (H_2S), which at high concentrations, can become quite inhibitory for microbial growth. H_2S is a very weak acid (pKa of 7 at 30°C) and therefore at neutral values sulfide is mainly present as H_2S (hydrogen sulfide or free sulfide) and HS^- (bisulfide). Hydrogen sulfide is the most toxic form of sulfide. The neutrality of H_2S-molecule allows its easy diffusion through the lipid cell membrane into cytoplasm, where it reacts with cell components. Sulfate is generally not toxic for anaerobic bacteria at concentrations up to 10 g.L^{-1}. Sulfite is very toxic for microorganisms and for that reason it is used as anti-bacterial agent, e.g., wine processing. The mechanism is not known precisely. Alcohols (methanol) are toxic for microorganisms at high concentrations, presumably due to the fact that they damage the cell membrane and due to inhibition of glycolytic enzymes. Methanol degradation by AB may result in the accumulation of acetate, which is toxic for microorganisms at higher concentrations. As with sulfide, unionised acetate (acetic acid) is considered the most toxic. Proteins and trace elements need to be available to microorganisms in order to facilitate growth. Bacteria compete for trace elements when these are limiting. Iron, cobalt, nickel, zinc and copper are trace elements that are necessary to maintain maximum growth. Microbials produce chelating agents. Methanol is a suitable substrate for the removal of inorganic sulfur compounds from off-gases in anaerobic bioreactors operated at 60°C or higher.

Electrons:

Electrons acceptors	Electrons donors
Sulfate	Glucose
Sulfite	Fructose
Thiosulfate	Formate
Nitrate	Acetate
	Propionate
	Butyrate
	Lactate
	Fumarate
	Succinate
	Methanol
	Butanol
	Isobutanol
	H_2/CO_2

Under thermophilic conditions, methanol can be used by several groups of microorganisms. The conversion of methanol to sulfate is energetically a more favourable process than the production of methane or acetate from methanol. The ability of microorganisms to survive extreme environmental conditions is partly determined by their capacity to protect DNA from destructive chemical and physical conditions. Some bacteria form endospores, which can resist extreme conditions. Spores may remain dormant for centuries and may even survive over geological time and through intersteller travel. Heat resistant spores are found among several species. Thermophilic, sulfate-reducing bacteria are found in a wide range of environments including hot springs, geothermal groundwater, fresh water, cold marine sediments, compost and manure, and anaerobic bioreactors. The deposition of sulfuroxyanions like sulfate, sulfite and thiosulfate by man cause severe envionmental problems like anaerobiosis of surface water and acid rain. In a biological process, the sulfuroxyanions are converted to the relatively high valued elemental sulfur. Sulfate reducers can use the methanol directly or indirectly via acetate, H_2/CO_2, or formate to produce sulfide.

FLUORIDATION

1ppm flourides result in the reduction of tooth decay. Fluorides inhibit the action of lipase. Fluoride inhibits neuromuscular activity. Fluoride administration to inmates decreased resistance to authority and induce physical deteriorization in the Soviet concentration camps during 1940. Fluorine may cause anoxia in the newborn and shorten the period of their survival. Fluoride was used in the German concentration camps to render the prisoners docile and inhibit the questioning of authority in 1942. The death rate is three times higher in areas with drinking water contained 0.4 ppm fluoride. Fluorides are general protoplasmic poisons, changing the permeability of the cell membrane by inhibiting certain enzymes. Fluorine compounds are used as insecticidal sprays for fruits and vegatables and the mining, and conversion of phosphate rock to superphosphate, which is used as a fertilizer. Fluorides are used in the smelting of many metals, such as steel and aluminum, and in the production of glass, enamel and brick. Uranium hexafluoride may have a marked central nervous system effect, with mental confusion, drowsiness and lassitude as the conspicuous features. Fluoride component is the causative factor. A by-product of aluminum manufacture is toxic sodium fluoride. Sodium fluoride is a highly toxic substance. Drinking water containing as little as 1.2 to 3.0 ppm of fluorine will cause such developmental disturbances in bones as osteosclerosis, spondylosis, and osteopetrosis, as well as goiter and other serious systemic disturbances. Beer was classified as a food and fluoride as a poison in 1945. Fluoride's affinity for magnesium and manganese ions enables it to deplete their availability for vital enzyme functions. Blood clotting may be affected by fluorides. Fluorides added to the water supplies may induce submission by damaging the brain and stem. There are instances in which difficulties in emotional adjustment are the primary result of alterations in the brain. Fluorides are violent poisons to all living tissue because of their precipitation of calcium. They cause fall of blood pressure, respiratory failure, and general paralysis. Continous ingestion of non-fatal doses causes permanent inhibition of growth. Fluorides lower haemoglobin and may cause irreversible loss of potassium from the red cells. Chronic intoxications resulting from prolonged intake of smaller amounts of fluorides include dental fluorosis. Fluoride also tends to accumulate in bones, leading to hypercalcification and brittleness. Ligaments and tendons also become calcified. Other effects include baldness in young men, accompanied by increased fluoride concentrations in hair and nails, anaemia and decreased blood clotting power due to the binding of calcium. Dysmenorrhea, alterations in growth and weight, lowered birth rate, high incidence of fractures, thyroid alterations and liver damage have been observed in regions of endemic fluorosis. There is a link between fluoridation and cancer. Fluoride increases cancer rate by 10% in only 13-17 years. The cancer death rate of the fluoridated areas is to far exceed the rate of the unfluoridated areas, 22% increase in cancer death rates. Chronic fluoride poisoning accelerates aging and degenerative disease. Fluoride reduces resistance in people to authority. Fluoridated toothpaste should not be used where the water supply is fluoridated. Fluoridation delays eruption of teeth. 25% to 50% of fluoride is deposited in the skeleton. Chronic fluoride intoxication abnormalities may occur after constant, or frequent exposure to fluorides. Fluoride is associated with stomatitis. Mottled teeth are teeth showing symptoms of fluorosis, and the enamel of mottled teeth is brittle and subject to mechanical injury, which is difficult or impossible to repair. Fluorosis occur in 10-20% of children. Fluorides produce primary damage by injuring the genetic material of the cells they enter. The deposit of fluorine on teeth and bones starts through the placenta as early as the embryo period, and then takes place through the mothers milk through the infancy period, and through food, as well as directly through the inside of the oral cavity. As fluorine is a known active enzyme poison, it is known to affect cell division (mitotic) in the foetus, resulting in anatomical anomalies (teratism). Fluoride at 1ppm concentration may cause a 25% decrease in activitiy of the enzyme succinic dehydrogenase in the liver. In the kidney this enzyme activity is reduced by 47.8% by fluoride at 1ppm in drinking water. This significant enzyme inhibition due to prolonged fluoride administration demonstrates the impairment of an important step in cellular metabolism. Fluoride inhibits the immune system to attack tumours (15-25%). Fluorides induce significant mitotic and meiotic chromosome alterations in tomato plants. 5-Fluoracil as a cancer treatment represents a case of classic overkill, which does more harm than the cancer itself. Fluoride acts to tie up magnesium-forming magnesium fluoride-an insoluble compound, which thus prevents the essential enzyme from using magnesium. As a result, mental processes are seriously interfered with, and nerve reactions throughout the body depressed, which plays a role in epileptic seizures and other convulsions. Water fluoridation accounts for only part of the daily fluoride consumption. Fluoride topical paste induces gum damage. Fluorides inhibit the repair of DNA damaged by radiation. Fluoride increases the incidence of Down syndrome, depresses the growth rate of human cells in culture, causes dermatitis, affects thyroid function, alters the concentration of metabolites in blood and urine, inhibits many important enzymes, affects reproduction, may incorporate itself into naturally occurring compounds to form others, is capable of producing cancer and giving rise to deformed babies, reduces fertility in female rats by >60%, and mimics the effects of some hormones-disturbing the hormonal balance. Fluoride lowers the immunobiological response of man against certain diseases, such as typhoid fever, anthrax, tuberculosis and Staphylococcus aureus. The activity of the human serum enzyme alkaline phosphatase is reduced by either dietary fluoride or fluoridated drinking water. Fluoride leaches out copper. There is potentially hazards amounts of lead in the toothpaste. The first symptoms of exposure to many toxic chemicals (e.g., fluorides) are not physiological, but psychological, and a number of individuals develop immediate symptoms of headache, burning in the throat, dizziness, nausea and vomiting to test solutions of <1 ppm. In the presence of aluminum in a concentration as small as 20 parts per billion, fluoride is able to cause an even larger increase in cAMP levels. cAMP inhibits the migration rate of white blood cells, as well as the ability of the WBC to destroy pathogenic organisms. The fluoride-induced increase in cAMP levels, might have effect on the immune system. Fluorides cause genetic damage in human blood cells. Fluorides in the body disrupt collagen synthesis. Fluoride stimulates superoxide production in resting white blood cells, seriously depressing the ability of WBC to destroy pathogenic agents. Superoxide in the bloodstream also gives rise to tissue damage and acceleration of the aging process. Fluoride at a concentration that stimulate vigourous superoxide production by the cells, abolishes phagocytosis. Intake levels 5 mg/day of fluoride will cause crippling deformities of the spine and major joints (skeletal fluorosis). Fluoride disrupts the synthesis of collagen in the body and leads to the breakdown of collagen in bone, tendon, muscle, skin cartilage, lung, kidney and trachea. It appears that fluoride disruption of collagen synthesis in cells responsible for laying down collagen leads these cells to try and compensate for their inability to put out intact collagen by producing larger quantities of imperfect collagen and/or non-collagenous protein. As little as 1ppm fluoride in the body interferes with collagen metabolism and leads to collagen breakdown, causing osteoporosis, bone cancer, brittle bones and teeth, and loss of connective tissue. Fluoride reacts strongly with the bonds that maintain the normal shapes of proteins in the body. By distorting the configuration of the body's own protein, the immune system attacks its own protein, resulting in an autoimmune or allergic response. Flruoides contribute to the development of an acquired immune deficiency syndrome, which is covered up by the media and medical community, to maintain a public focus on a viral cause for the problem, promoting the harmless HIV virus as the cause for AIDS, covering up the pharmo-chemical sensitization of the people. Fluorides (as in fluoridated water supplies, toothpaste,etc) stimulate granule formation and oxygen consumption in WBC. 1ppm fluoride in the water induces increase hydroxyproline and hydroxylysine levels in the blood and urine, as well as a decrease in skin and lung collagen levels. Fluoride confuses the human immune system and causes it to attack the body's own tissues, cumulatively resulting in accelerated aging. Less than 1ppm fluoride increases collagen formation by 50%, leading to formation of irregular bone. Fluoride causes genetic damage in human blood cells. Fluoride directly stimulates proliferation and alkaline phosphatase activity of bone forming cells. Fluoride induces morphological and neoplastic transformation, chromosome aberrations and unscheduled DNA synthesis. Fluoride toothpaste contains up to 1 mg/g of fluoride. A whole tube contains about 199 mg of fluoride-enough to kill a 25-pound child. Toothbrushing causes ingestion of 0.25 mg/day. Low levels of fluorides in the body delay WBC response to foreign agents. Intracellular fluorides can alter the kinetic properties of calcium currents in hippocampal neurons, which can affect behavior. Fluoride may cause genetic damage in bone marrow cells. Low levels of fluoride increase the incidence of melanotic tumours in living organisms from 12% to 100%. Bottled water may contain fluoride in significant amount. There is as much as a 50% increase in oral cancer rates in fluoridated areas. There is a link between oral precancerous growth and fluoride, as well as an increase in osteomas, osteosarcomas and hepatocholangiocarcinoma. There is a negative total fertility rate (TFR) as a result of fluoride in drinking water. Fluorides inhibit the adenylyl cyclase system in human spermatozoa. Fluorides lowers protein synthesis in osteoblasts. There is a connection between fluorides and non-ulcer dyspepsia, which could be due to prevalence of Heliobacter pylori infection as well as fluoride toxicity. Fluorides interfere with androgenesis and adversely impair the target organ structures. Fluorides may alter the configuration of cellular receptors, thereby inhibiting the action of testosterone in the same way that hormone-disrupting chemicals (chemicals that mimick oestogens) do. Fluorides affect the hippocampus, which is the central processor that integrates inputs from the environment, memory, and motivational stimuli to produce behavioral decisions and modify memory. The effects on behavior relate directly to plasma fluoride levels in the brain, and fluoride accumulation in the brain. Long-term accumulation of fluorides in the brain does occur, even with minimal doses. The behavioral changes occur from fluoride exposures are consistent with interrupted hippocampal development, with potential for motor dysfunction, IQ deficits and/or learning disabilities in humans. Substances that accumulate in brain tissue potentiate concerns about neurotoxic risks. Fluorides are cumulative in the brain. Fluorides interrupt normal brain development in babies and children, and fluorides affect the hippocampus, the central processor in the brain which integrates inputs from the environment, memory, and motivational stimuli to produce behavioral decisions and modify memory. Fluoridation undermines the capacity for people to make decisions and become motivated to act against abnormal environments (such as that present in tyranny). Fluorides in drinking water affect the nervous system directly without first causing skeletal deformations from fluorosis, neural effects occur at a very low dosage. The anti-obesity drug, dexfenfluramine can cause brain damage by altering the brain chemical serotonin. Fluoride was found to be an equivocal carcinogen, with a 5% increase in all types of cancers in fluoridated communities. Fluoride facts: 1- Fluoride is highly toxic. 2- Fluoride increases hip fractures. 3- Fluoride increases fertility dysfunction. 4- Infertility in women is increased with water fluoridation. 5- Fluoride increases fluorosis. 6- Fluoride is not effective in reducing tooth decay.

A I R

Unfortunately, little effort has been made to integrate all the available medical information. Air is made up of these chemical elements: nitrogen 78%, oxygen 20%, argon 1%, carbon dioxide 1/3% and other rare gasses 2/3%. These elements combine with other matter suspended in air to form molecules. An ion is formed if any of the molecular complexes acquire an electrical charge. The task of hygiene, apart from sanitation of the conditions of the external medium, is to work out measures to increase work capacity and speed up rehabilitation processes after work. Air ions are classified according to their size: small (molecules), medium (tiny particles), and large (dust). The small ions are the most mobile of the 3 categories and, for that reason play a central role in atmospheric influences upon the human system. Small ions, those electrically clusters of molecules of atmospheric gases, of both positive and negative polarity, are normally generated by radiation processes, radioactive fallout, cosmic (non solar) sources, natural radioactive materials in the atmosphere, and some aspects of solar radiations lying beyond the ultraviolet range, especially during solar disruptions. The air present in the pores of the soil and in larger subterranean cavities is continually ionised by external radiations in combination with local mineral deposits. The production of air ions varies not only in space, but also in time. The rate of ionisation is influenced by air pressure. As air pressure changes from day to day and from season to season, the rate of ionisation due to solar and cosmic sources also varies. The period of time over which small ions maintain their effective radiation (half-life), depends on the amount of pollutants in the air. The cleaner the air, the longer the half-life of small ions, especially on high mountains. Clean air means high concentration, providing there is an ionisation source present. The lower the visibility, the lower the small ion concentration; this is because low visibility means the presence of many large ions. In the vicinity of waterfalls there are up to 37,000 small negatively charged ions/cc of air. Coniferous trees give off a considerable amount of negative ions into the surrounding atmosphere. Foggy or dusty environments are low in small ion concentrations. Usually small air ions exist in a proportion of 5 positive to 4 negative ions. In the open country, in conditions of sunshine, 400-500 air ions/cm^3 are found. Values can go up to 1,000 in the mountains and down to 10 in crowded cities. Air ion ratio variations occur in the free atmosphere under normal conditions, the ratio depending on the strength and polarity of the environmental electrostatic field. The normal electrostatic polarity of an outdoor environment is that of a positive charge in the air space above the earth, and a negative charge at the surface of the earth itself. Between the two polarities of this electric atmospheric field, there is a continuous movement of ionic charges, both towards the positive air space above and the negative earth below. This ration remains fairly constant under normal conditions. But with a disturbed atmosphere, such as a thunderstorm, the polarity of the atmospheric field can be reversed, which correspondingly reverses the flow of ionic charges. The electrical conductivity of the atmosphere is related to the small ion concentration. Any reduction in small ion concentration results in a corresponding reduction in atmospheric conductivity. The air pollution is an anathema to small ions. Air pollution is a major cause of changing ion ratios in the environment, disturbed atmosphere being another. Ionisation of the air is a natural phenomenon. The ion density varies according to terrestrial and atmospheric conditions: radioactivity, barometric pressure, humidity, season, time of day, air movement, precipitation, altitude, intensity of ultra-violet and cosmic radiations, or air pollution. An ion is a molecule or a small group of molecules that become electrically charged when gaining or losing an electron. The addition of an electron produces a negative ion (-) while loss of an electron converts it to a positive (+) one. Naturally there is a delicate balance of ions. The relatively low natural ion levels do not permit systematic evaluation of their biologic properties. The effects of positive air ions are: 1- Reduction of the rate of ciliary beat, sometime progressing to complete stoppage, and a parallel reduction of the rate of mucus flow. 2- Contraction of the membraneous posterior wall of the trachea. 3- Drying of mucosal surface. 4- Induction in the cilia of a state of vulnerability to trauma. Negative ions increase ciliary activity and mucus flow above normal values. Negative air ions increase the rate at which serotonin (5-HT) is oxidised, probably through a cytochrome-linked reaction. Negative aero-ionisation should be carefully controlled in coronary artery disease. Artificially air is gradually gaining recognition as a valuable adjunct to other established methods of medical treatment. Ozone is a molecule composed of 3 atoms of oxygen. Two atoms of oxygen form the basic oxygen molecule-the oxygen we breathe that is essential to life. The third oxygen atom can detach from the ozone molecule, and re-attach to molecules of other substances, thereby altering their chemical composition. It is this ability to react with other substances that forms the basis of manufacturers' claims. The same chemical properties that allow high concentrations of ozone to react with organic material outside the body give it the ability to react with similar organic material within the body, and potentially cause harmful health consequences. Inhaled, ozone can damage the lungs. Relatively low amounts of ozone can cause chest pain, coughing, shortness of breath, and throat irritation. Ozone may also worsen chronic respiratory diseases such as asthma and compromise the ability of the body to fight respiratory infections. Exercise during exposure to ozone causes a greater amount of ozone to be inhaled, and increases the risk of harmful respiratory effects. Recovery from the harmful effects can occur following short-term exposure to low levels of ozone, but health effects may become more damaging and recovery less certain at higher levels or from longer exposures. Ozone is a toxic gas with vastly different chemical and toxicological properties from oxygen. Ozone heath effects and standards include: a- decreases in lung function, b- aggravation of asthma, c- throat irritation and cough, d- chest pain and shortness of breath, e- inflammation of lung tissue, and f- higher susceptibility to respiratory infection. Factors expected to increase risk and severity of ozone health effects are: 1- increase in ozone air concentration, 2- greater duration of exposure, 3- activities that raise the breathing rate (e.g., exercise), and 4- certain pre-existing lung diseases (e.g., asthma). Ozone in the upper atmosphere-referred to as stratospheric ozone helps filter out damaging ultraviolet radiation from the sun. Ozone in the atmosphere, which is the air we breathe can be harmful to the respiratory system. Harmful levels of ozone can be produced by the interaction of sunlight with certain chemicals emitted to the environment (e.g., automobile emissions and chemical emissions of industrial plants). For many of the chemicals commonly found in indoor environments, the reaction process with ozone may take months or years. For many of the chemicals with which ozone does readily react, the reaction can form a variety of harmful or irritating by-products. For example, chemicals from new carpet, reacting with ozone can produce a variety of aldehydes, and the total concentration of organic chemicals in the air increase rather than decrease. In addition to aldehydes, ozone may also increase indoor concentrations of formic acid, both of which can irritate the lungs if produced in sufficient amounts. Some of the potential by-products produced by ozone's reactions with other chemicals are themselves very reactive and capable of producing irritating and corrosive by-products. Ozone does not remove particles (e.g., dust and pollen) from the air, including the particles that cause most allergies. An ionizer is a device that disperses negatively (and/or positively) charged ions into the air. These ions attach to particles in the air giving them a negative (or positive) charge so that the particles may attach to nearby surfaces such as walls or furniture, or attach to one another and settle out of the air. Ionizers are less effective in removing particles of dust, tobacco smoke, pollen or fungal spores than either high efficiency particle filters or electrostatic precipitators. Ozone is not effective in reducing formaldehyde concentration, and not useful for odour removal in building ventilation systems. Ozone may react with acrolein, one of the many odorous and irritating chemicals found in secondhand tobacco smoke. Ozone does not effectively remove viruses, bacteria, mold, or other biological pollutants. Ozone have to be 5-10 times higher than public health standards (0.50-0.80 ppm) to decontaminate the air sufficiently to prevent survival and regeneration of the organisms once the ozone is removed. Even at high concentrations, ozone may have no effect on biological contaminants embedded in porous material such as duct lining or ceiling tiles. Ozone has been extensively used for water purification, but ozone chemistry in water is not the same as ozone chemistry in air. High concentrations of ozone in air, when people are not present, are sometimes used to help decontaminate an unoccupied space from certain chemical or biological contaminants or odours (e.g., fire restoration). Ozone can adversely affect indoor plants, and damage materials such as rubber, electrical wire coatings, and fabrics and art work containing susceptible dyes and pigments. Approaches to reducing indoor air pollution, in order of effectiveness, include: 1- Source control eliminates or control the sources of pollution. 2- Ventilation dilutes and exhaust pollutants through outdoor air ventilation. 3- Air cleaning removes pollutants through proven air cleaning methods. Source control is the most effective, which involves: minimizing the use of products and materials that cause indoor pollution, employing good hygiene practices to minimize biological contaminants (including the control of humidity and moisture, and occasional cleaning and disinfection of wet or moist surfaces), and using good housekeeping practices to control particles. The outdoor air ventilation approach is also effective and commonly employed. Ventilation methods include installing an exhaust fan close to the source of contaminants, increasing outdoor air flows in mechanical ventilation systems, and opening windows, especially when pollutant sources are in use. The air cleaning approach is not generally regarded as sufficient in itself, but is sometimes used to supplement source control and ventilation. Air filters, electronic particle air cleaners and ionizers are often used to remove airborne particles, and gas adsorbing material is sometimes used to remove gaseous contaminants when source control and ventilation are inadequate.

THE BIOLOGICAL EFFECTS OF AIR IONS

If anyone is exposed to a high level of negative atmosphere,e.g., a shower (supply of negative ions), within half an hour serotonin is excreted in the urine. Hydrotherapy is based on re-ionising the body, i.e., increasing the negative ion supply. Acupuncture points are ion absorption points, it enhances the body's ability to utilise oxygen properly and vitamins also have a similar effect. Clothing made of synthetic fabric increases ionic flow by causing friction with immobilised ionic particles in an electro-magnetically shielded environment. Under normal conditions the body balances the ionisation by means of nasal cycle e.g., taking the breath in through the left nostril negatively ionises the air and the right nostril positively ionises it. Running water also produces negative ions. Dimethylsufoxide (DMSO) (CH_3) acts biologically as a weak donor of electrons, when applied externally induces dramatic relief of pain. It has local analgesic properties and is absorbed through intact skin. Cortisone ($C_{21}H_{28}O_5$) is also a weak electron donor in tissues. Both negative and positive ions can: A- inhibit the growth of bacteria and fungi on solid media, B- exert a lethal effect on vegetative forms of bacteria suspended in small droplets of water, and C- reduce the viable amount of bacterial aerosols. Negative ions increase and positive ions decrease the rate of fibroblasts proliferation. Plants appear to benefit from increases in both positive and negative ionisation. Ionisation markedly increases the rate of growth of higher plants such as barley, oats and lettuce. Positive and negative ions expedite both the uptake of iron and its utilisation of the production of ion-containing enzymes. The ions stimulate the metabolism of the high-energy compound ATP in the chloroplast's and augment both nucleic acid metabolism and oxygen uptake. All of these phenomena are consistent with the observed ion-induced increase in growth rate. Ions of either charge increase synthesis of 3 enzymes (catalase, peroxidase and cytochrome C Oxidase). The site of action of air ions is the mucosa of the upper respiratory tract. Positive ions raise blood levels of serotonin (5-hydroxy tryptamine or 5-HT), while negative ions have the opposite effect. Small negative ions may stimulate, while small positive ions block the action of monamine oxidase (MAO), thus producing respectively a drop or rise in the concentration of free 5-HT in certain tissues and eliciting a corresponding physiological response. 5-HT may be the sole agent responsible for air-ion induced alteration of physiological function. The air in many locales near waterfalls for whatever reason, contains a high concentration of small air ions with a ratio of negative to positive ions being considerably greater than normal. 12-36 hours before the characteristic changes in wind, temperature and humidity, the total number of ions increase (from 1500 ions/cm^3 to 2600 ions/cm^3) and the ratio of positive to negative ions jumped from the normal 1.2 to 1.33. This early shift in ion density and ratio coincide with the onset of nervous and physical symptoms in weather sensitive people. Individuals of the serotonin hyperfunction syndrome suffer from insomnia, irritability, tension, migraine, amblyopia, oedema, palpitations, precordial pain, respiratory distress, hot flashes, tremor, chills, diarrhoea, polyuria, vertigo, etc. These patients display an increased output of serotonin in the urine and they experience relief when treated with negative ions or with serotonin blocking drugs. Air ion imbalance is the direct cause of the irritation syndrome, the positive air-ion-induced hypersecretion of serotonin. A dry throat, husky voice, headache, itch or obstructed nose and a reduction in maximum breathing capacity occur when exposed to nasal inhalation of positive ions in concentration of 3.2×10^4 ions/cm^3. Central heating and air conditioning, smoking, the usual household activities of dusting and cooking all combine to lower levels of small ions in indoor environments. Further, the static electricity generated by the widespread use of synthetic fibres in clothing and room furnishing as well as stray electric fields add a different dimension to the indoor climate, which is not conducive to the preservation of small air ions. The small physiologically active air ions readily combine with gaseous and particulate pollutants to form large (Langevin) ions that are considered physiologically inert. People travelling to work in polluted air, spending 8 hrs a day in offices or factories and living their leisure hours in urban dwellings inescapably breathe ion depleted air for substantial proportions of their lives. There is increasing evidence that this ion depletion leads to discomfort, enervation and lassitude and loss of physical and mental efficiency. This syndrome appears to develop quite apart from the direct toxic effects of the usual atmospheric pollutants. Any enclosed compartments with conditioned air such as a space capsule, are likely to be depleted of negative ions and have a considerable excess of positive ions and that prolonged stays in such an ion environment is detrimental. An excess of positive ions also affects the neurohormone adrenalin. The body responds to levels above 1000 ions/cc.

Some average sample readings of negative air ions taken in various locations:

By a waterfall:	50,000 ions/cc.
In the mountains:	5,000 ions/cc.
In the country:	1,500 ions/cc.
In an average modern office:	50 ions/cc.

NEGATIVE ION GENERATOR (How Ionizers Work)

The positive ions make us feel bad (TV screen produces positive ions). Air conditioning, crowds, air pollution (e.g., vehicle exhausts) and dust largely use up the biologically available negative ions in the air, leaving mainly positive ions. Positive ions are known to have an unpleasant effect on many people, producing headaches, anxiety, depression and breathing difficulties, as many hot dry winds that occur around the world are known to do (stress). The air's negative ions basically get used up by the first half hour of early morning traffic. Negative ion generator or ioniser are available since the 1930's and have been used throughout Europe, the Middle East, the Orient and the former Soviet Union in medical and general applications. A high concentration of small negative ions lowers infection rates in patients at risk, decreasing blood clotting and the need for pain killers while accelerating healing. The effect on asthma patients is particularly impressive, and even those without specific ailments would claim remarkable benefits in energy levels and well-being. Ions are atoms that have gained or lost an electron. Every molecule resembles a tiny solar system, with a nucleus of positively charged protons orbited by negatively charged electrons. The much lighter electrons are easily forced out of orbit, upsetting the stability of the molecule. When a molecule loses an electron, it becomes a positive ion, and when the displaced electron joins the orbit of another molecule, that once-stable molecule is now known as a negative ion. Friction caused by winds, rushing water and lightning will produce both positive and negative ions. In addition, radiation from the sun and from the earth's crust also result in the continuous formation of ions. The city air is not merely polluted by a collection of dust, smoke and other toxic substances, but also severely lacking in ions. But more seriously, the ratio of negative ions to positive is low. Covering the earth with roads, concrete and buildings, the earth's natural generation of ions is also inhibited. Air completely devoid of ions can not support life. Ions are necessary for absorption of oxygen needed to sustain life. The balance of positive to negative ions is crucial to optimum health and well-being. Increased levels of small negative ions have measurable beneficial effects. Victims of wind sickness are producing up to 1000% more serotonin than normal, serotonin-blocking drugs or negative ions provide fast relief. A small number of negative ions quickly kill the airborne bacteria. Atmosphere of high positive ions increase serotonin, leading to exhaustion and infection, while exposure to a high negative ion atmosphere reduces serotonin. The serotonin as a neurotransmitter carries nerve impulses throughout the body and is involved in the processes of sleep and immune function. Sick building syndrome, the indoor atmosphere of modern homes can be even more detrimental to health than smoggy city streets. Man-made environments, be the homes or workplaces, are rife with synthetic fibres and electronic equipment, plus heating and air conditioning units-all of which produce an abundance of positive ions. More than 60% of workers in new office buildings would complain of nasal congestion, frequent headaches or nausea, due to reduced negative ions from the air. Replacement of negative ion generators in selected areas of the interior, reduces complaints and dramatically decreases their number and frequency. The unprecedented rise in respiratory disease and especially asthma, may be related to constant exposure to dead, or ion-depleted air. Small negative ions enhance the function of the human immune system, increasing levels of IgA. Too many negative ions could prove as ineffective as too few. It was claimed that the ideal ioniser (e.g, ELANRA) should modulate frequencies down to as low as 7-10 Hz for relaxation and meditation. According to electro medicine, frequencies in the 7.83 Hz range may be the universally permeating clock frequencies or carriers on which mind or consciousness states can be impressed. Ioniser may be a possible alternative to treating psychiatric patients with powerful drugs. Negative ion concentration is measured per cubic centimetre of air. This ranges from almost zero in an air-conditioned city office to 50,000/c.c. near a large waterfall. Over open land the normal count is between 1,000 and 2,000/cc. The ELANRA may mimic this natural ideal. Humans in different circumstances, with different conditions, require different negative ion concentrations. The ELANRA boasts 144 different settings, and 9 different frequencies including the Schumann resonance, the Earth's brainwave. The negative ions can: 1- boost the immune system, 2- increase the body's ability to use the oxygen in the air, 3- increase the lung's ability to get rid of pollutants, 4- make it feel easier to breathe, 5- improve sleep, 6- reduce stress, 7- increase alertness, and 8- decrease platelet aggregation (blood clotting). Small negative air ions of oxygen, when breathed in, increase the body's production of IgA. IgA guards against viral and bacterial attachment and colonization of mucous membrane. Improved lung capacity and relief from asthma symptoms are achieved quite rapidly by exposure to small negative air ions. After a few session of exposure to negative ionisation (3-5 sittings) for just 40 min/day normal sleep is expected to return in 80% of insomnia patients; 75% of headache patients would have their pain disappear; over 50% of depressive patients would return to normal; and 100% of anxiety sufferers symptoms would disappear. Serotonin is an important brain chemical related to mood and stress. Many antidepressant drugs, including Prozac, alter brain metabolism of serotonin. So do negative ions, without the side-effects. Pollution has resulted in all sorts of physical, emotional, and mental ailments, including immune deficiency. The negative influences that ruin the immune systems everyday include chlorinated fluoridated water, lack of exercise, poor diet, junk food, artificial sweeteners, pollution, stress, positive ions, ect. Most medical practitioners are as unaware as their patients that something in the air can cause physical and mental distress in human beings. In the mountains, in forests and near moving water, friction between rapidly moving molecules in the atmosphere produces an abundance of tiny negatively charged ions of oxygen, ionisation also occurs due to natural radiation from the earth's surface. The smallest oxygen ions can be breathed in and enter the bloodstream via the alveoli, giving a relaxed and energized feeling. Large negative ions have no biological effect, while an overload of positively charged ions will raise the body's levels of the serotonin, leading to a variety of unpleasant symptoms. Serotonin plays a part in almost every function of the central nervous system. When the brain under/or over-produces serotonin the imbalance can be reflected in distressing symptoms. Abnormally high or low serotonin levels correlate with a range of conditions including depression, psychosis, sexual dysfunction, lethargy, migraine, low pain thresholds, learning disabilities, poor sleep patterns, asthma, aggression and even suicide. Histamine can be found in many parts of the body. The presence of too much histamine in the body may also cause many adverse reactions, similar to those associated with excess serotonin, including allergies and hay fever. The serotonin irritation syndrome is characterized by anxiety, headaches, palpitations and insomnia, exposure to negative ions is recommended as one form of treatment. Negative ionisation brings relief to asthma and allergy sufferers (85%), this dramatic effect applies to other diseases as well, including hay-fever, skin rashes, and the discomfort of sinus inflammation. MAO inhibitors, alters the metabolism of serotonin just as ionisation does. Ionisation is safe, drug-free treatment for depression. Treatment with high density negative ions of oxygen is indicated for depressive disorders characterized by reverse neuro-vegetative symptoms: hypersomnia, hyperphagia (with carbohydrate cravings) and fatigue. The headaches and mood swings some women experience at pre-menstrual times and during the menopause correlate with conditions already known to occur when serotonin levels are elevated. The hot flushes may be the result of the hypothalmus' failure to maintain a constant body temperature. Migraine headache, hemicrania can be treated by negative air ionisation. Ionisation should be recommend as a first line of treatment. Sexuality and reproduction can also benefit from ionisation. Exposure to negative oxygen ions boosts sexual interest and increases fertility, prevents miscarriage in those women most at risk, and stimulates lactation after birth. Excess serotonin inhibits the flow of information in the brain, interfering with the ability to learn. High levels of serotonin are almost always present in children with learning disabilities or attention deficit and hyperactivity disorder (ADHD). Learning disabilities and ADD are often found together and both are understood to be the result of poor communication between the right and left brain hemispheres. Exposure to negative ions increases processing ability (response between the hemispheres) in learning-disabled children. Reducing serotonin levels improve IQ and improve behaviour in autistic children. ADD may be a form of borderline autism. Ionisers have been used in hospitals to speed up healing and relieve pain throughout Europe, Japan and the former Soviet Union for many years. An ioniser by the bedside invariably results in deeper sleep, a refreshed feeling on awakening, and the ability to recall dreams. Unfortunately, a deficiency of negative ions is a feature of urban living. Electromagnetic fields from appliances, static electricity from synthetic fibres and air that is conditioned indoors or polluted by chemicals outdoors all serve to deplete negative air ions and alter the body's chemistry. Various diseases of man is influenced by atmospheric electricity. Atmospheric electricity depends upon the existence of gaseous ions in the air. Ion depletion and charge imbalance may play a significant role in a wide range of human ailments including respiratory infection in office workers and the malaise caused by weather conditions. Artificially generated air ions may prove valuable as a therapeutic modality in the treatment of burns, reparatory disorders, stomach ulcers and neurological disorders. In normal pollutant free air over land, there are 1500-4000 ions/cm^3. But negative ions are more mobile and Earth's surface has a negative charge, so negative ions are repelled from the Earth's surface. Thus the normal ratio of positive to negative ions is 1.2 to 1, respectively. Air ion formation begins when enough energy acts on a gaseous molecule to eject an electron. The displaced electron attaches itself to an adjacent molecule that becomes a negative ion, the original molecules then become a positive ions. Molecular collisions transfer the charge, so that positive charges come to reside on molecules with the lowest ionisation potential, while electrons are attracted to the species of greatest stability. Next, small numbers of molecules of water vapour, hydrogen and oxygen cluster about the ions to form small air ions. Certain small air ions unite with condensation nuclei and with most classes of air pollutants to form large or Langevin ions. In both cases the biological activity of the small air ions is lost. This is true also of the combination that occurs between small air ions of opposite charge. Further, ions like charge (unipolar ions) repel one another and tend to flow to enclosing surfaces where their ionic nature dissipates. Since they are small and carry a charge, they are deflected by electrical fields. All of these characteristics make it difficult to maintain high concentrations of small air ions and means that air ion densities are significantly altered by the indoor living and air pollution characteristic of urban life. In the case of air ions there is no disagreement about the disparate physical nature of air ions and non ionised gaseous molecules, but there is considerable reluctance to grant that this diversity is of biological significance. The maximal ion density one can attain in a closed atmosphere is approximately 1×10^6 ions/cm^3 of air. Air contains 2.7×10^{19} non ionised molecules/cm^3, so that the ratio of small ions to non ionised molecules is 1:27 trillion. Ions have a very brief life span and under the conditions ordinarily prevailing, attainable ion densities usually are considerable less than 1×10^6 ions/cm^3, making the final dilution in non ionised air greater by one or two orders of magnitude. Excitation or violent attraction or ejection of free electrons from the atoms is responsible for many changes of the target material. It has been found that molecules are held together stably by means of the chemical bonds formed through the sharing of electrons between neighbouring atoms. The explosive loss of an electron disrupts such bonds and converts a stable molecule into energy-rich radicals, which in their attempt to regain stable electronic configurations, may decompose several other molecules. An electric charge holds a varies number of molecules together, imparting to them certain specific characteristics, which are lost when the charge is removed.

BIOLOGIC CHANGES OF NEGATIVE ION THERAPY

All air pollutants, nuclear, industrial, and domestic result in increasing a positive charge in the surrounding atmosphere. Even the air in open country areas is predominantly positive because it probably received wind carried pollutants, which originated in distant industrial zones. One electric power station alone produces enough sulphur-dioxide (SO_2), with its affinity for positive electricity in the atmosphere to extend over vast areas of the earth's surface. At home, the viewing surface of a television set gives off into a room, electric emissions ranging in magnitude from about 7,000 volts to 11,000 volts. These emissions have the effect of producing a positive charge into the air and on the surface of all items within immediate range of the TV set. For an indoor environment, the artificial production of negative ions has been found to be an effective means of counteracting the effects of excessive positive air ions. The negative ionised air gas molecule that an ioniser is produce should enable the negatively charged air ion to be carried throughout the room, rather than be dissipated soon after its emission from the generator. A moisture envelope facilitates a more effective ion assimilation by the respiratory system and ultimate absorption of the ions into the bloodstream (Hydroscopic clean negative ions). Sick persons evidence a quicker biological response to the presence of negative ions than do healthy persons. The most important application of negative ion generator is its prophylactic aspect, preventing illness before it happens. The negative ion generator has both prophylactic and therapeutic applications. Anti-social behaviour, including some crimes, might have a basis in excess lead deposits retained in the body. Minor irritations caused occasionally by inhalation of ionised air, e.g., headaches, dizziness, dryness of nose and throat, and clogging of nasal passages, disappear readily after a few minutes in the outside air and without further after effects. Negatively ionised air increase CO_2 capacity of the blood plasma. Artificially ionised air is a potentially effective biologic factor, which, if properly harnessed, controlled, and utilised, may become a valuable adjunct to other forms of therapy. The inhalation of highly ionised air of negative polarity, induce considerable improvement in the general tone, cheerfulness, energy, good sleep and appetite. Negative ions can be produced by copying any of the ultraviolet (lamps), radioactive sources, etc. A bathroom shower provides a plentiful supply, but it is either too dangerous, too expensive or just impractical. The corona discharge technique is similar to lightning. A high voltage (but at extremely limited current, for safety) is applied to one or more needles. Electricity is a flow of individual electrons, and these electrons, supplied by the internal circuit, are pushed down the needle towards the point. The nearer they get to the point, the closer they become forced together. Electrons naturally repell each other, so as they reach the tip, the pressure becomes too much and they jump off, onto the nearest air molecule, turning it into an ion. Negative ions again repell each other, so they are driven from the needles as a gentle breeze, forming a dense cloud in front of the ionizer, which disperses in all directions into the room. The ions leaving the ionizer are small, high velocity ones. These are found to be most beneficial to health. If they collide with particles of smoke or pollution near the ionizer, they pass on their static charge. This particle is then strongly attracted to the nearest earthed surface, which could be a wall or the shelf on which the ionizer is placed. Out into the room the ions naturally begin to slow down. As they drift, pollutants such as dust, pollen, cigarette smoke and even vapourised substances like aerosol propellants and car fumes are attracted to and cluster around the ions. This has the effect of making the ion grow in size. There comes a point where it is too heavy to be carried in the air, so it falls to the ground. The smaller the particle, the harder it is for the immune systems to cope with. So ionizers excel at removing microscopic particles and at the same time they restore a vitality to the air. The quoted range of an ionizer is usually the distance at which it can maintain a certain concentration of ions. The standard figure is 1000 ions/cc of air could be the lower threshold level for health purposes, however most ionizers will have a cleaning effect over a greater distance. A badly designed ionizer may produce ozone, and with it nitrous oxide. These are toxic substances and can cause respiratory difficulties and stinging eyes. The maximum acceptable level of ozone is 0.1ppm (parts per million). Ozone is a very active form of oxygen, consisting of 3 atoms instead of the normal 2. Atoms such as hydrogen as H_2 and oxygen as O_2, when cling together in pairs, they form molecules. Under certain conditions these pairs of atoms can become separated. It takes quite a lot of energy to do this and usually happens when oxygen is subjected to high voltage electricity or high energy ultraviolet light (UVC), or by the energy of lightning or sunlight. Ozone is a blue gas-a very powerful oxidizing agent and extremely poisonous. It occupies a very small part of the atmosphere, mainly located about 30 km above the surface-the ozone layer. Here it prevents most of the ultraviolet radiation from reaching the earth where it could do great damage. Intense ultraviolet light from the sun hits the upper atmosphere. When the UV rays collide with oxygen they split the molecule into its 2 separate atoms. These atoms would dearly like to re-combine but have been driven apart by the impact. The next best thing they can do is to tag on to a nearby normal oxygen molecule, which now becomes O_3 instead of the usual O_2. The triple bond of ozone is quite weak and one atom needs to be shed to make it stable again. So when it passes near another single oxygen atom or another ozone molecule, the surplus (loose) atoms will break away and re-combine into oxygen pairs. The sun shines on the upper atmosphere reacting to form the ozone layer, and the energy absorbed by it all, reduces the intensity of the UV light to a level that life on the earth below can tolerate. In the upper atmosphere, there are very few other substances to react with, so the oxygen molecules just keep breaking down and re-combining with themselves (the exception being those aerosol pollutants that bind with the oxygen and actually break the cycle-thus causing the hole in the ozone layer). Things are quite different at ground level. The ozone layer is beneficial to us, but ozone itself is not. Normal oxygen is itself a very reactive gas, which easily combines with other substances. It supports combustion and oxidises (or breaks down) many materials. For example when oxygen combines with iron, then rust (iron oxide) is formed, and the green tarnish on copper is copper oxide. Cooking oil goes rancid and beer and wine go vinegary when oxidised. But ozone is far more reactive than oxygen. Because the third oxygen atom is so loosely bonded, it takes far less persuasion to break away and bind with other substances so it has far greater oxidising power. The bleach hydrogen peroxide is formed when ozone dissolves in water, and this is what happens when it comes into contact with moisture in eyes, nose and lungs. Ozone can also perish rubber and some plastics. The smell of ozone is the smell that comes out of the photocopier and laser printer. That's because they use static electricity to transfer the toner powder into an image on paper, and anything that uses high voltages in this way has a tendancy to produce ozone. Health and safety standards dictate that the levels must be kept very low, so office equipment is protected by special ozone filters. If these filters deteriorate and there is insufficient ventilation, soon the eyes start to sting and nose and throat become irritated. With building high voltage equipment it's very easy to generate ozone, as a by-product. A badly designed ionizer may produce ozone, and with it nitrous oxide. Also, the ionizer's emitter needles can deteriorate severely in the presence of ozone. In other words, if the ionizer generates ozone, it will need to have it's needles replaced quite often. Astrid ionizers are specifically designed not to generate ozone. The words natural de-odouriser or nature's de-odouriser usually means the ionizer gives off ozone. Ozone kills mould and bacteria by oxidising them, but also oxidises the cells in the body. Ozone generators are used industrially to fumigate buildings but all living things must be kept well away from the area. There are several means of artificially producing air ions, including corona discharge and tritium generators. These ion generators make it possible to re-establish optimal microclimatic conditions in living and working places. The ion depleted air in offices and factories lowers resistance to influenza and perhaps other infections, inhaling a mixture of air (4000 ions/cm^3) and with negative ions predominating, should increase resistance. The ratio of days lost because of respiratory illness should be reduced to 1 day in 3 months. Pain, restlessness and incidence of infection would be reduced and healing is promoted in burn patients treated for 1-1.5 hours a day in hospital, and out patients for 25-30 minutes, with negative ion concentrations as high as 10,000 ions/cm^3. Inhalation of negative ions increases the conversion of serotonin to 5-hydroxyindolacetic acid, with the relief of pain in burn patients treated with a high concentration of negative ions. Exposure to positive ions increases the respiratory rate and degree of bronchospasm in infants and adults with asthmatic (spastic) bronchitis, treatment with negative ions produces an opposite and therapeutic effect. The negative ion therapy terminates the spastic attack after a much shorter period than that is required by the conventional mode of treatment without adverse side effects common to the drug therapy. Negative ionisation increases parasympathetic nervous system activity, and only slightly increase sympathetic nervous system activity. The negative ions lowers activation of the psychiatric patients, and may treat psychoneurosis and anxiety syndromes. Sessions varies from fifteen minutes to two hours and the number of treatments from ten to twenty. The negative and positive ions affect the nervous system, the bloodstream, the respiratory system, and the endocrine glands. The negative ions are more mobile than the positive. Positive ions increase if a surface is heated. As a result, air in the neighbourhood of a functioning iron stove or central heating pipes is rich in positive air ions.

Ionosphere:
In the upper atmosphere the action of the sun's ultra violet rays, forms a thick band of ions called the ionosphere, which starts at 60 kilometres, and finishes around 100 kilometres, above the Earth. It is the ionosphere that reflects radio waves back to earth, allowing us to broadcast over long distances. The reason distant stations fade after dark is because the ions diminish as the sunlight disappears.
Aurora:
The Sun is constantly ejecting streams of charged particles into space. (This is known as the solar wind). As these particles approach the earth they are caught in its magnetic field and swept round to the North and South poles. Pouring into the atmosphere, they ionize its component gases in a spectacular display of glowing colours, called the Aurora.

Enclosed spaces in buildings, automobiles, airplanes, etc., are frequently over-saturated with positive ions. Positive ions are serotonin releasers, and that a local accumulation of serotonin in the trachea is the immediate cause of positive ion effects. Inhalation of positive ions produces swelling of the nasal mucosa. Like positive ion effects, the serotonin effects can be reversed by treatment with negative air ions. Both positive and negative ions in the lungs are taken up into the bloodstream, and whereas the erythrocytes of the blood pick up air oxygen, thrombocytes react to the positive ions releasing their neurohormone, serotonin. Treating patients for burns with negative air ions, increases mental alertness. Where there is a number of people, e.g., in an enclosed space, the number of negative small ions get used up quickly, leaving a predominance of positive ions. The concentration of positive ions tends to make people feel uncomfortable. The introduction of negative ions, improves substantially the environment wherein a group of people are congregated. Negative ionisation can replace the risky prophylactic use of anticoagulants after operations. Deep venous thrombosis (DVT) and pulmonary embolism (PE) are important cause of disability and death in hospitalised patients. Pharmacological methods are quite effective in preventing DVT and PE, but fatal cases can still happen. The late sequels of this disease-the postphlebitic syndrome (phleboedema, chronic venous insufficiency, varicose veins, ulceration and induration) represent an equal distressing situation. In almost 50% of the patients, the thrombotic process is clinically silent, and fatal embolism may be the first symptom of DVT and PE. Multiple factors are involved in the process, such as blood changes, which predispose to thrombosis or trigger thrombus formation, i.e., alterations in the properties of platelets and of the components of the coagulation and fibrinolytic systems, furthermore, changes in the vessel wall are directly redated to coagulation and fibrinolysis, mainly platelet adhesiveness and aggregation. Processes associated with tissue damage (operation, burns, fractures, cancer) leading to an increased platelet adhesiveness or increased platelet coagulant activity, or both, initiate the formation of platelet thrombi. Certain patients such as those undergoing open prostatectomy or total hip replacement, are a high risk for thromboembolism. The final outcome in such cases is determined by the balance between thrombogenic and fibrolytic mechanisms, vascular injury and reactions of the platelets being the main contributory causes of thrombosis. Patients with pulmonary embolism often have reduced arterial oxygen tension. Air ions are taken up by the respiratory tract and part of them reach the lungs. As they are mostly ionised oxygen and water aerosols they are taken up by the erythrocytes and thrombocytes at the alveoler site together with normal oxygen and water. Positive air ions release serotonin from the thrombocytes, negative-ones counter this effect, and, moreover, can inhibit platelet aggregation. As post-operative thromboembolism is mainly due to platelet aggregation, the mechanism of its prevention should be sought in the negative charge conveyed to the blood platelets by negative air ionisation, since negatively charged particles do not tend to agglomerate. Likewise, it has been claimed that the glucoproteids, which cost the cell membranes of all blood cells prevent agglomeration by virtue of their negative charge. Positive air ionisation provokes neurohormonal changes, especially serotonin release, which may precipitate thromboembolism. Electro-magnetic waves affect serotonin release just like air ionisation.

VITAMINS and MINERALS OF THE AIR

Vitamin is a composite of 2 separate words with different meanings to them: vita, comes from the Latin word for life, and an amine is any of various compounds derived from ammonia by replacement of hydrogen by one or more univalent hydrocarbon radicals. A vitamin is essentially a life gas/air, i.e., an air nutrient. The air in the woods or in gardens contains vitamins given off by plants. In each cubic meter of air there are several milligrams of volatile substances, including vitamins. Two of the nutrients emitted from the tree bark and needles of fragrant pine odoe, aere vitamin C and a form of bioflavonoid called proanthocyanidins. Because bioflavonoids and ascorbic acid often appear in combination with each other in plants, they form powerful antioxidant effects. Pycnogenol also is in the bark of the French maritime pine tree, which grows in the Les Landes pine forest along the Atlantic coast of southern France. These antioxidants are useful for controlling the deadly activity of a group of compounds within the body called free radicals. These molecules are lacking an electron and roam through the body at random, robbing normal molecules of their electrons. In doing so, free radicals create a great deal of havoc and mischief with the body's delicately structured biochemistry. What antioxidants do is to curb or check this destructive action. Vitamins prevent fatigue and increase work capacity (primarily vitamins C and B_1). Negatively ionised air has an anti-scurvy effect, and also having a favourable effect on reduced vitamin B_1 level. The urinary excretion of vitamin B_1 normally fluctuates in adults between 5 and 15 g/h. The content of the pyrotartaric acid permits to ascertain indirectly the vitamin B_1 metabolism. The simultaneous administration of vitamins and areoionisation produce an even greater reduction in the concentrations of pyrotartaric acid in the urine. Aeroionisation causes a slight reduction in the serum pyrotartaric. The taking of a vitamin complex simultaneously with aeroionisation, decreases the serum pyrotartaric acid still further. Under the effect of aeroionisation sessions the content of vitamin B_2 (riboflavin) increases. The administration of the vitamin complex including vitamin B_2 largely intensifies the normalising effect of aeroionisation and lowers the level of riboflavin content in the urine. Negatively ionised air has also a favourable effect on vitamin C metabolism. Hydrogen and oxygen are trace elements, which considered to be microminerals without which the body could not function and would readily perish in no time at all. Beyond them, however, are other minerals like calcium, magnesium, iron, potassium, phosphorus, and zinc, which the body depends upon for the normal maintenance of good health. Because of massive advertising by the health food industry and the single-mindedness of scientific opinion, consumers are led to believe that the best source for minerals is food and food supplements. Very little is ever said about water, let alone air, as being potential contributors of minerals. Tuberculosis sufferers are recommended to breath mountain air; asthma sufferers should prescribed a warm, dry, desert-like air; while emphysema patients and those experiencing chronic lung inflammation should put in an air reminiscent of a sea coast.

The problems arising from the local wind of depression:

Body pains, Sick headaches, Dizziness, Nausea, Variations in body salts - sodium, calcium and magnesium, Respiratory problems, Asthma, Higher incidence of heart attacks, Slower reaction time, Irritation, Exhaustion, Listlessness, Anxiety, and Depression.

BIOLOGICAL FIELDS AND STRESS

Acupunctural augmentation (increase) of ionic flows to counteract defective biological fields relieves pain because it is really within these fields that pain is experienced. Similarly, anaesthesia can be produced. Cellular disease is remedied therefore as cells begin to conform to overlying biological fields. Acupuncture needles in the mid-forearm cure headaches and needles in the big toe produce anaesthesia for dental extractions. Presumably the mechanism is due to their effects on and within the intricate, continually cognising, biological field. The body is a system of electric-magnetic energy. The way ions tend to behave in magnetic fields, one way or the other, depends upon the ionic charge, and the North/South orientation of the magnetic field. For example, red blood cells rotate on their flat axes in-vivo; in-vitro they will rotate if a magnetic field is applied. The direction of RBC rotation in-vitro reverses if the field is reversed. When radio energy (electro-magnetic waves) is applied to substances at proper frequencies, it causes the electrons to resonate. The frequency at which resonance occurs is also dependent upon the magnetic field strength. Each living cell appears to be a tiny battery generating its own current by chemical action, i.e. the Na/K pump. Extra-ordinarily active tissues are associated with more intensified fields and inactive tissues with diminished fields. There is definitely a relationship of diseased tissue to magnetic fields, so that when an oscillating magnetic pulse is applied at the right frequency and penetrating strength, cellular metabolism through correction of RBC rotation is re-established at the proper level. Biological fields may be affirmed to become defective during emotional or physical stress, which if prolonged, subsequently results in cellular disease. Electro-magnetic augmentation of ionic flows to counteract such defective biological fields, relieves pain, because it is really within such defective fields that pain is experienced. Similarly analgesia can be produced.

CHRONIC FATIGUE SYNDROME

The hypothesis that Chronic fatigue syndrome (CFS) is a psychosomatic illness has resulted in millions of ruined and destitute lives. The CFS victims could not collect insurance support or disability and descended into poverty. That hypothesis was probably an important part of the chief cause of death, which was suicide. CFS or myalgic encephalomyelitis (ME) is a disease characterised by several of the symptoms of impaired sleep, muscle twitching at night, extreme long lasting fatigue that gets much worse with exercise, loss of memory, disruption of the circadian rhythm, sore throat, muscle and joint aches, headache, cough, photophobia, night sweats depression that has much lower ACTH and cortisol secretion than typical depression, also a much lower secretion of growth hormone (results from over secretion of somatostatin in the hypothalamus), a failure for a 383 amino acid cortisol binding protein to decline under stress, lower secretion of DHEA (Dehydroepiandrosterone) that is the most abundant circulating steroid, there is a less rise of DHEA under ACTH stimulation compared to cortisol than normal, hypovolaemia, lymph node pain, low blood pressure upon standing (orthostatic intolerance), eye pain and fibromyalgia (muscle pain) as well as white spots on MRI brain scans and single-photon emission computed tomography (SPECT) scans, reduced blood flow to the part of the brain that controls the stomach muscles, loss of fingerprints in a third of the patients, changes in the body's hormones, increased sensitivity to glucocorticoid hormones, alterations in some of the immune enzymes, and a chronic low level activation of the immune system that last may be accounting for many of the non neurological symptoms, but all very variable, perhaps because different parts of the brain are attacked, perhaps because there is more than one species of virus involved, and perhaps because of strong affects from the large variety of secondary infections, which have been identified or even all three. The most consistent laboratory abnormality in patients with CFS is an extremely low erythrocyte sedimentation rate (ESR), which approaches zero (0-3 mm/h). Another consistent abnormality is an increased excretion of citrate in the urine. It has been suggested that this binds with and causes an increased excretion of magnesium. Citrus food and other acid foods like vinegar could be a precipitated factor. Women are much more often affected than men, because they complain more often than men and are more likely to be afflicted with emotional trauma, which is a triggering circumstance. Other names for the syndrome include chronic fatigue immune dysfunction syndrome (CFIDS) because the immune system is distorted, neuroendocrine immune dysfunction syndrome (NDS), myalgic encephalomyelitis (ME), and post viral fatigue syndrome (PVFS). 80% of fibromyalgia victims have CFS. Low molecular weight RNase L increased activity correlates well with severity of CFS symptoms but is normal in fibromyalgia, rheumatoid arthritis, lupus erythematosus, HIV, and depression. Viruses activate the 2-5-synthetase enzyme. This in turn converts ATP into 2-5 oligoadenylate and activates the RNase L enzyme, which degrades viral and single stranded RNA. Various protein kinase enzymes also become activated and elevated, which again inhibits both viral replication and protein synthesis. It has been suggested that environmental toxins in the presence of heat shock proteins can also activate this pathway. The stealth virus is thought to be capable of taking genetic sequences from bacteria and other hosts. Fragments of mycoplasma pathogen species have been found in CFS and fibromyalgia but they are probably opportunistic infections. However a high percentage of veterans with Gulf war syndrome are infected with Mycoplasmin fermentans especially and quite a few of their family members also become infected with symptoms of autism in their children. There is a benign treatment for mycoplasma and other intracellular pathogens using glutathione. Platelets are proposed to have an immune function, CFS could result from a defect in the platelet's immune function. A low molecular weight (37 kD) RNase-L, has been found in CFIDS (CFS) patients. It can be six times as destructive as the typical RNase-L. If the virus-fighting system is working normally, the normal form of RNase-L prevents the virus from reproducing. The rest of the immune system wipes out the virus, and then the entire immune system returns to normal levels, and the person recovers. In CFIDS, the RNase-L shifts to the more destructive form, and instead of deactivating, it stays active much longer, causing serious cellular metabolic dysfunction, which ultimately affects the liver. The cells can no longer produce essential enzymes, and without them, the liver can't do its job of detoxifying the body. Ciguatera poisons picked up by oceanic fish in the tropics have been linked to CFS, especially in Japan. This is a poison of many carbon rings generated by algae, which toxin can not be degraded by heat and which is thought to bind to sodium cell wall pumps. It remains in the body for a long time. Mannitol has been proposed as a treatment. Removals of mercury amalgam tooth fillings have resulted in significant improvement of CFS symptoms. Mercury release is especially high if amalgams are touched by gold or stainless steel. Fish contain unacceptable amounts of mercury. Aluminium has been found to be significantly higher in CFS than in normal people. Perhaps aluminium baking powder and pots and pans should be avoided. Half of women with endometriosis have allergies and even a greater per cent of those with allergies also has CFS. 7% of women with endometriosis have CFS and CFS is 100 times as common as in the female population and fibromyalgia is twice as common. Lupron makes fibromyalgia worse. CFS is associated with hypothyroidism. There is a correlation between dioxin (from living near an incinerator or a diet high in fish) and endometriosis. 50% of people who have fibromyalgia are sensitive to pollution/exhaust, cigarette smoke, gas/paint/solvent fumes, and perfumes. Chemical sensitivity may be operating through NO/peroxynitrite mechanism. Nitric oxide-mediated stimulation of neural transmitter release; peroxynitrite-mediated stimulation of post-synaptic NMDA sensitisation; peroxynitrite-mediated blood brain barrier permeabilization and NO inhibition of cytochrome P450 metabolism, all of them acting synergistically to create an extreme sensitisation. NMDA (N-methyl-D-aspartate) is a neuroreceptor in the brain and spinal cord for the neurotransmitter glutamate (the most important excitatory transmitter in the brain) may be involved in the toxic effects of excessive glutamate. NMDA is also an ion channel and is involved in chronic pain. Magnesium is known to lower NMDA sensitivity. Aspartame, an artificial sweetener, may damage the hypothalamus, which part of the brain controls steroids and seems to be defective in CFS. Aspartame degrades to the very poisonous methyl alcohol (methanol), poisonous because it can become formaldehyde. Aspartame is suspected to cause blindness, systemic lupus, cancer, and may mimics multiple sclerosis. The macula area of the retina and optic nerve fibers is highly reactive to the toxicity of methyl alcohol because of these fibers and nerve cells require from four to six times as much oxygen and nutrition and the visual pathways peripheral nerves. These fibers and nerves (Tods and Cones) can not go without oxygen for more than 90 seconds without some visual loss. Aspartame is broken down by the upper digestive tract and methyl alcohol is then absorbed into the blood vessels of the region, which travels through the bodies entire vascular system. This toxic effect causes extensive damage to the highly demanding metabolic requirements of the human visual pathway, ending with optic nerve atrophy as well as retinal starvation and visual loss. Other organ systems are affected, e.g., nerve damage to the arms, legs, central nerve impairment, resulting in impaired nerve function as well as neurosis and learning difficulties in school work as well as behaviour. Aspartame causes retinal detachments, floaters, blurred vision, blindness, etc. Aspartame causes macular degeneration. Eliminating monosodium glutamate (MSG) or MSG plus aspartame from the diet would resolve symptoms within months. Excitotoxins are molecules, such as MSG and aspartate, which act as excitatory neurotransmitters, and can lead to toxicity of nerves when used in excess. The elimination of MSG and other excitotoxins from the diets of patients with fibromyalgia offers a benign treatment option that has the potential for dramatic results in a subset of patients. Imidazolidinyl, a preservative in cosmetics, may cause disequilibrium of K, Na, and Ca channels. Poor nutrition and lack of exercise are contributing factors. A vegetarian diet using lots of raw vegetables has significantly improved the symptoms of fibromyalgia. Raw vegetables devoid of meat, but with vitamin B_{12} supplement, reduce the pain, and allow sleeping better, reduces weight, and reduces cholesterol at the end of 3 months, but does not improve exercise ability by the end of 3 months. Despite its benefits, none of the patients would chose to adhere to the diet. The pain gradually return as they drifted back to their previous omnivorous and probably junk food diet. A pervasive vitamin C deficiency probably exists in every society. Vitamin C cures viral infections. Increased potassium and magnesium should work almost as well. 5000-6000 mg of potassium/day and 460 mg of magnesium is recommended. Magnesium injections mute the symptoms significantly. The following medicines can help create Mg deficiency: benzthiazide, bumetanide, chlorothiazide, chlortrianisene, chlortetracycline, cholestyramine resin, conjugated oestrogens, corticosteroids, demeclocycline, diethylstilbestrol, digoxin, doxycycline, esterified oestrogens, ethacrynic acid, furosemide, hydrochlorothiazide, hydroflumethiazide, indapamide, methyclothiazide, metolazone, minocycline, oral contraceptives, oxytetracycline, penicillamine, polythiazide, quinethazone, tetracyclines, torsemide and trichlormethiazide. Magnesium is excreted in greater amounts during respiratory acidosis, so it may be desirable to sleep with the windows open at night or to install a carbon dioxide absorber in the bedroom. This suspicion is reinforced by the fact that the steroid hormone that stimulates acid excretion, 18-hydroxy deoxycorticosterone is secreted poorly during CFS, possibly a mechanism to allow an acid serum to activate immune enzymes. Also, respiratory alkalosis increases Mg excretion. Magnesium was found to be normal in the red cells in CFS patients and magnesium is normal in blood cells during a Mg deficiency as well, so red cell content can not be used in diagnosis. Chlorella pyrenoidos (an alga plant) supplements improve fibromyalgia. 10 g of dry chlorella furnishes 83 mg of potassium and 33 mg of magnesium. Eating a vegan diet including nori and chlorella algae doubles the vitamin B_{12} in the blood. Chlorella contains vitamin B_{12}. Chlorella also contains inositol, which is essential for some of the electrolyte pumps. A whole body (cell content) analysis of K has found that K averaged a little lower in CFS than the general population. Extremely low intakes of sodium chloride also increase potassium excretion. Lower cell K requires lower serum K for adequate nerve transmission, but the serum K does not drop correspondingly during a magnesium deficiency. Potassium is released from the cells by cold below 4°C. Low secretion of ACTH by the pituitary gland causes a lower secretion of 18-hydroxy deoxycorticosterone (18OH DOC) steroid. This in turn interferes with excretion of acid since the body probably uses that steroid to stimulate excretion of hydrogen ion. As a result aldosterone is decreased in order to prevent loss of K and the serum potassium rises. The cell fluid becomes more alkaline and the cell fluid pH (alkalinity) is the most reliable indicator of the intensity of muscle spasms. If muscle spasms are associated with chronic fatigue syndrome, it is possible that low intracellular calcium in that disease even though serum calcium is normal could be the reason. If Mg deficiency develops, half a year of Mg supplements can be required for complete normalisation of Mg and K-Na pumps. Magnesium supplements reduces leg cramps during pregnancy in 3 weeks without any change in serum. IL-1, IL-6, and TNF increase dramatically in the serum due to Mg deficiency as well as the neuropeptide substance P. The later causes histamine and PGE_2 to rise within 5 days. Changes in the immune system (such as any change in IgE antigen or rise in WBCs) during CFS that might occur could easily be augmented by a Mg deficiency. Agar seaweed is a rich source of Mg and contains 770 mg/100 g of dry weight. Vinegar may augment the pain. Potassium as the chloride may raise blood pressure. For CFS patients Mg injections may be necessary at first. Vitamin D is proposed as necessary for reabsorption of Mg in the kidneys. Half of fibromyalgia patients have vitamin D levels less than the amount that would stimulate parathyroid hormone. It is possible that CFS victims may have net bone resorption. Meals should be more than three times per day in smaller increments. Smaller meals would help prevent surges of K too high for those with weakened kidneys to handle efficiently as well as possibly increasing the useful cell retention by virtue of preventing the correction of high plasma K, which otherwise takes place by excretion in the urine and lower colon. Also, it would be advisable to drinking liquids 20 minutes later, liquids delay emptying of the stomach. Such a strategy helps prevent the bacterial overgrowth resulting from delayed emptying of the stomach. Correcting the choline deficiency using choline dihydrogen citrate along with vitamin C considerably improves myalgic encephalomyelitis (ME, probably fibromyalgia). Extra vitamin C creates less fatigue in healthy people. Vitamin B_{12} and folic acid are cofactors with choline. Because of very low vitamin B_{12} in the brain large injections is advisable. Melatonin is the hormone, which is thought to be central in circadian sleep patterns and has been shown to be helpful against depression caused by poor sleep habits. Linoleic acid (ω-6) may be deficient in myalgic encephalomyelitis (ME). The ratio of ω-6 to ω-3 oils should be one, but modern diets are much higher. ω-3 may possibly inhibit the immune system in excess but apparently not the white cell (T and B cells) functions. Low copper diet is associated with drop in the levels of enkephalins (the internally produced substances that provide pain relief and pleasure) that are produced in the brain. Depression has been relieved with copper supplements. There is a significant inverse relation between vitamin E and fatigue in CFS. NADH (nicotinamide adenine dinucleotide) has helped CFS's patients, SAM-e (S-adenosylmethionine) has reduced pain and depression in fibromyalgia, and 5-HTP (5 hydroxytryptophan) is thought to increase the brain's sleep and antidepression hormones. There have been encouraging improvements achieved in CFS victims with lifestyle changes including nutrition, alterations in intestinal bacterial flora, and removal of foods causing allergic reactions. Food elimination strategies may produce significant clinical responses in 50-80% of patients with particular benefits seen in gastrointestinal complaints, migraine, arthralgias, and recurrent upper respiratory tract infections including the sinuses and urinary tract infections. Weight gain during CFS could be largely from food sensitivity. Monosodium glutamate, cashew nuts, and acid foods such as vinegar or KCl augments headache. Enteric coated peppermint oil reduces intestinal bacterial overgrowth. Over training can precipitate CFS and exercise brings on a severe fatigue, which lasts for days. Short, mild treadmill exercise causes no obvious problem. Short periods of mild exercise across the day would be the preferred routine. Remove of electrolyte hormones such as aldosterone (which removal decreases K losses and Na retention) is probably an important part of the value of exercise. The poor tolerance to exercise in CFS could be due to cardiac infection with various herpes viruses, so it would be advisable to approach exercise cautiously. Hot shower or immersion in water at 30°C several times a week reduces cytokines that regulate the immune system, which are responsible for much of the problem. The most debilitating infirmity other than fatigue is loss of memory. CFS patients should carry maps with them showing the way home and notebooks with important information like phone numbers and grocery lists. This should help considerably. For those who have lost fingerprints a good ID should always be on them and perhaps name and phone number imprinted on their arm with a dye. Caffeine increases potassium excretion. The amino acid, arginine, accentuates the symptoms of herpes and may even trigger a resurgence of a dormant infection such as shingles (which disease is a resurgence of dormant chicken pox virus from nerves near the spine). Foods high in arginine are peanuts, cashews (peanuts are 50% higher than cashews but cashews are substantial), chocolate, and seeds other than the grass derived grain. Lysine helps to mute the effects of the herpes virus (such as chicken pox, shingles, infectious mononucleosis) significantly, reducing the occurrence (when taken routinely during the disease), severity, and healing time of herpes simplex virus. It probably does so by interfering with the absorption of arginine by the virus. NOS stimulates neural sensitisation by acting on arginine. Large lysine supplements may lead to relief from virus, and relief from CFS. CFS's patients should not be afraid to experiment with nutrients. The most common problem-causing foods or ingredients for the patients include corn, wheat, dairy food, citrus and sugar. Sugar can produce an allergy; sucrose and fructose can interfere considerably with copper metabolism. Irritable bowel symptoms are quite common in ME/CFS but it's worth screening for adult onset coeliac disease (AOCD), especially if the onset of ME/CFS is gradual. There is also increased risk of osteoporosis and small bowel adenocarcinoma. Millions of people eat things about which no records are kept, such as hydrogenated oils and additives. A 1998 medical report estimated that adverse reactions to prescription drugs kill about 106,000 Americans annually, roughly three times as many as are killed in automobile accidents. Fibromyalgia seems often to be made worst in hypertensive patients who are treated with ACE (angiotensin conversion enzyme) inhibitors and ACE receptor blockers (they induce muscle pain). Isotretinoin (brand name Roaccutane, for acne) can trigger an attack of chronic fatigue syndrome. CFS patients are very susceptible to adverse reactions from some anaesthetics and other medications and usually much smaller doses are indicated. Good jokes, camaraderie, and tactile approval (like hugs) will not cure the disease, but there is a good chance they will mute or distract some of the symptoms and make an eventual defeat of whatever infection is involved or become involved opportunistically a little more likely. Massage has been helpful for fibromyalgia, massage being the most helpful of physical manipulations, and also may have a placebo effect on the immune system. Emotional support for children is especially important. Staying warm enhances immunity and fear is well known to affect the immune hormones. Fear may be contributing to the lower K in CFS by increasing aldosterone as well. CFS and fibromyalgia are potentially extremely dangerous to society because of their severity and length of recovery time. 6.4% of patients may be triggered by a blood transfusion. By implementing all mitigating factors that have been discovered so far, CFS and fibromyalgia will be able to lead reasonably satisfactory lives of higher quality. Thyroid problems can cause fatigue. Endometriosis can be associated with hypothyroidism. Chronic fatigue syndrome, thyroid problems, and some kinds of porphyria seem to be more common in women of reproductive age, and are often exacerbated by premenstrual hormone changes. SLE is almost entirely a disease of women of child-bearing age. One possibility for this could be that women during this period harbour a peculiar flora, i.e., large numbers of gram-positive lactobacilli in the vagina. Numerous environmental, metabolic, infectious, immunologic, and psychiatric disturbances have been implicated in the many complaints of chronic fatigue syndrome (CFS). The latent tetany syndrome (LTS) parallels CFS in its neuromuscular and psychiatric manifestations, as well as in inner ear disturbances: vestibular in CFS and FM, as well as in LTS, and increased vulnerability to noise-induced deafness in LTS. Microvascular damage to the cochlea is seen in Mg deficiency, noise-induced deafness, and might be a factor in migraine and other severe headaches in both LTS and in CFS and FM. When the CFS was found to be associated with abnormal immunologic responses to infection, it was termed postinfectious neuromyasthenia, chronic virus infection, myalgic encephalomyelitis, chronic fatigue immune dysfunction syndrome (CFIDS), and fibromyalgia (FM). Chronic fatigue, weakness, depression and anxiety, sleep disturbances, paraesthesias and sensorineural hearing loss, as well as neuromuscular irritability and myalgias have long been known to respond to long-term Mg supplementation. Stress commonly precedes acute CFS events, which is another indication that Mg inadequacy might be a factor, because stress hormones cause Mg loss, and low Mg levels increase secretion of catecholamines. Muscle weakness, fatigue, and pain (aching and/or cramps with spasms or tetany) characterise LTS patients, in whom Mg deficiency has been identified. The sleep disorder of Mg deficiency in humans might be related to catecholamine excess, abnormal cerebral monoamines, histamine, and other neurotransmitters. Serotonin uptake is Mg-dependent. Mg repletion restores to normal both monoamine metabolites (brain dopamine levels and 5-hydroxy-indoleacetic acid) and the EEG, as well as decreasing vigilance wakefulness and increasing sleep. The cerebral monoamines, including norepinephrine and serotonin, homovanillic acid, and 5-hydroxy-indoleacetic acid, are affected by Mg deficiency. The structures most affected are hypothalamus, brain stem, corpus striatum, and structures that have important roles in maintenance of vigilance, as well as in various regulatory functions: neuroendocrine (hypothalamus), autonomic (brain stem) and motor (corpus striatum). The degree of sleep impairment, which leads to overwhelming daytime weariness, has been proposed as a contributory factor to the fatigue component of CFS and FM. Severe human Mg deficiency causes brain dysfunction, which include apathy, poor memory, confusion, disorientation and hallucinations, before coma or convulsions ensue. Alcoholism is known to be one of the important causes of Mg deficiency. Alcoholism is implicated in the learning defects of infants borne to alcoholic mothers, via the resultant Mg loss-induced abnormality in regulation of the NMDA receptor. Cognitive dysfunction that includes impaired attention, loss of ability to concentrate, mood changes, and memory loss is not uncommon in CFS. Low levels of Mg, have been considered in the pathogenesis of migraine. Mg has been reported effective prophylactically and therapeutically against migraine attacks. Mg inhibits histamine-induced bronchial constriction. High dietary Mg (182-654 mg) is associated with better lung function and decreased risk of airway hyper-reactivity and wheezing. Non-infectious rhinopathy, allergy, pseudoallergic vasomotor rhinitis, and asthma are associated with Mg deficiency. Influenza vaccine lowers RBC Mg. Familial differences in absorption and excretion of Mg and the high heritability of tissue Mg levels, which has been associated with the major histocompatability complex, may influence individual and familial variability in susceptibility to immunologic disorders. Emotional or physical stress precedes acute episodes of CFS and FM. The resultant increased catecholamine release, which increases Mg loss, can be a factor in CFS and FM. Low levels of Mg intensify the secretion of the catecholamines, thus increasing the risk of adverse effects of stress. The stress of loud noise causes Mg loss and intensifies the need for Mg. Increased catecholamines have been implicated in constriction of cochlear arteries with reduction of cochlear blood flow and of Mg levels in fluid around hair cells, with increased influx of Na and Ca and disturbance of energy metabolism of the inner ear. Mg deficiency-induced release of substance P induces expression of the endothelium-leukocyte adhesion molecule of the cochlear microvasculature, which further reduces cochlear blood flow. The deafness is induced both from the direct auditory trauma, and from the Mg deficit. The stria is the metabolic energy source of the cochlea. The apex of the cochlea is where the low frequency sounds are transduced; since that is where the blood supply is most distal, it is the area most vulnerable to impairment of blood flow. Low frequency deafness occurs frequently in post-menopausal women with CVD. Mg protects against development of CVD, especially that associated with microangiopathy that gives rise to cardiomyopathy. Microvascular lesions of the inner ear might similarly be contributed to by Mg deficiency.

EEG and EMG changes of CFS:

1- Sinusoidal slow waves generated by the reticulate neuronal hypersynchrony.
2- Reticular neuronal hypersynchrony generated by hyperapnoea (sinusoidal slow waves).
3- Neuromuscular hyperexcitability.
4- Sleep disorders: agitated sleep with frequent nocturnal awakenings, increased percentage and duration of light slow-wave sleep, and rapid, frequent changes of various stages of light slow wave sleep and of rapid eye movement (REM) sleep.

Cytokines involved in CFS:

Cytokine	Comment
Transforming growth factor beta (TGF- β)	Elevated serum level.
Interleukin 1 beta (IL-1 β)	Stimulated by lipopolysaccharide.
Interleukin 6 (IL-6)	Stimulated by phytohaemagglutinin.
Tumour necrosis factor-alpha (TNF-α)	Significantly increased in peripheral blood mononuclear cell.

Humoral and cellular immunology affected by Mg:

1- Mg inadequacy can cause histamine release (Cutaneous hyperaemia and inflammation).
2- Mg deficiency causes 10-fold increase in eosinophilia (Precede release of histamine and serotonin).
3- Mg deficiency depresses cell-mediated immunity.
4- Mg deficiency impairs phagocytic activity.
5- Mg deficiency impairs lymphocytic function.
6- Mg participates in the T-cell mediated immunity, which requires cooperation of the mononuclear phagocyte system and is involved in production of cytokines, ILs, INFs, TGF, and TNF
7- Mg deficiency increases plasma concentrations of inflammatory cytokines such as TNF, IL-I, IL-6, and the inflammatory neuropeptide substance P
8- Mg deficiency is associated with the decreased numbers and activity of natural killer (NK) or cytotoxic T-cells (may contribute to the postulated increased predilection to lymphoid and other cancers in CFS patients).

Oestrogen shifts blood Mg to bone, resulting in lowering of circulating Mg, especially in those with marginal Mg intakes. As serum Mg falls, the counteraction by Mg of the procoagulative effect of calcium cannot take effect. Thus, women on high oestrogen oral contraceptives, their tendency to develop thromboembolic events can be reduced by Mg supplements. Additional Mg-influenced substances that affect platelet aggregation, blood coagulation and vasoconstriction including those increased by Mg deficiency (thromboxane, endothelin) that increase risk, and those that are increased by optimal Mg levels and that decrease risk (prostacyclin and NO). Blood coagulation and vasoconstriction counteracted by Mg, might be the mechanisms by which Mg limits brain hypoxia. Excess serotonin is released from platelets of migraineurs; Mg reduces serotonin-induced spasms of cerebral arteries. Central neuronal hyperexcitability involves overactivity of the excitatory amino acids. Stimuli that activate the migraine attack evoke neuronal depolarisation, slow depolarisation shifts, and spreading suppression of spontaneous neuronal activity, possibly by glutamate and K^+-dependent mechanisms. Reduced gating of glutamatergic receptors may provide the link between the physiologic threshold for a migraine attack and the mechanisms of the attack itself by promoting glutamate hyperactivity, neuronal hyperexcitability, and susceptibility to glutamate-dependent spreading depression. Mg deficiency causes neuromuscular excitability. Mg deficiency-induced sustained release of histamine-which it has been suggested might function as a central neurotransmitter, which might contribute to increased neurologic irritability of severe Mg deficiency. Magnesium at physiologic levels blocks neuronal N-methyl-D-aspartate (NMDA). NMDA-receptors are normally activated by glutamate and/or aspartate, which are the principal neurotransmitters for excitatory synaptic transmission. This explains the anticonvulsive activity of Mg. The epileptiform activity of Mg deficiency is also blocked by other NMDA receptor antagonists. Excitatory amino acids are highly sensitive to extracellular free ionic Mg. Mg also decreases motor neurone responses evoked by norepinephrine, and by the inflammatory neuropeptide, e.g., substance P. Magnesium deficiency increases production of substance P, which is found in CNS neurons. Substance P may contribute to the irritability caused by Mg deficiency via its proinflammatory and free radical-releasing effects. Interpretation of measured Mg levels is difficult. Magnesium content in sublingual cells correlates well with cardiac tissue levels, with severe depression and with chronic disorders with characteristics of CFS, termed electromagnetic dysthymia. Histamine binds to sensory nerves to produce an afferent signal. Magnesium deficiency results both in early release of substance P and in subsequent histamine release. Magnesium deficiency is associated with free radical production. Mg and taurine (antioxidant amino acid) has been proposed as therapy for migraine. Magnesium should be combined with antioxidant therapy in CFS. Diagnosis of mitral valve prolapse is more common both in the disorder that is clearly associated with Mg deficiency and in CFS and FM. Analgesia are implicated as a risk factor in acquiring chronic fatigue syndrome (CFS), but not methylsulfonylmethane (MSM). Disturbed DC magnetic fields in the sleeping place confuses the body and can cause insomnia, chronic fatigue, and other sleep related problems. Uranium (U) is a nonessential element that is very abundant in rock, especially granite. It is present at widely varying levels in ground (drinking) water, root vegetables, and present in high phosphate fertilisers. The primary symptom of low level chronic U excess is chronic fatigue. Low thyroid function may also be an important factor in chronic fatigue and depression. Chronic fatigue syndrome patients display cortisol hyposecretion in saliva as well as plasma. T3 improves sleep patterns; decreases chronic fatigue, reduce weight, and increases exercise. Chemical syndromes include multiple chemical sensitivity (MCS), chronic fatigue syndrome (CFS), fibromyalgia (FM), and gulf war syndrome (GWS). Confusion has reigned since the symptoms-based descriptions of chronic fatigue syndrome, myalgic encephalopathy and multiple chemical sensitivity. The absence of objective abnormalities from physical findings, imaging or other tests meant diagnosis depended on complaints/symptoms. Disorders due to chemicals are frequent (chemical sensitivity-chronic fatigue). Sensitivity to chemicals has 15 names and designations that are so similar that they may/must describe the same phenomena. Many people are affected by chemicals spread indoors, but some experienced massive releases into the atmosphere. Chemicals impair human brain function. Often patients identify a specific chemical or a group that first bothered them but subsequently find many chemicals trigger ill effects. Concentrations of chemicals in air, absorbed into blood, fat and the brain and incorporated may be correlated with impaired CNS function. Therapeutic interventions, even avoidance should have record keeping and research to measure any effects of reducing exposures and body burdens of chemicals and drugs. Giving antioxidants and oxygen at the same time is irrational. Abnormal brain function in most people after several community wide chemical exposure's points to the brain as being primarily affected by chemicals. It modifies moods, stimulates (up regulates) or depresses (down regulates) functions and disturbs perceptions with unpleasant feelings. The human brain directs an orchestra of extremely diverse responses. The brain initiates rapid or deep breathing, fast or abnormal heart rhythms, dilates or contracts blood vessels, increases or decreases stomach and bowel activity (emptying versus stasis), and modifies vision, balance and the speed of response. The brain is generally or focally over stimulated to cause seizures, epilepsy or under-stimulated, sedated producing sleep and coma. Drugs that over stimulate, the convulsants including strychnine, picrotoxin and phenylenetetrazol have generalised effects as do amphetamines, LSD, PCP and OXY (Oxycontin). In contrast, sedative and narcotic (pain killing) drugs induce coma. The human lung is the ultimate absorptive organ. To facilitate intake of oxygen it has a minimal membrane thickness, 1/10,000 that of skin and the area of tennis court (80 m^2). Oxygen absorption can reach 4 liters per minute during exercise. Hydrocarbon solvents, hydrogen sulfide, ammonia, and chlorine enter blood and brain like oxygen. Inhaled ether causes unconsciousness in seconds. The brain controls all bodily operations and functions. Conscious, cortical impulses conducted to the limbic system and hypothalamus co-ordinate the autonomic nervous system (ANS) and influence metabolism, growth and reproduction via autonomic adrenergic and cholinergic (vagus) nerves and the ductless endocrine glands, thyroid and adrenal. The circuit for emotions consists of the amygdala, hippocampus and fornix and mamillary body, and the limbic structures. Through the cingulate gyrus, the frontal association cortex, parietal and temporal association cortices access limbic structures. Thus the cerebral cortex stimulates the limbic system to influence breathing, circulation, digestion, muscle tone and movements and endocrine functions. Chlorpyrifos, polychlorinated biphenyls and diesel exhaust impair the human brain. The high brain toxicity of chlorpyrifos and sprayed indoors to kill insects has been realised. It persists for many months on carpets, drapes, wall coverings and surfaces and occupants develop insidious neurobehavioural impairment and symptoms. A single high dose of chlorpyrifos cause cholinergic crises but less appreciated that insidious poisonings frequently present as chemical sensitivity and brain impairment. Pyrethroids, synthetic chrysanthemum insecticides that have been sprayed in airliner cabins to persist have poisoned flight attendants and people indoors at home with permanent effects. Similarly PCBs persist for years and heating transforms them into dibenzofurans increasing their neurotoxic potency 1,000- to 10,000-fold. Diesel exhaust is often pulled indoors and into cars and concentrated after absorption to cause insidious poisoning. Chemical brain injury include: Multiple chemical sensitivity, Chronic fatigue syndrome, Indoor air illness, sick building syndrome, Fibromyalgia, Depression, Asthma, Hyper reactive airways disorder, Gulf War syndrome, Organophosphates, Mould and mycotoxin disorders, Cacosmia: smell dysfunction, Chronic Lyme disease, Neuroendocrine hyperactivity, Degenerative brain diseases, PD-ALS, Traumatic brain injury, and ADHD (Attention deficit hyperactivity disorder). Specific chemicals as possible cause in indoor air include: Toluene diisocyanate, Trimellitic anhydride (TMA), Organophosphate insecticide spray residuals, Organochlorine-chlordane residuals, Pyrethroids residuals, Moulds and mycotoxins, Organic solvents; chlorinated ones (cacosmia), Formaldehyde, chlorine, and ammonia. Typical outdoor chemicals: Chlorine, hydrogen chloride, hydrogen sulfide, and diesel exhaust. Fibromyalgia (FM) and CFS are different diseases but closely related. Patients with these diseases have in common a decrease in corticotrophin releasing hormone (CRH), which controls cortisol output from the adrenals. Both have a decrease in levels of arginine vasopressin (AVP), a hormone that controls the ability of the body to release fluid. With a lack of AVP the patients would feel increasingly thirsty and have frequent urination with having to urinate about every 20-30 minutes. Both of these hormones are produced in an area of the brain called the supraoptic nucleus. Oxytocin (OT) is produced by the same nerve cells, which also have the capacity of making CRH and AVP. OT is a hormone produced in many parts of the body. In the brain, it is produced and released on a daily rhythm with its peak in the human brain occurring at around noon. OT is also produced in the posterior retina, in the pineal gland, thymus, pancreas, testicle, ovary, and adrenal glands. OT is one of the controlling factors of the microcirculation of the human body and brain. A decrease in OT can cause problems with decreased circulation in the extremities. Therefore, patients often complain of cold hands and feet, along with history of recurrent headaches. Oxytocin's ability to vasodilate the blood vessels is due to its capacity to stimulate the body's cells to produce nitric oxide (NO), a powerful vasodilator of the microcirculation. OT is released as a mother nurses her baby. Stimulation of the hormone release causes the mother to have an instinct to want to cuddle. As she nurses the child, her desire to cuddle intensifies. This same feeling can be experienced during intimacy. OT has the ability to increase libido. Stress can restrain the production of OT. OT seems to stimulate the ability of the brain to concentrate, contributes to mental alertness and improves memory. Patients lacking this hormone may find difficulty in concentrating, and feel like they are thinking in a fog. This has been noted in FM. OT can occupy multiple hormonal receptor sites in the body. An empty receptor for OT may potentially cause pain. Oxytocin has the ability to occupy not only its own receptor sites, but opiate (narcotic) receptor sites as well. Oxytocin has been given to humans to kill cancer pain, low back pain and bowel pain from irritable bowel syndrome. OT is produced in the posterior retina of the eye. A decrease in OT level can cause problems with intermittent blurring of vision. Visual disturbances in FM have been observed. OT is made in the pineal gland of the brain, as is melatonin, a hormone that enhances sleep. Insomnia is a sleep disorder frequently seen in patients with FM/CFIDS, and could indicate a deficiency in melatonin. Melatonin has the ability to activate the immune system, so the use of this product is usually contraindicated in the presence of autoimmune disease (e.g., lupus, and rheumatoid arthritis). The ovaries make OT. In the ovary, OT helps in the fine tuning of progesterone release. When patients are lacking OT, they may frequently complain of ovarian pain, even though pathology does not support the presence of either cysts or tumours. Ovulation function may be impaired with menstrual irregularity. Oxytocin is synthesised in the adrenal glands where it can stimulate or inhibit steroid production. Patients with a decreased OT level often complain of flank pain underneath the posterior ribs. Malfunction in the adrenal steroid production occurs in FM. OT is synthesised by the thymus gland. The thymus gland utilises OT to help process white blood cells, which help control autoimmunity. Normal levels of OT also help stimulate these cells into greater action. Women who nurse their children have a much lesser incidence of breast cancer. This hormone may be protective in its ability to prevent breast cancer, through its influences on the immune system. Oxytocin is produced by the pancreas. In the pancreas, OT stimulates the production of glucagon, a hormone that helps the intestines to relax. Therefore, in treating a patient with decreased levels of this hormone, may solve problems with increased intestinal spasms, secondary to a lack of glucagon production from the pancreas. OT can function as an antianxiety agent in the brain. It can also stimulate social behaviour. A lack of this hormone may be expressed as antisocial behaviour with some anxiety. Oxytocin can function as an antidepressant. Depression occurs in FM/CFIDS. OT can serve as a regulator of cardiovascular function and autonomic nervous system function. This explains why patients lacking this hormone have trouble controlling their blood pressure when going from a sitting to upright position, or when standing for a long period of time This is known as neurally mediated hypotension. They often complain of near syncope (light headed) and possible dizziness. A drop of OT levels in the brain leads to manifestation of baroreceptor malfunction. Oxytocin has the capacity to induce the body to mildly retain fluid. This is in part due to its physical and biological similarity to arginine vasopressin. AVP is a hormone that controls fluid metabolism, pain and memory. With a lack of OT, patients have increased thirst. They also have increased urinary output due to decreased ability to retain fluid.

NUTRITIONAL STRATEGIES FOR TREATING CFS

The development of chronic fatigue syndrome (CFS) may be due to deficiencies of various B vitamins, vitamin C, magnesium, sodium, zinc, L-tryptophan, L-carnitine, coenzyme Q_{10}, and essential fatty acids. Serum folate is highly correlated with the folate level of the CSF. While erythrocyte folate is usually a better indicator of folate deficiency, serum folate is a better indicator of CSF folate. A chronically low CSF folic acid level would be a reasonable basis for suspecting that brain folate could be diminished in CFS, causing impairment in brain function. Fatigue and depression, common findings in CFS, are also prominent features of folate deficiency. Folate deficiency can cause immunodepression. Substantially larger dosages of folate (10,000 μg of folate daily) have to be prescribed for a substantially longer period of time (3 month) for CFS patients. 30% of CFS patients have elevated methylmalonic acid, a urinary metabolite believed to be considerably more sensitive than serum vitamin B_{12} for diagnosing cobalamin deficiency. The levels of vitamin B_{12} in the CSF of CFS patients significantly correlate with measures of fatigability and neurasthenia. Fatigue and depression are features common to both disorders. A substantial amount of vitamin B_{12} appears to be necessary to relieve the symptoms of CFS. Vitamin B_{12} seems to have substantial analgesic properties. Patients with vertebral pain syndromes, degenerative neuropathies, and cancer would notice excellent pain relief with injections of 5,000 to 10,000 μg daily. The improved feelings of well-being in CFS patients following vitamin B_{12} supplementation could be at least partly due to the analgesic effect of the vitamin when administered at pharmacologic dosages. CFS symptoms are associated with an increased percentage of abnormally shaped erythrocytes (non-discocytes). 40-100% of CFS patients' erythrocytes are grossly deformed and can be identified as rigid stomatocytes and dimpled spherocytes. Erythrocytes normally measure 8 microns in diameter, while the diameter of the vessels through which they flow may be only 3 microns. Loss of the normal biconcave form impairs the ability of erythrocytes to change shape in order to traverse the microcirculation. The result is a reduction in blood flow on the microcirculatory level, causing an oxygen deficit and an accumulation of by-products of cellular respiration. This pathophysiological change may explain why CFS patients often present with symptoms referable to multiple organ systems. Vitamin B_{12} administration may relieve CFS symptoms by reversing the erythrocyte abnormalities leading to improved tissue oxygenation. A dose of 2,500-5,000 μg cyanocobalamin (subcutaneous or IM) every 2-3 days would eventually lead to response with an increase in energy, stamina, or well-being, usually within 2-3 weeks of treatment in 50-80% of the patients. Excellent pain relief is expected with injections of 5,000-10,000 μg daily (neuropathies and cancer). CFS patients have elevated urinary concentrations of 5-hydroxy-indoleacetic acid, the major metabolite of the neurotransmitter serotonin. The concentrations of 5-hydroxy-indoleacetic acid would return to normal following NADH supplementation. Nicotinamide adenine dinucleotide (NADH) is the reduced coenzyme form of the vitamin niacin. Other B vitamins for which there is evidence of reduced nutriture in CFS include riboflavin, thiamine, and pyridoxine. Marginal deficiency of vitamin C may cause fatigue, lassitude, and depression, which responds to supplementation. Since vitamin C deficiency causes capillary fragility, perhaps the best method of assaying vitamin C stores is to perform the Rumpel-Leede test in which a tourniquet is applied to the arm for five minutes to see whether petechiae appear. Ascorbic acid appears to exert a substantial analgesic effect at pharmacologic dosages. Vitamin C supplementation may also share with vitamin B_{12} the ability to reverse erythrocyte membrane abnormalities seen in CFS and thus improve capillary blood flow. Fifteen minutes postinfusion of vitamin C (15 g), over 80% of the membrane abnormalities would disappear. Vitamin C supplementation also bolsters immune responses including increased neutrophil motility and chemotaxis, increased immunoglobulin levels, and increased lymphocyte blastogenesis in response to mitogens. Vitamin C enhances interferon activity (antiviral activity). Stress hormones, including both catecholamines and corticoids, can promote a reduction in tissue Mg levels. Many of the symptoms and findings in CFS resemble those of Mg deficiency. Magnesium supplementation should be combined with malic acid, since malate plays an important role in energy metabolism; specifically the generation of mitochondrial ATP. Treating primary fibromyalgia patients for an average of 8 weeks with 200-600 mg magnesium and 1,200-2,400 mg malate daily may decrease significantly the mean tender point index. While pain from fibromyalgia appears to respond in about 2 days, fatigue may take 2 weeks to respond. Almost 40% of patients with CFS may show improvement after starting supplementation. Neurally-mediated hypotension, a term that refers to an abnormal neurocardiogenic reflex in individuals with structurally normal hearts, is a common cause of recurrent lightheadedness and fainting. When venous pooling during long sitting or standing causes a reduced ventricular preload, susceptible people respond with an increased catecholamine response, resulting in augmented inotropic activity and excessive stimulation of mechanoreceptors in the left ventricle. This causes an exaggerated parasympathetic response, resulting in vasodilation, bradycardia, hypotension, and possibly syncope. After the episode, fatigue is prominent and may last for an extensive period of time. Neurally-mediated hypotension is a common finding in chronic fatigue syndrome.

Symptoms associated with inadequate sodium intake include undue fatigue after moderate exertion, lassitude, headache, sleeplessness, and inability to concentrate. Zinc is another mineral often marginally deficient in CFS. Zinc deficiency can cause immunodepression and produce muscle pain and fatigue. Leukonychia, refers to white spots on the fingernails, may be a sign of marginal zinc deficiency and is correlated with frequent feelings of drowsiness. Giving normal subjects with no evidence of zinc deficiency, 135 mg of zinc daily for 15 days would increase isokinetic strength and isometric endurance in their leg muscles. L-tryptophan is depressed in the plasma of 80% of CFS patients, suggesting that brain serotonin levels may be depressed. Tryptophan is the dietary precursor of serotonin, a neurotransmitter intimately connected with mood. A low tryptophan diet may cause relapse in recovering depressives, while low tryptophan concentrations may rise when depression remits. Tryptophan supplementation provides a mild degree of analgesia and may be especially effective for the subset of chronic pain patients with a disorder of serotonergic transmission. After 3 months of 100 mg TID of 5-hydroxytryptophan, a metabolite of tryptophan and immediate precursor of serotonin, 50% of the patients with CFS are expected to have a fair-to-good degree of overall improvement, with highly significant improvements in fatigue, number of tender points, pain intensity, anxiety, and sleep quality. There is enhanced degradation of tryptophan in infectious diseases, possibly due to the increased formation of γ-interferon during activation of cell-mediated immunity. L-carnitine and its esters prevent toxic accumulations of fatty acids in the cellular cytoplasm, and of acyl CoA in the mitochondria, while providing acetyl CoA for mitochondrial energy production. Carnitine deficiency may impair mitochondrial function, resulting in generalised fatigue along with myalgia, muscle weakness, and malaise following physical exertion. CFS patients may suffer from a clinically-relevant carnitine deficiency. There is significant decrease in serum acylcarnitine, with increased ratio of acylcarnitine to free carnitine in CFS patients. In CFS, serum carnitine levels appear to be a biochemical marker for both symptom severity and ability to function. But, only one-third of CFS patients is carnitine responder (1 gm TID/qid, up to 6 gm a day), serum levels of L-carnitine won't predict who would respond. Mononuclear cell carnitine may be a better predictor of carnitine response than serum carnitine. CoQ_{10} facilitates cellular respiration. 80% of CFS patients are deficient in CoQ_{10}, which further decrease following mild exercise or over the course of normal daytime activity. 90% will have reduction and/or disappearance of clinical symptoms, and 85% will have decreased post-exercise fatigue when giving 100 mg CoQ_{10} daily or more for 3 months. Low levels of essential fatty acids (EFAs) appear to be a common finding in chronic fatigue syndrome. Viruses, as part of their attack strategy, may reduce the ability of cells to make 6-desaturated EFAs while interferon requires 6-desaturated EFAs in order to exert its antiviral effects. Changes in EFA metabolites could cause the immune, endocrine, and sympathetic nervous system dysfunctions, which is seen in CFS. The formation of prostaglandin E1, for example, can be enhanced by increasing intake of ω-6 fatty acids. This prostaglandin improves erythrocyte membrane fluidity and filterability; i.e., the ability of erythrocytes to pass through a small membrane filter. Supplementation with both evening primrose oil, a source of ω-6 fatty acids, and fish oils, a source of ω-3 fatty acids, has been shown to improve erythrocyte filterability. CFS is characterised by a cluster of symptoms that often defies diagnosis and treatment. CFS may be caused by a weakened immune system. There are immunologic alterations in CFS patients. Some of the symptoms of chronic fatigue syndrome so closely resemble those of an autoimmune illness that CFS has frequently been mistaken for both multiple sclerosis and lupus. 50-80% of all CFS patients suffers from allergies. OPCs are as much as 20 times more potent than vitamin C and 50 times more potent than vitamin E in their ability to provide antioxidant protection. OPCs have the ability to inhibit all 3 phases of the cancer process: initiation, promotion and progression. OPCs in the membranes of grape seeds have antioxidant and anti-inflammatory effects. Mast cells circulate throughout the body and they can release their inflammatory mediators everywhere. Degranulation of mast cells not only occurs in the final stage of capillary upheavals, but in each and every case of inflammation and allergy. Histamine is produced in 2 ways: 1- under the influence of hyaluronidase in mast cells; 2- by decarboxylation of the amino acid histidine. OPCs hinder the HD enzyme and thereby inhibits (prevents) the production of histamine. The anti-inflammatory and anti-allergic activity of OPC is based on their antioxidant activity. Free radicals activate the release of histamine and other mediators and they are also produced by these mediators. This is how free radicals form essential components in the development and perpetuation of inflammations and allergic reactions.

DIETARY INTERVENTION

Muscle comprises the greatest proportion of non-water body weight. Bone has a smaller relative growth rate than either fat or muscle, so the ratio of muscle to bone increases as the weight increases. The effect of sex is primarily on fat composition. Fats in the body serve five primary functions: 1- Provide the largest store of potential energy. 2- Serve as a cushion for the protection of vital organs. 3- Provide insulation from the thermal stress of cold environments. 4- Allow for the absorption of the fat soluble vitamins A, D, E, K and many other phytonutrients, such as β-carotene. 5- Act as appetite depressors. Fats are transported in the blood as chylomicrons, lipoproteins and free fatty acids. The chylomicrons contain triglycerides (fatty acids and glycerol) and small amounts of cholesterol and phospholipids (lecithins). After absorbing fats into the cells, most blood fat consists of lipoproteins (a combination of triglycerides, cholesterol and phospholipids) combined with protein. Fats are stored in the body exclusively as triglycerides. Fat anabolism, include: 1- Triglycerides are composed of fatty acids and glycerol. They can be made from excess sugar or protein and are stored mainly in fat cells. Potentially a huge amount of body fat can be stored in this way. The formation of triglycerides is under dietary and hormonal control. 2- Excess carbohydrates and or unstabilized high glycaemic (high insulin response) carbohydrates can be readily converted to triglycerides and stored as body fat. Fat cells are economical and very efficient in energy storage. They contain 2.25 times the energy level of glucose. It is impossible to lose body weight as purely fat shrinkage without some loss of lean body mass. This is why adding supplementation to target fat energy utilisation and to maintain muscle mass is essential. Obesity is the storing of surplus feed energy as fat both within and around body tissues. Obesity is the result of excess energy for maintenance and growth. Obese individuals are fed full high-energy rations feeding program. Fat Catabolism: 1- Hydrolysis of fats (mostly in liver cells) into fatty acids and glycerol. 2- Glycerol is converted to glyceraldehyde-3-phosphate, which enters the glycolytic pathway and eventually the Krebs' cycle to produce ATP. 3- Fatty acids are converted by β-oxidation to acetyl-CoA. When fat catabolism is accelerated, acetyl-CoA condenses to form ketone bodies (ketogenesis); this occurs mainly in the liver; the largest proportion of ketones enters blood from liver cells to be transported to the tissues for oxidation to CO_2 and H_2O via Krebs' cycle. 4- The rate of fat breakdown is inversely related to the rate of carbohydrate catabolism, i.e., more fat is broken down when less carbohydrates are available for energy. 5- The stabilising of carbohydrate metabolism can increase the fat catabolism rate, excessive carbohydrates or high insulin response (high glycaemic) carbohydrates can increase the amount of fat stored. Many nutrients, when combined with a low-calorie diet and exercise, can enhance fat loss and gains in lean mass (muscle). The nutrients that speed fat-burning can also help save muscle from being lost. Dieting or energy restriction causes the body to shed a combination of fat and some muscle tissue. Most carbohydrates absorbed are used to generate glucose, a small amount are anabolized to storage glucose or glycogen found mostly in muscle cells and the liver. Carbohydrates at proper dietary levels protect lean mass (muscle) from being broken down, and increase the amount of fat that is burned for energy; thus increasing the lean (muscle) to fat ratio. A balanced relationship between insulin and glucagon maximises the generation usable energy from glucose. The glycaemic factor or index: The more rapidly a carbohydrate can be converted to blood sugar (glucose) the faster and with greater efficiency can this carbohydrate be converted to fat. The reverse is also true-that the less rapidly a carbohydrate is converted to blood sugar (glucose) the less likely it is that the carbohydrate will be converted to fat. Many common feed carbohydrates have a high glycaemic factor e.g., corn, wheat, oats, sweet feeds (quick acting carbohydrates). Dietary proteins maintain and integrate all body tissues and cells. Proteins are present in all cells and are utilised to make hair, skin, hooves, hormones, enzymes, structural proteins, muscle, immune compounds, cell-membranes and internal cellular material. Acid-base balance is regulated by proteins, actin and myosin allow for normal muscular contraction. Proteins are involved in the transfer of genetic information. Amino acids singularly or in combination are involved in energy production, hormonal balance and a variety of other metabolic regulations. Proteins are used rapidly and have a high turnover rate. There are basically 2 types of protein (nitrogen) balance, negative nitrogen balance occurs when tissue is breaking down faster than it can be replaced, causing a drop in muscle mass and a lowering of the lean to fat ratio. Positive nitrogen balance, which is the protein build up occurs at a faster rate than tissue is breaking down. In protein metabolism, anabolism (or building) is primary and catabolism (burning) is secondary, in carbohydrate and fat metabolism the opposite is true. Growth and development is highly dependent on the endocrine system. The thyroid gland plays a major role in controlling energy expenditure and determining the basal metabolic rate (BMR). The thyroid gland requires more than just iodine to function optimally. The vitamins A, B_2, C, E; the minerals potassium, magnesium, copper, zinc; the amino acid tyrosine; and possibly the trace element rubidium are all involved in healthy thyroid function. Inadequate thyroid function can also depress liver function where a large part of fat metabolism is activated. Insulin increases cellular uptake of amino acids. This increases protein synthesis and minimises the use of amino acids for energy production. Insulin enhances the functional thyroid status. An over stimulation of insulin release increases fat production markedly and very quickly. This results in a lowered lean to fat ratio because of the increase in body fat. Growth hormone (GH) or somatotropin promotes the growth of cells and tissues by stimulating protein synthesis by: 1- increasing amino acid transport into the cells, and 2- stimulation of RNA synthesis. GH stimulates the release of fat for the production of energy, increase the rate of lipolysis and inhibit the lipogenesis. Physical exercise, sleep, caloric restriction, low blood sugar and certain amino acids can stimulate the release of growth hormone. Growth hormone's primary synergists are insulin, testosterone and the minerals magnesium, zinc and potassium. The primary antagonist to growth hormone is the stress hormone cortisol produced by the adrenals in response to any physiologically stressful event. Testosterone can increase the uptake of amino acids into muscle. Testosterone affects nuclear DNA, which produces new RNA followed by new enzymes and new proteins that generate cellular changes for growth and development. Testosterone's chief synergists are GH, insulin and the minerals zinc and cobalt, also the vitamins A, E, C, folic acid, B-complex and vitamin D. Maintaining healthy testicular function is paramount for the benefits of healthy testosterone synthesis and release. In the body after water the largest percentage of tissue composition is protein. Most of the protein's amino acids can be synthesised by the body. However, the ten essential amino acids that cannot be made must be ingested through high quality dietary proteins. The essential amino acids are: 1- Isoleucine, 2- Leucine, 3- Lysine, 4- Methionine, 5- Phenylalanine, 6- Threonine, 7- Tryptophan, 8- Valine, 9- Histidine, and 10- Arginine. High quality protein hydrolysates (predigested to a combination of single amino acids, di- and tri-peptides) are much better absorbed and utilised than either intact proteins or single amino acid combinations. The branched-chain amino acids leucine, isoleucine and valine make up one third of muscle protein. During exercise they are used up rapidly. They can also increase protein synthesis and prevent muscle proteins from breaking down. They may also prevent testosterone levels from falling during exercise and increase testosterone levels after training. The amino acid glycine can increase muscular function. The rate of glycogen release from the liver to supply carbohydrate based energy from the breakdown of glycogen to glucose is enhanced by glycine. The amino acids arginine and ornithine have similar metabolic properties. Both are also very effective detoxifiers of ammonia. Improved immune function, an increased ability for muscular contraction, improved recovery from stress and an increased capillary blood flow are also benefits of arginine and ornithine. The amino acids lysine and methionine have importance in reducing free radical stress and in stepping up the lipolysis rate. This combination supplies methyl groups, increases the synthesis of lipolysis compounds, detoxifies the system of metabolic waste and environmental toxins and prevents lipogenesis in the liver. The peptides of high quality protein hyrolysates increase positive nitrogen balance, facilitate protein synthesis and maintain a normal, healthy immune function, and maintaining and increasing a high lean mass to body fat ratio. Glutamine is the most abundant amino acid in muscle and is a main transporter of nitrogen waste. A strong relationship exists between glutamine and the branched chain amino acids to prevent muscle breakdown and support muscle growth. L-carnitine is a dipeptide made from the amino acid lysine. Its synthesis occurs primarily in the liver and kidneys, and requires adequate amounts of the vitamins B_6, C, niacin along with iron and the amino acid methionine. L-carnitine transports fatty acids across mitochondrial membranes where the fatty acids can be burned for energy. Carnitine aids in the removal of waste products from the mitochondria and increases the rate of fat oxidation in the liver. L-carnitine aids the oxidation of pyruvate and the branched chain amino acids in Krebs' cycle, and prevents the build up of certain fatty acid complexes that destabilise muscle cell membranes. L-carnitine reduces lactic acid in the muscle (a major cause of fatigue). L-carnitine enters the mitochondria may spare some of the antioxidants such as vitamins C, E and selenium; thus minimising the damaging effects of free radical exposure. γ-Linoleic acid (GLA) is made in the body from one of the essential fatty acids-linoleic acid. GLA is very important in the fat metabolising process. GLA serves as a precursor to PGE_1, which exert an anti-inflammatory influenced in the tissues. Adequate levels of the vitamins B_6, C, biotin and B_3 along with the mineral zinc and magnesium are required for the conversion of linoleic acid to GLA. Lipotropics prevent the accumulation of fat-most often in the liver, such as choline, inositol, methionine, betaine, lipase and a variety of plant compounds including silymarin and lipoic acid. Natural lecithin production can increase from the use of lipotropics, which protects the liver from environmental toxicity. Improvement of liver function provides a more effective activation of certain vitamins. The lipotropics aid in eliminating the metabolic waste products of protein metabolism and may improve disease resistance. Soy lecithin is a high phosphatide component of soybeans. It is very rich in the lipotropic choline as phosphatidyl choline or PC and also contains phosphatidyl inositol and phosphatidyl ethanolamine. This high choline and other lipotropic content aids in a fat clearing effect in the liver by emulsifying fats and making fat digestion easier. Lecithin keeps cholesterol and triglyceride fats soluble in the body for easy transport. Lecithin is a rich source of the essential fatty acid, linoleic acid, from which GLA is made. Cell membranes have a high lecithin content especially nerve tissue and the lecithin phosphatides are involved in cell-membrane stabilisation. Keeping the liver clear of excess fats is crucial to maximising liver function and improving the metabolism. Free radicals are toxic byproducts of oxygen metabolism. Free radical damage includes damage to the muscle cells and a subsequent decrease in the metabolism of the mitochondria. The B complex are involved in a variety of functions including: 1- Metabolism of carbohydrates and fats for energy. 2- Increasing cellular energy production by improving mitochondrial efficiency. 3- Increase the use of glycogen for energy. 4- Involved in protein and amino acid metabolism. 5- Involved in the formation of fatty acids and glucose. 6- Involved in the formation of steroid hormones and all new proteins including red blood cells, bone marrow cells, the lining of the intestinal tract and haemoglobin. 7- Involved in the formation of all new cells. The fat-soluble vitamins are: 1- Essential for vision, cellular growth, the health of skin, mucous membranes and normal immune function. 2- Essential for bone health and mineral balance. 3- Involved in the normal clotting processes of the blood. 4- Act as antioxidants in fatty tissue and other fats. 5- Compliments vitamin C as an antioxidant. Only 1% of calcium is involved in metabolic functions-among them: 1- Required for normal blood clotting. 2- Controls the excitability of nerves and muscles. 3- Stabilises cell membranes. 4- Involved in the regulation of hormone release. 5- Regulates cell division. Calcium levels are controlled both hormonally and nutritionally. Calcium is stored in the bones at a 2:1 ration with phosphorus. Enzymes basically act as catalysts to increase the rate of physiological reactions with a minimum of energy expenditure. By increasing the speed and the efficiency of digestion, proteins, carbohydrates, fats and fiber are broken down more completely, which improve bio-availability to body cells. Doctors must give their patient a goal they can meet. 30 minutes or more of moderate-intensity physical activity (walking and stair climbing) on a daily bases is essential. Very modest physical activity can reduce the risk of heart disease by 45% over those who are completely sedentary. Zone dieting is aimed to avoid ketosis, which may lead to loss of muscle weight even with high protein diets. Zone dieting is aimed to promote the desired hormonal response throughout the day, including insulin, glucagon, and eicosanoids. The goal of Zone dieting is to eat the correct balance of carbohydrates and protein so that glucagon is produced instead of insulin, because glucagon promotes the formation of good eicosanoids while insulin promotes the formation of bad eicosanoids that cause all sorts of diseases. To reach the Zone, one must consume foods in a protein-to-carbohydrate ratio of 0.75, which is 3 g of protein for every 4 g of carbohydrate, at every meal and snack, and without allowing more than 5 hrs between meals or snacks. The Zone diet not only limits fat intake, but also advises against the use of saturated fat. Monounsaturated fats have effect on eicosanoid production and preferred in this type of diet. The Zone's calculation of protein intake is based on lean body mass and activity level, which is scientifically invalid and will restrict protein intake too much for almost everyone who is limiting carbohydrate in the diet. The popularity of the Zone diet could be due to individuals using the recommended ratios but not limiting their intake to the calculated portions of carbohydrate, protein and fat. Whether or not the control of eicosanoid production by specific nutrient intake provides all the effects claimed is not at all scientifically proven. Zone diet is complex for most persons to follow with any consistency. In order for vitamins to be utilised by the body, they must first be converted into their active coenzyme forms.

EATING DISORDERS

Anorexia nervosa is an eating disorder characterised by excess control-a morbid fear of obesity leads the sufferer to try and limit or reduce their weight by excessive dieting, exercising, vomiting, purging and use of diuretics. Sufferers are typically >15% below the average weight for their height/sex/age and typically have amenorrhoea (if female) or low libido (if male). 1-2% of female teenagers are anorexic. Sugar is added to most processed foods. Overconsumption of refined food causes the saccharin disease. Organ changes from the saccharin disease results in diabetes, peptic ulcer, constipation, and other disorders. Carbohydrate-rich/protein-poor diet may influence the action of serotonin (brain chemical). Bread, backed potatoes, white rice, bananas and most cold breakfast cereals have a high glycaemic index. They cause a fast rise in blood sugar. This increases insulin production, which induces hypoglycaemia, making the individual feels starved. Baked potatoes have a higher glycaemic index than pure sugar. Beans, lentils, whole grains and protein-enriched pasta release glucose slower. The glycaemic factor may be the most promising area in nutrition. Glycaemic index may increase the risk of developing insulin resistance. The body compensates by producing abnormally large amounts of insulin, which increases the risk of diabetes mellitus, heart disease etc. Low glycaemic index foods reduce hunger, e.g., high-fiber foods, whole grains, less-processed foods, protein-enriched pasta etc. Eating disorders are not uncommon. Between 0.5% and 1% of teens develop anorexia nervosa (AN); almost 5% of older adolescents and young adult women will develop bulimia nervosa (BN). Binge/purge behaviour is very common, with some surveys finding an incidence as high as 25% in the older adolescents and young adult women surveyed. The pathogenesis of eating disorders could be a combination of biological, psychological, and societal factors. Biological factors include an imbalance of neurotransmitters, particularly serotonin (5-HT) and serotonin precursors. Precipitating psychosocial factors include difficulty with developmental tasks, puberty changes, and peer group issues. Teens should always be questioned about the use of supplements. Some teen athletes also may use caffeine as a supplement. The only supplement that adolescent athletes may need is a multivitamin, along with a modest increase in protein intake (meat has 1 g of protein/ounce). Growth hormone exerts its effect on growth by producing another class of hormone called insulin-like growth factor-1 (IGF-1), formerly known as somatomedin-C. Tests of growth hormone function include both GH and IGF-1 levels, since both are decreased in growth hormone deficiency. Normal IGF-1 levels can be as low as zero in younger children, although testing at times of natural "peaks", such as 20 minutes after awakening from sleep or following 20 minutes of vigorous exercise, may be helpful in obtaining a peak level. IGF binding proteins (IGFBP-3) are age-dependent and normal ranges are narrower. A decrease in haematocrit is a fairly late event in iron deficiency and a low or low-normal haematocrit will not occur until iron stores have been depleted. The glycaemic index (GI) ranks foods according to their potential to increase blood glucose levels. A high-carbohydrate diet subjects people to an oral glucose tolerance test every day; white bread (GI 100), modern starchy foods have a high GI (>70), whereas "traditional" starchy foods, such as barley, legumes/beans, pastas, and whole-grain cereal have a low GI (<55). The ranking of GI in foods is the same in different people; all people will have a higher response to foods with a high GI than to foods with a lower GI. Low-GI foods are more satiating and more effectively delay the return of hunger than high-GI foods. Other blood glucose determinants include the nature of CHO; the quantity of CHOs, fat, protein, and other nutrients present in foods; gastric emptying time; and circadian/diurnal variations. The rate of gastric emptying affects postprandial blood glucose (PPBG). The time of day the meal is consumed is another variable. A meal that causes one result in the morning may cause a much higher result after lunch or dinner.

DYSLIPIDAEMIA

Cholesterol is an essential component of cell membranes and a metabolic precursor of bile acids and steroid hormones (e.g., adrenocortical and sex hormones). It is obtained from the diet and synthesised in the liver, intestinal mucosa, and other cells. Secondary causes of hypercholesterolaemia and lipoprotein abnormalities include poorly controlled diabetes mellitus, hypothyroidism, nephrotic syndrome, dysproteinaemia, obstructive liver disease, drug therapy (e.g., cyclosporine, glucocorticoids), and alcoholism. Atherosclerosis begins with accumulation of lipoproteins (primarily LDL) within the inner layer of the arterial wall, where they no longer come in contact with antioxidants and other constituents in the bloodstream. Chemical modification (particularly oxidation) of lipoproteins leads to a local inflammatory reaction involving macrophages, which ingest oxidised lipoproteins and form foam cells. Accumulation of foam cells contributes to fatty lesion formation. Reverse cholesterol transport out of the tissues mediated by HDL may also occur. Over time, fatty lesions progress to fibrous plaques. Fissures may develop in a plaque, exposing the underlying tissues to platelets and other constituents of blood. Platelet adhesion, activation, and aggregation lead to thrombus formation, partially or completely occluding the vessel lumen and causing clinical symptoms of CHD (e.g., myocardial ischaemia or infarction). Advantages of testing for total cholesterol over a complete fasting lipoprotein profile include greater availability of the test, lower cost, and lack of a requirement that the patient fast before the test. Fasting defined as nothing by mouth with caloric value in the preceding 9-12 hours. Dyslipidaemia is an important risk factor in the context of cardiovascular disease, and appropriate intervention can have a significant impact on clinical outcomes. Cholesterol is one of the major lipid particles in the body; the other is triglyceride (TG). Both of these particles serve important functions; however as insoluble molecules, they must be transported in the blood in complexes known as lipoproteins. Plasma lipoproteins are composed of a core of TG and cholesterol ester, enveloped by a surface coat of phospholipid, unesterified (free) cholesterol, and special proteins called apolipoproteins (or apoproteins). Measurements of VLDL and LDL are based on a fasting TG level and are valid only when the TG level is <400 mg/dl (5 mmol/l). This explains the requirement for a fasting blood sample to determine blood lipid levels. For cell membranes, cholesterol provides an essential stabilising function and facilitates membrane transport. Cholesterol is required for biosynthesis of hormones and is a precursor of adrenal and sex hormones. Cholesterol is also required for the synthesis of bile acids; bile acids are crucial for the absorption of dietary fat in the small intestine. Triglycerides (TGs) offer an ideal means of energy storage and production; TGs are stored in adipose cells and are light in weight relative to their energy content when compared to glycogen stores in liver or muscle. Cardiac and skeletal muscle cells extract TG transported in circulating lipoproteins and converted to fatty acids and glycerol through lipolysis. Fatty acids are a major energy source for muscle cells. They are broken down in an β-oxidation reaction and then enter the Krebs cycle with subsequent production of ATP. In a number of cells, such as muscle and liver cells, fatty acids can also be converted into glucose through gluconeogenesis. TGs are insoluble particles and must be packaged into lipoproteins in order to circulate in the plasma, from sites of synthesis or absorption to sites of use. The core of the lipoprotein, containing cholesterol ester and TG, is nonpolar and hydrophobic, and the outer layer of the lipoprotein particle (containing free cholesterol, phospholipid, and specific apolipoproteins), is polarised, permitting the lipoprotein particles to be transported in the circulation. Apolipoproteins (apo) such as apoB, apoC and apoE, coat lipoprotein particles and serve a number of functions including the transport of lipids in the blood and recognition of lipoprotein particles by enzymes, which process or remove lipids from the lipoprotein particles. For example, apoC-II activates the enzyme lipoprotein lipase (LPL), which removes TG from lipoprotein particles such as chylomicrons and VLDL. Chylomicrons are produced in the intestinal lumen following the absorption of digested fat. They are the largest lipoprotein and are rich in TG. Because of their particle size, chylomicrons scatter more light and may cause the serum to take on a cloudy appearance after meals or in patients with dyslipidaemic syndromes characterised by the inability of catabolic chylomicrons and TG to rich lipoproteins. Chylomicrons are transported in the blood to tissues such as skeletal muscle, fat, and the liver. The capillary beds of these tissues contain high concentrations of LPL. Lipoprotein lipase (LPL) hydrolyses TG in the chylomicrons into free-fatty acids that are either oxidised by the muscle cells to generate energy, stored in adipose tissue, oxidised in the liver, or used in hepatic VLDL synthesis. Once the chylomicrons have been processed by LPL, the TG-depleted chylomicron is called a remnant particle, which is then transported to the liver for further processing. VLDL contains a high concentration of TG. VLDL is synthesised from free-fatty acids (FFAs) formed in the catabolism of chylomicrons in the liver, or from endogenous production of TG. Reverse cholesterol transport refers to the process by which cholesterol is removed from the tissues and returned to the liver. HDL is the key lipoprotein involved in reverse cholesterol transport and the transfer of cholesteryl esters between lipoproteins. High-density lipoproteins (HDL) is formed through a maturation process whereby precursor particles (nascent HDL) secreted by the liver and intestine proceed through a series of conversions (HDL cycle) to attract cholesterol from cell membranes and free cholesterol to the core of the HDL particle. There are subclasses of HDL particles, including HDL2 and HDL3. Possible mechanisms include the action of cholesteryl ester transfer protein, which transforms HDL into a TG-rich particle that interacts with hepatic-triglyceride lipase. Cholesterol ester-rich HDL may also be taken up directly by the receptors in the liver. Another mechanism may be that cholesterol esters are delivered directly to the liver for uptake without catabolism of the HDL cholesterol particle.

Non-HDL cholesterol = Total cholesterol - HDL

Non-HDL cholesterol therefore encompasses a broader indication of cardiovascular disease risk. Dyslipidaemia is an abnormality within the lipid profile, encompassing a variety of disorders relating to elevations in total cholesterol, LDL, or TG, or conversely, lower levels of HDL. Dyslipidaemia may present as a single disorder affecting only one lipoprotein parameter, or may represent a combination of lipoprotein abnormalities, such as elevated TG and low HDL. Dyslipidaemia may be the result of over-production or lack of clearance of the lipoprotein particles, or related to other defects in the apolipoproteins or enzyme deficiencies. Primary dyslipidaemia refers to a genetic defect in the lipid metabolism, e.g., familial hypercholesterolaemia. Secondary dyslipidaemia may be attributed to another cause. For example, environmental factors (such as a diet rich in saturated fat or a sedentary lifestyle), diseases (such as diabetes, hypothyroidism, obstructive liver disease), and medications (such as thiazide diuretics, progestins, or anabolic steroids) may result in a secondary dyslipidaemia, which could be potentially resolved by correcting the underlying condition. Nutrition therapy remains the cornerstone of dyslipidaemia management. Dietary changes, combined with regular exercise and weight management, collectively are known as therapeutic lifestyle changes (TLC). Cholesterol is part of a lipoprotein, or transport molecule. Cholesterol is a waxy, fat-like substance that is an important component of the cell membrane and is used in steroid hormone synthesis. Cholesterol is insoluble in plasma, as is triglyceride (TG). They are both encapsulated in lipoprotein molecules for transport in the blood. Lipoproteins are macromolecules that contain cholesterol esters and TG contained in a core and surrounded by phospholipids, nonesterified cholesterol, and apoproteins. Lipoproteins carrying TG have lower density than lipoproteins carrying cholesterol. There are 5 types of lipoproteins: 1- Chylomicrons; are the largest lipoprotein molecule; synthesised (in wall of small intestine). They are composed of exogenous TG and cholesterol (absorbed from the GI tract). They Transport dietary TG to fat and muscle cells, then transport cholesterol to liver. 2- Very-low-density lipoprotein (VLDL); is composed of large amount of TG and lesser cholesterol esters produced in the liver. VLDL transports endogenous TG to fat and muscle cells. 3- Intermediate-density lipoprotein (IDL); is rich in cholesterol. They are recycled in the liver after returning from the body, into VDL, or in the vascular compartment into LDL. 4- LDL; is composed of Cholesterol. They are the main carriers of cholesterol. LDL circulates until removed by receptor-mediated pathway or non-receptor-dependent pathway. 5- HDL is rich in surface phospholipids, low in cholesterol. HDL facilitates clearance of cholesterol from atheromatous plaque to the liver where it is excreted. HDL inhibits cellular uptake of LDL. Other chemical risk factors include fibrinogen, high-sensitivity C-reactive protein (hs-CRP), lipoprotein α (Lp-α), apoprotein B, and homocysteine. Fibrinogen is active in the clotting process; it can be decreased by exercise and by the intake of niacin. hs-CRP, a marker for inflammation, is increased by use of HRT. Measurement of hs-CRP is nonspecific, since smoking, collagen vascular disease, and other conditions will falsely elevate it. Statin drugs and niacin decrease hs-CRP levels. Lp-α is structurally related to LDL, can attach to circulating LDL and damage the endothelium. Lp-α concentration is not affected by diet or exercise, but can be decreased with niacin therapy or HRT. Apoprotein B is a component of the LDL molecule, and is known to be highly atherogenic. Apoprotein B testing reflects the density of LDL particles in the bloodstream. The high density and small size of the LDL molecule enhance the ability of LDL to pass through arteriolar lumen and cause damage. Apoprotein B can be reduced by niacin or HMG-CoA reductase (statin) therapy. Homocysteine is another marker for inflammation and correlates directly with adverse coronary events. Elevated homocysteine can especially reflect the endothelial damage seen in patients with diabetes. Folic acid (found in fresh fruits, vegetables, and other foods rich in B vitamins and fiber) reduces homocysteine levels. LDL molecules range in size and density in the bloodstream. Small LDL molecules are allowed rapid entry into the arterial walls where they deposit their cholesterol. Small LDL molecules more easily oxidise, intensifying the atherosclerotic activity of LDL. Small LDL trait is associated with reduced levels of HDL2b. Small LDL trait is associated with decreased uptake of LDL by receptors in the liver, allowing longer exposure of the arterial wall to LDL. Small LDL trait is associated with increased insulin resistance, which contributes to coronary artery disease (CAD) risk, both independently and through development of diabetes. Diabetes mellitus increases the risk of CHD 2- to 4-fold. Therapeutic lifestyle changes (TLC) as a nonpharmacologic approaches considered effective for lipid management include dietary modifications, weight loss or control, aerobic exercise, moderate alcohol consumption, and smoking cessation. The dietary recommendations include reduction of saturated fats to less than 7% of total calories and cholesterol to less than 200 mg/d, increasing consumption of plant stanols/sterols to 2 g/d, and increasing intake of viscous (soluble) fiber to 10-25 g/d. Physical activity is an essential element in managing hypercholesterolaemia. Obesity is not listed as a risk factor for CHD because it acts indirectly through other risk factors, such as diabetes mellitus, hyperlipidaemia, and hypertension. Restricting alcohol intake, because alcohol contributes calories and increases serum triglyceride concentrations in many people. If after at least 6 months of dietary therapy and exercise the reduction in LDL-C levels is inadequate, the addition of drug therapy to dietary therapy should be considered. Measurement of LDL-C levels 4 to 6 weeks and 3 months after initiation of antilipaemic drug therapy is recommended.

INTERVENTION FOR CANCER

In some cases of cancer, surgery is a cure, in others it is only palliative. Elderly patient should be treated as aggressively as possible but may require additional protective support. An understanding of the pathobiology of individual tumours is fundamental to the surgical treatment of cancer. The goal of all cancer therapy is to completely eradicate the tumour with the least amount of disruption or damage to normal tissue and organ function as possible. Interdisciplinary collaboration and treatment planning are necessary to select the most effective treatment method for cancer. Tumour that are slow growing and that consists of cells with prolonged cell cycles lend themselves best to surgical treatment, as they are more likely to be confined locally. Cancers can be insidious and progress as an asymptomatic lesion too small to be detected for long periods of time. A surgical procedure intended to be curative must involve resection of the entire tumour mass and normal tissue surrounding the tumour to ensure a margin of safety for removal of all cancer cells. The initial operation performed for removal of a cancer has a better chance for success than a subsequent operation performed for recurrence. Micrometastases may be present in 60% of individuals by the time a tumour is large enough to be detected clinically. A metastatic lesion can be found at a great distance from its primary lesion. The metastasis may have mutated so far from the primary site that the primary pathology is never found. Surgical resection of the metastatic lesions can result in long-term care without the primary source ever being discovered. Superficial and encapsulated tumours are more easily resected than those that are embedded in inaccessible or delicate tissue or those that have invaded tissues in multiple directions. The histological diagnosis is essential to select effective treatment because each type of cancer responds differently to therapy. Surgical diagnostic techniques are used to procure cells or tissue specimens for histopathologic examination. The biopsy specimen should contain both normal cells and tumour cells for comparison, it should be intact and not crushed or contaminated, and it should be labelled and preserved properly for complete evaluation. Only positive biopsy findings are definitive, negative biopsy means no cancer or the specimen is not representative of the tumour. Important principles of biopsy include: 1- minimising dissection, 2- maintaining adequate haemostasis to avoid inadvertent tumour spread. The use of incisional, excisional, aspiration, or core biopsy depends on tumour size, location, and growth pattern. Possible complications following any biopsy are pain, bleeding, haematoma, infection, dehiscence, and tumour cell seeding. Fine-needle aspiration or biopsy is the procedure of choice when there is a high index of suspicion for malignancy and the lesion is both accessible and solid. Needle biopsies are well tolerated by the patient, result in a small amount of trauma to tissue, and cause minimal manipulation of the tumour. Haematoma and infection are potential complications of needle biopsy. Core biopsies are indicated to confirm malignancy for nonsurgical treatment options. Regional biopsy involves obtaining several samples of tissue from different locations within a tumour or within a diseased organ. Stereotactic biopsy uses radiographic images to create three-dimensional images of a suspected neoplasm. Surgical procedures are selected for the precise diagnosis of cancer and for defining the stage of the disease. Before recommending a surgical resection of the tumour for cure, the physician undertakes a search for evidence of distant metastasis. A debulking surgical procedure can be used to reduce tumour burden prior to initiating systemic therapy. The decision-making process for creating a treatment plan for cancer depends on: 1- the natural history, 2- the clinical stage, 3- goals of treatment, and 4- indications and risks for all modalities of treatments or combined treatment. Curable desire should be expected throughout the surgical course. Radical surgical resections are performed when the tumour is surgically accessible and there is hope that the tumour can be resected en bloc, along with the necessary local or regional tissues and lymphatics. A radical resection may include the primary tumour and the regional lymph nodes surrounding the area. The en bloc regional lymph node dissection is critical to preventing local tumour recurrence. Surgical resections alter an individual's body image as well as the structure and function of the body. Indications for extensive radical surgery include primary tumours that grow slowly, have a wide infiltration, are large. Surgery is local therapy and thus is limited in what it can achieve as a treatment modality for cancer. Combination or adjuvant treatments are used to improve the rates of cure and disease-free survival. Radiotherapy combined with surgery for local and regional control of the tumour. Chemotherapy provides systemic control of micrometastases and distant metastases. Cytoreductive surgery is used to debulk or reduce the tumour mass to a size in which combination therapy can be most effective. Resection of the metastatic lesion is not indicated if there is evidence of additional metastatic disease or if the metastatic lesion is particularly aggressive or inaccessible. Palliative surgical procedures are aimed at controlling the cancer and improving the quality of life for the individual with cancer, even when all the cancer cannot be removed. The goal of palliative surgery is to relieve suffering and minimise the symptoms of the disease. Several surgical techniques are used for palliation of cancer such as fulguration, electrocoagulation, lasers, photodynamic therapy, shunts, and bone stabilisation procedures. Palliative surgery is useful in relieving suffering caused by an obstructive process. The patient must also tolerate anaesthesia. Surgery may be used to decompress vital structures (laminectomy), or to help in the control of pain. Surgery can be used for rehabilitation, e.g., breast construction following mastectomy, facial reconstruction after head and neck surgery, skin grafts, various implants, microvascular surgery, allografts, and autogenous reconstruction. Reconstructive procedures can be done immediately following resection or can be delayed several days or years. Rehabilitation potential is considered before initiation of primary treatment. Rehabilitative teaching and counselling generally is begun before primary surgical treatment is initiated. Rehabilitation is sometimes necessary for achieving the highest possible level of functioning. Nonanaemic patients can donate up to 6 units of blood prior to surgery. Blood may be donated from 42 days to 72 hours prior to surgery. Surgery accelerates basic energy requirements up to 1.5 times normal. The nutritionally debilitated individual with cancer is a poor surgical candidate and likely to experience severe postoperative complications unless their nutritional status is dealt with aggressively preoperatively. Elevated clotting factors and shortened partial thromboplastin and prothrombin times may occur in patients with cancer. Surgical procedures may become necessary during active radiotherapy or chemotherapy treatment cycles, e.g., insertion of vascular access device. Most chemotherapy acts by interfering with protein synthesis.

Phases of radiation effect:

Phase	Tissues most at risk
Acute effect phase: 1- Hours to days. 2- Proliferating cells are more radiosensitive than quiescent cells. 3- Brisk reactions heal completely with residual damage.	Bladder, bone marrow, colon, oesophagus, lymph nodes, oral mucosa, ovary, testis, salivary gland, skin, small bowel, stomach, and vagina.
Subacute effect phase: 1- Weeks to few months. 2- Clinical significant damage.	Brain, heart, kidney, liver, lung, and spinal cord.
Late effect phase: 1- Manifest late effects. 2- Injury to vasoconnective tissue and parenchymal cells. 3- Occurs in tissues with low cell turnover. 4- Dependent on fractionation, treatment volume, total radiation dose.	Bile ducts, bone, brain, breast, cartilage, lymph tissue, pancreas, pituitary, and thyroid.

The direct radiation effect occurs within the cell, e.g., DNA and RNA damage. Types of DNA damage include 1- change or loss of a base (thymine, adenine, guanine, or cytosine), 2- breakage of the hydrogen bond between the 2 chains of the DNA molecules, 3- breaks in one or both chains of the DNA molecules, and 4- cross-linking of the chains after breakage. The indirect radiation effect occurs when ionisation takes place in the medium surrounding the molecular structures within the cell. Radiation absorbed by the water molecules results in the formation of a free radical when an electron is knocked out of orbit surrounding the ion. Radiation can cause damage to proteins, carbohydrates, and enzymes within the cell. Alterations in the permeability of the cell membrane, may contribute to the ultimate effects of radiation at the cellular level. Radiosensitivity may be maximised during the M and G_2 phases of the cell cycle borders. Radiation doses must be measured through different mediums to accurately determine patient dose. The cell cycle is made of 5 phases G_1, S, G_2, M, G_0. The growth fraction is the portion of cells actively cycling compared to the entire population. Following mitosis, a cell can: 1- enter resting state (G_0) and re-enter the cycle at some later time (stem cells), or 2- enter the G_1 phase and continue to cycle. Synthesis of RNA and proteins occurs predominantly in the G_1 phase. Synthesis (S) phase, is when DNA is being replicated and is a relatively short period compared with the overall time a cell is cycling. The G_2 phase is typically brief, occurring after DNA synthesis and just before cell division. Mitosis, or cell division, ensues during the M phase, resulting in 2 identical daughter cells. The cycling time is the time from mitosis to mitosis. Cells that have left the cycle to enter G_0 are considered to be in a resting or dormant phase. These cells can actively synthesise RNA and proteins and differentiate, they are resistant to the cytotoxic effects of chemotherapy. Tumour cells are characterised by loss of controlled cell division. Pulmonary infiltrates in an immunocompromised patient can represent infection, metastatic disease, recurrence of the primary cancer, or chemotherapy related pulmonary toxicity. 5-20% of patients receiving chemotherapy will have pulmonary disease. History of drug exposure, subjective complaints (dyspnoea, nonproductive cough, and fever), and objective evidence of a significant deterioration from baseline (PFT, ABGs, and X-ray) may represent chemotherapy pulmonary toxicities. Many antineoplastic agents are associated with pulmonary damage. Pathobiology of chemotherapy pulmonary toxicities includes: 1- imbalance of the oxidant-antioxidant system, 2- direct cytotoxic injury to the lung parenchymal cells, 3- capillary leak syndrome, and 4- a hypersensitivity reaction. Chemotherapy can directly injure alveolar endothelial cells and can cause type I epithelial cell necrosis. Increased collagen deposition can result in severe irreversible pulmonary fibrosis. Bleomycin stimulates fibroblast proliferation and directly increases collagen deposition. The capillary leak syndrome, or noncardiogenic pulmonary oedema is rare. Mitomycin C induces haemolytic uraemic syndrome (part of it is the capillary leak syndrome). Mitomycin-induced lung injury may be the result of OFRs effect on the endothelium. Chemotherapeutic agents may stimulate chemokines production, which is responsible for attracting various inflammatory cells (neutrophils and lymphocytes). Lymphocytes and macrophages may then in turn recognise the chemotherapeutic agent as a foreign antigen and elicit a hypersensitivity reaction. The reaction of the type II cells may cause the nuclei to look atypical. The direct cytotoxic changes start as endothelial blebs in the alveolar capillaries associated with interstitial oedema. As the lesions progress, electron microscopy demonstrates a decrease in the number of type I pneumocytes and a proliferation of type II pneumocytes. A lymphocytic predominance occurs in the bronchoalveolar lavage fluid when diffuse alveolar damage is present. Pulmonary injury may present acutely with respiratory insufficiency and ARDS. Chemotherapy-induced pulmonary toxicity includes: bronchospasm, pleural effusion, bronchiolitis obliterans, and pulmonary veno-occlusive disease. High FiO_2 by itself can generate ROS that, in turn cause a cascade of damage to the alveoli and lung parenchyma. Oxygen therapy may act in a synergistic manner with chemotherapeutic agents to injure the lung. Concomitant illnesses may predispose the patient to drug-induced toxicity. Tobacco use appears to sensitise lung tissue to the effects of chemotherapy. Chest x-ray has a low specificity and sensitivity for chemotherapy-induced pulmonary toxicity. Hypersensitivity pneumonitis most often presents as a diffuse acinar infiltrate, pleural effusion are more common in this setting. The synergistic effect of radiation and bleomycin may be manifested as pulmonary toxicity in 2 forms: 1- bleomycin sensitises the tissue to radiotherapy (when bleomycin is given prior to radiotherapy), and 2- bleomycin increases the risk of developing pulmonary toxicity in patients who have had prior radiotherapy (42%). Noncardiogenic pulmonary oedema develops within 18-48 hrs of exposure to high FiO_2 in patients with a history of bleomycin administration in the prior 6-12 months. Bleomycin has more toxic effects on the lung when used in lymphoma patients. Renal impairment increases the risk of bleomycin toxicity. Once pulmonary toxicity occurs, the overall mortality approaches 50%. Mitomycin C-induced pulmonary toxicity occurs in 2-12% of cases, manifested as thrombotic microangiopathy, in which fibrin and platelet thrombi are deposited in pulmonary arterioles and capillaries. Combination of chemotherapy with cyclophosphamide as part of a protocol increases the incidence of pulmonary toxicity. Methotrexate accumulates in the lungs, initiating an immunologic reaction, in which lymphocyte predominance with an increase in helper T cells or an increase in suppressor T cells. Methotrexate-induced pulmonary toxicity injury may be direct cytotoxic effects on the lung. Acute pneumonitis, chemical pleuritis, chronic progressive pulmonary fibrosis, and noncardiogenic pulmonary oedema are clinical presentations of Methotrexate-induced pulmonary toxicity. Chemical pleuritis occurs in 9% of cases and presented as sudden-onset pleuritic chest pain lasting 3-5 days. Methotrexate-induced pulmonary toxicity x-ray feature is bibasilar reticulonodular pneumonitis. IL2 activates lymphocytes causing increased permeability of the endothelial cells with marked mononuclear cell infiltrate in the interstitium and perivascular areas. Concomitant therapy and preexisting lung disease are associated risk factors for IL2 pulmonary toxicity. Mortality rate from IL2-induced pulmonary toxicity is 2%. Intervention for chemotherapy-induced pulmonary toxicity is by discontinuation of the drug, antioxidants, and corticosteroids. Stomatitis is defined as a chemotherapy-related toxicity manifesting as oral cavity ulcers that interfere with one's ability to eat. Sucralfate, an oral, nonabsorbable aluminium salt of sulphated sucrose, has long been used for the treatment of gastrointestinal ulcerations. Its mechanism of action include: 1 it forms a protective mucosal barrier by binding to positively charged proteins on the eroded mucosal surface, and 2- it stimulates production of reparative mucus via prostaglandin synthesis and upregulation of the local circulation. Sucralfate is a relatively inexpensive and well tolerated approach that may be used as a treatment for stomatitis/mucositis. Glutamine, a nonessential amino acid, is used as an energy source for intestinal epithelial integrity. It has been shown previously to reduce the extent of intestinal injury from radiation and chemotherapy in laboratory rats. Oral glutamine has also been shown to reduce the degree of stomatitis in human subjects treated with chemotherapy, and also inhibit radiation-induced mucositis. Povidone-iodine mouth rinse is nonirritating and nontoxic, reduces the incidence, severity, and duration of chemoradiotherapy-induced stomatitis/mucositis. Azelastine, a free-radical scavenger with other antioxidants (vit. C, E, glutathione, and CoQ_{10}) delay the onset of mucositis development and reduce the severity of mucositis. Granulocyte-macrophage colony-stimulating factor (GM-CSF) affects mature leukocytes, macrophages, and dendritic cells as well as fibroblasts and keratinocytes, resulting in more rapid resolution of cuts, burns, skin grafts, and leg ulcers. The use of GM-CSF in high-dose chemotherapy and stem cell transplantation seems to be associated with a lower risk of stomatitis. Subcutaneous GM-CSF given during radiation therapy reduces the severity of acute mucositis. Significant number of patients would complain of fever, bone pain, and skin reactions at the GM-CSF injection site. There is no difference in lowering risk of stomatitis between either sucralfate or sucralfate plus subcutaneous GM-CSF. Most tumours in human persist for months to years without neovascularization until a subset of tumours cells acquires an angiogenic phenotype. Tumour cells may over-express angiogenic factors, may mobilise angiogenic protein from the ECM, may recruit host cells such as macrophages, which then produce their own angiogenic processes. Upregulation of angiogenic factors, is not sufficient in itself for a tumour cell to become angiogenic. Newly formed capillaries have fragmented basement membranes and are leaky, making them more penetrable by tumour cells than mature vessels. In normal tissues endothelial cells are quiescent, whereas in tumours they are activated and proliferating. Antiangiogenesis interfere with a wide range of biological processes involved in growth, migration, and differentiation of blood vessels. The vascular targeting provides rapid destruction and cell death of the vessels, infarcting large areas of the tumour, which require the identification of target molecules that are present at sufficient density on the surface of vascular endothelial cells in solid tumour but absent from endothelial cells in normal tissues. Angiogenesis is critical for the growth of solid tumours and for cell shedding from the primary tumour and the metastatic spread to distant sites. Almost 50% of all cancers have already metastasised by the time of diagnosis. Of all tumour cells that reach the circulation, only a small percentage (<0.01%) will ultimately succeed in initiating metastatic colonies, the majority of tumour cells being eliminated by random events. Evidence from the use of genetic markers suggests that the metastatic subpopulation dominate the primary tumour early in its growth. Some genetic changes lead to unrestrained growth, and additional genetic changes are required for invasion and metastasis. By differentiating invasion from metastasis, sequential steps in the metastatic cascade have been identified and studied separately. Basement membrane is a dense matrix (collagen, glycoproteins and proteoglycans) not normally allowing for passive cell traversal. Invasion of the basement membrane is an active process requiring: a- attachment to the basement membrane, b- matrix dissolution, and c- migration. Attachment is mediated by cell surface receptors of the integrin and non-integrin types. Degradative enzymes are secreted by the tumour cells producing a localised zone of lysis immediately adjacent to the cell surface. Proteolysis plays a crucial role not only during invasion at the primary site but also during intravasation, extravasation and successful invasion and establishment of metastases in distant organs. In the area of matrix proteolysis the tumour cell must migrate through basement membrane and stroma. This cell movement is directional, mediated by ligands binding to the cell surface and inducing a coordinated mobilisation of cytoskeletal elements. Cytokines, which regulate random tumour cell motility include autocrine motility factor (AMF), scatter factor, and autotaxin (ATX). The level of AMF in the urine of bladder cancer patients has been shown to be associated with invasion levels, stage and grade of disease. ATX stimulates both random and directed motility. Its activity appears to be receptor mediated. The action of proteinases is counterbalanced by natural proteinase inhibitors produced either by the tumour cell itself or by the host. The access of tumour cells to the lumen of blood vessel is facilitated within the primary tumour. Because of structural vascular defects the newly formed vessels are abnormally leaky and tumour cells either pass through junctions between endothelial cells or directly traverse the endothelial cells themselves (intracellular passage), which is related to the surface area of tumour vessels. During their passage through the circulation tumour cells use interactions with themselves (homotypic) as well as interactions with blood cells (heterotypic) in order to increase metastatic capacity. Homotypic interaction positively affects implantation in the microcirculation at the metastatic site. By interacting with platelets and fibrin, tumour cells become more insensitive to shear forces and immune effectors cells. Inhibitors of platelet aggregation inhibit metastasis. Heterotypic interaction also leads to coagulation abnormalities with increased turnover of constituents of the clotting system. Haemostatic abnormalities indicative of disseminated intravascular coagulation (DIC) are commonly detected in patients with various metastatic malignancies. Tumour cells also activate the platelets. The platelets store lytic enzymes, serotonin and growth factors as well as arachidonic acid (AA) metabolites known to increase metastasis. The ability to generate proteolytic and adhesive properties is probably essential for invasion, implantation and angiogenesis. Cancer screening programs claim to result in a reduction of patient mortality of up to 30%, for certain types of malignancy where primary tumours grow to a relatively large size before metastatic disseminations. In other tumour systems, 90% of metastases appear to be present when the age of the primary tumour is 43 doubling times equals to 6 mm size. In 1889 Paget proposed the seed and soil hypothesis. Tumour cells adhere to junctional regions between endothelial cells (basal lamina) and cause endothelial cell retraction. There is a correlation between the levels of membrane-bound PKC activity and haematogenous metastasising abilities of tumour. Angiogenesis is a complex event starts with the stimulation of resting endothelial cells in the parent vessel to degrade the basement membrane. This is followed by endothelial cell migration into the perivascular stroma and by formation of a capillary sprout. Proteinase inhibitors block both endothelial cell invasion and tumour cell invasion. During the final steps of dissemination the release of proteolytic enzymes (hydrolases, collagenases, cathepsins, and plasminogen activators) and active tumour cell motility completes metastasis formation. mRNA levels of nm23 are dramatically reduced in tumour cells of high metastatic capacity. Loss of nm23 RNA is strongly associated with poor survival. It is assumed that nm23 is involved in signal transduction of cell-cell communication and thereby plays a critical role in normal tissue development. By affecting microtubule assembly, nm23-like NDP kinases may regulate cellular functions such as mitotic spindle formation and cell locomotion. This would also explain the high degree of aneuploidy (genetic instability) observed in metastatic tumours due to aberrant mitosis. Treatment strategies for tumour invasion should include: 1- inhibition of tumour cell invasiveness, 2- suppression of angiogenesis, 3-

enhancement of local stromal reaction, 4- enhancement of tumour cell immunogenicity/activation of the immune system, 5- inhibition of tumour cell attachment, 6- antibody-mediated selective tumour cell killing. The median metastasis growth duration is 18 doubling times or 3.7 years before metastases become clinically detectable. Strategies to block tumour angiogenesis, invasion, and metastasis include: A- Inhibition of angiogenesis: α-interferon was the first angiogenesis inhibitor to be tested in clinical trials. Antiangiogenic therapy strategies include: 1- the inhibition of basic fibroblast growth factor secretion by tumour cells; 2- the use of an antibody to block the interaction of an angiogenic factor with its receptor; 3- antiendothelial angiogenesis inhibitors; 4- collagenase inhibitors, such medroxyprogestrone; 5- inhibitors of basement-membrane turnover; 6- angiostatic steroids, which exert their effects by: a) increasing the dissolution of the basement-membrane components specifically in growing capillary vessels, b) inhibit endothelial cell migration and proliferation, and c) reduce plasminogen-activator activity. An angiostatic steroid does not require heparin for its potentiation; 7- fungus-derived angiogenesis inhibitors, AMG 1470 reduces tumour growth with little or no toxicity; 8- inhibitors of arachidonic acid metabolism, different synthetic derivatives of these inhibitors display different activity on tubulogenesis and angiogenesis; 9- platelet factor 4, binds and neutralises heparin. Thrombospondin is another heparin-binding protein. Both have antiangiogenic activity; 10- D-penicillamine and gold thiomalate, inhibit endothelial cell proliferation, also gold thiomalate inhibits the release of angiogenic stimuli by macrophages; 11- vitamin D3 analogues; 12- interferon, α-interferon was the first angiogenesis inhibitor to be tested in clinical trials. The mechanism of action of interferons is due to: a) inhibitory effect on endothelial cell migration, and b) inhibition of angiogenesis. B- Inhibition of invasion: The invasive capacity of cancer cells is necessary not only for migration from the primary site but also for the formation of metastatic secondary tumours.

Classification of angiogenic inhibitors:

A- Inhibitors and endogenous antagonists of angiogenic growth:
    1- Antibodies to bFGF.
    2- Antibodies to VEGF.
    3- Angiogenin antagonist.
    4- Suramin and analogues.
    5- Sulfonic derivatives of disamycin A.
    6- Recombinant platelet factor 4.
    7- Angiostatin.
    8- Retinoids.
    9- Vitamin D and analogues.

B- Inhibitors of endothelial cell growth and/or migration:
    1- AGM-1470 (angioinhibin or TNP-470).
    2- Quinoline-3-carboxamide (linomide).
    3- Tamoxifen.
    4- D-Penicillamine.
    5- Genistein.
    6- Thalidomide.
    7- IFNs.
    8- IL12.

C- Angiogenic inhibitors targeting the basement membrane and ECM:
    1- Antibody anti-$\alpha v \beta_3$.
    2- Metalloproteinase inhibitors (batimastat).
    3- Minocycline.
    4- PAI-1.
    5- Aurintricarboxylic acid.

D- Potential new antiangiogenic therapeutics:
    1- NOS inhibitor.
    2- Thrombospondin antagonist.
    3- fps/fes inhibitor.

For successful invasion to occur, tumour cells must be capable of initiating the following steps: a) attachment, mediated by cell-surface receptors recognising membrane glycoproteins; b) matrix dissolution, achieved by secretion of degradative enzymes; and c) cell migration, mediated by ligands binding to the cell surface and inducing a coordinated mobilisation of cytoskeletal elements. Attachment to the extracellular matrix, migration, and adherence to other cells are modulated by cell-adhesion molecules (CAMs). Degenerative enzymes, producing a localised zone of lysis adjacent to the cell surface. Cell migration into the region of the matrix modified by proteolysis has been shown to be directional, and cytokines regulating tumour-cell motility have been identified. Inhibition of tumour-cell adhesion: Targeting of an adhesion molecule with a monoclonal antibody is expected to result in a highly specific inhibitory effect on binding to one substrate only. Selectins present on endothelial cells and platelets, are ligands for carbohydrate structures present on leukocytes. They play an important role during physiological processes such as inflammation. Increased expression of E-selectin, induced by cytokines, results in stronger adhesion of leukocytes to endothelial cells. Tumour cells could utilise selectin-carbohydrate interaction when they adhere at metastatic sites. Tumour cells aggregate with platelets through the binding of P-selectin on platelets to the carbohydrates on tumour cells. Tumour cells platelet aggregation is an accepted mechanism that facilitates entrapment in capillary tubes. Glycosphingolipids on tumour-cell surfaces play an important role in metastatic spread through a possible interaction with adhesive events and, perhaps, tumour-cell migration. Inhibition of proteolysis: Following the attachment of tumour cells to the basement membrane, proteolytic enzymes released or induced by the tumour cells digest the matrix and thus promote invasion. Tumour-cell binding to laminin, through its receptors, enhances the production of type IV collagenase in tumour cells. Antibody-mediated inhibition of urokinase binding to the cell-surface receptors reduces invasion significantly. ITI (inter-α-trypsin inhibitor) is the precursor of UTI (urinary trypsin inhibitor), which inhibits trypsin, chemotrypsin, plasmin, elastase, and cathepsins B and L. The invasive capacity of cancer is partially inhibited by an anti-uPA (urokinase-type plasminogen activator) antibody as well as by a specific inhibitor of cysteine proteases. Type IV collagenase levels have been positively correlated with increasing histopathological grade in human prostate cancer tissues. The action of metalloproteinases is counterbalanced by natural inhibitors such as TIMP (tissue inhibitor of metalloproteinase), a glycoprotein secreted by fibroblasts, endothelial cells, and chondrocytes.

Endogenous mediators of angiogenesis:

Angiogenic stimulators	Angiogenic inhibitors
FGFs	Angiostatin
VEGF	IFN
Angiotenin	PF4
TGFα	Thrombospondin
	Prolactin
TNFα	bFGF soluble receptor
Platelet-derived endothelial cell growth factor	TGFβ
TGFβ	TIMPs
Placental growth factor	Placental proliferin-related peptide
IL8	Glioma-derived angiogenesis inhibitory factor
Hepatocyte growth factor	
PDGF	
GSF	
PGE_2, PGE_1	

Upon transfection of the TIMP gene into tumour cells with a high-level expression of metalloproteinases, a striking inhibitory effect on local invasion and partial suppression of haematogenous metastasis is noticed. Modulation of TIMP activity might influence progression in some tumour systems. Inhibition of tumour-cell migration: A tumour-invasion-inhibitor factor (IIF-2) appears to bind to the cell surface, eliminating the invasive ability of the cells. IIF-2 does not affect tumour-cell adhesion, degradation of the extracellular matrix, or cell growth. Autocrine motility factor (AMF) is completely inhibited by IIF-2. An albumin conjugate of IIF-2 can inhibit colonization. AMF in the urine of bladder-cancer patients has been associated with invasion. C- Inhibition of metastasis: Products of the AA cascade have been associated with inhibition and promotion of malignancy. Tumourigenic phorbol esters have been shown to stimulate the production of PGE_2 and PGF_2, which can then be inhibited by treatment with cyclooxygenase inhibitor such as indomethacin. Inhibitors of prostaglandin (PG) synthesis that lead to the inhibition of cancer progression may be exerted by their inhibitory effect on platelet aggregation. Intravascular aggregation between platelets and tumour cells might promote metastatic spread by protecting tumour cells from the external milieu. Aspirin (cyclooxygenase inhibitor, inhibiting PG synthesis) was tested, but no effect was demonstrated. The reduction of PA activity by angiostatic steroids not only interferes with angiogenesis but might also directly inhibit tumour-cell invasion. A proteolytic cascade inhibited by PAs results in basement-membrane dissolution. Inhibitors of AA metabolism display antiangiogenic activity and simultaneously interfere with the process of initiation and promotion of malignancy. Inhibition of PG synthesis (metabolic products of AA) might interfere with heterotypic intravasal aggregation between platelets and tumour cells, which might facilitate metastatic dissemination. It appears likely that different tumours will respond differently to antiangiogenic therapy. The nm23 suppressor gene may interfere with the transforming growth factor (TGF)-β signal-transduction pathway for the stimulation of metastatic dissemination. The mechanism by which nm23 regulates metastasis may therefore involve the suppression of angiogenesis. Vitamin A and its analogues (retinoids) are actively studied as chemopreventive as well as therapeutic agents. Retinoids act as transcriptional regulators and generally inhibit growth and promote epithelial cell differentiation. The synthetic retinoid n-4-hydroxyphenyl retinamide (4HPR) has been shown to suppress chemical carcinogenesis in the bladder and prostate, part of its antitumour activity is due to the inhibition of angiogenesis.

ANTIOXIDANTS FOR CANCER CHEMOTHERAPY, AND CANCER CHEMOPREVENTION

Dietary supplementation with antioxidants may provide a safe and effective means of enhancing the response to cancer chemotherapy. Quality of life (QoL) of patients after chemotherapy may be improved by dietary supplementation with antioxidants that reduce or prevent chemotherapy-induced side effects. Free radicals and other reactive oxygen species (ROS) are essential for life, because they are involved in cell signalling and are used by phagocytes for their bacteriocidal action. ROS are also produced in all respiring organisms as a consequence of mitochondrial respiration, which consumes oxygen in the process of generating ATP by the coupling of electron transport and oxidative phosphorylation. Nonessential production of ROS, i.e., oxidative stress, can also be induced by exogenous factors such as drugs and environmental toxins. Oxidative stress is potentially harmful to cells, and ROS are implicated in the aetiology and progression of many disease processes including cancer. Oxidative stress reduces the rate of cell proliferation, and that occurring during chemotherapy may interfere with the cytotoxic effects of antineoplastic drugs, which depend on rapid proliferation of cancer cells for optimal activity. Antioxidants detoxify ROS and may enhance the anticancer effects of chemotherapy. ROS cause or contribute to certain side effects that are common to many anticancer drugs, such as gastrointestinal toxicity and mutagenesis. ROS also contribute to side effects that occur only with individual agents, such as doxorubicin-induced cardiotoxicity, cisplatin-induced nephrotoxicity, and bleomycin-induced pulmonary fibrosis. Antioxidants can reduce or prevent many of these side effects, alopecia and myelosuppression are not prevented by antioxidants. Under conditions of excessive oxidative stress, however, cellular antioxidants are depleted and ROS can damage cellular components and interfere with critical cellular activities. The cell cycle consists of a presynthetic phase (G_1), which precedes DNA synthesis, the phase of DNA synthesis (S), an interval that follows DNA synthesis (G_2), and mitosis (M), during which the G_2 cell with a double complement of DNA divides into two daughter cells. As the length of the cell cycle increases, the duration of the G_1 phase increases, whereas the durations of the S, G_2, and M phases remain relatively constant. The length of the cell cycle can be influenced by many factors, e.g., oxidative stress. During oxidative stress, the excessive production of ROS results in lipid peroxidation. Because the rate of DNA synthesis and the rate of cell proliferation, of cancer cells and normal cells, are inversely related to the degree of lipid peroxidation, oxidative stress prolongs the G_1 phase or may result in cells entering the G_0 phase. Antineoplastic agents act by blocking the synthesis of DNA precursors, damaging the integrity of DNA, or interfering with DNA replication, separation of the two double helices after replication, or the function of the mitotic spindle. Tumour cells in the nonproliferative G_0 state are little affected by anticancer drugs and can re-enter the division cycle after chemotherapy is completed, resulting in recurrence of disease. Tumour cells with prolonged G_1 phases are resistant to the cytotoxic effects of antineoplastic agents.

Major ROS of oxidative stress:

Free radicals		Nonradical oxygen species	
Hydroxyl radical	HO.	Hydrogen peroxide	H_2O_2
Superoxide radical	O_2^-	Singlet oxygen	$^1O_2$
Peroxyl	ROO.		
Alkoxyl	RO.		

Phases of the cell cycle:

Phase	Description	Phase-specific drugs
G_1	Preparation for DNA synthesis and cell division.	
S	DNA synthesis:	Methotrexate. Cytosine arabinoside.
G_2	Preparation for mitosis	
M	Mitosis (prophase, metaphase, anaphase, telophase):	Vincristine. Vinblastine. Paclitaxel.
G_0	A nonproliferative phase	

Anticancer drugs are cytotoxic only when tumour cells are proliferating rapidly. This rapid proliferation provides a mechanism whereby oxidative stress that slows or arrests cell growth, interferes with chemotherapeutic effectiveness. This rapid proliferation explains why slow-growing tumours, such as lung and colon carcinoma, (many cells in the G_0 phase for prolonged periods of time) are relatively unresponsive to chemotherapy. Cancer cells have highly evolved protective mechanisms to prevent lipid peroxidation so that rapid cell proliferation can occur. The primary mechanism whereby cancer cells prevent lipid peroxidation consists of a marked increase in vitamin E relative to the peroxidizable moieties (methylene groups) of the PUFA in their biological membranes. Tumour cells also have relatively low levels of the components of the NADPH-cytochrome P-450 electron transport chain, which results in less favourable conditions for the initiation and propagation of lipid peroxidation. Excessive oxidative stress may overcome even the highly evolved protective mechanisms of cancer cells, resulting in inhibition of cancer cell proliferation and interference with the cytotoxic activity of antineoplastic agents. Cancer imparts an oxidative stress on the host organism. There is a rapid elevation of lipid peroxidation products and impaired antioxidant status with various tumours. The impairment of antioxidant status may be due, in part, to poor nutrition. Tumour cells are able to overcome the level of oxidative stress that is induced by cancer and maintain their rapid rate of proliferation (highly effective antioxidant systems). Cancer cells-induced oxidative stress accelerates the neoplastic process by playing a role in the promotion and progression of cancer (high level of ROS in the host organism). Cancer chemotherapy results in a much greater degree of oxidative stress than that induced by cancer itself. This is illustrated by: 1- the elevation of lipid peroxidation products, 2- the reduction of total radical-trapping capacity of blood plasma, the marked reduction in plasma levels of antioxidants such as vitamin E, vitamin C, and β-carotene, and 3- the marked reduction of tissue glutathione (GSH) levels that occurs during chemotherapy with essentially all antineoplastic drugs. The high level of oxidative stress during chemotherapy may overcome the antioxidant defences of cancer cells, resulting in lipid peroxides that reduce or halt cancer cell proliferation and interfere with antineoplastic activity. ROS are not involved in the mechanism of action of most anticancer drugs. The scission of DNA by bleomycin involves an activated species of oxygen; it also requires the presence of an electron donor, a function that can be fulfilled by certain antioxidants. Oxidative stress resulting from chemotherapy-induced ROS contributes to or is responsible for many adverse effects of the treatment, including impairing antineoplastic activity. ROS have been implicated, in the aetiology of doxorubicin-induced cardiotoxicity, bleomycin-induced pulmonary fibrosis, and nephrotoxicity, neurotoxicity, and ototoxicity resulting from cisplatin administration. Chemotherapy causes a generalised mucositis of the GIT, which contributes to nausea and vomiting and causes diarrhoea, stomatitis, decreased resistance to the penetration of pathogens through the GI barrier resulting in an increased risk of infection (bacterial translocation), enhanced absorption of potentially toxic macromolecular substances, and impaired absorption of nutrients. Also, the mucositis is due to the cytotoxic effect of antineoplastic agents on the rapidly proliferating cells of the GI mucosa. A delayed side effect of chemotherapy is the development of secondary malignancies, most commonly acute leukaemias, which usually occur with a latency of 4-5 years after treatment of the original cancer. Secondary malignancies are frequently more refractory to treatment than the original. ROS and lipid peroxidation products are mutagenic and function as the mediator of the carcinogenic activity of many substances. ROS generated during chemotherapy play a significant role in the development of secondary malignancies. Antioxidants prevent the mutagenic and carcinogenic effects of antineoplastic agents. Myelosuppression and alopecia result from the cytotoxic effect of chemotherapy on bone marrow progenitor cells and cells of hair follicles. Vitamin E may enhance the myelosuppression resulting from the administration of doxorubicin. Vitamin E resides in the lipid domain of biological membranes and plasma lipoprotein, where it prevents lipid peroxidation of PUFA. Vitamin E is the most important antioxidant for preventing lipid peroxidation. Vitamin E reverses the inhibitory effect of lipid peroxidation on cell proliferation. Vitamin E enhances the cytotoxic effect of several anticancer drugs, including 5-fluorouracil (5-FU), doxorubicin, vincristine, dacarbazine, cisplatin, and tamoxifen. Vitamin E may enhance antineoplastic activity by inducing apoptosis via inhibition of protein tyrosine kinases (PTKs). Vitamin E is a protective agent against the cardiotoxicity of doxorubicin. Chronic cardiotoxicity, which is dose dependent, results in the development of a cardiomyopathy and congestive heart failure (CHF) that is unresponsive to digitalis. Vitamin E prevents doxorubicin-induced lipid peroxidation, cardiac lipid peroxidation and the elevation of serum lipid peroxides. Parenteral administration of 20-4,100 IU/kg vitamin E reduces the acute cardiotoxicity of doxorubicin, but not chronic doxorubicin-induced cardiotoxicity. Vitamin E reduces bleomycin-induced chromosomal breakage (secondary tumours). Vitamin E at 1,667 IU/kg also reduces cisplatin-induced nephrotoxicity. There is dysregulation of cytoplasmic calcium during doxorubicin administration. 200 IU of vitamin E intramuscularly 6 hours before treatment and nifedipine (60 mg/day PO) for 2 days before treatment may be protective against acute doxorubicin-induced cardiotoxicity. Vitamin E (2,000 IU/m^2/day PO), offers no protection from the development of chronic doxorubicin-induced cardiotoxicity. Intensive topical treatment with vitamin E (400 IU/ml, 1 ml twice daily) may facilitate the healing of chemotherapy-induced stomatitis. Vitamin C protects against lipid peroxidation by scavenging ROS in the aqueous phase before they can initiate lipid peroxidation. Vitamin C enhances the cytotoxic activity of doxorubicin, cisplatin, paclitaxel, dacarbazine, 5-FU, and bleomycin. Vitamin C increases drug accumulation and partially reverses vincristine resistance of human non-small-cell lung cancer cells. 2,000 mg/kg/day (in animal) of vitamin C prevented doxorubicin-induced lipid peroxidation and reduced the acute cardiotoxic effect of doxorubicin. Vitamin C protects against chemotherapy-induced mutagens. Supplementation with 1,000 mg of vitamin C significantly reduces the chromosomal damage in human. CoQ_{10} (Ubiquinone) is an indispensable cofactor in the electron transport chain of mitochondria, functioning as an electron carrier between the enzyme complexes of the respiratory chain. CoQ_{10} is a lipid-soluble antioxidant that scavenges lipid radicals within biological membranes. CoQ_{10} within mitochondria have an important role in preventing lipid peroxidative damage of mitochondrial membranes. CoQ_{10} prevents chronic doxorubicin-induced cardiotoxicity. Chronic cardiotoxicity mechanism involves the production of oxidising agents through an iron-dependent process and free radicals by doxorubicin, which results in mitochondrial lipid peroxidation within myocardial cells. Other effects of doxorubicin on the mitochondria of cardiac myocytes include: 1) reduction of the CoQ_{10} content of mitochondrial membranes, 2) inhibition of respiratory chain CoQ_{10}-dependent enzymes that interferes with the aerobic generation of ATP, and 3) inhibition of mitochondrial biosynthesis of CoQ_{10}. These effects of doxorubicin on CoQ_{10} biosynthesis and function may explain the acute and chronic forms of doxorubicin-induced cardiotoxicity. A single dose of doxorubicin can induce acute doxorubicin-induced cardiotoxicity (reversible side effect). Acute doxorubicin-induced cardiotoxicity may be due to inhibition of respiratory enzymes by doxorubicin, which may result from competition between CoQ_{10} and doxorubicin for the coenzyme's enzymatic sites, because both compounds contain a quinone group. Enzyme inhibition may also result from oxidation of CoQ_{10} by doxorubicin or doxorubicin-induced ROS, thus preventing CoQ_{10} from functioning as a cofactor for electron transport. Inhibition of the respiratory generation of ATP by doxorubicin would interfere with the electrophysiological activity of the heart. Chronic doxorubicin-induced cardiotoxicity may be attributable to depletion of mitochondrial CoQ_{10} resulting from inhibition of CoQ_{10} biosynthesis by doxorubicin. CoQ_{10} depletion disrupts electron transport and mitochondrial respiratory bioenergetics, and ultimately lead to loss of mitochondrial integrity and necrosis of cardiac myocytes. Mitochondrial degeneration is the earliest and most prominent ultrastructural change associated with the chronic cardiomyopathy induced by doxorubicin. Supplementation with CoQ_{10} (1-15 mg/kg/day) reverses doxorubicin-induced enzyme inhibition and restores the ATP-generating capacity of mitochondria. Vitamin E and vitamin C are able to counter doxorubicin-induced oxidative stress, thus preserving mitochondrial CoQ_{10} for its role in the redox reactions of the electron transport chain. Like vitamin E, CoQ_{10} effectively reduces doxorubicin-induced plasma and tissue lipid peroxidation. CoQ_{10} does not interfere with the antitumour activity of doxorubicin, nor does it prevent doxorubicin-induced myelosuppression. CoQ_{10} does not interfere with doxorubicin-induced suppression of DNA synthesis. CoQ_{10} is safe and effective means of reducing or preventing the development of the cardiomyopathy associated with chronic administration of anthracyclines. β-Carotene is one of >600 carotenoids that are produced by microorganisms and plants. β-Carotene is lipid-soluble antioxidant, exists as a mixture of cis- and trans-isomers. Cis β-Carotene is a more effective antioxidant than is trans β-Carotene. β-Carotene (5-50 mg/kg) reduces the genotoxicity of cyclophosphamide, mitomycin C, methyl methane-sulfonate, and bleomycin. Carotene reduces the rate of tumour induction by chemotherapy. Glutathione (GSH), a tripeptide of glutamic acid, cysteine, and glycine (GluCysGly), is the major water-soluble antioxidant in the cytoplasm, nuclei, and mitochondria of cells. The antioxidant functions of GSH require GSH peroxidase, which exists in several forms. Reduction of oxidised GSH (GSH disulfide), which is produced by reactions involving GSH peroxidase, requires GSH reductase. GSH is not transported into cells. In order for circulating GSH to increase intracellular GSH concentrations, it must first be hydrolysed to Glu and CysGly, which are subsequently transported into the cell and serve as substrates for GSH synthesis. Thus GSH administered orally or parenterally and that produced by the liver and released into the circulation enhance tissue levels of GSH by providing a source of its constituent amino acids. GSH monoesters are well absorbed after oral administration, are readily transported to cells and then hydrolysed to GSH and the corresponding alcohol. Higher cellular levels of GSH result from oral administration of GSH monoesters than from oral administration of comparable doses of GSH. GSH contains a thiol (sulfhydryl, a -SH moiety) group. Thiols are strongly nucleophilic and form stable covalent compounds with electrophilic compounds, such as the platinum coordination complexes cisplatin and carboplatin. Formation of the thiol-platinum complex inactivates the antineoplastic agent, which blocks not only its ROS-generating activity but also its cytotoxic effects. GSH and cisplatin are cleared rapidly from the circulation. Separation of GSH and cisplatin administration by a short period of time prevents the intravascular inactivation of cisplatin. GSH is taken up far more by normal cells (especially kidney cells) than by cancer cells, since only normal cells possess the enzyme (γ-glutamyl transpeptidase) that hydrolyses GSH to Glu and CysGly and transports these products into the cell. This difference between normal cells and cancer cells accounts for selective protection of normal cells by GSH. GSH or GSH mono-ethyl ester protect against free radical-induced hepatic venoocclusive disease secondary to high doses of BCNU or cyclophosphamide, without altering the myelosuppressive or antineoplastic activities of the antineoplastic drugs. The expected degree of neurotoxicity or nephrotoxicity is diminished and the dose of cisplatin can be increased to 175% of the standard dose before the expected degree of nephrotoxicity occur, when intravenous infusion of GSH (1.5-3.0 g/m^2 over 15 minutes) is giving shortly before administration of cisplatin. The decrease in creatinine clearance, the development of peripheral neuropathy, and the severity of hair loss are significantly reduced in the patients receiving GSH. The rapid and selective uptake of GSH by normal cells is most likely accounting for the lack of its interference with the antineoplastic activity of cisplatin. The clinical response to cisplatin is improved by administration of GSH.

Administration of GSH with cisplatin may provide a means of improving chemotherapeutic responsiveness by allowing dose escalation or a greater number of chemotherapy cycles. N-acetylcysteine (NAC) is a nucleophilic thiol compound, a free radical scavenger, and is well absorbed after oral administration and readily transported into cells, where it is deacetylated. NAC provides an intracellular source of cysteine, a substrate for GSH synthesis. NAC (like GSH) can form stable covalent compounds with electrophilic alkylating agents and platinum coordination complexes, which inactivate the anticancer drugs. NAC blocks the cytotoxicity and ROS-generating capability of cisplatin. NAC prevents haemorrhagic cystitis that results from administration of cyclophosphamide or ifosfamide. Haemorrhagic cystitis results from the toxic effect of acrolein, a metabolic product of cyclophosphamide and ifosfamide, on the bladder mucosa. The mechanism whereby NAC prevents this toxicity may be prevention of the intracellular depletion of antioxidants, such as GSH, by acrolein. Concomitant administration of NAC with cyclophosphamide or ifosfamide does not impair antineoplastic activity, because both anticancer drugs are inactive until they are metabolised by the liver to their phosphoramide mustard metabolites. Maintenance of cellular antioxidant systems prevents the acute form but not the chronic form of cardiotoxicity resulting from doxorubicin administration. NAC (50-2,000 mg/kg), by means of free radical scavenging or by enhancing intracellular levels of GSH, protects against acute doxorubicin-induced cardiotoxicity without interfering with the antitumour activity of the drug, but not against chronic doxorubicin-induced cardiotoxicity. NAC reduces cisplatin-induced nephrotoxicity by formation of an NAC-cisplatin complex, and protects against bleomycin-induced genotoxicity. There is no apparent impact of NAC on the development of myelosuppression or the antitumour response to ifosfamide. NAC can provide some protection from the haemorrhagic cystitis induced by ifosfamide. NAC may be protective against the acute but not the chronic form of doxorubicin-induced cardiotoxicity. Glutamine is an essential amino acid that does not possess antioxidant activity, but it serves as a source of glutamate for GSH synthesis, thus supporting cellular antioxidant systems. Glutamine is a primary fuel source for the rapidly proliferating enterocytes of the GIT. Glutamine supplementation during chemotherapy reduces GI injury that results from administration of antineoplastic drugs, especially the severe mucositis that results from treatment with antimetabolites such as 5-FU and methotrexate. Glutamine enhances the antitumour effectiveness of methotrexate, by increasing the intracellular tumour concentration of methotrexate. Oral glutamine, but not intravenous glutamine reduces the bacteraemia and mucosal injury associated with methotrexate-induced enterocolitis. Glutamine, administered by intragastric infusion, accelerates healing of the gut mucosa in rats receiving 5-FU. Glutamine is a GI protectant. Oral (30 g/day), but not intravenous, glutamine attenuates the increased GI permeability that results from chemotherapy (cisplatin and 5-FU) plus radiation therapy. Glutamine (swish-and-swallow) significantly reduces the duration and the severity of stomatitis with chemotherapy regimens. The severity and duration of diarrhoea resulting from chemotherapy with an anthracycline, etoposide, and cytosine arabinoside were significantly reduced in patients by oral administration of 6 g of glutamine 3 times daily beginning 3 days before initiation of chemotherapy and continuing for an average of 18 days. Inorganic selenium does not have antioxidant properties. Selenium has an important role in cellular antioxidant defences as a necessary component of selenoproteins. Selenium is incorporated into selenoproteins as selenocysteine. The GSH peroxidases are the best-characterised selenoproteins, but other circulating selenoproteins also have antioxidant functions. Organic and inorganic selenium protects against cisplatin-induced nephrotoxicity, without apparent inhibition of the antineoplastic activity of cisplatin. Also selenium administration allows for higher doses of cisplatin to be used. Selenium administration reduces cisplatin-induced myelosuppression. Selenium, with chemical properties similar to those of sulfur, can bind with platinum and inactivate the antineoplastic platinum coordination complexes. Selenium inhibits doxorubicin-induced decreases in myocardial vitamin E and GSH peroxidase levels and to reduce changes in myocardial function that are consistent with acute doxorubicin-induced cardiotoxicity. Selenium enhances cellular antioxidant systems. Selenium does not protect against chronic doxorubicin-induced cardiotoxicity. Selenium administration is associated with less leukopenia, less use of granulocyte colony-stimulating factor, and less need for blood transfusions. Soybeans contain a number of isoflavones, including genistein and daidzein. Soybean isoflavones enhance cellular antioxidant status by scavenging ROS and by increasing the activity of antioxidant enzymes including GSH peroxidase, GSH reductase, and superoxide dismutase. Genistein abrogates chemical-and ligand-induced generation of ROS. Genistein may enhance the antineoplastic effects of cancer chemotherapy. Genistein is an inhibitor of topoisomerase I and II. Genistein has been shown to induce apoptosis in a number of tumour cell lines (inhibit PTKs). Genistein also inhibits the binding of ATP to its binding site on the enzyme. Administration of genistein during chemotherapy with doxorubicin or etoposide may enhance antineoplastic activity. Enhanced antineoplastic activity may also result from inhibition of topoisomerase I by genistein resulting in arrest of the cell cycle in the S phase. Genistein has well-established antioestrogenic property. Genistein may be of benefit not only during the initial chemotherapeutic management of breast cancer, but also during long-term treatment with the antioestrogenic tamoxifen (synergism). Genistein enhances the accumulation of cisplatin and doxorubicin in accumulation-defective resistant cancer cells and increases the accumulation of cisplatin in non-resistant cancer cells. Soybean products reduce methotrexate-induced damage of the GI mucosa, effects that can be attributed to antioxidant isoflavones and the glutamate content of soy protein. Daidzein has a cytostatic effect on human gastric cancer cells (arrest of the cell cycle in the G1 phase). Daidzein does not inhibit topoisomerase II. Quercetin is ubiquitous polyphenolic flavonoid. Quercetin is antioxidant. Quercetin inhibits topoisomerase II activity, it enhances topoisomerase II-dependent DNA cleavage and the formation of stable protein-DNA cleavage complexes. Quercetin inhibits topoisomerase II-catalysed ATP hydrolysis. Quercetin inhibits several PTKs. The standard recommendation for daily vitamins intake are based on the amount required to avoid deficiency diseases and not necessarily the amount for optimal health. Cancer prevention includes: 1- primary prevention, e.g., avoiding cancer-causing substances, or supplementation with dietary protective agents. 2- Secondary prevention, e.g., detection and removal of benign lesions, or screening. Tobacco smoking and alcohol are synergistic and associated with cancer at various sites. Whole dietary regimens directed at tumour inhibition have involved lowering dietary fat content, increasing intake of fibre, fruit, and vegetables, and the administration of specific micronutrients, including most of the naturally occurring nutrients. When studying natural dietary regimens, the preclinical studies and phase I toxicology studies can generally be omitted. Malnourished cancer patients have increased risks of morbidity and mortality when undergoing major surgical procedures. Weight loss may be defined as loss of >10% of pre-illness weight. No single measurement is sensitive and specific for the identification of malnutrition. The nutritional risk index may be assessed by monitoring the albumin level and degree of preoperative unintended weight loss. Malnutrition is generally defined by a score of 100 or less.

The nutritional risk index = 1.59 X serum albumin level (g/l) + 0.417 X (current weight/usual weight) x 100

Patients who are unable to eat for an indefinite period of time (generally 7-10 days) require nutritional support. Patients suffering major trauma or undergoing bone marrow transplantation benefit from supplemental nutrition.

Antineoplastic Agents

Drug	Mechanism of Action	Toxicities
Most of the agents induced-toxicities include myelosuppression, alopecia, nausea, vomiting, anorexia, and secondary malignancies.		
1- Alkylating Agents:		
A- Cyclophosphamide	Alkylation of DNA	Cystitis GI ulceration
B- Ifosfamide	Alkylation of DNA	Cystitis CNS disturbances.
C- Busulfan	Alkylation of DNA.	1- Pulmonary fibrosis. 2- Hepatic venoocclusion.
D- BCNU (N,N-bis(2-chloroethyl)-N-nitrosourea)	Alkylation of DNA.	1- Pulmonary fibrosis. 2- Hepatotoxicity.
E- Dacarbazine	Alkylation of DNA.	hepatotoxicity
2- Platinum Coordination Complexes:		
A- Cisplatin	DNA cross-linking	1- Nephrotoxicity. 2- Peripheral neuropathy. 3- Ototoxicity.
B- Carboplatin	DNA cross-linking.	1- Nephrotoxicity. 2- Peripheral neuropathy. 3- Ototoxicity.
3- Antimetabolites:		
A- Methotrexate	Dihydrofolate reductase inhibition.	Mucositis.
B- 5-Fluorouracil	Thymidylate synthase inhibition.	Mucositis
C- Antimitotic Drugs		
D- Vincristine	Disruption of microtubules.	Peripheral neuropathy.
E- Vinblastine	Disruption of microtubules.	Peripheral neuropathy.
F- Paclitaxel	Disruption of microtubules.	1- Peripheral neuropathy. 2- Mucositis.
4- Antibiotics:		
A- Doxorubicin	1- DNA intercalation. 2- Topoisomerase II inhibition.	Cardiomyopathy
B- Bleomycin.	DNA fragmentation.	1- Pulmonary fibrosis. 2- Cutaneous toxicity.
5- Epipodophyllotoxins:		
Etoposide	Topoisomerase II inhibition	1- Mucositis. 2- Hepatotoxicity.

Patients with immunosuppression may benefit from nutritional manipulation. Severe malnutrition is defined as having weight losses >10-15% or albumin levels <2.8 g/dl, will benefit from preoperative and postoperative nutrition. Nutritional support of the surgical oncology patient is possible on many different levels (parenteral and enteral). Additional indications for nutritional support may include the support of patients undergoing chemotherapy or radiotherapy who would otherwise be unable to complete the course of treatment without nutritional supplementation. Immunosuppression is a major factor in the cancer patient. Tumour burden, malnutrition, and surgery, with its attendant anaesthesia and blood transfusion, all depress the immune system. Patients with cancer have depressed both cellular and humoral immune functions. T-cell proliferative responses to both mitogen and alloantigen are reduced with increasing tumour burden. Patients with advanced disease often lack a delayed type hypersensitivity reaction to skin recall antigens (i.e., anergy), the absence of this response being predictive of increased sepsis and mortality. T-cell activation is the primary abnormality in the delayed type hypersensitivity response. Cytotoxic T-lymphocytes have a decreased cytotoxicity to autologous tumour. Operative therapy intended to eradicate tumour occasionally promotes metastatic growth. Laparotomy abrogates the antitumour effect of lymphocyte activated killer cells (LAK). Perioperative blood transfusion is associated with poorer prognosis. Arginine is a dibasic amino acid with secretagogue properties (induces the secretion of growth factor, insulin, glucagon, prolactin, and somatostatin). Supplemental dietary arginine has thymotrophic properties and enhances the responsiveness of thymic lymphocyte. Arginine augments cellular immunity.

Dietary supplements

Antioxidant	Activities
Vitamin E	1- Chain-breaking lipid-soluble antioxidant.
	2- Inhibitor of protein tyrosine kinases (PTKs).
Vitamin C	Water-soluble antioxidant.
Coenzyme Q_{10}	1- Lipid-soluble antioxidant.
	2- Cofactor for electron transport.
β-Carotene	Lipid-soluble antioxidant.
Glutathione	Water-soluble antioxidant.
Glutamine	1- Cellular fuel for enterocytes and lymphocytes.
	2- Source of glutamate for glutathione synthesis.
Selenium	Incorporated into selenoproteins.
N-acetylcysteine (NAC)	1- Water-soluble antioxidant.
	2- Source of cysteine for glutathione synthesis.
Genistein	1- Antioxidant.
	2- Inhibitor of topoisomerase I and II.
	3- Inhibitor of protein tyrosine kinases (PTKs).
Quercetin	3- Antioxidant.
	2- Inhibitor of topoisomerase II.
	3- Inhibitor of protein tyrosine kinases.
Daidzein	Antioxidant.

Chemoprevention in uro-oncology:

Vitamin E	Prostate
DHEA	Prostate
Finastride	Prostate
NAC	Bladder
4-HPR	Bladder, prostate
NSAIDs	Bladder
Oltipraz	Bladder, prostate.

Transfusion reactions:

A- Immediate:
    1- Acute haemolytic transfusion reaction.
    2- Bacterial contamination, sepsis, septic shock
    3- Circulatory overload.
    4- Air metabolism
    5- Citrate toxicity
    6- Hypocalcaemia
    7- Hyperkalaemia
    8- Hypothermia
    9- Iron overload
    10- Respiratory distress
    11- Febrile reactions, chills
    12- Allergy, urticaria, anaphylaxis
B- Delayed:
    1- Delayed haemolytic transfusion reaction
    2- Graft-vs-host disease
    3- Infection, hepatitis (A, B, C), retrovirus, cytomegalovirus, HIV, human T-cell lymphotrophic virus type 1, parasites, malaria, babesiosis.
    4- Alloimmunisation.
    5- Bacterial contamination.

NUTRITION, WASTING, INVOLUNTARY WEIGHT LOSS (IWL) AND PROTEIN-ENERGY MALNUTRITION (PEM)

Protein-energy malnutrition (PEM) is defined as a pathologic state characterised by inadequate intake of energy and protein, which the most common form of malnutrition in the acutely injured, chronically ill, or incapacitated populations. There is a high prevalence of PEM in chronic wound patients, and it is a cause of poor healing and the development of wounds such as pressure ulcers. Initial symptoms of fatigue (cell-energy crisis) are followed by a progressive weight loss, including both fat and lean mass. The magnitude of lean mass loss, however, is what produces the morbidity and mortality of PEM. The definition of significant weight loss is a loss of 10% of body weight over a 6-month period. Weight loss itself is used as a medical quality-assurance marker for the status of nutrition in patient populations at risk. Starvation is the pathologic process whereby there is inadequate nutrient intake to meet demands, if prolonged, will result in malnutrition. The host response to illness, injury, or infection is an amplification of the flight or fight reaction. The initial insult leads to local and generalised inflammation and to the activation of an abnormal hormonal response, characterised by a marked increase in catecholamines and other stress hormones. This response produces a hypermetabolic-catabolic state. The degree of hypermetabolism and catabolism is dependent on both the degree of injury and the host response to injury. The hormonally induced metabolic response produces a marked increase in energy demands and change in nutrient use, with 50% coming from fat, 30% from carbohydrates, and 20% (or more) from protein. An energy deficit is common. The increased use of protein for fuel is counter to normal nutrient partitioning principles and rapidly depletes lean body mass. The body's protein is contained in lean body mass, mostly as skeletal muscle. Lean body mass is 50% to 60% muscle mass by weight and the rest is bone and tendon. Protein makes up the critical cell structure in muscle, viscera, red cells, and connective tissue. Enzymes that direct metabolism and antibodies that maintain immune functions are also proteins. It is the loss of body protein, not fat loss that produces the complications of malnutrition; protein synthesis is essential for any tissue repair. Skin is composed primarily of the protein collagen. Stored body fat is used primarily as a reservoir for energy. The size of the fat depot is controlled by both genetic and environmental stimuli, such as diet. Excess nutrients, especially carbohydrates, will expand the depot, while inadequate intake will decrease depot size.

Definition of significant involuntary weight loss:

1- Involuntary weight loss producing a significant health risk.	≥ 5% weight loss in 30 days. ≥ 7.5% weight loss in 90 days. 10% weight loss in 180 days.
2- Involuntary weight loss in association with any stress or comorbid factors.	May complicate many disease states.

Common causes of involuntary weight loss and PEM:

Cause	Criteria
Acute injury or disease process:	1- Other stressors, such as pain and anxiety, can lead to the same end point. 2- Increased nutrient losses due to gastrointestinal disease. 3- Lack of any adaptive or protective responses. 4- The degree of PEM exceeds the degree of weight loss, as lean mass is not protected. 5- Often the insult or "stress" resolves, but the weight loss and PEM are never corrected in the recovery phase.
Inadequate nutrient intake (quality and quantity):	Very common in the elderly, those with disabilities, those with lack of appetite from chronic illness, those with mental illness, and those with poverty.

Types of lipids include: 1- Free fatty acids (fat). 2- Phospholipids. 3- Lecithin. 4- Cholesterol. 5- Lipoproteins. Resistance exercise, which is muscle activity against a force (e.g., lifting objects, weight-bearing activity, and a weight-lifting program), leads to an anabolic stimulus that increases the protein synthesis of muscle. A 40% loss of total lean mass is fatal. Decreased muscle mass and strength is called sarcopenia, which is a Greek term for losing flesh. Sarcopenia is caused by inactivity and poor nutrition. It leads to weakness, disability, skin ulcers, and infections; leads to increased body fat caused by decreased activity and low metabolic rate; and leads to metabolic abnormalities, including diabetes, which leads to decreased quality of life (QoL). Conditions Associated with development of PEM include: 1- Catabolic illness, "the stress response" (e.g., trauma, surgery, wounds, infection, and corticosteroids). 2- Involuntary weight loss exceeding 10% of ideal, for any reason. 3- Chronic illnesses (e.g., diabetes, cancer, mental impairment, arthritis, and renal failure). 4- Increased nutritional losses; open wounds, enteral fistulas. 5- Intestinal tract diseases impairing absorption. Energy is required for all metabolic activities, including protein synthesis, and is rapidly adjusted according to need. Any hypermetabolic state increases demands, and the less efficient use of nutrients for energy results in the nutrient recycling at the level of pyruvate that reverts back to glucose, capturing only a small portion of the potential energy. The byproduct is heat. Therefore, more nutrients are used to meet the demands. Protein is metabolised into amino acids and peptides. With normal anabolic hormone activity, most of the protein byproducts are used for protein synthesis, not for energy. The amino acid profile found in vegetables is very different from the human profile and therefore deficient in key amino acids, while the profile of amino acids in egg albumin, milk, and meat protein provides the necessary protein substrate required by humans. With an inadequate anabolic drive, up to 30% of consumed protein ends up being used for energy. Nutrient partitioning is also disrupted with the activation of the "stress response" or "flight-fight response". In this case, protein is inappropriately used for fuel and the lean mass loss increases in response to catabolism. The insult leads to the release of inflammatory mediators, which activates an abnormal hormonal response due to neuroactivation and leads to a marked increase in catecholamines and other hormones, which produces a hypermetabolic-catabolic state. The degree of hypermetabolism and catabolism is largely dependent on the degree of injury or infection.

Effect of injury on metabolic demands:

Insult	Increase in Metabolic rate above basal.
Starvation	-10-0%
Elective operation	25-50%
Pneumonia	50-70%
Long bone fracture	50-70%
Multiple blunt trauma	50-70%
Infection	50-70%
Head injury	50-70%
Thermal injury: A- 10% body surface area	25%
B- 20% body surface area	50%
C- 30% body surface area	70%
D- 40% body surface area	85%

Terminology of protein energy malnutrition (PEM):

Term	Definition
Energy	The capacity to do work.
Energy production	Defined in terms of standard energy units produced per time.
Energy consumption	Energy used/time (e.g., kcal/hour or ml O_2/minute).
Kilocalorie (kcal)	Standard measure of the quantity of energy obtained from nutrients (often referred to as a calorie)
Metabolism	Sum (body) total of all chemical reactions required for cell function-an energy requiring process.
Hypermetabolism	Increase in metabolic rate above normal.
Anabolism	Constructive metabolism or new tissue formation with protein synthesis.
Catabolism	Destructive metabolism or tissue degradation with protein breakdown.

Causes of inadequate nutrient intake (quantity or quality) include: 1- Aging. 2- Mental illness. 3- Alcoholism. 4- Drug addiction. 5- Avoidance of specified food groups (meat, eggs, milk, fruits and vegetables, grains). 6- Poor dentition. 7- Food idiosyncrasies. 8- Poverty, isolation. 9- Anorexia (from disease process, drugs, emotional problems). 10- Recent weight loss or gain. 11- Inappropriate food choices from lack of information. The stress response can progress to multiple-organ dysfunction (MOD) with loss of body protein and direct cell injury by oxidants and other mediators. Critical illness caused by severe trauma, infection, or a wound will activate the stress response, as will an elective surgical procedure. Metabolic abnormalities due to the stress response of injury include: 1- Increased catabolic hormones (cortisol and catechols). 2- Decreased anabolic hormones (human growth hormone and testosterone). 3- Marked increase in metabolic rate. 4- Sustained increase in body temperature. 5- Marked increase in glucose demands and liver gluconeogenesis. 6- Rapid skeletal muscle breakdown with amino acid use as energy source (counter to normal nutrient channelling). 7- Lack of ketosis, indicating that fat is not the major calorie source. 8- Unresponsiveness of catabolism to nutrient intake. Increased glucose production is produced by protein breakdown and converting amino acids, predominantly alanine, into carbon skeletons, which are then transformed in the liver to glucose, resulting in a net protein loss. The aging population, especially if socially isolated or when living in a chronic care facility, is very prone to involuntary weight loss resulting in complications of lost lean mass. The use of RDA values for energy, protein, and micronutrients are based on what is needed to maintain function, not restore a deficit. A 50% increase in all values is needed to restore losses; otherwise, involuntary weight loss and PEM occur. The specific nutrient requirements for the elderly are higher than for the younger population. The elderly requires increased nutritional intake to avoid losing lean mass, bone, calcium, cognition, and developing chronic illness such as adult onset diabetes. Increased nutritional needs of the elderly include: 1- Calories at least 25% greater than standard RDA table to maintain activity. 2- Protein 1-1.2 g/kg/day to maintain lean mass as protein synthesis is less efficient. 3- Calcium and vitamin D (due to increased losses and decreased intake) in order to decrease osteoporosis. 4- Vitamin B complex, folic acid, to counteract increased homocysteine and its cardiovascular effects. An impaired or aging GIT is less efficient at absorbing nutrients. In addition, disease states such as inflammatory bowel disease, chronic diarrhoea, and medication-induced intestinal disorders all lead to increased losses requiring increased intake. Poor absorption occurs most frequently in the elderly with drug-induced gastrointestinal disorders. Prior bowel resections will also impede absorption, increasing the risk of PEM. Complications of involuntary weight loss (IWL) and PEM include: 1- progressive disability, 2- decreased activity, 3- discomfort, 4- decreased appetite, 5- impaired immune function, and 6- formation of chronic wounds. Complications begin with a decreased activity level and poor nutrition, which accelerates further weight and lean mass loss, which results in depression, poor QoL, and the progression of a downward spiral. Pneumonia often results and is a major cause of death in the PEM population. Malnutrition is a metabolic disorder, and the diagnosis depends on the history, physical exam, and biochemical markers; the biochemical markers are the most sensitive indicators. Because this assessment is not an exact science, there are a variety of different scales used for defining the degree of malnutrition. Probable physical findings include: 1- Unintentional loss of body weight. 2- Loss of subcutaneous fat, evidenced by loose skin, especially on extremities. 3- Muscle wasting, usually first evidenced by quadriceps wasting. 4- Presence of peripheral oedema, in the absence of recognised cardiac disease or circulatory disorder. 5- Poor healing of chronic wounds or pressure sores. 6- Glossitis, cracking at edges of mouth. 7- Chronic infections. 8- Listlessness, apathy. Serum albumin is a common indicator of the patient's protein stores. But because albumin has a half-life of about 20 days, and large amounts are stored in the body, a patient may already be malnourished before serum albumin levels drop. Serum albumin below 3.5 g/dl is considered low and a level below 2.5 g/dl indicates seriously deficient protein stores. Serum transferrin responds more readily than serum albumin to acute changes in protein status. Serum transferrin has a shorter half-life (8-10 days) and smaller body stores than albumin. A serum transferrin level below 200 mg/dl is considered low and below 100 mg/dl is considered severe. Serum cholesterol can indicate malnutrition if it is below 150 mg/dl. Decreased T-lymphocyte cell function indicates protein store depletion.

Objectives to restoring lost lean mass:

1- To eliminate the catabolic state.
2- To restore sufficient nutrient intake to meet current energy and protein needs (calorie intake up to a 100% increase; protein intake 1.5 g/kg/day to 2 g/kg/day).
3- To increase energy or calorie intake to about 50% above daily needs to begin the process of weight and lean mass gain during recovery.
4- To increase protein intake to 2 times RDA (0.8 g/kg/day; i.e., to 1.5 g/kg/day) to allow for restoration of lost lean mass during recovery.
5- To increase anabolic stimulation (which is abnormally low with PEM) to direct the substrate from protein intake into protein synthesis (and restore normal nutrient partitioning).
6- To avoid replacement of lost lean mass with fat gain.
7- To use exercise (mainly resistance exercise) to increase the bodies' anabolic drive to more rapidly regain lean mass.
8- To consider use of exogenous anabolic hormones to increase net protein synthesis.

The metabolic effects of growth hormone (HGH) include: 1- Increased nitrogen retention, protein synthesis. 2- Increased cell amino acid influx, decreased efflux. 3- Decreased urea formation. 4- Increased insulin-like growth factor (IGF) levels. 5- Increased fat oxidation, decreased catabolism. 6- Increased metabolic rate (10%). 7- Insulin resistance, hyperglycaemia. To accomplish the objectives to restoring lost lean mass, the following components may be required: 1- energy or caloric requirements; 2- protein requirements; 3- nutrient mix; and 4- micronutrient requirements. Optimal nutrition is essential to keep up with the increased calorie demands and to decrease the rate of catabolism. Nutrient mix is typically 50-60% carbohydrate (CHO), 25% fat, and 20-25% protein, with the rate of onset, osmolarity of solution, and rate of progression being dependent on gastrointestinal tolerance. The increased protein intake will decrease the net nitrogen losses by increasing the amino acid flow into the protein synthesis channel. The sudden restoration of nutrient intake, especially carbohydrates, in severe malnutrition will suddenly reactivate a number of dormant metabolic pathways. The sudden availability of CHO will exceed downregulated cell demands, necessitating energy for fat production and leading to a further energy deficit. In addition, there will be a shift of the previously depleted electrolytes, potassium, phosphorous, and magnesium, back into cells, resulting in potentially severe hypokalaemia, hypomagnesaemia, and hypophosphataemia. These compounds must be returned into the intracellular compartment. A dose 5-10 times the RDA of micronutrients is usually suggested until the PEM is corrected. Water is an essential nutrient. It is required to transport nutrients and remove by-products from cell metabolism as well as to maintain cardiovascular stability. Dehydration is a major problem in patients with the "stress response" due to increased evaporative losses and established PEM due to both lack of intake and increased losses. In the elderly the aging kidneys are not able to concentrate urine efficiently. Causes of dehydration include: 1- reduced thirst sensation, 2- reduced intake throughout the day, 3- limited access, 4- increased kidney loss due to aging kidney, 5- lack of replacement after increased gastrointestinal losses, and 6- losses due to medications, especially diuretics. The use of anabolic steroids in medicine: 1- Used for hypogonadism debilitating disease (1950s). 2- Abuse in strength sports (1950s). 3- Used for anaemia, osteoporosis, and to protect bone marrow from radiation (1960s). 4- Used for lean mass loss and for wound healing (1960s, 1970s). 5- Increased abuse resulting in decreased clinical use (1980s). 6- Resurgence of use in 1990s for AIDS and now burns and wounds.

Micronutrients (Some essential nutrients):

Compound	Metabolic Function
1- Organic Compounds:	
A- Fat-soluble Vitamins:	
1- Vitamin A (retinol).	1- Synthesis of rhodopsin, epithelial cell.
	2- Bone growth
	3- Inflammatory stimulant.
	4- Wound healing
2- Vitamin E.	Antioxidant in cell membranes.
3- Vitamin D.	Regulation of calcium metabolism.
4- Vitamin K.	Activates blood clotting factors under II, VII, IX, and X.
5- β carotene	Provitamin A, potent antioxidant at membrane lipid.
B- Water-soluble Vitamins:	
1- Thiamine (vitamin B_1)	Oxidative decarboxylation.
2- Riboflavin (vitamin B_2)	Electron transfer during oxidative phosphorylation.
3- Niacin (vitamin B_3)	1- Nicotinamide-adenine dinucleotide (NAD).
	2- Electron transfer reactions.
4- Pantothenic acid	Part of coenzyme A.
5- Biotin	Carbon dioxide transfer reactions.
6- Pyridoxine (vitamin B_6).	Transamination and decarboxylation reactions.
7- Folic acid.	One carbon transfer reaction.
8- Vitamin B_{12} (cobalamin).	1- Production of methionine.
	2- Coenzyme A reactions.
9- Ascorbic acid (vitamin C).	1- Antioxidant in cytosol.
	2- Collagen. Synthesis.
	3- Carnitine production.
2- Inorganic compounds:	
Microminerals:	
1- Chromium	Use of glucose and insulin potentiates insulin action.
2- Cobalt	Required for vitamin B_{12} synthesis
3- Copper	1- Connective tissue developments through collagen cross linking.
	2- Constituent of superoxide dismutase and of the scavenger ceruplasmin.
4- Iodine	Thyroid hormones
5- Iron	1- For haemoglobin and oxygen transport
	2- Electron transfer in oxidative phosphorylation
	3- Constituent of catase.
6- Manganese	1- Procollagen ground substance formation.
	2- Brain function.
	3- Neuromuscular function.
	4- Fatty acid synthesis.
	5- Constituent of superoxide dismutase in mitochondria.
7- Molybdenum	1- Metabolism of purines, pyrimidines.
	2- Redox reactions.
8- Selenium	1- Antioxidant and need for fat metabolism.
	2- Constituent of glutathione peroxidase.
9- Zinc	1- Energy metabolism
	2- Collagen formation
	3- Protein synthesis
	4- Epithelium proliferation
	5- Constituent of superoxide dismutase in cytosol.
3- Glutamine	Substrate for endogenous glutathione.

The action of all anabolic agents currently in clinical use is twofold: 1- To drive amino acids into the protein synthesis channel in the cell. The metabolic pathway used may be different for different drugs, but the outcome is increased protein. 2- Anticatabolic, because all of these hormones (analogues) appear to decrease protein degradation, possibly by blocking cell cortisol receptors. In the absence of a sufficient anabolic activity, the protein synthesis channel pathway is underused and the excess energy is stored as fat. The plasma HGH level is decreased after severe injury, sepsis, or chronic illness, thereby decreasing normal anabolic activity. Although the half-life of HGH is only a few hours, the effect on protein synthesis persists for more than 24 hours. Potential complications of HGH include: 1- Insulin resistance (hyperglycaemia). 2- Fluid retention. 3- Hypermetabolism. 4- May increase mortality rate. Testosterone levels are decreased immediately after severe trauma or critical illness and throughout the recovery period, resulting in decreasing anabolic activity during a period of catabolism. Endogenous levels of testosterone decrease with adult age (or chronic illness). Male muscle 100 times more responsive than female muscle to testosterone. High doses given exogenously will increase anabolism but it has a short half life. Testosterone is metabolised by the liver. Large-muscle group exercise is most effective, and, in a patient population in whom catabolism is pronounced, an aggressive, early program of resistance exercise that continues through the recovery phase is of major importance as an additional anabolic stimulus. Effect of resistance exercise includes: 1- Increased muscle fiber stretch and tension leads to microtears. 2- Muscle responds with a marked increase in protein synthesis from a local and systemic anabolic stimulus. 3- An increased production of endogenous HGH is a component of the anabolism. 4- Muscle cell responds by increased protein deposition, increasing mass. 5- This process reduces lean mass loss during "stress" and increases rate of gain during recovery. 6- Increased energy and protein substrate required. The metabolic effects of exercise include: 1- Increases muscle blood flow in order to increase substrate. 2- Improves glucose use due to increased muscle insulin sensitivity. 3- Increases use of glucose for energy for muscle work and protein synthesis, decreasing glucose conversion to fat. 4- Weight gain favours lean mass over fat mass. Daily fluid intake: A- 30-35 ml/kg body weight. B- Minimum of 1500 ml/day. C- 1-1.5 ml/calorie consumed. D- Replacement of added losses from disease or medications. Weight loss is associated with decreased QoL. Malnutrition per se has adverse clinical consequences. It is difficult to describe malnutrition precisely, though it may be obvious visually. The reason for this difficulty is that there are many aspects of "nutrition", including both macronutrients and micronutrients: 1- Macronutrient status reflects the total mass of the body, while micronutrient status reflects the efficiency of the body's cellular functions. 2- Micronutrient deficiency may exist without macronutrient deficiency, while macronutrient deficiency almost always has associated micronutrient deficiencies. The concept of malnutrition also has static and dynamic features, such as chronic stable malnutrition versus progressive tissue depletion: 1- Body weight varies within 3% over time in a stable, healthy adult. 2- In the classic study of controlled semistarvation in normal volunteers, physical performance began to decline after a cumulative weight loss of >10% of prestudy weight. 3- Subacute weight loss of >20% is associated with increased risk of hospitalisation. Starvation is defined as deprivation of food, either voluntary or involuntary, that leads to weight loss. Malabsorption can be viewed as starvation associated with the additional depletion of electrolytes and water. The effects of starvation can be reversed by providing food. Cachexia is characterised by a disproportionate loss of lean body mass, which results from specific alterations in intermediary metabolism. Feeding is not sufficient to reverse the effects of cachexia. The major operational difference between starvation and cachexia is that the nutritional effects of starvation can be reversed by appropriate feeding while the nutritional changes in cachexia are not reversed by feeding. Body weight may be an inaccurate measure of nutritional status in the presence of clinical disease. The body cell mass is a key tissue compartment. It consists of cells in the muscles and organs, plus circulating cells. This is the part of the body that consumes oxygen and produces carbon dioxide. The body cell mass is the compartment in which all metabolic activity occurs. The most obvious change in body shape is an increase in waist size and thinning of the extremities, often with increased prominence of the veins in the arms and legs. Facial changes include increased wrinkling of the skin and a loss of fat lateral to the nasolabial folds. Women may experience significant breast enlargement and a marked decrease in the size of the thighs as they lose weight. Redistribution of fat is common, with decreased amounts of subcutaneous adipose tissue and increased amounts of visceral adipose tissue in both men and women. Several metabolic abnormalities, especially hypertriglyceridaemia may observed in immunocompromised patients. There are many possible pathogenic mechanisms that might contribute to the observed body composition and metabolic changes in immunocompromised. These include alterations in energy balance, alterations in the body's systemic inflammatory response, possible autoimmune phenomena, and alterations in endocrine function, especially alterations in hypothalamus-pituitary-adrenal axis activity. Low levels of zinc impair taste and olfaction. Selenium deficiency induces cardiomyopathy. Antioxidant deficiency potentially is important, oxidative stress may promote, for example HIV replication through interactions with NFκB. Quality of life-physical performance-correlate significantly with body cell mass, and with the degree of immune depletion. Thus, the level of body cell mass is related to a physical function. Possible causes for altered food intake are multifactorial and include oropharyngeal and oesophageal pathology, psychosocial and economic factors, fatigue, focal or diffuse neurological diseases, and anorexia due to medications, malabsorption, or systemic infection. Many immunocompromised patients (HIV) fear embarrassment of stool incontinence or diarrhoea (either acutely or chronically) and therefore make a conscious effort to avoid eating in order to reduce stool frequency. Hypertriglyceridaemia and decreased serum cholesterol concentrations are common metabolic alterations. Deficiencies in endogenous anabolic activity may promote protein depletion. Decreased food intake, not hypermetabolism, is the most important predictor of weight loss. Effective treatment of a disease complication promoting malnutrition may lead to repletion in the absence of nutritional support. Megestrol acetate (Megace) is a derivative of progesterone that was first evaluated for the treatment of metastatic breast cancer. Appetite stimulation is not a substitute for aggressive diagnosis and management of specific disease complications. Appetite stimulation therapy will fail in the event of an untreated systemic infection. The major side-effects of these agents is related to suppression of the pituitary secretion of gonadotropins. Women receiving megestrol acetate may develop reversible amenorrhoea and men may develop reversible erectile dysfunction (ED). Weight gain due to megestrol acetate is in the form of fat and water, and not lean body mass. Suppression of testosterone release may be one underlying factor for this result. Hypogonadism may be associated with muscle wasting and a mood disorder in the absence of significant weight loss. A functioning immune system is better than any supportive therapy. There are a number of factors that contribute to the continuing increase in the morbidity and mortality associated with involuntary weight loss and Protein-energy malnutrition (PEM): 1- The increase in the number of patients at highest risk for these syndromes; specifically, the elderly, the disabled, and the chronically ill populations suffering with wounds. 2- The increase in patients with involuntary weight loss and PEM is the rapid transition of patients from the acute care setting to either rehabilitation or home care (or chronic care) settings. Both injury and infection activate the "stress response", which leads to pronounced catabolism and an inevitable loss of lean body mass (LBM), especially muscle. The elderly, the disabled, and the chronically ill suffer more frequently from these traumatic insults or infections, due to the aging process, osteoporosis, loss of muscle strength, and immune impairment. Skin is a bilayer organ whose functions are essential for survival. The epidermis is the outer thinner layer of the skin, composed mainly of epithelial cells. The basal or deepest epidermal cells are continually dividing and migrating toward the surface to replace lost surface cells (e.g., after an injury). The same type of regenerating epidermal cells is found in hair follicles and other skin appendages, which are anchored in the dermis. As the cells mature and migrate to the surface, they form keratin, which becomes an effective barrier to environmental hazards such as infection and to excess water evaporation. The dermis is a dynamic layer of thick connective tissue. Fibronectin is a key fibroblast-derived signal protein for the orchestration of healing. Cell types of the epidermis include keratinocytes, epithelial cells, Langerhans' cells. Cell types of the dermis include fibroblasts and macrophages. Epithelial cells make up the majority of the epidermis. Fibroblasts of mesenchymal origin are normally present in the dermis and produce dermal replacement components. Fibroblast products include: 1- collagen (type 1), 2- matrix proteins (fibronectin, tenascin, others), 3- proteoglycans, glycosaminoglycans, hyaluronic acid, and other matrix components, and 4- cytokines and other growth stimulants. There are 5 major interrelated and overlapping components to the healing process: 1- inflammation, 2- cellular proliferation, 3- connective tissue formation, 4- wound contraction, and 5- wound remodelling. The most common causes of tissue hypoxia are: 1- decrease in systemic blood volume and O_2 delivery, 2- decrease in O_2 saturation of haemoglobin, 3- eschar on the wound surface, and 4- surface exudates or infection consuming local oxygen. Pressure ulcers are initiated by excessive compression of soft tissues, which decreases blood flow and leads to tissue ischaemia, followed by necrosis. A secondary infection is very common, accelerating tissue damage. Invariably, one or more comorbid factors are present, which increases the probability of continued wound breakdown. Other factors that increase the risk of developing pressure sores are increasing age, skin thinning, weight loss, and vascular insufficiency. Once skin destruction occurs, it proceeds rapidly at a pace suggestive of bacterial enzymatic digestion. Bacteria allow many pressure sores to become deeper by further inducing the inflammatory response and resultant enzymatic digestion of normal tissue. Denervated tissue also appears to be much more susceptible to bacterial infection. Patients with weight loss who are malnourished have oedema, and skin breakdown occurs readily. They are at higher risk for infection as well as nonhealing ulcers. More than 95% of pressure sores occur below the umbilicus. For supine patients, pressure sores occur over the sacrum and posterior heel. Patients positioned on their side commonly develop ischial pressure sores, and ambulatory patients with diabetes develop pressure ulcerations over their metatarsal.

Skin growth factors:

Cell proliferation	Epithelial, endothelial, and fibroblast
Cell migration	Epithelial, endothelial, fibroblast, white blood cells.
Structure formation	Capillaries, epidermis.
Production of proteins	Collagen, matrix proteins, keratin.

The wound remodelling:

1- Increasing collagen crosslinking, resulting in increased strength.
2- Action of collagenase to begin breaking down excess collagen accumulation.
3- Regression of the lush network of surface capillaries as metabolic demands diminish.
4- Decreasing proteoglycan and, in turn, wound water content.

Conditions associated with development of protein-energy malnutrition (PEM) include: 1- Catabolic illness: the stress response (e.g., trauma, surgery, wounds, infection, corticosteroids). 2- Involuntary weight loss from any cause (exceeding 10% of ideal body weight). 3- Chronic illness (e.g., diabetes, cancer, renal failure). 4- Increased nutritional losses from intestinal disease, surgery. 5- Intestinal tract diseases impairing absorption. With a loss of lean body mass (LBM) <10%, healing of the wound takes priority, using available protein substrate. As LBM is decreased, more consumed protein is used to restore LBM with less being available to the wound. The wound healing rate then decreases until LBM is restored. With a loss of LBM exceeding 20% of total, spontaneous wounds can develop due to the thinning of skin from lost collagen. The most common precipitating cause of PEM is an acute injury or illness leading to a "stress response" with resulting hypermetabolism and increased catabolism. This abnormal metabolic response is aggravated by an increase in catecholamines, cortisol and glucagon and a decrease in anabolic hormones, testosterone, and growth hormone. Increased local metabolic activity and cellular work is required at the site of injury. The host must reabsorb damaged and devitalized tissue and then the tissues must be repaired. The wound consumes large quantities of energy during the healing process both by the large population of inflammatory cells and by the fibroblasts' production of collagen and matrix. Pyruvate and fatty acids recycle back to glucose and fat instead of being metabolised to CO_2 and H_2O. This results in less ATP, and a wasted increase of energy in the form of heat production. A catabolic insult (e.g., a hip fracture) results in the "stress response" being activated. The rate of LBM loss can reach 1 to 2 pounds a day, depending on the degree of catabolism and nutritional support. The recovery phase is entered when the catabolic insult has resolved and the metabolic rate is returning to normal. The presence of a chronic illness can rapidly lead to PEM that continues to worsen, especially in the presence of a wound. Chronic illness leading to PEM and impaired wound healing include cardiovascular disease, chronic obstructive pulmonary disease, diabetes, renal dysfunction, morbid obesity, alcoholism, substance abuse, chronic infection, chronic musculoskeletal problems (e.g., osteoporosis), and age-related frailty. With increasing age, there is often a gradual decrease in body muscle mass and a thinning of the dermis layer of the skin due to protein loss. This process is called degenerative aging and is usually found in those 70 years of age and older.

Skin layers:

Characteristics	Function
Epidermis:	
1- Protection from environmental insults.	
2- Ability to regenerate to regenerate every 2 weeks.	
	1- Protection from desiccation.
	2- Protection from bacterial entry.
	3- Protection from toxins.
	4- Fluid balance (avoid excessive evaporation loss).
	5- Neurosensory.
	6- Social interaction.
Dermis:	
1- Provides durability, flexibility of skin.	
2- Faster for all the components required for replication and repair of epidermis and dermis.	
3- Scaffolding for cell migration and the conduit for nutrient delivery.	
	1- Protection from trauma (elasticity, durability).
	2- Fluid balance (regulation of skin blood flow).
	3- Thermoregulation (control of skin blood flow).
	4- Growth factors and contact direction for epidermal replication and dermal repair.

Components of healing improved by moisture:

Epithelialization:	
	1- Keratinocyte proliferation.
	2- Keratinocyte migration.
	3- Keratinocyte differentiation.
Fibroplasia:	
	1- Fibroblast proliferation.
	2- Collagen matrix synthesis.
Angiogenesis:	
	1- Endothelial cell proliferation.
	2- New blood vessel formation.

Inflammatory phase (immediate onset):

Components	Abnormalities
1- Clotting of bleeders (1-5 minutes).	1- Process suppressed, inadequate O_2 for host defenses, corticosteroids stopping the process.
2- Increased blood flow (20-30 minutes).	2- Process accentuated, tissue damage, excessive inflammation.
3- Increased oxygen in wound.	
4- Antibodies released in wound.	
5- Increased neutrophils (bacterial killing).	
6- Increased macrophages.	

The thinning of skin and lost muscle puts the elderly at increased risk for pressure sores. The lost tissue protein is due to a combination of poor nutrition, muscle inactivity, and decreased anabolic activity. The degenerative aging process can be controlled with good nutrition and exercise. Diabetes Mellitus prevents wound healing through a number of mechanisms including increased blood sugar, decreased activity with LBM loss, PEM, and impaired local immune defenses. Progressive loss of LBM and replacement of fat mass is a common problem for people with diabetes. Uncontrolled blood glucose impairs blood flow through the critical small vessels at the wound surface by impeding red blood cell permeability. The increased glucose makes the cell wall rigid, thereby impeding flow. In addition, the haemoglobin release of oxygen is impaired, resulting in an oxygen and nutrient deficit in the healing wound. PEM is commonly seen in diabetics. Endogenous anabolic activity provided by human growth hormone and insulin are also decreased. Impaired local immune defenses are due to decreased blood flow and decreased neutrophil killing, both secondary to high glucose levels. The treatment for both is increasing muscle mass and controlling blood sugar. The increased muscle mass will more effectively control the blood sugar, decrease fat deposition, and increase endogenous anabolism, which will help the wound healing process. PEM is defined as inadequate intake of energy and protein to meet bodily needs. The prevalence of PEM in hospitals and chronic care facilities ranges from 20% to 50%, with an even higher incidence in those patients who also have a chronic nonhealing wound. Signs suggesting of PEM include nutritional deficiencies, including muscle wasting or weakness, dermatitis; ulceration of mucous membranes, delayed wound healing, CNS depression, glossitis, and congestive heart failure. Serum albumin is a common indicator of the patient's protein stores. But because albumin has a half-life of about 20 days, and large amounts are stored in the body, a patient may already be malnourished before serum albumin levels drop. Serum transferrin is a more accurate indicator of protein stores. Serum transferrin has a shorter half-life (8-10 days) and smaller body stores than albumin.

Complications relative to loss of lean body mass (LBM):

LBM loss	Mortality	Complications
10%	10%	Impaired immunity. Increased infection.
20%	30%	1- Decreased healing. 2- Weakness. 3- Infection.
30%	50%	1- Too weak to sit. 2- Pressure sore. 3- Pneumonia. 4- No healing.
40%	100%	Death (pneumonia).

Conditions associated with PEM:

	Incidence	Cause
Severe trauma, burn, and infection	80%	Catabolic response to the insult is characteristic of this degree of insult.
Spinal cord	50%	1- Acute losses caused by response to injury and immobility. 2- Chronic losses caused by ongoing catabolism and decreased anabolism.
Chronic care	≥25%	1- Weight loss is often the reason for lost nontreated PEM. 2- Comorbid factors increase risk of PEM. 3- Wound presence often further increases the risk factor.
Chronic wounds	≥50%	1- Wound will increase catabolic activity and increase the degree of PEM. 2- Have ongoing needs for increased energy, protein. 3- Often have chronic wound due to PEM.

The stages of pressure ulcers are:

Stage 1	Nonblanchable erythema of intact skin indicative of early skin ulceration.
Stage 2	Partial-thickness skin loss involving epidermis and variable portions of dermis.
Stage 3	Full-thickness skin loss also involving damage to subcutaneous tissue.
Stage 4	Full-thickness skin loss with extensive necrosis of underlying fat, muscles, and bone.

Total lymphocyte cell count is decreased with PEM. One of the most common markers for PEM is involuntary weight loss (unintentional weight loss). Defined as 5% loss of body weight in 30 days, 7% loss in 3 months, or 10% loss in 6 months, this amount of weight loss may produce a significant health risk. In acute wounds, the healing rate will be impaired when losses of LBM exceed 10% of the total. Healing will nearly cease with LBM losses approaching 20%. The acute wound then becomes a nonhealing wound. Chronic wounds, mainly pressure sores, begin to develop as a result of involuntary weight and LBM losses exceeding 20% of total. The objectives of correction of PEM and altered healing are a multifactorial process including: 1- Control the catabolic state. 2- Restore sufficient nutrient intake to meet current energy and protein needs. 3- Increase energy or calorie intake to about 50% above daily needs to begin the process of weight and LBM gain. 4- Maintain protein intake at 1.5 g/kg/day to allow for restoration of lost LBM. 5- Increase anabolic stimulation (which is abnormally low with PEM) directs the protein intake into protein synthesis (and restore normal nutrient partitioning). 6- Avoid replacement of lost LBM with fat gain. 7- Use exercise (mainly resistance exercise) to increase the body's anabolic drive. 8- Consider use of exogenous anabolic hormones to increase net protein synthesis. The micronutrients are vitamins and minerals. The majority of the calorie or energy needs (the energy compartment) come from the absorption and breakdown of carbohydrates and fat. Excess carbohydrates and fat are converted to fat as a reserve depot. Protein consumed is used mainly for protein synthesis. Protein consumed is digested and absorbed as peptides and amino acids to be used mainly for protein synthesis, thereby restoring and maintaining the LBM compartment (i.e., muscle, skin, the immune system, and wound healing). The elderly break down more protein daily and therefore need more protein to be available for the added protein synthesis needed to keep up with losses. Consumed protein is diverted to the LBM compartment by the endogenous anabolic activity of the body, which includes hormones such as testosterone and growth hormone and muscle activity. Only about 5% of protein intake is used for energy or fat (unless there is a very low endogenous anabolic activity, in which more protein will be used for calories). As previously described, 20% to 30% of body protein is used for gluconeogenesis during the "stress response", leading to a rapid loss of LBM. More protein is required, as is more anabolic activity, to control the degree of protein loss.

Local factors that impede healing:

Inadequate nutrients	1- Inadequate energy. 2- Inadequate protein. 3- Inadequate anabolic stimulus.
Tissue hypoxia	1- Low blood flow. 2- Eschar or exudates. 3- Consumption of O_2.
Tissue desiccation	1- Occurs with open wound. 2- Impedes epithelial migration. 3- Risk of wound conversion.
Wound exudates	1- Due to impaired local defense. 2- Exposure to microbes in the environment. 3- Increased inflammation-induced injury.
Wound trauma	1- Environmental insult. 2- Use of toxic chemicals. 3- Traumatic dressing changes.

Categories of patients who are at high risk for developing PEM:

High risk Cause	Feature	Intervention
Acutely injured/infected patient: 1- Marked hypermetabolism, stress-induced metabolism. 2-Inactivity leading to muscle loss. 3- Decreased endogenous anabolic activity.	1- Rapid loss of lean body mass. 2- Micronutrient depletion.	1- Early aggressive initiation of nutrition support (preferably oral) of high-protein, high-energy, micronutrient-rich diet. 2- Nutrient supplements may be required. 1- Early initiation of physical rehabilitation. 2- Appropriate nitrogen and caloric intake Use of anabolic agents to decrease lean mass loss and more rapidly restore lean mass.
Age-Related Frailty: 1- Poor nutrition due to disability, chewing and swallowing problems. 2- Inactivity leading to muscle and further inactivity (cycle of inactivity).	Significant psychosocial issues.	1-Aggressive restoration of high-protein, high-energy, micronutrient-rich diet. 2- Early use of ancillary support services. Resistance exercise program.
Immobility (bedridden or wheelchair-bound individuals): 1- Poor nutrition due to disability, chewing and swallowing problems. 2- Inactivity leading to muscle and further inactivity (cycle of inactivity). 3- Decreased endogenous anabolic activity.	Significant psychosocial issues.	1-Aggressive restoration of high-protein, high-energy, micronutrient-rich diet. 2- Early use of ancillary support services. Appropriate physical rehabilitation. Use of anabolic agents to restore lost weight and lean mass.
Diabetes Mellitus: 1-Decreased capillary blood flow from elevated glucose. 2- Protein malnutrition and decreased muscle mass. 3- Decreased insulin and anabolic hormones.		Careful monitoring and control of glucose, education. Increase protein intake to 1.5 g/kg/day. 1- Increase exercise. 2- Adequate insulin. 3- Anabolic agents.

Interventions for mild to moderate PEM (weight loss <15%): 1- Replacement of an existing loss requires a protein intake of 1.5 g/kg/day. 2- Caloric requirements about 30 kcal/kg/day. 3- Begin with multiple small meals using combination of easy-to-digest food and high-energy, high-protein liquid supplements. 4- Carbohydrates should make up 55% to 60% of calories, some simple sugar but mostly complex carbohydrates. 5- Fat should make up 25% of calories. 6- Protein should make up 20% of calories; use proteins of high biologic value, such as milk, eggs, fish, and meat. 7- Water: needs at least 1 ml of water/calorie. 8- Consider additional use of anabolic hormone to assist in restoring lean body mass rather than gaining fat. . Interventions for severe PEM (weight loss ≥20%): 1- Because of disability, anorexia, and high morbidity risk, enteral +/- parenteral nutrition is indicated. 2- Use of tube feeding is preferred until PEM is beginning to resolve. 3- Intravenous feedings, if needed. 4- Energy requirements and protein requirements are the same as moderate PEM, as patient cannot process any more nutrients until the recovery process is underway. 5- Consider anabolic agent once patient is tolerating nutrient intake goal. Nutritional support should begin within 48 hrs. The optimum route is the enteral route either using food and a nutrient supplement or the use of nasogastric tube feeding. The parenteral route is used if the enteral approach does not adequately meet needs. Water is an essential element for survival and a deficiency leads to a variety of complications. Patients with PEM have a greater risk of inadequate water intake due to a variety of reasons.

Route of nutrient delivery:

	Enteral	Parental
Advantage	1- Improves nutrient use, especially protein. 2- Protects gut mucosa.	1- Can initiate nutrition if gut not functional. 2- Does not require patient cooperation.
Disadvantage	1- Aspiration. 2- Impaired absorption. 3- Requires patient cooperation.	1- High infection risk (line sepsis). 2- Other catheter complications.
Timing	As soon as possible to avoid gut function problems.	1- Postresuscitation (usually day 3). 2- If needed to supplement enteral route.
Approach	1- Eating plus nutrient supplements area. 2- Tube feeding.	1- Central line through nonburn area. 2- Peripheral line for more dilute.

Systemic factors impeding healing:

Inadequate wound blood volume (hypovolaemia and oxygen delivery):
    1- Hypovolaemia.
    2- Dehydration.
Loss of body protein (lean body mass):
    1- Caused by ongoing catabolism.
    2- Caused by decreased anabolism.
    3- Caused by inadequate protein intake.
    4- Decreased amino acids in the wound.
Inadequate nutrition:
    1- Inadequate energy to wound for protein synthesis.
    2- Inadequate amino acids for healing.
Systemic infection:
    1- Increases catabolism.
    2- Blood shunted from wound.
    3- Collagenolysis.
Uncontrolled stress response:
    1- Inadequate energy due to hypermetabolism.
    2- Catabolism.
    3- Excess generalised inflammation.

Markers of PEM:

Index	Mild degree	Moderate degree	Severe degree
Albumin (not an early marker)	2.8-3.5 g/dl	2.1-2.7 g/dl	<2.1 g/dl
Transferrin	151-200 mg/dl	100-150 mg/dl	<100 mg/dl
Total lymphocyte count.	12000-15000/mm^3.	800-1199 mm^3.	<800 mm^3.

Factors associated with dehydration in PEM include: 1- Reduced thirst sensation. 2- Reduced intake throughout the day. Limited access. 3- Increased kidney loss due to the aging kidney. 4- Lack of replacement after increased gastrointestinal losses. 5- Losses from medications, especially diuretics. Anabolic hormones may attenuate the catabolic stimulus during stress, restore lean mass loss more rapidly, and restore normal nutrient partitioning such that protein consumed is not converted to energy and weight gained is not fat mass. The rationale for the addition of anabolic agents is to correct LBM loss more rapidly and improve healing. The actions of anabolic hormones include: 1- Anticatabolic, by decreasing loss of amino acids from the protein synthesis pathway. 2- Anabolic, by increasing the rate of protein synthesis. The metabolic effects HGH include: 1- Increased nitrogen retention, protein synthesis. 2- Increased cell amino acid influx, decreased efflux. 3- Decreased urea formation. 4- Increased insulin-like growth factor-1 (IGF-1) levels. 5- Increased fat oxidation, decreased catabolism. 6- Increased metabolic rate (10%). Potential Complications of HGH: A- Insulin resistance (hyperglycaemia). B- Fluid retention (usually self-limiting). C- Hypermetabolism. D- Increased mortality rate in certain critical care population. Resistance exercise, defined as muscle movement against a resistance such as weight lifting, is a potent anabolic stimulus. The mechanism remains unclear. Resistance exercise diminishes the degree of protein loss. Large-muscle exercise should be given priority. Lack of resistance exercise, as occurs with bed rest, a fixed splint, or general inactivity, will lead to muscle atrophy (net catabolism) in addition to that caused by a wound. Daily fluid intake: A- 25-30 ml/kg body weight. B- Minimum of 1500 ml. C- 1-1.5 ml/calorie consumed. D- Replacement of added losses from disease or medications.

RESPIRATORY DISEASE

Catabolism-induced malnutrition and involuntary weight loss (IWL) can lead to both acute and chronic respiratory failure. Adequate energy and protein synthesis is critical to cell function. Any potentially life-threatening impairment is related to the loss of lean body or protein mass due to the stress response to injury, infection, or inflammation. The catabolic hypermetabolic state, known as the stress response is responsible for impaired respiratory functions as well as abnormal body composition, leading to significant morbidity and mortality. Prevention of the problem is much easier than correction. The relationship of respiratory distress and PEM include: 1- The degree of respiratory dysfunction will become more severe in the presence of PEM and ongoing catabolism. 2- Major PEM-induced respiratory complications will accentuate the catabolic state. 3- The degree of catabolism and PEM will become more severe with respiratory failure. 4- Respiratory failure also decreases endogenous anabolic activity, impairing restoration of lost body protein even with good nutrition. 5- Restoring lost protein in the presence of respiratory disease is much more difficult than maintaining total body protein. 6- Processing the macronutrients to correct PEM is energy demanding. 7- Existing severe respiratory disease impedes the ability to correct PEM. Acute respiratory illness activates the stress response, which increases energy demands, increases protein catabolism, and increases the need for micronutrients. The loss of lean mass in a catabolic state leads to loss of chest wall and diaphragm muscle, and immune function and infection control. An increase in host anabolic activity will further decrease the loss of body protein, which occurs with a catabolic insult as seen with respiratory disease. Involuntary weight loss (IWL) is loss of 10% of actual body weight over a 6-month period or less, which produces a significant health risk. IWL is often an associated factor in malnutrition, as well as in injured or ill patients. The degree of morbidity, including poor wound healing and development of ongoing respiratory disease, increases as the degree of weight loss increases. Protein-energy malnutrition (PEM) is defined by inadequate intake of energy, and protein malnutrition is the most common form in the acutely injured, chronically ill, or incapacitated respiratory dysfunction populations. The rate of body weight and lean mass loss in a catabolic state is 5- to 10-fold faster than the rate of restoration. This problem is the result of a marked increase in catabolism relative to anabolism with any significant insult. Body composition can be divided into a fat and fat-free (or lean mass) component. All body protein is present in the lean mass compartment. Lean mass contains the body's protein content. The majority of the protein in the lean body mass is in the skeletal muscle mass. The rest comprises skin protein, immune defences, visceral organ structure, and so on. Protein is responsible for the metabolic activity and body structure essential for survival. The chest wall and diaphragm muscles are components of the lean body mass. Maintaining lean mass involve: 1- Intense genetic drive to maintain essential protein stores. 2- Anabolic hormones, which increase during periods of inadequate energy to preserve protein stores. 3- Continuous cell signalling to activate protein synthesis when needed. 4- Adequate protein intake to meet demands. The loss of protein stores or lean mass and inadequate energy result in significant morbidity and mortality. Elderly or disabled patients, and those with chronic illness, especially respiratory disease, do not tolerate any further loss of body protein. The most common problem caused by catabolism is progressive weakness and the development of decreased muscle activity. Complications of IWL/PEM correlate with the percent loss from ideal or normal lean mass. Often acute and chronic respiratory dysfunction patients already have lean body mass (LBM) loss as a result of the insult or infection causing the respiratory disease. A decrease in immune defences is the first complication noted, followed by impaired healing and weakness, both significant components of any respiratory dysfunction. Complications of IWL and PEM include: 1- Increasing disability (decreased activity, discomfort, decreased appetite, and progressing PEM). 2- Impaired lung function both acute and chronic. 3- Weakness. 4- Increased infection. 5- Progression of existing lung disorders. 6- Nonhealing wounds. 7- Poor quality of life (QoL). Acute and chronic respiratory deficiency invariably includes a catabolic state. All acute lung disorders produce a stress response, as do the frequent infections in chronic obstructive pulmonary disease (COPD). Pneumonia is a major cause of death in the PEM population. Common causes of catabolism, IWL, and PEM include: 1- Acute injury or a disease process where increased nutrient requirements and wasting of body protein are characteristics. 2- The presence of other stress, such as pain and anxiety. 3- The lack of any adaptive or protective responses. 4- Inadequate nutrient intake, both quality and quantity, which is very common with aging, disability, lack of appetite from chronic illness, mental illness, and poverty. Common conditions associated with development of PEM include: 1- Catabolic illness: the stress response, e.g., trauma, surgery, wounds, infection. 2- Acute or chronic respiratory disease. 3- Inflammation-induced disease. 4- Neuromuscular disorder. 5- Elderly. 6- Chronic illness. Typically, 60% of calories come from consumed carbohydrates, 30% from fat and only 5% to 10% from protein. The rest of the protein consumed is used for protein synthesis. Loss of lean mass would result in weakness, infection, and other complications impeding the musculoskeletal functions necessary to restore nutrition intake. The stress response defines the activation of the genetically preprogrammed fight-flight response as a result of some bodily insult. The response is essentially an activation of the hypothalamus-pituitary-adrenal axis (HPA) to release catecholamines and cortisol (and other stress hormones). The hormonal response and the subsequent inflammatory response to injury or infection are maladaptive, meaning there is no attempt to conserve energy or protein, but rather an all-out attempt to generate energy very rapidly from protein stores, at the expense of lean mass. A catabolic insult is any physical (or psychological) threat, i.e., injury, infection that activates the stress response. The response is a rapid and persistent increase in energy demands and use of protein (through the use of the amino acids alanine and glutamine) for fuel via conversion to glucose. Even nutrient intake will not reverse the process, until the insult, i.e., a wound/emphysema, has resolved. Until then a catabolic state persists. The hypermetabolic, catabolic response to stress is maladaptive as energy demands increase but production is very inefficient, leading to heat (wasted energy). Protein stores are consumed, especially the important amino acid glutamine. From a healthy state it takes approximately 2 weeks for a patient to progress to significant PEM with a catabolic insult. However, in the presence of malnutrition, it takes only a few days. The catabolism caused by a fracture or GIT condition even modest can increase an underlying lung disorder in an already compromised elder. ARDS caused by multiorgan trauma will activate the stress response and will accentuate the catabolic state. Inflammation is a significant component of virtually all acute or acute on chronic respiratory disease. The inflammatory mediators such as the proinflammatory cytokines, the oxidants, and proteases, all cause catabolism or protein degradation. Noninfectious inflammatory lung disorders that can cause acute respiratory failure include: 1- ARDS, 2- idiopathic pulmonary fibrosis, 3- sarcoid alveolitis, 4- Goodpasture's syndrome, 5- hypersensitivity pneumonitis, 6- acute eosinophilic pneumonia, and 7- systemic lupus erythematosus.

Stress response major metabolic abnormalities:

1- Increased catabolic hormones (cortisol and catechols).
2- Decreased anabolic hormones (human growth hormone and testosterone).
3- Marked increase in metabolic rate.
4- Marked increase in conversion of amino acids to glucose through liver gluconeogenesis.
5- Rapid skeletal muscle breakdown with amino acid use as an energy source.
6- Abnormal nutrient channelling.
7- Lack of ketosis, indicating that fat is not the major calorie source.
8- Unresponsiveness of catabolism to nutrient intake.
9- Inflammatory mediator-induced catabolism (oxidant and cytokine-induced catabolism).
10- Muscle cachexia from injury or illness.

Secretory products of activated alveolar macrophages:

Enzymes	Arachidonic acid	Cytokines	Oxygen metabolites
1- Elastase	Protease inhibitors	1- IL-1	1- Hydrogen peroxide
2- Plasminogen activator		2- TNF	2- Superoxide anion
3- Acid hydrolase		3- IL6	3- Singlet oxygen
			4- Hydroxyl radical

During early postinjury or infectious conditions, the initial cytokine response to such insults likely mediates beneficial protective signalling of the immune system. Prolonged production of these tissue cytokines sustains some metabolic effects of the hypercatabolic state. The proinflammatory cytokine peptides are produced by diverse cell types of both myeloid and nonmyeloid origin. These proteins may function by autocrine or systemic mechanisms of action. They produce local tissue responses at very low concentration but exert systemic effects in higher concentrations. These cytokines cause metabolic changes that lead to a catabolic state, which will result in a loss of protein stores as well as decreased use of fat for fuel. An increase in body temperature and a hypermetabolic state are produced. Tumour necrosis factor (cachexin) and other cytokines release oxidants and proteases that also produce protein degradation. Effect of proinflammatory cytokines (IL1, TNF, IL6) include: 1- altered metabolism (catabolism, hypermetabolism, and fat deposition in cells), 2- increased body temperature, and 3- activation and release of mediators (oxidants, proteases). The proinflammatory cytokines are released when any lung inflammatory process occurs, any added inflammation can result in respiratory failure in addition to systemic catabolism. Oxygen radicals are unstable metabolites of oxygen that have an unpaired electron, making them potent oxidising agents and toxic to cells. Hydroxyl anion (OH) appears to be the most damaging, and the cause of many of the oxygen radical lung injuries. A number of endogenous antioxidants remove the minute quantities of oxidants normally produced. Increased oxygen radical production or decreased antioxidants can overwhelm the antioxidant system. Tissue damage, especially to the lung, will be produced by these excess oxygen radicals via several mechanisms. Direct cell damage occurs, since the radicals oxidise the lipid moiety of the cell membrane. The resulting lipid peroxidation alters membrane function. Lung lipid peroxidation is known to occur in several forms of ARDS, including sepsis. The lung interstitial matrix is also altered, in particular, the crucial hyaluronic acid framework, which is fragmented by the OH radical. The process leads to loss of integrity of the tissue matrix, increasing the ease of oedema formation. Fragmentation of the lung basement membrane is known to occur with oxidant injury. Other direct effects include the generation of a potent chemotactic lipid from arachidonic acid, as well as stimulation of prostanoid and free lipid peroxide production, which, in turn, can affect distant tissues. An important indirect effect is the neutralisation of various lung antiproteases, in particular, $\alpha2$ antitrypsin, which, in turn, increases the risk of local protease damage. Oxidants are most important when lung injury develops with the onset of lung inflammation, or any acute lung process in established chronic lung disease. There is a decrease in the protective lung antioxidants through increased use and lack of production, in severe lung disease or any lung disease with concomitant PEM. Proteases, released from activated neutrophils or macrophages, will also attack normal lung tissue. Antiprotease activity must be maintained. Excess oxidants impede these antiproteases, leading to cell damage. Both antioxidants and proteases are catabolic, i.e., degrade tissue protein. Oxidants, especially OH, alter arachidonic acids (AAH) to produce AAO_2 and arachidonic acid-free radicals, or via superoxidant (O_2^-), a chemotactic lipid that attracts more inflammatory cells to the lung. Hydrogen peroxide, a hydroxide radical, both disrupts hyaluronic acid in the lung interstitium and deactivates antiproteases, leading to more protease damage. All oxidants increase thromboxane production (TxA_2), a potent vasoconstrictor. Proteases are enzymes that degrade protein, are released from white cells in response to a stimulus usually initiated by cytokines. PEM decreases the level of these protective proteins. Neutrophil proteases include: 1- neutral protease (elastase, cathepsin G, and collagenases), and 2- acid hydrolase (cathepsin B and D, β-glucuronidase, and β-glucosamines). Severe cerebral injury can lead to a series of respiratory disorders. In addition, severe cerebral damage results in a marked hypermetabolic catabolic state comparable to a severe burn. The increase in metabolic rate varies from a 50% to 100% increase above normal. Marked catabolism is also seen due both to brain damage via cytokine release and activation of the stress response, as well as through disuse atrophy if coma is present. The protein loss is extremely difficult to control even with a protein intake of 2 g/kg/day. The head-injury-induced hypoxia in the absence of oedema appears to be due to marked alteration in ventilation/perfusion matching mediated by the CNS. Pneumonia is seen as a result of an PEM-induced immune deficiency state and increased risk of aspiration. Any patient with a severe cerebral injury develops a profound catabolic state comparable to a burn of 50% of the body. Despite the presence of coma, catabolism and energy demands are very high. Also present to a significant degree are systemic inflammatory-induced catabolism and a neuromyopathic state. Even with maximum nutrition, it is difficult to maintain or restore body weight. Spinal cord injury results in an acute stress response and a catabolic state followed by disuse atrophy. There is also a well-documented long-term increase in energy demands as well as a chronic hypo-anabolic state. The risk of PEM is very high. If the injury is in the cervical spine, a long-term process of respiratory dysfunction can exist, especially recurrent pneumonias. The nutritional solution is the same as for any acute and chronic catabolic state (meeting energy demands while providing high-protein, high-micronutrient intake). There are 2 catabolic processes that occur by the use of paralytic agents in the management of severe respiratory failure, especially ARDS and asthma: 1- The first is disuse atrophy, which can be severe, as the patients are also catabolic due to the underlying respiratory disease process. 2- The second is a persistent neuromuscular blockade with evidence of weakness and atrophy of mainly type 2 fibers, including muscle necrosis. The process is most prominent in asthmatic patients on corticosteroids and paralytic agents. The catabolism is prominent in all muscle groups including the chest wall (a pound of muscle losses per day). Polyneuropathy of prolonged critical illness leads to a severe persistent catabolic state, which is poorly understood and lasts longer than the cortical illness but usually resolves eventually. There is significant protein loss due to disuse atrophy in combination with the underlying stress response evident in a ventilator-bound patient. Corticosteroids are used to treat a number of respiratory and nonrespiratory disorders. These agents produce catabolism, the degree being: 1- dependent on dose, 2- dependent on duration of treatment, and 3- dependent on the status of nutrition and anabolic activity. Many patients with chronic disorders treated with corticosteroids are also at high risk for PEM and decreased anabolic activity. Anabolic steroids reverse the catabolic response to corticosteroids but do not reverse the anti-inflammatory effects of the corticosteroids. The anabolic steroids are considered to be both anabolic and anticatabolic. Weight loss is used as a medical quality assurance marker for adequacy of nutrition in patient populations at risk. The degree of morbidity, including poor wound healing and development of pressure sores increases as the degree of weight loss increases.

Usual weight - Present weight X 100 = Percent of weight change

The loss of weight is only a marker of the risk of PEM as the type of weight lost, i.e., degree of lean mass versus fat and the degree of nutritional deficiency, defines the degree of protein-energy malnutrition. Malnutrition is a metabolic disorder; the diagnosis depends on a combination of findings on history and physical examination as well as biochemical markers. The correction of IWL and PEM with acute respiratory disease is multifactorial with time to correction requiring a much longer period than time to development. The objectives are: 1- Eliminate the catabolic state. 2- Restore sufficient nutrient intake to meet demands. 3- Restore antioxidants and micronutrients. 4- Increase anabolic activity to restore nutrient partitioning. 5- Avoid increasing only fat. 6- Employ resistance exercise. The BMR is the amount of energy expended at complete rest, shortly after awakening, and in a fasting state for 12-18 hrs. The BMR depends on age, sex, and body size and correlates roughly with body surface area. It is proportional to lean body mass, not to fat. Resting energy expenditure, which is often used synonymously with BMR, represents the amount of energy expended 2 hrs after a meal under conditions of rest and thermal neutrality. It is generally 10% higher than the BMR. The metabolic rate (energy demands) increases 20% after elective surgery and 110% after a severe burn. A wound, an infection, or traumatic injury will fall between these 2 extremes.

Energy expenditure = BMR X stress factor X activity factor

In most cases, the caloric mix is 60% carbohydrates, 20% fat, and 20% to 25% protein. If the respiratory quotient increases to 1 or above reflecting excess CO_2 production, then carbohydrate (CHO) intake needs to be decreased and fat needs to be increased. Protein must remain at 1.5 to 2 g/KgBW. A healthy adult requires about 0.8 g of protein/KgBW/day or about 60-70 g of protein to maintain homeostasis, i.e., tissue synthesis equals tissue breakdown. Stressed patients need more protein, in the range of 1.5-2.0 g of protein/KgBW/day. Urinary nitrogen losses after injury and illness are measured with an increased degree of stress. Nitrogen content is used as a marker for protein; (6.25 g of protein is equal to 1 g of nitrogen). Nitrogen balance studies, such as a 24-hr urinary urea nitrogen (UUN) measurement, that compare nitrogen intake with nitrogen excretion can be helpful in determining needs by at least matching losses with intake. Nutritionally depleted nonstressed patients also require at least 1.5 g/kg/day to restore lost body protein. Stressed, depleted patients usually cannot metabolise more than 2 g/kg/day of protein unless an anabolic agent (oxandrolone) is added that can override the catabolic stimulus. Micronutrients are essential for cellular function. Marked deficiencies in key micronutrients occur during the severe stress response as a result of increased losses, increased consumption during metabolism, and inadequate replacement. A deficiency state of micronutrient further amplifies stress, metabolic derangements, and ongoing catabolism or PEM. The micronutrients include key amino acids, such as glutamine and arginine, organic compounds (vitamins), and inorganic compounds (trace minerals). Because measurement of levels is difficult, prevention of a deficiency is accomplished only by providing increased intake. Deficiency states can lead to severe morbidity because adequate amounts of these key components are essential for minimising complications. Oxidant injury is a well-recognised component of acute and chronic respiratory distress. The replacement of antioxidants needs to begin in the early part of any acute illness with lung involvement as endogenous antioxidants are rapidly depleted. The antioxidant glutathione is made from glutamine and cysteines.

Physical Findings of PEM:

1- Loss of body weight.
2- Loss of subcutaneous fat evidenced by loose skin especially on extremities.
3- Muscle wasting usually first evidenced with quadriceps wasting.
4- Presence of peripheral oedema (in absence of recognised cardiac disease).
5- Glossitis, cracking at edges of mouth.
6- Hair loss, lack of luster.
7- Chronic infections.
8- Poor healing; chronic wounds, pressure sores.
9- Listless, apathetic.
10- Recurrent pulmonary infections.

The voluntary intake of food alone is insufficient for the increased energy and protein demands of the acute and chronic respiratory failure populations with PEM. Criteria for selecting nutrient supplement: 1- sufficient protein content, 2- sufficient caloric content, 3- quantity of carbohydrates, 4- quality of the contained nutrients, 5- route of administration, i.e., taken orally or tube feeding, and 6- palatability. Water is an essential element for survival and a deficiency leads to a multitude of complications. Water is a very essential nutrient, as water is required to transport nutrients and remove by-products from cell metabolism in addition to its roll in maintaining cardiovascular stability. Dehydration is a major problem in patients with a stress response due to increased evaporative losses. Dehydration is a major problem with established PEM due to both lack of intake and increased losses, especially in the elderly due to the inability of aging kidneys to concentrate urine. These problems, combined with a decreased thirst sensation, result in inadequate hydration. More water is needed to process and excrete the necessary increased calorie and protein intake. Causes of dehydration: 1- Reduced thirst sensation. 2- Reduced intake throughout the day. 3- Limited access. 4- Increased kidney loss due to the aging kidney. 5- Lack of replacement after increased gastrointestinal losses. 6- Losses from medications, especially diuretics. The daily fluid intake recommendations include: A) 30-35 ml/KgBW, B) minimum of 1500 ml/day, C) 1-1.5 ml/calorie consumed plus replacement of added losses from disease or medications. Corticosteroid catabolism can be reversed by an anabolic steroid (oxandrolone) without impeding the corticosteroid anti-inflammatory activity. In acute and chronic respiratory failure, excess carbohydrate (CHO) provided over that efficiently metabolised would lead to an increase in CO_2 production, leading to an increase in ventilation demands. An excess CHO can be monitored by the respiratory quotient (CO_2 produced/O_2 consumed). Once the respiratory quotient (RQ) reaches or exceeds 1, CHO intake needs to be decreased.

Stress factors:

Minor injury	1.2
Minor surgery	1.2
Clean wound	1.2
Bone fracture	1.3
Infected wound	1.3
Respiratory diseases	1.5
Major trauma	1.5
Major infection	1.5
Severe burn	2
Combination	2

Antioxidants:

Parts of the cell	Antioxidants act to prevent oxidant damage
Cell membranes	CoQ_{10}, vitamin E, β-carotene, and catalase.
Cytosole	Vitamin C, GSH-GSSG, GSHPx, Cu/Zn superoxide dismutase, and catalase.
Mitochondria	Vitamin C, CoQ_{10}, Mn, SOD, GSHPx, and GSH-GSSG.
Lysosomes	Vitamin C, vitamin E, and β-carotene.

Antioxidant replacement in stress disorders:

Antioxidant	Daily dose
Vitamin C	1-2 g
Vitamin E	400-1000 IU
β-carotene	50-100 mg
Copper	2-3 mg
Manganese	20-50 mg
Selenium	100-150 mg
Zinc	50-75 mg

An increase in fat calories can be provided instead. A decrease in the serum intracellular electrolytes, phosphate, magnesium, calcium, and potassium can lead to muscle weakness, including respiratory muscle function. Improvement in muscle function occurs with replacement, recognising that a low serum value means a marked decrease in intracellular levels. Nutritional restoration in existing PEM causes a rapid shift of the electrolytes into the cell, resulting in a rapid decrease in serum levels. This process, if unrecognised, will lead to a sudden decrease in muscle function. Appropriate electrolyte replacement will correct the muscle weakness. Oxidant damage is common in both acute and chronic lung disease and a rapid decrease in endogenous antioxidants, due to use and lack of adequate replacement. Adequate antioxidant replacement is essential. Corticosteroids are catabolic. The anabolic steroid (oxandrolone) prevents the catabolic response in corticosteroid-dependent patients with burns or wounds. The benefit of an androgen would be an increase in protein synthesis negating the protein loss from the corticosteroids. Anticatabolic therapy is aimed at attenuating the degree of protein loss in the presence of hypercatabolism. The anticatabolic agents include: 1- Nutrients (protein/peptides, glutamine), 2- antioxidants, 3- anabolic hormones, anxiolytics/pain medications (for psychological stress). The focus of anticatabolic therapy is to decrease the degree of protein degradation. Glutathione is a prominent antioxidant in the lung, especially in the airway mucosa. There is a well-recognised decrease in glutathione in acute and chronic respiratory disease due in large part to a lack of glutamine and cysteine. The decreased glutathione corresponds to airway injury. The psychological stress response is seen after any injury. Stress generates a hypermetabolic catabolic state through the same neuropathway initiated by physical stress. Psychological stress: 1- activates hypothalamus-pituitary-adrenal (HPA) axis, 2- increases catechols, 3- causes hypermetabolism, 4- causes immune dysfunction, 5- decreases wound blood flow, and 6- is correctable with adequate pain/stress management. Emotional structures are located mainly in the limbic system. The hypothalamus, in particular, with its connections to both central and peripheral nervous structures and neuroendocrine integrated systems, is concerned with the organisation of motivated behavioural and endocrine responses. In humans, mental stress is known to induce pronounced and reproducible activation of the sympathoadrenal system, with elevation of plasma epinephrine and norepinephrine concentrations and subsequent metabolic consequences identical to those produced by a wound insult. In both the catabolic and the recovery phase of respiratory disease, endogenous anabolic activity remains depressed especially in elderly patients, in those with chronic illness, and in patients with preexisting involuntary weight loss. Adequacy of substrate (1.5 g/kg/day protein) may not be sufficient to jump-start restoration of lean body mass. Protein intake ends up in the production pathway for energy in the absence of adequate endogenous anabolic hormone activity. The action of all anabolic agents is twofold: 1- First, amino acids are driven into the protein synthesis channel in the cell through action of cell surface receptors on all the tissues in the lean mass compartment. 2- The second action is anticatabolic. All anabolic agents appear to decrease protein degradation, possibly by blocking cell cortisol receptors.

Nutritional problems in acute and chronic respiratory disease:

Acute respiratory disease:	
A- Excess CO_2 production with excess carbohydrate (CHO) calories:	1- monitor respiratory quotient (RQ) to avoid a value of 1 or greater 2- increase fat calories
B- Respiratory muscle weakness with:	1- Hypophosphataemia 2- Hypomagnesaemia 3- Hypokalaemia 4- Hypocalcaemia
C- Lung damage with low antioxidants	
Chronic respiratory disease:	
A- Excess carbohydrate (CHO) more of a problem than in acute disease.	
B- Respiratory muscle weakness with electrolyte problems.	
C- Lung dysfunction improves with:	1- Vitamin C, niacin. 2- Correcting low copper, zinc.
D- Increased O_2 demands and CO_2 production with too rigorous repletion of nutrition.	1- Need to begin slower in spite of severe PEM.

Anabolic Hormones

1- Attenuate the catabolic stimulus during stress.
2- More rapidly restore lean mass loss.
3- Restore normal nutrient partitioning such that protein consumed is not converted to energy and weight gained is not fat mass.
4- Anticatabolic by decreasing loss of amino acids from the protein synthesis pathway.
5- Anabolic by increasing the rate of protein synthesis.
6- Anticatabolic by eliminating net catabolism with corticosteroid use.

In the recovery phase after acute or chronic respiratory illness or any other insult, endogenous anabolic activity remains depressed, especially in the cases of the elderly, of those with a chronic illness (COPD), and of those who have simply experienced an involuntary weight loss. The rate of protein synthesis is dependent on: 1- adequate substrate, namely, amino acids and peptides as a nitrogen source; and 2- overall anabolic drive relative to the catabolic stimulus to further increase protein synthesis. Anabolic hormones increases lean mass and correct PEM in the following: 1- acute respiratory disease (failure to wean), 2- chronic obstructive pulmonary disease, 3- elderly, frail population, 4- chronic dialysis population, 5- AIDS patients with weight loss, 6- major burns, acute phase and recovery phase, 7- elderly with osteoporosis to restore and maintain bone calcium, and 8- chronic nonhealing wounds. Human growth hormone (HGH) is normally produced by the pituitary gland (0.8 mg/day) and is a potent endogenous anabolic hormone. HGH binds to specific cell receptors, which leads to a host of metabolic effects, some due to direct hormone activity on tissues, especially in the liver. Other effects are due to the release of insulin-like growth factor-1 (IGF-1), which has potent wound-healing effects. The metabolic effects of HGH spare protein while increasing fat for fuel. A major complication is its anti-insulin effect and hyperglycaemia. Testosterone levels are decreased immediately after severe trauma or critical illness and throughout the recovery period, decreasing anabolic activity during a period of catabolism. Exogenously administered (oral or parenteral) testosterone is rapidly metabolised in the liver, resulting in a half-life of approximately 10 minutes, which is not practical for clinical use. Slow-release testosterone can be injected, but androgenic effects leading to hirsutism, hypersexualism, and mood changes may occur. Testosterone characteristics: 1- Produced mainly by testis in males and by adrenal gland in females. 2- Conversion of adrenal precursor androstenedione to testosterone in liver. 3- Both androgenic and anabolic activity via androgenic receptors. 4- Modest androgenic activity. 5- Endogenous levels decrease with adult age (or chronic illness). 6- Male muscle 100 times more responsive than female muscle to testosterone. 7- High doses given exogenously will increase anabolism but short half-life. 8- Metabolised by the liver. Megestrol acetate decreases lean body mass (LBM) as a result of its anti-testosterone effects. The predominant weight gain is fat and water. Megestrol acetate characteristics: A- effective appetite stimulant, B- 85% of weight gain is fat due to progestational steroid effect, C- can produce hypogonadism, D- decreases lean mass gain, and E- not beneficial for lean mass gain. Anabolic steroids are class of drugs that include testosterone and its derivatives. In 1940 it was reported that testosterone can stimulate production of RBCs, and anaemia was the early indication for testosterone use. In the 1950s, testosterone was used to treat debilitating disease and osteoporosis because of its anabolic properties. The mechanism of action of testosterone is its action on cell androgenic receptors (defined in the 1960s), which leads to cell amino acid influx in lean mass tissue and increased protein synthesis. Cells with a high androgenic receptor density include skin fibroblasts and skeletal muscle myocytes. The lower the ratio of androgenic to anabolic the better the quality of testosterone analogue, which indicates very little masculinizing effects, compared with a very potent anabolic effect. All testosterone analogs are cleared primarily by the liver, except oxandrolone. The anabolic steroid oxandrolone promotes weight gain following weight loss due to a variety of conditions, maintains weight in the face of ongoing IWL, and offsets the protein catabolism associated with the chronic use of corticosteroids. It is cleared primarily by the kidney, markedly decreasing any hepatotoxicity. Oxandrolone is a 17-β-hydroxy-17-α-methyl ester of testosterone, which is cleared primarily by the kidney.

Oxandrolone characteristics:

1- Oral 17 α analogue.
2- Metabolised mainly by kidney.
3- Not hepatotoxic.
4- Readily absorbed by intestine with 99% bioavailability.
5- Biologic life of 9-12 hours.
6- Highly anabolic.
7- Minimal androgenic effects.
8- Increases nitrogen retention and protein synthesis.
9- Acts on all androgenic receptors.
10- Infrequent effect on glucose metabolism.

Hepatotoxicity is minimal, even at doses >20 mg/day. Oxandrolone is the only steroid in which a carbon atom within the phenanthrene nucleus has been replaced by oxygen. This alteration appears to be responsible for its potent anabolic activity, which is 5 to 10 times that of methyltestosterone. Its androgenic effect is considerably less than testosterone. Oxandrolone is given orally, with 99% bioavailability. It is protein-bound in plasma with a biologic half-life of 9 hours. Oxandrolone metabolic effects are comparable to HGH in that protein synthesis increases, with no anti-insulin effects. Contraindications include the presence of hypercalcaemia or the presence of a tumour with androgenic receptors, namely, prostate cancer or male breast cancer. Exercise, through resistance training, not only increases the rate of anabolism but also increases the percent of weight gain, which is lean mass. The mechanism remains unclear. The musculoskeletal disability after weight loss and PEM, even in the aged or chronically ill population, can be significantly improved when exercise is added to nutrition along with an anabolic agent. Components of respiratory function include: 1- Adequate lung parenchymal function. 2- Normal chest wall mechanics. 3- Normal respiratory drive. 4- Adequate lung immune defences. 5- Work of breathing. Ventilation/perfusion matching is equally dependent on lung parenchyma and chest wall mechanics. The adequacy of oxygen exchange is most dependent on lung parenchymal function. The adequacy of carbon dioxide removal is very dependent on lung mechanics and ventilatory drive. Alveolar ventilation is most dependent on ventilatory drive and chest wall mechanics. Chest wall mechanics is dependent on the integrity of the bone and muscle activity as well as the state of the alveolar parenchyma. Respiratory dysfunction refers to lung function that is not normal. The first to fail is likely to be the immune defences followed by chest wall function, and then, less frequently, ventilatory drive. In the presence of a catabolic state, the components of respiratory function can fail quite rapidly, especially if a preexisting PEM is also present. Impaired arterial oxygenation is the first function to become abnormal. Causes of respiratory dysfunction include: 1- Parenchymal dysfunction (impaired gas exchange). 2- Common impaired chest wall function. 3- Impaired immune defences. 4- Impaired ventilatory drive. Acute lung injury can result from a simple localised lung infection (pneumonia) or from a systemic process that leads to diffuse alveolar damage as seen with the acute respiratory distress syndrome (ARDS). Most acute respiratory illnesses when combined with cough and/or dyspnea, oral intake is generally poor. Patients with severe lung injury may require endotracheal intubation and mechanically assisted ventilation, which precludes adequate oral intake. Acute lung failure often occurs in the setting of sepsis or trauma, conditions associated with a hypercatabolic state. The negative nitrogen balance leads to decreased respiratory muscle strength because of protein catabolism, diminished ventilatory drive, and altered immune function. Acute lung injury can be a catabolic insult resulting in an autodestructive process. COPD is repeated lung infections with a constant increase in energy demands as well as repeated catabolic states. COPD with acute respiratory failure is the most complex disorder to nutritionally manage. The failure of any or all of the components required for respiratory function will lead to acute or chronic respiratory failure. Components of chest wall function are: 1- Integrity of the ribs and sternum. 2- Strength of intercostals and accessory muscles. 3- Integrity of the neuromuscular junctions. 4- Diaphragm strength and function. Any alterations in function of any of these chest wall components will alter respiratory function. The traumatic injury leads to a hypermetabolic, catabolic state with increased demands for gas exchange in the presence of impaired chest wall motion. Also, the host defences are impaired due to inability to cough and clear secretions. Neuromyopathy is disuse atrophy from paralytic agents, spinal cord injury, and complete ventilatory support. The respiratory muscles are subject to fatigue from an imbalance between demands and capabilities. Muscle fatigue is the inability of respiratory muscle to attain the adequate force required to generate its prior force. Muscle weakness is the chronic inability of the respiratory muscles to attain the adequate force required to meet ventilatory demands. Both fatigue and weakness have important relationships to nutrition. Loss of over 20% of ideal body weight, moderate to severe PEM, results in a marked reduction (35% to 40%) in maximum inspiratory and expiratory pressure as well as a measurable decrease in diaphragm thickness. Poor nutritional status may result in a myopathy of the remaining respiratory muscles.

Progressive respiratory disease:

Cause	Intervention
A- Acutely injured or infected patient: 1- Marked hypermetabolism with energy deficit: stress-induced metabolism leading to rapid lean mass loss or micronutrient depletion. 2- Immobilisation stress (inactivity leading to muscle loss). 3- Decreased endogenous anabolic activity.	1- Early aggressive initiation of high-protein, high-energy, micronutrient-rich diet. 2- Early initiation of physical rehabilitation. 3- Anabolic agents to decrease lean mass loss and more rapidly restore lean mass.
B- Chronic obstructive pulmonary disease: 1- Poor nutrition due to anorexia and increased nutrient needs caused by increased work of breathing. 2- Decreased oxygen saturation results in altered cellular metabolism increasing inefficient anaerobic pathways. 3- Fatigue and muscle wasting secondary to dyspnea and inactivity.	Aggressive restoration of high-protein, high-energy, micronutrient-rich diet. Substitute lipid calories for carbohydrate calories to reduce the respiratory quotient. Resistance exercise to prevent further atrophy.
C- Age-related frailty: 1- Poor nutrition due to disability, chewing and swallowing problems, and significant psychosocial issues. 2- Immobilisation stress (inactivity leading to muscle loss and further inactivity "cycle of inactivity"). 3- Decreased endogenous anabolic activity.	High-protein, high-energy, micronutrient-rich diet. Resistance exercise program. Anabolic agents to restore lost weight and lean mass.
D- Acute respiratory failure 1- Marked hypermetabolism and catabolism due both to a focus of injury and infection and also lung inflammation. 2- Possible inactivity on ventilator. 3- Possible use of corticosteroids. 4- Decreased net anabolic activity.	1- Increased calories, protein, micronutrients. 2- Avoid excess carbohydrates. Weaning to work chest wall. Increase anabolic activity. Anabolic agents to decrease losses and increase protein gains.

Resistance exercise:

1- Increased muscle fiber stretch and tension leads to microtears.
2- Muscle responds with a marked increase in protein synthesis from a local and systemic anabolic stimulus.
3- An increased production of endogenous HGH is a component of the anabolism.
4- Muscle cell responds by increased protein deposition increasing mass.
5- This process reduces lean mass loss during "stress" and increases rate of gain during recovery.
6- Increased energy and protein substrate required.
7- Weight gain is mainly lean mass.
8- Increases use of glucose, thereby improving insulin resistance of diabetes.

Severe respiratory muscle weakness leads to ventilatory failure (hypercapnic respiratory failure). Fatigue results when the energy demand of the muscle exceeds the energy supply. COPD subjects are predisposed to heightened respiratory-muscle energy requirements based on an increased ventilatory demand, high inspiratory workload, and reduced mechanical efficiency. The reversal of muscle fatigue generally requires a period of rest for recovery of muscle function. Respiratory muscle fatigue is best viewed as a continuous process. The process begins when the respiratory muscles are subjected to an excessive mechanical load, leading to a series of changes within the neuromuscular command chain. Electrolyte abnormalities including hypokalaemia, hypophosphataemia, hypocalcaemia, and hypomagnesaemia are recognised to affect respiratory muscle function adversely. Hypophosphataemia is a particularly common complication of COPD. PEM results in a progressive reduction in body weight and skeletal muscle mass. PEM also results in biochemical changes within skeletal muscle, which are independent of changes in muscle fiber dimension. Abnormal respiratory muscle cellular energy metabolism is characteristic of COPD patients with respiratory failure and may relate to malnutrition. There is reduced energy stores in the form of ATP (adenosine triphosphate) and phosphocreatine, which adds to the loss of muscle power. Early aggressive nutrition support will gradually correct the PEM and existing muscle weakness. Antioxidant restoration is also essential. PEM is associated with decrease in hypoxic and hypercarbic (or decreased pH) ventilatory drive. PEM patients also become hypometabolic due to cachexia and a chronic energy deficit. Therefore, oxygen demands and CO_2 production are lower than normal. The respiratory system has a number of important immune defences, which depend on chest wall function and several are immune defences of the lung itself. Systemic immune defences are also very important. The severity of all aspects of host defences increases with increasing severity of PEM. Catabolism and PEM-induced lung immune dysfunction: 1- Altered function of the nasopharyngeal filter, especially if endotracheal tube present. 2- Impaired cough due to weakness. 3- Impaired mucociliary clearance (impaired cough, energy deficit). 4- Oropharyngeal and tracheobronchial colonisation with pathogens (impaired cough, immunosuppression). 5- Impaired cellular and humoral immunity. 6- Immunosuppression from malnutrition. 7- Inability to contain infection. The presence of a catabolic state and/or moderate to severe PEM and IWL requires at least a 50% increase in calories and high protein intake. Increased micronutrients are essential because of increased needs and increased losses. In addition, increased zinc, copper, vitamin C and vitamin A, glutamine, and arginine. The respiratory muscles normally provide the force required for inspiration. Expiratory flow is driven by the elastic recoil of the respiratory system (passive). Normally, <5% of the total oxygen consumption is used for the work of breathing; although this value may approach 50% under conditions of respiratory disease.

Impaired chest wall function:

Impaired bone structure with fractures:	1- Produces catabolic state. 2- Impairs mechanics. 3- Impairs lung defences.
Impaired muscle function due to:	A- Current catabolic state and weakness. B- Inadequate nutrient intake to meet current demands or corrects PEM. C- Inadequate anabolic drive to restore muscle mass. D- Use of corticosteroids (further catabolic insult). E- Disuse atrophy.

Causes of respiratory muscle failure in COPD:

1- Nutritional status.
2- Infection.
3- Drug therapy (sedatives, paralytics).
4- Hypoxaemia.
5- Hypercarbia.
6- Electrolyte abnormalities (hypocalcaemia, hypokalaemia, hypophosphataemia, hypomagnesaemia).
7- Corticosteroid use.

All components of respiratory function affect work of breathing. Any stiffness of the lung or of the chest wall will significantly increase work. Also, work will be increased with increased cough, increased respiratory rate, abdominal distention, etc. If the increased oxygen demands cannot be met and muscle fatigue occurs, respiratory distress will follow. Higher energy consumption by the respiratory muscles to meet the demands of compromised respiratory function produces a hypermetabolic state and leads to progressive weight loss when output exceeds caloric intake. Anabolic therapy added to nutrition can more rapidly restore a previous loss of lean mass to avoid a rapid fatigue from increased work. The diffuse lung injury (ARDS) remains a major cause of mortality and morbidity. ARDS is part of an inflammation-induced systemic disease state that can evolve to multiple system organ failure (MSOF). MSOF is the leading cause of death in posttrauma or surgical patient disease. ARDS from shock, infection, and trauma is both a primary pulmonary process and a component of a generalised inflammatory reaction. The pathobiology is nearly identical (hypoxaemia, increased shunt, decreased compliance, and an increase in lung water). There is a significant parenchymal, interstitial and airways inflammatory component to the disease process responsible for the increased lung stiffness. The systemic disorder initiating lung inflammation also activates the stress response to injury or infection. Therefore, a maladaptive hormonal environment and diffuse inflammation lead to a progressive catabolism and hypermetabolism. The initial response is the onset of a catabolic state as a result of a systemic insult. The first phase of the lung inflammation is some mild hypoxaemia and tachypnoea, which may be due to a CNS response because a respiratory alkalosis is usually present. It is likely that this early lung inflammation phase is very common and resolves along with the initiating insult before developing any lung dysfunction. If lung inflammation increases and if the WBCs are now activated by cytokines to release toxic mediators, a phase 2 (or progression of the) lung injury develops. The inflammatory mediators-cytokines, oxidants, and proteases produce ongoing lung damage. The second phase begins within 12-24 hrs of the early symptoms and is characterised by physiologic and pathologic evidence of lung injury. Mediator damage results in injury to the circulatory side of the alveolar capillary membrane. Hypoxaemia is now evident, along with a continuing tachypnoea and increasing shunt. Radiographic evidence of early parenchymal change is noted by peripheral patchy lung opacities (fluffy infiltrates) appearing initially in dependent lung fields (ARDS). An autodestructive lung inflammatory response is present. The majority of phase 2 lung processes also resolve if the underlying systemic insult is corrected. If the systemic insult worsens or if the inflammatory lung response increases, then an established ARDS results, which has a mortality rate of >50%. The third phase is characterised by acute respiratory failure necessitating mechanical ventilation. The severely-injured lung causes ongoing stress response and hormonal maladaptation. Although the mortality rate at this stage is high, complete resolution can occur over the subsequent weeks if both the systemic and lung process can be turned off. If not, MSOF can occur due to the visceral organ response to the release of inflammatory mediators from the lung, resulting in sepsis syndrome.

Lung immune defences:

A- Nasopharyngeal filter:
    1- Mechanical filter removing particles.
    2- Immune defences (cellular and humoral).
    3- White cells to kill microbes.
B- Oropharynx preventing pathogen overgrowth:
    1- Mechanical clearance (cough).
    2- Immune defences (cellular and humoral).
C- Airway macrociliary activity:
    1- Microbes and particles to upper airway for removal.
    2- Mechanical (ciliary movement).
D- Tracheobronchial immune defences:
    1- Mechanical (cough).
    2- Cellular and humoral immunity.
E- Cough reflex to remove mucus and microbes:
    1- Which have been in the airways.
    2- Mechanical (chest wall activity).
F- Lung parenchymal cellular and humoral immunity:
    1- Alveolar macrophages that are residents of the lung.
    2- Lymphocytes (T and B).
    3- Macrophages and neutrophils that migrate to the lung to assist lung white cells.
    4- Humoral defences produced by the resident cells (antibodies, complement, and opsonins).
G- Lung containment of infection or injury:
    1- Prevention of spread of infection by using cellular, humoral defences, and immune defences.
    2- Dependent on cough (chest wall) and mucociliary function.

Respiratory failure and catabolic state:

1- A systemic catabolic insult produces acute respiratory failure (ARDS, pneumonia):
    A- Lung inflammatory disease (ARDS).
    B- Impaired lung immunity (pneumonia).
2- Acute respiratory failure causing catabolism and its complications:
    A- ARDS leading to MSOF.
    B- Chronic recurrent pneumonia.
3- Catabolic insult with COPD and PEM causing respiratory failure.

As the lung macrophages are activated and intense inflammation is established, cytokines and mediators released from the lung injure systemic organs and amplify the hypermetabolic catabolic state. With time the lung exudates become lung fibrosis, much more difficult to resolve, and single organ failure evolves to MSOF. Mortality rate at this stage exceeds 90%. Chronic lung disease is either obstructive or restrictive. Obstructive disease is the most common form, characterised by airflow obstruction mainly during expiration, e.g., asthma, emphysema, cystic fibrosis, and chronic bronchitis. Restrictive disease is characterised by a decrease in lung compliance or an impairment in lung expansion, e.g., pulmonary lung fibrosis, infiltrative disease, collagen vascular, amyloid, and sarcoid. The major chronic restrictive diseases can be divided into lung, chest wall, and diaphragm. Lung disorders include diseases such as pulmonary fibrosis from prior acute disease or an idiopathic cause, cystic fibrosis, and a variety of collagen vascular diseases. Chest wall dysfunction includes mainly muscle weakness. However, chest wall deformity also fits into the category, including that caused by obesity, which impairs chest movement. Obstructive pulmonary disease is characterised by the lung parenchyma is hyperinflated and the diaphragm is flattened. Obstructive lung disease can be classified into: 1- reversible type (bronchial asthma, other acute obstructive disorders), or 2- irreversible type (obstructive emphysema, chronic bronchitis, chronic bronchiolitis, cystic fibrosis). The initial cause of asthma is increased airway smooth muscle contraction-caused by a variety of stimuli including anxiety brought on by inhaling an allergen and inhalation of cold air, which activates the smooth muscle. Bronchial inflammation is characteristic and the predominant cells are mast cells (macrophages) and eosinophils. Recurrent episodes of an acute increase in airway resistance markedly increase energy expenditures. Asthma attacks are characterised by the catabolism of inflammation and infection. The airways are extremely sensitive to arachidonic acid and its metabolites. Corticosteroids, both local and systemic, are frequently used; the latter, leads to further catabolism and lean mass loss. The major predisposing factor of obstructive emphysema is smoking. There is a subgroup of this disorder caused by environmental hazards. In addition, there is the untreated emphysema caused by α1 antitrypsin globulin deficiency. The principal change is the loss of lung elastic recoil due to destruction of lung elastic properties likely due to activation of a variety of proteolytic enzymes including elastase. A hyperinflated lung and a flattened diaphragm are typically seen. There is a marked increase in the work of breathing, requiring excess calories to avoid weight loss. Episodes of acute respiratory failure are common, especially in the later stages, due mainly to infection as lung defense mechanisms, especially cough, are severely compromised. Catabolism and weight-loss-induced protein loss are common. Each acute episode produces a catabolic state, which can lead to progressive losses of lean mass, with markedly increased risk of further infection and further impairment in lung function. Chronic bronchitis is characterised by chronic secretions of excess mucin by the bronchial tree. Cigarette smoking is the major initiating factor. Recurrent infections are not uncommon, leading to an almost constant catabolic state. Lung status is very dependent on maintenance of the muscles of respiration. The addition of corticosteroids further increases net catabolism. Cystic fibrosis is associated with constant increased energy and frequent catabolic states. This genetically transmitted disorder involves a number of organs including lung, pancreas, and liver. The lung dysfunction characterised by thick inspirational secretions resembles a COPD. The increase in work of breathing results in an increase in energy demands by 20% to 50% above normal. There are several naturally occurring antioxidants (superoxide dismutase, catalase, and glutathione) present in the airways to protect against inhaled oxidants. Vitamin A, ceruloplasmin, copper, methionine, and vitamins E and C may also protect against oxidant-induced destruction. The amount of oxidants in large amounts of smoke may still exceed these defences. Dietary antioxidants such as vitamin C and retinols may limit the destruction of lung tissue by proteases and protect against development or progression of COPD. An improvement in QoL and decreased complications is expected when optimising nutrition. PEM and IWL are common in patients with COPD, with incidences ranging from 20% to 60%. Patients with IWL who are >10% heavier than their ideal weight have a 5-fold increase in mortality compared with COPD patients without significant weight loss. Mechanism of weight loss: 1- inadequate dietary intake (anorexia, dyspnoea), 2- increased energy demands (increased work of breathing), 3- increased catabolism from cytokines and oxidants, and 4- impaired cardiopulmonary functions impeding nutrient delivery and processing. As the severity of the COPD disorders increases, a hypermetabolic catabolic state and PEM frequently develop. The catabolic proinflammatory cytokine, TNF (cachexin), is chronically elevated even without infections. Repeated pulmonary infections also cause increased protein losses. This patient group has difficulty meeting goals due to both pulmonary problems (dyspnoea) and gastrointestinal problems (bloating, early satiety, and anorexia). The latter appear to be due to the characteristic flattening of the diaphragm impinging the abdominal cavity. Restoring lost body weight using nutrient supplements (and anabolic steroids if indicated) and increased antioxidants has been shown to improve respiratory muscle function.

Summary of ARDS (inflammatory lung disease):

Phase I	1- Early lung changes caused by systemic focus of injury and inflammation with onset of a catabolic state (systemic insult initiating lung inflammation). 2- Early phase: intense lung congestion and inflammation.
Phase II	1- Onset of increased lung parenchymal inflammation, exudates, and infiltrates (days 1-4). Lung damage caused by inflammation. 2- Increased alveolar consolidation and capillary permeability begins to impair gas exchange, increasing short fraction.
Phase III	1- Acute progressive ARDS causing catabolism (days 2-10). 2- Intense lung inflammation produces large amounts of cytokines and mediators, which cause systemic damage (sepsis syndrome). 3- Characterised by: A- Diffuse lung infiltrates and air bronchograms. B- Catabolic inflammatory state as well as a maladaptive hormonal response.
Phase IV	1- Pulmonary fibrosis or pneumonia (≥10 days). 2- Systemic catabolic state caused by the ARDS. 3- Multisystem organ failure (MOF). 4- Lung fibrosis and recurrent pneumonia (sepsis). 5- The lung becomes the organ of sepsis with cytokine release and frequent pneumonias. 6- Late phase: macrophage infiltration and increased collagen. Macrophages release inflammatory cytokines that increase the inflammatory and maladaptive hormonal-induced systemic catabolic state.

Nutrition for COPD:

	Stable	Stressed
Protein	1.5 g/kg	2 g/kg
Micronutrients	5-10 times RDA	5-10 RDA
Pancreatic enzymes	As needed	As needed
Antioxidants	Increase	Increase
Anabolic agents	Not indicated	Indicated to retain/restore lean mass.
Calories	30 cal/kg	35 cal/kg

Nutrition without weight does not improve function. Acute-on-chronic respiratory failure (ACRF) is the exacerbations of chronic ventilatory failure, often requiring ICU admission and usually occurring in patients with COPD. Precipitating factors include: 1- infection (bacterial, viral), 2- cardiac disease, 3- environmental, 4- medications (β-blocking agents, narcotics/sedatives), and 5- respiratory muscle fatigue. The aim of nutritional therapy in the acute setting is to maintain calorie energy demands and correct obvious deficiencies. Caloric intake that exceeds energy demand results in heightened carbon dioxide production and increased ventilatory requirements. Excess energy intake is generally converted to fat stores without significant expansion of lean tissue mass during short-term nutritional repletion. Anabolic agents may be of benefit. A loss of lean body mass and the presence of IWL and PEM clearly lead to deterioration of respiratory function. Both acute respiratory distress and COPD produce an increase in energy expenditure and either a continuous or intermittent increase in net catabolism (body protein loss). Catabolism, IWL, and PEM accelerate both acute and chronic respiratory disease. ARDS and COPD lead to increased energy, protein, and micronutrient demands. IWL and PEM are more difficult to correct than prevent in the presence of respiratory failure. An early diagnosis of a hypermetabolic, catabolic state is essential. Early optimisation of nutrition, both macro- and micro-nutrients including antioxidants, is essential to preventing this autodestructive cycle. A hypercatabolic state is produced from any stress response to injury or infection, any inflammatory state, as well as a host of neuromuscular disorders including use of corticosteroids. A net protein loss is likely with any respiratory disorder, and increased energy demands.

DIETARY INTERVENTION FOR CHRONIC DISEASE PREVENTION

Most leading causes of adult deaths in developed nations including coronary heart disease (CHD), cancer, and stroke are influenced by diet. Other diseases associated with significant morbidity, including hypertension, diabetes mellitus, obesity, and osteoporosis, are also closely linked with dietary intake, as an aetiologic or exacerbating factor. Low-density lipoprotein (LDL) is the major cholesterol-carrying lipoprotein particle in circulation. It is also the major atherogenic lipoprotein in humans, with serum concentration of LDL-cholesterol (LDL-C) positively associated with the risk of CHD. For every 1 mg/dl increase in LDL-C, the risk of CHD increases by 1-2%. The oxidation of LDL specifically, the peroxidation of polyunsaturated fatty acids (PUFAs) in LDL particles is atherogenic. High-density lipoprotein (HDL) is involved in reverse cholesterol transport, carrying cholesterol from peripheral tissues to the liver for metabolism or excretion. HDL is an antiatherogenic lipoprotein; a 1 mg/dl increase in HDL-C concentration is associated with a 2-3% decrease in the risk of CHD. Triglyceride (triacylglycerol, TG) is the main form of dietary fat, and is positively associated with the risk of CHD. Hypertriglyceridaemia is associated with low HDL-C concentration and increased blood coagulability. Dietary fat is not a single entity but consists of various fatty acids and cholesterol. A diet high in SFAs (e.g., Palmitic acid) will raise LDL-C, and to a lesser extent HDL-C. Lauric and myristic acids (found in tropical oils) also raise cholesterol concentration. Stearic acid (the primary fatty acid in chocolate) has not been shown to elevate cholesterol concentration significantly. Polyunsaturated fatty acids are classified as ω-6 or ω-3 fatty acids, depending on the location of the first double bond. The primary ω-6 fatty acid is linoleic acid, an essential fatty acid found in plant oils such as corn oil. Replacing saturated fats in the diet with unsaturated fats may ultimately have the same effect on CHD risk as replacing fat calories with those from carbohydrates. The main dietary ω-3 PUFA is linolenic acid, found in soybean, canola, and fish oils. Linolenic acid reduces plasma TG concentration. Linolenic acid is elongated in the body to eicosapentaenoic acid (EPA) and docosahexaenoic acid (DHA). These long-chain fatty acids may decrease the risk of thrombosis and lower blood pressure. Eating fish at least once a week reduces the risk of sudden cardiac death by half compared with men who consumed fish less than once per month. Fish oil has high susceptibility to peroxidation. The only nutritionally important monounsaturated fatty acids (MUFA) is oleic acid, abundant in olive and canola oils and also in nuts. Liberal amounts of olive oil as the primary source of fat result in an equivalent drop in LDL-C, but also an increase in HDL-C. Monounsaturated fatty acids may decrease serum glucose and TG concentrations in type II diabetics. Monounsaturated fatty acids may decrease the susceptibility of LDL to oxidative modification. First-pressed extra virgin olive oil contains appreciable amounts of polyphenolic compounds, which may inhibit the oxidation of LDL. Extra virgin olive oil contains other antioxidants, such as tocopherols and flavonoids, which may account for some of its purported beneficial effects. Trans-monounsaturated fatty acids (TFAs) are produced through the hydrogenation of PUFAs, converting them from liquid oils into semisolid fats for use in commercially prepared foods. Hydrogenation results in the formation of trans double bonds, and the resulting TFAs (margarines, snacks, crackers, cookies, and in fast-food service fats and oils) resemble SFAs in conformation. TFAs increase LDL-C to levels similar to those produced by SFAs, but also decrease HDL-C concentration, potentially making them more atherogenic than SFAs. Trans-monounsaturated fatty acids have been associated with an increased risk of CHD. Dietary cholesterol (primarily egg yolks, fatty meats, and dairy products) has less effect on serum cholesterol than do SFAs or PUFAs. Dietary fat may be related to the development of other chronic diseases and certain cancers, including breast, colon, prostate, and pancreas. Any connection between fat and cancer may actually be due to the confounding effect of energy. Dietary fiber refers to plant materials that are not readily digestible by enzymes in the human gut. Water-soluble fibers include pectins, gums, mucilages, and some hemicelluloses. Insoluble fibers include cellulose and other hemicelluloses. The consumption of insoluble fiber, especially cellulose, tends to normalise large bowel function, decreasing transit time in constipation and increasing it in certain conditions characterised by chronic diarrhoea. Dietary cellulose increases stool weight and bulk but can result in dry stools. The addition of a soluble fiber such as pectin increases the water content of stools, providing for easier evacuation. The anticarcinogenic effects previously attributed to fiber may in fact be due to the effects of the nonfiber constituents of fruits and vegetables with possible anticancer properties, such as antioxidant vitamins and minerals. High fiber intake is associated with lower energy intake, which may also decrease cancer risk independently and confound the relation between fiber and colorectal cancer. Most soluble fibers reduce plasma total cholesterol and LDL-C concentrations. High fiber intake may have a beneficial effect on blood pressure. Water-soluble fiber, especially gel-forming pectins and gums, blunts the postprandial glycaemic and insulinaemic responses in normal and diabetic individuals and may have a role in the prevention and treatment of diabetes mellitus and insulin resistance. High intake of fiber has been associated with a reduced risk of cancer of hormonal tissues, including the breast and ovary. Because some of the purported benefits of high-fiber foods may be due to components other than fiber, high-fiber foods, rather than isolated fiber, should be recommended whenever possible. Pectin is found in fruits and vegetables, especially apples, oranges, strawberries, and carrots. Gums are found in oat bran, barley, and legumes. Whole-wheat flour, wheat bran, and vegetables are good sources of cellulose, while hemicellulose is found in wheat bran and whole grains. Consumption of liberal amounts of water should be encouraged when adding fiber to prevent faecal impaction. High-fiber foods should be added gradually to the diet to minimise gastrointestinal symptoms, which may include flatulence, borborygmus, cramps, and diarrhoea. The relationship between sodium (Na) and blood pressure is not a simple one. 30-50% of hypertensive patients and 15-33% of normotensive individuals are salt-sensitive. Only about 5-10% of sodium consumed occurs naturally in foods (primarily in meat). The rest is added during cooking or at the table (20-40%) or is consumed in processed foods (30-70%). Foods associated with excessive saltiness (e.g., potato chips) often contain much less Na than foods with hidden Na, such as bread and canned vegetables and soups. An increase in potassium (K) intake may provide some protection from the deleterious effects of a high-sodium diet in humans. Dietary potassium supplementation lowers blood pressure in established hypertension. The liberal intake of potassium-rich fresh fruits and vegetables (including bananas, orange juice, prunes, and dried beans) should be encouraged in otherwise healthy individuals. Processed fruits and vegetables are generally depleted of K and higher in Na compared with their fresh counterparts, resulting in much lower K-to-Na ratios. Calcium (Ca) is an integral component of bone in the form of hydroxyapatite, a crystalline structure of calcium phosphate in an organic matrix of collagenous protein. Optimal calcium intake (1,200-1,500 mg/day) in childhood and young adulthood is crucial to achieving peak adult bone mass. Inadequate Ca intake is associated with reduced bone mass in men and optimal Ca intake should also be emphasised in men. The DRI (Dietary reference intake) for men above the age of 50 and postmenopausal women is 1,200 mg/day. Deficient Ca status in adults older than 65 years is common due to decreased Ca intake and absorption and decreased vitamin D intake and synthesis. Increased Ca intake in this age group may also reduce bone loss and decrease the incidence of fractures. Dietary Ca recommendations for the prevention of osteoporosis would likely be adequate for blood pressure lowering. High Ca intake may decrease the risk of colorectal cancer. Low Ca intake is associated with increased risk of ischaemic stroke. High Ca intake may prevent nephrolithiasis, possibly by maintaining a lower rate of parathyroid hormone secretion and reducing absorption of oxalate. Nonfat or low-fat (1% fat) dairy products such as milk and yoghurt are excellent sources of Ca due to their high Ca content.

Lipodystrophy syndrome vs wasting syndrome:

Parameter	Lipodystrophy syndrome	Wasting syndrome
Body weight	Increase or decrease	Decrease
Body fat	Peripheral decrease, and central increase	Decrease, all areas.
Lean body mass	Unchanged	Decrease
Total cholesterol	Increase	Decrease
VLDL cholesterol	Increase	Decrease
LDL cholesterol	Increase	Decrease
HDL cholesterol	Decrease	Decrease
Triglycerides	Increase	Increase
Diabetes/insulin resistance	Yes	No
Coronary artery disease	Yes	No

Calcium supplements (in the form of carbonate or citrate) may be necessary to ensure adequate intake in cases of inadequate intake from foods. Optimal Ca utilisation is dependent on adequate vitamin D status. Up to 20 µg/day (800 IU/day) improves Ca balance and may help reduce the risk of fractures in the elderly. Excessive vitamin D intake (≥50 µg/day) can result in hypercalcaemia and hypercalciuria and should be avoided. Fish, such as salmon and sardines, contain substantial amounts of vitamin D. Supplements may be necessary when dietary intake and sun exposure are limited. The postsecretory oxidation of LDL by oxygen free radicals (OFR) results in the unregulated uptake of cholesterol in arterial walls, accelerating the atherosclerotic process. Cumulative oxidative damage to DNA may be associated with the rise in cancer risk that occurs with increasing age. Excess production of free radicals may contribute to tissue damage in rheumatoid arthritis, inflammatory bowel diseases (including Crohn's disease and ulcerative colitis), cataracts, macular degeneration, and neurodegenerative diseases and may be a major contributor to the aging process. Being lipid-soluble, α-tocopherol is present in LDL particles and inhibits the oxidation of LDL particles. There is a significant reduction in the incidence of cardiac events in men and women taking high-dose vitamin E supplements (200-400 mg/day). Patients should be advised to increase their consumption of healthful foods, including fruits and vegetables, rather than specific nutrients, such as antioxidant vitamins. Patients should be advised to eat at least 5-7 servings of fruits and vegetables each day. Excess caloric intake over time without a corresponding increase in energy expenditure results in the accumulation of excess body fat (obesity). Weight loss results in modest reductions in blood pressure. Abdominal obesity is associated with glucose intolerance and insulin resistance and is a risk factor for type 2 diabetes mellitus. Weight loss improves glucose tolerance and decreases insulin resistance, reducing the risk and severity of diabetes. Obesity increases the risk of gallbladder disease (including cholelithiasis), especially in women. High caloric intake is associated with an increased prevalence of cancers of the breast, cervix, endometrium, ovary, gallbladder, colon cancer, and prostate. Higher intakes of alcohol are associated with deleterious effects, including increased risk of hypertension, cardiomyopathy, and haemorrhagic stroke. More than 3 drinks/day is associated with a rise in blood pressure, and at >4 drinks/day, the average increase is 5-6 mmHg in systolic pressure and 2-4 mmHg in diastolic pressure.

Criteria of lipodystrophy syndrome or metabolic syndrome X:

1- Loss of peripheral fat tissue.
2- Increased central fat tissue (abdominal girth).
3- Visible subcutaneous veins.
4- Loss of temple, nasolabial, and cheek fat pads.
5- Buffalo hump or bull neck.
6- Breast hypertrophy.
7- Enlarged supraclavicular fat pads.
8- Bilateral symmetric lipomatosis (enlarged axillary fat pads).
9- Decreased thigh size.
10- Hyperlipidaemia (cholesterol, triglycerides).
11- Diabetes.
12- Peripheral insulin resistance.
13- Hyperinsulinism.
14- Hyperuricaemia.
15- Low serum testosterone.
16- Increased C peptide.
17- Increased free fatty acids (FFAs).
18- Coronary artery disease, ischaemic heart disease.
19- Hypertension.

Lipoproteinlipase:

Tissue	Role
Muscle (cardiac, skeletal)	Energy provision
White adipose tissue	Triglyceride storage
Brown adipose tissue	Thermogenesis
Lactating breast	Milk triglyceride synthesis
Lung	Surfactant synthesis
Brain	Phospholipid and glycolipid synthesis

Excess ethanol consumption is also associated with increased plasma TG concentration. Ethanol consumption is associated with an increased risk of certain malignancies, including oropharyngeal, laryngeal, and oesophageal cancer. Alcohol intake is positively associated with the risk of breast cancer in a dose-response relation. Excessive ethanol consumption may increase the risk of colorectal cancer and its precursor lesion, adenomatous polyps. Alcohol intake is a risk factor for osteoporosis. Excessive ethanol consumption is a well-established risk factor for liver disease, including fatty liver, alcoholic hepatitis, and cirrhosis, and is associated with increased mortality from pancreatitis and gastritis. Moderate elevations of plasma homocysteine, a sulfur-containing amino acid is with increased risk of peripheral vascular and cerebrovascular disease and CHD. Folic acid is required for the methylation reaction that converts homocysteine to methionine, another amino acid involved in one-carbon transfer reactions. Reduced plasma concentrations of folic acid can result in homocystinaemia. Folic acid, a water-soluble vitamin required for several one-carbon transfer reactions. Folic acid supplementation decreases plasma homocysteine concentrations, an effect that may be enhanced with the addition of vitamin B_{12}. Folic acid deficiency is related to an increased prevalence of precancerous dysplastic changes in cervical and colorectal epithelium. Adequate folic acid intake (400 μg/day) should be ensured through diet (especially fresh, dark-green leafy vegetables) and, if necessary, supplements.

CACHEXIA IN IMMUNOCOMPROMISED PATIENTS

More than 60% of HIV-positive persons are presented with PEM and vitamin and mineral deficit. This leads to progressive physical-metabolic wasting (wasting syndrome/cachexia) and increased susceptibility to opportunistic infections and drug toxicity. Protein energy malnutrition (PEM) is, either in association with other diseases or as a proper disease in itself, the first stage of cachexia. An insufficient intake of food is the first cause of malnutrition, and infectious diseases are the second. Malnutrition progresses to cachexia, its final stage, in oncologic and infectious diseases. Cachexia defines a state of general wasting associated with individual, social, or medical causes. Cachexia is caused by a series of factors, such as anorexia, sickness, dysgeusia, poor alimentary habits, and disturbances of digestion and absorption linked to digestive tract infections. All of these factors cause progressive weight loss, lipid store and muscular body mass loss, and negative nitrogen balance, with depletion of circulating and visceral proteins, in some cases advancing to African slim disease (African famine figure). Malnutrition and cachexia is either a weight loss of more than 10% of the ideal or usual weight or a body weight that falls below the 15th percentile for persons of the same age and height. Under normal physiologic conditions, the central nervous system balances the quantity of assumed and consumed calories through hunger and satiety, thereby maintaining body weight. The human body consumes energy in 3 ways: 1- calories burned for basal metabolism (REE), 2- calories burned for absorption of nutrients (dynamic specific action [DSA]), and 3- calories burned for physical activity. A rapid weight loss (over days) reflects a predominant reduction of the hydric mass of the body, while a reduction of tissue mass takes longer (weeks or months). In the case of infectious diseases, neoplasms, burns, traumas, and operations, a hypercatabolism develops via similar metabolic responses (e.g., acute phase response [APR]) and leads to self-cannibalism and cachexia. During serious illness, normal nourishment is not sufficient to balance the intense catabolism and nitrogen loss caused by the pathologic process. In such a condition, glucose, lipid, and protein reserves are burned in order to cover energy needs (self-cannibalism). If this phenomenon lasts for a long time, the effect can be fatal: the loss of one third of body proteins (2 kg of protein or 8 kg of cellular mass) in a person of average weight causes death in <1 month. The degree of cachexia itself is a factor of mortality independent of its cause. At death, the cellular mass is reduced on average to 54% of normal, corresponding to a body weight that is 66% of the ideal. In voluntary or forced starvation, death occurs when the body weight is reduced to approximately one third of ideal weight. During the septic period, starvation does not activate the energy-saving mechanism occurring in absence of infection, so that caloric withdrawal causes a further protein loss, which can reach 90 g/d. In sepsis, the basal energy requirement increases because of hyperthermia induced by cytokines and futile metabolic cycles. Hypertriglyceridaemia starts mainly through the inhibition of lipoprotein lipase (LPL), an enzyme that regulates the clearance of plasma triglycerides and energy production from lipids. LPL is inhibited by TNF and estradiol, while it is stimulated by progestins. An increase of liver lipogenesis is associated with increased production of very low-density lipoprotein (VLDL). In the fat cell, LPL inhibition causes a reduction of lipogenesis with consequent lipolysis. These effects can be avoided by administering prostaglandin synthesis inhibitors. The increase of plasma triglycerides observed in cachexia is due not only to the LPL inhibition and clearance but also to an increase of VLDL synthesis from free fatty acids (FFA) resulting from the lipolysis of peripheral fat. The liver synthesis of VLDL starts from FFA mobilised from peripheral fat, re-esterified to triglycerides by the liver, and secreted as VLDL (energy-wasting futile metabolic cycle). Fatty acids mobilised from peripheral fat reach the liver, where they elude oxidation. They are then re-esterified to triglycerides and pass into the blood circulation as VLDL. These lipoproteins return in the adipose tissue to be hydrolysed again to fatty acids by LPL. They are stored in fat cells as triglycerides. This vicious circle, which is caused by TNF, IL-1, IL-6, and IFN-α, does not produce energy (in the form of ATP) but rather consumes it, thus contributing to weight loss. In fat cells, an increase in cAMP activates a protein kinase, which in turn activates, by phosphorylation, a triglyceride lipase. Fat tissue stores and mobilises available energy quickly, and also produces hormone-like substances. One of these is called leptin, which regulates hypothalamic centers of hunger and satiety. 60% of adipose tissue is composed of adipocytes, the rest being blood cells, pericytes, preadipocytes, and fibroblasts. In the human embryo, fat tissue is observed from the second trimester of pregnancy, and its localisation remains constant in foetal and adult life. At birth, the fatty tissue is approximately 14% of body weight, which varies according to the mother's general condition. During the first year, adipocytes undergo hypertrophy and then replicate until full body development is completed. Only 2% of mature adipocytes can undergo mitosis with appropriate stimulation. Therefore, adipocyte hypertrophies, rather than an increase in cell number, seems responsible for diffuse or localised increases in fatty masses. Adipose tissue distribution depends on genetic and hormonal factors. In women, fat tissue is mainly localised to the thighs, with a central redistribution occurring after menopause. In men, fat tissue has a mainly visceral distribution. An excess visceral fat storage is associated with hyperinsulinaemia, diabetes, hyperlipidaemia, hypertension, decreased glucose tolerance, and increased cardiovascular disease risk. Metabolic activity in adipose tissue varies according to different regional distributions. Omental fat is more susceptible to catecholamines than epigastric fat because its adipocytes are bigger and contain more β_3-receptors. Since an increased lipolytic hydrolysis takes place in the visceral fat, a flow of FFA is directly conveyed to the liver. Hepatic VLDL synthesis increases, whereas uptake of insulin by the liver decreases. Chronic hyperinsulinaemia induces a peripheral insulin resistance, increasing diabetes risk. Subcutaneous thigh fat is less susceptible to catecholamines and has a higher LPL index, leading to a greater liposynthetic activity. The quantity of cortisol and androgen receptors is greater in visceral than in subcutaneous fat. As a consequence, LPL stimulation is stronger. Androgen stimulation of β-receptors increases hormone-sensitive lipase activity. The main fat disturbance in cachexia is a decrease in cholesterol levels, an increase in triglyceride levels, and a global loss of fat together with lean tissue in any body area. These patterns can be clearly distinguished from lipodystrophy. Female patients, especially, are psychologically distressed by such changes in their body shape. These changes include an increase in abdominal girth caused by excess fat deposits in the buttocks and abdomen; loss of buccal, parotid, Bichat, and preauricular fat pads, resulting in a cachectic appearance, with thinning and increased wrinkling of the face, prominent zygomata, and sunken eyeballs; and thinning of the arms and legs with prominence of subcutaneous veins (peripheral lipodystrophy). TNF inhibits and progestin derivatives stimulate LPL, the key enzyme of lipid metabolism. Inhibition of CRABP-1 and cytochrome P-450 3A isoform results in decreased cell differentiation and in apoptosis with reduced triglyceride storage and release. In normal conditions, retinoic acid is transformed to cis-9-retinoic acid after binding to CRABP-1 in peripheral adipocytes, which activates retinoic X receptor (RXR). Early and aggressive nutritional treatment improves the prognosis of various pathologic conditions. Hypercaloric-hyperproteinic diets represent the easiest way to increase body weight. The parameters used most for determining the start of therapy are predisease weight loss greater than 10%; albumin value below 3.5 g/dl; and reductions in transferrin, creatinine/weight index, and vitamins and microelements, as well as changes in other indices of nutrition. Nutritional intervention must be prescribed at all stages of HIV infection; during the asymptomatic period, counselling may be sufficient to promote a balanced protein-caloric diet for weight maintenance and prevention of PEM and vitamin/mineral deficiencies. Later, when the disease progresses, enteral and/or parenteral nutrition is often necessary to provide adequate dietary support. When signs of malnutrition appear in a patient, suitable nutritional treatment is advised. In primary malnutrition, a combined dietary-pharmacologic treatment has little or only transient effect. Secondary malnutrition, on the other hand, can be corrected by medical-dietary intervention. Lipids must be prescribed to avoid essential fatty acid deficit. Such energy substrates as polyunsaturated fatty acids (PUFAs), ornithine-ketoglutarate acid, medium chain triglycerides (MCTs), short chain fatty acids, and glutamine can be used to modulate the different stages of cachexia. The administration of lipids has little effect on nitrogen loss if not combined with glucose and proteins. Dextrose infusion inhibits protein catabolism: 150-200 g of dextrose (3 L at 5%) reduces the urinary loss of nitrogen by half. Hepatic neoglycogenesis is blocked, but the oxidative deamination is not. At least 600 g of dextrose is necessary to block this catabolism also and to reduce the urinary loss of nitrogen to 25%. In order to maintain glycaemia under 150 g/dl, insulin must be administered. A reduced supply of ω-6 PUFAs and an increased supply of ω-3 PUFAs may modulate the inflammatory process and cytokine production. Cachexia and tumour growth rate can be reduced by replacing a portion of carbohydrates with lipid derivatives of flaxseed oil at 50% of the calories normally from carbohydrates. Even neoplastic patients with a weight loss greater than 32% recover their weight with isocaloric diets in which the energy is supplied by MCTs at 70%. This effect seems to be due to eicosapentaenoic acid (EPA). This acid, together with docosahexaenoic acid, is the main component of flaxseed oil. In contrast to other fatty acids of the ω-3 and ω-6 series, EPA is a direct suppressant of lipid mobilisation factor and counteracts weight loss, lipolysis, and protein catabolism as well as tumour growth. In order to block protein loss, it is necessary to administer amino acids in a quantity ranging from 1.5-2 g/kg/d, as well as a quota (<0.7 g/kg) of essential amino acids. Positive effects may be obtained also through the use of appetite-stimulating drugs such as cyproheptadine, medroxyprogesterone acetate (MAP), megestrol acetate, IGF-1, corticosteroids, and growth hormones. Administration of flaxseed oil and/or vegetable-derived PUFAs, along with L-carnitine, in order to increase β-oxidative processes of long-chain fatty acids and replace saturated fats with polyunsaturated fats may be considered. Physical exercise is prescribed to partly compensate for the reduction of peripheral fat tissue by increasing muscle mass. Growth hormone reduces abdominal fat without affecting peripheral fat loss and lipids. Androgenic anabolic steroids (oxandrolone, nandrolone decanoate) increase muscular body mass without causing changes in lipids and body fat. In combined disorders (high cholesterol and high triglyceride levels), statins and fibrates together may control lipid metabolism, but they may also cause muscle damage (rhabdomyolysis). Severe malnutrition increases morbidity, susceptibility to infection, and the need for hospitalisation or custodial care. Malnutrition manifested as cachexia and wasting, is a common clinical problem in patients with AIDS. Malnutrition adversely effects the immune function, accelerates disease progression, and diminishes quality of life (QoL). Cachexia or wasting is defined as a state of ill health, malnutrition, emaciation, and generalised weakness. Weight loss and abnormalities in carbohydrate, protein, and lipid metabolism are characteristic in cachexia, resulting in a persistent negative energy balance. A ten percent weight loss from original body weight can be diagnostic for HIV-related wasting syndrome. Weight loss or wasting is secondary to a complicating process rather than HIV infection, and that not all patients who were HIV+ve lost weight as their disease progressed. The body cell mass (the metabolically active compartment of the body) is depleted to a greater extent than the total body weight in HIV-related wasting syndrome. There is a greater degree of wasting in body cell mass than in body fat content. A stressed state rather than a starved state seems to exist in HIV-related wasting syndrome, since the latter leads to a loss of body fat.

Drugs' intervention for cachexia:

A- Hormone:
    1- IGF-1
    2- Growth hormone
    3- Steroids (nandrolone decanoate, dexamethasone, prednisone, methylprednisolone, progestin derivatives as medroxyprogesterone acetate, megestrol acetate).
B- Anticytokines:
    1- Hydralazine sulfate (Reduces PEP-carboxykinase).
    2- Pentoxyphylline (Increases cAMP).
    3- Amrinone (Reduces phosphodiesterase).
    4- Dobutamine (β-agonist) (Increases cAMP).
    5- Thalidomide (Reduces TNF).
    6- NSAIDs, e.g., indomethacin, ibuprofen, aspirin (Reduces PG synthesis).
    7- Antibodies against TNF, LPS, IL-1.
C- Antiserotonin:
    Cyproheptadine.

Malnutrition may be a complication associated with AIDS, rather than an inevitable consequence of the disease. The aetiology of malnutrition in HIV infection is multifactorial. The primary mechanisms for involuntary weight loss may include decreased nutrient intake, malabsorption, metabolic disturbances, and increased protein energy requirements. Diminished nutrient intake may often result in anorexia; which may be due to mechanical obstruction such as disorders of the oral cavity, pharynx, or oesophagus, aphthous stomatitis, herpetic mucositis or oesophagitis, oesophageal cytomegalovirus infection, oral or oesophageal candidiasis, oral herpes virus, and bulky oral pharyngeal Kaposi's sarcoma or non-Hodgkin's lymphoma. Neurological disorders, psychological complications such as organic brain syndrome and drugs may also contribute to decreased oral intake. Malabsorption is often the result of intestinal microvilli damage. Metabolic disturbances include hypermetabolism due to disease-induced alterations in cytokine and thyroid hormone utilisation, fever due to systemic infections (e.g., Pneumocystis carinii pneumonia), and protein wasting. An alteration in protein metabolism is observed in patients with HIV infection, with a marked loss in skeletal muscle protein and a shift in hepatic protein synthesis from transport proteins (i.e., albumin) to acute-phase reactant proteins (i.e., α[1]-acid glycoprotein). Acute-phase reactant proteins are responsible for augmenting local tissue defences, clearing the interstitial tissues of cellular debris, and limiting the systemic effects of local inflammatory responses. These proteins consume large amounts of energy that are derived from skeletal muscle essential amino acids. Consequently, when AIDS patients develop chronic infections, severe muscle wasting develops. Testosterone deficiency has also been linked to protein deficiencies in male AIDS patients. Secondary causes of involuntary weight loss include drug-induced vomiting and diarrhoea and infectious diarrhoea (e.g., Clostridium difficile-induced colitis, Cryptosporidia, Isospora, Microsporidia, and Mycobacterium avium-intracellulare). HIV infection increases oxidative stress, which may in turn lead to faster progression of HIV disease. Oxidative stress may lead to progression of HIV disease due to impairment of immune function, enhancement of HIV replication, or enhancement of apoptosis. HIV replication is enhanced by oxidative stress, possibly via activation of NFκB. Apoptosis may contribute to the depletion of CD4 T cells in HIV infected persons. Hydrogen peroxide and TNF-α induce apoptosis in HIV infected cells, while antioxidants such as NAC inhibits apoptosis. Vitamin A consists of preformed vitamin A (retinol) found in animal foods, as well as certain carotenoids primarily β-carotene found mainly in vegetables and fruits, which metabolised to retinol. Vitamin A enhances humoral immune function, as well as cellular immune function including CD4 cells and natural killer cells, a decrease in cells expressing IL2R, and increased delayed-type hypersensitivity. Retinoic acid inhibits and enhances HIV replication. Carotenoid halocynthiaxanthin inhibits the reverse transcriptase enzyme. Vitamin A deficiency is associated with an increased risk of progression of HIV disease. Vitamin E is the major lipid soluble antioxidant present in cellular membranes that provides protection against lipid peroxidation. Vitamin E may increase immune response to antigens, improve host resistance against challenge with microorganisms, and enhance B and T cell lymphocyte functions, as well as phagocytic function. Vitamin E supplementation improves cell-mediated immune function including an increase in delayed type hypersensitivity. Supplementation with vitamin E is associated with increased production of INF-γ and IL-2, enhanced splenocyte proliferation and natural killer cell activity, and normalisation of levels of certain cytokines such as IL-6 and TNF-α. High serum vitamin E concentration is associated with a significantly decreased risk of progression to AIDS. High intake or levels of vitamin E may be beneficial in slowing progression of HIV disease. Vitamin C is a water-soluble vitamin and can function as both an intracellular and extracellular antioxidant. Vitamin C deficiency results in depressed cell-mediated immune response. T and B lymphocyte proliferative responses increase with vitamin C supplementation. Vitamin C inhibits replication in acutely and chronically infected T cells and to inhibit HIV reactivation in T cells stimulated by TNF-α. High vitamin C intake has a protective effect on progression of HIV disease. Selenium is an essential component of the antioxidant enzyme glutathione peroxidase, which reduces organic hydroperoxides and H_2O_2, and oxidises glutathione to glutathione disulphide. Selenium deficiency inhibits nonspecific immune function, humoral immunity, cellular immunity, and resistance to infection. Selenium supplementation enhances these immune functions, as well as resistance to infection (hepatitis infections). Only selenium deficiency was significantly associated with AIDS-related mortality. Supplementation with multiple vitamins and mineral is associated with improved immune function. Multivitamin supplement decreases risk of progression of HIV disease. Multivitamin supplementation is associated with a significant increase in CD4, CD8, and CD3 cell counts; and with a significantly decreased risk of foetal death and other pregnancy outcomes and a significant increases in haemoglobin levels. Low serum vitamin B_{12} levels are associated with an increased risk of subsequent progression to AIDS. Dietary intake of vitamin B complex inversely associated with progression to AIDS. Vitamin B complex is associated with a decreased risk of progression of HIV disease. Zinc intake is positively associated with an increased risk of progression to AIDS or death. High doses of zinc may impair immune function. NAC is a precursor of glutathione, an intracellular tripeptide that reduces ROS via its oxidation to glutathione disulphide. NAC scavenges free radicals directly. Glutathione plays a critical role in T cell function including enhancement of proliferation, inhibition of oxidative stress-induced apoptosis, and enhancement of cytotoxicity. NAC enhances T cell function and growth, and increases CD4 T cell counts in healthy subjects. NAC inhibits transcription and replication of HIV. NAC inhibits apoptosis of lymphocytes. NAC reduces mortality among HIV infected individuals. Oxidative stress caused by HIV accelerates progression of HIV disease.

NUTRITIONAL SUPPORT IN CRITICALLY ILL PATIENTS

Medium-chain fatty acids contain 8-10 carbon atoms. They do not require carnitine for entry into mitochondria to undergo oxidation. They can serve as an energy source even during marked stress. There are 3 branched-branched amino acids: leucine, isoleucine, and valine. They serve as important fuel for skeletal muscle, especially during stress. They promote protein synthesis, reduce protein degradation, and serve as substrates for gluconeogenesis. Their metabolism occurs in skeletal muscle. This increases their usefulness in the presence of liver dysfunction. Branched-chain amino acids improve nitrogen balance. 45% branched-chain amino acids enrichment is considered most optimal for nitrogen sparing and protein synthesis. Glutamine is a necessary component for protein and nucleotide synthesis and is an important fuel substrate for most rapidly dividing cells, including those of GIT, pancreas, pulmonary alveoli, and WBCs. The use of glutamine as a fuel by these tissues is increased greatly during times of stress. Oxidation of glutamine via the Krebs cycle yields 30 mmol ATP per mole glutamine, comparing with 36 mmol ATP produced from glucose oxidation. Short-chain fatty acids (acetate, propionate, and butyrate) are produced in the colon by fermentation of dietary carbohydrates and fiber polysaccharides and are absorbed readily through the colonocyte brush border. Short-chain fatty acids are readily metabolised by the colonocyte as fuel, and only a small amount enters the blood. Short-chain fatty acids may help maintain the colonic mucosal barrier to passage of bacteria and their toxins. Free radicals play an important role in the pathogenesis of many human diseases, including systemic lupus erythematosus, emphysema, ARDS, cirrhosis, stroke, inflammatory bowel disease, asthma, multiple sclerosis and arthritis of either immunologic or traumatic aetiology, and the transformation of normal cells to cancerous cells. The lack of specificity of free radical (generated by WBCs when they encounter and kill bacteria) killing results in normal tissue injury as well as bacterial killing leading to the common signs of inflammation. Iron can be readily mobilised by superoxide from ferritin (its storage protein):

$$O_2^- + \text{Ferritin-Fe}^{3+} \longrightarrow O_2 + \text{Ferritin} + Fe^{2+}$$

The resulting ferrous iron can react with pre-existing lipid peroxide to initiate a chain reaction of oxidation of membrane phospholipids. When oxidative stress is severe enough, the natural cellular defences, consisting of the antioxidant enzymes superoxide dismutase, glutathione peroxidase, phospholipid hydroperoxide glutathione peroxidase, and catalase, may be overwhelmed or exhausted, leading to lethal injury. Supplemental iron should not be administered during critical illness unless absolutely necessary. The iron supplementation in diet may lead to pre-existing excessively high iron stores in older patients, exposing them to much greater risk of oxidant injury during severe stress than comparably stressed younger patients or patients who have had chronic blood loss. Albumin binds free ferrous ion, reducing the potential risk of chain-reaction oxidant injury. Vitamin E can combine with propagating lipid radicals to form very stable vitamin E radicals, breaking the chain reaction. The vitamin E radicals extends into the aqueous interface of the membrane, where it can be reduced back to functional vitamin E by vitamin C. Arginine is a potent secretagogue for growth hormone, prolactin, insulin, and glucagon. It is a major source of nitrous and nitric oxides, both of which are important as mediators of vascular dilation, hepatic protein synthesis, and electron transport in mitochondria. Arginine stimulates various immunological functions (T-cell-mediated immunity). Arginine reduces protein catabolism during stress. Glutamine plays a major role in interorgan nitrogen transport in normal and stressful states. With the increased metabolic demands of inflammation or injury, the consumption of glutamine as a fuel may exceed the increased release from skeletal muscle and serum glutamine concentrations can fall dramatically. Glutamine deficiency is associated with progressive intestinal atrophy, resulting in decreased mucosal thickness and weight, loss of brush border enzymes, villous disruption, and bacterial invasion. Glutamic acid or glutamate appears to have effects similar to glutamine, although not as potent. Dietary nucleotides are essential for cell-mediated immunity and helper/inducer T-lymphocyte function. ω-3 fatty acids compete with ω-6 fatty acids for the production of the eicosanoids-PGs, TX, prostacyclin, and leukotrienes-producing a 3-series of eicosanoids. Production of the 3-series results in decreased production of IL-1 (responsible for fever, anorexia, anaerobic metabolism, and increased endothelial permeability), decreased production of TNF (responsible for endothelial adhesion, fever, and augment the catabolic response), and decreased production of IL-6 (which stimulates acute phase protein synthesis). ω-3 eicosanoids are not immunosuppressive, exerting much less suppression of proliferative immune responses, antibody and lymphokinin production, and cell-mediated cytolysis.

SURGICAL METABOLISM AND NUTRITION

Glutamine is nonessential, neutral amino acids that can be synthesised by virtually all tissues. Following injury, operation, sepsis, and other catabolic events, intracellular glutamine stores may decrease by >50% and plasma levels by 25%. Glutamine is avidly consumed by replicating cells, such as fibroblasts, lymphocytes, tumour cells, and intestinal epithelial cells. Glutamine accounts for a major portion of the amino acids released by muscle as a result of proteolysis. GIT uptake of glutamine occurs in the epithelial cells of the small bowel villi. Glutamine is the major energy substrate for the gut. Efficient functioning of the immune system depends upon a balance of eicosanoid production between the ω-6 and ω-3 PUFA. Diet high in ω-6 fatty acids suppresses immune function by inhibiting mitogenesis due to increased PGE_2 synthesis, which inhibits T cell proliferation. The administration of additional ω-3 PUFAs negates this effect. Selenium is part of the enzyme glutathione peroxidase. A decrease in the activity of this enzyme leads to peroxidation of membrane lipids, resulting in elevated concentrations of pentane in expired air. Manganese is the cofactor for the metalloenzymes pyruvate carboxylase and manganese-superoxide dismutase, being involved in the initial step in gluconeogenesis and in cellular antioxidant capability. Zinc is a cofactor for a number of metalloenzymes involved in carbohydrate, fat, amino acids, and nucleic acid metabolism. Chromium forms a complex with a small peptide containing nicotinic acid to produce glucose tolerance factor (GTF). GTF facilitates the binding of insulin to membrane receptors. Copper is a component of a number of metalloenzymes, including cytochrome oxidase (the terminal enzyme of the electron transport chain), dopamine hydroxylase, and lysyl oxidase. Fat-soluble vitamins participate in immune function and wound healing. After absorption they are delivered to the tissue in chylomicrons and stored in large quantities in the liver (A and K) or subcutaneous tissue and skin (D and E). Water-soluble vitamins serve as cofactors that facilitate reactions involved in the generation and transfer of energy and in amino acids and nucleic acid metabolism. Because of their limited storage, water soluble vitamin deficiencies are common. After an overnight fast, liver glycogen is rapidly depleted because of a fall in insulin and a rise in glucagon levels in plasma. There is an increase in hepatic gluconeogenesis from amino acids derived from the breakdown of muscle protein. The release of amino acids from muscle is regulated by insulin, which stimulates amino acid uptake, polyribosome formation, and protein synthesis. The fall in insulin along with a rise in plasma glucagon levels results in an increase in the concentration of cAMP in adipose tissue, which stimulates a hormone-sensitive lipase to hydrolyse triglycerides and release free fatty acids (FFA). Gluconeogenesis and FFA mobilisation require the presence of ambient cortisol and thyroid hormone. During starvation, the body attempts to conserve substrate by recycling metabolic intermediates. The haematopoietic system utilises glucose anaerobically and increases lactate production. Lactate is recycled back to glucose in the liver via the glucogenic Cori cycle. The glycerol released during peripheral triglyceride hydrolysis is converted into glucose via gluconeogenesis. Alanine and glutamine are the preferred substrates for hepatic gluconeogenesis from amino acids, and they contribute 75% of the amino acid carbon for glucose production. Alanine and glutamine constitute 75% of the amino acids released from skeletal muscle during starvation. Glutamine is taken up by the small bowel, transmitted to form additional alanine, and released into the portal circulation. These amino acids plus glucose participate in the glucose-alanine/glutamine-BCAA cycle, which shuttles amino groups and carbon from muscle to liver for conversion into glucose. Gluconeogenesis from amino acids results in a urinary nitrogen excretion of 8-12 g/d, predominantly as urea, which is equivalent to a loss of 340 g/d of lean tissue. At this rate, 35% of the lean body mass would be lost in 1 month, a uniformly fatal amount. Starvation can be survived for 2-3 months, as long as water is available. The effects of elective operations and trauma differ from that simple starvation due to activation of neural and endocrine systems, which accelerates the loss of lean tissue and inhibits adaptation. This heightened neuroendocrine secretion (ACTH, TSH, ADH, norepinephrine, epinephrine, aldosterone, insulin, glucagon, and growth hormone) produces: 1-

peripheral lipolysis, by the synergistic activation of hormone-sensitive lipase by glucagon, epinephrine, cortisol, and thyroid hormone; 2- accelerates catabolism, consisting of a rise in proteolysis, stimulated by cortisol; and 3- decreases peripheral glucose uptake due to insulin antagonism by growth hormone and epinephrine. In sepsis, the REE rises 50-80% above control values, and urinary nitrogen excretion reaches 20-30 g/d, equivalent to a median survival of 10 days without nutritional input. The glucose, amino acids, and FFA levels increase more than in trauma. Hepatic protein synthesis is stimulated, with both enhanced secretion of export and accumulation of structural protein. There is intense FFA oxidation. Septic patients have an abnormal plasma amino acid pattern similar to that of patients with liver failure. Gluconeogenesis stops in terminal sepsis together with further increase in plasma amino acids and fall in glucose concentration. The REE increases by 20-30% in patients with extreme cachexia. Patients with cancer retain nitrogen despite losses in most lean tissue. Synthesis, catabolism, and turnover of body protein are all increased, but the change in catabolism is greatest. There is anaerobic glucose metabolism in the neoplastic tissue, with susceptibility to lactic acidosis. There is impaired glucose tolerance, elevated glucose turnover rates, enhanced Cori cycle activity, and gluconeogenesis from alanine.

OBESITY

Obesity is far more related to carbohydrate consumption than fat intake. Corn contains <5% fat, it is almost 90% carbohydrate. A high carbohydrate, reduced fat diets promote weight gain. Insulin lowers the concentration of glucose in blood by inhibiting hepatic glucose production and by stimulating the uptake of glucose by muscle and adipose tissue. Insulin inhibits lipolysis, stimulates fatty acid synthesis and decreases the hepatic concentration of carnitine (carnitine shuttles fatty acids into mitochondria in most cells for use as ATP energy fuel). Insulin stimulates the fat cells to take up fat and sugar from the blood and store it away as body fat, especially in the middle of the body, within the abdomen and around the vital organs. Overweight people tend to have hyperinsulinaemia that promotes lipogenesis. Eating a candy bar or drinking a soft drink will normally raise blood sugar and insulin, within minutes. While starch may be slightly slower to raise blood sugar and insulin, the modern industrialised starches, such as white flour and finely ground corn meal, used to make pasta, bread, cakes, corn chips, crackers, cookies, etc., are digested and absorbed almost as quickly as simple sugar foods. Glucagon is secreted by the pancreas. The secretion of glucagon is regulated by dietary glucose, insulin, amino acids, and fatty acids. Glucose is a potent glucagon inhibitor. Insulin lowers high blood sugar; glucagon raises low blood sugar. Residing on the surface of the fat cells are 2 enzymes (both regulated by insulin and glucagon) that are responsible for herding fat into or out of the fat cells: 1- lipoprotein lipase (LPL), transports fatty acids into the fat cell and keeps them there; 2- hormone-sensitive lipase (HSL), does just the opposite, it releases the fat from fat cells into the blood (fat gets shuttled into the mitochondria for oxidation, where it is completely burned for cellular energy). Insulin stimulates the activity of LPL, the fat-storage enzyme, and glucagon inhibits it; glucagon stimulates HSL, and insulin inhibits it. LPL is highly sensitive to variations in the metabolic state, rapidly increased by oral glucose, by high carbohydrate diet and after usual meals. The chief dietary stimulant for insulin release is carbohydrate; the chief stimulant for glucagon release is protein. Even on a high carbohydrate/fat diet, untreated Type I diabetic (no secreting insulin) will continually lose fat (and muscle, as well), and may even lose 30-40 pounds in a month. One sucrose (white sugar) molecule is one glucose bonded to one fructose. Fructose decreases the ATP content of the liver. The liver is the chief metabolic organ of the body, uses 12% of the body's total ATP. Anything that seriously lowers liver ATP is by definition a metabolic poison. Fructose is incorporated into blood triglycerides more rapidly than is glucose. Unrefined starches (especially vegetables) will tend to cause less hyperinsulinaemia than sugar-rich foods such as candy, cake, pie, doughnuts, soft drinks, sports drinks, etc., as well as natural sugar foods such as dates, figs, dried pineapple, etc. The 4 chief pillars of aging include excess insulin, excess cortisol, excess blood glucose, and excess free radicals. Syndrome X, involves the strong clustering of hypertension, insulin resistance, hyperinsulinaemia, hypertriglyceridaemia, glucose intolerance, obesity, low HDL cholesterol and heart disease. Hyperinsulinaemia and insulin resistance are the common denominator of the syndrome, even-smaller amounts of sugar lead to ever-higher blood glucose levels, i.e., glucose intolerance. Insulin causes sodium retention with consequent water retention, hence the hypertension connection. Insulin promotes fat storage in fat cells (obesity). Insulin induces hypertriglyceridaemia. The arterial wall is an insulin-sensitive tissue. Insulin promotes proliferation of arterial smooth muscle cells (plaque formation, a beginning phase of atherosclerotic) and enhances lipid synthesis and low-density lipoprotein (LDL) receptor activity. Insulin-injecting diabetics typically develop atherosclerosis 10-20 years earlier than non-insulin-injecting diabetics. Hyperinsulinaemia is the more likely root cause of all 4 conditions: obesity, glucose intolerance, high triglycerides and hypertension. Excess cortisol and excess blood glucose are intimately tied to excess insulin. A consequence of obesity is the development of insulin resistance as weight is gained. Most cells can burn either fat or glucose for fuel, but the brain (under non-fasting conditions) can only burn glucose and typically needs 400-500 calories/day of glucose, i.e., about one half the normal total circulating blood sugar. The brain doesn't need insulin to absorb glucose. Glucagon is the primary hormone that should raise blood sugar to adequately feed hungry brain. Insulin acts as a glucagon release-inhibiting paracrine hormone, especially at high concentrations. So then the body releases cortisol. Cortisol comes to the brain's rescue in 2 ways: 1- Increases gluconeogenesis, the making of glucose by breaking down proteins from skin, muscle and organ tissue and converting them to glucose in the liver. 2- Causes a moderate decrease in the rate of glucose utilisation by cells everywhere in the body (cortisol causes insulin resistance). So, the first 3 pillars of aging excess insulin, cortisol and blood glucose are all interlocking and mutually enhancing. Cortisol contributes mightily to obesity. Adrenal corticosteroids play a role in the development of hypothalamic obesity, gold thioglucose obesity, and dietary obesity. Cortisol is also secreted to raise blood sugar in those who frequently skip meals, are fasting, practice starvation dieting, or are under severe stress. Hormone (Latin, meaning chemical-messenger) is the first messenger. Insulin acts through the second messengers inositol triphosphate (IP_3) and diacylglycerol (DAG). Perhaps the commonest second messenger, however, is cyclic AMP (cAMP). Many hormones utilise cAMP as a second messenger, including calcitonin, chorionic gonadotropin, corticotrophin, epinephrine, FSH, glucagon, LH, lipotrophin, melanocyte-stimulating hormone (MSH), norepinephrine, parathyroid hormone, TSH, and vasopressin. Insulin and glucagon have opposing second messengers: IP_3/DAG vs. cAMP. Adenylate cyclase enzyme produces the cAMP second messenger inside the cell. Insulin opposes cAMP production by adenylate cyclase. Insulin is one of the few hormones (cortisol being the other major one), which increases with age. Most other hormones, such as thyroid, DHEA, testosterone, oestrogen, growth hormone, etc., decrease with age. Hyperinsulinaemia tends to distort the overall hormone interactions, and promote low fidelity hormonal communication. Hyperinsulinaemia tends to damage the entire metabolism, because the sum total of the myriad biochemical reactions in our trillions of cells is under the control of tightly synchronised and integrated hormonal interactions. Autocrine eicosanoids, unlike endocrine hormones, are not secreted by glands, nor do they travel through the bloodstream to reach distant target tissues. Rather they are continuously being produced, in minute quantities, at the local cellular level, and last for seconds. Eicosanoids are powerful local biological response modifiers, or feedback modulators, helping to coordinate/fine-tune cellular reactions. Prostaglandins (PG) of the one-series, derived from the fatty acid γ-linolenic acid (GLA), are generally considered good PGs, while PGs of the two-series (PG2) are considered bad PGs-at least when present beyond some bare minimum necessary levels. PG2s are derived from the fatty acid arachidonic acid (AA), which in turn can either be made from GLA or get preformed from the diet. PGs modulate intracellular cAMP levels. PGE are those most implicated in adipose tissue regulation. PGE_1 stimulates adenylate cyclase, resulting increase in cAMP production that ultimately leads to accelerated lipolysis. PGE_2 has an inhibitory effect on adenylate cyclase resulting in a decrease of intracellular cAMP. PGE_1 plays a role in insulin secretion and glucose tolerance. The pancreatic β-cell regulation of insulin release is influenced by PGE_1. PGE_1 inhibits insulin secretion, perhaps by normalising insulin receptor sensitivity. Low levels of PGE_1 have been found in diabetics. Dietary/nutrient influence over PGE_1/PGE_2 includes: 1- Increasing the effectiveness of the conversion of cis-linoleic acid (a fatty acid common to many vegetable oils) into GLA. 2- Influencing the fate of the GLA metabolite dihomo-γ-linolenic acid (DGLA). DGLA can end up either as "good" PGE_1 or "bad" PGE_2, depending on whether or not the conversion of DGLA to AA is successfully blocked. 3- Restricting the dietary intake of preformed AA. Cis-linoleic acid (cLA) is the chief polyunsaturated fatty acid found in most vegetable oils, such as sunflower, safflower, corn, soy and sesame oils. The conversion of cLA to GLA is catalysed/controlled by the activity of the enzyme δ-6-desaturase (D6D). The activity of D6D can be blocked by a host of factors: 1- trans-fatty acids (common in hydrogenated oils, margarine's and shortenings), 2- high saturated fat intake, 3- cholesterol, 4- deficiencies of zinc, pyridoxine (vitamin B_6), or magnesium, 5- diabetes (i.e., severe insulin deficiency), 6- excessive alcohol intake, 7- aging, 8- oncogenic viruses, 9- chemical carcinogens, and 10- ionising radiation. Avoiding hydrogenated oil/margarine-based food products; eating only low-fat meat, poultry and dairy products; minimising alcohol intake; avoiding chemical additive-containing processed/manufactured (junk) foods; and taking supplements of zinc (15 mg/day), vitamin B_6 (10-50 mg/day) and magnesium (200-500 mg/day), will tend to maximise D6D activity, at least somewhat increasing conversion of cLA to GLA. Vitamin B_6 may also aid the conversion of GLA to DGLA for conversion to cAMP-enhancing PGE_1. Vitamin C and niacin (vitamin B_3) are needed to convert DGLA to PGE_1; so supplements of vitamin C (300-500 mg/day, minimum) and vitamin B_3 (50-100 mg/day) may also aid PGE_1 formation. Supplements of preformed GLA from evening primrose oil, borage oil, or blackcurrant oil may be helpful. DGLA can be converted to AA by the enzyme D5D. The primary activator of D5D is insulin. The primary hormonal suppressor of D5D is glucagon, and the fish-oil fatty acid EPA (eicosapentaenoic acid) is also a significant inhibitor of D5D. Feedlot beef, etc., is rich in AA; low-fat range-fed beef, poultry, etc., is low in AA, and contains some EPA. Growth hormone and insulin are both anabolic. Growth hormone promotes fat burning/loss, while insulin opposes fat burning and promotes fat gain. Increased insulin levels and decreased growth hormone levels are characteristic of obesity. PGE_1 suppresses insulin release; PGE_1 increases pituitary growth hormone release. Growth hormone-releasing hormone requires adequate pituitary cAMP levels to perform its growth hormone-releasing. Lowering insulin through a low-carbohydrate diet combined with GLA/EPA supplements to enhance PGE_1/cAMP levels may restore age-declining GH function. Growth hormone helps to build muscle mass when combined with testosterone. Obesity/high carbohydrate diet-elevated insulin inhibits the testosterone-producing activity of FSH/LH. In both men and women, testosterone may be converted to oestrogen through an aromatase enzyme. The aromatase enzyme exists and functions primarily in body fat. Oestrogen is a powerful pro-fat hormone, deposits in the fat of the breasts, subcutaneous tissues, the buttocks and thighs. Insulin, oestrogen and cortisol are the 3 primary pro-fat hormones of the human body. Severe chronic stress is a threat to normal male testosterone levels. Both testosterone and cortisol are made from the precursor protohormone pregnenolone. Normal daily male testosterone production is 5 mg, while 10-20 mg of cortisol is produced daily under non-stressed life conditions. The amount of cortisol produced under stress may double, stealing scarce pregnenolone needed for (decreasing with age) testosterone production. Cortisol is extremely pro-fat, and is the chief agent of muscle catabolism. The late twentieth century Western epidemic of obesity is as much due to widespread chronic hypokinesis (too little bodily movement), as it is to the carbohydrate/caloric excess typical of modern humans. Exercise decreases storage fat rather than LBM (lean body mass), whereas dietary interventions, i.e., dieting, tend to reduce both body fat and LBM. Exercise increases insulin sensitivity and decreases insulin resistance.

Trace element tests:

Symbol	Name	Functions	Laboratory Tests
Zn	Zinc	Forms an essential part of many enzymes (e.g., carbonic anhydrase, important in carbon dioxide metabolism), and plays an important role in protein synthesis and in cell division.	Plasma Zn, CRP, Leukocyte Zn.
Cu	Copper	Component of various proteins, including caeruloplasmin, erythrocuprein, cytochrome c oxidase, tyrosinase, etc.	Plasma Cu, plasma caeruloplasmin, CRP.
Se	Selenium	Constituent of the enzyme glutathione peroxidase, may be closely associated with vitamin E.	Plasma Se, RBC glutathione peroxidase, whole blood Se.
Fe	Iron	Essential constituent of haemoglobin, cytochrome, and other components of respiratory enzyme systems. Its chief functions are in the transport of oxygen to tissues (haemoglobin) and in cellular oxidation mechanisms.	Ferritin, CRP.
Mn	Manganese	Occurs throughout the body, concentrated in the mitochondria, chiefly in the pituitary, liver, pancreas, kidney, and bone. It is essential for the synthesis of mucopolysaccharides and activates a number of enzymes.	Plasma or whole blood Mn, LFT.
Cr	Chromium	Plays a role in glucose metabolism.	Plasma Cr, glucose tolerance.
Mo	Molybdenum	Component of the enzyme xanthine oxidase, aldehyde oxidase, and nitrate reductase.	Urine xanthine/hypoxanthine, urine sulphate.
I	Iodide	Incorporated into thyroid hormones.	TFT.
F	Fluoride	Structure of bone and teeth.	Urine F.

Actively exercising muscles may take in up to 30 times more blood sugar than they do when at rest, and this cellular uptake of glucose occurs without insulin. Walking provides the body with an alternative method to remove excess glucose from the bloodstream without the usual need for insulin secretion. Taking a brisk long walk 30-60 minutes after a large meal may help blunt the otherwise inevitable massive insulin surge large (carbohydrate-rich) meals normally induce. Salt increases plasma glucose and insulin response to starchy foods. Lean protein-low fat (ideally range-fed, organic) beef, lamb, chicken, turkey, fish etc., increases glucagon. Chromium picolinate (200-400 µg, three times daily) enhances insulin responsiveness (60% increase in insulin binding), and markedly enhances glucose and leucine uptake. DHEA, (10-50 mg, daily), high dose vitamin C (500-1000 mg 3-4 times daily), Gerovital-H3 (100 mg, daily), and Dilantin (Phenytoin) 25-50 mg at bedtime, may all help lower elevated cortisol levels. The large insulin releases generated by such carbo-bingeing preferentially increase tryptophan/serotonin in the brain, temporarily reducing anxiety and depression in such people. 5HTP (300 mg 3 times daily before meals) may help reduce weight by reducing total caloric intake, especially by reducing carbohydrate intake, thus lessening hyperinsulinaemia/insulin resistance. Also, 1000-1500 mg L-tryptophan at bedtime, or 50-100 mg 5HTP before meals may reduce carbohydrate-craving and intake. Panoply of antioxidants, e.g., 100-400 IU vitamin E, 500-2000 mg vitamin C, 200 µg selenium, 50-300 mg α-lipoic acid, 600 mg N-acetylcysteine (NAC), 2 mg copper as copper sebacate (SOD-mimetic), 50-100 mg grape seed extract/pycnogenol, 300-500 mg silymarin, 60-300 mg CoQ_{10}, 5-20 mg NADH, 10-100 mg, B_1, B_2, B_3, 50-200 mg B_5, 1-10 mg biotin, 200-500 mg magnesium etc., may help protect the essential fat burning furnaces. In a healthy human, storage fat is at a minimum and sooner or later all fat-dietary, body-manufactured, and storage fat ends up as fuel for the furnace, i.e., the trillions of mitochondrial. The mitochondria generates massive amounts of free radicals in producing ATP, they are rich in easily rancidified polyunsaturated fatty acids. Carnitine (1 g/day) is the shuttle molecule that escorts fatty acids into mitochondria where they are then oxidised. ALC (acetyl-L-carnitine), 1 g/day may also be a useful mitochondrial regenerator. The mitochondria become progressively deformed and dysfunctional with aging. Caffeine taken with a meal may induce increased thermogenesis. It may increase resting metabolic rate (60-70% of total daily energy consumption). Caffeine preadministration 45-60 minutes before exercise spares liver/muscle glycogen and enhances fatty acid burning in humans. Caffeine taken after at least an 8 hr fast, i.e., in the morning after arising, may be especially effective when combined with a 40-60 minute brisk walk enhances burning of stored body fat. The population may be subdivided into 4 subgroups: never overweight, pre-overweight, preclinical overweight, and clinical overweight. Fat women and men suffer social stigmatisation that worsens their QoL. The social disapproval is the major pressures for weight loss in women. Obesity is a reflection of increased enlarged fat stores, in both subcutaneous and visceral fat deposits. The secretory products of the enlarged fat cells produce most of the pathogenic changes that result in the complications associated with obesity; the rest are a consequence of the fat mass per se. The treatments for diabetes, hypertension, and heart disease are better than those available for obesity. Effective treatment for obesity can have a major impact on reducing the risk of developing comorbidities. There is an incidence rate of 8.5% for new cases of diabetes in patients who do not lose weight. To achieve the maximal reduction in the incidence of diabetes, weight loss needs to exceed 12%. Amphetamine as appetite suppressant is addictive. In the treatment of obesity, a plateau in body weight is often perceived as a therapeutic failure for the weight-loss drug. Aortic regurgitant lesions occur in up to 25% of the patients treated with the combination of fenfluramine and phentermine. There is a central controller transducer governing the search for and acquisition of food, as well as modulates the subsequent disposal of food once inside the body. And, there is a control system that ingests, digests, absorbs, transports, stores, metabolises, and excretes waste from the ingested food. The metabolic mixture oxidised by the body is related to the types of foods eaten, to the adaptive capacity of the body, and rate of energy expenditure. To maintain energy balance requires that the mix of fuels eaten be oxidised. A high rate of carbohydrate oxidation predicts future weight gain. When carbohydrate oxidation exceeds carbohydrate intake, the body needs carbohydrate to replace the limited stores. Since fatty acids cannot be converted to carbohydrate, amino acids are converted to carbohydrate equivalents that mobilise fat stores. Physical activity gradually declines with age, and maintaining a regular exercise program is difficult for many people, particularly as they get older. Adapting to a change from a lower- to a higher-fat diet takes time and can be accelerated by increasing exercise. Thyroid hormones increase energy expenditure and reduce body fat, but also increase the loss of lean body mass and bone calcium. Dinitrophenol, a thermogenic drug that uncouples oxidative phosphorylation, produces weight loss but has the highly undesirable side effect of producing cataracts and neuropathy. Insulin plays an important role in the activation of lipogenesis in fat cells. It also inhibits lipolysis and is involved in fat cell differentiation. Fat cell differentiation is a multistep process that requires a number of factors in addition to insulin. One of the early steps involves fatty acids interacting with the peroxisome proliferator-activated receptor (PPAR-γ). This PPAR-γ forms a heterodimer with the retinoid X receptor (RXR) to initiate the process of fat cell differentiation. Activation of the process increases fat cell differentiation, and inhibition of RXR inhibits this process. One clinical report suggests that defects in the gene for PPAR-γ may be related to obesity. Enzymes involved in fat metabolism are also important in obesity. The oral and nasal cavities are the first lines of exposure to food. Taste and smell can produce important positive and negative feedback signals. Taste and smell receptors for polyunsaturated fatty acids exist. Cholecystokinin (CCK), gastrin-releasing peptide, neuromedin B, and bombesin are gastrointestinal peptides that reduce food intake. Pancreatic peptides also modulate feeding. Both glucagon and glucagon-like peptide-1 (GLP-1) reduce food intake. Nutrients may also alter food intake peripherally by decreasing food intake. Leptin is a cytokine, derived primarily from fat cells, but also from the placenta and possibly the stomach, reduces food intake and increases the activity of the thermogenic components of the sympathetic nervous system. Leptin production is stimulated by insulin and glucocorticoids, and inhibited by β-adrenergic stimulation. The circulating levels of leptin are correlated with the level of body fat. This peptide exerts its effects through leptin receptors. Modulation of neurons in the arcuate nucleus by leptin results in the following: 1- Reduced secretion of neuropeptide Y (NPY). 2- Reduced secretion agouti-related protein (AGRP). 3- Increased secretion of proopiomelanocortin (POMC), the precursor of α-melanocyte-stimulating hormone (α-MSH), which reduces food intake. 4- Increased secretion of the peptide product of cocaine-amphetamine regulated transcript (CART). A brief dip in the circulating level of glucose precedes the onset of eating in more than 50% of meals consumed by humans. When this dip is blocked, food intake is delayed. The dip follows a small increase in insulin, suggesting an interdependent relationship between these 2 signals. Serotonin receptors modulate both the quantity of food and macronutrient selection. Stimulation of these receptors in the paraventricular nucleus reduces fat intake with little or no effect on the intake of protein or carbohydrate. There are 7 families of serotonin receptors (5-hydroxytryptamine [5-HT]). Noradrenergic receptors within the paraventricular hypothalamus (PVN) also modulate feeding. Norpinephrine can increase or decrease food intake depending on the type of adrenergic receptor it acts upon in the brain. Some of the α_1-adrenergic receptor agonists used to treat benign prostatic enlargement (BPE) are associated with weight gain. Stimulation of β_2- and β_3- adrenergic receptors in the brain decreases food intake. Neuropeptide Y is among the most potent stimulators of feeding. Its synthesis and release is modulated by insulin, leptin, and starvation. The melanocortin-receptor system is an important control point for feeding. Melanin concentrating hormone (MCH) is produced by neurons in the lateral hypothalamus. The opioid receptors modulate feeding, and modulate fat intake. The motor system for food acquisition, the endocrine system, and the autonomic nervous system are the major efferent control systems involved in acquiring food and regulating body fat stores. Among the endocrine controls are growth hormone, thyroid hormone, gonadal steroids (testosterone and oestrogens), glucocorticoids, and insulin. Testosterone increases lean mass relative to fat, and oestrogen has the opposite effect. Testosterone levels fall when human males grow older, with a corresponding increase in visceral and total body fat and a decrease in lean body mass. Growth hormone increases energy expenditure and increases the loss of fat. Growth hormone also reduces visceral fat more than total fat. Testosterone and anabolic steroids in males can lower visceral fat relative to total body fat, suggesting selective effects on different fat deposits. Testosterone may be associated with the development of prostatic hypertrophy and prostatic cancer. Excess growth hormone is associated with enhanced risk of cardiovascular disease. Adrenal glucocorticoids play an important role in the neuroendocrine control of food intake and energy expenditure. Glucocorticoids are critical for the development and maintenance of obesity. In humans, excess production of glucocorticoids produces modest obesity, and destruction of the adrenal glands is associated with loss of body fat. Using insulin or drugs that increase insulin secretion to treat diabetes leads to greater increases in body fat compared with other forms of diabetes treatment. Most of the resulting defects in obesity are caused by the metabolic consequences of an increased production of fatty acids and peptides due to the enlargement of fat cells. Other sequelae are the result of the increased mass of fat. Morbidity associated with enlarged fat cells and their excretory products include diabetes mellitus, heart disease, hypertension, gall bladder disease, and some forms of cancer. Sleep apnoea is a serious problem among overweight individuals and is more common in men than women. Intermittent airway obstruction at night leads to fitful sleeping and lowering of oxygenation. To compensate, there is increased sleepiness during the day and hypoventilation. The increased mass of fat in the pharyngeal area partially explains this complication of obesity. Treatment consists of weight loss and/or the use of intermittent positive-airway-pressure masks at night. Excess weight increases the risk of damaging joints (osteoarthritis). Weight gain during pregnancy, and the effect of pregnancy on subsequent weight gain, are important events in the weight gain history of women. Weight gain and changes in fat distribution occur after menopause. The decline in circulating levels of oestrogen and progesterone alters fat cell biology so that central fat deposition increases. Central or abdominal fat deposition is an important determinant of cardiovascular risk. Oestrogen replacement therapy does not prevent the weight gain. After ages 55-64 years, relative weight remains stable and then begins to decline, in men. Body weight varies throughout the day as food is eaten and metabolised. Body weight also varies from day to day, week to week, and over longer intervals. Weight cycling refers to the ups and downs in weight that often occur in people who diet, lose weight, stop dieting, and regain the weight they lost and sometimes more. A return of body weight to the normal range with no weight gain thereafter is rarely achieved and is unrealistic for most patients. The realistic goal is a loss of 5-15% of body weight. As the body fat accumulates, so too does the likelihood of developing certain cancers. The triggering factor may not just be the extra fat per se, but the related to imbalanced blood sugar mechanisms. The greater the amount of body fat a person has, the greater the risk of developing colorectal cancer. Excess body fat also makes an individual more prone to develop hyperinsulinaemia, a condition that arises when the body's decreasing sensitivity to insulin causes it to release chronically high levels of insulin in response. Individuals with high glucose and insulin levels, measured at fasting and 2 hrs after an oral glucose challenge, have >2-fold risk of developing colorectal cancer. Even individuals with moderate dysglycaemia are at increased risk. Increased waist circumference is also independently associated with colorectal cancer, more than doubling the rate of incidence. Insulin is a growth factor in the colon. It can stimulate each cancer cell to split into 2 new cells (mitosis), which fuel tumour growth. Insulin-like-growth factor-1 (IGF-1), a peptide in the body that shares a structural similarity to human insulin, is also thought to be another important player in this process. High levels of IGF-1 have been linked with colon cancer, prostate cancer, and breast cancer. The metabolic dysglycaemia profile provides advanced early warning of metabolic obesity parameters linked to the development of diabetes, accelerated aging, CVD, certain cancers, and other degenerative conditions. Comprehensive assessment measures fasting and 2-hour post-challenge levels of insulin and glucose tolerance, fasting blood assays of IGF-1, haemoglobin A1c, and fructosamine, and salivary assessment of bioavailable DHEA and cortisol. Obesity may be related to abnormal levels of dopamine, and exercise may increase dopamine. β-Blockers results in insulin resistance, which may aggravate existing diabetes and elicit diabetes in predisposed patients. Overweight and obesity are frequently complicated with hypertension and angina pectoris, which are often treated with β-blockers. The consequence may be aggravation of hypertension, insulin resistance and other atherogenic factors. Insulin resistance (IR) may result from the lack of adipocyte hormones (such as leptin) and increased metabolite (such as triglyceride) levels in nonadipose tissue. Adipose tissue in obesity becomes refractory to suppression of fat mobilization by insulin, i.e., high insulin interferes with conversion of fat (triglydcerides) in fat cells to energy, and also to the normal acute stimulatory effect of insulin on activation of lipoprotein lipase (involved in fat storage). There is increased intracellular insulin concentrations in obese patients with and without Type 2 diabetes, and these conditions are associated with a significant impairment of insulin receptor processing. Polycystic ovary syndrome (PCOS) [may cause obesity] is the most common endocrine disorder in women of reproductive age. Insulin resistance plays a significant role both as a cause and result of the syndrome. IR in PCOS seems to involve a postbinding defect in the insulin receptor and/or in the receptor signal transduction. Obese/insulin-resistant subjects are characterized by endothelial dysfunction and endothelial resistance to insulin's effect on enhancement of endothelium-dependent vasodilation. This endothelial dysfunction could contribute to the increased risk of atherosclerosis in obese IR subjects. Leptin levels are increased in obesity and may play a role in development of insulin resistance and NIDDM.

ASPARTAME

When aspartame is ingested with a carbohydrate rich meal the usual physiologic increase in tryptophan is blocked, while brain phenylalanine and tyrosine concentrations are increased. These changes in amino acid neurotransmitter precursors may alter indoleamine/catecholamine balance, and thus have a profound effect on mood and cognition resulting in depressed mood, anxiety, dizziness, panic attacks, nausea, irritability, impairment of memory and concentration. Bipolar is an affective disorder in which the patient exhibits both manic and depressive episodes. Aspartame has 50% phenylalanine, which as an isolate is neurotoxic and goes directly into the brain. Phenylalanine lowers the seizure threshold. Phenylalanine also depletes serotonin, which triggers manic depression, anxiety, hallucinations, panic attacks, insomnia, paranoia, mood swings and even suicidal tendencies. There are marked neuropsychologic changes when dopamine-serotonin balance is altered. Decreased brain serotonin has been associated with insomnia, depression, anxiety, panic attacks, hallucinations, suicidal attempts, hostility and psychopathic conditions. Aspartame inhibits the carbohydrate-induced synthesis of serotonin. Serotonin is an important component of the feedback system, which helps limit the consumption of carbohydrate to appropriate levels by blunting the carbohydrate craving. The amino acid tyrosine, derived from phenylalanine, reduces the amount of tryptophan that can cross the blood-brain barrier for utilisation in serotonin production. Altered brain dopamine concentrations in mental illness is relevant. The brain oedema and vascular stasis due to chronic methanol intake (10% of aspartame) could contribute to the neuropsychiatric manifestations of aspartame reactors. Aspartame could have an impact on catecholamine and indolamine metabolism. Depletion of tryptophan by aspartame consumption in nonalcoholics can induce behavioural changes. Aspartame destroys the brain, especially in the areas of learning. Aspartic acid, which is 40% of aspartame, is an excitotoxin. Aspartame also triggers brain haemorrhage, stroke and aneurysms. Several aspartame reactors have bipolar depression, wherein the depression would alternate with periods of manic behavioural characterised by intense excitement and overactivity. As many as 2 million Americas suffer manic-depression. Women outnumber men with unipolar depression by 2:1. Patients with unipolar depression have significantly higher local cerebral metabolic rates for glucose, but these are reduced in bipolar depression. Accordingly, a further reduction of the cerebral metabolic rate caused by aspartame or its by-products might precipitate clinical depression in pre-disposed individuals. A subset of manic-depression may be linked to a dominant gene on the tip of the short arm of chromosome 11. The tyrosine hydroxylase gene, which cascades the synthesis of dopamine, exists in this region. Cyclic antidepressants cause more behavioural disturbances in bipolar depression. The frequency and severity of depression, headache, nervousness, difficulty in remembering, insomnia, fatigue and malaise are striking among patients with a history of depression after ingesting aspartame. Individuals without a history of depression may not manifest frequent or severe symptoms when given aspartame initially. The decrease of serotonin by aspartame administration has considerable relevance to depression and other psychological problems. The rapid lowering of brain serotonin can precipitate clinical depressive symptoms in untreated individuals vulnerable to major depression. Aspartame decreases the availability of L-tryptophan (a precursor of serotonin) and alters its balance with norepinephrine, another important neurotransmitter. Aspartame may be likened to a lesion in the lateral hypothalamus causing depression, other psychiatric problems, and eating disorders. Aspartame interacts with antidepressants. Aspartame ruins female sexual response and induces male sexual dysfunction. Aspartame disrupts foetal development by aborting it or inducing defects. Aspartame damages DNA. Aspartame is estimated to be in more than 9000 products. Aspartame constitutes part of paediatric prescription drugs like paediatric penicillin, augmentin, chewable tylenol, chewable benadryl, paediatric vitamins, etc.

INTEGRATIVE BIOCHEMICAL ISSUES OF DIABETES

Diabetes mellitus refers to excessive urination and sugary urine, but it is now often diagnosed in people who neither urinate excessively nor pass glucose in the urine, on the basis of a high level of glucose in the blood. Many other signs (abnormal mucopolysaccharide metabolism with thickening of basement membranes, leakage of albumin through capillary walls and into the urine, a high level of free fatty acids in the blood, insensitivity of tissues to insulin, or reduced sensitivity of the β cells to glucose) are considered diagnostic by others, who believe that the worst aspects of the disease can be prevented if they can be diagnosed early and take preventive measures. Insulin use constitutes a serious health problem. Diabetes reduces the symptoms of asthma, which get worse when insulin is given. Diabetic women have intellectually precocious children. Complications of diabetes are complications of insulin treatment. Insulin was introduced into medicine in the 1920s. Diabetes may develop shortly after a severe viral infection. Intense sickness and a high fever (and high doses of drugs given to treat the sickness) can cause very high levels of glucose in the blood, and even glucose in the urine, but this is a fairly well recognised consequence of stress. High doses of cortisone (prednisone, etc.) typically cause elevated glucose levels. Glucose sensing and glucose responses may involve integration of multiple organ systems, including the brain and pancreas, implying that a multisystem approach to therapeutic management of type 2 diabetes may be necessary to achieve euglycaemia. The brain has a role in regulating the effects of the hormones of energy balance. The hypothalamus has receptors for both insulin and leptin. Insulin and leptin signal the hypothalamus, in both normal and insulin-resistant states. Plasma free fatty acid (FFA) concentrations are strongly correlated with the development of insulin resistance, which in turn is a strong predictor for the development of type 2 diabetes. Increased plasma FFA concentrations are associated with the accumulation of triglycerides in extra-adipose tissues, and their accumulation within muscle cells is an especially good predictor of type 2 diabetes. Intramyocellular lipid accumulation results in insulin resistance. Free fatty acids might initiate a metabolic cascade, resulting in the inhibition of phosphofructokinase and the accumulation of glucose-6-phosphate within muscle cells; the latter would downregulate glucose transport. FFAs induce a decrease in intramyocellular glucose, which suggests that FFAs have an inhibitory effect directly on glucose transport. FFA infusions inhibit the activity of the enzyme phosphoinositide-3OH kinase (PI-3K), a necessary element in one key pathway of insulin-stimulated glucose transport into muscle cells. The lipids decrease insulin receptor substrate-1 (IRS-1) tyrosine phosphorylation and increase phosphorylation of serine 301, events known to inhibit the transduction of insulin signals beyond IRS-1 to PI-3K. The activation by intracellular fatty acyl-CoA of some isoforms of the serine kinase protein kinase C (especially PKC-θ) might play a role, and the inhibitory PKC signal cascade includes IκB kinase β (IKKβ). IKKβ activation and subsequent serine phosphorylation of IRS-1 can be blocked by salicylates. Short-term, high-dose (7 g/day) aspirin treatment acutely improve insulin sensitivity (increased glucose disposal and decreased hepatic glucose production) in patients with type 2 diabetes. Long-term treatment with leptin can improve glycaemia by decreasing hepatic steatosis and improving insulin sensitivity in the liver.

Insulin action on metabolic pathways in non-neural tissues:
  Insulin → IRS → PI-3K → activation of protein kinase B (PKB).

The classic PI-3K pathway is activated in the hypothalamus in response to insulin. Leptin also utilises the same PI-3K signalling pathway as insulin. The activated PI-3K pathway and attenuated hyperglycaemia without changing in plasma insulin levels does not change the body fat. Insulin and leptin signal integration in the hypothalamus for blood glucose homeostasis, imply a nutritional feedback loop whose afferent arc involves peripheral nutrient changes, release of insulin and leptin, and hypothalamic signalling by insulin and leptin, and whose efferent arc results in properly regulated nutrient metabolism and switching off of insulin and leptin. Decreasing insulin receptor expression in the arcuate nucleus results in defective suppression of hepatic glucose production, as well as hyperphagia and increased fat mass. The 2 major types of neurons in the arcuate nucleus-the orexigenic neuropeptide Y/agouti-related peptide (NPY/AGRP) neurons and the anorexigenic pro-opiomelanocortin/cocaine-amphetamine related transcript (POMC/CART) neurons, both receive signals from insulin, resulting in reciprocal transcriptional events. Insulin action results in decreased NPY/AGRP gene transcription and increased POMC/CART transcription. The electrochemical correlates of these events are hyperpolarisation of the NPY/AGRP neurons and depolarisation of the POMC/CART neurons. The glucose-mediated insulin secretion from pancreatic β cells involves the closure of ATP-sensitive potassium (K^+-ATP) channels, leading to depolarisation of the β cells. There are also K^+-ATP channels in hypothalamic neurons. In these glucose-sensitive neurons, leptin and insulin activate K conductance and hyperpolarize the cells. Glycogen could be synthesised and stored in neurons, and could play a role as an energy source for neurons during periods of glucose deprivation. Brain glucose is almost entirely derived from plasma glucose. The increments in brain glucose that result from increases in plasma glucose are not dependent on insulin action. The brain has the potential to store energy by converting glucose to glycogen. Both glycogen synthase and glycogen phosphorylase are found in glial cells and can be modulated by the usual physiologic regulators. The brain synthesises and stores glycogen as a reserve supply of energy in anticipation of hypoglycaemia. Most glucose entry into the brain is independent of insulin action. The 55 kDa isoform of GLUT1 is present in endothelial cells and could facilitate transfer of glucose from the lumen of brain capillaries into these cells. The 45 kDa isoform of GLUT1 is present in glial cells and the choroid plexus, while GLUT3 is present in neurons; these transporters could be responsible for glucose import into these respective cells. Diabetes induces no significant change in GLUT1 concentration or glucose uptake into different brain regions. Chronic hypoglycaemia is associated with an increase in GLUT1mRNA expression, with increase glucose transport across the blood-brain barrier. In the brain a tight coupling exists between neuronal synaptic activity and glucose metabolism. Astrocytes constitute a large fraction of the bulk of brain tissue, and since they are opposed to both blood vessels and neurons. Astrocytes can take up glucose in a concentration-dependent manner, and this process can be stimulated by glutamate. This process is mediated by classic glutamate receptor and glutamate transporters. Glutamate transporters also facilitate sodium influx. Glutamate increases the intracellular Na concentration, and this activates Na/K-ATPase-dependent signals. A functional consequence of this is to increase intracellular lactate. There are lactate transporters in both astrocytes and neurons. So, the synaptic activity releases glutamate, which increases Na flux into astrocytes, resulting in activation of Na/K-ATPase, increased glycolysis, production of lactate, and transfer of lactate into adjacent neurons as an alternative source of energy. Type 2 diabetes is a heterogeneous disorder with varying degrees of insulin resistance and insulin secretion. There is a progressive impairment in pancreatic islet function during the course of the disease. The β-cell mass is reduced in type 2 diabetes, which progressively leads to β-cell failure. Type 2 diabetes progresses from a stage of normal glucose tolerance through prediabetes to overt type 2 diabetes. This progression is associated with minimal changes in the degree of insulin resistance; however, insulin secretion is progressively blunted with transition from prediabetes to overt diabetes. Thus, β-cell failure or dysfunction is inherently associated with type 2 diabetes and may precede the onset of hyperglycaemia. The therapeutic interventions to maintain euglycaemia during the prediabetes stage may protect β cells from failing. The secretion of insulin from pancreatic β cells is a complex process involving the integration and interaction of multiple external and internal stimuli. Thus, nutrients, hormones, neurotransmitters, and drugs all activate or inhibit insulin release. Normally, glucose induces a biphasic pattern of insulin release. First-phase insulin release occurs within the first few minutes after exposure to an elevated glucose level; this is followed by a more enduring second phase of insulin release. The first-phase insulin secretion is lost in patients with type 2 diabetes.

Sequence of glucose-induced insulin secretion:

1- Glucose is transported into β cells through facilitated diffusion of GLUT2 glucose transporters.
2- Intracellular glucose is metabolised to ATP.
3- Elevation in the ATP/ADP ratio induces closure of cell-surface ATP-sensitive K^+ (KATP) channels, leading to cell membrane depolarisation.
4- Cell-surface voltage-dependent Ca^{2+} channels (VDCC) are opened, facilitating extracellular Ca^{2+} influx into the β cell.
5- A rise in free cytosolic Ca^{2+} triggers the exocytosis of insulin.

The KATP channels serve as the transducer of a glucose-generated metabolic signal (i.e., ATP) to cell electrical activity (membrane depolarisation). Thus, like neurons, β cells are electrically excitable and capable of generating Ca^{2+} action potentials that are important in synchronising islet cell activity and insulin release. KATP channels are the targets for sulfonylureas, which are commonly prescribed oral agents in the treatment of type 2 diabetes. The sulfonylureas, like glucose, induce closure of KATP channels and stimulate insulin secretion. The β-cell KATP channel is a complex octameric unit of 2 different proteins: the sulfonylurea receptor (SUR-1) and an inward rectifier (Kir 6.2). The sulfonylurea receptor belongs to a superfamily of ATP-binding cassette proteins and contains the binding site for sulfonylurea drugs and nucleotides. KATP channels are present in other tissues of the body, including heart (SUR-2A/Kir 6.2), smooth muscle (SUR-2B/Kir 6.2), and brain (SUR-1/Kir 6.2). Glucose sensing in the brain during hypoglycaemia may be mediated by KATP channels located in brain hypothalamic neurons. Mutations in SUR-1 have been identified in humans and are associated with hyperinsulinism (persistent hyperinsulinaemic hypoglycaemia of infancy). The incretins are another set of factors that are important hormonal regulators of insulin secretion. The incretins are polypeptide hormones released in the gut after a meal that potentiate insulin secretion in a glucose-dependent manner. The incretins promote insulin secretion without accompanying hypoglycaemia (a common complication of sulfonylurea treatment). The incretins act by activating Gs (G-protein that activates adenylyl cyclase) to increase cAMP in β cells. cAMP, like ATP, is an important signal that regulates insulin release. The main mechanism of action of cAMP is by activation of protein kinase A (PKA) that, in turn, phosphorylates other substrates to turn on (or off) vital cell functions. A novel protein, cAMP-GEF II, a cAMP sensor (cAMPS) that forms a complex with other intracellular proteins (Rim2 and Rab3) directly regulate insulin exocytosis. cAMP can directly promote exocytosis of insulin granules without activation of PKA (i.e., PKA-independent pathway), and thereby provide additional molecular targets for therapeutic intervention. Extracellular Ca^{2+} influx through L-type voltage-dependent Ca^{2+} channels (VDCC) raises free cytoplasmic Ca^{2+} levels and triggers insulin secretion. The structure of the VDCC is complex and consists of 5 subunits: α1, α2, β, γ, and δ units. The α subunit constitutes the ion-conducting pore, whereas the other units serve a regulatory role. A novel protein, Kir-GEM, inhibits α ionic activity and prevents cell-surface expression of α subunits. In the presence of Ca^{2+}, Kir-GEM binds to the β isoform, and this interaction interferes in the trafficking or translocation of α subunits to the plasma membrane. The potential therapeutic role of Kir-GEM lies in the inhibitory effects on VDCC activity that may serve to protect β cells from overstimulation and subsequent failure, which is part of the disease aetiology of type 2 diabetes. KATP channels of hypothalamic neurons may be important in glucose sensing. The hypothalamus responds to lowering of glucose by activation (opening) of KATP channels, which results in autonomic signalling to promote glucagon secretion from α cells. Therapeutic intervention using oral hypoglycaemic agents in the prediabetes stage may potentially prevent or delay progression to overt diabetes. Stress and aging make cells less responsive in many ways by damaging their ability to produce energy and to adapt. The polyunsaturated fats are universally toxic to the energy producing system, and act as a misleading signal channelling cellular adaptation down certain self-defeating pathways. Diabetes is one of the terminal diseases that can be caused by the polyunsaturated vegetable oils. Coconut oil, in diabetes as in other degenerative diseases, is highly protective. The oral contraceptive pill may produce signs of diabetes, including decreased glucose tolerance. Free fatty acids can block the Krebs cycle, with the result of relative insulin resistance. The oestrogen used to treat menopause causes an increase in free fatty acids. Oestrogen's effect is mediated by growth hormone. Women are much more likely than men to develop diabetes. Free unsaturated fatty acids inhibit mitochondrial respiration, and free linoleic and linolenic acids act as intracellular regulators, stimulating the protein kinase C (PKC) system, which is also stimulated by oestrogen and the (cancer promoting) phorbol esters. There is the association between breast implants and scleroderma. Silicone functions as an adjuvant, making exposure to irritants, solvents or infections more harmful. Oestrogen promotes collagen formation, and changes in the connective tissue are deeply associated with the processes of stress and aging. Cushing's syndrome usually involves hyperglycaemia. Intense and/or prolonged stress can damage the insulin-secreting cells in the pancreas. One of the problems associated with diabetes is atherosclerosis, though now there is more emphasis on fatty degeneration. Other blood vessel problems include hypertension, and poor circulation in general, leading to gangrene of the feet, erectile dysfunction (ED), and degeneration of the retina. In muscles, and probably in other tissues of diabetics, capillaries are more widely spaced, as if the basal oxidative requirement were lower than normal. However, mitochondria contain more respiratory enzymes, as if to partly compensate for the poor delivery of oxygen to the cells. Osteoporosis or osteopenia is a common complication of diabetes, and seems to be associated with the calcification of soft tissues. The stressed heart becomes rigid and unable to contract completely, or to relax completely. Excess calcium enters cells, and fatty acids are mobilised both locally and systemically, and both of these tend to damage the mitochondria. In diabetes, fatty acids are mobilised and oxidised instead of glucose, and calcium enters cells, increasing their rigidity and preventing relaxation of muscles in blood vessels. The result seems to be formation of insoluble soap deposits in cells, blocking many metabolic processes. Calcium and iron tend to be deposited together in devitalised tissues. The lack of energy increases the amount of calcium in a cell, and stimulation or excitation does the same, creating or exaggerating a deficiency of energy. In hypothyroidism, many (if not all) tissues are very easily damaged. Since glucose is needed by liver cells to produce the active (T3) form of thyroid, diabetes almost by definition will produce hypothyroidism, since in diabetes glucose can't be absorbed efficiently by cells. Cell damage caused by the excitotoxins, glutamic and aspartic acids, requires both stimulation, and difficulty in maintaining adequate energy production. This combination leads to both calcium uptake and lipid peroxidation. When cells are de-energised, they tend to activate iron by chemical reduction, producing lipid peroxidation. This increase in the iron concentration suggests that there has been prolonged injury (oxidative stress) to the cell, with increased production of the haeme group, which binds iron. Iron is a factor in inflammation. A decreased blood supply predisposes an organ to calcification. In diabetes, a characteristic feature is that the blood supply is relatively remote from cells in muscle and skin, so the oxygen and nutrients have to diffuse farther than in normal individuals, and the ATP level of cells is characteristically lower than normal. In blood cells, both RBCs and WBCs are probably more rigid in diabetes, because of lower ATP production, and higher intracellular Ca and Na. Magnesium in the cell is largely associated with ATP, as the complex Mg-ATP. When ATP is used or converted to ADP, this lower-energy substance associates with calcium, as Ca-ADP. In a hypothyroid state, the energy charge can be depleted by stress, causing cells to lose Mg. ATP is less stable when it isn't complexed with Mg, so the stress-induced loss of Mg makes the cell more susceptible to stress, by acting as a chronic background stimulation, forcing the cell to replace the ATP, which is lost because of its instability. In this state, the cell takes up an excess of Ca. The energy deficit in diabetes produces an alarm state, causing increased production of adrenaline and cortisol (pro-inflammatory hormones). Adrenaline mobilises fat from storage, and the free fatty acids create a chronic problem involving: 1- blocked ATP production, 2- activation of the protein kinase C system (increasing tension in blood vessels), 3- inhibition of thyroid function with its energetic, hormonal, and tissue-structure consequences, 4- availability of fats for prostaglandin synthesis, and 5- possibly a direct effect on clot dissolving, besides the PAI-1 (plasminogen activator inhibitor) effect seen in diabetes. Oestrogen has many pro-clotting effects, and one of them is a decreased activity of vascular plasminogen activator (PA). Increased entry of calcium into cells is complexly related to increased exposure to unsaturated fatty acids, decreased energy, and lipid peroxidation. Osteoporosis, calcification of soft tissues and high blood pressure are promoted by multiple stresses, hypothyroidism, and Mg deficiency. Calcium overload of cells can't be avoided by avoiding dietary Ca, because the bones provide a reservoir from which Ca is easily drawn during stress. Calcium supplement can temporarily help prevent muscle cramps, because it makes Mg more available to the muscles. To stop a disease that involves abnormal calcification or contraction of muscle, increase the consumption of Mg, and to cause cells to absorb and retain the Mg, increase the thyroid function. The use of coconut oil provides energy to stabilise blood sugar while protecting mitochondria and the thyroid system from the harmful effects of unsaturated fats. Unsaturated (pork) fat permits diabetes to develop, sugar is slightly protective, and coconut oil is very protective against the form of diabetes caused by a poison. A low protein diet increases sensitivity to diabetes. Cysteine, glutathione, and thioglycolic acid (antioxidants) are protective against diabetes. The chelator of metals, BAL (British anti-lewisite) protects against diabetes. The oxidizable unsaturated fats are involved in the process of producing diabetes. The unsaturated oils suppress the thyroid. Coconut oil increases the metabolic rate, apparently by normalising thyroid function. Hypothyroidism involves deposition of mucopolysaccharides in tissues, increased permeability of capillaries with leakage of albumin out of the blood, elevated adrenaline that can lead to increased production of cortisol, decreased testosterone production, high risk of heart and circulatory disease, including a tendency to ulceration of the extremities, and osteoporosis, all of which are recognised complications of diabetes. High safflower oil diet induces diabetes. The hormone patterns associated with obesity such as low thyroid can increase both oestrogen and cortisol, which support the formation of fat, and the fat cells can become a chronic source of oestrogen synthesis. Oestrogen and the polyunsaturated fatty acids (PUFA), linoleic and linolenic acid, alike activate the protein kinase C (PKC) system of the cell. Many of the functions of PUFA are similar to the functions of oestrogen. Oestrogen increases secretion of growth hormone (GH). GH causes an increase in free fatty acids in the blood. Oestrogen promotes iron retention, so it sets the stage for oxidative stress. Both oestrogen and PUFA promote the entry of calcium into the cell. In diabetes, there is a generalised excess activation of the PKC system. Starch stimulates the appetite, promotes fat synthesis by stimulating insulin secretion, and sometimes increases the growth of bacteria that produce toxins. It is often associated with allergens, whole starch grains can be persorbed from the intestine directly into the blood stream where they may block arterioles, causing widely distributed nests of cell-death. Dieticians urge the use of complex carbohydrates (starch) instead of sugar. Starch is rapidly digested and absorbed, raises blood glucose faster than sucrose (half fructose, half glucose) does. Brewer's yeast has been used successfully to treat diabetes. Besides its high B-vitamin and protein content, yeast is an unusual food that should be sparingly used, because of its high phosphorous/calcium ratio, high K to Na ratio, and high oestrogen content. The insulin-producing β cells of the pancreas have oestrogen receptors. DHEA regenerates β cells in the pancreatic islets. Basic anti-aging diet is the best diet for prevention and treatment of diabetes, scleroderma, and the various connective tissue diseases, e.g., high protein, low unsaturated fats, low iron, and high antioxidant consumption, with moderate or low starch consumption. Amino acids, especially in eggs, stimulate insulin secretion, and that this can cause hypoglycaemia, which in turn causes cortisol secretion. Eating fruit (or other carbohydrate), coconut oil, and salt at the same meal will decrease this effect of the protein. Magnesium carbonate and Epsom salts can also be useful and safe supplements, except when the synthetic material causes an allergic bowel reaction. Insulin resistance in subjects with NIDDM impairs the ability of insulin to stimulate Mg as well as glucose uptake. Low ATP with high respiration would suggest uncoupling; unsaturated fatty acids that are known uncouplers of respiration from energy production. β-blockers increase the efficiency of energy expenditure during ordinary physical activity by increasing the utilisation of carbohydrate and by decreasing the utilisation of fat. Oxidative stress may play an important role in the pathogenesis of diabetic complications. Visceral adiposity is associated with skeletal muscle insulin resistance but this is not due to glucose-FFA substrate competition. Individuals with visceral obesity have reduced post absorptive FFA utilisation by muscle. The pathobiology of mitochondrial diabetes mellitus is probably a delayed insulin secretion due to an impaired mitochondrial ATP production in consequence of the mtDNA defect. Generally, individuals inheriting these mitochondrial diseases are relatively normal in early life, develop symptoms during childhood, mid-life, or old age depending on the severity of the mutation; and then undergo a progressive decline. This particular kind of diabetes, which is combined with deafness in 60% of the patients, involves a variant mitochondrial gene and occurs in about 1.5% of diabetics. Oestrogen probably via GH increases FFAs, and adrenaline that is elevated in hypothyroidism increases the release of FFAs from storage. Free fatty acids impair mitochondrial energy production. Skin capillary circulation severely impaired in toes of patients with IDDM, with and without late diabetic complications. Glucocorticoids may play a role in the maintenance and/or production of insulin resistance produced by high-fat feeding. Normal aging is characterised by resistance to insulin-mediated glucose uptake. Testosterone production is attenuated by PUFAs. The inhibitory action of EPA on testosterone production is reversible when the PUFA is stopped. Hyperglycaemia causes excess urinary calcium and phosphorus excretion in patients with NIDDM. In response to urinary Ca loss, parathyroid hormone (PTH) secretion is mildly stimulated. Bone formation seems to be suppressed in the hyperglycaemic state in spite of increased PTH secretion. Hypothyroid individuals may have high cortisol. In diabetes epinephrine is increased in striatum, hippocampus and hypothalamus, norepinephrine is increased in hypothalamus and decreased in pons and medulla. Reduced concentrations of testosterone and adrenal C-19 steroid precursors are associated with increased body fatness rather than with excess visceral fat accumulation. The early morning rise in the insulin demand is related to the increased early morning cortisol secretion and to the nocturnal peaks of growth hormone concentration. Excess of plasminogen activator inhibitor-1 (PAI-1) the main regulator of the fibrinolytic system, is closely associated to other components of the insulin resistance syndrome, namely, excessive body weight, high waist to hip ratio, elevated blood pressure, hyperinsulinaemia and hypertriglyceridaemia. TNF-α derived from adipose tissues might be involved in the induction of peripheral insulin resistance. High glucose concentration increases transendothelial permeation of albumin. There is a reduced vasodilatory capacity in diabetes, and especially in patients who are leaking albumin.

LIPID DISORDERS

Cholesterol can be synthesised by most cells of the body, primarily the liver and intestines, and it is obtained from the diet in foods of animal origin. Because cholesterol is not synthesised in plants, the major sources of dietary cholesterol are egg yolks, red meats and liver. Increased LDL cholesterol has been shown to result in endothelial dysfunction and increased oxidation by macrophages with incorporation of the oxidised cholesterol into the coronary vessel wall. Therefore, high levels of oxidised cholesterol directly causes and accelerates cardiovascular disease by promoting narrow, occlusive blood vessels. The oxidation of the LDL cholesterol is a critical step in atherogenesis. Red Yeast rice extract is a traditional food prepared by yeast fermentation and directly inhibits the enzyme HMG-CoA reductase. A 2,400 mg red yeast rice extract per day is required in order to reduce cholesterol by almost 20%. Unfortunately, Red Yeast rice extract causes the reduced biosynthesis of CoQ_{10}. This is because CoQ_{10} is made from a metabolite of mevalonate, a metabolite blocked by red yeast rice extracts. Artichoke leaf extact is rich in luteolin and glycosides, which support cholesterol metabolism. A standardised extract of Guggul (4.0% guggulsterones), an Ayurvedic herb, supports the actions of Red Yeast. α-Lipoic acid, grape seed and taurine provide antioxidant protection against free radical damage, including oxidation of LDL cholesterol. Flavonoids have an excellent effect on hypercholesterolaemia and other disorders including hypertension and vascular fragility. Some flavonoids are poorly absorbed and subject to intestinal degradation, e.g., Quercetin has only 20% absorption, but they are nonetheless conditionally essential in heart disease. The utilisation of nutrients to counter cardiovascular disease is extremely powerful, nutritional therapies are more effective than coronary artery bypass grafting and balloon angioplasty. Familial hypercholesterolaemia (FH) is caused by poor delivery of cholesterol into the cell resulting in a faulty HMG-CoA feedback loop. Without proper levels of cholesterol entering the cell, this results in the cell synthesising more cholesterol. This is not an uncommon disorder; it is estimated that 5% of myocardial infarctions suffered by individuals under 60 is caused by FH. Evidence suggests that lovastatin's success may stem from its ability to block a protein associated with the cancer cells, not from its knack for reducing cholesterol in the patients' blood. The type of carbohydrate consumed can have an effect on the cholesterol HDL. Specifically, it appears that the carbs found in beans have a lower glycaemic index than those in potatoes, and that low-glycaemic-index carbohydrates raise plasma concentrations of HDL cholesterol. High concentrations of HDL cholesterol are associated with decreased risk of cardiovascular disease. Atherosclerosis is a disease of the larger arteries. It is a progressive disease beginning in childhood with clinical manifestations developing in middle to late adulthood. In the presence of hyperlipidaemia, endothelial cells express cell-specific adhesion molecules (e.g., vascular cell adhesion molecule-1 [VCAM-1]), which promote the recruitment of monocytes. As LDL-cholesterol enters the subendothelial space, circulating monocytes attach to the surface of the intact endothelium. As monocytes attach to the surface of the endothelium they slip between the endothelial cells and accumulate in the subendothelial space. Once in the subendothelial space, monocytes undergo activation and differentiation to become macrophage cells, which actively take up LDL-cholesterol via a distinct receptor designated the scavenger or acetyl-LDL receptor. Modification of LDL involves the peroxidation of polyunsaturated fatty acids (PUFA) present in LDL, which is augmented by low concentrations of heavy metals such as copper. Antioxidants such as probucol, vitamin E, and β-carotene appear to counter this conversion. As macrophage cells ingest this modified lipid, they are converted to foam cells, the initial lesion of atherosclerosis. Oxidized LDL is a potent chemotactic factor and attracts monocytes into the subintimal space. Oxidized LDL inhibits macrophage movement, thus trapping them within the subintimal space. Lipid peroxidation also causes direct damage to endothelial cells allowing greater entry of circulating monocytes or LDL particles into the subendothelial space. Macrophages also produce chemotactic factors (e.g., IL1), oxygen metabolites (e.g., cytotoxic superoxide anions), and growth factors (e.g., PDGF). The fatty streak results in little or no obstruction of the affected artery and causes no clinical manifestations of coronary heart disease (CHD). Fatty streaks progress to fibrous plaques through the release of growth factors from macrophages, smooth muscle cells, or endothelial cells. These growth factors promote the migration of smooth muscle cells from the media into the fatty streak until it becomes a dominant cell type and a major source of connective tissue matrix within fibrous plaques and later lesions. Fibrous plaques contain large numbers of intimal smooth muscle cells, lipid-laden macrophages, and T lymphocytes. Fibrous plaques are usually raised and protrude into the arterial lumen and may compromise blood flow. As foam cells accumulate and stretch the endothelium and cytotoxic substances are released by macrophage cells and modified LDL particles, the endothelium may be damaged and expose subendothelial tissue. This triggers the adherence of platelets and possibly the formation of a clot. Adhering platelets are activated to release a number of vasoactive, chemotactic, and growth factors from their granules, which contribute to the formation of atherosclerosis. Platelet derived growth factor (PDGF) is one of the substances released and is both chemotactic and mitogenic. PDGF may induce smooth muscle cell migration from the media into the intima, as well as stimulate smooth muscle proliferation. When plaques fissure, there is platelet adhesion and activation with subsequent thrombus formation. Fissures may heal, sealing thrombi within the plaque. The fissuring-rehealing process leads to advanced, complicated lesions. The synthesis and metabolism of lipoproteins and apolipoproteins play a major role in the regulation of serum lipids and the risk of developing atherosclerosis. The major plasma lipids are cholesterol, triglycerides, and phospholipids. Cholesterol maintains the integrity of cell membranes and is an essential precursor for steroid hormone and bile acid formation. Cholesterol consumed in the diet is unesterified and is hydrophobic but not readily absorbed. Incorporation of small amounts of unesterified dietary cholesterol and cholesterol previously excreted in the bile into mixed micelles containing bile acids, free fatty acids, and monoglycerides is critical for transfer of cholesterol into mucosal cells. Cholesterol synthesis occurs in most cells of the body; however, the majority occurs in the liver and intestinal mucosa. The rate of synthesis is greatest during the night and lowest during the day. Cholesterol biosynthesis involves several condensation reactions beginning with acetyl-CoA in the mitochondria. The rate-limiting step in the pathway is the conversion of 3-hydroxy-3-methylglutaryl-coenzyme A (HMG-CoA) to mevalonate via the enzyme HMG-CoA reductase. HMG-CoA reductase activity is regulated by the amount of cholesterol in the cells. Unesterified cholesterol can be utilized for cell membrane production or esterified with a fatty acid by the enzyme acyl coenzyme A:acyltransferase (ACAT) and stored as cholesterol ester. Plasma cholesterol concentrations are regulated by intestinal cholesterol absorption, endogenous hepatic cholesterol synthesis, hepatic-free cholesterol, and bile acid excretion. When hepatic cell cholesterol concentrations increase, there is an increase in hepatic secretion of cholesterol and triglyceride-rich particles and a decrease in receptor-mediated removal of cholesterol-rich particles. Triglycerides consist of free fatty acids (FFAs) and glycerol and are an important source of stored energy.

Common plasma lipoprotein classes:

Class	Density g/ml	Diameter nm	Predominant lipids
Chylomicrons	0.93	75-1200	Dietary triglycerides
Very-low-density lipoproteins (VLDL)	0.93-1.006	30-80	Endogenous triglycrides
Intermediate-density lipoproteins (IDL)	1.006-1.019	25-35	Endogenous triglycerides, Cholesteryl esters
Low-density lipoproteins (LDL)	1.019-1.063	18-25	Endogenous cholesteryl esters
High-density lipoproteins (HDL_2)	1.063-1.125	9-12	Endogenous cholesteryl esters
High-density lipoprotein (HDL_3)	1.125-1.210	5-9	Endogenous cholesteryl esters

Severe elevations in plasma triglycerides (>1000 mg/dl) increase the risk of pancreatitis, whereas borderline elevations (200-400 mg/dl) may play a role in the development of atherosclerosis. Dietary triglycerides are absorbed in the duodenum and proximal jejunum as monoglycerides and FFAs following partial hydrolysis by gut lipases. Reconversion of these elements into triglycerides occurs in the mucosal cells. Triglycerides also are formed in the liver from accumulated or synthesized fatty acids. Triglycerides from the intestinal mucosa and liver are transported to adipose tissues for storage. Triglyceride lipolysis releases FFAs, which are transported by albumin to various tissues and utilized for energy. Lipolysis is stimulated by glucagon, catecholamines, and hormones including corticotropin, and inhibited by insulin and PGE_1. Phospholipids are essential for cellular function and lipid transport. The major phospholipid in the body is lecithin. Phosphatidylcholine, a major component of phospholipids, is synthesized primarily by microsomal enzymes. The major lipoproteins are chylomicrons, very-low-density lipoprotein (VLDL), intermediate-density lipoprotein (IDL), LDL, HDL, and lipoprotein(a) (Lp[a]). They are high molecular weight, water-soluble particles responsible for the transport of water-insoluble cholesterol and triglycerides in plasma. The surface of each lipoprotein is made up of apolipoproteins and unesterified cholesterol integrated between a phospholipid monolayer. The inner core is made up of cholesterol esters and triglycerides. The width of the surface layer is similar in all lipoproteins, but the core size varies. Triglycerides make up the majority of the inner core of chylomicrons and VLDL, whereas the inner cores of LDL and HDL are composed mainly of cholesterol esters. Lipoprotein(a) is comparable in composition to LDL, but also contains apolipoprotein(a). The major structural components of lipoproteins are apolipoproteins. Apolipoproteins are amphopathic; the hydrophobic amino acids of the apolipoprotein form one side of the helix and interact with the fatty acid portion of the phospholipids, whereas the hydrophilic amino acids interact with the polar region of phospholipids. Apolipoprotein A-I appears to be the major structural protein involved in HDL synthesis. Apolipoprotein A-I is also an important activator of lecithin:cholesterol acyltransferase (LCAT), which catalyzes the esterification of free cholesterol to cholesterol ester. Apolipoprotein B-48 is required for the secretion of chylomicrons from the intestines. Apolipoprotein B-100 is a major structural protein of VLDL, IDL, and LDL, and is required for hepatic VLDL secretion. The C apolipoproteins are found on the surface of VLDL and chylomicron particles. Apolipoprotein C-II is a cofactor for the activation of the enzyme lipoprotein lipase (LPL), which is responsible for the breakdown of triglycerides into FFAs and monoglycerides. Deficiencies in apolipoprotein C-II can lead to severe hypertriglyceridaemia. Apolipoprotein E is a polymorphic apolipoprotein. Dietary cholesterol and triglycerides absorbed in the intestines are incorporated into chylomicrons. As chylomicrons pass through the capillaries, their core triglycerides are hydrolyzed by LPL, located on the endothelial cell walls of capillary beds, and hepatic lipase (HL), located on the endothelial cell walls of the liver. This results in the release of free fatty acids (FFAs) that are quickly taken up by adipose tissue, converted back to triglycerides, and stored. Endogenously synthesized cholesterol and triglycerides are packaged into VLDL particles and secreted into plasma. The triglycerides in VLDL particles are hydrolyzed via the lipases in a manner similar to chylomicrons. LDL is the major cholesterol carrier in the body. 60% to 80% of LDL particles are removed via hepatic and peripheral B/E receptors. Increased hepatic uptake of VLDL, IDL, or LDL reduces B/E receptor production and suppresses the de novo intracellular cholesterol production. HDL is important to the removal of cholesterol from peripheral tissues. Cholesterol esters in HDL are transferred to other lipoproteins through the action of cholesterol ester transfer protein (CETP) for removal by the liver, or the cholesterol in HDL may be removed directly by the liver if it contains sufficient apolipoprotein E. The cholesterol esters that are transferred during this process are exchanged for triglycerides. Lipid metabolism is an intricate process and literally hundreds of steps could malfunction to produce lipid disorders. Familial hypercholesterolaemia (FH) is caused by autosomal dominant inheritance of defective genes for LDL receptors from one or both parents. Accelerated cardiovascular and cerebrovascular diseases often develop in these patients before the age of 20 years. Severe accumulation of LDL-cholesterol 4-6 times higher than normal (LDL-C >500 mg/dl) occurs as a result of decreased LDL-cholesterol clearance. Triglyceride and HDL-cholesterol concentrations are usually normal. Clinically, patients with FH can be identified by: 1- Severe hypercholesterolaemia (elevated total and LDL-C concentrations) early in life. 2- Strong family history of premature coronary heart disease (CHD). 3- Tendon xanthomas (caused by cholesterol deposition), especially the Achilles tendon and extensor tendons of the hands. The prevalence of familial defective apolipoprotein B-100 disorder is less than that of heterozygous FH, estimated to occur in up to 1/1000 population.

Indications for further evaluation in adults include:

Patients	Risk
Total cholesterol equal to or greater than 240 mg/dl.	2-fold increased risk of CHD
Total cholesterol between 200 and 239 mg/dl with two or more CHD risk factors.	2-fold increased risk of CHD
HDL-cholesterol less than 35 mg/dl.	2-fold increased risk of CHD
Established CHD or other atherosclerotic disease.	5-7-fold increased risk of CHD

Polygenic hypercholesterolaemia is the most common cause of mild to moderately elevated LDL-cholesterol and remains one of the least well understood lipid disorders. Environmental (e.g., dietary) and inheritance patterns appear to be related to this hypercholesterolaemic disorder. More than 95% of persons with hypercholesterolaemia have the polygenic form. Familial combined hyperlipidaemia (FCHL) is more common than FH, occurring in 1-2% of the population and accounting for 15-25% of patients with hypertriglyceridaemia. Patients with FCHL tend to be obese, hypertensive, develop premature atherosclerosis in the fourth or fifth decade of life, and have a strong family history of premature CHD. Elevations in both cholesterol and triglycerides and family history of premature CHD and dyslipidaemias. Familial hypoalphalipoproteinaemia is characterised by low HDL-cholesterol concentrations (<35mg/dl), occur in 3-5% of the general population with or without mild to moderate elevations in LDL-cholesterol and in 25-30% of patients hospitalized for acute myocardial infarction (MI). Familial dysbetalipoproteinaemia is a genetic disorder resulting from an apolipoprotein $E_{2/2}$ phenotype and causing plasma accumulation of VLDL, chylomicron remnants, and IDL particles. The clinical presentation of this dyslipidaemia may include both hypercholesterolaemia and hypertriglyceridaemia, palmar and/or tuberoeruptive xanthomas, and premature atherosclerotic disease. Cholesterol and triglyceride concentrations are usually between 300 and 600 mg/dl and the VLDL-cholesterol:triglyceride ratio is usually >0.3. Secondary dyslipidaemia should be considered and excluded when evaluating individuals with dyslipidaemia. The most common secondary causes of hypercholesterolaemia are diabetes mellitus, hypothyroidism, and drugs. Accuracy is the agreement between the test result from a given device or laboratory and that obtained in a standardized laboratory and is expressed as a mean % bias. Precision is the degree to which a test result can be replicated by multiple runs from the same sample and is expressed as the coefficient of variation (CV). Clinicians should know the accuracy and precision performance of the laboratory they use to properly interpret results. Accuracy and precision are derived from group results and do not refer to individual patient results. To limit the number of misclassifications, cholesterol measurements within 5% of the total cholesterol or within 9% of the LDL-cholesterol decision point should be repeated at least two or more times before making classification and treatment decisions. Normally, cholesterol has a diurnal pattern; concentrations are generally lower at night and higher during the day. Cholesterol concentrations vary from day to day, usually by only 2-3%, but in some patients by as much as 40-50%. This natural variability is another reason for testing cholesterol several times before making a treatment decision. Patients who do not fit into any of the above categories should be given dietary advice and total and HDL-cholesterol should be remeasured every 5 years. Patients should fast (i.e., nothing by mouth except water) for at least 9 and preferably 12-14 hours prior to the evaluation to clear exogenous triglycerides from the bloodstream. The cholesterol carried in VLDL and LDL particles is estimated from the following calculations:

VLDL = Triglycerides/5
LDL = Total cholesterol - (HDL + Triglycerides/5)

These formulas are accurate when the triglyceride concentration is <400mg/dl. When values are above 400 mg/dl, LDL-cholesterol must be measured directly. LDL-cholesterol obtained from smokers has more lipid peroxidation products than that from nonsmokers. This oxidative modification of LDL-cholesterol may occur either directly by oxidants in smoke or indirectly by endogenous oxidants induced by activation of cells in the airway system. Smokers are deficient in endogenous antioxidants. Smoking increases adrenergic activity resulting in increased vasoconstriction. Toxins in cigarette smoke also may directly injure the endothelial lining of the arterial wall leading to smooth muscle cell proliferation, platelet aggregation, and thrombogenesis. The increased CHD risk associated with smoking declines within 2-5 years after smoking cessation. In hypertension, increased arterial pressure and altered blood flow characteristics may cause focal endothelial injury thereby inducing smooth muscle cell growth in underlying arterial tissue. Hypertension-induced endothelial injury also may lead to thrombus formation in the late phases of atherogenesis. Other mediators of increased blood pressure, such as catecholamines and angiotensin, may induce cellular changes, which accelerate atherogenesis. HDL-cholesterol is inversely associated with CHD risk. A concentration of less than 35 mg/dl is considered a positive CHD risk factor whereas concentrations higher than 60 mg/dl is a negative risk factor. Circulating HDL-cholesterol accumulates and traps cholesterol from extrahepatic tissues and delivers it to the liver for excretion, i.e., reverse cholesterol transport. Thus, a major role for HDL is to act as a scavenger of cholesterol from tissues, including the arterial wall. HDL may prevent excessive exposure of the arterial wall to the postprandial lipids in the circulation, which may be atherogenic. HDL also may have antioxidant activity, possibly due to the exchange of lipid peroxidation products between the lipoproteins. Increased HDL-cholesterol concentrations positively correlate with normalizing endothelial function in atherosclerotic coronary arteries. Diabetes increases CHD risk about 3-fold in men and 6-fold in women. In diabetics, LDL composition is abnormal, resulting in a LDL particle that is smaller, more cholesterol dense, and more easily oxidized. Diabetics have altered endothelial function with impaired response to NO and enhanced secretion of endothelin; mitogenic effects of insulin; impaired fibrinolytic activity; and enhanced platelet adhesiveness and aggregability. Atherosclerosis is associated with a 5- to 7-fold higher risk of MI. Goals of therapy are based on the level of the patient's CHD risk, which is based on LDL-cholesterol, concurrent risk factors, and presence of atherosclerotic vascular disease. These goals are not rigid, absolute endpoints of therapy but provide a framework for making treatment decisions. Lifestyle habits that may be causing an elevation in lipid concentrations and that may lead to the development of other CHD risk factors must be addressed effectively. Aggressive lifestyle modification and drug therapies should be initiated to significantly lower cholesterol concentrations toward the goal of therapy. Before initiating any intervention, a thorough assessment of the patient's lifestyle is needed with particular attention to weight, dietary and exercise habits, alcohol intake, and smoking history. Assessment of the patient's understanding of his or her health, ability to learn new information, and social and financial factors that may affect the ability to accept or afford therapeutic interventions is necessary. The goal is to achieve realistic, gradual, and steady weight loss. Very-low-calorie weight-loss diets (500-800 calories/d) should be avoided. A vegetarian diet can be effective if the diet is nutritionally balanced. However, ultra low-fat diets (<10% total fat), like very low-calorie diets, are difficult to sustain, may not be nutritionally balanced, and can be potentially detrimental. Patients who markedly increase carbohydrate intake to replace fat may experience a temporary increase in triglycerides and/or a reduction in high-density lipoprotein (HDL) cholesterol. Patients with normal or low cholesterol who implement an ultra-low-fat diet as a preventive measure may find that both LDL-cholesterol and HDL-cholesterol concentrations fall. A minimum of 6 months is needed to adequately test patient adherence and lipid response to any diet program. Cholesterol is found only in foods of animal origin. Saturated fat and cholesterol often are found in the food supply together. Egg yolks, organ meats, veal, and shellfish are high in cholesterol but low in saturated fat. Coconut and palm kernel oils contain no cholesterol but are high in saturated fat and should be avoided. Fish, skinless chicken and turkey, and lean cuts of beef or other meats should be limited to 5 ounces daily (142 g). Egg whites are high in protein and contain no fat or cholesterol and are recommended for persons on fat- and cholesterol-restricted diets. Saturated fats contain myristic acid, found in cholesterol-rich foods such as butter fat, whole milk, ice cream, and cheese. Myristic acid contributes more than any other fat to increasing LDL-cholesterol concentrations. Palmitic acid found in coconut oil, palm kernel oil, and animal fats (especially beef and pork), also contributes significantly to increasing LDL-cholesterol concentrations. Stearic acid is a common component of cocoa butter, beef, and other foods, but contributes little to LDL-cholesterol elevation. Polyunsaturated fats (PUFA) reduce total and LDL-cholesterol but, like all fats, they are calorie dense (9 Kcal/g). The primary ω-6 fatty acid is linoleic acid, derived from vegetable sources such as safflower, sunflower, soybean, and corn oils. ω-3 fatty acids, found in some fish oils, are metabolic breakdown products of linolenic acid. ω-3 fatty acids lower triglyceride concentrations by decreasing very-low-density lipoprotein (VLDL) synthesis and increasing VLDL catabolism, but they have no consistent beneficial effect on LDL-cholesterol.

Contraindications to major lipid-modifying drugs:

Drug	Factor
Bile acid sequestran	1- Chronic constipation. 2- Vitamin K deficiency/hypoprothrombinaemia. 3- Complete biliary obstruction. 4- Hypersensitivity. 5- Hypertriglyceridaemia. 6- Phenylalanine accumulation (e.g., phenylketonuria). 7- Concurrent administration of drugs (thyroid hormones, digoxin, warfarin, thiazide diuretics, amiodarone, propranolol, iron salts, tetracycline, phenobarbital, phenylbutazone, penicillin G, and loperamide).
Niacin	1- Active peptic ulcer disease. 2- Active liver disease. 3- Arterial bleeding. 4- Noninsulin-dependent diabetes mellitus (relative). 5- Gout (relative). 6- Concurrent administration of HMG-CoA reductase inhibitors (Increased risk of myositis and rhabdomyolysis.).
HMG-CoA reductase inhibitor	1- Alcohol abuse. 2- Active liver disease. 3- History of hypersensitivity. 4- Pregnant or lactating women. 5- Women of childbearing potential not practicing birth control. 6- Concurrent therapy with cyclosporine, gemfibrozil, niacin (Increased risk of myositis and rhabdomyolysis).

HDL-cholesterol may decrease with fish oil in some people. ω-6 fatty acids are converted to arachidonic acid (AA), which is acted on by cyclooxygenase to yield two prostaglandins (PG), thromboxane A_2 (TXA_2) and prostaglandin I_2 (PGI_2 or prostacyclin). ω-3 fatty acids yield TXA_2 (biologically inert) and PGI_3 (a prostacyclin analog with potent vasodilator activity and platelet aggregation inhibitory actions). Fish oils also may cause hyperglycaemia in patients with diabetes. Monounsaturated fatty acids lower LDL-cholesterol; they also may increase and do not appear to decrease HDL-cholesterol. Oleic acid found in olive oil, canola oil (rapeseed), and some varieties of sunflower seeds and oils are common sources of monounsaturated fat. Most nuts are high in monounsaturated fat (except walnuts, which contain more polyunsaturated fat). Hydrogenated fats (trans fatty acids), like saturated fats, raise LDL-cholesterol concentrations. Trans fatty acids may raise the risk of CHD. Soluble fiber products such as psyllium, oat bran, and gums may limit absorption and bioavailability of certain nutrients as well as produce uncomfortable GI side effects. Most patients need help in translating dietary recommendations into practical terms and concepts. Patients who fail to have an adequate response can be classified as: 1- Poor patient adherence to diet despite intensive and prolonged counseling. 2- Poor patient adherence to diet in the first few months but eventually better. Time, patience, and long-term support are critically important. 3- High LDL-cholesterol concentrations that are resistant to dietary modification. 4- LDL-cholesterol concentrations are so high that diet therapy is insufficient to reach target goals. Patients with abdominal (truncal) obesity often overproduce VLDL particles, thereby raising triglyceride and possibly LDL-cholesterol concentrations. Abdominal obesity is associated strongly with insulin resistance and hyperinsulinemia that may lead to CHD. Regular physical exercise may reduce triglycerides, raise HDL-cholesterol, promote weight loss or maintenance of desired weight, lower blood pressure, improve diabetes control, and cause favourable changes in coronary blood flow. Regular aerobic exercise (e.g., brisk walking, jogging, swimming, bicycling, and tennis) should raise the heart rate for 20-30 minutes. There is strong inverse association between wine intake and CHD but less benefit from other alcohol sources. The effects of smoking on LDL-cholesterol concentrations are minimal. In some patients, LDL-cholesterol may drop slightly when they stop smoking; HDL-cholesterol also may increase slightly. Other patients may gain weight and show a paradoxical increase in LDL-cholesterol concentrations. Appropriate diet planning and physical activity may counter any negative LDL-cholesterol changes. The benefit of cholesterol-lowering treatment is most evident in patients with multiple risk factors. There is a lack of improvement in overall survival in the studies supporting the benefit of lowering cholesterol. Niacin is associated with 29% reduction in nonfatal MI and 11% reduction in total mortality. The decision to initiate lipid-modifying drug therapy generally should be delayed until lifestyle modification has been pursued for at least 6 months. The majority of patients with lipid abnormalities should invest in a good trial of lifestyle changes before considering the use of drugs that are expensive and more likely to cause side effects. The decision to use drugs is usually a long-term, probably life-long, commitment. Bile acid sequestrants, niacin, and HMG-CoA reductase inhibitors are the major drugs for managing patients with hypercholesterolaemia because they exert the most potent LDL-cholesterol-lowering effects of the drugs approved for lipid modification. Bile acid sequestrants may be the safest medications because they are unabsorbed systemically; however, HMG-CoA reductase inhibitors have established a clear record of safety with nearly a decade of use. Niacin is the least expensive of the 3 agents and has the most potent triglyceride-lowering and HDL-raising effects. Combination therapy using low doses of 2 drugs that have complementary effects can often reduce LDL-cholesterol to a greater extent with less cost and fewer side effects in patients with familial hypercholesterolaemia or severe polygenic hypercholesterolaemia. The combination of a bile acid sequestrant with an HMG-CoA reductase inhibitor may lower LDL by 45-64%. The combination of a bile acid sequestrant with niacin reduces LDL-cholesterol by 48%. It is important to employ good patient selection, close monitoring, and patient education whenever they are used. Patients with elevated cholesterol and triglyceride concentrations should be approached somewhat differently than patients with hypercholesterolaemia alone although the goal of therapy (lower LDL-cholesterol) is the same. A border-line hypertriglyceridaemia (i.e., 200-400 mg/dl) with hypercholesterolaemia signals a relatively high risk of CHD. These patients often have other CHD risk factors including hypertension, diabetes or hyperinsulinaemia, truncal obesity, and low HDL-cholesterol.

Causes of hypertriglyceridaemia:

> Primary:
>   1- Abnormal chylomicron and VLDL assembly.
>   2- Apo CII deficiency, Apo E deficiency or abnormality.
>   3- (Type III), hepatic lipase deficiency, LPL deficiency.
>   4- (Type I), VLDL overproduction.
> Secondary:
>   1- Disease (Burns, Cushing's disease, diabetes mellitus, hypercalcaemia, hyperuricaemia, hypothyroidism, liver disease, menopause, multiple myeloma, MI, obesity, renal disease, sepsis, SLE, trauma).
>   2- Drug (Alcohol, blockers, bile acid sequestrants, oestrogen, oral contraceptives, glucocorticoids, nicotine, retinoids, thiazides).

They also have elevated VLDL-cholesterol and high concentrations of small, cholesterol-dense LDL particles. A common lipid disorder that presents with combined hyperlipidaemia is familial combined hyperlipidaemia. Combined hyperlipidaemia patients have an impaired ability to clear VLDL particles from the bloodstream, resulting in an accumulation of VLDL particles and elevated cholesterol and triglyceride concentrations. The primary aim of therapy in patients with combined hyperlipidaemia is to lower LDL-cholesterol to goal while reducing triglycerides to <200 mg/dl and VLDL-cholesterol to <40 mg/dl as well as raising HDL-cholesterol concentrations (>35 mg/dl) if necessary. Diet, weight reduction, and exercise are important lifestyle modifications in these patients. If control is not achieved with this approach alone, drug therapy may be considered. The first-line drugs for this disorder are niacin and HMG-CoA reductase inhibitors. Niacin (2000-3000 mg daily) substantially lowers triglycerides (and thereby VLDL-cholesterol) and reduces LDL-cholesterol 15-25%; usually, HDL-cholesterol increases 15-30%. Because they raise triglyceride concentrations, bile acid sequestrants are not recommended in patients with combined hyperlipidaemia except in combination with a triglyceride-lowering agent such as niacin or gemfibrozil. The higher the triglyceride concentration prior to treatment, the more likely gemfibrozil is to raise LDL-cholesterol. The increase in LDL-cholesterol with gemfibrozil is not desirable and may negate its other beneficial lipid effects. Symptoms of myalgia must be monitored and creatine kinase (CK) levels measured if symptoms appear. Hypertriglyceridaemia alone may not indicate an increased risk of CHD. Patients with very high triglycerides (>1000 mg/dl) have an increased risk of hepatomegaly, splenomegaly, hepatic steatosis, and pancreatitis (but not CHD) and are candidates for triglyceride-lowering diet and drug treatment. Patients whose fasting serum triglycerides are <1000 mg/dl are at a lower immediate risk for pancreatitis. The triglyceride-lowering drugs of choice are niacin and gemfibrozil. Niacin is indicated in patients without diabetes; gemfibrozil is the primary choice in patients with diabetes because it does not increase glucose concentrations. Gemfibrozil (1200 mg/d) may reduce triglyceride concentrations by 70-90% in patients with severe hypertriglyceridaemia. Drug therapy should be considered only after lifestyle modification has been pursued for 6 months unless severe cholesterol elevations or severe atherosclerotic disease is present. Bile acid sequestrants are nonabsorbable resins, which exchange chloride ions for bile acids and other anions in the intestinal lumen. This activity interrupts the enterohepatic recirculation of bile acids, increases their faecal excretion, and depletes the cholesterol pool in the hepatocyte. The hepato-cellular depletion of cholesterol stimulates upregulation of surface low-density lipoprotein (LDL) receptors, which increase the uptake of cholesterol-carrying LDL and intermediate-density lipoprotein (IDL) particles from the circulation. But, depletion of the cholesterol pool stimulates in vivo cholesterol synthesis, partly negating the cholesterol-lowering efficacy of these drugs. Bile acid sequestrants lower apolipoprotein B concentrations but generally to a lesser degree than LDL-cholesterol. Bile acid sequestrants increase hepatic very-low-density lipoprotein (VLDL) production and thereby increase blood triglyceride concentrations (up to 15%). Bile acid sequestrants reduce CHD events. Because bile acid sequestrants are not absorbed systemically, adverse effects generally are limited to the GIT. Common adverse effects include abdominal bloating, belching, flatulence, abdominal pain, heartburn, nausea, vomiting, haemorrhoids, and constipation. Bile acid sequestrants may bind to medications (e.g., digoxin, thyroxine, thiazide diuretics, iron, fat-soluble vitamins, folic acid, and warfarin) and decrease their absorption or entero-hepatic circulation. Cholestyramine and colestipol are indicated as adjunctive therapy to diet for the reduction of elevated serum cholesterol in patients with primary hypercholesterolaemia who do not respond adequately to diet therapy alone. Bile acid sequestrants are contraindicated in patients with biliary obstruction or in patients who have shown hypersensitivity to the components. Niacin influences both the lipid composition of lipoproteins and overall plasma lipid concentrations. Niacin reduces hepatic VLDL synthesis, resulting in decreased LDL-cholesterol and IDL-cholesterol, the intermediate products of VLDL metabolism. Reduced VLDL synthesis also results in lower concentrations of cholesterol and triglycerides carried by this particle. VLDL particle size also is decreased. Niacin may inhibit lipolysis in adipose tissue, which decreases fatty acid release, triglyceride synthesis, and, ultimately, plasma triglyceride concentrations. Niacin may directly inhibit LDL synthesis, an effect that appears to be dose related. Niacin reduces apolipoprotein B concentrations to an extent similar to LDL-cholesterol. Niacin is the only major lipid-modifying drug known to reduce Lp(a), but the significance of this has not been shown. Sustained-release niacin has modest effects on HDL-cholesterol (5-15% increase) and triglycerides (20-40% reduction), and on LDL-cholesterol (25-50% reduction). The latter effect, however, may be caused by an undesirable toxic effect on the liver. Immediate-release niacin (<3 g/d) reduces LDL-cholesterol concentrations by 15-25% and triglycerides by 40-50% with maximum doses; HDL-cholesterol concentrations are increased by 25-35%. Niacin has been used for the treatment of hyperlipidaemia since the 1950s. Lipid modification with niacin may have a long-term beneficial effect. Niacin reduces the mortality and morbidity associated with lipid disorders by interfering with the deposition and possibly enhanced removal of cholesterol from atherosclerotic lesions. On average, 25-50% of patients titrated to 3 g or more per day may discontinue therapy because of intolerable side effects. Flushing of the face and trunk with initial use and/or pruritus of the trunk. Niacin increases prostaglandin activity; ingestion of one 325-mg aspirin tablet 30 min prior to the morning dose decreases the symptoms. Use of sustained-release niacin preparations may decrease this incidence. Acanthosis nigricans (hyperpigmentation, especially in the underarm and groin areas) also may occur. Niacin competes for elimination with uric acid by renal tubular secretion and may cause a slight increase of uric acid in 10% of patients. In a rare cases, the elevation may be great and lead to acute gouty arthritis. The mechanism by which niacin increases serum glucose is unknown. Pancreatic β-cell sensitivity may be altered; alternatively, the increase in free fatty acids resulting from niacin therapy may result in increased utilization of this substrate in preference to glucose. Diabetes mellitus is a relative contraindication for niacin therapy. Although loss of glycaemic control may be encountered, it usually can be corrected by increasing the dose of hypoglycaemic balance/therapy. The resultant hyperinsulinaemia may increase the risk of development of atherosclerotic disease. However, results of the largest clinical study that included diabetic dyslipidaemic patients treated with niacin showed no increase in mortality after 6 years of treatment and a reduction in mortality 9 years after the completion of the study. Immediate-release niacin rarely is associated with hepato-toxicity whereas sustained-release niacin may produce hepatotoxic effects in up to 50% of patients. Hepatotoxicity appears to be dose related; most cases are associated with sustained-release niacin in doses of 2000 mg or more per day. Discontinuation of the drug declines symptoms within 2 weeks and laboratory abnormalities resolve within 1-4 months. Nausea and abdominal pain may occur in 10-30% of patients taking niacin, which may be minimized by taking niacin with meals. Niacin induces steal coronary in patients with unstable angina, which may worsen angina. The HMG-CoA (3-hydroxy-3-methylglutaryl-coenzymeA) reductase inhibitors competitively inhibit HMG-CoA reductase, the enzyme responsible for the conversion of HMG-CoA to mevalonate, the rate-limiting step in cholesterol biosynthesis. HMG-CoA reductase inhibitors are specific and competitive inhibitors of the HMG-CoA enzyme. The loss of cholesterol production up-regulates the cellular membrane expression of LDL receptors, which constitute the major removal mechanism of circulating cholesterol-rich LDL in the body. LDL receptors also are involved in the removal of VLDL and IDL particles and may play a role in the modest increases in HDL particles. Cimetidine inhibits the oxidative pathways of lovastatin and simvastatin, P450-dependent pathways may serve as the basis for some of the drug-drug interactions. GI complaints (e.g., diarrhoea, abdominal pain, constipation, and flatulence) and headache are the most commonly reported adverse effects of HMG-CoA reductase inhibitors. Higher concentrations of statins in cerebral spinal fluid (CSF) may result in insomnia and effects on daytime performance. Nevertheless, the overall problem of sleep disturbances is small.

Changes in liver function tests (LFTs) and the development of myopathy are the most clinically important adverse effects of HMG-CoA reductase inhibitors (1%). Dosage reduction or withdrawal of the HMG-CoA reductase inhibitor reverses these abnormalities. Myopathy, defined as muscle pain and soreness plus elevations in creatine kinase (CK) to 10 times the upper limit of normal, is rare (0.1%). The risk of myopathy may increase when HMG-CoA reductase inhibitors are combined with erythromycin, niacin, gemfibrozil, or cyclosporine. The incidence with concurrent erythromycin and niacin appears to be very low, whereas the risk of myopathy with the combination of HMG-CoA reductase inhibitors and cyclosporine is highest, and may approach an incidence of 33%. Myopathy may develop as early as one week and as long as two years after initiation of therapy. Fibric acids augment triglyceride and VLDL-cholesterol clearance, decrease synthesis of triglycerides and apolipoprotein B, mobilizing cholesterol from tissues, and decreasing release of FFAs. These agents also may inhibit hepatic VLDL production by decreasing lipolysis of stored triglycerides and reducing fatty acid uptake to the liver. Fibric acids increase HDL-cholesterol production. Gemfibrozil elimination half-life is 1.5 hrs but its biologic half-life is considerably longer as the drug undergoes enterohepatic recirculation and is reabsorbed in the GI tract. Triglycerides are reduced from 30-50% with fibric acid therapy. Total cholesterol is only modestly reduced (5-20%) in patients without elevated triglycerides. The effect of fibric acids on LDL-cholesterol is variable. Nonlipid effects of clofibrate include a reduction in platelet adhesiveness and plasma fibrinogen levels, increased antidiuretic hormone release from the posterior pituitary (clofibrate has been used to manage diabetes insipidus), diminution of arginine-stimulated insulin and glucagon release from the pancreas and, in diabetic patients, a drop in serum insulin and fasting blood glucose levels. Evidence for the effectiveness of clofibrate in reducing deaths due to CHD is lacking. The safety of gemfibrozil in patients with CHD is questionable and, thus, the drug is not recommended for use in these patients. Malignant gastric neoplasm and cholelithiasis may occur in patients given clofibrate. Fibric acids produce GI adverse effects including epigastric pain, nausea, vomiting, dyspepsia, diarrhoea, flatulence, and constipation. Fibric acids may occasionally cause skin rash, alopecia, weight gain, headache, impotence, decreased libido, and muscle cramps. Cholestatic jaundice, flu-like syndrome consisting of severe muscular cramps, stiffness, and weakness due to myositis may occur. Rhabdomyolysis has been reported in a few cases, especially in patients with renal dysfunction. Increased gastric malignancy and increased cholelithiasis have occurred in patients treated with clofibrate. Fenofibrate lowers triglycerides to a similar degree as other fibric acid derivatives but it may lower LDL-cholesterol to a greater extent. Fenofibrate should be considered adjunctive therapy for the treatment of very high triglyceride concentrations (i.e., >1000 mg/dl) in adults who are at risk for pancreatitis and who are not responsive to dietary restrictions. A conservative approach, couched with common sense, should be advocated. Older patients should not be treated if their life expectancy is not long enough to benefit from cholesterol-lowering therapy (i.e., at least 1 y), they have serious concurrent illnesses, or their treatment regimen would be unduly complicated or compromised by the addition of cholesterol-lowering therapies. Many elderly patients already are following a diet deficient in protein, minerals, and vitamins and so are vulnerable to further restrictions. Also, many elderly patients have become very health conscious and may exceed recommendations, which could lead to additional problems: 1- Severe restriction of total fat intake may result in calorie-deficient diets. 2- Replacing fat with carbohydrates can cause weight gain and promote glucose intolerance. 3- Restricting the intake of milk and milk products (trying to avoid rather than replace whole-milk products) can lead to calcium deficiency.

HOMOCYSTEINE

Homocysteine is a sulfur-containing amino acid. Homocysteine prevents collagen from properly cross-linking to form a stable bone matrix. Elevated homocysteine levels also may be associated with spina bifida, rheumatoid arthritis and some cancers. Deficiencies of certain B-complex vitamins B_6, B_{12} and folic acid can lead to a build up of an undesired amino acid in the blood (homocysteine). Homocysteine causes the fibrous and fibrocalcific plaques in children with homocystinuria. Regardless of the cholesterol level, elevated homocysteine level is associated with increased risk of arteriosclerosis and coronary heart disease (independent risk factor). Homocysteine causes coronary heart disease (CHD) by injuring the endothelial lining of the coronary arteries and by thickening of the wall of the arteries, regardless of the level of cholesterol in the blood. Homocysteine interferes with the way cells use oxygen, resulting in a build-up of damaging free radicals. These reactive chemical forms can oxidise LDL, producing oxycholesterols and oxidised fats and proteins within developing plaques. Homocysteine stimulates growth of smooth muscle cells, causing deposition of extracellular matrix (ECM) and collagen, which causes a thickening and hardening of artery walls. When clots form in a coronary artery the blood supply to the heart itself is reduced or stopped, i.e., coronary thrombosis, which results in the death of heart tissue (myocardial infarction). Homocysteine in its reactive form, homocysteine thiolactone, affects the reactivity of platelets and is extremely active in causing platelet aggregation, which can lead to clot formation. Also, homocysteine causes the binding of lipoprotein (a) [Lp(a)] to fibrin in very low concentrations. Lipoprotein(a) has also been linked to the thrombotic events in coronary heart disease. Homocysteine is involved in other important clotting factors including protein C, factor VII, factor XII, and other clotting factors. The production of fibrous tissues of sulfated glycosaminoglycans and the destruction of elastin fibers in the wall of the artery lead to the formation of fibrous plaques. The lesions in the arteries accumulate a great deal of fat and cholesterol, producing fibrolipid plaques. The majority of patients (85%) who have severe CHD and arteriosclerosis have cholesterol levels in the normal range (<250 mg/dl). The normal level of homocysteine in the blood for a middle aged man is 8-12 µmol/l, so 50% increase in this would be up to 17 µmol/l. This level of homocysteine in the blood is associated with an increased risk of development of myocardial infarction. There are many different factors that are related to elevation of the homocysteine level, include: 1- Dietary imbalance between too much methionine from dietary protein and too little of vitamin B_6, vitamin B_{12} and folic acid. 2- Genetic factors or inherited factors, as many as 1:8 of the population in general carries hidden genetic defect in reductase enzyme that causes them to require more folic acid than normal people would require to prevent elevation of homocysteine levels. 3- Aging process, over the age of about 60 the homocysteine level increases about 1 µmol/l for every ten years of age. 4- Hormonal factor, premenopausal woman has a level of about 2 µmol/l lower than men of the same age. Normal women have homocysteine levels of 6-10 µmol/l, compared to normal men of the same age who have 8-12. After the menopause, the level of homocysteine rises in the blood to approach that of men of the same age. In chronic partial thyroid hormone deficiency, the homocysteine level can rise and lead to increased risk of vascular disease. 5- Cigarette smoke, smokers have higher level of homocysteine than non-smokers. 6- Exercise, those who exercise strenuously have lower level of homocysteine than those who are sedentary. 7- Drugs, such as methotrexate, nitrous oxide, and azaribine can elevate the homocysteine level. 350-400 µg a day of folic acid is required to keep the homocysteine level in the normal range. In the elderly, absorptive problems may cause a marginal vitamin B_{12} deficiency. In the case of vitamin B_6, there is perhaps a little less certainty about the exact figure. 30-40% of elderly has an intake that is significantly low, in the range of 1.6-1.7 mg/day. History of chronic nutritional depletion and nutritional abuse, such as chronic alcoholic, is an indication for higher levels of these vitamins. These vitamins are non-toxic. B-vitamins are water-soluble and any excess vitamin beyond the body's requirement is excreted quite rapidly in the urine. Vitamin B_{12} is quite safe. Vitamin B_6 in dose of 50-200 mg/day is safe. Folic acid in the range of 0.5-1 mg/day is probably entirely safe. Carbon monoxide of cigarette smoke combines with vitamin B_6 inactivating pyridoxamine phosphate, one of the active co-enzymes for many different enzymes in the body. Smokers tend to have lower level of vitamin B_6, and a higher level of vitamin B_6 is indicated. Choline-deficiency is associated with increased homocysteine level in the blood. There is an alternative pathway for conversion of homocysteine to methionine in man involving betaine. This methyl donor function of betaine has a supplementary action in lowering elevated homocysteine levels both in chronic renal failure and also in patients with homocystinuria. Controlling homocysteine levels may be successful in prevention of vascular disease and the catastrophic complications of vascular disease, heart attack, stroke and gangrene. About half of patients with homocystinuria respond quite well to vitamin B_6 in large doses. Vitamin B_6 may decrease the risk of thrombosis in vascular complications. Patients with carpal tunnel syndrome respond well to doses of vitamin B_6 in the range of 50-200 mg/day over a period of years, and also associated with 75% reduction in the risk of angina pectoris and myocardial infarction. Children with the inherited disease homocystinuria, have arteriosclerosis. The original index case was published in 1933. Children with homocystinuria have increased risk of thrombosis and rapidly progressive arteriosclerosis. Homocysteine is damaging to the arterial wall. The elevation of the homocysteine level in the blood damages the arteries regardless of any particular enzyme defect, e.g., methylenetetrahydrofolate reductase deficiency, cobalamin C disease. The cholesterol level is normal in these children. Patients with CHD have higher level of homocysteine in their blood following an oral dose of methionine, compared with controls without coronary heart disease. In humans, homocysteine is involved in the increased risk of neural tube defects in children that are born to mothers who are deficient in folic acid. Folate-deficient mothers tend to have higher level of homocysteine in their blood, and the amniotic fluid has higher level of homocysteine than in those in mothers who have normal folate intake. In cancer, there is an abnormality of methionine and homocysteine processing in cancer cells. Malignant cells have a very specific abnormality of homocysteine thiolactone metabolism. Malignant cells are unable to convert homocysteine thiolactone to sulfate. A compound formed between homocysteine thiolactone and vitamin A acid (retinoic acid), called thioretinamide, is anticarcinogenic. Furthermore, thioretinamide forms an additional complex with vitamin B_{12}, a substance known as thioretinaco. This compound is also anticarcinogenic. The activation of thioretinaco may occur through ozone oxidation of the sulfur atoms of homocysteine. This oxidation reaction may cause it to be highly effective anti-cancer compound. Homocysteine levels are elevated in rheumatoid arthritis. These patients also have an increased requirement of vitamin B_6, although vitamin B_6 therapy does not ameliorate their arthritis. Trimethylglycine (TMG), also known as betaine-not betaine hydrochloride-is a chemical compound and good source of methyl groups. A methyl group is a molecule consisting of one carbon atom and three hydrogen atoms (CH_3). Methylation is a naturally occurring process in which methyl groups are attached to substances in the body to either protect or transform them. For example, when attached to DNA, methyl groups appear protective, preventing mutated genes from expressing themselves. Methyl groups also convert homocysteine, a toxic amino acid that can cause cardiovascular disease, to methionine, a beneficial amino acid. In addition, methylation influences levels of S-adenosylmethionine (SAM-e), which protects the liver and is an antidepressant. These substances, along with folic acid, B_6, and, to a lesser extent, choline all contain methyl donor groups. These nutrients, along with TMG, are part of a chemical contingent in the body that works against cancer, heart and neurological diseases, and nearly every age-related disorder. Medium to high levels of plasma homocysteine are associated with increased risk for heart disease, cerebrovascular disease and peripheral artery disease. Daily treatment with TMG, folic acid and choline normalise homocysteine levels. The recommended TMG dose is 500-1,000 mg/day-roughly equivalent to what a diet high in broccoli, spinach or beets would provide. The methylation-aging connection, differs from antioxidant theories. Free radicals are created during oxidation and are believed to trigger cellular deterioration. On the other hand, oxidation is also necessary for metabolising food and driving normal body functions such as fighting viruses and bacteria. In contrast, methylation is not crucial for normal body functions. Homocysteine is a naturally occurring, sulfur containing amino acid, formed from methionine, an essential amino acid, via transmethylation. Homocysteine may be irreversibly metabolised to cysteine or can be converted back to methionine by transfer a methyl group from a derivative of folic acid in a reaction that involves vitamin B_{12}. Homocysteine level in blood is an independent risk factor for atherosclerotic vascular disease affecting the coronary, cerebral, and peripheral arteries. Homocysteine occurs in plasma as the free thiol (free homocysteine, 1%), its symmetrical disulfide (homocysteine) asymmetrical disulfide and conjugated with protein through disulfide linkage. The bulk of plasma homocysteine thus occurs in conjugated form, making it inaccessible to common analytical techniques. Sample must therefore be treated with reacting agent before analysis to liberate homocysteine as the free thiol. Elevated homocysteine levels, has inverse relationship to the plasma levels of both folate and vitamin B_{12}. Dietary supplementation with folic acid B_{12} and B_6 reduces the plasma homocysteine levels by about 30% in almost all subjects.

Homocysteine contribution:

1- Homocysteine prevents small arteries from dilating (vasoconstriction), thereby making them more susceptible to obstruction by clot or plaque.
2- Homocysteine changes coagulation factor levels, which encourages blood clot formation.
3- Homocysteine generates superoxide and hydrogen peroxide, both of which have been linked to damage the endothelial lining of arterial vessels.
4- Homocysteine causes the smooth muscle cells that support the arterial wall to multiply (proliferation), which is part of the narrowing process.
5- Homocysteine thiolactone, causes platelets to aggregate, which is part of the clotting process.
6- The active form of excess homocysteine reacts with LDL to form LDL-homocysteine thiolactone aggregates. These are taken up by macrophages, which in turn promotes atherothrombosis.

HYPOGLYCAEMIA

Hypoglycaemia is not a medically recognised term, except in connection with diabetes and various other diseases. Both hypoglycaemia and diabetes share insulin resistance (IR). Cerebral hypoglycaemia, cerebral diabetes, and brain diabetes are terms pointing to abnormal glucose transport systems across the membranes of brain cells. During the dynamic phase of hypoglycaemia in response to the rapidly falling blood sugar, serum adrenaline levels rises. This physiological mechanism is involved in increasing serum glucose by promoting gluconeogenesis. Hypoglycaemia is characterised by abnormally low blood sugar (glucose). More severe hypoglycaemia can lead to reduced glucose supply to brain, resulting in irritability, confusion, dizziness, fatigue, weakness, visual abnormalities and comas. Dietary factors, medications, organ (pancreas, pituitary gland, adrenal gland, and liver) malfunction, and alcohol are possible causes of hypoglycaemia. The body's response to hypoglycaemia is by releasing epinephrine (adrenaline), which can cause sweating, nervousness, hunger, faintness, palpitations, hypothermia and headaches. Consuming large amounts of refined carbohydrates causes the pancreas to release excessive amounts of insulin, which greatly decreases blood glucose by promoting the cellular utilisation of glucose. Eating more frequent and smaller meals of vegetables, low glycaemic index (GI) fruit and lean proteins is recommended. Adequate amounts of chromium, zinc, magnesium and other nutrients are also required for blood sugar balance. Regular aerobic exercise (walking, jogging, swimming, etc.) helps to maintain blood sugar levels. The most common type of hypoglycaemia, known as functional low blood sugar. Excessive stimulation of pancreatic insulin production is due to the intake of large amounts of refined carbohydrates (sugar and white flour), caffeine, and other stimulants. The pancreas fails to cut back its production of insulin after responding to a raised blood sugar (glucose) level after intake of food, i.e., sustained high insulin production. The nervous system is very sensitive to fluctuation in blood sugar, so that the most common symptoms of hypoglycaemia are nervousness, irritability, exhaustion, depression, and headaches. Hypoglycaemia is a chronic condition, which requires special treatment, with good nutrition program a healthy prognosis is expected. Stress is often blamed as the root cause for anxiety, depression and fatigue, but although stress can make any problem worse, the source of such problems is often physical in nature. Hypoglycaemia is one of the major physical causes. Hypoglycaemia is the body's inability to properly regulate blood sugar levels, causing the level of sugar in the blood to be too low or to fall too rapidly. Hypoglycaemia is not a disease, it is a condition, and, in most cases, it is fully reversible. The more common type of hypoglycaemia-called functional, reactive, or fasting. Some types of hypoglycaemia are caused by a tumour or other physical damage to a gland. The glandular imbalances that result, as the glands struggle to regulate the sugar level, cause their own symptoms-especially high adrenaline, which is usually perceived as anxiety or panic, but, in some cases, can lead to violence. This has something to do with domestic violence and street crime.

The most frequent problems with the GTT:

1- The test is run for <6 hrs (a 3 hr test certainly can't catch a drop at the 5.5 hr point).
2- The test measures glucose level but fails to measure insulin and adrenaline, the blood sugar may be holding up because the adrenal gland is dumping huge amounts of adrenaline.
3- Glucose alone can't tell the full story.
4- The symptoms are not carefully observed during the test.
5- The lowest glucose level is important, but the rate of drop is just as important.

Hypoglycaemia symptoms:

Symptoms	Comment
A- Mental:	
1- Anxiety	Ranging from constant worry to panic attacks.
2- Phobias	Claustrophobia, agoraphobia, acrophobia, etc.
3- Nervousness	
4- Restlessness	
5- Irritability	
6- Depression	Especially with females
7- Violent outbursts	Especially with males
8- Obsessive Compulsive Behaviour	
9- Forgetfulness	May be due to choline/inositol deficiency.
10- Inability to concentrate	
11- Unsocial, asocial, anti-Social behaviour.	
12- Crying spells	
13- Nightmares and night terrors.	Terror can continue after waken up. Waken in a cold sweat, pressure on the chest, or inability to breathe.
B- Physical:	
1- Headaches	Especially if a meal is missed.
2- Tachycardia	Due to high adrenaline.
3- Fatigue	Weakness, rubbery legs. Tremor or trembling of arm, leg, or whole body (outside or inside).
4- Twitching, jerking, or cramping of a leg muscle cramping.	May be just calcium or magnesium deficiency or food allergy response.
5- Waking after 2-3 hrs sleep.	
6- Tinnitus-ringing in the ear.	Due to high insulin in about 70 % of tinnitus cases.
7- Abnormal weight	Too high or too low.
8- Compulsive craving for sweets, colas, coffee, alcohol.	
9- Lack of appetite.	
10- Prolapsed mitral valve.	
11- Crawling sensations on skin	
12- Fainting	
13- Blurred vision	
14- Smothering spells-gasping for breath.	
15- Red blotches on skin or circular arcs of red skin.	
16- Lack of sexual drive.	
17- Chest pain-severe, but normal ECG.	
18- Bright light or loud sounds intolerance.	
19- Joint pains	Worse in the early morning after waking, and get better after being up and around a full day.

Symptoms such as weakness and mental fog begin as a result of hypoglycaemia. The body responds to the emergency by dumping adrenaline into the system. More symptoms follow from the high adrenaline, such as palpitation, anxiety, etc. Typical hypoglycaemia symptoms are not only directly from low blood glucose but also from the glandular imbalances that result, especially high adrenaline. The simplest working definition of anxiety is the way the person perceives high adrenaline. If adrenaline is moderately high for too long, individuals feel anxious and wonder why. This is called free-floating anxiety. If, on the other hand, adrenaline shoots up to very high value rapidly, and then decreases rapidly, the anxiety is brief but intense. This is called panic attack. With repeated emergencies the body learns to dump higher and higher amounts of adrenaline at the slightest hint of an emergency. The adrenal gland puts out about 60 different hormones; repeated requests for adrenaline dumps will affect all the others. This lead to adrenaline dumping and anxiety, it also leads to hormonal imbalances. Glucose tolerance test (GTT) is unreliable as it is frequently done in an inadequate way or is misinterpreted. Hypoglycaemia is best diagnosed by its symptoms. The intervention for hypoglycaemia is to reverse the condition; the glands must be allowed to recover. This is done by eliminating all foods and beverages that deliver sugar rapidly. Stress makes all problems worse. Handling hypoglycaemia should consist of diet, and nutritional supplements. Hypoglycaemics should avoid the following: 1- Sugar including sucrose, fructose, raw sugar (sugar + dirt), brown sugar (sugar, dyed brown), corn syrup, dried cane juice, raisin juice, etc., molasses, malt, malted barley, even maple syrup and honey.

2- Hydrogenated and partially hydrogenated oils including vegetable shortening and margarine. These are hormone imbalancers. 3- White wheat flour and white rice-nutrients that prevent heart disease and cancer are removed from these products to give longer shelf life. Fiber is also removed; therefore the starch rapidly converts to sugar. 4- Peanuts and corn-high mould sources, which tax the immune system. 5- Avoid fruit juices. 6- Avoid skim milk. Xanthine oxidase makes milk a bad choice for all. The harmful enzyme xanthine oxidase is deactivated when yoghurt and cheeses are made. 7- Avoid artificial sweeteners (substitute for sugar), e.g., aspartame, they make recovery for hypoglycaemics much more difficult and are a major health hazard. High level vitamin C is recommended, which is the primary support for the adrenal gland. Comorbid depression appears to be a significant risk factor for relapse among recovering alcoholics. Untreated depression may increase risk for relapse to problem drinking. Sleep variables, anxiety, and depression are possible markers of relapse in persons treated for alcoholism. Up to 91% of the alcohol impaired drivers may complain of at least some acute symptoms of depression. The biochemical mechanisms that may contribute to alcohol craving include the stress response of the hypothalamic-pituitary-adrenal axis (HPAA), the endogenous opiate β-endorphin system, neurotransmitter synthesis and release, hypoglycaemia, and nutrient deficiencies. Chronic alcoholism is diabetogenic in susceptible individuals. Alcoholism is associated with hypoglycaemia. Glucose intolerance in alcoholic patients is a common finding, which occurs in the presence or absence of liver damage. Glucose metabolism in alcoholics in a withdrawal state may be disturbed by impaired insulin secretion and insulin resistance. Alcohol dependency is accompanied by zinc content decrease in the hippocampus. Alcoholics may be deficient in zinc, an essential coenzyme in alcohol dehydrogenase. Zinc improves gastric alcohol dehydrogenase activity in alcoholics. Another probable aetiology of low blood sugar is alcohol-induced inhibition of gluconeogenesis along with starvation.

Dietary recommendations:

1- Divide daily intake of food into at least 6 small meals (breakfast, snack, lunch, snack, dinner, and snack).
2- Eat a high-fat, protein-rich diet, and high fiber, complex carbohydrate separately.
3- Minimise sugar intake. Eat sweets only at special meals, in small amounts, and only as part of the meal.
4- Avoid caffeine. Caffeine is found in coffee, chocolate, tea and colas.
5- Reduce alcohol intake and avoid alcohol entirely on an empty stomach.
6- Exercise is important.

The prolonged hypoglycaemia caused cortical damage simulating ischaemic brain damage. Acute alcohol consumption brings on hypoglycaemia in many individuals, since it inhibits gluconeogenesis from lactate and amino acids. A defect in central serotonin metabolism may manifest itself in poor impulse control leading to attempts at suicide, violence towards others, and Type II alcohol abuse. Clinical anxiety may be associated with an elevated blood lactate level and increased lactate to pyruvate ratio. The lactate to pyruvate ratio is increased by alcohol, caffeine and sugar, and decreased by deficiencies of niacin, thiamine and magnesium. Normalising an elevated blood lactate level or an increased lactate to pyruvate ration may relieve anxiety. All of those dysfunctions together with hyperinsulinism can greatly enhance the risk of atherosclerotic vascular disease. Attention deficit and hyperactivity (ADHD) disorder is prevalent among obese patients and highest in those with extreme obesity. The causes for the comorbidity are unknown, but may involve brain dopamine or insulin receptor activity. Controlling blood glucose sugar levels is essential in order to help starve the cancer and bolster the immune system. Mannoheptulose inhibits tumour cell glucose uptake and also inhibits glucokinase, the enzyme used in glycolysis. Partial replacement of complex digestible carbohydrates with monounsaturated fatty acids (avocado as one of its main sources) in the diet of patients with non-insulin-dependent diabetes mellitus improves the lipid profile favourably, maintains an adequate glycaemic control. Endogenous depression is associated with lowered glucose utilisation rate and with insulin resistance. A deficiency of folate and vitamin B_{12} lead to pseudodementia. ω-3 fatty acids contained in fish oil may benefit people with bipolar disorder. 10-30% of depressed patients, mostly bipolar, develops a therapy-resistant illness. Bipolar patients can benefit from lecithin, which contains 10-20% phosphatidylcholine, the precursor of the neurotransmitter acetylcholine. Pantothenic acid (vitamin B_5 required as coenzyme in conversion) treats mania. 9.6 g of ω-3 fatty acids from fish oil per day in addition to other medications significantly improves bipolar patients. Bipolar disorder is associated with alterations in the metabolism of cytosolic, choline-containing compounds in the anterior cingulate cortex. Moderate hypoglycaemia may evoke a significant stress response, behavioural changes, and alterations in cerebral blood flow and metabolism. Both hypo- and hyper-glycaemia may produce neurologic changes. The insulin receptor becomes resistant to insulin, which requires chromium to function properly. The putative antidepressant effects of chromium could be accounted for by enhancement of insulin utilisation and related increases in tryptophan availability in the CNS, and/or by chromium's effects on norepinephrine release. Chromium may potentiate antidepressant pharmacotherapy for dysthymic disorder. Chromium deficiency is associated with anxiety, fatigue, glucose intolerance, growth impairment, and hypercholesterolaemia, and chromium toxicity is associated with dermatitis, GI ulcers, renal impairment, and liver impairment. Sugar consumption causes urinary chromium excretion. Chromium improves glucose tolerance and lipid metabolism. The lipophilic antioxidant and mitochondrial respiratory chain redox coupler, coenzyme Q_{10} (CoQ_{10}), has the potential to improve energy production in mitochondria by by-passing defective components in the respiratory chain as well as by reducing the effects of oxidative stress. A high ratio of copper/zinc may cause zinc deficiency and could lead to violent behaviour. Dark skin individuals have high Cu levels. Related to the Cu needed in the production of melanin skin pigmentation. The highest concentration of copper is found in the brain and liver. Copper (Cu) is found in all other tissues in varying amounts, and about 50% of the total copper content of the body is found in the bones and muscles. It is essential in the production of collagen and the neurotransmitter noradrenaline. It plays a role in the production of the skin pigment melanin by converting the amino acid tyrosine. It is involved in production of haemoglobin. Copper influences iron absorption and mobilisation from the liver and other tissue stores. Absorption of the mineral is increased by acids and inhibited by calcium. Copper bracelets reduced pain and inflammation. Superoxide dismutase (SOD) a zinc and copper or manganese containing enzyme, which reacts with superoxide radicals to convert them to less dangerous. It is the fifth most common protein in the human body. All organisms not killed by air contain SOD. Intracellular cytoplasmic SOD contains Zn and Cu, mitochondrial SOD contains Zn and Mg. Superoxide radicals are implicated in arthritis and cataract formation. Social adversity may be a risk factor for depression, by increasing stress induced cortisol secretion, which impairs serotonin (5-HT) neurotransmission. Chronic fatigue syndrome patients display cortisol hyposecretion in saliva as well as plasma. Increased cortisol secretion, altered cortisol metabolism, and/or increased tissue sensitivity to cortisol may link insulin resistance, hypertension, and obesity. Patients with glucose intolerance have enhanced central and peripheral sensitivity to glucocorticoids. The clinical syndrome of glucocorticoid excess (Cushing's syndrome) is associated with glucose intolerance, obesity and hypertension. By opposing the actions of insulin, glucocorticoids could contribute to insulin resistance and its association with other cardiovascular risk factors. Turmeric or curcumin reduces the blood sugar. Depressive patients have decreased glucose tolerance. One of the predictive factors of treatment-resistant depression is the syndrome of relative insulin resistance. Stress is commonly associated with a variety of psychiatric conditions, including major depression, and with chronic medical conditions, including diabetes and insulin resistance. Insulin resistance has been associated with individuals diagnosed with depression. Diabetics have an increased risk of depression. Insulin activity plays a role in serotonergic activity by increasing the influx of tryptophan into the brain. This increased influx of tryptophan has been shown to result in an increase in serotonin synthesis. It may be possible to treat depression by increasing insulin activity. The antioxidant α-lipoic acid increases insulin sensitivity. α-Lipoic acid should be used as an adjunct treatment for depression. There is a functional state of insulin resistance during major depressive illness and a more generalised biological disturbance in some depressed patients. Depressed patients would have high basal glucose levels, great cumulative glucose responses after the GTT and large cumulative insulin responses after the GTT. Insulin is transported through the blood-brain barrier (BBB) and influences brain function via widely distributed insulin receptors on neurons. These receptors are particularly dense on catecholaminergic synaptic terminals, while effects are variable dependent on brain region. Insulin promotes central catecholaminergic activity, perhaps by inhibiting synaptic re-uptake of norepinephrine. Psychological stress may lead to increased cortisol levels, interfering with serotonin synthesis causing endogenous depression. There is a highly significant correlation between sugar consumption and the annual rate of depression; a correlation does not necessarily imply aetiology. Obese subjects with psychiatric manifestations frequently have mild hypercortisolism, while carefully screened obese subjects with no such manifestations are eucortisolaemic. The former may have stress-induced glucocorticoid-mediated visceral obesity and metabolic syndrome manifestations. Hypersecretion of cortisol as well as the presence of cortisol receptors in the brain is suggested as the pathway for monoamine change. Hypoglycaemics should consider digestive enzymes; Hypo- and a-chlorhydria may be found in 19.2% of these patients and hyperchlorhydria in 15.4% of them. There is altered serotonergic activity in aggressive and impulsive behaviours in substance abusers. Many addicted people may have an actual flaw in the way they process sugar and carbohydrates. This flaw in metabolization causes an addict to respond to sugar as if it were an alcohol and to white flour products as if it were sugar. Substances like sugar, by creating insulin and rapidly penetrating the cell wall, actually alter the permeability of the cell. There is a link between brain chemistry and food addictions. Serotonin is a calming, analgesic-like substance, which is secreted in response to carbohydrate and sugar consumption. Sugar addiction may be a misguided attempt to replenish serotonin in the system. Exercise can reduce a prediabetic (hypoglycaemic) condition. Exercise increases dopamine release and raises the number of dopamine receptors. Hypoglycaemics should be advised not to drink alcohol. Alcohol results in a dose-related elevation in insulin levels with unaltered blood glucose and free fatty acid responses in NIDDM, which points to an aggravation of insulin resistance.

DYSAUTONOMIAS

Dysautonomia is the change in the autonomic nervous system function that adversely affects health. The changes range from transient, occasional episodes of natural mediated hypotension to progressive neurodegenerative diseases; from disorders in which altered autonomic function plays a primary pathophysiologic role to disorders in which it worsen an independent pathologic state; and from mechanically straight forward to mysterious and controversial entities. In chronic autonomic failure (pure autonomic failure, multiple system atrophy, or autonomic failure in Parkinson disease), orthostatic hypotension reflects sympathetic neurocirculatory failure from sympathetic denervation or deranged reflexive regulation of sympathetic outflows. Humans absolutely require a functionally intact sympathetic nervous system to tolerate the non-emergency behaviour of simply standing up. Dysautonomia refers to a condition in which altered autonomic function adversely affects health, ranging from transient episodes in otherwise healthy people to progressive neurodegenerative diseases. Dihydroxyphenylglycol (DHPG) is the main neuronal metabolite of neuroepinephrine, which is produced by the action of monoamine oxidase on norepinephrine in sympathetic axoplasm. Axoplasmic norepinephrine has 2 sources including leakage from storage vesicles and reuptake after exocytotic release. Simultaneous assessments of norepinephrine, L-dopa, and DHPG spillovers provide information about related but different aspects of sympathetic noradrenergic function. L-dopa is the precursor of catecholamines and the immediate product of the rate-limiting step in catecholamine biosynthesis. Entry (spillover) into the bloodstream reflects both loss of neuroepinephrine from vesicles by leakage and re-uptake of norepinephrine. Most chronic autonomic failure occurs as a consequence of disease processes, e.g., toxic agents or medications. Primary chronic autonomic failure in adults includes pure autonomic failure, multiple system atrophy, and autonomic failure associated with Parkinson disease. Multiple system atrophy can include Parkinsonian features, making it difficult to distinguish the 2 conditions. Distress, by increasing sympathetic and adrenomedullary outflows, can trigger morbid or even mortal cardiovascular events. In acute mental stress responses, sympathetic nervous system activation preferentially targets the heart, providing a straightforward mechanism for precipitation of myocardial infarction or ventricular arrhythmias in the presence of fixed coronary artery stenosis. During panic attack, the amplitude of bursts of sympathetic nerve firing increases markedly, accompanied by increased adrenomedullary secretion of epinephrine. Epinephrine may be released into the cardiac venous drainage, even when an attack is not occurring. Some patients with panic disorder describe severe, crushing precordial chest pain that resembles angina pectoris. ECG changes can indicate myocardial ischaemia in these patients. Congestive heart failure is associated with increased sympathetic nervous system outflows, which adversely affect clinical outcome. The failing heart is sympathetically denervated. Chronic fatigue syndrome CFS is characterised by new, unexplained fatigue that lasts at least 6 months, is not relieved by rest, and has no clear cause. The syndrome is associated with 4 or more new symptoms, such as memory or concentration problems, sore throat, tender lymphadenopathy, myalgia, arthralgia, headache, unrefreshing sleep, and postexertional malaise. Chronic fatigue syndrome is a sporadic illness with occasional, poorly understood geographic clusters. Women are affected 2-3 times as often as men. Young, middle-aged persons are most often affected. 10 to 1000 per 100000 persons in USA may have CFS. CFS may be a form of dysautonomia. When evaluated by prolonged head-up tilting at a 70° angle, >60% of patients with CFS have abnormal blood pressure or pulse responses, with sudden hypotension or severe bradycardia or tachycardia, which is accompanied by a decreased level of consciousness (neurally mediated hypotension). CFS is a fairly common, incompletely understood disorder that overlaps clinically with dysautonomias. The brain affects the heart. Emotions-related alterations in cardiovascular function might cause or contribute to disease. 40% of patients with untreated essential hypertension have chronically increased cardiac and renal spillover of norepinephrine and increased rates of efferent sympathetic nerve firing in the outflow to the skeletal muscle vasculature. The sympathetic activation originates within the CNS and seems to be driven by noradrenergic projections from the brainstem to the forebrain. Deficiency of the plasma membrane norepinephrine transporter can produce orthostatic tachycardia by amplifying delivery to norepinephrine to its receptors in the heart. The 3 most commonly used nonpharmacologic therapies to reduce blood pressure (calorie restriction, weight loss, and exercise training) tend to inhibit sympathetic nervous system outflows. In normal individuals, peroneal muscle sympathetic activity approximately doubles during orthostatic stress, with an approximate doubling of the plasma norepinephrine concentration. Orthostatic hypotension and orthostatic intolerance are not synonymous. Orthostatic intolerance is common, and generally occurs in young patients (15-45 y), but delayed orthostatic hypotension can occur. In elderly individuals, orthostatic intolerance can be a manifestation of cerebral hypoperfusion from carotid disease. They report dizziness, visual changes, head and neck discomfort, poor concentration while standing, fatigue while standing (as well as other times), palpitations, tremor, anxiety, pre-syncope, and occasionally syncope. There is a preganglionic lesion in multiple system atrophy and postganglionic lesion in Parkinson disease with autonomic failure. A patient with orthostatic hypotension from dysregulation of sympathetic outflows might be at increased risk for acute hypertension from herbal remedies, such as ma-huang and yohimbe bark, which release norepinephrine.

UBIQUINONE

Ubiquinones are a group of homologous quinones that are widely distributed in animals, plants and microorganisms. They consist of benzoquinone nucleus and an isoprenoid side-chain, the length of the side-chain differing in different species. The predominant ubiquinone in man has a side-chain of 10 isoprenoid units and is known as coenzyme Q_{10} (CoQ_{10}). The biosynthesis of CoQ_{10} is a multistage process, which can be divided into 3 steps: 1- synthesis of the benzoquinone nucleus from tyrosine, 2- formation of isoprenoid side-chain from acetyl-CoA via the mevalonate pathway, and 3- the condensation of these 2 structures by the enzyme transprenyl transferase. CoQ_{10} synthesis appears to be initiated in the rough endoplasmic reticulum (RER), and the final condensation reaction takes place in the Golgi apparatus. An important step regulating CoQ_{10} synthesis involves hydroxymethylglutaryl (HMG)-CoA reductase, an enzyme common to the cholesterol biosynthetic pathway. The major function of CoQ_{10} is to act as an electron carrier in the mitochondrial respiratory chain (MRC), where it serves to transport electrons from complex I (NADH:ubiquinone reductase) and complex II (succinate:ubiquinone reductase) to complex III (ubiquinone:cytochrome C reductase). CoQ_{10} also serves to transfer electrons liberated from the β-oxidation of fatty acids to complex III of MRC. During electron transport, each of the 2 carbonyl groups on the CoQ_{10} benzoquinone ring accepts an electron: first, one electron is accepted to form the semiquinone anion ($CoQ_{10}H$); this then accept a second electron to form ubiquinol $CoQ_{10}H_2$. CoQ_{10} may also translocate protons from the mitochondrial matrix to the intermembrane space, contributing to the energy conservation at coupling site 2 of the respiratory chain. CoQ_{10} has a role in uncoupling protein from brown adipose tissue, which by their regulated transmembrane proton transport is able to uncouple mitochondrial oxidative phosphorylation, producing heat rather than ATP. A small decrease in CoQ_{10} concentration may depress ATP production and cause organ dysfunction. CoQ_{10} is antioxidant, protecting the cell from free-radical-induced oxidation, which has been attributed to $CoQ_{10}H_2$. The ratio $CoQ_{10}H_2/CoQ_{10}$ has been proposed as a possible marker of oxidative stress. $CoQ_{10}H_2$ is closely linked to vitamin E and serves to generate the reduced (active) α-tocopherol form of the vitamin. CoQ_{10} is involved in the regulation of membrane fluidity. CoQ_{10} is present in most tissues of the body, mostly as ubiquinol ($CoQ_{10}H_2$), except in brain and lung, where CoQ_{10} predominates (67% and 65% of total, respectively), which reflects higher oxidative stress in these tissues. Cytosolic NADPH:ubiquinol reductase activity accounts for the preponderance of non-mitochondrial reduction of CoQ_{10}. The highest concentrations of total CoQ_{10} (CoQ_{10} and $CoQ_{10}H_2$) are found in the heart, kidney and liver, with tissue concentrations of 0.114, 0.066 and 0.055 µg/mg wet weight of tissue, respectively. Within the cell, CoQ_{10} is distributed within the ER (0.150 µg/mg protein), the Golgi body (2.620 µg/mg), lysosomes (1.860 µg/mg), peroxisomes (0.290 µg/mg) and plasma membrane (0.740 µg/mg) in addition to the mitochondria (1.860 µg/mg). Exogenous CoQ_{10} elevates blood concentrations. Diminished concentrations of CoQ_{10} are associated with wide range of disorders and conditions. CoQ_{10} deficiencies: A- Primary: 1- mevalonate kinase deficiency, and 2- mitochondrial encephalomyopathy. B- Others: 1- cardiovascular disease, 2- phenylketonuria, 3- cancer, and 4- neurodegenerative diseases. Patients with mevalonate kinase (MK) deficiency present with mevalonic aciduria (MVA), which has a broad range of clinical symptoms, including psychomotor retardation, ataxia, cerebral atrophy and myopathy. Mitochondrial encephalomyopathies present as a heterogenous group of disorders characterized by morphological, biochemical and genetic abnormalities of mitochondria. Decreased levels of muscle CoQ_{10} in patients with mitochondrial encephalomyopathy may be associated with oxidative loss of CoQ_{10} as a result of a deficiency in the respiratory chain. Recurrent episodes of myoglobinuria and a CNS dysfunction since early childhood are suggestive of severe muscle CoQ_{10} deficiency associated with mitochondrial encephalomyopathy. Decreased activities of the endogenous CoQ_{10}-dependent complexes II III (succinate and cytochrome C reductases) and I III (NADH and cytochrome C reductases) in muscle mitochondrial preparations. There is a tissue-specific deficiency in the CoQ_{10} biosynthetic pathways of muscle and brain. Marked clinical improvement is expected after oral supplementation of CoQ_{10}, presented as improved muscle function. The CoQ_{10} deficiency may be the result of other mechanisms such as accelerated tissue catabolism, or enzyme deficiency in the biosynthetic pathway, e.g., deficiency in the transprenyl transferase. Hypothermia of neonatal CoQ_{10} deficiency may reflect the essential role of CoQ_{10} plays in thermogenesis in brown adipose tissue. Multiple organ failure in conjunction with hypothermia and lactic acidosis may suggest CoQ_{10} deficiency. The higher energy requirements of the heart make it vulnerable to deficits in mitochondrial energy metabolism. This may explain the high incidence of cardiomyopathy in patients with deficiencies of enzymes in the β-oxidation of fatty acids and/or the respiratory chain. A small decrease in CoQ_{10} concentration could compromise mitochondrial energy production, impairing cardiac function. The lower CoQ_{10} levels in the heart disease may reduce the capacity of the cellular antioxidative defence system. Phenylalanine inhibits HMG-CoA reductase activity. CoQ_{10} deficiency may have a role in the pathophysiology of phenylketonuria (PKU). Reduced plasma CoQ_{10} have reported in breast cancer, myeloma, lymphoma and lung cancer. Decreased plasma and tissue CoQ_{10} concentrations in cancer patients might be due to increased consumption of this molecule by ROS. Increased metabolic demand of tumour cells causes increased extraction of CoQ_{10} from the blood. CoQ_{10}

levels are decreased in Parkinson's disease patients. Anti-parkinsonian drugs such as deprenyl lowers CoQ_{10} levels. There are low levels in platelet mitochondria and a significant lower ratio of reduced $CoQ_{10}H_2$ to CoQ_{10} in platelets from untreated Parkinson's disease patients. This altered redox state is a consequence of the decreased activity in the MRC complex I, or increased oxidative stress as a result of loss of reduced glutathione in Parkinson's disease patients. Statins are HMG-CoA reductase inhibitors of fungal or synthetic origin, which are used to lower serum LDL-cholesterol in patients with hypercholesterolaemia. Decreased serum CoQ_{10} concentrations are accompanied by an elevated lactate/pyruvate ratio, a finding that was interpreted in terms of statin-induced mitochondrial dysfunction. LDL is the carrier of CoQ_{10} in the circulation. Statin lowers liver and heart CoQ_{10}. Primary muscle CoQ_{10} deficiency can result in elevated serum creatine kinase, with recurrent episodes of myoglobinuria. Myopathy and rhabdomyolysis linked to treatment with statin may result from a deficiency in muscle CoQ_{10}. These myotoxic effects are dose-related, and could be prevented by co-administration of CoQ_{10}. Idebenone has more efficient cellular uptake and relative ease with which it crosses the blood-brain barrier when compared with CoQ_{10}. β-Blockers inhibit CoQ_{10}-dependent enzymes. A serum CoQ_{10} concentrations of at least 2.5 μg/ml is needed for any tangible clinical benefits to be perceived in CHF patients, and the beneficial clinical effects may not be apparent for several months, depending on the severity of the disease. There are concern that CoQ_{10} may reduce the efficacy of warfarin. The cardioprotective effect of CoQ_{10} is thought to be mediated by its ability to act as antioxidant rather than to replenish depleted myocardial levels. CoQ_{10} may be therapeutic in the treatment of Parkinson's' disease, with greatest effect being observed at 1200 mg/d. Clinical assessment of CoQ_{10} status is generally based on plasma measurements (0.5-2 μmol/l). The plasma CoQ_{10} concentration is highly dependent on serum lipid concentrations. In common with α-tocopherol, CoQ_{10} is carried in the circulation by lipoproteins, 58% of total CoQ_{10} is associated with LDL, 26% with HDL, and 16% with other lipoproteins. Plasma CoQ_{10} is higher in men than in women, and, due to long half-life (>24 h) of CoQ_{10} in the circulation, dietary intake significantly influences plasma concentrations, contributing up to 25% of the total amount. Conditions in which the overall energy metabolism is elevated (hyperthyroidism, endurance exercise) are likely to lead to decreased plasma concentrations of CoQ_{10} because of increased metabolic demand from tissues. CoQ_{10} concentrations in tissues increase from birth until the age of 20 y and then decline to concentrations found at birth, which could be due to decrease in metabolic demand of tissues with age.

RHABDOMYOLYSIS

Rhabdomyolysis is a syndrome in which acute injury to skeletal muscle can lead to both renal failure and compartment syndromes. It can be induced by traumatic and non-traumatic causes. Electrolyte disturbances that can produce rhabdomyolysis include hyperosmolar states, especially with marked hyperglycaemia or hyponatremia, severe hypokalemia, and hypophosphataemia, which is commonly seen in alcoholics. Hyponatremia is often an unrecognised initiating factor of rhabdomyolysis. Hyponatremia and hypotonicity can initiate skeletal muscle injury and rhabdomyolysis. Rhabdomyolysis may be caused by severe isotonic hyponatremia, such as in TURP. The pathogenetic mechanism causing the rhabdomyolysis in hyponatremia is controversial, but may involve abnormality of sodium:calcium exchange across the muscle cell. The reduced rate of entry of sodium ions reduces the exchange for calcium ions outwardly so that calcium accumulates inside the cell. When intracellular calcium ions reach a critical concentration, they activate neutral proteases and lipases that serve to destroy the cell. Rhabdomyolysis, myoglobinuria, and renal failure have been known to follow massive crush injury. The syndrome is often associated with hypocalcaemia caused by a shift of EC Ca^{2+} into injured muscle. Sarcolemmic Na^+-K^+-ATPase activity is impaired in damaged muscle. Attenuating the activity of this ionic pump would diminish the extrusion of Na^+ from the sarcoplasm and interfere indirectly with the efflux of Ca^{2+} from the cell. The cytosolic free Ca^{2+} level increases, activating neutral proteases that would disrupt myofibrils and trigger muscle damage in exertional, ischaemic, traumatic, metabolic, or toxic rhabdomyolysis. Stretching increases the leakiness of membrane to Ca^{2+} ions in skeletal muscle cells and nerve cells. The intramuscular pressure in patients with compressed or wedged limb injury may increase to >240 mmHg, causing rhabdomyolysis initially independent of ischemia. In such patients, full-blown rhabdomyolysis and anterior tibial compartment syndrome may occur despite normal arterial pedal pulses and warm skin. In injured muscle, intramuscular pressure may exceed arterial blood pressure within minutes after trauma. Cell swelling results from the accumulation of intracellular solutes, an increase in the leakiness of membrane, and a reduction in active ionic extrusion. When intracompartmental pressure exceeds arteriolar-perfusion pressure, it obliterates the circulation to the affected region, causing muscle tamponade and myoneuronal ischaemic damage within hours. Hours after rescue of survivors and decompression of the limbs, when serum urea and creatinine are still within normal limits, a dangerous degree of hyperkalaemia, hypocalcaemia, hyperphosphataemia, hyperuricemia, and metabolic acidosis may already be present. Iron pigments derived from muscle and blood may also play a part in catalysing the formation of OFRs. Except for patients who die of some other cause, such as crush or drug injury, mortality from rhabdomyolysis is below 5%, with death usually caused by a complication, such as sepsis. Rhabdomyolysis (striped muscle lysis) is a syndrome of injury to skeletal muscle that causes release of its contents, resulting in an increase in serum CK-MM (100% muscle isoenzyme) of at least 5 times the upper limit of normal. Elevation of serum creatine kinase (CK) is 90 times normal. Serum myoglobin rises before CK to a mean of 20 times normal, and serum aldolase, LDH and SGOT increase to a mean of 10 times normal. Myoglobin precipitates in and injury the renal tubule, which is aggravated by hypotension, hypovolaemia, or acid urine. High-voltage exposure, prehospital cardiac arrest, full-thickness burns, and compartment syndrome are associated with myoglobinuria. Electrical injuries account for 3-4% of all admissions to burn units. ARF resulting from rhabdomyolysis-induced myoglobinuria is a primary cause of death in most patients who die in the days after the initial electrocution. The incidence of myoglobinuria after electrical injury is 14-42% and may be as high as 75-100% after high-voltage exposures (>1,000 volts). Acute renal failure develops rapidly in patients with myoglobinuria, partly due to the associated severe Hypovolaemia and acidosis. The mortality of ARF after electrical injury is 50-100%. With immediate, aggressive fluid resuscitation to correct Hypovolaemia and acidosis, the incidence of ARF in patients with myoglobinuria can be reduced from 40% to 10%. The rhabdomyolysis syndrome is defined by the triad of elevated CK, o-tolidine positive urine, and pigmented granular casts in the urine. When 2 or more predictors are present, obtaining CK levels is not warranted for the sole purpose of diagnosing rhabdomyolysis. In rhabdomyolysis resulting from blunt injury, low venous bicarbonate levels (<17 mmol/l) in patients with myoglobinuria are predictive of ARF. In nontraumatic rhabdomyolysis, the myoglobin clearance rate has been shown to be a better predictor of ARF than the simple presence of myoglobinuria itself. Life-threatening myoglobinuric ARF can be prevented by early aggressive management of myoglobinuria after electrical trauma.

Flow of solutes and water in rhabdomyolysis:

Flux	Solutes/water	Consequences
1- Influx from EC compartment into muscle cells:	H_2O, Na^+, Cl^-, Ca^{2+}.	• Hypovolaemia and haemodynamic shock. • Pre-renal and later ARF. • Hypocalcaemia. • Aggravated hyperkalaemic cardiotoxicity. • Increased cytosolic calcium. • Activation of cytotoxic proteases.
2- Efflux from damaged muscle cells:	K^+	• Hyperkalaemia and cardiotoxicity aggravated by hypocalcaemia and hypotension.
	Purines from disintegrating cell nuclei.	• Hyperuricemia. • Nephrotoxicity.
	Phosphate	• Hyperphosphataemia. • Aggravation of hypocalcaemia.
	Lactic and organic acids	• Metabolic acidosis and aciduria.
	Myoglobin	• Nephrotoxicity with coexisting oliguria, aciduria, and uricosuria.
	Thromboplastin	• DIC.
	Creatine kinase	• Extreme elevation of PCK.
	Creatinine	• increased serum creatinine.

The risk factors, which are high-voltage exposure, prehospital cardiac arrest, full-thickness burns, and compartment syndrome, can be used to guide urine myoglobin testing and treatment. A myoglobinuria prediction rule can be used to screen out patients with insignificant risk and to identify those at high risk who should be presumptively treated while a more definitive urine myoglobin test is being performed. Serum CK levels do not contribute any additional value. Treatment of rhabdomyolysis consists of removing the precipitating cause and administering sufficient intravenous fluids, furosemide, and mannitol to maximise urine output, and sodium bicarbonate to alkalinise the urine. Alkalinization of the urine may be protective, presumably because haemoglobin and myoglobin are more soluble in an alkaline solution and the formation of casts is therefore retarded. Acute hyperphosphataemia, hyperuricemia, and the formation of thrombi in the glomerular capillary tufts due to DIC, are other factors inducing ARF in rhabdomyolysis. Patients who survive the crush syndrome and ARF ultimately recover completely. Forced mannitol-alkaline diuresis therapy for prophylaxis against hyperkalaemia and ARF should be undertaken. IV hypotonic sodium chloride and sodium bicarbonate (NaCl, 110 mmol/l; chloride, 70 mmol/l; and bicarbonate, 40 mmol/l) in 5% glucose solution to which 10 g of mannitol/l is added in a 20% solution; should infused at rate of 12 L/day in adult of 75 kg, forcing diuresis of 8 L/day and maintaining urinary pH above 6.5, until myoglobinuria disappears. The infusion of bicarbonate may be gradually discontinued after 36 hrs. Loop diuretic acidify the urine, which is a disadvantage. Also mannitol may induce or aggravate the ARF.

CAERULOPLASMIN, COPPER AND NUTRITION

A primordial single-domain cupredoxin may be evolved into the multidomain copper oxidases, and the differences reside primarily in insertions and deletions at junctions between secondary-structure elements. Tyrosine and threonine may have some role in electron transfer. Copper seems to be found in a very limited subset of structures; zinc and iron have a much wider variety of environments in proteins. The prion protein PrPc is a glycoprotein of unknown function normally found in neurons and glia. It is involved in diseases such as bovine spongiform encephalopathy (BSE), scrapie and Creutzfeldt-Jakob disease. Prion diseases are characterised by neuronal degeneration, gliosis and accumulation of PrPSc. A fragment of human PrP consisting of amino acids 106-126 is toxic to neurons. This toxic effect requires the presence of microglia, which respond to PrP106-126 by increasing their oxygen radical production. Caeruloplasmin (CP) is a protein known to show sequence identity to factor VIII. CP is a major copper transporting plasma protein. CP, the main copper transport glycoprotein found in the blood, delivers its copper to intracellular proteins via a plasma membrane receptor protein. Cells are able to extract copper atoms from caeruloplasmin and transport the copper to the cytosol. Caeruloplasmin (CP) is a copper-binding protein in vertebrate plasma. The CP gene was mapped to human chromosome 3 and long arm of chromosome 8. Caeruloplasmin is an abundant α 2-serum glycoprotein that contains 95% of the copper found in the plasma of vertebrate species. Coagulation factor V is a high molecular weight plasma glycoprotein that participates as a cofactor in the conversion of prothrombin to thrombin by factor Xa. The coding region includes 651 amino acids from the carboxyl terminus that constitute the light chain of human factor Va and 287 amino acids that are part of the connecting region of the protein. The amino acid sequence of the light chain is homologous (40%) with the carboxyl-terminal fragment (Mr, 73,000) of human factor VIII. Both fragments have a similar domain structure that includes a single ceruloplasmin-related domain followed by two C domains. The carboxyl terminus of the connecting region, however, shows no significant amino acid sequence homology with factor VIII. It is very acidic and contains a number of potential N-linked glycosylation sites. It also contains about 20 tandem repeats of nine amino acids. The positions of the three free cysteines of factor VIII are the same as three of the four cysteines present in ceruloplasmin. However, the positions of the free cysteines in factor VIII and caeruloplasmin are not conserved in factor V. Hereditary caeruloplasmin deficiency with haemosiderosis (aceruloplasminaemia) is a new disease characterised by systemic haemosiderosis, diabetes mellitus, neurological abnormalities and pigment degeneration of the retina. Loss of the ferroxidase activity of caeruloplasmin results in systemic iron deposition and tissue damage. Neuroimaging studies reveal iron deposition in basal ganglia and in the red and dentate nuclei. Cerebellar ataxia, extrapyramidal signs and dementia develop after middle age. Sequence analysis of the cDNA of caeruloplasmin from this patient revealed an insertion of adenine in exon 3; this produced a premature stop codon. CP is expressed mainly in the liver, and the lung is another major site of CP synthesis. CP mRNA is found in airway epithelium and in the ductal cells of the submucosal glands. The airway epithelial cells are the major source of CP in the lung fluid and support ceruloplasmin's critical role in host defense against oxidative damage and infection in the lung. Oxidation of lipids and lipoproteins by macrophages is an important event during atherogenesis. Activation of monocytic cells results in the release of multiple oxidant species and consequent oxidation of LDL. Ceruloplasmin, a copper-containing acute phase reactant, is secreted by activated monocytic cells, and that the protein has an important role in LDL oxidation by these cells. CP exhibits oxidant activity under the appropriate conditions. Cellular factors in addition to ceruloplasmin, possibly active oxygen species and/or lipoxygenases, are essential and act synergistically with caeruloplasmin to oxidise LDL. CP may play role in pulmonary injury or repair. The lung is a prominent site of caeruloplasmin gene expression during inflammation and hyperoxia. Exposure to 95% O_2 results in a five- to six-fold induction of caeruloplasmin mRNA in lung tissue within 46 h, reaching maximum values at 86 h. CP biosynthesis is associated with growth and differentiation in non-hepatic tissues. CP gene expression is localised to the epithelium lining the mammary gland alveolar ducts. The mammary gland is a prominent site of extrahepatic caeruloplasmin gene expression. The lung is the predominant extrahepatic site of CP gene expression during foetal development; CP may play a role in lung development or pulmonary antioxidant defense. Caeruloplasmin is an oxidant with a high affinity for oxygen. The transplasma membrane electron transport increases cell growth and thymidine incorporation. CP may act as a terminal oxidase for ferrous iron or ascorbate to stimulate transplasma membrane electron transport. The four-electron transfer from caeruloplasmin to oxygen to form water will prevent peroxide formation at the cell surface. Alternatively, superoxide formation inside the cell or membrane could employ the superoxide dismutase function of caeruloplasmin to produce peroxide. Either mechanism stimulates cell growth by external oxidants. Chloride affects the catalytic efficiency of human CP. Human CP is, in the plasma, under control of this anion. Cartilage matrix glycoprotein (CMGP) is a disulfide-bonded 550,000 D protein, which is synthesised by chondrocytes and ciliary epithelial cells.

Neither CP production nor copper uptake is regulated by intracellular copper levels. Copper active sites of different types appear to be in close contacts within the caeruloplasmin molecule. Intracellular copper transport is impaired in Wilson disease and Menkes disease. Some of copper oxidases include ferroxidase, azurin, plastocyanin, superoxide dismutase, tyrosinase and hemocyanin. A mutation in the CP gene is associated with systemic haemosiderosis in humans, which is associated with excessive iron deposition mainly in the brain, liver and pancreas. CP protective role against oxidising agents may prevent peroxidation of lipids from the smooth muscle cell membrane. Low-density lipoprotein (LDL) is oxidised by cellular and noncellular mechanisms, both leading to an increased binding to collagen. HDL inhibits copper-catalysed oxidation of low-density lipoproteins (LDL). CP inhibits the reaction of superoxide radicals generation as a result of Cu interaction with -SH groups of RBC membrane; the effect is more pronounced than the effect of catalase or superoxide dismutase. CP reception on RBC leads to membrane protection from superoxide and hydroxyl radicals, and represents a more complex process. Wilson's disease is an autosomal recessive, inherited disorder of copper metabolism. In normal individuals, copper homeostasis is controlled by the balance between intestinal absorption of dietary copper and hepatic excretion of excess copper in bile. In Wilson's disease, hepatic copper is neither excreted in bile nor incorporated into caeruloplasmin and copper accumulates to toxic levels. The Wilson's disease gene (WND) encodes a putative copper-transporting protein that is expressed almost exclusively in the liver. The predicted structure of the protein product is that of a P-type ATPase with striking homology to bacterial copper transporters and the gene product of another inherited disorder of copper metabolism, Menkes' disease. Clinical manifestations of Wilson's disease occur indirectly after the release of copper from the liver with subsequent damage to the brain (neuropsychiatric presentation) and other organs. Management of Wilson's disease involves decreasing excess levels of copper accumulated in the liver, brain, and other organs. Copper chelation therapy, to increase urinary excretion of copper, is the mainstay of treatment. In addition, oral zinc therapy may be useful at decreasing absorption of dietary copper and rendering tissue copper nontoxic, by increasing the formation of complexes with copper-binding proteins. Liver transplantation can be necessary for individuals with acute hepatic failure or complications of cirrhosis. Gene therapy may evolve in the future; however, medical management is effective in most patients. Transition metals such as copper are known to initiate free radical formation and lipid peroxidation. The effects of Cu^{2+} can be inhibited by a number of lipophilic antioxidants, including probucol, vitamin E, butylated hydroxytoluene, and a 21-aminosteroid. The plasma membrane of eukaryotic cells contains an NADH oxidase, which can transfer electrons across the membrane. This oxidase is controlled by hormones, growth factors and other ligands that bind to receptors in the plasma membrane. Oncogenes also affect activity of the oxidase. The oxidants activate growth-related signals such as cytosolic alkalinization and calcium mobilisation. Antiproliferative agents such as adriamycin and retinoic acid inhibit the plasma membrane electron transport. Flavin, Coenzyme Q and an iron chelate on the cell surface are apparent electron carriers for the transmembrane electron transport. CoQ_{10} stimulates cell growth, and Coenzyme Q analogs such as capsaicin and chloroquine reversibly inhibit both growth and transmembrane electron transport. The ligand-activated oxidase in the plasma membrane introduces a new basis for control of signal transduction in cells. The redox state of the quinone in the oxidase is proposed to control tyrosine kinase either by generation of H_2O_2 or redox-induced conformational change. Degenerative diseases (aneurysms, slipped disc, haemorrhoids, emphysema, and arthritis) are among the most destructive and painful conditions. Elastin diseases are extremely dangerous. Copper is below optimum in a large number of individuals. Haemodialysis patients have low copper and zinc serum level. There are several dozen enzymes and hormones containing or affecting copper. When insulin is injected in massive dose, wild swings in other hormones develop, especially 18hydroxy deoxycorticosterone (to excrete hydrogen ion), and cortisol. Zn helps heal diarrhoea by interfering with copper absorption by the bacteria. Shellfish uses haemocyanin (copper pigment) instead of iron to transport oxygen. Squid has fairly large fraction of the copper in the skin. Cadmium causes changes similar to a copper deficiency. Copper tends to mute the toxic effects of cadmium, sliver, and lead. Oysters are very high in cadmium, lead, and arsenic in some polluted areas. Shrimp can also have cadmium. Shrimp is good source of zinc. Spirulina seawoods is a very good source of copper (6 mg/100 g). Vegetables low in starch have about 1 mg/pound copper. Legumes have a range the same as most livers, as do some oil seeds. Cereal grains from which the germ has been removed lose up to 45% of copper. A dried apricot has exactly the same mineral content as it had directly from the field. Leafy vegetables tend to be higher than starchy vegetables. Copper bracelets are a rather ineffective remedy, but have small measurable effect on rheumatoid arthritis (RA), especially in summer. Milk induces copper deficiency. Increased CVD is associated with milk. Milk may be a greater risk factor than smoking cigarettes. All cheeses are included in this category. Human milk has 4-6 times as much copper as cow's milk, and 20-25% of it is ceruloplasmin. Cows depend on microorganisms to digest their food, so they probably must prevent intestinal disease with more urgency. Methionine may double the net absorption of copper in humans. Copper absorption is tripled or more in the intestines in the presence of Na. Citric, lactic, acetic and malic acids can solubilise copper, which may be why fruits improve copper retention. Apples increase copper retention. Eating large of zinc interferes with absorption of copper as does iron and antacids. Eating large amounts of vitamin C may interfere with utilisation of copper within the body. Phytates decrease copper absorption. Hemicellulose is inhibiting, while pectin and intact cellulose are inactive. Sulfide (sulfur and molasses) inhibits copper absorption. Molybdenum causes symptoms of a copper deficiency even though the liver copper remains high. Purified soybean protein can reduce copper absorption (90%). Cholesterol may lower liver copper. Too much copper is toxic (>200 mg). Cadmium causes a great increase of cellular copper toxicity. Chronic copper toxicity can cause loss of weight, hypertension, ED, inability to excrete K resulting in night-time muscle cramps, and oedema (related to disruption of K channels). Most of these symptoms may arise from a concurrent zinc deficiency, as copper interferes with zinc absorption. There are 4 K channels (Shaw, Shab, Shal, and Shaker). With the exception of Shaker, all of the channels contain 4 zinc elements ring in the intracellular part of channel. Zinc has a role in brain function. Zinc enhances learning in undernourished children. Zinc inhibits prostate growth. Zinc deficiency during a copper deficiency is more damaging for some enzymes than too much. Increased copper intake by menstruating women can exaggerate the symptoms of multiple sclerosis (MS) by virtue of interfering with zinc (Zn). Low Zn levels result in deficient CuZu-superoxide dismutase (CuZnSOD), which in turn leads to increased levels of superoxide. Menstruating women have low Mg and vitamin B_6 levels. Vitamin B_6 moderates intracellular NO production and extracellular Mg is required for NO release from the cell, so that a deficiency of these nutrients results in increased NO production in the cell and reduced release from the cell. NO combines with superoxide to form peroxynitrite, an extremely powerful free radical that leads to the myelin damage of MS. Iron (Fe), molybdenum (Mo) and cadmium (Cd) accumulation also increase superoxide production. Since vitamin D is paramount for Mg absorption, the much-reduced exposure to sunlight in the higher latitudes may account for the higher incidence in these areas. Vitamin B_2 is a co-factor for xanthine oxidase and its deficiency exacerbates the low levels of uric acid caused by high Cu levels, resulting in myelin degeneration. Selenium (Se) and vitamin E prevent lipid peroxidation and EPA and DHA upregulate CuZnSOD. Supplementation with 100 mg Mg, 25 mg vitamin B_6, 15 mg zinc, 400 IU vitamin E and D, 200 µg Se, 180 mg EPA and 120 mg DHA/day for young individuals (14-16 y of age) may prevent multiple sclerosis (MS). Selenium deficiency exaggerates copper toxicity. Obstructive jaundice may cause Cu toxicity. Diabetics are more efficient at absorbing copper and may have a narrow safe range. Copper combined with a wide range of chelating agents has been recommended for RA. When Lysyl oxidase activity increases, blood pressure does also. Elastin has a fairly high turnover rate and Lysyl oxidase has a half-life of 16 hrs. A normal body contains about 100 mg of copper. During copper repletion (5-6 mg/day), zinc needs to be increased by 7 times when a routine intake is established. 3-6 mg copper/day would recover the condition within 30 days. It is possible that growth of funguses is enhanced by free copper. Growth is enhanced by externally applied copper. Large amounts of copper can be toxic. Chronic potassium deficiency causes alterations in the intracellular free amino acids, interference by sodium with enzymes inside the cell as a result of the increased Na there, alterations of the K-Na regulating hormone patterns that then affect other physiological processes, or some combination of these. Lysyl oxidase is the enzyme that cross-links collagen and elastin connecting tissue. Copper, largely tied up as protein, enters the stomach, and there and in the upper intestine, the proteins other than those entering from the bile are degraded. The bile is the mean of excretion of copper in adults. Loss in sweat is usually negligible, as are losses in urine. The copper is moved across cell walls possibly associated with certain amino acids. It may be α aminoisobutyric acid that is involved since this amino acid behaves the opposite of other amino acids from cortisol. The copper moves past a metallothionein barrier inside the cells into the serum. Both copper and zinc increase the metallothionein barrier. The copper in the serum is carried largely complexed to albumin and histidine to the liver. The liver rapidly removes it and stores it until such time as unknown hormones cause the liver to release caeruloplasmin (which protein contains copper) to the target cells for general purposes, as well as unbound copper when under stress. Adrenaline (epinephrine) stimulates caeruloplasmin release 150% as well as free copper and may be the stress hormone for copper. Cortisol does not directly mobilise copper in stress. Caeruloplasmin has a half-life of 130 hours. Elastin makes up the vertebrate disks above the sacroiliac, the blood vessels, much of the skin, the lungs, and the bronchial tubes of all vertebrates except the jawless fishes. Emphysema can be produced in animals by a copper deficiency. Emphysema seems to have an elastin defect greater than can be explained by cross-linking alone. Copper deficiency can cause diseases affected by elastin tissue strength. The lysyl oxidase is secreted normally, but its activity is reduced. Oestrogen increases the efficiency of copper absorption. Dilated superficial veins (varicose veins) may be associated with copper deficiency. Elastin is about as flexible as a rubber band and can stretch to two times its length. Collagen is about 1000 times stiffer. A healthy artery requires about 1000 mmHg or 10 times the normal mean blood pressure in order to rupture. There is a serious defect with elastin connecting tissue in Marfan syndrome. Marfan syndrome is probably a genetic defect, but there is a strong probability that it either operates through the copper physiology or is greatly affected by a deficiency. Tyrosinase incorporates tyrosine into melanin pigment and is the reason why copper deficient sheep fail to pigment and rats. Human gray hair may also be related to such mechanism. The pigment loss could be due to cell death. Relieving the deficiency is not necessarily bringing back the pigment. Neutropenia is the earliest symptom in copper deficient babies. The immune system is very sensitive to adequate copper. Interleukin-2 is reduced as a result of copper deficiency. Neutrophils are reduced in numbers as well as function and superoxide anion production is 60% less. The copper deficient spleens show little growth during an infection. Supplements or copper rich foods should be used for babies with extreme care, as should be formula made from water out of copper plumbing (which can contribute 0.8 mg/day to adult intake). Babies have 19 mg total copper at term, half of it in the liver. New borne babies have 230 PPM in the liver, which compares to 35 PPM in an adult. 3 mg/day might overwhelm a baby in a short time if continued. Babies must have a mechanism for retarding absorption through the intestines during excess copper. Cytochrome C oxidase is a fundamental enzyme in the body's handling of oxygen. It contains 2 copper atoms and 2 iron atoms. Several brain neurotransmitters such as dopamine and norepinephrine are formed and catabolized by copper enzymes such as tyrosine hydroxylase and dopamine-β-hydroxylase. The brain other than the cerebellum and hypothalamus have these transmitters decreased 30-60% in various sectors by a copper deficiency. There is raised copper in the CSF in Parkinson's disease. Copper is thought to increase perception of red and green colour. Epilepsy may be caused in babies by a copper deficiency. Copper deficient embryos cause increased genetic defects (chromosome abnormalities). Oxidative damage to DNA is produced by reduced superoxide dismutase (SOD), which is an enzyme that degrades superoxides. Control of free radicals by SOD is the protective role of copper (ceruloplasmin) against inflammation. Copper deficiency in inflammation probably operates through the prostaglandin hormone system as well. Diabetics probably absorb copper twice more than normal people do. Diabetics may have a narrow safe range of intake. The pancreas can be irreversibly destroyed by a copper deficiency, but not the isles of Langerhan. There is somewhat of negative correlation between copper in drinking water and onset of juvenile diabetes. Juvenile diabetes is associated with lower than normal levels of reduced glutathione, caeruloplasmin oxidase activity, zinc, copper and sodium. It may be possible that adequate copper could help prevent insulin dependant diabetes. Gastric ulcers can be reduced (40%) by several copper organic molecule complexes and lesions induced by indomethacin as well (90%). Low copper levels mute the immune system. Recurrent diarrhoea is often observed in a copper deficit. Scurvy like bone changes is a long-term result of copper deficiency, probably caused by failure of bone collagen to cross-link. Copper supplements increase bone density in women. Copper deficiency increases cholesterol in the blood stream. A histidine-induced cholesterol rise is abolished by copper supplements. High zinc to copper intake ratio is an important part of this. Too low zinc status during a copper deficiency can be even more damaging to the heart. Adding copper without zinc can make the situation worse. The rise in cholesterol and triglycerides has been attributed to 40% or more reduction in lipoprotein lipase. This may be an adaptation to provide extra cholesterol for the arterial wall with deposits in order to help protect them against rupture by decreasing their internal diameter for the stress on the walls, which is directly proportional to the radius. The average cholesterol intake has not varied more than 5% in the last 100 years. Non caeruloplasmin copper may signal increase of cholesterol. Statins have not prevented deaths, cholesterol in food is not correlated well with heart conditions, and the cholesterol level is normal in the average heart attack victim. The network of connective tissue in the heart fails after a copper deficiency, and this could contribute some to heart failure in those affected. Potassium wasting infectious disease is the most likely reason for severe K deficiency, not nutritional failure. Resisting infection is an extremely important function of the body. The immune system is considerably weakened by inadequate copper. Shutting down creation of enzyme systems is not immediately essential to immunity such as cortisol does is one way to increase availability of copper. 11 Deoxycorticosterone (DOC) is a hormone probably used by the body to regulate Na and K when intake of both of them is high. It declines during a deficiency of K and Na. It stimulates collagen synthesis and would thus tend to cancel cortisol's effect during diarrhoea. The DOC effect is probably accentuated by low Na. The effects of muted cross-linking by cortisol drop are especially serious for elastin tissue because the disordered rubbery organisation of elastin depends entirely on the cross-linking for strength. Lysyl oxidase oxidises the amino group in lysine, which is common amino acid in elastin. The aldehyde that forms spontaneously combines with adjacent amine and aldehyde groups to form strong covalent bonds, and thus join together the fairly small protein precursor molecules. The same thing happens for collagen and bone also, but collagen in tendons has many less cross links, probably made possible by collagen's greater length and more ordered structure, which permits numerous weak hydrogen bonds to be effective. The immune system generates superoxide in order to help killing bacteria. Normally the copper catalysed superoxide dismutase (SOD) enzyme destroys superoxide radicals, which are derived from WBCs (neutrophils, eosinophils, and macrophages), as fast as the radicals form. This enzyme declines during infection and is undoubtedly used by the body to help defeat serum infections. Superoxide degrades the synovial fluids by depolymerizing hyaluronic acid and possibly collagen as well as bacteria. Decline of SOD has been proposed as one of the mechanisms accounting for some of the symptoms of rheumatoid arthritis, which if so, is an indication that this enzyme is indeed tied to the potassium enzyme systems. Superoxide dismutase is low in children with rheumatoid arthritis. Injections are said to be beneficial in osteoarthritis. Glucocorticoids (steroids oxygenated in the 17-carbon position) help the body to resist infection, by altering processes that increase pathogens' growth or adverse effects. Glucocorticoid mobilisation for fight or flight is an adjunct made possible because most processes that resist infection are an antithesis for fight or flight (HPA axis). Cortisol may be for intestinal disease and corticosterone for serum disease. Release of caeruloplasmin copper transport protein from the liver is useful for both situations and is therefore controlled by a different hormone, epinephrine, for fight or flight. The kidneys have twice as much caeruloplasmin as liver and, the kidneys may synthesise ceruloplasmin. Potassium loss is the most serious aspect of intestinal diseases, so the electrolyte capabilities of cortisol, but not corticosterone, are oriented around conserving K by migration into the cells upon decline of cortisol. Cortisol, but not corticosterone, has its secretion from the adrenal cortex markedly reduced by low serum K. Sodium, water, glucose, amino acids, chloride, hydrogen ion, WBC activity, copper enzymes, and numerous other hormones and enzymes are controlled by cortisol such as to survive during virulent intestinal disease.

POTASSIUM DEFICIENCY

The most serious aspect of diarrhoeas is wasting K. Cortisol is reduced during K deficiency, and this reduction accounts for many of the symptoms of rheumatoid arthritis (RA). Cortisol shuts down most of the copper enzymes when it declines so that excretion of copper is increased and Lysyl oxidase inhibited. Potassium loss force cAMP to excrete water into the intestinal lumen. Adrenal's cortisol secretion is inhibited by low serum K. Endotoxins bacterial diseases force the body to secrete cortisol by increasing ACTH and this is probably an adaptation by the bacteria to force the body to inhibit the immune system. Glucosteroids response-modifying factor (GRMF) secreted by T-cell then prevents the cortisol from having full effect on WBC other than suppresser cells and thus raises the set point, as does IL-1. Interleukin-1 also stimulates cortisol secretion, as does TNF (cachectin). Cortisol shuts down the production of copper-containing enzymes such as Lysyl oxidase and SOD. Lysyl oxidase catalyses the formation of cross-links in all connecting tissue including elastin. Elastin makes up the main strength of normal blood vessels, and has a rapid turn over; this is the most serious problem in arthritis. Caeruloplasmin carries copper to the immune system during infection. Arthritic individuals have a lower whole body K. RBCs have a higher K content than normal during RA. Dehydration reduces the blood volume during diarrhoea. Workers of potash mines have 25% lower incidence of heart disease than surrounding populations. Heart disease is prevalent in RA. Grapefruit or K-losing diuretics increase K loss. Hydrogen ion interfere with K excretion, this could the mechanism where eating vinegar or cherries is efficacious in RA. Finnish men who work in copper mines have little arthritis or susceptibility to infection. The high milk diet along with frequent saunas may be why Finns have the highest rates of arthritis in the world, since milk is the poorest source of copper and perspiration loses K. Milk has been shown to have a high statistical correlation with CVD, and may be greater than smoking, which disease in turn is correlated with RA. The Massai of Africa have a higher rate of RA than the surrounding tribes. The Massai use a lot of milk as well as very few vegetables. Eating a lot of shellfish or liver should reduce those symptoms related to copper deficiency since they are the richest sources. Wet heart disease of beriberi can not materialise when K is deficient. If KCl is dissolved in fruit juice it tastes good and reduces the danger of intestinal injury that even slow release enteric tablets may present. Unboiled, unfrozen, uncanned vegetables low in starch are the richest sources of K. Other high K food include celery or bamboo shoots as Effinger. However, recovery from K deficiency will be slower, since K is not associated with chloride and would take weeks or months longer. Potassium deficiency can arise from diarrhoea, processed food, reliance on grain or fatty foods, psychic stress stimulation of aldosterone, stress stimulation of cortisol, diuretics, licorice, grapefruit, profuse perspiration, excessive vomiting, eating Na-bicarbonate, hyperventilation, laxatives, enemas, shock from burns or injury, hostile or fearful emotions, and very high or very low Na intake. A chronic K deficiency causes degenerative disease. Potassium makes up 70% of the positive ions in the cells. The cell is essentially a little bag of K salts. During a K deficiency, K migrates out of the cell and causes the cell fluid to become acidic (lower pH). Enzyme systems are often sensitive to acidity, this drift toward acidity could easily be the cause of some of the symptoms from K deficiency (e.g., arthritis). The kidneys have enzymes that make ammonium ion, using glutamine as a precursor. Phosphate-dependent glutaminase enzyme splits off an ammonium ion. Then in a second step, glutamine dehydrogenase enzyme splits off another ammonium ion from the glutamate, which had resulted from the first step to form α-ketoglutarate. This enables the kidneys to excrete more acid, which interferes with K excretion. The above enzymes become more active when the cell's fluid becomes more acidic. Glutamine is an essential amino acid. This may be an adaptation primarily for the purpose of conserving K even though ammonium excretion may be directly related to H^+ (acid or low pH) and not to K concentration. Potassium depletion has been shown to increase ammonia production by the kidneys from both glutamine and glutamate, decrease glutamate conversion to glutamine, and increase ammonium removal from aspartate. The ammonium ion and K have the same charge and size, and they are handled at the same site in the kidneys. Hydrogen ion and K may compete at the same site in the kidney's distal tubules. It is possible that 18-hydroxy deoxycorticosterone steroid is the hormone that regulates the handling of ammonium. Potassium excretion is quite sensitive to H^+ concentration (acidity). Potassium deficiency puts an increased drain on glutamine, and would presumably be disadvantageous to someone not getting enough protein. It also seems likely that eating baked goods, which have been risen with sodium bicarbonate, or stomach antacids would worsen a deficiency. There is a possibility that fruits that contain acids, which acids can be absorbed but not metabolised would have a conserving effect on K. Cherries have a beneficial effect on arthritis. When a neuron decides to fire, the cell wall suddenly becomes permeable to Na^+, and sodium ions near the cell wall suddenly move into the cell, followed a microsecond later by a flow of thousands of potassium ions in the opposite direction. Half the metabolic energy supplied to nerve cells is required to move the Na back out of the cell in order to recharge it. For this system to work the K in the plasma has to be kept as close as possible to 187 mg/l (4.8 mEq/l). If it rises above 400 (9 or 10 mEq) or falls below about 80 (2 mEq or more) death is almost certain from failure of the nerves leading to vital organs to fire. Rising above 400 is the greatest risk because excessive loading of plasma is quite possible from supplements, metabolic shock, and various hormone failures. Potassium is activator for several enzyme systems. The amount needed for activation is usually about 40 mg/l. It is probable that no other enzymes apart from d-amino isobutyric acid (which is permanently disrupted at a cellular level) are inactivated directly by low K whole body count (cell content). Healthy collagen ranks with steel in strength of individual fibers. Healthy bone, which is essentially an ossified connective tissue, has a strength that approaches that of cast iron. The strength of connecting tissue and its ability to regenerate are of considerable importance.

THYROID DYSFUNCTION

Thyroid disease is caused by deficiencies of key minerals and that there are critical steps of mineral metabolism that need to be working for normal thyroid metabolism. Several minerals are necessary for the production of thyroid hormone (T4), such as iodine, iron, manganese, zinc, copper, chromium, selenium, cobalt, and possibly other ultratrace minerals. Selenium is necessary for formation of the deiodinase enzyme, which converts T4 into T3. Copper seems to be necessary for suppressing the production of immune system malfunctions that cause autoimmune Graves' disease, and appears to have other critical functions in preventing hyperthyroidism. Potassium, sodium, lithium, calcium, and magnesium regulate the passage of minerals, other nutrients, and T3, through cell membranes. Imbalances of these gateway minerals can limit T4 production by interfering with the transport of minerals into the thyroid cells and can also limit the amount of T3, which gets into the body's cells, thereby limiting the rate of metabolism. Vitamins, proteins, and fats, which work with the minerals, need to be present for the minerals to work properly to perform normal endocrine functions. The sodium-potassium channels, the lithium-sodium counter transport system, and the calcium channels all depend upon the ratios of minerals rather than their absolute concentrations. Hyperthyroidism and Graves' are the result of nutritional deficiencies and imbalances, mainly two key mineral relationships: 1- zinc/cadmium/iron/copper, and 2- sodium/potassium/calcium/magnesium. Copper is depleted by various methods: 1- excess zinc or cadmium can deplete copper; and 2- various vitamin deficiencies, such as biotin, PABA, pantothenic acid, B_2, niacin, or B_1, can prevent copper from being utilised correctly. Therefore copper can be replenished by supplementation (5-8 mg/day); removing excess cadmium (smoking, chocolate, coffee, excess green leafy vegetables, etc.); removing excess zinc (stop taking multiple vitamin/mineral supplements, limit high zinc/low copper meats, limit B_6 that assists zinc metabolism), and by limiting iron and the nutrients that help iron metabolism (B_{12}, manganese, folic acid). In hyperthyroidism, excess Na and Ca deplete K and Mg. Deficiencies of K and Mg causes wide-ranging problems. Magnesium deficiency causes rapid heart rate and K deficiency causes weak or irregular heart rate. Magnesium deficiency prevents the heart muscles from going through a complete relaxation phase so the heart rate accelerates. Potassium deficiency also causes water to enter the cells and stay there, causing the body to swell up with oedema. If the patient is oedematous, probably due to K deficient, so provide more K than Mg. If the patient is losing weight, then probably Mg is more deficient than K, so provide more magnesium than potassium. The deficiencies of K and Mg explain why Na and Ca (dairy products, etc.) aggravate hyperthyroidism symptoms, since sodium especially depletes K and Ca especially depletes Mg. Zinc depletes K and Cd depletes Mg. Copper assists magnesium metabolism and potassium seems essential to enable copper enter the cells. The key deficiencies are copper and its large collection of assisting nutrients, magnesium, and potassium. The daily requirement of K is about 3000 mg. It also helps to increase high potassium foods like bananas (400 mg in one) and potatoes (500 mg in one), but these are high glycaemic. Excess K can deplete Mg (tachycardia), while excess Mg can deplete K (arrhythmia). Other nutrients may need to be added (iron, selenium, chromium, zinc, B_{12}, etc.) to boost thyroid production up to normal. Potassium deficiency is a critical precursor of thyroid disease. The initial effects of long-term K deficiency cause hypothyroidism but when the deficiency gets severe, hyperthyroidism results. Of the five minerals involved in the cellular transport of nutrients, K seems to be the most likely to be deficient in thyroid disease. Deficiencies of K decrease the ability of nutrients and hormones to enter the cells by disrupting the Na/K transport system. This results in mineral deficiencies, which then cause abnormal thyroid function, and can cause other symptoms like causing the cells to accumulate water, resulting in cellular and bodily oedema. This oedema or weight gain is seen in both hypothyroidism and hyperthyroidism. Potassium deficiency decreases copper transport into the cells and this results in copper-deficiency anaemia, since the iron no longer has enough copper to form an adequate amount of haemoglobin. Intake of iron further depresses copper levels resulting in more hyper symptoms when iron, manganese, or cobalt are ingested. Copper and Mg transport into the cells reduction causes the various symptoms associated with hyperthyroidism. Magnesium deficiency makes the individual very intolerant of calcium intake, since Ca is a natural antagonist of Mg and is the promoter of muscular contraction. Magnesium supplementation can relieve many of the symptoms of thyroid disease. A high ratio of Na to K may favour the transport of Ca into the cells at the expense of Mg. This pushes the heart rate higher because of the subsequent high Ca/Mg ratio. Magnesium supplementation helps, but perhaps the key to increasing Mg is the use of K supplements along with Mg supplements. Supplementation with copper and the nutrients that help copper metabolism, control hyperthyroidism in the long run better. Thyroid disease begins with K deficiency, or rather too high ratio of Na to K over a long period of time. This result in the gradual increase in Cu and Mg deficiencies, and the serious symptoms of hyperthyroidism are the result of problems created when these nutrients become deficient. Adrenal hormones cortisol and aldosterone, which are increased during stress, stimulate K excretion. The synthesis of muscle protein requires K. The excessive use of salt deplete K. Coffee and sugar increase the excretion of K from the body. Decreasing or eliminating coffee and sugars, including fruits, helps hyperthyroidism recovery. Physical activity increases K excretion. Extensive physical exertion for 3 hours a day can dissipate from 700-800 mg of K, which radiates out from sweat. A slow pounding heart rate indicates K deficiency, while a fast heart rate indicates Mg deficiency. Potassium constitutes 5% of the total mineral content of the body. Potassium is not a trace mineral, but a major mineral. The body normally has a lot of K and requires a large intake each day to maintain adequate levels. The amount of K in the average person's daily diet is estimated at 2000-6000 mg, with an estimated minimum requirement of 2500 mg. Hypothyroidism is the result of long-term K deficiency and hyperthyroidism is the result of severe long-term K deficiency. Potassium itself is not going to relieve the major symptoms of thyroid disease; it just opens the cellular channels for the important minerals to get in so they can perform their function. The major effects of hypothyroidism on health were summed up more than 60 years ago. The pathobiology of atherosclerotic heart disease was basically solved before the Second World War, i.e., respiratory defect. Many other diseases are now known to be caused by respiratory defects. Inflammation, stress, immunodeficiency, autoimmunity, developmental and degenerative diseases, and aging, all involve significantly abnormal oxidative processes. Just brief oxygen deprivation, e.g., hypoxia, triggers processes that lead to lipid peroxidation, producing a chain of other oxidative reactions when oxygen is restored, i.e., reperfusion syndrome. The only effective way to stop lipid peroxidation is to restore normal respiration. By the mid-1930s, it was generally known that hypothyroidism causes the cholesterol level in the blood to increase; hypercholesterolaemia was a diagnostic sign of hypothyroidism. By the 1960s, the protein-bound iodine (PBI) test was proven to be irrelevant to the diagnosis of hypothyroidism. Triiodothyronine (T3), the active thyroid hormone, in the blood can be measured with reasonable accuracy (using radioimmunoassay [RIA]), and this single test corresponds well to the metabolic rate. Thyroxine is considered hydrophobic substance, which associate with proteins, cells, and lipoproteins in the blood, rather than dissolving in the water. Thyroxine contains some polar groups that, in the right (industrial or laboratory) conditions, can make it slightly water-soluble. This makes it a little different from progesterone, which is simply and thoroughly hydrophobic, though the term free hormone is often applied to progesterone, as it is to thyroid. When red cells are broken up, they are found to contain progesterone at about twice the concentration of the serum. In the serum, 40-80% of the progesterone is bound on albumin. Albumin easily delivers its progesterone load into tissues. Progesterone, like cholesterol, can be carried on/in the lipoproteins, in moderate quantities. This leaves a very small fraction to be bound to the steroid binding globulin. The thyroid hormones associate with 3 types of simple proteins in the serum: transthyretin (prealbumin), thyroid binding globulin, and albumin. A very significant amount is also associated with various serum lipoproteins, including HDL, LDL, and VLDL. A very large portion of the thyroid in the blood is associated with the RBCs. When laboratories measure the hormones in the serum only, they have already thrown out about 95% of the thyroid hormone that the blood contained. The T3 is strongly associated with the cells' cytoplasmic proteins, but to move rapidly between the proteins inside the cells and other proteins outside the cells. Huge amount of T3 bound to albumin is taken up by the liver. The specific binding of T3 to albumin alters the protein's electrical properties, changing the way the albumin interacts with cells and other proteins. Albumin becomes electrically more positive when it binds the hormone; this would make the albumin enter cells more easily. Giving up its T3 to the cell, it would become more negative, making it tend to leave the cell. This active role of albumin in helping cells take up T3 might account for its increased uptake by even fewer RBCs. This could also account for the favourable prognosis associated with higher levels of serum albumin in various diseases. The idea of measuring the free hormone is that it supposedly represents the biologically active hormone, but in fact it is easier to measure the biological effects than it is to measure this hypothetical entity. Proteins have a great affinity for fats, and fats for proteins; even soluble proteins, such as serum albumin, often have interiors that are extremely fat loving. The structural proteins of cell membranes are not dissolved in water, i.e., insoluble proteins. Thyroid, progesterone, and oestrogen have many immediate effects that change the cell's functions long before genes could be activated. Transthyretin, carrying the thyroid hormone, enters the cell's mitochondria and nucleus. In the nucleus, it immediately causes generalised changes in the structure of chromosomes, as if preparing the cell for major adaptive changes. Respiratory activation is immediate in the mitochondria, but as respiration is stimulated, everything in the cell responds, including the genes that support respiratory metabolism. Active transport and membrane pumps are ideas that seem necessary to people who haven't studied the complex forces that operate at phase boundaries, such as the boundary between a cell and its environment. Diffusion, codiffusion, and absorption describe the situation adequately. T3 is used faster than T4, removing it from the blood more quickly than it enters from the thyroid gland itself. A natural glandular balance may be more appropriate to supplement than pure thyroxine. Most T3 is produced from T4 in the liver, not in the brain. An excess of thyroxine, in a tissue that doesn't convert it rapidly to T3, has an antithyroid action. Increasing thyroxine dose for hypothyroid women, worse their symptoms. The brain concentrates T3 from the serum, and may have a concentration 6 times higher than the serum, and it can achieve a higher concentration of T3 than T4. It takes up and concentrates T3, while tending to expel T4. More active metabolism probably keeps the blood ratio of T3 to T4 relatively high, with the liver consuming T4 at about the same rate that T3 is used. T3 has a short half-life, and it should be taken frequently (e.g., few µg/h). Since it restores respiration and metabolic efficiency very quickly, it isn't usually necessary to take it every hour or two, but until normal temperature and pulse have been achieved and stabilised, sometimes it's necessary to take it four or more times during the day. An effective way to use supplements is to take a combination T4-T3 dose, e.g., 40 µg of T4 and 10 µg of T3 once a day, and to use a few µg of T3 at other times in the day. Keeping a 14-day chart of pulse rate and temperature allows the patient to see whether the dose is producing the desired response. If the figures aren't increasing at all after a few days, the dose can be increased, until a gradual daily increment can be seen, moving toward the goal at the rate of about 1/14 per day. Arthritis, irregularities of growth, wasting, obesity, a variety of abnormalities of the hair and skin, carotenaemia, amenorrhoea, tendency to miscarry, infertility in males and females, insomnia or somnolence, emphysema, various heart diseases, psychosis, dementia, poor memory, anxiety, cold extremities, anaemia, and many other problems were known reasons to suspect hypothyroidism. The thyroid hormone produced in the thyroid gland (T4) is not primarily responsible for speeding up the metabolism. It is simply the raw material that the tissues of the body use to make the active hormone (T3). By far, most of the active thyroid hormone is produced outside the thyroid gland, in the tissues of the body. Many patients have normal thyroid blood tests and yet still have low body temperatures and classic thyroid symptoms that respond quickly and completely to proper T3 therapy and often remain improved even after the treatment has been discontinued. Hypothyroidism can be fatal. Wilson's thyroid syndrome is not immediately fatal, and T4 doesn't help very much. Wilson's thyroid syndrome is a very common and treatable condition using T3 therapy. Wilson's thyroid syndrome is probably more common than all other thyroid system disorders combined. T3 improves sleep patterns; decreases chronic fatigue, reduce weight, and increases exercise. Complete absence of any physical ailments, symptoms, or complaints, and the general sense of overall well-being is highly expected with liothyronine.

NUTRITIONAL AND METABOLIC EFFECTS OF ALCOHOLISM

Hepatic protein trafficking is perturbed by ethanol consumption. TGF-α shares 30-40% amino acid sequence homology with EGF, and like EGF is a potent hepatocyte mitogen. TGF-α and EGF bind to EGFR. This binding activates the intrinsic receptor tyrosine kinase activity and leads to receptor autophosphorylation, internalisation, and downregulation. Chronic ethanol impairs the ability TGF-α to stimulate the autophosphorylation of the EGFR in hepatocytes. This resulting in altered signal transduction and to impaired reparative and regenerative processes in the liver. Chronic ethanol consumption (10 g/KgBW) significantly decreases the rate of respiration, P/O ratio, and respiratory control ratio (RCR). Chronic ethanol consumption results in marked alteration in both the function and morphology of hepatic mitochondria. Chronic alcohol consumption leads to distinctive alterations in the enzymatic activity and the content of certain proteins of the mitochondrial inner membrane, especially cytochrome aa_3, cytochrome b, NADH dehydrogenase, ATPase, and cytochrome oxidase. Chronic alcoholism is associated with increased risk of fracture and incidence of osteoporosis. Chronic alcohol consumption results in significant osteopenia and osteoporosis. Alcohol reduces the number of osteoblasts and number of active osteoblasts. Alcohol decreases osteocalcin levels, suggesting decreased osteoblastic activity. Alcohol inhibits osteoblastic cell proliferation; decreases trabecular bone volume, osteoblast number, mineral apposition, and bone formation rate with increased mineralisation lag time. Alcohol may affect bone resorption, may cause decoupling of resorption and formation. 30% of alcoholics develop liver disease. An alcoholic may replace up to 60% of the daily calorie intake by alcohol. Vitamin B complex deficit may cause encephalopathy, peripheral neuropathy, and cardiac disturbances. Vitamin A deficit may cause visual and gonadal alterations and zinc deficiency may lead to immunologic changes. Alcoholic hypoglycaemia is the better-known alteration of carbohydrate metabolism induced by alcohol. However, excessive alcohol intake also induces glucose intolerance. The diabetogenic effect of alcohol is probably due to an inhibition of insulin secretion and, less probably, to a lower peripheral insulin sensitivity. The global effect of alcohol is to induce protein loss. Intestinal protein absorption is inhibited and urinary nitrogen excretion increases. Alcohol intake leads to negative nitrogen balance, in spite of adequate protein intake. These effects of alcohol persist during the first week of abstinence in alcoholics. Alcohol possibly acts as a direct toxin on muscle proteins, generating a muscle damage that is observed in up to 50% of alcoholics. The catabolic effect of alcoholism could also be due to the effects of lymphokines on muscle. Alcoholism induces cytokine mRNA expression and secretion. Alcohol intake inhibits lipolysis and reduces free fatty acid (FFA) levels. Paradoxically, the inhibition of lipolysis induced by alcohol is associated with an increase in arterial ketone body levels, probably due to the conversion of acetate, derived from alcohol oxidation, to acetoacetate and D-3-hydroxybutyrate. The ketone body that increases preferentially is D-3-hydroxybutyrate, because during alcohol oxidation, NADH:NAD ratio increases and this change in redox potential leads to the accumulation of reduced metabolites. In clinical situations, the diagnosis of alcoholic ketosis is difficult, because most reactive strips for ketones detect acetoacetate but not D-3-hydroxybutyrate. Serum triacylglycerol levels also increase as a consequence of alcohol intake and they rapidly return to normal values during abstinence. Triacylglycerol clearance by lipoprotein lipase is not altered by alcohol, but the clearance of chylomicron remnants is reduced and there is an increase in very low-density lipoprotein secretion by the liver. Moderate alcohol consumption is associated with increased oxidability of LDLs, which may contribute to the formation of fatty streaks in arteries. Fatty acid metabolism is also modified by alcohol, which induces changes in fatty acid composition of lipids in several tissues. Alcohol ingestion may also promote the formation of esterification products of fatty acids and alcohol, which could be mediators of end-organ damage. Liver cell necrosis in alcoholic liver disease occurs predominantly in the centrolobular zone (zone 3 of Rappaport), where oxygen availability is limited. In this zone, CYP2E1 is expressed with greater intensity and more ethanol is metabolised, therefore increasing cell oxygen requirements and thus generating a relative hypoxic condition. Activation of Kupffer cells by endotoxin may also contribute to the hypermetabolic state induced by alcohol intake and thus to pericentral hypoxia. The increased liver oxygen consumption is partially compensated by an increased hepatic blood flow caused by alcohol, which does not suffice to compensate the higher oxygen consumption and hypoxia. Chronic alcohol administration increases malondialdehyde levels, conjugated dienes in microsomal membrane lipids, and peroxidation products such as 4-hydroxynonenal in liver cell microsomes. The increased peroxidation seen after alcohol administration could be due to the generation of free radicals during its metabolization, or to a reduction in antioxidant systems. Alcohol acts as an antioxidant. Alcohol oxidation by mixed-function oxidases of CYP2E1 may generate free radicals that can unveil a peroxidative process. Peroxidative phenomena are potentiated by iron, metal that is accumulated in the liver of alcoholics. Alcohol oxidation by alcohol dehydrogenase also increases peroxidative pressure. These radicals are converted to hydrogen peroxide by superoxide dismutase, which is destroyed by glutathione peroxidase, and also hydroxyl radicals. Acute and chronic alcohol intake lower liver-reduced glutathione levels. Glutathione plays an important role in the defense against free radical attack, since it acts as a free radical scavenger and as a regenerator of α-tocopherol. Iron overload enhances alcohol-induced glutathione depletion in alcoholic. Glutathione depletion increases the induction of apoptotic changes and alteration of membrane function in hepatocytes. This alteration is associated with changes in mitochondrial transmembrane potential and mitochondrial dysfunction. The depletion of glutathione is associated with functional abnormalities of these organelles and is more pronounced in perivenular mitochondria. In alcoholics there is an inverse relationship between the hepatic levels of the vitamin E and the degree of peroxidative damage. There is a defect in vitamin E transport across the cell membrane. Free radicals directly damage lipid components of the cell membrane. Certain peroxidation products, such as hydroxyethyl radicals, may form immunogenic adducts that can produce damage through cell-mediated cytotoxicity. Free radicals may produce mitochondrial damage, reducing fatty acid β-oxidation and thereby contributing to the genesis of liver steatosis. The inhibition of lipolysis induced by alcohol is directly dependent on acetate. This lower lipid mobilisation could contribute to liver steatosis. Acetate may also inhibit muscle glucose oxidation contributing to the glucose intolerance of alcoholics. Liver failure leads to undernutrition and wasting due to series of causes that range from anorexia to specific metabolic alterations, which provoke a hypercatabolic state. Liver pathologic changes are of greater magnitude in obese alcoholics. Alcoholic hepatitis or cirrhosis is at least twice as frequent in subjects with a weight over 120% of standard than in subjects with normal weight. Obesity is an independent predictor of alcoholic liver disease. Insulin inhibits CYP2E1, therefore its lower levels in alcoholics could have pathogenic implications. Proinflammatory cytokines (TNF and IL-1) have a pathogenic importance in alcoholic liver disease. Alcohol inhibits lymphokine secretion by monocytes. Vitamin E may have an immunostimulating effect, it reverts alcohol-induced inhibition of thymocyte and splenocyte secretion of TNF and IL-6. Kupffer cell activation can be caused by oxidative stress or by a higher endotoxin absorption in the intestine. Lymphokines may cause liver damage stimulating stellate cells to generate fibrogenesis, increasing the migration of inflammatory cells to hepatic parenchyma, or altering microcirculation. IL-1 stimulates fibroblasts, and hepatic lipogenesis is stimulated both by IL-1 and TNF. Adipose tissue expresses mRNA for TNF synthesis, and this expression is directly proportional to the degree of adiposity in non-alcoholic subjects. There is a positive correlation between the secretion of IL-1 and the percentage of adipose tissue of these subjects. Obese individuals may be secreting higher amounts of lymphokines, which contribute to the genesis of liver damage. Obese and diabetic subjects have a higher degree of CYP2E1 induction than their lean counterparts, which probably due to a stabilisation of microsomal proteins. CYP2E1 is induced in human obese subjects. Acetone, acetoacetate, and β-hydroxybutyrate levels, which increase with fasting and in alcoholism, all induce the microsomal system. When the diet contains predominantly saturated fatty acids, no liver damage appears in alcoholic, but when polyunsaturated fatty acids (PUFAs) predominate, liver necrosis unveils. The mechanisms for a greater liver-damaging capacity of polyunsaturated fatty acids are a higher induction of CYP2E1, a greater peroxidative capacity of these fatty acids, or an increased generation of thromboxanes (TXs) from AA. These TXs may cause damage through their vasoconstrictor effects and platelet aggregation, aggravating liver tissue hypoxia caused by alcohol. A constant feature of the reduction of AA levels in liver lipids is a decrease in the arachidonic/linoleic acid ratio. This reduction may be due to an inhibition of 5 and 6 desaturates caused by alcohol, with a consequent decrease in the synthesis of AA. Hydroxylated derivatives from AA can also be formed, which effects, e.g., inhibition of Na-K-ATPase, could be of pathogenic importance in alcoholic liver disease. Saturated fatty acids in liver of alcoholics or obese subjects could be incorporated to triacylglycerols less efficiently, alcoholics and obese subjects have higher levels of FFAs in the liver. Unsaturated fatty acids predominated in the triacylglycerols fractions, while saturated fatty acids predominated in the FFA fractions. Free fatty acids may be hepatotoxic and contribute to the liver damage of obese individuals and alcoholics. Ethyl alcohol (a term used interchangeably with ethanol or alcohol) is a colourless organic liquid that has well characterised psychophysical and mood-altering effects. It is an important dietary component, but unfortunately it is subject to misuse as its consumption in excessive quantities may lead to dependency or tissue damage. Clinically, alcohol-related harm is detectable at a consumption rate beginning from 30 g/d. It is estimated that in Western societies or societies undergoing Westernization, over 10% of the general adult population may be classified as misusers, and there is growing concern that even adolescents, teenagers, and women may be consuming ethanol in dangerous amounts. Neurologic tissues or liver are predominantly affected by alcohol, leading, e.g., Wernicke's encephalopathy or cirrhosis, respectively. The disease states include alcoholic muscle disease, osteoporosis, and osteopenia, alcoholic cardiomyopathy and bowel abnormalities. The intestinal changes will precipitate malabsorption and motility disturbances, leading to a compromised nutritional state. In alcohol misusers, the corresponding incidence of abnormalities in skeletal muscle, GIT, and bone organ systems are each between 30-60%. The most abundant component of mammalian tissue, apart from water, is protein. In muscle, proteins maintain the structural characteristics of cells (i.e., cytoskeletal proteins), provide mechanical and locomotor activity (i.e., through contractile proteins) and catalyse reactions for processes such as energy coupling (i.e., enzymes). In some organs, constituent (i.e., non-export) proteins act as a source of metabolic fuel. In catabolic phases, the degradation rate of protein will exceed the rate of protein synthesis, e.g., alcoholic myopathy, osteopathy, and intestinal atrophy. In growth or hypertrophy, tissue protein synthesis exceeds protein degradation, e.g., alcoholic cirrhosis is related to an increased concentration of hepatic collagen. The fall in protein content can occur as a result of: 1- increases in both the rate of protein synthesis and breakdown, 2- increases in the rate of protein breakdown alone, 3- decreases in the rate of protein synthesis alone, and 4- decreases in both protein synthesis and breakdown. In all these situations the rate of breakdown will exceed the rate of synthesis. The synthesis is measured as a fractional rate, i.e., the fractional rate of protein synthesis is defined as the percentage of the tissue protein pool renewed each day; i.e., %/d. Approximately 20% of whole-body protein synthesis can be ascribed to the liver, and similar contributions are derived from either the gastrointestinal tract, skeletal muscle, or combined skin and bone. The heart is only a minor contributor to whole-body protein synthesis, but nevertheless is central for metabolic integrity: even moderate changes in heart muscle protein content affect contractile activity. An average of 56 g of ethanol per day has a total caloric intake that is 16% greater than that of non-drinkers (identical body mass indices of 26). At high ethanol intakes, there is inefficient utilisation of calories derived from ethanol. At even higher ethanol intakes, there are even reductions in the weights of alcoholics, engendered by loss of lean tissue. Ethanol and/or acetaldehyde impairs protein deposition and, as a consequence, reduces the associated cytoplasmic contents, which are predominantly water (water is 70% of tissue weight). Thus, there is a cascade effect: loss of tissue protein is accompanied by loss of water, minerals, and electrolytes. Ethanol dosage as 30, 40, or 60% of total calories does not appear to enhance thermal energy losses, but increases nitrogen excretion (nitrogen, uric acid, and urea and lose weight) with associated mineral losses. Losses of calories as ethanol in urine (50 Kcal/d) and respiration (50 Kcal/d) together comprised only 8% of total ethanol intake or 4% of total ingested energy. Alcohol-induced skeletal myopathy is characterised by muscle weakness, frequent falls, and difficulties in gait. Affected patients may lose up to 30% of their muscle mass. Approximately half to two-thirds of all chronic alcoholics are affected. Myofibrillar protein breakdown is reduced in alcoholics with histologically proven myopathy compared to those without myopathy. Skeletal muscle total RNA is reduced in response to ethanol exposure, which contributes to impaired protein synthesis. Chronic alcoholism causes disturbances in cardiac myofibrillary architecture and related parameters of contractility. Reduced plasma albumin concentrations are often seen in severe liver disease. They are often associated with a fall in the concentrations of plasma albumin, probably as a result of perturbed synthesis or the inability of the liver to secrete albumin into the circulation.

THE ELEMENTS AND INFLAMMATORY RESPONSE

The acute phase response to injury or infection involving systemic physiological and biochemical alterations, and is associated with alteration in dynamics of many trace elements, particularly iron, zinc, and copper. The main physiological components include fever, increased metabolic rate, and leukocytosis. The biochemical changes include increased oxidation of fat and carbohydrate, increased transfer of amino acids from skeletal muscle to the liver with the synthesis of hepatic acute-phase proteins, and alterations in trace element metabolism. The metabolic rate after elective surgery may increase by 10-20%, whereas in patients with severe sepsis, the metabolic rate may increase by 50%. Many patients who are seriously ill have multiple sources of loss trace elements. Patients with a fistula or diarrhoea lose large amounts of Cu, Mn, and Zn, and patients with severe burns have substantial losses of all elements through burn exudate. Dialysis leads to loss of significant amounts of all water-soluble nutrients. Zinc plays a central role in the enzyme-catalysed reactions of protein synthesis; Se is required for activity of glutathione peroxidase, a key enzyme in pathways for free radical scavenging. Serum Zn and Fe fall rapidly after the commencement of an operation, 2-4 h after skin incision until 12-24 h after. Zinc returns to normal within 4-5 days. Trace element binding proteins, albumin (for Zn) and transferrin (for Fe) fall during the acute phase response. Serum Cu rises after 1-2 days of the injury, reaching a peak concentration several days later. The fall in serum Fe results from the transfer of Fe from the Fe-transferrin complex in the plasma to lactoferrin released from leukocytes at the site of inflammation. Subsequently, Fe is taken up and bound to ferritin in the liver and spleen. IL-1 and TNF alter Fe dynamics by increasing the uptake from plasma and dropping half life. IL-1 and TNF may induce transferrin receptors in cells, which increase uptake from the plasma of Fe bound to transferrin. Once released from the transferrin receptor, Fe is bound to the tissue storage protein ferritin, which consists of 24 subunits of 2 types, H and L. Ferritin synthesis is increased as part of the acute phase reaction. Movement of Fe diverts labile plasma Fe and other labile intracellular Fe into storage form, thus reducing its availability within the plasma and withholding Fe from bacteria. The reduction in Fe availability may lead to concomitant reduction in the conversion of superoxide radicals to free hydroxyl radicals, thus reducing oxidative damage to membranes or DNA. If the changes persist, Fe transfer to bone marrow is reduced, leading to the anaemia of chronic illness. Most plasma Zn circulates bound to albumin (55-90%), a significant fall in albumin leads to a fall in Zn concentration. During an acute phase reaction, the drop in plasma Zn is greater than in albumin. IL-1 increases uptake of Zn by liver, bone marrow and thymus; whereas Zn is lost from bone, skin, and intestine. Endotoxaemia causes a fall in plasma Zn concentration associated with an increase in both the extracellular Zn and hepatic Zn pool. IL-1 and endotoxaemia induce the low molecular weight protein metallothionein. This cysteine-rich protein can be induced in many tissues of the body in response to heavy metals, corticosteroids, and cytokines. The effect of IL-1 is mediated via IL-6. TNF increases hepatic metallothionein synthesis through IL-6. Metallothionein synthesis is also induced by an increase in glucocorticoid activity. Glucocorticoid is secreted from the adrenal cortex under the control of ACTH. IL-1, IL-6 and TNF act on the hypothalamus to stimulate CRF release and hence increased glucocorticoid secretion. Glucocorticoids inhibit the release of IL-1 and TNF from macrophages. Metallothionein is a free radical scavenger, because it has numerous reduced sulphydryl groups. Metallothionein acts as a buffer in providing Zn when required for cell activities, e.g., increased metalloenzyme activity such as in the pathways of protein synthesis, which are stimulated during the acute phase response. Zinc plays an important role in stabilising cell membranes. Zinc is also an essential part of Zn-finger components of the DNA-binding transcription factors, which are important in controlling the selectivity of protein synthesis. Caeruloplasmin synthesis is induced by IL-1 and IL-6. Copper is necessary to provide the molecule with its ferrooxide activity, which may be beneficial in scavenging OFR generated from neutrophils and macrophages. Severe illness, whether acute or chronic, is associated with protein-energy malnutrition. Malnourished patients have a reduced acute phase protein response, although the plasma IL-6 changes may be only slightly altered compared with well-nourished individuals, suggesting an impaired hepatic response to IL-6. Malnourished patients have reduced cytokine production, which can be correlated with protein refeeding. Inadequate trace-element status impairs some components of the acute phase response. Severe Zn depletion reduces the metallothionein response to IL-1 and endotoxin. Copper deficiency leads to reduction in the caeruloplasmin response to IL-1 and an impaired response to caeruloplasmin and Cu-Zn superoxide dismutase to endotoxin and high O_2 concentration. Probably the most important single test for assessment of trace element status is a measure of the acute phase response, e.g., serum C-reactive protein concentration. Tests of plasma or whole-blood concentration, or urinary excretion, may not reflect whole-body status or the concentration within the metabolically active intracellular compartment. Bone marrow-stainable Fe probably the most accurate method of determining iron adequacy. Plasma haemoglobin and blood film indicates patients who require blood transfusion. Measurement of leukocyte Zn is a good index of whole-body Zn status. Measurement of serum creatinine (Cr) concentration requires meticulous sampling techniques to prevent contamination of the blood sample, and specialist analytical facilities. Creatinine excretion is mainly via urine; patients with renal failure have high plasma concentration. Magnesium is excreted primarily in bile.

IRON ASSOCIATED PROBLEMS

The regulation of iron is a central function of the immune system. Iron is a potentially toxic heavy metal. In excess, it can cause cancer, heart disease, and other illnesses. An excess of dietary iron contributes to the development of leukaemia and lymphatic cancers. In the 1960s when iron supplements were given to anaemic patients in Africa, there was a great increase in the death rate from infectious diseases, especially malaria. Just like lead, mercury, cadmium, nickel and other heavy metals, stored iron produces destructive free radicals. Excess iron is a crucial element in the transformation of stress into tissue damage by free radicals. Excess iron accelerates the accumulation of age-pigment and other signs of aging. Blood transfusions damage immunity, and excess iron has been suspected to be one of the causes for this. Excess iron together with minor stress can produce a form of scleroderma. Excess iron has a role in infectious diseases, degenerative brain diseases, such as Parkinson's, ALS (Lou Gehrig's disease), Huntington's chorea, and Alzheimer's disease, and lipofuscin (age pigment) related conditions such as skin aging, atherosclerosis, and cataracts. Iron tissue storage increases with ageing. Lipofuscin means fatty brown stuff, which is an oxidised mass of unsaturated fat and iron, formed by uncontrolled free radicals. The iron content of food has been identified as the major life-shortening factor, rather than the calories. Iron added to cereals may contribute to the incidence of leukaemia and cancers of the lymphatic tissues in children. The normal amount of dietary iron causes an increased susceptibility to infections even in children, and a subnormal amount of iron slows the aging process. Only a few milligrams of iron are lost each day in menstruation. Some women who menstruate can donate blood regularly without showing any tendency to become anaemic. Women absorb iron much more efficiently than men do. From a similar meal, women normally absorb three times as much iron as men. When pregnant, their higher oestrogen levels cause them to absorb about nine times as much as men. Every time a woman menstruates, she loses a little iron, so that by the age of 50 she is likely to have less iron stored in her tissues than a man does at the same age, but by the age of 65 women generally have as much excess iron in their tissues as men. Increase in jaundice of the newborn could be related iron supplement to the mother during pregnancy. Arsenic, or iron, or other toxic material stimulates the formation of RBCs, which indicates that the body responds to a variety of harmful factors by speeding its production of blood cells. Radiation has the same kind of stimulating effect, because growth is a natural reaction to injury. The optimal nutritional intake should consider the resistance to disease, longevity or rate of aging and even mental ability. An excess of iron, by destroying vitamin E and oxidising the unsaturated fats in RBCs, can contribute to haemolytic anaemia. In elderly, red cells break down faster, and are usually produced more slowly, increasing the tendency to anaemia, but additional iron tends to be more dangerous for older people. Anaemia in women is caused most often by hypothyroidism, or by various nutritional deficiencies. Oestrogen causes dilution of the blood, so that it is normal for females to have lower haemoglobin than males. The added iron will destroy vitamins in the food. Aluminium and iron react similarly in cells and are suspected causes of Alzheimer's disease. Coffee percolated in an aluminium pot contains a large amount of dissolved aluminium, because of coffee's acidity. Glass utensils are safe, and certain kinds of stainless steel are safe, because their iron is relatively insoluble. Teflon-coated pans are safe unless they are chipped. There are 2 main types of stainless steel, magnetic and nonmagnetic. The nonmagnetic form has a very high nickel content, and nickel is allergenic and carcinogenic. It is much more toxic than iron or aluminium. Cooking in an iron frying pan put iron into food, especially if the food is acidic, e.g., sauces. Iron destroys vitamin E, so vitamin E should be taken as a supplement. Vitamin E should be avoided at the same time as the iron-contaminated food been taken, because iron reacts with it in the stomach. Coffee, when taken with food, strongly inhibits the absorption of iron. Vitamin C stimulates the absorption of iron; so, drinking orange juice should be avoided at the same meal with iron-rich foods. A copper deficiency causes tissues to retain an excess of iron. Copper (Cu) is the crucial element for producing the colour in hair and skin, for maintaining the elasticity of skin and blood vessels, for protecting against certain types of free radical, and especially for allowing the proper O_2 utilisation for the production of biological energy. It is also necessary for the normal functioning of certain nerve cells (substantia nigra) whose degeneration is involved in Parkinson's disease. The shape and texture of hair, as well as its colour can change in a copper deficiency. Excess iron can block the absorption of Cu. With aging, the tissues lose Cu as they store excess iron. Drinking coffee with iron rich foods can reduce iron's toxic effects. Elevated hair iron may be found in smokers, X-ray technicians and individuals with certain forms of cancer. Aspirin protects against iron toxicity, clot formation, and reduces lipid peroxidation while blocking prostaglandin formation. Iron interacts with oestrogen and unsaturated fats in ways that can change restraint and adaptation into sudden self-destruction, apoptotic cell death.

SOME GASTROINTESTINAL TRACT PROBLEMS

Most problems with the GIT occur at the beginning, or the end. Disorders of the mouth, throat, oesophagus, stomach and duodenum are common. So are problems of the colon. Functional dyspepsia (FD) and irritable bowel syndrome (IBS) account for more than 1/2 of the GIT units workload. FD and IBS share some clinical features. There is some overlap in their underlying pathophysiology, with abnormalities of both motility and visceral sensitivity. Functional dyspepsia is recurrent (at least 12 weeks/y) epigastric pain with no evidence of organic disease and associated upper abdominal bloating, early satiety, nausea, vomiting, and feelings of fullness. FD may be divided into reflux-like, ulcer-like, dysmotility-like, and non-specific dyspepsia. FD is extremely common, affecting up to 25% of the population. This condition is very costly in both economic and social terms. Hypnotherapy (HT) is extremely effective in treating IBS, leading to long-term improvement of symptoms and quality of life (QoL). HT is commonly considered a purely psychotherapeutic intervention. HT has the capacity to normalise visceral sensitivity and modulate motility in GIT. HT may be more effective than medical treatment or supportive therapy in both short- and long-term management of FD. HT improves all aspects of symptomatology and QoL, and has considerable economic advantages. A standard course of HT requires up to 12 sessions and thus at face value appears rather expensive in the short-term. But, once treated, patient seldom needs any further intervention. Medically treated patients continue to consult much more than those managed with HT. The aim of supportive intervention: 1- to control for medication given, 2- to maximise response to treatment to the greatest extent possible, and 3- as a result of needing further supplies of medications. The general relaxation associated with hypnosis provides nonspecific psychotherapeutic response reducing anxiety. HT can influence GIT physiological function. FD may result from some modification of gastric motility, gastric accommodation, or visceral sensitivity. HT modifies gastric secretion. When toxic matter and undigested food, collected in the intestines as a result of bowel toxaemia, are absorbed from the bowels into the blood stream, the result is the leaky gut syndrome. The undigested molecules act as antigens, foreign substances that provoke an immune reaction. Many of these antigens are similar in structure to normal body components, and the antibodies produced to fight them can destroy healthy tissues. Nutritional depletion is associated with increased intestinal permeability and a decrease in villous height. Stress can increase gut permeability, increase ion secretion by a mechanism involving neural stimulation or mast cells, increase mucin release and deplete goblet cells. The gut is the likely source of the antigens causing inflammatory arthritis. Small intestinal passive permeability is increased in some patients with atopic eczema. The body's ability to absorb calcium is diminished by a lack of ultraviolet light from the sun. When the bowel is inflamed, toxins are absorbed. The natural bacterial endotoxin produces many of the same inflammatory effects as the food additive, carrageenan. Carrageenan produces inflammation and immunodeficiency, synergizing with oestrogen, endotoxin and unsaturated fatty acids. Liver damage leads to hormonal imbalance. Carrageenan-containing foods such as apple cider, hot dogs, most ice creams and prepared sauces and jellies. Carrageenan causes colitis and anaphylaxis in humans, and it is often present in baby formulas and a wide range of milk products. In the 1940s, carrageenan, a polysaccharide made from a type of seaweed, was recognised as a dangerous allergen. Carrageenan produces granulomas, immunodeficiency, arthritis, and other inflammations. Stress and anxiety sharply reduce the circulation of blood to the intestine and liver. Prolonged stress damages the ability of the intestinal cells to exclude large molecules. Local irritation and inflammation of the intestine also increase its permeability and decrease its ability to exclude harmful materials. Starch grains, or other hard particles, can be found in the blood, urine, and other fluids after they have been ingested. The incidence of several inflammatory diseases, e.g., Crohn's disease has been increased during the last 50 years, and at the same time, the incidence of several liver diseases has also been increasing. The permeability of the intestine that allows bacteria to enter the blood stream is very serious if the phagocytic cells are weakened. Carrageenan poisoning is one known cause of the disappearance of macrophages. Its powerful immunosuppression would tend to be superimposed onto the immunological damage that has been produced by radiation, unsaturated fats, and oestrogens. Carrageenan contributes to the disappearance of the liver enzymes (the cytochrome P-450 system) that detoxify drugs, hormones, and variety of other chemicals. Carrageenan enters even the intact, uninflamed gut, and damages both chemical defences and immunological defences. When it has produced inflammatory bowel damage, the amount absorbed will be greater, as will the absorption of bacterial endotoxin. Carrageenan and endotoxin synergize in many ways, including their effects on nitric oxide (NO), prostaglandins (PGs), toxic free radicals, and the defensive enzyme systems. Once the protective barrier-functions of the intestine and liver have been damaged, allergens and many materials with specific biological effects can enter the tissues. The polysaccharide components of connective tissue constitute a major part of our regulatory system for maintaining differentiated cell functioning and absorbed starches act as "false signals", with a great capacity for deranging cellular functioning. Carrageenan changes cellular function in complex ways, imitating changes seen in cancer. Since the bowel becomes inflamed in influenza, it is reasonable to think that some of the symptoms of the flu are produced by absorbed bowel toxins. While rheumatic fever and glomerulonephritis are caused by the antigens of streptococci, and systemic lupus erythematosus (SLE) is probably caused by the antigens of gram-positive lactobacilli found in the normal flora. Chronic fatigue syndrome, thyroid problems, and some kinds of porphyria seem to be more common in women of reproductive age, and are often exacerbated by premenstrual hormone changes. SLE is almost entirely a disease of women of childbearing age. One possibility for this could be that women during this period harbour a peculiar flora, i.e., large numbers of gram-positive lactobacilli in the vagina. Lactic acid is a metabolic burden, especially when combined with oestrogen excess. On a typical diet, tissues progressively accumulate linoleic acid, and this alters the structure of mitochondrial cardiolipin, which governs the response of the mitochondrial enzymes to the thyroid hormone. In the autoimmune diseases, such as lupus, there are typically antibodies to cardiolipin, as if the body were trying to reject its own tissues, which have been altered by the storage of linoleic acid. The altered mitochondrial function, which is involved in so many symptoms, can become part of a vicious circle, with endotoxin and oestrogen having central roles, once the stage has been set by the combination of diet, stress, and toxins. The premenstrual oestrogen-dominance usually leads progressively to higher prolactin and lower thyroid function. Oestrogen is closely associated with endotoxinaemia, and with histamine and nitric oxide (NO) formation, and with the whole range of inflammatory and autoimmune diseases. Anything that irritates the bowel, leading to increased endotoxin absorption, contributes to the same cluster of metabolic consequences. Charcoal, besides binding and removing toxins, is also a powerful catalyst for the oxidative destruction of many toxic chemicals. Persorption is a process in which relatively large particles pass through the intact wall of the intestine and enter the blood or lymphatic vessels. The colon is designed to extract water and some nutrients, store waste for convenient discharge, and check the waste products and other things not absorbed by the body for bacterial content. A clean, strong and well functioning colon is essential to maintaining optimal health. An unclean, weak and poorly functioning colon, is a breeding ground for disease, sickness and death. Serious health problems can arise if the colon is not properly cared for. Huge money is spent annually on laxatives. Good colon health is as much a function of the quality of food, as it's elimination status. Over time, the colon will lose the ability to process vital nutrients, absorb water and to eliminate faecal matter from the body (constipation). Constipation is a condition where the faecal matter is so tightly packed together, that bowel movements are infrequent, with difficulty and much straining. Old hardened faeces stick to the walls of the colon. The passage through which the faeces are forced to travel is much reduced, so that stools are narrow in diameter. The average person has 5-20 pounds of accumulated waste matter in their colon. Colon toxicity can be the underlying cause of many of health problems, such as: constipation, headaches, haemorrhoids, backaches, arthritis, allergies, diarrhoea, distended abdomen, bad breath/halitosis, skin problems, asthma, CFS, irritability, depression, prostate trouble, difficult weight loss, frequent colds, hypoglycaemia, insomnia, food cravings, abdominal gas, hypertension, foul body odour, and menstrual problems. The colon is selective in what it absorbs just as is the mouth. Most cases of colitis come from IBS. If the stress goes on long enough and enough damage is done to the colon, then it becomes colitis. The colon is lined with immune system testing stations. The colon is very important in electrolyte and water balance in the body. The bulk of the water flowing from the small intestine is absorbed in the ascending colon. The colon is not particularly resistant to bacterial invasions, and in fact, is the home of several pounds of friendly bacteria in healthful individuals. Humans do not survive well without this healthful internal flora. Chlorine, antibiotics and other agents can have serious effects on this flora and indirectly have serious effects on health. Many neurotransmitters (e.g., serotonin) are made in the digestive tract. The nervous system makes transmitters to respond to different situations. They are released into the blood stream. This sets the response stage of the nervous system. Changes of moods, nervousness or other vague changes in emotional states occur, but no specific sensations related to the organs involved. No two humans will have the same reaction to foods via production of neuro-stimulating chemicals or in the reaction to the production of those chemicals. From the beginning of the colon, in the caecum, to the end, in the rectum, there is a gradual reduction in the ability to handle acid. Acid in the beginning of the colon tends to be an irritation. The acid comes from the stomach or some other acid producing cells inappropriately located in the wrong place in the GIT. Acid almost instantly burn the colon. Stomach acid is not intended ever to reach the colon. Coffee, tea and other digestive stimulants (steak) also stimulate it. The body responds to stress in the short run by stopping or strongly slowing the rate of flow through the digestive tract. Food that should enter the mouth and depart the anus within 4 hours or so, can take days to get out. Complete constipation is a common response to acute stress. If the stress continues unabated for days or weeks, constipation forced on the digestive tract by the nervous system begins to break down. Patients with IBS have some neural degeneration in the ganglia of the myenteric plexus, associated with infiltrate of CD3-positive T lymphocytes, and longitude muscle hypertrophy. Intraepithelial lymphocyte (IEL) counts are above the normal upper limit (26/100 epithelial cells), which may be driven by luminal factors such as concomitant drugs. The inflammatory damage to the neural structures may be responsible of dysmotility and secondary longitude muscle hypertrophy. Elderly have a fewer mucosal-associated lymphocytes and a decreased response to infection. Possible mechanisms of IBS caused by inflammatory damage to the enteric nervous system include: 1- Bacterial infection damaging the mucosa, activating the immune response, and indirectly damaging the myenteric plexus. 2- Direct damage by neurotropic viruses such as Herpes zoster, Epstein-Barr virus, CMV. 3- An autoimmune attack initiated by exposure to cross-reacting antigens. Inflammatory bowel disease is associated with neural damage including nerve trunk hypertrophy, alteration in neuropeptide profile, and receptor expression. IBS can begin acutely after bacterial enteritis, e.g., Campylobacter infection, which has high incidence of postinfectious IBS. Chemically induced colitis induces infiltration of the myenteric plexus with eosinophils within 6 hrs, whereas the mild trauma activates macrophages within 3hrs. It also induces the production of the ICAM-1, recruits circulating blood monocytes, and induces epithelial cells to produce cytokines, especially IL-8. Inflammatory cells and altered permeability allows access of colonic bacterial products, especially endotoxin, which induces secondary responses including IL-1β production by glial cells surrounding afferent nerve terminals. Acute bacterial enteritis might initiate an inflammatory response in the myenteric plexus. Autoimmune damage to peripheral and autonomic nerves can develop after bacterial infection if the organism has a cross-reacting antigen such as Campylobacter, whose lipopolysaccharide mimics the GM1 ganglioside found in peripheral nerves. Neurotropic viruses can easily enter enteric nerves without significant mucosal damage. Herpes zoster is associated with damage to the myenteric plexus and the development of pseudo-obstruction in immunocompromised patients. There is residual viral DNA in the myenteric plexus and lamina propria in patients with chronic idiopathic pseudo-obstruction, e.g., CMV and EBV. Epstein-Barr virus is associated with acute cholinergic dysautonomia and marked gastrointestinal dysfunction. Cytomegalovirus infection after liver transplantation is associated with chronic gastrointestinal symptoms, including delayed gastric emptying. Ganglionitis, or inflammation limited to the myenteric plexus has been associated with pseudo-obstruction that can be paraneoplastic, autoimmune, or idiopathic in origin. The Hu protein expressed in cancer induces an immunological reaction (antibodies and cytotoxic T cells), which cross-reacts with and damages the myenteric plexus. Neurological dysfunction leads to functional obstruction and subsequent muscular hypertrophy is interesting, but equally, cytokines and growth factors induced by tissue damage may well be responsible. Disturbances of anal pressure activity, colonic motor function, small bowel motility, gastric motility, and visceral sensitivity are features of IBS. Increased numbers of mast cells both in terminal ileum and in the colon has also been noticed, as well as increased numbers of T cells and enterochromaffin cells in rectal mucosa a year after Campylobacter enteritis with persistent IBS symptoms. Acute inflammation can alter nerve function in the gut towards a state hyperalgesia. Histopathologic abnormalities of the proximal jejunum in patients with IBS include: 1- infiltration of lymphocytes in and around the nerves constituting the myenteric plexus, 2- increased number of intraepithelial lymphocytes, 3- neuron degeneration in the myenteric plexus, 4- longitudinal muscle hypertrophy, and 5 abnormalities in ICC numbers and/or size. Increased of intraepithelial lymphocytosis may be a hallmark of a hyperreactive gut in the same way that respiratory tract infection can lead to hyperreactive airways. Myenteric plexus denervation is associated with an increase in muscle thickness both in denervated and distant bowel segments. The portal-drained viscera (PDV), i.e., the intestine, pancreas, spleen, and stomach, have a high rate of both energy expenditure and protein synthesis. Absorbed amino acids are used for the synthesis of mucosal cellular proteins, for oxidative purposes, and for secretion of glycoproteins. Glycoproteins secreted into the intestinal lumen are either digested or reabsorbed, or they are passed to the colon where they are degraded, fermented, and lost from the body. Intestinal mucosa receives nutrients from 2 sources, the diet (brush border membrane) and the systemic circulation (basolateral membrane). 50% of the dietary amino acid intake is used by the PDV during a 24 hr period. In the first hours after the start of feeding, only 30% of the dietary protein intake reach the portal circulation and become available for whole-body growth, and 57% during the last 6 hrs of feeding. About 20% of intestinal amino acids utilisation are used for constitutive gut growth by the intestinal mucosa. Glycoprotein secretion and recycling is of particular importance with regard to the protective function of the gut in general and for the nutrition of threonine in particular. Structurally, the mucosa is protected by a complex net-work of glycoproteins (mucus), of which 2 mucins, MUC-2 and MUC-3, are important components. After 6 hrs of feeding 46% of lysine reach portal vein, which is derived from secreted protein recycling. Lysine is made available to the host through release via microbial protein breakdown and its subsequent intestinal uptake. Given the high rate of intestinal protein metabolism, it is essential that as the enterocytes are initially exposed to the diet, their requirements be met first. Intestinal energy production is largely derived from the oxidation of glutamate, glutamine, aspartate, and glucose. Some essential amino acids (lysine, leucine, and phenylalanine) are known to be used for oxidative purposes by the intestinal tissues as well, which is a nutritional loss. Phenylalanine oxidation is associated with its hydroxylation to tyrosine as a first step. All aspartate, glutamate, and glutamine are extracted by the GIT tissues (90% of dietary intake). Glutamate, together with cysteine and glycine serves as a precursor for the biosynthesis of glutathione and for mucosal nucleic acids. Intestinal tissues are nutritionally significant sites of de novo synthesis of alanine, arginine, glycine, and especially tyrosine. Massive gut resection renders arginine a fully essential amino acid. Tryptophan is a precursor for the biosynthesis of serotonin within the intestinal epithelium because this tissue is highly innervated. One major component of the imbalance of mucosal homeostasis in the pathogenesis of inflammatory bowel disease (IBD) is characterised by changes in cytokine production by macrophages and lymphocytes. IL-1β, TNF-α, IL-6, IL-12, and IFNG play an important role in sustained inflammatory responses. Elevation of expression levels of IL-1β, IL-6, TNF-α, and IFNG in the colonic tissue from patients with IBD has been shown. The expression of these proinflammatory cytokine genes is mainly regulated by the transcription factor NFκB. Inhibition of NFκB activation may be a promising target for the treatment of patients with IBD. Sulfasalazine, mesalamine, and corticosteroids inhibit activation of NFκB. Treatment with corticosteroids causes many undesirable side effects, such as impaired glucose tolerance, adrenal suppression, and increased risk of infection. Curcumin, widely used as a spice and responsible for the yellow colour of curry, is a natural product of plants obtained from curcuma longa linn (turmeric). Curcumin have a variety of pharmacologic effects including antitumour, anti-inflammatory, and anti-infectious activities. Curcumin inhibits NFκB in different type of cells. Curcumin blocks signal upstream of NFκB-inducing kinase and IκB kinase. Curcumin attenuates both wasting disease and colonic inflammation. Curcumin markedly reduces $CD4^+$ T-cell infiltration in the lamina propria. NFκB activation in colonic mucosa is suppressed by treatment with curcumin. Curcumin inhibits the expression of proinflammatory and inflammatory cytokine genes in the colonic mucosa. In humans, the pharmacologic safety of nontoxic consumption of curcumin is up to 100 mg/d. Curcumin is capable of inhibiting degradation of IκB and translocation of NFκB into the nuclei of the colonic epithelial cells and macrophages. By blocking NFκB activation, curcumin may inhibit early steps of inflammation and modulate up-regulation of multiple proinflammatory genes by interrupting the downstream inflammatory cascade. Curcumin is rapidly metabolised to glucuronide and sulfate conjugates that are excreted primarily in bile and to a lesser extent in urine. Luminal curcumin may have topical activity on colonic epithelial cells independent of systemic absorption. The chemopreventive activity of curcumin is related to inhibition of cyclooxygenase 2 (COX-2) expression via modulation of signalling pathways that regulate stability of NFκB sequestering protein IκB. Curcumin may inhibit the development of colorectal cancer in patients with IBD. Curcumin can significantly attenuate the colonic injury and inflammation induced by toxic substances. On average, 6 people in USA die every hour of colorectal cancer. Colonic mucosa experiences significant levels of spontaneous apoptosis of crypt cells. Spontaneous apoptosis may serve to remove genetically aberrant cells arising from the generally high rates of cellular proliferation in the intestinal epithelium. DNA-damaging toxins such as irradiation, chemotherapeutic agents, or toxic chemicals also induce higher rates of apoptosis. Adenomas are the precursors to the majority of large bowel malignancies. Adenomas arise from mucosa with higher rates of proliferation. Mucosal homeostasis requires a balance between proliferation and apoptosis. There is a substantial association between lower apoptotic activity and adenoma elsewhere in the large bowel, even in the absence of association between rectal mucosal proliferation and adenoma. There is a very strong inverse association between adenomas and apoptosis. Patients with adenomas have a generalised decreased level of apoptosis throughout their colons. NSAID may reduce risk through their inhibitory effects on PGs. Human tumours contain or produce large quantities of PGs. Prostaglandins synthesis is associated with tumour promotion and metastatic potential. Conventional NSAIDs and selective COX-2 inhibitors can lead to the regression of colorectal adenomas in patients with familial adenomatous polyposis coli (FAP). The protective effect of NSAIDs is a consequence of their exclusive effect on adenomatous tissue in which COX-2 is upregulated and not on normal mucosa.

CENESTHOPATHY

Somatopause is the decline in growth hormone (GH) level that occurs gradually from young adulthood throughout life, and it occurs in both sexes at roughly the same rate. This decline in GH leads to a decline in IGF-1, the hormone-like substance that is made in the liver in response to GH. The decline in IGF-1 also parallels the decline of all the attributes related to testosterone. Decreased IGF-1 levels leads to a decrease in skin thickness, bone density, aerobic capacity, and the healing rate of wounds. Reduction of GH (and IGF-1) leads to increase in body fat, waistline, waist to hip ratio (an indicator for risk of heart attack), LDL cholesterol, average days of illness, and hospitalisation rate. Age management emphasises preventive medicine first-focusing on the prevention of disease, slowing the aging process, and enhancing health. Hormone modulation is the second component of anti-aging medicine, and it is geared towards enhancing performance, as well as preventing disease. Lifestyle modification, sound nutritional practice, and careful monitoring are essential to achieve the maximum benefit from hormone modulation. Testosterone levels decline gradually in men, starting from age 30, and this decline continues throughout life. In women, levels decline precipitously at menopause, along with oestrogens and progesterone. In both sexes, along with this decline in testosterone, comes a decrease in libido, lean body mass, strength, energy, mood, sexual performance and mental acuity. Hormonal replacement does not change the total body weight. Lean body mass is increased significantly in men on growth hormone plus testosterone. Strength is increased primarily by testosterone. Aerobic capacity substantially is boosted by GH. But, the combination of GH and testosterone is additive. Men and women using GH reduce body fat by 14%. Men on both GH and testosterone would decrease body fat by 17-18%. LDL (bad cholesterol) is reduced by GH. Total cholesterol is reduced by GH, and the ratio of total cholesterol to HDL (coronary risk ratio) also declines, indicating less risk for heart attack. Testosterone in men lowers triglycerides and raises HDL cholesterol, both of which reduce risk for heart attack. These effects of testosterone in relation to cholesterol vary in women. Niacin (Vitamin B_3) raises HDL cholesterol. Continually monitor the testosterone and HDL level is essential. The PSA would be dropped in men on GH plus testosterone. Cognitive performance is difficult to attribute to a particular hormone, since the majority would be making lifestyle modifications and adding a variety of nutriceuticals, some of which are designed to enhance memory and cognitive function. Growth hormone makes sleep better and awaken refreshed with more energy and improved aerobic capacity. Average bone density increases over the course of a year. The skin becomes thicker and smoother with fewer wrinkles. Spider veins also tend to decrease as a virtue of thickening of the skin. The cholesterol profile usually improves as LDL cholesterol generally is reduced with the use of GH. Finally, there is an enhanced feeling of well being, often described as mood elevation. The goal for laboratory measuring is to see the testosterone level rise to the upper normal for men. The levels should be checked at least every 3 months. The dose of GH is based on a patient's age, sex, weight, IGF-1 level, his or her response to therapy, and the affordability. Growth hormone is administered by subcutaneous injection, using a tiny 30-gauge needle (6 mornings a week). Monitoring of IGF-1 level, as with other indicators, is critical. Insulin-like growth factor (IGF)-I, a mitogenic and antiapoptotic peptide, can affect the proliferation of epithelial cells. There is no association between IGF-I concentrations and breast-cancer risk in postmenopausal women. There is a positive relation between circulating IGF-I concentrations and risk of breast cancer among premenopausal women. Plasma IGF-I concentrations may be useful in the identification of women at high risk of breast cancer and in the development of risk reduction strategies. IGF binding protein 3 (IGFBP-3) may contribute to the senescent growth arrest of these cells. The decline in levels of IGFBP-3 with age is not as great as that of IGF-I, resulting in a relative excess of IGFBP-3. GH release is decreased with obesity, especially with intra-abdominal adiposity. Adiposity is associated with a reduction in the frequency of GH secretory bursts and a significant shortening of the circulatory half-life of GH, both of which reduce integrated daily plasma GH concentrations. Because both ageing and obesity are associated with elevation of plasma insulin levels, one possible mechanism for the above effects is an increase in insulin action on hypothalamic and pituitary IGF-I receptors, resulting in enhanced feedback inhibition of GH, and hence IGF-I, secretion. Sustained programmes of moderate intensity resistance exercise and high intensity endurance training in elderly persons do not increase IGF-I levels, despite substantial improvements in body fat and lean masses and in aerobic capacity. Oral clinical oestrogen doses, exerts a direct inhibitory effect on hepatic IGF-I synthesis, resulting in reduced circulating IGF-I levels and disinhibition of GH secretion. Oral tamoxifen also decreases baseline IGF-I Levels. Ageing and pathological GH deficiency are associated with reduced muscle and bone mass, increased total and intra-abdominal fat, dyslipidaemia, increased frequency of coronary heart disease and hypertension, and increased cardiovascular mortality. The changes in body composition and endocrine-metabolic function are partly reversible after GH administration to elderly or GH-deficient patients. Ageing is associated with increased insulin resistance and a rise in fasting glucose levels, whereas GH deficiency is typically accompanied by enhanced insulin sensitivity and episodes of fasting hypoglycaemia. By comparison with GH-deficient patients, the elderly require lower doses of GH replacement and are more prone to hyperglycaemia and other side effects. The initial insulin-like effects of GH treatment, such as a fall in plasma FFAs, which have been described in GH-deficient patients, have not been reported in the elderly. Waist-to-hip ration rises with age. Over the adult lifespan total body fat increases by 18% in men and 12% in women, in part due to decreased physical activity. This accumulation is greater in the abdominal-visceral than the subcutaneous fat compartment. The reciprocal relationship between adiposity and GH secretion enhances body fat deposition, with adverse metabolic sequelae. A low plasma IGF-I level correlates with indices of malnutrition such as reduced serum albumin, transferring, cholesterol, triceps skin-fold thickness, body weight and body mass index, but not with age, energy status or functional capacity. Low IGF-I values are strongly predictive of future life-threatening complications such as severe infections, which may reflect nutritional intake and hepatic function rather than GH secretion. Rates of skeletal depletion vary widely among individuals so that some people do not become significantly osteopenic with advancing age. Bone mineral loss is a predictable accompaniment of ageing in both women and men. Peak skeletal mass in women is achieved at about age 35, after which bone mass falls at a rate of 0.5-1%/year until the menopause, subsequent to which the rate of loss accelerates to 2-3%/year for the next 5-10 years and continues at a slower rate thereafter. Peak bone mass in men is about 25% higher than in women and occurs at about age 35-40, after which there is a continued steady age-related loss of skeletal mass of about 0.3% per year. The development of clinically evident osteoporosis depends upon multiple variables, including peak adult bone mass, level of physical activity, genetic and dietary factors, and endocrine, autocrine and paracrine influences on bone. Serum level of IGFBP-3 is positively correlated with bone mineral density and negatively correlated with age in osteoporotic patients. Sleep apnoea may increase with advancing age. Patients with obstructive sleep apnoea have diminished slow wave sleep and reduced nocturnal GH release and treatment with continuous positive airway pressure increases the quantity of slow wave sleep and rapid eye movement (REM) sleep, with improvement in GH secretion. There is a physiological, age-related decline in spontaneous GH release and IGF-I levels, which begins in the third decade and continues into advanced old age, so that GH and IGF-I levels in the elderly approach those of younger adults with pathological GH deficiency. Acromegaly is usually caused by a GH-secreting pituitary adenoma. Somatic growth and metabolic dysfunction occur subsequent to unrestrained GH secretion and elevated insulin-like growth factor (IGF)-I and IGF-binding protein (IGFBP)-3 levels. Classic clinical features of acromegaly include acral overgrowth, sweating, headaches, menstrual disturbances, and glucose intolerance. Well-documented clinical risks of long-term tissue exposure to uncontrolled GH hypersecretion include cardiac disease and hypertension, diabetes, respiratory disorders, joint disease, and neuropathy. 60% of acromegaly patients succumb to cardiovascular disease; 25% of patients, the cause of death is attributed to malignancy. Suppression of GH to less than 1 ng/ml, during an oral glucose tolerance test, and normalisation of IGF-I levels portend a favourable mortality outcome. Excess GH, by inducing both $IGFBP_3$ and IGF-I levels, promotes dysregulated cell growth balance characterised by dynamic signals for cell apoptosis vs. cell growth advantage. Peripheral tissue somatic growth and metabolic dysfunction are caused by direct effects of GH on peripheral receptors, impact of hepatic-derived circulating and paracrine IGF-I, and also the impact of elevated circulating $IGFPB_3$ levels. Elevated IGF-I bioactivity and activation of the IGF-I receptor are associated with cell proliferation and growth advantage, whereas $IGFBP_3$ bioactivity promotes an apoptotic advantage. $IGFBP_3$ inhibits IGF-I-induced prostate cancer cell growth, and breast cancer cells are diverted into an apoptotic phase by $IGFPB_3$. In acromegaly, the activated IGF-I receptor accounts for increased kidney, heart, or acral bony tissue functional cell mass. Elevated $IGFBP_3$ accounts for an enhanced cell removal process and apoptosis. Pathologically elevated GH results in peripheral tissue exposure to both excessive growth-promoting and growth-arresting influences. GH and IGF-I transform lymphocytes, and also induce cell proliferation. IGF-I receptor mass is increased in neoplastic tissues, and the activated IGF-IR also mediates cell transformation. Several growth factors and inactivated tumour-suppressing genes also stimulate IGF-IR synthesis. There may be a permissive, rather than an initiating, role for IGF-I in tumourigenesis. Hypophysectomy has protective or palliative role for patients with neoplasia. Somatostatin administration lowers IGF-I levels. There is evidence indicate a role for the GH-IGF-I axis in mediating both physiologic and pathologic cell growth and tissue hypertrophy. Cancer in a patient with acromegaly and uncontrolled GH levels may likely be more aggressive, with potentially increased cancer-associated morbidity and mortality. Uncontrolled acromegaly may provide a growth advantage to concurrently occurring neoplasms in these patients. 15% of deaths in acromegaly are attributable to malignancies, which is lower than would be expected from the general population. There is no clear evidence for enhanced de novo cancer initiation in acromegaly and, as yet, no direct proven causal relationship of acromegaly with malignant disease. There is a potential role of GH in the progression of congestive heart failure. Endothelial dysfunction is a prominent feature of CHF. GH plays a role in vascular reactivity. A 3 month treatment with GH corrects endothelial dysfunction and improves non-endothelium-dependent vasodilation in patients with CHF. Growth hormone (GH) administration to patients with chronic heart failure (CHF) corrects their vascular dysfunction. IGF-I is believed to have both neurotrophic and neuroprotective effects. The wide distribution of binding sites for IGF-I in the brain, with a high density in the hippocampus, may reflect the psychological effects of IGF-I on the CNS as well as the vulnerability of specific brain areas to GH and IGF-I deficiency. There is a cognitive deficit in GH-deficient individuals. GH and IGF-I can cross the blood-brain barrier, but local synthesis of IGF-I and, to a lesser degree, GH also occurs in most areas of the brain. GH receptors are found throughout the brain, but are mainly concentrated on the choroid plexus, pituitary, hippocampus, putamen, and hypothalamus. In the choroid plexus, GHRs may serve as a system to transport GH across the blood-brain barrier. The high number of GHRs in the hippocampus (2-4 fold higher than in most other areas of the brain) suggests that GH play a relatively important role in the hippocampal functions. The number of GHRs declines with age throughout the brain. The IGF-I receptors are relatively concentrated in the hippocampus and parahippocampal areas, but they are also present in the amygdala, cerebellum, and cortex in humans. In humans, no alterations in the density of cortical IGF-IRs could be shown in connection with aging or Alzheimer's disease. IGF-I can stimulate the viability and function of different neuronal cell types such as cortical, cerebellar and hypothalamic neurons, and contributes to neuronal DNA synthesis as well as myelination. IGF-I can protect hippocampal neurons against β-amyloid-induced toxicity, while increased depositions of this peptide in the CNS have been associated with the development of Alzheimer's disease. We must focus on the prevention of disease, slowing the aging process, and enhancing health. Hormone modulation should aim towards enhancing performance, as well as preventing disease. Lifestyle modification, sound nutritional practice, and careful monitoring are essential to achieve the maximum benefit from hormone modulation. Testosterone levels decline gradually in men, starting from age 30, and continue throughout life. In women, levels decline precipitously at menopause, along with oestrogens and progesterone. In both sexes, the result is a decrease in libido, lean body mass, strength, energy, mood, sexual performance and mental acuity. Somatopause is the decline in growth hormone level that occurs gradually from young adulthood throughout life, in both sexes at roughly the same rate. This decline in growth hormone leads to a decline in IGF-1, the hormone that is made in the liver in response to growth hormone. Lowered IGF-1 levels induces decrease in skin thickness, bone density, aerobic capacity, and the healing rate of wounds. On the other hand body fat, waistline, waist to hip ratio (an indicator for risk of heart attack), LDL cholesterol, average days of illness, and hospitalisation rate increased. Lean body mass increases more in men on GH plus testosterone, than on men who are on either of those hormones alone. Strength is increased primarily by testosterone. GH has little or no effect on strength by itself. Aerobic capacity is boosted by growth hormone, testosterone improves aerobic capacity so slightly, and the combination of both hormones is additive. Men on testosterone reduce body fat by 3-5%. Men and women on GH the body fat is reduced by 14%. Both hormones are additive, men decrease body fat by 17-18%. Testosterone lowers triglycerides and raises HDL cholesterol, both of which reduce risk for heart attack. Testosterone may lower HDL in women, potentially increasing risk for heart attack. GH and testosterone decrease diastolic blood pressure in men. GH plus testosterone do not induce prostate complications and do not increase prostate symptoms, but lower the PSA level. The characteristic syndrome adult GH deficiency (GHD) consisting of decreased mood and well-being, with alteration in body composition and substrate metabolism. GH replacement results in anabolic effects on bone metabolism. GH has a biphasic effect: 1- an initial predominance of bone resorption, and 2- stimulation of bone formation leads to a net gain in bone mass after 12-24 months of treatment. Adults with hypopituitarism have reduced life expectancy, with a greater than two-fold increase in mortality from cardiovascular disease. Long-standing GHD predisposes to the development of premature atherosclerosis. Cardiac function is impaired in GHD and GH replacement reverses these deficits, and these benefits are sustained 3 years after commencement of GH therapy. Exercise performance capacity improves after 6 months of GH replacement. GH replacement results in significant improvements in QoL and psychological well being in patients with GHD. Those with the greatest psychological morbidity benefit most from GH treatment. Most adverse effects associated with GH replacement in GHD are related to the anti-natriuretic effect of GH.

Clinical features of GHD:

Symptoms	Signs
1- Abnormal body composition. 2- Reduced lean body mass. 3- Increased abdominal adiposity. 4- Reduced strength and exercise capacity. 5- Impaired psychological well being. 6- Depressed mood. 7- Reduced vitality and energy. 8- Emotional lability. 9- Impaired self-control. 10- Anxiety. 11- Increased social isolation.	1- Overweight, with predominantly central (abdominal) adiposity. 2- Thin, dry skin; cool peripheries; poor venous access. 3- Reduced muscle strength. 4- Reduction in exercise performance. 5- Depression, labile emotions.

Inflammation plays an important role in atherosclerosis, and inflammatory markers are predictive of cardiovascular events. Long-term growth hormone replacement in men reduces levels of inflammatory cardiovascular risk markers, decrease central fat, and increases lipoprotein(α) and glucose levels without affecting lipid levels. Low free IGF-I and high IGFBP-1 levels are associated with a decreased self-reported quality of health, but are not related to physical disability in the elderly. Circulating total IGF-I levels may reflect an underlying biological process that influences cognitive decline. GH administration increases energy expenditure, independent of changes in lean body mass, in healthy, obese, and GH-deficient subjects. Some of the clinical features of fibromyalgia (FM) resemble the ones described in the adult GH-deficiency syndrome. Fibromyalgia (FM) is a painful syndrome of nonarticular origin, characterised by fatigue and widespread musculoskeletal pain, tiredness, and sleep disturbances, without any other objective findings on examination. GH has more than one target organ but its primary target is the liver, where it causes the formation and release of insulin-like growth factor (IGF [somatomedin C]). IGF is a by-product of growth hormone, and may be responsible for most of the anabolic effects of the hormone itself. IGF levels are fairly constant in the blood and can be measured more easily than GH. Measuring serum IGF assesses the amount of circulating GH in the body. GH and IGF start to decrease sometime after age 15-20 and continue to do so quite rapidly. GH deficient patients have almost 50% higher rate of death from heart disease than expected. Taking GH raises IGF-1 levels in the blood. Low thyroid levels cause decreased body temperature, increased cholesterol, and increased body fat. Low thyroid levels can contribute to a subjective feeling of sluggishness and low energy as well as depression. Supplementation of thyroid hormone is easy and inexpensive. The goal is to restore T_3 and T_4 to their natural ratio and blood concentrations. My mission is to keep patients mentally sharp, physically fit, and disease resistant. Insulin and cortisol are the two hormones that promote aging and promote degenerative disease. Side effects of growth hormone are minor, and are related to fluid retention. They are resolved by decreasing the dose. Men on both GH and testosterone experience, on average, a decrease in PSA. GH increases lean body mass and reduces LDL, improves psychological well being and lowers the risk of atherosclerosis with reduce risk for stroke in patients treated with GH for ten years. Human growth hormone is taken to raise the level of IGF-1. Side effects in women include weight gain (27%), which is possibly due to the increase in lean mass. Men would notice 80%

improved libido, 60% better sexual performance, 70% improved sense of well-being, 65% better productivity, 73% greater strength and increased lean muscle mass, 63% reduced body fat, 50% better sleep pattern, 45% improved memory, 55% better skin tone, 45% improved facial appearance, 53% decreased frequency of illness, and 35% relief from chronic pain. Women would notice 91% improved sense of well-being, 77% sleeping better and have more energy, 73% greater strength, 68% more productivity and better athletic performance, 68% reduced body fat, 59% better memory, 59% decreased frequency of illness, 62% better libido, 55% better sexual performance, 62% improved skin tone, 55% better facial appearance, 45% increased lean muscle mass. A common side effect of testosterone is mild decrease in testicle size (25%). Minor joint discomfort may sometimes occur in 20% of patients with growth hormone, and is easily remedied by stopping for about a week, and resuming at a lower dose. GH benefits include: 1- increase in libido, 2- decrease in body fat, 3- increase in lean muscle, 4- increase in bone density, 5- increase in skin thickness, 6- decrease in skin wrinkling, 7- improved cholesterol profile, 8- faster wound healing with lower infection rate, 9- decrease in hospitalisation rate by 50%, 10- decrease in sick days from work, 11- increase in exercise capacity, 12- decrease in diastolic blood pressure, 13- decrease in waist/hip ratio, 14- increase in renal blood blow, 15- increase in feeling of well being/improved socialisation, 16- strengthen immune system, 17- may improve memory, 18- may improve cognitive function, 19- may help hair regrowth, and 20- may reduce spider veins. Reduce body fat and build lean muscle without exercise: 1- boost energy levels, 2- increases sexual function in both women and men, 3- acts as a natural sleep aid, 4- stimulates hair growth and restores colour, 5- boosts the immune system, 6- restores skin strength and elasticity, 7- smoothes out wrinkles and cellulite, 8- strengthens nails, 9- improve vision and memory, 10- burns fat, 11- increases muscle mass, 12- minimises risk of heart disease, 13- lowers blood pressure, 14- improves blood cholesterol profiles, 15- promotes oxygen intake, 16- aids in osteoporosis prevention, and 17- turn back the body's biological time clock 10-20 years.

The effects of GH replacement in the adult with GHD:

A- Body composition
- 1- Increased lean body mass.
- 2- Decreased fat mass.
- 3- Increased bone mineral density (BMD).
- 4- Increased plasma volume and total body water.

B- Cardiovascular system
- 1- Increased ventricular mass.
- 2- Increased cardiac output.
- 3- Decrease peripheral vascular resistance.
- 4- Increased preload.
- 5- Increased red blood cell mass.

C- Muscle strength and exercise performance
- 1- Increased muscle mass and strength.
- 2- Increased exercise capacity.

D- Metabolism
- 1- Increased resting energy expenditure.
- 2- Increased protein synthesis.
- 3- Transient increases in insulin resistance.
- 4- Decreased total and low-density lipoprotein (LDL) cholesterol.

E- Psychological well-being and quality of life:
- 1- Improved mood and energy levels.
- 2- Improved self-perceived well being.

Growth Hormone (GH) is a polypeptide hormone composed of a long chain of 191 amino acids. Physiologically, GH is secreted by the anterior pituitary gland. The hypothalamus initiates GH secretion by secreting growth hormone releasing hormone (GHRH); at the same time it stops secreting a growth hormone inhibitory hormone called somatostatin. The bursts of GH release occur primarily during deep stages of sleep, such as stage 3 and stage 4. Once released in the blood, GH is very short lived. It is generally completely metabolised and gone within half-hour. During that time, however, it manages to reach the liver and many other cells in the body, and induce them to make another polypeptide hormone IGF-1. IGF-1 travels around to the various tissues of the body to effect most of the benefits that attributed to GH. The secretion of GH is regulated by biofeedback loop. IGF-1 also feeds back on the pituitary and hypothalamus to help control GH secretion. The growth hormone can be used for the treatment of AIDS wasting syndrome. This is the condition of weakness, fatigue, and loss of muscle mass in AIDS patients. The growth hormone is used for short bowel syndrome. Menopause is the condition in women whereby the ovaries atrophy and cease to produce the sex hormones oestrogen, progesterone and testosterone. Somatopause signifies the gradual decline in growth hormone production by the adult pituitary gland in both men and women, which begins at the age 30 and continues at a steady rate throughout life. Somatopause is accompanied by deterioration in the structure and functional capacity of the body, which is ultimately devastating to the human condition. The reason for the morning injection has to do with the biofeedback mechanism for GH. Most of the natural pituitary GH secretion occurs at night during deep stages of sleep. Injecting GH at night raises the serum level of GH precisely during the time the pituitary is scheduled to become active. This high serum level of GH from the injection can suppress the natural pituitary function by negative feedback. For the most part, the pituitary has completed its function and is at rest by 5 a.m. Therefore injecting after awakening in the morning results in injecting "on top of the peak" of endogenous GH, so as not to suppress the pituitary. By the time the pituitary is ready again for its night-time activity, the GH given in the morning injection would have been completely metabolised. This eliminates the risk of pituitary suppression. Side effects of growth hormone are generally mild and are largely associated with salt and water retention associated with a vague feeling of puffiness. Carpal tunnel syndrome is a fluid retention, which accumulate in the closed carpal tunnel compartment of the wrist, compressing the median nerve. This results in numbness and tingling in the palm and fingers. These side effects are easily remedied by abstaining from GH for about a week, and then resuming the treatment with a 20% dose reduction. Older patients are more subject to side effects and are generally started at a low dose of GH than younger adults. Another potential side effect of GH is the elevation of blood sugar. Growth hormone mobilises body fat, causing the fat cells to breakdown and release free fatty acids into the bloodstream. These FFAs are energy molecules, which can be taken up by organs to be used for energy. When the muscles are consuming FFAs as a fuel, they are far less interested in sugar, therefore they tend to resist the effects of insulin, and extract less sugar from the blood. At the same time, growth hormone can increase glucose output from the liver to the blood. This combination of effects can raise blood sugar and raise insulin levels, neither of which is good. This is only a problem in people who eat a diet high in sugar and starch, and do little exercise. Any growth hormone program must include proper nutrition and exercise with emphasis on a low glycaemic diet. Testosterone supplementation (T) in men improves strength and increases de novo protein synthesis as well as muscle mass. T has also been shown to decrease body fat and particularly visceral body fat, increase libido in normal men and increase libido and sexual performance in hypogonadal men. Mood is also improved with T in both hypogonadal and older community dwelling men. Higher endogenous testosterone has been correlated in many studies with a reduction in a number of cardiovascular risk factors, among them lower-blood pressure, total cholesterol (TC), LDL-cholesterol (LDL), triglycerides (TG), visceral body fat, waist-hip ratio (WHR), serum insulin, fasting and post-prandial glucose, higher HDL-cholesterol (HDL) and greater insulin sensitivity. Levels of dehydroepiandrosterone sulfate (DHEAS) correlate inversely with depression and supplementation with DHEAS results in improvement. DHEAS also directly stimulates immune responsiveness in humans. In men, DHEAS may improve body composition. DHEAS levels have been shown to correlate inversely with hippocampal atrophy in the elderly. DHEAS may improve memory in depressed patients of middle and older age. The growth hormone deficiency syndrome (GHDS) is a condition associated with weight gain, abnormal body composition (increased fat mass and decreased lean body mass), decreased bone mass, an atherogenic lipid profile, and increased cardiovascular risk. Patient's with the GHDS generally have increased fat mass and decreased lean body mass compared with sex-, age-, height-, and weight-matched controls. Obesity and, especially, increased central adiposity are common in persons with this condition, as are insulin resistance, dyslipidaemia, and other metabolic, inflammatory, and vascular factors associated with accelerated atherogenesis. Some of these patients may have hypothalamic abnormalities, resulting in increased appetite and body fat. Patients treated with excess thyroid hormone are at increased risk for atrial fibrillation and osteopenia. Undertreated patients would be more likely to be obese and have an increased level of low-density lipoprotein cholesterol. Androgen levels are decreased in women with hypopituitarism. DHEA benefits body composition and bone mineral density in normal elderly patients with a low baseline level of DHEA sulfate. Statins can reduce markers of inflammation. Treatment with growth hormone has proven benefits for body composition, surrogate markers for CVD, and bone health. GH provides better living with prolonged survival. A diet of low glycaemic index (LGID) lowers LDL and TG, raise HDL, improve both glycaemic control and insulin sensitivity in non-insulin dependent diabetes, and reduce coronary artery disease (CAD) risk in women. Physical exercise is capable of increasing muscle mass decreasing body fat and lowering the waist-hip ratio (WHR). Patients on a daily restriction of 25 kcal/kg and GH would lose more fat and become more insulin sensitive than those on the same diet without GH. IGF-1 level should not exceeds 360 ng/ml (upper normal for the 39 to 54 year old adult), with the absolute upper limit of IGF-1 at 360 ng/ml. Arthralgia is the second most common at 20% with swelling of hands or feet at 10%. These are well known side effects of GH related to its antinatruretic effect and are dose dependent; they are easily reversed by abstaining from GH for 1 week and resuming at a dose reduced by 15%. Weight gain occurs in 15% of patients. This fits with the antinatruretic effect of GH but is likely also related to the anabolic effect of T. Nipple discomfort is likely related to the rise in estradiol (E2) that often accompanied the rise in T level. It was easily controlled by bringing E2 back down to the upper normal range with a very low dose aromatase inhibitor. There is a marked individual variability in biochemical and body composition responses to GH replacement. Men have a greater body composition response to GH therapy than women, and younger patients with low levels of GH-binding protein have the greatest responses. Human growth hormone (HGH) is a large fragile protein molecule with a molecular weight of 20,000. Molecules as large as HGH cannot be absorbed into the body across skin or mucous membranes. Even insulin, a molecule only half the size of

HGH, and of similar type and construction, cannot be absorbed in this manner. Given this limitation, HGH is digested, or broken down into simpler compounds if it is not injected. HGH is a very fragile molecule; it is dependent upon the retention of a precise complete amino acid sequence and three-dimensional structure, with some parts of the molecule necessarily linked to others. It works on cell receptors, so, even if the chemical formula remains the same, any change in shape blocks HGH activity. The second way to attempt to augment the amount of a hormone in circulation is to try to stimulate the body in such a way as to increase the amount of hormone that is produced by the body itself, or to enhance the function of hormone that has already been produced. A secretagogue is a substance that stimulates the pituitary gland to produce HGH. Some compounds (e.g., amino acid molecules) have been shown to stimulate the production of significant amounts of HGH. The pituitary gland may be stimulated in a too broad manner, and levels of other hormones that are not desirable may also raise, especially cortisol. Cortisol antagonises the beneficial effects of HGH action. Secretagogues in general do not demonstrate a certain HGH response in every individual. They seem to have their greatest effectiveness in young (under age 35) and highly athletic people, and require high dose to generate a meaningful response (35%) with 10-30% increases in IGF-1 level. With advancing age there is a progressively smaller response rate and magnitude of response. Even patented drugs must disclose what they contain so appropriate decisions can be made regarding safety. Rule number one with secretagogues should be that any marketing company must tell the consumer exactly what he or she is ingesting. Most secretagogue products on the market would be HGH precursors. Response rates can be as low as 10%, with only a small actual rise in HGH levels. Chronic excess of GH and IGF-1 cause prostate overgrowth and further phenomena of rearrangement, but not prostate cancer. Benign prostate hyperplasia may occur in 58% of the acromegalics and 26.6% of the controls. The prevalence of structural abnormalities, including calcifications, nodules, cysts, and vesicle inflammation. Aging is associated with reduced GH, IGF-I, and sex steroid axis activity and with increased abdominal fat. The most central findings in both GH deficiency in adults and the metabolic syndrome are abdominal/visceral obesity and insulin resistance. Striking similarities exists between the metabolic syndrome (syndrome X or primary insulin resistance syndrome) and untreated GH deficiency in adults. The metabolic syndrome is associated with multiple endocrine abnormalities. They include increased cortisol secretion, blunted secretion of gonadotrophins and sex steroids, and abnormalities in the GH/IGF-I axis. With increased adiposity, GH secretion is blunted with a decrease in the mass of GH secreted per burst but without any major impact on GH secretory burst frequency. The serum levels of IGF-I are inversely related to the percentage of body fat (%BF). Insulin sensitivity does not decrease on testosterone replacement therapy of male subjects with idiopathic hypogonadotrophic hypogonadism. Testosterone replacement is associated with decrease in other cardiovascular risk factors. Long term testosterone replacement to date appears to be safe and effective means of treating hypogonadal elderly males, provided that frequent follow-up blood tests and examinations are performed. Drugs that deplete melatonin include β-blockers, NSAIDs, steroids, nicotine, alcohol, caffeine, sleep aids and anti-anxiety medications. Fluoxetine (Prozac) may lower melatonin levels. Low melatonin may contribute to insomnia, sleep/wake disorders, or PMS. Some forms of depression are associated with low melatonin levels. Low levels have also been implicated in increased risk for coronary heart disease. A disturbance in the circadian rhythm of melatonin may influence other hormones such as thyroid, testosterone, and oestrogen. Melatonin influences other vital functions including cardiovascular and antioxidant protection, endocrine function, immune regulation and body temperature. A growing body of research links the balance of androgens, progesterone, and sex-hormone-binding-globulin (SHBG) and their metabolites with primary mechanisms of bone turnover, lipid metabolism, cardiac function, cognitive and emotional health, immune function, as well as hormone-dependent diseases, such as breast and endometrial cancers and lupus. Estradiol is the major oestrogen secreted by the active ovaries. It may also arise from adrenal and peripheral adipose sources via enzymatic action (aromatase) on the androgens androstenedione and testosterone. Oestrogens stimulate growth and development of tissues related to female reproduction such as the breasts, vagina, and uterus. Some vasodilatory and bone/cartilage stimulating effects are evident too. Human growth hormone (GH) seems to have hyperadaptive effects (i.e., an increase in small bowel length and function per unit length). GH combined with a glutamine- and fiber-modified diet results in a 30% improvement in carbohydrate and protein absorption in SBS. The result is increased in body weight, a significant improvement in electrolyte absorption, and a decreased gastric emptying rate. Treatment with low-dose GH increases net intestinal macronutrient absorption and lean body mass without major side effects in home parenteral nutrition (HPN)-dependent patients with short bowel syndrome (SBS) having permanent intestinal failure and having reached their maximal physiologic level of intestinal adaptation. There is a synergistic effect with glutamine-supplemented parenteral nutrition and GH. GH increases plasma glutamine. A hyperphagic diet is rich in complex carbohydrates and contains sufficient glutamine (i.e., 6-8% of dietary protein). A hyperphagic diet may be a condition for optimal intestinal adaptation and a prerequistic for pharmacologic adaptation under treatment with GH in SBS in the absence of malnutrition. Human malnutrition induces resistance to treatment with GH that is biologically expressed by a decrease in basal and GH-stimulated IGF-1. Normal levels of IGF-1 reflect an absence of severe malnutrition. Increased serum IGF-1 level is associated with both significantly increased IGFBP-3 and decreased GHBP levels. Hyperphagic diet and the absence of malnutrition are needed for an optimal hyperadaptive intestinal response under an appropriate dosage of GH (i.e., without any clinically significant adverse effects). Citrulline is a reliable biological marker of the enterocyte mass in patients with SBS. Vitamin C is known to help regulate the activity of brain chemicals (catecholamines), improve circulation, and reduce stress hormone levels and anxiety. Vitamin C is a strong promoter of sexual activity. Vitamin C increases frequency in penile-vaginal intercourse but not any other types of sexual activity.

TESTOSTERONE

The chemical structure of testosterone is $CH_{19}H_{28}O_2$, and the molecular weight is 288.42 g/Mol. The incidence of CaP rises at an age while testosterone levels fall. Low serum testosterone levels are associated with a high incidence in prostate cancer (CaP). High levels stimulate growth of the prostate with possible risk of malignancy. Increased transcriptional activity caused by a genetic defect of the androgen receptor (AR) leads to an increased androgen activity at the cellular level. Low testosterone levels induce the formation of a proto-oncogene (c-Met) in the membrane of prostate cells. In human, a higher expression of the c-Met proto-oncogene on the prostate cancer cells is associated with a higher degree of invasiveness. There is a trend toward androgen deficiency in aging males, which is presented by age-related disturbances in memory, muscle mass, and strength. Also loss of libido and impotence, and osteopenia are testosterone loss effects. The consequences of testosterone deficiency are severe and in some cases dangerous and debilitating. Disturbances in balance and declines in maximal oxygen uptake capacity (VO_2max) may also be related to declines in testosterone levels. Changes in food intake may also be effects of testosterone loss. Testosterone replacement has a role to play in improving the quality of life in older men. Testosterone levels correlate more strongly with libido effects than with erections. Men treated with sildenafil (Viagra) do not obtain an adequate erection if their testosterone level is low, but will respond to testosterone treatment. Testosterone seems to be required for the last stage of the erection, possibly because of the hormone's effect on nitric oxide synthase. There are strong relationships between bioavailable testosterone and performance on a number of different memory tests. Both bioavailable and total testosterone levels correlated extremely well with functional status. Testosterone deficiency causes a decline in muscle mass, as well as sarcopenia, and frailty, with numerous interactions between these effects. Mortality from hip fractures is higher in men with low testosterone. Deficiency is associated with minimal trauma hip fracture. There is a decline in haemoglobin of 1-2 g/dl in ageing men, which is most likely related to the decline in testosterone. Testosterone supplementation increases haematocrite approximately 1%. There is no clinical evidence that the risk of either prostate cancer or benign prostatic enlargement (BPE) increases with testosterone replacement. Both BPE and prostate cancer may decline, but not to a statistically significant degree. The lower the level of free testosterone in an individual, the more likely the chance to have coronary artery disease. Testosterone improves exercise-induced ST depression. Testosterone relaxes the coronary arteries by liberating nitric oxide (NO), an effect very similar to that of oestrogen. Testosterone relieves angina in 77% of subjects, and leads to positive lipid studies (variable), suggesting improvement in myocardial ischaemia. Symptoms suggesting testosterone deficiency include: 1- decrease in libido or sex drive, 2- lack of energy, 3- decrease in strength or endurance, 4- weight loss, 5- decreased enjoyment of life, 5- sad or grumpy, 6- erections is less strong, 7- a recent deterioration in the ability to play sports, 8- falling asleep after dinner, and 9- recent deterioration in work performance. In 1935, testosterone purification became possible. In 1940 and 1944, the first American published reports of testosterone use for "the male climacteric" were published. From then until the 1970's, testosterone was given according to subjective criteria and with generalised dosing regimens. Once testosterone levels could be measured, it became possible to more accurately assess individual levels as an aid to implementing and precisely monitoring therapy. After age 30, testosterone levels may decline an average of 2% a year. This decline in testosterone production is consistent with a decline in Leydig cell (testosterone producing) number in the testes and decreasing activity of the enzymes that produce testosterone. There is also a diminished response to pituitary signals that normally initiate testosterone production and diminished coordination of the release of the pituitary signals that are produced, decreasing any chance for the testes to continue a normal pattern of testosterone secretion. Sex hormone-binding globulin (SHBG) levels increase with age. Increasing SHBG levels reduce free testosterone to a greater extent than the reduction seen in total testosterone. Thus, less total testosterone production in conjunction with increasing binding protein levels act in tandem to synergistically depress free/functional testosterone levels. The age-related decline in testosterone levels is associated with: 1- A decline in muscle mass and strength. Diminishing testosterone levels directly correlate with a decrease in the synthesis rate of muscle proteins, formation of contractile structures, and the force generating capabilities of muscle cells. Declines in muscle mass are also correlated with increased risk for falls and fractures. 2- Increase in body fat mass (abdominal fat and pectoral fat). Sometimes, gynaecomastia may occur. Decreases in testosterone are also associated with increasing levels of leptin. Leptin is a peptide hormone produced by fat cells, and its circulating levels are directly reflective of an individual's fat mass. 3- Decrease of bone mass. Up to 30% of men aged 60 and over may become osteoporotic. 1 in 6 will fracture a hip at some point in her/his life.

Between age 60 and 80, an unsupplemented woman show a 50% reduction in her original bone mineral density, and 1 in 4 will suffer a vertebral or hip fracture. 4- Decline in sex drive and libido, which precedes declines in actual performance. 5- Increased frequency of erectile dysfunction in men, and diminished sexual response and pleasure in women. 6- Decreased sense of overall wellbeing, perception of energy level, and vigour. 7- Decline in stamina and exertional performance. Performance-minded individuals, such as business executives and people whose careers demand multi-tasking or complex problem solving skills would suffer significantly. 8- Decline in cognitive skills, concentration and memory. 9- Coronary artery disease and cholesterol derangement. The goal of testosterone replacement therapy is to minimise, prevent, or reverse the affects of our age related decline. For males, testosterone level of 260-1,000 ng/dl is given as the normal laboratory range from men aged 20-70. For females, this range is 15-70 ng/dl. Free testosterone levels average approximately 2% of the total, 50-210 pg/ml for men and 1-10pg/ml for women. Free testosterone is the slightly more valuable of the two, as it reflects the amount of testosterone available to perform useful work at any one moment. A more accurate approach would be to use the normal range seen from age 30-35 (approximately 700-900 ng/dl for men, and 50-70 ng/dl for women), and try to maintain these levels over time rather than let them continue to decline. Prostate specific antigen (PSA) measurement must accompany testosterone levels at the time of an initial evaluation in order to screen for any pre-existing prostate disease, to direct any prerequisite work-up of elevated level that may be associated with prostate disease, and to be used as a baseline for future program follow-up. Other studies, such as thyroid hormones, growth hormone (HGH), luteinizing hormone (LH), dehydroepiandrosterone (DHEA), blood count, lipid profiles, and other laboratory and metabolic markers-such as body composition and bone density-all play roles in maximising a testosterone replacement therapy. Once therapy is initiated, follow up levels for testosterone and some of these other markers must be monitored over time in order to assure adequate safety and maximise utility. There are several different modes of testosterone delivery, but the best method varies from individual to individual, and is dependent upon several factors. Optimally, a testosterone delivery method should be clinically effective in correcting the signs and symptoms of testosterone decline and produce predictable and reproducible physiologic levels of testosterone and estradiol. Testosterone can be converted to estradiol by an aromatase enzyme. Testosterone is available directly in oral, injectable, topical, and implantable formulations; and may also be supplemented indirectly by the administration of human chorionic gonadotropin, which stimulates testosterone production by the testes. There are no recommended oral testosterone formulations. Oral testosterone and androgens such as Fluoxymesterone, methyltestosterone, oxandrolone, or danazol are available for clinical use, but are not appropriate for long-term testosterone replacement therapy. Their use is specific for certain disorders and must be used with great caution as they can cause an increase in liver enzymes, blockage of liver drainage pathways, direct liver damage, and even liver tumours. They also dramatically raise serum LDL cholesterol, decrease HDL cholesterol, and have been associated with increased risk of myocardial infarction and stroke. Testosterone undecanoate is a testosterone compound given in an oral base that is taken up by the lymph ducts in the intestines and is able to bypass the liver, thus minimising the typical side effects. Men do not convert DHEA into meaningful levels of testosterone, but women may. Thus, improving testosterone levels in women is to optimise DHEA levels and re-check testosterone after 5-7 weeks then use that value as the definitive criteria for instituting testosterone therapy. The injectable form of testosterone is not associated with the above-mentioned undesirable effects of oral androgen administration and is available in a formulation-testosterone cypionate-that allows a relatively long biological effect time and typically requires a dosage interval of only once each week. This method requires an intramuscular self-injection and provides testosterone, for a relatively long period of time and is low in cost. Delivering testosterone this way has a 100% success rate in providing usable hormone. Testosterone pellets require a surgical procedure for implantation and removal, and once they are placed, do not allow a means for tailoring dosage based on an individual's response. Topical testosterone placement formulations allow testosterone absorption through the skin. This method raises circulating levels of DHT and increasing the exposure of prostate and hair follicle cells to DHT rather than testosterone, which is not as active in these cells and not as well taken up. Testosterone patches have also been associated with other minor disadvantages, include: low obtainable maximum serum testosterone levels, difficulties with the area of skin required to apply creams to achieve therapeutic levels in men, and local skin reactions. Mild to moderate reactions occurs in as many as 50% of men using some formulations of the skin patch, which may produce a 30-50% failure rate. The very small amounts of testosterone cream required to raise testosterone levels in women have not been associated with these problems. Patches may seem more user friendly compared to injections, but their use is limited due to the above concerns. Administration of HCG, therefore, can mimic the effect of LH and increase an individual's testosterone production without directly administering testosterone. In men who still have a functional LH/testosterone control loop, this way of raising testosterone is the most physiologic, and is not associated with testicular atrophy. HCG can be administered daily in small doses via a subcutaneous injection, or given twice weekly via the same route. While direct injection of testosterone has a 100% success rate, there is approximately a 10-15% failure rate seen in individuals using HCG. With normal aging, the testicles will at some point stop responding to the LH signal from the pituitary, this is usually associated with a rise in LH levels.

Starting testosterone therapy is only the first step in a replacement program. Continued monitoring is the hallmark of a truly safe and successful program. A rise in testosterone produces an exaggerated rise in estradiol levels, if this occurs, it can blunt or negate the beneficial effects of testosterone and may even produce gynaecomastia. If estradiol levels rise too high, it is important to recognise this and adjust testosterone dose or implement therapy to block the conversion of testosterone to excess estradiol. It is also important to follow other metabolic markers that may be associated with testosterone therapy. Haemoglobin and haematocrite may rise with testosterone therapy and must be monitored over time as well. If testosterone levels are kept too high, we may see derangement in cholesterol metabolism and fluid retention, potentially exacerbating high blood pressure or causing oedema. Overly elevated testosterone levels may also hasten the onset of prostatic hyperplasia. Testosterone replacement is potentially beneficial to aging men, particularly in the areas of bone density and body composition. The magnitude and longevity of the beneficial effects are not known, however. The frail elderly (age >55 y) individual may be more vulnerable to adverse effects from treatment and these potential risks and benefits must be weighed. In general, parenteral testosterone therapy in older men results in a decline in serum levels of total cholesterol and low-density lipoprotein (LDL) cholesterol, and may decline HDL cholesterol. In general, higher serum testosterone levels correlate with lower metabolic cardiovascular risk factors, including higher high-density lipoprotein (HDL) cholesterol levels, lower blood pressure, and lower levels of plasma fibrinogen, fasting insulin, and lipoprotein.

The biologic effects of androgens:

Target	Effects
Reproductive organs	1- Stimulation of prenatal differentiation and pubertal development of the testes, penis, epididymis, seminal vesicles, and prostate. 2- Maintenance of these structures. 3- Initiation and maintenance of spermatogenesis.
Sexual function and behaviour	1- In human self-assessed aggression is not clear. 2- Key role in stimulating and maintaining sexual function in men.
Muscle	1- Increase nitrogen balance, LBM, and body weight. 2- Increases the size of the muscle cells with little effect on their number.
Skin and hair	1- Increases sebum production with acne as a possible consequence. 2- Male hair pattern. 3- Hair follicles can metabolise to DHT.
Liver	1- Increases synthesis of clotting factors, hepatic triglyceride lipase, sialic acid, a-1-antitrypsin and haptoglobin. 2- Decreases production of SHBG, other hormone binding proteins, transferrin and fibrinogen.
Lipids	Decreases HDL plasma concentrations.
Bone	Stimulates the proliferation of bone cell.
Haematology	Stimulates erythropoietin production by the kidneys. Suppressive effects on both humoral and cellular immune responses. Antiinflammatory effect.

Gynaecomastia or tenderness occasionally occurs in <2% of subjects. Liver toxicity has not been seen with the types of parenteral testosterone administered in clinical studies to older men. Testosterone may improve concentration, lift depression, and restore the overall sense of wellbeing most women enjoy until they reach perimenopause. Menopausal women are living with vaginal dryness, diminishing clitoral sensitivity, difficulty reaching orgasm, infrequent intercourse, and a dysfunctional partner as a result of steroid starvation. Testosterone is the igniter of both sexual fantasy and function. A woman who was not lacking in testosterone before she started hormone replacement therapy may become deficient during the course of therapy. In addition to improving sexual function, testosterone raises circulating levels of free oestrogen in the bloodstream. The women taking testosterone would have significant increases in sexual sensation and desire and more frequent intercourse after 4 weeks and again at 8 weeks after starting therapy. Oestrogen given alone triggers the production of more SHBG, which then binds up more testosterone. Testosterone cream can be rubbed directly on the clitoris to enhance sexual sensation. Decreased testosterone levels in men are associated with decreased haematocrite, diminishing body and facial hair. Decreased salivary testosterone levels may be seen in primary and/or secondary hypogonadism, hypothyroidism, or in obesity with a body mass index of 30 or greater. Decreased salivary testosterone levels may also result from increased sex hormone-binding globulin (SHBG), especially in older men. Elevated SHBG may be due to increased oestrogen levels, which will raise SHBG levels. Lower testosterone levels with a blunting of the normal diurnal rhythm may occur with aging and in testicular failure. Suppression of the circadian rhythm of testosterone in normal adult males taking glucocorticosteroids is also documented. Plasma testosterone levels are significantly decreased in heroin addicts. Heroin users have impaired insulin secretion, which is not dependent on β-cell exhaustion. Effective immune and stress responses are dependent on adrenal cortisol reserve, which may provide an explanation for the heroin addict's vulnerability to AIDS and other infectious diseases. Chronic heroin addiction may produce a change in the rate of hepatic extraction of insulin. Heroin addicts maintained on high dosage methadone (80-150 mg/day) also have depressed testosterone levels. An inverse relationship between methadone dosage and plasma testosterone occurs during methadone detoxification. Heroin addicts have abnormal sexual hormones, which can contribute to infertility. Methadone addiction produces a metabolic state similar to insulin-resistant (IR). Both heroin and methadone addiction may alter glucose metabolism, which stress the findings of similarities between opiate addicts and non-insulin dependent diabetics. Drug addicts have high copper levels. Copper and bromine are significantly high, while zinc, iron, manganese, calcium, sulfur phosphorus, potassium, and chlorine are significantly low in proportion to the period of heroin intake. Drug addiction can block the reabsorption of dopamine, thereby increase dopamine, which is experienced as a high by drug addicts. Marijuana affects hormones including reduction in testosterone, (related to sex drive and aggression), cortisol etc.

CARDIOVASCULAR EFFECTS OF TESTOSTERONE

Coronary artery disease (CAD) ratio of male to female is 3:1. Combined oestrogen and progesterone HRT have no benefit and may have deleterious effects. The physiology of testosterone is complex. Because testosterone is converted to oestrogen by the aromatase enzyme in adipose tissue, an individual's concentration is affected by body habitus and weight. Men with CAD have significantly lower concentration of bioavailable testosterone than men with normal coronary angiogram and the prevalence of hypogonadism in a population of men with CAD is about twice that in the general population. Hypotestosteronaemia is associated with an atherogenic lipid profile (high low density lipoprotein, low high density lipoprotein, high triglycerides), high fibrinogen and hypercoagulable state, an increase in insulin resistance and hyperinsulinaemia, and higher systolic and diastolic blood pressure. Testosterone has direct vasoactive properties. Testosterone vasodilatory effect is independent of an intact endothelium. It is caused by a direct effect on the vascular smooth muscle, by either an effect on K or Ca channels. In man, testosterone causes a dose dependent vasodilation. When testosterone is instilled into the left coronary artery, vasodilation ensues and coronary flow increases. Acute administration of intravenous testosterone improves exercise tolerance and reduces angina threshold in men with CAD. Testosterone has anabolic effects on bone marrow causing an increase in haematocrite. The incidence of men with hypogonadism increases as age progresses. A relative low blood concentration of testosterone in the older man may have effects on atherosclerosis and explain the higher incidence of CAD.

ANDROPATHY

Soya supplements reduce serum testosterone and improve markers of oxidative stress. Soya supplements could protect against prostatic disease and atherosclerosis. Throughout life the steroid milieu of prostatic cells is under constant change, and interactions between the cells and their environment result in dynamic alterations in cell biology. Prostatic stromal hyperplasia occurs due to the increased oestrogen to androgen ratio, seen with declining androgen concentration after the age of 50. Soybeans are a rich source of many biologically active components, including isoflavones. Flavonoids are a large group of polyphenolic compounds found in many fruits, vegetables and beverages such as tea and wine. Over 4000 flavonoids have been identified and they are divided into several groups according to their chemical structure, including flavonols (quercetin and kaempherol), flavanols (the catechins), flavones (apigenin) and isoflavones (daidzein and genistein). Phytoestrogens are compounds such as isoflavones and lignans, which are weak oestrogens. Phytoestrogens compete with endogenous oestrogens for oestrogen receptor sites, acting as partial agonists. Phytoestrogens stimulate the liver to produce sex hormone binding globulin (SHBG) and thus indirectly reduce the amount of biologically active oestrogen. Isoflavonoids inhibit the enzymes 5α-reductase and 17β-hydroxysteroid dehydrogenase (both of which are required for androgen synthesis), inhibit angiogenesis, DNA topoisomerases and tyrosine-specific protein kinases. Testosterone is the most potent androgen secreted by the testis; the others (androstenedione and dehydroepiandrosterone) are produced in reduced concentration. Flavone is a strong inhibitor of cycloxygenase and has free radical scavenging properties. Flavonoid interaction with prostaglandin synthetase may be through antioxidant or free radical scavenging, the antioxidant activity being related to inhibition of lipoxygenase. Oxidative modification of LDL is an important step in the development of the atherosclerotic plaque. Oxidation of LDL is a process initiated and propagated by free radicals. All the cells of the vessel wall, including endothelial cells, smooth muscle cells, macrophages and lymphocytes, can modify LDL. Transition metal ion-mediated generation of hydroxyl radicals, production of reactive oxygen species (ROS) by enzymes such as myeloperoxidase and lipoxygenase, and direct modification by reactive nitrogen species (RNS). Soya can protect LDL against oxidation by myeloperoxidase. Soya supplements protect against LDL oxidation by copper. Myeloperoxidase (MPO), a haeme protein, is secreted from leukocytes in response to an inflammatory stimulus. The enzyme utilises hydrogen peroxide released from activated leukocytes, generating a variety of ROS including hydroxyl radical and singlet oxygen. MPO gives rise to 2 main radical species, hypochlorous acid and tyrosyl radical, which both are potently pro-oxidant and cause modification of LDL lipids and protein. MPO is a mediator of oxidative damage to lipoproteins in the arterial intima. Soya flour can inhibit MPO-mediated oxidation of LDL. Penile erection is a haemodynamic process, involving increased arterial inflow and restricted venous outflow from the penis, coordinated with corpus cavernous smooth muscle relaxation. The process is generally accepted to be under neuroregulatory control and involves the cholinergic, adrenergic and nonadrenergic noncholinergic neuroeffector systems. Diverse mediators have been proposed to function as nonadrenergic noncholinergic neurotransmitters, including neuropeptides (vasoactive intestinal peptide, calcitonin gene-related peptide, substance P), purines (adenosine, ATP) and other factors, including decarboxylated amino acids, histamine, serotonin, prostaglandins and bradykinin. A fundamental principle of erection physiology is that the corporal smooth muscle musculature of penis must undergo relaxation for physiological penile erection to occur. Nitric oxide (NO) exerts a significant role in the physiology of the penis, operating chiefly as the principal mediator of erectile function. Nitric oxide synthase (NOS) have also been localised to lumbosacral pathways that are implicated in governing erectile function. The autonomic innervation of the penis and locally NO operates as a postganglionic neurotransmitter of nonadrenergic noncholinergic penile erection. Upon its synthesis and release from nerves terminals within the erectile tissue of the penis, NO diffuses to neighbouring vascular and trabecular smooth musculatures of the penis whereby it activates guanylate cyclase present in smooth muscle cells to produce cGMP. This increased intracellular accumulation of cGMP is perceived to cause corporeal smooth muscle relaxation via biochemical cascade. A putative mechanism involves cGMP dependent protein kinase dephosphorylation of myosin light chains in corporeal smooth muscle cells (directly or as a consequence of lowering intracellular calcium stores). NO activates Na/K-ATPase in human corporeal smooth musculature. The smooth muscle component of erectile tissue could conceivably generate NO. NO exerts a direct feedback inhibition of NOS activity by interacting with haeme moiety of the enzyme. Oxygen tension has a major role in NO mediate penile erections since low oxygen states inhibit NO synthesis. Normal high levels of tissue superoxide dismutase protect NO from destruction by superoxide anions. Androgen deprivation reduces NOS content, activity and erectile responses. Acetylcholine and bradykinin stimulate endothelial NOS pathways to generate NO mediated smooth muscle relaxation. Vasoactive intestinal peptide (VIP) may also exert a relaxant effect on corporeal smooth musculature via the NO-cGMP pathway. VIP has been found to co-localise with NOS in human penile neurons, and inhibitors of NOS or guanylate cyclase VIP induce relaxation of corporeal tissue. Diabetes mellitus has been associated with impaired NOS dependent erectile mechanisms. Hypercholesterolaemia impairs endothelial NOS dependent mechanisms. The NO pathway is androgen dependent. Aging phenomena correlates with altered NO synthesis and erectile responses. Radiation reduces the number of penile NOS containing nerves. Erectile tissue degenerative changes include decomposition of neural elements, smooth muscle cell cytoplasmic vacuolisation and endothelial disintegration, which are frequently associated with hypoxia or acidosis. NO production critically requires the presence of oxygen and physiological hydrogen ion concentrations. A defective NO-cGMP pathway could also compound pathological changes via blood flow disturbances in the penis. Excess of NO could produce pathological changes in the penis, consistent with its cytotoxic potential as a free radical source. By interacting with specific molecular targets, NO is able to damage cells in many ways, such as inhibiting ATP production, disrupting DNA synthesis or inducing direct toxic effects involving mechanisms that remain unclear. There are several isoenzymes of phosphodiesterase (PDE) exist, each with different properties and tissue distributions, PDE1 through to PDE6. Impotence is an indiscriminate term used by the lay public and the general medical community to describe a variety of problems with male sex function: libido, penile erection, ejaculation, or orgasm. Impotentia erigendi means the inability to have penile erection, impotentia coeundi is the inability of the male to complete the sexual act, and impotentia generandi means the inability to produce. Erectile dysfunction (ED), is the inability to achieve or maintain erection to sufficient rigidity and duration to permit satisfactory sexual performance. Erectile dysfunction disturbs not only sexual life but it also impairs quality of life for the patient and spouse. Erection is a complex neurophysiologic event regulated by compliance of CSM. Tumescence is initiated by relaxation of vascular tone and a drop in resistance to arterial flow that bathes the cavernous tissues in highly oxygenated arterial blood. In 1970, it was believed that 80-90% of all ED was psychologically induced. Non-physical causes can be found in 50% of cases. Testosterone enhances libido, frequency of sexual acts and sleep-related erections. Sexual activity can be resumed after MI when the patient can climb 2 flights of stairs without cardiac symptoms. Post-MI patients have no greater risk of a MI following sexual activity than does a healthy individual. Nitroglycerine tablets may be taken before sexual activity to avoid anginal pain. ED is strongly age-associated, with estimated prevalence rates of 39% in men 40 years of age and 67% in those 70 years of age. For men 40-70 years of age, 10% report complete erectile dysfunction and 25% report moderate ED. Since the time that sildenafil became commercially available as a prescription medication for the treatment of ED, there has been extensive media coverage of sildenafil in both the medical and lay press. Psychological causes of ED include: 1- Expression, many men has great difficulty expressing their feelings, both verbally and sexually. 2- Performance, books, videos do not point out that sexual expression and even degrees of pleasure are very individualised. 3- The man taking charge of and orchestrating sexual performance. 4- James Bond, the notion that a man is always ready, willing, and able to engage in sexual activity. Television is constantly promoting the image of this type of man. We need to be more realistics about our expectations; a man can be interested or concerned with issues other than sexual intercourse. 5- All physical contact must leads to sex? 6- Sex equals intercourse? Sex defined as how we view ourselves as male or female, i.e. gender identification. Intercourse is the physical union of the genitalia, often ending in an orgasm for one or both partners. 7- An erection, is it necessary for sexual activity. 8- Intensity, indicates that as the pleasurable feelings one experiences through foreplay accelerate, the ultimate pleasure is achieved through intercourse. 9- Sex should be natural and spontaneous, we must learn about sexual functioning and how to experience the greatest pleasure before it can be truly enjoyed. Nonpsychological causes of ED include: 1- Vascular, anything that interfere with blood flow into or out of the penis results in erectile dysfunction: A- Atherosclerosis of the pelvic blood vessels interferes with blood flowing into the corpora cavernosa. Today's stressful lifestyles and poor dietary habits are contributing to an increase in the incidence of vascular problems at an earlier age. B- Hypertension decreases blood flow, or effect of the used antihypertensive medications. Change in the medication may bring about the desired results but this needs medical consultation. C- Cerebrovascular disease and cerebrovascular accidents, due to decrease in cerebral cortical output. D- Aortic aneurysm. E- Others, such as, Leriche's syndrome, sickle cell anaemia, leukaemia, and Hodgkin's disease. F- Pelvic steal syndrome, occurs when exercise such as coital movement in the missionary position required blood, is shunted into the lower extremities, diverting it from the pelvic arteries. G- Venous or cavernous leak due to lack of venous compression. Blood flowing into the arteries may be sufficient to achieve an erection, but there is no enough pressure exerted on the veins to compress them and prevent the blood from leaving the area. K- Congenital anomalies. L- Fibrosis of the sinusoids due to aging, Peyronie's disease, traumatic lesion of the artery. M- Inadequate production of cavernous neurotransmitters, such as vasoactive intestinal polypeptide (VIP). N- Nicotine, caffeine, vasoconstrictors. Smokers have lower penile blood flow than nonsmokers do. Even past history of heavy smoking is related. 2- Neurological causes: multiple sclerosis, Parkinson's disease, peripheral neuropathies, sympathectomy, SCI, spina bifida, electroshock therapy, temporal lobe lesions, tabes dorsalis, cerebral palsy, myasthenia gravis, amyotrophic lateral sclerosis. 3- Endocrine causes: A- Diabetes mellitus, due to retarded ejaculation, or lack of tactile sensation. B- Pituitary disorders, such as hypothalamus-pituitary-gonadal abnormalities: pituitary tumours, hypopituitarism-hypogonadism (hypoandrogenism), hyperprolactinaemia (the testosterone level is lower than normal range of 350-800 ng/ml), pituitary trauma or surgery, drugs, renal disease, haemodialysis, or idiopathic. C- Primary gonadal failure. D- Thyroid abnormalities, hyperthyroidism. 40-60% of men with hyperthyroidism are impotent, an increase in the plasma free estradiol/plasma free testosterone ratio could be a contributing factor. E- Female hormones treatment. F- Obesity, 160% or more of the ideal weight. There is significant decrease in plasma testosterone and increase in estradiol levels. Testosterone returns to normal after weight loss G- Primary hypogonadism, such as, Klinefelter's syndrome. K- Secondary hypogonadism, e.g., mumps, orchitis, radiation, cytotoxic chemotherapy, vascular disease or surgery. M- Adrenal abnormalities: Cushing's syndrome and Addison's disease. 4- Cardiorespiratory, atherosclerosis, myocardial infarction, angina, coronary insufficiency, CHF, medications, emphysema, shortness of breath and psychological. 5- Anatomic abnormalities of the penis. 6- Medications and drug abuse may affect libido, erection, ejaculation, orgasm, or detumescence. 7- Surgical interventions: A-P resection, cystectomy, radical prostatectomy, pelvic lymphadenectomy, renal transplantation, sympathectomy, transection of spinal cord, TURP, hernia repair, heart bypass. 8- Genitourinary causes, prostatitis. Peyronie's disease causes a decrease in blood flow into one or both corporal bodies causing a bent or crooked penis. Prolonged priapism, due to tissue damage resulting from blood clotting in the vascular spaces that leads to fibrosis or necrosis. 9- Unusual causes: a) long-distance bicycling, b) riding stationary bicycle, c) direct sport injury to the penis, and d) amputation. Bicycling causes bilateral compression of the pudendal nerves or constriction of blood supply to the penis.

FERTILITY DYSFUNCTION (FD)

Infertility (Impotentia Generandi) in a couple is defined as the inability to achieve conception despite one year of frequent unprotected intercourse. The fertility rate in a couple is influenced by several factors, such as the age of the female partner; exposure to sexually transmitted disease (STD) and to environmental and medical toxins; coexistent disease states. Infertility has the potential to affect every aspect of the man's life. Every effort should be made to alleviate the sense of powerlessness that accompanies infertility. The couple may feel a sense of obligation to the doctor to succeed. Genetic defects have been divided into 3 major categories: Mendelian disorders, chromosome disorders and multifactorial disorders. Multifactorial disorders are believed to involve interactions between genes and environmental factors. Genetic defects are responsible for a variety of clinical presentations of male infertility, from GnRH deficiency to spermatogenic failure to obstructive azoospermia. Infertility remains unexplained in 20% of cases because numerous reproductive defects are undetectable with current methods. Infertility treatment choices often seem more difficult than other clinical decisions because evidence from randomised clinical trials is scarce, and there may be high costs and unwanted adverse effects. The process to explore all options is time consuming but not atypical for clinical decisions for long-standing disorders in which psychosocial and economic factors strongly influence the choices to be made. If a pregnancy does not occur by 3 years, persistent infertility is very likely in the absence of medical intervention. 20% of infertility cases are attributed to male factors, 38% to female factors, 27% have causal factors identified in both partners, and 15% can not be satisfactorily attributed to either partner. Up to 10% of healthy men are infertile. Male infertility is due to heterogeneous group of disorders. Recognisable causes account for 30-50%. The remainder are idiopathic. The coexistence of infertility and psychosocial distress is supported by sound scientific evidence. Psychological distress may influence the female reproductive system at various levels: through the autonomic nervous system, the endocrine system, and the immune system. Stress-induced changes in sexual behaviour constitute a more obvious possible mechanism of action. Psychological distress may be a risk factor for reduced fertility among women with long menstrual cycles. Causes of testicular atrophy: 1- trauma, 2- testicular torsion, 3- hypopituitarism, 5- cryptorchidism, 6- Klinefelter's syndrome (47, XXY), 7- alcoholism and cirrhosis, 8- infection, e.g., mumps, orchitis, gonococcal epididymitis, 9- malnutrition and cachexia, and 10- radiation. Causes of male fertility dysfunction (FD): 1- Hypothalamic pituitary disease (secondary hypogonadism), 1-2%; 2- Testicular disease (primary hypogonadism), 30-40%, 3- Posttesticular defects (disorders of sperm transport), 10-20%, and 4- Nonclassifiable, 40-50%.

Normal plasma levels for pituitary and gonadal hormones in men:

Hormone	SI units
Testosterone, total	10.4-38.2 nmol/l
Testosterone, free	173-729 pmol/l
DHT	0.9-2.6 nmol/l
Androstenedione	1.7-6.9 nmol/l
Estradiol	55-150 pmol/l
Estrone	55.5-240 pmol/l
FSH	2-15 IU/l
LH	2-15 IU/l
Prolactin	4-18 µg/l

Conditions for conception to occur:

1- The testes must have normal spermatogenesis.
2- The spermatozoa must complete their maturation.
3- The ducts for sperm transport must be patent.
4- The prostate and seminal vesicles must supply adequate amounts of seminal fluid.
5- The coital technique must enable the male partner to deposit his semen near the female cervix.
6- The spermatozoa must be able to penetrate the cervical mucus and reach the uterine tubes.
7- The spermatozoa must undergo capacitation and the acrosome reaction, fuse with the oolemma, and be incorporated into the ooplasm.

CIGARETTE SMOKE

Polycyclic aromatic hydrocarbons (PAH) are environmental pollutants present in the air, water, and foods. Benzo(a)pyrene belongs to the family of polycyclic aromatic hydrocarbons; it is produced mainly by the combustion of fossil fuels, but also is present in charcoal-broiled foods and in cigarette smoke. Benzo(a)pyrene is highly mutagenic and carcinogenic. It is metabolised by enzyme systems (oxidases) to reactive hydrophilic intermediates arising from epoxidation. The major diol epoxide (DE-I) binds covalently to the 2-amino group of DNA guanosine and forms adducts designated BPDE-I-dG-DNA. These adducts are premutational lesions in guanosine nucleosides, that, if not repaired, constitute a potential source of carcinogenic damage. There is evidence for a direct causative link between BPDE-I-DNA adducts and lung cancer gene (p53) mutational spots. Cigarette smoking is associated with a reduction in sperm quality and viability. Polycyclic aromatic hydrocarbons, including benzo(a)pyrene, are recognised as major environmental pollutants. Their continuous release into the ambient air from the combustion of fossil fuels and vegetation, and the rather long half-life of benzo(a)pyrene (10 weeks), allow its accumulation in the gonads and their fluids. This leads to continuous exposure of germ cells to carcinogenic agents in the nonsmokers as a result of the ubiquitous nature of benzo(a)pyrene. Benzo(a)pyrene, a constituent of tobacco smoke decreases testicular weight, induces atrophy of seminiferous tubules, and alters spermatogenesis. Smoking-related adducts are likely to arise from oxidative damage. Cigarette smoke contains oxygen-derived species (O_2^-, H_2O_2, and OH.) that imbalance the prooxidant-to-antioxidant ratio in the blood plasma and semen in smokers, leading to oxidative stress. Accumulated oxidative stress from smoking is known to induce mitochondrial and nuclear DNA damage in human somatic cells. BPDE-DNA adducts in sperm, occurring in association with seminal plasma cotinine levels. Oxidative stress from smoking also may lead to alterations in chromosome numbers in the gametes as a result of interference with the assembly and function of the meiotic spindle. In oocytes, there is increased frequencies of chromosome diploidy (46 chromosomes instead of 23) in smokers compared with nonsmokers. The risk of germ cell damage from cigarette smoking, and from other environmental pollutants, has been postulated to be greater in males than in females. Biomarkers of physiologic damage to male germ cells from environmental pollutants include decreases in the number of sperm produced, the quality of the sperm, and the capacity of the sperm to penetrate and fertilise the egg. Sperm with DNA fragmentation are negatively associated with both fertilisation rates and seminal parameters. This may be one cause of the conception delay found in epidemiologic studies of smokers. Cigarette smoke alters levels of hormones that are involved in spermatogenesis. Cigarette smoking alone is associated with lowered semen quality including sperm density, total sperm count, total number of motile sperm, and motile sperm concentration. Drinking >4 cups of coffee and smoking >20 cigarettes per day has been found to increase the percent of dead sperm and decrease sperm motility. Nicotine exposure leads to atrophy of the testis and impaired spermatogenesis. Polonium-210 and α-emitting radioelement components of tobacco smoke are capable of damaging DNA and have been detected at higher concentrations in the semen of smokers.

ANTIOXIDANTS

Spermatozoa produce relatively low levels of reactive oxygen species (ROS). High levels of ROS are needed to affect sperm motility. As many as 25% of semen samples from an unselected population of men attended an infertility clinic produced significant levels of ROS. There is an inverse correlation between the percentage of motile spermatozoa and the level of ROS. Fatty acid peroxides generated by ROS attack on cell membrane phospholipids are associated with loss of mammalian sperm motility and decreases capacity for sperm-oocyte fusion. The degradation products of these lipid peroxides (hydroxyalkenals and malonaldehyde) are highly toxic to spermatozoa and cause an irreversible loss of motility. ROS induce reversible axonemal damage and sperm immobilisation mostly due to depletion and cellular ATP and insufficient axonemal protein phosphorylation. Hydrogen peroxide is the most toxic ROS towards human spermatozoa. Deficient spermatozoa form high levels of ROS. Membrane lipid peroxidation has been correlated with midpiece morphological defects and abnormal morphology. ROS produced by damaged spermatozoa could affect the function of normal spermatozoa. The high generation of ROS by damaged or morphologically abnormal spermatozoa is often associated with lowered motility and decreased sperm functions. A loss of motility occurs in 34% of the cases. Only a third of the ROS produced by spermatozoa is released outside the cell. NADPH oxidase at the level of the sperm membrane is responsible for the increased ROS formation in the semen of infertile patients. The activity of the sperm diaphorase (an NADH-dependent oxidoreductase) located in the middle piece of spermatozoa and integrated in the mitochondrial respiratory system is increased in semen of infertile patients. PMN found in semen produce different levels of ROS according to their level of activation. Activated PMN could be a threat for spermatozoa, especially in cases of epididymo-orchitis where spermatozoa are in contact with PMN for long periods of time, in a very restricted space and in the absence of the high ROS scavenging capacity of seminal plasma. Human spermatozoa and seminal plasma posses antioxidant system to scavenge ROS and prevent ROS related cellular damage. Superoxide dismutase and glutathione peroxidase/reductase enzymatic antioxidant systems exist in human spermatozoa. Catalase exists in human semen. Decreased ROS scavenging capacity of spermatozoa or seminal plasma is responsible for increased production of ROS in the semen of infertile patients. All aerobic organisms require oxygen for life. However, although it is an essential element, the metabolites of oxygen such as superoxide (O_2^-), hydrogen peroxide (H_2O_2), and the hydroxyl radical (OH^-) are capable of adversely modifying cell functions and mechanisms, ultimately endangering the survival of the cell. These ROS have been implicated as a major contributory factor in male infertility. Human sperm are particularly sensitive to free radical assault because of their high content of polyunsaturated fatty acids (PUFA) and lack of DNA repair mechanisms. The production of ROS is strongly associated with the loss of sperm motility. Almost 40% of infertile men have detectable amounts of ROS in their semen; there is no ROS activity in the semen of fertile men. Somatic cells contain antioxidants within their cytoplasm. However, sperm lose most of their cytoplasm during their maturation and, therefore, lack the endogenous repair mechanisms and enzymatic defences observed in other cell types. Sperms are protected from oxidative insult by seminal plasma, which contains an abundance of antioxidant enzymes such as superoxide dismutase and catalase, which remove key ROS such as O_2^- and H_2O_2, and scavengers such as albumin and taurine. Seminal plasma also contains crucial chain-breaking antioxidants such as urates, ascorbate, and thiol groups. Seminal plasma from infertile men has a significantly lower total antioxidant capacity than that from fertile men. Human spermatozoa and seminal plasma posses antioxidant system to scavenge ROS and prevent ROS related cellular damage. Superoxide dismutase and glutathione peroxidase/reductase enzymatic antioxidant systems exist in human spermatozoa. Catalase exists in human semen. Decreased ROS scavenging capacity of spermatozoa or seminal plasma is responsible for increased production of ROS in the semen of infertile patients. In vitro fertilisation with or without intracytoplasmic sperm injection is the most successful therapy for male factor infertility. During routine sperm preparation for these procedures, sperm are removed from their seminal plasma to concentrate the subpopulation with the best morphology and motility. During this process, sperms are deprived of their antioxidant protection and left more vulnerable to oxidative insult. The centrifugation of neat semen, leave sperm in close contact with leukocytes and other defective or damaged sperm. These are both sources of free radicals. Depriving sperm of the antioxidant protection available in seminal plasma leads to damage to DNA, and that protection can be provided by supplementing preparation media with antioxidants such as ascorbate and α-tocopherol even when no free radical activity can be detected. Excessive levels of ascorbate are thought to be associated with the inhibition of ovarian steroidogenesis, a decline in fertility, and an increased likelihood of abortion. Ascorbate is the first line of antioxidant defence and is the only endogenous antioxidant in blood and seminal plasma that can competitively protect the lipoproteins from detectable peroxidative damage induced by aqueous peroxyl radicals. It is the principal antioxidant in the seminal plasma of fertile men, contributing up to 65% of the total chain-breaking antioxidant capacity. Ascorbate concentration in seminal plasma is 10 times greater than in blood plasma (364 M compared with 40 M, respectively). Ascorbate actually is secreted from seminal vesicles during ejaculation. It has been suggested that dietary ascorbate supplementation can protect human sperm from endogenous oxidative DNA damage, which in turn could affect sperm quality, potentially increasing the risk of genetic defects in offspring. Ascorbate is known to have a genotoxic effect in phage, bacterial, and mammalian cells, and it is known to induce DNA single-strand breaks in human lymphocytes, neonatal fibroblasts, and a T-cell leukaemia cell line. Ascorbate is widely available as a food supplement at doses of 500 mg or more per day. However, Ascorbate causes an increase in 8-oxo-adenine, a potentially mutagenic lesion, in DNA isolated from lymphocytes after in vitro supplementation at 500 mg daily. At doses of <500 mg, the antioxidant properties of ascorbate appeared to prevail. Ascorbate in the range of 20-600 M affects sperm motility adversely in both normozoospermic and asthenozoospermic samples. The percentage of progressive motility, velocity of sperm movement, and sperm head oscillations are significantly reduced. The presence of ascorbate provides complete protection from induced DNA damage from x-ray radiation. Ascorbate has the ability to promote the release of transition metals from proteins, to redox cycle iron and copper to form oxygen radicals, and therefore may act as a prooxidant. α-Tocopherol is classified as a chain-breaking antioxidant because of its ability to break the lipoperoxidative chain reaction through its interaction with lipid peroxyl and alkoxyl radicals. It is a powerful antioxidant and provides mammalian cells some protection from oxidative attack. Oral administration of vitamin E improves significantly the function of human sperm. α-Tocopherol is present in small but consistent quantities in seminal plasma (0.08-0.9 mol/l). However, these trace amounts may be adequate, because ascorbate can regenerate α-tocopherol. α-Tocopherol and ascorbate work synergistically to protect against lipid peroxidation, with ascorbate recycling α-tocopherol, allowing it to function again as a free radical chain-breaker. There is an improvement in DNA integrity after α-tocopherol supplementation. Ascorbate and α-tocopherol when added to sperm preparation media, they appear to have a greater detrimental effect on sperm motility and sperm DNA integrity than when either is added alone. Antioxidant supplementation, although beneficial to sperm DNA integrity and thus useful in the preparation of sperm samples for intracytoplasmic sperm injection, is not useful in the preparation of sperm samples for IVF, where motility is essential.

VARICOCELE

Varicocele is the most common treatable cause of infertility in men. Although the exact pathophysiology of varicocele is unknown, it impairs normal testicular function by elevating scrotal temperature via the reflux of warm abdominal blood through incompetent valves of the spermatic veins. These vascular lesions are divided into 3 grades based on physical findings. Several pathobiological mechanisms may be involved in causing sperm dysfunction in patients with varicocele. Increased oxidation due to the enhanced formation of noxious and cytotoxic oxidants. The majority of men with varicocele have an abnormal spermiogram, and varicocele repair improves sperm quality and subsequently increases pregnancy rates. Mechanisms of infertility in varicocele: 1- Induction of testicular hypoxia by venous stasis and small vessel occlusion leading to Leydig cell and germinal cell dysfunction. 2- Retrograde flow of adrenal and renal metabolites from the renal vein down the left internal spermatic vein. 3- Elevation in scrotal temperature. 4- Depression of gonadotropin or androgen secretion, which may change the endocrine environment to which both testes are exposed. 5- Excessive formation of ROS, with oxygen metabolite most damaging to human spermatozoa being H_2O_2. ROS initiate the peroxidation of unsaturated fatty acids in the sperm plasma membrane resulting in a loss of motility and the capacity of sperm-oocyte fusion. ROS can be generated by the spermatozoa themselves or phagocytic leukocytes. 6- Nitric oxide (NO) is capable of inhibiting human sperm motility. NO can reduce ATP and thus decrease sperm motility. NO derives from several cells of the male genital system such as phagocytes, endothelial cells and smooth muscle cells. Peroxynitrite is a noxious oxidant formed by a rapid reaction between NO and superoxide. Nitric oxide synthase (NOS) and xanthine oxidase activities in blood of varicocele veins are greater than those in peripheral blood, resulting in a dramatic increase in the rate of NO, peroxynitrite and S-nitrosothiol release within the varicocele veins. There is a 25 fold increase of NO rate production in the varicocele vein. This NO may derive from the cells of the varicocele vein (endothelial or smooth muscle cells) or from the NOS that has been diffused out of the varicocele vein. NO may also be produced by the testis and accumulate within the spermatic vein. NO and superoxide rapidly react together to form peroxynitrite, an unstable species that at normal pH is protonated to peroxynitrous acid, which spontaneously decomposes to nitrite and hydroxyl radicals in 20-30% yield. The remaining peroxynitrite acid is directly isomerized to nitrite. Peroxynitrite is an important mediator of free radical species with strong oxidising properties towards biological molecules, including protein and nonprotein sulfhydrates, deoxyribonucleic acid and membrane phospholipids. Peroxynitrite also formed from nitrates free or protein associated tyrosine and other phenolics via a metal catalysed reaction, and/or via intermediate formation of tyrosyl radicals and nitrite. Peroxynitrite may be responsible for sperm dysfunction in varicocele patients, causing lipid peroxidation with subsequent changes of the physicochemical characteristics of plasma membrane on sperm cells. Varicocele veins contain dramatically increased S-nitrosothiols compared to the peripheral veins. Varicocele blood serum contains a constitutive form of the enzyme NOS, which converts L-arginine into NO in the presence of calcium/calmodulin. NOS activity in varicocele serum is 8 fold greater than in peripheral veins. Xanthine oxidase activity in varicocele blood serum is 7 fold than that in peripheral veins. Peroxynitrite ($ONOO^-$) reacts with bicarbonate (HCO_3^-) of human erythrocytes according to the reaction:

$$ONOO^- + HCO_3^- \rightarrow HCO_3 . + NO_2 . + OH^-$$

Bicarbonate radicals (HCO_3.) oxidise luminol, and other aromatic and heterocylic molecules. Alternatively, peroxynitrite may peroxidise bicarbonate to peroxybicarbonate, another strong oxidising species. Superoxide dismutase catalyses the nutrition by peroxynitrite of a wide range of phenolics, including tyrosine in proteins. The erythrocytes are particularly susceptible to oxidative damage as a result of the high polysaturated fatty acid content of the membranes and the cellular concentration of oxygen and haemoglobin. ROS and fatty acid peroxides generated by ROS exert their action on spermatozoa by increasing lipid peroxidation and depletion of ATP. Moreover, the degradation products of these fatty acids peroxides may also be toxic to spermatozoa. NO decrease sperm motility by a mechanism involving inhibition of cellular respiration resulting in depletion of sperm ATP.

IMMUNOLOGIC INFERTILITY

In addition to its role as a barrier, the immune system has the ability to destroy, remember, and diversify in response to its interaction with a nonself or "foreign" object or antigen. Because the antigen must be presented to lymphocytes and recognised by surface receptors in association with Class II major histocompatibility antigens or major histocompatibility complex (MHC) molecules to incite a response, cell-mediated immunity is restricted to antigens within a certain size range and only to those that are presented by or on other cells. The actual magnitude of the response is modulated by cytokines or secreted molecules, as well as by subsets of lymphocytes that may augment (T-helper) or suppress (T-suppressor) the response. The mechanisms of eliminating the antigen include: 1- direct contact, 2- antibody linkage, and 3- opsonization. Depends on soluble mediators called antibodies, complements cell-mediated immunity in the recognition and effector phase of antigen processing. Passive and active vaccination theories based on the body's ability to accept or generate immunoglobulin with exquisite specificity. Immunoglobulins are glycoprotein molecules that are secreted by plasma cells (B lymphocytes), they consist of 2 light chains and 2 heavy chains that define constant (Fc) and variable (Fab) regions. The variable region determines its antigen binding specificity or idiotype. The humoral immune response is therefore better suited for free floating extracellular pathogens, such as most bacteria, extracellular parasites, and some viruses. Once bound, pathogens are exposed to a variety of fates: 1- the antibody Fc region may activate complement and results in cell lysis; 2- the antibody may link the pathogen to other cytolytic immune cells and cause opsonization and phagocytosis; 3- the antibody antigen complex may initiate, mast cell degranulation and inflammatory response; 4- the interaction of pathogen and antibody may lead to lymphocyte binding and direct cell specific killing. Cytokines, firsts described in 1972, there are now over 30 mediators in this family of immune molecules. They are: 1- tightly regulated; 2- nonimunoglobulin effector molecules of different sizes; 3- secreted by various lymphoid cells; and 4- regulating the response level, differentiation, and proliferation of cells in an immune response. Cytokines act in a paracrine or autocrine manner to exert local tissue effects. Sperm developing within the testis posses antigens considered foreign by the immune system, yet normally no immune response is generated. In embryonic life, the developing immune system acquires, a tolerance to all "self" antigens in the body, enabling it to differentiate these antigens from nonself antigens by the time birth occurs. At puberty, differentiating germ cells begin to express new antigens as the mature from spermatogonia to mature sperm, and it is these antigens that the immune system has not seen before. Mature spermatozoa are foreign to a man's immune system throughout adulthood. But autoimmune state, in which the humoral and cell-mediated arms of the immune system produce antibodies and sensitised lymphocytes against sperm, does not normally occur. The reason for immunologic unresponsiveness to sperm in the normal male is unclear. In the interstitial spaces between seminiferous tubules macrophages and various lymphocytes exist along with extensive lymphatic channels and blood vessels. No immune cells are found within the somniferous tubule. This has led to the concept of a blood-testis barrier, in which the germinal tissue is separated, by morphologic and humoral barriers from the interstitial tissue. The barrier consists of muscle-like myoid cells that line the outer surface of the seminiferous tubule, Sertoli cell tight junctions, and various immunosuppressive agents within the testis. Developing along with sperm at puberty are specialised tight junctions between Sertoli cells within the tubule. These specialised attachments form one of the tightest known epithelial cell junctions and so comprise the major component of the blood-testis barrier. The rete testis and epididymis, are weaker physical barriers in these anatomic locations. Blood-testis barrier cannot completely confine these antigens (sperms) at the level of the rete testis, efferent ductules, and epididymis. Lymphocytes found in these locations, predominantly T suppressor cells may actively suppress any immune reaction. Vasectomy or trauma would induce a pathologic autoimmune reaction. Cytokines and other humoral mediators of the immune response may contribute to tolerance within the testis. IFN-γ, soluble Fc receptor and TGF-β may indeed be active in the testicular environment. An inverse correlation of serum levels of antisperm antibodies and testosterone, suggest that testosterone may act to suppress the immune response through T suppressor cell induction. Extratesticular mechanisms, such as immune modulators are known to exist in seminal plasma and may be important in the induction of a tolerant state once sperm are delivered to a second foreign environment, the female reproductive tract. Tolerance to sperm may be HLA related, in that genetic links may predispose an individual to autoimmune reactions to sperm. Immunogenic testicular antigens types are: 1) testis-specific antigens are the equivalent of organ-specific antigens and consist of any antigen in the testis that can induce an orchitis; 2) aspermatogenic autoantigens are a subset of testis-specific antigens that induce an immune response resulting specifically in destruction of germ cells and decreased sperm production. Only germ cells express aspermatogenic autoantigens. Passing through the epididymis, where motility is gained and maturation occurs, modulates sperm antigens. Exposure to seminal plasma results in further modification and adsorption of antigens, making the fully nature spermatozoa antigenically rich. Antibodies to sperm are found in 3-12% of men who undergo evaluation for infertility, located in serum, seminal plasma, and bound to sperm. Serum antibodies are considered clinically less important than sperm-bound antibodies. Seminal plasma antibodies may or may not be sperm-bound. Antibody production may be classified, according to the stimulus for their generation, into: a) primary response, implies an unknown, inflammatory, or infectious; b) secondary response, indicates obstruction as the reason for antibody presence. Only the IgM, IgG, and IgA types have been found to be directed against sperm antigens. The IgM is confined to serum and only rarely found in organs or secretions of the male genital tract. IgM antibodies are not routinely measured in detection systems. Sperm IgG antibodies are probably derived from local production in the genital tract or from transudation from the bloodstream. Only 1% of serum IgG content is found in male genital tract secretions, but often sperm-bound IgG can be measured in the absence of assayable serum IgG antibody levels. Breaches in the blood-testis barrier, overwhelming inoculations with sperm, antigens, defect in active immunosuppression may all account for pathologic antibody production. The blood-testis barrier may be damaged under conditions, which are associated with infertility and ASA, such as testicular trauma, torsion, biopsy, vasectomy, orchitis, cancer, cryptorchidism, and varicocele, with exposure of previously isolated sperm antigen to the immune system. In posttesticular obstruction or surgical disease of the genital tract, such as vasectomy, congenital vas absence and vasovasostomy, large inoculations of the immune system with sperm antigens can occur, overwhelming any pre-existing immunosuppressive state and resulting in antisperm antibodies (ASA) production. As a convenience, a normal antigenic "leak" may lead to a pathologic autoimmune response instead of the normal state of tolerance. Cytokine deficiency may reduce the recruitment of T suppressor cells to the "leaky" areas or alter the state of the normal nonrecognition of sperm antigens. Regardless of the mechanism, an immune response to sperm antigens involves recognition of the particular antigen as nonself in conjunction with the MHC system. Such antigens are presented by an antigen-presenting cell (APC) to T or B lymphocytes but are recognised only when presented along with MHC Class II or Ia molecules on the APC and lymphocyte surface. Once an immune response is stimulated, T helper cells stimulate B cells to become activated as plasma cells, and specific ASA are generated to these antigens. Once generated, antibodies directed against sperm may result in infertility by a variety of mechanisms: A- disturbances in sperm transport; and B- disruptions in proper gamete interaction. If >50% of sperm are antibody bound, then reduced penetration can result. The Fc region of the IgA molecule binds receptors in the cervical mucus and impairs sperm motility, when sperm contact the cervical mucus, forward motility changes into a "shaking" motion ineffective for penetration. Sperm-directed antibodies may also cause sperm to autoagglutinate or clump, which equally inhibits cervical passage. Immunoglobulins of the IgG class unlike IgA, do not appear to mediate this phenomenon. The phagocytosis of antibody-coated sperm within the uterine cavity has been reported. The release of follicular fluid, rich in complement components, may facilitate sperm opsonization and destruction by macrophages, either near the ovum, within the fallopian tubes or within the endometrial cavity. By blocking normal sperm binding to the oocyte zona pellucida, ASA may affect fertility at the level of gamete interaction. Essential for fertilisation, sperm-zona binding via specific receptors could be blocked by steric interference from attached sperm antibodies. ASAs may also disrupt gamete interaction by inhibiting sperm penetration of the oocyte once the zona barrier is breached. Poor penetration (<20%) was seen in 78% of cases with antibody coated sperm and in 18% of those using uncoated sperm. Antibody induces spontaneous abortion. Certain disease states can potentially increase the risk for the development of ASA, e.g., obstructive, infectious, physical, thermal or genetic in nature. 40-80% of men with active or chronic prostatitis may have serum ASA present. Physical injury to one testis has been shown to effect the contralateral testis, presumably by an immunologic mechanism, but ASA production is more difficult to demonstrate. Genetic may influence the development of ASA in some men. Only semen can carry an inhibitory effect to subsequent reproductive events. 200 IU of vitamin E can improve fertility by 30%.

Pre-testicular causes of male infertility

Disorder	Genetic mechanism	Diagnosis
Kallmann's syndrome	X-linked recessive, defect in KALIG-1 gene on Xp22.3 chromosome	Anosmia, craniofacial asymmetry, cleft palate, delayed puberty, small testes.
Prader-Willi syndrome	Cytogenetic deletion of chromosome 15q12 on paternally inherited chromosome	Obesity, mental retardation, hypotonia, small hands and feet.
Bardet-Biedl syndrome	Autosomal recessive inheritance, gene mapped to chromosome 16q21	Obesity, mental retardation, retinitis pigmentosa, polydactyly
Cerebellar ataxia with hypogonadotrophic hypogonadism	Parental consanguinity, autosomal recessive inheritance	Disturbance of speech and gait, lack of secondary sexual characteristics, decreased libido, small firm testes.
Sickle cell anaemia	Autosomal recessive inheritance, defect in gene for β-globulin chain of haemoglobin,	Anaemia, sickling of erythrocytes at low oxygen tension. Sickle cell test, haemoglobin electrophoresis.
β-Thalassemia	Autosomal dominant inheritance, defect in gene for β-globulin chain of haemoglobin.	Anaemia, iron overload, haemoglobin electrophoresis.

INTERVENTION FOR MALE FERTILITY DYSFUNCTION

Male fertility dysfunction is a multifactorial disease process with a number of potential contributing causes. 6% of adult males may have fertility dysfunction (FD). Male sperm counts are declining, and environmental factors, such as pesticides, exogenous oestrogens, and heavy metals may have negative impact on spermatogenesis. Carnitine, arginine, zinc, selenium, vitamin B_{12}, vitamin C, vitamin E, glutathione, coenzyme Q_{10}, and other nutritional therapies improve sperm counts and sperm motility. Chlamydia can reside in the epididymis and vas deferens, affecting sperm development and fertility (50%). The presence of anti-sperm antibodies may indicate an undiagnosed infection, and is estimated to be a relative cause of FD in 3-7% of cases. There is a significant correlation between β-carotene levels and antibody titers, suggesting dietary antioxidants are involved in mediating immune function in the male reproductive system. There is a 20% drop in volume and a substantial 58% decline in sperm production in the last 50 years. There may be environmental reasons for deteriorating sperm quality, including occupational exposure to various chemicals, heat, radiation, and heavy metals, as well as exposure to environmental oestrogens and pesticides. Lifestyle risk factors are also significant, including cigarette smoking, alcohol consumption, chronic stress, and nutritional deficiencies. Exogenous oestrogens impact foetal development by inhibiting the development of Sertoli cells, which determine the lifelong capacity for sperm production. Circulating oestrogens also inhibit enzymes involved in testosterone synthesis and may directly affect testosterone production. Diethylstilbestrol (DES) was prescribed from 1945 to 1971 to millions of women during pregnancy. Male offspring from those women had a higher incidence of developmental problems of the reproductive tract, as well as diminished sperm volume and sperm count. Synthetic oestrogens are still widely used in the livestock, poultry, and dairy industries. Men wishing to improve their fertility and sperm quality probably should avoid hormone-containing dairy products and meats and opt instead for organic or hormone-free foods. Organochloride compounds have oestrogenic effects within the body. Chemicals such as dioxin, DDT, and PCBs are known to interfere with spermatogenesis. Low levels of DDT cause degeneration in sperm production, a decrease in the total number of sperm, and reduce number of Leydig cells. DDT acts as a hormonal disrupter, damaging the seminiferous epithelium and lowering local testosterone levels. Adequate intake of essential fatty acids is important to ensure proper membrane fluidity and energy production in sperm cells. High dietary intake of hydrogenated oils, especially cottonseed oil has a negative impact on sperm cell function. Cottonseed oil contains toxic pesticide residues, and high levels of the chemical gossypol, which can interfere with spermatogenesis. Dietary aflatoxin increases the number of abnormal sperm by 50%. Infertile males have 40% higher hair mercury levels than fertile males. Occupational exposure to lead causes a significant decrease in male FD. Cigarette smoking is associated with decreased sperm count, alterations in motility, and an overall increase in the number of abnormal sperm. Seminal cadmium levels are significantly increased, especially in those smoking more than one pack per day. Nicotine can alter the function of the hypothalamic-pituitary axis, affecting growth hormone, cortisol, vasopressin, and oxytocin release, which then inhibits the release of LH and prolactin. Cigarette smokers have higher levels of circulating estradiol and decreased levels of LH, FSH, and prolactin than non-smokers, all of which potentially impact spermatogenesis. Smokers with low prolactin levels have defects in sperm motility. The main function of carnitine in the epididymis is to provide an energetic substrate for spermatozoa. Carnitine contributes directly to sperm motility and may be involved in the successful maturation of sperm. Epididymal sperm use fatty acid oxidation as their main source of energy metabolism, and thus tend to concentrate carnitine while in the epididymis, as carnitine is necessary for transport of fatty acids into the mitochondria. Reduced levels of carnitine lower fatty acid concentrations within the mitochondria, which leads to decreased energy production and potential alterations in sperm motility. There is a direct correlation between semen carnitine content and sperm motility. There is a positive correlation between free L-carnitine and both sperm count and the number of motile sperm/ml. 3 g/day of oral L-carnitine for 4 months would have a positive effect on sperm motility, and increases total number of spermatozoa per ejaculate. Arginine is a precursor in the synthesis of putrescine, spermidine, and spermine, which are thought to be essential to sperm motility. 4 g/day of oral arginine for 3 months, or 80 ml of a 10% arginine HCl solution for 6 months have a significant effect on sperm count and motility (74%) without any side effects. Zinc is a trace mineral essential for normal functioning of the male reproductive system. Numerous biochemical mechanisms are zinc dependent, including more than 200 enzymes in the body. Zinc deficiency is associated with decreased testosterone levels and sperm count. Zinc levels are generally lower in infertile men; supplemental zinc may prove helpful in treating male FD. Polyunsaturated fatty acids and phospholipids are key constituents in the sperm cell membrane and are highly susceptible to oxidative damage. Sperm produce controlled concentrations of reactive oxygen species, such as the superoxide anion, hydrogen peroxide, and nitric oxide, which are needed for fertilisation. High levels of free radicals can damage sperm cells directly. The concentration of ascorbic acid in seminal plasma directly reflects dietary intake, and lower levels of vitamin C may lead to FD and increased damage to the sperm's genetic material. In healthy men, daily intake of low 5 mg of vitamin C/day is associated with 50% reduction in seminal plasma levels of vitamin C, with a 91% increase in sperm with DNA damage. 1000 mg ascorbic acid significantly improves sperm quality, 140% increase in sperm count, and significant reductions in the number of agglutinated sperm. Lipid peroxidation in the seminal plasma and spermatozoa can be estimated by malondialdehyde (MDA) concentrations. Oral supplementation with vitamin E significantly decreases MDA concentration and improve sperm motility, resulting in a 21% pregnancy occurrence, and improves sperm function in the zona binding assay, therefore enhancing the ability of the sperm to penetrate the egg in vitro. The combination of vitamin E and selenium significantly increases sperm motility and the overall percentage of normal spermatozoa. Glutathione have a positive effect on sperm motility. Selenium and glutathione are essential to the formation of phospholipid hydroperoxide glutathione peroxidase, an enzyme present in spermatids that becomes a structural protein comprising over 50% of the mitochondrial capsule in the mid-piece of mature spermatozoa. In sperm cells, CoQ_{10} is concentrated in the mitochondrial mid-piece, where it is involved in energy production. CoQ_{10} also functions as an antioxidant, preventing lipid peroxidation of sperm membranes. CoQ_{10} significantly improves fertilisation rate. Vitamin B_{12} is important in cellular replication, especially for the synthesis of RNA and DNA, and deficiency states have been associated with decreased sperm count and motility. Methylcobalamin (1,500 µg/day) increases the standard sperm parameters by 60%. Ginseng enhances sperm motility, promotes increased sperm formation and testosterone levels, enhances erectile capacity and protect against atrophy and testicular damage induced by dioxin. The herb Pygeum africanum may also be an effective therapy for male infertility, especially in cases of diminished prostatic secretions. Pygeum extracts have been shown to increase alkaline phosphatase activity, which helps maintain the appropriate pH of seminal fluid, and increases total prostatic secretions. Sperm motility is partly determined by the pH of the prostatic fluid. Pygeum africanum may have a role in promoting and maintaining optimal sperm motility. 50% of cases of infertility are traced to the female, whose chances of fertility dysfunction (FD) increase after age 30. The normal menstrual cycle is usually every 26-35 days and abnormal cycles can adversely affect reproduction but can also be corrected. 40% of cases of FD is traced to the male who, ordinarily, may remain fertile throughout most of his life (30-40% due to low sperm count or varicose veins in the testicles). A 50% decrease in sperm counts in the last century is documented and blamed on polychlorinated biphenyls (PCBs) imitating oestrogen in food, air and water; 10% of cases of FD are traced to a combination of male and female FD; 5% from unknown factors. Fertility may be corrected as result of vitamin herbal supplements such as dong quai, vitamins B_6, zinc, liquid chlorophyll, evening primrose oil and a vegetarian diet. The endocrine glands, which secrete and control hormones, depend on a correct supply of nutrients, especially trace minerals. Nutritional deficiencies and harmful chemicals can harm the eggs and sperm. Substances associated with reduce fertility or cause complete sterility, include: tobacco, caffeine, alcohol, marijuana, prescription drugs, workplace or other environmental hazards such as lead, pesticides such as DDT and DCBs, radiation (including X-rays), anaesthetic, polystyrene, xylene, some solvents, benzene, heavy metals (including arsenic, boron, cadmium, lead, manganese and mercury), fyrol (a flame retardant used on mattresses, pillows and auto seat covers and which can be found in the seminal fluid, and a proven mutagen that can cause genetic damage and birth defects), food additives (including BHA, BHT, nitrates, nitrites, oxystearin, MSG and growth hormones often fed to commercially raised animals). Pyrroloquinoline quinone (PQQ) appears to play an important role in fertility. Good sources of PQQ include parsley, green tea, green peppers, and kiwi fruit.

Primary intervention for fertility dysfunction:

Intervention	Comment
1- Herbs	
Dong quai	1- Rich in vitamin E, cobalt and iron.
	2- Takes months of regular use to benefit.
	3- For female FD
Red raspberry leaves	1- Vitamin C, A, B complex, D, E, iron, phosphorous, manganese and calcium.
	2- For female FD.
Black cohosh	1- Hormone balancing herb.
	2- It contains triterpenes and flavonoids that suppress the secretion of luteinizing hormone and has a mild but significant oestrogenic effects.
	3- Excess consumption may cause headaches.
Alfalfa	1- Contains high amounts of β-carotene and trace minerals for glandular nutrition
	2- Chlorophyll content helps detoxify the blood.
Kelp	1- Rich in trace minerals, contains significant amounts of iodine, calcium and potassium.
	2- Iodine is helpful for hypothyroidism and elevated prolactin levels.
Ho shou wu	Beneficial effects on fertility and ovulation.
False unicorn	1- Strong reputations for promoting fertility.
	2- A uterine tonic and diuretic.
	3- Effective for menstrual or ovarian pain and dysfunction.
Damiana	Increases fertility and sexual desire in both males and females.
Wild Yam	1- Contains steroid-like compounds, which are easily converted into sex hormones in the body, triggering the release of FSH that stimulates the ovaries to release an egg.
	2- High yam consumption appears to stimulate release of more than one ovum each month.
Blessed Thistle	Hormone balancer used for general female problems.
Liquid chlorophyll	Regulates menstruation.
Evening primrose oil	Contains EFAs for PGs, which induce ovulation.
Astragalas	1- Stimulates the immune function.
	2- Significantly increases sperm motility (150%)
Chaste tree berry	1- Ability to raise progesterone levels and lowers oestrogen levels.
	2- It influences these hormones by acting on the pituitary gland.
Siberian ginseng	1- General tonic.
	2- Aphrodisiac properties that enhance and increase male sexual function.
	3- Stimulates testosterone production via production of LH.
Sarsaparilla	1- Male and female tonic.
	2- Aphrodisiac.
	3- May contain hormones testosterone, progesterone and cortin.
	4- Has been used to produce a synthetic of testosterone.
Saw palmetto	1- Male tonic.
	2- Aphrodisiac.
	3- Have oestrogenic compounds.
	4- Reduces symptoms of BPE.
Pumpkin seed	Rich in zinc (vital to healthy functioning of the male reproductive system).
Bee pollen	It is produced by the male part of flowering plants.
	It contains all the necessary vitamins and ten essential amino acids necessary for a complete protein as well as enzymes and coenzymes.
	Improves sperm production in men and menstrual problems in women.
	Allergy could be a problem.
2- Vitamins and minerals	
Vitamin B_6	1- Linked to the female reproductive process.
	2- Contraceptive pills almost completely eliminates it.
	3- Increases progesterone concentrations.
	4- Deficiency leads to a build-up of oestrogen in the system and the ovary responds by shutting down its progesterone production, which can lead to abortion and luteal phase defects.
	5- Lowers prolactin, too much prolactin can prevent pregnancy.
Vitamin E	1- Improves sperm's ability to impregnate.
	2- Prevents abortion by developing a more healthy uterine wall and increasing health of the placenta.
Vitamin C	1- Prevent sperm clumping and increases sperm motility, viability and number.
	2- Deficiency is associated with increased levels of DNA damage.
	3- Clomiphene works better for women with vitamin E.
Vitamins B-complex	1- Play a central role in healthy nervous system function and hormone balance.
	2- Deficiency can create an excess of oestrogen and excess oestrogen can further deplete B vitamins.
Vitamin A	For healthy sperm production.
Bioflavonoids	Flavone corrected non-traumatic uterine bleeding within 3 menstrual cycles.
Selenium	1- Half of the selenium in the male body is found in semen.
	2- Important as zinc for healthy sperm.
	3- Deficiency is associated with sexual dysfunction and infertility.
ZINC	1- Normalise deficient sperm counts and sperm motility.
	2- Marginal zinc deficiency can cause sperm counts to drop.
3- Prescription drugs:	
Metformin	1- Used to treat obesity.
	2- May cure infertility that is associated with polycystic ovary disease-even in people without diabetes.
Clomiphene	1- 75% of women expected to achieve ovulation.
	2- Only 40% may become pregnant.
	3- Significant side effects.
	4- Higher incidence of genetic abnormalities.

ANDROPAUSE

Aging male patients who complain of hot flashes, night sweats, depression, and/or sexual dysfunction may be experiencing andropause. After age 30 years, the mean serum testosterone level generally decreases by 1-2% per year (this decline can vary widely). About half of men older than 50 could have a low early morning bioavailable testosterone level (from <60 ng/dl to <130 ng/dl). Men ages 40-70 years, 52% would have minimal, moderate, or complete impotence. 95% of this androgen is produced by Leydig cells in the testes, in response to stimulation by LH from the anterior pituitary; the remaining 5% is produced in the adrenal glands. Plasma testosterone concentrations fluctuate due to its pulsatile secretion and to its diurnal rhythm, usually peaking at about 8 AM and falling to low levels at about 8 PM. Normally, free testosterone ranges between 9.0 and 30 ng/dl; total testosterone, between 300 and 1,200 ng/dl. The decline of testosterone (hypotestosteronaemia) leads to decreased libido, erectile dysfunction, decreased spermatogenesis, fatigue, depression, confusion, and hot flashes and night sweats. Persistently low testosterone levels can also be detrimental to organ systems other than the reproductive system (e.g., leading to osteoporosis and/or anaemia). The consequences of decreased testosterone parallel certain signs and symptoms of aging (sexual dysfunction, decreased bone and muscle mass, and decreased muscle strength). Additionally, age-related declines in growth hormone production and melatonin secretion may contribute to changes associated with testosterone deficiency. Primary hypogonadism is usually the result of damage directly to the testes through a congenital defect or trauma. Age-related hormone decline may be considered primary hypogonadism if testosterone falls to pathologic levels and the patient becomes symptomatic. A delayed diagnosis of Klinefelter's syndrome (a chromosomal abnormality) can account for andropause, karyotyping can confirm or rule out suspected cases. Trauma to the testes can be physical, chemical, or infectious. Physical trauma includes direct blunt or crushing injury that damages the substance of the glands and/or blocks the blood supply to the gonads. Even if only one testicle is directly involved in spermatic cord torsion, both are subject to serious damage. Chemicals that can be toxic to the testes include chemotherapeutic agents (although these are more damaging to sperm production than to testosterone production), the antifungal drug ketoconazole, and long-term glucocorticoids. Use of spironolactone, cimetidine, phenytoin, and/or flutamide has been associated with hypogonadism. Radiation therapy can be damaging to the testes. The most common source of infectious injury to the testes is the postpubertal mumps orchitis. Chronic renal failure and other systemic diseases are associated with male hypogonadism. Secondary hypogonadism (hypogonadotropic) is caused by a disturbance in hypothalamic-pituitary function, possibly the result of a disease or defect of the anterior pituitary. Secondary hypogonadism may also be attributed to chronic illness, malnutrition, anorexia, or obesity. In case of obesity the adipose tissue contains high concentrations of aromatase, which converts androgens to oestrogens; this condition can be reversed by weight loss. Hormonal decline occurs very gradually in men. Testosterone deficiency is not the most common cause of impotence; other causes include atherosclerosis, diabetes, hypertension, and use of certain medications. Clinical findings alone, however, are not sufficient to differentiate between age-related andropause (a diagnosis of exclusion) and adult-onset disease states. Differential diagnosis for andropause: 1- Pituitary adenoma, 2- Testicular cancer, 3- Klinefelter's syndrome, 4- Chemotherapy/radiation therapy, 5- Surgery (testicular, hernia, abdominal lymph node dissection), 6- Generalised vascular disease caused by diabetes, heavy smoking, dyslipidaemia, etc., and 7- Mumps orchitis. Signs and symptoms of andropause include arthralgia and stiffness, decreased bone density, decreased libido, decreased muscle mass, decreased strength, decreased visuospatial skills, depression, erectile dysfunction, fatigue, hot flashes, increased irritability, lethargy, mood swings, nervousness, and sleep disturbances. Careful patient screening and follow-up are important, however, as testosterone replacement has been associated with several serious adverse effects. Testosterone, converted to dihydrotestosterone in the prostate, is directly linked to benign prostatic enlargement, to increased levels of prostate-specific antigen (PSA), and to the growth of prostate cancer. Prostate cancer is an absolute contraindication for testosterone replacement; occult prostate cancer may be exacerbated by testosterone. Testosterone may lower levels of high-density lipoprotein cholesterol. Testosterone is associated with a decrease in total cholesterol, low-density lipoprotein cholesterol, and lipoprotein(α), all potentially cardioprotective changes. Thus, at present, it appears possible that the total impact of testosterone on cardiovascular risk factors in older men may be more beneficial than detrimental. Sleep apnoea may worsen in patients who are given testosterone. Exogenous testosterone may also put elderly men at increased risk of erythrocytosis. Other potential adverse effects are hepatotoxicity (with oral methyltestosterone), gynaecomastia, infertility, and aggressive behaviour (especially when physiologic doses are exceeded). The therapeutic goals are to diminish or relieve the adverse effects of hypotestosteronaemia and to restore libido, sexual function, and a sense of well-being. Hormonal supplementation also appears to improve bone mass, muscle mass, and strength in older men. The circadian cycle produced by normally functioning testes is best approximated by testosterone transdermal patches or gel. Other options include oral formulations and long- and short-acting intramuscular preparations. Applied at bedtime, both transdermal patches and testosterone gel achieve peak androgen levels in the early morning. The principal drawback of transdermal patches or gel is expense; skin irritation, and perspiration may make the patch difficult to apply. Oral products currently available are all rapidly metabolised by the liver. This makes achieving satisfactory (circadian-like) serum androgen levels difficult. Prolonged use may lead to hepatotoxicity and to unsatisfactory alterations in lipid profiles. Though not comfortable, the intramuscular injection is the most cost-effective and most commonly used method of testosterone replacement therapy. Long-acting intramuscular preparations achieve a peak serum level at approximately 72 hrs postinjection and decline thereafter, with possibly subclinical hormone levels by the time of the next injection (2-4 weeks later). Short-acting preparations have a short half-life (12-24 hrs) and must be administered every other day to achieve satisfactory serum testosterone levels; they are rarely used. Total and free serum testosterone levels should be checked at one month, then every 6-12 months; at 3 months, lipid levels and haematocrite should be checked. Every 6 months, PSA levels should be checked, and treatment efficacy and adverse effects reviewed. Low testosterone levels may be common in middle-aged men with treatment-resistant major depressive disorder, perhaps because chronic depressive symptoms lead to blunting of the hypothalamic-pituitary-gonadal axis or, possibly, because of effects of the antidepressant medications themselves. Testosterone gel may improve psychological aspects of depression, as reflected in the depressed mood, guilt, and psychological anxiety. Testosterone gel may produce antidepressant effects in the large and probably under-recognised population of depressed men with low testosterone levels. Potential risk of testosterone supplementation is development of paranoid symptoms, especially in subjects taking tricyclic antidepressants. Testosterone supplementation may produce antidepressant effects in men. Normal range of morning testosterone is 270-1070 ng/dl. Perioperative hGH treatment of younger patients undergoing major abdominal surgery preserves limb lean tissue mass, increases postoperative muscular strength, and reduces long-term postoperative fatigue. Aging is associated with reduced GH, IGF-1, and sex steroid axis activity and with increased abdominal fat. GH administration increases IGF-1 levels in women and men, with the increment in IGF-1 levels being higher in men. In women, neither GH, hormone replacement therapy, nor GH together with hormone replacement therapy alters total abdominal area, S.C fat, or visceral fat significantly. In contrast, in men, administration of GH and GH together with testosterone decreases total abdominal area by 3.9% and 3.8%, respectively, and S.C fat decreases by 10% after GH, and by 14% after GH together with testosterone. In healthy older individuals, GH and/or sex hormone administration (after 6 months) elicits a sexually dimorphic response on S.C abdominal fat. Insulin sensitivity improves more in men with relatively low testosterone values. Testosterone treatment of middle-aged abdominally obese men gives beneficial effects on well-being and the cardiovascular and diabetes risk profile. The mechanisms involved in these changes might act either via effects on visceral fat accumulation, followed by metabolic improvements, and/or via direct effects on muscle insulin sensitivity. As we age our bodies natural GH production decreases, and many of the effects of aging are seen as a result. Replacing human growth hormone in IGF-1 deficient adults, can significantly eliminate these symptoms, reverse the biological effects of aging, reduce body fat, increase lean muscle mass, strengthen the heart and improve sexual performance. One year of continual GH therapy reverses 10-20 years of age decline. Human growth hormone is an endocrine hormone that is produced by the anterior portion of the pituitary gland. It is made up of 191 amino acids. The production of GH decreases with ageing. Virtually every system in the human body may be dependent on hGH for proper functioning. Growth hormone peaks during adolescence and decreases dramatically thereafter. At age 40, GH production is only 40% of what it was at age 20. One year continual use of hGH increases lean muscle mass by 9%, reduction of 14% in body fat occurs in just 6 months of hGH use. hGH replacement therapy increases energy and endurance levels significantly, which improves exercise and athletic performance. hGH replacement therapy is the only anti-aging treatment that makes people look and feel younger. Changes in skin texture are one of the earliest changes seen in treatment. The decline of the male and female libido is directly related to the age-related declines in hGH and testosterone in the body. hGH and/or testosterone replacement therapy improves sexual function (potency and frequency) in 75% of men and women. Testosterone begins to decline in men at about age 25. Androgens, especially testosterone has a number of effects on muscles, bones, the central nervous system, bone marrow, and the prostate and sexual function. Testosterone can restore muscle tone and improve stamina. Testosterone can restore healthy sexual excitement and desire, which in turn results in an improvement in mood and overall well being. Testosterone is responsible for the sex drive for both men and women. Testosterone supplementation results in increased muscle strength, muscle size, increased energy level, decreased fat and increased desire and endurance for exercise. Natural testosterone results in a decreased cholesterol and increased HDL. The method of choice is injection. Natural testosterone gel is applied to the skin. When testosterone production is reduced, men tend to undergo a drop in sexual desire and performance. They may also experience depression, fatigue, loss of motivation, and osteoporosis. The size and strength of their muscles may diminish and their body hair may become sparse. These symptoms are not specific to testosterone deficiency; however, some men with hypogonadism often don't recognise that they have a medical problem that is treatable. Elderly men who suffer from co-morbid conditions such as malnutrition, heart disease and diabetes mellitus may have a more severe decline in testosterone than do healthy men. A full endocrine evaluation is necessary before embarking on any endocrine intervention. The goal is a balanced 30 y.o physiology: 1- Optimal brain function, 2- High energy including sexual energy and performance, 3- A strong immune system to lower the risk of disease, and 4- Body composition, with normal muscle/fat ratios, not only for appearance but also, for health.

PROSTATIC DISEASES

The prostate is composed of smooth muscle, fibrous tissue, epithelium, glandular lumen and blood vessels lined by endothelium. The proportion of the various components differs for each individual prostate. Benign prostatic enlargement (BPE) is excessive growth of the glandular stromal elements. The TUR surgical approach has dominated prostatic surgery because of the high success rate and low morbidity. 20-25% of patients is expected to be unsatisfied with long-term outcome, with a reoperation rate of 15% during an 8-year period. McNeal dismissed the concept of prostatic lobes and described 3 anatomical prostatic zones: 1- The peripheral zone, which compromises around two thirds of the gland. 2- The transition zone, which compromises 10% of the gland in 2 symmetrical lobes alongside the prostatic urethra, and is separated from the rest of the gland by the fibro-muscular stroma. 3- The central zone, compromises one quarter of the volume of the gland, found mainly at the base of the prostate with the tissue surrounding the ejaculatory ducts. BPE is a non-malignant neoplasm of the prostatic stroma and epithelial tissue that causes enlargement of the prostate gland. The gland can reach up to 10 times the normal adult size in severe cases. BPE is an age-related condition. Despite suffering as a result of their symptoms, many men are reluctant to seek treatment, fear of surgery and its potential side-effects may be one of the reasons for not seeking help. There is considerable regional variation in bladder wall blood flow and oxygenation. The bladder perfusion and oxygen tension are greater at the base than at the dome whether the bladder is empty, filling, or contracting. Bladder distension and contraction, especially against a closed bladder neck, induces significant ischaemia and hypoxia of the bladder wall. In the empty bladder, blood flow and oxygen tension in the bladder base is greater than at the dome with and without outlet obstruction. Bladder filling causes a significant decrease in bladder wall blood flow and oxygen tension with or without outlet obstruction. Strong bladder contractions are associated with a significant drop in bladder wall perfusion and bladder oxygenation in obstruction or non-obstruction. Maximal bladder distension is accompanied by high intravesical pressure places the bladder, especially the dome, at significant risk of ischaemic injury. The effect of hypoxia is reversible. The amount of energy that a certain bladder can generate is limited, there is biphasic nature of the bladder contraction: 1- Initial phasic contractile response, which determines the pressure response, it is related to the intracellular ATP concentration. 2- Plateau phase, determines the ability to empty which may be linked to active mitochondrial respiration. Bladder outlet obstruction (BOO) has shown to cause a marked increase in anaerobic metabolism in the bladder. In obstruction the bladder ability to empty is affected not the ability to generate pressure. If the bladder runs out of energy before it is empty, the detrusor contraction fades away prematurely and a certain amount of residual urine is the consequence of this decompensation. Many elderly suffer from loss of bladder distensibility and urinary concentrating ability. The bladder outlet obstruction induces significant bladder hypertrophy and contractile dysfunction. Degenerative alterations in the contractile apparatus and neural innervation and alterations of bladder metabolism, are the likely functional defects. The urinary bladder metabolises glucose as its major energy source and relies on metabolic energy to support contraction. Glucose metabolised to lactate rather than incorporated into glycogen, in the bladder. The obstructed bladder tissues has: 1- a significant reduction both in glucose consumption and CO_2 generation, and 2- a decrease in aerobic metabolism and increase in anaerobic metabolism, may indicate that either reduced mitochondrial function or presence of fewer mitochondria in the obstructed tissues, and 3- a decreased oxidative metabolism. Although anaerobic metabolism is capable of producing the phasic contraction of the urinary bladder, oxidative metabolism is required for bladder emptying, which in turn is related to the ability of the bladder to maintain increase in intravesical pressure (tonic phase of contraction). The urothelium is affected in urinary overload and increased urothelial mitotic activity (borderline between the submucosal and the muscular layers) after obstructive distension and hyperdiuresis. The action on the urothelium is directly on the basal cell layer. Partial bladder ischaemia also induces a rapid bladder growth response. The growth of the bladder is due to cell division (hyperplasia) within certain compartments of the bladder as well as enlargement (hypertrophy) of select cellular components. An important concept of the compensated phase of bladder hypertrophy is its reversibility. The decompensated phase is characterised by the loss of functional emptying ability. The decompensated phase of the bladder is not reversible. The action of polypeptide growth factors is mediated by their binding to specific receptor protein on the surface of individual cells. Through this binding, they are able to initiate cellular actions as diverse as proliferation, hypertrophy and apoptosis. During the phase of bladder response (hypertrophy), there is a dramatic increase in the bladder's expression of bFGF in conjunction with a decrease in the bladder's endogenous expression of TGF-β. The bFGF is an effective mitogen for fibroblasts, endothelial cells and even some types of epithelial cells as a stimulant of smooth muscle cell development. TGF-β is an inhibitor of cellular growth processes, especially for epithelial cells. When the obstruction is removed, the expression of these 2 growth factors returns to normal, and so does the size and cellular content of the hypertrophied (compensated) bladder. It is most likely that EGF induces extensive replication of urothelial cells in the earliest stages of the hypertrophic response to BOO. EGF always present in bladder urine being synthesised and secreted by the kidney. Urinary EGF cannot normally access these basal cell receptors because the outlet layers of the urothelium (deficient of EGFR) forms a tight barrier that prevents access of urinary proteins and solutes to the basal layer. The bladder overdistension that accompanies partial outlet obstruction stretches and breaches the outer urothelial barrier so that large urinary solutes and proteins such as EGF might now penetrate to the basal cell layer. The stretched urothelial lining in an overdistended bladder would be regenerated when an effective growth factor (EGF) in the urine was able to penetrate to the deepest layer of this tissue and stimulate the regrowth of a barrier layer of urothelial cells. Growth would cease once the barrier had been regenerated. Early molecular response to BOO would be highly characteristic of a stressed tissue undergoing a growth process, includes: 1- highly induced expression of a heat shock gene products (hsp-70) in conjunction with the induced expression of certain proto-oncogenes (cmyc and Ha-ras), and 2- reduced expression of TGF-β. Cyclical repetition of ischaemia would be expected to have a very detrimental effect on the bladder, resulting in the long-term loss of highly oxygen-dependent components (smooth muscle) and the development of ischemic fibrosis. Partial denervation of the bladder induces a significant increase (50%) in bladder weight. Both obstruction (which increases the functional load of the detrusor muscle cells) and denervation (which produces bladder paralysis) are known to induce hypertrophy of the detrusor smooth muscle cells. The obstructed bladders increase about tenfold after 7 weeks. At corresponding times the denervated bladders weigh about 4-6 times more than normal. Increased weight is due to: 1) increased amount of extracellular proteins, 2) the total amount of collagen is significantly higher after 10 days, and 3) hypertrophy of smooth muscle cells. The denervated bladder is paralysed and cannot void actively. The bladder rapidly becomes overdistended. The prostate gland lies on the levator muscle as if in a sling. The growth of the prostate can extend posteriorly towards the rectum, anteriorly towards the urethra or superiorly towards the bladder. Growth towards the rectum has no adverse physiological consequences. Growth towards the urethra can result in compression causing an infra-vesicle obstruction. Growth towards the bladder results in pressure on the trigone, producing irritative symptoms. Obstruction of the urethra is a purely mechanical effect. The adenoma extends forwards and distorts the urethra so that it appears to be curved. The response of the bladder to this obstruction is governed by its vesico-elasticity (compliance) and its contractility (expel its content). In the normal bladder, collagen represents half of the detrusor mass. If there is an increase in the collagen to muscle ratio the bladder becomes more fibrous and its compliance is reduced. The possible causes of increased collagen content include chronic infection and age. Collagen present in all tissues increases with time. To overcome obstruction, micturition pressure increases and the smooth muscle fibres of the bladder hypertrophy.

The BOO temporarily causes:

1- Bladder overdistension.
2- Reduces the integrity of the urothelial barrier.
3- Reduction in blood flow to the bladder (ischaemia), which may be the stressful condition underlying all aspects in the remodelling of the bladder.
4- Elevation of vascular endothelial growth factor (VEGF).

Muscle hypertrophy gives rise to trabeculatae folds and microscopically is associated with hyperplasia of the urothelium and fibroblasts. BPE may also have a direct effect on the trigone. When hyperplasia occurs, the prostate presses into the trigone, initially only slightly, and then increasingly until it totally compresses the bladder neck anteriorly and the genital tract posteriorly. If the median lobe is particularly large, urine can be physically trapped behind it, which explains the development of post-micturition residue. The bladder responses to obstruction initially by compensated phase. It struggles to combat the obstruction and chronic irritation, leading to an increase in basal tone on filling. Eventually, the bladder enters the decompensated phase in which contraction becomes inadequate, the bladder distends and its sensitivity is reduced. Muscle failure results from collagen infiltration, dissociating the muscle cells. There are 2 distinct reactions of the bladder to obstruction, the high pressure conflict bladder or the distended low pressure bladder. Urinary disturbances in BPE result from an inadequate balance between bladder contractility and urethral resistance. BPE is frequently accompanied by hypervascularization in Retzius' space but particularly in the prostatic pedicles. Bladder instability is found in 43-86% of subjects with BPE. The pathobiology of bladder decompensation includes: 1- progressive denervation, 2- selective dysfunction of the sarcoplasmic reticulum's ability to store and release calcium, and 3- selective mitochondrial dysfunction. The tissue concentrations of choline acetyl transferase decrease by 50% in obstructed tissue, which is metabolic evidence of cholinergic denervation. There is a selective decrease in microsomal calcium ATPase, results in loss of calcium regulatory function of the sarcoplasmic reticulum. There is no decrease in the activity of myosin ATPase. The decreased oxidative metabolism of glucose and pyruvate, the decreased activities of citrate synthase and malate dehydrogenase, and the decreased rate of ATP synthesis provide metabolic proof of mitochondrial dysfunction. Partial outlet obstruction results in bladder hypertrophy, bladder hypertrophy induces localised ischaemia/hypoxia, ischaemia/hypoxia induces a net release of calcium from the sarcoplasmic reticulum, high intracellular calcium activates calpain (calcium-activated protease), phospholipase A_2 and results in the generation of free radicals and the subsequent activation of lipid peroxidases. Nerve membranes, the sarcoplasmic reticulum, and mitochondrial membranes are selective targets for calpain, phospholipase A_2 and free-radical-stimulated lipid peroxidation. Pygeum africanum (Tadenan) protect the bladder against the development of contractile and metabolic dysfunctions induced by partial outlet obstruction. Pygeum africanum protects the bladder against the reduced mitochondrial and sarcoplasmic reticulum function. Specifically, the bladder citrate synthase and microsomal calcium ATPase activities of the Tadenan. Phytotherapy of the prostate, extracts of plants and herbs are widely used to treat symptomatic BPE. Various mechanisms of action have been suggested, including the inhibition of PG synthetase, lowering of cholesterol, inhibition of 5α-reductase, blocking of androgen binding to receptors and α-adrenergic blockade. The essential fatty acid, γ-linolenic acid (GLA), present in abundance in the flower of the evening primrose and the borage plant, is a powerful 5α-reductase inhibitor. Flavonoids, plants, herbs, grains and some fungi contain phyto-oestrogens. There are 3 major chemical types of phytoestrogens in plants: flavones, isoflavones and coumestans. Flavone is strong inhibitor of COX. At different substrate concentrations flavonoids may behave as either inhibitors or stimulators of COX. The stimulatory activity may depend on the free-radical scavenging properties of flavonoids. Flavonoid interaction with PG synthase may be linked to their anti-oxidant properties. The reduction of hydroxyperoxy groups of the intermediate endo-peroxide PGG_2 to PGH_2 by hydroxyperoxidase, results in the formation of hydroxyl radical, which may inactivate COX by feedback. The anti-inflammatory activity of some free-radical scavengers is by their acting as free-radical traps for the H_2O_2 and O_2^- produced during phagocytosis. The free radical scavenging activity of flavonoids in the inhibition and stimulation of PG synthetase has not been established. Flavones and flavonones are potent inhibitors of the lipoxygenase (LOX) enzymes. The antioxidant properties of flavonoids may be related to their inhibitory activity on LOX. Flavonoids inhibit COX and/or LOX pathways with differences in potency and selectivity intermediate between indomethacin and aspirin. Some flavonoids enhance the local generation of PGI_2, a potent vasodilator, and, by scavenging platelet-derived free radicals, may be able to protect NO from destruction by superoxide anions in the vascular endothelium covered with platelet thrombi. Both COX and LOX-derived products of AA metabolism can generate free radical species, which promote mutations, chromosomal aberration, cytotoxicity, carcinogenesis and cellular degeneration with ageing. These effects may be inhibited with flavone. There are 2 distinct 5α-reductase isoenzyme, type 1 and type 2. Type 1 isoenzyme predominates in the liver and in the skin, while both isoforms are expressed in the prostate. The extract of plant Serenosa repens is a potent 5α-reductase inhibitor. The active principle of the root of Hypoxis rooperi, sitosterol, is a sterol mixture in which the phytosetrol is a glycoside bond (β-sitosterol-β-D-glucoside), claimed to be both COX and LOX inhibitor. The PG content (PGE_2 and PGE_{2a}) of adenomatous prostatic tissue is significantly reduced by oral sitosterol. Tadenan (Pygeum africanum), are used clinically for the treatment of BPE, it is consist mainly of 3-β-sitostenone, 3-β-sitosterol glucoside, β-sitostanol, oleanolic acid and small proportion of long-chain fatty acid from C-12 (oleic) to C-22 (heptadecanoic).

Mechanism of action of Saw palmetto (Permixon):

1- Saw palmetto potential mechanisms include 5α-reductase inhibition, adrenergic receptor antagonism and intraprostatic AR blockade.
2- Saw palmetto is a non-competitive inhibitor of type I 5α-reductase and uncompetitively inhibit the type II isoenzyme.
3- Saw palmetto inhibits DHT binding to the cytosolic AR in prostate cells.
4- Saw palmetto inhibits the nuclear oestrogen receptors.
5- Saw palmetto has anti-inflammatory effects in the prostate via inhibition of AA metabolites and ROS produced by neutrophils.
6- Saw palmetto inhibits FGF induced prostatic epithelial proliferation.
7- Saw palmetto has α-adrenergic inhibitory property. Saw palmetto has antioedematous effects.
8- Saw palmetto modulates prolactin induced prostatic growth by receptor signal transduction.

Permixon (Saw palmetto) is a lipid-sterol extract of Serenoa repens (LSE-Sr), consists of free (90%) and esterified (7%) long-chain fatty acids and also contains several sterols, notably β-sitosterol, campesterol and stigmasterol, flavonoids and other substances. Permixon possess a peripheral anti-androgen effect by inhibiting 5α-reductase, 3-ketosteroid reductase and receptor binding of androgens. Permixon is a patent non-competitive and uncompetitive inhibitor of both the type 1 and 2 5α-reductase, respectively. LSE-Sr competitively blocks the translocation of cytosolic oestrogen and androgen receptors to the nuclei. LSE-Sr decreases progesterone receptors. 320 mg/day permixon inhibits 5α-reductase 3 times more strongly than finasteride (5 mg/day). Permixon inhibit 17β-dehydrogenase, the enzyme responsible for converting androstenedione from the adrenals and testes into testosterone. Permixon inhibits phospholipase A_2 and 5-lipoxygenase enzymes, blocking the release of free AA, the precursor fatty acid of the pro-inflammatory eicosanoid metabolites PGs and leukotrienes. Permixon does not have the undesirable side-effects of ED or loss of libido. Saw palmetto improves flow rate (FR) and symptoms in men with LUTS. The use of saw palmetto in patients with BPE is safe with no recognised adverse effects. Saw palmetto has no effect on serum PSA. Cernilton (Secale cereale) is a mixture of extracts, pharmacologically active fractions, a water-soluble fraction (T60) and a fat-soluble fraction (GBX). T60 acts on α-adrenergic receptors while GBX relaxes the external sphincter and in combination may relax both the internal and external sphincter muscle.

GBX inhibits 5α-reductase activity in the epithelium and stroma of the prostate. Cernilton inhibits the testosterone stimulated growth of the prostate, increases tissue and serum zinc levels. Cernilton has a beneficial effect in chronic non-bacterial prostatitis. Despite an abundance of studies and publications, there are still many controversies concerning the diagnosis and management of prostate cancer (CaP). The prostate gland compromises one anterior lobe, one posterior lobe, one median lobe, and two lateral lobes. The prostate gland consists of 3 zones: central, transition and peripheral. The prostate secretes substances that help maintain the motility and viability of sperm. BPE involves tissue growth within the transition zone, while cancer derives from glandular tissue in the peripheral zone. CaP is second to lung cancer as a cause of cancer related mortality in men. Almost 30% of men above 50 have evidence of malignant cells within the gland, but only 9.5% will have a clinical disease with a risk of mortality at 2.9%. About 45% of men older than 75 years have autopsy evidence of CaP. The overall lifetime risk of invasive CaP is 15% and an average of 6.4%/year chance to develop CaP in patients aged between 60-80 years. A man with a first degree relative (father or brother) has a 16% risk to develop CaP compared with 8% for men with a negative family history. There is increased risk among printers, painters, rubber workers, textile workers, mechanics, loggers, ship fitters, farmers, and drug and chemical workers. The cause of the disease is unknown. Patients are often asymptomatic at the time of diagnosis. Signs and symptoms of CaP may range from genitourinary distress to features of advanced metastatic disease. Signs of distant spread often results from bone metastases in the hips, spine, or legs and include pain in the back, hips, or perineal area, fractures, and joint stiffness. Some patients with minimal capsular invasion harbour cancer outside the prostate. Vitamin E can significantly inhibits the high-fat diet induced progression of human CaP. Oxidative pathways are important in prostate cancer progression and vitamin E may have beneficial preventive properties. Vitamin D is an important determinant of prostate cancer risk. Men with the most active form of the receptor have a lower frequency of CaP. Vitamin D may have a role in chemoprevention of CaP. Suramin may reduce PSA expression in CaP lines without affecting tumour growth. There are higher death rates from prostate cancer in current cigarette smokers; smoking may adversely affect survival in prostate cancer patients. Tobacco seems to have an antioestrogen effect that could alter the production, metabolism, and degradation of oestrogen and decrease the oestrogen available in the prostate. Cigarette smoking increases the production of DHT, circulating levels of androgen and the ratio of androgens to oestrogen at the target cell, which may contribute to the progression of CaP. Cigarette smoking increases OFRs, induces ischaemia and DNA mutations. No material association between cigarette smoking and prostate cancer incidence or mortality was found in CaP. Seminal vesicle invasion is not associated with a uniformly poor prognosis. Epidemiologically bladder and prostate cancers are predominantly diseases of male individuals with bladder cancer almost 4 times more prevalent in and prostate cancer exclusively limited to the male population. Anatomically the bladder and prostate have the same embryological origin and they are intimately related spatially. There are also similarities at molecular level. The rate of bladder cancer in patients with prostate cancer is 18 times higher, and the rate of prostate cancer in those with bladder cancer is 19 times higher. With up to a 70% rate of CaP in patients with bladder cancer. Prostate cancer is an epidemic world-wide disease. The incidence of clinical CaP has increased dramatically in past years due to greater awareness of the male population and increased use of screening tests. The main environmental risk factors are linked with a western life-style, e.g. a high intake of dietary fat, which is thought to play an important role in the multistep carcinogenesis of prostate cancer. Prostate cancer is a disease of old men, more of whom die with the disease than from the disease. For a 50-year-old, the life risk of developing cancer of the prostate is about 42%, the risk of developing the disease clinically is 6.1% at the age of 60 years, and the risk of dying from the disease is 2.9%. In men older than 55 years, the incidence of CaP increases by 65.6%, while the mortality due to CaP increases by 14%. Mortality rate of CaP is 5.1% per year. Prostate cancer increases faster with age than any other major cancer, and with lengthening longevity, the burden of illness of CaP will increase in the future. There is threefold increase in risk in first degree relatives of men died of CaP. Men with positive family history of CaP, after the age of 40 years, should undergo a DRE and serum PSA monitoring yearly. Familial prostate cancer has 2 important factors: 1- early age at onset, 2- the number of multiple affected family members. The estimated cumulative risk of developing prostate cancer is 88% for carriers as compared with 5% for non-carriers. Low animal fat and high fiber content essentially vegetables and fruit, may relate to the lower incidence of clinical CaP in the East than the West. Animal saturated fat intake: 1- may act as the environmental promotor responsible in part for the extreme variation in CaP progression, 2- is positively associated with the number of deaths from prostate cancer, and 3- is linked to the changes in risk that follow migration. A high legume content (beans, soyabean, lentils, and chick peas), health benefit is probably related to non-nutrient components such as isoflavonoids and lingans, particularly found in soyabean, tofu, grains, flaxseed, rapeseed, whole cereals, and asparagus. These foods are metabolised by the normal microflora to produce diphenolic weak oestrogens such as lignan enterolactone and the isoflavonoid genistein. Weak oestrogen agonistic or antagonistic properties like weakly oestrogenic tamoxifen used in the management of breast cancer. A high fibre and low fat intakes are more likely to protect men from prostatic cancer. Vitamin A and its precursor β-carotene are essential for normal epithelial cell proliferation and differentiation. Vitamin A, present in plants as β-carotene, reduces the risk of CaP. Vitamin A intake from animal sources increases the risk CaP. Individuals with low serum retenoid levels (a precursor of vitamin A) or a low dietary intake of foods containing retenoids have an increased risk of CaP. Retinoic acid mediates a downregulation of saturated fatty acid and an upregulation of unsaturated fatty acid in humans with CaP. The most important source of vitamin D is photosynthetic reaction of 7-dehydrocholesterol with sun light in the skin. Black skin, the protective effect of ultraviolet light mediated by vitamin D, is inhibited by cutaneous melanin. Blacks absorb less ultraviolet light and have lower vitamin D levels. Ageing, is accompanied by low serum levels of vitamin D. Diet that is high in fish oil is rich in vitamin D. Rubber and tyre workers are faced with a wide variety of environment hazards from the complex manufacturing processes involved in this industry. Farmers are at increasing risk of CaP, although pesticide exposure is only weakly applied. Zinc is necessary trace element in multiple intracellular metabolic pathways; for example, polymerases require zinc for replication and repair of DNA and RNA. In men with CaP, the prostate gland levels of zinc are significantly lower than in the normal prostate gland. Cadmium increases the risk of CaP by interacting with zinc. Hormones are known to have stimulatory effects on the development and growth of prostate epithelium cells and smooth muscle cells, and presumed to play an important role in the development of CaP. The level of testosterone in neoplastic tissue has been found to be higher than in the hyperplastic prostate, although the levels of DHT are higher in the hyperplastic gland. Total testosterone serum levels are generally decreased with high-fibre and low fat diets. Prolactin and oestrogens may also play a defined role in the prostate metabolism. There is a lower incidence of CaP in cirrhotic (3.3%) than non-cirrhotic (9%) patients, suggesting that prolonged exposure to high oestrogen levels could prevent or delay the development of CaP. There are a number of similarities between CaP and BPE: 1- the parallel increase in prevalence with age, 2- both require androgen for growth and development, and 3- both may respond to anti-androgen treatment regimens. BPE is present in around 80% of the patients with CaP. Death rate for prostate cancer is 3.7 times greater in BPE patients. The relative risk of developing CaP is calculated as 0.88-5.1 in patients previously treated for BPE. Alcohol, tobacco, and obesity have been reported to influence serum testosterone levels. Alcohol has an acute depressant effect on testosterone levels, and a small non-significant increase in CaP risk is evident. Cigarette smokers have been reported to have higher testosterone levels than non-smokers. Obese men have been reported to have lower circulating testosterone levels. Although the overall metastatic CaP population has decreased during the past 15 years, there is still no apparent decrease in mortality from CaP. The development and maintenance of prostatic differentiation is androgen-driven via the foetal mesenchyme and adult stroma, without a requirement for epithelial androgen receptor (AR). Growth and morphogenetic effects of androgens are believed to be mediated by growth factors acting between the epithelial and mesenchymal/stromal components of the glands. Muscle bands are thickest at the proximal, and thinnest at the distal ends of the prostatic ducts. The prostatic epithelium to fibroblastic stromal cells at the distal tips may be responsible for the continued low levels of proliferation found in these regions into adulthood. Expression of α-actin is followed by vinculin, myosin, desmin, and laminin. These markers are localised to the developing smooth muscle sheaths and are not expressed in the intraductal tissue of the prostate or the lamina propria of the seminal vesicle (SV). Genetic damage is a prerequisite for cancer development. Genetic damage alone is not sufficient to account for most cases of cancer growth and progression, only 9%. Prostate cancer consists of a heterogeneous population of cells of differing malignant potential that can be derived from multiple progenitor "stem" cell clones. Growth, differentiation and progression of cancer depend not only on intrinsic genetic changes in the cancer cells, but are also closely associated with the host organ's microenvironment. Agents for prevention and treatment would have to target both tumour epithelium and host microenvironment (such as tumour stroma, angiogenesis, and host immunity). Interaction between susceptibility genes (genetic) and environmental exposure (epigenetic) cause prostate cancer to develop and progress. At the cellular level, epigenetic factors may play a critical role affecting CaP growth, differentiation and metastasis. Prostate cancer is characterised by unpredictable biologic behaviour, and long latency before diagnosis. Undiagnosed microscopic foci of prostate cancer exist in 30-50% of men over the age of 50 and in nearly half of men in their eighth and ninth decades of life. It is estimated that 9 of 10 prostate cancers will remain clinically silent, and that <1% of men with histologically identifiable CaP actually die of it. At autopsy, the incidence of histologic CaP of men in their 30's and 40's are 27% and 34%, respectively. 15-20% of untreated early stage patients will develop metastatic disease within 10 years. CaP invariably becomes hormone resistant, most patients relapse at a median of 18 months. Molecular mutations correspond with histologic changes, as the tumour becomes more aggressive. The carcinogenic initiation of CaP appears to be a frequent event occurring in nearly one-third of all men over 45 years of age. Inherited and noninherited forms of CaP can share common genetic lesions. Men with two to three first-degree relatives affected with CaP have a five fold and 11-fold increased risk, respectively. A cell deficient in glutathione transferase production lacks one means of detoxifying potential carcinogens, found both in small latent foci and regions of advanced CaP. The mutation involves the inactivation of a prostate glutathione-s-transferase gene is via methylation (adding of a methyl group). The methylation in CaP suggests novel chemopreventive strategies to increase the activity of other functional glutathione transferases. The

Glutathione-s-transferase gene methylation could be valuable in molecular staging. Histologically CaP is heterogeneous with only a proportion having completed the process required to produce a clinically aggressive cancer. The natural history of CaP is unpredictable. Preneoplastic lesions of CaP are: 1- atypical adenomatous hyperplasia (AAH). Its role is unclear, and 2- prostatic intraepithelial neoplasia (PIN). Androgens are essential for development, differentiation as well as maintenance of morphology and secretory function in normal prostate. Androgens are necessary but not sufficient for sustained proliferation of neoplastic prostate cells. No oncogene has been correlated definitively with the initiation or progression of CaP. Growth factors are proteins that may be very important not only in normal but also in abnormal prostate growth, particularly in the early stages of neoplastic transformation. Hypoxia promotes tumour aggressiveness and metastases. Tumour cells appear to adapt to hypoxia by secreting angiogenetic factors. Hypoxia-induced factor 1 (HIF-1), a transcription factor involved in the cellular response to hypoxia, leads to increased transcription of VEGF, a key marker for angiogenesis. Hypoxic regions exist in CaP. Increasing levels of hypoxia in prostatic tissue correlates with increasing clinical stage and patient age. Oxidative stress results in damage to cellular structures and is linked to cancer. Antioxidant enzymes are endogenous proteins that work in combination to protect cells from ROS damage. ROS are physiologic by-products capable of directly injuring cells. Oxidative damage is diminished by antioxidant vitamins, non-vitamin A carotenoids, and trace elements such as selenium. Antioxidant enzyme expression is lower in PIN and CaP than in basal cells of benign prostatic epithelium. Antioxidant enzyme imbalance and oxidative stress alter cellular redox, arresting cell proliferation and growth by cell cycle arrest due to an activated p53 protein, inducing apoptosis, activating signal transduction pathways, activating transcription factors such as Fos, Jun, and NFκB and increasing mitochondrial activity. Antioxidant defense mechanisms, including ROS detoxification enzyme activity, decline with age. The antioxidant activity of selenium and vitamin E may account for their ability to prevent CaP. The risk for CaP seems to be reduced by certain antioxidant compounds (vitamins E and A, and selenium). Antioxidant enzymes and oxidative damage products are modulated in metastatic compared to primary CaP. There is a different oxidative metabolism in malignant prostate epithelium than normal prostatic epithelial cells. Metastatic lesions from primary CaP have higher levels of manganese superoxide dismutase and nuclear oxidative damage products than primary tumours. Glutathione (GSH) maintains an optimum cellular redox potential. Chemical depletion, physical efflux from the cell, or intracellular redistribution of this thiol antioxidant is associated with the onset of apoptosis. Thiol depletion can be used as an effective means of activating caspase-3 in both androgen sensitive and insensitive CaP cells. Direct activation of this effector caspase may serve as a useful strategy for inducing apoptosis in CaP cells. Antioxidants modulate human prostate cancer cell proliferation by altering apoptosis. Vitamin E triggers apoptosis in human CaP cells but not normal prostate cells, and modulates Fas signalling. Many human CaP cells have escaped the apoptotic effects of natural regulators of cell growth such as TGFβ-1 and TNF. Antioxidants modulate human CaP cell proliferation by altering apoptosis in dividing cells. Calcitonin-related peptides have been found in the human prostate, and calcitonin (CT) and calcitonin gene-related peptide (CGRP) have been demonstrated in subpopulations of neuroendocrine (NE) cells. CT and CGRP are present in NE cells of the human prostate. Calcitonin levels are significantly reduced in BPH, in parallel with a decreased number of CT-immunoreactive NE cells, and no significant changes in tissue levels of CGRP. Eicosanoids modulate the interaction of tumour cells with various host components in cancer metastasis. Their synthesis involves the release of arachidonic acid (AA) from cellular phospholipids by phospholipase A_2 (PLA2), followed by metabolism by cyclooxygenases (COXs) and lipooxygenases (LOXs). PGE2 production is necessary for rendering the cells invasive-permissive but not sufficient for inducing invasiveness. Prostaglandins (PGs) and COX enzymes may be involved in the initiation and/or the promotion of carcinogenesis. COX is a rate-limiting enzyme in PG synthesis because of its rapid autoinactivation. COX-1 is present in most tissues and is involved in the physiologic production of PGs for maintaining normal homeostasis. COX-2 is induced by mitogens, cytokines, and growth factors. COX-2 is responsible for PGs produced in inflammatory sites. COX is the first oxidase in the process of PG production from AA. All other PGs and TX are synthesised from PGH_2. COX-1 acts to maintain cellular homeostasis. COX-2 activity is very low in the normal state. COX-2 is modulated by the degree of CaP differentiation. COX-2 is high in the blood vessels and stromal tissues of CaP. COX-2 gene is up-regulated in several cancer tissues. COX-2 is induced in CaP. PGE_1 and PGE_2 (produced by COX-2) have a role in the proliferation of the malignant cells and metastasis of CaP. Inhibition of COX-2 development may lead to inhibition of the proliferation and metastasis CaP, and inhibition of prostate carcinogenesis. COX disturbing the balance between matrix metalloproteinases (MMPs) and tissue inhibitors of metalloproteinases (TIMPs) in prostate cancer cells. COX-2 selective inhibitors have a role in the prevention and therapy of prostate cancer invasion. COX-2, an inducible enzyme, which catalyses the formation of prostaglandins from arachidonic acid, is expressed in CaP cell. Selective COX-2 inhibitor suppresses CaP cell growth. Tumour growth suppression is achieved by a combination of direct induction of tumour cell apoptosis and down regulation of tumour VEGF with decreased angiogenesis. Nonsteroidal antiinflammatory drugs inhibiting COX enzyme activity in both its constitutive (COX-1) and inducible (COX-2) isoforms. Human CaP cells generate COX-2, and that COX-2 might play an important role in the proliferation of prostate carcinoma cells. Inhibition of COX-2 development may lead not only to inhibition of the proliferation and metastasis of prostate carcinoma but also to the inhibition of prostate carcinogenesis. There is a very weak expression of COX-1 in CaP cell. COX-2 is an inducer of angiogenesis of new blood vessels. The arachidonate 12-lipoxygenase, an enzyme metabolising AA to form 12(S)-hydroxyeicosatetraenoic acid (HETE), has a role in prostate cancer progression. As prostate cancer reaches a more advanced stage, the level of 12-lipoxygenase expression is increased. Overexpression of 12-lipoxygenase in human prostate cancer cells stimulates angiogenesis and tumour growth. The inhibitor of 12-lipoxygenase is effective against metastatic prostate tumour growth, and the inhibition of 12-lipoxygenase is related with the reduction of tumour angiogenesis. Prostate cancer patients experiencing a relapse in disease often express high serum TNF-α levels. Many androgen-insensitive CaP cells are TNF-α-insensitive because of the expression of antiapoptotic genes as part of the NF-κB family of transcription factors. NF-κB stimulates gene transcription when expressed in the nucleus; however, in resting cells, this nuclear import is prevented by association with the cytoplasmic inhibitor IκBα. This cytoplasmic retention of NF-κB is uncoupled by many extracellular signals including low levels of TNF-α. During normal cell activation, nuclear translocation of NF-κB is preceded by phosphorylation and degradation of IκBα. When phosphorylation is blocked, IκBα remains intact, thereby blocking NF-κB translocation to the nucleus and subsequent activation of antiapoptotic genes that cause TNF-α insensitivity. Inhibition of NF-κB selectively sensitises previously insensitive CaP cells to TNF-α, and decreases IL-6 production by TNF-α. TNF-α may play a role in the initiation of an androgen-independent state in CaP through its ability to inhibit AR sensitivity in CaP. IL-6 stimulates osteoclastic bone resorption. IL-6 appears to play a role in the development of bone metastases from CaP. PTHrP 1-34 stimulates IL-6 production. IL-6 is an enhancer/helper factor rather than a primary bone resorption factor in pathologic conditions associated with increased bone resorption. In tissues with AR, androgen is converted to dihydrotestosterone (DHT) in the cytol. Dihydrotestosterone is 7β-hydroxy-5α-androstan-3-one, a powerful androgenic hormone, formed in peripheral tissue by the action of the enzyme 5α-reductase on testosterone, it is thought to be the essential androgen responsible for somatic virilization during embryogenesis, for development of most male secondary characteristics at puberty, and for adult male sexual function. DHT is 3 times more potent than testosterone; testosterone is 5-10 times more potent than adrenal androgens. DHT is the key androgen in the human prostate responsible for prostate growth, the defect in the male pseudohermaphrodite syndrome associated with familial pseudovaginal perineoscrotal hypospadias involves a deficiency of 5α-reductase enzyme and thus a lack of DHT. Children with the disorder have ambiguous genitalia and under-developed prostate glands with puberty and a resultant rise in serum testosterone, masculinization occurs including penile enlargement and scrotal rugation but the prostate glands fails to develop and CaP does not occur. Androgens synthesised in both the adrenal cortex and testis ranges 3-10 mg/ml, of which 2% is present unbounded and therefore biologically active form. Testosterone entering the prostate directly or after release from the plasma steroid hormone-binding protein is either free to bind to its receptor, locating almost exclusively to the luminal epithelial cells, or is metabolised.

The digital rectal examination (DRE) serious limitations:

1- access is possible only to part of the gland.
2- a lesion could be located in an area that is not palpable.
3- a lesion may be too small to palpate.
4- only disease affecting the gland itself is detectable.
5- distant disease would not be discernible with this method.

Only one third of prostatic cancers become clinically manifested and one in three patients die of the disease. The natural history of prostatic carcinoma is unpredictable. In cases of relapse after adequate endocrine therapy, 90% of patients die from their disease within 2 years. In advanced tumours various degrees of differentiation are found. Undifferentiated tumour cells would have lost their potency to respond to androgen deprivation and their growth activity is responsible for progression towards androgen-independent disease. Prostatic atrophy (PA) is one of the most frequent mimics of prostatic adenocarcinoma. It occurs almost exclusively in the peripheral zone of the gland and gained importance with the increasing use of needle biopsies for the detection of prostatic carcinoma. PA is characterised by the presence of generalised atherosclerosis and benign or malignant nephrosclerosis. Fibrosis of the stroma may or may not be present in simple and hyperplastic atrophy. Hyperplastic atrophy associated with fibrosis of the stroma is the histologic subtype that most frequently mimics adenocarcinoma. PA increases with age, and ischaemia caused by local intense arteriosclerosis seems to be a potential factor for its aetiopathogenesis. PA is probably not a premalignant lesion. CaP has a constantly evolving histologic appearance with a general trend toward dedifferentiation with time. The Gleason scale evaluates the differentiation of the cells of the tumour and its growth pattern to determine how aggressive it may be, which assigns a grade of 2-10 to a tumour. Grading is important because there is a correlation with survival time. The small cell carcinoma of the prostate is not hormonally sensitive. Tumour markers are substances that present in the body in concentration, which is related to the presence of a tumour. Sensitivity measures elevated tumour marker level in the presence of particular tumour. Specificity measures proportion of patients without tumour with normal marker levels, i.e., true negative. Positive predictive value is the percentage of positive results of tumour marker, which is true positive. Predictive value is the conditional probability that a clinical test result correctly identifies a patient as having or not having a disease, i.e., the predictive value of a positive test (positive predictive value) is the probability that a person with a positive test is a true positive (i.e., does have the disease) and the predictive value of a negative test (negative predictive value) is the probability that a person with a negative test does not have the disease.

Variant of CaP:

Adenocarcinoma and associated tumours.
Adenoid cystic carcinoma/basal cell carcinoma.
Adenocarcinoma with neuroendocrine cells.
Adenocarcinoma with oncocytic features.
Ductal carcinoma (endometrioid carcinoma).
Lymphoepithelioma-like carcinoma.
Mucinous carcinoma.
Neuroendocrine carcinoma.
Sarcomatoid carcinoma.
Signet ring cell carcinoma.
Squamous and adenosquamous carcinoma.

The predictive value of a screening test is determined by the sensitivity and specificity of the test, and by the prevalence of the condition for which the test is used. The ideal tumour marker should have 100% sensitivity and 100% specificity. Tumour markers are used in screening, diagnosis, as prognostic indicator, to monitor therapy, and for early diagnosis of relapse. Tumour markers may have a high sensitivity in patients with advanced cancer but most have a low sensitivity in patients with early stage cancer. Metastasis is highly selective and consists of a series of sequential, interrelated steps that include growth, vascularization, invasion, and survival in the circulation, adhesion, extravasation, and proliferation at the distinct site. Cancers can enzymatically produce antiangiogenic factors (angiostatin) from precursor proteins, that has the ability to inhibit angiogenesis. There is only 1% chance to die from CaP. We can not predict which cancer will be aggressive. Most patients will suffer the side effects of therapy. 40% of patients with PSA 4-9.9 µg/ml will already have tumour spread outside the prostate. More PSA is protein bound in patients with CaP than BPE. Not all PSA >20 µg/ml have distant metastases.

PSA levels vary with age:

Age	PSA values
40-49 years	0-2.5 ng/dl
50-59 years	0-3.5 ng/dl
60-69 years	0-4.5 ng/dl
70-79 years	0-6.5 ng/dl

Lymph nodes metastases are associated with elevated PSA. The PSA has been termed the best marker by oncology in the early 1990s. In 1971, the PSA was identified in the seminal plasma, initially was called γ-semino protein, it has been used as a forensic marker in rape cases. In 1979, PSA was isolated from human serum. PSA is a glycoprotein consisting of 240 amino acids with enzymatic activity (MW 33,000-34,000), produced by epithelial cells in prostate glands. PSA production is androgen dependent property of cloacogenic glandular epithelium and has been demonstrated in urachal remnants, anal glands and periurethral glands, it is produced by normal and malignant prostatic cells in men and Skene's glands in women. PSA is produced by other organs in the body, e.g., breast, pancreas, etc. Free PSA half life is 12 hrs and is filtered by the glomeruli. Complex PSA half life is 2.2-3.2 days, and is thought to be cleared by the liver as it is too large to be filtered by the kidney. Several weeks may be necessary for PSA to become undetectable after radical prostatectomy. PSA is abundant in the prostate 10,000 times higher than other organ. Men must abstain from ejaculation for 42 hr prior to PSA sampling. Free PSA increases above the expected biological variation in 48% of men 1 hr after DRE (90%). Prostate needle biopsy causes a significant increase in free and total PSA that remains elevated for at least 1 week. Men with BPE have increased urinary PSA. Urinary PSA and serum-to-urinary PSA ratio provides separation between BPE and carcinoma in patient aged 50 year with PSA range of 4-10 ng/ml. An elevated PSA is not diagnostic of CaP. It is likely that PSA production in women is due to other cognate steroid hormones, such as progestin or glucocorticoid. Conditions and procedures associated with increased level of PSA are: 1- Benign prostatic enlargement. 2- Manipulation of the prostate, PSA is marginally affected by DRE that had little clinical impact. More vigorous prostate manipulation such as prostatic massage, TURP and prostate biopsy may cause more striking elevations in the serum PSA, remaining elevated from 2 weeks to 6 weeks. 3- Cystoscopy. 4- Prostatic infarction. 5- Chronic prostatitis. 6- Prostate cancer, is due to the increased cellularity of proliferating CaP, not to increased rates of PSA production by individual CaP cells. The value of PSA measurement in screening for CaP has been controversial since a high proportion of patients with BPE has abnormal PSA levels. The initial hyperplasia (BPH) has abnormal PSA levels. PSA level is insufficiently sensitive to be used alone as a screening test for CaP. Age related changes in prostate gland growth, provides increased sensitivity in young men and increased specificity in older men, and eliminates the need for volume measurements or density calculations. The ultimate value of this approach is that it decreases the number of biopsies in older patients with PSA values >4 ng/ml and that it may increase the rate of cancer detection in younger men with PSA values <4 ng/ml. The best results are seen when all 3 evaluations are performed. Screening should include both DRE and PSA measurement, with ultrasound added if either result is positive. It is uncertain whether early detection has any effect on survival time, and early asymptomatic patients should be treated or simply watched. The goal of annual evaluation is to detect tumours when they are small and confined to the gland and therefore curable. PSA levels should decline to zero after a radical prostatectomy in patients with disease confined to the gland and normal or undetectable levels after curative radiation therapy. Failure of PSA concentration to fall after definitive therapy usually indicates residual disease. Elevation after an initial fall to undetectable levels may be the earliest indication of recurrent disease. The rate of fall of PSA values after radiation therapy is dependent on the initial value, gland volume, and degree of BPE present, quantity of cancer left. PSA is exocrine-secreted protein, which develops its physiological function in the seminal plasma, high concentration of PSA is found in seminal fluid and urine. The presence of urinary PSA was first described in 1985. PSA is not excreted by human kidney. It is excreted by urachal remnants and periurethral glands. It is mixed in to the urine by passage through the prostatic urethra; it is higher in the first portion of voided urine than in mid stream. PSA urine concentration in urine after radical prostatectomy seems to be of some value in the follow up. Patients with small tumour but extensive boney metastases have high serum and low urinary PSA levels, whereas patients with locally advanced tumours without distant metastases have moderate serum and high urinary concentrations. The determination of PSA in serum and urine might allow distinguishing between local or distant progression of prostate cancer.

The positive predictive values in CaP are:

Screening methods	Values
DRE.	22-36%
TRUS.	15-41%
PSA >4 ng/dl.	22-35%
PSA >10 ng/dl.	65-67%
Abnormal DRE & TRUS.	37-61%
Abnormal TRUS & PSA.	33-52%
Abnormal DRE, TRUS & PSA.	62-74%
Normal DRE & PSA, positive ultrasound.	4-9%

Prostatic acid phosphatase (PAP) is the prostatic isoenzyme of the widely distributed enzymatic acid phosphatase. It is a glycoprotein that hydrolyses protein tyrosine phosphate esters into inorganic phosphate ions and alcohol. PAP has considerable cross-reactivity with other acid phosphatase isoenzymes of blood cells, bone, breast, kidney, and liver. PAP is valuable in identifying prostatic epithelium. PSA and PAP are secreted mostly but not exclusively by prostatic epithelium under androgenic regulation. Extraprostatic sources are in pancreas and salivary glands. The acid phosphatases are group of enzymes capable of hydrolysing esters of orthophosphoric acid in an acid medium resulting in the splitting of O-P bounds with the release of phosphoric acid. Acid phosphatases were first demonstrated in erythrocytes in 1924, in urine 1925, and in spleen and liver in 1934. Human ejaculate and the prostate are rich in acid phosphatase similar to that of urine. Although acid phosphate has its highest concentration in the cells and secretions of the prostate gland, this enzyme is also present in erythrocytes, leukocytes, platelets, liver, spleen, kidney, and other tissues. The acid phosphatases are widely distributed in all human body fluids and tissue. Histochemical studies indicate that the tissue enzyme activity is located mainly in glandular epithelium and that much of the acid phosphatase is of lysosomal origin. The enzyme activity is highest in prostate, with variable activity in cancers of the breast, stomach, and colon and low activity in cancers of the thyroid, kidney, and ovary. Acid phosphatase was the first tumour marker to be measured in the blood and more than 50 years have passed since an elevation of serum acid phosphatase concentration was observed in patient with CaP. Cells from many common neoplasms seed bone marrow via the blood stream as an early event. The subsequent growth of these cells into clinically significant metastatic lesions is associated with their ability to stimulate bone resorption through osteoclasts and macrophages or through a direct action on bone. The products of bone resorption (e.g., growth factors), act on the tumour cells to stimulate the expression of properties that promote their metastatic competence. Both bone marrow and bone include dynamic cell populations. Bone marrow is a source of mitogenic peptides that can stimulate the growth of cancer cells. Within the bone microcompartment, metastases are more frequent at sites of red marrow, where vascular sinusoids are lined by endothelial cells. Bone sites with CaP metastases display distinctive osteoblastic reactions, characterised by high bone turnover rates with increased osteoid surface, osteoid volume, and mineralisation rates. CaP cells can produce peptides with selective mitogenic activity for osteoblasts, and these cytokines may stimulate the proliferation of osteoblasts. Most cancer cells have ameboid properties and move by pseudopodial extension. Cancer cell migration and chemotaxis contribute to metastasis by promoting tumour invasion and extravasation. Bone resorption may occur in the absence of metastases and may be more widespread leading to osteopenia. Osteomalacia may be a manifestation of tumours either when bone formation is accelerated and mineralisation does not keep pace or when tumours overproduce a phosphaturic substance. Osteomalacia may occur when bone formation induced by CaP is so rapid that mineralisation lags behind. Osteoblast growth factors activity as well as enzymatic activity, are possible mechanisms. CaP is the most common neoplasm that is associated with osteoblastic lesions. All tumours metastatic lesions to bone are space-occupying lesions, all induce bone resorption in addition to bone formation. The predominant skeletal response to CaP is the production of new woven bone as a consequence of osteoblastic activation. Cancer may induce personal and family pressures about difficult medical decisions and the uncertainty of the prognosis. There are often concerns about money; employment and the travelling involved for medical treatment. Together with fatigue, nausea, pain and the specific ill effects of cancer, it is not surprising that there are repercussions for sexual health.

Problems with PAP:

1- Interference by acid phosphatase from other tissues.
2- Diurnal/random fluctuation.
3- Influence by prostatic manipulation (DRE in particular).
4- Requirement for proper collection and processing of specimen because of enzyme instability.

Marital difficulties do not occur in previously happy couples, although difficulties for couples already in some conflict can be exacerbated. Surgery is the most dramatic therapy to influence sexual health. Prostatitis is a clinical syndrome manifest as a result of prostatic inflammation. Inflammation in the absence of bacterial infection is more common. Prostatodynia, is a condition presented with similar features as prostatitis, but in which there is neither infection nor inflammation. Prostatitis is associated with an alteration in the concentration of prostate-derived seminal factors. Bacteria in sperm are associated with decreased mobility of sperm, but they could not link this to changes in prostatic constituents. Infertile men with bacteriospermia have a high pH, decreased seminal fluid volume and concentrations of citric acid, fructose, and acid phosphatase. Citric acid concentration is best parameter in discriminating those with prostatitis. A worsening in the quality of sex life has been reported in up to 25% of men with BPE. Retrograde ejaculation occurs in >73% of patients who had TURP, similar to open prostatectomy 77%. Postoperative ED occurs in 3-35% of patients. Effect of CaP on sexual function includes erectile dysfunction, ejaculatory disturbance, and loss of libido. The progression from PIN to early latent CaP is thought to require 10 years or more, with clinically significant cancer occurring some 3-5 years later. CaP is ideal for chemoprevention because of the features of the disease, including a high prevalence, long latency time, hormone dependency, and the availability of a marker, the availability of precursor lesion (PIN). The risk of CaP may be 70% greater in men with high energy intake. Animal dietary fat may be converted to androgens, with a resultant increased androgenic stimulation of the prostate that increases risk of hormonally induced tumours. High fat diet is associated with increased production of both androgens and oestrogens. There is increased risk of CaP in men with higher intakes of α-linolenic acid; α-linolenic acid is present essentially in red meat, butter and vegetable oils (soya bean oil, rapeseed oil). Polyunsaturated fats have damaging effect on DNA and other cell components, affecting immune defences, tissue invasiveness and tumour metastatic spread. Polyunsaturated fats alter 5α-reductase activity. ω-3 fatty acids inhibits CaP cells. ω-3 fatty acids obtained essentially from fatty fish and eicosanoid synthesis inhibitors block CaP invasion by regulating tumour cell proteolytic enzyme activity. ω-6 PUFAs stimulate tumour development. Carotenoids are a group of complex unsaturated hydrocarbons occurring as pigments in plants, e.g., carrots, and tomatoes. Some carotenoids are precursors of vitamin A whereas lycopene and astaxanthin have a different structure that is not convertible into vitamin A. Carotenoids have antioxidants potential, especially lycopene. Lycopene may interfere with IGF1 receptor signalling and cell cycle progression. Intake of tomato-based foods may be associated with a reduced risk of CaP. Vitamin E prevents the oxidation and peroxidation of membrane phospholipids. Vitamin E suuccinate triggers apoptosis in CaP cell, but not normal prostate cells. Vitamin E inhibits high-fat diet-promoted growth of CaP cell. α-Tecopherol together with β-carotene decreases the incidence of CaP (32% decrease). β-Carotene alone is associated with 23% and 15% higher incidence and mortality rates, respectively. Vitamin A (retinol) high dose-related side effects, including hepatotoxicity, CNS changes and mucocutaneous dryness. Dietary vitamin D is obtained from a plant sterol (ergosterol) or from milk. Specific vitamin D receptor genotypes are associated with an aggressive CaP phenotype. Low serum calcium levels stimulate the secretion of parathyroid hormone, which promotes the conversion of vitamin D into 1,25 $(OH)_2D$.

Conditions that can cause false-positive elevation of serum AP:

1- Manipulation of the prostate, DRE, biopsy/surgery, endoscopy, catheterisation
2- Prostatic disease, infarction, prostatitis, urinary retention, BPH.
3- Other malignancies, i.e., carcinoma with hepatic/skeletal metastasis, multiple myeloma, polycythaemia vera, Hodgkin's disease, Hairy cell leukaemia.
4- Kidney disease, i.e., chronic glomerulonephritis, Gout.
5- Skeletal disease, Paget's disease, osteosarcoma, osteogenesis imperfecta.
6- Liver disease, hepatitis, cirrhosis.
7- Other diseases.

Dietary fat and CaP:

Dietary constituent	Source	Mechanism	Effect
Total fat	Red meat/vegetable fat	Increases synthesis of androgen and oestrogen.	Promotion.
Saturated fat	Red meat, milk.	Increases synthesis of androgen.	Promotion.
Mono/polyunsaturated fats α-linolenic acid	Corn oil, soya bean oil, red meat.	DNA damage, increases cell proliferation.	Promotion.
ω-3 fatty acids Eicosapentanoic acid	Fatty fish.	Proteolytic enzyme inhibition.	Inhibition.
Cholesterol	Egg, red meat.		Insufficient

Vitamins and CaP:

Dietary constituent	Source	Mechanism	Effect
β-carotene	Carrots.	Unclear.	Insufficient.
Lycopene	Tomatoes.	? IGFR interference.	Inhibition.
Vitamin C	Fruit and vegetables.	Antioxidant	Insufficient.
Retinoids	Vegetables.	Inhibits cell proliferation	
Vitamin E (α-tocopherol)	Cotton seed oil, lettuce, watercress, hemp seed oil.	Inhibition of tumour progression (prolongs latent phase)	Inhibition.
Vitamin D	Vegetable, milk.	Decreases cell proliferation Increases cell differentiation.	Inhibition.

Selenium is a key component of a number of functional selenoproteins required for normal health, such as glutathione peroxidase enzyme antioxidants. Lipid peroxides impair membrane structure and function. These lipids are unstable end products, which decompose to produce free radicals and cytotoxic aldehydes with promotion of damage to DNA, i.e., they have carcinogenic properties. Selenium reduces overall cancer incidence by 37% and that of the CaP by 50%. High dose of selenium is hepatotoxic and toxic to CNS. Isoflavonoids are compound or plant pigments found in legumes. Genistein and diadzein are the major isoflavones shown to inhibit the growth of CaP. The lignan enterolactone and the soya-derived isoflavone genistein are inhibitors of several steroid-metabolising enzymes such as aromatase or 5α-reductase. Genistein inhibits angiogenesis through inhibition of the proliferating cell growth. Genistein induces apoptosis in CaP cell. Genistein decreases PSAmRNA, protein expression and secretion. Genistein inhibits cell proliferation independent of PSA signalling pathways. CaP was inversely associated with estimated consumption of nuts, oil seeds, fish, and soya products. Ligands act as weak oestrogens, interfering with the biological activity of normal oestrogens and inhibiting their synthesis. They also inhibit enzymes of cell growth and angiogenesis and act as antioxidants. Biological effects of phytoesterogen presents in isoflavonoids (genistein), flavonoids (quercetin), and ligand (enterolactone), include: 1- Antioestrogenic activity (nuclear receptor binding competition). 2- Stimulation of sex hormone binding globulin (SHBG) synthesis. 3- Decreases free testosterone plasma level. 4- 5α-reductase inhibition. 5- 17β-hydroxystenoid dehydrogense inhibition. 6- Tyrosine specific protein kinase inhibition. 7- FGF2 inhibition. Chemoprevention consists of the administration of agents aimed at preventing or delaying the progression of cancer. Chemoprevention agents are usually intended for relatively healthy subjects. Chemoprevention might be considered in patient with rising PSA after radical surgery. The rate of increase of PSA significantly improved in the nutritional supplement group, with a delay in PSA rise of 8 weeks with a 6 week course of supplements.

Elements and CaP:

Substance	Source	Mechanism	Effect
Selenium	Bread, cereals, fish, meat.	Antioxidant, apoptosis inducer, catalase enhancer, cytochrome P450 modifier, immunostimulant.	Inhibition.
Isoflavonoids	Peas, beans, soya beans.	Angiogenesis inhibition, antioxidant, apoptosis inducer, oncogene expression inhibition, EGFR inhibition, anti-oestrogen, ODC synthesis inhibition, decreases cholesterol and LDH.	Inhibition.
Fenretinide (4HPR)	Carrots.	Antiproliferative, apoptosis inducer, angiogenesis inhibition, cellular differentiation, IGF1 inhibition, immunostimulation, ODC synthesis inhibition, protein kinase C inhibition.	Inhibition.
Zinc	Meat.		Insufficient.
Calcium	Milk, cheese.	Vitamin D synthesis inhibition.	Promotion.

Chemoprevention for CaP:

Target	Advantages	Disadvantages
Healthy men	Applicable.	Expensive. May require biopsy at end of study to establish the status.
Strong family history. HGPIN.	Applicable.	Finding may not be applicable to general population.
Chemoactive cancer/incidental CaP.	Applicable. Ability to evaluate pathology specimen.	Would require subsequent biopsy. Finding may not be applicable to general population.

GYNOPATHY

Insulin resistance (IR) has been increasingly implicated in the pathogenesis of gout. Patients with depression have impaired insulin sensitivity and resultant hyperinsulinaemia and other abnormalities can be resolved after recovery from depression. The half-life of fats in human adipose tissue is about 600 days, i.e.; significant amounts of previously consumed oils will still be present up to 4 years after they have been removed from the diet. A solvent containing double bonds (e.g., soy oil or other oil containing PUFA) would very likely prevent the close association between vitamin E and ubiquinone, which is necessary for charge-transfer to occur. Electronic activation is the most important feature of the living state, and specific electronic interaction between vitamin E and ubiquinone plays an important role in the respiratory function of ubiquinone. Ubiquinone is part of the electron transport chain that can leak electrons, which might be one of the ways in which vitamin E can prevent the formation of toxic free-radicals. Unsaturated oils interfere with this very specific but delicate bond, which could explain, at least partly, their toxicity for mitochondria. Unsaturated fats inhibit thyroid function, increase lipid peroxides, and induce azoospermia. People who eat fish heads (or other animal heads) generally consume the thyroid gland, as well as the brain. The brain is the body's richest source of cholesterol, which, with adequate thyroid hormone and vitamin A, is converted into the steroid hormones pregnenolone, progesterone, and DHEA, in proportion to the quantity circulating in blood in LDL. The active thyroid hormone is also concentrated many-fold in the brain. The brain is also the richest source of the water-insoluble (hydrophobic) steroid hormones. DHEA (dehydroepiandrosterone) is low in individuals who are susceptible to heart disease or cancer, and all three of these steroids have a broad spectrum of protective actions. Thyroid hormone, vitamin A, and cholesterol, which are used to produce the protective steroids, have a similarly broad range of protective effects, even when used singly. A high level of serum cholesterol is practically diagnostic of hypothyroidism. Hypothyroid individuals are susceptible to infections, heart disease, and cancer. Toxic effects of fish oil may include testicular degeneration, softening of the brain, muscle damage, and spontaneous cancer. This toxic effects of fish oil result from an induced vitamin E deficiency. Lipolysis is associated with peroxidation of the fatty acids released. Enrichment of the tissues with highly unsaturated fatty acids results in an increase in lipid peroxidation even in the presence of normal concentrations of vitamin E. The unsaturated fats are associated with aging, lipofuscin, and oestrogen. Unsaturated fats have special properties of adsorption, and are more soluble in water than the saturated fats. The movement and modulation of proteins and nucleic acids might require these special properties. The high concentration of unsaturated fats in mitochondria suggests that they are required for mitochondrial structure, or function, or regulation, or reproduction. Thyroxin is among the structural antioxidants. The oxygen-sparing effects of progesterone would make it appropriate to include it among the structural antioxidants. A fat-free liver extract cures EFA deficiency. Diets rich in protein increase the requirement for vitamin B_6, which is a co-factor of transaminases.

Mankind are eating about 100 times more EFA than they should, because most of the food animals are fed large amounts of grains and soybeans. Saturated fats are dominant in tropical plants and in warm-blooded animals, which relates to the stability of these oils at high temperatures. Coconut oil that had been stored at room temperature for a year was found to have no measurable rancidity. Since growing coconuts often experience temperatures around 100°F, ordinary room temperature isn't an oxidative challenge. Fish oil or safflower oil can't be stored long at room temperature, and at 98°F, the spontaneous oxidation is very fast. In 1962, it was found that unsaturated fatty acids are directly toxic to mitochondria. Stress increases the amount of free fatty acids circulating in the blood, as well as lipid peroxides. Hypoxia increases the intracellular concentration of free fatty acids. Stress liberates even local tissue fats in the heart, and systematic drug treatment, including antioxidants, can stop the enlargement of stress-induced infarctions. The cardiac necrosis caused by unsaturated fats (linolenic acid, in particular) can be prevented by a cocoa butter supplement. A certain proportion of saturated fat appears to be necessary for stability of the mitochondria. The EFAs decrease the phosphorylation efficiency, the amount of usable energy produced by cellular respiration. Increased serum EPA from an EPA supplement occurs with butter. Palm oils have antithrombotic effect, in relation to platelet aggregation. Platelet aggregation is enhanced by sunflowerseed oil, but palm oil tends to decrease it. Vitamin E facilitates clot removal, by activating proteolytic enzymes. Unsaturated fats inhibit proteolytic enzymes in the blood. The equilibrium between clotting and clot dissolution is especially important in the veins, where blood moves more slowly, and spends more time. The slower blood flows the greater its predisposition to clotting. The intrinsic process, leading to fibrin production, is slow, taking up to a minute or more to occur. Thrombosis as a result of stasis occurs in the venous circulation; typically in the legs where venous return is slowest. Many thousands of small thrombi are formed each day in the lower body. These pass via the vena cava into the lungs where thrombolysis occurs. Vitamin E and oestrogen act in opposite directions on the clot-removing enzymes. Oestrogen increases blood lipids, and increases the incidence of strokes and heart attacks. Oestrogen's effect on clotting is very complex, since it increases the ratio of unsaturated to saturated fatty acids in the body, and increases the tendency of blood to pool in the large veins, in addition to its direct effects on the clotting factors. Unsaturated fats are immunosuppressive. The anti-inflammatory effect of ω-3 fatty acids (fish oil) may be related to the suppression of IL-1 and TNF. The suppression of these anti-tumour immune factors persists after the fish oil treatment is stopped. Stress and hypoxia can cause cells to take up large amounts of fatty acids. Cortisols kill white blood cells (which can be inhibited by extra glucose), which is an important part of its immunosuppressive effect, and this killing is mediated by causing the cells to take up unsaturated fats.

The similar effects of oestrogen and of polyunsaturated fats (PUFA):

1- Antagonism to vitamin E and thyroid, to respiration and proteolysis.
2- Promotion of lipofuscin formation and of clot formation.
3- Promotion of seizure activity.
4- Impairment of brain development and learning.
5- Involvement in positive or negative regulation of cell division, depending on cell type.

Unsaturated fats cause degranulation of mast cells. The short-chain fatty acids normally produced by bacteria in the bowel have a local anti-inflammatory action. Several aspects of the immune system are improved by short-chain saturated fats. Their anti-histamine action is important, because of histamines immunosuppressive effects. The neutrophil proteinases and plasma antiproteinases are important in the evolution of tissue damage. Unsaturated fats may block the action of cytotoxic cells. The inhibition of proteolytic enzymes by unsaturated fats will act at many sites: 1- digestion of protein, 2- digestion of clots, 3- digestion of the colloid in the thyroid gland that releases the hormones, 4- the activity of white cells, and 5- the normal digestion of cytoplasmic proteins involved in maintaining a steady state as new proteins are formed and added to the cytoplasm. The inhibition of the destruction of intracellular proteins would shift the balance toward growth. Cancer cells have a high level of unsaturated fats, yet they have a low level of lipid peroxidation; lipid peroxidation inhibits growth, and is often mentioned as a normal growth-restraining factor. A diet lacking fats prevent the development of spontaneous tumours. The unsaturated fats are essential for the development of tumours. Tumours secrete a factor that mobilises fats from storage. Saturated fats, coconut oil and butter, for example do not promote tumour growth. Olive oil is not a strong tumour promoter, but may have a slightly permissive effect on tumour growth. The carcinogenic action of unsaturated fats can be offset by adding thyroid hormone. Cystine also increases the tumour incidence. In hyperthyroidism, the ability to quickly oxidise larger amounts of the toxic oils would likely have a protective effect, preventing storage and subsequent peroxidation, and reducing the oils ability to synergies with oestrogen. Consumption of unsaturated fat is associated with both skin aging and with the sensitivity of the skin to ultraviolet damage, Ultraviolet light-induced skin cancer is mediated by unsaturated fats and lipid peroxidation. Butter and coconut oil contain significant amounts of the short and medium-chain saturated fatty acids, which are very easily metabolised, inhibit the release of histamine, promote differentiation of cancer cells, tend to counteract the stress-induced proteins, decrease the expression of prolactin receptors, and promote the expression of the T3 receptor. In the FFA form, the unsaturated fats are toxic to the mitochondria, but cancer cells are famous for their compensatory glycolysis. α-Linolenic acid ester has a stimulating effect on breast cancer. Skin fibroblasts have a specific preference for oleic acid, over polyunsaturated fat. Connective tissue cells have a low propensity to take up unsaturated fats. Breast cancer cells have a high affinity for fats, which shows a selective toxicity of oils for cancer cells. The use of unsaturated fats in cancer chemotherapy may explain their tendency to cause pulmonary embolism, their suppression of immunity including factors specifically involved in cancer resistance, and their carcinogenicity. Excessive unsaturated dietary fats interfere with learning and behaviour as a result of lipid peroxidation; some of the effects can be reduced with antioxidants. Lipid peroxidation is involved in seizures. Violent criminals have neurological defects (e.g., 90%), which may reflect unusual patterns of brain lipids. Stress or additional cortisone, which, by blocking the use of glucose, forces cells to take up more fat, causes accelerated aging of the brain. Dietary linoleic acid may be required for the development of alcoholic liver damage. Omitting cholesterol entirely from the diet may cause leakage of amino-transferase enzymes, which is similar to the effects of the presence of linoleic acid with ethanol. Dietary coconut oil causes decreased fat synthesis and storage, when compared with diets containing unsaturated fats. The short-chain fats in coconut oil may improve tissue response to the thyroid hormone (T3), and its low content of unsaturated fats might allow a more nearly optimal function of the thyroid gland and of mitochondria. The presence of palmitate in the lung surfactant phospholipids suggests that maternal overload with unsaturated fats might interfere with the formation of these important substances, causing breathing problems in the newborn. The bone-calcium mobilising effect of prostaglandins suggests that dietary fats might affect osteoporosis; the absence of osteoporosis in some tropical populations might relate to their consumption of coconut oil and other saturated tropical oils. Soy steroids can be converted by bowel bacteria into oestrogens. The role of PUFA in reproduction might be similar to that of oestrogen, i.e., the promotion of uterine and breast cell proliferation, water uptake, etc. Polyunsaturated fats are nearly ubiquitous. They are essential for cancer, and that they have other properties, which cause them to be toxic at certain levels. If PUFA are really essential for reproduction, unsaturated vegetable oils could temporarily be added to the diet when reproduction is desired. Low in PUFA, might prolong our characteristically human condition of delayed reproductive maturity.

MENOPAUSE

The concentration of a hormone in the blood doesn't directly represent the concentration in the various organs. The amount of oestrogen in tissue is decreased when progesterone is abundant. In the absence of progesterone, tissues retain oestrogen even when there is little oestrogen circulating in the blood. Melatonin decreases sharply at puberty when oestrogen increases, and then it decreases again at menopause. Prolactin (stimulated by oestrogen) increases around puberty, and instead of decreasing at menopause, it often increases, and its increase is associated with osteoporosis and other age-related symptoms. Oestrogen is produced in many tissues by the enzyme aromatase, even in the breast and endometrium. Aromatase increases with aging. Oestrogen is inactivated, mainly in the liver and brain, by being made water-soluble by the attachment of glucuronic acid and/or sulfuric acid. Oestrogen's concentration in a particular tissue depends on: 1- its affinity or binding strength for components of that tissue, relative to its affinity for the blood; 2- the activity in that tissue of the aromatase enzyme, which converts androgens to oestrogen; 3- the activity of the glucuronidase enzyme, which converts water-soluble oestrogen glucuronides into the oil soluble active forms of oestrogen; and 4- the sulfatases and several other enzymes that modify the activity and solubility of the oestrogens. Oestrogen receptors, which bind oestrogens in cells, are inactivated by progesterone, and activated by many physical and chemical conditions. Inflammation activates β-glucuronidase, and anti-inflammatory substances such as aspirin reduce many of oestrogen's effects. Free hormone is the hormone that isn't bound to a transporting protein, with the more or less explicit idea that it is dissolved in the water of the plasma or extracellular fluid (ECF). Albumin and other proteins enter cells more or less freely, depending on prevailing conditions. Within the blood, progesterone and thyroid hormone (T3) are much more concentrated in the red blood cells than in the serum. The oxygen atoms, and especially the phenolic group of oestrogen, slightly reduce the hormones' affinity for simple oils, but they interact with other polar or aromatic groups, giving oestrogen the ability to bind more strongly and specifically with some proteins and other molecules. Enzymes that catalyse oestrogen's oxidation-reduction actions are among the specific oestrogen-binding proteins. Many proteins and lipoproteins bind steroids, but some intracellular proteins bind them so strongly. Progesterone prevents the tissue from concentrating oestrogen. When progesterone is low, the tissues may contain 20-30 times more oestrogen than the plasma in non-pregnant women. In aging, the sharply decreased progesterone production creates a situation resembling the follicular phase of the menstrual cycle, allowing tissues to concentrate oestrogen even when the serum oestrogen may be low. Besides the relatively direct actions of progesterone on the oestrogen receptors, keeping their concentration low, and its indirect action by preventing prolactin from stimulating the formation of oestrogen receptors, there are many other processes that can increase or decrease the tissue concentration of oestrogen, and many of these influences change with aging. There are 2 kinds of enzyme that produce oestrogen. Aromatase converts male hormones into oestrogen. β-Glucuronidase converts the inactive oestrogen-glucuronides into active oestrogen. The healthy liver inactivates practically all the oestrogen that reaches it, mostly by combining it with the sugar acid, glucuronic acid. This makes the oestrogen water soluble, and it is quickly eliminated in the urine. But when it passes through inflamed tissue, these tissues contain large amounts of β-glucuronidase, which will remove the glucuronic acid, leaving the pure oestrogen to accumulate in the tissue. Liver impairment decreases its ability to excrete oestrogen, and oestrogen contributes to a variety of liver diseases. Hypothyroidism prevents the liver from attaching glucuronic acid to oestrogen, and so increases the body's retention of oestrogen, which in turn impairs the thyroid gland's ability to secrete thyroid hormone. Hypothyroidism often results from nutritional protein deficiency. Fat and the skin are major sources of oestrogen, especially in older people. The activity of aromatase increases with aging, and under the influence of prolactin, cortisol, prostaglandin, and the pituitary hormones, FSH (follicle stimulating hormone) and growth hormone. It is inhibited by progesterone, thyroid, aspirin, and high altitude. Aromatase can produce oestrogen in fat cells, fibroblasts, smooth muscle cells, breast and uterine tissue, pancreas, liver, brain, bone, skin, etc. Aromatase in mammary tissue appears to increase oestrogen receptors and cause breast neoplasia, independently of ovarian oestrogen. Oophorectomy causes the tissue aromatases to increase. The loss of progesterone and ovarian androgens is probably responsible for this generalised increase in the formation of oestrogen. In the brain, aromatase increases under the influence of oestrogen treatment. Sulfatase is another enzyme that releases oestrogen in tissues, and its activity is inhibited by anti-oestrogenic hormones. Progesterone inhibits the release or activation of β-glucuronidase, which increases with aging. Glucuronic acid also tends to inhibit the intracellular release of oestrogen by β-glucuronidase. Glucaric acid inhibits β-glucuronidase. The decline of the antioestrogenic factors in aging, combined with the increase of pro-oestrogenic factors such as cortisol, prolactin and FSH, occurs in both men and women. During the reproductive years, women's cyclic production of large amounts of progesterone probably retards their aging enough to account for their greater longevity. Childbearing also has a residual anti-oestrogenic effect and is associated with increased longevity. General aging contributes to the specific changes that lead to menopause, but fertility may be prolonged to a much greater age by preventing excitotoxic exhaustion of the hypothalamic nerves. Old sperms have been implicated in some birth defects. More frequent intercourse involves fresher sperms. When ovaries have been treated with X-rays to destroy their ability to ovulate, they have been found to produce more oestrogen than before. Ovulation is one thing, and the production of hormones is another thing. Both the brain and the skin are sources of steroid hormones, and it is possible that the death of skin cells and neurons is one factor in the age-related decline in the sex steroids. The cell's environment, the signals and substances and energy it receives, is complex, and involved in the aging process. High oestrogen is the cause of early puberty, a high cancer incidence, and a relatively short life. The occurrence of menopause at an early age in women is associated with a greater risk of death from all causes, including strokes and coronary heart disease. Ovarian aging (age-related decline in fertility) is an indicator of general aging. Oestrogen accelerates aging of the CNS, destroying the nerves, which regulate the pituitary gonadotropins, and causing ovarian failure and infertility. Infertility that developed at middle age is caused by a high rate of O_2 consumption in the uterus, causing the O_2 needed by the developing embryo to be consumed by uterine tissues, and causing suffocation of the embryo. This is the central mechanism by which the oestrogen-containing contraceptives work: at any stage of pregnancy, a sufficient dose of oestrogen kills the embryo. Menopause is the first missed period associated with suddenly increased bone loss, nervous symptoms such as depression, insomnia, and flushing. The onset of menopause corresponds to the failure to produce progesterone, while oestrogen is produced at normal levels. This results in a great functional excess of oestrogen, because it is no longer opposed by progesterone. Four years are required for the monthly oestrogen excess to disappear. Bone loss sets in immediately when progesterone fails because cortisol then is able to dominate, causing bone catabolism; progesterone normally protects against cortisol. Oestrogen dominance promotes mitosis of the prolactin-secreting cells of the pituitary, and that prolactin causes osteoporosis; by age 50, most people have some degree of tumefaction of the prolactin-secreting part of the pituitary. But oestrogen dominance (or progesterone deficiency) also clearly obstructs thyroid secretion, and thyroid governs the rate of bone metabolism and repair. Correcting the thyroid and progesterone should take care of the cortisol/prolactin/osteoporosis problem. The menopausal pituitary hormones, high levels of LH and FSH, are produced because the regulatory nerves in the hypothalamus have lost their sensitivity to oestrogen, not because oestrogen is deficient. The nerves are desensitised by their cumulative exposure to oestrogen. The mechanism by which oestrogen desensitises and kills brain cells is by the excitotoxic process, in which the excitatory transmitter glutamic acid is allowed to exhaust the nerve cells. Glutamic acid, or aspartic acid, or aspartame can cause brain damage and reproductive failure. Cortisol also activates the excitotoxic system, in other brain cells, causing stress-induced atrophy of those cells. Progesterone and pregnenolone are recognised as inhibitors of this excitotoxic process. Oestrogen's stress-mimicking action also probably tends to increase the secretion of LH, which can be corrected by supplementing progesterone and thyroid. Oestrogen creates the same conditions as occur in the shock phase of the stress reaction. Also shock, in a potential vicious circle, can increase the level of oestrogen. Oestrogen stimulates the adrenal glands, independently of the pituitary's ACTH. This can increase the production of adrenal androgens, leading to hirsutism, and other male traits, including anabolic effects. Oestrogen's nerve-exciting action lowers seizure thresholds; premenstrual epilepsy is probably another acute sign of the neurotoxicity of oestrogen. When fatigue and lethargy are associated with aging, the brain stimulating action of oestrogen can make a woman feel that she has more energy. Oestrogen inhibits one of the enzymic routes for inactivating brain amines, and so it has more general effects on the brain than just the glutamate system. This generalised effect on brain amines is more like the effects of cocaine or amphetamine. The reason for the menopausal progesterone deficiency is a complex of stress-related causes. Free radicals, e.g., from iron in the corpus luteum, interfere with progesterone synthesis, as do prolactin, ACTH, oestrogen, cortisol, carotene, and an imbalance of gonadotropins. A deficiency of thyroid hormone, vitamin A, and LDL-cholesterol can also prevent the synthesis of progesterone. The effect of an intra-uterine irritant, e.g., IUD, is to signal the ovary to suppress progesterone production, to prevent pregnancy while there is a problem in the uterus. ACTH suppresses progesterone synthesis, which is similar to prevent pregnancy during stress. Since progesterone and pregnenolone protect brain cells against the excitotoxins, anything that chronically lowers the body's progesterone level tends to accelerate the oestrogen-induced excitotoxic of brain cells death. Chronic constipation, and anxiety, which decreases blood circulation in the intestine, can increase the liver's exposure to endotoxin. Endotoxin (like intense physical activity) causes the oestrogen concentration of the blood to rise. Diets that speed intestinal peristalsis might be expected to postpone menopause. Penicillin treatment, probably by lowering endotoxin production, decreases oestrogen and cortisone, while increasing progesterone. The same effect can be achieved by eating raw carrots (especially with coconut oil/olive oil dressing) every day, to reduce the amount of bacterial toxins absorbed, and to help in the excretion of oestrogen. Long hours of daylight increase progesterone production, and long hours of darkness are stressful. Annually, our total hours of day and night are the same regardless of latitude, but different ways of living, levels of artificial illumination, etc., have a strong influence on our hormones. Prolonged exposure to light may delay some aspects of aging. Aging may be prevented or delayed by protecting against the excitotoxins. Minimising oestrogen and cortisone with optimal thyroid activity, and maximising pregnenolone and progesterone to prevent excitotoxic cell fatigue, can be done easily. A diet low in iron and unsaturated fats protects the respiratory system from the damaging effects of excessive excitation; and-since pregnenolone is formed in the mitochondrion-also helps to prevent the loss of these hormones.

YOUTH-ASSOCIATED HORMONES

Endotoxin formed in the bowel can chunk respiration and cause hormone imbalances contributing to instability of the nerves. Optimise bowel flora, e.g., with a carrot salad; a dressing of vinegar, coconut oil and olive oil, carried into the intestine by the carrot fiber, suppresses bacterial growth while stimulating healing of the wall of the intestine. The carrot salad improves the ratio of progesterone to oestrogen and cortisol, and so is as appropriate for epilepsy as for premenstrual syndrome, insomnia, or arthritis. 60 years ago, progesterone was found to be the main hormone produced by the ovaries. It is the most protective hormone the body produces, and the large amounts that are produced during pregnancy result from the developing baby's need for protection from the stressful environment. Normally, the brain contains a very high concentration of progesterone, reflecting its protective function for that most important organ. The thymus gland, the key organ of the immune system, is also profoundly dependent on progesterone. Progesterone is the basic hormone of adaptation and of resistance to stress. The adrenal glands use it to produce their anti-stress hormones, and when there is enough progesterone, they don't have to produce the potentially harmful cortisone. Progesterone deficiency is associated with increased production of cortisone. Excessive cortisone causes osteoporosis, aging of the skin, damage to brain cells, and the accumulation of fat, especially on the back and abdomen. Progesterone relieves anxiety, improves memory, protects brain cells, and even prevents epileptic seizures. It promotes respiration, and has been used to correct emphysema. It prevents bulging veins by increasing the tone of blood vessels, and improves the efficiency of the heart. It reverses many of the signs of aging in the skin, and promotes healthy bone growth. It relieves arthritis, and helps a variety of immunological problems. Progesterone dissolved in vitamin E is absorbed very efficiently, and distributed quickly to all of the tissues. If a woman has ovaries, progesterone helps them to produce both progesterone and oestrogen as needed, and also helps to restore normal functioning of the thyroid and other glands. If her ovaries have been removed, progesterone should be taken consistently to replace the lost supply. A progesterone deficiency is associated with increased susceptibility to cancer, and progesterone has been used to treat some types of cancer. While men do naturally produce progesterone, and can sometimes benefit from using it, it is not a male hormone. Used alone, progesterone often makes it unnecessary to use oestrogen for hot flashes or insomnia, or other symptoms of menopause. When dissolved in vitamin E, progesterone begins entering the blood stream almost as soon as it contacts any membrane, such as the lips, tongue, gums, or palate, but when it is swallowed, it continues to be absorbed as part of the digestive process. When taken with food, its absorption occurs at the same rate as the digestion and absorption of the food. Progesterone, because of its normal anaesthetic function (prevents the pain of labour when its level is adequate), directly calms nerves, and in this way suppresses many of the excitotoxic processes. It has direct effects on mitochondria, promoting energy production, and it facilitates thyroid hormone functions. Progesterone promotes the elimination of oestrogen from tissues, and is a diuretic in several benign ways, which are compatible with maintenance of blood volume. It antagonises the mineralocorticoids and the glucocorticoids, both of which promote seizures. In the 1940s, pregnenolone was tested in people who were sick or under stress, and it was found to have a wide range of beneficial actions. The side effects of some of the synthetic hormones were so awful, e.g., synthetic varieties of cortisone can destroy immunity, and can cause osteoporosis, diabetes, and rapid aging, with loss of pigment in the skin and hair, and extreme thinning of the skin. After the age of 40 or 45, it seems that everyone lives in a state of continuous stress, just as a normal part of aging, which is associated with decreased ability to produce an abundance of pregnenolone. Pregnenolone improves the memory and general performance. Pregnenolone is one of the major hormones in the brain. It is produced by certain brain cells, as well as being absorbed into the brain from the blood. It protects brain cells from injury caused by fatigue, and an adequate amount has a calming effect on the emotions, which is part of the reason that it protects us from the stress response that leads to an excessive production of cortisone. Pregnenolone has a face-lifting action, by improving circulation to the skin, and by an actual contraction of myoepithelial cells in the skin. A similar effect can improve joint mobility in arthritis, tissue elasticity in the lungs, and even eyesight. Pregnenolone is protective of fibrous tissues. Pregnenolone is largely converted into two other youth-associated protective hormones, progesterone and DHEA. At the age of 30, both men and women produce roughly 30-50 mg. of pregnenolone daily. When taken orally, even in the powdered form, it is absorbed fairly well. One dose of 300 mg keeps acting for about a week, as absorption continues along the intestine, and as it is recycled in the body. It improves the body's ability to produce its own pregnenolone. It tends to improve function of the thyroid and other glands. DHEA (dehydroepiandrosterone) is produced in the brain (from pregnenolone), but it is probably produced in other organs, including the skin. The brain contains a much higher concentration of DHEA than the blood does. Elderly produce only about 5% as much as youth do. This is about the same decrease that occurs with progesterone and pregnenolone. Protection against the toxic actions of other specialised hormones, e.g., cortisone, is a major function of DHEA and the other youth-associated hormones. Starvation, aging, and stress cause the skin to become thin and fragile. An excess of cortisone-whether it is from medical treatment, or from stress, aging, or malnutrition-does the same thing. Material from the skin is dissolved to provide nutrition for the more essential organs. Other organs, such as the muscles and bones, dissolve more slowly, but just as destructively, under the continued influence of cortisone. DHEA blocks these destructive effects of cortisone, and actively restores the normal growth and repair processes to those organs, strengthening the skin and bones and other organs. DHEA protects against cancer. DHEA regenerates the insulin-producing cells of the pancreas. Brewer's yeast and DHEA improve the sugar metabolism in DM. In diabetes, very little sugar enters the cells, so fatigue is a problem. DHEA stimulates cells to absorb sugar and to burn it, so it increases the general energy level and helps to prevent obesity. Young people produce about 12-15 mg of DHEA/day. DHEA is decreased by about 2 mg/day for every decade after the age of 30. DHEA, like the thyroid hormone, increases the heat production and ability to burn calories. At the age of 50, up to 4 mg of DHEA/day restore the level of DHEA in the blood to a youthful level. Excess DHEA is converted into oestrogen or testosterone, and large amounts of those sex hormones can disturb the function of the thymus gland (shrink) and the liver (enlarge). DHEA should be used with caution; DHEA can cause changes in glial cells resembling those seen in the aging brain. Supplements of pregnenolone and thyroid may be the safest way to optimise DHEA production.

THE MUSCULO-SKELETAL HEALTH

Vitamin D insufficiency/deficiency is a major problem in the elderly population. 30% of bone mass is influenced by exogenous factors (with nutrition clearly being one of the key factors open to modification). Vitamin D status is affected by the aging process, and inversely related to parathyroid hormone (PTH). Vitamin D (in combination with PTH) plays a crucial role in the regulation of calcium (Ca) and phosphate (P) metabolism and promotes Ca absorption from the gut and kidney tubules. There is high risk of bone fractures once serum 25(OH)D value is <30 nmol/l, and serum PTH is in the range of 4.9 pmol/l. 1.25 g of Ca and 400 IU of cholecalciferol for 8 weeks, results in suppression of PTH and an increase in 25(OH)D and serum Ca levels. Alfacalcidol (ALF), an active vitamin D analogue supplementation inhibits bone resorption markers, causes modulation of cytokines, and reduces pain in subjects with osteopenia. The skeleton plays an important role as a source of buffer contributing to both the preservation of the body's pH and defense of the system against acid-base disorders. Reduced extracellular pH directly stimulates osteoclasts to resorb bone. Almost all of the bone mineral release that occurs in response to acidosis results from osteoclast activation, which increases resorption pit formation in bone (with the organic matrix being destroyed at the same time). A small drop in pH causes a tremendous burst in bone resorption. The role of bone in acid-base balance is complex, the skeleton is a giant ion exchange column loaded with an alkali buffer, as 80% of body carbonate, 80% of body citrate, and 35% of body sodium are contained in solution within the hydration shell of bone and are released in response to metabolic acid. The intake of acid is a way of everyday life, and animal proteins and cereals are rich sources of phosphoric and sulphuric acid, which are recognised as acid ash foods. The concept of relation between diet and osteoporosis is going back to 1968. For 1 mEq of acid/day, 2 mEq of Ca/day is required to buffer this amount of fixed acid, this account for a 15% loss of inorganic bone in an average individual over a decade. The more acid precursors consumed, the greater the degree of systemic acidity, and with increasing age, overall renal function declines and acidity increases. Thus, humans become more acidic as they age. An increase in fruit and vegetable (alkali-forming foods) intake from 3.6 to 9.5 daily servings decreases urinary Ca excretion from 157 mg/24 hrs to 110 mg/24 hrs, and reduces acid load. Insufficient intake of dietary protein is implicated in the pathogenesis of osteoporosis, and protein supplementation improves the clinical outcomes of hip fractures. Isocaloric protein undernutrition is associated with alterations of somatotropic and gonadotropic axes and results in decreased bone mineral density (BMD) and bone strength of cancellous and cortical bone. Protein undernutrition, with normal calorie intake, reduces IGF-1 levels independently of growth hormone secretion. The resistance of both IGF-1 and growth hormone in target organs is associated with early depression of bone formation. The inhibition of the sex steroid secretion may explain, in part, the increase in bone resorption. Lower protein intake is significantly associated with greater bone loss at the spine and femur skeletal sites. Anorexia nervosa is associated with increased fracture risk (persisted for >10 years), whereas bulimia nervosa is associated with a borderline significant increase in fracture. Low bone mass is linked with eating disorders, especially those conditions that are associated with amenorrhoea. Weight loss is associated with increased bone resorption. Supplementation of vitamin D with Ca is effective in normalising both the low vitamin D status and hyperparathyroidism, which is associated with increased bone resorption. FGF-2, a modulator of cartilage and bone growth and differentiation, is expressed and regulated in osteoblastic cells. FGF-2 mediates bone formation. There is some evidence that compounds such as growth hormone or insulin-like growth factor I (IGF-I) can increase periosteal apposition, enlarging bone and increasing its bending strength. IGF-1 increases periosteal apposition, whereas pamidronate reduces endocortical resorption. IGF-I increases the external diameter of the midshaft tibia and femoral neck. The phytoestrogens (Genistein) have gained popularity as low-dose oestrogens for prevention of osteoporosis. Pre-osteoblastic cells treated with genistein form mineralised bone nodules and express the oestrogen receptors α and β. Isoflavonoid phytoestrogen may act through these receptors to stimulate bone formation and inhibit bone resorption. Genistein stimulates osteoblastogenesis and inhibits adipogenesis via oestrogen receptor-dependent pathways. Free radicals may increase bone resorption. Bone loss is prevented by α-tocopherol, which may be a simple intervention to help prevent oestrogen deficiency-induced bone loss. Loss of renal mass combined to metabolic acidosis and uraemic retention (increased serum uric acid) reduces renal production and degradation of calcitriol, and decreases activity of 1α-hydroxylase. Uraemic solute retention is responsible for changes in calcitriol metabolism, resulting in a net decrease of blood calcitriol levels. There is decrease concentration of the vitamin D receptor (VDR) and decrease in VDR content of parathyroid glands in renal failure. Once calcitriol is complexed with the VDR, binding occurs with the DNA of VDREs (vitamin D response elements). Monocytes and lymphocytes contain several calcitriol-regulated genes. Altered production and metabolization of calcitriol, altered expression of VDR and altered binding properties of hormone receptor complex to DNA may contribute to calcitriol resistance in renal failure. During pregnancy, a woman's ability to retain dietary calcium and iron increases, and the baby seems to be susceptible to overloading. A normal baby doesn't need dietary iron for several months. Iron overload, age pigment, and calcification of soft tissues are so commonly associated with old age, that it is important to recognise that the same cluster occurs at the other extreme of (young) age, and that respiratory limitations characterise both of these periods of life. Calcium (Ca) functions as a regulatory trigger in many cell processes, including cell stimulation and cell death. Calcium has a tendency to be deposited with iron in damaged tissue. Albumin has unusual affinity for metal ions. The oxygen delivery system and the immune system evolved together, recycling iron in a tightly controlled system. Macrophages are involved in the massive turnover of haemoglobin, and osteoclasts, which suggests that iron and calcium are handled in analogous ways. In a family with marble-bone disease, or osteopetrosis, it was found that their RBCs lacked one form of the carbonic anhydrase enzyme, and that as a result, their body fluids retained abnormally high concentrations of CO_2. An excess of carbon dioxide (CO_2) may be dissolving bones and causing osteoporosis or osteopenia, instead of osteopetrosis. The thyroid hormone is responsible for the CO_2 produced in respiration. Chronic hypothyroidism causes osteopenia, which is significant because women (as a result of oestrogen's effects on the thyroid) are much more likely than men to have hypothyroidism, and that, relative to men, women in general are osteopenic, i.e., they have more delicate skeletons than men do. Hypothyroid women are likely to have small bones and excessive cortisol production. Thyroxin is the thyroid-suppressive precursor material, it doesn't correct hypothyroidism. Hypothyroid women are likely to have cancer, osteoporosis, obesity, etc. Triiodothyronine (T3), the active form of thyroid hormone, does contribute to bone formation. Glucocorticoids cause a negative calcium balance, leading to osteoporosis; the thyroid hormone and progesterone oppose those hormones, protecting against osteoporosis. Hip fractures, like cancer, strokes, and heart disease are strongly associated with old age. Insulin is the main factor promoting fat storage, and it is anabolic for bone. The loss of bone mass coincides with the onset of clinical diabetes. Since excess cortisol can cause both high blood sugar and bone loss, when diabetes is defined on the basis of high blood sugar, it will often involve high blood sugar caused by excess cortisol, and there will be Ca loss. The onset of progesterone deficiency is coinciding with bone loss, which reflects the importance of progesterone's antagonism to cortisol. Progesterone (but not estrone, estradiol, testosterone, or androstenedione) is significantly lower in those losing bone mass most rapidly. Around the age of 50, when bone loss is increasing, progesterone and thyroid are likely to be deficient, and cortisol and prolactin are likely to be increased. Prolactin contributes directly to bone loss, and is likely to be one of the factors that contribute to decreased progesterone production. Oestrogen tends to cause increased secretion of prolactin and the glucocorticoids, which cause bone loss, but it also promotes insulin secretion, which tends to prevent bone loss. All of these factors are associated with increased cancer risk. Thyroid and progesterone, unlike oestrogen, stimulate bone building, and are associated with a decreased risk of cancer. Thyroid and progesterone have anti-degenerative effects, protecting the bones, joints, brain, immune system, heart, blood vessels, breasts, etc. Healthy high energy metabolism requires the exclusion of most Ca from cells, and when Ca enters the stimulated or deenergized cell, it is likely to trigger a series of reactions that lower energy production, interfering with oxidative metabolism. During aging, both calcium and iron tend to accumulate and they both seem to have an affinity for similar locations, and they both tend to displace Cu. Elastin is a protein, the units of which are probably bound together by Cu atoms. In old age, elastin is one of the first substances to calcify, e.g., in the elastic layers of arteries, causing them to lose elasticity, and to harden into almost bone-like tubes. In the heart and kidneys, the mitochondria (rich in copper-enzymes) are often the location showing the earliest calcification, e.g., when Mg is deficient. Certain proteins have higher than average affinity for copper, iron, and calcium. Copper (Cu) atoms bind the protein molecules into relatively elastic systems. Calcium forms the link between adhesive proteins. Atherosclerosis might represent a protective process. As a defensive reaction, binding iron and destroying unsaturated fatty acids, and by this detoxifying action, possibly protecting against calcification and destruction of elastin. Macrophages are the characteristic foam cells convert polyunsaturated oils into age pigment, which accounts for the depletion of those fats in the plaques. Damaged mitochondria start the process of pathological calcification in the heart and kidneys, and other tissues that are respiratorily stressed. Healthy respiration, producing CO_2, is needed to keep Ca outside the cell. This efficient defense system may facilitate the deposition of Ca in suitable places depending on specific protein binding. Carbon dioxide and bicarbonate are formed in the mitochondria, and the steady outward flow of the bicarbonate anion would facilitate the elimination of Ca from the mitochondria. The affinity of CO_2 for the amine groups on proteins (as in the formation of carbamino haemoglobin, which changes the shape of the protein) could change the affinity of collagen or other proteins for Ca. ATP and CO_2 are closely associated, because both are produced in respiration. Hormones such as progesterone also act as cardinal adsorbents, regulating the affinity of proteins for salts and other molecules. Cells have many proteins with variable affinity for Ca; e.g., in muscle, the endoplasmic reticulum, releases and then sequesters Ca to control contraction and relaxation. This Ca-binding system is backed up by and is spatially in close association with that of the mitochondrion. The earliest deposition of crystals on implanted material is calcium carbonate. In newly formed bone, the phosphate content is low, and increases with maturity. While mature bone has an apatite-like ratio of Ca and phosphate, newly calcified bone is very deficient in phosphate. During nucleation of crystals, chemical interaction between CO_2 and amino groups (e.g., amino acids, protein, or ammonia) removes the CO_2 from solution, and the carbamino acid formed becomes a bound anion with which Ca can form a salt. The divalent calcium (Ca^{2+}) forms a link between the monovalent carbamino acid and another anion. Linking with carbonate (CO_3^{2-}), one valence would be free to continue the salt-chain. This chemistry is compatible with the bone formation. The uncoupled oxidative phosphorylation is a subtle thermogenesis, which suggests that stimulated production of CO_2 is the factor that stimulates calcification. DHEA, which powerfully stimulates bone formation, is (like thyroid and progesterone) thermogenic. Stress-related abnormalities of muscle metabolism (e.g., high lactate formation) are consistent with hypothyroidism. During malignant hyperthermia, there is slow removal of Ca from the contractile apparatus of the muscles. Hypertonicity of muscles, various degrees of myopathy and rigidity, and uncoupling of oxidative phosphorylation occur in these patients. Lactic acidosis suggests that mitochondrial respiration is defective in the patients. Besides the sensitivity to anaesthetics, the muscles of these patients are abnormally sensitive to caffeine and elevated extracellular potassium. During surgery, artificial ventilation, combined with stress, toxic anaesthetics, and any extra-mitochondrial oxidation that might be occurring (NADH-oxidase, which produces no CO_2), make relative hyperventilation a plausible explanation for the development of hyperthermia. Hyperventilation can cause muscle contraction. Free intracellular Ca is the trigger for muscle contraction (and Mg is an important factor in relaxation). Capillary tone, similarly, is increased by hyperventilation, and relaxed by CO_2.

The muscle-relaxing effect of CO_2 shows that the binding of intracellular Ca is promoted by CO_2, as well as by ATP. A relaxed muscle and a strong bone are characterised by bound Ca. Activation of the sympathetic nervous system promotes hyperventilation. This means that hypothyroidism, with high adrenaline (resulting from a tendency toward hypoglycaemia because of inefficient use of glucose and O_2), predisposes to hyperventilation. Muscle stiffness, muscle soreness and weakness, and osteoporosis all seem to be consequences of inadequate respiration, allowing lactic acid to be produced instead of CO_2. Insomnia, hyperactivity, anxiety, and many chronic brain conditions also show evidence of defective respiration, e.g., either slow consumption of glucose or the formation of lactic acid, both of which are common consequences of hypothyroidism. Abnormal Ca regulation may be involved in epilepsy. The combination of supplements of T3, Mg, progesterone and pregnenolone can restore normal respiration, and it seems clear that this should normalise Ca metabolism, decreasing the calcification of soft tissues, increasing the calcification of bones, and improving the efficiency of muscles and nerves. Magnesium, like carbonate, is a component of newly formed bone. Avoiding polyunsaturated vegetable oils is important for protecting respiration; some of the prostaglandins (PGs) they produce are implicated in osteoporosis, but more generally, they antagonise thyroid function and they can interfere with Ca control. Polyunsaturated dietary oils are associated with bone-toxicity.

OESTROGENS

Soot containing polycyclic aromatic hydrocarbons, is both oestrogenic and carcinogenic. Phenolics and chlorinated hydrocarbons are significantly oestrogenic, and that many oestrogenic herbicides, pesticides, and industrial by-products persist in the environment, causing infertility, deformed reproductive organs, tumours, and other biological defects, including immunodeficiency. Natural oestrogens, from human urine, enter the rivers from sewage. Many tons of synthetic and pharmaceutical oestrogens, administered to menopausal women in quantities much larger than their bodies ever produced metabolically, are being added to the rivers. Weak oestrogens in the environment may become hundreds of times more estrogenic by synergistic interactions (e.g., chemicals sprayed on forests and added on lawns and gardens), i.e., combinations of natural, medical, dietary, and environmental oestrogens. Man-made endocrine-disrupting chemicals range across all continents and oceans. Because of their persistence in the body, they can be passed from generation to generation. The endocrine system is sensitive to perturbation and a target for disturbance. Endocrine-disrupting chemicals can undermine neurological and behavioural development and subsequent potential of individuals. DES (diethyl stilbestrol) is associated with cancer, abortion, blood clots, infertility, and deformities in the children, and even in the grandchildren. DES is a weak oestrogen; it doesn't compete with natural oestrogens for the oestrogen receptors. Estriol prevents implantation and destroys the blastocyst. The anti-progestational activity of estriol and estradiol are almost the same. The conversion of estradiol to other oestrogens occurs mainly in the liver, in the non-pregnant woman, as does the further metabolism of the oestrogens into glucuronides and sulfates. The hormonal conditions leading to and associated with breast cancer all affect the liver and its metabolic systems. The hydroxylating enzymes are also affected by toxins. Hypothyroidism (low T3), low progesterone, pregnenolone, DHEA, etiocholanolone, and high prolactin, growth hormone, and cortisol are associated with the chronic high oestrogen and breast cancer pathobiology, and modify the liver's regulatory ability. Estetrol, which has one more hydroxyl group than estriol, is a more sensitive and reliable indicator of foetal morbidity than estriol during toxaemic pregnancies, because it starts to decrease earlier, or decreases more, than estriol. There are 4 hormonal features in women with breast cancer: 1- diminished androgen production, 2- luteal inadequacy, 3- increased 16-hydroxylation of estradiol, and 4- increased prolactin. The 16-hydroxylation converts estradiol into estriol. The tumour have about a six times higher concentration of estriol-sulfate than liver or muscle of breast cancer patient. Myocardial infarction in men is associated with elevated conversion of estradiol to estriol. Oestrogens and phytoestrogens cause deformities in the genitals, feminisation of men, and anatomical changes in the brain as well as functional masculinization of the female brain. The effects of the phytoestrogens are very complex, because they modify the sensitivity of cells to natural oestrogens, and also modify the metabolism of oestrogens, with the result of the effects on tissues can be either pro- or anti-oestrogenic. The flavonoids, naringenin, quercetin and kaempherol modify the metabolism of estradiol, causing increased bioavailability of both estrone and estradiol. Phytoestrogens may regulate the plants' interactions with other organisms. Their biochemistry in animals is much more complicated than that of natural oestrogens. Phytoestrogens are so often associated with other food toxins-antithyroid factors, inhibitors of digestive enzymes, immunosuppressants, etc.-that the avoidance of certain foods is desirable. Soybeans inhibit the thyroid function. Asia has endemic hypothyroidism, and beans are widely associated with hypothyroidism. Oestrogen's effects, ranging from shock to cancer, all seem to relate to an interference with the use of oxygen. Oestrogen and high altitude have something in common, namely, oxygen deprivation. A large increase in the consumption of oestrogenic soy products or the use of oestrogen is associated with osteoporosis. Women using oestrogen have an 80% higher rate of breast cancer. Progesterone can promote bone rebuilding. Oestrogen is a stress-promoting and age-promoting hormone. The catabolic glucocorticoids tend to increase with aging. Progesterone is anticatabolic hormone that should be used to prevent stress-induced atrophy of skin, bones, brain, etc. Excess prolactin can cause osteoporosis. Oestrogen promotes the secretion of prolactin, and can cause hyperprolactinaemia. Women have a higher incidence of osteoporosis than men do. Young women have thinner more delicate bones than young men. Menstrual irregularities, and luteal defects, that involve relatively high oestrogen and low progesterone, increase bone loss. Insulin stimulates bone growth. Oestrogen increases the level of FFAs in the blood, i.e., antagonises insulin (insulin decreases the level of FFAs), and the fatty acids themselves strongly oppose the effects of insulin. Between the ages of 20 and 40, there is a considerable increase in the blood level of oestrogen in women. However, bone loss begins around the age of 23, and progresses through the years when oestrogen levels are rising. Osteoarthritis, which involves degeneration of the bones around joints, is strongly associated with high levels of oestrogen. Oestrogen causes the retention of Ca by soft tissues. Calcium accumulates in the soft tissues; normally in aging and stress. Oestrogen reduces the activity of osteoclasts. The osteoclast is a type of phagocytic cell, and is considered to be a macrophage. Calcium retention by soft tissues is a marker of tissue aging, tissue damage, excitotoxicity, and degeneration. The toxic effects of excessive intracellular Ca (decreased respiration and increased excitation) are opposed by Mg. Oestrogen supply is associated with Mg deficiency, which can be an important factor in osteoporosis. Oestrogen changes the fat and water content of tissues. Ultrasound measurements may provide very accurate measurements of bone density, without the fat and water artefacts that can produce misleading results in the X-ray procedure. Tumour necrosis factor (TNF) kills only a few cancers, but it damages every organ of the body, usually causing the tissues to waste away. Other names, lymphotoxin and cachectin, reflect its toxic actions on healthy tissues. The level of TNF increases with aging. Both men and women lose minerals from their bones at the rate of about 1% per year. During aging, as their bones get thinner, men's oestrogen levels keep rising. Muscle loss occurs at about the rate of 1% per year. Elderly have weaker muscles, and are more likely to injure themselves in a fall because their muscles don't react as well. Women's muscles are normally smaller than men's, and oestrogen contributes significantly to these differences. TNF can produce very rapid loss of tissue including bone, and in general, it rises with aging. Low doses of oestrogen increase TNF, high doses decrease TNF. TNF is produced by endotoxin, and oestrogen increases the amount of endotoxin in the blood. Oestrogen, alone can stimulate the production of TNF. Lactic acid and unsaturated fats and hypoxia can stimulate increased formation of TNF. Oestrogen increases production of nitric oxide (NO) systemically, and NO can stimulate TNF formation. TNF causes cells to take up too much Ca, which makes them hypermetabolic before killing them. TNF increases formation of NO and carbon monoxide (CO), blocking respiration. TNF can cause a 19.5-fold increased in the enzyme that produces CO, which blocks respiration. Tamoxifen, which has some oestrogenic effects, including the inhibition of osteoclasts, can kill osteoclasts when the dose is high enough. Some types of dementia, such as Alzheimer's disease, involve a life-long process of degeneration of the brain, with an inflammatory component, that probably makes them comparable to osteoporosis and muscle wasting. The microglia, which are similar to macrophages, and the astrocytes, can produce TNF. Using aspirin regularly is associated with a low incidence of dementia. Aspirin inhibits the formation of TNF, and aspirin has been found to retard bone loss. Menopause is the result of prolonged exposure to oestrogen. Overactivity of the pituitary leads to many other features of aging. The links between oestrogen and TNF appear to be essential factors in aging and its diseases. Each of these substances has its constructive, but limited, place in normal physiology, but as excitatory factors, they must operate within the appropriate constraints. Adequate oxygen, a generous supply of carbon dioxide, saturated fats, thyroid, and progesterone restrain TNF, while optimising other cytokines and immune functions, including thymic protection. The respiratory production of energy and CO_2, and the respiratory defect in which lactic acid is produced, correspond to successful adaptation, and to stressful/excitotoxic maladaptation, respectively. Unsaturated fats, iron, and lactic acid are closely related to the actions and regulation of TNF, and therefore they strongly influence the nature of stress and the rate of aging. Stimulation in the presence of O_2 produces CO_2, allowing cells to excrete calcium and to deposit it in bones, but stimulation in the absence of O_2 produces lactic acid and causes cellular Ca uptake. The destructive effects such as multiple organ failure/congestive heart failure/shock-lung, etc., apparently involve arachidonic acid (AA) and its metabolites, which are based on the so-called essential fatty acids (EFAs). When O_2 and the correct nutrients are available, the hypermetabolism produced by TNF could be reparative, rather than destructive. Saturated fats, carbon dioxide, progesterone, and thyroid hormone together restore stability to a system that has been stimulated beyond its capacity to adapt without injury. The combination of hypoglycaemia with elevation of cortisone probably accounts for the nocturnal incidence of seizures. Seizures can be caused by lack of glucose, lack of oxygen, vitamin B_6 deficiency, and magnesium deficiency. They are more likely to occur during the night, during puberty, premenstrually, during pregnancy, during the first year of life, and can be triggered by hyperventilation, running, strong emotions, or unusual sensory stimulation. Water retention and low sodium increase susceptibility to seizures. A metabolic disturbance, especially if combined with intestinal irritation, may cause fits. Low albumin, high prealbumin, low Mg and high Ca all suggest hypothyroidism. Problems with the bowel, liver, and sex hormones are highly associated with hypothyroidism, both as causes and as effects. Water intoxication increased susceptibility to seizures even in normal individuals. Someone with hyponatraemia would be more susceptible to induction of a seizure by excessive water intake. Hypothyroid patients tend to lose Na easily, and unopposed oestrogen increases water retention, without an equivalent Na retention, so low thyroid, high oestrogen people have 2 of the conditions (oedema and hyponatraemia) known to predispose to seizures. Hypoglycaemia and high adrenaline predominate during in the early pre-dawn hours. Hypoglycaemia, like oxygen deprivation, is enough to cause convulsions during those hours. Progesterone and thyroid promote normal energy production, and their deficiency causes a tendency toward hypoglycaemia, oedema and instability of nerves. The most popular anticonvulsant drugs are both neurotoxic and teratogenic, that is, they damage the patient's brain, and greatly increase the incidence of birth defects. Stress can cause brain damage, as well as other diseases. Excitotoxicity is the harmful cellular effect (death/injury) caused by an excitatory transmitter such as glutamate or aspartate acting on a cell whose energetic reserves aren't adequate to sustain the level of activity provoked by the transmitter. As Ca enters cells, K leaves, and enzymes are activated, producing FFAs (e.g., linoleic and arachidonic) and prostaglandins, which activate other processes, including lipid peroxidation and free radical production. Protein kinase C (promoted by unsaturated fats and oestrogen) facilitates the release of excitatory amino acids. Oestrogen supports acetylcholine release, which leads to increased extracellular K and excitatory amino acids. Oestrogen also stimulates the production of free radicals. Calcium, free radicals, and unsaturated free fatty acids impair energy production, decreasing the ability to regulate K and Ca. The increased oestrogen associated with seizures is associated with reduced serum Ca. Ammonia is produced by stimulated nerves, and normally its elimination helps to eliminate and control the excitotoxic amino acids, glutamate and aspartate. The production of urea consumes aspartic acid, converting it to fumaric acid, but this requires CO_2, produced by normal mitochondrial function. Seizures can be induced by hypoxia. A deficiency of CO_2 would reduce the delivery of O_2 to the brain by constricting blood vessels and changing haemoglobin's affinity for O_2 (limiting CO_2 production), and the failure to consume aspartate (in urea synthesis) and glutamate (as α-ketoglutarate) and aspartate (as oxaloacetate) in the Krebs cycle, means that as energy becomes deficient, excitation tends to be promoted. Excitotoxicity is probably involved in insomnia. Balloonists and mountain climbers at extremely high elevations suffer from severe insomnia. Since excitation can promote the toxic forms of oxidation, many surprising substances turn out to have an antioxidant function. Magnesium, Na (balancing Ca and K), thyroid and progesterone (increasing energy production), and in some situations, CO_2. Aspirin, by inhibiting prostaglandin synthesis (and maybe other mechanisms) often lowers free radical production. Adenosine seems to have a variety of antioxidant functions, and one mechanism seems to be its function as an antiexcitatory transmitter. One of oestrogen's excitant actions on the brain may involve its antagonism to adenosine. Albumin serves to protect respiration, by binding FFAs. Oestrogen blocks the liver's ability to produce albumin, and increases the level of circulating FFAs. Free fatty acids cause brain oedema, which could be another aspect of oestrogen's contribution to seizure susceptibility. Magnesium deficiency is a factor that increases susceptibility to seizures. Hypothyroidism reduces the ability of cells to retain Mg. Thyroid hormone keeps oestrogen and adrenal hormones low, and increases production of progesterone and pregnenolone. It facilitates retention of Mg and of Na. Thyroid hormone, progesterone, and a high quality protein diet generally correct the epilepsy problem, together with minimising the consumption of the unsaturated fats. Vitamin E is likely to reduce the intensity and frequency of seizures. Coconut oil lowers the requirement for vitamin E, and reduces the toxicity of the unsaturated fats, favouring effective respiration and improving thyroid hormone and progesterone production. When the brain loses its oxygen supply, consciousness is lost immediately, before there is much decrease in the ATP concentration. While ATP constitutes a kind of reservoir of cellular energy, the flow of CO_2 through the brain cell is almost the mirror image of the flow of O_2. Oxygen scarcity leads directly to CO_2 scarcity. The sensitive state, consciousness, might require the presence of CO_2 as well as ATP, to sustain a co-operative, semi-stable, state of the cytoplasmic proteins.

MOODS DISORDERS

Lifetime rates of depression are nearly twice as high among women as they are among men. Women are more susceptible to depression due to certain potentials including increased genetic predisposition, increased exposure to stressful life events, and modulation of the neuroendocrine system in response to fluctuating gonadal hormones. The neuroendocrine molecule 17-β estradiol plays a role in depressive disorders among women. Rates of depression among females correspond with menarche (oestrogen level rise sharply). Oestrogen and progesterone interact with CNS at the level of neurotransmitters, especially serotonin. SSRI are efficacious in attenuating mood symptoms in the context of premenstrual syndrome (PMS) and premenstrual dystrophic disorder (PMDD), during the second half of the menstrual cycle or continuously over the course of the menstrual cycle. Variations in the levels of reproductive hormones also occur during pregnancy and birth, and abnormal levels of these hormones may play a role in the onset of postpartum depression. Psychosocial stressors such as lack of social support, marital discord, and a poor relationship may also play a role in postpartum depression. Perimenopausal women with hot flushes are significantly more likely to be depressed than are perimenopausal women without hot flushes. A lifetime history of depression may lead to an earlier decline in ovarian function and transition to perimenopause. Premenopausal women with a history of depression are 1.2 times more likely to enter perimenopausal early, and 3 times more likely to have an early perimenopause if they were depressed and using antidepressants. Oestrogen treatment is associated with a decrease in hot flushes, sweating, sleep difficulties, and headaches among menopausal women. The hormonal therapy may not be beneficial for depressive symptoms among postmenopausal women. Women with bipolar disorder who are childbearing age warrants special considerations because of the impact of the female reproductive cycle on the course of bipolar illness. Mood disorders are more than one and a half times as prevent among women as they are among men. The lifetime prevalence of major depression is twice as high in women as it is among men. Many women with bipolar disorder are misdiagnosed with major depression. Misdiagnosis of bipolar disorder for major depression may be the result of 2 factors: 1- women with bipolar disorder tend to present with major depressive episodes more often and earlier than men do, and 2- women with bipolar disorder tend to display a later onset of mania, as compared with men. Women are more likely to suffer rapid cycling bipolar. There is a higher prevalence of hypothyroidism in women and antidepressant-induced switching or cycle acceleration among bipolar disorder patients misdiagnosed as suffering from unipolar depression. Lithium is less effective for rapid cycling bipolar disorder characterised by mixed states. Antimanic agents are teratogenics. Carbamazine, oxcarbazepine, and topiramate are associated with oral contraceptive failure due to their induction of cytochrome P450 enzymes and associated metabolism of oestrogen and progesterone. Gabapentin, lamotrigine, and valproate are associated with oral contraceptive failure. There is no perfect solution, and no decision is risk-free. Certain atypical antipsychotics affect the hormone prolactin and may lead to hyperprolactinaemia. Risperidone tends to elevate prolactin levels, whereas clozapine, olanzapine, quetiapine, and ziprasidone do not. Clozapine and risperidone are associated with significant weight gain among patients and may predispose mothers to obesity and hyperglycaemia, which in turn increase the risk of congenital malformations in the child. Lithium may be associated with an increased risk of cardiovascular malformations, such as Ebstein's anomaly (.05%). Valproic acid (anticonvulsant) is associated with an increased risk of neural tube defects, when used in the first trimester may lead to spina bifida (3-8%) and increased long-term behavioural sequelae. Carbamazepine is associated with a 0.5% risk of spina bifida as well as elevated risks for craniofacial anomalies and microcephaly. Calcium channel blockers (verapamil) and ω-3 fatty acids may be options for the treatment of bipolar disorder during pregnancy. Sexual dysfunction commonly occurs in the context of major depressive disorder. Sexual dysfunction is not a symptom of major depressive disorder per se; decreased sexual desire and arousal may be characteristics associated with depression-related anhedonia. Sexual dysfunction is common among individuals with major depressive disorder. 40% of men and 50% of women would suffer decreased sexual interest; and 40-50% of reduced levels of arousal. Serotonin reuptake inhibitors (SRIs)-induced sexual dysfunction ranges from 30% to 70%. Intolerable side effects may be one reason that patients are noncompliant with antidepressant treatment. Premature discontinuation is associated with high rates of relapse and recurrence of depression. The sexual response cycle consists of 4 phases: desire, arousal, orgasm, and resolution. The phases of the sexual response cycle are affected by reproductive hormones and neurotransmitters. Oestrogen, testosterone, and progesterone promote sexual desire; dopamine promotes desire and arousal, and norepinephrine promotes arousal. Prolactin inhibits arousal, and oxytocin promotes orgasm. Serotonin, in contrast to most of these other molecules, appears to have a negative impact on the desire and arousal phases of the sexual response cycle, and this seems to occur through its inhibition of dopamine and norepinephrine. Serotonin also appears to exert peripheral effects on sexual functioning by decreasing sensation and by inhibiting nitric oxide (NO). The serotonergic system, therefore, may contribute to various sexual problems across the sexual response cycle: The primary sexual disorders, such as hypoactive sexual desire disorder, as well as secondary causes, such as psychiatric disorders (e.g., depression) and endocrine disorders (e.g., diabetes mellitus, which may cause neurologic and/or vascular complications). Situational and psychosocial stressors (e.g., job changes), and the use of substances exert a negative impact upon sexual functioning, such as psychotropic medication and drug abuse, e.g., alcohol. Antidepressant-induced sexual dysfunction is common, but only 14.2% of depressed patients taking selective SRIs (SSRIs) for depression spontaneously report sexual complaints. Risk factors for sexual dysfunction include: 1- age (50 years old or older), 2- having less than a college education, 3- not being employed full-time, tobacco use (6-20 times per day), 3- a prior history of antidepressant-induced sexual dysfunction, 4- a history of little or no sexual enjoyment, and 5- considering sexual functioning as not or only somewhat important. Drug holidays may provide relief from SSRI-induced sexual dysfunction, but, may result in SSRI discontinuation symptoms after 1-2 days or encourage medication noncompliance. There is significantly high rate of sexual dysfunction with both duloxetine and paroxetine. There is a relationship between depression and ischaemic heart disease (IHD). Patients with depressed affect and/or moderate/severe hopelessness have a much higher risk for both fatal and nonfatal IHD. Depressive-like symptomatology may be a risk factor for the subsequent development of IHD. Depression is also very common in patients with CAD (23%). The rate of coronary deaths among the nondepressed patients is 3% at 6 months and 6% at 18 months. Among the depressed patients, the incidence of coronary death is 16.5% at 6 months and 20% at 18 months. Depression increases the rate of coronary death within 6 months of the initial MI. Depression increases the risk of developing MI by 4.54, and even dysphoria increases the odds ratio for developing MI by over 2-fold. Barbiturates, phenothiazines, and lithium are associated with an increased MI risk, whereas tricyclic antidepressants (TCAs) are not. Depression is a significant independent risk factor for CAD (1.5- to 2-fold), which is similar to tobacco and smoking. Depression is associated with a number of alterations, including hyperactivity of the HPA axis and increased sympathetic activity. Functional abnormalities in platelet function are one of the key mediators of clot formation. Platelets have neuroectodermal origin and so are of the same lineage as central serotonergic neurons. Depression is associated with reduced heart rate variability, a known predictor for poor outcome in CAD. Tricyclic drugs are particularly dangerous in patients with heart disease, as they reduce heart rate variability. Reductions in heart rate variability may result in heart failure, conduction abnormalities or arrhythmia in the diseased heart. Depression is an independent risk factor for the development of IHD. Depression in the general population may range from 2% to 15%, while it is much higher in patients with diabetes (9% to 27%), stroke (22% to 50%), or cancer (18% to 39%). There is at least a 2-fold increased risk for developing depression in diabetic patients (type 1 or type 2); and among children and adolescents with type 1 diabetes, there is at least 2- to 3-fold increase in the prevalence of depression. The key symptoms of both diseases overlap, including insomnia, fatigue, and changes in appetite. Comorbid depression increases the incidence of diabetic complications and is associated with poor adherence to diabetic medication, exercise, and diet, as well as poor glycaemic control. Depression itself is also often associated with elevated glucocorticoids, which may further complicate diabetes. There may be elevated glucose levels and insulin resistance in depressed patients following an oral glucose tolerance test (GTT). Depression may increase the risk for subsequently developing diabetes by nearly 2-fold. Depression may further complicate diabetes. Medications used to treat depression do not further worsen the condition. Monoamine oxidase inhibitors and TCAs in combination with sulfonylureas may induce hypoglycaemia, and TCAs and mirtazapine may increase weight gain and $HbA1_c$ levels. Fluoxetine and presumably SSRIs in general produce an increase in insulin sensitivity, and sertraline lowers $HbA1_c$ levels. TCAs and SSRIs are also useful treatments for painful neuropathy. Treating depression in diabetic patients is associated with better QoL and improved glycaemic control. Depression is common in stroke patients (15-25%) and is particularly common in females. The location of stroke may not affect the incidence of poststroke depression, although it is possible that specific left-frontal cortical strokes may specifically increase the rate of depression. Poststroke depression is associated with a lack of recovery in daily activities, and with a reduced 10-year survival. 60% of nondepressed patients are expected to be alive 10 years after their initial stroke, while only 30% of depressed patients do. Depression is very prevalent in cancer, with rates in some forms of cancer (pancreatic, oropharyngeal) as high as 40% to 50%. Chronic depression may also be a risk factor for developing cancer. The immune system and nervous system interact, immune activations resemble many of the cardinal symptoms (i.e., sickness behaviour). Cytokines, which are powerful modulators of the immune system and released after infection, can produce anhedonia, malaise, weakness, social withdrawal, anorexia, hypersomnia, and hyperalgesia, which are all reminiscent of the major symptoms in depression. Patients with major depression have elevated serum concentrations of IL-6, a powerful proinflammatory cytokine. The use of interferons is associated with a number of debilitating side effects including flu-like symptoms, fatigue, nausea, anorexia, vomiting, and, in some cases, mood changes, cognitive changes, and/or psychosis. Paroxetine is associated with significantly higher compliance and reduce depression. Depression hinders compliance, increases symptom burden, and also diminishes survival in both cancer and stroke. Depression is a systemic illness that adversely affects the entire body. Depression is associated with a decreased number of circulating lymphocytes including T cells, B cells, and particularly natural killer cells, which are involved in the detection and removal of tumour cells. Age, hospitalisation, melancholia, and alcoholism are associated with the altered immune function seen in depressed patients. There is increased production of proinflammatory cytokines in major depression. Elevated levels of IL-6 occur in treatment-resistant depression. Cytokines may contribute to depression. Patients with inflammatory disease show an increased risk of depression, although this of course does not suggest causality. The sickness behaviour-induced by IL-6 is associated with sleep disorders, anorexia, and anhedonia. These symptoms mimic many of the cardinal features of depression. Sickness behaviour is not inherently maladaptive in response to an acute inflammatory process as it lowers energy usage and may help limit the spread of an infection. Cytokines are predominantly released peripherally in response to inflammatory processes. Cytokines are detected by the vagus nerve, which then releases cytokines within the CNS. The actively transported cytokines also cross the blood brain barrier (BBB). Cytokines are elevated in students prior to stressful exams, which correlates with self reported stress. Tamoxifen may increase risk of developing depression in breast cancer patients (27%). Adverse childhood experiences may increase the risk of adulthood depression, and alterations in the hypothalamic-pituitary-adrenal (HPA) axis function may be related. Dysfunction of the HPA axis is also one of the most consistent neurobiological findings in major depression. Early life stress may produce HPA dysfunction as well as induce behavioural changes reminiscent of the cardinal signs of depression. Early adverse life events are associated with long-lasting alterations in HPA axis function, which predisposes to psychopathology. Although depression is common in the postpartum period, treatment with antidepressants is often complicated by fear that the drugs may affect infants during breast-feeding. Bright lights for seasonal affective disorder (SAD), dawn simulation for SAD, and bright lights for non-SAD depression may be efficacious, with effect sizes equivalent to those from antidepressant pharmacotherapy trials. Duloxetine is effective in the treatment of major depressive disorder (MDD) in elderly patients. Some adolescents with treatment-resistant depression (TRD) may benefit from having bupropion added to their SSRI regimen. Drugs that deplete melatonin include β-blockers, NSAIDs, steroids, nicotine, alcohol, caffeine, sleep aids and anti-anxiety medications. Fluoxetine (Prozac) may lower melatonin levels. Low melatonin may contribute to insomnia, sleep/wake disorders, or PMS. Some forms of depression are associated with low melatonin levels. Low levels have also been implicated in increased risk for coronary heart disease (CHD). Melatonin influences other vital functions including cardiovascular and antioxidant protection, endocrine function, immune regulation and body temperature. Rapid, shallow breathing increases muscle tension. Tense muscles could, in turn, increase the likelihood of repetitive stress injuries, such as carpal tunnel syndrome. Individuals breathe faster and more shallowly when they type on a keyboard or use a mouse. Low melatonin levels occur in bulimia or neuralgia and in women with fibromyalgia; replacement reduces pain, sleeping disorders, and depression in fibromyalgia and bulimia. Lithium carbonate administration in bipolar disorder can cause folic acid deficiency. Calcium levels may be decreased in manic patients, ω-6 essential fatty acids may reduce certain side effects of lithium, excess vanadium may cause mania and can be reduced by ascorbic acid (nutritional precursors of neurotransmitters), L-phenylalanine, phosphatidylcholine (precursor of acetylcholine), L-tryptophan (precursor of serotonin) may be effective. Safflower oil may reverse lithium toxicity (tremor and ataxia), folic acid supplementation may improve condition, lithium affects sodium metabolism. Weight gain is a frequent adverse effect associated with lithium use. Leptin is an adipocyte hormone, regulating food intake and energy balance providing the hypothalamus with information on the amount of body fat. Leptin may be associated with lithium-induced weight gain. All memory systems are impaired during acute hypoglycaemia, with working memory and delayed memory being particularly susceptible. Hypoglycaemia is common in insulin-treated diabetics. Insulin resistance is correlated to both depression and sleep alterations. Overweight is related to NSSD (nonspecific symptoms of depression), sleep alterations, and hormonal changes. Eating a meal that is right for the individual's metabolic type should produce marked and lasting improvement in the energy, mental capacities, emotional well-being, and leave the individual feeling well-satisfied for several hours.

APPENDICES

APPENDIX 1: ABBREVIATIONS AND DEFINITIONS

α-AE	Alpha-amidating enzyme.
ΔΨm	The mitochondrial inner membrane potential.
α-TO·	α-Tocopheroxyl radical.
α-TOH	α-Tocopherol.
4-OH-TEMPO	4-hydroxy-2,2,6,6-tetramethyl-piperidine-1-oxyl.
^{99m}Tc-DMSA	Technetium-99m-dimercaptosuccinic acid.
AA	Arachidonic acid.
AAA	Aromatic amino acids.
ALA	α-lipoic acid.
AAST	American association of the surgery of trauma.
ABC	ATP-binding cassette.
Abdominal reflex	Contractions of the abdominal muscles on scratching of the abdominal wall.
ACE	Angiotensin-converting enzyme.
ACH	Adrenal congenital hyperplasia.
ACS	Abdominal compartment syndrome.
AD	Autonomic dysreflexia.
ADCC	Antibody-dependent cell-mediated cytotoxicity.
Adduct	Addition product of alkylating agent and biological molecule
ADHD	Attention deficit and hyperactivity disorder.
ADHs	Alcohol dehydrogenases.
Adjuvant	Assisting or aiding. A substance that aids another.
ADMA	Asymmetric dimethylarginine.
AdoHcy	The thioether S-adenosylmethionine.
Ado-Met	The nucleoside S-adenosylmethionine.
AGEs	Advanced glycation end products.
AGNII	Angiotensin II.
AGRP	Agouti-related protein.
AFGF	acidic FGF.
AICD	Activation-induced cell death.
AIDS	Acquired immunodeficiency syndrome.
AIF1	Apoptosis inducing factor 1.
AIS	Androgen insensitivity syndromes.
Al	Aluminium
Alkoxy radical	RO·
Alkylation	Formation of a hydrid molecule by covalent bond formation between a radical molecule and another, less reactive molecule.
ALA	5-aminolevulinic acid.
ALA	α-lipoic acid.
ALA-D	5-aminolevulinic acid (ALA) dehydrogenase.
ALARA	As Low As Reasonably Achievable.
ALA-S	5-aminolevulinic acid (ALA) synthase.
ALC	Acetyl-L-Carnitine.
ALD	Alcohol-induced liver disease.
ALEs	Advanced lipoxidation end products.
ALI	Acute lung injury.
ALLs	Acute lymphocytic leukemias.
AMD	Age-related macular degeneration.
AMF	Autocrine motility factor.
Ammonia	NH_3.

Ammonium	NH_4.
AMP	Adenosine monophosphate.
AMPA	α-amino-3-hydroxy-5-methyl-4-isoxazolepropionic acid.
Amygdalin	A cyanogenetic glycoside found in seeds and other plant parts of members of the family Rosaceae, e.g., bitter almonds. It is split by enzymatic hydrolysis into glucose, benzaldehyde, and hydrocyanic acid. The term is sometimes used interchangeably with Laetrile.
Anal reflex	Contraction of the anal sphincter on scratching or other irritation of the skin of the anus.
ANCA	Anti-neutrophile cytoplasmic antibody.
Anoikis	The disruption of epithelial cell interaction with the ECM induced apoptosis.
ANP	Atrial natriuretic peptide.
ANT	Adenine nucleotide translocase.
ANV	Actual nightly voids.
AO	Amine oxidase MADH, methylamine dehydrogenase
AAO	Ascorbic acid oxidase.
AOPP	Advanced oxidation protein products.
AOS	Activated oxygen species.
AP-1	Activator protein 1.
APACHE	Acute physiological and chronic health evaluation.
Apatite	Any of a group of minerals with the general formula $10Ca^{2+}$; $6PO_4^{3-}$; X^- where X is a monovalent anion such as a chloride, carbonate, fluoride, or hydroxyl ion, when it contains a hydroxyl ion the compound is hydroxyapatite.
APC	Activated Protein C.
APC	Antigen-presenting cell.
Apo-B	Apolipoprotein B.
Apoptosome	cyto c, Apaf-1, and procaspase-9.
Apostat	Apoptotic stimulus.
APP	Amyloid precursor protein.
APRF	Acute phase response factor.
APS	Adenosine-5'-phosphosulfate.
APTT	Activated partial thromboplastin time.
AQP2	Aquaporin-2.
AR	Androgen receptor.
ARE	Antioxidant responsive element.
ARF	Acute renal failure.
ARNT	Aryl hydrocarbon receptor nuclear translocator.
ART	Assisted reproduction techniques.
ASP	Apoptosis-specific protein.
ASA	American Society of Anesthesiologists.
ASA	American Standards Association.
ASA	American Surgical association.
ASA	Antisperm antibodies.
AT	Antithrombin-III.
AT	Ataxia telangiectasia.
ATC	Around the clock.
ATF2	Activating transcription factor 2.
ATII	Alveolar type II. Type II pneumocytes.
Atherogenesis	the formation of atheromatous lesion in the intima.
ATM	Ataxia talangiectasia.
Auscultation	The act of listening to sounds within the body. Mediate a. auscultation performed by the aid of an instrument (stethoscope) interposed between the ear and the part being examined.
Auto-oxidation	The phenomenon of combining directly with oxygen at ordinary temperature, without catalysis. Spontaneous, non-catalysed oxidation of a compound
AVP	Arginine-vasopressin.
Az	Azurin.

B	Boron.
β_2-m	β_2-microglobulin.
BBB	The blood-brain barrier.
Babiniski's reflex	Dorsiflexion of the big toe on stimulating the sole of the foot, normal in infants but in others a sign of a lesion in the CNS, particularly pyramidal tract.
BAL	Blood Alcohol Level.
Ballottement	A tossing about. Ballottement is a palpatory manoeuvre to test for a floating object. Renal b. palpation of the kidney by pressing one hand into the abdominal wall while the other hand makes quick thrusts forward from behind so as to throw the kidney against the anterior hand.
BCAA	Branched chain amino acids.
BCG	Bacillus Calmette-Guerin.
Bcl	B cell lymphoma.
BCR	Break point cluster region gene.
BDNF	Brain-derived neurotrophic factor.
BDZ	Benzodiazepine.
Be	Beryllium.
Beckwith-Wiedemann's syndrome	A congenital autosomal dominant syndrome with variable expressivity characterised by exomphalos, macroglossia, and gigantism, often associated with viceromegaly, adrenocortical cytomegaly, and dysplasia of the renal medulla.
BFGF	basic FGF.
BFGF	basic fibroblast growth factor.
BFU-E	Burst-forming unit-erythroid.
BH	Bcl-2 homology domains.
BH_4	Tetrahydrobioprotein.
BHLH	Basic Helix-loop-Helix Proteins.
Bi	Bismuth.
Bioassay	Determination of the active power of a substance by noting its effect on a live animal or an isolated organ preparation as compared with the effect of a standard preparation
Biotransformation	Biochemical processing of a coompound, which results in decreased or increased toxicity.
BK	Berkelium = an element of atomic number 97, atomic weight 247, produced by bombardment of the isotope of americium of atom weight 241 by helium ions.
Bladder reflex	Reflex contracting and emptying of the bladder in response to filling, the first step in micturating reflex it can be voluntarily inhibited by impulses from the brain in persons with normal neurological function.
BLMG	Buccal mucosa graft.
BMC	Bone marrow cell.
BMD	Bone mineral density.
BMI	Body mass index.
BMPs	Bone morphogenetic proteins.
BMR	Basal metabolic rate.
Bp	Base pairs.
BPIP	Bacterial permeability-increasing protein.
Bracken fern	Poisonous plant in veterinary medicine, is used as a forge for cattle and sheep in certain parts of the world and also as a component of vegetable salad in certain countries. It causes bone marrow hyperplasia, leading to death. The plant produces severe intoxication due to enzymatic destruction of thiamine by thiaminase.
BS	Bulbospongiosus.
BSE	Bovine spongioform encephalopathy.
BUF	Bacterial ureteroplegic factor.
Bulbospongiosus (bulbocavernosus) (penile) reflex	Contraction of the bulbospongiosus muscle in response to a tap on the dorsum of the penis.
Bufotenin	5HT derivative.
BXO	Balanitis Xerotica Obliterans.

C. neoformans	Cryptococcus neoformans.
C4bBP	C4b-binding protein.
CABG	Coronary artery bypass graft.
CABP	Chronic abacterial prostatitis.
CAD	Caspase-activated deoxyribonuclease.
CAD	Coronary artery disease.
CAH	Congenital adrenal hyperplasia.
CAIV	Carbonic anhydrase IV.
CAK	Ceramide activated protein kinase.
cak	Cyclin activating kinase.
cAMP	Cyclic 3' 5'-adenosine monophosphate.
CAM	Cell-adhesion molecules.
CA-MRSA	Community-acquired MRSA.
CAN	Cardiovascular autonomic neuropathy.
CaP	Carcinoma of the prostate, or prostate cancer.
CAP37	Cationic antimicrobial protein.
CAPK	Ceramide activated protein kinase.
CaR	The Ca^{2+}_o-sensing regulator. The extracellular calcium-sensing receptor.
Carbene	Compound with electron pair of a carbon atom readily available for bond formation.
Carbonyl	The divalent radical CO=, e.g., carbonyl chloride (phosgene $COCl_2$), nickel carbonyl (gaseous industrial product). Carbonyl also is the divalent group C=O, occuring in compounds such as aldehyde, ketones, carboxylic acid, and esters.
CARD	Caspase recruitment domain.
CART	Cocaine-amphetamine regulated transcript.
Caspase	Cysteine aspartic acid-specific proteases.
Caveolae	Flask-shaped sacrolemmal invaginations of fairly uniform size that are reported to be present in all smooth muscle cells.
CBAVD	Cystic fibrosis/congenital bilateral absence of the vas deferens.
CBG	Corticosteroid-binding globulin.
CBP	Chronic bacterial prostatitis.
CBP	Continuous blood purification.
CBP	CREB-binding protein.
CBT	Behaviour therapy.
CCl_4	Carbon tetrachloride.
CD95	Also called Fas or Apo-1.
Cd	Cadmium (bivalent metal).
CD	Cluster of differentiation.
CD	Conventional dosing.
CDC genes	Cell division cycle genes.
cdk	Cyclin dependent kinase
cdkI	Cyclin dependent kinase inhibitor.
CDKs	Cyclin-dependent kinases.
CDT	Cadeveric donor transplantations.
CED	Caenorhabditis elegans.
CEF	Cyclophosphamide, epirubicin and fluorouracil.
Centrifugal	Moving away from a center.
Centripetal	Moving toward a centre.
CEOOH	Cholesterylester hydroperoxide.
CETP	Cholesterol ester transfer protein.
CFIDS	Chronic fatigue immune dysfunction syndrome.
CFTR	Cystic fibrosis transmembrane conductance regulator.
CFU	Colony forming unit.
CFU-E	Colony-forming unit, erythroid.

CFU-GEMM	Colony-forming unit-granulocyte/erythroid/monocyte/megakaryocyte.
CFUs/ml	CFUs per mililiter of bacteria in the urine.
CGD	Chronic granulomatous disease.
CGH	Comparative genomic hybridisation.
CGMP	Cyclic 3',5'-guanosine monophosphate.
CGRP	Calcitonin gene-related peptide.
Ch-4S	Chondroitin 4-sulphate.
Ch-6S	Chondroitin 6-sulphate.
CHD	Coronary heart disease.
CHDF	Continuous haemodiafiltration.
Chelate	To combine with a metal in complexes compound in which a metallic ion is sequestered and firmly bound into a ring within the chelating molecule.
Chemical bond	Force holding 2 atoms in place and resisting their separation.
Chemotaxis	Cellular migration guided by a chemical concentration gradient.
CHF	Continuous hemifiltration.
CHIP28	Channel-forming integral protein of 28 kDa.
CI	Continuous infusion.
CIAP	Cellular inhibitor of apoptosis
CIC	Clean intermittent catheterisation.
CIEP	Counter immunoelectrophoresis.
CISC	Clean intermittent self-catheterisation.
CISCA	Cyclophosphamide, doxorubicin, cisplatin.
Cisternae	Flat sacs.
CK	Creatine kinase.
CK-aAb	CK-auto-antibodies.
CL	Chemiluminescence.
CM	Calmodulin.
CM	Cisplatin, methotrexate.
CMGP	Cartilage matrix glycoprotein.
CML	Carboxymethyllysine.
CMLs	Chronic myelogenous leukemias.
$CMRO_2$	Cerebral metabolic rate for oxygen.
CMO	Cetyl myristoleate.
CMV	Methotrexate, vinblastine, cisplatin.
CNF	Ciliary neurotrophic factor.
CNTF	Ciliary Neurotrophic factor.
COD	Calcium oxalate dihydrate.
COM	Calcium oxalate monohydrate.
Complete response	Complete disappearance of all evidence of the tumour.
COOH	Carboxyl group.
COP	Cardiac output.
COPD	Chronic obstructive pulmonary disease.
COPs	Cytosolic coat proteins.
$CoQ_{10}^{\cdot -}$	Ubisemiquinone-10 radical.
$CoQ_{10}H_2$	Ubiquinol-10.
CoQ	CoQ_{10}.
COX	Cyclo-oxygenase.
COX	Cytochrome c oxidase.
CP	Caeruloplasmin
CP	Cisplatin.
CPLA2	Cytosolic phospholipase A_2.
CR	Complete response.
CR	Computed radiography.
CrCl	Creatinine clearance.

C-reactive protein	Named after it reacted with the pneumococcal C-polysaccharide during acute pneumococcal pneumonia.
Creatine kinase	An Mg^{++}-activated transferase enzyme, catalyses the phosphorylation of creatine by ATP to form phosphocreatine. This reaction stores ATP as phosphocreatine in muscle and brain tissue. It occurs in 3 isoenzyme, 2 M (muscle), B (brain). CK_1 (BB) is found primarily in brain, CK_2 (MB) primarily in cardiac muscle, and CK_3 (MM) primarily in skeletal muscle.
CREB	cAMP response element binding protein.
Cremastric (Geijel's) (hypogastric) reflex	Stimulation of the skin on the front and inner side of the thigh retracts the testis on the same side. Indicates integrity of the first lumber nerve segment of the spinal cord or its root, absence indicates damage of the first lumber nerve segment or its root or lesion of the corticospinal tract.
CRF	Chronic renal failure.
CRP	C-reactive protein.
CS	Chondroitin sulfate.
CSF	Cerebrospinal fluid.
CT1	Cardiotrophin-1.
CTGF	Connective-tissue growth factor.
CTF	CAAAT transcription factor.
CTL	Cytotoxic T-lymphocytes.
CTP	Cystidine triphosphate.
CTR	Cough transmission ratio.
CTRF	Corticotropin-releasing hormone.
Cu	Copper.
CU	Cresta urethralis.
Cu^+	Cuprous, copper in its monovalent form.
Cu^{2+}	Cupric, copper in its divalent form in aqueous solution..
cyto c	Cytochrome c.
Cytochrome	Any class of haemoproteins whose principal biologic function is electron transport by virtue of a reversible valency change of its heme iron; cytochromes, are distinguished according to their prosthetic group as a, b, c, and d. Cytochrome b_5 reductase, a flavoprotein that transfers electrons from NADH or NADPH to cytochrome b_5, as in the liver microsomal desaturation of fatty acids. Cytochrome c oxidase, a copper-containing cytochrome of the a type which receive electrons from cytochrome c and transfers them to oxygen, enabling the oxygen to combine with hydrogen ions to form water, also called c. aa_3.
Cytokine	A generic term from nonantibody proteins released by one cell population on contact with specific antigen, which act as intra-cellular mediators in the generation of an immune response.
Cytosol	The liquid medium of the cytoplasm, i.e., cytoplasm minus organelles and nonmembranous insoluble components.
DAF	Decay-accelerating factor.
DAG	Diacylglycerol.
Dalton	Atomic mass unit.
Dartos (scrotal) reflex	vermicular contractions of the dartos muscle when a cold or stroking stimulus is applied to the perineum.
DAT	Dopamine transporter.
DBH	Dopamine β-hydroxylase
DcRs	Decoy receptors.
Deconditioning	A change in CV function after prolonged periods of weightlessness, probably related to a shift of a quantity of blood from the lower limbs to the thorax, resulting in reflex diuresis and a reduction of blood volume.
DED	Death effector domain.
dentritic cells (DC)	A heterogeneous group of nonphagocytic lymph node constituents comprising follicular dentritic cells of the germinal centres, interdigitating cells of the deep cortex, and veil cells of the afferent lymph and lymphatic sinuses, all of which have an irregular shape with numerous branching processes and an inconspicuous complement of cell organelles. DC are bone marrow derived

	professional antigen presenting cells (APC).
DES	Diethylstilbestrol, a nonsteroid oestrogen.
DFMO	Difluoromethylornithine.
DHA	Dehydroascorbic acid.
DHAES	Dehydroepiandrosterone sulfate.
DHEA	Dehydroepiandrosterone.
DHIC	Detrusor hyperactivity with impaired contractility.
DHPG	Dihydroxyphenylglycol.
DIC	Disseminated intravascular coagulation.
Diradical	Highly reactive molecule consisting of 2 atoms, each with an unpaired electron, e.g., oxygen, chlorine.
Dismutation	Oxidation -reduction reaction between more than one molecule of the same substance.
Disequilibrium syndrome	a group of symptoms occurring during or following hemodialysis or peritoneal dialysis, resulting from an osmotic shift of water into the brain, usually there is headache and less often nausea, muscle cramps, nervous irritability, drowsiness, and convulsions.
DKA	Diabetic ketoacidosis.
DM-Mt3243	diabetes mellitus due to 3243 bp mitochondrial tRNA mutation.
DMSO	Dimethyl sulfoxide.
DNA-PK	DNA-dependent protein kinase
DNA	Deoxyribonucleic acid.
DNIC	Dinitrosyl-iron complexes.
DNP	Dorsal nerve of the penis.
DNR	Do not resuscitate.
DO	Detrusor overactivity.
Do_2	Oxygen delivery.
Do_2I	Oxygen delivery index.
Dopa	3,4 dihydroxyphenylalanine.
Dorsal (erector spinae) reflex	contraction of the back muscle in response to stimulation of the skin over the erector spinal muscle.
DPL	Diagnostic peritoneal lavage.
Dpp	Dipeptide permease.
DRE	Digital rectal examination.
DRG	Dorsal root ganglion.
DS	Dermatan sulphate.
DSA	Digital subtraction angiography.
DSB	DNA double-strand breakage.
DMSO	Dimethyl sulfoxide.
DTH	Delayed type hypersensitivity responses.
Dysesthesia	Impairment of any sense, especially of that of touch, unpleasant abnormal sensation produced by normal stimuli.
Dysontogenesis	Defective embryonic development.
Dysontogenetic	Characterised of dysontogenesis.
Dystocia	Abnormal or difficult labour.
EAA	Essential amino acids.
EAA	Excitatory amino acid neurotransmitters.
EAS	External anal sphincter.
ECD	Extracellular domain.
ECF	Extracellular fluid.
ECM	Extracellular matrix.
ECP	Eosinophil cationic protein.
ECU	Environmental control unit.
ED	Endogenous defences.
EDN	Eosinophil derived neurotoxin.
EDS	Excessive daytime sleepiness.

EDTA	Ethylenediamine tetraacetic acid.
EFA	Essential fatty acid.
EGF	Epidermal growth factor.
ELAM	Endothelial leukocyte-adhesion molecule.
Electrolytes	solutes that forms ions in the solution and conduct electricity.
ELISA	Enzyme-linked immunosorbent assay. ELISA describes an enzyme-based immunoassay method which is useful for measuring antigen concentrations.
ELR	Glutamic acid-leucine-arginine.
Emla	Eutectic of local anaesthetics.
EMT	Epithelial-mesenchymal transition.
ENA-78	Epithelial-derived neutrophil attractant-78.
ENFs	Epidermal nerve fibers.
EP	Endogenous pyrogen.
EPA	Eicosapentaenoic acid.
EPCR	Protein C receptor.
Epigastric reflex	contraction of the abdominal muscles caused by stimulation of the skin of the epigastric or over the 5th and 6th intercostal spaces near the axilla.
epitope	An antigenic determinant of known structure.
EPN	Emphysematous pyelonephritis.
EPO	Eosinophil peroxidase.
EPO	Erythropoietin.
EPS	Expression of prostatic secretion.
EPS	Extraprostatic space.
EPS	Extracellular polymeric substances.
ER	Oestrogen receptor.
ERK	Extracellular signal-regulated kinase.
ERT kinase	EGF receptor threonine kinase.
ESPs	Extracellular signalling proteins.
ESR	Erythrocyte sedimentation rate.
ESRF	Endstage renal failure.
ETC	electron transport chain.
ETF	Electron transferring flavoprotein.
ETS	Enviornmental tobacco smoke.
EUA	Examination under anaesthesia.
Eukaryote	An organism whose cells have a true nucleus.
FAA	Free fatty acids.
Fab	Variable heavy chains.
FADD	Fas-associated death domain, also called Mot 1.
Fas	Fatty acid synthetase.
Fas	(CD95/APO-1) A cell surface receptor directly responsible for triggering cell death by apoptosis
FasL	Fas ligand.
Fc	Constant heavy chains.
FD	Functional dyspepsia.
FDPs	Fibrin degradation products.
Fe	Ferrum (iron).
Fe^{2+}	Ferrous.
Fe^{3+}	Ferric.
Fe^{3+}-O_2-Fe^{2+}	Perferryl radicals.
Ferric	Containing iron in its plus-three oxidation state, Fe^{3+}.
Ferrous	Containing iron in its plus-two oxidation state, Fe^{2+}.
Ferrum	Iron, Fe.
FH	Familial hypercholesterolaemia.
FIo_2	Fractional inspired oxygen.

FLIPs	FADD-like ICE inhibitory proteins.
Flow cytometry	Is a powerful analytic technique in which individual cells can be simultaneously analysed for several parameters, including size and granularity, as well as, the expression of surface and intracellular markers defined by fluorescent antibodies.
FNA	Fine needle aspiration.
FRSs	Free radical scavengers.
FSH	Follicle-stimulating hormone.
FTD	Frontotemporal dementia.
FWD	Free water deficit.
GAGs	Glycosaminoglycans.
GAP	Glans-approximation procedure.
GALT	Gut-associated lymphoid tissue.
GAPs	GTPase activating proteins.
GAS	Group A Streptococcus pyogenes.
GBM	Glomerular basement membrane.
GC	Glucocorticoids.
GCAP	Germ cell-specific alkaline phosphatase.
GCP	Granulocyte chemotactic protein.
GCSF	Granulocyte-colony-stimulating factor.
GCT	Germ cell tumour.
Geijel's (inguinal) reflex	Reflex in the female corresponding to the cremasteric reflex in the male, on stroking of the inner anterior aspect of the upper thigh there is a contraction of the muscular fibers at the upper edge of Poupart's ligament.
Genotype	The alleles present at one or more specific foci.
GFAT	Glutamine:fructose-6-phosphate-amidotransferase.
GFN	The genitofemoral nerve.
GHB	γ-hydroxybutyric acid.
GIFT	Gamete intrafallopian transfer.
GISA	Glycopeptide intermediate-susceptible S aureus.
GIT	Gastrointestinal tract.
GLA	γ-linolenic acid.
GLIP	Glycerol intrinsic protein.
GlucN-6-P	Glucosamine-6-phosphate.
Glut-1	Human erythrocyte glucose transporter.
GM-CSF	Granulocyte-macrophage colony stimulating factor.
GnRH	Gonadotropin-releasing hormone.
GPI	Glycerophosphatidylinositol.
GO	Galactose oxidase.
Gonadorelin	Synthetic luteinizing hormone-releasing hormone, structurally identical to the natural hormone.
GOT	Glutamate oxaloacetate transaminase.
GPI	Glycosyl phosphatidylinisitol.
GR	Glucocorticoid receptor.
GRMF	Glucosteroids response-modifying factor.
GRAP	Glanular reconstruction and preputioplasty.
GRO	Growth regulated oncogene.
GRP	Gastrin-releasing peptide.
Gram's staining	a staining procedure devised by Gram in which microorganisms are stained with crystal violet, treated with 1:15 dilution of Lugol's iodine; decolorized with ethanol or ethanol acetone, and counterstained with a contrasting dye, usually safranin. Those microorganisms that retain the crystal violet stain are said to be gram-positive, and those that lose the crystal violet stain by decolorization but stain with the counterstain are said to be gram-negative.
Gram-negative	losing the stain or decolorized by alcohol in Gram's method of staining, a primary characteristic of bacteria having a cell wall surface more complex in chemical deposition than do the gram-positive.

Gram-positive	retaining the stain or resisting decolorization by alcohol in Gram's method of staining, a primary characteristic of bacteria whose cell wall is composed of peptidoglycan and teichoic acid.
Gray (Gy)	A unit of absorbed radiation dose equal to 100 rads.
Grb2	Growth factor receptor-binding protein 2.
GSH	Glutathione (reduced).
GSI	Genuine stress incontinence.
GSLs	Glycosphingolipids.
GSNO	S-nitrosoglutathione.
GSSG	Glutathione (oxidised).
GSSH	Glutathione disulfide.
GT	Glucose transporters.
GTF	Glucose tolerance factor.
GTP	Guanosine triphosphate.
GTS	Growing teratoma syndrome.
GTT	Glucose tolerance tests.
HA	Hyaluronic acid.
Haemiacidrin	Solution containing citric and gluconic acids, magnesium hydroxycarbonate, magnesium acid citrate and calcium carbonate.
Haemin	Ferric heme. Heme is used as a generic expression denoting no particular iron valence state.
HA-MRSA	Hospital-acquired MRSA.
HbA1c	Haemoglobin A1c.
H-B-EGF	Heparin-binding epidermal growth factor-like growth factor.
Hc	Haemocyanin.
H/R	Hypoxia/reoxygenation.
H_2O_2	Hydrogen peroxide.
H_2O	Water.
H_4 bioprotein	Tetrahydrobioprotein.
H_2S	Hydrogen sulfide.
H_2SO_3	Sulfurous acid.
HCAs	Heterocyclic amines.
hCG	Human chorionic gonadotropin.
HCO_3^-	Bicarbonate, is a proton acceptor (base).
H_2CO_3	Carbonic acid, is a proton donor (acid).
HDL	High-density lipoprotein.
Hg	Mercury.
HFEMF	High frequency electromagnetic fields.
HGF	Hepatocyte growth factor.
HHNK	Hyperosmolar hyperglycaemic nonkitotic coma.
HIT	Heparin-induced thrombocytopenia.
hKLK2	Human glandular kallikrein-1.
HLA	Human leucocyte antigen.
hMG	Human menopausal gonadotropin.
HMT	Histamine-N-methyltransferase.
HMW	High-molecular-weight.
HMW-AOPP	High molecular weight-AOPP.
HNO_2	Nitrous acid.
HOCl	Hypochlorous acid.
HOCM	High osmolar contrast media.
Homolysis	Lysis of a cell by extracts of the same type of tissue.
HOONO	Peroxynitrous acid.
HOX	Hypohalous acids.
HPA	Hypothalamus-Pituitary-Adrenal axis.
HPAA	hypothalamic pituitary adrenal axis.
hPAP	Human prostatic acid phosphatase.
HPMC	The human peritoneal mesothelial cell.

HPN	Home parentral nutrition.
HPP-CFC	High proliferative potential colony-forming cell.
HPS	Hepatopulmonary syndrome.
HPT	Hypothalamus-Pituitary-Thymus axis.
h, hr, hrs	Hour, hours.
HREs	Hormone response elements.
Hrs.	Hydrophobic regions.
HRV	Heart rate variability.
HS	Heparan sulphate.
HSC	Haematopoietic stem cell.
HSF	Heat shock transcription factors.
5-HT1A	5-hydroxytryptamine$_{1A}$.
HTGp	Prostate specific transglutaminase.
HTL	Human T-cell lymphotropic virus type I.
HUS	Haemolytic uremic syndrome.
Hydrazine	A colourless, gaseous diamine, H_2N-NH_2, also any member of a group of its substitution derivatives. Also called diamide.
Hydrogen peroxide	H_2O_2
Hydrogen sulfide	H_2S
8-iso-PGF$_{2\alpha}$	8-iso-prostaglandin $F_{2\alpha}$ or 8-epi-prostaglandin $F_{2\alpha}$
IADSA	Intraarterial digital subtraction angiogram.
IBD	Inflammatory bowel disease.
IBS	Irritable bowel syndrome.
I-CAM1	Intracellular adhesion molecule 1.
I/R injury	Ischaemia/reperfusion injury.
IA	Intra-arterial.
IAP	Intra-abdominal pressure.
IAP	Inhibitor of apoptosis proteins.
IBT	Immunobead rosette test.
IC	Ischiocavernosus.
IC	Interstitial cystitis.
ICAM	Intercellular adhesion molecule.
ICE	IL1β converting enzyme, is a cysteine protease that converts prointerleukin-1β to its active form by cleaving it at aspartate residues.
ICM	Intra-celluar matrix (cytosol).
ICS	Immunocytochemical staining.
I.C.S	International Continence Society.
ICSI	Intracytoplasmic sperm injection.
ICU	Intensive care unit.
IDDM	Insulin-dependent diabetes mellitus.
IEMA	Monoclonal immunoenzymeteric assay.
IERG	Immediate early response genes.
IES	Invaginated extraprostatic space.
IF	Intrinisic factor.
IFN	Interferon.
IGF1	Insulin-like growth factor 1.
IGIF	IFN-γ inducing factor.
IIR	Inflammatory immune response.
IKK	Inhibitor of κB (I-κB) kinase complex.
IL	Interleukin.
IMP	Inosine monophosphate.

IMPs	Intramembranous particles.
iNOS	Inducible nitric oxide synthase.
INR	International normalised ratio.
Intracrine	A type of hormone function in which a regulatory factor acts within the cell that synthesises it by binding to intracellular receptors.
ionophore	Any molecule, as drug, that increases the permeability of cell membranes to a specific ion.
IMPs	Intramembranous particles.
IP-10	Interferon-γ inducible protein-10.
IP6	Inositol Hexaphosphate.
IR	Insulin receptor.
IR	Insulin resistance.
IRMA	Monoclonal immunoradiometric assay.
ISD	Intrinsic sphincter deficiency.
ISH	In situ hybridisation.
ISI	International sensitive index.
IT	Immunotherapy.
ITP	Inosine triphosphate.
IUI	Intrauterine insemination.
IVDSA	Intravenous digital subtraction angiogram.
IVF	In vitro fertilisation.
IVU	Intravenous urography.
IWL	Involuntary weight loss.
Jaffe reaction	Measurement of PRA, plasma electrolytes and creatinine.
Jak	Janus family.
JNK	c-Jun N-terminal kinase, or SAPK.
K	Potassium.
K⁺	Potassium ion.
KA	kang ai.
KALIG	Kallmann's syndrome interval gene.
kD	kilo Dalton.
KDSM	Keratinizing desquamative squamous metaplasia.
KGF	Keratinocyte growth factor.
kilobase, kb	A unit used designating the length of a nucleic acid sequence. 7 kb means a sequence 7000 nucleotides long.
Kilodalton, kda, kDa, kD, kd	A unit of mass, being 1000 daltons.
Kir	**K** channel, **i**nward **r**ectifier.
Knudson's two-mutation theory of cancer	In order for a cell to become transformed, the functions of a critical growth control gene must be inactivated by mutations, or by epigenetic events that affect both copies of the gene.
Kocher's (testicular compression) reflex	contraction of the abdominal muscle on compression of the testicle.
KS	Keratan sulfate.
L	carbon-centred lipid radical
LAB	Lactic acid bacteria.
LAD	Leukocyte-adhesion-deficiency syndrome
Laetrile	Trademark for l-mandelonitrile-b-glucuronic acid, derived by hydrolysis of amygdalin and oxidation of the resulting l-mandelonitrile-b-glucoside; it is alleged to have antineoplastic properties. The term is sometimes used interchangeably with amygdalin.
LAK	Lymphokine-activated killer.
LAMPs	Lysosome-associated membrane proteins.
LAPP	Leech antiplatelet protein.
LBM	Lean body mass.
LCP	Lineage-committed progenitors.

LD_{50}	Median lethal dose.
LDL	Low-density lipoprotein.
LH	Luteinizing hormone.
LHRH	Luteinizing hormone-releasing hormone.
LIF	Leukaemia inhibiting factor.
Lineage	Descent traced down from or back to a common ancestor.
Lithotomy position	The patient in dorsal decubitus with hips and knees flexed and the thigh abducted and externally rotated.
LMA	Longitudinal muscle of the anus.
LMNL	lower motor neurone lesions.
LMWA	Low molecular weight antioxidant.
LMW-AOPP	Low molecular weight-AOPP.
LMWCr's	Low molecular weight chromium compounds.
LOO·	Lipid peroxyl radical.
LOOH	Lipid hydroperoxide.
LOCM	Low osmolar contrast media.
LOX	Lipoxygenase.
LPB	LPS-binding protein.
LPL	Lipoprotein lipase.
LPS	Lipopolysaccharide.
LRDT	Living related donor renal transplantations.
LUTO	Lower urinary tract obstruction.
M	Molality = the number of moles of solute dissolved in 1,000 g of solvent.
M	Molarity = the number of moles of solute in exactly 1 litre of solution.
M-6-P	Mannose-6-phosphate.
MAP	Mean arterial pressure.
MAP	Mitogen-activated ptotein.
MAP-2 kinase	Microtubule associated protein-2 kinase.
MAPK	Mitogen-activated protein kinase.
MAPKK kinases	MAPK kinase kinases.
MAR	Mixed agglutination reaction.
Masked state	State of ecological illness in which sensitivities to chemical or food is obscured by addiction to them; patient can become unmasked through avoiding the offending food or chemical.
Mass (Riddochs) reflex	In severe spinal cord injury, stimulation below the level of the lesion produces flexion reflexes to the lower extremity, evacuation of the bowel and bladder, and sweating of the skin below the level of the lesion.
MBP	Major basic protein.
MBP kinase	Myelin basic protein kinase.
MCO	Metal catalysed oxidation.
MCP	Monocyte chemoattractant protein.
MCP-1	Macrophage chemoattractant protein 1.
MCS	Multiple chemical sensitivity.
MDF	Myocardial depressant factor.
MDH	Malate dehydrogenase.
mdr or MDR	multidrug resistance gene.
ME	Myalgic encephalomyelitis.
MELAS	Mitochondrial encephalopathy, lactic acidosis, and stroke-like episodes.
Menaquinone	Any of a series of compounds in which the phytyl side chain of phytonadione (vit K_1) is replaced by a side chain of prenyl units and which have vitamin K activity, they are synthesised in gram-positive bacteria. Called also farnoquinone and vit K_2.
Merism	The repetition of parts in an organism so as to form a regular pattern.
MET	Mesenchymal-epithelial transition.
MetHb	Methaemoglobin.
MFO	Mixed-function oxidation.

MGF	Mast cell growth factor.
mGy	MilliGrays.
MHCs	myosin heavy chain subunits.
Micturation reflex	Any of the reflexes necessary for effortless evacuation of urine and subconscious maintenance of continence, vesical contraction following distention of the bladder, vesical contraction evoked by urethral flow, vesical contraction evoked by proximal urethral distension, relaxation of the urethra resulting from running liquid in the urethra, distention of the bladder resulting in relaxation of the external sphincter, relaxation of the proximal urethral smooth muscle by distension of the bladder and vesical contraction related to running liquid through the urethra.
MIC	Minimum inhibitory concentration.
MIG	Monokine induced by IFNγ.
MIP	Major intrinsic protein.
MIP-1α	Macrophage inflammatory protein-1α.
MIS	Mullerian inhibiting substance.
MIWC	Mercurial-insensitive water channel.
Mixed-function oxidase system (MFO)	Enzymatic detoxifying system to render lipid-soluble molecules more water-soluble for excretion.
MLCK	Myosin light chain kinase.
MLCP	Myosin light-chain phosphatase.
MLNs	Mesenteric lymph nodes.
MLR	Mixed lymphocyte reaction.
MM	Malignant melanoma.
MMD	Mitochondrial membrane depolarization.
MMP	Matrix metalloproteinase.
Mn	Manganese.
MNV	Mean nuclear volume.
MØ	The macrophage lineage.
Modelling	Experiments stimulating natural phenomena under conditions.
MODS	Multiple organ dysfunction syndrome.
MOF	Multiple organ failure.
Molal	Containing one mole of solute/kg of solvent. molal refers to the weight of the solvent, molar to the volume of solvent.
Mole, mol	The amount of substance (in a system) that contains as many elementary entities (atoms, ions, molecules, or radicals) as there are carbon atoms in 12 grams of carbon-12(^{12}C), or that amount of chemical compound whose mass in grams is equivalent to its formula mass.
MMSE	Mini-mental state examination.
MPAP	Mean pulmonary arterial pressure.
MPF	Mutation promotion factor.
MPIF	Monocyte/macrophage procoagulant inducing factor.
MPO	Myeloperoxidase.
MPS	Mononuclear phagocyte system.
MPT	Membrane permeability transition.
MRNA	Messenger ribonucleic acid.
MRSA	Methicillin-resistant Staphylococcus aureus.
MS	Multiple sclerosis.
MSC	Mechanosensitive ion channels.
MSG	Monosodium glutamate.
MSH	Melanocyte-stimulating hormone.
MSLT	Multiple sleep latency test.
MSM	Methylsulfonylmethane, also known as dimethyl sulfone.
MT	Metallothionein.
MtDNA	Mitochondrial DNA.
MTHFR	Methylenetetrahydrofolate reductase.
MTIR	Multiple total internal reflection.

MUFA	Monounsaturated fatty acids.
Mutagen	A chemical or physical agent that induces or increases genetic mutations by causing changes in DNA.
MVAC	Methotrexate, vinblastine, doxorubicin, and cysplatin.
MW	Microwaves.
MW	Molecular weight.
Na	Sodium
Na^+	Sodium ion.
NAC	N-acetyl cysteine.
NAD^+	Nicotinamide adenine dinucleotide (oxidised form).
NADH	Dihydronicotinamide adenine dinucleotide (reduced form).
$NADP^+$	Nicotinamide adenine dinucleotide phosphate (oxidised form).
NADPH	Dihydronicotinamide adenine dinucleotide phosphate (reduced form).
NANC	Noradrenergic, noncholinergic.
NAP	Neutrophil activating protein.
NAT2	N-acetyltransferase 2.
NCAM	Neural cell adhesion molecule.
NDI	Nephrogenic diabetes insipidus.
NDNA	Nuclear DNA.
NDP kinases	Nucleoside diphosphate kinases family.
NDS	Neuroendocrine immune dysfunction syndrome.
ND:YAG	Neodymium:yttrium-aluminum-garnet.
NE	Norepinephrine.
NEAA	Non-essential amino acids.
NED	No evidence of disease.
NEFA	Nonesterified fatty acid.
Neoadjuvant	A term used to describe preliminary cancer therapy. Chemotherapy administered before cystectomy or definitive RT.
NF1	Neurofibromatosis type-1 gene.
NF-IL6	Nuclear factor of IL-6.
NFκB	Nuclear factor kappa B.
NFR	Not for resuscitation.
NGF	Nerve growth factor.
Ni	Nickel.
NIDDM	Non-insulin-dependent diabetes mellitus.
NIK	NFκB-inducing kinase.
NIR	Nitrite reductase.
NIRO	Near-infrared oxygen monitoring
Nitrate	NH_3^-.
Nitrite	NH_2^-.
NK	Natural killer cells.
NM	Nanomolar.
NMDA	N-methyl-D-aspartate.
NMR	Nuclear magnetic resonance.
NO_2	Nitrogen dioxide.
NO^{2-}	Nitronium ion.
N_2O_3	Dinitrogen trioxide.
N_2O_4	Dinitogen tetroxide.
NOS	Nitric oxide synthase, which is a cytochrome. P450 type haemoprotein. . Type-1 NOS (neuronal, nNOS), type-3 NOS (endothelial, eNOS), and type-2 or inducible NOS (iNOS).
NPTR	Nocturnal penile tumescence and rigidity.
NPY	Neuropeptide Y.
NRP	Non-radical product.
NRPs	Nuclear regulatory proteins.

NRTK	Non-receptor tyrosine kinase.
NSE	Neuron specific enolase.
NTA	Nitrilotriacetic acid
NTS	Nucleus tractus solitarius.
NUV	Nocturnal urine volume.
6-OHDA	6-Hydroxydopamine.
8OhdG	8-hydroxy-2'-deoxyguanosine.
O_2	Oxygen.
$^1O_2$	Singlet oxygen.
O_2^-	Superoxide anion radical.
O_2^{2-}	Peroxide anion.
·OH	The hydroxyl radical.
OCl⁻	The hypochlorite ion.
ODC	Orotidine 5'-phosphate decarboxylase.
OFRs	Oxygen free radicals.
ONOO⁻	Peroxynitrite.
OPG	Osteoprotagan.
OPG	Osteoprotegerin.
ORFs	Open reading frames.
Orphan receptor	Receptor whose ligands have not been identified yet.
OSA	Obstructive sleep apnoea.
OSCE	Objective Structured Clinical Examination.
OSLER	Objective Structured Long Examination Record.
OSM	Oncostatin M.
o-tolidine	A compound related to benzidine and formerly used in testing occult blood.
Oxide	Any compound oxygen with an element or radical.
OXPHOS	Oxidative phosphorylation.
Ozonide	Product of ozone addition to a small molecule (unsaturated hydrocarbon).
PABA	Para-aminobenzoic acid.
PAF	Platelet activating factor.
PAF	Prostatic antibacterial factor.
PAF1	Plasminogen-activator inhibitor 1.
PAHs	Polycyclic aromatic hydrocarbons.
PAI-1	Plasminogen activator inhibitor.
Palpation	The application of the palm of the fingers with light pressure to the surface of the body for the purpose of determining the consistence of the parts beneath in physical diagnosis.
PARP	Poly(ADP-ribose)polymerase.
Partial response	Decrease by 50% in the volume of tumour.
PASD	The prostate-specific antigen denisty.
Pb	Lead.
PC	Phosphatidylcholine.
PCA	Procoagulant activity.
PCBs	Polychlorinated biphenyls.
PCi	Picocurie.
PCM	Pubococcygeus muscle.
PCMP	Posterior portion of pubococcygeus muscle.
PCN	Percutaneous nephrostomy.
PCNA	Proliferating cell nuclear antigen.
PCNL	Percutaneous nephrostolithotomy.
PCOS	Polycystic ovary syndrome.
PC-PLC	Phosphocholine-phospholipase C.
PCR	Polymerase chain reaction.
Pcr	Phosphocreatine.
PDE	Phosphodiesterase.

PDGF	Platelet-derived growth factor.
PDN	Painful diabetic neuropathy.
PDV	Portal-drained viscera.
PE	Phosphatidylethanolamine.
PE	Polyethylene.
PE	Prostatic epithelium.
PEM	Protein energy malnutrition.
PEP	Phosphoenolpyruvate.
Peroxidation	Propagative, oxidative breakdown of fatty acids in biological membranes.
Peroxyl radical	ROO·
Peroxynitrite	ONOO⁻
PEST	pro-glu-ser-thr.
PET	Positron emission tomography.
PF4	Platelet factor 4.
Pfu	Plaque forming units.
PGAl	Phosphoglyceraldehyde.
PGs	Proteoglycan.
PGFs	Peptide growth factors.
PGHS	Prostaglandin H synthase.
PGP	Protein gene product.
Ph⁺	Philadelphia+ chromosome.

Phase I studies evaluate toxicity and dose-schedule. They are usually done in patients who have failed other therapies. Any response is an added bonus.

Phase II studies evaluate a specific dose or schedule against a specific tumour.

Phase III studies compare the current best therapy against the experimental treatment. For some tumours, there is no best treatment and the effects are measured against appropriate controls. Phase III studies may also identify equally effective regimens that may have less toxicity.

Phenotype	The expression of a single gene or gene pair.
Pheromone	A substance secreted to the outside of the body by an individual and perceived by a second individual of the same species, releasing a specific reaction of behaviour in the percipient
PHN	Postherpetic neuralgia.
Phospholamban	22 kilodalton membrane-bound polypeptide of the sacroplasmic reticulum.
PI-3K	Phosphatidylinositol-3-kinase.
P_i	Inorganic phosphate.
PI_3-kinase	Phosphatidylinositol 3-kinase.
PKA	Protein kinase A.
PKC	Protein kinase C.
PKG	Protein kinase G.
PLAP	Placental alkaline phosphatase.
Plastic	Tending to build up tissues or to restore a lost part.
Plasticity	The quality of being plastic or conformable, the ability of the embryonic cells to alter in conformity with the immediate environment.
PLC-g	Phospholipase C-g.
PLGF	Placental growth factor.
PLMS	Periodic limb movements in sleep.
PLP	Pyridoxal-phosphate.
PLS	Pathogenesis, location and degree of symptoms, and symptoms as prognostic factors.
PMA	Phorbol myristate acetate.
PMDS	Persistent Mullerian-duct syndrome.
PMNs	Polymorphonucleocytes.
PMØ	Peritoneal macrophages.
PMS	Pre-menstrual syndrome.
PND	Percutaneous nephrostomy drainage.
Pol II	RNA polymerase II.

Polyene	A chemical compound in which there are several conjugated double bonds.
Polyol	An alcohol containing more than 2 hydroxyl groups, e.g., sugar alcohols, inositol.
POMC	Proopiomelanocortin.
Ppb	Parts per billion.
Ppi	Inorganic pyrophosphate.
PPIase	Peptidyl-prolyl cis-trans isomerase.
Ppm	Parts per million.
PQQ	Pyrrolo-quinoline quinone.
PR	Partial response.
PR	Persistent positive cytology.
PR	per-rectal examination.
PR	Progesterone receptor.
PRA	plasma renin activity.
Prehn's sign	Pain due to torsion of spermatic cord increases when scrotum is gently lifted up onto the symphysis pubis. Pain due to epididymoorchitis is relieved.
PRNP	Prion protein.
Progression	Growth of the tumour (as new lesions) in spite of therapy.
PROS	Partially reduced oxygen species.
PS	Phosphatidyl serine.
PS	Prostatic stroma.
PSP	Photostimulable phosphor.
PSPs	Protein serine phosphatases.
PTSD	Post-traumatic syndrome.
PT	The mitochondrial permeability transition.
PTH	Parathyroid hormone
PTHrP	Parathyroid hormone-related peptide.
PTCL	Peripheral T-cell lymphoma.
PTFE	Polytetrafluoroethylene.
PTK	Protein tyrosine kinase.
Pto_2	Oxygen tension.
PTP1C	Protein phosphatase-1C.
PTPases	Protein tyrosine phosphatases.
PTR	Pressure transmission ratio.
PU	Polyurethane.
PUFAs	Polyunsaturated fatty acids.
PUJO	Pelviureteric junction obstruction, as UPJO.
PUL	Pubourethral ligament.
PUSM	Periurethral striated muscle.
PVC	Polyvinyl chloride.
PVFS	Post viral fatigue syndrome.
PVN	Paraventricular nucleus of the hypothalamus.
PVR	Postvoiding residual of urine.
PVRI	Pulmonary vascular resistance index.
QFIA	Quantitative fluorescence image analysis.
QoL	Quality of life.
QST	Quantitative sensory testing.
QUIN	Quinolinic acid.
Quinone	Any benzene derivative in which two hydrogen atoms are replaced by two oxygen atoms.
RA	Rheumatoid arthritis.
Rad	A unit of measurement of the absorbed dose of ionising radiation, it corresponds to an energy transfer of 100 ergs/gm of any absorbing material. The biological effect of 1 rad depends on the tissue exposed to radiation.
RAR	Retinoic acid receptor.
RasGAP	GTPase activating protein.
RAST	Radio-allergo-absorbent.

RB	Retinoblastoma.
RBC	Red blood cells.
RCOs	Reactive carbonyl compounds.
RDA	Recommended daily dose allowances.
Redox cycling molecule	One that can cyclically accept electrons and donate them to oxygen or other activated oxygen species, thereby generating oxidant molecular species.
REE	Resting energy expenture.
REM	Rapid eye movement.
Reno-intestinal reflex	Inhibition of the mobility of the intestine resulting from renal irritation.
Reno-renal reflex	A reflex pain or anuria in a sound kidney in cases in which the other kidney is diseased.
RER	Rough endoplasmic reticulum.
RF	Radio-frequency radiation.
RFLP	Restriction fragment length polymorphism.
RGD	Arg-Gly-Asp, a recognition sequence on many ligands of integrins.
Rheology	The science of the deformation and flow of matter, such as the flow of blood through the heart and blood vessels.
RIA	Radioimmunoassay.
RIP	Receptor-interacting protein.
RLF	Replication licensing factor
RMR	Resting metabolic rate.
RNH-Cl	Chloramines.
R-NHI	Endogenous amines.
RNI	Reactive nitrogen intermediate.
RO	Reverse osmosis.
ROC	Receiver operating characteristic.
ROI	Reactive oxygen intermediates.
ROM	Reactive oxygen metabolites.
ROO$^-$	Peroxyl radical.
ROR	Reactive oxygen radicals.
ROS	Reactive oxygen species.
RPNLD	Retroperitoneal lymph node dissection.
Rrs	Response rates.
RSK-kinase	Ribosomal S6 protein kinase.
RSNO	S-nitrosothiols.
RT	Radiotherapy.
RTLFs	Respiratory tract lining fluids.
RT-PCR	Transcriptase-polymerase chain reaction.
RVD	Regulatory cell volume decrease.
RVI	Regulatory cell volume increase.
RXR	Retinoid X receptor.
SAA	Serum amyloid A.
SAM-e	S-adenosymethionine.
Sanative	Having a tendency to heal, curative.
Sanogenesis	The science dealing with sanative, i.e., the mechanisms of prevention and elimination of the pathological process, and mechanisms responsible for compensation and recovery of disordered functions.
SAPK	Stress activated protein kinase, or JNK.
SAS	Sleep apnoea syndrome.
Sb	Antimony
SBR	Small bowel resection.
SBS	Small bowel syndrome.
SCAMPS	Secretory carrier membrane proteins.
SCC	Squamous cell carcinoma.
SCF	Stem cell factor.

SCFA	Small-chain fatty acids.
Schiff's reagent	A reagent for testing for the presence of aldehydes. German chemist.
Scission	In chemistry, the splitting of a molecule into two or more simpler molecules. Splitting, fission.
Scr	Serum creatinine.
SDH	Sorbitol dehydrogenase.
SDF	Stromal cell-protein.
Se	Selenium.
Senescence	The process or condition of growing old, especially the condition resulting from the transitions and accumulations of the deleterious aging processes.
SERMs	Selective oestrogen receptor modulators.
Ser-P	Serine phosphorylation.
Sexual reflex	Reflex of erection and ejaculation produced by stimulation of the genitals.
SF-1	Steroidogenic factor 1.
SG	Specific gravity.
SGC	Soluble guanylate cyclase.
SH	Sulfhydryl.
SH2 domains	Scr homolgy domain 2.
SHBG	Sex hormone binding globulin.
SI	Stress incontinence.
SIgA, S-IgA	Secretory immunoglobulin A.
SIRS	Systemic inflammatory response syndrome.
SK	Streptokinase.
Sle$^\alpha$	Sialyl Lewise$^\alpha$.
Sle^x	Sialyl Lewis^x.
SM	Sphingomyelin.
Smase	Sphingomyelinase.
SMCs	Smooth muscle cells.
Sn	Inorganic tin.
SPN	Sacral parasympathetic nucleus.
SO$_2^-$	Superoxide radical.
SOD	Superoxide dismutase.
SP	Single positive cells.
SP	Substance P.
SPA	Surfactant protein-A.
SPECT	Single photon emission computed tomography.
SPF	S-phase promoting factors.
SR	Sarcoplasmic reticulum.
SRB	Sulfate reducing bacteria.
SSB	Single-strand breaks (DNA).
STAT or Stat proteins	Signal tranducers and activators of transcription.
Steinstrasse	The stream of tiny stone fragments that may fill the ureter after lithotripsy-and may obstruct it.
SSRIs	Selective serotonin reuptake inhibitors.
STE	Stryneric theroplastic elastomer.
STF	Semineferous tubule fluid.
Stochastic	Able to conjecture skilfully; arrived at by skilful conjecturing; random.
Sucralfate	Sucrose aluminum sulfonate.
Sulfurous acid	H_2SO_3
Suprapubic (supraumbilical) (epigastric) reflex	Stroking the abdomen above Pouparts ligament cause deviation of the linea alba toward the side that is stroked.
SVRI	Systemic vascular resistance index.
SWS	Slow-wave sleep.
Synthetase	Ligase.

TAFI	Thrombin activateable fibrinolysis inhibitor.
TAP	Tunica albuginea plication.
Tc	Technetium.
^{99m}Tc	Technetium-99m.
^{99m}Tc-DTPA	Technetium-99m diethylenetriaminepentaacetic acid.
^{99m}Tc-MAG3	Technetium-99m mercaptoacetyltriglycine.
TCA cycle	Tricarboxylic acid cycle (Krebs cycle).
TCGF	T cell growth factor.
TCR	T-cell antigen receptor.
TCRE	Transcervical resection of endometrium.
TCSF	Tumour collegenase-stimulating factor.
TD	Tardive dyskinesia.
TDF	Testis determining gene or factor.
TeBG	Testosterone-binding globulin.
TEE	Total energy expenditure.
Terpene	Any hydrocarbon of the formula $C_{10}H_{16}$, derivable chiefly from essential oils, resins, and other vegetable aromatic products. They may be acyclic, bicyclic, or monocyclic, and differ somewhat in physical properties.
TF	Tissue factor.
Tf	Transferrin.
TFAs	Trans-monounsaturated fatty acids.
TFPI	Tissue factor pathway inhibitor.
TG	Triglycerides.
TGFβ	Transforming growth factor-β.
TGN	Trans-Golgi network.
TSGs	Tumour suppressor genes.
TH	Tyrosine hydroxylase.
TH-cells	T-helper cells.
The lateral spermatic ligament	is that portion of the trasversalis fascia enveloping the spermatic vessels and fixing them to the lateral abdominal wall. This structure is revealed only by separation of the spermatic vessels from the posterior peritoneum.
TIF	Testicular interstitial fluid.
TIL	Tumour-infiltrating leukocytes.
TIPS	Transjugular intrahepatic portosystemic shunt.
TLC	Therapeutic lifestyle changes.
TM	Thrombomodulin.
TMA	Thrombotic microangiopathy.
TIMP	Tissue inhibitors of metalloproteinases.
t_{max}	Maximum elimination rate.
TMP	Tocopherol-mediated peroxidation.
TMP	Trimethoprim.
TMP-SMX	Trimethoprim-Sulpha-methoxazole.
TNAP	Tissue-non-specific alkaline phosphatase.
TNF	Tumour necrosis factor.
TNFR	Tumour necrosis factor receptor.
TNG	Trans-Golgi network.
Topa	3,4,6 trihydroxyphenylalanine.
Torr	A unit of pressure equal to 1 mmHg to within one part in 7 million. A unit of pressure equal to 1/760 atmosphere.
Tp	T-precursor lymphocyte.
TPA	Tissue plasminogen activator.
TPN	Total parenteral nutrition.
TR55	TNFR subunit p55.

TRP	Tubular reabsorption of phosphate.
TRADD	TNF-receptor-associated through death domain.
TRAF2	The ring finger protein.
TRAF2	TNFR-associated factor-2.
Trendelenburg's position	One in which the patient is supine on the table or bed, the head of which is titled downward 30-40 degrees, and the table or bed angulated beneath the knees.
Trihalomethanes	Collective term for halogenated aliphatic hydrocarbons, which often contaminate water supplies.
TRPM-2	Testosterone-repressed prostatic message-2.
TSGs	Tumour suppressor genes.
TTP	Thrombotic thrombocytopenic purpura.
TUEP	Transurethral electrovaporization.
TUIP	Transurethral incision of the prostate.
TULIP	Transurethral laser incision of the prostate.
Tumefaction	A swelling, a state of being swollen.
TUMT	Transurethral microwave therapy.
TUNA	Transurethral needle ablation.
TUR	Transurethral resection.
TURBN	Transurethral resection of the bladder neck.
TURP	Transurethral resection of the prostate.
TUVP	Transurethral vaporisation of the prostate.
TVOCs	Toxic volatile organic hydrocarbons.
U	Uranium.
Ubiquinol	The form of ubiquinone when reduced by two electrons.
Ubiquinone	Coenzyme Q.
UCA	Urinary concentrating ability.
U/E	Urea and electrolyte.
UGE	Urogenital epithelium.
UGM	Urogenital sinus mesenchyme.
UGS	Urogenital sinus.
UIC	Uninhibited contraction.
UK	Urokinase.
UMNL	Upper motor neurone lesions.
uPA	Urokinase type plasminogen activator.
UPJO	Ureteropelvic junction obstruction, as PUJO.
Urea	$CO(NH_2)_2$
Urea cycle	A serious of metabolic reactions, occurring in the liver, by which ammonia is converted to urea using cyclically regenerated ornithine as a carrier.
Ureolysis	The decomposition of urea into CO_2 and NH_3.
Urokinase	An enzyme found in the urine, it is elaborated by the parenchymal cells of the human kidney and functions as a plasminogen activator. It is used as a thrombolytic (fibrinolytic) agent.
Uromucoid	Tamm-Horsfall glycoprotein.
US	Ultrasound scan.
UTI	Urinary tract infection.
UTP	Uridine triphosphate.
UV	Ultraviolet.
VAC	Volume-activated channels.
VAP	Ventilator-associated pneumonia.
VB	Voided bladder.
VCAM	Vascular cell-adhesion molecule.
VCAM1	Vascular cell adhesion molecule 1.
VDAC	Voltage-dependent anion channel.
VEGF	Vascular endothelial growth factor.
VILI	Ventilator-induced lung injury.
VIP	Vasoactive intestinal peptide.

VIP21	Caveolin.
VLAP	Visual laser ablation of the prostate.
VLDL	Very-low-density lipoprotein.
V_{O_2}	Oxygen consumption.
VOC	volatile organic compounds.
VPF	Vascular permeability factor.
VRE	Vancomycin-resistant Enterococcus.
vs.	Versus.
VSMC	Vascular smooth muscle cells.
VUJ	Ureterovesical junction.
VUR	Vesicoureteral reflux.
VZV	Varicella zoster virus.
WBC	White blood cells.
WCH-CD	Water channel expressed selectivity in kidney collecting duct apical membrane.
WHO	World health organisation.
X	Halide.
Xenobiotic	A foreign chemical to the biological system.
ZCE	Zone of critical elasticity.
ZIFT	Zygote intrafallopian transfer.
Zn	Zinc.

APPENDIX 2: LABORATORY TESTS

Haematology:

	Conventional Units	SI Units
Erythrocytes		
Males	4.6–6.2 million/mm^3	4.6–6.2 × 10^{12}/L
Females	4.2–5.4 million/mm^3	4.2–5.4 × 10^{12}/L
Children (varies with age)	4.5–5.1 million/mm^3	4.5–5.1 × 10^{12}/L
Leukocytes, total	4500–11,000/mm^3	4.5–11.0 × 10^9/L
Leukocytes, differential counts		
Myelocytes	0%	0/L
Band neutrophils	3–5%	150–400 × 10^6/L
Segmented neutrophils	54–62%	3000–5800 × 10^6/L
Lymphocytes	25–33%	1500–3000 × 10^6/L
Monocytes	3–7%	300–500 × 10^6/L
Eosinophils	1–3%	50–250 × 10^6/L
Basophils	0–1%	15–50 × 10^6/L
Platelets	150,000–400,000/mm^3	150–400 × 10^9/L
Reticulocytes	25,000–75,000/mm^3 (0.5–1.5% of erythrocytes)	25–75 × 10^9/L
Coagulation tests		
Bleeding time (template)	2.75–8.0 min	2.75–8.0 min
Coagulation time (glass tube)	5–15 min	5–15 min
D-Dimer	<0.5 mg/mL	<0.5 mg/L
Factor VIII and other coagulation factors	50–150% of normal	0.5–1.5 of normal
Fibrin split products	<10 µg/mL	<10 mg/L
Fibrinogen	200–400 mg/dL	2.0–4.0 g/L
Partial thromboplastin time, activated (aPTT)	20–35 s	20–35 s
Prothrombin time (PT)	12.0–14.0 s	12.0–14.0 s
Coombs' test		
Direct	Negative	Negative
Indirect	Negative	Negative
Corpuscular values of erythrocytes		
Mean corpuscular hemoglobin (MCH)	26–34 pg/cell	26–34 pg/cell
Mean corpuscular volume	80–96 mm^3	80–96 fL
Mean corpuscular haemoglobin concentration (MCHC)	32–36 g/dL	320–360 g/L
Haptoglobin	20–165 mg/dL	0.20–1.65 g/L
Hematocrit		
Males	40–54 mL/dL	0.40–0.54
Females	37–47 mL/dL	0.37–0.47
Newborns	49–54 mL/dL	0.49–0.54
Children (varies with age)	35–49 mL/dL	0.35–0.49
Haemoglobin:		
Males	13.0–18.0 g/dL	8.1–11.2 mmol/L
Females	12.0–16.0 g/dL	7.4–9.9 mmol/L
Newborns	16.5–19.5 g/dL	10.2–12.1 mmol/L
Children (varies with age)	11.2–16.5 g/dL	7.0–10.2 mmol/L
Haemoglobin, foetal	<1.0% of total	<0.01 of total
Haemoglobin A$_{1c}$	3–5% of total	0.03–0.05 of total
Haemoglobin A$_2$	1.5–3.0% of total	0.015–0.03 of total
Haemoglobin, plasma	0.0–5.0 mg/dL	0.0–3.2 mmol/L
Methaemoglobin	30–130 mg/dL	19–80 mmol/L
Sedimentation rate (ESR):		
Wintrobe:		
Males	0–5 mm/h	0–5 mm/h
Females	0–15 mm/h	0–15 mm/h
Westergren:		
Males	0–15 mm/h	0–15 mm/h
Females	0–20 mm/h	0–20 mm/h

Biochemistry (Blood, Serum, and Plasma):

	Conventional Units	SI Units
Acetoacetate plus acetone		
Qualitative	Negative	Negative
Quantitative	0.3–2.0 mg/dL	30–200 μmol/L
Acid phosphatase, serum (thymolphthalein monophosphate substrate)	0.1–0.6 U/L	0.1–0.6 U/L
ACTH (see Corticotropin)		
Alanine aminotransferase (ALT) serum (SGPT)	1–45 U/L	1–45 U/L
Albumin, serum	3.3–5.2 g/dL	33–52 g/L
Aldolase, serum	0.0–7.0 U/L	0.0–7.0 U/L
Aldosterone, plasma		
Standing	5–30 ng/dL	140–830 pmol/L
Recumbent	3–10 ng/dL	80–275 pmol/L
Alkaline phosphatase (ALP), serum		
Adult	35–150 U/L	35–150 U/L
Adolescent	100–500 U/L	100–500 U/L
Child	100–350 U/L	100–350 U/L
Ammonia nitrogen, plasma	10–50 μmol/L	10–50 μmol/L
Amylase, serum	25–125 U/L	25–125 U/L
Anion gap, serum, calculated	8–16 mEq/L	8–16 mmol/L
Ascorbic acid, blood	0.4–1.5 mg/dL	23–85 μmol/L
Aspartate aminotransferase (AST) serum (SGOT)	1–36 U/L	1–36 U/L
Base excess, arterial blood, calculated	0 ± 2 mEq/L	0 ± 2 mmol/L
Bicarbonate		
Venous plasma	23–29 mEq/L	23–29 mmol/L
Arterial blood	21–27 mEq/L	21–27 mmol/L
Bile acids, serum	0.3–3.0 mg/dL	0.8–7.6 μmol/L
Bilirubin, serum		
Conjugated	0.1–0.4 mg/dL	1.7–6.8 μmol/L
Total	0.3–1.1 mg/dL	5.1–19.0 μmol/L
Caeruloplasmin, serum	23–44 mg/dL	230–440 mg/L
Calcium, serum	8.4–10.6 mg/dL	2.10–2.65 mmol/L
Calcium, ionized, serum	4.25–5.25 mg/dL	1.05–1.30 mmol/L
Carbon dioxide, total, serum or plasma	24–31 mEq/L	24–31 mmol/L
Carbon dioxide tension (PCO_2), blood	35–45 mm Hg	35–45 mm Hg
β-carotene, serum	60–260 μg/dL	1.1–8.6 μmol/L
Chloride, serum or plasma	96–106 mEq/L	96–106 mmol/L
Cholesterol, serum or ethylenediaminetetraacetic acid (EDTA) plasma		
Desirable range	<200 mg/dL	<5.20 mmol/L
Low-density lipoprotein (LDL) cholesterol	60–180 mg/dL	1.55–4.65 mmol/L
High-density lipoprotein (HDL) cholesterol	30–80 mg/dL	0.80–2.05 mmol/L
Copper	70–140 μg/dL	11–22 μmol/L
Corticotropin (ACTH), plasma, 8 AM	10–80 pg/mL	10–80 pg/mL
2–18 pmol/L	2–18 pmol/L	2–18 pmol/L
Cortisol, plasma		
8 AM	6–23 μg/dL	170–630 nmol/L
4 PM	3–15 μg/dL	80–410 nmol/L
10 PM	<50% of 8 AM value	<50% of 8 AM value
Creatine, serum		
Males	0.2–0.5 mg/dL	15–40 μmol/L
Females	0.3–0.9 mg/dL	25–70 μmol/L
Creatine kinase (CK), serum		
Males	55–170 U/L	55–170 U/L
Females	30–135 U/L	30–135 U/L
Creatine kinase MB isoenzyme, serum	<5% of total CK activity	
	<5.0 ng/mL by immunoassay	
Creatinine, serum	0.6–1.2 mg/dL	50–110 μmol/L
Estradiol-17b, adult		
Males	35–240 pmol/L	35–240 pmol/L
Females		
Follicular	30–100 pg/mL	110–370 pmol/L
Ovulatory	200–400 pg/mL	730–1470 pmol/L
Luteal	50–140 pg/mL	180–510 pmol/L
Ferritin, serum	20–200 ng/mL	20–200 μg/L
Fibrinogen, plasma	200–400 mg/dL	2.0–4.0 g/L
Folate		
Serum	3–18 ng/mL	6.8–41 nmol/L
Erythrocytes	145–540 ng/mL	330–1220 nmol/L
Follicle-stimulating hormone (FSH), plasma		
Males	4–25 mU/mL	4–25 U/L
Females, premenopausal	4–30 mU/mL	4–30 U/L
Females, postmenopausal	40–250 mU/mL	40–250 U/L
Gamma-glutamyltransferase (GGT), serum	5–40 U/L	5–40 U/L
Gastrin, fasting, serum	0–100 pg/mL	0–100 mg/L
Glucose, fasting, plasma or serum	70–115 mg/dL	3.9–6.4 nmol/L
Growth hormone (hGH), plasma, adult, fasting	0–6 ng/mL	0–6 ng/L
Haptoglobin, serum	20–165 mg/dL	0.20–1.65 gm/L
Iron, serum	75–175 μg/dL	13–31 μmol/L
Iron binding capacity, serum		
Total	250–410 μg/dL	45–73 μmol/L
Saturation	20–55%	0.20–0.55
Lactate		
Venous whole blood	5.0–20.0 mg/dL	0.6–2.2 mmol/L
Arterial whole blood	5.0–15.0 mg/dL	0.6–1.7 mmol/L
Lactate dehydrogenase (LD), serum	110–220 U/L	110–220 U/L
Lipase, serum	10–140 U/L	10–140 U/L

Biochemistry (Blood, Serum, and Plasma) Continued:

	Conventional Units	SI Units
Lutropin (LH), serum		
Males	1–9 U/L	1–9 U/L
Females		
Follicular phase	2–10 U/L	2–10 U/L
Midcycle peak	15–65 U/L	15–65 U/L
Luteal phase	1–12 U/L	1–12 U/L
Postmenopausal	12–65 U/L	12–65 U/L
Magnesium, serum	1.3–2.1 mg/dL	0.65–1.05 mmol/L
Osmolality	275–295 mOsm/kg water	275–295 mOsm/kg water
Oxygen, blood, arterial, room air		
Partial pressure (PaO_2)	80–100 mmHg	80–100 mmHg
Saturation (SaO_2)	95–98%	95–98%
pH, arterial blood	7.35–7.45	7.35–7.45
Phosphate, inorganic, serum		
Adult	3.0–4.5 mg/dL	1.0–1.5 mmol/L
Child	4.0–7.0 mg/dL	1.3–2.3 mmol/L
Potassium		
Serum	3.5–5.0 mEq/L	3.5–5.0 mmol/L
Plasma	3.5–4.5 mEq/L	3.5–4.5 mmol/L
Progesterone, serum, adult		
Males	0.0–0.4 ng/mL	0.0–1.3 mmol/L
Females		
Follicular phase	0.1–1.5 ng/mL	0.3–4.8 mmol/L
Luteal phase	2.5–28.0 ng/mL	8.0–89.0 mmol/L
Prolactin, serum		
Males	1.0–15.0 ng/mL	1.0–15.0 µg/L
Females	1.0–20.0 ng/mL	1.0–20.0 µg/L
Protein, serum, electrophoresis		
Total	6.0–8.0 g/dL	60–80 g/L
Albumin	3.5–5.5 g/dL	35–55 g/L
Globulins		
α1	0.2–0.4 g/dL	2.0–4.0 g/L
α2	0.5–0.9 g/dL	5.0–9.0 g/L
β	0.6–1.1 g/dL	6.0–11.0 g/L
γ	0.7–1.7 g/dL	7.0–17.0 g/L
Pyruvate, blood	0.3–0.9 mg/dL	0.03–0.10 mmol/L
Rheumatoid factor	0.0–30.0 IU/mL	0.0–30.0 kIU/L
Sodium, serum or plasma	135–145 mEq/L	135–145 mmol/L
Testosterone, plasma		
Males, adult	300–1200 ng/dL	10.4–41.6 nmol/L
Females, adult	20–75 ng/dL	0.7–2.6 nmol/L
Pregnant females	40–200 ng/dL	1.4–6.9 nmol/L
Thyroglobulin	3–42 ng/mL	3–42 µg/L
Thyrotropin (hTSH), serum	0.4–4.8 µIU/mL	0.4–4.8 mIU/L
Thyrotropin-releasing hormone (TRH)	5–60 pg/mL	5–60 ng/L
Thyroxine (FT_4), free, serum	0.9–2.1 ng/dL	12–27 pmol/L
Thyroxine (T_4), serum	4.5–12.0 µg/dL	58–154 nmol/L
Thyroxine-binding globulin (TBG)	15.0–34.0 µg/mL	15.0–34.0 mg/L
Transferrin	250–430 mg/dL	2.5–4.3 g/L
Triglycerides, serum, 12-h fast	40–150 mg/dL	0.4–1.5 g/L
Triiodothyronine (T_3), serum	70–190 ng/dL	1.1–2.9 nmol/L
Triiodothyronine uptake, resin (T_3RU)	25–38%	0.25–0.38
Urate		
Males	2.5–8.0 mg/dL	150–480 µmol/L
Females	2.2–7.0 mg/dL	130–420 µmol/L
Urea, serum or plasma	24–49 mg/dL	4.0–8.2 nmol/L
Urea nitrogen, serum or plasma	11–23 mg/dL	8.0–16.4 nmol/L
Viscosity, serum	1.4–1.8 X water	1.4–1.8 X water
Vitamin A, serum	20–80 µg/dL	0.70–2.80 µmol/L
Vitamin B12, serum	180–900 pg/mL	133–664 pmol/L

Therapeutic Drug Monitoring (Serum)

	Therapeutic range	Toxic concentrations	Proprietary name(s)
Analgesics			
Acetaminophen	10–20 µg/mL	>250 µg/mL	Tylenol, Datril
Salicylate	100–250 µg/mL	>300 µg/mL	Aspirin, Bufferin
Antibiotics			
Amikacin	25–30 µg/mL	Peak >35 µg/mL	Amikin
		Trough >10 µg/mL	
Gentamicin	5–10 µg/mL	Peak >10 µg/mL	Garamycin
		Trough >2 µg/mL	
Tobramycin	5–10 µg/mL	Peak >10 µg/mL	Nebcin
		Trough >2 µg/mL	
Vancomycin	5–35 µg/mL	Peak >40 µg/mL	Vancocin
		Trough >10 µg/mL	
Anticonvulsants	5–12 µg/mL	>15 µg/mL	Tegretol
Ethosuximide	40–100 µg/mL	>150 µg/mL	Zarontin
Phenobarbital	15–40 µg/mL	40–100 ng/mL (varies widely)	Luminal
Phenytoin	10–20 µg/mL	>20 µg/mL	Dilantin
Primidone	5–12 µg/mL	>15 µg/mL	Mysoline
Valproic acid	50–100 µg/mL	>100 µg/mL	Depakene
Antineoplastics and Immunosuppressives			
Cyclosporine	50–400 ng/mL	>400 ng/mL	Sandimmune
Methotrexate, high dose, 48-h	Variable	>1 mmol/L 48 h after dose	
Tacrolimus (FK-506), whole blood	3–10 mg/L	>15 mg/L	Prograf
Bronchodilators and Respiratory Stimulants			
Caffeine	3–15 ng/mL	>30 ng/mL	
Theophylline (aminophylline)	10–20 µg/mL	>20 µg/mL	Elixophyllin, Quibron
Cardiovascular Drugs			
Amiodarone (obtain specimen more than 8 h after last dose)	1.0–2.0 µg/mL	>2.0 µg/mL	Cordarone
Digitoxin (obtain specimen 12–24 h after last dose)	15–25 ng/mL	>35 ng/mL	Crystodigin
Digoxin	0.8–2.0 ng/mL	>2.4 ng/mL	Lanoxin
Disopyramide	2–5 µg/mL	>7 µg/mL	Norpace
Flecainide	0.2–1.0 ng/mL	>1 ng/mL	Tambocor
Lidocaine	1.5–5.0 µg/mL	>6 µg/mL	Xylocaine
Mexiletine	0.7–2.0 ng/mL	>2 ng/mL	Mexitil
Procainamide	4–10 µg/mL	>12 µg/mL	Pronestyl
Procainamide plus N-acetyl-p-aminophenol (NAPA)	8–30 µg/mL	>30 µg/mL	
Propranolol	50–100 ng/mL	Variable	Inderal
Quinidine	2–5 µg/mL	>6 µg/mL	Cardioquin, Quinaglute
Tocainide	4–10 ng/mL	>10 ng/mL	Tonocard
Psychopharmacologic Drugs			
Amitriptyline	120–150 ng/mL	>500 ng/mL	Elavil, Triavil
Bupropion	25–100 ng/mL	Not applicable	Wellbutrin
Desipramine	150–300 ng/mL	>500 ng/mL	Norpramin
Imipramine	125–250 ng/mL	>400 ng/mL	Tofranil
Lithium (obtain specimen 12 h after last dose)	0.6–1.5 mEq/L	>1.5 mEq/L	Lithobid
Nortriptyline	50–150 ng/mL	>500 ng/mL	Aventyl, Pamelor

Urine Biochemistry

	Conventional Units	SI Units
Acetone and acetoacetate, qualitative	Negative	Negative
Albumin		
Qualitative	Negative	Negative
Quantitative	10–100 mg/24 h	0.15–1.5 µmol/d
Aldosterone	3–20 mg/24 h	8.3–55 nmol/d
δ-Aminolevulinic acid (δ-ALA)	1.3–7.0 mg/24 h	10–53 µmol/d
Amylase	<17 U/h	<17 U/h
Amylase/creatinine clearance ratio	0.01–0.04	0.01–0.04
Bilirubin, qualitative	Negative	Negative
Calcium (regular diet)	<250 mg/24 h	<6.3 nmol/d
Catecholamines		
Epinephrine	<10 mg/24 h	<55 nmol/d
Norepinephrine	<100 µg/24 h	<590 nmol/d
Total free catecholamines	4–126 mg/24 h	24–745 nmol/d
Total metanephrines	0.1–1.6 mg/24 h	0.5–8.1 µmol/d
Chloride (varies with intake)	110–250 mEq/24 h	110–250 mmol/d
Copper	0–50 µg/24 h	0.0–0.80 µmol/d
Cortisol, free	10–100 µg/24 h	27.6–276 nmol/d
Creatine		
Males	0–40 mg/24 h	0.0–0.30 mmol/d
Females	0–80 mg/24 h	0.0–0.60 mmol/d
Creatinine	15–25 mg/kg/24 h	0.13–0.22 mmol/kg/d
Creatinine clearance (endogenous)		
Males	110–150 mL/min/1.73 m^2	110–150 mL/min/1.73 m^2
Females	105–132 mL/min/1.73 m^2	105–132 mL/min/1.73 m^2
Cystine or cysteine	Negative	Negative
Dehydroepiandrosterone (DHEA)		
Males	0.2–2.0 mg/24 h	0.7–6.9 µmol/d
Females	0.2–1.8 mg/24 h	0.7–6.2 µmol/d
Oestrogens, total		
Males	4–25 µg/24 h	14–90 nmol/d
Females	5–100 µg/24 h	18–360 nmol/d
Glucose (as reducing substance)	<250 mg/24 h	<250 mg/d
Haemoglobin and myoglobin, qualitative	Negative	Negative
Homogentisic acid, qualitative	Negative	Negative
17-Ketogenic steroids		
Males	5–23 mg/24 h	17–80 µmol/d
Females	3–15 mg/24 h	10–52 µmol/d
17-Hydroxycorticosteroids		
Males	3–9 mg/24 h	8.3–25 µmol/d
Females	2–8 mg/24 h	5.5–22 µmol/d
5-Hydroxyindoleacetic acid		
Qualitative	Negative	Negative
Quantitative	2–6 mg/24 h	10–31 µmol/d
17-Ketosteroids		
Males	8–22 mg/24 h	28–76 µmol/d
Females	6–15 mg/24 h	21–52 µmol/d
Magnesium	6–10 mEq/24 h	3–5 mmol/d
Metanephrines	0.05–1.2 ng/mg creatinine	0.03–0.70 mmol/mmol creatinine
Osmolality	38–1400 mOsm/kg water	38–1400 mOsm/kg water
PH	4.6–8.0	4.6–8.0
Phenylpyruvic acid, qualitative	Negative	Negative
Phosphate	0.4–1.3 g/24 h	13–42 mmol/d
Porphobilinogen		
Qualitative	Negative	Negative
Quantitative	<2 mg/24 h	<9 µmol/d
Porphyrins		
Coproporphyrin	50–250 µg/24 h	77–380 nmol/d
Uroporphyrin	10–30 µg/24 h	12–36 nmol/d
Potassium	25–125 mEq/24 h	25–125 mmol/d
Pregnanediol		
Males	0.0–1.9 mg/24 h	0.0–6.0 µmol/d
Females		
Proliferative phase	0.0–2.6 mg/24 h	0.0–8.0 µmol/d
Luteal phase	2.6–10.6 mg/24 h	8–33 µmol/d
Postmenopausal	0.2–1.0 mg/24 h	0.6–3.1 µmol/d
Pregnanetriol	0.0–2.5 mg/24 h	0.0–7.4 µmol/d
Protein, total		
Qualitative	Negative	Negative
Quantitative	10–150 mg/24 h	10–150 mg/d
Protein/creatinine ratio	<0.2	<0.2
Sodium (regular diet)	60–260 mEq/24 h	60–260 mmol/d
Specific gravity		
Random specimen	1.003–1.030	1.003–1.030
24-hour collection	1.015–1.025	1.015–1.025
Urate (regular diet)	250–750 mg/24 h	1.5–4.4 mmol/d
Urobilinogen	0.5–4.0 mg/24 h	0.6–6.8 µmol/d
Vanillylmandelic acid (VMA)	1.0–8.0 mg/24 h	5–40 µmol/d

Toxic Substances

	Conventional Units	SI Units
Arsenic, urine	<130 µg/24 h	<1.7 µmol/d
Bromides, serum, inorganic	<100 mg/dL	<10 mmol/L
Toxic symptoms	140–1000 mg/dL	14–100 mmol/L
Carboxyhemoglobin, blood: Saturation		
Urban environment	<5%	<0.05
Smokers	<12%	<0.12
Symptoms		
Headache	>15%	>0.15
Nausea and vomiting	>25%	>0.25
Potentially lethal	>50%	>0.50
Ethanol, blood	<0.05 mg/dL (<0.005%)	<1.0 mmol/L
Intoxication	>100 mg/dL (>0.1%)	>22 mmol/L
Marked intoxication	300–400 mg/dL (0.3–0.4%)	65–87 mmol/L
Alcoholic stupor	400–500 mg/dL (0.4–0.5%)	87–109 mmol/L
Coma	>500 mg/dL (>0.5%)	>109 mmol/L
Lead, blood		
Adults	<25 µg/dL	<1.2 µmol/L
Children	<15 µg/dL	<0.7 µmol/L
Lead, urine	<80 µg/24 h	<0.4 µmol/d
Mercury, urine	<30 µg/24 h	<150 nmol/d

Semen Analysis

	Conventional Units	SI Units
Volume	2–5 mL	2–5 mL
Liquefaction	Complete in 15 min	Complete in 15 min
PH	7.2–8.0	7.2–8.0
Leukocytes	Occasional or absent	Occasional or absent
Spermatozoa		
Count	60–150 × 10^6/mL	60–150 × 10^6/mL
Motility	>80% motile	>0.80 motile
Morphology	80–90% normal forms	>0.80–0.90 normal forms
Fructose	>150 mg/dL	>8.33 mmol/L

Immunologic Function

	Conventional Units	SI Units
Complement, Serum		
C3	85–175 mg/dL	0.85–1.75 gm/L
C4	15–45 mg/dL	150–450 mg/L
Total hemolytic (CH_{50})	150–250 U/mL	150–250 U/mL
Immunoglobulins, Serum, Adult		
IgG	640–1350 mg/dL	6.4–13.5 g/L
IgA	70–310 mg/dL	0.70–3.1 g/L
IgM	90–350 mg/dL	0.90–3.5 g/L
IgD	0.0–6.0 mg/dL	0.0–60 mg/L
IgE	0.0–430 ng/dL	0.0–430 mg/L

Liquid measure - Imperial [(fl)oz]

	Meteric (ml)
1/6 (fl)oz	4.924 ml
1/3 (fl)oz	9.857 ml
1/2 (fl)oz	14.79 ml
1 (fl)oz	29.51 ml
2 (fl)oz	59.15 ml
4 (fl)oz	118.3 ml
8 (fl)oz	236.6 ml
16 (fl)oz	478.2 ml
32 (fl)oz	946.4 ml
1/2 gal [64 (fl)oz]	1893 ml
1 gal [128 (fl)oz]	3785 ml
5 gal [640 (fl)oz]	18930 ml

Dry weight - Imperial (oz)

	Meteric (g)
1 oz	28.35 g
2 oz	56.7 g
4 oz	113.4 g
8 oz	226.8 g
16 oz (1 lb)	453.6 g
80 oz (5 lb)	2268 g (2.268 kg)
160 oz (10 lb)	4536 g (4.536 kg)
320 oz (20 lb)	9072 g (9.072 kg)

Accuracy is the agreement between the test result from a given device or laboratory and that obtained in a standardized laboratory and is expressed as a mean % bias. Precision is the degree to which a test result can be replicated by multiple runs from the same sample and is expressed as the coefficient of variation (CV). Clinicians should know the accuracy and precision performance of the laboratory they use to properly interpret results.

APPENDIX 3: E Numbers

E100	Curcumin, turmeric [Colouring]
E101	Riboflavin (Vitamin B2), formerly called lactoflavin (Vitamin G) [Colouring] [likely to be GM]
E101a	Riboflavin-5'-Phosphate [Colouring] [likely to be GM]
E102	Tartrazine, FD&C Yellow 5 [Colouring] [possible allergic reaction]
E103	Chrysoine Resorcinol [Colouring]
E104	Quinoline Yellow [Colouring] [possible allergic reaction]
E105	Fast Yellow AB [Colouring]
E106	Riboflavin-5-Sodium Phosphate [Colouring]
E107	Yellow 2G [Colouring] [possible allergic reaction]
E110	Sunset Yellow FCF, Orange Yellow S, FD&C Yellow 6 [Colouring] [possible allergic reaction]
E111	Orange GGN [Colouring]
E120	Cochineal, Carminic acid, Carmines, Natural Red 4 [Colouring] [possible allergic reaction] [animal origin]
E121	Orcein, Orchil [Colouring]
E122	Carmoisine, Azorubine [Colouring] [possible allergic reaction]
E123	Amaranth, FD&C Red 2 [Colouring] [possible allergic reaction]
E124	Ponceau 4R, Cochineal Red A, Brilliant Scarlet 4R [Colouring] [possible allergic reaction]
E125	Scarlet GN [Colouring]
E126	Ponceau 6R [Colouring]
E127	Erythrosine, FD&C Red 3 [Colouring] [possible allergic reaction]
E128	Red 2G [Colouring] [possible allergic reaction]
E129	Allura Red AC, FD&C Red 40 [Colouring] [possible allergic reaction]
E130	Indanthrene blue RS [Colouring]
E131	Patent Blue V [Colouring] [possible allergic reaction]
E132	Indigo carmine, Indigotine, FD&C Blue 2 [Colouring] [possible allergic reaction]
E133	Brilliant Blue FCF, FD&C Blue 1 [Colouring] [possible allergic reaction]
E140	Chlorophylls and Chlorophyllins: (i) Chlorophylls (ii) Chlorophyllins [Colouring]
E141	Copper complexes of chlorophylls and chlorophyllins (i) Copper complexes of chlorophylls (ii) Copper complexes of chlorophyllins [Colouring]
E142	Greens S [Colouring] [possible allergic reaction]
E150a	Plain Caramel [Colouring] [likely to be GM]
E150b	Caustic sulphite caramel [Colouring] [likely to be GM]
E150c	Ammonia caramel [Colouring] [likely to be GM]
E150d	Sulphite ammonia caramel [Colouring] [likely to be GM]
E151	Black PN, Brilliant Black BN [Colouring] [possible allergic reaction]
E152	Black 7984 [Colouring] [possible allergic reaction]
E153	Carbon black, Vegetable carbon [Colouring] [likely to be GM] [possibly of animal origin]
E154	Brown FK, Kipper Brown [Colouring] [possible allergic reaction]
E155	Brown HT, Chocolate brown HT [Colouring] [possible allergic reaction]
E160a	α-carotene, β-carotene, γ-carotene [Colouring] [possibly animal content]
E160b	Annatto, bixin, norbixin [Colouring] [possible allergic reaction]
E160c	Capsanthin, capsorubin, Paprika extract [Colouring]
E160d	Lycopene [Colouring] [possibly GM]
E160e	β-apo-8'-carotenal (C 30) [Colouring]
E160f	Ethyl ester of β-apo-8'-carotenic acid (C 30) [Colouring]
E161a	Flavoxanthin [Colouring]
E161b	Lutein [Colouring]
E161c	Cryptoaxanthin [Colouring] [likely to be GM]
E161d	Rubixanthin [Colouring]
E161e	Violaxanthin [Colouring]
E161f	Rhodoxanthin [Colouring]
E161g	Canthaxanthin [Colouring] [possibly of animal origin]
E162	Beetroot Red, Betanin [Colouring]
E163	Anthocyanins [Colouring]
E170	Calcium carbonate, Chalk [Colouring]
E171	Titanium dioxide [Colouring]
E172	Iron oxides and hydroxides [Colouring]
E173	Aluminium [Colouring]
E174	Silver [Colouring]
E175	Gold [Colouring]
E180	Pigment Rubine, Lithol Rubine BK [Colouring]
E181	Tannin [Colouring]
E200	Sorbic acid [Preservative]
E201	Sodium sorbate [Preservative]
E202	Potassium sorbate [Preservative]
E203	Calcium sorbate [Preservative]
E210	Benzoic acid [Preservative] [possible allergic reaction]
E211	Sodium benzoate [Preservative] [possible allergic reaction]
E212	Potassium benzoate [Preservative] [possible allergic reaction]
E213	Calcium benzoate [Preservative] [possible allergic reaction]
E214	Ethyl para-hydroxybenzoate [Preservative] [possible allergic reaction]
E215	Sodium ethyl para-hydroxybenzoate [Preservative] [possible allergic reaction]
E216	Propyl para-hydroxybenzoate [Preservative] [possible allergic reaction]
E217	Sodium propyl para-hydroxybenzoate [Preservative] [possible allergic reaction]
E218	Methyl para-hydroxybenzoate [Preservative] [possible allergic reaction]
E219	Sodium methyl para-hydroxybenzoate [Preservative] [possible allergic reaction]
E220	Sulphur dioxide [Preservative] [possible allergic reaction]
E221	Sodium sulphite [Preservative] [possible allergic reaction]
E222	Sodium hydrogen sulphite [Preservative] [possible allergic reaction]
E223	Sodium metabisulphite [Preservative] [possible allergic reaction]
E224	Potassium metabisulphite [Preservative] [possible allergic reaction]
E226	Calcium sulphite [Preservative] [possible allergic reaction]
E227	Calcium hydrogen sulphite [Preservative] [Firming Agent] [possible allergic reaction]
E228	Potassium hydrogen sulphite [Preservative] [possible allergic reaction]
E230	Biphenyl, diphenyl [Preservative]
E231	Orthophenyl phenol [Preservative]
E232	Sodium orthophenyl phenol [Preservative]
E233	Thiabendazole [Preservative]
E234	Nisin [Preservative]
E235	Natamycin, Pimaracin [Preservative]
E236	Formic acid [Preservative]
E237	Sodium formiate [Preservative]
E238	Calcium formiate [Preservative]
E239	Hexamethylene tetramine, Hexamine [Preservative]
E240	Formaldehyde [Preservative]
E242	Dimethyl dicarbonate [Preservative]
E249	Potassium nitrite [Preservative]
E250	Sodium nitrite [Preservative]
E251	Sodium nitrate, Saltpetre [Preservative]
E252	Potassium nitrate (Saltpetre) [Preservative] [possibly of animal origin]
E260	Acetic acid [Preservative] [Acidity regulator]

E261	Potassium acetate [Preservative] [Acidity regulator]
E262	Sodium acetates (i) Sodium acetate (ii) Sodium hydrogen acetate (sodium diacetate) [Preservative] [Acidity regulator]
E263	Calcium acetate [Preservative] [Acidity regulator]
E264	Ammonium acetate [Preservative]
E270	Lactic acid [Preservative] [Acid] [Antioxidant] [possibly of animal origin]
E280	Propionic acid [Preservative]
E281	Sodium propionate [Preservative]
E282	Calcium propionate [Preservative] [possible allergic reaction]
E283	Potassium propionate [Preservative]
E284	Boric acid [Preservative]
E285	Sodium tetraborate (borax) [Preservative]
E290	Carbon dioxide [Acidity regulator]
E296	Malic acid [Acid] [Acidity regulator]
E297	Fumaric acid [Acidity regulator]
E300	Ascorbic acid (Vitamin C) [Antioxidant]
E301	Sodium ascorbate [Antioxidant]
E302	Calcium ascorbate [Antioxidant]
E303	Potassium ascorbate [Antioxidant]
E304	Fatty acid esters of ascorbic acid (i) Ascorbyl palmitate (ii) Ascorbyl stearate [Antioxidant]
E306	Tocopherol-rich extract (natural) [Antioxidant] [possibly GM]
E307	α-tocopherol (synthetic) [Antioxidant] [possibly GM]
E308	γ-tocopherol (synthetic) [Antioxidant] [possibly GM]
E309	δ-tocopherol (synthetic) [Antioxidant] [possibly GM]
E310	Propyl gallate [Antioxidant] [possible allergic reaction]
E311	Octyl gallate [Antioxidant] [possible allergic reaction]
E312	Dodecyl gallate [Antioxidant] [possible allergic reaction]
E315	Erythorbic acid [Antioxidant]
E316	Sodium erythorbate [Antioxidant]
E317	Erythorbin acid [Antioxidant]
E318	Sodium erythorbin [Antioxidant]
E319	Butylhydroxinon [Antioxidant]
E320	Butylated hydroxyanisole (BHA) [Antioxidant]
E321	Butylated hydroxytoluene (BHT) [Antioxidant] [possible allergic reaction]
E322	Lecithin [Emulsifier][likely to be GM] [possibly of animal origin]
E325	Sodium lactate [Antioxidant] [possibly of animal origin]
E326	Potassium lactate [Antioxidant] [Acidity regulator] [possibly of animal origin]
E327	Calcium lactate [Antioxidant] [possibly of animal origin]
E329	Magnesium lactate [Antioxidant]
E330	Citric acid [Antioxidant]
E331	Sodium citrates (i) Monosodium citrate (ii) Disodium citrate (iii) Trisodium citrate [Antioxidant]
E332	Potassium citrates (i) Monopotassium citrate (ii) Tripotassium citrate [Antioxidant]
E333	Calcium citrates (i) Monocalcium citrate (ii) Dicalcium citrate (iii) Tricalcium citrate [Acidity regulator] [Firming Agent]
E334	Tartaric acid (L(+)-) [Acid] [Antioxidant]
E335	Sodium tartrates (i) Monosodium tartrate (ii) Disodium tartrate [Antioxidant]
E336	Potassium tartrates (i) Monopotassium tartrate (cream of tartar) (ii) Dipotassium tartrate [Antioxidant]
E337	Sodium potassium tartrate [Antioxidant]
E338	Phosphoric acid [Antioxidant]
E339	Sodium phosphates (i) Monosodium phosphate (ii) Disodium phosphate (iii) Trisodium phosphate [Antioxidant]
E340	Potassium phosphates (i) Monopotassium phosphate (ii) Dipotassium phosphate (iii) Tripotassium phosphate [Antioxidant]
E341	Calcium phosphates (i) Monocalcium phosphate (ii) Dicalcium phosphate (iii) Tricalcium phosphate [Anti-caking agent] [Firming Agent]
E343	Magnesium phosphates (i) monomagnesium phosphate (ii) Dimagnesium phosphate [Anti-caking agent]
E350	Sodium malates (i) Sodium malate (ii) Sodium hydrogen malate [Acidity regulator]
E351	Potassium malate [Acidity regulator]
E352	Calcium malates (i) Calcium malate (ii) Calcium hydrogen malate [Acidity regulator]
E353	Metatartaric acid [Emulsifier]
E354	Calcium tartrate [Emulsifier]
E355	Adipic acid [Acidity regulator]
E356	Sodium adipate [Acidity regulator]
E357	Potassium adipate [Acidity regulator]
E363	Succinic acid [Acidity regulator]
E365	Sodium fumarate [Acidity regulator]
E366	Potassium fumarate [Acidity regulator]
E367	Calcium fumarate [Acidity regulator]
E370	1,4-Heptonolactone [Acidity regulator]
E375	Nicotinic acid, Niacin, Nicotinamide [Colour retention agent] [possible allergic reaction]
E380	Triammonium citrate [Acidity regulator]
E381	Ammoniumferrocitrate [Acidity regulator]
E385	Calcium disodium ethylene diamine tetra-acetate (Calcium disodium EDTA) [Antioxidant]
E400	Alginic acid [Thickener] [Stabiliser] [Gelling agent] [Emulsifier]
E401	Sodium alginate [Thickener] [Stabiliser] [Gelling agent] [Emulsifier]
E402	Potassium alginate [Thickener] [Stabiliser] [Gelling agent] [Emulsifier]
E403	Ammonium alginate [Thickener] [Stabiliser] [Emulsifier]
E404	Calcium alginate [Thickener] [Stabiliser] [Gelling agent] [Emulsifier]
E405	Propane-1,2-diol alginate (Propylene glycol alginate) [Thickener] [Stabiliser] [Emulsifier]
E406	Agar [Thickener] [Gelling agent] [Emulsifier]
E407	Carrageenan [Thickener] [Stabiliser] [Gelling agent] [Emulsifier] [possible allergic reaction]
E407a	Processed eucheuma seaweed [Thickener] [Stabiliser] [Gelling agent] [Emulsifier]
E410	Locust bean gum (Carob gum) [Thickener] [Stabiliser] [Gelling agent] [Emulsifier]
E412	Guar gum [Thickener] [Stabiliser]
E413	Tragacanth [Thickener] [Stabiliser] [Emulsifier] [possible allergic reaction]
E414	Acacia gum (gum arabic) [Thickener] [Stabiliser] [Emulsifier] [possible allergic reaction]
E415	Xanthan gum [Thickener] [Stabiliser] [possibly GM]
E416	Karaya gum [Thickener] [Stabiliser] [Emulsifier] [possible allergic reaction]
E417	Tara gum [Thickener] [Stabiliser]
E418	Gellan gum [Thickener] [Stabiliser] [Emulsifier]
E420	Sorbitol (i) Sorbitol (ii) Sorbitol syrup [Emulsifier] [Sweetener] [Humectant]
E421	Mannitol [Anti-caking agent] [Sweetener]
E422	Glycerol [Emulsifier] [Sweetener] [possibly of animal origin]
E425	Konjac (i) Konjac gum (ii) Konjac glucomannane [Emulsifier] E430 Polyoxyethylene (8) stearate [Emulsifier] [Stabiliser] [possible allergic reaction] [possibly of animal origin]
E431	Polyoxyethylene (40) stearate [Emulsifier] [possibly of animal origin]
E432	Polyoxyethylene (20) sorbitan monolaurate (polysorbate 20) [Emulsifier] [possibly of animal origin]
E433	Polyoxyethylene (20) sorbitan monooleate (polysorbate 80) [Emulsifier] [possibly of animal origin]
E434	Polyoxyethylene (20) sorbitan monopalmitate (polysorbate 40) [Emulsifier] [possibly of animal origin]
E435	Polyoxyethylene (20) sorbitan monostearate (polysorbate 60) [Emulsifier] [possibly of animal origin]
E436	Polyoxyethylene (20) sorbitan tristearate (polysorbate 65) [Emulsifier] [possibly of animal origin]
E440	Pectins (i) pectin (ii) amidated pectin [Emulsifier]
E441	Gelatine [Emulsifier] [Gelling agent] [animal origin]
E442	Ammonium phosphatides [Emulsifier] [possibly of animal origin]
E444	Sucrose acetate isobutyrate [Emulsifier]
E445	Glycerol esters of wood rosins [Emulsifier]
E450	Diphosphates (i) Disodium diphosphate (ii) Trisodium diphosphate(iii) Tetrasodium diphosphate (iv) Dipotassium diphosphate (v) Tetrapotassium diphosphate (vi) Dicalcium diphosphate (vii) Calcium dihydrogen diphosphate [Emulsifier]
E451	Triphosphates (i) Pentasodium triphosphate (ii) Pentapotassium triphosphate [Emulsifier]
E452	Polyphosphates (i) Sodium polyphosphates (ii) Potassium polyphosphates (iii) Sodium calcium polyphosphate (iv) Calcium polyphophates [Emulsifier]

E459	β-cyclodextrine [Emulsifier]
E460	Cellulose (i) Microcrystalline cellulose (ii) Powdered cellulose [Emulsifier]
E461	Methyl cellulose [Emulsifier]
E462	Ethyl cellulose [Emulsifier]
E463	Hydroxy propyl cellulose [Emulsifier]
E464	Hydroxy propyl methyl cellulose [Emulsifier]
E465	Ethyl methyl cellulose [Emulsifier]
E466	Carboxy methyl cellulose, Sodium carboxy methyl cellulose [Emulsifier]
E467	Ethyl hydroxyethyl cellulose [Emulsifier]
E468	Crosslinked sodium carboxy methyl cellulose [Emulsifier]
E469	Enzymically hydrolysed carboxymethylcellulose [Emulsifier]
E470a	Sodium, potassium and calcium salts of fatty acids [Emulsifier] [Anti-caking agent] [possibly of animal origin]
E470b	Magnesium salts of fatty acids [Emulsifier] [Anti-caking agent] [possibly of animal origin]
E471	Mono- and diglycerides of fatty acids (glyceryl monostearate, glyceryl distearate) [Emulsifier] [likely to be GM] [possibly of animal origin]
E472a	Acetic acid esters of mono- and diglycerides of fatty acids [Emulsifier] [possibly GM] [possibly of animal origin]
E472b	Lactic acid esters of mono- and diglycerides of fatty acids [Emulsifier] [possibly of animal origin]
E472c	Citric acid esters of mono- and diglycerides of fatty acids [Emulsifier] [possibly of animal origin]
E472d	Tartaric acid esters of mono- and diglycerides of fatty acids [Emulsifier] [possibly of animal origin]
E472e	Mono- and diacetyl tartaric acid esters of mono- and diglycerides of fatty acids [Emulsifier] [possibly of animal origin]
E472f	Mixed acetic and tartaric acid esters of mono- and diglycerides of fatty acids [Emulsifier] [possibly of animal origin]
E473	Sucrose esters of fatty acids [Emulsifier] [possibly GM] [possibly of animal origin]
E474	Sucroglycerides [Emulsifier] [possibly of animal origin]
E475	Polyglycerol esters of fatty acids [Emulsifier] [possibly GM] [possibly of animal origin]
E476	Polyglycerol polyricinoleate [Emulsifier] [possibly GM] [possibly of animal origin]
E477	Propane-1, 2-diol esters of fatty acids, propylene glycol esters of fatty acids [Emulsifier] [possibly GM] [possibly of animal origin]
E478	Lactylated fatty acid esters of glycerol and propane-1 [Emulsifier] [possibly of animal origin]
E479b	Thermally oxidized soya bean oil interacted with mono- and diglycerides of fatty acids [Emulsifier] [likely to be GM] [possibly of animal origin]
E481	Sodium stearoyl-2-lactylate [Emulsifier] [possibly of animal origin]
E482	Calcium stearoyl-2-lactylate [Emulsifier] [possibly of animal origin]
E483	Stearyl tartrate [Emulsifier] [possibly of animal origin]
E491	Sorbitan monostearate [Emulsifier] [possibly GM] [possibly of animal origin]
E492	Sorbitan tristearate [Emulsifier] [possibly of animal origin]
E493	Sorbitan monolaurate [Emulsifier] [possibly of animal origin]
E494	Sorbitan monooleate [Emulsifier] [possibly of animal origin]
E495	Sorbitan monopalmitate [Emulsifier] [possibly of animal origin]
E500	Sodium carbonates (i) Sodium carbonate (ii) Sodium hydrogen carbonate (Bicarbonate of soda) (iii) Sodium sesquicarbonate [Acidity regulator] [Raising Agent]
E501	Potassium carbonates (i) Potassium carbonate (ii) Potassium hydrogen carbonate [Acidity regulator]
E503	Ammonium carbonates (i) Ammonium carbonate (ii) Ammonium hydrogen carbonate [Acidity regulator]
E504	Magnesium carbonates (i) Magnesium carbonate (ii) Magnesium hydroxide carbonate (syn. Magnesium hydrogen carbonate) [Acidity regulator] [Anti-caking agent]
E507	Hydrochloric acid [Acid]
E508	Potassium chloride [Acidity regulator]
E509	Calcium chloride [Acidity regulator]
E510	Ammonium chloride, ammonia solution [Acidity regulator]
E511	Magnesium chloride [Acidity regulator]
E512	Stannous chloride [Acidity regulator]
E513	Sulphuric acid [Acid]
E514	Sodium sulphates (i) Sodium sulphate (ii) Sodium hydrogen sulphate [Acidity regulator]
E515	Potassium sulphates (i) Potassium sulphate (ii) Potassium hydrogen sulphate [Acidity regulator]
E516	Calcium sulphate [Acidity regulator] [Firming Agent]
E517	Ammonium sulphate [Acidity regulator]
E518	Magnesium sulphate, Epsom salts [Acidity regulator]
E519	Copper sulphate [Acidity regulator]
E520	Aluminium sulphate [Acidity regulator]
E521	Aluminium sodium sulphate [Acidity regulator]
E522	Aluminium potassium sulphate [Acidity regulator]
E523	Aluminium ammonium sulphate [Acidity regulator]
E524	Sodium hydroxide [Acidity regulator]
E525	Potassium hydroxide [Acidity regulator]
E526	Calcium hydroxide [Acidity regulator] [Firming Agent]
E527	Ammonium hydroxide [Acidity regulator]
E528	Magnesium hydroxide [Acidity regulator]
E529	Calcium oxide [Acidity regulator]
E530	Magnesium oxide [Acidity regulator]
E535	Sodium ferrocyanide [Acidity regulator] [Anti-caking agent]
E536	Potassium ferrocyanide [Acidity regulator]
E538	Calcium ferrocyanide [Acidity regulator]
E540	Dicalcium diphosphate [Acidity regulator] [Emulsifier]
E541	Sodium aluminium phosphate, acidic [Emulsifier]
E542	Bone phosphate [Anti-caking agent] [animal origin]
E543	Calcium sodium polyphosphate
E544	Calcium polyphosphate [Emulsifier]
E545	Ammonium polyphosphate [Emulsifier]
E550	Sodium silicate
E551	Silicon dioxide (Silica) [Emulsifier] [Anti-caking agent]
E552	Calcium silicate [Anti-caking agent]
E553a	(i) Magnesium silicate (ii) Magnesium trisilicate [Anti-caking agent]
E553b	Talc [Anti-caking agent] [possible allergic reaction]
E554	Sodium aluminium silicate [Anti-caking agent]
E555	Potassium aluminium silicate [Anti-caking agent]
E556	Calcium aluminium silicate [Anti-caking agent]
E558	Bentonite [Anti-caking agent]
E559	Aluminium silicate (Kaolin) [Anti-caking agent]
E570	Stearic acid (Fatty acid) [Anti-caking agent] [possibly GM] [possibly of animal origin]
E572	Magnesium stearate, calcium stearate [Emulsifier] [Anti-caking agent] [possibly GM] [possibly of animal origin]
E574	Gluconic acid [Acidity regulator]
E575	Glucono-δ-lactone [Acidity regulator] [Sequestrant]
E576	Sodium gluconate [Sequestrant]
E577	Potassium gluconate [Sequestrant]
E578	Calcium gluconate [Firming Agent]
E579	Ferrous gluconate [Colouring]
E585	Ferrous lactate [Colouring] [possibly of animal origin]
E620	Glutamic acid [Flavour enhancer] [possible reaction] [possibly GM]
E621	Monosodium glutamate [Flavour enhancer] [possible reaction] [likely to be GM]
E622	Monopotassium glutamate [Flavour enhancer] [possible reaction] [possibly GM]
E623	Calcium diglutamate [Flavour enhancer] [possible reaction] [possibly GM]
E624	Monoammonium glutamate [Flavour enhancer] [possible reaction] [possibly GM]
E625	Magnesium diglutamate [Flavour enhancer] [possible reaction] [possibly GM]
E626	Guanylic acid [Flavour enhancer]
E627	Disodium guanylate, sodium guanylate [Flavour enhancer]
E628	Dipotassium guanylate [Flavour enhancer]
E629	Calcium guanylate [Flavour enhancer]
E630	Inosinic acid [Flavour enhancer]
E631	Disodium inosinate [Flavour enhancer] [possibly of animal origin]

E632	Dipotassium inosinate [Flavour enhancer]
E633	Calcium inosinate [Flavour enhancer]
E634	Calcium 5'-ribonucleotides [Flavour enhancer]
E635	Disodium 5'-ribonucleotides [Flavour enhancer] [possibly of animal origin]
E636	Maltol [Flavour enhancer]
E637	Ethyl maltol [Flavour enhancer]
E640	Glycine and its sodium salt [Flavour enhancer] [possibly of animal origin]
E650	Zinc acetate [Flavour enhancer]
E900	Dimethyl polysiloxane [Anti-foaming agent] [Anti-caking agent]
E901	Beeswax, white and yellow [Glazing agent][possible allergic reaction] [animal origin]
E902	Candelilla wax [Glazing agent]
E903	Carnauba wax [Glazing agent] [possible allergic reaction]
E904	Shellac [Glazing agent] [animal origin]
E905	Microcrystalline wax [Glazing agent]
E907	Crystalline wax [Glazing agent]
E910	L-cysteine [animal origin]
E912	Montanic acid esters
E913	Lanolin, sheep wool grease [Glazing agent] [animal origin]
E914	Oxidized polyethylene wax [Glazing agent]
E915	Esters of Colophane [Glazing agent]
E920	L-cysteine hydrochloride [Improving agent] [animal origin]
E921	L-cysteine hydrochloride monohydrate [Improving agent] [animal origin]
E924	Potassium bromate [Improving agent]
E925	Chlorine [Preservative] [Bleach]
E926	Chlorine dioxide [Preservative] [Bleach]
E927b	Carbamide [Improving agent]
E928	Benzole peroxide [Improving agent]
E938	Argon [Packaging gas]
E939	Helium [Packaging gas]
E941	Nitrogen [Packaging gas]
E942	Nitrous oxide [Propellant]
E943a	Butane [Propellant]
E943b	Iso-butane [Propellant]
E944	Propane [Propellant]
E948	Oxygen [Packaging gas]
E949	Hydrogen
E950	Acesulfame K [Sweetener]
E951	Aspartame [Sweetener] [possible reaction]
E952	Cyclamic acid and its Na and Ca salts [Sweetener]
E953	Isomalt [Sweetener]
E954	Saccharin and its Na, K and Ca salts [Sweetener]
E957	Thaumatin [Sweetener] [Flavour enhancer]
E959	Neohesperidine DC [Sweetener]
E965	Maltitol (i) Maltitol (ii) Maltitol syrup [Sweetener] [Stabiliser] [Humectant]
E966	Lactitol [Sweetener] [animal origin]
E967	Xylitol [Sweetener]
E999	Quillaia extract [Foaming agent]
E1103	Invertase [Preservative]
E1105	Lysozyme [Preservative]
E1200	Polydextrose [Stabiliser] [Thickening agent] [Humectant] [Carrier]
E1201	Polyvinylpyrrolidone [Stabiliser]
E1202	Polyvinylpolypyrrolidone [Carrier] [Stabiliser]
E1400	Dextrin [Stabiliser] [Thickening agent]
E1401	Modified starch [Stabiliser] [Thickening agent]
E1402	Alkaline modified starch [Stabiliser] [Thickening agent]
E1403	Bleached starch [Stabiliser] [Thickening agent]
E1404	Oxidized starch [Emulsifier] [Thickening agent]
E1410	Monostarch phosphate [Stabiliser] [Thickening agent]
E1412	Distarch phosphate [Stabiliser] [Thickening agent]
E1413	Phosphated distarch phosphate [Stabiliser] [Thickening agent]
E1414	Acetylated distarch phosphate [Emulsifier] [Thickening agent]
E1420	Acetylated starch, mono starch acetate [Stabiliser] [Thickening agent]
E1421	Acetylated starch, mono starch acetate [Stabiliser] [Thickening agent]
E1422	Acetylated distarch adipate [Stabiliser] [Thickening agent]
E1430	Distarch glycerine [Stabiliser] [Thickening agent]
E1440	Hydroxy propyl starch [Emulsifier] [Thickening agent]
E1441	Hydroxy propyl distarch glycerine [Stabiliser] [Thickening agent]
E1442	Hydroxy propyl distarch phosphate [Stabiliser] [Thickening agent]
E1450	Starch sodium octenyl succinate [Emulsifier] [Stabiliser] [Thickening agent]
E1451	Acetylated oxidised starch [Emulsifier] [Thickening agent]
E1505	Triethyl citrate [Foam Stabiliser]
E1510	Ethanon
E1518	Glyceryl triacetate (triacetin) [Humectant]
E1520	Propylene glycol [Humectant]

APPENDIX 4: Carcinogens, Known to be a human carcinogen:

Aflatoxins
Alcoholic Beverage Consumption
4-Aminobiphenyl
Analgesic Mixtures Containing Phenacetin (See Phenacetin and Analgesic Mixtures Containing Phenacetin)
Arsenic Compounds, Inorganic
Asbestos
Azathioprine
Benzene
Benzidine (See Benzidine and Dyes Metabolized to Benzidine)
Beryllium and Beryllium Compounds
1,3-Butadiene
1,4-Butanediol Dimethylsulfonate (Myleran)
Cadmium and Cadmium Compounds
Chlorambucil
1-(2-Chloroethyl)-3-(4-methylcyclohexyl)-1-nitrosourea (McCCNU)
bis(Chloromethyl) Ether and Technical-Grade Chloromethyl Methyl Ether
Chromium Hexavalent Compounds
Coal Tar Pitches (See Coal Tars and Coal Tar Pitches)
Coal Tars (See Coal Tars and Coal Tar Pitches)
Coke Oven Emissions
Cyclophosphamide
Cyclosporin A (Ciclosporin)
Diethylstilbestrol
Dyes Metabolized to Benzidine (See Benzidine and Dyes Metabolized to Benzidine)
Environmental Tobacco Smoke (See Tobacco Related Exposures)
Erionite
Estrogens, Steroidal
Ethylene Oxide
Melphalan
Methoxsalen with Ultraviolet A Therapy (PUVA)
Mineral Oils (Untreated and Mildly Treated)
Mustard Gas
2-Naphthylamine
Nickel Compounds (See Metallic Nickel and Nickel Compounds)
Radon
Silica, Crystalline (Respirable Size)
Smokeless Tobacco (See Tobacco Related Exposures)
Solar Radiation (See Ultraviolet Radiation Related Exposures)
Soots
Strong Inorganic Acid Mists Containing Sulfuric Acid
Sunlamps or Sunbeds, Exposure to (See Ultraviolet Radiation Related Exposures)
Tamoxifen 223
2,3,7,8-Tetrachlorodibenzo-p-dioxin (TCDD); "Dioxin"
Thiotepa
Thorium Dioxide
Tobacco Smoking (See Tobacco Related Exposures)
Vinyl Chloride
Ultraviolet Radiation, Broad Spectrum UV Radiation (See Ultraviolet Radiation Related Exposures)
Wood Dust

APPENDIX 5: Carcinogens, anticipated to be a human carcinogen.

Acetaldehyde
2-Acetylaminofluorene
Acrylamide
Acrylonitrile
Adriamycin (Doxorubicin Hydrochloride)
2-Aminoanthraquinone
o-Aminoazotoluene
1-Amino-2-methylanthraquinone
2-Amino-3-methylimidazo[4,5-f]quinoline (IQ)
Amitrole
o-Anisidine Hydrochloride
Azacitidine (5-Azacytidine, 5-AzaC)
Benz[a]anthracene (See Polycyclic Aromatic Hydrocarbons)
Benzo[b]fluoranthene (See Polycyclic Aromatic Hydrocarbons)
Benzo[j]fluoranthene (See Polycyclic Aromatic Hydrocarbons)
Benzo[k]fluoranthene (See Polycyclic Aromatic Hydrocarbons)
Benzo[a]pyrene (See Polycyclic Aromatic Hydrocarbons)
Benzotrichloride
Bromodichloromethane
2,2-bis-(Bromoethyl)-1,3-propanediol (Technical Grade)
Butylated Hydroxyanisole (BHA)
Carbon Tetrachloride
Ceramic Fibers (Respirable Size)
Chloramphenicol
Chlorendic Acid
Chlorinated Paraffins (C_{12}, 60% Chlorine)
1-(2-Chloroethyl)-3-cyclohexyl-1-nitrosourea
bis(Chloroethyl) nitrosourea
Chloroform
3-Chloro-2-methylpropene
4-Chloro-o-phenylenediamine
Chloroprene
p-Chloro-o-toluidine and p-Chloro-o-toluidine Hydrochloride (See p-Chloro-o-toluidine and p-Chloro-o-toluidine Hydrochloride)
Chlorozotocin
C.I. Basic Red 9 Monohydrochloride
Cisplatin
p-Cresidine
Cupferron
Dacarbazine
Danthron (1,8-Dihydroxyanthraquinone)
2,4-Diaminoanisole Sulfate
2,4-Diaminotoluene
Dibenz[a,h]acridine (See Polycyclic Aromatic Hydrocarbons)
Dibenz[a,j]acridine (See Polycyclic Aromatic Hydrocarbons)
Dibenz[a,h]anthracene (See Polycyclic Aromatic Hydrocarbons)
7H-Dibenzo[c,g]carbazole (See Polycyclic Aromatic Hydrocarbons)
Dibenzo[a,e]pyrene (See Polycyclic Aromatic Hydrocarbons)
Dibenzo[a,h]pyrene (See Polycyclic Aromatic Hydrocarbons)
Dibenzo[a,i]pyrene (See Polycyclic Aromatic Hydrocarbons)
Dibenzo[a,l]pyrene (See Polycyclic Aromatic Hydrocarbons)
1,2-Dibromo-3-chloropropane
1,2-Dibromoethane (Ethylene Dibromide)
2,3-Dibromo-1-propanol
tris(2,3-Dibromopropyl) Phosphate
1,4-Dichlorobenzene
3,3'-Dichlorobenzidine and 3,3'-Dichlorobenzidine Dihydrochloride (See 3,3'-Dichlorobenzidine and 3,3'-Dichlorobenzidine Dihydrochloride)
Dichlorodiphenyltrichloroethane (DDT)
1,2-Dichloroethane (Ethylene Dichloride)
Dichloromethane (Methylene Chloride)
1,3-Dichloropropene (Technical Grade)
Diepoxybutane
Diesel Exhaust Particulates
Diethyl Sulfate
Diglycidyl Resorcinol Ether
3,3'-Dimethoxybenzidine (See 3,3'-Dimethoxybenzidine and Dyes Metabolized to 3,3'-Dimethoxybenzidine)
4-Dimethylaminoazobenzene
3,3'-Dimethylbenzidine (See 3,3'-Dimethylbenzidine and Dyes Metabolized to 3,3'-Dimethylbenzidine)
Dimethylcarbamoyl Chloride
1,1-Dimethylhydrazine
Dimethyl Sulfate
Dimethylvinyl Chloride
1,6-Dinitropyrene (See Nitroarenes [selected])
1,8-Dinitropyrene (See Nitroarenes [selected])
1,4-Dioxane
Disperse Blue 1
Dyes Metabolized to 3,3'-Dimethoxybenzidine (See 3,3'-Dimethoxybenzidine and Dyes Metabolized to 3,3'-Dimethoxybenzidine)
Dyes Metabolized to 3,3'-Dimethylbenzidine (See 3,3'-Dimethylbenzidine and Dyes Metabolized to 3,3'-Dimethylbenzidine)
Epichlorohydrin
Ethylene Thiourea
di(2-Ethylhexyl) Phthalate
Ethyl Methanesulfonate
Formaldehyde (Gas)
Furan
Glasswool (Respirable Size)
Glycidol
Hexachlorobenzene
Hexachlorocyclohexane Isomoers
Hexachloroethane
Hexamethylphosphoramide
Hydrazine and Hydrazine Sulfate (See Hydrazine and Hydrazine Sulfate)
Hydrazobenzene
Indeno[1,2,3-cd]pyrene (See Polycyclic Aromatic Hydrocarbons)
Iron Dextran Complex
Isoprene
Kepone (Chlordecone)
Lead Acetate (See Lead Acetate and Lead Phosphate)
Lead Phosphate (See Lead Acetate and Lead Phosphate)

Carcinogens, anticipated to be a human carcinogen (Continued)

Lindane and Other Hexachlorocyclohexane Isomers
2-Methylaziridine (Propylenimine)
5-Methylchrysene (See Polycyclic Aromatic Hydrocarbons)
4,4'-Methylenebis(2-chloroaniline)
4-4'-Methylenebis(N,N-dimethyl)benzenamine
4,4'-Methylenedianiline and 4,4'-Methylenedianiline Dihydrochloride (See 4,4'-Methylenedianiline and its Dihydrochloride Salt)
Methyleugenol
Methyl Methanesulfonate
N-Methyl-N'-nitro-N-nitrosoguanidine
Metronidazole
Michler's Ketone [4,4'-(Dimethylamino)benzophenone]
Mirex
Nickel (Metallic) (See Nickel and Nickel Compounds)
Nitrilotriacetic Acid
o-Nitroanisole
6-Nitrochrysene (See Nitroarenes [selected])
Nitrofen (2,4-Dichlorophenyl-p-nitrophenyl ether)
Nitrogen Mustard Hydrochloride
2-Nitropropane
1-Nitropyrene (See Nitroarenes [selected])
4-Nitropyrene (See Nitroarenes [selected])
N-Nitrosodi-n-butylamine
N-Nitrosodiethanolamine
N-Nitrosodiethylamine
N-Nitrosodimethylamine
N-Nitrosodi-n-propylamine
N-Nitroso-N-ethylurea
4-(N-Nitrosomethylamino)-1-(3-pyridyl)-1-butanone
N-Nitroso-N-methylurea
N-Nitrosomethylvinylamine
N-Nitrosomorpholine
N-Nitrosonornicotine
N-Nitrosopiperidine
N-Nitrosopyrrolidine
N-Nitrososarcosine
Norethisterone
Ochratoxin A
4,4'-Oxydianiline
Oxymetholone
Phenacetin (See Phenacetin and Analgesic Mixtures Containing Phenacetin)
Phenazopyridine Hydrochloride
Phenolphthalein
Phenoxybenzamine Hydrochloride
Phenytoin
Polybrominated Biphenyls (PBBs)
Polychlorinated Biphenyls (PCBs)
Polycyclic Aromatic Hydrocarbons (PAHs)
Procarbazine Hydrochloride
Progesterone
1,3-Propane Sultone
β-Propiolactone
Propylene Oxide
Propylthiouracil
Reserpine
Safrole
Selenium Sulfide
Streptozotocin
Styrene-7,8-oxide
Sulfallate
Tetrachloroethylene (Perchloroethylene)
Tetrafluoroethylene
Tetranitromethane
Thioacetamide
Thiourea
Toluene Diisocyanate
o-Toluidine and o-Toluidine Hydrochloride(See o-Toluidine and o-Toluidine Hydrochloride)
Toxaphene
Trichloroethylene
2,4,6-Trichlorophenol
1,2,3-Trichloropropane
Ultraviolet A Radiation (See Ultraviolet Radiation Related Exposure)
Ultraviolet B Radiation (See Ultraviolet Radiation Related Exposure)
Ultraviolet C Radiation (See Ultraviolet Radiation Related Exposure)
Urethane
Vinyl Bromide
4-Vinyl-1-cyclohexene Diepoxide
Vinyl Fluoride

Urotext
Urology and health
www.urotext.com

I N D E X

A

Abdominal pain, 60, 230, 248, 251, 322, 390, 452, 466, 491, 629
Absorption, 8, 13, 237, 320-323, 364, 409, 492, 586, 635, 694
ACE inhibitor, 142, 247, 251, 365, 387
Acetaldehyde, 82, 147, 160, 161, 241, 407, 543, 544, 649, 738
Acetaminophen, 82, 83, 254, 256, 286, 403, 404, 448, 730
Acetate, 30, 127, 223, 298, 308, 362, 441, 509, 547, 648, 738
Acetyl L-carnitine, 408, 455, 517
Acetylcholine, 37, 118, 316-319, 343, 451-456, 528, 666, 702
Acid-base balance, 161, 221, 225, 234, 334, 365, 506, 567, 695
Acidosis, 67, 121, 133-138, 230-235, 365, 425, 511, 534, 666
Acne, 321, 322, 452, 491, 561, 664
Acupuncture, 257, 311, 551, 557
Adaptation, 10, 16, 48, 51-53, 142, 189, 201, 210, 237, 282, 325, 356, 365, 424, 441, 483, 526, 612, 643, 651, 661, 693, 698
Addiction, 147, 325, 326, 458, 459, 585, 635, 664, 665, 716
Adduct, 82, 83, 148, 214, 247, 274, 305, 393, 543, 648, 668, 704
Adenocarcinoma, 224, 315, 338, 348, 561, 584, 685
ADH (Antidiuretic hormone), 30, 125, 130, 132, 133, 151, 159, 161, 199, 200, 230, 291, 324, 348, 612
Adolescent, 149, 158, 333, 468, 569, 649, 701, 702, 728
Adrenal cortex, 10, 131-133, 298, 348, 363, 366, 644, 650, 684
Adrenal gland, 11, 159, 171, 224, 308, 314, 318, 365, 410, 443, 446, 457, 484, 517, 524-526, 599, 617, 632-634, 675, 693
Adrenal insufficiency, 230, 286, 365, 524, 525, 526, 531
Adriamycin, 70, 107, 187, 343, 372, 388, 403, 641, 738
Aerobic, 8, 27, 49, 117, 184-188, 261, 385, 401, 530, 655-658
Aflatoxins, 271, 429, 737
Ageing, 70, 211, 227, 304, 374, 384, 468, 500, 656, 682
Aging, 210, 212, 215, 217, 219, 221, 224, 226, 234, 237, 375, 420, 498, 585, 631, 661, 666, 693
Air, 462, 545, 549, 552, 557, 583
ALA (α-lipoic acid), 60, 61, 251, 256, 416, 419, 454, 513, 704
Alcohol, 82, 147, 150, 152, 156, 157, 160, 161, 162, 175, 262, 310, 324, 339, 399, 403, 441, 462, 510, 522, 543, 607, 627, 628, 634, 647, 682, 704, 706
Alcoholism, 151, 160, 561, 585, 634, 648
Aldehyde, 33, 186, 243, 328, 334, 513, 643, 707
Aldosterone, 131, 133, 137, 362, 365, 424, 728, 731
Alkaline phosphatase (ALP), 11, 109, 228, 323, 548, 674, 728
Alkalosis, 67, 133, 134, 136, 137, 161, 199, 230-232, 425, 559
Alkoxyl, 65, 147, 148, 419, 578, 670
Alkylating, 185, 377, 581, 582, 704
Alkylation, 401, 582, 704
Allergic, 89, 117, 273, 314, 352, 412, 459, 495, 507, 560, 733
Allergy, 8, 117, 182, 313, 371, 443, 459, 484, 553, 583, 633, 675
Allopurinol, 72, 293, 333, 383, 523
Almond, 262, 311, 335, 336, 343, 444, 449, 457, 459, 538, 539
Alzheimer's disease (AD), 85, 423, 331, 405, 535-539
Amino acids, 11, 27, 126, 322, 353, 567, 622
Ammonia, 139, 147, 401, 431, 699, 704, 728, 733
Ammonium, 705, 734, 735
Amygdalin, 336, 705
Amyloidosis, 289, 291
Anabolic steroid, 596, 600
Anaemia, 10, 182, 192, 288, 289, 320, 452, 651, 673
Anaerobic, 19, 27, 56, 186, 208
Analgesia, 159, 253, 256, 282, 325, 360, 558, 563
Aneurysm, 667
Angiostatin, 279, 576, 577
Angiotensin I (AGNI), 123, 124, 125, 132, 139, 704
Angiotensin II (AGNII), 123, 124, 125, 132, 139, 247, 704
Angiotensin-converting enzyme (ACE), 360
Angiotensinogen, 191
Anion gap, 232, 728
Anorexia nervosa (AN), 569, 695
Anoxia, 56, 66, 67, 79, 81, 84, 85, 167, 409, 547
Antibiotics, 14, 15, 34, 82, 117, 165, 172, 186, 201-204, 209, 233, 309, 319, 378, 402, 415, 462, 484, 498, 582, 653, 730
Antigen, 88, 487
Apoptosis, 44, 46, 59, 96, 97, 99, 100, 102-106, 108, 110, 112, 114, 267, 268, 296, 360, 381, 391, 520, 610, 704

Arachidonic acid (AA), 34, 171, 177, 449, 595, 704
Arginine, 114, 222, 344, 351, 360, 395, 464, 568, 583, 611, 674
Aromatic hydrocarbon (PAH), 148, 149, 269, 271, 274, 528, 529, 668, 697, 738, 739,
Arteriosclerosis, 80, 179, 328, 340, 452, 499, 513, 517, 630, 684
Arthritis, 65, 85, 86, 220, 226, 286, 308, 329, 362, 419, 443, 449, 452, 499, 527, 535, 540, 584, 611, 635, 647, 653, 693, 694
Ascorbate, 20, 66, 70, 77, 143, 148, 169, 188, 196, 235, 244, 256, 257, 273, 379, 380-384, 402, 410-417, 420, 441, 670
Aspartame, 256, 325, 515, 559, 618, 736
Aspirin, 26, 226, 229, 256, 259, 286, 308, 336, 413, 540, 577, 651, 698, 730
Asthma, 380, 486, 557, 564, 603
Atherosclerosis, 235, 405, 503, 529, 570, 623, 626, 667, 696
Athletes, 345, 346, 358, 395, 454, 569
ATN (acute tubular necrosis), 171, 172, 497
ATP, 18, 19, 20, 21, 22, 23, 26, 27, 28, 30, 37, 41, 45, 46, 49, 50, 53, 54, 55, 56, 58, 64, 66, 68, 69, 79, 97, 98, 102, 104, 121, 126, 138, 142, 144, 155, 161, 167, 169, 176, 204, 214, 232, 234, 252, 256, 263, 270, 299, 309, 320, 326, 335, 350, 351, 367, 376, 384, 386, 387, 388, 389, 397, 399, 403, 407, 408, 414, 417, 419, 421, 422, 426, 452, 455, 478, 480, 481, 482, 490, 513, 514, 519, 539, 543, 544, 546, 552, 558, 565, 567, 570, 578, 580, 590, 601, 608, 611, 613, 615, 620, 637, 666, 669, 670, 671, 678, 679, 696, 699, 709
ATP depletion, 45, 54, 55, 56, 66, 104, 270, 519
ATPase, 19, 20, 27, 32, 54, 56, 60, 71, 126, 142, 364, 426, 641, 647, 679
Autism, 311, 321, 502
Autoimmune, 8, 101, 108, 162, 182, 212, 226, 238, 249, 289, 312, 330, 374, 412, 443, 468, 479, 501, 511, 588, 654, 672
Autonomic dysreflexia (AD), 179, 180, 181

B

B cell, 87, 97, 120, 157, 165, 182, 216, 218, 238, 475, 560, 672, 701, 706
Bacteria, 49, 50, 53, 86, 190, 203, 208, 294, 297, 462, 493, 547, 589, 687
Bacteroides, 204, 209, 233, 353
bcl$_2$, 97, 98, 104, 105, 106, 108
bFGF, 106, 279, 297, 529, 534, 576, 577, 678
Bicarbonate, 134, 671, 713, 728, 735
Bile, 29, 231, 313, 328, 573, 627, 628, 728
Bilirubin, 261, 273, 728, 731
Biological water, 152
Biopsy, 70, 75, 120, 263, 275, 336, 511, 572, 672, 685, 686, 689
Biotin, 257, 259, 452, 458, 463, 546, 587
Biotransformation, 13, 15, 83, 202, 706
Bisphosphonates, 220, 283
Bladder outlet obstruction (BOO), 678
Bleeding, 320, 452, 727
Blood, 122, 150, 154, 165, 176, 180, 182, 229, 250, 282, 286, 292, 336, 387, 397, 482, 487, 488, 506, 526, 533, 548, 562, 573, 593, 651, 667, 706, 728, 729
Blood flow, 250, 667
Blood pressure, 180, 229, 526
Blood transfusion, 165, 651
Bone marrow, 87, 227, 650, 687, 706
Bone metastases, 281, 282, 480, 681, 684
Bone mineral density (BMD), 220, 221, 527, 659, 695
Brain, 342, 405, 420, 445, 533, 534, 542, 573, 587, 607, 619
Breast, 264, 291, 308, 607, 691
Bromine, 418, 665
Buccal mucosa, 12, 146, 706
Burns, 11, 230, 320, 367, 409, 448, 525, 554, 574, 599, 628, 659
Butter, 260, 444, 448, 450, 690

C

Cachexia, 192, 284, 588, 608, 609
Cadmium (Cd), 147, 331, 418, 430, 489, 641, 682, 707, 737
Caeruloplasmin, 199, 319, 640, 644, 650, 708, 728
Caffeine, 256, 330, 331, 338, 468, 474, 506, 536, 560, 615, 634
Calcitonin, 138, 158, 230, 349, 683, 708

Calcium, 42, 57, 67, 76, 137, 160, 167, 169, 177, 205, 230, 289, 309, 318, 337, 338, 418, 427, 452, 458, 463, 464, 537, 568, 586, 606, 621, 689, 695, 698, 700, 708, 728, 731, 733, 734, 735, 736
Calcium oxalate, 327, 361, 368, 410, 429, 708
Calcium phosphate, 137, 312, 337, 361, 606, 734
Calmodulin, 75, 76, 104, 140, 183, 205, 303, 304, 347, 430, 671
cAMP, 28, 39, 40, 42, 43, 47, 54, 72, 104, 116, 118, 130, 140, 170, 183, 205, 218, 256, 286, 296, 299, 304, 330, 346, 347, 348, 350, 422, 430, 453, 548, 608, 610, 612, 614, 620, 644, 707, 709
Cancer, 98, 263, 264, 268, 269, 288, 290, 308, 335, 352, 372, 375, 377, 396, 399, 412, 429, 477, 480, 483, 532, 541, 575, 578, 582, 657, 687, 690
Candida, 184, 186, 203, 209, 315, 447, 543
CaP (prostate cancer), 44, 112-114, 224, 272, 273, 281, 282, 287, 300, 377, 662, 681, 684-689, 707
Capillary, 42, 167, 240, 296, 696
Caprylic acid, 442
Carbene, 707
Carbohydrate, 11, 21, 30, 90, 120, 162, 187, 200, 225, 258-262, 311-314, 339, 364, 416, 448, 553, 597-599, 612-616, 661
Carbon chain, 152, 357, 397, 441, 442, 449
Carcinogenesis, 114, 267-273, 276, 310, 316, 327, 338, 341, 358, 370-372, 393, 417, 429-431, 532, 577, 580, 683, 684
Carnitine, 30, 257, 445, 452, 566, 568, 587, 615, 673
Carotene, 8, 66, 71, 148, 269, 273, 309, 343, 373, 375-385, 390, 417, 456, 483, 580, 587, 610, 673, 681, 688, 693, 728, 733
Carpal tunnel syndrome ((CTS), 322, 660
Carrot, 373, 693
Catecholamines, 51, 79, 139, 146, 150, 163-166, 171, 195, 319, 387, 405, 524, 561, 565, 585, 595, 608, 624, 636, 662, 731
Catheter, 175, 180, 181, 409, 593
Cell cycle, 22, 34, 49, 57, 100-104, 112, 210, 223, 266, 270, 276, 356, 372, 391-394, 425, 430, 572, 578, 581, 683, 687
Cell division, 23, 59, 107, 143, 223, 267, 300, 425, 532, 573, 615
Cell volume, 31, 46-48, 97, 268
Ceramide, 33, 98, 104, 105, 155, 183, 315, 707
cGMP, 39, 54, 72-75, 141, 205, 239, 301, 330, 347, 350, 666
Chelation, 327, 344, 369, 484, 498-500, 641
Chemical bond, 74, 531, 554, 708
Chemical sensitivity, 482, 559
Chemotaxis, 43, 68, 69, 78, 89, 94, 118, 120, 147, 156, 165, 171, 184, 192-194, 293, 377, 413, 565, 687, 708
Chemotherapy, 96, 113, 288, 572, 573, 579, 676, 718
Cherries, 260, 645
Children, 398, 488, 493, 495, 532, 546, 631, 684, 727, 732
Chlamydia, 184, 207, 447, 673
Chlorella, 341, 500, 559
Chloride, 130, 199, 205, 362, 640, 728, 731, 737, 738
Chlorophyll, 341, 498, 675
Cholesterol, 31, 86, 298, 315, 343, 374, 377, 382, 389, 441, 448, 537, 570, 571, 584, 623, 625, 626, 642, 688, 707, 728
Choline, 319, 507
Chromium, 257, 259, 441, 451, 452, 463, 464, 513, 587, 612, 615, 634, 737
Chronic fatigue syndrome (CFS), 56, 325, 360, 413, 558-566, 636, 636, 653
Chrysin, 341, 454
Chymotrypsin, 335, 336
Cisplatin, 289, 582, 708, 738
Citrate, 234, 404, 429, 583
Citrulline, 62, 76, 239, 351, 396, 398, 401, 662
CJD, 536
c-jun, 22, 48, 52, 97, 183, 190, 235, 247, 271, 357
CLA (Conjugated linoleic acid), 274, 323, 337, 451, 456
Clinical trials, 9
Clostridium, 111, 204, 209, 221, 481, 610
Coagulopathy, 166, 320
Cobalt, 418, 464, 587
Coconut, 444, 445, 621, 626, 690, 699
Coffee, 147, 331, 646, 651, 653
Collagen, 84, 96, 199, 294, 305, 371, 587, 589, 590, 642, 679
Colon, 42, 264, 282, 308, 653
Constipation, 181, 209, 230, 241, 254, 313, 329, 330, 452, 472, 492, 495, 511, 569, 605, 627, 629, 630, 653, 693
Cooking, 530, 555, 651
Copper, 73, 270, 319, 320, 321, 339, 353, 363, 418, 452, 456, 463, 464, 507, 509, 516, 537, 587, 598, 612, 615, 635, 640, 645, 650, 651, 665, 696, 709, 728, 731, 733, 735

CoQ_{10}, 69, 148, 169, 255, 259, 309, 340, 346, 376, 377, 381, 383, 384, 386, 387, 388, 390, 408, 416, 454, 517, 537, 566, 574, 579, 598, 615, 623, 635, 637, 641, 674, 708
Corticosteroids, 119, 142, 286, 338, 380, 524, 526, 596, 598, 603
Cortisol, 198, 317, 525, 614, 642, 644, 661, 692, 728, 731
Cortisone, 199, 443, 551
Creatinine, 125, 225, 229, 274, 351, 639, 651, 708, 728, 731
Creatinine clearance, 229, 708, 731
CT scan, 263, 277, 282
Curcumin, 255, 353, 371, 511, 635, 655, 733
Cyanide, 19, 20, 63, 73, 114, 147, 336
Cyclin dependent kinase (cdk), 102, 103, 119, 276, 357, 707
Cyclophosphamide, 582, 707, 708, 737
Cyclosporine, 56, 730
Cysteine, 78, 79, 328, 402, 406, 491, 622, 707
Cystine, 329, 330, 361, 404, 406, 690, 731
Cystitis, 151, 180, 197, 581, 582
Cystoscopy, 186, 686
Cytochrome, 20, 50, 81, 82, 113, 114, 115, 154, 169, 320, 520, 521, 643, 708, 709
Cytochrome C, 20, 27, 34, 55, 69, 99, 113-115, 142, 169, 186, 214, 320, 358, 475, 514, 520-552, 615, 637, 643, 708, 709
Cytochrome P450, 60, 62, 81-83, 149, 151, 168, 202, 227, 286, 298, 324, 370, 384, 402-404, 516, 542, 559, 629, 689, 700
Cytokines, 24, 87, 88, 89, 90, 91, 97, 106, 116, 120, 123, 157, 165, 166, 174, 188, 191, 194, 239, 262, 270, 293, 487, 524, 562, 575, 595, 671, 701

D

Dairy, 233, 262, 316, 323, 337, 338, 358, 388, 441, 455, 489, 506, 507, 534, 561, 605, 606, 614, 645, 673
Deep venous thrombosis (DVT), 556
Deficiencies, 8, 410, 430, 518, 588, 630, 645
Degeneration, 10, 16, 121, 210, 248, 251, 272, 313, 314, 317, 373, 390, 403, 449, 463, 500, 530, 580, 651, 673, 680, 698
Degenerative disease, 70, 114, 215, 385, 389, 403, 418, 443, 450, 459, 462, 470, 513, 527, 541, 548, 621, 641, 446, 658
Dehydroepiandrosterone (DHEA), 224, 537, 558, 710, 731
Dementia, 219, 221, 317, 323, 343, 391, 405-408, 456, 485, 490, 492, 499, 5506, 15-518, 534-541, 640, 698
Depression, 222, 223, 226, 325, 454, 456, 466, 503, 510, 535, 557, 560, 564, 633, 658, 670, 700
Deterioration, 109, 175, 217, 284, 419, 481, 536, 573, 632, 662
Detoxification, 8, 15, 71, 149, 210, 211, 249, 272, 289, 318, 339, 384, 402, 417, 451, 457, 463, 482-502, 507, 522, 665, 683
Dexamethasone, 102, 107, 116, 268, 285, 287, 466, 610
Diabetes insipidus, 133
Diabetes mellitus, 235, 237, 238, 245, 510, 571, 618, 629, 666
Dialysis, 78, 115, 230, 255, 287, 320, 361, 490, 507, 599, 650
Diarrhoea, 231, 321, 363, 413, 452
Diethylstilbestrol (DES), 559, 673, 697, 710, 737
Digestion, 8, 23, 97, 114, 172, 200, 260-263, 294, 311-316, 322, 336, 418, 442-445, 495, 511, 543, 563, 589, 608, 690, 694
Dihydrotestosterone (DHT), 219, 297, 300, 348, 663, 668, 684
Dioxin, 509, 559, 673, 674, 737
Dioxygen, 21, 64, 70
Diradical, 64, 710
Disease, 8, 378, 463, 506, 628
Dismutation, 70, 187, 189, 239, 376, 478, 710
Diuresis, 122, 130, 133, 134, 139, 142, 149, 164, 412, 639, 709
DMSO, 50, 329, 551, 710
DNA, 18, 20, 22, 23, 25, 28, 37, 39, 44, 46, 49, 50, 51, 56, 57, 59, 62, 64, 65, 69, 71, 72, 74, 77, 82, 84, 88, 89, 95, 96, 97, 98, 99, 102, 103, 104, 106, 108, 110, 111, 112, 114, 119, 120, 142, 149, 152, 155, 164, 168, 185, 188, 196, 204, 210, 211, 214, 215, 218, 224, 227, 239, 240, 241, 242, 243, 246, 247, 263, 266, 268, 269, 270, 271, 272, 274, 276, 284, 291, 292, 300, 305, 309, 319, 322, 324, 328, 338, 340, 341, 343, 346, 348, 349, 350, 351, 354, 356, 357, 365, 371, 372, 375, 376, 378, 380, 385, 389, 392, 395, 404, 407, 417, 418, 421, 425, 429, 440, 444, 447, 458, 475, 480, 482, 489, 495, 498, 516, 519, 522, 524, 529, 531, 532, 538, 544, 547, 548, 567, 573, 578, 579, 582, 606, 618, 631, 643, 650, 654, 657, 665, 668, 669, 674, 675, 681, 687, 688, 695, 710, 717, 718, 723
Dobutamine, 610
Dopamine, 57, 123, 125, 320, 325, 407, 458, 468, 474, 480, 516, 542, 709
Duodenum, 13, 43, 62, 208, 237, 336, 349, 364, 624, 652

Dysautonomia, 179, 636, 654

E

E. coli (Escherichia coli), 50, 204, 209, 311
Echinacea, 312
Eicosanoids, 28, 69, 123, 140-142, 171, 177, 255, 316, 377, 443, 456, 466, 569, 612, 614, 683
Eicosapentaenoic acid (EPA), 256, 285, 312, 381, 394, 448, 456, 459, 465, 466, 498, 538, 605, 609, 614, 622, 642, 690
ELANRA, 553
Electrolytes, 47, 161, 230, 291, 294, 338, 366, 395, 429, 586, 588, 598, 649, 711
Electron, 27, 66, 169, 320, 376, 384, 429, 587, 711
Emphysema, 10, 148, 320, 420, 490, 557, 595, 611, 642, 693
Endometriosis, 559, 561
Endotoxin, 46, 85, 162, 198, 293, 366, 540-542, 648-654, 693
Energy, 18, 19, 27, 261, 284, 314, 432, 539, 584, 585, 587, 593, 597, 607
Enterobacter, 209
Enzyme, 20, 35, 82, 83, 228, 280, 334, 401, 580, 644
Eosinophil, 94, 118, 119, 710, 711
Epilepsy, 159, 410, 445, 457, 496, 542, 563, 643, 693, 697, 699
Epithelial cell, 32, 46, 87, 112, 119, 184, 589
Erectile dysfunction (ED), 249, 666
Erythrocytes, 79, 240, 385, 415, 429, 565, 727, 728
Erythropoietin, 51, 53, 225, 287-289, 644
Essential fatty acid (EFA), 256, 443, 449, 451, 456, 566, 689
Estradiol, 44, 85, 158, 220-2224, 298, 660-668, 674, 696, 699
Eukaryotic cell, 19, 32, 182-185, 187, 189, 207, 426, 475, 641
Evening primrose oil (EPO), 256, 312, 675
Exercise, 86, 144, 146, 259, 261, 339, 403, 526, 534, 550, 600, 615, 631, 634, 635, 658

F

FAD, 18, 19, 21, 27, 28, 30, 50, 194, 350, 414
Familial hypercholesterolaemia (FH), 623, 625
Fasting, 258, 329, 377, 570
Fat, 165, 314, 510, 545, 567, 592, 608, 615, 692
Fatigue, 254, 284, 458, 565, 601, 633
Fatty acid, 28, 30, 269, 309, 442, 450, 567, 570, 587, 608, 648, 669, 711, 734, 735
Fatty degeneration, 30, 443, 463, 621
Fenton, 51, 65, 71, 168, 269, 270, 522
Fertility, 87, 258, 298, 309, 322, 375, 483, 548, 667-675, 692
Fever, 72, 87, 110, 151, 165, 166, 171, 192-196, 200, 229, 285, 291, 314, 399, 400, 413, 475, 507, 548, 574, 610, 618, 652
Fiber, 210, 237, 308-312, 327, 377, 464, 519, 587, 601, 634, 693
Fibrosis, 176, 191, 211, 228, 294, 303, 317, 400, 521, 667, 684
Fimbriae, 184, 200, 201, 331
Fish oil, 256, 285, 326, 442, 444, 459, 538, 605, 627, 634, 690
Flavin, 50, 69, 76, 186, 196, 641
Flavon, 340, 341, 369, 376, 623, 665, 680
Flavone, 454, 665, 675, 680
Flavonoid, 330, 342, 369, 370, 376, 454, 582, 665, 680
Flavoprotein, 19, 26, 27, 29, 71, 76, 113, 214, 233, 392, 709
Flaxseed oil, 323, 449, 609
Fluconazole, 202, 203
Fluorescent, 332, 492
Fluorine, 436, 463, 537, 547, 548
Folate, 259, 309, 338, 537, 565, 728
Folic acid, 10, 452, 458, 463, 465, 507, 537, 571, 587, 607, 631
Formaldehyde, 147, 485, 493, 564, 733, 738
Free radical, 49, 64, 66, 67, 147, 154, 169, 177, 210, 227, 238, 240, 247, 256, 268, 293, 312, 320, 374, 385, 412, 413, 416, 419, 429, 432, 444, 454, 480, 482, 516, 540, 566, 568, 578, 611, 632, 648, 693, 695, 712
Fructose, 56, 237, 246, 457, 544, 547, 613, 732
Fruits, 311, 316, 333
FSH, 44, 287, 298, 348, 515, 614, 668, 674, 675, 692, 712, 728
Functional dyspepsia (FD), 652
Functional medicine, 8
Furan, 738
Furosemide, 47, 121, 122, 141, 142, 230, 424, 425, 559, 639

G

Garlic, 343, 344, 500

G-CSF, 87, 115, 218
Gene therapy, 113, 498, 641
Germ cell, 218, 266, 478, 668, 669, 671, 672, 712
GFR, 15, 24, 80, 121, 123, 125, 127, 128, 131, 133, 134, 141, 149, 225, 227, 234, 245, 263, 264, 282, 287, 424
GIFT, 712
Ginkogo Biloba, 256, 257
GIT, 11, 12, 14, 63, 72, 102, 107, 111, 134, 138, 151, 161, 171, 205, 231, 232, 233, 234, 261, 282, 284, 287, 308, 329, 338, 349, 352, 353, 364, 386, 396, 423, 443, 451, 463, 466, 483, 485, 490, 501, 503, 512, 527, 579, 586, 595, 611, 612, 629, 649, 652, 712
Glomerulosclerosis, 142, 247, 389
Glucocorticoids, 89, 103, 120, 146, 164, 192, 193, 198, 617, 622, 643, 650, 696, 712
Glucose, 28, 39, 50, 53, 77, 79, 122, 126, 145, 235, 236, 238, 240, 241, 242, 243, 249, 259, 311, 318, 362, 414, 418, 421, 428, 451, 508, 511, 517, 537, 547, 613, 619, 620, 633, 678, 713, 728, 731
Glucosteroids response-modifying factor (GRMF), 198, 644
Glutamate, 136, 176, 195, 350, 398, 405, 427, 515, 542, 620, 654, 712
Glutamine, 135, 136, 165, 345, 359, 395, 397, 458, 568, 574, 581, 583, 587, 611, 612, 644, 712
Glutathione, 34, 65, 71, 79, 114, 154, 187, 210, 228, 245, 247, 269, 289, 292, 350, 369, 376, 379, 392, 397, 398, 401, 404, 405, 409, 415, 417, 421, 537, 544, 580, 583, 598, 611, 648, 674, 683, 713
Glycaemic, 84, 251, 259, 262, 285, 309, 512, 517, 567, 569, 606, 623, 629, 632, 634, 645, 660, 701
Glycosaminoglycans (GAG), 108, 254, 339, 589, 631, 712
GM-CSF, 87, 90, 91, 92, 116, 184, 216, 217, 246, 297, 316, 412, 574, 712
GnRH, 109, 298, 667, 712
Gout, 360, 627, 688
Green tea, 310, 331, 340, 353, 368, 371, 372, 399, 451, 511, 675
Growth factors, 47, 68, 108, 116, 123, 240, 242, 247, 263, 279, 280, 294, 295, 297, 541, 590, 683
Growth hormone (GH), 44, 128, 138, 158, 347, 397, 567, 569, 609, 610, 614, 617, 656, 657, 660, 677, 728
Gut, 309, 396, 400
Gynaecomastia, 662, 664, 676

H

H_2, 70, 160, 238, 286, 321, 409, 414, 440, 547, 555
Haber-Weiss, 71, 77, 85, 115, 168, 187, 188, 189, 197, 389
Haematuria, 149, 490, 491
Haemodialysis, 334, 390, 641
Haemoglobin, 10, 66, 72, 74, 85, 162, 289, 292, 344, 498, 664, 713, 727, 731
Haemorrhagic cystitis, 581
Haemostasis, 45, 178, 235, 276, 293, 384, 572
Hair, 491, 597, 664
hCG, 298, 348, 713
HCO_3^-, 47, 67, 134, 136, 231, 232, 233, 234, 314, 671, 713
Health, 9, 224, 304, 483, 527, 555
Heart disease, 85, 146, 224, 257, 308, 337, 367, 372-375, 408, 443-447, 500, 514, 569, 605, 630-637, 644, 692, 700-702
Heart failure, 11, 86, 132, 134, 255, 328, 385-388, 490, 501, 508, 512, 579, 591, 636, 643, 657, 698
Heat, 52, 56, 84, 85, 291, 356, 412, 450, 547, 714
Heat shock proteins (HSP), 48, 52, 56, 81, 84, 85, 164, 187, 291, 292, 412, 679
Heavy metal, 317, 360, 410, 462, 487, 497, 506, 526
Hemp, 450, 688
Heparin, 10, 16, 119, 229, 241, 244, 277, 293, 296, 360, 361, 498, 499, 576, 317
Hepatitis C, 8, 58, 157, 404
Hepatocyte growth factor (HGF), 39, 247, 265, 279, 357, 577
Hepatorenal syndrome, 47, 161, 232
Hepatotoxicity, 15, 154, 312, 334, 370, 582, 600, 629, 676, 688
High-density lipoprotein (HDL), 146, 147, 160-164, 236, 258, 322, 340, 374, 389, 429, 447, 451, 509-512, 570, 571, 605, 613, 623-630, 641, 655, 656, 658, 663, 664, 677, 713, 728
Histamine, 118, 159, 174, 181, 229, 286, 314, 352, 467, 495, 507, 553, 563, 566
HIV, 34, 100, 106, 157, 182, 186, 187, 188, 218, 368, 375, 445, 466, 506, 514, 515, 549, 558, 583, 588, 609, 610

Homeostasis, 16, 36, 42, 53, 71, 89-93, 100, 104, 111, 134, 137, 141, 176, 190, 211, 229, 268, 288, 292, 315-317, 319, 332, 353, 376, 396, 422, 473, 502, 522-526, 597, 619, 655, 683
Homocysteine, 78, 114, 198, 339, 496, 516, 537, 571, 630, 632
Hormone, 34, 39, 347, 349, 610, 614, 655, 659, 668, 675, 714
Hydrochloric acid (HCL), 8, 81, 208, 228, 231, 261, 313, 314, 322, 340, 459, 544, 674, 735
Hydrogen peroxide, 64, 65, 71, 81, 115, 169, 185, 238, 389, 419, 493, 578, 595, 596, 610, 669, 713, 714
Hydrogenation, 450, 605
Hydrophilic, 12, 14, 79, 83, 151-153, 207, 237, 320, 346, 423, 431, 434-436, 438, 442, 624, 668
Hydrophobic, 33- 39, 41, 42, 44, 57, 99, 102, 152-154, 203, 335, 346, 431-440, 570, 624, 646, 689, 714
Hydroxyl, 64, 71, 77, 82, 115, 189, 214, 319, 389, 419, 578, 595
Hypercalcaemia, 36, 43, 55, 109, 110, 221, 230, 281-283, 287, 288, 291, 428, 499, 600, 606, 628
Hyperkalaemia, 202, 225, 230, 255, 360-365, 583, 638, 639
Hypermagnesaemia, 230
Hypernatraemia, 225, 229, 365
Hyperoxaluria, 32
Hyperoxia, 51, 115, 116, 171, 293, 381, 640
Hyperphosphataemia, 231, 638, 639
Hypersensitivity, 11, 24, 215, 219, 241, 329, 334, 395, 459, 472, 473, 493, 532, 543, 544, 573, 574, 583, 595, 610, 627, 629
Hypertension, 72, 77, 87, 141, 151, 160-167, 170-181, 225-227, 236, 256-259, 301, 337, 360-371, 388, 419, 449, 487-490, 497, 508-512, 528, 605-608, 621, 656, 626, 656, 667, 676
Hyperthermia, 96, 112, 285, 399, 487, 608, 696
Hyperthyroidism, 211, 221, 228, 347, 428, 638, 645, 667, 690
Hypertrophy, 118, 121, 161, 234, 247, 528, 617, 653, 657, 678
Hypoglycaemia, 67, 163, 251, 288, 291, 326, 446, 456, 517, 632, 633, 699, 702
Hyponatraemia, 162, 192, 225, 229-232, 254, 288, 361, 699
Hypothalamus, 89, 132, 179, 200, 228, 285, 297, 301, 347-349, 470-474, 514-518, 558-563, 619-622, 643, 657, 692, 702
Hypothermia, 320, 583, 632, 637
Hypothyroidism, 334, 347, 456, 506, 535, 540, 622, 646, 692, 697
Hypoxia, 24, 51, 52, 79, 139, 171, 281, 289, 443, 501, 522, 683, 690, 713

I

IgA, 89, 90, 199, 203, 208, 215, 261, 360, 400, 526, 553, 672, 732
IgE, 87, 89, 90, 117, 120, 199, 203, 486, 497, 560, 732
IGF (Insulin-like growth factor), 39, 103, 106, 119, 159, 191, 192, 219-222, 240-247, 251, 252, 271, 284, 309, 338, 345-349, 569, 586, 593, 599, 609, 610, 617, 655-661, 677, 695
IgG, 87, 89, 90, 173, 203, 215, 244, 386, 430, 487, 672, 732
IL1, 86, 88, 89, 90, 91, 92, 95, 116, 165, 170, 171, 181, 183, 187, 189, 190, 191, 193, 197, 201, 238, 240, 283, 286, 487, 523, 595, 623
IL2, 22, 25, 87-92, 102, 165, 193, 197, 219, 574
IL6, 86-91, 107, 116, 165, 170, 173, 184, 190, 191, 193, 219, 244, 284, 286, 375, 476, 480, 487, 521, 523, 595
Immunoglobulin, 37, 68, 87, 91-93, 105, 165, 199, 208, 215, 216, 278, 280, 413, 565, 671
Immunotherapy, 92, 479, 482, 715
Impotence, 666
INFγ, 87, 477, 479, 610
Infection, 157, 205, 583, 585, 591, 602
Infertility, 549, 667, 692
Inflammation, 85, 89, 193, 201, 374, 498, 541, 595, 646, 658, 687, 691
Inositol, 257, 327, 458, 509, 715
Internal pollution, 462
Interstitial nephritis, 497
Involuntary weight loss (IWL), 584, 586, 594, 597, 600, 602
Iodine, 418, 452, 456, 463, 587, 675
Ipriflavone, 310, 339
Iron, 59, 61, 62, 63, 65, 67, 82, 190, 269, 288, 320, 378, 380, 403, 405, 418, 452, 458, 463, 464, 488, 499, 516, 522, 537, 547, 583, 587, 611, 615, 621, 638, 642, 648, 651, 695, 711, 728, 733, 738
Irritable bowel syndrome (IBS), 503, 652-654, 714
Ischaemia, 33, 54, 56, 66, 68, 85, 108, 167, 169, 176, 249, 255, 384, 714

Itraconazole, 84, 202, 203

J

Juvenile, 509, 643

K

Ketoconazole, 202
KGF, 265, 295, 296, 715
Kidney, 42, 260, 491, 533, 688
Klebsiella, 209
Krebs cycle, 18, 19, 20, 27, 28, 30, 127, 143, 399, 411, 458, 570, 611, 621, 699, 724
Kryptopyrrole, 321, 464, 502, 506, 5087, 508

L

Laboratory, 8, 125, 226, 272, 335, 493, 506, 615, 656, 727-732
Lactobacillus, 204, 209, 233, 337
Lactoferrin, 185, 193, 338
Laetrile, 336, 705, 715
L-arginine, 62, 72, 75, 196, 239, 303, 344, 345, 350, 671
Lauric acid, 445-448
Lead (Pb), 360, 429, 488, 499, 719, 732, 738
Lean body mass (LBM), 358, 584, 589-594, 606, 655, 658, 664
Lecithin, 228, 309, 451, 568, 584, 734
Lemon, 527, 545
Leukaemia, 10, 24, 190, 264, 291, 368, 410, 492, 532, 651, 670
Leukocytes, 24, 46, 68, 70, 87, 114, 140, 167, 177, 194-196, 278, 295, 376, 411, 475, 496, 574, 577, 650, 686, 727, 732
Leukotriene, 69, 120, 196, 324, 390, 401, 448
Leydig cell, 297, 298, 662, 670, 673, 675
L-glutamine, 208, 396, 406, 513, 526
L-glutathione, 491
LH, 36, 287, 298, 301, 348, 378, 382, 515, 614, 663, 668, 674, 675, 692, 716, 729
LHRH, 298, 716
LIF, 87, 91, 190, 191, 286, 716
Light, 51, 52, 65, 223, 251, 269, 309, 311, 371, 407, 442, 450, 467, 468, 482, 496, 530-532, 555, 633, 652, 682, 690, 693
Lindane, 444, 739
Linolenic acid, 448, 605
Lipid peroxidation, 31, 80, 82, 176, 240, 385, 445, 477, 540, 623, 674, 691
Lipoprotein, 35, 258, 451, 570, 624, 630, 716
Lithium, 232, 459, 700, 730
Liver, 162, 343, 347, 381, 403, 406, 420, 449, 476, 493, 521, 641, 648, 652, 664, 688, 692
Longevity, 210, 223, 227, 342, 429, 651, 664, 681, 692
Low-density lipoprotein (LDL), 34, 40, 80, 103, 146-163, 240, 257-259, 298, 305, 324, 340, 369, 373-390, 498-513, 570-572, 605-607, 623-632, 641, 655-666, 689, 693, 716, 728
Lung, 264, 279, 282, 291, 293, 402, 596, 599, 603, 604, 607
Lutein, 373, 375, 376, 733
Lyme disease, 506, 564
Lymphocyte, 79, 88, 95, 111, 116, 120, 156, 170, 193, 215-219, 278, 334, 359, 412, 476, 483, 565, 586, 610, 653, 671, 672
Lysine, 77, 560, 568, 654

M

Macrophages, 46, 74, 86, 87, 90, 97, 103, 118, 170, 182, 196, 242, 244, 271, 294, 295, 603, 604, 623, 695
Magnesium, 309, 318, 337, 339, 340, 362, 380, 422, 426, 429, 452, 458, 463, 501, 537, 559, 563, 565, 621, 645, 651, 697, 699, 729, 731, 734, 735
Malnourished, 161, 582, 650
Malnutrition, 221, 399, 463, 503, 582, 584, 586, 588, 596, 608, 609, 610
Manganese, 339, 343, 418, 452, 463, 587, 598, 612, 615, 717
Masked state, 716
Mast cell, 90, 118, 171, 195, 514, 566, 717
Matrix, 20, 193, 223, 271, 278, 294, 477, 717
M-CSF, 87, 220, 271
Meat, 314, 354, 361, 441, 689
Melatonin, 219, 229, 285, 326, 350, 458, 526, 532, 538, 560, 564, 661, 691, 702
Mental illness, 456, 464, 470, 471, 502-505, 584, 585, 594, 618

Mercury (Hg), 147, 488, 497, 501, 559, 713, 732
Metabolic acidosis, 63, 231, 234, 363, 424, 639
Metal, 73, 115, 148, 169, 192, 211, 238, 280, 292, 312, 327, 378, 406, 485-489, 495-500, 502, 506, 513, 648, 651, 671, 695
Metalloproteinase (MMP), 88, 112, 191-193, 223, 295, 372, 477, 512, 576, 577
Metastasis, 92, 103, 106, 114, 157, 263, 265, 270, 277-283, 289, 310, 336, 375, 400, 475-477, 572-577, 682-688
Methionine, 78, 273, 340, 353, 465, 507, 568, 641
Methotrexate, 574, 578, 582, 708, 718, 730
Milk, 210, 260, 314, 325, 337, 353, 359, 458, 459, 545, 607, 641, 644, 689
Mineral, 43, 108, 158, 220, 259, 295, 312, 313, 337-339, 410, 432, 463, 490, 500, 539, 566, 608, 645-649, 674, 695, 737
Mitochondria, 22, 28, 51, 56, 58, 61, 64, 84, 98, 142, 155, 168, 169, 214, 418, 422, 426, 520, 598
Mitomycin C, 574
Molybdenum, 333, 463, 464, 587, 615, 642
Monocyte, 116, 271, 404, 716, 717
Monounsaturated fatty acid, 449, 605, 627, 718
Mood, 151, 660, 700
MRI, 9, 282, 291, 536, 558
MSM, 329, 339, 360, 394, 500, 546, 563, 717
Mucin, 209, 276, 379, 380, 604, 652
Multiple sclerosis (MS), 470, 517, 566, 611, 642
Muscle, 20, 145, 229, 234, 274, 359, 428, 441, 561, 566, 586, 587, 595, 597, 601, 607, 659, 664, 679, 682, 697, 698
Mutagen, 83, 675, 718
Mutation, 32, 41, 77, 100-103, 108, 111-114, 117, 143, 214, 215, 246, 263, 266-272, 278, 355, 371, 389, 537, 622, 641, 682
Mycobacterium tuberculosis, 117
Myosin, 56, 75, 161, 177, 204, 208, 303, 520, 567, 666, 679, 682

N

NAC (N-acetyl cysteine), 57, 216, 255, 256, 257, 259, 376, 380, 394, 397, 405, 406, 412, 415, 501, 514, 536, 581, 583, 610, 615, 718
NADH, 18, 19, 20, 21, 26, 27, 30, 50, 51, 54, 59, 63, 64, 70, 79, 143, 151, 159, 169, 194, 214, 237, 238, 245, 247, 332, 376, 378, 384, 387, 407, 408, 411, 413, 523, 543, 560, 565, 615, 637, 641, 647, 709, 718
NADPH, 20, 29, 48, 63, 65, 70, 71, 76, 77, 79, 105, 115, 149, 159, 167, 194, 196, 237, 238, 245, 247, 249, 292, 350, 369, 376, 377, 384, 392, 402, 411, 416, 521, 637, 669, 709, 718
Necrosis, 13, 69, 97, 105, 110-112, 171, 268, 519-521, 648, 690
Negative ion, 494, 550-556
Neoplasia, 38, 39, 46, 94, 263, 288, 338, 377, 657, 683, 692
Nephron, 32, 80, 121, 138-142, 225, 234, 255, 337, 364, 424
Neuropeptide, 205, 251, 252, 302, 466, 560, 616, 663, 654, 718
Niacin, 257, 309, 320, 342, 452, 463, 587, 627, 628, 656, 734
Nickel, 149, 492, 718, 737, 739
Nicotine, 146, 149, 457, 490, 528, 667, 669, 673
Nicotinic acid, 504, 734
Nitrate, 76, 168, 547, 718
Nitric oxide (NO), 50, 57, 62-69, 72-76, 90, 113, 121-123, 125, 146-149, 166, 167, 183-189, 194-198, 205, 267-269, 301-305, 419, 520-524, 559, 626, 653, 666, 670, 680, 698, 718
Nitric oxide synthase (NOS), 34, 57, 74-76, 122, 166, 174, 184, 195-197, 244, 267, 301-304, 350, 522, 576, 666, 670, 671
Nitriloside, 335, 337
Nitroglycerine, 64, 72, 75, 203, 666
Nitrous oxide, 73, 462, 555, 556, 631, 736
NMDA (N-methyl-D-aspartate), 559
NSAIDs, 15, 82, 111, 122, 229, 251, 254, 255, 282, 284, 286, 310, 326, 360, 395, 583, 610, 655, 661, 702
Nucleus, 18, 25, 43, 97, 113, 120, 152, 164, 174, 182, 199, 216, 267, 200, 327, 387, 418, 466, 475, 519, 552, 600, 647, 684
Nutrient, 8, 123, 189, 201, 236, 284, 308, 316-318, 321, 374-376, 413, 452-4555, 584-586, 590-595, 601, 610, 614, 619
Nutrition, 8, 237, 308, 329, 378, 396, 410, 449, 462, 485, 503, 559, 571, 584, 590-596, 599-602, 604, 632, 660, 671, 694
Nutritional deficiencies, 232, 233, 288, 411, 591, 651, 673, 674

O

Obesity, 86, 162, 235, 237, 444, 509, 567, 571, 607, 613, 616, 648, 660, 667, 673
Oedema, 10, 117, 132, 168, 210, 285, 362, 449, 501, 552, 699

Oestrogen, 37, 109, 158, 160, 178, 228, 298, 336, 337, 344, 539, 562, 615, 617, 621, 642, 651, 653, 664, 690, 691, 696, 697, 699, 711
Oil, 262, 328, 361, 386, 436, 441, 442, 444, 450, 489, 544, 614, 641, 688, 689, 691
Olive oil, 86, 309, 310, 444, 445, 448, 450, 489, 605, 627, 690
Oncogene, 103, 106, 112, 267, 279, 286, 421, 429, 683, 689
Organic, 48, 65, 142, 149, 232, 233, 330, 342, 484, 492, 545, 564, 581, 587
Orthomolecular medicine, 311
Orthomolecularism, 417, 462
Osmolality, 99, 168, 227, 229, 261, 729, 731
Osteoblast, 43, 109, 110, 158, 159, 220, 253, 287, 647
Osteoclast, 43, 108, 109, 158, 220, 235, 252, 281, 348, 695, 698
Osteopenia, 221, 527, 621, 647, 699, 660, 662, 687, 694, 695
Overweight, 613, 617, 658, 702
Oxandrolone, 597, 598, 600, 609, 663
Oxidation, 26, 63, 78, 105, 185, 238, 246, 374, 376, 378, 381, 611, 640, 665, 710
Oxidative stress, 49, 53, 65, 71, 77, 81, 115, 150, 155, 250, 269, 291, 345, 369, 374, 378, 381, 403, 405, 419, 522, 578, 579, 610, 622, 668, 683
OXPHOS, 58, 69, 403, 719
Oxygen free radicals (OFRs), 70, 77, 139, 140, 167, 169, 171-173, 182, 187, 201, 211, 267, 301, 378, 381, 403, 415, 416, 574, 638, 681
Oxytocin, 302, 348, 564, 674, 700
Ozone, 80, 379, 462, 533, 550, 555

P

p53, 26, 96, 98, 100, 101, 102, 103, 104, 107, 108, 112, 114, 168, 212, 214, 266, 268, 269, 271, 276, 278, 281, 357, 391, 475, 483, 521, 668, 683
P5P, 322, 328, 502
Pacemaker, 467, 532
Paclitaxel, 113, 578, 582
Pain, 177, 226, 254, 282, 290, 304, 360, 415, 556, 721
Pancreas, 259, 349, 511
Pantothenic acid, 340, 452, 463, 464, 587, 634
Papillary necrosis, 122, 141, 255
Parasites, 11, 201, 312, 330, 459, 478, 488, 502, 533, 583, 671
Parathyroid hormone (PTH), 39, 43, 137, 138, 157, 158, 348, 220, 287, 337, 348, 349, 424, 425, 622, 694
Parkinson's disease (PD), 57, 344, 372, 405, 407, 516, 517, 638
Pathobiology, 16, 75, 245, 249, 573
PDGF, 37, 39, 52, 92, 95, 119, 146, 191, 194, 238, 242, 244, 247, 271, 294, 295, 296, 315, 534, 577, 623, 720
Peanut, 260, 262, 311, 343, 388, 445, 449
Pectin, 202, 353, 459, 545, 605, 606, 642, 734
Pentoxifylline, 256, 285, 521
Peroxidation, 71, 77, 177, 185, 213, 239, 268, 328, 370, 373, 381, 390, 405, 426, 445, 605, 623, 648, 670, 687-690, 720
Peroxidative, 155, 384, 386, 389, 426, 648, 649, 670
Peroxyl, 147, 148, 185, 268, 377, 378, 383, 578, 670, 720, 722
Pesticide, 330, 485, 487, 544, 673, 682
Phagocyte, 77, 92, 120, 182-194, 197, 562
Phenobarbital, 15, 60, 203, 627, 730
Phenol, 309, 485, 503, 733
Phenylalanine, 457, 464, 508, 526, 568, 618, 627, 637, 654
Phosphate, 138, 440, 639, 729, 731, 733, 738
Phospholipids, 38, 69, 153, 382, 450, 584, 624
Phosphorus, 418, 463, 464
PMS, 321, 322, 326, 450, 454, 485, 506, 661, 699, 720
Polarised, 31-33, 40, 99, 277, 570
Pollution, 10, 117, 308, 353, 380, 462, 488-490, 551-555, 559
Potassium (K), 34, 40, 55, 67, 97, 127, 133, 143, 167, 199, 229-231, 283, 331, 359-367, 425, 463, 509, 559, 598, 642-646
Pregnancy, 10, 30, 171, 237, 290, 323, 336, 366, 410, 415, 462, 509, 541, 608, 617, 668, 673, 674, 692, 693, 695, 698-700
Pregnenolone, 298, 455, 526, 540, 694
Premature ejaculation, 304
Progesterone, 391, 540, 646, 691, 693, 696, 697, 721, 729, 739
Prolactin, 298, 301, 349, 507, 577, 668, 682, 691, 696, 700, 729
Prostaglandin, 121, 169, 170, 376, 720
Prostate cancer (CaP), 281, 283, 288, 676, 681, 686
Prostate specific antigen (PSA), 224, 273, 280, 283, 300, 323, 656, 658, 663, 676, 677, 680, 681, 685, 686, 688, 689
Prostatic acid phosphatase (PAP), 686, 687

Prostatitis, 667, 673, 681, 686, 687, 688
Prostatodynia, 687
Protein, 15, 20, 23, 31, 38, 45, 49, 55, 98, 105, 116, 119, 140, 148, 166, 183, 191, 195, 210, 212, 221, 238, 246, 313, 318, 320, 335, 358, 382, 391, 439, 457, 463, 464, 475, 493, 584, 585, 586, 587, 592, 594, 595, 597, 599, 604, 608, 695, 699, 705, 711, 720, 721, 729, 731
Protein kinase C (PKC), 38, 40, 104, 105, 138, 150, 153, 160, 205, 217, 237, 243-247, 250, 264, 270, 271, 304, 317, 327, 346-348, 389, 392, 393, 427, 575, 619, 621, 622
Protein-energy malnutrition (PEM), 584-586, 588-605, 608, 609
Proteus, 209
Pseudomonas, 95, 185, 198, 205, 209, 409, 480
PUFA, 29, 65, 239, 315, 337, 380, 381, 390, 448, 536, 579, 605, 612, 622, 623, 627, 669, 689, 690
Pulmonary oedema, 68, 81, 115-117, 175, 381, 389, 574
Pumpkin, 324, 675
Pyridoxine, 322, 325, 430, 452, 457, 458, 504, 587

Q

QoL (Quality of life), 17, 224, 284, 287, 288, 289, 318, 335, 388, 421, 472, 506, 578, 584, 586, 588, 594, 604, 609, 615, 652, 658, 701, 721
Quercetin, 330, 368, 369, 508, 582, 583, 623
Quinone, 721

R

Radiation, 64, 107, 112, 113, 267, 396, 444, 474, 530, 531, 573, 651, 666, 676, 737, 739
Radiation therapy, 267, 676
Radiotherapy, 572, 722
Rancidity, 65, 371, 417, 444, 445, 690
Ras oncogene, 47, 269
RDA, 221, 227, 375, 376, 429, 462, 585, 586, 604, 722
Receptor, 183, 201, 206, 346, 719
Redox, 52, 78, 194, 273, 376, 402, 587, 722
Reduction, 35, 63, 67, 121, 151, 161, 182, 227, 251, 273, 275, 322, 358, 370, 376, 377, 487, 550, 580, 655, 658, 679
Regeneration, 19, 21, 51, 56, 175, 193, 212, 249-253, 267, 294, 321, 350, 370, 379, 384, 401, 413, 519, 522, 540-542, 551
Renal insufficiency (RI), 24, 142, 225, 230, 234, 235, 255, 363, 457, 500
Renal papillary necrosis, 255
Renin, 139, 141, 287
Reoxygenation, 52, 54, 70-72, 79, 81, 167-169, 420, 522
Resistance, 199, 222, 258, 345, 584, 592, 593, 601
Respiratory acidosis, 136, 231, 232, 360, 559
Resveratrol, 255, 310, 339
Retrograde ejaculation, 249, 687
Rheumatoid arthritis (RA), 11, 218, 226, 261-263, 404, 641, 642, 644
Riboflavin, 215, 452, 458, 463, 587, 733
Rice, 260, 328, 359
Rifampicin, 83
RNA, 21, 22, 23, 25, 53, 57, 69, 77, 96, 107, 199, 263, 266, 272, 275, 300, 319, 322, 329, 338, 341, 351, 356, 357, 370, 396, 406, 418, 425, 430, 440, 447, 478, 489, 540, 558, 567, 573, 649, 674, 682, 720

S

Salad, 314, 353, 544, 545, 693
Salmonella, 200, 205, 206, 445
SAM-e (S-adenosylmethionine), 340, 350, 560
Sarcopenia, 221, 584
Saturated fatty acid, 337, 442, 446-450, 540, 649, 682, 690
Schistosoma, 269
Schizophrenia, 405, 457, 504, 510
Scission, 65, 376, 391, 393, 579, 723
Scurvy, 257, 331, 336, 376, 408, 411, 412, 415, 417, 462, 504, 505, 557, 643
Seaweed, 312, 398, 560, 652
Selenium, 257, 273, 308, 376, 379, 380, 391, 404, 452, 463, 502, 581, 583, 587, 588, 598, 611, 612, 615, 642, 645, 674, 675, 688, 689, 723, 739
Semen, 407, 668-670, 673, 674, 732
Seminal vesicle, 681

Sepsis, 55, 57, 74, 75, 90, 165, 169-173, 185, 187, 189, 196, 222, 231, 288, 291, 326, 395-397, 401, 404, 523-525, 583, 587, 593, 600, 603, 604, 608, 612, 613, 628, 638, 650
Serotonin, 293, 324, 349, 454, 458, 466, 503, 553, 561, 616, 618, 635, 700
Sertoli cell, 44, 297, 298, 672, 673
SIADH, 288
Silicone, 621
Silymarin, 369, 370, 521, 568, 615
Singlet oxygen, 419, 578, 595, 719
Skin, 59, 150, 225, 264, 282, 491, 584, 589, 590, 622, 664, 691
Smog, 462, 490
Smooth muscle, 349, 390, 723
SOD (superoxide dismutase), 21, 65, 70, 71, 77, 115, 168, 169, 185, 187, 210, 239, 256, 292, 293, 320, 379, 404, 405, 413, 478, 502, 517, 522, 598, 635, 643, 644, 723
Sodium (Na), 47, 73, 85, 113, 122, 131, 198, 361, 365, 418, 463, 472, 548, 644, 718, 729, 731, 733, 734, 735
Solvent, 126, 128, 147, 156, 423, 431, 434, 438, 439, 484-486, 496, 527, 535, 559, 689
Somatopause, 655, 657, 659
Soy, 260, 309, 343, 359, 368, 568, 691
Sperm, 484, 669, 671, 674
Spermatozoa, 669, 732
Sphingolipids, 33, 155, 315, 316, 521
Spinal cord injury (SCI), 175
St John's Wort, 459
Staphylococcus, 186, 188, 190, 209, 368, 447, 474, 548, 717
Starch, 544, 622, 652, 736
STD, 667
Stress, 49, 84, 85, 151, 200, 226, 375, 398, 441, 514, 521, 526, 534, 542, 561, 564, 565, 595, 598, 621, 632, 652, 690, 699, 722, 723
Stroke, 86, 146, 226, 310, 368, 400, 427, 503, 534, 605-607, 701
Sugar, 236, 325, 334, 442, 463, 506, 561, 569, 633
Sulfite, 362, 543, 546, 547
Sulfonamide, 402, 546
Sulfur, 329, 363, 374, 418, 543, 544, 546
Sulfur dioxide, 363, 543, 544, 546
Superoxide, 65, 73, 74, 75, 77, 115, 168, 177, 183, 184, 194, 239, 273, 292, 411, 419, 476, 549, 578, 595, 635, 643, 669, 671, 719, 723
Supplement, 8, 200, 222, 239, 262, 273, 318, 322, 345, 399, 408, 417, 453, 500, 513, 551, 569, 597, 611, 647, 670, 689, 690
Suramin, 576, 681
Syndrome, 258, 291, 405, 613

T

T cell, 37, 38, 58, 87, 93, 94, 101, 111, 120, 157, 170, 182, 188, 206, 210, 215, 217, 220, 238, 244, 283, 399, 429, 448, 475, 519, 574, 610, 612, 654, 701, 724
Taurine, 257, 259, 328, 345, 400, 458, 546
TCR, 38, 88, 217, 724
Tea, 330, 368
Telomerase, 57, 218, 223, 274, 275, 478
Test, 9, 10, 157, 226, 259, 321, 334, 353, 428, 466, 471, 488, 493, 502, 525, 548, 570, 625, 633, 646, 673, 686, 701, 727
Testis, 42, 282, 724
Testosterone, 223, 224, 297, 298, 301, 567, 587, 599, 610, 616, 622, 655, 658, 660, 662, 665, 668, 676, 684, 729
Theanine, 371, 372
Thiamine, 452, 458, 463, 538, 587
Thiol, 30, 66, 73-80, 113, 198, 376, 406, 491, 516-521, 580, 683
Thrombocytopenia, 288, 289, 484
Thyroid, 338, 347, 348, 457, 561, 587, 616, 645, 667, 689, 696, 699
Tiredness, 290, 363, 452, 466, 468, 470, 496, 503, 658
TMG, 340, 465, 507, 631
TNF (Tumour necrosis factor), 26, 33, 51, 56, 59, 69, 76, 85, 86, 88, 89, 90, 91, 92, 95, 97, 100, 103, 107, 109, 113, 115, 116, 118, 144, 155, 157, 162, 165, 166, 170, 171, 174, 182, 183, 186, 188, 189, 190, 191, 193, 194, 197, 200, 201, 206, 208, 210, 216, 218, 220, 238, 240, 242, 243, 244, 246, 247, 256, 257, 262, 267, 270, 279, 281, 284, 286, 287, 288, 290, 294, 315, 316, 357, 362, 371, 375, 379, 399, 476, 487, 509, 519, 523, 524, 560, 562, 577, 595, 604, 608, 610, 612, 623, 644, 648, 650, 654, 683, 690, 698, 724
TNM classification, 263

Tobacco, 81, 86, 146, 154, 269, 332, 374, 462, 528, 574, 582, 681, 737
Tocotrienol, 375, 448, 516
Tolerance, 83, 105, 146, 153, 187, 237, 254, 257, 322, 388, 409, 419, 428, 455, 490, 507, 542, 586, 607, 612-621, 672, 701
Topoisomerase, 426, 582
Toxic, 83, 139, 147, 311, 399, 411, 450, 459, 462, 463, 474, 484, 485, 487, 493, 494, 535, 689, 725, 730, 732
Toxicity, 369, 401, 431, 463
Toxin, 114, 117, 185, 187, 190, 200, 204-209, 369, 388, 410, 448, 474, 488, 514, 547, 559, 648
TP53, 26, 97
Trace element, 310, 320, 386, 394, 418, 483, 484, 487, 547, 557, 567, 615, 650, 682, 683
Transcription, 23, 119, 190, 223
Transcription factor, 119, 223
Trans-fatty acids, 442, 530, 614
Transferrin, 61, 191, 593, 724, 729
Trans-monounsaturated fatty acids (TFAs), 605
Trauma, 172, 173, 293, 301, 474, 676
Trichloroethylene, 485, 739
Tricyclic antidepressants (TCAs), 251-254, 466, 472, 677, 701
Trimethoprim, 202, 203, 233, 361
Trophoblast, 278, 287, 335-337
Tropical, 261, 444, 450, 541, 605, 690, 691
Trypsin, 174, 191, 212, 335, 336, 359, 577
Tumour, 26, 33, 46, 91, 106, 200, 263, 266, 269, 270, 276, 277, 278, 279, 280, 283, 336, 401, 413, 425, 478, 562, 572, 573, 577, 578, 579, 582, 595, 683, 685, 698, 724, 725
Tumour suppressor gene, 266, 724, 725
TUNEL, 110
TURP, 638, 667, 686, 687, 725
Tyrosine, 35, 36, 52, 75, 85, 183, 190, 198, 349, 409, 480, 526, 640, 688, 724

U

Ultrasound, 496, 511, 686, 698, 725
Unrefined, 259, 308, 451, 613
Unsaturated fatty acid, 29, 66, 185, 315, 442, 449, 540, 649, 696
Urate, 24, 128, 292, 361, 379, 729, 731
Urea, 114, 122, 125, 128, 137, 725, 729
Urgency, 197, 199, 331, 362, 641
Uric acid, 10, 23-26, 70, 71, 128, 143, 269, 291, 331, 333, 360-362, 379, 380, 399, 490, 504, 505, 629, 642, 649, 691, 695
Urine, 8, 15, 23, 27, 66, 83, 78, 91, 120, 127-137, 141, 152, 161, 181, 200, 227, 230, 285, 309, 319, 332, 363-366, 412-415, 487-489, 496, 500, 508, 526, 587, 615, 655, 679, 697, 731
Urokinase, 191, 193, 280, 295, 310, 577, 725
Uterus, 42, 140, 170, 171, 264, 377, 409, 507, 661, 692, 693

V

Valsalva manoeuvre, 222
Vanadium, 256, 258, 508
Varicocele, 670
Vasectomy, 672
Vasoactive intestinal peptide (VIP), 164, 301-303, 349, 666
Vasopressin, 32-43, 125, 150, 164, 348, 422, 565, 614, 674, 705
Vegetable, 274, 373, 386, 464, 540, 688, 733
Vegetarian, 10, 134, 209, 277, 376, 455, 559, 626, 674
Vesicles, 11, 35, 57, 108, 145, 146, 195, 264, 347, 402, 428, 664

Vinegar, 527, 560
Vitamin A, 223, 257, 376, 409, 417, 452, 456, 463, 538, 577, 587, 604, 610, 647, 675, 681, 688, 729
Vitamin B complex, 586, 611, 647
Vitamin B_{12}, 227, 452, 456, 458, 463, 505, 538, 560, 565, 587, 631, 674, 729
Vitamin B_{15}, 336
Vitamin B_{17}, 336
Vitamin B_5, 453, 463
Vitamin B_6, 78, 322, 328, 339, 352, 430, 453, 463, 507, 542, 614, 631, 642, 675
Vitamin C, 8, 78, 177, 200, 218, 239, 245, 257, 269, 273, 292, 308, 339, 344, 346, 375, 376, 377, 378, 380, 383, 390, 391, 408, 410, 413, 415, 417, 441, 452, 458, 463, 508, 518, 539, 559, 565, 579, 583, 598, 599, 610, 614, 651, 662, 675, 688, 734
Vitamin D, 93, 137, 159, 221, 275, 339, 362, 452, 463, 539, 560, 576, 587, 681, 688, 689, 694
Vitamin E, 66, 148, 177, 188, 218, 238, 245, 256, 257, 259, 273, 292, 308, 353, 374, 376, 377, 379, 381, 384, 389, 390, 397, 417, 430, 452, 463, 464, 516, 536, 583, 587, 598, 610, 611, 648, 651, 675, 681, 687, 688, 690, 699
Vitamin K, 339, 452, 463, 587, 627
Vitamins, 339, 417, 453, 456, 458, 503, 526, 557, 587, 645, 675, 688

W

Warfarin, 84, 226
Water, 42, 122, 129, 132, 142, 152, 431, 462, 494, 533, 548, 586, 592, 597, 698, 713, 726
Water channel, 42, 726
Weight loss, 222, 259, 284, 582, 584, 588, 591, 596, 607, 609, 695
Whey protein, 210, 345, 358, 359
Wound healing, 51, 52, 140, 147, 166, 192, 212, 240, 272, 291, 293, 317, 395, 410, 449, 462, 511, 529, 587-596, 612, 659

X

X chromosome, 44, 297, 300
Xanthine, 23-25, 69-71, 140, 159, 169, 333, 453, 615, 634, 671
Xanthine dehydrogenase (XD), 24, 25, 67, 70, 167, 292, 333
Xanthine oxidase (XO), 23-25, 55, 66-72, 79, 140, 154, 169, 186, 250, 292, 333, 369, 383, 522, 615, 634, 642, 670, 671
Xanthinuria, 24, 25, 333
Xenobiotic, 15, 380, 404, 484, 726
X-ray, 263, 337, 530, 573, 651, 698
XX, 297

Y

Y chromosome, 297
Yeast, 394, 479, 623

Z

Zeaxanthin, 373, 376
Zinc, 219, 257, 309, 320, 322, 339, 353, 361, 418, 430, 452, 463, 464, 465, 489, 502, 506, 509, 526, 539, 566, 587, 598, 611, 612, 615, 634, 642, 645, 650, 674, 682, 689, 726, 736

Lightning Source UK Ltd.
Milton Keynes UK
18 September 2009

143896UK00001B/3/A